Food Microbiology

FUNDAMENTALS AND FRONTIERS

Third Edition

Food Microbiology

FUNDAMENTALS AND FRONTIERS

Third Edition

EDITED BY

Michael P. Doyle and Larry R. Beuchat

Center for Food Safety, The University of Georgia, Griffin, GA 30223-1797

ASM PRESS   WASHINGTON, D.C.

Cover images: *Fusarium graminearum* macroconidia (courtesy Lloyd B. Bullerman, University of Nebraska, Lincoln); Norwalk virus (courtesy C. Büchen-Osmond, ICTVdB Management, Columbia University, http://www.ncbi.nlm.nih.gov/ICTVdb/ICTVdB/00.012.htm); *Giardia lamblia* (courtesy Jeffrey M. Farber, Health Canada, Ottawa); *Campylobacter jejuni* biofilm on stainless steel (courtesy John Austin and Greg Sanders, Bureau of Microbial Hazards, Health Canada, Ottawa); background, *Listeria monocytogenes* biofilm on glass, stained with fluorescent wheat germ agglutinin (courtesy John Austin and Greg Sanders).

Copyright © 2007   ASM Press
American Society for Microbiology
1752 N Street, N.W.
Washington, DC 20036-2804

Library of Congress Cataloging-in-Publication Data

Food microbiology : fundamentals and frontiers / editors, Michael P.
   Doyle and Larry R. Beuchat.—3rd ed.
      p. ; cm.
   Includes bibliographical references and index.
   ISBN-13: 978-1-55581-407-6
   ISBN-10: 1-55581-407-7
   1. Food—Microbiology.   I. Doyle, Michael P. II. Beuchat, Larry R.
   [DNLM: 1. Food Microbiology.   QW 85 F6855 2007]
QR115.F654 2007
664.001′579—dc22

                                                                         2006033548

10   9   8   7   6   5   4   3   2   1

Address editorial correspondence to: ASM Press, 1752 N St., N.W., Washington, DC
20036-2904, U.S.A.

Send orders to: ASM Press, P.O. Box 605, Herndon, VA 20172, U.S.A.
Phone: 800-546-2416; 703-661-1593
Fax: 703-661-1501
Email: Books@asmusa.org
Online: estore.asm.org

Contents

X. Advanced Techniques in Food Microbiology

Contributors

TIMOTHY J. BARRETT
Foodborne and Diarrheal Diseases Branch, Centers for Disease Control and
Prevention, MS-C03, 1600 Clifton Rd., Atlanta, GA 30333

DANE BERNARD
Keystone Foods LLC, 5 Tower Bridge, 300 Barr Harbor Dr., Suite 600,
West Conshohocken, PA 19428

LARRY R. BEUCHAT
Center for Food Safety, University of Georgia, 1109 Experiment St., Griffin,
GA 30223-1797

GREGORY A. BOHACH
Dept. of Microbiology, Molecular Biology, and Biochemistry, College of
Agricultural and Life Sciences, Agricultural Sciences Bldg. Room 47,
University of Idaho, Moscow, ID 83844

FREDERICK BREIDT, JR.
U.S. Dept. of Agriculture, Agricultural Research Service, Dept. of Food Science,
322 Schaub Hall, Box 7624, North Carolina State University, Raleigh, NC
27695-7624

PAUL BROWN
Senior Investigator [retired], National Institutes of Health, Bethesda, MD 20814

R. L. BUCHANAN
Center for Food Safety and Applied Nutrition, Food and Drug Administration,
5100 Paint Branch Pkwy., College Park, MD 20740

LLOYD B. BULLERMAN
Dept. of Food Science and Technology, 322 Food Industry Complex, East
Campus, University of Nebraska, Lincoln, NE 68583-0919

FRANK F. BUSTA
National Center for Food Protection and Defense, University of Minnesota, 925
Delaware St., SE, Minneapolis, MN 55455

DIDIER CABANES
Group of Molecular Microbiology, Institute for Molecular and Cell Biology,
Rua do Campo Alegre 823, 4150-180 Porto, Portugal

IAIN CAMPBELL
International Centre for Brewing and Distilling, Heriot-Watt University,
Edinburgh EH10 4AS, United Kingdom

FREDERIC CARLIN
Institut National de la Recherche Agronomique, UMR408, Sécurité et Qualité
des Produits d'Origine Végétale, Site Agroparc, 84914 Avignon cedex 9, France

MICHAEL L. CHIKINDAS
Dept. of Food Science, Rutgers-The State University of New Jersey, New
Brunswick, NJ 08901-8520

ASHOK K. CHOPRA
Dept. of Microbiology and Immunology, University of Texas Medical Branch,
Medical Research Bldg., 3.142H, 301 University Blvd., Galveston,
TX 77555-1070

PASCALE COSSART
Institut Pasteur, Unité des Interactions Bactéries-Cellules; Inserm, U604; and
INRA, USC2020, Paris, F-75015, France

MICHELLE D. DANYLUK
Dept. of Food Science and Human Nutrition, University of Florida, IFAS,
Citrus Research and Education Center, Lake Alfred, FL 33850-2299

JEAN-YVES D'AOUST
Food Directorate, Health Products & Food Branch, Health Canada, Sir F. G.
Banting Research Centre, Postal Locator 22.04.A2, Tunney's Pasture, Ottawa,
Ontario, Canada K1A 0K9

P. MICHAEL DAVIDSON
Dept. of Food Science and Technology, 2605 River Rd., University of Tennessee,
Knoxville, TN 37996-4591

LINDA A. DETWILER
Center for Public and Corporate Veterinary Medicine, Virginia-Maryland
Regional College of Veterinary Medicine, 8075 Greenmead Dr., College Park,
MD 20742

ILENYS DÍAZ-MUÑIZ
U.S. Dept. of Agriculture, Agricultural Research Service, Dept. of Food Science,
322 Schaub Hall, Box 7624, North Carolina State University, Raleigh, NC
27695-7624

MICHAEL P. DOYLE
Center for Food Safety, University of Georgia, Griffin, GA 30223

DORIS H. D'SOUZA
Dept. of Food Science & Technology, Institute of Agriculture, University of
Tennessee, Knoxville, TN 37996-4591

TRI DUONG
Genomic Sciences Program, Dept. of Food Science, North Carolina State
University, Raleigh, NC 27695-7624

JEFFREY M. FARBER
Bureau of Microbial Hazards, Health Products and Food Branch, Food
Directorate, Health Canada, Sir F. G. Banting Research Centre, P/L2204A2, 251
Sir Frederick Banting Driveway, Ottawa, Ontario, K1A 0K9, Canada

JÓSEF FARKAS
Dept. of Refrigeration and Livestock Products' Technology, Faculty of Food
Science, Corvinus University of Budapest, Ménesi út 45, H-1118 Budapest,
Hungary

PETER FENG
Division of Microbiological Studies, U.S. Food and Drug Administration,
HFS-516, CFSAN, 5100 Paint Branch Pkwy., College Park, MD 20740-3835

GRAHAM H. FLEET
School of Chemical Sciences and Engineering, The University of New South
Wales, Sydney, New South Wales 2052, Australia

JOSEPH F. FRANK
Dept. of Food Science and Technology, University of Georgia, Athens,
GA 30602-7610

CRISTI L. GALINDO
Dept. of Microbiology and Immunology, University of Texas Medical Branch,
301 University Blvd., Galveston, TX 77555-1070

H. RAY GAMBLE
National Research Council, 500 Fifth St., NW, Washington, DC 20001

PETER GERNER-SMIDT
Foodborne and Diarrheal Diseases Branch, Centers for Disease Control and
Prevention, 1600 Clifton Rd., Atlanta, GA 30333

PER EINAR GRANUM
Dept. of Food Safety and Infection Biology, Norwegian School of Veterinary
Science, P.O. Box 8146 Dep., N-0033 Oslo, Norway

LINDA J. HARRIS
Dept. of Food Science and Technology, University of California, One Shields
Ave., Davis, CA 95616-8598

EUGENE G. HAYUNGA
National Institute of Child Health and Human Development, National Institutes
of Health, Bethesda, MD 20892-7510

CRAIG W. HEDBERG
Divison of Environmental Health Sciences, School of Public Health, University
of Minnesota, Room 1242 Mayo Bldg., 420 Delaware St. S.E., Minneapolis,
MN 55455

AILSA D. HOCKING
Food Science Australia, Riverside Life Sciences Centre, P.O. Box 52, North
Ryde, New South Wales 1670, Australia

LEE-ANN JAYKUS
Dept. of Food Science, College of Agriculture and Life Sciences, North Carolina State University, Raleigh, NC 27695-7624

ERIC A. JOHNSON
Food Research Institute, University of Wisconsin-Madison, Madison, WI 53706

MARK E. JOHNSON
Center for Dairy Research, Dept. of Food Science, University of Wisconsin-Madison, Madison, WI 53706-1565

JAMES B. KAPER
Center for Vaccine Development, Dept. of Microbiology and Immunology, University of Maryland School of Medicine, 685 West Baltimore St., Baltimore, MD 21201

JIMMY T. KEETON
Dept. of Animal Science, Room 338 Kleberg Animal and Food Science Center, Texas A&M University, College Station, TX 77843-2471

SHAUN P. KENNEDY
National Center for Food Protection and Defense, University of Minnesota, 925 Delaware St., SE, Minneapolis, MN 55455

CHARLES W. KIM
Center for Infectious Diseases, Health Sciences Center, State University of New York at Stony Brook, Stony Brook, NY 11794

TODD R. KLAENHAMMER
Genomic Sciences Program, Dept. of Food Science, 339 Schaub Hall, Box 7624, North Carolina State University, Raleigh, NC 27695-7624

KEITH A. LAMPEL
Center for Food Safety and Applied Nutrition, Food and Drug Administration, 5100 Paint Branch Pkwy., College Park, MD 20740

RAQUEL F. LENATI
Bureau of Microbial Hazards, Health Products and Food Branch, Food Directorate, Health Canada, Sir F. G. Banting Research Centre, P/L2204A2, 251 Sir Frederick Banting Driveway, Ottawa, Ontario, K1A 0K9, Canada

ALEX S. LOPEZ
6621 Creeping Thyme St., Las Vegas, NV 89148

DOUGLAS L. MARSHALL
Dept. of Food Science, Nutrition, and Health Promotion, Room 110 Herzer, Stone Blvd., Box 9805, Mississippi State University, Mississippi State, MS 39762-9805

KARL R. MATTHEWS
Dept. of Food Science, School of Environmental and Biological Sciences, Rutgers, The State University of New Jersey, 65 Dudley Rd., New Brunswick, NJ 08901-8520

ANTHONY T. MAURELLI
Dept. of Microbiology and Immunology, F. Hébert School of Medicine,

Uniformed Services University of the Health Sciences, 4301 Jones Bridge Rd., Bethesda, MD 20814-4799

JOHN MAURER
Dept. of Population Health, University of Georgia, Athens, GA 30602

BRUCE A. MCCLANE
Dept. of Molecular Genetics and Biochemistry, University of Pittsburgh School of Medicine, E1240 Biomedical Science Tower, Pittsburgh, PA 15261-2072

JENNIFER CLEVELAND MCENTIRE
Institute of Food Technologists, 1025 Connecticut Ave., NW, Suite 503, Washington, DC 20036

ROGER F. MCFEETERS
U.S. Dept. of Agriculture, Agricultural Research Service, Dept. of Food Science, 322 Schaub Hall, Box 7624, North Carolina State University, Raleigh, NC 27695-7624

JIANGHONG MENG
Dept. of Nutrition and Food Science, University of Maryland, College Park, MD 20742

KENNETH B. MILLER
Nutrition and Natural Product Sciences, Technical Center, Hershey Foods Corp., 1025 Reese Ave., Hershey, PA 17033-0805

CHRISTINE L. MOE
Hubert Dept. of Global Health, Rollins School of Public Health, Emory University, Atlanta, GA 30322

THOMAS J. MONTVILLE
Dept. of Food Science, School of Environmental and Biological Sciences, Rutgers, The State University of New Jersey, 65 Dudley Rd., New Brunswick, NJ 08901-8520

IRVING NACHAMKIN
Dept. of Pathology and Laboratory Medicine, University of Pennsylvania School of Medicine, 4th Floor, Gates Building, 3400 Spruce St., Philadelphia, PA 19104-4283

M. J. ROBERT NOUT
Dept. of Agrotechnology and Food Sciences, Wageningen University, Bloemenweg 2, 6703HD Wageningen, The Netherlands

GEORGE-JOHN E. NYCHAS
Laboratory of Food Microbiology and Biotechnology of Foods, Dept. Food Science and Technology, Agricultural University of Athens, Iera Odos 75, Athens 11855, Greece

JAMES D. OLIVER
Dept. of Biology, University of North Carolina at Charlotte, 9201 University City Blvd., Charlotte, NC 28223

YNES R. ORTEGA
Center for Food Safety, University of Georgia, 1109 Experiment St., Griffin, GA 30223-1797

FRANCO J. PAGOTTO
Bureau of Microbial Hazards, Health Products and Food Branch, Food
Directorate, Health Canada, Sir F. G. Banting Research Centre, P/L2204A2,
251 Sir Frederick Banting Driveway, Ottawa, Ontario, K1A 0K9, Canada

ERIKA PFEILER
Genomic Sciences Program, Dept. of Food Science, North Carolina State
University, Raleigh, NC 27695-7624

MERLE D. PIERSON
Dept. of Food Science and Technology, Virginia Polytechnic Institute and State
University, Blacksburg, VA 24061

JOHN I. PITT
Food Science Australia, P.O. Box 52, North Ryde, New South Wales 1670,
Australia

STEVEN C. RICKE
Dept. of Food Science, University of Arkansas, 2650 North Young Ave.,
Fayetteville, AR 72704-4605

ROY M. ROBINS-BROWNE
Dept. of Microbiology and Immunology, University of Melbourne, Victoria
3010, and Murdoch Children's Research Institute, Royal Children's Hospital,
Parkville, Victoria 3052, Australia

PRABIR K. SARKAR
Dept. of Botany, University of North Bengal, Siliguri 734013, India

VIRGINIA N. SCOTT
GMA/FPA, 1350 I St., NW, Suite 300, Washington, DC 20005

KEUN SEOK SEO
Dept. of Microbiology, Molecular Biology, and Biochemistry, University of
Idaho, Moscow, ID 83844

PETER SETLOW
Dept. of Molecular, Microbial and Structural Biology, University of Connecticut
Health Center, Farmington, CT 06030-3305

L. MICHELE SMOOT
Silliker, Inc., 2057 Builders Place, Columbus, OH 43204

JOHN N. SOFOS
Dept. of Animal Sciences, Colorado State University, 1171 Campus Delivery,
Fort Collins, CO 80523-1171

WILLIAM H. SPERBER
Corporate Food Safety and Regulatory Affairs, Cargill, Inc., 15407 McGinty
Rd. West, Wayzata, MN 55391-9300

JAMES L. STEELE
Dept. of Food Science, University of Wisconsin-Madison, 1605 Linden Dr.,
Madison, WI 53706-1565

BALA SWAMINATHAN
Centers for Disease Control and Prevention, 1600 Clifton Rd., MS-C03,
Atlanta, GA 30333

T. MATTHEW TAYLOR
Dept. of Food Science and Technology, 2605 River Rd., University of Tennessee, Knoxville, TN 37996-4591

STERLING S. THOMPSON
Microbiology Research & Services, Technical Center, Hershey Foods Corp., 1025 Reese Ave., Hershey, PA 17033-0805

R. C. WHITING
Center for Food Safety and Applied Nutrition, Food and Drug Administration, 5100 Paint Branch Pkwy., College Park, MD 20740

IRENE ZABALA DIAZ
Departmento Experimental de Biologia, F.E.C.-L.U.Z., Av. Goajira, Bloque A-1, Maracaibo, Edo. Zulia, 4001, Venezuela

DANTE S. ZARLENGA
U.S. Dept. of Agriculture, Agricultural Research Service, 10300 Baltimore Ave., Beltsville, MD 20705

WEI ZHANG
National Center for Food Safety and Technology, Illinois Institute of Technology, 6502 South Archer Rd., Summit, IL 60501

SHAOHUA ZHAO
Division of Animal and Food Microbiology, Center for Veterinary Medicine/ Office of Research, Food & Drug Administration, Laurel, MD 20708

TONG ZHAO
Center for Food Safety, University of Georgia, Griffin, GA 30223

DON L. ZINK
U.S. Food and Drug Administration, 5100 Paint Branch Parkway, College Park, MD 20740

Preface to the Third Edition

The field of food microbiology is among the most diverse of the areas of study within the discipline of microbiology. Its scope encompasses a wide variety of microorganisms including spoilage, probiotic, fermentative, and pathogenic bacteria, molds, yeasts, viruses, prions, and parasites; a diverse composition of foods and beverages; a broad spectrum of environmental factors that influence microbial survival and growth; and a multitude of research approaches that range from very applied studies of survival and growth of foodborne microorganisms to basic studies of the mechanisms of pathogenicity of disease-causing, foodborne microorganisms.

Several excellent books address many different aspects of food microbiology. The purpose of *Food Microbiology: Fundamentals and Frontiers* is to complement these books by providing new, state-of-the-science information that emphasizes the molecular and mechanistic aspects of food microbiology, and not to dwell on other aspects well covered in introductory food microbiology texts. The third edition provides new information regarding recent advances in all aspects of food microbiology. Major revisions have been made to chapters addressing foodborne pathogens, which is an area exploding with new findings. New chapters on *Enterobacter sakazakii*, prions, genomics and proteomics, and molecular source tracking/molecular subtyping appear in the new third edition. Chapters focused on nuts and cereals and on biodefense also have been added.

This advanced reference text fulfills the need of research microbiologists, graduate students, and professors of food microbiology courses for an in-depth treatment of food microbiology. It provides current, definitive, factual material written by experts on each subject. The book is written at a level which presupposes a general background in microbiology and biochemistry needed to understand the "how and why" of food microbiology at a basic scientific level.

The book is composed of 10 major sections that address each of the major areas of the field. "Factors of Special Significance to Food Microbiology" provides a perspective on and description of the basic principles that affect the growth, survival, and death of microorganisms, coverage of bacterial spores, the use of indicator microorganisms and microbiological criteria, and biodefense. "Microbial Spoilage of Foods and Public Health Concerns" covers the principles of spoilage, dominant microorganisms, and spoilage patterns for each of four major food categories. The 14 chapters in the "Foodborne Pathogenic Bacteria" section provide a current molecular understanding of foodborne bacterial pathogens in the context of their pathogenic mechanisms, tolerance to preservation methods, and underlying epidemiology as well as basic information about metabolic characteristics of each microorganism, symptoms of illness, and common food reservoirs. Similar perspectives are given by chapters in the sections "Mycotoxigenic Molds," "Viruses," "Prions," and "Foodborne and Waterborne Parasites."

"Preservatives and Preservation Methods" presents information on mechanisms, models, and kinetics in three chapters which elucidate physical, chemical, and biological methods of food preservation. The "Fermentations and Beneficial Microorganisms" section emphasizes the genetics and physiology of microorganisms involved in fermentation of foods and beverages. The influence of fermentation on product characteristics is examined, and the benefits of probiotics and prebiotics in promoting health are presented.

Rapid, genetic, and immunological methods for detecting foodborne microorganisms, predictive modeling and quantitative risk assessment, hazard analysis and critical control points, and molecular source tracking and subtyping are key issues to the future of food microbiology. Hence, it is appropriate that these topics are covered in the closing section, "Advanced Techniques in Food Microbiology."

We are grateful to all of our coauthors for their dedication to producing a book that is at the cutting edge of food microbiology, and to the reviewers whose critical evaluations enabled us to fine tune each chapter.

Michael P. Doyle
Larry R. Beuchat

Factors of Special Significance to Food Microbiology

I

Food Microbiology: Fundamentals and Frontiers, 3rd Ed.
Edited by M. P. Doyle and L. R. Beuchat
© 2007 ASM Press, Washington, D.C.

Thomas J. Montville
Karl R. Matthews

Growth, Survival, and Death of Microbes in Foods

1

Food microbiologists must understand microbiology and food systems and be able to integrate them to solve problems in complex food ecosystems. This chapter addresses this in three parts by (i) examining foods as ecosystems and discussing intrinsic and extrinsic environmental factors that control bacterial growth, (ii) explaining first-order or pseudo-first-order kinetics which govern the log phase of microbial growth and many types of lethality, and (iii) focusing on physiology and metabolism of foodborne microbes. The ability of bacteria to use different biochemical pathways which generate different amounts of ATP influences their ability to grow under adverse conditions in foods. The generation and utilization of energy, "bioenergetics," quorum sensing, and the ability to grow as biofilms are critically important to growth in food. The last section of this chapter reviews the limitations of classical microbiology.

FOOD ECOSYSTEMS, HOMEOSTASIS, AND HURDLE TECHNOLOGY

Foods as Ecosystems

Foods are complex ecosystems. Ecosystems are each composed of the environment and the organisms that live in it. The food environment is composed of intrinsic factors inherent to the food (i.e., pH, water activity, and nutrients) and extrinsic factors external to it (i.e., temperature, gaseous environment, the presence of other bacteria). When intrinsic and extrinsic factors are manipulated to preserve food, food preservation can be viewed as "the ecology of zero growth" (15).

When applied to microbiology, ecology can be defined as "the study of the interactions between the chemical, physical, and structural aspects of a niche and the composition of its specific microbial population" (99). "Interactions" highlights the multivariable nature of ecosystems. Computer modeling can be very helpful in understanding the complex relationship among the bacteria and the multiple environmental parameters in foods. A complete set of reviews about food ecosystems has been published by the Society for Applied Bacteriology (14).

Foods can be heterogeneous on a micrometer scale. Heterogeneity and its associated gradients of pH, oxygen, nutrients, water activity etc. are key ecological factors in foods (15). Foods may contain several distinct microenvironments. This is well illustrated by the food poisoning outbreaks in "aerobic" foods caused by the

Thomas J. Montville and Karl R. Matthews, Dept. of Food Science, School of Environmental and Biological Sciences, Rutgers, The State University of New Jersey, New Brunswick, NJ 08901-8520.

"obligate anaerobe" *Clostridium botulinum*. Growth of *C. botulinum* in foods such as potatoes and sautéed onions exposed to air has caused botulism outbreaks (85). The oxygen in these foods is driven out during cooking and diffuses back in so slowly that, while the surface layer is aerobic, the bulk of the product remains anaerobic.

Intrinsic Factors that Influence Microbial Growth

Those factors inherent to the food are "intrinsic" factors. These include natural food compounds that stimulate or retard microbial growth, added preservatives, the oxidation-reduction potential, water activity, and pH. Most of these factors are covered separately in the chapters on physical and chemical methods of food preservation. The influence of pH is particularly important and covered in some depth below.

Intracellular pH (pH_i) must be maintained above some critical pH_i at which intracellular proteins become irreversibly denatured. Three progressively more stringent mechanisms, the "homeostatic response," the "acid tolerance response," and the synthesis of "acid shock" protein maintain a pH_i consistent with viability. These have been studied most extensively in *Salmonella enterica* serovar Typhimurium (53, 54, 135).

The "homeostatic" response helps cells maintain their pH_i in mildly acidic (external pH [pH_o] > 6.0) conditions. The homeostatic response maintains pH_i by allosterically modulating the activity of proton pumps, antiports, and symports to increase the rate at which protons are expelled from the cytoplasm. The homeostatic mechanism is constitutive and functions in the presence of protein synthesis inhibitors. The proton-translocating F_0F_1 ATPase described in the bioenergetics section (below) is especially important in regulating pH_i.

The "acid tolerance response" (ATR) is triggered by a pH_o of 5.5 to 6.0 (52, 53) and maintains a pH_i of >5.0 at pH_o values as low as 4.0. Optimal pH for triggering the ATR response varies by organism (74). In *Listeria monocytogenes*, ATR appears to involve the membrane-bound F_0F_1 ATPase proton pump (18, 90). In enterobacteria at least four regulatory systems, an alternative sigma factor, a two-component signal transduction system (PhoPQ), the major iron-regulatory protein Fur, and Ada (involved in adaptive response to alkylating agents), are involved with acid survival (10). These systems may be activated depending on whether the stress is from an inorganic or organic acid (9). Loss of the gene encoding the general transcription factor σ^B in *L. monocytogenes* diminishes acid tolerance but has no effect on virulence in a mouse model (143). Induced ATR in *Escherichia coli* O157: H7 alters the expression of 86 genes, of which 6 are

important for low-pH survival (1). The ATR response differs for log-phase and stationary-phase cells. In *Salmonella* serovar Typhimurium, OmpR is critical to stationary-phase ATR but not to the log-phase ATR (2, 3). In addition, the ATR response of *Salmonella* serovars can differ; *Salmonella enterica* serovar Typhi is 200 to 2,000 times more susceptible to lethal acidity than *Salmonella* serovar Typhimurium. Acid-adapted salmonellae have increased resistance to a low-pH gastric environment, which may increase virulence (56).

The ATR may confer cross-protection to other environmental stressors. Acid adaptation increases heat and freeze-thaw resistance of *Escherichia coli* O157:H7 (78). The exposure of *Salmonella* serovar Typhimurium cells to pH 5.8 for a few cell doublings induces 12 proteins, represses 6 proteins, and renders the cells less sensitive to sodium chloride and heat (81). Exposure of *S. enterica* serovar Typhimurium to short-chain fatty acids increases acid resistance (77). Following exposure to nisin, survival of acid-adapted *L. monocytogenes* is approximately 10-fold greater than that of nonadapted cells (139). Acid-adapted *L. monocytogenes* has increased resistance against heat shock, osmotic stress, alcohol stress (118), and nisin (19). Acid adaptation of *E. coli* O157:H7 enhances thermotolerance (41).

The third way that cells regulate pH_i, the synthesis of acid-shock proteins, is triggered by pH_o from 3.0 to 5.0. Acid-shock proteins are a set of *trans*-acting regulatory proteins. The majority of acid-induced proteins in *L. monocytogenes* are common for the responses to acid adaptation and acid stress (118), but some are unique. Three stationary phase-dependent acid resistance systems protect *E. coli* O157:H7 under extreme acid (pH 2.5 or less). These include the oxidative or glucose-repressed system, the glutamate decarboxylase system, and the arginine decarboxylase system (30). DNA-binding proteins (Dps) interact with DNA to form stable complexes which protect the DNA from acid-mediated damage (33). Survival of an *E. coli* O157:H7 *dps* mutant is significantly less (4 log CFU/ml reduction) than the parent strain (1 log CFU/ml reduction) after acid (pH 1.8) exposure.

External pH (pH_o) can also regulate the expression of genes governing proton transport, amino acid degradation, adaptation to acidic or basic conditions, and even virulence (113). The expression of the *Yersinia enterocolitica inv* gene in laboratory media at 23°C but not at 37°C seems paradoxical, since its expression is required for infection of warm-blooded animals. However, at the pH of the small intestine (5.5), the *inv* gene is expressed at 37°C (117). The *yst* gene, which codes for a heat-stable enterotoxin in *Y. enterocolitica*, is regulated similarly (94). The *toxR* gene, which controls expression of cholera toxin in *Vibrio cholerae*, is regulated in part by pH (101). In *Salmonella*, exposure to low pH enhances survival in macrophages. *Salmonella enterica* serovar Dublin virulence genes are induced by low pH (127). Exposure of *Salmonella enterica* serovar Enteritidis to pH 10 or 1.5% trisodium phosphate significantly increases thermotolerance (122).

Extrinsic Factors that Influence Microbial Growth

Temperature and gas composition are the primary extrinsic factors influencing microbial growth. Controlled and modified atmospheres are covered in depth in chapter 32. The influence of temperature on microbial growth and physiology cannot be overemphasized. While the influence of temperature on growth kinetics is obvious and covered here in some detail, the influence of temperature on gene expression is equally important. Cells grown at refrigerated temperature express different genes and are physiologically different than those grown at ambient temperature. Later chapters provide organism-specific detail about the way temperature regulates phenotypes ranging from motility to virulence.

A "rule of thumb" in chemistry suggests that reaction rates double with every 10°C increase in temperature. This simplifying assumption is valid for bacterial growth rates only over a limited range of organism-dependent temperatures (Fig. 1.1). Bacteria are classified as psychrophiles, psychrotrophs, mesophiles, and thermophiles according to the way in which temperature influences their growth.

Both psychrophiles and psychrotrophs grow, albeit slowly, at 0°C. True psychrophiles have optimum growth rates at 15°C and cannot grow above 25°C. Psychrotrophs, such as *L. monocytogenes* and *C. botulinum* type E, have optima of ~25°C and cannot grow above 40°C. Because these foodborne pathogens, and even some mesophilic *Staphylococcus aureus* strains, can grow at <10°C, conventional refrigeration cannot ensure the safety of a food (116). Additional barriers to microbial growth should be incorporated into refrigerated foods containing no other inhibitors (102).

Several metabolic capabilities are important for growth in the cold. Homeoviscous adaptation enables cells to maintain membrane fluidity at low temperatures. As temperature decreases, cells synthesize increasing amounts of mono- and diunsaturated fatty acids (36, 124). The "kinks" caused by the double bonds prevent tight packing of the fatty acids into a more crystalline array. The accumulation of compatible solutes at low temperatures (70) is analogous to their accumulation under conditions of low water activity, as discussed in chapters 32 and 33. The membrane's physical state can regulate the expression of genes, particularly those that respond to temperature (140). The production of cold shock proteins (CSPs) contributes to an organism's ability to grow at low temperatures. CSPs appear to function as RNA chaperones, minimizing the folding of mRNA, thereby facilitating the translation process. *Streptococcus thermophilus* CSPs are maximally expressed at 20°C. Northern blot analysis revealed a ninefold induction of *csp* mRNA and that its regulation takes place at the transcriptional level (147). Pretreatment at 20°C increases survival approximately 1,000-fold compared to nonadapted cells. *E. coli* CSPs are categorized into two groups. Class I proteins are expressed at low levels at 37°C and increase dramatically after shift to low

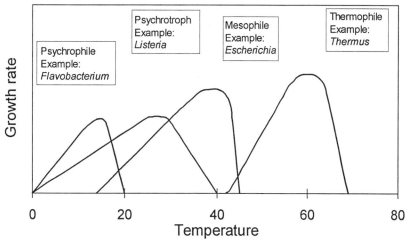

Figure 1.1 Relative growth rates of bacteria at different temperatures.

temperature. Class II CSPs increase only a fewfold after downshift in temperature (134). Cold-shocking *L. monocytogenes* from 37 to 5°C induces 12 CSPs with molecular weights ranging from 48,000 to 14,000 (8). Expression of the *fri* gene, encoding ferritin, protects *L. monocytogenes* against multiple stresses including cold and heat shock (43). Exposure of *E. coli* O157:H7 to cold stress decreases its acid tolerance (44).

Temperature regulates the expression of virulence genes in several pathogens. The expression of 16 proteins on seven operons on the *Y. enterocolitica* virulence plasmid is high at 37°C, weak at 22°C, and undetectable at 4°C (126). Similarly, the gene(s) required for virulence of *Shigella* spp. is expressed at 37°C, but not at 30°C. The expression of genes required for *L. monocytogenes* virulence is also temperature regulated (79). Cells grown at 4, 25, and 37°C all synthesize internalin, a protein required for penetration of the host cell. Cells grown at 37°C, but not those grown at 4 or 25°C, are hemolytic. However, the hemolytic activity is restored during the infection process (35). Temperature influences expression of *Vibrio cholerae toxT* and *toxR* genes essential for cholera toxin production. Maximal expression occurs at 30°C, whereas at 37°C expression is significantly decreased or abolished (101, 123). In enterohemorrhagic *E. coli*, temperature modulates transcription of the *esp* genes; synthesis of Esp proteins is enhanced when bacteria are grown at 37°C. Esp proteins are required for signal transduction events leading to the formation of the attaching and effacing lesions linked to virulence (11).

The growth temperature can influence a cell's thermal sensitivity. *L. monocytogenes* cells preheated at 48°C have increased thermal resistance (50). Holding listeria at 48°C for 2 h in sausages increases their *D* values at 64°C by 2.4-fold. This thermotolerance is maintained for 24 h at 4°C (48). Subjecting *E. coli* O157:H7 cells to sublethal heating at 46°C increases their *D* value at 60°C by 1.5-fold. Two proteins, putatively GroEL and DnaK, increase following heat shock (67). The role of heat shock proteins in increased thermal resistance is discussed in more depth in chapter 32. In short, the heat shock response and regulated synthesis of heat shock proteins (HSPs) in gram-negative bacteria can differ markedly from gram-positive bacteria. Many HSPs are molecular chaperones (e.g., DnaK and GroEL) or ATP-dependent proteases (e.g., Lon and ClpAP) and function in protein folding, assembly, transport, and repair under stress and nonstress conditions (136, 148). Shock proteins synthesized in response to one stressor may provide cross-protection against other stressors (81). Exposing *Bacillus subtilis* to mild heat stress enables the organism to survive not only otherwise-lethal temperatures

but also exposure to toxic concentrations of NaCl (141). Heat adapted (50°C for 45 min) listeriae are more resistant to acid shock (118). Similarly, sublethal heat treatment of *E. coli* O157:H7 cells increases their tolerance to acidic conditions (142).

THE IMPORTANCE OF FIRST-ORDER KINETICS

Growth Kinetics

Food microbiology is concerned with all four phases of microbial growth. Growth curves showing the lag, exponential logarithmic or log, stationary, and death phases of a culture are normally plotted as the number of cells on a logarithmic scale or $\log_{10}$ cell number versus time. These plots represent the states of microbial populations rather than individual microbes. Thus, both the lag phase and stationary phase of growth represent periods when the growth rate equals the death rate to produce no net change in cell numbers.

During the lag phase, cells adjust to their new environment by inducing or repressing enzyme synthesis and activity, initiating replication of DNA, and, in the case of spores, differentiating into vegetative cells (see chapter 3). The length of the lag phase depends on temperature, the inoculum size (larger inocula usually have shorter lag phases), and the physiological history of the organism. If actively growing cells are inoculated into an identical fresh medium at the same temperature, the lag phase may vanish. Conversely, these factors can be manipulated to extend the lag phase beyond the time where some other food quality attribute (such as proteolysis or browning) becomes unacceptable. Foods are generally considered microbially safe if obvious spoilage precedes microbial growth. However, "spoiled" is a subjective and culturally biased concept. It is safer to create conditions that prevent growth altogether.

During the exponential or log phase of growth, bacteria reproduce by binary fission. One cell divides into two cells, which divide into four cells, which divide into eight cells, etc. Thus, during exponential growth, first-order reaction kinetics can be used to describe the change in cell numbers. Food microbiologists often use doubling times as the kinetic constant to describe the rate of logarithmic growth. Doubling times (t_d), which are also referred to as "generation" times (t_{gen}), are related to classical kinetic constants as shown in Table 1.1.

The influence of different parameters on a food's final microbial load can be illustrated by manipulating the equations in Table 1.1. Equation 1a states that the number of organisms (N) at any time is directly proportional to the initial number of organisms (N_0). Thus, decreasing the initial microbial load 10-fold will reduce

Table 1.1 First-order kinetics can be used to describe exponential growth and inactivation

Growth[a]	Thermal inactivation[b]	Irradiation[c]
1a. $N = N_0 e^{\mu t}$	1b. $N = N_0 e^{-kt}$	1c. $N = N_0 e^{-Ds/Do}$
2a. $2.3\log(N/N_0) = \mu \Delta t$	2b. $2.3\log(N/N_0) = -(k\Delta t)$	
3a. $\Delta t = [2.3\log(N/N_0)]/\mu$	3b. $\Delta t = -[2.3\log(N/N_0)]/k$	
4a. $t_d = 0.693/\mu$	4b. $D = 2.3/k$	
	5b. $E_a = \dfrac{2.3RT_1T_2}{z} \times \dfrac{9}{5}$	

[a] N, cell number (CFU/g); N_0, initial cell number (CFU/g); t, time (h); μ, specific growth rate (h^{-1}); t_d, doubling time (h).
[b] k, rate constant (h^{-1}); D, decimal reduction time (h) at a constant temperature; E_a, activation energy (kcal/mol); T_1T_2, reference temperature and test temperature (Kelvin); z, degrees required to change D value by a factor of 10.
[c] Do, rate constant (h^{-1}); Ds, dose (grays).

the cell number at any time by 10-fold, although at extended times, the population from the lower inoculum may reach the same final number. Because the instantaneous specific growth rate (μ) and time are in the power function of the equation, they have more marked effects on N. Consider a food where $N_0 = 1 \times 10^4$ CFU/g and $\mu = 0.2$ h^{-1} at 37°C. After 24 h, the cell number would be 1.2×10^6 CFU/g. Reducing the initial number by 10-fold will reduce the number after 24 h 10-fold to 1.2×10^5 CFU/g. However, reducing the temperature from 37 to 7°C has a much more profound effect. If one makes the simplifying assumption that the growth rate decreases two-fold with every 10°C decrease in temperature, then μ will be decreased eightfold to 0.025 h^{-1} at 7°C. When equation 1a is solved using these values (i.e., $N = 10^4 e^{0.025 \times 24}$), then N at 24 h is 1.8×10^4 CFU/g. Both time and temperature have much greater influence over the final cell number than does the initial microbial load.

Equation 3a can be used to determine how long it will take a microbial population to reach a certain level. Consider the case of ground meat manufactured with an N_0 of 1×10^4 CFU/g. How long can it be held at 7°C before reaching a level of 10^8 CFU/g? According to equation 3, $t = [2.3(\log 10^8/10^4)]/0.025$ or 368 h.

Food microbiologists frequently use doubling times (t_d) to describe growth rates of foodborne microbes. The relationship between t_d and μ is more obvious if equation 2a is written using natural logs (i.e., $\ln[N/N_0] = \mu \Delta t$) and solved for the condition where t is equal to t_d and N is equal to $2N_0$. Since the natural log of 2 is 0.693, the solution for equation 2a is $0.693/\mu = t_d$ (equation 4a). The average rate constant k, defined as the number of generations per unit time (i.e., $1/t_{gen}$), is also used by applied microbiologists. The instantaneous growth rate constant is related to k by the equation $\mu = 0.693k$. Both rate constants characterize populations in the exponen-

tial phase of growth. Some typical specific growth rates and doubling times are given in Table 1.2.

Death Kinetics

The killing of microbes by energy input (equations 1b, 1c), acid, bacteriocins, and other lethal agents is often governed by first-order kinetics. If one knows the initial microbial number, the first-order rate constant, and the time of exposure, one can predict the number of viable cells remaining. In food microbiology, the D value (decimal reduction time; amount of time required to reduce N_0 by 90% at a constant temperature) is the most frequently used kinetic constant. The use of D values in thermobacteriology is covered in more depth in chapter 32. D values are inversely proportional to the rate constant k as shown in equation 4b. Both D and k values are defined for a given temperature. The relationship between k and T is related to the activation energy E_a as determined by the Arrhenius equation, $k = s^{-E_a/RT}$, where s is the frequency constant, R is the ideal gas constant, and T is degrees Kelvin. In thermobacteriology, the relationship between D and T is given by the z value. The z value is defined as the number of degrees Fahrenheit required to change the D value by a factor of 10. The z value is related to the E_a

Table 1.2 Representative specific growth rates and doubling times of microorganisms

Microorganism	μ (h^{-1})	t^d (h)
Bacteria		
Optimal conditions	2.3	0.3
Limited nutrients	0.20	3.46
Psychrotroph, 5°C	0.023	30
Molds, optimal	0.1–0.03	6.9–23

by the equation $z = 2.3RT_1T_2/E_a \times (9/5)$, where T_1 and T_2 are actual and reference temperatures. A z value of 18°F equals an E_a of about 40 kcal/mol.

MICROBIAL PHYSIOLOGY AND METABOLISM

The Second Law of Thermodynamics dictates that all things progress to the state of maximum randomness in the absence of energy input. Since life is an ordered process, all living things must generate energy. Foodborne bacteria do this by oxidizing reduced compounds. Oxidation only occurs in a chemical couple where the oxidation of one compound is linked to the reduction of another. In the case of aerobic bacteria, the initial carbon source, glucose, is oxidized to carbon dioxide, oxygen is reduced to water, and 38 ATP are generated. Most of the ATP is generated through oxidative phosphorylation in the electron transport chain. In oxidative phosphorylation, the energy of the electrochemical gradient generated when oxygen is used as the terminal electron acceptor drives the formation of a high-energy bond between inorganic phosphate and an adenine nucleotide. Anaerobic bacteria, which lack functional electron transport chains, must reduce an internal compound through the process of fermentation and generate only 1 or 2 mol of ATP per mol of hexose catabolized. In this case, ATP is formed by substrate-level phosphorylation and the phosphate group is transferred from an organic compound to the adenine nucleotide.

Glycolytic Pathways—Carbon Flow and Substrate Level Phosphorylation

Embden-Meyerhof-Parnas Pathway

The most commonly used pathway for glucose catabolism (glycolysis) is the Embden-Meyerhof-Parnas (EMP) pathway (Fig. 1.2). In many organisms, the pathway is bidirectional (i.e., amphibolic) and synthesizes glucose, glycogen, or starch. The overall rate of glycolysis is regulated by the activity of phosphofructokinase. This enzyme converts fructose-6-phosphate to fructose-1,6-bisphosphate. Phosphofructokinase activity is subject to allosteric regulation, where the binding of AMP or ATP at one site inhibits or stimulates (respectively) the phosphorylation of fructose-6-phosphate at the enzyme's active site. Fructose-1,6-bisphosphate activates lactate dehydrogenase (see below) so that the flow of carbon to pyruvate is tightly linked to the regeneration of NAD when pyruvate is reduced to lactic acid.

Another key enzyme of the EMP pathway is aldolase. The ultimate fermentation end products generated by the catabolism of pentoses and hexoses are partially determined by which enzyme converts the sugars to smaller units. Aldolase cleaves one molecule of fructose-1,6-bisphosphate to two three-carbon units: dihydroxyacetone phosphate and glyceraldehyde-3-phosphate. Other glycolytic pathways use keto-deoxyphosphogluconate (KDPG) aldolase to make two three-carbon units or phosphoketolase to produce one two-carbon compound and one three-carbon unit. Substrate-level phosphorylation generates a net gain of two ATP when 1,3-diphosphoglycerate and phosphoenolpyruvate donate phosphoryl groups to ADP.

Entner-Doudoroff Pathway

The Entner-Doudoroff pathway is an alternate glycolytic pathway that yields one ATP per molecule of glucose and diverts one three-carbon unit to biosynthetic pathways. In aerobes that use this pathway, such as *Pseudomonas* species, the difference between forming one ATP by this pathway versus the two ATP formed by the EMP pathway is inconsequential compared to the 34 ATP formed from oxidative phosphorylation. In the Entner-Doudoroff pathway, glucose is converted to 2-keto-3-deoxy-6-phosphogluconate. The enzyme KDPG aldolase cleaves this to one molecule of pyruvate (directly, without the generation of an ATP) and one molecule of 3-phosphoglyceraldehyde. The 3-phosphoglyceraldehyde is then catabolized by the same enzymes used in the EMP pathway with the generation of one ATP by substrate-level phosphorylation using phosphoenol pyruvate as the phosphoryl group donor.

Heterofermentative Catabolism

Heterofermentative bacteria, such as *Leuconostoc* and some lactobacilli, have neither aldolases nor KDPG aldolase. The heterofermentative pathway is based on pentose catabolism. The pentose can be obtained by transport into the cell or by intracellular decarboxylation of hexoses. In either case, the pentose is converted to xylulose-5-phosphate with ribulose-5-phosphate as an intermediate. The xylulose-5-phosphate is cleaved by phosphoketolase to a glyceraldehyde-3-phosphate and a two-carbon unit which can be converted to acetaldehyde, acetate, or ethanol. Although this pathway yields only one ATP, it offers cells a competitive advantage by allowing them to utilize pentoses which homolactic organisms cannot catabolize.

Homofermentative Catabolism

Homofermentative bacteria in the genera *Lactococcus* and *Pediococcus* and some *Lactobacillus* species produce lactic acid as the sole fermentation product. The

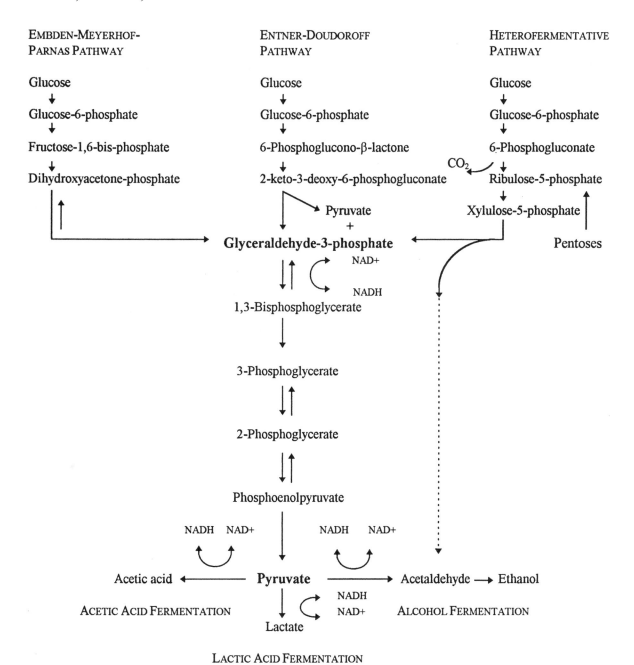

Figure 1.2 Major catabolic pathways used by foodborne bacteria.

EMP pathway is used to produce pyruvate, which is then reduced by lactate dehydrogenase, forming lactic acid and regenerating NAD. Some *Lactobacillus* species, such as *Lactobacillus plantarum* (137), are characterized as "facultatively heterofermentative." Hexoses are their preferred carbon source and are metabolized by the homofermentative pathway. If only pentoses are available, the cell shifts to a heterofermentative mode. When grown at low hexose concentrations, these bacteria do not make enough fructose-1,6-bisphosphate to activate their lactate dehydrogenase. This also causes them to shift to heterofermentative catabolism.

The Tricarboxylic Acid Cycle
The tricarboxylic acid (TCA) cycle links glycolytic pathways to respiration. It generates $NADH_2$ and $FADH_2$ as substrates for oxidative phosphorylation while providing additional ATP through substrate level phosphorylation.

With each turn of the TCA cycle, 2 pyruvate + 2 ADP + 2 FAD + 8 NAD → 6 CO_2 + 2 ATP + 2 $FADH_2$ + 8 NADH. Succinic acid, oxaloxlate, and α-ketoglutarate link the TCA cycle to amino acid biosynthesis. The TCA cycle is used by all aerobes, but some anaerobes lack all of the enzymes required to have a functional TCA cycle.

The tricarboxylic acid cycle is also the basis for two industrial fermentations important to the food industry. The microbial production of citric acid by *Aspergillus niger* and *Aspergillus wenti* and of glutamic acid by *Corynebacterium glutamicum* depends on mutations that affect α-ketoglutarate dehydrogenase and cause TCA intermediates to accumulate.

Aerobes, Anaerobes, the Regeneration of NAD, and Respiration

The flow of carbon to pyruvate always consumes NAD, which must be regenerated for continued catabolism. When $NADH_2$ is oxidized to NAD, another compound must be reduced, i.e., serve as an electron acceptor. Aerobes having electron transport chains use molecular oxygen as the terminal electron acceptor during oxidative phosphorylation. As electrons travel down the electron transport chain, protons are pumped out, forming a proton gradient across the membrane. This proton gradient can be converted to ATP by the action of the F_0F_1 ATPase. Oxidation of $NAD(P)H_2$ yields three ATP. Oxidation of $FADH_2$ yields two ATP. ATP and NADH are thus, in a sense, interconvertible. Sulfur and nitrite can also serve as terminal electron acceptors in "anaerobic respiration."

Anaerobes, in contrast, have a fermentative metabolism. Fermentations oxidize carbohydrates in the absence of an external electron acceptor. The final electron acceptor is an organic compound produced during carbohydrate catabolism. In the most obvious case, pyruvic acid is the terminal electron acceptor when it accepts an electron from NADH and is reduced to lactic acid. Some anaerobes are aerotolerant and can generate more energy in the presence of low levels of oxygen than in its absence. For example, some lactic acid bacteria have inducible NADH oxidases that regenerate NAD by reducing molecular oxygen to H_2O_2 (137). This spares the use of pyruvate as an electron acceptor and allows it to be converted to acetic acid with the generation of an additional ATP. These lactic acid bacteria have an NADH peroxidase which detoxifies the H_2O_2. Obligate anaerobes cannot detoxify H_2O_2 and die when exposed to air.

Bioenergetics

All catabolic pathways generate energy with which the bacteria can perform useful work. Energy generation and utilization are critical to microbial life. Several excellent reviews (68, 88) and books (63, 104) on bioenergetics provide depth and clarity. The preceding section on microbial biochemistry stressed the role of ATP in the cell's energy economy, but transmembrane gradients of other compounds play an equally important role. Transmembrane gradients release energy when one compound moves from high concentration to low concentration (i.e., "with the gradient"). This energy can be coupled to the transport of a second compound from a low concentration to a high concentration (i.e., "against the gradient").

According to Mitchell's chemiosmotic theory, the proton motive force (PMF) has two components. An electrical component, the membrane potential ($\Delta\Psi$), represents the charge potential across the membrane. The transmembrane pH gradient (ΔpH) is the second component. Together, these constitute the PMF, as stated by the equation PMF = $\Delta\Psi - z\Delta pH$. In this equation, z is equal to 2.3 RT/F, R is the gas constant, T is the absolute temperature, and F is the Faraday constant. The factor z converts the pH gradient into millivolts and has a value of 59 mV at 25°C. The PMF is defined as being interior negative and alkaline, resulting in a negative value. (In the equation above, $z\Delta pH$ is not being subtracted from $\Delta\Psi$, but it makes this negative term more negative.) There also is some interconversion of the $\Delta\Psi$ and the ΔpH components of PMF. If, for example, the ΔpH component decreases when an organism is transferred to a more neutral environment, the cell compensates by increasing $\Delta\Psi$ so that the total PMF remains relatively constant. PMF values can be as high as –200 mV for aerobes, or in the range of –100 mV to –150 mV for anaerobes. Protein phosphorylation, flagellar synthesis and rotation, reversed electron transfer, and protein transport use PMF as an energy source (63).

PMF is generated by several mechanisms (Fig. 1.3). The translocation of protons down the electrochemical gradient during respiration generates a proton gradient. The oxidation of NADH is accompanied by the export of enough protons to make three ATP. The proton gradient is converted to ATP by the F_0F_1 ATPase when it is driven in the direction of ATP synthesis (88). The bacterial F_0F_1 ATPase is nearly identical to chloroplast and mitochondrial F_0F_1 ATPases.

The F_0F_1 ATPase is reversible. Aerobes use it to convert PMF to ATP. In anaerobes, it converts ATP to PMF. Maintaining internal pH homeostasis may be the principal role of the F_0F_1 ATPase in anaerobes (68). Internal pH not only influences the activity of cytoplasmic enzymes, but it also regulates the expression of genes responsible for functions ranging from amino acid degradation to virulence (113). Anaerobes deacidify their cytoplasms

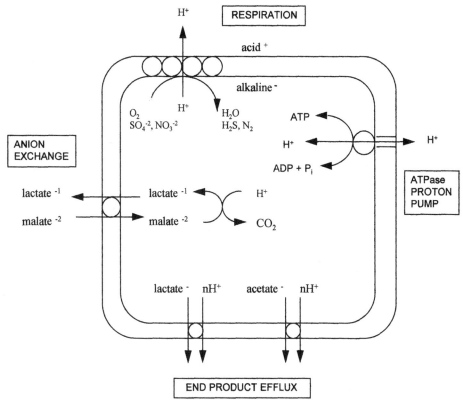

Figure 1.3 Proton motive force can be generated by respiration, ATP hydrolysis, end-product efflux, or anion exchange mechanisms. Modified from reference 137.

using the F_0F_1 ATPase to pump protons out. The proton pumping is driven by ATP hydrolysis. Some of the energy lost from ATP hydrolysis can be recovered if the resultant proton gradient is used to perform useful work, such as transport (see below). Most bacteria maintain their internal pH (pH_i) near neutrality, but lactic acid bacteria can tolerate lower pH_i values and expend less ATP on pH homeostasis. Acid-induced death is the direct result of an excessively low pH_i.

Given their limited capacity for ATP generation, it is not unexpected that some lactic acid bacteria can also generate ΔpH by ATP-independent mechanisms. The electropositive excretion of protons with acidic end products (93) has been demonstrated for lactate and acetate. For example, under some conditions, *Lb. plantarum* excretes three protons per molecule of acetate, thus sparing one ATP (137). The antiport (see below) exchange of precursor and product in anion degrading systems, such as the malate^{2-}:lactate^{1-} exchange of the malolactic fermentation, might contribute to the generation of $\Delta\Psi$ (88).

Bacteria have evolved several mechanisms to achieve similar ends. The accumulation of compounds against a gradient (i.e., transport) is work and requires energy. In the case of primary transport systems and group translocation, this work is done by phosphoryl group transfer (Fig. 1.4). Secondary transport systems are fueled by the energy stored in the gradients which make up the PMF.

Cell Signaling and Quorum Sensing

Introduction

The explosion of papers on cell-to-cell communication gives new perspectives to many food microbiology issues. Many reviews of quorum sensing (6, 8, 95, 128) and signal transduction (64, 72, 75, 119, 131) provide the details. Both mechanisms turn on genes that would be superfluous to isolated cells, but which are advantageous to large populations. Cellular communication is by two main mechanisms:

- The two-component signal transduction system is composed of a membrane-spanning histidine kinase sensor and a response regulator protein. Three-component systems are used extensively by lactic acid bacteria, which can excrete small, often antimicrobial, peptides as the

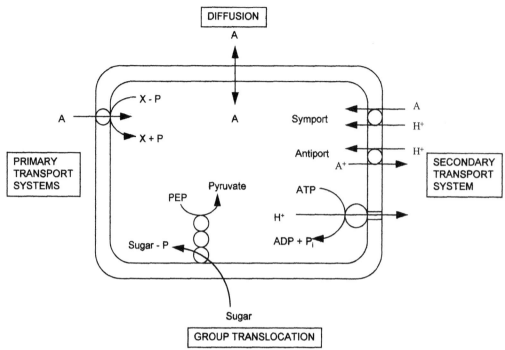

Figure 1.4   Transport can be at the direct expense of high-energy phosphate bonds or can be linked to the proton gradient of the proton motive force.

autoinducer. Although signal transduction is used for quorum sensing, not all signal transduction is related to quorum sensing.

- The quorum sensing homoserine lactone system is based on autoinducer diffusion across membranes. When a threshold concentration is reached, the autoinducers interact with intracellular regulator proteins to modulate gene transcription. Strictly speaking, "quorum sensing controlled behaviors are those that only occur when bacteria are present in high cell numbers" (6).

Signal Transduction

The two-component "signal transduction" system consists of a histidine kinase receptor and a response regulator. An extracellular "trigger" molecule binds at the N terminus on the "out" side of an integral membrane protein kinase. The protein kinase transduces the signal across the membrane through a conformational change to increase the kinase activity at its C-terminal cytoplasmic side. The increased kinase activity phosphorylates a response regulator protein. The phosphorylated response regulator protein can modulate gene expression, enzymatic activity, or flagellar rotation (64). Signal molecules of gram-positive bacteria are usually small posttranslationally processed peptide signals. Lactic acid bacteria use a three-component system. The signal is a small peptide coded for by structural genes on the operon. It

is excreted, sometimes after posttranslational modification. When the peptide reaches a certain extracellular concentration, it binds to a specific receptor, transduces a signal to phosphorylate a response regulator, and upregulates its own synthesis (119, 131). This has been studied extensively for the lantibiotic nisin (72).

Quorum Sensing

"Quorum sensing" is fundamentally different from signal transduction. Rather than acting on a transmembrane protein, the signal compound diffuses across the membrane and binds to a regulator protein that affects transcription of a regulon(s) to elicit a cellular response (7, 144). The signal compound is made by a gene product of the same regulon and, hence, is autoinduced (6).

In gram-negative bacteria, N-acyl homoserine lactones (abbreviated in the literature as both AHLs and HSLs) generally act as signaling molecules (132). These are referred to as AI-1 (autoinducer 1) and are synthesized by AHL synthase, encoded by the *luxI* gene. AHLs obtain their species specificity from their differing acyl side chains. AI-2, originally thought to be unique to *Vibrio*, is a furanosyl-bromide diester product of LuxS. LuxS is encoded by *luxS*, which is also involved in the synthesis of the newly discovered AI-3. At high concentration, these molecules bind to and activate a transcriptional activator which in turn induces target gene expression.

In most bacteria, signaling molecules are at their greatest concentration during stationary phase; however, for *E. coli* O157:H7 and *Salmonella* serovar Typhimurium, quorum sensing is critical for regulating behavior in the prestationary phase of growth (132). Moreover, unlike other gram-negative bacteria, the inhibitory activity does not require transcription or translation to be effective.

Examples of Cell Signaling in Foodborne Microbes

Given the current excitement about quorum sensing, it is tempting to speculate that it has a role in spoilage (66). However, many studies use a positive bioluminescence response in the *Vibrio harveyi* or other detection systems as evidence of quorum sensing, without actually isolating the autoinducing compound or identifying a regulated phenotype. Such signals have been detected in a variety of organisms in a variety of foods. Cloak et al. (34) used an AI-2 detecting system to find positive signals from *Campylobacter*, *Salmonella*, and *E. coli* O157:H7 in broth, chicken soup, and milk. All of these produced a response, but in amounts that varied by 1,000-fold. Conceptually similar studies (60) detected positive signals in bean sprouts, vacuum-packed beef, fish fillet, and turkey where the microbiota contains *Enterobacteriaceae*, *Pseudomonas*, *Aeromonas*, *Shewanella*, and *Photobacterium*. Other foods, and preservatives such as sodium benzoate and sodium propionate, could inhibit the bioluminescence response (84).

A more rigorous study casts doubt on the linkage of quorum sensing and spoilage. Bruhn and coworkers (25) used thin-layer chromatography and mass spectrometry to demonstrate that the bioluminescence-inducing compound in five samples of commercial vacuum-packaged meat was *N*-3-oxo-hexanoyl homoserine lactone. However, meat spoiled at the same rate whether it was inoculated with wild-type strains or AHL synthase knockout mutants. Furthermore, addition of halogenated furones (quorum-sensing inhibitors) did not influence spoilage, leading to the conclusion that quorum sensing does not regulate spoilage in vacuum-packed meat.

There are many phenotypes where a role for cell signaling has been established at a genetic level (Table 1.3). Many of these effects are pleiotropic. The discovery of autoinduction in bacteriocin-producing lactic acid bacteria explains the hitherto puzzling loss of bacteriocin production by cells which still have the requisite genes (and the ability to coax it back by adding supernatants from normally producing cultures) and the fact that some strains produce bacteriocins on agar but not liquid media.

Caveats

Fuqua and Greenberg (57) caution that communication requires not only the sending of a signal, but also receiving and acting on the information. AI-2 certainly meets these criteria in *V. harveyi*, but *V. harveyi*'s ability to receive and respond to AI-2 produced by other species does not prove that the other species hear and respond

Table 1.3 Examples of quorum sensing in food microbiology

Organism	Signal system	Phenotype	Genetic involvement	Reference(s)
L. monocytogenes	Signal transduction, two component	Growth at low temperature and high salt concentration	*kdpE*, *orfX* (RsbQ homolog)	24
		Virulence	*pclA*, *hly*, *actA*, *inlA*, host *srcFR* (encodes kinase that acts on actin)	138
S. aureus	Signal transduction, two component	Pleiotropic effects on cytotoxins, enterotoxins, proteases	*agr* (accessory gene regulator locus) activated by RAP signaling peptide	73, 106, 114
Lactic acid bacteria	Signal transduction, three component	Bacteriocin production	*cln* locus in *Carnobacterium piscicola*, *pln* locus in *Lb. plantarum*, *nis* locus in *Lactobacillus lactis*	25, 72, 75
S. enterica serovar Typhimurium	Quorum sensing, AI-2	"Fitness" in chickens	Pleiotropic effect of *luxS*	22
Enteropathogenic *E. coli*	Quorum sensing, AI-2	Flagella, formation of attachment and effacement lesions	Pleiotropic (?) effect of *luxS*, *qse* (quorum sensing regulator), *ee* (enterocyte effacement) locus	114, 125
V. cholerae	Quorum sensing, AI-1, AI-2, other	Virulence	Activates virulence regulon by repressing *hapR*	96

to the signal. One must identify the target of AI-2 in the organism that produces it before attributing its role to quorum sensing.

Winzer et al. (144) propose that four strict criteria must be met to confirm quorum sensing in a given organism:

- Production of the signal compound is specific to an event.
- The signal accumulates extracellularly and is recognized by a specific receptor.
- A specific response is generated after the signal reaches a threshold concentration.
- The response goes beyond metabolism or detoxification of the signal compound.

Furthermore, they suggest that the widespread production of AI-2 by many microbes may be related to its role as a by-product of the activated methyl cycle. Ribosyl homocysteine is cleaved by LuxS or a homologous ribosyl homocysteinase to homocysteine and (after another conversion) methyl hydroxyfuranone. This compound is toxic to the cell and can be converted to AI-2, which is excreted (34). Thus, in some cells, AI-2 production may be a mechanism for excreting a toxic substance rather than a means of cellular communication.

Biofilms

Cells in biofilms are more resistant to heat, chemicals, and sanitizers. *L. monocytogenes* reduction by treatment with a combination of sodium hypochlorite and heat is approximately 100 times lower in biofilms than for free cells (55). Increased chemical resistance is attributed to the very slow growth rates of cells in biofilms and not a diffusional barrier created by the biomatrix (37). Indeed, cells in the nutrient-deplete interior of the microcolony may be in the "viable but nonculturable" (VNC) state. Reviews on biofilms in the food industry (29, 149) pragmatically emphasize the importance of cleaning prior to sanitizing process equipment. True biofilms take days to weeks to reach equilibrium. Proper cleaning insures that the cells in the nascent biofilm can be reached by sanitizers. Trisodium phosphate is effective towards *E. coli* O157:H7, *Campylobacter jejuni*, and *Salmonella* serovar Typhimurium free and biofilm cells (130). Newer methods for control of biofilms include superhigh magnetic fields, ultrasound treatment, and high pulsed electric fields (76). The design of equipment with smooth, highly polished surfaces also impedes biofilm formation by making the initial adsorption step more difficult.

To suggest that more research is needed about biofilms would be a gross understatement. Although planktonic (i.e., free, single) cells are easy to study, and pure culture is the foundation of microbiology as we know it, "in all natural habitats studied to date bacteria prefer to reproduce on any available surface rather than in the liquid phase" (29). Furthermore, biofilms exist as communities of microbial species embedded in a biopolymer matrix. Biofilms are heterogeneous in time and space, frequently appearing as collections of mushroom-shaped microcolonies with moving water channels between them (29, 149). Foodborne pathogens *E. coli* O157:H7, *L. monocytogenes*, *Y. enterocolitica*, and *C. jejuni* form biofilms on food surfaces and food contact equipment, leading to serious health problems and economic losses due to spoilage of food (76).

Biofilm formation is a multistep process. First, the solid surface undergoes a conditioning process that allows cells to be adsorbed by weak reversible electrostatic forces. Biopolymer formation follows rapidly and anchors these cells. The synthesis of the matrix polymer may be upregulated by quorum sensing when the local concentration of cells increases by adsorption. The microcolonies have defined boundaries which allow fluid channels to run through the biomatrix. This requires higher-level differentiation, quorum sensing, or some kind of cell-to-cell communication to prevent undifferentiated growth from filling in these channels which bring nutrients and remove wastes. Costerton (37) paints a vivid picture of this system, concluding that "the highly structured biofilm mode of growth provide[s] bacteria with a measure of homeostasis, a primitive circulatory system, a framework for the development of...specialized cell functions,...[and] protection from antimicrobial agents."

In *Pseudomonas aeruginosa*, transcription of alginate biosynthetic genes is activated by response regulators which increase synthesis of a sigma-like factor which regulates transcription of the *algD* promoter (20). The *algD* promoter regulates virtually all of the alginate biosynthetic operon. This system also contains an alginate lyase that disperses cells when the environment threatens communal life. Quorum sensing may also be involved, since *P. aeruginosa* mutants lacking the signal molecule 3-oxo-C_{12}-HSL form a biofilm that is thinner and lacks the three-dimensional structure of the parent (39).

Homeostasis and Hurdle Technology

Instead of setting one parameter to the extreme limit for growth, hurdle technology "deoptimizes" a variety of factors (80). For example, a limiting water activity of 0.85 or a limiting pH of 4.6 prevents the growth of foodborne pathogens. Hurdle technology might obtain similar inhibition at pH 5.2 and a water activity of 0.92. Hurdle technology assaults multiple homeostatic processes (58). In acidic conditions, cells use energy to pump out protons. In low-water-activity environments, cells use energy to accumulate compatible solutes. Maintenance of membrane fluidity also requires energy. When the energy needed for biosynthesis is diverted into maintenance of homeostasis, cell growth is inhibited. When homeostatic energy demands exceed the cell's energy-producing capacity, the cell dies. Hurdle technology can encompass the use of antimicrobial agents (e.g., nisin) and technology including the use of ozone and the application of irradiation in conjunction with shifts in pH and water activity to inhibit microbial growth (87, 108).

LIMITATIONS OF CLASSICAL MICROBIOLOGY

Limitations of Plate Counts

All methods based on the plate count and pure culture microbiology have limitations. The "plate count" is based on the assumptions that every cell forms one colony and that every colony originates from one cell. The ability of a given cell to form a colony depends on many factors including the physiological state of the cell, the medium used for enumeration, the incubation temperature and time, and the number of cells present. Table 1.4 (83) illustrates these points by providing *D* values at 55°C for *L. monocytogenes* with different thermal histories (heat-shocked for 10 min at 80°C or not heat treated) on selective (McBride's) or nonselective (TSAY [Tryptic Soy Agar + 0.5% yeast extract]) media under aerobic or anaerobic atmospheres. The *D* values for *E. coli* O157:H7 are affected similarly (100). Injured cells and cells that are "viable, but nonculturable" pose special problems as discussed below.

Injury

Microorganisms may be injured by sublethal levels of stressors such as heat, radiation, acid, or sanitizers. Freezing at –20°C for 24 h can injure ~99% of an *E. coli* O157:H7 population (62). The type of food

influences both injury and subsequent recovery. Injury is characterized by decreased resistance to selective agents or by increased nutritional requirements (65). Molecular events associated with injury are complex and are still being defined. Injury is influenced by time, temperature, concentration of the injurious agent, strain of the target pathogen, and experimental methodology. For example, while a standard sanitizer test indicates that several sanitizers kill listeriae, viable cells can be recovered using listeria repair broth (121). The degree of injury decreases and the extent of lethality increases as the time and sanitizer concentration increase. For example, *L. monocytogenes* cells grown at <28°C undergo a 3- to 4-log kill when exposed to 52°C. However, if grown at 37 or 42°C, there is little death, but 2 to 3 logs of injury when heated to 52°C (129).

Data illustrating injury are shown in Fig. 1.5. Cells subjected to a mild stress are plated on a rich nonselective medium and a selective medium containing 6% NaCl. The difference between the numbers of colonies on each medium represents injured cells. (If 10^7 CFU/ml of a population are enumerated on the nonselective medium and 10^4 CFU/ml can grow on the selective medium, then 10^3 CFU/ml are injured.) Specialized enumeration media are often required because growth and gene expression of an organism cultured on a nonselective medium can be quite different when cultured on selective medium (26, 59, 71, 103).

Microbial injury is important to food safety for several reasons. (i) If injured cells are classified as dead during the determination of thermal resistance, the thermal sensitivity will be overestimated and the *D* values will be errantly low. (ii) Injured cells that escape detection at the time of postprocessing sampling may repair before consumption and present a safety or spoilage problem. Heat treatments can be optimized to consider injury of surviving bacteria (86). (iii) The "selective agent" may be common food ingredients such as salt, organic acids, humectants, or even suboptimal temperature.

Injury in spores is even more complex. The many biochemical steps of sporulation, germination, and outgrowth explained in chapter 2 provide a plethora of targets which can be damaged. Thermal injury is the most well-studied form of injury and can occur during extrusion as well as during conventional thermal processing (82). Spores can also be injured by chemicals and irradiation. DNA, RNA, enzymes, and membranes may be damaged during thermal injury. Irradiation-induced injury of spores is primarily caused by single-strand breaks in DNA and is also manifested by increased sensitivity to pH, salt, and heat (49). *Rec* systems can repair

Table 1.4 Influence of thermal history and enumeration protocols on experimentally determined *D* values at 55°C for *L. monocytogenes* (83)

	D_{55} values (min)			
	TSAY medium		McBride's medium	
	Heat shock		Heat shock	
Atmosphere	+	−	+	−
Aerobic	18.7	8.8	9.5	6.6
Anaerobic	26.4	12.0	No growth	

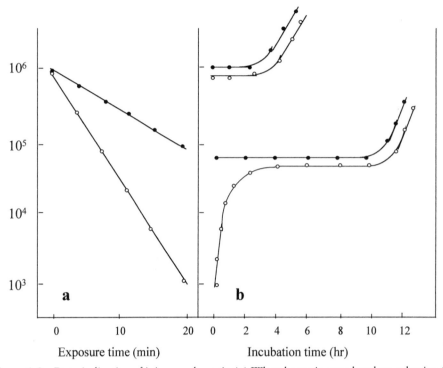

Exposure time (min)   Incubation time (hr)

Figure 1.5   Data indicative of injury and repair. (a) When bacteria are plated on selective (○) or nonselective (●) media during exposure to some stressor (e.g., heat), the decrease in CFU on a nonselective medium represents the true lethality, while the difference between the values obtained on each medium is defined as "injury." (b) During "repair," resistance to selective agents is regained, and the value obtained on the selective medium approaches that of the nonselective medium. Unstressed controls are shown at the top of panel b. Modified and redrawn from reference 27.

injury caused by single-strand DNA breaks. Radiation-induced heat sensitivity is caused by damage to the spore cortex peptidoglycan and can last for weeks to months (51, 52).

Vegetative cells injured by heat, freezing, and detergents usually leak intracellular constituents from damaged membranes. Membrane integrity is reestablished during repair (13). Osmoprotectants can prevent or minimize freeze injury in *L. monocytogenes* (45, 46). Oxygen toxicity also causes injury. Recovery of injured cells is often enhanced by adding peroxide detoxifying agents such as catalase or pyruvate to the recovery medium or by excluding oxygen through the use of anaerobic incubation conditions or adding Oxyrase® (which enzymatically reduces oxygen) to the recovery medium. "Repair" is the process by which cells recover from injury. Repair requires de novo synthesis of RNA and protein (23) and often appears as an extended lag phase. The extent and rate of repair are influenced by environmental factors. *L. monocytogenes* organisms injured at 55°C for 20 min start to repair immediately at 37°C and are completely recovered by 9 h (92). Heat-injured *L. monocytogenes*

organisms do not replicate in milk at 4°C. Repair at 4°C is delayed for 8 to 10 days, and full recovery requires 16 to 19 days (38).

Viable but Nonculturable

Salmonella, *Campylobacter*, *Escherichia*, *Shigella*, and *Vibrio* species, and other genera, can exist in a state where they are viable but cannot be cultured (109). This differentiation of vegetative cells into a dormant "viable but nonculturable" (VNC) state is a survival strategy for many nonsporulating species. The VNC state is morphologically different from that of the "normal" vegetative cell. During the transition to the VNC state, rod-shaped cells shrink and become small spherical bodies which are not spores (71, 105). Changes in membrane fatty acid composition occur in *Vibrio* during entry into the VNC state (40, 146). It takes from 2 days to several weeks for an entire population of vegetative cells to become VNC (105).

The viability of VNC cells can be demonstrated through cytological methods (5, 69). The structural integrity of the bacterial cytoplasmic membrane can be determined by the permeability of cells to fluorescent nucleic acid stains (42). Bacteria with intact cell membranes stain fluorescent green, whereas bacteria with damaged membranes stain fluorescent red (21). Iodonitrotetrazolium violet can also identify VNC cells. Respiring cells reduce iodonitrotetrazolium violet to form an insoluble compound detectable by microscopic observation (105). Unculturable (<10 CFU/ml) *Salmonella enteritis* populations starved at 7°C have been quantified as 10^4 viable cells per ml using these methods (31). Experimental data (105) in Fig. 1.6 illustrate a *Vibrio vulnificus* population that appears to have died off (i.e., gone through a 6-log reduction in CFU/ml) although >10^5 per ml are quantified as viable.

VNC cells can also be identified by their substrate-responsive metabolism. When VNC cells are incubated with yeast extract (as a nutrient) and nalidixic acid or ciprofloxacin (inhibitor of cell division), their elongation can be quantified microscopically. This method detects VNC cells of *L. monocytogenes* (12). Other methods have been used to demonstrate the VNC state for *Streptococcus faecalis*, *Micrococcus flavus*, and *B. subtilis* (28).

Powerful new methods for detecting VNC cells are being developed as understanding of bacteria at the molecular level and techniques for genetic manipulation advance. The detection of specific RNA by reverse transcriptase PCR is one such method (115). Alternatively, reporter genes via green fluorescent protein-tagged and Lux-tagged methods are used to identify cells synthesizing proteins (32). A buoyant-density gradient method also allows detection of viable and VNC cells (145).

Because the VNC state is most often induced by nutrient limitation in aquatic environments, it might appear irrelevant to the nutrient-rich milieu of food. However, the VNC state can also be induced by changes in salt concentration, exposure to hypochlorite, and shifts in temperature (94, 112). *V. vulnificus* populations shifted to refrigeration temperatures are still lethal to mice when the entire population of 10^5 viable cells becomes nonculturable (<0.04 CFU/ml) (110). The bacteria resuscitate in the mice and can be cultured postmortem. *E. coli* and *Salmonella* serovar Typhimurium entered a VNC state following chlorination of waste water (111). Although the pathogens could not be resuscitated, they may still present a public health hazard. *Salmonella* serovar Typhimurium DT104 held at 5°C for 235 days entered into and was recovered from the VNC state (61). Temperature changes can induce the VNC state. When starved at 4 or 30°C for more than a month, *V. harveyi* become VNC at 4°C but remained culturable at 30°C. In contrast, *E. coli* cells entered the VNC state at 30°C, but die at 4°C (50). Foodborne pathogens in nutritionally rich media can become VNC when shifted to refrigerated temperature (91, 105, 110). This has chilling implications for the safety of refrigerated foods.

Resuscitation of VNC cells is demonstrated by an increase in culturability that is not accompanied by an increase in the total cell numbers. The return to culturability can be induced by temperature shifts or the gradual return of nutrients. The same population of bacteria can go through multiple cycles of the VNC and culturable states in the absence of growth (105). No one specific

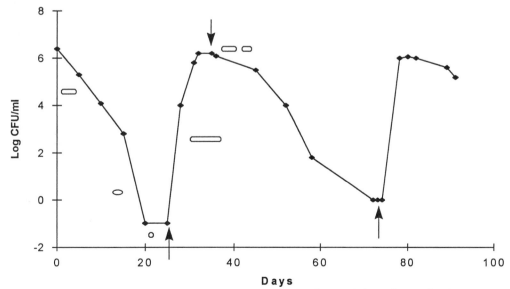

Figure 1.6 Data showing changes of plate count and cell morphology during development of the VNC state induced by temperature downshifts (at time 0 and ↓) and resuscitation by temperature upshifts (↑). Reprinted from reference 105 with permission.

factor can be identified as preventing the resuscitation of VNC cells. Amending media with catalase or sodium pyruvate restores culturability of VNC cells of *E. coli* O157 and *Vibrio parahaemolyticus* (97, 98), suggesting that the transfer of cells to nutrient-rich media initiates rapid production of superoxide and free radicals. Catalase hydrolyzes H_2O_2. Sodium pyruvate degrades H_2O_2 through decarboxylation of the α-keto acid to form acetic acid and CO_2 (47). VNC campylobacters have been resuscitated when injected into fertilized chicken eggs and incubated at 37°C for 48 h (133).

Increased awareness of the VNC state should lead to a reexamination of our concept of "viability," our dependence on "enrichment culture" to isolate pathogens, and our reliance on cultural methods to monitor microbes in the environment (4, 91, 120). The mechanisms of VNC formation, the mechanisms which make VNC cells resistant to environmental stress, and the mechanisms that signal resuscitation are largely unknown. The relationship between viability and culturability may need to be redefined and, indeed, has spawned extensive discussion (5, 69, 89, 107). Some investigators question the ability of bacteria to enter or resuscitate from a VNC state and caution blind acceptance of methods developed to determine viability (5, 16, 17). Clearly, more research is needed on "viable but nonculturable" foodborne pathogens.

CONCLUSION

Microbial growth in foods is a complex process governed by genetic, biochemical, and environmental factors. Much of what we "know" about foodborne microbes must be held with the detached objectivity required of an unproven hypothesis. Developments in molecular biology and microbial ecology will change or deepen our perspective about the growth of microbes in foods. Some of these developments are detailed in this book.

References

1. **Arnold, C. N., J. McElhanon, A. Lee, R. Leonhart, and D. A. Siegele.** 2001. Global analysis of *Escherichia coli* gene expression during the acetate-induced acid tolerance response. *J. Bacteriol.* **183:**2178–2186.

2. **Bang, I. S., B. H. Kim, J. W. Foster, and Y. K. Park.** 2000. *ompR* regulates the stationary-phase acid tolerance response of *Salmonella enterica* serovar Typhimurium. *J. Bacteriol.* **182:**2245–2252.

3. **Bang, I. S., J. P. Audia, Y. K. Park, and J. W. Foster.** 2002. Autoinduction of the *ompR* response regulator by acid shock and control of the *Salmonella enterica* acid tolerance response. *Mol. Microbiol.* **44:**1235–1250.

4. **Barer, M. R., L. T. Gribbon, C. R. Harwood, and C. E. Nwaguh.** 1993. The viable but non-culturable hypothesis and medical bacteriology. *Rev. Med. Microbiol.* **4:**183–191.

5. **Barer, M. R., and C. R. Harwood.** 1999. Bacterial viability and culturability. *Adv. Microb. Physiol.* **41:**93–137.

6. **Bassler, B. L.** 2002. Small talk: cell-to-cell communication in bacteria. *Cell* **109:**421–424.

7. **Bauer, W. D., U. Mathesius, and M. Teplitski.** 2005. Eukaryotes deal with quorum sensing. *ASM News* **71:** 129–135.

8. **Bayles, D. O., B. A. Annous, and B. J. Wilkinson.** 1996. Cold stress proteins induced in *Listeria monocytogenes* in response to temperature downshock and growth at low temperatures. *Appl. Environ. Microbiol.* **62:**1116–1119.

9. **Bearson, B. L., L. Wilson, and J. W. Foster.** 1998. A low pH-inducible, PhoPQ-dependent acid tolerance response protects *Salmonella typhimurium* against inorganic acid stress. *J. Bacteriol.* **180:**2409–2417.

10. **Bearson, S., B. Bearson, and J. W. Foster.** 1997. Acid stress responses in enterobacteria. *FEMS Microbiol. Lett.* **147:**173–180.

11. **Beltrametti, F., A. U. Kresse, and C. A. Guzman.** 1999. Transcriptional regulation of *esp* genes of enterohemorrhagic *Escherichia coli*. *J. Bacteriol.* **181:**3409–3418.

12. **Besnard, V., M. Federighi, and J. M. Cappelier.** 2000. Development of a direct viable count procedure for the investigation of VBNC state in *Listeria monocytogenes*. *Lett. Appl. Microbiol.* **31:**77–81.

13. **Beuchat, L. R.** 1978. Injury and repair of gram-negative bacteria with special consideration of the involvement of the cytoplasmic membrane. *Adv. Appl. Microbiol.* **23:**219–243.

14. **Board, R. G., D. Jones, R. G. Kroll, and G. L. Pettipher** (ed). 1992. Ecosystems: microbes: food. *J. Appl. Bacteriol.* **73:**1S–178S.

15. **Boddy, L., and J. W. T. Wimpenny.** 1992. Ecological concepts in food microbiology. *J. Appl. Bacteriol.* **73:** 23S–38S.

16. **Bogosian, G., N. D. Aardema, E. V. Bourneauf, P. J. Morris, and J. P. O'Neil.** 2000. Recovery of hydrogen peroxide-sensitive culturable cells of *Vibrio vulnificus* gives the appearance of resuscitation from a viable but nonculturable state. *J. Bacteriol.* **182:**5070–5075.

17. **Bogosian, G., P. J. L. Morris, and J. P. O'Neil.** 1998. A mixed culture recovery method indicates that enteric bacteria do not enter the viable but nonculturable state. *Appl. Environ. Microbiol.* **64:**1736–1742.

18. **Bonnet, M.** 2005. Acid tolerance response of *Listeria monocytogenes*: bioenergetics and mechanisms of resistance to the antimicrobial nisin. Ph.D. dissertation. The Graduate School, Rutgers, the State University of New Jersey, New Brunswick.

19. **Bonnet, M., and T. J. Montville.** 2005. Acid-tolerant *Listeria monocytogenes* persist in a model food system fermented with nisin-producing bacteria. *Lett. Appl. Microbiol.* **40:**237–242.

20. **Boyd, A., and A. M. Chakrabarty.** 1995. *Pseudomonas aeruginosa* biofilms: role of the alginate exopolysaccharide. *J. Ind. Microbiol.* **15:**162–168.

21. **Bouwer, P., and T. Abee.** 2000. Assessment of viability of microorganisms employing fluorescence techniques. *Int. J. Food Microbiol.* **55:**193–200.

22. Brandl, M. T., W. G. Miller, A. H. Bates, and R. E. Mandrell. 2005. Production of autoinducer-2 in *Salmonella enterica* serovar Thompson contributes to its fitness in chickens, but not in cilantro leaf surfaces. *Appl. Environ. Microbiol.* **71:**2653–2662.

23. Brock, T. D., and M.T. Madigan. 1988. *Biology of Microorganisms*, p. 793–795. Prentice Hall, Englewood Cliffs, N.J.

24. Brondsted, L., B. H. Kallipolitis, H. J. Ingmer, and S. Knochel. 2003. KdpE and a putative RsbQ homologue contribute to growth of *Listeria monocytogenes* at high osmolarity and low temperature. *FEMS Microbiol. Lett.* **219:**233–239.

25. Bruhn, J. B., A. B. Christensen, L. R. Flodgaard, K. F. Nielsen, T. O. Larson, M. Givskov, and L. Gram. 2004. Presence of acylated homoserine lactones (AHLs) of AHL-producing bacteria in meat and potential role of AHL in spoilage of meat. *Appl. Environ. Microbiol.* **70:**4293–4302.

26. Bull, M. K., M. M. Hayman, C. M. Stewart, E. A. Szabo, and S. J. Knabel. 2005. Effect of prior growth temperature, type of enrichment medium, and temperature and time of storage on recovery of *Listeria monocytogenes* following high pressure processing of milk. *Int. J. Food Microbiol.* **101:**53–61.

27. Busta, F. F. 1978. Introduction to injury and repair of microbial cells. *Adv. Appl. Microbiol.* **23:**195–201.

28. Byrd, J. J., H.-S. Xu, and R. R. Colwell. 1991. Viable but nonculturable bacteria in drinking water. *Appl. Env. Microbiol.* **57:**875–878.

29. Carpentier, B., and O. Cerf. 1993. Biofilms and their consequences, with particular reference to hygiene in the food industry. *J. Appl. Bacteriol.* **75:**499–511.

30. Castanie-Cornet, M., T. A. Penfound, D. Smith, J. F. Elliot, and J. W. Foster. 1999. Control of acid resistance in *Escherichia coli*. *J. Bacteriol.* **181:**3525–3535.

31. Chmielewski, R., and J. F. Frank. 1995. Formation of viable but nonculturable *Salmonella* during starvation in chemically defined solutions. *Lett. Appl. Microbiol.* **20:**380–384.

32. Cho, J. C., and S. J. Kim. 1999. Green fluorescent protein-based direct viable count to verify a viable but non-culturable state of *Salmonella typhi* in environmental samples. *J. Microbiol. Methods* **36:**227–235.

33. Choi, S. H., D. J. Baumler, and C. W. Kasper. 2000. Contribution of *dps* to acid stress tolerance and oxidative stress tolerance in *Escherichia coli* O157:H7. *Appl. Environ. Microbiol.* **66:**3911–3916.

34. Cloak, O. M., B. T. Solow, C. Briggs, C. Y. Chen, and P. M. Fratamico. 2002. Quorum sensing and production of antoinducer-2 in *Campylobacter* spp., *E. coli* O157:H7, and *Salmonella enterica* serovar Typhimurium in foods. *Appl. Environ. Microbiol.* **68:**4666–4671.

35. Conte, M. P., C. Longhi, G. Petrone, M. Polidoro, P. Valenti, and L. Seganti. 1994. *Listeria monocytogenes* infection of Caco-2 cells: role of growth temperature. *Res. Microbiol.* **145:**677–682.

36. Cossins, A. R., and M. Sinensky. 1984. Adaptation of membranes to temperature, pressure and exogenous lipids, p. 1–20. *In* M. Shinitzky (ed.), *Physiology of Membrane Fluidity*. CRC Press, Boca Raton, Fla.

37. Costerton, J. W. 1995. Overview of microbial biofilms. *J. Ind. Microbiol.* **15:**137–140.

38. Crawford, R. W., C. M. Belizeau, T. J. Poeler, C. W. Donnelly, and U. K. Bunning. 1989. Comparative recovery of uninjured and heat-injured *Listeria monocytogenes* cells from bovine milk. *Appl. Environ. Microbiol.* **55:**1490–1494.

39. Davies, D. G., M. R. Parsek, J. P. Pearson, B. H. Iglewski, J. W. Costeron, and E. P. Greenberg. 1997. The involvement of cell-to-cell signals in the development of a bacterial biofilm. *Science.* **280:**295–298.

40. Day, A. P., and J. D. Oliver. 2004. Changes in membrane fatty acid composition during entry of *Vibrio vulnificus* into a viable but nonculturable state. *J. Microbiol.* **42:**69–73.

41. Duffy, G., D. C. Riordan, J. J. Sheridan, J. E. Call, R. C. Whiting, I. S. Blair, and D. A. McDowell. 2000. Effect of pH on survival, thermotolerance, and verotoxin production of *Escherichia coli* O157:H7 during simulated fermentation and storage. *J. Food Prot.* **63:**12–18.

42. Duncan, S., L. A. Glover, K. Killham, and J. I. Prosser. 1994. Luminescence-based detection of activity of starved and viable but nonculturable bacteria. *Appl. Environ. Microbiol.* **60:**1308–1316.

43. Dussurget, O., E. Dumas, C. Archambaud, I. Chafsey, C. Chambon, M. Hebraud, and P. Cossart. 2005. *Listeria monocytogenes* ferritin protects against multiple stresses and is required for virulence. *FEMS Microbiol. Lett.* **250:**253–261.

44. Elhanafi, D., B. Leenanon, W. Bang, and M. A. Drake. 2004. Impact of cold and cold-acid stress on poststress tolerance and virulence factor expression of *Escherichia coli* O157:H7. *J. Food Prot.* **67:**19–26.

45. El-Kest, S. E., and E. H. Marth. 1991. Injury and death of frozen *Listeria monocytogenes* as affected by glycerol and milk components. *J. Dairy Sci.* **74:**1201–1208.

46. El-Kest, S. E., and E. H. Marth. 1991. Strains and suspending menstrua as factors affecting death and injury of *Listeria monocytogenes* during freezing and frozen storage. *J. Dairy Sci.* **74:**1209–1213.

47. Elstner, E. F., and A. Heupel. 1973. On the decarboxylation of α-keto acid by isolated chloroplasts. *Biochim. Biophys. Acta* **352:**182–188.

48. Farber, J. M., and B. E. Brown. 1990. Effect of prior heat shock on heat resistance of *Listeria monocytogenes* in meat. *Appl. Environ. Microbiol.* **56:**1584–1587.

49. Farkas, J. 1994. Tolerance of spores to ionizing radiation: mechanisms of inactivation, injury, and repair. *J. Appl. Bacteriol.* **76:**81S–90S.

50. Fedio, W. M., and H. Jackson. 1989. Effect of tempering on the heat resistance of *Listeria monocytogenes*. *Lett. Appl. Microbiol.* **9:**157–160.

51. Feeherry, F. E., D. T. Munsey, and D. B. Rowley. 1987. Thermal inactivation and injury of *Bacillus stearothermophilus* spores. *Appl. Environ. Microbiol.* **53:**365–370.

52. Foegoding, P. M., and F. F. Busta. 1981. Bacterial spore injury—an update. *J. Food Prot.* **44:**776–786.

53. Foster, J. W., and H. K. Hall. 1991. Inducible pH homeostasis and the acid tolerance response of *Salmonella typhimurium*. *J. Bacteriol.* **173:**5129–5135.

54. Foster, J. W., Y. K. Park, L. S. Bang, K. Karem, H. Betts, H. K. Hall, and E. Shaw. 1994. Regulatory circuits involved with pH-regulated gene expression in *Salmonella typhimurium*. *Microbiology* **140**:341–352.

55. Frank, J. F., and R. A. Koffi. 1990. Surface-adherent growth of *Listeria monocytogenes* is associated with increased resistance to surfactant sanitizers and heat. *J. Food Prot.* **48**:740–742.

56. Fratamico, P. M. 2003. Tolerance to stress and ability of acid-adapted and non-acid-adapted *Salmonella enterica* serovar Typhimurium DT104 to invade and survive in mammalian cells in vitro. *J. Food Prot.* **66**:1115–1125.

57. Fuqua, C., and E. P. Greenberg. 1998. Cell-to-cell communication in *Escherichia coli* and *Salmonella typhimurium*: they may be talking, but who's listening? *Proc. Natl. Acad. Sci. USA* **95**:6571–6572.

58. Gould, G. W. 1995. Homeostatic mechanisms during food preservation by combined methods, p. 397–410. *In* G. V. Barbosa-Canovas and J. Welti-Chanes (ed.), *Food Preservation by Moisture Control*. Technomic Publishing Co., Inc., Lancaster, Pa.

59. Gracia, K. S., and J. L. McKillip. 2004. A review of conventional detection and enumeration methods for pathogenic bacteria in food. *Can. J. Microbiol.* **50**:883–890.

60. Gram, L., L. Ravin, M. Rasch, J. B. Bruhn, A. B. Christensen, and M. Givskov. 2002. Food spoilage—interactions between food spoilage bacteria. *Int. J. Food Microbiol.* **78**:79–97.

61. Gupte, A. R., C. L. Rezende, and S. W. Joseph. 2003. Induction and resuscitation of viable but nonculturable *Salmonella enterica* serovar Typhimurium DT104. *Appl. Environ. Microbiol.* **69**:6669–6675.

62. Hara-Kudo, Y., M. Ifedo, H. Kodaka, H. Nakagawa, K. Goto, T. Masuda, H. Konuma, T. Kojima, and S. Kumagai. 2000. Selective enrichment with resuscitation step for isolation of freeze-injured *Escherichia coli* O157:H7 from foods. *Appl. Environ. Mircobiol.* **66**:2866–2872.

63. Harold, F. M. 1981. *The Vital Force: a Study of Bioenergetics*. W. H. Freeman & Company, New York, N.Y.

64. Hellingwerf, K. J., W. C. Crieland, M. J. T. de Mattos, W. D. Hoff, R. Kort, D. T. Verhamme, and C. Avignone-Rosa. 1998. Current topics in signal transduction in bacteria. *Antonie Leeuwenhoek* **74**:211–227.

65. International Commission on Microbiological Specifications for Foods. 1980. Injury and its effect on recovery, p. 205–214. *In Microbial Ecology of Foods*, vol. 1. *Factors Affecting Life and Death of Microorganisms*. Academic Press, Inc., New York, N.Y.

66. Jay, J. M., J. P. Vilai, and M. E. Huges. 2003. Profile and activity of bacteria biota of ground beef held from freshness to spoilage at 5–7°C. *Int. J. Food Microbiol.* **81**:105–111.

67. Juneja, V. K., P. G. Klein, and B. S. Marmer. 1998. Heat shock and thermotolerance of *Escherichia coli* O157:H7 in a model beef gravy system and ground beef. *J. Appl. Microbiol.* **84**:677–684.

68. Kashket, E. R. 1987. Bioenergetics of lactic acid bacteria: cytoplasmic pH and osmotolerances. *FEMS Microbiol. Rev.* **46**:233–244.

69. Kell, D. B., A. S. Kaprelyants, D. H. Weichart, C. R. Harwood, and M. R. Barer. 1998. Viability and activity in readily culturable bacteria: a review and discussion of the practical issues. *Antonie Leeuwenhoek* **73**:169–187.

70. Ko, R., L. T. Smith, and G. M. Smith. 1994. Glycine betaine confers enhanced osmotolerance and cryotolerance in *Listeria monocytogenes*. *J. Bacteriol.* **176**:426–431.

71. Kobayashi, H., T. Miyamoto, Y. Hashimoto, M. Kirki, A. Motomatsu, K. Honjoh, and M. Iio. 2005. Identification of factors involved in recovery of heat-injured *Salmonella enteritidis*. *J. Food Prot.* **68**:932–941.

72. Konings, W. N., J. Kok, O. P. Kuipers, and B. Poolman. 2000. Lactic acid bacteria: the bugs of the new millennium. *Curr. Opin. Microbiol.* **3**:276–282.

73. Korem, M., A. S. Sheoran, Y. Gov, S. Tzipori, I. Borovok, and N. Balahan. 2003. Characterization of RAP, a quorum sensing activator of *Staphylococcus aureus*. *FEMS Microbiol. Lett.* **223**:165–175.

74. Koutsoumanis, K. P., and J. N. Sofos. 2004. Comparative acid stress response of *Listeria monocytogenes*, *Escherichia coli* O157:H7 and *Salmonella* Typhimurium after habituation at different pH conditions. *Lett. Appl. Microbiol.* **38**:321–326.

75. Kuipers, O. P., P. G. G. A. de Ruyter, M. Kleerebezem, and W. M. de Vos. 1998. Quorum sensing-controlled gene expression in lactic acid bacteria. *J. Biotechnol.* **64**:15–21.

76. Kumar, C. G., and S. K. Anand. 1998. Significance of microbial biofilms in food industry: a review. *Int. J. Food Microbiol.* **42**:9–27.

77. Kwon, Y. M., and S. C. Ricke. 1998. Induction of acid resistance of *Salmonella typhimurium* by exposure to short-chain fatty acids. *Appl. Environ. Mircobiol.* **64**:3458–3463.

78. Leenanon, B., and M. A. Drake. 2001. Acid stress, starvation, and cold stress affect poststress behavior of *Escherichia coli* O157:H7 and nonpathogenic *Escherichia coli*. *J. Food Prot.* **64**:970–974.

79. Leimeister-Wachter, M., E. Domann, and T. Chakraborty. 1992. The expression of virulence genes in *Listeria monocytogenes* is thermoregulated. *J. Bacteriol.* **174**:947–952.

80. Leistner, L. 1994. Principles and applications of hurdle technology, p. 1–21. *In* G. W. Gould (ed.), *New Methods of Food Preservation*. Blackie Academic and Professional, Glasgow, Scotland.

81. Leyer, G. J., and E. A. Johnson. 1993. Acid adaptation induces cross-protection against environmental stresses in *Salmonella typhimurium*. *Appl. Environ. Microbiol.* **59**:1842–1847.

82. Likimani, T. A., and J. N. Sofos. 1990. Bacterial spore injury during extrusion cooking of corn/soybean mixtures. *Int. J. Food Microbiol.* **11**:243–249.

83. Linton, R. H., J. B. Webster, M. D. Pierson, J. R. Bishop, and C. R. Hackney. 1992. The effect of sublethal heat shock and growth atmosphere on the heat resistance of *Listeria monocytogenes* Scott A. *J. Food Prot.* **55**:84–87.

84. Liu, L., M. E. Hume, and S. D. Pillai. 2004. Autoinducer-2-like activity associated with foods and its interaction with food additives. *J. Food Prot.* **67**:1457–1462.

85. Lund, B. M. 1992. Ecosystems in vegetable foods. *J. Appl. Bacteriol.* 73:115S–126S.

86. Mafart, P. 2000. Taking injuries of surviving bacteria into account for optimizing heat treatments. *Int. J. Food Microbiol.* 55:175–179.

87. Mahapatra, A. K., K. Muthukumarappan, and J. L. Julson. 2005. Application of ozone, bacteriocins and irradiation in food processing: a review. *Cit. Rev. Food Sci. Nutr.* 45:447–461.

88. Maloney, P. C. 1990. Microbes and membrane biology. *FEMS Microbiol. Rev.* 87:91–102.

89. McDougald, D., S. A. Rice, D. Weichart, and S. Kjelleberg. 1998. Nonculturability: adaptation or debilitation? *FEMS Microbiol. Ecol.* 25:1–9.

90. McEntire, J. C., G. M. Carman, and T. J. Montville. 2004. Increased ATPase activity is responsible for acid sensitivity of nisin-resistant *Listeria monocytogenes* ATCC 700302. *Appl. Environ. Microbiol.* 70:2717–2721.

91. McKay, A. M. 1992. Viable but non-culturable forms of potentially pathogenic bacteria in water. *Lett. Appl. Microbiol.* 14:129–135.

92. Meyer, D. H., and C. W. Donnelly. 1992. Effect of incubation temperature on repair of heat-injured Listeria in milk. *J. Food Prot.* 55:579–582.

93. Michels, P. A. M., J. P. J. Michels, J. Boonstra, and W. L. Konings. 1979. Generation of an electrochemical proton gradient in bacteria by the excretion of metabolic end products. *FEMS Microbiol. Lett.* 5:357–364.

94. Mikulskis, A. V., I. Delor, V. H. Thi, and G. R. Cornelis. 1994. Regulation of *Yersinia enterocolitica* enterotoxin *yst* gene. Influence of growth phase, temperature, osmolarity, pH and bacterial host factors. *Mol. Microbiol.* 14:905–915.

95. Miller, M. B., and B. L. Bassler. 2001. Quorum sensing in bacteria. *Annu. Rev. Microbiol.* 55:165–169.

96. Miller, M. B., K. Skorupski, D. H. Lenz, R. K. Taylor, and B. L. Bassler. 2002. Parallel quorum sensing systems converge to regulate virulence in *Vibrio cholerae. Cell* 110:303–314.

97. Mizunoe, Y., S. N. Wai, A. Takade, and S. Yoshida. 1999. Restoration of culturability of starvation-stressed and low temperature-stressed *Escherichia coli* O157 cells by using H_2O_2-degrading compounds. *Arch. Microbiol.* 172:63–67.

98. Mizunoe, Y., S. N. Wai, T. Ishikawa, A. Takade, and S. Yoshida. 2000. Resuscitation of viable but nonculturable cells of *Vibrio parahaemolyticus* induced at low temperature under starvation. *FEMS Microbiol. Lett.* 186:115–120.

99. Mossel, D. A. A., and C. B. Struijk. 1992. The contribution of microbial ecology to management and monitoring of the safety, quality and acceptability (SQA) of foods. *J. Appl. Bacteriol.* 73:1S–22S.

100. Murano, E. A., and M. O. Pierson. 1993. Effect of heat shock and incubation atmosphere on injury and recovery of *Escherichia coli* O157:H7. *J. Food Prot.* 56:568–572.

101. Murley, Y. M., P. A. Carrol, K. Skorupski, R. K. Taylor, and S. B. Calderwood. 1999. Differential transcription of the *tcpPH* operon confers biotype-specific control of the *Vibrio cholerae* ToxR virulence regulon. *Infect. Immun.* 67:5117–5123.

102. National Food Processors Association. 1988. Factors to be considered in establishing good manufacturing practices for the production of refrigerated food. *Dairy Food Sanit.* 8:288–291.

103. Ngutter, C., and C. Donnelly. 2003. Nitrate-induced injury of *Listeria monocytogenes* and the effect of selective versus nonselective recovery procedures on its isolation from frankfurters. *J. Food Prot.* 66:2252–2257.

104. Nicholls, D. G., and S. J. Ferguson. 1992. *Bioenergetics 2.* Academic Press, San Diego, Calif.

105. Nilsson. L., J. D. Oliver, and S. Kjelleberg. 1991. Resuscitation of *Vibrio vulnificus* from the viable but nonculturable state. *J. Bacteriol.* 173:5054–5059.

106. Novick, R. P. 1999. Regulation of pathogenicity in *Staphylococcus aureus* by a peptide-based density-sensing system, p. 129–146. *In* G. M. Dunny and S. C. Winns (ed.), *Cell-Cell Signaling in Bacteria.* ASM Press, Washington, D.C.

107. Nystrom, T. 1998. To be or not to be: the ultimate decision of the growth-arrested bacterial cell. *FEMS Microbiol. Rev.* 21:283–290.

108. Olasupo, N. A., D. J. Fitzgerald, A. Narbad, and M. J. Gasson. 2004. Inhibition of *Bacillus subtilis* and *Listeria innocua* by nisin in combination with some naturally occurring organic compounds. *J. Food Prot.* 67:596–600.

109. Oliver, J. D. 2005. The viable but nonculturable state in bacteria. *J. Microbiol.* 43:93–100.

110. Oliver, J. D., and R. Bocklan. 1995. In vivo resuscitation, and virulence towards mice, of viable but nonculturable cells of *Vibrio vulnificus. Appl. Environ. Microbiol.* 61:2620–2623.

111. Oliver, J. D., M. Dagher, and K. Linden. 2005. Induction of *Escherichia coli* and *Salmonella typhimurium* into the viable but nonculturable state following chlorination of wastewater. *J. Water Health* 3:249–257.

112. Oliver, J. D., F. Hite, D. McDougald, N. L. Andon, and L. M. Simpson. 1995. Entry into, and resuscitation from, the viable but nonculturable state by *Vibrio vulnificus* in an estuarine environment. *Appl. Environ. Microbiol.* 61:2624–2630.

113. Olson, E. R. 1993. Influence of pH on bacterial gene expression. *Mol. Microbiol.* 8:5–14.

114. Otto, M. 2001. *Staphylococcus aureus* and *Staphylococcus epidermidis* peptide pheromones produced by accessory gene regulator *agr* system. *Peptides* 22:1603–1608.

115. Pai, S., J. K. Actor, E. Sepulveda, R. L. Hunter, and C. Jegannath. 2000. Identification of viable and non-viable *Mycobacterium tuberculosis* in mouse organs by directed RT-PCR for antigen 85B mRNA. *Microb. Pathog.* 28:335–342.

116. Palumbo, S. A. 1986. Is refrigeration enough to restrain foodborne pathogens? *J. Food Prot.* 49:1003–1009.

117. Pepe, J. C., J. L. Badger, and V. L. Miller. 1994. Growth phase and low pH affect the thermal regulation of the *Yersinia enterocolitica inv* gene. *Mol. Microbiol.* 11:123–135.

118. Phan-Thanh, L., F. Mahouin, and S. Alige. 2000. Acid responses of *Listeria monocytogenes. Int. J. Food Microbiol.* 55:121–126.

119. Quadri, L. E. N. 2002. Regulation of antimicrobial peptide production by autoinducer mediated quorum sensing in lactic acid bacteria. *Antonie Leeuwenhoek* **83:**133–145.

120. Rollins, D. M., and R. R. Colwell. 1986. Viable but nonculturable stage of *Campylobacter jejuni* and its role in survival in the natural aquatic environment. *Appl. Environ. Microbiol.* **52:**531–538.

121. Sallam, S. S., and C. W. Donnelly. 1992. Destruction, injury and repair of *Listeria* species exposed to sanitizing compounds. *J. Food Prot.* **59:**771–776.

122. Sampathkumar, B., G. G. Khachatourians, and D. R. Korber. 2004. Treatment of *Salmonella enterica* serovar Enteritidis with a sublethal concentration of trisodium phosphate or alkaline pH induces thermotolerance. *Appl. Environ. Microbiol.* **70:**4613–4620.

123. Schuhmacker, D. A., and K. E. Klose. 1999. Environmental signals modulate ToxT-dependent virulence factor expression in *Vibrio cholerae. J. Bacteriol.* **181:**1508–1514.

124. Sinensky, M. 1974. Homeoviscous adaptation—a homeostatic process that regulates the viscosity of membrane lipids in *Escherichia coli. Proc. Natl. Acad. Sci. USA* **71:**522–525.

125. Sircili, M. P., M. Matthews, L. R. Trabulsi, and V. Sperandio. 2004. Modulation of enteropathogenic *Escherichia coli* virulence by quorum sensing. *Infect. Immun.* **72:**2329–2337.

126. Skurnik, M. 1985. Expression of antigens encoded by the virulence plasmid of *Yersinia enterocolitica* under different growth conditions. *Infect. Immun.* **47:**183–190.

127. Slonczewski, J. L., and J. W. Foster. 1999. pH-regulated genes and survival at extreme pH. *In* F. C. Neidhardt et al. (ed.), *Escherichia coli and Salmonella: Cellular and Molecular Biology.* ASM Press, Washington, D.C.

128. Smith, J. L., P. M. Fratamico, and J. S. Novak. 2004. Quorum sensing: a primer for food microbiologists. *J. Food Prot.* **67:**1053–1070.

129. Smith, J. L., B. S. Marmer, and R. C. Benedict. 1991. Influence of growth temperature on injury and death of *Listeria monocytogenes* Scott A during a mild heat treatment. *J. Food Prot.* **54:**166–169.

130. Somers, E. B., J. L. Schoeni, and A. C. L. Wong. 1994. Effect of trisodium phosphate on biofilm and planktonic cells of *Campylobacter jejuni, Escherichia coli* O157:H7, *Listeria monocytogenes,* and *Salmonella typhimurium. Int. J. Food Microbiol.* **22:**269–276.

131. Sturme, M. H. L., M. Kleerebezem, J. Nakayama, A. D. L. Akkermans, E. E. Vaughan, and W. M. de Vos. 2002. Cell-to-cell communication by autoinducing peptides in gram-positive bacteria. *Antonie Leeuwenhoek* **81:**233–243.

132. Surette, M. G., and B. L. Bassler. 1998. Quorum sensing in *Escherichia coli* and *Salmonella typhimurium. Proc. Natl. Acad. Sci. USA* **95:**7046–7050.

133. Talibart, R., M. Denis, A. Castillo, J. M. Cappelier, and G. Ermel. 2000. Survival and recovery of viable but noncultivable forms of *Campylobacter* in aqueous microcosm. *Int. J. Food Microbiol.* **55:**263–267.

134. Thieringer, H. A., P. G. Jones, and M. Inouye. 1998. Cold shock and adaptation. *Bioessays* **20:**49–57.

135. Tiwari, R. P., N. Sachdeva, and G. S. Hoondal. 2004. Adaptive acid tolerance response in *Salmonella enterica* serovar Typhimurium and *Salmonella enterica* serovar Typhi. *J. Basic Microbiol.* **2:**137–146.

136. Tomoyasu, T., A. Takaya, T. Sasaki, T. Nagase, R. Kikuno, M. Morioka, and T. Yamamoto. 2003. A new heat shock gene, *agsA,* which encodes a small chaperone involved in suppressing protein aggregation in *Salmonella enterica* serovar Typhimurium. *J. Bacteriol.* **185:**6331–6339.

137. Tseng, C.-P., and T. J. Montville. 1993. Metabolic regulation of end product distribution in lactobacilli: causes and consequences. *Biotechnol. Prog.* **9:**113–121.

138. Van Langendonck, N., P. Velge, and E. Bottreau. 1998. Host cell protein tyrosine kinases are activated during the entry of *Listeria monocytogenes. FEMS Microbiol. Lett.* **162:**169–176.

139. Van Schaik, W., C. G. Gahan, and C. Hill. 1999. Acid-adapted *Listeria monocytogenes* displays enhanced tolerance against the lantibiotics nisin and lactin 3147. *J. Food Prot.* **62:**536–539.

140. Vigh, V., B. Maresca, and J. L. Harwood. 1998. Does the membrane's physical state control the expression of heat shock and other genes? *Trends Biochem. Sci.* **23:**369–372.

141. Volker, U., H. Mach, R. Schmid, and M. Hecker. 1992. Stress proteins and cross-protection by heat shock and salt stress in *Bacillus subtilis. J. Gen. Microbiol.* **138:**2125–2135.

142. Wang, G., and M. P. Doyle. 1998. Heat shock response enhances acid tolerance of *Escherichia coli* O157:H7. *Lett. Appl. Microbiol.* **26:**31–34.

143. Wiedmann, M., T. J. Arvik, R. J. Hurley, and K. J. Boor. 1998. General stress transcription factor σ^B and its role in acid tolerance and virulence of *Listeria monocytogenes. J. Bacteriol.* **180:**3650–3656.

144. Winzer, K., K. R. Hardic, and P. Williams. 2002. Bacterial cell-to-cell communication: sorry, can't talk now—gone to lunch! *Curr. Opin. Microbiol.* **5:**216–222.

145. Wolffs, P., B. Norling, J. Hoorfar, M. Griffiths, and P. Radstrom. 2005. Quantification of *Campylobacter* spp. in chicken rinse samples by using flotation prior to real-time PCR. *Appl. Environ. Microbiol.* **71:**5759–5764.

146. Wong, H. C., C. T. Shen, C. N. Chang, Y. S. Lee, and J. D. Oliver. 2004. Biochemical and virulence characterization of viable but nonculturable cells of *Vibrio parahaemolyticus. J. Food Prot.* **67:**2430–2305.

147. Wouters, J. A., F. M. Rombouts, W. M. deVos, O. P. Kuipers, and T. Abee. 1999. Cold shock proteins and low-temperature response of *Streptococcus thermophilus* CNRZ302. *Appl. Environ. Microbiol.* **65:**4436–4442.

148. Yura, T., and K. Nakahigashi. 1999. Regulation of the heat-shock response. *Curr. Opin. Microbiol.* **2:**153–158.

149. Zottola, E. A., and K. C. Sasahara. 1994. Microbial biofilms in the food processing industry—should they be a concern? *Int. J. Food Microbiol.* **23:**125–148.

Food Microbiology: Fundamentals and Frontiers, 3rd Ed.
Edited by M. P. Doyle and L. R. Beuchat
© 2007 ASM Press, Washington, D.C.

Jennifer Cleveland McEntire
Thomas J. Montville

Antimicrobial Resistance

2

The specter of antibiotic-resistant microbes is an international concern due to the public health implications of a postantibiotic era. Consumers worry about the increased difficulty in treating bacterial infections when the organism is antibiotic resistant, but few understand that the greatest risk factor for infection with an antibiotic-resistant pathogen is having recently taken antibiotics. Difficult-to-treat hospital-acquired infections caused by antibiotic-resistant microbes such as methicillin-resistant *Staphylococcus aureus* (MRSA) and vancomycin-resistant enterococci have captured public attention. In 2003, more than a quarter of nosocomial enterococcal infections were resistant to vancomycin, and almost 60% of nosocomially acquired *S. aureus* infections were MRSA. This represents a 12% and 11% increase in percentage of resistant infections compared to the previous 5-year averages, respectively (11). There are also economic implications; in the mid-1990s, an estimated $4 billion of the >$7 billion in human antibiotic sales targeted nosocomial infections caused by antibiotic-resistant bacteria (33). The Centers for Disease Control and Prevention and the medical community are promoting changes in antibiotic-prescribing practices and encouraging prudent use of antibiotics to decrease the selective pressure that increases the prevalence of antibiotic-resistant microbes. These recommendations include renewed emphasis on preventing the spread of infection and targeting of specific antibiotics for specific pathogens to limit the use of broad-spectrum antibiotics (http://www.cdc.gov/drugresistance/healthcare/ltc/12steps_ltc.htm).

ANTIMICROBIAL USE IN FOOD AND PRODUCTION AGRICULTURE

Antibiotics

Antibiotics usage extends beyond human medicinal purposes. Antibiotics are used in food production to treat infected animals, prevent infection of potentially exposed animals, and promote growth. The growth-promoting function of antibiotics is only partly understood and is the subject of debate (24, 52). Many argue that the widespread use of antibiotics at subtherapeutic levels exacerbates the selective pressure that enriches microbial populations for antibiotic-resistant bacteria and suggest that growth promotion and disease prevention could be achieved

---

Jennifer Cleveland McEntire, Institute of Food Technologists, Washington, D.C. 20036. **Thomas J. Montville,** Dept. of Food Science, Cook College, Rutgers, the State University of New Jersey, New Brunswick, NJ 08901-8520.

through nonantibiotic alternatives and novel management practices (48). As a result of the debate, many European countries have discontinued the administration of antibiotics for growth promotion. Denmark banned the use of antibiotics for growth promotion in food animal production in an effort to decrease the prevalence of antibiotic-resistant human pathogens. While the total use of antibiotics in animals in Denmark decreased 30% from 1997 (before the ban) to 2004, there has been a 68% increase in the use of antibiotics for therapeutic purposes during the same period (http://www.who.int/salmsurv/en/Expertsreportgrowthpromoterdenmark.pdf). This makes it difficult to assess the "success" of the ban.

A 2002 meeting of the World Health Organization assessing the impact of the Danish ban on growth-promoting antibiotics found that resistance of enterococci to those agents decreased; *Enterococcus faecium* resistance to virginiamycin decreased from approximately 60% to 10% in broiler meat between 1997 (the peak of virginiamycin use) and 1999 (when there was no virginiamycin use). The incidence of *Salmonella*, *Campylobacter*, and *Yersinia* infections in humans was not affected (http://www.who.int/salmsurv/en/Expertsreportgrowthpromoterdenmark.pdf).

Agricultural antibiotics usage extends beyond animal use, though presumably to a much lesser extent. In the United States, fruit plants may be sprayed with oxytetracycline and streptomycin. There is very limited use of antibiotics in U.S. aquaculture. Only sulfadimethoxine/ormetoprim, sulfamerizine, and oxytetracycline are approved for use, although sulfamerizine is no longer commercially available. The extent of antibiotic use in other countries, including by seafood and produce exporters, is unknown and presumed to be much higher. There are no surveillance systems for antibiotic-resistant bacteria on crops; however, it appears that the transfer of antibiotic resistance determinants between plant pathogens and human pathogens does not occur under natural conditions (39).

Food Antimicrobials

The number of cases of many types of bacterial foodborne illness has steadily decreased over the past 10 years (12). These changes are likely due to the effective interventions of food scientists and increased consumer awareness of safe food handling practices. While there has been progress in limiting contamination of food by pathogens, some contamination is inevitable and must be controlled. Microorganisms are constantly responding, evolving and adapting to their environments and food safety interventions, so maintenance of food safety requires continuous effort.

It is illegal for antibiotics to be applied to or present on finished food products. Therefore, food antimicrobials are used to control bacterial growth (of both pathogens and spoilage organisms). Many food antimicrobials, such as nisin, lysozyme, lactoferrin, essential oils, and organic acids, are derived from natural sources, but even if their source is microbial, they are not classified as antibiotics (19).

Sanitizers and disinfectants are used throughout food production, from the farm to the consumers' homes. Resistance to food antimicrobials and sanitizers is much more difficult to characterize and quantify than antibiotic resistance, since standardized methods have not been developed and surveillance systems are not in place. When resistance has been documented, such as resistance of *Listeria monocytogenes* and *S. aureus* to quaternary ammonium compounds (QACs), the level where "resistance" is observed is severalfold lower than what is used industrially (32). While these bacteria may be classified as "resistant," they are likely still susceptible to QACs at normal usage levels.

Overview of Antimicrobial Uses in Food Production

Microorganisms encounter a multitude of stresses during the journey from farm to fork. On the farm, antibiotics are used in animals, on fruit trees to a limited extent, and in aquaculture. Sanitizers and disinfectants are often used in these environments as well. During food processing, raw meats may be decontaminated with a variety of physical processes or acidic rinses. Fruits and vegetables may be washed in chlorine rinses and/or packed in modified-atmosphere environments. To decrease spoilage and, in many cases, prevent pathogen growth, antimicrobials may be added to foods. Food processing plants are cleaned frequently using agents such as QACs and other sanitizers. Antimicrobials such as triclosan are used in the food plants and are now being introduced into many consumer products. Ultimately, the consumer may wash, heat, or otherwise treat the product before consumption. In the event of bacterial foodborne illness, the consumer may be treated with antibiotics.

RESISTANCE TERMINOLOGY

"Antimicrobial resistance" is a broad term with many meanings. It describes the response of a multitude of microbes to a variety of agents by many different mechanisms. It is useful to distinguish among antibiotic resistance, food antimicrobial resistance, and resistance to sanitizers and disinfectants.

Antibiotic Resistance

There are several ways to define antibiotic resistance (21). Broadly, antibiotic resistance refers to the "temporary or permanent ability of an organism and its progeny to remain viable and/or multiply under conditions that would destroy or inhibit other members of the strain" (14). Resistance to antibiotics is generally genetically based, and progeny of antibiotic-resistant bacteria are also resistant. While antibiotic resistance can result from spontaneous mutation in the absence of antibiotic use (53), most experts and government agencies agree that the increasing prevalence of antibiotic resistance is directly related to increased antibiotic use, both prudent and frivolous, in clinical and agricultural settings. Estimates of the combined total use of antibiotics, including human and animal use, vary widely. In 1989, the Institute of Medicine estimated a total yearly antibiotic usage of 50 million pounds. The Animal Health Institute estimated that about 20 million pounds of antibiotics were administered to animals in 2003 (http://www.ahi.org/Documents/Antibioticuse2003.pdf). It appears that, on a mass basis, agricultural practices, including therapeutic treatment, contribute between 25 and 50% of the antibiotics introduced into the biosphere.

Food Antimicrobial Resistance

Unlike antibiotic resistance, for which mechanisms of resistance (and methodology to determine resistance) are well established, "resistance" to food antimicrobials, while well studied, remains poorly understood. There is no standard definition, or threshold, to characterize a microbe as resistant to a specific food antimicrobial, sanitizer, or disinfectant. Differences in methods also hamper the comparison of data. Another complicating factor is that, in many cases, "resistance" manifests as a temporary adaptation that is not displayed by subsequent generations. Sublethal stresses may also lead to selection of stress-resistant survivors which may become stress-hardened or cross-protected and more difficult to control. Sublethally stressed pathogens may be even more virulent. In general, stress-adapted or -hardened pathogens may have greater potential for survival and growth in foods and in sublethal environments. Thus, traditional antimicrobial hurdles may be inadequate for pathogen control and may lower the infective dose of some strains (2, 49).

Sanitizer and Disinfectant Resistance

Sanitizers and disinfectants are used in all areas of food production, and antibacterial agents are increasingly being added to consumer products such as soaps. The only class of industrial sanitizers for which laboratory resistance has been documented is quaternary ammonium compounds. However, their recommended usage is typically at concentrations greatly exceeding the levels of observed "resistance," making "resistance" of little practical relevance. The incorporation of triclosan into many consumer products has raised the concern that bacteria may become triclosan resistant and that triclosan use may select for bacteria resistant to other antibiotics. Several studies have shown that this is an incorrect hypothesis (1).

A fundamental understanding of how the various agents work and an understanding of the mechanisms of resistance will enable food scientists to be better informed about the potential for cross-resistance, whereby microbes are resistant both to food processing agents or techniques and antibiotics. In some cases, however, the mechanisms that confer resistance to one agent may result in increased sensitivity to another agent.

GENETIC CAUSES OF RESISTANCE

For the purpose of this chapter, "resistance" refers to a genetic difference between wild-type and resistant strains of the same microbe. In contrast, "innate resistance" refers to the noninhibition of a microbe that has never been sensitive to a particular antimicrobial. In some cases, this type of resistance may be due to differences in cell wall composition. For example, the cell wall structure of gram-negative bacteria makes them innately "more resistant" to penicillin than are gram-positive bacteria. While innate resistance has major implications, it is outside the scope of this chapter. "Adaptation" refers to an inducible, transient phenotype.

Resistance Mechanisms

The ways by which microbes protect themselves from inhibitory agents vary and can be due to multiple mechanisms. The mechanisms of antibiotic resistance include efflux pumps, modification of the cellular target, changes in membrane permeability, and degradation of the antibiotic (22). Table 2.1 illustrates the parallels between the mechanisms of antibiotic resistance and mechanisms of organic acid and bacteriocin resistance.

Active export contributes to bacterial resistance to many agents, including antibiotics, but has no corresponding mechanism for bacteriocins (which do not accumulate intracellularly). As detailed in chapter 33, the intracellular accumulation of protons from organic acids acidifies the cytoplasm, inhibiting many metabolic processes. The cell tries to maintain pH homeostasis by pumping these protons out of the cytoplasm using ATP as an energy source. The resultant energy depletion is a

Table 2.1   Examples of resistance mechanisms

Mechanism	Action	Antibiotic(s)	Organic acids	Bacteriocin(s)
Export	Specific	Tetracycline	F_0F_1 ATPase pumps out protons, anions accumulate intracellularly	Not applicable, bacteriocins not in cytoplasm
Destruction	Specific or general	β-Lactamases	Not applicable	Protease, specific "bacteriocinase"
Modification	Specific	Methylation of aminoglycosides	Not applicable	Dehydroreductases can inactivate lantibiotics like nisin
Altered receptors	Specific	Penicillin binding proteins	No receptor required	Reported for nisin
Membrane composition	General	Altered membranes in resistant *E. coli* and bacilli	May affect permeability	Demonstrated for nisin resistance

major factor in the inhibition caused by organic acids. Tetracycline resistance in *Escherichia coli* is at least partially due to an energy-dependent efflux mechanism (40). A similar mechanism has been implicated in *E. coli* fluoroquinolone resistance (16, 17, 29). The genes for multiple antibiotic resistance in *Pseudomonas aeruginosa* might be on an efflux operon (46).

Enzymatic degradation or modification of antibiotics is common. The primary mechanism of resistance to β-lactam antibiotics (penicillins and cephalosporins) is via hydrolysis of the β-lactam ring (10). Chloramphenicol is inactivated by the acetylation catalyzed by the chloramphenicol acetyltransferase of resistant microbes (22). Resistant pathogens modify aminoglycosides using methylases, acetyltransferases, nucleotidyltransferases, and phosphotransferases (51). Enzymatic degradation of preservatives can be specialized or general but would be different from the enzymes that inactivate antibiotics. For example, some bacteria metabolize citric acid, rendering it ineffective in their presence. Many proteases inactivate bacteriocins in a nonspecific fashion.

The most common form of intrinsic resistance to antibiotics is altered membranes: the outer membrane of gram-negative bacteria acts as a strong permeability barrier to many antibiotics (44). Decreased chloramphenicol permeability may be a cause of resistance in *Haemophilus influenzae* and *Pseudomonas cepacia* (8, 9). In gram-positive *Staphylococcus epidermidis*, glycopeptide antibiotic resistance may result from overproduction of glycopeptide-binding sites within the cell wall peptidoglycan (50). *Bacillus subtilis* mutants resistant to the protonophore carbonyl cyanide *m*-chlorophenylhydrazone (CCCP) have altered membranes with reduced amounts of C16 fatty acids and increased ratios of iso:nteiso branches. The CCCP-resistant mutants are cross-resistant to other inhibitors including 2,4-dinitrophenol, tributylin, and neomycin (26).

Antibiotics Select for Existing Mutations

Microbes can become resistant by mutation or by acquiring a resistance gene from another microbe. The use of antibiotics does not cause resistance, it selects for resistant mutants already present in the population. Mutations which result in antibiotic resistance arise spontaneously. They may be caused by errors in DNA synthesis, chemical changes induced by mutagens, or incorrect repair of radiation-induced single-strand breaks. Mutations occur at the same rate in the presence and in the absence of antibiotics. The frequency of mutation is 10^{-9} per bacterial generation (20). Had antibiotics never been used, there could still be low levels of antibiotic resistance due to these normal, expected mutations. Approximately 0.3% of isolates from deep-sea cold-seep sediments produced β-lactamases, the enzymes responsible for conferring resistance to penicillin, even though penicillin has presumably never been introduced into this environment (53). Microbes may acquire genes from other cells which result in resistance. This gene transfer can be among populations of the same strain or between different strains, species, or even genera. The intergeneric transfer of resistant genes has important implications; although antibiotic resistance in foodborne pathogens may not have clinical significance, if it is transferred to other species it may impede clinical treatment of their diseases. These mechanisms of genetic transfer, including plasmids, integrons, and transposons, are not unique to resistance genes and are covered in chapter 44.

If all antibiotic use was suddenly discontinued, would antibiotic resistance in bacteria disappear or be maintained? Until recently, it was thought that, in the absence of the selective pressure caused by the antibiotic, antibiotic-resistant genetic determinants responsible for resistance would be lost and the bacteria would revert to an antibiotic sensitivity. However, in a study of a remote Bolivian community where antibiotics have not knowingly been used in agriculture and are rarely used by humans, antibiotic-resistant bacteria were isolated from 73% of the population's fecal samples (4). Possible reasons for the high level of resistance in the absence of antibiotics include unknown selective advantages unrelated to antibiotic resistance, selective pressure due to exposure of heavy metals which coselects for antibiotic resistance, and the presence of naturally occurring antibiotic-producing strains in the ecosystem.

Whether antibiotic resistance results from chromosomal mutations or gene acquisition, there is often an associated "fitness cost." In other words, the ability of a cell to resist an antibiotic often comes at the expense of another cellular function. For example, ribosomal mutations which prevent antibiotics from binding to their ribosomal target may inhibit protein synthesis (6). If antibiotic resistance is no longer advantageous, the cell would benefit from reverting to the wild type by allowing that protein synthesis to proceed unhindered. However, reversion does not always happen. In these cases, compensatory mutations alleviate the fitness cost, and the resistance phenotype is maintained (35). In the case of Denmark, where the discontinuation of antibiotic growth promoters decreased overall antibiotic use, the prevalence of antibiotic-resistant enterococci did decrease, counter to the prediction that resistance determinants would be maintained (http://www.who.int/salmsurv/en/Expertsreportgrowthpromoterdenmark.pdf). To explain these seemingly opposite scenarios, it is again necessary to examine the mechanism of resistance in the host organism; the maintenance of resistance when the underlying cause is ribosomal may be very different compared to resistance due to another mechanism. The "fitness cost" of some mechanisms of resistance may be less than others, making those traits less likely to be lost, or there may be numerous possibilities for compensatory mutations. In other cases, the burden of maintaining a plasmid, for example, may outweigh the benefits of continued expression of that genetic information.

ADAPTATION TO FOOD ANTIMICROBIALS

Bacteria, yeast, and molds may be innately resistant to certain food antimicrobials, but sensitive organisms do not typically mutate or acquire resistance to food antimicrobials. However, exposure to subinhibitory antimicrobial levels may cause a temporary adaptation, so that subsequent exposure to normally lethal levels is less effective. Bacterial adaptation is the term used to describe temporary phenotypic changes in response to stress. Adaptation is more common for food antimicrobial agents than antibiotics. New genetic material is not required for bacteria to adapt; instead, the stressor quickly activates certain existing pathways and mechanisms to produce a physiological response that helps the microbe withstand the stressor. In contrast to the antibiotic selection for resistance, for adaptation, the antimicrobial causes the observed resistance.

Some mechanisms for adaptation are known. The synthesis of stress response proteins, including heat shock proteins, is triggered by low levels of the stress. These stress response proteins protect the cell from subsequent related or unrelated stresses. In many cases, these proteins serve a general protective function. For example, exposure to low levels of acid may activate the stress response system, providing those cells protection from subsequent heat exposure as well as acid exposure (3). Stress response proteins are discussed in chapter 1 of this volume.

There are conflicting data on the persistence of adapted phenotypes. The duration of the phenotype may depend on the underlying mechanism. For example, in the acid tolerance response (chapter 1), stress response proteins are quickly produced, but few studies have demonstrated their persistence over time and with subsequent generations. Quorum sensing (see chapter 1) allows cells to communicate with each other by secreting signaling molecules. If older cells signal to newer cells that a stress was encountered, cells can exchange information in a nongenetic fashion. The result would be that future generations of cells would exhibit the altered phenotype in the absence of the stressor.

ANTIMICROBIAL RESISTANCE IN THE REAL WORLD

Selective Pressure from Farm to Fork

It is difficult to apply resistance prevalence trend data from across the farm-to-fork continuum to foodborne illness. The data provided to the National Antibiotic Resistance Monitoring System (NARMS) by the Centers for Disease Control and Prevention (CDC) monitor antibiotic resistance in human infections. While the data are broken down by microorganisms, they do not attribute the infection to specific vehicles, such as foods. The NARMS data collected by the U. S. Department of Agriculture track resistance rates in food animals, but not during the many steps between the animal, the food, its consumption, and foodborne illness (Fig. 2.1). Slaughter techniques and contamination rates of the raw product, consumer handling, especially storage and cooking, and human dose response are all variables that make it difficult to correlate on-farm resistance rates with human illness.

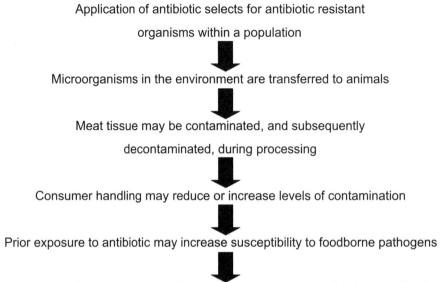

Application of antibiotic selects for antibiotic resistant organisms within a population

Microorganisms in the environment are transferred to animals

Meat tissue may be contaminated, and subsequently decontaminated, during processing

Consumer handling may reduce or increase levels of contamination

Prior exposure to antibiotic may increase susceptibility to foodborne pathogens

Treatment failure may result if bacteria are resistant to antibiotic prescribed

Figure 2.1   Flow of antibiotic-resistant and susceptible microorganisms from farm to fork.

In 2002, NARMS was expanded to include surveillance by the Food and Drug Administration (FDA) of antibiotic-resistant *Salmonella*, *Campylobacter*, *E. coli*, and *Enterococcus* in raw retail meats. These data provide insight into the transfer of antibiotic-resistant organisms from the animal to the meat, but since meat is generally not consumed raw, NARMS will still be unable to provide a direct relationship between antibiotic-resistant bacteria and human foodborne illness. There are currently no monitoring systems or sets of standardized, validated methods to track resistance to other antimicrobials used in food, to sanitizers, or to disinfectants. The transient nature of adaptation makes it difficult to quantify and track.

An examination of Fig. 2.1, using meat as an example, illustrates that there are several points along the farm-to-table pathway where interventions can influence the selection and dissemination of antibiotic-resistant bacteria. Decreasing the use of antibiotics, particularly in humans and agriculture, will decrease the selection of antibiotic-resistant bacteria; however, as noted, a low level of resistant mutants cannot be eliminated. Food scientists have no control over this point in the process. Good sanitation practices, along with management practices including utilization of vaccines and competitive exclusion, may limit the transfer and colonization of animals by bacteria, both antibiotic resistant and susceptible. The main points of influence of food microbiologists begin during slaughter and processing, when techniques and processes, such as carcass decontamination, can reduce the concentration

and limit the spread of organisms to meat tissue. Handling, packaging, and storage conditions en route to retail distribution are also influenced by food microbiologists. Consumers also play a role in general food safety, through their handling and cooking of these products. The presence of pathogens in the consumed product may result in foodborne illness. If the consumer had recently taken antibiotics, he or she could be more susceptible to infection by a pathogen. In the case of antibiotic-resistant bacteria, the treatment of illness could be compromised.

Antibiotic Resistance in Specific Foodborne Pathogens

While it is useful for food microbiologists to consider resistance with respect to specific foodborne pathogens and their contribution to foodborne illness, the limitations of this narrow approach must be acknowledged. This approach neglects the public health importance of antibiotic-resistant bacteria that are not foodborne, such as those involved in nosocomial infections. This limited perspective does not consider the environmental impact of antibiotic resistance, such as the changes in soil or aquatic microbiota, or the potential horizontal transfer of resistance determinants from nonpathogens to pathogens. Examining the present status of antibiotic resistance in foodborne pathogens cannot predict future responses to food safety interventions. Microorganisms evolve quickly and might develop new mechanisms of adaptation or resistance to novel interventions. The information provided below on resistance in specific foodborne pathogens must be considered with these constraints in mind.

Salmonella

The variety of serotypes that make up the genus *Salmonella* makes it extremely difficult to generalize about antibiotic resistance in *Salmonella*. The serovars of *Salmonella* responsible for foodborne illnesses vary widely with time, without respect to antibiotic resistance. For example, while the incidence of foodborne illness due to *Salmonella enterica* serovar Typhimurium decreased 40% between 1996 to 1998 and 2004, *Salmonella enterica* serovar Newport-related foodborne illnesses increased 40% (12). Changes in serovar prevalence should be considered when examining trends in antibiotic resistance, to put the potential impact on human health in proper perspective.

Salmonella serovar Typhimurium DT104 is resistant to at least five antibiotics and has been associated with consumption of dairy and beef products (15, 56). This chromosomally mediated resistance results from integrons. Over the past several years, *Salmonella* serovar Newport MDR-AmpC, which is resistant to at least four additional antibiotics, has received much attention due to its epidemic spread. It appears that serovar Newport MDR-AmpC isolates are from a clonal population.

Campylobacter

More than a million cases of campylobacteriosis are estimated to occur in the United States each year (41). While the bacterium is typically associated with poultry, outbreaks are most often associated with produce, possibly as a result of cross-contamination from raw poultry (31). In cases in which infection by *Campylobacter* is treated with antibiotics, erythromycin or a fluoroquinolone is most commonly prescribed. Fluoroquinolones are also used in animal agriculture. The FDA deemed that fluoroquinolone use in poultry could increase the prevalence of fluoroquinolone-resistant *Campylobacter* and compromise the treatment of humans. After lengthy court proceedings, the FDA withdrew the approval of the fluoroquinolone enrofloxacin for use in poultry (23).

Shigella

When considering antibiotic-resistant *Shigella*, it is important to understand that humans, not food animals, are hosts for this pathogen. This means that cases of shigellosis are not due to carriage of the pathogen in the food animal or contamination of crops with manure from food animals. Human fecal contamination, either on the farm or during food preparation, is the root cause of shigellosis. The infection is classified as a foodborne illness because food is often the vehicle for *Shigella*. Both sanitation practices and antibiotic use vary from country to country. Today's global food supply warrants examination of imported food for contamination with antibiotic-resistant *Shigella*.

Listeria monocytogenes

Infection with *L. monocytogenes* is typically treated with a combination of ampicillin and aminoglycoside. Rates of antibiotic resistance in this pathogen have not been studied to the same extent as in other pathogens such as *Salmonella* and *Campylobacter*. Because it is considered widely distributed in the environment, is not strictly a zoonotic organism, and has not been considered problematic in terms of antibiotic resistance, it is not included in any NARMS surveys. Studies to date have found little to no antibiotic resistance in *L. monocytogenes*, with the exception of those by Prazak et al. (47), who determined that 95% of isolates from Texan cabbage, environmental, or water samples were resistant to two or more antibiotics, with 85% of those resistant to penicillin.

The documented plasmid-mediated resistance of some *L. monocytogenes* strains to QACs (27) is of more immediate concern to food microbiologists. The prevalence of QAC-resistant *L. monocytogenes* is much higher in isolates from food processing facilities than in clinical isolates. While QAC resistance could increase prevalence of *L. monocytogenes* in the processing plant, these strains or serotypes may not be the dominant causes of human listeriosis.

Isolates of *L. monocytogenes* resistant to the bacteriocin nisin (discussed in further detail in chapter 21) have been studied and characterized. While the genetic determinant of resistance is currently unknown, the strains typically have a more rigid cell membrane that may physically impair attachment or insertion of nisin molecules (36, 42). Nisin is approved for use in some foods to prevent against clostridial outgrowth, not to control *L. monocytogenes*. It is not known whether nisin resistance results in increased survival of the bacterium in foods treated with nisin and increases the potential of foodborne illness.

Escherichia coli

E. coli may be one of the more publicly recognizable foodborne pathogens, although only certain pathogenic types, such as O157:H7, cause illness. Most *E. coli* strains are nonpathogenic and are part of the natural microbiota of many animals, including humans. Human infection by pathogenic, enterohemorrhagic *E. coli* is not typically treated with antibiotics. This makes antibiotic resistance in these strains practically irrelevant from a public health standpoint. When animal and human clinical isolates of *E. coli* O157:H7 have been tested for antibiotic susceptibility, resistance against tetracycline, streptomycin, and sulfamethoxazole is most commonly found.

While innate resistance is generally outside the scope of this chapter, there are noteworthy characteristics of *E. coli* O157:H7. A major outbreak of *E. coli* O157:H7 was associated with unpasteurized apple juice, an acidic product in which conventional knowledge about generic *E. coli* deemed the organism would not survive. The enhanced ability of the strain to survive acidic conditions is an innate quality. Interestingly, compared to many other strains of *E. coli*, serotype O157:H7 isolates have a slightly lower maximum growth temperature.

Staphylococcus aureus

Extracellular toxin production by large cell numbers of *S. aureus* causes foodborne illness; ingestion of the bacteria themselves does not. There has been considerable publicity in the medical community and among the general public regarding MRSA. MRSA typically causes nosocomial infection, not foodborne illness. Since there is no recognized increase in staphylococcal enterotoxin production by MRSA, while this pathogen is of great clinical significance its antibiotic resistance has no influence on staphylococcal food poisoning. Some strains of *S. aureus* are resistant to QACs, due either to plasmid acquisition or chromosomal alteration that results in efflux of the sanitizer (25). While this may enable the pathogen to persist in the food processing environment, most cases of foodborne illness related to *S. aureus* are related to postprocessing contamination by human contact, making the industrial relevance of QAC resistance among *S. aureus* strains questionable.

Fermentative Microorganisms

Fermentative microorganisms, such as lactic acid bacteria, are used in the production of many food products and are not known to be pathogenic. Therefore, there is no direct relationship between antibiotic resistance in these microbes and foodborne illness. However, these microorganisms may harbor resistance determinants, and the use of these microbes in food production may provide the opportunity for these determinants to spread to other microorganisms, including pathogens. For example, genetic analysis of clinical *L. monocytogenes* isolates led to the hypothesis that streptomycin resistance had been transferred from enterococci and/or streptococci (13).

Yeasts and Molds

Spoilage microorganisms are a major concern in food microbiology. The effectiveness of agents to control them must be maintained to ensure a reasonable shelf life for foods. Many yeasts and molds are innately resistant to sorbates and benzoates. When certain yeast strains have been exposed in laboratory studies to subinhibitory concentrations of sorbic or benzoic acid, they were more resistant to higher levels of the acids than were nonexposed cells (7, 58).

CROSS-RESISTANCE AND COSELECTION

The ability of a bacterial strain to display resistance to more than one antibiotic can be due to cross-resistance or coselection. In cross-resistance, this multidrug resistance may be due to the same mechanism. For example, an alteration in a cellular receptor may make a cell resistant to all antibiotics that act at that target site. In cases of coselection, the microbe may be resistant to a number of different antibiotics, each with distinct mechanisms of action. The different resistance determinants are generally genetically linked.

The multiple antibiotic resistance (*mar*) operon is of special interest in food microbiology. This global regulator can be expressed by a number of pathogens, including *E. coli* O157:H7 and *Salmonella enterica* serovar Enteritidis. Activation of the operon in serovar Enteritidis by chlorine, sodium nitrite, sodium benzoate, or acetic acid, all of which are used by the food industry, causes increased resistance to antibiotics such as tetracycline and ciprofloxacin.

A few studies have examined the response of antibiotic-resistant bacteria to food processing treatments such as heat. Antibiotic-resistant *Salmonella* and *L. monocytogenes* are as sensitive to heat as their susceptible counterparts (3, 57). In contrast, there is a relationship between ATR and heat resistance; activation of ATR confers increased protection against heat (3). This can be explained by the ATR mechanism. The induction of stress response proteins (due to exposure to acid) results in increased heat resistance, for which stress response proteins also play a role. This exemplifies the need to examine situations of cross-resistance on a case-by-case basis; broad assumptions about cross-resistance cannot be made without an understanding of the mechanism(s).

Only one study (30) has linked nisin resistance with antibiotic resistance. *S. aureus* nisin-resistant isolates were resistant to antibiotics. The minimal inhibitory concentrations for the antibiotics increased as much as 30-fold in the nisin-resistant strains. However, similar studies with *L. monocytogenes* revealed no significant increase in resistance to antibiotics (18, 37). Nisin-resistant *L. monocytogenes* and *Clostridium botulinum* are more sensitive to food preservatives such as low pH, salt, sodium nitrite, and potassium sorbate (36). Their increased acid sensitivity is caused by increased ATPase activity, which decreases intracellular energy stores (38).

Forthcoming data from the NARMS retail meat survey should reveal whether there is any relationship between antibiotic-resistant pathogens harbored by animals and contamination rates in the meat products. Higher contamination rates by pathogens in general may correspond with increased likelihood of illness, and contamination by antibiotic-resistant pathogens may affect the likelihood of an individual becoming ill (especially if the consumer was recently treated with an antibiotic) and may affect treatment of the illness with antibiotics.

The relationship between antibiotic resistance and increased virulence has not been fully elucidated (28), but there are reports that antibiotic-resistant bacteria are somewhat more virulent than susceptible strains (5). With respect to ATR, the enhanced ability of acid-adapted bacteria to survive passage through the stomach may increase the likelihood of intestinal infection, in addition to any direct relationship between activation of the ATR and virulence (59).

OPPORTUNITIES AND CONCERNS

An understanding of mechanisms that result in resistance and adaptation can enable food microbiologists to develop effective intervention strategies to improve the overall safety of foods. While exposure to sublethal stresses may result in stress adaptation, concurrent exposure to multiple stresses may cause cell death due to metabolic exhaustion, as bacteria try to maintain their cellular integrity and homeostatic balance. This is commonly referred to as "hurdle technology" (34). This approach should decrease resistance prevalence, because a microorganism would have to be resistant to multiple stressors to survive. Again, understanding the mechanisms of resistance will aid in the selection of the specific "hurdles" to minimize potential resistance.

In other cases, instead of sublethal exposure resulting in cross-resistance to other agents (such as the activation of the stress response proteins), the cell's susceptibility to other stresses may be enhanced. For example, after exposure to alkali cleaning solutions, four of five strains of L. monocytogenes were as sensitive or more sensitive to heat than unexposed cells, and all were more sensitive to free chlorine, benzalkonium chloride, and cetylpyridiunium chloride, components of sanitizers, than were controls (55).

A mechanistic understanding may also identify opportunities of collateral sensitivity, where the cellular changes resulting in resistance leave the cell more vulnerable to other types of antimicrobial agents. For example, several studies suggest that nisin resistance results in physiological changes that decrease resistance to other food safety interventions such as heat (20), acid, salt, sorbates, and nitrite (36). This was also true for some cephalosporins, which are normally ineffective against L. monocytogenes. A nisin-resistant strain was highly sensitive to expanded-spectrum and broad-spectrum cephalosporins at concentrations by which the wild-type strain was unaffected (39). The converse may also be true; i.e., resistance to some antibiotics increases sensitivity to nisin. A penicillin-resistant S. aureus mutant was 50-fold more sensitive to nisin (54). Bacillus licheniformis, which is resistant to the bacitracin it produces, is highly sensitive to detergents, likely due to a specific membrane change (45).

CONCLUSIONS

Antibiotics play a vital role in improving human, animal, and plant health. However, their effectiveness is endangered when their targets become resistant. The human medical community has begun reducing frivolous usage of antibiotics to decrease the selective pressure that generates antibiotic-resistant microorganisms. Similar efforts are ongoing in animal agriculture through prudent-use practices. Since sick animals must be treated, the complete elimination of animal antibiotic use is not practical and may be counterproductive. Efforts to minimize the agricultural use of antibiotics must be balanced to decrease the emergence of antibiotic-resistant microbes while not increasing the level of pathogens associated with food products, as this could result in an overall net increase in foodborne disease.

Not all manifestations of resistance result in increased foodborne illness or affect the ability to treat it. However, from a broader ecological perspective, resistance may still raise questions. For example, how might antibiotic resistance in commensal organisms and soil microflora affect foodborne illness? Is antibiotic resistance of no concern in foodborne pathogens whose illnesses are not treated with antibiotics, such as E. coli O157:H7? What role can food microbiologists play in combating the emergence of antibiotic-resistant pathogens?

Food microbiologists have little influence over the use of antibiotics in animal production, veterinary medicine, or human medicine which provides selective pressure that increases the prevalence of antibiotic-resistant microbes. Food microbiologists do use bacteriostatic and bactericidal agents such as food antimicrobials, sanitizers, and disinfectants and must be aware of the prevalence and impact of microbial resistance to these agents.

Standardized methods to assess resistance to food antimicrobial agents and sanitizers need to be developed. There is a continuing need to determine what combinations of microorganisms and antimicrobial agents represent the greatest resistance challenges for food safety. This information is difficult to ascertain, because FoodNet data do not distinguish foodborne illnesses caused by antibiotic-resistant and antibiotic-susceptible microorganisms, and food attribution data are generally lacking. With a better understanding of the resistance profiles of microorganisms causing illnesses, and information on the farm-to-table path of these microbes, we may be able to alter practices along that continuum and design intelligent approaches to counter resistance. This must be a continuing effort, since microorganisms are constantly evolving and finding new ways to resist antibiotics.

Finally, antibiotic-resistant bacteria appear to respond to food preservation interventions similarly to their susceptible counterparts. Therefore, interventions against pathogens in foods will be equally effective for antibiotic-resistant and -susceptible cells. The main preventive measure food microbiologists can take to counter antibiotic-resistant organisms is to continue to develop and implement effective interventions to improve the overall safety of foods.

References

1. **Aiello, A. E., B. Marshall, S. B. Levy, P. Della-Latta, S. X. Lin, and E. Larson.** 2005. Antibacterial cleaning products and drug resistance. *Emerg. Infect. Dis.* **11:**1565–1570.

2. **Archer, D. L.** 1996. Preservation microbiology and safety: evidence that stress enhances virulence and triggers adaptive mutations. *Trends Food Sci. Technol.* **7:**91–95.

3. **Bacon, R. T., J. R. Ransom, J. N. Sofos, P. A. Kendall, K. E. Belk, and G. C. Smith.** 2003. Thermal inactivation of susceptible and multiantimicrobial-resistant *Salmonella* strains grown in the absence or presence of glucose. *Appl. Environ. Microbiol.* **69:**4123–4128.

4. **Bartoloni, A., F. Bartalesi, A. Mantella, E. Dell'Amico, M. Roselli, M. Strohmeyer, H. G. Barahona, V. P. Barron, F. Paradisi, and G. M. Rossolini.** 2004. High prevalence of acquired antimicrobial resistance unrelated to heavy antimicrobial consumption. *J. Infect. Dis.* **189:**1291–1294.

5. **Barza, M., and K. Travers.** 2002. Excess infections due to antimicrobial resistance: the "attributable fraction". *Clin. Infect. Dis.* **34:**S126–S130.

6. **Bilgin, N., F. Claesens, H. Pahverk, and M. Ehrenberg.** 1992. Kinetic properties of *Escherichia coli* ribosomes with altered forms of S12. *J. Mol. Biol.* **224:**1011–1027.

7. **Bills, S., L. Restaino, and L. M. Lenovich.** 1982. Growth response of an osmotolerant sorbate-resistant yeast, *Saccharomyces rouxii*, at different sucrose and sorbate levels. *J. Food Prot.* **45:**1120–1124.

8. **Burns, J. L., P. M. Mendelman, J. Levy, T. L. Stull, and A. L. Smith.** 1985. A permeability barrier as a mechanism of chloramphenicol resistance in *Haemophilus influenzae*. *Antimicrob. Agents Chemother.* **27:**46–54.

9. **Burns, J. L., L. A. Hedin, and D. M. Lien.** 1989. Chloramphenicol resistance in *Pseudomonas cepacia* because of decreased permeability. *Antimicrob. Agents Chemother.* **33:**136–141.

10. **Bush, K., and R. B. Sykes.** 1984. Interaction of β-lactam antibiotics with β-lactamases as a cause for resistance, p. 131. *In* L. E. Bryan (ed.), *Antimicrobial Drug Resistance.* Academic Press, Orlando, Fla.

11. **Centers for Disease Control and Prevention.** 2004. National Nosocomial Infection Surveillance (NNIS) System report. Data summary from January 1992–June 2004. Issued October 2004. *Am. J. Infect. Control* **32:**470–485.

12. **Centers for Disease Control and Prevention.** 2005. Preliminary FoodNet data on the incidence of infection with pathogens transmitted commonly through food—10 sites, United States, 2004. *Morb. Mortal. Wkly. Rep.* **54:**352–356.

13. **Charpentier, E., G. Gerbaud, C. Jacquet, J. Rocourt, and P. Courvalin.** 1995. Incidence of antibiotic resistance in *Listeria* species. *J. Infect. Dis.* **172:**277–281.

14. **Cloete, T. E.** 2003. Resistance mechanisms of bacteria to antimicrobial compounds. *Int. Biodeterior. Biodegradation* **51:**277–282.

15. **Cody, S. H., S. L. Abbott, A. A. Marfin, B. Schulz, P. Wagner, K. Robbins, J. C. Mohle-Boetani, and D. J. Vugia.** 1999. Two outbreaks of multidrug-resistant *Salmonella* serotype Typhimurium DT104 infections linked to raw-milk cheese in northern California. *JAMA* **281:**1805–1810.

16. **Cohen, S. P., D. C. Hooper, J. S. Wolfson, K. S. Souza, L. M. McMurry, and S. B. Levy.** 1988. Endogenous active efflux of norfloxacin in susceptible *Escherichia coli*. *Antimicrob. Agents Chemother.* **32:**1187–1191.

17. **Cohen, S. P., L. M. McMurry, D. C. Hooper, J. S. Wolfson, and S. B. Levy.** 1989. Cross-resistance to fluoroquinolones in multiple-antibiotic-resistant (Mar) *Escherichia coli* selected by tetracycline or chloramphenicol: decreased drug accumulation associated with membrane changes in addition to OmpF reduction. *Antimicrob. Agents Chemother.* **33:**1318–1325.

18. **Crandall, A. D., and T. J. Montville.** 1998. Nisin resistance in *Listeria monocytogenes* ATCC 700302 is a complex phenotype. *Appl. Environ. Microbiol.* **64:**231–237.

19. **Davidson, P. M., and M. A. Harrison.** 2002. Resistance and adaptation to food antimicrobials, sanitizers, and other process controls. *Food Technol.* **56:**69–78.

20. **Davies, J.** 1994. Inactivation of antibiotics and the dissemination of resistance genes. *Science* **264:**375–382.

21. **Davison, H. C., J. C. Low, and M. E. Woolhouse.** 2000. What is antibiotic resistance and how can we measure it? *Trends Microbiol.* **8:**554–559.

22. **Dever, L. A., and R. S. Dermody.** 1991. Mechanisms of bacterial resistance to antibiotics. *Arch. Intern. Med.* **151:**886–895.

23. **Food and Drug Administration.** 2005. *Animal Drugs, Feeds, and Related Products; Enrofloxacin for Poultry; Withdrawal of Approval of New Animal Drug Application*, p. 44048–44049. Docket no. 2000N-1571, OC 2005194. FR doc. 05-15223. 1 August 2005.

24. Gaskins, H. R., C. T. Collier, and D. B. Anderson. 2002. Antibiotics as growth promotants: mode of action. *Anim. Biotechnol.* 13:29–42.

25. Gillespie, M. T., B. R. Lyon, and R. A. Skurray. 1989. Gentamicin and antiseptic resistance in epidemic methicillin-resistant *Staphylococcus aureus. Lancet* i:503.

26. Guffanti, A. A., S. Clejan, L. H. Falk, D. B. Hicks, and T. A. Krulwich. 1987. Isolation and characterization of uncoupler-resistant mutants of *Bacillus subtilis. J. Bacteriol.* 169:4469–4478.

27. Heir, E., B. A. Lindstedt, O. J. Rotterud, T. Vardund, G. Kapperud, and T. Nesbakken. 2004. Molecular epidemiology and disinfectant susceptibility of *Listeria monocytogenes* from meat processing plants and human infections. *Int. J. Food Microbiol.* 96:85–96.

28. Helms, M., P. Vastrup, P. Gerner-Smidt, and K. Molbak. 2002. Excess mortality associated with antimicrobial drug-resistant *Salmonella typhimurium. Emerg. Infect. Dis.* 8:490–495.

29. Hooper, D. C., J. S. Wolfson, K. S. Souza, E. Y. Ng, G. L. McHugh, and M. N. Swartz. 1989. Mechanisms of quinolone resistance in *Escherichia coli*: characterization of *nfxB* and *cfxB*, two mutant resistance loci decreasing norfloxacin accumulation. *Antimicrob. Agents Chemother.* 33:283–290.

30. Hossack, D. J. N., M. C. Bird, and A. A. Fowler. 1983. The effects of nisin on the sensitivity of microorganisms to antibiotics and other chemotherapeutic agents, p. 425–433. *In* M. Woodbine (ed.), *Antimicrobials and Agriculture.* Butterworth, London, United Kingdom.

31. Institute of Food Technologists. 2001. *Analysis and Evaluation of Preventive Control Measures for the Control and Reduction/Elimination of Microbial Hazards on Fresh and Fresh-Cut Produce.* Institute of Food Technologists, Chicago, Ill.

32. Institute of Food Technologists. 2006. Antimicrobial resistance: implication for the food system. *Comprehensive Rev. Food Sci. Food Safety* S(3).

33. John, J. F., Jr., and N. O. Fishman. 1997. Programmatic role of the infectious diseases physician in controlling antimicrobial costs in the hospital. *Clin. Infect. Dis.* 24:471–485.

34. Lestner, L., and G. W. Gould. 2002. *Multiple Hurdle Technologies.* Kluwer Academic, New York, N.Y.

35. Levin, B. R., V. Perrot, and N. Walter. 2000. Compensatory mutations, antibiotic resistance and the population genetics of adaptive evolution in bacteria. *Genetics* 154:985–997.

36. Mazzotta, A. S., K. Modi, M. L. Chikindas, and T. J. Montville. 2000. Inhibition of nisin-resistant (Nis[r]) *Listeria monocytogenes* and Nis[r] *Clostridium botulinum* by common food preservatives. *J. Food Sci.* 65:888–890.

37. McEntire, J. C. 2003. Relationship between nisin resistance and acid sensitivity of *Listeria monocytogenes.* Ph.D. dissertation. Rutgers University, New Brunswick, N.J.

38. McEntire, J. C., G. M. Carman, and T. J. Montville. 2004. Increased ATPase activity is responsible for acid sensitivity of nisin-resistant *Listeria monocytogenes* ATCC 700302. *Appl. Environ. Microbiol.* 70:2717–2721.

39. McManus, P. S., V. O. Stockwell, G. W. Sundin, and A. L. Jones. 2002. Antibiotic use in plant agriculture. *Annu. Rev. Phytopathol.* 40:443–465.

40. McMurray, L., R. E. Petrucci, Jr., and S. B. Levy. 1980. Active efflux of tetracycline encoded by four genetically different tetracycline resistance determinants in *Escherichia coli. Proc. Natl. Acad. Sci. USA* 77:3974–3977.

41. Mead, P. S., L. Slutsker, V. Dietz, L. F. McCaig, J. S. Bresee, C. Shapiro, P. M. Griffin, and R. V. Tauxe. 1999. Food-related illness and death in the United States. *Emerg. Infect. Dis.* 5:607–625.

42. Ming, X., and M. Daeschel. 1993. Nisin resistance of food-borne bacteria and the specific resistance responses of *Listeria monocytogenes* Scott A. *J. Food Prot.* 56:944–948.

43. Modi, K., M. L. Chikindas, and T. J. Montville. 2000. Sensitivity of nisin-resistant *Listeria monocytogenes* to heat and the synergistic action of heat and nisin. *Lett. Appl. Bacteriol.* 30:249–253.

44. Nikaido, H., and M. Vaara. 1985. Molecular basis of bacterial outer membrane permeability. *Microbiol. Rev.* 49:1–32.

45. Podlesek, Z., A. Comino, B. Herzog-Velikonja, and M. Grabnar. 2000. The role of the bacitracin ABC transporter in bacitracin resistance and collateral detergent sensitivity. *FEMS Microbiol. Lett.* 188:103–106.

46. Poole, K., K. Krebes, C. McNally, and S. Neshat. 1993. Multiple antibiotic resistance in *Pseudomonas aeruginosa*: evidence for involvement of an efflux operon. *J. Bacteriol.* 175:7363–7372.

47. Prazak, M. A., E. A. Murano, I. Mercado, and G. R. Acuff. 2002. Antimicrobial resistance of *Listeria monocytogenes* isolated from various cabbage farms and packing sheds in Texas. *J. Food Prot.* 65:1796–1799.

48. Rosen, G. D. 2003. Pronutrient antibiotic replacement standard discussed. *Feedstuffs* 75:11–13, 16.

49. Samelis, J., and J. N. Sofos. 2003. Strategies to control stress-adapted pathogens and provide safe foods, p. 303–351. *In* A. E. Yousef and V. K. Juneja (ed.), *Microbial Adaptation to Stress and Safety of New-Generation Foods.* CRC Press, Inc., Boca Raton, Fla.

50. Sanyal, D., and D. Greenwood. 1993. An electron microscope study of glycopeptide antibiotic-resistant strains of *Staphylococcus epidermidis. J. Med. Microbiol.* 39:204–210.

51. Shaw, K. J., P. N. Rather, R. S. Hare, and G. H. Miller. 1993. Molecular genetics of aminoglycoside resistance genes and familial relationship of the aminoglycoside-modifying enzymes. *Microbiol. Rev.* 57:138–163.

52. Shryock, T. 2000. Growth promotion and feed antibiotics, p. 735–743. *In* J. F. Prescott, J. D. Baggot, and R. D. Walker (ed.), *Antimicrobial Therapy in Veterinary Medicine.* Iowa State Press, Ames.

53. Song, J. S., J. H. Jeon, J. H. Lee, S. H. Jeong, B. C. Jeong, S.-J. Kim, J.-H. Lee, and S. H. Lee. 2005. Molecular characterization of TEM-type β-lactamases identified in cold-seep sediments of Edison Seamount (South of Lihir Island, Papua New Guinea). *J. Microbiol.* 43:172–178.

54. Szybalski, W. 1953. Genetic studies on microbial cross resistance to toxic agents. II. Cross resistance of *Micrococcus*

pyogenes var. *aureus* to thirty-four antimicrobial agents. *Antibiot. Chemother.* 3:1095–1103.

55. Taormina, P. J., and L. R. Beuchat. 2002. Survival of *Listeria monocytogenes* in commercial food-processing equipment cleaning solutions and subsequent sensitivity to sanitizers and heat. *J. Appl. Microbiol.* **92**:71–80.

56. Villar, R. G., M. D. Macek, S. Simons, P. S. Hayes, M. J. Goldoft, J. H. Lewis, L. L. Rowan, D. Hursh, M. Patnode, and P. S. Mead. 1999. Investigation of multidrug-resistant *Salmonella* serotype Typhimurium DT104 infections linked to raw-milk cheese in Washington State. *JAMA* **281**:1811–1816.

57. Walsh, D., J. J. Sheridan, G. Duffy, I. S. Blair, D. A. McDowell, and D. Harrington. 2001. Thermal resistance of wild-type and antibiotic-resistant *Listeria monocytogenes* in meat and potato substrates. *J. Appl. Microbiol.* **90**:555–560.

58. Warth, A. D. 1988. Effect of benzoic acid on growth yield of yeasts differing in their resistance to preservatives. *Appl. Environ. Microbiol.* **54**:2091–2095.

59. Wilmes-Riesenberg, M. R., B. Bearson, J. W. Foster, and R. Curtis III. 1996. Role of the acid tolerance response in virulence of *Salmonella typhimurium*. *Infect. Immun.* **64**:1085–1092.

Food Microbiology: Fundamentals and Frontiers, 3rd Ed.
Edited by M. P. Doyle and L. R. Beuchat
© 2007 ASM Press, Washington, D.C.

Peter Setlow
Eric A. Johnson

3

Spores and Their Significance

Members of the gram-positive *Bacillus* and *Clostridium* spp. and some closely related genera respond to slowed growth or starvation by initiating the process of sporulation. Spores formed by certain species in the genera *Alicyclobacillus*, *Bacillus*, *Clostridium*, *Desulfotomaculum*, *Paenibacillus*, and *Sporolactobacillus* present practical problems in food microbiology. The molecular biology of sporulation and spore resistance has been extensively studied in *Bacillus subtilis* (38, 48, 92, 93, 105, 110, 111, 131–133, 135, 136). This chapter describes the fundamental basis of sporulation and problems that spores present to the food industry.

The first obvious morphological event in sporulation is an unequal cell division. This creates the smaller prespore or forespore compartment and the larger mother cell compartment. As sporulation proceeds, the mother cell engulfs the forespore, resulting in a forespore within a mother cell, and eventually the mother cell lyses (7, 28, 48). Since the spore is formed within the mother cell, it is termed an endospore.

Throughout sporulation gene expression is ordered not only temporally but also spatially, as some genes are expressed only in the mother cell or the forespore (48, 110, 111). The pattern of gene expression during sporulation is controlled by the ordered synthesis and/or activation of five new sigma (σ; specificity) factors for RNA polymerase and a number of DNA binding proteins (48, 110). As sporulation proceeds there are striking morphological and biochemical changes in the developing spore. It becomes encased in two and sometimes three novel layers: a large layer of peptidoglycan (PG) termed the spore cortex, whose structure differs from that of growing cell PG; a number of layers of spore coats; and, in spores of some species, an exosporium. Both the coats and exosporium contain proteins unique to spores (119, 154). The spore's central region or core also accumulates a huge depot (≥10% of spore dry weight) of pyridine-2,6-dicarboxylic acid (dipicolinic acid [DPA]) (Fig. 3.1), as well as a large amount of divalent cations, and the core loses much water (38). The developing forespore also synthesizes a large amount of small acid-soluble proteins (SASP), some of which coat the spore chromosome and protect the DNA from damage (25, 92, 93, 131–133, 135, 136). As a result of these and other changes, the spore becomes metabolically dormant and resistant to many harsh conditions including heat, radiation, and chemicals (38, 92, 135).

Peter Setlow, Dept. of Biochemistry, University of Connecticut Health Center, Farmington, CT 06030-3305. Eric A. Johnson, Food Research Institute, University of Wisconsin-Madison, Madison, WI 53706.

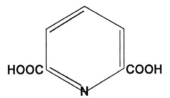

Figure 3.1 Structure of dipicolinic acid. Note that at physiological pH both carboxyl groups will be ionized.

Despite the spore's extreme dormancy, if given the appropriate stimulus (often a nutrient such as a sugar or amino acid), the spore can return to life via spore germination (86, 92, 101). Within minutes of exposure to a germinant, spores lose their unique characteristics, including loss of DPA by excretion, loss of the cortex and SASP by degradation, and loss of spore resistance. Completion of germination allows progression into outgrowth, when metabolism of endogenous and exogenous compounds begins and macromolecular synthesis is initiated. Eventually the outgrowing spore is converted back into a growing cell.

Detailed study of sporulation, spores, and spore germination and outgrowth has been motivated by a number of factors, one of which is the attraction of this model developmental system. This attraction has led to a detailed understanding of mechanisms that modulate gene expression during sporulation, as well as factors involved in spore resistance. Another motivating factor is the major role played by sporeformers in food spoilage and foodborne diseases; related to this factor is the recent widespread recognition of the potential for the use of spores of *Bacillus anthracis* as an agent of bioterrorism. While a tremendous amount of knowledge has been gained in studies motivated by these disparate factors, the amount of cross talk between individuals working on either basic or applied aspects of spore research has never been optimal. Thus, one purpose of this chapter is to highlight the state of knowledge of molecular mechanisms of sporulation, spore resistance and dormancy, and spore germination and outgrowth. Hopefully, it will provide a counterpoint to more applied aspects of this system. This review will focus on molecular mechanisms, most of which have been examined in *B. subtilis*. This bacterium is neither an important pathogen nor an important agent of food spoilage. However, its natural transformability, as well as an abundance of molecular biological and genetic information, including the complete sequence of its genome, has made *B. subtilis* the bacterium of choice for mechanistic studies on sporulation, spore germination, and spore resistance. Although much less mechanistic work has been carried out with other sporeformers, these studies indicate that the mechanisms regulating gene expression during sporulation, bringing about spore resistance and dormancy, and involved in spore germination and outgrowth are probably similar to those in *B. subtilis*. Determination of the genome sequences of many other gram-positive sporeformers is at or near completion, including data for at least seven *Bacillus* species, five *Clostridium* species, and two other species closely related to *Bacillus*. Comparison of these sequence data with those for *B. subtilis* indicates that there is a tremendous degree of conservation among genes involved in sporulation, spore resistance, and spore germination and outgrowth.

SPORULATION
Distribution of Sporeformers
The sporulating bacteria discussed in this chapter form heat-resistant endospores that contain DPA and are refractile or phase bright under phase-contrast microscopy. Most studies on sporulation, spores, and spore germination have been carried out with species of either the aerobic bacilli or the anaerobic clostridia. However, members of other closely related genera form similar spores, among these being *Geobacillus*, *Paenibacillus*, *Sporosarcina*, *Sporolactobacillus*, and *Thermoactinomyces* species (4, 90, 138).

Induction of Sporulation
The mechanism for inducing sporulation in the laboratory is limitation for one or more nutrients, by either exhaustion of nutrients during growth or shifting cells from a rich to a poor medium. Addition of an inhibitor (decoyinine) of guanine nucleotide biosynthesis also leads to sporulation. Although these laboratory methods cause the great majority of cells in a culture to sporulate, this may not be how sporulation is induced in nature. Even cells in a growing culture produce a small but finite number of spores, with the percentage of spores increasing as the culture's growth rate decreases (21).

Massive sporulation of a cell population generally takes place only when cells enter stationary phase. However, sporulation is not an obligatory outcome of entry into stationary phase. Indeed, many early events in stationary phase are attempts to access new sources of nutrients such that sporulation either will not take place or is at least delayed. These stationary-phase events include the following: (i) synthesis and secretion of degradative enzymes such as amylases and proteases (109, 140); (ii) synthesis and secretion of antibiotics such as gramicidin or bacitracin (109, 140); (iii) in some species, synthesis and release of protein toxins active against insects (3) or animals (84); (iv) development of motility (96); (v) killing and cannabalism of sister cells in the population (42); and (vi) in a few species (i.e., *B. subtilis*), development of genetic competence (26) (Fig. 3.2). Although these phenomena are not necessary for sporulation, they are often regulated by mechanisms that modulate gene expression during sporulation (26, 109). Like sporulation, competence is also regulated in a cell-density-dependent manner via the secretion of small peptides termed competence pheromones (26, 73).

There are also drastic changes in the metabolism of the stationary-phase cell, with these changes extending into sporulation. Some of these changes include catabolism of polymers like poly-β-hydroxybutyrate formed in vegetative growth and initiation of oxidative metabolism due to synthesis of tricarboxylic acid (TCA) cycle enzymes (140). Many of these latter enzymes are not present in

the developing forespore. Consequently, the mother cell and forespore have different metabolic capabilities.

Progression into sporulation generally requires regulatory signals that lead to the derepression of genes expressed only in stationary phase, although usually not the products of genes encoding degradative enzymes, antibiotics, toxins, and proteins involved in motility or competence (109, 141). However, induction of sporulation does require completion of chromosome replication, repair of DNA damage, and induction of synthesis of TCA cycle enzymes (109, 122, 141). Entry into sporulation is also increased in cells growing at high cell density by small molecules secreted into the growth medium (73).

Morphological, Biochemical, and Physiological Changes during Sporulation

Sporulation is divided into seven stages based on morphological characteristics of cells throughout the

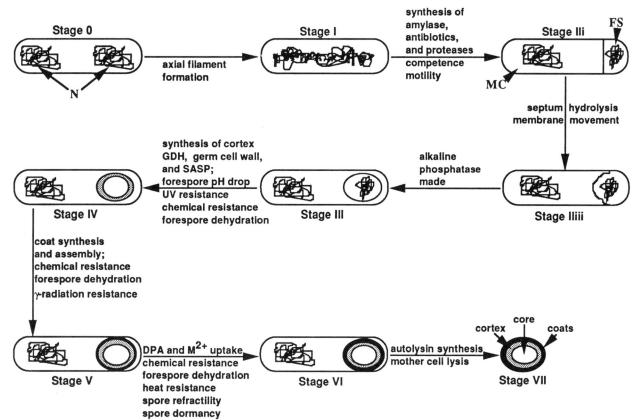

Figure 3.2 Morphological, biochemical, and physiological changes during sporulation of a rod-shaped *Bacillus* cell. In stage 0, a cell with two nucleoids (N) is shown; in stage IIi the mother cell and forespore are designated MC and FS, respectively. Note that the forespore nucleoid is more condensed than that in the mother cell. Stage IIii is not shown in this scheme, and the forespore nucleoid is not shown after stage III for clarity. The time of some biochemical and physiological events, such as forespore dehydration and acquisition of types of resistance to different chemicals (all lumped together as "chemical resistance"), stretches over a number of stages. The data for this figure are taken from references 28 and 136.

developmental process, which can take as little as 8 h (Fig. 3.2). Growing cells are in stage 0. Older literature had suggested that the first morphological feature unique to sporulation was the presence of the two nucleoids in an axial filament that could be observed in electron micrographs; this was defined as stage I. Because of concerns that this axial filament was an artifact of preparation and because no single mutant has been found that is blocked in the stage 0/stage I transition, stage I has often been ignored. However, there is more recent evidence that stage I is a discrete stage in sporulation (110).

The first morphological feature of sporulation that can easily be seen in the light microscope is the formation of an asymmetric septum dividing the sporulating cell into the mother cell and forespore compartments (110). Despite the isolation of many genes involved in division septum formation (30), the mechanism guiding the asymmetric placement of the sporulation septum is not clear. Immediately after asymmetric septation, only ca. 30% of the chromosome destined for the spore is present in the forespore; the remainder is subsequently transferred into the forespore, and then both mother cell and forespore compartments contain complete and apparently identical single chromosomes (130). However, the forespore chromosome is initially more condensed than the mother cell chromosome (130). A biochemical marker for late stage II is the synthesis of alkaline phosphatase. During stage II, genes in the two compartments of the sporulating cell may be expressed differentially.

Following septum formation, the mother cell membrane engulfs the forespore, surrounding the forespore with two complete membranes, the inner and outer forespore or spore membranes (see below). These two membranes have opposite polarities (28). There are a number of changes that occur in the transition from stage II to stage III, leading to the subdivision of stage II into three substages, two of which are shown in Fig. 3.2.

In the transition from stage III to stage IV, a large PG structure termed the cortex is laid down between the inner and outer forespore membranes. The cortex PG has a structure similar to that of cell wall PG, but with a number of differences (described below). The spore's germ cell wall is made at about the same time as the cortex but appears to have the same structure as growing cell wall PG (113). During the stage III to stage IV transition, the forespore also synthesizes glucose dehydrogenase and SASP; a number of the SASP are involved in spore resistance. The developing forespore acquires full UV resistance and some chemical resistance at this time, and the forespore chromosome adopts a ringlike shape due to the binding of specific SASP (110, 131). Late in stage III the forespore pH falls by 1 to 1.5 units and forespore dehydration begins.

In the stage IV-to-stage V transition, the proteinaceous coat layers are laid down outside the outer forespore membrane (154). The coats of spores of *B. anthracis* and *B. subtilis* contain ≥ 30 proteins, almost all of which are unique to spores (68, 71, 154). The reason for the plethora of coat proteins is not clear, as many *B. subtilis* coat proteins can be lost with no apparent phenotypic effect. Forespore γ-radiation resistance begins to be acquired during this period, as does further chemical resistance, and forespore dehydration continues (131). During the stage V-to-stage VI transition, the spore core's depot of DPA is accumulated following DPA synthesis in the mother cell. DPA uptake is paralleled by uptake of divalent cations, predominantly Ca^{2+}, but much Mg^{2+} and Mn^{2+} as well (38). The great majority of these cations are in the spore core, presumably in a 1:1 complex with DPA. The precise state of these compounds in the spore core is not known, although the amount of DPA accumulated exceeds its solubility. During this period the spore's central region or core undergoes the final process of dehydration. Because of the high ratio of solids to water in the spore core at this stage, the spore appears bright in phase-contrast microscopy. It has been suggested that the spore core is in a glasslike state (1), but this suggestion remains controversial (74). Because of permeability changes in the spore membranes, most likely the inner membrane, the spore at this stage stains poorly, if at all, with common bacteriological stains. The spore also becomes metabolically dormant during this period and acquires further γ-radiation and chemical resistance (131). Finally, in the transition to stage VII, autolysins are produced in the mother cell, resulting in its lysis and release of the spore.

The seven stages are particularly useful in characterizing various asporogenous or *spo* mutants that have no defect in growth but are blocked in a particular stage in sporulation. These *spo* mutants are given an added designation denoting the stage in which they are blocked. Thus, *spo0* and *spoII* mutants are blocked in stages 0 and II, respectively. The various stages have also allowed correlation of biochemical changes with morphological changes (Fig. 3.2). However, the sporulation scheme outlined above is an oversimplification for several reasons. First, since the various stages are intermediates in a continuous developmental process, it is probably misleading to think of the stages as discrete entities. Second, the scheme is only for rod-shaped bacteria that sporulate without terminal swelling of the forespore compartment. There are many sporeformers in which the forespore compartment swells considerably and the mother cell elongates (138). Some sporeformers also grow as cocci (138).

Regulation of Gene Expression during Sporulation

Much of our knowledge of the regulation of gene expression during sporulation has been derived predominantly from analysis of *spo* mutants in *B. subtilis*. Asporogeny can be caused by mutations in any 1 of more than 75 distinct genetic loci (111). The identification of biochemical or physiological markers for various stages of sporulation provides another major aid in understanding gene regulation during sporulation (Fig. 3.2). Analysis of these markers in *spo* mutants indicates that sporulation is primarily a linear process of sequential events, although there are many checkpoints that coordinate development in the two cell types. In general, *spo0* mutants exhibit no sporulation-specific events, *spoII* mutants exhibit only stage 0-specific events, etc.

While analysis of *spo* mutants has furnished a broad outline of the regulation of gene expression during sporulation, molecular genetic technology has provided a detailed understanding of this process (48, 110, 111). It was initially hypothesized that sporulation requires transcription of many new genes, and this has been elegantly demonstrated by a variety of techniques, most recently using microarray technology (27, 145, 164). It was further suggested that changes in transcription during sporulation were modulated by changes in the specificity of the cell's RNA polymerase due to synthesis of new σ factors. This suggestion has been expanded over the past 20 years, as changes in gene expression during sporulation require modulation of transcription by many mechanisms including the following: (i) synthesis of RNA polymerase σ factors with altered promoter specificity; (ii) activation or inactivation of σ factors by covalent and noncovalent modification; (iii) regulatory communication to coordinate development in the mother cell and forespore; (iv) synthesis of DNA binding proteins to activate or repress transcription; and (v) degradation of DNA binding proteins and σ factors (48, 110, 111). This information has provided a detailed picture of the control of gene expression during sporulation that is striking not only in its complexity but in the redundancy of its control mechanisms. The following discussion of these control mechanisms has been simplified and concentrates on major regulatory gene products.

Initiation of sporulation requires expression of *spo0* gene products at a significant level in vegetative cells (110). The most important of these *spo0* gene products is Spo0A. This protein is the response regulator half of a two-component regulatory system. These signal transduction systems transmit signals, often by binding to DNA and affecting transcription, when an aspartyl residue in the response regulator is phosphorylated by a sensor kinase (105). In growing cells, Spo0A is primarily in the dephosphorylated state. However, under conditions that initiate sporulation, the level of Spo0A rises, as does its degree of phosphorylation, the latter through action of multiple sensor kinases. In *B. subtilis* the majority of phosphate on Spo0A is derived via a phosphorelay initiated by phosphorylation of Spo0F (Fig. 3.3 and 3.4). The phosphate is then transferred from Spo0F~P to Spo0A by Spo0B. There are two major kinases, KinA and KinB, that initiate the phosphorelay (57, 105). Mutations in either *kinA* or *kinB* have little or no effect on sporulation, but a *kinA kinB* double mutant is asporogenous. There are also three other minor kinases (57, 105). In addition to the kinases there are a number of phosphatases that can dephosphorylate Spo0A~P, either directly or through dephosphorylation of Spo0F~P (Fig. 3.3). This multiplicity of kinases and phosphatases allows a variety of environmental signals to be integrated to determine the intracellular level of Spo0A~P, and a threshold concentration of Spo0A~P is needed to initiate sporulation, as Spo0A and Spo0A~P modulate the expression of ≥100 genes (35, 36). A number of regulators of the kinases and phosphatases that modulate the level of Spo0A~P have been identified (35, 36, 57, 105, 110, 122). Since overexpression of any of the kinases that generate Spo0A~P triggers sporulation in growing cells, these kinases may be held inactive by some metabolite whose level decreases on entry into stationary phase/sporulation (36).

The phosphorylation of Spo0A increases its affinity for binding to sites upstream of several key genes, although different sites have different affinities for Spo0A~P (36). The *abrB* gene has a high affinity for Spo0A~P, and binding of Spo0A~P decreases *abrB* transcription. Since AbrB is labile, a decrease in *abrB* transcription rapidly

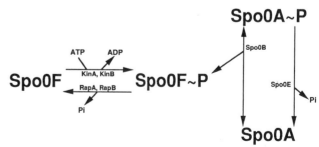

Figure 3.3 Some gene products and reactions that affect levels of Spo0A~P. Spo0E is a phosphatase that acts on Spo0A~P; RapA and RapB are phosphatases that act on Spo0F~P (105, 114, 161).

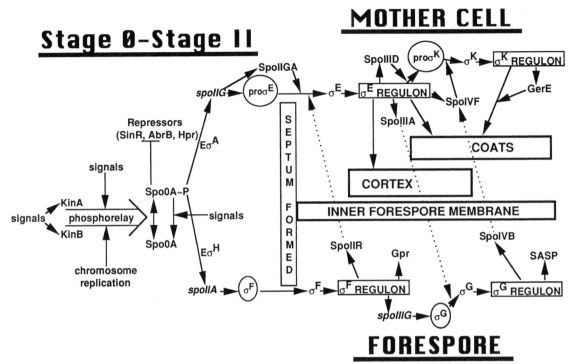

Figure 3.4   Regulation of gene expression during sporulation. The effect of Spo0A~P on repressors is negative; other effects of regulatory molecules on reactions are generally positive, although the effect of signals may be positive or negative. The enclosure of the pro-σ factors and σ factors denotes that at this time these factors are inactive. This figure is adapted from that in reference 136.

decreases the intracellular AbrB concentration. AbrB is a repressor of a number of genes normally expressed in the stage 0 to stage II transition, including *spo0H*, which encodes σ^H. AbrB is also an activator for synthesis of a second repressor, termed Hpr, that represses additional stage 0-expressed genes, especially the genes for several proteases. In addition to AbrB and Hpr, there is a third repressor, termed SinR, that represses other genes expressed in the stage 0 to stage II transition, including those of the *spoIIG* and *spoIIA* operons. SinR action is blocked by synthesis of a protein termed SinI that binds to SinR and blocks its action. Synthesis of SinI is stimulated by Spo0A~P and probably repressed by AbrB and Hpr. There is also a fourth repressor, CodY, that represses many genes whose expression is needed early in stationary phase/sporulation, including genes whose products are involved in competence, motility, and the TCA cycle (88, 118). CodY also represses genes required for sporulation including *spo0A*, and increased levels of total Spo0A are a hallmark of sporulation (35, 36, 88, 118). The CodY corepressor is GTP, and the following facts have long been known: (i) a decrease in the cell's GTP pool takes place early in stationary phase/

sporulation; and (ii) an artificial reduction in the GTP pool can induce sporulation in growing cells. The consequences of all this regulation are that early in the stage 0-to-stage II transition, genes normally repressed during vegetative growth are derepressed through a decrease in GTP levels and an increase in the level of Spo0A~P (Fig. 3.4). Important among these stage 0-to-stage II genes is *spo0H*, which encodes σ^H. There is some transcription of *spo0H* during vegetative growth by RNA polymerase containing the cell's main sigma factor, σ^A. However, levels of active σ^H in growing cells are kept low, probably by a posttranscriptional mechanism, but in sporulation, increased levels of σ^H promote transcription by Eσ^H (110). As Spo0A~P levels increase, concentrations become sufficient to bind the promoters of the *spoIIG* and *spoIIA* operons which have been derepressed by removal of SinR as a SinI-SinR complex (110). The binding of Spo0A~P stimulates transcription of these genes by RNA polymerase containing σ^A or σ^H (109, 110). The *spoIIA* operon is transcribed by Eσ^H and encodes three proteins, with the third cistron (*spoIIAC*) encoding another sigma factor termed σ^F. The *spoIIAB* gene encodes an inhibitor of σ^F function, whereas *spoIIAA* encodes an antagonist of SpoIIAB (48, 110). Prior to septation, σ^F is inactive in the sporulating cell due to its interaction with SpoIIAB. Interaction of SpoIIAB with σ^F rather than with SpoIIAA is promoted by the high ATP/ADP ratio in the preseptation sporulating cell; phosphorylation of SpoIIAA by SpoIIAB also prevents SpoIIAA-SpoIIAB interaction (48, 110). The *spoIIG* operon is transcribed by Eσ^A and encodes two proteins, with the second cistron, *spoIIGB*, encoding another σ factor for RNA polymerase, σ^E (48, 110, 111). Unlike σ^F, σ^E is synthesized as an inactive precursor termed pro-σ^E. The first gene of the *spoIIG* operon (*spoIIGA*) is the protease responsible for processing pro-σ^E to σ^E. Action of SpoIIGA on pro-σ^E requires gene expression under σ^F control in the forespore (Fig. 3.4). Although *spoIIG* is transcribed before septum formation, σ^E is not generated until after septation, and then only in the mother cell (48, 110). σ^F is also produced before septum formation and is maintained in an inactive state by interaction with SpoIIAB as noted above. However, following septation, σ^F becomes active in the forespore, as SpoIIAB leaves σ^F and interacts with SpoIIAA. This process is promoted largely by dephosphorylation of SpoIIAA~P by the SpoIIE phosphatase that resides in the sporulation septum at this time (16, 169), but it is not clear how σ^F activity is confined to the forespore. Eσ^F transcribes a number of genes in the forespore including *rsfA*, *spoIIR*, *gpr*, and *spoIIIG* (48, 110, 111, 164) (Fig. 3.4). RsfA is a DNA binding protein that modifies the specificity of Eσ^F (164). SpoIIR promotes pro-σ^E processing by SpoIIGA in the mother cell, although the mechanism of action of SpoIIR is not clear. SpoIIR can promote pro-σ^E processing not only in the mother cell, but also in the forespore (48, 110). However, pro-σ^E is normally not present in the forespore shortly after septation. The requirement for *spoIIR* transcription in the forespore for pro-σ^E processing in the mother cell is the first of a number of examples of cross talk between the two compartments of the sporulating cell that coordinate the development of the mother cell and forespore (48, 110, 111). Among other Eσ^F-transcribed genes, *gpr* encodes a protease that acts on SASP in the first minute of spore germination and *spoIIIG* encodes another σ factor, termed σ^G, that is responsible for the bulk of forespore-specific transcription (111). However, σ^G is not active following its synthesis in stage III. Instead, some mother cell-specific event(s), including transcription of the *spoIIIA* operon by Eσ^E, is required for generation of active Eσ^G (48, 110) (Fig. 3.3 and 3.4). This is the second example of cross talk between mother cell and forespore, although the mechanism for regulation of σ^G activity in the forespore is not clear (129). The generation of σ^E only in the mother cell and the synthesis of σ^G only in the forespore now establish compartment-specific transcription during sporulation.

As noted above, Eσ^G transcribes a number of genes expressed only in the forespore. As the third example of regulatory cross talk, one such gene (*spoIVB*) is responsible for communicating with the mother cell and coordinating gene expression in this compartment (48, 110, 111) (Fig. 3.4) (see below). The *ssp* genes are a large set of genes transcribed by Eσ^G. These genes encode the SASP that are major protein components of the spore core (see below) (131, 132). Transcription of genes by Eσ^G is modulated by the DNA binding protein SpoVT, and the *spoVT* gene is transcribed by Eσ^G (164).

In contrast to Eσ^G, Eσ^E transcribes genes only in the mother cell, including genes needed for cortex biosynthesis and some genes (*cot* genes) needed for coat formation and assembly (27, 37, 111). Eσ^E also transcribes *spoIIID*, which encodes a DNA binding protein that modulates Eσ^E action, resulting in transcription of different classes of Eσ^E-dependent genes. One such Eσ^E-dependent gene that requires SpoIIID for transcription is *sigK*, which encodes a final new σ factor termed σ^K (48, 110, 111). In *B. subtilis*, the *sigK* gene has an intervening sequence that is removed only from the mother cell genome by a recombinase. Expression of the recombinase is regulated such that generation of an intact *sigK* gene just precedes *sigK* transcription. However, this intervening sequence is absent in the *sigK* genes of other *Bacillus* species (48, 110). As is the case with σ^E, σ^K is synthesized as an inactive pro-σ^K and is processed proteolytically about 1 h after its synthesis (48, 110). Conversion of pro-σ^K to σ^K in the mother cell requires expression of one σ^G-controlled gene (*spoIVB*) in the forespore (the third example of cross talk) and participation of the σ^E-controlled *spoIVF* operon in the mother cell (48, 110, 111) (Fig. 3.4). SpoIVFB is the protease that processes pro-σ^K, and SpoIVFA is an inhibitor of SpoIVFB function. Eσ^K also transcribes *sigK* (in conjunction with SpoIIID), the gene for DPA synthase, other *cot* genes, and *gerE*, which encodes a DNA binding protein that modulates Eσ^K activity, causing a change in the pattern of *cot* gene expression (48, 110, 111) (Fig. 3.4). Most knowledge of gene expression during sporulation ends at this point, including information about genes that may be required for mother cell lysis. However, the gene for one autolytic enzyme (CwlC) is transcribed by Eσ^K (111).

The preceding picture of regulation of gene expression during sporulation is derived predominantly from studies with *B. subtilis*. Is the picture similar in bacteria from other species or genera, including *Clostridium*? Although there is no definitive answer to this question, genomic sequence data indicate that most key regulatory genes such as *spo0A*, *spoIIGA*, *spoIIGB*, *spoIIIG*, *sigK*, etc. are present in other sporeformers, including *Clostridium* spp. The striking conservation of both the sequences of the encoded proteins and the organization of these genes strongly suggests that regulation of sporulation across species is very similar.

THE SPORE

Spore Structure

The structure of the dormant spore is quite different from that of a growing cell (Fig. 3.5), as many spore structures, including the exosporium and coats, have no counterparts in growing cells. The outermost spore layer, the exosporium, varies significantly in size between species and, while very large in spores of *Bacillus cereus*, *B. anthracis*, *Bacillus thuringiensis*, and some *Clostridium* species, may not be present in spores of species such as *B. subtilis* (146, 153). The exosporium is composed of lipid, carbohydrate, and protein, and the proteins are unique to spores (119, 144, 153, 163). However, the function of the exosporium is not known. Underlying the exosporium are the spore coats. In *B. subtilis* spores, a number of distinct coat layers are seen in electron micrographs (154). The coats are composed primarily of protein, and there are ≥30 different proteins in the *B. subtilis*

spore coat, all of which are unique to spores (68). Some coat proteins appear to function in the assembly of the coat structure, but specific roles for most coat proteins are not known. Posttranslational modification of some coat proteins, in particular formation of dityrosine and γ-glutamyl-lysine cross-links, may also be important in spore coat structure (116, 154). While most coat proteins have no known function, as a whole the coats protect spore PG from attack by lytic enzymes, the spore's inner layers against many chemicals, and the spore itself against digestion by predatory protozoa (see below). However, the spore coats play no significant role in maintenance of spore resistance to heat or radiation (38, 131). Underlying the spore coats is the outer forespore membrane. This is a functional membrane in the developing forespore but may not be functional in the dormant spore (101). The protein composition of this membrane is different from that of the inner forespore membrane.

Underlying the outer forespore membrane is the spore's cortex. This large PG layer is structurally similar to cell wall PG, but with several differences (113). Cortical PG contains diaminopimelic acid, even if vegetative cell PG contains lysine, and ca. 65% of the muramic acid residues in cortical PG lack peptide residues. Although some muramic acid residues contain a single D-alanine, much is present as muramic acid-δ-lactam (MAL), a

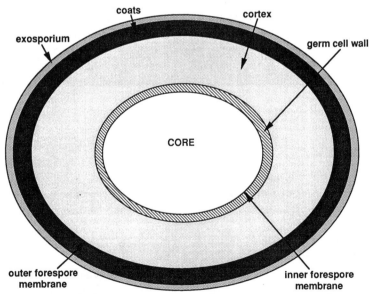

Figure 3.5 Structure of a dormant spore. The various structures are not drawn precisely to scale, especially the exosporium, whose size varies tremendously between spores of different species. The relative size of the germ cell wall is also generally smaller than that shown. The positions of the inner and outer forespore membranes, between the core and the germ cell wall and between the cortex and coats, respectively, are also noted.

compound not present in growing cell PG. A number of genes involved in cortex synthesis have been identified and characterized, and the pathway for MAL synthesis during sporulation is fairly well understood (37, 39, 113). While the timing of cortex biosynthesis and cross-linking during sporulation has been determined (113), the distribution of cross-links throughout the three-dimensional cortical volume is not known. This is unfortunate, as the spore cortex is likely responsible for the dehydration of the spore core and thus for spore dormancy and much spore resistance (see below). Between the cortex and the inner forespore membrane is the germ cell wall. The structure of this PG layer appears identical to that of vegetative cell PG (113).

The next structure, the inner forespore membrane, is a functional membrane and is an extremely strong permeability barrier slowing the entry of almost all molecules, including perhaps even water, into the spore core (17, 165). A lipid probe in this membrane is largely immobile, suggesting that this membrane has a relatively "frozen" structure (19). However, the inner membrane's phospholipid content is similar to that of growing cells (28). The volume surrounded by the inner forespore membrane appears smaller than expected based on this membrane's phospholipid content. However, the core's volume expands significantly upon completion of spore germination and in the absence of membrane synthesis (19). A lipid probe in the inner membrane also becomes fully mobile at this time as well.

Finally, the central region or core contains the spore's DNA, ribosomes, and most enzymes, as well as the DPA and most divalent cations. There are also many unique gene products in the dormant spore, including the large SASP pool (ca. 10% of spore protein), much of which are bound to spore DNA (131) (see below). A notable feature of the spore core is its low water content (38). While vegetative cells have ca. 4 g of water per g of dry weight, the spore core has only 0.4 to 1 g of water per g of dry weight. The core's low water content likely plays a major role in spore dormancy and in spore resistance to a variety of agents (38, 131). The core's low water content is also likely the reason for the immobility of ions and protein in the spore core (18). In contrast to the low water content in the spore core, the water content in other regions of the spore is similar to that in growing cells (38).

Spore Macromolecules

Some proteins in the spore coats and core are not present in the vegetative cell. The SASP are particularly noteworthy, and some of these proteins play a major role in spore resistance (131, 132). SASP are synthesized in the forespore during stage III of sporulation, when their coding genes are transcribed primarily by $E\sigma^G$; SASP are located almost exclusively in the spore core (25). There are three kinds of SASP in *Bacillus* spp., termed γ-type, α/β-type, and minor. There are at least nine minor SASP in *B. subtilis* spores with sizes ranging from 34 to 71 amino acids (25). The function of these minor proteins is generally unclear. The 75- to 105-residue γ-type SASP make up ca. 5% of spore protein and do not bind to any other spore macromolecule. Their only known function is to be degraded during outgrowth, providing amino acids for metabolism and protein synthesis. γ-Type SASP degradation is initiated by the sequence-specific protease GPR, the product of the *gpr* gene. The γ-type SASP are encoded by a single gene in *Bacillus* spp. However, the γ-type SASP are not present in spores of *Clostridium* species. The sequences of γ-type SASP are not homologous to other proteins in currently available databases, and these proteins' sequences have diverged significantly during evolution. The α/β-type SASP make up 3 to 5% of total spore protein and are named for the two major proteins of this type in *B. subtilis* spores. These proteins are found in spores of *Clostridium* species (25, 131, 132). In contrast to γ-type SASP, α/β-type SASP are coded for by up to seven genes. All of these genes are expressed in parallel, although in most species two proteins (major α/β-type SASP) are expressed at high levels and the remainder at much lower levels. However, all α/β-type SASP have similar properties in vitro and in vivo (see below). The amino acid sequences of these small (60- to 75-residue) proteins are highly conserved both within and across species, although they too have no sequence homology to other proteins in current databases (132). The α/β-type SASP are also degraded to amino acids in the first minutes of spore outgrowth in a process initiated by GPR.

The α/β-type SASP are DNA binding proteins, both in vivo and in vitro (25, 131, 132). In vivo these proteins saturate the spore chromosome and convert it into a ringlike toroidal structure (25, 34). The binding of α/β-type SASP provides much of the spore DNA's resistance to various treatments (see below). Binding of α/β-type SASP to DNA is not sequence specific but exhibits a preference for G/C-rich regions, although A/T-rich regions are bound. The proteins bind to the outside of the DNA helix, interacting primarily with the sugar phosphate backbone. This binding alters the DNA structure, although the precise structure of the α/β-type SASP-DNA complex is unknown (34), and provides resistance to chemical and enzymatic cleavage of the DNA backbone, alters the DNA's UV photochemistry, and greatly slows DNA depurination (see below). The spore nucleoid also contains HBsu, the major protein found on the vegetative cell nucleoid; this protein covers ca. 5% of spore DNA and modulates the effect of α/β-type SASP on spore DNA properties (121).

A number of proteins present in vegetative cells are absent from spores. These include amino acid and nucleotide biosynthetic enzymes that are degraded during sporulation and resynthesized during spore outgrowth (101, 136). Other proteins, including enzymes of amino acid and carbohydrate catabolism, are present at similar levels in spores and cells. Spores also contain enzymes needed for RNA and protein synthesis and many DNA repair enzymes. However, at least one protein needed for initiation of DNA replication may be absent. Enzyme activities common to both cell and spore proteins are invariably the same gene product. In general, spore and cell rRNAs and tRNAs are similar if not identical, although much tRNA in spores lacks the 3′-terminal A residue and little if any spore tRNA is aminoacylated. However, spores lack most if not all functional mRNA. Spore DNA appears identical to growing cell DNA but has a different structure due to different associated proteins as noted above.

Spore Small Molecules

Spores differ from growing cells in their content of small molecules (Table 3.1), which are located in the spore core. The small amount of spore core water and the huge depot of DPA and divalent cations have already been noted (Table 3.1). The ions in the spore core are immobile as is at least one normally mobile protein (18), consistent with a dearth of free water in spores. The pH in the spore core is 1 to 1.5 units lower than that in a growing cell or the mother cell compartment (Table 3.1). Spores of most species accumulate a large depot of 3-phosphoglyceric acid (3PGA) (Table 3.1) shortly after and in response to the forespore pH decrease (101, 131). However, spores have little if any of the common "high-energy" compounds found in growing cells, including deoxynucleoside triphosphates, ribonucleoside triphosphates, reduced pyridine nucleotides, and acyl-coenzyme A (CoA), although the "low-energy" forms of these compounds are present (Table 3.1). The high-energy forms of these latter compounds are lost from the forespore late in sporulation (Fig. 3.2). Much of the CoA in spores is in disulfide linkage, some as a CoA disulfide and some linked to protein (Table 3.1). The function of these disulfides is not known, but they are reduced early in spore outgrowth. In addition to lacking many amino acid biosynthetic enzymes, spores have low levels of most free amino acids but often have high levels of glutamic acid (Table 3.1).

Spore Dormancy

Spores are metabolically dormant, catalyzing no metabolism of endogenous or exogenous compounds. The major

Table 3.1 Small molecules in cells and spores of *Bacillus* species

Molecule	Content (mmol/g dry wt) in:	
	Cells[a]	Spores[b]
ATP	3.6	≤0.005
ADP	1	0.2
AMP	1	1.2–1.3
Deoxynucleotides	0.59[c]	<0.025[d]
NADH	0.35	<0.002[e]
NAD	1.95	0.11[e]
NADPH	0.52	<0.001[e]
NADP	0.44	0.018[e]
Acyl-CoA	0.6	<0.01[e]
CoASH[f]	0.7	0.26[e]
CoASSX[g]	<0.1	0.54[e]
3-PGA	<0.2	5–18
Glutamic acid	38	24–30
DPA	<0.1	410–470
Ca²⁺		380–916
Mg²⁺		86–120
Mn²⁺		27–56
H⁺	7.6–8.1[h]	6.3–6.9[h]

[a] Values for *B. megaterium* in mid-log phase are from references 131 and 136.
[b] Values are the range from spores of *B. cereus*, *B. subtilis*, and *B. megaterium* and are from references 131 and 136.
[c] Value is the total of all four deoxynucleoside triphosphates.
[d] Value is the sum of all four deoxynucleotides.
[e] Values are for *B. megaterium* only.
[f] CoASH, free CoA.
[g] CoASSX, CoA in disulfide linkage to CoA or a protein.
[h] Values are expressed as pH and are the range in *B. cereus*, *B. megaterium*, and *B. subtilis*.

cause of this dormancy is undoubtedly the low water content of the spore core, which likely precludes protein mobility and enzyme action (18, 38, 131). A reflection of this dormancy is that the spore core contains at least two enzyme-substrate pairs which are stable for months to years but which interact in the first 15 to 30 min of spore outgrowth, resulting in substrate degradation. These two enzyme-substrate pairs are 3PGA-phosphoglycerate mutase (PGM) and SASP-GPR (131). Although special regulatory mechanisms other than dehydration stabilize 3PGA and SASP in the developing forespore despite the presence of PGM and GPR, the lack of PGM and GPR action on their substrates in dormant spores may again be due to spore core dehydration.

SPORE RESISTANCE

The spore's metabolic dormancy is undoubtedly one factor in its ability to survive long periods in the absence of nutrients. A second factor is the spore's extreme resistance to potentially lethal treatments including heat, radiation, chemicals, and desiccation (38, 92, 132, 135), as spores are much more resistant than vegetative cells to a variety of killing treatments. Table 3.2 presents representative data for growing cells and spores of *B. subtilis*. Note, however, that some bacteria form much more resistant spores than does *B. subtilis*. Spore resistance is due to a variety of factors, with spore core dehydration and

Table 3.2 Killing and mutagenesis of spores and cells of *B. subtilis* by various treatments[a]

A. Freeze-drying

No. of freeze-drying cycles	Survival (%)		
	Cells[b]	Wild-type spores	α⁻β⁻ spores[c]
1	2		
3		100 (<0.5)	7 (14)

B. 10% Hydrogen peroxide[a]

Time of treatment (min)	Survival (%)		
	Cells[b]	Wild-type spores	α⁻β⁻ spores[c]
2.5	0.3	92	
5		88	50
10			10 (14)
20		50	0.1
60		6 (≤0.5)	

C. UV[d]

	Dose to kill 90% of the population (J/m²)		
	Cells[b]	Wild-type spores	α⁻β⁻ spores[c]
	40	315	25

D. Moist heat

Treatment temp (°C)	D value		
	Cells[b]	Wild-type spores	α⁻β⁻ spores[c]
95		14 min (≤0.5)	
85		360 min (≤0.5)	15 min (13)
65	<15 s	105 h	10 h
22		2.5 yr (≤0.5)	2.8 mo (18)

E. Dry heat

Treatment temp (°C)	D value		
	Cells[b]	Wild-type spores	α⁻β⁻ spores[c]
120		33 min (12)	
90	5 min		5.5 min (12)

[a] Data taken from references 32, 131, and 136. Values in parentheses are the percentages of survivors with asporogenous or auxotrophic mutations when spores undergo 30 to 99% killing.
[b] Cells in the log phase of growth. Similar results have been obtained with wild-type and α⁻β⁻ cells.
[c] These spores lack the two major α/β-type SASP and thus ca. 80% of the α/β-type SASP pool.
[d] UV irradiation with light predominantly at 254 nm.

α/β-type SASP involved in many types of resistance, whereas the impermeability of the spore's inner membrane may be involved in only one. Since different, sometimes multiple factors contribute to different types of spore resistance, it is not surprising that spore resistance to different treatments is acquired at different times in sporulation (Fig. 3.2). In recent years, some mechanisms of spore resistance to heat, UV, and chemicals have been elucidated. The following discussion of spore resistance concentrates on *B. subtilis* because of the detailed mechanistic data available for this bacterium. However, studies with spores of other bacteria, in particular on the mechanism of heat resistance, have indicated that factors involved in resistance of *B. subtilis* spores are also involved in resistance of spores of other species and genera. However, the relative importance of particular factors in spore resistance may vary significantly between species.

Spore Freezing and Desiccation Resistance
Growing bacteria are killed somewhat during freezing and even more during desiccation, unless special precautions are taken. The precise mechanism(s) of this killing is not clear, but one cause may be DNA damage. In contrast to the sensitivity of growing cells to freeze-drying, spores are resistant to multiple cycles of freeze-drying (92) (Table 3.2). The α/β-type SASP are one cause of spore resistance to freeze-drying by preventing DNA damage. Spores lacking these proteins (termed α⁻β⁻ spores) are much more sensitive to killing by freeze-drying than are wild-type spores (Table 3.2), with the killing of α⁻β⁻ spores due in large part to DNA damage. The spore's DPA is another component of spore desiccation resistance, as DPA-less spores are more sensitive to desiccation than their DPA-replete counterparts, and killing of DPA-less spores by desiccation is also due to DNA damage (S. Atluri, B. Setlow, R. Kitchell, K. Koziol-Dube, and P. Setlow, unpublished results). Accumulation of carbohydrates such as trehalose is often important in the resistance of yeast or other fungal spores to freezing or desiccation (20). However, bacterial spores do not accumulate such sugars (136).

Spore Pressure Resistance
Spores are much more resistant to high pressures (>50 megapascals [MPa]) than are growing cells (92). In general, spores are killed more rapidly at lower pressures (50 to 300 MPa) than at higher pressures (400 to 600 MPa) (92). This apparent anomaly is because spore killing by pressure requires spore germination and then spore killing. The most effective way to kill pressure-germinated spores is by heat, and thus pressure treatments are often carried out at elevated temperatures. However, spore

germination by lower pressures proceeds by the activation of the spore's nutrient germinant receptors (see below) (9, 99, 168), and the action of nutrient germinant receptors decreases at elevated temperatures. In contrast, germination promoted by higher pressures does not require the nutrient germinant receptors but may involve a change in the permeability of the spore's inner membrane allowing DPA release (99, 168; E. P. Black, J. Wei, K. Koziol-Dube, D. E. Cortezzo, D. G. Hoover, and P. Setlow, unpublished results); the rate of this process actually increases as the temperature is raised (Black, Wei, Koziol-Dube, Cortezzo, Hoover, and Setlow, unpublished). Even if the high pressure-germinated spore goes no further in spore germination, the now DPA-less spore is much more heat sensitive than is the initial dormant spore (100).

Spore γ-Radiation Resistance

Spores are more resistant to γ-radiation than are growing cells, and γ-radiation resistance is acquired 1 to 2 h before acquisition of heat resistance (92). Although SASP are not involved, the factors involved in spore γ-radiation resistance are not known (132). The low water content in the spore core would be expected to provide protection against γ-radiation. However, no analysis has correlated the degree of spore core dehydration with spore γ-radiation resistance. One impediment to understanding spore γ-radiation resistance is the lack of knowledge of the lethal damage, presumably to spore DNA, caused by γ-radiation.

Spore UV Radiation Resistance

Spores of many species are 7 to 50 times more resistant than are growing cells to UV radiation at 254 nm (93, 131–133) (Table 3.2), the wavelength giving maximal killing. Spores are also more resistant than growing cells at longer and shorter UV wavelengths. UV resistance is acquired by the developing forespore ca. 2 h before acquisition of heat resistance (Fig. 3.2), in parallel with synthesis of α/β-type SASP. The α/β-type SASP are essential for spore UV resistance but spore coats, cortex, and core dehydration are not.

The major reason for spore UV resistance is the different UV photochemistry of DNA in spores and in growing cells. The major photoproducts formed upon UV irradiation of growing cells or purified DNA are cyclobutane-type dimers between adjacent pyrimidines (93). The most abundant of these are between adjacent thymine residues (TT) (Fig. 3.6A), with smaller amounts formed between adjacent cytosine and thymine residues (CT) or adjacent cytosine residues (CC). In addition, UV irradiation of cells or purified DNA generates various

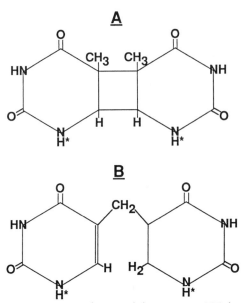

Figure 3.6 Structures of (A) cyclobutane-type TT dimer and (B) 5-thyminyl-5,6-dihydrothymine adduct (spore photoproduct). The positions of the hydrogens noted by the asterisks are the locations of the glycosylic bond in DNA.

6,4-photoproducts (64PPs) also formed between adjacent pyrimidines (93, 133). All these photoproducts can be lethal as well as mutagenic. In contrast to the UV photochemistry of purified DNA or DNA in growing cells, UV irradiation of spores generates few cyclobutane-type dimers and 64PPs, but rather large amounts of a thyminyl-thymine adduct initially termed "spore photoproduct" (SP) (93, 133) (Fig. 3.6B). The yield of SP as a function of UV fluence in spores is similar to the yield of TT as a function of UV fluence in growing cells, and SP is a potentially lethal photoproduct (132, 133). Hence, the difference in UV photochemistry between DNA in growing cells and spores is insufficient to explain spore UV resistance, as there must be a difference in the capacity of cells and spores to repair TT and SP. Indeed, spores have at least two mechanisms for SP repair, both of which operate in the first minutes of spore outgrowth. One mechanism is via the excision repair system that repairs TT and other lesions in growing cells (133). Spores lacking this repair system are two- to threefold more sensitive to UV than are wild-type spores (93, 133). The second repair system, unique to both spores and SP, monomerizes SP to two thymines without excision of the lesion and is very error-free (93, 133, 135). Spores lacking this SP-specific repair system are 5- to 10-fold more UV sensitive than are wild-type spores. Spores lacking both repair systems are 20- to 40-fold more UV sensitive than are wild-type spores (93, 133, 135). SP-specific repair is catalyzed by the spore photoproduct lyase (Spl), an iron-sulfur protein that uses S-adenosylmethionine (SAM) as a cofactor. Spl is a member of the "radical SAM" family of enzymes that use an [Fe-S] center plus SAM to generate a catalytic adenosyl radical (93, 133). In contrast to enzymes of excision repair that are present in growing cells and spores, Spl is present only in the developing forespore (93, 133).

The major factor causing the altered UV photochemistry of spore DNA is the saturation of DNA with α/β-type SASP; all α/β-type SASP appear relatively interchangeable in this (93, 131–133). Spores lacking ca. 80% of these proteins ($\alpha^-\beta^-$ spores) are more UV sensitive than are growing cells (Table 3.2), and UV irradiation of $\alpha^-\beta^-$ spores generates significant amounts of TT and 64PPs and reduced amounts of SP (22). UV irradiation of DNA saturated with any of a number of purified α/β-type SASP generates SP and no TT or 64PPs. The yields of SP and other photoproducts as a function of UV fluence in these complexes are ca. 10-fold lower than the yields in spores (22, 23). This difference is due to the spore's DPA that acts as a photosensitizer. Binding of α/β-type SASP to DNA in vitro also blocks formation of CT and CC, as well as 64PPs. The change in the UV photochemistry of DNA upon binding of α/β-type SASP strongly indicates that the DNA in this complex is in an altered structure, although the precise nature of this altered structure is unknown (34).

During spore outgrowth, α/β-type SASP are degraded to amino acids that support much protein synthesis. SASP degradation is initiated by the sequence-specific protease termed GPR (131–133). Because of the photosensitizing action of DPA, spores early in outgrowth are more UV resistant than are dormant spores due to release of DPA prior to significant SASP degradation. However, as SASP degradation proceeds, this elevated UV resistance falls to that of the vegetative cell.

Spore Chemical Resistance

Spores are more resistant than growing cells to a variety of chemical compounds, including glutaraldehyde, oxidizing agents (Table 3.2), phenols, formaldehyde chloroform, octanol, alkylating agents such as ethylene oxide, iodine, and detergents, as well as to pH extremes and lytic enzymes such as lysozyme (81, 92, 135). Resistance of spores to these agents is acquired at different times in sporulation (Fig. 3.2). For enzymes such as lysozyme, spores that lack much coat protein due to chemical treatment or mutation are sensitive to cortex degradation by lytic enzymes, and coats protect spores against digestion by predatory protozoa (70). Coat-defective spores are more sensitive to many chemicals including oxidizing agents, perhaps because coat proteins serve as a "reactive armor" that inactivates toxic chemicals before they reach more-sensitive targets farther within the spore. The very slow rate of passage of hydrophilic molecules across the spore's inner membrane also plays an important role in spore resistance to many chemicals (17). A number of chemicals, including formaldehyde and alkylating agents, kill spores at least in part by DNA damage, whereas others, including many oxidizing agents, do not (17, 92, 135). This latter group of chemicals often inactivates the spore's germination apparatus. However, this is likely not the mechanism for spore killing, as these chemicals appear to damage the spore's inner membrane, such that even if the treated spores are germinated artificially, the inner membrane ruptures, leading to spore death (17, 92, 135).

Some detailed information is available about the points raised above for oxidizing agents such as peroxides. Peroxides kill cells by several mechanisms, but a major one is the generation of hydroxyl radicals that cause mutagenic or lethal DNA damage (135). However, spore killing by hydrogen peroxide or organic hydroperoxides such as t-butylhydroperoxide (tBHP) is not accompanied by significant DNA damage or mutagenesis (Table 3.2) (135). Consequently, DNA in spores must be well protected against damage by these agents. One reason for the high resistance of spores to peroxides may be that spore coats form a protective barrier (135). It is likely that the spore's inner membrane is relatively impermeable to these hydrophilic compounds (17), and the decreased core water content also plays a role in spore peroxide resistance (135). However, these mechanisms do not protect spore DNA against these agents, which is accomplished through the saturation of the spore chromosome with α/β-type SASP (135). In contrast to wild-type spores, which are extremely resistant to hydrogen peroxide and tBHP, $\alpha^-\beta^-$ spores are much more sensitive (92, 135) (Table 3.2). While there is no increase in mutations among survivors of wild-type spores killed 90 to 99% by hydrogen peroxide or tBHP, 5 to 15% of the survivors of $\alpha^-\beta^-$ spores treated similarly have obvious mutations (92, 135) (Table 3.2). DNA from $\alpha^-\beta^-$ spores killed by hydrogen peroxide also has a high frequency of single-strand breaks, whereas DNA from wild-type spores killed similarly exhibits no such damage (135). These data suggest that in wild-type spores, the saturation of DNA with α/β-type SASP provides such good protection against DNA damage that spore killing by peroxides is by other mechanisms. However, in $\alpha^-\beta^-$ spores, the rates of DNA damage caused by peroxides are greatly increased such that DNA damage is a significant cause of spore death. Supporting this simple model is that one component of spore hydrogen peroxide resistance is acquired during sporulation in parallel with accumulation of α/β-type SASP. This component of resistance is not acquired during sporulation of an α$^-$β$^-$ strain. Saturation of DNA with α/β-type SASP also blocks hydrogen peroxide cleavage of the DNA backbone in vitro (135). However, α/β-type SASP binding to DNA is not the only factor in spore hydrogen peroxide resistance. Indeed, one additional component of resistance is acquired during sporulation at about the time of final spore core dehydration (135). This may reflect roles in hydrogen peroxide resistance for spore core dehydration or the low permeability of the spore's inner membrane, or that the loss of reduced pyridine nucleotides from the developing spore (which also occurs at this time) blocks extensive production of hydroxyl radicals through the Fenton reaction (17, 135). Unlike the situation in growing cells, enzymes such as catalases and alkylhydroperoxide reductases and the DNA binding protein MrgA play no role in spore resistance to peroxides (135).

Spore Heat Resistance

Heat resistance, probably the spore resistance most familiar to food microbiologists, has great implications for the food industry and is probably the most studied spore resistance property (38, 135). Spore heat resistance is remarkable, as spores of many species can withstand 100°C for many minutes. Heat resistance is often quantified as a D_t value, which is the time in minutes at temperature (t) needed to kill 90% of a cell or spore population. Generally, D values for spores at a temperature of $t + 40$°C are approximately equal to those for their vegetative cell counterparts at temperature t. The extended survival of spores at elevated temperatures is paralleled by even longer survival times at lower temperatures. Spore D values increase 4- to 10-fold for each 10°C decrease in temperature (38). Consequently,

a spore with a D value at 90°C ($D_{90°C}$) of 30 min may have a $D_{20°C}$ value of many years.

One deficiency in our understanding of the spore heat resistance mechanism is the identity of the target(s) whose damage results in heat killing of spores. This target is not spore DNA, as spore heat killing is associated with neither DNA damage nor mutagenesis (32) (Table 3.2). A protein may be the target of spore heat killing (135), but such a protein has not been identified. Sublethal heat treatment can also damage spores in some way, with this damage being repairable during spore germination and outgrowth (50), but the nature of this damage is unknown. In contrast to our lack of knowledge about the mechanism(s) of heat killing, there is much more information on factors which modulate spore heat resistance, as discussed below.

Sporulation Temperature

Elevated sporulation temperatures increase spore resistance to moist heat (8, 38, 83). Indeed, spores of thermophiles generally have much higher heat resistance than spores of mesophiles. Since spore macromolecules are generally identical to cell macromolecules, spore macromolecules are not intrinsically heat resistant. Presumably, the total macromolecular content of spores from thermophiles is more heat stable than that from mesophiles, accounting for the higher heat resistance of spores of thermophiles. However, spores of one strain are also more heat resistant when prepared at higher temperatures (Table 3.3). This is probably not due to temperature-related changes in total macromolecular composition and may be due to the reduced core water content in spores prepared at higher temperatures (8) (Table 3.3). However, how higher sporulation temperatures cause reduced spore core water contents is not known. Elevated sporulation temperatures do not alter spore resistance to

Table 3.3 Heat resistance of *B. subtilis* spores prepared at different temperatures with different ions and with or without α/β-type SASP[a]

Spore	Prepn temp (°C)	Mineralization	H$_2$O (g/g of spore core wet wt)	$D_{100°C}$
B. subtilis	50	Native	0.335	45
	37	Ca^{2+}	0.425	37
	37	Native	0.50	8.9
	37	H$^+$	0.571	2.7
	20	Native	0.55	4.9
B. subtilis 168 wild type	37	Native	0.37	360[b]
B. subtilis 168 α$^-$β$^-$	37	Native	0.37	15[b]

[a] Data from references 8 and 32.
[b] $D_{85°C}$.

dry heat or UV radiation but do increase spore resistance to a number of chemicals (83, 135). This latter effect may be due to changes in levels of spore coat proteins as a function of the sporulation temperature (83).

In growing bacteria, adaptation to heat stress involves the proteins of the heat shock response. The levels of these proteins would be expected to increase with increasing sporulation temperature. Indeed, a heat shock during sporulation increases the heat resistance of the resultant spores (85). However, heat shock proteins play no role in spore heat resistance (92, 135).

α/β-Type SASP

The killing of spores by moist heat is not by DNA damage, as neither general mutagenesis (Table 3.2) nor DNA damage accompanies spore killing (131, 132, 135), even though high temperatures would be expected to cause DNA depurination. Therefore, spore DNA is remarkably well protected against heat damage such that inactivation of spores by moist heat is due to mechanisms other than DNA damage. The major cause of spore DNA protection against moist-heat damage appears to be the saturation of spore DNA by α/β-type SASP, and $\alpha^- \beta^-$ spores of *B. subtilis* have D values 5 to 10% of those for wild-type spores (Tables 3.2 and 3.3). In addition, while moist-heat killing of wild-type spores generates <1% obvious mutations in survivors, such killing of $\alpha^- \beta^-$ spores produces ca. 20% obvious mutations among survivors (Table 3.2). Killing of $\alpha^- \beta^-$ spores by moist heat is also accompanied by a large amount of DNA damage. This DNA damage includes abasic sites that may be the primary heat-induced lesion, as well as single-strand breaks that may be generated by secondary cleavage reactions at abasic sites (135, 136). As would be predicted from these latter findings, α/β-type SASP slow the rate of DNA depurination in solution at least 20-fold (135).

Wild-type spores are much more resistant to dry heat than to moist heat, as D values are 2 to 3 orders of magnitude higher in dry versus moist spores (Table 3.2) (132). Wild-type spores also exhibit a high level of mutagenesis accompanying killing by dry heat (132, 135) (Table 3.2), and this mutagenesis is associated with generation of DNA damage (92). $\alpha^- \beta^-$ spores are more sensitive to dry heat than are wild-type spores (with survivors of dry-heat killing of $\alpha^- \beta^-$ spores also exhibiting a high percentage of mutations) and exhibit heat resistance similar to that of dry vegetative cells (135) (Table 3.2). These findings suggest the following: (i) α/β-type SASP are a major factor increasing the dry-heat resistance of wild-type spores and (ii) α/β-type SASP provide significant DNA protection against dry-heat damage; but (iii) dry spores are so well protected against mechanisms of heat

killing other than DNA damage that at elevated temperatures DNA damage does eventually kill dry spores. In support of these findings, α/β-type SASP slow the depurination caused by dry-heat treatment of purified DNA. However, in contrast to the ≥20-fold decrease in DNA depurination in solution caused by α/β-type SASP these proteins only slow dry-heat-induced depurination ca. 3-fold (135).

Spore Mineralization

Spores accumulate large amounts of divalent cations late in sporulation, approximately in parallel with DPA. DPA was initially thought to be important in causing spore heat resistance. However, this role of DPA was called into question by the isolation of mutants that formed DPA-less but heat-resistant spores (100, 135). Unfortunately, the mutations in these latter strains have not been characterized, and there may have been multiple mutations. Analyses of mutations in *B. subtilis* that eliminate DPA synthetase have revealed that DPA-less spores are much less moist-heat resistant than are their DPA-replete brethren (100). Much of the latter effect is likely due to the increased core water content of DPA-less spores (see below). However, DPA also provides significant protection to spore DNA against damage by either dry or moist heat when α/β-type SASP are absent (Atluri, Setlow, Kitchell, Koziol-Dube, and Setlow, unpublished). DPA also appears to be required for dormant spore stability, as DPA-less spores germinate spontaneously, although they can be stabilized by other mutations (100).

The amount and type of mineral ions accumulated also affect spore heat resistance (8, 38). Analyses of spores of several species from which mineral ions have been removed by titration with acid and the spores then back-titrated with a mineral hydroxide give the order of spore heat resistance with different cations as $H^+ < Na^+ < K^+ < Mg^{2+} < Mn^{2+} < Ca^{2+} <$ untreated. Despite the clear role for mineral ions in spore heat resistance, the mechanism of this effect remains unclear. Alteration of spore mineralization can alter spore core water content (Table 3.3), which presumably causes a significant effect on heat resistance. However, mineralization may also affect spore heat resistance independently of its effects on core water content (Atluri, Setlow, Kitchell, Koziol-Dube, and Setlow, unpublished) (Table 3.3).

Spore Core Water Content

Low core water content is a major factor contributing to spore heat resistance. Dehydration of the spore begins in the stage III to stage IV transition and continues through stages IV and V, with final dehydration taking place approximately in parallel with acquisition of full spore heat resistance (38, 135). Synthesis of the spore cortex is essential both for effecting this dehydration and for maintaining the dehydrated state of the spore core. This is undoubtedly due to the ability of PG to change its volume markedly upon changes in ionic strength and/or pH. If an expansion in cortex volume is restricted to one direction, i.e., towards the spore core, core water would be extruded via mechanical action. Although the precise mechanism of this process is unclear, the important role of the cortex in heat resistance is shown by the inverse correlation between spore heat resistance and the volume occupied by the spore cortex relative to that of the core (38). This correlates not only across species, but also in a single species in which spore cortex biosynthesis has been altered by mutation (113). Presumably, the volume of the spore cortex influences the degree of core dehydration, and the amount of cortex and presumably its mechanical strength are crucial to the spore's ability to maintain core dehydration during heat treatment. Another factor determining spore core water content is DPA, as DPA-less spores have higher core water levels than otherwise identical DPA-replete spores (100). Presumably the core volume normally occupied by DPA is occupied by water in DPA-less spores. As noted above, DPA-less spores are much more sensitive to moist heat than are DPA-replete spores.

Studies of spores from a large number of species have revealed a good correlation between spore core water content and moist-heat resistance over a 20-fold range of D values (8, 38) (Fig. 3.7). However, at the extremes of core water contents, D values vary widely, presumably reflecting the importance of other factors such as sporulation temperature, cortex structure, etc., in modulating spore moist-heat resistance (38). One value that is unfortunately missing from analyses of spore water content is the amount of core water that is free water. Studies of protein and ion movement in spores have indicated that these molecules are immobile (18), consistent with a dearth of free water in the core. Presumably, the low water content in the spore core causes heat resistance as well as long-term spore survival by slowing water-driven chemical reactions such as DNA depurination and protein deamidation. A low water content also stabilizes proteins against denaturation by restricting their molecular motion. It would be informative to know the precise amount of water associated with spore core macromolecules in order to calculate the degree of their stabilization by the low core water content.

DNA Repair

As noted above, DNA repair is essential for spore resistance to UV and is also essential for spore resistance to

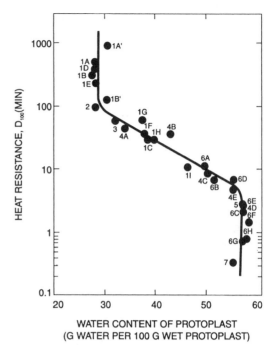

Figure 3.7 Correlation of spore heat resistance and protoplast (core) water content of lysozyme-sensitive spore types from seven *Bacillus* species that vary in thermal adaptation and mineralization. Reprinted from Gerhardt and Marquis (38) with permission. The numbers refer to spores of various species: 1, *G. stearothermophilus*; 2, "*Bacillus caldolyticus*"; 3, *Bacillus coagulans*; 4, *B. subtilis*; 5, *B. thuringiensis*; 6, *B. cereus*; and 7, *Bacillus macquariensis*. The letters denote the sporulation temperature or the mineralization of the spores of various species as described in the original publication.

chemicals that kill spores by DNA damage (135). However, spore resistance to moist heat is not altered by loss of DNA repair functions, including those mediated by RecA (92). This is not surprising since moist heat does not kill wild-type spores by DNA damage. However, spore resistance to dry heat, a treatment that does kill spores by DNA damage, is markedly decreased by the loss of various DNA repair activities, including RecA and other proteins (92, 125, 164).

SPORE ACTIVATION, GERMINATION, AND OUTGROWTH

Activation

Although spores are metabolically dormant and can remain in this state for many years, if given the proper stimulus they return to active metabolism within minutes through the process of spore germination (Fig. 3.8). A spore population will often germinate more rapidly

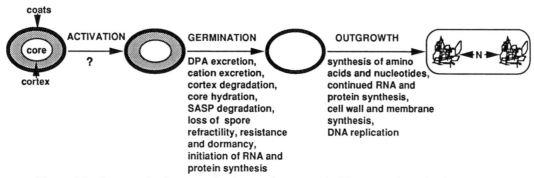

Figure 3.8 Spore activation, germination, and outgrowth. The events in activation are not known, hence the question mark. The loss of the spore cortex and the hydration and swelling of the core are shown in the germinated spore. The figure is adapted from Fig. 3 in reference 134.

and completely if activated prior to addition of a germinant (101, 134). However, the requirement for activation varies widely among spores of different species. A number of agents cause spore activation, including low pH and some chemicals, although the most widely used agent is sublethal heat. The precise changes induced by spore activation are not clear, although in some species, the activation process is reversible.

Germination

The precise period encompassed by spore germination, as distinguished from subsequent spore outgrowth, has been given a number of different definitions. For the purposes of this review, we consider spore germination as those events that take place without an immediate need for metabolic energy. During germination, a resistant dormant spore with a cortex and a large pool of DPA and mineral ions is transformed into a sensitive germinated spore in which the cortex has been degraded, DPA and most mineral ions have been excreted, and the core water content has become similar to that in a growing cell (101, 134). These changes can take place in the absence of metabolism (55, 101, 134). However, conversion of this germinated spore into a growing cell via outgrowth requires exogenous nutrients.

The initiation of spore germination can be triggered by a variety of compounds, including nucleosides, amino acids, sugars, salts, CaDPA, and long-chain alkylamines (101, 134). CaDPA and alkylamines are relatively universal germinants, whereas the specific nutrients that act as germinants vary from species to species. For nutrient germinants, metabolism of the germinant is not required. The stereospecificity for nutrient germinants (e.g., L-alanine is a germinant, whereas D-alanine often inhibits germination) strongly suggests that these molecules interact with specific proteins (86, 101, 134).

Analysis in *B. subtilis* has indicated that these proteins are the products of one of three tricistronic *ger* operons, *gerA*, *gerB*, and *gerK*. These operons are expressed in the developing forespore during sporulation, and mutations in any cistron of these operons result in spores altered in initiating spore germination in response to one or more, but not all, nutrient germinants. Spores lacking all three of these operons germinate extremely poorly in all nutrients (101, 134). It appears likely that the products of any one operon form a complex which functions as a nutrient germinant receptor (53, 86, 101, 134), and these receptors are located in the spore's inner membrane (86, 134). In addition to the tricistronic *ger* operons noted above, there are several other *ger* genes in *B. subtilis* in which mutations alter spore germination (52, 86, 101, 134). For some of these additional *ger* genes the function is known, whereas for others it is not. In general, all *ger* genes identified in *B. subtilis* have counterparts in other *Bacillus* species, and some, in particular genes encoding nutrient germinant receptors, are also in *Clostridium* species.

The earliest biochemical events in germination following mixing of nutrient germinants with spores are the release of protons and other monovalent cations and probably Zn^{2+} (101). Releases of DPA and associated divalent cations are next and are associated with some uptake of water by the spore core; cortex degradation and further core water uptake follow (101, 134). These events require large changes in the permeability of the inner forespore membrane. However, neither the control nor the mechanism of these changes is well understood, although proteins encoded by the *spoVA* operon may be involved in DPA excretion (157, 162). The changes accompanying spore germination may be extremely rapid, in as little as 30 s to 2 min for an individual spore. However, for a spore population the time can be much longer, as individual spores initiate germination after very different lag times. Indeed, some spores in a population appear superdormant and may require special conditions or treatment to induce their germination (101). In *B. subtilis* the release of DPA and other ions and their replacement by water increases the core water content from 35 to 45% of wet weight (101, 134). This change reduces spore moist-heat resistance significantly but has little effect on resistance to other agents and does not allow resumption of enzyme action or protein mobility (18, 101, 134). The release of DPA and other ions and their replacement by water have been termed stage I of germination, and germination can be halted after stage I by mutations that block degradation of the spore cortex.

Initiation of cortex degradation is the crucial next event in germination, as it allows a rapid two- to three-fold increase in spore core volume through water uptake once the cortex that restricts core expansion is removed. This is termed stage II of spore germination and leads to spore outgrowth. The importance of cortex lysis in spore germination has focused attention on cortex-lytic (lysozymelike) enzymes (CLEs). A number of CLEs have been isolated from spores of different species, including CwlJ, SleB, SleC, SleL, and SleM (80, 86, 101, 134). These enzymes generally are located in the spore's outer layers, are active only on intact spore cortex or cortex fragments, and appear to be specific for peptidoglycan containing MAL. This is consistent with the lack of degradation of the cortex upon initiation of germination of spores of a *B. subtilis cwlD* mutant, as CwlD is required for formation of MAL (39, 113). The specificity of CLEs prevents these enzymes from hydrolyzing the germ cell wall during spore germination.

The function of the various CLEs in vivo has been studied best in *B. subtilis*, where CwlJ and SleB play redundant but overlapping roles in cortex degradation (80, 101, 134). However, a *cwlJ sleB* mutant completes stage I of spore germination, indicating that the stage I events do not require cortex degradation (134). SleB appears to be a lytic transglycosylase, but the bond specificity of CwlJ has not been established (6). Both CwlJ and SleB are present in spores in a mature, potentially active form and must therefore be regulated in some fashion. While regulation of SleB is not understood, CwlJ activation requires exposure to the CaDPA that is released in stage I of germination (98). CwlJ can also be activated by high concentrations (tens of mM) of exogenous CaDPA, and this is the reason that CaDPA acts as a germinant (98). *Clostridium perfringens* spores contain a lytic enzyme, SleC, which is present in spores as a zymogen that is activated by proteolysis early in spore germination (80).

A similar enzyme has been suggested to be present in spores of at least one *Bacillus* sp., and there are reports of inhibition of spore germination by protease inhibitors (101). However, the target for these protease inhibitors is not known, and no mutations in proteases or other genes that might encode CLEs have been identified as affecting germination of *B. subtilis* spores.

Outgrowth

Following the completion of stage II of spore germination, core hydration has returned to that of a growing cell and protein mobility is restored (18, 101, 134). Consequently, enzymatic reactions begin in the spore core (101, 134). These include utilization of the spore's depot of 3PGA to generate ATP and NADH, degradation of SASP initiated by GPR and completed by peptidases, catabolism of some of the amino acids produced by SASP degradation, and initiation of catabolism of exogenous compounds. Carbohydrate metabolism early in outgrowth uses glycolysis and/or the hexose monophosphate shunt but stops at acetate as the spore lacks a number of enzymes of the TCA cycle. In *Bacillus megaterium*, endogenous energy reserves can support ATP production during the first 10 to 15 min of outgrowth, but no longer.

RNA synthesis is initiated in the first minutes of outgrowth, using nucleotides stored in the spore or generated by the breakdown of preexisting spore RNA. The identity of the first RNAs made at this time is not clear, but they are probably mRNA. The precise structure of the RNA polymerase acting early in outgrowth is also not clear. Protein synthesis begins shortly after RNA synthesis. All the components of the protein-synthesizing machinery stored in the dormant spore appear functional, with the exception of some tRNA lacking the 3'-terminal adenosine residue. However, this tRNA is repaired by tRNA nucleotidyltransferase early in outgrowth. As was the case with RNA synthesis early in outgrowth, endogenous amino acids derived largely from SASP breakdown support most protein synthesis in the first 20 to 30 min of outgrowth. Despite the utilization of endogenous reserves early in outgrowth, completion of outgrowth, a process that can take as little as 90 min, requires exogenous nutrients (101, 134). During outgrowth, the spore regains the ability to synthesize amino acids, nucleotides, and other small molecules due to synthesis of biosynthetic enzymes at defined times. However, the regulation of gene expression during spore outgrowth has not been thoroughly studied.

DNA replication is not generally initiated until at least 60 min after the start of outgrowth. DNA repair can occur well before DNA replicative synthesis, and spores contain deoxynucleoside triphosphates by the first minutes of outgrowth (101, 134). However, the importance of DNA repair in the first minutes of outgrowth, other than repair of UV damage, has not been well studied. During spore outgrowth, the volume of the outgrowing spore continues to increase, requiring the synthesis of membrane and cell wall components.

One question that remains intriguing is whether there are genes that are needed for outgrowth but no other stage of growth. A number of mutations affecting outgrowth (termed *out* mutants) have been isolated in *B. subtilis*, and most have been identified (101, 128). While the functions of some of these genes have been established, to date none has been shown to function solely in outgrowth.

PRACTICAL PROBLEMS OF SPORES IN THE FOOD INDUSTRY

Sporeforming bacteria and heat-resistant fungi pose specific problems for the food industry. Three species of sporeformers, *Clostridium botulinum*, *C. perfringens*, and *B. cereus*, are well known to produce toxins that can cause illness in humans and animals (40, 47, 49, 166), and many species of sporeformers cause spoilage of foods (56, 87, 148). Some strains of *Clostridium butyricum* and *Clostridium baratii* can also produce botulinal neurotoxin, but fortunately these isolates are rare and their spores appear to have a lower heat resistance than their nontoxigenic counterparts (47, 59, 60). Sporeformers causing foodborne illness and spoilage are particularly important in low-acid foods (equilibrium pH, ≥4.6) packaged in cans, bottles, pouches, or other hermetically sealed containers ("canned" foods), which are processed by heat (148). Certain sporeformers also cause various types of spoilage of high-acid foods (equilibrium pH, <4.6) (12, 87). Psychrotrophic sporeformers have also been increasingly recognized to cause spoilage of refrigerated foods (87). Fungi that produce heat-resistant ascospores are also an important cause of spoilage of acidic foods and beverages such as fruits and fruit products (112, 156).

Food spoilage by sporeforming bacteria was discovered by Pasteur during investigation of butyric acid fermentation in wines (13). Pasteur was able to isolate a bacterium he termed *Vibrion butyrique*, which is probably the same microorganism now classified as *C. butyricum*, the type species of the genus *Clostridium*. Endospores were discovered independently by Ferdinand Cohn and Robert Koch in 1876, soon after Pasteur had made microbiology famous (64). Investigations by Pasteur and Koch led to the association of microbial activity with the safety and quality of foods. During investigation of the anthrax bacillus, *B. anthracis*, the famous Koch's postulates were born; these proved that a disease is caused by a specific microorganism and also led to the development of pure culture techniques. Diseases and spoilage problems caused by sporeformers have historically been associated with foods that are thermally processed, because heat selects for survival and subsequent growth of sporeforming microorganisms.

The process of appertizing, or preserving food sterilized by heat in a hermetically sealed container, was invented in the late 1700s by Appert (2), who believed that the elimination of air was responsible for the long shelf lives of thermally processed foods. The empirical use of thermal processing gradually developed into modern-day thermal-processing industries. In the late 1800s and early 1900s, several North American and European scientists were instrumental in developing scientific principles to ensure the safety and prevent the spoilage of thermally processed foods (41). As a result of these studies, thermal processing of foods in hermetically sealed containers became an important industry (115, 124). In early studies of canned food spoilage, Prescott and Underwood at the Massachusetts Institute of Technology and Russell at the University of Wisconsin found that endospore-forming bacilli caused the spoilage of thermally processed clams, lobsters, and corn (115, 124). The classic studies of Esty and Meyer (31) in California provided definitive values of the heat resistance of *C. botulinum* type A and B spores and helped rescue the United States canning industry from its near demise from botulism in commercially canned olives and other foods. Quantitative thermal processes, understanding of spore heat resistance, aseptic processing, and implementation of HACCP (Hazard Analysis-Critical Control Point) concepts also resulted from developments in the canned food industry.

Low-Acid Canned Foods

The U.S. Food and Drug Administration (FDA) and the U.S. Department of Agriculture (USDA) Food Safety Inspection Service define a low-acid canned food as one with a finished equilibrium pH of ≥4.6 and a water activity (a_w) of greater than 0.85. The regulations regarding thermal processing of canned foods are described in the U.S. Code of Federal Regulations (21 CFR, parts 108 to 114). The USDA has regulatory oversight of products that contain at least 3% raw red meat or 2% cooked poultry, and the FDA regulates other products. A description of the thermal process, including the facility, equipment, and formulations, must be filed with the proper governmental agency prior to commercial processing. The processor must report process deviations or instances of spoilage to the FDA or the USDA.

Low-acid foods are packaged in hermetically sealed containers, often cans or glass jars but also plastic pouches and other types of containers. These "cans" (as collectively defined in this chapter) containing low-acid food products must be processed by heat to achieve commercial sterility (107, 126), which is a condition achieved by application of heat that inactivates microorganisms of public health significance, as well as any microorganisms of non-health significance capable of reproducing in the food under normal nonrefrigerated conditions of storage and distribution. Commercial sterility is an empirical term to indicate a low level of microbial survival and provision of shelf stability (107, 126). Preservation procedures such as acidification or lowering of water activity by brining or other means can be used to attain commercial sterility. These preservation procedures are often combined with a reduced heat treatment.

In canned low-acid foods, the primary goal is to inactivate spores of *C. botulinum*, because this bacterium has the highest heat resistance of microbial foodborne pathogens. The degree of heat treatment applied varies considerably according to the class of food, spore numbers, pH, storage conditions, and other factors. For example, canned low-acid vegetables and uncured meats usually receive a $12D$ process (see below) or "botulinum cook." Lesser heat treatments are applied to shelf-stable canned cured meats as well as to foods with reduced water activity or other antimicrobial factors that inhibit growth of sporeforming bacteria. In practice, the spore content of ingredients and the cleanliness of the cannery environment are of major importance for successful heat treatment of low-acid canned foods (87). Certain foods and food ingredients such as mushrooms, potatoes, spices, sugars, and starches may contain large numbers of spores of *C. botulinum* and other sporeformers, which can be monitored for their spore input to a process. Honey and certain other foods can harbor *C. botulinum* spores and should not be fed to infants of less than 1 year of age because only a few spores may be sufficient to cause infant botulism (151).

Researchers in the early and mid-1900s described thermal processes to prevent botulism and spoilage in canned foods (reviewed in references 41 and 148). Pflug (106–108) refined the semilogarithmic microbial destruction model and designed a strategy to achieve the required heat process F_T value. He also introduced the concept of the probability of a nonsterile unit (PNSU) on a one-container basis. This reasoning logically explains the traditional $12D$ term, which designates the time required in a thermal process for a 12-log reduction of

C. botulinum spores. In thermal processing, two values have historically been used to describe the thermal inactivation of an organism: the D value is the time required for a 1-log reduction of a microorganism, and the z value is the temperature change required to change the D value by a factor of 10 or 1 $\log_{10}$ (148). While the D value represents the resistance of a microorganism to a specific temperature, the z value represents the relative resistance of a microorganism to inactivation at different temperatures (77, 148). The thermal process for the 10^{-11} to 10^{-13} level of probability of a botulism incident occurring will depend on the initial number of *C. botulinum* spores present in a container. This value can be quite high, for example, 10^4 for a container of mushrooms, or very low, such as 10^{-1} or less for a container of a meat product (45). Pflug (106–108) emphasized that the thermal processing industry should prioritize process design to protect against (i) public health hazard (botulism) from *C. botulinum* spores, (ii) spoilage from mesophilic sporeforming microorganisms, and (iii) spoilage from thermophilic microorganisms in containers stored in warm climates or environments. Generally, low-acid foods in hermetically sealed containers are heated to achieve 3 to 6 min at a temperature of 121°C (250°F) or equivalent at the center or most heat-impermeable region of a food. This ensures inactivation of the most heat-resistant *C. botulinum* spores with a $D_{121°C (250°F)}$ of 0.21 min and a z of 10°C (18°F). Economic spoilage is also avoided by achieving a ca. $5D$ killing of mesophilic spores that typically have a $D_{121°C}$ of ca. 1 min (Table 3.4). Foods that are distributed in warm climates of tropical or desert areas require a particularly severe thermal treatment of ca. 20 min at 121°C to achieve a $5D$ killing of *Clostridium thermosaccharolyticum*, *Geobacillus stearothermophilus*, and *Desulfotomaculum nigrificans* because these organisms have a $D_{121°C}$ of ca. 3 to 4 min (Table 3.4). Such severe treatment can have a detrimental impact on nutrient content and organoleptic qualities, but it assures a shelf-stable food. Predictive models have been proposed to simulate growth of proteolytic *C. botulinum* during cooling of cooked meat (61).

The concept of semilogarithmic or first-order kinetics of heat inactivation of spores and the use of D and z values have been carefully reevaluated (103, 104). It was concluded that most populations of spores differ in heat sensitivities and that non-log-linear death kinetics more accurately explain spore inactivation. Analysis using non-log kinetics can explain the tailing and curving that is commonly observed in heat inactivation of spore populations. The kinetics of heat inactivation of spore populations closely followed a cumulative Weibull distribution, i.e., $\log S = b(T)^{tn(T)}$, where S is the survival

Table 3.4 Heat resistance of sporeformers of importance in foods[a]

Organism	Approx. D value[b] at temp:						
	80°C	85°C	90°C	95°C	100°C	110°C	120°C
Spores of public health significance							
Group I *C. botulinum* types A and B			50		7–30	1–3	0.1–0.2
Group II *C. botulinum* type B		1–30	0.1–3	0.03–2			
C. botulinum type E	0.3–3				0.01		
B. cereus					3–200	0.03–2.4	
C. perfringens			3–145		0.3–18	2.3–5.2	
Mesophilic aerobes							
B. subtilis					7–70	6.9	0.5
B. licheniformis					13.5	0.5	
B. megaterium					1		
Bacillus polymyxa			4–5		0.1–0.5		
Bacillus thermoacidurans			11–30		2–3		
A. acidoterrestris			16	2.6			
Thermophilic aerobes							
G. stearothermophilus					100–1,600		1–6
Bacillus coagulans					20–300		2–3
Mesophilic anaerobes							
C. butyricum	4–5	0.4–0.8					
Clostridium sporogenes					80–100	21	0.1–1.5
Clostridium tyrobutyricum	13						
Thermophilic anaerobes							
D. nigrificans					<480		2–3
C. thermosaccharolyticum				400		3–4	

[a] Data are from references 89 and 136.
[b] At pH ca. 7 and a_w of >0.95.

ratio and and $b(T)$ and $n(T)$ are temperature-dependent coefficients (104).

Heat treatments are now commonly applied to aseptically processed low-acid canned foods, whereby commercially sterilized cooled product is filled into presterilized containers followed by aseptic hermetic sealing with a closure in a sterile environment. This technology was initially used for commercial sterilization of milk and creams in the 1950s and then encompassed that of other food products such as soups, eggnog, cheese spreads, sour cream dips, puddings, and high-acid products such as fruit and vegetable drinks. Aseptic processing and packaging systems have the potential to reduce energy, packaging, and distribution costs. New developments in food preservation such as ohmic heating, high pressure, UV light, and sound-inactivation technologies have also renewed interest in the mechanisms of spore, cell, and enzyme inactivation (reviewed in reference 117).

Spores of *C. botulinum* can survive for long duration in high-acid foods with pH values ≤4.6 (40). Growth of *C. botulinum* and outbreaks of botulism in high-acid foods have been reported. Some outbreaks in high-acid foods were associated with inadvertent increases in pH, such as by growth of molds with catabolism of organic acids (40).

Bacteriology of Sporeformers of Public Health Significance

Three species of sporeformers, *C. botulinum*, *C. perfringens*, and *B. cereus*, are well known to cause foodborne illness (Table 3.5). Certain other species of *Bacillus* such as *Bacillus licheniformis*, *B. subtilis*, and *Bacillus pumilus* have also been reported to sporadically cause foodborne disease (104), and rare strains of *C. butyricum* and *C. baratii* produce type E and F botulinal toxins, respectively (47, 59). Devastating incidences of intestinal anthrax caused by ingestion of contaminated raw or poorly cooked meat have been reported (46). Spores of *B. anthracis* have recently been of considerable concern as agents of bioterrorism (29, 54).

The principal microbial hazard in heat-processed foods and in minimally processed refrigerated foods is *C. botulinum*. The genus *Clostridium* consists of gram-positive, anaerobic, sporeforming bacilli that obtain energy by

Table 3.5   Growth requirements of sporeformers of public health significance

Organism	Inhibitory condition			Temp range for growth (°C)
	Minimum pH	NaCl concn (%)	Minimum a_w	
Group I *C. botulinum*	4.6	10	0.94	10–50
Group II *C. botulinum*	5.0	5	0.97	3.3–45
B. cereus	4.35–4.9	ca. 10	0.91–0.95	5–50
C. perfringens	5.0	ca. 7	0.95–0.97	15–50

fermentation (14, 47, 49, 60). The species *C. botulinum* is a heterogeneous collection of strains that differ widely in genetic relatedness and phenotypic properties but which all have the property of producing a characteristic neurotoxin of extraordinary potency (60, 152). *C. botulinum* and other pathogenic bacteria produce spores that swell the mother sporangium, giving a "tennis racket" or club-shaped appearance (Fig. 3.9). Spores of *C. botulinum* types B and E frequently possess an exosporium (146), and type E characteristically produces appendages (Fig. 3.10). *C. botulinum* is commonly divided into four physiological groups (I through IV) on the basis of phenotypic properties (60, 139). Group I (strongly proteolytic strains producing neurotoxin types A, B, and F) and group II (nonproteolytic strains producing neurotoxin serotypes B, E, and F) are the two groups of concern in food safety (40, 79, 139). Strains in groups I and II have certain properties that affect their ability to grow in foods. The spores of group II have considerably less heat resistance than group I spores, but they can grow and produce toxin at refrigerator temperatures. The ability of *C. botulinum* to grow at low temperatures has generated considerable concern that refrigerated food products could lead to botulism outbreaks (40). It was recommended for refrigerated products with a shelf life of greater than 5 days and that receive a heat-treatment killing less than 6 $\log_{10}$ of psychrotrophic spores of *C. botulinum* that additional intrinsic preservation factors should be included to ensure the botulinal safety (40, 43).

Determination of the genomic sequence of several clostridia, including *C. botulinum* serotype A (ATCC 3502 Hall strain) (Sanger Centre, unpublished), has revealed unexpected metabolic, pathogenic, and spore-related features (102). In contrast to *Bacillus* species, *C. botulinum* lacks genes for the proteins Spo0F and Spo0B involved in sporulation (167). These findings suggest that initiation and regulation of sporulation are controlled

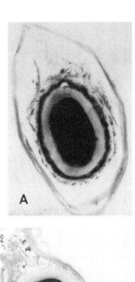

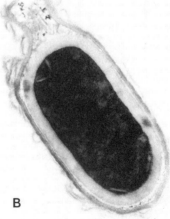

Figure 3.10   Electron micrographs of *C. botulinum* type B (A) and type E (B) showing characteristic exosporium in types B and E and appendages in type E. Micrographs courtesy of Philipp Gerhardt from spores produced in E.A.J.'s laboratory.

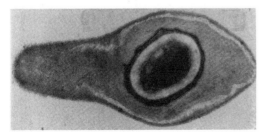

Figure 3.9   Transmission electron micrograph (×50,000) of a longitudinal section through a spore and sporangium of *C. botulinum* type A, showing the characteristic club-shaped morphology.

by mechanisms distinct from those involved in *Bacillus* species. Bioinformatic analyses of the genome sequence should also reveal characteristics underlying spore resistance and properties governing growth in foods.

C. perfringens is widespread in soils and is a normal resident of the intestinal tracts of humans and certain animals (47). *C. perfringens* can grow extremely rapidly in high-protein foods such as meats that have been cooked to eliminate competitors and are inadequately cooled, allowing it to produce in the intestinal tract an enterotoxin that causes diarrheal disease (chapter 19). It also produces a variety of other extracellular toxins and degradative enzymes, but these are mainly of significance in gas gangrene and diseases in animals (142). *C. perfringens* differs from many other clostridia in being nonmotile, reducing nitrate, and carrying out a stormy fermentation of lactose in milk. Contributing to the ability of *C. perfringens* to cause foodborne illness are its ubiquitous distribution in foods and food environments, the formation of resistant endospores that survive cooking of foods, and an extremely rapid growth rate in warm foods (6 to 9 min at 43 to 45°C) (chapter 19).

C. perfringens is the cause of a relatively common type of foodborne illness in the United States and in several other countries where surveillance has been conducted. Many foods contain spores of *C. perfringens* (chapter 19). In addition to vegetables and fruits that acquire spores from soil, foods of animal origin are contaminated during slaughtering with spores in the environment or spores residing in the intestinal tract. Surveys have revealed that the incidence of *C. perfringens* in raw meat samples is high, often 40 to 50%, depending on the sample size and detection methods (chapter 19). Accordingly, most foodborne outbreaks due to *C. perfringens* are associated with meat products. Dried foods such as spices are a common source of *C. perfringens* and other sporeformers (82, 124).

Sporulation of *C. perfringens* is often difficult to obtain in many laboratory media and in many foods, and Duncan-Strong or related complex media are generally used. When spores do occur, they are large, oval, and centrally or subterminally located and swell the cells. The optimum temperature for growth of vegetative cells is ca. 43 to 45°C (109 to 113°F), and in rich media or in certain foods at the optimum temperature for growth, doubling times as short as 6 to 9 min have been observed. Due to the rapid growth of *C. perfringens* in many foods, cell numbers sufficient to cause foodborne illness can be produced rapidly (chapter 19). Germination of spores and growth of cells can occur at up to 50°C (ca. 122°F). *C. perfringens* does not generally grow below 15°C (68°F), and true psychrotrophic strains of the bacterium have not

been isolated. *C. perfringens* can grow over a pH range of 5.0 to 9.0 or 8.5, and the optimum is ca. 6.5 (chapter 19). At temperatures below 45°C (113°F), some strains will grow at pH 5. As with most bacteria, organic acids such as acetate, lactate, and citrate are much more effective than are mineral acids for inhibiting the growth of *C. perfringens*. Growth of most strains is also inhibited by 5 to 6% sodium chloride (Table 3.5). Sodium nitrite in cured meat products also inhibits growth (chapter 19). Conditions for growth of *C. perfringens* in many foods have been reviewed (56) (chapter 19).

The heat resistance of *C. perfringens* spores is strain dependent and varies considerably. In general, two classes of heat sensitivity are common. Heat-resistant spores have $D_{90°C}$ ($D_{194°F}$) values of 15 to 145 min and z values of 9 to 16°C (16 to 29°F), compared to heat-sensitive spores which have $D_{90°C}$ values of 3 to 5 min and z values of 6 to 8°C (11 to 14°F). The spores of the heat-resistant class generally require a heat shock of 75 to 100°C (167 to 212°F) for 5 to 20 min in order to germinate. The basis of the wide variation in heat resistance is not currently understood. The spores of both classes may survive cooking of foods and may be stimulated by heat shock for germination during the heating procedures. Both classes can cause diarrheal foodborne illness, although it would be expected that the heat-resistant forms would be a more frequent cause of illness.

Foodborne illness caused by *C. perfringens* nearly always involves temperature abuse of a cooked food, and the great majority of foodborne illnesses caused by *C. perfringens* could be avoided if cooked foods were eaten immediately after cooking or rapidly chilled and reheated before consumption to inactivate vegetative cells. The objective in prevention of foodborne illnesses is to limit the multiplication of vegetative cells in the food. Since the spores are widespread and are resistant to heat, they will often survive the cooking procedure, germinate, and rapidly outgrow to large vegetative-cell populations if the rate of cooling is inadequate. This property has led to USDA performance standards for cooling in the production of certain ready-to-eat meat and poultry products (chapter 19). The USDA-Food Safety Inspection Service draft guidelines state that cooked meat products should be cooled at a rate adequate to prevent a 1-$\log_{10}$ increase of *C. perfringens* cell counts (chapter 19). If foods are not rapidly cooled, they should be held at 60°C (140°F) or higher. For cooled foods, the maximum internal temperature should not remain between 130°F and 80°F for more than 1.5 h, and not between 80°F and 40°F for more than 5 h. For products that receive a pasteurization step, the entire process must not allow greater than 1 $\log_{10}$ growth in the product during processing and cooling. Predictive models have been proposed for evaluation of *C. perfringens* growth and inactivation (chapter 19).

The genome sequence for *C. perfringens* strain 13 has been reported (137). The genome of strain 13 possesses interesting features regarding metabolism, spore formation, resistance properties, and pathogenicity. The genome sequence could serve as a framework for analysis of properties pertaining to growth and survival in foods, and for rational design of food safety systems.

The genus *Bacillus* contains only two species, *B. anthracis* and *B. cereus*, that are recognized as definitive human pathogens. *B. cereus* can produce a heat-labile enterotoxin causing diarrheal illness and a heat-stable toxin giving an emetic response in humans (chapter 20). Generally the bacterium must grow to very high numbers ($>10^6$/g of food) to cause human illness. *B. cereus* is closely related to *B. megaterium*, *B. thuringiensis*, and *B. anthracis*, but *B. cereus* can be distinguished from these species by biochemical tests and the absence of toxin crystals. Other bacilli including *B. licheniformis*, *B. subtilis*, and *B. pumilis* have been reported to cause foodborne-disease outbreaks, primarily in the United Kingdom (12, 44).

B. cereus spores occur widely in foods and are commonly found in milk, cereals, starches, herbs, spices, and other dried foodstuffs. They are also frequently found on the surfaces of meats and poultry, probably because of soil or dust contamination. Investigators in Sweden reported isolation of *B. cereus* from 47.8% of 3,888 different food samples. In the United Kingdom, *B. cereus* was isolated from 98/108 (91%) of rice samples. The bacterium causes spoilage of raw and unpasteurized milk (15), and foods containing dried milks, such as infant formulas, may possess fairly high numbers of spores or cells.

The bacterium grows over the temperature range of approximately 10° to 48°C (50° to 118.4°F), with an optimum of 28° to 35°C (82.4° to 95°F) (Table 3.5). Psychrotrophic strains that produce enterotoxin in milk have been isolated (14, 160). The doubling time at the optimum temperature in a nutritious medium is 18 to 27 min. Several strains can grow slowly in sodium chloride concentrations of 7.5%. The minimum water activity for growth is 0.95. The bacterium grows over a pH range of approximately 4.9 to 9.3, but these environmental limits for growth are dependent on water activity, temperature, and other interrelated parameters.

Spores of *B. cereus* are ellipsoidal and central to subterminal and do not distend the sporangium. Spore germination can occur over the temperature range of 8° to 30°C (46.4° to 86°F). Spores from strains associated with foodborne illness had a heat resistance $D_{95°C}$ (203°F) of

ca. 24 min. Other strains were shown to have a wider range of heat resistance. It has been suggested that strains involved in foodborne illness have higher heat resistances and therefore are more apt to survive cooking. Spores are hydrophobic and attach to food-contact surfaces (51).

Since *B. cereus* is widespread in nature and survives extended storage in dried food products, it is not practical to eliminate low numbers of spores from foods. Control against foodborne illness should be directed at preventing germination of spores and preventing multiplication of large populations of the bacterium. Cooked foods should rapidly and efficiently be cooled to less than 7°C (45°F), or maintained above 60°C (140°F), and should be thoroughly reheated before serving.

B. anthracis has rarely been associated with gastrointestinally mediated anthrax (54), generally resulting from occupational exposure such as in the tanning of hides, but gastrointestinal anthrax has also been reported from consumption of spore-contaminated meat (29, 54). In late 2001, the deliberate release of *B. anthracis* spores revealed that the pathogen could also cause disease and death as a bioterrorist agent, by means including dissemination in foods (29, 54). The resistance properties of spores of *B. anthracis* have been reviewed and should provide a framework for control (29, 120, 143), although little is known regarding resistance in various foods.

Heat Resistance of *C. botulinum* Spores

Group I *C. botulinum* type A and B strains can produce spores of remarkable heat resistance and are the most important sporeformers in public health safety of thermally processed foods. The classic investigation on the heat resistance of *C. botulinum* spores was carried out by Esty and Meyer (31) in California as a result of commercial outbreaks of botulism in canned olives and certain other canned vegetables. They examined 109 type A and B strains at five heating temperatures over the range 100° to 120°C (212° to 248°F). They found that the inactivation rate was logarithmic between 100° and 120°C and depended on the spore concentration, the pH, and the heating menstruum. Esty and Meyer determined that 0.15 M phosphate buffer (Sorensen's buffer), pH 7.0, gave the most consistent heat inactivation results, and their use of a standardized system enables comparisons of heat resistance by researchers today. The use of a reproducible system is valuable in periodically determining the heat resistance of new spore crops. Extrapolating the data of Esty and Meyer gives a maximum value for $D_{121°C}$ of 0.21 min for *C. botulinum* type A and B spores in phosphate buffer (158). The thermal-processing industries have used

$D_{121°C}$ as a standard in calculating process requirements. Proteolytic *C. botulinum* type F spores have a heat resistance $D_{98.9°C}$ of 12.2 to 23.2 min and a $D_{110°C}$ of 1.45 to 1.82 min (118), which is much lower than that of type A spores. Spores of nonproteolytic *C. botulinum* types B and E have much lower heat resistance than proteolytic type A and B strains. The type E spores have a $D_{70°C (158°F)}$ varying from 29 to 33 min and a $D_{80°C (176°F)}$ from 0.3 to 3 min, depending on the strains. The *z* value ranged from 13 to 15°F. These values are comparable to those of Ohye and Scott (95), who obtained $D_{80°C}$ values of 3.3 and 0.4 min for spores of two type E strains. Spores of nonproteolytic *C. botulinum* type B have heat resistance considerably higher than that of type E spores. Scott and Bernard (127) determined that the $D_{82.2°C}$ of spores of nonproteolytic type B strains ranged from 1.5 to 32.3 min as compared with a $D_{82.2°C}$ of 0.33 min for a type E strain. Media containing lysozyme can significantly enhance recovery of group II *C. botulinum* spores because lysozyme substitutes for spore lytic enzymes that are inactivated by heat (76). *D* values at 85° and 95°C were 100 and 4.4 min, respectively, for spores of strain 17B and 45.6 and 2.8 min for spores of strain Beluga E on medium containing lysozyme. The thermal resistance of *C. botulinum* spores is strongly dependent on environmental and recovery conditions. Heat resistance is markedly affected by acidity (62). Esty and Meyer found that spores had maximum resistance at pH 6.3 and 6.9, and resistance decreased markedly at pH values below 5 or above 9. Increased levels of sodium chloride or sucrose and decreased a_w increase the heat resistance of *C. botulinum* spores (149). Sugiyama (150) found that spores grown in media containing fatty acids increased their heat resistance. *C. botulinum* spores coated in oil were more resistant to heat (139). In common with *Bacillus* spp., sporulation of *C. botulinum* at higher temperatures results in spore crops with greater heat resistance (159).

Little is known of the compositional factors contributing to heat resistance of *C. botulinum* spores. The metal composition of purified group I *C. botulinum* spores is different from spores of *Bacillus* (66). The minerals required for sporulation and mechanisms of heat resistance of *C. botulinum* and *Bacillus* spp. are probably not the same (65, 66). Unlike *Bacillus* spp., which require manganese for sporulation, *C. botulinum* type B sporulation was enhanced by zinc and inhibited by copper (65). During sporulation, *C. botulinum* accumulated relatively high concentrations of transition metals, particularly zinc (ca. 1% of cell dry weight) and iron and copper (0.05 to 1%). Spores containing increased contents of iron or copper were more rapidly inactivated by heat than were native spores or spores containing increased manganese or zinc

(66). In the anaerobic growth environment of clostridia, transition metals would be expected to have important roles in the sporulation and resistance properties of spores. Metals that undergo redox changes, including copper, iron, and manganese, tend to precipitate as the hydroxides or oxides in aerobic environments, but they are more soluble and biologically available at low redox potentials. The mechanisms by which iron and copper accelerate heat inactivation and zinc and manganese protect *C. botulinum* spores against thermal energy have not been elucidated. Iron and copper are redox-active transition metals and may catalyze hydrolytic reactions and may also spontaneously react with oxygen, generating toxic oxygen species that cause mutations in DNA (75). Manganese and zinc ions can associate with nucleic acids, and manganese ions can provide protection against heat denaturation and can also effectively scavenge free radicals in biological systems (14). Studies have also indicated that heat inactivation of *C. botulinum* spores is accelerated in modified gas atmospheres (67).

Spores of *C. botulinum* groups I and II are highly resistant to γ-irradiation compared with vegetative cells of most microorganisms, and it is probably not practical to inactivate them in foods by irradiation. *C. botulinum* spores have a *D* of 0.1 to 0.45 Mrad (2.0 to 4.5 kGy). *C. botulinum* spore γ-radiation resistance varies depending on the type; proteolytic types A, B, and F are most resistant (102). *C. botulinum* spores are also highly resistant to ethylene oxide but are inactivated by halogen sanitizers and by hydrogen peroxide (67). Hydrogen peroxide is commonly used for sanitizing surfaces in aseptic packaging, and halogen sanitizers are used in cannery cooling waters. Alternatives to hydrogen peroxide such as peracetic acid are receiving renewed consideration due to the deleterious effects of hydrogen peroxide on packaging equipment and materials (11). The assessment of biocides and food preservatives in sporicidal efficacy has recently been reviewed (69, 123).

Incidence of Foodborne Illness Caused by *C. botulinum*

The epidemiology of botulism has been thoroughly reviewed (33, 40, 43), and only aspects pertaining to spore survival and outgrowth are presented in this chapter. Fortunately, the incidence of botulism in commercial foods is very low. It has been estimated that about 30 billion cans, bottles, and pouches of low-acid foods are consumed in the United States each year. From 1940, when heat-processing principles were firmly established, through 1975, fewer than 10 botulism outbreaks and fewer than four deaths were caused by inadequately commercially canned foods in the United States (78, 91).

From 1971 through 1982, however, botulinum toxin was detected in several commercial canned foods such as mushrooms, salmon, soups, peppers, tuna fish, beef stew, and tomatoes. Survival of spores and toxin production during this period were caused mainly by underprocessing or by container leakage following processing. The detection of botulinal toxin in canned mushrooms and in canned salmon in the 1970s and 1980s prompted United States regulatory agencies to recommend that chlorine or sanitizers be used in cooling water. Botulism has been transmitted via several commercial low-acid foods including chopped garlic in oil, cheese sauce, bean dip, and clam chowder (40, 47, 59). These incidences resulted from temperature abuse of products labeled "keep refrigerated" and the absence of inhibitory conditions other than temperature. Although botulism from commercial products has been quite rare due to excellent control during thermal processing, botulism from tainted home-canned and improperly fermented products occurs relatively frequently throughout the world (58). The heat resistance of *C. botulinum* spores is often not understood by home canners, and 20 to 30 botulism cases occur each year in the United States with a current case-fatality rate of about 10%. Botulism is more common in some countries such as Poland and China, where improper home canning of meats and poor fermentation of soybean curd occur relatively frequently (59). In recent years, the United States and certain other countries have seen a resurgence of botulism in restaurant-prepared foods, most often caused by poor temperature control of the prepared foods (59).

Inadvertent temperature abuse of foods has resulted in botulism outbreaks. Botulism has occurred from potatoes that were wrapped in foil, baked, and then held at room temperature until they were used for preparing salads (40, 59). During cooking, vegetative organisms are killed but the spores of *C. botulinum* survive and grow in the anaerobic environment created by wrapping the potatoes in foil (40). In April 1994, an outbreak of botulism in Texas affected 23 individuals, 17 of whom were hospitalized (40, 59). This was the largest botulism outbreak in the United States since 1983. The food vehicle was a potato-based dip (skordalia), which was prepared using foil-wrapped potatoes that were left at room temperature after baking. These examples illustrate ways in which changes in food processing, elimination of antimicrobials, and relying solely on refrigeration can result in incidences of botulism. Changes in formulation, processing, or packaging of food products can lead to botulism. To ensure the botulinogenic safety of a food with potential of supporting *C. botulinum* growth, it is recommended that laboratory challenge tests be performed.

Guidelines have been recommended for such challenge studies (24, 40, 94).

HACCP, FSO, and Prevention of Foodborne Disease by Sporeformers

The safety of thermally processed low-acid foods is enhanced by application of risk management programs including Hazard Analysis and Critical Control Point (HACCP) and Food Safety Objectives (FSO) (147). These systems entail a systematic and quantitative risk assessment program to assure the safety of foods. They were designed to have strict control over all aspects of the safety of food production including raw materials, processing methods, the food plant environment, personnel, storage, and distribution. In practice, the identification of potential hazards for a given process and meticulous control of critical control points are required. Methods for HACCP and FSO quality assurance programs for thermally processed foods have been described previously (147).

Spoilage of Acid and Low-Acid Canned and Vacuum-Packaged Foods by Sporeformers

Thermally processed low-acid foods receive a heat treatment adequate to kill spores of *C. botulinum* but not sufficient to kill more heat-resistant spores of mesophiles and thermophiles. Acid and acidified foods with an equilibrium pH of ≤4.6 are not processed sufficiently to inactivate all spores, since most species of sporeformers do not grow under acid conditions and inactivation of all spores would be detrimental to food quality and nutritional composition. Certain foods such as cured meats and hams do not receive a thermal process sufficient to inactivate sporeformers and thus must be kept under refrigerated conditions for microbial stability. These classes of foods present opportunities for the growth of sporeformers that do not present a public health hazard but which can cause economic spoilage (12, 87). Most of the spoilers and their characteristic spoilage patterns have been recognized for many years (87, 148). In recent years, however, spoilage of vegetable and fruit products by *Alicyclobacillus acidoterrestris* has captured the interest of many microbiologists due to its ability to grow under highly acidic conditions and cause spoilage at low spore levels (97). Psychrotrophic clostridia have also been increasingly observed to spoil vacuum-packaged chilled meats, and certain strains can produce botulinal toxin (63, 89).

In practice, the inherent spore contamination of foods and food ingredients and of the cannery environment contributes to spoilage problems. Dry ingredients such

Table 3.6 Spoilage of canned foods by sporeformers[a]

Type of spoilage	pH	Major sporeformers responsible	Spoilage defects
Flat-sour	≥5.3	*B. coagulans, B. stearothermophilus*	No gas, pH lowered. May have abnormal odor and cloudy liquor.
Thermophilic anaerobe	≥4.8	*C. thermosaccharolyticum*	Can swells, may burst. Anaerobic anaerobe and products give sour, fermented, or butyric odor. Typical foods are spinach, corn.
Sulfide spoilage	≥5.3	*D. nigrificans, Clostridium bifermentans*	Hydrogen sulfide produced, giving rotten egg odor. Iron sulfide precipitate gives blackened appearance. Typical foods are corn, peas.
Putrefractive anaerobe	≥4.8	*Clostridium sporogenes*	Plentiful gas. Disgusting putrid odor. pH often increased. Typical foods are corn, asparagus.
Psychrotrophic clostridia	>4.6		Spoilage of vacuum-packaged chilled meats. Production of gas, off flavors and odors, discoloration.
Aerobic sporeformers	≥4.8	*Bacillus* spp.	Gas usually absent except for cured meats; milk is coagulated. Typical foods are milk, meats, beets.
Butyric spoilage	≥4.0	*C. butyricum, Clostridium tertium*	Gas, acetic and butyric odor. Typical foods are tomatoes, peas, olives, cucumbers.
Acid spoilage	≥4.2	*Bacillus thermoacidurans*	Flat (*Bacillus*) or gas (butyric anaerobes). Off odors depend on organism. Common foods are tomatoes, tomato products, other fruits.
	<4	*A. acidoterrestris*	Flat spoilage with off flavors. Most common in fruit juices and acid vegetables, and also reported to spoil iced tea.

[a]Data are from references 89, 115, and 136.

as sugar, starches, flours, and spices often contain high levels of sporeformers (82). Spore populations can also accumulate in a food plant, such as thermophilic spores on heated equipment and saccharolytic clostridia in plants processing sugar-rich foods such as fruits.

The principal spoilage microorganisms and spoilage manifestations are presented in Table 3.6. The principal classes of sporeformers causing spoilage are thermophilic flat-sour organisms, thermophilic anaerobes not producing hydrogen sulfide, thermophilic anaerobes forming hydrogen sulfide, putrefactive anaerobes, facultative *Bacillus* mesophiles, butyric clostridia, lactobacilli, and heat-resistant molds and yeasts (12, 87). *A. acidoterrestris* and psychrotrophic clostridia have recently been implicated in the spoilage of meat and fruit products, respectively. Practical control of these bacteria includes the following: monitoring of raw foods entering the cannery, particularly sugars, starches, spices, onions, mushrooms, and dried foods, to limit the initial spore load in a food product; adequate thermal processing depending on subsequent storage and distribution conditions; rapid cooling of products; chlorination of cooling water; and implementing and maintaining good manufacturing practices within the food plant.

Certain species of sporeforming psychrophiles have the ability to spoil refrigerated foods and have caused spoilage of meats and dairy products in recent years. Psychrophilic strains of *Bacillus* have been isolated from spoiled dairy products (15). Psychrophilic clostridia have also been associated with the spoilage of meats (72).

Heat-resistant fungi are economically spoilers of acidic foods, particularly fruit products (112, 156). While most filamentous fungi and yeasts are killed by heating for a few minutes at 60 to 75°C, heat-resistant fungi produce thick-walled ascospores that survive heating at ca. 85°C for 5 min. The most common genera of heat-resistant fungi causing spoilage are *Byssochlamys*, *Neosartorya*, *Talaromyces*, and *Eupenicillium* (112, 156). Certain heat-resistant fungi also produce toxic secondary metabolites collectively referrred to as mycotoxins (156). To prevent spoilage of heat-treated foods, raw materials should be screened for heat-resistant fungi, and strict Good Manufacturing Practices and sanitation programs should be followed during processing. Additionally, manipulation of water activity and oxygen tension and the application of antimycotic agents can be used to prevent fungal growth.

Modeling Growth of Sporeformers in Foods

Abundant information exists on the behavior of microorganisms in foods, and it is often useful to generate statistical models to quantify safety risks in foods. Two general types of models have mainly been used in food microbiology: (i) those analyzing experimental growth and survival data using simple and higher-order polynominals, and (ii) theoretical models derived from basic scientific principles and computer analysis, used to predict microbial survival. Statistical models can be particularly useful to define important variables and predict microbial behavior in advance of practical testing (94). Certain companies and institutions are beginning to use models involving neural nets. These are valuable because they are adaptive and improve in precision and predictive capability over time. An example of a model used extensively in the dairy industry is that of Tanaka et al. (155) for preventing growth of *C. botulinum* in processed cheese. Although models can provide guidelines for food safety, assurance of sterility generally needs to be ascertained by laboratory challenge studies.

CONCLUSION

The scientific investigation of sporeformers has greatly contributed to the development of microbiology for the enhancement of food safety and quality. The fundamental understanding of sporulation in *B. subtilis* provides an elegant model of cellular differentiation. Advances in the understanding of the mechanisms of spore heat resistance have contributed to a greater knowledge of dormancy and the ecological success of sporeformers. The remarkable resistance properties of spores and their impact on human disease, particularly botulism, tetanus, and anthrax, have led to the development of microbiology and its importance in medicine and industry. Bacterial and fungal spores present specific problems to the food industry including the cause of foodborne illness and spoilage of foods. A recent consideration of sporeformers as disease agents is their deliberate release as bioterrorist agents in aerosols or in foods.

Work in the laboratory of P.S. has been supported by the Army Research Office, the National Institutes of Health (NIH) (GM19698), and the USDA. Research in the laboratory of E.J. has been supported by the FDA, USDA, NIH, and the industrial sponsors of the Food Research Institute.

References

1. **Ablett, A. H., P. J. Lillford, and D. R. Martin.** 1999. Glass formation and dormancy in bacterial spores. *Int. J. Food Sci. Technol.* **34**:59–69.

2. **Appert, N.** 1810. L'Art de conserver pendant plusieurs anees toutes les substances animales et vegetales. *In* S. A. Goldblith, M. A. Joslyn, and J. T. R. Nickerson (ed.), *Introduction to the Thermal Processing of Foods.* 1961. AVI Publishing Co., Westport, Conn.

3. **Aronson, A. I.** 1993. Insecticidal toxins, p. 953–964. *In* A. L. Sonenshein, J. A. Hoch, and R. Losick (ed.), *Bacillus subtilis and Other Gram-Positive Bacteria: Biochemistry, Physiology, and Molecular Genetics.* American Society for Microbiology, Washington, D.C.

4. **Ash, C., F. G. Priest, and M. D. Collins.** 1993. Molecular identification of rRNA group 3 bacilli (Ash, Farrow, Wallbanks and Collins) using a PCR probe test. Proposal for the creation of a new genus *Paenibacillus. Antonie Leeuwenhoek* **64**:253–260.

5. Reference deleted.

6. **Atrih, A., and S. J. Foster.** 2001. *In vivo* roles of the germination-specific lytic enzymes of *Bacillus subtilis* 168. *Microbiology* **147**:2925–2932.

7. **Barak, I., and A. J. Wilkinson.** 2005. Where asymmetry in gene expression originates. *Mol. Microbiol.* **57**:611–620.

8. **Beaman, T. C., and P. Gerhardt.** 1986. Heat resistance of bacterial spores correlated with protoplast dehydration, mineralization, and thermal adaptation. *Appl. Environ. Microbiol.* **52**:1242–1246.

9. **Black, E. P., K. Koziol-Dube, D. Guan, J. Wei, D. E. Cortezzo, D. G. Hoover, and P. Setlow.** 2005. Factors influencing the germination of *Bacillus subtilis* spores via the activation of nutrient receptors by high pressure. *Appl. Environ. Microbiol.* **71**:5879–5887.

10. Reference deleted.

11. **Blackistone, B., R. Chuyate, D. Kautter, Jr., J. Charbonneau, and K. Suit.** 1999. Efficacy of oxonia active against selective spore formers. *J. Food Prot.* **62**:262–267.

12. **Brown, K. L.** 2000. Control of bacterial spores. *Br. Med. Bull.* **56**:158–171.

13. **Bulloch, W.** 1938. *The History of Bacteriology.* Oxford University Press, Oxford, England.

14. **Cato, E. P., W. L. George, and S. M. Finegold.** 1986. The genus *Clostridium*, p. 1141–1200. *In* H. A. Sneath, N. S. Mair, and M. E. Sharpe (ed.), *Bergey's Manual of Systematic Bacteriology*, vol. 2. Williams & Wilkins, Baltimore, Md.

15. **Champagne, C. P., R. R. Laing, D. Roy, A. A. Mafu, and M. W. Griffiths.** 1994. Psychrotrophs in dairy products: their effects and their control. *Crit. Rev. Food Sci. Nutr.* **34**:1–30.

16. **Clarkson, J., I. D. Campbell, and M. D. Yudkin.** 2004. Efficient regulation of σF, the first sporulation-specific sigma factor in *B. subtilis. J. Mol. Biol.* **342**:1187–1195.

17. **Cortezzo, D. E., and P. Setlow.** 2005. Analysis of factors that influence the sensitivity of spores of *Bacillus subtilis* to DNA damaging chemicals. *J. Appl. Microbiol.* **98**:606–617.

18. **Cowan, A. E., D. E. Koppel, B. Setlow, and P. Setlow.** 2003. A soluble protein is immobile in dormant spores of *Bacillus subtilis* but is mobile in germinated spores: implications for spore dormancy. *Proc. Natl. Acad. Sci. USA* **100**:4209–4214.

19. **Cowan, A. E., E. M. Olivastro, D. E. Koppel, C. A. Loshon, B. Setlow, and P. Setlow.** 2004. Lipids in the inner membrane of dormant spores of *Bacillus* species are immobile. *Proc. Natl. Acad. Sci. USA* **101**:7733–7738.

20. **Crowe, J. H., E. A. Hoekstra, and L. M. Crowe.** 1992. Anhydrobiosis. *Annu. Rev. Physiol.* **54**:579–599.

21. Dawes, I. W., and J. Mandelstam. 1970. Sporulation of *Bacillus subtilis* in continuous culture. *J. Bacteriol.* 103:529–535.

22. Douki, T., B. Setlow, and P. Setlow. 2005. Effects of the binding of α/β-type small, acid-soluble spore proteins on the photochemistry of DNA in spores of *Bacillus subtilis* and in vitro. *Photochem. Photobiol.* 81:163–169.

23. Douki, T., B. Setlow, and P. Setlow. 2005. Photosensitization of DNA by dipicolinic acid, a major component of spores of *Bacillus* species. *Photochem. Photobiol. Sci.* 4:591–597.

24. Doyle, M. P. 1991. Evaluating the potential risk from extended shelf-life refrigerated foods by *Clostridium botulinum* inoculation studies. *Food Technol.* 45:154–156.

25. Driks, A. 2002. Proteins of the spore core and coat, p. 527–535. *In* A. L. Sonenshein, J. A. Hoch, and R. Losick (ed.), *Bacillus subtilis and Its Closest Relatives: from Genes to Cells.* ASM Press, Washington, D.C.

26. Dubnau, D., and C. M. Lovett, Jr. 2002. Transformation and recombination, p. 453–471. *In* A. L. Sonenshein, J. A. Hoch, and R. Losick (ed.), *Bacillus subtilis and Its Closest Relatives: from Genes to Cells.* ASM Press, Washington, D.C.

27. Eichenberger, P., M. Fujita, S. T. Jensen, E. M. Conlon, D. Z. Rudner, S. T. Wang, C. Ferguson, T. Sato, J. S. Liu, and R. Losick. 2005. The program of gene transcription for a single differentiating cell type during sporulation in *Bacillus subtilis*. *PLoS Biol.* 2:e328.

28. Ellar, D. J. 1978. Spore specific structures and their functions, p. 295–325. *In* R. Y. Stanier, H. J. Rogers, and J. B. Ward (ed.), *Relations between Structure and Function in the Prokaryotic Cell.* Cambridge University Press, London, England.

29. Erickson, M. C., and J. L. Kornacki. 2003. *Bacillus anthracis*: current knowledge in relation to contamination of food. *J. Food Prot.* 66:691–699.

30. Errington, J. 2001. Septation and chromosome segregation during sporulation in *Bacillus subtilis*. *Curr. Opin. Microbiol.* 4:660–666.

31. Esty, J. R., and K. F. Meyer. 1922. The heat resistance of the spores of *Bacillus botulinus* and allied anaerobes. *J. Infect. Dis.* 31:650–663.

32. Fairhead, H., B. Setlow, and P. Setlow. 1993. Prevention of DNA damage in spores and in vitro by small, acid-soluble proteins from *Bacillus* species. *J. Bacteriol.* 175:1367–1374.

33. Franciosa, G., P. Aureli, and R. Schechter. 2003. *Clostridium botulinum*, p. 61–89. *In* M. D. Bier and J. W. Miliotis (ed.), *International Handbook of Foodborne Pathogens.* Marcel Dekker, New York, N.Y.

34. Frenkiel-Krispin, D., R. Sack, J. Englander, E. Shimoni, M. Eisenstein, E. Bullitt, R. Horowitz-Scherer, C. S. Hayes, P. Setlow, A. Minsky, and S. G. Wolf. 2004. Structure of the DNA-SspC complex: implications for DNA packaging, protection, and repair in bacterial spores. *J. Bacteriol.* 186:3525–3530.

35. Fujita, M., J. E. González-Pastor, and R. Losick. 2005. High- and low-threshold genes in the Spo0A regulon of *Bacillus subtilis*. *J. Bacteriol.* 187:1357–1368.

36. Fujita, M., and R. Losick. 2005. Evidence that entry into sporulation in *Bacillus subtilis* is governed by a gradual increase in the level and activity of the master regulator Spo0A. *Genes Dev.* 19:2236–2244.

37. Fukushima, T., H. Yamamoto, A. Atrih, S. J. Foster, and J. Sekiguchi. 2002. A polysaccharide deacetylase gene (*pdaA*) is required for germination and for production of muramic δ-lactam residues in the spore cortex of *Bacillus subtilis*. *J. Bacteriol.* 184:6007–6015.

38. Gerhardt, P., and R. E. Marquis. 1989. Spore thermoresistance mechanisms, p. 17–63. *In* I. Smith, R. Slepecky, and P. Setlow (ed.), *Regulation of Procaryotic Development.* American Society for Microbiology, Washington, D.C.

39. Gilmore, M. E., D. Bandyopadhyay, A. M. Dean, S. D. Linnstaedt, and D. L. Popham. 2004. Production of muramic δ-lactam in *Bacillus subtilis* spore peptidoglycan. *J. Bacteriol.* 186:80–89.

40. Glass, K. G., and E. A. Johnson. 2001. Formulating low-acid foods for botulinal safety, p. 323–350. *In* V. K. Juneja and J. N. Sofos (ed.), *Control of Foodborne Organisms.* Marcel Dekker, New York, N.Y.

41. Goldblith, S. A., M. A. Joslyn, and J. T. R. Nickerson. 1961. *An Anthology of Food Science*, vol. 1. *Introduction to the Thermal Processing of Foods.* AVI Publishing, Westport, Conn.

42. Gonzalez-Pastor, J. E., E. C. Hobbs, and R. Losick. 2003. Cannibalism by sporulating bacteria. *Science* 301:510–513.

43. Gould, G. W. 1999. Sous vide foods: conclusions of an ECFF botulinum working party. *Food Control* 10:47–51.

44. Granum, P. E., and T. C. Baird-Parker. 2000. *Bacillus* species, p. 1029–1039. *In* B. M. Lund, T. C. Baird-Parker, and G. W. Gould (ed.). *The Microbiological Safety and Quality of Food*, vol. II. Aspen Publishers, Gaithersburg, Md.

45. Greenberg, R. A., R. B. Tompkin, B. O. Blade, R. S. Kittaka, and A. Anelis. 1966. Incidence of mesophilic spores in raw pork, beef, and chicken in processing plants in the United States and Canada. *Appl. Microbiol.* 14:789–793.

46. Guillemin, J. 1999. *Anthrax. The Investigation of a Deadly Outbreak.* University of California Press, Berkeley, Calif.

47. Hatheway, C. L., and E. A. Johnson. 1998. *Clostridium*: the spore-bearing anaerobes, p. 732–782. *In* W. J. Hausler and M. Sussman (ed.), *Topley and Wilson's Microbiology and Microbial Infections*, 9th ed., vol. 3. Edward Arnold, London, England.

48. Hilbert, D. W., and P. J. Piggot. 2004. Compartmentalization of gene expression during *Bacillus subtilis* spore formation. *Microbiol. Mol. Biol. Rev.* 68:234–262.

49. Hippe, H., J. R. Andreesen, and G. Gottschalk. 1992. The genus *Clostridium* – nonmedical, p. 1800–1866. *In* A. Balows, H. G. Truper, M. Dworkin, W. Harder, and K. H. Schleifer (ed.), *The Prokaryotes*, 2nd ed., vol. II. Springer Verlag, New York, N.Y.

50. Hurst, A. 1983. Injury, p. 255–274. *In* A. Hurst and G. W. Gould (ed.), *The Bacterial Spore*, vol. 2. Academic Press, London, England.

51. Husmark, U., and U. Ronner. 1992. The influence of hydrophobic, electrostatic and morphologic properties on the adhesion of *Bacillus* spores. *Biofouling* 5:335–344.

52. Igarashi, T., B. Setlow, M. Paidhungat, and P. Setlow. 2004. Analysis of the effects of a *gerF* (*lgt*) mutation on

the germination of spores of *Bacillus subtilis*. *J. Bacteriol.* **186**:2984–2991.

53. Igarashi, T., and P. Setlow. 2005. Interaction between individual protein components of the GerA and GerB nutrient receptors that trigger germination of *Bacillus subtilis* spores. *J. Bacteriol.* **187**:2513–2518.

54. Inglesby, T. V., D. A. Henderson, J. G. Bartlett, M. S. Ascher, E. Eitzen, A. M. Friedlander, J. Hauer, J. McDade, M. T. Osterholm, T. O'Toole, G. Parker, T. M. Perl, P. K. Russell, and K. Tonat. For the Working Group on Civilian Biodefense. 1999. Anthrax as a biological weapon. Medical and public health management. *JAMA* **281**:1735–1745.

55. Ingram, M. 1969. Sporeformers as food spoilage organisms, p. 549–610. *In* G. W. Gould and A. Hurst (ed.), *The Bacterial Spore*. Academic Press, London, England.

56. International Commission on Microbiological Specifications for Foods. 1996. *Microorganisms in Foods. 5. Characteristics of Microbial Pathogens*. Blackie Academic & Professional, London, England.

57. Jiang, M., W. Shao, M. Perego, and J. A. Hoch. 2000. Multiple histidine kinases regulate entry into stationary phase and sporulation in *Bacillus subtilis*. *Mol. Microbiol.* **38**:535–542.

58. Johnson, E. A. 1991. Microbiological safety of fermented foods, p. 135–169. *In* J. G. Zeikus and E. A. Johnson (ed.), *Mixed Cultures in Biotechnology*. McGraw Hill, New York, N.Y.

59. Johnson, E. A., and M. C. Goodnough. 1998. Botulism, p. 724–741. *In* W. J. Hausler and M. Sussman (ed.), *Topley and Wilson's Microbiology and Microbial Infections*, 9th ed., vol. 3. Edward Arnold, London, England.

60. Johnson, E. A. 2006. *Clostridium botulinum* and *Clostridium tetani*, p. 1035–1088. *In* S. P. Borriello, P. R. Murray, and G. Funke (ed.), *Topley and Wilson's Microbiology and Microbial Infections*, 8th ed. Hodder Arnold, London, England.

61. Juneja, V. K., and H. M. Marks. 1999. Proteolytic *Clostridium botulinum* growth at 12–48°C simulating the cooling of cooked meat: development of a predictive model. *Food Microbiol.* **16**:583–592.

62. Juneja, V. K., B. S. Marmer, J. G. Phillips, and A. J. Miller. 1995. Influence of the intrinsic properties of food on thermal inactivation of spores of nonproteolytic *Clostridium botulinum*: development of a predictive model. *J. Food Safety* **15**:349–364.

63. Kalinowski, R. M., and R. B. Tompkin. 1999. Psychrotrophic clostridia causing spoilage in cooked meat and poultry products. *J. Food Prot.* **62**:766–772.

64. Keynan, A., and N. Sandler. 1984. Spore research in historical perspective, p. 1–48. *In* A. Hurst and G. W. Gould (ed.), *The Bacterial Spore*, vol. 2. Academic Press, London, England.

65. Kihm, D. J., M. T. Hutton, J. H. Hanlin, and E. A. Johnson. 1988. Zinc stimulates sporulation in *Clostridium botulinum* 113B. *Curr. Microbiol.* **17**:193–198.

66. Kihm, D. J., M. T. Hutton, J. H. Hanlin, and E. A. Johnson. 1990. Influence of transition metals added during sporulation on heat resistance of *Clostridium botulinum* 113B spores. *Appl. Environ. Microbiol.* **56**:681–685.

67. Kihm, D. J., and E. A. Johnson. 1990. Hydrogen gas accelerates thermal inactivation of *Clostridium botulinum* spores. *Appl. Microbiol. Biotechnol.* **33**:705–708.

68. Kim, H., M. Hahn, P. Grabowski, D. C. McPherson, M. M. Otte, R. Wang, C. C. Ferguson, P. Eichenberger, and A. Driks. 2006. The *Bacillus subtilis* spore coat interaction network. *Mol. Microbiol.* **59**:487–502.

69. Kim, J., and P. M. Foegeding. 1993. Principles of control, p. 121–176. *In* A. H. W. Hauschild and K. L. Dodds (ed.), *Clostridium botulinum. Ecology and Control in Foods*. Marcel Dekker, New York, N.Y.

70. Klobutcher, L. A., K. Ragkousi, and P. Setlow. 2006. The *Bacillus subtilis* spore coat provides "eat resistance" during phagosomal predation by the protozoan *Tetrahymena thermophila*. *Proc. Natl. Acad. Sci. USA* **103**:165–170.

71. Lai, E. M., N. D. Phadke, M. T. Kachman, R. Giorno, S. Vazquez, J. A. Vazquez, J. R. Maddock, and A. Driks. 2003. Proteomic analysis of the spore coats of *Bacillus subtilis* and *Bacillus anthracis*. *J. Bacteriol.* **185**:1443–1454.

72. Lawson, P., R. H. Dainty, N. Kristiansen, J. Berg, and M. D. Collins. 1994. Characterization of a psychrotrophic *Clostridium* causing spoilage in vacuum-packed cooked pork: description of *Clostridium algidicarnis* sp. nov. *Lett. Appl. Microbiol.* **19**:153–157.

73. Lazazzera, B., T. Palmer, J. Quisel, and A. D. Grossman. 1999. Cell density control of gene expression and development in *Bacillus subtilis*, p. 27–46. *In* G. M. Dunny and S. C. Winans (ed.), *Cell-Cell Signaling in Bacteria*. American Society for Microbiology, Washington, D.C.

74. Leuschner, R. G. K., and P. J. Lillford. 2003. Thermal properties of bacterial spores and biopolymers. *Int. J. Food Microbiol.* **87**:8–14.

75. Loeb, L. A., E. A. James, A. M. Waltersdorph, and S. J. Klebanoff. 1988. Mutagenesis by the autoxidation of iron with isolated DNA. *Proc. Natl. Acad. Sci. USA* **85**:3918–3922.

76. Lund, B. M., and M. W. Peck. 1994. Heat resistance and recovery of spores of nonproteolytic *Clostridium botulinum* in relation to refrigerated, processed foods with extended shelf-life. *J. Appl. Bacteriol. Symp.* **76**:115S–128S.

77. Lund, D. 1975. Thermal processing, p. 31–92. *In* M. Karel, O. R. Fennema, and D. B. Lund (ed.), *Principles of Food Science. Part II. Physical Principles of Food Preservation*. Marcel Dekker, New York, N.Y.

78. Lynt, R. K., D. A. Kautter, and R. B. Read, Jr. 1975. Botulism in commercially canned foods. *J. Milk Food Technol.* **38**:546–550.

79. Lynt, R. K., D. A. Kautter, and H. M. Solomon. 1982. Differences and similarities among proteolytic strains of *Clostridium botulinum* types A, B, E and F: a review. *J. Food Prot.* **45**:466–474.

80. Makino, S., and R. Moriyama. 2002. Hydrolysis of cortex peptidoglycan during bacterial spore germination. *Med. Sci. Monit.* **8**:RA119–RA127.

81. McDonnell, G., and A. D. Russell. 1999. Antiseptics and disinfectants: activity, action and resistance. *Clin. Microbiol. Rev.* **12**:147–179.

82. McKee, L. H. 1995. Microbial contamination of spices and herbs: a review. *Lebensm.-Wiss. Technol.* **28**:1–11.

83. Melly, E., P. C. Genest, M. E. Gilmore, S. Little, D. L. Popham, A. Driks, and P. Setlow. 2002. Analysis of the properties of spores of *Bacillus subtilis* prepared at different temperatures. *J. Appl. Microbiol.* **92:**1105–1115.

84. Moayeri, M., and S. H. Leppla. 2004. The roles of anthrax toxin in pathogenesis. *Curr. Opin. Microbiol.* **7:**19–24.

85 Mohavedi, S., and W. M. Waites. 2000. A two-dimensional protein gel electrophoresis study of the heat stress response of *Bacillus subtilis* cells during sporulation. *J. Bacteriol.* **182:**4758–4763.

86. Moir, A., B. M. Corfe, and J. Behravan. 2002. Spore germination. *Cell. Mol. Life Sci.* **59:**403–409.

87. Moir, C. J. (ed.). 2001. *Spoilage of Processed Foods: Cause and Diagnosis.* Australian Institute of Food Science and Technology, NSW Branch, Waterloo DC, New South Wales, Australia.

88. Molle, V., Y. Nakaura, R. P. Shivers, H. Yamaguchi, R. Losick, Y. Fujita, and A. L. Sonenshein. 2003. Additional targets of the *Bacillus subtilis* global regulator CodY identified by chromatin immunoprecipitation and genome-wide transcript analysis. *J. Bacteriol.* **185:**1911–1922.

89. Moorhead, S. M., and R. G. Bell. 1999. Psychrotrophic clostridia mediated gas and botulinal toxin production in vacuum-packed chilled meat. *Lett. Appl. Microbiol.* **28:**108–112.

90. Nazina, T. N., T. P. Tourova, A. B. Poltaraus, E. V. Novikova, A. A. Grigoryan, A. E. Ivanova, A. M. Lysenko, V. V. Petrunyaka, G. A. Osipov, S. S. Belyaev, and M. V. Ivanov. 2001. Taxonomic study of aerobic thermophilic bacilli: descriptions of *Geobacillus subterraneus* gen nov., sp. nov. and *Geobacillus uzenensis* sp. nov. from petroleum reservoirs and transfer of *Bacillus stearothermophilus*, *Bacillus thermocatenulatus*, *Bacillus thermoleovorans*, *Bacillus kaustophilus*, *Bacillus thermodenitrificans* to *Geobacillus* as the new combinations G. *stearothermophilus*, G. *thermocatenulatus*, G. *thermoleovorans*, G. *kaustophilus*, G. *thermoglucosidasius* and G. *thermodenitrificans*. *Int. J. Syst. Evol. Microbiol.* **51:**433–446.

91. NFPA/CMI Container Integrity Task Force, Microbiological Assessment Group Report. 1984. Botulism risk from post-processing contamination of commercially canned foods in metal containers. *J. Food Prot.* **47:**801–816.

92. Nicholson, W. L., N. Munakata, G. Horneck, H. J. Melosh, and P. Setlow. 2000. Resistance of *Bacillus* endospores to extreme terrestrial and extraterrestrial environments. *Microbiol. Mol. Biol. Rev.* **64:**548–572.

93. Nicholson, W. L., A. C. Schuerger, and P. Setlow. 2005. The solar UV environment and bacterial spore UV resistance: considerations for Earth-to-Mars transport by natural processes and human spaceflight. *Mutat. Res.* **571:**248–264.

94. Notermans, S., P. in't Veld, T. Wijtzes, and G. C. Mead. 1993. A user's guide to microbial challenge testing for ensuring the safety and stability of food products. *Food Microbiol.* **10:**145–157.

95. Ohye, D. F., and W. J. Scott. 1957. Studies in the physiology of *Clostridium botulinum* type E. *Aust. J. Biol. Sci.* **10:**85–94.

96. Ordal, G. W., L. Marquez-Magana, and M. J. Chamberlin. 1993. Motility and chemotaxis, p. 765–784. In A. L. Sonenshein, J. A. Hoch, and R. Losick (ed.), *Bacillus subtilis and Other Gram-Positive Bacteria: Biochemistry, Physiology, and Molecular Genetics.* American Society for Microbiology, Washington, D.C.

97. Orr, R. V., and L. R. Beuchat. 2000. Efficacy of disinfectants in killing of spores of *Alicycobacillus acidoterrestris* and performance of media supporting colony development by survivors. *J. Food Prot.* **63:**1117–1122.

98. Paidhungat, M., K. Ragkousi, and P. Setlow. 2001. Genetic requirements for induction of germination of spores of *Bacillus subtilis* by Ca²⁺-dipicolinate. *J. Bacteriol.* **183:**4886–4893.

99. Paidhungat, M., B. Setlow, W. B. Daniels, D. Hoover, E. Papafragkou, and P. Setlow. 2002. Mechanisms of initiation of germination of spores of *Bacillus subtilis* by pressure. *Appl. Environ. Microbiol.* **68:**3172–3175.

100. Paidhungat, M., B. Setlow, A. Driks, and P. Setlow. 2000. Characterization of spores of *Bacillus subtilis* which lack dipicolinic acid. *J. Bacteriol.* **182:**5505–5512.

101. Paidhungat, M., and P. Setlow. 2002. Spore germination and outgrowth, p. 537–548. In A. L. Sonenshein, J. A. Hoch, and R. Losick (ed.), *Bacillus subtilis and Its Closest Relatives: from Genes to Cells.* ASM Press, Washington, D.C.

102. Parades, C. J., K. V. Alasker, and E. T. Papsoutsakis. 2005. A comparative genomic view of clostridial sporulation and physiology. *Nat. Rev. Microbiol.* **3:**969–978.

103. Peleg, M., and M. B. Cole. 1998. Reinterpretation of microbial survival curves. *Crit. Rev. Food Sci.* **38:**353–380.

104. Peleg, M., and M. B. Cole. 2000. Estimating the survival of *Clostridium botulinum* spores during heat treatments. *J. Food Prot.* **63:**190–195.

105. Perego, M., and J. A. Hoch. 2002. Two-component systems, phosphorelays, and regulation of their activities by phosphatases, p. 473–482. In A. L. Sonenshein, J. A. Hoch, and R. Losick (ed.), *Bacillus subtilis and Its Closest Relatives: from Genes to Cells.* ASM Press, Washington, D.C.

106. Pflug, I. J. 1987. Endpoint of a preservation process. *J. Food Prot.* **50:**347–351.

107. Pflug, I. J. 1987. Factors important in determining the heat process value, F_T, for low acid canned foods. *J. Food Prot.* **50:**528–533.

108. Pflug, I. J. 1987. Calculating F_T-values for heat preservation of shelf-stable, low acid canned foods using the straight-line semilogarithmic model. *J. Food Prot.* **50:**608–615.

109. Phillips, Z. E., and M. A. Strauch. 2002. *Bacillus subtilis* sporulation and stationary phase gene expression. *Cell. Mol. Life Sci.* **59:**392–402.

110. Piggot, P. J., and D. W. Hilbert. 2004. Sporulation of *Bacillus subtilis.* *Curr. Opin. Microbiol.* **7:**579–586.

111. Piggot, P. J., and R. Losick. 2002. Sporulation genes and intercompartmental regulation, p. 483–518. In A. L. Sonenshein, J. A. Hoch, and R. Losick (ed.), *Bacillus subtilis and Its Closest Relatives: from Genes to Cells.* ASM Press, Washington, D.C.

112. Pitt, J. I., and A. D. Hocking (ed). 1997. *Fungi and Food Spoilage*, 2nd ed. Blackie Academic & Professional, London, England.

113. **Popham, D. L.** 2002. Specialized peptidoglycan of the bacterial endospore: the inner wall of the lockbox. *Cell. Mol. Life Sci.* **59:**426–433.

114. **Pottahil, M., and B. A. Lazazzera.** 2003. The extracellular Phr peptide-Rap phosphatase signaling circuit of *Bacillus subtilis. Front. Biosci.* **8:**32–45.

115. **Prescott, S. C., and W. L. Underwood.** 1897. Microorganisms and sterilizing processes in the canning industries. *Technol. Q.* **10:**183–199.

116. **Ragkousi, K., and P. Setlow.** 2004. Transglutaminase-mediated cross-linking of GerQ in the coats of *Bacillus subtilis* spores. *J. Bacteriol.* **186:**5567–5575.

117. **Rahman, M. S. (ed.).** 1999. *Handbook of Food Preservation.* Marcel-Dekker, New York, N.Y.

118. **Ratnayake-Lecamwasam, M., P. Serror, K. W. Wong, and A. L. Sonenshein.** 2001. *Bacillus subtilis* CodY represses early-stationary-phase genes by sensing GTP levels. *Genes Dev.* **15:**1093–1103.

119. **Redmond, C., L. W. Baillie, S. Hibbs, A. J. Moir, and A. Moir.** 2004. Identification of proteins in the exosporium of *Bacillus anthracis. Microbiology* **150:**355–363.

120. **Rice, E. W., N. J. Adcock, M. Sivaganesan, and L. J. Rose.** 2005. Inactivation of spores of *Bacillus anthracis* Sterne, *Bacillus cereus*, and *Bacillus thuringiensis* by chlorination. *Appl. Environ. Microbiol.* **71:**5587–5589.

121. **Ross, M. A., and P. Setlow.** 2000. The *Bacillus subtilis* HBsu protein modifies the effects of α/β-type small, acid-soluble spore proteins on DNA. *J. Bacteriol.* **182:**1942–1948.

122. **Rowland, S. L., W. F. Burkholder, K. A. Cunningham, M. W. Maciejewski, A. D. Grossman, and G. F. King.** 2004. Structure and mechanism of action of Sda, an inhibitor of the histidine kinases that regulate initiation of sporulation in *Bacillus subtilis. Mol. Cell* **13:**689–701.

123. **Russell, A. D.** 1998. Assessment of sporicidal efficacy. *Int. Biodeterior. Biodegradation* **41:**281–287.

124. **Russell, H. L.** 1896. Gaseous fermentations in the canning industry, p. 227–231. *In Twelfth Annual Report of the Agricultural Experiment Station of the University of Wisconsin.* University of Wisconsin, Madison, Wis.

125. **Salas-Pacheco, J. M., B. Setlow, P. Setlow, and M. Pedraza-Reyes.** 2005. Role of Nfo (YqfS) and ExoA apurinic/apyrimidinic endonucleases in protecting *Bacillus subtilis* spores from DNA damage. *J. Bacteriol.* **187:**7374–7381.

126. **Schmitt, H. P.** 1966. Commercial sterility in canned foods, its meaning and determination. *Assoc. Food Drug Off. U.S. Q. Bull.* **30:**141–151.

127. **Scott, V. N., and D. T. Bernard.** 1982. Heat resistance of spores of non-proteolytic type B *Clostridium botulinum. J. Food Prot.* **45:**909–912.

128. **Scotti, C., M. Piatti, A. Cuzzoni, P. Perani, A. Tognoni, G. Grandi, A. Galizzi, and A. M. Albertini.** 1993. A *Bacillus subtilis* large ORF coding for a polypeptide highly similar to polyketide synthases. *Gene* **130:**65–71.

129. **Serrano, M., A. Neves, C. M. Soares, C. P. Moran, Jr., and A. O. Henriques.** 2004. Role of the anti-sigma factor SpoIIAB in regulation of σ^G during *Bacillus subtilis* sporulation. *J. Bacteriol.* **186:**4000–4013.

130. **Setlow, P.** 1993. DNA structure, spore formation, and spore properties, p. 181–194. *In* P. J. Piggot, P. Youngman, and C. P. Moran, Jr. (ed.), *Regulation of Bacterial Differentiation.* American Society for Microbiology, Washington, D.C.

131. **Setlow, P.** 1994. Mechanisms which contribute to the long-term survival of spores of *Bacillus* species. *J. Appl. Bacteriol.* **176:**49S–60S.

132. **Setlow, P.** 1995. Mechanisms for the prevention of damage to the DNA in spores of *Bacillus* species. *Annu. Rev. Microbiol.* **49:**29–54.

133. **Setlow, P.** 2001. Resistance of spores of *Bacillus* species to ultraviolet light. *Environ. Mol. Mutagen.* **38:**97–104.

134. **Setlow, P.** 2003. Spore germination. *Curr. Opin. Microbiol.* **6:**550–556.

135. **Setlow, P.** 2006. Spores of *Bacillus subtilis*: their resistance to and killing by radiation, heat and chemicals. *J. Appl. Microbiol.* **101:**514–525.

136. **Setlow, P., and E. A. Johnson.** 2001. Spores and their significance, p. 33–70. *In* M. P. Doyle, L. R. Beuchat, and T. J. Montville (ed.), *Food Microbiology: Fundamentals and Frontiers*, 2nd ed. ASM Press, Washington, D.C.

137. **Shimizu, T., K. Ohtani, H. Hirakawa, K. Ohshima, A. Yamashita, T. Shiba, N. Ogasawara, M. Hattori, S. Kuhara, and H. Hayashi.** 2002. Complete genome sequence of *Clostridium perfringens*, an anaerobic flesh-eater. *Proc. Natl. Acad. Sci. USA* **99:**996–1001.

138. **Slepecky, R. A., and E. R. Leadbetter.** 1994. Ecology and relationships of endospore-forming bacteria: changing perspectives, p. 195–206. *In* P. J. Piggot, C. P. Moran, Jr., and P. Youngman (ed.), *Regulation of Bacterial Differentiation.* American Society for Microbiology, Washington, D.C.

139. **Smith, L. D. S., and H. Sugiyama.** 1988. *Botulism. The Organism, Its Toxins, the Disease*, 2nd ed. Charles C Thomas, Springfield, Ill.

140. **Sonenshein, A. L.** 2000. Endospore-forming bacteria—an overview, p. 133–150. *In* L. J. Shimkets and Y. V. Brun (ed.), *Prokaryotic Development.* American Society for Microbiology, Washington, D.C.

141. **Sonenshein, A. L.** 2000. Control of sporulation initiation in *Bacillus subtilis. Curr. Opin. Microbiol.* **3:**561–566.

142. **Songer, J. G.** 1996. Clostridial enteric diseases of domestic animals. *Clin. Microbiol. Rev.* **9:**216–234.

143. **Spotts Whitney, E. A., M. E. Beatty, T. H. Taylor, Jr., R. Weyant, J. Sobel, M. J. Arduino, and D. A. Ashford.** 2003. Inactivation of *Bacillus anthracis* spores. *Emerg. Infect. Dis.* **9:**623–627.

144. **Steichen, C. T., J. F. Kearney, and C. L. Turnbough, Jr.** 2005. Characterization of the exosporium basal layer protein BxpB of *Bacillus anthracis. J. Bacteriol.* **187:**5868–5876.

145. **Steil, L., M. Serrano, A. O. Henriques, and U. Volker.** 2005. Genome-wide analysis of temporally regulated and compartment-specific gene expression in sporulating cells of *Bacillus subtilis. Microbiology* **151:**399–420.

146. **Stevenson, K. E., and R. H. Vaughn.** 1972. Exosporium formation in sporulating cells of *Clostridium botulinum* 78A. *J. Bacteriol.* **112:**618–621.

147. Stringer, M. 2005. Summary report. Food safety objectives - role in microbiological food safety management. *Food Control* **16:**775–794.

148. Stumbo, C. R. 1973. *Thermobacteriology in Food Processing*, 2nd ed. Academic Press, New York, N.Y.

149. Sugiyama, H. 1951. Studies on factors affecting the heat resistance of spores of *Clostridium botulinum*. *J. Bacteriol.* **62:**81–96.

150. Sugiyama, H. 1952. Effect of fatty acids on the heat resistance of *Clostridium botulinum* spores. *Bacteriol. Rev.* **16:**125–126.

151. Sugiyama, H. 1986. Mouse models for infant botulism, p. 73–91. *In* O. Zak and M. A. Sande (ed.), *Experimental Models in Antimicrobial Chemotherapy*, vol. 2. Academic Press, New York, N.Y.

152. Sugiyama, H. 1980. *Clostridium botulinum* neurotoxin. *Microbiol. Rev.* **44:**419–448.

153. Sylvestre, P., E. Couture-Tosi, and M. Mock. 2002. A collagen-like surface glycoprotein is a structural component of the *Bacillus anthracis* exosporium. *Mol. Microbiol.* **45:**169–178.

154. Takamatsu, H., and K. Watabe. 2002. Assembly and genetics of spore protective structures. *Cell. Mol. Life Sci.* **59:**434–444.

155. Tanaka, N., E. Traisman, P. Plantinga, L. Finn, W. Flom, L. Meske, and J. Guffisberg. 1986. Evaluation of factors involved in antibotulinal properties of pasteurized process cheese spreads. *J. Food Prot.* **49:**526–531.

156. Tournas, V. 1994. Heat-resistant fungi of importance to the food and beverage industry. *Crit. Rev. Microbiol.* **20:**243–263.

157. Tovar-Rojo, F., M. Chander, B. Setlow, and P. Setlow. 2002. The products of the *spoVA* operon are involved in dipicolinic acid uptake into developing spores of *Bacillus subtilis*. *J. Bacteriol.* **184:**584–587.

158. Townsend, C. T., J. R. Esty, and F. C. Baselt. 1938. Heat-resistance studies on spores of putrefactive anaerobes in relation to the determination of safe processes for canned foods. *Food Res.* **3:**323–346.

159. Trent, J. D., M. Gabrielson, B. Jensen, J. Neuhard, and J. Olsen. 1994. Acquired thermotolerance and heat shock proteins in thermophiles from the three phylogenetic domains. *J. Bacteriol.* **176:**6148–6152.

160. Van Netton, P., A. Van de Moosdijk, P. Van de Hoensel, D. A. A. Mossel, and I. Perales. 1990. Psychrotrophic strains of *Bacillus cereus* producing enterotoxin. *J. Appl. Bacteriol.* **69:**73–79.

161. Veening, J. W., L. W. Hamoen, and O. P. Kuipers. 2005. Phosphatases modulate bistable sporulation gene expression pattern in *Bacillus subtilis*. *Mol. Microbiol.* **56:**1481–1494.

162. Vepachedu, V. R., and P. Setlow. 2005. Localization of SpoVAD to the inner membrane of spores of *Bacillus subtilis*. *J. Bacteriol.* **187:**5677–5682.

163. Waller, L. N., M. J. Stump, K. F. Fox, W. M. Harley, A. Fox, G. C. Stewart, and M. Shahgholi. 2005. Identification of a second collagen-like glycoprotein produced by *Bacillus anthracis* and demonstration of spore-specific sugars. *J. Bacteriol.* **187:**4592–4597.

164. Wang, S. T., B. Setlow, E. M. Conlon, J. L. Lyon, D. Imamura, T. Sato, P. Setlow, R. Losick, and P. Eichenberger. 2006. The forespore line of gene expression in *Bacillus subtilis*. *J. Mol. Biol.* **358:**16–37.

165. Westphal, A. J., P. B. Price, T. J. Leighton, and K. E. Wheeler. 2003. Kinetics of size changes of individual *Bacillus thuringiensis* spores in response to changes in relative humidity. *Proc. Natl. Acad. Sci. USA* **100:**3461–3466.

166. Willis, A. T. 1969. *Clostridia of Wound Infection*. Butterworths, London, England.

167. Wörner, K., H. Szurmant, C. Chiang, and J. A. Hoch. 2006. Phosphorylation and functional analysis of the sporulation initiation factor Spo0A from *Clostridium botulinum*. *Mol. Microbiol.* **59:**1000–1012.

168. Wuytack, E. Y., J. Soons, F. Poschet, and C. W. Michiels. 2000. Comparative study of pressure- and nutrient-induced germination of *Bacillus subtilis* spores. *Appl. Environ. Microbiol.* **66:**257–261.

169. Yudkin, M. D., and J. Clarkson. 2005. Differential gene expression in genetically identical sister cells: the initiation of sporulation in *Bacillus subtilis*. *Mol. Microbiol.* **56:**578–589.

Food Microbiology: Fundamentals and Frontiers, 3rd Ed.
Edited by M. P. Doyle and L. R. Beuchat
© 2007 ASM Press, Washington, D.C.

Merle D. Pierson
Don L. Zink
L. Michele Smoot

4

Indicator Microorganisms and Microbiological Criteria

INTRODUCTION

Purpose of Microbiological Criteria

Microbiological criteria are used to distinguish between acceptable and unacceptable products or between acceptable and unacceptable food processing and handling practices. The numbers and types of microorganisms present in or on a food product may be used to judge the microbiological safety and quality of that product. Safety is determined by the presence or absence of pathogenic microorganisms or their toxins, the number of pathogens, and the expected control or destruction of these agents. The level of spoilage microorganisms reflects the microbiological quality, or wholesomeness, of a food product as well as the effectiveness of measures used to control or destroy such microorganisms. Indicator organisms may be used to assess either the microbiological quality or food safety risk of a food. Indicator organisms used to assess microbiological safety of a food require that a positive relationship between the occurrence of the indicator organism and the likely presence of a pathogen or toxin has been established. Specifically, microbiological criteria are used to assess the following: (i) the safety of food, (ii) adherence to good manufacturing practices (GMPs),

(iii) the keeping quality (shelf life) of certain perishable foods, and (iv) the utility (suitability) of a food or ingredient for a particular purpose (42). When appropriately applied, microbiological criteria can be a useful means for ensuring the safety and quality of foods, which in turn elevates consumer confidence. Microbiological criteria provide the food industry and regulatory agencies with guidelines for control of food processing systems and are an underlying component of any critical control point that addresses a microbiological hazard in Hazard Analysis and Critical Control Points (HACCP) systems. In addition, internationally accepted criteria can advance free trade through standardization of food safety/quality requirements.

Need To Establish Microbiological Criteria

A microbiological criterion should be established and implemented only when there is a need and when it can be shown to be both effective and practical. There are many factors to consider when establishing meaningful microbiological criteria (42), including:

- evidence of a hazard to health based on epidemiological data or a hazard analysis

Merle D. Pierson, Dept. of Food Science and Technology, Virginia Polytechnic Institute and State University, Blacksburg, VA 24061. **Don L. Zink,** U.S. Food and Drug Administration, 5100 Paint Branch Parkway, College Park, MD 20740. **L. Michele Smoot,** Silliker, Inc., 2057 Builders Place, Columbus, Ohio 43204.

- the nature of the natural and commonly acquired microflora of the food and the ability of the food to support microbial growth
- the effect of processing on the microflora of the food
- the potential for microbial contamination and/or growth during processing, handling, storage, and distribution
- the category of consumers at risk
- the state in which the food is distributed
- the potential for abuse at the consumer level
- spoilage potential, utility, and GMPs
- the manner in which the food is prepared for ultimate consumption
- reliability of methods available to detect and/or quantify the microorganism(s) and toxin(s) of concern
- the costs/benefits associated with the application of the criterion

Definitions

The National Research Council of the National Academy of Sciences (NAS) addressed the issue of microbiological criteria in their 1985 report entitled *An Evaluation of the Role of Microbiological Criteria for Foods and Food Ingredients*. In that text, it was established that a microbiological criterion will stipulate that a type of microorganism, group of microorganisms, or toxin produced by a microorganism must either not be present at all, be present in only a limited number of samples, or be present as no less than a specified number or amount in a given quantity of a food or food ingredient. In addition, a microbiological criterion should include the following information (42):

- a statement describing the identity of the food or food ingredient
- a statement identifying the contaminant of concern
- an analytical method to be used for the detection, enumeration, or quantification of the contaminant of concern
- a sampling plan
- microbiological limits considered appropriate to the food and commensurate with the sampling plan

Criteria may be either mandatory or advisory. A mandatory criterion is one that may not be exceeded, and food that does not meet the specified limit is required to be subjected to some action, such as rejection, destruction, reprocessing, or diversion. An advisory criterion permits acceptability judgments to be made, and it should serve as an alert to deficiencies in processing, distribution, storage, or marketing. For application purposes, the three categories of criteria that are applied include standards, guidelines, and specifications. The following definitions were recommended in the National Research Council's Subcommittee on Microbiological Criteria for Foods and Food Ingredients (42). *Codex Alimentarius* (12) does not provide specific definitions; however, they are implied in the Application of Microbiological Criteria section of their criteria principles document.

- Standard: A microbiological criterion that is part of a law, ordinance, or administrative regulation. A standard is a mandatory criterion. Failure to comply constitutes a violation of the law, ordinance, or regulation and will be subject to the enforcement policy of the regulatory agency having jurisdiction (40). ...a criterion contained in a Codex Alimentarius standard. Wherever possible it should contain limits only for pathogenic microorganisms of public health significance in the food concerned. Limits for non-pathogenic microorganisms may be necessary when the methods of detection for the pathogens of concern are cumbersome or unreliable. Standards based on fixed numbers of nonpathogenic microorganisms may result in the recall or down grading of otherwise wholesome food. To minimize the effect of this approach penalty provisions could be applied when a lot is rejected. Such penalties would result in suspension in the privilege to process food only after repeated violations occur over a specified time period (10).
- Guideline: A microbiological criterion often used by the food industry or regulatory agency to monitor a manufacturing process. Guidelines function as alert mechanisms to signal whether microbiological conditions prevailing at critical control points or in the finished product are within the normal range. Hence, they are used to assess processing efficiency at critical control points and conformity with Good Manufacturing Practices. A microbiological guideline is advisory (42). ...is intended to increase assurance that the provisions of hygienic significance in the Code have been met. It may include microorganisms which are not of direct public health significance (12).
- Specifications: A microbiological criterion that is used as a purchase requirement whereby conformance becomes a condition of purchase between buyer and vendor of a food ingredient. A microbiological specification may be advisory or mandatory (42). ...is applied at the establishment at a specified point during or after processing to monitor hygiene. It is intended to guide the manufacturer and is not intended for official control purposes (12). The Codex use of "specification" only refers to end products and does not include raw

materials, ingredients, or foods in contractual agreements between two parties.

Who Establishes Microbiological Criteria?

Different scientific organizations have been involved in developing general principles for the application of microbiological criteria by regulatory agencies and the food industry. The scientific organizations which have influenced the U.S. food industry the most include the Joint Food and Agricultural Organization (FAO) and World Health Organization (WHO) *Codex Alimentarius* International Food Standards Program, the International Commission on Microbiological Specifications for Foods (ICMSF), the U.S. NAS, and the U.S. National Advisory Committee on Microbiological Criteria for Foods. The *Codex Alimentarius* program first formulated "General Principles for the Establishment and Application of Microbiological Criteria" in 1981 (12). These microbiological principles have been well accepted internationally. In 1984, the NAS National Research Council Subcommittee on Microbiological Criteria for Foods and Food Ingredients formulated general principles for the application of microbiological criteria to food and food ingredients as requested by four U.S. regulatory agencies (42). The Codex Committee published a revision to their earlier document "General Principles for the Establishment and Application of Microbiological Criteria for Foods" (17). *Codex Alimentarius* contains standards for all principal foods in the forms (i.e., processed, semiprocessed, and raw) in which they are delivered to the consumer. Fresh perishable commodities not traded internationally are excluded from these standards (36, 46). When establishing microbiological criteria for those foods intended for international trade, materials provided by the ICMSF (25, 28, 29) and the *Codex Alimentarius* (12) should be consulted.

The World Trade Organization provides a framework for ensuring fair trade and harmonizing standards and import requirements on food traded, through the Agreements on Sanitary and Phytosanitary Measures and Technical Barriers to Trade. Countries are required to base their standards on science, to base programs on risk analysis methodologies, and to develop ways of achieving equivalence between methods of inspection, analysis, and certification between trading countries (36, 46). The World Trade Organization recommends the use of standards, guidelines, and recommendations developed by *Codex Alimentarius* to facilitate harmonization of standards. Improvements in the development and execution of microbiological criteria will continue to be made with international acceptance and widespread implementation

of HACCP, a science-based preventative system for food control. Many countries have mandated HACCP requirements and have established specific HACCP requirements for public sectors of their domestic food industries (24).

SAMPLING PLANS

Introduction

A sampling plan includes both the sampling procedure and the decision criteria regarding the disposition of a lot of product based on the results of the sampling plan. To examine a food for the presence of microorganisms, a representative sample is examined by defined methods. A lot is that quantity of product produced, handled, and stored within a specified time period under uniform conditions. Since it is impractical to conduct microbiological analysis on the entire lot, statistical concepts of population probability and sampling must be used to determine the number and size of sample units from the lot and to provide conclusions drawn from the analytical results. The sampling plan is designed so that inferior lots, within a specified level of confidence, are rejected. Detailed information regarding statistical concepts of population probabilities and sampling, choice of sampling procedures, decision criteria, and practical aspects of application as applied to microorganisms in food can be found in publications by the ICMSF (26, 29).

A simplified example of a sampling plan described in the aforementioned publication is provided below. Assume 10 samples were taken and analyzed for the presence of a particular microorganism. Based on the decision criterion, only a certain number of the sample units could be positive for the presence of that microorganism for the lot to still be considered acceptable. If in the criterion the maximum allowable positive units had been set at 2 ($c = 2$), then a positive result for more than 2 of the 10 sample units ($n = 10$) would result in rejection of the lot. Ideally, the decision criterion is established to accept lots that are of the desired quality and reject lots that are not. However, because the entire lot is not examined, there is always the risk that an acceptable lot may be rejected or an unacceptable lot may be accepted. The more samples that are examined or the larger the value of n, the lower the risk of making an incorrect decision about the lot quality. However, as n increases, sampling becomes more time consuming and costly. Generally, a compromise is made between the size of n and the level of risk that is acceptable.

The level of risk that is associated with a particular sampling plan can be determined by an operating characteristic curve, as seen in Fig. 4.1. The two vertical scales in

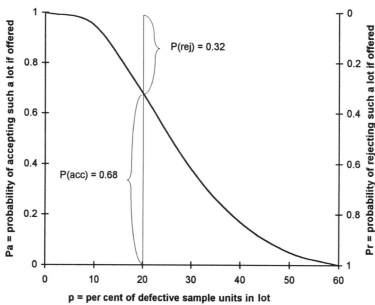

Figure 4.1 Operating characteristic curve for $n = 10$, $c = 2$. From reference 26 with permission.

the graph show (i) the probability of acceptance, P_a, or the ratio of the number of times that the results will indicate a lot should be accepted to the number of times a lot of the given quality is sampled for a decision and (ii) the probability of rejection, P_r, or the ratio of the number of times that the result will indicate a lot should be rejected to the number of times a lot of that given quality is sampled for a decision. The horizontal axis indicates the percent of defective sample units (p) that are in the lot. This probability is usually expressed as percentages of "defectives." The operating characteristic curve in Fig. 4.1, in which $n = 10$ and $c = 2$, shows that as p increases, P_a will decrease. In other words, if the probability that a test unit will yield a positive result is great, then the probability that the lot will be accepted becomes small. The risks to be considered include those for the consumers or buyers and those for the vendors or producers. The vendors' risk is that probability that a lot of acceptable quality is rejected, while the probability that a lot of defective quality is accepted is referred to as the consumers' risk. The level of risk that a vendor is willing to accept relative to himself and the consumer is then used to design the sampling plan.

Types of Sampling Plans

Sampling plans are divided into two main categories: variables and attributes. A variables plan depends on the frequency distribution of microorganisms in the food. For correct application of a variables plan the microorganisms must be distributed in a log normal fashion (i.e., counts transformed to logarithms are normally distributed).

When the food is from a common source and it is known to be produced/processed under uniform conditions, log normal distribution of the microbes present is reasonably assumed (32, 33). Attributes sampling is the preferred plan when microorganisms are not homogeneously distributed throughout the food or when the target microorganism is present at low levels, which is often the case with pathogenic microorganisms. Attributes plans are also widely used to determine the acceptance or rejection of product at ports or other points of entry because there is little or no knowledge of how the food was processed or past performance records are not available. Attributes sampling plans may also be used to monitor performance relative to accepted GMPs. Attributes sampling, however, is not appropriate when there is no defined lot or when random sampling is not possible, as might occur when monitoring cleaning practices.

Variables Sampling Plans

Because attributes sampling plans are widely used over variables sampling, this text will not cover variables sampling in detail. For more detailed information on variables sampling plans, the reader is referred to ICMSF publications (26, 29) and reports from Kilsby (32), Kilsby et al. (33), and Malcolm (37).

Attributes Sampling Plans

Two-Class Plans

The two-class attributes sampling plan assigns the concentration of microorganisms of the sample units tested to a particular attribute class depending on whether the microbiological counts are above or below some preset concentration, represented by the letter m. The decision criterion is based on (i) the number of sample units tested, n, and (ii) the maximum allowable number of sample units yielding unsatisfactory test results, c. For example, when $n = 5$ and $c = 2$ in a two-class sampling plan designed to make a presence/absence decision on the lot (i.e., $m = 0$), the lot is rejected if more than two of the five sample units tested are positive. As n increases for the set number c, the stringency of the sampling plan also increases. Conversely, for a set sample size n, as c increases, the stringency of the sampling plan decreases, allowing for a higher probability of accepting, P_a, food lots of a given quality. Two-class plans are applied most often in qualitative (semiquantitative) pathogen testing, where the results are expressed as the presence or absence of the specific pathogen per sample weight analyzed.

Three-Class Plans

Three-class sampling plans use the concentration of microorganisms in the sample units to determine levels of quality and/or safety. Counts above a preset concentration M for any of the n sample units tested are considered unacceptable, and the lot is rejected. The level of the test organism acceptable in the food is denoted by m. This concentration in a three-class attribute plan separates acceptable lots (i.e., counts less than m) from marginally acceptable lots (i.e., counts greater than m but not exceeding M). Counts above m and up to and including M are not desirable but the lot can be accepted provided the n number of samples that exceed m is no greater than the preset number, c. Thus, in a three-class sampling plan, the food lot will be rejected if any one of the sample units exceeds M or if the number of sample units with contamination levels above m exceeds c. Similar to the two-class sampling plan, the stringency of the three-class sampling plan is also dependent on the two numbers denoted by n and c. The larger the value of n for a given value of c, the better the food quality must be to have the same chance of passing, and vice versa. From n and c it is then possible to find the probability of acceptance, P_a, for a food lot of a given microbiological quality.

ESTABLISHING LIMITS

Microbiological limits, as defined in a criterion, represent the level above which action is required. Levels should be realistic and should be determined based on knowledge of the raw materials and the effects of processing, product handling, storage, and end use of the product. Limits should also take into account the likelihood of uneven distribution of microorganisms in the food, the inherent variability of the analytical procedure, the risk associated with the microorganisms, and the conditions under which the food is expected to be handled and consumed. Microbiological limits should include sample weight to be analyzed, method reference, and confidence limits of the referenced method where applicable.

The shelf life of a perishable product is often determined by the number of microorganisms initially present. As a general rule, a food containing a large population of spoilage microbes will have a shorter shelf life than the same food containing fewer numbers of the same spoilage microorganisms. However, the relationship between total counts and shelf life is not absolute. Some types of microorganisms have a greater impact on the organoleptic characteristics of a food than others due to the presence of different enzymes acting upon the food constituents. In addition to the effect of certain levels and/or types of spoilage microorganisms, changes in perceptible quality characteristics will also vary depending on the food and the conditions of storage, such as temperature and gaseous atmosphere. All of these parameters need to be considered when establishing limits for the microbiological criteria used to determine product quality and/or shelf life.

Foods produced and stored under GMPs may be expected to have a different microbiological profile than those foods produced and stored under poor conditions. The use of poor-quality materials, improper handling, or unsanitary conditions may result in higher bacterial counts in the finished product. However, low counts in the finished product do not necessarily mean that the product was produced under acceptable GMPs. Processing steps such as heat treatments, fermentation, freezing, or frozen storage can reduce the counts of bacteria that have resulted from noncompliance with GMPs. Other products, such as ground beef, may normally contain high microbial counts even under the best conditions of manufacture due to the growth of psychrotrophic bacteria during refrigeration.

Limits set for microbiological criteria used to assess adherence to GMPs require a working knowledge of the types and levels of microorganisms present at the different processing steps to establish a relationship between the microbiology of the food and adherence to GMPs.

Microbiological criteria are also an underlying component of any critical control point in an HACCP system that addresses a microbiological hazard (9). In an HACCP system, a critical limit is set for each critical control point identified for an operation. A critical limit is the maximum and/or minimum value to which a biological, chemical, or physical parameter must be controlled at a

CCP to prevent, eliminate, or reduce to an acceptable level the occurrence of a food safety hazard. Limits assigned to microbiological criteria are not necessarily microbiological in nature but are often a physical or chemical attribute. For example, high-acid foods not thermally processed are required to have a pH of less than 4.6 or a water activity of less than 0.85 to prevent outgrowth of *Clostridium botulinum* spores and botulinum toxin production.

Setting limits for a microbiological criterion allows for a decision to be made about the operation of a critical step in a process. The criterion may be either absolute or performance based. An absolute criterion is one in which a specific upper level or frequency of a microorganism, group of microorganisms, or product of microbial metabolism has been established. A negative result for *Escherichia coli* O157:H7 in 25 g of ground beef is an example of a critical limit established for an absolute microbiological criterion. Alternatively, a performance criterion refers to a specified change that a process is expected to exert on the level or frequency of a microorganism or microbial metabolite (9). For example, low-acid canned foods are required to be heat treated at a specific temperature and time to reduce the number of *C. botulinum* spores by a factor of at least 10^{12}.

The critical limits associated with CCPs dictate the level of stringency of a food production process. For a limit to be effective, it must be well understood how that limit is related to the microbiological criterion, how the criterion is related to the microbiological hazard identified in the hazard analysis, and societies' and regulatory agencies' expectations. The use of quantitative risk assessment techniques to scientifically determine the probability of occurrence and severity of known human exposure to foodborne hazards is highly advocated. The process consists of (i) hazard identification, (ii) hazard characterization, (iii) exposure assessment, and (iv) risk characterization (34). Though quantitative risk assessment techniques are well established for chemical agents, their application to food safety microbiology is relatively new. The application of risk assessment principles to biological agents has been complicated by the inherent differences between chemical and biological agents. Levels of microorganisms are often not uniformly distributed and are subject to change during processing and storage. Although systematic application of risk analysis methodology is still needed to establish microbiological criteria, quantitative data on microbial hazards may be substituted with current qualitative data within this same framework to facilitate establishment of microbiological limits.

The concept of a Food Safety Objective (FSO) can also be helpful in establishing limits. The ICMSF defines an FSO as "the maximum frequency and/or concentration of a microbial hazard in a food considered tolerable for consumer protection" (29). FSOs are used to translate a particular public health objective, such as a 50% reduction in the number of cases of listeriosis, into a microbiological or food safety criterion that can be measured and used by food producers and government agencies.

Examples of FSOs include the following:

- The number of *Listeria monocytogenes* organisms in ready-to-eat foods may not exceed 100 CFU per g.
- *Salmonella* spp. must not be detected in powdered infant formula in 60 25-g subsamples of the production lot.
- Aflatoxin concentration in peanuts should not exceed 15 µg/kg (56).

FSOs must not be confused with microbiological criteria. An FSO states the level of control that is needed to achieve a public health objective. FSOs are not applied to individual lots of product and do specify sampling plans or testing methods. The level of control required by an FSO can be achieved through proper application of HACCP, GMPs, good agricultural practices, and proper application of performance criteria. FSOs should be quantitative and measurable by some means, such as testing or performance criteria. Hence, the FSO is a means of communicating the stringency with which food safety control systems must operate.

The ICMSF (29) lists the uses of FSOs as follows:

- translate a public health goal into a measurable level of control upon which food processes can be designed so the resulting food will be acceptable
- validate food processing operations to ensure they will meet the expected level of control
- assess the acceptability of a food operation by control authorities or other auditors
- highlight food safety concerns, separate from quality and other concerns
- force change in a food commodity and improve its safety
- serve as the basis for establishing microbiological criteria for individual lots or consignments of food when the source or conditions of manufacture are uncertain

INDICATORS OF MICROBIOLOGICAL QUALITY

Introduction

Estimation of a product for indicator microorganisms can provide simple, reliable, and rapid information about process failure, postprocessing contamination, contamination from the environment, and the general level of hygiene under which the food was processed and stored. However, these methods cannot replace examination for specific pathogens where suitable methods exist and where such testing is appropriate, but they usually provide information in a shorter time than that required for isolation and identification of specific microorganisms or pathogens.

Indicator Microorganisms

Indicator microorganisms can be used in microbiological criteria. These criteria might be used to address existing product quality or to predict shelf life of the food. Some examples of indicators and the products in which they are used are shown in Table 4.1.

Those microorganisms listed in Table 4.1 are the primary spoilage microbes corresponding to each specific product. Loss of quality in other products may not be limited to one microorganism but rather to a variety of microorganisms due to the unrestricted environment of the food. In those types of products it is often more practical to determine the counts of groups of microorganisms most likely to cause spoilage in that particular food.

Aerobic plate count (APC) or standard plate count (SPC) is commonly used to determine "total" numbers of microorganisms in a food product. By modifying the environment of incubation or the medium used, APC can be used to preferentially screen for groups of microorganisms such as those that are anaerobic, thermoduric, mesophilic, psychrophilic, thermophilic, proteolytic, and lipolytic. APC may be a component of microbiological criteria assessing product quality when that criterion is used to (i) monitor foods for compliance with standards or guidelines set by various regulatory agencies, (ii) monitor foods for compliance with purchase specifications, and (iii) monitor adherence to GMPs (42).

Table 4.1 Organisms highly correlated with product quality[a]

Organisms	Products
Acetobacter spp.	Fresh cider
Bacillus spp.	Bread dough
Byssochlamys spp.	Canned fruits
Clostridium spp.	Hard cheeses
Flat-sour spores	Canned vegetables
Lactic acid bacteria	Beers, wines
Lactococcus lactis	Raw milk (refrigerated)
Leuconostoc mesenteroides	Sugar (during refining)
Pectinatus cerevisiiphilus	Beers
"*Pseudomonas putrefaciens*"	Butter
Yeasts	Fruit juice concentrates
Zygosaccharomyces bailii	Mayonnaise, salad dressing

[a] Reprinted with permission from reference 30.

A microbiological criterion as a specification is used to determine the usefulness of a food or food ingredient for a particular purpose. For example, specifications are set for thermophilic spores in sugar and spices intended for use in the canning industry. Lots of sugar failing to meet specifications may not be suitable for use in the low-acid canning but could be diverted for other uses. The APC of refrigerated perishable foods such as milk, meat, poultry, and fish may be used to indicate the condition of equipment and utensils used, as well as the time/temperature profile of the storage and distribution of the food.

When evaluating results of APCs for a particular food, it is important to remember that (i) APCs only measure live cells and therefore would not be of value, for example, to determine the quality of raw materials used for a heat-processed food; (ii) APCs are of little value in assessing organoleptic quality since high microbial counts generally must be present prior to organoleptic quality loss; and (iii) because different bacteria vary in their biochemical activities, quality loss may also occur at low total counts, depending on the predominant microbes present. With any food, specific causes of unexpected high counts can be identified by examination of samples at control points and by plant inspection. Reliable interpretation of the APC of a food requires knowledge of the expected microbial population at the point in the process or distribution at which the sample is collected. If counts are higher than expected, this will point to the need to determine why there has been a violation of the criterion. A review by Silliker (47) provides a detailed discussion of the use of total counts as an index of sanitary quality, organoleptic quality, and safety.

The direct microscopic count (DMC) is used to give an estimate of both viable and nonviable cells in samples containing a large number of microorganisms (i.e., $>10^5$ CFU/ml). Considering that the DMC does not differentiate between live and dead cells (unless a fluorescent dye such as acridine orange is employed) and it requires that the total cell count exceed 10^5 cells/ml, use of the DMC is of limited value as part of microbiological criteria for quality issues. The use of DMC as part of microbiological criteria for foods or ingredients is restricted to a few products such as raw, non-grade A milk, dried milks, liquid and frozen eggs, and dried eggs (42).

Other methods commonly used to indicate quality of different food products include the Howard mold count, yeast and mold count, heat-resistant mold count, and thermophilic spore count. The Howard mold count is used to detect the presence of moldy material in canned fruit and tomato products (21) as well as to evaluate the sanitary condition of processing machinery in vegetable canneries (19). Yeasts and molds frequently become predominant on foods when conditions for bacterial growth are less favorable. Therefore, they can potentially be a problem in fermented dairy products, fruits, fruit beverages, and soft drinks. Yeasts and mold counts are used as part of the microbiological standards of various dairy products such as cottage cheese and frozen cream (51) and sugar (43). Heat-resistant molds, such as *Byssochlamys fulva* and *Aspergillus fisheri*, that may survive the thermal processes applied to fruit and fruit products may need limits in purchase specifications for ingredients such as fruit concentrates. Concern for thermophilic spores in ingredients used in the canning industry is related to their ability to cause defects in foods held at elevated temperatures because of inadequate cooling and/or storage at too-high temperatures. Purchase specifications and verification criteria are often used for thermophilic spore counts in ingredients intended for use in low-acid, heat-processed canned foods.

Metabolic Products

In certain cases, bacterial populations in a food product can also be estimated by testing for metabolic products produced by the microorganisms present in the food. When a correlation is established between the presence of a metabolic product and product quality loss, tests for the metabolite may be a part of a microbiological criterion.

An example of the use of metabolic products as part of a microbiological criterion is the organoleptic evaluation of imported shrimp. Trained personnel are able to classify the degree of decomposition (i.e., quality loss) into one of three classes through organoleptic examination. The shrimp are placed into one of the following quality classes: class 1, passable; class 2, decomposed (slight but definite); or class 3, decomposed (advanced). Limits of acceptability of a lot are based on the number of shrimp in a sample that are placed into each of the three classes (4). Other commodities in which organoleptic examination is used to determine quality deterioration include raw milk, meat, poultry, and fish and other seafoods. The food industry also uses these examinations to classify certain foods into quality grades. For additional information about organoleptic examination of foods, the reader is referred to the work of Amerine et al. (3) and Larmond (35).

Other examples of metabolic products used to assess product quality are listed in Table 4.2.

INDICATORS OF FOODBORNE PATHOGENS AND TOXINS

Introduction

Microbiological criteria as they apply to product safety should only be developed when the application of a criterion can reduce or eliminate a potential foodborne

Table 4.2 Some microbial metabolic products that correlate with food quality[a]

Microbial metabolite(s)	Applicable food product(s)
Cadaverine and putrescine	Vacuum-packaged beef
Diacetyl	Frozen juice concentrate
Ethanol	Apple juice, fishery products
Histamine	Canned tuna
Lactic acid	Canned vegetables
Trimethylamine	Fish
Total volatile bases, total volatile nitrogen	Seafoods
Volatile fatty acids	Butter, cream

[a] Reprinted with permission from reference 30.

hazard. Each food type should be carefully evaluated through risk assessment to determine the potential hazards and their significance to consumers. When a food is repeatedly implicated as a vehicle in foodborne disease outbreaks, application of microbiological criteria may be useful. Public health officials and the dairy industry responded to widespread outbreaks of milk-borne disease occurring in the early 1900s in the United States. By imposing controls on milk production, developing safe and effective pasteurization procedures, and setting microbiological criteria, the safety of commercial milk supplies was greatly improved. Epidemiologic evidence alone, however, does not necessitate imposing microbiological criteria. The criteria should only be applied when their use results in a safer food (1).

Food products frequently subject to contamination by harmful microorganisms, such as shellfish, may benefit from the application of microbiological criteria. The National Shellfish Sanitation Program utilizes microbiological criteria in this manner to prevent use of shellfish from polluted waters which may contain various intestinal pathogens (2). Depending on the type and level of contamination anticipated, imposition of microbiological criteria may or may not be justified. Foods contaminated with pathogens that do not have the opportunity to grow to levels that would potentially result in a health hazard do not warrant microbiological criteria. Though fresh vegetables are often contaminated with small numbers of *C. botulinum*, *Clostridium perfringens*, and *Bacillus cereus* cells, epidemiologic evidence indicates that this contamination presents no health hazard. Hence, imposing microbiological criteria would not be beneficial regarding these microorganisms. However, criteria may be appropriate for enteric pathogens on produce, since there have been several outbreaks of foodborne illness resulting from fresh produce contaminated with enteric pathogens (7, 50).

Often food processors alter the intrinsic or extrinsic parameters of a food (nutrients, pH, water activity, inhibitory chemicals, gaseous atmosphere, temperature of storage, and the presence of competing microbes) to prevent growth of undesirable microorganisms. If control over one or more of these parameters is lost, then there may be a risk of a health hazard. For example, in the manufacture of cheese or fermented sausage, a lactic acid starter culture is relied upon to produce acid quickly enough to inhibit the growth of *Staphylococcus aureus* to cell numbers that would be potentially harmful. Process CCPs, such as the rate of acid formation, are implemented to prevent growth or contamination of the food by harmful microorganisms and to assure control of the process is maintained (44).

Depending on the pathogen, low levels of the microorganism in the food product may or may not be of concern. Some microorganisms have such a low infective dose that their mere presence in a food presents a significant public health risk. For such microbes, the concern is not whether the pathogen is able to grow in the food but that the microorganism could survive for any length of time in the food. Foods having intrinsic or extrinsic factors sufficient to prevent survival of pathogens or toxigenic microorganisms of concern may not be candidates for microbiological criteria related to safety. For example, the acidity of certain foods, such as fermented meat products, might be assumed sufficient for pathogen control. In fact, the growth and toxin production of

S. aureus might be prevented; however, enteric pathogens such as *E. coli* O157:H7 could survive and result in a product unsafe for consumption.

An important and sometimes overlooked consideration when evaluating the potential microbiological risks associated with a food is the consumer. More rigid microbiological requirements may be needed if the food is intended for use by infants, the elderly, or immunocompromised people because they are more susceptible to infectious agents than are healthy adults. The sampling plan specified in a microbiological criterion should be appropriate to the hazard expected to be associated with the food, the consumer of the food, and the severity of the illness. The hazard associated with a food is determined by (i) the type of microorganism expected to be encountered and (ii) the expected conditions of handling and consumption after sampling. A more stringent sampling plan is desired for products expected to contain higher degrees of hazards. The ICMSF (26, 29) proposed a system for classification of foods according to risk into 15 hazard categories called cases, with suggested appropriate sampling plans, as shown in Table 4.3.

The stringency of sampling plans for foods is based either on the hazard to the consumer from pathogenic microorganisms and their toxins or toxic metabolites or on the potential for quality deterioration to an unacceptable state, and it should take into account the types of microorganisms present and their numbers (26, 29).

Table 4.3 Plan stringency (case) in relation to degree of health hazard and conditions of use[a]

Type of hazard	Conditions in which food is expected to be handled and consumed after sampling, in the usual course of events[b]		
No direct health hazard	Increased shelf life	No change	Reduced shelf life
Utility (e.g., general contamination, reduced shelf life, and incipient spoilage)	Case 1 Three-class $n = 5, c = 3$	Case 2 Three-class $n = 5, c = 2$	Case 3 Three-class $n = 5, c = 1$
Health hazard	Reduced hazard	No change	Increased hazard
Low, indirect (indicator)	Case 4 Three-class $n = 5, c = 3$	Case 5 Three-class $n = 5, c = 2$	Case 6 Three-class $n = 5, c = 1$
Moderate, direct, limited spread	Case 7 Three-class $n = 5, c = 2$	Case 8 Three-class $n = 5, c = 1$	Case 9 Three-class $n = 5, c = 0$
Serious, incapacitating but not usually life threatening; sequelae are rare, moderate duration	Case 10 Two-class $n = 5, c = 0$	Case 11 Two-class $n = 10, c = 0$	Case 12 Two-class $n = 20, c = 0$
Severe, for the general population or restricted populations, causing life-threatening or substantial chronic sequelae or illness of long duration	Case 13 Two-class $n = 15, c = 0$	Case 14 Two-class $n = 30, c = 0$	Case 15 Two-class $n = 60, c = 0$

[a] Reprinted with permission from reference 29.
[b] More stringent plans would generally be used for sensitive foods destined for susceptible populations.

Foodborne pathogens can be grouped into one of three categories based on the severity of the potential hazard (i.e., severe hazards, moderate hazards with potentially extensive spread, and moderate hazards with limited spread) as shown in Table 4.4. Pathogens with the potential for extensive spread are often initially associated with specific foods; however, secondary spread to other foods commonly occurs from environmental contamination and cross-contamination within processing plants and food preparation areas, including homes. An example is fresh beef that is contaminated with *E. coli* O157:H7. One or a few contaminated piece(s) of meat can lead to widespread contamination of product during processing, such as grinding to produce ground beef. There can also be cross-contamination if the fresh beef is improperly stored with ready-to-eat foods. Microbial pathogens in the lowest risk group (moderate hazards, limited spread) are found in many foods, usually in small numbers. Generally, illness is caused only when ingested foods contain large numbers of the pathogen, e.g., *C. perfringens*, or have at some time contained large enough numbers to produce sufficient toxin to cause illness, e.g., *S. aureus*. Outbreaks are usually restricted to consumers of a particular meal or a particular kind of food (42). A summary of those vehicles associated with outbreaks of foodborne diseases, and the microorganisms involved, that occurred in the United States from 1977 to 1984 has been prepared by Bryan (8) and from 1983 to 1987 by Bean et al. (6).

Table 4.4 Hazardous microorganisms and parasites grouped on the basis of risk severity[a]

I. Severe hazards
   Clostridium botulinum types A, B, E, and F
   Shigella dysenteriae
   Salmonella enterica serovar Typhi, *Salmonella paratyphi* A and B
   Enterohemorrhagic *Escherichia coli* (EHEC)
   Hepatitis A and E
   Brucella abortis, Brucella suis
   Vibrio cholerae O1
   Vibrio vulnificus
   Taenias solium

II. Moderate hazards: potentially extensive spread[b]
   Listeria monocytogenes
   Salmonella spp.
   Shigella spp.
   Other enterovirulent *Escherichia coli* (EEC)
   Streptococcus pyogenes
   Rotavirus
   Norwalk virus group
   Entamoeba histolytica
   Diphyllobothrium latum
   Ascaris lumbricoides
   Cryptosporidium parvum

III. Moderate hazards: limited spread
   Bacillus cereus
   Campylobacter jejuni
   Clostridium perfringens
   Staphylococcus aureus
   Vibrio cholerae, non-O1
   Vibrio parahaemolyticus
   Yersinia entercolitica
   Giardia lamblia
   Taenia saginata

---

[a] Reprinted with permission from reference 45.
[b] Although classified as moderate hazards, complications and sequelae may be severe in certain susceptible populations.

Indicator Microorganisms

Microbiological criteria for food safety may use tests for indicator microorganisms which suggest the possibility of a microbial hazard. *E. coli* in drinking water, for example, indicates possible fecal contamination and, therefore, the potential presence of enteric pathogens. Jay (30) suggested that an indicator used to assess food safety should ideally meet the following criteria:

- be easily and rapidly detectable
- be easily distinguishable from other members of the food flora
- have a history of constant association with the pathogen whose presence it is to indicate
- always be present when the pathogen of concern is present
- be a microorganism whose number ideally should correlate with those of the pathogen of concern
- possess growth requirements and a growth rate equaling those of the pathogen
- have a die-off rate that at least parallels that of the pathogen and, ideally, persist slightly longer than the pathogen of concern
- be absent from foods that are free of the pathogen except perhaps at certain minimum numbers

Buttiaux and Mossel (11) suggested additional criteria for fecal indicators used in food safety. They include the following:

- Ideally the bacteria selected should demonstrate specificity, occurring only in intestinal environments.
- They should occur in very high numbers in feces so as to be encountered in high dilutions.
- They should possess a high resistance to the external environment, the pollution of which is to be assessed.
- They should permit relatively easy and fully reliable detection even when present in low numbers.

With the exception of *Salmonella* and *S. aureus*, most tests for assuring safety use indicator microorganisms rather than direct tests for the specific hazard. An overview of some of the more common indicator microorganisms used for assuring food safety is given below.

Fecal Coliforms and *E. coli*

Fecal coliforms, including *E. coli*, are easily destroyed by heat, and cell numbers may decline during freezing and frozen storage of foods. Microbiological criteria involving *E. coli* may be useful in those cases where it is desirable to determine whether fecal contamination may have occurred. However, criteria involving *E. coli* are generally not useful for detecting likely fecal contamination for foods that have been processed sufficiently to destroy this bacterium. *E. coli* may exist in the food processing plant environment and recontaminate processed foods. Hence, *E. coli* in such processed foods does not necessarily indicate fecal contamination. Contamination of a food with *E. coli* implies a risk that other enteric pathogens may be present in the food. Fecal coliform bacteria are used as a component of microbiological standards to monitor the wholesomeness of shellfish and the quality of shellfish-growing waters (23, 54). The purpose is to reduce the risk of harvesting shellfish from waters polluted with fecal material. The fecal coliforms have a higher probability of including microbes of fecal origin than do coliforms which are comprised of bacteria of both fecal and nonfecal origin. However, many fecal coliform bacteria are not *E. coli* and are indigenous to vegetation and plant materials and so are not of fecal origin. Hence, fecal coliforms are not a reliable indicator of fecal contamination. Like *E. coli*, fecal coliforms can become established on equipment and utensils in the food processing environment and contaminate processed foods. Hence, fecal coliforms do not necessarily indicate fecal contamination in processed foods. At present, *E. coli* is the most widely used indicator of fecal contamination. The failure to detect *E. coli* in a food, however, does not assure the absence of enteric pathogens (23, 48). In many raw foods of animal origin, small numbers of *E. coli* can be expected because of the close association of these foods with the animal environment and the likelihood of contamination of carcasses from fecal material, hides, or feathers during slaughter-dressing procedures. Rapid direct plating methods for *E. coli* now exist; hence, it would be better to use measurement of *E. coli* rather than fecal coliforms as a component of microbiological criteria for foods.

The presence of *E. coli* in a heat-processed food indicates either process failure or, more commonly, postprocessing contamination from equipment or employees or from contact with contaminated raw foods. In the case of refrigerated ready-to-eat products such as shrimp and crabmeat, coliforms are recommended as indicators of process integrity with regard to reintroduction of pathogens from environmental sources and maintenance of adequate refrigeration (10). The source of coliforms in these types of products after thermal processing is usually the processing environment, resulting from inadequate sanitation procedures and/or temperature control. Coliforms were recommended over *E. coli* and APCs, because coliforms are often present in higher numbers than *E. coli* and the levels of coliforms do not increase over time when the product is stored properly.

Enterococci

Sources of enterococci include fecal material, from both warm-blooded and cold-blooded animals, and plants (39). Enterococci differ from coliforms in that they are salt tolerant (grow in the presence of 6.5% NaCl) and are relatively resistant to freezing (30). Certain enterococci (*E. faecalis* and *E. faecium*) are also relatively heat resistant and may survive the usual milk pasteurization temperatures. Enterococci can establish themselves and persist in the food processing establishment for long periods. Because many foods contain small numbers of enterococci, a thorough understanding of the role and significance of enterococci in a food is required before any significance can be attached to their presence and population numbers. Enterococci counts have few useful applications in microbiological criteria for food safety. This indicator bacterium may be applicable in specific cases to identify poor manufacturing practices.

Metabolic Products

Some microbiological criteria related to safety rely on tests for metabolites to indicate a potential hazard rather than direct tests for pathogenic or indicator microorganisms. Examples of metabolites as a component of microbiological criteria include (i) tests for thermonuclease (thermostable deoxyribonuclease) in foods containing or suspected of containing $\geq 10^6$ *S. aureus* cells per ml or g and (ii) illumination of grains under UV light to detect the presence of aflatoxin produced by *Aspergillus* species (30, 38).

APPLICATION AND SPECIFIC PROPOSALS FOR MICROBIOLOGICAL CRITERIA FOR FOOD AND FOOD INGREDIENTS

The application of useful microbiological criteria should address the following issues: (i) the sensitivity of the food product(s) relative to safety and quality, (ii) the need for a microbiological standard(s) and/or guideline(s), (iii) assessment of information necessary for establishment

Table 4.5 Microbiological criteria for verification of cooked, ready-to-eat shrimp and cooked, ready-to-eat crabmeat[a]

Microorganism	Criteria Shrimp	Criteria Crabmeat	Explanation
Salmonella spp.	$n = 30$ $c = 0$ $m = M = 0$	$n = 30$ $c = 0$ $m = M = 0$	Analytical unit = 25 g
Listeria monocytogenes	$n = 5$ $c = 0$ $m = M = 0$	$n = 5$ $c = 0$ $m = M = 0$	Sample unit = 50 g; analytical unit = 25 g through compositing of 5-g portions from 5 sample units
Staphylococcus aureus	$n = 5$ $c = 2$ $m = 50/g$ $M = 500/g$	$n = 5$ $c = 2$ $m = 100/g$ $M = 1,000/g$	
Thermal tolerant coliforms[b]	$n = 5$ $c = 2$ $m = 100/g$ $M = 1,000/g$	$n = 5$ $c = 2$ $m = 500/g$ $M = 5,000/g$	

[a] Reprinted with permission from reference 10.
[b] "Thermal tolerant coliforms" used in lieu of the more traditional designation, "fecal coliforms."

of a criterion if one seems to be indicated, and (iv) where the criterion should be applied (40). In this section, examples of the application of microbiological criteria to various foods and food ingredients are presented. There are no general criteria suitable for all food product groups. The relevant background literature that relates to the quality and safety of a specific food product should be consulted prior to implementing microbiological criteria. Such information may be found in ICMSF publications, industry codes of practice, legislation, and peer reviewed technical publications. Suggested microbiological limits for a wide range of products and product groups were recently published as a guidance document for all involved in producing, using, and interpreting microbiological criteria in the food and catering industries (49).

The utilization of microbiological criteria, as stated previously, may be either mandatory (standards) or advisory (guidelines, specifications). One obvious example of a mandatory criterion (standard) as it relates to food safety is the "zero tolerance" established for *Salmonella* in all ready-to-eat (RTE) foods. The U.S. Department of Agriculture (USDA)/Food Safety and Inspection Service (FSIS) has mandated a "zero tolerance" for *E. coli* O157:H7 in fresh ground beef. Since the emergence of *L. monocytogenes* as a foodborne pathogen, the U.S. FDA and USDA have also applied a "zero tolerance" for this bacterium in all RTE foods. Absence of *L. monocytogenes* is one of the four microbiological criteria included for verification for cooked RTE shrimp and crabmeat, as shown in Table 4.5. However, there is considerable scientific debate as to whether a "zero tolerance" is warranted for *L. monocytogenes* (18). The ICMSF (25) has recommended that the international community allow for a tolerance of *L. monocytogenes* as outlined in Table 4.6.

Table 4.6 Summary of cases and sampling plans for *Listeria monocytogenes*[a]

Intended consumer	Health hazard	Conditions in which food is expected to be handled and consumed after sampling, in the usual course of events Reduce degree of hazard	Cause no change in hazard	May increase hazard
Normal individuals	Moderate, direct, potentially extensive spread	Case 10 $n = 5$ $\leq 100/g$	Case 11 $n = 10$ $\leq 100/g$	Case 12 $n = 20$ $\leq 100/g$
Highly susceptible individuals	Severe, direct	Case 13 $n = 15$ $< 1/375$ g	Case 14 $n = 30$ $< 1/750$ g	Case 15 $n = 60$ $< 1/1,500$ g

[a] Reprinted with permission from reference 25.

These recommendations are an example of advisory criteria. Other advisory criteria which have been established to address the safety of foods are shown in Table 4.7.

Mandatory criteria in the form of standards have also been employed for quality issues. Presented in Table 4.8 are examples of foods and food ingredients for which federal, state, and local, as well as international, microbiological standards have been developed.

Some of the advisory criteria that have been developed and applied to foods for quality monitoring are presented in Table 4.7.

The USDA/FSIS coordinates a pathogen reduction program intended to reduce the level of pathogenic microorganisms in meat and poultry products (53). As part of this program, microbiological testing of carcasses, ground meat, and poultry for the presence of *Salmonella*

Table 4.7 Examples of various food products for which advisory microbiological criteria have been established

Product category	Test parameters	Case	Plan class	n	c	Limit per g m	M	Reference
Roast beef	*Salmonella* sp.	12	2	20	0	0		26
Pâté	*Salmonella* sp.	12	2	20	0	0		
Raw chicken	APC	1	3	5	3	5×10^5	10^7	26
Cooked poultry, frozen, ready to eat	*S. aureus*	8	3	5	1	10^3	10^4	26
Cooked poultry, frozen, to be reheated	*S. aureus*	8	3	5	1	10^3	10^4	
	Salmonella sp.	10	2	5	0	0		
Chocolate/confectionery	*Salmonella* sp.	11	2	10^a	0	0		26
Dried milk	APC	2	3	5	2	3×10^4	3×10^5	26
	Coliforms	5	3	5	1	10	10^2	
	Salmonella sp.[b]	10	2	5	0	0		
	(normal routine)	11	2	10	0	0		
		12	2	20	0	0		
	Salmonella sp.[b]	10	2	15	0	0		
	(high-risk	11	2	30	0	0		
	populations)	12	2	60	0	0		
Fresh cheese[c]	*S. aureus*			5	2	10^2	10^3	13
	Coliforms			5	2	10^2	10^3	
Soft cheese[c]	*S. aureus*			5	2	10^2	10^3	13
Pasteurized liquid, frozen, and dried egg products	APC	2	3	5	2	5×10^4	10^6	26
	Coliforms	5	3	5	2	10^3	10^3	
	Salmonella sp.[a]	10	2	5	0	0		
	(normal routine)	11	2	10	0	0		
		12	2	20	0	0		
	Salmonella sp.[a]	10	2	15	0	0		
	(high-risk	11	2	30	0	0		
	populations)	12	2	60	0	0		
Fresh and frozen fish; to be cooked before eating	APC	1	3	5	3	5×10^3	10^7	26
	E. coli		4		3	11	500	
	Salmonella sp.[d]	10	2	5	0	0		
	V. parahaemolyticus[d]	7	3	5	2	10^2	10^3	
	S. aureus[d]	7	3	5	2	10^3	10^4	
Coconut	*Salmonella* sp.	1	3	5	3	5×10^5	10^7	26
	(growth not expected)	11	2	10	0	0		
	(growth expected)	12	2	20	0	0		

[a] The 25-g analytical unit may be composited.
[b] The case is to be chosen based on whether the hazard is expected to be reduced, unchanged, or increased.
[c] Requirements only for fresh and soft cheese made from pasteurized milk.
[d] For fish derived from inshore or inland waters of doubtful bacteriological quality or for instances in which fish are to be eaten raw, additional tests may be desirable.

Table 4.8 Examples of various food products for which mandatory microbiological criteria have been established

Product category	Test parameter(s)	Comment	Reference
United States			
Dairy products			
Raw milk	Aerobic plate count	Recommendations of U.S. Public Health Service (USPHS)	55
Grade A pasteurized	Aerobic plate count Coliforms	USPHS	
Grade A pasteurized (cultured)	Aerobic plate count Coliforms	USPHS	55
Dry milk (whole)	Standard plate count Coliforms	USPHS	55
Dry milk (nonfat)	Standard plate count Coliforms	Standards of Agricultural Marketing Service (USDA)	48
Frozen desserts	Standard plate count Coliforms	USPHS	55
Starch and sugars	Total thermophilic spore count Flat-sour spores Thermophilic anaerobic spores Sulfide spoilage spores	National Canners Assoc. (NFPA)	41
Breaded shrimp	Aerobic plate count *E. coli* *S. aureus*	FDA Compliance Policy Guide	22
International			
Caseins and caseinates	Total plate count Thermophilic organisms Coliforms	Europe	42
Natural mineral waters	Aerobic mesophilic count Coliforms *E. coli* Fecal streptococci Sporulating sulfite-reducing anaerobes *Pseudomonas aeruginosa*, parasites and pathogenic organisms	Codex	42
Hot meals served by airlines	*E. coli* *S. aureus* *B. cereus* *C. perfringens* *Salmonella* sp.	Europe	5
Tomato juice	Mold count	Canada	42
Fish protein	Total plate count *E. coli*	Canada	42
Gelatin	Total plate count Coliform *Salmonella* sp.	Canada	42

would be required. The results of daily *Salmonella* testing for its presence or absence are evaluated over a specific time period using a statistical procedure known as the "moving-window sum" to determine whether the process is in control. The presence of *Salmonella* is compared to a target frequency by examining a group of consecutive days (the window) of a specific duration in relation to a prespecified acceptable limit. The microbiological targets are summarized in Table 4.9.

If the number of positive samples is less than or equal to the acceptable limit, the process is in control. As each new sampling day is added, the window moves 1 day. Changing the size of the window, the acceptable limit, or the number of samples to be taken each day can alter the

Table 4.9 Moving-sum rules for meat and poultry commodities[a]

Commodity	Target (% positive *Salmonella*)	Window size (days)	Acceptable limit[b]
Steer/heifers	1	82	1
Cow/bulls	1	82	1
Raw ground beef	4	38	2
Fresh sausage	12	19	3
Turkeys	18	15	3
Hogs	18	17	4
Broilers	25	16	5

[a] Reprinted from reference 52.
[b] There is approximately an 80% probability of meeting the acceptable limit when the process % positive equals the target.

stringency and sensitivity of the evaluation. The current acceptable limits are based on the criterion that there is an 80% probability that the plant is actually exceeding the target value if it exceeds the acceptable limit. As in any statistical approach, one needs to develop a program that (i) has a low probability of exceeding the limit when the producer is meeting the target and (ii) has a high probability of exceeding the limit when the producer is not meeting the target. The 80% level was deemed to be a reasonable balance between the two, thus providing reasonable decision criteria. This verification procedure gives meat and poultry producers an opportunity to measure their process performance in relation to baseline data and national targets for pathogen reduction. Establishments whose production process exceeds the national targets are required to reevaluate their processing controls and, with FSIS oversight, initiate corrective actions (52).

CURRENT STATUS

The ICMSF recommends a series of steps be taken to manage microbiological hazards for foods intended for international trade (27). These steps include conducting a risk assessment and an assessment of risk management options, establishing an FSO, and confirming that the FSO is achievable by application of GMPs and HACCP. This stepwise approach is based on the following Codex documents: "Principles and Guidelines for the Control of Microbiological Risk Assessment" (15), "General Principles of Food Safety Management" (20), "General Principles of Food Hygiene" (16), and "Hazard Analysis and Critical Control Points (HACCP) System and Guidelines for Its Application" (14).

Industry and regulatory authorities may use performance criteria, process criteria, and end product criteria associated with the food to meet the FSO. For foods in international commerce, FSOs provide an objective basis to identify what an importing country is willing to accept in relation to the microbiological safety of its food supply and to assess, or demonstrate, the equivalence of the microbiological outcome of different control measures (31). The use of FSOs should facilitate the harmonization of international trade where practices between countries differ but both practices provide safe food products.

SUMMARY

Microbiological criteria are most effectively applied as part of quality assurance programs in which HACCP and other prerequisite programs are in place. Here, criteria provide a means of determining the effectiveness of those control measures used to eliminate, reduce, or control the presence, survival, and growth of microorganisms. The establishment and application of microbiological criteria for foods will continue to evolve, as new pathogens emerge and new food vehicles are implicated as a result of the changing industrial ecology of food production and consumption. Further development of quantitative risk assessment techniques will facilitate the establishment of microbiological criteria for foods in international trade and guidelines for national standards and policies.

References

1. **Acuff, G. R.** 1993. Microbiological criteria, p. A6.01–A6.07. *In Proceedings of the World Congress on Meat and Poultry Inspection.* October 10–14, 1993. Food Safety and Inspection Service, U.S. Department of Agriculture, Washington, D.C.

2. **Ahmed, F. E. (ed.).** 1991. *Seafood Safety.* National Academy Press, Washington, D.C.

3. **Amerine, M. A., R. M. Pangborn, and E. B. Roessler.** 1965. *Principles of Sensory Evaluation of Food.* Academic Press, New York, N.Y.

4. **Anonymous.** 1979. *Shrimp Decomposition Workshop.* National Shrimp Breaders and Processors Association, National Fisheries Institute, and U.S. Food and Drug Administration, Tampa, Fla.

5. **Association of European Airlines.** 1996. *Hygiene Guidelines, Routine Microbiological Standards for Aircraft Ready Food.* Association of European Airlines, Brussels, Belgium.

6. **Bean, N. H., P. M. Griffin, J. S. Goulding, and C. B. Ivey.** 1990. Foodborne disease outbreaks, 5-year summary, 1983–1987. *J. Food Prot.* **53:**711–728.

7. **Beuchat, L. R.** 1996. Pathogenic microorganisms associated with fresh produce. *J. Food Prot.* **59:**204–216.

8. **Bryan, F. L.** 1988. Risks associated with vehicles of foodborne pathogens and toxins. *J. Food Prot.* **51:**498–508.

9. Buchanan, R. L. 1995. The role of microbiological criteria and risk assessment in HACCP. *Food Microbiol.* **12**:421–424.

10. Buchanan, R. L. 1991. Microbiological criteria for cooked, ready-to-eat shrimp and crabmeat. *Food Technol.* **45**: 157–160.

11. Buttiaux, R., and D. A. A. Mossel. 1961. The significance of various organisms of faecal origin in foods and drinking water. *J. Appl. Bacteriol.* **24**:353–364.

12. Codex Alimentarius Commission. 1981. *Report of the 17th Session of the Codex Committee on Food Hygiene.* Alinorm 81/13. Food and Agriculture Organization, Rome, Italy.

13. Codex Alimentarius Commission. 1993. *Report of the 20th Session of the Codex Commission on Food Hygiene.* Alinorm 93/13A. Food and Agriculture Organization, Rome, Italy.

14. Codex Alimentarius Commission. 1997. Joint FAO/WHO Food Standards Programme, Codex Committee on Food Hygiene, Supplement to Volume 1B-1997. *Hazard Analysis and Critical Control Point (HACCP) System and Guidelines for Its Application.* Annex to CAC/RCP 1-1969, Rev. 3. Food and Agriculture Organization, Rome, Italy.

15. Codex Alimentarius Commission. 1997. Joint FAO/WHO Food Standards Programme, Codex Committee on Food Hygiene. *Proposed Draft Principles and Guidelines for the Conduct of Microbiological Risk Assessment.* CX/FH 97/4. Food and Agriculture Organization, Rome, Italy.

16. Codex Alimentarius Commission. 1997. Joint FAO/WHO Food Standards Programme, Codex Committee on Food Hygiene, Supplement to Volume 1B-1997. *Recommended International Code of Practices, General Principles of Food Hygiene.* CAC/RCP 1-1969, Rev. 3. Food and Agriculture Organization, Rome, Italy.

17. Codex Alimentarius Commission. 1997. Joint FAO/WHO Food Standards Programme, Codex Committee on Food Hygiene, Supplement to Volume 1B-1997. *Principles for the Establishment and Application of Microbiological Criteria for Foods.* CAC/GL 21-997. Food and Agriculture Organization, Rome, Italy.

18. Doyle, M. P. 1991. Should regulatory agencies reconsider the policy of zero-tolerance of *Listeria monocytogenes* in all ready-to-eat foods? *Food Safety Notebook* **2**:98.

19. Eisenberg, W. V., and S. M. Cichowicz. 1977. Machinery mold-indicator organism in food. *Food Technol.* **31**:52–56.

20. Food and Agriculture Organization of the United Nations/World Health Organization (FAO/WHO). 1997. *Risk Management and Food Safety. Report of the Joint FAO/WHO Consultation, Rome, Italy.* FAO food and nutrition paper 65. Food and Agriculture Organization of the United Nations/World Health Organization, Rome, Italy.

21. Food and Drug Administration. 1978. *The Food Defect Action Levels.* Food and Drug Administration, Washington, D.C.

22. Food and Drug Administration. 1989. Raw breaded shrimp—microbiological criteria for evaluating compliance with current good manufacturing practice regulations, (CPG 7108.25) chapter 8. *In Compliance Policy Guides.* Food and Drug Administration, Washington, D.C.

23. Hackney, C. R., and M. D. Pierson (ed.). 1994. *Environmental Indicators and Shellfish Safety.* Chapman and Hall, New York, N.Y.

24. Hathaway, S. 1999. Management of food safety in international trade. *Food Control* **10**:247–253.

25. International Commission on Microbiological Specifications for Foods (ICMSF). 1994. Choice of sampling plan and criteria for *Listeria monocytogenes. Int. J. Food Microbiol.* **22**:89–96.

26. International Commission on Microbiological Specifications for Foods (ICMSF). 1986. *Microorganisms in Foods 2. Sampling for Microbiological Analysis: Principles and Applications,* 2nd ed. University of Toronto Press, Toronto, Ontario, Canada.

27. International Commission on Microbiological Specifications for Foods (ICMSF). 1997. Establishment of microbiological safety criteria for foods in international trade. *World Health Stat. Q.* **50**:119–123.

28. International Commission on Microbiological Specifications for Foods (ICMSF). 1998. *Microorganisms in Foods 6: Microbial Ecology of Food Commodities.* Blackie, London, England.

29. International Commission on Microbiological Specifications for Foods (ICMSF). 2002. *Microorganisms in Foods 7: Microbiological Testing in Food Safety Management.* Kluwer Academic/Plenum Publishers, New York, N.Y.

30. Jay, J. M. 2000. *Modern Food Microbiology,* 6th ed. Aspen Pub., Gaithersburg, Md.

31. Jouve, J.-L. 1999. Establishment of food safety objectives. *Food Control* **10**:303–305.

32. Kilsby, D. 1982. Sampling schemes and limits, p. 387–421, *In* M. H. Brown (ed.), *Meat Microbiology.* Applied Science Publishers, London, United Kingdom.

33. Kilsby, D., L. J. Aspinall, and A. C. Baird-Parker. 1979. A system for setting numerical microbiological specifications for foods. *J. Appl. Bacteriol.* **46**:591–599.

34. Lammerding, A. M. 1997. An overview of microbial food safety risk assessment. *J. Food Prot.* **60**:1420–1425.

35. Larmond, E. 1977. *Laboratory Methods for Sensory Evaluation of Food.* Pub. No. 1937. Research Branch, Canada Department of Agriculture, Ottawa, Quebec, Canada.

36. Lupien, J. R., and M. F. Kenny. 1998. Tolerance limits and methodology: effect on international trade. *J. Food Prot.* **61**:1571–1578.

37. Malcolm, S. 1984. A note on the use of the non-central t-distribution in setting numerical specifications for foods. *J. Appl. Bacteriol.* **57**:175–177.

38. Marshall, R. T. (ed.). 1992. *Standard Methods for the Examination of Dairy Products,* 16th ed. American Public Health Association, Washington, D.C.

39. Mundt, J. O. 1970. Lactic acid bacteria associated with raw plant food materials. *J. Milk Food Technol.* **33**:550–553.

40. National Advisory Committee on Microbiological Criteria for Foods. 1993. Generic HACCP for raw beef. *Food Microbiol.* **10**:449–488.

41. National Canners Association. 1968. *Laboratory Manual for Food Canners and Processors,* vol. 1. AVI Publishing, Westport, Conn.

42. National Research Council. 1985. *An Evaluation of the Role of Microbiological Criteria for Foods and Food Ingredients.* National Academic Press, Washington, D.C.

43. **National Soft Drink Association.** 1975. *Quality Specifications and Test Procedures for "Bottler's Granulated and Liquid Sugar."* National Soft Drink Association, Washington, D.C.

44. **Pierson, M. D.** 1996. Critical limits, significance and determination of critical limits, p. 72–78. *In Proceedings of the 2nd Australian HACCP Conference.* Sydney, Australia.

45. **Pierson, M. D., and D. A. Corlett, Jr. (ed.).** 1992. *HACCP: Principles and Applications.* Van Nostrand Reinhold, New York, N.Y.

46. **Randell, A. W., and A. J. Whitehead.** 1997. Codex Alimentarius: food quality and safety standards for international trade. *Rev. Sci. Tech. Off. Int. Epizoot.* **16:**313–321.

47. **Silliker, J. H.** 1963. Total counts as indexes of food quality, p. 102–112. *In* L. W. Slanetz, C. O. Chichester, A. R. Gaufin, and Z. J. Ordal (ed.), *Microbiological Quality of Foods.* Academic Press, New York, N.Y.

48. **Silliker, J. H., and D. A. Gabis.** 1976. ICMSF method studies. VII. Indicator tests as substitutes for direct testing of dried foods and feeds for *Salmonella. Can. J. Microbiol.* **22:**971–974.

49. **Stannard, C.** 1997. Development and use of microbiological criteria for foods. *Food Sci. Technol. Today* **11:**137–177.

50. **Tauxe, R., H. Kruse, C. Hedberg, M. Potter, J. Madden, and K. Wachsmuth.** 1997. Microbial hazards and emerging issues associated with produce: a preliminary report to the National Advisory Committee on Microbiological Criteria for Foods. *J. Food Prot.* **60:**1400–1408.

51. **U.S. Department of Agriculture.** 1975. General specifications for approved dairy plants and standards for grades of dairy products. *Fed. Regist.* **40:**47910–47940.

52. **U.S. Department of Agriculture.** 1995. Moving sum procedures for microbial testing in meat and poultry establishments. Science and Technology Program, Food Safety and Inspection Service, U.S. Department of Agriculture, Washington, D.C.

53. **U.S. Department of Agriculture-Food Safety and Inspection Service (USDA-FSIS).** Pathogen reduction; hazard analysis and critical control point (HACCP) systems; proposed rule. *Fed. Regist.* **60:**6774–6889.

54. **U.S. Department of Health, Education and Welfare.** 1965. *National Shellfish Sanitation Program, Manual Operations. Part 1. Sanitation of Shellfish Growing Areas.* U.S. Government Printing Office, Washington, D.C.

55. **U.S. Public Health Service/Food and Drug Administration.** 1978. *Grade A Pasteurized Milk Ordinance. 1978 Recommendations.* PHS/FDA publication no. 229. U.S. Government Printing Office, Washington, D.C.

56. **van Schothorst, M.** 1998. Principles for the establishment of microbiological food safety objectives and related control measures. *Food Control* **9:**379–384.

Food Microbiology: Fundamentals and Frontiers, 3rd Ed.
Edited by M. P. Doyle and L. R. Beuchat
© 2007 ASM Press, Washington, D.C.

Shaun P. Kennedy
Frank F. Busta

5

Biosecurity: Food Protection and Defense

Many food-related habits from prior generations and practices of ancient cultures, from certain religious dietary guidelines to a preference for brewed or boiled fluids, have some basis in preventing foodborne illness. Yet the problem of food safety, as we currently understand it, is a relatively modern concern. In many parts of the world, food security—access to sufficient calories—is a more dominant issue. Developed societies have moved from having great concern over food security to enjoying a relative abundance of food. With this relative abundance, however, the possibility of becoming ill from the food itself has become a significant concern. Sometimes an overall shift in relative health risks and the abundance of food resulting from improvements in the food supply chain drive new foodborne illness risks. This is especially true where the desire for quick-to-prepare, ready-to-eat foods such as fresh fruits and vegetables has resulted in consumers being exposed to a broader array of potentially naturally contaminated foods from around the globe.

The intentional contamination of food to cause harm, and consequently the concern for food biosecurity, also has a very long history. Historical examples of the use of contaminated food and water as weapons include the Athenians' contamination of drinking water for the city of Kirrha of the Amphictyonic League with the plant root of helleborous (Christmas rose) in 590 to 600 B.C., reportedly causing severe gastrointestinal illness that rendered the city defenseless for the ensuing attack. Contamination of wine with mandagora by the Carthaginian General Maharbal and various historic instances of plague-infested animal/human bodies dumped into water supplies continued as examples of intentional food contamination from the Roman times forward (55). During World War II the Japanese Army experimented with the use of food for the delivery of pathogens such as *Bacillus anthracis*, *Shigella* spp., *Vibrio cholerae*, *Salmonella enterica* serovar Paratyphi, and *Yersinia pestis* (23, 39). It is only very recently, however, that protecting food from intentional contamination has become a significant concern on regional, national, and international levels. While vulnerabilities were identified well beforehand (66), the terrorist attacks in the United States on 11 September 2001 significantly elevated concerns about food system biosecurity (32). The intersection of incredibly efficient, highly integrated food supply systems with individuals or organizations potentially using those very systems to inflict public health or economic harm on a massive scale

Shaun P. Kennedy and Frank F. Busta, National Center for Food Protection and Defense, University of Minnesota, 925 Delaware Street, SE, Minneapolis, MN 55455.

is the new face of food system biosecurity. This chapter examines the risk of intentional food contamination and provides an overview of current methods for evaluating risk and defining appropriate interventions.

HISTORICAL PERSPECTIVE

There are many intentional food contamination incidents, including those cited above, that involve the use of diseased animals or toxic plants or microorganisms as a malicious action. These have been generally localized, often based on specific military objectives and not based on any detailed understanding of the microbial contaminant introduced. But the rising number of contamination events over the last 30 years suggests that there are increasingly more individuals and organizations that see the food system as a target of opportunity for attack. A survey of some of the documented intentional contamination events is provided in Table 5.1.

Most of these events are still somewhat narrow in scope, such as the numerous instances of using pesticides/poisons in China or the use of *Salmonella* by the Rajneeshee cult in Oregon. They unfortunately, however, illustrate how foods can be used to effectively deliver biological or chemical agents resulting in significant morbidity and/or mortality. Some events are on a slightly larger scale, such as the intentional contamination of ground meat with the pesticide nicotine sulfate. To date, however, unintentional

Table 5.1   A survey of intentional food contamination events

Yr	Country(ies)	Morbidity	Mortality	Event description
1964	United States	47	0	Hepatitis contamination of potato salad by an enlisted soldier (44)
1970	Canada	4	0	Roommate contamination of food items (63)
1978	Belgium, France, The Netherlands, Germany, and the United Kingdom	5	0	Mercury-contaminated Jaffa oranges from Israel by "The Arab Revolutionary Group" (57)
1984	United States	751	0	*Salmonella enterica* serovar Typhimurium contamination of multiple salad bars by the Rajneeshees cult (76)
1987	Philippines	140	19	Contaminated water provided to participants in a "fun run" (64)
1994	Tajikistan	53	15	Cyanide-contaminated champagne intended for Russian soldiers (64)
1996	United States	13	0	*Shigella dysenteriae*-laced pastries provided to laboratory coworkers (50)
1997	China	44	0	Insecticide-contaminated pork provided to villagers by another resident (57)
1998	Japan	60	4	Cyanide contamination of food items at a festival (64)
1999	China	148	0	Nitric acid contamination of donkey soup by competitors to the food vendor (57)
1999	China	48	0	Rat poison contamination of sweet rolls by competitors to the food vendor (62)
2000	Afghanistan	60	2	Contaminated food at a religious school (60)
2000	Canada	27	0	Arsenic contamination of coffee in a vending machine (57)
2000	United States	34	0	Rat poison contamination of school lunch salsa by classmates (62)
2000	India	2	2	Cyanide-contaminated liquor (62)
2001	China	120	0	Rat poison-contaminated noodles by a competitor to the food vendor (60)
2002	China	92	0	Rat poison-contaminated school lunch (57)
2002	Zimbabwe	47	7	Pesticide-contaminated tea at a church group (57)
2002	Russia	60	0	*Salmonella enterica* serovar Typhimurium-contaminated school meal(s) (57)
2002	China	400	41	Rat poison-contaminated breakfast foods by a competitor to the food vendor (57)
2002	United States	111	0	Nicotine sulfate-contaminated ground beef by a worker (10)
2003	United States	16	1	Arsenic-contaminated coffee and snacks at a church group by a parishioner (57)
2003	China	241	0	Rat poison-contaminated school breakfasts (57)

food contamination events provide the more striking examples of how certain foods can impact a significant number of consumers in a geographically dispersed manner. For example, a 1994 outbreak of *Salmonella enterica* serovar Enteritidis infections resulting from accidental contamination of ice cream in Minnesota involved an estimated 224,000 individuals (41), and a 1985 outbreak of *Salmonella enterica* serovar Typhimurium infections associated with milk in Illinois involved 60,000 consumers (67). Such illustrations of the potential scope of a food system contamination event, combined with the knowledge that far more aggressive microorganisms, microbially derived toxins, or nonmicrobial agents could be used, drive the current emphasis on food system protection and defense. While biosecurity is generally associated with only biological contaminants, the tools required to effectively prevent intentional contamination with biological or nonbiological agents are substantially the same in most cases.

INTENTIONAL VERSUS UNINTENTIONAL CONTAMINATION

Both intentional and unintentional food contamination are of concern when the contamination can result in illness due to the lack of any further processing steps, including home preparation, to eliminate the deleterious impact of the contamination. This generally means that the contaminant is stable in the food, survives final preparation, and is not organoleptically obvious, thus providing no flavor or other sensory attributes to suggest the presence of the contaminant. The subsequent foodborne illness results in either morbidity or mortality, along with a concomitant economic impact. For either intentional or unintentional contamination, the risk management control strategy includes identifying food/contaminant combinations of potential risk/vulnerability and then inserting controls to reduce the risk/vulnerability. The controls could be inserted at any point, from preharvest or preslaughter inputs through the point of consumption. A recent case in Taiwan where high dioxin levels in a pasture resulted in high dioxin content in milk illustrates how even environmental inputs can be of concern (52). For control of the more traditional biological contaminants of concern for food safety, the Hazard Analysis-Critical Control Point (HACCP) system (see chapter 46) is just one example of the approaches that are in use. The difference between intentional and unintentional contamination begins with selection of the actual contaminants and then includes where and how they are introduced, to what foods they are introduced, the level of introduction, and other considerations.

In traditional food safety, foodborne illness results from an overall system risk management failure which enables the introduction, survival, and growth of the contaminant to populations high enough to cause harm. In most cases, food safety issues arise primarily from equipment, process, or operator failure. This could be because not all reasonably feasible risks were identified, resulting in no attempt to provide an intervention to prevent the risk. It could also be due to insufficient interventions or controls to manage an identified risk or equipment, process, or operator failure. Extensive efforts undertaken across the food industry, academia, and government to identify all reasonably feasible risks and then develop control strategies have dramatically improved food safety, minimizing most anticipatable accidental contamination events. However, failures can and do occur at any point in the food system, e.g., a refrigeration unit not holding temperature and thus allowing microbial growth, poultry process water cross-contamination, undercooked meat or poultry at retail or in the consumer's home, and consumers eating known high-risk foods such as unpasteurized milk or raw oysters.

Foodborne illness from intentional contamination results not from system failure, but instead from intentional attacks on a system that defeat the in-place food safety controls. One scenario would be if the controls and detection strategies in place for natural contamination could have identified and contained the intentional contamination, but the strategies were actively overridden or bypassed. More worrisome, however, are intentional contamination attacks that succeed because the contamination "could not happen," e.g., the agent is not normally present or of any realistic concern. Such scenarios provide no reasonable incentive to firms for incurring the incremental costs to prevent them, increasing the potential for such an intentional contamination to cause harm. Metal detectors, magnets, and screens are designed to catch low probability, accidental contamination and keep it out of the food supply. These same tools are also useful for catching intentional contamination events such as those with metal, glass, or other particulate agents (75). Intervention strategies for events that do not normally occur would be considered economically unviable in most cases, and in many others they may not exist. If the microorganism or contaminant is not normally present at any stage in the preharvest-through-consumption system, there is very little published work on inactivation or detection strategies. Earlier there was also little justification for such work, but concerns about food system terrorism have elevated in recent years. The World Health Organization report, "Terrorist Threats to Food: Guidance for Establishing and Strengthening Prevention and Response Systems" (84), and the U.S. Government Accounting Office report, "Food Processing Security" (31), provide a national and international review of the reasons for concern. Homeland Security Presidential Directive 9, "Defense of United States Agriculture and Food" (12), enacted 30 January 2004, elevated the food and agriculture sector to the status of a critical U.S. infrastructure in need of significant efforts to protect it from intentional harm. These and other events of the last several years have initiated investigations into how to counter intentional contamination of food and water with pathogens and other agents that could cause significant and catastrophic morbidity and mortality.

The microbial pathogens and toxins of most concern are listed in Table 5.2, along with some of the potential chemical agents of concern provided for comparison; several references are available that go into significant detail on the agents of concern and illness progression (3, 29, 35, 48, 72, 74) and food protection more broadly (65). This chapter will focus specifically on biological contaminants, although a number of chemicals can also be of concern. Some very virulent biological agents, e.g., smallpox virus or Ebola virus, are not listed in Table 5.2, given their very limited stability outside of their host as naturally expressed (38), and thus the difficulty of using them to contaminate food. In addition, should one stabilize such viruses, their efficiency as a weapon via delivery vehicles other than food makes their use for food contamination less likely. Many of the conventional microbial food pathogens are also not included on this list as they are covered in other chapters. In addition, while many food pathogens are readily accessible, if introduced into foods where they are a traditional safety concern, the quality and safety control systems in place would generally be sufficient to minimize the risk of illness.

When introduced into foods with which they are not normally associated, or introduced in the production system at a point after standard processing steps that would inactivate them, conventional food pathogens pose a more significant risk. If the food/pathogen combination is not typical, the quality and safety systems for that food may not adequately mitigate the risk. For example, the detection system may not work for the food/pathogen combination, or the postcontamination processing may not be sufficient to inactivate the pathogen. In addition, uncommon associations of illness, pathogen, outbreak profile, and food could slow the public health system response as it tries to find the source of the contamination (78), a possibility demonstrated by the misdiagnosis of the cause of a botulism outbreak in 1963 (49). In addition to slow epidemiologic attribution, the potential for delayed diagnosis of the disease itself due to the current unfamiliarity of many physicians with the disease presentation has been reported (25). The above notwithstanding, however, the likely result of conventional food pathogens used for intentional contamination is high morbidity and relatively lower mortality compared to the pathogens of greatest concern. Conventional pathogens are therefore not the primary focus of efforts to prevent catastrophic public health consequences through intentional contamination of food. It should be noted, however, that conventional foodborne pathogens could be used to great effect in causing significant economic harm if, through an intentional contamination, the event results in significant loss of public confidence in a segment of the food system or in the government's ability to protect the public.

Microorganisms and their toxins not normally associated with food, especially those listed as select agents by the Centers for Disease Control and Prevention (21), are of greatest concern. Many of these agents are more difficult to produce at the level of purity and in populations or volumes necessary to result in a significant contamination of the food system relative to some of the conventional foodborne pathogens. In addition, these agents are now so tightly controlled and regulated that obtaining significant quantities can be difficult. However, several characteristics make both select agents and genetically engineered microorganisms potentially more likely pathogens for intentional contamination of the food system. Such microorganisms and their toxins are significantly more infective or toxic than typical foodborne pathogens, requiring much less material for a successful attack. There are typically no inactivation processes or detection systems developed and validated for detecting or quantitating them in foods. Consequently, the pathogen could be consumed at infective/toxic doses without detection. At first disease presentation to health care providers, the initial diagnosis of potential select agents could be more limited than expected (28), with some studies suggesting as low as 10% identified, due to lack of familiarity with the disease presentation (25). Outbreak investigations where bioterrorism was considered have had lag times in Centers for Disease Control notification of 2 to 27 days of problem identification (5). In the early stages of an outbreak investigation, the fact that these agents are not generally associated with food could mean that public health officials may have to evaluate a wide range of possible sources of the causative agent beyond food, and this would dramatically slow any response. This combination of high infectivity and toxicity, unproven inactivation and detection systems, and uncertain public health system response, along with the shock value of the agents themselves, makes select

Table 5.2 Potential food contamination agents

Agent	LD_{50} [a]	Heat sensitivity	Solubility of toxin	Organoleptic characteristics
Microbiological agents and their toxins				
Alpha amanitin (59)	1.0 µg/kg of body weight in mice, oral (59)	255°C MP[b], thermally stable at normal cooking temperatures (42)	Water, alcohols (59)	NA[c]
Bacillus anthracis	>10^8 spores in rabbits, oral (53)	Heat stable up to 159°C for 2 h (14)	Water, alcohols	No odor/taste (20)
Botulinum neurotoxin	0.001 µg/kg in mice, oral (27)	Anaerobic spores destroyed by boiling 10 min; moist heat at 120°C for 10 min destroys (27)	Water (40)	Current food contaminant
Clostridium perfringens epsilon toxin	0.1–5.0 µg/kg in mice, oral (27)	Freezing causes loss of viability	Water	Current food contaminant
Diacetoxyscirpenol	120 µg/kg in mice, oral (59)	Heat stable, MP 161–162°C (59)	Moderately polar solvents (59)	NA
Francisella tularensis	10^8 CFU in humans, oral (27)	Heat sensitive, killed by heating at 55°C for 10 min (27)	Water	NA
Shigatoxin-producing *E. coli*	0.002 µg/kg in mice, oral (27)	Optimum growth 37°C, viable growth range 7–8 to 46°C	Water (14)	Current food contaminant
Staphylococcal enterotoxin	1–25 µg/kg in humans, oral (11)	Heat stable for minutes at 100°C and resistant to freezing (27)	Water (14)	
T-2 toxin	1,210 µg/kg in mice, oral (27)	MP of 151–152°C, thermally stable at normal cooking temperatures and times; decomposes at 816°C/30 min (27)	Soluble in lower alcohols and polar solvents (59)	No odor/taste (54)
Yersinia pestis	100 CFU, oral (13)	Inactivated at 55°C/15 min (27)	Water (14)	Prior contamination
Other agents for comparison				
Abrin	0.04 µg/kg in mice, oral (27)	Heat stable to 60°C for 30 min (59)	Slightly soluble in water (20)	NA
Aconitine	100 µg/kg in mice oral (27)	MP 204°C (59)	Water at 310 µg/ml (59)	
Conotoxins	5 µg/kg in mice, oral (27)	NA	Water (14)	No odor/taste (59)
Cyanide	30 µg/kg in rats, oral (22)	Heating can release irritating or toxic gases; MP −13.4°C (hydrogen cyanide); MP 563.7°C (sodium) (22)	Water at 48 g/100 ml at 10°C (22)	Faint bitter-almond odor; odorless when dry, slight odor of HCN when moist (22)
Digoxin	177.8 µg/kg in humans, oral; 76.7 µg/kg in mice, oral (59)	Heat stable, decomposes at 235°C (59)	Insoluble in water, soluble in dilute alcohols (59)	Odorless, bitter taste (42)
Fluoroacetic acid	5 µg/kg in rats, oral; 7 µg/kg in mice, oral (59)	MP 35.2°C, boiling point 165°C (59)	Soluble in water at 50 µg/ml at 25°C	Odorless powder (59)
Nicotine sulfate	85.5 µg/kg in mice, oral; 83.0 µg/kg in rats, oral (59)	Decomposes on heating, producing toxic fumes (42)	Water, alcohol	None for crystals, aqueous solution has tobacco/fishy odor (59)
Organophosphates (as a class)	150 µg/kg in rats, oral; 600 µg/kg in mice, oral (59)	Varies by specific toxin but 35–36°C is general MP and generally thermally stable to cooking (59)	Water, 55 µg/ml at 20°C (59)	Range of no odor to pungent, garliclike (59)

(Continued)

Table 5.2 Potential food contamination agents *(Continued)*

Agent	LD_{50}[b]	Heat sensitivity	Solubility of toxin	Organoleptic characteristics
Picrotoxin	150 µg/kg in rats, oral (59)	MP 203°C and thermally stable to cooking (59)	Water at 3,030 µg/ml ambient, 20% boiling (59)	Odorless, bitter taste (22)
Ricin	3–5 µg/kg in mice, oral (27)	Stable but detoxifies at 80°C (10 min) and 50°C (50 min) (59)	Water (59)	No odor/taste (59)
Saxitoxin	263 µg/kg in mice, oral (59)	Heat stable, cooking does not remove (59)	Water and methanol (59)	No odor/taste (59)
Strychnine	2,350 µg/kg in rats, oral; 20 µg/kg in mice, oral (59)	MP 275–285°C (59)	Water at 20% (59)	Odorless; pure material has a bitter metallic taste and bitter taste in solution at 1.4 µg/ml (22)
Tetrodotoxin	8.0 µg/kg in mice, oral; 200 µg/kg in cats, oral (27)	Darkens above 220°C without decomposing (59)	Soluble in diluted acetic acid, slightly soluble in water (59)	No odor (59)

[a] LD_{50}, 50% lethal dose.
[b] MP, melting point.
[c] NA, not applicable.

agents and genetically engineered microorganisms more attractive as terrorism agents, even given the significantly higher technical expertise required to utilize them.

Because the list of potential agents is extensive and varied, evaluation of intentional food contamination vulnerabilities requires a different set of considerations than preventing accidental contamination. First, there is the overall compatibility of the microorganism with the food, i.e., its technical attractiveness. If the pathogen is stable in the food matrix and it is not inactivated by conventional processing, it could be of concern. This includes food and microorganism combinations where the food supports growth of the microorganism. In addition, the ease with which the microorganism can be mixed into the food product and how readily it can be dispersed and remain visually unidentifiable are important considerations. If the population of the microorganism necessary to provide an infective dose turns a clear fluid product turbid, for example, that food/microorganism combination is not a likely scenario. Just as important is the organoleptic impact of the microorganism or contaminant on the food item. Negative public health consequences would be avoided if consumers would not eat sufficient product due to noticeable differences in expected organoleptic quality, in much the same way as conventional spoilage by microorganisms renders food inedible, thereby possibly preventing illness that could be caused by pathogens that had also grown in the food. In the case of intentional nicotine sulfate contamination of ground beef in Michigan, the spicy

flavor impact of the contaminant was one of the ways in which the incident was uncovered (9). Similar early detection would be expected for foods with off flavors or any other organoleptic shift.

The potential risk based on characteristics of the specific food, i.e., the relative attractiveness it might have as a vehicle for delivery of the microbial agent based on compatibility factors, is an another important consideration. Large batch sizes, thorough mixing after contamination, and the absence of any postcontamination terminal treatment are factors that increase vulnerability due to the amount of product that could be contaminated with a harmful dose of the pathogen. The speed of distribution of the product and the rate of consumer consumption also have a direct impact on the severity of the public health consequences from the intentional contamination. The fear and rage impact of the event are also impacted by any preferential consumption by more vulnerable populations, especially children (73).

Beyond such specific technical considerations, another aspect of the food that could increase its vulnerability for intentional contamination is the lingering potential impact on consumer confidence and the economy. If the food or commodity is targeted in a way that maximizes consumer concerns about the near- or long-term ability to protect that food item from future attacks, then recovery from the attack would be slower and therefore result in a more vulnerable food or commodity. Since it is unlikely that any attack would be significant enough to negatively impact the total availability of food, any economic impacts due to consumption changes are at the firm, manufacturer, item, commodity, or category level, reflecting a shift of purchasing patterns. The $200 million loss to the Chilean grape industry resulting from intentional contamination of grapes with cyanide in 1989 (33) and the deepening of the bankruptcy challenges of a restaurant chain following an outbreak of hepatitis associated with eating food items containing green onions at the restaurant (61) are recent examples of the economic impacts of food incidents at the sector or firm level. Similar to any natural disaster, there are also real economic losses in lives and future productivity resulting from the attack itself, and these could be substantial. There also could be significant trade implications if the attack resulted in a restriction of trade due to trading partners' concerns about the safety of the food type, something more commonly considered in animal trade and the official animal disease status, e.g., the status of foot-and-mouth disease in a given country. Intentional contamination of a particular food becomes more attractive as a target as the magnitude of various economic consequences increases. It is important to note that an intentional contamination event could result in no direct public health consequences and yet result in significant or catastrophic economic harm.

INTENTIONAL FOOD CONTAMINATION RISK MANAGEMENT

Intervention strategies to prevent attacks on the food system suffer from a different challenge than traditional food safety efforts in that making decisions on investments and efficacy is hampered by the lack of a natural occurrence of the event (46). In such cases stakeholders are understandably reluctant to invest to protect against such events (6). Addressing the economic models that could be useful is beyond the scope of this chapter, but there are technical considerations on how to evaluate intervention strategies that are useful to review. Although potentially difficult to validate, measuring the success of an intervention to prevent terrorism can be achieved in the absence of any ability to demonstrate failed attacks. The general approach is to evaluate the degree to which the intervention reduces the vulnerability of each subsystem within the overall infrastructure of concern (83). In the critical infrastructure of the food system, there are multiple approaches currently being used to assess vulnerability or risk at all levels, from the single-unit level to national infrastructure and global supply systems. Independent of the risk assessment approach, it is relatively straightforward to evaluate the success of a particular

intervention. First, one analyzes the vulnerability of the particular section or operation within the food system prior to applying the intervention. Then, either after deploying the intervention or in a theoretical construct, the section or operation in the food system is analyzed again. If the vulnerability of the food system section or operation is reduced, then the intervention can be considered successful. This does not, however, answer whether or not the intervention is economically justified, which represents another entirely new set of analyses. Given their widespread use by federal and state agencies, Operational Risk Management (ORM) and CARVER (Criticality, Accessibility, Recuperability, Vulnerability, Effect, and Recognizability; see below) variations are worth reviewing as probable risk assessment tools for food system operations. In addition, there are many tools developed by the private sector for either risk assessment or development of food system protection plans that could be of use (1, 58, 77).

Operational Risk Management

ORM is a tool that originated from efforts in the U.S. National Aeronautics and Space Administration (NASA) and the U.S. Department of Defense (DoD) to reduce the risks of failure of aircraft, space missions, and weapons. It is still actively used today by units such as the U.S. Naval Safety Center (81) and was adopted by the U.S. Food and Drug Administration—Center for Food Safety and Nutrition for early food system risk assessments (31). In total, ORM is a five-step process for identifying and managing risks: (i) identify the hazards; (ii) assess the potential consequence of the hazards; (iii) determine which risks to manage with which interventions; (iv) implement the interventions; and (v) assess the success of the interventions and modify as necessary.

Traditionally, in ORM, risk is defined as a function of the severity of the failure and the probability of the failure. As noted above, since intentional contamination is not a stochastic event, there is not a predictive function for estimating the probability of intentional contamination events. For food defense, probability can best be considered as the probability of success if an appropriately skilled person or group tried to contaminate the food system. The severity and probability can be evaluated on separate scales. For any food item, commodity, facility, unit operation, or other definable subset of the food system, one can conduct this analysis and come up with a location on an assessment grid which compares the severity against the probability, to allow focusing on those events with the highest combination of both. The basic scales used in ORM as they could be applied to the food system are noted in Table 5.3.

Table 5.3 Operational risk management semantic scales

Severity scale	Probability scale
Very high (catastrophic public health impact, high morbidity and mortality)	Very high (system continuously vulnerable)
High (significant public health impact, primarily morbidity, some mortality)	High (system regularly vulnerable)
Medium (some morbidity, no mortality)	Medium (system sporadically vulnerable)
Low (no real impact)	Low (system seldom vulnerable) Very low (system rarely vulnerable)

Table 5.4 Operational risk management: grid for categorizing[a]

Severity	Probability				
	Very high	High	Medium	Low	Very low
Very high	++++	++++	++++	+++	+
High	++++	+++	+++	++	+
Medium	+++	++	++	+	+
Low	++	+	+	+	+

[a] Grid locations represent varying risk levels.

The general approach is to select the food system(s) of concern, identify the points of vulnerability (hazards), and conduct a severity-versus-probability assessment for the vulnerability. Given the breadth of information necessary to complete an ORM risk assessment, it generally requires a team of experts on the food system or facility, with additional facilitators with detailed knowledge of the ORM approach and any threat information that might not be available to the users. This may include specific details on microbial or chemical contaminants of concern. Importantly, successful utilization of ORM does not require sensitive details on potential agents. Only general characteristics need to be introduced, such as a determination of how much of a contaminant could reasonably be added at a particular stage of the food supply chain. Depending on the result, the facilitator or expert can narrow down the scope of the potential vulnerability. Through a simple charting or scoring exercise, one can then evaluate each of the vulnerabilities and rank them as to those of greatest risk of allowing a catastrophic event. Intervention efforts can then be prioritized for implementation to reduce the system vulnerability. The ORM process then comes full circle with the implementation of the chosen intervention(s), evaluating the success of the intervention, and repeating the risk assessment process. This cycle is repeated until the resulting risk reaches an acceptable level. A representation of the ranking grid that could be used for this purpose is presented in Table 5.4.

CARVER+Shock

CARVER, and its further refinement CARVER+Shock, is another strategy for completing a risk assessment that is a tool now in use by both the FDA and the U.S. Department of Agriculture (USDA) in analyzing points of vulnerability (19, 24). CARVER+Shock is adapted from DoD evaluation systems that were developed to identify targets of greatest effect on an enemy as well as those of greatest vulnerability (43, 47). CARVER+Shock risk assessment is composed of seven elements which are all included to evaluate the vulnerability of a system by analyzing each node within a system. Modified to fit the concerns presented by intentional food contamination, these can be defined as follows:

- Criticality: the degree to which the public health or economic consequences are nationally significant. High scores equate to catastrophic morbidity, mortality, or economic harm.
- Accessibility: physical access to the target; the ability of the perpetrator to gain access to the point of contamination and escape undetected.
- Recuperability: overall system resiliency as measured by the time required to bring the system back into operation, with low scores for only days to recover and high scores for recovery going a year or longer.
- Vulnerability: attack feasibility as viewed by the potential for a successful attack. This includes both the ability to introduce enough of the material of concern to cause harm and the potential for subsequent processing to reduce that risk.
- Effect: direct loss from the attack as defined by the fraction of the food system that has been impacted by the attack.
- Recognizability: ease of target identification is a measure of the degree of specialized knowledge needed in order to identify the point for the intentional contamination.
- Shock: combined health, economic, and psychological impact of the attack, which is a measure of the overall impact. Importantly, the economic and psychological impacts of an attack may not require any morbidity or mortality if they result in a substantial lack of public confidence in the food system or the government.

To apply CARVER+Shock to a food system, the first step is to define the scales to use for each of the rating elements. In selecting the scales, it is very important to avoid potential overweighting of any individual component. The scales developed by the FDA and USDA implementation have been modified from the original DoD scales to render them more applicable to the food system. To allow for sufficient discrimination, 10-point (1 to 10) semantic scales are used for each of the rating categories. For the FDA/USDA evaluations, the 10-point scale is aggregated into five semantic rating groups, giving the evaluator the option of a "high" and a "low" score for each semantic rating. Semantic scales used by the FDA are shown in Table 5.5 (17).

Using these scales or a different set of semantic scales, the next step is to evaluate the food facility, system, or operation in order to identify unique nodes or portions of the system. These nodes, when taken in total, should represent the entirety of the food system, while they individually represent known or likely differences in vulnerability. For the model system depicted in Fig. 5.1, there would be 32 initial nodes for evaluation. Each of the unit operations shown, such as blending or air injection, and each transfer between unit operations represent unique nodes for risk assessment.

Each node of a system is then evaluated against each of the CARVER+Shock semantic scales, and a score for each element is assigned. As noted for the utilization of ORM, a team of experts is generally required to complete this evaluation. A composite score is compiled for the food system or for the section of the food system that is under consideration. These ratings allow for a rational comparison of vulnerabilities across portions of a food system, with interventions or countermeasures designed to address those nodes with the highest scores, indicating a greater vulnerability. While generally utilized to assess a specific food system or facility, the same strategy could be utilized to compare across multiple food systems to identify those with the greatest need for interventions or countermeasures. The selection of interventions, implementation, and assessment could then be conducted in an ORM-like process.

Assessing the overall risk through ORM only is simpler than with CARVER+Shock in that it has only two rating elements for ranking risk, severity and probability, and it requires less training. Conversely, however, it may also require greater expertise in the specific food system under consideration. In some cases a desirable strategy may be to use CARVER+Shock to identify the food system, facility/operation, or node of greatest concern and then allow those who work within that system or facility/operation to use ORM to minimize specific vulnerabilities.

Food System Interventions

Independent of the system used to identify which food system, facility, or operation is vulnerable to intentional contamination, the first levels of interventions are basic security considerations. Standard perimeter and access control, often referred to as "guns, gates, and guards" (10, 68, 79), represents the most basic level, but by necessity interventions need to go beyond that to include facility- and operation-specific physical and human security considerations. Important for their successful implementation, any intervention needs to have a clear technical justification (56). Many of these are already normal practice for most food operations, such as restricting access to locations within a facility to those with a need for access in order to complete their task. Human resource functions become much more important because dealing with disgruntled workers and screening out intentional terrorists attempting to become employees with access to points of high vulnerability are important and necessary interventions. A major challenge in implementing appropriate human resources restrictions is ensuring their compliance with personal privacy and civil rights regulations. There are many other facility- or operation-specific interventions that may be considered that are independent of a specific microbiological or chemical agent. Many of these are detailed in the various guidance documents developed by the FDA and the USDA to assist food systems firms in developing a food security system (18, 34).

When considering defense against contamination with specific microbial agents that may be of unique concern to a specific food item, the characterization of the food itself, the process and safety/quality systems already in place, and final retail/food service/consumer handling practices must also be considered. If the microbial agent has a specific sensitivity to pH, the product may be able to be modified to shift the pH enough to reduce the long-term survivability of the agent. The same could be true for other physiochemical characteristics of the food. More likely, however, is that a food process modification could be introduced to reduce or eliminate the activity of the microbial pathogen so that even if it is introduced into the product, it will not result in morbidity or mortality. Pasteurization is routinely used to control normally prevalent or anticipated accidental or natural contamination of some types of foods. Either the current pasteurization processes, or more aggressive versions, could be applied to certain highly vulnerable foods to mitigate any potential harm from their use to deliver specific agents. This obviously requires that the organoleptic quality of the food not be compromised through the process. In the United States, a majority of the dairy industry has done this already for some of their

Table 5.5 CARVER+Shock semantic scales

Criticality: introduction of threat agents into food at this location would have significant health or economic impact. Example metrics are as follows.

Criticality criteria	Scale
Loss of >10,000 lives OR loss of >$100 billion	9–10
Loss of life is 1,000–10,000 OR loss of $10 billion–$100 billion	7–8
Loss of life 100–1,000 OR loss of $1 billion–$10 billion	5–6
Loss of life <100 OR loss of <$1 billion	3–4
No loss of life OR loss <$100 million	1–2

Accessibility: a target is accessible when an attacker can reach the target to conduct the attack and egress the target undetected. Example metrics are as follows.

Accessibility criteria	Scale
Easily accessible (e.g., target is outside building and no perimeter fence); limited physical or human barriers or observation; attacker has relatively unlimited access to the target; attack can be carried out using medium or large volumes of contaminant without undue concern of detection; multiple sources of information concerning the facility and the target are easily available	9–10
Accessible (e.g., target is inside building, but in unsecured part of facility); human observation and physical barriers limited; attacker has access to the target for 1 h or less; attack can be carried out with moderate to large volumes of contaminant but requires the use of stealth; only limited specific information is available on the facility and the target	7–8
Partially accessible (e.g., inside building, but in a relatively unsecured, but busy, part of facility); under constant possible human observation; some physical barriers may be present; contaminant must be disguised, and time limitations are significant; only general, nonspecific information is available on the facility and the target	5–6
Hardly accessible (e.g., inside building in a secured part of facility); human observation and physical barriers with an established means of detection; access generally restricted to operators or authorized persons; contaminant must be disguised and time limitations are extreme; limited general information available on the facility and the target	3–4
Not accessible; physical barriers, alarms, and human observation; defined means of intervention in place; attacker can access target for less than 5 min with all equipment carried in pockets; no useful publicly available information concerning the target.	1–2

Recuperability: a target's recuperability is measured in the time it will take for the specific facility to recover productivity. Example metrics are as follows.

Recuperability criteria	Scale
>1 yr	9–10
6 mo–1 yr	7–8
3–6 mo	5–6
1–3 mo	3–4
<1 mo	1–2

Vulnerability: a measure of the ease with which threat agents can be introduced in quantities sufficient to achieve the attacker's purpose once the target has been reached. Example metrics are as follows.

Vulnerability criteria	Scale
Target characteristics allow for easy introduction of sufficient agents to achieve aim	9–10
Target characteristics almost always allow for introduction of sufficient agents to achieve aim	7–8
Target characteristics allow 30–60% probability that sufficient agents can be added to achieve aim	5–6
Target characteristics allow moderate probability (10–30%) that sufficient agents can be added to achieve aim	3–4
Target characteristics allow low probability (<10%) that sufficient agents can be added to achieve aim	1–2

Effect: effect is a measure of the percentage of system productivity (total production capacity) damaged by an attack at a single facility. Example metrics are as follows.

Effect criteria	Scale
>50% of the system's production impacted	9–10
25–50% of the system's production impacted	7–8
10–25% of the system's production impacted	5–6
1–10% of the system's production impacted	3–4
<1% of the system's production impacted	1–2

(Continued)

Table 5.5   *(Continued)*

Recognizability: a target's recognizability is the degree to which it can be identified by an attacker without confusion with other targets or components. Example metrics are as follows.

Recognizability criteria	Scale
The target is clearly recognizable and requires little or no training for recognition.	9–10
The target is easily recognizable and requires only a small amount of training for recognition.	7–8
The target is difficult to recognize or might be confused with other targets or target components and requires some training for recognition	5–6
The target is difficult to recognize. It is easily confused with other targets or components and requires extensive training for recognition	3–4
The target cannot be recognized under any conditions, except by experts	1–2

Shock: shock is the combined measure of the national health, psychological, and collateral economic impacts of a successful attack. Example metrics are as follows.

Shock	Scale
Target has major historical, cultural, religious, or other symbolic importance; loss of >10,000 lives; major impact on sensitive subpopulations, e.g., children or the elderly; national economic impact >$100 billion	9–10
Target has high historical, cultural, religious, or other symbolic importance; loss of 1,000–10,000 lives; significant impact on sensitive subpopulations, e.g., children or the elderly; national economic impact $10 billion–$100 billion	7–8
Target has moderate historical, cultural, religious, or other symbolic importance; loss of life 100–1,000; moderate impact on sensitive subpopulations, e.g., children or the elderly; national economic impact $1 billion–$10 billion	5–6
Target has little historical, cultural, religious, or other symbolic importance; loss of life <100; small impact on sensitive subpopulations, e.g., children or the elderly; national economic impact $100 million–$1 billion	3–4
Target has no historical, cultural, religious, or other symbolic importance; loss of life <10; no impact on sensitive subpopulations, e.g., children or the elderly; national economic impact <$100 million	1–2

products to reduce the vulnerability of those foods (30). Similar interventions could be applied to other foods to mitigate the vulnerability, including other processing interventions, e.g., high-pressure processing and ionizing radiation (51). In all cases, however, it is important to design the food system such that intentional contamination after the insertion of additional interventions is extremely unlikely or readily detectable/traceable (45) so that the food does not become a more attractive target than it was originally.

Detection and diagnostics, both new technology platforms, and integration of new platforms into new systems are fields of intense interest and activity in the broadly defined area of homeland security. As applied to the food and agriculture critical infrastructure, detection and diagnostic tools have the potential to prevent or contain intentional contamination events. Just as in the case of food safety, however, detection and diagnostic systems have limitations. The traditional challenges of sampling strategies, preanalytical processing, and time to detection are amplified while concerns about false positives and false negatives are elevated when applied to detecting contamination events that should not happen. This requires a new approach on how to evaluate the utility of any detection intervention, both with regard to its feasibility to contain the unlikely event and the cost of implementation. An overview of technologies currently available for use in analyzing food items for potential contamination can be found in recent articles (2, 4, 7, 8, 15, 16, 36, 69–71).

The following presents an approach for evaluating detection and diagnostic system intervention options that takes the principles of "detect to warn," "detect to treat," and "detect to recover" described by former Rear Admiral Richard Danzig (26) and translates them into a mental construct for use in the food system of detect to prevent, detect to protect, and detect to respond or recover:

"Detect to prevent" means an overall strategy that enables positive confirmation of contamination before the finished food item leaves the facility in order to eliminate any chance of public health consequences. This can be at the farm level, such as field-packed produce, all the way through to food processing facilities or large-scale commissaries, and at any point in between. The firm incurs the economic cost of the detection strategy, any further economic impact from actual contamination events being limited to disposal of any contaminated product.

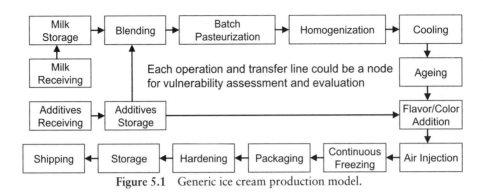

Figure 5.1 Generic ice cream production model.

"Detect to protect" still intends to prevent any public health consequences from intentional food contamination, but after the food has left the facility of concern. While the time frame varies greatly based on the food item, there is a lag time from when food leaves any harvest, processing, or handling facility until it is purchased for consumption by consumers. If the detection system can enable the supply chain to maintain control and prevent the sale of contaminated products, then the public health and primary economic consequences are averted. The firm still has an economic loss, both direct, from moving product through the supply chain, lost sales, and disposal costs, and indirect, through customer and consumer loss of confidence for not preventing or containing the contamination before distribution.

"Detect to recover" is the strategy to enable rapid identification of intentional contamination that has moved into consumer consumption in order to quickly contain the event and minimize its impact. Here the economic consequences are far more significant and there are important public health consequences. In many ways this is similar to how current foodborne illness response systems work. When food safety issues are not identified prior to distribution and consumption, resulting in a foodborne illness outbreak, our public health and food systems work to identify and contain the outbreak while learning from it to prevent future outbreaks. If the agent is present at an infective or lethal level, however, the consequences from an intentional event would be far more dire than those from historic natural foodborne illness outbreaks. In this case the economic losses go beyond the firm or impacted commodity to include losses due to illness and overall macroeconomic impacts.

To illustrate how these three approaches can be applied to the food system, refer to Fig. 5.2, a generic process model for egg processing incorporated into the USDA-FSIS Model Food Security Plan for Egg Processing Facilities guidance document (80). While detection

technologies could be deployed during or after each unit operation shown, such as storage or grading, the arrows indicate potentially preferred points for inserting a detection intervention. The decision of where to insert a detection intervention is based on factors such as the ease of obtaining a representative sample (balance tanks better than grading), the potential risk of contamination at that point (washing versus breaking), and proximity to the terminus of a set of sequenced operations.

To use the above diagram to design a detection strategy for any food item, one needs additional information beyond the identified points of potential detection. The time to detection, specificity, and sensitivity of the technique need to be combined with an understanding of the velocity of the food item from that point forward so that the detection intervention can be categorized as a detect-to-prevent, detect-to-protect, or detect-to-recover strategy. In some cases the most practical strategy for the use of detection technologies could be a rapid technique for identifying suspect food items soon enough to allow diversion prior to final loss of supply chain owner control, followed by a highly specific technique to ensure that the suspect contaminant is present. These technical challenges are in addition to the difficulties in establishing sampling strategies and acquiring samples. While the likelihood of higher levels of the agent from intentional contamination, as compared to unintentional contamination, with a foodborne pathogen reduces sampling uncertainty challenges, this likelihood does not eliminate such challenges (82). Whatever detection strategy is selected, however, the economic considerations will also be paramount. The cost of the detection system implementation itself could be a significant barrier to industry adoption and could require regulatory actions to drive use of the technologies. Beyond those costs are the costs of false positives, also referred to as "detect to regret." These costs drive the need for total detection system false-positive rates to be significantly lower than

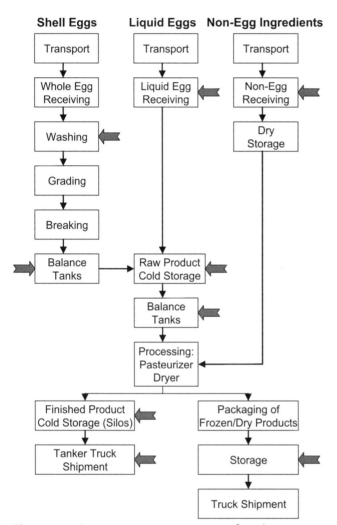

Figure 5.2 Generic egg processing process flow diagram.

occurring microbial pathogens cannot anticipate and protect against individuals committed to using the food system as a means of economic or public health terrorism. Approaches to harden the food system against intentional contamination and enhance its inherent biosecurity are being adapted from ORM, CARVER, CARVER+Shock, and other systems from military and law enforcement experience. Countering the potential risks posed by intentional use of microbiological pathogens requires a continual reassessment and optimization of the food system. While similar to the continual improvement approach dictated by food safety systems to prevent unintentional microbiological contamination of food products, food system biosecurity represents an additional set of challenges. These include recognizing that a successful attack on the food system would not necessarily result in any direct morbidity or mortality. These challenges cover the full range from basic microbiology to food system supply-chain management to legal protection of potentially sensitive private-sector information. Ensuring that the biosecurity of the food system is not compromised will require that all aspects are addressed, with food microbiology being a necessary, but not the only, element.

References

1. **AIB International.** 2006. *Food Security Assessment.* AIB International, Manhattan, Kans. [Online.] http://www.aibon line.org/audits/foodsafety/foodsecurityaudits/. Accessed 23 February 2006.

2. **Akbulut, D., K. A. Grant, and J. McLauchlin.** 2004. Development and application of real-time PCR assays to detect fragments of the *Clostridium botulinum* types A, B, and E neurotoxin genes for investigation of human foodborne and infant botulism. *Foodborne Pathog. Dis.* **1:**247–257.

3. **Arnon, S., R. Schechter, T. V. Inglesby, D. A. Henderson, J. G. Bartlett, M. S. Ascher, E. Eitzen, A. D. Fine, J. Hauer, M. Layton, S. Lillibridge, M. T. Osterholm, T. O'Toole, G. Parker, T. M. Perl, P. K. Russell, D. L. Swerdlow, and K. Tonat.** 2001. Botulinum toxin as a biological weapon. Medical and public health management. *JAMA* **285:**1059–1070.

4. **Arranz, I., W. R. G. Baeyens, G. Van Der Weken, S. De Saeger, and C. Van Peteghem.** 2004. Review: HPLC determination of fumonisin mycotoxins. *Crit. Rev. Food Sci. Nutr.* **44:**195–203.

5. **Ashford, D. A., R. M. Kaiser, M. E. Bales, K. Shutt, A. Patrawalla, A. McShan, J. W. Tappero, B. A. Perkins, and A. L. Dannenberg.** 2003. Planning against biological terrorism: lessons from outbreak investigations. *Emerg. Infect. Dis.* **9:**515–519.

6. **Bazerman, M. H., and M. D. Watkins.** 2004. *Predictable Surprises: The Disasters You Should Have Seen Coming and How to Prevent Them.* Harvard Business School Press, Cambridge, Mass.

for many of the individual detection technologies now available. The scope of food production is such that if the scale of detection desired is the fluid bulk truck, for example, that a false-positive rate of 1/100,000 would result in nearly 40 false positives a year in something like fluid milk. If that false positive could not be rapidly confirmed, it would result in the investigation of 40 potential terrorism events every year.

In summary, ensuring that the United States and the world benefit from the overarching goal of an abundant, nutritious, and safe food system is more complex now that intentional contamination is a very real and credible threat. The overall food system risk management system that has developed over the last few decades serves as a sound and necessary foundation for approaches that will protect the food system from intentional abuse, but it is not sufficient. HACCP and other systems to control naturally

7. Bennett, R. W. 2005. Staphylococcal enterotoxin and its rapid identification in foods by enzyme-linked immunosorbent assay-based methodology. *J. Food Prot.* **68:**1264–1270.

8. Blyn, L. B. 2006. Biosensors and food protection. *Food Technol.* **60:**36–41.

9. Boulton, M., M. Stanbury, D. Wade, J. Tilden, D. Bryan, J. Payne, and B. Eisenga. 2003. Nicotine poisoning after ingestion of contaminated ground beef. *Morb. Mortal. Wkly. Rep.* **52:**413–416.

10. Brooks, L. F. 2004. Testimony, Committee on Government Reform, Subcommittee on National Security, Emerging Threats, and International Relations. *Congressional Record.*

11. Burrows, W. D., and S. E. Renner. 1999. Biological warfare agents as threats to potable water. *Environ. Health Perspect.* **107:**975–984.

12. Bush, G. W. 2004. Homeland Security Presidential Directive/HSPD-9: Defense of United States Agriculture and Food. The White House, Washington, D.C. [Online.] http://www.whitehouse.gov/news/releases/2004/02/20040203-2.html.

13. Carus, W. S. 2001. *Bioterrorism and Biocrimes: The Illicit Use of Biological Agents Since 1900.* Center for Counterproliferation Research, National Defense University, Washington, D.C.

14. CBWInfo. 2006. Factsheets on Chemical and Biological Warfare Agents. CBWInfo. [Online.] http://www.cbwinfo.com/intro.html. Accessed November 2006.

15. Cebula, T., E. W. Brown, S. A. Jackson, M. K. Mammel, A. Mukherjee, and J. E. LeClerc. 2005. Molecular identification of foodborne pathogens. *Expert Rev. Mol. Diagn.* **5:**432–445.

16. Cebula, T. A., S. A. Jackson, E. W. Brown, B. Goswami, and J. E. LeClerc. 2005. Chips and SNPs, bugs and thugs: a molecular sleuthing perspective. *J. Food Prot.* **68:**1271–1284.

17. Center for Food Safety and Applied Nutrition. 2005. Scales in use as of 11/30/2005 provided by CFSAN. Center for Food Safety and Applied Nutrition, College Park, Md.

18. Center for Food Safety and Applied Nutrition. 2005. Food defense and terrorism. Center for Food Safety and Applied Nutrition. [Online.] http://www.cfsan.fda.gov/~dms/fsterr.html. Accessed 20 November 2005.

19. Center for Food Safety and Applied Nutrition. 2005. SPPA questions and answers. Center for Food Safety and Applied Nutrition. [Online.] http://www.cfsan.fda.gov/~dms/agroter4.html. Accessed 23 February 2006.

20. Centers for Disease Control and Prevention. 2005. Bioterrorism Agents/Diseases. Centers for Disease Control and Prevention, Atlanta, Ga. [Online.] http://www.bt.cdc.gov/agent/agentlist.asp. Accessed 20 November 2005.

21. Centers for Disease Control and Prevention. 2006. HHS and USDA Select Agents and Toxins: 7 CFR Part 331, 9 CFR Part 121, and 42 CFR Part 73. Centers for Disease Control and Prevention. [Online.] http://www.cdc.gov/od/sap/docs/salist.pdf. Accessed 20 February 2005.

22. Centers for Disease Control and Prevention, Agency for Toxic Substances and Disease Registry. 2006. [Online.] http://www.atsdr.cdc.gov/.

23. Christopher, G. W., T. J. Cieslak, J. A. Pavlin, and E. M. Eitzen. 1997. Biological warfare: a historical perspective. *JAMA* **278:**412–417.

24. Conner, C. F. 2006. Testimony of the Honorable Charles F. Conner, Deputy Secretary, United States Department of Agriculture Before the U.S. Senate Committee on Agriculture, Nutrition, and Forestry Subcommittee on Research, Nutrition and General Legislation. USDA. [Online.] http://www.usda.gov/homelandsecurity/DepSecTest_jan09.pdf. Accessed 1 March 2006.

25. Cosgrove, S. E., T. M. Perl, X. Song, and S. D. Sisson. 2005. Ability of physicians to diagnose and manage illness due to category A bioterrorism agents. *Arch. Intern. Med.* **165:**2002–2006.

26. Danzig, R. 2003. *Catastrophic Bioterrorism - What Is To Be Done?* Center for Technology and National Security Policy, National Defense University, Washington, D.C.

27. Darling, R. G., and J. B. Woods. 2004. *Medical Management of Biological Casualties Handbook,* 5th ed. USAMRIID, Fort Detrick, Md.

28. Dembek, Z. F., R. L. Buckman, S. K. Fowler, and J. L. Hadler. 2003. Missed sentinel case of naturally occurring pneumonic tularemia outbreak: lessons for detection of bioterrorism. *J. Am. Board Fam. Pract.* **16:**339–342.

29. Dennis, D. T., T. V. Inglesby, D. A. Henderson, J. G. Bartlett, M. S. Ascher, E. Eitzen, A. D. Fine, A. M. Friedlander, J. Hauer, M. Layton, S. R. Lillibridge, J. E. McDade, M. T. Osterholm, T. O'Toole, G. Parker, T. M. Perl, P. K. Russell, and K. Tonat. 2001. Tularemia as a biological weapon: medical and public health management. *JAMA* **285:**2763–2773.

30. Detlefsen, C. 2005. Dairy industry vigilant in addressing food security. *Cheese Market News* **25:**1.

31. Dyckman, L. J. 2003. *Food-Processing Security: Voluntary Efforts Are Under Way, but Federal Agencies Cannot Fully Assess Their Implementation.* Publication no. 03-342. GAO, Washington, D.C.

32. Dyckman, L. J. 2003. *Bioterrorism: A Threat to Agriculture and the Food Supply.* Publication no. 04-259T. GAO, Washington, D.C.

33. Food and Drug Administration. 2005. An Introduction to Food Security Awareness. U.S. Food and Drug Administration, Rockville, Md. [Online.] http://www.fda.gov/ora/training/orau/FoodSecurity/startpage.html. Accessed 20 November 2005.

34. Food Safety and Inspection Service. 2005. Food Security and Emergency Preparedness Security Guidelines. Food Safety and Inspection Service, USDA, Washington, D.C. [Online.] http://www.fsis.usda.gov/Food_Security_&_Emergency_Preparedness/Security_Guidelines/index.asp. Accessed 20 November 2005.

35. Franz, D. R., P. B. Jahrling, A. M. Friedlander, D. J. McClain, D. L. Hoover, R. W. Bryne, J. A. Pavlin, G. W. Christopher, and E. M. Eitzen. 1997. Clinical recognition and management of patients exposed to biological warfare agents. *JAMA* **278:**399–411.

36. Garber, E. A., R. M. Eppley, M. E. Stack, M. A. McLaughlin, and D. L. Park. 2005. Feasibility of immunodiagnostic devices for the detection of ricin, amanitin, and T-2 toxin in food. *J. Food Prot.* **68:**1294–1301.

37. Reference deleted.

38. Harper, G. J. 1961. Airborne microorganisms: survival test with four viruses. *J. Hyg.* **59:**479–486.

39. Harris, S. H. 2003. Japanese biomedical experimentation during the World-War-II era, p. 463–506. *In* D. E. Lounsbury and R. F. Bellamy (ed.), *Military Medical Ethics*, vol. II. Office of the Surgeon General, Department of the Army, Washington, D.C.

40. Hauschild, A. H. W., and K. L. Dodds. 1993. *Clostridium botulinum. Ecology and Control in Foods.* Marcel Dekker, Inc., New York, N.Y.

41. Hennessy, T. W., C. W. Hedberg, L. Slutsker, K. E. White, J. M. Besser-Wiek, M. E. Moen, J. Feldman, W. W. Coleman, L. M. Edmonson, K. L. MacDonald, and M. T. Osterholm. 1996. A national outbreak of *Salmonella enteritidis* infections from ice cream. *N. Engl. J. Med.* **334**:1281–1286.

42. IPCS INCHEM. 2005. Chemical Safety Information from Intergovernmental Organizations. The International Programme on Chemical Safety. [Online.] http://www.inchem. org. Accessed 20 November 2005.

43. Jones, T. S. 2005. *NAVMC Directive 3500.86.* United States Marine Corps, Washington, D.C.

44. Joseph, P. R., J. D. Millar, and D. A. Henderson. 1965. An outbreak of hepatitis traced to food contamination. *N. Engl. J. Med.* **273**:188–194.

45. Jotcham, R. 2005. Authentication, antitamper, and track-and-trace technology options to protect foods. *J. Food Prot.* **68**:1314–1317.

46. Kahn, A. S., D. L. Swerdlow, and D. D. Juranek. 2001. Precautions against biological and chemical terrorism directed at food and water supplies. *Public Health Rep.* **116**:3–14.

47. Kavanaugh, J. J. 1996. *Air Force Handbook 31-302.* United States Air Force, Washington, D.C.

48. Klietmann, W. F., and K. L. Ruoff. 2001. Bioterrorism: implications for the clinical microbiologist. *Clin. Microbiol. Rev.* **14**:364–381.

49. Koenig, M. G., A. Spickard, M. A. Cardella, and D. E. Rogers. 1964. Clinical and laboratory observations on type E botulism in man. *Medicine* **43**:517–545.

50. Kolavic, S. A., A. Kimura, S. L. Simons, L. Slutsker, S. Barth, and C. E. Haley. 1997. An outbreak of *Shigella dysenteriae* type 2 among laboratory workers due to intentional food contamination. *JAMA* **278**:396–398.

51. Lado, B. H., and A. E. Yousef. 2002. Alternative food-preservation technologies: efficacy and mechanisms. *Microbes Infect.* **4**:433–440.

52. Lee, T. 2005. Dioxin in local cows' milk meets with EU standard, EQPF says. *China Post* **2005**:9–28.

53. Lincoln, R. E., J. S. Walker, F. Klein, A. J. Rosenwald, and W. I. Jones, Jr. 1967. Value of field data for extrapolation in anthrax. *Fed. Proc.* **26**:1558–1562.

54. Locasto, D., M. Allswede, and T. M. Stein. 2005. CBRNE-T-2 Mycotoxins. http://www.emedicine.com. Accessed 20 November 2005.

55. Mayor, A. 2004. Poison waters, deadly vapors, p. 99–118. Sweet sabotage, p. 145–169. *In Greek Fire, Poison Arrows & Scorpion Bombs: Biological and Chemical Warfare in the Ancient World.* The Overlook Press, Peter Mayer Publishers, Inc., Woodstock, N.Y.

56. Miller, A. J., C. L. Hielman, S. Droby, and N. Paster. 2005. Science and technology based countermeasures to foodborne terrorism: introduction. *J. Food Prot.* **68**:1253–1255.

57. Mohtadi, H., and A. Murshid. 2006. A global chronology of incidents of chemical, biological, radioactive and nuclear attacks: 1950–2005. [Online.] http://www.ncfpd. umn.edu/files/GlobalChron.pdf.

58. National Infrastructure Institute, Center for Infrastructure Expertise. 2006. CARVER2. [Online.] http://www.ni2cie. org/CARVER2.asp. Accessed 23 February 2006.

59. National Library of Medicine. 2005. Toxicology Data Network. National Library of Medicine, National Institutes of Health, Bethesda, Md. [Online.] http://www.toxnet.nlm. nih.gov. Accessed 20 November 2005.

60. National Memorial Institute for the Prevention of Terrorism. 2005. MIPT Terrorism Knowledge Base. [Online.] http:// www.tkb.org/Home.jsp. Accessed 10 November 2006.

61. Oltmanns, R. 2004. Puts & Calls: A Tale of Two Vegetables (OK, a Veggie and a Fruit). Corporate Communications Key In Times of Crisis. *Crisis Manager,* 1 September 2004 [Online.] http://www.bernsteincrisismanagement. com/nl/crisismgr040901.html.

62. Pate, J., and G. Cameron. 2001. Covert Biological Weapons Attacks against Agricultural Targets: Assessing the Impact against U.S. Agriculture. BCSIA Discussion Paper 2001–9, ESDP Discussion Paper ESDP-2001-05. John F. Kennedy School of Government, Harvard University, Cambridge, Mass.

63. Phillis, J. A., J. Harrold, G. V. Whiteman, and L. Perelmutter. 1972. Pulmonary infiltrates, asthma and eosinophilia due to *Ascaris suum* infestation in man. *N. Engl. J. Med.* **286**:965–970.

64. Purver, R. 1995. *Chemical and Biological Terrorism: the Threat According to the Open Literature.* Canadian Security Intelligence Service, Ottawa, Canada.

65. Rasco, B. A., and G. E. Bledsoe. 2005. *Bioterrorism and Food Safety.* CRC Press, Boca Raton, Fla.

66. Robertson, R. E. 1999. *Food Safety: Agencies Should Further Test Plans for Responding to Deliberate Contamination.* Publication no. RCED-00-3. GAO, Washington, D.C.

67. Ryan, C. A., M. K. Nickels, N. T. Hargrett-Bean, M. E. Potter, T. Endo, L. Mayer, C. W. Langkop, C. Gibson, R. C. McDonald, R. T. Kenney, N. D. Puhr, P. J. McDonnell, R. J. Martin, M. L. Cohen, and P. A. Blake. 1987. Massive outbreak of antimicrobial-resistant salmonellosis traced to pasteurized milk. *JAMA* **258**:3269–3274.

68. Salerno, R. M., and J. G. Koelm. 2002. *Biological Laboratory and Transportation Security and the Biological Weapons Convention.* SAND no. 2002-1067P. Sandia National Laboratories, Albuquerque, N.M.

69. Sapsford, K. E., C. R. Taitt, N. Loo, and F. S. Ligler. 2005. Biosensor detection of botulinum toxoid A and staphylococcal enterotoxin B in food. *Appl. Environ. Microbiol.* **71**:5590–5592.

70. Sharma, S. K., B. S. Eblen, R. L. Bull, D. H. Burr, and R. C. Whiting. 2005. Evaluation of lateral-flow *Clostridium botulinum* neurotoxin detection kits for food analysis. *Appl. Environ. Microbiol.* **71**:3935–3941.

71. Sharma, S. K., and R. C. Whiting. 2005. Methods for detection of *Clostridium botulinum* toxin in foods. *J. Food Prot.* **68**:1256–1263.

72. Sidell, F. R., E. T. Takafuji, and D. R. Franz. 1997. *Medical Aspects of Chemical and Biological Warfare.* Office of the Surgeon General, Department of the Army, Washington, D.C.

73. **Slovic, P.** 1987. Perception of risk. *Science* **236**:280–285.

74. **Stark, A.-A.** 2005. Threat assessment of mycotoxins as weapons: molecular mechanisms of acute toxicity. *J. Food Prot.* **68**:1285–1293.

75. **Stier, R.** 2003. The dirty dozen: ways to reduce the 12 biggest foreign materials problems. *Food Safety* **03**:6–7.

76. **Torok, T. J., R. V. Tauxe, R. P. Wise, J. R. Livengood, R. Sokolow, S. Mauvais, K. A. Birkness, M. Skeels, J. Horan, and L. R. Foster.** 1997. A large community outbreak of salmonellosis caused by intentional contamination of restaurant salad bars. *JAMA* **278**:389–395.

77. **Trap-It Security, Inc.** http://www.trap-it.com/ (Accessed February 23, 2006).

78. **Treadwell, T. A., D. Koo, K. Kuker, and A. S. Kahn.** 2003. Epidemiologic clues to bioterrorism. *Morb. Mortal. Wkly. Rep.* **118**:92–98.

79. **Tucker, J. B.** 2003. Biosecurity: limiting terrorist access to deadly pathogens. *Peaceworks* no. 52.

80. **USDA, Food Safety and Inspection Service.** 2005. Model Food Security Plan for Egg Processing Facilities. USDA, Food Safety and Inspection Service. [Online.] http://www.fsis.usda.gov/PDF/Model_FoodSec_Plan_Eggs.pdf. Accessed 15 November 2005.

81. **U. S. Navy.** 2005. *Naval Safety Center.* U.S. Navy. [Online.] http://www.safetycenter.navy.mil/orm/default.htm. Accessed 12 October 2005.

82. **Whitaker, T. B., and A. S. Johansson.** 2005. Sampling uncertainties for the detection of chemical agents in complex food matrices. *J. Food Prot.* **68**:1306–1313.

83. **Woodbury, G.** 2005. Measuring prevention. *Homeland Security Affairs* **1**:1–9.

84. **World Health Organization.** 2002. Terrorist Threats to Food: Guidance for Establishing and Strengthening Prevention and Response Systems. World Health Organization, Geneva, Switzerland. [Online.] http://www.who.int/foodsafety/publications/general/en/terrorist.pdf.

Microbial Spoilage and Public Health Concerns

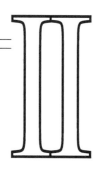

Food Microbiology: Fundamentals and Frontiers, 3rd Ed.
Edited by M. P. Doyle and L. R. Beuchat
© 2007 ASM Press, Washington, D.C.

George-John E. Nychas
Douglas L. Marshall
John N. Sofos

Meat, Poultry, and Seafood

6

A muscle food, including meat, poultry, and seafood, is described as spoiled when it is considered unacceptable by consumers based on its sensory characteristics. Specific sensory characteristics that become unacceptable in a spoiled food include appearance, flavor, and texture. Spoilage occurs when the formation of off flavors, off odors, discoloration, or slime or any other changes in physical appearance or chemical characteristics make the food unacceptable to the consumer (133). Changes in muscle food characteristics are due to native or microbial enzymatic activity or to other chemical reactions (6). A food is described as spoiled based on subjective judgments by the consumer, which may be influenced by cultural and economic considerations and background, as well as the sensory acuity of the individual and the intensity of the change (133). However, when spoilage progresses, most consumers would agree that gross discoloration, strong off odors, and the development of slime would constitute spoilage (99).

Spoilage changes associated with muscle foods may vary with the microbial association, muscle type (e.g., high or low pH, thigh, breast, enzyme activity, etc.), product composition (e.g., content of sugars, lipids, etc.), and storage environment (e.g., temperature, gas composition of packs, etc.). A spoiled food is not necessarily an unsafe food, and therefore, spoilage is considered only an economic loss and it is not generally regulated. Thus, spoilage of foods often does not receive the attention given to microbial changes leading to foodborne illness. Economic losses as well as the food wasted and loss of consumer confidence due to spoilage, however, are of major significance. Thus, it is important to understand the causes and mechanisms of spoilage in order to minimize these losses and provide a food supply of high quality and adequate shelf life.

ECOLOGY OF THE SPOILAGE MICROFLORA OF MUSCLE FOODS

Origin and Types of the Microflora in Muscle Foods

It is generally accepted (47, 101, 129) that bacteria are absent, undetectable, or at extremely low populations in muscle tissues (i.e., red meat and the tissues of poultry, finfish, and shellfish) of healthy live food animals.

George-John E. Nychas, Dept. of Food Science and Technology, Agricultural University of Athens, Laboratory of Food Microbiology and Biotechnology of Foods, Iera Odos 75, Athens 11855, Greece. Douglas L. Marshall, Dept. of Food Science, Nutrition, and Health Promotion, Mississippi State University, Room 110 Herzer, Stone Blvd., Mississippi State, Mississippi 39762-9805. John N. Sofos, Dept. of Animal Sciences, Colorado State University, 1171 Campus Delivery, Fort Collins, Colorado 80523-1171.

As the inherent protective barriers (skins, hides, scales, and shells) and natural antimicrobial defense mechanisms (lysozyme and antimicrobial peptides) of the live animal are destroyed at slaughter, the resulting meat becomes exposed to increasing levels of contamination and, depending on various extrinsic parameters (temperature, packaging, processing method, etc.), may undergo rapid microbial decay. Unless effectively controlled, the slaughtering process may lead to extensive contamination of the exposed cut surfaces of muscle tissue with a vast array of gram-negative and gram-positive bacteria as well as fungi (Table 6.1). Sources of these microorganisms (129, 153, 156, 233) include the external surfaces of the animal and the gastrointestinal tract, as well as the environment with which the animal had contact at some time before or during slaughter (Table 6.2).

Among bacteria, genera in the family *Enterobacteriaceae*, *Photobacterium phosphoreum*, *Shewanella* (*Alteromonas*) *putrefaciens*, *Brochothrix thermosphacta*, *Pseudomonas* spp., *Aeromonas* spp., and lactic acid bacteria have been found to be major contributors to muscle food spoilage, depending on the product type and the conditions surrounding the product (Table 6.3). A comparative analysis of the 16S rRNA of lactic acid bacteria reveals reliable phylogenetic relationships among muscle food contaminants. On this basis, at least 17 major lines were identified (124). Among these, the most important genera of lactic acid bacteria were *Carnobacterium*, *Lactobacillus*, and *Weissella* (41, 42), followed by *Brochothrix*, *Kurthia*, and *Listeria*. Micrococci and staphylococci share common habitats such as the skin of the animals, while *Bacillus* and *Clostridium* are the most important endospore-forming genera with respect to muscle systems. Representatives of the genera *Brevibacterium* (DNA, 60 to 70 mol% G+C), *Corynebacterium* (51 to 65 mol% G+C), and *Propionibacterium* (66 to 67 mol% G+C) together with *Bifidobacterium* form part of the actinomycete branch of bacteria (124) that may be present in meat.

It needs to be noted that the identification and characterization of gram-negative bacteria have not been studied in depth. This is especially evident with *Pseudomonas*, a genus with high heterogeneity and biodiversity within species and/or subspecies. Furthermore, many members of the pseudomonad group have no clear taxonomic status or natural relationships with other genera (184, 210–212). For both reasons mentioned above, conventional phenotypic methods offer limited classification results (37, 183). On the other hand, molecular methods have proven to be powerful tools not only for identification at the species level but also for strain characterization (10, 137, 149). It needs to be noted that although molecular fingerprinting methods have been successfully applied to bacteria of

medical interest or foodborne pathogens in epidemiological studies, they are not always efficient in tracking a particular strain in the food environment (44).

High numbers of bacteria are present on the hide, hair, hooves, and feathers of red meat and poultry animals, as well as in their gastrointestinal tract. Microorganisms on the hide include asymptotic bacterial pathogens, as well as nonpathogenic bacteria, yeasts, and molds. They are normally associated with the microflora of the skin or with species present in fecal material or soil (47, 101). The population and composition of this microflora are influenced by environmental conditions; for example, wet or muddy hides may contain larger populations of bacteria indigenous to soil, while contamination of the hide with fecal material may increase the proportion of microorganisms of fecal origin (47, 101, 115, 233).

Fish, being poikilothermic, possess a microflora influenced by the temperature of the water and by the microflora of the bottom sediment of the area of catch (Table 6.2). Fish caught on a line may have lower bacterial counts than fish that are trawled by dragging a net along the bottom; the trawl net drags through the bottom sediment, which usually has high counts of microorganisms (114, 128). Unlike other crustacean shellfish (i.e., lobster, crab, and crayfish) that are kept alive until preparation for consumption, shrimp die soon after harvesting. Decomposition of shrimp starts soon after death and involves bacteria on the surface that originate from the marine environment or are introduced during handling and washing (85, 187). Molluscan shellfish (i.e., oysters, clams, scallops, and mussels) are sessile and filter feeders, and therefore their microflora depends greatly on the quality of the water in which they reside, the quality of wash water, and other factors (135).

The type and extent of microbial contamination on muscle foods immediately after harvest will also depend on various washing and decontamination treatments that may be applied during the dressing process (94, 153, 235).

Effects of Carcass or Meat Decontamination

Chemical (chlorine, organic acids, inorganic phosphates, proteins, oxidizers, etc.) or physical (knife-trimming, cold or hot water, vacuum, and/or steam) agents and combinations of two or more agents simultaneously or in sequence are applied as carcass decontamination/sanitization treatments in the United States (153, 235). It is worth noting that no chemical or physical decontamination/sanitization treatments of meat and poultry carcasses are allowed by European Union regulations at any stage of production and processing, including live animals (before slaughter), carcasses, primal cuts, and final

Table 6.1 Genera of microorganisms commonly found on meats, poultry, and seafood[a]

| Microorganism | Gram reaction | Occurrence in type of muscle food[b] | | | |
| | | Meat and poultry | | | |
		Fresh	Processed	Vacuum packaged	Fish
Bacteria					
Achromobacter	−	X			
Acinetobacter	−	XX	X	X	X
Aeromonas	−	XX	X	X	X
Alcaligenes	−	X			X
Alteromonas	−	X	X		X
Arthrobacter	−/+	X	X		
Bacillus	+	X	X	XX	X
Brochothrix	+	X	X		
Campylobacter	−	X			
Carnobacterium	+	X		XX	
Chromobacterium	−	X			X
Citrobacter	−	X			
Clostridium	+	X			
Corynebacterium	+	X	X	X	X
Cytophaga					X
Enterobacter	−	X	X	X	X
Enterococcus	+	XX	X	XX	X
Escherichia	−	X			
Flavobacterium	−	X			X
Hafnia	−	X	X		
Halobacterium	−				X
Janthinobacterium	−		X		
Klebsiella	−	X			
Kluyvera	−	X			
Kocuria	+	X	X	X	
Kurthia	+	X		X	
Lactobacillus	+	X	XX	XX	X
Lactococcus	+	X			
Leuconostoc	+	X	X	X	
Listeria	+	X	X		
Microbacterium	+	X	X	X	X
Micrococcus	+	X	X	X	
Moraxella	−	XX			X
Morganella	−				X
Paenibacillus	+	X	X		
Pantoea	−	X			
Photobacterium	−				X
Proteus	−	X			
Providencia	−	X	X	X	
Pseudomonas	−	XX	X		XX
Shewanella	−	X	X	X	
Staphylococcus	+	X	X	X	X
Streptococcus	+	X	X		X
Vibrio	−	X			
Weissella	+	X	X	X	
Yersinia	−	X		X	

(Continued)

Table 6.1   Genera of microorganisms commonly found on meats, poultry, and seafood[a] *(Continued)*

| | | Occurrence in type of muscle food[b] | | | |
| | | Meat and poultry | | | |
Microorganism	Gram reaction	Fresh	Processed	Vacuum packaged	Fish
Yeasts					
Candida		XX	X		
Cryptococcus		X			
Debaryomyces		X	XX		
Hansenula		X			
Pichia		X			
Rhodotorula		X			
Saccharomyces			X		
Torulopsis		XX			
Trichosporon		X	X		
Molds					
Alternaria		X	X		
Acremonium		X			
Aspergillus		X	XX		
Aureobasidium		X			
Botrytis			X		
Cladosporium		XX	X		
Chrysosporium		X			
Fusarium		X	X		
Geotrichum		XX	X		
Monascus		X			
Monilia		X	X		
Mucor		XX	X		
Neurospora		X			
Penicillium		X	XX		
Rhizopus		XX	X		
Scopulariopsis			X		
Sporotrichum		XX			
Thamnidium		XX	X		

[a] Based on references 47, 62, 81, 135, 152, 156, 171, 239, 252.
[b] X, known to occur; XX, most frequently isolated.

products (15, 153). In general, most decontamination/sanitization technologies in the United States are applied to carcasses immediately after hide removal but before evisceration, as well as at the end of the dressing process or before carcass chilling. Recently, decontamination immediately before transfer of chilled carcasses to the cutting room for boning has been advocated. Proposed or applied interventions for reduction of contamination on carcasses are based on treatments with water or steam (e.g., hot water, steam pasteurization, steam-vacuum) at various temperatures and pressures, and chemical solutions, mostly organic acids (e.g., lactic, acetic, and citric), as well as chlorine or chlorine dioxide, trisodium phosphate, acidified sodium chlorite, peroxyacetic acid, cetylpyridinium chloride, hydrogen peroxide, ozone, and

protein compounds such as lactoferrin (15, 126, 153, 234–237). The effectiveness of these treatments in reducing microbial contamination is affected by a number of factors, including water pressure, temperature, chemicals used and their concentration, duration of exposure (which depends on speed of slaughter and length of the application chamber), method of application, and time or stage of application during slaughter and processing (234, 235, 237). Decontamination procedures also have been used for fish (128, 247). In the United States, decontamination systems are approved by the Food Safety and Inspection Service of the U.S. Department of Agriculture (USDA) if the agents used (i) are "generally recognized as safe," (ii) do not create an "adulterant" situation, (iii) do not create labeling issues (i.e., added ingredients),

Table 6.2 Sources of microbial contamination of muscle foods[a]

Source of contamination	Source of contamination of:			
	Red meat	Poultry	Finfish	Shellfish
Hide	Y			
Hair	Y			
Skin, feathers, and feet		Y		
Hooves	Y			
Method of harvesting			Y	Y
Fishing vessel			Y	Y
Ice or seawater			Y	Y
Nets and rough handling			Y	
Marine environment water and bottom sediment			Y	Y
Gastrointestinal tract	Y	Y	Y	
Hides (direct contact or aerosols)	Y	Y		
Litter and feces	Y	Y		
Scalding	Y	Y		
Dehairing machines	Y			
Defeathering equipment		Y		
Singeing	Y			
Evisceration (viscera)	Y	Y	Y	
Processing environment for muscles (floors, walls, contact surfaces, knives, workers' hands)	Y	Y	Y	Y
Fabrication of muscles (surfaces, hands, grinding equipment, trimming, etc.)	Y	Y	Y	Y
Further processing (ingredients, spices, salt, etc.)	Y	Y	Y	Y

[a] Based on references 41, 101, 114, 128, 156, and 238. Y, yes.

and (iv) can be supported with scientific studies as being effective (234–236).

Despite the generally accepted effectiveness of decontamination technologies in reducing numbers and prevalence of pathogenic and/or spoilage bacteria on meat and poultry carcasses, there are a number of concerns associated with their use. These are (i) potential spreading and redistribution of bacteria over the carcass or penetration into the tissue (237) and (ii) potential development of stress-resistant pathogenic bacteria (219–222, 235, 237, 270–273). The concern for development of stress-resistant pathogens can be attributed to the stress hardening phenomenon, which refers to the increased tolerance of a pathogen to a specific stress after adaptation through continuous exposure to the same or a different sublethal stressing environment (168, 219, 220, 235). It should be noted, however, that no published studies have examined such concerns in spoilage bacteria. Nevertheless, decontamination interventions should be evaluated not only for contributions to food safety improvement but also for potential development of stress-resistant pathogens and potential consequences of the modification of the natural microbial spoilage association after decontamination. Studies should also examine the consequences of

the imposed stress on characteristics of surviving spoilage microbes and the resulting modified competition patterns among survivors during product storage. Proposed strategies to control stress resistance of bacteria involve the continued application of lethal concentrations of antimicrobials or optimization of decontamination interventions, in type, intensity, and sequence, to maximize microbial destruction and minimize resistance development (220, 228, 235).

Development of the Microbial Association

Although muscle foods may be contaminated with a wide range of microbes (Table 6.3), their spoilage in developed countries is caused by relatively few of these microorganisms that become dominant through selection during storage and develop a microbial association. It is evident from Table 6.4 that cold storage and the gaseous composition surrounding the tissue determine the composition of the dominating microflora. As noted above, selective factors favor the growth of particular microorganisms and, as a consequence, a characteristic and specific microbial association develops and is present at the time of spoilage, leading to its characteristic spoilage features. For example, with the advent of supermarkets in the late 1950s,

Table 6.3 Bacteria associated with raw chilled muscle products[a]

Gram-negative bacteria	Gram-positive bacteria
Aerobes	Catalase reaction weak
Neisseriaceae	*Brochothrix thermosphacta*
Psychrobacter immobilis	*Kurthia zopfii*
Psychrobacter phenylpyruvica	*Staphylococcus* spp.
Acinetobacter spp.	*Clostridium estertheticum*
A. lwoffii	*Clostridium frigidicarnis*
A. johnsonii	*Clostridium casigenes*
	Clostridium algidixylanolyticum sp. nov.
Pseudomonadaceae	
Pseudomonas rRNA homology	Catalase reaction negative
Group 1 (*Pseudomonas fluorescens*)	*Lactobacillus* spp.
Biovars I, II, III, IV, V (includes 7 clusters)	*L. sakei*
P. lundensis	*L. curvatus*
P. fragi	*L. bavaricus*
P. putida	*Carnobacterium* spp.
	C. divergens
Facultative anaerobes	*C. piscicola*
Shewanella putrefaciens	*Leuconostoc* spp.
S. baltica	*L. carnosum*
S. oneidensis	*L. gelidum*
Photobacterium phosphoreum	*L. amelibiosum*
Enterobacteriaceae	*L. mesenteroides* subsp. *mesenteroides*
Serratia spp.	*Pediococcus* spp.
S. marcescens	*Weissella hellenica*
S. liquefaciens	*Lactococcus raffinolactis*
Citrobacter spp.	
C. freundii	
C. koseri	
Providencia aerogenes	
Enterobacter spp.	
E. aerogenes	
E. cloacae	
E. agglomerans	
Hafnia alvei	
Kluyvera spp.	
Morganella morganii	
Pantoea agglomerans	
Raoultella planticola (Klebsiella	
pneumoniae)	
Vibrionaceae	

[a] Based on references 8, 28, 29, 81, 90, 124, 204, 239, and 252.

storage of meat aerobically at cold temperatures and in high relative-humidity environments became a major selective factor for *Pseudomonas* to dominate as the main spoilage genus. Gram-positive bacteria (e.g., lactic acid bacteria and *Brochothrix thermosphacta*) and gram-negative bacteria (e.g., *Photobacterium phosphoreum* and *S. putrefaciens*) did so in chilled fish and meat stored in modified-atmosphere packaging (MAP). A schematic presentation of microbial associations is given in Fig. 6.1.

Studies focused on the contribution of yeasts to the spoilage of muscle foods have attracted little attention even though yeasts may be common contaminants. Yeasts do not outgrow bacteria on muscle tissue and muscle products unless a bacteriostatic agent is included in specific products (e.g., British fresh sausages) or the product is stored for extended periods in cold and dry environments. Acidic decontamination of meat may lead to changes in the dominating microbial association and

Table 6.4 Specific spoilage microflora dominating on fresh meat stored at 0 to 4°C under different gas atmospheres[a]

Gas composition	Organisms on indicated type of muscle food	
	Meat and poultry	Fish
Air	*Pseudomonas* spp.	*Shewanella putrefaciens, Pseudomonas* spp.
>50% CO_2 with O_2	*Brochothrix thermosphacta*	*B. thermosphacta, S. putrefaciens*
50% CO_2	*Enterobacteriaceae*, lactic acid bacteria	*Photobacterium phosphoreum*, lactic acid bacteria
50% CO_2 with O_2	*B. thermosphacta*, lactic acid bacteria	*P. phosphoreum*, lactic acid bacteria, *B. thermosphacta*
100% CO_2	Lactic acid bacteria	Lactic acid bacteria
Vacuum packaged	*Pseudomonas* spp., *B. thermosphacta*	*Pseudomonas* spp.

[a] Based on references 113, 114, 124, 129, 197, 202, and 239.

if combined with long-term storage may lead to development of yeasts (153, 219, 220, 235).

Bacterial Attachment

As mentioned above, edible muscle tissues of healthy animals are generally sterile prior to processing, and contaminating microorganisms are usually found only on the surfaces of whole muscle. Therefore, spoilage of meat, poultry, and seafood is generally a consequence of microbial growth, which follows attachment of microbial cells and colonization of muscle surfaces (9, 69, 87). An exception is with certain live shellfish that are consumed with their intestinal tracts intact; in this case some contaminating microorganisms are encased within the edible tissue.

Two stages are involved in the attachment of bacteria on biotic (e.g., muscle tissue) or abiotic (e.g., processing or fabrication equipment) surfaces. The first is a loose, reversible sorption that may be due to van der Waals forces or other physicochemical factors (9, 84, 87). One of the factors that influence this type of attachment is the number of bacteria in the water (film) phase (38, 87, 265). The second stage consists of an irreversible attachment of cells to surfaces, or biofilm formation, involving the production of an extracellular polysaccharide layer—a polymer matrix known as a glycocalyx (45, 67). Factors such as surface characteristics, properties of the substratum, physiological stage and cell surface characteristics, and cell motility may also influence bacterial attachment to surfaces (49, 64, 68, 93). Recent studies have shown that for a given microorganism in a defined model system, the flow rate of the growth medium or the carbon source provided can drastically alter the structure and function of communities that form colonies or biofilms (240, 241).

The attachment of bacteria on biotic or abiotic surfaces to form colonies or biofilms, although mainly unpredictable (26), may be influenced by (i) interventions applied during processing, (ii) the adaptation properties of the individual bacterial cells, and (iii) the ability of other bacteria present to attach or influence attachment to surfaces (96, 156, 175, 250). On the other hand, Farber and Idziak (84) concluded that minimal competition occurs among meat spoilage bacteria during attachment to beef longissimus dorsi muscle. Chung et al. (38) observed an absence of significant competition in the attachment of spoilage and pathogenic bacteria to fat and lean beef surfaces. Differences in the rates of attachment of certain bacterial strains may influence the composition of the initial microbial flora; for example, *Pseudomonas* has been reported to attach more rapidly to meat surfaces than do various other types of spoilage bacteria (32, 88). Pre-evisceration spray-washing of carcasses in the United States is commonly applied immediately after hide removal with the objective of removing initial microbial contamination before strong attachment on the carcass surface (235, 237).

Ephemeral and Specific Spoilage Association

Studies have established that spoilage is caused only by an ephemeral fraction of the initial microbial association (127, 202). This concept has contributed significantly to our understanding of meat and seafood spoilage. The microbial associations developing on muscle tissues stored aerobically at cold temperatures are characterized by an oxidative metabolism. The gram-negative bacteria that spoil meat are either aerobes or facultative anaerobes. *Pseudomonas* (*Pseudomonas fragi, P. fluorescens, P. putida*, and *P. lundensis*), *Shewanella putrefaciens*, and *Photobacterium phosphoreum* were found to be the dominant species on muscle tissue (beef, lamb, pork, poultry, and fish) stored at cold temperatures (113, 114, 239) (Table 6.4). *Brochothrix thermosphacta* and cold-tolerant *Enterobacteriaceae* (e.g., *Hafnia alvei*,

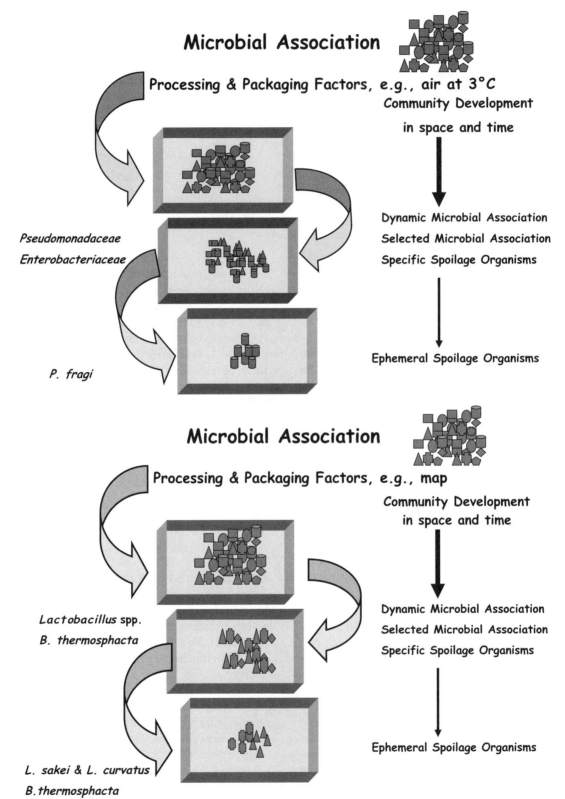

Figure 6.1 Examples of development of microbial association in muscle foods.

Serratia liquefaciens, and *Enterobacter agglomerans*) also occur on chilled muscle foods stored aerobically (18, 73, 113, 202), but in terms of population numbers they do not contribute to the microbial associations. Lactic acid bacteria, although they have been detected in the aerobic spoilage flora of chilled meat, are not considered to be important in spoilage except possibly for lamb (70, 124). Both lactic acid bacteria and *B. thermosphacta* are the main, if not the most important, causes of spoilage characterized by muscle souring. The other distinct type of muscle spoilage is characterized by putrefaction and is related to proteolytic activity and off odor production by gram-negative bacteria that dominate under aerobic conditions (202). Other common spoilage conditions and causative bacteria are listed in Table 6.5.

In general, the metabolic activity of the ephemeral microbial association, which prevails in a muscle ecosystem under certain aerobic conditions, or is generally introduced during processing, leads to the manifestation of changes that are characterized as spoilage of meat.

Table 6.5 Common defects in meat products and causal bacteria[a]

Defect	Meat product(s)	Organism(s)
Slime	Meats	*Pseudomonas, Lactobacillus, Enterococcus, Weissella, Brochothrix*
H_2O_2 greening	Meats	*Weissella, Leuconostoc, Enterococcus, Lactobacillus*
H_2S greening	Vacuum-packaged meats	*Shewanella*
H_2S production	Cured meats	*Vibrio, Enterobacteriaceae*
Sulfide odor	Vacuum-packaged meats	*Clostridium, Hafnia*
Cabbage odor	Bacon	*Providencia*
Potato odor	Ham	*Burkholderia, Pseudomonas*
Putrefaction	Ham	*Enterobacteriaceae, Proteus*
Bone taint	Whole meats	*Clostridium, Enterococcus*
Bone taint	Bacon	*Proteus, Vibrio*
Pocket taint	Bacon	*Vibrio, Alcaligenes, Proteus*
Internal taint	Ham	*Providencia*
Souring	Ham	Lactic acid bacteria, *Enterococcus, Micrococcus, Bacillus, Clostridium*

[a] From reference 171.

These undesirable spoilage changes are related to the type, composition, and population of the microbial association and the type and availability of energy substrates in meat. Indeed, the type and the extent of spoilage are governed by the availability of low-molecular-weight compounds (e.g., glucose and lactate) existing in meat (202); eventual muscle food changes and subsequent overt spoilage are due to catabolism of nitrogenous compounds as well as secondary metabolic reactions.

Interactions among Spoilage Bacteria

The selection of the microbial association and subsequent chemical changes that occur during muscle spoilage depend on selected and applied environmental conditions, and on microbial interactions (202, 254, 267). Although this concept has only been partly exploited in muscle food microbiology, it may be important in understanding spoilage, as it was found that there are interactions among spoilage bacteria (Tables 6.3, 6.4, and 6.5). Indeed, studies have shown that competition for nutrients (e.g., glucose and protein nitrogen) or chemical elements (e.g., iron), metabiosis (i.e., production of a favorable environment), and cell-to-cell communication (i.e., quorum sensing) can also affect the physiological attributes of microorganisms under the imposed ecological determinants (71, 73, 112, 161, 198, 204, 254, 267). This is, for example, the case with *Pseudomonas* spp. that can either inhibit or enhance the growth of *S. putrefaciens* or *Listeria monocytogenes*, respectively. This is achievable due to the ability of *Pseudomonas* to utilize glucose and/or to produce siderophores at higher rates than *S. putrefaciens*, or to provide compounds (protein hydrolysates) in the case of *L. monocytogenes* (170, 257). Furthermore, Worm et al. (267) concluded that in proteolytic communities, data on growth and nitrogen availability showed that protein hydrolysates were available to both the proteolytic and nonproteolytic strains. This interaction can be the major factor governing the development of a spoilage microflora. So far, studies with single-culture or coculture model systems (254, 267) have been helpful in simplifying the natural food ecosystem. Indeed, this approach (single-culture or coculture model systems) permits a better understanding of the mechanisms through which development of potential ephemeral spoilage organisms is affected by possible interactive behaviors, and may help identify the responsible metabolite(s) which may be further used as a unique chemical spoilage index. However, it should be noted that chemical changes (e.g., production of off odors) occurring in naturally contaminated fish (150) and meat (Table 6.5) differ significantly from those on sterile muscle tissue inoculated with ephemeral spoilage organisms (150, 255). As a consequence, metabolic interactions among spoilage bacteria and other members of muscle bacterial communities should be studied in situ in order to have a complete "picture" of muscle spoilage.

Bacterial Interactions with Nutrients

The contribution of specific nutrients to bacterial behavior has been shown in single-culture or coculture studies. The cardinal compound, glucose, was found to be metabolized more rapidly by the obligate aerobic strains of pseudomonads than by the facultative anaerobic strains of *B. thermosphacta* and oxidative (mostly gram-negative) strains of *S. putrefaciens* (254). The coculture of pseudomonads with either *Shewanella* or *B. thermosphacta* did not affect growth rates, although an acceleration of glucose consumption was evident. It was concluded that pseudomonads could play a syntrophic role for the former (254). This observation is of great importance since *B. thermosphacta* has a much greater spoilage potential than lactobacilli and can be important in both aerobic and anaerobic spoilage of muscle foods. This bacterium utilizes glucose and glutamate but not other amino acids during aerobic incubation (Table 6.6).

Worm et al. (267) reported that the mobilization of protein for growth was severely depressed when proteolytic

bacteria constituted a minor part of the model community. This conclusion agrees well with findings for lactic acid bacteria growing in milk, where a significant reduction in growth and in milk acidification occurred when the proteolytic strain constituted less than 10 to 50% of a batch culture inoculum (140, 266). The competition had drastic consequences for the proteolytic activity of the model community, as the proteolytic *P. fluorescens* strain ON2 was a better competitor for protein hydrolysates than was the nonproteolytic strain DF57. It appeared that *P. fluorescens* ON2 was able to utilize hydrolysates derived from a low level of proteinase activity, whereas nitrogen limitation (and growth arrest) of *P. fluorescens* DF57 was only relieved once higher extracellular proteinase activity had developed. These observations substantiate the complex relation between measured enzymatic activity (e.g., proteinase) and the formation as well as the rate consumption of hydrolysates (263). It was concluded that proteinase activity easily becomes the rate-limiting step for bacterial growth on protein substrates when nonproteolytic bacteria are dominating the system. The proteolytic capacity of a bacterial community is then tightly linked to the abundance and competitive success of the proteinase-producing strain(s) (263, 267).

Table 6.6 Order of substrate utilization during growth of major muscle spoilage bacteria[a]

Substrate	Aerobic					Anaerobic[b]				
	A	B	C	D	E	A	B	C	D	E
Glucose	1	1	1	1	1	1	1	1	1	1
Glucose-6-phosphate	2			2					2	
Lactate	3	2		3						
Pyruvate	4	3				2[c]				
Gluconate	5	4				2[c]				
Gluconate 6-phosphate	6									
Proprionate		5								
Formate							1			
Ethanol		6								
Acetate		7				2				
Amino acids	7	8	2	4		2	1[c]		3	2
Serine and cysteine							1[c]			
Creatine	8									
Creatinine	9									
Citrate	10									
Aspartate	11									
Glutamate	12					2[c]				
Ribose			3							
Glycerol			4							
Lipids										

[a] Modified from references 80, 202, and 239. In the order, 1 is first and 12 is last. A, *Pseudomonas* spp.; B, *Shewanella putrefaciens*; C, *Brochothrix thermosphacta*; D, *Enterobacter* spp.; E, lactic acid bacteria.
[b] Under oxygen limitation and/or CO_2 inhibition.
[c] No specific order is given.

Components of Muscle Tissue Supporting Microbial Growth

Status of Energy Sources

Detailed reviews of the composition and spoilage processes of postmortem muscle tissues have been published for red meat, poultry, and fish (31, 98, 99, 131, 132, 135, 158, 164, 194, 195, 231). It is well established (106) that water, proteins, and fat are the main constituents of muscles (Table 6.7), while minor components include carbohydrates such as glycogen and intermediate glycolytic products (i.e., glucose, glucose-6-phosphate, and lactate) (Table 6.8). Cessation of muscle cell respiration brings ATP synthesis to an end, and glycolysis leads to accumulation of lactic acid and a concomitant decrease in pH. The final postmortem pH and the residual glycogen content of tissues are influenced by the initial amounts of glycogen in the muscle. In tissue with a high initial glycogen content (e.g., red muscle and breast muscle), the pH may decrease to 5.5 to 5.9 before enzymatic activity associated with glycolysis ends due to an inability to maintain sufficient concentrations of ATP. When small amounts of glycogen (e.g., in pork muscle, leg muscle, and finfish) are initially present, there is a direct correlation between glycogen content and final pH (6.0 to 6.7) (31, 132, 176, 177, 191); however, tissue with a lower ultimate pH always contains some residual glycogen when glycolytic activity ceases, since ATP concentration is the limiting factor. In addition to glycogen, glycolytic intermediates such as glucose-6-phosphate and glucose are reduced to low levels following rigor (106) (Table 6.9). The decrease in muscle pH and accumulation of various metabolites following rigor facilitate denaturation of some proteins. The release of proteolytic enzymes, such as cathepsins from lysosomes, results in a small amount of protein breakdown (106, 164), yielding soluble low-molecular-weight compounds constituting 1.2 to 3.5% of muscle tissue.

Table 6.7 Proximate composition of lean muscle tissue[a]

Muscle food	Water (%)	Protein (%)	Lipid (%)	Ash (%)	Energy (kJ/100 g)
Beef	71–73	20–22	5–8	0.7	168
Pork	71	22	6	0.8	ND
Shellfish	73	20	5.2	ND[b]	127
Lamb	74	20	5	0.6	ND
Chicken	76	21	3	1	ND
Cod	81	18	0.7	1	96
Mackerel	64	19	14	1	199

[a] Modified from references 128 and 160.
[b] ND, not determined.

Table 6.8 Average chemical composition of typical adult mammalian muscle and fish after rigor mortis and before commencement of decomposition postmortem[a]

Component	Wet wt (%)	
	Meat	Fish
Water	75.0	70.0
Protein	19.0	19.0
Myofibrillar	11.5	15.0
Sarcoplasmic	5.5	3.5
Connective tissue and organelle	2.0	0.5
Lipid	2.5	1.0
Carbohydrate and lactic acid	1.2	0.8
Lactic acid	0.90	0.45
Glucose-6-phosphate	0.15	NAD[b]
Glycogen	0.10	NAD
Glucose, traces of other glycolytic intermediates	0.05	NAD
Miscellaneous soluble nonprotein substances	2.3	NAD
Nitrogenous	1.65	
Creatine	0.55	
Inosine	0.30	
Monophosphate ATP, AMP	0.10	
Amino acids	0.35	
Carnosine, anserine	0.35	
Inorganic	0.65	
Vitamins	Traces	0.2

[a] Adapted from references 132 and 164.
[b] NAD, no available data.

The composition of raw poultry muscle tissue varies with age, sex, anatomy, and species (31). For example, concentrations of glucose and lactate in breast muscle have been reported to be 98 and 670 mg/100 g, respectively, while corresponding amounts in thigh muscle were 8 and 360 mg/100 g (197). Similarly, the composition of fish muscle may fluctuate widely depending on size, season, fishing grounds, and diet in the case of aquacultured species (231). Fat in poultry is not distributed throughout the muscle tissue as in red meat and fish, but is mostly present beneath the skin of the abdominal cavity. According to Shewan (231), non-protein-soluble components constitute approximately 1.5% of fish muscle and their composition and concentration vary with species, while within species these components may vary with size, season, and fishing ground. Such components consist of sugars, minerals, vitamins, and nonprotein nitrogen compounds

Table 6.9 Main low-molecular-weight components of beef and fish pre- and post-rigor mortis[a]

Component	Concn (mg/100g)			
	Meat		Finfish	
	Pre	Post	Pre	Post
Creatine phosphate	300	NF[b]	9.3[e]	0.2[e]
Creatine	450	650	ND	ND
Betaine	NF	ND[c]	NAD	100[f]
ATP	300	NAD[d]	6.5[e]	0.2[e]
IMP	20	300[e]	ND	ND
Glycogen	1,000	100	220	40
Glucose	500	100	220	40
Glucose-6-phosphate	100	200	21	32
Lactic acid	100	900	100	400
pH	7.2	5.5	7.3	6.5
Free amino acids	200	350	NAD	250
Trimethylamine oxide	ND	ND	ND	350–1,000
Carnosine and anserine	300	300	ND	100[g]

[a] From references 128, 135, and 202.
[b] NF, not found.
[c] ND, not determined.
[d] NAD, no available data.
[e] In micromoles per gram.
[f] In some fish.
[g] Cod.

such as free amino acids, ammonia, trimethylamine oxide, creatine, taurine, anserine, uric acid, betaine, carnosine, and histamine (30, 135, 231). Elasmobranchs (e.g., sharks and rays) contain about twice as high a concentration of soluble components as do other fish (128).

Surprisingly, despite the large amount of proteins and lipids reported for muscles, the amounts of other compounds (e.g., low-molecular-weight compounds shown in Tables 6.8 and 6.9) are also sufficient to support massive development of a diverse microcosm on meat (100). Indeed, several studies have shown that bacteria grow on red meat, fish, poultry, and muscle at the expense of one or more of the low-molecular-weight soluble components (150, 197). Changes that occur during spoilage take place in the aqueous phase of meat (203), which contains glucose, lactic acid, amino acids, nucleotides, urea, and water-soluble proteins. These substrates are catabolized by almost all bacteria in muscle food microflora (71, 72, 97, 201).

Glucose: the Key Energy Source
Glucose is the preferred energy substrate and the first to be used by various microorganisms growing on muscle foods. The order in which various substrates are attacked by the main spoilage bacteria (e.g., pseudomonads, *Enterobacteriaceae*, *B. thermosphacta*, and lactic acid bacteria) under aerobic and anaerobic conditions is shown in Table 6.6. Under aerobic conditions none of the

bacteria is known to cease growth because of substrate exhaustion at the muscle surface; oxygen availability, however, has been suggested to be a limiting factor. The rate of growth and the type of metabolic activity of *S. putrefaciens*, *B. thermosphacta*, and *Pseudomonas* spp., when cultured individually or in all possible combinations in gel cassette systems supplemented with 0.1% glucose at 5°C, were influenced by and depended on the combination of the tested bacteria (150, 254). The pseudomonads predominate because of their higher growth rates and greater affinity for oxygen, and as a consequence greater catabolism of glucose, over the other muscle spoilage bacteria (102, 103). For example, *Pseudomonas* spp. showed a greater rate of consumption of glucose and lactate than did *Lactobacillus* spp. in beef samples stored unpackaged or packaged under vacuum in highly permeable oxygen film (255). Other studies have shown that the transport capacity and oxidation rate of glucose in microbial cells, as well as microbial antagonism, could contribute also to the dominance of *Pseudomonas* in muscle foods stored under aerobic conditions (182, 202, 254).

Lactate
Lactate is almost exclusively the second energy source utilized by the microbial association of muscle under both aerobic and anaerobic conditions (71, 72, 103, 182) (Table 6.6). However, the preferential use of glucose over lactate and amino acids has attracted considerable attention (71,

83, 97, 161, 163, 197). Available data suggest a sequential use of glucose followed by lactate. Molin (182), however, observed that lactate is used by *P. fragi* in broth culture in the presence of glucose under both aerobic and oxygen-limiting conditions. Similar results were reported with various muscles or models mimicking muscle foods (e.g., gel cassette model, beef strip loins, poultry, and fish) naturally contaminated or inoculated with *Lactobacillus* spp. and stored aerobically or under oxygen limitation or anaerobic conditions (17, 57, 71, 142, 200, 203, 254, 255). The general conclusion drawn from these studies was that the lactate concentration decreased following glucose utilization in muscle samples stored aerobically, in vacuum packages, or under other MAP conditions. It was evident also that the rate of glucose and lactate utilization in vacuum- or MAP-stored samples was less than that in samples stored aerobically.

Amino Acids

In principle, amino acids are the third main energy pool for bacteria in muscle foods. Many workers have relied on changes in free amino acids in meat in their attempts to determine if protein degradation had occurred (61, 176, 177). One would expect that the amino acid content would remain constant until shortly before the onset of spoilage due to glucose exhaustion (52, 100) or until bacterial numbers reached 7 to 8 log CFU/g (53). Subsequent to that, amino acids would decline before rising again sharply when proteolysis begins. However, this scenario is not in agreement with the results reported by Newton and Rigg (192). Storage of beef (199), pork (225, 226), poultry skin and fillets (205), and fish (150) under aerobic conditions has revealed that the sum of the free amino acid and water-soluble protein content increases during storage, which is associated with increases in colony counts. Nychas and Arkoudelos (199) and Nychas and Tassou (205) showed that this increase occurred in meat samples with relatively high concentrations of glucose. Moreover, the rate of free amino acid increase under aerobic conditions was greater than that occurring under MAP conditions. These observations could be of commercial importance since spoilage is most frequently associated only with post-glucose utilization of amino acids by pseudomonads (100).

CHEMICAL CHANGES IN THE MUSCLE ECOSYSTEM

Substrate Conversion to Spoilage Compounds

It is established that the end products of microbial metabolism in muscle tissue are associated with spoilage defects such as off flavors, foul odors, slime, and changes in color (Table 6.5). The production of metabolic compounds that can eventually spoil muscle tissues is the outcome of substrate interactions with the developed microbial association and with the indigenous muscle enzymes as well as of nonenzymatic chemical reactions and physical changes. Since microbial association is dependent on the imposed environmental conditions (e.g., packaging) (Table 6.10) and on microbial competition, the observed chemical changes may be considered an expression of the muscle ecosystem in which two extreme situations are possible. The first scenario that could lead to changes in the ecosystem is related to competition among facultatively anaerobic gram-positive bacteria, while the other would be competition among members of the aerobic gram-negative microflora. These two possibilities as well as the effect of indigenous enzymes and chemical/physical changes are discussed below.

Indigenous Enzymes Versus Microbial Activity

Glycolytic enzymes indigenous to muscle tissues participate in the postmortem glycolysis that ceases when the ultimate postmortem pH reaches 5.4 to 5.5. Church et al. (39) reported that endogenous meat enzymes may be responsible, at least in part, for D-lactic acid production, since similar concentrations were found in slices of meat taken at progressively increasing depths from the meat surface. If D-lactic acid production had been due solely to microbial metabolism, higher levels would have been expected at the surface, with increasingly lower levels at increasing depth from the meat surface. However, Meyns et al. (181) found levels of lactic acid of 10 mg/100 g at a depth of 1 cm and 40 mg/100 g at 5 cm in beef stored in vacuum packages, providing further support for the hypothesis that endogenous meat enzymes produce D-lactic acid during the early stages of storage. The concentration of D-lactic acid at a given storage time has also been found to be independent of the packaging atmosphere and of differences in lactic acid bacteria. This supports the concept that D-lactic acid production is due mainly to meat (muscle) metabolism. The absence of D-lactic acid in freshly slaughtered meat of different species and its gradual increase during storage have also been reported (17, 72, 161, 203). Others have shown that when poultry or beef meat was inoculated with homofermentative or heterofermentative lactic acid bacteria in vacuum packages or in 100% CO_2 and then stored at 10°C, the concentration of D-lactic acid was always higher in inoculated samples than in uninoculated product (17, 203). Thus, although present, the contribution of muscle enzymes to spoilage is minor compared to the enzymatic action of microbial flora (205, 255). Further

Table 6.10 Distribution of microflora of beef knuckles vacuum packaged and stored for 21 days at 0 to 2°C in packages with oxygen transmission rates ranging from 1 to 400 cc/m^2/24 h[a]

Microbe(s)	Distribution (%) at different O$_2$ transmission rates					
	1 cc	10 cc	12 cc	13 cc	30 cc	400 cc[b]
Micrococcus					2.6	
Lactobacillus spp.						
L. coryniformis				6.1		
L. plantarum	11.9		5.6	15.1	14.0	
L. cellobiosus	45.7	57.0	65.7	5.5	6.6	3.9
All *Lactobacillus* spp.	57.6	57.0	71.3	26.7	20.6	3.9
Leuconostoc spp.						
L. mesenteroides	1.1	38.3	15.0	39.8	51.3	12.4
L. paramesenteroides	35.9		12.5	28.5	6.7	
All *Leuconostoc* spp.	37.0	38.3	27.5	68.3	58.0	12.4
Streptococcus		2.3				
Brochothrix thermosphacta			1.0	0.1	4.6	3.8
Coryneform bacteria	1.1			1.5		
Staphylococcus				1.8		
Moraxella-Acinetobacter		0.3	0.2	0.1		
Flavobacterium	2.2					
Pseudomonas		1.6		1.5	14.2	79.9
Erwinia herbicola	1.1					
Aeromonas	1.0	0.3				
All gram-negative rods	4.3	2.2	0.2	1.6	14.2	79.9

[a] From references 133 and 224. Values are averages of three knuckles.
[b] Values are for microflora at 14 days; storage of knuckles packaged and stored in this film was not extended beyond 14 days.

investigation using surface-sterilized meat inoculated with or without lactic acid bacteria should clarify the contribution of this group to D-lactic acid formation.

The activity of indigenous proteolytic and lipolytic enzymes may not be adequate to affect meat spoilage. Indeed, this was evident from the results of studies where the hypothesis that a similar pattern of protein breakdown could be expected in poultry naturally contaminated or inoculated with pseudomonads when stored under different temperature and packaging conditions. This hypothesis was not confirmed (205). It is well established that microbiological activity is by far the most important factor influencing changes that cause proteolytic spoilage in muscle foods (150, 202). However, a clear exception to this rule has been observed in crustaceans, where the endogenous tissue enzymes in the hepatopancreas cause rapid postmortem muscle breakdown that is independent of microbial proteases (46). This is the primary reason that lobsters, crabs, and crayfish are kept alive after harvesting. Death of these animals after harvesting results in rapid liquefaction of edible tissues.

The need for a critical microbial population density of 7.8 log CFU per cm^2 or g (97, 100) for proteolysis to become sensorially evident (e.g., slime formation and sulfur or ammonia odors) in various muscles has been disputed (151, 205, 243). The perception that this population can be used as an indication of bacterial proteolytic spoilage in muscle foods should have only a utilitarian role rather than a fundamental significance. An example where high microbial populations do not equate to spoilage is the case of bilgewater fish, in which recently harvested fish come into contact with heavily contaminated holding tank water. In this case, the fish are inoculated with high numbers of bacteria that can potentially cause spoilage but the edible flesh is very fresh. According to Sutherland (243), microbial numbers per se are not always "a guide to the extent of spoilage," but microbial activity is extremely important (e.g., bacteria producing proteolytic and lipolytic enzymes). This is in agreement with other studies, in which proteolysis was evident even during very early stages of storage, regardless of populations in the microbial association and the presence of low-molecular-weight compounds (e.g., glucose). The proteolytic activity that occurs in muscle tissue under these conditions is exclusively attributed to microbial activity (23–25, 64, 205, 225, 226). Both pseudomonads and lactic acid bacteria have been found to contribute to muscle proteolysis (205). Proteolytic activity can lead to penetration of pseudomonads into meat (107, 119), gaining an ecological advantage through penetration because they then have access to a new niche with available nutrient resources for exploitation, which would not be accessible or available to nonproteolytic or less proteolytic bacteria. Soluble sarcoplasmic proteins have been reported to probably be the initial substrate for proteolytic attack in muscle foods (121, 122, 136).

While many spoilage microorganisms are able to produce lipases, limitations in existing methodologies have led to conclusions of limited or inhibited production of lipases in the presence of carbohydrates, lipids, and proteins in the substrate (1, 2, 12, 89, 158). On the other hand, oxidative rancidity of muscle fat occurs when oxygen reacts with unsaturated fatty acids during storage and stable compounds such as aldehydes, ketones, and short-chain fatty acids are produced, resulting in the eventual development of rancid off flavors and odors (98). Autoxidation, independent of microbial activity, occurs in muscle foods stored in aerobic environments, and its rate is influenced by the presence of unsaturated fatty acids in the fat (89). Autoxidation of fat is of particular importance in the deterioration of fatty fish and pork, which contain highly unsaturated lipids (158). The phospholipid component of muscle tissue membranes is also rich in unsaturated fatty acids that are susceptible to oxidation.

Concluding, it is the total microbial activity or growth per se, rather than the activity of microbial enzymes and accumulation of metabolic by-products, that characterizes the various types of muscle food spoilage (23–25, 151). Thus, it is important that the chemistry of spoilage is discussed in the context of two distinct storage scenarios (aerobic versus anaerobic), and such a discussion follows.

Chemical Changes under Aerobic Conditions

Chemical Changes by Gram-Negative Bacteria

Pseudomonads

The key chemical changes associated with the metabolic attributes of pseudomonads have been studied extensively in broth and in model systems such as meat extracts and gel cassettes (71, 97, 182, 254). The rationale for these studies is based on the importance of the organism in the spoilage of muscle tissues stored under cold and aerobic conditions. The metabolic idiosyncrasies of pseudomonads are summarized here. D-Glucose and L- and D-lactic acid have been found to be metabolized sequentially by pseudomonads (71), while D-glucose is preferred over DL-lactate (Table 6.6). This metabolic sequence observed in the muscle ecosystems is well documented (75, 150, 201). The extracellular oxidation of glucose and glucose-6-phosphate causes a transient accumulation of D-gluconate and pyruvate and an increase in the concentration of 6-phosphogluconate (78, 149). These important observations (201) led to a proposed method for control of microbial activity in muscle foods by addition of glucose and its transformation to gluconate (100, 161, 230). This is based on (i) a decrease in pH by the accumulation of oxidative products, (ii) a transient pool of gluconate and the inability of other organisms of the association to catabolize it, and (iii) postponement of catabolism of creatine, creatinine, and other amino acids, which would have as a consequence an increase of pH due to release of ammonia (201).

Other volatile compounds found in spoiled muscle foods are listed in Table 6.10. Indeed, odors of by-products such as sulfides and methyl esters are usually the first manifestation of spoilage of chilled meat, poultry, and seafoods stored under aerobic conditions (53–55, 59, 61, 77); the compounds involved are mostly products of amino acids. It was mentioned previously that meat is usually considered sensorially spoiled when bacterial populations exceed 8 log CFU/cm^2 or the glucose/gluconate concentration at the meat surface is reduced to undetectable levels (97, 201). The timescale for production of these odors is also consistent with the established restriction of amino acid metabolism until after the depletion of glucose/gluconate or lactate on the meat surface (70, 97, 98). Pseudomonads, particularly *P. fragi*, are the major and possibly the sole producer of ethyl esters in aerobically stored meat (53, 77, 176, 177). Other bacteria responsible for these volatile compounds are *S. putrefaciens*, *Proteus*, *Citrobacter*, *Hafnia*, and *Serratia* (178). In addition to the formation of various malodorous compounds, the release of large amounts of ammonia also contributes to the development of spoilage odors (50, 60, 61). Schmitt and Schmidt-Lorenz (225, 226) found that the concentration of ammonia increased in aerobically stored samples of broiler skin whose microflora was dominated by pseudomonads. About half of the pseudomonads and *Enterobacteriaceae* produce ammonia in a medium having a chemical composition similar to that of chicken skin (176, 177). Concentrations of four of the volatile compounds, acetone, methyl ethyl ketone, dimethyl sulfide, and dimethyl disulfide (Table 6.11), increase continuously during aerobic storage of ground beef at 5, 10, or 20°C (242). Hydrogen sulfide, another potential indicator of spoilage, is not produced by pseudomonads, while dimethyl sulfide is not produced by the *Enterobacteriaceae* (53). Hydrogen sulfide and ammonia are formed as a result of the conversion of cysteine to pyruvate by the enzyme cysteine desulfhydrase (98). Hydrogen sulfide combines with the muscle pigment to give a green discoloration. Putrescine, cadaverine, histamine, tyramine, spermine, and spermidine were found to be present in ground pork, beef, poultry, and fish stored at cold temperatures (77, 81, 138, 139, 155, 165, 187, 225, 226). Cadaverine was the major biogenic amine in poultry stored either aerobically or in vacuum packages. Schmitt and Schmidt-Lorenz (225) reported that putrescine and cadaverine, which are detectable at colony counts of 5 log CFU/cm², could indicate onset of spoilage in poultry. Pure culture experiments proved that pseudomonads were the major source of putrescine, while the *Enterobacteriaceae* produced mostly cadaverine.

Several pseudomonad-like bacteria have been isolated from shrimp (6) and found to be prolific foul-odor producers. *Chryseomonas luteola*, *Serratia marcescens*, *Pseudomonas fluorescens*, and *Brevundimonas* spp. had the greatest odor-producing potential and formed trimethylamine, a volatile compound notably associated with the "fishy" odor of spoiled seafoods. The compound is produced by enzymatic reduction of trimethylamine oxide, a natural constituent of seafood muscle. These bacteria also produced an abundance of sulfur compounds that included methanethiol (garbage odor), dimethyl disulfide (onion odor), thiophene (skunky odor), and dimethyltrisulfide (cat urine odor).

Other aroma compounds produced by these bacteria included isovaleric acid (sweaty foot odor) and butyric acid (baby vomit odor). *Chryseomonas luteola* produced the highest intensity off odors of the spoilage bacteria (6). The U.S. Food and Drug Administration currently uses indole (mothball or tar odor) as an indicator of the microbial quality of shrimp. This compound is a microbial metabolite of L-tryptophan. Other useful seafood spoilage indicators include the biogenic amines histamine, putrescine, and cadaverine.

Enterobacteriaceae

If the meat ecosystem favors their growth, genera in the family *Enterobacteriaceae* may also be important in muscle food spoilage. Conditions allowing growth of *Enterobacteriaceae* include limited oxygen and low temperature (Table 6.4). They utilize mainly glucose and glucose-6-phosphate as carbon sources (Table 6.6), while exhaustion of these substrates is followed by amino acid degradation (100, 103). Thus, members of this family produce ammonia and volatile sulfides, including hydrogen sulfide and malodorous amines, from amino acid metabolism (105, 120). *Enterobacteriaceae* and *B. thermosphacta* do not produce esters in pure culture, while acids and alcohols are among their end products. The production of branch chain esters (Table 6.10) could be due to the possibility of pseudomonads catalyzing the interaction of excreted products or to the possibility that they are formed by direct chemical interaction (53, 77). Inoculation experiments with *Enterobacteriaceae* and *B. thermosphacta* showed an initial increase in the levels of acetoin and diacetyl, which are often detected at the same time as the esters (53). As pseudomonads catabolize acetoin and diacetyl, the concentrations of both diminish with time (183).

Chemical Changes by Gram-Positive Bacteria

In general, gram-positive bacteria, especially the lactic acid bacteria, are unimportant contaminants of muscle foods stored under aerobic conditions (50, 196, 199). *B. thermosphacta* may have some importance (5, 75, 245) in the spoilage of pork, lamb, and fish, particularly on fatty surfaces. Physiological attributes (e.g., the end products of metabolism) and factors influencing the role of glucose and oxygen limitation (e.g., pH and temperature) on lactic acid bacteria and *B. thermosphacta* isolated from muscle tissues have been studied in model systems (meat extracts, gel cassettes, and sterile muscle blocks) (Tables 6.12 and 6.13). *B. thermosphacta* has a much greater spoilage potential than lactobacilli and can be important in both aerobic and anaerobic

Table 6.11 Factors and precursors affecting the production of malodorous end products of gram-negative bacteria (e.g., *Pseudomonas* spp., *Shewanella putrefaciens*, and *Moraxella*) inoculated in broth, a sterile model system, and naturally spoiled muscle[a]

End product	Meat and poultry	Fish	Factor(s)	Precursor(s)
Sulfur compounds				
Sulfides	+	+	Temp and sub-	Cysteine, cystine, methionine
Dimethylsulfide	+	+	strate (glucose)	Methanethiol, methionine
Dimethyldisulfite	+	+	limitation	Methionine
Methyl mercaptan	+	+		NAD
Methanethiol	+	+		Methionine
Hydrogen sulfide	−/+[b]	+	High pH	Cystine, cysteine
Dimethyltrisulfide	+	+	NAD[d]	Methionine, methanothiol
Esters				
Methyl esters (acetate)	+	+	Glucose (l[e])	NAD
Ethyl esters (acetate)	+	+	Glucose (l)	NAD
Ketones				
Acetone	+	+	NAD	NAD
2-Butanone	+	+	NAD	NAD
Acetoin and diacetyl	+/−[c]	+	NAD	NAD
Aromatic hydrocarbons				
Diethylbenzene	+	+	NAD	NAD
Trimethylbenzene	+	+	NAD	NAD
Toluene	+	+	NAD	NAD
Aliphatic hydrocarbons				
Hexane	+	+	NAD	NAD
2,4-Dimethylhexane and methylheptone	+	+	NAD	NAD
Aldehydes				
(2-methylbutanal)	+	+	NAD	Isoleucine
Alcohols				
Methanol	+	+	NAD	NAD
Ethanol	+	+	NAD	NAD
2-Methylpropanol	+	+	NAD	Valine
2-Methylbutanol	+	+	NAD	Isoleucine
3-Methylbutanol	NAD	+	NAD	Leucine
Other compounds				
Ammonia	+	+	Glucose (l)	Amino acids
Trimethylamine	+	+	NAD	Trimethylamine oxide

[a] From references 129, 147, 150, 166, 202, and 254.
[b] Production only by *Shewanella putrefaciens*.
[c] These compounds decreased during storage.
[d] NAD, no available data.
[e] l, low concentration of glucose.

spoilage of muscle foods. This bacterium utilizes glucose and glutamate but no other amino acid during aerobic incubation (103), and it produces a mixture of end products (Table 6.13). The assimilation of glucose and production of formic and acetic acids in a model system (gel cassette) are affected by the presence of other spoilage bacteria (e.g., pseudomonads and *Shewanella* spp.) (255).

Table 6.12 Factors and precursors affecting the maximum formation of end products of homofermentative lactic acid bacteria (*Lactobacillus*, *Leuconostoc*, and *Carnobacterium*) that were inoculated in broth, a sterile model system, and naturally spoiled meat[a]

End product	Beef and poultry	Fish	Factor(s)	Precursor(s)
Aerobic storage				
L-Lactic acid	+	+	NAD	Glucose
D-Lactic acid	+	ND	NAD	Glucose
Acetic acid	+	+	Glucose (l), O_2 (h), E	Glucose, lactate, pyruvate
Acetoin and diacetyl	+	+	pH (l), glucose (h)	Pyruvate
Hydrogen peroxide	+	NAD	NAD	NAD
Formic acid	+	+	NAD	Glucose, acetic acid
Ethanol	+	+	NAD	Glucose
Various gaseous conditions				
L-Lactic acid	+	+	NAD	Glucose
D-Lactic acid	+	+	NAD	Glucose
Acetic acid	+	+	Glucose (l), O_2 (h), E	Glucose, lactate, pyruvate
Acetoin	+	NAD	pH (l)	Pyruvate
Formic acid	+	+	NAD	Glucose, acetic acid
Ethanol	+	+	NAD	NAD

[a] From references 43, 48, 202, 255, and 256. h, high oxygen; l, low concentration of glucose; E, appropriate enzymes (iLDH, NADH peroxidase, lactate, or pyruvate oxidase); NAD, no available data; ND, not detected.

Chemical Changes under Oxygen Limitation or Anoxic Conditions

Facultatively anaerobic food ecosystems differ from aerobic ecosystems in that products of fermentation (e.g., lactic and acetic acids) that accumulate and cause spoilage may also have antimicrobial activity. Chemical changes, such as an increase in concentrations of D-lactic and acetic acids, may offer a reliable tool for monitoring the quality of muscle foods (172, 202). Acetic acid, as a product of further oxidation of lactic acid, has been used for the construction of models to evaluate quality in poultry, fish, and beef (52, 112, 141, 142). Homofermentative or heterofermentative types of metabolism and their ecological determinants are of importance (17, 142). The microbial metabolites detected in naturally contaminated samples of chilled meat stored under vacuum and in MAP are shown in Table 6.12. Under these conditions, the putrid odors associated with storage in air are replaced by relatively inoffensive sour/acid odors. Such odors have been assumed to arise from the acidic end products of glucose fermentation, which is the primary source of energy for microbial growth (97, 99, 104). However, the production of such off odors is difficult to explain in terms of accumulation of acetic, isobutanoic, L-isopentanol, and D-lactic acids because the amounts are relatively small compared to the amount of endogenous L-lactic acid of normal pH muscle (51, 65). The dairy/cheesy odors found in beef stored in gas mixtures with CO_2 were produced by *B. thermosphacta* and lactic acid bacteria, both of which can produce diacetyl or acetoin and alcohols (57, 58, 255).

Chemical Changes by Gram-Negative Bacteria

Since gram-negative bacteria such as pseudomonads are inhibited in reduced-oxygen environments, the spoilage of muscle foods stored under oxygen limitation and carbon dioxide-enriched atmospheres is due to undefined actions of lactic acid bacteria and/or *B. thermosphacta*, *S. putrefaciens*, and *P. phosphoreum*, the last two in fish (62, 112, 202, 255). However, the presence of sulfur compounds, such as propyl esters and 3-methylbutanol compounds, as well as the production of formic and acetic acids (Table 6.11) lead again to the crucial question as to whether this inhibition is due to carbon dioxide enrichment or oxygen limitation. It is well known that pseudomonads are very sensitive to carbon dioxide at low storage temperatures, which increase the solubility of this gas (274). The high affinity of pseudomonads for oxygen could be the main reason for their preponderance in muscle foods. Indeed, it was reported that under oxygen-limiting conditions, although the order of substrate utilization remains the same with the exception of glucose and lactate, this group can use alternative carbon sources (71, 182, 255). The production of tyramine, putrescine, and cadaverine has also been attributed to lactic acid bacteria in meat stored under vacuum or MAP conditions (56, 165, 232).

Table 6.13 End products formed by *Brochothrix thermosphacta* in naturally spoiled meat, poultry, and fish or in model muscle systems (e.g., broth or gel cassette)[a]

End product	M/P	F	Factor(s)	Precursor(s)
Aerobically				
Acetoin	+	NA	Glucose (h), pH (h/l), T (h/l)	Glucose (MJ), alanine (MN), diacetyl
Acetic acid	+	+	Glucose (h), pH (h/l), T (h/l)	Glucose (MJ), alanine (MN)
L-Lactic acid	NP	+	T (h), pH (h), O_2 (l)	Glucose
Formic acid	+	+	T (h), pH (h)	Glucose
Ethanol	+	NA	T (h), glucose	NAD
CO_2	+	NA	NAD	Glucose
Isobutyric acid	+	NAD	Glucose (l), T (l), pH (h)	Valine, leucine
Isovaleric acid	+	NAD	Glucose (l), T (l), pH (h)	Valine, leucine
2-Methylbutyric acid	+	NAD	Glucose (l), pH (h)	Isoleucine
3-Methylbutanol	+	NAD	Glucose (h), pH (l)	NAD
2-Methylbutanol	NA			NAD
2-Methylbutanol	NA			Isoleucine
3-Methylbutanol	NA			Leucine
2,3-Butanediol	+	NAD	Glucose (h), T (h/l)	Diacetyl
Diacetyl	+	NAD		
			NAD	NAD
2-Methylpropanol	+		Glucose (h)	Valine
2-Methylpropanal	NA		NAD	Valine
Free fatty acids	NT	NAD	Glucose (l), pH, O_2, T (h)	Meat fat
In different gaseous atmospheres				
L-Lactic acid	+	+	Glucose (h), pH (h), T (ns)	Glucose
Acetic acid	+	+	O_2 (h), glucose (l)	Glucose
Ethanol	+	NAD	T (h), pH (h)	NAD
Formic	+	+	T (h), pH (h)	NAD

[a] From references 74, 75, 150, 202, and 254. M, naturally spoiled meat; P, poultry; F, fish; h, high pH, concentration of glucose, or storage temperature; l, low pH, concentration of glucose, or storage temperature; h/l, contradictory results; NS, not significant factor; MJ, major contribution; MN, minor contribution; NP, no production under strictly aerobic conditions; ND, not determined; NT, not tested; NA, not analyzed; NAD, no available data; T, temperature.

Chemical Changes by Gram-Positive Bacteria

Lactic Acid Bacteria

Studies on food spoilage with pure cultures in sterile food model systems as well as with natural ecosystems have found changes in lactate concentration due to either production or assimilation of this compound. For example, Nassos et al. (188–190), Drosinos et al. (73), Koutsoumanis and Nychas (150), and Tsigarida and Nychas (255) found an increase in lactate concentrations during storage of muscle foods under anoxic conditions and recommended its use as a spoilage indicator. It is important to distinguish between D- and L-lactate, which is accomplished by high-performance liquid chromatography (HPLC) (188–190). For example, it is known that D-lactate decreases in muscle foods during storage under aerobic or MAP conditions (73, 199, 202, 204). When the acid profile of water-soluble compounds was analyzed by HPLC (162, 203), it was confirmed that the chromatographic area of lactic acid did not change significantly, compared to changes found with an enzymatic method of analysis used in these studies (75, 141, 203). This could be due to the fact that D-lactate was formed during storage. Indeed, when D- and L-lactate were analyzed enzymatically, it was found that the amount of L-lactate decreased while that of D-lactate increased during storage of meat under different conditions (202). Similar results have been reported in studies on muscle foods stored under conditions of 100% CO_2, 100% N_2, or 20%:80% CO_2-O_2 or in vacuum packages (203). D-Lactate, however, is not a product of *B. thermosphacta* metabolism or endogenous anaerobic glycolysis because in both cases only L-lactate is produced (11, 123, 146, 209). Therefore, the increase in concentration of D-lactate is due to metabolism by lactic acid bacteria, particularly *Carnobacterium*, *Leuconostoc*, or *Weissella*, which generate D-, L-, or DL-lactate (41, 42, 145, 146).

Production of L or D forms of lactate depends on the presence of D-nLDH and/or L-nLDH (specific NAD$^+$-dependent lactate dehydrogenases). A few lactic acid bacteria (e.g., *Lactobacillus curvatus* and *Lactobacillus sakei*) produce a racemase which converts L-lactic acid to D-lactic acid (92). L-Lactic acid induces the racemase, which results in a mixture of D- and L-lactic acids. Generally, L-lactic acid is the major form produced in the early growth phase, while D-lactic acid follows during the late exponential to the stationary phase (92).

Ordonez et al. (209) reported that no consistent patterns were obtained for L-lactic acid concentrations in pork packed in 20% CO_2–80% air and 20% CO_2–80% O_2, whereas the D-lactic acid concentration was reported to increase along with counts for lactobacilli. D-Lactic acid was not detected in samples during the initial stages of storage; however, low levels were detected after 5 days of storage under both atmospheres, while after 20 days levels had risen to 12 to 18 mg/100 g of meat. Similar patterns have also been reported by others (72, 161, 203), who found that D-lactate was produced in beef (ground), lamb, chicken, pork, and dry ham samples stored under vacuum or MAP conditions. The production of D-lactic acid has been attributed to heterofermentative lactic acid bacteria (72, 209), although in poultry meat a higher concentration of this acid was always associated with samples inoculated with the homofermentative *Lactobacillus plantarum* rather than with the heterofermentative *Weisella minor* (202). Similar results have been reported by Borch and Agerhem (17) for inoculated beef slices.

The increase of acetate in beef, pork, and fish stored under different vacuum or MAP conditions could be attributed to a shift from homo- to heterofermentative metabolism of lactic acid bacteria (17, 51, 65, 142, 150, 209). The production of acetate from lactic acid is affected by environmental factors such as pO_2, pH, and glucose limitation (4, 13, 14, 145, 146). Indeed, the type of energy source (glucose or galactose), glucose limitation, degree of aeration, concentration of lactate dehydrogenase (iLDH), NADH peroxidase or fructose 1,6-diphosphate, and stereospecificity of NAD-independent flavin-containing lactate dehydrogenase, lactate oxidase, or pyruvate oxidase may all influence the conversion of lactate or pyruvate to acetate (4, 19, 40, 92, 145, 185, 186, 218, 227, 248, 253). Nychas et al. (203) reported that alcohols, particularly ethanol and propanol, appear to be the most promising compounds as indicators of spoilage in meat and meat products stored under vacuum or MAP conditions. As mentioned above, ethanol could be a fermentation by-product of the heterofermentative leuconostocs and carnobacteria, or a product of homofermentative lactic acid bacteria when environmental stress conditions cause a shift in metabolism to heterofermentative (145, 146). Results from laboratory broth culture experiments suggest that carnobacteria may also produce formic acid in meat stored under vacuum or MAP conditions (125, 246).

Brochothrix thermosphacta

Studies with fresh fish and model meat systems (gel cassettes) stored under various gaseous conditions showed that the metabolic products arising from *B. thermosphacta* are different from those produced under strictly aerobic conditions (75, 255). For example, when the oxygen tension is low (<0.2 μM oxygen), the main metabolic end products of *B. thermosphacta* are L-lactate and ethanol (11, 19, 117, 123). There was no production of acetic acid, D-lactic acid, 2,3-butanediol, isovaleric acid, isobutyric acid, or acetoin in broth samples flushed with gases other than oxygen (11, 123), but excessive amounts of acetic acid were produced in fish supplemented with glucose (75). Formic acid was among the end products regardless of the gaseous storage atmosphere used. It was suggested by Hitchener et al. (123) that glucose metabolism by this bacterium could be through (i) the Embeden-Meyerhof glycolytic pathway, in which after the conversion of glucose to two molecules of pyruvate, the latter compound is metabolized to lactate and/or ethanol plus carbon dioxide; (ii) conversion of glucose via 6-phosphogluconate and pentose phosphate to equimolar amounts of lactate, ethanol, and carbon dioxide; or (iii) the Entner-Doudoroff pathway, in which glucose is converted via 6-phosphogluconate and 2-keto-3-deoxy-6-phosphogluconate and 2-keto-3-deoxy-6-phosphogluconate is converted to pyruvate. *Brochothrix thermosphacta* behaves as a heterofermentative bacterium under glucose-limited conditions.

Quality and Safety of Reduced-Oxygen-Packaged Muscle Foods

Limiting growth of spoilage microorganisms using reduced-oxygen packaging may create an environment conducive to pathogen growth and toxin production before there is evidence of spoilage (258). For example, in reduced-oxygen-packaged products, where refrigeration is the sole barrier to outgrowth of nonproteolytic *Clostridium botulinum* and spores have not been destroyed (e.g., vacuum-packaged raw fish, unpasteurized crayfish, or crab meat), the temperature must be maintained at 3.3°C or below from packing to consumption to prevent growth and toxin formation. Temperature control by processors is usually possible; however, transportation, retail, and home storage conditions may be inadequate. The use of time temperature integrators or antimicrobial agents may offer appropriate control strategies throughout distribution. Alternatively, products may be frozen to prevent pathogen growth.

Vacuum-packaged refrigerated fresh and cooked meat and poultry products have been found to undergo an unusual type of spoilage due to growth of psychrotrophic spoilage *Clostridium* spp. (27, 143, 144). These clostridia (*C. laramiense, C. difficile, C. beijerinckii,* and *C. lituseburense*) can grow, sporulate, and germinate at 0°C or below. This type of sporadic spoilage has been detected in products of normal pH during storage at 2°C or lower and is characterized by an initial pinkish red color which becomes green and is associated with large amounts of hydrogen sulfide (known as "blown pack") and purge during storage when the meat undergoes proteolysis (133). Prediction or prevention of this type of spoilage is still difficult.

SPOILAGE OF SPECIFIC MUSCLE TISSUES

Spoilage of Adipose Tissue

The spoilage process of adipose tissue is similar to that for lean muscle tissue; the adipose tissue, however, is used more efficiently as a substrate for microbial growth when it is emulsified and there is a water phase present (133). Since lipolytic enzymes are not produced by microorganisms until carbohydrates are exhausted, fatty tissue spoilage does not depend on the ability of the microflora to produce lipases (195). In addition, amounts of soluble components (e.g., glucose, glycolytic intermediates, and amino acids) are smaller in adipose than in muscle tissue, and soluble nutrients from the underlying tissue are not readily replenished by diffusion (98, 195). While the spoilage process and rate of growth of spoilage bacteria are similar on adipose and muscle tissues, spoilage odors are associated with lower numbers of bacteria on adipose tissue, where most available glucose is depleted when populations reach 6 log CFU/cm². In addition to containing limited amounts of carbohydrates, adipose tissue has less lactic acid than does muscle tissue, and therefore the surface pH is higher (i.e., approaching 7.0). As a result, spoilage of adipose tissue has been compared to that of dark, firm, dry (DFD) meat (194, 195) and may involve growth of *S. putrefaciens*. The growth rates of psychrotrophic bacteria such as *H. alvei, S. liquefaciens,* and *L. plantarum* have been reported to be higher on fat than on lean beef and pork tissues, but there was little difference in growth rates of other bacteria (259). Generally, spoilage of fat before lean tissue is unlikely, considering the presence of muscle tissue purge in cut vacuum-packaged meats and restricted bacterial growth on dry fatty carcass surfaces (133, 195).

DFD Meats

Excessive stress or exercise before slaughter is associated with increased animal metabolism that can deplete levels of muscle glycogen and lead to production of less lactic acid and a higher ultimate pH following slaughter. Muscles with a pH of >6.0 appear darker than normal red meat because the higher muscle respiration rate reduces the depth of oxygen penetration and, therefore, reduces the level of oxymyoglobin formation (118). This condition is known as DFD muscle or meat, and although it is more common in beef, it may also be detected in muscles of pigs and other meat animals (133). In addition to a higher ultimate pH resulting from lower lactic acid content, DFD tissue is also deficient in glucose and glycolytic intermediates (191), and spoilage is faster than that of meat with a normal postmortem pH. The absence of glucose in the tissues allows faster degradation of amino acids by pseudomonads, and spoilage occurs at lower bacterial cell densities (>6 log CFU/cm²) than in normal meat (191).

Storage of DFD meat in vacuum packages or in MAP also results in rapid spoilage, typically associated with green discolorations (105). The higher pH and absence of glucose and glucose-6-phosphate may allow *S. liquefaciens* and *S. putrefaciens* to compete successfully with the normally dominant lactic acid bacteria and to comprise a significant portion of the spoilage microflora (105, 213). The green discoloration observed in DFD meat is associated with formation of hydrogen sulfide from cysteine or glutathione by *S. putrefaciens* (105). The hydrogen sulfide reacts with myoglobin to form green sulfmyoglobin (193). The presence of even low numbers of *S. liquefaciens* organisms may be associated with spoilage odors on DFD meat because this organism produces small amounts of hydrogen sulfide that may be insufficient to cause green discoloration (105, 191, 213).

PSE Meat

Pale, soft, exudative (PSE) muscle tissue or meat is a condition that may occur in pork and poultry, and to a lesser extent in beef. It develops when accelerated postmortem glycolysis decreases the muscle pH to its ultimate value while the muscle temperature is still high (98, 118). The condition affects 5 to 20% of pig carcasses and is characterized by the development of a pale color, soft texture, and exudation of fluids from the muscle (98, 118). Development of PSE is directly associated with porcine stress syndrome, which may lead to death of animals exposed to mild stress, or to malignant hyperthermia, in which death may be caused by exposure to certain anesthetics (118). There is debate as to whether PSE meat spoils more slowly than meat with normal pH values. Rey et al. (215) reported that the rate of bacterial growth was higher in DFD beef and lower in PSE pork, suggesting that muscle pH may affect growth of spoilage microorganisms under various conditions. Gill (98) concluded that even though an ultimate pH of 5.1 or below has been reported for PSE pork, there is little difference in the ultimate pH and chemical composition of PSE and normal meat. Since the concentrations of soluble low-molecular-weight compounds should be similar in PSE and normal meat, the development and characteristics of spoilage would likely be similar (98, 133).

Ground and Comminuted Products

The reduced shelf life of comminuted muscle foods has been attributed to higher levels of initial contamination originating from lower-quality raw materials or trimmings used in grinding and to cross-contamination and microbial spreading during processing, as well as to the physical effects of the comminution process on the muscle tissue. Grinding or comminution ruptures tissue cells and releases fluids and nutrients that are easily used by bacteria. Additionally, bacteria that have been restricted to the surface of meats are distributed throughout the mass, resulting in cross-contamination (131, 132, 244). Although initial bacterial populations are greater, the types of microorganisms and resulting spoilage resemble those on intact muscle tissue. The dominant microflora on the surface of aerobically stored comminuted products consists of *Pseudomonas*, *Acinetobacter*, and *Moraxella* species, while lactic acid bacteria may dominate in the interior due to the limited availability of oxygen (131). When present, contaminants such as *Aeromonas* spp. and genera within the Enterobacteriaceae are more frequently found on comminuted than on intact tissue products (133).

As with other comminuted fresh meats, fresh sausage also spoils rapidly, since the addition of salt and spices is not sufficient to delay microbial growth. Microorganisms dominating in fresh sausage spoilage are similar to those present in other fresh meats; however, spoilage microflora of refrigerated pork sausage may also include *B. thermosphacta* (131, 133, 244).

Cooked Products

Exposure of muscle foods to heat during cooking results in destruction of vegetative bacterial cells, while bacterial spores may survive. Spoilage during storage of perishable, cooked, uncured meats is the result of microflora surviving heat processing or of postcooking contaminants. Dominant spoilage microorganisms in such products may include psychrotrophic micrococci, streptococci, lactobacilli, and *B. thermosphacta* (131). Cooked-meat recontamination becomes a concern in products handled and stored after cooking or exposed to abusive temperatures for extended times during serving. Where recontamination is minimal, nonproteolytic bacteria usually dominate and result in development of sour odors. Recontamination with proteolytic microorganisms results in putrid odors due to breakdown of amino acids (132). Spoilage of canned muscle food products is usually due to spoiled raw materials, inadequate thermal processing allowing survival of heat-resistant mesophilic sporeformers, slow cooling or storage at high temperatures that allow proliferation of thermophilic sporeformers, or reintroduction of microorganisms from postprocessing leakage (131, 133).

Processed Products

Microbial spoilage of perishable processed muscle foods is influenced by the nature of the raw materials, ingredients used in the formulation, type of processing, and conditions of storage. Spoilage differs among cured, heat-processed, fermented, or dried products of various a_ws and pHs that are stored aerobically or under MAP conditions. Detailed reviews of the microbiology of cured meats are provided by Tompkin (249), Ingram and Simonsen (131), Kraft (158), Gardner (91), and the International Commission on Microbiological Specifications for Foods (132).

Spoilage defects of frankfurters, bologna, sausage, and luncheon meats may be classified as slimy, souring, and greening (133, 135). Slimy spoilage is usually confined to outside product surfaces and may start developing as discrete colonies which then expand to form a uniform gray layer of slime. Formation of slime usually requires the presence of moisture, and microorganisms implicated in its development include yeasts, *Lactobacillus*, *Enterococcus*, and *B. thermosphacta* (133, 135). Souring usually occurs on product surfaces under casings, where lactobacilli, enterococci, and *B. thermosphacta* metabolize lactose and other sugars to produce organic acids. In contrast to greening of fresh meats, which may be a result of hydrogen sulfide production by certain bacteria, the same defect in cured meat products may develop in the presence of hydrogen peroxide, which may form when surfaces of vacuum or MAP products are exposed to air. Hydrogen peroxide may then react with nitrosohemochrome to produce choleglobin, which has a greenish color (116). The absence of muscle catalase (breaks down H_2O_2), through heat inactivation during the cooking process, and growth of catalase-negative bacteria may result in greening of cured meat products that contain nitrite. As a result, and because of the low oxidation/reduction potential in the meat interior, hydrogen peroxide may accumulate. As with surface greening, reaction of the hydrogen peroxide with nitrosohemochrome may result in the production of a green pigment. *Lactobacillus viridescens* is the most common cause of this type of greening, but species of *Streptococcus* and *Leuconostoc* may also produce this defect. The presence of these bacteria often is a result of poor sanitation before or following processing (116, 131, 133, 135).

The reduced a_w of cured meats inhibits spoilage by gram-negative psychrotrophic bacteria at refrigeration temperatures and prevents putrefaction (131). Cured meats, however, are spoiled by lactobacilli or micrococci, which tolerate low a_w (251). Aerobic micrococci predominate on cured meats stored aerobically, while lactobacilli predominate in the spoilage microflora of products stored in vacuum packaging or MAP (130). If sucrose is present in the formulation of cured meats, a slimy dextran layer may be formed through the activity of *Leuconostoc* species or other bacteria such as *L. viridescens*. *B. thermosphacta* may also be involved in the spoilage of these products (249). Vacuum-packaged bacon spoils at rates similar to those of other cured products, and its spoilage involves the same microorganisms. It has been reported that different spoilage characteristics may develop on fat and lean tissue of aerobically stored bacon (130); a smoked fishy flavor may develop on the lean tissue of bacon, while fatty tissues may develop a rancid or cheesy flavor. The low a_w, the presence of nitrite, and exposure to smoke provide stability in dry-cured products by suppressing bacterial growth; their spoilage involves mainly yeasts and molds if products are stored in humid environments (249). Dry-cured meats may be spoiled if microbial growth occurs before proper and adequate salt penetration, curing, and drying (249). Dried meat products remain stable when properly prepared and stored. To restrict the growth of some species of bacteria, yeasts, and molds that can grow at low a_w, the water content of dried meat products must be reduced to about 20%. To inhibit the growth of xerotolerant molds, the water content must be further reduced to about 15% (131). Microbial spoilage of these products should not occur unless exposure to high relative humidity or other high-moisture conditions results in an uptake of moisture. Addition of antifungal agents may be valuable to retard growth of spoilage fungi in these products (133, 135).

Biogenic Amines

Risks to consumers from consumption of fresh muscle foods and muscle food products may be due to the presence of pathogens and biogenic amines. Although pathogenic microorganisms are not considered part of the spoilage association per se, their occurrence in muscle food products is possible due to their presence in animals and the environment.

Production of biogenic amines by the natural microbial flora may be an issue of concern in stored muscle foods (223). Amines have been detected in fresh meat and some fish (primarily scombroid species such as tuna, mahimahi, and mackerel) stored under aerobic or vacuum or MAP conditions (Table 6.14). The formation of these biogenic amines during storage can lead to human illness known as scombroid poisoning, which is characterized as a severe and sometimes fatal allergic reaction that occurs shortly after consumption of contaminated products. Among the amines, levels of histamine, putrescine, and cadaverine may show a constant increase during storage. Concentrations of spermine, spermidine, and tryptamine usually remain

Table 6.14 Production of biogenic amines by muscle microbial flora in muscle foods and broths[a]

| Biogenic amine | Bacteria | Storage condition | | Factor(s) |
		Temp (°C)	Medium or pack	
Putrescine	*Hafnia alvei, Serratia liquefaciens, Shewanella putrefaciens*	1	VP[b]	pH, ornithine (arginine)
		1	Fish broth	utilization
Cadaverine	*H. alvei, S. liquefaciens, S. putrefaciens*	1	VP	pH, lysine utilization
		1		
Histamine	*Proteus morganii, Klebsiella pneumoniae, H. alvei, Aeromonas hydrophila, Morganella morganii, Photobacterium phosphoreum, S. putrefaciens*	1.7	Fish in VP	Temp, pH, histidine
		2.1	Fish in VP, fish broth	utilization
Spermine				pH, spermidine
Spermidine				pH, agmatine, arginine
Tyramine	*Lactobacillus* sp., *L. carnis, L. divergens, Enterococcus faecalis*	1	VP	
		20	Air[c]	pH
Tryptamine				pH

[a] From references 20, 34, 56, 81, 147, 167, 169, 202, and 261.
[b] VP, vacuum pack.
[c] Aerobic storage.

steady, while a small increase in amounts of tyramine may be observed early in the storage period (56, 165, 223). Since lactic acid bacteria and *B. thermosphacta* do not produce amines, the formation of these compounds has been attributed to *Enterobacteriaceae*; however, tyramine can also be formed by some strains of the genus *Lactobacillus*. Proper sanitation, proper storage temperature, and storage time limitation should minimize human health problems associated with biogenic amines in muscle foods.

EVALUATION, MODELING, AND PREDICTION OF QUALITY IN MUSCLE FOODS

Muscle Food Quality

Quality is generally a subjective and sometimes elusive term. Freshness and safety of muscle food products are generally considered the most important contributors to quality. It is therefore crucial to have valid methods to monitor freshness and quality (3, 260). Indeed, methods should be valid for application by industry and consumers in order to obtain reliable information on freshness status when merchandising and purchasing products. The meat, poultry, and seafood industries need methods to determine the type of processing needed for raw materials and also to predict the remaining shelf life of their products. Inspection authorities need reliable methods for regulatory purposes, while the wholesale and retail

sectors need valid methods to ensure the quality of their products. It is important to understand that differences between species, variation in geographical origin, and, most importantly, changes in preservation technologies (e.g., introduction of new packaging and processing methods) may alter spoilage patterns of muscle products. Therefore, the role of scientists is to suggest reliable and valid methods to meet the requirements of retailers, consumers, and regulatory authorities.

Currently, sensory and microbiological analyses are most often used to evaluate the freshness or spoilage of meat, poultry, and seafood. The disadvantage of sensory analysis, which is probably the most acceptable and appropriate method, is its reliance on highly trained panelists, which is necessary to minimize subjectivity but makes it costly and unattractive for routine analysis (52, 134). Microbiological analyses, at least in their traditional form (total viable counts), are often misleading, and scientists have shown that it is more meaningful to measure either the specific spoilage organisms or even the ephemeral spoilage organisms when this is feasible (Fig. 6.1 and 6.2). Unfortunately, microbiological analyses are lengthy, costly, and destructive to test products; therefore, efforts have been made to replace both microbiological and sensory analyses with biochemical changes in muscle foods, e.g., based on nucleotide catabolism (*K* value), production of amines, ammonia, trimethylamine, and sulfur compounds, or physical methods (e.g., the Torrymeter, Freshtester, or RT Freshmeter), which provide

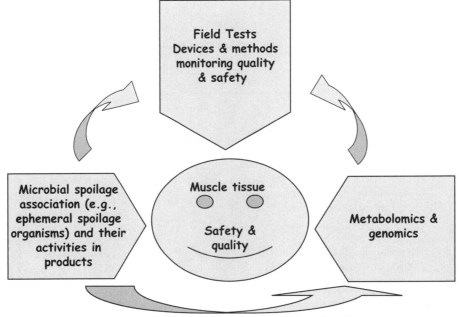

Figure 6.2 Future trends for the evaluation of spoilage.

information that can be indicative of microbial growth on fish and meat (108, 109, 206, 208).

Chemical/Biochemical Approach

The correlation between microbial growth and chemical changes during spoilage has been continuously recognized as a means of revealing indicators that may be useful for quantifying muscle tissue quality as well as the degree of spoilage (52, 79, 112, 134, 150, 202). The ideal indicator should (i) be absent or at least at low levels in fresh and unspoiled muscle tissue, (ii) increase with storage time, (iii) be produced by the microbial flora dominating during spoilage, and (iv) have good correlation with sensory testing (134). As a consequence, identification of such an ideal compound has proven to be a difficult task. There are two types of potential metabolites for such use: (i) those which are specific (e.g., gluconate) to certain microorganisms (e.g., pseudomonads) and (ii) those that are common (e.g., acetic or lactic acid) to different members of the microbial association of muscle foods. In both cases, however, spoilage information provided may be incorrect. Additionally, the imposed interventions (e.g., packaging, biopreservation, and temperature of storage), sophisticated testing procedures needed (e.g., instrumentation and highly educated personnel), and lack of a sound definition for spoilage make the task of evaluating spoilage objectively (i.e., identifying spoilage indicators) difficult.

Recently some interesting analytical approaches have been forwarded for the rapid and quantitative monitoring of meat spoilage (Table 6.15). These include biosensors

Table 6.15 Compounds potentially useful for assessment of shelf life of raw meat and fish under different packaging conditions[a]

Compound(s)	Test(s)	Packaging condition(s)	Red meat and poultry	Finfish
Glucose	Enzymatic kit	Air, VP,[b] MAP	Y[c]	Y
Acetate	Enzymatic kit, HPLC	VP, MAP	Y	Y
Gluconate	Enzymatic kit	Air, VP, MAP	Y	
Total lactate	HPLC	VP, MAP	Y	
D-Lactate	Enzymatic kit	VP, MAP	Y	Y
Ethanol	Enzymatic kit, GLC[d]	VP, MAP	Y	
Free amino acids	Chromatometric	Air	Y	
Ammonia	Enzymatic, colorimetric	Air	Y	Y
Acetone, methyl ethyl ketone, dimethyl sulfide, dimethyldisulfide, hydrogen sulfide	GLC, GC/MS,[e] sulfur selective detector	VP, MAP	Y	Y
Diacetyl, acetoin	Colorimetric	VP, MAP	Y	
Biogenic amines	HPLC, sensors, enzymatic test, GLC, enzyme electrodes, test strips	Air, VP, MAP	Y	Y
Diamines	Amperometric electrodes (enzymatic systems)	Air	Y	
Microbial activity	Enzymatic/resazurin	Air	Y	Y
Volatiles (odors)	Electronic noses, PTR[f]-MS (chemical sensors)	Air, VP, MAP	Y	Y
Proteolysis (amides, amines, etc.)	FT-IR, NIR,[g] MIR[h]	Air, VP, MAP	Y	Y
Torymeter				Y
Microbial activity	Impedance—capacitance	Air	Y	Y
ATP/IMP	HPLC	Air		Y
K values	HPLC	Air		Y
Lipid oxidation	HPLC, GC, spectroscopy	Air, VP		Y
TMA[i]	HPLC, GC, sensors, enzymatic	Air		Y

[a] Based on references 12, 16, 18, 33, 35, 36, 79, 80, 108, 109, 148, 159, 172, 173, 180, 204, 207, 208, 217, 261, 262, 264, 267–269.
[b] VP, vacuum packaged.
[c] Y, yes.
[d] GLC, gas-liquid chromatography.
[e] GC/MS, gas chromatography/mass spectroscopy.
[f] PTR, proton transfer reaction–mass spectroscopy.
[g] NIR, near-infrared.
[h] MIR, mid-infrared.
[i] TMA, trimethylamine.

Table 6.16 Software available for modeling microbial changes[a]

GrowthPredictor (United Kingdom)—http://www.ifr.ac.uk/Safety/GrowthPredictor/ (based on data previously used in the FoodMicromodel software; 18 models for growth of pathogenic bacteria; available free of charge)

Sym'Previus—http://www.symprevius.net (French predictive microbiology application software under development)

Seafood Spoilage and Safety Predictor (shelf life of seafoods and growth of specific spoilage organisms; *Listeria monocytogenes* in cold-smoked salmon) (63)

Food Spoilage and *Escherichia coli* Predictors (under development) (212)

Safety Monitoring and Assurance System (151, 154) (Greek predictive microbiology application software under development); software is based on kinetic data of spoilage bacteria derived from fish, meat, and milk in situ

Pathogen Modelling Program (United States)—http://www.arserrc.gov/mfs/pathogen.htm (37 models of growth, survival, and inactivation; frequently updated [version 7.0]); available free of charge during the last 15 years; ~5,000 downloads per year

[a] Based on references 63, 154, and 211 and http://www.arserrc.gov/mfs/pathogen.htm and http://www.ifr.ac.uk/Safety/GrowthPredictor/.

(enzymatic reactor systems), electronic noses (array of sensors), and Fourier transform infrared (FT-IR) analysis. For example, FT-IR spectroscopy, in combination with appropriate machine learning methods, presents a novel method for the quantitative detection of food spoilage. Further, the development of an "expert system" that can automatically classify the sensorial input into a "diagnosis" based on extracted preprocessing features is necessary before widespread adoption is possible. The application of advanced statistical methods (discriminant function analysis, clustering algorithms, and chemometrics) and intelligent methodologies (neural—artificial or not—networks, fuzzy logic, evolutionary algorithms, and genetic programming) may be used for qualitative and quantitative indices in parallel with unsupervised or supervised learning algorithms (7, 80, 110, 111, 229).

Regardless of the methodology used for the quantitative evaluation of spoilage and safety for control purposes, factors such as (i) food structure and physicochemical parameters (e.g., type, concentration and nutrient availability, and diffusivity), (ii) microbial competition and physiological state of bacterial cells as well as effects of dynamic storage (e.g., fluctuation of temperature, vacuum packaging or MAP, and film permeability) conditions, and (iii) understanding microbial ecology mechanisms (e.g., bacterial communication and deciphering the function of genes, i.e., genomics) of growth and survival of established and emerging pathogens and spoilage bacteria in stressful food environments should be taken into account (Fig. 6.2).

An understanding of the origin of metabolites (metabolomics) involved in spoilage (i.e., responsible organism and substrate), the genes that regulate their production at the cell level (genomics and proteomics), and the effect of muscle characteristics on the rate and type of metabolite formation is of great importance (Fig. 6.2). Application

of rapid analytical methodologies based on this knowledge will be of benefit to the industry by providing valid methods to ensure the freshness and safety of their products and in the case of disputes between buyers and sellers, regulatory or inspection authorities (e.g., reliable methods for control purposes), and consumers (e.g., reliable and transparent indications of the safety and quality status of muscle tissue in retail and until consumed).

Models Describing Spoilage and Safety

The statement and the question of Box and Draper (22) that "All models are wrong. The question is, how wrong do they have to be not to be useful?" does not discourage scientists from working in this field. Mathematical models have been developed for predicting the effects of temperature, a_w, pH, and carbon dioxide on the growth and survival of spoilage and pathogenic bacteria such as *B. thermosphacta*, lactic acid bacteria, *P. phosphoreum*, pseudomonads, *S. putrefaciens*, *Listeria monocytogenes*, and *Salmonella* in muscle foods (21, 66, 76, 82, 86, 157, 174, 179, 216). However, the application of these models has not been focused on monitoring muscle food quality per se, but mainly as a management tool for shelf life and safety prediction and as a scientific tool to gain insight into muscle food spoilage.

Different models (kinetic or stochastic) have been developed for various muscle products, but still, an accurate prediction of shelf life is not attainable (95, 152, 154, 214, 259, 260). There are limited successfully validated models for predicting the growth of ephemeral spoilage organisms that have been included in application software, and this has facilitated prediction of food shelf life under constant and dynamic temperature storage conditions (Table 6.16). Measurement of enzyme synthesis and activity could be used to estimate shelf life, offering a completely different modeling approach (243).

References

1. **Alford, J. A., and L. E. Elliott.** 1960. Lipolytic activity of microorganisms at low and intermediate temperatures. I. Action of *Pseudomonas fluorescens* on lard. *Food Res.* 25:296–303.

2. **Alford, J. A., J. L. Smith, and H. D. Lilly.** 1971. Relationship of microbial activity to changes in lipids in foods. *J. Appl. Bacteriol.* 34:133–146.

3. **Anonymous.** 1997. The need for methods to evaluate fish freshness, p. 1–396. *In Research and Industry Proceedings of the Final Meeting of the Concerted Action—Evaluation of Fish Freshness. Methods To Determine the Freshness of Fish in Research and Industry.* International Institute of Refrigeration, Paris, France.

4. **Axelsson, L. T.** 1993. Lactic acid bacteria: classification and physiology, p. 1–63. *In* S. Salminen and A. Von Wright (ed.), *Lactic Acid Bacteria.* Marcel Dekker Inc., New York, N.Y.

5. **Barlow, J., and A. G. Kitchell.** 1966. A note on the spoilage of prepacked lamb chops by *Microbacterium thermosphactum. J. Appl. Bacteriol.* 29:185–188.

6. **Bazemore, R., S. G. Fu, Y. Yoon, and D. Marshall.** 2003. Major causes of shrimp spoilage and methods for assessment, p. 223–234. *In* A. M. Rimando and K. K. Schrader (ed.), *Off-Flavors in Aquaculture.* ACS Symposium Series no. 848. American Chemical Society, Washington, D.C.

7. **Beavis, R. C., S. M. Colby, R. Goodacre, P. B. Harrington, J. P. Reilly, S. Sokolow, and C. W. Wilkerson.** 2000. Artificial intelligence and expert systems in mass spectrometry, p. 11558–11597. *In* R. A. Meyers (ed.), *Encyclopedia of Analytical Chemistry.* John Wiley & Sons, Chichester, United Kingdom.

8. **Bell, G. R., S. M. Moorhead, and D. R. Broda.** 2001. Influence of heat shrink treatment on the onset of clostridial "blown pack" spoilage of vacuum pack chilled meat. *Food Res. Int.* 34:271–275.

9. **Benedict, R. C.** 1988. Microbial attachment to meat surfaces. *Reciprocal Meat Conf. Proc.* 41:1–6.

10. **Bennasar, A., J. Guasp, and J. Lalucat.** 1998. Molecular methods for the detection and identification of *Pseudomonas stutzeri* in pure culture and environmental samples. *Microb. Ecol.* 35:22–53.

11. **Blickstad, E., and G. Molin.** 1984. Growth and end-product formation in fermenter cultures of *Brochothrix thermosphacta* ATCC 11509^T and two psychrotrophic *Lactobacillus* spp. in different gaseous atmospheres. *J. Appl. Bacteriol.* 57:213–220.

12. **Blixt, Y., and E. Borch.** 1999. Using an electronic nose for determining the spoilage of vacuum-packaged beef. *Int. J. Food Microbiol.* 46:123–134.

13. **Bobillo, M., and V. M. Marshall.** 1991. Effect of salt and culture aeration on lactate and acetate production by *Lactobacillus plantarum. Food Microbiol.* 8:153–160.

14. **Bobillo, M., and V. M. Marshall.** 1992. Effect of acidic pH and salt on acid end-products by *Lactobacillus plantarum* in aerated, glucose-limited continuous culture. *J. Appl. Bacteriol.* 73:67–70.

15. **Bolder, R. D.** 1997. Decontamination of meat and poultry carcasses. *Trends Food Sci. Technol.* 8:221–227.

16. **Boothe, D. H., and J. W. Arnold.** 2002. Electronic nose analysis of volatile compounds from poultry meat samples, fresh and after refrigerated storage. *J. Sci. Food Agric.* 82:315–322.

17. **Borch, E., and H. Agerhem.** 1992. Chemical, microbial and sensory changes during the anaerobic cold storage of beef inoculated with a homofermentative *Lactobacillus* sp. or a *Leuconostoc* sp. *Int. J. Food Microbiol.* 15:99–108.

18. **Borch, E., M.-L. Kant-Muermans, and Y. Blixt.** 1996. Bacterial spoilage of meat and cured meat products. *Int. J. Food Microbiol.* 33:103–120.

19. **Borch, E., and G. Molin.** 1989. The aerobic growth and product formation of *Lactobacillus, Leuconostoc, Brochothrix* and *Carnobacterium* in batch cultures. *Appl. Microbiol. Biotechnol.* 30:81–88.

20. **Bover-Cid, S., T. Hernandez-Jover, M. J. Miguelez-Arrizado, and M. C. Vidal-Carou.** 2003. Contribution of contaminant enterobacteria and lactic acid bacteria to biogenic amine accumulation in spontaneous fermentation of pork sausages. *Eur. Food Res. Technol.* 216:477–482.

21. **Bovill, R. A., J. Bew, and J. Baranyi.** 2001. Measurements and predictions of growth for *Listeria monocytogenes* and *Salmonella* during fluctuating temperature. II. Rapidly changing temperatures. *Int. J. Food Microbiol.* 67:131–137.

22. **Box, G. E. P., and N. R. Draper.** 1987. *Empirical Model Building and Response Surfaces.* John Wiley and Sons, New York, N.Y.

23. **Braun, P., and J. P. Sutherland.** 2003. Predictive modelling of growth and measurement of enzymatic synthesis and activity by a cocktail of *Pseudomonas* spp., *Shewanella putrefaciens* and *Acinetobacter* sp. *Int. J. Food Microbiol.* 86:271–282.

24. **Braun, P., and J. P. Sutherland.** 2004. Predictive modelling of growth and measurement of enzymatic synthesis and activity by a cocktail of *Brochothrix thermosphacta. Int. J. Food Microbiol.* 95:169–175.

25. **Braun, P., and J. P. Sutherland.** 2005. Predictive modelling of growth and measurement of enzymatic synthesis and activity by a cocktail of selected Enterobacteriaceae and *Aeromonas hydrophila. Int. J. Food Microbiol.* 105:257–266.

26. **Bridson, E. Y., and G. W. Gould.** 2000. Quantal microbiology. *Lett. Appl. Microbiol.* 30:95–98.

27. **Broda, D. M., K. M. De Lacy, and R. G. Bell.** 1998. Influence of culture media on the recovery of psychrotrophic Clostridium spp. associated with the spoilage of vacuum-packed chilled meats. *Int. J. Food Microbiol.* 39:69–78.

28. **Broda, D. M., P. A. Lawson, R. G. Bell, and D. R. Musgrave.** 1999. *Clostridium frigidicarnis* sp. nov., a psychrotolerant bacterium associated with 'blown pack' spoilage of vacuum-packed meats. *Int. J. Syst. Bacteriol.* 49:1539–1550.

29. **Broda, D. M., D. J. Saul, R. G. Bell, and D. R. Musgrave.** 2000. *Clostridium algidixylanolyticum* sp. nov., a psychrotolerant, xylan-degrading, spore-forming bacterium. *Int. J. Syst. Evol. Microbiol.* 50:623–631.

30. **Brown, W. D.** 1986. Fish muscle as food, p. 405–451. *In* P. J. Bechtel (ed.), *Muscle as Food.* Academic Press, Orlando, Fla.

31. Bryan, F. L. 1980. Poultry and poultry meat products, p. 410–469. *In* J. H. Silliker, R. P. Elliot, A. C. Baird-Parker, F. L. Bryan, J. H. B. Christian, D. S. Clark, J. C. Olsen, Jr., and T. A. Roberts (ed.), *Microbial Ecology of Foods*, vol. 2. *Food Commodities*. Academic Press, New York, N.Y.

32. Butler, J. L., J. C. Stewart, C. Vanderzant, Z. L. Carpenter, and G. C. Smith. 1979. Attachment of microorganisms to pork skin and surfaces of beef and lamb carcasses. *J. Food Prot.* 42:401–406.

33. Byrne, L., K. T. Lau, and D. Diamond. 2003. Development of pH sensitive films for monitoring spoilage volatiles released into packaged fish headspace. *Ir. J. Agric. Food Res.* 42:119–129.

34. Chaouqy, N. E., E. I. Marakchi, and A. Zekhnini. 2005. Bacteria active in the spoilage of anchovy (*Engraulis encrasicholus*) stored in ice and at ambient temperature. *Sci. Aliment.* 25:129–146.

35. Chemnitius, G. C., and U. Bilitewski. 2000. Development of screen-printed electrodes for the estimation of fish quality. *Sensors Actuators* B 67:254–260.

36. Chemnitius, G. C., M. Suzuki, I. Kimiyasu, J. Kimura, I. Karube, and R. Schmid. 1992. Thin-film polyamine biosensor: substrate specificity and application to fish freshness determination. *Anal. Chim. Acta* 263:93–100.

37. Christensen, H., M. Boye, L. K. Poulsen, and O. F. Rasmussen. 1994. Analysis of fluorescent pseudomonads based on 23S ribosomal DNA sequences. *Appl. Environ. Microbiol.* 60:2196–2199.

38. Chung, K.-T., J. S. Dickson, and J. D. Crouse. 1989. Attachment and proliferation of bacteria on meat. *J. Food Prot.* 52:173–177.

39. Church, P. N., A. R. Davies, A. Slade, R. J. Hart, and P. A. Gibbs. 1992. Improving the safety and quality of meat and meat products by modified atmosphere and assessment by novel methods. *FLAIR 89055 Interim 2nd Year Report*, EEC DGXII, Brussels, Belgium.

40. Cogan, J. F., D. Walsh, and S. Condon. 1989. Impact of aeration on the metabolic end-products formed from glucose and galactose by *Streptococcus lactis*. *J. Appl. Bacteriol.* 66:77–84.

41. Collins, M. D., J. A. E. Farrow, B. A. Phillips, S. Ferusu, and D. Jones. 1987. Classification of *Lactobacillus divergens*, *Lactobacillus piscicola*, and some catalase-negative, asporogenous, rod-shaped bacteria from poultry in a new genus, *Carnobacterium*. *Int. J. Syst. Bacteriol.* 37:310–316.

42. Collins, M. D., J. Samelis, J. Metaxopoulos, and S. Wallbanks. 1993. Taxonomic studies on some leuconostoc-like organisms from fermented sausages: description of a new genus *Weissella* for the *Leuconostoc paramesenteroides* group of species. *J. Appl. Bacteriol.* 75:595–603.

43. Condon, S. 1987. Responses of lactic acid bacteria to oxygen. *FEMS Microbiol. Rev.* 46:269–280.

44. Costas, M., H. B. Olmes, S. L. W. On, and D. E. Stead. 1992. Identification of medically important *Pseudomonas* species using computerized methods. *Appl. Environ. Microbiol.* 58:1–18.

45. Costerson, J. W., R. T. Irvin, and K.-J. Cheng. 1981. The bacterial glycocalyx in nature and disease. *Annu. Rev. Microbiol.* 35:299–324.

46. Cotton, L. N., and D. L. Marshall. 1998. Rapid impediometric method to determine crustacean food freshness, p. 147–160. *In* M. H. Tunick, S. A. Palumbo, and P. M. Fratamico (ed.), *New Techniques in the Analysis of Foods*. Plenum Publishing Corp., New York, N.Y.

47. Cox, N. A., L. J. Richardson, J. S. Bailey, D. E. Cosby, J. A. Cason, and M. T. Musgrove. 2005. Bacterial contamination of poultry as a risk to human health, p. 21–43. *In* J. Mead (ed.), *Food Safety Control in the Poultry Industry*. CRC Woodhead Publishing Limited, New York, N.Y.

48. Cselovszky, J., G. Wolf, and W. P. Hammes. 1992. Production of formate, acetate and succinate by anaerobic fermentation of *Lactobacillus pentosus* in the presence of citrate. *Appl. Microbiol. Biotechnol.* 37:94–97.

49. Cutter, L. A., P. M. van Schie, and M. Fletcher. 2003. Adhesion of anaerobic microorganisms to solid surfaces and the effect of sequential attachment on adhesion characteristics. *Biofouling* 19:9–18.

50. Dainty, R., and B. M. Mackey. 1992. The relationship between the phenotypic properties of bacteria from chill-stored meat and spoilage process, p. 103S–114S. *In* R. G. Board, D. Jones, R. G. Kroll, and G. L. Pettipher (ed.), *Society for Applied Bacteriology Symposium Series No. 21*. Blackwell Scientific Publications, Oxford, United Kingdom.

51. Dainty, R. H. 1981. Volatile fatty acid detected in vacuum packed beef during storage at chill temperatures, p. 688–690. *In Proceedings of 27th Meeting of European Meat Workers*. Vienna, Austria.

52. Dainty, R. H. 1996. Chemical/biochemical detection of spoilage. *Int. J. Food Microbiol.* 33:19–34.

53. Dainty, R. H., R. A. Edwards, and C. M. Hibbard. 1985. Time course of volatile compound formation during refrigerated storage of naturally contaminated beef in air. *J. Appl. Bacteriol.* 59:303–309.

54. Dainty, R. H., R. A. Edwards, and C. M. Hibbard. 1989. Spoilage of vacuum-packed beef by a *Clostridium* sp. *J. Sci. Food Agric.* 49:473–486.

55. Dainty, R. H., R. A. Edwards, C. M. Hibbard, and J. J. Marnewick. 1989. Volatile compounds associated with microbial growth on normal and high pH beef stored at chill temperatures. *J. Appl. Bacteriol.* 66:281–289.

56. Dainty, R. H., R. A. Edwards, C. M. Hibbard, and C. V. Ramantanis. 1987. Amines in fresh beef of normal pH and the role of bacteria in changes in concentration observed during storage in vacuum packs at chill temperature. *J. Appl. Bacteriol.* 63:427–434.

57. Dainty, R. H., and C. M. Hibbard. 1980. Aerobic metabolism of *Brochothrix thermosphacta* growing on meat surfaces and in laboratory media. *J. Appl. Bacteriol.* 48:387–396.

58. Dainty, R. H., and C. M. Hibbard. 1983. Precursors of the major end products of aerobic metabolism of *Brochothrix thermosphacta*. *J. Appl. Bacteriol.* 55:387–396.

59. Dainty, R. H., and F. J. K. Hoffman. 1983. The influence of glucose concentration and culture incubation time on end-product formation during aerobic growth of *Brochothrix thermosphacta*. *J. Appl. Bacteriol.* 55:233–239.

60. Dainty, R. H., B. G. Shaw, K. A. De Boer, and S. J. Scheps. 1975. Protein changes caused by bacterial growth on beef. *J. Appl. Bacteriol.* 39:73–81.

61. **Dainty, R. H., B. G. Shaw, and T. A. Roberts.** 1983. Microbial and chemical changes in chill-stored red meat, p. 151–178. *In* T. A. Roberts and F. A. Skinner (ed.), *Food Microbiology: Advances and Prospects.* Academic Press, London, United Kingdom.

62. **Dalgaard, P.** 2000. Fresh and lightly preserved seafood, p. 110–139. *In* C. M. D. Man and A. A. Jones (ed.), *Shelf-Life Evaluation of Foods.* Aspen Publishers Inc., London, United Kingdom.

63. **Dalgaard, P., P. Buch, and S. Silberg.** 2002. Seafood spoilage predictor—development and distribution of a product specific application software. *Int. J. Food Microbiol.* **73:**343–349.

64. **Dens, E. J., and J. F. Van Impe.** 2003. Modelling applied to foods: predictive microbiology for solid food systems, p. 475–506. *In* P. Zeuthen and L. Bogh-Sorensen (ed.), *Food Preservation Techniques.* CRC/Woodhead Publishing Limited, Cambridge, United Kingdom.

65. **de Pablo, B., M. A. Asensio, B. Sanz, and J. A. Ordonez.** 1989. The D(–) lactic acid and acetoin/diacetyl as potential indicators of the microbial quality of vacuum-packed pork and meat products. *J. Appl. Bacteriol.* **66:**185–190.

66. **Devlieghere, F., A. H. Geeraerd, K. J. Versyck, B. Vandewaetere, J. Van Impe, and J. Debevere.** 2001. Growth of *Listeria monocytogenes* in modified atmosphere packed cooked meat products: a predictive model. *Food Microbiol.* **18:**53–66.

67. **Dewanti, R., and A. C. L. Wong.** 1995. Influence of culture conditions on biofilm formation by *Escherichia coli* O157:H7. *Int. J. Food Microbiol.* **26:**147–164.

68. **Dickson, J. S.** 1991. Attachment of *Salmonella typhimurium* and *Listeria monocytogenes* to beef tissue: effects of inoculum level, growth temperature and bacterial culture age. *Food Microbiol.* **8:**143–151.

69. **Dickson, J. S., and M. E. Anderson.** 1992. Microbiological decontamination of food animal carcasses by washing and sanitizing systems: a review. *J. Food Prot.* **55:**133–140.

70. **Drosinos, E. H.** 1994. Microbial associations of minced lamb and their ecophysiological attributes. Ph.D. thesis. University of Bath, Bath, United Kingdom.

71. **Drosinos, E. H., and R. G. Board.** 1994. Metabolic activities of pseudomonads in batch cultures in extract of minced lamb. *J. Appl. Bacteriol.* **77:**613–620.

72. **Drosinos, E. H., and R. G. Board.** 1995. Attributes of microbial associations of meat growing as xenic batch cultures in a meat juice at 4°C. *Int. J. Food Microbiol.* **26:**279–293.

73. **Drosinos, E. H., K. Lampropoulou, E. Mitre, and G.-J. E. Nychas.** 1997. Attributes of fresh gilt-head seabream (*Sparus aurata*) fillets treated with potassium sorbate, sodium gluconate and stored under a modified atmosphere at 0–1°C. *J. Appl. Microbiol.* **83:**569–575.

74. **Drosinos, E. H., and G.-J. E. Nychas.** 1996. *Brochothrix thermosphacta*, the climax micro-organism on Greek fish tsipoura (*Sparus aurata*) and gopa (*Boops boops*) stored under a modified atmosphere at 0–4°C. *Ital. J. Food Sci. Technol.* **8:**323–330.

75. **Drosinos, E. H., and G.-J. E. Nychas.** 1997. Production of acetate and lactate in relation to glucose content during

modified atmosphere storage of gilt-head seabream (*Sparus aurata*) at 0+1°C. *Food Res. Int.* **30:**711–717.

76. **Duffy, L. L., P. B. Vanderlinde, and F. H. Grau.** 1994. Growth of *Listeria monocytogenes* on vacuum-packed cooked meats: effect of pH, a_w, nitrite and ascorbate. *Int. J. Food Microbiol.* **23:**377–390.

77. **Edwards, R. A., R. H. Dainty, and C. M. Hibbard.** 1987. Volatile compounds produced by meat pseudomonads and related reference strains during growth in air at chill temperatures. *J. Appl. Bacteriol.* **62:**403–412.

78. **Eisenberg, R. C., S. J. Butters, S. C. Quay, and S. B. Friedman.** 1974. Glucose uptake and phosphorylation in *Pseudomonas fluorescens*. *J. Bacteriol.* **120:**147–153.

79. **Ellis, D. I., D. Broadhurst, D. B. Kell, J. J. Rowland, and R. Goodacre.** 2002. Rapid and quantitative detection of the microbial spoilage of meat by Fourier transform infrared spectroscopy and machine learning. *Appl. Environ. Microbiol.* **68:**2822–2828.

80. **Ellis, D. I., and R. Goodacre.** 2001. Rapid and quantitative detection of the microbial spoilage of muscle foods: current status and future trends. *Trends Food Sci. Technol.* **12:**414–424.

81. **Emborg, J., B. G. Laursen, and P. Dalgaard.** 2005. Significant histamine formation in tuna (*Thunnus albacares*) at 2°C—effect of vacuum- and modified atmosphere-packaging on psychrotolerant bacteria. *Int. J. Food Microbiol.* **101:**263–279.

82. **Farber, J. M., Y. Cai, and W. H. Ross.** 1996. Predictive modeling of the growth of *Listeria monocytogenes* in CO_2 environments. *Int. J. Food Microbiol.* **32:**133–144.

83. **Farber, J. M., and E. S. Idziak.** 1982. Detection of glucose oxidation products in chilled fresh beef undergoing spoilage. *Appl. Environ. Microbiol.* **44:**521–524.

84. **Farber, J. M., and E. S. Idziak.** 1984. Attachment of psychrotrophic meat spoilage bacteria to muscle surfaces. *J. Food Prot.* **47:**92–95.

85. **Feiger, E. A., and A. F. Novak.** 1961. Microbiology of shellfish deterioration, p. 561–611. *In* G. Borgstrom (ed.), *Fish as Food*, vol. 1. *Production, Biochemistry, and Microbiology.* Academic Press, New York, N.Y.

86. **Fernandez, P. S., S. M. George, C. S. Sills, and M. W. Peck.** 1997. Predictive model of the effect of CO_2, pH, temperature, and NaCl on the growth and survival of foodborne pathogenic bacteria. *Int. J. Food Microbiol.* **37:**37–45.

87. **Firstenberg-Eden, R.** 1981. Attachment of bacteria to meat surfaces: a review. *J. Food Prot.* **44:**602–607.

88. **Firstenberg-Eden, R., S. Notermans, and M. Van Schothorst.** 1978. Attachment of certain bacterial strains to chicken and beef meat. *J. Food Safety* **1:**217–228.

89. **Freeman, L. R., G. J. Silverman, P. Angelini, C. Merritt, Jr., and W. B. Esselen.** 1976. Volatiles produced by microorganisms isolated from refrigerated chicken at spoilage. *Appl. Environ. Microbiol.* **32:**222–231.

90. **Garcia-Lopez, M. L., M. Prieto, and A. Otero.** 1998. The physiological attributes of Gram-negative bacteria associated with spoilage of meat and meat products, p. 1–34. *In* R. G. Board and A. R. Davies (ed.), *The Microbiology of Meat and Poultry.* Blackie Academic and Professional, London, United Kingdom.

91. **Gardner, G. A.** 1982. Microbiology of processing: bacon and ham, p. 129–178. *In* M. H. Brown (ed.), *Meat Microbiology*. Applied Science Publishers, Inc., New York, N.Y.

92. **Garvie, E. I.** 1980. Bacterial lactate dehydrogenase. *Microbiol. Rev.* **44:**106–139.

93. **Geesey, G. G.** 2001. Bacterial behavior at surfaces. *Curr. Microbiol.* **4:**296–300.

94. **Geornaras, I., and J. N. Sofos.** 2005. Combining physical and chemical decontamination interventions for meat, p. 433–460. *In* J. N. Sofos (ed.), *Improving the Safety of Fresh Meat*. CRC/Woodhead Publishing Limited, Cambridge, United Kingdom.

95. **Giannakourou, M., K. Koutsoumanis, G.-J. E. Nychas, and P. S. Taoukis.** 2001. Development and assessment of an intelligent shelf life decision system (SLDS) for quality optimization of the food chill chain. *J. Food Prot.* **64:**1051–1057.

96. **Giaouris, E., N. Chorianopoulos, and G.-J. E. Nychas.** 2005. Effect of temperature, pH, and water activity on biofilm formation by *Salmonella enterica* Enteritidis PT4 on stainless steel surfaces as indicated by the bead vortexing method and conductance measurements. *J. Food Prot.* **68:**2149–2154.

97. **Gill, C. O.** 1976. Substrate limitation of bacterial growth at meat surfaces. *J. Appl. Bacteriol.* **41:**401–410.

98. **Gill, C. O.** 1982. Microbial interaction with meats, p. 225–264. *In* M. H. Brown (ed.), *Meat Microbiology*. Applied Science Publishers, Inc., New York, N.Y.

99. **Gill, C. O.** 1983. Meat spoilage and evaluation of the potential storage life of fresh meat. *J. Food Prot.* **46:**444–452.

100. **Gill, C. O.** 1986. The control of microbial spoilage in fresh meats, p. 49–88. *In* A. M. Pearson and T. R. Dutson (ed.), *Advances in Meat Research*, vol. 2. *Meat and Poultry Microbiology*. AVI Publishing Co., Inc., Westport, Conn.

101. **Gill, C. O.** 2005. Sources of microbial contamination at slaughtering plants, p. 231–243. *In* J. N. Sofos (ed.), *Improving the Safety of Fresh Meat*. CRC/Woodhead Publishing Limited, Cambridge, United Kingdom.

102. **Gill, C. O., and G. Molin.** 1991. Modified atmospheres and vacuum packaging, p. 172–199. *In* N. J. Russell and G. W. Gould (ed.), *Food Preservatives*. Blackie, Glasgow, United Kingdom.

103. **Gill, C. O., and K. G. Newton.** 1977. The development of aerobic spoilage flora on meat stored at chill temperatures. *J. Appl. Microbiol.* **43:**189–195.

104. **Gill, C. O., and K. G. Newton.** 1978. The ecology of bacterial spoilage of fresh meat at chill temperatures. *Meat Sci.* **2:**207–217.

105. **Gill, C. O., and K. G. Newton.** 1979. Spoilage of vacuum-packaged dark, firm, dry meat at chill temperatures. *Appl. Environ. Microbiol.* **37:**362–364.

106. **Gill, C. O., and K. G. Newton.** 1982. Effect of lactic acid concentration on growth on meat of gram-negative psychrotrophs from a meatworks. *Appl. Environ. Microbiol.* **43:**284–288.

107. **Gill, C. O., and N. Penney.** 1977. Penetration of bacteria into meat. *Appl. Environ. Microbiol.* **33:**1284–1286.

108. **Gill, T. A.** 1992. Chemical and biochemical indices in seafood quality, p. 377–387. *In* H. H. Huss, M. Jacobsen, and J. Liston (ed.), *Quality Assurance in the Fish Industry*. Elsevier, Amsterdam, The Netherlands.

109. **Gill, T. A.** 1997. Advanced analytical tools in seafood science. *Dev. Food Sci.* **38:**479–490.

110. **Goodacre, R.** 2000. Applications of artificial neural networks to the analysis of multivariate data, p. 123–152. *In* H. M. Cartwright (ed.), *Intelligent Data Analysis in Science: a Handbook*. Oxford University Press, Oxford, United Kingdom.

111. **Goodacre, R., S. Vaidyanathan, W. B. Dunn, G. G. Harrigan, and D. B. Kell.** 2004. Metabolomics by numbers: acquiring and understanding global metabolite data. *Trends Biotechnol.* **22:**245–252.

112. **Gram, L., and P. Dalgaard.** 2002. Fish spoilage bacteria—problems and solutions. *Curr. Opin. Biotechnol.* **13:**262–266.

113. **Gram, L., and H. H. Huss.** 1996. Microbiological spoilage of fish and fish products. *Int. J. Food Microbiol.* **33:**121–137.

114. **Gram, L., and H. H. Huss.** 2000. Fresh and processed fish and shellfish, p. 472–506. *In* B. M. Lund, A. C. Baird-Parker, and G. W. Gould (ed.), *The Microbiological Safety and Quality of Foods*. Aspen Publishers, Inc., Gaithersburg, Md.

115. **Grandin, T.** 2005. Animal welfare and food safety at the slaughter plant, p. 244–258. *In* J. N. Sofos (ed.), *Improving the Safety of Fresh Meat*. CRC/Woodhead Publishing Limited, Cambridge, United Kingdom.

116. **Grant, G. F., A. R. McCurdy, and A. D. Osborne.** 1988. Bacterial greening in cured meats: a review. *Can. Inst. Food Sci. Technol. J.* **21:**50–56.

117. **Grau, F. H.** 1980. Inhibition of the anaerobic growth of *Brochothrix thermosphacta* by lactic acid. *Appl. Environ. Microbiol.* **40:**433–436.

118. **Greaser, M. L.** 1986. Conversion of muscle to meat, p. 37–102. *In* P. J. Bechtel (ed.), *Muscle as Food*. Academic Press, Inc., New York, N.Y.

119. **Gupta, L. K., and Y. Nagamohini.** 1992. Penetration of poultry meat by *Pseudomonas* and *Lactobacillus* spp. *World J. Microbiol. Biotechnol.* **8:**212–213.

120. **Hanna, M. O., D. L. Zink, Z. L. Carpenter, and C. Vanderzant.** 1976. *Yersinia enterocolitica*-like organisms from vacuum-packaged beef and lamb. *J. Food Sci.* **41:**1254–1256.

121. **Hasegawa, T., A. M. Pearson, J. F. Price, and R. V. Lechowich.** 1970. Action of bacterial growth on the sarcoplasmic and urea-soluble proteins from muscle. I. Effects of *Clostridium perfringens*, *Salmonella enteritidis*, *Achromobacter liquefaciens*, *Streptococcus faecalis* and *Kurthia zopfii*. *Appl. Microbiol.* **20:**117–122.

122. **Hasegawa, T., A. M. Pearson, J. H. Rampton, and R. V. Lechowich.** 1970. Effect of microbial growth upon sarcoplasmic and urea-soluble proteins from muscle. *J. Food Sci.* **35:**720–724.

123. **Hitchener, B. J., A. F. Egan, and P. J. Rogers.** 1979. Energetics of *Microbacterium thermosphactum* in glucose-limited continuous culture. *Appl. Environ. Microbiol.* **37:**1047–1052.

124. Holzapfel, W. H. 1998. The Gram-positive bacteria associated with meat and meat products, p. 35–84. *In* R. G. Board and A. R. Davies (ed.), *The Microbiology of Meat and Poultry*. Blackie Academic and Professional, London, United Kingdom.

125. Holzapfel, W. H., and E. S. Gerber. 1983. *Lactobacillus divergens* sp. nov., a new homofermentative *Lactobacillus* species producing L(+) lactate. *Syst. Appl. Microbiol.* **4**:522–534.

126. Huffman, N. M. 2002. Current and future technologies for the decontamination of carcasses and fresh meat. *Meat Sci.* **62**:285–294.

127. Huis in't Veld, J. H. J. 1996. Microbial and biochemical spoilage of foods: an overview. *Int. J. Food Microbiol.* **33**:1–18.

128. Huss, H. H. 1995. Assurance of seafood quality. FAO Fisheries technical paper no. 334. United Nations Food and Agriculture Organization, Rome, Italy.

129. Huss, H. H., P. Dalgaard, and L. Gram. 1997. Microbiology of fish and fish products. *Dev. Food Sci.* **38**:413–430.

130. Ingram, M., and R. H. Dainty. 1971. Changes caused by microbes in spoilage of meats. *J. Appl. Bacteriol.* **34**:21–39.

131. Ingram, M., and B. Simonsen. 1980. Meats and meat products, p. 333–409. *In* J. H. Silliker, R. P. Elliot, A. C. Baird-Parker, F. L. Bryan, J. H. B. Christian, D. S. Clark, J. C. Olsen, Jr., and T. A. Roberts (ed.), *Microbial Ecology of Foods*, vol. 2. *Food Commodities*. Academic Press, New York, N.Y.

132. **International Commission on Microbiological Specifications for Foods.** 1998. *Microorganisms in Foods 6, Microbial Ecology of Food Commodities*. Blackie Academic and Professional, London, United Kingdom.

133. Jackson, T. C., D. L. Marshall, G. R. Acuff, and J. S. Dickson. 2001. Meat, poultry, and seafood, p. 91–110. *In* M. P. Doyle, L. R. Beuchat, and T. J. Montville (ed.), *Food Microbiology: Fundamentals and Frontiers*, 2nd ed. ASM Press, Washington, D.C.

134. Jay, J. M. 1986. Microbial spoilage indicators and metabolites, p. 219–240. *In* M. D. Pierson and N. J. Sterm (ed.), *Foodborne Microorganisms and Their Toxins: Developing Methodology*. Marcel Dekker Inc., Basel, Switzerland.

135. Jay, J. M. 2000. *Modern Food Microbiology*, 6th ed. Aspen Publishers, Gaithersburg, Md.

136. Jay, J. M., and L. A. Shelef. 1976. Effect of microorganisms on meat proteins at low temperatures. *J. Agric. Food Chem.* **24**:1113–1116.

137. Johansen, K., S. Andersen, and C. Jacobsen. 1996. Phenotypic and genotypic characterization of phenanthrene-degrading fluorescent *Pseudomonas* biovars. *Appl. Environ. Microbiol.* **62**:3818–3825.

138. Jørgensen, L. V. 2000. Spoilage and safety of cold-smoked salmon. Ph.D. thesis. Danish Institute for Fisheries Research, Lyngby, Denmark.

139. Jørgensen, L. V., H. H. Huss, and P. Dalgaard. 2000. The effect of biogenic amine production by single bacterial cultures and metabiosis on cold-smoked salmon. *J. Appl. Microbiol.* **89**:920–934.

140. Juillard, V., and J. Richard. 1994. Mixed cultures in milk of a proteinase-positive and a proteinase-negative variant of *Lactococcus lactis* subsp. *lactis*; influence of initial percentage of proteinase-positive cells on the growth parameters of each strain and on the rate of acidification. *Lait* **74**:3–12.

141. Kakouri, A., E. Drosinos, and G.-J. E. Nychas. 1997. Storage of Mediterranean fresh fish (*Boops boops*, and *Sparus aurata*) under modified atmospheres or vacuum at 3 and 10°C. *Dev. Food Sci.* **38**:171–178.

142. Kakouri, A., and G.-J. E. Nychas. 1994. Storage of poultry meat under modified atmospheres or vacuum packs; possible role of microbial metabolites as indicator of spoilage. *J. Appl. Bacteriol.* **76**:163–172.

143. Kalchayanand, N., B. Ray, and R. A. Field. 1993. Characteristics of psychrotrophic *Clostridium laramie* causing spoilage of vacuum-packaged refrigerated fresh and roasted beef. *J. Food Prot.* **56**:13–17.

144. Kalinowski, R. M., and R. B. Tompkin. 1999. Psychrotrophic clostridia causing spoilage in cooked meat and poultry products. *J. Food Prot.* **62**:766–772.

145. Kandler, O. 1983. Carbohydrate metabolism in lactic acid bacteria. *Antonie Leeuwenhoek* **49**:209–224.

146. Kandler, O., and N. Weiss. 1986. Genus *Lactobacillus*, p. 1208–1234. *In* P. H. A. Sneath, N. S. Mair, M. E. Sharpe, and J. G. Holt (ed.), *Bergey's Manual of Systematic Bacteriology*. Williams and Wilkins, Baltimore, Md.

147. Karpas, Z., B. Tilman, R. Gdalevsky, and A. Lorber. 2002. Determination of volatile biogenic amines in muscle food products by ion mobility spectrometry. *Anal. Chim. Acta* **463**:155–163.

148. Karube, I., I. Satoh, Y. Araki, S. Suzuki, and H. Yamada. 1980. Monoamine oxidase electrode in freshness testing of meat. *Enzyme Microb. Technol.* **2**:117–120.

149. Kersters, K., W. Ludwig, M. Vancanneyt, P. De Vos, M. Gillis, and K. H. Schleifer. 1996. Recent changes in the classification of pseudomonads: an overview. *Syst. Appl. Microbiol.* **19**:465–477.

150. Koutsoumanis, K., and G.-J. E. Nychas. 1999. Chemical and sensory changes associated with microbial flora of Mediterranean boque (*Boops boops*) stored aerobically at 0, 3, 7, and 10°C. *Appl. Environ. Microbiol.* **65**:698–706.

151. Koutsoumanis, K., A. Stamatiou, P. Skandamis, and G.-J. E. Nychas. 2006. Development of a microbial model for the combined effect of temperature and pH on spoilage of ground meat and validation of the model under dynamic temperature conditions. *Appl. Environ. Microbiol.* **72**:124–134.

152. Koutsoumanis, K. P. 2001. Predictive modeling of the shelf life of fish under nonisothermal conditions. *Appl. Environ. Microbiol.* **67**:1821–1829.

153. Koutsoumanis, K. P., I. Geornaras, and J. N. Sofos. 2006. Microbiology of land muscle food. *In* Y. H. Hui, *Handbook of Food Science*. Marcel Dekker Inc., New York, N.Y.

154. Koutsoumanis, K. P., M. Giannakourou, P. S. Taoukis, and G.-J. E. Nychas. 2002. Application of SLDS (shelf life decision system) to marine cultured fish quality. *Int. J. Food Microbiol.* **73**:375–382.

155. **Koutsoumanis, K. P., K. Lambropoulou, and G.-J. E. Nychas.** 1999. Biogenic and sensory changes associated with the microbial flora of Mediterranean gilt-head seabream (*Sparus aurata*) stored aerobically at 0, 8, and 15°C. *J. Food Prot.* **62**:392–402.

156. **Koutsoumanis, K. P., and J. N. Sofos.** 2004. Microbial contamination of carcasses and cuts, p. 727–737. *In* W. K. Jensens (ed.), *Encyclopedia of Meat Sciences.* Elsevier Academic Press, Amsterdam, The Netherlands.

157. **Koutsoumanis, K. P., and P. Taoukis.** 2005. Meat safety, refrigerated storage and transport: modeling and management, p. 503–561. *In* J. N. Sofos (ed.), *Improving the Safety of Fresh Meat.* Woodhead/Publishing, Ltd, CRC Press, Cambridge, United Kingdom.

158. **Kraft, A. A.** 1992. *Psychrotrophic Bacteria in Foods*: Disease and Spoilage. CRC Press, Inc., Boca Raton, Fla.

159. **Kuda, T., C. Matsumoto, and T. Yano.** 2002. Changes in acid and alkaline phosphatase activities during the spoilage of raw muscle from horse mackerel *Trachurus japonicus* and gurnard *Lepidotriga microptera. Food Chem.* **76**:443–447.

160. **Lambert, A. D., J. P. Smith, and K. L. Dodds.** 1991. Shelf life extension and microbiological safety of fresh meat—a review. *Food Microbiol.* **8**:267–297.

161. **Lambropoulou, K., E. H. Drosinos, and G.-J. E. Nychas.** 1996. The effect of the addition of glucose to normal and high pH meat. *Int. J. Food Microbiol.* **30**:281–291.

162. **Lambropoulou, K. A.** 1995. The role of glucose in meat. M.Sc. thesis. University of Humberside, Humberside, United Kingdom.

163. **Lasta, J. A., N. Pensel, M. Masana, H. R. Rodriguez, and P. T. Garcia.** 1995. Microbial growth and biochemical changes on naturally contaminated chilled-beef subcutaneous adipose tissue stored aerobically. *Meat Sci.* **39**:149–158.

164. **Lawrie, R. A.** 1985. *Meat Science*, 4th ed. Pergamon Press, New York, N.Y.

165. **Lehane, L., and J. Olley.** 2000. Histamine fish poisoning revisited. *Int. J. Food Microbiol.* **58**:1–37.

166. **Lopez-Caballero, M. E., J. A. Sanchez-Fernandez, and A. Moral.** 2001. Growth and metabolic activity of *Shewanella putrefaciens* maintained under different CO_2 and O_2 concentrations. *Int. J. Food Microbiol.* **64**:277–287.

167. **Lopez-Caballero, M. E., M. D. A. Torres, J. A. Sanchez-Fernandez, and A. Moral.** 2002. *Photobacterium phosphoreum* isolated as a luminescent colony from spoiled fish, cultured in model system under controlled atmospheres. *Eur. Food Res. Technol.* **215**:390–395.

168. **Lou, Y. Q., and A. E. Yousef.** 1997. Adaptation to sublethal environmental stresses protects *Listeria monocytogenes* against lethal preservation factors. *Appl. Environ. Microbiol.* **63**:1252–1255.

169. **Luten, J. B., W. Bouquet, L. A. J. Seuren, M. M. Burggraaf, G. Riekwel-Booy, P. Durand, M. Etienne, J. P. Gouyou, A. Landrein, A. Ritchie, M. Leclerq, and R. Guinet.** 1992. Biogenic amines in fishery products: standardization methods within EC, p. 427–439. *In* H. H. Huss, M. Jakobsen, and J. Liston (ed.), *Quality Assurance in the Fish Industry.* Elsevier, Amsterdam, The Netherlands.

170. **Marshall, D. L., L. S. Andrews, J. H. Wells, and A. J. Farr.** 1992. Influence of modified atmosphere packaging on the competitive growth of *Listeria monocytogenes* and *Pseudomonas fluorescens* on precooked chicken. *Food Microbiol.* **9**:303–309.

171. **Marshall, D. L., and M. F. A. Bal'a.** 2001. Microbiology of meats, p. 149–169. *In* Y. H. Hui, W. K. Nip, R. W. Rogers, and O. A. Young (ed.), *Meat Science and Applications.* Marcel Dekker, Inc., New York, N.Y.

172. **Marshall, D. L., and P. L. W. Lehigh.** 1993. Nobody's nose knows. *Chemtech* **23**:38–42.

173. **Mayr, D., R. Margesin, E. Klingsbichel, E. Hartungen, D. Jenewein, F. Skinner, and T. D. Mark.** 2003. Rapid detection of meat spoilage by measuring volatile organic compounds by using proton transfer reaction mass spectrometry. *Appl. Environ. Microbiol.* **69**:4697–4705.

174. **McClure, P. J., C. Blackburn, M. B. Cole, P. S. Curtis, J. E. Jones, J. D. Legan, I. D. Ogden, K. M. W. Peck, T. A. Roberts, J. P. Sutherland, and S. J. Walker.** 1994. Modeling the growth, survival and death of microorganisms in foods: the UK food micromodel approach. *Int. J. Food Microbiol.* **34**:265–275.

175. **McEldowney, S., and M. Fletcher.** 1987. Adhesion of bacteria from mixed cell suspension to solid surfaces. *Arch. Microbiol.* **148**:57–62.

176. **McMeekin, T. A.** 1975. Spoilage association of chicken breast muscle. *Appl. Microbiol.* **29**:44–47.

177. **McMeekin, T. A.** 1977. Spoilage association of chicken leg muscle. *Appl. Microbiol.* **33**:1244–1246.

178. **McMeekin, T. A.** 1982. Microbial spoilage of meats, p. 1–40. *In* R. Davies (ed.), *Developments in Food Microbiology—1.* Applied Science Publishers, London, United Kingdom.

179. **McMeekin, T. A., J. N. Olley, T. Ross, and D. R. Ratkowsky.** 1993. *Predictive Microbiology—Theory and Application*, p. 339. Research Studies Press Ltd., Taunton, United Kingdom.

180. **Metcalfe, A. M., and D. L. Marshall.** 2004. Capacitance method to determine the microbiological quality of raw shrimp (*Penaeus setiferus*). *Food Microbiol.* **21**: 361–364.

181. **Meyns, B., N. Begazo, and W. Schmidt-Lorenz.** 1992. Concentration changes of glucose, glycogen, L(+) and D(−) lactic acid during microbial spoilage of beef. *Mitt. Geb. Lebensmittelunters. Hyg.* **83**:121–125.

182. **Molin, G.** 1985. Mixed carbon source utilization of meat-spoiling *Pseudomonas fragi* 72 in relation to oxygen limitation and carbon dioxide inhibition. *Appl. Environ. Microbiol.* **49**:1442–1447.

183. **Molin, G., and A. Tenstrom.** 1986. Phenotypically based taxonomy of psychrotrophic *Pseudomonas* isolated from spoiled meat, water, and soil. *Int. J. Syst. Bacteriol.* **36**:257–274.

184. **Moore, W. E. C., E. P. Cato, and L. V. H. Moore.** 1985. Index of bacterial and yeast nomenclatural changes published in the *International Journal of Systematic Bacteriology* since the 1980s. Approved lists of bacterial names (1 January 1980 to 1 January 1985). *Int. J. Syst. Bacteriol.* **35**:382–407.

185. **Murphy, M. G., and S. Condon.** 1984. Comparison of aerobic and anaerobic growth of *Lactobacillus plantarum* in a glucose medium. *Arch. Microbiol.* **138:** 49–53.

186. **Murphy, M. G., and S. Condon.** 1984. Correlation of oxygen utilization and H_2O_2 accumulation with oxygen induced enzymes in *Lactobacillus plantarum* cultures. *Arch. Microbiol.* **138:**44–48.

187. **Nakamura, M., Y. Wada, H. Sawaya, and T. Kawabata.** 1979. Polyamine content in fresh and processed pork. *J. Food Sci.* **44:**515–517, 523.

188. **Nassos, P. S., A. D. King, Jr., and A. E. Stafford.** 1983. Relationship between lactic acid concentration and bacterial spoilage in ground beef. *Appl. Environ. Microbiol.* **46:**894–900.

189. **Nassos, P. S., A. D. King, Jr., and A. E. Stafford.** 1985. Lactic acid concentration and microbial spoilage in anaerobically and aerobically stored ground beef. *J. Food Sci.* **50:**710–712, 715.

190. **Nassos, P. S., A. D. King, Jr., and A. E. Stafford.** 1988. Lactic acid concentration as an indicator of acceptability in refrigerated or freeze-thawed ground beef. *Appl. Environ. Microbiol.* **54:**822–823.

191. **Newton, K. G., and C. O. Gill.** 1980. The microbiology of DFD fresh meats: a review. *Meat Sci.* **5:**223–232.

192. **Newton, K. G., and W. J. Rigg.** 1979. The effect of film permeability on the storage life and microbiology of vacuum packaged meat. *J. Appl. Bacteriol.* **47:**433–441.

193. **Nichol, D. J., M. K. Shaw, and D. A. Ledward.** 1970. Hydrogen sulfide production by bacteria and sulfmyoglobin formation in prepacked chilled beef. *Appl. Microbiol.* **19:**937–939.

194. **Nottingham, P. M.** 1982. Microbiology of carcass meats, p. 13–65. *In* M. H. Brown (ed.), *Meat Microbiology.* Applied Science Publishers Ltd., New York, N.Y.

195. **Nottingham, P. M., C. O. Gill, and K. G. Newton.** 1981. Spoilage at fat surfaces of meat, p. 183–190. *In* T. A. Roberts, G. Hobbs, J. H. B. Christian, and N. Skovgaard (ed.), *Psychrotrophic Microorganisms in Spoilage and Pathogenicity.* Academic Press, New York, N.Y.

196. **Nychas, G.-J., A. Robinson, and R. G. Board.** 1992. Microbiological and physico-chemical evaluation of ground beef from retail shops. *Fleisch. Int.* **1:**49–53.

197. **Nychas, G.-J. E.** 1994. Modified atmosphere packaging of meats, p. 417–436. *In* R. P. Singh and F. A. R. Oliveira (ed.), *Minimal Processing of Foods and Process Optimization, an Interface.* CRC Press, London, United Kingdom.

198. **Nychas, G.-J. E.** 2005. Bacterial talks—in which language? p. 38–39. *In Proceedings of the 1st Panhellenic Conference on Biotechnology of Foods, March 31–April 2.* The Association of Greek Chemists, Athens, Greece. (In Greek.)

199. **Nychas, G.-J. E., and J. S. Arkoudelos.** 1990. Microbiological and physico-chemical changes in minced meat under carbon dioxide, nitrogen or air at 3°C. *Int. J. Food Sci. Technol.* **25:**389–398.

200. **Nychas, G.-J. E., and J. S. Arkoudelos.** 1991. The influence of *Brochothrix thermosphacta* on the quality of minced meat. *Agric. Res.* **15:**103–115.

201. **Nychas, G.-J. E., V. M. Dillon, and R. G. Board.** 1988. Glucose the key substrate in the microbiological changes occurring in meat and certain meat products. *Biotechnol. Appl. Biochem.* **10:**203–231.

202. **Nychas, G.-J. E., E. H. Drosinos, and R. G. Board.** 1998. Chemical changes in stored meat, p. 288–326. *In* R. G. Board and A. R. Davies (ed.), *The Microbiology of Meat and Poultry.* Blackie Academic and Professional, London, United Kingdom.

203. **Nychas, G.-J. E., P. A. Gibbs, R. G. Board, and J. J. Sheridan.** 1994. Improving the safety and quality of meat and meat products by modified atmosphere and assessment by novel methods. FLAIR proposal no. 89055, contract no. AGRF/0024 (SCP), final report. EU, DGXII, Brussels, Belgium.

204. **Nychas, G.-J. E., and P. Skandamis.** 2005. Fresh meat spoilage and modified atmosphere packaging (MAP), p. 461–502. *In* J. N. Sofos (ed.), *Improving the Safety of Fresh Meat.* CRC/Woodhead Publishing Limited, Cambridge, United Kingdom.

205. **Nychas, G.-J. E., and C. C. Tassou.** 1997. Spoilage process and proteolysis in chicken as noted by HPLC method. *J. Sci. Food Agric.* **74:**199–208.

206. **Oehlenschläger, J.** 1992. Evaluation of some well established and some underrated indices for the determination of freshness and/or spoilage of ice stored wet fish, p. 339–350. *In* H. H. Huss, M. Jacobsen, and J. Liston (ed.), *Quality Assurance in the Fish Industry.* Elsevier, Amsterdam, The Netherlands.

207. **Olafsdottir, G., A. Hognadottir, E. Martinsdottir, and H. Jonsdottir.** 2000. Application of an electric nose to predict total volatile bases in capelin (*Mallotus villosus*) for fishmeal production. *J. Agric. Food Chem.* **48:**2353–2359.

208. **Olafsdottir, G., E. Martinsdottir, and E. H. Jonsson.** 1997. Gas sensor and GC measurements of volatile compounds in capelin. *Dev. Food Sci.* **38:**507–520.

209. **Ordonez, J. A., B. de Pablo, B. P. de Castro, M. A. Asensio, and B. Sanz.** 1991. Selected chemical and microbiological changes in refrigerated pork stored in carbon dioxide and oxygen enriched atmospheres. *J. Agric. Food Chem.* **39:**668–672.

210. **Palleroni, N. J.** 1984. Genus I. *Pseudomonas,* p. 141–199. *In* N. R. Krieg and J. G. Holt (ed.), *Bergey's Manual of Systematic Bacteriology.* Williams and Wilkins Co., Baltimore, Md.

211. **Palleroni, N. J.** 1992. Introduction to the family Pseudomonadaceae, p. 3071–3085. *In* A. Balows, H. G. Trüper, M. Dworkin, W. Harder, and K.-H. Schleifer (ed.), *The Prokaryotes,* 2nd ed. Springer-Verlag, New York, N.Y.

212. **Palleroni, N. J.** 1992. Human- and animal-pathogenic pseudomonads, p. 3086–3103. *In* A. Balows, H. G. Trüper, M. Dworkin, W. Harder, and K.-H. Schleifer (ed.), *The Prokaryotes,* 2nd ed. Springer-Verlag, New York, N.Y.

213. **Patterson, J. T., and P. A. Gibbs.** 1977. Incidence and spoilage potential of isolates from vacuum-packaged meat of high pH value. *J. Appl. Bacteriol.* **43:**25–38.

214. **Rasmussen, S. K. J., T. Ross, and T. McMeekin.** 2002. A process risk model for the shelf life of Atlantic salmon fillets. *Int. J. Food Microbiol.* **73:**47–60.

215. **Rey, C. R., A. A. Kraft, D. G. Topel, F. C. Parrish, Jr., and D. K. Hotchkiss.** 1976. Microbiology of pale, dark and normal pork. *J. Food Sci.* **41:**111–116.

216. **Ross, T.** 1996. Indices for performance evaluation of predictive models in food microbiology. *J. Appl. Bacteriol.* **81:**501–508.

217. **Saby, C., T. V. Nguyen, and J. H. T. Luong.** 2004. An electrochemical flow analysis system for putrescine using immobilized putrescine oxidase and horseradish peroxidase. *Electroanalysis* **16:**260–267.

218. **Sakamoto, M., and K. Komagata.** 1996. Aerobic growth of and activities of NADH oxidase and NADH peroxidase in lactic acid bacteria. *J. Ferment. Bioeng.* **82:**210–216.

219. **Samelis, J., and J. N. Sofos.** 2003. Strategies to control stress-adapted pathogens, p. 303–351. *In* A. E. Yousef and V. K. Juneja (ed.), *Microbial Stress Adaptation and Food Safety.* CRC Press, Boca Raton, Fla.

220. **Samelis, J., and J. N. Sofos.** 2003. Organic acids, p. 98–120. *In* S. Roller (ed.), *Natural Antimicrobials for the Minimal Processing of Foods.* Woodhead Publishing Limited, Cambridge, United Kingdom.

221. **Samelis, J., J. N. Sofos, P. A. Kendall, and G. C. Smith.** 2001. Fate of *Escherichia coli* O157:H7, *Salmonella* Typhimurium DT 104, and *Listeria monocytogenes* in fresh meat decontamination fluids at 4 and 10°C. *J. Food Prot.* **64:**950–957.

222. **Samelis, J., J. N. Sofos, P. A. Kendall, and G. C. Smith.** 2001. Influence of the natural microbial flora on the acid tolerance response of *Listeria monocytogenes* in a model system of fresh meat decontamination fluids. *Appl. Environ. Microbiol.* **67:**2410–2420.

223. **Santos, M. H. S.** 1996. Biogenic amines: their importance in foods. *Int. J. Food Microbiol.* **29:**213–231.

224. **Savell, J. W., D. B. Griffin, C. W. Dill, G. R. Acuff, and C. Vanderzant.** 1986. Effect of film oxygen transmission rate on lean color and microbiological characteristics of vacuum-packaged beef knuckles. *J. Food Prot.* **49:**917–919.

225. **Schmitt, R. E., and W. Schmidt-Lorenz.** 1992. Formation of ammonia and amines during microbial spoilage of refrigerated broilers. *Lebensm.-Wiss. Technol.* **25:**6–10.

226. **Schmitt, R. E., and W. Schmidt-Lorenz.** 1992. Degradation of amino acids and protein changes during microbial spoilage of chilled unpacked and packed chicken carcasses. *Lebensm.-Wiss. Technol.* **25:**11–20.

227. **Sedewitz, B., K. H. Schleifer, and F. Gotz.** 1984. Physiological role of pyruvate oxidase in the aerobic metabolism of *Lactobacillus plantarum. J. Bacteriol.* **160:**462–465.

228. **Shadbolt, C., T. Ross, and T. A. McMeekin.** 2001. Differentiation of the effects of lethal pH and water activity: food safety implications. *Lett. Appl. Microbiol.* **32:**99–102.

229. **Shaw, A. D., M. K. Winson, A. M. Woodward, A. C. McGovern, H. M. Davey, N. Kaderbhai, D. Broadhurst, R. J. Gilbert, J. Taylor, E. Timmins, B. K. Alsberg, J. J.**

Rowland, R. Goodacre, and D. B. Kell. 1999. Rapid analysis of high-dimensional bioprocesses using multivariate spectroscopies and advanced chemometrics. *Adv. Biochem. Eng. Biotechnol.* **66:**83–114.

230. **Shelef, L. A.** 1977. Effect of glucose on the bacterial spoilage of beef. *J. Food Sci.* **42:**1172–1175.

231. **Shewan, J. M.** 1961. The microbiology of sea-water fish, p. 487–560. *In* G. Borgstrom (ed.), *Fish as Food*, vol. 1. *Production, Biochemistry, and Microbiology.* Academic Press, New York, N.Y.

232. **Smith, J. S., P. B. Kenney, C. L. Kastner, and M. M. Moore.** 1993. Biogenic amine formation in fresh vacuum-packaged beef during storage at 1°C for 120 days. *J. Food Prot.* **56:**497–500, 532.

233. **Sofos, J. N.** 1994. Microbial growth and its control in meat poultry and fish, p. 359–403. *In* A. M. Pearson and T. R. Dutson (ed.), *Quality Attributes and Their Measurements in Meat, Poultry and Fish Products.* Blackie Academic and Professional, Glasgow, United Kingdom.

234. **Sofos, J. N.** 2002. Approaches to pre-harvest food safety assurance, p. 23–48. *In* F. J. M. Smulders and J. D. Collins (ed.), *Food Safety Assurance and Veterinary Public Health*, vol. 1. *Food Safety Assurance in the Pre-Harvest Phase.* Wageningen Academic Publishers, Wageningen, The Netherlands.

235. **Sofos, J. N.** 2005. *Improving the Safety of Fresh Meat.* CRC/Woodhead Publishing, Limited, Cambridge, United Kingdom.

236. **Sofos, J. N., K. E. Belk, and G. C. Smith.** 1999. Processes to reduce contamination with pathogenic microorganisms in meat, p. 596–605. *In Proceedings of the 45th International Congress of Meat Science and Technology.* Japan Society for Meat Science and Technology, Yokohama, Japan.

237. **Sofos, J. N., and G. C. Smith.** 1998. Non-acid meat decontamination technologies: model studies and commercial applications. *Int. J. Food Microbiol.* **44:**171–188.

238. **Sofos, J. N., and G. C. Smith.** 1999. Animal, carcass and meat hygiene to enhance food safety. *In EOLSS Encyclopedia.* EOLSS Publishers Ltd., Oxford, United Kingdom. [Online.] www.eolss.net

239. **Stanbridge, L. H., and A. R. Davis.** 1998. The microbiology of chill-stored meat, p.174–219. *In* R. G. Board and A. R. Davies (ed.), *The Microbiology of Meat and Poultry.* Blackie Academic and Professional, London, United Kingdom.

240. **Stoodley, P., I. Dodds, J. D. Boyle, and H. M. Lappin-Scott.** 1999. Influence of hydrodynamics and nutrients on biofilm structure. *J. Appl. Microbiol.* **85:**19S–28S.

241. **Stoodley, P., A. Jacobsen, and B. C. Dunsmore.** 2001. The influence of fluid shear and AICI3 on the material properties of *Pseudomonas aeruginosa* PAO1 and *Desulfovibrio* sp. EX265 biofilms. *Water Sci. Technol.* **43:**113–120.

242. **Stutz, H. K., G. J. Silverman, P. Angelini, and R. E. Levin.** 1991. Bacteria and volatile compounds associated with ground beef spoilage. *J. Food Sci.* **56:**1147–1153.

243. **Sutherland, J.** 2003. Modelling food spoilage, p. 451–474. *In* P. Zeuthen and L. Bogh-Sorensen (ed.), *Food*

Preservation Techniques. CRC Woodhead Publishing Limited, Cambridge, United Kingdom.

244. **Sutherland, J. P., and A. Varnam.** 1982. Fresh meat processing, p. 103–128. *In* M. H. Brown (ed.), *Meat Microbiology*. Applied Science Publishers Ltd., New York, N.Y.

245. **Talon, R., M. Paron, D. Bauchart, F. Duboisset, and M.-C. Montel.** 1992. Lipolytic activity of *Brochothrix thermosphacta* on natural triglycerides. *Lett. Appl. Microbiol.* **14**:153–157.

246. **Tassou, C., V. Aletras, and G.-J. E. Nychas.** 1996. The use of HPLC to monitor changes in the organic acid profile extracted from poultry stored under different storage conditions, p. 496–499. *In Proceedings of the 17th National Chemistry Conference*. University of Patras, Patras, Greece.

247. **Tassou, C. C., K. Lambropoulou, and G.-J. E. Nychas.** 2004. Effect of pre-storage treatments and storage conditions on the survival of *Salmonella* Enteritidis PT4 and *Listeria monocytogenes* on fresh marine and freshwater aquaculture fish. *J. Food Prot.* **67**:193–198.

248. **Thomas, T. D., D. C. Ellwood, and V. M. C. Longyear.** 1979. Change from homo- to heterolactic fermentation by *Streptococcus lactis* resulting from glucose limitation in anaerobic chemostat cultures. *J. Bacteriol.* **138**:109–117.

249. **Tompkin, R. B.** 1986. Microbiology of ready-to-eat meat and poultry products, p. 89–121. *In* A. M. Pearson and T. R. Dutson (ed.), *Advances in Meat Research*, vol. 2. *Meat and Poultry Microbiology*. AVI Publishing Co., Inc., Westport, Conn.

250. **Tremoulet, F., O. Duche, A. Namane, B. Martinie, and J.-C. Labadie.** 2002. Comparison of protein patterns of *Listeria monocytogenes* grown in biofilm or in planktonic mode by proteomic analysis. *FEMS Microbiol. Lett.* **210**:25–31.

251. **Troller, J. A.** 1979. Food spoilage by microorganisms tolerating low-a_w environments. *Food Technol.* **33**(1):72–75.

252. **Tryfinopoulou, P., E. Tsakalidou, and G.-J. E. Nychas.** 2002. Characterization of *Pseudomonas* spp. associated with spoilage of gilt-head seabream stored under various conditions. *Appl. Environ. Microbiol.* **68**:65–72.

253. **Tseng, C. P., and T. J. Montville.** 1990. Enzyme activities affecting end-product distribution by *Lactobacillus plantarum* in response to changes in pH and O_2. *Appl. Environ. Microbiol.* **56**:2761–2763.

254. **Tsigarida, E., I. S. Boziaris, and G.-J. E. Nychas.** 2003. Bacterial synergism or antagonism in a gel cassette system. *Appl. Environ. Microbiol.* **69**:7204–7209.

255. **Tsigarida, E., and G.-J. E. Nychas.** 2001. Ecophysiological attributes of a *Lactobacillus* sp. and a *Pseudomonas* sp. on sterile beef fillets in relation to storage temperature and film permeability. *J. Appl. Microbiol.* **90**:696–705.

256. **Tsigarida, E., and G.-J. E. Nychas.** 2006. Effect of high-barrier packaging films with different oxygen transmission rates on the growth of Lactobacillus sp. on meat fillets. *J. Food Prot.* **69**:943–947.

257. **Tsigarida, E., P. N. Skandamis, and G.-J. E. Nychas.** 2000. Behaviour of *Listeria monocytogenes* and autochthonous flora on meat stored under aerobic, vacuum and modified atmosphere packaging conditions with or without the presence of oregano essential oil at 5°C. *J. Appl. Microbiol.* **89**:901–909.

258. **U.S. Food and Drug Administration, Center for Food Safety & Applied Nutrition.** 2001. *Clostridium botulinum* toxin formation. *In Fish and Fisheries Products Hazards and Controls Guidance*, 3rd ed. [Online.] http://www.cfsan.fda.gov/~comm/haccp4m.html.

259. **Vanderzant, C., J. W. Savell, M. O. Hanna, and V. Potluri.** 1986. A comparison of growth of individual meat bacteria on the lean and fatty tissue of beef, pork and lamb. *J. Food Sci.* **51**:5–8, 11.

260. **Van Impe, J. F. M., and K. Bernaerts.** 2000. *Predictive Modelling in Foods—Conference Proceedings*. KULeuven/BioTec, Leuven, Belgium.

261. **Veciana-Nogués, M. T., A. Marine-Font, and M. C. Vidal-Carou.** 1997. Biogenic amines as hygienic quality indicators of tuna. Relationships with microbial counts, ATP-related compounds, volatile amines, and organoleptic changes. *J. Agric. Food Chem.* **45**:2036–2041.

262. **Ventitanarayanan, K. S., C. Faustman, T. Hoagland, and B. W. Berry.** 1997. Estimation of spoilage bacteria load on meat by fluorescein diacetate hydrolysis or resazurin reduction. *J. Food Sci.* **62**:601–604.

263. **Vetter, Y. A., J. W. Deming, P. A. Jumars, and B. B. Krieger-Brockett.** 1998. A predictive model of bacterial foraging by means of freely released extracellular enzymes. *Microb. Ecol.* **36**:75–92.

264. **Vinci, G., and M. L. Antonelli.** 2002. Biogenic amines: quality index of freshness in red and white meat. *Food Control* **13**:519–524.

265. **Wilson, P. D. G., T. F. Brocklehurst, S. Arino, D. Thuault, M. Jakobsen, M. Lange, J. Farkas, J. W. T. Wimpenny, and J. F. Van Impe.** 2002. Modelling microbial growth in structured foods: towards a unique approach. *Int. J. Food Microbiol.* **73**:275–289.

266. **Winkel, S. A., and G. H. Richardson.** 1984. Cell mass and acid production of proteinase-positive and proteinase-negative lactic cultures in buffered non-fat milk. *J. Dairy Sci.* **67**:2856–2859.

267. **Worm, J., L. E. Jensen, T. S. Hansen, M. Sondergaard, and O. Nybroe.** 2000. Interactions between proteolytic and non-proteolytic *Pseudomonas fluorescens* affect protein degradation in a model community. *FEMS Microbiol. Ecol.* **32**:103–109.

268. **Yano, Y., K. Yokoyama, and I. Karube.** 1996. Evaluation of meat spoilage using a chemiluminescence-flow injection analysis system based on immobilized putrescine oxidase and photodiode. *Lebensm.-Wiss. Technol.* **29**:498–502.

269. **Yano, Y., K. Yokoyama, E. Tamiya, and I. Karube.** 1996. Direct evaluation of meat spoilage and the progress of aging using biosensors. *Anal. Chim. Acta* **320**:269–276.

270. **Yuk, H. G., and D. L. Marshall.** 2003. Heat adaptation alters *Escherichia coli* O157:H7 membrane lipid composition and verotoxin production. *Appl. Environ. Microbiol.* **69**:5115–5119.

271. **Yuk, H. G., and D. L. Marshall.** 2004. Adaptation of *Escherichia coli* O157:H7 to pH alters membrane lipid composition, verotoxin secretion, and resistance to simulated gastric fluid acid. *Appl. Environ. Microbiol.* **70:**3500–3505.

272. **Yuk, H. G., and D. L. Marshall.** 2005. Influence of acetic, citric, and lactic acid on *Escherichia coli* O157:H7 membrane lipid composition, verotoxin secretion, and acid resistance in simulated gastric fluid. *J. Food Prot.* **68:**673–679.

273. **Yuk, H. G., and D. L. Marshall.** 2006. Effect of trisodium phosphate adaptation on changes in membrane lipid composition, verotoxin secretion, and acid resistance of *Escherichia coli* O157:H7 in simulated gastric fluid. *Int. J. Food Microbiol.* **106:**39–44.

274. **Zhao, Y. Y., J. H. Wells, and D. L. Marshall.** 1992. Description of log phase growth for selected microorganisms during modified atmosphere storage. *J. Food Process Eng.* **15:**299–317.

Food Microbiology: Fundamentals and Frontiers, 3rd Ed.
Edited by M. P. Doyle and L. R. Beuchat
© 2007 ASM Press, Washington, D.C.

Joseph F. Frank

Milk and Dairy Products

7

Being both highly perishable and nutritious, milk has since prehistoric times been subject to a variety of preservation treatments. Modern dairy processing utilizes pasteurization, heat sterilization, fermentation, dehydration, refrigeration, and freezing as preservation treatments. The result of these, when combined with component separation processes (i.e., churning, filtration, and coagulation), is an assortment of dairy foods having vastly different tastes and textures and a complex variety of spoilage microflora. Some defects of milk and cheese caused by microorganisms are listed in Tables 7.1 and 7.2. In this chapter, the discussion of spoilage is organized by types of microorganisms associated with various defects. These include gram-negative psychrotrophic microorganisms, coliform and lactic acid bacteria, spore-forming bacteria, and yeasts and molds. The major objective of this chapter is to describe the interactions of these microorganisms with dairy foods that lead to commonly encountered product defects.

MILK AND DAIRY PRODUCTS AS GROWTH MEDIA

Milk

Milk is a good growth medium for many microorganisms because of its high water activity, near neutral pH, and available nutrients. Milk, however, is not an ideal growth medium since, for example, the addition of yeast extract or protein hydrolysates often increases growth rates. Table 7.3 lists the major nutritional components of milk and their normal concentrations. These components consist of lactose, fat, protein, minerals, and various nonprotein nitrogenous compounds. Many microorganisms cannot utilize lactose and therefore must rely on proteolysis or lipolysis to obtain carbon and energy. In addition, freshly collected raw milk contains various growth inhibitors that decrease in effectiveness with storage.

Carbon and Nitrogen Availability

Carbon sources in milk include lactose, protein, and fat. The citrate in milk can be utilized by many microorganisms but is not present in a sufficient amount to support significant growth. A sufficient amount of glucose is present in milk to allow initiation of growth by some microorganisms, but for fermentative microorganisms to continue growth they must have a lactose transport and metabolism system. These are described in chapter 35. Other spoilage microorganisms may oxidize lactose to lactobionic acid.

Joseph F. Frank, Dept. of Food Science and Technology, University of Georgia, Athens, GA 30602-7610.

Table 7.1 Some defects of fluid milk which result from microbial growth

Defect	Associated microorganisms	Types of enzyme	Metabolic product(s)	Reference(s)
Bitter flavor	Psychrotrophic bacteria, *Bacillus*	Protease, peptidase	Bitter peptides	76, 82
Rancid flavor	Psychrotrophic bacteria	Lipase	Free fatty acids	101
Fruity flavor	Psychrotrophic bacteria	Esterase	Ethyl esters	87
Coagulation	*Bacillus* spp.	Protease	Casein destabilization	76
Sour flavor	Lactic acid bacteria	Glycolytic	Lactic, acetic acids	38, 100
Malty flavor	Lactic acid bacteria	Oxidase	3-Methylbutanal	79
Ropy texture	Lactic acid bacteria	Polymerase	Exopolysaccharides	14

Few spoilage microorganisms utilize milk fat as a carbon or energy source. This is because the milk fat globules are surrounded by a protective membrane composed of glycoproteins, lipoproteins, and phospholipids. Milk fat is available for microbial metabolism only if the globule membrane is physically damaged or enzymatically degraded (2).

Caseins are present in the form of highly hydrated micelles and are readily susceptible to proteolysis. Whey proteins (β-lactoglobulin, α-lactalbumin, serum albumin, and immunoglobulins) remain soluble in the milk after precipitation of casein. They are less susceptible than caseins to microbial proteolysis. Milk contains nonprotein nitrogenous compounds such as urea, peptides, and amino acids that are readily available for microbial utilization (Table 7.3). These compounds are present in insufficient quantity to support the extensive growth required for spoilage.

Minerals and Micronutrients

Milk is a good source of B vitamins and minerals. The major salt cations and anions present in milk are listed in Table 7.3. Although milk contains many trace mineral nutrients such as iron, cobalt, copper, and molybdenum, some of these, such as iron, may not be present in a readily usable form. Supplementation of milk with trace elements may be necessary to achieve maximum microbial growth rates. Milk also contains the growth stimulant orotic acid (a metabolic precursor for pyrimidines).

Natural Inhibitors

The major microbial inhibitors in raw milk are lactoferrin and the lactoperoxidase system. Natural inhibitors of lesser importance include lysozyme, specific immunoglobulins, and folate and vitamin B_{12} binding systems. Lactoferrin, a glycoprotein, acts as an antimicrobial agent

Table 7.2 Some defects of cheese which result from microbial growth

Defect(s)	Associated microorganisms	Metabolic product(s)	Reference
Open texture, fissures	Heterofermentative lactobacilli	Carbon dioxide	61
Early gas	Coliforms, yeasts	Carbon dioxide, hydrogen	70
Late gas	*Clostridium* spp.	Carbon dioxide, hydrogen	26
Rancidity	Psychrotrophic bacteria	Free fatty acids	64
Fruity flavor	Lactic acid bacteria	Ethyl esters	9
White crystalline surface deposits	*Lactobacillus* spp.	Excessive D-lactate	89
Pink discoloration	*Lb. delbrueckii* subsp. *bulgaricus*	High redox potential	99

Table 7.3 Approximate concentrations of some nutritional components of milk[a]

Component	Amt[b]
Water	87.3
Lactose	4.8
Fat	3.7
Casein	2.6
Whey protein	0.6
Salt cations	
Sodium	58
Potassium	140
Calcium	118
Magnesium	12
Salt anions	
Citrate	176
Chloride	104
Phosphorus	74
NPN[c]	
Total NPN	296
Urea N	42
Peptide N	32
Amino acid N	44
Creatine N	25

[a] Adapted from reference 54.
[b] The amounts for the first five components are in grams per 100 g, those for the salt cations and anions are in milligrams per 100 g, and those for the nitrogen are in milligrams per liter.
[c] NPN, nonprotein nitrogen.

by binding iron. Human milk contains over 2 mg of lactoferrin per ml, but it is of lesser importance in cow's milk, which contains only 20 to 200 µg/ml (71). Psychrotrophic aerobes are inhibited by lactoferrin, but the presence of citrate in cow's milk limits its effectiveness, as the citrate competes with lactoferrin for binding the iron (8).

The most effective microbial inhibitor in cow's milk is the lactoperoxidase system, which consists of lactoperoxidase, thiocyanate, and hydrogen peroxide. Lactoperoxidase catalyzes the oxidation of thiocyanate and simultaneous reduction of hydrogen peroxide, resulting in the accumulation of hypothiocyanite ($OSCN^-$). Two mechanisms for this reaction are illustrated in Fig. 7.1. Hypothiocyanite oxidizes sulfhydryl groups of proteins, resulting in enzyme inactivation and structural damage to the microbial cytoplasmic membrane (117). Lactoperoxidase and thiocyanate are present in milk during synthesis, whereas hydrogen peroxide is formed in milk when oxygen is metabolized by lactic acid bacteria. Hydrogen peroxide is the limiting substrate of the reaction, so the effective use of this inhibitor system for preserving milk

1a. $2SCN^- + H_2O_2 + 2H^+ \xrightarrow[\text{(lactoperoxidase)}]{} (SCN)_2 + H_2O$

1b. $(SCN)_2 + H_2O \xrightarrow{} HOSCN + SCN^- + H^+$

1c. $HOSCN \rightleftharpoons H^+ + OSCN^-$

2. $SCN^- + H_2O_2 \xrightarrow[\text{(lactoperoxidase)}]{} OSCN^- + H_2O$

Figure 7.1 Two mechanisms for generation of hypothiocyanite ($OSCN^-$) inhibitor in milk. Adapted from reference 117.

relies on adding a hydrogen peroxide-generating system or, less effectively, hydrogen peroxide to the milk (28). Lactic acid bacteria, coliforms, and various pathogens are inhibited by this system (117).

Effect of Heat Treatments

The minimum required heat treatment for milk to be sold for fluid consumption in the United States is 72°C for 15 s for continuous systems and 63°C for 30 min for batch processes, though most processors use higher temperatures and longer holding times. Pasteurization affects the growth rate of the spoilage microflora by destroying inhibitor systems (117), so postpasteurization contaminants may be able to grow more rapidly in pasteurized milk than in the raw product. More severe heat treatments affect microbial growth by increasing available nitrogen through protein hydrolysis and by liberating inhibitory sulfhydryl compounds. Lactoperoxidase is partially inactivated by normal pasteurization treatments (117).

Dairy Products

Dairy products provide substantially different growth environments than fluid milk because these products have nutrients added, removed, or concentrated or have a lower pH or a_w. The composition, pH, and a_w of selected dairy products are presented in Table 7.4. Butter is a water-in-oil emulsion, so microorganisms are trapped within serum droplets. If butter is salted, the mean salt content of the water droplets will be 6 to 8%, sufficient to inhibit gram-negative spoilage organisms that could grow during refrigeration. However, individual droplets will have a significantly higher or lower salt content if the salt is not uniformly distributed during manufacture. This can result in psychrotrophic bacteria growing in droplets of low salt content. Unsalted butter is usually made from acidified cream and relies on low pH and refrigeration for preservation.

Table 7.4 Approximate compositions, pHs, and water activities of selected dairy products[a]

Product	Component (g/100 g)				a_w	pH
	Water	Fat	Protein	Carbohydrate		
Butter	16.0	81.0	3.6	0.06		6.3
Cheddar cheese	37.0	32.8	24.9	1.3	0.90–0.95	5.2
Swiss cheese	37.2	27.4	28.4	3.4		5.6
Nonfat dried milk	3.2	0.8	36.2	52.0	0.2	
Evaporated skim milk	79.4	0.2	7.5	11.3	0.93–0.98	
Yogurt	89	1.7	3.5	5.1		4.3

[a] Compiled from references 5, 7, 21, and 56.

PSYCHROTROPHIC SPOILAGE

Preservation of fluid milk relies on effective sanitation, timely marketing, heat treatment, and refrigeration. Raw milk is rapidly cooled after collection and is kept cold until pasteurized, after which it is kept cold until consumption. There is often sufficient time between milk collection and consumption for psychrotrophic bacteria to grow. Flavor defects can result from this growth. Pasteurized milk often has a shelf life of 16 to 22 days; however, contamination of the contents of a container with even one rapidly growing psychrotrophic microorganism can reduce the shelf life to 10 days or less. Growth of psychrotrophic bacteria in raw milk can lead to defects in products made from that milk because of residual enzyme activity.

Psychrotrophic Bacteria in Milk

Psychrotrophic bacteria which spoil raw and pasteurized milk are primarily aerobic gram-negative rods in the family *Pseudomonadaceae*. It is typical that 65 to 70% of psychrotrophic isolates from raw milk are in the *Pseudomonas* genus (41). Although representatives of other genera, including *Aeromonas*, *Listeria*, *Staphylococcus*, *Enterococcus*, and the family *Enterobacteriaceae*, may be present in raw milk and increase in number during storage, they are usually outgrown by the gram-negative obligate aerobes when milk is held at its typical 3 to 7°C storage temperature (60). The psychrotrophic spoilage microflora of milk is generally proteolytic, with many isolates able to produce extracellular lipases, phospholipase, and other hydrolytic enzymes, but unable to utilize lactose. The bacterium most often associated with flavor defects in refrigerated milk is *Pseudomonas fluorescens* (31), with *Pseudomonas fragi*, *Pseudomonas putida*, and *Pseudomonas lundensis* (106) also commonly encountered. Jaspe et al. (52) observed that incubating raw milk for 3 days at 7°C selected for a population of *Pseudomonas* spp. which had a 10-fold-higher growth rate at 7°C and a lower growth rate at 21°C than the initial *Pseudomonas* population. These strains also exhibited

1,000-fold-greater proteolytic activity and 280-fold-greater lipolytic activity than the initial *Pseudomonas* population. Psychrotrophic bacteria commonly found in raw milk are inactivated by pasteurization.

Sources of Psychrotrophic Bacteria in Milk

Soil, water, animals, and plant material constitute the natural habitat of psychrotrophic bacteria found in milk (22). Plant materials, such as grass and hay used for animal feed, may contain over 10^8 psychrotrophs per gram (107). Psychrotrophic bacteria can be isolated in low numbers from water used on the dairy farms (108). Use of this water to clean and rinse milking equipment provides a direct means for their entry into milk. Psychrotrophic bacteria isolated from water are often active producers of extracellular enzymes and grow rapidly in refrigerated milk (22). Consequently, water is an important source of milk spoilage bacteria. The teat and udder area of the cow can harbor high levels of psychrotrophic bacteria, even after washing and sanitizing (80). These psychrotrophs probably originate from soil.

Milking equipment, utensils, and storage tanks are the major source of psychrotrophic contamination of raw milk (22). Proper cleaning and sanitizing procedures can effectively reduce contamination from these sources. Milk residues on unclean equipment provide a growth niche for psychrotrophic bacteria which enter milking machines, pipelines, and holding tanks with water rinses or milk. Milking equipment is generally fabricated from stainless steel; however, for some parts, rubber or other nonmetal materials must be used. Rubber materials are difficult to sanitize, since only moderate use results in the formation of microscopic cracks. Bacteria attached to these parts are difficult to inactivate by chemical sanitization (81).

Pasteurized milk products become contaminated with psychrotrophic bacteria by exposure to contaminated equipment or air. Schröder (98) determined that filling equipment was most often the source of psychrotrophs in packaged milk. Although only low levels of psychrotrophic bacteria are found in air, only one viable cell per container is required to spoil the product. Aseptic packaging eliminates this low level of contamination and therefore is used to extend the shelf life of pasteurized milk.

Growth Characteristics and Defects

Generation times in milk of the most rapidly growing psychrotrophic *Pseudomonas* spp. isolated from raw milk are 8 to 12 h at 3°C and 5.5 to 10.5 h at 3 to 5°C (104). These growth rates are sufficient to cause spoilage within 5 days if the milk initially contains only one cell per milliliter. However, most psychrotrophic pseudomonads present in raw milk grow much more slowly, causing refrigerated milk to spoil in 10 to 20 days.

Defects of fluid milk associated with the growth of psychrotrophic bacteria are related to the production of extracellular enzymes. Sufficient enzyme to cause defects is usually present when the population of psychrotrophs reaches 10^6 to 10^7 CFU/ml (32). Bitter and putrid flavors and coagulation result from proteolysis. Rancid and fruity flavors result from lipolysis. The production of extracellular enzymes by psychrotrophic bacteria in raw milk also has implications for the quality of products produced from that milk.

Proteases

Factors Affecting Protease Production

P. fluorescens and other psychrotrophs found in milk generally produce extracellular protease during the late exponential and stationary phases of growth (42, 93, 113), probably due to the release of preformed enzyme from the cells (42). The temperature for optimum production of protease by psychrotrophic *Pseudomonas* spp. is lower than the temperature for the optimum rate of growth (73). Relatively large amounts of protease are produced at temperatures as low as 5°C (72). Raw milk held at 2°C exhibits little proteolysis after 10 days of storage (45).

Since *Pseudomonas* spp. are obligate aerobes, it is expected that oxygen is required for protease synthesis. Myhara and Skura (84) reported optimum protease production for *P. fragi* in a medium containing 7.4 μg of dissolved oxygen per ml, slightly less than the 8.4 μg/ml contained in saturated water.

The effect of calcium and iron ions on protease production by *Pseudomonas* spp. is relevant to dairy spoilage. McKellar and Cholette (74) reported that in the absence of ionic calcium, an inactive precursor to proteinase was produced by *P. fluorescens*. This precursor could not be activated, indicating that calcium is required to stabilize the enzyme. Iron, which may be at a growth-limiting concentration in milk, will repress protease production by *Pseudomonas* spp. when added to milk (33). Maximum protease production will occur only if iron is growth limiting, though this may be an indirect effect of reduced cytochrome synthesis and decreased energy levels (73). Pyoverdine, an iron-chelating pigment produced by *P. fluorescens*, stimulates protease production (75).

Most evidence indicates that *P. fluorescens* regulates protease production as a means to provide carbon to the cell rather than amino acids for protein synthesis (32). Protease production is induced by various low-molecular-weight protein degradation products and is subject to end product and catabolite repression. Asparagine is the most effective inducer, and citric acid is an effective inhibitor of synthesis (73). Protease production by *P. fluorescens* in raw milk is preceded by the depletion of glucose, galactose, lactate, glutamine, and glutamic acid. Supplementation of milk with these compounds delays or inhibits protease production (53).

Properties of Proteases from Psychrotrophic *Pseudomonas*

Fox et al. (36) summarized properties of proteinases produced by *P. fluorescens*. Properties most relevant to dairy product spoilage include temperature optima from 30 to 45°C, with significant activity at 4°C and pH optima near neutrality or at alkaline conditions; all are metalloenzymes containing either Zn^{2+} or Ca^{2+}. Perhaps the most important technological characteristic of these enzymes is their heat stability. Decimal reduction times at 140°C range from 50 to 200 s, sufficient to retain significant activity after ultrahigh-temperature (UHT) milk processing (58). This heat stability is surprising, considering that the enzymes are produced and active at refrigeration temperatures. Another unexpected property of these enzymes is their susceptibility to autodegradation at 55 to 60°C, apparently due to an unfolding of the protein chain into a more sensitive conformation (102).

Protease-Induced Product Defects

Proteases of psychrotrophic bacteria cause product defects either at the time they are produced in the product or as a result of enzyme surviving a heat process. Most investigators have observed that these proteases preferentially hydrolyze κ-casein, although some show preference for β- or $α_{s-1}$-casein (23). Degradation of casein results in the liberation of bitter peptides. Bitterness is a common off flavor in pasteurized milk that has been subject to postpasteurization contamination with psychrotrophic bacteria. Continued proteolysis results in putrid off flavors associated with lower-molecular-weight degradation products such as ammonia, amines, and sulfides. Bitterness in UHT (commercially sterile) milk develops when sufficient psychrotrophic bacterial growth occurs in raw milk (estimated at 10^5 to 10^7 CFU/ml) to leave residual enzyme after heat treatment (82). Low-level protease activity in UHT milk can also result in coagulation or sediment formation. UHT milk appears to be more sensitive to protease-induced defects than raw milk, probably a result of either heat-induced changes in casein micelle structure or heat inactivation of protease inhibitors (88).

The effect of proteases of psychrotrophic origin on quality of cheese and cultured products is minimal, because the combination of low pH and low storage temperature inhibits their activity (62). In addition, proteases are removed with the whey fraction during cheese manufacture. However, growth of proteolytic bacteria in raw milk lowers cheese yield because proteolytic products of casein degradation are lost to the whey rather than becoming part of the cheese (118). Proteases can also alter the functional characteristics of milk powders (17).

Lipases

Psychrotrophic *P. fluorescens* isolated from milk often produces extracellular lipase in addition to protease. Other commonly found lipase-producing psychrotrophs include *P. fragi* and *Pseudomonas aeruginosa*.

Factors Affecting Lipase Production

Lipases of psychrotrophic pseudomonads, like proteases, are produced in the late log or stationary phase of growth (1, 73). As with protease, optimal synthesis of lipase generally occurs below the optimum temperature for growth. For example, Andersson (3) reported optimum production of lipase at 8°C by a *P. fluorescens* strain which exhibited optimum growth at 20°C. Milk is an excellent medium for lipase production by pseudomonads. Lipase production requires an organic nitrogen source; however, some amino acids repress lipase production, especially those that serve as nitrogen but not carbon sources (33).

Ionic calcium is required for lipase activity, as activity is inhibited by EDTA, which chelates the calcium (1). *P. fluorescens* can produce lipase in calcium-free media, although the amount produced is less than in the presence of calcium (74). Lipase production by *P. fluorescens*, to a greater degree than protease production, is stimulated by limiting iron availability (75). Supplementation of milk with iron delays the onset of lipase production by *P. fluorescens* in raw milk (34).

Although there is conflicting evidence, most reports indicate that lipase production is subject to catabolite repression (73, 101). Lipases are produced by *P. fluorescens* and *P. fragi* in the absence of triglycerides. The presence of triglycerides can either inhibit or stimulate production of lipase, depending on the specific strain, triglyceride concentration, and growth conditions. Surface-active agents such as polysaccharides and lecithin often stimulate the release of lipase from the cell surface (101).

Properties of Lipase from Psychrotrophic *Pseudomonas* spp.

The properties of lipases produced by *Pseudomonas* spp. have been summarized by Stead (101) and Fox et al. (36). Temperatures for optimal activity of these lipases range from 22 to 70°C, with most between 30 and 40°C. The optimal pH for activity is from 7.0 to 9.0, with most of the organisms having optima between 7.5 and 8.5. *P. fluorescens* lipase in milk is active at refrigeration temperatures, and significant activity remains at subfreezing temperatures and at low a_w (4).

The heat stability of these lipases is similar to that of the proteases. Decimal reduction times at 140°C range from 48 to 437 s (58), sufficient to provide residual activity after UHT treatment. Most of these lipases are also subject to accelerated irreversible inactivation at temperatures less than 100°C (59). Histidine and $MgCl_2$ were required for the low-temperature inactivation of a heat-stable lipase isolated from *P. fluorescens* (20).

Lipase-Induced Product Defects

The triglycerides in raw milk are present in globules that are protected from enzymatic degradation by a membrane. Milk becomes susceptible to lipolysis if this membrane is disrupted by excessive shear force (from pumping, agitation, and freezing). Raw milk contains a mammalian lipase (milk lipase) which will rapidly act on the fat if the globule membrane is disrupted. Most cases of rancidity in milk are a result of this process, rather than from the growth of lipase-producing microorganisms. Phospholipase C and protease produced by psychrotrophic bacteria can degrade the fat globule membrane, resulting in the enhancement of milk lipase activity (2, 19). Milk lipase is heat labile, so most milk products (other than raw-milk cheeses) will not have residual activity. Sufficient bacterial lipase can be produced in raw milk to cause defects in products manufactured from that milk. Since residual activities are usually low, and the reaction environment less than optimum, usually only products with long storage times or high storage temperatures are affected. This includes

UHT milk, some cheeses, butter, and whole-milk powder. Lipase-producing bacteria can also recontaminate pasteurized milk and cream and grow in these products during refrigerated storage. The rancid flavor and odor resulting from lipase action are usually from the liberation of C_4 to C_8 fatty acids. Fatty acids of higher molecular weight produce a flavor described as soapy. Low levels of unsaturated fatty acids liberated by enzymatic activity may be oxidized to ketones and aldehydes to produce oxidized or "cardboardy" off flavor (27). *P. fragi* produces a fruity off flavor in milk by esterifying free fatty acids with ethanol (87). Ethyl butyrate and ethyl hexanoate are the esters formed at the largest amounts, with low levels of ethyl esters of acetate, propionate, and isovalerate also produced. Low levels of ethanol in milk, present as a result of microbial activity, stimulate ester production. Residual activity from heat-stable microbial lipases can cause off flavors in UHT milk, but lipase-induced defects are not as common as those resulting from microbial protease (82).

Rancid defect in butter may result from growth of lipolytic microorganisms during storage, residual heat-stable microbial lipase originating from the growth of psychrotrophic bacteria in the milk or cream, or native milk lipase activity in the raw milk or cream. When butter is manufactured from rancid cream, low-molecular-weight free fatty acids are removed with the watery portion of the cream (buttermilk), so the resulting butter will have not a typical rancid flavor, but a less pronounced soapy off flavor associated with C_{10} to C_{12} free fatty acids. However, the typical odor of rancid butter is associated with lower-molecular-weight fatty acids (C_4 to C_8). Microbial lipases present in butter will exhibit activity even if the product is stored at $-10°C$ (85). Growth of psychrotrophic bacteria in butter occurs only if the product is made from sweet rather than ripened (sour) cream. Sweet cream butter is preserved by salt and refrigeration. Butter is a water-in-fat emulsion, so moisture and salt will not equilibrate during storage. Consequently, if salt and moisture are not evenly distributed in the product during manufacture, then lipolytic psychrotrophs will have pockets of high a_w in which to grow (27).

Cheese is more susceptible to defects caused by bacterial lipases than those caused by proteases because lipases, unlike most proteases, are concentrated along with the fat in the curd. The acidic environment of most cheeses limits lipase activity. Some cheeses, such as Camembert and Brie, increase in pH to near neutrality during ripening. Camembert is, in fact, susceptible to defects associated with microbial lipase (30). More acidic cheeses, e.g., Cheddar, are susceptible if cured for several months or if large amounts of lipase are present (63).

Law et al. (64) reported that Cheddar cheese made from milk containing *P. fluorescens* at more than 10^6 CFU/ml before pasteurization developed rancid flavor in 6 to 8 months.

Whole-milk powder containing bacterial lipase may develop rancidity, and low-fat milk-derived powders may contain residual lipase which becomes active when these products are used in fat-containing food formulations (17, 101).

Control of Product Defects Associated with Psychrotrophic Bacteria

Raw Milk
Preventing product defects which result from growth of psychrotrophic bacteria in raw milk involves limiting contamination levels, rapid cooling immediately after milking, and maintenance of cold storage temperatures. Limiting the populations of bacteria primarily involves cleaning, sanitizing, and drying cows' teats and udders before milking, and using cleaned and sanitized equipment. Removal of residual milk solids from milk contact surfaces is critical to psychrotroph control, since these residues protect cells from the action of chemical sanitizers and provide nutrients for growth. Subsequent growth over a period of days results in a biofilm which, in addition to containing high numbers of bacteria, is highly resistant to chemical sanitizers (37). Rapid cooling of milk after collection is important because contamination of the product with psychrotrophic bacteria is unavoidable. As previously indicated, psychrotrophic activity in milk is inhibited at $2°C$, but freezing of milk causes disruption of the fat globule membrane, making it highly susceptible to lipolysis. Therefore, the challenge of farm storage systems is to rapidly cool milk to as low a temperature as possible while avoiding ice formation. Proteolysis in raw milk can be inhibited by addition of carbon dioxide (67).

Pasteurized Products
Preventing contamination of pasteurized dairy products with psychrotrophic bacteria is primarily a matter of equipment cleaning and sanitation, although airborne psychrotrophs may also limit product shelf life. Even when filling equipment is effectively cleaned and sanitized, it can still become a source of psychrotrophic microorganisms that accumulate during normal hours of continuous use (98). These microorganisms probably enter the filler through the vacuum system or from containers. Complete elimination of psychrotrophic microorganisms from products is best achieved by using aseptic packaging technologies.

SPOILAGE BY FERMENTATIVE NONSPOREFORMERS

Spoilage of milk and dairy products resulting from growth of acid-producing fermentative bacteria occurs when storage temperatures are sufficiently high for these microorganisms to outgrow psychrotrophic bacteria, or when the product composition is inhibitory to gram-negative aerobic organisms. For example, the presence of lactic acid in fluid milk is a good indication that the product was exposed to an unacceptably high storage temperature that allowed growth of lactic acid bacteria. Fermented dairy foods, though manufactured using lactic acid bacteria, can be spoiled by the growth of "wild" lactic acid bacteria that produce unwanted gas, off flavors, or appearance defects. Fluid milk, cheese, and cultured milks are the major dairy products susceptible to spoilage by non-spore-forming fermentative bacteria.

Non-spore-forming bacteria responsible for fermentative spoilage of dairy products are mostly in either the lactic acid-producing or coliform group. Genera of lactic acid bacteria involved in spoilage of milk and fermented products include *Lactococcus, Lactobacillus, Leuconostoc, Enterococcus, Pediococcus,* and *Streptococcus* (16). Coliforms can spoil milk, but this is seldom a problem since they are usually outgrown by either the lactic acid or psychrotrophic bacteria. Coliform-induced spoilage is more common in some cheese varieties. Members of the *Enterobacter* and *Klebsiella* genera are often spoilage coliforms of concern.

Sources of Fermentative Spoilage Bacteria

Lactic acid-producing bacteria are normal inhabitants of the cow's teat. Lactic acid bacteria are also associated with silage and other animal feeds and feces. Coliform bacteria are present on udder skin as a result of fecal contamination, so ineffective cleaning of this area before milking will contribute to high coliform populations in milk. Coliform bacteria in raw milk are also associated with inadequately cleaned milking equipment (10).

Defects of Fluid Milk Products

The most common fermentative defect in fluid milk products is souring caused by the growth of lactic acid bacteria. Lactic acid by itself has a clean, pleasant acid flavor and no odor. The unpleasant "sour" odor and taste of spoiled milk are a result of small amounts of acetic and propionic acids (100). Sour odor can be detected before a noticeable acid flavor develops. For discussion of lactic acid production in milk, see chapter 35. Other defects may occur in combination with acid production. A malty flavor results from growth of *Lactococcus lactis* subsp. *lactis* biovar *maltigenes*. This strain is unique among

lactococci in its ability to produce 2-methylpropanal, 3-methylbutanal, and the corresponding alcohols (79). The aldehydes are produced by decarboxylation of α-ketoisocaproic and α-ketoisovaleric acids. These keto-acids are also concurrently used to synthesize leucine and valine by transamination with glutamic acid. Alcohols corresponding to the aldehydes are formed by the action of alcohol dehydrogenase in the presence of $NADH_2$. Malty flavor is primarily from 3-methylbutanal.

Another defect associated with growth of lactic acid bacteria in milk is "ropy" texture. Most dairy-associated species of lactic acid bacteria have strains that produce exocellular polymers which cause the ropy defect (14). Some of these strains are used to produce high-viscosity fermented products such as yogurt and Scandinavian ropy milk (villi, skyr). The defect in noncultured fluid milk products is usually caused by growth of specific strains of lactococci. The polymer produced by these organisms is a polysaccharide containing glucose and galactose with small amounts of mannose, rhamnose, and pentose (15).

Defects in Cheese

Lactic Acid Bacteria

Some strains of lactic acid bacteria produce flavor and appearance defects in cheese. Lactobacilli are a normal part of the dominant microflora of aged Cheddar cheese. If heterofermentative lactobacilli predominate, the cheese is prone to develop an "open" texture or fissures, a result of gas production during aging (61). Off flavors are also associated with the growth of these organisms (112). Gassy defects in aged Cheddar cheese are more often associated with growth of lactobacilli than with growth of coliforms, yeasts, or sporeformers. The use of elevated ripening temperatures for Cheddar cheese, e.g., 15 rather than 8°C, encourages growth of heterofermentative lactobacilli but not that of non-lactic acid bacteria (25). This phenomenon limits the use of high-temperature storage to accelerate ripening. *Lactobacillus brevis* and *Lactobacillus casei* subsp. *pseudoplantarum* have been associated with gas production in retail Mozzarella cheese (50). *Lactobacillus casei* subsp. *casei* produces a soft body defect in Mozzarella cheese (51). The softened cheese cannot be readily sliced or grated and does not melt properly.

Some cheese varieties occasionally exhibit a pink discoloration. Pink spots in Swiss-type varieties result from the growth of pigmented strains of propionibacteria. In Italian cheese varieties, a pink discoloration may occur either in a band near the surface or throughout the whole cheese. This defect is associated with strains of *Lactobacillus delbrueckii* subsp. *bulgaricus* that fail to lower the redox potential of the cheese (99). Another common defect of aged Cheddar cheese is the appearance of white crystalline deposits on the surface. Although they do not affect flavor, these deposits reduce consumer acceptability. Rengpipat and Johnson (89) observed an atypical strain of a facultatively heterofermentative *Lactobacillus* associated with the deposits. This strain produces an unusually large amount of D-lactic acid during cheese aging, resulting in the formation of insoluble calcium lactate crystals, the primary component of the white deposits. *Lactobacillus casei* subsp. *alactosus* and *Lactobacillus casei* subsp. *rhamnosus* have been associated with the development of a phenolic flavor in Cheddar cheese, described as being similar to horse urine (50). The flavor develops after 2 to 6 months of aging.

Fruity off flavor in Cheddar cheese is usually not caused by growth of psychrotrophic bacteria, as it is in milk, but rather is a result of growth of lactic acid bacteria (usually *Lactococcus* spp.) which produce esterase. Fruity-flavored cheeses contain high levels of ethanol, a substrate for esterification (9). The major esters contributing to fruity flavor in cheese are ethyl hexanoate and ethyl butyrate.

Coliform Bacteria

Coliform bacteria were recognized as causing a gassy defect in Cheddar and related cheese varieties as early as 1885 (94). If present, they grow during the cheese manufacture process or shortly thereafter, producing an "early gas" defect. In hard cheeses, such as Cheddar, this defect occurs when lactic acid fermentation fails to rapidly lower pH, or when highly contaminated raw milk is used. Cheese varieties in which acid production is purposely delayed by washing the curds are highly susceptible to coliform growth (40). Soft, mold-ripened cheeses, such as Camembert, have an increase in pH during ripening, with a resulting susceptibility to coliform growth (39, 95). Gas formation in retail Mozzarella cheese has been associated with growth of *Klebsiella pneumoniae* (70). Coliform growth in retail cheese is often manifested as a swelling of the plastic package. Large populations (10^7 CFU/g) of coliforms are needed to produce a gassy defect.

Defects in Fermented Milk Products

Fermented milk products such as cultured buttermilk, sour cream, and cottage cheese rely on diacetyl produced during fermentation for their typical "buttery" flavor and aroma. These products lose consumer appeal when this flavor is lost due to reduction of diacetyl to acetoin and 2,3-butanediol (38). Lactococci capable of growing at

7°C may produce sufficient diacetyl reductase to destroy diacetyl in cultured milks (48). Other psychrotrophic contaminants in cultured milks, including yeasts and coliforms, may also be involved in diacetyl reduction (114).

Control of Defects Caused by Lactic Acid and Coliform Bacteria

Defects in fluid milk caused by coliforms and lactic acid bacteria are controlled by good sanitation practices during milking, maintaining raw milk at temperatures below 7°C, pasteurization, and refrigeration of pasteurized products. These microorganisms seldom grow to significant levels in refrigerated pasteurized milk because of their low growth rates compared to those of psychrotrophic bacteria. Control of coliform growth in cheese is achieved by using pasteurized milk, encouraging rapid fermentation of lactose, temperature and salt control (77), and good sanitation during manufacture. Controlling defects produced by undesirable lactic acid bacteria in cheese and fermented milks is more difficult, since growth of lactic acid bacteria must be encouraged during manufacture, and the products often provide suitable growth environments. Undesirable strains of lactic acid bacteria are readily isolated from the manufacturing environment, so their control requires attention to plant cleanliness and protecting the product during manufacture.

SPORE-FORMING BACTERIA

Spoilage by spore-forming bacteria can occur in low-acid fluid milk products that are preserved by substerilization heat treatments and packaged with little chance for recontamination with vegetative cells. Products in this category include aseptically packaged milk and cream, and sweetened and unsweetened concentrated canned milks. Nonaseptic packaged refrigerated fluid milk may spoil due to growth of psychrotrophic *Bacillus cereus*, *Bacillus mycoides*, and *Bacillus polymyxa* in the absence of more rapidly growing gram-negative psychrotrophs (44, 106). Hard cheeses, especially those with low interior salt concentrations, are also susceptible to spoilage by spore-forming bacteria.

Spore-forming bacteria that spoil dairy products usually originate in the raw milk. Populations present in raw milk are generally quite low (<5,000 CFU/ml), and the occurrence of a sporeformer-induced defect does not always correlate with initial numbers of sporeformers in the raw product (78). This is because products prone to support sporeformer growth are stored for sufficiently long periods that outgrowth of small numbers of cells can eventually cause a defect. Spore-forming bacteria in raw milk are predominantly *Bacillus* spp., with *Bacillus licheniformis*, *B. pallidus*, *B. cereus*, *B. subtilis*, and *B. megaterium* commonly isolated (69, 97). *Clostridium* spp. are present in raw milk at such low levels that enrichment and most-probable-number techniques must be used for quantification (91). Populations of spore-forming bacteria in raw milk vary seasonally. Growth of these bacteria in silage contributes to high numbers of spores in raw milk (105).

Defects in Fluid Milk Products

Pasteurized milk packaged under conditions that limit recontamination can spoil due to the growth of psychrotrophic *B. cereus*. This topic has been reviewed by Meer et al. (76). Psychrotrophic *B. cereus* is present in over 80% of raw milk samples. There is also evidence that psychrotrophic *Bacillus* spp. are introduced into the milk at the processing plant as postpasteurization contaminants (43). Psychrotrophic *B. cereus* can reach populations exceeding 10^6 CFU/ml in milk held for 14 days at 7°C, although lower growth is more common (78). The defect is described as sweet curdling, since it first appears as coagulation without significant acid or off flavor being formed. Coagulation is caused by a chymosin-like protease (18). Eventually, the enzyme degrades casein sufficiently to produce a bitter-flavored product. Growth may become visible as "buttons" at the bottom of the carton; these buttons are actually bacterial colonies. Psychrotrophic *Bacillus* spp. other than *B. cereus* are also capable of spoiling heat-treated milk. Cromie et al. (24) observed that psychrotrophic *Bacillus circulans* was the predominant spoilage organism in aseptically packaged heat-treated milk. This microorganism produces acid from lactose, giving the milk a sour flavor. *Bacillus mycoides* is another psychrotrophic sporeformer frequently isolated in milk (86).

Most bacterial spores present in raw milk are moderately heat labile and destroyed by UHT treatments. The major heat-resistant species in milk is *Geobacillus stearothermophilus* (formerly *Bacillus stearothermophilus*) (83). *Bacillus sporothermodurans* and *Paenibacillus lactis* have been isolated from UHT milk (96).

Defects in Canned Condensed Milk

Canned condensed milk may be either sweetened with sucrose and glucose to lower the a_w or left unsweetened. The unsweetened product must be sterilized by heat treatment, whereas the sweetened product has sufficiently low a_w to inhibit spore germination. Defects associated with growth of surviving spore-forming organisms in this product have been described by Carić (13). "Sweet coagulation" is caused by growth of *Bacillus coagulans*, *G. stearothermophilus*, or *B. cereus*. This defect is similar to the sweet-curdling defect caused by psychrotrophic *B. cereus* in pasteurized milk. Protein destruction, in addition to curdling, can also occur and is usually caused by growth of *B. subtilis* or *B. licheniformis*. Swelling or bursting of cans can be caused by growth of *Clostridium sporogenes*. "Flat sour" defect (acidification without gas production) can result from growth of *G. stearothermophilus*, *B. licheniformis*, *B. coagulans*, *Bacillus macerans*, and *B. subtilis* (55).

Control of Sporeformer-Associated Defects in Fluid Products

Methods for controlling growth of sporeformers in fluid products mainly involve the use of appropriate heat treatments. UHT treatments produce products microbiologically stable at room temperature. However, when sub-UHT heat treatments are more severe than that required for pasteurization, the shelf life of cream and milk can actually decrease, a phenomenon attributed to spore activation (83). A practical means to prevent sporeformers from spoiling nonfermented liquid dairy products given sub-UHT heat treatments has not been developed.

Defects in Cheese

The major defect in cheese caused by spore-forming bacteria is gas formation resulting from growth of clostridia, often *Clostridium tyrobutyricum* and *Clostridium beijerinckii* and occasionally from growth of *Clostridium sporogenes* and *Clostridium butyricum* (65). This defect is often called "late gas" because it occurs after the cheese has aged for several weeks. Emmental, Swiss, Gouda, and Edam cheeses are most often affected because of their relatively high pH and moisture content and low interior salt levels. Late gas defect results from the fermentation of lactate to butyric acid, acetic acid, carbon dioxide, and hydrogen gas. Populations of *C. tyrobutyricum* spores of less than one per milliliter of milk can produce the defect, because the spores are concentrated in the cheese curd during manufacture (92). The number of spores required to cause late gas in 9-kg wheels of rinded Swiss cheese was estimated at >100 per liter of raw milk (26). The presence of *C. tyrobutyricum* spores in milk has been traced to the consumption of contaminated silage, which increases levels in cow feces (26). Contaminated silage generally has a high pH that allows growth of clostridia.

Control of Sporeformer-Associated Defects in Cheese

Ideally, control of late gas defect would occur at the farm by instituting feeding and management practices that would reduce the number of spores entering the milk supply (46). In practice, this approach has not achieved the required results, so cheese manufacturers have tried to control the defect by removing spores from the milk at the plant, or inhibiting their growth in the cheese (103). Numbers of bacterial spores can be reduced in milk by a centrifugation process known as bactofugation (56). Bacteriocins produced by lactic acid bacteria may provide a highly specific means of inhibiting anaerobic spore germination (109).

YEASTS AND MOLDS

Growth of yeasts and molds is a common cause of spoilage of fermented dairy products, because these microorganisms are able to grow well at low pH. Yeast spoilage is manifest as a fruity or yeasty odor and/or gas formation. Cured cheeses, when properly made, have small amounts of lactose, thus limiting the potential for yeast growth. Cultured milks, such as yogurt and buttermilk, and fresh cheeses, such as cottage, normally contain sufficient lactose to support yeast growth. "Fermented/yeasty" flavor was observed in Cheddar cheese spoiled by growth of a *Candida* sp. and was associated with elevated ethanol, ethyl acetate, and ethyl butyrate (49). The affected cheese had a high moisture content (associated with low starter activity, and therefore high residual lactose) and low salt content, which contributed to allowing yeast growth. Yeast spoilage can also occur in dairy foods with low a_w, such as sweetened condensed milk and butter. The most common yeasts present in dairy products are *Kluyveromyces marxianus* and *Debaryomyces hansenii* (the teleomorph) and their asporogenous counterparts (the anamorph), *Candida* species, and *Zygosaccharomyces microellipsoides* (12, 35, 57). Also prevalent are *Rhodotorula mucilaginosa*, *Yarrowia lipolytica*, *Torulospora*, and *Pichia* (90, 116). Fermented dairy products provide a highly specialized ecological niche for yeasts, selecting for those that can utilize lactose or lactic acid and that tolerate high salt concentrations (35). Yeasts able to produce proteolytic or lipolytic enzymes may also have a selective advantage for growth in dairy products.

Mold Spoilage of Cheese

Growth of spoilage molds on cheese is a problem which still has significance, though it dates back to prehistory. The most common molds found on cheese are *Penicillium* spp. (11, 110), with others occasionally found, including *Aspergillus*, *Alternaria*, *Mucor*, *Fusarium*, *Cladosporium*, *Geotrichum*, and *Hormodendrum*. Penicillia commonly isolated from processed cheese include

Penicillium roqueforti, *P. cyclopium*, *P. viridicatum*, and *P. crustosum* (111). Vacuum-packaged cured cheese supports the growth of *Cladosporium cladosporioides*, *Penicillium commune*, *Penicillium glabrum*, *Cladosporium herbarum*, and *Phoma glomerata* (6, 47).

Controlling Mold Spoilage

Yeasts and molds that spoil dairy products can usually be isolated in the processing plant on packaging equipment, in the air, in salt brine, on manufacturing equipment, and in the general environment (floors, walls, ventilation ducts, etc.). Successful control efforts must start with limiting exposure of products to these sources. Most mold spores do not survive pasteurization (29). If the initial contamination level is limited, strategies to inhibit growth are more likely to succeed. These include packaging to reduce oxygen (and/or increase carbon dioxide), cold storage, and the use of antimycotic chemicals such as sorbate, propionate, and natamycin (pimaricin). Added liquid smoke is also a potent mold inhibitor (115). None of these control measures is completely effective. Vacuum-packaged cheese is susceptible to thread mold defect, where the fungi grow in the wrinkles of the plastic film (47). Some molds are resistant to antimycotic additives. Sorbate-resistant molds are commonly isolated from sorbate-treated cheese, but not from untreated cheese (66). Some *Penicillium* spp. not only are resistant to sorbate but also will degrade it by decarboxylation, producing 1,3-pentadiene (68). This imparts a kerosene-like odor to the cheese. Some *Mucor* spp. degrade sorbate to 4-hexenol, and some *Geotrichum* spp. degrade it to 4-hexenoic acid. Sorbate can also be used as a carbon source or be oxidized to carbon dioxide and water (66). The ability of some molds to degrade sorbate explains why cheeses with high levels of mold contamination are not effectively preserved by this additive.

References

1. Abad, P., A. Villafafila, J. D. Frias, and C. Rodriguez-Fernandez. 1993. Extracellular lipolytic activity from *Pseudomonas fluorescens* biovar I (*Pseudomonas fluorescens* NC1). *Milchwissenschaft* **48**:680–683.

2. Alkanhal, H. A., J. F. Frank, and G. L. Christen. 1985. Microbial protease and phospholipase C stimulate lipolysis of washed cream. *J. Dairy Sci.* **68**:3162–3170.

3. Andersson, R. E. 1980. Lipase production, lipolysis and formation of volatile compounds by *Pseudomonas fluorescens* in fat containing media. *J. Food Sci.* **45**:1694–1701.

4. Andersson, R. E. 1980. Microbial lipolysis at low temperatures. *Appl. Environ. Microbiol.* **39**:36–40.

5. Banwart, G. J. 1981. *Basic Food Microbiology*. AVI Publishing Co., Westport, Conn.

6. **Basilico, J. C., M. Z. deBasilico, C. Chiericatti, and C. G. Vinderola.** 2001. Characterization and control of thread mould in cheese. *Lett. Appl. Microbiol.* **32**:419–423.

7. **Bassette, R., and J. S. Acosta.** 1988. Composition of milk products, p. 39–79. *In* N. P. Wong (ed.), *Fundamentals of Dairy Chemistry.* Van Nostrand Reinhold Co., New York, N.Y.

8. **Batish, V. K., H. Chander, K. C. Zumdegni, K. L. Bhatia, and R. S. Singh.** 1988. Antibacterial activity of lactoferrin against some common food-borne pathogenic organisms. *Aust. J. Dairy Technol.* **43**:16–18.

9. **Bills, D. D., M. E. Morgan, L. M. Reddy, and E. A. Day.** 1965. Identification of compounds responsible for fruit flavor defect of experimental Cheddar cheeses. *J. Dairy Sci.* **48**:1168–1170.

10. **Bramley, A. J., and C. H. McKinnon.** 1990. The microbiology of raw milk, p. 163–208. *In* R. K. Robinson (ed.), *Dairy Microbiology*, vol. 1. Elsevier Applied Science, New York, N.Y.

11. **Bullerman, L. B., and F. J. Olivigni.** 1974. Mycotoxin producing potential of molds isolated from Cheddar cheese. *J. Food Sci.* **39**:1166–1168.

12. **Cappa, F., and P. F. Cocconcilli.** 2001. Identification of fungi from dairy products by means of 18S rRNA analysis. *Int. J. Food Microbiol.* **69**:157–169.

13. **Carić, M.** 1994. Concentrated and dried dairy products. VCH Publishers, Inc., New York, N.Y.

14. **Cerning, J.** 1990. Exocellular polysaccharides produced by lactic acid bacteria. *FEMS Microbiol. Rev.* **87**:113–130.

15. **Cerning, J., C. Bouillanne, M. Landon, and M. Desmazeaud.** 1992. Isolation and characterization of exopolysaccharides from slime-forming mesophilic lactic acid bacteria. *J. Dairy Sci.* **75**:692–699.

16. **Chapman, H. R., and M. E. Sharpe.** 1990. Microbiology of cheese, p. 203–289. *In* R. K. Robinson (ed.), *Dairy Microbiology*, vol. 2. Elsevier Applied Science, New York, N.Y.

17. **Chen, L., R. M. Daniel, and T. Coolbear.** 2003. Detection and impact of protease and lipase activities in milk and milk powders. *Int. Dairy J.* **13**:255–275.

18. **Choudhery, A. K., and E. M. Mikolajcik.** 1971. Activity of *Bacillus cereus* proteinases in milk. *J. Dairy Sci.* **53**:363–366.

19. **Chrisope, G. L., and R. T. Marshall.** 1976. Combined action of lipase and microbial phospholipase C on a model fat emulsion and raw milk. *J. Dairy Sci.* **59**:2024–2030.

20. **Christen, G. L., and R. T. Marshall.** 1985. Effect of histidine on thermostability of lipase and protease of *Pseudomonas fluorescens* 27. *J. Dairy Sci.* **68**:594–604.

21. **Christian, J. H. B.** 1980. Reduced water activity, p. 70–91. *In Microbial Ecology of Foods*, vol. 1. Academic Press, New York, N.Y.

22. **Cousin, M. A.** 1982. Presence and activity of psychrotrophic microorganisms in milk and dairy products: a review. *J. Food Prot.* **45**:172–207.

23. **Cousin, M. A.** 1989. Physical and biochemical effects of milk components, p. 205–225. *In* R. C. McKellar (ed.), *Enzymes of Psychrotrophs in Raw Food*. CRC Press, Inc., Boca Raton, Fla.

24. **Cromie, S. J., T. W. Dommett, and D. Schmidt.** 1989. Changes in the microflora of milk with different pasteurization and storage conditions and aseptic packaging. *Aust. J. Dairy Technol.* **44**:74–77.

25. **Cromie, S. J., J. E. Giles, and J. R. Dulley.** 1987. Effect of elevated ripening temperatures on the microflora of Cheddar cheese. *J. Dairy Res.* **54**:69–76.

26. **Dasgupta, A. R., and R. R. Hull.** 1989. Late blowing of Swiss cheese: incidence of *Clostridium tyrobutyricum* in manufacturing milk. *Aust. J. Dairy Technol.* **44**:82–87.

27. **Deeth, H. C., and C. H. Fitz-Gerald.** 1983. Lipolytic enzymes and hydrolytic rancidity in milk and milk products, p. 195–239. *In* P. F. Fox (ed.), *Developments in Dairy Chemistry, part II*. Applied Science, London, England.

28. **Dionysius, D. A., P. A. Grieve, and A. C. Vos.** 1992. Studies on the lactoperoxidase system: reaction kinetics and antibacterial activity using two methods for hydrogen peroxide generation. *J. Appl. Bacteriol.* **72**:146–153.

29. **Doyle, M. P., and E. H. Marth.** 1975. Thermal inactivation of conidia from *Aspergillus flavus* and *Aspergillus parasiticus*. I. Effects of moist heat, age of conidia, and sporulation medium. *J. Milk Food Technol.* **38**:678–682.

30. **Dumont, J. P., G. Delespaul, B. Miquot, and J. Adda.** 1977. Influence des bactéries psychrotrophs sur les qualités organoleptiques de fromages à pâte molle. *Lait* **57**:619–630.

31. **Ewings, K. N., R. E. O'Conner, and G. E. Mitchell.** 1984. Proteolytic microflora of refrigerated raw milk in South East Queensland. *Aust. J. Dairy Technol.* **39**:65–68.

32. **Fairbairn, D. J., and B. A. Law.** 1987. The effect of nitrogen and carbon sources on proteinase production by *Pseudomonas fluorescens*. *J. Appl. Bacteriol.* **62**:105–113.

33. **Fernandez, L., J. A. Alvarez, P. Palacios, and C. San Jose.** 1992. Proteolytic and lipolytic activities of *Pseudomonas fluorescens* grown in raw milk with variable iron content. *Milchwissenschaft* **47**:160–163.

34. **Fernandez, L., C. San Jose, and R. C. McKellar.** 1990. Repression of *Pseudomonas fluorescens* extracellular lipase secretion by arginine. *J. Dairy Res.* **57**:69–78.

35. **Fleet, G. H.** 1990. Yeasts in dairy products. *J. Appl. Bacteriol.* **68**:199–211.

36. **Fox, P. F., P. Power, and T. M. Cogan.** 1989. Isolation and molecular characteristics, p. 57–120. *In* R. C. McKellar (ed.), *Enzymes of Psychrotrophs in Raw Food*. CRC Press, Inc., Boca Raton, Fla.

37. **Frank, J. F., and R. A. Koffi.** 1990. Surface-adherent growth of *Listeria monocytogenes* is associated with increased resistance to surfactant sanitizers and heat. *J. Food Prot.* **53**:560–564.

38. **Frank, J. F., and E. H. Marth.** 1988. Fermentations, p. 656–738. *In* N. P. Wong (ed.), *Fundamentals of Dairy Chemistry*, 3rd ed. Van Nostrand Reinhold Co., New York, N.Y.

39. **Frank, J. F., E. H. Marth, and N. F. Olson.** 1977. Survival of enteropathogenic and nonpathogenic *Escherichia coli* during the manufacture of Camembert cheese. *J. Food Prot.* **40**:835–842.

40. **Frank, J. F., E. H. Marth, and N. F. Olson.** 1978. Behavior of enteropathogenic *Escherichia coli* during manufacture and ripening of brick cheese. *J. Food Prot.* **41**:111–115.

41. **Garcia, M. L., B. Sanz, P. Garcia-Collia, and J. A. Ordonez.** 1989. Activity and thermostability of the extracellular

lipases and proteinases from pseudomonads isolated from raw milk. *Milchwissenschaft* 44:547–560.

42. **Griffiths, M. W.** 1989. Effect of temperature and milk fat on extracellular enzyme synthesis by psychrotrophic bacteria during growth in milk. *Milchwissenschaft* 44:539–543.

43. **Griffiths, M. W., and J. D. Phillips.** 1990. Incidence, source and some properties of psychrotrophic *Bacillus* spp. found in raw and pasteurized milk. *J. Soc. Dairy Technol.* **43:** 62–70.

44. **Hanson, M. L., W. L. Wendorff, and K. B. Houck.** 2005. Effect of heat treatment of milk on activation of *Bacillus* spores. *J. Food Prot.* 68:1484–1486.

45. **Haryani, S., N. Datta, A. J. Elliot, and H. C. Deeth.** 2003. Production of proteinases by psychrotrophic bacteria in raw milk stored at low temperature. *Aust. J. Dairy Technol.* 58:15–20.

46. **Herlin, A. H., and A. Christansson.** 1993. Cheese-blowing anaerobic spores in bulk milk from loose-housed and tied dairy cows. *Milchwissenschaft* 48:686–689.

47. **Hocking, A. D., and M. Faedo.** 1992. Fungi causing thread mould spoilage of vacuum packaged Cheddar cheese during maturation. *Int. J. Food Microbiol.* **16:**123–130.

48. **Hogarty, S. L., and J. F. Frank.** 1982. Low-temperature activity of lactic streptococci isolated from cultured buttermilk. *J. Food Prot.* 43:1208–1211.

49. **Horwood, J. F., W. Stark, and R. R. Hull.** 1987. A "fermented, yeasty" flavour defect in Cheddar cheese. *Aust. J. Dairy Technol.* 42:25–26.

50. **Hull, R., S. Toyne, I. Haynes, and F. Lehman.** 1992. Thermoduric bacteria: a re-emerging problem in cheesemaking. *Aust. J. Dairy Technol.* 47:91–94.

51. **Hull, R. R., A. V. Roberts, and J. J. Mayes.** 1983. The association of *Lactobacillus casei* with a soft-body defect in commercial Mozzarella cheese. *Aust. J. Dairy Technol.* 22:78–80.

52. **Jaspe, A., P. Oviedo, L. Fernandez, P. Palacios, and C. Sanjose.** 1995. Cooling raw milk: change in the spoilage potential of contaminating *Pseudomonas*. *J. Food Prot.* 58:915–921.

53. **Jaspe, A., P. Palacios, P. Matias, L. Fernandez, and C. Sanjose.** 1994. Proteinase activity of *Pseudomonas fluorescens* grown in cold milk supplemented with nitrogen and carbon sources. *J. Dairy Sci.* 77:923–929.

54. **Jenness, R.** 1988. Composition of milk, p. 1–38. *In* N. P. Wong (ed.), *Fundamentals of Dairy Chemistry.* Van Nostrand Reinhold Co., New York, N.Y.

55. **Kalogridou-Vassiliadou, D.** 1992. Biochemical activities of *Bacillus* species isolated from flat sour evaporated milk. *J. Dairy Sci.* 75:2681–2686.

56. **Kosikowski, F. V.** 1982. *Cheese and Fermented Milk Foods,* 2nd ed. F. V. Kosikowski and Associates, Brooktondale, N.Y.

57. **Kosse, D., H. Seiler, R. Amann, W. Ludwig, and S. Scherer.** 1997. Identification of yoghurt-spoiling yeasts with 18s rRNA oligonucleotide probes. *Syst. Appl. Microbiol.* 20:468–480.

58. **Kroll, S.** 1989. Thermal stability, p. 121–152. *In* R. C. McKellar (ed.), *Enzymes of Psychrotrophs in Raw Food.* CRC Press, Inc., Boca Raton, Fla.

59. **Kumura, H., K. Mikawa, and Z. Saito.** 1993. Influence of milk proteins on the thermostability of the lipase from *Pseudomonas fluorescens* 33. *J. Dairy Sci.* 76:2164–2167.

60. **Lafarge, V., J.-C. Ogier, V. Girard, V. Maladen, J.-Y. Leveau, A. Gruss, and A. Delacroix-Buchet.** 2004. Raw cow milk bacterial population shifts attributable to refrigeration. *Appl. Environ. Microbiol.* 70:5644–5650.

61. **Lalaye, L. C., R. E. Simard, B.-H. Lee, R. A. Holley, and R. N. Giroux.** 1987. Involvement of heterofermentative lactobacilli in development of open texture in cheeses. *J. Food Prot.* 50:1009–1012.

62. **Law, B. A.** 1979. Reviews of the progress of dairy science: enzymes of psychrotrophic bacteria and their effects on milk and milk products. *J. Dairy Res.* 46:573–588.

63. **Law, B. A., C. M. Cousins, M. E. Sharpe, and F. L. Davies.** 1979. Psychrotrophs and their effects on milk and dairy products, p. 137–152. *In* A. D. Russell and R. Fuller (ed.), *Cold Tolerant Microbes in Spoilage and the Environment.* Academic Press, New York, N.Y.

64. **Law, B. A., M. E. Sharpe, and H. R. Chapman.** 1976. Effect of lipolytic Gram negative psychrotrophs in stored milk on the development of rancidity in Cheddar cheese. *J. Dairy Res.* 43:459–468.

65. **Le Bourhis, A.-G., K. Saunier, J. Doré, J.-P. Carlier, J.-F. Chamba, M.-R. Popoff, and J.-L. Tholozan.** 2005. Development and validation of PCR primers to assess the diversity of *Clostridium* spp. in cheese by temporal temperature gradient gel electrophoresis. *Appl. Environ. Microbiol.* 71:29–38.

66. **Liewen, M. B., and E. H. Marth.** 1985. Growth and inhibition of microorganisms in the presence of sorbic acid: a review. *J. Food Prot.* 48:364–375.

67. **Ma, Y., D. M. Barbano, and M. Santos.** 2003. Effect of CO_2 addition to raw milk on proteolysis and lipolysis at 4°C. *J. Dairy Sci.* 86:1616–1631.

68. **Marth, E. H., C. M. Capp, L. Hasenzahl, H. W. Jackson, and R. V. Hussong.** 1966. Degradation of potassium sorbate by *Penicillium* species. *J. Dairy Sci.* 49:1197–1205.

69. **Martin, J. H., D. P. Stahly, W. J. Harper, and I. A. Gould.** 1962. Sporeforming microorganisms in selected milk supplies. *Proc. XVI Int. Dairy Congr.* C:295–304.

70. **Massa, S., F. Gardini, M. Sinigaglia, and M. E. Guerzoni.** 1992. *Klebsiella pneumoniae* as a spoilage organism in Mozzarella cheese. *J. Dairy Sci.* 75:1411–1414.

71. **Masson, P. L., and J. F. Heremans.** 1971. Lactoferrin in milk from different species. *Comp. Biochem. Physiol.* 39B:119–129.

72. **McKellar, R. C.** 1982. Factors influencing the production of extracellular proteinase by *Pseudomonas fluorescens.* *J. Appl. Bacteriol.* 53:305–316.

73. **McKellar, R. C.** 1989. Regulation and control of synthesis, p. 153–172. *In* R. C. McKellar (ed.), *Enzymes of Psychrotrophs in Raw Food.* CRC Press, Inc., Boca Raton, Fla.

74. **McKellar, R. C., and H. Cholette.** 1986. Possible role of calcium in the formation of active extracellular proteinase by *Pseudomonas fluorescens.* *J. Appl. Bacteriol.* 60:37–44.

75. **McKellar, R. C., K. Shamsuzzaman, C. San Jose, and H. Cholette.** 1987. Influence of iron (iii) and pyoverdine, a siderophore produced by *Pseudomonas fluorescens* B52,

on its extracellular proteinase and lipase production. *Arch. Microbiol.* **147:**225–230.

76. Meer, R. R., J. Baker, F. W. Bodyfelt, and M. W. Griffiths. 1991. Psychrotrophic *Bacillus* spp. in fluid milk products: a review. *J. Food Prot.* **54:**969–979.

77. Melilli, C., D. M. Barbano, M. Cacamo, M. A. Calvo, G. Schembari, and G. Licitra. 2004. Influence of brine concentration, brine temperature, and presalting on early gas defect in raw milk pasta filata cheese. *J. Dairy Sci.* **87:**3648–3657.

78. Mikolojcik, E. M., and N. T. Simon. 1978. Heat resistant psychrotrophic bacteria in raw milk and their growth at 7°C. *J. Food Prot.* **41:**93–95.

79. Morgan, M. E. 1976. The chemistry of some microbially induced flavor defects in milk and dairy foods. *Biotechnol. Bioeng.* **18:**953–965.

80. Morse, P. M., H. Jackson, C. H. McNaughton, A. G. Leggatt, G. B. Landerkin, and C. K. Johns. 1968. Investigation of factors contributing to the bacteria count of bulk tank milk. II. Bacteria in milk from individual cows. *J. Dairy Sci.* **51:**1188–1191.

81. Mosteller, T. M., and J. R. Bishop. 1993. Sanitizer efficacy against attached bacteria in milk biofilm. *J. Food Prot.* **56:**34–41.

82. Mottar, J. F. 1989. Effect on the quality of dairy products, p. 227–243. *In* R. C. McKellar (ed.), *Enzymes of Psychrotrophs in Raw Food.* CRC Press, Inc., Boca Raton, Fla.

83. Muir, D. D. 1989. The microbiology of heat treated fluid milk products, p. 209–270. *In* R. K. Robinson (ed.), *Dairy Microbiology*, vol. 1. Elsevier Applied Science, New York, N.Y.

84. Myhara, R. M., and B. Skura. 1990. Centroid search optimization of cultural conditions affecting the production of extracellular proteinase by *Pseudomonas fragi* ATCC 4973. *J. Appl. Bacteriol.* **69:**530–538.

85. Nashif, S. A., and F. E. Nelson. 1953. The lipase of *Pseudomonas fragi*. III. Enzyme action in cream and butter. *J. Dairy Sci.* **36:**481–488.

86. Phillips, J. D., and M. W. Griffiths. 1986. Factors contributing to the seasonal variation of *Bacillus* species in pasteurized products. *J. Appl. Bacteriol.* **61:**275–285.

87. Reddy, M. C., D. D. Bills, R. C. Lindsay, and L. M. Libbey. 1968. Ester production by *Pseudomonas fragi*. I. Identification and quantification of some esters produced in milk cultures. *J. Dairy Sci.* **51:**656–659.

88. Reimerdes, E. H. 1982. Changes in the proteins of raw milk during storage, p. 271. *In* P. F. Fox (ed.), *Developments in Dairy Chemistry*, part I. Applied Science, London, England.

89. Rengpipat, S., and E. A. Johnson. 1989. Characterization of a Lactobacillus strain producing white crystals on cheddar cheese. *Appl. Environ. Microbiol.* **56:**2579–2582.

90. Rohm, H., F. Eliskases-Lechner, and M. Bräuer. 1992. Diversity of yeasts in selected dairy products. *J. Appl. Bacteriol.* **72:**370–376.

91. Rosen, B., U. Merin, and I. Rosenthal. 1989. Evaluation of clostridia in raw milk. *Milchwissenschaft* **44:**356–357.

92. Rosen, B., G. Popel, and I. Rosenthal. 1990. The affinity of *Clostridium tyrobutyricum* to casein in raw milk. *Milchwissenschaft* **45:**152–154.

93. Rowe, M. T. 1990. Growth and extracellular enzyme production by psychrotrophic bacteria in raw milk stored at low temperature. *Milchwissenschaft* **45:**495–499.

94. Russell, H. L. 1885. Gas producing bacteria and the relation of the same to cheese, p. 139–150. *In Wisconsin Agriculture Experimental Station 12th Annual Report.*

95. Rutzinski, J. L., E. H. Marth, and N. F. Olson. 1979. Behavior of *Enterobacter aerogenes* and *Hafnia* species during the manufacture and ripening of Camembert cheese. *J. Food Prot.* **42:**790–793.

96. Scheldeman, P., K. Goossens, M. Rodriguez-Diaz, A. Pil, J. Goris, L. Herman, P. De Vos, N. A. Logan, and M. Heyndrickx. 2004. *Paenibacillus lactis* sp. nov., isolated from raw and heat-treated milk. *Int. J. Syst. Evol. Microbiol.* **54:**885–891.

97. Scheldeman, P., A. Pil, L. Herman, P. De Vos, and M. Heyndrickx. 2005. Incidence and diversity of potentially highly heat-resistant spores isolated at dairy farms. *Appl. Environ. Microbiol.* **71:**1480–1494.

98. Schröder, M. J. A. 1984. Origins and levels of post pasteurization contamination of milk in the dairy and their effects of keeping quality. *J. Dairy Res.* **51:**59–67.

99. Shannon, E. L., N. F. Olson, and J. H. von Elbe. 1969. Effect of lactic starter culture on pink discoloration and oxidation-reduction potential in Italian cheese. *J. Dairy Sci.* **52:**1567–1561.

100. Shipe, W. F., R. Bassette, D. D. Deane, W. L. Dinkley, E. G. Hammond, W. J. Harper, D. H. Klein, M. E. Morgan, J. H. Nelson, and R. A. Scanlan. 1978. Off flavors in milk: nomenclature, standards, and bibliography. *J. Dairy Sci.* **61:**856–869.

101. Stead, D. 1986. Microbial lipases: their characteristics, role in food spoilage and industrial uses. *J. Dairy Res.* **53:**481–505.

102. Stepaniak, L., E. Zakrzewski, and T. Sorhaug. 1991. Inactivation of heat-stable proteinase from *Pseudomonas fluorescens* P1 at pH 4.5 and 56EC. *Milchwissenschaft* **46:**139–142.

103. Su, Y. C., and S. C. Ingham. 2000. Influence of milk centrifugation, brining and ripening conditions in preventing gas formation by *Clostridium* spp. in Gouda cheese. *Int. J. Food Microbiol.* **54:**147–154.

104. Suhren, G. 1989. Producer microorganisms, p. 3–34. In R. C. McKellar (ed.), *Enzymes of Psychrotrophs in Raw Food.* CRC Press, Inc., Boca Raton, Fla.

105. te Giffel, M. C., A. Wagendorp, A. Herrewegh, and F. Driehuis. 2002. Bacterial spores in silage and raw milk. *Antonie Leeuwenhoek* **81:**625–630.

106. Ternstrom, A., M. A. Lindberg, and G. Molin. 1993. Classification of the spoilage flora of raw and pasteurized bovine milk, with special reference to *Pseudomonas* and *Bacillus*. *J. Appl. Bacteriol.* **75:**25–34.

107. Thomas, S. B. 1966. Sources, incidence, and significance of psychrotrophic bacteria in milk. *Milchwissenschaft* **21:**270–275.

108. Thomas, S. B., and B. F. Thomas. 1973. Psychrotrophic bacteria in refrigerated bulk-collected raw milk. Part I. *Dairy Ind.* **38:**11–15.

109. Thualt, D., E. Beliard, J. Je Guern, and C.-M. Bourgeois. 1991. Inhibition of *Clostridium tyrobutyricum* by

bacteriocin-like substances produced by lactic acid bacteria. *J. Dairy Sci.* **74:**1145–1150.

110. **Torrey, G. S., and E. H. Marth.** 1977. Isolation and toxicity of molds from foods stored in homes. *J. Food Prot.* **40:**187–190.

111. **Tsai, W.-Y. J., M. B. Liewen, and L. Bullerman.** 1988. Toxicity and sorbate sensitivity of molds isolated from surplus commodity cheese. *J. Food Prot.* **51:**457–462.

112. **Turner, K. W., and T. D. Thomas.** 1980. Lactose fermentation in Cheddar cheese and the effect of salt. *N. Z. J. Dairy Sci. Technol.* **15:**265–276.

113. **Vilafafila, A., J. D. Frias, P. Abad, and C. Rodriguez-Fernandez.** 1993. Extracellular proteinase activity from psychrotrophic *Pseudomonas fluorescens* biovar 1 (*Ps. fluorescens* NC1). *Milchwissenschaft* **48:**435–438.

114. **Wang, J. J., and J. F. Frank.** 1981. Characterization of psychrotrophic bacterial contamination of commercial buttermilk. *J. Dairy Sci.* **64:**2154–2160.

115. **Wendorff, W. L., W. E. Riha, and E. Muehlenkamp.** 1993. Growth of molds on cheese treated with heat or liquid smoke. *J. Food Prot.* **56:**963–966.

116. **Westall, S., and O. Liltenborg.** 1998. Spoilage yeasts of decorated soft cheese packed in modified atmosphere. *Food Microbiol.* **15:**243–249.

117. **Wolfson, L. M., and S. S. Sumner.** 1993. Antibacterial activity of the lactoperoxidase system: a review. *J. Food Prot.* **56:**887–892.

118. **Yan, L., B. E. Langlois, J. O'Leary, and C. Hicks.** 1983. Effect of storage conditions of grade A raw milk on proteolysis and cheese yield. *Milchwissenschaft* **38:**715–719.

Food Microbiology: Fundamentals and Frontiers, 3rd Ed.
Edited by M. P. Doyle and L. R. Beuchat
© 2007 ASM Press, Washington, D.C.

Frédéric Carlin

Fruits and Vegetables

8

Fresh fruits and vegetables are an extraordinary dietary source of nutrients, micronutrients, vitamins, and fiber for humans, and an essential basic raw material for the food industry. They are living organs detached from their parent plants and have a high water content, which contributes to their natural fragility. In addition, fruits and vegetables are widely exposed to microbial contamination through contact with soil, dust, and water, and by handling at harvest or during postharvest processing, thereby establishing conditions that may lead to spoilage and loss of quality.

A single produce such as tomato is cultivated under many different climates and at many latitudes from its tropical area of origin to the colder Nordic countries. Giving an exhaustive account of microbial spoilage of tomato, as well as many other types of produce, presents some difficulties. Moreover, fruits and vegetables cover many different species and include many different plant organs at varied stages of physiological maturity in their consumed forms. Reviewing microbial spoilage of fruits and vegetables is a real challenge, and the literature on this topic is particularly rich. This chapter focuses on the origin, description, and control of bacterial and fungal spoilage of fruits and vegetables. Table 8.1 presents examples of some important spoilage molds and bacteria, their host vegetables or fruits, and symptoms of infection. More extensive descriptions of fruit and vegetable spoilage can be found elsewhere (*The Commercial Storage of Fruits, Vegetables, and Florist and Nursery Stocks*, available at http://www.ba.ars.usda.gov/hb66) (3, 26, 89, 105, 106).

MAIN CHARACTERISTICS OF FRUITS AND VEGETABLES AND THEIR ROLES IN MICROBIAL SPOILAGE

Fruits and vegetables are the edible parts of plant organs of very diverse nature: leaves (lettuce and cabbage, for instance), stems (leek and asparagus), flowers (artichoke, cauliflower, and broccoli), roots (beet, carrot, and turnip), bulbs (garlic and onion), tubers (potato), and fruits in their botanical meaning, e.g., simple fruit such as tomato, cucumber, and pepper, stone fruit such as peach, seed fruit such as apple, multiple fruit such as pineapple, aggregate fruit such as raspberry, and fruits consumed in their immature form such as green beans. Plant tissues in fruits and vegetables consist of an assemblage of cells surrounded by a pectic and cellulosic cell

Frédéric Carlin, INRA, UMR408, Sécurité et Qualité des Produits d'Origine Végétale, Avignon, F-84914, France, and University of Avignon, Avignon, F-84029, France.

Table 8.1 Important microbial agents of postharvest spoilage of fruits and vegetables[a]

Spoilage agent	Type of postharvest disease or spoilage	Produce affected	Biology
Alternaria alternata and other *Alternaria* spp.	Black rot, black spots, dark lesions	Cucurbit, solanaceous vegetables, green bean, brassica, potato, citrus, persimmon, mango, pome fruits	Stem-end pathogen. Penetration by flower or stem scars. Infection may remain quiescent.
Botrytis cinerea	Soft rot covered with gray mold	Cucurbit, solanaceous vegetables, green bean, pea, brassica, artichoke, celery, lettuce, chicory, onion, garlic, carrot, citrus, apple, strawberry, raspberry	Wide spectrum. Infection before or after harvest through damaged or senescent tissue. Favored by wet conditions. May spread into neighboring fruits, causing "nesting." Possible growth even at low temperatures.
Colletotrichum musae	Anthracnose; dark circular spots on ripening fruits	Banana	Quiescent infection until fruit ripening.
Colletotrichum gloeosporioides and other *Colletotrichum* spp.	Anthracnose; lesion on the skin. Dark spots, sunken lesions.	Cucurbit, solanaceous vegetables, green bean, avocado, apple, mango	Quiescent infections. May form appressoria on the plant cuticle. Development of decay during fruit maturation.
Geotrichum candidum	Sour rot	Cucurbit, carrot, citrus, tomato	Soil pathogen. Transmission by insects. Wound pathogen.
Monilinia spp.	Brown rot. Brown spots, white molds in concentric circles.	Apple, stone fruits	Survival in winter on mummified fruits. Infection may remain quiescent on immature fruits.
Penicillium spp.	Blue mold, green mold covering lesions or rot	Cucurbit, onion, garlic, grape, apple, citrus	Wound pathogen. Colored spores at the center of the lesions. Slow development at low temperature. May spread from fruits to fruits.
Rhizopus spp.	Soft, very wet rot. Development of profuse mycelium with spore heads turning black.	Cucurbit, solanaceous vegetables, green bean, pea, sweet potato, stone fruits, papaya	Ubiquitous. Infection at, or after harvest by wound, or by contact with soil or infected produce. Rapid decay above 20°C.
Erwinia carotovora	Soft rot	Cucurbit, solanaceous vegetables, brassica, asparagus, celery, lettuce, chicory, carrot	Infection by wounds, scars, and lenticels before or after harvest, and by contact with decaying vegetable. Favored by wet conditions and temperatures of 24–30°C.

[a] Adapted from references 3, 89, and 105.

wall organized in a network, and a middle lamella rich in pectin cementing together cell walls (53, 117). Both the water content of the cell vacuole and the cell wall organization and composition contribute to edible plant tissue firmness. The outer parts of fruits and vegetables are characterized by layers varying in thickness according to the type of produce, and include a hydrophobic cuticle consisting of cutin and wax covering an epidermis made of a layer of cells and eventually a layer of cork cells (45). Fruit and vegetable surfaces can be interrupted by natural openings, e.g., stomata or lenticels involved in respiration and transpiration of the plant organs, trichomes, cracks, wounds caused by insects, mechanical

injury, and other stress assaults, and scars resulting from detachment from plants.

Once fruits and vegetables have detached from the plant, their physiological activity continues and is negatively correlated with shelf life. Commodities, such as potato, onion, garlic, carrot, cantaloupe, and watermelon, with a low respiration rate (<10 ml of CO_2/kg/h at 5°C) can be stored for longer periods than those such as Brussels sprouts, spinach, broccoli, asparagus, or mushroom, with a high respiration rate (>20 ml of CO_2/kg/h at 5°C) (59). For "climacteric" fruits (e.g., apple, tomato, avocado, and banana), several biochemical changes associated with natural respiration occur and, triggered by autocatalytic production of the plant hormone ethylene, will lead to horticultural maturity (102). Nonclimacteric commodities (e.g., strawberry, citrus, grape, and cherry) are picked at their horticultural maturity, a stage of development corresponding to the prerequisites for utilization by consumers. The senescence process will lead, after horticultural maturity, to irreversible changes in structure and metabolism of the organs, and finally to deterioration (92, 104). Postharvest moisture loss due to respiration causes structural damage in fruits such as apples and wilting in leafy vegetables. A decrease in acidity and softening is followed by cell wall degradation. Degradation of phenolic compounds during ripening of climacteric fruits is among the physiological changes that may favor the invasion and growth of spoilage microorganisms.

Fruits and vegetables have a high water content and contain significant amounts of nutrients essential for microbial growth (Table 8.2). The main limitation preventing growth of most bacteria is the low pH of fruits (as low as 2.0 in some *Citrus* species). Spoilage of low-pH fruits is restricted to molds and yeasts, which are more tolerant than bacteria to high acidity. The favorable effects of basic nutrients in produce tissues may be balanced by compounds known for their antimicrobial activity, such as phenolic compounds and tannins in many

fruits and vegetables, tomatine (a saponin in tomato), sulfur-derived compounds in the *Alliaceae* (onion and garlic), or terpenoids in carrot (3, 16, 19).

ORIGIN OF CONTAMINATION

Fruits and vegetables harbor a wide range of microbial contaminants. Populations of the aerobic mesophilic bacteria ranging from 10^2 to 10^8 CFU/g of produce and bacterial contamination at levels of 10^5 to 10^6 CFU/g have been reported for fruits (76, 89). Numbers of yeasts and molds (although CFU quantification is probably not appropriate because of the possible presence of extensive mycelia or fruiting structures) ranging between 10^1 and 10^6 CFU/g on vegetables and between 10^3 and 10^7 CFU/g on fruits have been reported (23, 76, 89). These variations reflect the diversity of conditions prevailing during cultivation and postharvest storage and, to some extent, methods used for enumeration (89). With few exceptions, gram-negative species of bacteria are dominant on vegetables and basidiomycetous yeasts are among the major species most frequently found on fruits (23, 89). Postharvest spoilage microorganisms do not seem to include a dominating species in sound fresh fruits and vegetables (89). The dominating species of bacteria and

Table 8.2 Approximate pH values and water, protein, and sugar contents of some fresh fruits and vegetables[a]

Fruit or vegetable	pH	Water (g/100 g)	Protein (g/100 g, fresh wt)	Sugars (g/100 g, fresh wt)
Asparagus	5.0–6.1	93.2	2.2	1.9
Beans (lima)	5.4–6.5	70.2	6.8	1.5
Broccoli	6.5	89.3	2.8	1.7
Carrot	4.9–6.3	88.3	0.93	4.5
Cauliflower	6.0–6.7	91.9	2.0	2.4
Corn (sweet)	5.9–7.3	76.0	3.2	3.2
Lettuce	6.0–6.4	95.6	0.9	1.7
Onion	5.0–5.8	88.5	0.9	4.3
Pepper (red)	5.3–5.8	92.0	1.0	4.2
Potato tuber	5.6–6.2	81.6	1.7	1.2
Spinach	5.1–6.8	91.4	2.9	0.4
Squash	5.0–5.4	94.6	1.2	2.2
Tomato (ripe)	3.4–4.7	94.5	0.88	2.6
Apple	2.9–3.3	85.6	0.26	10.4
Banana	4.5–5.2	74.9	1.1	2.4
Grape	3.4–4.5	80.5	0.72	15.0
Lime	1.8–2.0	88.3	0.70	1.7
Melon (cantaloupe)	6.2–6.5	90.2	0.84	7.8
Orange	3.6–4.3	86.8	0.94	9.4

[a] From reference 75 and U.S. Department of Agriculture, Agricultural Research Service, 2005, National Nutrient Database for Standard Reference, release 18. Nutrient Data Laboratory home page, available at http://www.nal.usda.gov/fnic/foodcomp.

yeasts are generally not known for their ability to cause decay on fresh produce, and when detected, spoilage microorganisms represent only a low proportion of the normal microflora.

Postharvest spoilage microorganisms may take many different routes to contaminate fruits and vegetables. Seeds, including vegetable seeds, have been shown to be the primary source of postharvest diseases such as *Colletotrichum* infections on pepper, onion bulb neck rot caused by *Botrytis* spp., potato tuber soft rot caused by *Erwinia* spp., and potato tuber gangrene caused by *Phoma* spp. (89). Rain water is a significant vehicle of microorganisms from plant to plant, plant to soil, and soil to plant through splashing (72). Irrigation water from different sources may contain pectolytic bacteria (100). Soil and its components (rhizosphere and plant debris) are the natural reservoirs of spoilage bacteria such as *Bacillus* and *Clostridium* (74) and molds such as *Sclerotinia* spp. (63) and *Rhizoctonia solani* (106), and facilitate survival, in particular in plant debris or on fruits and vegetables in contact with the rhizosphere, as shown for *Erwinia carotovora* (93). Air and wind disperse spores or fruiting bodies of molds (38), leaves and microorganisms adhering to leaves, and aerosol particles that may contain bacteria, presuming the acquisition of some resistance to desiccation (72). Postharvest handling of produce has been shown to be a cause of contamination with spoilage microorganisms (89). Immersion tank solutions have been suspected to be a cause of redistribution of *Phialophora malorum* on pears (109). Wooden boxes previously used to store carrots may contain many species of molds pathogenic to carrot, including *Rhizoctonia carotae*, *Sclerotinia sclerotiorum*, and *Botrytis cinerea*, which can cause lesions on sound carrots (64). The common postharvest spoilage molds *B. cinerea* and *Penicillium* spp. can be found at multiple sites in production areas, e.g., on fruits, in orchard litter and soil, in orchard air, on packing lines, and in cold storage air; however, in these cases, contamination on the fruit surface is critical to further decay (68). The inoculum can consist of spores or conidia, mycelium, or sclerotia. Mycelium is generally infectious. Germination of spores and sclerotia generally depends on water, the presence of nutrients at least in small quantities that may have leached from fruit and vegetable tissues, or juice released from damaged tissues in wounds (3, 89).

The time elapsed between contact of microorganisms with the surface of the plant organ and initiation of spoilage is extremely variable. Anyway, this process presumes the establishment of some sort of colonization on the organ surface. Fluorescent pseudomonads, including pectolytic strains, form bacterial communities aggregated in a matrix of exopolymers and assimilated to form biofilms (11, 86). Adhesion of molds to plant surfaces involves very specific interactions which can involve lectins, hydrophobic contact with the plant cuticle, or secreted adhesives (113). Fungal agents of storage diseases may then colonize a few cells in a limited area of plant tissue, followed by a delay before becoming active under specific circumstances. This period without growth is known as quiescence. Infection during the quiescent period can be symptomless, as that of germinated spores of *Colletotrichum* spp. on various types of produce, or can result in visible but nonexpanding symptoms, as in ghost spot of tomato caused by *B. cinerea* (95). The quiescent period can be observed while fruits are still attached to the parent plant. For instance, initial infection of gray mold on strawberry or grape occurs at blooming and remains dormant until fruit formation or during postharvest storage (3).

Spoilage involves to some extent the penetration of microbial cells into plant tissue. The cuticle barrier can be compromised by many postharvest spoilage microorganisms, e.g., *B. cinerea* on cucumber and tomato fruits, *Colletotrichum* spp. on tomato and bell pepper (89), or *S. sclerotiorum* on carrot (63). Despite evidence of cutinase activity of some fungal pathogens, the actual contribution of these enzymes to tissue penetration remains controversial, some mutants unable to produce cutinases still being pathogenic (3, 45). The formation at the tip of the germ tube is an organ called the appressorium, a structure used by fungal pathogens to press against and attach to plant surfaces in preparation for infection (24, 82) and which contributes, for example, to the development of ghost spots caused by *B. cinerea* on tomato fruits, to anthracnose caused by *Colletotrichum* in pepper and by *Colletotrichum musae* in banana (3, 89), or to spoilage of carrot caused by *S. sclerotiorum* (63). In the case of bananas, appressoria formed on preharvest fruits remain quiescent until harvest and fruit maturation to eventually penetrate into tissue and cause spoilage. Germination and appressorium formation by *C. musae* and *Colletotrichum gloeosporioides* may be induced by ethylene produced by the infected commodity (41).

Fruits and vegetables also offer a large diversity of natural or accidental openings which can serve as ports of entry for penetration of postharvest spoilage microorganisms. Adverse conditions (e.g., wind, frost, and contact between fruits and limbs) in orchards, vineyards, and fields, as well as harvesting and postharvest handling involving mechanical or human interventions, can result in wounding of fruits caused by stems of other fruits or abrasion during transport (83). Up to 14% of hand-picked pears and up to 30% of apples may present wounds after harvest (1, 108). Wounds are a common site of penetration of postharvest spoilage microorganisms and are critical, for instance, for infection of tomatoes by *Rhizopus stolonifer*, carrot by a range of postharvest pathogens, and plums by *Monilinia fructicola* (52, 73, 89). Natural openings, such as dead tissues at the blossom end, stem scars on apple and citrus, and the calyx of tomato, eggplant, and bell pepper, are also potential sites for microorganisms to penetrate tissues (83, 89). Soft-rot *Erwinia* spp. have been shown to penetrate into potato tissue through lenticels under certain conditions (hydrostatic pressure) (6). A higher density of lenticels on various apple cultivars is correlated with higher susceptibility to infection with *Penicillium* spp. (1). The presence of free water in which bacteria may be suspended can favor infiltration into plant tissues and their internalization. This process can be enhanced by hydrostatic pressure created by immersing warm fruits in cold wash water. Internalization and infiltration have been well described for human pathogens, but also occur for spoilage microorganisms (83).

CAUSING THE DISEASE OR THE SPOILAGE

Specific factors associated with virulence of microorganisms are the primary mechanisms at the origin of postharvest spoilage of fruits and vegetables. Their role is usually evident when comparing, within species, certain strains able to cause spoilage to those unable to cause spoilage. A comparison of different strains of *Pseudomonas fluorescens* has shown, for instance, that the production of a biosurfactant (a peptidolipid named viscosin) is a key factor in the decay of broccoli florets caused by the bacterium, while pectolytic activity is necessary but not sufficient to cause spoilage (50, 51). The aggressiveness of *Mycocentrospora acerina* is related to production of pectinase and glucanase (67). Molecular biology techniques enable the production of mutants with specific characteristics and tests to determine if these characteristics (and their genes) are necessary for microorganisms to cause spoilage. This approach has enabled researchers to demonstrate the role of pectinases in pathogenesis of soft-rot bacteria. Mutants of the soft-rot bacterium *Erwinia chrysanthemi* that do not produce pectin methylesterase or a specific isoenzyme of pectate lyase have a reduced ability to macerate potato tubers, whereas the production of another pectate lyase does not seem to be necessary for tissue maceration (4). These enzymes also have different implications in pathogenesis. Pectic enzymes are clearly involved in postharvest spoilage of fruits and vegetables by bacteria as well as by molds. Purified pectin-degrading enzymes are able to macerate plant tissue and cause cell death without involving other enzymes or toxic factors. Pectinases break down pectic components in the middle

lamella and cell wall, resulting in tissue maceration, loss of rigidity of plant tissue, and irreversible cell damage. Polygalacturonase, pectin esterase, and pectin lyase are the main types of pectinases that act at different sites on the D-galacturonic acid chain constituting the basic structure of pectic compounds (47). Pectolytic bacteria and molds generally produce several types of pectinases, not always playing the same role (4). Eight isozymes of polygalacturonase are produced by *Phomopsis cucurbitae*, a quiescent infection mold, during postharvest infection and decay of cantaloupe, and their activity changes during maturation of the fruit (121). An endopolygalacturonase mutant of *Alternaria citri* loses its ability to cause black rot on citrus, while a mutant of *Alternaria alternata* for the same function is still able to cause brown spot on citrus (55). Pectic enzymes are necessary for *A. citri* to progress from the pedicel in the central axis of the citrus fruit to the sac juice containing nutrients (56).

Some postharvest spoilage bacteria and molds also produce toxins. Some molds, in particular *Alternaria* spp., have very specific host-parasite interactions (30). Mycotoxins are also known for their toxicity to humans or animals. Implications of the nonspecific toxins produced by molds causing postharvest spoilage are highly variable (3). For instance, production of patulin by *Penicillium expansum* does not appear to be involved in postharvest spoilage of apples during storage. In contrast, the production of oxalic acid has been shown to occur in a number of postharvest diseases caused by *Sclerotium rolfsii* or *S. sclerotiorum* (63, 97).

In some instances, spoilage can be fully opportunistic. The physiological activity of minimally processed vegetables is markedly changed by processing and storage conditions, e.g., modified-atmosphere packaging (88). The lactic acid bacterium *Leuconostoc mesenteroides* is not known as a plant pathogen, but it has been shown to be strongly associated with the spoilage of shredded carrots. In this situation spoilage is thought to be due to a shift toward anaerobic metabolism in modified atmospheres with concentrations of CO_2 in excess and/or to low concentrations of O_2 that induce toxicity to carrot cells and leakage of electrolytes and nutrients used by the saprophyte (15). On minimally processed green leafy salads, in contrast, spoilage can be explained by the development of pectolytic fluorescent pseudomonads (90).

DEFENSE REACTIONS

Fruits and vegetables offer a range of barriers to infection by postharvest spoilage microorganisms. Some are preformed or constitutive in the plant organ. The cuticle barrier is the most external barrier to penetration.

Removal of waxes from the cuticle has been shown to increase the vulnerability of pepper fruits to infection by *Colletotrichum capsici* and *C. gloeosporioides*, and of cabbage by *B. cinerea* (89). The cuticle thickness is correlated with the resistance of tomato fruit or grape berries to *B. cinerea* and of peaches to *Monilinia fructigena* (3, 43). In many types of produce the barrier effect of the cuticle is reinforced by epidermis or periderm tissues, which can be relatively thick structures such as the rind of citrus and melons. Enhanced resistance to infection linked to these structures can be explained by their higher resistance to crack formation and therefore to penetration by spoilage microorganisms, larger amounts of protective material to be degraded, and lower diffusion or access to water and nutrients required for the infection process (3).

Preformed antimicrobial compounds may also be involved in plant resistance, although demonstration of their actual effects on resistance is relatively difficult to achieve because of difficulties in assessing inhibitory activity and in correlating changes in concentrations with decay development, as critically underpinned by Prusky (95). These compounds can be extremely diverse, as fruits and vegetables cover a wide range of plant families. The presence of the phenolic compounds catechol and protocatehuic acid has a role in the resistance of onion to *Colletotrichum circinans*, and the presence of the alkaloidal saponin tomatine has a role in the resistance of tomato to *B. cinerea* (3, 78). Decreases in concentrations of 5,12-*cis*-heptadecenyl resorcinol and 5-pentadecenyl resorcinol in the skin of mango fruit during ripening are related to an increase in susceptibility to *A. alternata*, lower decreases occurring in the most resistant cultivars (95). Quiescent infections of *C. gloeosporioides* may be regulated by preformed epicatechin acting as an inhibitor of lipoxygenase activity and consequently delaying degradation of an antifungal diene present in unripe avocado fruit (95). A monoterpene aldehyde, citral, present in particular in the oil cavities of citrus albedo, is thought to be involved in the resistance of young mature green lemons to *Penicillium* (95). The resistance of carrot roots to infection by *Mycocentrospora acerina* could be attributed to falcarindiol, a polyacetylene compound which accumulates in the peridermis and the pericyclic parenchyma of the root tissue at concentrations 50-fold higher than in the core xylem parenchyma, and which has a pronounced inhibitory effect on molds (44).

A wide range of defensive barriers may also be formed in reaction to an infection. Pathogens are known to produce elicitors which stimulate these reactions of defense (48). Plant tissues may produce small molecules of varied nature, known as phytoalexins (78). Bramley's

Seedling apples produce benzoic acid in response to infection by *Nectria galligena* (110), and the resistance of carrot to *B. cinerea* has been attributed to a coumarin, 6-methoxymellein (46). Structural changes such as accumulation of lignin or suberin or development of callus have also been observed as defensive reactions (99), in particular in carrot, potato tuber, and pear (5, 70, 108). Polygalacturonase-inhibiting proteins produced by plants and acting against endopolygalacturonases of plant-pathogenic molds that cause wall degradation and tissue maceration have been detected in a range of fruits and vegetables, including apple, pear, grape, raspberry, onion, and pepper (25). An esterase produced during the interaction between mature peppers and *C. gloeosporioides* has been shown to inhibit the formation of the fungal appressoria, and therefore decay (61). Production of reactive oxygen species (including H_2O_2) is induced after infection of pepper by *C. gloeosporioides*. The direct inhibitory effect of these species on the pathogen still remains unclear, but they likely are at the origin of activation of the phenylpropanoid pathway and accumulation of the antifungal compound diene, both implicated in resistance (7). Other mechanisms implicating enzymatic activities (chitinases and lipoxygenases) or accumulation of hydroxyproline-rich proteins have been proposed (3, 89). Interactions between plants and pathogens are a very fertile research area, and novel defense mechanisms are regularly discovered.

CONTROLLING SPOILAGE

Effects of Temperature, Relative Humidity, and Modified Atmosphere

Postharvest control of temperature, relative humidity, and composition of the gaseous atmosphere aims at reducing the physiological activity of fruits and vegetables by delaying ripening and senescence, consequently prolonging the shelf life. These environmental factors may act by giving less opportunity to the pathogen to develop by retaining the integrity of the plant organ and by directly inhibiting microbial growth. The most suitable temperature, relative humidity, and modified atmosphere for preserving the quality of most fruits and vegetables are now relatively well established and have been extensively reviewed (3, 87, 101). A decrease in temperature by 10°C reduces the respiratory activity by two- to fourfold (59), and a temperature close to 0°C is recommended for most commodities, with the exception of those of tropical origin and a few temperate produce which suffer from physiological disorders (chilling injuries) when stored at refrigeration temperatures. For these produce, the optimal storage temperature is close to 10°C. While growth of many postharvest spoilage microorganisms is still possible at very low refrigeration temperature, their rate of development is reduced. In addition, some major spoilage microorganisms are inhibited at refrigeration temperatures at which produce is often stored. The bacterium *Erwinia carotovora* subsp. *carotovora* cannot grow at temperatures below 6°C, and the molds *Rhizopus stolonifer*, *Phytophtora infestans*, and *Aspergillus niger* cannot grow below 2 to 5, 4 and 11°C, respectively (3, 66, 74).

Modified atmospheres, used in combination to chill storage, also reduce the physiological activity of fresh produce. For instance, reducing the O_2 concentration from 21 to 2.5% during storage of broccoli florets results in approximately a 50% decrease in respiration rate (59). Recommended CO_2 concentrations for produce storage rarely exceed 10%, and 1 to 5% O_2 is tolerated (59, 60, 101). Exposure to higher CO_2 (lower O_2) may result in physiological disorders, leading to a loss in quality. As reviewed by El Goorani and Sommer (37), either a reduction in O_2 or an increase in CO_2 delays in vitro growth of many postharvest pathogens, without complete inhibition. Modified atmospheres reduce microbial spoilage of fruits and vegetables in many instances, although some spoilage microorganisms are not directly inhibited. For example, controlled-atmosphere storage of apples at 1°C in an atmosphere containing 5% CO_2 and 3% O_2 prevents the development of lesions due to *Pezicula alba*, while in vitro, this gaseous atmosphere has no effect (10). Generally, high concentrations of CO_2 or low concentrations of O_2, often less than concentrations tolerated by the plant organs, are needed for a significant reduction of in vitro growth. The lower susceptibility to postharvest pathogens of fruits and vegetables stored under controlled atmospheres is mainly due to delayed senescence (37). However, modified atmospheres can increase the extent of diseases in potato tubers, carrot, and other root crops. Reduced physiological activity slows wound healing, giving pathogens additional time to establish infections. Under extreme conditions (anoxia), strict anaerobes such as pectolytic soft-rotting clostridia may cause spoilage, as shown on potato, for instance (74).

Water loss is a consequence of respiratory activity and is highly detrimental to quality. A 10% weight loss (much less for leafy vegetables) makes produce unacceptable to consumers (8). Storage at relative humidity higher than 90% is recommended for most commodities, with a few exceptions, e.g., garlic and onion (87). High humidity also increases the availability of nutrients to spoilage microorganisms. Despite this, storage under high humidity is generally in favor of the produce. Storage of cabbage, celery, leek, and carrot at 98 to 100% relative humidity instead of 90 to 95% relative humidity results in lower losses caused by decay (114). Thorne (111) observed that spoilage due to *R. stolonifer* only occurs on carrots that have lost more than 3 to 8% of their fresh weight.

Physical Treatments

The possibility of using ionizing radiation to extend the shelf life of fruits and vegetables has been studied since the 1950s (3). Postharvest spoilage bacteria and fungi are sensitive to ionizing radiation. Doses lower than 4 kGy reduce 1,000-fold the germination of major postharvest pathogens such as *Penicillium* spp., *Monilinia fructicola*, *R. stolonifer*, and *Alternaria* spp. Decimal reduction doses for *Pseudomonas* and other gramnegative bacteria are approximately 0.2 kGy (3, 91). However, it appears that some fruits and vegetables are adversely affected by doses necessary to inactivate some postharvest pathogens (81). Treatment with ionizing radiation is consequently useful to lower initial contamination or to inhibit growth of postharvest pathogens and to prolong the development of disease, but not for complete elimination of microbial contaminants. In the case of strawberry, a 2-kGy dose prolongs the shelf life by several days (3).

Prestorage heat treatment has been used with some success to reduce postharvest spoilage of a wide range of fresh produce, including green pepper, apple, and citrus (39, 103). Two main types of applications can be distinguished: short-term exposure to heat, from a few seconds at 60 to 62°C to 60 to 120 min at about 45°C, by immersion in or rinsing and spraying with hot water; or long-term exposure, also known as "curing" (for a few hours to several days), mainly in hot air (103). Two kinds of effects can be observed, either direct effects on microbial contaminants or indirect effects on the treated fruit or vegetable. Treatment of tomatoes for 3 days at 38°C before storage at 20°C for 7 days results in inhibition of *B. cinerea* without alteration of quality (40). Short-time exposure at 56°C delays germination of *Penicillium digitatum*, and treatment at 59°C and 62°C inhibits germination (94). A similar level of sensitivity to heat has been observed for *M. fructicola*, *B. cinerea*, *Cladosporium herbarum*, *R. stolonifer*, *A. alternata*, and *Penicillium expansum* (3). These treatments also reduce 1,000-fold the natural epiphytic microflora on citrus. On whole citrus, hot water treatment of fruits artificially inoculated with *P. digitatum* has been shown to result in less than 20% decay after 4 days at 24°C, while all control fruits are spoiled (94).

Chemical Treatments

Fungicides

Synthetic antimicrobial chemicals are still widely applied to fruits and vegetables after harvest, with substantial differences among countries, according to approval by legal authorities, as complements to modifications of storage environments or when modifications are not possible. Their spectrum of activity and possibilities of applications have been reviewed elsewhere (31, 32, 80) and in chapter 33. These chemical compounds have different modes of action. For instance, benzimidazoles have a systemic action beneath the surface of the host as well as antisporulant properties, and they inhibit tubulin assembly and therefore mitosis (22). Imalazil belongs to a family of systemic fungicides that inhibits the biosynthesis of ergosterol, an essential compound in the membrane of fungal cells (3). Dicarboximides alter the osmotic adaptation capabilities of molds (118). Concerns about toxicity limit official authorization for use of fungicides on defined commodities and create some negative attitudes among consumers. In addition, the occurrence of fungal resistance restricts possible applications of fungicides. Alternative chemicals such as compounds generally recognized as safe or natural compounds, e.g., acetaldehyde, hexanal, and essential oils, have also shown antimicrobial properties suitable for postharvest control of phytopathogens (112). The presence of the essential oil carvone in the storage atmosphere has been shown to reduce, for instance, decay of potato tubers inoculated with *Phoma exigua* and *Fusarium sulphureum* but not with *Fusarium solani* (49).

Decontamination

Decontamination aims at reducing the number of microbial contaminants on the surface of fruits and vegetables, thereby prolonging the time required to develop spoilage. Various chemicals have been tested for possible application to fruits and vegetables. Hypochlorous acid (HClO), chlorine dioxide (ClO$_2$), ozone, and hydrogen peroxide are among the most extensively studied chemicals (9). Limitations in their action, in particular that of chlorine, may be due to various factors: the natural hydrophobicity of plant surfaces, inaccessibility of microorganisms within plant tissue, and possible neutralization of the decontaminating agent upon contact with fruit and vegetable tissue components. Chlorine at concentrations that will cause several decimal reductions of pathogens in pure water is often slightly more efficient than washing produce with pure water (9, 88). Reduction in microbial populations is generally 1 to 2 log CFU/g. However, despite these limitations, application of disinfectants may

result in better retention of quality during storage (89). For example, prestorage chlorine application at suitable concentrations and for appropriate exposure times results in significant reductions of black spot disease caused by *A. alternata* on persimmon (96) and spoilage of nectarines and plums caused by *Monilinia laxa* (79). Sanitizers are also useful to prevent buildup of microorganisms on equipment and in wash water, and to avoid dissemination from contaminated material.

Triggering Defense Reactions

As stated above, fruits and vegetables, as any plant organ, have intrinsic defense mechanisms that respond to postharvest microbial infection. The induction of these mechanisms may be systemic, as shown for carrot roots, in which inoculation with *B. cinerea* and *S. sclerotiorum* near the root tip reduces infection after subsequent inoculation with these molds near the crown (84). A range of biotic and abiotic factors that induce defense reactions have been shown to improve quality retention during storage. Appropriate fertilization of plants with calcium or direct application of a calcium solution on the fruit and vegetable to be stored increases the calcium content in tissue. These treatments have been successful in controlling the growth of *P. expansum* on apples and soft rot caused by *E. carotovora* subsp. *atroseptica* on potatoes. The effect of calcium is attributed to a reinforcement of pectin bonds, making the cell wall more resistant to pectic enzymes, or a direct inhibitory effect of calcium on the enzymes themselves (21). Accelerating healing of wounds by suberization of parenchyma cells as a result of exposure at moderately high, ambient, or low temperatures is a common procedure to reduce postharvest spoilage of carrot, potato, sweet potato, onion, yam, and pear (17, 40, 73, 89). Heat treatment may also induce structural changes in the epicuticular wax (resulting in fewer cracks and therefore offering a mechanical barrier to spoilage microorganisms), delay ripening, and inhibit antifungal activity, phytoalexin production, or chitinase and glucanase activities which play a role in degradation (103).

Chitosan, a compound derived from chitin, has antimicrobial properties when used, for instance, to coat fresh fruits and vegetables for the purpose of regulating gas and moisture exchange (27, 115). In addition to damaging fungal hyphae, chitosan also reduces the ability of fungal pectic enzymes to macerate plant tissue by a direct effect on enzymes and indirect effects by inhibiting the production of oxalic and fumaric acids by pathogens. Chitosan may also inhibit host-specific toxin production, phytoalexin production, and structural changes in the cell wall, as shown in tomatoes infected by the black mold rot mold, *A. alternata*, or on bell pepper fruit infected by *B. cinerea* (33, 34, 98). Stimulation of antifungal hydrolyses by chitinases and glucanases has also been reported (115).

Plant regulators are involved in the development and response of produce to environmental stresses. Jasmonates in particular may play a role in plant defense responses to microbial attacks, and their postharvest application on fruits is effective in controlling spoilage molds such as *M. fructicola* and *P. expansum* on peaches and *P. digitatum* on grapefruit (29, 119). This positive effect has been attributed to induced resistance of the fruit, in particular to higher activities of chitinase, glucanase, phenylalanine ammonia lyase, and peroxidase (119). Similar activities or accumulation of antimicrobial compounds related to disease resistance are induced in peach, grapefruit, and carrot after exposure to UV-C (28, 35, 85). However, UV-C has little activity against *B. cinerea* inoculated on wounded bell pepper, indicating that wounded tissues are protected against exposure to irradiation, while complete inhibition has been observed in vitro (85). In this case and in grapefruit inoculated with *P. digitatum*, the induction of defense reactions was the likely cause of a lower rate of decay (28, 85).

Resistance of Cultivars

Cultivars within a plant species may exhibit differences in susceptibility to infection by postharvest pathogens. Laboratory experiments and practical experience have enabled a classification of potato cultivars according to their susceptibility to *Phoma exigua*, which causes gangrene (73). Tests done using artificial inoculation of produce show some differences in susceptibility among cultivars, e.g., susceptibility of broccoli to *Pseudomonas marginalis* (14), onion to *Aspergillus niger* and *Botrytis allii* (62, 71), pepper to *C. gloeosporioides* (77), and sweet potato to *Rhizopus* spp. (18). Resistance of cultivars to spoilage is usually only delayed, not prevented (89). Evaluation of cultivars may depend on testing procedures: the ranking of potato cultivars for resistance against *E. carotovora* subsp. *atroseptica*, for example, depends on the inoculation method (2). The resistance of cultivars to postharvest spoilage microorganisms may be the result of several complex factors and interactions. A low number of pores and thickness of external grape berry skin, for example, are positively correlated with resistance to *B. cinerea*, both factors likely giving a general protection against fungal attack (43). In contrast, among 12 apple cultivars tested for resistance against *B. cinerea*, *P. expansum*, *Mucor piriformis*, and *Pezicula malicorticis*, no single cultivar was the most resistant to all pathogens and each cultivar that was the most resistant to one pathogen was also the most

susceptible to one of the other pathogens. It has been suggested that Golden Delicious cultivar apples, despite their fragile epidermis, have a lower probability of wounding than Granny Smith cultivar apples because of the particularity of their pedicels, stiffer and shorter in the latter case (107). Increased firmness in transgenic tomatoes with reduced levels of polygalactoronase results in increased resistance to *Geotrichum candidum* and *R. stolonifer*, while tomatoes with suppressed expression of ripening-related expansin, with a similar increased firmness, are not more resistant to *B. cinerea* and *A. alternata* (12, 65).

Biological Control

Biological control refers to the application of microbial antagonists of postharvest spoilage microorganisms, in particular on wounds, which are a natural site of entry for pathogens. Biological control is based on the selection of naturally occurring microorganisms particularly adapted to surviving or growing on fruit and vegetable surfaces. Many mechanisms are probably involved in interactions between antagonists and pathogens (57). These may include antibiosis, such as the production of pyrrolnitrin by *Pseudomonas cepacia*, shown to control blue mold rot caused by *P. expansum* and gray mold rot caused by *B. cinerea* on pome fruits (57), or *Fusarium sambucinum* on potato (13). Competition for space and nutrients is a seductive hypothesis to account for the antagonistic effect of some biological agents, in particular on wound sites, which are often rich in nutrients. For instance, the yeast-like *Aureobasidium pullulans* depletes amino acids in vitro and inhibits germination of *P. expansum* in apple juice (58). In addition to this possible competition for space and nutrients, the antagonist *A. pullulans* induces apple β-1,3-glucanase, chitinase, and peroxidase activity, which controls decay caused by *B. cinerea* and *P. expansum* (54). A similar induction process has been shown with the antagonist yeast *Candida saitoana* (36). Induction of these defense reactions could explain the effect of antagonists against postharvest pathogens. Some yeasts also show a strong attachment to the hyphae of postharvest spoilage molds and production of cell wall-degrading enzymes (116). Some of these biocontrol agents, in particular strains of the yeast *Candida oleophila* and of the bacterium *Pseudomonas syringae*, are registered as biopesticides in the United States (42).

The diversity of biocontrol methods illustrates the diversity of possibilities based on a direct control of pathogen contamination or of its development, indirect control through delayed senescence and preserved integrity of the organ host, or induced defense mechanisms. Under practical conditions, delay of spoilage depends on combinations of several factors: prevention of wounding during harvest and postharvest handling, cold storage together with modified atmospheres, and application of fungicides for many commodities, e.g., apples and pears, which can be successfully stored for several months and therefore offered for sale to the consumer long after harvest. Recent research has also revealed the possibility of other combinations, such as postharvest heat treatment and use of antagonists on apples followed by cold storage to control *P. expansum* and *Colletotrichum acutatum* (20, 69). In the near future, integrated approaches, combining control along the production chain at various preharvest and postharvest stages, will probably be increasingly used. Preharvest applications, for instance, of calcium, growth regulators, and a fungicide, combined with postharvest heat treatment, antagonist yeast treatment, and another fungicide application, have been shown to be successful in controlling *P. digitatum* and other decay microorganisms on mandarin oranges (120).

CONCLUSION

Spoilage of fruits and vegetables is the result of complex interactions between a living plant organ and its microflora, and therefore deals with plant pathology and plant physiology as much as with food microbiology. Control of postharvest spoilage microorganisms largely accounts for these interactions. Improving the quality retention of fruits and vegetables will raise for a long time some exciting and difficult questions about microbial ecology impacting the survival and development of microorganisms in the environment, in particular during production, and on the plant organ to be stored and later consumed.

The global exchange of fruits and vegetables will likely not decrease. At present, millions of tons of fruits and vegetables are crossing seas, oceans, and continents, from the southern to the northern hemisphere, and from tropical to temperate zones. Developing countries increasingly play a role in this world market. Further development of minimally processed fruits and vegetables will bring new questions: how to maintain the quality of processed produce, which requires prevention of spoilage, when they are often heavily stressed and natural defenses of the intact tissues have been overwhelmed. Consumer demand for high-quality fruits and vegetables produced under environmentally friendly conditions will probably not decrease. Finding solutions to historical problems associated with preservation of fruits and vegetables against infection and spoilage by bacteria and fungi will be in actuality the challenge for the near future.

References

1. **Amiri, A., and G. Bompeix.** 2005. Diversity and population dynamics of *Penicillium* spp. on apples in pre- and postharvest environments: consequences for decay development. *Plant Pathol.* 54:74–81.

2. **Bain, R. A., and C. M. Perombelon.** 1988. Methods of testing potato cultivars for resistance to soft rot of tubers caused by *Erwinia carotovora* subsp. *atroseptica*. *Plant Pathol.* 37:431–437.

3. **Barkai-Golan, R.** 2001. *Postharvest Diseases of Fruits and Vegetables. Development and Control.* Elsevier, Amsterdam, The Netherlands.

4. **Barras, F., F. Vangijsegem, and A. K. Chatterjee.** 1994. Extracellular enzymes and pathogenesis of soft-rot *Erwinia. Annu. Rev. Phytopathol.* 32:201–234.

5. **Bartz, J. A., and J. W. Eckert.** 1987. Bacterial diseases of vegetable crops after harvest, p. 351–376. *In* J. Weichmann (ed.), *Postharvest Physiology of Vegetables.* Marcel Dekker, New York, N.Y.

6. **Bartz, J. A., and A. Kelman.** 1985. Infiltration of lenticels of potato tubers by *Erwinia carotovora* pv. *carotovora* under hydrostatic pressure in relation to bacterial soft rot. *Plant Dis.* 69:69–74.

7. **Beno-Moualem, D., and D. Prusky.** 2000. Early events during quiescent infection development by *Colletotrichum gloeosporioides* in unripe avocado fruits. *Phytopathology* 90:553–559.

8. **Ben-Yehoshua, S.** 1987. Transpiration, water stress, and gas exchange, p. 113–170. *In* J. Weichmann (ed.), *Postharvest Physiology of Vegetables.* Marcel Dekker, New York, N.Y.

9. **Beuchat, L. R.** 1998. Surface decontamination of fruits and vegetables eaten raw. WHO/FSF/FOS/98.2. [Online.] http://www/who.int/foodsafety/publications/fs-management/en/surface-decon.pdf.

10. **Bompeix, G.** 1978. The comparative development of *Pezicula alba* and *P. malicortis* on apples and in vitro (air and controlled atmosphere). *Phytopathol. Z.* 91:97–109.

11. **Boureau, T., M. A. Jacques, R. Berruyer, Y. Dessaux, H. Dominguez, and C. E. Morris.** 2004. Comparison of the phenotypes and genotypes of biofilm and solitary epiphytic bacterial populations on broad-leaved endive. *Microb. Ecol.* 47:87–95.

12. **Brummell, D. A., W. J. Howie, C. Ma, and P. Dunsmuir.** 2002. Postharvest fruit quality of transgenic tomatoes suppressed in expression of a ripening-related expansin. *Postharv. Biol. Technol.* 25:209–220.

13. **Burkhead, K. D., D. A. Schisler, and P. J. Slininger.** 1994. Pyrrolnitrin production by biological control agent *Pseudomonas cepacia* B37w in culture and in colonized wounds of potatoes. *Appl. Environ. Microbiol.* 60:2031–2039.

14. **Canaday, C. H., J. E. Wyatt, and J. A. Mullins.** 1991. Resistance in broccoli to bacterial soft rot caused by *Pseudomonas marginalis* and fluorescent *Pseudomonas* species. *Plant Dis.* 75:715–720.

15. **Carlin, F., C. Nguyen-The, Y. Chambroy, and M. Reich.** 1990. Effects of controlled atmospheres on microbial spoilage, electrolyte leakage and sugar content of 'fresh-ready-to-use' grated carrots. *Int. J. Food Sci. Technol.* 25:110–119.

16. Chung, K. T., T. Y. Wong, C. I. Wei, Y. W. Huang, and Y. Lin. 1998. Tannins and human health: a review. *Crit. Rev. Food Sci. Nutr.* **38**:421–464.

17. Clark, C. A. 1992. Postharvest diseases of sweet potatoes and their control. *Postharv. News Inf.* **3**:75N–79N.

18. Clark, C. A., and M. W. Hoy. 1994. Identification of resistance in sweet potato to *Rhizopus* soft rot using two inoculation methods. *Plant Dis.* **78**:1078–1082.

19. Conner, D. E. 1993. Naturally occurring compounds, p. 441–468. *In* P. M. Davidson and A. L. Brannen (ed.), *Antimicrobials in Foods*. Marcel Dekker, New York, N.Y.

20. Conway, W. S., B. Leverentz, W. J. Janisiewicz, R. A. Saftner, and M. J. Camp. 2005. Improving biocontrol using antagonist mixtures with heat and/or sodium bicarbonate to control postharvest decay of apple fruit. *Postharv. Biol. Technol.* **36**:235–244.

21. Conway, W. S., C. E. Sams, and A. Kelman. 1994. Enhancing the natural resistance of plant tissues to postharvest diseases through calcium applications. *HortScience* **29**:751–754.

22. Davidse, L. C. 1986. Benzimidazole fungicides: mechanism of action and biological impact. *Annu. Rev. Phytopathol.* **24**:43–65.

23. Deak, T. 1991. Foodborne yeasts. *Adv. Appl. Microbiol.* **36**:179–278.

24. Dean, R. A. 1997. Signal pathways and appressorium morphogenesis. *Annu. Rev. Phytopathol.* **35**:211–234.

25. De Lorenzo, G., R. D'Ovidio, and F. Cervone. 2001. The role of polygalacturonase-inhibiting proteins (PGIPs) in defense against pathogenic fungi. *Annu. Rev. Phytopathol.* **39**:313–335.

26. Dennis, C. 1983. *Post-Harvest Pathology of Fruits and Vegetables*. Academic Press, London, United Kingdom.

27. Devlieghere, F., A. Vermeulen, and J. Debevere. 2004. Chitosan: antimicrobial activity, interactions with food components and applicability as a coating on fruit and vegetables. *Food Microbiol.* **21**:703–714.

28. Droby, S., E. Chalutz, B. Horev, L. Cohen, V. Gaba, C. L. Wilson, and M. Wisniewski. 1993. Factors affecting UV-induced resistance in grapefruit against the green mold decay caused by *Penicillium digitatum*. *Plant Pathol.* **42**:418–424.

29. Droby, S., R. Porat, L. Cohen, B. Weiss, B. Shapiro, S. Philosoph-Hadas, and S. Meir. 1999. Suppressing green mold decay in grapefruit with postharvest jasmonate application. *J. Am. Soc. Hortic. Sci.* **124**:184–188.

30. Durbin, R. D. 1983. The biochemistry of fungal and bacterial toxins and their modes of action, p. 137–162. *In* J. A. Callow (ed.), *Biochemical Plant Pathology*. John Wiley & Sons, Chichester, United Kingdom.

31. Eckert, J. W., and J. M. Ogawa. 1985. The chemical control of postharvest diseases: subtropical and tropical fruits. *Annu. Rev. Phytopathol.* **23**:421–454.

32. Eckert, J. W., and J. M. Ogawa. 1988. The chemical control of postharvest diseases: deciduous fruits, berries, vegetables, and root/tuber crops. *Annu. Rev. Phytopathol.* **26**:433–469.

33. El Ghaouth, A., J. Arul, C. Wilson, and N. Benhamou. 1994. Ultrastructural and cytochemical aspects of the effect of chitosan on decay of bell pepper fruit. *Physiol. Mol. Plant Pathol.* **44**:417–432.

34. El Ghaouth, A., J. Arul, C. Wilson, and N. Benhamou. 1997. Biochemical and cytochemical aspects of the interactions of chitosan and *Botrytis cinerea* in bell pepper fruit. *Postharv. Biol. Technol.* **12**:183–194.

35. El Ghaouth, A., C. L. Wilson, and A. M. Callahan. 2003. Induction of chitinase, beta-1,3-glucanase, and phenylalanine ammonia lyase in peach fruit by UV-C treatment. *Phytopathology* **93**:349–355.

36. El Ghaouth, A., C. L. Wilson, and M. Wisniewski. 2003. Control of postharvest decay of apple fruit with *Candida saitoana* and induction of defense responses. *Phytopathology* **93**:344–348.

37. El Goorani, M. A., and N. F. Sommer. 1981. Effects of modified atmospheres on postharvest pathogens of fruits and vegetables. *Hortic. Rev.* **3**:412–461.

38. Elliot, M. A., and N. J. Talbot. 2004. Building filaments in the air: aerial morphogenesis in bacteria and fungi. *Curr. Opin. Microbiol.* **7**:594–601.

39. Fallik, E. 2004. Prestorage hot water treatments (immersion, rinsing and brushing). *Postharv. Biol. Technol.* **32**:125–134.

40. Fallik, E., J. Klein, S. Gringerg, E. Lomaniec, S. Lurie, and A. Lalazar. 1993. Effect of postharvest heat treatment of tomatoes on fruit ripening and decay caused by *Botrytis cinerea*. *Plant Dis.* **77**:985–988.

41. Flaishman, M. A., and P. E. Kolattukudy. 1994. Timing of fungal invasion using hosts ripening hormone as a signal. *Proc. Natl. Acad. Sci. USA* **91**:6579–6583.

42. Fravel, D. R. 2005. Commercialization and implementation of biocontrol. *Annu. Rev. Phytopathol.* **43**:337–359.

43. Gabler, F. M., J. L. Smilanick, M. Mansour, D. W. Ramming, and B. E. Mackey. 2003. Correlations of morphological, anatomical, and chemical features of grape berries with resistance to *Botrytis cinerea*. *Phytopathology* **93**:1263–1273.

44. Garrod, B., B. G. Lewis, and D. T. Coxon. 1978. *cis*-Heptadeca-1,9-diene-4,6-diyne-3,8-diol, an antifungal polyacetylene from carrot root tissue. *Physiol. Plant Pathol.* **13**:241–246.

45. Glenn, G. M., B.-S. Chiou, S. Imam, D. Wood, and W. Orts. 2005. Role of cuticles in produce quality and preservation, p. 19–53. *In* O. Lamikanra, S. Iman, and D. Ukuku (ed.), *Produce Degradation—Pathways and Prevention*. Taylor and Francis Group, Boca Raton, Fla.

46. Goodliffe, J. P., and J. B. Heale. 1978. The role of 6-methoxy mellein in the resistance and susceptibility of carrot root tissue to the cold-storage pathogen *Botrytis cinerea*. *Physiol. Plant Pathol.* **12**:27–43.

46a. Gross, K. C., C. Y. Wang, and M. Saltveit (ed.). 2004. The commercial storage of fruits, vegetables, and florist and nursery stocks. Draft version of revised USDA Agriculture Handbook. [Online.] http://www.ba.ars.usda.gov/hb66. Accessed November 2006.

47. Gummadi, S. N., and T. Panda. 2003. Purification and biochemical properties of microbial pectinases—a review. *Process Biochem.* **38**:987–996.

48. Hahn, M. G. 1996. Microbial elicitors and their receptors in plants. *Annu. Rev. Phytopathol.* **34**:387–412.

49. Hartmans, K. J., P. Diepenhorst, W. Bakker, and L. G. M. Gorris. 1995. The use of carvone in agriculture: sprout

suppression of potatoes and antifungal activity against potato tuber and other plant diseases. *Ind. Crops Prod.* **4**:3–13.

50. Hildebrand, P. D. 1989. Surfactant-like characteristics and identity of bacteria associated with broccoli head rot in Atlantic Canada. *Can. J. Plant Pathol.* **11**:205–214.

51. Hildebrand, P. D., P. G. Braun, K. B. McRae, and X. Lu. 1998. Role of the biosurfactant viscosin in broccoli head rot caused by a pectolytic strain of *Pseudomonas fluorescens*. *Can. J. Plant Pathol.* **20**:296–303.

52. Hong, C. X., T. J. Michailides, and B. A. Holtz. 1998. Effects of wounding, inoculum density, and biological control agents on postharvest brown rot of stone fruits. *Plant Dis.* **82**:1210–1216.

53. Imam, S., J. Shey, D. Wood, G. Glenn, B.-S. Chiou, M. Inglesby, C. Ludvik, A. Klamczynski, and W. Orts. 2005. Structure and function of complex carbohydrates in produce and their degradation process, p. 563–597. *In* O. Lamikanra, S. Iman, and D. Ukuku (ed.), *Produce Degradation—Pathways and Prevention.* Taylor and Francis Group, Boca Raton, Fla.

54. Ippolito, A., A. El Ghaouth, C. L. Wilson, and M. Wisniewski. 2000. Control of postharvest decay of apple fruit by *Aureobasidium pullulans* and induction of defense responses. *Postharv. Biol. Technol.* **19**:265–272.

55. Isshiki, A., K. Akimitsu, M. Yamamoto, and H. Yamamoto. 2001. Endopolygalacturonase is essential for citrus black rot caused by *Alternaria citri* but not brown spot caused by *Alternaria alternata*. *Mol. Plant-Microbe Interact.* **14**:749–757.

56. Isshiki, A., K. Ohtani, M. Kyo, H. Yamamoto, and K. Akimitsu. 2003. Green fluorescent detection of fungal colonization and endopolygalacturonase gene expression in the interaction of *Alternaria citri* with citrus. *Phytopathology* **93**:768–773.

57. Janisiewicz, W. J., and L. Korsten. 2002. Biological control of postharvest diseases of fruits. *Annu. Rev. Phytopathol.* **40**:411–441.

58. Janisiewicz, W. J., T. J. Tworkoski, and C. Sharer. 2000. Characterizing the mechanism of biological control of postharvest diseases on fruits with a simple method to study competition for nutrients. *Phytopathology* **90**:1196–1200.

59. Kader, A. A. 1987. Respiration and gas exchange of vegetables, p. 25–43. *In* J. Weichmann (ed.), *Postharvest Physiology of Vegetables.* Marcel Dekker, New York, N.Y.

60. Kader, A. A., D. Zagory, and E. L. Kerbel. 1989. Modified atmosphere packaging of fruits and vegetables. *Crit. Rev. Food Sci. Nutr.* **28**:1–30.

61. Kim, Y. S., H. H. Lee, M. K. Ko, C. E. Song, C. Y. Bae, Y. H. Lee, and B. J. Oh. 2001. Inhibition of fungal appressorium formation by pepper (*Capsicum annuum*) esterase. *Mol. Plant-Microbe Interact.* **14**:80–85.

62. Ko, S. S., J. W. Huang, J. F. Wang, S. Shanmugasundaram, and W. N. Chang. 2002. Evaluation of onion cultivars for resistance to *Aspergillus niger*, the causal agent of black mold. *J. Am. Soc. Hortic. Sci.* **127**:697–702.

63. Kora, C., M. R. McDonald, and G. J. Boland. 2003. Sclerotinia rot of carrot—an example of phenological adaptation and bicyclic development by *Sclerotinia sclerotiorum*. *Plant Dis.* **87**:456–470.

64. Kora, C., M. R. McDonald, and G. J. Boland. 2005. Occurrence of fungal pathogens of carrots on wooden boxes used for storage. *Plant Pathol.* **54**:665–670.

65. Kramer, M., R. Sanders, H. Bolkan, C. Waters, R. E. Sheeny, and W. R. Hiatt. 1992. Postharvest evaluation of transgenic tomatoes with reduced levels of polygalacturonase: processing, firmness and disease resistance. *Postharv. Biol. Technol.* **1**:241–255.

66. Lacey, J. 1989. Pre- and post-harvest ecology of fungi causing spoilage of foods and other stored products. *J. Appl. Bacteriol. Symp. Suppl.* **67**:11S–25S.

67. Le Cam, B., P. Massiot, and F. Rouxel. 1994. Cell-wall polysaccharide-degrading enzymes produced by isolates of *Mycocentrospora acerina* differing in aggressiveness on carrot. *Physiol. Mol. Plant Pathol.* **44**:187–198.

68. Lennox, C. L., R. A. Spotts, and L. A. Cervantes. 2003. Populations of *Botrytis cinerea* and *Penicillium* spp. on pear fruit, and in orchards and packinghouses, and their relationship to postharvest decay. *Plant Dis.* **87**:639–644.

69. Leverentz, B., W. S. Conway, W. J. Janisiewicz, R. A. Saftner, and M. J. Camp. 2003. Effect of combining MCP treatment, heat treatment, and biocontrol on the reduction of postharvest decay of "Golden Delicious" apples. *Postharv. Biol. Technol.* **27**:221–233.

70. Lewis, B. G., and B. Garrod. 1983. Carrots, p. 103–124. *In* C. Dennis (ed.), *Post-Harvest Pathology of Fruits and Vegetables.* Academic Press, London, United Kingdom.

71. Lin, M. W., J. F. Watson, and J. R. Baggett. 1995. Inheritance of resistance to neck-rot disease incited by *Botrytis allii* in bulb onions. *J. Am. Soc. Hortic. Sci.* **120**:297–299.

72. Lindow, S. E. 1996. Role of immigration and other processes in determining epiphytic bacterial populations. Implications for disease management, p. 155–168. *In* C. E. Morris, P. C. Nicot, and C. Nguyen-The (ed.), *Aerial Plant Surface Microbiology.* Plenum Press, New York, N.Y.

73. Logan, C. 1983. Potatoes, p. 179–217. *In* C. Dennis (ed.), *Post-Harvest Pathology of Fruits and Vegetables.* Academic Press, London, United Kingdom.

74. Lund, B. M. 1983. Bacterial spoilage, p. 219–257. *In* C. Dennis (ed.), *Post-Harvest Pathology of Fruits and Vegetables.* Academic Press, London, United Kingdom.

75. Lund, B. M. 1992. Ecosystems in vegetable foods. *J. Appl. Bacteriol. Symp. Suppl.* **73**:115S–126S.

76. Lund, B. M., and A. L. Snowdon. 2000. Fresh and processed fruits, p. 738–758. *In* B. M. Lund, A. C. Baird-Parker, and G. W. Gould (ed.), *The Microbiological Quality and Safety of Food*, vol. I. Aspen Publishers, Gaithersburg, Md.

77. Manandhar, J. B., G. L. Hartman, and T. C. Wang. 1995. Anthracnose development on pepper fruits inoculated with *Colletotrichum gloeosporioides*. *Plant Dis.* **79**:380–383.

78. Mansfield, J. W. 1983. Antimicrobial compounds, p. 237–265. *In* J. A. Callow (ed.), *Biochemical Plant Pathology.* John Wiley & Sons, Chichester, United Kingdom.

79. Mari, M., T. Cembali, E. Baraldi, and L. Casalini. 1999. Peracetic acid and chlorine dioxide for postharvest

control of *Monilinia laxa* in stone fruits. *Plant Dis.* 83: 773–776.

80. **Mastovska, K.** 2005. Role of pesticides in produce production, preservation quality, and safety, p. 341–378. *In* O. Lamikanra, S. Iman, and D. Ukuku (ed.), *Produce Degradation—Pathways and Prevention.* Taylor and Francis Group, Boca Raton, Fla.

81. **Maxie, E. C., N. F. Sommer, and F. G. Mitchell.** 1971. Infeasibility of irradiating fresh fruits and vegetables. *HortScience* **6**:202–204.

82. **Mendgen, K., M. Hahn, and H. Deising.** 1996. Morphogenesis and mechanisms of penetration by plant pathogenic fungi. *Annu. Rev. Phytopathol.* **34**:367–386.

83. **Mendonca, A.** 2005. Bacterial infiltration and internalization in fruits and vegetables, p. 441–482. *In* O. Lamikanra, S. Iman, and D. Ukuku (ed.), *Produce Degradation—Pathways and Prevention.* Taylor and Francis Group, Boca Raton, Fla.

84. **Mercier, J., and J. Arul.** 1993. Induction of systemic-disease resistance in carrot roots by pre-inoculation with storage pathogens. *Can. J. Plant Pathol.* **15**:281–283.

85. **Mercier, J., M. Baka, B. Reddy, R. Corcuff, and J. Arul.** 2001. Shortwave ultraviolet irradiation for control of decay caused by *Botrytis cinerea* in bell pepper: induced resistance and germicidal effects. *J. Am. Soc. Hortic. Sci.* **126**:128–133.

86. **Morris, C. E., and J. M. Monier.** 2003. The ecological significance of biofilm formation by plant-associated bacteria. *Annu. Rev. Phytopathol.* **41**:429–453.

87. **Morris, J. R., and P. L. Brady.** 2005. Temperature effects on produce degradation, p. 599–647. *In* O. Lamikanra, S. Iman, and D. Ukuku (ed.), *Produce Degradation—Pathways and Prevention.* Taylor and Francis Group, Boca Raton, Fla.

88. **Nguyen-The, C., and F. Carlin.** 1994. The microbiology of minimally processed fresh fruits and vegetables. *Crit. Rev. Food Sci. Nutr.* **34**:371–401.

89. **Nguyen-The, C., and F. Carlin.** 2000. Fresh and processed vegetables, p. 620–684. *In* B. M. Lund, A. C. Baird-Parker, and G. W. Gould (ed.), *The Microbiological Quality and Safety of Food.* Aspen Publishers, Gaithersburg, Md.

90. **Nguyen-The, C., and J.-P. Prunier.** 1989. Involvement of pseudomonads in deterioration of ready-to-use salads. *Int. J. Food Sci. Technol.* **24**:47–58.

91. **Patterson, M. F., and P. Loaharanu.** 2000. Irradiation, p. 65–88. *In* B. M. Lund, A. C. Baird-Parker, and G. W. Gould (ed.), *The Microbiological Quality and Safety of Food.* Aspen Publishers, Gaithersburg, Md.

92. **Paull, R. E.** 1992. Postharvest senescence and physiology of leafy vegetables. *Postharv. News Inf.* **3**:11N–20N.

93. **Perombelon, M. C. M., and A. Kelman.** 1980. Ecology of the soft rot erwinias. *Annu. Rev. Phytopathol.* **18**:361–387.

94. **Porat, R., A. Daus, B. Weiss, L. Cohen, E. Fallik, and S. Droby.** 2000. Reduction of postharvest decay in organic citrus fruit by a short hot water brushing treatment. *Postharv. Biol. Technol.* **18**:151–157.

95. **Prusky, D.** 1996. Pathogen quiescence in postharvest diseases. *Annu. Rev. Phytopathol.* **34**:413–434.

96. **Prusky, D., D. Eshel, I. Kobiler, N. Yakoby, D. Beno-Moualem, M. Ackerman, Y. Zuthji, and R. Ben Arie.** 2001. Postharvest chlorine treatments for the control of the persimmon black spot disease caused by *Alternaria alternata. Postharv. Biol. Technol.* **22**:271–277.

97. **Punja, Z. K.** 1985. The biology, ecology and control of *Sclerotium rolfsii. Annu. Rev. Phytopathol.* **23**:97–127.

98. **Reddy, M. V. B., P. Angers, F. Castaigne, and J. Arul.** 2000. Chitosan effects on blackmold rot and pathogenic factors produced by *Alternaria alternata* in postharvest tomatoes. *J. Am. Soc. Hortic. Sci.* **125**:742–747.

99. **Ride, J. P.** 1983. Cell walls and other structural barriers in defence, p. 215–236. *In* J. A. Callow (ed.), *Biochemical Plant Pathology.* John Wiley & Sons, Chichester, United Kingdom.

100. **Robinson, I., and R. P. Adams.** 1978. Ultra-violet treatment of contaminated irrigation water and its effect on the bacteriological quality of celery at harvest. *J. Appl. Bacteriol.* **45**:83–90.

101. **Saltveit, M. E.** 2003. Is it possible to find an optimal controlled atmosphere? *Postharv. Biol. Technol.* **27**:3–13.

102. **Saltveit, M. E.** 2004. Ethylene effects. *In* K. C. Gross, C. Y. Yang, and M. Saltveit (ed.), *The Commercial Storage of Fruits, Vegetables, and Florist and Nursery Stocks.* Draft version of revised *USDA Agriculture Handbook.* [Online.] http://www.ba.ars.usda.gov/hb66. Accessed November 2005.

103. **Schirra, M., G. D'hallewin, S. Ben-Yehoshua, and E. Fallik.** 2000. Host-pathogen interactions modulated by heat treatment. *Postharv. Biol. Technol.* **21**:71–85.

104. **Shewfelt, R. L.** 1986. Postharvest treatment for extending the shelf life of fruits and vegetables. *Food Technol.* **40**(5):71–80, 89.

105. **Snowdon, A. L.** 1990. *A Colour Atlas of Post-Harvest Diseases and Disorders of Fruits and Vegetables*, vol. 1. *General Introduction and Fruits.* Wolfe Scientific, London, United Kingdom.

106. **Snowdon, A. L.** 1991. *A Colour Atlas of Post-Harvest Diseases and Disorders of Fruits and Vegetables*, vol. 2. *Vegetables.* Wolfe Scientific, London, United Kingdom.

107. **Spotts, R. A., L. A. Cervantes, and E. A. Mielke.** 1999. Variability in postharvest decay among apple cultivars. *Plant Dis.* **83**:1051–1054.

108. **Spotts, R. A., P. G. Sanderson, C. L. Lennox, D. Sugar, and L. A. Cervantes.** 1998. Wounding, wound healing and staining of mature pear fruit. *Postharv. Biol. Technol.* **13**:27–36.

109. **Sugar, D., and R. A. Spotts.** 1993. Dispersal of inoculum of *Phialophora malorum* in pear orchards and inoculum redistribution in pear immersion tanks. *Plant Dis.* **77**: 47–49.

110. **Swinburne, T. R., and A. E. Brown.** 1975. The biosynthesis of benzoic acid in Bramley's seedling apples infected by *Nectria galligena* Bres. *Physiol. Plant Pathol.* **6**:259–264.

111. **Thorne, S. N.** 1972. Studies of the behaviour of stored carrots with respect to their invasion by *Rhizopus stolonifer* Lind. *J. Food Technol.* **7**:139–151.

112. **Tripathi, P., and N. K. Dubey.** 2004. Exploitation of natural products as an alternative strategy to control

postharvest fungal rotting of fruit and vegetables. *Postharv. Biol. Technol.* **32**:235–245.

113. **Tucker, S. L., and N. J. Talbot.** 2001. Surface attachment and pre-penetration stage development by plant pathogenic fungi. *Annu. Rev. Phytopathol.* **39**:385–417.

114. **van den Berg, L.** 1987. Water vapor pressure, p. 203–230. *In* J. Weichmann (ed.), *Postharvest Physiology of Vegetables*. Marcel Dekker, New York, N.Y.

115. **Wilson, C. L., A. Elghaouth, E. Chalutz, S. Droby, C. Stevens, J. Y. Lu, V. Khan, and J. Arul.** 1994. Potential of induced resistance to control postharvest diseases of fruits and vegetables. *Plant Dis.* **78**:837–844.

116. **Wisniewski, M., C. Biles, S. Droby, R. McLaughlin, C. Wilson, and E. Chalutz.** 1991. Mode of action of the postharvest biocontrol yeast, *Pichia guilliermondii*. 1. Characterization of attachment to *Botrytis cinerea*. *Physiol. Mol. Plant Pathol.* **39**:245–258.

117. **Wood, D., S. Imam, G. P. Sabellano, W. Orts, and G. M. Glenn.** 2005. Microstructure of produce degradation, p. 529–561. *In* O. Lamikanra, S. Iman, and D. Ukuku (ed.), *Produce Degradation—Pathways and Prevention*. Taylor and Francis Group, Boca Raton, Fla.

118. **Yamaguchi, I., and M. Fujimura.** 2005. Recent topics on action mechanisms of fungicides. *J. Pestic. Sci.* **30**:67–74.

119. **Yao, H. J., and S. P. Tian.** 2005. Effects of a biocontrol agent and methyl jasmonate on postharvest diseases of peach fruit and the possible mechanisms involved. *J. Appl. Microbiol.* **98**:941–950.

120. **Yildiz, F., P. Kinay, M. Yildiz, F. Sen, and I. Karacali.** 2005. Effects of preharvest applications of CaCl$_2$, 2, 4-D and benomyl and postharvest hot water, yeast and fungicide treatments on development of decay on Satsuma mandarins. *J. Phytopathol.* **153**:94–98.

121. **Zhang, J. X., B. D. Bruton, and C. L. Biles.** 1997. Polygalacturonase isozymes produced by *Phomopsis cucurbitae* in relation to postharvest decay of cantaloupe fruit. *Phytopathology* **87**:1020–1025.

Food Microbiology: Fundamentals and Frontiers, 3rd Ed.
Edited by M. P. Doyle and L. R. Beuchat
© 2007 ASM Press, Washington, D.C.

Michelle D. Danyluk
Linda J. Harris
William H. Sperber

9

Nuts and Cereals

The culinary definition of nuts is very broad and includes botanically defined nuts (e.g., acorn, chestnut, and filbert), seeds (e.g., Brazil nut, cashew, pignolia or pine nut, pumpkin, sesame, and sunflower), legumes (e.g., peanut), and drupes (e.g., almond, coconut, macadamia, pecan, pistachio, and walnut) (1, 40, 63). Nuts are grown all over the world in temperate, subtropical, and tropical zones. The most economically important nut worldwide is the peanut or groundnut (*Arachis hypogaea* L.). Other economically important crops include almond (*Prunus dulcis*), Brazil (*Bertholletia excelsa*), cashew (*Anacardium occidentale* L.), chestnut (*Castanea* spp.), coconut (*Cocos nucifera* L.), filbert or hazelnut (*Corylus* spp.), macadamia (*Macadamia* spp.), pecan (*Carya illinoinensis*), pistachio (*Pistacia vera* L.), and walnut (*Juglans* spp.).

Nuts may be sold in the shell or shelled. In addition to being eaten out of hand, nuts are used as ingredients in a wide variety of baked goods and confectionery, dairy, and snack foods, and are sold whole, chopped, diced, slivered, ground, or as a flour or paste. Detailed descriptions of a wide variety of nuts and their production, harvesting, processing, and microbiology can be found in a number of published works (63, 85).

Cereal grains are the most important agricultural products in the world. Since their first cultivation in prehistoric times, cereal grains have sustained the development of civilizations. Over the course of millennia, good agricultural practices have been refined in order to protect the security, safety, and quality of grain supplies. The most important cereal crops today, in terms of quantities produced, are wheat, rice, and corn. These cereals are produced in roughly equal quantities. The majority of the grain supply is consumed by humans. About one-fifth is fed to animals.

Many of the same intrinsic and extrinsic factors influence the survival and growth of spoilage and disease-causing microorganisms on nuts and cereals. The major intrinsic factor is water activity (a_w). All microorganisms have an a_w limit below which they will not grow, regardless of all other factors that may influence metabolic activities. At a_w values above which growth can occur, temperature is the major extrinsic factor influencing proliferation as well as toxin production. This chapter presents an overview of the behavior of microorganisms on nuts, cereals, and products produced from them, with particular emphasis on describing conditions that permit

Michelle D. Danyluk and **Linda J. Harris,** Dept. of Food Science and Technology, University of California, One Shields Ave., Davis, CA 95616-8598. **William H. Sperber,** Corporate Food Safety and Regulatory Affairs, Cargill, Inc., 15407 McGinty Road West, Wayzata, MN 55391-9300.

or inhibit growth and treatments that can be used for their control or elimination.

MICROBIOLOGY OF NUTS

Harvesting, Processing, and Storage of Nuts

The edible portion of a nut may be referred to as a seed, kernel, or meat. Seed coats or pellicles are present on all nuts and develop from tissues originally surrounding the ovule. The seed coats may be as thin as paper (e.g., peanuts) or thick and hard (e.g., coconuts). All nuts have a more or less rigid outer casing or shell. Some nuts (e.g., almond, coconut, pecan, pistachio, and walnut) also have a hull, which is the pulpy tissue that provides a protective outer covering for the shell.

Worldwide, production, harvesting, and processing techniques for nuts range from highly mechanized to labor-intensive, and methods vary significantly for the various types of nuts (63, 85). Many tree nuts are harvested by knocking the nut to the ground or onto a tarp, either by hand or mechanically. Some nuts are partially dried on the tree and then on the orchard floor or elsewhere on the ground, taking advantage of solar heat and low humidity conditions. In other cases, mechanical drying is used alone or in conjunction with natural drying. For example, almonds are generally dried in the orchard after they have been shaken to the ground. They are harvested by sweeping from the orchard floor, and the hull and often the shell are removed in a facility that is usually separate from further processing steps. At the processing facility almond kernels are sized and then stored for various times prior to final processing and packaging. Pecans are often mechanically dried before being stored in the shell; they are wetted before being cracked to soften the shell, thereby minimizing breakage and facilitating clean removal of kernel halves. The moist hull of the walnut is removed in a wet process prior to drying. Pistachios, particularly those grown in the United States, are harvested directly into trailers; the hulls are removed shortly thereafter in a wet process before the nuts are dried mechanically to a stable moisture content for storage.

Natural Microflora

Although production and processing vary for each type of nut, they are generally dried to an a_w of less than 0.70, with corresponding moisture contents ranging from 3.8% (macadamia) to 12.1% (chestnut) (7). The low a_w prevents the growth of microorganisms, resulting in microbiologically stable products. Populations of bacteria, yeasts, and molds present on raw nuts vary greatly and depend on growing, harvesting, and handling practices. Because nuts often come in contact with the soil

during harvest (tree nuts) or growth (peanuts), their microflora (type and population) is greatly influenced by the microorganisms present in the soil. For almonds harvested onto canvas tarps, aerobic plate and yeast and mold counts have been reported to be significantly lower than counts on almonds harvested from the ground (41). Higher microbial populations are also associated with the amount of foreign material present, the amount of insect damage, and shell integrity.

The thick shells of most tree nuts provide an important and effective barrier to microbial penetration, and the presence of a hull further reduces the risk of microbial invasion. The internal surface of a dry, intact nut picked from the tree is thought to be virtually sterile (16, 36, 51). Hull or shell dehiscence (splitting) can occur on the tree or after harvest. Different varieties within a single nut type may have widely differing shell thicknesses. Thin and porous shells are easily damaged and any damage to the shell during harvest, caused either naturally by birds, other vertebrates, or insects or by other means, will reduce the protection the shell provides (41). The shell may crack along the suture during wetting (49) or drying, providing another means for microbial entry (41, 51). Coconut water leaked from cracked shells will support the growth of microorganisms, which can lead to contamination of the meat prior to processing (65).

Cover crops may be used between rows of trees in orchards. The presence of grass or other cover crops in pecan orchards increases the likelihood of invasion of the kernel by molds (85). Grazing domestic animals on these cover crops is practiced in some regions, which increases the risk of nut contamination by fecal microorganisms, especially if the nuts drop to the ground before or during harvest (49, 65).

Moisture content of nuts in the field, particularly during harvest, will have a direct influence on the microflora. Nuts may come in contact with water in the orchard when they drop prematurely to the ground while the trees are still being irrigated or when rain falls during or after harvest. Nuts grown in humid areas are more likely to become contaminated with molds than are those grown in drier regions (85), and rainfall just prior to or during harvest can cause molding of almonds (39) and pecans (6) on the tree as well as on the ground. The dry almond hull swells upon exposure to water, and movement of *Salmonella* through intact almond hulls and shells to the kernel has been demonstrated when an intact hull and shell are immersed in water (19, 79). Pecan shells crack along the suture line when held in water for 48 h or more, and the cracks remain after nuts have dried (49). Walnuts with deteriorated hulls and shells are contaminated with *Escherichia coli* to a greater extent than those with intact hulls and shells when soaked in a buffered suspension of *E. coli* (51).

The initial microflora of peanuts, which develop beneath the soil surface, originates from the soil. Mold contamination of the pods is inevitable. Growing peanut crops in the same field year after year has been shown to increase mold populations in peanuts (57). A higher incidence of molds and pod cracking occurs following drought stress, which may be caused by either low rainfall or overplanting, and is more common in sandy soils. Systemic invasion by *Aspergillus flavus* throughout the peanut plant has been demonstrated under experimental conditions (58) and may provide a means by which peanut kernels can become contaminated. Recently, peanut lines resistant to fungal invasion have been identified (52). Resistance factors include increased tannin content, amino acid composition, waxy surface, permeability, and cell structure arrangement (52).

The aspergilli, especially *A. flavus*, are ubiquitous invaders of nuts (58) and can produce the mycotoxin aflatoxin. Additional fungal invaders include *Alternaria*, other species of *Aspergillus*, *Acremonium*, *Chaetomium*, *Cladosporium*, *Fusarium*, *Eurotium*, *Mucor*, *Paecilomyces*, *Penicillium*, *Phialophora*, *Phomopsis*, *Rhizopus*, *Tricothecium*, and *Trichosporon* (6, 16, 27, 30, 39, 58, 59, 87). Bacteria isolated from nuts include the genera *Achromobacter*, *Acinetobacter*, *Bacillus*, *Brevibacterium*, *Clostridium*, *Corynebacterium*, *Flavobacterium*, *Lactobacillus*, *Leuconostoc*, *Microbacterium*, *Micrococcus*, *Pseudomonas*, *Staphylococcus*, *Streptococcus*, and *Xanthomonas*, as well as members of the *Enterobacteriaceae* (33, 36, 41, 85).

Microbiological Spoilage

Spoilage of nuts is primarily due to oxidative rancidity of the fats. Microbial spoilage, which is less common, can be completely controlled by maintaining an appropriate a_w. Water activity values of less than 0.70 essentially eliminate bacterial and fungal growth in nuts. Molds may grow on nuts that are damp, and for almonds, bacteria have been shown to be capable of growing in wet hulls and shells (79) and in wet areas of the processing environment (25). Condensate may form on nuts that are cooled too quickly or are removed from refrigerated storage and placed in a warm, humid environment. Every effort should be made to protect nuts from moisture in areas in which nuts are processed and stored.

Safety Considerations

Data on the incidence of bacterial foodborne pathogens in nuts are limited to *Salmonella*. An ongoing unpublished survey of 6,728 100-g samples of almonds (after hulling and shelling but before further processing) harvested in California demonstrated a low background level of *Salmonella* (an average incidence of 0.85% over 4 years) (20). Approximately 28 different serovars were isolated from 65 positive lots of nuts. *Salmonella* was rarely recovered from retested samples; when the pathogen was detected, it was usually at a level of <3 MPN (most probable number)/100 g and occasionally 2 to 3 MPN/100 g (L. J. Harris, unpublished data).

Recently, *Salmonella* was isolated from 11 of 117 (9.4%) sesame seed samples and ready-to-eat sesame seed products (halva [a sesame seed paste-based candy, also known as halvah or halavah] and sesame seed paste [tahini]) collected from retail and delicatessen stores in Germany (12). Six samples contained multidrug-resistant *Salmonella enterica* serovar Typhimurium DT 104, which was associated with an international outbreak of salmonellosis ongoing at the time the products were collected (Table 9.1). Even when those six samples are not considered, the overall isolation rate for *Salmonella* was greater than 4%. The National Enteric Pathogen Surveillance Scheme in Australia reported the isolation of 17 different *Salmonella* serovars from sesame seeds and sesame seed products, including hummus (chickpeas with tahini), tahini, and halva tested on 30 occasions between 1985 and 2001 (55). *Salmonella* has also been isolated in macadamia nut (75), walnut (62), cashew and Brazil nut kernels (27), and dried coconut meat (65).

Nuts are not commonly associated with outbreaks of illness caused by microorganisms; however, salmonellosis has been associated with consumption of a wide variety of nuts and nut products (Table 9.1). In some outbreaks of salmonellosis, microbial analyses during traceback investigations have identified high numbers of positive samples (usually 25-g samples), but low numbers of *Salmonella* organisms per g of products tested. Recalled, unopened packages of in-shell peanuts ("dry-flavored" and roasted) were positive for *Salmonella* in, on average, 28 of 72 samples (38%) tested in Australia, Canada, and the United Kingdom (42). Populations of *Salmonella* determined in Australia in the same products ranged from <0.03 to approximately 2 MPN/g (42). Populations of less than three *Salmonella* organisms per g (three samples) and four *Salmonella* organisms per g (one sample) were reported for opened and unopened jars of peanut butter involved in a 1996 outbreak of salmonellosis (67).

Recalled lots of almonds were positive in 84% of 100-g samples tested; positive lots had populations of <0.3 MPN/g upon initial testing and 6 to 9 MPN/100 g upon retesting using larger sample sizes (20). Populations detected during outbreak investigations suggest that low

Table 9.1 Outbreaks of illness associated with the consumption of nuts

Nut	Product	Pathogen	Year	Outbreak location(s)	Reference(s)
Almond	Raw whole	*Salmonella* serovar Enteritidis PT 30	2000–2001	Canada, United States	34
	Raw whole	*Salmonella* serovar Enteritidis PT 9c	2004	Canada, United States	15
Coconut	Desiccated	*Salmonella* serovar Typhi, *Salmonella* serovar Senftenberg, and possibly other serovars	1953	Australia	88
	Desiccated	*Salmonella* serovar Java PT Dundee	1999	United Kingdom	54
	Milk	*Vibrio cholerae*	1991	United States	14, 76
Hazelnut	Conserve (for yogurt)	*Clostridium botulinum*	1989	United Kingdom	56
Peanut	Canned	*Clostridium botulinum*	1986	Taiwan	17
	Savory snack	*Salmonella* serovar Agona PT 15	1994–1995	United Kingdom, Israel	38, 69
	Peanut butter	*Salmonella* serovar Mbandaka	1996	Australia	67
	Flavored or roasted in the shell	*Salmonella* serovar Stanley and *Salmonella* serovar Newport	2001	Australia, Canada, United Kingdom	42
Sesame seed	Halva	*Salmonella* serovar Typhimurium DT 104	2001	Australia, Sweden, Norway, Germany, United Kingdom	55

numbers of *Salmonella* organisms in these products may have been sufficient to cause illness. Low infectious doses of salmonellae have also been reported for a number of other dried foods (22).

Inadequate temperature control played a role in an outbreak of cholera linked to the consumption of coconut milk (76). Frozen coconut milk was not heated to a temperature sufficient to kill *Vibrio cholerae*, and extended storage times at room temperature following thawing allowed for growth of the pathogen to infectious levels. Other *Vibrio* species, *Aeromonas* spp., and three serovars of *Salmonella* were isolated from the same lot of coconut milk and are further indications of a lack of process sanitation and insufficient heat treatment.

Inadequate processing was a possible reason for outbreaks of botulism implicating canned peanuts (17). A combination of high pH (5.0 to 5.5), a formulation change from sugar to aspartame, and insufficient heating enabled survival and outgrowth of *Clostridium botulinum* spores in a hazelnut conserve used for flavoring yogurt (56). *C. botulinum* is capable of growing and producing toxin in aerobically stored (but not anaerobically stored) peanut spread when the a_w is 0.96 or above (18). Molds growing on the surface of the spread stored under aerobic conditions were thought to provide microenvironments suitable for toxin formation. Under anaerobic conditions, growth of lactic acid bacteria lowered the

pH, effectively inhibiting the growth of *C. botulinum*. An a_w of 0.94 or less prevented growth of *C. botulinum*, regardless of atmospheric conditions. Conditions supporting growth of *C. botulinum* also allowed growth of lactic acid bacteria and molds, and the spreads were judged inedible before toxin was detected. Commercially manufactured peanut butter has an a_w of 0.70 or below, which prevents the growth of microorganisms. The potential for microbial growth should be evaluated when formulating unconventional nut spreads or other nut-based products with higher a_w.

Mycotoxigenic molds may be introduced into the nut while it is on the tree or in the ground or during harvest and storage. The presence of these molds is not a significant problem unless sufficient moisture becomes available to permit growth. Once the nut is dried to an a_w of 0.70 or below, growth is inhibited and toxin production is prevented. Aflatoxins, produced by *A. flavus*, *Aspergillus parasiticus*, and *Aspergillus nomius*, are the most common mycotoxins found in nuts and nut products. Aflatoxins are mainly produced under warm conditions and thus are more likely to be present in nuts grown in countries with warm temperate, subtropical, or tropical climates. Agricultural practices can also influence contamination levels. Peanuts, tree nuts that remain on the ground after maturity (allowing for wetting and drying cycles, split hulls, or shells that expose the kernel), and damaged nuts that allow for mold entry are most susceptible to contamination (85). See chapter 24 for more detailed information on aflatoxins.

Effects of Processing and Storage

Current processes for removing hulls and shells differ for each kind of nut and lead to various risks of cross contamination between the hull and/or shell and the kernel. Soils are partially removed during harvest by forced air or sifting, but significant amounts may be brought into storage and processing facilities and can contaminate equipment and nuts during processing.

Kokal and Thorpe (44) determined the incidence of *E. coli* on soft-shelled almonds at growth, production, harvest, hulling, and shelling stages. A low percentage (1%) of almonds on the tree were positive. The proportion of positive almonds gradually increased with each successive harvest step (shaking trees to detach nuts, sweeping, pickup, storage, weighing, and precleaning), up to 40% prior to hulling. During subsequent separate hulling and shelling steps, the percentage of *E. coli*-positive almonds fell to 13 to 17%. Most almonds are hulled and shelled in a single-step process that exposes the kernel directly to the outer part of the hull, further increasing the potential for soil contamination at this stage.

When hulls and shells of almonds and macadamia nuts, for example, are mechanically removed from the kernel under dry conditions, large volumes of fine particulate matter are generated. This dust, composed of soil, hulls, and shells, is extremely difficult to eliminate from the processing environment and can contribute significantly to the contamination of kernels. The addition of water or aqueous quaternary ammonium compounds (200 to 1,000 µg/ml) to almond "dust" can result in rapid growth of *Salmonella* and significant increases in the aerobic plate counts (25). As with all dry-processing environments, appropriate care should be used when cleaning and sanitizing equipment to ensure that rapid and complete drying occurs. The use of isopropyl alcohol-based sanitizers on almond hulling and shelling equipment has the dual benefit of decreasing microbial populations in the presence of a high organic load and drying rapidly on equipment surfaces (25).

A wet process is used to remove the hull of some nuts (e.g., pecan, pistachio, and walnut). The shells of pecans and some other nuts are conditioned or softened by soaking in water or by humidifying with water sprays, steaming, or holding under high relative humidity. In some cases the water may be chlorinated, but in dump or soak tanks the high organic load makes it difficult to maintain residual free chlorine at microbicidal concentrations. Water can be a significant source of cross contamination of nuts if the microbial populations are not controlled by chemical or physical means. However, there are few published data on these potential sources of contamination. Pecan meats from nuts with visibly intact shells did not become contaminated when soaked in lactose broth containing *E. coli* (49) or in water containing *Salmonella* (9). Although liquid can be absorbed at the base and apex of intact pecan shells, the pecan packing tissue (material surrounding the kernel) either inhibited growth or caused a reduction in the number of *Salmonella* organisms, depending upon the concentration of the tissue in broth (9). The authors hypothesized that this effect was due to tannins and polyphenolic compounds present in the packing tissue. Kokal (43) similarly hypothesized that tannins present in walnut skin might be responsible for observed reductions of *E. coli* when inoculated nuts were stored at room temperature.

Shelled or in-shell nuts may be fumigated for insect control and stored for up to a year before further processing and shipping (35). Long-term storage temperatures range from 4°C to ambient, but after processing and packaging, nuts are generally distributed and displayed at the retail level at ambient temperature. Consumers may store purchased almonds at ambient, refrigerated, or frozen temperatures for an additional 1 to 2 years (2).

Few data are available on the survival of microorganisms on nuts or nut products, but in general, survival is remarkably good. Generic *E. coli* or *Salmonella* inoculated into nuts or nut products were detected for several months to years (9, 13, 41, 43, 45, 77). Significantly higher numbers of *Salmonella* organisms survived on pecan halves, in-shell pecans (9), and almonds (77) and in peanut butter (13) stored under refrigerated conditions compared to ambient or elevated temperatures. A temperature effect was not observed when *Salmonella* was inoculated into halva (45).

Storage of almond kernels at -20 ± 2, 4 ± 2, and $23 \pm 3°C$ resulted in average calculated reductions of *Salmonella* serovar Enteritidis phage type (PT) 30 of 0, 0, and 0.25 log CFU/almond per month, respectively (77). On kernels stored at $35 \pm 2°C$, a biphasic survival curve was observed, and calculated reductions were 1.1 log CFU/almond per month from days 0 to 59 and 0 log CFU/almond per month from days 59 to 170 (77) (Fig. 9.1). Long-term survival was independent of initial inoculum level, but survival during initial drying was impacted significantly by the methods used to culture cells to prepare the inoculum. Collectively, these data indicate that *Salmonella* can survive on contaminated nuts and nut products throughout their typical 1- to 2-year shelf life.

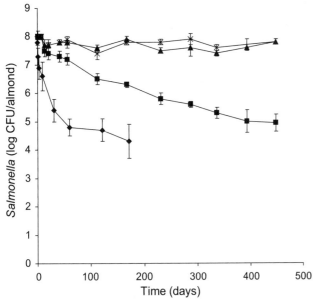

Figure 9.1 Survival during storage at 35 ± 2°C (♦), 23 ± 3°C (■), 4 ± 2°C (▲), and −20 ± 2°C (×) of *Salmonella* serovar Enteritidis PT 30 inoculated onto almonds at a population of approximately 7 log CFU/almond. Almonds stored at 35°C were not analyzed after 170 days (77).

Processing steps that involve a thermal treatment play a dual role of altering the texture and appearance of nuts and reducing microbial contamination. Some products are exposed to several sequential thermal processing steps. For example, to prepare roasted blanched almond slices or slivers, the almonds are first exposed to hot water or steam (blanching) to loosen the pellicle. After the pellicle is mechanically removed, the almonds are dried by exposure to relatively mild heat. To prepare the almonds for slicing, they may be soaked in water and then heated to increase pliability; after slicing or chopping, the pieces may be exposed to dry heat to roast or darken the kernel. In addition to hot air roasting, many nuts are roasted by immersion in hot oil, which is also considered to be a dry heat process. Alternatively, nuts may be moistened by water or steam or sprayed with a saturated salt solution prior to roasting. Each of these pretreatments will have an impact on microbial reduction during heating, although published nut-specific data are generally lacking in this area.

Microorganisms in dry environments and under dry processing conditions are significantly more heat resistant than when suspended in liquids or largely aqueous foods (61). A 5-log reduction of *Salmonella* serovar Enteritidis PT 30 or *Salmonella* serovar Senftenberg 775W on inoculated almonds was achieved within 2 min

when almonds were immersed in hot water at 88°C (78), but a temperature of 127°C was needed to achieve a similar reduction when almonds were exposed to hot oil for the same time (24). Typical industry practices for blanching almonds are to expose them to 88 to 99°C for 5 to 1 min, respectively, which is sufficient to achieve greater than a 5-log CFU/g reduction of *Salmonella*. A reduction of several log units of *Salmonella* on coconut pieces heated to 80°C for 5 min has been observed, without deleterious effects on the sensory quality of shredded product (65). Coconut meat is routinely pasteurized before or after shredding (85).

For oil roasting, exposure at 138 to 149°C for 3 to 12 min is required to obtain the desired level of roast in almonds; this exposure would also achieve substantial reductions of *Salmonella*. Dry air heating is significantly less efficient than oil roasting in reducing microbial contamination. Several factors influence this efficacy, including bed depth, air temperature and velocity, chamber humidity, and nut moisture. Almonds tolerate higher temperatures than do many other nuts such as coconut, walnut, macadamia, and Brazil (65). These nuts darken more rapidly than almonds, and it is not known what level of microbial reduction is achieved during normal roasting processes.

Propylene oxide (C_3H_6O), also known as methyloirane or propene oxide, is currently registered in the United States as a fumigant to reduce the populations of bacteria, yeasts, and molds in a variety of dry foods, including cocoa, processed spices, edible gums, and tree nuts. It is used routinely to reduce microbial populations in bulk tree nuts and their products without impacting sensory quality. The lethality of propylene oxide treatment is dependent upon concentration, exposure time, exposure temperature, and humidity (5, 21). Under commercial conditions, a 5-log or greater reduction of *Salmonella* serovar Enteritidis PT 30 is achieved when inoculated almonds are prewarmed to 30°C and exposed to propylene oxide at a concentration of 0.5 kg/m^3 for 4 h (21). However, this level of reduction was only observed consistently when almonds were held for 5 days postexposure. Aerobic plate counts declined by 0 to 1.1 log CFU/g during the same treatment conditions.

MICROBIOLOGY OF GRAINS AND MILLED CEREAL GRAINS

Cereal grains are subject to microbiological contamination and proliferation while growing in the field. Postharvest microbiological growth is limited by good storage practices, and much of the microflora is removed during the milling process. Microbiological growth in prepared cereal products can lead to spoilage or foodborne illness when well-established control measures are not used.

Spoilage of Pre- and Postharvest Grains

Cereal grains accumulate a large and varied microflora during growth in the field. Molds, or "field fungi," in the genera *Alternaria*, *Aspergillus*, *Cladosporium*, *Fusarium*, and *Helminthosporium* commonly grow on grain with an a_w of 0.90 or higher, a level that corresponds to a moisture content of about 20% or higher. Climatic conditions have a major effect on the amount of mold growth that occurs on cereal grains. The combination of below-normal temperatures and above-normal precipitation and relative humidity will foster excessive mold growth, particularly in wheat and barley crops. Under these conditions, field fungi can damage the grain, even to the point of total crop loss (48, 64). Less invasive mold infections can damage grain quality and sometimes result in the production of mycotoxins. Quality indices such as test weight/bushel and the 1,000-kernel weight test can be used to assess the soundness of grain (4).

Modern grain harvesting and storage practices prevent further mold growth. Crops are cut, threshed, and winnowed by mechanical harvesters in order to separate the grain from chaff. Mechanical dryers are often used to reduce the moisture content of grain before storage. Depending on the size of the global grain reserves, any particular crop could be stored for 1 year or longer before being processed for consumption. To protect quality during this period, grain must always be stored in bins and conveyed in vessels that exclude water, birds, insects, and rodents (48, 64).

Grains that are dried to 12 to 14% moisture will not mold during storage, as long as they are kept dry. Inadequately dried grains will support the growth of "storage fungi" belonging to the genera *Penicillium*, *Rhizopus*, *Mucor*, and *Aspergillus*. While it is not practical to store grains with a moisture content higher than 15% for long periods, fungistatic agents such as propionic acid, formaldehyde, and acetic acid, or combinations thereof, are used to prevent mold and yeast growth during shorter-term storage of high-moisture grains to be used as animal feed (48, 64).

A wide variety of bacteria, including lactic acid bacteria, micrococci, bacilli, and some enteric bacteria, can grow on cereal crops. Some species of enteric bacteria are normal plant saprophytes, and their presence in grain or milled products is not related to fecal contamination. Additionally, crops and stored grains are contaminated by microorganisms from dust, birds, rodents, insects, and other environmental sources. Some levels of insects or insect fragments and rodent hairs and excreta are almost certain to be present in the grain supply. Therefore, the U.S. Food and Drug Administration (FDA) was granted the authority to "establish maximum levels of natural or unavoidable defects in foods for human use that present no health hazard" (80). Maximum acceptable levels, called food defect action levels, associated with insect and rodent contamination of grain and cereal products have been established for wheat, wheat flour, corn meal, macaroni and noodle products, and popcorn (Table 9.2).

Effect of Milling on Microbiological Quality

Before milling, aspiration and screening steps not only clean the grain but also reduce the grain microflora. A significant further reduction of the microflora occurs when the bran is removed. The grain is usually sprayed with water and stored in tempering bins for 6 to 18 h before milling. The amount of water added is just enough

Table 9.2 U.S. FDA defect action levels for insect and rodent contamination of grains and cereal products[a]

Product	Defect	Action level
Wheat	Insect	Avg of 32 or more insect-damaged kernels/100 g
	Rodent	Avg of 9 mg or more of rodent excreta pellets and/or pellet fragments/kg
Wheat flour	Insect	Avg of 75 or more insect fragments/50 g
	Rodent	Avg of 1 or more rodent hairs/50 g
Corn meal	Insect	Avg of 1 or more whole insects/50 g or avg of 25 or more insect fragments/25 g
	Rodent	Avg of 1 or more rodent hairs/25 g or avg of 1 or more rodent excreta pellets/50 g
Macaroni and noodle products	Insect	Avg of 225 insect fragments or more/225 g in 6 or more subsamples
	Rodent	Avg of 4.5 rodent hairs or more/225 g in 6 or more subsamples
Popcorn	Rodent	1 or more rodent excreta pellets are found in 1 or more subsamples, and 1 or more rodent hairs are found in 2 or more other subsamples; or 2 or more rodent hairs/pound and rodent hair are found in 50% or more of the subsamples; or 20 or more gnawed grains/pound and rodent hair are found in 50% or more of the subsamples

[a] From reference 80.

to moisten the surface of the grain. Tempering simultaneously toughens the bran and softens the endosperm, thereby facilitating bran removal and endosperm crushing in the first milling step (4).

While many millers add up to 300 µg of calcium hypochlorite per ml to the temper water, the utility of this addition is unknown. The hypochlorite has no apparent antimicrobial effect, as it is quickly neutralized by the grain and its dust (46). In wheat that was tempered for 16 h with and without 200 µg of calcium hypochlorite per ml in the temper water, no reduction in total plate or coliform counts occurred in the hypochlorite-treated samples compared to the untreated samples (W. H. Sperber, unpublished data).

The dry milling process generally removes 90 to 99% of the microflora originally present on the grain. Flours with the lowest bran, or ash, content typically have the greatest reduction in microflora. Hesseltine (31) found counts (geometric means) of 5.4 log CFU/g in 53 lots of wheat. The flours milled from these lots had a mean count of 3.9 log CFU/g. In an attempt to reduce these counts, five wheat samples were heated to 60°C for periods ranging from 1 to 4 h. The heated wheat samples had a mean count of 4.6 log CFU/g, while the derived flours had a mean count of 2.4 log CFU/g (31). Kamphuis et al. (37) found that mold counts on corn were reduced from 3.8 to 3.5 log CFU/g during grinding.

Other than aspiration and screening, there is no step in dry milling that will further reduce the microbial load of milled cereal grains. Therefore, milled grains should be considered raw agricultural commodities, just as grains are considered. As introduced above, this consideration has been incorporated into U.S. federal regulations for several grains and cereal products (80) (Table 9.2).

The mean counts of several indicator microorganisms detected in a large number of wheat flour samples indicate the excellent microbiological quality of flour, despite its status as a raw agricultural commodity (Table 9.3). The upper limits detected for each indicator microorganism exceed the mean counts by at least a factor of 10, with the broadest range occurring in the enumeration of molds.

Table 9.3 Microbiological profile of wheat flour[a]

Microorganisms	No. of samples	Geometric mean (CFU/g)	Upper limit (CFU/g)
Total count	4,888	15,000	300,000
Yeasts	1,744	120	1,200
Molds	5,211	800	50,000
Coliforms	2,951	200	2,000

[a] Adapted from reference 73.

Table 9.4 Geometric mean values of combined yeast and mold counts in several milled cereal grains[a]

Type of flour/meal	No. of samples	Yeasts and molds (CFU/g)
Wheat	48	1,500
Rye	24	4,200
Buckwheat	24	19,000
Corn	48	130,000

[a] Adapted from reference 8.

At normal moisture levels of 12 to 14%, flours will not support microbial growth. The first group of microorganisms that can grow when the moisture content of flour is increased, by either condensation or other water addition, are xerophilic molds and yeasts. Therefore, the dry-milling industry has come to consider excessive mold or yeast counts per gram of flour to be indicative of moisture contamination in the flour handling system. Excessive counts trigger equipment inspections focused on detecting and eliminating sources of moisture (73). Yeast and mold counts vary considerably, depending upon the type of flour (Table 9.4).

Coliform counts in wheat flour have been reported to average 200/g, but they range as high as 2,000/g (73) (Table 9.3). Thinking that higher coliform counts represent an insanitary condition, many food processors expect their flour suppliers to not exceed a maximum coliform limit. While the coliform group of bacteria was adopted almost 100 years ago as an indicator of fecal contamination in drinking water and pasteurized dairy products, it is unsuited for a similar use in milled cereal grains. The same is true for fresh fruits and vegetables (see chapter 8). A number of coliform species in the genera *Erwinia*, *Pantoea*, *Serratia*, *Klebsiella*, and *Enterobacter* grow predominantly on plants and are usually not found in fecal matter. Growth of these particular coliforms on wheat plants will ensure their carryover into wheat flour, thereby negating any potential utility of a coliform specification for wheat flour or other milled cereal grains (73).

The incidence of salmonellae in 4,360 samples of wheat flour was found to be about 1% (73). This is not unexpected, as flour is a raw agricultural commodity. The occasional presence of salmonellae in flour is not necessarily considered a public health hazard, as most flour is used in products that are baked or cooked, processes that easily kill salmonellae. Moreover, there have been no cases of salmonellosis attributed to milled cereal grains or their conventional products.

Some uses of flour, e.g., incorporation into dry beverage mixtures or dry infant foods, do not have a process step before consumption that is lethal to salmonellae. Therefore, attempts have been made to reduce or eliminate salmonellae in flours by dry heating at temperatures as high as 60°C. However, prolonged heating above 55°C may degrade flour protein functionality (4). When a compounded beverage formulation is similarly heated, its nutritional attributes may be reduced (10).

At a_ws of 0.40 to 0.60, D values for *Salmonella* inactivation average about 1 h in the temperature range of 69 to 77°C, with a z value of about 30°C (3). Storing corn flour at 49°C for 24 h results in a 3-log CFU/g reduction of salmonellae (83). Microwave treatment has been shown to reduce salmonellae in dry corn-soymilk blends, but nutrient degradation occurs at higher temperatures (11). Soft wheat flour intended for use in cakes may be treated with chlorine gas. This treatment produces a whiter flour, but no evidence of microbicidal activity in this application has been reported (4).

Milled cereal grains are produced by a dry-milling process. The resulting flours can be further wet processed, or "hydroprocessed," in order to separate the gluten and starch components. Lactobacilli dominate the wet-processing system, substantially reducing the pH and thereby preventing the growth of nonlactic microflora originally present in the flour. Microorganisms such as salmonellae will not increase during wet processing. Therefore, products such as wheat gluten and starch will have essentially the same microbiological pathogen profile as the original wheat flour from which they are derived (Sperber, unpublished data).

Spoilage of Cereal Products

The baking of dough products simultaneously reduces the microflora and moisture content, thereby limiting the types of microorganisms that could cause spoilage. Even though mold spores and vegetative microbial cells are easily killed during baking, the predominant cause of baked-product spoilage is mold growth. The surfaces of baked products may be contaminated with airborne mold spores during the relatively long product cooling period between baking and packaging (74).

Many bakery products contain one or more fungistatic agents that retard mold growth. One of the most effective food-grade fungistatic agents is potassium sorbate, typically used at concentrations of 1,000 to 3,000 µg/g. However, because it inhibits yeast growth and metabolism, the use of potassium sorbate is limited to chemically leavened products. Yeast-leavened products are typically protected from mold spoilage by the use of calcium propionate at concentrations of 2,000 to 8,000 µg/g. Bakery products are relatively easily stabilized against rapid mold spoilage by an interaction of preservative factors. Reduced a_w and

pH values, residual ethanol content from yeast fermentation, and pasteurization during baking enhance the effectiveness of chemical preservatives (29, 48, 50).

Confectionery components of baked goods, e.g., fruit or cream fillings and icings, are vulnerable to yeast spoilage. This spoilage can be effectively controlled by use of one or more of the following preservative techniques: a_w reduction, pH reduction, and addition of a chemical preservative. Potassium sorbate is particularly effective in this application when used at 1,000 to 2,000 µg/g (48).

Bakery products with a moist interior have long been known to be vulnerable to "rope" spoilage. Rope is the result of extracellular capsular material production by *Bacillus subtilis*. Endospores that survive baking will grow under favorable conditions to very high numbers of vegetative cells and produce rope spoilage. Ropy products have a melon-like odor and a stringy, mucilaginous appearance when pulled apart. Early bakers were encouraged to avoid rope spoilage by adding acetic acid to the dough before baking. The fungistatic agents currently used in bakery products to retard mold spoilage have a limited bacteriostatic activity against *B. subtilis* (26, 86).

Refrigerated dough products are created by packing unbaked, chemically leavened doughs into containers that become hermetically sealed as the doughs proof. These raw doughs are then baked by the consumer. There are two general types of refrigerated dough products— breadstuffs which have a_w values above 0.90 and cookie doughs which have an a_w of ca. 0.80. The refrigeration temperature, reduced a_w, and high carbon dioxide pressure of the breadstuff doughs prevent the growth of almost all microorganisms. Microbial spoilage only occurs after temperature abuse and is almost exclusively caused by *Lactobacillus plantarum* or *Leuconostoc mesenteroides* (32). The only microorganisms that can grow in refrigerated cookie doughs are osmotolerant yeasts, and then only after abuse at ambient temperatures or higher. Bacterial growth is prevented by the low a_w of cookie dough. Xerophilic mold growth is prevented by carbon dioxide content in the dough (Sperber, unpublished data).

In the past several decades, considerable progress has been made to minimize rope spoilage in baked goods and lactic spoilage of refrigerated dough products. The principal factors for improvement in both cases are the improved sanitary design and improved cleaning and sanitation of the dough-handling equipment (48; Sperber, unpublished data). These advances have proven to be more practical and effective than earlier management attempts to control spoilage and safety principally by the use of microbiological specifications on the raw materials.

Mold spoilage of some types of specialty bakery products is retarded by the use of carbon dioxide and/or oxygen scavengers (70), or by the combined use of headspace carbon dioxide and/or ethyl alcohol (84). The use of headspace gases to inhibit mold growth requires the use of packaging materials with very low oxygen transmission rates, and the packages must be well sealed to prevent leakage of the gases.

Some bakery products can be protected against mold spoilage by heating them inside a hermetically sealed, moisture-impermeable package. The internally generated steam originating from moisture in these products is sufficient to kill mold spores on the product surface, thereby preventing mold growth indefinitely. A number of heat sources, including conventional and microwave ovens and infrared bulbs, can be used in this process. The generated steam later equilibrates with the product. This process is suitable for both refrigerated and shelf-stable baked goods. It is more effective and less costly than modified-atmosphere packaging (Sperber, unpublished data).

Food Safety Considerations for Cereal Products

The reduced a_w (0.94 or less) of most cereal products prevents the growth of many microorganisms that can cause foodborne illness. Specifically, *C. botulinum*, *Clostridium perfringens*, *E. coli*, and salmonellae are prevented from growing at this reduced a_w (72). Even though its minimum a_w for growth is 0.94, proteolytic strains of *C. botulinum* have received special attention, even in bakery products, because of the severity of the illness they cause and because their spores survive the baking process (71). Denny et al. (23) reported that *C. botulinum* cannot produce toxin in canned bread at an a_w of 0.95 or below.

A major new product category, cooked, refrigerated pasta products, can sometimes present a substantial safety challenge. These products are made both with and without a variety of fillings, and the pasta and filling components sometimes have a_ws above 0.95. The products are packaged in hermetically sealed containers with very low headspace oxygen in order to prevent mold growth. The resulting anaerobic condition favors the growth of *C. botulinum*. Additionally, cooking eliminates much of the competitive microflora, thereby facilitating *C. botulinum* growth if temperature abuse occurs. Glass and Doyle (28) showed that *C. botulinum* cannot produce toxin in refrigerated pasta, even when temperature abused at 30°C, when the a_w is 0.94 or below. A survey of commercially available refrigerated, filled pasta products revealed that some are capable of supporting *C. botulinum* growth and toxin production when they are abused at 30°C (66). Some of these commercial products

had a_w values as high as 0.983. Refrigerated pasta products such as these need to be formulated to prevent overt food safety hazards should temperature abuse occur. This product design activity must be an integral part of the manufacturer's hazard analysis and critical control point plan. If practical food safety measures, such as the control of pH and a_w limits, cannot be incorporated to protect products from overt food safety hazards during extended temperature abuse, the products should not be produced.

Analogous situations have been demonstrated to exist with cream- or custard-filled pastries, nonfruit pies, and very moist cakes. These products can present opportunities for *Staphylococcus aureus* and salmonellae to grow to populations sufficient to cause illness. Use of the "hurdle effect," i.e., combinations of preservative factors such as a_w reduction, pH reduction, and preservatives, can successfully eliminate the potential hazards posed by these pathogens when products are marketed at ambient temperatures (48, 68).

Historically, commercially produced dried pasta products were vulnerable to the growth of *S. aureus* and/or salmonellae during dough mixing, extrusion, and drying. Some strains of *S. aureus* have been found to be capable of growth in pasta doughs with a_ws as low as 0.86, though enterotoxin production is generally inhibited at a_ws of 0.90 to 0.94 (82). Staphylococci die within several weeks in dried pasta, while salmonellae can persist for more than 1 year (60). However, the potential safety and public health hazards presented by these two pathogens in contaminated dry pasta were found to be exactly the opposite at the time of pasta preparation and consumption. Surviving salmonellae in pasta would be easily killed during cooking, presenting no public health hazard. In contrast, the dead *S. aureus* cells could leave behind their thermostable enterotoxins, which would not be destroyed by cooking pasta, thereby remaining as a potential cause of illness (47).

Two advances have essentially eliminated the potential for growth of *S. aureus* in pasta dough. The first advance involves the use of modern pasta dryers which reduce the pasta a_w to less than 0.86 within 4 h, compared to periods up to 24 h with older-design pasta dryers. The second advance involves a very simple step taken by the mixer operator. Commercial pasta production involves continuous dough mixing under vacuum, in which mixers are operated continuously for days or weeks. Dough buildup in vacuum mixers is known to create a good environment for growth of and toxin production by *S. aureus*. In one case, *A. flavus* was found to have grown and produced aflatoxin in dough buildups. Once-daily elimination of the dough buildup has been validated to eliminate both of these potential foodborne hazards during continuous production periods of up to 30 days (Sperber, unpublished data).

Many outbreaks of illness caused by *Bacillus cereus* have been associated with consumption of fried rice (53). In these cases, rice was cooked and then stored at ambient temperature up to 24 h before being prepared and served as fried rice. Under these conditions, *B. cereus* can grow rapidly and produce a heat-stable emetic toxin. This potential hazard can be eliminated simply by storing the cooked rice at refrigeration temperatures before frying.

In 2005, an outbreak of salmonellosis was associated with the consumption of cake batter ice cream. Since other ice cream flavors produced by the same manufacturer were not involved in reported cases, it was thought that some component of the cake batter may have been responsible for the illnesses. The cake batter had not been baked or pasteurized before its addition to the ice cream mix. As a result, the U.S. FDA issued a warning against the use of atypical uncooked ingredients in ready-to-eat foods (81).

References

1. **Alden, L.** 2005. The cook's thesaurus: nuts. [Online.] http://www.foodsubs.com/Nuts.html. Accessed 6 October 2005.

2. **Almond Board of California.** 2005. Industry resources: almonds from bloom to market. [Online.] http://www.almondboard.com/Resources/content.cfm?ItemNumber = 637. Accessed 5 October 2005.

3. **Archer, J., E. T. Jervis, J. Bird, and J. E. Gaze.** 1998. Heat resistance of *Salmonella weltevreden* in low-moisture environments. *J. Food Prot.* 61:969–973.

4. **Atwell, W. A.** 2001. *Wheat Flour.* Eagan Press, St. Paul, Minn.

5. **Beuchat, L. R.** 1973. *Escherichia coli* on pecans: survival under various storage conditions and disinfection with propylene oxide. *J. Food Sci.* 38:1063–1066.

6. **Beuchat, L. R.** 1975. Incidence of molds on pecan nuts at different points during harvesting. *Appl. Microbiol.* 29:852–854.

7. **Beuchat, L. R.** 1978. Relationship of water activity to moisture content in tree nuts. *J. Food Sci.* 43:754–755, 758.

8. **Beuchat, L. R.** 1992. Enumeration of fungi in grain flours and meals as influenced by settling time in diluent and by the recovery medium. *J. Food Prot.* 55:899–901.

9. **Beuchat, L. R., and E. K. Heaton.** 1975. *Salmonella* survival on pecans as influenced by processing and storage conditions. *Appl. Microbiol.* 29:795–801.

10. **Bookwalter, G. N., R. J. Bothast, W. F. Kwolek, and M. R. Gumbmann.** 1980. Nutritional stability of corn-soy-milk blends after dry heating to destroy salmonellae. *J. Food Sci.* 45:975–980.

11. **Bookwalter, G. N., T. P. Shukla, and W. F. Kwolek.** 1982. Microwave processing to destroy *Salmonella* in corn-soy-milk blends and effect on product quality. *J. Food Sci.* 47:1683–1686.

12. **Brockmann, S. O., I. Piechotowski, and P. Kimmig.** 2004. *Salmonella* in sesame seed products. *J. Food Prot.* 67:178–180.

13. **Burnett, S. L., E. R. Gehm, W. R. Weissinger, and L. R. Beuchat.** 2000. Survival of *Salmonella* in peanut butter and peanut butter spread. *J. Appl. Microbiol.* 89:472–477.

14. **Centers for Disease Control and Prevention.** 1991. Cholera associated with imported frozen coconut milk—Maryland, 1991. MMWR dispatch 13 December 1991. [Online.] http://www.cdc.gov/mmwr/preview/mmwrhtml/00015726.htm.

15. **Centers for Disease Control and Prevention.** 2004. Outbreak of *Salmonella* serotype Enteritidis infections associated with raw almonds—United States and Canada, 2003–2004. MMWR dispatch 4 June 2004. [Online.] http://www.cdc.gov/mmwr/preview/mmwrhtml/mm53d604a1.htm.

16. **Chipley, R. J., and E. K. Heaton.** 1971. Microbial flora of pecan meat. *Appl. Microbiol.* 22:252–253.

17. **Chou, J. H., P. H. Hwang, and M. D. Malison.** 1988. An outbreak of type A foodborne botulism in Taiwan due to commercially preserved peanuts. *Int. J. Epidemiol.* 17:899–902.

18. **Clavero, M. R. S., R. E. Brackett, L. R. Beuchat, and M. P. Doyle.** 2000. Influence of water activity and storage conditions on survival and growth of proteolytic *Clostridium botulinum* in peanut spread. *Food Microbiol.* 17:53–61.

19. **Danyluk, M. D., M. T. Brandl, and L. J. Harris.** 2005. Migration of *Salmonella* Enteritidis PT 30 through almond hulls and shells, p. 158. *Progr. Abstr. Book, 92nd Annu. Meet., Int. Assoc. Food Prot.*

20. **Danyluk, M. D., L. J. Harris, and D. W. Schaffner.** 2006. Monte Carlo simulations assessing the risk of salmonellosis from consumption of almonds. *J. Food Prot.* 69:1594–1599.

21. **Danyluk, M. D., A. R. Uesugi, and L. J. Harris.** 2005. Survival of *Salmonella* Enteritidis PT 30 on inoculated almonds after commercial fumigation with propylene oxide. *J. Food Prot.* 68:1613–1622.

22. **D'Aoust, J.-Y., J. Maurer, and J. S. Bailey.** 2001. *Salmonella* species, p. 141–178. *In* M. P. Doyle, L. R. Beuchat, and T. J. Montville (ed.), *Food Microbiology: Fundamentals and Frontiers*, 2nd ed. ASM Press, Washington, D.C.

23. **Denny, C. B., D. J. Goeke, Jr., and R. Sternberg.** 1969. Inoculation tests of *Clostridium botulinum* in canned breads with special reference to water activity. Research publication no. 4-69. National Canners Association, Washington, D.C.

24. **Du, W.-X., and L. J. Harris.** 2005. Survival of *Salmonella* Enteritidis PT 30 on almonds after exposure to hot oil, p. 109. *Progr. Abstr. Book, 92nd Annu. Meet., Int. Assoc. Food Prot.*

25. **Du, W.-X., and L. J. Harris.** 2005. Evaluation of the efficacy of aqueous and alcohol-based quaternary ammonium sanitizers for reducing *Salmonella* in dusts generated in almond hulling and shelling facilities. Poster 89E-17, presented at Annual Meeting of the Institute of Food Technologists, 16 to 20 July 2005, New Orleans, La. [Online.] http://ift.confex.com/ift/2005/techprogram/paper_30016.htm.

26. **Fisher, E. A., and P. Halton.** 1928. A study of "rope" in bread. *Cereal Chem.* 5:192–208.

27. Freire, F. C. O., and L. Offord. 2002. Bacterial and yeast counts in Brazilian commodities and spices. *Braz. J. Microbiol.* **33:**145–148.

28. Glass, K. A., and M. P. Doyle. 1991. Relationship between water activity of fresh pasta and toxin production by proteolytic *Clostridium botulinum*. *J. Food Prot.* **54:**162–165.

29. Guynot, M. E., A. J. Ramos, V. Sanchis, and S. Marin. 2005. Study of benzoate, propionate, and sorbate salts as mould spoilage inhibitors on intermediate moisture bakery products of low pH (4.5–5.5). *Int. J. Food Microbiol.* **101:**161–168.

30. Hao, D. Y.-Y., E. K. Heaton, and L. R. Beuchat. 1989. Microbial, compositional, and other quality characteristics of pecan kernels stored at –20°C for 25 years. *J. Food Sci.* **54:**472–474.

31. Hesseltine, C. W. 1968. Flour and wheat: research on their microbiological flora. *Baker's Dig.* **42:**40–46.

32. Hesseltine, C. W., R. R. Graves, R. Rogers, and H. R. Burmeister. 1969. Aerobic and facultative microflora of fresh and spoiled refrigerated dough products. *Appl. Microbiol.* **18:**848–853.

33. Hyndman, J. B. 1963. Comparison of enterococci and coliform microorganisms in commercially produced pecan nut meats. *Appl. Microbiol.* **11:**268–272.

34. Isaacs, S., J. Aramini, B. Ceibin, J. A. Farrar, R. Ahmed, D. Middleton, A. U. Chandran, L. J. Harris, M. Howes, E. Chan, A. S. Pichette, K. Campbell, A. Gupta, L. J. Lior, M. Pearce, C. Clark, F. Rodgers, F. Jamieson, I. Brophy, and A. Ellis. 2005. An international outbreak of salmonellosis associated with raw almonds contaminated with a rare phage type of *Salmonella* Enteritidis. *J. Food Prot.* **68:**191–198.

35. Kader, A. A. 1996. In-plant storage, p. 274–277. *In* W. C. Micke (ed.), *Almond Production Manual*. Division of Agriculture and Natural Resources, University of California, Oakland.

36. Kajs, T. M., R. Hagenmaier, C. Vanderzant, and K. F. Mattil. 1976. Microbiological evaluation of coconut and coconut products. *J. Food Sci.* **41:**352–356.

37. Kamphuis, H. J., M. I. Van der Horst, R. A. Sampson, F. M. Rombouts, and S. Notermans. 1992. Mycological condition of maize products. *Int. J. Food Microbiol.* **16:**237–245.

38. Killalea, D., L. R. Ward, D. Roberts, J. de Louvois, F. Sufi, J. M. Stuart, P. G. Wall, M. Susman, M. Schwieger, P. J. Sanderson, I. S. T. Fisher, P. S. Mead, O. N. Gill, C. L. R. Bartlett, and B. Rowe. 1996. International epidemiological and microbiological study of outbreak of *Salmonella* Agona infection from a ready to eat savoury snack. I. England and Wales and the United States. *Br. Med. J.* **313:**1105–1107.

39. King, A. D., Jr., W. U. Halbrook, G. Fuller, and L. C. Whitehand. 1983. Almond nutmeat moisture and water activity and its influence on fungal flora and seed composition. *J. Food Sci.* **48:**615–617.

40. King, A. D., Jr., and T. Jones. 2001. Nut meats, p. 561–563. *In* F. P. Downes and K. Ito (ed.), *Compendium of Methods for the Microbiological Examination of Foods*. American Public Health Association, Washington, D.C.

41. King, A. D., Jr., M. J. Miller, and L. C. Eldridge. 1970. Almond harvesting, processing, and microbial flora. *Appl. Microbiol.* **20:**208–214.

42. Kirk, M. D., C. L. Little, M. Lem, M. Fyfe, D. Genobile, A. Tan, J. Threlfall, A. Paccagenella, D. Lightfoot, H. Lyi, L. McIntyre, L. Ward, D. J. Brown, S. Surnam, and I. S. T. Fisher. 2004. An outbreak due to peanuts in their shell caused by *Salmonella enterica* serotypes Stanley and Newport—sharing molecular information to solve international outbreaks. *Epidemiol. Infect.* **132:**571–577.

43. Kokal, D. 1965. Viability of *Escherichia coli* on English walnut meats (*Juglans regia*). *J. Food Sci.* **30:**325–332.

44. Kokal, D., and D. W. Thorpe. 1969. Occurrence of *Escherichia coli* in almonds of nonpareil variety. *Food Technol.* **23:**93–98.

45. Kotzekidou, P. 1998. Microbial stability and fate of *Salmonella* Enteritidis in halva, a low-moisture confection. *J. Food Prot.* **61:**181–185.

46. Kurtzman, C. P., and C. W. Hesseltine. 1970. Chlorine tolerance of microorganisms found in wheat and flour. *Cereal Chem.* **47:**244–246.

47. Lee, W. H., C. L. Staples, and J. C. Olson, Jr. 1975. *Staphylococcus aureus* growth and survival in macaroni dough and the persistence of enterotoxins in the dried products. *J. Food Sci.* **40:**119–120.

48. Legan, J. D. 2000. Cereals and cereal products, p. 759–783. *In* B. M. Lund, T. C. Baird-Parker, and G. W. Gould (ed.), *The Microbiological Safety and Quality of Food*. Aspen Publishers, Inc., Gaithersburg, Md.

49. Marcus, K. A., and H. J. Amling. 1973. *Escherichia coli* field contamination of pecan nuts. *Appl. Microbiol.* **26:**279–281.

50. Melnick, D., H. W. Vahlteich, and A. Hackett. 1956. Sorbic acid as a fungistatic agent for foods. XI. Effectiveness of sorbic acid in protecting cakes. *Food Res.* **21:**133–146.

51. Meyer, L. T., and R. H. Vaughn. 1969. Incidence of *Escherichia coli* in black walnut meats. *Appl. Microbiol.* **18:**925–931.

52. Mixon, A. C. 1980. Potential for aflatoxin contamination in peanuts (*Arachis hypogaea* L.) before and after harvest—a review. *J. Environ. Qual.* **9:**344–349.

53. Mortimer, P. R., and G. McCann. 1974. Food-poisoning episodes associated with *Bacillus cereus* in fried rice. *Lancet* **i:**1043–1045.

54. O'Brien, S., S. Brustin, G. Duckworth, and L. Ward. 1999. *Salmonella* Java phage type Dundee—rise in cases in England: update. [Online.] http://www.eurosurveillance.org/ew/1999/990318.asp. Accessed 11 December 2003.

55. O'Grady, K. A., J. Powling, A. Tan, M. Valcanis, D. Lightfoot, J. Gregory, K. Lalor, R. Guy, B. Ingle, R. Andrews, S. Crerar, and R. Stafford. 2001. Salmonella Typhimurium DT104—Australia, Europe. Archive no. 20010822.1980. [Online.] http://www.promedmail.org. Accessed 18 October 2004.

56. O'Mahony, M., E. Mitchell, R. J. Gilbert, N. D. Hutchinson, N. T. Begg, J. C. Rodhouse, and J. E. Morris. 1990. An outbreak of foodborne botulism associated with contaminated hazelnut yoghurt. *Epidemiol. Infect.* **104:**385–395.

57. Pettit, R. E., and R. A. Taber. 1968. Factors influencing aflatoxin accumulation in peanut kernels and the associated mycoflora. *Appl. Microbiol.* **16:**1230–1234.

58. Pitt, J. I., S. K. Dyer, and S. McCammon. 1991. Systemic invasion of developing peanut plants by *Aspergillus flavus*. *Lett. Appl. Microbiol.* **13**:16–20.

59. Pitt, J. I., A. D. Hocking, K. Bhudhasamia, B. F. Miscamble, K. A. Wheeler, and P. Tanboon-Ek. 1993. The normal microflora of commodities from Thailand. 1. Nuts and oilseeds. *Int. J. Food Microbiol.* **20**:211–216.

60. Rayman, M. K., J.-Y. D'Aoust, B. Aris, C. Maishmert, and R. Wasik. 1979. Survival of microorganisms in stored pasta. *J. Food Prot.* **42**:330–334.

61. Riemann, H. 1968. Effect of water activity on the heat resistance of *Salmonella* in "dry" materials. *Appl. Microbiol.* **16**:1621–1622.

62. Riyaz-Ul-Hassan, S., V. Verma, A. Malik, and G. N. Qazi. 2003. Microbiological quality of walnut kernels and apple juice concentrate. *World J. Microbiol. Biotechnol.* **19**:845–850.

63. Rosengarten, F., Jr. 1984. *The Book of Edible Nuts*. Dover Publications, Inc., Mineola, N.Y.

64. Sauer, D. B., R. A. Meronuck, and C. M. Christenson. 1992. Microflora, p. 313–340. *In* D. B. Sauer (ed.), *Storage of Cereal Grains and Their Products*, 4th ed. American Association of Cereal Chemists, St. Paul, Minn.

65. Schaffner, C. P., K. Mosbach, V. C. Bibit, and C. H. Watson. 1967. Coconut and *Salmonella* infection. *Appl. Microbiol.* **15**:471–475.

66. Schebor, C., and J. Chirife. 2000. A survey of water activity and pH values in fresh pasta packed under modified atmosphere manufactured in Argentina and Uruguay. *J. Food Prot.* **63**:965–969.

67. Scheil, W., S. Cameron, C. Dalton, C. Murray, and D. Wilson. 1998. A South Australian *Salmonella* Mbandaka outbreak investigation using a database to select controls. *Aust. N. Z. J. Public Health* **22**:536–539.

68. Shelf Stable Bakery Product Subcommittee. 2005. Voluntary protocol for the safety of an unrefrigerated pumpkin pie product during shelf and use life. *Am. Inst. Baking Tech. Bull.* **27**(5).

69. Shohat, T., M. S. Green, D. Merom, O. N. Gill, A. Reisfeld, A. Matas, D. Blau, N. Gal, and P. E. Slater. 1996. International epidemiological and microbiological study of outbreak of *Salmonella agona* infection from a ready to eat savoury snack. II. Israel. *Br. Med. J.* **313**:1107–1109.

70. Smith, J. P., B. Ooraikul, W. J. Koersen, E. D. Jackson, and R. A. Lawrence. 1986. Novel approach to oxygen control in modified atmosphere packaging of bakery products. *Food Microbiol.* **3**:315–320.

71. Sperber, W. H. 1982. Requirements of *Clostridium botulinum* for growth and toxin production. *Food Technol.* **36**:89–94.

72. Sperber, W. H. 1983. Influence of water activity on foodborne bacteria—a review. *J. Food Prot.* **46**:142–150.

73. Sperber, W. H. 2003. Microbiology of milled cereal grains. *Int. Assoc. Operative Millers Tech. Bull.* **3**:7929–7931.

74. Spicher, G. 1985. On the question of dough product hygiene. Third communication. The microbiological quality of currently available commercial dough products. *Getreide Mehl Brot* **39**:212–215.

75. St. Clair, V. J., and M. M. Klenk. 1990. Performance of three methods for the rapid identification of *Salmonella* in naturally contaminated foods and feeds. *J. Food Prot.* **53**:161–164.

76. Taylor, J. L., J. Tuttle, T. Pramukul, K. O'Brien, T. J. Barrett, B. Jolbaito, Y. L. Lim, D. J. Vugia, J. G. Morris, Jr., R. V. Tauxe, and D. M. Dwyer. 1993. An outbreak of cholera in Maryland associated with imported commercial frozen fresh coconut milk. *J. Infect. Dis.* **167**:1330–1335.

77. Uesugi, A. R., M. D. Danyluk, and L. J. Harris. 2006. Survival of *Salmonella* Enteritidis phage type 30 on inoculated almonds stored at –20, 4, 23, and 35°C. *J. Food Prot.* **69**:1851–1857.

78. Uesugi, A. R., and L. J. Harris. 2005. Survival of *Salmonella* Enteritidis PT 30 on almonds after exposure to hot water, p. 109. *Progr. Abstr. Book, 92nd Annu. Meet., Int. Assoc. Food Prot.*

79. Uesugi, A. R., and L. J. Harris. 2006. Growth of *Salmonella* Enteritidis phage type 30 in almond hull and shell slurries and survival in drying almond hulls. *J. Food Prot.* **69**:712–718.

80. U.S. Food and Drug Administration. 2005. The food defect action levels. [Online.] http://www.cfsan.fda.gov/~dms/dalbook.html.

81. U.S. Food and Drug Administration. 19 August 2005. Bulletin to the food service and retail food store industry regarding cake batter ice cream and similar products. [Online.] http://www.cfsan.fda.gov/~ear/ret-batt.html.

82. Valik, L., and F. Görner. 1993. Growth of *Staphylococcus aureus* in pasta in relation to its water activity. *Int. J. Food Microbiol.* **20**:45–48.

83. VanCauwenberge, J. E., R. J. Bothast, and W. F. Kwolek. 1981. Thermal inactivation of eight *Salmonella* serotypes in dry corn flour. *Appl. Environ. Microbiol.* **42**:688–691.

84. Vora, H. M., and J. S. Sidhu. 1987. Effect of varying concentrations of ethyl alcohol and carbon dioxide on the shelf life of bread. *Chem. Mikrobiol. Technol. Lebensm.* **11**:56–59.

85. Wareing, P. W., L. Nicolaides, and D. R. Twiddy. 2000. Nuts and nut products, p. 919–940. *In* B. M. Lund, T. C. Baird-Parker, and G. W. Gould (ed.), *The Microbiological Safety and Quality of Food*. Aspen Publishers, Inc., Gaithersburg, Md.

86. Watkins, E. J. 1906. Ropiness in flour and bread and its detection and prevention. *J. Soc. Chem. Ind.* **25**:350–355.

87. Wells, J. M. 1980. Toxigenic fungi isolated from late-season pecans. *J. Food Saf.* **4**:213–220.

88. Wilson, M. M., and E. F. Mackenzie. 1955. Typhoid fever and salmonellosis due to consumption of infected desiccated coconut. *J. Appl. Bacteriol.* **18**:510–521.

Foodborne
Pathogenic Bacteria

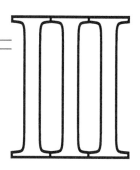

Food Microbiology: Fundamentals and Frontiers, 3rd Ed.
Edited by M. P. Doyle and L. R. Beuchat
© 2007 ASM Press, Washington, D.C.

Jean-Yves D'Aoust
John Maurer

Salmonella Species[†]

10

In the early 19th century, clinical pathologists in France first documented the association of human intestinal ulceration with a contagious agent, with the disease later being identified as typhoid fever. Further investigations by European workers led to the isolation and characterization of the typhoid bacillus responsible for typhoid fever and to the development of a serodiagnostic test for the detection of this serious human disease agent (116, 284). Differential clinical and serological traits were used subsequently to identify the closely related paratyphoid organisms. In the United States, contemporary work by Salmon and Smith in 1885 led to the isolation of *Bacillus cholerae-suis,* now known as *Salmonella enterica* serovar Choleraesuis, from swine suffering from hog cholera (284). During the first quarter of the 20th century, great advances occurred in the serological detection of somatic (O) and flagellar (H) antigens within the *Salmonella* group, a generic term coined by Lignières in 1900 (284). An antigenic scheme for the classification of salmonellae was first proposed by White in 1926 and subsequently expanded by Kauffmann in 1941 into the Kauffmann-White scheme, which currently includes more than 2,541 serovars (371).

[†]This chapter was erroneously credited to J. Stan Bailey and John J. Maurer in the contents of *Food Microbiology: Fundamentals and Frontiers,* 2nd ed.

CHARACTERISTICS OF THE ORGANISM

Taxonomy

Salmonella spp. are facultatively anaerobic gram-negative rod-shaped bacteria belonging to the family *Enterobacteriaceae.* Although members of this genus are motile by peritrichous flagella, nonflagellated variants, such as *Salmonella* serovar Pullorum and *Salmonella* serovar Gallinarum, and nonmotile strains resulting from dysfunctional flagella do occur. Salmonellae are chemo-organotrophic, with the ability to metabolize nutrients by both respiratory and fermentative pathways. The bacteria grow optimally at 37°C and catabolize D-glucose and other carbohydrates with the production of acid and gas. Salmonellae are oxidase negative and catalase positive, grow on citrate as a sole carbon source, generally produce hydrogen sulfide, decarboxylate lysine and ornithine, and do not hydrolyze urea. Many of these traits formed the basis for the presumptive biochemical identification of *Salmonella* isolates. According to a contemporary definition, a typical *Salmonella* isolate would produce acid and gas from glucose in triple sugar iron (TSI) agar medium and would not utilize lactose or sucrose in TSI or differential plating media, such as brilliant green, xylose lysine desoxycholate, and Hektoen

Jean-Yves D'Aoust, Food Directorate, Health Products & Food Branch, Health Canada, Sir F. G. Banting Research Centre, Postal Locator 22.04.A2, Tunney's Pasture, Ottawa, Ontario, Canada K1A 0K9. **John Maurer,** Department of Population Health, University of Georgia, Athens, GA 30602.

enteric agars. Additionally, typical salmonellae readily produce an alkaline reaction from the decarboxylation of lysine to cadaverine in lysine iron agar, generate hydrogen sulfide gas in TSI and lysine iron media, and fail to hydrolyze urea (14, 121). The dynamics of genetic variability arising from bacterial mutations and conjugative intra- and intergeneric exchanges of plasmids encoding determinant biochemical traits continue to reduce the proportion of typical *Salmonella* biotypes. From the early studies of Le Minor and colleagues confirming that *Salmonella* utilization of lactose and sucrose was plasmid mediated (286, 287), many studies have since emphasized the occurrence of Lac⁺ and/or Suc⁺ biotypes in clinical specimens and food materials. This situation is of public health concern because biochemically atypical salmonellae could easily escape detection on disaccharide-dependent plating media which are commonly used in hospital and food industry laboratories. Bismuth sulfite agar remains a medium of choice for isolating salmonellae because, in addition to its high level of selectivity, it responds solely and most effectively to the production of extremely low levels of hydrogen sulfide gas (116). The diagnostic hurdles engendered by the changing patterns of disaccharide utilization by *Salmonella* are being confounded further by the increasing occurrence of biotypes that cannot decarboxylate lysine, that possess urease activity, that produce indole, and that readily grow in the presence of KCN. Clearly, the recognition of *Salmonella* as a biochemically homogeneous group of microorganisms is becoming obsolete. The one serovar-one species concept is untenable because most serovars cannot be separated by biochemical tests. The situation has led to a reassessment of the diagnostic value of these and other biochemical traits and to their likely replacement with molecular technologies targeted at the identification of stable genetic loci and/or their products that are unique to the *Salmonella* genus.

The nomenclature of the *Salmonella* group has progressed through a succession of taxonomic schemes based on biochemical and serological characteristics and on principles of numerical taxonomy and DNA homology (Table 10.1). In the early development of taxonomic schemes, determinant biochemical reactions were used to separate salmonellae into subgroups. The Kauffmann-White scheme was the first attempt to systematically classify salmonellae using these scientific parameters. This major undertaking culminated in the identification of five biochemically defined subgenera (I to V) wherein individual serovars were afforded species status (256). Subgenus III included members of *S. arizonae*. Subsequently, a three-species nomenclatural system was proposed, using 16 discriminating tests to identify *S. typhi* (single serovar), *S. choleraesuis* (single serovar), and *S. enteritidis* (all other salmonella serovars). The latter scheme recognized members of the *Arizona* group as a distinct genus (146). A defining development in *Salmonella* taxonomy occurred in 1973, when Crosa et al. (107) determined by DNA-DNA hybridization that all serotypes and subgenera of *Salmonella* and the *Arizona* group were related at the species level. The single exception to this observation was the subsequent description of *S. bongori*, previously known as subgenus V, as a distinct species. Another

Table 10.1 Taxonomic schemes for *Salmonella* spp.

Diagnostic basis	Salient features	Serovar designation	Reference
Biochemical	Five subgenera (I to V) Serovar = species status *S. arizonae*	*S. typhimurium*	256
Biochemical	Three species (*S. typhi, S. choleraesuis, S. enteritidis*) *Arizona* = separate genus	*S. enteritidis* serovar Typhimurium	146
Phenetic/DNA homology	Single species (*S. choleraesuis*) Seven subspecies (choleraesuis, salamae, arizonae, diarizonae, houtenae, bongori, indica) Type strain = *S. choleraesuis*	*S. choleraesuis* subsp. *choleraesuis* serovar Typhimurium	288
Phenetic/DNA homology	Single species (*S. enterica*) Seven subspecies (see above) Type strain = serovar Typhimurium LT 2	*S. enterica* subsp. *enterica* serovar Typhimurium or *Salmonella* serovar Typhimurium	285
Multilocus enzyme electrophoresis	Two species (*S. enterica* [six subspecies] and *S. bongori*)		389

system based on numerical taxonomy and DNA relatedness proposed a single species (*S. choleraesuis*) consisting of seven subspecies (288). In numerical taxonomy, a statistical comparison of morphological and biochemical attributes of strains (phenetic analysis) measures the taxonomic proximity of test strains and allows for their separation into distinct taxons. In DNA homology, a high degree of hybridization of *Salmonella* reference DNAs with extracts of test strains confirms the genetic relatedness of nucleic acid reactants and supports the inclusion of the test microorganisms in the *Salmonella* genus.

A subsequent modification of the former scheme, while retaining the names of the seven recognized subspecies, changed the type species from *S. choleraesuis* to *S. enterica* (285). The following nomenclature is now widely used. The genus *Salmonella* consists of two species, each of which contains multiple serovars (Table 10.2). The two species are *S. enterica*, the type species, and *S. bongori*, which was formerly known as subspecies V. *S. enterica* is divided into six subspecies, which are referred to by Roman numerals and subspecies names (I, *S. enterica* subsp. *enterica*; II, *S. enterica* subsp. *salamae*; IIIa, *S. enterica* subsp. *arizonae*; IIIb, *S. enterica* subsp. *diarizonae*; IV, *S. enterica* subspecies *houtenae*; and VI, *S. enterica* subsp. *indica*) (68, 389). *S. enterica* subspecies are differentiated on the basis of biochemical traits and genomic relatedness.

The biochemical identification of foodborne and clinical *Salmonella* isolates is generally coupled to serological confirmation, a complex and labor-intensive technique involving the agglutination of bacterial surface antigens with *Salmonella*-specific antibodies. These include somatic (O) lipopolysaccharides (LPS) on the external surface of the bacterial outer membrane, flagellin (H) antigens associated with the peritrichous flagella, and the capsular (Vi) antigen, which occurs only in serovars Typhi, Paratyphi C, and Dublin (284). The heat-stable somatic (O) antigens are classified as major or minor antigens. The former category consists of antigens such as the somatic factors O:4 and O:3, which are specific determinants for the somatic groups B and E, respectively. In contrast, minor somatic antigenic components, such as O:12, are nondiscriminatory, as evidenced by their presence in different somatic groups. Smooth (S) variants relate to strains with well-developed serotypic LPS that readily agglutinate with specific antibodies, whereas rough (R) variants exhibit incomplete LPS antigens resulting in weak or no agglutination with *Salmonella* somatic antibodies. Flagellar (H) antigens are heat-labile proteins, and individual *Salmonella* strains may produce one (monophasic) or two (diphasic) sets of flagellar antigens. Although serovars such as serovar Dublin produce a single set of flagellar (H) antigens, most serovars can alternately elaborate two sets of antigens, i.e., phase 1 and phase 2 antigens. These homologous surface antigens are chromosomally encoded by the H_1 (phase 1) and H_2 (phase 2) genes and transcribed under the control of the vh_2 locus (284). Capsular (K) antigens commonly encountered in members of the *Enterobacteriaceae* are limited to the Vi antigen in the *Salmonella* genus. Thermal solubilization of the Vi antigen is necessary for the immunological identification of underlying serotypic LPS.

Serological testing procedures aim to derive the complete antigenic formulas of individual *Salmonella* isolates. We use serovar Infantis (6,7:r:1,5) as a working example. Commercially available polyvalent somatic antisera each consist of a mixture of antibodies specific for a limited number of major antigens, e.g., polyvalent B antiserum (BD Difco, Becton Dickinson and Company, Sparks, Md.) recognizes somatic (O) groups C_1, C_2, F, G, and H. Following a positive agglutination result with polyvalent B antiserum, single-group antisera representing the five somatic groups included in the polyvalent B reagent would be used to identify the serogroup of the isolate. The test isolate would react with the C_1 group antiserum, indicating that antigens 6 and 7 are present. Flagellar (H) antigens would then be determined by broth agglutination reactions, using polyvalent H antisera or the Spicer-Edwards series of antisera. In the former assay, a positive agglutination reaction with one of the five polyvalent antisera (Poly A to E; Becton Dickinson and Company) would lead to testing with single-factor antisera to specifically identify the phase 1 and/or phase 2 flagellar antigens present. Agglutination in Poly C flagellar antiserum and subsequent reaction of the isolate with single-group H antisera would confirm the presence of the r antigen (phase 1). The empirical antigenic formula of the isolate would then be 6,7:r. Phase reversal in semisolid agar supplemented with r antiserum would

Table 10.2 Species within the *Salmonella* genus[a]

Salmonella species and subspecies	No. of serovars
S. enterica subsp. *enterica* (I)	1,504
S. enterica subsp. *salamae* (II)	502
S. enterica subsp. *arizonae* (IIIa)	95
S. enterica subsp. *diarizonae* (IIIb)	333
S. enterica subsp. *houtenae* (IV)	72
S. enterica subsp. *indica* (VI)	13
S. bongori (V).	22
Total	2,541

[a] From references 68 and 371.

immobilize phase 1 salmonellae at or near the point of inoculation, thereby facilitating the recovery of phase 2 cells from the edge of the zone of migration. Serological testing of phase 2 cells with Poly E and 1-complex antisera would confirm the presence of the flagellar 1 factor. Confirmation of the flagellar 5 antigen with single-factor antiserum would yield the final antigenic formula 6,7: r:1,5, which corresponds to serovar Infantis. A similar analytical approach would be used with the Spicer-Edwards polyvalent H antisera, where the identification of flagellar (H) antigens would arise from the pattern of agglutination reactions among the four Spicer-Edwards antisera and with three additional polyvalent antisera consisting of the L, 1, and e,n complexes.

Physiology

Growth

Salmonella spp. consist of resilient microorganisms that readily adapt to extreme environmental conditions. Some *Salmonella* strains can grow at elevated temperatures ($\leq$54°C), and others exhibit psychrotrophic properties in their ability to grow in foods stored at 2°C to 4°C (113) (Table 10.3). Moreover, preconditioning of cells to low temperatures can markedly increase the growth and survival of salmonellae in refrigerated food products (5). Such growth characteristics raise concerns about the efficacy of chill temperatures to ensure food safety through bacteriostasis. These concerns are further heightened by

the widespread refrigerated storage of foods packaged under vacuum or a modified atmosphere to prolong shelf life. Gaseous mixtures consisting of 60 to 80% (vol/vol) CO_2 with various proportions of N_2 and/or O_2 can inhibit the growth of aerobic spoilage microorganisms, such as *Pseudomonas* spp., without promoting the growth of *Salmonella* spp. (118). However, the proliferation of salmonellae in inoculated raw minced beef and cooked crab meat stored at 8°C to 11°C under modified atmospheres containing low levels of CO_2 (20 to 50% [vol/vol]) warrants caution in the general application of this novel processing technology (54, 242).

Studies on the maximum temperature for growth of *Salmonella* spp. in foods are generally lacking. Notwithstanding an early report of growth of salmonellae in inoculated custard and chicken à la king at 45.6°C (15), more recent evidence indicates that prolonged exposure of mesophilic strains to thermal stress conditions results in mutants of serovar Typhimurium capable of growth at 54°C (137). Although the mechanism of this phenomenon has yet to be elucidated, preliminary findings indicate that two separate mutations enable serovar Typhimurium to grow actively at 48°C (*ttl*) and 54°C (*mth*).

The physiological adaptability of *Salmonella* spp. is further demonstrated by their ability to proliferate at pH values ranging from 4.5 to 9.5, with an optimum pH for growth of 6.5 to 7.5 (Table 10.3). It is well established that the bacteriostatic or antibacterial effects of acidic conditions are acidulant dependent (116). Of the many

Table 10.3 Physiological limits for growth of *Salmonella* spp. in foods and bacteriological media

Parameter	Limit Minimum	Maximum	Product	Serovar	Reference
Temperature (°C)	2.0 (24 h)		Minced beef[a]	Typhimurium	82
	2.0 (2 days)		Minced chicken[b]	Typhimurium	39
	4.0 ($\leq$10 days)		Shell eggs[b]	Enteritidis	263
		54.0[c]	Agar medium	Typhimurium	137
pH	3.99[d]		Tomatoes	Infantis	36
	4.05[e]		Liquid medium	Anatum Tennessee Senftenberg	92
		9.5	Egg washwater[b]	Typhimurium	230
a_w	0.93[f]		Rehydrated dried soup[b]	Oranienburg	456

[a] Naturally contaminated.
[b] Artificially contaminated.
[c] Mutants selected to grow at elevated temperature.
[d] Growth within 24 h at 22°C.
[e] Acidified with HCl or citric acid; growth within 24 h at 30°C.
[f] Growth within 3 days at 30°C.

organic acids produced by starter cultures in meat, dairy, and other fermented foods and of the various organic and inorganic acids used in product acidification, propionic and acetic acids are more bactericidal than the common food-associated lactic and citric acids. Interestingly, the antibacterial action of organic acids decreases with increasing length of the fatty acid chain (116). Early research on the propensity for growth of *Salmonella* in acidic environments revealed that wild-type strains preconditioned on pH gradient plates could grow in liquid and solid media at considerably lower pH values than those tolerated by the parent strains (238). These findings raise concerns regarding the safety of fermented foods, such as cured sausages and fermented raw milk products, where the progressive starter culture-dependent acidification of fermented foods could provide a favorable environment for the elevation of endogenous salmonellae to a state of increased acid tolerance. The growth and/or enhanced survival of salmonellae during the fermentative process would result in a contaminated ready-to-eat product. Leyer and Johnson (293) demonstrated the increased survival of acid-adapted *Salmonella* spp. in fermented milk and during storage (5°C) of Cheddar, Swiss, and mozzarella cheese derived from the milk. The presence of acid-tolerant salmonellae in such foods further heightens the level of public health risk because this acquired physiological trait could minimize the antimicrobial activity of gastric acidity (pH 2.5) and promote the survival of *Salmonella* within the acidic cytoplasm of mononuclear and polynuclear phagocytes of the human host (117). Bearson et al. (48) identified four regulator genes, of which two, *ada* and *phoP*, are associated with virulence and two, *rpoS* and *fur*, are associated with the development of acid tolerance. The complexity of the acid tolerance stress response suggests that acid tolerance and virulence are interrelated.

High salt concentrations have long been recognized for their ability to extend the shelf life of foods by inhibiting the growth of endogenous microflora (367). This bacteriostatic effect, which can also engender cell death, results from a dramatic decrease in water activity (a_w) and from bacterial plasmolysis commensurate with the hypertonicity of the suspending medium. Studies have revealed that foods with water activity values of <0.93 do not support the growth of salmonellae (116) (Table 10.3). Although *Salmonella* is generally inhibited in the presence of 3 to 4% NaCl, bacterial salt tolerance increases with increasing temperature in the range of 10 to 30°C. However, the latter phenomenon is associated with a protracted log phase and a decreased rate of growth. Evidence further suggests that the magnitude of this adaptive response is food and serovar specific (116).

Anaerobiosis can potentiate greater salt tolerance in *Salmonella* and raises concerns regarding the safety of modified atmosphere- and vacuum-packaged foods that contain high levels of salt (17).

The pH, salt concentration, and temperature of the microenvironment can exert profound effects on the growth kinetics of *Salmonella* spp. Several studies revealed the increased ability of salmonellae to grow under acidic (pH, ≤5.0) conditions or in environments of high salinity (≥2% NaCl) with increasing temperature (116, 151, 444). Similar research on the interrelationship between pH and NaCl at 20 to 30°C has underscored the dominance of the medium pH on the growth of *Salmonella* (444). Interestingly, the presence of salt in acidified foods can reduce the antibacterial action of organic acids as low concentrations of NaCl or KCl stimulate the growth of serovar Enteritidis in broth medium acidified to pH 5.19 with acetic acid (379). Such findings and other reports on the enhanced salt-dependent survival of salmonellae in rennet whey (pH 4.8 to 5.6) and mayonnaise (445) indicate that low levels of salt can undermine the preservative action of organic acids and potentially compromise the safety of fermented and acidified foods. Although the mechanisms for these salt phenomena remain elusive, the role of salinity in restoring cellular homeostasis may be linked to the Na⁺/K⁺-proton antiport systems (163, 164, 353).

The study of interactive forces generated by temperature, pH, and salt on the growth and survival of *Salmonella* has led to the development and application of mathematical models to predict the fate of salmonellae in foods. In this approach, the growth, survival, or inactivation of a target microorganism under various conditions of pH, NaCl, temperature, or other environmental factors of interest is laboriously characterized in laboratory media. The generated data are then used to derive mathematical models that depict the response of the microorganism to different combinations of environmental factors (179). Predictive models are not without limitations. For example, the use of a model to predict the growth of salmonellae in a food whose salt content lies beyond the range of values originally studied for the derivation of the mathematical model would likely lead to erroneous conclusions. Moreover, models are generally based on the behavior of a few *Salmonella* strains under selected environmental conditions. The physiological diversity among the large number of serovars (Table 10.2) may unduly challenge the reliability of models in predicting bacterial growth responses. Additionally, the lot-to-lot variations in the composition of a given food and in the types and numbers of background microflora are variables that could seriously undermine the predictive capability of mathematical models. A special issue on predictive modeling published in the *International Journal of Food Microbiology* (vol. 23, no. 3/4, 1994) provides invaluable information on the merits and limitations of the modeling approach for *Salmonella* and other foodborne bacterial pathogens and should be consulted for an in-depth review of the subject.

Survival

Although the potential growth of foodborne *Salmonella* spp. is of primary importance in safety assessments, the propensity for these pathogens to persist in hostile environments further heightens public health concerns. The survival of *Salmonella* for prolonged periods of time in foods stored at freezer and ambient temperatures is well documented (116). It is noteworthy that the composition of the freezing menstruum, the kinetics of the freezing process, the physiological state of foodborne salmonellae, and the serovar-specific responses to extreme temperatures determine the fate of salmonellae during freezer storage of foods (101). The viability of salmonellae in dry foods stored at $\geq 25°C$ decreases with increasing storage temperature and with increasing moisture content (102, 116).

Heat is widely used in food manufacturing processes to control the bacterial quality and safety of end products. Factors that potentiate the greater heat resistance of *Salmonella* and other foodborne bacterial pathogens in food ingredients and finished products have been studied extensively (102). Although the heat resistance of *Salmonella* spp. increases as the a_w of the heating menstruum decreases, detailed studies have revealed that the nature of solutes used to alter the a_w of the heating menstruum plays a determinant role in the level of acquired heat resistance (116). For example, the heating of serovar Typhimurium in menstrua adjusted to an a_w of 0.90 with sucrose and glycerol conferred different levels of heat resistance, as evidenced by $D_{57.2}$ values of 40 to 55 min and 1.8 to 8.3 min, respectively (191). The *D*-value represents the amount of time required to effect a 90% decrease (1.0 log_{10} reduction) in the number of viable cells upon heating at a specified constant temperature. Other important features associated with this adaptive response include the greater heat resistance of salmonellae grown in nutritionally rich than in minimal media, of cells derived from stationary- rather than logarithmic-phase cultures, and of salmonellae previously stored in a dry environment (191, 268, 343). The ability of *Salmonella* to acquire greater heat resistance following exposure to sublethal temperatures is equally notable. The phenomenon stems from a rapid adaptation of the bacterium to rising temperatures in the

microenvironment to a level of enhanced thermotolerance quite distinct from that described by conventional time-temperature curves of thermal lethality. This adaptive response has potentially serious implications in the safety of thermal processes that expose or maintain food products at marginally lethal temperatures. Exposure of salmonellae to sublethal temperatures ($\leq 50°C$) for 15 to 30 min enhances their heat resistance through a rapid chloramphenicol-sensitive synthesis of heat shock proteins (240, 303, 304). Changes in the fatty acid composition of cell membranes in heat-stressed *Salmonella* cells to provide a greater proportion of saturated membrane phospholipids reduce the fluidity of the bacterial cell membranes with an attendant increase in membrane resistance to heat damage (240). The likelihood that other protective cellular functions are triggered by heat shock stimuli cannot be discounted.

The complexities of the foregoing considerations can best be illustrated by the following scenario involving *Salmonella* contamination of a chocolate confectionery product. Experience has shown that the survival of salmonellae in dry-roasted cocoa beans can lead to contamination of in-line and finished products. Thermal inactivation of salmonellae in molten chocolate is most difficult because the time-temperature conditions that would be required to effectively eliminate the pathogen in this sucrose-containing product of low a_w would likely result in an organoleptically unacceptable product. The problem is further compounded by the ability of salmonellae to survive for many years in the finished product when stored at ambient temperature (114). Clearly, effective decontamination of raw cocoa beans and stringent in-plant control measures to prevent cross-contamination of inline products are of prime importance to this food industry.

Brief exposure of serovar Typhimurium cells to mild acidic environments of pH 5.5 to 6.0 (preshock) followed by exposure of the adapted cells to a pH of ≤ 4.5 (acid shock) triggers a complex acid tolerance response (ATR) that potentiates the survival of the microorganism in extreme acidic environments (pH 3.0 to 4.0). The response translates into an induced synthesis of 43 acid shock and outer membrane proteins, a reduced growth rate, and pH homeostasis, as demonstrated by the bacterial maintenance of internal pH values of 7.0 to 7.1 and 5.0 to 5.5 upon sequential exposure of cells to external pHs of 5.0 and 3.3, respectively (164, 228). In the ATR, a bacterial Mg^{2+}-dependent proton-translocating ATPase encoded by the *atp* operon plays an important role in maintaining the cellular pH at ≥ 5.0 through the energy-dependent transport of intracellular protons to the cell exterior (164). Other transmembrane mechanisms that putatively operate in pH homeostasis include the H^+-coupled ion transport systems (antiport) for the intra- and extracellular transfer of K^+, Na^+, and H^+ ions, the electron transport chain-dependent efflux of H^+, and transport systems committed to the symport of H^+ and solutes (353). The Fe^{2+}-binding regulatory protein encoded by the *fur* (ferric uptake regulator) gene also impacts bacterial acid tolerance, as evidenced by the inability of *fur* mutant strains to survive under highly acidic conditions (175). Acid-induced activation of amino acid decarboxylases in *Salmonella* provides an additional protective mechanism whereby cadaverine and putrescine, from the enzymatic breakdown of lysine and ornithine, respectively, potentiate acid neutralization and enhanced bacterial survival (353). Further characterization of the *Salmonella* response to acid stress has led to the identification of two additional protective mechanisms that operate in salmonellae in the stationary phase of growth and which are distinct from the previously discussed ATR, which prevails in log-phase cells (164, 280). One of these pH-dependent responses, designated stationary-phase ATR, provides greater acid resistance than the log-phase ATR and is induced at pHs of <5.5 and functions maximally at pH 4.3. The stationary-phase ATR, which induces the synthesis of only 15 acid shock proteins, is not affected by mutations in the *atp* and *fur* genes and is *rpoS* independent (280). The induction of the remaining acid-protective mechanism associated with *Salmonella* in the stationary phase is independent of the external pH and dependent on the alternative sigma factor (σ^S) encoded by the *rpoS* locus. The mechanism seemingly reinforces the ability of stationary-phase cells to survive under hostile environmental conditions. We have thus seen that three possibly overlapping cellular systems confer acid tolerance in *Salmonella* spp. These include (i) the pH-dependent, *rpoS*-independent log-phase ATR, (ii) the pH-dependent and *rpoS*-independent stationary-phase ATR, and (iii) pH-independent, *rpoS*-dependent stationary-phase acid resistance. These systems likely operate in the acidic environments that prevail in fermented and acidified foods and in phagocytic cells of the infected host.

Acid stress can also trigger enhanced bacterial resistance to other adverse environmental conditions. The growth of serovar Typhimurium at pH 5.8 engendered increased thermal resistance at 50°C, enhanced tolerance to high osmotic stress (2.5 M NaCl) ascribed to the induced synthesis of the OmpC outer membrane proteins, a greater surface hydrophobicity, and increased resistance to the antibacterial lactoperoxidase system and surface-active agents such as crystal violet and polymyxin B (293).

Reservoirs

The widespread occurrence of *Salmonella* spp. in the natural environment, coupled with the intensive husbandry practices used in the meat, fish, and shellfish industries and the recycling of offals and inedible raw materials into animal feeds, has favored the continued prominence of this human bacterial pathogen in the global food chain (116, 120). Of the many sectors within the meat industry, poultry meat and eggs remain a predominant reservoir of *Salmonella* spp. in many countries and tend to overshadow the importance of other meats, such as pork, beef, and mutton, as potential vehicles of infection (116, 209, 210, 452, 488). In an effort to actively address the problem of *Salmonella* spp. in meat products, the USDA Food Safety Inspection Service (FSIS) published the "Final Rule on Pathogen Reduction and Hazard Analysis and Critical Control Point (HACCP) Systems" in July 1996. This Final Rule requires the meat and poultry industries to implement HACCP plans in all plants and requires them to systematically sample final products for the indicator organism *Escherichia coli* biotype I. At the same time, FSIS conducts *Salmonella* spp. testing to verify that the implemented HACCP is helping to control *Salmonella* spp. on finished products. Baseline studies prior to the implementation of the Final Rule indicated a *Salmonella* spp. contamination rate of about 24% for U.S. broiler chickens. In 1999, the rate of *Salmonella* spp. contamination on processed broiler chickens had been reduced to about 11%. Major reductions were also observed with other meat animals. A gradual annual increase in *Salmonella* prevalence rates in chickens has been observed since 1999, to a level of approximately 16% in 2005. Future improvement in the *Salmonella* status of meat animals hinges on coordinated and enduring efforts from all sectors of the meat industry towards the implementation of stringent control measures at the farm level and within the processing, distribution, and retailing sectors (488).

The continuing pandemic of human serovar Enteritidis phage type 4 (Europe) and phage type 8 (North America) infections associated with the consumption of raw or lightly cooked shell eggs and egg-containing products further emphasizes the importance of poultry as a vehicle of human salmonellosis and the need for sustained and stringent bacteriological control of poultry husbandry practices. This egg-related pandemic is of particular concern because the problem arises from transovarian transmission of the infective agent into the interior of the egg prior to shell deposition. The viability of internalized serovar Enteritidis cells remains unaffected by the egg surface sanitizing practices currently applied in egg-grading stations. The pandemic engenders significant public health and societal costs and major economic losses to the poultry industry as a result of depopulation of infected layer flocks and mandatory pasteurization of their shell eggs. Vaccination of breeder and egg-laying flocks, together with improved on-farm biosecurity inverventions, has led in recent years to major reductions in the prevalence of serovar Enteritidis in eggs and to reductions in eggborne outbreaks of salmonellosis in the United Kingdom and the United States.

The rapid depletion of feral stocks of fish and shellfish in recent years has greatly increased the importance of the international aquaculture industry as an alternative source for these popular food items. The high-density farming conditions required to maximize biological yields and to satisfy growing market demands open gateways to the widespread infection of species reared in earthen ponds and other unprotected facilities that are continuously exposed to environmental contamination. It is noteworthy that many of the currently available aquaculture products originate from the Asiatic, African, and South American continents. The use of raw meat scraps and offals, of night soil potentially contaminated with typhoid and paratyphoid salmonellae, and of Salmonella-contaminated animal feeds and feces is not uncommon in these geographic areas. Clearly, such husbandry practices favor widespread bacterial contamination during rearing (120). The human health risk associated with the consumption of lightly cooked aquaculture fish or shellfish or a raw sushi dish potentially contaminated with Salmonella spp. cannot be minimized. Moreover, aquaculture products from developing countries could be contaminated with highly resistant salmonellae arising from the use of subtherapeutic levels of antibiotics in rearing ponds to prevent disease and to promote the growth of aquaculture species.

Fruits and vegetables have gained notoriety in recent years as vehicles of human salmonellosis. The situation has developed from the increased global export of fresh and dehydrated fruits and vegetables from countries that enjoy tropical and subtropical climates. The prevailing hygienic conditions during the production, harvesting, and distribution of products in these countries do not always meet minimum standards and will facilitate product contamination. More specifically, the fertilization of crops with untreated sludge or sewage effluents potentially contaminated with antibiotic-resistant Salmonella spp., the irrigation of garden plots and fields and the washing of fruits and vegetables with contaminated waters, the repeated handling of product by local workers, and the propensity for environmental contamination of spices and other condiments during drying in unprotected facilities identify weak production links that undermine food safety. Critical control interventions such as field irrigation with treated effluents,

fertilization of soils with fully composted animal wastes, washing of fruits and vegetables with potable and/or bactericidal wash waters, education of local workers on the hygienic handling of fresh produce, and greater protection of products from environmental contamination during all phases of production and marketing would markedly enhance the bacterial quality and safety of fresh ready-to-eat products (58, 67, 123, 339).

Foodborne Outbreaks

National epidemiologic registries continue to underscore the importance of Salmonella spp. as a leading cause of foodborne bacterial illnesses in humans, where reported incidents of foodborne salmonellosis tend to dwarf those associated with other foodborne pathogens (Table 10.4). Although the incidence of human foodborne salmonellosis seemingly is considerably greater in some countries, national disease statistics should be considered with caution because of differences in the inclusivity and refinement of national reporting systems, in the ability of health agencies to conduct comprehensive and timely epidemiological investigations, and in the reliability of clinical and food analysis data. Worldwide trends in the incidence of human foodborne salmonellosis are revealing (Table 10.5). Several countries have reported decreases in the annual number of incidents in recent years, whereas other countries report little change in the occurrence of foodborne salmonellosis. Major outbreaks of foodborne salmonellosis in the last few decades are of interest because they underscore the multiplicity of foods and Salmonella serovars that have been implicated in human illness (see Tables 10.6 to 10.11).

Dairy products have been incriminated in large outbreaks of human salmonellosis (Table 10.6). Outbreaks of salmonellosis in Australia and Scotland recall the persistence of bacterial disease outbreaks in the United States from the consumption of raw fluid milk, which many consider a health fetish (376). In 1984, Canada experienced a large outbreak of no fewer than 2,700 confirmed cases of serovar Typhimurium PT10 infection from the consumption of Cheddar cheese manufactured from heat-treated and pasteurized milk. Manual override of the flow diversion valve reportedly led to the entry of raw milk into vats of thermized and pasteurized cheese milk (115). The following year witnessed the largest outbreak of salmonellosis in the United States where 16,284 confirmed cases of illness were associated with the consumption of pasteurized fluid milk (278, 402). Although the cause of this outbreak was never ascertained, a cross-connection between raw and pasteurized milk lines seemingly was at fault. In 1994, a major outbreak of foodborne salmonellosis in the United States

Table 10.4 Human epidemiology of foodborne bacterial pathogens

Country	Year(s)	Mean no. of reported incidents[a]								Reference
		Salmonella spp.	Staphylococcus aureus	Clostridium perfringens	Campylobacter spp.	Bacillus cereus	E. coli	Shigella spp.	Listeria monocytogenes	
Australia	1995–2000	15.0	NA	6.0	1.2	NA	NA	NA	1.0	112
Canada	2000	42.0	3.0	17.0	26.0	2.0	27.0[b]	6.0	3.0	211
Denmark	1998	27.0	4.0	5.0	0.0	3.0	0.0	5.0	NA	494
England and Wales	2000	35.0	0.0	4.0	8.0	0.0	6.0	1.0	NA	495
Finland	2000	5.0	1.0	7.0	0.0	2.0	0.0	1.0	0.0	495
France	2000	163.0	45.0	8.0	NA	13.0	NA	2.0	NA	495
Germany	2000	69.0	0.0	1.0	1.0	2.0	1.0[c]	0.0	NA	495
Hungary	2000	155.0	9.0	1.0	17.0	2.0	0.0	0.0	0.0	495
Italy	2000	113.0	9.0	6.0	1.0	1.0	NA	NA	0.0	495
Japan	2002	465.0	72.0	37.0	447.0	NA	96.0	NA	NA	28
The Netherlands	2000	8.0	1.0	8.0	0.0	14.0	1.0[c]	NA	NA	495
Romania	1998	23.0	6.0	1.0	NA	0.0	0.0	1.0	NA	494
Scotland	2000	6.0	0.0	1.0	0.0	NA	1.0	NA	NA	495
Spain	1998	554.0	36.0	22.0	1.0	4.0	12.0	3.0	NA	494
Sweden	2000	11.0	3.0	0.0	4.0	1.0	2.0[d]	0.0	NA	495
United States	1997	60.0	9.0	6.0	2.0	4.0	8.0[c]	10.0	NA	352

[a] Mean annual number of reported incidents, where applicable. NA, not available.
[b] Generic and enterohemorrhagic E. coli O157:H7.
[c] Generic reporting only.
[d] Enterotoxigenic E. coli.

Table 10.5 Trends in human foodborne salmonellosis

Country	Annual no. of reported incidents[a]								No. of cases[b]	References
	1985	1987	1989	1991	1993	1995	1998	2000		
Austria	124	151	440	963	922	873	880	NA	80.8–92.4	489, 490, 494
Bulgaria	15	10	12	13	12	14	NA	7	2.1–31.9	489, 490, 494, 495
Canada	59	53	51	28	43	27	NA	NA	0.5–3.8	122, 209, 210
Denmark	12	5	5	11	7	17	27	NA	27.0–94.6	489, 490, 494
England and Wales	372	421	935	936	136	90	57	35	22.0–45.0	489, 490, 494, 495
Finland	11	3	2	4	3	8	1	5	44.9–153.2	489, 490, 494, 495
France	7	178	462	477	200	185	267	163	21.9–31.9	489, 490, 494, 495
Germany	—[c]	—[c]	23	62	150	117	108	69	103.5–242.7	489, 490, 494, 495
Hungary	116	122	131	110	158	266	269	155	18.0–88.9	489, 490, 494, 495
Japan	NA	90	146	159	NA	NA	757	518	2.9–8.2	28, 442
Poland	380	690	709	625[d]	398	365	325	295	44.1–93.9	489, 490, 494, 495
Scotland	133	180	151	115	NA	NA	11	6	4.0–46.0	489, 490, 494, 495
Sweden	8	7	7	15	8	5	4	11	0.2–7.3	489, 490, 494, 495
United States	79	52	117	122	68	90	NA	NA	1.1–4.3	47, 83, 352

[a] NA, not available.
[b] Range of cases per 100,000 population within the review period (1985–2000).
[c] German unification occurred in 1989.
[d] Estimated from graphic representation of data.

was associated with ice cream contaminated with serovar Enteritidis. The transportation of pasteurized ice cream mix in an unsanitized truck that had previously carried raw eggs was identified as the source of contamination (20, 222). In 1998, a nationwide outbreak of serovar Enteritidis PT8 in Canada was associated with contaminated Cheddar cheese included in retail luncheon packages consumed primarily by children.

In the last decade, fresh fruits and vegetables figured prominently as vehicles of human salmonellosis (Table 10.7). Many factors contributed to this disquieting situation, including the perishable nature of these products, which generally preempts their analysis for human

bacterial pathogens prior to marketing. Moreover, the growing popularity of convenience foods, such as pre-cut and prepackaged produce, introduces new public health concerns because damaged plant tissues release nutrients and provide a favorable matrix for bacterial proliferation. Repeated outbreaks of human salmonellosis from fresh tomatoes, lettuce, mixed salads, bean and alfalfa sprouts, raw almonds, cantaloupes, and orange juice (Table 10.7) underscore the major challenge to the produce industry and to government regulatory agencies in the implementation and consistent application of stringent on-farm pathogen control measures. *Salmonella* contamination of fruits and vegetables could arise

Table 10.6 Examples of major foodborne outbreaks of human salmonellosis from dairy products

Year	Country	Vehicle	Serovar	No.		Reference(s)
				Cases[a]	Deaths	
1973	Trinidad	Milk powder	Derby	3,000[b]	NS[c]	475
1976	Australia	Raw milk	Typhimurium PT9	>500	NS[c]	414
1981	Scotland	Raw milk	Typhimurium PT204	654	2	95
1984	Canada	Cheddar cheese	Typhimurium PT10	2,700	0	115
1985	United States	Pasteurized milk	Typhimurium	16,284	7	278
1994	United States	Ice cream	Enteritidis PT8	740	0	20, 222
1998	Canada	Cheddar cheese	Enteritidis PT8	700	0	388

[a] Confirmed cases, unless stated otherwise.
[b] Estimated number of cases.
[c] NS, not specified.

Table 10.7 Examples of major foodborne outbreaks of human salmonellosis from fruits and vegetables

Year(s)	Country	Vehicle	Serovar	No. Cases[a]	No. Deaths	Reference(s)
1981	The Netherlands	Salad base	Indiana	600[b]	0	49
1984	United States	Salad bars	Typhimurium	751	0	453
1991	United States and Canada	Cantaloupes	Poona	>400	NS[c]	166
1991	Germany	Fruit soup	Enteritidis	600	NS	178
1993	United States	Tomatoes	Montevideo	100	0	212
1994	Finland and Sweden	Alfalfa sprouts	Bovismorbificans	492	0	370
1995	United States	Orange juice	Hartford Gaminara	62	0	83, 359
1996	United States	Alfalfa sprouts	Montevideo Meleagridis	481	1	331
1998	United States	Toasted oat cereal	Agona	209	0	83
1999	Australia	Orange juice	Typhimurium	427	NS	37
1999	Canada	Alfalfa sprouts	Paratyphi B (Java)	>53	NS	193
1999	Japan	Peanut sauce	Enteritidis PT1	644	NS	251
1999	United States and Canada	Orange juice	Muenchen	>220	0	63
1999	United States	Tomatoes	Baildon	86	3	110
1999	United States and Brazil	Mangoes	Newport	78	2	419
2000	United States	Orange juice	Enteritidis	>74	0	77
2000	United States	Mung bean sprouts	Enteritidis	>45	0	22
2000	The Netherlands	Bean sprouts	Enteritidis PT4b	12	0	465
2000	Europe (five countries)	Lettuce	Typhimurium PT204b	>392	0	105
2000–2001	Canada and United States	Raw almonds	Enteritidis PT30	168	0	245
2000–2002	United States and Canada	Cantaloupe	Poona	155	1	13
2001	United States	Alfalfa sprouts	Kottbus	31	NS	485
2001	Canada and Australia	Shandong peanuts	Stanley	93	NS	23
2001	Europe and Australia	Halva	Typhimurium DT104	>70	NS	24, 25
2002	United States	Roma tomatoes	Javiana	159	0	455
2002–2003	Australia and New Zealand	Tahini	Montevideo	68	NS	460
2003	Germany	Aniseed herbal tea	Agona	42	0	271
2003–2004	United States and Canada	Raw almonds	Enteritidis	29	0	257
2004	United States and Canada	Roma tomatoes	Braenderup	125	0	100
			Javiana	390	0	
			Typhimurium	27	0	
			Anatum	5	0	
			Thompson	4	0	
			Muenchen	4	0	
2004	United Kingdom	Lettuce	Newport	>372	0	183
2005	Austria	Mixed salad	Enteritidis PT21	85	0	410
2005	Finland	Iceberg lettuce	Typhimurium var. Copenhagen DT104b	60	0	441

[a] Confirmed cases, unless stated otherwise.
[b] Estimated number of cases.
[c] NS, not specified.

from the entry of pathogens through scar tissues, from the natural uptake of pathogens through root systems, from surface contamination of flowering plants and subsequent entrapment of the pathogen during embryogenesis of the fruit or vegetable, and from the transfer of surface contaminants onto edible plant tissues during slicing or into the juice of freshly pressed fruits (205, 297, 339). The human health risk is increased further by the propensity of salmonellae to actively grow on cut tomatoes and melons, in precut salad mixes, and on fresh produce moisturized during retail display at ambient temperatures (122). Worldwide reports of the presence of *Salmonella* spp.

in sesame seed products, notably tahini and halva, and attendant recalls of adulterated products imported primarily from Mid-Eastern countries in the last 2 decades (122, 496) finally culminated in confirmed international outbreaks of serovar Typhimurium DT104 and serovar Montevideo in 2001 and 2002, respectively (Table 10.7). Scant reports of human salmonellosis from the consumption of tahini or halva likely stem from single or few cases of illness among family members that are not reported to health agencies. The rash of bacterial disease outbreaks from the consumption of sprouted seeds in recent years (46, 340) likely resulted from the germination of internally contaminated seeds at ambient temperatures under high-moisture conditions and the concurrent growth of salmonellae on the leguminous sprouts. *Salmonella* spp. are most likely localized under the seed coat, as suggested by the difficulty in isolating salmonellae from intact seeds and the need to germinate seeds for successful cultural isolation of the pathogen. The inability of bactericidal agents to effectively eliminate salmonellae from intact seeds further points to an internal localization of *Salmonella* spp. (46, 340).

The ubiquity of *Salmonella* spp. in the natural environment, the intense husbandry practices associated with the rearing of meat animals, the provision of *Salmonella*-contaminated animal feeds to reared species, and the propensity for the cross-contamination of animals in feed lots, in holding pens, and during animal slaughter and processing of carcasses are factors that could contribute to the presence of salmonellae in raw meat products (122). In 1984, liver pâté manufactured in France was incriminated as the cause of no fewer than 506 and 250 cases of infection with serovar Goldcoast in France and in England and Wales, respectively (Table 10.8). Localization of the pathogen within the external gelatin layer of the pâté suggested that product adulteration resulted from product handling by plant workers (64). In the same year, gelatin glaze was incriminated in another international outbreak of 866 cases of serovar Enteritidis PT4 infection among airline passengers, flight crews, ground personnel, and food catering staff who had consumed glazed canapés and appetizers prepared for a major European airline. The incident was tentatively ascribed to cross-contamination of the aspic glaze by infected food handlers in the flight kitchen. Fortunately, no serious airline mishaps resulted from this episode other than flight delays and the grounding of a Concorde aircraft in Washington, D.C., because the entire crew was incapacitated by salmonellosis and a replacement crew was not readily available. In 1997, numerous participants at a church fundraising dinner were infected with serovar Heidelberg, where undercooking and slow cooling of stuffed ham and the use of a single unsanitized mechanical meat slicer contributed to the magnitude of the outbreak. A large protracted outbreak of serovar Bovismorbificans PT24 in Germany was putatively ascribed to the

Table 10.8 Examples of major foodborne outbreaks of human salmonellosis from meats and meat products

Year(s)	Country	Vehicle	Serovar	No. Cases[a]	No. Deaths	Reference(s)
1984	France and England	Liver pâté	Goldcoast	756	0	64, 358
1984	International	Aspic glaze	Enteritidis PT4	866	2	75
1997	United States	Stuffed ham	Heidelberg	746	1	50
2000	United States	Hamburger buns/ infected worker	Thompson	55	0	266
2001	United States	Pork	Uganda	24	NS[b]	253
2003	United States	Ground beef	Typhimurium DT104	58	0	127
2004	United States	Ground beef	Typhimurium	>31	0	30
2004	Germany	Raw minced pork	Give	115	NS	250
2004	United States	Roast beef	*Salmonella* spp.	28	NS	109
2004–2005	Germany	Raw pork	Bovismorbificans PT24	402	1	185
2005	Canada	Deli meats	Typhimurium PTU302	55	0	31
2005	The Netherlands	Imported raw beef	Typhimurium PT104	165	0	269
2005	Honduras	Cooked chicken	*Salmonella* spp.	>600	NS	161
2005	Spain	Cooked chicken	Hadar	2,138	1	289
2005	United States	Cooked turkey	Enteritidis	>304	1	162
2005	Canada	Roast beef	*Salmonella* spp.	155	0	29
2005	England	Kebab	Enteritidis PT1	195	NS	12

[a] Confirmed cases, unless stated otherwise.
[b] NS, not specified.

Table 10.9 Examples of major foodborne outbreaks of human salmonellosis from fish and fish products

YEAR	Country	Vehicle	Serovar	No. Cases[a]	No. Deaths	Reference
1988	Japan	Cuttlefish	Champaign	330	0	350
1999	Japan	Dried squid	*Salmonella* spp.	>453	0	21
1999	Japan	Cuttlefish chips	Chester Oranienburg	≥1,500	NS[b]	458
2001	Norway and Sweden	Fish	Livingstone	60	3	197

[a] Confirmed cases, unless stated otherwise.
[b] NS, not specified.

consumption of raw minced pork and fermented raw pork sausages. In 2005, undercooked chicken was incriminated as the vehicle of infection of 600 participants in a political campaign in Honduras. In the same year, more than 2,000 human cases of serovar Hadar PT2 infection resulted from the consumption of a single nationally distributed brand of roasted, vacuum-packaged chicken in Spain; the incident resulted in the hospitalization of more than 200 infected patients and in one fatality. In 2005, undercooked turkey served to restaurant patrons resulted in more than 300 cases of serovar Enteritidis infection in the United States.

Major outbreaks of salmonellosis associated with fish and shellfish occurred mainly in Japan, where marine species are a food staple (Table 10.9). The consumption of cuttlefish, a cephalopod mollusc, which had been left to thaw at room temperature for up to 30 h and then boiled

for only a short period of time was incriminated in 330 infections of serovar Champaign in children. In 1999, more than 400 cases of salmonellosis in Japan were associated with dried squid from a processing plant whose supply of well water was contaminated. In the same year, cuttlefish chip snacks contaminated with serovar Chester and serovar Oranienburg were implicated in an outbreak involving more than 1,500 human cases of illness in Japan. Consumption of fish gratin manufactured in Sweden and marketed as a frozen product in Norway and Sweden resulted in 60 cases of serovar Livingstone infection and three fatalities in medically challenged elderly persons. Indirect evidence suggested that the egg powder ingredient for the fish gratin was the source of contamination.

The ongoing pandemic of egg-borne serovar Enteritidis continues to impact the incidence of foodborne outbreaks worldwide (Table 10.10). A large Swedish outbreak of

Table 10.10 Examples of major foodborne outbreaks of human salmonellosis from eggs and egg products

YEAR	Country	Vehicle	Serovar	No. Cases[a]	No. Deaths	Reference
1974	United States	Potato salad	Newport	3,400[b]	0	235
1976	Spain	Egg salad	Typhimurium	702	6	16
1977	Sweden	Mustard dressing	Enteritidis PT4	2,865	0	218
1987	People's Republic of China	Egg drink	Typhimurium	1,113	NS[c]	499
1988	Japan	Cooked eggs	*Salmonella* spp.	10,476	NS	326
1993	France	Mayonnaise	Enteritidis	751	0	392
2001	United States	Tuna salad with eggs	Enteritidis	688	0	136
2001	Latvia	Cake/raw egg sauce	Enteritidis PT4	19	0	248
2002	Spain	Custard-filled pastry	Enteritidis PT6	1,435	0	78
2002	England	Bakery products	Enteritidis PT14b	>150	1	468
2003	England, Wales, and Scotland	Egg sandwiches	Bareilly	186	NS	103
2003	Australia	Raw egg mayonnaise	*Salmonella* spp.	>106	1	160
2003	United States	Egg salad kit	Typhimurium	18	0	258
2004	People's Republic of China	Cake/raw egg topping	Enteritidis	197	NS	298
2005	England	Imported shell eggs	Enteritidis PT6	68	0	99

[a] Confirmed cases, unless stated otherwise.
[b] Estimated number of cases.
[c] NS, not specified.

serovar Enteritidis PT4 in 1977 was attributed to the consumption of a mayonnaise dressing prepared in a central kitchen and distributed to school cafeterias in Stockholm. Although no fatalities were associated with this episode, 50 of the 80 patients requiring hospitalization suffered from rheumatoid sequelae. In 1993, school cafeterias in Douai, France, were implicated in a large outbreak of serovar Enteritidis infections among 751 schoolchildren, teachers, and support staff who had consumed tuna salad prepared with raw egg mayonnaise. In 2001, consumption of tuna salad and hardboiled eggs contaminated with serovar Enteritidis PT2, PT13a, and PT23 infected 688 inmates in four prison facilities in South Carolina. In the following year, pastries filled with vanilla cream were incriminated as the vehicle of 1,435 cases of serovar Enteritidis PT6 infection in the province of Gerona, Spain. Cross-contamination of the supply bakery by fresh shell eggs was identified as the probable cause of the outbreak. In 2002, the use of fresh shell eggs in the preparation of uncooked bakery products in Cheshire, England, led to more than 150 cases of infection with serovar Enteritidis PT14b. More recently, a cake pastry topped with a mixture containing raw eggs resulted in 197 human cases of serovar Enteritidis infection in the People's Republic of China. In 2005, several catering services in northeast England were associated with a total of 68 cases of serovar Enteritidis PT6 infection. Shell eggs imported from The Netherlands were suspected as the source of infection. Other serovars have been involved in eggborne outbreaks. In 1974, temperature abuse of egg-containing potato salad served at an outdoor barbecue led to an estimated 3,400 human cases of serovar Newport infection, where cross-contamination of the salad by an infected food handler was suspected. In 1987–1988, large outbreaks of serovar Typhimurium and of *Salmonella* spp.

linked to an egg drink and to a cooked egg dish were reported in China and Japan, respectively. Prepackaged egg sandwiches were identified as the vehicle of serovar Bareilly infections in 186 consumers from England, Wales, and Scotland. Meat rolls prepared with raw eggs were implicated in a restaurant outbreak of more than 100 cases of salmonellosis in Melbourne, Australia.

In 1973, milk chocolate manufactured in Canada was implicated in more than 200 cases of serovar Eastbourne infection in Canada and the United States (Table 10.11). Dry roasting of imported cocoa beans and environmental contamination of the manufacturing plant with dust from the raw cocoa bean storage rooms contributed to the adulteration of finished products. Additional episodes of human salmonellosis involving chocolate products are noteworthy. In 1982, 245 cases of serovar Napoli infection resulted from the consumption of chocolate bars imported from Italy. In 1987, Norway experienced an outbreak of 349 cases of serovar Typhimurium infection associated with a variety of domestic chocolate products manufactured by a single plant in Trondheim. Chocolate bars manufactured by this company and exported to Finland were also incriminated as the cause of an additional 12 cases of salmonellosis. More recently, chocolate products from a large firm in Germany were identified as the vehicle of 439 cases of serovar Oranienburg infection in Germany and of additional cases in several European countries and, probably, Canada. In 1993, paprika imported from South America was incriminated as the contaminated ingredient used in the manufacture of potato chips distributed in Germany. The consumption of foods likely cross-contaminated by raw poultry in a Florida restaurant kitchen affected no fewer than 850 diners and staff, who suffered gastrointestinal symptoms from serovar Newport. In 2001, more than 250 salmonellosis patients, including 194 children, in northern

Table 10.11 Examples of major foodborne outbreaks of human salmonellosis from other products

| Year(s) | Country | Vehicle | *Salmonella* serovar | No. | | Reference(s) |
				Cases[a]	Deaths	
1973	Canada and United States	Chocolate	Eastbourne	217	0	104, 113
1982	England and Wales	Chocolate	Napoli	245	0	182
1987	Norway and Finland	Chocolate	Typhimurium	361	0	255
1993	Germany	Paprika chips	Saintpaul	>670	0	283
			Javiana			
			Rubislaw			
1995	United States	Restaurant foods	Newport	850	0	19
2000–2001	Germany and International	Chocolate	Oranienburg	>439	NS[b]	476
2001	Romania	Pastries	*Salmonella* spp.	>250	0	159

[a] Confirmed cases, unless stated otherwise.
[b] NS, not specified.

Romania were infected following the consumption of *Salmonella*-tainted pastries.

CHARACTERISTICS OF THE DISEASE

Symptoms and Treatment

Human *Salmonella* infections can lead to several clinical conditions, including enteric (typhoid) fever, uncomplicated enterocolitis, and systemic infections by nontyphoid microorganisms. Enteric fever is a serious human disease associated with the typhoid and paratyphoid strains, which are particularly well adapted for invasion and survival within host tissues. Clinical manifestations of enteric fever appear after a period of incubation ranging from 7 to 28 days and may include diarrhea, prolonged and spiking fever, abdominal pain, headache, and prostration (117). Diagnosis of the disease relies on the isolation of the infective agent from blood or urine samples in the early stages of the disease or from stools after the onset of clinical symptoms (116). An asymptomatic chronic carrier state commonly follows the acute phase of the disease. The treatment of enteric fever is based on supportive therapy and/or the administration of chloramphenicol, ampicillin, or trimethoprim-sulfamethoxazole. Marked global increases in the resistance of typhoid and paratyphoid microorganisms to these antibacterial drugs have greatly undermined their efficacy in human therapy. The problem is particularly serious in developing countries, where multiply antibiotic-resistant salmonellae are frequently implicated in outbreaks of enteric fever and are the cause of unusually high fatality rates (80, 382). The widespread use of fluoroquinolones in these countries is hampered by the high cost of these drugs.

In response to the current era of antibiotic-resistant superbugs, there is a growing interest worldwide in the use of phage therapy for the clinical management of severe human infections with multiple antibiotic-resistant bacterial pathogens and for veterinary purposes (431). The potential of bacteriophages for the treatment of human diseases was first recognized by Félix d'Herelle in 1919. Subsequently, the Elivia Institute of Bacteriophage, Microbiology and Virology (Tbilisi, Soviet Republic of Georgia) promoted the use of bacteriophage therapy in several European countries. Prophylactic and therapeutic bacteriophage preparations are currently marketed in several European countries. In 2006, the U.S. Food and Drug Administration approved a bacteriophage mixture marketed by Intralytix Inc. (Baltimore, Md.) for the control of *Listeria monocytogenes* in deli meats and in other ready-to-eat foods. The potentially high affinity and specificity of a phage for a selected bacterial target is attractive and could provide a useful adjunct to antimicrobial agents for the effective treatment of tenacious human bacterial infections without disrupting the microbiological balance in the intestinal tract of the host.

Human infections with nontyphoid *Salmonella* commonly result in enterocolitis, which appears 8 to 72 h after contact with the invasive pathogen. The clinical condition is generally self-limiting, and remission of the characteristic nonbloody diarrheal stools and abdominal pain usually occurs within 5 days of the onset of symptoms. The successful treatment of uncomplicated cases of enterocolitis may only require supportive therapy, such as fluid and electrolyte replacement. The use of antibiotics in such episodes is contraindicated because it tends to prolong the carrier state and the intermittent excretion of salmonellae (116). This asymptomatic persistence of salmonellae in the gut likely results from a marked antibiotic-dependent repression of the native gut microflora, which normally competes with *Salmonella* for nutrients and intestinal binding sites. Human infections with nontyphoid strains can also degenerate into systemic infections and precipitate various chronic conditions. In addition to serovars Dublin and Choleraesuis, which exhibit a predilection towards septicemia, similarly high levels of virulence have been observed with other nontyphoid strains. Preexisting physiological, anatomical, and immunological disorders in human hosts can also favor severe and protracted illness through the inability of host defense mechanisms to respond effectively to the presence of invasive salmonellae (116). More frequent reports in recent years on the chronic and debilitating sequelae of nontyphoid systemic infections are of concern because this emerging medical pattern may be linked to increased levels of virulence among nontyphoid salmonellae, increased susceptibility of human populations to chronic bacterial diseases, or synergy between both factors (91).

Salmonella-induced chronic conditions, such as aseptic reactive arthritis, Reiter's syndrome, and ankylosing spondylitis, are noteworthy. Bacterial prerequisites for the onset of these chronic diseases include the ability of the bacterial strain to infect mucosal surfaces, the presence of outer membrane LPS, and a propensity to invade host cells (74, 423). Recent evidence suggests that these arthropathies may be linked to a genetic predisposition in individuals that carry the class I HLA-B27 histocompatibility antigen. Other antigens encoded by the HLA-B locus, such as the B7, B22, B40, and B60 antigens, which serologically cross-react with B27 antiserum, have also been associated with reactive arthritis (74, 313, 445). It is disconcerting that therapeutic eradication of a human *Salmonella* infection in the intestinal tract does not preclude the subsequent onset of chronic rheumatoid diseases in a distal, noninfected limb (445).

The major global impact of typhoid and paratyphoid salmonellae on human health led to the early development of parenteral vaccines consisting of heat-, alcohol-, or acetone-killed cells (135). The phenol-heat-killed typhoid vaccine, which continues to be used widely for protective immunity, can precipitate adverse reactions, including fever, headache, pain, and swelling at the site of injection. Although the need for immunogenic preparations against nontyphoid *Salmonella* could be argued, the development of prophylactics against the multiplicity of serovars and bacterial surface antigens in the global ecosystem, the rapid succession of serovars in human populations and in reservoirs of infection, and the unpredictable pathogenicity of infective strains have dampened research interest and hampered significant advances in this field of clinical sciences. Nevertheless, the pandemic of poultry- and eggborne serovar Enteritidis infections that continue to afflict consumers in many countries underscores the potential benefit of specific vaccines for human and veterinary applications (116, 311, 397).

Live attenuated vaccines continue to generate great research interest because such preparations induce strong and durable humoral and cell-mediated responses in vaccinees (86, 111). The identification of bacterial genes wherein mutations lead to the attenuation of the carrier strain has greatly accelerated the development of live oral vaccines (86, 135). The avirulent mutant strain of serovar Typhi (Ty21a) is a product of chemical mutagenesis for which the underlying molecular basis for attenuation is poorly understood. The absence of Vi antigens in the Ty21a strain and mutation in the *galE* locus, which significantly reduces the number of biosynthetic LPS enzymes, have been proposed as determinants for attenuation (86, 135). A mutation in the *rpoS* gene also contributes to the safety of this strain for humans (395). The *rpoS* gene, a component of the *Salmonella* virulence arsenal, engages the synthesis of many protective proteins in response to environmental stress conditions. The *rpoS* mutation in Ty21a suggestively is linked to a single nucleotide change in genomic DNA and results in a frameshift and in gene transcription into an unstable RNA polymerase designated the σ^s factor (395). The Ty21a vaccine is manufactured in the form of enteric capsules by the Swiss Serum and Vaccine Institute (Berne, Switzerland) under the trade name Vivotif Berna. The preparation, which has successfully met the challenge of intense field testing in Chile and Egypt (60 and 96% efficacies, respectively), currently enjoys wide usage in European and North American medical communities. Immunization results from the administration of three capsules on alternate days and provides up to 3 years of protection.

An injectable vaccine prepared from the Vi capsular polysaccharide antigen of serovar Typhi was released by Pasteur Mérieux Sérums et Vaccins (Lyon, France) under the trade name Typhim Vi. A single dose of phenol-preserved antigen provides a rapid rise in serum levels of Vi antibodies and confers protection for up to 3 years. In the presence of an active typhoid infection, the Vi antibodies facilitate the bactericidal action of serum complement (classical pathway). Two large-scale clinical trials, in Nepal and Eastern Transvaal (South Africa), confirmed the vaccine efficacy, which ranged from 65 to 75%, with low levels of adverse reactions (259). In contrast to the Ty21a vaccine, whose heterogeneous molecular composition predisposes recipients to stronger and varied secondary reactions, the use of a single, chemically defined antigen in Typhim Vi minimizes the risks of undesirable side effects. Developmental research may soon culminate in the marketing of new typhoid vaccines exhibiting greater levels of immunogenicity, biological safety, and storage stability (135). For example, a temperature-sensitive (*ts*) strain of serovar Typhi that grows optimally at 29°C but whose proliferation ceases at 37°C shows great potential as an immunogen in a novel live oral typhoid vaccine (53). Research at the National Institutes of Health (Bethesda, Md.) has also yielded promising results on the development of a conjugate vaccine consisting of O-acetylated pectin linked to a tetanus toxoid that raises the level of typhoid antibodies in mice (435). This preparation is of particular interest as a human prophylactic because it is strongly immunogenic, is divalent, and carries no endotoxic moiety. Although it is beyond the scope of the present dissertation, another rapidly evolving and productive field of research involves the use of attenuated *Salmonella* strains as vectors for the delivery of heterologous antigens to mammalian immune systems (86, 135). For example, the administration of a live attenuated *Salmonella* strain harboring the gene encoding fragment C of the tetanus toxin would provoke a potent immunological response against the carrier *Salmonella* strain and the tetanus toxin (135). Clearly, the field of vaccine development has witnessed great advances in recent years, and such success will undoubtedly continue as new knowledge on the molecular complexities of bacterial attenuation facilitates the delivery of defined, effective, and stable prophylactic preparations. The subject of live attenuated vaccines has been reviewed thoroughly (418). A U.S. Food and Drug Administration (FDA)-approved (1999) live avirulent serovar Typhimurium vaccine has been developed (207) for use in broiler and breeder chickens, whereas serovar

Enteritidis vaccines are used extensively in the European egg-layer industry to abate the problem of serovar Enteritidis infections.

Antibiotic Resistance

Antibiotic resistance in *Salmonella* spp. has been reported since the early 1960s (73, 466), when most of the reported resistance was to a single antibiotic (73, 88). However, in the mid-1970s, multiple-drug-resistant (MDR) *Salmonella* emerged. In 1979, a clone of serovar Typhimurium with chromosomally encoded and integron-mediated resistance to chloramphenicol, streptomycin, sulfonamides, and tetracycline moved through the food chain from calves to humans (399). Other MDR clones were subsequently identified, with some of these clones appearing to be geographically localized, whereas others, such as serovar Typhimurium DT204, DT204c, and DT193, were encountered in several countries (399, 447, 448). The appearance of serovar Typhimurium DT104 in the late 1980s raised major concerns because of its common pentaresistance to ampicillin, chloramphenicol, streptomycin, sulfonamides, and tetracycline. Resistance of this phagovar to additional antibiotics such as trimethoprim, nalidixic acid, and spectinomycin, coupled with a reduced sensitivity to ciprofloxacin, has been reported (449, 451). Although DT104 was originally detected in cattle (300, 471), this clone was later identified in other livestock species, companion animals, and wildlife (498). DT104 has been at the forefront of discussions concerning the impact of agricultural use of antibiotics on public health because of its dual resistance to chloramphenicol and the veterinary analog florfenicol, which is currently approved for use in cattle. While MDR serovar Typhimurium DT104 continues to cause serious foodborne outbreaks worldwide (219), an MDR serovar Newport has emerged which is resistant to β-lactams, the β-lactam inhibitor clavulanic acid, and cephalosporin as well as several other classes of antibiotics (27, 152). While DT104 was still susceptible to antibiotics generally prescribed to treat gastrointestinal illnesses, with a few notable exceptions (412), emerging MDR *Salmonella* strains were resistant to broad-spectrum cephalosporins currently prescribed to treat *Salmonella* infections in children (27, 152, 206). Like DT104, these MDR *Salmonella* strains, most notably serovar Newport strains, were first reported in cattle (27, 152, 206) and have since been found in other animal species (472, 483, 484, 502). Why the cephalosporin resistance phenotype appears to be more widespread among MDR *S. enterica* serovars other than DT104 may be attributed to where the antibiotic resistance locus resides in the genome,

i.e., in the plasmid versus the chromosome, as discussed in more detail below. It has been argued, with DT104 (190), that the emergence of MDR serovar Newport is linked to the veterinary use of antibiotics, specifically ceftiofur, a cephalosporin currently approved for use in cattle and poultry (152). Since many of the MDR serovar Newport isolates are also resistant to florfenicol, early use of this antibiotic may have contributed to the emergence of this MDR *Salmonella* serovar (131).

With the exception of quinolone/fluoroquinolone resistance, antibiotic resistance in *Salmonella* is attributed to the acquisition of foreign genes that encode enzymes to destroy, chemically inactivate, or "pump" the noxious drug out of the bacterial cell or provide an alternative pathway to the one targeted by the antibiotic. Depending on the drug resistance gene, the enzyme's specificity ranges from narrow to broad. These antibiotic resistance genes often reside on mobile genetic elements, including plasmids, transposons, and integrons that can potentially ferry the resistance from a commensal to the pathogen (261). One genetic element in particular, the integron, appears to be especially culpable in the development and dissemination of MDR for many microorganisms, including *Salmonella*. An integron is a genetic element capable of capturing, combining, or swapping a large assortment of antibiotic resistance genes through recognition of a 59-bp element (be) present in "selected" genes by its recombinase IntI and then integrating the captured gene into its resident integration site, *attI*. This genetic element can create a tandem of antibiotic resistance genes (430). There have been as many as seven gene cassettes reported for a single integration site of an integron (336). There are at least eight classes of integrons in nature, with the two newest classes coming from analyses of environmental DNA (345). Integrons are widespread in nature (336, 345) and, as reported in one study, constitute a "sizeable" reservoir for antibiotic resistance genes (336). Among the eight classes of integrons, class 1 integrons are the most studied and prevalent among human and veterinary pathogens (192, 215, 415). In addition to the basic genetic structure outlined above, the class 1 integrons also possess a truncated quaternary ammonium resistance gene, *qacΔE* (361), and the sulfonamide resistance gene *sul1*, 3' of the *attI* site and its resident drug resistance cassette(s) (430). The resistance genes associated with class 1 integrons confer resistance to a diverse array of antibiotics and disinfectants, including aminoglycosides, β-lactams, chloramphenicol, macrolides, quaternary ammonium, and trimethoprim (155). Tetracycline resistance genes are the only drug resistance genes that have not been identified among the myriad of class 1 integron cassettes. Class 1 integrons occur in various *S. enterica* serovars; have been detected in serovars isolated from humans (346, 383), animals (192, 383, 502), and foods (478); and possess a vast assortment of drug resistance genes (346, 383, 502). The integron itself is not mobile, but it often resides in mobile plasmids or transposons, such as the mercury resistance transposon Tn*21*, which can "hop" into conjugative/mobile plasmids (295).

The class 1 integron is also the backbone of the MDR genomic island, SGI1, which is present in serovar Typhimurium DT104. Two integrons flank the phenicol resistance gene and tetracycline resistance gene (*floR* and *tetG*, respectively) (65). The integrons themselves possess the antibiotic resistance genes *aadA2* and *pse-1*, which encode resistance to streptomycin and ampicillin, respectively. Besides serovar Typhimurium DT104, different antibiotic resistance cassettes have been identified in either integrons or *attI* integration sites for the genomic island SGI1 (290). At one point in its evolution, this MDR locus appears to have been part of a conjugative plasmid that integrated into the *Salmonella* chromosome. Subsequent deletions and rearrangements within the integrated plasmid resulted in the genomic island SGI1, presented

in Fig. 10.1 (65). The SGI1 genomic island, specifically the MDR locus, has been reported in other serovar Typhimurium phage types and *S. enterica* serovars isolated from many animal species (65, 133, 134, 290, 332). Once thought unique to MDR *Salmonella*, a segment of SGI1 was recently discovered within a 20-kb MDR genomic island in *Acinetobacter baumannii* (165), and the unusual phenicol resistance gene *floR* has also been identified in other gram-negative veterinary and human pathogens (94, 260, 407, 477). *floR* was first described for the fish pathogen *Photobacterium dansalae* in Japan, where it resides in a conjugative plasmid (264). The same phenicol resistance gene is also present in a mobile genomic island of MDR *Vibrio cholerae* (229). Detailed genetic analysis of the *V. cholerae* MDR genomic island revealed that this locus is a phage/plasmid chimera: fusing plasmid conjugation genes, without a plasmid *ori*, with *int*, *xis*, and integration sequences of a λ-like phage (45). The DT104 SGI1 locus also has *int* and *xis* genes of a site-specific transposon and excises from the chromosome as a circular intermediate, but unlike the *V. cholerae* locus, it requires a helper plasmid for mobilization (132).

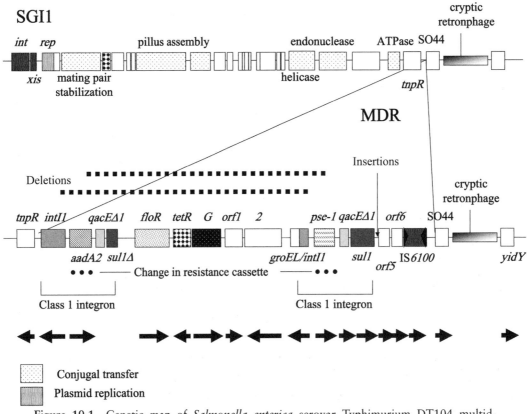

Figure 10.1 Genetic map of *Salmonella enterica* serovar Typhimurium DT104 multidrug resistance locus, SGI1. Regions of SGI1 with deletions, insertions, or an alternate gene cassette(s) within the class 1 integron are indicated within the figure (65, 132–134).

Extended-spectrum cephalosporin-resistant *S. enterica* serovars have become a significant public health concern due to their resistance to cephalosporins, which are currently prescribed to treat salmonellosis in children (152). There are four functional groups of enzymes that confer resistance to β-lactams, cephalosporins, monobactams, and carbapenems through cleavage of each antibiotic's β-lactam ring (76). Serovar Newport resistance to β-lactams, the β-lactamase inhibitor clavulanic acid, and cephalosporins is attributed to a plasmid-borne cephamycinase, Bla$_{CMY-2}$, also referred to as AmpC. *ampC* is often part of the normal repertoire of genes of *Enterobacteriaceae*, with the exception of the genus *Salmonella*, in which the gene is absent (56). While the gene is silent, microbes develop resistance to cephalosporins through either induction with cefoxitin or ampicillin or antibiotic selection of mutants with point mutations in the promoter of the resident gene (76). The serovar Newport *bla*$_{CMY-2}$ gene is homologous to the *ampC* gene of *Citrobacter freundii* (483). At some point in its evolution, a chromosome-borne *bla*$_{CMY-2}$ gene was captured by mobile conjugative plasmids. The serovar Newport *ampC* gene is upstream of a transposon (181), which may explain the mechanism behind its evolution from a chromosomal origin to its present plasmid-borne status. In addition to its occurrence in serovar Newport isolates, the plasmid-borne *ampC* gene has been identified in other *S. enterica* serovars (472, 483, 484, 501) and in *E. coli* strains (407, 484, 501) isolated from many animal species (407, 483, 484, 501). Due to the spread of these plasmid-borne cephamycinases, bacterial infections have become more difficult to treat, even in the veterinary setting (407, 425, 472). These bacterial infections are especially problematic because in addition to cephalosporin resistance, plasmids have also acquired other antibiotic resistance genes (407, 472) via the plasmid's resident integron(s) that confer onto its bacterial host resistance to aminoglycosides, phenicols, sulfonamide, tetracycline, and trimethoprim (407). While most extended-spectrum cephalosporin resistance in *Salmonella* is due to *bla*$_{CMY-2}$, there have been reports of other mechanisms of resistance, including extended-spectrum β-lactamases, which confer broad resistance to many β-lactams and cephalosporins but remain sensitive to the β-lactam inhibitor clavulanic acid (384, 462, 501). Since many of these antibiotic resistance genes reside on mobile genetic elements, some with fairly broad host ranges, we will likely continue to encounter MDR *Salmonella* strains with different antibiotic resistance phenotypes appearing in diverse genetic backgrounds.

Antibiotic usage in agriculture and its impact on human health have long been a contentious issue, since the release of the Swann Report in 1969 (434), and have been reviewed and debated ever since (11, 243, 244, 341, 342, 491, 493). At the center of this debate is whether veterinary usage of antibiotics is responsible for the emergence of MDR in foodborne pathogens and for the therapeutic failure of antibiotics in treating bacterial infections. Disease, disease prevention, government regulations, practicality, and economics are often the most important factors that influence antibiotic usage in food animal production today. The use of different antibiotics in veterinary medicine is predicated on the pharmacokinetics of the drug for the targeted animal species, on treatment cost and efficacy, and on drug residues in animal tissues, as well as on the susceptibility of the pathogen to the antibiotic (377). The development of antibiotic resistance led the American Veterinary Medical Association to adopt judicious usage guidelines to reduce its spread (454). In several countries, government regulations also prohibit or limit the use of certain antibiotics to specific food animal species, mostly to avoid tissue residues that could prove harmful to human health (e.g., chloramphenicol has been associated with aplastic anemia) (181).

There are far fewer antibiotics available for agriculture than for human medicine. However, there is concern over whether the usage of veterinary analogs to the drugs currently prescribed in human medicine (e.g., the veterinary analog enrofloxacin versus the human drug ciprofloxacin) will lead to cross-resistance in foodborne pathogens to the human drugs. This prospect has led the U.S. FDA to reassess and develop a new framework for evaluating currently approved or new veterinary antibiotics (158), and to the European Union's ban of growth-promoting antibiotics in food animals (81). Several antimicrobial susceptibility monitoring programs are currently in place to examine longitudinally the susceptibility trends among foodborne pathogens and commensal bacteria isolated from both food animals and humans (309, 318). Several studies have attempted to correlate the rise in antibiotic resistance with the approval of a drug for use in an animal species (96, 424), and there are several examples where actual antibiotic usage on the farm appeared to correlate with the decline in antibiotic susceptibility among farm and human isolates (239, 270, 291, 312, 463). However, there are also several studies that have failed to link antibiotic usage with the development of antibiotic-resistant microbes in animals (147, 262). There are also examples where the emergence of drug-resistant microbes predated regulatory approval or the use of an antibiotic in an animal species (57, 260, 385). While there are examples where antibiotic susceptibilities have increased with discontinuation of antibiotic usage (270), there are also examples where drug resistance still persists despite the discontinued use of a drug in food animals (260, 337). In sorting through and making sense of these studies with their contradictory results, we often forget that we are dealing with an open system in which bacteria and, more importantly, the drug resistance genes themselves know no boundaries. The antibiotic resistance genes currently present in bacterial pathogens may originate from antibiotic-producing bacteria present in soil (125, 267, 310, 480). The microbiota present in the environment may serve as hosts or "reservoirs" for the antibiotic resistance genes acquired by bacterial pathogens (126, 337, 394). The likelihood that a microorganism develops resistance, persists, and spreads involves more than the selective pressure of antibiotic use alone, and could also result from a complex interaction of genes, ecosystems, and the environment (55, 486). Understanding all the arguments surrounding this debate is complex and beyond the scope of this review. Readers are referred to several government and scientific reports on this topic (11, 243, 244, 341, 342, 434, 491, 493, 497).

With recent events, this continued debate may become moot as governments continue to restrict or ban antibiotic use in agriculture, as evident from the European Union's ban of antibiotic growth promoters and the FDA's repeal of its approval for the veterinary use of enrofloxacin in food animal species. Market forces have also had an impact, as some companies within the poultry industry have refrained from therapeutic and prophylactic use of antibiotics. The challenges facing veterinarians and the food industry relate to keeping animals healthy (81), especially with few or no alternatives to antibiotics for treating or preventing bacterial infections, while still producing safe products for consumers (401).

Infectious Dose

It is well established that newborns, infants, the elderly, and immunocompromised individuals are more susceptible to *Salmonella* infections than are healthy adults (116). The incompletely developed immune system in newborns and infants, the frequently weak and/or delayed immunological responses in the elderly and debilitated persons, and the generally low gastric acid production in infants and seniors facilitate the intestinal colonization and systemic spread of salmonellae in these segments of the population (62, 117). Antibiotic treatment of subjects before their encounter with *Salmonella* enhances bacterial virulence through an antibiotic-mediated clearance of native gut microflora, which reduces the level of bacterial competition for nutrients and attachment sites in the intestinal tract of the host (231, 426).

Detailed investigations of foodborne outbreaks have indicated that the ingestion of only a few *Salmonella* cells can be infectious (Table 10.12). Although early reports showed that large numbers of salmonellae inoculated into eggnog and fed to human volunteers produced overt disease (317), more recent evidence suggests that 1 to 10 cells can constitute a human infectious dose (115, 255). Determinant factors in salmonellosis are not limited to the immunological heterogeneity within human populations and to the virulence of infecting strains but may also include the chemical composition of incriminated food vehicles. A common denominator of the foods associated with low infectious doses (Table 10.12) is the high fat content in chocolate (cocoa butter), cheese (milk fat), and meat (animal fat). It has been suggested that entrapment of salmonellae within hydrophobic lipid micelles would afford protection against the bactericidal action of gastric acidity. Following bile-mediated dispersion of the lipid moieties in the duodenum, the viable salmonellae would resume their infectious course in search of suitable points of attachment in the lower portion of the small intestine (colonization). The rapid emptying of gastric contents could also provide an alternate mechanism for the successful infection of susceptible hosts. The swift passage of a liquid bolus through an empty stomach would minimize bacterial exposure to gastric acidity and sustain the migration of viable salmonellae in the intestinal tract (62).

Table 10.12 Human infectious doses of *Salmonella*[a]

Food	Serovar	Infectious dose (CFU)	Reference
Eggnog	Meleagridis	10^4–10^7	317
	Anatum	10^5–10^7	
Goat cheese	Zanzibar	10^5–10^{11}	390
Carmine dye	Cubana	10^4	277
Imitation ice cream	Typhimurium	10^4	33
Chocolate	Eastbourne	10^2	113
Hamburger	Newport	10^1–10^2	157
Cheddar cheese	Heidelberg	10^2	156
Chocolate	Napoli	10^1–10^2	194
Cheddar cheese	Typhimurium	10^0–10^1	115
Chocolate	Typhimurium	$\leq 10^1$	255
Paprika potato chips	Saintpaul	$\leq 4.5 \times 10^1$	283
	Javiana		
	Rubislaw		
Alfalfa sprouts	Newport	$\leq 4.6 \times 10^2$	1
Ice cream	Enteritidis	$\leq 2.8 \times 10^1$	470

[a] Adapted from reference 116.

Publications on the dynamics of human *Salmonella* infections are of singular interest (188, 189). An in-depth epidemiologic study of a large outbreak of serovar Typhimurium involving chicken served to delegates at a medical conference revealed that the clinical course in patients was directly related to the number of ingested salmonellae (189). The incubation period for the onset of symptoms was inversely related to the infectious dose. Patients with short (≤22 h) periods of incubation suffered more frequent diarrheal bowel movements, higher maximum body temperatures, greater persistence of clinical symptoms, and a greater frequency of hospitalization. Interestingly, no association between the ages of infected individuals and the length of the incubation period was noted. Similar findings were reported in retrospective dose-response studies of foodborne salmonellosis (189, 324).

The compelling evidence that ingestion of a few *Salmonella* cells can develop into a variety of clinical conditions and deteriorate into septicemia and even death underscores the unpredictable pathogenicity of this large and heterogeneous group of human bacterial pathogens. Food producers, processors, and distributors should recognize that low levels of salmonellae in a finished food product could lead to serious public health consequences (116).

PATHOGENICITY AND VIRULENCE FACTORS

Specific and Nonspecific Human Responses

The presence of viable salmonellae in the human intestinal tract confirms the successful evasion of nonspecific host defenses by ingested bacteria. Antibacterial lactoperoxidase in saliva, gastric acidity, mucoid secretions from intestinal goblet cells, intestinal peristalsis, and sloughing of luminal epithelial cells synergistically oppose bacterial colonization of the intestinal mucosa. In addition to these constitutive hurdles to bacterial infection, the antibacterial action of nonspecific phagocytic cells (neutrophils, macrophages, and monocytes) coupled with the immune responses associated with specific T and B lymphocytes, the epitheliolymphoid tissues (Peyer's patches), and the classical or alternative pathways for complement inactivation of invasive pathogens mount a formidable defense against the systemic spread of *Salmonella*.

The human diarrheagenic response to foodborne salmonellosis results from the migration of the pathogen in the oral cavity to intestinal tissues and mesenteric lymph follicles (enterocolitis). The event coincides with extensive leukocyte influx into the infected tissues, increased

mucous secretion by goblet cells, and mucosal inflammation triggered by the leukocytic release of prostaglandins. The last occurrence also activates the adenyl cyclase in intestinal epithelial cells, resulting in increased fluid secretion into the intestinal lumen (154, 369). The failure of host defense systems to hold invasive *Salmonella* in check can degenerate into septicemia and other chronic clinical conditions.

Genetics of *Salmonella* Virulence

The complete annotated sequence of the serovar Typhimurium genome was first published in 2001 (315), and since that time, the genomes for *Salmonella* serovars Typhi, Paratyphi A, and Choleraesuis have been annotated and published (89, 316, 360). In genomic comparisons between serovar Typhimurium and *E. coli* K-12, ~25% of the 4,552 genes of *Salmonella* are absent from the *E. coli* genome, and of these 1,002 *Salmonella* genes, ~50% are common to the four published *S. enterica* serovars Typhimurium, Typhi, Paratyphi A, and Choleraesuis. Further genomic comparisons between *E. coli* and *Salmonella* revealed that many of the "*Salmonella*-specific" genes are clustered together at focal points or loci within the bacterial chromosome into "islands" (Fig. 10.2 and 10.3). These genomic islands vary in size, generally map next to tRNA genes, often have remnants of phage genes at their borders, and usually have a GC content lower than the cognizant GC content for most metabolic and housekeeping genes, suggesting that horizontal gene transfer was involved in the evolution of *Salmonella* spp. (411). Many of these genomic islands in *Salmonella enterica* encode important functions that are essential to the pathogen's virulence (Table 10.13), and these loci define this genus and species (45). In addition to identifying putative virulence genes, genomic comparisons between *E. coli* and *Salmonella* also highlight metabolic differences that exist between these two bacteria and identify those genes that may provide *Salmonella* with a competitive advantage in the gastrointestinal tract. For example, while fimbriae are believed to be important contributors to *Salmonella* colonization of animal hosts (464), the ability of the bacterium to utilize, as an energy source, the hydrogen present in the gastrointestinal tract also appears vital for *Salmonella* colonization and virulence (305). It remains to be determined if other metabolic pathways, e.g., propanediol utilization, play a similar role.

While there are many genetic differences between *E. coli* and *Salmonella*, there is also significant genetic variability within the species *S. enterica* (84, 374), in which 2 to 8% of the serovar Typhimurium LT2 complement of 4,500 genes is absent in one or more *S. enterica* serovars.

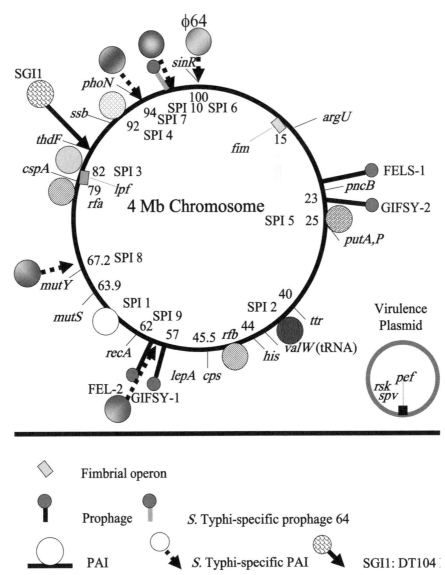

Figure 10.2 Genetic map of *S. enterica* virulence loci (65a, 315, 316, 408a). Genomic islands are depicted on the chromosome relative to the positions of housekeeping genes or specific tRNAs. All 10 *Salmonella* pathogenicity islands, including SPI6 to SPI10, which are only present in *Salmonella* serovars associated with enteric fever, are illustrated as circles, with serovar Typhi genomic islands attached to the map via arrows with dotted lines. Prophage genomes are also included in the map and are designated by the "lollipop" symbol shown. Squares denote adhesins/fimbrial operons. More detailed genetic organization of SPI1-5 is given in Fig. 10.3.

This genetic variability is due in part to the virulence genes resident on mobile genetic elements, i.e., phages, plasmids, and transposons, and the distribution of these mobile genetic elements among *S. enterica* serovars (3, 38, 69, 140). Sequencing of the *S. enterica* Typhimurium LT2 genome has revealed the presence of four complete prophage genomes within the bacterial chromosome (315). Within these prophage genomes are ancillary, nonphage genes, which are believed to contribute to some aspect of *Salmonella* pathogenesis (Fig. 10.4) (69). The circulation of these prophages and their resident virulence-associated genes vary among and within *S. enterica* serovars (84, 374), as does the distribution of the individual phage-associated virulence genes among the different lysogenic phages that infect *Salmonella* (325). Further characterization of the prophages resident in *Salmonella*

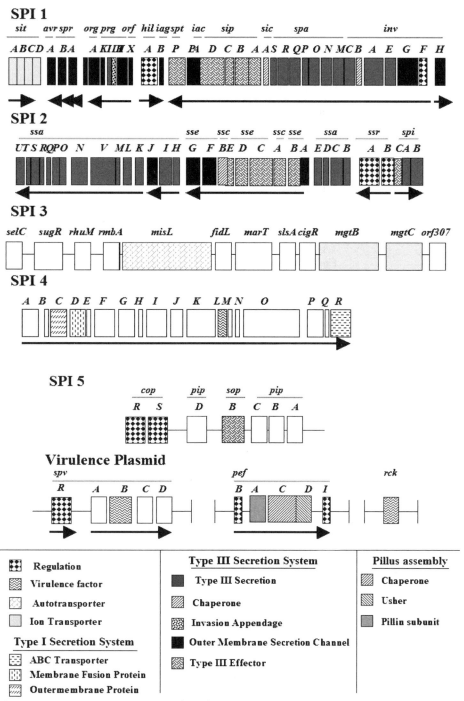

Figure 10.3 Genetic organization of virulence genes present in SPI1-5 and virulence plasmid. The genes and their organization into operons are shown, with arrows demarcating operons and their directions of transcription. The known functions of genes are depicted in this illustration, with shading or pattern designating the functions as described in the key at the bottom of the figure. Those genes with no known function are shown as white boxes. Type III effectors are bacterial proteins injected into host cell cytoplasm via the type III secretion system.

Table 10.13 *Salmonella* genomic islands, prophages, and virulence plasmid

Locus	Regulation	Function(s)	Distribution	Reference(s)
rfa/waa	*waaH*	Distal core LPS synthesis; serum survival	*S. enterica*; *S. bongori*	411a
rfb/wba		O-antigen biosynthesis; serum survival	*S. enterica*[a]; *S. bongori*[a]	411a
SPI1	*phoP/Q*; *ompR/envZ*; *sirA*	Cell invasion (enteritis)	*S. enterica*[b]; *S. bongori*	2, 10, 51, 347
SPI2	*phoP/Q*; *ompR/envZ*; *rpoS*	Macrophage survival; cell invasion	*S. enterica*	279, 360
SPI3		Macrophage survival	*S. enterica*[b]; *S. bongori*	10, 61, 360
SPI4	*waaH?*; *sirA*	Macrophage survival?; colonization of mammals		3, 330
SPI5	*sirA*	Fluid secretion (enteritis)	*S. enterica*[b]; *S. bongori*	3, 10, 360
SPI6		Fimbrial operons *safA–D*, *tcsA–R*	*S. enterica* "enteric fever" serovars[c]	360, 374
SPI7		Vi antigen; type IV pili	*S. enterica* "enteric fever" serovars[d]	84, 360, 374
SPI8		Bacteriocin immunity	*S. enterica* "enteric fever" serovars	360
SPI9		Type I secretion; RTX homologue	*S. enterica* "enteric fever" serovars	360
SPI10		Fimbrial operon *sefA–R*	*S. enterica* "enteric fever" serovars	360
SGI1		DT104 MDR locus	*S. enterica* serovars[c]	65, 132, 134
Virulence plasmid	*rpoS*	Proliferation within reticuloendothelial system	*S. enterica* serovars[e]	84, 275, 374

[a] Diversity in genetic composition of operon.
[b] Genetic variability within locus among *Salmonella* serovars.
[c] Present in *S. enterica* serovar Choleraesuis.
[d] Present in *S. enterica* serovars Typhi, Paratyphi, and Dublin.
[e] Variable distribution among *Salmonella* serovars.

has identified new virulence genes among different *S. enterica* serovars (38) and phage types (404). While the genomic islands, SPI1 to SPI5, are conserved in *S. enterica*, there is genetic variability within several of these loci among the serovars (10). Recent completion of the

serovar Typhi genome identified five additional pathogenicity islands, SPI6 to SPI10 (Fig. 10.2; Table 10.13), that appear to be unique to serovar Typhi and other serovars associated with enteric fever in humans (374). Genome comparisons between serovar Typhimurium and serovar

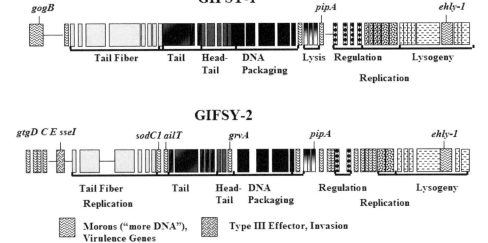

Figure 10.4 Genetic maps of serovar Typhimurium GIFSY prophages. The detailed genetic maps presented in this figure highlight putative virulence genes within the genomes of two lambda-like prophages present in the *S. enterica* serovar Typhimurium LT2 genome (69). One of these genes, *sseI*, is injected into the host cytoplasm via the type III secretion system present in SPI1.

Typhi have also identified new fimbrial operons, bringing the total number of distinctive fimbrial operons to 20 (140). Several of these operons are variable in their distribution among *S. enterica* serovars, and some appear to be unique to specific serovar or phylogenetic clades, possibly explaining the predilection of specific *Salmonella* serovars for certain animal species (44). There are also additional, subtle genetic differences between *Salmonella* serovars. In detailed genetic analysis of the 4,323 open reading frames, there were 210 pseudogenes identified, including several mutations in genes tied to *Salmonella* pathogenesis (360). Pseudogenes arise from point mutations, insertions, and deletions within the open reading frame that result in nonsense or frameshift mutations. Recent analysis of serovar Paratyphi A, which is associated with enteric fever, revealed that 173 of its ~4,400 genes are pseudogenes, 30 of which are also present in serovar Typhi. Seven of the 30 pseudogenes common to both serovars Typhi and Paratyphi are in loci that contribute to gastroenteritis (316). The serovar Choleraesuis genome has 151 pseudogenes, including several genes involved in colonization, chemotaxis, and motility (89).

The virulence and pathogenesis of *Salmonella* spp. have evolved such that acquisition of mobile genetic elements, subtle point mutations, and single-nucleotide insertions and deletions in key genes can have a profound impact on the behavior of *Salmonella* spp. in vivo. For a serovar with as broad a host range as that of serovar Typhimurium, the complement of genes required for colonization varies depending on the animal host, e.g., avian versus mammalian species (330). The differences between broad-host-range and host-adapted *S. enterica* serovars may be due to subtle changes in gene expression (329, 482) more than to the acquisition or loss of genes (143, 316, 373, 374). Subtle mutations in promoters and global regulators may also explain differences in pathogen behaviors that occur within bacterial populations (274, 396). While microarray-based genomic analyses have provided useful insights into the evolution of *Salmonella* (373), more detailed sequence-based genomic and proteomic studies (143) should identify the subtle genomic differences among serovars that contribute to host adaptation and virulence.

In *Salmonella* pathogenesis, a wide array of virulence and cellular repair genes are coordinately regulated as *Salmonella* travels from the intestinal tract into tissues and the bloodstream, turning genes on/off as they are needed in the animal host. Table 10.13 lists key global regulatory elements, most notably *phoP/Q* and *rpoS*, that repress or activate these key virulence loci in *Salmonella*. Many genomic islands are part of a regulon(s) involved in scavenging limiting nutrients or adapting to environments detrimental to the survival of *Salmonella*. A "regulon" is a regulatory circuitry that has evolved where common promoter and regulatory sequences are recognized by a single repressor, activator, or alternative sigma factor for genes with disparate functions but a common goal, i.e., survival (299). The genes that make up this regulon are often physically separate from one another, existing as discrete loci scattered throughout the bacterial chromosome. Disrupting this coordinated, spatial regulation at either the transcriptional (319) or posttranscriptional level (8, 439, 440) profoundly affects the virulence of the bacterium. However, many of these global regulatory elements are also responsible for maintaining homeostasis and regulating several key cellular repair and detoxification enzymes. Therefore, it is difficult to determine whether virulence attenuation is attributed to reduced pathology or physiological crippling of the pathogen. Certain details concerning regulation of key virulence genes are presented in discussions of *Salmonella* pathogenesis. For a synopsis of gene regulation as it relates to *Salmonella* virulence, physiology, and food safety, see the recent review by Maurer and Lee (314).

Attachment and Invasion

The establishment of a human *Salmonella* infection depends on the ability of the bacterium to attach to (colonization) and enter (invasion) intestinal columnar epithelial cells (enterocytes) and specialized M cells overlying Peyer's patches (Fig. 10.5). Salmonellae must successfully compete with indigenous gut microflora for suitable attachment sites on the luminal surface of the intestinal wall and evade capture by secretory immunoglobulin A that may also be present on the surfaces of epithelial cells. Colonization of enterocytes arises, in part, from the interaction of bacterial type 1 (mannose-sensitive) or type 3 (mannose-resistant) fimbriae, surface adhesins, nonfimbriate (mannose-resistant) hemagglutinins, or enterocyte-induced polypeptides with host glycoprotein receptors located on the microvilli or glycocalyx of the intestinal surface (117, 369). Although the role of *Salmonella* motility in the adherence and invasion processes is uncertain, motility may not be essential for bacterial internalization and probably would simply increase the frequency of productive contacts between the pathogen and its targeted epithelial cell (172). An authoritative review on the morphology of interactions between salmonellae and enterocytes, M cells, and the immune apparatus in the mucosal and submucosal layers of the host intestine is recommended as additional reading (369). Fine-structural studies have also revealed that proteinaceous appendages develop on the surfaces of salmonellae upon contact with

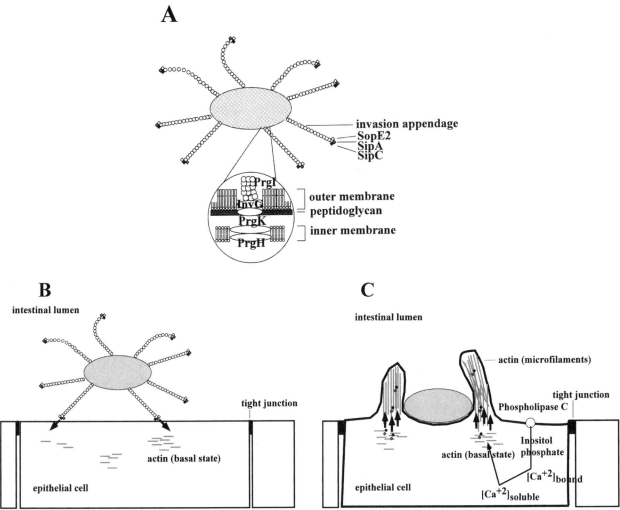

Figure 10.5　Mechanism for *Salmonella* invasion of host epithelial cells.

epithelial cells (172, 187). These invasion appendages are 0.3 to 1.0 nm long and ca. 60 nm in diameter, which is considerably thicker than flagella (ca. 20 nm) and type 1 fimbriae (ca. 7 nm). The assembly of these *Salmonella* appendages is energy dependent but independent of de novo bacterial protein synthesis (98, 187). The appendages are short-lived and are shed concomitant with the appearance of membrane ruffles on colonized epithelial cells (187). Following bacterial attachment, signal transduction between the pathogen and host cell culminates in energy-dependent *Salmonella* invasion of enterocytes and M cells (177, 187). Although bacterial protein synthesis is not a prerequisite for the onset of invasion, its prolonged inhibition compromises the ability of salmonellae to enter cultured epithelial cells (302). The role of the *inv* ("invasion") locus in *Salmonella* spp. is of singular importance in the invasion mechanism because it triggers

two profound changes in enterocytes and M cells, namely, a Ca^{2+} influx and cytoskeleton rearrangement in the targeted host cells. The cytoskeleton of eukaryotic cells is a highly organized network of contractile and supportive filaments consisting of actin or other protein elements that define the apical-basal polarity of intestinal epithelial cells and control the movement of intracellular organelles (154, 369). Adherent *Salmonella* cells promote the influx of luminal Ca^{2+} into the mammalian cell, an important signal that influences actin polymerization (400). Salmonellae deliver the invasion proteins SipA, SipB, SipC, SptP, SopE2, and SopB to eukaryotic cells (97, 98). These proteins, in turn, cause polymerization of host cell actin into microfilaments in the vicinity of the invading pathogen (171, 186). *Salmonella* invasion proteins SipA and SipC act as catalysts in the nucleation and subsequent polymerization of F-actin into microfilaments (208, 503,

504). Another invasion protein, SopE2, activates a signal transduction cascade involved in cytoskeletal rearrangement (427). Histologically, *Salmonella* invasion is seen as an evagination of the apical cytoplasm of epithelial cells, i.e., membrane ruffling, around the adherent salmonellae that subsequently mediates the pinocytotic uptake of the bacterial pathogen (252).

We have thus far seen that *Salmonella* attaches to the microvilli of the intestinal epithelium, an intercellular contact that activates salmonellae to the production of transient, proteinaceous appendages deemed essential for *Salmonella* invasion. These events are closely followed by actin polymerization and the formation of membrane ruffles. The complete internalization of *Salmonella* into the epithelial cell precipitates a reversion of actin microfilaments and membrane ruffles to their basal states. This final event may be mediated by the *Salmonella* tyrosine phosphatase SptP, a protein that causes disruption of the actin cytoskeleton and the disappearance of stress fibers (170). *Salmonella* invasion of the epithelial cells also has profound effects on the cell's physiology. Epithelial cells produce cytokines, including chemoattractants, in response to the invading bacteria (139). The invading salmonellae also cause apoptosis (programmed cell death) of epithelial cells (265), neutrophils, and macrophages (227, 328, 393). Diarrhea associated with salmonellae now appears to be a response to bacterial invasion of enterocytes and M cells instead of the action of a putative enterotoxin, Stn (457, 473).

Salmonella invasion genes are organized into contiguous and functionally related loci, SPI1, located within a 40- to 50-kb segment of chromosomal DNA (i.e., pathogenicity island) that encodes determinant factors for the facilitated entry of salmonellae into host cells (196, 322) (Fig. 10.3; Table 10.13). The *inv* pathogenicity island is a multigenic locus consisting of 30 genes. Many of these genes encode enzymes and transcriptional activators responsible for the regulation, expression, and translocation of important effectors, such as SipA, to the surfaces of host cells. Most of these genes encode the type III secretion components involved in the assembly of the invasion appendages and the export and insertion of *Salmonella* invasion proteins into the epithelial cell (171, 237). This invasion-associated type III secretion apparatus structurally resembles the flagellar basal bodies of gram-negative bacteria (237). The characterization of a second *Salmonella* pathogenicity island, SPI2, has identified additional genes necessary for the export of a membrane fusion protein, SipB (223). SPI1 invasion genes are evolutionarily conserved among *Salmonella* spp. (66), making these genes useful targets for PCR-based detection of *Salmonella* in foods (87). Although this locus is not essential for the salmonellae

to traverse the intestinal barrier and spread systemically (334), it does appear to be one of several factors that mediate enteritis associated with *Salmonella* infection (457).

Environmental factors such as high osmolarity and low pO$_2$ enhance bacterial invasiveness by altering the superhelicity of chromosomal DNA, which consequently impacts the level of transcription of invasion-related genes such as *invA* (149, 177). There are several key transcriptional regulatory elements involved in the regulation of the *inv* locus (3, 51, 124, 141). In fact, *Salmonella* coordinately regulates its genes as it develops from a localized to a systemic infection (217, 417) (Table 10.13).

Growth and Survival within Host Cells

In contrast to several bacterial pathogens, such as *Yersinia*, *Shigella*, and enteroinvasive *E. coli*, which replicate within the cytoplasm of host cells, salmonellae are confined to endocytotic vacuoles wherein bacterial replication begins within hours following internalization (177). The infected vacuoles move from the apical to the basal pole of the host cell, where *Salmonella* are released into the lamina propria (246, 369). During their proliferation within epithelial vacuoles, *Salmonella* cells also induce the formation of stable filamentous structures (Sif) within the epithelial cytosol (176). These organelles, which contain lysosomal membrane glycoproteins and acid phosphatase, are seemingly connected to the infected vacuoles. The formation of these filaments requires viable salmonellae within the membranous vacuoles and is blocked by inhibitors of vacuolar acidification. Although the role of these induced filamentous structures is poorly understood, they may be necessary for the intravacuolar replication of *Salmonella* spp. A bacterial surface mechanism has tentatively been identified that facilitates the migration of salmonellae into deeper layers of tissues upon their release into the lamina propria. The demonstrated ability of thin aggregate fimbriae on the outer surfaces of *Salmonella* cells to bind host plasminogen and the tissue-type plasminogen activator could markedly increase the invasiveness of infecting strains. It is likely that the zymogen would be converted to its proteolytic (plasmin) form on the bacterial surface, thereby providing salmonellae with an effective tool for breaching host tissue barriers and facilitating transcytosis into deeper tissues (420).

The systemic migration of *Salmonella* exposes the bacterium to phagocytosis by macrophages, monocytes, and polymorphonuclear leukocytes and to the antibacterial conditions that prevail in the cytoplasm of these host defense cells (117). Survival of the bacterial cell within the hostile confines of the macrophage or neutrophil determines the host's fate. Whether the host develops enteric fever from *Salmonella* infection is determined partly by the genetics of the host (421) and of salmonellae (221, 413). Intracellular pathogens have developed different strategies to survive within professional phagocytic cells, which include (i) escaping the phagosomes (41), (ii) inhibiting acidification of the phagosomes (234), (iii) preventing phagosome-lysosome fusion (32), and (iv) withstanding the toxic environment of phagolysosome (391). Although there is evidence to support *Salmonella*'s ability to prevent acidification of the phagosomes (6, 7), fusion with the lysosome (71, 459), or maturation of the phagolysosome (386), *Salmonella* is a hardy bacterium that can survive and replicate within the bactericidal milieu of the acidified phagolysosome (351, 387). For example, the membrane oxidase-dependent formation of toxic oxygen products, such as singlet oxygen, superoxide anion, hydrogen peroxide, and hydroxyl radicals, during the oxidative metabolic burst of phagocytes is countered by the protective activities of several bacterial enzymes, including superoxide dismutase, peroxidase, and catalase. *Salmonella* synthesizes 30 bacterial proteins in response to the toxic oxygen products of macrophages. Of these, nine are dependent on the synthesis of a nucleic acid activator encoded by the chromosomal *oxyR* locus that induces the transcription of genetic loci responsible for the synthesis of protective proteins (372).

The *phoP/phoQ* regulon is a two-component transcriptional regulator system that enables *Salmonella* to survive within the hostile environment of phagocytes, notably the high acidity within phagolysosomes (a host cell construct of an infected phagosome with incorporated lysosomal granules) and the release of antibacterial defensins by phagocytic cells (320, 372). Defensins, which also occur in epithelial cells, are small, nonspecific cationic peptides that inactivate salmonellae by inserting into the outer bacterial membrane, thereby creating transmembrane channels that increase bacterial permeability to ions and precipitate cell death (153). The antidefensin gene product(s) encoded by the *Salmonella phoP/phoQ* system appears to involve chemical alteration of the LPS core (461), which makes the salmonellae resistant to neutrophil antimicrobial peptides (416), as well as the production of a cell surface protease that degrades the defensin (199). The *phoP/phoQ* locus is part of a transcriptional cascade of activators/repressors (199, 203) that are necessary for the expression of a subset of genes essential to the survival of the bacterial cell within the macrophage (48, 199, 203). *phoQ* is a kinase that reportedly senses the hostile phagolysosomal environment by means of a short chain of 20 of its 487 amino acid residues that extends into the periplasmic space of the *Salmonella* cell envelope. Half of the amino acids in this periplasmic chain are acidic and form an anionic box that functions as a receptor for the cationic defensins or as

a sensor of highly acidic conditions in the phagocyte. An interaction between a signal and the *phoQ* kinase anionic box triggers the autophosphorylation of the kinase sensor at a histidine residue, followed by a kinase-dependent transfer of phosphate to an aspartate residue in the amino terminus of the PhoP transcriptional activator protein (320, 321). The phosphorylated PhoP protein activates the transcription of several *phoP*-activated genes (*pag* genes), including *pagA*, *pagB*, *pagC*, *psiD*, and the *phoN* locus, which encodes the periplasmic nonspecific acid phosphatase. The PhoP-activated genes *pagA* and *psiD* are not involved in *Salmonella* virulence. The *Salmonella pagC* gene encodes an outer membrane protein that promotes survival within macrophages. Although the amino acid sequence of PagC is similar to that of the *ail* gene product in *Yersinia enterocolitica*, this structural homology between both gene products does not extend to functional homology, as evidenced by the loss of invasiveness in *ail* but not *pagC* mutants (372). The transcription of *pagC* within the macrophage is maximal at pH 4.9, and its protein product reportedly contributes to serum resistance but affords no protection against defensins or highly acidic environments (320, 467). Interestingly, transcription of the *pag* genes is induced within acidified *Salmonella*-infected phagocytes but not in infected epithelial vacuoles (7). The PhoP-activated gene *pagB* is part of an operon that includes polymyxin resistance locus *prmAB* (306) genes that encode a two-component regulatory system (398) similar to the *phoP/phoQ*- and *ompR/envZ*-encoded sensor kinase-activators involved in the regulation of multiple factors (40, 48, 180, 422), including virulence (130, 321). The *prmAB* locus appears to regulate the *pmrH-FIJKLM* operon, whose genes are necessary for the synthesis of 4-aminoarabinose lipid A. This modified LPS confers resistance to polymyxin and defensins (204). In addition to the 5 previously mentioned *phoP*-activated genes, 13 more positively regulated loci (*pagD* to *pagP*) have been identified (52). These additional *pag* genes do not contribute to the *phoP/phoQ*-dependent resistance to defensins, whereas the *pagD*, *pagJ*, *pagK*, and *pagM* genes participate in mouse virulence and in bacterial survival in macrophages.

In contrast to the *pag* genes that are expressed under adverse environmental conditions, such as low pH, nutrient deficiency, and stationary phase, the *phoP/phoQ* system also regulates the expression of *phoP*-repressed genes (*prg*), which are induced under nonstress conditions (51). Stated differently, conditions that activate *pag* genes generally repress *prg* expression. The *prgH* operon (SPI1) plays a determinant role in the export of bacterial proteins necessary for invasion of epithelial cells (51, 362). Other proteins arising from the transcription of *prg* genes are required for the diffuse membrane ruffling of macrophages, increased pinocytosis, and formation within macrophages of spacious phagosomes wherein slow acidification favors greater bacterial survival and attendant virulence (6). It is noteworthy that only a limited number of *phoP*-activated genes contribute to *Salmonella* virulence (467) and that mutations in this regulon (*pagC*, *-D*, *-J*, *-K*, and *-M*) result in phenotypes exhibiting decreased survival within macrophages, increased susceptibility to acidic pH, serum complement (*pagC*), and defensins, and reduced invasiveness of epithelial cells (*prg H*).

As with invasion, several bacterial genes important for survival and proliferation in macrophages have been identified as part of pathogenicity islands in the *Salmonella* chromosome (61, 223). The SPI2 pathogenicity island, discussed earlier with reference to invasion, encodes a factor(s) necessary for intracellular survival and growth in epithelial cells and macrophages (93, 224). *Salmonella* produces an effector, SpiC, that is exported through the type III secretion system encoded by SPI2 to the host cell cytosol. This bacterial protein inhibits endosome-endosome fusion and prevents fusion between the phagosome and lysosome (459). A third pathogenicity island, SPI3, is also required for intracellular growth of *Salmonella* within a macrophage (60). The *mgtCB* genes, which are present in SPI3, are required for growth under magnesium-limiting conditions and essential for intracellular growth (60). The other genes identified in this locus do not contribute to *Salmonella* pathogenesis in a mouse typhoid model. Their functions are currently unknown (61).

Other Virulence Factors

The virulence of *Salmonella* spp., as reflected in the ability to cause acute and chronic diseases in humans and in a variety of animal hosts, stems from structural and physiological attributes that act synergistically or independently in promoting bacterial adhesion and invasiveness (116, 117). The ATR is an example of the physiological responsiveness of salmonellae to adverse environmental conditions, which concurrently potentiates greater bacterial virulence in hosts and greater acid tolerance in fermented foods. In addition to the *inv* locus and the *phoP/phoQ* system, which encode determinants for *Salmonella* invasiveness and resistance to the antibacterial conditions within phagocytes, respectively, several other virulence factors impact *Salmonella* pathogenicity.

Virulence Plasmids

Virulence plasmids are large cytoplasmic DNA structures that replicate in synchrony with the bacterial chromosome, contain many virulence loci ranging from 30 to 60 MDa, and occur with a frequency of one to two copies per chromosome (200). The presence of virulence plasmids within the *Salmonella* genus is limited and has been confirmed for serovars Typhimurium, Dublin, Gallinarum-Pullorum, Enteritidis, Choleraesuis, and Abortusovis (200). The absence of a virulence plasmid in the host-adapted and highly infectious serovar Typhi is noteworthy (117, 200). Although limited in their distribution in nature, virulence plasmids are self-transmissible (4). Gene products from the transcription of a virulence plasmid potentiate systemic spread and infection of extraintestinal tissues but exert no effect on *Salmonella* adhesion and invasion of epithelial and M cells (116). More specifically, it is suggested that virulence plasmids enable carrier strains to multiply rapidly within host cells and overwhelm host defense mechanisms (275, 438). These plasmids also confer on salmonellae the abilities to induce lysis of macrophages and to illicit an inflammatory response (198) and enteritis in the animal host (294).

The *Salmonella* plasmid virulence (*spv*) region consists of a gene cluster that encodes products for the prolific growth of salmonellae in host reticuloendothelial tissues. This genetic entity was formerly identified as *mka* (mouse killing agent), *mkf* (mouse killing factor), or *vir* (virulence) (372). Signals that trigger *spv* transcription include the hostile environment within host phagocytes, iron limitation, elevated temperatures, low pH, and nutrient deprivation associated with the stationary phase of growth (216, 438). The *spv* regulon approximately contains 8.0 kb and at least five genes (*spvR*, *-A*, *-B*, *-C*, and *-D*) and two principal promoters, one for the *spvR* locus and another for the *spvABCD* transcriptional unit (202). These loci are transcribed in a single direction under the combined regulatory action of *spvR* and a chromosomal *rpoS*-encoded sigma (σ^S) factor (216). The sigma factor, a subunit within the RNA polymerase, recognizes the *spvR* promoter and activates transcription of the *spv* regulon (275). The *spvR*-encoded protein positively regulates the expression of the downstream *spvABCD* transcriptional unit through the *spvA* promoter sequence (200, 433). SpvB functions as an ADP-ribosyltransferase (356) that ADP-ribosylates F-actin and blocks its polymerization into filaments. The nucleotide sequences of the *spv* loci are highly conserved within the *Salmonella* genus (433, 438). Mutations in this region strongly attenuate or inactivate the ability of salmonellae to establish deep-seated infections (201, 294). Minor differences in the nucleotide sequences of the *spvR* genes in serovars Dublin and Typhimurium markedly alter the capacity of the *spvR* gene to induce the *spvA* promoter and transcription of the *spvABCD* operon. These findings provide some insight on the molecular basis for the comparatively greater virulence of serovar Dublin in humans (438). The *Salmonella* virulence plasmid also contains a fimbrial operon (*pef*) that encodes an adhesin involved in colonization of the small intestine and enteropathogenicity of the bacterium (43, 169). The distribution of this and other adhesins may explain the narrow versus broad host ranges exhibited by salmonellae (44).

Iron Acquisition

Siderophores are yet another facet of the *Salmonella* virulence armamentarium. These elements retrieve essential iron from host tissues to drive key cellular functions, such as the electron transport chain, and enzymes associated with iron cofactors (106). To this end, *Salmonella* must compete with host transferrin, lactoferrin, and ferritin ligands for available iron (106, 117). For example, transferrin scavenges tissue fluids for Fe^{3+} ions to form Fe^{3+}-transferrin complexes that bind to surface host cell receptors. Upon internalization, the complexes dissociate and the released Fe^{3+} is complexed with ferritin for intracellular storage. In response to a limited availability of Fe^{3+} in host tissue, *Salmonella* sequesters Fe^{3+} ions by means of a high-affinity phenolate enterochelin (also designated enterobactin), consisting of a cyclic trimer of dihydroxybenzoic acid and L-serine, and a low-affinity hydroxamate aerobactin chelator, an anabolic product derived from one citrate and two lysine (one hydroxylated and one acetylated) residues (338, 372). The *fur* (ferric uptake regulator) gene regulates the synthesis of these bacterial siderophores (106). *Salmonella* binding of trivalent iron proceeds with the interaction of a ferri-siderophore complex with an outer membrane protein receptor induced in response to limiting concentrations of intracellular iron. The complex is then transposed into the bacterial cytoplasm, where the ferric moiety is reduced to the ferrous state. The low affinity of the siderophore for Fe^{2+} results in the release of the divalent ion into the bacterial cytoplasm for subsequent use in key metabolic functions. It is notable that the degree of *Salmonella* virulence is directly related to the enterochelin content of the infecting strain. Experimental insight into this relationship follows from the reduced virulence of auxotrophic strains deficient in the aromatic pathway, which is responsible for the formation of dihydroxybenzoic acid, the precursor of enterochelin (116). Siderophores are not the only mechanism by which *Salmonella* can acquire iron from its host. A new *fur*-regulated *sitABCD* operon was identified within SPI1 that can compete with chelators like 2,2′-dipyridyl for iron and influence *Salmonella* growth in vivo (249, 505). These iron transport proteins have considerable homology to the *yfe* ABC iron transport system in *Yersinia pestis*. The

sitABCD operon in SPI1 (Fig. 10.3) is evolutionarily conserved among the salmonellae (505).

Toxins

Diarrheagenic enterotoxin is a putative *Salmonella* virulence factor. The release of toxin into the cytoplasm of infected host cells precipitates an activation of adenyl cyclase localized in the epithelial cell membrane and a marked increase in the cytoplasmic concentration of cyclic AMP in host cells. The concurrent fluid exsorption into the intestinal lumen results from a net secretion of Cl^- ions into the crypt regions of the intestinal mucosa and depressed Na^+ absorption at the level of the intestinal villi (117, 428). Enterotoxigenicity is a virulence phenotype that prevails in *Salmonella* serovars, including serovar Typhi, and which is expressed within hours following bacterial contact with the targeted host cells (116, 150).

One *Salmonella* enterotoxin is a thermolabile protein with a molecular mass of 90 to 110 kDa (117). The narrow pH range (6.0 to 8.0) for active enterotoxin suggests that the delivery of a functional toxin may be impaired or inhibited in acidic phagolysosomes (pH 4.5 to 5.0) and in epithelial endosomes (pH 5.0 to 6.5). The ability of enterotoxin to bind to GM_1 ganglioside receptors on host cell surfaces remains a controversial issue (116, 117, 150, 380). The *Salmonella* enterotoxin is encoded by a 6.3-kb chromosomal gene (*stx*) which regulates the synthesis of three proteins, of 45, 26, and 12 kDa (90).

In addition to enterotoxin, *Salmonella* strains generally elaborate a thermolabile cytotoxic protein which is localized in the bacterial outer membrane (34, 117). The cytotoxin is not inactivated with antisera raised against Shiga toxin or *E. coli* Shiga toxins 1 and 2, as determined in green monkey kidney (Vero) and HeLa cell bioassays (34, 129). Maximum production of cytotoxin in laboratory media occurs at pH 7.0 and 37°C during the early stationary phase of growth. Hostile environments such as acidic pH and elevated (42°C) temperature induce an extracellular release of toxin, possibly as a result of induced bacterial lysis (129). The virulence attribute of cytotoxin stems from its inhibition of protein synthesis and lysis of host cells, thereby promoting the dissemination of viable salmonellae into host tissues (117, 272). The ability of added Ca^{2+} ions to block the cytotoxin-dependent disruption of cell monolayers suggests that cytotoxin may function as a chelator of divalent cations that normally contribute to the structural integrity of monolayers (363). This cytotoxin may actually be one of the *Salmonella* invasion proteins reported to activate apoptosis in epithelial and phagocytic cell types (227, 265, 328, 393).

The *Salmonella* pathogenicity island SPI5 has been identified and appears to be responsible for the pathogen's ability to induce diarrhea (Fig. 10.3). This locus is conserved among salmonellae. Several genes were identified, including *pipA, -B, -C, -D,* and *sopB*, that were necessary for *Salmonella* to cause intestinal secretion and an inflammatory response but not essential for the bacterium to cause systemic infection (487). SopB appears to function as an inositol phosphate phosphatase and hydrolyzes phosphatidylinositol 3,4,5-triphosphate, an inhibitor of Ca^{2+}-dependent chloride secretion. A mutation affecting the phosphatase activity of SopB diminished the ability of *Salmonella* to cause diarrhea (347). The SPI1 locus has also emerged as an important factor in the enteropathogenicity of *Salmonella* (457, 473). This evidence questions the role, if any, of the enterotoxin in diarrhea (473). Two new toxins have recently been identified as part of mobile genetic elements in *S. enterica* (38, 404). Their role in pathogenesis has not yet been elucidated.

Vi Antigen, LPS, and Porins

Lastly, three virulence determinants located within or on the external surface of the *Salmonella* outer membrane are discussed briefly here. The capsular polysaccharide Vi (virulence) antigen occurs in most strains of serovar Typhi, in a few strains of serovar Paratyphi C, and rarely in serovar Dublin (372). The Vi antigen significantly increases the virulence of carrier strains by inhibiting the opsonization of the C3b host complement factor to surface LPS, a critical event in the induction of macrophage phagocytosis of invasive salmonellae (117). Two genes (*viaA* and *-B*) are associated with the formation of the Vi antigen. However, the *viaB* locus, which encodes no fewer than six proteins, appears to be the key genetic element in the control of Vi synthesis, as evidenced by the presence of *viaA* in members of the *Enterobacteriaceae* and in *Salmonella* spp. that do not express the Vi antigen (372). Recent data further indicate that the *ompR* locus also impacts Vi capsule formation in serovar Typhi (365).

The lengths of serotypic LPS that protrude from the bacterial outer membrane not only define the rough (short LPS) and smooth (long LPS) phenotypes but also play an important role in repelling the potentially lytic attack of the host complement system (117). In effect, the LPS of smooth variants sterically hinder the stable insertion of the C5b-9 complement factor into the inner cytoplasmic membrane that would precipitate bacteriolysis. The general inability of the short LPS in rough variants to protect against the C5b-9 lytic insertion renders such variants more susceptible to lysis and, consequently, less virulent (117). The primary carbohydrate composition of LPS can

also affect the level of serum complement activation. For example, an isogenic pair of *Salmonella* spp. containing either the somatic B (factors 4 and 12) or C (factors 6 and 7) LPS antigens activated serum complement at low and high rates, respectively (409). Correspondingly, serogroup B exhibits greater virulence than does serogroup C.

Porins are outer membrane proteins that function as transmembrane (outer membrane) channels in regulating the influx of nutrients, antibiotics, and other small molecular species (108, 344). To date, four porins, encoded by the *ompF, ompC, ompD,* and *phoE* genes of serovar Typhimurium, have been described (344). Gene transcription of the *ompF* and *ompC* loci is induced by changes in the microenvironment. Low osmolarity, low nutrient availability, and low temperature can trigger transcription of the *ompF* gene, with concomitant repression of the *ompC* locus. Conversely, favorable environmental conditions induce transcription of the *ompC* gene and repression of the *ompF* locus (108, 327, 344). This is analogous to the previously described *phoP/phoQ*-dependent activation of *pag* genes under adverse conditions, with concomitant repression of *prg* genes (51). The transmembrane channels formed by OmpF and OmpC are 1.1 to 1.3 nm in diameter and consist of trimers formed from monomeric subunits. EnvZ activates OmpR through a phosphorylation-dependent mechanism, whereas the activated *ompR* locus regulates the transcription of *ompF* and *ompC* (108, 327). This mechanism recalls the previously discussed *phoP/phoQ* regulon, where phosphorylation of the *phoP* gene product by the *phoQ*-encoded kinase activated the transcription of *pag* genes. Mutations in the *ompR/envZ* (designated *ompB*) regulon dramatically attenuate the virulence of carrier strains. In contrast, mutations in the *ompC, -F,* or *-D* locus exert little effect on *Salmonella* virulence (130). However, mutations in both *ompC* and *ompF* reduced the virulence of *Salmonella* spp. in mice and protected vaccinated mice upon challenge with wild-type serovar Typhimurium (85). Like *phoP/phoQ*, the two-component *ompB* regulatory system controls a series of genes important to the physiology (40, 180, 422) and pathogenesis of *Salmonella* spp. (279, 323). Interestingly, reports on the ability of *Salmonella* porins to elicit immunological host defense responses suggest that porins are of limited importance in *Salmonella* pathogenicity (117, 436). Recent evidence indicates that the *rck* locus (resistance to complement killing), which is located on the *Salmonella* virulence plasmid, encodes gene products that protect both smooth and rough variants against the lytic C5b-9 complement factor (213, 214). Nucleotide sequencing of *rck* and amino acid analysis of the Rck protein revealed homology with the *phoP/phoQ*-dependent *pagC* locus and gene product (214).

CONCLUSION

Salmonella spp. continue to be a leading cause of foodborne bacterial illnesses. The situation has endured because of the widespread occurrence of salmonellae in the natural environment and their prevalence in many sectors of the global food chain (116). Intense animal husbandry practices and difficulties in controlling the spread of *Salmonella* in vertically integrated meat and poultry production and processing industries contribute to the situation of raw poultry and meats being principal vehicles of human foodborne salmonellosis. The use of prophylactic doses of medically important antibiotics for reared animal species may promote on-farm selection of antibiotic-resistant strains and increase the human health risks associated with handling and consumption of contaminated meat products. A similar scenario can also be drawn for the aquaculture industry, in which intensive rearing of fish and shellfish in generally unprotected facilities together with the use of antibiotic treatment is common practice. The use of fluoroquinolones as prophylactic drugs in the agricultural and aquaculture industries is of concern because such an approach may undermine the clinical efficacy of these novel and invaluable drugs (116, 117). Reports on the acquired resistance of foodborne *Salmonella* spp. and *Campylobacter* spp. to fluoroquinolones are already available (116). Moreover, the propensity for bacterial cross-resistance to fluoroquinolones adds yet another cynical dimension to this situation (195, 247). The importance of *Salmonella* vehicles other than raw poultry and meats and derived products cannot be minimized. Cases of human salmonellosis associated with the consumption of fresh fruits and vegetables, spices, chocolate, and milk products (Tables 10.6 to 10.11) reiterate the importance of sanitary practices during the harvesting, processing, and distribution of raw foods and food ingredients (114, 120).

The arsenal of virulence factors that enable *Salmonella* spp. to evade the various antibacterial host defense mechanisms is remarkable yet disconcerting. The physiological adaptability of salmonellae to hostile conditions in the natural environment safeguards their survival and infectious potential (Table 10.3). In addition, the pathogen benefits from chromosome-, phage-, and plasmid-carried virulence determinants that provide for its facilitated attachment and invasion of host cells (*inv* and *prgH*) (Fig. 10.5); resistance to intraphagocyte acidity (*phoP/phoQ*), complement lysis (Vi, LPS, *phoP/phoQ*, and *rck*), and antibacterial substances (*ompF*, *ompC*, *ompD*, and *phoE*); widespread invasion of deep host tissues (*spv*); competition for available Fe^{3+} (siderophores); and induction of diarrhea.

The problem of *Salmonella* in the global food chain and its current and projected repercussions on human health are cause for concern. Changes in agricultural and aquaculture practices are needed to reduce the incidence of human foodborne salmonellosis. The foregoing considerations on the pathogenicity and human infectious dose for foodborne salmonellosis underscore the primacy of sensitive and cost-effective methods for the detection of salmonellae in raw and processed foods. Although a detailed discussion on the diagnostic principles and performance of cultural and commercially available rapid molecular methods is beyond the scope of this chapter, a brief overview of the operating characteristics and sensitivity of PCR technology, which is rapidly gaining application in food microbiology, is warranted. PCR technology has made significant advances in diagnostic microbiology, becoming an important tool for the rapid detection of bacterial, viral, and protozoan pathogens (138, 335, 355). While it is unlikely that PCR techniques will ever completely replace culture-based methods for detection, this novel technology, when appropriately utilized, can help to make important and timely decisions within the food microbiology laboratory (406). Important developments in the processing of samples for PCR analysis (282), increased sensitivities of PCR techniques (354), shortened detection times (307), and standardized use in foods (307) have been reported. With recent advances in PCR technologies, real-time PCR and PCR–enzyme-linked immunosorbent assay may provide the user with a qualitative (yes/no) answer as well as a quantitative assessment of pathogen loads in foods (233, 296). When the cell number of pathogens in the assay sample is low (<100), the ability to accurately quantitate cells may be limited. Most PCR-based tests for *Salmonella* are targeted towards the detection of *invA*, a gene that is conserved and unique to this genus (381). Other genes with a more limited distribution in *S. enterica* have also been examined for the ability to detect specific *Salmonella* serovars (e.g., serovar Enteritidis [127]), phage types (225), or MDR *Salmonella* strains (79). PCR has also been examined as a possible tool for serotyping *S. enterica* by targeting a gene(s), gene combinations, or sequences responsible for the antigenic variability in *S. enterica* (226, 301). As a molecular typing tool, PCR has also proven useful for identifying strain differences (432) where other molecular tools, such as pulsed-field gel electrophoresis (PFGE), fail (236). In these typing approaches, primers target repetitive elements (enterobacterial repetitive intergenic consensus [ERIC PCR] [474]) or short 10-mer oligonucleotides (random amplified polymorphic DNA

[RAPD] or arbitrary-primer PCR [236]) that randomly bind within the bacterial genome, resulting in distinctive DNA patterns. However, unlike PFGE, ERIC and RAPD PCR assays have not been standardized to allow for lab-to-lab comparisons of isolates. Recently, a new PCR typing tool, multilocus sequence typing (MLST), has been examined as a possible replacement for PFGE for subtyping bacterial isolates (403). MLST is a DNA sequence-based typing tool that targets about seven genes for PCR amplification followed by sequencing to identify allele differences within a bacterial population (145, 432). Genetic relatedness among clinical isolates can be inferred from a collective comparison of all genes typed by this method. The advantages of this methodology over PFGE include the fact that costs associated with DNA sequencing have been considerably reduced to make it affordable, no additional specialized equipment or software is required for a laboratory already set up for PCR, and there is at least one website for users to compare their results against a database of information deposited by users (2). The challenge is to identify the set of genes that exhibit enough sequence diversity to be included in MLST for organism X and to obtain agreement by the scientific community on the genes selected for MLST (148, 273, 432). Other molecular diagnostic tools on the horizon include microarrays, a nucleic acid hybridization technology for the concurrent detection of a myriad of genes on a single platform. Detection of a specific gene(s) or organism is based on the selective capture of fluorescently labeled DNA or RNA to its corresponding, complementary 50-mer oligonucleotide or PCR amplicon at a preselected position on a nylon membrane or glass slide (168). Presently, this technology is experimental, used to study evolution (84), genetic diversity (374), and gene expression (378) in *S. enterica*, but it has great promise as a tool that can be adapted to the diagnostic or food microbiology laboratory (469).

We gratefully thank Stan Bailey, USDA-ARS, Russell Research Center, Athens, Ga., for his insightful comments and valued input to this chapter.

References

1. **Aabo, S., and D. L. Baggesen.** 1997. Growth of *Salmonella* Newport in naturally contaminated alfalfa sprouts and estimation of infectious dose in a Danish *Salmonella* Newport outbreak due to alfalfa sprouts, p. 425–426. *In Proceedings of Salmonella and Salmonellosis*, Ploufragan, France.

2. **Achtman, M.** 2006. MLST databases at the MPI für Infektionsbiologie. http://web.mpiib-berlin.mpg.de/mlst/. Accessed 8 February 2006.

3. **Ahmer, B. M. M., J. van Reeuwijk, P. R. Watson, T. S. Wallis, and F. Heffron.** 1999. *Salmonella* SirA is a global regulator of genes mediating enteropathogenesis. *Mol. Microbiol.* 31:971–982.

4. **Ahmer, B. M., M. Tran, and F. Heffron.** 1999. The virulence plasmid of *Salmonella typhimurium* is self-transmissible. *J. Bacteriol.* 181:1364–1368.

5. **Airoldi, A. A., and E. A. Zottola.** 1988. Growth and survival of *Salmonella typhimurium* at low temperature in nutrient deficient media. *J. Food Sci.* 53:1511–1513.

6. **Alpuche-Aranda, C. M., E. L. Racoussin, J. A. Swanson, and S. I. Miller.** 1994. *Salmonella* stimulate macrophage macropinocytosis and persist within spacious phagosomes. *J. Exp. Med.* 179:601–608.

7. **Alpuche-Aranda, C. M., J. A. Swanson, W. P. Loomis, and S. I. Miller.** 1992. *Salmonella typhimurium* activates virulence gene transcription within acidified macrophage phagosomes. *Proc. Natl. Acad. Sci. USA* 89:10079–10083.

8. **Altier, C., M. Suyemoto, and S. D. Lawhon.** 2000. Regulation of *Salmonella enterica* serovar Typhimurium invasion genes by *csrA*. *Infect. Immun.* 68:6790–6797.

9. **Altmeyer, R. M., J. K. McNern, J. C. Bossio, I. Rosenshine, B. B. Finlay, and J. E. Galan.** 1993. Cloning and molecular characterization of a gene involved in *Salmonella* adherence and invasion of cultured epithelial cells. *Mol. Microbiol.* 7:89–98.

10. **Amavisit, P., D. Lightfoot, G. F. Browning, and P. F. Markham.** 2003. Variation between pathogenic serovars within *Salmonella* pathogenicity islands. *J. Bacteriol.* 185:3624–3635.

11. **American Society for Microbiology.** 1995. *Report of the ASM Task Force on Antibiotic Resistance.* ASM Press, Washington, D.C.

12. **Anaraki, S., I. Giraudon, and S. Cathcart.** 2005. Large outbreak of *Salmonella* Enteritidis in north east London. *Eurosurveill. Wkly. Rep.* 10:050317.

13. **Anderson, S. M., L. Verchick, R. Sowadsky, B. Sun, R. Civen, J. C. Mohle-Boetani, S. B. Werner, M. Starr, S. Abbott, M. Gutierrez, M. Palumbo, J. Farrar, P. Shillam, E. Umland, M. Tanuz, M. Sewell, J. Cato, W. Keene, M. Goldoft, J. Hofmann, P. Waller, C. Braden, G. Djomand, M. Reller, and W. Chege.** 2002. Multistate outbreaks of *Salmonella* serotype Poona infections associated with eating cantaloupe from Mexico—United States and Canada, 2000–2002. *Morb. Mortal. Wkly. Rep.* 51:1044–1047.

14. **Andrews, W. H., V. R. Bruce, G. A. June, P. Sherrod, T. S. Hammack, and R. M. Amaguana.** 1995. *Salmonella*, chapter 5. *In Bacteriological Analytical Manual (BAM)*, 8th ed. AOAC International, Arlington, Va.

15. **Angelotti, R., M. J. Foter, and K. H. Lewis.** 1961. Time-temperature effects on salmonellae and staphylococci in foods. *Am. J. Public Health* 51:76–88.

16. **Anonymous.** 1977. *Aviation Catering. Report of Working Group.* World Health Organization (WHO) Regional Office for Europe, Copenhagen, Denmark.

17. **Anonymous.** 1986. Microbiological safety of vacuum-packed, high salt foods questioned. *Food Chem. News* 28:30–31.

18. **Anonymous.** 1995. Aquaculture cited by ASM as problem for antibiotic resistance. *Food Chem. News* 37:20–21.

19. **Anonymous.** 1995. Florida *Salmonella* outbreak caused changes in state inspections. *Food Chem. News* 37:27.

20. **Anonymous.** 1995. Ice-cream firm reaches tentative *Salmonella* case agreement. *Food Chem. News* 36:53.

21. **Anonymous.** 25 April 1999. Salmonellosis, dried squid—Japan (Tokyo). *Mainichi Daily News.*

22. **Anonymous.** 2000. Salmonellosis outbreak associated with raw mung bean sprouts. News release 22-00, April. California Department of Health Services, Sacramento, Calif.

23. **Anonymous.** 2001. *Salmonella* Stanley and *Salmonella* Newport in imported peanuts—international outbreak. *W. H. O. Newsl.* **70**:5.

24. **Anonymous.** 2001. International outbreak of *Salmonella* Typhimurium DT104 due to contaminated sesame seed products. Update from Germany, United Kingdom and Norway. *Eurosurveill. Wkly. Rep.* **5**:010816.

25. **Anonymous.** 2001. *Salmonella* Typhimurium definitive type 104 in halva. *Commun. Dis. Wkly. Rep.* **11**(34), 3 August.

26. Reference deleted.

27. **Anonymous.** 2002. Outbreak of multidrug-resistant *Salmonella* Newport—United States, January-April 2002. *Morb. Mortal. Wkly. Rep.* **51**:545–548.

28. **Anonymous.** 2003. Salmonellosis in Japan as of June 2003. *Infect. Agents Surveill. Rep. Jpn.* **24**:179–180.

29. **Anonymous.** 7 June 2005. Roast beef source of *Salmonella* that sickened 155 at Mother's Day brunch. *Canadian Press Wire.*

30. **Anonymous.** 2006. Multistate outbreak of *Salmonella* Typhimurium infections associated with eating ground beef—United States, 2004. *Morb. Mortal. Wkly. Rep.* **55**:180–182.

31. **Anonymous.** 2006. Outbreak of *Salmonella* Typhimurium phage type U302 in Ontario, spring 2005. *Can. Commun. Dis. Rep.* **32**:75–82.

32. **Arenas, G. N., A. S. Staskevich, A. Aballay, and L. S. Mayorga.** 2000. Intracellular trafficking of *Brucella abortus* in J774 macrophages. *Infect. Immun.* **68**:4255–4263.

33. **Armstrong, R. W., T. Fodor, G. T. Curlin, A. B. Cohen, G. K. Morris, W. T. Martin, and J. Feldman.** 1970. Epidemic *Salmonella* gastroenteritis due to contaminated imitation ice cream. *Am. J. Epidemiol.* **91**:300–307.

34. **Ashkenazi, S., T. G. Cleary, B. E. Murray, A. Wanger, and L. K. Pickering.** 1988. Quantitative analysis and partial characterization of cytotoxin production by *Salmonella* strains. *Infect. Immun.* **56**:3089–3094.

35. **Asperilla, M. A., R. A. Smego, Jr., and L. K. Scott.** 1990. Quinolone antibiotics in the treatment of *Salmonella* infections. *Rev. Infect. Dis.* **12**:873–889.

36. **Asplund, K., and E. Nurmi.** 1991. The growth of salmonellae in tomatoes. *Int. J. Food Microbiol.* **13**:177–182.

37. **Australian Health Commission.** 25 March 1999. End to *Salmonella* outbreak. Press release (Adelaide).

38. **Bacciu, D., G. Falchi, A. Spazziani, L. Bossi, G. Marogna, G. S. Leori, S. Rubino, and S. Uzzau.** 2004. Transposition of the heat-stable toxin *astA* gene into a gifsy-2-related prophage of *Salmonella enterica* serovar Abortusovis. *J. Bacteriol.* **186**:4568–4574.

39. **Baker, R. C., R. A. Qureshi, and J. H. Hotchkins.** 1986. Effect of an elevated level of carbon dioxide containing atmosphere on the growth of spoilage and pathogenic bacteria. *Poult. Sci.* **65**:729–737.

40. **Bang, I. S., B. H. Kim, J. W. Foster, and Y. K. Park.** 2000. OmpR regulates the stationary-phase acid tolerance response of *Salmonella enterica* serovar Typhimurium. *J. Bacteriol.* **182**:2245–2252.

41. **Barry, R. A., H. G. Bouwer, D. A. Portnoy, and D. J. Hinrichs.** 1992. Pathogenicity and immunogenicity of *Listeria monocytogenes* small-plaque mutants defective for intracellular growth and cell-to-cell spread. *Infect. Immun.* **60**:1625–1632.

42. **Baumler, A. J., A. J. Gilde, R. M. Tsolis, A. W. M. Van Der Velden, B. M. M. Ahmer, and F. Heffron.** 1997. Contribution of horizontal gene transfer and deletion events to development of distinctive patterns of fimbrial operons during evolution of *Salmonella* serotypes. *J. Bacteriol.* **179**:317–322.

43. **Baumler, A. J., R. M. Tsolis, F. A. Bowe, J. G. Kusters, S. Hoffmann, and F. Heffron.** 1996. The *pef* fimbrial operon of *Salmonella typhimurium* mediates adhesion to murine small intestine and is necessary for fluid accumulation. *Infect. Immun.* **64**:61–68.

44. **Baumler, A. J., R. M. Tsolis, T. A. Ficht, and L. G. Adams.** 1998. Evolution of host adaptation in *Salmonella enterica*. *Infect. Immun.* **66**:4579–4587.

45. **Beaber, J. W., B. Hochhut, and M. K. Waldor.** 2002. Genomic and functional analyses of SXT, an integrating antibiotic resistance gene transfer element derived from *Vibrio cholerae. J. Bacteriol.* **184**:4259–4269.

46. **Beales, N.** 2004. *Review of the Microbiological Risks Associated with Sprouted Seeds.* Report no. 41. Campden and Chorleywood Food Research Association Group, Chipping Campden, Gloucestershire, United Kingdom.

47. **Bean, N. H., J. S. Goulding, C. Lao, and F. J. Angulo.** 1996. Surveillance for foodborne disease outbreaks—United States, 1988–1992. *Morb. Mortal. Wkly. Rep.* **45**:1–66.

48. **Bearson, B. L., L. Wilson, and J. W. Foster.** 1998. A low pH-inducible, PhoPQ-dependent acid tolerance response protects *Salmonella typhimurium* against inorganic acid stress. *J. Bacteriol.* **180**:2409–2417.

49. **Beckers, H. J., M. S. M. Daniels-Bosman, A. Ament, J. Daenen, A. W. J. Hanekamp, P. Knipschild, A. H. H. Schuurmann, and H. Bijkerk.** 1985. Two outbreaks of salmonellosis caused by *Salmonella indiana*. A survey of the European Summit outbreak and its consequences. *Int. J. Food Microbiol.* **2**:185–195.

50. **Beers, A.** 1997. Maryland church dinner *Salmonella* outbreak blamed on faulty food preparation. *Food Chem. News* **39**:19–20.

51. **Behlau, I., and S. I. Miller.** 1993. A PhoP-repressed gene promotes *Salmonella typhimurium* invasion of epithelial cells. *J. Bacteriol.* **175**:4475–4484.

52. **Belden, W. J., and S. I. Miller.** 1994. Further characterization of the PhoP regulon: identification of new PhoP-activated virulence loci. *Infect. Immun.* **62**:5095–5101.

53. **Bellanti, J. A., B. J. Zeligs, S. Vetro, Y. H. Pung, S. Luccioli, M. J. Malavasic, A. M. Hooke, T. R. Ubertini, R. Vanni, and L. Nencioni.** 1993. Studies of safety, infectivity and immunogenicity of a new temperature-sensitive (ts) 51-1 strain of *Salmonella typhi* as a new live oral typhoid fever vaccine candidate. *Vaccine* II:587–590.

54. **Bergis, H., G. Poumeyrol, and A. Beaufort.** 1994. Etude de développement de la flore saprophyte et de *Salmonella* dans les viandes hachées conditionnées sous atmosphère modifiée. *Sci. Aliments* **14:**217–228.

55. **Bergstrom, C. T., M. Lipsitch, and B. R. Levin.** 2000. Natural selection, infectious transfer and the existence conditions for bacterial plasmids. *Genetics* **155:**1505–1519.

56. **Bergstrom, S., F. P. Lindberg, O. Olsson, and S. Normark.** 1983. Comparison of the overlapping *frd* and *ampC* operons of *Escherichia coli* with the corresponding DNA sequences in other gram-negative bacteria. *J. Bacteriol.* **155:**1297–1305.

57. **Besser, T. E., M. Goldoft, L. C. Pritchett, R. Khakhria, D. D. Hancock, D. H. Rice, J. M. Gay, W. Johnson, and C. C. Gay.** 2000. Multiresistant *Salmonella typhimurium* DT104 infections of humans and domestic animals in the Pacific Northwest of the United States. *Epidemiol. Infect.* **124:**193–200.

58. **Beuchat, L. R., B. V. Nail, B. B. Adler, and M. R. S. Clavero.** 1998. Efficacy of spray application of chlorinated water in killing pathogenic bacteria on raw apples, tomatoes and lettuce. *J. Food Prot.* **61:**1305–1311.

59. **Björnerot, L., A. Franklin, and E. Tysen.** 1996. Usage of antibacterial and antiparasitic drugs in animals in Sweden between 1988 and 1993. *Vet. Rec.* **21:**282–286.

60. **Blanc-Potard, A. B., and E. A. Groisman.** 1997. The *Salmonella selC* locus contains a pathogenicity island mediating intramacrophage survival. *EMBO J.* **16:**5376–5385.

61. **Blanc-Potard, A. B., F. Solomon, J. Kayser, and E. A. Groisman.** 1999. The SPI-3 pathogenicity island of *Salmonella enterica. J. Bacteriol.* **181:**998–1004.

62. **Blaser, M. J., and L. S. Newman.** 1982. A review of human salmonellosis. 1. Infective dose. *Rev. Infect. Dis.* **4:**1096–1106.

63. **Boase, J., S. Lipsky, P. Simani, S. Smith, C. Skilton, S. Greenman, S. Harrison, J. Duchin, M. Samadpour, R. Gautom, S. Lankford, T. Harris, K. Ly, D. Green, J. Kobyashi, E. DeBess, T. McGivern, S. Mauvais, V. Balan, D. Fleming, K. Sanchez, P. D. Vertz, J. C. Mohle-Boetani, D. Seuring, J. H. Goddard, S. A. Bidol, J. Bender, C. M. Sewell, I. N. Vold, L. Marengo, and J. Archer.** 1999. Outbreak of *Salmonella* serotype Muenchen infections associated with unpasteurized orange juice—United States and Canada. *Morb. Mortal. Wkly. Rep.* **48:**582–585.

64. **Bouvet, E., C. Jestin, and R. Ancelle.** 1986. Importance of exported cases of salmonellosis in the revelation of an epidemic, p.303. *In Proceedings of the Second World Congress on Foodborne Infections and Intoxications, Berlin, Germany.*

65. **Boyd, D., G. A. Peters, A. Cloeckaert, K. S. Boumedine, E. Chaslus-Dancla, H. Imberechts, and M. R. Mulvey.** 2001. Complete nucleotide sequence of a 43-kilobase genomic island associated with the multidrug resistance region of *Salmonella enterica* serovar Typhimurium DT104 and its identification in phage type DT120 and serovar Agona. *J. Bacteriol.* **183:**5725–5732.

65a. **Boyd, D. A., G. A. Peters, L. Ng, and M. R. Mulvey.** 2000. Partial characterization of a genomic island associated with the multidrug resistance region of *Salmonella enterica* DT104. *FEMS Microbiol. Lett.* **189:**285–291.

66. **Boyd, E. F., J. Li, H. Ochman, and R. K. Selander.** 1997. Comparative genetics of the *inv-spa* invasion gene complex of *Salmonella enterica. J. Bacteriol.* **179:**1985–1991.

67. **Brackett, R. E.** 1992. Shelf stability and safety of fresh produce as influenced by sanitation and disinfection. *J. Food Prot.* **55:**808–814.

68. **Brenner, F. W., R. G. Villar, F. J. Angulo, R. Tauxe, and B. Swaminathan.** 2000. *Salmonella* nomenclature. *J. Clin. Microbiol.* **38:**2465–2467.

69. **Brussow, H., C. Canchaya, and W. D. Hardt.** 2004. Phages and the evolution of bacterial pathogens: from genomic rearrangements to lysogenic conversion. *Microbiol. Mol. Biol. Rev.* **68:**560–602.

70. **Bryan, J. P., H. Rocha, and W. M. Scheld.** 1986. Problems in salmonellosis: rationale for clinical trials with newer beta-lactam agents and quinolones. *Rev. Infect. Dis.* **8:**189–203.

71. **Buchmeier, N. A., and F. Heffron.** 1991. Inhibition of macrophage phagosome-lysosome fusion by *Salmonella typhimurium. Infect. Immun.* **59:**2232–2238.

72. **Bulletin épidémiologique annuel.** 1999. *Epidémiologie des Maladies Infectieuses en France.* Situation en 1997 et tendances évolutives récentes. Réseau National de Santé Publique, Saint-Maurice, France.

73. **Bulling, E., R. Stephan, and V. Sebek.** 1973. The development of antibiotic resistance among *Salmonella* bacteria of animal origin in the Federal Republic of Germany and West Berlin: 1st communication: a comparison between the years of 1961 and 1970–1971. *Zentbl. Bakteriol. Mikrobiol. Hyg. Abt. 1. Orig. A* **225:**245–256.

74. **Bunning, V. K., R. B. Raybourne, and D. L. Archer.** 1988. Foodborne enterobacterial pathogens and rheumatoid disease. *J. Appl. Bacteriol. Symp.* **17**(Suppl.):87S–107S.

75. **Burslem, C. D., M. J. Kelly, and F. S. Preston.** 1990. Food poisoning—a major threat to airline operations. *J. Soc. Occup. Med.* **40:**97–100.

76. **Bush, K.** 2001. New beta-lactamases in gram-negative bacteria: diversity and impact on the selection of antimicrobial therapy. *Clin. Infect. Dis.* **32:**1085–1089.

77. **Butler, M. A.** 2000. *Salmonella* outbreak leads to juice recall in Western states. *Food Chem. News* **42:**19–20.

78. **Camps, N., A. Dominguez, M. Cy, M. Perez, J. Pardos, T. Llobet, M. A. Usera, L. Salleras, and the Working Group for the Investigation of the Outbreak of Salmonellosis in Torroella de Montgri.** 2005. A foodborne outbreak of *Salmonella* infection due to overproduction of egg-containing foods for a festival. *Epidemiol. Infect.* **133:**817–822.

79. **Carlson, S. A., L. F. Bolton, C. E. Briggs, H. S. Hurd, V. K. Sharma, P. J. Fedorka-Cray, and B. D. Jones.** 1999. Detection of multiresistant *Salmonella typhimurium* DT104 using multiplex and fluorogenic PCR. *Mol. Cell. Probes* **13:**213–222.

80. **Carmen Palomino, W., R. Lucia Aguad, L. Manuel Rodriguez, G. Graciela Cofre, and J. Villanueva.** 1986. Clinical course of infections by *Salmonella typhi*, paratyphus A and paratyphus B in relation to the sensitivity of the etiological agent to chloramphenicol. *Rev. Méd. Chile* **114:**919–927.

81. **Casewell, M., C. Friis, E. Marco, P. McMullin, and I. Phillips.** 2003. The European ban on growth-promoting antibiotics and emerging consequences for human and animal health. *J. Antimicrob. Chemother.* **52:**159–161.

82. Catsaras, M., and D. Grebot. 1984. Multiplication des *Salmonella* dans la viande hachée. *Bull. Acad. Vét. Fr.* 57:501–512.

83. Centers for Disease Control and Prevention. 2000. 1999 surveillance results. [Online.] www.cdc.gov/ncidod/dbmd/foodnet.

84. Chan, K., S. Baker, C. C. Kim, C. S. Detweiler, G. Dougan, and S. Falkow. 2003. Genomic comparison of *Salmonella enterica* serovars and *Salmonella bongori* by use of an *S. enterica* serovar Typhimurium DNA microarray. *J. Bacteriol.* 185:553–563.

85. Chatfield, S. N., C. J. Dorman, C. Hayward, and G. Dougan. 1991. Role of *ompR*-dependent genes in *Salmonella typhimurium* virulence: mutants deficient in both *ompC* and *ompF* are attenuated in vivo. *Infect. Immun.* 59:449–452.

86. Chatfield, S., M. Roberts, P. Londono, I. Cropley, G. Douce, and G. Dougan. 1993. The development of oral vaccines based on live attenuated *Salmonella* strains. *FEMS Immunol. Med. Microbiol.* 7:1–8.

87. Chen, S., A. Yee, M. Griffiths, C. Larkin, C. T. Yamashiro, R. Behari, C. Paszko-Kolva, K. Rahn, and S. A. De Grandis. 1997. The evaluation of a fluorogenic polymerase chain reaction assay for the detection of *Salmonella* species in food commodities. *Int. J. Food Microbiol.* 35:239–250.

88. Cherubin, C. E. 1981. Antibiotic resistance of *Salmonella* in Europe and the United States. *Rev. Infect. Dis.* 3:1105–1125.

89. Chiu, C. H., P. Tang, C. Chu, S. Hu, Q. Bao, J. Yu, Y. Y. Chou, H. S. Wang, and Y. S. Lee. 2005. The genome sequence of *Salmonella enterica* serovar Choleraesuis, a highly invasive and resistant zoonotic pathogen. *Nucleic Acids Res.* 33:1690–1698.

90. Chopra, A. K., C. W. Houston, J. W. Peterson, R. Prasad, and J. J. Mekalanos. 1987. Cloning and expression of the *Salmonella* enterotoxin gene. *J. Bacteriol.* 169:5095–5100.

91. Christmann, D., T. Staub, and Y. Hansmann. 1992. Manifestations extra-digestives des salmonelloses. *Méd. Mal. Infect.* 22:289–298.

92. Chung, K. C., and J. M. Goepfert. 1970. Growth of *Salmonella* at low pH. *J. Food Sci.* 35:326–328.

93. Cirillo, D. M., R. H. Valdivia, D. M. Monack, and S. Falkow. 1998. Macrophage-dependent induction of the *Salmonella* pathogenicity island 2 type III secretion system and its role in intracellular survival. *Mol. Microbiol.* 30:175–188.

94. Cloeckaert, A., S. Baucheron, and E. Chaslus-Dancla. 2001. Nonenzymatic chloramphenicol resistance mediated by IncC plasmid R55 is encoded by a *floR* gene variant. *Antimicrob. Agents Chemother.* 45:2381–2382.

95. Cohen, D. R., I. A. Porter, T. M. S. Reid, J. C. M. Sharp, G. I. Forbes, and G. M. Paterson. 1983. A cost benefit study of milk-borne salmonellosis. *J. Hyg.* 91:17–23.

96. Cohen, M. L., and R. V. Tauxe. 1986. Drug-resistant *Salmonella* in the United States: an epidemiologic perspective. *Science* 234:964–969.

97. Collazo C. M., and J. E. Galan. 1997. The invasion-associated type III system of *Salmonella typhimurium* directs the translocation of Sip proteins into the host cell. *Mol. Microbiol.* 24:747–756.

98. Collazo, C. M., M. K. Zierler, and J. E. Galan. 1995. Functional analysis of the *Salmonella typhimurium* invasion genes *inv*I and *inv*J and identification of a target of the protein secretion apparatus encoded in the *inv* locus. *Mol. Microbiol.* 15:25–38.

99. Communicable Disease Report (CDR) Weekly. 2005. Outbreaks of infection with *Salmonella* Enteritidis PT6 infection in the northeast of England associated with eggs. *CDR Wkly.* 15, no. 41 (13 October 2005).

100. Corby, S., V. Lanni, V. Kistler, V. Dato, A. Weltman, C. Yozviak, K. Waller, K. Nalluswami, M. Moll, J. Lockett, S. Montgomery, M. Lynch, C. Braden, S. K. Gupta, and A. DuBois. 2005. Outbreaks of *Salmonella* infections associated with eating roma tomatoes—United States and Canada, 2004. *Morb. Mortal. Wkly. Rep.* 54:325–328.

101. Corry, J. E. L. 1971. *The Water Relations and Heat Resistance of Microorganisms*, p. 1–42. Scientific and technical survey no. 73. The British Food Manufacturing Industries Research Association, Leatherhead, Surrey, United Kingdom.

102. Corry, J. E. L. 1976. The safety of intermediate moisture foods with respect to *Salmonella*, p. 215–238. *In* R. Davies, G. G. Birch, and K. J. Parker (ed.), *Intermediate Moisture Foods*. Applied Science Publishers Ltd., London, United Kingdom.

103. Cowden, J., S. O'Brien, B. Adak, et al. 2003. Outbreak of *Salmonella* Bareilly in Great Britain. Results from the case-control study. *Eurosurveill. Wkly. Rep.* 7:031031.

104. Craven, P. C., D. C. Mackel, W. B. Baine, W. H. Barker, E. J. Gangarosa, M. Goldfield, H. Rosenfeld, R. Altman, G. Lachapelle, J. W. Davies, and R. C. Swanson. 1975. International outbreak of *Salmonella eastbourne* infection traced to contaminated chocolate. *Lancet* i:788–793.

105. Crook, P. D., J. F. Aguilera, E. J. Threlfall, S. J. O'Brien, G. Sigmunddsdottir, D. Wilson, I. S. T. Fisher, A. Ammon, H. Briem, J. M. Cowden, M. E. Locking, H. Tschäpe, W. van Pelt, L. R. Ward, and M. A. Widdowson. 2003. A European outbreak of *Salmonella enterica* serotype Typhimurium definitive phage type 204b in 2000. *Clin. Microbiol. Infect.* 9:839–845.

106. Crosa, J. H. 1989. Genetics and molecular biology of siderophore-mediated iron transport in bacteria. *Microbiol. Rev.* 53:517–530.

107. Crosa, J. H., D. J. Brenner, W. H. Ewing, and S. Falkow. 1973. Molecular relationships among the salmonellae. *J. Bacteriol.* 115:307–315.

108. Csonka, L. N. 1989. Physiological and genetic responses of bacteria to osmotic stress. *Microbiol. Rev.* 53:121–147.

109. Cukan, A. 13 October 2005. *Caregiving: Food Safety Is a Big Concern*. United Press International.

110. Cummings, K., E. Barrett, J. C. Mohle-Boetani, J. T. Brooks, J. Farrar, T. Hunt, A. Fiore, K. Komatsu, S. B. Werner, and L. Slutsker. 2001. A multistate outbreak of *Salmonella enterica* serotype Baildon associated with domestic raw tomatoes. *Emerg. Infect. Dis.* 7:1046–1048.

111. Curtiss, R., III, S. M. Kelly, and J. O. Hassan. 1993. Live oral avirulent *Salmonella* vaccines. *Vet. Microbiol.* 37:397–405.

112. Dalton, C. B., J. Gregory, M. D. Kirk, R. J. Stafford, R. Givney, E. Kraa, and D. Gould. 2004. Foodborne disease outbreaks in Australia. *Commun. Dis. Intell.* 28:211–224.

113. D'Aoust, J.-Y., B. J. Aris, P. Thisdele, A. Durante, N. Brisson, D. Dragon, G. Lachapelle, M. Johnston, and R. Laidley. 1975. *Salmonella eastbourne* outbreak associated with chocolate. *Can. Inst. Food Sci. Technol. J.* 8:181–184.

114. D'Aoust, J.-Y. 1977. *Salmonella* and the chocolate industry. A review. *J. Food Prot.* 40:718–727.

115. D'Aoust, J.-Y., D. W. Warburton, and A. M. Sewell. 1985. *Salmonella typhimurium* phage-type 10 from cheddar cheese implicated in a major Canadian foodborne outbreak. *J. Food Prot.* 48:1062–1066.

116. D'Aoust, J.-Y. 1989. *Salmonella*, p. 327–445. *In* M. P. Doyle (ed.), *Foodborne Bacterial Pathogens*. Marcel Dekker, New York, N.Y.

117. D'Aoust, J.-Y. 1991. Pathogenicity of foodborne *Salmonella*. *Int. J. Food Microbiol.* 12:17–40.

118. D'Aoust, J.-Y. 1991. Psychrotrophy and foodborne *Salmonella*. *Int. J. Food Microbiol.* 13:207–216.

119. D'Aoust, J.-Y., A. M. Sewell, E. Daley, and P. Greco. 1992. Antibiotic resistance of agricultural and foodborne *Salmonella* isolates in Canada: 1986–1989. *J. Food Prot.* 55:428–434.

120. D'Aoust, J.-Y. 1994. *Salmonella* and the international food trade. *Int. J. Food Microbiol.* 24:11–31.

121. D'Aoust, J.-Y., and U. Purvis. 1998. Isolation and identification of *Salmonella* from foods. MFHPB-20. Health Protection Branch, Health Canada, Ottawa, Ontario, Canada.

122. D'Aoust, J.-Y. 2000. *Salmonella*, p. 1233–1299. *In* B. M. Lund, A. C. Baird-Parker, and G. W. Gould (ed.), *The Microbiological Safety and Quality of Food*. Aspen Publishers, Gaithersburg, Md.

123. D'Aoust, J.-Y. 2001. Foodborne salmonellosis: current international concerns. *Food Safety* 7:10–17, 51.

124. Darwin, K. H., and V. L. Miller. 1999. InvF is required for expression of genes encoding proteins secreted by the SPI1 type III secretion apparatus in *Salmonella typhimurium*. *J. Bacteriol.* 181:4949–4954.

125. Davies, J. 1994. Inactivation of antibiotics and the dissemination of resistance genes. *Science* 264:375–382.

126. D'Costa, V. M., K. M. McGrann, D. W. Hughes, and G. D. Wright. 2006. Sampling the antibiotic resistome. *Science* 311:374–377.

127. Dechet, A. M., E. Scallan, K. Gensheimer, R. Hoekstra, J. Gunderman-King, J. Lockett, D. Wrigley, W. Chege, J. Sobel, and Multistate Working Group. 2006. Outbreak of multidrug-resistant *Salmonella enterica* serotype Typhimurium definitive type 104 infection linked to commercial ground beef, northeastern United States, 2003–2004. *Clin. Infect. Dis.* 42:747–752.

128. De Medici, D., L. Croci, E. Delibato, S. Di Pasquale, E. Filetici, and L. Toti. 2003. Evaluation of DNA extraction methods for use in combination with SYBR green I real-time PCR to detect *Salmonella enterica* serotype Enteritidis in poultry. *Appl. Environ. Microbiol.* 69:3456–3461.

129. Dewanti, R., and M. P. Doyle. 1992. Influence of cultural conditions on cytotoxin production by *Salmonella enteritidis*. *J. Food Prot.* 55:28–33.

130. Dorman, C. J., S. Chatfield, C. F. Higgins, C. Hayward, and G. Dougan. 1989. Characterization of porin and *ompR* mutants of a virulent strain of *Salmonella typhimurium*: *ompR* mutants are attenuated in vivo. *Infect. Immun.* 57:2136–2140.

131. Doublet, B., A. Carattoli, J. M. Whichard, D. G. White, S. Baucheron, E. Chaslus-Dancla, and A. Cloeckaert. 2004. Plasmid-mediated florfenicol and ceftriaxone resistance encoded by the *floR* and *bla*(CMY-2) genes in *Salmonella enterica* serovars Typhimurium and Newport isolated in the United States. *FEMS Microbiol. Lett.* 233:301–305.

132. Doublet, B., D. Boyd, M. R. Mulvey, and A. Cloeckaert. 2005. The *Salmonella* genomic island 1 is an integrative mobilizable element. *Mol. Microbiol.* 55:1911–1924.

133. Doublet, B., P. Butaye, H. Imberechts, D. Boyd, M. R. Mulvey, E. Chaslus-Dancla, and A. Cloeckaert. 2004. *Salmonella* genomic island 1 multidrug resistance gene clusters in *Salmonella enterica* serovar Agona isolated in Belgium in 1992 to 2002. *Antimicrob. Agents Chemother.* 48:2510–2517.

134. Doublet, B., R. Lailler, D. Meunier, A. Brisabois, D. Boyd, M. R. Mulvey, E. Chaslus-Dancla, and A. Cloeckaert. 2003. Variant *Salmonella* genomic island 1 antibiotic resistance gene cluster in *Salmonella enterica* serovar Albany. *Emerg. Infect. Dis.* 9:585–591.

135. Dougan, G. 1994. The molecular basis for the virulence of bacterial pathogens: implications for oral vaccine development. *Microbiology* 140:215–224.

136. Drociuk, D., S. Carnasale, G. Elliot, L. J. Bell, J. J. Gibson, L. Wolf, D. Briggs, B. Jenkins, J. M. Maillard, M. Huddle, F. Virgin, C. Braden, P. Srikantiah, A. Stoica, and T. Chiller. 2003. Outbreaks of *Salmonella* serotype Enteritidis infection associated with eating shell eggs—United States, 1999–2001. *Morb. Mortal. Wkly. Rep.* 51:1149–1151.

137. Droffner, M. L., and N. Yamamoto. 1992. Procedure for isolation of *Escherichia*, *Salmonella*, and *Pseudomonas* mutants capable of growth at the refractory temperature of 54°C. *J. Microbiol. Methods* 14:201–206.

138. D'Souza, D. H., and L. Jaykus. 2006. Molecular approaches for the detection of foodborne viral pathogens, p. 91–118. *In* J. Maurer (ed.), *PCR Methods in Foods*. Springer, New York, N.Y.

139. Eckmann, L., J. R. Smith, M. P. Housley, M. B. Dwinell, and M. F. Kagnoff. 2000. Analysis by high-density cDNA arrays of altered gene expression in human intestinal epithelial cells in response to infection with the invasive enteric bacteria *Salmonella*. *J. Biol. Chem.* 275:14084–14094.

140. Edwards, R. A., G. J. Olsen, and S. R. Maloy. 2002. Comparative genomics of closely related salmonellae. *Trends Microbiol.* 10:94–99.

141. Eichelberg, K., and J. E. Galan. 1999. Differential regulation of *Salmonella typhimurium* type III secreted proteins by pathogenicity island 1-encoded transcriptional activators InvF and HilA. *Infect. Immun.* 67:4099–4105.

142. Eichelberg, K., C. C. Ginocchio, and J. E. Galan. 1994. Molecular and functional characterization of the *Salmonella typhimurium* invasion genes *invB* and *invC*: homology of *invC* to the F_0F_1 ATPase family of proteins. *J. Bacteriol.* 176:4501–4510.

143. Encheva, V., R. Wait, S. E. Gharbia, S. Begum, and H. N. Shah. 2005. Proteome analysis of serovars Typhimurium and Pullorum of *Salmonella enterica* subspecies I. *BMC Microbiol.* 5:42.

144. **Endtz, H. P., G. J. Ruijs, B. van Klingeren, W. H. Jansen, T. van der Reyden, and R. P. Mouton.** 1991. Quinolone resistance in *Campylobacter* isolated from man and poultry following the introduction of fluoroquinolones in veterinary medicine. *J. Antimicrob. Chemother.* **27:**199–208.

145. **Enright, M. C., and B. G. Spratt.** 1999. Multilocus sequence typing. *Trends Microbiol.* **7:**482–487.

146. **Ewing, W. H.** 1972. The nomenclature of *Salmonella*, its usage, and definitions for the three species. *Can. J. Microbiol.* **18:**1629–1637.

147. **Fairchild, A. S., J. L. Smith, U. Idris, J. Lu, S. Sanchez, L. B. Purvis, C. Hofacre, and M. D. Lee.** 2005. Effects of orally administered tetracycline on the intestinal community structure of chickens and on *tet* determinant carriage by commensal bacteria and *Campylobacter jejuni*. *Appl. Environ. Microbiol.* **71:**5865–5872.

148. **Fakhr, M. K., L. K. Nolan, and C. M. Logue.** 2005. Multilocus sequence typing lacks the discriminatory ability of pulsed-field gel electrophoresis for typing *Salmonella enterica* serovar Typhimurium. *J. Clin. Microbiol.* **43:**2215–2219.

149. **Falkow, S., R. R. Isberg, and D. A. Portnoy.** 1992. The interaction of bacteria with mammalian cells. *Annu. Rev. Cell Biol.* **8:**333–363.

150. **Fernandez, M., J. Sierra-Madero, H. de la Vega, M. Vazquez, Y. Lopez-Vidal, G. M. Ruiz-Palacios, and E. Calva.** 1988. Molecular cloning of a *Salmonella typhi* LT-like enterotoxin gene. *Mol. Microbiol.* **2:**821–825.

151. **Ferreira, M. A. S. S., and B. M. Lund.** 1987. The influence of pH and temperature on initiation of growth of *Salmonella* spp. *Lett. Appl. Microbiol.* **5:**67–70.

152. **Fey, P. D., T. J. Safranek, M. E. Rupp, E. F. Dunne, E. Ribot, P. C. Iwen, P. A. Bradford, F. J. Angulo, and S. H. Hinrichs.** 2000. Ceftriaxone-resistant salmonella infection acquired by a child from cattle. *N. Engl. J. Med.* **342:**1242–1249.

153. **Fields, P. I., E. A. Groisman, and F. Herron.** 1989. A *Salmonella* locus that controls resistance to microbicidal proteins from phagocytic cells. *Science* **243:**1059–1060.

154. **Finlay, B. B.** 1994. Molecular and cellular mechanisms of *Salmonella* pathogenesis. *Curr. Top. Microbiol. Immunol.* **192:**163–185.

155. **Fluit, A. C., and F. J. Schmitz.** 1999. Class 1 integrons, gene cassettes, mobility, and epidemiology. *Eur. J. Clin. Microbiol. Infect. Dis.* **18:**761–770.

156. **Fontaine, R. E., M. L. Cohen, W. T. Martin, and T. M. Vernon.** 1980. Epidemic salmonellosis from cheddar cheese: surveillance and prevention. *Am. J. Epidemiol.* **111:**247–253.

157. **Fontaine, R. E., S. Arnon, W. T. Martin, T. M. Vernon, E. J. Gangarosa, J. J. Farmer, A. B. Moran, J. H. Silliker, and D. L. Decker.** 1978. Raw hamburger: an interstate common source of human salmonellosis. *Am. J. Epidemiol.* **107:**36–45.

158. **Food and Drug Administration.** March 1999. *A Proposed Framework for Evaluating and Assuring the Human Safety of the Microbial Effects of Antimicrobial New Animal Drugs Intended for Use in Food-Producing Animals.* U. S. Food and Drug Administration, Washington, D.C.

159. **Food Safety Net—Canada.** 27 October 2001. Romanian kids sickened by *Salmonella*. www.foodsafetynetwork.ca.

160. **Food Safety Net—Canada.** 17 January 2003. Gastro cases top 100 as likely cause found. www.foodsafetynetwork.ca.

161. **Food Safety Net—Canada.** 28 September 2005. Salmonellosis, political gathering—Honduras, Atlantida. www.foodsafetynetwork.ca.

162. **Food Safety Net—Canada.** 3 June 2005. Salmonellosis, foodborne, fatal—USA (South Carolina): turkey. www.foodsafetynetwork.ca.

163. **Foster, J. W., and B. Bearson.** 1994. Acid-sensitive mutants of *Salmonella typhimurium* identified through a dinitrophenol lethal screening strategy. *J. Bacteriol.* **176:**2596–2602.

164. **Foster, J. W., and H. K. Hall.** 1991. Inducible pH homeostasis and the acid tolerance response of *Salmonella typhimurium*. *J. Bacteriol.* **173:**5129–5135.

165. **Fournier, P. E., D. Vallenet, V. Barbe, S. Audic, H. Ogata, L. Poirel, H. Richet, C. Robert, S. Mangenot, C. Abergel, P. Nordmann, J. Weissenbach, D. Raoult, and J. M. Claverie.** 2006. Comparative genomics of multidrug resistance in *Acinetobacter baumannii*. *PLoS Genet.* **2:**e7.

166. **Francis, B. J., J. V. Altamirano, M. G. Stobierski, W. Hall, B. Robinson, S. Dietrich, R. Martin, F. Downes, K. R. Wilcox, C. Hedberg, R. Wood, M. Osterholm, G. Genese, M. J. Hung, S. Paul, K. C. Spitalny, C. Whalen, and J. Spika.** 1991. Multistate outbreak of *Salmonella poona* infections—United States and Canada, 1991. *Morb. Mortal. Wkly. Rep.* **40:**549–552.

167. **Franklin, A.** 1997. Current status of antibiotic resistance in animal production in Sweden, p. 229–235. *In Report of a WHO Meeting.* WHO/EMC/ZOO/97.4. World Health Organization, Geneva, Switzerland.

168. **Freeman, W. M., D. J. Robertson, and K. E. Vrana.** 2000. Fundamentals of DNA hybridization arrays for gene expression analysis. *Biotechnology* **29:**1042–1055.

169. **Friedrich, M. J., N. E. Kinsey, J. Vila, and R. J. Kadner.** 1993. Nucleotide sequence of a 13.9 kb segment of the 90 kb virulence plasmid of *Salmonella typhimurium*: the presence of fimbrial biosynthetic genes. *Mol. Microbiol.* **8:**543–558.

170. **Fu, Y., and J. E. Galan.** 1998. The *Salmonella typhimurium* tyrosine phosphatase SptP is translocated into host cells and disrupts the actin cytoskeleton. *Mol. Microbiol.* **27:**359–368.

171. **Galan, J. E., and D. Zhou.** 2000. Striking a balance: modulation of the actin cytoskeleton by *Salmonella*. *Proc. Natl. Acad. Sci. USA* **95:**8754–8761.

172. **Galan, J. E., and C. Ginocchio.** 1994. The molecular genetic bases of *Salmonella* entry into mammalian cells. *Biochem. Soc. Trans.* **22:**301–306.

173. **Galan, J. E., and R. Curtiss III.** 1991. Distribution of the *invA*, *-B*, *-C*, and *-D* genes of *Salmonella typhimurium* among other *Salmonella* serovars: *invA* mutants of *Salmonella typhi* are deficient for entry into mammalian cells. *Infect. Immun.* **59:**2901–2908.

174. **Galan, J. E., C. Ginocchio, and P. Costeas.** 1992. Molecular and functional characterization of the *Salmonella* invasion gene *invA*: homology of InvA to members of a new protein family. *J. Bacteriol.* **174:**4338–4349.

175. Garcia-del Portillo, F., J. W. Foster, and B. B. Finlay. 1993. Role of acid tolerance response genes in *Salmonella typhimurium* virulence. *Infect. Immun.* **61**:4489–4492.

176. Garcia-del Portillo, F., M. B. Zwick, K. Y. Leung, and B. B. Finlay. 1994. Intracellular replication of *Salmonella* within epithelial cells is associated with filamentous structures containing lysosomal membrane glycoproteins. *Infect. Agents Dis.* **2**:227–231.

177. Garcia-del-Portillo, F., and B. B. Finlay. 1994. Invasion and intracellular proliferation of *Salmonella* within non-phagocytic cells. *Microbiologia SEM* **10**:229–238.

178. Geiss, H. K., I. Ehrhard, A. Rösen-Wolff, H. G. Sonntag, J. Pratsch, A. Wirth, D. Krüger, I. Knollmann-Schanbacher, H. Kühn, and C. Treiber-Klötzer. 1993. Foodborne outbreak of a *Salmonella enteritidis* epidemic in a major pharmaceutical company. *Gesundheits-w.* **55**:127–132.

179. Gibson, A. M., N. Bratchell, and T. A. Roberts. 1988. Predicting microbial growth: growth responses of salmonellae in laboratory medium as affected by pH, sodium chloride and storage temperature. *Int. J. Food Microbiol.* **6**:155–178.

180. Gibson, M. M., E. M. Ellis, K. A. Graeme-Cook, and C. F. Higgins. 1987. OmpR and EnvZ are pleiotropic regulatory proteins: positive regulation of the tripeptide permease (*tppB*) of *Salmonella typhimurium*. *Mol. Gen. Genet.* **207**:120–129.

181. Giles, W. P., A. K. Benson, M. E. Olson, R. W. Hutkins, J. M. Whichard, P. L. Winokur, and P. D. Fey. 2004. DNA sequence analysis of regions surrounding bla_{CMY-2} from multiple *Salmonella* plasmid backbones. *Antimicrob. Agents Chemother.* **48**:2845–2852.

182. Gill, O. N., P. N. Socket, C. L. R. Bartlett, M. S. B. Vaile, B. Rowe, R. J. Gilbert, C. Dulake, H. C. Murrell, and S. Salmaso. 1983. Outbreak of *Salmonella* Napoli infection caused by contaminated chocolate bars. *Lancet* i:574–577.

183. Gillespie, I. 2004. Outbreak of Salmonella Newport infection associated with lettuce in the U. K. *Eurosurveill. Rep.* **8**:041007.

184. Gilmore, A. 1986. Chloramphenicol and the politics of health. *Can. Med. Assoc. J.* **134**:423–435.

185. Gilsdorf, A., A. Jansen, K. Alpers, H. Dieckmann, U. van Treeck, A. M. Hauri, G. Fell, M. Littmann, P. Rautenberg, R. Prager, W. Rabsch, P. Roggentin, A. Schroeter, A. Miko, E. Bartelt, J. Bräunig, and A. Ammon. 2005. A nationwide outbreak of *Salmonella* Bovismorbificans PT24, Germany, December 2004–March 2005. *Eurosurveill. Wkly. Rep.* **10**:050324.

186. Ginocchio, C., J. Pace, and J. E. Galan. 1992. Identification and molecular characterization of a *Salmonella typhimurium* gene involved in triggering the internalization of salmonellae into cultured epithelial cells. *Proc. Natl. Acad. Sci. USA* **89**:5976–5980.

187. Ginocchio, C. C., S. B. Olmsted, C. L. Wells, and J. E. Galan. 1994. Contact with epithelial cells induces the formation of surface appendages on *Salmonella typhimurium*. *Cell* **76**:717–724.

188. Glynn, J. R., and D. J. Bradley. 1992. The relationship of infecting dose and severity of disease in reported outbreaks of *Salmonella* infections. *Epidemiol. Infect.* **109**:371–388.

189. Glynn, J. R., and S. R. Palmer. 1992. Incubation period, severity of disease, and infecting dose: evidence from a *Salmonella* outbreak. *Am. J. Epidemiol.* **136**:1369–1377.

190. Glynn, M. K., C. Bopp, W. Dewitt, P. Dabney, M. Mokhtar, and F. J. Angulo. 1998. Emergence of multidrug-resistant *Salmonella enterica* serotype Typhimurium DT104 infections in the United States. *N. Engl. J. Med.* **338**:1333–1338.

191. Goepfert, J. M., I. K. Iskander, and C. H. Amundson. 1970. Relation of the heat resistance of salmonellae to the water activity of the environment. *Appl. Microbiol.* **19**:429–433.

192. Goldstein, C., M. D. Lee, S. Sanchez, C. R. Hudson, B. Phillips, B. Register, M. Grady, C. Liebert, A. O. Summers, D. G. White, and J. J. Maurer. 2001. Incidence of class 1 and 2 integrases in clinical and normal flora bacteria from livestock, companion animals, and exotics. *Antimicrob. Agents Chemother.* **45**:723–726.

193. Gordenker, A. 1999. Common seed source identified in Canadian outbreak. *Food Chem. News* **41**:23.

194. Greenwood, M. H., and W. L. Hooper. 1983. Chocolate bars contaminated with *Salmonella napoli*: an infectivity study. *Br. Med. J.* **286**:1394.

195. Griggs, D. J., M. C. Hall, Y. F. Jin, and L. J. V. Piddock. 1994. Quinolone resistance in veterinary isolates of *Salmonella*. *J. Antimicrob. Chemother.* **33**:1173–1189.

196. Groisman, E. A., and H. Ochman. 1993. Cognate gene clusters govern invasion of host epithelial cells by *Salmonella typhimurium* and *Shigella flexneri*. *EMBO J.* **12**:3779–3787.

197. Guerin, P. J., B. De Jong, E. Heir, V. Hasseltvedt, G. Kapperud, K. Styrmo, B. Gondrosen, J. Lassen, Y. Andersson, and P. Aavitsland. 2004. Outbreak of *Salmonella* Livingstone infection in Norway and Sweden due to contaminated processed fish products. *Epidemiol. Infect.* **132**:889–895.

198. Guilloteau, L. A., T. S. Wallis, A. V. Gautier, S. McIntyre, D. J. Platt, and A. J. Lax. 1996. The *Salmonella* virulence plasmid enhances *Salmonella*-induced lysis of macrophages and influences inflammatory responses. *Infect. Immun.* **64**:3385–3393.

199. Guina, T., E. C. Yi, H. Wang, M. Hackett, and S. I. Miller. 2000. A *phoP*-regulated outer membrane protease of *Salmonella enterica* serovar Typhimurium promotes resistance to alpha-helical antimicrobial peptides. *J. Bacteriol.* **182**:4077–4086.

200. Guiney, D. G., F. C. Fang, M. Krause, and S. Libby. 1994. Plasmid-mediated virulence genes in non-typhoid *Salmonella* serovars. *FEMS Microbiol. Lett.* **124**:1–10.

201. Gulig, P. A., T. J. Doyle, J. A. Hughes, and H. Matsui. 1998. Analysis of host cells associated with the *spv*-mediated increased intracellular growth rate of *Salmonella typhimurium* in mice. *Infect. Immun.* **66**:2471–2485.

202. Gulig, P. A., H. Danbar, D. G. Guiney, A. J. Lax, F. Norel, and M. Rhen. 1993. Molecular analysis of *spv* virulence genes of the *Salmonella* virulence plasmids. *Mol. Microbiol.* **7**:825–830.

203. Gunn, J. S., and S. I. Miller. 1996. PhoP-PhoQ activates transcription of *prmAB*, encoding a two-component regulatory system involved in *Salmonella typhimurium*

antimicrobial peptide resistance. *J. Bacteriol.* **178:**6857–6864.

204. **Gunn, J. S., S. S. Ryan, J. C. van Velkinburgh, R. K. Ernst, and S. I. Miller.** 2000. Genetic and functional analysis of a PmrA-PmrB-regulated locus necessary for lipopolysaccharide modification, antimicrobial peptide resistance, and oral virulence of *Salmonella enterica* serovar Typhimurium. *Infect. Immun.* **68:**6139–6146.

205. **Guo, X., J. Chen, R. E. Brackett, and L. R. Beuchat.** 2001. Survival of *Salmonella* on and in tomato plants from the time of inoculation at flowering and early stages of fruit development through fruit ripening. *Appl. Environ. Microbiol.* **76:**4760–4764.

206. **Gupta, A., J. Fontana, C. Crowe, B. Bolstorff, A. Stout, S. van Duyne, M. P. Hoekstra, J. M. Whichard, T. J. Barrett, F. J. Angulo, and The National Antimicrobial Resistance Monitoring System PulseNet Working Group.** 2003. Emergence of multidrug-resistant *Salmonella enterica* serotype Newport infections resistant to expanded-spectrum cephalosporins in the United States. *J. Infect. Dis.* **188:**1707–1716.

207. **Hassan, J. O., and R. Curtiss.** 1994. Development and evaluation of an experimental vaccination program using a live avirulent *Salmonella typhimurium* strain to protect immunized chickens against challenge with homologous and heterologous *Salmonella* serotypes. *Infect. Immun.* **62:**5519–5527.

208. **Hayward, R. D., and V. Koronakis.** 1999. Direct nucleation and bundling of actin by the SipC protein of invasive *Salmonella*. *EMBO J.* **18:**4926–4934.

209. **Health Canada.** 1991. *Foodborne and Waterborne Disease in Canada. Annual Summary 1985–86.* Polyscience Publications, Morin Heights, Québec, Canada.

210. **Health Canada.** 2000. *Foodborne and Waterborne Disease in Canada. Annual Summary 1994–95.* Polyscience Publications, Laval, Québec, Canada.

211. **Health Canada.** 2006. *Provincial and Territorial Enteric Outbreaks in Canada, 1996–2003.* Foodborne, Waterborne and Zoonotic Infections Division, Centre for Infectious Disease Prevention and Control, Public Health Agency of Canada (information kindly provided by C. Tinga, J. Valcour, and K. Doré).

212. **Hedberg, C. W., F. J. Angulo, K. E. White, C. W. Langkop, W. L. Schell, M. G. Stobierski, A. Schuchat, J. M. Besser, S. Dietrich, L. Helsel, P. M. Griffin, J. W. McFarland, and M. T. Osterholm.** 1999. Outbreaks of salmonellosis associated with eating uncooked tomatoes: implications for public health. *Epidemiol. Infect.* **122:**385–393.

213. **Hefferman, E. J., J. Harwood, J. Fierer, and D. Guiney.** 1992. The *Salmonella typhimurium* virulence plasmid complement resistance gene *rck* is homologous to a family of virulence-related outer membrane protein genes, including *pagC* and *ail*. *J. Bacteriol.* **174:**84–91.

214. **Hefferman, E. J., L. Wu, J. Louie, S. Okamoto, J. Fierer, and D. G. Guiney.** 1994. Specificity of the complement resistance and cell association phenotypes encoded by the outer membrane protein genes *rck* from *Salmonella typhimurium* and *ail* from *Yersinia enterocolitica*. *Infect. Immun.* **62:**5183–5186.

215. **Heir, E., B. A. Lindstedt, T. M. Leegaard, E. Gjernes, and G. Kapperud.** 2004. Prevalence and characterization of

integrons in blood culture *Enterobacteriaceae* and gastrointestinal *Escherichia coli* in Norway and reporting of a novel class 1 integron-located lincosamide resistance gene. *Ann. Clin. Microbiol. Antimicrob.* **3:**12.

216. **Heiskanen, P., S. Taira, and M. Rhen.** 1994. Role of *rpoS* in the regulation of *Salmonella* plasmid virulence (*spv*) genes. *FEMS Microbiol. Lett.* **123:**125–130.

217. **Heithoff, D. M., C. P. Conner, P. C. Hanna, S. M. Julio, U. Hentschel, and M. J. Mahan.** 1997. Bacterial infection as assessed by in vivo gene expression. *Proc. Natl. Acad. Sci. USA* **94:**934–939.

218. **Hellström, L.** 1980. Food-transmitted *S. enteritidis* epidemic in 28 schools, p. 397–400. *In Proceedings of the World Congress on Foodborne Infections and Intoxications*, Berlin, Germany.

219. **Helms, M., S. Ethelberg, K. Molbak, and the DT104 Study Group.** 2005. International *Salmonella* Typhimurium DT104 infections, 1992–2001. *Emerg. Infect. Dis.* **11:**859–867.

220. **Helmuth, R., and D. Protz.** 1997. How to modify conditions limiting resistance in bacteria in animals and other reservoirs. *Clin. Infect. Dis.* **24:**8136–8138.

221. **Henderson, S. C., D. I. Bounous, and M. D. Lee.** 1999. Early events in the pathogenesis of avian salmonellosis. *Infect. Immun.* **67:**3580–3586.

222. **Henkel, J.** 1995. Ice cream linked to *Salmonella* outbreak. *FDA Consumer* **29:**30–31.

223. **Hensel, M., J. E. Shea, B. Raupach, D. Monack, S. Falkow, C. Gleeson, T. Kubo, and D. W. Holden.** 1997. Functional analysis of *ssaJ* and the *ssaK/U* operon, 13 genes encoding components of the type III secretion apparatus of *Salmonella* pathogenicity island 2. *Mol. Microbiol.* **24:**155–167.

224. **Hensel, M., J. E. Shea, S. R. Waterman, R. Mundy, T. Nikolaus, G. Banks, A. Vazquez-Torres, C. Gleeson, F. C. Fang, and D. W. Holden.** 1998. Genes encoding putative effector proteins of the type III secretion system of *Salmonella* pathogenicity island 2 are required for bacterial virulence and proliferation in macrophages. *Mol. Microbiol.* **30:**163–174.

225. **Hermans, A. P., T. Abee, M. H. Zwietering, and H. J. Aarts.** 2005. Identification of novel *Salmonella enterica* serovar Typhimurium DT104-specific prophage and nonprophage chromosomal sequences among serovar Typhimurium isolates by genomic subtractive hybridization. *Appl. Environ. Microbiol.* **71:**4979–4985.

226. **Herrera-Leon, S., J. R. McQuiston, M. A. Usera, P. I. Fields, J. Garaizar, and M. A. Echeita.** 2004. Multiplex PCR for distinguishing the most common phase-1 flagellar antigens of *Salmonella* spp. *J. Clin. Microbiol.* **42:**2581–2586.

227. **Hersh, D., D. M. Monack, M. R. Smith, N. Ghori, S. Falkow, and A. Zychlinsky.** 1999. The *Salmonella* invasin SipB induces macrophage apoptosis by binding to caspase-1. *Proc. Natl. Acad. Sci. USA* **96:**2396–2401.

228. **Hickey, E. W., and I. N. Hirshfield.** 1990. Low pH-induced effects of patterns of protein synthesis and on internal pH in *Escherichia coli* and *Salmonella typhimurium*. *Appl. Environ. Microbiol.* **56:**1038–1045.

229. **Hochhut, B., Y. Lotfi, D. Mazel, S. M. Faruque, R. Woodgate, and M. K. Waldor.** 2001. Molecular analysis

of antibiotic resistance gene clusters in *Vibrio cholerae* O139 and O1 SXT constins. *Antimicrob. Agents Chemother.* 45:2991–3000.

230. Holley, R. A., and M. Proulx. 1986. Use of egg washwater pH to prevent survival of *Salmonella* at moderate temperatures. *Poult. Sci.* 65:922–928.

231. Holmberg, S. D., M. T. Osterholm, K. A. Senger, and M. L. Cohen. 1984. Drug-resistant *Salmonella* from animal fed antimicrobials. *N. Engl. J. Med.* 311:617–622.

232. Holt, J., D. Propes, C. Patterson, T. Bannerman, L. Nicholson, M. Bundesen, E. Salehi, M. DiOrio, C. Kirchner, R. Tedrick, R. Duffy, and J. Mazurek. 2003. Multistate outbreak of *Salmonella* Typhimurium infections associated with drinking unpasteurized milk—Illinois, Indiana, Ohio and Tennessee, 2002–2003. *Morb. Mortal. Wkly. Rep.* 52:613–615.

233. Hong, Y., M. Berrang, T. Liu, C. Hofacre, S. Sanchez, L. Wang, and J. J. Maurer. 2003. Rapid detection of *Campylobacter coli*, *C. jejuni*, and *Salmonella enterica* on poultry carcasses using PCR–enzyme-linked immunosorbent assay. *Appl. Environ. Microbiol.* 69:3492–3499.

234. Horwitz, M. A., and F. R. Maxfield. 1984. *Legionella pneumophila* inhibits acidification of its phagosome in human monocytes. *J. Cell Biol.* 99:1936–1943.

235. Horwitz, M. A., R. A. Pollard, M. H. Merson, and S. M. Martin. 1977. A large outbreak of foodborne salmonellosis on the Navajo nation Indian reservation, epidemiology and secondary transmission. *Am. J. Public Health* 67:1071–1076.

236. Hudson, C. R., M. Garcia, R. K. Gast, and J. J. Maurer. 2001. Determination of close genetic relatedness of the major *Salmonella enteritidis* phage types by pulsed-field gel electrophoresis and DNA sequence analysis of several *Salmonella* virulence genes. *Avian Dis.* 45:875–886.

237. Hueck, C. J. 1998. Type III secretion systems in bacterial pathogens of animals and plants. *Microbiol. Mol. Biol. Rev.* 62:379–433.

238. Huhtanen, C. N. 1975. Use of pH gradient plates for increasing the acid tolerance of salmonellae. *Appl. Microbiol.* 29:309–312.

239. Humphrey, T. J., F. Jorgensen, J. A. Frost, H. Wadda, G. Domingue, N. C. Elviss, D. J. Griggs, and L. J. Piddock. 2005. Prevalence and subtypes of ciprofloxacin-resistant *Campylobacter* spp. in commercial poultry flocks before, during, and after treatment with fluoroquinolones. *Antimicrob. Agents Chemother.* 49:690–698.

240. Humphrey, T. J., N. P. Richardson, K. M. Statton, and R. J. Rowbury. 1993. Effects of temperature shift on acid and heat tolerance in *Salmonella enteritidis* phage type 4. *Appl. Environ. Microbiol.* 59:3120–3122.

241. Inami, G. B., and S. E. Moler. 1999. Detection and isolation of *Salmonella* from naturally contaminated alfalfa seeds following an outbreak investigation. *J. Food Prot.* 62:662–664.

242. Ingham, S. C., R. A. Alford, and A. P. McCown. 1990. Comparative growth rates of *Salmonella typhimurium* and *Pseudomonas fragi* on cooked crab meat stored under air and modified atmosphere. *J. Food Prot.* 53:566–567.

243. Institute of Medicine. 1988. *Report of a Study. Human Health Risks with the Subtherapeutic Use of Penicillin or Tetracyclines in Animal Feed*, p. 1–216. National Academy Press, Washington, D.C.

244. Institute of Medicine. 1998. *Antimicrobial Resistance: Issues and Options. Forum on Emerging Infection.* National Academy Press, Washington, D.C.

245. Isaacs, S., J. Arimini, B. Cieben, et al. 2005. An international outbreak of salmonellosis associated with raw almonds contaminated with a rare phage type of *Salmonella* Enteritidis. *J. Food Prot.* 68:191–198.

246. Isberg, R. R., and G. T. V. Nhieu. 1994. Two mammalian cell internalization strategies used by pathogenic bacteria. *Annu. Rev. Genet.* 27:395–422.

247. Jacobs-Reitsma, W. F., P. M. F. J. Koenraad, N. M. Bolder, and R. W. A. W. Mulder. 1994. In vitro susceptibility of *Campylobacter* and *Salmonella* isolates from broilers to quinolones, ampicillin, tetracycline, and erythromycin. *Vet. Q.* 16:206–208.

248. Jalava, K., J. Perevoscikovs, A. Siitonen, et al. 2002. *Salmonella* Enteritidis PT 4 infections among a group of Finns visiting Riga: effective collaboration between Latvian and Finnish authorities resolved an outbreak. *EpiNorth J.* 2(1).

249. Janakiraman, A., and J. M. Slauch. 2000. The putative iron transport system SitABCD encoded on SPI1 is required for full virulence of *Salmonella typhimurium*. *Mol. Microbiol.* 35:1146–1155.

250. Jansen, A., C. Frank, R. Prager, H. Oppermann, and K. Stark. 2005. Nation-wide outbreak of *Salmonella* Give in Germany, 2004. *Z. Gastroenterol.* 43:707–713.

251. Japan Ministry of Health and Welfare. 1999. *Departmental Health Report*. Japan Ministry of Health and Welfare, Tokyo, Japan.

252. Jones, B. D., H. F. Paterson, A. Hall, and S. Falkow. 1993. *Salmonella typhimurium* induces membrane ruffling by a growth factor-receptor-independent mechanism. *Proc. Natl. Acad. Sci. USA* 90:10390–10394.

253. Jones, R. C., V. Reddy, L. Kornstein, J. R. Fernandez, F. Stavinsky, A. Agasan, and S. I. Gerber. 2004. *Salmonella enterica* serotype Uganda infection in New York City and Chicago. *Emerg. Infect. Dis.* 10:1665–1667.

254. Kaniga, K., J. C. Bossio, and J. E. Galan. 1994. The *Salmonella typhimurium* invasion genes *invF* and *invG* encode homologues of the AraC and PulD family of proteins. *Mol. Microbiol.* 13:555–568.

255. Kapperud, G., S. Gustavsen, I. Hellesnes, A. H. Hansen, J. Lassen, J. Hirn, M. Jahkola, M. A. Montenegro, and R. Helmuth. 1990. Outbreak of *Salmonella typhimurium* infection traced to contaminated chocolate and caused by a strain lacking the 60-megadalton virulence plasmid. *J. Clin. Microbiol.* 28:2597–2601.

256. Kauffmann, F. 1966. *The Bacteriology of* Enterobacteriaceae. Munksgaard, Copenhagen, Denmark.

257. Keady, S., G. Briggs, J. Farrar, J. C. Mohle-Boetani, J. O'Connell, S. B. Werner, D. Anderson, L. Tenglesen, S. Bidols, B. Albanese, C. Gordan, E. DeBess, J. Hatch, W. E. Keene, M. Plantenga, J. Tierheimer, A. L. Hackmann, C. E. Rinehardt, C. H. Sandt, A. Ingram, S. Hansen, S. Hurt, M. Poulson, R. Pallipamu, J. Wiclund, C. Braden, J. Lockett, S. van Duyne, A. Dechet, and C. Smelser. 2004. Outbreak of *Salmonella* serotype Enteritidis infections associated with raw almonds—United States

and Canada, 2003–2004. *Morb. Mortal. Wkly. Rep.* 53:484–487.

258. **Keene, W. E., K. Hedberg, P. Cieslak, S. Schafer, and E. Dechet.** 2004. *Salmonella* serotype Typhimurium outbreak associated with commercially processed egg salad—Oregon, 2003. *Morb. Mortal. Wkly. Rep.* 53:1132–1134.

259. **Keitel, W. A., N. L. Bond, J. M. Zahradnik, T. A. Cramton, and J. B. Robbins.** 1994. Clinical and serological responses following primary and booster immunization with *Salmonella typhi* Vi capsular polysaccharide vaccines. *Vaccine* 12:195–199.

260. **Keyes, K., C. Hudson, J. J. Maurer, S. G. Thayer, D. G. White, and M. D. Lee.** 2000. Detection of florfenicol resistance genes in *Escherichia coli* isolated from sick chickens. *Antimicrob. Agents Chemother.* 44:421–424.

261. **Keyes, K., M. D. Lee, and J. J. Maurer.** 2003. Antibiotics: mode of action, mechanism of resistance and transfer, p. 45–46. *In* M. Torrence and R. Isaacson (ed.), *Current Topics in Food Safety in Animal Agriculture.* Iowa State University Press, Ames, Iowa.

262. **Khachatryan, A. R., D. D. Hancock, T. E. Besser, and D. R. Call.** 2004. Role of calf-adapted *Escherichia coli* in maintenance of antimicrobial drug resistance in dairy calves. *Appl. Environ. Microbiol.* 70:752–757.

263. **Kim, C. J., D. A. Emery, H. Rinke, K. V. Nagaraja, and D. A. Halvorson.** 1989. Effect of time and temperature on growth of *Salmonella enteritidis* in experimentally inoculated eggs. *Avian Dis.* 33:735–742.

264. **Kim, E.-H., and T. Aoki.** 1996. Sequence analysis of the florfenicol resistance gene encoded in the transferable R-plasmid of a fish pathogen, *Pasteurella piscicida*. *Microbiol. Immunol.* 40:665–669.

265. **Kim, J. M., L. Eckmann, T. C. Savidge, D. C. Lowe, T. Witthoft, and M. F. Kagnoff.** 1998. Apoptosis of human intestinal epithelial cells after bacterial invasion. *J. Clin. Investig.* 102:1815–1823.

266. **Kimura, A. C., M. S. Palumbo, H. Meyers, S. Abbott, R. Rodriguez, and S. B. Werner.** 2005. A multistate outbreak of *Salmonella* serotype Thompson infection from commercially distributed bread contaminated by an ill food handler. *Epidemiol. Infect.* 133:823–828.

267. **Kimura, H., M. Izawa, and Y. Sumino.** 1996. Molecular analysis of the gene cluster involved in cephalosporin biosynthesis from *Lysobacter lactamgenus* YK90. *Appl. Microbiol. Biotechnol.* 44:589–596.

268. **Kirby, R. M., and R. Davies.** 1990. Survival of dehydrated cells of *Salmonella typhimurium* LT 2 at high temperatures. *J. Appl. Microbiol.* 68:241–246.

269. **Kivi, M., W. van Pelt, D. Notermans, A. van de Giessen, W. Wannet, and A. Bosman.** 2005. Large outbreak of *Salmonella* Typhimurium DT 104, The Netherlands, September-November, 2005. *Eurosurveill. Wkly.* 10:051201.

270. **Klare, I., D. Badstubner, C. Konstabel, G. Bohme, H. Claus, and W. Witte.** 1999. Decreased incidence of VanA-type vancomycin-resistant enterococci isolated from poultry meat and from fecal samples of humans in the community after discontinuation of avoparcin usage in animal husbandry. *Microb. Drug Resist.* 5:45–52.

271. **Koch, J., A. Schrauder, K. Alpers, D. Werber, C. Frank, R. Prager, W. Rabsch, S. Broll, F. Feil, P. Roggentin, J.** Bockemühl, H. Tschäpe, A. Ammon, and K. Stark. 2005. *Salmonella* Agona outbreak from contaminated aniseed, Germany. *Emerg. Infect. Dis.* 11:1124–1127.

272. **Koo, F. C. W., J. W. Peterson, C. W. Houston, and N. C. Molina.** 1984. Pathogenesis of experimental salmonellosis: inhibition of protein synthesis by cytotoxin. *Infect. Immun.* 43:93–100.

273. **Kotetishvili, M., O. C. Stine, A. Kreger, J. G. Morris, Jr., and A. Sulakvelidze.** 2002. Multilocus sequence typing for characterization of clinical and environmental *Salmonella* strains. *J. Clin. Microbiol.* 40:1626–1635.

274. **Koutsolioutsou, A., E. A. Martins, D. G. White, S. B. Levy, and B. Demple.** 2001. A *soxRS*-constitutive mutation contributing to antibiotic resistance in a clinical isolate of *Salmonella enterica* (serovar Typhimurium). *Antimicrob. Agents Chemother.* 45:38–43.

275. **Kowarz, L., C. Coynault, V. Robbe-Saule, and F. Norel.** 1994. The *Salmonella typhimurium katF* (*rpoS*) gene: cloning, nucleotide sequence, and regulation of *spvR* and *spvABCD* virulence plasmid genes. *J. Bacteriol.* 176:6852–6860.

276. Reference deleted.

277. **Lang, D. J., L. J. Kunz, A. R. Martin, S. A. Schroeder, and L. A. Thomson.** 1967. Carmine as a source of nosocomial salmonellosis. *N. Engl. J. Med.* 276:829–832.

278. **Lecos, C.** 1986. Of microbes and milk: probing America's worst *Salmonella* outbreak. *Dairy Food Sanit.* 6:136–140.

279. **Lee, A. K., C. S. Detweiler, and S. Falkow.** 2000. OmpR regulates the two-component system SsrA-SsrB in *Salmonella* pathogenicity island 2. *J. Bacteriol.* 182:771–781.

280. **Lee, I. S., J. L. Slonczewski, and J. W. Foster.** 1994. A low-pH-inducible, stationary-phase acid tolerance response in *Salmonella typhimurium*. *J. Bacteriol.* 176:1422–1426.

281. **Lee, L. A., N. D. Puhr, E. K. Maloney, N. H. Bean, and R. V. Tauxe.** 1994. Increase in antimicrobial-resistant *Salmonella* infections in the United States, 1989–1990. *J. Infect. Dis.* 170:128–134.

282. **Lee, M. D., and A. Fairchild.** 2006. Sample preparation for PCR, p. 41–50. *In* J. Maurer (ed.), *PCR Methods in Foods.* Springer, New York, N.Y.

283. **Lehmacher, A., J. Bockemühl, and S. Aleksic.** 1995. A nationwide outbreak of human salmonellosis in Germany due to contaminated paprika and paprika-powdered potato chips. *Epidemiol. Infect.* 115:501–511.

284. **Le Minor, L.** 1981. The genus *Salmonella*, p. 1148–1159. *In* M. P. Starr, H. Stolp, H. G. Truper, A. Balows, and H. G. Schlegel (ed.), *The Prokaryotes.* Springer-Verlag, New York, N.Y.

285. **Le Minor, L., and M. Y. Popoff.** 1987. Request for an opinion. Designation of *Salmonella enterica* sp. nov., nom. rev., as the type and only species of the genus *Salmonella*. *Int. J. Syst. Bacteriol.* 37:465–468.

286. **Le Minor, L., C. Coynault, and G. Pessoa.** 1974. Déterminisme plasmidique du caractère atypique "lactose positif" de souches de *S. typhimurium* et de *S. oranienburg* isolées au Brésil lors d'épidémies de 1971 à 1973. *Ann. Microbiol.* 125A:261–285.

287. **Le Minor, L., C. Coynault, R. Rhode, B. Rowe, and S. Aleksic.** 1973. Localisation plasmidique de déterminant

génétique du caractère atypique "saccharose +" des *Salmonella. Ann. Microbiol.* **124B**:295–306.

288. **Le Minor, L., M. Y. Popoff, B. Laurent, and D. Hermant.** 1986. Individualisation d'une septième sous-espèce de *Salmonella: S. choleraesuis* subsp. *indica* subsp. nov. *Ann. Inst. Pasteur Microbiol.* **137B**:211–217.

289. **Lenglet, A.** 2005. E-alert 9 August: over 2000 cases so far in *Salmonella* Hadar outbreak in Spain associated with consumption of pre-cooked chicken, July-August, 2005. *Eurosurveill. Wkly. Rep.* **10**:050811.

290. **Levings, R. S., D. Lightfoot, S. R. Partridge, R. M. Hall, and S. P. Djordjevic.** 2005. The genomic island SGI1, containing the multiple antibiotic resistance region of *Salmonella enterica* serovar Typhimurium DT104 or variants of it, is widely distributed in other *S. enterica* serovars. *J. Bacteriol.* **187**:4401–4409.

291. **Levy, S. B., G. B. FitzGerald, and A. B. Macone.** 1976. Changes in intestinal flora of farm personnel after introduction of a tetracycline-supplemented feed on a farm. *N. Engl. J. Med.* **295**:583–588.

292. **Leyer, G. J., and E. A. Johnson.** 1992. Acid adaptation promotes survival of *Salmonella* spp. in cheese. *Appl. Environ. Microbiol.* **58**:2075–2080.

293. **Leyer, G. J., and E. A. Johnson.** 1993. Acid adaptation induces cross-protection against environmental stresses in *Salmonella typhimurium. Appl. Environ. Microbiol.* **59**:1842–1847.

294. **Libby, S. J., L. G. Adams, T. A. Ficht, C. Allen, H. A. Whitford, N. A. Buchmeier, S. Bossie, and D. G. Guiney.** 1997. The *spv* genes on the *Salmonella dublin* virulence plasmid are required for severe enteritis and systemic infection in the natural host. *Infect. Immun.* **65**:1786–1792.

295. **Liebert, C. A., R. M. Hall, and A. O. Summers.** 1999. Transposon Tn*21*, flagship of the floating genome. *Microbiol. Mol. Biol. Rev.* **63**:507–522.

296. **Liming, S. H., and A. A. Bhagwat.** 2004. Application of a molecular beacon-real-time PCR technology to detect *Salmonella* species contaminating fruits and vegetables. *Int. J. Food Microbiol.* **95**:177–187.

297. **Lin, C., and C. Wei.** 1997. Transfer of *Salmonella* Montevideo onto the interior surfaces of tomatoes by cutting. *J. Food Prot.* **60**:858–863.

298. **Liu, L., H. F. He, C. F. Dai, L. H. Liang, T. Li, L. H. Li, H. M. Luo, and R. Fontaine.** 2006. Salmonellosis outbreak among factory workers—Huizhou, Guangdong Province, China, July 2004. *Morb. Mortal. Wkly. Rep.* **55**(Suppl. 01):35–38.

299. **Loewen, P. C., and R. Hengge-Aronis.** 1994. The role of the sigma factor S (KatF) in bacterial global regulation. *Annu. Rev. Microbiol.* **48**:53–80.

300. **Low, J. C., G. Hopkins, T. King, and D. Munro.** 1996. Antibiotic resistant *Salmonella typhimurium* DT104 in cattle. *Vet. Rec.* **138**:650–651.

301. **Luk, J. M., U. Kongmuang, P. R. Reeves, and A. A. Lindberg.** 1993. Selective amplification of abequose and paratose synthase genes (*rfb*) by polymerase chain reaction for identification of *Salmonella* major serogroups (A, B, C2, and D). *J. Clin. Microbiol.* **31**:2118–2123.

302. **MacBeth, K. J., and C. A. Lee.** 1993. Prolonged inhibition of bacterial protein synthesis abolishes *Salmonella* invasion. *Infect. Immun.* **61**:1544–1546.

303. **Mackey, B. M., and C. Derrick.** 1990. Heat shock protein synthesis and thermotolerance in *Salmonella typhimurium. J. Appl. Bacteriol.* **69**:373–383.

304. **Mackey, B. M., and C. M. Derrick.** 1986. Elevation of the heat resistance of *Salmonella typhimurium* by sublethal heat shock. *J. Appl. Bacteriol.* **61**:389–393.

305. **Maier, R. J., A. Olczak, S. Maier, S. Soni, and J. Gunn.** 2004. Respiratory hydrogen use by *Salmonella enterica* serovar Typhimurium is essential for virulence. *Infect. Immun.* **72**:6294–6299.

306. **Makela, P. H., M. Sarvas, S. Calcagno, and K. Lounatmaa.** 1978. Isolation and characterization of polymyxin-resistant mutants of *Salmonella typhimurium. FEMS Microbiol. Lett.* **3**:323–326.

307. **Malorny, B., E. Paccassoni, P. Fach, C. Bunge, A. Martin, and R. Helmuth.** 2004. Diagnostic real-time PCR for detection of *Salmonella* in food. *Appl. Environ. Microbiol.* **70**:7046–7052.

308. **Malorny, B., P. T. Tassios, P. Radstrom, N. Cook, M. Wagner, and J. Hoorfar.** 2003. Standardization of diagnostic PCR for the detection of foodborne pathogens. *Int. J. Food Microbiol.* **83**:39–48.

309. **Marano, N. N., S. Rossiter, K. Stamey, K. Joyce, T. J. Barrett, L. K. Tollefson, and F. J. Angulo.** 2000. The National Antimicrobial Resistance Monitoring System (NARMS) for enteric bacteria, 1996–1999: surveillance for action. *J. Am. Vet. Med. Assoc.* **217**:1829–1830.

310. **Marshall, C. G., I. A. Lessard, I. Park, and G. D. Wright.** 1998. Glycopeptide antibiotic resistance genes in glycopeptide-producing organisms. *Antimicrob. Agents Chemother.* **42**:2215–2220.

311. **Mason, J.** 1994. *Salmonella enteritidis* control programs in the United States. *Int. J. Food Microbiol.* **21**:155–169.

312. **Mathew, A. G., K. N. Garner, P. D. Ebner, A. M. Saxton, R. E. Clift, and S. Liamthong.** 2005. Effects of antibiotic use in sows on resistance of *E. coli* and *Salmonella enterica* Typhimurium in their offspring. *Foodborne Pathog. Dis.* **2**:212–220.

313. **Mattila, L., M. Leirisalo-Repo, S. Koskimies, K. Granfors, and A. Sütonen.** 1994. Reactive arthritis following an outbreak of *Salmonella* infection in Finland. *Br. J. Rheumatol.* **33**:1136–1141.

314. **Maurer, J. J., and M. D. Lee.** 2006. *Salmonella*: virulence, stress response, and resistance, p. 215–239. *In* M. Griffiths (ed.), *Understanding Pathogen Behaviour: Virulence, Stress Response and Resistance.* Woodhead Publishers Limited, Cambridge, United Kingdom.

315. **McClelland, M., K. E. Sanderson, J. Spieth, S. W. Clifton, P. Latreille, L. Courtney, S. Porwollik, J. Ali, M. Dante, F. Du, S. Hou, D. Layman, S. Leonard, C. Nguyen, K. Scott, A. Holmes, N. Grewal, E. Mulvaney, E. Ryan, H. Sun, L. Florea, W. Miller, T. Stoneking, M. Nhan, R. Waterston, and R. K. Wilson.** 2001. Complete genome sequence of *Salmonella enterica* serovar Typhimurium LT2. *Nature* **413**:852–856.

316. **McClelland, M., K. E. Sanderson, S. W. Clifton, P. Latreille, S. Porwollik, A. Sabo, R. Meyer, T. Bieri, P. Ozersky, M. McLellan, C. R. Harkins, C. Wang, C. Nguyen, A. Berghoff, G. Elliott, S. Kohlberg, C. Strong, F. Du, J.**

Carter, C. Kremizki, D. Layman, S. Leonard, H. Sun, L. Fulton, W. Nash, T. Miner, P. Minx, K. Delehaunty, C. Fronick, V. Magrini, M. Nhan, W. Warren, L. Florea, J. Spieth, and R. K. Wilson. 2004. Comparison of genome degradation in Paratyphi A and Typhi, human-restricted serovars of *Salmonella enterica* that cause typhoid. *Nat. Genet.* **36**:1268–1274.

317. McCullough, N. B., and C. W. Eisele. 1951. Experimental human salmonellosis. 1. Pathogenicity of strains of *Salmonella meleagridis* and *Salmonella anatum* obtained from spray-dried whole egg. *J. Infect. Dis.* **88**:278–290.

318. Mevius, D. J., M. J. Sprenger, and H. C. Wegener. 1999. EU conference 'The Microbial Threat.' *Int. J. Antimicrob. Agents* **11**:101–105.

319. Miller, S. I., and J. J. Mekalanos. 1990. Constitutive expression of the *phoP* regulon attenuates *Salmonella* virulence and survival within macrophages. *J. Bacteriol.* **172**:2485–2490.

320. Miller, S. I. 1991. PhoP/PhoQ: macrophage-specific modulators of *Salmonella* virulence. *Mol. Microbiol.* **5**:2073–2078.

321. Miller, S. I., A. M. Kukral, and J. J. Mekalanos. 1989. A two-component regulatory system (phoP/phoQ) controls *Salmonella typhimurium* virulence. *Proc. Natl. Acad. Sci. USA* **86**:5054–5058.

322. Mills, D. M., V. Bajaj, and C. A. Lee. 1995. A 40 kb chromosomal fragment encoding *Salmonella typhimurium* invasion genes is absent from the corresponding region of the *Escherichia coli* K-12 chromosome. *Mol. Microbiol.* **15**:749–759.

323. Mills, S. D., S. R. Ruschkowski, M. A. Stein, and B. B. Finlay. 1998. Trafficking of porin-deficient *Salmonella typhimurium* mutants inside HeLa cells: *ompR* and *envZ* mutants are defective for the formation of *Salmonella*-induced filaments. *Infect. Immun.* **66**:1806–1811.

324. Mintz, E. D., M. L. Cartter, J. L. Hadler, J. T. Wassell, J. A. Zingeser, and R. V. Tauxe. 1994. Dose-response effects in an outbreak of *Salmonella enteritidis*. *Epidemiol. Infect.* **112**:13–23.

325. Mirold, S., W. Rabsch, H. Tschäpe, and W. D. Hardt. 2001. Transfer of the *Salmonella* type III effector *sopE* between unrelated phage families. *J. Mol. Biol.* **312**:7–16.

326. Miyagawa, S., and A. Miki. 1992. The epidemiological data of food poisoning in 1991. *Food Sanit. Res.* **42**:78–104.

327. Mizuno, T., and S. Mizushima. 1990. Signal transduction and gene regulation through the phosphorylation of two regulatory components: the molecular basis for the osmotic regulation of the porin genes. *Mol. Microbiol.* **4**:1077–1082.

328. Monack, D. M., B. Raupach, A. E. Hromockyj, and S. Falkow. 1996. *Salmonella typhimurium* invasion induces apoptosis in infected macrophages. *Proc. Natl. Acad. Sci. USA* **93**:9833–9838.

329. Monsieurs, P., S. De Keersmaecker, W. W. Navarre, M. W. Bader, F. De Smet, M. McClelland, F. C. Fang, B. De Moor, J. Vanderleyden, and K. Marchal. 2005. Comparison of the PhoPQ regulon in *Escherichia coli* and *Salmonella typhimurium*. *J. Mol. Evol.* **60**:462–474.

330. Morgan, E., J. D. Campbell, S. C. Rowe, J. Bispham, M. P. Stevens, A. J. Bowen, P. A. Barrow, D. J. Maskell, and T. S. Wallis. 2004. Identification of host-specific colonization factors of *Salmonella enterica* serovar Typhimurium. *Mol. Microbiol.* **54**:994–1010.

331. Mouzin, E., S. B. Werner, R. G. Bryant, et al. 1997. When a health food becomes a hazard: a large outbreak of salmonellosis associated with alfalfa sprouts—California. Presented at Epidemic Intelligence Service Conference, Centers for Disease Control and Prevention, Atlanta, Ga.

332. Mulvey, M. R., D. Boyd, A. Cloeckaert, R. Ahmed, L. K. Ng, and Provincial Public Health Laboratories. 2004. Emergence of multidrug-resistant *Salmonella* Paratyphi B dT+, Canada. *Emerg. Infect. Dis.* **10**:1307–1310.

333. Murray, B. E. 1986. Resistance of *Shigella*, *Salmonella*, and other selected enteric pathogens to antimicrobial agents. *Rev. Infect. Dis.* **8**(Suppl. 2):S172–S181.

334. Murray, R. A., and C. A. Lee. 2000. Invasion genes are not required for *Salmonella enterica* serovar Typhimurium to breach the intestinal epithelium: evidence that *Salmonella* pathogenicity island 1 has alternative functions during infection. *Infect. Immun.* **68**:5050–5055.

335. Mustapha, A., and Y. Li. 2006. Molecular detection of foodborne pathogens, p. 69–90. *In* J. Maurer (ed.), *PCR Methods in Foods*. Springer, New York, N.Y.

336. Naas, T., Y. Mikami, T. Imai, L. Poirel, and P. Nordmann. 2001. Characterization of In53, a class 1 plasmid- and composite transposon-located integron of *Escherichia coli* which carries an unusual array of gene cassettes. *J. Bacteriol.* **183**:235–249.

337. Nandi, S., J. J. Maurer, C. Hofacre, and A. O. Summers. 2004. Gram-positive bacteria are a major reservoir of class 1 antibiotic resistance integrons in poultry litter. *Proc. Natl. Acad. Sci. USA* **101**:7118–7122.

338. Nassif, X., and P. Sansonetti. 1987. Les systèmes bactériens de captation du fer: leur rôle dans la virulence. *Bull. Inst. Pasteur* **85**:307–327.

339. National Advisory Committee on Microbiological Criteria for Foods. 1999. Microbiological safety evaluations and recommendations on fresh produce. *Food Control* **10**:117–143.

340. National Advisory Committee on Microbiological Criteria for Foods. 1999. Microbiological safety evaluations and recommendations on sprouted seeds. *Int. J. Food Microbiol.* **52**:123–153.

341. National Research Council. 1980. *Effects on Human Health of Subtherapeutic Use of Antimicrobials in Animal Feeds*. National Academy Press, Washington, D.C.

342. National Research Council. 1999. *The Use of Drugs in Food Animals: Benefits and Risks*. National Research Council, Food and Nutrition Board, Institute of Medicine. National Academy Press, Washington, D.C.

343. Ng, H., H. G. Bayne, and J. A. Garibaldi. 1969. Heat resistance of *Salmonella*: the uniqueness of *Salmonella senftenberg* 775W. *Appl. Microbiol.* **17**:78–82.

344. Nguyen Van, J. C., and L. Gutman. 1994. Résistance aux antibiotiques par diminution de la perméabilité chez les bactéries à Gram négatif. *Presse Méd.* **23**:522–531.

345. Nield, B. S., A. J. Holmes, M. R. Gillings, G. D. Recchia, B. C. Mabbutt, K. M. Nevalainen, and H. W. Stokes.

2001. Recovery of new integron classes from environmental DNA. *FEMS Microbiol. Lett.* **195**:59–65.

346. **Nogrady, N., I. Gado, A. Toth, and J. Paszti.** 2005. Antibiotic resistance and class 1 integron patterns of non-typhoidal human *Salmonella* serotypes isolated in Hungary in 2002 and 2003. *Int. J. Antimicrob. Agents* **26**:126–132.

347. **Norris, F. A., M. P. Wilson, T. S. Wallis, E. E. Galyov, and P. W. Majerus.** 1998. SopB, a protein required for virulence of *Salmonella dublin*, is an inositol phosphate phosphatase. *Proc. Natl. Acad. Sci. USA* **95**:14057–14059.

348. **Notermans, D. W., W. van Pelt, M. Kivi, A. W. van de Giessen, W. J. Wannet, and A. Bosman.** 2005. Sharp increase of infections with Salmonella serotype Typhimurium DT104 in The Netherlands. *Ned. Tijdschr. Geneeskd.* **149**:2992–2994.

349. **Novick, R. P.** 1981. The development and spread of antibiotic-resistant bacteria as a consequence of feeding antibiotics to livestock. *Ann. N.Y. Acad. Sci.* **368**:23–59.

350. **Ogawa, H., H. Tokunou, M. Sasaki, T. Kishimoto, and K. Tamura.** 1991. An outbreak of bacterial food poisoning caused by roast cuttlefish "yaki-ika" contaminated with *Salmonella* spp. Champaign. *Jpn. J. Food Microbiol.* **7**:151–157.

351. **Oh, Y., C. Alpuche-Aranda, E. Berthiaume, T. Jinks, S. I. Miller, and J. A. Swanson.** 1996. Rapid and complete fusion of macrophage lysosomes with phagosomes containing *Salmonella typhimurium*. *Infect. Immun.* **64**:3877–3883.

352. **Olsen, S. J., L. C. MacKinon, J. S. Goulding, N. H. Bean, and L. Slutsker.** 2000. Surveillance for foodborne disease outbreaks—United States, 1993–1997. *Morb. Mortal. Wkly. Rep.* **49**(SS01):1–51.

353. **Olson, E. R.** 1993. Influence of pH on bacterial gene expression. *Mol. Microbiol.* **8**:5–14.

354. **Olsvik, O., T. Popovic, E. Skjerve, K. S. Cudjoe, E. Hornes, J. Ugelstad, and M. Uhlen.** 1994. Magnetic separation techniques in diagnostic microbiology. *Clin. Microbiol. Rev.* **7**:43–54.

355. **Ortega, Y.** 2006. Molecular tools for the identification of foodborne parasites, p. 119–146. *In* J. Maurer (ed.), *PCR Methods in Foods.* Springer, New York, N.Y.

356. **Otto, H., D. Tezcan-Merdol, R. Girisch, F. Haag, M. Rhen, and F. Koch-Nolte.** 2000. The *spvB* gene product of the *Salmonella enterica* virulence plasmid is a mono(ADP-ribosyl)transferase. *Mol. Microbiol.* **37**:1106–1115.

357. **Pacer, R. E., J. S. Spika, M. C. Thurmond, N. Hargrett-Bean, and M. E. Potter.** 1989. Prevalence of *Salmonella* and multiple antimicrobial-resistant *Salmonella* in California dairies. *J. Am. Vet. Med. Assoc.* **195**:59–63.

358. **Palmer, S. R., and B. Rowe.** 1986. Trends in *Salmonella* infections. *Public Health Lab. Serv. Microbiol. Dig.* **3**:18–21.

359. **Parish, M. E.** 1998. Coliforms, *Escherichia coli* and *Salmonella* serovars associated with a citrus-processing facility implicated in a salmonellosis outbreak. *J. Food Prot.* **61**:280–284.

360. **Parkhill, J., G. Dougan, K. D. James, N. R. Thomson, D. Pickard, J. Wain, C. Churcher, K. L. Mungall, S. D.** Bentley, M. T. Holden, M. Sebaihia, S. Baker, D. Basham, K. Brooks, T. Chillingworth, P. Connerton, A. Cronin, P. Davis, R. M. Davies, L. Dowd, N. White, J. Farrar, T. Feltwell, N. Hamlin, A. Haque, T. T. Hien, S. Holroyd, K. Jagels, A. Krogh, T. S. Larsen, S. Leather, S. Moule, P. O'Gaora, C. Parry, M. Quail, K. Rutherford, M. Simmonds, J. Skelton, K. Stevens, S. Whitehead, and B. G. Barrell. 2001. Complete genome sequence of a multiple drug resistant *Salmonella enterica* serovar Typhi CT18. *Nature* **413**:848–852.

361. **Paulsen, I. T., T. G. Littlejohn, P. Radstrom, L. Sundstrom, O. Skold, G. Swedberg, and R. A. Skurray.** 1993. The 3′ conserved segment of integrons contains a gene associated with multidrug resistance to antiseptics and disinfectants. *Antimicrob. Agents Chemother.* **37**:761–768.

362. **Pengues, D. A., M. J. Hantman, I. Behlau, and S. I. Miller.** 1995. PhoP/PhoQ transcriptional repression of *Salmonella typhimurium* invasion genes: evidence for a role in protein secretion. *Mol. Microbiol.* **17**:169–181.

363. **Peterson, J. W., and D. W. Niesel.** 1988. Enhancement by calcium of the invasiveness of *Salmonella* for HeLa cell monolayers. *Rev. Infect. Dis.* **10**:S319–S322.

364. **Pezzi, G. H., I. M. Gallardo, S. M. Ontanon, et al.** 1995. Vigilancia de brotes de infecciones e intoxica ciones de origen alimentario, Espana, ans 1994 (excluye brotes hidricos). *Bol. Epidemiol. Seml.* **3**:293–299.

365. **Pickard, D., J. Li, M. Roberts, D. Maskell, D. Hone, M. Levine, G. Dougan, and S. Chatfield.** 1994. Characterization of defined *ompR* mutants of *Salmonella typhi*: *ompR* is involved in the regulation of Vi polysaccharide expression. *Infect. Immun.* **62**:3984–3993.

366. **Piddock, L. J. V., C. Wray, I. McLaren, and R. Wise.** 1990. Quinolone resistance in *Salmonella* species: veterinary pointers. *Lancet* **336**:125.

367. **Pivnick, H.** 1980. Curing salts and related materials, p. 136–159. *In* J. H. Silliker, R. P. Elliott, A. C. Baird-Parker, F. L. Bryan, J. H. B. Christian, D. S. Clark, J. C. Olson, Jr., and T. A. Roberts (ed.), *Microbial Ecology of Foods*, vol. 1. *Factors Affecting Life and Death of Microorganisms.* Academic Press, New York, N.Y.

368. **Pohl, P., Y. Glupczynskj, M. Marin, G. van Robaeys, P. Lintermans, and M. Couturier.** 1993. Replicon typing characterization of plasmids encoding resistance to gentamicin and apramycin in *Escherichia coli* and *Salmonella typhimurium* isolated from human and animal sources in Belgium. *Epidemiol. Infect.* **III**:229–238.

369. **Polotsky, Y., E. Dragunsky, and T. Khavkin.** 1994. Morphologic evaluation of the pathogenesis of bacterial enteric infections. *Crit. Rev. Microbiol.* **20**:161–208.

370. **Pönkä, A., Y. Andersson, A. Sütonen, B. de Jong, M. Jahkola, O. Haikala, A. Kuhmonen, and P. Pakkala.** 1995. *Salmonella* in alfalfa sprouts. *Lancet* **345**:462–463.

371. **Popoff, M. Y., J. Bockemuhl, and L. L. Gheesling.** 2004. Supplement 2002 (no. 46) to the Kauffmann-White scheme. *Res. Microbiol.* **155**:568–570.

372. **Popoff, M. Y., and F. Norel.** 1992. Bases moléculaires de la pathogénicité des *Salmonella. Méd. Mal. Infect.* **22**:310–324.

373. **Porwollik, S., C. A. Santiviago, P. Cheng, L. Florea, and M. McClelland.** 2005. Differences in gene content

between *Salmonella enterica* serovar Enteritidis isolates and comparison to closely related serovars Gallinarum and Dublin. *J. Bacteriol.* **187**:6545–6555.

374. **Porwollik, S., E. F. Boyd, C. Choy, P. Cheng, L. Florea, E. Proctor, and M. McClelland.** 2004. Characterization of *Salmonella enterica* subspecies I genovars by use of microarrays. *J. Bacteriol.* **186**:5883–5898.

375. **Porwollik, S., R. M. Wong, and M. McClelland.** 2002. Evolutionary genomics of *Salmonella*: gene acquisitions revealed by microarray analysis. *Proc. Natl. Acad. Sci USA* **99**:8956–8961.

376. **Potter, M. E., A. F. Kaufmann, P. A. Blake, and R. A. Feldman.** 1984. Unpasteurized milk. The hazards of a health fetish. *JAMA* **252**:2048–2052.

377. **Prescott, J. F., J. D. Baggot, and R. D. Walker.** 2000. *Antimicrobial Therapy in Veterinary Medicine*, 3rd ed. Iowa State University Press, Ames, Iowa.

378. **Prouty, A. M., I. E. Brodsky, S. Falkow, and J. S. Gunn.** 2004. Bile-salt-mediated induction of antimicrobial and bile resistance in *Salmonella typhimurium*. *Microbiology* **150**:775–783.

379. **Radford, S. A., and R. G. Board.** 1995. The influence of sodium chloride and pH on the growth of *Salmonella enteritidis* PT 4. *Lett. Appl. Microbiol.* **20**:11–13.

380. **Rahman, H., V. B. Singh, and V. D. Sharma.** 1994. Purification and characterization of enterotoxic moiety present in cell-free culture supernatant of *Salmonella typhimurium*. *Vet. Microbiol.* **39**:245–254.

381. **Rahn, K., S. A. Grandis, R. C. Clarke, S. A. McEwen, J. E. Galan, C. Ginocchio, R. Curtiss III, and C. L. Gyles.** 1992. Amplification of *invA* gene sequences of *Salmonella typhimurium* by polymerase chain reaction as a specific method of detection of *Salmonella*. *Mol. Cell Probes* **6**:271–279.

382. **Rajajee, S., T. B. Anandi, S. Subha, and B. R. Vatsala.** 1995. Patterns of resistant *Salmonella typhi* infection in infants. *J. Trop. Pediatr.* **41**:52–54.

383. **Randall, L. P., S. W. Cooles, M. K. Osborn, L. J. Piddock, and M. J. Woodward.** 2004. Antibiotic resistance genes, integrons and multiple antibiotic resistance in thirty-five serotypes of *Salmonella enterica* isolated from humans and animals in the UK. *J. Antimicrob. Chemother.* **53**:208–216.

384. **Rankin, S. C., J. M. Whichard, K. Joyce, L. Stephens, K. O'Shea, H. Aceto, D. S. Munro, and C. E. Benson.** 2005. Detection of a *bla*(SHV) extended-spectrum betalactamase in *Salmonella enterica* serovar Newport MDR-AmpC. *J. Clin. Microbiol.* **43**:5792–5793.

385. **Rasmussen, B. A., K. Bush, D. Keeney, Y. Yang, R. Hare, C. O'Gara, and A. A. Medeiros.** 1996. Characterization of IMI-1 beta-lactamase, a class A carbapenemhydrolyzing enzyme from *Enterobacter cloacae*. *Antimicrob. Agents Chemother.* **40**:2080–2086.

386. **Rathman, M., L. P. Barker, and S. Falkow.** 1997. The unique trafficking pattern of *Salmonella typhimurium*-containing phagosomes in murine macrophages is independent of the mechanism of bacterial entry. *Infect. Immun.* **65**:1475–1485.

387. **Rathman, M., M. D. Sjaastad, and S. Falkow.** 1996. Acidification of phagosomes containing *Salmonella*

typhimurium in murine macrophages. *Infect. Immun.* **64**:2765–2773.

388. **Ratman, S., F. Stratton, C. O'Keefe, A. Roberts, R. Coates, M. Yetman, S. Squires, R. Khakhria, and J. Hockin.** 1999. *Salmonella* Enteritidis outbreak due to contaminated cheese—Newfoundland. *Can. Commun. Dis. Rep.* **25**:17–20.

389. **Reeves, M. W., G. M. Evins, A. A.Heiba, B. D. Plikaytis, and J. J. Farmer III.** 1989. Clonal nature of *Salmonella typhi* and its genetic relatedness to other salmonellae as shown by multilocus enzyme electrophoresis, and proposal of *Salmonella bongori* comb. nov. *J. Clin. Microbiol.* **27**:313–320.

390. **Reitler, R., D. Yarom, and R. Seligmann.** 1960. The enhancing effect of staphylococcal enterotoxin on *Salmonella* infection. *Med. Off.* **104**:181.

391. **Rhoades, E. R., and H. J. Ullrich.** 2000. How to establish a lasting relationship with your host: lessons learned from *Mycobacterium* spp. *Immunol. Cell Biol.* **78**:301–310.

392. **Richard, F., E. Pons, B. Lelore, V. Bleuze, B. Grandbastien, C. Collinet, R. Mathis, J.-F. Diependale, and P. Legrand.** 1994. Toxi-infection alimentaire collective du 8 juin 1993 à Douai. *Bull. Epidemiol. Hebdo.* **3**:9–11.

393. **Richter-Dahlfors, A., A. M. J. Buchan, and B. B. Finlay.** 1997. Murine salmonellosis studied by confocal microscopy: *Salmonella typhimurium* resides intracellularly inside macrophages and exerts a cytotoxic effect on phagocytes in vivo. *J. Exp. Med.* **186**:569–580.

394. **Riesenfeld, C. S., R. M. Goodman, and J. Handelsman.** 2004. Uncultured soil bacteria are a reservoir of new antibiotic resistance genes. *Environ. Microbiol.* **6**:981–989.

395. **Robbe-Saule, V., C. Coynault, and F. Norel.** 1995. The live oral typhoid vaccine Ty21a is a rpoS mutant and is susceptible to various environmental stresses. *FEMS Microbiol. Lett.* **126**:171–176.

396. **Robbe-Saule, V., G. Algorta, I. Rouilhac, and F. Norel.** 2003. Characterization of the RpoS status of clinical isolates of *Salmonella enterica*. *Appl. Environ. Microbiol.* **69**:4352–4358.

397. **Roberts, J. A., and P. N. Sockett.** 1994. The socioeconomic impact of human *Salmonella enteritidis* infection. *Int. J. Food Microbiol.* **21**:117–129.

398. **Roland, K. L., L. E. Martin, C. R. Esther, and J. K. Spitznagel.** 1993. Spontaneous *prmA* mutants of *Salmonella typhimurium* LT2 define a new two-component regulatory system with a possible role in virulence. *J. Bacteriol.* **175**:4154–4164.

399. **Rowe, B., E. J. Threlfall, L. R. Ward, and A. S. Ashley.** 1979. International spread of multiresistant strains of *Salmonella typhimurium* phage types 204 and 193 from Britain to Europe. *Vet. Rec.* **105**:468–469.

400. **Ruschkowski, S., I. Rosenshine, and B. B. Finlay.** 1992. *Salmonella typhimurium* induces an inositol phosphate flux in infected epithelial cells. *FEMS Microbiol. Lett.* **95**:121–126.

401. **Russell, S. M.** 2003. The effect of air sacculitis on bird weights, uniformity, fecal contamination, processing errors, and populations of *Campylobacter* spp. and *Escherichia coli*. *Poult. Sci.* **82**:1326–1331.

402. Ryan, C. A., M. K. Nickels, N. T. Hargrett-Bean, M. E. Potter, T. Endo, L. Mayer, C. W. Langkop, C. Gibson, R. C. McDonald, R. T. Kenney, N. D. Puhr, P. J. McDonnell, R. J. Martin, M. L. Cohen, and P. A. Blake. 1987. Massive outbreak of antimicrobial-resistant salmonellosis traced to pasteurized milk. *JAMA* **258:**3269–3274.

403. Sails, A. D., B. Swaminathan, and P. I. Fields. 2003. Utility of multilocus sequence typing as an epidemiological tool for investigation of outbreaks of gastroenteritis caused by *Campylobacter jejuni*. *J. Clin. Microbiol.* **41:**4733–4739.

404. Saitoh, M., K. Tanaka, K. Nishimori, S. Makino, T. Kanno, R. Ishihara, S. Hatama, R. Kitano, M. Kishima, T. Sameshima, M. Akiba, M. Nakazawa, Y. Yokomizo, and I. Uchida. 2005. The *artAB* genes encode a putative ADP-ribosyltransferase toxin homologue associated with *Salmonella enterica* serovar Typhimurium DT104. *Microbiology* **151:**3089–3096.

405. Sameshima, T., H. Ito, I. Uchida, H. Danbara, and N. Terakado. 1993. A conjugative plasmid pTE195 coding for drug resistance and virulence phenotypes from *Salmonella naestved* strain of calf origin. *Vet. Microbiol.* **36:**197–203.

406. Sanchez, S. 2006. Making PCR a normal routine of the food microbiology lab, p. 51–68. *In* J. Maurer (ed.), *PCR Methods in Foods.* Springer, New York, N.Y.

407. Sanchez, S., M. A. McCrackin Stevenson, C. R. Hudson, M. Maier, T. Buffington, Q. Dam, and J. J. Maurer. 2002. Characterization of multi-drug-resistant *Escherichia coli* associated with nosocomial infections in dogs. *J. Clin. Microbiol.* **40:**3586–3595.

408. Sandaa, R. A., V. L. Torsvik, and J. Goksoyr. 1992. Transferable drug resistance in bacteria from fish-farm sediments. *Can. J. Microbiol.* **38:**1061–1065.

408a. Sanderson, K. E., A. Hessel, S. Liu, and K. E. Rudd. 1996. The genetic map of *Salmonella typhimurium*, edition VIII, p. 1903–1999. *In* F. C. Neidhardt et al. (ed.), Escherichia coli *and* Salmonella: *Cellular and Molecular Biology*, 2nd ed. ASM Press, Washington, D.C.

409. Saxen, H., I. Reima, and P. H. Mäkelä. 1987. Alternative complement pathway activation by *Salmonella* O polysaccharide as a virulence determinant in the mouse. *Microb. Pathog.* **2:**15–28.

410. Schmid, D., S. Schandl, A. M. Pichler, C. Kornschober, C. Berghold, A. Beranek, G. Neubauer, M. Neuhold-Wassermann, W. Schwender, A. Klauber, A. Deutz, P. Pless, and F. Allerberger. 2006. *Salmonella* Enteritidis phage type 21 outbreak in Austria, 2005. *Eurosurveill. Mon. Rep.* **11:**67–69.

411. Schmidt, H., and M. Hensel. 2004. Pathogenicity islands in bacterial pathogenesis. *Clin. Microbiol. Rev.* **17:**14–56.

411a. Schnaitman, C. A., and J. D. Klena. 1993. Genetics of lipopolysaccharide synthesis in enteric bacteria. *Microbiol. Rev.* **57:**655–682.

412. Schroeter, A., B. Hoog, and R. Helmuth. 2004. Resistance of *Salmonella* isolates in Germany. *J. Vet. Med. B.* **51:**389–392.

413. Schwan, W. R., X. Huang, L. Hu, and D. J. Kopecko. 2000. Differential bacterial survival, replication, and apoptosis-inducing ability of *Salmonella* serovars

within human and murine macrophages. *Infect. Immun.* **68:**1005–1013.

414. Seglenieks, Z., and S. Dixon. 1977. Outbreak of milk-borne *Salmonella* gastroenteritis—South Australia. *Morb. Mortal. Wkly. Rep.* **26:**127.

415. Seward, R. J. 1999. Detection of integrons in worldwide nosocomial isolates of *Acinetobacter* spp. *Clin. Microbiol. Infect.* **5:**308–318.

416. Shafer, W. M., L. E. Martin, and J. K. Spitznagel. 1984. Cationic antimicrobial proteins isolated from human neutrophil granulocytes in the presence of diisopropyl fluorophosphates. *Infect. Immun.* **45:**29–35.

417. Shea, J. E., M. Hensel, C. Gleeson, and D. W. Holden. 1996. Identification of a virulence locus encoding a second type III secretion system in *Salmonella typhimurium*. *Proc. Natl. Acad. Sci. USA* **93:**2593–2597.

418. Sirard, J.-C., F. Niedergang, and J.-P. Kraehenbuhl. 1999. Live attenuated *Salmonella*: a paradigm of mucosal vaccines. *Immunol. Rev.* **171:**5–26.

419. Sivapalasingam, S., E. Barrett, A. Kumura, S. van Duyne, W. De Witt, M. Ying, A. Frisch, Q. Phan, E. Gould, P. Shillam, V. Reddy, T. Cooper, M. Hoekstra, C. Higgins, J. P. Sanders, R. V. Tauxe, and L. Slutsker. 2003. A multistate outbreak of *Salmonella enterica* serotype Newport infection linked to mango consumption: impact of water-dip disinfestation technology. *Clin. Infect. Dis.* **37:**1585–1590.

420. Sjöbring, U., G. Pohl, and A. Olsen. 1994. Plasminogen, adsorbed by *Escherichia coli* expressing curli or by *Salmonella enteritidis* expressing thin aggregative fimbriae, can be activated by simultaneously captured tissue-type plasminogen activator (t-PA). *Mol. Microbiol.* **14:**443–452.

421. Skamene, E., E. Schurr, and P. Gros. 1998. Infection genomics: *nramp1* as a major determinant of natural resistance to intracellular infections. *Annu. Rev. Med.* **49:**275–287.

422. Slauch, J. M., S. Garrett, D. E. Jackson, and T. J. Silhavy. 1988. EnvZ functions through OmpR to control porin gene expression in *Escherichia coli* K-12. *J. Bacteriol.* **170:**439–441.

423. Smith, J. L. 1994. Arthritis and foodborne bacteria. *J. Food Prot.* **57:**935–941.

424. Smith, K. E., J. M. Besser, C. W. Hedberg, F. T. Leano, J. B. Bender, J. H. Wicklund, B. P. Johnson, K. A. Moore, and M. T. Osterholm. 1999. Quinolone-resistant *Campylobacter jejuni* infections in Minnesota, 1992–1998. *N. Engl. J. Med.* **340:**1525–1532.

425. Song, W., E. S. Moland, N. D. Hanson, J. S. Lewis, J. H. Jorgensen, and K. S. Thomson. 2005. Failure of cefepime therapy in treatment of *Klebsiella pneumoniae* bacteremia. *J. Clin. Microbiol.* **43:**4891–4894.

426. Spika, J. S., S. H. Waterman, G. W. Soo Hoo, M. E. St. Louis, R. E. Pacer, S. M. James, M. L. Bissett, L. W. Mayer, J. Y. Chiu, B. Hall, K. Greene, M. E. Potter, M. L. Cohen, and P. A. Blake. 1987. Chloramphenicol-resistant *Salmonella newport* traced through hamburger to dairy farms. *N. Engl. J. Med.* **316:**565–570.

427. Stender, S., A. Friebel, S. Linder, M. Rohde, S. Mirold, and W. D. Hardt. 2000. Identification of SopE2 from

Salmonella typhimurium, a conserved guanine nucleotide exchange factor for Cdc42 of the host cell. *Mol. Microbiol.* **36**:1206–1221.

428. Stephen, J., T. S. Wallis, W. G. Starkey, D. C. A. Candy, M. P. Osborne, and S. Haddon. 1985. Salmonellosis: in retrospect and prospect. *In Microbial Toxins and Diarrhoeal Disease Ciba Fndn. Symp.* **112**:175–192.

429. Stickney, R. R. 1990. A global overview of aquaculture production. *Food Rev. Int.* **6**:299–315.

430. Stokes, H. W., and R. M. Hall. 1989. A novel family of potentially mobile DNA elements encoding site-specific gene integration functions: integrons. *Mol. Microbiol.* **3**:1669–1683.

431. Stone, R. 2002. Stalin's forgotten cure. *Science* **298**:728–731.

432. Sukhnanand, S., S. Alcaine, L. D. Warnick, W. L. Su, J. Hof, M. P. Craver, P. McDonough, K. J. Boor, and M. Wiedmann. 2005. DNA sequence-based subtyping and evolutionary analysis of selected *Salmonella enterica* serotypes. *J. Clin. Microbiol.* **43**:3688–3698.

433. Suzuki, S., K. Komase, H. Matsui, A. Abe, K. Kawahara, Y. Tamura, M. Kijima, H. Danbara, M. Nakamura, and S. Sato. 1994. Virulence region of plasmid pN2001 of *Salmonella enteritidis. Microbiology* **140**:1307–1318.

434. Swann, M. M. 1969. *Report of the Joint Committee on the Use of Antibiotics in Animal Husbandry and Veterinary Medicine.* Cmnd. 4190. Her Majesty's Stationery Office, London, United Kingdom.

435. Szu, S. C., S. Bystricky, M. Hinojosa-Ahumada, W. Egan, and J. B. Robbins. 1994. Synthesis and some immunologic properties of an O-acetyl pectin [poly(1→4)-α-D-GalpA]–protein conjugate as a vaccine for typhoid fever. *Infect. Immun.* **62**:5545–5549.

436. Tabaraie, B., B. K. Sharma, P. R. Sharma, R. Sehgal, and N. K. Ganguly. 1994. Stimulation of macrophage oxygen free radical production and lymphocyte blastogenic response by immunization with porins. *Microbiol. Immunol.* **38**:561–565.

437. Tacket, C. O., L. B. Dominguez, H. J. Fisher, and M. L. Cohen. 1985. An outbreak of multiple-drug-resistant *Salmonella* enteritis from raw milk. *JAMA* **253**:2058–2060.

438. Taira, S., P. Heiskanen, R. Hurme, H. Heikkilä, P. Riikonen, and M. Rhen. 1995. Evidence for functional polymorphism of the *spvR* gene regulating virulence gene expression in *Salmonella. Mol. Gen. Genet.* **246**:437–444.

439. Takaya, A., M. Suzuki, H. Matsui, T. Tomoyasu, H. Sahinami, A. Nakane, and T. Yamamoto. 2003. Lon, a stress-induced ATP-dependent protease, is critically important for systemic *Salmonella enterica* serovar Typhimurium infection of mice. *Infect. Immun.* **71**:690–696.

440. Takaya, A., T. Tomoasu, A. Tokumitsu, M. Morioka, and T. Yamamoto. 2002. The ATP-dependent Lon protease of *Salmonella enterica* serovar Typhimurium regulates invasion and expression of genes carried on *Salmonella* pathogenicity island 1. *J. Bacteriol.* **184**:224–232.

441. Takkinen, J., U. M. Nakari, T. Johansson, T. Niskanen, A. Siitonen, and M. Kuusi. 2005. A nationwide outbreak of multiresistant *Salmonella* Typhimurium var. Copenhagen DT 104B infection in Finland due to contaminated lettuce from Spain, May 2005. *Eurosurveill. Rep.* **10**:050630.

442. Tanaka, N. 1993. Food hygiene in Japan—Japanese food hygiene regulations and food poisoning incidence. *Dairy Food Environ. Sanit.* **13**:152–156.

443. Taormina, P. M., L. R. Beuchat, and L. Slutsker. 1999. Infections associated with eating seed sprouts: an international concern. *Emerg. Infect. Dis.* **5**:626–634.

444. Thomas, L. V., J. W. T. Wimpenny, and A. C. Peters. 1992. Testing multiple variables on the growth of a mixed inoculum of *Salmonella* strains using gradient plates. *Int. J. Food Microbiol.* **15**:165–175.

445. Thomson, G. T. D., D. A. DeRubeis, M. A. Hodge, C. Rajanayagam, and R. D. Inman. 1995. Post-*Salmonella* reactive arthritis: late clinical sequelae in point source cohort. *Am. J. Med.* **98**:13–21.

446. Threlfall, E. J. 1992. Antibiotics and the selection of food-borne pathogens. *J. Appl. Bacteriol. Symp.* 73 (Suppl.):96S–102S.

447. Threlfall, E. J., B. Rowe, J. L. Ferbuson, and L. R. Ward. 1986. Characterization of plasmids conferring resistance to gentamicin and apramycin in strains of *Salmonella typhimurium* phage type 204c isolated in Britain. *J. Hyg.* (Cambridge) **97**:419–426.

448. Threlfall, E. J., B. Rowe, J. L. Ferguson, and L. R. Ward. 1985. Increasing incidence of resistance to gentamicin and related aminoglycosides in *Salmonella typhimurium* phage type 204c in England, Wales and Scotland. *Vet. Rec.* **117**:355–357.

449. Threlfall, E. J., F. J. Angulo, and P. G. Wall. 1998. Ciprofloxacin-resistant *Salmonella typhimurium* DT104. *Vet. Rec.* **142**:255.

450. Threlfall, E. J., M. D. Hampton, H. Chart, and B. Rowe. 1994. Identification of a conjugative plasmid carrying antibiotic resistance and *Salmonella* plasmid virulence (*spv*) genes in epidemic strains of *Salmonella typhimurium* phage type 193. *Lett. Appl. Microbiol.* **18**:82–85.

451. Threlfall, E. J., L. R. Ward, J. A. Skinner, and B. Rowe. 1997. Increase in multiple antibiotic resistance in non-typhoidal salmonellas from humans in England and Wales: a comparison of data for 1994 and 1996. *Microb. Drug Resist.* **3**:263–266.

452. Todd, E. C. D. 1994. Surveillance of foodborne disease, p. 461–536. *In* Y. H. Hui, J. R. Gorham, K. D. Murrell, and D. O. Cliver (ed.), *Foodborne Disease Handbook. Diseases Caused by Bacteria,* vol. 1. Marcel Dekker, New York, N.Y.

453. Torok, T. J., R. V. Tauxe, R. P. Wise, J. R. Livengood, R. Sokolow, S. Mauvais, K. A. Birkness, M. R. Skeels, J. M. Horan, and L. R. Foster. 1997. A large community outbreak of salmonellosis caused by intentional contamination of restaurant salad bars. *JAMA* **278**:389–395.

454. Torrence, M. E. 2001. Activities to address antimicrobial resistance in the United States. *Prev. Vet. Med.* **51**:37–49.

455. Toth, B., D. Bodager, R. M. Hammond, S. Stenzel, J. K. Adams, T. Kass-Hout, R. M. Hoekstra, P. S. Mead, and P. Srikantiah. 2002. Outbreak of *Salmonella* serotype Javiana infections—Orlando, Florida, June 2002. *Morb. Mortal. Wkly. Rep.* **51**:683–684.

456. Troller, J. A. 1986. Water relations of foodborne bacterial pathogens—an updated review. *J. Food Prot.* **49:**656–670.

457. Tsolis, R. M., L. G. Adams, T. A. Ficht, and A. J. Baumler. 1999. Contribution of *Salmonella typhimurium* virulence factors to diarrheal disease in calves. *Infect. Immun.* **67:**4879–4885.

458. Tsujii, H., and K. Hamada. 1999. Outbreak of salmonellosis caused by ingestion of cuttlefish chips contaminated by both *Salmonella* Chester and *Salmonella* Oranienburg. *Jpn. J. Infect. Dis.* **52:**138–139.

459. Uchiya, K., M. A. Barbieri, K. Funato, A. H. Shah, P. D. Stahl, and E. A. Groisman. 1999. A *Salmonella* virulence protein that inhibits cellular trafficking. *EMBO J.* **18:**3924–3933.

460. Unicomb, L. E., G. Simmons, T. Merritt, J. Gregory, C. Nicol, P. Jelfs, M. Kirk, A. Tan, R. Thomson, J. Adamopoulos, C. L. Little, A. Currie, and C. B. Dalton. 2005. Sesame seed products contaminated with *Salmonella*: three outbreaks associated with tahini. *Epidemiol. Infect.* **133:**1065–1072.

461. Vaara, M., T. Vaara, M. Jenson, I. Helander, M. Nurminen, E. T. Rietschel, and P. H. Makelä. 1981. Characterization of the lipopolysaccharide from the polymixin-resistant *pmrA* mutants of *Salmonella typhimurium*. *FEBS Lett.* **129:**145–149.

462. Vahaboglu, H., S. Dodanli, C. Eroglu, R. Ozturk, G. Soyletir, I. Yildirim, and V. Avkan. 1996. Characterization of multiple-antibiotic-resistant *Salmonella typhimurium* strains: molecular epidemiology of PER-1-producing isolates and evidence for nosocomial plasmid exchange by a clone. *J. Clin. Microbiol.* **34:**2942–2946.

463. van den Bogaard, A. E., R. Willems, N. London, J. Top, and E. E. Stobberingh. 2002. Antibiotic resistance of faecal enterococci in poultry, poultry farmers and poultry slaughterers. *J. Antimicrob. Chemother.* **49:**497–505.

464. van der Velden, A. W., A. J. Baumler, R. M. Tsolis, and F. Heffron. 1998. Multiple fimbrial adhesins are required for full virulence of *Salmonella typhimurium* in mice. *Infect. Immun.* **66:**2803–2808.

465. Van Duynhoven, Y. T. H. P., M. Widdowson, C. M. de Jager, T. Fernandes, S. Neppelenbroek, W. van den Brandhof, W. J. B. Wannet, J. A. van Kooij, H. J. M. Rietveld, and W. van Pelt. 2002. *Salmonella enterica* serotype Enteritidis phage type 4b outbreak associated with bean sprouts. *Emerg. Infect. Dis.* **8:**440–443.

466. Van Leeuwen, W. J., J. D. A. van Embden, P. A. M. Guinee, E. H. Kampelmacher, A. Manten, M. van Schothorst, and C. E. Voogd. 1979. Decrease of drug resistance in *Salmonella* in The Netherlands. *Antimicrob. Agents Chemother.* **16:**237–239.

467. Vescovi, E. G., F. C. Soncini, and E. A. Groisman. 1994. The role of the PhoP/PhoQ regulon in *Salmonella* virulence. *Res. Microbiol.* **145:**473–480.

468. von Radowitz., J. 15 October 2002. PA News article.

469. Vora, G. J., C. E. Meador, D. A. Stenger, and J. D. Andreadis. 2004. Nucleic acid amplification strategies for DNA microarray-based pathogen detection. *Appl. Environ. Microbiol.* **70:**3047–3054.

470. Vought, K. J., and S. R. Tatini. 1998. *Salmonella enteritidis* contamination of ice cream associated with a 1994 multistate outbreak. *J. Food Prot.* **61:**5–10.

471. Wall, P. G., D. Morgan, K. Lamden, M. Ryan, M. Griffin, E. J. Threlfall, L. R. Ward, and B. Rowe. 1994. A case control study of infection with an epidemic strain of multiple resistant *Salmonella typhimurium* DT104 in England and Wales. *Commun. Dis. Rep.* **4:**R130-R135.

472. Ward, M. P., T. H. Brady, L. L. Couëtil, K. Liljebjelke, J. J. Maurer, and C. C. Wu. 2005. Investigation and control of an outbreak of salmonellosis caused by multidrug-resistant *Salmonella* Typhimurium in a population of hospitalized horses. *Vet. Microbiol.* **107:**233–240.

473. Watson, P. R., E. E. Galyov, S. M. Paulin, P. W. Jones, and T. S. Wallis. 1998. Mutation of *invH*, but not *stn*, reduces *Salmonella*-induced enteritis in cattle. *Infect. Immun.* **66:**1432–1438.

474. Weigel, R. M., B. Qiao, B. Teferedegne, D. K. Suh, D. A. Barber, R. E. Isaacson, and B. A. White. 2004. Comparison of pulsed field gel electrophoresis and repetitive sequence polymerase chain reaction as genotyping methods for detection of genetic diversity and inferring transmission of *Salmonella*. *Vet. Microbiol.* **100:**205–217.

475. Weissman, J. B., R. M. A. D. Deen, M. Williams, N. Swanton, and S. Ali. 1977. An island-wide epidemic of salmonellosis in Trinidad traced to contaminated powdered milk. *West Indian Med. J.* **26:**135–143.

476. Werber, D., J. Dreesman, F. Feil, U. van Treeck, G. Fell, S. Ethelberg, A. M. Hauri, P. Roggentin, I. S. Fisher, S. C. Behnke, E. Bartelt, E. Weise, A. Ellis, A. Siitonen, Y. Andersson, H. Tschäpe, M. H. Kramer, and A. Ammon. 2005. International outbreak of *Salmonella* Oranienburg due to German chocolate. *BMC Infect. Dis.* **5:**7.

477. White, D. G., C. R. Hudson, J. J. Maurer, S. Ayers, S. Zhao, M. D. Lee, L. F. Bolton, T. Foley, and J. Sherwood. 2000. Characterization of chloramphenicol and florfenicol resistance in bovine pathogenic *Escherichia coli*. *J. Clin. Microbiol.* **38:**4593–4598.

478. White, D. G., S. Zhao, R. Sudler, S. Ayers, S. Friedman, S. Chen, P. F. McDermott, S. McDermott, D. D. Wagner, and J. Meng. 2001. The isolation of antibiotic-resistant salmonella from retail ground meats. *N. Engl. J. Med.* **345:**1147–1154.

479. Wiedemann, B., and P. Heisig. 1994. Mechanisms of quinolone resistance. *Infection* **22**(Suppl. 2):S73-S79.

480. Wiener, P., S. Egan, A. S. Huddleston, and E. M. Wellington. 1998. Evidence for transfer of antibiotic-resistance genes in soil populations of streptomycetes. *Mol. Ecol.* **7:**1205–1216.

481. Wierup, M. 1997. Ten years without antibiotic growth promoters—results from Sweden with special reference to production results, alternative disease preventive methods and the usage of antibacterial drugs, p. 229–235. *In Report of a WHO Meeting.* WHO/EMC/ZOO/97.4. World Health Organization, Geneva, Switzerland.

482. Winfield, M. D., and E. A. Groisman. 2004. Phenotypic differences between *Salmonella* and *Escherichia coli* resulting from the disparate regulation of homologous genes. *Proc. Natl. Acad. Sci. USA* **101:**17162–17167.

483. Winokur, P. L., A. Brueggemann, D. L. DeSalvo, L. Hoffmann, M. D. Apley, E. K. Uhlenhopp, M. A. Pfaller, and G. V. Doern. 2000. Animal and human multidrug-resistant, cephalosporin-resistant *Salmonella* isolates expressing a plasmid-mediated CMY-2 AmpC beta-lactamase. *Antimicrob. Agents Chemother.* 44:2777–2783.

484. Winokur, P. L., D. L. Vonstein, L. J. Hoffman, E. K. Uhlenhopp, and G. V. Doern. 2001. Evidence for transfer of CMY-2 AmpC beta-lactamase plasmids between *Escherichia coli* and *Salmonella* isolates from food animals and humans. *Antimicrob. Agents Chemother.* 45:2716–2722.

485. Winthrop, K. L., M. S. Palumbo, J. A. Farra, J. C. Mohle-Boetani, S. Abbot, M. E. Beatty, G. Inami, and S. B. Serner. 2003. Alfalfa sprouts and *Salmonella* Kottbus infection: a multistate outbreak following inadequate seed disinfection with heat and chlorine. *J. Food Prot.* 66:13–17.

486. Witte, W. 2000. Ecological impact of antibiotic use in animals on different complex microflora: environment. *Int. J. Antimicrob. Agents* 14:321–325.

487. Wood, M. W., M. A. Jones, P. R. Watson, S. Hedges, T. S. Wallis, and E. E. Galyov. 1998. Identification of a pathogenicity island required for *Salmonella* enteropathogenicity. *Mol. Microbiol.* 29:883–891.

488. World Health Organization. 1988. *Salmonellosis Control: the Role of Animal and Product Hygiene.* Technical Report Series 774. World Health Organization, Geneva, Switzerland.

489. World Health Organization. 1992. *WHO Surveillance Programme for Control of Foodborne Infections and Intoxications in Europe. Fifth Report (1985–1989).* Institute of Veterinary Mecidine, Robert von Ostertag Institute, Berlin, Germany.

490. World Health Organization. 1995. *WHO Surveillance Programme for Control of Foodborne Infections and Intoxications in Europe. Sixth Report (1990–1992).* Federal Institute for Health Protection of Consumers and Veterinary Medicine, Berlin, Germany.

491. World Health Organization. 1997. *The Medical Impact of the Use of Antimicrobials in Food Animals: Report of a WHO Meeting, Berlin, Germany.* Document no. WHO/E. C./ZOO/97.4. World Health Organization, Geneva, Switzerland.

492. World Health Organization. 1998. Use of quinolones in food animals and potential impact on human health, p. 1–24. *In Report of a WHO Meeting.* WHO/EMC/ZDI/98.10. World Health Organization, Geneva, Switzerland.

493. World Health Organization. 2000. *WHO Issues New Recommendations To Protect Human Health from Antimicrobial Use in Food Animals.* [Online.] www.who.int.

494. World Health Organization. 2001. *WHO Surveillance Programme for Control of Foodborne Infections and Intoxications in Europe. Seventh Report (1993–1998).* Federal Institute for Health Protection of Consumers and Veterinary Medicine, Berlin, Germany.

495. World Health Organization. 2003. *WHO Surveillance Programme for Control of Foodborne Infections and Intoxications in Europe. Eighth Report (1999–2000).* WHO Regional Office for Europe, Copenhagen, Denmark.

496. World Health Organization. 2004. German study looks at prevalence of *Salmonella* in sesame seed products. *WHO Newsl.* 80:6.

497. Wray, C., and A. Wray. 2000. *Salmonella in Domestic Animals.* CABI, Oxon, United Kingdom.

498. Wray, C., R. W. Hedges, K. P. Shannon, and D. E. Bradley. 1986. Apramycin and gentamicin resistance in *Escherichia coli* and salmonellas isolated from farm animals. *J. Hyg.* (Cambridge) 97:445–456.

499. Ye, X. L., C. C. Yan, H. H. Xie, X. P. Tan, Y. Z. Wang, and L. M. Ye. 1990. An outbreak of food poisoning due to *Salmonella typhimurium* in the People's Republic of China. *J. Diarrheal Dis. Res.* 8:97–98.

500. Young, H. K. 1994. Do nonclinical uses of antibiotics make a difference? *Infect. Control Hosp. Epidemiol.* 15:484–487.

501. Zhao, S., D. G. White, P. F. McDermott, S. Friedman, L. English, S. Ayers, J. Meng, J. J. Maurer, R. Holland, and R. D. Walker. 2001. Identification and expression of cephamycinase bla_{CMY} genes in *Escherichia coli* and *Salmonella* isolated from animals and food. *Antimicrob. Agents Chemother.* 45:3647–3650.

502. Zhao, S., P. J. Fedorka-Cray, S. Friedman, P. F. McDermott, R. D. Walker, S. Qaiyumi, S. L. Foley, S. K. Hubert, S. Ayers, L. English, D. A. Dargatz, B. Salamone, and D. G. White. 2005. Characterization of *Salmonella* Typhimurium of animal origin obtained from the National Antimicrobial Resistance Monitoring System. *Foodborne Pathog. Dis.* 2:169–181.

503. Zhou, D., M. S. Mooseker, and J. E. Galan. 1999. An invasion-associated *Salmonella* protein modulates the actin-bundling activity of plastin. *Proc. Natl. Acad. Sci. USA* 96:10176–10181.

504. Zhou, D., M. S. Mooseker, and J. E. Galan. 1999. Role of the *S. typhimurium* actin-binding protein SipA in bacterial internalization. *Science* 283:2092–2095.

505. Zhou, D., W. D. Hardt, and J. E. Galan. 1999. *Salmonella typhimurium* encodes a putative iron transport system within the centisome 63 pathogenicity island. *Infect. Immun.* 67:1974–1981.

Food Microbiology: Fundamentals and Frontiers, 3rd Ed.
Edited by M. P. Doyle and L. R. Beuchat
© 2007 ASM Press, Washington, D.C.

Irving Nachamkin

Campylobacter jejuni

11

CHARACTERISTICS OF THE ORGANISM

Campylobacter jejuni subsp. *jejuni* (hereafter referred to as *C. jejuni*) is one of many species within the genus *Campylobacter*, family *Campylobacteraceae* (62). Since its likely initial description by Theodor Escherich in 1886 (16) and subsequent isolation by Jones et al. in 1931 (36), many investigators have become interested in studying the taxonomy and clinical importance of campylobacters and have greatly expanded the number of genera and species associated with this group of bacteria.

Three genera, *Campylobacter*, *Arcobacter*, and *Sulfurospirillum*, are included in the family *Campylobacteraceae* (62, 89). The family *Campylobacteraceae* includes at present 17 species within the genus *Campylobacter*, 6 species in the genus *Arcobacter*, and 6 species in the genus *Sulfurospirillum* (Table 11.1). *Campylobacter* is the type genus within the family *Campylobacteraceae*. Campylobacters are gram-negative, highly motile, S-shaped, curved or thin spiral rods and vary in size from 0.2 to 0.9 μm wide and 0.5 to 5 μm long. A unique feature of most campylobacters is their requirement for a microaerobic environment for optimal growth. It is now apparent that some *Campylobacter* species also require increased hydrogen (ca. 6%) for microaerobic growth. The genomes of

Campylobacter species vary in size from approximately 1.68 Mb to 1.77 Mb and 29.64 to 34.54% G+C (18, 64). A detailed description of the taxonomy, clinical significance, and methods for isolation and identification from clinical materials has been published (17).

ENVIRONMENTAL SUSCEPTIBILITY

Campylobacter jejuni is susceptible to a variety of environmental conditions that make it unlikely to survive for long periods of time outside the host. The bacterium does not grow at temperatures below 30°C, is microaerobic, and is sensitive to drying, high-oxygen conditions, and low pH (13). *Campylobacter jejuni* does not possess genes involved in cold-shock protein responses, and the inability to grow at low temperatures may be due to the absence of these protective proteins (63). *Campylobacter* spp. produce a number of enzymes that inactivate reactive oxygen intermediates, such as superoxide dismutase, alkyl hydroperoxide reductase, and catalase, and the enzymes protect the bacteria from the products of oxygen exposure (63).

The decimal reduction time for campylobacters varies and depends on the food source and temperature

Irving Nachamkin, Dept. of Pathology and Laboratory Medicine, University of Pennsylvania School of Medicine, Philadelphia, PA 19104-4283.

Table 11.1 Reservoirs and disease-associated species in the family *Campylobacteraceae*[a]

Bacterium	Reservoirs	Disease, sequelae, or comments	
		Human	Animals
C. jejuni subsp. *jejuni*	Human, cattle, wild birds, poultry, pets	Diarrhea, systemic illness, GBS,[b] RA[c]	Diarrhea, abortion
C. jejuni subsp. *doylei*	Humans	Diarrhea, bloodstream	
C. fetus subsp. *fetus*	Cattle, sheep	Abortion, systemic illness, diarrhea	Abortion
C. fetus subsp. *venerealis*	Cattle		Infertility
C. coli	Pigs, birds, poultry, cats	Diarrhea	
C. lari	Birds, dogs	Diarrhea	
C. upsaliensis	Domestic pets, poultry	Diarrhea, bloodstream	Diarrhea
C. hyointestinalis subsp. *hyointestinalis*	Cattle, pigs, hamsters, deer	Rare, proctitis, diarrhea	Proliferative enteritis
C. hyointestinalis subsp. *lawsonii*	Pigs		
C. mucosalis	Pigs	Rare, diarrhea	Proliferative enteritis
C. sputorum biovar sputorum	Humans	Oral cavity, abscesses	Genital tract of bulls, abortion in sheep
C. sputorum biovar paraureolyticus	Cattle	Diarrhea	
C. sputorum biovar faecalis	Cattle, sheep		Enteritis
C. lanienae	Pigs, cattle		
C. insulaenigrae	Marine mammals		
C. hominis	Humans		
C. concisus	Humans	Periodontal disease	
C. curvus	Humans	Periodontal disease	
C. rectus	Humans	Periodontal disease, pulmonary infections	
C. showae	Humans	Periodontal disease	
C. helveticus	Cats, dogs		Diarrhea
C. gracilis	Humans	Infections of head, neck, other sites	
A. butzleri	Cattle, pigs, poultry	Diarrhea, other	Diarrhea, abortion
A. cryaerophilus	Cattle, sheep, pigs, poultry	Diarrhea, bacteremia	Abortion
A. skirrowii	Cattle, sheep, pigs, poultry		Abortion, diarrhea; isolated from genital tract of bulls
A. nitrofigilis	Plants		
A. cibarius	Poultry		

[a] From references 62, 79, 90, 91, and 98.
[b] GBS, Guillain-Barré syndrome.
[c] RA, reactive arthritis.

(Table 11.2). Hence, the bacterium should not survive in food products brought to adequate cooking temperatures. Campylobacters are susceptible to gamma irradiation (1 kGy), but the rate of killing is dependent on the type of food being processed. Irradiation is less effective for frozen materials than for refrigerated or room temperature meats. Early-log-phase cells are more susceptible than cells grown to log or stationary phase (70). *Campylobacter* spp. are more radiation sensitive than other foodborne pathogens such as salmonellae and *Listeria monocytogenes*, and irradiation treatment that is effective against the latter set of bacteria should be sufficient to kill *Campylobacter* spp. as well (65).

Disinfectants such as sodium hypochlorite (Clorox), *o*-phenylphenol (Amphyl), iodine-polyvinylpyrrolidine (Betadine), alkylbenzyl dimethylammonium chloride (Zephiran), glutaraldehyde (Cidex), formaldehyde, and ethanol have antibacterial activity at commonly used concentrations (93). Ascorbic acid at 0.05% inhibits campylobacters and is bactericidal at 0.09% (32).

Table 11.2 *D* values determined in different foods[a]

Food	Temp (°C)	*D*-value range (min)
Skim milk	48	7.2–12.8
	55	0.74–1.0
Red meat	50	5.9–6.3
	60	<1
Ground chicken	49	20
	57	<1

[a] Data from reference 32.

Campylobacter spp. are susceptible to low pH and are killed readily at pH 2.3 (9). Campylobacters remain viable and multiply in bile at 37°C and survive better in feces, milk, water, and urine held at 4°C than in material held at 25°C. Multidrug efflux systems operative in *C. jejuni* likely play a role in resistance to bile (63). The maximum periods of viability of *Campylobacter* spp. at 4°C were 3 weeks in feces, 4 weeks in water, and 5 weeks in urine (9). Freezing reduces the cell numbers of *Campylobacter* in contaminated poultry, but even after freezing to –20°C, low levels of *Campylobacter* can be recovered (32).

RESERVOIRS AND FOODBORNE OUTBREAKS

C. jejuni is zoonotic, with many animals serving as reservoirs for human disease. Reservoirs for *C. jejuni* include poultry (chickens, turkeys, ducks, and geese), livestock (cows, pigs, sheep, and goats), wild birds, domestic animals, a variety of wild mammals, rodents, shellfish, and occasionally, contaminated produce (50) (Table 11.3). Campylobacters are frequently isolated from water, and water supplies have been sources of infection in some

Table 11.3 Isolation of *Campylobacter* from different food sources[a]

Food source	Avg % positive (range)
Chicken	33–53 (3–98)
Turkey	56 (1–94)
Duck, goose, pheasant, game hen, guinea fowl	32 (3–100)
Cattle	45 (2–97)
Beef	6 (1–54)
Swine	27 (2–100)
Sheep, lamb, goat	33 (0–92)
Seafood	16 (0–52)
Produce	1 (0–22)

[a] Data from reference 50.

reported outbreaks. *C. jejuni* can remain dormant in water in a state that has been termed "viable but nonculturable" (75); that is, under unfavorable conditions, the bacteria essentially remain dormant and cannot be easily recovered on artificial media. Under favorable conditions, campylobacters are able to multiply (48). The role of these forms as a source of infection for humans is not clear (50).

Over 30,000 cases of campylobacteriosis have been reported in approximately 900 outbreaks worldwide from 1978 to 2003 (Fig. 11.1) (50). In the United States, outbreaks are infrequent; from 1978 to 1996, there were 111 outbreaks of *Campylobacter* enteritis reported to the Centers for Disease Control and Prevention (CDC), affecting 9,913 individuals (21). The vehicles of *Campylobacter* outbreaks have changed over the past 2 decades. Water and unpasteurized milk were responsible for over one-half of outbreaks between 1978 and 1987, whereas other foods accounted for over 80% of outbreaks between 1988 and 1996 (21). In a 10-year review of outbreaks from 1981 to 1990, 20 outbreaks occurred that affected 1,013 individuals who drank raw milk. The attack rate was 45%. At least one outbreak occurred each year, and most of the outbreaks occurred in children who had gone on field trips to dairy farms (97). From 2000 to 2003, there were 47 outbreaks reported to the CDC, affecting 1,075 individuals. The most recent milkborne outbreaks were reported from Wisconsin in 2001 (28) and Michigan in 2003 (http://www.cdc.gov/foodborneoutbreaks/us_outb/fbo2003/summary03.htm).

The seasonal distribution of outbreaks is somewhat different from that of sporadic cases. Milkborne and

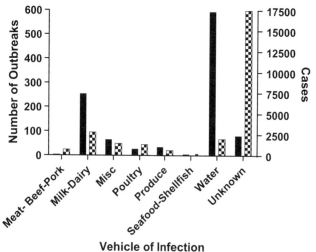

Figure 11.1 Foodborne and waterborne outbreaks of *Campylobacter* infections, 1978 to 2003. Solid bars, number of cases; checked bars, number of outbreaks. Adapted from reference 50.

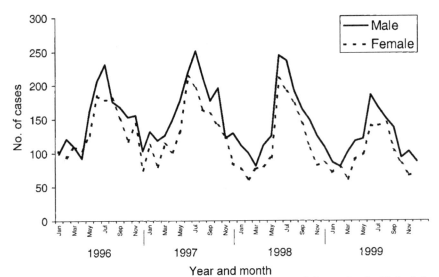

Figure 11.2 Season distribution of sporadic cases of *Campylobacter* in the United States, 1996–1999, Original FoodNet Sites, CDC (77).

waterborne outbreaks tend to occur in the spring and fall but do not occur frequently in the summer months (21). Sporadic cases are usually most frequent during the late spring and summer (Fig. 11.2) (77).

In contrast to the relatively low occurrence of outbreaks caused by campylobacters, these organisms have been identified in many studies as being among the most common causes of sporadic bacterial enteritis in the United States (21, 77). In 1995, the CDC, U.S. Department of Agriculture, and Food and Drug Administration developed an active surveillance system for foodborne diseases, including those caused by *Campylobacter*, called the Foodborne Diseases Active Surveillance Network, also known as FoodNet. According to a variety of data including FoodNet incidence data, U.S. Census Bureau data, and data from other sources, there are 1.4 million *Campylobacter* infections annually in the United States by current estimates (77), down from earlier estimates of 2.4 million (21). Sporadic cases occur more often during the summer months and usually follow ingestion of improperly handled or cooked food, primarily poultry products (21, 77, 81). Other exposures include drinking raw milk and contaminated surface water, overseas travel, and contact with domestic pets (Table 11.4).

CHARACTERISTICS OF DISEASE

C. jejuni and *C. coli*

Most *Campylobacter* species are associated with lower gastrointestinal tract infection; however, extraintestinal infections are common with some species, such as *C. fetus* subsp. *fetus* (hereafter referred to as *C. fetus*), and sequelae of *Campylobacter* infection are being more frequently recognized. *C. jejuni* and *C. coli* have been recognized since the 1970s as agents of gastrointestinal tract infection. Campylobacters have been identified as common causes of sporadic bacterial enteritis in the United States (20).

C. jejuni and *C. coli* are the most common *Campylobacter* species associated with diarrheal illness and are clinically indistinguishable. Because *C. jejuni* is not readily

Table 11.4 Risk factors for *Campylobacter* infection based on population attributable fraction identified in a U.S. case-control study, 1998–1999[a]

Risk factor	Population attributable fraction (95% confidence interval)
Consumption of chicken, restaurant prepared	24 (17–31)
Consumption of nonpoultry, restaurant prepared	21 (13–30)
Contact with farm animals	4 (1–7)
Contact with domestic puppies	5 (3–7)
Consumption of turkey, restaurant prepared	4 (1–6)
Consumption of undercooked chicken	3 (1–6)
Consumption of raw seafood	3 (0.3–5)
Consumption of raw milk	1.5 (0.4–3)

[a] Data from reference 20.

distinguished from *C. coli* in clinical laboratories, the precise ratio of *C. jejuni* to *C. coli* as causes of gastrointestinal infection is not known. *C. coli* is estimated to account for 5 to 10% of campylobacters isolated from patients with campylobacteriosis, but this number may vary in different parts of the world (60).

Patients who develop campylobacter infection may be asymptomatic or have mild to severe illness. Typical symptoms and signs include fever, abdominal cramping, and diarrhea (with or without blood or fecal leukocytes) that lasts several days to more than 1 week. Symptomatic infections are usually self-limited, but relapses may occur in 5 to 10% of untreated patients. *Campylobacter* infection may occasionally present as acute appendicitis and result in patients undergoing unnecessary surgery. Extraintestinal infections and sequelae do occur and include bacteremia, bursitis, urinary tract infection, meningitis, endocarditis, peritonitis, erythema nodosum, pancreatitis, abortion and neonatal sepsis, reactive arthritis, and Guillain-Barré Syndrome (GBS). Deaths directly attributable to *C. jejuni* infection have been reported but rarely occur (21).

C. jejuni and *C. coli* are susceptible to a variety of antimicrobial agents, including macrolides, fluoroquinolones, aminoglycosides, chloramphenicol, and tetracycline. Erythromycin has been the drug of choice for treating *C. jejuni* gastrointestinal tract infections, but ciprofloxacin is a good alternative drug. Early therapy of *Campylobacter* infection with erythromycin or ciprofloxacin is effective in eliminating the campylobacters from stool and may also reduce the duration of symptoms associated with infection (8).

C. jejuni is generally susceptible to erythromycin, with resistance rates of less than 5%. Rates of erythromycin resistance in *C. coli* vary considerably, with up to 80% of strains having resistance in some studies. Although ciprofloxacin has been effective in treating *Campylobacter* infections, fluoroquinolone resistance during therapy has been reported. Significant rates of primary fluoroquinolone resistance in *C. jejuni* are now well documented, nationally approximately 18% but maybe significantly higher locally (15, 39, 54). Patients infected with fluoroquinolone-resistant *C. jejuni* may have prolonged illness, and thus, alternative therapy should be used (57).

Other *Campylobacter* and *Arcobacter* Species

Campylobacter species other than *C. jejuni* have been isolated with increasing frequency from a variety of human infections. Improvements in culture methods and recognition that some species required increased hydrogen concentrations in the microaerobic environment have contributed to this increased isolation.

C. fetus is associated with a variety of localized infections (gastroenteritis) and systemic infections (bacteremia), but patients with underlying diseases are particularly at risk for infection with *C. fetus* (8). Patients may develop septic arthritis, abscesses, meningitis, endocarditis, mycotic aneurysm, septic abortion, salpingitis, thrombophlebitis, and peritonitis (8). Gastroenteritis without systemic involvement occurs more frequently than previously thought and is probably underestimated because the bacterium may not grow well at 42°C on some common selective media for stool culture (92). *C. fetus* subsp. *venerealis* is a rare cause of human infection and causes bovine venereal campylobacteriosis and is a cause of bovine infertility (88).

Campylobacter upsaliensis is infrequently isolated from routine stool cultures because it is susceptible to antimicrobial agents in common selective media and requires alternative methods for isolation, such as filtration. *C. upsaliensis* causes diarrhea and bacteremia and is also associated with canine and feline gastroenteritis (17). *C. lari* has occasionally been reported from humans with bacteremia and gastrointestinal and urinary tract infections and was identified as responsible for a waterborne outbreak of infection (17).

The role of other *Campylobacter* species in causing human disease is based mostly on case reports or small series, and so their pathogenic role has not been determined. The species isolated from patients with gastrointestinal and/or localized infections include *C. jejuni* subsp. *doylei*, *C. hyointestinalis*, *C. concisus*, *C. sputorum*, *C. curvus*, and *C. gracilis* (17). Several species are more associated with periodontal infections including *C. rectus*, *C. curvus*, and *C. showae* (17). A more extensive review on the clinical significance of non-*C. jejuni*/*C. coli* species was published previously (43).

Arcobacter

Arcobacter species are increasingly recognized in human infections and have been underappreciated because of the use of suboptimal culture conditions. Only two species have been associated with human infection, *Arcobacter butzleri* and *A. cryaerophilus*. *A. butzleri* was reported to be the fourth most common *Campylobacter*-like organism isolated from patients with diarrhea (92) and has also been isolated from patients with bacteremia, endocarditis, peritonitis, and diarrhea (43, 92). *Arcobacter cryaerophilus* has also been isolated from patients with bacteremia and diarrhea (43, 92). The other non-human-associated species are *A. nitrofigilis*, *A. cibarius*, and *A. halophilus* (17).

EPIDEMIOLOGIC SUBTYPING SYSTEMS USEFUL FOR INVESTIGATING FOODBORNE ILLNESSES

Numerous typing systems for *Campylobacter* epidemiologic studies have been described and vary in complexity and ability to discriminate between strains. Biotyping, phage typing, and serotyping are common phenotypic methods used in epidemiologic studies (58, 66). Serotyping is probably the most frequently used phenotypic system. The heat-labile serotyping scheme described by Lior et al. (45) detects over 100 serotypes of *C. jejuni*, *C. coli*, and *C. lari*. Bacterial surface antigens and flagellar antigens are the serodeterminants for this serotyping system (1). The heat-stable Penner (HS) serotyping scheme (67) detects 60 types of *C. jejuni* and *C. coli* (66) and detects a *Campylobacter* capsular polysaccharide (38).

Molecular approaches to typing have become routine for epidemiologic studies of *Campylobacter* infections (40). Methods include restriction endonuclease analysis, ribotyping, PCR-based techniques, pulsed-field gel electrophoresis of macrorestricted chromosomal DNA (PFGE), and amplified fragment length polymorphism. A standardized protocol of PFGE has been used by the PulseNet national surveillance system at CDC (61, 73). DNA sequence-based typing systems such as multilocus sequence typing have been found to be useful for campylobacter investigations (12). Microarray technology using genomic sequence data has also been developed as a potential typing system (40). Unfortunately, no single method can be recommended for outbreak investigations, and to determine strain relatedness, a combination of methods such as serotyping and a molecular method may be required (58).

INFECTIVE DOSE AND SUSCEPTIBLE POPULATIONS

Campylobacter jejuni is susceptible to low pH, and hence, the gastric environment is sufficient to kill most campylobacters (9). The infective dose of *C. jejuni*, however, does not appear to be high, with <1,000 organisms being capable of causing illness (7). The only controlled study to examine the infective dose of *C. jejuni* was conducted by Black and colleagues (7). In a study using two different strains of *C. jejuni*, only 18% of human volunteers became ill when infected with 10^8 CFU of strain A3249; however, 46% of the volunteers became ill when infected with another strain, 81-176. In an interesting experiment, Robinson ingested 500 CFU of *C. jejuni* in milk, which resulted in abdominal cramps and nonbloody diarrhea occurring 4 days after ingestion and lasting 3 days (74). Although not studied directly, there appeared to be a dose-related effect on both rate of infection and severity of illness in individuals involved in an outbreak of *Campylobacter* infection after ingesting raw milk (10).

Young children and young adults, 20 to 40 years of age, have the highest incidence of sporadic infections in the United States (77). The incidence of *Campylobacter* infection in developing countries may be orders of magnitude higher than in the United States (85, 86). In contrast to developed countries, campylobacters are frequently isolated from individuals who may or may not have diarrheal disease. Most symptomatic infections occur in infancy and early childhood, and the incidence decreases with age (11, 86, 87). Age-related increases in humoral immune responses to *Campylobacter* antigens are associated with a decrease in symptomatic illness (47, 87). Travelers to developing countries may acquire *Campylobacter* infection, with isolation rates from 0 to 39% reported in different studies (86).

Bacteremia reportedly occurs at a rate of 1.5 per 1,000 intestinal infections, with the highest rate in the elderly (80). Persistent diarrheal illness and bacteremia may occur in immunocompromised hosts, such as in patients with human immunodeficiency virus infection or hypogammaglobulinemia, and are difficult to treat (8).

VIRULENCE FACTORS AND MECHANISMS OF PATHOGENICITY

Despite decades of research, relatively little is known about the mechanism by which *C. jejuni* causes human disease. *Campylobacter jejuni* can cause an enterotoxigenic-like illness with loose or watery diarrhea or an inflammatory colitis with fever and the presence of fecal blood and leukocytes and occasionally bacteremia that suggests an invasive mechanism of disease. A major problem in elucidating the pathogenesis of *Campylobacter* infection has been the lack of suitable animal models (35, 99).

Cell Association and Invasion

Campylobacter interacts with eukaryotic cells in a variety of ways and involves the interaction of surface structures as well as potential secretion and glycosylation systems that affect the cellular response to *Campylobacter*. Surface structures that appear to be involved with this interaction include lipooligosaccharide (LOS), capsular polysaccharide, outer membrane proteins, and flagella (33, 37, 82). Both N-linked and O-linked protein glycosylation pathways are present in *Campylobacter* and may affect host-cell interactions, including adherence, invasion, and colonization (82, 83). Like other bacterial pathogens, *Campylobacter* must first establish itself in the intestinal tract, with motility and chemotaxis being important factors in this initial interaction with the host (30). Inducible proteins are expressed by *Campylobacter* upon contact with eukaryotic cells, but the roles of many of these proteins are not known (72). *C. jejuni* harbors a virulence plasmid, pVir, that contains several open reading frames with similarity to type IV secretion system genes and may be further involved in host-cell interactions (42). *Campylobacter* likely induces host cell cytoskeletal rearrangements, similar to other enteric pathogens, on contact with the cell in order to induce cellular invasion (30).

Flagella and Motility

Campylobacter species are motile and have a single polar, unsheathed flagellum at one or both ends. Motility and flagella are important determinants for the invasion-translocation process. Motility and *flaA* are essential for colonization (25, 56, 95). Motility mutants are frequently isolated in mutagenesis studies and point to the importance of this factor in virulence (29). Chemotactic motility is an important factor for the pathogenesis of *Campylobacter* and other bacteria, producing a number of chemoreceptors that are under investigation (29).

Toxins

Investigations on the cytolethal distending toxin (CDT) have received increased attention over the past few years, and the biology of this toxin was recently reviewed by Pickett and Lee (68). First reported by Johnson and Lior in 1988, the role of CDT (34) in pathogenesis is unknown, but the toxin is involved in blocking the cell cycle. *cdt* genes were characterized by Pickett et al., and three adjacent genes, *cdtA*, *cdtB*, and *cdtC*, were determined to encode this toxin and were similar to genes that encode *Escherichia coli* CDT proteins (69). Most strains of *C. jejuni* have *cdt* genes, and these genes are present in other *Campylobacter* species. In addition to blocking the cell cycle, this toxin may have a role in inducing pro-inflammatory cytokines (19).

Other toxic activities of *Campylobacter* have been previously described (94). The completion of the genome sequence of *C. jejuni* NCTC 11168 by Parkhill and colleagues has enabled a reexamination of the presence of putative toxin genes (64). Genes *cdtA* through *cdtC* were identified in the genome sequence; however, cholera-like toxin genes were notably absent (64). Genes with similarity to those encoding contact-dependent hemolysins present in pathogenic *Serpulina* and *Mycobacterium* species, integral membrane protein with a hemolysin domain, and phospholipase (*pldA*) also were identified in NCTC

11168 (64). Genes for other putative toxins, such as a cholera-like enterotoxin (94), that have been reported in *Campylobacter* were not detected in NCTC 11168 or other strains that were recently sequenced (18).

Other Factors

Environmentally regulated gene expression is an important area of research of *Campylobacter*. Campylobacters are microaerobic and for *C. jejuni* have higher temperatures for optimal growth, 42°C, than other bacterial pathogens. Understanding the impact of environmental signals on the growth, metabolism, and pathogenicity of *Campylobacter* will have a major impact on the ability to control campylobacters in the environment and food chain. Such pathways include response to iron, oxidative stress, temperature regulation including cold and heat shock responses, and starvation. Multidrug efflux systems are present in *Campylobacter* and may be important in antimicrobial resistance and resistance to bile and affect colonization (44). The environmental regulatory pathways of *Campylobacter* were recently reviewed by Park (63).

Autoimmune Sequelae

It is now clearly established that *Campylobacter* infection is a major trigger of GBS, an acute, immune-mediated paralytic disorder affecting the peripheral nervous system (52).

It is estimated that the annual incidence of GBS preceded by *C. jejuni* infection ranges from 0.17 to 0.51 case per 100,000 population and accounts for 425 to 1,272 cases per year in the United States (51). The pathogenesis of GBS induced by *C. jejuni* is not entirely clear, but molecular mimicry between bacterial LOS and relevant target epitopes in peripheral nerve tissue is a major mechanism of *Campylobacter*-induced GBS (96).

Campylobacter infection has been recognized as the most common identifiable event preceding GBS and has been estimated to occur in up to 40% of GBS patients (31). Although serologic and culture studies revealed that some patients with GBS had evidence of infection, an important study by Kuroki and colleagues solidified the association of *Campylobacter* and GBS (41). In a study of Japanese patients with GBS, Kuroki et al. isolated *C. jejuni* from 14 of 46 GBS patients (30.4%) compared with only 6 (1.2%) of 503 in a healthy control population. By use of serotyping to characterize the isolates, 10 of 12 available isolates were found to have the same HS serotype, HS:19. This serotype, however, occurred in only 1.7% of 1,150 *C. jejuni* isolates from patients with uncomplicated gastrointestinal infection. Serotypes HS:19 and HS:41 have been shown to be predominant GBS-associated serotypes in several studies (23, 84). However, studies from European investigators have not observed these predominant serotypes in GBS patients (14).

Structural studies have revealed that strains of serotype HS:19 as well as certain other serotypes (HS:4 and HS:1) of *Campylobacter* have core LOS structures that mimic ganglioside structures such as GM1 and GD1a, a component of motor neurons (2, 3, 100). Ganglioside-epitopes are produced by strains involved in GBS as well as by enteritis-associated isolates (55). *Campylobacter* organisms exhibit a high degree of variation in their LOS structure, and GBS strains appear to be restricted to certain LOS classes (22, 24). The genetics of ganglioside-epitope expression in *C. jejuni* is only now being understood. *C. jejuni* produces a number of sugar transferases and sialyltransferases involved in sialyation of the outer core LOS (22). A variety of antiganglioside antibodies are produced by patients with GBS, and these antibodies likely play an important role in the pathogenesis of the disease (96). Host genetic factors such as HLA type clearly play an important role in the development of disease (46).

IMMUNITY

Protective immunity appears after infection with *Campylobacter* species and is likely antibody mediated. Black et al. (7) determined that rechallenge of homologous strains 28 days after the initial volunteer challenge resulted in protection against illness but not necessarily against colonization by *C. jejuni*. Blaser et al. (10) revealed that 76% of acutely exposed individuals who had not been exposed previously to raw milk became acutely ill, compared with none of 10 individuals who were regular milk drinkers and who drank the implicated milk. Humoral immunity is likely to be an important component of protective immunity, as suggested by studies on persistent infection in immunocompromised hosts with human immunodeficiency virus infection or hypogammaglobulinemia (8).

In developing countries where *Campylobacter* infections are endemic, immunity to *Campylobacter* infection appears to be age dependent. In a cohort of Mexican children (11), the ratio of symptomatic to asymptomatic infection decreased from infancy to 6 years of age. An inverse relationship between serum antiflagellin antibodies and diarrheal illness in another cohort of children also supports the role of antibodies in protective immunity (47). Production of cytokines during intestinal infection may play some role in immunity and immune responses. Using a mouse colonization model, Baqar et al. (6)

determined that oral administration of interleukin-5 and interleukin-6 reduced the level of gut colonization with *C. jejuni*.

Flagellin is an important immunogen during *Campylobacter* infection (26), and antibodies against this protein correlate to some degree with protective immunity. Breast-feeding provides protection against *Campylobacter* infection in developing countries (47, 49, 53, 76). Antibodies against flagellin present in breast milk appear to be associated with protection of infants against infection (53).

Several researchers are working toward developing a vaccine for *Campylobacter* infection. Baqar et al. (5) tested an oral whole-cell killed vaccine coadministered with *E. coli* heat-labile enterotoxin as an immunoadjuvant to rhesus monkeys. The vaccine elicited both humoral and cellular responses; however, the ability to protect against infection was not studied. Killed whole-cell vaccine provided colonization protection in a mouse model (4). Guerry et al. (27) described the production of a *recA* mutant of *C. jejuni* 81-176 that colonized rabbits and induced colonization protection. Whether this strain was attenuated is not known, but *recA* mutants of other bacteria such as *Salmonella enterica* serovar Typhimurium are avirulent (27). Strategies for human vaccine development were reviewed by Scott and Tribble (78).

Other strategies to prevent the transmission of infection to humans include improved hygiene practices during broiler production, such as decontamination of water supplies, use of competitive exclusion flora which may prevent *C. jejuni* colonization of young chicks, and immunological approaches through the use of animal vaccines (59).

CONCLUSIONS

Campylobacter species are among the most important of human bacterial enteric pathogens, yet little is known about how these intriguing bacteria cause disease. Difficult challenges remain in identifying suitable in vitro and in vivo models of infection that will enable investigators to study *Campylobacter* species at both the biological and genetic levels. Elucidation of the genetic basis for pathogenesis of *Campylobacter* infection is still in an early stage; however, with the completion of the *C. jejuni* genome project and additional genome sequencing studies, a rich source of information is now available to study this intriguing pathogen. Finally, *Campylobacter* has received renewed attention by governmental agencies, which has already led to infusion of resources for research in the future (71).

References

1. Alm, R. A., P. Guerry, M. E. Power, H. Lior, and T. J. Trust. 1991. Analysis of the role of flagella in the heat-labile Lior serotyping scheme of thermophilic campylobacters by mutant allele exchange. *J. Clin. Microbiol.* **29:**2438–2445.

2. Aspinall, G. O., A. G. McDonald, H. Pang, L. A. Kurjanczyk, and J. L. Penner. 1994. Lipopolysaccharides of *Campylobacter jejuni* serotype O:19: structures of core oligosaccharide regions from the serostrain and two bacterial isolates from patients with the Guillain-Barré syndrome. *Biochemistry* **33:**241–249.

3. Aspinall, G. O., A. G. McDonald, T. S. Raju, H. Pang, L. A. Kurjanczyk, J. L. Penner, and A. P. Moran. 1993. Chemical structure of the core region of *Campylobacter jejuni* serotype O:2 lipopolysaccharide. *Eur. J. Biochem.* **213:**1029–1037.

4. Baqar, S., L. A. Applebee, and A. L. Bourgeois. 1995. Immunogenicity and protective efficacy of a prototype *Campylobacter* killed whole-cell vaccine in mice. *Infect. Immun.* **63:**3731–3735.

5. Baqar, S., A. L. Bourgeois, P. J. Schultheiss, R. I. Walker, D. M. Rollins, R. L. Haberberger, and O. R. Pavlovskis. 1995. Safety and immunogenicity of a prototype oral whole-cell killed *Campylobacter* vaccine administered with a mucosal adjuvant in non-human primates. *Vaccine* **13:**22–28.

6. Baqar, S., N. D. Pacheco, and F. M. Rollwagen. 1993. Modulation of mucosal immunity against *Campylobacter jejuni* by orally administered cytokines. *Antimicrob. Agents Chemother.* **37:**2688–2692.

7. Black, R. E., M. M. Levine, M. L. Clements, T. P. Hughs, and M. J. Blaser. 1988. Experimental *Campylobacter jejuni* infections in humans. *J. Infect. Dis.* **157:**472–480.

8. Blaser, M. J., and B. M. Allos. 2005. *Campylobacter jejuni* and related species, p. 2548–2557. *In* G. L. Mandell, J. E. Bennett, and R. Dolin (ed.), *Principles and Practice of Infectious Diseases.* Elsevier Churchill Livingstone, Philadelphia, Pa.

9. Blaser, M. J., H. L. Hardesty, B. Powers, and W. L. Wang. 1980. Survival of *Campylobacter fetus* subsp. *jejuni* in biological milieus. *J. Clin. Microbiol.* **11:**309–313.

10. Blaser, M. J., E. Sazie, and P. Williams. 1987. The influence of immunity on raw milk-associated *Campylobacter* infection. *JAMA* **257:**43–46.

11. Calva, J. J., G. M. Ruiz-Palacios, A. B. Lopez-Vidal, A. Ramos, and R. Bojalil. 1988. Cohort study of intestinal infection with *Campylobacter* in Mexican children. *Lancet* **i:**503–505.

12. Dingle, K. E., F. M. Colles, D. R. A. Wareing, R. Ure, A. J. Fox, F. E. Bolton, H. J. Bootsma, R. J. L. Willems, R. Urwin, and M. C. J. Maiden. 2001. Multilocus sequence typing system for *Campylobacter jejuni. J. Clin. Microbiol.* **39:**14–23.

13. Doyle, M. P., and D. M. Jones. 1992. Food-borne transmission and antibiotic resistance of *Campylobacter jejuni*, p. 45–48. *In* I. Nachamkin, M. J. Blaser, and L. S. Tompkins (ed.), Campylobacter jejuni: *Current Status and Future Trends.* American Society for Microbiology, Washington, D.C.

14. Endtz, H. P., C. W. Ang, N. Van Den Braak, B. Duim, A. Rigter, L. J. Price, D. L. Woodward, F. G. Rodgers, W. M. Johnson, J. A. Wagenaar, B. C. Jacobs, H. A. Verbrugh, and A. van Belkum. 2000. Molecular characterization of *Campylobacter jejuni* from patients with Guillain-Barre and Miller Fisher syndromes. *J. Clin. Microbiol.* **38:**2297–2301.

15. Engberg, J., F. M. Aarestrup, D. E. Taylor, P. Gerner-Smidt, and I. Nachamkin. 2001. Quinolone and macrolide resistance in *Campylobacter jejuni* and *C. coli*: resistance mechanisms and trends in human isolates. *Emerg. Infect. Dis.* **7:**24–34.

16. Escherich, T. 1886. Beitrage zur Kenntniss der Darmbacterien. III. Ueber das Vorkommen von Bironen im Darmcanal und den Stuhlgangen der Sauglinge [Articles adding to the knowledge of intestinal bacteria. III. On the existence of vibrios in the intestines and feces of babies]. *Munch. Med. Wochenschr.* **33:**815–817.

17. Fitzgerald, C., and I. Nachamkin. 2007. *Campylobacter* and *Arcobacter. In* P. R. Murray, E. J. Baron, J. H. Jorgensen, M. L. Landry, and M. A. Pfaller (ed.), *Manual of Clinical Microbiology.* ASM Press, Washington, D.C.

18. Fouts, D. E., E. F. Mongodin, R. E. Mandrell, W. G. Miller, D. A. Rasko, J. Ravel, L. M. Brinkac, R. T. DeBoy, C. T. Parker, S. C. Daugherty, R. J. Dodson, A. S. Durkin, R. Madupu, S. A. Sullivan, J. U. Shetty, M. A. Ayodeji, A. Shvartsveyn, M. C. Schatz, J. H. Badger, C. M. Fraser, and K. E. Nelson. 2005. Major structural differences and novel potential virulence mechanisms from the genomes of multiple *Campylobacter* species. *PLoS Biol.* **3:**72–85.

19. Fox, J. G., A. B. Rogers, M. T. Whary, Z. Ge, N. S. Taylor, S. Xu, B. H. Horvitz, and S. E. Erdman. 2004. Gastroenteritis in NF-kB deficient mice is produced with wild-type *Campylobacter jejuni* but not with *C. jejuni* lacking cytolethal distending toxin despite persistent colonization with both strains. *Infect. Immun.* **72:**1116–1125.

20. Friedman, C. R., R. M. Hoekstra, M. Samuel, R. Marcus, J. Bender, B. Shiferaw, S. Reddy, S. Ajuja, D. L. Helfrick, F. Hardnett, M. Carter, B. Anderson, and R. V. Tauxe. 2004. Risk factors for sporadic *Campylobacter* infection in the United States. A case-control study in FoodNet sites. *Clin. Infect. Dis.* **38(Suppl. 3):**285–296.

21. Friedman, C. R., J. Neimann, H. C. Wegener, and R. V. Tauxe. 2000. Epidemiology of *Campylobacter jejuni* infections in the United States and other industrialized nations, p. 121–138. *In* I. Nachamkin and M. J. Blaser (ed.), Campylobacter, 2nd ed. ASM Press, Washington, D.C.

22. Gilbert, M., P. C. R. Godschalk, C. T. Parker, H. Ph. Endtz, and W. W. Wakarchuk. 2005. Genetic basis for the variation in the lipooligosaccharide outer core of *Campylobacter jejuni* and possible association of glycotransferase genes with post-infectious neuropathies, p. 219–248. *In* J. M. Ketley and M. E. Konkel (ed.), Campylobacter: *Molecular and Cellular Biology.* Horizon Bioscience, Norfolk, United Kingdom.

23. Goddard, E. A., A. J. Lastovica, and A. C. Argent. 1997. *Campylobacter* O:41 isolation in Guillain-Barré syndrome. *Arch. Dis. Child.* **76:**526–528.

24. Godschalk, P. C. R., A. P. Heikema, M. Gilbert, T. Komagamine, C. W. Ang, J. Glerum, D. Brochu, J. Li, N. Yuki,

B. C. Jacobs, A. van Belkum, and H. P. Endtz. 2004. The crucial role of *Campylobacter jejuni* genes in anti-ganglioside antibody induction in Guillain-Barre syndrome. *J. Clin. Investig.* **114:**1659–1665.

25. Grant, C. C. R., M. E. Konkel, W. Cieplak, and L. S. Tompkins. 1993. Role of flagella in adherence, internalization, and translocation of *Campylobacter jejuni* in nonpolarized and polarized epithelial cells. *Infect. Immun.* **61:**1764–1771.

26. Guerry, P., R. A. Alm, C. Szymanski, and T. J. Trust. 2000. Structure, function and antigenicity of *Campylobacter* flagella, p. 405–421. *In* I. Nachamkin and M. J. Blaser (ed.), Campylobacter, 2nd ed. ASM Press, Washington, D.C.

27. Guerry, P., P. M. Pope, D. H. Burr, J. Leifer, S. W. Joseph, and A. L. Bourgeois. 1994. Development and characterization of *recA* mutants of *Campylobacter jejuni* for inclusion in vaccines. *Infect. Immun.* **62:**426–432.

28. Harrington, P., J. Archer, J. P. Davis, D. R. Croft, and J. K. Varma. 2002. Outbreak of *Campylobacter jejuni* infections associated with drinking unpasteurized milk procured through a cow-leasing program, Wisconsin, 2001. *Morb. Mortal. Wkly. Rep.* **51:**548–549.

29. Hendrixson, D. R., B. J. Akerley, and V. J. DiRita. 2001. Transposon mutagenesis of *Campylobacter jejuni* identifies a bipartite energy taxis system required for motility. *Mol. Microbiol.* **40:**214–224.

30. Hu, L., and D. J. Kopecko. 2005. Invasion, p. 369–383. *In* J. M. Ketley and M. E. Konkel (ed.), Campylobacter: *Molecular and Cellular Biology.* Horizon Bioscience, Norfolk, United Kingdom.

31. Hughes, R. A. C., and J. H. Rees. 1997. Clinical and epidemiologic features of Guillain-Barré syndrome. *J. Infect. Dis.* **176**(Suppl. 2):S92–S98.

32. Jacobs-Reitsma, W. 2000. *Campylobacter* in the food supply, p. 467–481. *In* I. Nachamkin and M. J. Blaser (ed.), Campylobacter, 2nd ed. ASM Press, Washington, D.C.

33. Jagannathan, A., and C. Penn. 2005. Motility, p. 331–347. *In* J. M. Ketley and M. E. Konkel (ed.), Campylobacter: *Molecular and Cellular Biology.* Horizon Bioscience, Norfolk, United Kingdom.

34. Johnson, W. M., and H. Lior. 1988. A new heat-labile cytolethal distending toxin (CLDT) produced by *Campylobacter* spp. *Microb. Pathog.* **4:**115–126.

35. Jones, F. R., S. Baqar, A. Gozalo, G. Nunez, N. Espinoza, S. M. Reyes, M. Salazar, R. Meza, C. K. Porter, and S. E. Walz. 2006. New world monkey *Aotus nancymae* as a model for *Campylobacter jejuni* infection and immunity. *Infect. Immun.* **74:**790–793.

36. Jones, F. S., M. Orcutt, and R. B. Little. 1931. Vibrios (*Vibrio jejuni*, n.sp.) associated with intestinal disorders of cows and calves. *J. Exp. Med.* **53:**853–864.

37. Karlyshev, A. V., O. L. Champion, G. W. P. Joshua, and B. W. Wren. 2005. The polysaccharide capsule of *Campylobacter jejuni*, p. 249–258. *In* J. M. Ketley and M. E. Konkel (ed.), Campylobacter: *Molecular and Cellular Biology.* Horizon Bioscience, Norfolk, United Kingdom.

38. Karlyshev, A. V., D. Linton, N. A. Gregson, A. J. Lastovica, and B. W. Wren. 2000. Genetic and biochemical evidence of a *Campylobacter jejuni* capsular polysaccharide that accounts for Penner serotype specificity. *Mol. Microbiol.* **35:**529–541.

39. Kassenborg, H. D., K. E. Smith, D. J. Vugia, T. Rabatsky-Ehr, M. R. Bates, M. A. Carter, N. B. Dumas, M. P. Cassidy, N. Marano, R. V. Tauxe, and F. J. Angulo. 2004. Fluoroquinolone-resistant *Campylobacter* infections: eating poultry outside of the home and foreign travel are risk factors. *Clin. Infect. Dis.* **38**(Suppl. 3):279–284.

40. Klena, J. D., and M. E. Konkel. 2005. Methods for epidemiologic analysis of *Campylobacter jejuni*, p. 165–179. *In* J. M. Ketley and M. E. Konkel (ed.), Campylobacter: *Molecular and Cellular Biology.* Horizon Bioscience, Norfolk, United Kingdom.

41. Kuroki, S., T. Saida, M. Nukina, T. Haruta, M. Yoshioka, Y. Kobayashi, and H. Nakanishi. 1993. *Campylobacter jejuni* strains from patients with Guillain-Barre syndrome belong mostly to Penner serogroup 19 and contain B-N-acetylglucosamine residues. *Ann. Neurol.* **33:**243–247.

42. Larsen, J. C., and P. Guerry. 2005. Plasmids of *Campylobacter jejuni* 81–176, p. 181–192. *In* J. M. Ketley and M. E. Konkel (ed.), Campylobacter: *Molecular and Cellular Biology.* Horizon Bioscience, Norfolk, United Kingdom.

43. Lastovica, A. J., and M. B. Skirrow. 2000. Clinical significance of *Campylobacter* and related species other than *Campylobacter jejuni*, p. 89–121. *In* I. Nachamkin and M. J. Blaser (ed.), Campylobacter, 2nd ed. ASM Press, Washington, D.C.

44. Lin, J., M. Akiba, and Q. Zhang. 2005. Multidrug efflux systems in *Campylobacter*, p. 205–218. *In* J. M. Ketley and M. E. Konkel (ed.), Campylobacter: *Molecular and Cellular Biology.* Horizon Bioscience, Norfolk, United Kingdom.

45. Lior, H., D. L. Woodward, J. A. Edgar, L. J. Laroche, and P. Gill. 1982. Serotyping of *Campylobacter jejuni* by slide agglutination based on heat-labile antigenic factors. *J. Clin. Microbiol.* **15:**761–768.

46. Magira, E. E., M. Papaioakim, I. Nachamkin, A. K. Asbury, C. Y. Li, T. W. Ho, J. W. Griffin, G. M. McKhann, and D. S. Monos. 2003. Differential distribution of HLA-DQb/DRb epitopes in the two forms of Guillain-Barre syndrome, acute motor axonal neuropathy (AMAN) and acute inflammatory demyelinating polyneuropathy (AIDP): identification of DQβ epitopes associated with susceptibilty to and protection from AIDP. *J. Immunol.* **170:**3074–3080.

47. Martin, P. M. V., J. Mathiot, J. Ipero, M. Kirimat, A. J. Georges, and M. C. Georges-Courbot. 1989. Immune response to *Campylobacter jejuni* and *Campylobacter coli* in a cohort of children from birth to 2 years of age. *Infect. Immun.* **57:**2542–2546.

48. Medema, G. J., F. M. Schets, A. W. van de Giessen, and A. H. Gavelaar. 1992. Lack of colonization of 1 day old chicks by viable, non-culturable *Campylobacter jejuni*. *J. Appl. Bacteriol.* **72:**512–516.

49. Megraud, F., G. Boudraa, K. Bessaoud, S. Bensid, F. Dabis, R. Soltana, and M. Touhami. 1990. Incidence of *Campylobacter* infection in infants in western Algeria and the possible protective role of breast feeding. *Epidemiol. Infect.* **105:**73–78.

50. Miller, W. G., and R. E. Mandrell. 2005. Prevalence of *Campylobacter* in the food and water supply: incidence,

outbreaks, isolation and detection, p. 101–163. *In* J. Ketley and M. E. Konkel (ed.), Campylobacter: *Molecular and Cellular Biology*. Horizon Bioscience, Norfolk, United Kingdom.

51. **Mishu, B., and M. J. Blaser.** 1993. Role of infection due to *Campylobacter jejuni* in the initiation of Guillain-Barré syndrome. *Clin. Infect. Dis.* **17:**104–108.

52. **Nachamkin, I., B. M. Allos, and T. W. Ho.** 1998. *Campylobacter* and Guillain-Barré syndrome. *Clin. Microbiol. Rev.* **11:**555–567.

53. **Nachamkin, I., S. H. Fischer, X. H. Yang, O. Benitez, and A. Cravioto.** 1994. Immunoglobulin A antibodies directed against *Campylobacter jejuni* flagellin present in breastmilk. *Epidemiol. Infect.* **112:**359–365.

54. **Nachamkin, I., B. S. Ung, and M. Li.** 2002. Increasing fluoroquinolone resistance in *Campylobacter jejuni*, Pennsylvania, USA, 1982–2001. *Emerg. Infect. Dis.* **8:**1501–1503.

55. **Nachamkin, I., H. Ung, A. P. Moran, D. Yoo, M. M. Prendergast, M. A. Nicholson, K. Sheikh, T. W. Ho, A. K. Asbury, G. M. McKhann, and J. W. Griffin.** 1999. Ganglioside GM1 mimicry in *Campylobacter* strains from sporadic infections in the United States. *J. Infect. Dis.* **179:**1183–1189.

56. **Nachamkin, I., X. H. Yang, and N. J. Stern.** 1993. Role of *Campylobacter jejuni* flagella as colonization factors for three-day-old chicks: analysis with flagellar mutants. *Appl. Environ. Microbiol.* **59:**1269–1273.

57. **Nelson, J. M., K. E. Smith, D. J. Vugia, T. Rabatsky-Ehr, S. D. Segler, H. D. Kassenborg, S. M. Zansky, K. Joyce, N. Marano, R. M. Koekstra, and F. J. Angulo.** 2004. Prolonged diarrhea due to ciprofloxacin-resistant *Campylobacter* infection. *J. Infect. Dis.* **190:**1150–1157.

58. **Newell, D. G., J. A. Frost, B. Duim, J. A. Wagenaar, R. H. Madden, J. van der Plas, and S. L. W. On.** 2000. New developments in the subtyping of *Campylobacter* species, p. 27–44. *In* I. Nachamkin and M. J. Blaser (ed.), Campylobacter, 2nd ed. ASM Press, Washington, D.C.

59. **Newell, D. G., and J. A. Wagenaar.** 2000. Poultry infections and their control at the farm level, p. 497–509. *In* I. Nachamkin and M. J. Blaser (ed.), Campylobacter, 2nd ed. ASM Press, Washington, D.C.

60. **Oberhelman, R. A., and D. N. Taylor.** 2000. *Campylobacter* infections in developing countries, p. 139–153. *In* I. Nachamkin and M. J. Blaser (ed.), Campylobacter, 2nd ed. ASM Press, Washington, D.C.

61. **Olsen, S. J., G. R. Hansen, L. Bartlett, C. Fitzgerald, A. Sonder, R. Manjrekar, T. Riggs, J. Kim, R. Flahart, G. Pezzino, and D. L. Swerdlow.** 2001. An outbreak of *Campylobacter jejuni* infections associated with food handler contamination: the use of pulsed-field gel electrophoresis. *J. Infect. Dis.* **183:**164–167.

62. **On, S. L. W.** 2005. Taxonomy, phylogeny, and methods for the identification of *Campylobacter* species, p. 13–42. *In* J. M. Ketley and M. E. Konkel (ed.), Campylobacter: *Molecular and Cellular Biology*. Horizon Bioscience, Norfolk, United Kingdom.

63. **Park, S. F.** 2005. *Campylobacter jejuni* stress responses during survival in the food chain and colonization, p. 311–330. *In* J. M. Ketley and M. E. Konkel (ed.), *Campylobacter*: *Molecular and Cellular Biology*. Horizon Bioscience, Norfolk, United Kingdom.

64. **Parkhill, J., B. W. Wren, K. Mungall, J. M. Ketley, C. Churcher, D. Basham, T. Chillingworth, R. M. Davies, T. Feltwell, S. Holroyd, K. Jagels, A. V. Karlyshev, S. Moule, M. J. Pallen, C. W. Penn, M. A. Quail, M. A. Rajandream, K. M. Rutherford, A. H. M. van Vliet, S. Whitehead, and B. G. Barrell.** 2000. The genome sequence of the foodborne pathogen *Campylobacter jejuni* reveals hypervariable sequences. *Nature* **403:**665–668.

65. **Patterson, M. F.** 1995. Sensitivity of *Campylobacter* spp. to irradiation in poultry meat. *Lett. Appl. Microbiol.* **20:**338–340.

66. **Patton, C. M., and I. K. Wachsmuth.** 1992. Typing schemes: are current methods useful?, p. 110–128. *In* I. Nachamkin, M. J. Blaser, and L. S. Tompkins (ed.), Campylobacter jejuni: *Current Status and Future Trends*. American Society for Microbiology, Washington, D.C.

67. **Penner, J. L., and J. N. Hennessy.** 1980. Passive hemagglutination technique for serotyping *Campylobacter fetus* subsp. *jejuni* on the basis of soluble heat-stable antigens. *J. Clin. Microbiol.* **12:**732–737.

68. **Pickett, C. L., and R. B. Lee.** 2005. Cytolethal distending toxin, p. 385–395. *In* J. M. Ketley and M. E. Konkel (ed.), Campylobacter: *Molecular and Cellular Biology*. Horizon Bioscience, Norfolk, United Kingdom.

69. **Pickett, C. L., E. C. Pesci, D. L. Cottle, G. Russell, N. Erdem, and H. Zeytin.** 1996. Prevalence of cytolethal distending toxin production in *Campylobacter jejuni* and relatedness of *Campylobacter cdtB* genes. *Infect. Immun.* **64:**2070–2078.

70. **Radomyski, T., E. A. Murano, D. G. Olson, and P. S. Murano.** 1994. Eliminaton of pathogens of significance in food by low-dose irradiation: a review. *J. Food Prot.* **57:**73–86.

71. **Ransom, G. M., B. Kaplan, A. M. McNamara, and I. K. Wachsmuth.** 2000. *Campylobacter* prevention and control: the USDA-Food Safety and Inspection Service role and new food safety approaches, p. 511–528. *In* I. Nachamkin and M. J. Blaser (ed.), Campylobacter, 2nd ed. ASM Press, Washington, D.C.

72. **Raphael, B. H., M. R. Monteville, J. D. Klena, L. A. Joens, and M. E. Konkel.** 2005. Interactions of *Campylobacter jejuni* with non-professional phagocytic cells, p. 397–413. *In* J. M. Ketley and M. E. Konkel (ed.), Campylobacter: *Molecular and Cellular Biology*. Horizon Bioscience, Norfolk, United Kingdom.

73. **Ribot, E. M., C. Fitzgerald, K. Kubota, B. Swaminathan, and T. J. Barrett.** 2001. Rapid pulsed-field gel electrophoresis protocol for subtyping *Campylobacter jejuni*. *J. Clin. Microbiol.* **39:**1889–1894.

74. **Robinson, D. A.** 1981. Infective dose of *Campylobacter jejuni* in milk. *Br. Med. J.* **282:**1584.

75. **Rollins, D. M., and R. R. Colwell.** 1986. Viable but nonculturable stage of *Campylobacter jejuni* and its role in the survival in the natural aquatic environment. *Appl. Environ. Microbiol.* **52:**531–538.

76. **Ruiz-Palacios, G. M., J. J. Calva, L. K. Pickering, Y. Lopez-Vidal, P. Volkow, H. Pezzarossi, and M. S. West.** 1990. Protection of breast-fed infants against *Campylobacter* diarrhea by antibodies in human milk. *J. Pediatr.* **116:**707–713.

77. **Samuel, M. C., D. J. Vugia, S. Shallow, R. Marcus, S. Segler, T. McGivern, H. Kassenborg, K. Reilly, M. Kennedy,**

F. Angulo, and R. V. Tauxe. 2004. Epidemiology of sporadic *Campylobacter* infection in the United States and declining trend in incidence, FoodNet 1996–1999. *Clin. Infect. Dis.* **38**(Suppl. 3):165–174.

78. **Scott, D. A., and D. R. Tribble.** 2000. Protection against *Campylobacter* infection and vaccine development, p. 303–319. *In* I. Nachamkin and M. J. Blaser (ed.), Campylobacter. ASM Press, Washington, D.C.

79. **Skirrow, M. B.** 1994. Diseases due to *Campylobacter, Helicobacter* and related bacteria. *J. Comp. Pathol.* **111**:113–149.

80. **Skirrow, M. B., D. M. Jones, E. Sutcliffe, and J. Benjamin.** 1993. *Campylobacter* bacteremia in England and Wales, 1981–1991. *Epidemiol. Infect.* **110**:567–573.

81. **Stern, N. J.** 1992. Reservoirs for *Campylobacter jejuni* and approaches for intervention in poultry, p. 49–60. *In* I. Nachamkin, M. J. Blaser, and L. S. Tompkins (ed.), Campylobacter jejuni: *Current Status and Future Trends.* American Society for Microbiology, Washington, D.C.

82. **Szymanski, C. M., S. Goon, B. Allan, and P. Guerry.** 2005. Protein glycosylation in *Campylobacter*, p. 259–273. *In* J. M. Ketley and M. E. Konkel (ed.), *Campylobacter: Molecular and Cellular Biology.* Horizon Bioscience, Norfolk, United Kingdom.

83. **Szymanski, C. M., and B. W. Wren.** 2005. Protein glycosylation in bacterial mucosal pathogens. *Nat. Rev. Microbiol.* **3**:225–237.

84. **Takahashi, M., M. Koga, K. Yokoyama, and N. Yuki.** 2005. Epidemiology of *Campylobacter jejuni* isolated from patients with Guillain-Barre and Fisher syndromes in Japan. *J. Clin. Microbiol.* **43**:335–339.

85. **Tauxe, R. V.** 1992. Epidemiology of *Campylobacter jejuni* infections in the United States and other industrialized nations, p. 9–19. *In* I. Nachamkin, M. J. Blaser, and L. S. Tompkins (ed.), Campylobacter jejuni: *Current Status and Future Trends.* American Society for Microbiology, Washington, D.C.

86. **Taylor, D. N.** 1992. *Campylobacter* infections in developing countries, p. 20–30. *In* I. Nachamkin, M. J. Blaser, and L. S. Tompkins (ed.), Campylobacter jejuni: *Current Status and Future Trends.* American Society for Microbiology, Washington, D.C.

87. **Taylor, D. N., P. Echeverria, C. Pitarangsi, J. Seriwatana, L. Bodhidatta, and M. J. Blaser.** 1988. Influence of strain characteristics and immunity on the epidemiology of *Campylobacter* infections in Thailand. *J. Clin. Microbiol.* **26**:863–868.

88. **Thompson, S. A., and M. J. Blaser.** 2000. Pathogenesis of *Campylobacter fetus* infections, p. 321–347. *In* I. Nachamkin and M. J. Blaser (ed.), Campylobacter, 2nd ed. ASM Press, Washington, D.C.

89. **Vandamme, P.** 2000. Taxonomy of the family *Campylobacteraceae*, p. 3–26. *In* I. Nachamkin and M. J. Blaser (ed.), Campylobacter, 2nd ed. ASM Press, Washington, D.C.

90. **Vandamme, P., M. I. Daneshvar, F. E. Dewhirst, B. J. Paster, K. Kersters, H. Goossens, and C. W. Moss.** 1995. Chemotaxonomic analyses of *Bacteroides gracilis* and *Bacteroides ureolyticus* and reclassification of *B. gracilis* as *Campylobacter gracilis* comb. nov. *Int. J. Syst. Bacteriol.* **45**:145–152.

91. **Vandamme, P., L. J. VanDoorn, S. T. Alrashid, W. G. V. Quint, J. VanderPlas, V. L. Chan, and S. L. W. On.** 1997. *Campylobacter hyoilei* Alderton et al. 1995 and *Campylobacter coli* Veron and Chatelain 1973 are subjective synonyms. *Int. J. Syst. Bacteriol.* **47**:1055–1060.

92. **Vandenberg, O., A. Dediste, K. Houf, S. Ibekwem, H. Souayah, S. Cadranel, N. Douat, G. Zissis, J.-P. Butzler, and P. Vandamme.** 2004. *Arcobacter* species in humans. *Emerg. Infect. Dis.* **10**:1863–1867.

93. **Wang, W. L., B. W. Powers, N. W. Luechtefeld, and M. J. Blaser.** 1980. Effects of disinfectants on *Campylobacter jejuni. Appl. Environ. Microbiol.* **45**:1202–1205.

94. **Wassenaar, T. M.** 1997. Toxin production by *Campylobacter* spp. *Clin. Microbiol. Rev.* **10**:466–476.

95. **Wassenaar, T. M., N. M. Bleumink-Pluym, and B. A. van der Zeijst.** 1991. Inactivation of *Campylobacter jejuni* flagellin genes by homologous recombination demonstrates that flaA but not flaB is required for invasion. *EMBO J.* **10**:2055–2061.

96. **Willison, H. J.** 2005. The immunobiology of Guillain-Barre syndromes. *J. Peripher. Nerv. Syst.* **10**:94–112.

97. **Wood, R. C., K. L. MacDonald, and M. T. Osterholm.** 1992. Campylobacter enteritis outbreaks associated with drinking raw milk during youth activities. A 10-year review of outbreaks in the United States. *JAMA* **268**:3228–3230.

98. **Young, V. B., and L. S. Mansfield.** 2005. *Campylobacter* infection—clinical context, p. 1–12. *In* J. M. Ketley and M. E. Konkel (ed.), Campylobacter: *Molecular and Cellular Biology.* Horizon Bioscience, Norfolk, United Kingdom.

99. **Young, V. B., D. B. Schauer, and J. G. Fox.** 2000. Animal models of *Campylobacter* infection, p. 287–301. *In* I. Nachamkin and M. J. Blaser (ed.), Campylobacter, 2nd ed. ASM Press, Washington, D.C.

100. **Yuki, N., T. Taki, F. Inagaki, T. Kasama, M. Takahashi, K. Saito, S. Handa, and T. Miyatakes.** 1993. A bacterium lipopolysaccharide that elicits Guillain-Barre syndrome has a GM1 ganglioside structure. *J. Exp. Med.* **178**:1771–1775.

Food Microbiology: Fundamentals and Frontiers, 3rd Ed.
Edited by M. P. Doyle and L. R. Beuchat
© 2007 ASM Press, Washington, D.C.

Jianghong Meng
Michael P. Doyle
Tong Zhao
Shaohua Zhao

Enterohemorrhagic *Escherichia coli*

12

Escherichia coli is a common part of the normal facultative anaerobic microflora in the intestinal tract of humans and warm-blooded animals. These commensal *E. coli* strains rarely cause disease except in immunocompromised hosts or when the normal gastrointestinal barriers are breached. However, several strains have acquired specific virulence attributes that enable them to cause a broad spectrum of diseases including diarrheal disease, urinary tract infections, and sepsis and meningitis.

E. coli isolates are serologically differentiated based on three major surface antigens, which enable serotyping: the O (somatic), H (flagella), and K (capsule) antigens. A total of 173 O antigens, 56 H antigens, and 103 K antigens have been identified to date (2). It is considered necessary only to determine the O and the H antigens, not the K antigens, to serotype strains of *E. coli* associated with diarrheal disease. The O antigen identifies the serogroup of a strain, and the H antigen identifies its serotype. The application of serotyping to isolates associated with diarrheal disease has shown that specific serogroups often fall into one category of diarrheagenic *E. coli*. However, some serogroups such as O55, O111, O126, and O128 appear in more than one category.

Diarrheagenic *E. coli* isolates are categorized into specific groups (pathotypes) based on virulence properties, mechanisms of pathogenicity, clinical syndromes, and distinct O:H serotypes. These categories include enteropathogenic *E. coli* (EPEC), enterotoxigenic *E. coli* (ETEC), enteroinvasive *E. coli* (EIEC), diffuse-adhering *E. coli* (DAEC), enteroaggregative *E. coli* (EAEC), and enterohemorrhagic *E. coli* (EHEC). This chapter focuses on EHEC, which among the *E. coli* strains that cause foodborne illness in the United States is the most significant group based on frequency and severity of illness. More information on other diarrheagenic *E. coli* strains is available in several review articles (27, 39).

EPEC

EPEC was the first pathotype of *E. coli* to be described and can cause severe diarrhea in infants, especially in developing countries. EPEC previously was also associated with outbreaks of diarrhea in nurseries in developed countries. The major O serogroups associated with illness include O55, O86, O111ab, O119, O125ac, O126, O127, O128ab, and O142. Humans are an important

Jianghong Meng, Dept. of Nutrition and Food Science, University of Maryland, College Park, MD 20742. **Michael P. Doyle and Tong Zhao,** Center for Food Safety, University of Georgia, Griffin, GA 30223. **Shaohua Zhao,** Division of Animal and Food Microbiology, Center for Veterinary Medicine/Office of Research, Food & Drug Administration, Laurel, MD 20708.

reservoir. The original definition of EPEC is "diarrheagenic *E. coli* belonging to serogroups epidemiologically incriminated as pathogens but whose pathogenic mechanism has not been proven to be related to either enterotoxins, or *Shigella*-like invasiveness." However, EPEC have been determined to induce attaching and effacing (A/E) lesions in cells to which they adhere and can invade epithelial cells.

ETEC

ETEC are a major cause of infantile diarrhea in developing countries. They are also the agents most frequently responsible for travelers' diarrhea. ETEC colonize the proximal small intestine by fimbrial colonization factors (e.g., CFA/I and CFA/II) and produce heat-labile or heat-stable enterotoxin that elicits fluid accumulation and a diarrheal response. The most frequent ETEC serogroups include O6, O8, O15, O20, O25, O27, O63, O78, O85, O115, O128ac, O148, O159, and O167. Humans are the principal reservoir of ETEC strains that cause human illness.

EIEC

EIEC cause nonbloody diarrhea and dysentery similar to that caused by *Shigella* spp. by invading and multiplying within colonic epithelial cells. As for *Shigella*, the invasive capacity of EIEC is associated with the presence of a large plasmid (ca. 140 MDa) which encodes several outer membrane proteins involved in invasiveness. The antigenicity of these outer membrane proteins and the O antigens of EIEC are closely related. The principal site of bacterial localization is the colon, where EIEC invade and proliferate in epithelial cells, causing cell death. Humans are a major reservoir, and the serogroups most frequently associated with illness include O28ac, O29, O112, O124, O136, O143, O144, O152, O164, and O167. Among these serogroups, O124 is the serogroup most commonly encountered.

DAEC

DAEC have been associated with diarrhea primarily in young children who are older than infants. The relative risk of DAEC-associated diarrhea increases with age from 1 year to 5 years. The basis for this age-related infection is unknown. DAEC are most commonly of serogroups O1, O2, O21, and O75. Typical symptoms of DAEC infection are mild diarrhea without blood or fecal leukocytes, and DAEC strains produce a characteristic diffuse-adherent pattern of attachment to HEp-2 or HeLa cell lines. DAEC generally do not elaborate heat-labile, heat-stable, or elevated levels of Shiga toxin, nor do they possess EPEC adherence factor plasmids or invade epithelial cells.

EAEC

EAEC recently have been associated with persistent diarrhea in infants and children in several countries worldwide. These organisms are uniquely different from the other types of pathogenic *E. coli* because of their ability to produce a characteristic pattern of aggregative adherence on HEp-2 cells. EAEC adhere in an appearance of stacked bricks to the surface of HEp-2 cells. Serogroups associated with EAEC include O3, O15, O44, O77, O86, O92, O111, and O127. A gene probe derived from a plasmid associated with EAEC strains has been developed to identify *E. coli* of this type; however, more epidemiologic information is needed to elucidate the significance of EAEC as an agent of diarrheal disease.

EHEC

EHEC were first recognized as human pathogens in 1982 when *E. coli* O157:H7 was identified as the cause of two outbreaks of hemorrhagic colitis. Since then, many other serotypes of *E. coli* such as O26, O111, and sorbitol-fermenting O157:NM also have been associated with cases of hemorrhagic colitis and have been classified as EHEC. However, serotype O157:H7 is the predominant cause of EHEC-associated disease in the United States and many other countries. All EHEC produce factors cytotoxic to African green monkey kidney (Vero) cells and hence have been named verotoxins (VT) or Shiga toxins (Stx) because of similarity to the Shiga toxin produced by *Shigella dysenteriae* type 1 (42). Production of Shiga toxins by *E. coli* O157:H7 was first reported by Johnson et al. (26). Karmali et al. (28) subsequently associated Shiga toxin-producing *E. coli* infections with a severe and sometimes fatal condition, hemolytic uremic syndrome (HUS). *E. coli* organisms of many different serotypes subsequently were determined to produce Shiga toxins; hence, they have been named Shiga toxin-producing *E. coli* (STEC). More than 600 serotypes of STEC have been identified, including approximately 160 O serogroups and 50 H types, and the list continues to grow (see the VTEC table compiled by K. A. Bettelheim at http://www.microbionet.com.au/vtectable.htm). However, only those strains that cause hemorrhagic colitis are considered to be EHEC, and there are at least 130 EHEC serotypes that have been recovered from human patients (Table 12.1). Major non-O157 EHEC

Table 12.1 Serotypes of non-O157 Shiga toxin-producing *E. coli* recovered from patients with hemorrhagic colitis and/or HUS[a]

Sero-group	H type	Sero-group	H type	Sero-group	H type	Sero-group	H type	Sero-group	H type	Sero-group	H type	Sero-group	H type
O1	—[b]	O15	—	O68	4	O100	32	O112ac	—	O134	25	O?	—
	7	O18	—	O69	—	O101	—	O113	21	O137	41		11
O2	6	O20	7	O70	35	O103	2	O118	—	O145	—	OR	—
	7		19	O73	34		18		12		25		4
	29	O22	5	O75	5		21		16		28		9
O4	—		8	O76	7		25		30	O146	8		11
	5	O23	0	O77	—	O104	—	O119	2		21		16
	10		7	O79	7		2		5		28		25
O5	—		16	O83	1		21		6	O153	2		49
O6	—	O25	2	O84	—	O105	18	O121	10		25	OX3	—
	2	O26	—	O86	—	O105ac	18		19	O163	19		2
	4		11	O91	—	O107	27	O125	—	O165	—		21
O8	2	O45	2		10	O111	—	O126	27		25	OX174	2
	9	O46	31		21		2	O127	21	O168	—		21
	19	O48	21		40		7	O128	—	O172	—	OX177	—
	21	O50	—	O92	—		8		2	O173	2	OX181	49
O9	—		7		33		11		7	O174	—		
O11	2	O55	6	O98	—	O111ac	—	O128ab	2		2		
O14	—		10	O100	25	O112ab	2		45		21		

[a] Not a comprehensive listing.

[b] —, nonmotile.

serogroups identified in the United States include O26, O111, O103, O111, O121, O45, and O145 (6). Since *E. coli* O157:H7 is the most common serotype of the EHEC and because more is known about this serotype than other serotypes of EHEC, this chapter focuses on *E. coli* O157:H7. Non-O157 EHEC are included where information is available.

CHARACTERISTICS OF *E. COLI* O157:H7 AND NON-O157 EHEC

E. coli O157:H7 was first identified as a foodborne pathogen in 1982. There had been prior isolation of the organism, identified retrospectively among isolates at the Centers for Disease Control and Prevention (CDC); the isolate was from a Californian woman with bloody diarrhea in 1975 (23). In addition to production of Shiga toxin(s), most strains of *E. coli* O157:H7 also possess several characteristics uncommon to most other *E. coli* organisms: inability to grow well, if at all, at temperatures ≥44.5°C in *E. coli* broth, inability to ferment sorbitol within 24 h, inability to produce β-glucuronidase (i.e., inability to hydrolyze 4-methylumbelliferyl-D-glucuronide [MUG]), possession of a pathogenicity island known as the locus of enterocyte effacement (LEE), and carriage of a 60-MDa plasmid. Non-O157 EHEC do not share the previously described growth and metabolic

characteristics, although they all produce Shiga toxin(s), and most contain LEE and the large plasmid.

Acid Resistance

Foodborne pathogens must pass through an acidic gastric barrier with pH values as low as 1.5 to 2.5 to cause infections in humans. Some enteric pathogens such as *Vibrio cholerae* use an "assault tactic" that involves large numbers of infecting cells, in the hope that a few will survive and gain entrance to the intestine. *E. coli* O157:H7, however, has effective mechanisms in resisting extreme acid stress. Three systems in EHEC are involved in acid resistance, including an acid-induced oxidative, an acid-induced arginine-dependent, and a glutamate-dependent system (31). The oxidative system is less effective in protecting the bacterium from acid stress than the arginine-dependent and glutamate-dependent systems. The alternate sigma factor RpoS is required for oxidative acid tolerance but is only partially involved with the other two systems. Once induced, the acid resistance state can persist for a prolonged period of time (≥28 days) at refrigeration temperature. More detailed information on acid resistance can be found in a review article by Foster (17).

The minimum pH for *E. coli* O157:H7 growth is 4.0 to 4.5 but is dependent on the interaction of pH with other growth factors. Studies on inactivation of *E. coli*

O157:H7 with organic acid sprays on beef using acetic, citric, or lactic acid at concentrations up to 1.5% revealed that E. coli O157:H7 populations were not appreciably affected by any of the treatments (4). E. coli O157:H7, when inoculated at high populations, survived fermentation, drying, and storage of fermented sausage (pH 4.5) for up to 2 months at 4°C (22), in mayonnaise (pH 3.6 to 3.9) for 5 to 7 weeks at 5°C and for 1 to 3 weeks at 20°C (61), and in apple cider (pH 3.6 to 4.0) for 10 to 31 days or 2 to 3 days at 8 or 25°C (62), respectively. More importantly, induction of acid resistance in E. coli O157:H7 also can increase tolerance to other environmental stresses. Acid resistance-induced cells also have increased tolerance to heating, radiation, and antimicrobials.

Antibiotic Resistance

Initially, when E. coli O157:H7 was first associated with human illness, the pathogen was susceptible to most antibiotics affecting gram-negative bacteria (29). However, more recent studies indicate a trend toward increasing resistance to antibiotics among E. coli O157:H7 isolates (37, 51). E. coli O157:H7 strains isolated from humans, animals, and food have developed resistance to multiple antibiotics, with streptomycin-sulfisoxazole-tetracycline being the most common resistance profile. Non-O157 EHEC strains isolated from humans and animals also have acquired antibiotic resistance, and some are resistant to multiple antibiotics commonly used in human and veterinary medicine (50). However, antibiotic resistance among EHEC and STEC was low compared to non-STEC E. coli strains (51). Antibiotic-resistant EHEC strains can possess a selective advantage over other bacteria colonizing the gastrointestinal tract of animals that are treated with antibiotics (therapeutically or subtherapeutically). Hence, antibiotic-resistant EHEC strains may become the predominant E. coli present under antibiotic selective pressure and subsequently be more prevalent in fecal excretions.

Inactivation by Heat and Irradiation

Studies on the thermal sensitivity of E. coli O157:H7 in ground beef revealed that the pathogen has no unusual resistance to heat, with D values at 57.2, 60, 62.8, and 64.3°C of 4.5, 0.75, 0.4, and 0.16 min, respectively (14). Heating ground beef sufficiently to kill typical strains of Salmonella also kills E. coli O157:H7 (Table 12.2). The presence of fat protects E. coli O157:H7 in ground beef, with D values for lean (2.0% fat) and fatty (30.5% fat) ground beef of 4.1 and 5.3 min at 57.2°C, respectively, and 0.3 and 0.5 min at 62.8°C, respectively (32). Pasteurization of milk (72°C, 0.27 min) is an effective treatment that kills more than

Table 12.2 Comparison of D values for E. coli O157:H7 and Salmonella spp. in ground beef

Temp (°C)	D value (min)		
	E. coli O157:H7		
	30.5% fat	17–20% fat	Salmonella spp.
51.7	115.5	ND	54.3
57.2	5.3	4.5	5.43
62.8	0.47	0.40	0.54

10^4 E. coli O157:H7 per ml (11). Proper heating of foods of animal origin, e.g., heating foods to an internal temperature of at least 68.3°C for several seconds is an important critical control point to ensure inactivation of E. coli O157:H7.

The use of irradiation to eliminate many types of foodborne pathogens in food has been approved by many countries. Unlike many other processing technologies, irradiation at dosages that kill enteric foodborne pathogens retains the raw character of foods. In the United States, an irradiation dose of 4.5 kGy is approved for refrigerated, and 7.5 kGy for frozen, raw ground beef. D_{10} values for E. coli O157:H7 in raw ground beef patties range from 0.241 to 0.307 kGy, depending on temperature, with D_{10} values significantly higher for patties irradiated at –16°C than at 4°C (9). Hence, an irradiation dose of 1.5 kGy should be sufficient to eliminate E. coli O157:H7 at the cell numbers likely to occur in ground beef.

SOURCES OF E. COLI O157:H7 AND NON-O157 EHEC

Cattle

Undercooked ground beef and, less frequently, unpasteurized milk are well-recognized vehicles associated with outbreaks of E. coli O157:H7 infection; hence, cattle have been the focus of many studies on their role as a reservoir of E. coli O157:H7.

Detection of E. coli O157:H7 and STEC on Farms

The first reported isolation of E. coli O157:H7 from cattle was from a <3-week-old calf with colibacillosis in Argentina in 1977 (44). Prevalence rates of STEC as high as 60% have been found in bovine herds in many countries, but in most cases, the rates range from 10 to 25%. The isolation rates of E. coli O157:H7 are much lower than those of non-O157 STEC. A study on STEC carriage by dairy cattle on farms in Canada revealed that 36% of cows and 57% of calves were STEC positive in all of 80 herds tested (59). Of these, only seven animals (0.45%) on four farms (5%) were positive for E. coli

O157:H7. A 2002 USDA national study revealed that 38.5% of dairy farms had at least one *E. coli* O157:H7-positive cow and that 4.3% of individual cows were *E. coli* O157 positive. *E. coli* O157:H7 levels in calf feces range from detectable by enrichment culture (but <10^2 CFU/g) to 10^5 CFU/g. Young, weaned animals more frequently carry *E. coli* O157:H7 than adult cattle. Studies in some locations indicate an increased prevalence of *E. coli* O157:H7 in cattle during warmer months of the year, which correlates with the seasonal variation in human disease. For example, a 1997 study in England revealed that *E. coli* O157:H7 was isolated from 38% of cattle presented for slaughter in the spring but from only 4.8% of cattle during the winter months. A U.S. study also revealed a high prevalence of *E. coli* O157:H7 in beef cattle presented for slaughter from July to August 1999 (15). *E. coli* O157:H7 was present in 28% (91 of 327) of fecal samples and on 11% (38 of 355) of hide samples. All but 2 of the 30 lots of cattle or carcasses tested were positive for *E. coli* O157:H7.

Factors Associated with Bovine Carriage of *E. coli* O157:H7

Several tentative associations between fecal shedding of *E. coli* O157:H7 and feed or environmental factors have been made from epidemiologic studies of dairy herds. For example, some calf starter feed regimes or environmental factors and feed components such as whole cottonseed were associated with reduced prevalence of *E. coli* O157:H7. In contrast, grouping calves before weaning, sharing among calves feeding utensils without sanitation, and early feeding of grain were associated with increased carriage of *E. coli* O157:H7. Controversial results have been obtained from studies on the effect of grain and hay feeding on *E. coli* O157:H7 carriage by cattle. An early study suggested that grain feeding increased both the number and degree of acid resistance of *E. coli* compared with hay feeding (13). However, a later study determined that hay-fed cattle fecally shed *E. coli* O157:H7 longer than grain-fed animals, and the pathogen was equally acid resistant in the bovine host irrespective of diet (24).

Environmental factors such as water and feed sources or farm management practices such as manure handling may play important roles in influencing the prevalence of *E. coli* O157:H7 on dairy farms. The pathogen was frequently found in water troughs on farms. Several studies have revealed that *E. coli* O157:H7 can survive for weeks or months in bovine feces and water. Commercial feeds often contained detectable *E. coli*, indicating widespread fecal contamination, although *E. coli* O157:H7 was only infrequently detected.

Susceptibility of cattle to subsequent gastrointestinal colonization with *E. coli* O157:H7 is largely a function of age. Young animals are more likely to be positive than older ones of the same herd. For many cattle, *E. coli* O157:H7 is transiently carried in the gastrointestinal tract and is intermittently excreted for a few weeks to months in young calves and heifers. Carriage of more than one strain of *E. coli* O157:H7 has been described. Other influences that disrupt the normal flora of the gastrointestinal tract such as temporary withholding of feed can alter the pattern of fecal shedding. Fasting calves for 48 h can increase the populations of *E. coli* O157:H7 in the gastrointestinal tract of some animals.

Cattle Model for Infection of *E. coli* O157:H7

E. coli O157:H7 is not a pathogen of weaned calves and adult cattle; hence, animals that carry the pathogen are not ill. However, there is evidence that *E. coli* O157:H7 can cause diarrhea and A/E lesions in neonatal calves. The initial sites of localization of *E. coli* O157:H7 in cattle are the forestomachs (rumen, omasum, and reticulum); however, the principal site of colonization appears to be a dense region of lymphoid follicles on the mucosal surface of the terminal rectum (40).

Domestic Animals and Wildlife

Although cattle are thought to be the primary source of STEC in the food chain, STEC have also been isolated from other domestic animals and wildlife such as sheep, goats, deer, dogs, horses, swine, cats, chickens (3), water buffaloes (18), wild birds, rodents (41), and guanacos (38). Prevalence of *E. coli* O157:H7 and STEC in sheep is generally higher than in other animals. A 6-month study of healthy, naturally infected ewes revealed that fecal shedding of the pathogen was transient and seasonal, with 31% of sheep positive in June, 5.7% positive in August, and none in November (30). In a survey of seven animal species in Germany, STEC were isolated most frequently from sheep (66.6%), goats (56.1%), and cattle (21.1%) (3), with lower prevalence rates in chickens (0.1%), pigs (7.5%), cats (13.8%), and dogs (4.8%).

Humans

Fecal shedding of *E. coli* O157:H7 by patients with hemorrhagic colitis or HUS usually lasts for no more than 13 to 21 days following onset of symptoms. However, in some instances, the pathogen can be excreted in feces for many weeks. A child infected during a day care center-associated outbreak continued to excrete the pathogen for 62 days after the onset of diarrhea (43).

Studies of persons living on dairy farms to determine carriage of *E. coli* O157:H7 by farm families revealed elevated antibody titers against the surface antigens of *E. coli* O157; however, the pathogen was not isolated from feces. An asymptomatic long-term carrier state has not been identified. The significance of fecal carriage of *E. coli* O157:H7 by humans is the potential for person-to-person dissemination of the pathogen, a situation which has been observed repeatedly in outbreak settings. A contributing factor to person-to-person transmission of the pathogen is its extraordinarily low infectious dose, estimated at <100 cells; possibly as few as 10 can produce illness in highly susceptible populations. Inadequate attention to personal hygiene, especially after using the bathroom, can transfer the pathogen to other persons through contaminated hands, resulting in secondary transmission.

DISEASE OUTBREAKS

Geographic Distribution

E. coli O157:H7 has been the cause of many major outbreaks of severe illness worldwide. At least 30 countries on six continents have reported *E. coli* O157:H7 infection in humans. In the United States, 350 outbreaks of *E. coli* O157:H7 infection were documented in 49 states, accounting for 8,598 cases from 1982 to 2002 (Table 12.3) (47). Among the cases, there were 1,493 (17.4%) hospitalizations, 354 (4.1%) cases of HUS, and 40 (0.4%) deaths. Outbreak size ranged from 2 to 781. The number of reported outbreaks increased since 1982 and peaked in 2000 with 46 (Fig. 12.1). This dramatic increase may be due in part to improved recognition of *E. coli* O157:H7 infection following the publicity of a large multistate outbreak in the western United States in 1993. The median outbreak size, however, declined from 1982 to 2002. Of the 326 outbreaks involving a single state, Minnesota reported the most (43 outbreaks), followed by Washington (27 outbreaks), New York (22 outbreaks), California (18 outbreaks), and Oregon (18 outbreaks).

The precise incidence of *E. coli* O157:H7 foodborne illness in the United States is not known because infected persons presenting mild or no symptoms and persons with nonbloody diarrhea are less likely to seek medical attention than patients with bloody diarrhea; hence, such cases would not be reported. Foodborne Diseases Active Surveillance Network (FoodNet; http://www.cdc. gov/ncidod/dbmd/foodnet/) reports that the annual rate of *E. coli* O157:H7 infection at several surveillance sites in the United States has ranged from 0.9 to 2.8 cases per 100,000 population. Between 2000 to 2004 there

has been a decline in *E. coli* O157:H7 infections from 2.0 cases per 100,000 to 0.9 case per 100,000, respectively. In 1999, the CDC estimated that *E. coli* O157:H7 causes 73,480 illnesses and 61 deaths annually in the United States and non-O157 STEC account for an additional 37,740 cases with 30 deaths (34). Eighty-five percent of these cases are attributed to foodborne transmission. However, with the dramatic reduction in cases reported by FoodNet during the past 5 years, these estimates are likely to be revised downward.

Large outbreaks of *E. coli* O157:H7 infections involving hundreds of cases also have been reported in Canada, Japan, and the United Kingdom. The largest multiple outbreaks reported worldwide occurred in May to December 1996 in Japan, involving more than 11,000 reported cases. In the same year, 21 elderly people died in a large outbreak involving 501 cases in central Scotland. Although *E. coli* O157:H7 is still the predominant serotype of EHEC in the United States, Canada, the United Kingdom, and Japan, an increasing number of outbreaks and sporadic cases related to EHEC of serotypes other than O157:H7 have been reported. A large epidemic involving several thousand cases of *E. coli* O157:NM infection occurred in Swaziland and South Africa, following consumption of contaminated surface water. In continental Europe, Australia, and Latin America, non-O157 EHEC infections are more common than *E. coli* O157:H7 infections. Details of many reported foodborne and waterborne outbreaks of EHEC infections are provided in Table 12.4. There are no distinguishing biochemical phenotypes for non-O157 EHEC, making screening for these bacteria problematic and labor-intensive, and for this reason many clinical laboratories do only limited testing for them. Therefore, the prevalence of non-O157 EHEC infections is likely underestimated.

Seasonality of *E. coli* O157:H7

Outbreaks and clusters of *E. coli* O157:H7 infections peak during the warmest months of the year. Approximately 89% of 350 outbreaks reported in the United States occurred from May to November (47). FoodNet data indicate the same trend. The reasons for this seasonal pattern are unknown but may include (i) an increased prevalence of the pathogen in cattle or other livestock or vehicles of transmission during the summer, (ii) greater human exposure to ground beef or other *E. coli* O157:H7 contaminated foods during the "cook-out" months, and/or (iii) greater improper handling (temperature abuse and cross-contamination) or incomplete cooking of products such as ground beef during warm months than other months.

Table 12.3 Outbreaks and cases of *E. coli* O157 infection by transmission route, 1982–2002[a]

Transmission route	Outbreaks			Outbreak size	Cases		
	No.	% of total	% of foodborne	Median (range)	No.	% of total	% of foodborne
Ground beef	75	21	41	8 (2–732)	1,760	20	33
Unknown food vehicle	42	12	23	8 (2–86)	646	8	12
Produce	38	11	21	20 (2–736)	1,794	21	34
Other beef	11	3	6	17 (2–323)	563	7	11
Other food vehicle	10	3	5	15 (2–47)	206	2	4
Dairy product	7	2	4	8 (2–202)	300	3	6
Subtotal, foodborne	183	52		11 (2–736)	5,269	61	
Unknown transmission route	74	21		4 (2–140)	812	9	
Person-to-person	50	14		7 (2–63)	651	8	
Recreational water	21	6		8 (2–45)	280	3	
Animal contact	11	3		5 (2–111)	319	4	
Drinking water	10	3		26 (2–781)	1,265	15	
Laboratory related	1	<1		2	2	<1	
Subtotal, other routes	167	48		5 (2–781)	3,329	39	
Total	350				8,598		

[a] Reproduced from reference 47.

Age of Patients

All age groups can be infected by *E. coli* O157:H7, but the very young and the elderly most frequently experience severe illness with complications. HUS usually occurs in children rather than adults. Population-based studies have suggested that the highest age-specific incidence of *E. coli* O157:H7 infection occurs in children of age 2 to 10 years. The high rate of infection in this age group likely is the result of increased exposure to contaminated foods, contaminated environments, and infected animals, more opportunities for person-to-person spread between infected children with relatively undeveloped hygiene skills, and the lack of adequate protective antibodies to Shiga toxins.

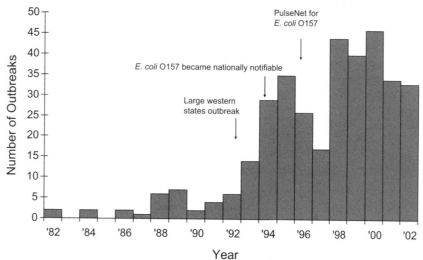

Figure 12.1 Number of *E. coli* O157:H7 outbreaks in the United States by year, 1982–2002 (*n* = 350), reproduced from reference 47.

Table 12.4 Representative foodborne and waterborne outbreaks of *E. coli* O157:H7 and other EHEC infections[a]

Yr	Mo	Location[b]	No. of cases/no. of deaths	Setting	Vehicle
1982	2	Oregon	26	Community	Ground beef
1982	5	Michigan	21	Community	Ground beef
1985		Canada	73/17	Nursing home	Sandwiches
1987	6	Utah	51	Custodial institution	Ground beef/person-to-person
1988	10	Minnesota	54	School	Precooked ground beef
1989	12	Missouri	243	Community	Water
1990	7	North Dakota	65	Community	Roast beef
1991	11	Massachusetts	23	Community	Apple cider
1991	7	Oregon	21	Community	Swimming water
1992[c]		France	>4	Community	Goat cheese
1992	12	Oregon	9	Community	Raw milk
1993	1	California, Idaho, Nevada, Washington	732/4	Restaurant	Ground beef
1993	3	Oregon	47	Restaurant	Mayonnaise?
1993	7	Washington	16	Church picnic	Pea salad
1993	8	Oregon	27	Restaurant	Cantaloupe
1994[d]	2	Montana	18	Community	Milk
1994	11	Washington, California	19	Home	Salami
1995[e]	2	Adelaide, Australia	>200	Community	Semidry sausage
1995	10	Kansas	21	Wedding	Punch/fruit salad
1995	11	Oregon	11	Home	Venison jerky
1995	7	Montana	74	Community	Leaf lettuce
1995	9	Maine	37	Camp	Lettuce
1996[f]		Komatsu, Japan	126	School	Luncheon
1996	5, 6	Connecticut, Illinois	47	Community	Mesclun lettuce
1996	7	Osaka, Japan	7,966/3	Community	White radish sprouts
1996	10	California, Washington, Colorado	71/1	Community	Apple juice
1996	11	Central Scotland, United Kingdom	501/21	Community	Cooked meat
1997	5	Illinois	3	School	Ice cream bar
1997	6	Michigan, Virginia	108	Community	Alfalfa sprouts
1997	11	Wisconsin	13	Church banquet	Meatballs/coleslaw
1998	6	Wisconsin	63	Community	Cheese curds
1998	6	Wyoming	114	Community	Water
1998	7	North Carolina	142	Restaurant	Coleslaw
1998	7	California	28	Prison	Milk
1998	8	New York	11	Deli	Macaroni salad
1998	9	California	20	Church	Cake
1999	7	Ohio	18	Restaurant	Coleslaw
1999[g]	7	Texas	56	Camp	Salad bar
1999[h]	7	Connecticut	11	Community	Lake water
1999	8	New York	900/2	Fair	Well water
1999	10	Ohio, Indiana	47	Community	Lettuce
2000	11	Iowa, Minnesota, Wisconsin	52	Community	Ground beef
2002	7	10 states	38	Community	Ground beef
2002	8	Washington	32	Camp	Romaine lettuce
2005	9, 10	Minnesota	23	Community	Prepackaged lettuce
2006	8–10	26 states, Canada	206/3	Community	Prepackaged fresh spinach

[a] *E. coli* O157:H7 unless otherwise noted.
[b] State of the United States unless otherwise noted.
[c] *E. coli* O119.
[d] *E. coli* O104:H21; bloody diarrhea was not observed.
[e] *E. coli* O111:NM.
[f] *E. coli* O118:H2.
[g] *E. coli* O111:H8.
[h] *E. coli* O121:H19.

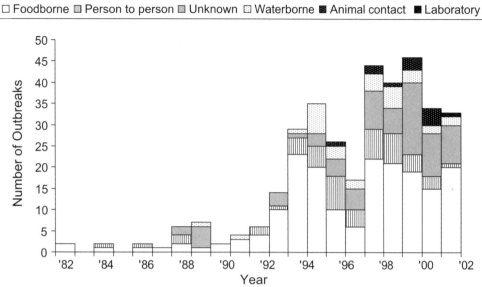

Figure 12.2 Transmission routes of *E. coli* O157:H7 outbreaks in the United States by year, 1982–2002, reproduced from reference 47.

Locations of Outbreaks

Reported foodborne outbreaks have most frequently occurred in communities (29%), restaurants or food facilities (28%), and schools (16%) (47). Among 51 restaurant and food facility outbreaks, 22 (43%) were in chain establishments and 29 were in single food service establishments.

Transmission of *E. coli* O157:H7

Food remained the predominant transmission route (Fig. 12.2), accounting for 52% of 350 outbreaks and

61% of 8,598 outbreak-related cases from 1982 to 2002 (47). A variety of foods have been identified as vehicles of *E. coli* O157:H7 infections, although ground beef has been the most frequent food vehicle (Fig. 12.3). Examples of other foods that have been implicated in outbreaks include roast beef, cooked meats, venison meat and jerky, salami, raw milk, pasteurized milk, yogurt, cheese, ice cream bars, lettuce, unpasteurized apple cider or juice, cantaloupe, handling potatoes, radish sprouts, alfalfa sprouts, fruit or vegetable salad, and cake (36). Among 183 foodborne outbreaks, the food vehicle in

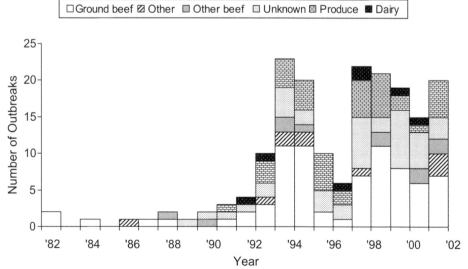

Figure 12.3 Vehicles of foodborne *E. coli* O157:H7 outbreaks in the United States by year, 1982–2002, reproduced from reference 47.

75 (41%) was ground beef, in 42 (23%) was unknown, in 38 (21%) was produce, in 11 (6%) was other beef, in 10 (5%) was other foods, and in 7 (4%) was dairy products (Table 12.3).

The route of *E. coli* O157:H7 transmission for many outbreaks (74; 21%) was unknown. Outbreaks attributed to transmission by person-to-person contact (50; 14%), water (31; 9%), animal contact (11; 3%), and laboratory exposure (1; 0.3%) have also been reported (Table 12.3). An outbreak investigation reported that a petting zoo-associated *E. coli* O157:H7 infection subsequently caused secondary transmission, asymptomatic infection, and prolonged shedding in children in the classroom (12). In contrast to *E. coli* O157:H7 outbreaks in which a food is most often identified as a vehicle, the modes of transmission of most outbreaks caused by non-O157 EHEC are unknown (6, 25). Only a few outbreaks of non-O157 EHEC have been clearly associated with foods and/or water (Table 12.4).

Examples of Foodborne and Waterborne Outbreaks

The Original Outbreaks

The first documented outbreak of *E. coli* O157:H7 infection occurred in the state of Oregon in 1982, with 26 cases and 19 persons hospitalized. All patients had bloody diarrhea and severe abdominal pain. The median age was 28 years, with a range of 8 to 76 years. The duration of illness ranged from 2 to 9 days, with a median of 4 days. This outbreak was associated with eating undercooked hamburgers from fast-food restaurants of a specific chain. *E. coli* O157:H7 was recovered from stools of patients. A second outbreak followed 3 months later and was associated with the same fast-food restaurant chain in Michigan, with 21 cases and 14 persons hospitalized. The median age was 17 years, with a range of 4 to 58 years. Contaminated hamburgers again were implicated as the vehicle, and *E. coli* O157:H7 was isolated both from patients and from a frozen ground beef patty. That *E. coli* O157:H7, a heretofore unknown human pathogen, was the causative agent was established by its association with the food and recovery of the bacterium with identical microbiologic characteristics from both the patients and the meat from the implicated supplier.

Large Multistate Outbreak

A large multistate outbreak of *E. coli* O157:H7 infection in the United States occurred in Washington, Idaho, California, and Nevada in early 1993. Approximately 90% of primary cases were associated with eating at a single fast-food restaurant chain (Chain A), from which *E. coli* O157:H7 was isolated from hamburger

patties. Transmission was amplified by secondary spread (48 patients in Washington alone) via person-to-person transmission. In total, 731 cases were identified, with 629 in Washington, 13 in Idaho, 57 in Las Vegas, Nev., and 34 in Southern California. The median age of patients was 11 years, with a range of 4 months to 88 years. One hundred seventy-eight persons were hospitalized, 56 developed HUS, and 4 died. Because neither specific laboratory testing nor surveillance for *E. coli* O157:H7 was carried out for earlier cases in Nevada, Idaho, and California, the outbreak went unrecognized until a sharp increase in cases of HUS was identified and investigated in the state of Washington.

The outbreak resulted because of insufficient cooking of hamburgers by Chain A restaurants. Epidemiologic investigation revealed that 10 of 16 hamburgers cooked according to Chain A's cooking procedures in Washington State had internal temperatures below 60°C, which was substantially less than the minimum internal temperature of 68.3°C required by the state of Washington. Cooking patties to an internal temperature of 68.3°C for several seconds would have been sufficient to kill the low populations of *E. coli* O157:H7 detected in the contaminated ground beef.

Outbreaks Associated with Produce

Produce-associated outbreaks of *E. coli* O157:H7 infection were first reported in 1991, and produce has remained a prominent food vehicle (Fig. 12.3). Raw vegetables, particularly lettuce and alfalfa and vegetable sprouts, have been implicated in several outbreaks of *E. coli* O157:H7 infection in North America, Europe, and Japan. In May, 1996, a mesclun mix of organic lettuce was associated with a multistate outbreak in which 47 cases were identified in Illinois and Connecticut. Isolates of *E. coli* O157:H7 from patients in the two states were indistinguishable by DNA subtyping. Traceback studies implicated one grower as the likely source of the contaminated lettuce and revealed that cattle were present in the vicinity of the lettuce-growing and -processing areas.

Between May and December 1996, multiple outbreaks of *E. coli* O157:H7 infection occurred in Japan, involving 11,826 cases and 12 deaths. The largest outbreak affected 7,892 schoolchildren and 74 teachers and staff in Osaka in July 1996, among which 606 individuals were hospitalized, 106 had HUS, and 3 died. Epidemiologic investigations revealed that white radish sprouts were the vehicle of transmission.

Apple Cider and Apple Juice Outbreaks

The first confirmed outbreak of *E. coli* O157:H7 infection associated with apple cider occurred in

Massachusetts in 1991, involving 23 cases. In 1996, three outbreaks of *E. coli* O157:H7 infection associated with unpasteurized apple juice or cider were reported in the United States. The largest of the three occurred in three western states (California, Colorado, and Washington) and British Columbia, Canada, with 71 confirmed cases and one death. *E. coli* O157:H7 was isolated from the implicated apple juice. An outbreak also occurred in Connecticut, with 14 confirmed cases. Contamination of apples by manure was the suspected source of *E. coli* O157:H7 in several of the outbreaks. Using apple drops (i.e., apples picked up from the ground) for making apple cider was a common practice, and apples can become contaminated by resting on soil contaminated with manure. Apples also can become contaminated if transported or stored in areas that contain manure or are treated with contaminated water. Investigation of the 1991 outbreak in Massachusetts revealed that the implicated cider press processor also raised cattle that grazed in a field adjacent to the cider mill. Fecal droppings from deer also were found in the orchard where apples used to make the cider were harvested.

Waterborne Outbreaks

Reported waterborne outbreaks of *E. coli* O157:H7 infection have increased substantially in recent years, being associated with swimming water, drinking water, well water, and ice. Investigations of lake-associated outbreaks revealed that in some instances the water was likely contaminated with *E. coli* O157:H7 by toddlers defecating while swimming and that swallowing lake water was subsequently identified as the risk factor. A 1995 outbreak in Illinois involved 12 children ranging in age from 2 to 12 years. Although *E. coli* O157:H7 was not recovered from water samples, high levels of *E. coli* were detected, indicating likely fecal contamination. A large waterborne outbreak of *E. coli* O157:H7 among attendees of a county fair in New York occurred in August 1999. More than 900 persons were infected, of which 65 were hospitalized and two died. Unchlorinated well water used to make beverages and ice was the vehicle, and *E. coli* O157:H7 was isolated from samples of well water.

Waterborne outbreaks of *E. coli* O157 infections also have been reported in other locations of the world. Drinking water, which was probably contaminated with bovine feces, was implicated in outbreaks in Scotland and southern Africa. *E. coli* O157:NM was isolated from water associated with the latter outbreak. Outbreaks associated with drinking contaminated well water also have been reported in Japan.

Outbreaks of EHEC O111 Infection

Several outbreaks of EHEC *E. coli* O111 infection have been reported worldwide; however, an outbreak in early 1995 in South Australia was one in which the vehicle of transmission was identified. Twenty-three cases of HUS among children >12 years of age (median age, 4 years) were reported after consumption of an uncooked, semidry fermented sausage product. Of 10 sausage samples obtained from the homes of nine patients (eight homes), 8 were positive for Shiga toxin genes by PCR, and *E. coli* O111:NM was isolated from 4 of these samples. Eighteen (39%) of 47 additional sausage samples from the same manufacturer and retail stores were PCR positive; three yielded *E. coli* O111:NM. In June 1999, an outbreak of *E. coli* O111:H8 involving 58 cases occurred at a teenage cheerleading camp in Texas. Contaminated ice was the implicated vehicle.

CHARACTERISTICS OF DISEASE

The spectrum of human illness of *E. coli* O157:H7 infection includes nonbloody diarrhea, hemorrhagic colitis, and HUS. Some persons may be infected but asymptomatic, but typically for a short period of time (<3 weeks). Ingestion of the bacterium is followed typically by a 3- to 4-day incubation period (range, 2 to 12 days), during which colonization of the large bowel occurs. Illness begins with nonbloody diarrhea and severe abdominal cramps for 1 to 2 days and then progresses in the second or third day of illness to bloody diarrhea that lasts for 4 to 10 days (1, 52). Many outbreak investigations revealed that more than 90% of microbiologically documented cases of diarrhea caused by *E. coli* O157:H7 were frank blood in the stools, but in some outbreaks there have been reports of 30% of cases with nonbloody diarrhea. Symptoms usually resolve after 1 week, but about 6% of patients progress to HUS, one-half of whom require dialysis, and 75% require transfusions of erythrocytes and/or platelets. The case-fatality rate from *E. coli* O157:H7 infection is about 1%. Similar symptoms have been observed in infections with non-O157 EHEC.

HUS largely affects children, for whom it is the leading cause of acute renal failure. The risk that a child younger than 10 years with a diagnosed *E. coli* O157:H7 infection will develop HUS is about 15% (52). The syndrome is characterized by a triad of features: acute renal insufficiency, microangiopathic hemolytic anemia, and thrombocytopenia. Significant pathological changes include swelling of endothelial cells, widened subendothelial regions, and hypertrophied mesangial cells between glomerular capillaries. These changes combine to narrow the lumina of the glomerular capillaries and afferent arterioles and result in thrombosis of the arteriolar and glomerular microcirculation. Complete obstruction of renal microvessels can produce glomerular and tubular necrosis, with an increased probability of subsequent hypertension or renal failure.

INFECTIOUS DOSE

Retrospective analyses of foods associated with outbreaks of EHEC infection revealed that the infectious dose is very low. For example, between 0.3 and 15 CFU of *E. coli* O157:H7 isolate per g was enumerated in lots of frozen ground beef patties associated with a 1993 multistate outbreak in the western United States. Similarly, 0.3 to 0.4 *E. coli* O157:H7 isolate per g was detected in several intact packages of salami that were associated with a foodborne outbreak. These data suggest that the infectious dose of *E. coli* O157:H7 may be fewer than 100 cells. In an outbreak of *E. coli* O111:NM infection in Australia, the implicated salami was estimated to contain fewer than one cell per 10 g. Additional evidence for a low infectious dose is the capability for person-to-person and waterborne transmission of EHEC infection.

MECHANISMS OF PATHOGENICITY

Significant virulence factors associated with the pathogenicity of EHEC have been identified based on histopathology of tissues of HUS and hemorrhagic colitis patients, studies with tissue culture and animal models, and studies using genetic approaches. A general body of knowledge of the pathogenicity of EHEC has been developed and indicates that the bacteria cause disease by their ability to adhere to the host cell membrane and colonize the large intestine, and then producing one or more Stxs.

Attaching and Effacing

Most studies addressing the pathogenesis of EHEC have focused on elucidating the mechanisms of adherence and colonization. By adhering to intestinal epithelial cells, EHEC subvert cytoskeletal processes to produce a histopathological feature known as an A/E lesion (Fig. 12.4). *E. coli* O157:H7 produces an A/E lesion in the large intestine similar to that induced by EPEC, which in contrast occurs predominantly in the small intestine. The A/E lesion is characterized by intimate attachment of the bacteria to the plasma membranes of the host epithelial cells, localized destruction of the brush border microvilli, and assembly of highly organized pedestal-like actin structures (21). The molecular genetics of the A/E lesion was first studied in EPEC and subsequently in *E. coli* O157:H7. All proteins associated with the

formation of A/E lesions identified to date are encoded on a chromosomal pathogenicity island known as the LEE (Fig. 12.5). These include structural components of a type III secretion system (TTSS), an outer membrane adhesin (intimin), and translocated intimin receptor (Tir) and other effector proteins. The complete DNA sequences of the LEE from EPEC strain E2348/69 and EHEC O157:H7 strain EDL 933 have been determined (16, 46). The EHEC LEE is larger, containing ca. 43 kb

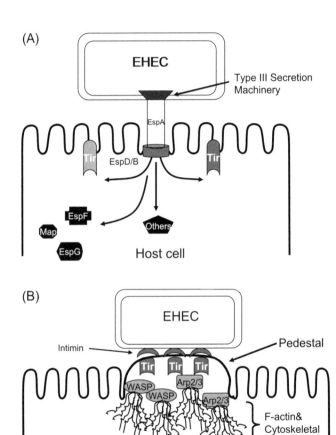

Figure 12.4 Schematic illustration of A/E lesion formation in EHEC, modified based on references 8 and 56. (A) A/E translocation of effector proteins through TTSS that forms a pore through the membranes of EHEC. EHEC translocate a number of proteins: EspB and EspD, which form a translocon in the plasma membrane; the cytoplasmic proteins EspF, EspG, and Map; the translocated intimin receptor Tir, which inserts into the plasma membrane; and other unidentified effectors. (B) Formation of EHEC pedestal. EHEC intimately attaches to the host cell through intimin-Tir binding. The binding triggers the formation of actin-rich pedestals beneath adherent bacteria after Wiskott-Aldrich syndrome protein (WASP) and the heptameric actin-related protein Arp2/3 are recruited to the pedestal tip.

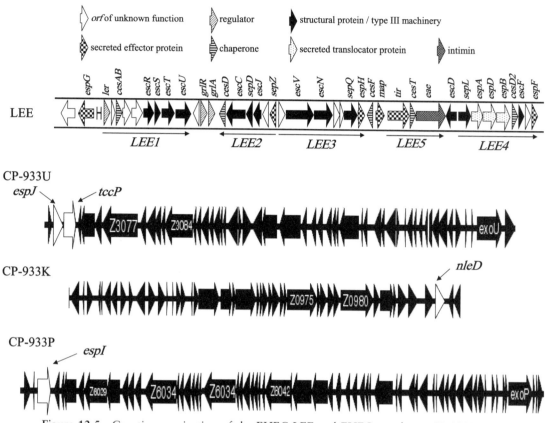

Figure 12.5 Genetic organization of the EHEC LEE and EHEC prophages CP-933U, CP-933K, and CP-933P, reproduced from reference 21.

compared with ca. 35 kb for EPEC. Most of the size difference is due to a 7.5-kb putative prophage present on the EHEC LEE. The G+C contents of the EPEC LEE and EHEC LEE are 38.3 and 39.6%, respectively, which are significantly lower than the 50.8% G+C content of the total *E. coli* genome. This suggests that EPEC and EHEC may have acquired this pathogenicity island by horizontal transfer from another species.

TTSS

TTSS are associated with the virulence of many gram-negative bacterial pathogens. The TTSS apparatus (Fig. 12.6) is a complex "needle and syringe" structure that is assembled from the products of approximately 20 genes in LEE (21). The system is used by EHEC to directly translocate virulence factors from the bacteria into the targeted host cells in a single step.

Many components are broadly conserved among virulence TTSS. In EHEC, Esc (mnemonic for *E. coli* secretion) C and EscV are the main components of the outer and inner membrane ring structures, respectively. EscJ, a lipoprotein spanning the periplasm, serves as a

bridge between the inner and outer membrane protein rings. The EHEC TTSS needle is composed of a single protein, EscF, that forms a projection channel required for TTSS-dependent protein secretion. The unique feature of the EHEC TTSS is the presence of a filamentous extension to the needle complex-associated EscF needle, called the Esp (for *E. coli* secreted protein) A filament. The EspA filament is a polymer of the translocator protein EspA, which is the sole constituent of these hollow filamentous conduits. Polymerization of the EspA filaments is mediated by coil-coil interactions between EspA subunits. EspA filaments bind directly to the needle protein EscF.

The EspA filaments establish a transient link between the bacterium and the host cell, which enables effector protein translocation; therefore, they are important adhesion factors. Following effector protein translocation, EspA filaments and the needle complex are eliminated from the bacterial cell surface to allow intimate bacterial attachment through intimin-Tir interactions.

Effector proteins are delivered to the host cell cytoplasm from the extremity of the EspA filament through

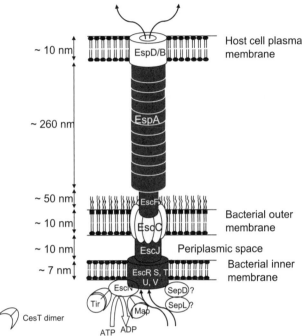

Figure 12.6 TTSS apparatus of EHEC. The basal body of the TTSS is composed of the secretin EscC, the inner membrane proteins EscR, EscS, EscT, EscU, and EscV, and the EscJ lipoprotein, which connects the inner and outer membrane ring structures. EscF constitutes the needle structure, whereas EspA subunits polymerize to form the EspA filament. EspB and EspD form the translocation pore in the host cell plasma membrane, connecting the bacteria with the eukaryotic cell via EspA filaments. The cytoplasmic ATPase EscN provides the energy to the system by hydrolyzing ATP molecules into ADP. SepD and SepL have been represented as cytoplasmic components of the TTSS. (Reproduced from reference 21.)

a translocation pore formed in the plasma membrane of the host cell by the translocator proteins EspB and EspD (Fig. 12.4A) (21). Additional proteins, Sep (mnemonic for secretion of EPEC protein) L and SepD, also play a role in the formation of the translocation apparatus. SepL is a soluble cytoplasmic protein which interacts with SepD. These proteins could be involved in the "switch" from secretion of translocator proteins to secretion of effector proteins through the type III machinery.

Intimin

Intimin is a 94-kDa outer membrane protein encoded by *eae* (*E. coli* attaching and effacing). The *eae* genes of pathogenic *E. coli* present a considerable heterogeneity in their 3′ end that encodes the C-terminal 280 amino acids (Int280) involved in binding to enterocytes and Tir, and the corresponding changes in the amino acid sequence also represent antigenic variations. Based on the sequence and antigenic differences, more than 10 distinct intimin

types have been identified and classified (55), with α, β, ε, and γ being the main intimin types (45). Intimin α is generally found in EPEC, whereas ε and γ are closely associated with EHEC, and β is present in both EPEC and EHEC. *E. coli* O157:H7 produces intimin γ.

Intimin is exported via the general secretory pathway to the periplasm, where it is inserted into the outer membrane by a putative autotransport mechanism. Intimin has two functional regions: the highly conserved N-terminal region is inserted into the bacterial outer membrane, forming a β-barrel-like structure, and mediates dimerization. The variable Int280 extends from the bacterium and interacts with receptors in the host cell plasma membrane. Int280 comprises three globular domains: two immunoglobulin-like domains and a C-type lectin-like domain; the latter, together with the last immunoglobulin-like domain, contains the receptor-binding site. Intimin binds, in addition to Tir, a receptor encoded by the host cell. Two potential host cell-carried intimin receptors (Hir) have so far been identified, β_1 integrin and nucleolin. Interaction of intimin with host cells stimulates production of microvillus-like processes.

Effector Proteins

Numerous effector proteins have been identified in EHEC and are translocated into the host cell via the LEE-encoded TTSS (Table 12.5) (21). Several effector proteins including Tir (transmembrane intimin receptor), Map (mitochondrion-associated protein), EspF, EspG, EspH, SepZ, and EspB are encoded in LEE, whereas others such as Cif (cycle inhibiting factor), EspI, EspJ, and TccP (Tir-cytoskeleton coupling protein) are encoded in prophages.

Tir localizes to the host cell plasma membrane. Tir contains two membrane-spanning transmembrane domains and forms a hairpin-like structure with both its C and N termini located within the host cell and the region between the two transmembrane domains forming an extracellular loop, exposed on the surface of the cell, which interacts with intimin. Like intimin in the bacterial outer membrane, plasma membrane-bound Tir is a dimer. The two Tir-binding domains interact with Tir molecules belonging to different Tir dimers. This binding pattern generates a reticular array-like conformation that clusters Tir under adherent bacteria. Tir intracellular amino and carboxy termini interact with a number of focal adhesion and cytoskeletal proteins, linking the extracellular bacterium to the host cell cytoskeleton. These interactions lead to the formation of actin-rich pedestals beneath adherent bacteria after Wiskott-Aldrich syndrome protein (WASP) and the heptameric actin-related protein Arp2/3 are recruited to the pedestal tip (Fig. 12.4B). Descriptions of

Table 12.5 List of effector proteins of EHEC[a]

Effector protein	Molecular mass (kDa)	Localization in the genome	Localization in host cells	Function(s)
Tir	56–68	LEE	Tip of pedestal	Intimate attachment; A/E lesion formation; actin polymerization
Map	20	LEE	Mitochondria	Disruption of TER and mitochondrial membrane potential; filopodium formation
EspF	20–31	LEE	Mitochondria	Disruption of TER and mitochondrial membrane potential; induction at cell death
EspG	44	LEE	Pedestal/colocalization with tubulin	Destruction of microtubule network
EspH	21	LEE	Underneath microcolony	Cytoskeleton modulation
SepZ	9.5	LEE	Unknown	Unknown
EspB	32	LEE	Bacterial attachment site	Cytoskeleton modulation
Cif	32	λ prophage	Unknown	Cyclomodulin activity
EspI/NleA	54	CP-933P	Golgi	Unknown
EspJ	25	CP-933U	Unknown	Regulation of clearance dynamics in vivo
TccP	42.4	CP-933U	Tip of pedestal	A/E lesion formation; actin polymerization
NleB	39	O island 122/CP-933K	Unknown	Unknown
NleC	40	CP-933K	Unknown	Unknown
NleD	28	CP-933K	Unknown	Colonization in calves
NleE	27	O island 122	Unknown	Unknown
NleF	24	O island 71	Unknown	Unknown
NleH	33	nt 2281573–2282484d/ CP-933K	Unknown	Unknown

[a] Data from reference 21.

other effector factors are provided in Table 12.3. Additional information on effector proteins is available in a review article by Garmendia et al. (21).

Other Adhesins

In addition to intimin, genomic sequence analysis has shown that at least 12 regions encode putative adhesins in EHEC (55). One of these regions contains genes closely related to the long polar fimbrial (*lpf*) operon of *Salmonella enterica* serovar Typhimurium. The EHEC *lpf* operon increases fimbria expression and adherence to tissue culture cells. A nonfimbrial factor encoded outside the LEE in adherence, termed *efa* (EHEC factor for adherence), has also been proposed to influence bacterial adhesion and intestinal colonization. A large gene with significant homology to *efa* is known as *toxB*, which is located in the EHEC pO157 plasmid (see below). In addition, several EHEC proteins act as mediators of adherence but without an obvious role in pathogenesis, including Iha (*V. cholerae* Irg homologue adhesin), Cah (calcium-binding antigen 43 homologue), and OmpA (outer membrane protein A). Toma et al (54) reported that the prevalence of these putative adhesins is associated mainly with serotypes and not with the source of isolates.

60-MDa Plasmid (pO157)

E. coli O157:H7 isolates possess a plasmid (pO157) of approximately 60 MDa (unrelated to the 60-MDa plasmid present in EPEC) that contains DNA sequences common to plasmids present in other serotypes of EHEC isolated from patients with hemorrhagic colitis. Because of the consistent association of the 60-MDa plasmid with several EHEC serotypes, the plasmid has been implicated in the pathogenesis of EHEC infections. However, its exact role in virulence has not been fully elucidated.

Based on DNA sequence analysis, pO157 is a 92-kb F-like plasmid composed of segments of putative virulence genes in a framework of replication and maintenance regions, with seven insertion sequence elements located largely at the boundaries of the virulence segments. There are 100 open reading frames, of which 19 have been previously sequenced and implicated as potential virulence genes, including those encoding a potential adhesin (ToxB), EHEC-hemolysin, a serine protease (EspP), a catalase, and the StcE protein. ToxB shares sequence similarity with the EPEC LifA and Efa-1. The term EHEC-hemolysin is used to distinguish it from α-hemolysin, to which it is related but not identical. EHEC-hemolysin belongs to the repeats-in-toxin family of exoproteins. Four gene products of the *hlyCABD* operon encode a pore-forming cytolysin and its secretion apparatus. Toxicity results from the insertion of HlyA into the cytoplasmic membrane of target mammalian cells, thereby disrupting permeability. The EHEC catalase-peroxidase is encoded by *katP*, whose product is a bifunctional periplasmic enzyme that protects the bacterium against oxidative stress, a possible defense strategy of mammalian cells during bacterial infection.

Shiga Toxins

EHEC produce one or two Stxs. The nomenclature of the Stx family and their important characteristics are listed in Table 12.6. Molecular studies on Stx1 from different *E. coli* strains revealed that Stx1 is either completely identical to the Stx of *S. dysenteriae* type 1 or differs by only one amino acid. However, during the last decade, several variants of Stx1 have been described. Some minor variants have 99% nucleotide sequence homology with *stx*$_1$ of phage 933J. A more substantial deviation of Stx1 was observed in an ovine strain OX3:H8 131/3, and subsequently among human isolates (60). It differs from Stx1 of phage 933J by nine amino acids within the A subunit and three amino acids within the B subunit, and is designated as Stx1c. Another Stx1 variant (Stx1d) was recently identified in STEC ONT:H19 of bovine origin, showing difference from Stx1 by 20 amino acids in the A subunit and 7 amino acids in the B subunit (7).

Unlike Stx1, toxins of the Stx2 group are not neutralized by antiserum raised against Stx and do not cross-hybridize with Stx1-specific DNA probes. There is sequence and antigenic variation within toxins of the Stx2 family produced by *E. coli* O157:H7 and other STEC. At least 11 variants of Stx2 have been identified, including Stx2, Stx2c (Stx2vh-a and Stx2vh-b), Stx2d (Stx2d-OX3a and Stx2d-Ount), Stx2e, Stx2f, and Stx2g (5, 20). The Stx2c subgroup is approximately 97% related to the amino acid sequence of the B subunits of Stx2, whereas the A subunit of Stx2c shares 98 to 100% amino acid sequence homology with Stx2. Stx2e is associated with edema disease that principally occurs in piglets and shares 93 and 84% amino acid sequence homology with the A and B subunits, respectively, of Stx2. Hence, the Stx2-related toxins have only partial serological reactivity with anti-Stx2 serum. Stx2f and Stx2g of STEC strains isolated from feral pigeons and cattle waste water have been described recently (20, 49).

Structure of the Stx Family

Shiga toxins are a holotoxin composed of a single enzymatic A subunit of approximately 32 kDa in association with a pentamer of receptor-binding B subunits of 7.7 kDa (42). The Stx A subunit can be split by trypsin into an enzymatic A1 fragment (approximately 27 kDa) and a carboxyl terminal A2 fragment (approximately 4 kDa) that links A1 to the B subunits. The A1 and A2

Table 12.6 Nomenclature and biological characteristics of Shiga toxins (Stx)[a]

		Biological characteristics						
		% Nucleotide sequence homology to *stx*		% Nucleotide sequence homology to *stx*$_2$			Activated by	
Nomenclature	Genetic loci	A subunit	B subunit	A subunit	B subunit	Receptor	intestinal mucus	Disease
Stx	Chromosome	N/A	N/A			Gb3	No	Human diarrhea, HC[b], HUS
Stx1	Phage	99	100			Gb3	No	Human diarrhea, HC, HUS
Stx1c	Chromosome	97	96			Unknown	Unknown	Human and sheep?
Stx1d	Unknown	93	92			Unknown	Unknown	Cattle?
Stx2	Phage			N/A	N/A	Gb3	No	Human diarrhea, HC, HUS
Stx2c	Phage			100	97	Gb3	No	Human diarrhea, HC, HUS
Stx2d	Phage			99	97	Gb3	Yes	Human diarrhea, HC, HUS
Stx2e	Chromosome			93	84	Gb4	No	Pig edema disease
Stx2f	Unknown			63	57	Unknown	Unknown	Pigeon
Stx2g	Unknown			94	91	Unknown	Unknown	Bovine

[a] From references 7, 35, and 60; N/A, not applicable.
[b] Hemorrhagic colitis.

subunits remain linked by a single disulfide bond until the enzymatic fragment is released and enters the cytosol of a susceptible mammalian cell. Each B subunit is composed of six antiparallel strands forming a closed barrel capped by a single helix between strands 3 and 4. The A subunit lies on the side of the B subunit pentamer, nearest to the C-terminal end of the B-subunit helices. The A subunit interacts with the B-subunit pentamer through a hydrophobic helix which extends to one-half of the 2.0-nm length of the pore in the B pentamer. This pore is lined by the hydrophobic side chains of the B-subunit helices. The A subunit also interacts with the B subunit via a four-stranded mixed sheet composed of residues of both the A2 and A1 fragments.

Genetics of Stxs

While most stx_1 operons share a great deal of homology, there is considerable heterogeneity in the stx_2 family. Unlike other Stx2 whose genes are located on bacteriophage that integrate into the chromosome, Stx of *S. dysenteriae* type 1, Stx1c, and Stx2e are encoded by chromosomal genes (58, 60). A sequence comparison of the growing stx_2 family indicates that genetic recombination among the B subunit genes, rather than base substitutions, has given rise to the variants of Stx2 present in human and animal strains of *E. coli* (19). However, the operons for every member of the Stx subgroups are organized identically; the A and B subunit genes are arranged in tandem and separated by a 12- to 15-nucleotide gap in between. The operons are transcribed from a promoter which is located 5' to the A subunit gene, and each gene is preceded by a putative ribosome-binding site. The existence of an independent promoter for the B subunit genes has been suggested. The holotoxin stoichiometry suggests that expression of the A and B subunit genes is differentially regulated, permitting overproduction of the B polypeptides. Finally, Stx and Stx1 production is negatively regulated at the transcriptional level by an iron-Fur protein corepressor complex which binds at the stx_1 promoter but is unaffected by temperature, whereas Stx2 production is neither iron nor temperature regulated.

Receptors

All members of the Stx family bind to globoseries glycolipids on the eukaryotic cell surface; Stx, Stx1, Stx2, Stx2c, and Stx2d bind to glycolipid globotriaosylceramide (Gb_3), whereas Stx2e primarily binds to glycolipid globotetraosylceramide (Gb_4) (10). The alteration of binding specificity between Stx2e and the rest of the Stx family is related to carbohydrate specificity of receptors (33). The amino acid composition of B subunits of Stx2 and Stx2e differ at only 11 positions, yet Stx2e binds

primarily to Gb_4, whereas Stx2 binds only to Gb_3. High-affinity binding also depends on multivalent presentation of the carbohydrate, as would be provided by glycolipids in a membrane. The affinity of Stx1 for Gb_3 isoforms is influenced by fatty acyl chain length and by its level of saturation. Stx1 binds preferentially to Gb_3 containing C20:1 fatty acid, whereas Stx2c prefers Gb_3 containing C18:1 fatty acid. The basis for these findings may be related to the ability of different Gb_3 isoforms to present multivalent sugar-binding sites in the optimal orientation and position at the membrane surface. It is also possible that different fatty acyl groups affect the conformation of individual receptor epitopes on the sugar.

Mode of Action of the Stxs

Shiga toxins act by inhibiting protein synthesis. Each of the B subunits is capable of binding with high affinity to an unusual disaccharide linkage (galactose 1-4 galactose) in the terminal trisaccharide sequence of Gb_3 (or Gb_4) (53). Following binding to the glycolipid receptor, the toxin is endocytosed from clathrin-coated pits and transferred first to the trans-Golgi network and subsequently to the endoplasmic reticulum and nuclear envelope. While it appears that transfer of the toxin to the Golgi apparatus is essential for intoxication, the mechanism of entry of the A subunit from the endosome to the cytosol, and particularly the role of the B subunit in the process, remains unclear. In the cytosol, the A subunit undergoes partial proteolysis and splits into a 27-kDa active intracellular enzyme (A1) and a 4-kDa fragment (A2) bridged by a disulfide bond. Although the entire toxin is necessary for its toxic effect on whole cells, the A1 subunit is capable of cleaving the *N*-glycoside bond in one adenosine position of the 28S rRNA that comprises 60S ribosomal subunits (48). This elimination of a single adenine nucleotide inhibits the elongation factor-dependent binding to ribosomes of aminoacyl-bound transfer RNA molecules. Peptide chain elongation is truncated and overall protein synthesis is suppressed, resulting in cell death.

Role of Stxs in Disease

The precise role of Stxs in mediating colonic disease, HUS, and neurological disorders has not been fully elucidated. There is no satisfactory animal model for hemorrhagic colitis or HUS, and the severity of disease precludes study of experimental infections in humans. Therefore, our present understanding of the role of Stxs in causing disease is obtained from a combination of studies, including histopathology of diseased human tissues, animal models, and endothelial tissue culture cells. Results of recent studies support the concept that Stxs contribute to pathogenesis by directly damaging vascular endothelial cells in certain organs, thereby disrupting the homeostatic properties of these cells.

The involvement of Stx in enterocolitis was demonstrated when fluid accumulation and histological damage occurred after purified Stx was injected into ligated rabbit intestinal loops. The fluid secretion may be due to the selective killing of absorptive villus tip intestinal epithelial cells by Stx. However, intravenous administration of Stx to rabbits can produce nonbloody diarrhea, suggesting that other mechanisms of diarrhea are possible. Studies with genetically mutated STEC strains also indicate that Stx has a role in intestinal disease, but the significance of Stx in provoking a diarrheal response differs depending on the animal model used.

Many epidemiologic studies have identified a correlation between enteric infection with *E. coli* O157:H7 and development of HUS in humans. Histopathologic examination of kidney tissue from HUS patients revealed profound structural alterations in the glomeruli, the basic filtration unit of the kidney (35). Glomerular endothelial cells were swollen and were often detached from the glomerular basement membrane. Hence, subendothelial matrix components may be exposed and serve as sites for platelet adherence and activation. The damage caused by Stxs is often not limited to the glomeruli. Arteriolar damage, involving internal cell proliferation, fibrin thrombus deposition, and perivascular inflammation, occurs (57). Cortical necrosis also occurs in a small number of HUS cases. In addition, human glomerular endothelial cells are sensitive to the direct cytotoxic action of bacterial endotoxin. Endotoxin in the presence of Stxs also can activate macrophage and polymorphonuclear neutrophils to synthesize and release cytokines, superoxide radicals, or proteinases and amplify endothelial cell damage.

Neurological symptoms in patients and experimental animals infected with *E. coli* O157:H7 also have been described, and may be caused by secondary neuron disturbances that result from endothelial cell damage by Stxs. Studies in mice perorally administered an *E. coli* O157:H7 strain revealed that Stx2v impaired the blood-brain barrier and damaged neuron fibers, resulting in death. The presence of the toxin in neurons was verified by immunoelectron microscopy.

Epidemiologic studies also have revealed that *E. coli* O157:H7 strains isolated from patients with hemorrhagic colitis usually produce both Stx1 and Stx2 or Stx2 only; isolates producing only Stx1 are uncommon. Patients infected with *E. coli* O157:H7 producing only Stx2 or in combination with Stx1 were more likely to develop serious renal or circulatory complications than patients infected with enterohemorrhagic *E. coli* strains producing Stx1 only.

CONCLUDING REMARKS

The serious nature of the symptoms of hemorrhagic colitis and HUS caused by *E. coli* O157:H7 places this pathogen in a category apart from most other foodborne pathogens, which typically cause only mild symptoms. The severity of the illness it causes combined with its apparent low infectious dose (<100 cells) qualifies *E. coli* O157:H7 to be among the most serious of known foodborne pathogens. Although the pathogen has been isolated from a variety of domestic animals and wildlife, cattle are a major reservoir of *E. coli* O157:H7, with undercooked ground beef being the single most frequently implicated vehicle of transmission. An important feature of this pathogen is its acid tolerance. Outbreaks have been associated with consumption of contaminated high acid foods, including apple juice and fermented dry salami. Many other foods and recreational and drinking water also have been identified as vehicles of transmission of *E. coli* O157:H7 infections. Significant progress has been made towards understanding the mechanisms of pathogenicity of *E. coli* O157:H7 in the last decade; production of one or more Shiga toxins and A/E adherence are important virulence factors.

Enterohemorrhagic *E. coli* other than O157:H7 have been increasingly associated with cases of HUS. Hundreds of non-O157 STEC serotypes have been isolated from humans, but not all of these serotypes have been shown to cause illness. Some STEC may have a low potential to cause HUS; other non-O157 STEC isolates, found in healthy individuals, may not be pathogens. Furthermore, multiple STEC serotypes have been isolated from a single patient. The contribution of each serotype to the pathogenesis of disease is difficult to determine. *E. coli* O157:H7 is still by far the most important serotype of STEC in North America. Isolation of non-O157:H7 STEC requires techniques not generally used in clinical laboratories; hence, these bacteria are infrequently sought or detected in routine practice. Recognition of non-O157 EHEC strains in foodborne illness necessitates identification of serotypes of EHEC other than O157:H7 in persons with bloody diarrhea and/or HUS and preferably in implicated food. The increased availability in clinical laboratories of techniques such as testing for Stxs or their genes and identification of other virulence markers unique for EHEC should enhance the detection of disease attributable to non-O157 EHEC.

References

1. **Besser, R. E., P. M. Griffin, and L. Slutsker.** 1999. *Escherichia coli* O157:H7 gastroenteritis and the hemolytic uremic syndrome: an emerging infectious disease. *Annu. Rev. Med.* **50:**355–367.

2. Bettelheim, K. A., and G. H. Thomas. 2005. *E. coli* as pathogens. http://ecoli.bham.ac.uk/path/sero.html. [Online.] Accessed 15 November 2005.

3. Beutin, L., D. Geier, H. Steinruck, S. Zimmermann, and F. Scheutz. 1993. Prevalence and some properties of verotoxin (Shiga-like toxin)-producing *Escherichia coli* in seven different species of healthy domestic animals. *J. Clin. Microbiol.* **31:**2483–2488.

4. Brackett, R., Y. Hao, and M. Doyle. 1994. Ineffectiveness of hot acid sprays to decontaminate *Escherichia coli* O157: H7 on beef. *J. Food Prot.* **57:**198–203.

5. Brett, K. N., M. A. Hornitzky, K. A. Bettelheim, M. J. Walker, and S. P. Djordjevic. 2003. Bovine non-O157 Shiga toxin 2-containing *Escherichia coli* isolates commonly possess stx2-EDL933 and/or stx2vhb subtypes. *J. Clin. Microbiol.* **41:**2716–2722.

6. Brooks, J. T., E. G. Sowers, J. G. Wells, K. D. Greene, P. M. Griffin, R. M. Hoekstra, and N. A. Strockbine. 2005. Non-O157 Shiga toxin-producing *Escherichia coli* infections in the United States, 1983–2002. *J. Infect. Dis.* **192:**1422–1429.

7. Burk, C., R. Dietrich, G. Acar, M. Moravek, M. Bulte, and E. Martlbauer. 2003. Identification and characterization of a new variant of Shiga toxin 1 in *Escherichia coli* ONT: H19 of bovine origin. *J. Clin. Microbiol.* **41:**2106–2112.

8. Campellone, K. G., and J. M. Leong. 2003. Tails of two Tirs: actin pedestal formation by enteropathogenic *E. coli* and enterohemorrhagic *E. coli* O157:H7. *Curr. Opin. Microbiol.* **6:**82–90.

9. Clavero, M., J. Monk, L. Beuchat, M. Doyle, and R. Brackett. 1994. Inactivation of *Escherichia coli* O157: H7, salmonellae, and *Campylobacter jejuni* in raw ground beef by gamma irradiation. *Appl. Environ. Microbiol.* **60:**2069–2075.

10. Cohen, A., G. E. Hannigan, B. R. Williams, and C. A. Lingwood. 1987. Roles of globotriosyl- and galabiosylceramide in verotoxin binding and high affinity interferon receptor. *J. Biol. Chem.* **262:**17088–17091.

11. D'Aoust, J., C. Park, R. Szabo, E. Todd, B. Emmons, and R. McKellar. 1988. Thermal inactivation of *Campylobacter* species, *Yersinia enterocolitica*, and hemorrhagic *Escherichia coli* O157:H7 in fluid milk. *J. Dairy Sci.* **71:**3230–3236.

12. David, S. T., L. MacDougall, K. Louie, L. McIntyre, A. M. Paccagnella, S. Schleicher, and A. Hamade. 2004. Petting zoo-associated *Escherichia coli* O157:H7—secondary transmission, asymptomatic infection, and prolonged shedding in the classroom. *Can. Commun. Dis. Rep.* **30:**173–180.

13. Diez-Gonzalez, F., T. R. Callaway, M. G. Kizoulis, and J. B. Russell. 1998. Grain feeding and the dissemination of acid-resistant *Escherichia coli* from cattle. *Science* **281:**1666–1668.

14. Doyle, M. P., and J. L. Schoeni. 1984. Survival and growth characteristics of *Escherichia coli* associated with hemorrhagic colitis. *Appl. Environ. Microbiol.* **48:**855–856.

15. Elder, R. O., J. E. Keen, G. R. Siragusa, G. A. Barkocy-Gallagher, M. Koohmaraie, and W. W. Laegreid. 2000. Correlation of enterohemorrhagic *Escherichia coli* O157 prevalence in feces, hides, and carcasses of beef cattle during processing. *Proc. Natl. Acad. Sci. USA* **97:**2999–3003.

16. Elliott, S. J., L. A. Wainwright, T. K. McDaniel, K. G. Jarvis, Y. K. Deng, L. C. Lai, B. P. McNamara, M. S. Donnenberg, and J. B. Kaper. 1998. The complete sequence of the locus of enterocyte effacement (LEE) from enteropathogenic Escherichia coli E2348/69. *Mol. Microbiol.* **28:**1–4.

17. Foster, J. W. 2004. *Escherichia coli* acid resistance: tales of an amateur acidophile. *Nat. Rev. Microbiol.* **2:**898–907.

18. Galiero, G., G. Conedera, D. Alfano, and A. Caprioli. 2005. Isolation of verocytotoxin-producing *Escherichia coli* O157 from water buffaloes (*Bubalus bubalis*) in southern Italy. *Vet. Rec.* **156:**382–383.

19. Gannon, V. P., C. Teerling, S. A. Masri, and C. L. Gyles. 1990. Molecular cloning and nucleotide sequence of another variant of the *Escherichia coli* Shiga-like toxin II family. *J. Gen. Microbiol.* **136:**1125–1135.

20. Garcia-Aljaro, C., M. Muniesa, J. E. Blanco, M. Blanco, J. Blanco, J. Jofre, and A. R. Blanch. 2005. Characterization of Shiga toxin-producing *Escherichia coli* isolated from aquatic environments. *FEMS Microbiol. Lett.* **246:**55–65.

21. Garmendia, J., G. Frankel, and V. F. Crepin. 2005. Enteropathogenic and enterohemorrhagic *Escherichia coli* infections: translocation, translocation, translocation. *Infect. Immun.* **73:**2573–2585.

22. Glass, K., J. Loeffelholz, J. Ford, and M. Doyle. 1992. Fate of *Escherichia coli* O157:H7 as affected by pH or sodium chloride in fermented, dry sausage. *Appl. Environ. Microbiol.* **58:**2513–2516.

23. Griffin, P. M., and R. V. Tauxe. 1991. The epidemiology of infections caused by *Escherichia coli* O157:H7, other enterohemorrhagic *E. coli*, and the associated hemolytic uremic syndrome. *Epidemiol. Rev.* **13:**60–98.

24. Hovde, C. J., P. R. Austin, K. A. Cloud, C. J. Williams, and C. W. Hunt. 1999. Effect of cattle diet on *Escherichia coli* O157:H7 acid resistance. *Appl. Environ. Microbiol.* **65:**3233–3235.

25. Johnson, R., R. Clarke, J. Wilson, S. Read, K. Rahn, S. Renwick, K. Sandhu, D. Alves, M. Karmali, H. Lior, S. McEwen, J. Spika, and C. Gyles. 1996. Growing concerns and recent outbreaks involving non-O157:H7 serotypes of verotoxigenic *Escherichia coli*. *J. Food Prot.* **59:**1112–1122.

26. Johnson, W. M., H. Lior, and G. S. Bezanson. 1983. Cytotoxic *Escherichia coli* O157:H7 associated with haemorrhagic colitis in Canada. *Lancet* **i:**76.

27. Kaper, J. B., J. P. Nataro, and H. L. Mobley. 2004. Pathogenic *Escherichia coli*. *Nat. Rev. Microbiol.* **2:**123–140.

28. Karmali, M. A., M. Petric, C. Lim, P. C. Fleming, G. S. Arbus, and H. Lior. 1985. The association between idiopathic hemolytic uremic syndrome and infection by verotoxin-producing *Escherichia coli*. *J. Infect. Dis.* **151:**775–782.

29. Kim, H. H., M. Samadpour, L. Grimm, C. R. Clausen, T. E. Besser, M. Baylor, J. M. Kobayashi, M. A. Neill, F. D. Schoenknecht, and P. I. Tarr. 1994. Characteristics of antibiotic-resistant *Escherichia coli* O157:H7 in Washington State, 1984–1991. *J. Infect. Dis.* **170:**1606–1609.

30. Kudva, I. T., P. G. Hatfield, and C. J. Hovde. 1996. *Escherichia coli* O157:H7 in microbial flora of sheep. *J. Clin. Microbiol.* **34:**431–433.

31. Lin, J., M. Smith, K. Chapin, H. Baik, G. Bennett, and J. Foster. 1996. Mechanisms of acid resistance in entero-hemorrhagic *Escherichia coli*. *Appl. Environ. Microbiol.* 62:3094–3100.

32. Line, J., A. Fain, A. Moran, L. Martin, R. Lechowich, J. Carosella, and W. Brown. 1991. Lethality of heat to *Escherichia coli* O157:H7: D-value and z-value determinations in ground beef. *J. Food Prot.* 54:762–766.

33. Lingwood, C. A. 1996. Role of verotoxin receptors in pathogenesis. *Trends Microbiol.* 4:147–153.

34. Mead, P. S., L. Slutsker, V. Dietz, L. F. McCaig, J. S. Bresee, C. Shapiro, P. M. Griffin, and R. V. Tauxe. 1999. Food-related illness and death in the United States. *Emerg. Infect. Dis.* 5:607–625.

35. Melton-Celsa, A., and A. O'Brien. 1998. Structure, biology, and relative toxicity of Shiga toxin family members for cells and animals, p. 121–128. *In* J. Kaper and A. O'Brien (ed.), Escherichia coli *O157:H7 and Other Shiga Toxin-Producing* E. coli *Strains.* ASM Press, Washington, D.C.

36. Meng, J., and M. P. Doyle. 1998. Microbiology of Shiga toxin-producing *Escherichia coli* in foods, p. 92–111. *In* J. Kaper and A. O'Brien (ed.), Escherichia coli *O157:H7 and Other Shiga Toxin-Producing* E. coli *Strains.* ASM Press, Washington, D.C.

37. Meng, J., S. Zhao, M. Doyle, and S. Joseph. 1998. Anti-biotic resistance of *Escherichia coli* O157:H7 and O157:NM isolated from animals, food and humans. *J. Food Prot.* 61:1511–1514.

38. Mercado, E. C., S. M. Rodriguez, A. M. Elizondo, G. Marcoppido, and V. Parreno. 2004. Isolation of Shiga toxin-producing *Escherichia coli* from a South American camelid (Lama guanicoe) with diarrhea. *J. Clin. Microbiol.* 42:4809–4811.

39. Nataro, J. P., and J. B. Kaper. 1998. Diarrheagenic *Escherichia coli*. *Clin. Microbiol. Rev.* 11:142–201.

40. Naylor, S. W., J. C. Low, T. E. Besser, A. Mahajan, G. J. Gunn, M. C. Pearce, I. J. McKendrick, D. G. Smith, and D. L. Gally. 2003. Lymphoid follicle-dense mucosa at the terminal rectum is the principal site of colonization of enterohemorrhagic *Escherichia coli* O157:H7 in the bovine host. *Infect. Immun.* 71:1505–1512.

41. Nielsen, E. M., M. N. Skov, J. J. Madsen, J. Lodal, J. B. Jespersen, and D. L. Baggesen. 2004. Verocytotoxin-producing *Escherichia coli* in wild birds and rodents in close proximity to farms. *Appl. Environ. Microbiol.* 70:6944–6947.

42. O'Brien, A. D., V. L. Tesh, A. Donohue-Rolfe, M. P. Jackson, S. Olsnes, K. Sandvig, A. A. Lindberg, and G. T. Keusch. 1992. Shiga toxin: biochemistry, genetics, mode of action, and role in pathogenesis. *Curr. Top. Microbiol. Immunol.* 180:65–94.

43. Orr, P., D. Milley, D. Colby, and M. Fast. 1994. Prolonged fecal excretion of verotoxin-producing *Escherichia coli* following diarrheal illness. *Clin. Infect. Dis.* 19:796–797.

44. Orskov, F., I. Orskov, and J. A. Villar. 1987. Cattle as reservoir of verotoxin-producing *Escherichia coli* O157:H7. *Lancet* ii:276.

45. Oswald, E., H. Schmidt, S. Morabito, H. Karch, O. Marches, and A. Caprioli. 2000. Typing of intimin genes in human and animal enterohemorrhagic and enteropatho-genic *Escherichia coli*: characterization of a new intimin variant. *Infect. Immun.* 68:64–71.

46. Perna, N. T., G. F. Mayhew, G. Posfai, S. Elliott, M. S. Donnenberg, J. B. Kaper, and F. R. Blattner. 1998. Molecular evolution of a pathogenicity island from entero-hemorrhagic *Escherichia coli* O157:H7. *Infect. Immun.* 66:3810–3817.

47. Rangel, J. M., P. H. Sparling, C. Crowe, P. M. Griffin, and D. L. Swerdlow. 2005. Epidemiology of *Escherichia coli* O157:H7 outbreaks, United States, 1982–2002. *Emerg. Infect. Dis.* 11:603–609.

48. Sandvig, K., and B. van Deurs. 1996. Endocytosis, intra-cellular transport, and cytotoxic action of Shiga toxin and ricin. *Physiol. Rev.* 76:949–966.

49. Schmidt, H., J. Scheef, S. Morabito, A. Caprioli, L. H. Wieler, and H. Karch. 2000. A new Shiga toxin 2 variant (Stx2f) from *Escherichia coli* isolated from pigeons. *Appl. Environ. Microbiol.* 66:1205–1208.

50. Schroeder, C. M., J. Meng, S. Zhao, C. DebRoy, J. Torcolini, C. Zhao, P. F. McDermott, D. D. Wagner, R. D. Walker, and D. G. White. 2002. Antimicrobial resistance of *Escherichia coli* O26, O103, O111, O128, and O145 from animals and humans. *Emerg. Infect. Dis.* 8:1409–1414.

51. Schroeder, C. M., C. Zhao, C. DebRoy, J. Torcolini, S. Zhao, D. G. White, D. D. Wagner, P. F. McDermott, R. D. Walker, and J. Meng. 2002. Antimicrobial resistance of *Escherichia coli* O157 isolated from humans, cattle, swine, and food. *Appl. Environ. Microbiol.* 68:576–581.

52. Tarr, P. I., C. A. Gordon, and W. L. Chandler. 2005. Shiga-toxin-producing *Escherichia coli* and haemolytic uraemic syndrome. *Lancet* 365:1073–1086.

53. Tesh, V. L., and A. D. O'Brien. 1991. The pathogenic mechanisms of Shiga toxin and the Shiga-like toxins. *Mol. Microbiol.* 5:1817–1822.

54. Toma, C., E. Martinez Espinosa, T. Song, E. Miliwebsky, I. Chinen, S. Iyoda, M. Iwanaga, and M. Rivas. 2004. Dis-tribution of putative adhesins in different seropathotypes of Shiga toxin-producing *Escherichia coli*. *J. Clin. Micro-biol.* 42:4937–4946.

55. Torres, A. G., X. Zhou, and J. B. Kaper. 2005. Adherence of diarrheagenic *Escherichia coli* strains to epithelial cells. *Infect. Immun.* 73:18–29.

56. Vallance, B. A., and B. B. Finlay. 2000. Exploitation of host cells by enteropathogenic *Escherichia coli*. *Proc. Natl. Acad. Sci. USA* 97:8799–8806.

57. van Setten, P. A., L. A. Monnens, R. G. Verstraten, L. P. van den Heuvel, and V. W. van Hinsbergh. 1996. Effects of verocytotoxin-1 on nonadherent human monocytes: binding characteristics, protein synthesis, and induction of cytokine release. *Blood* 88:174–183.

58. Weinstein, D. L., M. P. Jackson, J. E. Samuel, R. K. Holmes, and A. D. O'Brien. 1988. Cloning and sequencing of a Shiga-like toxin type II variant from *Escherichia coli* strain responsible for edema disease of swine. *J. Bacteriol.* 170:4223–4230.

59. Wilson, J. B., R. C. Clarke, S. A. Renwick, K. Rahn, R. P. Johnson, M. A. Karmali, H. Lior, D. Alves, C. L. Gyles, K. S. Sandhu, S. A. McEwen, and J. S. Spika. 1996.

Vero cytotoxigenic *Escherichia coli* infection in dairy farm families. *J. Infect. Dis.* **174:**1021–1027.

60. **Zhang, W., M. Bielaszewska, T. Kuczius, and H. Karch.** 2002. Identification, characterization, and distribution of a Shiga toxin 1 gene variant (*stx*₁c) in *Escherichia coli* strains isolated from humans. *J. Clin. Microbiol.* **40:** 1441–1446.

61. **Zhao, T., and M. P. Doyle.** 1994. Fate of enterohemorrhagic *Escherichia coli* O157:H7 in commercial mayonnaise. *J. Food Prot.* **57:**780–783.

62. **Zhao, T., M. P. Doyle, and R. Besser.** 1993. Fate of enterohemorrhagic *Escherichia coli* O157:H7 in apple cider with and without preservatives. *Appl. Environ. Microbiol.* **59:**2526–2530.

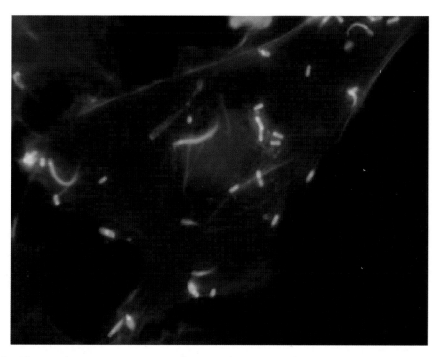

Color Plate 1 (chapter 21) Directional actin polymerization by *L. monocytogenes*. *L. monocytogenes* isolates were processed for triple-labeling fluorescence microscopy, 5 h after starting the infection of Vero cells. Bacteria (red) were visualized with a polyclonal anti-*Listeria* antibody, actin (green) with phalloidin, and cell nuclei (blue) with DAPI (4′,6′-diamidino-2-phenylindole). Magnification, ×100.

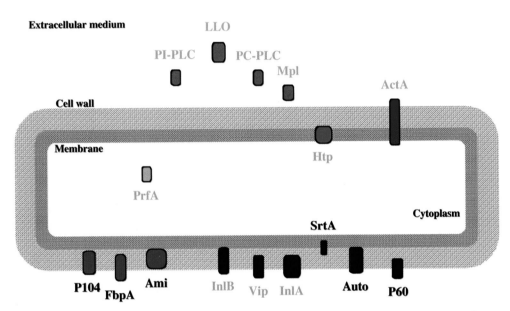

Color Plate 2 (chapter 21) Schematic representation of the *L. monocytogenes* virulence factors. The localization of factors implicated in adhesion (green), entry (blue), escape from the phagosome and intracellular growth (red), and intracytoplasmic movement and cell-to-cell spreading (purple) are indicated. The names of factors whose expression is regulated by PrfA are in orange.

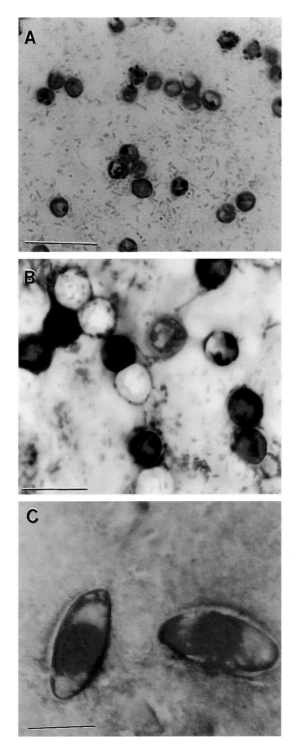

Color Plate 3 (chapter 31)   Acid-fast staining. (A) *C. parvum*. (B) *C. cayetanensis*. (C) *I. belli*. (Bars, 20 µm.)

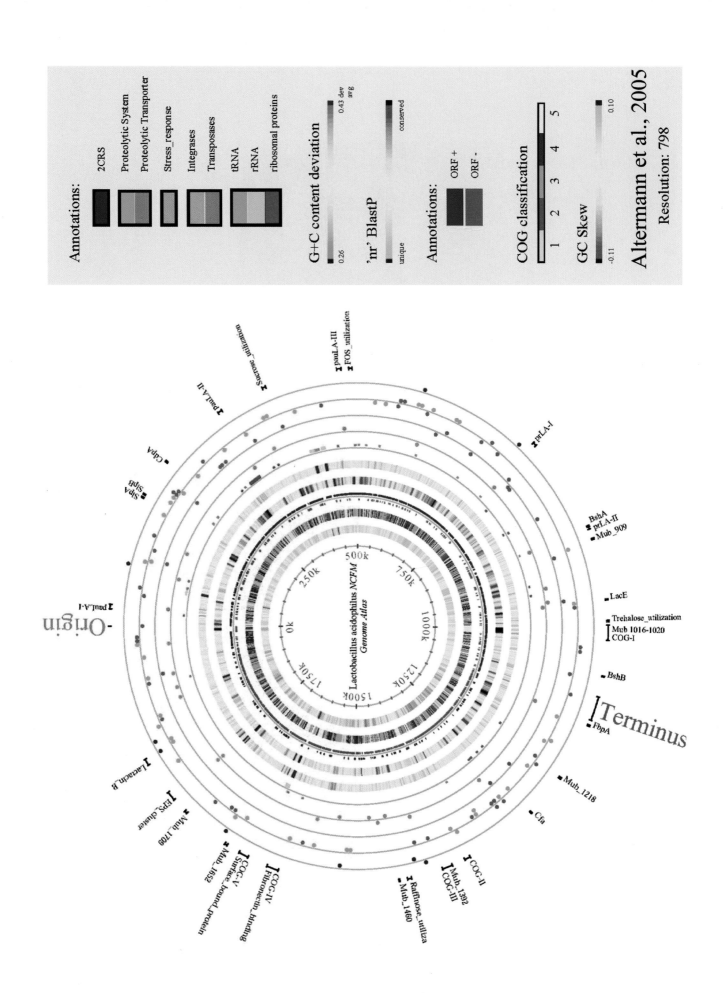

Annotations:

2CRS
Proteolytic System
Proteolytic Transporter
Stress_response
Integrases
Transposases
tRNA
rRNA
ribosomal proteins

G+C content deviation

0.26   0.43 dev   avg

'nr' BlastP

unique   conserved

Annotations:

ORF +
ORF -

COG classification

1   2   3   4   5

GC Skew

-0.11   0.10

Altermann et al., 2005
Resolution: 798

Sucrose_utilization
pauLA-III
FOS_utilization
pauLA-II
ClpA
SlpA
SlpB
pauLA-I
Origin
prLA-I
BshA
prLA-II
Mub_909
LacE
Trehalose_utilization
Mub 1016-1020
COG-I
BshB
Terminus
FbpA
Mub_1218
Cfa
Mub_1392
COG-II
COG-III
Raffinose_utiliza
Mub_1460
Fibronectin_binding
COG-IV
COG-V
Surface_bound_protein
Mub_1652
Mub_1709
EPS_cluster
Lacacin_B

Lactobacillus acidophilus NCFM
Genome Atlas

0k
250k
500k
750k
1000k
1250k
1500k
1750k

Color Plate 4 (chapter 42) Genome atlas of *L. acidophilus* NCFM. The atlas represents a circular view of the complete genome sequence of *L. acidophilus* NCFM. The key on the right describes the single circles in the top-down-outermost-innermost direction, as follows. Circle 1 (innermost), GC-skew. Circle 2, COG classification. Predicted open reading frames (ORFs) were analyzed using the COG database and grouped into five major categories: 1, information storage and processing; 2, cellular processes and signaling; 3, metabolism; 4, poorly characterized; 5, ORFs with uncharacterized COGs or no COG assignment. Circle 3, ORF orientation. ORFs in the sense orientation (ORF+) are shown in blue; ORFs in the antisense orientation (ORF−) are shown in red. Circle 4, BLAST similarities. Deduced amino acid sequences compared against the nonredundant (nr) database using gapped BLASTP (12). Regions in blue represent unique proteins in NCFM, whereas highly conserved features are shown in red. The degree of color saturation corresponds to the level of similarity. Circle 5, G+C content deviation. Deviations from the average G+C content are shown in either green (low-GC spike) or orange (high-GC spike). A boxfilter was applied to visualize contiguous regions of low or high deviations. Circle 6, ribosomal machinery. tRNAs, rRNAs, and ribosomal proteins are shown as green, cyan, and red lines, respectively. Clusters of proteins are represented as colored boxes to maintain readability. Circle 7, mobile elements. Predicted transposases are shown as light purple dots, and phage-related integrases are shown as orange dots. Circle 8, stress response. Genes involved in the general stress response, including chaperones, and genes involved in heat shock, DNA repair, and pH regulation are shown as dark purple dots. Circle 9, peptide and amino acid utilization. Proteases and peptidases are shown as green dots, and non-sugar-related transporters are shown as light blue dots. Circle 10 (outermost), two-component regulators (2CRS). Each 2CRS is represented as a brown dot, consisting of a response regulator and a histidine kinase. In circles 7 to 10 each full dot represents one predicted ORF and stacked dots represent clusters of ORFs. Selected features representing single ORFs and ORF clusters are shown outside of circle 10 with bars indicating their absolute size. The origin and terminus of DNA replication are identified in green and red, respectively. Other features: SlpA and SlpB (S-layer proteins), CdpA (cell division protein [50]), sugar utilization (sucrose, fructo-oligosaccharide, trehalose, and raffinose), LacE (phosphotransferase system [PTS]-sugar transporter), BshA and BshB (bile salt hydrolases), Mub-909 to Mub-1709 (mucus-binding proteins; the numbers correspond to the La numbering scheme), FbpA (fibronectin-binding protein), Cfa (cyclopropane fatty acid synthase), Fibronectin_binding (fibronectin-binding protein cluster), EPS_cluster (exopolysaccharides), Lactacin_B (bacteriocin), pauLA-I to pauLA-III (potential autonomous units), and prLA-I and prLA-II (phage remnants). Reprinted from reference 3.

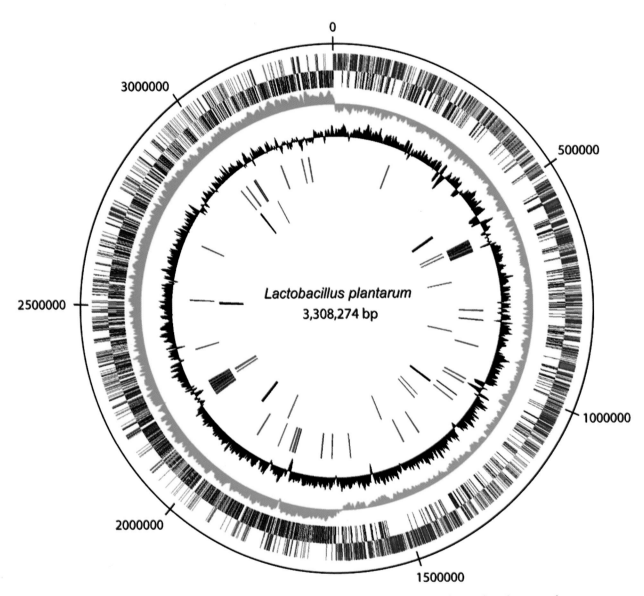

Color Plate 5 (chapter 44) Genome atlas of *L. plantarum* WCFS1. The predicted origin of replication is shown at the top. The outer to inner circles show (i) positive-strand ORFs (red); (ii) negative-strand ORFs (blue); (iii) GC-skew (green); (iv) G+C content (black); (v) prophage-related functions (green) and *IS*-like elements (purple); and (vi) rDNA operons (black) and tRNA-encoding genes (red). Reprinted from reference 38 with permission.

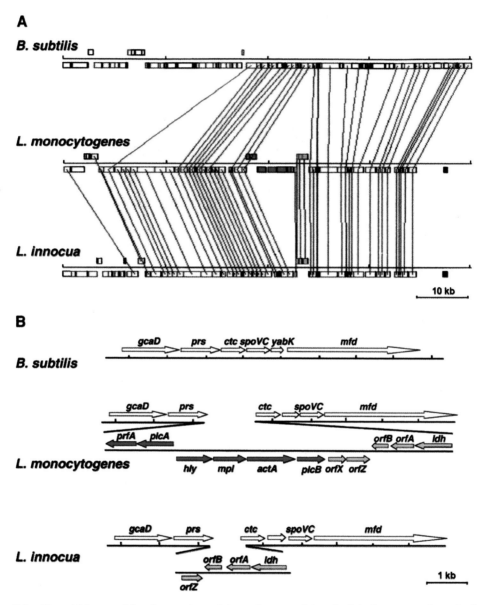

A

B. subtilis

L. monocytogenes

L. innocua

10 kb

B

B. subtilis

gcaD *prs* *ctc* *spoVC* *yabK* *mfd*

L. monocytogenes

gcaD *prs* *ctc* *spoVC* *mfd*

prfA *plcA* *orfB* *orfA* *ldh*

hly *mpl* *actA* *plcB* *orfX* *orfZ*

L. innocua

gcaD *prs* *ctc* *spoVC* *mfd*

orfB *orfA* *ldh*

orfZ

1 kb

Color Plate 6 (chapter 44)   Comparison of the region containing the "virulence gene cluster" of *L. monocytogenes* and the homologous region of the *L. innocua* and *Bacillus subtilis* genomes. Open blue boxes and arrows indicate orthologs among the three genomes; solid red boxes and arrows indicate the virulence gene cluster; solid yellow boxes and arrows indicate genes absent from *B. subtilis*. (A) Scheme generated by GenomeScout (LION Bioscience). (B) Enlargement of the region containing the virulence gene cluster. Reprinted from reference 31 with permission.

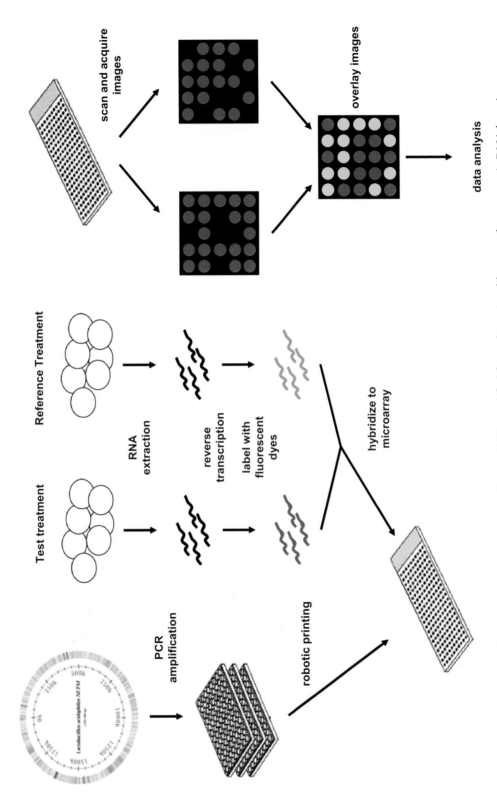

Color Plate 7 (chapter 44)   Microarrays. Probes are PCR amplified from clones in a library or from genomic DNA by using gene-specific primers. Individual amplicons are purified and spotted onto glass slides. Total RNA is labeled from both a test and reference sample by using fluorescent dyes and allowed to hybridize to the probes on the array. The array is then visualized using a laser scanner that generates color images, which are overlaid and compared for intensity and source. Adapted from reference 26 with permission.

Food Microbiology: Fundamentals and Frontiers, 3rd Ed.
Edited by M. P. Doyle and L. R. Beuchat
© 2007 ASM Press, Washington, D.C.

Franco J. Pagotto
Raquel F. Lenati
Jeffrey M. Farber

Enterobacter sakazakii

13

The first report of a "yellow-pigmented coliform" as the causative agent in a case of septicemia in an infant dates back to 1929 (93). In 1961, a yellow-pigmented coliform was reported as the causative agent in two cases of terminal neonatal meningitis that occurred in 1958 in St. Albans, England (116). Another report described a case of meningitis caused by *Enterobacter sakazakii* in a child from Denmark who had presented with severe mental and neurological sequelae (51). The name *E. sakazakii* was recorded for the first time in 1977 (11), and finally in 1980, *E. sakazakii* was distinguished from *Enterobacter cloacae* and became a new species named after the Japanese microbiologist Riichi Sakazaki, in honor of his contribution to the current understanding of enteric bacteriology.

The first documented link of *E. sakazakii* to powdered infant formula (PIF) as a causative factor of infection was reported by Clark et al. in 1990 (15), who confirmed the link between *E. sakazakii* isolated from patients and from PIF in two hospitals by using plasmid analysis, antibiograms, chromosomal restriction endonuclease analysis, ribotyping, and multilocus enzyme electrophoresis. To date, there have been approximately 80 reported cases of *E. sakazakii* infections resulting in 19 deaths, described

in 11 different countries and often associated with neonates and children from 3 days to 4 years of age. While only nine cases of adult infections from *E. sakazakii* have been documented (21, 39, 50, 61), there have been other reports of the isolation of *E. sakazakii* from adults.

E. sakazakii is an emerging foodborne pathogen that has increasingly raised interest among the scientific community, health care providers, and the food industry since the early 1980s. However, there is still a lack of information on this bacterium, including its natural habitat, mechanisms of virulence, and dose response for humans. In 2002, the International Commission for Microbiological Specifications for Foods (ICMSF) (42) classified *E. sakazakii* as a severe hazard for restricted populations, causing life-threatening or substantial chronic sequelae or illness of long duration, with the risk populations being newborns and immunocompromised infants. In 2004, the Food and Agriculture Organization of the United Nations (FAO) and the World Health Organization (WHO) jointly held an expert meeting on *E. sakazakii* and other microorganisms of concern in PIF (28), aiming to gather information for revising the 1979 *Recommended International Code of Hygienic Practice for Foods for Infants and*

Franco J. Pagotto and Raquel F. Lenati, Bureau of Microbial Hazards, Health Products and Food Branch, Food Directorate, Health Canada, Sir F.G. Banting Research Centre, P/L2204A2, Ottawa, Ontario, K1A 0L2, Canada. Jeffrey M. Farber, Bureau of Microbial Hazards, Health Products and Food Branch, Food Directorate, Health Canada, Sir F.G. Banting Research Centre, P/L2203G3, Ottawa, Ontario, K1A 0L2, Canada.

Children. More recently, in 2005, a *Proposed Draft Recommended International Code of Hygienic Infant Formula* was released, intended to replace the 1979 Code (27).

The WHO/FAO expert meeting of 2004 agreed on a list of recommendations to the scientific community and infant formula manufacturers focusing on the need for a better understanding of *E. sakazakii* and potentially other microorganisms that could be found in infant formula, including (i) the use of internationally validated detection and molecular typing methods for *E. sakazakii*; (ii) investigation and reporting of sources and vehicles of infection by *E. sakazakii* including the establishment of a laboratory-based network; and (iii) a better understanding of the ecology, taxonomy, virulence, and other characteristics of this emerging pathogen. The aim of this chapter is to provide a review of the basic biology, ecology, pathogenicity, and epidemiology of this emerging opportunistic foodborne pathogen.

CHARACTERISTICS OF THE ORGANISM

E. sakazakii belongs to the genus *Enterobacter* and, like most species in the genus *Enterobacter*, is considered an opportunistic pathogen. *E. sakazakii* organisms are gram-negative, oxidase-negative, non-sporeforming, non-acid-fast, straight, rod-shaped bacteria, having dimensions of 0.3 to 1.0 by 1.0 to 6.0 μm. The bacterium is motile by peritrichous flagella, nonhalophilic, and facultatively anaerobic and grows over a wide range of temperatures (6 to 45°C). Prior to the early 1980s, when it became a new species, *E. sakazakii* was referred to as a yellow-pigmented *E. cloacae*. However, the two species were found to be significantly different based on a low percentage of DNA-DNA homology (32 to 49%), biochemical profiles, antimicrobial susceptibility, and pigment production (10, 11, 26). Initially, major biochemical differences among *E. sakazakii* and other *Enterobacter* species were noted, such as the former's inability to ferment D-sorbitol, its positive α-glucosidase activity, and pigment production. The production of an extracellular DNase was shown in subsequent studies to not be uniform among *E. sakazakii* isolates, demonstrating the differential ability to ferment D-sorbitol and, rarely, to be α-glucosidase negative. Table 13.1 summarizes the biochemical characteristics of *E. sakazakii* according to the second edition of Bergey's manual (31), while Table 13.2 shows the

Table 13.1 Biochemical characterization of *E. sakazakii*[a]

Characteristic		Acid from:		Utilization of:			
Motility (36°C)	+	Adonitol	−	*cis*-Aconitate	+	Maltitol	+
Yellow pigment	+	L-Arabinose	+	*trans*-Aconitate	d	D-Melibiose	+
Urea hydrolysis	−	D-Arabitol	−	Adonitol	−	1-O-Methyl-α-galactoside	+
Indole production	D	Cellobiose	+	4-Aminobutyrate	+	1-O-Methyl-D-glucose	−
β-Xylosidase test	+	Sucrose	+	5-Amonivalerate	−	1-O-Methyl-α-D-glucoside	+
Methyl red	−	Dulcitol	−	D-Arabitol	−	Mucate	−
Voges-Proskauer	+	*meso*-Erythritol	−	Benzoate	−	Palatinose	+
Growth in KCN	+	Glycerol	−	Citrate	+	Phenylacetate	−
Gelatin hydrolysis (22°C)	−	*myo*-Inositol	(+)	*m*-Coumarate	−	L-Proline	+
DNase (25°C)	(+)	Maltose	+	Dulcitol	d	Protocatechuate	−
Lysine decarboxylase	−	D-Mannitol	+	L-Fucose	−	Putrescine	+
Arginine dehydrolase	+	Melibiose	+	Gentisate	−	Quinate	−
Ornithine decarboxylase	+	α-Methylglucosidase	+	Histamine	−	D-Raffinose	+
Phenylalanine deaminase	D	Raffinose	+	3-Hydroxybenzoate	−	L-Rhamnose	+
Glucose dehydrogenase	+	L-Rhamnose	+	4-Hydroxybenzoate	−	D-Saccharate	−
Gluconate dehydrogenase	−	Salicin	+	3-Hydroxybutyrate	−	D-Sorbitol	−
Growth at 41°C	+	D-Sorbitol	−	*myo*-Inositol	d	Sucrose	+
Esculin hydrolysis	+	Trehalose	+	5-Ketogluconate	−	D-Tagatose	−
Acetate	−	D-Xylose	+	2-Ketoglutarate	−	*meso*-Tartrate	−
		Lactose	+	Lactose	+	Tricarballylate	−
		Mucate	+	Lactulose	+	Tryptamine	−
				D-Lyxose	−	D-Turanose	d
				D-Malate	(d)	L-Tyrosine	−
				Malonate	(d)		

[a] +, 90 to 100% positive in 1 to 2 days; (+), 90 to 100% positive in 1 to 4 days; −, 90 to 100% negative in 4 days; d, positive or negative in 1 to 4 days; (d), positive or negative in 3 to 4 days.
[b] Modified from Garrity et al. (31).

Table 13.2 Biochemical profiles useful for differentiation of *E. sakazakii* and *E. cloacae*[a]

Test	E. sakazakii	E. cloacae
Yellow pigmentation	+	−
Production of oxidase	−	+
Production of α-glucosidase	+	−
Production of Tween 80 esterase	+	−
Production of phosphoamidase	−	+
Delayed production of DNase (25°C)	+	−
Urea hydrolysis	+	−
Acid from D-sorbitol	−	+
Utilization of 4-aminobutyrate	+	−
Utilization of D-saccharate	−	+
Utilization of D-sorbitol	−	+

[a] Modified from Lehner and Stephan (67) and Garrity et al. (31).

biochemical profiles which can be used to differentiate between *E. sakazakii* and *E. cloacae* (67).

Colonies of *E. sakazakii* on standard laboratory growth media such as Trypticase soy agar (TSA) measure 2 to 3 mm and 1 to 1.5 mm in diameter at 36 and 25°C, respectively, after overnight incubation. *E. sakazakii* produces colonies with distinct morphologies; one colony type is described as being dry, matte, and leathery or rubbery, retracting to the agar when touched, and having little biomass adhering to an inoculation loop. The second type is moist and glossy and easy to remove from the agar with a loop (36). Conditions influencing the production of an exopolysaccharide by *E. sakazakii* have been studied (103), with distinguishing features being its high viscosity and gel formation. Maximum amounts of exopolysaccharide produced by *E. sakazakii* were obtained with a carbon:nitrogen ratio of 20.2:1 at 27°C in media supplemented with glucose. Other bacteria such as *Enterobacter aerogenes*, *Xanthomonas campestris*, and *Arthrobacter viscosus* have also been previously shown to have enhanced EPS production in the presence of glucose. Harris and Oriel (38) have registered a patent for the production of gums from *E. sakazakii*, hypothesizing that in addition to being a frictional drag reduction agent in aqueous systems, the heteropolysaccharide could be used as a suspending, thickening, or stabilizing agent (38).

Interestingly, the two morphologies described for *E. sakazakii* cells may be lost upon repeated subculturing. The yellow pigment production is also unstable and, similar to colony morphology, can revert from one state to another upon subculturing. Other features that may be related to the heteropolysaccharide production are the sediment formation when the organism is grown in static liquid media and the increase of growth medium viscosity when it is grown with agitation.

INITIAL DATA ON PHYLOGENETIC CHARACTERIZATION

Being an emerging pathogen, there is a large amount of information that remains unknown about *E. sakazakii*. Nevertheless, advances have been made in the molecular characterization of *E. sakazakii* by using amplification and sequencing of the 16S rRNA gene, pulsed-field gel electrophoresis (PFGE), ribotyping, and plasmid typing (85).

Clementino et al. (16) were one of the first research groups to use tRNA intergenic spacer (tDNA-PCR) and 16S-23S internal transcribed spacer (ITS)-PCR for the characterization of *E. cloacae*. In their study, specific and reproducible patterns were obtained for *E. sakazakii* strain ATCC 29004. However, the study aimed at differentiating *E. cloacae*, and only a single *E. sakazakii* strain was used in their analyses.

The first PCR-based application for *E. sakazakii* was described by Keyser et al. (56). They used the full-length 16S rRNA gene from an *E. sakazakii* type strain to develop a detection system for upflow anaerobic sludge blankets. Subsequently, Lehner et al. (68) reevaluated this approach, overcame some of the specificity aspects, and generated more 16S rRNA sequences which led to the description of two strain lineages. They then developed a 16S rRNA gene-specific PCR method to distinguish *E. sakazakii* strains belonging to both lineages.

Iversen et al. (49) used partial 16S rRNA, along with *hsp60* sequences, to investigate the phylogenetic relationship among 126 *E. sakazakii* strains. Their study reinforced the polyphyletic nature of *Enterobacter* spp., as did a previous study based on the *gyrB* gene (20). Iversen et al. (49) identified four clusters and reported that a substantial amount of taxonomic heterogeneity existed within the species. They were able to group most strains in one cluster and hypothesized that the other three clusters could possibly represent new lineages within the species. Interestingly, they reported similarity among an *E. sakazakii* type strain and *Citrobacter koseri*. Equally important, their work demonstrated that current methods based on biochemical profiling (e.g., API 20E or ID 32E) did not always align with the 16S rRNA-based approach; we have reached a similar conclusion in our laboratory.

Liu et al. (73) reported on the use of a PCR-oligonucleotide array for the detection of *E. sakazakii* from infant formula. Using phylogenetic analyses of the ITS region of *E. sakazakii*, the group developed two pairs of species-specific primers that amplified a 282- and 251-bp fragment of the ITS G-operon and ITS IA-operon, respectively. Ten oligonucleotide features (i.e., probes) were designed to target these ITS products by using a nylon membrane array.

A 5' nuclease real-time PCR for the specific identification of *E. sakazakii* isolates has been developed (74). In this method, the 16S rRNA gene was targeted, along with a coamplified internal amplification control (IAC) in order to monitor false-negative results. Researchers concluded that based on the use of an IAC, their assay may be potentially suitable for the rapid detection of *E. sakazakii* in foods, and they reinforced the importance of using an IAC in diagnostic PCRs to decrease false negatives caused by PCR inhibitors, technical errors, or equipment malfunction. Their proposed method is 3 to 4 days faster than culture and isolation methods. If one were to start with a putative positive on a plate, their method would yield a result in approximately 2 h, compared to 2 days using the traditional biochemical methods.

In an epidemiological study done in 1990, Clark et al. (15) used plasmid profiling to analyze 32 *E. sakazakii* isolates, 27 from a hospital outside the United States and 5 from a hospital in the United States. Among the 27 isolates, 24 were infant formula and 3 were patient isolates, while of the 5 U.S. isolates, 1 was an infant formula isolate, 1 was isolated from a blender used for infant formula preparation, and 3 were isolated from patients. Of the 27 isolates, 26 (3 from patients and 23 from infant formula) had the same plasmid profile with four bands of approximately 3.2, 42, 70, and 85 MDa in size. The five isolates from the United States had the same plasmid profile, giving three bands of 3.2, 29, and 75 MDa.

Recently, we have undertaken a comprehensive phenotypic and genotypic characterization of over 200 strains of *E. sakazakii* collected from around the world. Using 1 kb of PCR-amplified 16S rRNA gene sequences, phylogenetic analyses indicated that there is a >95% sequence identity among isolates from environmental, food, and clinical sources (71). These same isolates were typed with PFGE and automated ribotyping. Using the restriction endonuclease XbaI, 119 different pulsotypes were generated from 145 strains, of which the two largest clusters comprised five isolates each. The five strains forming one of the clusters had the same ribotype. In addition to one cluster in common with PFGE, the ribotyping technique generated two other clusters containing five strains each. These three larger clusters were part of the 110 ribotypes generated with the restriction enzyme EcoRI. While both PFGE and ribotyping are techniques presently used for epidemiological studies during interlaboratory network investigations of outbreaks or isolated cases of bacterial infections, the congruence between the two approaches was found to be only 20% among the strains analyzed in our laboratory. PFGE seemed to better discriminate strains according to their sources, grouping strains isolated from the same outbreak and separating some that were known to be of different origin.

SUSCEPTIBILITY TO PHYSICAL AND CHEMICAL TREATMENTS

Temperature

Studies focusing on thermotolerance have reported on D and z values for *E. sakazakii* grown in reconstituted PIF (9, 48, 87). Most studies report $D_{54°C}$ values varying from 10.2 to 16.4 min, $D_{58°C}$ from 2.4 to 4.2 min, $D_{60°C}$ values of 1.1 min, and $D_{62°C}$ ranging from 0.2 to 0.4 min. The z values fall in the range of 5.6 to 5.8°C, with one study citing a z value of 3.1°C (87). Extrapolating from these collective data, the decimal reduction time at 71.2°C would be 0.7 s and the minimum high-temperature-short-treatment pasteurization process of 15 s at 71.7°C should theoretically result in a 21-log reduction in viable counts of the organism, indicating that pasteurized milk should be free of *E. sakazakii* after a pasteurization step. This would suggest that PIF testing positive for this bacterium are most likely contaminated after the various heating steps which are used in the process.

Recently, Williams et al. (121) used a top-down proteomics approach to look at the proteins expressed by 12 strains of *E. sakazakii* with the aim of identifying proteins expressed in a particular group of strains. Using strains resistant to higher temperatures, based on their previous work on the thermal inactivation of *E. sakazakii* in reconstituted PIF (22), they looked for markers of thermal tolerance. Applying liquid chromatography mass spectrometry, they obtained protein expression profiles for the 12 strains and found a putative marker common among thermotolerant *E. sakazakii* strains. Interestingly, this protein was homologous to a protein in the bacterium *Methylobacillus flagellatus* KT, an obligate methylotroph that grows at relatively high temperatures. This same protein has not been found in genera closely related to *E. sakazakii*, such as *Escherichia coli* and *Salmonella enterica*.

The effects of microwave heating on *E. sakazakii* strains have also been evaluated. In a study by Kindle et al. (57), five strains of *E. sakazakii* were suspended in reconstituted PIF at an initial concentration of 5 log CFU/ml and then placed in a microwave oven until the first signs of boiling, reaching temperatures of approximately 82 to 93°C. After the microwave treatment, four of the five samples of reconstituted infant formula tested negative for *E. sakazakii*, while one of the samples contained 20 CFU/ml. The researchers hypothesized that differences in infant formula composition could possibly account for different rates of inactivation of *E. sakazakii*.

There has been little work done on the ability of *E. sakazakii* to survive in frozen foods. Our laboratory has shown that the organism can be frozen in reconstituted PIF for over 6 months without any decrease of viable cell counts (70).

Guidelines for preparing and manipulating reconstituted infant formula and fortified breast milk in the household and health care settings in order to manage the risk of *E. sakazakii* growth are starting to be released by groups such as the American Dietetic Association (ADA) (94), the European Society for Pediatric Gastroenterology Hepatology and Nutrition (ESPGHAN) (3), the U.S. Food and Drug Administration (FDA) (114), Health Canada (40), and the New Zealand Ministry of Health (88). In addition, in order to evaluate current recommendations regarding "hang times" (i.e., the amount of time a formula is kept at room temperature in the feeding bag and accompanying lines during enteral tube feeding), Telang et al. (112) assessed total bacterial growth in fortified breast milk. Most guidelines recommend that when nutritionally appropriate, sterile liquid formula products should be used in preference to formulas reconstituted from nonsterile PIF. A summary of some other current recommendations with regard to the handling of PIF and fortified breast milk can be found in Table 13.3.

Water Activity (a_w)

The major food product linked to cases of *E. sakazakii* infection is PIF. It is known that the organism can survive for a long period of time in PIF (23) as well as in spices (46). However, the exact mechanism(s) that this organism uses to survive in these very dry environments (e.g., PIF average $a_w = 0.2$) is unknown.

It has been previously demonstrated that some bacteria increase osmolarity by the intracellular accumulation of ions (e.g., K^+) and compatible solutes such as proline, glycine betaine, and trehalose (55). When the drying process causes osmotic stress for the bacteria, polyhydroxyl compounds such as trehalose can be used by the organism to replace the shell of water around macromolecules, thus protecting cells from damage. The role of trehalose accumulation in the cells during the drying process and the osmotic tolerance of *E. sakazakii* was studied by Breeuwer et al. (9). They demonstrated that stationary-phase cells of *E. sakazakii* are more resistant to drying than are exponential-phase cells. Further work in this area done with *E. coli* has demonstrated that the trehalose synthesis genes are induced by stationary-phase sigma factor rpoS (41, 62, 63) and that the exposure of the organism to dry conditions results in permeability

changes caused by the influx of trehalose. Nonstressed stationary-phase *E. coli* cells did not accumulate trehalose, similar to the situation observed with nonstressed stationary-phase cells of *E. sakazakii*. The trehalose concentration in nonstressed *E. sakazakii* stationary-phase cells was ca. 0.040 μmol/mg of protein, while in dried stationary-phase cells, the concentration increased to 0.23 μmol/mg of protein. It would appear that dry stress of *E. sakazakii* stationary-phase cells could be a prerequisite for its accumulation of trehalose. The higher synthesis of trehalose and the induction of periplasmic trehalose in stationary-phase cells may stimulate the transport of nonfunctional trehalose to the periplasm, where it is converted to glucose. This cycle was described in a study with *E. coli*, where mutant strains defective in periplasmic trehalose accumulated large amounts of trehalose in the medium (111).

Breeuwer et al. (9) demonstrated that, after air drying *E. sakazakii* stationary-phase cell cultures in an incubator at 25°C and air humidity of 20.7%, initial populations of 9 log CFU/ml decreased only 1 to 1.5 logs when resuspended from the air-dried state 46 days later, and 1.5 to 2.5 logs when air dried and maintained for the same period of time at 45°C. Similar results were obtained for air-dried cells maintained in desiccators with saturated solutions of LiCl (a_w, 0.113), potassium acetate (a_w, 0.225), or magnesium nitrate (a_w, 0.529). Not surprisingly, exponential-phase cells of *E. sakazakii* were more sensitive to drying and decreased in greater numbers than did stationary-phase cells when exposed to a low a_w. In the same study, Breeuwer et al. (9) demonstrated that *E. sakazakii* survived better in dry environments than did *Salmonella* and *E. coli*. Therefore, the colonization of postpasteurization environments by *E. sakazakii* could represent a greater concern in dry products that become contaminated during ingredient mixing, filling, or packaging. The greater survival of air-dried *E. sakazakii* at elevated temperatures such as 45 to 47°C represents another advantage of this bacterium in warm and dry environments surrounding the areas of drying equipment in food factories.

Edelson-Mammel et al. (23) studied the long-term survival of *E. sakazakii* in PIF by inoculation of approximately 6 log CFU of *E. sakazakii* per ml into reconstituted PIF. The inoculated samples were stored at room temperature in a closed screw-cap bottle for a period of >1.5 years. Levels of *E. sakazakii* cells declined 2.5 logs during the initial 5 months of storage, at a rate of approximately 0.5 logs per month, with another 0.5-log decline observed in the following year reaching a final count of 3 log CFU/ml at the end of the 1.5-year study. This study demonstrated that *E. sakazakii* can survive

Table 13.3 Guidelines for reconstituted PIF and human breast milk powder-fortified preparation and handling

Reconstituted PIF	Human breast milk + powdered BMF[a]	Reference(s)
Refrigerate prepared formula to 4°C (40°F) within 1 h of preparation. Single-use bottles and nipples (except specialty products not available as single use) are recommended. Use only chilled, sterile water for IF preparation. Use equipment that keeps PIF chilled at 4°C (40°F) to safely transport it to the patient unit. Autoclaving or a thermal process is recommended for PIF preparation equipment and utensils. Discard any formula left in the bottle after the feed.	Store expressed breast milk at ≤20°C (−4°F). Each mother's expressed breast milk must be stored in a separate container. BMF should be stored in the refrigerator at 2–4°C (35–40°F) and should be used within 24 h. Fortifiers must be measured accurately, using aseptic techniques.	94
Reconstitute small volume of formula for each feeding to reduce the quantity and "hang time" at room temperature before consumption. Due to differences in PIF preparation among hospitals, each facility should identify and follow procedures appropriate to minimize microbial growth in reconstituted PIF. Minimize the "hang time" (i.e., the amount of time a PIF is at room temperature in the feeding bag and accompanying lines during enteral tube feeding), with no hang time exceeding 4 h.		94, 114
Preparation of PIF should be done in a laminar flow hood. Rehydrate PIF with sterile water. Follow recommendations of ADA.		40
Use PIF powder within 4 weeks of opening the can. Sterilize bottles. Prepare only the amount for baby's next feeding, and prepare it close to feeding time. Use warmed formula within 20 min. Discard any reconstituted formula left in the bottle after the feeding. Discard reconstituted formula left out of the refrigerator (4°C) for more than 4 h.		88
	Colony counts of aerobic mesophilic bacteria and *E. sakazakii* did not increase significantly in either unfortified or fortified breast milk samples over 6 h. Supplementing BMF with up to 1.44 mg iron/14 kcal does not compromise antibacterial activity of human milk held at room temperature for 6 h. Support ADA recommendations.	112

[a] BMF, human breast milk with fortifiers.

for extended periods in a dry environment such as PIF. Further studies are required to address the survival of *E. sakazakii* in infant formula stored at higher temperatures and moisture levels such as in tropical countries, preferably with naturally contaminated product.

Biofilm Formation

Contaminated food handling equipment such as brushes, blenders, and spoons has been linked to *E. sakazakii* neonatal infections (5, 105), indicating that the organism can survive on materials such as metal and plastics, perhaps through biofilm formation. Iversen et al. (48) used a biofilm formation assay to demonstrate that *E. sakazakii* adhered better to silicon, latex (8 × 10³ CFU/cm²),

and polycarbonate than stainless steel (50 CFU/cm²). It was also observed that encapsulated strains formed a denser biofilm than did nonencapsulated strains. It may be worthwhile to see if outbreak strains of *E. sakazakii* are significantly more encapsulated than food or environmental isolates and how this function is regulated.

Lehner et al. (66) worked with 56 *E. sakazakii* strains in another study of biofilm formation to investigate the putative formation of cellulose as one of the components of the biofilm. The group also investigated other factors that could potentially facilitate survival of *E. sakazakii* in the environment, such as adherence to hydrophilic and hydrophobic surfaces, production of intracellular polysaccharides and the presence of cell-to-cell signaling molecules. Pellicle and flock formation was observed in 21 and 44 of the strains grown in Luria-Bertani and brain heart infusion (BHI) broth, respectively, indicating that growth media may play a role in the ability of *E. sakazakii* to form biofilms. Interestingly, 12 isolates did not form any pellicles or flocks under either growth condition. A calcofluor white stain assay indicated the presence of cellulose on the fibrils of the extracellular matrices, possibly suggesting that cellulose was present as an extracellular compound when the biofilm was formed. In addition, different surfaces such as glass and polyvinyl chloride were investigated for *E. sakazakii* adherence; 23 of 56 isolates were able to adhere to glass surfaces in shaken cultures, and 33 strains formed a liquid-solid interface between the polyvinyl chloride surface and the liquid medium observed after 1% crystal violet staining. High-performance liquid chromatography analysis was used to determine the composition of the polysaccharide, which was made up of glucose, galactose, fructose, and glucuronic acid. Thin-layer chromatography analysis of culture cell-free supernatants performed on ethyl acetate demonstrated the presence of 3-oxo-C6-homoserine lactone (HSL) and 3-oxo-C8-HSL, two different types of acylated HSLs. This would suggest the potential ability of *E. sakazakii* to produce cell-to-cell signaling molecules. The significance of the latter finding in *E. sakazakii* biofilm formation is unknown at present.

The ability of gram-negative bacteria to coordinate colonization and associate with other cells by intercellular communication systems triggered by low-molecular-weight signaling compounds has been previously investigated (106, 120). *N*-Acyl derivatives of HSL (acyl HSLs) are examples of signaling molecules previously described in gram-negative bacteria. Modulation of the physiological processes controlled by acyl HSLs as well as by non-acyl HSL-mediated systems depends on the cell density and on the growth phase of the organisms. Gram et al. (33) have studied the production of acylated HSLs in *Enterobacteriaceae* strains in naturally contaminated as well as artificially inoculated foods at concentrations of 5 to 7 log CFU/g, suggesting that acylated HSLs could be implicated in regulating phenotypes important in food spoilage and thus could possibly play a role in food quality as well as food safety.

Chemical Inactivation

E. sakazakii has been isolated from acidic food products such as fermented bread and beverages (average pH of 4.0) (18, 32, 84). Nair et al. (82) studied the antibacterial properties of caprylic acid on *E. sakazakii* in reconstituted PIF. Caprylic acid is a natural eight-carbon chain fatty acid present in human breast milk and has GRAS

(generally recognized as safe) status in the United States. In their study, Nair et al. (82) inoculated reconstituted PIF with 6 log CFU of *E. sakazakii* per ml, containing monocaprylin at a final concentration of 25 or 50 mM, with samples being incubated at 4, 8, 23, or 37°C. At concentrations of 50 mM, monocaprylin rapidly inactivated *E. sakazakii* at 37°C, reducing its population to undetectable levels even after enrichment, while at 25 mM monocaprylin, *E. sakazakii* populations were reduced by 4.5 logs after 24 h. At room temperature, the number of *E. sakazakii* decreased by 5 and 4 logs in the presence of 50 and 25 mM monocaprylin, respectively, in 1 h. *E. sakazakii* remained at <10 CFU/ml after 24 h in the presence of 50 mM monocaprylin but increased in number from 2 to 4.5 logs CFU/ml at the end of 24 h in the presence of 25 mM monocaprylin. At 4 and 8°C, both concentrations of monocaprylin were effective in inactivating *E. sakazakii* in the reconstituted PIF. In the United States, monoglycerides are currently added to infant formula as emulsifiers. Therefore, the incorporation of monocaprylin as an antimicrobial ingredient in PIF is potentially feasible.

Competitive Exclusion/Probiotics

Telang et al. (112) evaluated the growth of resident aerobic mesophilic flora on artificially inoculated *E. sakazakii* in fresh human breast milk, fresh human breast milk with fortifiers, and reconstituted PIF. Samples were inoculated with 2 or 3 log CFU/ml of *E. sakazakii* and were maintained at 22°C with sampling every 2 h until hour 6. Total counts of *E. sakazakii* in both fortified and nonfortified human breast milk as well as in infant formula increased less than 1 log and were not significantly different over the 6-h period. In contrast, Chan (13) demonstrated a greater increase of total bacterial growth in fortified human breast milk as compared to nonfortified breast milk. Chan suggested that the high iron content in breast milk fortifier formulations could play a role in enhancing the growth of bacteria in the fortified breast milk, due to the inhibition of lactoferrin in high-iron environments. Lactoferrin is a well-known iron-binding glycoprotein which is present in human breast milk and is believed to have antibacterial properties.

The survival and growth of *E. sakazakii* in infant rice cereal were investigated by Richards et al. (100). Water, apple juice, milk, or liquid infant formula was used to reconstitute infant rice cereal prior to inoculation with a 10-strain cocktail of *E. sakazakii* at populations of 0.27, 0.93, and 9.3 CFU/ml, followed by incubation at 4, 12, 21, or 30°C for up to 72 h. Growth was not observed in the samples reconstituted with apple juice at any temperature, nor in the samples rehydrated with water, milk, or formula incubated at 4°C. The lag time observed for growth in cereal reconstituted with water, milk, or formula increased as the incubation temperature decreased (lag time$_{12°C}$ > lag time$_{21°C}$ > lag time$_{30°C}$). In addition, it was also found that once the bacterial population reached its maximum cell density (8 log CFU/ml) numbers decreased, in some cases to nondetectable levels during subsequent storage, concurrent with a decrease in pH.

Our laboratory recently completed a study on the growth of *E. sakazakii* in reconstituted PIF and human breast milk with and without fortification at 10, 23, and 37°C (69). In this study, reconstituted PIF, human breast milk, or human breast milk with fortifiers was inoculated with approximately 10 CFU of clinical, environmental, or food isolates of *E. sakazakii* per ml, incubated at 10, 23, or 37°C, and sampled periodically. A total of 105 growth curves were modeled using the modified Gompertz model from which estimated lag and generation time parameters were derived (69). At 10°C, growth was observed in 14 of the 23 curves and, interestingly, *E. sakazakii* did not grow in fortified breast milk at 10°C. At 23 and 37°C, growth occurred in all experiments, with longer generation times being observed at 23°C. Lower generation time parameters were seen in PIF than in human breast milk or human breast milk with fortifiers.

NICHES AND RESERVOIRS

The natural habitat of *E. sakazakii* remains unknown. In recent years, efforts have focused on detecting this organism in a wide variety of environments. Not surprisingly, like other members of the family *Enterobacteriaceae*, *E. sakazakii* has been isolated from environmental samples such as water, dust, soil, plant materials, mud, and even household vacuum cleaner bags, indicating that its ecological niche is quite diverse. It also supports the notion that its reservoir is likely an environmental source. Interestingly, new vectors and sources of contamination such as flies and rodents (37, 53, 60) have been recently reported. Although Muytjens and Kollee (77) were not able to isolate *E. sakazakii* from cattle, cattle milk, domesticated animals, rodents, bird dung, grain, rotting wood, mud, soil, or surface water, several other reports have described the isolation of *E. sakazakii* from various sources such as food manufacturing facilities and food products (e.g., powdered milk, cereal, chocolate, potato flour, pasta, and spices), households, water samples, dust, and grass silage (36, 52, 53, 76, 118). In a survey of water springs in Spain, which included waters from hypothermal (<30°C), mesothermal (30 to 40°C), and hyperthermal (>40°C) sources, 10 of 31 species

isolated from hyperthermal springs were identified as *E. sakazakii* (76).

E. sakazakii has been recovered from clinical specimens such as cerebrospinal fluid (CSF), blood, sputum, throat, nose, stool, gut, skin, wounds, bone marrow, eye, ear, stomach aspirates, anal swabs, and the breast abscess of infected patients (7, 30, 39, 58, 59, 65, 75, 79, 81, 98, 101, 107, 117, 122). Farmer et al. (26) reported the isolation of *E. sakazakii* from the respiratory tract of 29 patients in one hospital over a 7-month time period. *E. sakazakii* has also been shown to survive in clinical settings over extended periods of time. For example, based on ribotyping, Nazarowec-White and Farber (85) reported the persistence of one isolate of *E. sakazakii* in the same hospital over an 11-year period.

Foods such as milk powders, cheese products, and dry food ingredients such as herbs and spices and rice seeds (17) have also been reported to contain *E. sakazakii*; however, these foods have never been linked to human illnesses. In addition to its ability to survive in low a$_w$ environments, *E. sakazakii* can also survive in acidic food products. For example, it has been isolated from a Saudi Arabian fermented bread called Khamir (pH 3.8) (32); from two traditional drinks consumed in Jordan, one with a final pH of 8.6 and the other with a final pH of 2.8 (32); and from a fermented cassava product (pH 4.4) known as Attiéké, popular in African countries (18).

The major food commodity associated with *E. sakazakii* infection in neonates is PIF (90, 96, 105, 117). This bacterium was isolated for the first time from an unopened can of dried milk by Farmer et al. in 1980 (26). Muytjens et al. (80) isolated *Enterobacteriaceae* from 52.5% of 141 infant formula cans from 35 countries, with *E. sakazakii* being detected in 20 of 141 (14.2%) samples from 13 of the 35 countries. Leuschner and Bew (72) isolated *E. sakazakii* from 8 of 58 (13.8%) samples of infant formula from 11 countries. Iversen et al. (46) surveyed 82 samples of PIF and 404 other food products for the presence of *Enterobacteriaceae*. *E. sakazakii* was isolated from two of the formulas, 5 of 49 (10.2%) dried infant foods, 3 of 72 (4.1%) milk powders, 2 of 62 (3.2%) cheese products, and various dry food ingredients, including 40 of 122 (37.8%) herbs and spices. Soriano et al. (109) isolated *E. sakazakii* from raw lettuce from a survey of products consumed in restaurants in Spain. Leclercq et al. (64), in a study done to compare fecal coliform agar and violet red bile lactose agar for the enumeration of fecal coliform in foods, isolated *E. sakazakii* from cheese, minced beef, sausage meat, and vegetables. Sprouts (alfalfa and mung bean) have been reported twice as a source of *E. sakazakii* (19, 102). In our laboratory, we isolated this bacterium from crab meat (70), using buffered peptone water and mLST as the enrichment and selective enrichment broths, respectively. Using the most probable number (MPN) to enumerate the microorganism, we estimated an E. sakazakii load of >4 log CFU/g. The detection of E. sakazakii in PIF that meets the current microbiological standards has led policy makers in many countries to propose new microbiological criteria for PIF products (Table 13.4). A list of the various foods from which E. sakazakii has been isolated can be seen in Table 13.5.

PIF was introduced as a replacement for human breast milk more than 50 years ago. The powder form constitutes over 80% of the infant formula used worldwide. In the United States, between powder and start powder, the consumption of PIF in 2005 was greater than 6 billion 8-oz. bottles in addition to 133 million 8-oz. bottles of follow-on powder (89). The powdered form has advantages over the liquid form, in terms of both cost and storage. However, while the liquid form is sterile, the powder may contain low levels of microorganisms. The understanding of the behavior of E. sakazakii cells in dry products is a key element to be considered in the evaluation of potential treatments for inactivating E. sakazakii and other pathogens in PIF (28). Current processing technology is unable to completely eliminate the potential for microbial contamination in PIF without affecting its organoleptic and nutritional requirements. Based on currently available knowledge, sterilization of the final product in its dry form in cans or sachets seems possible only using irradiation. However, the doses that are likely to be required

to inactivate E. sakazakii do not appear to be feasible due to organoleptic deterioration of the product. A number of other technologies, such as ultrahigh pressure and magnetic fields, combined with other potential hurdles, may be potential candidates in the future.

ISOLATION METHODS

One of the most commonly used methods for isolation of E. sakazakii from PIF was described by the FDA (115). Developed in 2002, it is based on the isolation method described by Muytjens et al. in 1988 (80). The FDA method was developed for the isolation and enumeration of E. sakazakii from PIF. It consists of preenrichment of the infant formula powder in sterile distilled water overnight at 36°C, followed by selective enrichment in Enterobacteriaceae enrichment broth overnight at 36°C. Samples are surface plated (100 μl) onto violet red bile glucose (VRBG) agar and incubated overnight at 36°C. Presumptive colonies are selected and streaked onto TSA, and after 48 to 72 h of incubation at 25°C, yellow colonies are selected for biochemical tests using API 20E test strips.

Kandhai et al. (53) analyzed 152 environmental samples by using isolation methods with and without an enrichment step. Approximately 65% of the samples were tested without being enriched, 35% were tested after enrichment, and 18% were tested with both methods for comparative purposes. The enrichment used was buffered peptone water, incubated for 20 to 24 h at 36°C, followed by streaking onto VRBG agar media. When the enrichment step was not used, samples were streaked directly onto VRBG. Presumptive E. sakazakii colonies were selected and streaked onto TSA media and incubated for 48 h at room temperature. Yellow colonies were put through the API 20E kit as well as tested for a-glucosidase activity. Ribotyping was also done on presumptive positive isolates. From this study, the researchers concluded that E. sakazakii could be isolated from environmental samples with or without the enrichment step and that, combined, yellow pigmentation and α-glucosidase positive activity would be sufficient for E. sakazakii identification. The method proposed by Kandhai et al. (53) is shorter than that of the FDA and could be useful in routine screening of environmental samples from food manufacturing sites.

Much work has been done looking into the development of rapid, sensitive, and accurate methods for isolation of E. sakazakii. In 2004, Leuschner and Bew (72) described a modified version of the FDA method for E. sakazakii, which was validated in 16 laboratories from eight European countries. The modification suggested was the use of nutrient agar supplemented with 4-methylumbelliferyl-α-D-glucosidase (NA-α-MUG). The preenrichment and selective enrichment steps were similar

Table 13.4 Microbiological criteria to be applied to PIF[a,b]

Bacterium	n	c	m	M (per g)	Class plan[c]
Mesophilic aerobic bacteria[d]	5	2	10^3/g	10^4/g	3
Enterobacteriaceae[e]	10	2	0/g	NA[f]	2
Salmonella	60	0	0/25g	NA	2
E. sakazakii	30	0	0/10g	NA	2

[a] ISO methods are to be used for all determinations listed Abbreviations: n = number of sample units examined from a lot; c, number of samples being accepted between m and M; m, microbial limit separating good quality from marginally acceptable quality; M, microbial limit separating marginally acceptable quality from defective quality.

[b] Available at ftp://ftp.fao.org/codex/ccfh38/fh38_07e.pdf. This new format is to be discussed at the 38th session of the CODEX Committee on Food Hygiene (joint FAO/WHO Food Standards Programme), December 4–9, 2006, Houston, Tex.

[c] Class plan 2: two classes of limits (m) involved: ≤m and >m, where the maximum number of sample(s) yielding unsatisfactory testing results is represented by c. Class plan 3: m reflects the upper limit of GMP (Good Manufacturing Practices) and M indicates the limit beyond which the level of contamination is unacceptable (27).

[d] The proposed criteria for mesophilic aerobic bacteria are reflective of Good Manufacturing Practices and do not include nonpathogenic microorganisms that may be intentionally added, such as probiotics.

[e] Reductions in the levels of Enterobacteriaceae in PIF will lead to lower populations of E. sakazakii.

[f] NA, not applicable.

Table 13.5   Foods from which *Enterobacter sakazakii* has been isolated

Food type	Incidence/levels	Comments	Reference(s)
Powdered milk		First reported isolation from powdered milk	26
PIF		Czechoslovakia	96
PIF	<1 CFU/g from 14.2% of the products from 13 of the 35 countries	Survey of 141 milk substitutes obtained in 35 countries	80
Fermentation of sorghum bread (khamir)	1×10^4 CFU/g prefermentation, decreasing to (<1 CFU/ml) after 2×24-h fermentation	Prefermentation pH, 6.77; final product pH, 3.93	32
PIF		Plasmid profiles done by Clark et al. (15) showed similar profiles for infant formula and patient isolates.	15, 105
PIF		22 infant formula isolates had biotypes, antibiograms, and plasmid profiles identical to those of isolates from 4 infants, reported by Clark et al. (15).	7, 15
PIF		*E. sakazakii* was isolated from a blender used to prepare formula.	90
PIF		Isolates from stomach aspirate, anal swab, and/or blood sample for 6 of the 12 neonates were reported to be similar to isolate from infant formula (by PCR).	117
PIF		First outbreak associated with infant formula. There was a total of 9 cases, 2 infected and 7 colonized infants in a NICU[a].	119
Raw lettuce	1/40 samples	Restaurant in Spain	109
Raw rice	20 *E. sakazakii* isolates/one lot	Farms in Philippines	17
PIF		Isolated from blender after an infant case in a hospital in Jerusalem	8
Cheese, minced beef, sausage meat, and vegetables		France	64
Sous and tamarind drinks	6.3×10^2 CFU/ml in sous drink; 2 CFU/ml in tamarind drink	Sous drink pH, 8.6; tamarind drink pH, 2.8	84
Fermented cassava (attiéké)	Inoculum (2.3×10^6 CFU/g); grated pulp (1.8×10^5 CFU/ml); fermented pulp (2.3×10^2)	Inoculum pH, 5.0; grated pulp pH, 6.2; fermented pulp pH, 4.4	18
PIF, dried infant foods, milk powders, cheese products, dry food ingredients (herbs and spices)	2/82 (infant formula), 5/49 (dried infant foods), 3/72 (milk powders), 2/62 (cheese products), 40/122 (herbs and spices)		46
	10/74 +ve[b] for *E. sakazakii* using 25-g samples for testing	Indonesia	25
Mung bean sprouts	2×10^2 to 2×10^7 CFU of *E. sakazakii* per g among *Enterobacter* spp. in 25% of 300 sprouts tested		102
Alfalfa sprouts	8% +ve (5/60)	Mexico City	19
Crab meat	>1.1×10^4 CFU/g	MPN method used to enumerate	70

[a] NICU, neonatal intensive care unit.
[b] +ve, positive results for food matrices tested for *E. sakazakii*.

to previously described methods. Selective enrichment broths were streaked in parallel onto VRBG and NA-α-MUG plates. On NA-α-MUG plates, *E. sakazakii* colonies are yellow under normal light and show blue-violet fluorescence under UV light. The authors recommended the use of both agar media for the presumptive detection of *E. sakazakii*.

Another recently developed chromogenic medium, Oh and Kang (OK) agar medium, uses 4-methyl-umbelliferyl-α-D-glucoside as a selective marker in a differential medium for *E. sakazakii* (91). The fluorogenic substrate was added to VRBG, tryptone bile agar, and TSA media. On OK medium, *E. sakazakii* shows a strong fluorogenic characteristic, which clearly distinguishes it from other microorganisms under UV light. All presumptive *E. sakazakii*-positive fluorescent colonies were confirmed to be *E. sakazakii* by API 20E strips.

Iversen et al. (47) also described a new chromogenic medium named Druggan-Forsythe-Iversen (DFI), based on detecting the presence of the enzyme α-glucosidase by using the substrate 5-bromo-4-chloro-3-indolyl-α-D-glucopyranoside (XaGlc). *E. sakazakii* hydrolyzes this substrate to an indigo pigment, producing blue-green colonies. The authors reported on a comparative study of the new medium with the FDA method and showed that 95 clinical and food isolates of *E. sakazakii* were detected on DFI agar 2 days sooner than the current FDA method. The characteristics of 148 strains representing 17 genera of the *Enterobacteriaceae* family other than *E. sakazakii* were also compared using the two methods. Only a few isolates of *Escherichia vulneris*, *Pantoea* spp., and *C. koseri* strains gave false-positive results on DFI agar. A few α-glucosidase-positive strains were identified as *Pantoea* spp. based on API 20E biochemical profiles but had a higher percentage of identification than *E. sakazakii* by use of the ID32E biochemical test strip. It was concluded that the DFI medium enables the detection of *E. sakazakii* within mixed cultures of *Enterobacteriaceae*, something not feasible with VRBG agar, a general *Enterobacteriaceae* selective medium.

A new, promising selective agar, the *E. sakazakii* Plating Medium (ESPM), has recently been described (99). Its innovative characteristic is the use of two chromogenic substrates that are *E. sakazakii* specific, carbohydrates that are not metabolized by *E. sakazakii*, and a pH indicator to indicate the presence of other bacteria that ferment these sources of sugar, diminishing the possibility of false-positive results. A rapid 6-h screening medium consisting of a bisugar medium containing sucrose and melibiose was also developed. Presumptive colonies on ESPM can be streaked to the bisugar plate (99).

Guillaume-Gentil et al. (34) described a method for the detection and identification of *E. sakazakii* from environmental samples. Their method includes enrichment using mLST broth and an incubation temperature of 45°C for 22 to 24 h, followed by streaking onto TSA containing bile salts (TSAB). The group suggested exposure to light during incubation of the TSAB plates at 37°C, in order to observe the characteristic yellow colonies, which were confirmed using API 20E and α-glucosidase activity tests. This method was superior (40% positives) to the "reference" method (enrichment in buffered peptone water followed by isolation on VRBG agar), which yielded 26% positives from a total of 192 environmental samples.

A real-time PCR method for the detection of *E. sakazakii* in infant formula, in which the specific target sequence is within the macromolecular synthesis (MMS) operon, has also been developed (104). The MMS operon consists of three genes, *fpsU*, *dnaG*, and *rpoD*, which are involved in the synthesis of DNA, RNA, and protein. The *dnaG-rpoD* intergenic sequence differs in length and primary sequence between species. The 5′-nuclease-based amplifications are often more specific than standard PCR amplifications due to a requirement for 100% homology between probe and template, in addition to primer-template specificity. In contrast to the FDA method, the authors reported that their method takes 2 days to complete. In addition, their method had a sensitivity of 100 CFU/ml of broth and could detect contamination levels as low as 0.6 CFU/g of PIF.

Recently, Gurtler and Beuchat (35) compared the ability of spiral plating and ecometric techniques to recover stressed cells of *E. sakazakii*. Five stress conditions were used to test the survival of *E. sakazakii* cells on TSA supplemented with 0.1% pyruvate (TSAP, a nonselective control medium); Leuschner, Baird, Donald, and Cox (LBDC) agar (a differential, nonselective medium); OK agar; fecal coliform agar (FCA); DFI medium; VRBG agar; and *Enterobacteriaceae* enrichment (EE) agar. *E. sakazakii* cells were stressed using heat (55°C for 5 min), freeze-thawing (−20°C for 24 h, thawed, frozen again at −20°C for 2 h, and thawed), acidification at pH 3.54, alkaline conditions (pH of 11.25), and desiccation in PIF (a_w of, 0.25; 21°C for 31 days). The study revealed that TSAP > LBDC > FCA > OK, VRBG > DFI > EE in terms of spiral plating recovery for heat-, freezing-, acid-, and alkaline-stressed cells, as well as unstressed control cells. Desiccation stress, however, gave different results, with TSAP = LBDC = FCA = OK > DFI = VRBG = EE. ESPM agar was included in the recovery study using the ecometric technique, with better recovery being observed. The media were evaluated as best to worst for recovering stressed cells of *E. sakazakii*, as follows: TSAP = LBDC > FCA > ESPM = VRBG = OK > DFI = EE.

The study by Gurtler and Beuchat (35) highlights the fact that while some of the newer differential and chromogenic media are useful when using pure cultures of *E. sakazakii*, variability in recovery does occur. Interestingly, in this study, spiral plating was considered better than the ecometric technique when attempting to evaluate the performance of the recovery media. Equally noteworthy is the relatively poor performance of EE agar, further emphasizing the need for better isolation methodologies. Table 13.6 summarizes the current methods used to detect and/or enumerate *E. sakazakii*.

Table 13.6 Summary of isolation and enumeration methods for *Enterobacter sakazakii*

Format	Selective broth	Selective agar	Confirmation method	Duration[a]	Reference
Biochemical	EE broth at 37°C	VRBG	API 20E	7 days	115
	mLST broth at 45°C	Chromogenic *E. sakazakii* medium (XaGlc + dimethylformamide)	Oxidase, L-ornithine, L-arginine, D-sorbitol, L-rhamnose, D-sucrose, D-melibiose, amygdaline, and citrate	7 days	43
	BPW[b] at 37°C	VRBG	API 20E, α-glucosidase	5–6 days	53
	EE broth at 37°C	VRBG, nutrient agar-α-MUG (fluorescence under UV light)	API 20E	4 days	72
	N/A[c]	OK media (with 4-methyl-umbelliferyl-α-D-glucoside)	API 20E	N/A	91
	N/A	DFI (Oxoid) (XaGlc)	API 20E, ID 32E	N/A	47
	N/A	ESPM, pH indicator for fermentation of sugars not fermented by *E. sakazakii*	N/A	N/A	99
	ESSB at 37°C	ESIA chromogenic medium	N/A	2 days	AES Labs
	mLST broth at 45°C	TSA + bile salts	Yellow colonies confirmed using API 20E and α-glucosidase activity	3–5 days	34
DNA					
16s rRNA	N/A	N/A	N/A	4 h	68
Real-time PCR	EE broth	VRBG agar	MMS operon (*fpsU*, *dnaG* and *rpoD*)	2 days	104
PCR-oligo array	mLST + van, at 44°C, 20 h	BHI, 5 h, followed by DNA extraction	ITS G and ITS IA-operon	2 days	73
16s rRNA and IAC	N/A	N/A	N/A	2 h	74
BAX PCR	mLST + van (45°C, 20–22 h)	BHI, 3 h, 37°C	PCR result on machine, isolation from enrichment broths to NA	2 days	Dupont Qualicon

[a] Duration is approximate and includes confirmation using biochemical tests such as the API 20E.
[b] BPW, buffered peptone water.
[c] N/A, not available.

The International Organization for Standardization (ISO) (43) is working on the development of a methodology for the isolation of *E. sakazakii* from milk and milk products, which includes the use of the selective enrichment broth mLST incubated at 45°C and containing vancomycin (10 µg/ml) as well as a higher concentration of NaCl (34 g/liter), in addition to a chromogenic agar medium.

SYMPTOMS, AT-RISK POPULATIONS, AND FOODBORNE OUTBREAKS

E. sakazakii has been associated mainly with necrotizing enterocolitis, septicemia, and meningitis. Neurological sequelae are commonly reported and include brain abscess and infarction, ventricle compartmentalization due to necrosis of brain tissue and liquefaction of white cerebral matter, and cranial cystic changes, as well as hemorrhagic and nonhemorrhagic intercerebral infarctions leading to cystic encephalomalacia (12, 44, 59, 110, 122).

Although *E. sakazakii* has caused disease in all age groups, on the basis of the age distribution of reported cases it was deduced that the group at particular risk is infants less than 1 year old (28). Among infants, the immunocompromised and neonates younger than 28 days are considered to be at greatest risk; low-birth-weight (LBW) neonates, weighing less than 2.5 kg at birth, could possibly be at even greater risk. However, a survey by Stoll et al. (110) identified only one case of *E. sakazakii* sepsis among 10,660 LBW infants, suggesting that outside an epidemic situation, LBW infants are likely not a group at high risk for *E. sakazakii* infection. Sondheimer et al. (108) found that premature or term neonates secrete less gastric acid than older infants, a potentially important factor contributing to the increased survival of *E. sakazakii* during its passage through the stomach and then into the intestine. An additional group at risk in developing countries is infants of human immunodeficiency virus-positive mothers, considering that they may be fed solely PIF in order to prevent the documented transmission of the virus from human immunodeficiency virus-positive mothers to infants through breast-feeding (28).

The first reported cases of *E. sakazakii* occurred in 1958 in England, and since then, cases of neonatal and infant infections associated with this bacterium have been reported and further described in many regions of the world including Belgium, Canada, Denmark, France, Germany, Greece, Iceland, Israel, The Netherlands, Spain, the United Kingdom, New Zealand, and nine states in the United States. A summary of the reported cases of *E. sakazakii* infections reported from 1958 until 2004, including symptoms and outcomes, is given in Table 13.7.

Most of the *E. sakazakii* infections that have been reported have occurred in developed nations. The lack of reports from developing countries may be due to lack of awareness of the organism and methods to isolate it, rather than the absence of illness related to this microorganism. Because of the general lack of awareness of this organism, combined with poor existing methodologies, there are likely to be a number of unreported or misdiagnosed cases, even in developed countries.

Cases within adult populations have been reported, although with a much smaller incidence. Only nine cases of *E. sakazakii* infection have been reported among adults, all of whom were immunocompromised (21, 24, 39, 50, 61, 97).

VIRULENCE FACTORS AND PATHOGENICITY

In general, the most frequently cited risk factors for the acquisition of an *Enterobacter* infection are severe debilitating underlying illness and the prior use of antimicrobial agents, factors that facilitate the colonization of *Enterobacter* spp. in the human intestinal tract. The immature neonatal immune system may increase the risk of acquiring an *E. sakazakii* infection. However, it is not known exactly what host and environmental factors need to be present in order to cause infection in neonates.

Since *E. sakazakii* is a recently recognized human pathogen, not much research has been conducted on this organism, and therefore, little is known about its pathogenesis. Keller et al. (54) have indicated that the virulence factors of *E. sakazakii* (exotoxins, aerobactin, and hemagglutinin) may be similar to those of *E. cloacae*.

Our laboratory has demonstrated the organism's potential for producing cytotoxins and/or enterotoxins by using Vero, CHO, and Y-1 cell lines and a suckling mice assay, respectively (92).

No additional data on dose-response models for *E. sakazakii* in humans have become available since the FAO/WHO meeting in 2004 (28). Studies done with suckling mice suggest that a large number of cells may be required to cause infection in healthy neonates (92). Iversen et al. (45) hypothesized a minimum infectious dose of 1,000 CFU, based on a comparison with *E. coli* O157:H7 and *Listeria monocytogenes* serotype 4b. They further surmised that based on an average number of 0.36 to 66 CFU of *E. sakazakii*/100 g, previously reported (80, 86) to be found in unopened cans of PIF and the recommended average portion of 18 g of powder reconstituted for one single feeding, 14 generations would be needed to reach 1,000 CFU. This would take average times of 7 h at 37°C, 17.9 h at 21°C, 1.7 days at 18°C, and 7.9 days at 10°C, according to the generation times that the researchers observed in growth studies. In summary, therefore, PIF containing the levels of *E. sakazakii* previously reported is unlikely to cause illness unless there is temperature abuse, contamination during handling (15), long periods of storage, and/or a combination of these factors.

Little is known about the virulence factors and pathogenicity of *E. sakazakii*, with the first real attempt stemming from work in our laboratory. A toxin assay done in our laboratory was the first describing putative virulence factors of *E. sakazakii* (92). Supernatants free of bacterial cells and boiled for 20 min did not have any effects except on Vero cells, for which there was an observed decrease in cytopathic effect. Interestingly, not all *E. sakazakii* isolates tested demonstrated cytopathic effects against the cell lines and some of the strains that tested positive in the enterotoxin assay done with suckling mice did not cause any cytopathic effects in the cell lines tested. It is possible that the results obtained in these experiments indicate differences in receptor-toxin binding or denote a complex regulation system. Alternatively, it is possible that *E. sakazakii* produces more than one cytotoxin or that horizontal gene transfer events involving the transfer of cytotoxin (and for related genes) occurs only in certain isolates. These results may help to explain why some infants are reported to be colonized by *E. sakazakii* but do not develop any symptoms. In the same study, intraperitoneal injection infectivity assays in suckling mice showed that all of the tested strains of *E. sakazakii* were lethal to the animals at a dose of 8 log CFU. Death occurred within 3 days after dosing and typically within 24 to 48 h. Two of the clinical *E. sakazakii* isolates tested

Table 13.7 Summary of *E. sakazakii* cases and outbreaks in infants reported in the literature

Date and location	No. of cases and outcome	Disease symptoms/organs affected	Comments	Reference
1958, St. Albans, England	2 infants	White cranial matter resulting in degeneration into soft hemorrhagic mass; *E. sakazakii* isolated from brain, CSF, liver, and marrow swabs	Isolate sensitive to chloramphenicol and streptomycin	116
1965, Denmark	1 infant, 2 days old; recovered at 4 months of age, yet experienced extreme mental impairment.	Meningitis; *E. sakazakii* isolated from CSF; blood, feces, and throat cultures tested negative.	Treatment with streptomycin, chloramphenicol, ampicillin, sulfadiazine, sulfadimidine, sulfamerazine, and sulfacombin	51
1979, Macon, G.	1 infant, 7 days old	First reported case of nonmeningital bacteremia caused by *E. sakazakii*	Ampicillin treatment	75
1981, Indianapolis, Ind.	1 infant, 5 weeks old; recovery, after treatment; severely developmentally delayed	Increase in head's circumference; meningoencephalitis and cerebral ventricular compartmentation; bulging fontanels and grand mal seizures	Ampicillin and gentamicin treatments	58
1983, The Netherlands	8 cases in a 6-year period	Meningitis; 2 of 8 patients experiencing necrotizing enterocolitis; *E. sakazakii* isolated from the blood and CSF of all patients	*E. sakazakii* isolates from CSF indistinguishable from isolates recovered from PIF and utensils used to prepare the formula	81
1985, Athens, Greece	1 infant, 3 days old; other infants were colonized but not infected by *E. sakazakii*.	Septicemia caused by *E. sakazakii* commingled with *K. pneumoniae*	Resistant to ampicillin, netilmicin, cefotaxime, and amikacin	4
1986–1987, Reykjavik, Iceland	3 infants; *E. sakazakii* recovered from urine, groin, and anal swabs of a 3-day-old asymptomatic child	1st infant became mentally retarded and quadriplegic after recovery; 2nd child had Down's syndrome and died of complications from the infection 5 days after birth; 3rd child developed a seizure disorder and was moderately delayed in all developmental areas.	Infants were fed PIF. *E. sakazakii* was recovered from five packages of infant formula; 22 of 23 isolates from infant formula were identical in biotype, antibiotic profile, and plasmid profile to the 3 neonatal strains.	7
1987, Boston, Mass., and New Orleans, La.	2 infants, 4 weeks and 8 days old	Meningitis, necessitating ventricular shunts; cerebral destruction, developmental damage, and severe neurologic complications	The patients were never geographically proximate.	122
1988, Memphis, Tenn.	4 preterm neonates	Bacteremia, septicemia, urinary tract infection, abdominal distension, and bloody diarrhea or stool	Blender tested positive for *E. sakazakii* and *E. cloacae*. Isolates from infant formula and isolates from infants had the same plasmid and multilocus enzyme profile.	105
1990, Baltimore, Md.	1 infant, 6 months old	Septicemia following small bowel complications (exploratory laparotomy and a gastrostomy tube); blood cultures were positive for both *E. sakazakii* and *Leuconostoc mesenteroides*; stool cultures were negative for both bacteria.	Resistant to vancomycin and ampicillin. The blender used to rehydrate the powdered infant formula was heavily contaminated with both bacteria.	90
1990, Cincinnati, Ohio	1 infant, 2 days old	Blood culture tested positive for the presence of *E. sakazakii*; hemorrhage, abscess, and brain infarction	Ampicillin and cefotaxime treatment	30

(Continued)

Table 13.7 (Continued)

Date and location	No. of cases and outcome	Disease symptoms/organs affected	Comments	Reference
1995/1996, Boston, Mass.	5 cases within 1 year period; individuals of 3, 39, 73, 76, and 82 years of age	All patients had potentially immunosuppressive illnesses under treatment; the 2 youngest patients survived infection.	Gentamicin, cefotaxime, ceftazidime, cefuroxime axetil, and clindamycin treatment	61
1998, Belgium	12 infants; 4 patients required operative treatment; 2 patients (twins) died within 3 weeks of each other.	Blood, anal swabs, and stomach aspirates of 6 of the 12 patients were positive for *E. sakazakii*; 2 strains with two differing morphologies were isolated from 1 patient.	All patients were fed infant formula prior to the illness; a survey of unopened cans of PIF yielded 14 *E. sakazakii* isolates.	117
2000, Winston Salem, N.C.	1 premature infant, 6 days old; no apparent neurological or developmental deficits after 5 weeks	Brain abscess, high fever, irritability, and seizure activity; abnormal cerebritis-like indicators in the frontal lobe of the brain; draining of purulent fluid via craniotomy; *E. sakazakii* was isolated from blood and CSF but not from urine.	Ampicillin, cefotaxime, and bactrim treatment; 1st reported case of *E. sakazakii* being isolated directly from a drained cranial abcess	12
1993, 1995, 1997–2000, Israel	2 underweight infants, 1 being premature, 4 and 9 days of age, recovered after treatment.	Meningitis, seizures, infarction, liquefaction, and cavitation of the brain; ventriculoperitoneal shunt	Cefotaxime treatment; infants were fed PIF.	5
	3 infants were colonized but not infected.	*E. sakazakii* was recovered from CSF and blood.	Antibiotic treatment did not eliminate *E. sakazakii* from colonized asymptomatic patients; blender used to prepare infant formula tested positive for *E. sakazakii*.	
2001, Knoxville, Tenn.	49 infants in a NICU[a] screened—10 infants tested positive for *E. sakazakii*.	7 of the 10 *E. sakazakii*-positive neonates were colonized but not infected; source of the bacterium was traced to PIF.		119
2003, Brazil	14-day-old breast-fed infant girl	Failed ampicillin and ceftriaxone treatment; meningitis and death	Appears to be first report of transmission of *E. sakazakii* via human breast milk.	6
2004, France	9 infants	2 deaths	Linked to infant formula; recall of the implicated infant formula by manufacturer; common source and distribution of infant formula	1
2004, New Zealand	1 premature infant; Waikato NICU	Meningitis and death	Linked to infant formula	88

[a] NICU, neonatal intensive care unit.

in the study, one being positive and the other negative in the in vitro assay using cell lines, had the lowest minimum lethal intraperitoneal doses, but were nonlethal at oral doses. This could indicate that some *E. sakazakii* strains may contain virulence factors that would allow them to survive passage through the stomach and/or translocate across the intestinal wall. Iversen et al. (46) found *E. sakazakii* to be toxic to N2a neuroblastoma cells using the 3-(4,5-dimethylthiazol-2-yl)-2,5-diphenyl tetrazolium bromide (MTT) cell proliferation assay. In the same

unpublished study, the researchers suggested that protease, phosphatase, and lipase activities may contribute to host cell death during an *E. sakazakii* infection.

E. sakazakii infections in newborns were at one time suspected of occurring via passage of the organism through the mother's birth canal, similar to newborn infections caused by other pathogens transmitted from mother to child (75, 113). However, this hypothesis lost favor based on the occurrence of *E. sakazakii* infections in neonates born by Caesarean section (5, 77, 81) and in neonates testing negative for bacterial infection at birth who later developed *E. sakazakii* infections.

The low-acidity conditions existing in the upper gastrointestinal tract of neonates could facilitate the survival of *E. sakazakii* as it passes through the stomach and then into the small intestine. Iversen et al. (49) suggested that similarly to the pneumococci, *Haemophilus*, and meningococci, the major pathogens associated with meningitis in infants, *E. sakazakii* could also present a developmental dependence on access to the central nervous system. In addition, although the route by which most pathogens enter the CSF and cause meningitis is not established, the authors suggested that due to paracellular and transcellular mechanisms that may induce permeability of the blood-brain barrier, the choroids plexus is the most likely entry site.

Willis and Robinson (122) described two cases of *E. sakazakii* infections involving cerebral infarctions followed by the development of cystic lesions, symptoms which are similar to those caused by *Citrobacter diversus*. Based on previous work by Foreman et al. (29), which described possible cyst formation involving a sequence of vasculitis, necrosis, and liquefaction of the cerebral white matter possibly misdiagnosed as abscesses, Willis and Robinson (122) suggested a similar ability of *E. sakazakii* and *C. diversus* to induce a cascade of events leading to a high rate of cyst formation. Farmer et al. (26) had previously shown that *C. koseri* is 50% related to

E. sakazakii by DNA-DNA hybridization, while Iversen et al. (49) compared *Citrobacter* species and *E. sakazakii* 16S rRNA gene sequences and found that *E. sakazakii* was 97.8% similar to *C. koseri*.

A recent review of 46 cases of invasive infections caused by *E. sakazakii* in infants focused on the identification of host risk factors and disease course (8a). Interestingly, using meningitis and bacteremia alone to characterize the disease, invasive infant cases were divided into two groups based on gestational age and birthweight.

Currently, we are looking at a number of nonprimate animal models to gain a better understanding of the mechanisms of pathogenesis and dose response of *E. sakazakii*. Figure 13.1 illustrates our current understanding of the pathogenesis of *E. sakazakii*.

ANTIBIOTIC RESISTANCE

Exposing microorganisms to a wide array and concentrations of antimicrobials potentially leads to an increase in the resistance to antibiotics available for human treatment. Several studies have described the resistance of *Enterobacter* isolates to quinolones, β-lactams, and trimethoprim-sulfamethoxazole (36, 95). Bacterial meningitis demands prompt and effective treatment, and knowledge of antibiotic susceptibility is of great

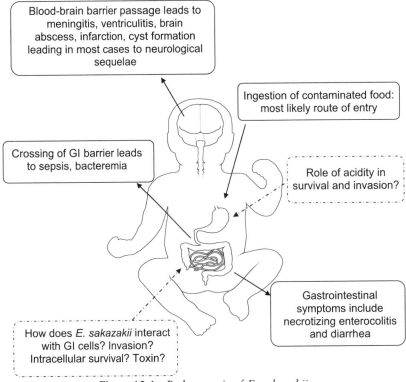

Figure 13.1 Pathogenesis of *E. sakazakii*.

importance regarding neonatal *E. sakazakii* infections and their treatments. *E. sakazakii* has been reported as being more sensitive than other *Enterobacter* spp. to some antibiotics including the aminoglycosides, ureido-penicillins, ampicillin, and carboxypenicillins (2, 36, 85, 122). Muytjens et al. (78) tested the antibiotic suscepti-bility of 195 *E. sakazakii* isolates from various sources such as the respiratory tract, digestive tract, CSF, super-ficial wounds, urine, upper respiratory tract, blood, and utensils against 29 antimicrobials and reported *E. sakazakii* to be the most susceptible among the eight *Enterobacter* spp.tested. Concentrations of 24 of the 29 antibiotics necessary to inhibit at least 90% of the strains were ≤8 µg/ml, which was twofold lower than that required for inhibiting *E. cloacae*. Antibiotics having MICs greater than 8 µg/ml were chloramphenicol (16 µg/ml), cefaloridin (16 µg/ml), cefsulodin (32 µg/ml), cephalothin (>128 µg/ml), and sulfamethoxazole (>128 µg/ml).

Although Lai (61) has reported successful treatment of the first known cases of *E. sakazakii* infections with ampicillin in combination with gentamicin or chloram-phenicol, and Willis and Robinson (122) described the ampicillin-gentamicin treatment as the "gold standard" for *E. sakazakii* treatments, recent reports have described resistance of *E. sakazakii* to the β-lactams, as well as to the gold standard antibiotics. *Enterobacter* spp. are known to be prolific in terms of their ability to inactivate broad-spectrum penicillins and cephalosporins, and this ability appears to be on the increase (14, 24, 61). Pitout et al. (95) tested eight strains of *E. sakazakii* for susceptibil-ity against ampicillin, ampicillin-sulbactam, amoxacillin-clavulanic acid, ticarcillin, ticarcillin-clavulanic acid, piperacillin, piperacillin-tazobactam, aztreonam, cepha-lothin, cefazolin, cefoxitin, cefotaxime, ceftriaxone, ceftazidime, cefepime, and imipenem. In this study, dere-pressed mutant phenotype strains were resistant to all antimicrobials except to imipenem, whereas the wild-type strain (possessing an inducible Bush group 1 β-lactamase) or basal strain (not expressing or expressing in low levels a Bush group 1 β-lactamase) was sensitive to all antimicro-bials tested. One wild-type strain was resistant to cefoxi-tin. Block et al. (8) have also reported β-lactamase activity in *E. sakazakii* isolates recovered from six neonatal and childhood infections. The enzyme was believed to be most likely part of Bush group 1 β-lactamase. Lai (61) and Weir (119) have indicated the possibility of using carbapenems or the newer broad-spectrum cephalosporins together with an aminoglycoside or trimethoprim-sulfamethoxazole to treat *E. sakazakii* meningitis.

Interestingly, Muytjens et al. (81) reported that the *E. sakazakii* strains isolated from their study were sus-ceptible in vitro to ampicillin, gentamicin, chloram-phenicol, and kanamycin. However, six of eight patients responded poorly and died after being treated with these antimicrobials. Arseni et al. (4) also reported the resis-tance to ampicillin treatment, among others (netilmicin, cefotaxime, and amikacin), in a neonatal case of septice-mia caused by *E. sakazakii* and possibly *K. pneumoniae*. In contrast, Naqvi et al. (83) successfully treated an *E. sakazakii* infection in one patient by using cefotaxime, and Willis and Robinson (122) were able to effectively treat two cases of *E. sakazakii*-induced neonatal men-ingitis with moxalactam after observing that ampicillin-gentamicin therapy was unresponsive.

Kleiman et al. (58) reported the resistance of an *E. sakazakii* isolate from a 5-week-old female with meningo-encephalitis to cephalothin (MIC = 16 µg/ml). Nazarowec-White and Farber (85) tested the antibiotic resistance of 17 strains of *E. sakazakii* and found four antibiotic sus-ceptibility patterns (antibiograms). The largest cluster contained five strains that were resistant to sulfisoxazole and cephalothin and susceptible to ampicillin, cefotaxime, chloramphenicol, gentamicin, kanamycin, polymyxin-B, trimethoprim-sulfamethoxazole, tetracycline, and strepto-mycin. The authors concluded that when compared to ribo-typing, PFGE, and random amplified polymorphic DNA, antibiogram patterns were the least discriminatory to dis-tinguish bacterial strains. Farmer et al. (26) reported on the MICs of 12 antibiotics, namely, ampicillin, carbenicillin, cephalothin, amikacin, gentamicin, kanamycin, tobramy-cin, chloramphenicol, sulfisoxazole, tetracycline, nalidixic acid, nitrofurantoin, and trimethoprim-sulfamethoxazole, against 10 *E. sakazakii* strains. They also tested the anti-biotic resistance of 24 strains of *E. sakazakii* against 12 antibiotics, showing that 100% of the strains tested were susceptible to gentamicin, kanamycin, chloramphenicol, and ampicillin while none of them were susceptible to penicillin and only 13% were susceptible to cephalothin. Although vancomycin is not commonly used to treat infec-tions caused by *E. sakazakii*, it is used at a concentration of 10 µg/ml in the selective broth mLST.

CONCLUSIONS

E. sakazakii has become a growing concern for govern-ment regulatory agencies, health care providers (spe-cially those in neonatal intensive care units), and PIF manufacturers. In the past few years, great efforts have been undertaken by regulatory agencies and the scien-tific community in order to acquire new knowledge on this emerging opportunistic pathogen. In the next 3 to 5 years, our understanding of this organism will increase tremendously, and as a result, science- and evidence-based policies and guidelines will likely be developed to better control this organism and reduce the number of *E. sakazakii* outbreaks related to PIF.

References

1. Acoignard, B., V. Vaillant, J. P. Vincent, A. Lefièche, P. Mariani-Kurkdjian, C. Bernet, F. L'Hériteau, H. Sénéchal, P. Grimont, E. Bingen, and J. C. Desenclos. 2006. Infections sévères à *Enterobacter sakazakii* chez des nouveaunés ayant consommé une préparation en poudre pour nourrissons, France, octobre-décembre 2004. *Bull. Epidemiol. Hebd.* **2-3:**10–13.

2. Adamson, D. H., and J. R. Rogers. 1981. *Enterobacter sakazakii* meningitis with sepsis. *Clin. Microbiol. Newsl.* **3:**19–20.

3. Agostoni, C., I. Axelsson, O. Goulet, B. Koletzko, K. F. Michaelsen, J. W. Puntis, J. Rigo, R. Shamir, H. Szajewska, D. Turck, Y. Vandenplas, and L. T. Weaver. 2004. Preparation and handling of powdered infant formula: a commentary by the ESPGHAN Committee on Nutrition. *J. Pediatr. Gastroenterol. Nutr.* **39:**320–322.

4. Arseni, A., E. Malamou-Ladas, C. Koutsia, M. Xanthou, and E. Trikka. 1987. Outbreak of colonization of neonates with Enterobacter sakazakii. *J. Hosp. Infect.* **9:**143–150.

5. Bar-Oz, B., A. Preminger, O. Peleg, C. Block, and I. Arad. 2001. *Enterobacter sakazakii* infection in the newborn. *Acta Paediatr.* **90:**356–358.

6. Barreira, E. R., D. Costa de Souza, P. de Freitas Góis, and J. C. Fernandes. 2003. Meningite por *Enterobacter sakazakii* em recém-nascido: relato de caso. *Pediatria* (São Paulo) **25:**65–70.

7. Biering, G., S. Karlsson, N. C. Clark, K. E. Jonsdottir, P. Ludvigsson, and O. Steingrimsson. 1989. Three cases of neonatal meningitis caused by *Enterobacter sakazakii* in powdered milk. *J. Clin. Microbiol.* **27:**2054–2056.

8. Block, C., O. Peleg, N. Minster, B. Bar-Oz, A. Simhon, I. Arad, and M. Shapiro. 2002. Cluster of neonatal infections in Jerusalem due to unusual biochemical variant of *Enterobacter sakazakii. Eur. J. Clin. Microbiol. Infect. Dis.* **21:**613–616.

8a. Bowen, A. B., and C. R. Braden. 2006. Invasive *Enterobacter sakazakii* disease in infants. *Emerg. Infect. Dis.* [Serial online.] http://www.cdc.gov/ncidod/EID/vol12no08/05-1509.htm.

9. Breeuwer, P., A. Lardeau, M. Peterz, and H. M. Joosten. 2003. Desiccation and heat tolerance of *Enterobacter sakazakii. J. Appl. Microbiol.* **95:**967–973.

10. Brenner, D. J. 1974. DNA reassociation for the clinical differentiation of enteric bacteria. *Public Health Lab.* **32:**118–130.

11. Brenner, D. J., J. J. Farmer III, F. W. Hickman, M. A. Asbury, and A. G. Steigerwalt. 1977. *Taxonomic and Nomenclature Changes in Enterobacteriaceae.* Centers for Disease Control and Prevention, Atlanta, Ga.

12. Burdette, J. H., and C. Santos. 2000. *Enterobacter sakazakii* brain abscess in the neonate: the importance of neuroradiologic imaging. *Pediatr. Radiol.* **30:**33–34.

13. Chan, G. M. 2003. Effects of powdered human milk fortifiers on the antibacterial actions of human milk. *J. Perinatol.* **23:**620–623.

14. Chow, J. W., M. J. Fine, D. M. Shlaes, J. P. Quinn, D. C. Hooper, M. P. Johnson, R. Ramphal, M. M. Wagener, D. K. Miyashiro, and V. L. Yu. 1991. *Enterobacter* bacteremia:

clinical features and emergence of antibiotic resistance during therapy. *Ann. Intern. Med.* **115:**585–590.

15. Clark, N. C., B. C. Hill, C. M. O'Hara, O. Steingrimsson, and R. C. Cooksey. 1990. Epidemiologic typing of *Enterobacter sakazakii* in two neonatal nosocomial outbreaks. *Diagn. Microbiol. Infect. Dis.* **13:**467–472.

16. Clementino, M. M., I. de Filippis, C. R. Nascimento, R. Branquinho, C. L. Rocha, and O. B. Martins. 2001. PCR analyses of tRNA intergenic spacer, 16S-23S internal transcribed spacer, and randomly amplified polymorphic DNA reveal inter- and intraspecific relationships of *Enterobacter cloacae* strains. *J. Clin. Microbiol.* **39:**3865–3870.

17. Cottyn, B., E. Regalado, B. Lanoot, M. De Cleene, T. W. Mew, and J. Swings. 2001. Bacterial populations associated with rice seed in the tropical environment. *Phytopathology* **91:**282–292.

18. Coulin, P., Z. Farah, J. Assanvo, H. Spillmann, and Z. Puhan. 2006. Characterisation of the microflora of attieke, a fermented cassava product, during traditional small-scale preparation. *Int. J. Food Microbiol.* **106:**131–136. (First published 5 October 2005.)

19. Cruz, A. C., E. Fernandez, E. Salinas, P. Ramirez, C. Montiel, and C. A. Eslaval. 2004. Characterization of *Enterobacter sakazakii* isolated from different sources, abstr. Q-051. *Abstr. 104th Gen. Mtg. Am. Soc. Microbiol.,* 23–27 May, New Orleans, La.

20. Dauga, C. 2002. Evolution of the gyrB gene and the molecular phylogeny of *Enterobacteriaceae*: a model molecule for molecular systematic studies. *Int. J. Syst. Evol. Microbiol.* **52:**531–547.

21. Dennison, S. K., and J. Morris. 2002. Multiresistant *Enterobacter sakazakii* wound infection in an adult. *Infect. Med.* **19:**533–535.

22. Edelson-Mammel, S. G., and R. L. Buchanan. 2004. Thermal inactivation of *Enterobacter sakazakii* in rehydrated infant formula. *J. Food Prot.* **67:**60–63.

23. Edelson-Mammel, S. G., M. K. Porteous, and R. L. Buchanan. 2005. Survival of *Enterobacter sakazakii* in a dehydrated powdered infant formula. *J. Food Prot.* **68:**1900–1902.

24. Emery, C. L., and L. A. Weymouth. 1997. Detection and clinical significance of extended-spectrum beta-lactamases in a tertiary-care medical center. *J. Clin. Microbiol.* **35:**2061–2067.

25. Estuningsih, E. 2004. *Present Work and Future Research Needs on* Enterobacter sakazakii. FAO/WHO regional conference on food safety for Asia and Pacific, Seremban, Malaysia, 24–27 May.

26. Farmer, J. J., M. A. Asbury, F. W. Hickman, D. J. Brenner, and the Enterobacteriaceae Study Group. 1980. *Enterobacter sakazakii,* new species of *Enterobacteriaceae* isolated from clinical specimens. *Int. J. Syst. Bacteriol.* **30:**569–584.

27. Food and Agriculture Organization. 1994. *Codex Alimentarius: Code of Hygienic Practice for Foods for Infants and Children.* Food and Agriculture Organization of the United Nations, Rome, Italy.

28. Food and Agriculture Organization/World Health Organization. 2004. *Joint FAO/WHO Workshop on* Enterobacter sakazakii *and Other Microorganisms in Powdered Infant*

Formula. Geneva, 2–5 February, 2004. Food and Agriculture Organization/World Health Organization. [Online.] http://www.who.int/foodsafety/publications/micro/mra6/en/.

29. **Foreman, S. D., E. E. Smith, N. J. Ryan, and G. R. Hogan.** 1984. Neonatal *Citrobacter* meningitis: pathogenesis of cerebral abscess formation. *Ann. Neurol.* **16:**655–659.

30. **Gallagher, P. G., and W. S. Ball.** 1991. Cerebral infarctions due to CNS infection with *Enterobacter sakazakii. Pediatr. Radiol.* **21:**135–136.

31. **Garrity, G., D. J. Brenner, N. Krigg, and J. T. Staley.** 2005. *Bergey's Manual of Systematic Bacteriology,* vol. II. *The Proteobacteria,* Part B. *The Gammaproteobacteria.* Springer, New York, N.Y.

32. **Gassem, M. A.** 1999. Study of the micro-organisms associated with the fermented bread (khamir) produced from sorghum in Gizan region, Saudi Arabia. *J. Appl. Microbiol.* **86:**221–225.

33. **Gram, L., A. B. Christensen, L. Ravn, S. Molin, and M. Givskov.** 1999. Production of acylated homoserine lactones by psychrotrophic members of the *Enterobacteriaceae* isolated from foods. *Appl. Environ. Microbiol.* **65:**3458–3463.

34. **Guillaume-Gentil, O., V. Sonnard, M. C. Kandhai, J. D. Marugg, and H. Joosten.** 2005. A simple and rapid cultural method for detection of *Enterobacter sakazakii* in environmental samples. *J. Food Prot.* **68:**64–69.

35. **Gurtler, J. B., and L. R. Beuchat.** 2005. Performance of media for recovering stressed cells of *Enterobacter sakazakii* as determined using spiral plating and ecometric techniques. *Appl. Environ. Microbiol.* **71:**7661–7669.

36. **Gurtler, J. B., J. L. Kornacki, and L. R. Beuchat.** 2005. *Enterobacter sakazakii:* a coliform of increased concern to infant health. *Int. J. Food Microbiol.* **104:**1–34.

37. **Hamilton, J. V., M. J. Lehane, and H. R. Braig.** 2003. Isolation of *Enterobacter sakazakii* from midgut of *Stomoxys calcitrans. Emerg. Infect. Dis.* **9:**1355–1356.

38. **Harris, L. S., and P. J. Oriel.** 21 February 1989. Heteropolysaccharide produced by *Enterobacter sakazakii.* U.S. patent 4,806,636.

39. **Hawkins, R. E., C. R. Lissner, and J. P. Sanford.** 1991. *Enterobacter sakazakii* bacteremia in an adult. *South. Med. J.* **84:**793–795.

40. **Health Canada (Health Professional Advisory).** 2002. *Enterobacter sakazakii* infection and powdered infant formulas. [Online.] http://www.hc-sc.gc.ca/fn-an/securit/ill-intox/esakazakii/enterobacter_sakazakii_e.html. Accessed 3 February 2006.

40a. **Health Protection Agency.** 2004. *Klebsiella, Enterobacter, Serratia,* and *Citrobacter* spp bacteraemia, England, Wales, and Northern Ireland: 2003. *CDR Wkly.* 20 May 2004, vol. 14, no. 21. [Serial online.] http://www.hpa.org.uk/cdr/archives/2004/bact_2104.pdf.

41. **Hengge-Aronis, R., W. Klein, R. Lange, M. Rimmele, and W. Boos.** 1991. Trehalose synthesis genes are controlled by the putative sigma factor encoded by *rpoS* and are involved in stationary-phase thermotolerance in *Escherichia coli. J. Bacteriol.* **173:**7918–7924.

42. **International Commission on Microbiological Specifications for Foods.** 2002. *Microorganisms in Foods,* vol. 7. *Microbiological Testing in Food Safety Management,* p. 128–130. Kluwer Academic/Plenum Publishers, New York, N.Y.

43. **International Organisation for Standardization.** 2006. *Milk and Milk Products—Detection of* Enterobacter sakazakii(TC 34/SC; ISO Standards). Document ISO/TS 22964:2006. ISO, Geneva, Switzerland.

44. **Iversen, C., and S. Forsythe.** 2003. Risk profile of *Enterobacter sakazakii,* an emergent pathogen associated with infant milk formula. *Trends Food Sci. Technol.* **14:**443–454.

45. **Iversen, C., N. Lazar Adler, and S. J. Forsythe.** 2004. Virulence factors of *Enterobacter sakazakii,* abstr. P-108. *Abstr. 104th Gen. Mtg. Am. Soc. Microbiol.,* 23–27 May, New Orleans, La.

46. **Iversen, C., J. Caubilla-Baron, and S. Forysthe.** 2004. Isolation of *Enterobacter sakazakii, Enterobacteriaceae,* and other microbial contaminants from powdered infant formula milk and related products, abstr. P-004. *Abstr. 104th Gen. Mtg. Am. Soc. Microbiol.,* 23–27 May, New Orleans, La.

47. **Iversen, C., P. Druggan, and S. Forsythe.** 2004. A selective differential medium for *Enterobacter sakazakii,* a preliminary study. *Int. J. Food Microbiol.* **96:**133–139.

48. **Iversen, C., M. Lane, and S. J. Forsythe.** 2004. The growth profile, thermotolerance and biofilm formation of *Enterobacter sakazakii* grown in infant formula milk. *Lett. Appl. Microbiol.* **38:**378–382.

49. **Iversen, C., M. Waddington, S. L. On, and S. Forsythe.** 2004. Identification and phylogeny of *Enterobacter sakazakii* relative to *Enterobacter* and *Citrobacter* species. *J. Clin. Microbiol.* **42:**5368–5370.

50. **Jimenez, E. B., and C. Gimenez.** 1982. Septic shock due to *Enterobacter sakazakii. Clin. Microbiol. Newsl.* **4:**30.

51. **Joker, R. N., T. Norholm, and K. E. Siboni.** 1965. A case of neonatal meningitis caused by a yellow *Enterobacter. Danish Med. Bull.* **12:**128–130.

52. **Kandhai, M. C., M. W. Reij, L. G. Gorris, O. Guillaume-Gentil, and M. van Schothorst.** 2004. Occurrence of *Enterobacter sakazakii* in food production environments and households. *Lancet* **363:**39–40.

53. **Kandhai, M. C., M. W. Reij, K. van Puyvelde, O. Guillaume-Gentil, R. R. Beumer, and M. van Schothorst.** 2004. A new protocol for the detection of *Enterobacter sakazakii* applied to environmental samples. *J. Food Prot.* **67:**1267–1270.

54. **Keller, R., M. A. Pedroso, R. Ritchman, and R. M. Silva.** 1998. Occurrence of virulence associated properties in *Enterobacter cloacae. Infect. Immun.* **66:**645–649.

55. **Kempf, B., and E. Bremer.** 1998. Uptake and synthesis of compatible solutes as microbial stress responses to high-osmolality environments. *Arch. Microbiol.* **170:**319–330.

56. **Keyser, M., R. C. Witthuhn, L. C. Ronquest, and T. J. Britz.** 2003. Treatment of winery effluent with upflow anaerobic sludge blanket (UASB) granular sludges enriched with *Enterobacter sakazakii. Biotechnol. Lett.* **25:**1893–1898.

57. **Kindle, G., A. Busse, D. Kampa, U. Meyer-Koenig, and F. D. Daschner.** 1996. Killing activity of microwaves in milk. *J. Hosp. Infect.* **33:**273–278.

58. Kleiman, M. B., S. D. Allen, P. Neal, and J. Reynolds. 1981. Meningoencephalitis and compartmentalization of the cerebral ventricles caused by *Enterobacter sakazakii*. *J. Clin. Microbiol.* **14**:352–354.

59. Kline, M. W. 1988. Pathogenesis of brain abscess caused by *Citrobacter diversus* or *Enterobacter sakazakii*. *Pediatr. Infect. Dis. J.* **7**:891–892.

60. Kuzina, L. V., J. J. Peloquin, D. C. Vacek, and T. A. Miller. 2001. Isolation and identification of bacteria associated with adult laboratory Mexican fruit flies, *Anastrepha ludens* (Diptera: Tephritidae). *Curr. Microbiol.* **42**:290–294.

61. Lai, K. K. 2001. *Enterobacter sakazakii* infections among neonates, infants, children, and adults. Case reports and a review of the literature. *Medicine* (Baltimore) **80**:113–122.

62. Lange, R., and R. Hengge-Aronis. 1991. Growth phase-regulated expression of *bolA* and morphology of stationary-phase *Escherichia coli* cells are controlled by the novel sigma factor σ^s. *J. Bacteriol.* **173**:4474–4481.

63. Lange, R., and R. Hengge-Aronis. 1991. Identification of a central regulator of stationary-phase gene expression in *Escherichia coli*. *Mol. Microbiol.* **5**:49–59.

64. Leclercq, A., C. Wanegue, and P. Baylac. 2002. Comparison of fecal coliform agar and violet red bile lactose agar for fecal coliform enumeration in foods. *Appl. Environ. Microbiol.* **68**:1631–1638.

65. Lecour, H., A. Seara, J. Cordeiro, and M. Miranda. 1989. Treatment of childhood bacterial meningitis. *Infection* **17**:343–346.

66. Lehner, A., K. Riedel, L. Eberl, P. Breeuwer, B. Diep, and R. Stephan. 2005. Biofilm formation, extracellular polysaccharide production, and cell-to-cell signaling in various *Enterobacter sakazakii* strains: aspects promoting environmental persistence. *J. Food Prot.* **68**:2287–2294.

67. Lehner, A., and R. Stephan. 2004. Microbiological, epidemiological, and food safety aspects of *Enterobacter sakazakii*. *J. Food Prot.* **67**:2850–2857.

68. Lehner, A., T. Tasara, and R. Stephan. 2004. 16S rRNA gene based analysis of *Enterobacter sakazakii* strains from different sources and development of a PCR assay for identification. *BMC Microbiol.* **4**:43.

69. Lenati, R., K. Hebert, J. Farber, and F. Pagotto. The growth and survival of *Enterobacter sakazakii* in fortified and non-fortified human breast milk and reconstituted powdered infant formula. Unpublished data.

70. Lenati, R., K. Hebert, J. Farber, and F. Pagotto. Survival of *Enterobacter sakazakii* isolates in frozen crab meat and in reconstituted powdered infant formula. Unpublished data.

71. Lenati, R., K. Hébert, Y. Kou, S. McIlwham, K. Tyler, J. Farber, and F. Pagotto. A comprehensive phenotypic and genotypic characterization of 220 strains of *Enterobacter sakazakii*. Unpublished data.

72. Leuschner, R. G., and J. Bew. 2004. A medium for the presumptive detection of *Enterobacter sakazakii* in infant formula: interlaboratory study. *J. AOAC Int.* **87**:604–613.

73. Liu, Y., Q. Gao, X. Zhang, Y. Hou, J. Yang, and X. Huang. 2006. PCR and oligonucleotide array for detection of *Enterobacter sakazakii* in infant formula. *Mol. Cell. Probes* **1**:11–17.

74. Malorny, B., and M. Wagner. 2005. Detection of *Enterobacter sakazakii* strains by real-time PCR. *J. Food Prot.* **68**:1623–1627.

75. Monroe, P. W., and W. L. Tift. 1979. Bacteremia associated with *Enterobacter sakazakii* (yellow, pigmented *Enterobacter cloacae*). *J. Clin. Microbiol.* **10**:850–851.

76. Mosso, M. A., M. C. de la Rosa, C. Vivar, and M. R. Medina. 1994. Heterotrophic bacterial populations in the mineral waters of thermal springs in Spain. *J. Appl. Bacteriol.* **77**:370–381.

77. Muytjens, H. L., and L. A. Kollee. 1990. *Enterobacter sakazakii* meningitis in neonates: causative role of formula? *Pediatr. Infect. Dis. J.* **9**:372–373.

78. Muytjens, H. L., and J. van der Ros-van de Repe. 1986. Comparative in vitro susceptibilities of eight *Enterobacter* species, with special reference to *Enterobacter sakazakii*. *Antimicrob. Agents Chemother.* **29**:367–370.

79. Muytjens, H. L., and L. A. Kollee. 1982. Neonatal meningitis due to *Enterobacter sakazakii*. *Tijdschr. Kindergeneeskd.* **50**:110–112.

80. Muytjens, H. L., H. Roelofs-Willemse, and G. H. Jaspar. 1988. Quality of powdered substitutes for breast milk with regard to members of the family *Enterobacteriaceae*. *J. Clin. Microbiol.* **26**:743–746.

81. Muytjens, H. L., H. C. Zanen, H. J. Sonderkamp, L. A. Kollee, I. K. Wachsmuth, and J. J. Farmer III. 1983. Analysis of eight cases of neonatal meningitis and sepsis due to *Enterobacter sakazakii*. *J. Clin. Microbiol.* **18**:115–120.

82. Nair, M. K., J. Joy, and K. S. Venkitanarayanan. 2004. Inactivation of *Enterobacter sakazakii* in reconstituted infant formula by monocaprylin. *J. Food Prot.* **67**:2815–2819.

83. Naqvi, S. H., M. A. Maxwell, and L. M. Dunkle. 1985. Cefotaxime therapy of neonatal gram-negative bacillary meningitis. *Pediatr. Infect. Dis.* **4**:499–502.

84. Nassereddin, R. A., and M. I. Yamani. 2005. Microbiological quality of sous and tamarind, traditional drinks consumed in Jordan. *J. Food Prot.* **68**:773–777.

85. Nazarowec-White, M., and J. M. Farber. 1999. Phenotypic and genotypic typing of food and clinical isolates of *Enterobacter sakazakii*. *J. Med. Microbiol.* **48**:559–567.

86. Nazarowec-White, M., and J. M. Farber. 1997. *Enterobacter sakazakii*: a review. *Int. J. Food Microbiol.* **34**:103–113.

87. Nazarowec-White, M., and J. M. Farber. 1997. Thermal resistance of *Enterobacter sakazakii* in reconstituted dried-infant formula. *Lett. Appl. Microbiol.* **24**:9–13.

88. New Zealand Ministry of Health. April 5, 2005. *News and Issues*: E. sakazakii *Meningitis To Become a Notifiable Disease*. [Online.] www.moh.govt.nz/moh.nsf/bfc540e5ac1abe02cc256e7d0082eede/d6081bec1d116521cc256fd90074d556?OpenDocument#1. Accessed 1 February 2006.

89. Nielsen, A. C. Personal communication.

90. Noriega, F. R., K. L. Kotloff, M. A. Martin, and R. S. Schwalbe. 1990. Nosocomial bacteremia caused by *Enterobacter sakazakii* and *Leuconostoc mesenteroides* resulting from extrinsic contamination of infant formula. *Pediatr. Infect. Dis. J.* **9**:447–449.

91. Oh, S. W., and D. H. Kang. 2004. Fluorogenic selective and differential medium for isolation of *Enterobacter sakazakii*. *Appl. Environ. Microbiol.* **70**:5692–5694.

92. **Pagotto, F. J., M. Nazarowec-White, S. Bidawid, and J. M. Farber.** 2003. *Enterobacter sakazakii*: infectivity and enterotoxin production *in vitro* and *in vivo*. *J. Food Prot.* **66:**370–375.

93. **Pangalos, G.** 1929. Sur un bacille chromogene isole par hemoculture. *C. R. Soc. Biol.* **100:**1097.

94. **Pediatric Nutrition Practice Group of the American Dietetic Association.** 2004. *Infant Feedings: Guidelines for Preparation of Formula and Breastmilk in Health Care Facilities.* [Online.] http://www.eatright.org/cps/rde/xchg/ada/hs.xsl/nutrition_5441_ENU_HTML.htm. Accessed 14 February 2006.

95. **Pitout, J. D., E. S. Moland, C. C. Sanders, K. S. Thomson, and S. R. Fitzsimmons.** 1997. β-Lactamases and detection of β-lactam resistance in *Enterobacter* spp. *Antimicrob. Agents Chemother.* **41:**35–39.

96. **Postupa, R., and E. Aldova.** 1984. *Enterobacter sakazakii*: a Tween 80 esterase-positive representative of the genus *Enterobacter* isolated from powdered milk specimens. *J. Hyg. Epidemiol. Microbiol. Immunol.* **28:**435–440.

97. **Pribyl, C., R. Salzer, J. Beskin, R. J. Haddad, B. Pollock, R. Beville, B. Holmes, and W. J. Mogabgab.** 1985. Aztreonam in the treatment of serious orthopaedic infections. *Am. J. Med.* **78:**51–56.

98. **Reina, J., F. Parras, J. Gil, F. Salva, and P. Alomar.** 1989. Human infections caused by *Enterobacter sakazakii*. Microbiologic considerations. *Enferm. Infecc. Microbiol. Clin.* **7:**147–150.

99. **Restaino, L., E. W. Frampton, W. C. Lionberg, and R. J. Becker.** 2006. A chromogenic plating medium for the isolation and identification of *Enterobacter sakazakii* from foods, food ingredients, and environmental sources. *J. Food Prot.* **69:**315–322.

100. **Richards, G. M., J. B. Gurtler, and L. R. Beuchat.** 2005. Survival and growth of *Enterobacter sakazakii* in infant rice cereal reconstituted with water, milk, liquid infant formula, or apple juice. *J. Appl. Microbiol.* **99:**844–850.

101. **Ries, M., D. Harms, and J. Scharf.** 1994. Multiple cerebral infarcts with resulting multicystic encephalomalacia in a premature infant with *Enterobacter sakazakii* meningitis. *Klin. Padiatr.* **206:**184–186.

102. **Robertson, L. F., G. S. Johannessen, B. K. Gjerde, and S. Loncarevic.** 2002. Microbiological analysis of seed sprouts in Norway. *Int. J. Food Microbiol.* **75:**119–126.

103. **Scheepe-Leberkuhne, M., and F. Wagner.** 1986. Optimization and preliminary characterization of an exopolysaccharide synthesized by *Enterobacter sakazakii*. *Biotechnol. Lett.* **8:**695–700.

104. **Seo, K. H., and R. E. Brackett.** 2005. Rapid, specific detection of *Enterobacter sakazakii* in infant formula using a real-time PCR assay. *J. Food Prot.* **68:**59–63.

105. **Simmons, B. P., M. S. Gelfand, M. Haas, L. Metts, and J. Ferguson.** 1989. *Enterobacter sakazakii* infections in neonates associated with intrinsic contamination of a powdered infant formula. *Infect. Control Hosp. Epidemiol.* **10:**398–401.

106. **Smith, J. L., P. M. Fratamico, and S. J. Novak.** 2004. Quorum sensing: a primer for microbiologists. *J. Food Prot.* **67:**1063–1070.

107. **Sogaard, P., and P. Kjaeldgaard.** 1986. Two isolations of enteric group 69 from human clinical specimens. *Acta Pathol. Microbiol. Immunol. Scand. Sect.* B **94:**365–367.

108. **Sondheimer, J., D. Clark, and E. Gervaise.** 1985. Continuous gastric pH measurement in young and older healthy preterm infants receiving formula and clear liquid feedings. *J. Pediatr. Gastroenterol. Nutr.* **4:**352–355.

109. **Soriano, J. M., H. Rico, J. C. Molto, and J. Manes.** 2001. Incidence of microbial flora in lettuce meat and Spanish potato omelette from restaurants. *Food Microbiol.* **18:**159–163.

110. **Stoll, B. J., N. Hansen, A. A. Fanaroff, and J. A. Lemons.** 2004. *Enterobacter sakazakii* is a rare cause of neonatal septicemia or meningitis in VLBW infants. *J. Pediatr.* **144:**821–823.

111. **Styrvold, O. B., and A. R. Strom.** 1991. Synthesis, accumulation, and excretion of trehalose in osmotically stressed *Escherichia coli* K-12 strains: influence of amber suppressors and function of the periplasmic trehalase. *J. Bacteriol.* **173:**1187–1192.

112. **Telang, S., C. L. Berseth, P. W. Ferguson, J. M. Kinder, M. Deroin, and B. W. Petschow.** 2005. Fortifying fresh human milk with commercial powdered human milk fortifiers does not affect bacterial growth during 6 hours at room temperature. *J. Am. Diet. Assoc.* **105:**1567–1572.

113. **Tift, W. L.** 1977. Group B streptococcal infections in the neonate. *J. Med. Assoc. Ga.* **66:**703–705.

114. **United States Food and Drug Administration (Center for Food Safety and Applied Nutrition Office of Nutritional Products, Labeling and Dietary Supplements).** April 11, 2002; revised October 10, 2002. Health professionals letter on *Enterobacter sakazakii* infections associated with use of powdered (dry) infant formulas in neonatal intensive care units. U.S. FDA, Rockville, Md.

115. **United States Food and Drug Administration.** 2002. Isolation and enumeration of *Enterobacter sakazakii* from dehydrated powdered infant formula. [Online.] www.cfsan.fda.gov/~comm/mmesakaz.html. Accessed 23 October 2005.

116. **Urmenyi, A. M. C., and A. W. Franklin.** 1961. Neonatal death from pigmented coliform infection. *Lancet* **i:**313–315.

117. **van Acker, J., F. de Smet, G. Muyldermans, A. Bougatef, A. Naessens, and S. Lauwers.** 2001. Outbreak of necrotizing enterocolitis associated with *Enterobacter sakazakii* in powdered milk formula. *J. Clin. Microbiol.* **39:**293–297.

118. **Van Os, M., P. G. Van Wikeselaar, and S. F. Spoelstra.** 1996. Formation of biogenic amines in well fermented grass silages. *J. Agric. Sci. Cambridge* **127:**97–107.

119. **Weir, E.** 2002. Powdered infant formula and fatal infection with *Enterobacter sakazakii*. *CMAJ* **166:**1570.

120. **Whitehead, N. A., A. M. L. Barnard, H. Slater, N. J. L. Simpson, and G. P. C. Salmond.** 2001. Quorum-sensing in gram-negative bacteria. *FEMS Microbiol. Rev.* **25:**365–404.

121. **Williams, T. L., S. R. Monday, S. Edelson-Mammel, R. Buchanan, and S. M. Musser.** 2005. A top-down proteomics approach for differentiating thermal resistant strains of *Enterobacter sakazakii*. *Proteomics* **5:**4161–4169.

122. **Willis, J., and J. E. Robinson.** 1988. *Enterobacter sakazakii* meningitis in neonates. *Pediatr. Infect. Dis. J.* **7:**196–199.

Food Microbiology: Fundamentals and Frontiers, 3rd Ed.
Edited by M. P. Doyle and L. R. Beuchat
© 2007 ASM Press, Washington, D.C.

Roy M. Robins-Browne

Yersinia enterocolitica

14

CHARACTERISTICS OF THE ORGANISM

The genus *Yersinia* comprises 11 species within the family *Enterobacteriaceae* (Table 14.1) (15, 244) and includes three well-characterized pathogens of mammals, one of fish, and several other species whose etiologic role in disease is uncertain (for a review of the latter, see reference 212). The four known pathogenic species are *Yersinia pestis*, the causative agent of bubonic and pneumonic plague; *Y. pseudotuberculosis*, a rodent pathogen which occasionally causes mesenteric lymphadenitis, septicemia, and immune-mediated diseases in humans; *Y. ruckeri*, the cause of enteric redmouth disease in salmonids and other freshwater fish; and *Y. enterocolitica*, a versatile intestinal pathogen which is the most prevalent *Yersinia* species amongst humans.

Y. pestis is transmitted to its host via flea bites or respiratory aerosols, whereas *Y. pseudotuberculosis* and *Y. enterocolitica* are foodborne pathogens. Nevertheless, these three species share a number of essential virulence determinants which enable them to overcome the innate defenses of their hosts. Analogs of these virulence determinants occur in several other enterobacteria, such as enteropathogenic and enterohemorrhagic *Escherichia*

coli and *Salmonella* and *Shigella* species, as well as in various other pathogens of animals (e.g., *Pseudomonas aeruginosa* and *Bordetella* species) and plants (e.g., *Erwinia amylovora*, *Xanthomonas campestris*, and *Pseudomonas syringae*), thus providing evidence for horizontal transfer of virulence genes between diverse bacterial pathogens.

Classification

Y. enterocolitica first emerged as a human pathogen during the 1930s (21, 212). It shares between 10 and 30% DNA homology with other genera in the *Enterobacteriaceae* and is approximately 50% related to *Y. pseudotuberculosis* and *Y. pestis*. The last two species share >90% DNA homology, with genetic analysis suggesting that *Y. pestis* is a clone of *Y. pseudotuberculosis* which evolved some 1,500 to 20,000 years ago, shortly before the first known pandemics of human plague (2).

Y. enterocolitica is a heterogenous species, being divisible into a large number of subgroups, chiefly according to biochemical activity and lipopolysaccharide (LPS) O antigens (Tables 14.2 and 14.3). Biotyping is based on the ability of *Y. enterocolitica* to metabolize selected

Roy M. Robins-Browne, Dept. of Microbiology and Immunology, University of Melbourne, Victoria 3010, and Murdoch Children's Research Institute, Royal Children's Hospital, Parkville, Victoria 3052, Australia.

Table 14.1 Some biochemical tests used to differentiate *Yersinia* species[a]

Test	Result[b]											
			Y. enterocolitica									
	Y. aldovae	*Y. bercovieri*	Biotypes 1–4	Biotype 5	*Y. frederiksenii*	*Y. intermedia*	*Y. kristensenii*	*Y. mollaretii*	*Y. pestis*	*Y. pseudotuberculosis*	*Y. rohdei*	*Y. ruckeri*
Indole	–	–	D	–	+	+	D	–	–	–	–	–
Voges-Proskauer	+	–	+	+[c]	D	+	–	–	–	–	–	–
Citrate (Simmons)	D	–	–	–	D	+	–	–	–	–	+	–
L-Ornithine	+	+	+	–	+	+	+	+	–	–	+	+
Mucate, acid	D	+	–	–	D	D	+[c]	+	–	–	–	–
Pyrazinamidase	+	+	D	–	+	+	+	+	–	–[c]	+	ND
Sucrose	–	+	+	D	+	+	–	+	–	–	+	–
Cellobiose	–	+	+	+	+	+	+	+	–	–	+	–
L-Rhamnose	+	–	–	–	+	+	–	–	–	+	–	–
Melibiose	–	–	–	–	–	+	–	–	D	+	D	–
L-Sorbose	–	–	D	D	+	+	+	+	–	–	ND	ND
L-Fucose	D	+	D	–	+	D	D	–	ND	–	ND	ND

[a] Adapted from reference 246.
[b] +, positive; –, negative; D, different reactions; ND, not determined.
[c] Some reactions may be delayed or weakly positive.

Table 14.2 Biotyping scheme for *Y. enterocolitica*[a]

Test	Reaction of biotype[b]					
	1A	1B	2	3	4	5
Lipase (Tween hydrolysis)	+	+	−	−	−	−
Esculin hydrolysis	D	−	−	−	−	−
Indole production	+	+	(+), −	−	−	
D-Xylose fermentation	+	+	+	+	−	D
Voges-Proskauer reaction	+	+	+	+	+	(+)
Trehalose fermentation	+	+	+	+	+	−
Nitrate reduction	+	+	+	+	+	−
Pyrazinamidase	+	−	−	−	−	−
β-D-Glucosidase	+	−	−	−	−	−
Proline peptidase	D	−	−	−	−	−

[a] Adapted from reference 247.
[b] +, positive; (+), delayed positive; −, negative; D, different reactions.

organic substrates and provides a convenient means to subdivide the species into subtypes of variable clinical and epidemiological significance (Tables 14.2 and 14.3) (245). Most primary pathogenic strains from humans and domestic animals fall within biotypes 1B, 2, 3, 4, and 5. In contrast, *Y. enterocolitica* strains of biotype 1A are commonly obtained from terrestrial and freshwater ecosystems. For this reason, they are often referred to as environmental strains, although some of them may be responsible for intestinal infections (221).

The most frequent *Y. enterocolitica* biotype obtained from human clinical material worldwide is biotype 4.

Table 14.3 Relationship between O serotype and pathogenicity of *Y. enterocolitica* and related species

Species and biotype	Serotype(s)[a]
Y. enterocolitica	
1A	O:4; O:5; O:6,30; O:6,31; O:7,8; O:7,13; O:10; O:14; O:16; O:21; O:22; O:25; O:37; O:41,42; O:46; O:47; O:57; NT
1B	**O:4,32; O:8; O:13a,13b;** O:16; **O:18; O:20; O:21;** O:25; O:41,42; NT
2	**O:5,27; O:9;** O:27
3	**O:1,2,3; O:3; O:5,27**
4	**O:3**
5	**O:2,3**
Y. bercovieri	O:8; O:10; O;58,16; NT
Y. frederiksenii	O:3; O:16; O:35; O:38; O:44: NT
Y. intermedia	O:17; O:21,46; O:35; O:37; O:40; O:48; O:52; O:55; NT
Y. kristensenii	O:11; O:12,25; O:12,26; O:16; O:16,29; O:28,50; O:46; O:52; O:59; O:61; NT
Y. mollaretii	O:3, O:6,30; O:7,13; O:59: O:62,22; NT

[a] NT, not typable. Serotypes which include strains considered to be primary pathogens are shown in bold.

Biotype 1B yersiniae are usually isolated from patients in the United States and are referred to as "American" strains, although they have also been found in a number of countries in Europe, Africa, Asia, and Australasia. Although not common anywhere, biotype 1B yersiniae are inherently more virulent for mice (and probably for humans) than strains in the other pathogenic categories and have been identified as the cause of several food-borne outbreaks of yersiniosis in the United States.

Serotyping of *Y. enterocolitica*, based on LPS surface O antigens, coincides to some extent with biotyping (Table 14.3) and provides a useful additional tool to subdivide this species in a way that relates to pathologic significance (243). Serotype O:3 is the variety most frequently isolated from humans, and almost all of these isolates are biotype 4. Other serotypes commonly obtained from humans, particularly in northern Europe, include O:9 (biotype 2) and O:5,27 (biotype 2 or 3). The usefulness of serotyping is limited to some extent by the fact that the overwhelming majority of human infections are due to strains of serotype O:3 and by the presence of cross-reacting O antigens in *Y. enterocolitica* strains of variable pathological and epidemiological significance. In addition, some bacteria that were originally allocated to O serotypes of *Y. enterocolitica* were later reclassified as separate species (243).

At least 18 flagellar (H) antigens of *Y. enterocolitica*, designated by lowercase letters (a,b; b,c; b,c,e,f,k; m; etc.), have also been identified. Although there is little overlap between the H antigens of *Y. enterocolitica* sensu stricto and those of related species, antigenic characterization of isolates by complete O and H serotyping is seldom attempted (243).

Other schemes for subtyping *Yersinia* species include bacteriophage typing, multienzyme electrophoresis, multilocus sequence typing, and the demonstration of restriction fragment length polymorphisms of chromosomal and plasmid DNA (111, 124). These techniques can be used to determine the relatedness of different isolates and to facilitate epidemiological investigations of outbreaks or to trace the sources of sporadic infections (72, 238).

Susceptibility and Tolerance

Y. enterocolitica is unusual amongst pathogenic enterobacteria in that it is psychrotrophic, as evidenced by its ability to replicate at temperatures below 4°C. The doubling time for this organism at the optimum growth temperature (approximately 28 to 30°C) is about 34 min, which increases to 1 h at 22°C, 5 h at 7°C, and approximately 40 h at 1°C (193). *Y. enterocolitica* readily withstands freezing and can survive in frozen foods for extended periods, even after repeated freezing and thawing (227). Studies of the ability of *Y. enterocolitica* to survive and grow in artificially contaminated foods under various storage conditions have shown that it generally survives better at room temperature and refrigeration temperatures than at intermediate temperatures. *Y. enterocolitica* persists longer in cooked foods than in raw foods, probably due to an increased availability of nutrients in cooked foods and the fact that the presence of other psychrotrophic bacteria in unprocessed food may restrict yersinia growth (193). The number of viable *Y. enterocolitica* cells may increase more than a millionfold on cooked beef or pork within 24 h at 25°C or within 10 days at 7°C (89). *Y. enterocolitica* can grow at refrigeration temperatures in vacuum-packed meat, boiled eggs, boiled fish, pasteurized liquid eggs, pasteurized whole milk, cottage cheese, and tofu (soybean curd) (63, 193). Proliferation also occurs in refrigerated seafoods, such as oysters, raw shrimp, and cooked crab meat, but at a lower rate than in pork or beef (160). Bacteria may also persist for extended periods in refrigerated vegetables and cottage cheese, particularly in the presence of chicken meat (199). The psychrotrophic nature of *Y. enterocolitica* also poses problems for the blood transfusion industry, largely because of its ability to proliferate and release endotoxin in blood products stored at 4°C without manifestly altering their appearance (9).

Y. enterocolitica and *Y. pseudotuberculosis* are able to grow over a pH range from approximately pH 4 to 10, with an optimum pH of approximately 7.6 (189). They tolerate alkaline conditions extremely well, but their acid tolerance is less pronounced and depends on the acidulant used, the environmental temperature, the composition of the medium, and the growth phase of the bacteria (4, 239). The acid tolerance of *Y. enterocolitica*

is enhanced by the production of urease, which hydrolyzes urea to release ammonia that elevates the cytoplasmic pH (55).

Y. enterocolitica and *Y. pseudotuberculosis* are susceptible to heat and are readily destroyed by pasteurization at 71.8°C for 18 s or 62.8°C for 30 min (51, 227). Exposure of surface-contaminated meat to hot water (80°C) for 10 to 20 s reduces bacterial viability at least 99.9% (208). *Y. enterocolitica* is also susceptible to ionizing and ultraviolet irradiation (29, 58), high hydrostatic pressure (249), and sodium nitrate and nitrite added to food (52). It is relatively resistant to sodium nitrate and nitrite in solution, however, and can also tolerate NaCl at concentrations of up to 5% (52, 210). *Y. enterocolitica* is generally susceptible to organic acids, such as lactic, citric, and acetic acids, and to chlorine (60, 239). However, some resistance to chlorine occurs in yersiniae grown under conditions that approximate natural aquatic environments or cocultivated with predatory aquatic protozoa (91, 122).

CHARACTERISTICS OF INFECTION

Infections with *Y. enterocolitica* typically manifest as nonspecific, self-limiting diarrhea but may give rise to a variety of suppurative and autoimmune complications (Table 14.4), with the risk of these determined partly by host factors, in particular age and underlying immune status.

Acute Infection

Y. enterocolitica enters the gastrointestinal tract after ingestion of contaminated food or water. The median infective dose for humans is not known but is likely to exceed 10^4 CFU. Gastric acid appears to be a significant barrier to infection with *Y. enterocolitica*, and for individuals with gastric hypoacidity, the infectious dose may be lower (55, 66).

Most symptomatic infections with *Y. enterocolitica* occur in children, especially in those younger than 5 years of age. In these patients, yersiniosis presents as diarrhea, often accompanied by low-grade fever and abdominal pain (101, 138). The character of the diarrhea varies from watery to mucoid. A small proportion of children (generally <10%) have frankly bloody stools. Children with *Y. enterocolitica*-induced diarrhea often complain of abdominal pain and headache. Sore throat is a frequent accompaniment and may dominate the clinical picture in older patients (215). The illness typically lasts from a few days to 3 weeks, although some patients develop chronic enterocolitis which may persist for several months (186). Occasionally, acute enteritis

Table 14.4 Clinical manifestations of infections with *Y. enterocolitica*

Common manifestations
  Diarrhea (gastroenteritis), especially in young children
  Enterocolitis
  Pseudoappendicitis syndrome due to terminal ileitis, acute mesenteric lymphadenitis
  Pharyngitis
  Postinfection autoimmune sequelae
    Arthritis, especially associated with HLA-B27
    Erythema nodosum
    Uveitis, associated with HLA-B27
    Glomerulonephritis (uncommon)
    Myocarditis (uncommon)
    Thyroiditis (uncertain)

Less common manifestations
  Septicemia
  Visceral abscesses, for example, in liver, spleen, and lungs
  Skin infection (pustules, wound infection, pyomyositis, etc.)
  Pneumonia
  Endocarditis
  Osteomyelitis
  Peritonitis
  Meningitis
  Intussusception
  Eye infections (conjunctivitis, panophthalmitis)

progresses to intestinal ulceration and perforation or to ileocolic intussusception, toxic megacolon, or mesenteric vein thrombosis (47). On rare occasions, patients may present with peritonitis in the absence of intestinal perforation (172).

In children older than 5 years of age and in adolescents, acute yersiniosis often presents as a pseudoappendicular syndrome due to acute inflammation of the terminal ileum or the mesenteric lymph nodes. The usual features of this syndrome are abdominal pain and tenderness localized to the right lower quadrant. These symptoms are usually accompanied by fever, with little or no diarrhea. The importance of this form of the disease lies in its close resemblance to appendicitis (47). Of those patients with this syndrome who undergo surgical treatment, approximately 60 to 80% have terminal ileitis, with or without mesenteric adenitis, and a normal or slightly inflamed appendix (167, 234). *Y. enterocolitica* may be cultured from the distal ileum and the mesenteric lymph nodes. The pseudoappendicular syndrome appears to be more frequent for patients infected with the relatively more virulent strains of *Y. enterocolitica*, notably strains of biotype 1B. *Y. enterocolitica* is rarely found in patients with true appendicitis (234).

Although *Y. enterocolitica* is seldom isolated from extraintestinal sites, there appears to be no tissue in which it will not grow. In adults, pharyngitis, sometimes with cervical lymphadenitis, may dominate the clinical presentation (215). Focal disease in the absence of obvious bacteremia may present as cellulitis, subcutaneous abscess, pyomyositis, suppurative lymphadenitis, septic arthritis, osteomyelitis, urinary tract infection, renal abscess, sinusitis, pneumonia, lung abscess, or empyema (47).

Bacteremia is a rare complication of infection, except in patients who are immunocompromised or in an iron-overloaded state (66). Factors which predispose patients to the development of *Yersinia* bacteremia include immunosuppression, blood dyscrasias, malnutrition, chronic renal failure, cirrhosis, alcoholism, diabetes mellitus, and acute and chronic iron overload states, particularly when managed by chelation therapy with desferrioxamine B (47). Bacteremic dissemination of *Y. enterocolitica* may lead to various manifestations, including splenic, hepatic, and lung abscesses, catheter-associated infections, osteomyelitis, panophthalmitis, endocarditis, mycotic aneurysm, and meningitis (47). *Yersinia* bacteremia is reported to have a case fatality rate of between 30 and 60%.

Bacteremia may also result from direct inoculation of *Y. enterocolitica* into the circulation during blood transfusion (25, 128). Indeed, *Y. enterocolitica* is one of the most important causes of fatal bacteremia following transfusion with packed red blood cells or platelets (25). Patients infused with contaminated blood may develop symptoms of a severe transfusion reaction minutes to hours after exposure, depending on the number of bacteria and the amount of endotoxin administered with the blood (9). The varieties of *Y. enterocolitica* responsible for transfusion-acquired yersiniosis are the same serobiotypes as those associated with enteric infections. The probable sources of these infections are blood donors with low-grade, subclinical bacteremia. A small number of bacteria in donated blood can increase during storage at refrigeration temperatures without manifestly altering the appearance of the blood (224).

Autoimmune Complications

Although most episodes of yersiniosis remit spontaneously without long-term sequelae, infections with *Y. enterocolitica* are noteworthy for a variety of immunological complications, such as reactive arthritis, erythema nodosum, uveitis, glomerulonephritis, carditis, and thyroiditis, which have been reported to follow acute infection (47). Of these, reactive arthritis is the most widely recognized (6, 20, 132). This manifestation of yersiniosis is infrequent before the age of 10 years and occurs most often in Scandinavian countries, where serotype O:3 strains and the human leukocyte antigen HLA-B27 are especially prevalent. Men and women are affected equally. Arthritis typically follows the onset of diarrhea or the pseudoappendicular syndrome by 1 to 2 weeks. The joints most commonly involved are the knees, ankles, toes, tarsal joints, fingers, wrists, and elbows. Synovial fluid from affected joints contains large numbers of inflammatory cells, principally polymorphonuclear leukocytes, and is invariably sterile, although it generally contains bacterial antigens (77). The duration of arthritis is typically less than 3 months, and the long-term prognosis is good, although some patients may have symptoms that persist for several years (38, 99). Many patients with arthritis also have extra-articular symptoms, including urethritis, uveitis, and erythema nodosum (20). *Y. enterocolitica*-induced erythema nodosum occurs predominantly in women and is not associated with HLA-B27. Other autoimmune complications of yersiniosis, including Reiter's syndrome, uveitis, acute proliferative glomerulonephritis, collagenous colitis, and rheum-like carditis, have been reported, mostly from Scandinavian countries (127). Yersiniosis has also been linked to various thyroid disorders, including Graves' disease hyperthyroidism, nontoxic goiter, and Hashimoto's thyroiditis, although the causative role of yersiniae in these conditions is uncertain (225). In Japan, *Y. pseudotuberculosis* has been implicated in the etiology of Kawasaki's disease (229).

RESERVOIRS

Infections with *Yersinia* species are zoonoses. The subgroups of *Y. enterocolitica* which commonly occur in humans also occur in domestic animals, whereas those which are infrequent in humans generally infect wild rodents. *Y. enterocolitica* can occupy a broad range of environments and has been isolated from the intestinal tracts of many different mammalian species as well as from birds, frogs, fish, flies, fleas, crabs, and oysters (21, 47, 229).

Foods that may harbor *Y. enterocolitica* include pork, beef, lamb, poultry, and dairy products, such as milk, cream, and ice cream (63, 117, 193). *Y. enterocolitica* is also commonly found in a variety of terrestrial and freshwater ecosystems, including soil, vegetation, lakes, rivers, wells, and streams, and can persist for extended periods in soil, vegetation, streams, lakes, wells, and spring water, particularly at low environmental temperatures (34, 117). Most environmental isolates of *Y. enterocolitica* lack markers of bacterial virulence and are of uncertain significance for human or animal health (149).

Although *Y. enterocolitica* has been recovered from a variety of wild and domesticated animals, pigs are the only animal species from which *Y. enterocolitica* strains of biotype 4 and serotype O:3 (the variety most commonly associated with human disease) have been isolated with any degree of frequency (117). Pigs may also carry *Y. enterocolitica* strains of serotypes O:9 and O:5,27, particularly in regions where human infections with these varieties are comparatively common. In countries with a high incidence of human yersiniosis, *Y. enterocolitica* is commonly isolated from pigs at slaughterhouses (8, 70, 85). The tissue most frequently culture positive at slaughter is the tonsils, which appears to be the preferred site of *Y. enterocolitica* infection in pigs. Other tissues which frequently yield yersiniae include the tongue, cecum, rectum, feces, and gut-associated lymphoid tissue. *Y. enterocolitica* is seldom isolated from meat offered for retail sale, however, apart from pork tongue (117, 193), although standard methods of bacterial isolation and detection may underestimate the true incidence of contamination (71, 152).

Although some domesticated farm animals, notably sheep and cattle, may suffer symptoms as a result of infection with *Y enterocolitica* (206, 207), in most cases the biotypes and serotypes of these bacteria differ from those responsible for human infection, indicating a lack of transmission of these particular bacteria between animals and humans. In contrast, individual isolates of *Y. enterocolitica* from pigs and humans are indistinguishable from each other in terms of serotype, biotype, restriction fragment length polymorphisms of chromosomal and plasmid DNA, and carriage of virulence determinants (117). Further evidence that pigs are a significant reservoir of human infections is provided by epidemiological studies pointing to the ingestion of raw or undercooked pork as a major risk factor for the acquisition of yersiniosis (157, 220). Infection also occurs after handling of contaminated chitterlings (pig intestine), particularly by children (116, 129).

Food animals are seldom infected with biotype 1B strains of *Y. enterocolitica*, whose reservoir remains unknown (193) but appears likely to be wild rodents (93). The relatively low incidence of human yersiniosis caused by these strains, despite their comparatively high virulence, points to a lack of significant contact between their reservoir and humans.

FOODBORNE OUTBREAKS

Considering the widespread occurrence of *Y. enterocolitica* in nature and its ability to colonize food animals, to persist within animals and the environment, and to proliferate at refrigeration temperatures, outbreaks of yersiniosis are surprisingly uncommon. Most foodborne outbreaks for which a source has been identified have been traced to milk (Table 14.5). Since *Y. enterocolitica* is killed by proper pasteurization, infection results from the consumption of raw milk or milk that is contaminated after pasteurization (59, 192). During the mid-1970s, two outbreaks of yersiniosis caused by *Y. enterocolitica* O:5,27 occurred amongst 138 Canadian schoolchildren who had consumed raw milk, but the organism was not recovered from the suspected source (119). In 1976, serotype O:8, biotype 1B *Y enterocolitica* was responsible for an outbreak in New York State which affected 217 people, 38 of whom were culture positive (18). The source of infection was chocolate-flavored milk, which evidently became contaminated after pasteurization.

In 1981, an outbreak of infection with *Y. enterocolitica* O:8 affected 35% of 455 individuals at a diet camp in New York State (150). Seven patients were hospitalized as a result of infection, five of whom underwent appendectomy. The source of the infection was reconstituted powdered milk and/or chow mein, which probably became contaminated during preparation by an infected food handler. In 1982, 172 cases of infection with *Y. enterocolitica* O:13a,13b occurred in an area which included parts of Tennessee, Arkansas, and Mississippi (216). The suspected source was pasteurized milk which may have become contaminated with pig manure during transport (59).

More recently, three unrelated outbreaks of infection with *Y. enterocolitica* O:3 affecting infants and children in Atlanta, Chicago, and Tennessee between 1989 and 2002 were attributed to the transmission of bacteria from raw chitterlings to affected children via the hands of food handlers (33, 116, 129). Other foods which have been responsible for outbreaks of yersiniosis include "pork cheese" (a type of sausage prepared from chitterlings), bean sprouts, and tofu (47, 137, 214). In the outbreaks associated with bean sprouts and tofu, contaminated well or spring water was the probable source of the bacteria. Water was also the putative source of infection in a case of sporadic *Y. enterocolitica* bacteremia in a 75-year-old man in New York State (120) and in a small family outbreak in Ontario, Canada (223). Several outbreaks of presumed foodborne infection with *Y. enterocolitica* O:3 have also been reported from the United Kingdom and Japan, but in most cases the sources of these outbreaks were not identified.

MECHANISMS OF PATHOGENICITY

Yersinia enterocolitica is an invasive enteric pathogen whose virulence determinants have been the subject of intensive investigation (41, 92, 236), but not all strains of *Y. enterocolitica* are equally virulent (Table 14.6). *Y. enterocolitica* strains of biotypes 1B, 2, 3, 4, and 5 possess a panoply of virulence determinants, including a chromosomally encoded invasin and a ca. 70-kb virulence plasmid, termed pYV (plasmid for *Yersinia* virulence) (41, 167, 236). In addition, biotype 1B strains of *Y. enterocolitica* carry two pathogenicity islands, which are associated with enhanced virulence. All pYV-bearing

Table 14.5 Selected foodborne outbreaks of infections with *Y. enterocolitica*

Location	Yr	Mo	No. of cases	Serotype	Source[a]	Reference
Canada	1976	April	138	O:5,27	Raw milk (?)	119
New York	1976	September	38	O:8	Chocolate-flavored milk	18
Japan	1980	April	1,051	O:3	Milk	140
New York	1981	July	159	O:8	Powdered milk/chow mein	198
Washington	1981	December	50	O:8	Tofu/spring water	216
Pennsylvania	1982	February	16	O:8	Bean sprouts/well water	47
Southern United States	1982	June	172	O:13a,13b	Milk (?)	218
Hungary	1983	December	8	O:3	Pork "cheese" (sausage)	137
Georgia	1989	November	15	O:3	Pork chitterlings	130
Northeastern United States	1995	October	10	O:8	Pasteurized milk (?)	3

[a] (?), the bacteria were not isolated from the incriminated source.

Table 14.6 Characteristics of pathogenic subgroups of *Y. enterocolitica*

Subgroup	Capacity to invade epithelial cells in vitro	Biotype(s)	Virulence-associated determinants
Classical	High (lower if pYV present)	2, 3, 4, 5	Invasin, Ail, fibrillae (Myf), urease, heat-stable enterotoxin (Yst), LPS, and a virulence plasmid (pYV) encoding YadA, a type III secretory pathway, and translocator and effector proteins
	High (lower if pYV present)	1B	All of the above, together with the HPI, the YSA pathogenicity island, a type II secretory pathway (Yts), and phospholipase A (YplA)
Atypical	Low to moderate	1A	Products of genes homologous to those for insecticidal toxin complex

clones of *Y. enterocolitica* have the capacity to invade epithelial cells in large numbers in vitro, which distinguishes them from clones that never carry pYV (79, 175). Paradoxically, however, this highly invasive phenotype is not specified by genes within pYV and is maximally expressed by bacteria cured of pYV.

Until recently, weakly invasive, pYV-negative strains of *Y. enterocolitica*, most of which belong to biotype 1A, were regarded as avirulent because they do not carry pYV or any of the other well-characterized virulence-associated genes of this species (see below). There is now persuasive clinical, epidemiological, and experimental evidence, however, which shows that at least some of these strains may cause gastrointestinal symptoms clinically indistinguishable from those due to pYV-bearing strains (221). This is supported by the laboratory demonstration that biotype 1A strains fall into two categories, i.e., isolates recovered from symptomatic patients can penetrate epithelial cells in moderate numbers and resist killing by macrophages to a greater extent than biotype 1A strains obtained from nonclinical sources (78, 79, 200). Clinical and environmental biotype 1A strains of the same O serogroup also differ from each other genetically (185). A recent report indicated that the products of a gene complex resembling the insecticidal toxin complex genes first discovered in *Photorhabdus luminescens* may contribute to the virulence of some strains of biotype 1A by facilitating their persistence in vivo (222). Because the mechanisms by which biotype 1A strains cause disease are largely unknown, the remainder of this section focuses chiefly on the virulence determinants of the classical pathogenic, i.e., pYV-bearing, highly invasive strains of *Y. enterocolitica*.

Pathological Changes

Examination of surgical specimens from patients with yersiniosis reveals that *Y. enterocolitica* is an invasive pathogen which displays a tropism for lymphoid tissue (28, 39). Ulceration and necrosis of the intestinal epithelium are usual, with the distal ileum, in particular the epithelium overlying the intestinal lymphoid follicles (Peyer's patches), bearing the brunt of the infection. The mesenteric lymph nodes are frequently enlarged and contain focal areas of necrosis. Because investigations in volunteers are precluded by the risk of autoimmune sequelae, most information regarding the pathogenesis of yersiniosis in vivo has been obtained from animal models, in particular mice and rabbits (94, 134, 231). Although these animals are not the natural hosts of the serotypes of *Y. enterocolitica* that commonly infect humans, they have provided valuable insights into the probable pathogenesis of human disease. Nevertheless, some data derived from animal studies should be interpreted with caution, particularly where death is used as the end point of infection, since this is not the usual outcome of human infection (162).

After oral inoculation of mice with a virulent strain of *Y. enterocolitica*, most bacteria remain within the intestinal lumen, while a small number adhere to the mucosal epithelium, showing no particular preference for any cell type (83). However, invasion of the epithelium takes place almost exclusively through M (microfold) cells (Fig. 14.1) (83). M cells are specialized epithelial cells that overlie Peyer's patches, where they play a major role in antigen sampling (115, 125, 188). Studies with experimentally infected rabbits and pigs have shown that after penetrating the epithelium, *Y. enterocolitica* traverses the basement membrane to reach the gut-associated lymphoid tissue and the lamina propria, where it causes localized tissue destruction leading to the formation of microabscesses (Fig. 14.2) (134, 181). These lesions occur chiefly within intestinal crypts but may extend as far as the crypt-villus junction. *Y. enterocolitica* often

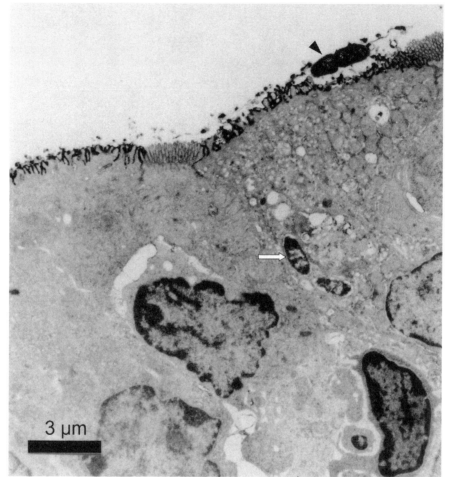

Figure 14.1 Transmission electron micrograph showing the initial interaction (arrowhead) and transport (arrow) of *Y. enterocolitica* through an intestinal M cell 60 min after inoculation into mouse ileum. (Reprinted with permission from reference 83.)

spreads via the lymph to the draining mesenteric lymph nodes, where it may also lead to microabscess formation. If the bacteria circumvent the lymph nodes to enter the bloodstream, they can disseminate to any organ, but they continue to show a tropism for lymphoid tissue by preferentially localizing in the reticuloendothelial tissues of the liver and spleen. Although *Y. enterocolitica* is often regarded as a facultative intracellular pathogen because of its innate resistance to killing by macrophages (54, 167), most of the bacteria observed in histological sections are located extracellularly (90). Nevertheless, macrophages may provide a niche for bacterial replication during the early stages of infection and serve as a vehicle for bacterial dissemination throughout the body (167). In mice, highly virulent (biotype 1B) strains of *Y. enterocolitica* can bypass the Peyer's patches and disseminate directly via the bloodstream (31).

Virulence Determinants

Chromosomal Determinants of Virulence

Invasin

All strains of *Y. enterocolitica* which carry pYV also produce a 91-kDa surface-expressed protein termed invasin. This outer membrane protein was first identified in *Y. pseudotuberculosis* as a 102-kDa protein product of the chromosomal *inv* gene (109). When introduced into an innocuous laboratory strain of *E. coli*, such as *E. coli* K-12, *inv* imbues the recipient with the ability to penetrate mammalian cells, including epithelial cells and macrophages (54, 109). Despite the difference in size of invasins from *Y. enterocolitica* and *Y. pseudotuberculosis*, the two proteins are functionally highly conserved.

The amino terminus of invasin is inserted in the bacterial outer membrane, while the carboxyl terminus is

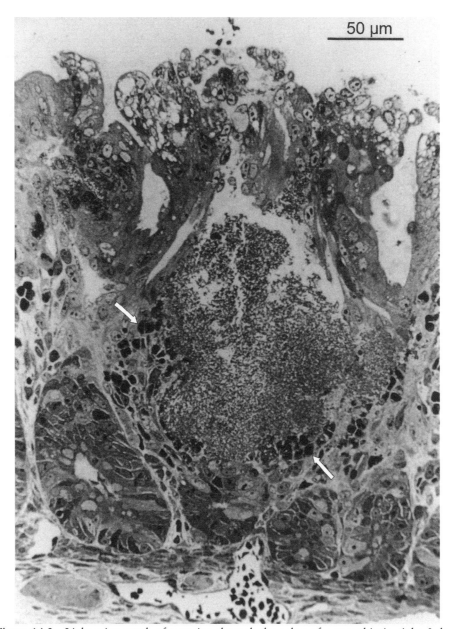

Figure 14.2 Light micrograph of a section through the colon of a gnotobiotic piglet 3 days after inoculation with a virulent strain of *Y. enterocolitica* O:3. Note the microabscess, comprising mostly bacteria, the surrounding inflammatory cells (arrows), and the disrupted epithelium with vacuolated and necrotic cells. An epoxy section stained with methylene blue is shown. (Reprinted with permission from reference 230.)

exposed on the surface, where it mediates binding to host cell integrins (187). The latter are heterodimeric transmembrane proteins which communicate extracellular signals to the cytoskeleton. Integrins comprise α and β subunits, which form the basis of their classification into families. The β_1 class integrins, which are the principal receptors for invasin, occur on many cell types, including epithelial cells, macrophages, T lymphocytes, and

Peyer's patch M cells (37). Their physiological role is to act as receptors for fibronectin, laminin, and related host proteins, which may bear conformational similarities to invasin (87). However, the affinity of invasin for integrins $\alpha_3\beta_1$, $\alpha_4\beta_1$, $\alpha_5\beta_1$, and $\alpha_6\beta_1$ is much greater than that of fibronectin. Accordingly, when invasin binds to these integrins, it causes them to cluster and initiate a sequence of events, including the activation of focal adhesion kinase (Fak) and the small GTPase Rac-1, which results in reorganization of the host cell cytoskeleton and internalization of the bacteria (108, 246). Inhibitors of actin polymerization and tyrosine kinases block invasin-mediated invasion (65, 184), indicating that the uptake of *Y. enterocolitica* by eukaryotic cells requires both an intact cytoskeleton and signal transduction pathways involving tyrosine phosphorylation. The internalization process is governed entirely by the host cell, because nonviable bacteria and even latex particles coated with invasin are internalized in the same way as living bacteria (110).

Although DNA sequences homologous to *inv* occur in all *Yersinia* species (except *Y. ruckeri*), this gene is functional only in *Y. pseudotuberculosis* and the classical pathogenic biotypes (1B through 5) of *Y. enterocolitica*, suggesting that invasin plays a key role in virulence (146). Nevertheless, although *inv* mutants of *Y. enterocolitica* show a pronounced reduction in the ability to invade epithelial cells in vitro, their virulence for orally inoculated mice is barely affected (162). There is evidence, however, that invasin contributes to gastrointestinal tract colonization of mice, as *inv* mutants of *Y. enterocolitica* show diminished translocation into Peyer's patches compared with that of wild-type strains (139, 162). The fact that *inv* mutants show only marginally reduced virulence for mice suggests that M cells may be able to internalize the bacteria to some extent in the absence of a specific stimulus or that *Y. enterocolitica* expresses supplementary factors, such as YadA (see below), which can compensate for the absence of invasin.

Ail

Classical pathogenic strains of *Y. enterocolitica* produce an outer membrane protein, unrelated to invasin, which also confers invasive ability on *E. coli* (144). This 17-kDa peptide is specified by a chromosomal *ail* (attachment-invasion) locus, so called because it mediates bacterial attachment to some cultured epithelial cell lines and invasion of others. In concert with YadA, a pYV-encoded protein, Ail may also allow yersiniae to persist extracellularly by protecting them from being killed by serum (17, 145). The *ail* gene occurs only in the classical pathogenic varieties of *Y. enterocolitica* (175) and is expressed optimally in vitro at 37°C (unlike *inv*, which is expressed optimally at 25°C, unless the pH is lowered to 5.5 [161]). Further circumstantial evidence supporting a role for Ail in virulence stems from the finding that *ail* mutants fail to adhere to or invade cultured cells when the *inv* gene is not expressed (107). Surprisingly, however, an *ail* mutant of *Y. enterocolitica* showed no reduction in virulence for perorally inoculated mice, indicating that Ail is not required to establish infection or even to cause systemic infection in these animals (241). However, since *Y. enterocolitica*

is inherently resistant to complement-mediated killing by murine serum (in contrast to its pronounced susceptibility to killing by human serum [158]), the mouse model may not be well suited for investigating the contribution of anticomplement factors to virulence.

Heat-Stable Enterotoxins

When first isolated from clinical material, most strains of *Y. enterocolitica* secrete a heat-stable enterotoxin, known as Yst (or Yst-a), which is reactive in infant mice (159). Yst is made up of 30 amino acids (Fig. 14.3). Its carboxyl terminus is homologous to those of heat-stable enterotoxins from enterotoxigenic *E. coli*, *Citrobacter freundii*, and non-O1 serotypes of *Vibrio cholerae* and also to that of guanylin, an intestinal paracrine hormone (Fig. 14.3) (7, 50, 84, 102, 171, 217–219). These polypeptides also share a common mechanism of action which involves binding to and activation of cell-associated guanylate cyclase, with subsequent elevation of intracellular concentrations of cyclic GMP (179). This in turn causes perturbation of fluid and electrolyte transport in intestinal absorptive cells, which may result in diarrhea.

Despite its similarity to known virulence factors and the fact that the production of Yst is by and large restricted to the classical pathogenic biotypes of *Y. enterocolitica* (57), the contribution of Yst to the pathogenesis of diarrhea associated with yersiniosis is uncertain. Doubts regarding the role of Yst in virulence stem from the following observations: (i) the toxin is generally not detectable in bacterial cultures incubated at temperatures above 30°C, (ii) the production of Yst has not been demonstrated in vivo, and (iii) strains of *Y. enterocolitica* that have spontaneously lost the ability to produce Yst retain full virulence for experimental animals (181, 190). On the other hand, Delor and Cornelis showed that a *yst* mutant of a serotype O:9 strain of *Y. enterocolitica* caused milder diarrhea in infant rabbits than did the wild-type strain (56). This observation suggests that the reason why *yst* (and *inv*) is not normally expressed at 37°C is that the conditions used to study the expression of this gene in vitro do not reflect those to which the bacteria are exposed in vivo. In this regard, the finding that *Y. enterocolitica* can produce Yst at 37°C if the bacterium is grown in media with an osmolarity and pH resembling those in the intestinal lumen is significant, although the precise stimulus of Yst synthesis in vivo is not yet known (143).

After repeated passage or prolonged storage, Yst-secreting strains of *Y. enterocolitica* frequently become toxin negative. This phenomenon is not caused by mutation of the *yst* gene but is due to silencing of this gene

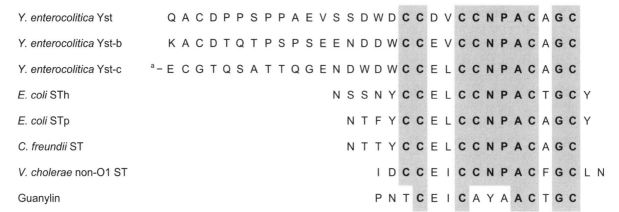

Figure 14.3 Amino acid sequences of the mature heat-stable enterotoxins produced by *Y. enterocolitica* (102, 173, 221), enterotoxigenic *E. coli* (human [STh] and porcine [STp] subtypes) (7, 219), *C. freundii* (84), and non-O1 *V. cholerae* (220) and of the intestinal hormone guanylin (50). Amino acid residues which are shaded are common to all seven peptides. The first 23 amino acids at the N terminus of the Yst-c mature toxin (denoted by a superscript "a") are not included in the sequence alignment.

by YmoA (*Yersinia* modulator). The latter is an 8-kDa, Hha-like protein which downregulates gene expression in yersiniae by binding to DNA and altering its topology (45). The expression of *yst* is also regulated by RpoS, an alternative sigma factor of RNA polymerase, which is involved in regulating the expression of a number of stationary-phase genes in other enterobacteria (104).

Toxins that resemble Yst in terms of heat stability and reactivity in infant mice, but with a different structure, molecular weight, and/or mechanism of action, have been detected in various *Yersinia* species, including biotype 1A strains of *Y. enterocolitica* and "avirulent" *Yersinia* species, such as *Y. bercovieri* and *Y. mollareti* (171, 180, 213, 252). Some of these toxins have been characterized, and the genes encoding their production have been cloned and sequenced (Fig. 14.3). The availability of diagnostic probes for the genes encoding Yst-b and Yst-c has indicated that biotype 1A *Y. enterocolitica* strains commonly carry the genes for Yst-b, whereas Yst-c, which was originally identified in a serotype O:3 strain of biotype 4, is rare (79, 171, 252).

Some strains of *Y. enterocolitica* may elaborate Yst or other enterotoxins over a wide range of temperatures, from 4°C to 37°C (118, 180). Since these toxins are relatively acid stable, they could resist inactivation by stomach acid, and thus conceivably cause food poisoning, if they were ingested preformed in food. In artificially inoculated foods, however, these toxins are synthesized optimally at 25°C during the stationary phase of bacterial growth (191). Accordingly, the storage conditions required for their production in food would generally result in spoilage,

making the possibility of ingestion of preformed toxin unlikely.

Myf

Most intestinal pathogens possess distinctive colonization factors on their surfaces which mediate their adherence to the intestinal epithelium. In noninvasive, enterotoxin-secreting bacteria, such as enterotoxigenic *E. coli*, these factors frequently take the form of surface fimbriae, which allow the bacteria to deliver their toxins close to epithelial cells whilst resisting removal by peristalsis (73). In enteroinvasive bacteria, surface adhesins may augment virulence by allowing the bacteria to home in on cells, such as M cells, which they preferentially invade. The key intestinal colonization factors of *Y. enterocolitica* are invasin and YadA (see below), which mediate binding to M cells. In addition, some strains of *Y. enterocolitica* produce a fimbrial adhesin, named Myf (for mucoid *Yersinia* fibrillae) because it bestows a mucoid appearance on bacterial colonies which express it (106). Myf proteins are narrow, flexible fimbriae which resemble CS3, an essential colonization factor of some human strains of enterotoxigenic *E. coli*. MyfA, the major structural subunit of Myf, shows some homology to the PapG protein of pyelonephritis-associated strains of *E. coli* and is 44% identical at the DNA level to the pH6 antigen of *Y. pseudotuberculosis* and *Y. pestis*, which also has a fibrillar structure and mediates thermoinducible binding of *Y. pseudotuberculosis* to tissue culture cells (135, 250). Like *ail* and *yst*, *myf* occurs predominantly in *Y. enterocolitica* strains of the classical pathogenic varieties commonly associated with disease (106). The pH6 antigen of *Y. pseudotuberculosis* is synthesized within the acidic phagolysosomes of macrophages and may play a role in the interaction between bacteria and phagocytic cells, although it does not appear to contribute to bacterial survival in these cells. Its main role in virulence may relate to its ability to mediate binding of bacteria to intestinal mucus before the bacteria make contact with epithelial cells (139). Although Myf may contribute to the colonizing ability of *Y. enterocolitica*, direct proof of this possibility is lacking.

LPS

As with other enterobacteria, *Y. enterocolitica* can be classified as smooth or rough depending on the amount of O side chain polysaccharide attached to the inner core region of the cell wall LPS. Synthesis of the O side chain by *Y. enterocolitica* is regulated by temperature, such that colonies are smooth when grown at temperatures below 30°C but rough at 37°C (204). O antigen mutants of a *Y. enterocolitica* serotype O:8 strain showed an impaired ability to colonize Peyer's patches, spleens, and livers of mice infected via different routes (14). These mutants also showed altered expression or function of other virulence-associated determinants, including YadA, Ail, phospholipase A (YplA), and flagellin, indicating that LPS may be required for the normal expression or function of other surface molecules or secretory systems. In addition, the outer core region of LPS plays a role in maintaining outer membrane integrity and may contribute to the resistance of *Y. enterocolitica* to bactericidal peptides in host tissues and macrophages (205). Smooth LPS may also enhance virulence by increasing bacterial hydrophilicity and thus facilitate bacterial passage through the mucous secretions which line the intestinal epithelium.

Flagella

When grown in vitro, *Y. enterocolitica* is motile at 25°C but not at 37°C. The expression of the genes encoding flagellin and invasin appears to be coordinately regulated in that mutants of *Y. enterocolitica* which are defective in the expression of invasin are hypermotile (11). In addition, a strain bearing a mutation in the master regulatory operon for flagellum biosynthesis, *flhDC*, showed enhanced production of pYV-encoded virulence proteins. Nevertheless, motility does not appear to contribute directly to the virulence of *Y. enterocolitica* for mice (105).

Phospholipase

Some isolates of *Y. enterocolitica* are hemolytic due to the production of phospholipase A (YplA). A strain of *Y. enterocolitica* in which the *yplA* gene encoding this enzyme was deleted showed diminished virulence for perorally inoculated mice (195). Interestingly, this mutant induced less inflammation and necrosis in intestinal and lymphoid tissues than did the wild type, suggesting that phospholipase contributes to microabscess formation by *Y. enterocolitica*. YplA is secreted by *Y. enterocolitica* via the same type III export apparatus that is used for flagellar proteins (242).

Iron Acquisition and the High-Pathogenicity Island (HPI)

Iron is an essential micronutrient of almost all bacteria. Despite the nutrient-rich environment provided to bacteria by mammalian tissues, the availability of iron in many extracellular locations is limited (247). This is because most extracellular iron is bound to high-affinity transport glycoproteins, such as transferrin and lactoferrin, or is incorporated into organic molecules. Several species of pathogenic bacteria produce low-molecular-weight, high-affinity iron chelators known as siderophores (151). These compounds are secreted by the bacteria into the surrounding medium, where they form a complex with ferric iron. The resultant ferri-siderophore complex then binds to specific receptors on the bacterial surface to be taken up by the cell. The observation that patients suffering from iron overload show increased susceptibility to severe infections with *Y. enterocolitica* suggested that the availability of iron in tissues may determine the outcome of yersiniosis (178).

Investigation of the relationship of yersiniae to iron has revealed that these bacteria employ a wide array of processes to acquire iron from inorganic and organic sources (164, 165). The fact that most clinical isolates of *Y. enterocolitica* do not produce siderophores accounts for their reliance on abnormally high concentrations of iron for growth in animals. Interestingly, however, the highly virulent biotype 1B strains, as well as *Y. pestis* and *Y. pseudotuberculosis*, carry genes for the biosynthesis, transport, and regulation of a 482-Da, catechol-containing siderophore known as yersiniabactin. The ca. 40-kbp *ybt* locus which contains these genes has a higher G+C content (57.5 mol%) than that of the *Y. enterocolitica* chromosome (47 mol%), is flanked on one side by an *asn* tRNA gene, and carries the gene for a putative integrase (30). These features, which are typical of a pathogenicity island, have led to the *ybt* locus also being known as the *Yersinia* HPI. The designation "high" alludes to the observation that bacteria which carry this locus are more virulent for mice infected perorally (median lethal dose, $<10^3$ CFU) than are strains which lack it (median lethal dose, typically $>10^6$ CFU).

Interestingly, the *Yersinia* HPI is also found in a variety of other enterobacteria, such as *E. coli*, *Klebsiella*, *Citrobacter*, and *Salmonella* spp., indicating the possible lateral transfer of these virulence genes between different species of pathogens (30, 169).

The HPI of serotype O:8 *Y. enterocolitica* contains 22 open reading frames within 43.4 kb, approximately 30.5 kb of which are conserved between *Y. enterocolitica*, *Y. pestis*, and *Y. pseudotuberculosis* (Fig. 14.4) (169). Synthesis of yersiniabactin requires *irp* (iron-regulated protein) genes *irp-1* to *irp-5*, as well as *ybtA*, which encodes an AraC-like regulator, whilst *irp-6* and *irp-7* are required for utilization of the ferri-yersiniabactin complex (26). The major receptor for this complex is a 65-kDa outer membrane protein named FyuA, which also serves as a receptor for pesticin, a bacteriocin produced by *Y. pestis* (95). Transport of ferri-yersiniabactin complexes across the cell wall of *Y. enterocolitica* resembles the analogous pathway in *E. coli* in that it is an energy-dependent process which requires TonB. The latter protein couples energy provided by inner membrane metabolism to outer membrane protein receptors, such as FyuA. TonB-, FyuA-, and yersiniabactin-deficient mutants of *Y. enterocolitica* all show reduced virulence for mice, presumably because of their limited capacity to acquire sufficient iron to grow in tissues (12). The fact that *aroA* is also required for yersiniabactin synthesis may partly explain the attenuation of *aroA* mutants of *Y. enterocolitica* for virulence in mice (23).

Although biotypes of *Y. enterocolitica* other than 1B do not produce yersiniabactin, they are able to acquire iron from a number of sources, including ferri-siderophore complexes in which the siderophore, such as desferrioxamine B, was synthesized by another microorganism (12, 177). The iron-desferrioxamine complex (known as ferrioxamine) binds to FoxA, a 76-kDa outer membrane protein, which shares 33% amino acid homology with FhuA, the high-affinity ferrichrome receptor of *E. coli* (13). The ability of *Y. enterocolitica* to acquire iron from ferrioxamine may have important clinical implications because desferrioxamine B is used therapeutically to reduce iron overload in patients with hemosiderosis and other forms of iron intoxication. When administered to patients, desferrioxamine B forms a ferri-siderophore complex, which *Y. enterocolitica* can utilize as a growth factor (176). Accordingly, if patients undergoing iron chelation therapy with desferrioxamine B become infected with *Y. enterocolitica*, the bacterium may be able to proliferate in tissues where, under normal circumstances, the poor availability of iron would limit its growth. Apart from its effects on microbial iron metabolism, desferrioxamine B may also increase susceptibility to systemic yersiniosis by interfering with host immune responses (10).

YSA Pathogenicity Island

Biotype 1B strains of *Y. enterocolitica* possess another pathogenicity island, comprising approximately 30 open reading frames, which encodes a type III secretion system (TTSS) known as the *Yersinia* secretion apparatus (YSA). This apparatus is distinct from the well-characterized pYV-encoded Ysc TTSS (see below). The YSA TTSS is required for the export of a set of at least 11 proteins called *Yersinia* secreted proteins (Ysps). These include YspA, -B, -C, and -D as well as YopE, YopN, and YopP, which are also exported by the Ysc TTSS (255). Although the contribution of Ysps to virulence is not known, YSA-defective mutants show a reduced ability to colonize the mouse intestine (235).

Biotype 1B strains also possess the genes for a type II protein secretion pathway which contributes to virulence but is absent from less virulent strains of this species (112). The protein(s) secreted by this apparatus is not known.

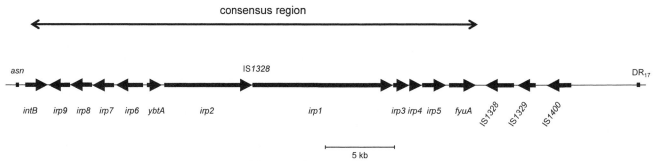

Figure 14.4 Representation of the HPI of *Y. enterocolitica* O:8 strain WA-C. Arrows indicate the positions of the open reading frames and the direction of transcription. The region that is conserved in *Y. pestis* and *Y. pseudotuberculosis* is indicated by a double-headed arrow. (Adapted from reference 169.)

Urease

All enteric pathogens must negotiate the acid barrier of the stomach to cause disease. In *Y. enterocolitica*, acid tolerance relies on the production of urease, which catalyzes the release of ammonia from urea and allows the bacterium to resist pHs as low as 2.5 (55, 256). Urease also contributes to the survival of *Y. enterocolitica* in host tissues, but the mechanism by which this occurs is not known (81).

The Virulence Plasmid (pYV)

All fully virulent strains of *Y. enterocolitica* biotypes 1B to 5 carry pYV, a ca. 70-kb plasmid which is highly conserved amongst *Y. enterocolitica*, *Y. pestis*, and *Y. pseudotuberculosis* (for reviews, see references 44 and 42). pYV specifies functions which interfere with innate immune responses, such as phagocytosis, complement activation, and the production of proinflammatory cytokines, thus allowing yersiniae to proliferate extracellularly in tissues (Table 14.7). In addition, some factors encoded by pYV may act on T and B cells directly to modify adaptive immune responses.

Yersiniae which carry pYV exhibit a distinctive phenotype, known as "calcium dependency" or "the low calcium response" because it manifests when the bacteria are grown in media containing low concentrations of Ca^{2+}. The principal features of this response are the cessation of bacterial growth after one or two generations and the appearance of at least 12 new proteins (Yops) on the bacterial surface or in the culture medium. Yops are characterized by their common mode of secretion and their regulation by a pYV-encoded regulator known as VirF. Yops are so named because they were once thought to be outer membrane proteins, but they are now known to be secreted by the bacteria via a type III secretory pathway encoded by pYV. The expression of pYV-encoded proteins in vitro imbues *Y. enterocolitica* with novel properties, such as autoagglutination, resistance to killing by normal human serum, and an ability to bind Congo red and crystal violet (22). Some of these characteristics have been exploited in the design of culture media, such as calcium-depleted agar containing Congo red (16, 62), to facilitate the isolation and identification of pYV-bearing yersiniae.

Several pYV plasmids have been sequenced fully (for example, see reference 209). The genes carried on these plasmids include those for (i) an outer membrane protein adhesin, YadA; (ii) a type III secretory apparatus (Ysc) whereby Yops are transported across the *Yersinia* cell wall; (iii) at least six distinct antihost effector Yops; (iv) a translocation apparatus comprising certain Yops which the effector Yops require to gain access to the host cell cytosol; and (v) factors for the regulation of Yop biosynthesis, secretion, and translocation (Table 14.7). Genes for the effector Yops are scattered around pYV, while those required for Yop secretion and translocation are clustered together (Fig. 14.5). Although Yops are highly conserved between *Yersinia* species, there is little

Table 14.7 Major pYV-encoded determinants of *Y. enterocolitica* and their roles in virulence

Determinant(s)	Contribution to virulence
YadA	Bacterial adhesion and invasion; reduction of opsonization by interfering with binding of complement proteins
Ysc complex (product of *virC*, *virG*, *virA*, and *virB*)	Type III secretory apparatus
Yops	
Effector Yops	
YopH	Protein tyrosine phosphatase which interferes with phagocytosis and other immunological responses by dephosphorylating focal adhesion kinase, etc.
YopE, YopT, YopO/YpkA[a]	Interfere with phagocytosis by disrupting Rho GTPases and the host cell cytoskeleton
YopP/YopJ[a]	Reduces inflammation by suppressing signaling via MAP kinase and NF-κB
YopM	Suppression of immune responses?
YopB, YopD, LcrV	Yop translocation; antihost effects; regulation of *yop* gene expression
Syc proteins SycE, -H, -D, -N, -T	Yop chaperones
VirF	Regulation of expression of *yop*, *virC*, and *yadA* genes

[a] Designation in *Y. pseudotuberculosis*.

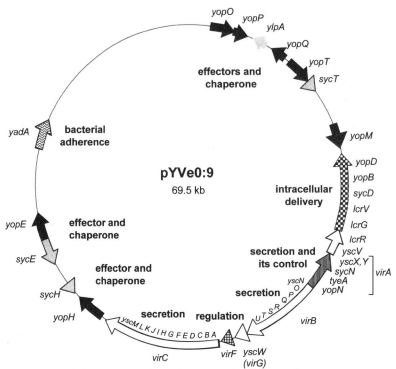

Figure 14.5 Map of the virulence plasmid, pYVe, of *Y. enterocolitica* serogroup O:9 showing the locations and directions of transcription (arrows) of the genes encoding (i) YadA; (ii) YlpA; (iii) YopB, -D, -E, -H, -M, -N, -O, -P, -Q, and -T and LcrV; (iv) the specific Yop chaperones SycD, -E, -H, and -T; (v) the secretion elements VirA, -B, -C, and -G; and (vi) the regulatory element VirF. (Adapted from reference 103.)

homology between the individual Yops of a single species. pYV also encodes YlpA, a 29-kDa lipoprotein related to the TraT proteins encoded by plasmids in various enterobacteria, and in *Y. enterocolitica* strains of biotypes 2 to 5 (but not biotype 1B or *Y. pseudotuberculosis*), it carries an operon which specifies resistance to arsenic (154).

YadA, a pYV-Encoded Adhesin

YadA, formerly known as Yop1 or P1, is a 44- to 47-kDa outer membrane protein which belongs to the family of autotransporter proteins (96). These proteins have a characteristic arrangement of functional domains, including an N-terminal signal sequence, an internal passenger domain, and a C-terminal translocator domain. YadA represents a subfamily of autotransporter proteins, termed oligomeric coiled-coil adhesins or trimeric autotransporters because they possess a short trimeric translocator domain at their C termini (46, 96). Individual YadA monomers aggregate in solution to form oligomers with an apparent molecular mass of about 200 kDa. On the bacterial surface, however, YadA forms trimers which appear as lollipop-shaped structures that envelop the entire outer membrane as a densely packed array (5).

YadA mediates bacterial adhesion to intestinal mucus and to certain extracellular matrix proteins, including collagen, laminin, and cellular fibronectin (203). These proteins in turn may bind to β_1 integrins on epithelial cells and stimulate bacterial internalization by endocytosis in a manner similar to that mediated by invasin. Thus, YadA may contribute to bacterial invasion, even though binding to integrins via matrix proteins generally does not stimulate bacterial uptake to the same extent as that mediated by invasin (19). YadA may also promote bacterial invasion by binding to integrins directly. Both YadA and invasin induce cell signaling via β_1 integrin binding to kinases, such as small GTPases and mitogen-activated protein (MAP) kinases (including p38, MEK1, and Jun N-terminal protein kinase), leading to the production of interleukin-8 (IL-8) (194). The latter is a proinflammatory cytokine which promotes chemotaxis and activation of polymorphonuclear leukocytes (PMNs).

Apart from its role as an adhesin and invasin, YadA also contributes to virulence by conveying resistance to complement-mediated opsonization. It achieves this by binding factor H, thereby reducing the deposition of C3b on the bacterial surface (36). As a result, YadA is associated with resistance of *Y. enterocolitica* to complement-mediated lysis and phagocytosis and with an ability to inhibit the respiratory burst of PMNs (35, 36). Given the pluripotential capacity of YadA to increase the likelihood of bacterial survival in host tissues, it is not surprising that YadA mutants of *Y. enterocolitica* show markedly reduced virulence for mice (163, 182). This is in contrast to YadA mutants of *Y. pseudotuberculosis*, which show no attenuation, and to *Y. pestis*, which is extremely virulent for mice but is naturally defective in YadA production due to a single-base-pair deletion resulting in a shift of the reading frame of the gene (88).

Yop Secretion and the Ysc Secretion Apparatus

Secretion of Yops occurs by the type III secretory pathway, a system which is utilized by several species of pathogenic bacteria to transfer proteins directly from their own cytoplasm to that of a host cell via a cytoplasmic bridge. TTSSs bypass the Sec-dependent, or general secretory, pathway by integrating transport across the inner and outer membranes in a temporally linked fashion (reviewed by Ghosh in reference 75). In contrast to proteins exported via the general secretory pathway, the N termini of type III secreted proteins show no resemblance to each other and are not cleaved during export. Another feature of type III secretion is the presence of structurally conserved chaperone proteins, which bind to individual secreted proteins and guide them to the secretion apparatus while preventing their premature interaction with other proteins. Chaperone proteins for Yops are denoted by the prefix Syc (for specific Yop chaperone) and include SycE (the chaperone for YopE), SycH (for YopH), SycD (for YopB and YopD), SycN (for YopN), and SycT (for YopT). The genes encoding these chaperones are located on pYV close to the corresponding *yop* genes (Fig. 14.5). Although they share no significant homology with each other, all Yop chaperones identified to date are low-molecular-mass (14- to 19-kDa) proteins with a C-terminal amphipathic α-helix and a pI of about 4.5 (44).

The passage of proteins through the TTSS is tightly controlled to ensure that only selected proteins can exit the bacterial cell via this pathway. Nevertheless, the signal by which the TTSS recognizes proteins destined for export is not known. Evidence has been presented for two apparently contradictory scenarios, namely, (i) that the secretion signal is contained within the noncleaved N terminus of the protein and (ii) that it is embodied

within the 5′ end of the mRNA encoding the protein (170). Although the weight of current opinion favors the idea that the signal for initial secretion is in the protein, it is possible that secretion may be cotranslational during the later stages of infection (42). Whichever is correct, it is clear that the secretion signal is contained within the first 15 to 20 amino acids or codons of the secreted protein.

The Ysc secretion apparatus itself is a paradigm of all type III secretory machines (44). The 29 genes encoding this apparatus are contained within four contiguous loci, called *virC* (comprising *yscABCDEFGHIJKLM*), *virG* (which encodes YscW), *virA* (encoding YopN, TyeA, SycN, YscX, YscY, YscV, and YscR/LcrD), and *virB* (comprising *yscNOPQRSTU*) (Fig. 14.5) (103). Ten Ysc proteins (YscD, -J, -L, -N, -Q, -R, -S, -T, -U, and -V) have counterparts in almost every type III secretion apparatus, including the TTSS for flagella.

YscD, -R, -S, -T, -U, and -V appear to span the inner membrane (Fig. 14.6) (31). YscN has ATP-binding motifs (Walker boxes A and B) resembling the catalytic subunit

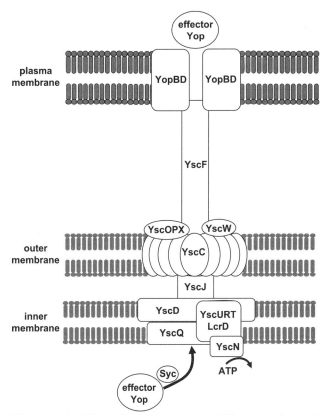

Figure 14.6 Schematic representation of Yop secretion and translocation by *Y. enterocolitica*. The major structural proteins of the secretory apparatus are shown in relation to their known or deduced locations in the cell wall. The effector Yop chaperone (Syc) and a translocation pore comprising YopB and YopD are also depicted.

of the F_0F_1 proton translocase and related ATPases and is predicted to energize the secretory apparatus (248). The results of yeast two- and three-hybrid experiments suggest that YscN may also form a complex with YscK, YscQ, and YscL (113). YscJ is a lipoprotein whose counterpart in enteropathogenic *E. coli*, EscJ, was recently shown to form a large ring structure with extensive grooves, ridges, and electrostatic features, suggesting that it functions as an essential platform for TTSS assembly (251). YscC also has a counterpart in most TTSSs, apart from that for flagella. It is an outer membrane protein of the secretin family and is envisaged to form a ring-like structure with an external diameter of 20 nm and an apparent central pore of around 5 nm (123). It is stabilized in the outer membrane by YscW, a lipoprotein product of the *virG* gene (123).

Several Ysc proteins, including YscO, -P, and -X, are secreted into the medium together with Yops. Because they are required for Yop secretion, however, they seem likely to constitute the most external part of the type III secretory apparatus (31). At least one Yop, YopN, appears to serve as a constituent protein of the type III secretory machine rather than as one of its substrates. Its role is evidently to plug the secretory channel together with TyeA, a surface protein which binds to YopN, to prevent the uncontrolled release of translocated Yops from the secretory channel (68).

Electron microscopic examination has shown that the distal portion of the Ysc TTSS comprises a "needle" formed by the polymerization of the 6-kDa YscF protein (100). Isolated needles have lengths of approximately 60 to 80 nm, widths of 6 to 7 nm, and a hollow center of around 2 nm. Apart from its role in virulence protein secretion, YscF may also contribute to the cell contact-dependent regulation of the TTSS (228).

Little is known about the actual mechanism of protein secretion via the TTSS, but it is envisaged that the Ysc apparatus serves as a hollow channel through which exported proteins traverse the inner membrane, peptidoglycan layer, and outer membrane in a single step. Whether proteins travel folded or unfolded is not known, but given the dimensions of the channel, it is likely that they travel at least partially unfolded (42).

Yop Translocation

The principal function of the Ysc TTSS is to inject (translocate) proteins into the cytosol of eukaryotic cells, where they act on specific targets (42, 236). Yops are classified into those which are translocated into host cells to exert an antihost action (effector Yops) and those which are primarily involved in the translocation process. There is evidence, however, that some Yops required for translocation may also serve as antihost effectors.

The transport of Yops from the bacterial cytoplasm via Ysc into the host cell cytosol via the translocation apparatus is envisaged to occur in one step from bacteria which are closely bound to the host cell (130). Two of the proteins required for translocation are YopB (42 kDa) and YopD (33 kDa), both of which have hydrophobic domains suggesting that they interact directly with host cell membranes (86, 153). YopB resembles members of the RTX toxin family and can form pores of approximately 2 nm in diameter in the plasma membranes of eukaryotic cells (153). YopD associates with YopB and appears to contribute to the formation of the pores (153). The observation that YopD is also translocated into cells, however, suggests that it may fulfill more than one function (69). YopB also has additional functions, including the stimulation of proinflammatory signaling in host cells (239).

YopB and YopD are encoded by the *lcrGV-sycD-yopBD* operon, which also encodes LcrV, LcrG, and SycD, the chaperone for YopB and YopD (Fig. 14.5). LcrG is needed for the efficient translocation of Yops, but its precise location is unknown. It also appears to act in concert with LcrV as a Yop regulator (131) and may assist the Syc/translocation channel to bind to a heparin-like receptor on host cells (24).

LcrV, also known as the V antigen, is a versatile Yop which is required for the formation of the translocation pore (42). It is also involved in regulating the expression of Ysc and Yops. Unlike other Yops, LcrV is secreted into the extracellular milieu in significant concentrations by *Y. pestis*, at least, both in vitro and in vivo (48). Some LcrV protein coats the bacterial surface, while free LcrV can exert profound immunological effects, such as inhibiting chemotaxis of PMNs and stimulating the release of the immunosuppressive cytokine IL-10 (48). The key role of LcrV in the virulence of *Y. pestis* is evidenced by the fact that antibodies to LcrV protect mice from infection with this species (48, 76, 234). The mechanism of action of these anti-LcrV antibodies is uncertain, but there is evidence that they opsonize the bacteria in preparation for phagocytosis by PMNs and that they interfere with formation of the translocation pore (48, 76). The relative contribution of LcrV to the virulence of *Y. enterocolitica* is less than that for *Y. pestis*.

Mechanism of Action of Effector Yops

The collective actions of the effector Yops are to inhibit bacterial uptake and killing by phagocytic cells, inhibit cytokine production, and induce apoptosis, thus allowing yersiniae to persist extracellularly in tissues. The effector Yops achieve these outcomes mostly by disrupting the proinflammatory signaling pathways that are activated in response to stimulation by invasin, YadA, and LPS. A detailed review of the mechanisms of action of individual Yops was published recently, and readers are referred to this source for more information (236).

Four Yops, YopH, YopE, YopT, and YopO, act synergistically to prevent phagocytosis (82). YopH (51 kDa) is a potent protein tyrosine phosphatase. In HeLa cells, YopH dephosphorylates focal adhesion kinase (Fak) and the focal adhesion proteins paxillin and p130^{cas} (166). Focal adhesions are sites where the integrin receptor acts as a transmembrane bridge between extracellular matrix proteins and intracellular signaling proteins. By targeting these proteins and others involved in integrin-mediated endocytosis, YopH interferes with the uptake of yersiniae by epithelial cells and professional phagocytes (238). YopH also affects other immune response pathways, for example, by preventing the production of macrophage chemoattractant protein 1, counteracting T- and B-cell activation by antigen, and reducing T-cell receptor-mediated production of IL-2 by T cells, amongst other activities (236).

YopE is a 25-kDa GTP-activating protein that targets RhoA, Rac-1, and Cdc42, all three of which are small GTPases that control specific cytoskeletal elements in host cells (238). YopE alters the conformations of these proteins to impair their signaling function. This leads to disruption of the actin cytoskeleton, which manifests as rounding and detachment of affected cells in culture. Through its action on Rho GTPases, YopE may also inhibit the production of proinflammatory cytokines by epithelial cells infected with yersiniae (237).

YopT also acts on Rho GTPases through its action as a cysteine protease which removes the lipid modification from RhoA, Rac-1, or Cdc42. Its effect on epithelial cells resembles that of YopE. Since YopT is not produced by serotype O:3 strains of *Y. enterocolitica*, however, and since a YopT mutant of a serotype O:8 strain of *Y. enterocolitica* was no less virulent than the wild-type strain, the contribution of YopT to virulence is uncertain (236).

YopO (an 82-kDa protein; known as YpkA in *Y. pseudotuberculosis*) is a serine/threonine kinase that binds to Rho GTPases. Like that of YopE and YopT, its action leads to the disruption of the actin cytoskeletons of cultured cells, but the way by which it achieves this is not known (238).

YopP (known as YopJ in *Y. pseudotuberculosis*) is a 32-kDa protein which is encoded by the same operon as YopO/YpkA. YopP/J is a cysteine protease which downregulates inflammatory responses by inhibiting signaling pathways that are controlled by MAP kinase and NF-κB. YopP/J binds directly to MAP kinase and prevents its activation (156). Its action on NF-κB is due to the ability of YopP/J to inhibit phosphorylation of IKKβ (156), preventing the degradation of IKKβ and retarding the translocation of NF-κB to the nucleus. The resultant inhibition of NF-κB, a transcription factor of central importance in activating inflammatory responses, has profound effects on cytokine production and induces macrophage apoptosis (255). The enzymic activity of YopP/J appears to involve the removal of ubiquitin or ubiquitin-like modifications from proteins, but its specific target(s) in cells has not been identified (156).

YopM does not exhibit any enzymic activity. This protein contains 12 to 20 repeats of a 19-residue leucine-rich repeat motif (133). After YopM is translocated into cells, it traffics to the nucleus by means of a vesicle-associated, microtubule-dependent pathway (201), but its intranuclear role, if any, has not been determined. YopM also forms a complex with two cytoplasmic kinases, RSK1 and PRK2, and activates them (141). Although the precise mechanism of action of YopM is not known, its contribution to virulence is evident from the fact that *yopM* mutants of yersinae are significantly attenuated. This may be due to the role that YopM plays in depleting NK cells and the proinflammatory cytokines IL-12, IL-18, and gamma interferon (121).

Regulation of Yop Production and Secretion

All Yops and YlpA are produced in vitro at 37°C (but not below 30°C) when the concentration of Ca^{2+} is sufficiently low to induce bacteriostasis. The mechanism of regulation by temperature involves VirF and DNA supercoiling (183). VirF is a pYV-encoded, DNA-binding protein of the AraC family of transcriptional regulators and is only active at 37°C (40). Other members of this family include VirF from *Shigella flexneri*, Rns from enterotoxigenic *E. coli*, and ExsA from *Pseudomonas aeruginosa*, all of which are involved in the regulation of expression of virulence in their respective bacteria. VirF plays a central role in the virulence of *Y. enterocolitica* by governing the transcriptional activation of *yadA*, *ylpA*, all of the *yop* genes, and the *virC* operon. Because these genes are coregulated, they have been named the Yop regulon (43). Transcription of *virF* itself is also regulated by temperature.

In vitro transcription of the *yop* genes, but not the *ysc* genes, *yadA*, or *virF*, is repressed by Ca^{2+} or by mutations in the Ysc secretion apparatus. Repression due to mutations in Ysc is due to feedback inhibition from the closed secretion apparatus (31). Repression by Ca^{2+} may occur in a similar manner, as supported by the observation that Ca^{2+} affects the interaction between YopN, TyeA, and LcrG, permitting them to bind together and plug the Ysc channel (64).

There is no doubt that Yops are produced in vivo because animals and humans infected with virulent *Yersinia* species develop antibodies to these proteins during the course of infection (Fig. 14.7). Although host temperature (37°C) is likely to be a key stimulus for the production of Yops in vivo, the signal equating to low Ca^{2+} is not known. Forsberg et al. (67) have shown that extracellular yersiniae produce Yops and are cytotoxic for HeLa cells even in the presence of millimolar concentrations of Ca^{2+}, provided that the bacteria can attach to the target cells. This observation indicates that in tissues, contact between *Y. enterocolitica* and host cells permits Yop release in the same way that Ca^{2+} depletion does in vitro.

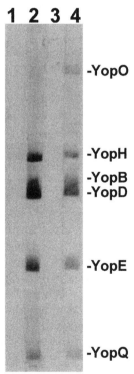

Figure 14.7 Antibody response of sheep infected with *Y. enterocolitica* or *Y. pseudotuberculosis* to Yops. Yops were prepared from *Y. enterocolitica* serogroup O:3, separated by polyacrylamide gel electrophoresis, transferred to a nitrocellulose membrane, and incubated with preimmune (lanes 1 and 3) or immune (lanes 2 and 4) sera from lambs with naturally acquired infection with pYV-bearing *Y. enterocolitica* (lanes 1 and 2) or *Y. pseudotuberculosis* (lanes 3 and 4). (Reprinted with permission from reference 174.)

Antihost Action of pYV

Strains of *Y. enterocolitica* which have been cured of pYV can colonize the intestinal tracts of experimental animals, penetrate Peyer's patches, and even travel to the mesenteric lymph nodes, indicating that pYV-encoded determinants are not required during the early stages of infection (167). After a time, however, pYV-negative strains are eliminated from the body, whereas pYV-bearing strains proliferate and spread to other tissues.

Macrophages and PMNs are major participants in the first line of defense against invading microbes by ingesting and destroying bacteria, producing and releasing proinflammatory cytokines, and priming T and B cells. The effects of pYV-encoded mediators on these cells account for the contribution of pYV to virulence. When *Y. enterocolitica* first penetrates the gut epithelium, before pYV-encoded proteins are expressed, the bacteria may be taken up by macrophages. They are able to survive within these cells, possibly due to the combined actions of OmpR and GrsA (167). OmpR is a regulator that mediates changes in gene expression in response to changes in osmolarity, whereas GrsA (also known as HtrA or DegP) is a heat shock-induced serine protease. Although Yops are not synthesized while the bacteria are located within host cells (114), yersiniae which survive their initial encounter with phagocytes and escape from these cells will express and translocate Yops once they reestablish contact with phagocytes. The translocated Yops then inhibit bacterial uptake, impair the respiratory burst, interfere with the production and release of several proinflammatory cytokines, and induce apoptosis. If pYV-bearing bacteria are preopsonized by antibodies, resistance to phagocytosis is mediated largely by YadA, whereas unopsonized bacteria resist ingestion by phagocytes due to the action of YopH, acting in concert with YopE, YopO, and YopT. YopH, -E, -O, and -T are also involved in inhibition of the respiratory burst, whereas inhibition of cytokine release and the induction of apoptosis are primarily due to YopP/YopJ.

pYV-bearing strains of *Y. enterocolitica* also resist ingestion and killing by PMNs, whereas plasmidless strains are killed (35). The principal virulence determinants involved in this process are YadA, which interferes with opsonization, and YopH and YopE, which inhibit the respiratory burst and retard phagocytosis (36). Yops also inhibit the production of receptor-dependent superoxide anions by granulocytes, but the mechanism by which this occurs is not known (240). Should pYV-bearing bacteria be ingested by PMNs, they can resist killing due to their relatively low susceptibility to the antimicrobial peptides produced by these cells. This resistance is attributable to YadA and smooth LPS (205).

Pathogenesis of *Yersinia*-Induced Autoimmunity

Arthritis

Following acute infection with pYV-bearing *Y. enterocolitica*, a proportion of patients develop autoimmune (reactive) arthritis. A similar syndrome may also occur after infections with *Campylobacter*, *Salmonella*, *Shigella*, or *Chlamydia* species (173, 233). The overwhelming majority of individuals with postinfective reactive arthritis caused by *Y. enterocolitica* carry the class I human leukocyte antigen HLA-B27 (198). In addition, HLA-B27-positive individuals have more severe arthritic symptoms and a more prolonged course than do individuals with different HLA antigens.

The pathogenesis of reactive arthritis is poorly understood (233). The synovial fluid from affected joints of patients with *Yersinia*-induced arthritis is culture negative but generally contains bacterial antigens, such as LPS and heat shock proteins, within inflammatory cells (74, 77). The presence of these antigens may reflect an impaired ability of HLA-B27-bearing phagocytes to eliminate them. LPS in synovial fluid may stimulate the local production of tumor necrosis factor alpha, which plays a key role in the pathogenesis of reactive arthritis (173).

Explanations for the link between yersiniosis and autoimmunity include antigen persistence, molecular mimicry, impaired immune responsiveness, and infection-induced presentation of normally cryptic cellular antigens (97, 225, 233). Although patients with reactive arthritis display higher levels of serum immunoglobulin A antibodies to *Yersinia* antigens than do individuals without arthritis, these antibodies are unlikely to contribute to the development of arthritis. Instead, they probably reflect enhanced stimulation of the mucosal immune system due to the persistence of bacterial antigens in the intestine or other tissues (53).

The observation that rats transgenic for HLA-B27 show a weaker cytotoxic T-cell response against *Y. pseudotuberculosis* than that of nontransgenic syngeneic rats provides a link between prolonged infection, antigen persistence, and HLA-B27 (61). Furthermore, HLA-B27-expressing monocytes show an impaired capacity to limit the intracellular replication of salmonellae (233).

Support for the molecular mimicry hypothesis comes from the demonstration of autoreactive T cells in the synovial tissues of patients with reactive arthritis (98). Among the bacterial antigens which may provoke autoreactivity are heat shock proteins, which are somewhat conserved among bacteria and mammals and share a number of antigenic determinants. Accordingly, an immune response to selected epitopes on bacterial heat shock proteins may lead to an autoimmune response at sites where bacterial antigens accumulate, including the joints of patients with reactive arthritis. In keeping with this suggestion is the observation that synovial fluid from patients with reactive arthritis may contain CD4+ major histocompatibility complex (MHC) class II-restricted T lymphocytes which recognize epitopes that are shared by a 60-kDa *Yersinia* heat shock protein and its human counterpart (98). Further evidence for the molecular mimicry hypothesis comes from the finding that an immunodominant epitope from GroEL, a heat shock protein of *Salmonella* serovar Typhimurium (which is almost identical to the homologous protein in *Y. enterocolitica*), can be presented by a class I major histocompatibility complex antigen in mice and can then be recognized by CD8+ cytotoxic T cells which cross-react with a peptide from a murine heat shock protein (136). Other bacterial antigens which have been claimed to contribute to autoimmunity via specific interactions with the immune system are YadA and the β-subunit of urease (80, 202). However, early suggestions that YadA shared epitopes with the peptide-binding groove of the HLA-B27 antigen have been discounted (126). The β-subunit of urease is a cationic protein which (i) is recognized by CD4+ T cells from patients with *Yersinia*-induced reactive arthritis and (ii) evokes arthritis when injected into the joints of rats (202). The finding that a urease-deficient mutant of *Y. enterocolitica* retained the capacity to induce arthritis in a rat model, however, casts doubt on the role of urease in this condition (81). There is also a suggestion that HLA-B27 itself may serve as an autoantigen in the pathogenesis of arthritis (173).

Y. enterocolitica and *Y. pseudotuberculosis* may also induce polyclonal T-cell stimulation by virtue of their ability to secrete toxins which resemble superantigens (1, 211). YpmA and its variant, YpmB, are 14- to 15-kDa proteins produced by approximately 20% of *Y. pseudotuberculosis* strains that activate human T cells of Vβ phenotypes 3, 9, 31.1, and 13.2 (32, 148). The *Y. enterocolitica* superantigen is not as well characterized as that of *Y. pseudotuberculosis*, but evidently it can stimulate T cells with Vβ phenotypes 3, 7, 8.1, 9, and 11 (211). Yersiniae may also provoke nonspecific immune stimulation when invasin binds to β₁ integrins on T lymphocytes, thus providing a costimulatory signal to these cells (27).

Thyroid Diseases

Y. enterocolitica has been implicated in the etiology of various thyroid disorders, including autoimmune thyroiditis and Graves' disease hyperthyroidism (225, 226).

The latter is an immunologic disorder mediated by auto-antibodies to the thyrotropin receptor. The chief link between *Y. enterocolitica* and thyroid diseases is that patients with these disorders frequently have elevated titers of serum agglutinins to *Y. enterocolitica* O:3 (197). Since there is no clear relationship between the incidence or geographic distribution of yersiniosis and that of auto-immune thyroid diseases, however, the presence of circulating antibodies to *Y. enterocolitica* in such patients is likely to reflect a fortuitous cross-reaction between *Yersinia* and thyroid antigens rather than a causal relationship (225). In addition, follow-up of patients many years after infection with *Y. enterocolitica* or *Y. pseudotuberculosis* has shown no increased frequency of thyroid disease, nor is there any evidence that hyperthyroidism is exacerbated by infection with *Y. enterocolitica* (225).

Although few studies have been conducted on the pathogenesis of *Yersinia*-induced erythema nodosum or glomerulonephritis, experiences with other infective agents suggest that these manifestations are caused by the deposition of immune complexes in affected organs.

SUMMARY AND CONCLUSIONS

Y. enterocolitica is a versatile foodborne pathogen with a remarkable ability to adapt to a wide range of environments within and outside its mammalian hosts. *Y. enterocolitica* typically accesses its hosts via food or water, in which it will have grown to stationary phase at ambient temperature. Under these circumstances, *Y. enterocolitica* expresses factors, such as urease, flagella, and smooth LPS, which facilitate its passage through the stomach and the mucous layer of the small intestine. Bacteria in this state may also carry Myf and invasin, which may promote adherence to and penetration of the dome epithelium overlying the Peyer's patches. The higher infectivity of *Y. enterocolitica* when grown at ambient temperature than that at 37°C may account for the small number of reports of human-to-human transmission of yersiniosis (155).

Once *Y. enterocolitica* begins to replicate in the body at 37°C, LPS becomes rough, and Ail and YadA appear on the bacterial surface. These factors may promote further invasion while protecting bacteria from complement-mediated opsonization. After a period in the host, which may include some time spent within professional phagocytes, yersiniae make contact with host cells in lymphoid tissue, where they synthesize and translocate the effector Yops, YopH, -E, -T, -O, and -P, which further frustrate the efforts of phagocytes to ingest and remove the bacteria. Subsequent bacterial replication may lead to tissue damage and the formation of microabscesses. If

strains of *Y. enterocolitica* which bear the HPI gain access to tissues where iron supplies would normally be growth limiting, they may produce yersiniabactin, which will enable them to acquire iron and allow bacterial replication to proceed. Eventually, the cycle is completed when the bacteria rupture through microabscesses in intestinal crypts to reenter the intestine and regain access to the environment. This well-defined life cycle of *Y. enterocolitica*, with its distinctive temperature-induced phases, is reminiscent of the flea-rat-flea cycle of *Y. pestis*.

Although much remains to be learned about *Y. enterocolitica*, investigations into the pathogenesis of yersiniosis to date have provided fascinating new insights into bacterial pathogenesis as a whole and into its genetic control. *Y. enterocolitica* and *Y. pseudotuberculosis* were the first invasive human pathogens for which plasmid-mediated virulence was documented (256), from which internalins (invasin, Ail, and YadA) were cloned and characterized (107, 147), for which the relationship between iron limitation and ferri-siderophore uptake assumed clinical significance (177), and in which type III protein secretion was identified (142). Future research in this area will no doubt lead to new and unexpected discoveries of bacterial strategies to evade immune responses that will further advance our understanding of the interface between microbes and their hosts.

References

1. Abe, J., M. Onimaru, S. Matsumoto, S. Noma, K. Baba, Y. Ito, T. Kohsaka, and T. Takeda. 1997. Clinical role for a superantigen in *Yersinia pseudotuberculosis* infection. *J. Clin. Investig.* **99:**1823–1830.

2. Achtman, M., K. Zurth, G. Morelli, G. Torrea, A. Guiyoule, and E. Carniel. 1999. *Yersinia pestis*, the cause of plague, is a recently emerged clone of *Yersinia pseudotuberculosis*. *Proc. Natl. Acad. Sci. USA* **96:**14043–14048.

3. Ackers, M. L., S. Schoenfeld, J. Markman, M. G. Smith, M. A. Nicholson, W. DeWitt, D. N. Cameron, P. M. Griffin, and L. Slutsker. 2000. An outbreak of *Yersinia enterocolitica* O:8 infections associated with pasteurized milk. *J. Infect. Dis.* **181:**1834–1837.

4. Adams, M. R., C. L. Little, and M. C. Easter. 1991. Modelling the effect of pH, acidulant and temperature on the growth rate of *Yersinia enterocolitica*. *J. Appl. Bacteriol.* **71:**65–71.

5. Aepfelbacher, M., R. Zumbihl, K. Ruckdeschel, C. A. Jacobi, C. Barz, and J. Heesemann. 1999. The tranquilizing injection of *Yersinia* proteins: a pathogen's strategy to resist host defense. *Biol. Chem.* **380:**795–802.

6. Ahvonen, P., K. Sievers, and K. Aho. 1969. Arthritis associated with *Yersinia enterocolitica*. *Acta Rheumatol. Scand.* **15:**232–255.

7. Aimoto, S., T. Takao, Y. Shimonishi, S. Hara, T. Takeda, Y. Takeda, and T. Miwatani. 1982. Amino-acid sequence of a heat-stable enterotoxin produced by human enterotoxigenic *Escherichia coli*. *Eur. J. Biochem.* **129:**257–263.

8. **Andersen, J. K., R. Sorensen, and M. Glensbjerg.** 1991. Aspects of the epidemiology of *Yersinia enterocolitica*: a review. *Int. J. Food Microbiol.* **13:**231–237.

9. **Arduino, M. J., L. A. Bland, M. A. Tipple, S. M. Aguero, M. S. Favero, and W. R. Jarvis.** 1989. Growth and endotoxin production of *Yersinia enterocolitica* and *Enterobacter agglomerans* in packed erythrocytes. *J. Clin. Microbiol.* **27:**1483–1485.

10. **Autenrieth, I. B., R. Reissbrodt, E. Saken, R. Berner, U. Vogel, W. Rabsch, and J. Heesemann.** 1994. Desferrioxamine-promoted virulence of *Yersinia enterocolitica* in mice depends on both desferrioxamine type and mouse strain. *J. Infect. Dis.* **169:**562–567.

11. **Badger, J. L., and V. L. Miller.** 1998. Expression of invasin and motility are coordinately regulated in *Yersinia enterocolitica*. *J. Bacteriol.* **180:**793–800.

12. **Baumler, A., R. Koebnik, I. Stojiljkovic, J. Heesemann, V. Braun, and K. Hantke.** 1993. Survey on newly characterized iron uptake systems of *Yersinia enterocolitica*. *Int. J. Med. Microbiol. Virol. Parasitol. Infect. Dis.* **278:**416–424.

13. **Baumler, A. J., and K. Hantke.** 1992. Ferrioxamine uptake in *Yersinia enterocolitica*: characterization of the receptor protein FoxA. *Mol. Microbiol.* **6:**1309–1321.

14. **Bengoechea, J. A., H. Najdenski, and M. Skurnik.** 2004. Lipopolysaccharide O antigen status of *Yersinia enterocolitica* O:8 is essential for virulence and absence of O antigen affects the expression of other *Yersinia* virulence factors. *Mol. Microbiol.* **52:**451–469.

15. **Bercovier, H., and H. H. Mollaret.** 1984. Genus XIV. *Yersinia* Van Loghem 1944, 15^AL, p. 498–506. *In* N. R. Krieg and J. G. Holt (ed.), *Bergey's Manual of Systematic Bacteriology*, vol. 1. Williams & Wilkins, Baltimore, Md.

16. **Bhaduri, S., B. Cottrell, and A. R. Pickard.** 1997. Use of a single procedure for selective enrichment, isolation, and identification of plasmid-bearing virulent *Yersinia enterocolitica* of various serotypes from pork samples. *Appl. Environ. Microbiol.* **63:**1657–1660.

17. **Biedzka-Sarek, M., R. Venho, and M. Skurnik.** 2005. Role of YadA, Ail, and lipopolysaccharide in serum resistance of *Yersinia enterocolitica* serotype O:3. *Infect. Immun.* **73:**2232–2244.

18. **Black, R. E., R. J. Jackson, T. Tsai, M. Medvesky, M. Shayegani, J. C. Feeley, K. I. E. MacLeod, and A. M. Wakelee.** 1978. Epidemic *Yersinia enterocolitica* infection due to contaminated chocolate milk. *N. Engl. J. Med.* **298:**76–79.

19. **Bliska, J. B., and S. Falkow.** 1994. Interplay between determinants of cellular entry and cellular disruption in the enteropathogenic *Yersinia*. *Curr. Opin. Infect. Dis.* **7:**323–328.

20. **Borg, A. A., J. Gray, and P. T. Dawes.** 1992. *Yersinia*-related arthritis in the United Kingdom. A report of 12 cases and review of the literature. *Q. J. Med.* **84:**575–582.

21. **Bottone, E. J.** 1977. *Yersinia enterocolitica*: a panoramic view of a charismatic microorganism. *Crit. Rev. Microbiol.* **5:**211–241.

22. **Bottone, E. J.** 1997. *Yersinia enterocolitica*: the charisma continues. *Clin. Microbiol. Rev.* **10:**257–276.

23. **Bowe, F., P. O'Gaora, D. Maskell, M. Cafferkey, and G. Dougan.** 1989. Virulence, persistence, and immunogenicity of *Yersinia enterocolitica* O:8 *aroA* mutants. *Infect. Immun.* **57:**3234–3236.

24. **Boyd, A. P., M. P. Sory, M. Iriarte, and G. R. Cornelis.** 1998. Heparin interferes with translocation of Yop proteins into HeLa cells and binds to LcrG, a regulatory component of the Yersinia Yop apparatus. *Mol. Microbiol.* **27:**425–436.

25. **Brecher, M. E., and S. N. Hay.** 2005. Bacterial contamination of blood components. *Clin. Microbiol. Rev.* **18:**195–204.

26. **Brem, D., C. Pelludat, A. Rakin, C. A. Jacobi, and J. Heesemann.** 2001. Functional analysis of yersiniabactin transport genes of *Yersinia enterocolitica*. *Microbiology* **147:**1115–1127.

27. **Brett, S. J., A. V. Mazurov, I. G. Charles, and J. P. Tite.** 1993. The invasin protein of *Yersinia* spp. provides costimulatory activity to human T cells through interaction with beta 1 integrins. *Eur. J. Immunol.* **23:**1608–1614.

28. **Brubaker, R. R.** 1991. Factors promoting acute and chronic diseases caused by yersiniae. *Clin. Microbiol. Rev.* **4:**309–324.

29. **Butler, R. C., V. Lund, and D. A. Carlson.** 1987. Susceptibility of *Campylobacter jejuni* and *Yersinia enterocolitica* to UV radiation. *Appl. Environ. Microbiol.* **53:**375–378.

30. **Carniel, E.** 2001. The *Yersinia* high-pathogenicity island: an iron-uptake island. *Microbes Infect.* **3:**561–569.

31. **Carniel, E., I. Autenrieth, G. Cornelis, H. Fukushima, F. Guinet, R. Isberg, J. Pham, M. Prentice, M. Simonet, M. Skurnik, and G. Wauters.** 22 November 2002, posting date. *Y. enterocolitica* and *Y. pseudotuberculosis*. *In* M. Dworkin, S. Falkow, E. Rosenberg, K.-H. Schleifer, and E. Stackebrandt (ed.), *The Prokaryotes: an Evolving Electronic Resource for the Microbiological Community*, 3rd ed., release 3.11. Springer-Verlag, New York, N.Y. [Online.] http://141.150.157.117:8080/prokPUB/chaprender/jsp/showchap.jsp?chapnum=338.

32. **Carnoy, C., H. Müeller-Alouf, S. Haentjens, and M. Simonet.** 1998. Polymorphism of *ypm*, *Yersinia pseudotuberculosis* superantigen encoding gene. *Zentrbl. Bakteriol.* **29**(Suppl.):397–398.

33. **Centers for Disease Control and Prevention.** 2003. *Yersinia enterocolitica* gastroenteritis among infants exposed to chitterlings—Chicago, Illinois, 2002. *Morb. Mortal. Wkly. Rep.* **52:**956–958.

34. **Chao, W. L., R. J. Ding, and R. S. Chen.** 1988. Survival of *Yersinia enterocolitica* in the environment. *Can. J. Microbiol.* **34:**753–756.

35. **China, B., B. T. N'Guyen, M. de Bruyere, and G. R. Cornelis.** 1994. Role of YadA in resistance of *Yersinia enterocolitica* to phagocytosis by human polymorphonuclear leukocytes. *Infect. Immun.* **62:**1275–1281.

36. **China, B., M. P. Sory, B. T. N'Guyen, M. de Bruyere, and G. R. Cornelis.** 1993. Role of the YadA protein in prevention of opsonization of *Yersinia enterocolitica* by C3b molecules. *Infect. Immun.* **61:**3129–3136.

37. **Clark, M. A., B. H. Hirst, and M. A. Jepson.** 1998. M-cell surface beta1 integrin expression and invasin-mediated targeting of *Yersinia pseudotuberculosis* to mouse Peyer's patch M cells. *Infect. Immun.* **66:**1237–1243.

38. **Colmegna, I., and L. R. Espinoza.** 2005. Recent advances in reactive arthritis. *Curr. Rheumatol. Rep.* **7:**201–207.

39. Cornelis, G., Y. Laroche, G. Balligand, M. P. Sory, and G. Wauters. 1987. *Yersinia enterocolitica*, a primary model for bacterial invasiveness. *Rev. Infect. Dis.* **9:**64–87.

40. Cornelis, G., C. Sluiters, C. L. de Rouvroit, and T. Michiels. 1989. Homology between *virF*, the transcriptional activator of the *Yersinia* virulence regulon, and AraC, the *Escherichia coli* arabinose operon regulator. *J. Bacteriol.* **171:**254–262.

41. Cornelis, G. R. 2002. The Yersinia Ysc-Yop 'type III' weaponry. *Nat. Rev. Mol. Cell Biol.* **3:**742–752.

42. Cornelis, G. R. 2002. *Yersinia* type III secretion: send in the effectors. *J. Cell Biol.* **158:**401–408.

43. Cornelis, G. R., T. Biot, C. Lambert de Rouvroit, T. Michiels, B. Mulder, C. Sluiters, M. P. Sory, M. Van Bouchaute, and J. C. Vanooteghem. 1989. The *Yersinia* yop regulon. *Mol. Microbiol.* **3:**1455–1459.

44. Cornelis, G. R., A. Boland, A. P. Boyd, C. Geuijen, M. Iriarte, C. Neyt, M.-P. Sory, and I. Stainier. 1998. The virulence plasmid of *Yersinia*, an antihost genome. *Microbiol. Mol. Biol. Rev.* **62:**1315–1352.

45. Cornelis, G. R., C. Sluiters, I. Delor, D. Geib, K. Kaniga, C. Lambert de Rouvroit, M. P. Sory, J. C. Vanooteghem, and T. Michiels. 1991. ymoA, a *Yersinia enterocolitica* chromosomal gene modulating the expression of virulence functions. *Mol. Microbiol.* **5:**1023–1034.

46. Cotter, S. E., N. K. Surana, and J. W. St. Geme III. 2005. Trimeric autotransporters: a distinct subfamily of autotransporter proteins. *Trends Microbiol.* **13:**199–205.

47. Cover, T. L., and R. C. Aber. 1989. *Yersinia enterocolitica*. *N. Engl. J. Med.* **321:**16–24.

48. Cowan, C., A. V. Philipovskiy, C. R. Wulff-Strobel, Z. Ye, and S. C. Straley. 2005. Anti-LcrV antibody inhibits delivery of Yops by *Yersinia pestis* KIM5 by directly promoting phagocytosis. *Infect. Immun.* **73:**6127–6137.

49. Reference deleted.

50. Currie, M. G., K. F. Fok, J. Kato, R. J. Moore, F. K. Hamra, K. L. Duffin, and C. E. Smith. 1992. Guanylin: an endogenous activator of intestinal guanylate cyclase. *Proc. Natl. Acad. Sci. USA* **89:**947–951.

51. D'Aoust, J. Y., C. E. Park, R. A. Szabo, E. C. Todd, D. B. Emmons, and R. C. McKellar. 1988. Thermal inactivation of *Campylobacter* species, *Yersinia enterocolitica*, and hemorrhagic *Escherichia coli* O157:H7 in fluid milk. *J. Dairy Sci.* **71:**3230–3236.

52. de Giusti, M., and E. de Vito. 1992. Inactivation of *Yersinia enterocolitica* by nitrite and nitrate in food. *Food Addit. Contam.* **9:**405–408.

53. de Koning, J., J. Heesemann, J. A. Hoogkamp-Korstanje, J. J. Festen, P. M. Houtman, and P. L. van Oijen. 1989. *Yersinia* in intestinal biopsy specimens from patients with seronegative spondyloarthropathy: correlation with specific serum IgA antibodies. *J. Infect. Dis.* **159:**109–112.

54. de Koning-Ward, T. F., T. Grant, F. Oppedisano, and R. M. Robins-Browne. 1998. Effect of bacterial invasion of macrophages on the outcome of assays to assess bacterium-macrophage interactions. *J. Immunol. Methods* **215:**39–44.

55. de Koning-Ward, T. F., and R. M. Robins-Browne. 1995. Contribution of urease to acid tolerance in *Yersinia enterocolitica*. *Infect. Immun.* **63:**3790–3795.

56. Delor, I., and G. R. Cornelis. 1992. Role of *Yersinia enterocolitica* Yst toxin in experimental infection of young rabbits. *Infect. Immun.* **60:**4269–4277.

57. Delor, I., A. Kaeckenbeeck, G. Wauters, and G. R. Cornelis. 1990. Nucleotide sequence of *yst*, the *Yersinia enterocolitica* gene encoding the heat-stable enterotoxin, and prevalence of the gene among pathogenic and nonpathogenic yersiniae. *Infect. Immun.* **58:**2983–2988.

58. Dion, P., R. Charbonneau, and C. Thibault. 1994. Effect of ionizing dose rate on the radioresistance of some food pathogenic bacteria. *Can. J. Microbiol.* **40:**369–374.

59. Doyle, M. P. 1990. Pathogenic *Escherichia coli*, *Yersinia enterocolitica*, and *Vibrio parahaemolyticus*. *Lancet* **336:**1111–1115.

60. Escudero, M. E., L. Velazquez, M. S. Di Genaro, and A. M. de Guzman. 1999. Effectiveness of various disinfectants in the elimination of *Yersinia enterocolitica* on fresh lettuce. *J. Food Prot.* **62:**665–669.

61. Falgarone, G., H. S. Blanchard, B. Riot, M. Simonet, and M. Breban. 1999. Cytotoxic T-cell-mediated response against *Yersinia pseudotuberculosis* in HLA-B27 transgenic rat. *Infect. Immun.* **67:**3773–3779.

62. Farmer, J. J., III, G. P. Carter, V. L. Miller, S. Falkow, and I. K. Wachsmuth. 1992. Pyrazinamidase, CR-MOX agar, salicin fermentation-esculin hydrolysis, and D-xylose fermentation for identifying pathogenic serotypes of *Yersinia enterocolitica*. *J. Clin. Microbiol.* **30:**2589–2594.

63. Feng, P., and S. D. Weagant. 1993. *Yersinia*, p. 427–460. *In* Y. H. Hui, J. R. Gorham, K. D. Murrell, and D. O. Cliver (ed.), *Foodborne Disease Handbook*, vol. 1. Marcel Dekker, New York, N.Y.

64. Ferracci, F., F. D. Schubot, D. S. Waugh, and G. V. Plano. 2005. Selection and characterization of *Yersinia pestis* YopN mutants that constitutively block Yop secretion. *Mol. Microbiol.* **57:**970–987.

65. Finlay, B. B., and S. Falkow. 1988. Comparison of the invasion strategies used by *Salmonella cholerae-suis*, *Shigella flexneri* and *Yersinia enterocolitica* to enter cultured animal cells: endosome acidification is not required for bacterial invasion or intracellular replication. *Biochimie* **70:**1089–1099.

66. Foberg, U., A. Fryden, E. Kihlstrom, K. Persson, and O. Weiland. 1986. *Yersinia enterocolitica* septicemia: clinical and microbiological aspects. *Scand. J. Infect. Dis.* **18:**269–279.

67. Forsberg, A., R. Rosqvist, and H. Wolf-Watz. 1994. Regulation and polarized transfer of the *Yersinia* outer proteins (Yops) involved in antiphagocytosis. *Trends Microbiol.* **2:**14–19.

68. Forsberg, A., A. M. Viitanen, M. Skurnik, and H. Wolf-Watz. 1991. The surface-located YopN protein is involved in calcium signal transduction in *Yersinia pseudotuberculosis*. *Mol. Microbiol.* **5:**977–986.

69. Francis, M. S., and H. Wolf-Watz. 1998. YopD of Yersinia pseudotuberculosis is translocated into the cytosol of HeLa epithelial cells: evidence of a structural domain necessary for translocation. *Mol. Microbiol.* **29:**799–813.

70. Fredriksson-Ahomaa, M., M. Bucher, C. Hank, A. Stolle, and H. Korkeala. 2001. High prevalence of Yersinia enterocolitica 4:O3 on pig offal in southern Germany: a slaughtering technique problem. *Syst. Appl. Microbiol.* **24:**457–463.

71. Fredriksson-Ahomaa, M., and H. Korkeala. 2003. Low occurrence of pathogenic *Yersinia enterocolitica* in clinical, food, and environmental samples: a methodological problem. *Clin. Microbiol. Rev.* **16:**220–229.

72. Fredriksson-Ahomaa, M., T. Korte, and H. Korkeala. 2000. Contamination of carcasses, offals, and the environment with *yadA*-positive *Yersinia enterocolitica* in a pig slaughterhouse. *J. Food Prot.* **63:**31–35.

73. Gaastra, W., and A. M. Svennerholm. 1996. Colonization factors of human enterotoxigenic *Escherichia coli* (ETEC). *Trends Microbiol.* **4:**444–452.

74. Gaston, J. S., C. Cox, and K. Granfors. 1999. Clinical and experimental evidence for persistent *Yersinia* infection in reactive arthritis. *Arthritis Rheum.* **42:**2239–2242.

75. Ghosh, P. 2004. Process of protein transport by the type III secretion system. *Microbiol. Mol. Biol. Rev.* **68:**771–795.

76. Goure, J., P. Broz, O. Attree, G. R. Cornelis, and I. Attree. 2005. Protective anti-V antibodies inhibit *Pseudomonas* and *Yersinia* translocon assembly within host membranes. *J. Infect. Dis.* **192:**218–225.

77. Granfors, K., S. Jalkanen, R. von Essen, R. Lahesmaa-Rantala, O. Isomaki, K. Pekkola-Heino, R. Merilahti-Palo, R. Saario, H. Isomaki, and A. Toivanen. 1989. *Yersinia* antigens in synovial-fluid cells from patients with reactive arthritis. *N. Engl. J. Med.* **320:**216–221.

78. Grant, T., V. Bennett-Wood, and R. M. Robins-Browne. 1999. Characterization of the interaction between *Yersinia enterocolitica* biotype 1A and phagocytes and epithelial cells in vitro. *Infect. Immun.* **67:**4367–4375.

79. Grant, T., V. Bennett-Wood, and R. M. Robins-Browne. 1998. Identification of virulence-associated characteristics in clinical isolates of *Yersinia enterocolitica* lacking classical virulence markers. *Infect. Immun.* **66:**1113–1120.

80. Gripenberg-Lerche, C., M. Skurnik, L. Zhang, K.-O. Söderström, and P. Toivanen. 1994. Role of YadA in arthritogenicity of *Yersinia enterocolitica* serotype O:8: experimental studies with rats. *Infect. Immun.* **62:**5568–5575.

81. Gripenberg-Lerche, C., L. Zhang, P. Ahtonen, P. Toivanen, and M. Skurnik. 2000. Construction of urease-negative mutants of *Yersinia enterocolitica* serotypes O:3 and O:8: role of urease in virulence and arthritogenicity. *Infect. Immun.* **68:**942–947.

82. Grosdent, N., I. Maridonneau-Parini, M. P. Sory, and G. R. Cornelis. 2002. Role of Yops and adhesins in resistance of *Yersinia enterocolitica* to phagocytosis. *Infect. Immun.* **70:**4165–4176.

83. Grützkau, A., C. Hanski, H. Hahn, and E. O. Riecken. 1990. Involvement of M cells in the bacterial invasion of Peyer's patches: a common mechanism shared by *Yersinia enterocolitica* and other enteroinvasive bacteria. *Gut* **31:**1011–1015.

84. Guarino, A., R. Giannella, and M. R. Thompson. 1989. *Citrobacter freundii* produces an 18-amino-acid heat-stable enterotoxin identical to the 18-amino-acid *Escherichia coli* heat-stable enterotoxin (STIa). *Infect. Immun.* **57:**649–652.

85. Gurtler, M., T. Alter, S. Kasimir, M. Linnebur, and K. Fehlhaber. 2005. Prevalence of *Yersinia enterocolitica* in fattening pigs. *J. Food Prot.* **68:**850–854.

86. Håkansson, S., T. Bergman, J. C. Vanooteghem, G. Cornelis, and H. Wolf-Watz. 1993. YopB and YopD constitute a novel class of *Yersinia* Yop proteins. *Infect. Immun.* **61:**71–80.

87. Hamburger, Z. A., M. S. Brown, R. R. Isberg, and P. J. Bjorkman. 1999. Crystal structure of invasin: a bacterial integrin-binding protein. *Science* **286:**291–295.

88. Han, Y. W., and V. L. Miller. 1997. Reevaluation of the virulence phenotype of the *inv yadA* double mutants of *Yersinia pseudotuberculosis. Infect. Immun.* **65:**327–330.

89. Hanna, M. O., J. C. Stewart, D. L. Zink, Z. L. Carpenter, and C. Vanderzant. 1977. Development of *Yersinia enterocolitica* on raw and cooked beef and pork at different temperatures. *J. Food Sci.* **42:**1180–1184.

90. Hanski, C., U. Kutschka, H. P. Schmoranzer, M. Naumann, A. Stallmach, H. Hahn, H. Menge, and E. O. Riecken. 1989. Immunohistochemical and electron microscopic study of interaction of *Yersinia enterocolitica* serotype O8 with intestinal mucosa during experimental enteritis. *Infect. Immun.* **57:**673–678.

91. Harakeh, M. S., J. D. Berg, J. C. Hoff, and A. Matin. 1985. Susceptibility of chemostat-grown *Yersinia enterocolitica* and *Klebsiella pneumoniae* to chlorine dioxide. *Appl. Environ. Microbiol.* **49:**69–72.

92. Hartland, E. L., and R. M. Robins-Browne. 1998. Infections with enteropathogenic *Yersinia* species: paradigms of bacterial pathogenesis. *Rev. Med. Microbiol.* **9:** 191–205.

93. Hayashidani, H., Y. Ohtomo, Y. Toyokawa, M. Saito, K. Kaneko, J. Kosuge, M. Kato, M. Ogawa, and G. Kapperud. 1995. Potential sources of sporadic human infection with *Yersinia enterocolitica* serovar O:8 in Aomori Prefecture, Japan. *J. Clin. Microbiol.* **33:**1253–1257.

94. Heesemann, J., K. Gaede, and I. B. Autenrieth. 1993. Experimental *Yersinia enterocolitica* infection in rodents: a model for human yersiniosis. *APMIS* **101:**417–429.

95. Heesemann, J., K. Hantke, T. Vocke, E. Saken, A. Rakin, I. Stojiljkovic, and R. Berner. 1993. Virulence of *Yersinia enterocolitica* is closely associated with siderophore production, expression of an iron-repressible outer membrane polypeptide of 65,000 Da and pesticin sensitivity. *Mol. Microbiol.* **8:**397–408.

96. Henderson, I. R., F. Navarro-Garcia, M. Desvaux, R. C. Fernandez, and D. a'Aldeen. 2004. Type V protein secretion pathway: the autotransporter story. *Microbiol. Mol. Biol. Rev.* **68:**692–744.

97. Hermann, E. 1993. T cells in reactive arthritis. *APMIS* **101:**177–186.

98. Hermann, E., D. T. Yu, K. H. Meyer zum Buschenfelde, and B. Fleischer. 1993. HLA-B27-restricted CD8 T cells derived from synovial fluids of patients with reactive arthritis and ankylosing spondylitis. *Lancet* **342:**646–650.

99. Herrlinger, J. D., and J. U. Asmussen. 1992. Long term prognosis in yersinia arthritis: clinical and serological findings. *Ann. Rheum. Dis.* **51:**1332–1334.

100. Hoiczyk, E., and G. Blobel. 2001. Polymerization of a single protein of the pathogen *Yersinia enterocolitica* into needles punctures eukaryotic cells. *Proc. Natl. Acad. Sci. USA* **98:**4669–4674.

101. **Hoogkamp-Korstanje, J. A. A., and V. M. M. Stolk-Engelaar.** 1995. *Yersinia enterocolitica* infection in children. *Pediatr. Infect. Dis.* J. **14:**771–775.

102. **Huang, X., K. Yoshino, H. Nakao, and T. Takeda.** 1997. Nucleotide sequence of a gene encoding the novel *Yersinia enterocolitica* heat-stable enterotoxin that includes a pro-region-like sequence in its mature toxin molecule. *Microb. Pathog.* **22:**89–97.

103. **Iriarte, M., and G. R. Cornelis.** 1999. Identification of SycN, YscX, and YscY, three new elements of the *Yersinia yop* virulon. *J. Bacteriol.* **181:**675–680.

104. **Iriarte, M., I. Stainier, and G. R. Cornelis.** 1995. The *rpoS* gene from *Yersinia enterocolitica* and its influence on expression of virulence factors. *Infect. Immun.* **63:**1840–1847.

105. **Iriarte, M., I. Stainier, A. V. Mikulskis, and G. R. Cornelis.** 1995. The *fliA* gene encoding σ^{28} in *Yersinia enterocolitica*. *J. Bacteriol.* **177:**2299–2304.

106. **Iriarte, M., J. C. Vanooteghem, I. Delor, R. Diaz, S. Knutton, and G. R. Cornelis.** 1993. The Myf fibrillae of *Yersinia enterocolitica*. *Mol. Microbiol.* **9:**507–520.

107. **Isberg, R. R.** 1990. Pathways for the penetration of enteroinvasive *Yersinia* into mammalian cells. *Mol. Biol. Med.* **7:**73–82.

108. **Isberg, R. R., and P. Barnes.** 2001. Subversion of integrins by enteropathogenic *Yersinia*. *J. Cell Sci.* **114:**21–28.

109. **Isberg, R. R., and S. Falkow.** 1985. A single genetic locus encoded by *Yersinia pseudotuberculosis* permits invasion of cultured animal cells by *Escherichia coli* K-12. *Nature* **317:**262–264.

110. **Isberg, R. R., and G. T. Van Nhieu.** 1994. Two mammalian cell internalization strategies used by pathogenic bacteria. *Annu. Rev. Genet.* **28:**395–422.

111. **Iteman, I., A. Guiyoule, and E. Carniel.** 1996. Comparison of three molecular methods for typing and subtyping pathogenic *Yersinia enterocolitica* strains. *J. Med. Microbiol.* **45:**48–56.

112. **Iwobi, A., J. Heesemann, E. Garcia, E. Igwe, C. Noelting, and A. Rakin.** 2003. Novel virulence-associated type II secretion system unique to high-pathogenicity *Yersinia enterocolitica*. *Infect. Immun.* **71:**1872–1879.

113. **Jackson, M. W., and G. V. Plano.** 2000. Interactions between type III secretion apparatus components from *Yersinia pestis* detected using the yeast two-hybrid system. *FEMS Microbiol. Lett.* **186:**85–90.

114. **Jacobi, C. A., A. Roggenkamp, A. Rakin, R. Zumbihl, L. Leitritz, and J. Heesemann.** 1998. In vitro and in vivo expression studies of *yopE* from *Yersinia enterocolitica* using the *gfp* reporter gene. *Mol. Microbiol.* **30:**865–882.

115. **Jepson, M. A., and M. A. Clark.** 1998. Studying M cells and their role in infection. *Trends Microbiol.* **9:**359–365.

116. **Jones, T. F.** 2003. From pig to pacifier: chitterling-associated yersiniosis outbreak among black infants. *Emerg. Infect. Dis.* **9:**1007–1009.

117. **Kapperud, G.** 1991. *Yersinia enterocolitica* in food hygiene. *Int. J. Food Microbiol.* **12:**53–65.

118. **Kapperud, G.** 1982. Enterotoxin production at 4°, 22°, and 37° among *Yersinia enterocolitica* and *Y. enterocolitica*-like bacteria. *APMIS* **90B:**185–189.

119. **Kasatiya, S. S.** 1976. *Yersinia enterocolitica* gastroenteritis outbreak—Montreal. *Can. Dis. Wkly. Rep.* **2:**73–74.

120. **Keet, E. E.** 1974. *Yersinia enterocolitica* septicemia: source of infection and incubation period identified. *N. Y. State J. Med.* **74:**2226–2229.

121. **Kerschen, E. J., D. A. Cohen, A. M. Kaplan, and S. C. Straley.** 2004. The plague virulence protein YopM targets the innate immune response by causing a global depletion of NK cells. *Infect. Immun.* **72:**4589–4602.

122. **King, C. H., E. B. Shotts, Jr., R. E. Wooley, and K. G. Porter.** 1988. Survival of coliforms and bacterial pathogens within protozoa during chlorination. *Appl. Environ. Microbiol.* **54:**3023–3033.

123. **Koster, M., W. Bitter, H. de Cock, A. Allaoui, G. R. Cornelis, and J. Tommassen.** 1997. The outer membrane component, YscC, of the Yop secretion machinery of Yersinia enterocolitica forms a ring-shaped multimeric complex. *Mol. Microbiol.* **26:**789–797.

124. **Kotetishvili, M., A. Kreger, G. Wauters, J. G. Morris, Jr., A. Sulakvelidze, and O. C. Stine.** 2005. Multilocus sequence typing for studying genetic relationships among *Yersinia* species. *J. Clin. Microbiol.* **43:**2674–2684.

125. **Kraehenbuhl, J. P., and M. R. Neutra.** 2000. Epithelial M cells: differentiation and function. *Annu. Rev. Cell Dev. Biol.* **16:**301–332.

126. **Lahesmaa, R., M. Skurnik, K. Granfors, T. Mottonen, R. Saario, A. Toivanen, and P. Toivanen.** 1992. Molecular mimicry in the pathogenesis of spondyloarthropathies. A critical appraisal of cross-reactivity between microbial antigens and HLA-B27. *Br. J. Rheumatol.* **31:**221–229.

127. **Larsen, J. H.** 1980. *Yersinia enterocolitica* infection and rheumatic diseases. *Scand. J. Rheumatol.* **9:**129–137.

128. **Leclercq, A., L. Martin, M. L. Vergnes, N. Ounnoughene, J. F. Laran, P. Giraud, and E. Carniel.** 2005. Fatal *Yersinia enterocolitica* biotype 4 serovar O:3 sepsis after red blood cell transfusion. *Transfusion* **45:**814–818.

129. **Lee, L. A., A. R. Gerber, D. R. Lonsway, J. D. Smith, G. P. Carter, N. D. Puhr, C. M. Parrish, R. K. Sikes, R. J. Finton, and R. V. Tauxe.** 1990. *Yersinia enterocolitica* O:3 infections in infants and children, associated with the household preparation of chitterlings. *N. Engl. J. Med.* **322:**984–987.

130. **Lee, V. T., D. M. Anderson, and O. Schneewind.** 1998. Targeting of *Yersinia* Yop proteins into the cytosol of HeLa cells: one-step translocation of YopE across bacterial and eukaryotic membranes is dependent on SycE chaperone. *Mol. Microbiol.* **28:**593–601.

131. **Lee, V. T., C. Tam, and O. Schneewind.** 2000. LcrV, a substrate for *Yersinia enterocolitica* type III secretion, is required for toxin targeting into the cytosol of HeLa cells. *J. Biol. Chem.* **275:**36869–36875.

132. **Leirisalo-Repo, M.** 1987. *Yersinia* arthritis. Acute clinical picture and long-term prognosis. *Contrib. Microbiol. Immunol.* **9:**145–154.

133. **Leung, K. Y., and S. C. Straley.** 1989. The *yopM* gene of *Yersinia pestis* encodes a released protein having homology with the human platelet surface protein GPIb alpha. *J. Bacteriol.* **171:**4623–4632.

134. **Lian, C. J., W. S. Hwang, J. K. Kelly, and C. H. Pai.** 1987. Invasiveness of *Yersinia enterocolitica* lacking the

virulence plasmid: an in-vivo study. *J. Med. Microbiol.* **24**:219–226.

135. Lindler, L. E., and B. D. Tall. 1993. *Yersinia pestis* pH 6 antigen forms fimbriae and is induced by intracellular association with macrophages. *Mol. Microbiol.* **8**:311–324.

136. Lo, W. F., A. S. Woods, A. DeCloux, R. J. Cotter, E. S. Metcalf, and M. J. Soloski. 2000. Molecular mimicry mediated by MHC class Ib molecules after infection with gram-negative pathogens. *Nat. Med.* **6**:215–218.

137. Marjai, E., M. Kalman, I. Kajary, A. Belteky, and M. Rodler. 1987. Isolation from food and characterization by virulence tests of *Yersinia enterocolitica* associated with an outbreak. *Acta Microbiol. Hung.* **34**:97–109.

138. Marks, M. I., C. H. Pai, L. Lafleur, L. Lackman, and O. Hammerberg. 1980. *Yersinia enterocolitica* gastroenteritis: a prospective study of clinical, bacteriologic, and epidemiologic features. *J. Pediatr.* **96**:26–31.

139. Marra, A., and R. R. Isberg. 1997. Invasin-dependent and invasin-independent pathways for translocation of *Yersinia pseudotuberculosis* across the Peyer's patch intestinal epithelium. *Infect. Immun.* **65**:3412–3421.

140. Maruyama, T. 1987. *Yersinia enterocolitica* infection in humans and isolation of the microorganism from pigs in Japan. *Contrib. Microbiol. Immunol.* **9**:48–55.

141. McDonald, C., P. O. Vacratsis, J. B. Bliska, and J. E. Dixon. 2003. The *Yersinia* virulence factor YopM forms a novel protein complex with two cellular kinases. *J. Biol. Chem.* **278**:18514–18523.

142. Michiels, T., P. Wattiau, R. Brasseur, J. M. Ruysschaert, and G. Cornelis. 1990. Secretion of Yop proteins by yersiniae. *Infect. Immun.* **58**:2840–2849.

143. Mikulskis, A. V., I. Delor, V. H. Thi, and G. R. Cornelis. 1994. Regulation of the *Yersinia enterocolitica* enterotoxin Yst gene. Influence of growth phase, temperature, osmolarity, pH and bacterial host factors. *Mol. Microbiol.* **14**:905–915.

144. Miller, V. L. 1992. *Yersinia* invasion genes and their products. *ASM News* **58**:26–33.

145. Miller, V. L., K. B. Beer, G. Heusipp, B. M. Young, and M. R. Wachtel. 2001. Identification of regions of Ail required for the invasion and serum resistance phenotypes. *Mol. Microbiol.* **41**:1053–1062.

146. Miller, V. L., and S. Falkow. 1988. Evidence for two genetic loci in *Yersinia enterocolitica* that can promote invasion of epithelial cells. *Infect. Immun.* **56**:1242–1248.

147. Miller, V. L., B. B. Finlay, and S. Falkow. 1988. Factors essential for the penetration of mammalian cells by *Yersinia. Curr. Top. Microbiol. Immunol.* **138**:15–39.

148. Miyoshi-Akiyama, T., W. Fujimaki, X. J. Yan, J. Yagi, K. Imanishi, H. Kato, K. Tomonari, and T. Uchiyama. 1997. Identification of murine T cells reactive with the bacterial superantigen *Yersinia pseudotuberculosis*-derived mitogen (YPM) and factors involved in YPM-induced toxicity in mice. *Microbiol. Immunol.* **41**:345–352.

149. Mollaret, H. H., H. Bercovier, and J. M. Alonso. 1979. Summary of the data received at the WHO Reference Centre for *Yersinia enterocolitica. Contrib. Microbiol. Immunol.* **5**:174–184.

150. Morse, D. L., M. Shayegani, and R. J. Gallo. 1984. Epidemiologic investigation of a *Yersinia* camp outbreak linked to a food handler. *Am. J. Public Health* **74**:589–592.

151. Neilands, J. B. 1981. Microbial iron compounds. *Annu. Rev. Biochem.* **50**:715–731.

152. Nesbakken, T., G. Kapperud, K. Dommarsnes, M. Skurnik, and E. Hornes. 1991. Comparative study of a DNA hybridization method and two isolation procedures for detection of *Yersinia enterocolitica* O:3 in naturally contaminated pork products. *Appl. Environ. Microbiol.* **57**:389–394.

153. Neyt, C., and G. R. Cornelis. 1999. Insertion of a Yop translocation pore into the macrophage plasma membrane by *Yersinia enterocolitica*: requirement for translocators YopB and YopD, but not LcrG. *Mol. Microbiol.* **33**:971–981.

154. Neyt, C., M. Iriarte, V. H. Thi, and G. R. Cornelis. 1997. Virulence and arsenic resistance in yersiniae. *J. Bacteriol.* **179**:612–619.

155. Nilehn, B. 1969. Studies on *Yersinia enterocolitica* with special reference to bacterial diagnosis and occurrence in human acute enteric disease. *Acta Pathol. Microbiol. Scand.* **206**(Suppl.):1–48.

156. Orth, K. 2002. Function of the *Yersinia* effector YopJ. *Curr. Opin. Microbiol.* **5**:38–43.

157. Ostroff, S. M., G. Kapperud, L. C. Hutwagner, T. Nesbakken, N. H. Bean, J. Lassen, and R. V. Tauxe. 1994. Sources of sporadic *Yersinia enterocolitica* infections in Norway: a prospective case-control study. *Epidemiol. Infect.* **112**:133–141.

158. Pai, C. H., and L. De Stephano. 1982. Serum resistance associated with virulence in *Yersinia enterocolitica. Infect. Immun.* **35**:605–611.

159. Pai, C. H., V. Mors, and S. Toma. 1978. Prevalence of enterotoxigenicity in human and nonhuman isolates of *Yersinia enterocolitica. Infect. Immun.* **22**:334–338.

160. Peixotto, S. S., G. Finne, M. O. Hanna, and C. Vanderzant. 1979. Presence, growth and survival of *Yersinia enterocolitica* in oyster, shrimp and crab. *J. Food Prot.* **42**:974–981.

161. Pepe, J. C., J. L. Badger, and V. L. Miller. 1994. Growth phase and low pH affect the thermal regulation of the *Yersinia enterocolitica inv* gene. *Mol. Microbiol.* **11**:123–135.

162. Pepe, J. C., and V. L. Miller. 1993. *Yersinia enterocolitica* invasin: a primary role in the initiation of infection. *Proc. Natl. Acad. Sci. USA* **90**:6473–6477.

163. Pepe, J. C., M. R. Wachtel, E. Wagar, and V. L. Miller. 1995. Pathogenesis of defined invasion mutants of *Yersinia enterocolitica* in a BALB/c mouse model of infection. *Infect. Immun.* **63**:4837–4848.

164. Perry, R. D. 1993. Acquisition and storage of inorganic iron and hemin by the yersiniae. *Trends Microbiol.* **1**:142–147.

165. Perry, R. D., and J. D. Fetherston. 1997. *Yersinia pestis*: etiologic agent of plague. *Clin. Microbiol. Rev.* **10**:35–66.

166. Persson, C., N. Carballeira, H. Wolf-Watz, and M. Fallman. 1997. The PTPase YopH inhibits uptake of *Yersinia*, tyrosine phosphorylation of p130Cas and FAK, and the associated accumulation of these proteins in peripheral focal adhesions. *EMBO J.* **16**:2307–2318.

167. Pujol, C., and J. B. Bliska. 2005. Turning *Yersinia* pathogenesis outside in: subversion of macrophage function by intracellular yersiniae. *Clin. Immunol.* **114:**216–226.

168. Puylaert, J. B., R. J. Vermeijden, S. D. van der Werf, L. Doornbos, and R. K. Koumans. 1989. Incidence and sonographic diagnosis of bacterial ileocaecitis masquerading as appendicitis. *Lancet* **ii:**84–86.

169. Rakin, A., C. Noelting, S. Schubert, and J. Heesemann. 1999. Common and specific characteristics of the high-pathogenicity island of *Yersinia enterocolitica*. *Infect. Immun.* **67:**5265–5274.

170. Ramamurthi, K. S., and O. Schneewind. 2003. Substrate recognition by the *Yersinia* type III protein secretion machinery. *Mol. Microbiol.* **50:**1095–1102.

171. Ramamurthy, T., K. Yoshino, X. Huang, G. B. Nair, E. Carniel, T. Maruyama, H. Fukushima, and T. Takeda. 1997. The novel heat-stable enterotoxin subtype gene (ystB) of *Yersinia enterocolitica*: nucleotide sequence and distribution of the yst genes. *Microb. Pathog.* **23:**189–200.

172. Reed, R. P., R. M. Robins-Browne, and M. L. Williams. 1997. *Yersinia enterocolitica* peritonitis. *Clin. Infect. Dis.* **25:**1468–1469.

173. Reveille, J. D., and F. C. Arnett. 2005. Spondyloarthritis: update on pathogenesis and management. *Am. J. Med.* **118:**592–603.

174. Robins-Browne, R. M., A. M. Bordun, and K. J. Slee. 1993. Serological response of sheep to plasmid-encoded proteins of *Yersinia* species following natural infection with *Y. enterocolitica* and *Y. pseudotuberculosis*. *J. Med. Microbiol.* **39:**268–272.

175. Robins-Browne, R. M., M. D. Miliotis, S. Cianciosi, V. L. Miller, S. Falkow, and J. G. Morris, Jr. 1989. Evaluation of DNA colony hybridization and other techniques for detection of virulence in *Yersinia* species. *J. Clin. Microbiol.* **27:**644–650.

176. Robins-Browne, R. M., and J. K. Prpic. 1985. Effects of iron and desferrioxamine on infections with *Yersinia enterocolitica*. *Infect. Immun.* **47:**774–779.

177. Robins-Browne, R. M., J. K. Prpic, and S. J. Stuart. 1987. Yersiniae and iron. A study in host-parasite relationships. *Contrib. Microbiol. Immunol.* **9:**254–258.

178. Robins-Browne, R. M., A. R. Rabson, and H. J. Koornhof. 1979. Generalised infection with *Yersinia enterocolitica* and the role of iron. *Contrib. Microbiol. Immunol.* **5:**277–282.

179. Robins-Browne, R. M., C. S. Still, M. D. Miliotis, and H. J. Koornhof. 1979. Mechanism of action of *Yersinia enterocolitica* enterotoxin. *Infect. Immun.* **25:**680–684.

180. Robins-Browne, R. M., T. Takeda, A. Fasano, A. M. Bordun, S. Dohi, H. Kasuga, G. Fang, V. Prado, R. L. Guerrant, et al. 1993. Assessment of enterotoxin production by *Yersinia enterocolitica* and identification of a novel heat-stable enterotoxin produced by a noninvasive *Y. enterocolitica* strain isolated from clinical material. *Infect. Immun.* **61:**764–767.

181. Robins-Browne, R. M., S. Tzipori, G. Gonis, J. Hayes, M. Withers, and J. K. Prpic. 1985. The pathogenesis of *Yersinia enterocolitica* infection in gnotobiotic piglets. *J. Med. Microbiol.* **19:**297–308.

182. Roggenkamp, A., H.-R. Neuberger, A. Flugel, T. Schmoll, and J. Heesemann. 1995. Substitution of two histidine residues in YadA protein of *Yersinia enterocolitica* abrogates collagen binding, cell adherence and mouse virulence. *Mol. Microbiol.* **16:**1207–1219.

183. Rohde, J. R., J. M. Fox, and S. A. Minnich. 1994. Thermoregulation in *Yersinia enterocolitica* is coincident with changes in DNA supercoiling. *Mol. Microbiol.* **12:**187–199.

184. Rosenshine, I., V. Duronio, and B. B. Finlay. 1992. Tyrosine protein kinase inhibitors block invasin-promoted bacterial uptake by epithelial cells. *Infect. Immun.* **60:**2211–2217.

185. Sachdeva, P., and J. S. Virdi. 2004. Repetitive elements sequence (REP/ERIC)-PCR based genotyping of clinical and environmental strains of *Yersinia enterocolitica* biotype 1A reveal existence of limited number of clonal groups. *FEMS Microbiol. Lett.* **240:**193–201.

186. Saebo, A., and J. Lassen. 1992. Acute and chronic gastrointestinal manifestations associated with *Yersinia enterocolitica* infection. A Norwegian 10-year follow-up study on 458 hospitalized patients. *Ann. Surg.* **215:**250–255.

187. Saltman, L. H., Y. Lu, E. M. Zaharias, and R. R. Isberg. 1996. A region of the *Yersinia pseudotuberculosis* invasin protein that contributes to high affinity binding to integrin receptors. *J. Biol. Chem.* **271:**23438–23444.

188. Sansonetti, P. J., and A. Phalipon. 1999. M cells as ports of entry for enteroinvasive pathogens: mechanisms of interaction, consequences for the disease process. *Semin. Immunol.* **11:**193–203.

189. Schiemann, D. A. 1980. *Yersinia enterocolitica*: observations on some growth characteristics and response to selective agents. *Can. J. Microbiol.* **26:**1232–1240.

190. Schiemann, D. A. 1981. An enterotoxin-negative strain of *Yersinia enterocolitica* serotype O:3 is capable of producing diarrhea in mice. *Infect. Immun.* **32:**571–574.

191. Schiemann, D. A. 1988. Examination of enterotoxin production at low temperatures by *Yersinia* spp. in culture media and foods. *J. Food Prot.* **51:**571–573.

192. Schiemann, D. A. 1987. *Yersinia enterocolitica* in milk and dairy products. *J. Dairy Sci.* **70:**383–391.

193. Schiemann, D. A. 1989. *Yersinia enterocolitica* and *Yersinia pseudotuberculosis*, p. 601–672. *In* M. P. Doyle (ed.), *Foodborne Bacterial Pathogens*. Marcel Dekker, New York, N.Y.

194. Schmid, Y., G. A. Grassl, O. T. Buhler, M. Skurnik, I. B. Autenrieth, and E. Bohn. 2004. *Yersinia enterocolitica* adhesin A induces production of interleukin-8 in epithelial cells. *Infect. Immun.* **72:**6780–6789.

195. Schmiel, D. H., E. Wagar, L. Karamanou, D. Weeks, and V. L. Miller. 1998. Phospholipase A of *Yersinia enterocolitica* contributes to pathogenesis in a mouse model. *Infect. Immun.* **66:**3941–3951.

196. Shayegani, M., D. Morse, I. DeForge, T. Root, L. M. Parsons, and P. S. Maupin. 1983. Microbiology of a major foodborne outbreak of gastroenteritis caused by *Yersinia enterocolitica* serogroup O:8. *J. Clin. Microbiol.* **17:**35–40.

197. **Shenkman, L., and E. J. Bottone.** 1976. Antibodies to *Yersinia enterocolitica* in thyroid disease. *Ann. Intern. Med.* 85:735–739.

198. **Simonet, M. L.** 1999. Enterobacteria in reactive arthritis: *Yersinia, Shigella,* and *Salmonella. Rev. Rhum. Engl. Ed.* 66:14S–18S.

199. **Sims, G. R., D. A. Glenister, T. F. Brocklehurst, and B. M. Lund.** 1989. Survival and growth of food poisoning bacteria following inoculation into cottage cheese varieties. *Int. J. Food Microbiol.* 9:173–195.

200. **Singh, I., and J. S. Virdi.** 2005. Interaction of *Yersinia enterocolitica* biotype 1A strains of diverse origin with cultured cells in vitro. *Jpn. J. Infect. Dis.* 58:31–33.

201. **Skrzypek, E., C. Cowan, and S. C. Straley.** 1998. Targeting of the *Yersinia pestis* YopM protein into HeLa cells and intracellular trafficking to the nucleus. *Mol. Microbiol.* 30:1051–1065.

202. **Skurnik, M., S. Batsford, A. Mertz, E. Schiltz, and P. Toivanen.** 1993. The putative arthritogenic cationic 19-kilodalton antigen of *Yersinia enterocolitica* is a urease beta-subunit. *Infect. Immun.* 61:2498–2504.

203. **Skurnik, M., Y. el Tahir, M. Saarinen, S. Jalkanen, and P. Toivanen.** 1994. YadA mediates specific binding of enteropathogenic *Yersinia enterocolitica* to human intestinal submucosa. *Infect. Immun.* 62:1252–1261.

204. **Skurnik, M., and P. Toivanen.** 1993. *Yersinia enterocolitica* lipopolysaccharide: genetics and virulence. *Trends Microbiol.* 1:148–152.

205. **Skurnik, M., R. Venho, J. A. Bengoechea, and I. Moriyon.** 1999. The lipopolysaccharide outer core of *Yersinia enterocolitica* serotype O:3 is required for virulence and plays a role in outer membrane integrity. *Mol. Microbiol.* 31:1443–1462.

206. **Slee, K. J., and C. Button.** 1990. Enteritis in sheep and goats due to *Yersinia enterocolitica* infection. *Aust. Vet. J.* 67:396–398.

207. **Slee, K. J., and N. W. Skilbeck.** 1992. Epidemiology of *Yersinia pseudotuberculosis* and *Y. enterocolitica* infections in sheep in Australia. *J. Clin. Microbiol.* 30:712–715.

208. **Smith, M. G.** 1992. Destruction of bacteria on fresh meat by hot water. *Epidemiol. Infect.* 109:491–496.

209. **Snellings, N. J., M. Popek, and L. E. Lindler.** 2001. Complete DNA sequence of *Yersinia enterocolitica* serotype O:8 low-calcium-response plasmid reveals a new virulence plasmid-associated replicon. *Infect. Immun.* 69:4627–4638.

210. **Stern, N. J., M. D. Pierson, and A. W. Kotula.** 1980. Effects of pH and sodium chloride on *Yersinia enterocolitica* growth at room and refrigeration temperatures. *J. Food Sci.* 45:64–67.

211. **Stuart, P. M., and J. G. Woodward.** 1992. *Yersinia enterocolitica* produces superantigenic activity. *J. Immunol.* 148:225–233.

212. **Sulakvelidze, A.** 2000. Yersiniae other than *Y. enterocolitica, Y. pseudotuberculosis,* and *Y. pestis*: the ignored species. *Microbes Infect.* 2:497–513.

213. **Sulakvelidze, A., A. Kreger, A. Joseph, R. M. Robins-Browne, A. Fasano, G. Wauters, N. Harnett, L. DeTolla, and J. G. Morris, Jr.** 1999. Production of enterotoxin by *Yersinia bercovieri,* a recently identified *Yersinia enterocolitica*-like species. *Infect. Immun.* 67:968–971.

214. **Tacket, C. O., J. Ballard, N. Harris, J. Allard, C. Nolan, T. Quan, and M. L. Cohen.** 1985. An outbreak of *Yersinia enterocolitica* infections caused by contaminated tofu (soybean curd). *Am. J. Epidemiol.* 121:705–711.

215. **Tacket, C. O., B. R. Davis, G. P. Carter, J. F. Randolph, and M. L. Cohen.** 1983. *Yersinia enterocolitica* pharyngitis. *Ann. Intern. Med.* 99:40–42.

216. **Tacket, C. O., J. P. Narain, R. Sattin, J. P. Lofgren, C. Konigsberg, Jr., R. C. Rendtorff, A. Rausa, B. R. Davis, and M. L. Cohen.** 1984. A multistate outbreak of infections caused by *Yersinia enterocolitica* transmitted by pasteurized milk. *JAMA* 251:483–486.

217. **Takao, T., T. Hitouji, S. Aimoto, Y. Shimonishi, S. Hara, T. Takeda, Y. Takeda, and T. Miwatani.** 1983. Amino acid sequence of a heat-stable enterotoxin isolated from enterotoxigenic *Escherichia coli* strain 18D. *FEBS Lett.* 152:1–5.

218. **Takao, T., Y. Shimonishi, M. Kobayashi, O. Nishimura, M. Arita, T. Takeda, T. Honda, and T. Miwatani.** 1985. Amino acid sequence of heat-stable enterotoxin produced by *Vibrio cholerae* non-O1. *FEBS Lett.* 193:250–254.

219. **Takao, T., N. Tominaga, S. Yoshimura, Y. Shimonishi, S. Hara, T. Inoue, and A. Miyama.** 1985. Isolation, primary structure and synthesis of heat-stable enterotoxin produced by *Yersinia enterocolitica. Eur. J. Biochem.* 152:199–206.

220. **Tauxe, R. V., J. Vandepitte, G. Wauters, S. M. Martin, V. Goossens, P. de Mol, R. Van Noyen, and G. Thiers.** 1987. *Yersinia enterocolitica* infections and pork: the missing link. *Lancet* i:1129–1132.

221. **Tennant, S. M., T. H. Grant, and R. M. Robins-Browne.** 2003. Pathogenicity of *Yersinia enterocolitica* biotype 1A. *FEMS Immunol. Med. Microbiol.* 38:127–137.

222. **Tennant, S. M., N. A. Skinner, A. Joe, and R. M. Robins-Browne.** 2005. Homologues of insecticidal toxin complex genes in *Yersinia enterocolitica* biotype 1A and their contribution to virulence. *Infect. Immun.* 73:6860–6867.

223. **Thompson, J. S., and M. J. Gravel.** 1986. Family outbreak of gastroenteritis due to *Yersinia enterocolitica* serotype O:3 from well water. *Can. J. Microbiol.* 32:700–701.

224. **Tipple, M. A., L. A. Bland, J. J. Murphy, M. J. Arduino, A. L. Panlilio, J. J. Farmer III, M. A. Tourault, C. R. Macpherson, J. E. Menitove, and A. J. Grindon.** 1990. Sepsis associated with transfusion of red cells contaminated with *Yersinia enterocolitica. Transfusion* 30:207–213.

225. **Toivanen, P., and A. Toivanen.** 1994. Does *Yersinia* induce autoimmunity? *Int. Arch. Allergy Immunol.* 104:107–111.

226. **Tomer, Y., and T. F. Davies.** 1993. Infection, thyroid disease, and autoimmunity. *Endocr. Rev.* 14:107–120.

227. **Toora, S., E. Budu-Amoako, R. F. Ablett, and J. Smith.** 1992. Effect of high-temperature short-time pasteurization, freezing and thawing and constant freezing, on the survival of *Yersinia enterocolitica* in milk. *J. Food Prot.* 55:803–805.

228. **Torruellas, J., M. W. Jackson, J. W. Pennock, and G. V. Plano.** 2005. The *Yersinia pestis* type III secretion needle plays a role in the regulation of Yop secretion. *Mol. Microbiol.* 57:1719–1733.

229. Tsubokura, M., K. Otsuki, K. Sato, M. Tanaka, T. Hongo, H. Fukushima, T. Maruyama, and M. Inoue. 1989. Special features of distribution of *Yersinia pseudotuberculosis* in Japan. *J. Clin. Microbiol.* **27:**790–791.

230. Tzipori, S., R. Robins-Browne, and J. K. Prpic. 1987. Studies on the role of virulence determinants of *Yersinia enterocolitica* in gnotobiotic piglets. *Contrib. Microbiol. Immunol.* **9:**233–238.

231. Une, T. 1977. Studies on the pathogenicity of *Y. enterocolitica*. I. Experimental infection in rabbits. *Microbiol. Immunol.* **21:**349–363.

232. Une, T., and R. R. Brubaker. 1984. Roles of V antigen in promoting virulence and immunity in yersiniae. *J. Immunol.* **133:**2226–2230.

233. Vahamiko, S., M. A. Penttinen, and K. Granfors. 2005. Aetiology and pathogenesis of reactive arthritis: role of non-antigen-presenting effects of HLA-B27. *Arthritis Res. Ther.* **7:**136–141.

234. Van Noyen, R., R. Selderslaghs, J. Bekaert, G. Wauters, and J. Vandepitte. 1991. Causative role of *Yersinia* and other enteric pathogens in the appendicular syndrome. *Eur. J. Clin. Microbiol. Infect. Dis.* **10:**735–741.

235. Venecia, K., and G. M. Young. 2005. Environmental regulation and virulence attributes of the Ysa type III secretion system of *Yersinia enterocolitica* biovar 1B. *Infect. Immun.* **73:**5961–5977.

236. Viboud, G. I., and J. B. Bliska. 2005. Yersinia outer proteins: role in modulation of host cell signaling responses and pathogenesis. *Annu. Rev. Microbiol.* **59:**69–89.

237. Viboud, G. I., S. S. So, M. B. Ryndak, and J. B. Bliska. 2003. Proinflammatory signalling stimulated by the type III translocation factor YopB is counteracted by multiple effectors in epithelial cells infected with *Yersinia pseudotuberculosis*. *Mol. Microbiol.* **47:**1305–1315.

238. Virdi, J. S., and P. Sachdeva. 2005. Molecular heterogeneity in *Yersinia enterocolitica* and '*Y. enterocolitica*-like' species: implications for epidemiology, typing and taxonomy. *FEMS Immunol. Med. Microbiol.* **45:**1–10.

239. Virto, R., D. Sanz, I. Alvarez, S. Condon, and J. Raso. 2005. Inactivation kinetics of *Yersinia enterocolitica* by citric and lactic acid at different temperatures. *Int. J. Food Microbiol.* **103:**251–257.

240. Visser, L. G., E. Seijmonsbergen, P. H. Nibbering, P. J. van den Broek, and R. van Furth. 1999. Yops of *Yersinia enterocolitica* inhibit receptor-dependent superoxide anion production by human granulocytes. *Infect. Immun.* **67:**1245–1250.

241. Wachtel, M. R., and V. L. Miller. 1995. In vitro and in vivo characterization of an *ail* mutant of *Yersinia enterocolitica*. *Infect. Immun.* **63:**2541–2548.

242. Warren, S. M., and G. M. Young. 2005. An aminoterminal secretion signal is required for YplA export by the Ysa, Ysc, and flagellar type III secretion systems of *Yersinia enterocolitica* biovar 1B. *J. Bacteriol.* **187:**6075–6083.

243. Wauters, G., S. Aleksic, J. Charlier, and G. Schulze. 1991. Somatic and flagellar antigens of *Yersinia enterocolitica* and related species. *Contrib. Microbiol. Immunol.* **12:**239–243.

244. Wauters, G., M. Janssens, A. G. Steigerwalt, and D. J. Brenner. 1988. *Yersinia mollaretii* sp. nov. and *Yersinia bercovieri* sp. nov., formerly called *Yersinia enterocolitica* biogroups 3A and 3B. *Int. J. Syst. Bacteriol.* **38:**424–429.

245. Wauters, G., K. Kandolo, and M. Janssens. 1987. Revised biogrouping scheme of *Yersinia enterocolitica*. *Contrib. Microbiol. Immunol.* **9:**14–21.

246. Weidow, C. L., D. S. Black, J. B. Bliska, and A. H. Bouton. 2000. CAS/Crk signalling mediates uptake of *Yersinia* into human epithelial cells. *Cell. Microbiol.* **2:**549–560.

247. Weinberg, E. D. 1984. Iron withholding: a defense against infection and neoplasia. *Physiol. Rev.* **64:**65–102.

248. Woestyn, S., A. Allaoui, P. Wattiau, and G. R. Cornelis. 1994. YscN, the putative energizer of the *Yersinia* Yop secretion machinery. *J. Bacteriol.* **176:**1561–1569.

249. Wuytack, E. Y., A. M. Diels, and C. W. Michiels. 2002. Bacterial inactivation by high-pressure homogenisation and high hydrostatic pressure. *Int. J. Food Microbiol.* **77:**205–212.

250. Yang, Y., and R. R. Isberg. 1997. Transcriptional regulation of the *Yersinia pseudotuberculosis* pH 6 antigen adhesin by two envelope-associated components. *Mol. Microbiol.* **24:**499–510.

251. Yip, C. K., T. G. Kimbrough, H. B. Felise, M. Vuckovic, N. A. Thomas, R. A. Pfuetzner, E. A. Frey, B. B. Finlay, S. I. Miller, and N. C. Strynadka. 2005. Structural characterization of the molecular platform for type III secretion system assembly. *Nature* **435:**702–707.

252. Yoshino, K., T. Takao, X. Huang, H. Murata, H. Nakao, T. Takeda, and Y. Shimonishi. 1995. Characterization of a highly toxic, large molecular size heat-stable enterotoxin produced by a clinical isolate of *Yersinia enterocolitica*. *FEBS Lett.* **362:**319–322.

253. Young, B. M., and G. M. Young. 2002. Evidence for targeting of Yop effectors by the chromosomally encoded Ysa type III secretion system of *Yersinia enterocolitica*. *J. Bacteriol.* **184:**5563–5571.

254. Young, G. M., D. Amid, and V. L. Miller. 1996. A bifunctional urease enhances survival of pathogenic *Yersinia enterocolitica* and *Morganella morganii* at low pH. *J. Bacteriol.* **178:**6487–6495.

255. Zhang, Y., A. T. Ting, K. B. Marcu, and J. B. Bliska. 2005. Inhibition of MAPK and NF-kappa B pathways is necessary for rapid apoptosis in macrophages infected with *Yersinia*. *J. Immunol.* **174:**7939–7949.

256. Zink, D. L., J. C. Feeley, J. G. Wells, C. Vanderzant, J. C. Vickery, W. D. Roof, and G. A. O'Donovan. 1980. Plasmid-mediated tissue invasiveness in *Yersinia enterocolitica*. *Nature* **283:**224–226.

Food Microbiology: Fundamentals and Frontiers, 3rd Ed.
Edited by M. P. Doyle and L. R. Beuchat
© 2007 ASM Press, Washington, D.C.

Keith A. Lampel
Anthony T. Maurelli

Shigella Species

15

Bacillary dysentery or shigellosis is caused by *Shigella* species. Dysentery was the term used by Hippocrates to describe an illness characterized by frequent passage of stools containing blood and mucus accompanied by painful abdominal cramps. Perhaps one of the greatest historical impacts of this disease has been its powerful influence in military operations. Long, protracted military campaigns and sieges almost always spawned epidemics of dysentery, causing large numbers of military and civilian casualties. With a low infectious dose required to cause disease coupled with oral transmission via fecally contaminated food and water, it is not surprising that dysentery caused by *Shigella* spp. follows in the wake of many natural (earthquakes, floods, and famine) and man-made (war) disasters. Apart from these special circumstances, shigellosis remains an important disease in developed countries as well as in underdeveloped countries.

Foodborne shigellosis is a neglected area of study on a global scale. Recent international efforts have directed focus on better coordination and awareness amongst the World Health Organization (WHO) and public regulatory agencies about the effects of foodborne illnesses and the pathogens that cause these diseases (40). The purpose of this chapter is to educate food microbiologists about the members of the *Shigella* spp. and the disease they cause. Modes of transmission and examples of recent foodborne outbreaks are also presented. Finally, our most current understanding of the genetics of *Shigella* pathogenesis, the genes involved in causing disease, and how they are regulated is discussed. Since no single review can be completely comprehensive, the reader is encouraged to refer to several excellent recent reviews for additional information (31, 44, 84, 105, 121).

CHARACTERISTICS OF THE ORGANISM

Classification and Biochemical Characteristics

There are four species of the genus *Shigella*, which are serologically grouped (43 serotypes) based on their somatic O antigens, including *Shigella dysenteriae* (group A), *S. flexneri* (group B), *S. boydii* (group C), and *S. sonnei* (group D). *S. dysenteriae* type 1 differs from the other serotypes in that it carries the genetic determinants for Shiga toxin, a potent cytotoxin. As members of the

Keith A. Lampel, Center for Food Safety and Applied Nutrition, Food and Drug Administration, 5100 Paint Branch Parkway, College Park, MD 20740. Anthony T. Maurelli, Department of Microbiology and Immunology, Uniformed Services University of the Health Sciences, F. Hébert School of Medicine, 4301 Jones Bridge Road, Bethesda, MD 20814-4799.

family *Enterobacteriaceae*, they are nearly genetically identical to the *Escherichieae* and are closely related to salmonellae (101). *Shigella* spp. are nonmotile, oxidase-negative, gram-negative rods. Some important biochemical characteristics that distinguish these bacteria from other enterics include their inability to ferment lactose (although some strains of *S. sonnei* may ferment lactose slowly) or to utilize citric acid as a sole carbon source. They do not produce H_2S, except for *S. flexneri* 6 and *S. boydii* serotypes 13 and 14, and do not produce gas from glucose. *Shigella* spp. are inhibited by potassium cyanide and do not synthesize lysine decarboxylase (43). Entero-invasive *Escherichia coli* (EIEC) strains have pathogenic and biochemical properties that are similar to those of *Shigella*. This similarity poses a problem in distinguishing these pathogens. For example, EIEC strains are nonmotile and are unable to ferment lactose. Some serotypes of EIEC also share identical O antigens with *Shigella* (117).

Shigella spp. are not particularly fastidious in their growth requirements, and in most cases, these organisms are routinely cultivated in the laboratory on artificial medium. Cultures of *Shigella* are easily isolated and grown from analytical samples, including water and clinical samples. In the latter case, *Shigella* isolates are present in fecal specimens in large numbers (10^3 to 10^9 per g of stool) during the acute phase of infection, and therefore identification is readily accomplished using culture media, biochemical analysis, and serological typing. *Shigella* organisms are shed and continue to be detected from convalescent patients (10^2 to 10^3 per g of stool) for weeks or longer after the initial infection. Isolation of *Shigella* at this stage of infection is more difficult because a selective enrichment broth for *Shigella* is not available, and therefore *Shigella* can be outgrown by resident bacterial fecal flora.

Isolation of *Shigella* from foods is not as facile as that from other sources. Foods have many different physical attributes that may affect the successful recovery of shigellae. These factors include composition, such as the fat content of the food; physical parameters, such as pH and salt; and the natural microbial flora of the food. In the last case, other microbes in a sample may overgrow *Shigella* during culture in broth media. The amount of time from the clinical report of a suspected outbreak to the analysis of the food samples can be considerable and therefore can lessen the chances of identifying the causative agent. The physiological state of shigellae present in the food is a contributing factor in the successful recovery of this pathogen. *Shigella* may be present in low numbers or in a poor physiological state in suspected food samples. Under these conditions, special enrichment procedures are required for successful detection of *Shigella* (7).

Shigella in Foods

Shigella spp. are not associated with any specific foods. Common foods that have been implicated in outbreaks caused by shigellae include potato salad, chicken salad, tossed salad, and shellfish. The locations where these types of *Shigella*-contaminated foods were served ranged from the home to restaurants, camps, picnics, schools, airlines, sorority houses, and military mess halls (128). In many cases, the source (food) was not determined. From 1983 to 1987, 2,397 foodborne outbreaks, including 54,453 cases, were reported to the Centers for Disease Control and Prevention (CDC). In only 38% of the cases was the source of the etiologic agent identified (9). Although epidemiologic investigations may strongly implicate a common food source, *Shigella* spp. are not often recovered and identified from foods by using standard bacteriological methods. Also, since shigellae are not commonly associated with any specific food, routine testing of foods to identify these pathogens is not usually performed.

The traditional approach to determine if foods are microbially contaminated in the processing plant is to assay the final product. There are several drawbacks to this approach (51), including the fact that current bacteriological methods are often time-consuming and laborious. An alternative to end-product testing is the Hazard Analysis and Critical Control Point (HACCP) (http://vm.cfsan.fda.gov/~lrd/haccp.html) system, which identifies certain points of the processing system that may be most vulnerable to microbial, chemical, and physical contamination (see chapter 46). This approach is a preventive program with less reliance on end-product testing. A system such as HACCP may be well suited for pathogenic bacteria such as *E. coli* O157:H7 and *Salmonella* spp., which are known to be associated with specific foods, e.g., with beef and poultry and with egg products, respectively.

In contrast, establishing specific critical control points for preventing *Shigella* contamination of foods is more challenging for the HACCP concept. This pathogen is usually introduced into foods by an infected person, such as a food handler with poor personal hygiene. In some cases, this may occur at the manufacturing site, but more often it happens at a point between the processing plant and the consumer. Another factor is that foods such as vegetables (e.g., lettuce) can be contaminated at the site of harvest and shipped contaminated directly to market. Although HACCP systems can effectively reduce or eliminate pathogen contamination of foods, pathogens that are not indigenous to but rather introduced into foods during food harvesting or preparation, such as *Shigella*, may not be controlled as effectively by HACCP systems.

Survival and Growth in Foods

Depending upon in vitro environmental conditions, *Shigella* spp. can survive in media with a pH range of 2 to 3 for several hours. Acid resistance is modulated by a sigma factor encoded by the *rpoS* (*katF*) gene (127). However, under acidic conditions, shigellae do not typically survive well, either in foods and or in stool samples. Among foods, studies using citric juices (orange, grape, and lemon), carbonated beverages, and wine revealed that some shigellae could be recovered after 1 to 6 days (58). In neutral-pH foods, such as butter or margarine, shigellae can be recovered after 100 days when stored frozen or at 6°C.

Shigella can survive a temperature range of –20°C to room temperature. Survival of shigellae is longer in foods stored frozen or at refrigeration temperatures than in those stored at room temperature. In foods such as salads with mayonnaise and some cheese products, *Shigella* can survive for 13 to 92 days. As an example, two strains of *Shigella boydii* 18, one from an outbreak in which contaminated bean salad was the vehicle and the other from the American Type Culture Collection, were inoculated into bean salad and stored at 4 or 23°C. No growth was observed at 4°C, but both strains survived well. However, at 23°C, a 2-log CFU/g increase occurred over 2 days and then the number of viable cells decreased (2). *Shigella* spp. can survive on dried surfaces for an extended period of time and in foods, e.g., shrimp, ice cream, and minced pork meat, stored frozen. Growth and survival rates of *Shigella* spp. are reduced in the presence of 3.8 to 5.2% NaCl at pH 4.8 to 5.0, 300 to 700 mg of $NaNO_2$/liter, and 0.5 to 1.5 mg of sodium hypochlorite (NaClO)/liter in water at 4°C. These bacteria are sensitive to ionizing radiation, with a reduction of 10^7 CFU at 3 kGy.

FOODBORNE OUTBREAKS

It is estimated that the number of foodborne illnesses in the United States is 6 million to 81 million annually (11, 89) and that the actual number of foodborne outbreaks may be greatly underreported (129). The adverse effect of foodborne pathogenesis on public health is reflected by its notable morbidity and mortality (111). In the United States, foodborne diseases are responsible for an estimated 5,000 to 9,000 deaths per year and for 325,000 hospitalizations. Although the CDC has reported that foodborne outbreaks of shigellosis in the United States

have declined recently, shigellosis continues to be a major public health concern. In compiling data from 1982 to 1997, which includes reported cases (from 1983 to 1992) and the number of cases from passive (1992 to 1997) and active surveillance via FoodNet (24; http://www.cdc.gov/ncidod/dbmd/foodnet), the CDC estimated that there were approximately 448,000 cases of shigellosis in the United States, making it the third leading cause of foodborne outbreaks by bacterial pathogens (89), and 600 deaths were attributed to *Shigella* (www.aphl.org/Conferences/PULSENET_UPDATE_MEETING_2004/files/2004/Wednesday/08-Jennings.pdf). Recent (2004) FoodNet data indicated that of a total of 15,806 laboratory-confirmed cases of infections, *Shigella* spp. were identified in 2,231 cases, the third largest number, behind *Salmonella* (6,464 cases) and *Campylobacter* (5,665 cases) (27). Worldwide, *Shigella* spp. cause endemic outbreaks in developing countries, whereas *S. dysenteriae* is responsible for large epidemics, often with high mortality rates. The WHO estimates that *Shigella* spp. are responsible for 164.7 million cases of shigellosis annually in developing countries and 1.5 million cases in industrialized countries (70). Approximately 700,000 people die each year from bloody diarrhea caused by shigellae. The WHO (*Guidelines for the Control of Shigellosis, including Epidemics Due to Shigella dysenteriae Type 1* [http://whqlibdoc.who.int/publications/2005/9241592330.pdf]), CDC, and other international agencies have instituted surveillance systems to monitor different aspects of the epidemiology of outbreaks caused by *Shigella*. Currently, CDC has the following six systems in place:

1. Public Health Laboratory Information System (http://www.cdc.gov/ncidod/dbmd/phlisdata). For information specifically for *Shigella* from this database, use http://www.cdc.gov/ncidod/dbmd/phlisdata/shigella.htm.
2. National Electronic Telecommunications System for Surveillance (http://www.cdc.gov/epo/dphsi/netss.htm). These data can be accessed at http://www.cdc.gov/mmwr or http://www.cdc.gov/mmwr/summary.html.
3. FoodNet is an active surveillance system that gathers data from over 300 laboratories throughout the country (http://www.cdc.gov/foodnet/annuals.htm).
4. National Molecular Subtyping Network for Foodborne Diseases Surveillance (PulseNet) uses pulsed-field gel electrophoresis (PFGE) patterns to create a database of DNA fingerprinting of several pathogens, including *Shigella* (http://www.cdc.gov/pulsenet).
5. National Antimicrobial Resistance Monitoring System monitors antimicrobial resistance of

not only *Shigella* but also other human bacterial pathogens (http://www.cdc.gov/narms).

6. Foodborne Outbreak Detection Unit (http://www.cdc.gov/epo/mmwr/preview/mmwrhtml/ss4901a1.htm).

The number of outbreaks due to each species of *Shigella* is located on the CDC website at http://www.cdc.gov/ncidod/dbmd/phlisdata/shigella.htm.

Transmission and Susceptible Populations

The primary means of transmission of *Shigella* is by the fecal-oral route through human feces. Most cases of shigellosis are caused by the ingestion of fecally contaminated food or water, and in the case of foods, the major factor for contamination is poor personal hygiene of food handlers. From infected carriers, this pathogen can spread by several routes, including food, fingers, feces, and flies. The last usually transmit the bacteria from fecal matter to foods. The highest incidence of shigellosis occurs during the warmer months of the year. Improper storage of contaminated foods is the second most common contributing factor to foodborne outbreaks of shigellosis (128). Other contributing factors are inadequate cooking, contaminated equipment, and food obtained from unsafe sources (9). To reduce the spread of shigellosis, infected food handlers should be monitored and not allowed to handle food until stool samples are negative for *Shigella*.

Shigella is a frank pathogen capable of causing disease in otherwise healthy individuals. Certain populations, however, may be more predisposed to infection due to the nature of transmission of the pathogen. The greatest frequency of shigellosis occurs among children of <4 to 6 years of age. In the United States, outbreaks of shigellosis and other diarrheal diseases in day care centers are increasing as more single-parent and two-parent working families use these facilities to care for their children (75, 106). Typical toddler behavior, such as oral exploration of the environment and inadequate personal hygiene habits, creates conditions ideally suited to transmission of bacterial, protozoal, and viral pathogens that are spread by fecal contamination. Transmission of *Shigella* in this population is very efficient, and the pathogen's low infectious dose increases the risk for shigellosis. Increased risk also extends to family contacts of day care attendees (142). Worldwide, epidemics of shigellosis in developing countries largely affect children between the ages of 1 and 4 years, and shigellosis accounts for nearly 10% of all diarrheal diseases in this age group. *Shigella dysenteriae* type 1 affects all age groups, most frequently in developing countries. In the United States,

S. sonnei accounts for approximately 85% of shigellosis cases, and *S. flexneri* accounts for approximately 15%. The influence of global travel on shigellosis (36) is evident by an increased incidence of an emerging *S. boydii* serotype in the United States among either travelers visiting other countries or individuals who had contact with foreign visitors (62).

Shigellosis can be endemic in a variety of institutional settings, such as prisons, mental hospitals, and nursing homes, where crowding or inadequate hygienic conditions create an environment for direct fecal-oral contamination. Crowded conditions and poor sanitation contribute to endemic shigellosis in developing countries as well.

When natural or man-made disasters destroy the sanitary waste treatment and water purification infrastructure, developed countries assume the conditions of developing countries. These conditions place the population at risk for diarrheal diseases, such as cholera and dysentery. Recent examples include famine and political upheaval in Somalia and the war in Bosnia-Herzegovina (75). Massive population displacement, e.g., refugees fleeing from Rwanda into Zaire in 1994, can also lead to explosive epidemics of diarrheal disease caused by *Vibrio cholerae* and *S. dysenteriae* type 1 (48).

The low infectious dose and oral route of transmission also make *Shigella* a potential biological weapon. The literature includes at least one report of deliberate contamination of food with *Shigella* that resulted in a dozen cases of shigellosis (69).

Reservoirs and Vehicles of Infection

Humans are the natural reservoir of *Shigella* infections, although several cases of transmission from monkeys have been reported (66). In one instance, three animal caretakers at a monkey house complained of having diarrhea. *S. flexneri* serotype 1b was isolated from stool samples from these employees, and further investigation revealed that four monkeys were shedding the identical serotype. The disease was spread by direct contact of the caretakers with excrement from the infected monkeys.

Asymptomatic carriers of *Shigella* may exacerbate the maintenance and spread of this pathogen in developing countries. Two studies, one in Bangladesh (56) and the other in Mexico (50), revealed that *Shigella* was present in stool samples from asymptomatic children under the age of 5 years. *Shigella* spp. were rarely found in infants under the age of 6 months.

Examples of Foodborne Outbreaks

In the cases illustrated below and in Table 15.1, it is apparent that a wide range of foods can be contaminated

Table 15.1 Examples of foodborne outbreaks caused by *Shigella* spp.

Year	Location; source of contamination[a]	Isolate	Reference
1986	Texas; shredded lettuce	*S. sonnei*	30
1987	Rainbow Family gathering; food handlers	*S. sonnei*	144
1988–1989	Monroe, NY; multiple sources	*S. sonnei*	145
1988	Outdoor music festival, Michigan; food handlers	*S. sonnei*	73
1988	Commercial airline; cold sandwiches	*S. sonnei*	53
1989	Cruise ship; potato salad	*S. flexneri*	76
1990	Operation Desert Shield (U.S. troops); fresh produce	*Shigella* spp.	57
1991	Alaska; moose soup	*S. sonnei*	46
1992–1993	Operation Restore Hope, Somalia, U.S. troops	*Shigella* spp.	124
1994	Europe; shredded lettuce from Spain	*S. sonnei*	64
1994	Midwest U.S.; green onions	*S. flexneri*	22
1994	Cruise ship	*S. flexneri*	23
1998	Fresh parsley	*S. sonnei*	25
2000	Bean dip	*S. sonnei*	26
2001	Tomato	*S. flexneri*	110

[a] The source of contamination is listed, when known.

with shigellae. Disease is caused by the ingestion of these contaminated foods and, in some instances, can subsequently lead to rapid dissemination in the population.

1987—Rainbow Family Gathering

At an annual gathering of the Rainbow Family, as many as half of the 12,700 people in attendance may have had shigellosis (144). *S. sonnei* was isolated from stool cultures from attendees. Spread of the pathogen most likely occurred by the fecal-oral route in a crowded environment by contamination of the food, water, or both. Subsequent outbreaks in three other states were due to attendees returning home and secondarily infecting other persons.

1989 and 1994—Shigellosis aboard Cruise Ships

In October 1989, 14% of passengers and 3% of crew members aboard a cruise ship reported gastrointestinal symptoms (76). A multiple antibiotic-resistant strain of *S. flexneri* serotype 2a was isolated from several ill passengers and crew. German potato salad was identified as the vehicle of infection. Contamination was introduced by infected food handlers, initially in the country where the food was originally prepared and secondarily by a member of the galley crew on the cruise ship. Another outbreak of shigellosis occurred in August 1994 on the cruise ship SS *Viking Serenade* (23). Thirty-seven percent (586) of passengers and 4% (30) of the crew reported having diarrhea, and one death occurred. *S. flexneri* 2a was isolated from patients, and spring onions were the suspected vehicle of infection.

1990—Operation Desert Shield

Diarrheal diseases during a military operation can be a major factor in reducing troop readiness. In Operation Desert Shield, enteric pathogens were isolated from 214 U.S. soldiers, and of those, 113 cases were diagnosed as shigellosis, with *S. sonnei* being the most prevalent species isolated (57). Shigellosis accounted for more time lost from military duties and was responsible for more severe morbidity than enterotoxigenic *E. coli*, the most common enteric pathogen isolated from U.S. troops in Saudi Arabia (57). The suspected source was contaminated fresh vegetables, specifically lettuce. Twelve heads of lettuce were examined, and enteric pathogens were isolated from all.

1991—Alaska Moose Soup

In September 1991, 25 people who participated in a gathering of local residents in Galena, Alaska, contracted shigellosis. Homemade moose soup was implicated as the vehicle of infection. One of five women who made the soup had reported that she had gastroenteritis before or at the time of preparing the soup. *S. sonnei* was isolated from one hospitalized patient.

1994—Contaminated Lettuce in Norway and the United Kingdom

One hundred ten culture-confirmed cases of shigellosis caused by *S. sonnei* were reported in an outbreak in Norway in 1994 (64). Iceberg lettuce from Spain which was served on a salad bar was suspected as the source of the Norwegian outbreak and was likely responsible for increases in shigellosis in other European countries, such as the United Kingdom (43) and Sweden. *S. sonnei* was isolated from patients from several Northwest European countries but was not isolated from any foods. Strong epidemiologic evidence indicated that the source of these outbreaks was imported lettuce.

1994—Green Onions

In the summer of 1994 in the Midwest United States, an outbreak of *S. flexneri* serotype 6 (mannitol negative) occurred (22). Although not confirmed, the suspected vehicle was Mexican green onions (scallions or spring onions). In June of that year, 17 people developed diarrhea at a church potluck meal in Indiana, with three confirmed cases of shigellosis. Twenty-nine people contracted shigellosis at an anniversary reception in Indiana in July. In Illinois, 26 culture-confirmed, mannitol-negative *S. flexneri* or *Shigella* sp. isolates were reported to the State Department of Health. Isolates of *S. flexneri* serotype 6 were also reported in Missouri, Minnesota, Wisconsin, Michigan, and Kentucky. Ingestion of green onions was strongly implicated as the source of shigellosis. An infected worker most likely contaminated the food at the time of harvest or packing.

1998—Fresh Parsley

During August 1998, several health departments in the United States and Canada reported many cases of shigellosis among people who ate at restaurants that served chopped, uncooked parsley. The causative agent was identified as *S. sonnei*, and based on PFGE subtyping of isolates, epidemiologic traceback, and data from other investigations, one farm in Mexico was implicated as the source of contaminated parsley. Further investigations revealed that the water supply used for chilling the parsley in a hydrocooler and making ice was unchlorinated and prone to bacterial contamination (25).

2000—Five-Layer Bean Dip

An outbreak of shigellosis caused by ingestion of contaminated five-layer (bean, salsa, guacamole, nacho cheese, and sour cream) party dip occurred in three West Coast states (26). *S. sonnei* was isolated from at least 30 patients and one layer (cheese) of the dip (144).

2001—Tomatoes

Several restaurants in the New York area purchased tomatoes that were overripe and bruised from one distributor. These tomatoes were identified by epidemiologic data, including the observation that only those restaurants that served these tomatoes were linked to a number of illnesses, with the tomatoes as the vehicle of infection. More than 880 people were ill, and *S. flexneri* 2a was isolated from several restaurant patrons as well as from ill and asymptomatic employees of these restaurants. None of the workers from the restaurants reported being ill before the tomatoes arrived. PFGE subtyping of *S. flexneri* isolates associated all restaurant patrons with this one outbreak (110).

One of the striking features about foodborne outbreaks caused by shigellae is that contamination of foods usually does not occur at the processing plant, but rather the source is usually a food handler. As evident from the examples described above, these incidents can occur in a variety of settings, ranging from small town gatherings and picnics to restaurants and cruise ships.

CHARACTERISTICS OF DISEASE

Clinical Presentation

Disease caused by *Shigella* spp. is distinguished from diseases caused by most of the other foodborne pathogens described in this volume in at least two important aspects, i.e., the production of bloody diarrhea or dysentery and the low infectious dose required to cause clinical symptoms. Bloody diarrhea refers to diarrhea where the stools contain visible red blood. Dysentery has the same meaning, but the passage of bloody mucoid stools is accompanied by severe abdominal and rectal pain, cramps, and fever. While abdominal pain and diarrhea are experienced by nearly all patients with shigellosis, fever occurs in about one-third of cases, and frank blood in the stools occurs in about 40% of cases (32).

The clinical picture of shigellosis ranges from a mild watery diarrhea to severe dysentery. All *Shigella* spp. can cause acute bloody diarrhea. The dysentery stage of the disease caused by *Shigella* spp. may or may not be preceded by watery diarrhea. This stage reflects the transient multiplication of bacteria as they pass through the small bowel. Jejunal secretions probably are not effectively reabsorbed in the colon due to transport abnormalities caused by bacterial invasion and destruction of the colonic mucosa (68). The dysentery stage of disease correlates with extensive bacterial colonization of the colonic mucosa. The bacteria invade the epithelial cells of the colon and spread from cell to cell but penetrate only as far as the lamina propria. Foci of individually infected cells produce microabscesses that coalesce, forming large abscesses and mucosal ulcerations. As the infection progresses, dead cells of the mucosal surface slough off, thus leading to the presence of blood, pus, and mucus in the stools.

The incubation period for shigellosis is 1 to 7 days, but the illness usually begins within 3 days. Strains of *S. dysenteriae* type 1 cause the most severe disease, whereas *S. sonnei* produces the mildest disease. *S. flexneri* and *S. boydii* infections can be either mild or severe. Despite the severity of the disease, shigellosis is self-limiting. If left untreated, clinical illness usually persists for 1 to 2 weeks (although it may last as long as a month), and the patient recovers.

Infectious Dose

As mentioned previously, an important aspect of *Shigella* pathogenesis is the extremely low 50% infectious dose, i.e., the experimentally determined oral dose required to cause disease in 50% of volunteers challenged with a virulent strain of the pathogen. The 50% infectious dose for *S. flexneri*, *S. sonnei*, and *S. dysenteriae* is approximately 5,000 cells. Volunteers become ill when doses as low as 200 cells are given (33). The low infectious dose of *Shigella* underscores the high communicability of bacillary dysentery and gives the disease great explosive potential for person-to-person spread as well as foodborne and waterborne outbreaks of diarrhea.

Complications

Shigellosis can be a very painful and incapacitating disease and is more likely to require hospitalization than other bacterial diarrheas. It is not usually life-threatening, and mortality is rare, except in malnourished children, immunocompromised individuals, and the elderly (13). However, complications arising from the disease include severe dehydration, intestinal perforation, toxic megacolon, septicemia, seizures, hemolytic-uremic syndrome (HUS), and reactive arthritis (12). HUS is a rare but potentially fatal complication associated with infection by *S. dysenteriae* 1 (108). The syndrome is characterized by hemolytic anemia, thrombocytopenia, and acute renal failure. Epidemiologic studies suggest that Shiga toxin produced by *S. dysenteriae* 1 may be the cause of HUS (77). This hypothesis is supported by the fact that HUS is also caused by strains of *E. coli* O157:H7, which produce high levels of Shiga toxin (65). It has been suggested that Shiga toxin may cause HUS by entering the bloodstream and damaging vascular endothelial cells such as those in the kidney (64, 77, 131). Reactive arthritis is a postinfection sequela to shigellosis that is strongly associated with individuals of the HLA-B27 histocompatibility group (126). The syndrome is comprised of three symptoms, urethritis, conjunctivitis, and arthritis, with the last being the most dominant symptom. Infections caused by several other gram-negative enteric pathogens can

also lead to this type of sterile inflammatory polyarthropathy (20).

Treatment and Prevention

Although stool fluid losses are not as massive as those with other bacterial diarrheas, the diarrhea associated with shigellosis, combined with water loss due to fever and decreased water intake due to anorexia, may result in severe dehydration (12). Oral fluid intake can generally replace fluid losses, although intravenous rehydration may be required for very young and elderly patients.

The antibiotic of choice for treatment of shigellosis is trimethoprim-sulfamethoxazole (32). However, there is some controversy regarding the use of antibiotics in treating shigellosis. Since the infection is self-limited in normally healthy patients and full recovery occurs without the use of antibiotics, drug therapy is usually not indicated. In addition, multiple drug resistance among isolates of *Shigella* is becoming more common. Clinical isolates resistant to sulfonamides, ampicillin, trimethoprim-sulfamethoxazole, tetracycline, chloramphenicol, and streptomycin have been found (14, 55). Extensive use of antibiotics selects for drug-resistant microbes, and therefore many believe that antimicrobial therapy for shigellosis should be reserved only for the most severely ill patients. On the other hand, there are persuasive public health arguments for the antibiotic management of shigellosis. Antibiotic treatment limits the duration of disease and shortens the period of fecal excretion of bacteria (52). Since an infected person or asymptomatic carrier can be an index case for person-to-person and food- and waterborne spread, antibiotic treatment of these individuals can be a significant public health tool to contain the spread of shigellosis. However, antibiotics are not a substitute for improved hygienic conditions to contain the secondary spread of shigellosis. The single most effective means of preventing secondary transmission is handwashing. Food handling and preparation are important processes that also deserve attention, and persons with diarrhea should be excluded from handling food.

Despite many years of intensive effort, an effective vaccine against shigellosis still has not been developed. Attenuated oral vaccine strains of *S. flexneri* are currently being tested. One such strain is SC602. This strain has mutations that block iron uptake and abolish the intracellular and intercellular motility phenotypes (8). A limited challenge study showed protection among persons treated with vaccines. However, one major drawback still to be resolved is the transient fever and mild diarrhea associated with administration of the vaccine (29). This study highlights one of the persistent problems impeding the development of a safe *Shigella* vaccine, i.e., designing a strain that can induce a protective immune response without producing unacceptable side effects. Other vaccines are currently in different stages of development (preclinical to phase III), including live attenuated *S. flexneri-S. sonnei* and *S. flexneri-S. dysenteriae* combinations (www.who.int/vaccine_research/documents/en/Status_Table_April05.pdf).

VIRULENCE FACTORS

Hallmarks of Virulence

Shigella spp. and EIEC are the principal agents of bacillary dysentery and, as such, belong to the group of enteric pathogens that cause disease by overt invasion of epithelial cells in the large intestine. The clinical symptoms of shigellosis can be attributed directly to the hallmarks of *Shigella* virulence, which include the abilities to induce diarrhea, invade epithelial cells of the intestine, multiply intracellularly, spread from cell to cell, and induce an inflammatory response.

Shigella cells colonize the small intestine only transiently and cause little tissue damage (112). The production of enterotoxins by the bacteria while they are in the small bowel likely results in the diarrhea that generally precedes the onset of dysentery (39, 97). The jejunal secretions elicited by these toxins may facilitate passage of the bacteria through the small intestine and into the colon, where they colonize and invade the epithelium.

Formal and colleagues established the essential role of epithelial cell invasion in *Shigella* pathogenesis in a landmark study that employed both in vitro tissue culture assays for invasion and animal models (71). Spontaneous colonial variants of *S. flexneri* 2a that are unable to invade epithelial cells in tissue culture do not cause disease in monkeys. Gene transfer studies using *E. coli* K-12 donors and *S. flexneri* 2a recipients established the third hallmark of *Shigella* virulence. An *S. flexneri* 2a recipient that inherited the *xyl-rha* region of the *E. coli* K-12 chromosome retained the ability to invade epithelial cells but has a reduced ability to multiply within these cells (38). This hybrid strain fails to cause fatal infection in an opium-treated guinea pig model and is unable to cause disease when fed to rhesus monkeys (42).

It is necessary but not sufficient for *Shigella* to be able to multiply within host epithelial cells after invasion. The bacterium must also be able to spread through the epithelial layer of the colon by cell-to-cell spread that does not require the bacterium to leave the intracellular environment and be reexposed to the intestinal lumen. Mutants of *Shigella* that are competent for invasion and multiplication but unable to spread between cells in this fashion have been isolated. Observation of these mutants established intracellular spread as the fourth hallmark of *Shigella* virulence, and they are discussed further below.

Along with the ability to colonize and cause disease, an intrinsic part of a bacterium's pathogenicity is its mechanism for regulating the expression of the genes involved in virulence. Virulence in *Shigella* spp. is regulated by the growth temperature. After growth at 37°C, virulent strains of *Shigella* are able to invade mammalian cells, but when cultivated at 30°C, they are noninvasive. This noninvasive phenotype is reversible by shifting the growth temperature to 37°C. The temperature change enables the bacteria to reexpress their virulence properties (82). Temperature regulation of virulence gene expression is a characteristic that *Shigella* shares with other human pathogens, such as *E. coli*, *Salmonella enterica* serovar Typhimurium, *Bordetella pertussis*, *Yersinia* spp., and *Listeria monocytogenes* (see reference 80 for a review). Regulation of gene expression in response to environmental temperature is a useful bacterial strategy. By sensing the ambient temperature of the mammalian host (e.g., 37°C for humans) to trigger gene expression, this strategy permits *Shigella* spp. to economize energy that would be expended on the synthesis of virulence products when the bacteria are outside the host. The system also permits the bacteria to coordinately regulate the expression of multiple unlinked genes that are required for the full virulence phenotype. Temperature regulation in *S. flexneri* 2a operates at the level of gene transcription and is mediated by both positive and negative transcription factors. A chromosomal gene, *virR* (*hns*), encodes a repressor of virulence gene expression (85), while two other genes, *virF* and *virB*, encode positive activators (1, 112). These genes are discussed in a later section. A more thorough treatment of virulence gene regulation in *Shigella* can be found in several review articles (31, 101).

Genetics

Virulence-Associated Plasmid Genes

Given the complexity of the interactions between host and pathogen, it is not surprising that *Shigella* virulence is multigenic, involving both chromosomal and plasmid-encoded genes (Table 15.2). Another landmark report on the pathogenicity of *Shigella* was the demonstration of the indispensable role of a large plasmid in invasion. A 180-kb plasmid in *S. sonnei* and a 220-kb plasmid in *S. flexneri* are essential for invasion (118, 119). Other *Shigella* spp., as well as strains of EIEC, contain homologous plasmids which are functionally interchangeable and share significant degrees of DNA homology (117). Hence, it is probable that the plasmids of *Shigella* and

Table 15.2 Virulence-associated loci of *Shigella*

Locus	Product	Role in virulence
Chromosomal loci		
cpxR	Response regulator of CpxA-CpxR two-component system	Activator of *virF*
iuc	Synthesis of aerobactin and receptor	Acquisition of iron in the host
rfa and *rfb*	Enzymes for core and O antigen biosynthesis	Correct polar localization of IcsA
set[a]	ShET1	Enterotoxin
she	Putative hemagglutinin and mucinase	Unknown
sodB	Superoxide dismutase	Inactivation of superoxide radicals; defense against oxygen-dependent killing in host
stx[b]	Shiga toxin	Destruction of vascular tissue
vacB (*rnr*)	Exoribonuclease RNase R	Posttranscriptional regulation of virulence gene expression
virR (*hns*)	Histone-like protein	Repressor of virulence gene expression
Plasmid-carried loci		
icsA (*virG*)	Cell-bound and secreted protein	Actin polymerization for intracellular motility and intercellular spread
ipaA	Secreted effector	Efficient invasion; binds to vinculin and promotes F-actin
ipaB	Secreted effector	Invasion; lysis of vacuole; induction of apoptosis
ipaC	Secreted effector	Invasion; induces cytoskeletal reorganization
ipaD	Secreted effector	Invasion; antisecretion plug (with IpaB); depolymerization
ipgC	17-kDa protein	Cytoplasmic chaperone for IpaB and IpaC; coactivator of MxiE
ipgB1	Secreted effector	Promotes membrane ruffling by activation of Rac1 and Cdc42
mxi/spa	20 proteins	Type III system for secretion of Ipa and other virulence proteins
mxiE	Transcriptional activator	AraC family postinvasion activator
ospD1	Secreted effector	Antiactivator of MxiE
ospG	Secreted effector	Targets ubiquitin-conjugating enzymes; down-regulates host innate immune response
icsB	Secreted effector	Shields IcsA from binding autophagy protein Atg5; protects against autophagy
sen	ShET2	Enterotoxin
virB	Transcriptional activator	Temperature regulation of virulence genes
virF	Transcriptional activator	Temperature regulation of virulence genes

[a] The *set* locus and production of ShET1 are found almost exclusively in *S. flexneri*.
[b] The *stx* locus and production of Shiga toxin are observed only in *S. dysenteriae* 1.

EIEC were derived from a common ancestor (116). The DNA sequence of the virulence plasmid of *S. flexneri* was the first to be published (19, 60, 138), and the virulence plasmids of *S. sonnei*, *S. dysenteriae* 1, and *S. boydii* have also been sequenced (59, 146).

A 37-kb region of the invasion plasmid of *S. flexneri* 2a contains all of the genes necessary to permit the bacterium to penetrate into tissue culture cells. This DNA segment was identified as the minimal region of the virulence plasmid necessary to allow a plasmid-cured derivative of *S. flexneri* (and *E. coli* K-12) to invade tissue culture cells (81). The region carries about 33 genes contained in two groups transcribed in opposite orientations (Fig. 15.1). These genes encode proteins that are components of the type III secretion system (T3SS) apparatus, effectors, translocators, transcription activators, and chaperones. Although a precise transcription map of these genes has not been defined, available evidence and the DNA sequence of the region suggest a multiple-operon organization.

The genes comprising the *ipaBCDA* (invasion plasmid antigens) cluster encode the immunodominant antigens detected in sera from convalescent patients and experimentally challenged monkeys (98). The *ipaBCD* genes have been demonstrated experimentally to be required for invasion of mammalian cells (91). An *ipaA* mutant is still invasive but has a 10-fold reduced ability compared to the wild-type strain (134). The Ipa products act as translocators by interacting with the host cell membrane subsequent to being shunted through the bacterial

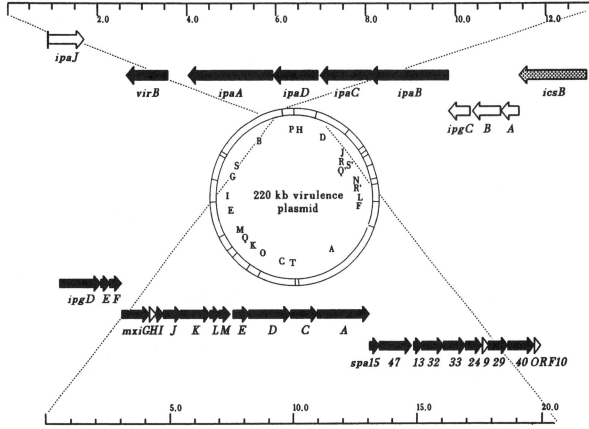

Figure 15.1 Map of virulence plasmid from *S. flexneri* 2a. A SalI restriction map of the 220-kb plasmid is shown in the center. Sections of SalI fragments B and P (top map) and fragments P, H, and D (bottom map) are expanded to illustrate the virulence loci carried in these regions. The expanded regions are contiguous and cover 32 kb. The open reading frame for *icsB* is separated from that of *ipgD* by 314 bp. The entire sequences of the virulence plasmids from *S. flexneri* 2a and 5 have been determined (19, 60, 138).

membrane through the T3SS. IpaB and IpaC (and probably IpaA) form a complex on the bacterial cell surface and are responsible for transducing the signal that leads to entry of *Shigella* into host cells via bacterium-directed phagocytosis (93). IpaD is also required for insertion of IpaBC into the membrane (107). However, when IpaB and IpaC are used to coat latex beads, they form a complex that promotes uptake of the beads by HeLa cells (90). This complex of Ipa proteins also binds to cell surface receptors such as α5-β1 integrins (140). Purified IpaC induces cytoskeletal reorganization via actin polymerization and depolymerization, including the formation of filopodia and lamellipodial extensions on permeabilized cells (135). IpaA binds to vinculin and promotes F-actin depolymerization (17, 134). This step is thought to facilitate reorganization of the host cell surface structures induced by contact with *Shigella* and to modulate bacterial entry. A recent overview of the interactions of

the gene products encoded by the large virulence plasmid has been published (105).

Although the Ipa proteins have no typical signal sequence, they are secreted into the extracellular medium. Contact of the bacterium with epithelial cells causes increased secretion of the cytoplasmic pool of Ipa products (5, 92). IpaD forms an antisecretion complex, or plug, with IpaB. Consequently, *ipaD* mutants are hypersecreters of Ipa products (92). *ipgC* (invasion plasmid gene) is required for invasion and acts as a cytoplasmic chaperone which prevents IpaB and IpaC from forming complexes while in the bacterial cytoplasm (93). In the absence of IpgC, IpaB and IpaC are rapidly degraded. IpgC also plays a role as a coactivator in virulence gene regulation (see below).

The product of *ipaB* has also been postulated to be the "contact hemolysin" which is responsible for lysis of the phagocytic vacuole minutes after entry of the bacterium into the host cell (54). The ability of *S. flexneri* to induce apoptosis in infected macrophages is an additional property assigned to IpaB (147).

The products of the *ipa* genes are actively secreted into the extracellular medium even though they contain no signal sequence for recognition by the usual gram-negative bacterial transport system. Ipa secretion is mediated by a T3SS pathway (121) and requires a dedicated apparatus composed of gene products from the *mxi/spa* loci (Fig. 15.1). The *mxi* (membrane expression of invasion plasmid antigens) genes comprise an operon that encodes several lipoproteins (MxiJ and MxiM), a transmembrane protein (MxiA), and proteins containing signal sequences (MxiD, MxiJ, and MxiM) (3–6). MxiH, MxiJ, MxiD, and MxiA share homology with proteins involved in the secretion of virulence proteins (Yops) in *Yersinia* spp. (3, 4, 6). MxiE, an AraC homologue, is a transcriptional activator that is involved in regulating the expression of a number of virulence plasmid genes (63, 86).

The *spa* (surface presentation of Ipa antigens) genes encode proteins that share significant homology with proteins involved in flagellar synthesis in *E. coli*, *Salmonella* serovar Typhimurium, *Bacillus subtilis*, and *Caulobacter crescentus* (120, 137). Included among these genes is *spa47*, which encodes a protein that probably functions as the energy-generating component of the secretion apparatus since it has sequence similarities with ATPases of the flagellar assembly machineries of other bacteria (137).

The proteins encoded by the *mxi* and *spa* genes comprise the *Shigella* T3SS (121). The secretion apparatus consists of the inner membrane proteins MxiG and MxiJ and an outer membrane protein, MxiD, that, along with MxiM, is proposed to form a bridge across the periplasmic space to contact MxiJ and MxiG (122). The secretion "needle" is composed of MxiH and, possibly, MxiI (16, 130). Spa47 plays a critical role in facilitating the transfer of proteins through the needle complex (61), while Spa32 is involved in controlling needle length (78). Spa33 associates with MxiG and MxiJ in the putative C ring of the secretion apparatus and controls the secretion of IpaB and IpaC (94, 123). Nonpolar null mutations in all of the *mxi/spa* genes tested to date result in a loss of the ability to secrete the Ipa proteins and a loss of invasive capacity for cultured cells. Hence, the T3SS is an essential component of *Shigella* virulence. The exception to this statement is *mxiE*, which encodes a transcriptional regulator of genes expressed postinvasion. MxiE is addressed below in the section on gene regulation.

Apart from the resemblance to genes involved in flagellar synthesis and the secretion of Yops, there is an even more striking similarity in both gene organization

and predicted protein sequence between the *mxi/spa* region of the *Shigella* virulence plasmid and a virulence-associated chromosomal region of *Salmonella* serovar Typhimurium (44, 49). The *Salmonella spa* region encodes homologues of the *Shigella spa* genes in the same gene order. Sequence identities between the protein homologues range as high as 86% (Spa9 versus SpaQ). The relatedness of the *spa* regions strongly suggests that these two human pathogens both evolved similar mechanisms for secretion of the virulence proteins required for signal transduction with the mammalian host. Plant pathogens such as *Erwinia carotovora*, *Xanthomonas campestris*, and *Pseudomonas solanacearum* also contain genes that encode homologues of the *mxi/spa*-encoded proteins (136). It is now recognized that the T3SS for transport of virulence proteins is a critical element of both plant and animal bacterial pathogenesis.

Secretion of the Ipa proteins via the Mxi/Spa apparatus is induced when *Shigella* cells contact the host cell (92, 140). Other compounds that mimic this signal and induce Ipa secretion are fibronectin, laminin, collagen type IV, Congo red, bile salts, and fetal bovine serum (104, 141). A similar phenomenon of contact-induced secretion is observed in other bacterial pathogens that utilize a T3SS pathway for extracellular transport of virulence effectors (121).

A new class of exported proteins was identified from the sequence of the virulence plasmid. Their genes are designated *osp* (outer *Shigella* protein) genes, and their functions are only just beginning to be determined (19). OspD1 negatively controls virulence gene expression by acting as an antiactivator (see below), and OspG is involved in down-regulation of the host innate immune response through interference with activation of the NF-κB pathway (67).

A plasmid-carried virulence gene which is unlinked to the 37-kb region shown in Fig. 15.1 is not required for invasion but is crucial for intra- and intercellular motility. This gene, known as *virG* or *icsA* (intracellular spread), encodes a protein that catalyzes the polymerization of actin in the cytoplasm of the infected host cell (15, 79). The IcsA protein is unusual in that it is expressed asymmetrically on the bacterial surface, being found at only one pole (47). The polymerization of actin monomers by IcsA forms a tail leading from the pole and provides the force that propels the bacterium through the cytoplasm. Hence, unipolar expression of IcsA imparts directionality of movement to the bacterium. Although the mechanism of unipolar localization of IcsA is unknown, it is dependent on the synthesis of a complete lipopolysaccharide (LPS) (114). LPS mutants of *S. flexneri* 2a express surface IcsA in a circumferential fashion, and while the protein is still capable of polymerizing actin, cell movement is restricted as the bacterium becomes encased in a shell of actin. A plasmid-encoded protease, SopA/IcsP, cleaves IcsA and is proposed to play a role in the unipolar localization of IcsA (35, 125). However, *E. coli* and plasmid-cured derivatives of *S. flexneri* transformed with a cloned *icsA* gene localize IcsA normally (115). These results suggest that IcsA localization does not require any other virulence plasmid-carried gene and that motifs within IcsA itself contain the information that directs the protein to the pole. The use of translational fusions between IcsA and green fluorescent protein defined the amino acids that are sufficient for polar localization and confirmed correct polar localization of IcsA expressed in *E. coli* as well as in *Salmonella* serovar Typhimurium, *Yersinia pseudotuberculosis*, and *Vibrio cholerae* (28). Surface-expressed IcsA is a trigger for autophagy. IcsB is a protein secreted via the *Shigella* T3SS that acts to shield IcsA from binding the autophagy protein Atg5 and thus protects the bacterium from being trapped and degraded by autophagy within the infected host cell (102).

Chromosomal Virulence Loci

In contrast to the genes of the virulence plasmid that are responsible for invasion of mammalian tissues, most of the chromosomal loci associated with *Shigella* virulence are involved in regulation or survival within the host. Mutations that alter O antigen and core synthesis or assembly lead to a "rough" phenotype and render *Shigella* avirulent. Synthesis of a complete LPS, which is crucial for correct unipolar localization of IcsA (see above), requires chromosomal loci such as *rfa* and *rfb*. In the case of *S. sonnei* and *S. dysenteriae* 1, plasmid-carried genes are also necessary for synthesis of the LPS O side chain (18).

Although *Shigella* and *E. coli* are very closely related at the genetic level, there are significant differences in these organisms beyond the presence of the virulence plasmid in *Shigella*. Pathogenicity islands, which are clusters of genes that have a role in virulence, have been identified in several virulent bacteria, including *Shigella*. These regions are typically large (20 to 200 kb), and in addition to virulence genes, they may have transposable elements, e.g., insertion sequence elements, and other modes of mobility, such as plasmid or bacteriophage genes, all indicating possible horizontal transfer of genetic information.

Two pathogenicity islands have been identified in the chromosome of *S. flexneri* (Table 15.2). *Shigella* pathogenicity island 1 (SHI-1) contains the *set* gene, which encodes an enterotoxin (39, 109). It is contained within the open reading frame of another gene, *she*, which encodes a protein with putative hemagglutinin and mucinase activities (109). The *iuc* locus, which contains the genes for aerobactin synthesis and transport, is present in SHI-2 (95, 139). Aerobactin is a hydroxamate siderophore that *S. flexneri* uses to scavenge iron. When the *iuc* locus is inactivated, aerobactin-deficient mutants retain the capacity to invade host cells but are altered in virulence, as measured in animal models. These results suggest that aerobactin synthesis is important for bacterial growth within the mammalian host (72, 96).

S. dysenteriae 1, like *E. coli* O157:H7, produces Shiga toxin, which has been implicated in producing some of the symptoms of HUS. Acquisition of the Shiga toxin genes, *stxA* and *stxB*, by *S. dysenteriae* has been postulated to be a consequence of a lysogenic bacteriophage. Over time, the phage became stably integrated due to a loss of genetic function of essential phage genes, most likely as a result of rearrangement and transposition events. Shiga toxin inhibits protein synthesis by inactivating rRNA (28S rRNA) in the 60S subunits of mammalian ribosomes and preventing elongation factor 2 from interacting with the ribosome (99). Shiga toxins are found infrequently in other *Shigella* spp. The Shiga toxin produced by *S. dysenteriae* and that elaborated by *E. coli* O157:H7 are nearly identical, differing from each other by just one amino acid.

A mutation in the *stx* locus does not alter the ability of the bacterium to invade epithelial cells or cause keratoconjunctivitis in the Sereny test. However, when tested in macaque monkeys, the mutant strain caused less vascular damage in the colonic tissue than did the toxin-producing parent (41). Hence, the production of Shiga toxin may account for the generally more severe infections caused by *S. dysenteriae* 1 than those caused by the other species of *Shigella*. Two other enterotoxins are produced by *Shigella*. The gene for *Shigella* enterotoxin 1 (ShET1) is carried chromosomally and is present in *S. flexneri* but is not usually found in other *Shigella* strains. ShET2 (*Shigella* entertoxin 2) is encoded by the virulence plasmid in many, but not all, *Shigella* serotypes.

Early genetic studies by Formal and colleagues (42) and recent reports that select genetic loci are not present in the *Shigella* chromosome but are present in the nonpathogenic *E. coli* chromosome (83) suggest that the absence of particular genes in *Shigella* has a direct consequence on its pathogenesis. The absence of these genetic determinants in *Shigella* spp. occurred by either deletion or mutation and is referred to as a "black hole" (83). When a gene complementary to a genetic locus that is absent in *Shigella* is introduced into the pathogen, the presence of the newly acquired gene product in *Shigella* may inhibit the action of a particular virulence factor. For instance, although lysine decarboxylase activity is present in >85% of *E. coli* strains, it is missing in all strains of *Shigella* spp. and EIEC. Lysine decarboxylase (encoded by the *cadA* gene) produces cadaverine from lysine. Cadaverine inhibits the action of the *Shigella* enterotoxins that are believed to be responsible for the diarrheal symptoms associated with shigellosis. Therefore, *cadA* may be considered an antivirulence gene for *Shigella* (83). Thus, in the evolution of *Shigella* as a human pathogen from a nonpathogenic *E. coli* lineage, the acquisition of the large virulence plasmid may also have involved the loss of specific incompatible genetic loci.

Virulence Gene Regulation

An important feature of *Shigella* pathogenesis is the ability of the bacterium to modulate the expression of its virulence genes in response to the growth temperature. Several activators and a repressor control the virulence regulon of *Shigella*, such that genes are turned on at 37°C and turned off at 30°C. The product of the chromosomal *virR* (*hns*) locus is a histone-like protein, H-NS, which behaves as a repressor of *Shigella* virulence gene expression in response to the growth temperature (85). Mutations in *virR* (*hns*) cause a deregulation of temperature control and result in the expression of genes in the virulence regulon even at the nonpermissive temperature of 30°C. The *virR* (*hns*) locus is allelic with regulatory loci in other enteric bacteria, and like *virR* (*hns*), these alleles act as repressors of their respective regulons (see the references cited in reference 31). Several different models to explain how VirR/H-NS acts as a transcriptional repressor have been proposed (31). However, because VirR/H-NS is involved in gene regulation in response to diverse environmental stimuli, such as osmolarity, pH, and temperature, a comprehensive model to explain its activity has been elusive.

H-NS binds to the promoter regions of two transcriptional activators, *virF* and *virB*, and blocks their transcription at 30°C (10, 37). At 37°C, the H-NS binding sites undergo a conformational transition and no longer bind H-NS, thus leading to increased transcription of the activator genes. One of the targets of VirR/H-NS regulation is the plasmid-carried transcriptional activator gene *virB* (133). The expression of genes in the *ipa* and *mxi/spa* clusters is dependent on VirB, and mutations in *virB* abolish the bacterium's ability to invade tissue culture cells (1, 21). Transcription of *virB* is dependent on the growth temperature and VirF (132). VirB is a DNA-binding protein and shares homology with the plasmid-partitioning proteins ParB of bacteriophage P1 and SopB of plasmid F. Purified VirB shows preferential binding to the intergenic *icsB-ipgD* region (Fig. 15.1) and displays a

similar preference for binding to the *spa* and *virA* structural gene promoter regions (88).

The product of the *virF* locus is a key element in temperature regulation of the *Shigella* virulence regulon. A helix-turn-helix motif in the carboxyl-terminal portion of VirF is characteristic of members of the AraC family of transcriptional activators (45). Consistent with its predicted role as a DNA-binding protein, VirF binds to sequences upstream of *virB* (133). The binding of VirF may act as an antagonist to binding by VirR/H-NS and thereby provides a mechanism for responding to temperature. The expression of *virF* is subject to temperature regulation by H-NS. Repression of *virF* occurs at a critical temperature of <32°C and takes place through the binding of H-NS at two sites within the *virF* promoter (37).

Activation of the T3SS leads to increased expression of a subset of virulence genes encoding secreted effectors. Expression of these genes is regulated by MxiE, a transcriptional activator of the AraC family (63, 86). MxiE-regulated promoters are preceded by a 17-bp MxiE box 33 to 49 bp upstream of the transcription start site (87). The activity of MxiE requires a coactivator, IpgC, the chaperone of IpaB and IpaC (86). Under nonsecretion conditions, IpgC is associated with IpaB and IpaC, whereas MxiE is associated with the T3SS substrate OspD1, which acts as an antiactivator. Hence, both the activator (MxiE) and coactivator (IpgC) are titrated and unavailable to interact with each other. When the secretion apparatus is activated, IpaB and IpaC are secreted and release IpgC. After OspD1 is secreted, MxiE is released and is free to bind IpgC and activate MxiE-controlled promoters (103). Expression profile analysis defined the following three classes of T3SS secreted substrates: (i) those that are controlled by VirB (expressed independently of secretion activity), (ii) those that are controlled by MxiE (expressed only under secretion conditions), and (iii) those that are controlled by both VirB and MxiE (expressed under nonsecretion conditions and induced under secretion conditions) (74).

CONCLUSIONS

While foodborne infections due to members of the *Shigella* spp. may not be as frequent as those caused by other foodborne pathogens, they have the potential for explosive spread due to the low infectious dose that can cause overt clinical disease. In addition, cases of bacillary dysentery frequently require medical attention (even hospitalization) and cause lost time from work because the severity and duration of symptoms can be incapacitating. There is no effective vaccine against dysentery caused by *Shigella*. These features, coupled with the wide geographical distribution of strains and the sensitivity of the human population, make *Shigella* a formidable public health threat.

Research on the genetics of Shigella *virulence in the laboratory of A.T.M. is supported by USUHS protocol H073KB and Public Health Service grant AI24656 from the National Institute of Allergy and Infectious Diseases. The opinions or assertions contained herein are the private ones of A.T.M. and are not to be construed as official or reflecting the views of the Department of Defense or the Uniformed Services University of the Health Sciences.*

References

1. **Adler, B., C. Sasakawa, T. Tobe, S. Makino, K. Komatsu, and M. Yoshikawa.** 1989. A dual transcriptional activation system for the 230 kb plasmid genes coding for virulence-associated antigens of *Shigella flexneri*. *Mol. Microbiol.* **3:**627–635.

2. **Agle, M. E., S. E. Martin, and H. P. Blaschek.** 2005. Survival of *Shigella boydii* 18 in bean salad. *J. Food Prot.* **68:**838–840.

3. **Allaoui, A., P. J. Sansonetti, and C. Parsot.** 1992. MxiJ, a lipoprotein involved in secretion of *Shigella* Ipa invasins, is homologous to YscJ, a secretion factor of the *Yersinia* Yop proteins. *J. Bacteriol.* **174:**7661–7669.

4. **Allaoui, A., P. J. Sansonetti, and C. Parsot.** 1993. MxiD, an outer membrane protein necessary for the secretion of the *Shigella flexneri* Ipa invasins. *Mol. Microbiol.* **7:**59–68.

5. **Andrews, G. P., A. E. Hromockyj, C. Coker, and A. T. Maurelli.** 1991. Two novel virulence loci, *mxiA* and *mxiB*, in *Shigella flexneri* 2a facilitate excretion of invasion plasmid antigen. *Infect. Immun.* **59:**1997–2005.

6. **Andrews, G. P., and A. T. Maurelli.** 1992. *mxiA* of *Shigella flexneri* 2a, which facilitates export of invasion plasmid antigens, encodes a homologue of the low-calcium response protein LcrD, of *Yersinia pestis*. *Infect. Immun.* **60:**3287–3295.

7. **Andrews, W. H.** 1989. Methods for recovering injured classical enteric pathogenic bacteria (*Salmonella, Shigella* and enteropathogenic *Escherichia coli*) from foods, p. 55–113. *In* B. Ray (ed.), *Injured Index and Pathogenic Bacteria: Occurrence and Detection in Foods, Water and Feeds*. CRC Press, Boca Raton, Fla.

8. **Barzu, S., A. Fontaine, P. Sansonetti, and A. Phalipon.** 1996. Induction of a local anti-IpaC antibody response in mice by use of a *Shigella flexneri* 2a vaccine candidate: implications for use of IpaC as a protein carrier. *Infect. Immun.* **64:**1190–1196.

9. **Bean, N. H., and P. M. Griffin.** 1990. Foodborne disease outbreaks in the United States, 1973–1987: pathogens, vehicles, and trends. *J. Food Prot.* **53:**804–817.

10. **Beloin, C., and C. J. Dorman.** 2003. An extended role for the nucleoid structuring protein H-NS in the virulence gene regulatory cascade of *Shigella flexneri*. *Mol. Microbiol.* **47:**825–838.

11. **Bennett, J. V., S. D. Holmberg, M. F. Rogers, and S. L. Solomon.** 1987. Infectious and parasitic diseases, p. 102–114. *In* R. W. Amlet and H. B. Dull (ed.), *Closing the Gap: the Burden of Unnecessary Illness*. Oxford Press, New York, N.Y.

12. **Bennish, M. L.** 1991. Potentially lethal complications of shigellosis. *Rev. Infect. Dis.* **13**(Suppl. 4):S319–S324.

13. **Bennish, M. L., J. R. Harris, B. J. Wojtynaik, and M. Struelens.** 1990. Death in shigellosis: incidence and risk factors in hospitalized patients. *J. Infect. Dis.* **161:**500–506.

14. **Bennish, M. L., and M. A. Salam.** 1992. Rethinking options for the treatment of shigellosis. *J. Antimicrob. Chemother.* **30:**243–247.

15. **Bernardini, M. L., J. Mounier, H. d'Hauteville, M. Coquis-Rondon, and P. J. Sansonetti.** 1989. Identification of *icsA*, a plasmid locus in *Shigella flexneri* that governs bacterial intra- and intercellular spread through interaction with F-actin. *Proc. Natl. Acad. Sci. USA* **86:**3867–3871.

16. **Blocker, A., N. Jouihri, E. Larquet, P. Gounon, F. Ebel, C. Parsot, P. Sansonetti, and A. Allaoui.** 2001. Structure and composition of the *Shigella flexneri* "needle complex," a part of its type III secreton. *Mol. Microbiol.* **39:**652–663.

17. **Bourdet-Sicard, R., M. Rudiger, B. M. Jockusch, P. Gounon, P. J. Sansonetti, and G. T. Nhieu.** 1999. Binding of the *Shigella* protein IpaA to vinculin induces F-actin depolymerization. *EMBO J.* **18:**5853–5862.

18. **Brahmbhatt, H. N., A. A. Lindberg, and K. N. Timmis.** 1992. *Shigella* lipopolysaccharide: structure, genetics, and vaccine development. *Curr. Top. Microbiol. Immunol.* **180:**45–64.

19. **Buchrieser, C., P. Glaser, C. Rusniok, H. Nedjari, H. d'Hauteville, F. Kunst, P. Sansonetti, and C. Parsot.** 2000. The virulence plasmid pWR100 and the repertoire of proteins secreted by the type III secretion apparatus of *Shigella flexneri*. *Mol. Microbiol.* **38:**760–771.

20. **Bunning, V. K., R. B. Raybourne, and D. L. Archer.** 1988. Foodborne enterobacterial pathogens and rheumatoid disease. *J. Appl. Bacteriol. Symp.* **65**(Suppl.):87S–107S.

21. **Buysse, J. M., M. M. Venkatesan, J. Mills, and E. V. Oaks.** 1990. Molecular characterization of a transacting, positive effector (*ipaR*) of invasion plasmid antigen synthesis in *Shigella flexneri* serotype 5. *Microb. Pathog.* **8:**197–211.

22. **Centers for Disease Control and Prevention.** 1994. Personal communication.

23. **Centers for Disease Control and Prevention.** 1994. Outbreak of *Shigella flexneri* 2a infections on a cruise ship. *Morb. Mortal. Wkly. Rep.* **43:**657.

24. **Centers for Disease Control and Prevention.** 1996. The Foodborne Diseases Active Surveillance Network. *Morb. Mortal. Wkly. Rep.* **46:**258–261.

25. **Centers for Disease Control and Prevention.** 1999. Outbreaks of *Shigella sonnei* infection associated with eating fresh parsley—United States and Canada, July–August, 1998. *Morb. Mortal. Wkly. Rep.* **48:**285–289.

26. **Centers for Disease Control and Prevention.** 2000. Outbreaks of *Shigella sonnei* infections associated with eating a nationally distributed dip—California, Oregon, and Washington, January 2000. *Morb. Mortal. Wkly. Rep.* **49:**60–61.

27. **Centers for Disease Control and Prevention.** 2005. Preliminary FoodNet data on the incidence of infections with pathogens transmitted commonly through food—10 sites, United States. *Morb. Mortal. Wkly. Rep.* **54:**352–356.

28. **Charles, M., M. Perez, J. H. Kobil, and M. B. Goldberg.** 2001. Polar targeting of *Shigella* virulence factor IcsA in

Enterobacteriaceae and *Vibrio. Proc. Natl. Acad. Sci. USA* 98:9871–9876.

29. Coster, T. S., C. W. Hoge, L. L. VanDeVerg, A. B. Hartman, E. V. Oaks, M. M. Venkatesan, D. Cohen, G. Robin, A. Fontaine-Thompson, P. J. Sansonetti, and T. L. Hale. 1999. Vaccination against shigellosis with attenuated *Shigella flexneri* 2a strain SC602. *Infect. Immun.* 67:3437–3443.

30. Davis, H., J. P. Taylor, J. N. Perdue, G. N. Stelma, Jr., J. M. Humphreys, Jr., R. Rowntree III, and K. D. Greene. 1988. A shigellosis outbreak traced to commercially distributed lettuce. *Am. J. Epidemiol.* 128:1312–1321.

31. Dorman, C. J., and M. E. Porter. 1998. The *Shigella* virulence gene regulatory cascade: a paradigm of bacterial gene control mechanisms. *Mol. Microbiol.* 29:677–684.

32. DuPont, H. L. 2005. *Shigella* species (bacillary dysentery), p. 2655–2661. *In* G. L. Mandell, J. E. Bennett, and R. Dolin (ed.), *Principles and Practice of Infectious Diseases.* Churchill Livingstone, New York, N.Y.

33. DuPont, H. L., M. M. Levine, R. B. Hornick, and S. B. Formal. 1989. Inoculum size in shigellosis and implications for expected mode of transmission. *J. Infect. Dis.* 159:1126–1128.

34. Edwards, P. R., and W. H. Ewing. 1972. *Identification of Enterobacteriaceae.* Burgess Publishing Co., Minneapolis, Minn.

35. Egile, C., H. d'Hauteville, C. Parsot, and P. J. Sansonetti. 1997. SopA, the outer membrane protease responsible for polar localization of IcsA in *Shigella flexneri. Mol. Microbiol.* 23:1063–1073.

36. Ekdahl, K., and Y. Andersson. 2005. The epidemiology of travel-associated shigellosis—regional risks, seasonality and serogroups. *J. Infect.* 51:222–229.

37. Falconi, M., B. Colonna, G. Prosseda, G. Micheli, and C. O. Gualerzi. 1998. Thermoregulation of *Shigella* and *Escherichia coli* EIEC pathogenicity. A temperature-dependent structural transition of DNA modulates accessibility of *virF* promoter to transcriptional repressor H-NS. *EMBO J.* 17:7033–7043.

38. Falkow, S., H. Schneider, L. Baron, and S. B. Formal. 1963. Virulence of *Escherichia-Shigella* genetic hybrids for the guinea pig. *J. Bacteriol.* 86:1251–1258.

39. Fasano, A., F. R. Noriega, D. R. Maneval, Jr., S. Chanasongcram, R. Russell, S. Guandalini, and M. M. Levine. 1995. *Shigella* enterotoxin 1: an enterotoxin of *Shigella flexneri* 2a active in rabbit small intestine in vivo and in vitro. *J. Clin. Investig.* 95:2853–2861.

40. Flint, J. A., Y. T. Van Duynhoven, F. J. Angulo, S. M. DeLong, P. Braun, M. Kirk, E. Scallan, M. Fitzgerald, G. K. Adak, P. Sockett, A. Ellis, G. Hall, N. Gargouri, H. Walke, and P. Braam. 2005. Estimating the burden of acute gastroenteritis, foodborne disease, and pathogens commonly transmitted by food: an international review. *Clin. Infect. Dis.* 41:698–704.

41. Fontaine, A., J. Arondel, and P. J. Sansonetti. 1988. Role of the Shiga toxin in the pathogenesis of bacillary dysentery studied by using a Tox– mutant of *Shigella dysenteriae* 1. *Infect. Immun.* 56:3099–3109.

42. Formal, S. B., E. H. LaBrec, T. H. Kent, and S. Falkow. 1965. Abortive intestinal infection with an *Escherichia coli-Shigella flexneri* hybrid strain. *J. Bacteriol.* 89:1374–1382.

43. Frost, J. A., M. B. McEvoy, C. A. Bentley, Y. Andersson, and B. Rowe. 1995. An outbreak of *Shigella sonnei* infection associated with consumption of iceberg lettuce. *Emerg. Infect. Dis.* 1:26–29.

44. Galán, J. E., and P. J. Sansonetti. 1996. Molecular and cellular bases of *Salmonella* and *Shigella* interactions with host cells, p. 2757–2773. *In* F. C. Neidhardt, R. Curtiss III, C. A. Gross, J. I. Ingraham, E. C. Lin, K. B. Low, Jr., B. Magasanik, W. Reznikoff, M. Riley, M. Schaechter, and H. E. Umbarger (ed.), Escherichia coli *and* Salmonella typhimurium: *Cellular and Molecular Biology*, 2nd ed. American Society for Microbiology, Washington, D.C.

45. Gallegos, M. T., C. Michan, and J. L. Ramos. 1993. The XylS/AraC family of regulators. *Nucleic Acids Res.* 21:807–810.

46. Gessner, B. D., and M. Beller. 1994. Moose soup shigellosis in Alaska. *West. J. Med.* 160:430–433.

47. Goldberg, M. B., O. Barzu, C. Parsot, and P. J. Sansonetti. 1993. Unipolar localization and ATPase activity of IcsA, a *Shigella flexneri* protein involved in intracellular movement. *J. Bacteriol.* 175:2189–2196.

48. Goma Epidemiology Group. 1995. Public health impact of Rwandan refugee crisis: what happened in Goma, Zaire, in July, 1994? *Lancet* 345:339–344.

49. Groisman, E. A., and H. Ochman. 1993. Cognate gene clusters govern invasion of host epithelial cells by *Salmonella typhimurium* and *Shigella flexneri. EMBO J.* 12:3779–3787.

50. Guerrero, L., J. J. Calva, A. L. Morrow, F. R. Velazquez, F. Tuz-Dzib, Y. Lopez-Vidal, H. Ortega, H. Arroyo, T. G. Cleary, L. K. Pickering, and G. M. Ruiz-Palacios. 1994. Asymptomatic *Shigella* infections in a cohort of Mexican children younger than two years of age. *Pediatr. Infect. Dis. J.* 13:597–602.

51. Hall, P. A. 1994. Scope for rapid microbiological methods in modern food production, p. 255–267. *In* P. Patel (ed.), *Rapid Analysis Techniques in Food Microbiology.* Blackie Academic & Professional, New York, N.Y.

52. Haltalin, K., J. Nelson, and R. Ring. 1967. Double-blind treatment study of shigellosis comparing ampicillin, sulfadiazone and placebo. *J. Pediatr.* 70:970–981.

53. Hedberg, C. W., W. C. Levine, K. E. White, R. H. Carlson, D. K. Winsor, D. N. Cameron, K. L. MacDonald, and M. T. Osterholm. 1992. An international foodborne outbreak of shigellosis associated with a commercial airline. *JAMA* 268:3208–3212.

54. High, N., J. Mounier, M. C. Prevost, and P. J. Sansonetti. 1992. IpaB of *Shigella flexneri* causes entry into epithelial cells and escape from the phagocytic vacuole. *EMBO J.* 11:1991–1999.

55. Hoge, C. W., J. M. Gambel, A. Srijan, C. Pitarangsi, and P. Echeverria. 1998. Trends in antibiotic resistance among diarrheal pathogens isolated in Thailand over 15 years. *Clin. Infect. Dis.* 26:341–345.

56. Hossain, M. A., K. Z. Hasan, and M. J. Albert. 1994. *Shigella* carriers among non-diarrhoeal children in an endemic area of shigellosis in Bangladesh. *Trop. Geogr. Med.* 46:40–42.

57. Hyams, K. C., A. L. Bourgeois, B. R. Merrell, P. Rozmajzl, J. Escamilla, S. A. Thornton, G. M. Wasserman, A. Burke,

P. Echeverria, K. Y. Green, A. Z. Kapikian, and J. N. Woody. 1991. Diarrheal disease during Operation Desert Shield. *N. Engl. J. Med.* **325**:1423–1428.

58. International Commission on Microbiological Specifications for Foods of the International Union of Biological Societies. 1996. *Shigella*, p. 280–298. *In Microorganisms in Foods*, vol 5. *Microbiological Specifications of Food Pathogens*. Blackie Academic & Professional, New York, N.Y.

59. Jiang, Y., F. Yang, X. Zhang, J. Yang, L. Chen, Y. Yan, H. Nie, Z. Xiong, J. Wang, J. Dong, Y. Xue, X. Xu, Y. Zhu, S. Chen, and Q. Jin. 2005. The complete sequence and analysis of the large virulence plasmid pSS of *Shigella sonnei*. *Plasmid* **54**:149–159.

60. Jin, Q., Z. Yuan, J. Xu, Y. Wang, Y. Shen, W. Lu, J. Wang, H. Liu, J. Yang, F. Yang, X. Zhang, J. Zhang, G. Yang, H. Wu, D. Qu, J. Dong, L. Sun, Y. Xue, A. Zhao, Y. Gao, J. Zhu, B. Kan, K. Ding, S. Chen, H. Cheng, Z. Yao, B. He, R. Chen, D. Ma, B. Qiang, Y. Wen, Y. Hou, and J. Yu. 2002. Genome sequence of *Shigella flexneri* 2a: insights into pathogenicity through comparison with genomes of *Escherichia coli* K12 and O157. *Nucleic Acids Res.* **30**:4432–4441.

61. Jouihri, N., M. P. Sory, A. L. Page, P. Gounon, C. Parsot, and A. Allaoui. 2003. MxiK and MxiN interact with the Spa47 ATPase and are required for transit of the needle components MxiH and MxiI, but not of Ipa proteins, through the type III secretion apparatus of *Shigella flexneri*. *Mol. Microbiol.* **49**:755–767.

62. Kalluri, P., K. C. Cummings, S. Abbott, G. B. Malcolm, K. Hutcheson, A. Beall, K. Joyce, C. Polyak, D. Woodward, R. Caldeira, F. Rodgers, E. D. Mintz, and N. Strockbine. 2004. Epidemiological features of a newly described serotype of *Shigella boydii*. *Epidemiol. Infect.* **132**:579–583.

63. Kane, C. D., R. Schuch, W. A. Day, Jr., and A. T. Maurelli. 2002. MxiE regulates intracellular expression of factors secreted by the *Shigella flexneri* 2a type III secretion system. *J. Bacteriol.* **184**:4409–4419.

64. Kapperud, G., L. M. Rorvik, V. Hasseltvedt, E. A. Hoiby, B. G. Iversen, K. Staveland, G. Johnsen, J. Leitao, H. Herikstad, Y. Andersson, G. Langeland, B. Gondrosen, and J. Lassen. 1995. Outbreak of *Shigella sonnei* infection traced to imported iceberg lettuce. *J. Clin. Microbiol.* **33**:609–614.

65. Karmali, M. A., M. Petric, C. Lim, P. C. Fleming, G. S. Arbus, and H. Lior. 1985. The association between idiopathic hemolytic syndrome and infection by verotoxin-producing *Escherichia coli*. *J. Infect. Dis.* **151**:775–782.

66. Kennedy, F. M., J. Astbury, J. R. Needham, and T. Cheasty. 1993. Shigellosis due to occupational contact with non-human primates. *Epidemiol. Infect.* **110**:247–257.

67. Kim, D. W., G. Lenzen, A. L. Page, P. Legrain, P. J. Sansonetti, and C. Parsot. 2005. The *Shigella flexneri* effector OspG interferes with innate immune responses by targeting ubiquitin-conjugating enzymes. *Proc. Natl. Acad. Sci. USA* **102**:14046–14051.

68. Kinsey, M. D., S. B. Formal, G. J. Dammin, and R. A. Giannella. 1976. Fluid and electrolyte transport in rhesus monkeys challenged intracecally with *Shigella flexneri* 2a. *Infect. Immun.* **14**:368–371.

69. Kolavic, S. A., A. Kimura, S. L. Simons, L. Slutsker, S. Barth, and C. E. Haley. 1997. An outbreak of *Shigella dysenteriae* type 2 among laboratory workers due to intentional food contamination. *JAMA* **278**:396–398.

70. Kotloff, K. L., J. P. Winickoff, B. Ivanoff, J. D. Clemens, D. L. Swerdlow, P. J. Sansonetti, G. K. Adak, and M. M. Levine. 1999. Global burden of *Shigella* infections: implications for vaccine development and implementation of control strategies. *Bull. W. H. O.* **77**:651–666.

71. LaBrec, E. H., H. Schneider, T. J. Magnani, and S. B. Formal. 1964. Epithelial cell penetration as an essential step in the pathogenesis of bacillary dysentery. *J. Bacteriol.* **88**:1503–1518.

72. Lawlor, K. M., P. A. Daskeleros, R. E. Robinson, and S. M. Payne. 1987. Virulence of iron-transport mutants of *Shigella flexneri* and utilization of host iron compounds. *Infect. Immun.* **55**:594–599.

73. Lee, L. A., S. M. Ostroff, H. B. McGee, D. R. Johnson, F. P. Downes, D. N. Cameron, N. H. Bean, and P. M. Griffin. 1991. An outbreak of shigellosis at an outdoor music festival. *Am. J. Epidemiol.* **133**:608–615.

74. Le Gall, T., M. Mavris, M. C. Martino, M. L. Bernardini, E. Denamur, and C. Parsot. 2005. Analysis of virulence plasmid gene expression defines three classes of effectors in the type III secretion system of *Shigella flexneri*. *Microbiology* **151**:951–962.

75. Levine, M. M., and O. S. Levine. 1994. Changes in human ecology and behavior in relation to the emergence of diarrheal diseases, including cholera. *Proc. Natl. Acad. Sci. USA* **91**:2390–2394.

76. Lew, J. F., D. L. Swerdlow, M. E. Dance, P. M. Griffin, C. A. Bopp, M. J. Gillenwater, T. Mercatante, and R. I. Glass. 1991. An outbreak of shigellosis aboard a cruise ship caused by a multiple-antibiotic-resistant strain of *Shigella flexneri*. *Am. J. Epidemiol.* **134**:413–420.

77. Lopez, E. L., M. Diaz, S. Grinstein, S. Dovoto, F. Mendila-Harzu, B. E. Murray, S. Ashkenazi, E. Rubeglio, M. Woloj, M. Vasquez, M. Turco, L. K. Pickering, and T. G. Cleary. 1989. Hemolytic uremic syndrome and diarrhea in Argentine children: the role of Shiga-like toxins. *J. Infect. Dis.* **160**:469–475.

78. Magdalena, J., A. Hachani, M. Chamekh, N. Jouihri, P. Gounon, A. Blocker, and A. Allaoui. 2002. Spa32 regulates a switch in substrate specificity of the type III secreton of *Shigella flexneri* from needle components to Ipa proteins. *J. Bacteriol.* **184**:3433–3441.

79. Makino, S., C. Sasakawa, K. Kamata, T. Kurata, and M. Yoshikawa. 1986. A genetic determinant required for continuous reinfection of adjacent cells on large plasmid in *Shigella flexneri* 2a. *Cell* **46**:551–555.

80. Maurelli, A. T. 1989. Temperature regulation of virulence genes in pathogenic bacteria: a general strategy for human pathogens? *Microb. Pathog.* **7**:1–10.

81. Maurelli, A. T., B. Baudry, H. d'Hauteville, T. L. Hale, and P. J. Sansonetti. 1985. Cloning of virulence plasmid DNA sequences involved in invasion of HeLa cells by *Shigella flexneri*. *Infect. Immun.* **49**:164–171.

82. Maurelli, A. T., B. Blackmon, and R. Curtiss III. 1984. Temperature-dependent expression of virulence genes in *Shigella* species. *Infect. Immun.* **43**:195–201.

83. Maurelli, A. T., R. E. Fernández, C. A. Bloch, C. K. Rode, and A. Fasano. 1998. "Black holes" and bacterial pathogenicity: a large genomic deletion that enhances the virulence of *Shigella* spp. and enteroinvasive *Escherichia coli*. *Proc. Natl. Acad. Sci. USA* **95:**3943–3948.

84. Maurelli, A. T., and K. A. Lampel. 2001. *Shigella*, p. 323–343. *In* Y. H. Hui, M. D. Pierson, and J. R. Gorham (ed.), *Foodborne Disease Handbook*, vol. 1, 2nd ed. Marcel Dekker Publishers, New York, N.Y.

85. Maurelli, A. T., and P. J. Sansonetti. 1988. Identification of a chromosomal gene controlling temperature regulated expression of *Shigella* virulence. *Proc. Natl. Acad. Sci. USA* **85:**2820–2824.

86. Mavris, M., A. L. Page, R. Tournebize, B. Demers, P. Sansonetti, and C. Parsot. 2002. Regulation of transcription by the activity of the *Shigella flexneri* type III secretion apparatus. *Mol. Microbiol.* **43:**1543–1553.

87. Mavris, M., P. J. Sansonetti, and C. Parsot. 2002. Identification of the *cis*-acting site involved in activation of promoters regulated by activity of the type III secretion apparatus in *Shigella flexneri*. *J. Bacteriol.* **184:**6751–6759.

88. McKenna, S., C. Beloin, and C. J. Dorman. 2003. In vitro DNA-binding properties of VirB, the *Shigella flexneri* virulence regulatory protein. *FEBS Lett.* **545:**183–187.

89. Mead, P. S., L. Slutsker, V. Dietz, L. F. McCaig, J. S. Bresee, C. Shapiro, P. M. Griffin, and R. V. Tauxe. 1999. Food-related illness and death in the United States. *Emerg. Infect. Dis.* **5:**607–625.

90. Ménard, R., M.-C. Prévost, P. Gounon, P. Sansonetti, and C. Dehio. 1996. The secreted Ipa complex of *Shigella flexneri* promotes entry into mammalian cells. *Proc. Natl. Acad. Sci. USA* **93:**1254–1258.

91. Ménard, R., P. J. Sansonetti, and C. Parsot. 1993. Nonpolar mutagenesis of the *ipa* genes defines IpaB, IpaC, and IpaD as effectors of *Shigella flexneri* entry into epithelial cells. *J. Bacteriol.* **175:**5899–5906.

92. Ménard, R., P. J. Sansonetti, and C. Parsot. 1994. The secretion of the *Shigella flexneri* Ipa invasins is induced by the epithelial cell and controlled by IpaB and IpaD. *EMBO J.* **13:**5293–5302.

93. Ménard, R., P. J. Sansonetti, C. Parsot, and T. Vasselon. 1994. The IpaB and IpaC invasins of *Shigella flexneri* associate in the extracellular medium and are partitioned in the cytoplasm by a specific chaperon. *Cell* **76:**829–839.

94. Morita-Ishihara, T., M. Ogawa, H. Sagara, M. Yoshida, E. Katayama, and C. Sasakawa. 2006. *Shigella* Spa33 is an essential C-ring component of type III secretion machinery. *J. Biol. Chem.* **281:**599–607.

95. Moss, J. E., T. J. Cardozo, A. Zychlinsky, and E. A. Groisman. 1999. The *selC*-associated SHI-2 pathogenicity island of *Shigella flexneri*. *Mol. Microbiol.* **33:**74–83.

96. Nassif, X., M. C. Mazert, J. Mounier, and P. J. Sansonetti. 1987. Evaluation with an *iuc*::Tn*10* mutant of the role of aerobactin production in the virulence of *Shigella flexneri*. *Infect. Immun.* **55:**1963–1969.

97. Nataro, J. P., J. Seriwatana, A. Fasano, D. R. Maneval, L. D. Guers, F. Noriega, F. Dubovsky, M. M. Levine, and J. G. Morris, Jr. 1995. Identification and cloning of a novel plasmid-encoded enterotoxin of enteroinvasive *Escherichia coli* and *Shigella* strains. *Infect. Immun.* **63:**4721–4728.

98. Oaks, E. V., T. L. Hale, and S. B. Formal. 1986. Serum immune response to *Shigella* protein antigens in rhesus monkeys and humans infected with *Shigella* spp. *Infect. Immun.* **53:**57–63.

99. O'Brien, A. D., and R. K. Holmes. 1995. Protein toxins of *Escherichia coli* and *Salmonella*. *In* F. C. Neidhardt, R. Curtiss III, C. A. Gross, J. I. Ingraham, E. C. Lin, K. B. Low, Jr., B. Magasanik, W. Reznikoff, M. Riley, M. Schaechter, and H. E. Umbarger (ed.), Escherichia coli *and* Salmonella typhimurium: *Cellular and Molecular Biology*, 2nd ed. American Society for Microbiology, Washington, D.C.

100. Ochman, H., T. S. Whittam, D. A. Caugant, and R. K. Selander. 1983. Enzyme polymorphism and genetic population structure in *Escherichia coli* and *Shigella*. *J. Gen. Microbiol.* **129:**2715–2726.

101. O'Connell, C. M. C., R. C. Sandlin, and A. T. Maurelli. 1995. Signal transduction and virulence gene regulation in *Shigella* spp.: temperature and (maybe) a whole lot more, p. 111–127. *In* R. Rappuoli (ed.), *Signal Transduction and Bacterial Virulence*. R.G. Landes Company, Austin, Tex.

102. Ogawa, M., T. Yoshimori, T. Suzuki, H. Sagara, N. Mizushima, and C. Sasakawa. 2005. Escape of intracellular *Shigella* from autophagy. *Science* **307:**727–731.

103. Parsot, C., E. Ageron, C. Penno, M. Mavris, K. Jamoussi, H. d'Hauteville, P. Sansonetti, and B. Demers. 2005. A secreted anti-activator, OspD1, and its chaperone, Spa15, are involved in the control of transcription by the type III secretion apparatus activity in *Shigella flexneri*. *Mol. Microbiol.* **56:**1627–1635.

104. Parsot, C., R. Ménard, P. Giunon, and P. J. Sansonetti. 1995. Enhanced secretion through the *Shigella flexneri* Mxi-Spa translocon leads to assembly of extracellular proteins into macromolecular structures. *Mol. Microbiol.* **16:**291–300.

105. Parsot, C., and P. J. Sansonetti. 1996. Invasion and the pathogenesis of *Shigella* infections. *Curr. Top. Microbiol. Immunol.* **209:**25–42.

106. Pickering, L. K., A. V. Bartlett, and W. E. Woodward. 1986. Acute infectious diarrhea among children in day-care: epidemiology and control. *Rev. Infect. Dis.* **8:**539–547.

107. Picking, W. L., H. Nishioka, P. D. Hearn, M. A. Baxter, A. T. Harrington, A. Blocker, and W. D. Picking. 2005. IpaD of *Shigella flexneri* is independently required for regulation of Ipa protein secretion and efficient insertion of IpaB and IpaC into host membranes. *Infect. Immun.* **73:**1432–1440.

108. Raghupathy, P., A. Date, J. C. M. Shastry, A. Sudarsanam, and M. Jadhav. 1978. Haemolytic-uraemic syndrome complicating shigella dysentery in south Indian children. *Br. Med. J.* **1:**1518–1521.

109. Rajakumar, K., C. Sasakawa, and B. Adler. 1997. Use of a novel approach, termed island probing, identifies the *Shigella flexneri* she pathogenicity island which encodes a homolog of the immunoglobulin A protease-like family of proteins. *Infect. Immun.* **65:**4606–4614.

110. Reller, M. E., J. M. Nelson, K. Mølbak, D. M. Ackman, D. J. Schoonmaker-Bopp, T. P. Root, and E. D. Mintz. 2006. A large, multiple-restaurant outbreak of infection with *Shigella flexneri* serotype 2a traced to tomatoes. *Clin. Infect. Dis.* **42**:163–169.

111. Roberts, T. 1989. Human illness costs of foodborne bacteria. *Am. J. Agric. Econ.* **71**:468–474.

112. Rout, W. R., S. B. Formal, R. A. Giannella, and G. J. Dammin. 1975. Pathophysiology of *Shigella* diarrhea in the rhesus monkey: intestinal transport, morphological, and bacteriological studies. *Gastroenterology* **68**:270–278.

113. Sakai, T., C. Sasakawa, and M. Yoshikawa. 1988. Expression of four virulence antigens of *Shigella flexneri* is positively regulated at the transcriptional level by the 30 kilodalton *virF* protein. *Mol. Microbiol.* **2**:589–597.

114. Sandlin, R. C., K. A. Lampel, S. P. Keasler, M. B. Goldberg, A. L. Stolzer, and A. T. Maurelli. 1995. Avirulence of rough mutants of *Shigella flexneri*: requirement of O-antigen for correct unipolar localization of IcsA in bacterial outer membrane. *Infect. Immun.* **63**:229–237.

115. Sandlin, R. C., and A. T. Maurelli. 1999. Establishment of unipolar localization of IcsA in *Shigella flexneri* 2a is not dependent on virulence plasmid determinants. *Infect. Immun.* **67**:350–356.

116. Sansonetti, P. J., H. d'Hauteville, C. Ecobichon, and C. Pourcel. 1983. Molecular comparison of virulence plasmids in *Shigella* and enteroinvasive *Escherichia coli*. *Ann. Microbiol.* (Paris) **134A**:295–318.

117. Sansonetti, P. J., T. L. Hale, and E. V. Oaks. 1985. Genetics of virulence in enteroinvasive *Escherichia coli*, p. 74–77. *In* D. Schlessinger (ed.), *Microbiology—1985*. American Society for Microbiology, Washington, D.C.

118. Sansonetti, P. J., D. J. Kopecko, and S. B. Formal. 1981. *Shigella sonnei* plasmids: evidence that a large plasmid is necessary for virulence. *Infect. Immun.* **34**:75–83.

119. Sansonetti, P. J., D. J. Kopecko, and S. B. Formal. 1982. Involvement of a plasmid in the invasive ability of *Shigella flexneri*. *Infect. Immun.* **35**:852–860.

120. Sasakawa, C., K. Komatsu, T. Tobe, T. Suzuki, and M. Yoshikawa. 1993. Eight genes in region 5 that form an operon are essential for invasion of epithelial cells by *Shigella flexneri* 2a. *J. Bacteriol.* **175**:2334–2346.

121. Schuch, R., and A. T. Maurelli. 2000. The type III secretion pathway: dictating the outcome of bacterial-host interactions, p. 203–223. *In* K. A. Brogden, J. A. Roth, T. B. Stanton, C. A. Bolin, F. C. Minion, and M. J. Wannemuehler (ed.), *Virulence Mechanisms of Bacterial Pathogens*, 3rd ed. ASM Press, Washington, D.C.

122. Schuch, R., and A. T. Maurelli. 2001. MxiM and MxiJ, base elements of the Mxi-Spa type III secretion system of *Shigella*, interact with and stabilize the MxiD secretin in the cell envelope. *J. Bacteriol.* **183**:6991–6998.

123. Schuch, R., and A. T. Maurelli. 2001. Spa33, a cell surface-associated subunit of the Mxi-Spa type III secretory pathway of *Shigella flexneri*, regulates Ipa protein traffic. *Infect. Immun.* **69**:2180–2189.

124. Sharp, T. W., S. A. Thornton, M. R. Wallace, R. F. Defraites, J. L. Sanchez, R. A. Batchelor, P. J. Rozmajzl, R. K. Hanson, P. Echeverria, A. Z. Kapikian, X. J.

Xiang, M. K. Estes, and J. P. Burans. 1995. Diarrheal disease among military personnel during Operation Restore Hope, Somalia, 1992–1993. *Am J. Trop. Med. Hyg.* **52**:188–193.

125. Shere, K. D., S. Sallustio, A. Manessis, T. G. D'Aversa, and M. B. Goldberg. 1997. Disruption of IcsP, the major *Shigella* protease that cleaves IcsA, accelerates actin-based motility. *Mol. Microbiol.* **25**:451–462.

126. Simon, D. G., R. A. Kaslow, J. Rosenbaum, R. L. Kaye, and A. Calin. 1981. Reiter's syndrome following epidemic shigellosis. *J. Rheumatol.* **8**:969–973.

127. Small, P., D. Blankenhorn, D. Welty, E. Zinser, and J. L. Slonczewski. 1994. Acid and base resistance in *Escherichia coli* and *Shigella flexneri*: role of *rpoS* and growth pH. *J. Bacteriol.* **176**:1729–1737.

128. Smith, J. L. 1987. *Shigella* as a foodborne pathogen. *J. Food Prot.* **50**:788–801.

129. Snyder, O. P. 1992. HACCP—an industry food safety self-control program. IV. *Dairy Food Environ. Sanit.* **12**:230–232.

130. Tamano, K., S. Aizawa, E. Katayama, T. Nonaka, S. Imajoh-Ohmi, A. Kuwae, S. Nagai, and C. Sasakawa. 2000. Supramolecular structure of the *Shigella* type III secretion machinery: the needle part is changeable in length and essential for delivery of effectors. *EMBO J.* **19**:3876–3887.

131. Tesh, V. L., and A. D. O'Brien. 1991. The pathogenic mechanisms of Shiga toxin and the Shiga-like toxins. *Mol. Microbiol.* **5**:1817–1822.

132. Tobe, T., S. Nagai, N. Okada, B. Adler, M. Yoshikawa, and C. Sasakawa. 1991. Temperature-regulated expression of invasion genes in *Shigella flexneri* is controlled through the transcriptional activation of the *virB* gene on the large plasmid. *Mol. Microbiol.* **5**:887–893.

133. Tobe, T., M. Yoshikawa, T. Mizuno, and C. Sasakawa. 1993. Transcriptional control of the invasion regulatory gene *virB* of *Shigella flexneri*: activation by VirF and repression by H-NS. *J. Bacteriol.* **175**:6142–6149.

134. Tran Van Nhieu, G., A. Ben-Ze'ev, and P. J. Sansonetti. 1997. Modulation of bacterial entry into epithelial cells by association between vinculin and the *Shigella* IpaA invasin. *EMBO J.* **16**:2717–2729.

135. Tran Van Nhieu, G., E. Caron, A. Hall, and P. J. Sansonetti. 1999. IpaC induces actin polymerization and filopodia formation during *Shigella* entry into epithelial cells. *EMBO J.* **18**:3249–3262.

136. Van Gijsegem, F., S. Genin, and C. Boucher. 1993. Conservation of secretion pathways for pathogenicity determinants of plant and animal bacteria. *Trends Microbiol.* **1**:175–180.

137. Venkatesan, M. M., J. M. Buysse, and E. V. Oaks. 1992. Surface presentation of *Shigella flexneri* invasion plasmid antigens requires the products of the *spa* locus. *J. Bacteriol.* **174**:1990–2001.

138. Venkatesan, M. M., M. B. Goldberg, D. J. Rose, E. J. Grotbeck, V. Burland, and F. R. Blattner. 2001. Complete DNA sequence and analysis of the large virulence plasmid of *Shigella flexneri*. *Infect. Immun.* **69**:3271–3285.

139. **Vokes, S. A., S. A. Reeves, A. G. Torres, and S. M. Payne.** 1999. The aerobactin iron transport system genes in *Shigella flexneri* are present within a pathogenicity island. *Mol. Microbiol.* **33**:63–73.

140. **Watarai, M., T. Tobe, M. Yoshikawa, and C. Sasakawa.** 1995. Contact of *Shigella* with host cells triggers release of Ipa invasins and is an essential function of invasiveness. *EMBO J.* **14**:2461–2470.

141. **Watarai, M., S. Funato, and C. Sasakawa.** 1996. Interaction of Ipa proteins of *Shigella flexneri* with α5-β1 integrin promotes entry of the bacteria into mammalian cells. *J. Exp. Med.* **183**:991–999.

142. **Weissman, J. B., A. Schmerler, P. Weiler, G. Filice, N. Godby, and I. Hansen.** 1974. The role of preschool children and day-care centers in the spread of shigellosis in urban communities. *J. Pediatr.* **84**:797–802.

143. **Wetherington, J., J. L. Bryant, K. A. Lampel, and J. M. Johnson.** 2000. PCR screening and isolation of *Shigella sonnei* from layered party dip. *Lab. Inf. Bull.* **16**:1–8.

144. **Wharton, M., R. A. Spiegel, J. M. Horan, R. V. Tauxe, J. G. Wells, N. Barg, J. Herndon, R. A. Meriwether, J. N. MacCormack, and R. H. Levine.** 1990. A large outbreak of antibiotic-resistant shigellosis at a mass gathering. *J. Infect. Dis.* **162**:1324–1328.

145. **Yagupsky, P., M. Loeffelholz, K. Bell, and M. A. Menegus.** 1991. Use of multiple markers for investigation of an epidemic of *Shigella sonnei* infections in Monroe County, New York. *J. Clin. Microbiol.* **29**:2850–2855.

146. **Yang, F., J. Yang, X. Zhang, L. Chen, Y. Jiang, Y. Yan, X. Tang, J. Wang, Z. Xiong, J. Dong, Y. Xue, Y. Zhu, X. Xu, L. Sun, S. Chen, H. Nie, J. Peng, J. Xu, Y. Wang, Z. Yuan, Y. Wen, Z. Yao, Y. Shen, B. Qiang, Y. Hou, J. Yu, and Q. Jin.** 2005. Genome dynamics and diversity of *Shigella* species, the etiologic agents of bacillary dysentery. *Nucleic Acids Res.* **33**:6445–6458.

147. **Zychlinsky, A., B. Kenny, R. Mènard, M. C. Prevost, I. B. Holland, and P. J. Sansonetti.** 1994. IpaB mediates macrophage apoptosis induced by *Shigella flexneri*. *Mol. Microbiol.* **11**:619–627.

Food Microbiology: Fundamentals and Frontiers, 3rd Ed.
Edited by M. P. Doyle and L. R. Beuchat
© 2007 ASM Press, Washington, D.C.

James D. Oliver
James B. Kaper

Vibrio Species

16

Whereas the 8th edition of *Bergey's Manual of Systematic Bacteriology* listed five *Vibrio* species, with two recognized as human pathogens, over 80 species have now been described (189), including at least 12 capable of causing infection in humans. A number of reviews on the pathogenic vibrios have appeared over the years (88, 132, 141, 149, 151), although with the exception of those of *Vibrio cholerae* and *V. parahaemolyticus*, relatively little is known of the virulence mechanisms they employ. Of the 12 human pathogens, 8 have been directly associated with foods, and these pathogens are the subject of this review.

INCIDENCE OF VIBRIOS IN SEAFOOD

One of the most consistent features of human vibrio infections is a recent history of seafood consumption. Vibrios, which are generally the predominant bacterial genus in estuarine waters, are associated with a great variety of seafoods. To cite a few studies, Gopal et al. (59) reported that shrimp samples from India had a mean of up to 4.4×10^4 CFU of *Vibrio* spp., with *V. alginolyticus*, *V. parahaemolyticus*, *V. vulnificus*, *V. fluvialis*, *V. mimicus*, and occasionally *V. cholerae* identified. A survey of frozen raw shrimp imported from Mexico, China, and Ecuador found over 63% to harbor *Vibrio* species, including *V. vulnificus* and *V. parahaemolyticus* (11). Buck (19) reported that 36 to 60% of finfish and shellfish sampled from supermarkets were contaminated with *Vibrio* spp., with *V. parahaemolyticus* and *V. alginolyticus* most commonly isolated. The recovery of vibrios from molluscan shellfish was substantially greater during the summer months. Lowry et al. (115) found that 100 and 67% of the raw oysters they examined contained *V. parahaemolyticus* and *V. vulnificus*, respectively. Interestingly, 50 and 25% of the cooked oysters tested also contained *V. parahaemolyticus* and *V. vulnificus*, respectively. In a study of the distribution of vibrios in oysters (*Crassostrea virginica*) originating from the coast of Brazil, Matté et al. (119) found, in order of incidence, *V. alginolyticus* (81%), *V. parahaemolyticus* (77%), *V. cholerae* non-O1 (31%), *V. fluvialis* (27%), *V. furnissii* (19%), *V. mimicus* (12%), and *V. vulnificus* (12%). DePaola et al. (44) reported up to 1.2×10^5 vibrios/g of oyster, with pathogenic strains in 21.8% of the 156 samples. Finally, Elhadi et al. (49), examining 768 samples of a variety of

James D. Oliver, Dept. of Biology, University of North Carolina at Charlotte, 9201 University City Blvd., Charlotte, NC 28223. **James B. Kaper**, Center for Vaccine Development, Division of Geographic Medicine, Dept. of Medicine, University of Maryland School of Medicine, 685 West Baltimore St., Baltimore, MD 21201.

seafoods from Malaysian markets, detected eight different pathogenic *Vibrio* spp., with all eight present in some shrimp and cockles. Their results indicate that seafoods in Malaysia are typically contaminated with potentially pathogenic vibrios regardless of the season.

The use of molecular methods to detect vibrios in shellfish is becoming more routine. For example, Panicker and Bej (155) described the use of real-time PCR for the specific detection of *V. vulnificus* in oysters and reported that ca. 10^3 CFU were detectable in oyster homogenates following a brief enrichment period.

ISOLATION

Various enrichment broths have been described previously for the isolation of vibrios, although alkaline peptone water (APW) remains the most commonly employed (147). These broths are frequently coupled with thiosulfate-citrate-bile salts-sucrose (TCBS) agar or other plating media selective for vibrios. By employing sucrose as a differentiating trait, 11 of the 12 human pathogenic vibrios can be separated on TCBS into 6 species which are generally sucrose positive (*V. cholerae*, *V. metschnikovii*, *V. fluvialis*, *V. furnissii*, *V. alginolyticus*, and *V. carchariae*) and 5 species which are generally sucrose negative (*V. mimicus*, *V. hollisae*, *V. damsela*, *V. parahaemolyticus*, and *V. vulnificus*). Although most of the vibrios grow well on TCBS agar, *V. hollisae* exhibits very poor to no growth on this medium, which is also the case for *V. damsela* when incubated at 37°C. Further, the determination of oxidase activity, a crucial test for distinguishing vibrios from the *Enterobacteriaceae*, may give erroneous results when colonies are taken from TCBS agar. Therefore, sucrose-positive colonies on

TCBS should be subcultured by heavy inoculation onto a nonselective medium, such as blood agar. For recent reviews of enrichment and plating media for the vibrios, see Oliver (147) and Harwood et al. (70).

IDENTIFICATION

The salient differentiating features of the eight human pathogenic vibrios associated with foods are shown in Table 16.1. Comments regarding the taxonomy of individual species are noted below. However, although determination of select phenotypic traits for the species-level identification of vibrios remains routine, identification based on this classic method has always been problematic for this genus. Some researchers have developed immunological assays in an attempt to overcome this problem, whereas most now employ molecular techniques as the primary tool to this end (70). For example, DNA probes targeting the thermolabile hemolysin (*tlh*) and the thermostable direct hemolysin (TDH) (*tdh*) genes are used to confirm total numbers of and numbers of pathogenic *V. parahaemolyticus* cells, respectively. Similarly, the hemolysin gene (*vvhA*) is used as a target specific for *V. vulnificus*, as are the 16S-23S rRNA intergenic spacer region and *ctx* for total numbers of and numbers of toxin-producing strains of *V. cholerae*, respectively.

EPIDEMIOLOGY

The numbers of cells of most *Vibrio* spp. in both surface waters and shellfish have a definite seasonal correlation, generally being greatest during the warm-weather months. Seasonality is most notable for *V. vulnificus* and

Table 16.1 Key differential characteristics of food-associated pathogenic *Vibrio* species[a]

| Test | % Positive for[b]: | | | | | | |
	V. cholerae	*V. mimicus*	*V. hollisae*	*V. fluvialis*[c]	*V. alginolyticus*	*V. parahae-molyticus*	*V. vulnificus*
Voges-Proskauer (1% NaCl)	75	9	0	0	95	0	0
Motility (36°C)	99	98	0	70–89	99	99	99
Acid production from:							
Sucrose	100	0	0	100	99	1	15
D-Mannitol	99	99	0	97	100	100	45
Cellobiose	8	0	0	30	3	5	99
Salicin	1	0	0	0	4	1	95

[a] Adapted from reference 2.
[b] Numbers indicate the percentage of strains that are positive after 48 h of incubation at 36°C (unless other conditions are indicated). Most of the positive reactions occur during the first 24 h.
[c] Includes *V. furnissi*, which differs from *V. fluvialis* primarily by production of gas in D-glucose.

V. parahaemolyticus but is not associated with some vibrios, such as *V. fluvialis*, which occur throughout the year.

Although there is considerable variation in the severities of the different vibrio diseases, and indeed even in those of infections caused by the same species, the most severely ill patients generally have preexisting underlying illnesses, with chronic liver disease being one of the most common. An exception to this generalization occurs with *V. cholerae* O1 and O139, which can readily cause disease in noncompromised individuals. In almost all cases, a recent history of consumption of seafood, especially raw oysters, is noted. *V. cholerae* O1 and O139 strains are also exceptional in having a broader vehicle range for infection, although seafood remains important in their transmission. In a report by Bonner et al. (16) covering infections caused by vibrio isolates recovered during a 10-year period in Alabama, 87% of the patients indicated a recent history of seawater-associated activity. Lowry et al. (115) determined that all 51 persons with diarrhea who had a *Vibrio* species present in a stool specimen had eaten raw or cooked seafood. Unfortunately, because vibrios are part of the normal estuarine microflora and not a result of fecal contamination, vibrio infections will not likely be controlled through shellfish sanitation programs. It is thus essential that raw seafood be adequately refrigerated or iced to prevent substantial bacterial growth.

INCIDENCE OF VIBRIO INFECTIONS

The routine use of TCBS agar for screening stool samples for vibrios suggests a low incidence of these infections. Hoge et al. (79), for example, obtained only 40 *Vibrio* isolates from 32 patients (during a 15-year survey of all stool specimens obtained in a hospital adjacent to the Chesapeake Bay in Maryland), for an overall incidence of 1.6 per 100,000 patients per year. Similarly, Magalhaes et al. (116) examined 3,250 diarrheal stool specimens received between 1989 and 1991 at a clinical laboratory in Brazil and, despite enrichment in APW prior to culture on TCBS, isolated vibrios from only 55 (1.7%) of the samples. Bonner et al. (16), who concluded that vibrios were not a major cause of infectious diarrhea, found only one case of vibrio gastroenteritis during the four years TCBS was used in a Gulf Coast survey. However, the routine use of TCBS for screening of stool and wound samples may be warranted in some instances. Desenclos et al. (46) reported an overall annual incidence rate for any vibrio illness to be 11.3 per million for raw-oyster-eating populations in Florida but only 2.2 per million for persons who did not eat raw oysters. These rates were greatly affected by several host factors. For example, the annual incidence

rate of vibrio illness for raw-oyster eaters with liver disease was 95.4 per million, compared to a rate of 9.2 per million for raw-oyster eaters without liver disease. Similarly, Levine et al. (111) suggested that, although the reported incidence rate of infections yielding *Vibrio* species is low (0.7 per 100,000) compared to that of infections yielding other enteric pathogens (e.g., 7.7 per 100,000 for *Shigella* species), the high proportion (45%) of patients with vibrio gastroenteritis who are hospitalized suggests that many milder infections may occur which are not reported. These authors also noted that 86% of patients with *V. fluvialis* and 35% of those with *V. parahaemolyticus* had bloody stools.

Lesmana et al. (108) reported that *Vibrio* spp. were isolated from more than 21% of diarrhea patients admitted to a community hospital in Jakarta, Indonesia. These species included *V. cholerae* O1 (49.5%), *V. parahaemolyticus* (30%), non-O1 *V. cholerae* (17%), and *V. fluvialis* (9%). Smaller numbers of *V. furnisii*, *V. metschnikovii*, *V. mimicus*, and *V. hollisae* strains were also isolated.

SUSCEPTIBILITY TO PHYSICAL AND CHEMICAL TREATMENTS

With the exception of those of *V. cholerae*, *V. parahaemolyticus*, and *V. vulnificus*, relatively little is known of the susceptibilities of vibrios to various food preservation methods. Summarized here are some of the studies relevant to this point; when detailed studies exist for the various vibrios, they are noted as each is described.

Cold

Although reports generally indicate that vibrios are sensitive to cold temperatures, seafoods have also been reported to be protective for vibrios at refrigeration temperatures. Wong et al. (202) isolated several psychrotrophic strains of *V. mimicus*, *V. fluvialis*, and *V. parahaemolyticus* from frozen seafoods and found these strains to survive well at 10, 4, and –30°C. *V. parahaemolyticus* may undergo an initial rapid reduction (ca. 99%) in survival when incubated on whole shrimp at 3, 7, 10, or −18°C, although survivors remained at the end of the 8-day study. *V. parahaemolyticus* can also survive storage in shellstock oysters for at least 3 weeks at 4°C and subsequently multiply after incubation at 35°C for 2 to 3 days. Similarly, numbers of cells of *V. parahaemolyticus* were reduced in cooked fish mince and surimi stored at 5°C for 48 h, but growth occurred when the product was held at 25°C. Cook (36) reported that *V. vulnificus* failed to grow in oysters held at 13°C and below, with significant increases in numbers of cells in oysters stored at 18°C and above. Such studies suggest that naturally occurring vibrios are able to multiply in unchilled shellstock oysters. Similarly, studies involving temperature abuse of octopus, cooked shrimp, and crabmeat all have documented growth of *V. parahaemolyticus* to very large numbers of cells when the meat was held for even short periods of time under improper refrigeration.

The persistence of *V. vulnificus* in oysters following freezing and storage at −20°C, with or without vacuum packaging, was dependent on the length of frozen-storage time for cells packaged without vacuuming, with a decrease from ca. 10^5 to ca. 10^1 CFU/g. Vacuum-packaged samples showed significantly lower concentrations of *V. vulnificus* than normally packaged samples over a 70-day study period (158). Further, individually quick frozen technology for oysters has advanced to a point where oysters with a shelf life of more than 18 months, and a reduction of numbers of *V. vulnificus* cells to nondetectable levels, have been routinely produced (75).

Heat

All of the vibrios are sensitive to heat, although a wide range of thermal inactivation rates have been reported. Heating of *V. parahaemolyticus* cells at 60, 80, or 100°C for 1 min is lethal for small (5×10^2) populations, although some cells survived heating at 60°C and even 80°C for 15 min when populations of 2×10^5 were used (194). Cook and Ruple (37) reported that decimal reduction times at 47°C averaged 78 s for the 52 strains of *V. vulnificus* examined. Thorough heating of shellfish to provide an internal temperature of at least 60°C for several minutes appears to be sufficient to eliminate the pathogenic vibrios (194). Cook and Ruple (37) found that heating oysters for 10 min in water at 50°C is sufficient to reduce *V. vulnificus* populations to nondetectable levels. Low-temperature pasteurization (e.g., 50°C for 10 min) of oysters reduces *V. vulnificus* populations by 6 logs (6). Nascumento et al. (138), studying an O1 strain of *V. cholerae* inoculated into white shrimp (*Penaeus schimitti*), reported that boiling results in total destruction within 1 to 2 min. The U.S. Food and Drug Administration (FDA) recommends steaming shellstock oysters, clams, and mussels for 4 to 9 min, frying shucked oysters for 10 min at 375°C, or baking oysters for 10 min at 450°C.

Irradiation

Doses of 3 kGy of gamma irradiation kill vibrios in frozen shrimp (164). High levels (>50 kilorads) of ^{60}Co have been used to eliminate *V. cholerae* from both fresh and frozen frog legs (172). The use of ionizing radiation for reducing levels of *V. vulnificus* cells in shellstock oysters has also been studied, with a dose of 1 kGy resulting in

more than a 5-log reduction and no mortality in the oysters. Higher doses, e.g., 1.5 kGy, completely inactivate *V. vulnificus* but result in increased oyster mortalities of up to 16% (47). Similar studies of *V. parahaemolyticus* in oysters revealed that doses of 3 kGy reduce pathogen levels by up to 6 logs with no oyster mortalities (82). The effects of low-dose gamma irradiation on vibrios in oysters have also been described by Andrews et al. (7).

High Pressure

Recently, the use of high hydrostatic pressure (30,000 to 50,000 lb/in^2) has been evaluated for the killing of vibrios in oysters (72). While such treatment results in the death of the animal, reduction in *V. vulnificus* levels of up to 6 logs has been observed after 10 min of treatment, and reduction of up to 9 logs has been observed for *V. parahaemolyticus* when oysters were treated for as little as 30 s (23, 93).

Miscellaneous Inhibitors

The bactericidal effects on *V. parahaemolyticus* of a large variety of dried spices, oils of several herbs, tomato sauce, and several organic acids have been reported previously, with many of these substances being highly toxic at low levels (167). Similarly, *V. vulnificus* is inactivated by several fruit and vegetable juices and a variety of spice extracts. *V. parahaemolyticus* is highly sensitive to as little as 50 parts of butylated hydroxyanisole per million and is inhibited by 0.1% sorbic acid. Of 10 GRAS (generally recognized as safe) compounds tested against *V. vulnificus*, only diacetyl was inhibitory to the pathogen when *V. vulnificus* was present in shellstock oysters (184). Vibrios are highly acid sensitive, although growth in media at as low as pH 4.8 has been reported for *V. parahaemolyticus*.

Depuration

Depuration, wherein filter-feeding bivalves purify themselves through the pumping of bacteria-free water through their tissues, is of considerable value in removing contaminating bacteria such as *Salmonella* species and *Escherichia coli* (20, 166). Similarly, oysters artificially infected with vibrios in the laboratory can effectively eliminate these added bacteria through this process (50, 62, 168). However, the many studies that have been done on depuration of oysters and clams show that this method, by itself, does not significantly reduce the naturally occurring *Vibrio* microflora present in these animals (84, 187). Motes et al. (137) have demonstrated that relaying oysters to off-shore waters with high salinities (30 to 34%) can reduce *V. vulnificus* levels from 10^3 to 10^4 to <10 MPN (most probable number)/g within 7 to 17 days.

V. CHOLERAE

V. cholerae O1 is the causative agent of cholera, one of the few foodborne diseases with epidemic and pandemic potential. V. cholerae has been well defined on the basis of biochemical tests and DNA homology studies, but as reviewed elsewhere (88), this species is not homogeneous with regard to pathogenic potential. Specifically, important distinctions within the species are made on the basis of production of cholera enterotoxin (cholera toxin, or Ctx), serogroup, and potential for epidemic spread. Until recently, the public health distinction was simple, that is, V. cholerae strains of the O1 serogroup which produced Ctx were associated with epidemic cholera and all other members of the species either were nonpathogenic or were only occasional pathogens. However, with the emergence of cholera due to strains of the O139 serogroup (see below), such previous distinctions are no longer valid. There are two serogroups, O1 and O139, that have been associated with epidemic disease, but there are also strains of these serogroups which do not produce Ctx, do not cause cholera, and are not involved in human disease. Conversely, there are occasional strains of serogroups other than O1 or O139 that are clearly pathogenic, either through the production of Ctx or through other virulence factors (see below); however, none of these other serogroups have caused large epidemics or pandemics. Therefore, in assessing the public health significance of an isolate of V. cholerae, there are two critical properties to be determined beyond the biochemical identification of the species as V. cholerae. The first of these properties is production of cholera toxin, which is the toxin that is responsible for severe, cholera-like disease in epidemic and sporadic forms. The second property is possession of the O1 or O139 antigen, which, since the actual determinant of epidemic and pandemic potential is not known, is at least a marker of such potential. The subject of cholera has been reviewed elsewhere, and readers are referred to these reviews for primary references, particularly from the older literature (88, 144). Unless otherwise stated, all information given about this species will apply to V. cholerae O1 or O139 strains capable of causing cholera.

Classification

V. cholerae O1

V. cholerae strains of the O1 serogroup that produce a Ctx have long been associated with epidemic and pandemic cholera. Strains isolated from environmental samples in areas without epidemics are usually Ctx negative and are considered to be nonpathogenic based on results from volunteer studies (see "Reservoirs" below). However, Ctx-negative V. cholerae O1 strains have been isolated in

occasional cases of diarrhea or extraintestinal infections. This serogroup can be further subdivided into serotypes of the O1 serogroup called Ogawa and Inaba. V. cholerae O1 can also be divided into two biotypes, classical and El Tor, which differ in several characteristics. The El Tor biotype is currently the most important biotype; strains of the classical biotype have not been isolated for several years.

V. cholerae Non-O1 and Non-O139

Nearly 200 serogroups of V. cholerae have been described. In recent years, until the emergence of the O139 serogroup, all isolates that were identified as V. cholerae on the basis of biochemical tests but that were negative for the O1 antigen were referred to as non-O1 V. cholerae. In earlier years, non-O1 V. cholerae strains were referred to as NCV (noncholera vibrios) or NAG (nonagglutinable) vibrios. The basis for serotyping of V. cholerae is the lipopolysaccharide (LPS) somatic antigen; H-antigens are not useful in serotyping. The great majority of these strains do not produce cholera toxin and are not associated with epidemic diarrhea (reviewed by Morris [131, 133]). These strains are occasionally isolated in cases of diarrhea (usually associated with consumption of shellfish) and have been isolated in a variety of extraintestinal infections. These strains are regularly found in estuarine environments, and infections due to these strains are commonly of environmental origin. Although most of these strains do not produce Ctx, some strains may produce other toxins (see below); however, for many strains of V. cholerae non-O1 and non-O139 isolated in cases of gastroenteritis, the pathogenic mechanisms are unknown. Strains of the O141 serogroup have been isolated in sporadic cases of severe diarrhea and produce Ctx and the toxin coregulated pilus (TCP) colonization factor typical of O1 and O139 strains (39) but also produce a novel type III secretion system (TTSS; see below).

V. cholerae O139 Bengal

The simple distinction between V. cholerae O1 and V. cholerae non-O1 was rendered obsolete in early 1993 when the first reports appeared of a new epidemic of severe, cholera-like disease emerging from eastern India and Bangladesh (reviewed by Albert [4]). Further investigations revealed that the causative bacterium did not belong to the O serogroups previously described for V. cholerae but to a new serogroup, which was given the designation O139 and the synonym Bengal in recognition of the origin of this strain. This pathogen appears to be a hybrid of the O1 strains and the non-O1 strains. In important virulence characteristics, such as clinical manifestations of infection and Ctx and TCP sequences, V. cholerae O139 is indistinguishable from typical El Tor

V. cholerae O1 strains (4). However, this bacterium does not produce the O1 LPS due to deletion of 22 kb of DNA necessary for production of the O1 antigen and insertion of a 35-kb region encoding the O139 antigen (35). Furthermore, like many strains of non-O1 *V. cholerae* and unlike *V. cholerae* O1, this bacterium produces a polysaccharide capsule (see below). When it initially appeared, the O139 serogroup replaced the O1 serogroup in some parts of southeast Asia and was feared to represent a new pandemic (the eighth pandemic) of cholera. However, few cases of O139 infection were reported beyond southeast Asia and cases of O1 infection became dominant again in this part of the world (5).

Isolation and Identification

TCBS is the most commonly employed isolation medium for *V. cholerae*, although this species can grow as lactose-negative colonies on MacConkey agar. Enrichment of specimens usually employs the nonselective APW, with plating after 6 to 8 h (to prevent overgrowth by other species) or overnight incubation (90). Suspected *V. cholerae* isolates can be transferred from primary isolation plates onto a standard series of biochemical media used for identification of *Enterobacteriaceae* and *Vibrionaceae*. Both conventional tube tests and commercially available enteric identification systems are suitable for identifying this species, although the accuracy of commercial identification kits can range from 50 to 97% (2). Several key characteristics for distinguishing *V. cholerae* from other species are given in Table 16.1.

The key confirmation for identification of *V. cholerae* O1 is agglutination in polyvalent antisera raised against the O1 antigen. Polyvalent antisera for *V. cholerae* O1 and O139 are commercially available and can be used in slide agglutination or coagglutination tests. Monoclonal-antibody-based, coagglutination tests suitable for testing isolated colonies or diarrheal stool samples for O1 or O139 are also available commercially (Cholera SMART and Bengal SMART; New Horizons Diagnostics Corp., Columbia, Md.). Oxidase-positive organisms (determined by using colonies grown on nonselective media) that agglutinate in O1 or O139 antisera can be reported presumptively as *V. cholerae* O1 or O139 and then forwarded to a public health reference laboratory for confirmation. Antisera for serogroups other than O1 or O139 are not commercially available.

Molecular Techniques

Nucleic acid probes are not routinely employed for the identification of *V. cholerae* due to the ease of identifying this species by conventional methods. Where DNA probes and PCR techniques have been extremely useful is in distinguishing those strains of *V. cholerae* that contain genes encoding cholera toxin (*ctx*) from those that do not contain these genes. This distinction is particularly important in examining environmental isolates of *V. cholerae* since most of these strains lack *ctx* sequences.

A number of DNA fragment probes and synthetic oligonucleotide probes have been developed to detect *ctx* sequences in isolated colonies (reviewed by Popovic et al. [160]), although PCR techniques are now more commonly employed (90). PCR has also been used to detect toxigenic *V. cholerae* O1 in food samples including fruit, vegetables, and shellfish specimens that have been seeded with *V. cholerae* O1 (99). PCR was used to investigate a small outbreak of cholera in New York due to crabs imported from Ecuador. Although the crabs implicated in the outbreak did not yield *V. cholerae* O1 after culture, the *ctx* PCR technique detected *ctx* sequences in one of four crab samples examined.

Subtyping of *V. cholerae* strains by using a variety of techniques such as restriction fragment length polymorphism analysis, pulsed-field gel electrophoresis (PFGE), ribotyping, and multilocus sequence typing has yielded significant insights into the molecular epidemiology of *V. cholerae*. For example, restriction fragment length polymorphism analysis revealed that a toxigenic O1 strain isolated from a cholera patient in Maryland was identical to isolates from Louisiana and Texas, which concurred with results of epidemiologic investigations showing that the crabs eaten by the Maryland patient were harvested along the Texas coast. PFGE has been used to differentiate strains of *V. cholerae* in many studies all over the world, and a study by the Centers for Disease Control and Prevention (CDC) revealed that PFGE could distinguish strains that were identical when examined by multilocus enzyme electrophoresis or ribotyping (24). Multilocus sequence typing using only three housekeeping genes (*gyrB*, *pgm*, and *recA*) was found to offer superior discriminating ability compared to PFGE (101). A very recent approach is to use a single multiplex PCR assay to simultaneously amplify 95 "diagnostic" regions from *V. cholerae* and other *Vibrio* species (e.g., species- or serogroup-specific genes and toxin genes, etc.) that are hybridized to a microarray containing these genes (195). This approach allows rapid and definitive inter- and intraspecies discrimination that can be helpful in epidemiologic, environmental, and health risk assessment surveillance.

Reservoirs

Environment

V. cholerae is part of the normal, free-living (autochthonous) bacterial flora in estuarine areas. Non-O1 and non-O139 strains are much more commonly isolated from the environment than are O1 strains, even in settings where epidemics occur in which fecal contamination of the environment may be expected. Outside of areas where epidemics occur (and away from areas that may have been contaminated by cholera patients), O1 environmental isolates are almost always Ctx negative. However, it is clear that Ctx-producing *V. cholerae* O1 can persist in the environment in the absence of known human disease. An environmental reservoir is the most likely explanation for the persistence of a single strain in the U.S. Gulf Coast for more than 20 years. Periodic introduction of such environmental isolates into the human population through ingestion of uncooked or undercooked shellfish appears to be responsible for isolated foci of endemic disease along the U.S. Gulf Coast and in Australia (15).

V. cholerae strains are capable of colonizing the surfaces of zooplankton such as copepods with 10^4 to 10^5 *V. cholerae* cells attached to a single copepod. Other aquatic biota, such as water hyacinths, filamentous green algae, and insects, have also been shown to be colonized by *V. cholerae* (reviewed by Butler and Camilli [22]). *V. cholerae* produces a chitinase and is able to bind to chitin, which is the principal component of crustacean shells; the bacterium can grow in media with chitin as the sole carbon source. Chitin was recently shown to induce natural competence in *V. cholerae*, thus suggesting that this species can acquire new genetic material by transformation during growth on chitin (125). The environmental reservoir is a major part of the *V. cholerae* life cycle where bacteria discharged from the human host reside in association with aquatic life forms until they are once again ingested by humans via contaminated water or food (165).

The persistence of *V. cholerae* within the environment may be facilitated by its ability to assume survival forms, including a viable but nonculturable (VBNC) state, biofilms, and a rugose survival form. The VBNC state is a dormant state in which *V. cholerae* is still viable but not culturable in conventional laboratory media (reviewed by Colwell and Huq [33] and Oliver [146, 148, 150]). In this dormant state, the cells are reduced in size and become ovoid. The continued viability of the nonculturable *V. cholerae* can be assessed by a direct viable count procedure in which cells are incubated in the presence of yeast extract and nalidixic acid and examined microscopically for cell elongation. The VBNC state can be induced in the laboratory by incubating a culture of *V. cholerae* in phosphate-buffered saline at 4°C for several days. Although these cells are not culturable with nonselective enrichment broth or plates, nonculturable *V. cholerae* O1 injected into ligated rabbit ileal loops or ingested by volunteers has yielded culturable *V. cholerae* O1 in intestinal contents or stool specimens (34). A recent microarray study found that VBNC *V. cholerae* still produces transcripts of virulence factor genes (195).

V. cholerae can also form biofilms that can enhance survival in the environment (198). A variety of factors including flagella, the type IV pilus mannose-sensitive hemagglutinin (MSHA), and quorum sensing (QS; see below) are involved in the formation of biofilms. Biofilms allow the bacterium to persist in association with biotic and abiotic surfaces and help to prevent predation by grazing protozoa (120). Furthermore, *V. cholerae* cells present in biofilms are much more resistant to killing by acid shock than are planktonic cells that are not in biofilms (209). This increased acid resistance may help the bacterium survive stomach acidity after ingestion by the next human host.

A phenomenon that is closely related to biofilms is the formation of a rugose or wrinkled colony appearance due to the production of an exopolysaccharide. Rugose colonies can form spontaneously in the laboratory at a frequency of about 1 in 1,000, a rate that can be increased under stress by certain culture conditions, and the process is reversible (reviewed by Kaper et al. [88]). When examined by light and electron microscopy, cells in a rugose culture are small and spherical and are embedded in an amorphous matrix material composed primarily of carbohydrate. In this state, the cells are protected against adverse environmental conditions. Notably, rugose variants survive in the presence of chlorine and other disinfectants and have enhanced capacity to form biofilms. Such rugose forms are nonetheless still capable of causing diarrhea in volunteers. The rugose polysaccharide produced by O1 El Tor strains is called VPS^{ETr}, and genes required for synthesis are clustered in a 30-kb region called the *vps* locus (207). Microarray analysis of rugose and smooth variants of the same strain implicated 124 differentially regulated genes encoding regulators, surface properties, and motility in the formation of rugose colonies (206).

Humans and Animals

Long-term carriage of *V. cholerae* in humans is extremely rare and is not considered to be significant in transmission of disease. However, short-term carriage of *V. cholerae* by humans is quite important in transmission of disease. Persons with acute cholera excrete 10^7 to 10^8 *V. cholerae* CFU per g of stool; for patients who have 5 to 10 liters of diarrheal stool, total output of *V. cholerae* can be in the range of 10^{11} to 10^{13} CFU. Even after cessation of symptoms, patients who have not been treated with antibiotics may continue to excrete vibrios for 1 to 2 weeks. Furthermore, a high percentage of persons infected with

V. cholerae in areas where infection is endemic have inapparent illness and can still excrete the bacterium, although excretion generally lasts for less than a week. Asymptomatic carriers are most commonly identified among members of households of persons with acute illness: in various studies, the rate of asymptomatic carriage in this group has ranged from 4% to almost 22%. There have also been studies indicating that *V. cholerae* O1 can be sporadically carried by household animals, including cows, dogs, and chickens, but no animal species consistently carries the bacterium.

In the 1970s it was widely accepted that asymptomatic and convalescent human (and possibly animal) carriers were the primary reservoir for cholera. With the recognition that *V. cholerae* can live and multiply in the environment, much greater attention has been given to identification and characterization of environmental reservoirs (reviewed by Colwell and Huq [33]). Nonetheless, Ctx-producing *V. cholerae* O1 (i.e., disease-causing strains) continues to be isolated almost exclusively from areas that have been contaminated by human feces or sewage from persons or groups of persons known to have had cholera. Similarly, in areas where the disease is endemic, such as Lima, Peru, rates of isolation from the environment correlate primarily with the degree of sewage contamination. However, even areas where cholera is not endemic may be contaminated by ballast water from ships originating in areas where cholera is endemic. For example, oysters in Mobile Bay, Alabama, contained toxigenic *V. cholerae* O1 strains with PFGE patterns that were identical to those of toxigenic *V. cholerae* strains isolated from ballast water from ships docked at Gulf of Mexico ports (136). A dynamic relationship between human and environmental sources of the bacterium is apparent, with carriage and amplification by human populations playing a critical role in epidemic spread of Ctx-producing *V. cholerae*. Environmental bacteriophages capable of lysing *V. cholerae* O1 or O139 are also involved in this relationship, and a recent 3-year study in Bangladesh revealed that the presence of cholera phages was inversely correlated with the occurrence of viable *V. cholerae* in the environment and the number of cholera cases in the local community (53).

Foodborne Outbreaks

The critical role of water in the transmission of cholera has been recognized for more than a century, ever since the London physician and epidemiologist John Snow showed in 1854 that illness was associated with consumption of water from a water system that drew its water from the Thames at a point below major sewage inflows. In developing countries, ingestion of contaminated water and food is probably the major vehicle for the transmission of cholera, whereas in developed countries, foodborne transmission is more important (58). Such distinctions are often difficult to make since contaminated water is frequently used in food preparation. For example, rice prepared with water contaminated with *V. cholerae* O1 has been implicated in outbreaks in Bangladesh as well as on the U.S. Gulf Coast. Fruit juices diluted with contaminated water and vegetables irrigated with untreated sewage have been associated with disease in South America. Seafood may acquire the bacterium from environmental sources and may serve as a vehicle in both endemic and epidemic disease, particularly if it is uncooked or only partially cooked.

The role of food in transmitting *V. cholerae* O1 has been extensively reviewed elsewhere (91, 130, 162), and the reader is referred to these reviews for more detail and primary references. The spectrum of food items implicated in the transmission of cholera includes crabs, shrimp, raw fish, mussels, cockles, squid, oysters, clams, rice, raw pork, millet gruel, cooked rice, street vendor food, frozen coconut milk, and raw vegetables and fruit (130). One shared characteristic of the implicated foods is their neutral or nearly neutral pHs. Hence, in investigating a suspected foodborne outbreak with many possible vehicles, one can eliminate the foods with an acid pH and concentrate on neutral or alkaline foods (91). This predilection for neutral foods was demonstrated in an epidemiologic study in West Africa, where boiled rice is commonly prepared in the morning, held unrefrigerated, and eaten with sauce at the midday and evening meals. In a case control study of illness, tomato sauces with a pH of 4.5 to 5.0 were protective against illness whereas less acidic sauces (pH 6.0 to 7.0) prepared from ground peanuts were associated with illness due to *V. cholerae* O1 (130). Alkaline conditions are employed in enrichment cultures of food and water samples wherein the use of APW (pH 8.5) followed by plating onto TCBS agar is a common isolation method (91). Survival and growth of *V. cholerae* O1 in food are also enhanced by low temperatures, high organic content, high moisture, and the absence of competing flora (130). Survival is increased when foods are cooked before contamination; cooking eliminates competing microbes and has also been suggested to destroy some heat-labile growth inhibitors and produce denatured proteins that the bacterium uses for growth (130).

As noted below, food buffers *V. cholerae* O1 against killing by gastric acid. Although many different food items can provide this buffering capacity, the protection provided by chitin is noteworthy because crustaceans are frequent vehicles of disease. In dilute hydrochloric acid solutions of approximately the same pH as human gastric acid, survival of *V. cholerae* absorbed to chitin was enhanced compared to that of *V. cholerae* in the absence of chitin (130).

In the United States, both domestically acquired and imported cases of cholera occur. For domestic cases, crabs, shrimp, and oysters have been the most frequently implicated vehicles, although the largest single outbreak (16 cases) was due to ingestion of contaminated rice. In this outbreak, which occurred on a Gulf Coast oil rig in 1981, cooked rice was moistened with water contaminated by human feces and then held for 8 h after cooking (130). Although most domestic cases occur in states bordering the Gulf Coast, seafood shipped from this area has caused disease in both Maryland and Colorado (130). The risk of imported cholera has greatly increased since the establishment of endemic cholera in South America in 1991. The largest such outbreak (75 cases) involved crab salad served on an airplane flying from Peru to California. A smaller outbreak of eight cases occurred in New Jersey due to crabs purchased in Ecuador and carried to the United States in an individual's luggage. Importation from Asia can also occur, even in commercially imported food. A small outbreak of four cases in Maryland was attributed to frozen coconut milk imported from Thailand which was subsequently used in a topping for a rice pudding.

Most cases of gastroenteritis caused by *V. cholerae* of serogroups other than O1 or O139 have been linked to the consumption of raw oysters (reviewed by Morris [131]). Both disease incidence and isolation rates for non-O1 and non-O139 serogroups from oysters are highest in the summer. Such strains were isolated from up to 14% of freshly harvested oysters in one study conducted by the U.S. FDA. Outside of the United States, outbreaks have also been linked to consumption of contaminated potatoes, chopped eggs, preprepared gelatin, vegetables, and meat samples (76, 131). As with *V. cholerae* O1, survival of non-O1 and non-O139 *V. cholerae* is enhanced in foods of alkaline pH.

Characteristics of Disease

The explosive, potentially fatal dehydrating diarrhea that is characteristic of cholera is actually seen in only a minority of persons infected with cholera toxin-producing *V. cholerae* O1 or O139. Most infections with *V. cholerae* O1 are mild or even asymptomatic. It has been estimated that 11% of patients with classical infections develop severe disease, compared with 2% of those with El Tor infections. An additional 5% of El Tor infections and 15% of classical infections result in moderate illness (reviewed by Kaper et al. [88]). Symptoms of persons infected with

V. cholerae O139 Bengal appear to be virtually identical to those of persons infected with O1 strains.

The incubation period for cholera can range from several hours to 5 days and is dependent in part on inoculum size. The onset of illness may be sudden, with profuse, watery diarrhea, or there may be premonitory symptoms such as anorexia, abdominal discomfort, and simple diarrhea. Initially the stool is brown with fecal matter, but soon the diarrhea develops a pale gray color with an inoffensive, slightly fishy odor. Mucus in the stool imparts the characteristic "rice water" appearance. Vomiting is often present, occurring a few hours after the onset of diarrhea.

In the most severe form of cholera, termed cholera gravis, the level of diarrhea may quickly reach 500 to 1,000 ml/h, leading rapidly to tachycardia, hypotension, and vascular collapse due to dehydration. Peripheral pulses may be absent, and blood pressure may be unobtainable. Skin turgor is poor, giving the skin a doughy consistency; the eyes are sunken; and hands and feet become wrinkled, as after long immersion in water ("washerwoman's hands"). Such severe dehydration can lead to death within hours of the onset of symptoms unless fluids and electrolytes are rapidly replaced. While cholera gravis is a striking clinical entity, milder illnesses are not readily differentiated from other cases of gastroenteritis in areas where cholera is endemic.

Gastroenteritis associated with *V. cholerae* non-O1 and non-O139 is generally of mild to moderate severity, although severe, cholera-like illness has also been seen occasionally. Besides nonbloody and occasionally bloody diarrhea, symptoms may include abdominal cramps and fever with nausea and vomiting occurring in a minority of patients (131). Non-O1, non-O139 *V. cholerae* strains are also frequently isolated in cases of extraintestinal infections such as septicemia, wound infections, and ear infections; these infections usually involve exposure to fresh or brackish water (131, 133). The case fatality rate of extraintestinal infections can exceed 50%, and individuals with preexisting liver disease are particularly at risk.

Infectious Dose and Susceptible Population

In healthy North American volunteers, doses of 10^{11} CFU of *V. cholerae* were required to consistently cause diarrhea when the inoculum was given in buffered saline (pH 7.2). When stomach acidity was neutralized with 2 g of sodium bicarbonate immediately prior to administration of the inoculum, attack rates of 90% were seen with an inoculum of 10^6. Food has a buffering capacity comparable to that of sodium bicarbonate. Ingestion of 10^6 vibrios with food such as fish and rice resulted in the same high attack rate (100%) as ingestion of this inoculum administered with a buffer (109). Further studies revealed that most volunteers who receive as few as 10^3 to 10^4 vibrios with a buffer develop diarrhea, although lower inocula correlate with longer incubation periods and diminished severity. The incubation time in volunteers between the ingestion of vibrios and the onset of diarrhea has ranged from 8 to 96 h. The inoculum size in naturally occurring infections is not known with certainty but is estimated to be in the range of 10^2 to 10^3 vibrios (58).

The volunteer data on the effect of a buffer on the infectious dose are consistent with epidemiologic data indicating that people who are achlorhydric because of surgery, medication (e.g., antacids), or other reasons are at increased risk for cholera. Individuals of blood group O are at increased risk for more severe cholera, as demonstrated for natural infection as well as experimental infection; the mechanism of this increased susceptibility is unknown. In addition to these factors, additional host factors, as yet poorly defined, play a role in susceptibility to cholera. The effect of host factors is illustrated by a study in which identical inocula caused 44 liters of diarrhea in one volunteer and little or no illness in other individuals (109).

Strains of non-O1, non-O139 *V. cholerae* have also been studied with volunteers. Of three non-Ctx-producing strains fed to volunteers, only one strain caused diarrhea. This strain produced a heat-stable enterotoxin (ST)-like toxin (see below) and caused diarrhea in 6 of 8 volunteers at doses of 10^6 to 10^9 vibrios after neutralization of stomach acid with sodium bicarbonate (131). The severity of disease was generally mild, but in one volunteer, diarrheal stool volume exceeded 5 liters.

Virulence Mechanisms

Infection due to *V. cholerae* O1 or O139 begins with the ingestion of food or water contaminated with the bacterium. After passage through the acid barrier of the stomach, vibrios colonize the epithelium of the small intestine by means of one or more adherence factors. Invasion into epithelial cells or the lamina propria does not occur. Production of cholera enterotoxin (and possibly other toxins) disrupts ion transport by intestinal epithelial cells. The subsequent loss of water and electrolytes leads to the severe diarrhea characteristic of cholera.

Cholera Toxin

Volunteer studies demonstrate that the massive, dehydrating diarrhea characteristic of cholera is induced by cholera enterotoxin, also referred to as cholera toxin, or Ctx. Cholera toxin is among the best characterized of bacterial toxins and has been extensively reviewed (42, 174). Ctx is a prototypic A-B subunit toxin where the B subunit serves to bind the holotoxin to the eukaryotic cell receptor and the A subunit possesses specific enzymatic activities, ADP-ribosyltransferase and NAD-glycohydrolase, that act intracellularly. Ctx consists of five identical B subunits and a single A subunit; neither of the subunits individually has significant secretogenic activity in animal or intact cell systems. The mature B subunit contains 103 amino acids with a subunit weight of 11.6 kDa. The mature A subunit has a mass of 27.2 kDa and is proteolytically cleaved to yield two polypeptide chains, a 195-residue A1 peptide of 21.8 kDa and a 45-residue A2 peptide of 5.4 kDa, which are linked by a disulfide bond. The *ctxA* and *ctxB* genes encoding the A and B subunits are carried by a filamentous bacteriophage (CTXM) which is capable of transducing *ctx* genes into nontoxigenic strains (reviewed by McLeod et al. [122]).

The receptor for Ctx is the ganglioside GM1, and the binding of toxin to epithelial cells is enhanced by a neuraminidase produced by *V. cholerae*. This 83-kDa enzyme catalyzes the conversion of higher-order gangliosides into GM1, thereby enhancing the binding of Ctx and leading to greater fluid secretion. Binding of the CtxB pentamer to GM1 in lipid rafts of the plasma membrane triggers toxin internalization via endocytic vesicles. The internalized holotoxin traffics through the trans-Golgi network and to the endoplasmic reticulum (reviewed by de Haan and Hirst [42]). Ultimately, the A1 peptide is translocated into the cell cytosol, where it irreversibly ADP-ribosylates the α subunit of the G_s protein. G proteins link many cell surface receptors to effector proteins at the plasma membrane, thereby regulating an extensive set of metabolic pathways. G_s is involved in regulation of the adenylate cyclase complex, which mediates the transformation of ATP into cyclic AMP (cAMP), a crucial intracellular messenger for a variety of cellular pathways. The Ctx A1 peptide catalyzes the transfer of the ADP-ribose moiety of NAD to a specific arginine residue in the $G_{s\alpha}$ protein, resulting in the activation of adenylate cyclase and subsequent increases in intracellular levels of cAMP. cAMP activates a cAMP-dependent protein kinase, protein kinase A (PKA), leading to protein phosphorylation of the major Cl^- channel in epithelial cells, the cystic fibrosis transmembrane conductance regulator (CFTR) (Fig. 16.1). Increased Cl^- secretion by intestinal crypt cells through the CFTR and decreased NaCl-coupled absorption by villus cells (by mechanisms that are not well understood) result in a transepithelial osmotic gradient that causes water flow into the lumen of the intestine. The massive volume of water overwhelms the absorptive capacity of the intestine, resulting in diarrhea.

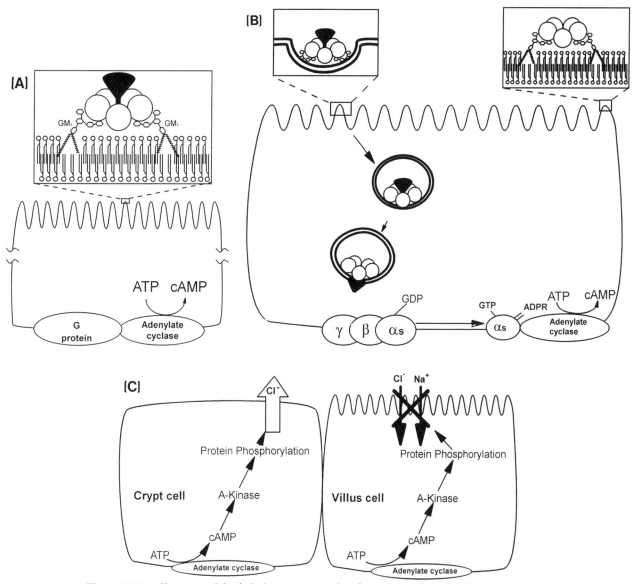

Figure 16.1 Classic model of cholera toxin mode of action involving cAMP. More recent evidence indicates that prostaglandins and the ENS are also involved in the response to cholera toxin (see the text for details). (A) Adenylate cyclase, located in the basolateral membrane of intestinal epithelial cells, is regulated by G proteins. Cholera toxin binds via the B subunit pentamer (shown as open circles with the A subunit as the inverted solid triangle) to the GM1 ganglioside receptor inserted into the lipid bilayer. (B) The toxin enters the cell via endosomes, and the A1 peptide ADP-ribosylates $G_{s\alpha}$ located in the basolateral membrane. (C) Increased cAMP activates PKA, leading to protein phosphorylation. In crypt cells, the protein phosphorylation leads to increased Cl^- secretion; in villus cells, it leads to decreased NaCl absorption. Adapted from reference 88.

The activation of adenylate cyclase leading to increased cAMP levels and activation of the CFTR via PKA is the classic mode of action of Ctx. However, there are additional effects of Ctx that may contribute to the secretory effects of cholera toxin, including production of prostaglandins and stimulation of the enteric nervous system (ENS) reviewed by Sears and Kaper [174] and de Haan and Hirst [42]). Ctx has been shown to activate platelet-activating factor, leading to activation of phospholipase A2 and subsequent accumulation of arachidonic acid and prostaglandins. A substantial portion of the secretogenic potential of Ctx can be blocked by addition of platelet-activating factor receptor antagonists and phospholipase A2 inhibitors. Consistent with these observations, cholera patients in the active secretory disease stage have elevated jejunal concentrations of prostaglandin E$_2$ compared to patients in the convalescent stage. The ENS plays an important role in normal intestinal secretion and absorption and has been implicated in the diarrheal response to *V. cholerae*. Ctx can stimulate the release of vasoactive intestinal peptide and serotonin (5-hydroxytryptamine), factors that can induce secretion via the ENS. Results from a variety of studies using receptor antagonists and ganglionic or neurotransmitter blockers as well as direct measurements of increased levels of serotonin (5-hydroxytryptamine) and vasoactive intestinal peptide in cholera patients support the role of the ENS in cholera.

Other Toxins Produced by *V. cholerae*

When the first recombinant *V. cholerae* vaccine strains with specific deletions of genes encoding Ctx were tested in volunteers, it was somewhat surprising that mild to moderate diarrhea was still seen in ca. 50% of volunteers (110). The diarrhea was not the severe, dehydrating diarrhea seen with wild-type strains, which can exceed 40 liters in volume, but a much milder diarrhea which ranged from 0.3 to 2.1 liters in volume. Some volunteers also experienced abdominal cramps, anorexia, and low-grade fever when fed Δ*ctx V. cholerae* strains. These results prompted a search for additional toxins produced by *V. cholerae*, and it is now known that *V. cholerae* produces a variety of extracellular products that have deleterious effects on eukaryotic cells (reviewed by Fullner [55]). The zonula occludens toxin (Zot) increases the permeability of the small intestinal mucosa by affecting the structure of the intercellular tight junction, or zonula occludens (54). Accessory cholera enterotoxin (Ace) causes fluid accumulation in rabbit ligated ileal loops and increases the potential difference in intestinal tissue mounted in Ussing chambers (192). The *ace* and *zot* genes are located immediately upstream of the *ctx* genes and are also believed to be components of the CTXM filamentous phage carrying the cholera toxin gene (122). The soluble hemagglutinin/protease (HA/P) is a zinc-dependent metalloprotease that is capable of nicking and activating the A subunit of Ctx as well as cleaving mucin, fibronectin, and lactoferrin. HA/P can perturb the barrier function of epithelial cells by digesting occludin and rearranging the distribution of ZO-1 in tight junctions (205). Hemolysin (also called cytolysin or El Tor hemolysin) was originally described and utilized to distinguish

between the El Tor (hemolytic) and classical (nonhemolytic) biotypes of *V. cholerae*. The purified hemolysin is cytolytic for a variety of erythrocytes and cultured mammalian cells, is rapidly lethal for mice, and causes bloody fluid accumulation in ligated rabbit ileal loops (178). The RtxA toxin is produced by O1 El Tor and O139 strains and is related to members of the RTX (repeats in toxin) family, a group which includes the hemolysin of uropathogenic *E. coli* and adenylate cyclase of *Bordetella pertussis*. The huge RtxA toxin, with a predicted unprocessed size of ca. 500 kDa, causes depolymerization of actin stress fibers and covalent cross-linking of cellular actin, and is responsible for the cytotoxicity of El Tor strains to HEp-2 cells (56).

The role of toxins other than cholera toxin in the pathogenesis of disease due to *V. cholerae* is largely unknown. These toxins clearly cannot cause cholera gravis because the diarrhea seen with Δ*ctx* strains presumably still producing these toxins is not the severe purging seen with wild-type *V. cholerae* strains. Δ*ctx V. cholerae* vaccine candidate strains lacking genes encoding Zot, Ace, hemolysin, or RtxA toxin still caused mild to moderate diarrhea as well as fever and abdominal cramps in volunteers. However, mutation of the *hap* gene encoding HA/P did reduce the reactogenicity seen with Δ*ctx V. cholerae* vaccine strains in volunteers (10). An El Tor strain with deletions of genes encoding Ctx, hemolysin, HA/P, and RtxA caused reduced pulmonary inflammation compared to the wild type when administered intranasally to mice (57) but the relevance to human intestinal infection is not clear. Toxins other than Ctx may contribute in part to the diarrhea and other symptoms seen with *V. cholerae* strains, perhaps serving as a secondary secretogenic mechanism when conditions for producing cholera toxin are not optimal.

Toxins of *V. cholerae* Non-O1 or Non-O139 Strains

Most non-O1 or non-O139 strains do not contain *ctx* genes but usually possess genes encoding the hemolysin, RtxA, and HA/P. Some strains of *V. cholerae* non-O1 or non-O139 (<10%) produce a 17-amino-acid heat-stable enterotoxin (designated NAG-ST for nonagglutinable *Vibrio* ST) that shows 50% sequence homology to the STa of enterotoxigenic *E. coli*. In a volunteer study, one subject who ingested a Ctx-negative *V. cholerae* non-O1 strain producing NAG-ST purged over 5 liters of diarrheal stool (131). The *tdh* gene encoding the TDH of *V. parahaemolyticus* (see below) has also been found in some *V. cholerae* non-O1 and non-O139 strains.

A recent genome sequence analysis of a clinical non-O1, non-O139 strain (O39 serogroup) that was particularly virulent in an animal model revealed the presence of a TTSS that is not present in O1 and O139 strains (48). This specialized secretion system mediates the translocation of toxins and other effector proteins from the bacterial cytoplasm into the mammalian cell and is an important virulence factor of several gram-negative pathogens, including *Salmonella*, *Shigella*, and *Yersinia* species, enteropathogenic *E. coli*, and *Pseudomonas* species. The TTSS found in *V. cholerae* is related to the TTSS2 of *V. parahaemolyticus* (see below) and was present in 6 of 12 *V. cholerae* non-O1, non-O139 serogroups tested, including all clinical isolates of the O141 serogroup.

Colonization Factors

TCP

TCP is the best characterized intestinal colonization factor of *V. cholerae*. The TCP consists of long filaments 7 nm in diameter which are laterally associated in bundles. The name of the pilus results from the fact that expression of the pilus is correlated with expression of cholera toxin (188). TCP belongs to the type IV family of pili and is the only colonization factor of *V. cholerae* whose importance in human disease has been proven. Volunteers ingesting *V. cholerae* strains with specific mutations in the *tcpA* gene did not experience diarrhea, and no vibrios were recovered from the stools of the volunteers. Direct binding of TCP to epithelial cells has not been demonstrated, and the role of this factor may be to mediate interbacterial aggregation and thereby facilitate intestinal colonization. Synthesis of TCP is complex, and up to 15 open reading frames are found in the *tcp* gene cluster. The *tcp* gene cluster along with *toxT* (see below) is carried on a 40-kb pathogenicity island called VPI (*Vibrio* pathogenicity island) that is present in all O1 and O139 clinical isolates and absent from nearly all environmental O1 and O139 isolates and most clinical non-O1, non-O139 isolates (89).

Other Potential Colonization Factors

A number of other potential colonization factors have been described for *V. cholerae*, including a mannose-fucose-resistant hemagglutinin, MSHA, and several outer membrane proteins, but their role in human intestinal colonization remains unproven. The MSHA is a type IV pilus that has been proven in volunteer studies not to be necessary for human colonization but does appear to be involved in biofilm formation and adherence to zooplankton (199). A 53-kDa protein, called GlpA, was recently implicated in attachment to both zooplankton and human epithelial cells by binding to a sugar present on both surfaces (96).

Motility and Flagella

V. cholerae strains are motile by means of a single, polar, sheathed flagellum. Motility is an important virulence property, with nonmotile, fully enterotoxinogenic mutants being diminished in virulence. In several animal and in vitro models, motile *V. cholerae* rapidly enters the mucus gel overlaying the intestinal epithelium and can be found in intervillus spaces within minutes to a few hours. The role of chemotaxis in *V. cholerae* infection is complex, and some chemotactic mutants show increased infectivity in infant mice. A model has been proposed wherein temporary downregulation of chemotaxis in wild-type *V. cholerae* enhances initial colonization specifically of the upper small intestine, after which chemotaxis is upregulated to allow penetration through the mucus (22).

LPS and Capsule

The LPS of *V. cholerae* O1 is the major protective antigen of this pathogen, and its importance in protective immunity greatly outweighs that of the cholera toxin. The importance of this antigen was seen in the initial outbreaks in India and Bangladesh when the O139 serogroup caused widespread disease in individuals who were presumably immune to the O1 serogroup. *V. cholerae* O1 is unencapsulated, but strains of *V. cholerae* O139 produce a polysaccharide capsule, which has also been termed an O-antigen capsule. Both O1 LPS and O139 capsule have been implicated in intestinal colonization, probably by multiple mechanisms involving direct mucosal adherence and increased intestinal survival in the presence of bile and other factors (139).

Adherence Factors of Non-O1, Non-O139 *V. cholerae*

Intestinal colonization factors of non-O1, non-O139 *V. cholerae* strains are poorly characterized. Only a minority of clinical non-O1, non-O139 strains, such as those belonging to the O141 serogroup, possess *tcp* genes. A variety of fimbria hemagglutinins have been described for these strains, but their role in intestinal adherence is unclear. Most non-O1, non-O139 isolates produce a polysaccharide capsule which in addition to potentially aiding intestinal adherence may facilitate the septicemia that often occurs with these strains.

Regulation

Regulation of virulence gene expression in *V. cholerae* is highly complex, and many regulatory systems, both distinct and overlapping, have been described previously. The complex regulation is understandable in light of the dramatically different environments—the human intestine and the aquatic environment—that make up the *V. cholerae* life cycle. Expression of several virulence genes in *V. cholerae* O1 and O139 is coordinately regulated so that multiple genes respond in a similar fashion to environmental conditions. A wide variety of techniques have been used to study gene expression, including recombinase-base in vivo expression technology and microarray analysis. The use of recombinase-base in vivo expression technology has identified in vivo-expressed genes and elucidated the temporally and spatially separable transcription events in the mouse small intestine (106). Microarray analysis has been applied to examine *V. cholerae* gene expression in rice water stools from cholera patients. Such an analysis suggests that vibrios in the stool are already modulating gene expression to prepare for entry into the aquatic environment and to create a hyperinfectious state that lowers the infectious dose for the next victim (127).

The very active research in this area continues to yield the discovery of novel regulatory systems as well as detailed characterization of well-established regulatory systems in this species. For example, a recently described regulatory system of *V. cholerae* uses cyclic dinucleotide 3′,5′-cyclic diguanylic acid as an intracellular signal to activate virulence genes and repress biofilm formation (reviewed by Camilli and Bassler [25]). The effect of iron on virulence and gene expression in this species has long been known, and recent microarray studies using a strain with a mutation in the iron-dependent negative regulator Fur have revealed the numerous genes involved in the response to this element (128). Space limitations do not permit a detailed discussion of regulation in *V. cholerae*, but two important systems that allow the pathogen to respond to its environment and regulate virulence gene expression will be briefly described.

ToxR Regulon

The ToxR regulon controls expression of several critical virulence factors and has been the most extensively characterized regulatory system of *V. cholerae* (reviewed by Reidl and Klose [165] and Krukonis and DiRita [104]). The key regulator of Ctx and TCP expression in this system is ToxT, a member of the AraC/XylS family of transcription regulators that is encoded by a gene carried on the VPI. ToxT binds upstream of *tcpA* and *ctxA* to activate expression of these genes. Expression of ToxT itself is modulated by ToxR, a 32-kDa transmembrane protein that senses environmental conditions. The 19-kDa transmembrane protein ToxS associates with ToxR and helps to assemble or stabilize ToxR monomers into a dimeric form. ToxRS also directly regulates expression of the OmpT and OmpU outer membrane proteins independently of ToxT. The importance of ToxR in human disease was demonstrated in volunteer studies wherein a ToxR mutant *V. cholerae* O1 strain was fed to volunteers who subsequently suffered no diarrheal symptoms and did not shed the strain in their stools.

Expression of *toxT* is also regulated by the TcpP and TcpH proteins that are encoded by genes located on the VPI upstream of *toxT*. TcpPH and ToxRS have sequence and functional similarities that permit the transmission of environmental signals to modulate expression of *toxT*. Expression of the *tcpPH* genes is in turn regulated by the AphAB proteins which are encoded by genes outside the VPI. Microarray analysis of *toxT*, *tcpPH*, and *toxRS* mutants revealed 13, 27, and 60 genes, respectively, that were transcriptionally repressed in the mutants compared to those in the wild type (13). Hence, the major virulence factors of *V. cholerae*, cholera toxin and TCP, are regulated in a cascade fashion in which AphAB controls expression of TcpPH, which acts together with ToxRS to activate expression of ToxT, which then activates expression of Ctx and TCP.

QS

QS is a density-dependent regulatory process whereby small molecules termed autoinducers (AIs) act as signaling molecules to activate transcription of genes. AIs produced by bacteria are detected and responded to by other members of a population to coordinate gene expression, thereby allowing unicellular microbes to act as multicellular organisms. When the population density is high, the AI level is also high, leading to expression of genes that would not be expressed at low densities. The most common scenario for QS in bacterial pathogens is the activation of virulence genes at high population densities (reviewed by Kaper and Sperandio [87]). *V. cholerae* also regulates expression of virulence genes by QS, but in this case, high densities lead to repression of virulence factors.

V. cholerae has three distinct QS systems (reviewed by Lenz et al. [107]). The CAI-1 and AI-2 QS systems produce and respond to distinct AI molecules, but both end up in a common pathway involving the response regulator LuxO. Phosphorylation of LuxO in response to either CAI-1 or AI-2 activates expression of four genes encoding small RNAs, ultimately leading to transcription of the master regulator HapR. HapR serves as a repressor of genes encoding Ctx, TCP, and biofilm formation via the AphA regulator (see above) and activates expression of HA/P. At low population densities, little HapR is made, thereby allowing expression of Ctx and Tcp and formation of biofilms, but at high population densities, expression of HapR leads to repression of these factors.

A third QS system has a distinct signaling pathway but ultimately ties into the LuxO/HapR signaling cascade. A model has been proposed (66) in which TCP and Ctx are expressed early in infection at low densities along with biofilm formation, which helps intestinal colonization. Later in the infection, when *V. cholerae* cells are at high densities, biofilm production ceases and HA/P production increases, thereby allowing the pathogen to exit the host and adapt to an environmental reservoir where expression of Ctx and TCP is not necessary. A similar QS system in which virulence factors are expressed at low densities and repressed at high densities is also seen with *V. parahaemolyticus* (see below).

V. MIMICUS

Prior to 1981, *V. mimicus* was known as sucrose-negative *V. cholerae* non-O1. This species was determined to be a distinctly different species on the basis of biochemical reactions and DNA hybridization studies, and the name *mimicus* was given because of its similarity to *V. cholerae* (41). This organism is isolated chiefly in cases of gastroenteritis, but occasional strains have been isolated in ear infections as well.

Reservoir

As with that of other *Vibrio* species, the reservoir of *V. mimicus* is the aquatic environment. One interesting study compared the ecologies of *V. mimicus* in the tropic, polluted environment of Bangladesh and the cleaner, more temperate climate of Okayama, Japan (30). This species was isolated from Bangladeshi waters throughout the year, whereas it was not isolated in Okayama when the water temperature decreased below 10°C. In Japan, *V. mimicus* was present both in freshwater and in brackish waters with a salinity optimum of 4 ppt; the bacterium was not recovered from waters with salinities of >10 ppt. Besides being present free in the water column, *V. mimicus* was also isolated from the roots of aquatic plants, sediments, and plankton at levels of up to 6 × 10^4 CFU per 100 g of plankton. *V. mimicus* has also been isolated from fish (100) and freshwater prawns (31).

Foodborne Outbreaks

Gastroenteritis due to *V. mimicus* has been associated only with consumption of seafood. In initial patient studies in the United States (175), consumption of raw oysters was significantly associated with disease due to this species; one patient reported eating only shrimp and crab, not oysters, in the week before the onset of disease. Cases are usually sporadic, rather than associated with common-source outbreaks, but three patients in Louisiana became ill with *V. mimicus* infection after attending a company banquet where the foods served included crawfish (175). Disease in Japan is associated with consumption of raw fish, and at least two outbreaks of seafood-borne disease have involved *V. mimicus* of serogroup O41 (100).

Characteristics of Disease

Disease due to *V. mimicus* is characterized by diarrhea, nausea, vomiting, and abdominal cramps in most patients. In a minority of patients, fever, headache, and bloody diarrhea also occur. In one group of patients, diarrhea lasted a median of 6 days, with a range of 1.5 to 10 days (175). In this study, disease was associated with consumption of seafood, particularly raw oysters, and the median interval from the time of consumption to the onset of illness was 24 h (range, 3 to 72 h). In a 1989 survey of vibrio infections on the Gulf Coast, *V. mimicus* was isolated in 4 cases of gastroenteritis, *V. parahaemolyticus* was isolated in 26 cases, and non-O1 *V. cholerae* was isolated in 18 cases (111).

Infectious Dose and Susceptible Population

There are no volunteer or epidemiologic data to enable estimation of an infectious dose for *V. mimicus*. There is not a particularly susceptible population, other than people who eat raw oysters. Most patients who develop gastroenteritis with *V. mimicus* were in good health before the onset of illness (175).

Virulence Factors

V. mimicus appears not to produce unique enterotoxins, but many strains produce toxins that were first described for other *Vibrio* species, including Ctx, TDH, Zot, and a heat-stable enterotoxin apparently identical to the NAG-ST produced by *V. cholerae* non-O1 and non-O139 strains. One study in Bangladesh (30) revealed that 10% of clinical isolates produced a Ctx-like toxin compared to less than 1% of environmental strains producing such a toxin. Another study in Japan revealed that 94% of clinical isolates produced TDH whereas none of the environmental strains produced this toxin.

There is little information about potential intestinal colonization factors of *V. mimicus*. Like *V. cholerae* non-O1 and non-O139, *V. mimicus* rarely, if ever, expresses TCP or possesses *tcp* genes. Pili and hemagglutinins have been reported for this species but have not been definitively linked to pathogenesis.

V. PARAHAEMOLYTICUS

V. parahaemolyticus is the leading bacterial cause of intestinal infections due to ingestion of seafood, usually raw fish or shellfish. It has been implicated in numerous outbreaks of diarrheal disease throughout the world ever since the first description of its involvement in a major outbreak of food poisoning in 1950. *V. parahaemolyticus* is also the *Vibrio* species most frequently isolated from clinical specimens in the United States (2). Between 1973 and 1998, a total of 40 outbreaks of *V. parahaemolyticus* infections in the United States were reported to the CDC, with over 1,000 persons involved (40).

Classification

Biochemical characteristics that distinguish *V. parahaemolyticus* from other *Vibrio* species are given in Table 16.1. Commercial identification systems vary in their abilities to correctly identify this and other *Vibrio* species, but addition of NaCl up to 2% to the test suspension diluent can improve the accuracy (2, 118). Conventional media for determining biochemical reactions should contain 2 to 3% NaCl (90). Although the core characteristics of *V. parahaemolyticus* are relatively consistent, as many previously described variant strains have been named as new species, strain variation can be seen with some traits, most notably with the production of urease by ca. 15% of strains. The *ure* gene is linked to the *trh* gene encoding a potential toxin of this bacterium (see below). In addition, some traits, such as H_2S production, are dependent on the medium or assay method employed, and caution must be exercised in their determination.

V. parahaemolyticus is serotyped according to both its somatic O- and capsular polysaccharide K-antigens, and there are presently 12 O (LPS)-antigens and over 70 K (acidic polysaccharide)-antigens recognized (90). Although many environmental and some clinical isolates are untypeable by using the K-antigen, most clinical strains can be classified according to their O type. Until the recent description of serotype O3:K6, there appeared to be no correlation between serotype and virulence. This serotype, however, has recently been involved in epidemics of gastroenteritis throughout the world and is now referred to as the pandemic strain of *V. parahaemolyticus*. Antigenic variants belonging to serotypes O4:K68 and O1:K untypeable have also emerged that are largely indistinguishable from the O3:K6 pandemic strain with regard to ribotyping, PFGE patterns, and other subtyping methods (32). A recent study of 178 U.S. isolates from environmental, food, and clinical sources revealed 27 different serotypes and 28 ribotypes, with most clinical isolates from outside the Pacific Coast being O3:K6 (44, 45).

A special consideration in the classification of *V. parahaemolyticus* is the ability of certain strains to produce a hemolysin, termed TDH, or Kanagawa hemolysin, which is correlated with virulence in this species (see

"Virulence Mechanisms" below). The production of this hemolysin was originally established in a study using a special blood agar called Wagatsuma agar in which beta-hemolytic strains were termed Kanagawa phenomenon positive (KP⁺) and nonhemolytic strains were termed KP⁻. However, this difficult-to-prepare medium (which requires freshly drawn human blood) has been supplanted by *tdh* gene probes and *tdh*-based PCR protocols. Detailed probe and PCR protocols are described in the online FDA *Bacteriological Analytical Manual* (90).

Reservoirs

V. parahaemolyticus occurs in estuarine waters throughout the world and is easily isolated from coastal waters of the entire United States, as well as from sediment, suspended particles, plankton, and a variety of fish and shellfish. The last include at least 30 different species, among them clams, oysters, lobster, scallops, shrimp, and crab. In a study conducted by the FDA, 86% of the 635 seafood samples examined were positive for this species. Counts for *V. parahaemolyticus* have been reported to be as high as 1,300 CFU/g of oyster tissue and 1,000 CFU/g of crabmeat, although levels of 10 CFU/g are more typical for seafood products. Hackney et al. (65), in a 3-year survey of 716 seafood samples obtained in North Carolina, found 46% to be positive for *V. parahaemolyticus*. Notably, most *V. parahaemolyticus*-positive samples were unshucked oysters (79% positive), unshucked clams (83% positive), unpeeled shrimp (60% positive), or live crabs (100% positive). Another study isolated *V. parahaemolyticus* from ca. 69 to 100% of the commercially obtained or cultured oysters, clams, and shrimp tested but only 42% of the crabs.

Hackney et al. (65), as well as others, have observed no correlation with fecal coliforms or other indicators, but *V. parahaemolyticus* levels have a definite seasonal variation, with samples analyzed in January and February often free of *V. parahaemolyticus*. Others have also reported the ecology of *V. parahaemolyticus* to be heavily influenced by water temperature, salinity, and association with certain plankton, with highest levels occurring in warmer months and in waters of intermediate salinity. The relationship between each of these parameters has been nicely demonstrated by Kaneko and Colwell (86) in their study of the occurrence of this species in Chesapeake Bay waters. *V. parahaemolyticus* has been occasionally isolated from freshwater sites but only at extremely low levels (<5 CFU/liter) and only during the warmest periods of the year. In a recent example of the impact of climate change on infectious diseases, rising water temperatures in the Gulf of Alaska over the past 25 years have been implicated in causing a summertime outbreak of gastroenteritis among 62 cruise ship passengers associated with consumption of Alaskan oysters (121).

KP⁺ strains of *V. parahaemolyticus* are of prime importance in human disease, although occasional KP⁻ strains have also been isolated from diarrheal stools. KP⁺ strains constitute a very small percentage (typically <1%) of the *V. parahaemolyticus* strains found in aquatic environments and seafoods. Thus, the simple isolation of this species from water or foodstuffs does not, in itself, indicate a health hazard (see "Virulence Mechanisms" below).

The isolation of *V. parahaemolyticus* from foodstuffs generally involves a preenrichment step. A variety of enrichment broths have been described previously, the most common of which is APW. APW provides superior recovery of *V. parahaemolyticus* from a variety of fish and shellfish, even when the samples have been chilled or frozen (147). Of the many plating media suggested for this species, TCBS remains the most commonly employed (147). The U.S. FDA *Bacteriological Analytical Manual* prescribes an overnight enrichment at 35 to 37°C in APW, from which a loopful is streaked onto TCBS agar to obtain isolated colonies of *V. parahaemolyticus*. TCBS is also commonly used for direct plating from stool samples (2). A chromogenic agar medium (CHROMagar Vibrio; CHROMagar Microbiology, Paris, France) on which *V. parahaemolyticus* produces purple colonies has recently been described as yielding improved recovery of this species from seafood (68). Most-probable-number and membrane filtration procedures have been described previously for enumerating CFU of this species from seafood specimens (90). DNA probe and PCR procedures have been described previously to detect the *tlh* gene (see below) to identify the isolate as *V. parahaemolyticus* and the *tdh* gene to identify pathogenic strains of this species (90).

Foodborne Outbreaks

Gastroenteritis with *V. parahaemolyticus* is almost exclusively associated with seafood which is consumed raw, inadequately cooked, or cooked but recontaminated. In Japan, *V. parahaemolyticus* is the major cause of foodborne illness, with as many as 70% of all bacterial food poisonings in that country in the 1960s attributable to this species (193). Outbreaks of *V. parahaemolyticus* gastroenteritis in the United States occurring between 1973 and 1987 have been summarized by Daniels et al. (40). Whereas most Japanese outbreaks involve fish, the U.S. outbreaks involved primarily crab, shrimp, lobster, and oysters. Indeed, of the 42 outbreaks described by Beuchat (12) and Bean and Griffin (9), shellfish were implicated in 33. The first major outbreak (320 persons ill) in the United States occurred in Maryland in 1971, a result of

improperly steamed crabs (38). Subsequent outbreaks occurred at all U.S. coasts and Hawaii. A very large outbreak of *V. parahaemolyticus* infection occurred during the summer of 1978 and affected 1,133 of 1,700 persons attending a dinner in Port Allen, La. (8). All stool isolates were KP⁺. The food implicated was boiled shrimp, which yielded cultures positive for *V. parahaemolyticus*. The raw shrimp had been purchased and shipped in standard wooden seafood boxes. They were boiled the morning of the dinner but returned to the same boxes in which they had been shipped. The warm shrimp were then transported 40 miles in an unrefrigerated truck to the site of the dinner and held an additional 7 to 8 h until being served that night. Another major outbreak (416 cases) was linked to consumption of raw oysters from Galveston Bay, Tex. (40).

In their survey of four Gulf Coast states, Levine et al. (111) found that *V. parahaemolyticus* was the most common cause of gastroenteritis (37% of 71 cases) in that area. Similarly, Desenclos et al. (46) found *V. parahaemolyticus* to be the second leading cause (over 26%) of gastroenteritis cases in those persons who had consumed raw oysters in Florida. In a 15-year survey of *Vibrio* infections reported by a hospital adjacent to the Chesapeake Bay, Hoge et al. (79) found 9 (>69%) of the 13 *Vibrio*-positive stool specimens to contain *V. parahaemolyticus* as the sole pathogen. In the largest survey reported to date, Hlady and Klontz (77) reported the epidemiology of 690 *Vibrio* infections in Florida during a 13-year period. They reported 68% of the gastroenteritis cases to be associated with raw oyster consumption.

Characteristics of Disease

V. parahaemolyticus has a remarkable ability for rapid growth, and generation times as short as 8 to 9 min at 37°C have been reported. Even in seafoods, generation times of 12 to 18 min have been demonstrated. As a result, *V. parahaemolyticus* strains have the ability to rapidly increase in numbers, both in vitro and in vivo, and this ability is evidenced in the characteristics of the disease that the bacterium produces. Symptoms reported in the Maryland outbreak cited above began 4 h to more than 30 h after food consumption, with a mean time of 23.6 h. The primary symptoms were diarrhea (100%) and abdominal cramps (86%), along with nausea and vomiting (26%) and fever (23%). Symptoms subsided in 3 to 5 days in most individuals, although they lasted 5 to 7 days in 30% and more than 7 days in another 20%. In the more severe cases, diarrhea was watery with mucus, blood, and tenesmus.

In the Louisiana outbreak of 1978, a mean incubation period of 16.7 h was reported, with a range of 3 to 76 h

(8). The duration was from less than 1 day to more than 8 days, with a mean of 4.6 days. Hospitalization was required for over 7% of the victims. Symptoms included diarrhea (95.1%), cramps (91.5%), weakness (90.2%), nausea (71.9%), chills (54.9%), headache (47.7%), fever (47.5%), and vomiting (12.2%). Ages of patients ranged from 13 to 78 years.

An extensive clinical study of 28 patients in Bangladesh demonstrated an acute inflammatory response evident in intestinal specimens obtained from both the small and large intestines (161). Occult or frank blood was detected in the stool specimens of 68% of the patients, and the inflammatory responses were more severe in patients infected with *V. parahaemolyticus* than in those infected with *V. cholerae* O1 or O139. *V. parahaemolyticus* can also cause extraintestinal infections. In the study of U.S. infections from 1973 to 1998, 34% of all *V. parahaemolyticus* infections were found to be wound infections and 5% were septicemia, compared to 59% for gastroenteritis (40). Of those patients with septicemia, 29% died, compared to 2% mortality associated with gastroenteritis.

Infectious Dose and Susceptible Population

Along with *V. cholerae*, *V. parahaemolyticus* is the only vibrio for which experimental evidence exists regarding the dosages required for initiation of gastroenteritis. Studies using human volunteers have shown that ingestion of 2×10^5 to 3×10^7 CFU of KP$^+$ cells leads to the rapid development of gastrointestinal disease. Conversely, volunteers receiving as many as 1.6×10^{10} CFU of KP$^-$ cells exhibited no signs of diarrheal disease (173). The KP$^+$ and KP$^-$ strains used in the volunteer studies were not isogenic, and so other factors besides the TDH may also contribute to the difference in infectious doses. It is not known whether the pandemic O3:K6 clone has a lower infectious dose than other serotypes. Based on typical numbers of *V. parahaemolyticus* present in fish and shellfish and the low incidence of KP$^+$ cells in these natural samples (see below), it would appear that a meal of raw shellfish would likely contain no greater than 10^4 CFU of KP$^+$ cells. Hence, for disease to result from consumption of contaminated food, it would appear that mishandling at temperatures allowing growth of the cells would be necessary (193). *V. parahaemolyticus* populations can increase by 3 logs in shellstock oysters held at 26°C for 24 h postharvest.

Virulence Mechanisms

The epidemiologic linkage between human virulence and the ability of *V. parahaemolyticus* isolates to produce the Kanagawa hemolysin (TDH) has long been established. In the initial study by Sakazaki et al. (171), 96.5% (2,655 of 2,720) of the strains isolated from human patients were KP$^+$ whereas only 1% (7 of 650) of environmental isolates were KP$^+$. Many subsequent studies have confirmed the rare frequency of KP$^+$ strains among environmental *V. parahaemolyticus* isolates (85), although some studies have shown higher frequencies (44). It is thought that a natural selection of KP$^+$ strains occurs in the intestinal tract and that KP$^-$ strains survive better in the estuarine environment (85).

The KP hemolysin is called TDH because of its heat stability, with only partial inactivity observed after 30 min at 100°C, and its direct lysis of erythrocytes, which occurs without additional substituents. The lytic activity on erythrocytes from a wide variety of animals was the initial focus of investigations into TDH activity (178). TDH forms pores in erythrocytes and planar lipid bilayers (69, 80), although it shows no significant homology to other pore-forming toxins. More recent studies have focused on the mechanisms by which TDH may cause diarrhea. By constructing isogenic mutants with mutations in the *tdh* gene encoding TDH, Nishibuchi et al. (142) determined that only the TDH$^+$ parent strain and not the *tdh* mutant was capable of inducing fluid accumulation in the rabbit ileal loop assay. Similar results were seen when culture supernatants of these strains were tested on rabbit ileal tissue mounted in Ussing chambers, a more sensitive measure of secretory activity. In this assay, the ability of TDH to alter ion transport in the intestinal tract was observed at nanogram levels, with no histological changes. The purified TDH protein induces chloride ion secretion in rabbit and human intestinal tissue mounted in Ussing chambers (163, 186). However, unlike cholera toxin, which acts through increased intracellular cAMP levels, TDH uses Ca^{2+} as an intracellular second messenger. This is the first bacterial enterotoxin for which the linkage between changes in intracellular calcium levels and secretory activity has been established.

KP$^+$ isolates usually contain two nonidentical copies of the *tdh* gene, but many KP$^-$ or weakly positive isolates contain only one gene copy. The predicted protein products of the two gene copies (*tdh*1 and *tdh*2) differ in seven amino acid residues, although the proteins are immunologically indistinguishable. Although both gene products contribute to the KP phenotype, >90% of the TDH protein can be attributed to high-level expression of the *tdh*2 gene (140). Expression of *tdh*2 but not that of *tdh*1 is under the control of a regulator similar to the ToxR of *V. cholerae*. The *tdh* genes are located on an 80-kb pathogenicity island on the small chromosome (117).

During one outbreak of gastroenteritis, KP⁻ isolates of *V. parahaemolyticus* produced a TDH-related hemolysin, termed TRH, but not TDH. The *trh* gene shows 69% identity to the *tdh2* gene, and the biological, immunological, and physicochemical characteristics of TRH are similar, but not identical, to those of TDH (143). A survey of 285 strains of *V. parahaemolyticus* revealed that not only were *tdh*-positive strains strongly associated with gastroenteritis but *trh*-positive strains were also so associated. In one study, *trh* was found in over 35% of 214 clinical strains, including 24% of those lacking the *tdh* gene (179). Purified TRH is less thermostable than TDH, being inactivated at 60°C for 10 min (178). Genes encoding a third, unrelated hemolysin, called thermolabile hemolysin (*tlh*), are found in all strains of *V. parahaemolyticus* and are often used as a species-specific gene probe (90). Similar to TDH, purified TRH induces Cl⁻ secretion in intestinal epithelial cells in a process involving increases in intracellular calcium levels (186). These results suggest that TRH may be an important virulence factor and possibly the cause of diarrhea in those patients from whom only KP⁻ strains of *V. parahaemolyticus* are isolated from stools. The *trh* gene is colocated with the *ure* gene encoding urease on a 15.7-kb pathogenicity island of G+C content (41%) lower than that of the total genome (46 to 47%) of this species (156). Hence, strains usually possess both *trh* and *ure* or lack both genes. As with *tdh*, environmental strains usually lack *trh* and *ure* (156). The function of urease in the pathogenesis of disease due to *ure⁺ V. parahaemolyticus* strains is unknown, although it may assist the survival of the pathogen during passage through the acidic environment of the stomach.

A surprising finding of the genome sequence analysis of a *V. parahaemolyticus* O3:K6 strain was the presence of two previously unknown TTSSs (117). Preliminary characterization of the functions of these two systems has shown that the system encoded on the large chromosome, TTSS1, is involved in cytotoxicity to HeLa cells, with cell death being mediated by apoptosis (154). TTSS1 is present in all *V. parahaemolyticus* strains—clinical and environmental, KP⁺ and KP⁻. TTSS2 is found only in KP⁺ strains (117) and is encoded on the same pathogenicity island in the small chromosome that contains *tdh*. Mutation of TTSS2 genes results in diminished enterotoxigenicity in rabbit ileal loops (157). TTSS1 is regulated by QS (73), but similar to that in *V. cholerae*, expression of the TTSS1 genes is repressed by QS at high densities. Since TTSS1 is present in all environmental and clinical strains, regulation by QS presumably plays a role in the environmental reservoir of *V. parahaemolyticus*.

Comparison of the O3:K6 pandemic clone to nonpandemic clones of this species has resulted in the identification of a protein biomarker that is unique to the former (200). This protein is a histone-like DNA-binding protein that is encoded on a 16-kb region present in O3:K6 strains but absent from nonpandemic clinical isolates. The presence of such a protein may cause altered regulation of genes encoding virulence factors and may account for the enhanced transmission or virulence of the pandemic clone. Strains of the O3:K6 pandemic clone also contain *tdh* but not *trh* or *ure* (32, 45).

The genome sequence revealed many other potential virulence factors, including genes that encode homologues of *E. coli* cytotoxic necrotizing factor, *Pseudomonas* exoenzyme T, an RTX toxin, adherence factors, and other factors (117). The contribution of these factors to human disease caused by *V. parahaemolyticus* is currently being investigated.

V. VULNIFICUS

Of all of the pathogenic vibrios, *V. vulnificus* is the most serious in the United States, alone responsible for 95% of all seafood-linked deaths in this country. Indeed, in Florida, this bacterium is the leading cause of reported deaths from all foodborne illness (78). Among that portion of the population which is at risk for infection by this bacterium, primary septicemia cases resulting from raw oyster consumption typically carry fatality rates of 50 to 60%. This is the highest death rate for any foodborne disease agent (124, 190). The CDC and FDA estimate that there are approximately 50 foodborne cases per year of *V. vulnificus* infection in the United States serious enough to be recognized by hospital personnel, although estimates of 17,500 to more than 41,000 total cases have been calculated (124, 190). The bacterium is unusual in being able to produce wound infections in addition to gastroenteritis and primary septicemias. Wound infections carry a 20 to 25% fatality rate, are also seawater and/or shellfish associated, and generally require surgical debridement or amputation of the affected tissue. The biology of *V. vulnificus*, as well as the clinical manifestations of both the primary septicemic and wound forms, has been reviewed previously (112, 149, 151, 183). Discussion here is limited to the foodborne (primary septicemic) form of infection caused by this bacterium.

Classification

The first detailed taxonomic study of this species was carried out by researchers at the CDC who in 1976 described 38 strains submitted by clinicians around the country. Originally termed the lactose-positive vibrio, its current name was suggested after a series of phenotypic and genetic studies by investigators in several laboratories (145). In 1982, a second biotype of *V. vulnificus* was described which can easily be differentiated from biotype 1 by its negative indole reaction. Biotype 2 strains are a major source of fatalities in eels but have been reported to lead to human infection in isolated instances. In 1999, a third biotype was described, which differs from biotype 1 and 2 isolates in being negative in the citrate and *o*-nitrophenyl-β-D-galactopyranoside tests and in lacking fermentation of salicin, cellobiose, and lactose. This is a fascinating bacterium which appears to be a genetic hybrid of the biotype 1 and 2 forms (14). Biotype 3 strains have to date been isolated only in Israel, and all cases were wound infections. In 98% of the 62 cases, *Tilapia* spp. were implicated as the source of these infections. The key differential traits for the three biotypes of *V. vulnificus* are shown in Table 16.2. The discussions here focus on the originally described biotype 1, which is the major human pathogen.

The isolation of *V. vulnificus* from blood samples is straightforward, as the bacterium grows readily on TCBS, MacConkey, and blood agars. Isolation and identification of environmental strains, on the other hand, have proven much more problematic. Vibrios tend to make up 50% or more of estuarine bacterial populations, and the great bulk of these vibrios have not been characterized. The phenotypic traits of this species have been fully described in a number of studies (2, 51, 90, 145). However, it is well recognized that there can be considerable variation in these traits, including lactose and sucrose fermentation, considered to be among the most important traits for identifying this species. This problem is exemplified by the isolation from a clinical sample of a bioluminescent strain of this species (153).

Various media have been proposed for isolating *V. vulnificus*, but the medium most commonly employed is colistin-polymyxin B-cellobiose agar or one of several derivatives

Table 16.2 Differentiation of the three biogroups of *V. vulnificus*[a]

Test	Result for biogroup[b]:		
	1	2	3
Ornithine decarboxylase	V	−	+
Indole production	+	−	+
Acid production from:			
D-Mannitol	V	−	−
D-Sorbitol	−	+	−
Cellobiose	+	+	−
Salicin	+	+	−

[a] Adapted from reference 2.
[b] See footnote *b* of Table 1. V, variable; −, negative; +, positive.

(147). Many studies have employed this medium for the isolation of *V. vulnificus* from oysters and clams, and Sun and Oliver (185) found that >80% of the 1,000 colonies obtained from oyster homogenates could be identified as this species. A similar conclusion was reported by Sloan et al. (182) following their study of five selective enrichment broths and two selective agar media for isolating *V. vulnificus* from oysters. Currently, the U.S. FDA *Bacteriological Analytical Manual* prescribes an 18- to 24-h enrichment at 35 ± 2°C in APW, from which a loopful is streaked onto mCPC or CC agar (derivatives of colistin-polymyxin B-cellobiose agar) to obtain isolated colonies of *V. vulnificus* (90).

The identification of this species is best made by employing a probe against its hemolysin (*vvhA*) gene (135), and the latest edition of the U.S. FDA *Bacteriological Analytical Manual* recommends the use of this probe (90). Several reports (e.g., reference 155) have described the use of multiplex PCR directed towards this and other gene targets for the detection of *V. vulnificus* in shellfish.

Susceptibility to Control Methods

Cook and Ruple (37) reported that those *V. vulnificus* cells naturally occurring in oysters underwent a time-dependent decrease in recovery when either shellstock oysters or shucked oyster meats were held at 4, 0, or −1.9°C. Cook (36) later found that, after harvest, *V. vulnificus* failed to grow within 30 h in shellstock oysters stored at 13°C or below. In oysters held at 18°C or higher for 12 or 30 h, however, *V. vulnificus* levels increased. The susceptibility of *V. vulnificus* to freezing, low-temperature pasteurization, high hydrostatic pressure, and ionizing radiation has been discussed in the introduction to this chapter.

Reservoirs

V. vulnificus is a widespread inhabitant of estuarine environments, having now been recovered from the Gulf, East, and Pacific Coasts of the United States and from around the world (90, 92, 149, 151). *V. vulnificus* has been isolated from oysters, crabs, clams, ark shells, tarbos, plankton, and seawater samples (145). DePaola et al. (43) have described its isolation from the intestinal tracts of a variety of bottom-feeding coastal fish and have suggested that such fish may represent a major reservoir of this species.

Most studies have reported a lack of correlation between the presence of *V. vulnificus* and the presence of fecal coliforms, but this may be environment dependent (159). As exemplified in Fig. 16.2, a strong correlation between water temperature and isolation of *Vibrio*

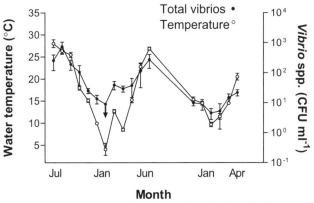

Figure 16.2 Correlation between culturability of *Vibrio* spp. from estuarine environments and water temperature. Reprinted with permission from Pfeffer et al. (159).

spp. has been observed by numerous investigators, in agreement with the findings of epidemiologic studies on infections caused by *V. vulnificus*. Because of this seasonality and the inability to isolate *V. vulnificus* from water or oysters when water temperatures are low, there has been considerable investigation of the apparent die-off of this species during cold-weather months. This situation is now attributed to a cold-induced VBNC state, wherein the cells remain viable but are no longer culturable on the routine media normally employed for their isolation. This phenomenon, which has now been demonstrated with at least 16 genera, has been the subject of several reviews (146, 148, 150). It is also possible, however, that *V. vulnificus* overwinters in certain environments. It is interesting in this context that DePaola et al. (43) reported its isolation during the winter from sheepshead fish at higher densities than that from sediment or seawater.

Two Genotypes of *V. vulnificus*

In no case has more than one person developed *V. vulnificus* infection following consumption of the same lot of oysters. Indeed, the consumption of raw oysters and subsequent development of disease in one family member while others exhibit no symptoms are most common. Epidemiologic studies are complicated by the fact that raw oysters are usually eaten whole, and hence rarely do there exist any remains of the implicated oyster to sample. Additional raw oysters from the same lot, or even the same serving, may remain, but studies have indicated that two oysters taken from the same estuarine location may have vastly different numbers of *V. vulnificus* cells. It is possible that certain methods, such as randomly amplified polymorphic DNA PCR, may allow molecular epidemiology to be used to aid in tracking the individual strains involved in an infection. Buchrieser

et al. (18), employing clamped homogeneous electric field gel electrophoresis to examine 118 strains of *V. vulnificus* isolated from three oysters, reported that no two isolates had the same profile. Indeed, studies from other laboratories using arbitrarily primed PCR, ribotyping, PFGE, and amplified fragment length polymorphisms all indicate that no two *V. vulnificus* isolates have the same chromosomal arrangement (197). On this point, it is interesting that *V. vulnificus*, like several other members of this genus, contains two distinct chromosomes (191).

As noted by Buchrieser et al. (18), whether all strains are capable of causing infection or whether only certain strains are pathogenic remains to be determined. However, an answer to this question now appears to be developing. Studies reported as early as 1994 suggested that the 16S rRNA of *V. vulnificus* exhibits significant sequence variation and that two distinct genotypes may exist. Subsequent sampling studies reported seasonal and environmental source variation that could be correlated to these two genotypes. The existence of several genotypes was further supported by results from Gutacker et al. (64), who used multilocus enzyme electrophoresis, randomly amplified polymorphic DNA PCR, and sequence analysis of the *recA* and *glnA* genes to study the three biotypes of *V. vulnificus*. More recently, a simple and rapid PCR procedure was used to show that the two genotypes show a strong correlation with the sources of their isolation. The great majority (90%) of the C genotype strains were from clinical samples, whereas 93% of the strains isolated from a variety of environmental sources were classified as E type (169). Hence, it is clear that two distinct genotypes of this pathogen exist, and it is likely that only one of these plays a significant role in initiating human infection.

Foodborne Outbreaks

Infections due to *V. vulnificus* show a pronounced correlation with water temperature, with most cases occurring during warm months. Further, nearly all cases of *V. vulnificus* infection result from consumption of raw oysters, and most of these infections result in primary septicemias. In a study of 422 infections occurring in 23 states, Shapiro et al. (177) reported that 96% of those patients developing primary septicemia had consumed raw oysters. Similar findings were reported by Strom and Paranjpye (183) and by Oliver (149) in a review of more than 500 cases.

In the only prospective study performed to determine the incidence of vibrios in symptomatic or asymptomatic infections among persons eating raw shellfish, Lowry et al. (115) found that 3 of the 479 persons tested had *V. vulnificus* present in their stools, although none were ill. Coupled with their finding that two-thirds of the raw oysters tested were culture positive for *V. vulnificus*, their study suggests that exposure to this species may be relatively high and underscores the need for at-risk persons to avoid raw seafood.

Characteristics of Disease

A number of major studies of infections caused by *V. vulnificus* have appeared. In a review of more than 100 cases of primary septicemia, Oliver (149) found the incubation period to range from 7 h to 10 days, with most patients becoming symptomatic within 36 h. The most significant symptoms included fever (94%), chills (86%), nausea (60%), and hypotension (systolic pressure, <85 mm; 43%). Although sometimes present, symptoms typical of gastroenteritis were not as common: abdominal pain (44%), vomiting (35%), and diarrhea (30%). An unusual symptom that generally (69%) occurred in these cases was the development of secondary lesions. These occurred most frequently on the legs, frequently developed into necrotizing fasciitis or vasculitis, and often necessitated surgical debridement or limb amputation. Hlady and Klontz (77) found that 94% of persons developing *Vibrio* infections in Florida were hospitalized, some for up to 43 days.

Surprisingly, gastrointestinal disease with associated diarrhea is relatively rare. Johnston et al. (83) were the first to provide evidence of such a syndrome, reporting on three males who presented with abdominal cramps; *V. vulnificus* was isolated from their diarrheal stools. All three had a history of alcohol abuse, routinely took antacids, and had eaten raw oysters during the week prior to their illness. Although symptoms in all three subsided without antibiotic treatment, diarrhea continued for a month in one case. Desenclos et al. (46), in their survey of *Vibrio* infections in raw-oyster eaters in Florida, reported eight cases of *V. vulnificus*-induced gastroenteritis, six of which involved consumption of raw oysters. Levine et al. (111) reported an additional three cases of gastroenteritis from *V. vulnificus*.

The time to death in the fatal cases of *V. vulnificus* infection varies considerably, ranging from 2 h posthospital admission to as long as 6 weeks (145). Most deaths occur within a few days. Hospital stays of several weeks are the norm for those surviving primary septicemic disease. To a great extent, survival depends on prompt antibiotic administration. Further, *V. vulnificus* is susceptible to most antibiotics, and a large variety of antibiotics have been used clinically.

Infectious Dose and Susceptible Population

In almost all *V. vulnificus* infections which follow ingestion of raw oysters, the patient has an underlying chronic disease (149). The most common of these is a liver- or blood-related disorder, with liver cirrhosis secondary to alcoholism or alcohol abuse being the most typical. These diseases typically result in elevated serum iron levels, and laboratory studies of experimental *V. vulnificus* infection have shown that elevated serum iron levels play a major role in this disease (203). Other risk factors include hematopoietic disorders, chronic renal disease, gastric disease, use of immunosuppressive agents, and diabetes.

The infectious dose of *V. vulnificus* is not known. Mouse models, however, offer some insight on this point. Wright et al. (203) observed that, when mice were injected with 16 μg of iron to produce iron overload in sera, the 50% lethal dose (LD_{50}) decreased from ca. 10^6 CFU to a single cell. In a variation of these studies, the administration of small amounts of CCl_4 to produce short-term liver necrosis was found to increase serum iron levels, with an inverse correlation between serum iron levels and LD_{50} observed. Such experimental data agree with the findings of epidemiologic studies indicating that liver damage and immunocompromising diseases are major underlying factors in the development of *V. vulnificus* infections and suggest that extremely low levels of this pathogen may be sufficient to initiate potentially fatal infections.

Although cases in children have been reported, infections tend to be in males (86.5% of those cases reviewed by Oliver [149]) whose average age exceeds 50 years. The reason for this gender specificity has now been determined; Merkel et al. (126) found that estrogen protects against the *V. vulnificus* endotoxin, a virulence factor critical to infection with *V. vulnificus* (see below).

Virulence Mechanisms

Capsule

The polysaccharide capsule which is produced by nearly all strains of *V. vulnificus* is essential to the bacterium's ability to initiate infection. Simpson et al. (181) examined 38 strains of *V. vulnificus*, including both clinical and environmental isolates and virulent and avirulent strains, and found that all virulent strains were of the opaque (encapsulated) colony type whereas isogenic cells taken from translucent (acapsular) colonies were avirulent (Fig. 16.3). It was further observed that encapsulated cells produce nonencapsulated cells and that this loss of capsule correlates with loss of virulence in otherwise isogenic strains. These authors subsequently found that only the encapsulated cells were able to utilize transferrin-bound iron and that only these cells were iron responsive, i.e., were virulent at an inoculum of 10^3 CFU in iron-overloaded mice. The primary reason

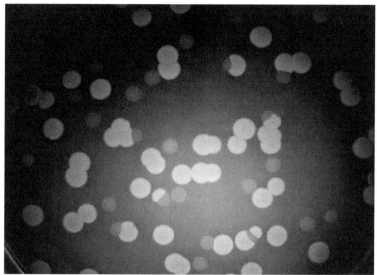

Figure 16.3 Morphologies of opaque (encapsulated) and translucent (nonencapsulated) colonies of *V. vulnificus*.

for the avirulence of translucent cells likely resides in the observation that the capsule allows these cells to resist phagocytosis.

In recent years, capsular phase variation has been described (27, 204), with intermediate colony morphotypes reported (170) which may result from downregulation of capsular polysaccharide. A recent paper (61) also reported on the existence of a rugose phenotypic variant of *V. vulnificus*. The significance of intermediate and rugose cells in human virulence has yet to be fully understood.

Simonson et al. (180) used polyclonal antibodies to study the capsule of *V. vulnificus* and found at least 10 distinct serotypes. All 10 serotypes were associated with human infection, with 34.6% of clinical strains being of type 2 or 4. In contrast, 84.6% of the typeable environmental strains were of serotype 3 or 5 and only 7.7% were of type 2 or 4. Similarly, Hayat et al. (71) reported that 19% of 21 clinical strains, but none of the 67 environmental isolates examined, agglutinated with antiserum prepared against a clinical isolate. Bush et al. (21) examined the sugars of these capsular polysaccharides from 120 *V. vulnificus* isolates and reported great diversity in capsular types, with no clear correlation of capsular types, genetic classification schemes, or pathogenic potential. Hence, whether capsular differences play a role in the epidemiology of *V. vulnificus* infections remains to be determined.

Iron

Elevated levels of iron in serum appear to be essential for the ability of *V. vulnificus* to multiply in the human host (145, 149, 183). The effect of elevated serum iron levels on reduction of LD_{50} in mice has been described above. Indeed, Wright et al. (203) have shown that normal human serum does not permit growth of *V. vulnificus*, suggesting that this bacterium may be able to produce septicemia only in those with elevated serum iron levels. Although *V. vulnificus* simultaneously produces both hydroxymate and phenolate siderophores, it is unable to compete with serum transferrin for iron, and this is likely a major factor in its inability to initiate infections in individuals with normal serum iron levels. A similar result has been reported for other iron-binding proteins, such as lactoferrin and ferritin. *V. vulnificus* is able to overcome the binding of haptoglobin to hemoglobin, however, and this may represent another aspect of the importance of iron in the pathogenesis of these infections.

LPS

The symptoms which occur during *V. vulnificus* septicemia, including fever, tissue edema, hemorrhage, and especially the significant hypotension, are those classically associated with gram-negative endotoxic shock. Thus, another product of *V. vulnificus* which may be critical to its virulence is the endotoxic LPS present in *V. vulnificus* cells. McPherson et al. (123) found that intravenous injections of extracted and partially purified LPS from *V. vulnificus* caused decreased arterial blood pressure in mice and rats within 10 min, which further declined, leading to decreased heart rate and death within 30 to 60 min. Subsequent studies found that an inhibitor of nitric oxide synthase (the LPS-induced enzyme responsible for the release of nitric oxide and subsequent host tissue damage) administered 10 min after LPS injection reversed this lethal effect. These results indicate that the classic symptoms of endotoxic shock observed following *V. vulnificus* infection are likely due to the stimulation by LPS of nitric oxide synthase and that inhibition of this enzyme is a possible treatment for the endotoxic shock produced by this pathogen. As noted above, the gender specificity of *V. vulnificus* infections appears to be due to an estrogen-induced protection of females against this endotoxin.

Simonson et al. (180), employing monoclonal antibodies, have identified five LPS serologic varieties in *V. vulnificus*. They found that 25% of the clinical isolates expressed LPS antigens 1 and/or 5 whereas only 0.3% of environmental isolates were of these serotypes. Although the chemical composition of the LPS extracted from one strain of *V. vulnificus* has been reported previously, nothing is known regarding the relative virulences of the different serotypes.

Other Toxins

In addition to the role of capsule, iron, and endotoxin in the pathogenesis of *V. vulnificus* infections, *V. vulnificus* produces a large number of extracellular compounds, including hemolysin, protease, elastase, collagenase, DNase, lipase, phospholipase, mucinase, chondroitin sulfatase, hyaluronidase, and fibrinolysin. To date, however, none of these putative virulence factors have definitively been shown to be involved in pathogenesis. The most studied of these is a powerful heat-stable hemolysin-cytotoxin which possesses cytolytic activity against a variety of mammalian erythrocytes, cytotoxic activity against CHO cells, vascular permeability activity against guinea pig skin, and lethal activity for mice. Production of the hemolysin, a metalloprotease, has recently been shown to be regulated by a transmembrane virulence regulator homologous to ToxRS of *V. cholerae*. The toxin has been purified and has a molecular weight of ca. 56,000 and an LD_{50} for mice of ca. 3 μg/kg of body weight following intravenous injection. It enhances vascular permeability through the release of bradykinin and can also specifically degrade type IV collagen, thereby destroying the basal membrane layer of capillary vessels. Interestingly, low-density lipoprotein inactivates this cytolysin through the oligomerization of the toxin monomer. However, hemolysin-cytolysin mutants show no reduction in the LD_{50} for mice following intraperitoneal injection (176), and the role of hemolysin-cytolysin as a virulence factor remains unclear. The reader is directed to a recent review on the hemolysins of *Vibrio* species (208) for more in-depth information on these putative virulence factors.

An elastolytic protease that lacks hemolytic activity but which degrades albumin, immunoglobulin G, elastin, and complement factors C3 and 4 has been described previously. Minimum lethal doses in mice, regardless of route of injection, were reported to be ca. 25 μg, with extensive hemorrhagic necrosis, edema, and muscle tissue destruction occurring. A broad-specificity metalloprotease is also produced, which cleaves several plasma proteins and in addition interferes with a number of blood-clotting functions (26).

Kim et al. (95) investigated the role of adenylate cyclase in the virulence of this species and reported that production of hemolysin and protease, motility, and cytotoxicity against HeLa cells were all decreased in a *cya* mutant. Further, colony morphology (due to capsule production) was modified in this mutant, and LD_{50} for mice increased 100-fold.

There is increasing indication that flagella and motility may play a part in the pathogenesis of this species. Recently, Kim et al. (94) used in vivo-induced antigen technology to identify *V. vulnificus* genes induced in serum taken from septicemic patients. They found proteins involved in chemotaxis, among other functions, to stand out as likely being important virulence gene products.

Although none of these many putative virulence factors are known to be essential to the virulence of *V. vulnificus*, they likely play a role in the pathogenesis of foodborne infections or may be essential for the wound infections also produced by this species. An excellent review of the molecular pathogenesis of this species has recently been published (63), and the reader is directed to that and other recent reviews on disease production by this species (141, 149, 183).

V. FLUVIALIS

Classification

V. fluvialis was originally described by Lee et al. (105) and referred to as group F. Subsequently, it was determined that this group was identical to that referred to as group EF-6 by the CDC, and further taxonomic study concluded that these isolates were a new species, *V. fluvialis*. Two biogroups of group F vibrios were originally described, of which biogroup I was anaerogenic and isolated from aquatic environments and in diarrheal cases, whereas biogroup II was aerogenic and not disease associated. Subsequent studies indicated that the aerogenic strains are a unique species, and these were reclassified (17) as *V. furnissii* (see next section).

The possibility of misidentifying *V. fluvialis* as *Aeromonas* spp. (especially *Aeromonas hydrophila*) exists as both are arginine dihydrolase positive. The simplest differentiation is the inability of *V. fluvialis*, being halophilic, to grow in media lacking NaCl. The lack of production of indole by *V. fluvialis* also differentiates this species from *Aeromonas* spp. (85).

Reservoirs

V. fluvialis has been frequently isolated from brackish and marine waters and sediments in the United States as well as other countries. It has also been reported from fish and shellfish from the Pacific Northwest and Gulf Coast. Wong et al. (201) reported that ca. 65 to 79% of the oysters, hard clams, and freshwater clams they examined harbored *V. fluvialis* but only 25% of the crabs and 6% of the shrimp. *V. fluvialis* has been found only rarely in freshwaters.

Foodborne Outbreaks

V. fluvialis was isolated from more than 500 patients with diarrheal stools at the Cholera Research Laboratory in Dacca, Bangladesh, during a 9-month period (81). Approximately half of the patients were less than 5 years of age. Since that outbreak, however, *V. fluvialis* has been reported only occasionally to be an enteric pathogen.

In the United States, Levine et al. (111) found that *V. fluvialis* accounted for 10% of the clinical cases in their survey of *Vibrio* infections along the Gulf Coast. All seven of these cases resulted in gastroenteritis, with three patients requiring hospitalization. Consumption of raw oysters was implicated in at least three of the seven cases, and shrimp was implicated in one other.

In the largest study of clinical cases of *V. fluvialis* infections occurring in the United States, Klontz and Desenclos (97) described 12 persons with gastroenteritis in Florida from whom this species was recovered between 1982 and 1988. Eight of the 10 from whom the species was isolated reported eating seafood during the week before becoming ill, with raw oysters implicated in five cases, shrimp in two, and cooked fish in one case. In a subsequent survey of vibriosis resulting from raw-oyster consumption in Florida during an 8-year period, Desenclos et al. (46) reported 5.6% of 125 gastroenteritis cases to be caused by *V. fluvialis*.

Characteristics of Disease

Gastroenteritis with *V. fluvialis* is similar to cholera, with diarrhea and vomiting (97%), moderate to severe dehydration (67%), abdominal pain (75%), and fever (35%) being common symptoms (81). Passage of 10 to 12 stools per day has been reported, with individual stool outputs of 0.5 to 7 liters. Diarrhea typically lasts from 16 h to more than 3 days. Stools collected during the original

Bangladesh outbreak revealed an average of 10^6 *V. fluvialis* cells per ml. A notable difference from cholera is the frequent occurrence of bloody stools in infections due to *V. fluvialis*. Huq et al. (81) reported that 75% of the stool samples they examined contained erythrocytes or leukocytes, and bloody stools were also reported in 86% of the seven cases reported by Levine et al. (111).

All 10 of the gastroenteritis patients reviewed by Klontz and Desenclos (97) had diarrhea (generally watery), with 2 to 20 stools per day (median of 7). Five patients reported at least one episode of bloody stools. Other symptoms included nausea, vomiting, and abdominal cramps. No patients had hypotension. The median age of the patients was 37 (range, 1 month to 67 years), with eight patients being male. The median incubation period for six of the patients who had eaten seafood was 39 h (range, 16 to 60 h). The median duration of the illness was 6 days (range, 1 to 60 days); five patients were hospitalized for a median of 4 days (range, 3 to 11 days).

Infectious Dose and Susceptible Population

The infectious dose for this species is not known. Persons developing gastroenteritis have been from 1 month old to >80 years old, and it does not appear that underlying disease plays a major role in the infections caused by this species. Klontz and Desenclos (97) reported that only 4 of the 10 patients they studied with *V. fluvialis* gastroenteritis had underlying medical conditions, which included diabetes, alcohol abuse, and ulcerative colitis. One patient had a history of cardiopulmonary disease and peptic ulcers and was taking antacids at the time of hospitalization. The patient described by Klontz et al. (98) did not have a history of alcoholism or liver disease but had extensive coronary artery disease.

Although successful resolution of infection without antibiotic therapy has been reported, antibiotic treatment, often along with intravenous fluids, is generally administered. Antibiotic therapy was not successful in at least one case.

Virulence Mechanisms

Most early studies on the virulence of *V. fluvialis* failed to detect evidence of an enterotoxin in this species. This failure may have been due to the growth conditions employed, as it was subsequently found that production of a heat-labile enterotoxin which induces fluid accumulation in the intestines of suckling mice is culture medium dependent. It has also been shown that, unlike clinical isolates, most environmental isolates do not induce fluid accumulation.

In a comprehensive study on the putative virulence factors produced by *V. fluvialis*, Chikahira and Hamada

(29) studied cell filtrates and extracts of 39 environmental and clinical isolates. An effect on CHO cells was found for 84 and 75% of the environmental and clinical isolates, respectively, and 45 and 12.5% caused cell elongation, respectively. The cell elongation factor was appreciably neutralized by anti-cholera toxin serum, but the cell killing factor was not. Live cultures of all human and half of the environmental isolates also caused fluid accumulation in rabbit ileal loops, an activity not neutralized by anti-cholera toxin serum. In addition, enzyme-linked immunosorbent assay studies indicated that 64% of the environmental and 67% of the human isolates produced cholera toxin, although none of the culture filtrates caused fluid accumulation in suckling mice. Culture filtrates from 15 of the 16 strains tested were found to be lethal for mice (one human isolate was not lethal), with death occurring within 20 min. Less than a third of the isolates were hemolytic for rabbit erythrocytes.

In addition to the above-described factors, Oliver et al. (152) reported that the strain of *V. fluvialis* that they studied produced elastase, mucinase, protease, lipase, lecithinase, chondroitin sulfatase, hyaluronidase, DNase, fibrinolysin, and a hemolysin. In this range of exoproducts, *V. fluvialis* is comparable to *V. vulnificus* and *V. mimicus*.

Han et al. (67) purified the hemolysin of *V. fluvialis* from culture supernatants and found that it is hemolytic for a variety of mammalian erythrocytes. The nucleotide sequence of the *vfh* gene encoding this toxin, as well as the physical nature of the 740-amino-acid protein which results, was also characterized. Kothary et al. (103) subsequently reported on this hemolysin (although a different molecular weight was derived) and noted that, in addition to lysing erythrocytes of eight different animal species, it is cytotoxic to CHO cells and elicits fluid accumulation in suckling mice. Most interesting was the finding that 14 of the first 20 N-terminal amino acid residues are identical to those of the El Tor hemolysin of *V. cholerae* and the heat-labile hemolysin of *V. mimicus*. Finally, Ahn et al. (3) recently reported on an iron-regulated hemin-binding outer membrane protein that affects hemolytic activity and oxidative stress response and a mutant of which had a significantly reduced LD_{50}.

V. FURNISSII

Classification

The taxonomy of *V. furnissii* has been extensively described by Brenner et al. (17) and Farmer et al. (52). Their studies included biochemical reactions and DNA-DNA hybridization to determine the relatedness of *V. furnissii* to *V. fluvialis*, as well as to *Aeromonas* and *Alteromonas*

species and other vibrios. This species is similar to *Aeromonas hydrophila*, from which it can easily be distinguished through its ability to grow in 6% NaCl. *V. furnissii* can be differentiated from *V. fluvialis* primarily by its production of gas from glucose (Table 16.1).

Reservoirs

V. furnissii has been isolated from river and estuarine waters and from marine molluscs and crustacea from throughout the world. Wong et al. (201) found ca. 7 to 12% of the oysters, clams, shrimp, and crabs they examined to harbor this species.

Foodborne Outbreaks

The largest documented occurrences of *V. furnissii* infection were in 1969, when this species was isolated in two outbreaks of acute gastroenteritis in American tourists returning from the Orient. In the first outbreak, 23 of 42 elderly passengers returning from Tokyo developed gastroenteritis; one woman died and two other persons required hospitalization. Food histories implicated shrimp and crab salad and/or the cocktail sauce served with the salad. *V. furnissii* was recovered from seven stool specimens, two of which also contained *V. parahaemolyticus*. The second outbreak affected 24 of 59 persons returning from Hong Kong. Nine persons were hospitalized. A food vehicle was not found, but *V. furnissii* was isolated from at least five fecal specimens. However, because several other potentially enteropathogenic bacteria were also isolated from these stool samples, an absolute causal role of *V. furnissii* could not be documented.

Characteristics of Disease

Symptoms described by Brenner et al. (17) for the gastroenteritis outbreaks described above included diarrhea (91 to 100%), abdominal cramps (79 to 100%), nausea (65 to 89%), and vomiting (39 to 78%). Fever was not reported. The onset of symptoms occurred between 5 and 20 h postexposure, with patients recovering within 24 h.

Infectious Dose and Susceptible Population

The infectious dose and susceptible population for *V. furnissii* are not known.

Virulence Mechanisms

Chikahira and Hamada (29) provided an extensive description of the toxic products produced by nine environmental strains of *V. furnissii*. Unlike those of strains of *V. fluvialis*, culture filtrates or cell extracts of only one strain of *V. furnissii* caused elongation of CHO cells. On the other hand, 100% cause cell death, an activity which Lockwood et al. (113) determined to be distinct from the elongation activity. Lockwood et al. found 86% of the *V. furnissii* strains to produce a factor cross-reacting with cholera toxin and 50% to cause fluid accumulation in rabbit ileal loops and to produce a hemolysin active against rabbit erythrocytes. The culture supernatants of all *V. furnissii* strains cause mouse lethality. A mutant deficient in phosphomannomutase had reduced virulence in mice and was more serum sensitive than the parent strain.

V. HOLLISAE

Classification

V. hollisae was described as a new species in 1982 (74). Of the original 16 strains obtained and characterized by the CDC, 15 were from stool samples or intestinal contents, and many of the corresponding patients had diarrhea. The phenotypic traits of this species are described in reference 74 and in Table 16.1. Based on 16S rRNA analysis, the placement of this species into a newly formed genus as *Grimontia hollisae* has recently been proposed (189). Vuddhakul et al. (196) recently described a series of DNA primers of value in identifying this species from environmental samples.

V. hollisae is unusual among the vibrios in its inability to grow on TCBS agar or MacConkey agar. As these two media are routinely employed for the examination of stool samples for vibrios, it is possible that infections caused by this species are missed in clinical labs. *V. hollisae* grows well on blood agar, however, and xylose-lysine-desoxycholate agar recovers this species. The use of the API-20E system has generally been reported to properly identify *V. hollisae*, although the system failed to identify one of the isolates described by Abbot and Janda (1).

Reservoirs

The distribution of *V. hollisae* is not well documented, although *V. hollisae* is likely a marine species and it appears that the organism prefers warmer marine waters (1).

Foodborne Outbreaks

According to Abbott and Janda (1), 30 cases of *V. hollisae* infection have been reported in the literature since the original description of this species. Most (87%) have been cases of gastroenteritis in adults, with 13% of infections causing extraintestinal disease. Three cases of septicemia associated with *V. hollisae* have been reported.

There appears to be a strong correlation between *V. hollisae* infections and consumption of raw seafood. Cases also have been reported following consumption of fried catfish and of dried and salted fish, suggesting an ability of the bacterium to survive these treatments.

In addition, Abbott and Janda (1) cited a case of *V. hollisae*-associated diarrhea that involved a 61-year-old male who denied recent travel or seafood consumption. This case suggests that additional vehicles may remain to be identified.

Levine et al. (111), in their study of 121 vibrio infections in four Gulf Coast states which occurred in 1989, reported 9 to be due to *V. hollisae*. Eight of these were cases of gastroenteritis, with one being a wound infection. Two of the patients required hospitalization. The most commonly consumed foods for the patients with gastroenteritis were raw oysters, with clams, crabs, and shrimp also implicated. Desenclos et al. (46), in a study of 333 cases of vibrio illnesses in adults in Florida over an 8-year period, found 32 to be due to *V. hollisae*. Of these, 20 (62.5%) followed ingestion of raw oysters. Although no description of these cases was included in their study, 17 of these 20 cases resulted in gastroenteritis, with another 3 resulting in septicemia. Morris et al. (134) described nine cases of diarrhea in which specimens were culture positive for *V. hollisae*, with no other enteric pathogen identified. All patients had diarrhea and abdominal pain, and all but one were hospitalized. Implicated were raw oysters or clams in six cases and raw shrimp in another. A seventh patient had eaten seafood but denied eating it raw. Abbott and Janda (1) have described two incidences of severe gastrointestinal disease caused by *V. hollisae*. In both cases, one in a 42-year-old male and the other in a 25-year-old female, ingestion of raw oysters preceded the infections. In one of the cases, steamed oysters and cooked crab were also consumed.

Although *V. hollisae* had previously been reported in a septicemia case (134), Lowry et al. (114) were the first to describe a case of septicemia in which *V. hollisae* alone was isolated from blood cultures. A 65-year-old male was admitted to the hospital with a 12-h history of night fever, vomiting, and abdominal pain. He had passed two loose stools. On the day before hospitalization, the patient ate fried Mississippi River catfish for lunch and again for dinner. His illness began at midnight, and upon admission *V. hollisae* was isolated from a blood culture. The patient was given antibiotics and was discharged 8 days after admission. This case is unusual not only because it involved septicemia but also because the infection occurred following consumption of a freshwater fish which had been fried. There was no history of exposure to saltwater or other seafood consumption by the patient. However, catfish are capable of adapting to low salinities, and the region of the Mississippi River where the fish was obtained often has salt concentrations of up to 0.5%. It is also possible that incomplete cooking or recontamination of the fish occurred. An additional case of foodborne septicemia has been reported which was unusual in that the vehicle appeared to be dried and salted fish obtained from a Southeast Asian food store and eaten uncooked.

In the first case reported in Europe, Gras-Rouzet et al. (60) described a case of gastroenteritis and bacteremia in a previously healthy 76-year-old man who ate cockles from Brittany, France.

Characteristics of Disease

Symptoms of gastroenteritis caused by *V. hollisae* are similar to those of gastroenteritis caused by non-O1 strains of *V. cholerae* (134) and typically include severe abdominal cramping, vomiting, fever, and watery diarrhea (1). In the cases reported by Morris et al. (134), the median duration of diarrhea was 1 day (range, 4 h to 13 days), with occasional reports of bloody diarrhea. Eight of the patients were admitted to the hospital (median duration of stay, 5 days; range, 2 to 9 days), but all recovered.

Infectious Dose and Susceptible Population

Although six of the nine gastroenteritis patients described by Morris et al. (134) were male, with an average age of 35 years (range, 31 to 59 years), other reports do not suggest any significant preference for gender or age (1), nor is a seasonality in the occurrence of infections apparent, with cases appearing in the winter as well as the summer. Most cases of gastroenteritis are in otherwise healthy individuals, although abnormal liver function secondary to alcohol abuse was present in one case reported by Morris et al. (134). In the septicemia case described by Lowry et al. (114), the patient had consumed a fifth of wine daily for 20 years and had previously undergone pancreatectomy and splenectomy as well as other surgical procedures. Unlike that in cases of gastroenteritis, underlying disease appears to be the norm for septicemic cases (1, 134).

V. hollisae is highly susceptible to most antimicrobial agents (1), and tobramycin and cefamandole were successfully used in a septicemic case (114).

Virulence Mechanisms

Gene sequences in *V. hollisae* which are homologous to those of the TDH gene of *V. parahaemolyticus* have been reported previously, although strain-to-strain variation apparently exists (140). The hemolysin has been purified and partially characterized and reported to be related to the *V. parahaemolyticus* TDH. The two toxins have also been found to have similar lethal toxicities in mice.

Intragastric administration of *V. hollisae* cells to infant mice has been reported to elicit intestinal fluid accumulation. An enterotoxin which elongates CHO cells and causes fluid accumulation in mice has been purified (102) and detected in extracts from infected mice and in culture fluids from various growth media.

A hydroxamate siderophore, identified as aerobactin, is produced by *V. hollisae* in response to iron limitation. This iron-binding protein may play a role in the ability of this species to cause infection, and the patient in one of the bacteremic cases had improperly dosed himself with ferrous sulfate as a supplement for chronic anemia, an act which may have exacerbated the disease process.

Miliotis et al. (129) have recently described the ability of *V. hollisae* to adhere to and invade cultured epithelial cells, with internalization involving both eukaryotic and prokaryotic factors. The authors suggested that, in addition to toxin production, the ability to invade epithelial cells is consistent with the invasive disease produced by *V. hollisae* in some patients (114).

V. ALGINOLYTICUS

Classification

V. alginolyticus was originally classified as a biotype of *V. parahaemolyticus*, and indeed the two are genetically quite similar. However, they can easily be differentiated phenotypically, most readily by the fermentation of sucrose by *V. alginolyticus*. Some differences in taxonomic traits between clinical and environmental isolates of *V. alginolyticus* have been suggested, but whether these are significant has not been demonstrated.

Reservoirs

V. alginolyticus strains have generally been reported to inhabit, often in high numbers, seawater and seafood taken from throughout the world (85). The bacterium is easily isolated from fish, clams, crabs, oysters, mussels, and shrimp, as well as water. Indeed, many surveys have reported this species to be one of the most commonly isolated of the vibrios. Several studies have reported a temperature correlation with its isolation, with numbers of cells greatest in the warm-water months.

Foodborne Outbreaks

Most *V. alginolyticus* infections are associated with the marine environment and generally remain superficial. Only rarely has *V. alginolyticus* been implicated as a food-associated pathogen. Indeed, this species can be isolated from 0.5% of healthy people in Japan, with no clinically associated intestinal disease evident. *V. alginolyticus* has been isolated from the rice water diarrheal stool of a female patient with acute enterocolitis and also from the trout roe she had consumed. *V. alginolyticus* has also been isolated from the blood of a leukemic 49-year-old female who had consumed raw oysters 1 week earlier. Desenclos et al. (46) reported a case of *V. alginolyticus*-induced gastroenteritis occurring in Florida during their 8-year survey but did not provide details. This case represented only one of the 333 bacteriologically confirmed *Vibrio* illnesses reported in that study. Chien et al. (28), however, have described a case of bacteremia in an immunocompromised patient who had eaten a large amount of seafood.

Characteristics of Disease

There are few reports in the literature describing the symptoms of gastroenteric disease caused by *V. alginolyticus*. The leukemic patient with *V. alginolyticus* infection was admitted to the hospital with confusion, shock, and anemia. She had a temperature of 40°C and a systolic blood pressure of 80 mm Hg. Despite antibiotic administration and treatment for shock, the patient died 12 days after admission. The exact role of *V. alginolyticus* in this case is unknown, however, as *Pseudomonas aeruginosa* was also isolated from the patient's blood and the presence of this pathogen was very possibly a factor in the fatal outcome.

Infectious Dose and Susceptible Population

Whereas extraintestinal infections are typically self-limiting and relatively mild, systemic infections are generally severe. Most such cases appear to involve patients who are immunocompromised due to severe burns or cancer. A history of alcohol abuse may also be important in the development of *V. alginolyticus* infections.

Virulence Mechanisms

Virtually nothing is known regarding the pathogenic mechanisms of *V. alginolyticus*. Oliver et al. (152) reported the production of lipase, lecithinase, chondroitin sulfatase, DNase, and hemolysin in the strain of *V. alginolyticus* that they examined but saw no evidence for the production of elastase, mucinase, protease, or hyaluronidase, nor were culture filtrates cytotoxic for CHO cells.

CONCLUSIONS

The presence of vibrios, especially *V. cholerae*, *V. vulnificus*, and *V. parahaemolyticus*, in foods represents a serious and growing public health hazard. These species, along with five other recognized human pathogenic vibrios present in foods, are found in estuarine waters and occur frequently in a variety of fish and shellfish. Studies routinely report 30 to 100% of fresh, frozen, or iced fish and shellfish to harbor vibrios, with *V. parahaemolyticus*,

non-O1 *V. cholerae*, *V. vulnificus*, *V. alginolyticus*, and *V. fluvialis* predominating. Most vibrios demonstrate seasonality both in their isolation from the environment and foodstuffs and in the infections they cause, both being generally greater during the warm-weather months. Although the overall incidence of infection with vibrios has been estimated to be relatively low, many milder infections likely occur which are not reported. Symptoms of infections with the various vibrios range from mild gastrointestinal upset to death; fatalities occur primarily with two species, *V. cholerae* and *V. vulnificus*. With some exceptions, e.g., those infected with *V. cholerae* O1 or O139, severely ill patients typically suffer preexisting underlying illness, with chronic liver disease being the most common.

Cholera produced by *V. cholerae* is one of the few foodborne diseases with epidemic potential. The seventh pandemic of cholera, which began in 1961 and continues as of this writing, has involved more than 100 countries, affected more than 3 million persons, and killed many thousands. As more has been learned about the pathogenesis of this bacterium in recent years, it has become clear that its pathogenic potential is quite heterogeneous. Although a number of toxins are elaborated by *V. cholerae*, it is evident that the production of cholera toxin is essential to disease production and that the possession of either the O1 or O139 antigen correlates with this potential. Disease due to *V. cholerae* often involves seafood, but a wide variety of nonacidic foods have been involved in disease transmission.

Along with *V. cholerae*, *V. parahaemolyticus* is the most studied of the pathogenic vibrios. Occurring in estuarine waters throughout the world, it is generally found in a wide variety of seafoods. In Japan, as many as 70% of all bacterial foodborne illnesses are attributable to this species. Outbreaks affecting over 1,100 people have occurred in the United States, and in all cases, seafood (especially shrimp and oysters) has been implicated. Disease production is generally limited to strains producing the so-called Kanagawa-positive hemolysin, or TDH, which interestingly is not produced by the great majority of environmental isolates.

Of all the pathogenic vibrios, *V. vulnificus* is the most serious in the United States, being responsible for 95% of all seafood-borne deaths in this country. Like *V. cholerae* and *V. parahaemolyticus*, *V. vulnificus* is found in estuarine waters and the molluscan shellfish that inhabit those environments. Development of human infection typically follows consumption of raw oysters taken from the Gulf Coast but is generally restricted to persons having certain underlying diseases such as liver cirrhosis. Fatality rates are approximately 60%. The recent finding that two distinct genotypes of biotype 1 *V. vulnificus* exist offers the possibility of a better understanding of the ecology and pathogenicity of this species.

Human infections with the remaining vibrios (*V. mimicus*, *V. fluvialis*, *V. furnissii*, *V. hollisae*, and *V. alginolyticus*) are less common and usually less severe, although deaths are occasionally reported. Seafood is the usual source of infection by these species.

Not enough is known of the susceptibilities of the vibrios to food preservation methods. Cold is often considered an effective defense against the proliferation of vibrios, although seafoods have been reported to be protective for some of the vibrios at refrigeration temperatures. Other vibrios appear to be able to increase in number at refrigeration temperatures. Thorough heating of shellfish appears to be adequate to kill all *Vibrio* spp. and is probably the only effective protective measure currently available.

References

1. Abbott, S. L., and J. M. Janda. 1994. Severe gastroenteritis associated with *Vibrio hollisae* infection: report of two cases and review. *Clin. Infect. Dis.* 18:310–312.

2. Abbott, S. L., J. M. Janda, J. A. Johnson, and J. J. Farmer III. 2006. *Vibrio* and related organisms, p. 723–733. *In* P. R. Murray, E. J. Baron, J. H. Jorgensen, M. L. Landry, and M. A. Pfaller (ed.), *Manual of Clinical Microbiology*, 9th ed. ASM Press, Washington, D.C.

3. Ahn, S.-H., J.-H. Han, J.-H. Lee, K.-J. Park, and I.-S. Kong. 2005. Identification of an iron-regulated hemin-binding outer membrane protein, HupO, in *Vibrio fluvialis*: effects on hemolytic activity and the oxidative stress response. *Infect. Immun.* 73:722–729.

4. Albert, M. J. 1994. *Vibrio cholerae* O139 Bengal. *J. Clin. Microbiol.* 32:2345–2349.

5. Albert, M. J., and G. B. Nair. 2005. *Vibrio cholerae* O139 Bengal—10 years on. *Rev. Med. Microbiol.* 16:135–143.

6. Andrews, L., D. L. Park, and Y.-P. Chen. 2000. Low temperature pasteurization to reduce the risk of *Vibrio* infections in raw-shellstock oysters. *Proceedings of the 25th Annual Meeting of the Seafood Science and Technology Society.* Seafood Science and Technology Society, Longboat Key, Fla.

7. Andrews, L., M. Jahncke, and K. Mallikarjunan. 2003. Low dose gamma irradiation to reduce pathogenic vibrios in live oysters (*Crassostrea virginica*). *J. Aquat. Food Prod. Technol.* 121:71–82.

8. Anonymous. 1978. *V. parahaemolyticus* foodborne outbreak—Louisiana. *Morb. Mortal. Wkly. Rep.* 27:345–346.

9. Bean, N. H., and P. M. Griffin. 1990. Foodborne disease outbreaks in the United States, 1973–1987: pathogens, vehicles, and trends. *J. Food Prot.* 53:804–817.

10. Benitez, J. A., L. Garcia, A. Silva, H. Garcia, R. Fando, B. Cedre, A. Perez, J. Campos, B. L. Rodriguez, J. L. Perez, T. Valmaseda, O. Perez, A. Perez, M. Ramirez, T. Ledon, M. D. Jidy, M. Lastre, L. Bravo, and G. Sierra. 1999.

11. Berry, T. M., D. L. Park, and D. V. Lightner. 1994. Comparison of the microbial quality of raw shrimp from China, Ecuador, or Mexico at both wholesale and retail levels. *J. Food Prot.* 57:150–153.

12. Beuchat, L. R. 1982. *Vibrio parahaemolyticus*: public health significance. *Food Technol.* 36(3):80–83, 92.

13. Bina, J., J. Zhu, M. Dziejman, S. Faruque, S. Calderwood, and J. Mekalanos. 2003. ToxR regulon of *Vibrio cholerae* and its expression in vibrios shed by cholera patients. *Proc. Natl. Acad. Sci. USA* 100:2801–2806.

14. Bisharat, N., D. I. Cohen, R. M. Harding, D. Falush, D. W. Crook, T. Peto, and M. C. Maiden. 2005. Hybrid *Vibrio vulnificus. Emerg. Infect. Dis.* 11:30–35.

15. Blake, P. A. 1994. Endemic cholera in Australia and the United States, p. 309–319. *In* I. K. Wachsmuth, P. A. Blake, and O. Olsvik (ed.), Vibrio cholerae *and Cholera: Molecular to Global Perspectives*. ASM Press, Washington, D.C.

16. Bonner, J. R., A. S. Coker, C. R. Berryman, and H. M. Pollock. 1983. Spectrum of *Vibrio* infections in a Gulf Coast community. *Ann. Intern. Med.* 99:464–469.

17. Brenner, D. J., F. W. Hickman-Brenner, J. V. Lee, A. G. Steigerwalt, G. R. Fanning, D. G. Hollis, J. J. Farmer III, R. E. Weaver, S. W. Joseph, and R. J. Seidler. 1983. *Vibrio furnissii* (formerly aerogenic biogroup of *Vibrio fluvialis*), a new species isolated from human feces and the environment. *J. Clin. Microbiol.* 18:816–824.

18. Buchrieser, C., V. V. Gangar, R. L. Murphree, M. L. Tamplin, and C. W. Kaspar. 1995. Multiple *Vibrio vulnificus* strains in oysters as demonstrated by clamped homogeneous electric field gel electrophoresis. *Appl. Environ. Microbiol.* 61:1163–1168.

19. Buck, J. D. 1998. Potentially pathogenic *Vibrio* spp. in market seafood and natural habitats from Southern New England and Florida. *J. Aquat. Food Prod. Technol.* 7:53–62.

20. Burkhardt, W., III, S. R. Rippey, and W. D. Watkins. 1992. Depuration rates of Northern quahogs *Mercenaria mercenaria* (Linnaeus, 1758) and Eastern oysters *Crassostrea virginica* (Gmelin, 1791) in ozone- and ultraviolet light-disinfected seawater systems. *J. Shellfish Res.* 11:105–109.

21. Bush, C. A., P. Patel, S. Gunawardena, J. Powell, A. Joseph, J. A. Johnson, and J. G. Morris. 1997. Classification of *Vibrio vulnificus* strains by the carbohydrate composition of their capsular polysaccharides. *Anal. Biochem.* 250:186–195.

22. Butler, S. M., and A. Camilli. 2005. Going against the grain: chemotaxis and infection in *Vibrio cholerae. Nat. Rev. Microbiol.* 3:611–620.

23. Calik, H., M. T. Morrissey, P. Reno, R. Adams, and H. An. 2000. The use of high hydrostatic pressure for reduction of vibrio in oysters. *Proceedings of the 25th Annual Meeting of the Seafood Science and Technology Society.* Seafood Science and Technology Society, Longboat Key, Fla.

24. Cameron, D. N., F. M. Khambaty, I. K. Wachsmuth, R. V. Tauxe, and T. J. Barrett. 1994. Molecular characterization of *Vibrio cholerae* O1 by pulsed-field gel electrophoresis. *J. Clin. Microbiol.* 32:1685–1690.

Preliminary assessment of the safety and immunogenicity of a new CTXF-negative, hemagglutinin/protease-defective El Tor strain as a cholera vaccine candidate. *Infect. Immun.* 67:539–545.

25. Camilli, A., and B. L. Bassler. 2006. Bacterial small-molecule signaling pathways. *Science* **311:**1113–1116.

26. Chang, A. L., H. Y. Kim, J. E. Park, P. Acharya, I.-S. Park, S. M. Yoon, H. J. You, K.-S. Ham, J. K. Park, and J. S. Lee. 2005. *Vibrio vulnificus* secretes a broad-specificity metalloprotease capable of interfering with blood homeostasis through prothrombin activation and fibrinolysis. *J. Bacteriol.* **187:**6909–6916.

27. Chatzidaki-Livanis, M., M. K. Jones, and A. C. Wright. 2006. Genetic variation in the *Vibrio vulnificus* group 1 capsular polysaccharide operon. *J. Bacteriol.* **188:**1987–1998.

28. Chien, J. Y., J. T. Shih, P. R. Hsueh, P. C. Yang, and K. T. Luh. 2002. *Vibrio alginolyticus* as the cause of pleural empyema and bacteremia in an immunocompromised patient. *Eur. J. Clin. Microbiol. Infect. Dis.* **21:**401–403.

29. Chikahira, M., and K. Hamada. 1988. Enterotoxigenic substances and other toxins produced by *Vibrio fluvialis* and *Vibrio furnissii. Jpn. J. Vet. Sci.* **50:**865–873.

30. Chowdhury, M. A. R., K. M. S. Aziz, B. A. Kay, and Z. Rahim. 1987. Toxin production by *Vibrio mimicus* strains isolated from human and environmental sources in Bangladesh. *J. Clin. Microbiol.* **25:**2200–2203.

31. Chowdhury, M. A. R., H. Yamanaka, S. Miyoshi, K. M. S. Aziz, and S. Shinoda. 1989. Ecology of *Vibrio mimicus* in aquatic environments. *Appl. Environ. Microbiol.* **55:**2073–2078.

32. Chowdhury, N. R., S. Chakraborty, T. Ramamurthy, M. Nishibuchi, S. Yamasaki, Y. Takeda, and G. B. Nair. 2000. Molecular evidence of clonal *Vibrio parahaemolyticus* pandemic strains. *Emerg. Infect. Dis.* **6:**631–636.

33. Colwell, R. R., and A. Huq. 1994. Vibrios in the environment: viable but nonculturable *Vibrio cholerae*, p. 117–133. *In* K. Wachsmuth, P. A. Blake, and O. Olsvik (ed.), Vibrio cholerae *and Cholera: Molecular to Global Perspectives.* ASM Press, Washington, D.C.

34. Colwell, R. R., M. L. Tamplin, P. R. Brayton, A. L. Gauzens, B. D. Tall, D. Herrington, M. M. Levine, S. Hall, A. Huq, and D. A. Sack. 1990. Environmental aspects of *Vibrio cholerae* in transmission of cholera, p. 327–343. *In* R. B. Sack and Y. Zinnaka (ed.), *Advances in Research on Cholera and Related Diarrheas*, vol. 7. KTK Scientific Publishers, Tokyo, Japan.

35. Comstock, L. E., J. A. Johnson, J. M. Michalski, J. G. Morris, Jr., and J. B. Kaper. 1996. Cloning and sequence of a region encoding surface polysaccharide of *Vibrio cholerae* O139 and characterization of the insertion site in the chromosome of *Vibrio cholerae* O1. *Mol. Microbiol.* **19:**815–826.

36. Cook, D. W. 1994. Effect of time and temperature on multiplication of *Vibrio vulnificus* in postharvest Gulf Coast shellstock oysters. *Appl. Environ. Microbiol.* **60:**3483–3484.

37. Cook, D. W., and A. D. Ruple. 1992. Cold storage and mild heat treatment as processing aids to reduce the numbers of *Vibrio vulnificus* in raw oysters. *J. Food Prot.* **55:**985–989.

38. Dadisman, T. A., R. Nelson, J. R. Molenda, and H. J. Garber. 1972. *Vibrio parahaemolyticus* gastroenteritis in Maryland. I. Clinical and epidemiological aspects. *Am. J. Epidemiol.* **96:**414–426.

39. Dalsgaard, A., O. Serichantalergs, A. Forslund, W. Lin, J. Mekalanos, E. Mintz, T. Shimada, and J. G. Wells. 2001. Clinical and environmental isolates of *Vibrio cholerae* serogroup O141 carry the CTX phage and the genes encoding the toxin-coregulated pili. *J. Clin. Microbiol.* **39:**4086–4092.

40. Daniels, N. A., L. MacKinnon, R. Bishop, S. Altekruse, B. Ray, R. M. Hammond, S. Thompson, S. Wilson, N. H. Bean, P. M. Griffin, and L. Slutsker. 2000. *Vibrio parahaemolyticus* infections in the United States, 1973–1998. *J. Infect. Dis.* **181:**1661–1666.

41. Davis, B. R., G. R. Fanning, J. M. Madden, A. G. Steigerwalt, H. B. Bradford, Jr., H. L. Smith, Jr., and D. J. Brenner. 1981. Characterization of biochemically atypical *Vibrio cholerae* strains and designation of a new pathogenic species, *Vibrio mimicus. J. Clin. Microbiol.* **14:**631–639.

42. de Haan, L., and T. R. Hirst. 2004. Cholera toxin: a paradigm for multi-functional engagement of cellular mechanisms. *Mol. Membr. Biol.* **21:**77–92.

43. DePaola, A., G. M. Capers, and D. Alexander. 1994. Densities of *Vibrio vulnificus* in the intestines of fish from the U.S. Gulf Coast. *Appl. Environ. Microbiol.* **60:**984–988.

44. DePaola, A., J. L. Norstrom, J. C. Bowers, J. G. Wells, and D. W. Cook. 2003. Seasonal abundance and total and pathogenic *Vibrio parahaemolyticus* in Alabama oysters. *Appl. Environ. Microbiol.* **69:**1521–1526.

45. DePaola, A., J. Ulaszek, C. A. Kaysner, B. J. Tenge, J. L. Nordstrom, J. Wells, N. Puhr, and S. M. Gendel. 2003. Molecular, serological, and virulence characteristics of *Vibrio parahaemolyticus* isolated from environmental, food, and clinical sources in North America and Asia. *Appl. Environ. Microbiol.* **69:**3999–4005.

46. Desenclos, J.-C. A., K. C. Klontz, L. E. Wolfe, and S. Hoecherl. 1991. The risk of *Vibrio* illness in the Florida raw oyster eating population, 1981–1988. *Am. J. Epidemiol.* **134:**290–297.

47. Dixon, W. D. 1992. The effects of gamma radiation (^{60}Co) upon shellstock oysters in terms of shelf life and bacterial reduction, including *Vibrio vulnificus* levels. M.S. thesis. University of Florida, Gainesville, Fla.

48. Dziejman, M., D. Serruto, V. C. Tam, D. Sturtevant, P. Diraphat, S. M. Faruque, M. H. Rahman, J. F. Heidelberg, J. Decker, L. Li, K. T. Montgomery, G. Grills, R. Kucherlapati, and J. J. Mekalanos. 2005. Genomic characterization of non-O1, non-O139 *Vibrio cholerae* reveals genes for a type III secretion system. *Proc. Natl. Acad. Sci. USA* **102:**3465–3470.

49. Elhadi, N., S. Radu, C.-H. Chen, and M. Nishibuchi. 2004. Prevalence of potentially pathogenic *Vibrio* species in the seafood marketed in Malaysia. *J. Food. Prot.* **67:**1469–1475.

50. Eyles, M. J., and G. R. Davey. 1984. Microbiology of commercial depuration of the Sydney rock oyster, *Crassostrea commercialis. J. Food Prot.* **47:**703–706.

51. Farmer, J. J., III, F. W. Hickman-Brenner, and M. T. Kelly. 1985. Vibrio, p. 282–301. *In* E. H. Lennette et al. (ed.), *Manual of Clinical Microbiology*, 4th ed. American Society for Microbiology, Washington, D.C.

52. Farmer, J. J., III, R. E. Weaver, S. W. Joseph, and R. J. Seidler. 1983. *Vibrio furnissii* (formerly aerogenic biogroup

of *Vibrio fluvialis*), a new species isolated from human feces and the environment. *J. Clin. Microbiol.* **18:**816–824.

53. Faruque, S. M., I. B. Naser, M. J. Islam, A. S. Faruque, A. N. Ghosh, G. B. Nair, D. A. Sack, and J. J. Mekalanos. 2005. Seasonal epidemics of cholera inversely correlate with the prevalence of environmental cholera phages. *Proc. Natl. Acad. Sci. USA* **102:**1702–1707.

54. Fasano, A., B. Baudry, D. W. Pumplin, S. S. Wasserman, B. D. Tall, J. M. Ketley, and J. B. Kaper. 1991. *Vibrio cholerae* produces a second enterotoxin, which affects intestinal tight junctions. *Proc. Natl. Acad. Sci. USA* **88:**5242–5246.

55. Fullner, K. J. 2003. Toxins of *Vibrio cholerae*: consensus and controversy, p. 481–502. *In* G. A. Hecht (ed.), *Microbial Pathogenesis and the Intestinal Epithelial Cell.* ASM Press, Washington, D.C.

56. Fullner, K. J., and J. J. Mekalanos. 2000. In vivo covalent cross-linking of cellular actin by the *Vibrio cholerae* RTX toxin. *EMBO J.* **19:**5315–5323.

57. Fullner, K. J., J. C. Boucher, M. A. Hanes, G. K. Haines, III, B. M. Meehan, C. Walchle, P. J. Sansonetti, and J. J. Mekalanos. 2002. The contribution of accessory toxins of *Vibrio cholerae* O1 El Tor to the proinflammatory response in a murine pulmonary cholera model. *J. Exp. Med.* **195:**1455–1462.

58. Glass, R. I., and R. E. Black. 1992. The epidemiology of cholera, p. 129–154. *In* D. Barua and W. B. Greenough III (ed.), *Cholera.* Plenum Medical Book Co., New York, N.Y.

59. Gopal, S., S. K. Otta, S. Kumar, I. Karunasagar, M. Nishibuchi, and I. Karunasagar. 2005. The occurrence of *Vibrio* species in tropical shrimp culture environments; implications for food safety. *Int. J. Food Microbiol.* **102:**151–159.

60. Gras-Rouzet, S., P. Y. Donnio, F. Juguet, P. Plessis, J. Minet, and J. L. Avril. 1996. First European case of gastroenteritis and bacteremia due to *Vibrio hollisae. Eur. J. Clin. Microbiol. Infect. Dis.* **15:**864–866.

61. Grau, B. L., M. C. Henk, and G. S. Pettis. 2005. High-frequency phase variation of *Vibrio vulnificus* 1003: isolation and characterization of a rugose phenotypic variant. *J. Bacteriol.* **187:**2519–2525.

62. Groubert, T. N., and J. D. Oliver. 1994. Interaction of *Vibrio vulnificus* and the Eastern oyster, *Crassostrea virginica. J. Food Prot.* **57:**224–228.

63. Gulig, P. A., K. L. Bourdage, and A. M. Starks. 2005. Molecular pathogenesis of *Vibrio vulnificus. J. Microbiol.* **43:**118–131.

64. Gutacker, M., N. Conza, C. Benagli, A. Pedroli, M. V. Bernasconi, L. Permin, R. Aznar, and J.-C. Piffaretti. 2003. Population genetics of *Vibrio vulnificus*: identification of two divisions and a distinct eel-pathogenic clone. *Appl. Environ. Microbiol.* **69:**3203–3212.

65. Hackney, C. R., B. Ray, and M. L. Speck. 1980. Incidence of *Vibrio parahaemolyticus* in and the microbiological quality of seafood in North Carolina. *J. Food Prot.* **43:**769–773.

66. Hammer, B. K., and B. L. Bassler. 2003. Quorum sensing controls biofilm formation in *Vibrio cholerae. Mol. Microbiol.* **50:**101–104.

67. Han, H.-H., J.-H. Lee, Y.-H. Choi, J.-H. Park, T.-J. Choi, and I.-S. Kong. 2002. Purification, characterization and molecular cloning of *Vibrio fluvialis* hemolysin. *Biochim. Biophys. Acta* **1599:**106–114.

68. Hara-Kudo, Y., T. Nishina, H. Nakagawa, H. Konuma, J. Hasegawa, and S. Kumagai. 2001. Improved method for detection of *Vibrio parahaemolyticus* in seafood. *Appl. Environ. Microbiol.* **67:**5819–5823.

69. Hardy, S. P., M. Nakano, and T. Iida. 2004. Single channel evidence for innate pore-formation by *Vibrio parahaemolyticus* thermostable direct haemolysin (TDH) in phospholipid bilayers. *FEMS Microbiol. Lett.* **240:**81–85.

70. Harwood, V. J., J. P. Gandhi, and A. C. Wright. 2004. Methods for isolation and confirmation of *Vibrio vulnificus* from oysters and environmental sources: a review. *J. Microbiol. Methods* **59:**301–316.

71. Hayat, U., G. P. Reddy, C. A. Bush, J. A. Johnson, A. C. Wright, and J. G. Morris, Jr. 1993. Capsular types of *Vibrio vulnificus*: an analysis of strains from clinical and environmental sources. *J. Infect. Dis.* **168:**758–762.

72. He, H., R. M. Adams, D. F. Farkas, and M. T. Morrissey. 2001. The use of high hydrostatic pressure to shuck oysters and extend shelf-life. *J. Shellfish Res.* **20:**1299–1300.

73. Henke, J. M., and B. L. Bassler. 2004. Quorum sensing regulates type III secretion in *Vibrio harveyi* and *Vibrio parahaemolyticus. J. Bacteriol.* **186:**3794–3805.

74. Hickman, F. W., J. J. Farmer III, D. G. Hollis, G. R. Fanning, A. G. Steigerwalt, R. E. Weaver, and D. J. Brenner. 1982. Identification of *Vibrio hollisae* sp. nov. from patients with diarrhea. *J. Clin. Microbiol.* **15:**395–401.

75. Hillman, C. 2000. Commercial pioneering of frozen oysters. *Proceedings of the 25th Annual Meeting of the Seafood Science and Technology Society.* Seafood Science and Technology Society, Longboat Key, Fla.

76. Hlady, W. G. 1997. *Vibrio* infections associated with raw oyster consumption in Florida, 1981–94. *J. Food Prot.* **60:**353–357.

77. Hlady, W. G., and K. C. Klontz. 1996. The epidemiology of *Vibrio* infections in Florida, 1981–1993. *J. Infect. Dis.* **173:**1176–1183.

78. Hlady, W. G., R. C. Mullen, and R. S. Hopkin. 1993. *Vibrio vulnificus* from raw oysters. Leading cause of reported deaths from foodborne illness in Florida. *J. Fla. Med. Assoc.* **80:**536–538.

79. Hoge, C. W., D. Watsky, R. N. Peeler, J. P. Libonati, E. Israel, and J. G. Morris, Jr. 1989. Epidemiology and spectrum in *Vibrio* infections in a Chesapeake Bay USA community. *J. Infect. Dis.* **160:**985–993.

80. Honda, T., Y. Ni, Y. Miwatani, T. Adachi, and J. Kim. 1992. The thermostable direct hemolysin of *Vibrio parahaemolyticus* is a pore-forming toxin. *Can. J. Microbiol.* **38:**1175–1180.

81. Huq, M. I., A. K. M. J. Alam, D. J. Brenner, and G. K. Morris. 1980. Isolation of *Vibrio*-like group, EF-6, from patients with diarrhea. *J. Clin. Microbiol.* **11:**621–624.

82. Jakabi, M., D. S. Gelli, J. C. Torre, M. A. B. Rodas, B. D. G. M. Franco, M. T. Destro, and M. Landgraf. 2003. Inactivation by ionizing radiation of *Salmonella* enteritidis, *Salmonella* infantis, and *Vibrio parahaemolyticus* in oysters (*Crassostrea brasiliana*). *J. Food Prot.* **66:**1025–1029.

83. **Johnston, J. M., S. F. Becker, and L. M. McFarland.** 1986. Gastroenteritis in patients with stool isolations of *Vibrio vulnificus. Am. J. Med.* 80:336–338.

84. **Jones, S. H., T. L. Howell, and K. R. O'Neill.** 1991. Differential elimination of indicator bacteria and pathogenic *Vibrio* sp. from Eastern oysters (*Crassostrea virginica gmelin*, 1791) in a commercial purification facility in Maine. *J. Shellfish Res.* 10:105–112.

85. **Joseph, S. W., R. R. Colwell, and J. B. Kaper.** 1982. *Vibrio parahaemolyticus* and related halophilic vibrios. *Crit. Rev. Microbiol.* 10:77–124.

86. **Kaneko, T., and R. R. Colwell.** 1978. The annual cycle of *Vibrio parahaemolyticus* in Chesapeake Bay. *Microb. Ecol.* 4:135–155.

87. **Kaper, J. B., and V. Sperandio.** 2005. Bacterial cell-to-cell signaling in the gastrointestinal tract. *Infect. Immun.* 73:3197–3209.

88. **Kaper, J. B., J. G. Morris, Jr., and M. M. Levine.** 1995. Cholera. *Clin. Microbiol. Rev.* 8:48–86.

89. **Karaolis, D. K. R., J. A. Johnson, C. C. Bailey, E. C. Boedeker, J. B. Kaper, and P. R. Reeves.** 1998. A *Vibrio cholerae* pathogenicity island associated with epidemic and pandemic strains. *Proc. Natl. Acad. Sci. USA* 95:3134–3139.

90. **Kaysner, C. A., and A. DePaola, Jr.** 2004. *Bacteriological Analytical Manual Online*, chapter 9. [Online.] U.S. Food and Drug Administration, Rockville, Md. http://www.cfsan.fda.gov/~ebam/bam-9.html.

91. **Kaysner, C. A., and W. E. Hill.** 1994. Toxigenic *Vibrio cholerae* O1 in food and water, p. 27–39. *In* I. K. Wachsmuth, P. A. Blake, and O. Olsvik (ed.), *Vibrio cholerae and Cholera: Molecular to Global Perspectives*. ASM Press, Washington, D.C.

92. **Kaysner, C. A., C. Abeyta, M. M. Wekell, A. DePaola, R. F. Stott, and J. M. Leitch.** 1987. Virulent strains of *Vibrio vulnificus* from estuaries of the U.S. West Coast. *Appl. Environ. Microbiol.* 53:1349–1351.

93. **Kilgen, M. B.** 2000. Processing controls for *Vibrio vulnificus* in raw oysters—commercial hydrostatic high pressure. *Proceedings of the 25th Annual Meeting of the Seafood Science and Technology Society*. Seafood Science and Technology Society, Longboat Key, Fla.

94. **Kim, R. K., S. E. Lee, C. M. Kim, S. Y. Kim, E. K. Shin, D. H. Shin, S. S. Chung, H. E. Choy, A. Progulske-Fox, J. D. Hillman, M. Handfield, and J. H. Rhee.** 2003. Characterization and pathogenic significance of *Vibrio vulnificus* antigens preferentially expressed in septicemic patients. *Infect. Immun.* 71:5461–5471.

95. **Kim, Y. R., S. Y. Kim, C. M. Kim, S. E. Lee, and J. H. Rhee.** 2005. Essential role of an adenylate cyclase in regulating *Vibrio vulnificus* virulence. *FEMS Microbiol. Lett.* 243:497–503.

96. **Kirn, T. J., B. A. Jude, and R. K. Taylor.** 2005. A colonization factor links *Vibrio cholerae* environmental survival and human infection. *Nature* 438:863–866.

97. **Klontz, K. C., and J.-C. A. Desenclos.** 1990. Clinical and epidemiological features of sporadic infections with *Vibrio fluvialis* in Florida, USA. *J. Diarrhoeal Dis. Res.* 8:1–2.

98. **Klontz, K. C., D. E. Cover, F. N. Hyman, and R. C. Mullen.** 1994. Fatal gastroenteritis due to *Vibrio fluvialis* and nonfatal bacteremia due to *Vibrio mimicus*: unusual vibrio infections in two patients. *Clin. Infect. Dis.* 19:541–542.

99. **Koch, W. H., W. L. Payne, B. A. Wentz, and T. A. Cebula.** 1993. Rapid polymerase chain reaction method for detection of *Vibrio cholerae* in foods. *Appl. Environ. Microbiol.* 59:556–560.

100. **Kodama, H., Y. Gyobu, N. Tokuman, H. Uetake, T. Shimada, and R. Sakazaki.** 1988. Ecology of non-O1 *Vibrio cholerae* and *Vibrio mimicus* in Toyama prefecture, p. 79–88. *In* N. Ohtomo and R. B. Sack (ed.), *Advances in Research on Cholera and Related Diarrheas*, vol. 6. KTK Scientific Publishers, Tokyo, Japan.

101. **Kotetishvili, M., O. C. Stine, Y. Chen, A. Kreger, A. Sulakvelidze, S. Sozhamannan, and J. G. Morris, Jr.** 2003. Multilocus sequence typing has better discriminatory ability for typing *Vibrio cholerae* than does pulsed-field gel electrophoresis and provides a measure of phylogenetic relatedness. *J. Clin. Microbiol.* 41:2191–2196.

102. **Kothary, M. H., E. F. Claverie, M. D. Miliotis, J. M. Madden, and S. H. Richardson.** 1995. Purification and characterization of a Chinese hamster ovary cell elongation factor of *Vibrio hollisae. Infect. Immun.* 63:2418–2423.

103. **Kothary, M. H., H. Lowman, B. A. McCardell, and B. D. Tall.** 2003. Purification and characterization of enterotoxigenic El Tor-like hemolysin produced by *Vibrio fluvialis. Infect. Immun.* 71:3213–3220.

104. **Krukonis, E. S., and V. J. DiRita.** 2003. From motility to virulence: sensing and responding to environmental signals in *Vibrio cholerae. Curr. Opin. Microbiol.* 6:186–190.

105. **Lee, J. V., P. Shread, and A. L. Furniss.** 1978. The taxonomy of group F organisms: relationships to *Vibrio* and *Aeromonas. J. Appl. Bacteriol.* 45:ix.

106. **Lee, S. H., D. L. Hava, M. K. Waldor, and A. Camilli.** 1999. Regulation and temporal expression patterns of *Vibrio cholerae* virulence genes during infection. *Cell* 99:625–634.

107. **Lenz, D. H., M. B. Miller, J. Zhu, R. V. Kulkarni, and B. L. Bassler.** 2005. CsrA and three redundant small RNAs regulate quorum sensing in *Vibrio cholerae. Mol. Microbiol.* 58:1186–1202.

108. **Lesmana, M., D. S. Subekti, P. Tjaniadi, C. H. Simanjuntak, N. H. Punjabi, J. R. Campbell, and B. A. Oyofo.** 2002. Spectrum of *vibrio* species associated with acute diarrhea in North Jakarta, Indonesia. *Diagn. Microbiol. Infect. Dis.* 43:91–97.

109. **Levine, M. M., R. E. Black, M. L. Clements, D. R. Nalin, L. Cisneros, and R. A. Finkelstein.** 1981. Volunteer studies in development of vaccines against cholera and enterotoxigenic *Escherichia coli*: a review, p. 443–459. *In* T. Holme, J. Holmgren, M. H. Merson, and R. Mollby (ed.), *Acute Enteric Infections in Children. New Prospects for Treatment and Prevention*. Elsevier/North-Holland Biomedical Press, Amsterdam, The Netherlands.

110. **Levine, M. M., J. B. Kaper, D. Herrington, G. Losonsky, J. G. Morris, M. L. Clements, R. E. Black, B. Tall, and R. Hall.** 1988. Volunteer studies of deletion mutants of *Vibrio cholerae* O1 prepared by recombinant techniques. *Infect. Immun.* 56:161–167.

111. **Levine, W. C., P. M. Griffin, and the Gulf Coast Vibrio Working Group.** 1993. *Vibrio* infections on the Gulf

Coast: results of first year of regional surveillance. *J. Infect. Dis.* **167**:479–483.

112. Linkous, D. A., and J. D. Oliver. 1999. Pathogenesis of *Vibrio vulnificus. FEMS Microbiol. Lett.* **174**:207–214.

113. Lockwood, D. E., A. S. Kreger, and S. H. Richardson. 1982. Detection of toxins produced by *Vibrio fluvialis. Infect. Immun.* **35**:702–708.

114. Lowry, P. W., L. M. McFarland, and H. K. Threefoot. 1986. *Vibrio hollisae* septicemia after consumption of catfish. *J. Infect. Dis.* **154**:730–731.

115. Lowry, P. W., L. M. McFarland, B. H. Peltier, N. C. Roberts, H. B. Bradford, J. L. Herndon, D. F. Stroup, J. B. Mathison, P. A. Blake, and R. A. Gunn. 1989. *Vibrio* gastroenteritis in Louisiana: a prospective study among attendees of a scientific congress in New Orleans. *J. Infect. Dis.* **160**:978–984.

116. Magalhaes, V., A. Castello Filho, M. Magalhaes, and T. T. Gomes. 1993. Laboratory evaluation on pathogenic potentialities of *Vibrio furnissii. Mem. Inst. Oswaldo Cruz Rio de Janeiro* **88**:593–597.

117. Makino, K., K. Oshima, K. Kurokawa, K. Yokoyama, T. Uda, K. Tagomori, Y. Iijima, M. Najima, M. Nakano, A. Yamashita, Y. Kubota, S. Kimura, T. Yasunaga, T. Honda, H. Shinagawa, M. Hattori, and T. Iida. 2003. Genome sequence of *Vibrio parahaemolyticus*: a pathogenic mechanism distinct from that of *V. cholerae. Lancet* **361**:743–749.

118. Martinez-Urtaza, J., A. Lozano-Leon, A. Vina-Feas, J. de Novoa, and O. Garcia-Martin. 2006. Differences in the API 20E biochemical patterns of clinical and environmental *Vibrio parahaemolyticus* isolates. *FEMS Microbiol. Lett.* **255**:75–81.

119. Matté, G. R., M. H. Matté, I. G. Rivera, and M. T. Martins. 1994. Distribution of pathogenic vibrios in oysters from a tropical region. *J. Food Prot.* **57**:870–873.

120. Matz, C., D. McDougald, A. M. Moreno, P. Y. Yung, F. H. Yildiz, and S. Kjelleberg. 2005. Biofilm formation and phenotypic variation enhance predation-driven persistence of *Vibrio cholerae. Proc. Natl. Acad. Sci. USA* **102**:16819–16824.

121. McLaughlin, J. B., A. DePaola, C. A. Bopp, K. A. Martinek, N. P. Napolilli, C. G. Allison, S. L. Murray, E. C. Thompson, M. M. Bird, and J. P. Middaugh. 2005. Outbreak of *Vibrio parahaemolyticus* gastroenteritis associated with Alaskan oysters. *N. Engl. J. Med.* **353**:1463–1470.

122. McLeod, S. M., H. H. Kimsey, B. M. Davis, and M. K. Waldor. 2005. CTXphi and *Vibrio cholerae*: exploring a newly recognized type of phage-host cell relationship. *Mol. Microbiol.* **57**:347–356.

123. McPherson, V. L., J. A. Watts, L. M. Simpson, and J. D. Oliver. 1991. Physiological effects of the lipopolysaccharide of *Vibrio vulnificus* on mice and rats. *Microbios* **67**:141–149.

124. Mead, P. S., L. Slutsker, V. Dietz, L. F. McCaig, J. S. Bresee, C. Shapiro, P. M. Griffin, and R. B. V. Tauxe. September–October, 1999, posting date. September 15, 1999. *Emerg. Infect. Dis.* **5**. [Online.] http://www.cdc.gov/ncidod/eid/vol5no5/mead.htm.

125. Meibom, K. L., M. Blokesch, N. A. Dolganov, C. Y. Wu, and G. K. Schoolnik. 2005. Chitin induces natural competence in *Vibrio cholerae. Science* **310**:1824–1827.

126. Merkel, S. M., S. Alexander, J. D. Oliver, and Y. M. Huet-Hudson. 2001. Essential role for estrogen in protection against *Vibrio vulnificus*-induced endotoxic shock. *Infect. Immun.* **69**:6119–6122.

127. Merrell, D. S., S. M. Butler, F. Qadri, N. A. Dolganov, A. Alam, M. B. Cohen, S. B. Calderwood, G. K. Schoolnik, and A. Camilli. 2002. Host-induced epidemic spread of the cholera bacterium. *Nature* **417**:642–645.

128. Mey, A. R., E. E. Wyckoff, V. Kanukurthy, C. R. Fisher, and S. M. Payne. 2005. Iron and fur regulation in *Vibrio cholerae* and the role of Fur in virulence. *Infect. Immun.* **73**:8167–8178.

129. Miliotis, M. D., B. D. Tall, and R. T. Gray. 1995. Adherence to and invasion of tissue culture cells by *Vibrio hollisae. Infect. Immun.* **63**:4959–4963.

130. Mintz, E. D., T. Popovic, and P. A. Blake. 1994. Transmission of *Vibrio cholerae* O1, p. 345–356. *In* I. K. Wachsmuth, P. A. Blake, and O. Olsvik (ed.), Vibrio cholerae *and Cholera: Molecular to Global Perspectives.* ASM Press, Washington, D.C.

131. Morris, J. G., Jr. 1990. Non-O group 1 *Vibrio cholerae*: a look at the epidemiology of an occasional pathogen. *Epidemiol. Rev.* **12**:179–191.

132. Morris, J. G., Jr. 1995. "Noncholera" *Vibrio* species, p. 671–685. *In* M. J. Blaser, P. D. Smith, J. I. Ravdin, H. B. Greenberg, and R. L. Guerrant (ed.), *Infections of the Gastrointestinal Tract.* M. Raven Press, Ltd., New York, N.Y.

133. Morris, J. G., Jr. 2003. Cholera and other types of vibriosis: a story of human pandemics and oysters on the half shell. *Clin. Infect. Dis.* **37**:272–280.

134. Morris, J. G., Jr., H. G. Miller, R. Wilson, C. O. Tacket, D. G. Hollis, F. W. Hickman, R. E. Weaver, and P. A. Blake. 1982. Illness caused by *Vibrio damsela* and *Vibrio hollisae. Lancet* **i**:1294–1296.

135. Morris, J. G., Jr., A. C. Wright, D. M. Roberts, P. K. Wood, L. M. Simpson, and J. D. Oliver. 1987. Identification of environmental *Vibrio vulnificus* isolates with a DNA probe for the cytotoxin-hemolysin gene. *Appl. Environ. Microbiol.* **53**:193–195.

136. Motes, M., A. DePaola, S. Zywno-Van Ginkel, and M. McPhearson. 1994. Occurrence of toxigenic *Vibrio cholerae* O1 in oysters in Mobile Bay, Alabama: an ecological investigation. *J. Food Prot.* **57**:975–980.

137. Motes, M. L., A. DePaola, D. W. Cook, J. E. Veazey, J. C. Hunsucker, W. E. Garthright, R. J. Blodgett, and S. Chirtel. 1998. Influence of water temperature and salinity on *Vibrio vulnificus* in Northern Gulf and Atlantic Coast oysters (*Crassostrea virginica*). *Appl. Environ. Microbiol.* **64**:1459–1465.

138. Nascumento, D. R., R. H. Vieira, H. B. Almeida, T. R. Patel, and S. T. Iaria. 1998. Survival of *Vibrio cholerae* O1 strains in shrimp subjected to freezing and boiling. *J. Food Prot.* **61**:1317–1320.

139. Nesper, J., S. Schild, C. M. Lauriano, A. Kraiss, K. E. Klose, and J. Reidl. 2002. Role of *Vibrio cholerae* O139 surface polysaccharides in intestinal colonization. *Infect. Immun.* **70**:5990–5996.

140. Nishibuchi, M., and J. B. Kaper. 1995. Thermostable direct hemolysin gene of *Vibrio parahaemolyticus*: a

virulence gene acquired by a marine bacterium. *Infect. Immun.* **63:**2093–2099.

141. **Nishibuchi, M., and A. DePaola.** 2005. *Vibrio* species, p. 251–271. *In* P. M. Fratamico, A. K. Bhunia, and J. L. Smith (ed.), *Food-Borne Pathogens: Microbiology and Molecular Biology.* Caister Academic Press, Norfolk, United Kingdom.

142. **Nishibuchi, M., A. Fasano, R. G. Russell, and J. B. Kaper.** 1992. Enterotoxigenicity of *Vibrio parahaemolyticus* with and without genes encoding thermostable direct hemolysin. *Infect. Immun.* **60:**3539–3545.

143. **Nishibuchi, M., T. Taniguchi, T. Misawa, V. Khaeomanee-iam, T. Honda, and T. Miwatani.** 1989. Cloning and nucleotide sequence of the gene (*trh*) encoding the hemolysin related to the thermostable direct hemolysin of *Vibrio parahaemolyticus. Infect. Immun.* **57:**2691–2697.

144. **Okeke, I. N., J. Eardley, C. C. Bailey, and J. B. Kaper.** 2001. *Vibrio cholerae,* p. 1191–1236. *In* M. Sussman (ed.), *Molecular Medical Microbiology.* Academic Press, London, United Kingdom.

145. **Oliver, J. D.** 1989. *Vibrio vulnificus,* p. 569–600. *In* M. P. Doyle (ed.), *Foodborne Bacterial Pathogens.* Marcel Dekker, New York, N.Y.

146. **Oliver, J. D.** 2000. Public health significance of viable but nonculturable bacteria, p. 277–300. *In* R. R. Colwell and D. J. Grimes (ed.), *Nonculturable Microorganisms in the Environment.* ASM Press, Washington, D.C.

147. **Oliver, J. D.** 2000. Culture media for the isolation and enumeration of pathogenic *Vibrio* species in foods and environmental samples. *In* J. E. L. Corry, G. D. W. Curtis, and R. M. Baird (ed.), *Culture Media for Food Microbiology,* 2nd ed. Elsevier Science, Amsterdam, The Netherlands.

148. **Oliver, J. D.** 2005. Viable but nonculturable bacteria in food environments. *In* P. M. Fratamico, A. K. Bhunia, and J. L. Smith (ed.), *Food-Borne Pathogens: Microbiology and Molecular Biology.* Caister Academic Press, Norfolk, United Kingdom.

149. **Oliver, J. D.** 2005. *Vibrio vulnificus,* p. 253–276. *In* S. Belkin and R. R. Colwell (ed.), *Oceans and Health: Pathogens in the Marine Environment.* Springer Science, New York, N.Y.

150. **Oliver, J. D.** 2005. The viable but nonculturable state in bacteria. *J. Microbiol.* **43:**93–100.

151. **Oliver, J. D.** 2006. *Vibrio vulnificus,* p. 349–366. *In* F. L. Thompson, B. Austin, and J. Swing (ed.), *Biology of Vibrios.* ASM Press, Washington, D.C.

152. **Oliver, J. D., M. B. Thomas, and J. Wear.** 1986. Production of extracellular enzymes and cytotoxicity by *Vibrio vulnificus. Diagn. Microbiol. Infect. Dis.* **5:**99–111.

153. **Oliver, J. D., D. M. Roberts, V. K. White, M. A. Dry, and L. M. Simpson.** 1986. Bioluminescence in a strain of the human bacterial pathogen *Vibrio vulnificus. Appl. Environ. Microbiol.* **52:**1209–1211.

154. **Ono, T., K. S. Park, M. Ueta, T. Iida, and T. Honda.** 2006. Identification of proteins secreted via *Vibrio parahaemolyticus* type III secretion system 1. *Infect. Immun.* **74:**1032–1042.

155. **Panicker, G., and A. K. Bej.** 2005. Real-time PCR detection of *Vibrio vulnificus* in oysters: comparison of oligonucleotide primers and probes targeting *vvhA. Appl. Environ. Microbiol.* **71:**5702–5709.

156. **Park, K. S., T. Iida, Y. Yamaichi, T. Oyagi, K. Yamamoto, and T. Honda.** 2000. Genetic characterization of DNA region containing the *trh* and *ure* genes of *Vibrio parahaemolyticus. Infect. Immun.* **68:**5742–5748.

157. **Park, K. S., T. Ono, M. Rokuda, M. H. Jang, K. Okada, T. Iida, and T. Honda.** 2004. Functional characterization of two type III secretion systems of *Vibrio parahaemolyticus. Infect. Immun.* **72:**6659–6665.

158. **Parker, R. W., E. M. Maurer, A. B. Childers, and D. H. Lewis.** 1994. Effect of frozen storage and vacuum-packaging on survival of *Vibrio vulnificus* in Gulf Coast oysters (*Crassostrea virginica*). *J. Food Prot.* **57:**604–606.

159. **Pfeffer, C. S., M. F. Hite, and J. D. Oliver.** 2003. The ecology of *Vibrio vulnificus* in estuarine waters of eastern North Carolina. *Appl. Environ. Microbiol.* **69:**3526–3531.

160. **Popovic, T., P. I. Fields, and O. Olsvik.** 1994. Detection of cholera toxin genes, p. 41–52. *In* I. K. Wachsmuth, P. A. Blake, and O. Olsvik (ed.), Vibrio cholerae *and Cholera.* American Society for Microbiology, Washington, D.C.

161. **Qadri, F., M. S. Alam, M. Nishibuchi, T. Rahman, N. H. Alam, J. Chisti, S. Kondo, J. Sugiyama, N. A. Bhuiyan, M. M. Mathan, D. A. Sack, and G. B. Nair.** 2003. Adaptive and inflammatory immune responses in patients infected with strains of *Vibrio parahaemolyticus. J. Infect. Dis.* **187:**1085–1096.

162. **Rabbani, G. H., and W. B. Greenough III.** 1999. Food as a vehicle of transmission of cholera. *J. Diarrhoeal Dis. Res.* **17:**1–9.

163. **Raimondi, D., J. P. Y. Kao, J. B. Kaper, S. Guandalini, and A. Fasano.** 1995. Calcium-dependent intestinal chloride secretion by *Vibrio parahaemolyticus* thermostable direct hemolysin in a rabbit model. *Gastroenterology* **109:**381–386.

164. **Rashid, H. O., H. Ito, and I. Ishigaki.** 1992. Distribution of pathogenic vibrios and other bacteria in imported frozen shrimps and their decontamination by gamma-irradiation. *World J. Microbiol. Biotechnol.* **8:**494–499.

165. **Reidl, J., and K. E. Klose.** 2002. *Vibrio cholerae* and cholera: out of the water and into the host. *FEMS Microbiol. Rev.* **26:**125–139.

166. **Richards, G. P.** 1988. Microbial purification of shellfish: a review of depuration and relaying. *J. Food Prot.* **51:**218–251.

167. **Robach, M. C., and C. S. Hickey.** 1978. Inhibition of *Vibrio parahaemolyticus* by sorbic acid in crab meat and flounder homogenates. *J. Food Prot.* **41:**699–702.

168. **Rodrick, G. E., K. R. Schneider, F. A. Steslow, N. J. Blake, and W. S. Otwell.** 1988. Uptake, fate and ultraviolet depuration of vibrios in *Mercenaria campechiensis. Mar. Technol. Soc. J.* **23:**21–26.

169. **Rosche, T. M., Y. Yano, and J. D. Oliver.** 2005. A rapid and simple PCR analysis indicates there are two subgroups of *Vibrio vulnificus* which correlate with clinical or environmental isolation. *Microbiol. Immunol.* **49:**381–389.

170. **Rosche, T. M., B. Smith, and J. D. Oliver.** 2006. Evidence for an intermediate colony morphology of *Vibrio vulnificus. Appl. Environ. Microbiol.* **72:**4356–4359.

171. Sakazaki, R., K. Tamura, T. Kato, Y. Obara, S. Yamai, and K. Hobo. 1968. Studies on the enteropathogenic, facultatively halophilic bacteria, *Vibrio parahaemolyticus*. III. Enteropathogenicity. *Jpn. J. Med. Sci. Biol.* 21:325–331.

172. Sang, F. C., M. E. Hugh-Jones, and H. V. Hagstad. 1987. Viability of *Vibrio cholerae* O1 on frog legs under frozen and refrigerated conditions and low dose radiation treatment. *J. Food Prot.* 50:662–664.

173. Sanyal, S. C., and P. C. Sen. 1974. Human volunteer study on the pathogenicity of *Vibrio parahaemolyticus*, p. 227–230. *In* T. Fujino, G. Sakaguchi, R. Sakazaki, and Y. Takeda (ed.), *International Symposium on* Vibrio parahaemolyticus. Saikon Publishing Co., Ltd., Tokyo, Japan.

174. Sears, C. L., and J. B. Kaper. 1996. Enteric bacterial toxins: mechanisms of action and linkage to intestinal secretion. *Microbiol. Rev.* 60:167–215.

175. Shandera, W. X., J. M. Johnston, B. R. Davis, and P. A. Blake. 1983. Disease from infection with *Vibrio mimicus*, a newly recognized *Vibrio* species. *Ann. Intern. Med.* 99:169–171.

176. Shao, C.-P., and L.-I. Hor. 2000. Metalloprotease is not essential for *Vibrio vulnificus* virulence in mice. *Infect. Immun.* 68:3569–3573.

177. Shapiro, R. L., S. Altekruse, L. Hutwagner, R. Bishop, R. Hammond, S. Wilson, B. Ray, S. Thompson, R. V. Tauxe, and P. M. Griffin. 1998. The role of Gulf Coast oysters harvested in warmer months in *Vibrio vulnificus* infections in the United States, 1988–1996. *J. Infect. Dis.* 178:752–759.

178. Shinoda, S. 1999. Haemolysins of *Vibrio cholerae* and other *Vibrio* species, p. 373–385. *In* J. E. Alouf and J. H. Freer (ed.), *The Comprehensive Sourcebook of Bacterial Protein Toxins*. Academic Press, London, United Kingdom.

179. Shirai, H., H. Ito, T. Kirayama, Y. Nakamoto, N. Nakabayashi, K. Kumagni, Y. Takeda, and M. Nishibuchi. 1990. Molecular epidemiologic evidence for association of thermostable direct hemolysin (TDH) and TDH-related hemolysin of *Vibrio parahaemolyticus* with gastroenteritis. *Infect. Immun.* 58:3568–3573.

180. Simonson, J. G., P. Danieu, A. B. Zuppardo, R. J. Siebeling, R. L. Murphree, and M. L. Tamplin. 1995. Distribution of capsular and lipopolysaccharide antigens among clinical and environmental *Vibrio vulnificus* isolates, abstr. B-286, p. 215. *Abstr. Annu. Meet. Am. Soc. Microbiol.* American Society for Microbiology, Washington, D.C.

181. Simpson, L. M., V. K. White, S. F. Zane, and J. D. Oliver. 1987. Correlation between virulence and colony morphology in *Vibrio vulnificus*. *Infect. Immun.* 55:269–272.

182. Sloan, E. M., C. J. Hagen, G. A. Lancette, J. T. Peeler, and J. N. Sofos. 1992. Comparison of five selective enrichment broths and two selective agars for recovery of *Vibrio vulnificus* from oysters. *J. Food Prot.* 55:356–359.

183. Strom, M. S., and R. N. Paranjpye. 2000. Epidemiology and pathogenesis of *Vibrio vulnificus*. *Microb. Infect.* 2:177–188.

184. Sun, Y., and J. D. Oliver. 1994. Effects of GRAS compounds on natural *Vibrio vulnificus* populations in oysters. *J. Food Prot.* 57:921–923.

185. Sun, Y., and J. D. Oliver. 1995. The value of CPC agar for the isolation of *Vibrio vulnificus* from oysters. *J. Food Prot.* 58:439–440.

186. Takahashi, A., Y. Sato, Y. Shiomi, V. V. Cantarelli, T. Iida, M. Lee, and T. Honda. 2000. Mechanisms of chloride secretion induced by thermostable direct haemolysin of *Vibrio parahaemolyticus* in human colonic tissue and a human intestinal epithelial cell line. *J. Med. Microbiol.* 49:801–810.

187. Tamplin, M. L., and G. M. Capers. 1992. Persistence of *Vibrio vulnificus* in tissues of Gulf Coast oysters, *Crassostrea virginica*, exposed to seawater disinfected with UV light. *Appl. Environ. Microbiol.* 58:1506–1510.

188. Taylor, R. K., V. L. Miller, D. B. Furlong, and J. J. Mekalanos. 1987. Use of *phoA* gene fusions to identify a pilus colonization factor coordinately regulated with cholera toxin. *Proc. Natl. Acad. Sci. USA* 84:2833–2837.

189. Thompson, F. L., T. Iida, and J. Swings. 2004. Biodiversity of vibrios. *Microbiol. Mol. Biol. Rev.* 68:403–431.

190. Todd, E. C. D. 1989. Preliminary estimates of costs of foodborne disease in the United States. *J. Food Prot.* 52:595–601.

191. Trucksis, M., J. Michalski, Y. K. Denk, and J. B. Kaper. 1998. The *Vibrio cholerae* genome contains two unique circular chromosomes. *Proc. Natl. Acad. Sci. USA* 95:14464–14469.

192. Trucksis, M., J. E. Galen, J. Michalski, A. Fasano, and J. B. Kaper. 1993. Accessory cholera enterotoxin (Ace), the third toxin of a *Vibrio cholerae* virulence cassette. *Proc. Natl. Acad. Sci. USA* 90:5267–5271.

193. Twedt, R. M. 1989. *Vibrio parahaemolyticus*, p. 543–568. *In* M. P. Doyle (ed.), *Foodborne Bacterial Pathogens*. Marcel Dekker, New York, N.Y.

194. Vanderzant, C., and R. Nickelson. 1972. Survival of *Vibrio parahaemolyticus* in shrimp tissue under various environmental conditions. *Appl. Microbiol.* 23:34–37.

195. Vora, G. J., C. E. Meador, M. M. Bird, C. A. Bopp, J. D. Andreadis, and D. A. Stenger. 2005. Microarray-based detection of genetic heterogeneity, antimicrobial resistance, and the viable but nonculturable state in human pathogenic *Vibrio* spp. *Proc. Natl. Acad. Sci. USA* 102:19109–19114.

196. Vuddhakul, V., T. Nakai, C. Matsumoto, T. Oh, T. Nishino, C.-H. Chen, M. Nishibuchi, and J. Okuda. 2000. Analysis of *gyrB* and *toxR* gene sequences of *Vibrio hollisae* and development of *gyrB*- and *toxR*-targeted PCR methods for isolation of *V. hollisae* from the environment and its identification. *Appl. Environ. Microbiol.* 66:3506–3514.

197. Warner, J. M., and J. D. Oliver. 1999. Randomly amplified polymorphic DNA analysis of clinical and environmental isolates of *Vibrio vulnificus* and other *Vibrio* species. *Appl. Environ. Microbiol.* 65:1141–1144.

198. Watnick, P. I., and R. Kolter. 1999. Steps in the development of a *Vibrio cholerae* El Tor biofilm. *Mol. Microbiol.* 34:586–595.

199. Watnick, P. I., K. J. Fullner, and R. Kolter. 1999. A role for the mannose-sensitive hemagglutinin in biofilm formation by *Vibrio cholerae* El Tor. *J. Bacteriol.* 181:3606–3609.

200. Williams, T. L., S. M. Musser, J. L. Nordstrom, A. DePaola, and S. R. Monday. 2004. Identification of a protein biomarker unique to the pandemic O3:K6 clone of *Vibrio parahaemolyticus. J. Clin. Microbiol.* **42:**1657–1665.

201. Wong, H.-C., S.-H. Ting, and W.-R. Shieh. 1992. Incidence of toxigenic vibrios in foods available in Taiwan. *J. Appl. Bacteriol.* **73:**197–202.

202. Wong, H.-C., L.-L. Chen, and C.-M. Yu. 1994. Survival of psychrotrophic *Vibrio mimicus*, *Vibrio fluvialis* and *Vibrio parahaemolyticus* in culture broth at low temperatures. *J. Food Prot.* **57:**607–610.

203. Wright, A. C., L. M. Simpson, and J. D. Oliver. 1981. Role of iron in the pathogenesis of *Vibrio vulnificus* infections. *Infect. Immun.* **34:**503–507.

204. Wright, A. C. J. L. Powell, M. K. Tanner, L. A. Ensor, A. B. Karpas, J. G. Morris, Jr., and M. B. Sztein. 1999. Differential expression of *Vibrio vulnificus* capsular polysaccharide. *Infect. Immun.* **67:**2250–2257.

205. Wu, Z., P. Nbom, and K.-E. Magnusson. 2000. Distinct effects of *Vibrio cholerae* haemagglutinin/protease on the structure and localization of the tight junction-associated proteins occludin and ZO-1. *Cell. Microbiol.* **2:**11–17.

206. Yildiz, F. H., X. S. Liu, A. Heydorn, and G. K. Schoolnik. 2004. Molecular analysis of rugosity in a *Vibrio cholerae* O1 El Tor phase variant. *Mol. Microbiol.* **53:**497–515.

207. Yildiz, F. H., and G. K. Schoolnik. 1999. *Vibrio cholerae* O1 El Tor: identification of a gene cluster required for the rugose colony type, exopolysaccharide production, chlorine resistance, and biofilm formation. *Proc. Natl. Acad. Sci. USA* **96:**4028–4033.

208. Zhang, X.-H., and B. Austin. 2005. A review. Haemolysins in *Vibrio* species. *J. Appl. Microbiol.* **98:**1011–1019.

209. Zhu, J., M. B. Miller, R. E. Vance, M. Dziejman, B. L. Bassler, and J. J. Mekalanos. 2002. Quorum-sensing regulators control virulence gene expression in *Vibrio cholerae. Proc. Natl. Acad. Sci. USA* **99:**3129–3134.

Food Microbiology: Fundamentals and Frontiers, 3rd Ed.
Edited by M. P. Doyle and L. R. Beuchat
© 2007 ASM Press, Washington, D.C.

Cristi L. Galindo
Ashok K. Chopra

Aeromonas and *Plesiomonas* Species

17

The genera *Aeromonas* and *Plesiomonas* comprise gram-negative, facultatively anaerobic, oxidase-positive, glucose-fermenting, rod-shaped bacteria that are generally motile. However, the species *Aeromonas salmonicida* and *A. media* are reported to be nonmotile. *Aeromonas* spp. (aeromonads) usually have a single polar flagellum, but up to 50% of isolates can produce numerous lateral flagella that are distinct from the polar flagellum. *Plesiomonas shigelloides* has two to seven polar flagella and may also produce lateral flagella (96). Both genera were previously classified in the family *Vibrionaceae*, and due to their prevalences in aquatic environments and their enteropathogenicities, they have traditionally been considered together. However, molecular genetic evidence (including that from 16S rRNA cataloguing, 5S rRNA sequencing, and rRNA-DNA hybridization) distinguishes these bacteria from each other and from *Vibrio* species, and they are currently classified as separate families in *Bergey's Manual of Systematic Bacteriology*. *Aeromonas* spp. form the *Aeromonadaceae* family, and *Plesiomonas shigelloides*, which is the only member of the *Plesiomonas* genus, has been placed in the family *Enterobacteriaceae* (20, 106, 139).

The histories of the two genera have been reviewed by Farmer et al. (56). Aeromonads were discovered more than 100 years ago, but the recognition of their role in human illness dates to 1954, when the bacteria were associated with the death of a 40-year-old Jamaican woman with acute fulminating metastatic myositis (33). In 1964, the first well-described case associating *Aeromonas* spp. with diarrhea was reported (191). The bacterium now known as *P. shigelloides* was first described in 1947 by Ferguson and Henderson, who isolated it from a fecal specimen (7). Members of both genera can cause serious gastroenteritis and extraintestinal infections, particularly in immunocompromised hosts, though much controversy existed earlier concerning enteropathogenicity due to a lack of conclusive human volunteer trials and animal models. However, the recent development of animal models and case control human studies clearly demonstrated the role of aeromonads in gastroenteritis (6, 198). Further, substantial clinical and microbiological evidence supports the epidemiologic evidence that certain *Aeromonas* spp. can cause gastroenteritis, and as such aeromonads have been listed by the Environmental Protection Agency on the Contaminant Candidate List for U.S. water supplies. Moreover, recent studies have identified previously unknown *Aeromonas* virulence mechanisms that may contribute significantly

Cristi L. Galindo and **Ashok K. Chopra**, Department of Microbiology and Immunology, University of Texas Medical Branch, Galveston, TX 77555.

to human pathogenicity. This chapter reviews existing knowledge of *Aeromonas* virulence factors and host responses and future directions that may provide a more detailed understanding of *Aeromonas*-associated human diseases. Much less is known about *P. shigelloides* and its possible virulence determinants than is known about *Aeromonas* spp., but potential virulence markers were recently described that may play a role in gastroenteritis caused by *P. shigelloides*.

CLASSIFICATION AND IDENTIFICATION

Aeromonas Species

The taxonomy of the *Aeromonas* genus is considered complex and has undergone constant change over the last decade (56, 100, 103, 106). The number of recognized species is now 17, including 3 recently identified species: *A. clicicola*, isolated from mosquitoes and drinking water; *A. simiae*, isolated from feces of healthy monkeys; and *A. molluscorum*, isolated from bivalve mollusks. *Aeromonas* spp. can be differentiated from *Vibrio* spp. by growth on 6% NaCl, which is positive for vibrios and negative for aeromonads. However, *Aeromonas* spp. can be biochemically confused with other bacteria, including *Vibrio* spp., and proposed phenotypic identification schemes typically involve a large number of biochemical tests (4, 106) with potentially low accuracy levels. Analysis of 16S rRNA and multilocus sequence analysis of housekeeping genes have been proposed as alternatives for more accurate identification of *Aeromonas* spp.

The genus *Aeromonas* was originally divided into a single mesophilic species, *A. hydrophila*, and a single psychrophilic species, *A. salmonicida*. Thus, in the older literature (and unfortunately in some more recent reports), the name *A. hydrophila* is used to include all aeromonads except the latter species, a fish pathogen. DNA hybridization technology revealed that there are at least 17 distinct hybridization groups (HGs), 8 of which are associated with human illness (182). More than 85% of gastroenteritis-associated isolates belong to three main species, *A. hydrophila* HG1, *A. caviae* HG4, and *A. veronii* biovar sobria HG8/10 (21, 80, 101, 122). In the discussion to follow, specific species names will be used only if the species have been identified by HG, either by genetic or by biochemical means. Otherwise the bacteria will be referred to as *Aeromonas* spp. or aeromonds to avoid confusion.

The taxonomy of the *Aeromonas* genus is still evolving. It is important that strains are accurately identified, as antibiotic susceptibilities and pathogenic mechanisms (see below) may vary between strains. Misidentification

of unusual *Aeromonas* species as members of the genus *Vibrio* is a continuing problem (1). Comprehensive biochemical typing schemes that will accurately identify all *Aeromonas* isolates, including all environmental strains, have been proposed, but the large numbers of tests (>40) are beyond the scope of most routine laboratories (111). There is a need for further work in this area to develop schemes that will accurately and easily identify all *Aeromonas* isolates. Until then, food isolates of *Aeromonas* species should be identified at least to a complex (*A. hydrophila*, *A. caviae*, or *A. veronii*) by using key phenotypic features (103).

Plesiomonas shigelloides

In contrast to the genus *Aeromonas*, *Plesiomonas* consists of a homogeneous species with a single DNA HG (*P. shigelloides*) which is separable from aeromonads by simple phenotypic tests: exoenzyme production and a range of carbohydrate fermentations (22, 56, 152), including myoinositol fermentation, decarboxylation of ornithine, and the L-histidine decarboxylase test (conventional Moeller's tube test), which may also be a potentially useful diagnostic feature for differentiating *P. shigelloides* from members of the family *Vibrionaceae* (166). More than 100 serovars of *P. shigelloides* have been described, several of which react with *Shigella* antisera. At present, 76 O- and 41 H-antigens are included in a proposed international typing scheme (8–10, 22), though not all *P. shigelloides* strains are groupable by using this system (103). Ampicillin-containing selective *Aeromonas* media are not suitable for the isolation of *P. shigelloides*. Suitable isolation media include xylose-sodium deoxycholate-citrate and inositol-brilliant green bile salts and *Plesiomonas* agars (22, 56, 94, 104, 152).

TOLERANCE OF OR SUSCEPTIBILITY TO PRESERVATION METHODS

Temperature, pH, and Salt Tolerance

Most aeromonads have an optimal growth temperature of 28°C; however, they constitute a very heterogeneous group and include strains with a very wide temperature growth range (<5 to 45°C) (127). Many strains can grow at refrigeration temperatures, and *Aeromonas* species present in foods of many types can increase 10- to 1,000-fold during 7 to 10 days of storage at 5°C (17, 29, 171). Psychrotrophic strains, which grow very rapidly, have also been described, including one with a theoretical minimum temperature for growth of −5.3°C that was isolated from goat milk (113, 186). Most strains isolated from clinical specimens can also grow at refrigeration temperatures (51, 114, 171), indicating that temperature alone is ineffective for controlling the growth of *Aeromonas* strains in foods (18, 112).

Most strains of *P. shigelloides* do not grow below 8°C, but at least one strain has been reported to grow at 0°C, and *P. shigelloides* may occur in the aquatic environments of cold climates (137, 152, 153). The optimal temperature for growth of *P. shigelloides*, however, is considered to be 38 to 39°C, with a maximum of around 45°C. About 25% of strains grow at 45°C (56, 152). Temperatures of 42 to 44°C have been recommended for the isolation of *P. shigelloides* from environmental specimens in which the presence of aeromonads and other organisms poses a problem (94).

Strains of *Aeromonas* spp. and *P. shigelloides* can be recovered from foods stored at −20°C for considerable periods (years) and are successfully stored in culture collections at −70°C. However, *Aeromonas* spp. are fairly sensitive to low pH (<5.5). Studies have revealed that *Aeromonas* spp. are unlikely to grow in foods with more than 3 to 3.5% (wt/wt) NaCl and pH values of below 6.0 when foods are stored at low temperature (112). Acetic, lactic, tartaric, citric, sulfuric, and hydrochloric acids, in that order, are effective at restricting growth (2), and polyphosphates can also control the growth of aeromonads in certain foods (170). *P. shigelloides* grows at pH 5 to 8 but may be especially susceptible to low pH. Cells are killed rapidly at pH 4 (22, 152). Tolerance to salt is similar to that of *Aeromonas* spp. (153).

Atmosphere, Heat, Irradiation, Disinfectants, and Chlorine

Overall, aeromonads grow as well anaerobically as they do aerobically, but they are more sensitive to $NaNO_2$ under anaerobic than aerobic conditions. *Aeromonas* growth under modified atmospheres depends on the nature and number of competing microflora, and the use of modified atmospheres to extend the shelf lives of packaged meats and fresh vegetables may enable aeromonads to grow to high populations (17, 18). The type of vegetable has more influence on the *Aeromonas* growth rate than the type of atmosphere present, with more rapid growth occurring on shredded endive and iceberg lettuce than on brussels sprouts or grated carrots (98). *Aeromonas* spp. contribute to the spoilage of many foods, but in other foods, such as milk, they can reach large populations (up to 10^8 CFU) without detectable organoleptic changes (119). *P. shigelloides*, on the other hand, is extremely susceptible to modified-atmosphere storage (80% CO_2, no O_2), with no growth under such conditions in cooked crayfish tails held at 11 and 14°C (95).

Aeromonas spp. are readily killed by either heat or irradiation (2, 163, 172, 185); however, some *Aeromonas* toxins are reportedly heat stable (56°C for up to 20 min to 100°C for up to 30 min) (36, 40). *Aeromonas* strains are susceptible to disinfectants, including chlorine, but recovery of aeromonads from chlorinated water supplies is commonly reported, possibly due to posttreatment recontamination, inactivation of chlorine by organic matter, the presence of very high initial numbers of aeromonads, or the persistence of the organism in a "viable but nonculturable" state following treatment (2, 112). A recently identified concern is that activated carbon filters, commonly used on household faucets to improve the quality and taste of tap water, may amplify the growth of *Aeromonas* spp. by promoting biofilm formation (35).

More research is needed to determine the effects of the factors discussed above on plesiomonads, including the ability of these organisms to survive and grow under storage conditions commonly used for many foods, especially those of aquatic origin. Studies to date indicate that adequately cooked foods will not contain viable *P. shigelloides* and proper refrigeration should control its growth in stored foods (153).

RESERVOIRS

Members of both genera are primarily aquatic organisms. The motile, mesophilic aeromonads are found in salt, fresh, stagnant, estuarine, and brackish water worldwide. They are also commonly present in drinking water, sinks, drain pipes, and household effluents, as well as in sewage-contaminated waters (14, 56). A significant correlation between organic matter content and total numbers of aeromonads in waters has been reported previously (13), and the distribution of species and the proportion of strains producing toxins can vary in different geographic regions. Clinical strains reflect differences in the distributions of environmental strains expressing virulence-associated properties (117, 118).

Motile aeromonads may associate with or colonize many water-dwelling plants and animals (e.g., healthy fish, leeches, and frogs). They are recognized as the causal agent of red-leg disease in amphibians and are responsible for diseases in reptiles, fish, shellfish, and snails. In many fish species, they cause hemorrhagic septicemia (106). They can be found as minor components of fecal flora of a wide variety of animals, including domestic animals used for food (pigs, cows, sheep, and poultry). They have been implicated in outbreaks of bovine abortion and diarrhea in piglets, as well as other infections in a variety of animals (77, 182). More recently, *Aeromonas* spp. were isolated from houseflies, mosquitoes, ticks, and even tree bark, a testament to the ubiquity and adaptive nature of these bacteria (4, 61, 156, 159, 160, 179, 180, 206–208, 214).

Aeromonads are not generally considered to be normal inhabitants of the gastrointestinal tract of humans. However, the fecal carriage rate can approach 3% in asymptomatic persons in temperate climates (109, 154) and 30% or more in the tropics or developing regions (109, 181). Recently, *Aeromonas* (*A. simiae*) was also isolated from the feces of nonhuman primates (*Macaca fascicularis*) (82). Aeromonads are widespread in foods and are readily isolated from meat, raw milk, poultry, fish, shellfish, and vegetables. They have also been detected in prepared foods, such as pasteurized milk, though less frequently (59, 112, 119). There are geographic differences in the species isolated from water, and some species are more highly associated with certain types of foods (29, 70, 78, 113, 162). *P. shigelloides* is found in fresh and estuarine waters, as well as seawater in warm weather (8–10, 56, 142). Its limited temperature range for growth influences the more frequent occurrence of *P. shigelloides* in tropical and subtropical climates than in other climates and also accounts for the seasonal variation (summer incidence) that occurs in its isolation from river water in temperate climates. However, *P. shigelloides* has been isolated from aquatic environments in cold climates, such as Northern Europe (128). It also has been isolated from warm- and cold-blooded animals, including dogs, cats, cattle, pigs, snakes, shellfish, and tropical fish, as well as healthy humans, particularly in developing countries (e.g., 24%, or 12 of 51 adults in Thailand) (12, 22). Few systematic studies of the prevalence of *P. shigelloides* in foods have been conducted; to date, *P. shigelloides* has been isolated predominantly from fish and seafood (56, 153).

CHARACTERISTICS OF DISEASE

There is a species-associated disease spectrum for *Aeromonas*. *A. veronii* biovar sobria (HG8/10) is more frequently associated with bacteremia than the other species. *A. schubertii* (HG12) is associated with aquatic wound infections but as yet has not been documented in association with gastroenteritis (30, 32). *A. veronii* biovar sobria also occurs in wound infections and in infections following the use of medicinal leeches (75, 204). *A. veronii* biovar sobria (HG8/10), *A. hydrophila* (HG1 and HG3), *A. caviae* (HG4), and to a lesser extent *A. veronii* biovar veronii (HG8/10), *A. trota* (HG14), and *A. jandaei* (HG9) are the species that have been associated with gastroenteritis (30, 31, 100, 106). *A. veronii* biovar sobria has been associated with the dysenteric presentation more frequently than the other species,

although studies of invasive *Aeromonas* gastroenteritis are few (130, 216). *A. caviae* (HG4) is most common in pediatric diarrhea (158). *A. media* (HG5) is also occasionally isolated from human feces, but its significance is not known (30, 100).

Gastroenteritis

Aeromonas- and *Plesiomonas*-associated cases of gastroenteritis are distinctly seasonal (sharp summer peak). Clinical manifestations of disease vary considerably for *Aeromonas*-associated gastroenteritis. Mild cases are self-limiting with watery diarrhea, and more severe cases can result in a *Shigella*-like dysenteric form, with blood and mucus in the stools, inflammatory exudate, fever, and abdominal pain (101). Although less common, chronic diarrheal episodes caused by aeromonads have also been described previously. In most of these cases, the episode lasted 7 to 10 days; however, a diarrheal episode exceeding 1 year's duration due to *A. caviae* or *A. hydrophila* infection has also been reported (99, 101). Acute, self-limited diarrhea is particularly problematic for children, and *Aeromonas* spp. have been detected worldwide as the only pathogen from as many as 20% of children suffering from diarrhea and from fewer than 2% of children without diarrhea (25, 39, 48, 50, 74, 93, 140, 165, 213). Older patients, on the other hand, are at higher risk for presenting with *Aeromonas*-mediated chronic enterocolitis (150). Aeromonads are highly prevalent enteric pathogens, and in recent years the incidence of gastroenteritis due to *Aeromonas* spp. has increased significantly (73). For the first half of 1991, for instance, the University of Iowa Hygienic Laboratory reported an incredible 224 cases of *Aeromonas* gastroenteritis (184). Additionally, *Aeromonas* spp. have emerged in some countries such as Japan and Finland as an important cause of traveler's diarrhea, though the severity of illness is typically milder than that of illness caused by *Escherichia coli* (79, 220).

Patients with stools positive for *P. shigelloides* also have either watery diarrhea or diarrhea with blood and mucus. The secretory form is usually reported to last from 1 to 7 days but can be prolonged (~3 weeks), with many (up to 30) bowel movements per day at the peak of the disease (22, 152). Holmberg et al. (91) determined that most patients (adults) from whom *P. shigelloides* was isolated had symptoms of invasive disease (bloody, mucus-containing stools with polymorphonuclear leukocytes). Patients often had severe abdominal cramps, vomiting, and some degree of dehydration. *P. shigelloides* has also been associated with traveler's diarrhea in the West (126) and especially in the Orient (136, 202).

The lack of suitable animal models which reproduce the features of *Aeromonas*- and *Plesiomonas*-associated diarrhea has made it difficult to establish definitively the enteropathogenicities of these bacteria. Additionally, the role of *Aeromonas* as an enteric pathogen was a subject of controversy, largely because of a 1985 report indicating that oral feeding with *Aeromonas* failed to induce diarrhea in human volunteers (155). However, it has become increasingly clear that not all *Aeromonas* strains cause gastroenteritis (129), and *Aeromonas* may have failed to cause diarrhea in human volunteers because strains of questionable suitability were used in the study or were cultivated on artificial media for an extended period of time, thereby losing virulence traits (155). Most of these strains were either nonenterotoxigenic in animal models or obtained from wound specimans and healthy individuals and, as shown by the lack of demonstrable fecal shedding by volunteers, apparently were without the necessary attributes to survive the gastrointestinal tract. However, recent animal model and case control human studies provided conclusive evidence of *Aeromonas*-associated gastroenteritis (6, 198). More human volunteer studies using highly virulent strains are risky because of increased recognition of the ability of *Aeromonas* to cause systemic infections, which are often fatal.

In strong support of the enteropathogenic potential of *Aeromonas* spp., a specific intestinal secretory immunoglobulin A (sIgA) response directed against the exoproteins in patients with naturally acquired *Aeromonas* diarrhea was reported previously (46). However, no sIgA titer increases were detected in 14 patients shedding *P. shigelloides*. Thus, although this study provided further evidence for the significance of *Aeromonas* species as pathogens in acute diarrhea, it raised questions about the role of *P. shigelloides*, at least in adults with traveler's diarrhea (105).

Extraintestinal Infections

Although extraintestinal infections with *Aeromonas* spp. and *P. shigelloides* are uncommon, they tend to be severe and often fatal, particularly in patients with sepsis and meningitis. *Aeromonas* extraintestinal diseases (peritonitis, endocarditis, pneumonia, conjunctivitis, and urinary tract infections) are more common and varied than those caused by *P. shigelloides* (cellulitis, arthritis, endophthalmitis, and cholecystitis) (22, 58, 109, 177). The gastrointestinal tract may be the source of some disseminating infections, although some may originate from infected wounds and trauma. *Aeromonas* spp. are well documented as causative agents of wound infections, usually linked to water-associated injuries or aquatic recreational activities (72).

PATHOGENICITY AND VIRULENCE FACTORS OF *AEROMONAS* SPECIES

The varied clinical picture of *Aeromonas* infections, and gastroenteric illness in particular, suggests that complex pathogenic mechanisms occur in aeromonads. For the different gastroenteric disease manifestations, it may be that strains possess multiple virulence factors in different combinations which may interact with different levels of the intestine. Depending on the species and strain, *Aeromonas* spp. produce an impressive array of virulence factors, including hemolysins, cytotonic and cytotoxic enterotoxins, proteases, lipases, leucocidins, endotoxin, adhesins, and the ability to form an S layer (66). Additionally, several new virulence factors have been recently described for aeromonads, including a type III secretion system (T3SS) and the glycolytic enzyme enolase (27, 197, 200, 221). Host responses to *Aeromonas* spp. and their various putative virulence factors also have been examined in more depth, the results of which have shed new light on the mechanism of *Aeromonas*-associated infections.

Aeromonas Surface Molecules

Many strains of *Aeromonas* possess a regularly arrayed surface layer (S layer) tethered to the bacterial cell surface via lipopolysaccharide (LPS) that allows bacteria to resist host defenses (108). Merino et al. (148) reported that *Aeromonas* spp. belonging to serogroup O:11 with an S layer resisted complement-mediated killing by impeding complement activation. However, serum resistance of *Aeromonas* strains lacking the S layer was due to their inability to form C5b or C5b-9. In subsequent studies, Merino et al. (147) determined that a 39-kDa outer membrane protein, which is bound to C1q, is not accessible in *Aeromonas* strains possessing an O-antigen and imparts serum resistance to the bacteria.

The O-antigen polysaccharide, which is composed of repeating oligosaccharide units, is covalently attached to the lipid A-core complex of the LPS and extends outward from the cell surface. The genus *Aeromonas* has been classified into 96 serogroups, and the O-antigen LPS has been shown to play an important role in adhesion to and colonization of host cells (149, 222). Some serogroups of *A. hydrophila* (O:11 and O:34) also possess capsular polysaccharide (222), which may confer serum resistance (5) and has also been shown to play a role in adherence to and invasion of host cells (144, 145).

Porin II was recently reported to be an important surface molecule involved in serum susceptibility and C1q binding in *Aeromonas* (164). Porin loss by different mechanisms may lead to serum resistance of *Aeromonas*, with a potential for the organism to develop antibiotic resistance, resulting in serious clinical problems.

Extracellular Enzymes

Most aeromonads elaborate a variety of extracellular enzymes: proteases, DNase, RNase, elastase, lecithinase, amylase, lipases, gelatinase, and chitinase. Several of these enzymes have now been cloned, as have two *exe* operons required for extracellular secretion of proteins by *Aeromonas* species (11, 38, 71, 143, 189). The roles of these enzymes in virulence have yet to be determined, but there is much evidence that correlates their production with increased pathogenicity.

Aeromonas spp. produce a wide range of proteases, which cause tissue damage and aid in establishing an infection by overcoming host defenses and by providing nutrients for cell proliferation (178). At least three types of proteases have been identified, which include heat-labile serine protease and heat-stable and EDTA-sensitive or -insensitive metalloproteases. Recently, additional proteases were described for *A. veronii* biovar sobria and *A. caviae* that may directly and indirectly play a role in bacterial virulence (157, 205). In addition, some aminopeptidases may have a specific function such as activation of cytotoxic enterotoxin (Act) and aerolysin. *Aeromonas* spp. produce glycerophospholipid:cholesterol acyltransferase (GCAT), which functions as a lipase or phospholipase and may cause erythrocyte lysis by digesting the plasma membranes of erythrocytes (178). Although a role of GCAT in the fish disease furunculosis has been suggested, the roles of GCAT and of proteases as virulence factors in humans are presently undefined. Lipases may be important for bacterial nutrition. They may also constitute virulence factors by interacting with human leukocytes or by affecting several immune system functions through free fatty acids generated by lipolytic activity.

Siderophores

Efficient mechanisms for iron acquisition from the host during an infection are considered essential for virulence. Siderophores are low-molecular-weight compounds with high affinities for various forms of iron. Mesophilic aeromonads produce either of two siderophores, enterobactin and amonabactin (28, 141). Enterobactin is produced by various gram-negative bacteria; amonabactin has so far been detected only in *Aeromonas* spp. Two biologically active forms of the latter siderophore exist. The type of siderophore produced by aeromonads varies with hybridization group. Amonabactin (*amo*A) and enterobactin (*aeb*C) biosynthetic genes have been cloned, which will facilitate

determination of the role(s) of these siderophores in *Aeromonas* virulence (15, 28, 141).

Cytotoxic and Cytolytic Toxins

A. hydrophila Cytotoxic Enterotoxin (Act)

Act was originally purified from a diarrheal isolate, SSU, of *A. hydrophila* (40–42) and was subsequently determined to possess a variety of biological activities, including hemolysis, cytotoxicity, and enterotoxicity, and to cause lethality in mice (57). In addition to the roles of Act as a cytotoxic enterotoxin and a factor increasing the lethality of *A. hydrophila* (HG1) for mice, studies from Chopra's group determined that Act inhibits the phagocytic ability of mouse phagocytes both in vitro and in vivo and stimulates the chemotactic activity of human leukocytes (57). Act also induces the production of proinflammatory cytokines, such as tumor necrosis factor alpha (TNF-α), interleukin-1 (IL-1), IL-6, macrophage inflammatory protein, and RANTES, in a macrophage cell line and IL-8 in colonic epithelial cells. These activities may contribute to the pathology observed in the small intestine following *Aeromonas* infection (45, 67, 188). Act also induces the production of prostaglandins and reactive oxygen species (45).

Subsequent studies by Galindo et al. on the effects of purified Act on macrophages and colonic epithelial cells revealed that Act induces extensive host cell signaling, including the influx and release of calcium from intracellular stores, activation of transcription factors (NF-κB, CREB, AP-1, and C/EBP-β), activation of all three major Mitogen-activated protein kinase pathways (p38, Jun N-terminal protein kinase, and extracellular signal-related kinase 1/2), transcription of multiple host genes, and phosphorylation of proteins involved in immune responses and inflammation (62–64, 67). Of particular note, Act was recently determined to cause classical, caspase-dependent apoptosis of murine macrophages and human colonic epithelial cells, which may contribute significantly to the pathology of *A. hydrophila* infections in humans (63, 67). Although Act causes the release of cytochrome *c* and apoptosis-inducing factor from mitochondria, experiments using bone marrow-derived murine macrophages in which one or both of the receptors for TNF-α (TNF receptor 1 [TNFR-1] and TNFR-2) were mutated (or deleted) revealed that Act-induced apoptosis requires the presence of TNFR-1. Act-induced apoptosis of host cells may explain the extensive tissue damage caused by aeromonads in vivo, although a direct correlation between apoptosis and human disease manifestation requires further study.

While it is clear that Act induces extensive host cell signaling, it is unknown how exactly the toxin exerts these effects. Act is a pore-forming toxin and binds cholesterol, but it is unclear whether the toxin mediates its effects via punching holes in host cell membranes, binding to cholesterol or a host cell target protein, or being internalized and interacting with cytosolic host proteins. If a host cell receptor for Act exists, it has yet to be discovered. However, a recent study indicated that Act could bind to nine different human proteins, based on protein microarrays. Two of these proteins (SNAP23 and galectin-3) were determined to be important for Act-induced apoptosis based on small interfering RNA transfection experiments, which is an important step toward delineating the mechanism of action of this toxin on host cells (65). However, additional studies are needed to determine if Act-associated apoptosis of host cells is directly caused by the interaction of Act with galectin-3 and SNAP23 or involves additional mechanisms. For instance, Act may bind to other host membrane proteins, potentially via association of the toxin with cholesterol-rich lipid rafts, and directly induce signal cascades. Alternatively, Act-induced alteration in host intracellular signaling may be the result of a rapid influx of calcium subsequent to the production of holes in host cell plasma membranes or via internalization of Act and interaction with host cytosolic factors.

Act is itself regulated by a variety of factors and is maximally expressed at 37°C and pH 7.0 (199). While the presence of Ca^{2+} increases *act* promoter activity, glucose and iron down regulate the promoter activity of the *act* gene. Sha et al. also determined that Act production is regulated by glucose-inhibited division gene A (*gidA*) and an iron-regulated ferric uptake regulatory (*fur*) gene (199). Increased expression of virulence factors, like Act, is a common pathogenic mechanism utilized by a variety of bacteria. Other *Aeromonas* iron-regulated genes include those that code for colicin receptor, heme receptor, and a ferric siderophore receptor, which are more highly expressed in vivo than in vitro (49).

Aerolysins

Aerolysin proteins are channel-forming proteins which produce a transmembrane channel that destroys cells by breaking their permeability barriers (37, 176, 212, 217, 218). After the toxin is activated by proteolysis of about 25 to 48 amino acids from the carboxy-terminal end, it binds to the receptor glycophorin on eukaryotic cells (69), oligermerizes, and then inserts itself into the host cell membrane (23, 37, 175, 217, 218). The resulting pores are ~1.5 nm in diameter and exhibit weak anion selectivity. This may explain the enterotoxic properties of the protein at low concentrations (37, 212). However, it also has been suggested that aerolysin does not cause cell death simply by allowing leakage of intracellular components.

One study has determined that channel formation leads to selective permeabilization of the target cell to small ions, such as potassium, which causes depolarization of the plasma membrane. This in turn results in vacuolation of the endoplasmic reticulum and selective alteration of the early biosynthetic membrane pathway by inhibiting transport of newly synthesized membrane proteins to the plasma membrane (3). More recently, aerolysin was shown to cause apoptosis of T lymphomas, which was attributed to channel formation. Like Act, aerolysin is cytotoxic, hemolytic, and lethal; however, few studies have been conducted with respect to the potential effects of this toxin on host cell signaling.

Hemolysins

Aeromonads produce at least two major classes of hemolysins, alpha and beta. Act- and aerolysin-like molecules (beta-hemolysins) are enterotoxigenic cytolysins and have been cloned from several *A. hydrophila* (HG1, HG2, and HG3) strains (86, 219). These toxins, termed hemolysin (HlyA), have been proposed to contribute to the virulence of aerolysin-positive strains and to account for some of the species-associated differences in *Aeromonas* virulence mechanisms (85, 219). The second class of hemolysins, alpha-hemolysin, is elaborated during the late stationary phase of growth and causes incomplete lysis of erythrocytes (double-zone lysis on blood agar). It is not expressed at temperatures above 30°C and has not been associated with enterotoxic properties. Studies conducted by Fujii et al. (59a) revealed that T84 cells stimulated with aerolysin-like hemolysin from *A. sobria* produce cyclic AMP in the medium, and that may lead to fluid secretion, possibly by channel insertion into the apical membrane and by activation of protein kinase C.

Cytotonic Enterotoxins

A variety of cytotonic enterotoxins have been described previously, and several have now been cloned. They produce increased levels of cyclic AMP and cause elongatioin of Chinese hamster ovary (CHO) cells. They are thought to act similarly to cholera toxin (CT) despite their differing molecular properties and various reactivities with cholera antitoxin. Several non-CT-reactive cytotonic enterotoxins have been described for *Aeromonas* spp., including a 15- to 20-kDa heat-stable protein (132) and a 44-kDa heat-labile protein (40, 41). Other cytotonic enterotoxins purified from various *Aeromonas* strains have cross-reactivity with CT (36, 158, 183, 195). Chopra et al. cloned the gene encoding Alt and subsequently purified the recombinant protein from *E. coli*. The recombinant Alt was slightly smaller (35 to 38 kDa) than the native Alt (44 kDa and above) purified from

A. hydrophila (HG1) (40, 44). At the amino acid level, Alt exhibited significant homology with lipase and phospholipase of *A. hydrophila* (HG1, HG2, and HG3), although it did not exhibit phospholipase activity (43, 143). This toxin increases Ca^{2+} levels in CHO cells through host cell phospholipase C activation (7). Further, active immunization of mice with Alt reduced the fluid secretory capacity of *A. hydrophila* (HG1) in a mouse ligated ileal loop model (43). In a subsequent study, the *alt* gene was detected by PCR in 75% (23 of 31) of food and water isolates, including the *A. hydrophila* (HG1, HG2, and HG3) strain implicated in an outbreak of foodborne illness following ingestion of raw, fermented fish (76).

A second cytotonic enterotoxin gene (designated *ast*) was also identified in the Alt⁺ *A. hydrophila* strain, and a 4.8-kb SalI-BamHI DNA fragment encoding this heat-stable toxin hybridized to a 3.5-kb BamHI DNA fragment of a plasmid that contained the *A. trota* heat-stable cytotonic enterotoxin gene (44). This toxin also elevated cyclic AMP levels in cells (7). *Aeromonas* spp., therefore, may produce different types of cytotonic enterotoxins that are functionally similar. However, the mechanism(s) involved in elevating second messengers (e.g., cyclic AMP) may be quite different. Cytotonic enterotoxins may be produced only by a minority of strains. Seidler et al. reported production of these enterotoxins in <6% (20 of 330) of strains (196), and Shimada et al. determined that a CT-like enterotoxin was produced by *Aeromonas* spp. in ~4.5% (8 of 179) of strains tested (203).

Recently (6), investigators compared the distributions of *ast* and *alt* toxin genes in over 250 aeromonads isolated from 1,735 children with diarrhea, 830 control children, and 120 randomly selected environmental water samples. Results revealed that the number of isolates positive for both the *alt* and *ast* genes was significantly higher for those obtained from diarrheal children than for those obtained from control children or the environment. Isolates that were positive for only the *ast* gene were more frequently found in environmental isolates, and isolates positive for only *alt* were not correlated with any single isolate type. These results suggest that the products of both the *alt* and *ast* genes may act synergistically to induce severe diarrhea.

Shiga-Like Toxins

Aeromonas infection has been associated with hemolytic-uremic syndrome (19, 55, 101, 190). Hemolytic-uremic syndrome is caused by Shiga toxin (Stx)-producing bacteria. Hence, it is possible that some strains of *Aeromonas* have acquired the ability to produce Shiga-like toxins. One study revealed Stx1 production in 3 of 28 (11%) clinical *Aeromonas* isolates and 1 of 11 (9%) environmental

isolates. Stx-producing bacteria were idenified on the basis of neutralization of cytotoxicity and PCR amplification of isolated DNA by using oligonucleotide primers from the *stx1* gene of *E. coli* O157:H7. The gene encoding Stx1 was found on a small 2.14-kb plasmid (81). Additional studies are needed to confirm these findings.

Invasins

The Sereny test for tissue invasiveness has been universally negative for aeromonads (56). Only a few studies have examined the ability of strains to penetrate epithelial cell lines, such as HEp-2 cells of epipharyngeal origin and the intestinal Caco-2 cells (130, 161, 216). Invasive ability has been reported most commonly for strains of *A. sobria* (*A. veronii* biovar sobria), followed by *A. hydrophila* (HG1, HG2, and HG3), particularly strains isolated from patients with symptoms of dysentery (130, 216). Invasive strains of *A. caviae* also have been described previously, however (201). The possible link between tissue culture cell invasiveness and dysentery needs to be confirmed with larger numbers of patients. The mechanisms of *Aeromonas* cell line adhesion and invasion remain to be elucidated. Invasin determinants are probably chromosomally located (130). Nishikawa et al. determined that the DNA of four invasive *Aeromonas* strains did not hybridize with probes to the genes associated with the attaching and effacing (*eae*) and invasion (*ipaB*) abilities of *E. coli* (161).

Filamentous Adhesins

Adhesion is likely an essential virulence factor for aeromonads which infect through mucosal surfaces or cause gastroenteric disease (110). Because receptors in the host are usually carbohydrate moities which may also be expressed on erythrocytes, hemagglutination has been used as a screening model to detect adhesins. The hemagglutinins of *Aeromonas* species are numerous, with some associated with filamentous appendages (fimbriae or pili) and others with outer membrane proteins. Some hemagglutination patterns (e.g., not inhibited by fucose, galactose, or mannose) correlate with diarrhea, but hemagglutination screening has not proved useful for the identification of potentially significant strains (24).

Aeromonas filamentous surface structures are diverse. Their number and type vary depending on the source of the isolate and bacterial growth conditions (87, 110, 120). Environmental strains of *A. veronii* biovar sobria are heavily piliated when isolated (120). Short, rigid pili are the predominant type expressed on heavily piliated aeromonads. These pili cause autoaggregation of bacteria but are not hemagglutinating and do not bind to intestinal cells (92). The role of the short, rigid pili in colonization is unknown. In contrast, many *Aeromonas* isolates (in particular strains of *A. veronii* biovar sobria and *A. caviae*) from diarrheal stools are poorly piliated (<10 per cell).

Long, thin, flexible pili have been purified from all *Aeromonas* species associated with diarrheal infection (88–90, 97, 121, 123). They form a family of type IV pili whose structural subunits share N-terminal amino acid homology and have molecular masses of 19 to 23 kDa. They have been designated bundle-forming pili (Bfp) because the bundles of pili that link cells are of this pilus type (124). There is substantial evidence that the Bfp are intestinal colonization factors (88–90, 97, 118, 121). Bfp also may promote colonization by forming bacterium-to-bacterium linkages (121, 124). By contrast, a second family of type IV pili, designated Tap pili, do not appear to be as significant as Bfp for *Aeromonas* intestinal colonization. However, their widespread conservation suggests that they do play an important role in the biology of the bacterium (60, 115).

Lateral flagella may facilitate intestinal colonization and biofilm formation at the intestinal mucosal surface by mediating swarming motility (83). They may also form bacterium-to-bacterium linkages to facilitate these processes (116, 125) as has been described previously for *Vibrio parahaemolyticus* (16).

Quorum Sensing

Quorum sensing, which is a mechanism for controlling gene expression in response to an expanding bacterial population, has been reported for *Aeromonas* spp. and is a subject of intensive investigation of many gram-negative bacteria (209). The quorum-sensing signal molecule belongs to the *N*-acylhomoserine lactone (AHL) family, and the signal generator proteins responsible for the synthesis of AHLs belong to the LuxI family. Accumulation of this molecule above a threshold concentration provides an indication that the minimum bacterial population size has been reached and that the appropriate target gene(s) should be activated via the LuxR family of transcriptional activators. The LuxR protein consists of two domains with an AHL binding site within the N-terminal end and a helix-turn-helix DNA binding motif within the C-terminal domain. It is plausible that the expression of various virulence factors of *Aeromonas* may be controlled by quorum sensing. The role of an AHL-dependent, quorum-sensing system, based on the LuxRI homolog AhyRI in *A. hydrophila*, has been reported previously (209).

The major signal molecule synthesized by the *ahyI* locus in *A. hydrophila* was *N*-(butanoyl)-L-homoserine lactone (BHL), also referred to as C4-HSL, with AHL synthesized in relatively smaller amounts (210). Downstream of the *ahyI* locus is a gene with homology to the *iciA* gene, an inhibitor of chromosome replication in *E. coli*, suggesting that in *Aeromonas* cell division may be linked to quorum sensing. Further, both AhyRI and BHL are required for the transcription of *ahyI*. In other bacteria, such as *Pseudomonas aeruginosa*, BHL is involved in the regulation of the secretion of multiple exoproducts, including elastase, hemolysin, chitinase, alkaline protease, cyanide, lectins, staphylolytic activity, pyocyanin, and the alternative stationary-phase sigma factor RpoS (210).

A. hydrophila produces both a serine protease and a metalloprotease, and there is evidence to suggest that their production may be regulated by quorum sensing (210, 215). The presence of the quorum-sensing autoinducer C4-HSL in *A. hydrophila* biofilm development has also been reported previously (210), and an *ahyI* mutant that could not produce C4-HSL failed to form mature biofilms (135). Bacteria in biofilms are more resistant to host defenses and antimicrobial agents and may express more virulent phenotypes as a result of gene activation through bacterial communication (quorum sensing) or gene transfer (209). More in-depth studies are needed to definitively establish the role of quorum sensing in *Aeromonas*-associated infections.

Type III Secretion System

A crucial fish virulence factor, *A. salmonicida* toxin (AexT), which is a homolog of *Pseudomonas aeruginosa* exoenzyme T/S (ExoT/S), produces cytotoxic effects on gonad cells of rainbow trout (26). ExoT/S is secreted by a T3SS, and secretion of AexT occurs only after contact with fish cells, which led to the discovery of a T3SS in *Aeromonas*. A total of 19 open reading frames were identified in *A. salmonicida* that code for the T3SS and exhibit homology with the T3SS of *Yersinia* species (27). Mutation in the *A. salmonicida ascV* gene, a homolog of the *yscV* gene in yersiniae and the product of which is a highly conserved inner membrane protein found in every known T3SS, resulted in no toxic effect on rainbow trout gonad cells. Likewise, mutation in the *aopB* and *acrV* genes of *A. salmonicida*, which are homologs of *Yersinia yopB* and *lcrV* genes, prevented translocation of AexT into host cells and hence resulted in a lack of cytotoxicity (27). In a fish isolate of *A. hydrophila*, 25 open reading frames coding for another T3SS were identified (221). Mutation in *aopB* and *aopD* (homolog of *yopD* in yersiniae) resulted in decreased cytotoxicity in carp epithelial cells and also prevented mortality in fish, possibly due to an increase in phagocytosis (221).

Sha et al. (200) identified and characterized a T3SS in a clinical isolate of *A. hydrophila* SSU, which suggested that a T3SS may be an important virulence factor of *A.*

hydrophila in humans. By using probes specific to the *ascF* and *ascG* genes of the T3SS, 84 clinical isolates of *Aeromonas* were examined for the presence of the T3SS. Fifty percent of the isolates possessed the hybridizing sequences, with a higher prevalence in *A. hydrophila* and *A. veronii* than in *A. caviae* (34). Mutation of a component of the *A. hydrophila* T3SS needle structure (*aopB*) resulted in less cytotoxicity to macrophages and colonic epithelial cells. By using single- and double-knockout mutants of *aopB* and *act* toxin, it was determined that the T3SS, but not Act, likely plays an important role in host cell cytotoxicity during early (within 2 h) bacterial-host cell interactions and that during the later stages of infection, Act contributes more significantly to *Aeromonas*-associated cell cytotoxicity than does the T3SS. More interestingly, deletion of the two virulence genes reduces production of lactones compared to that in wild-type bacteria, which suggests a positive correlation among the presence of the T3SS, Act toxin, and quorum sensing (200).

Other Virulence Factors

Recent studies mapped opsonophagocytosis resistance to *ftsE* and *ftsX* genes in *A. hydrophila* (146). FtsE and FtsX form a complex in the inner membrane that bears the characteristics of an ABC-type transporter involved in cell division. It is believed that an *A. hydrophila ftsE* mutation renders a filamentous phenotype at 37°C, which may interfere with opsonophagocytosis.

Another potential virulence factor of *A. hydrophila* is the glycolytic enzyme enolase. By using a murine peritoneal culture model, Sha et al. (197) identified via restriction fragment differential display PCR five genes of *A. hydrophila*, one of which was the enolase gene, that were differentially expressed under in vivo versus in vitro growth conditions. In addition to its role in glycolysis, enolase is expressed on the surface of activated immune cells, which contributes to intracellular matrix digestion and subsequent cell migration via plasminogen activation (173, 187). Enolase is also important in the pathogenesis of *Streptococcus pyogenes* (174). Secretion and surface expression of enolase in *A. hydrophila* and binding of enolase to human plasminogen have also been determined previously (197). It is possible that the increased expression of the enolase gene in vivo contributes to the pathogenesis of *A. hydrophila* infections, although additional studies are needed to definitively determine the role of enolase in bacterial virulence and human disease.

Superoxide dismutases (SODs) have also been implicated as potential *Aeromonas* virulence factors (47, 131). SODs are responsible for detoxifying superoxide anions generated inside phagocytic cells, thus allowing bacteria to survive inside the hostile environment of the host. Genes encoding SodA and SodB in *A. salmonicida* were sequenced (47), and SOD levels were higher in bacteria grown under in vivo conditions than in the organisms cultivated in vitro. Further, SOD levels were lower in avirulent versus virulent cultures of *A. salmonicida*. Studies of the SODs of *A. hydrophila* revealed that Fe-SOD was crucial for bacterial viability, as a mutation in the gene coding for SodB was lethal (131). The susceptibility of an Mn-SOD mutant to hydrogen peroxide was similar to that of the wild-type bacterium, indicating that this SOD was not involved in protection against intracellular superoxide. However, the survival of the Mn-SOD mutant was reduced compared to that of wild-type *A. hydrophila* when bacteria were exposed to hypoxanthine-xanthine oxidase, indicating a role for this enzyme against external superoxides (131).

DNA adenine methyltransferase (DAM) was recently identified as a potential virulence factor of *Aeromonas* spp. DAM methylates DNA at adenine residues in GATC sequences (169), which occurs after DNA replication on the parental strand, generating hemimethylated DNA that is distinct from the rest of the chromosomal DNA. The hemimethylated status of newly synthesized DNA provides a time frame during which cellular processes, such as DNA replication (133, 151, 192) and repair of mismatched bases, as well as alteration of gene expression (138) occur. Additionally, DAM appears to play a crucial role in modulating virulence gene expression in a broad range of bacterial pathogens (54, 84, 107, 134). The *dam* gene in diarrheal isolate SSU of *A. hydrophila* was recently characterized, and it was revealed that its overexpression attenuates bacterial virulence, including Act- and T3SS-associated cytotoxicity, motility, and virulence, in a mouse lethality model (52). More studies are needed to determine if DAM methylation plays a similar role in virulence regulation in other aeromonads.

PATHOGENICITY AND VIRULENCE FACTORS OF *PLESIOMONAS* SPECIES

There are no virulence factors that are widely accepted as being important for *Plesiomonas*-associated infections. No potential virulence factors are tested for routinely (56, 102, 168), and very few have been investigated in any detail.

A glycocalyx, which was detected on the outer surface of *P. shigelloides* by ruthenium red staining and transmission electron microscopy, may be involved in the attachment of plesiomonads to epithelial cells. Endotoxin may play a role in *Plesiomonas* virulence as it does for

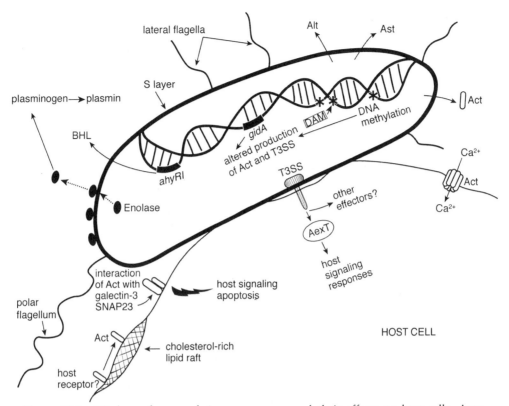

Figure 17.1 Virulence factors of *Aeromonas* spp. and their effects on host cells. Aeromonads produce a single polar flagellum and several lateral flagella which aid in colonization within the host. The S layer (shown as a black line surrounding the bacterial cell) contributes to pathogenicity by allowing the bacteria to evade complement killing. Accumulation of the quorum-sensing molecule, BHL, which is synthesized by the *ahyRI* locus, can lead to coordinated expression of virulence genes. Enolase is expressed on the bacterial cell surface and is also secreted, after which it binds to and activates plasminogen, leading to a fibrinolytic cascade and potential dissemination of the bacteria in the host. Other secreted virulence factors include the cytotonic toxins Alt (heat labile) and Ast (heat stable), which cause intestinal fluid secretion, and Act (cytotoxic enterotoxin), which exerts multiple effects on host cells. Act and the related aerolysin toxins (latter not shown) heptamerize and punch holes in host cell membranes, which leads to an influx of calcium. Act specifically can bind to galectin-3 and SNAP23, host cell molecules that are required for Act-induced host cell apoptosis. Additionally, Act binds to cholesterol, which may help aggregate the toxin on the cell surface to become internalized or bring it into contact with one or more host surface receptors through which it may signal. The newly described *Aeromonas* T3SS can directly inject pathogenicity factors, such as the AexT toxin, into host cells. The production of Act and components of the T3SS are regulated by glucose-inhibited division gene A (*gidA*) and DAM, which methylates specific bacterial DNA sequences. Only some of the crucial virulence factors produced by *Aeromonas* spp. are shown. Figure drawn by Erin Boyle.

other pathogens (22). Plesiomonads do produce extracellular enzymes, such as elastase, which may be involved in connective tissue degradation. However, whether strains produce toxins or not has been a controversial issue. Several investigators concluded that plesiomonads do not produce toxins because rabbit ileal loop and suckling mouse assays, cell culture assays (Y₁, Vero, and CHO), and rabbit skin permeability toxin assays all yielded neg-

ative results (22, 56, 152). However, the production of both heat-stable and heat-labile cytotoxic enterotoxins from all strains, irrespective of origin or serotype, has been reported previously (194). Iron regulates production of a factor that causes an elongation of CHO cells similar to that produced by *Vibrio cholerae* enterotoxin (68). More than 90% of 36 *Plesiomonas* strains tested produced a beta-hemolysin, as judged by the results of agar overlay and contact-dependent hemolysis assays. The hemolysin was cell associated and was produced at both 25 and 35°C (102). It may play a role in iron acquisition in vivo or have another, as yet unknown, role in gastrointestinal disease. Hemolytic and elastolytic activities of *P. shigelloides* are enhanced when bacteria are grown in an iron-depleted medium (193).

Culture filtrates from *P. shigelloides* isolated from water have been shown to cause intracellular vacuolation on mammalian cells and produce hemolytic and enterotoxic activities (53). Okawa et al. (167) isolated and purified a heat-stabile cytotoxin from a clinical diarrheal isolate of *P. shigelloides*. This toxin is a large (>600-kDa) complex of three LPS binding proteins and LPS. The protein portion specifically is both cytotoxic and enteropathogenic and therefore may play a role in *P. shigelloides* pathogenicity.

Little is known about intestinal colonization and mucus interactions by plesiomonads. Some isolates from clinical material reportedly adhere to HeLa, HEp-2, and INT 407 cells. However, adhesion is low compared with that of clinical *Aeromonas* isolates (214). Invasiveness in the Sereny test does not occur, but invasion of the cell lines described above occurs. Invasion of HeLa cells by freshly isolated strains from children with acute diarrhea was comparable with that by *Shigella sonnei* (22, 32, 56, 168, 194). More recently, 12 clinical isolates of *P. shigelloides* were shown, by transmission electron microscopy, to adhere to intact microvilli and enter differentiated Caco-2 intestinal epithelial cells (211). However, more studies are required to determine the specific factors involved in adherence to and invasion of host cells and the importance of these processes in *P. shigelloides*-associated gastroenteritis.

SUMMARY AND CONCLUSION

Aeromonas species are commonly isolated from foods and water. Molecular genetic studies have led to considerable advances in the taxonomy of these bacteria in recent years. Good typing schemes for accurately identifying clinical isolates now exist, but there is a need for additional studies to develop schemes that will identify all environmental isolates. *Aeromonas* spp. are the etiologic agents in a variety of extraintestinal infections involving immunocompetent as well as immunocompromised individuals. There is also strong evidence that some strains are enteropathogenic. Disease-causing strains appear to be only a possibly small subset among the diverse strains of *Aeromonas* spp. present in the environment.

There is still much to be learned about *Aeromonas* virulence determinants and how they combine to result in the virulent subsets within each *Aeromonas* species that cause disease. At present, it is not possible to identify the disease-causing strains because of our incomplete understanding of *Aeromonas* virulence mechanisms. Some of the confusion surrounding *Aeromonas* enterotoxins has been resolved, and several new virulence factors have been described. Major virulence factors and their effects on host cells are depicted in Fig. 17.1. As shown in this figure, the production of multiple pathogenic effectors enables some strains of *Aeromonas* to evade host defenses and modulate host cell signaling. The discovery of new virulence traits, such as a T3SS, enolase, and AexT as well as several regulatory genes, have provided new insights into the mechanisms that lead to human disease. It is likely that ongoing molecular genetic investigations will further elucidate the mechanisms of adhesion and invasiveness which will be useful in developing tests that can identify those strains which pose a risk to human health.

The case for *Plesiomonas*'s being an enteric pathogen is less convincing than that for *Aeromonas* spp. Although *Plesiomonas* has been isolated from diarrheal patients and has been incriminated in several large water- and foodborne outbreaks, no definite virulence mechanism has been identified in most strains associated with gastrointestinal infections. The negative results obtained from human volunteer studies and the lack of an intestinal sIgA response in infected patients cast further doubt on the enteropathogenicity of the species. Establishment of the role of *P. shigelloides* in human disease therefore awaits further studies.

Grant support from NIH/NIAID (AI41611) and the American Water Works Association Research Foundation is acknowledged. C.L.G. was a predoctoral fellow with funding from the National Science Foundation.

References

1. Abbott, S. L., L. S. Seli, M. Catino, Jr., M. A. Hartley, and J. M. Janda. 1998. Misidentification of unusual *Aeromonas* species as members of the genus *Vibrio*: a continuing problem. *J. Clin. Microbiol.* **36:**1103–1104.

2. Abeyta, Jr., C., S. A. Palumbo, and G. N. Stelma. 1994. *Aeromonas hydrophila* group, p. 1–27. *In* H. Hui, J. R. Groham, K. D. Murrel, and D. O. Cliver (ed.), *Foodborne Disease Handbook*, vol. 1. Marcel Dekker Inc., New York, N.Y.

3. Abrami, L., M. Fivaz, P. E. Glauser, R. G. Parton, and F. G. van der Goot. 1998. A pore-forming toxin interacts with a GPI-anchored protein and causes vacuolation of the endoplasmic reticulum. *J. Cell Biol.* **140:**525–540.

4. Aguilera-Arreola, M. G., C. Hernandez-Rodriguez, G. Zuniga, M. J. Figueras, and G. Castro-Escarpulli. 2005. *Aeromonas hydrophila* clinical and environmental ecotypes as revealed by genetic diversity and virulence genes. *FEMS Microbiol. Lett.* **242:**231–240.

5. **Aguillar, A., M. M. Marino, M. M. Nogueras, M. Regue, and J. M. Tomas.** 1999. Two genes from the capsule of *Aeromonas hydrophila* (serogroup O:34) confer serum resistance to *Escherichia coli* strains. *Res. Microbiol.* **150:**395–402.

6. **Albert, M. J., M. Ansaruzzaman, K. A. Talukder, A. K. Chopra, I. Kuhn, M. Rahman, A. S. Faruque, M. S. Islam, R. B. Sack, and R. Mollby.** 2000. Prevalence of enterotoxin genes in *Aeromonas* spp. isolated from children with diarrhea, healthy controls, and the environment. *J. Clin. Microbiol.* **38:**3785–3790.

7. **Albert, M. J., and A. K. Chopra.** 2000. Personal communication.

8. **Aldova, E.** 2000. New serovars of *Plesiomonas shigelloides*—1992–1998. *Cent. Eur. J. Public Health* **8:**150–151.

9. **Aldova, E.** 1997. New serovars of *Plesiomonas shigelloides*—1996. *Cent. Eur. J. Public Health* **5:**21–23.

10. **Aldova, E.** 1987. Serotyping of *Plesiomonas shigelloides* strains with our own antigenic scheme. An attempted epidemiological study. *Zentbl. Bakteriol. Mikrobiol. Hyg. A* **265:**253–262.

11. **Anguita, J., L. B. Rodriguez Aparicio, and G. Naharro.** 1993. Purification, gene cloning, amino acid sequence analysis, and expression of an extracellular lipase from an *Aeromonas hydrophila* human isolate. *Appl. Environ. Microbiol.* **59:**2411–2417.

12. **Arai, T., N. Ikejima, T. Itoh, S. Sakai, T. Shimada, and R. Sakazaki.** 1980. A survey of *Plesiomonas shigelloides* from aquatic environments, domestic animals, pets and humans. *J. Hyg.* (London) **84:**203–211.

13. **Araujo, R. M., R. M. Arribas, F. Lucena, and R. Pares.** 1989. Relation between *Aeromonas* and faecal coliforms in fresh waters. *J. Appl. Bacteriol.* **67:**213–217.

14. **Ashbolt, N. J., A. Ball, M. Dorsch, C. Turner, P. Cox, A. Chapman, and S. M. Kirov.** 1995. The identification and human health significance of environmental aeromonads. *Water Sci. Technol.* **31:**263–269.

15. **Barghouthi, S., S. M. Payne, J. E. Arceneaux, and B. R. Byers.** 1991. Cloning, mutagenesis, and nucleotide sequence of a siderophore biosynthetic gene (*amoA*) from *Aeromonas hydrophila*. *J. Bacteriol.* **173:**5121–5128.

16. **Belas, M. R., and R. R. Colwell.** 1982. Scanning electron microscope observation of the swarming phenomenon of *Vibrio parahaemolyticus*. *J. Bacteriol.* **150:**956–959.

17. **Berrang, M. E., R. E. Brackett, and L. R. Beuchat.** 1989. Growth of *Aeromonas hydrophila* on fresh vegetables stored under a controlled atmosphere. *Appl. Environ. Microbiol.* **55:**2167–2171.

18. **Beuchat, L. R.** 1991. Behavior of *Aeromonas* species at refrigeration temperatures. *Int. J. Food Microbiol.* **13:**217–224.

19. **Bogdanovic, R., M. Cobeljic, M. Markovic, V. Nikolic, M. Ognjanovic, L. Sarjanovic, and D. Makic.** 1991. Haemolytic-uraemic syndrome associated with *Aeromonas hydrophila* enterocolitis. *Pediatr. Nephrol.* **5:**293–295.

20. **Boone, D. R., R. W. Castenholz, and G. M. Garrity.** 2001. *Bergey's Manual of Systematic Bacteriology*, 2nd ed., vol. 2. Springer, New York, N.Y.

21. **Borrell, N., M. J. Figueras, and J. Guarro.** 1998. Phenotypic identification of *Aeromonas* genomospecies from clinical and environmental sources. *Can. J. Microbiol.* **44:**103–108.

22. **Brenden, R. A., M. A. Miller, and J. M. Janda.** 1988. Clinical disease spectrum and pathogenic factors associated with *Plesiomonas shigelloides* infections in humans. *Rev. Infect. Dis.* **10:**303–316.

23. **Buckley, J. T.** 1991. Secretion and mechanism of action of the hole-forming toxin aerolysin. *Experientia* **47:**418–419.

24. **Burke, V., M. Cooper, J. Robinson, M. Gracey, M. Lesmana, P. Echeverria, and J. M. Janda.** 1984. Hemagglutination patterns of *Aeromonas* spp. in relation to biotype and source. *J. Clin. Microbiol.* **19:**39–43.

25. **Burke, V., M. Gracey, J. Robinson, D. Peck, J. Beaman, and C. Bundell.** 1983. The microbiology of childhood gastroenteritis: *Aeromonas* species and other infective agents. *J. Infect. Dis.* **148:**68–74.

26. **Burr, S. E., K. Stuber, and J. Frey.** 2003. The ADP-ribosylating toxin, AexT, from *Aeromonas salmonicida* subsp. *salmonicida* is translocated via a type III secretion pathway. *J. Bacteriol.* **185:**6583–6591.

27. **Burr, S. E., T. Wahli, H. Segner, D. Pugovkin, and J. Frey.** 2003. Association of type III secretion genes with virulence of *Aeromonas salmonicida* subsp. salmonicida. *Dis. Aquat. Organ.* **57:**167–171.

28. **Byers, B. R., G. Massad, S. Barghouthi, and J. E. Arceneaux.** 1991. Iron acquisition and virulence in the motile aeromonads: siderophore-dependent and -independent systems. *Experientia* **47:**416–418.

29. **Callister, S. M., and W. A. Agger.** 1987. Enumeration and characterization of *Aeromonas hydrophila* and *Aeromonas caviae* isolated from grocery store produce. *Appl. Environ. Microbiol.* **53:**249–253.

30. **Carnahan, A. M., S. Behram, and S. W. Joseph.** 1991. Aerokey II: a flexible key for identifying clinical *Aeromonas* species. *J. Clin. Microbiol.* **29:**2843–2849.

31. **Carnahan, A. M., T. Chakraborty, G. R. Fanning, D. Verma, A. Ali, J. M. Janda, and S. W. Joseph.** 1991. *Aeromonas trota* sp. nov., an ampicillin-susceptible species isolated from clinical specimens. *J. Clin. Microbiol.* **29:**1206–1210.

32. **Carnahan, A. M., M. A. Marii, G. R. Fanning, M. A. Pass, and S. W. Joseph.** 1989. Characterization of *Aeromonas schubertii* strains recently isolated from traumatic wound infections. *J. Clin. Microbiol.* **27:**1826–1830.

33. **Caselitz, F. H.** 1996. How the *Aeromonas* story started in medical microbiology. *Med. Microbiol. Lett.* **5:**46–54.

34. **Chacon, M. R., L. Soler, E. A. Groisman, J. Guarro, and M. J. Figueras.** 2004. Type III secretion system genes in clinical *Aeromonas* isolates. *J. Clin. Microbiol.* **42:**1285–1287.

35. **Chaidez, C., and C. Gerba.** 2004. Comparison of the microbiologic quality of point-of-use (POU)-treated water and tap water. *Int. J. Environ. Health Res.* **14:**253–260.

36. **Chakraborty, T., M. A. Montenegro, S. C. Sanyal, R. Helmuth, E. Bulling, and K. N. Timmis.** 1984. Cloning of enterotoxin gene from *Aeromonas hydrophila* provides conclusive evidence of production of a cytotonic enterotoxin. *Infect. Immun.* **46:**435–441.

37. Chakraborty, T., A. Schmid, S. Notermans, and R. Benz. 1990. Aerolysin of *Aeromonas sobria*: evidence for formation of ion-permeable channels and comparison with alpha-toxin of *Staphylococcus aureus*. *Infect. Immun.* **58:** 2127–2132.

38. Chang, M. C., S. Y. Chang, S. L. Chen, and S. M. Chuang. 1992. Cloning and expression in *Escherichia coli* of the gene encoding an extracellular deoxyribonuclease (DNase) from *Aeromonas hydrophila*. *Gene* **122:**175–180.

39. Chatterjee, B. D., and K. N. Neogy. 1972. Studies on *Aeromonas* and *Plesiomonas* species isolated from cases of choleraic diarrhoea. *Indian J. Med. Res.* **60:**520–524.

40. Chopra, A. K., and C. W. Houston. 1999. Enterotoxins in *Aeromonas*-associated gastroenteritis. *Microbes Infect.* **1:** 1129–1137.

41. Chopra, A. K., and C. W. Houston. 1989. Purification and partial characterization of a cytotonic enterotoxin produced by *Aeromonas hydrophila*. *Can. J. Microbiol.* **35:**719–727.

42. Chopra, A. K., C. W. Houston, C. T. Genaux, J. D. Dixon, and A. Kurosky. 1986. Evidence for production of an enterotoxin and cholera toxin cross-reactive factor by *Aeromonas hydrophila*. *J. Clin. Microbiol.* **24:**661–664.

43. Chopra, A. K., J. W. Peterson, X. J. Xu, D. H. Coppenhaver, and C. W. Houston. 1996. Molecular and biochemical characterization of a heat-labile cytotonic enterotoxin from *Aeromonas hydrophila*. *Microb. Pathog.* **21:**357–377.

44. Chopra, A. K., R. Pham, and C. W. Houston. 1994. Cloning and expression of putative cytotonic enterotoxin-encoding genes from *Aeromonas hydrophila*. *Gene* **139:**87–91.

45. Chopra, A. K., X. Xu, D. Ribardo, M. Gonzalez, K. Kuhl, J. W. Peterson, and C. W. Houston. 2000. The cytotoxic enterotoxin of *Aeromonas hydrophila* induces proinflammatory cytokine production and activates arachidonic acid metabolism in macrophages. *Infect. Immun.* **68:**2808–2818.

46. Crivelli, C., A. Demarta, and R. Peduzzi. 2001. Intestinal secretory immunoglobulin A (sIgA) response to *Aeromonas* exoproteins in patients with naturally acquired *Aeromonas* diarrhea. *FEMS Immunol. Med. Microbiol.* **30:**31–35.

47. Dacanay, A., S. C. Johnson, R. Bjornsdottir, R. O. Ebanks, N. W. Ross, M. Reith, R. K. Singh, J. Hiu, and L. L. Brown. 2003. Molecular characterization and quantitative analysis of superoxide dismutases in virulent and avirulent strains of *Aeromonas salmonicida* subsp. *salmonicida*. *J. Bacteriol.* **185:**4336–4344.

48. Deodhar, L. P., K. Saraswathi, and A. Varudkar. 1991. *Aeromonas* spp. and their association with human diarrheal disease. *J. Clin. Microbiol.* **29:**853–856.

49. Ebanks, R. O., A. Dacanay, M. Goguen, D. M. Pinto, and N. W. Ross. 2004. Differential proteomic analysis of *Aeromonas salmonicida* outer membrane proteins in response to low iron and in vivo growth conditions. *Proteomics* **4:**1074–1085.

50. Eko, F. O., and S. J. Utsalo. 1989. Characterization and significance of *Aeromonas* spp. isolated from diarrhoeic stools in Nigeria. *J. Trop. Med. Hyg.* **92:**97–101.

51. Eley, A., I. Geary, and M. H. Wilcox. 1993. Growth of *Aeromonas* spp. at 4°C and related toxin production. *Lett. Appl. Microbiol.* **16:**36–39.

52. Erova, T. E., L. Pillai, A. A. Fadl, J. Sha, S. Wang, C. L. Galindo, and A. K. Chopra. 2006. DNA adenine methyltransferase alters virulence of *Aeromonas hydrophila*. *Infect. Immun.* **74:**410–424.

53. Falcon, R., G. V. Carbonell, P. M. Figueredo, F. Butiao, H. O. Saridakis, J. S. Pelayo, and T. Yano. 2003. Intracellular vacuolation induced by culture filtrates of *Plesiomonas shigelloides* isolated from environmental sources. *J. Appl. Microbiol.* **95:**273–278.

54. Falker, S., M. A. Schmidt, and G. Heusipp. 2005. DNA methylation in *Yersinia enterocolitica*: role of the DNA adenine methyltransferase in mismatch repair and regulation of virulence factors. *Microbiology* **151:**2291–2299.

55. Fang, J. S., J. B. Chen, W. J. Chen, and K. T. Hsu. 1999. Haemolytic-uraemic syndrome in an adult male with *Aeromonas hydrophila* enterocolitis. *Nephrol. Dial. Transplant.* **14:**439–440.

56. Farmer, J. J., III, M. J. Arduino, and F. W. Hickman-Brenner. 1992. The genera *Aeromonas* and *Plesiomonas*, p. 3012–3028. *In* H. G. Truper, M. Dworkin, W. Harder, and H. H. Schleifer (ed.), *The Prokaryotes*, 2nd ed. Springer, New York, N.Y.

57. Ferguson, M. R., X. J. Xu, C. W. Houston, J. W. Peterson, D. H. Coppenhaver, V. L. Popov, and A. K. Chopra. 1997. Hyperproduction, purification, and mechanism of action of the cytotoxic enterotoxin produced by *Aeromonas hydrophila*. *Infect. Immun.* **65:**4299–4308.

58. Freij, B. J. 1987. Extraintestinal *Aeromonas* and *Plesiomonas* infections in humans. *Experientia* **43:**359–360.

59. Fricker, C. R., and S. Tompsett. 1989. *Aeromonas* spp. in foods: a significant cause of food poisoning? *Int. J. Food Microbiol.* **9:**17–23.

59a. Fujii, Y., T. Nomura, R. Yokoyama, S. Shinoda, and K. Okamoto. 2003. Studies of the mechanism of action of the aerolysin-like hemolysin of *Aeromonas sobria* in stimulating T84 cells to produce CAMP. *Infect. Immun.* **71:**1557–1560.

60. Fullner, K. J., and J. J. Mekalanos. 1999. Genetic characterization of a new type IV-A pilus gene cluster found in both classical and El Tor biotypes of *Vibrio cholerae*. *Infect. Immun.* **67:**1393–1404.

61. Gajewska, J., A. Miszczyk, and Z. Markiewicz. 2004. Characteristics of bacterial strains inhabiting the wood of coniferous trees. *Pol. J. Microbiol.* **53:**283–286.

62. Galindo, C. L., A. A. Fadl, J. Sha, and A. K. Chopra. 2004. Microarray analysis of *Aeromonas hydrophila* cytotoxic enterotoxin-treated murine primary macrophages. *Infect. Immun.* **72:**5439–5445.

63. Galindo, C. L., A. A. Fadl, J. Sha, C. Gutierrez, V. L. Popov, I. Boldogh, B. B. Aggarwal, and A. K. Chopra. 2004. *Aeromonas hydrophila* cytotoxic enterotoxin activates mitogen-activated protein kinases and induces apoptosis in murine macrophages and human intestinal epithelial cells. *J. Biol. Chem.* **278:**37597–37612.

64. Galindo, C. L., A. A. Fadl, J. Sha, L. Pillai, C. Gutierrez, Jr., and A. K. Chopra. 2005. Microarray and proteomics analyses of human intestinal epithelial cells treated with the *Aeromonas hydrophila* cytotoxic enterotoxin. *Infect. Immun.* **73:**2628–2643.

65. Galindo, C. L., C. Gutierrez, Jr., and A. K. Chopra. 2006. Potential involvement of galectin-3 and SNAP23 in *Aeromonas hydrophila* cytotoxin enterotoxin-induced host cell apoptosis. *Microb. Pathog.* **40**:56–68.

66. Galindo, C. L., J. Sha, A. A. Fadl, L. Pillai, and A. K. Chopra. 2006. Host immune responses to *Aeromonas* virulence factors. *Curr. Immun. Rev.* **2**:13–26.

67. Galindo, C. L., J. Sha, D. A. Ribardo, A. A. Fadl, L. Pillai, and A. K. Chopra. 2003. Identification of *Aeromonas hydrophila* cytotoxic enterotoxin-induced genes in macrophages using microarrays. *J. Biol. Chem.* **278**:40198–40212.

68. Gardner, S. E., S. E. Fowlston, and W. L. George. 1990. Effect of iron on production of a possible virulence factor by *Plesiomonas shigelloides*. *J. Clin. Microbiol.* **28**:811–813.

69. Garland, W. J., and J. T. Buckley. 1988. The cytolytic toxin aerolysin must aggregate to disrupt erythrocytes, and aggregation is stimulated by human glycophorin. *Infect. Immun.* **56**:1249–1253.

70. Gobat, P. F., and T. Jemmi. 1993. Distribution of mesophilic *Aeromonas* species in raw and ready-to-eat fish and meat products in Switzerland. *Int. J. Food Microbiol.* **20**:117–120.

71. Gobius, K. S., and J. M. Pemberton. 1988. Molecular cloning, characterization, and nucleotide sequence of an extracellular amylase gene from *Aeromonas hydrophila*. *J. Bacteriol.* **170**:1325–1332.

72. Gold, W. L., and I. E. Salit. 1993. *Aeromonas hydrophila* infections of skin and soft tissue: report of 11 cases and review. *Clin. Infect. Dis.* **16**:69–74.

73. Gomez Campdera, J., P. Munoz, F. Lopez Prieto, R. Rodriguez Fernandez, M. Robles, M. Rodriguez Creixems, and E. Bouza Santiago. 1996. Gastroenteritis due to *Aeromonas* in pediatrics. *An. Esp. Pediatr.* **44**:548–552. (In Spanish.)

74. Gracey, M. 1988. Gastro-enteritis in Australian children: studies on the aetiology of acute diarrhoea. *Ann. Trop. Paediatr.* **8**:68–75.

75. Graf, J. 1999. Symbiosis of *Aeromonas veronii* biovar sobria and *Hirudo medicinalis*, the medicinal leech: a novel model for digestive tract associations. *Infect. Immun.* **67**:1–7.

76. Granum, P. E., K. O'Sullivan, J. M. Tomas, and O. Ormen. 1998. Possible virulence factors of *Aeromonas* spp. from food and water. *FEMS Immunol. Med. Microbiol.* **21**:131–137.

77. Grey, P. A., and S. M. Kirov. 1993. Adherence to HEp-2 cells and enteropathogenic potential of *Aeromonas* spp. *Epidemiol. Infect.* **110**:279–287.

78. Hanninen, M. L. 1993. Occurrence of *Aeromonas* spp. in samples of ground meat and chicken. *Int. J. Food Microbiol.* **18**:339–342.

79. Hanninen, M. L., S. Salmi, L. Mattila, R. Taipalinen, and A. Siitonen. 1995. Association of *Aeromonas* spp. with travellers' diarrhoea in Finland. *J. Med. Microbiol.* **42**:26–31.

80. Hanninen, M. L., and A. Siitonen. 1995. Distribution of *Aeromonas* phenospecies and genospecies among strains isolated from water, foods or from human clinical samples. *Epidemiol. Infect.* **115**:39–50.

81. Haque, Q. M., A. Sugiyama, Y. Iwade, Y. Midorikawa, and T. Yamauchi. 1996. Diarrheal and environmental isolates of *Aeromonas* spp. produce a toxin similar to Shiga-like toxin 1. *Curr. Microbiol.* **32**:239–245.

82. Harf-Monteil, C., A. L. Fleche, P. Riegel, G. Prevost, D. Bermond, P. A. Grimont, and H. Monteil. 2004. *Aeromonas simiae* sp. nov., isolated from monkey faeces. *Int. J. Syst. Evol. Microbiol.* **54**:481–485.

83. Harshey, R. M. 1994. Bees aren't the only ones: swarming in gram-negative bacteria. *Mol. Microbiol.* **13**:389–394.

84. Heithoff, D. M., R. L. Sinsheimer, D. A. Low, and M. J. Mahan. 1999. An essential role for DNA adenine methylation in bacterial virulence. *Science* **284**:967–970.

85. Heuzenroeder, M. W., C. Y. Wong, and R. L. Flower. 1999. Distribution of two hemolytic toxin genes in clinical and environmental isolates of *Aeromonas* spp.: correlation with virulence in a suckling mouse model. *FEMS Microbiol. Lett.* **174**:131–136.

86. Hirono, I., and T. Aoki. 1991. Nucleotide sequence and expression of an extracellular hemolysin gene of *Aeromonas hydrophila*. *Microb. Pathog.* **11**:189–197.

87. Ho, A. S., T. A. Mietzner, A. J. Smith, and G. K. Schoolnik. 1990. The pili of *Aeromonas hydrophila*: identification of an environmentally regulated "mini pilin." *J. Exp. Med.* **172**:795–806.

88. Hokama, A., Y. Honma, and N. Nakasone. 1990. Pili of an *Aeromonas hydrophila* strain as a possible colonization factor. *Microbiol. Immunol.* **34**:901–915.

89. Hokama, A., and M. Iwanaga. 1992. Purification and characterization of *Aeromonas sobria* Ae24 pili: a possible new colonization factor. *Microb. Pathog.* **13**:325–334.

90. Hokama, A., and M. Iwanaga. 1991. Purification and characterization of *Aeromonas sobria* pili, a possible colonization factor. *Infect. Immun.* **59**:3478–3483.

91. Holmberg, S. D., I. K. Wachsmuth, F. W. Hickman-Brenner, P. A. Blake, and J. J. Farmer III. 1986. *Plesiomonas* enteric infections in the United States. *Ann. Intern. Med.* **105**: 690–694.

92. Honma, Y., and N. Nakasone. 1990. Pili of *Aeromonas hydrophila*: purification, characterization, and biological role. *Microbiol. Immunol.* **34**:83–98.

93. Hossain, M. A., K. M. Rahman, S. M. Asna, Z. Rahim, T. Hussain, and M. R. Miah. 1992. Incidence of *Aeromonas* isolated from diarrhoeal children and study of some virulence factors in the isolates. *Bangladesh Med. Res. Counc. Bull.* **18**:61–67.

94. Huq, A., A. Akhtar, M. A. Chowdhury, and D. A. Sack. 1991. Optimal growth temperature for the isolation of *Plesiomonas shigelloides*, using various selective and differential agars. *Can. J. Microbiol.* **37**:800–802.

95. Ingham, S. C. 1990. Growth of *Aeromonas hydrophila* and *Plesiomonas shigelloides* on cooked crayfish tails during cold storage under air, vacuum, and a modified atmosphere. *J. Food Prot.* **53**:665–667.

96. Inoue, K., Y. Kosako, K. Suzuki, and T. Shimada. 1991. Peritrichous flagellation in *Plesiomonas shigelloides* strains. *Jpn. J. Med. Sci. Biol.* **44**:141–146.

97. Iwanaga, M., and A. Hokama. 1992. Characterization of *Aeromonas sobria* TAP13 pili: a possible new colonization factor. *J. Gen. Microbiol.* **138**:1913–1919.

98. Jacxsens, L., F. Devlieghere, P. Falcato, and J. Debevere. 1999. Behavior of *Listeria monocytogenes* and *Aeromonas* spp. on fresh-cut produce packaged under

equilibrium-modified atmosphere. *J. Food Prot.* **62**:1128–1135.

99. **Janda, J., and S. L. Abott.** 1996. Human pathogens, p. 151–173. *In* B. Austin, M. Altwegg, P. Gosling, and S. Joseph (ed.), *The Genus* Aeromonas. John Wiley & Sons, Chichester, England.

100. **Janda, J. M.** 1991. Recent advances in the study of the taxonomy, pathogenicity, and infectious syndromes associated with the genus *Aeromonas. Clin. Microbiol. Rev.* **4**:397–410.

101. **Janda, J. M., and S. L. Abbott.** 1998. Evolving concepts regarding the genus *Aeromonas*: an expanding panorama of species, disease presentations, and unanswered questions. *Clin. Infect. Dis.* **27**:332–344.

102. **Janda, J. M., and S. L. Abbott.** 1993. Expression of hemolytic activity by *Plesiomonas shigelloides. J. Clin. Microbiol.* **31**:1206–1208.

103. **Janda, J. M., and S. L. Abbott.** 1999. Unusual foodborne pathogens. *Listeria monocytogenes, Aeromonas, Plesiomonas*, and *Edwardsiella* species. *Clin. Lab. Med.* **19**:553–582.

104. **Jeppesen, C.** 1995. Media for *Aeromonas* spp., *Plesiomonas shigelloides* and *Pseudomonas* spp. from food and environment. *Int. J. Food Microbiol.* **26**:25–41.

105. **Jiang, Z. D., A. C. Nelson, J. J. Mathewson, C. D. Ericsson, and H. L. DuPont.** 1991. Intestinal secretory immune response to infection with *Aeromonas* species and *Plesiomonas shigelloides* among students from the United States in Mexico. *J. Infect. Dis.* **164**:979–982.

106. **Joseph, S. W., and A. Carnahan.** 1994. The isolation, identification and systematics of the motile *Aeromonas* species. *Ann. Rev. Fish. Dis.* **4**:315–343.

107. **Julio, S. M., D. M. Heithoff, D. Provenzano, K. E. Klose, R. L. Sinsheimer, D. A. Low, and M. J. Mahan.** 2001. DNA adenine methylase is essential for viability and plays a role in the pathogenesis of *Yersinia pseudotuberculosis* and *Vibrio cholerae. Infect. Immun.* **69**:7610–7615.

108. **Kay, W. W., and T. J. Trust.** 1991. Form and functions of the regular surface array (S-layer) of *Aeromonas salmonicida. Experientia* **47**:412–414.

109. **Kelly, K. A., J. M. Koehler, and L. R. Ashdown.** 1993. Spectrum of extraintestinal disease due to *Aeromonas* species in tropical Queensland, Australia. *Clin. Infect. Dis.* **16**:574–579.

110. **Kirov, S. M.** 1993. Adhesion and piliation of *Aeromonas* spp. *Med. Microbiol. Lett.* **2**:274–280.

111. **Kirov, S. M.** 1997. *Aeromonas.* AIFST (NSW Branch) Food Microbiology Group, Hobart, Tasmania, Australia.

112. **Kirov, S. M.** 1993. The public health significance of *Aeromonas* spp. in foods. *Int. J. Food Microbiol.* **20**:179–198.

113. **Kirov, S. M., M. J. Anderson, and T. A. McMeekin.** 1990. A note on *Aeromonas* spp. from chickens as possible food-borne pathogens. *J. Appl. Bacteriol.* **68**:327–334.

114. **Kirov, S. M., E. K. Ardestani, and L. J. Hayward.** 1993. The growth and expression of virulence factors at refrigeration temperature by *Aeromonas* strains isolated from foods. *Int. J. Food Microbiol.* **20**:159–168.

115. **Kirov, S. M., T. C. Barnett, C. M. Pepe, M. S. Strom, and M. J. Albert.** 2000. Investigation of the role of type IV *Aeromonas* pilus (Tap) in the pathogenesis of *Aeromonas* gastrointestinal infection. *Infect. Immun.* **68**:4040–4048.

116. **Kirov, S. M., M. Castrisios, and J. G. Shaw.** 2004. *Aeromonas* flagella (polar and lateral) are enterocyte adhesins that contribute to biofilm formation on surfaces. *Infect. Immun.* **72**:1939–1945.

117. **Kirov, S. M., and L. J. Hayward.** 1993. Virulence traits of *Aeromonas* in relation to species and geographic region. *Aust. J. Med. Sci.* **14**:54–58.

118. **Kirov, S. M., L. J. Hayward, and M. A. Nerrie.** 1995. Adhesion of *Aeromonas* sp. to cell lines used as models for intestinal adhesion. *Epidemiol. Infect.* **115**:465–473.

119. **Kirov, S. M., D. S. Hui, and L. J. Hayward.** 1993. Milk as a potential source of *Aeromonas* gastrointestinal infection. *J. Food Prot.* **56**:306–312.

120. **Kirov, S. M., I. Jacobs, L. J. Hayward, and R. H. Hapin.** 1995. Electron microscopic examination of factors influencing the expression of filamentous surface structures on clinical and environmental isolates of *Aeromonas veronii* biotype sobria. *Microbiol. Immunol.* **39**:329–338.

121. **Kirov, S. M., L. A. O'Donovan, and K. Sanderson.** 1999. Functional characterization of type IV pili expressed on diarrhea-associated isolates of *Aeromonas* species. *Infect. Immun.* **67**:5447–5454.

122. **Kirov, S. M., B. Rees, R. C. Wellock, J. M. Goldsmid, and A. D. Van Galen.** 1986. Virulence characteristics of *Aeromonas* spp. in relation to source and biotype. *J. Clin. Microbiol.* **24**:827–834.

123. **Kirov, S. M., and K. Sanderson.** 1995. *Aeromonas*: recognizing the enemy. *Todays Life Sci.* **7**:30–35.

124. **Kirov, S. M., and K. Sanderson.** 1996. Characterization of a type IV bundle-forming pilus (SFP) from a gastroenteritis-associated strain of *Aeromonas veronii* biovar sobria. *Microb. Pathog.* **21**:23–34.

125. **Kirov, S. M., B. C. Tassell, A. B. Semmler, L. A. O'Donovan, A. A. Rabaan, and J. G. Shaw.** 2002. Lateral flagella and swarming motility in *Aeromonas* species. *J. Bacteriol.* **184**:547–555.

126. **Knebel, U., N. Sloot, M. Eikenberg, H. Borsdorf, U. Hoffler, and J. F. Riemann.** 2001. Gastroenteritis due to *Plesiomonas shigelloides*—rare cases in the Western world. *Med. Klin.* (Munich) **96**:109–113. (In German.)

127. **Knochel, S.** 1990. Growth characteristics of motile *Aeromonas* spp. isolated from different environments. *Int. J. Food Microbiol.* **10**:235–244.

128. **Krovacek, K., L. M. Eriksson, C. Gonzalez-Rey, J. Rosinsky, and I. Ciznar.** 2000. Isolation, biochemical and serological characterisation of *Plesiomonas shigelloides* from freshwater in Northern Europe. *Comp. Immunol. Microbiol. Infect. Dis.* **23**:45–51.

129. **Kuhn, I., M. J. Albert, M. Ansaruzzaman, N. A. Bhuiyan, S. A. Alabi, M. S. Islam, P. K. Neogi, G. Huys, P. Janssen, K. Kersters, and R. Mollby.** 1997. Characterization of *Aeromonas* spp. isolated from humans with diarrhea, from healthy controls, and from surface water in Bangladesh. *J. Clin. Microbiol.* **35**:369–373.

130. Lawson, M. A., V. Burke, and B. J. Chang. 1985. Invasion of HEp-2 cells by fecal isolates of *Aeromonas hydrophila*. *Infect. Immun.* **47**:680–683.

131. Leclere, V., M. Bechet, and R. Blondeau. 2004. Functional significance of a periplasmic Mn-superoxide dismutase from *Aeromonas hydrophila*. *J. Appl. Microbiol.* **96**:828–833.

132. Ljungh, A., P. Eneroth, and T. Wadstrom. 1982. Cytotonic enterotoxin from *Aeromonas hydrophila*. *Toxicon* **20**:787–794.

133. Lobner-Olesen, A., M. G. Marinus, and F. G. Hansen. 2003. Role of SeqA and Dam in *Escherichia coli* gene expression: a global/microarray analysis. *Proc. Natl. Acad. Sci. USA* **100**:4672–4677.

134. Low, D. A., N. J. Weyand, and M. J. Mahan. 2001. Roles of DNA adenine methylation in regulating bacterial gene expression and virulence. *Infect. Immun.* **69**:7197–7204.

135. Lynch, M. J., S. Swift, D. F. Kirke, C. W. Keevil, C. E. Dodd, and P. Williams. 2002. The regulation of biofilm development by quorum sensing in *Aeromonas hydrophila*. *Environ. Microbiol.* **4**:18–28.

136. Maluping, R. P., C. R. Lavilla-Pitogo, A. DePaola, J. M. Janda, and K. Krovacek. 2004. Occurrence, characterisation and detection of potential virulence determinants of emerging aquatic bacterial pathogens from the Philippines and Thailand. *New Microbiol.* **27**:381–389.

137. Manorama, V. T., R. K. Agarwal, and S. C. Sanyal. 1983. Enterotoxins of *Plesiomonas shigelloides*: partial purification and characterization. *Toxicon* **3**:269–272.

138. Marinus, M. G. 1996. Methylation of DNA, p. 782–791. *In* F. C. Neidhardt, R. Curtiss III, J. L. Ingraham, E. C. C. Lin, K. B. Low, B. Magasnik, W. S. Reznikoff, M. Riley, M. Schaechter, and H. E. Umbarger (ed.), *Escherichia coli* and *Salmonella*: *Cellular and Molecular Biology*, 2nd ed. ASM Press, Washington, D.C.

139. Martinez-Murcia, A. J., S. Benlloch, and M. D. Collins. 1992. Phylogenetic interrelationships of members of the genera *Aeromonas* and *Plesiomonas* as determined by 16S ribosomal DNA sequencing: lack of congruence with results of DNA-DNA hybridizations. *Int. J. Syst. Bacteriol.* **42**:412–421.

140. Martinez-Silva, R., M. Guzmann-Urrego, and F. H. Caselitz. 1961. On the problem of the significance of *Aeromonas* strains in enteritis in infants. *Z. Tropenmed. Parasitol.* **12**:445–451. (In German.)

141. Massad, G., J. E. Arceneaux, and B. R. Byers. 1991. Acquisition of iron from host sources by mesophilic *Aeromonas* species. *J. Gen. Microbiol.* **137**(Pt. 2):237–241.

142. Medema, G., and C. Schets. 1993. Occurrence of *Plesiomonas shigelloides* in surface water: relationship with faecal pollution and trophic state. *Zentbl. Hyg. Umweltmed.* **194**:398–404.

143. Merino, S., A. Aguilar, M. M. Nogueras, M. Regue, S. Swift, and J. M. Tomas. 1999. Cloning, sequencing, and role in virulence of two phospholipases (A1 and C) from mesophilic *Aeromonas* sp. serogroup O:34. *Infect. Immun.* **67**:4008–4013.

144. Merino, S., A. Aguilar, X. Rubires, N. Abitiu, M. Regue, and J. M. Tomas. 1997. The role of the capsular polysaccharide of *Aeromonas hydrophila* serogroup O:34 in the adherence to and invasion of fish cell lines. *Res. Microbiol.* **148**:625–631.

145. Merino, S., A. Aguilar, X. Rubires, D. Simon-Pujol, F. Congregado, and J. M. Tomas. 1996. The role of the capsular polysaccharide of *Aeromonas salmonicida* in the adherence and invasion of fish cell lines. *FEMS Microbiol. Lett.* **142**:185–189.

146. Merino, S., M. Altarriba, R. Gavin, L. Izquierdo, and J. M. Tomas. 2001. The cell division genes (ftsE and X) of *Aeromonas hydrophila* and their relationship with opsonophagocytosis. *FEMS Microbiol. Lett.* **198**:183–188.

147. Merino, S., M. M. Nogueras, A. Aguilar, X. Rubires, S. Alberti, V. J. Benedi, and J. M. Tomas. 1998. Activation of the complement classical pathway (C1q binding) by mesophilic *Aeromonas hydrophila* outer membrane protein. *Infect. Immun.* **66**:3825–3831.

148. Merino, S., X. Rubires, A. Aguilar, S. Alberti, S. Hernandez-Alles, V. J. Benedi, and J. M. Tomas. 1996. Mesophilic *Aeromonas* sp. serogroup O:11 resistance to complement-mediated killing. *Infect. Immun.* **64**:5302–5309.

149. Merino, S., X. Rubires, A. Aguillar, J. F. Guillot, and J. M. Tomas. 1996. The role of the O-antigen lipopolysaccharide on the colonization in vivo of the germfree chicken gut by *Aeromonas hydrophila* serogroup O:34. *Microb. Pathog.* **20**:325–333.

150. Merino, S., X. Rubires, S. Knochel, and J. M. Tomas. 1995. Emerging pathogens: *Aeromonas* spp. *Int. J. Food Microbiol.* **28**:157–168.

151. Messer, W., and M. Noyer-Weidner. 1988. Timing and targeting: the biological functions of Dam methylation in *E. coli*. *Cell* **54**:735–737.

152. Miller, M. L., and J. A. Kohburger. 1985. *Plesiomonas shigelloides*: an opportunistic food and waterborne pathogen. *J. Food Prot.* **48**:449–457.

153. Miller, M. L., and J. A. Kohburger. 1986. Tolerance of *Plesiomonas shigelloides* to pH, sodium chloride and temperature. *J. Food Prot.* **49**:877–879.

154. Millership, S. E., S. R. Curnow, and B. Chattopadhyay. 1983. Faecal carriage rate of *Aeromonas hydrophila*. *J. Clin. Pathol.* **36**:920–923.

155. Morgan, D. R., P. C. Johnson, H. L. DuPont, T. K. Satterwhite, and L. V. Wood. 1985. Lack of correlation between known virulence properties of *Aeromonas hydrophila* and enteropathogenicity for humans. *Infect. Immun.* **50**:62–65.

156. Mourya, D. T., M. D. Gokhale, V. Pidiyar, P. V. Barde, M. Patole, A. C. Mishra, and Y. Shouche. 2002. Study of the effect of the midgut bacterial flora of Culex quinquefasciatus on the susceptibility of mosquitoes to Japanese encephalitis virus. *Acta Virol.* **46**:257–260.

157. Nakasone, N., C. Toma, T. Song, and M. Iwanaga. 2004. Purification and characterization of a novel metalloprotease isolated from *Aeromonas caviae*. *FEMS Microbiol. Lett.* **237**:127–132.

158. Namdari, H., and E. J. Bottone. 1991. *Aeromonas caviae*: ecologic adaptation in the intestinal tract of infants coupled to adherence and enterotoxin production as factors in enteropathogenicity. *Experientia* **47**:434–436.

159. Nayduch, D., A. Honko, G. P. Noblet, and F. Stutzenberger. 2001. Detection of *Aeromonas caviae* in the common housefly Musca domestica by culture and polymerase chain reaction. *Epidemiol. Infect.* **127**:561–566.

160. Nayduch, D., G. P. Noblet, and F. J. Stutzenberger. 2002. Vector potential of houseflies for the bacterium *Aeromonas caviae*. *Med. Vet. Entomol.* **16**:193–198.

161. Nishikawa, Y., A. Hase, J. Ogawasara, S. M. Scotland, H. R. Smith, and T. Kimura. 1994. Adhesion to and invasion of human colon carcinoma Caco-2 cells by *Aeromonas* strains. *J. Med. Microbiol.* **40**:55–61.

162. Nishikawa, Y., and T. Kishi. 1988. Isolation and characterization of motile *Aeromonas* from human, food and environmental specimens. *Epidemiol. Infect.* **101**:213–223.

163. Nishikawa, Y., J. Ogasawara, and T. Kimura. 1993. Heat and acid sensitivity of motile *Aeromonas*: a comparison with other food-poisoning bacteria. *Int. J. Food Microbiol.* **18**:271–278.

164. Nogueras, M. M., S. Merino, A. Aguilar, V. J. Benedi, and J. M. Tomas. 2000. Cloning, sequencing, and role in serum susceptibility of porin II from mesophilic *Aeromonas hydrophila*. *Infect. Immun.* **68**:1849–1854.

165. Nojimoto, I. T., C. S. Bezana, C. do Carmo, L. M. Valadao, and M. Carrijo Kde. 1997. The prevalence of *Aeromonas* spp. in the diarrheal feces of children under the age of 5 years in the city of Goiania, Goias in the 1995–1996 biennium. *Rev. Soc. Bras. Med. Trop.* **30**:385–388. (In Portuguese.)

166. O'Brien, M., and M. Tandy. 1994. L-Histidine decarboxylase: a new diagnostic test for *Plesiomonas shigelloides*. *Aust. Microbiol.* **15**:A–69.

167. Okawa, Y., Y. Ohtomo, H. Tsugawa, Y. Matsuda, H. Kobayashi, and T. Tsukamoto. 2004. Isolation and characterization of a cytotoxin produced by *Plesiomonas shigelloides* P-1 strain. *FEMS Microbiol. Lett.* **239**:125–130.

168. Olsvik, O., K. Wachsmuth, B. Kay, K. A. Birkness, A. Yi, and B. Sack. 1990. Laboratory observations on *Plesiomonas shigelloides* strains isolated from children with diarrhea in Peru. *J. Clin. Microbiol.* **28**:886–889.

169. Palmer, B. R., and M. G. Marinus. 1994. The *dam* and *dcm* strains of *Escherichia coli*—a review. *Gene* **143**:1–12.

170. Palumbo, S. A., J. E. Call, P. H. Cooke, and A. C. Williams. 1995. Effect of pyrophosphates and NaCl on *Aeromonas hydrophila* K144. *J. Food Safety* **15**:77–87.

171. Palumbo, S. A., A. C. Maxino, A. C. Williams, R. L. Buchanan, and D. W. Thayer. 1985. Starch-ampicillin agar for the quantitative detection of *Aeromonas hydrophila*. *Appl. Environ. Microbiol.* **50**:1027–1030.

172. Palumbo, S. A., A. C. Williams, R. L. Buchanan, and J. G. Philips. 1987. Thermal resistance of *Aeromonas hydrophila*. *J. Food Prot.* **50**:761–764.

173. Pancholi, V. 2001. Multifunctional alpha-enolase: its role in diseases. *Cell. Mol. Life Sci.* **58**:902–920.

174. Pancholi, V., and V. A. Fischetti. 1998. Alpha-enolase, a novel strong plasmin(ogen) binding protein on the surface of pathogenic streptococci. *J. Biol. Chem.* **273**:14503–14515.

175. Parker, M. W., J. T. Buckley, J. P. Postma, A. D. Tucker, K. Leonard, F. Pattus, and D. Tsernoglou. 1994. Structure of the *Aeromonas* toxin proaerolysin in its water-soluble and membrane-channel states. *Nature* **367**:292–295.

176. Parker, M. W., F. G. van der Goot, and J. T. Buckley. 1996. Aerolysin—the ins and outs of a model channel-forming toxin. *Mol. Microbiol.* **19**:205–212.

177. Parras, F., M. D. Diaz, J. Reina, S. Moreno, C. Guerrero, and E. Bouza. 1993. Meningitis due to *Aeromonas* species: case report and review. *Clin. Infect. Dis.* **17**:1058–1060.

178. Pemberton, J. M., S. P. Kidd, and R. Schmidt. 1997. Secreted enzymes of *Aeromonas*. *FEMS Microbiol. Lett.* **152**:1–10.

179. Pidiyar, V., A. Kaznowski, N. B. Narayan, M. Patole, and Y. S. Shouche. 2002. *Aeromonas culicicola* sp. nov., from the midgut of Culex quinquefasciatus. *Int. J. Syst. Evol. Microbiol.* **52**:1723–1728.

180. Pidiyar, V. J., K. Jangid, M. S. Patole, and Y. S. Shouche. 2004. Studies on cultured and uncultured microbiota of wild culex quinquefasciatus mosquito midgut based on 16S ribosomal RNA gene analysis. *Am. J. Trop. Med. Hyg.* **70**:597–603.

181. Pitarangsi, C., P. Echeverria, R. Whitmire, C. Tirapat, S. Formal, G. J. Dammin, and M. Tingtalapong. 1982. Enteropathogenicity of *Aeromonas hydrophila* and *Plesiomonas shigelloides*: prevalence among individuals with and without diarrhea in Thailand. *Infect. Immun.* **35**:666–673.

182. Popoff, M. 1984. Genus III. *Aeromonas* Kluyver and Van Niel 1936, p. 545–548. *In* N. R. Krieg and J. G. Holt (ed.), *Bergey's Manual of Systematic Bacteriology*, vol. 1. The Williams and Wilkins Co., Baltimore, Md.

183. Potomski, J., V. Burke, J. Robinson, D. Fumarola, and G. Miragliotta. 1987. *Aeromonas* cytotonic enterotoxin cross reactive with cholera toxin. *J. Med. Microbiol.* **23**:179–186.

184. Quinn, J. P. 1991. UHL enteric bacterial disease surveillance. *Lab. Hotline* **28**:2.

185. Radomyski, T., E. A. Murano, D. G. Olson, and P. S. Murano. 1994. Elimination of pathogens of significance in food by low dose radiation: a review. *J. Food Prot.* **57**:73–86.

186. Ratkowsky, D. A., J. Olley, T. A. McMeekin, and A. Ball. 1982. Relationship between temperature and growth rate of bacterial cultures. *J. Bacteriol.* **149**:1–5.

187. Redlitz, A., B. J. Fowler, E. F. Plow, and L. A. Miles. 1995. The role of an enolase-related molecule in plasminogen binding to cells. *Eur. J. Biochem.* **227**:407–415.

188. Ribardo, D. A., K. R. Kuhl, I. Boldogh, J. W. Peterson, C. W. Houston, and A. K. Chopra. 2002. Early cell signaling by the cytotoxic enterotoxin of *Aeromonas hydrophila* in macrophages. *Microb. Pathog.* **32**:149–163.

189. Rivero, O., J. Anguita, C. Paniagua, and G. Naharro. 1990. Molecular cloning and characterization of an extracellular protease gene from *Aeromonas hydrophila*. *J. Bacteriol.* **172**:3905–3908.

190. Robson, W. L., A. K. Leung, and C. L. Trevenen. 1992. Haemolytic-uraemic syndrome associated with *Aeromonas hydrophila* enterocolitis. *Pediatr. Nephrol.* **6**:221.

191. Rosner, R. 1964. *Aeromonas hydrophila* as the etiological agent in a case of severe gastroenteritis. *Am. J. Clin. Pathol.* **42:**402–404.

192. Russell, D. W., and N. D. Zinder. 1987. Hemimethylation prevents DNA replication in *E. coli. Cell* **50:**1071–1079.

193. Santos, J. A., C. J. Gonzalez, T. M. Lopez, A. Otero, and M. L. Garcia-Lopez. 1999. Hemolytic and elastolytic activities influenced by iron in *Plesiomonas shigelloides. J. Food Prot.* **62:**1475–1477.

194. Saraswathi, B., R. K. Agarwal, and S. C. Sanyal. 1983. Further studies on enteropathogenicity of *Plesiomonas shigelloides. Indian J. Med. Res.* **78:**12–18.

195. Schultz, A. J., and B. A. McCardell. 1988. DNA homology and immunological cross-reactivity between *Aeromonas hydrophila* cytotonic toxin and cholera toxin. *J. Clin. Microbiol.* **26:**57–61.

196. Seidler, R. J., D. A. Allen, H. Lockman, R. R. Colwell, S. W. Joseph, and O. P. Daily. 1980. Isolation, enumeration and characterization of *Aeromonas* from polluted waters encountered in diving operations. *Appl. Environ. Microbiol.* **39:**1010–1018.

197. Sha, J., C. L. Galindo, V. Pancholi, V. L. Popov, Y. Zhao, C. W. Houston, and A. K. Chopra. 2003. Differential expression of the enolase gene under in vivo versus in vitro growth conditions of *Aeromonas hydrophila. Microb. Pathog.* **34:**195–204.

198. Sha, J., E. V. Kozlova, and A. K. Chopra. 2002. Role of various enterotoxins in *Aeromonas hydrophila*-induced gastroenteritis: generation of enterotoxin gene-deficient mutants and evaluation of their enterotoxic activity. *Infect. Immun.* **70:**1924–1935.

199. Sha, J., E. V. Kozlova, A. A. Fadl, J. P. Olano, C. W. Houston, J. W. Peterson, and A. K. Chopra. 2004. Molecular characterization of a glucose-inhibited division gene, *gidA*, that regulates cytotoxic enterotoxin of *Aeromonas hydrophila. Infect. Immun.* **72:**1084–1095.

200. Sha, J., L. Pillai, A. A. Fadl, C. L. Galindo, T. E. Erova, and A. K. Chopra. 2005. The type III secretion system and cytotoxic enterotoxin alter the virulence of *Aeromonas hydrophila. Infect. Immun.* **73:**6446–6457.

201. Shaw, J. G., J. P. Thornley, L. Plalmer, and I. Geary. 1995. Invasion of tissue culture cells by *Aeromonas caviae. Med. Microbiol. Lett.* **4:**316–323.

202. Shigematsu, M., M. E. Kaufmann, A. Charlett, Y. Niho, and T. L. Pitt. 2000. An epidemiological study of *Plesiomonas shigelloides* diarrhoea among Japanese travellers. *Epidemiol. Infect.* **125:**523–530.

203. Shimada, T., R. Sakazaki, K. Horigome, Y. Uesaka, and K. Niwano. 1984. Production of cholera-like enterotoxin by *Aeromonas hydrophila. Jpn. J. Med. Sci. Biol.* **37:**141–144.

204. Snower, D. P., C. Ruef, A. P. Kuritza, and S. C. Edberg. 1989. *Aeromonas hydrophila* infection associated with the use of medicinal leeches. *J. Clin. Microbiol.* **27:**1421–1422.

205. Song, T., C. Toma, N. Nakasone, and M. Iwanaga. 2004. Aerolysin is activated by metalloprotease in *Aeromonas veronii* biovar sobria. *J. Med. Microbiol.* **53:**477–482.

206. Stojek, N. M., and J. Dutkiewicz. 2004. Studies on the occurrence of gram-negative bacteria in ticks: Ixodes ricinus as a potential vector of *Pasteurella. Ann. Agric. Environ. Med.* **11:**319–322.

207. Sukontason, K., M. Bunchoo, B. Khantawa, S. Piangjai, R. Methanitikorn, and Y. Rongsriyam. 2000. Mechanical carrier of bacterial enteric pathogens by *Chrysomya megacephala* (Diptera: Calliphoridae) in Chiang Mai, Thailand. *Southeast Asian J. Trop. Med. Public Health* **31**(Suppl. 1):157–161.

208. Sulaiman, S., M. Z. Othman, and A. H. Aziz. 2000. Isolations of enteric pathogens from synanthropic flies trapped in downtown Kuala Lumpur. *J. Vector Ecol.* **25:**90–93.

209. Swift, S., A. Karlyshev, L. Fish, E. Durant, M. Winson, S. Chhabra, P. Williams, S. Macintyre, and G. Stewart. 1997. Quorum sensing in *Aeromonas hydrophila* and *Aeromonas salmonicida*: identification of the LuxRI homologs AhyRI and AsaRI and their cognate N-acylhomoserine lactone signal molecules. *J. Bacteriol.* **179:**5271–5281.

210. Swift, S., M. J. Lynch, L. Fish, D. F. Kirke, J. M. Tomas, G. S. Stewart, and P. Williams. 1999. Quorum sensing-dependent regulation and blockade of exoprotease production in *Aeromonas hydrophila. Infect. Immun.* **67:**5192–5199.

211. Theodoropoulos, C., T. H. Wong, M. O'Brien, and D. Stenzel. 2001. *Plesiomonas shigelloides* enters polarized human intestinal Caco-2 cells in an in vitro model system. *Infect. Immun.* **69:**2260–2269.

212. van der Goot, F. G., F. Pattus, M. Parker, and J. T. Buckley. 1994. The cytolytic toxin aerolysin: from the soluble form to the transmembrane channel. *Toxicology* **87:**19–28.

213. Verenkar, M., V. Naik, S. Rodrigues, and I. Singh. 1995. *Aeromonas* species and *Plesiomonas shigelloides* in diarrhoea in Goa. *Indian J. Pathol. Microbiol.* **38:**169–171.

214. Vipond, R., I. R. Bricknell, E. Durant, T. J. Bowden, A. E. Ellis, M. Smith, and S. MacIntyre. 1998. Defined deletion mutants demonstrate that the major secreted toxins are not essential for the virulence of *Aeromonas salmonicida. Infect. Immun.* **66:**1990–1998.

215. Vivas, J., B. E. Razquin, P. Lopez-Fierro, G. Naharro, and A. Villena. 2004. Correlation between production of acyl homoserine lactones and proteases in an *Aeromonas hydrophila aroA* live vaccine. *Vet. Microbiol.* **101:**167–176.

216. Watson, I. M., J. O. Robinson, V. Burke, and M. Gracey. 1985. Invasiveness of *Aeromonas* spp. in relation to biotype, virulence factors, and clinical features. *J. Clin. Microbiol.* **22:**48–51.

217. Wilmsen, H. U., J. T. Buckley, and F. Pattus. 1991. Site-directed mutagenesis at histidines of aerolysin from *Aeromonas hydrophila*: a lipid planar bilayer study. *Mol. Microbiol.* **5:**2745–2751.

218. Wilmsen, H. U., K. R. Leonard, W. Tichelaar, J. T. Buckley, and F. Pattus. 1992. The aerolysin membrane channel is formed by heptamerization of the monomer. *EMBO J.* **11:**2457–2463.

219. **Wong, C. Y., M. W. Heuzenroeder, and R. L. Flower.** 1998. Inactivation of two haemolytic toxin genes in *Aeromonas hydrophila* attenuates virulence in a suckling mouse model. *Microbiology* **144**(Pt. 2):291–298.

220. **Yamada, S., S. Matsushita, S. Dejsirilert, and Y. Kudoh.** 1997. Incidence and clinical symptoms of *Aeromonas*-associated travellers' diarrhoea in Tokyo. *Epidemiol. Infect.* **119**:121–126.

221. **Yu, H. B., P. S. Rao, H. C. Lee, S. Vilches, S. Merino, J. M. Tomas, and K. Y. Leung.** 2004. A type III secretion system is required for *Aeromonas hydrophila* AH-1 pathogenesis. *Infect. Immun.* **72**:1248–1256.

222. **Zhang, Y. L., E. Arakawa, and K. Y. Leung.** 2002. Novel *Aeromonas hydrophila* PPD134/91 genes involved in O-antigen and capsule biosynthesis. *Infect. Immun.* **70**:2326–2335.

Food Microbiology: Fundamentals and Frontiers, 3rd Ed.
Edited by M. P. Doyle and L. R. Beuchat
© 2007 ASM Press, Washington, D.C.

Eric A. Johnson

Clostridium botulinum

18

Botulism is a neuroparalytic disease in humans and animals, resulting from the actions of neurotoxins produced by *Clostridium botulinum* and rare strains of *Clostridium butyricum* and *Clostridium baratii* (16, 41, 50, 51, 63). Foodborne botulism occurs following ingestion of botulinal neurotoxin preformed in foods. Botulism can also result from the growth and toxin production by *C. botulinum* in the intestine (infant botulism and adult intestinal botulism) and in wounds (wound botulism) (4, 16, 42, 100). An "undetermined classification" refers to cases of diagnosed botulism for which no plausible food vehicle or intestinal colonization by *C. botulinum* can be demonstrated (4, 100). Inhalational botulism is extremely rare, but a few human cases have been observed and it has been demonstrated to occur in nonhuman primates and animal models (6, 89, 120). Intentional botulism (bioterrorism) was documented in the mid-1900s, and concerns that could be associated with terrorist activities have recently reemerged (6, 15, 73), particularly through foods (1, 73, 121). Botulism outbreaks can have a dramatic impact on the populations in which they occur (6, 26), and outbreaks of animal botulism have periodically caused devastating losses of domestic and wild animals (26, 33, 74, 104).

Since the early 1900s, botulism has been a serious concern of the food industry and regulatory agencies because of the resistance properties of the pathogen, its ability to survive and grow in many foods, and the severity of the disease (12, 16, 26, 39, 80). Resistant endospores produced by *C. botulinum* are widely distributed in soils and contaminate many foods (24, 25, 52–54, 63, 81, 105). In improperly processed and preserved foods, the endospores can germinate and vegetative cells can proliferate to form botulinal neurotoxins, which cause botulism on ingestion. Botulinal neurotoxins are the most poisonous toxins known and are toxic by the oral route including by food consumption (63, 96, 107). Consequently, a major goal of the food industry and of regulatory agencies is to prevent survival of spores and proliferation of vegetative cells in foods that could support the production of botulinum toxin. Hence, certain food regulations and industry practices have been designed specifically to prevent growth of, and toxin formation by, *C. botulinum* (12, 53, 63, 84, 92). The importance of *C. botulinum* and its neurotoxins in food safety has contributed to unique research approaches in food microbiology (44, 93).

Eric A. Johnson, Department of Food Microbiology and Toxicology and Bacteriology, Food Research Institute, University of Wisconsin, 1925 Willow Dr., Madison, WI 53706.

HISTORICAL FEATURES OF *C. BOTULINUM* AND BOTULISM

Although anecdotal evidence suggests that botulism was recognized over 1,000 years ago during the reign of Emperor Leo VI of Byzantium (886–911 A.D.) (26, 104), the disease was first definitively described in humans in Germany by Müller (1735–93) and Justinius Kerner (1786–1862) (20, 26, 61, 80). During the 1800s in Germany, raw "blood" and "liver sausages" were associated with a disease characterized by muscle paralysis and suffocation. The disease was referred to as sausage "botulus" poisoning. Kerner showed experimentally that a toxin was produced within the sausage and that exclusion of air was required for the toxin formation (61). Botulism was often referred to as Kerner's disease following his investigations (61, 80). Subsequent to the investigations in Germany, a disease with similar symptoms was recorded in Russia and Denmark from the consumption of fish, termed "ichthyism" (26). Botulism has since been recognized as a disease that occurs worldwide from the consumption of a variety of toxin-containing foods (42, 44, 52, 63).

Although numerous theories were proposed for the cause of botulism, its etiology remained obscure until a series of experiments were conducted by the Belgian microbiologist Emile Pierre van Ermengem in the late 1890s (61, 118). In 1895, van Ermengem published his classic treatise describing the isolation of an anaerobic bacillus from a raw salted ham implicated in a botulism outbreak affecting 34 individuals (118). He showed that the bacillus produced a very potent toxin that was released into the medium. van Ermengem clearly established the etiology of botulism by isolation of *"Bacillus botulinus"* from the ham and the spleens and large intestines of persons who had died from the food poisoning. His success in isolating the anaerobe and demonstrating the extracellular nature of the toxin was a triumph in food microbiology. In 1897, Kempner (69) subsequently showed that van Ermengem's cultures produced a substance that on injection in an inactive form gave rise to an antitoxin in the blood of goats, which prevented death in animals exposed to the toxin. This provided the first evidence that antitoxin to botulinal neurotoxin could neutralize toxicity and prevent death (61).

van Ermengem's and Kempner's landmark investigations established several principles of botulism that remain valid today and form the cornerstone for understanding and control of the disease: (i) foodborne botulism is a true toxemia caused by toxins produced by *C. botulinum*; (ii) the toxin is produced in foods by a specific organism, *"Bacillus botulinus"*; (iii) the toxin is active by the oral route; (iv) the toxin is inactivated by heat and alkali but is stable in acidic conditions; (v) the toxin is not produced in food containing sufficient salt or acid; (vi) *C. botulinum* produces heat-resistant endospores; (vii) animals vary in their susceptibility to botulinal neurotoxins; and (viii) animals can develop immunity to the botulinal neurotoxins by exposure to inactive toxin or toxoid (61, 118). Investigation of subsequent outbreaks established that there were several types of *C. botulinum* (104). Some of these types produce paralytic diseases similar to human botulism, such as "limberneck" in chickens and wild birds and flaccid paralysis in monkeys, lions, cats, horses, cattle, dogs, and many other animals (26, 33, 74, 104, 105). Strains isolated from some outbreaks were more proteolytic than the van Ermengem isolate, digesting meat and milk protein, and formed a toxin that was not neutralized by the antitoxin to the van Ermengem strain (61, 104). Over time, seven serologically distinguishable types of botulinal neurotoxins (A to G) have been identified, although evidence has shown that variations of the toxins can exist within a serotype (43, 49, 106; unpublished data). *C. botulinum* strains that produce more than one serotype of toxin or that contain "silent" or unexpressed genes have been isolated with increasing frequency (41–43, 62).

BIOCHEMISTRY AND PHARMACOLOGY OF BOTULINAL NEUROTOXINS

The exceptional feature of the botulinogenic clostridia is the production of a characteristic neurotoxin of extraordinary potency for humans and animals (95, 110). Botulinal neurotoxins are the most potent toxins known, with estimated lethal human intravenous doses of 0.1 to 1 ng per kg of body weight (6, 95, 96). The 50% lethal dose (LD_{50}) for a 20-g Swiss-Webster (18- to 22-g) mouse is 7 to 10 pg for most purified botulinal neurotoxins (60, 95). The mouse bioassay is an extremely sensitive and consistent method in interlaboratory studies and is the "gold standard" for measuring the active levels of botulinal neurotoxins in foods and clinical samples (16, 96, 109, 117). Botulinal neurotoxin has the unusual feature among protein toxins of being extremely poisonous by the oral route, with an estimated LD of 0.1 to 1 µg per kg (6, 96). The oral toxicity varies depending on the serotypes of botulinal neurotoxin, the toxic food consumed, and the presence of food and alcohol in the intestinal tract, as well as other factors (6, 96).

Botulinal neurotoxins are proteins of ~150 kDa that naturally exist as components of progenitor toxin complexes (13, 21, 22, 95, 110), whereby the neurotoxin component is associated with nontoxic proteins and RNA (96, 110). The nontoxic proteins in the complexes have been demonstrated to provide protection during experimental manipulations, food processing, and passage through the gastrointestinal tract (95, 96). Botulinal neurotoxins are produced as single-chain molecules of ca. 150 kDa that achieve their characteristic high toxicities of 10^7 to 10^8 mouse $LD_{50}s$ (MLD_{50}) per mg by posttranslational proteolytic cleavage to form a dichain molecule composed of an L chain (~50 kDa) and a heavy (H) chain (~100 kDa) linked by a disulfide bond (19, 70, 110, 113).

Botulinal neurotoxin consists of three basic functional domains (70, 82, 113): (i) L chain, the catalytic domain that has endopeptidase activity on neuronal substrates; (ii) H_N, the translocation domain residing in the N-terminal region of the H chain; and (iii) H_C, the receptor-binding domain located in the C-terminal region of the H chain. The gene and amino acid sequences of botulinal neurotoxins have been determined for a limited number of *C. botulinum* strains (106; M. Jacobsen and E. A. Johnson, unpublished data). Recent studies have demonstrated that subtypes of botulinal neurotoxins occur within specific serotypes, in which the 150-kDa proteins differ by 3 to 12% in amino acid sequence and associated properties such as immunogenicity and neutralization by antitoxins (106). These findings indicate that different evolutionary lineages of botulinal neurotoxins exist within this toxin family with differing properties of importance to microbiology and medicine.

Botulinal neurotoxin enters the circulation from the intestinal tract in foodborne and intestinal botulism, or from wound infections in cases of wound botulism (6, 17, 82). Once in the circulation, the neurotoxin binds to and is internalized within target nerves, primarily in cholinergic nerve endings at the neuromuscular junction (17, 18, 82). Botulinal neurotoxins enter nerves by a multistep process: (i) binding to receptor proteins and lipid gangliosides; (ii) internalization with vesicles by endocytosis; (iii) translocation of the catalytic portion (L_c) into the nerve cytosol; and (iv) cleavage of neuronal substrates (components of the SNARE apparatus) by the L_c (82). This ultimately results in the inhibition of release of acetylcholine at the neuromuscular junction, preventing muscle activation and causing the characteristic flaccid paralysis of botulism (6, 18, 82). The cellular mechanisms of botulinal neurotoxin intoxication have been an area of intense investigation, and excellent reviews of this intricate process are available (71, 82).

CLINICAL ASPECTS

Since botulism is a true toxemia and botulinal neurotoxin is solely responsible for the illness, foodborne, infant, and wound botulism are clinically similar. The hallmark clinical symptoms of botulism are a bilateral and descending weakening and paralysis of skeletal muscles (6, 17, 18) (Fig. 18.1). The incubation time for onset of symptoms varies with the type of botulism, the botulinal neurotoxin serotype, and the quantity of neurotoxin that reaches target nerves. Foodborne botulism occurs following the consumption of food contaminated with preformed botulinal neurotoxin, and the vast majority of cases are caused by types A, B, and E, and rarely by type F (42, 45, 49, 100, 124). In foodborne botulism cases, the onset time is usually 12 to 36 h following consumption of the toxic food. The incubation period can be as short as 2 h when high quantities of toxin are ingested or as long as 2 to 14 days with serotypes B or E or ingestion of low quantities of botulinal neurotoxin (49, 63, 107, 124). Wound botulism usually has a relatively long incubation period of 4 to 14 days, reflecting the time for neurotoxigenic clostridia to colonize the wound and produce neurotoxin (107, 124). Infant botulism has been reported to have an incubation time from 6 to 8 h to several days (4, 6, 40), although extremely rapid onset has occasionally been reported (40). It is controversial whether rapid onset of fulminant botulism is a cause of sudden infant death syndrome (5, 40, 85).

In most cases of botulism, cranial nerves are first affected, particularly those innervating the eyes, and the first symptoms are blurred and double vision, dilated pupils, and drooping eyelids (6, 15, 18, 100) (Fig. 18.1).

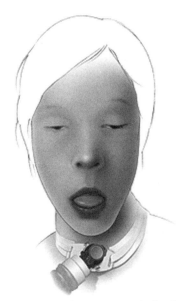

Figure 18.1 Portrayal of a person with the flaccid paralysis symptoms characteristic of botulism. Drawing prepared by James K. Archer, Centers for Disease Control and Prevention, Atlanta, Ga.

The eyes respond slowly to light in a darkened room (4). These abnormalities are followed by other symptoms in the cranium, including difficulty in swallowing, weakness of the neck and mouth, dysphagia (drooling and difficulty swallowing), and problems in speaking (6, 15, 18, 100). As the paralysis descends, weakness of the upper limbs and the torso occurs, and in severe cases muscles affecting respiration are weakened, and mechanical ventilation to prevent fatality by suffocation is required (6, 17, 18). Generally, the patient's hearing remains normal, consciousness is not lost, and the victim is cognizant of the progression of the disease. Other symptoms may occur including nausea or vomiting, dizziness or vertigo, diarrhea or constipation, dry mucous membranes in the mouth and throat, sore throat, and parasthesias, which may be related to botulinal neurotoxin or other toxins or pathogens in the contaminated foods. Infant botulism has certain distinct features including constipation, weakness and hypotonia, poor suck and feeding, weak cry, lack of head control, and cardiovascular abnormalities (hypotension, tachycardia) (4, 40). Certain cases of botulism have been associated with abnormalities of autonomic and sensory functions such as constipation, dry mouth, and difficulty in urinating (17, 18). A patient's awareness of weakening of muscle activity and ensuing paralysis can lead to considerable emotional distress including anxiety and depression. The most severe and long-lasting foodborne botulism generally occurs with type A (6, 17, 100). In general, the duration of illness follows the pattern A > B > F > E in animal models and in humans in foodborne botulism (6, 17, 34, 38, 100, 124). In infant botulism, nearly all cases are caused by types A and B, and A has a longer duration and causes more severe symptoms (4, 7, 40). The fatality rate from foodborne botulism has decreased from a range of 50 to 70% in the 1800s and 1900s to <10% in the past 20 to 30 years (16, 17, 49, 81). Convalescence and recovery from botulism are usually prolonged, requiring weeks to months depending on the serotype of toxin and quantity consumed. Recovery is usually complete, and patients regain full normal function (4, 17, 18).

Diagnosis of botulism includes clinical assessment of the initial visual disturbances and other cranial effects and detection of botulinal neurotoxin from appropriate specimens by mouse bioassay (6, 17, 49, 100). The definitive diagnosis is detection of botulinal neurotoxin in the patient's serum, stool, wound, vomitus, and/or suspect contaminated food (6, 17, 49, 100). However, detection of botulinal neurotoxin is often negative in serum and feces after 48 h of exposure (17), and foods may be unavailable for analysis. Electrodiagnostic testing can provide diagnosis of botulism and is particularly useful

for patients with clinical signs but whose samples test negative in the mouse bioassay (6, 17, 18). Electrophysiologic testing is also useful for differential diagnosis of other causes of acute flaccid paralysis syndromes such as Guillain-Barré syndrome, myasthenia gravis, tick paralysis, and Lambert-Eaton syndrome (6, 17, 18). Guidelines for electrodiagnostic testing for botulism have been outlined (17, 18).

Currently there is no treatment for botulism except for passive administration of antibodies (available from the CDC and some State Health Departments) at early stages in the disease before botulinal neurotoxin has begun internalization into nerves (6, 7, 16, 17, 49). Antitoxin can effectively neutralize unbound toxin in the circulatory system, but it does not prevent the disease once receptor binding and the internalization process of botulinal neurotoxin are underway (46). Therefore, antitoxin should be administered as soon as possible. The major treatment of botulism is supportive nursing care, with specific attention given to respiratory ability and the need for mechanical ventilation (6, 17, 18). Nasogastric or parenteral nutritional support may also be required. A treatment called human botulism immune globulin intravenous (Baby-BIG) has been prepared from immunized human donors for treatment of infants with botulism (7). These antibody preparations have been demonstrated to reduce the severity of botulism in infants and reduce the duration of hospital stay (7). In this extensive study, infants treated with BIG-IV had a decreased duration of hospital stay from 6.4 to 2.0 weeks and from 2.2 to 1.5 weeks for type A and type B botulism, respectively (7). Information on the drug can be obtained from the website www.infantbotulism.org, supported by the California Department of Health Services.

ECOLOGY OF BOTULINOGENIC CLOSTRIDIA

Botulinogenic clostridia are widely dispersed in nature owing to their ability to form resistant endospores (24, 53, 105). They have been isolated mainly from two principal habitats: soils, including sediments within lakes and oceans, and the intestinal tracts of animals (but not healthy humans) (24, 51, 53, 63). The prevalence of C. botulinum worldwide varies according to the geographical region, which is probably related to the physical and chemical composition of the soil and microflora present (33, 53, 80). Spores are readily dispersed in dust and aerosols and thus frequently occur in many environments and in foods (24, 25, 51, 63). Neurotoxigenic clostridia are saprophytic and do not have an obligatory relationship with an animal host (51).

Geographical regions with a high prevalence of *C. botulinum* spores generally experience a higher incidence of botulism than regions containing fewer spores (24, 25, 53, 63). The highest numbers of cases of recorded botulism are from Poland, China, France, the United States, and Russia, and these countries have regions with high levels of *C. botulinum* spores (24, 25, 53, 63). *C. botulinum* spores appear to have a relatively low prevalence in several countries including Great Britain, Sweden, The Netherlands, Switzerland, Austria, Greece, Australia, New Zealand, Mexico, and countries in the South American and African continents (24, 53, 63). Type B (nonproteolytic) and type E spores are relatively common in regions of Europe and Scandinavia. Type E, associated with marine and freshwater foods, is the predominant cause of botulism in cooler aquatic regions, including coastal regions of Canada, Alaska, and northern Japan, Scandinavia, and regions of Russia (52, 53, 63). *C. botulinum* serotype A is most prevalent in the western continental United States, and type A spores also have a high incidence in regions of China and Argentina. *C. botulinum* type B (proteolytic) is most prevalent in eastern U.S. soils (52, 53, 63). Twenty-four percent of the soils tested in the United States were found to harbor *C. botulinum* spores (24). The incidence of spores was higher in sediments and soils in and near Lake Michigan and along the North American Pacific coast (24). The relative incidences of *C. botulinum* spores in soils and sediments are similar in analogous regions of Canada, Central and South America, Europe, and Asia (24, 53).

C. botulinum spores have also been associated with a variety of foods, generally at low concentrations (25, 53, 63). The incidence and type of spores in various raw foods usually reflect the prevalence of the spores in the geographical origin of the foods. For example, fish are often contaminated with *C. botulinum* spores due to the prevalence of spores in many marine and freshwater coastal environments. Foods harvested from continental soils such as vegetables and fruits often contain type A and B spores (24, 53, 63). Certain foods are rarely contaminated with *C. botulinum* spores including many types of meats such as raw poultry, beef, pork, and dairy products (25, 53). However, spores can be inadvertently added, particularly through supplementation with fish, vegetables, and dry ingredients such as dehydrated vegetables and spices. Processing conditions and plant hygiene also affect the contamination of foods by *C. botulinum* spores. When new processing techniques are implemented in food production, the process must be carefully evaluated to verify that it will not increase the risk of *C. botulinum* growth in foods. In addition to processing considerations, the formulation of foods, taking

into consideration intrinsic and extrinsic factors (including packaging), is critical for controlling *C. botulinum* growth and toxin production in raw and minimally processed low-acid foods (44, 92). This aspect is discussed in more detail below.

EPIDEMIOLOGY OF FOODBORNE BOTULISM

As described in the previous section, foodborne botulism occurs mainly in clustered geographic regions of the world (25, 42, 52, 63). In many countries it is very rare, although the actual incidence of botulism is undoubtedly greater than is reported. It is likely that mild cases are not diagnosed and patients are not admitted for treatment. Some countries currently do not have adequate public health facilities for botulism diagnosis (for an example, see reference 119). Since it is a rare disease, botulism is often misdiagnosed as another neurologic disorder (6, 17, 18, 40). Nonetheless, in the United States and certain other countries with capable public health facilities, the characteristic paralytic symptoms and records of release of antitoxin for treatment probably make hospitalized botulism one of the most accurately reported foodborne diseases (16).

In his pioneering ecological surveys and studies of *C. botulinum*, K. F. Meyer and associates concluded that the risk of botulism is greater in geographic regions that have a high incidence of type A, B, and E spores (80, 81). This has been supported by other studies of *C. botulinum* spore incidence, generally following a botulism outbreak. Current epidemiological and microbiologic evidence indicates that types A, B, and E (and rarely F and C) are responsible for human botulism. Since their discovery in the mid-1980s (42, 49, 50), *C. baratii* and *C. butyricum*, which produce botulinal neurotoxins, have been recognized as causes of infant botulism (42, 50). Botulinogenic *C. butyricum* has recently been associated with foodborne outbreaks (116).

The primary regions of the world with reports of human foodborne botulism are East Asia (China and Japan), North America, some countries in Europe (Poland, Germany, France, Italy, Spain, Portugal, Denmark, and Norway), the Middle East (Iran), Latin America, Russia, and South Africa (52, 63). Cases of hospitalized human botulism appear to be rare in the United Kingdom, although notable outbreaks have occurred including the Loch Maree tragedy, the Birmingham incident, and the hazelnut yogurt outbreak (63). Human botulism is also rare in Africa, Australia, Israel, Taiwan, Greece, New Zealand, India, Mexico, and several South American countries (52). However, these conclusions regarding the epidemiology of botulism require careful consideration with an increasing global food supply and regional and worldwide trade and travel.

In the United States, botulism has occurred primarily due to home-prepared foods (16, 100, 107), although restaurant-associated foods have been associated with several outbreaks during the past 2 decades (63, 76, 100, 107). Botulism outbreaks in the United States reached a peak during the 1930s as procedures for diagnosis were developed and then declined due to vigorous preventive measures in the commercial canned food industry and the development of guidelines for home canning (16, 80, 81). From 1899 to 1949, there were 1,281 cases and 830 deaths; from 1950 to 1977, there were 678 cases and 169 deaths; from 1978 to 1993, there were 423 cases and 31 deaths in the United States; and from 1990 to 2000, there were 263 cases and 11 deaths (16, 80, 81, 92). Currently, approximately 25 to 50 foodborne botulism cases are diagnosed annually (100, 107). In the United States from 1899 to the early 1990s, approximately 60% of foodborne botulism cases were caused by type A toxin, 18% by type B, and 22% by type E (16). Botulism is endemic in the Pacific Coast states and in Alaska. During the period 1990–2000, 39% of the cases occurred in Alaska and were caused by type E toxin in traditional Alaska Native home-prepared fermented fish and sea mammals (100, 107). In the lower 49 states, the most common vehicles were home-canned vegetables (107). Nine percent of the cases were due to two restaurant-associated outbreaks, each affecting 25 individuals (107). During the period 1981–2002 in the United States, of 1,269 reported cases, 13 adult type F botulism cases caused by toxigenic *C. baratii* were diagnosed (45). In only one case was a food vehicle identified, and *C. baratii* was isolated from leftovers of spaghetti and tuna (45). Toxigenic *C. butyricum* has recently been reported to have caused foodborne botulism in India (116).

Foods involved in mainland U.S. botulism are usually home-canned fruits and vegetables followed in incidence by home-prepared meats, fish, and other miscellaneous foods (100, 107). Of 182 outbreaks identified with foods during 1971–89, 137 were caused by fruits and vegetables, 15 by meats, 13 by fish, and 17 by other foods including mixed vehicles. As expected, vegetables harvested from soils have caused botulism, particularly onions, garlic, and potatoes (92, 100, 107). These outbreaks have mainly been caused by improper preservation procedures such as storing garlic pieces in oil and baked potatoes in foil at ambient conditions (2, 84, 108). Such handling procedures created an anaerobic environment for growth and toxin production by *C. botulinum*. Although it is clear that *C. botulinum* spores are found on certain

vegetables, surveys have suggested that the numbers of spores on foods are quite low (25, 58). *C. botulinum* spores are rarely found in commercially produced meats; however, they are often found in some varieties of fish.

Botulism from commercial and restaurant-prepared foods has unexpectedly occurred with changes in food handling and preservation practices. Contributing factors to several of these outbreaks have been described (2, 8, 12, 42, 44, 53, 100, 107, 108). An excellent example is the poor process control and maintenance of chilled distribution of smoked whitefish from the Great Lakes, which resulted in a resurgence of botulism in the 1960s (39, 53). In the 1970s, changes in packaging procedures and underprocessing led to botulinal toxin production by *C. botulinum* in canned mushrooms prepared by seven U.S. commercial producers (75). Production of foods with minimal or no preservatives and relying primarily on refrigeration for controlling *C. botulinum* has also led to outbreaks of botulism (44, 92). Increased consumption of foods in restaurants and from food service establishments in the United States and many other countries has had an impact on foodborne botulism (44, 63, 76, 100, 107). In 2001, an outbreak of type A botulism affecting 15 individuals occurred at a church supper from consumption of chili (67). In 1993, an outbreak of foodborne botulism affecting 17 persons occurred in Texas from eating skordalia dip prepared from baked potatoes cooked in foil and stored at ambient temperature following cooking (2). A botulism outbreak affecting eight persons involved a commercial cheese sauce that was properly processed commercially but then was left opened in a food service establishment without proper refrigeration. Type A spores introduced during handling grew and produced botulinum neurotoxin (115). In July 1998, a large botulism outbreak reportedly affecting 1,400 persons and causing 19 deaths resulted from consumption of spoiled meat and poultry products, including paté and a processed meat called "casher" (44). Poor hygiene in the processing plant and inadequate refrigeration in distribution contributed to the outbreak. In 1997, an outbreak of botulism affecting 27 patients and causing one death resulted from eating locally made cheese in Iran (44). A type A botulism outbreak that affected nine bus drivers in Argentina was caused by a restaurant-prepared Argentine meat roll (*matambre*) (119). A summary of select botulism outbreaks involving commercially processed or restaurant-prepared foods that have occurred in various regions of the world has been reviewed (44). These outbreaks highlight the need for implementation of safe food-processing and -handling procedures, a rigorous inspection program, and education to prevent future outbreaks of botulism.

The only food that has been definitively associated with infant botulism is honey (4, 40). In California, approximately 30% of infant botulism cases were associated with honey consumption, but this has decreased to less than 10% with increased public awareness. In Europe, honey consumption has been associated with nearly 50% of the cases of infant botulism (40). Analyses of botulism-implicated honeys have revealed that C. botulinum spore numbers can be quite low, as few as 5 to 70 spores per g. As with many other foods produced today, commercial honeys are often blends from several worldwide geographic regions. There have also been several reported incidences in which honey containing botulinum spores was fed to infants that did not contract botulism (4, 85). Type B infant botulism was diagnosed in a 5-month-old female baby, and infant formula was initially thought to be the vehicle, but epidemiologic analysis using physiological tests and pulsed-field gel electrophoresis did not support the conclusion that the outbreak vehicle was unopened formula (64). A recent study described dust as a vehicle for C. botulinum spores in infant botulism (85).

Botulism contracted from eating commercially prepared foods has been extremely rare in the United States and many other countries (12, 16, 42, 44, 75, 92, 100). The current good safety record for commercial foods is due, in large part, to the diligence of food manufacturers in formulating, processing, and controlling temperature during the distribution of foods (44, 63, 92). Other contributing factors include the low incidence of botulinal spores in prepared foods (52, 53, 105), competition with spoilage organisms, and consumption of foods before toxin production can occur.

MICROBIOLOGICAL CHARACTERISTICS OF BOTULINOGENIC CLOSTRIDIA

The genus Clostridium is a diverse assemblage of anaerobic, spore-bearing Eubacteria with a gram-positive cell wall structure (14, 31, 51, 63, 65). Clostridia are strict anaerobes and obtain energy by fermentation (14, 31, 63). In toxigenic species, the spores are generally wider than the vegetative organisms in which they are formed, imparting spindle shapes, the characteristic clostridial forms (51, 57, 65, 66). In culture, neurotoxigenic clostridia typically grow as large rod-shaped bacteria and often form filaments or chains. The vegetative cells are often curved, their sides are parallel, and their ends are rounded (51, 57, 65, 66).

The defining feature of botulinogenic clostridia is that they produce botulinal neurotoxin. The currently recognized species are C. botulinum, Clostridium argentinense

(type G), and rare strains of C. butyricum and C. baratii (50, 65, 66). Seven neurotoxin serotypes (A, B, C_1, D, E, F, and G) are currently recognized, which are distinguished by toxicity neutralization using serotype-specific antitoxins raised against purified neurotoxins. A Clostridium strain termed RKG that presumptively produces a botulinum-like neurotoxin has recently been reported (23). Since the genes encoding botulinal neurotoxin and the complex proteins appear to be associated with mobile elements in many botulinogenic clostridia (60, 63, 66), it seems likely that normally nontoxigenic clostridia can acquire the genes for toxin formation.

Certain features are valuable in the initial evaluation of C. botulinum cultures in addition to demonstration of botulinum neurotoxin. All strains are anaerobic, gram-positive sporeforming rods (49, 51, 57, 65, 66). The Gram reaction may be weak or appear negative in cultures 24 h or older (57, 65, 66). Spores are generally present in rich media, but their microscopic presence may require several days to 1 to 2 weeks of incubation. The lipase reaction exhibiting a pearly film surrounding the colonies on egg yolk agar is characteristic of all serotypes of C. botulinum except type G and C. baratii and C. butyricum (49, 66). Botulinogenic clostridia are extremely heterogeneous in their physiology, and significant differences can occur in several characteristics including heat stability of spores, temperature range of growth, acid and salt tolerance, proteolytic activity, and substrate utilization (Table 18.1).

Studies from several laboratories have revealed that C. botulinum can be divided into four physiologically different groups (groups I to IV), and this grouping has been widely accepted (14, 42, 49, 57, 63). These groups are genetically distinct as determined by various methods, including DNA hybridization, pulsed-field gel electrophoresis, multilocus enzyme electrophoresis, sequencing of genes encoding rRNA, and gene arrangements within the toxin gene clusters (42, 60, 61). With the identification of toxigenic C. butyricum and C. baratii, Groups V and VI have been suggested (42, 49). These physiological groupings have also been related to metabolic and physiological characteristics, including nutritional requirements (49, 51, 57, 123), resistance to salt and acidity and other environmental and food components (9, 44, 53, 63, 87), spore heat resistance and germination properties (35, 42, 44, 50, 53, 59, 75; also see chapter 3), tolerance of high-pressure treatment (76, 90, 91), resistance to chemicals and sanitizers (53, 59), resistance or tolerance to air and modified atmospheres (44, 53, 63, 72, 122), minimum growth temperature (32, 53, 105), end product formation (16, 42, 49, 57), neurotoxin gene cluster composition and arrangement, and neurotoxin

Table 18.1 Groupings and relevant growth and resistance properties of botulinogenic clostridia[a]

	Group					
	(*C. botulinum*)			IV (*C. argentinense*)	V (*C. butyricum*)	VI (*C. baratii*)
Properties	I	II	III			
Neurotoxin types	A, B, F	B, E, F	C, D	G	E	F
Growth temp						
Minimum	10°C	3°C	15°C	12°C	10°C	20°C
Optimum	35–40°C	18–25°C	35–40°C	35–40°C	30–37°C	30–40°C
Maximum	48°C	45°C	NA	45°C	~40°C	NA
Minimum pH	4.6	5.0	NA	NA	~3.6	NA
Inhibitory a_w	0.94	0.97	NA	NA	~5.0	NA
Inhibitory NaCl (NA)	10%	5%	3%	>3%	NA	5.0
$D_{100°C}$ of spores	~25 min	<0.1 min	NA	NA	NA	NA
$D_{121°C}$ of spores	0.21 min	<0.005 min				

[a] Data are for a limited number of strains. NA, insufficient data available. Inhibitory factors interact; therefore, the inhibitory values may be affected by sublethal combinations of factors. *C. sporogenes*, often considered to be a surrogate for Group I *C. botulinum*, has $D_{121°C}$ values ranging from 0.5 to 6 min (53, 63).

expression (13, 21, 22, 62). The four groups also have distinctive surface antigen relationships (49, 51) and differ in the host range of bacteriophages and antibacterial activity of bacteriocins (boticins) (62, 68, 105). Characteristics of botulinogenic clostridia, particularly pertaining to growth and survival in foods, are summarized in Table 18.1.

Neurotoxigenic clostridia that produce more than one serotype of botulinal toxin or that carry genes for unexpressed toxin genes have been isolated (21, 41–43, 63). Neurotoxin formation in *C. botulinum* is unstable in some serotypes and strains (62, 63, 96), and the genes of the neurotoxin gene complex can be associated with mobile elements in certain serotypes (21, 42, 51, 62, 105). Bacteria resembling *C. baratii* and *C. butyricum* that produce botulinum-like neurotoxins were originally isolated from infants with botulism (42, 50, 51), and neurotoxigenic *C. butyricum* and *C. baratii* have recently been isolated from wider geographical regions and associated with foodborne as well as infant botulism (42, 50, 62, 63, 116).

Botulinal neurotoxin production is affected by a variety of factors including temperature, amino acid and peptone composition and concentration, glucose concentration, and several other factors. In group I, arginine and glucose decrease toxin production and activation, while in *C. botulinum* type E (group II), tryptophan decreases toxin formation (reviewed in reference 62). Botulinal neurotoxin in type A is formed during late exponential phase and into early stationary phase (21). Following synthesis, the single-chain neurotoxin is activated to the dichain form during stationary phase in proteolytic strains (21, 62). The genetic regulation of botulinal toxin

production is a relatively new area of investigation, largely due to the lack of genetic methods for its study. Genetic tools for elucidating genes involved in toxin formation and activation as well as other traits in *C. botulinum* and related neurotoxigenic clostridia are under development. An interesting feature of the genetic regulation of botulinal toxin formation is that it is positively regulated by a potentially new class of RNA polymerase sigma factors (for a review, see reference 31). Toxin gene regulation in *C. botulinum* and other toxin-forming clostridia has been reviewed (31, 62). As genetic tools become available, much more information will be elucidated regarding the genes involved in botulinal toxin biosynthesis, proteolytic activation, complex formation, and other properties such as stability, secretion, and function in *C. botulinum*.

Prevention of Growth and Toxin Formation in Foods

The type of botulism that can most readily be prevented is foodborne botulism. Prevention of infant, adult intestinal, and wound botulism relies currently on minimizing exposure to spores, but because up to 90% of spore sources of infant botulism are currently unknown (honey and dust being exceptions) (4, 24, 25, 42, 55, 85, 105), preventing exposure to spores is not practical in most instances. In contrast, the primary factors that affect survival of spores during processing of foods and subsequent growth of *C. botulinum* therein have been well studied and models have been developed for control, resulting in the ability to design and implement appropriate food formulation and processing principles for prevention of botulism (9, 10, 44, 53, 79, 97, 103). Empirical testing results of the potential for *C. botulinum* to grow and produce botulinal neurotoxin in a variety of foods have also been extensively summarized to assist the food industry in designing and producing botulism-safe foods (58).

In summary, prevention of botulinal neurotoxin formation in foods can be achieved by (i) avoiding contamination of foods by spores; (ii) inactivating spores that are present in foods; (iii) preventing spores from germination and vegetative cell growth resulting in botulinal neurotoxin formation; and (iv) inactivation of botulinal neurotoxins in food. Practically, it is difficult to prevent contamination of many foods by spores due to their widespread distribution and the possibility of contamination of foods during harvesting, formulation, and processing. Consequently, considerable research and development efforts have been devoted to spore inactivation and preventing growth in foods (see above) (for reviews, see references 44, 53, 102, and 111).

Spore Inactivation

Thermal processing methods for inactivation of *C. botulinum* spores have been described in chapter 3 and are only briefly summarized in this section. The excellent safety record of botulism in commercial low-acid canned foods attests to the efficacy of industry practices (16, 44, 52, 75, 81, 86). Most foodborne botulism occurs from home-canned or -prepared and home-fermented foods (16, 105, 107). Guidelines to prevent botulism in home-prepared foods can be obtained from the CDC, FDA, USDA, local health departments, and many university outreach programs.

Thermal processing is of considerable importance in the food industry for inactivation of *C. botulinum*. Heat processes used for inactivation of *C. botulinum* spores in low-acid canned foods also inactivate other foodborne microbes of public health concern. Group I *C. botulinum* spores (serotype A and proteolytic strains of serotypes B and F) have a much higher heat resistance than group II spores (type E and nonproteolytic strains of serotypes B and F) (Table 18.1). Spores from nonproteolytic type B and F strains generally have higher heat resistances than serotype E (chapter 3). The molecular basis for differences among the different groups of botulinogenic clostridia has not been elucidated, but with sequencing of the genome of *C. botulinum* type A (Sanger Centre [see "Genomics of *C. botulinum*" below]), it is anticipated that genes controlling sporulation and resistance properties will be identified (88, 125; Sanger Centre). The greatest heat resistance occurs in certain strains of *C. botulinum* serotype A, for which a $D_{121°C}$ (250°F) of 0.21 min is often used by canning and packaging

industries as a cardinal value. Interestingly, in *C. sporogenes*, which is used as a surrogate for *C. botulinum*, spores can have a $D_{121°C}$ of 0.5 to 6 min (see chapter 3). This large difference in thermal resistance and certain other physiological characteristics brings into question the use of *C. sporogenes* as a surrogate in the development of processing procedures and for safety testing of formulations. An ideal nontoxigenic surrogate would be derived from *C. botulinum* deletion of the toxin gene cluster as was accomplished for strain 62A (60, 62).

Inactivation of *C. botulinum* spores by ionizing irradiation has been investigated in buffers and in foods (53; see chapter 3). In general, bacterial spores are 5 to 15 times more resistant than their corresponding vegetative cells. The kinetics of inactivation initially show a soft shoulder, followed by a logarithmic rate of inactivation (53; chapter 3). Spores from proteolytic strains were reported to have *D* values in the range of 0.2 to 0.45 Mrads (2.0 to 4.5 kGy) at −50 to −10°C. In the range of −200 to +50°C, *D* values decreased by approximately 1 krad per °C (53; chapter 3). In contrast to thermal inactivation, type E spores were not significantly more sensitive than type A spores. The presence of oxygen and other factors such as pH appear to affect inactivation and recovery, but the mechanisms are not known (53; chapter 3). From a practical food-processing perspective, it appears unlikely that ionizing irradiation would satisfactorily destroy *C. botulinum* spores at the relatively high concentrations associated with foods.

Potential food preservation methods under development including high-pressure, pulsed electric fields, ohmic heating, and light and sound treatments are in preliminary stages of evaluation for spores, particularly *C. botulinum* spores (53; see also chapter 3). Conducting research with *C. botulinum* requires a CDC and/or USDA-registered select agent facility (16, 77), and very few laboratories with capabilities for applying the aforementioned processes to botulinum spores are available. The limited reports evaluating high-pressure processing have indicated that some degree of inactivation does occur, particularly when combined with heat (78, 90, 91). There is a need for critical and careful evaluation of advanced physical preservation techniques for *Clostridium* and *Bacillus* endospores. It should be kept in mind that when a physical process is introduced into a food production process, inadvertent events can occur that may enable growth and toxin formation by *C. botulinum* (see "Growth Requirements of *C. botulinum* in Foods" below).

A limited number of chemicals and gases have been used for spore inactivation. Chlorine and related compounds are among the most effective chemicals for destruction of spores (59; chapter 3). Pretreatment of spores with chlorine may also sensitize *C. botulinum* spores to other inactivation treatments such as heat. The effectiveness of chlorine is dependent on the concentration and form of chlorine used, the presence of organic material, pH, duration of treatment, temperature, and other factors. Group I *C. botulinum* spores, which have the greatest heat resistance, are also more resistant to chlorine. Since chlorine dioxide is less susceptible to sequestration by organic materials, its use can have advantages over sodium hypochlorite. For decontamination of relatively clean surfaces, solutions of 100 to 200 mg of hypochlorite per liter for 2 min is sufficient for *C. botulinum* spore inactivation. In water used for cooling heat-treated cans, 1 to 2 µg of hypochlorite per ml is recommended (86). The absence of hypochlorite in cooling water probably contributed to one of the largest and most economically devastating episodes of botulism in recent history, which was caused by canned salmon (75, 86).

C. botulinum spores can also be inactivated by hydrogen peroxide. For sporicidal activity, relatively high concentrations (ca. 35%) are required (see chapter 3). Hydrogen peroxide is most stable at pH 3.5 to 4.0 and loses activity as the pH increases. High temperatures also increase the activity. Hydrogen peroxide has utility for sterilizing aseptic processing materials and certain packaging systems. Ozone at 5 to 6 mg/liter with an exposure time of 32 min can also inactivate *C. botulinum* spores (53).

Botulinal Neurotoxin Inactivation

The stability of botulinal neurotoxins in foods and clinical samples is an important factor contributing to the botulinal safety of foods, but neurotoxin inactivation has not been well studied (102). Inactivation of botulinal neurotoxin toxicity is dependent on serotype and the type of protein complex in which the neurotoxin is present. In general, botulinal neurotoxins are labile to heat, alkali, chlorine, and certain other physical and chemical treatments (53, 96, 102). Although botulinal neurotoxins are heat sensitive (102, 118), they are more stable to heat under acidic conditions, especially at pH values of 3.5 to 5, with organic acids, proteins, and certain ions such as Ca^{2+} and Mg^{2+} (53, 102). Heating at 70°C for 1 h or 80°C for 30 min or boiling for 5 min inactivated toxicity in buffers and foods (102). However, the rate of thermal inactivation does not follow a classic log-linear kinetics associated with first-order inactivation kinetics, and considerable tailing has been observed during heat inactivation of botulinal neurotoxin inactivation in various buffers and foods (53, 102), particularly at lower processing temperatures. The biphasic curves observed complicate the use of traditional D values to model

thermal inactivation of botulinal neurotoxin, and it has been proposed that heat resistance be expressed as the time required for inactivation to concentration values that are below the threshold for toxicity of the toxin (53). The basis for a small percentage of botulinal neurotoxin having increased apparent heat resistance is unknown, and it has been suggested that it could be due to the formation of protein complexes, aggregation, changes in physical shape, or renaturation on cooling. At least theoretically, the tailing observed could affect the degree of inactivation of high levels of botulinal neurotoxin during pasteurization processes used for various foods such as milk and the possibility for retention of toxicity has led to speculation on the adequacy of current time and temperature conditions used in commercial pasteurization procedures to absolutely ensure elimination of the neurotoxin (1, 121).

The stability of botulinal neurotoxins in drinking and lake water depends on the presence of organic matter, higher acidity, and the presence of hypochlorite (102). Botulinal neurotoxin was stable in distilled water for 7 days, and stability was enhanced at pH values of 4 to 5 and by the presence of proteins. Botulinal neurotoxins are sensitive to chlorine, and the quantities of chlorine generally used in drinking water are adequate for inactivation. Contaminated objects or surfaces can be decontaminated with 0.1 to 0.5% sodium hypochlorite solution. For routine sanitation and cleansing of surfaces, exposure to 0.1% hypochlorite (commercial bleach is about 5.25% hypochlorite) is adequate, and for higher levels of contamination 0.5% should be used with 20- to 30-min exposure, followed by rinsing with distilled water. For decontamination of spills, the toxin solution on a surface should be covered with absorbent material to prevent aerosolization. Botulinal neurotoxin A toxin complex is inactivated by alkali at pH values of >9.5 to 10 (ca. 0.1 M NaOH) (53, 69). Limited studies have suggested that botulinal neurotoxin is inactivated by 250-ppm ozone (53).

In general, botulinal neurotoxins are not affected by freezing, especially in the presence of proteins and organic acids at pH values of 5 to 6.5. The notable exception to stabilization by organic acids during freezing is acetate buffer, in which botulinum type A toxin is entirely inactivated (96). Botulinal neurotoxins were not significantly inactivated by gamma irradiation from a ^{60}Co source (102). Drying under vacuum or lyophilization can stabilize botulinum toxins particularly in the presence of proteins such as gelatin or albumin and excipients including trehalose (63, 96). Overall, the effects of physical and chemical treatments on botulinal neurotoxin and its protein complexes have not been thoroughly studied, although such information could be used to enhance the safety of foods.

Growth Requirements of C. *botulinum* in Foods

The primary factors that control growth of C. *botulinum* in foods are temperature, pH and acidity, water activity (a_w), redox potential, nutrient sufficiency, the presence of antimicrobials, and competitive microflora (Table 18.2) (44, 53, 54, 58, 63, 69). Compilations of growth and neurotoxin formation by C. *botulinum* in various foods and media are available (44, 54, 58). Critical values for growth of botulinogenic clostridia in foods are presented in Table 18.1. In practice, these limit values apply to relatively few strains, and most strains require slightly less stringent values for growth and toxin formation to occur. In foods formulated to be botulism-safe, inhibition relies on the inhibitory activities of combinations of factors (44, 53, 54, 111). Among the most important inhibitory factors are acidic pH and water activity. The critical pH for botulism safety of foods is pH 4.6, below which C. *botulinum* outgrowth of vegetative cells does not occur. Food products with pH values of ≤4.6 are termed high-acid foods and have an excellent record of botulism safety (87). Low-acid foods are those having an equilibrium pH of >4.6 and a water activity (a_w) of >0.85. These foods require suitable processing or preservation procedures to prevent survival or growth of C. *botulinum* (44, 53). In refrigerated, minimally processed foods in which growth of group II C. *botulinum* is of concern, a pH of 5.0 is sufficient to prevent germination and growth. Short-chain organic acids (e.g., lactic acid, acetate, or malic acid) with pK_a values in the range of 3.0 to 5.0 are generally used as acidulants as these are more inhibitory than mineral acids at equivalent pH

values (44). When acids are used as acidulants, adequate time and mixing are required for diffusion of the acids throughout the foods, occasionally several hours to days. If the foods are not to be subsequently processed, then these acidulated foods should be refrigerated (≤40°F) until equilibrium is reached.

The classification of foods into high- and low-acid categories has been a useful criterion for ensuring the botulism safety of most foods. However, botulism outbreaks have been attributed to high-acid foods, particularly in tomato products (16, 53, 63, 81, 111). Certain conditions can enable toxin formation in high-acid foods including inadequate penetration of acids into larger pieces of foods, hence creating microenvironments of higher pH. A second important factor has been termed metabiosis, whereby yeasts, molds, or bacteria can metabolize acids and other components such as proteins and raise the pH to allow C. *botulinum* to grow (53, 111). Fungal mats can form at the surface of certain foods, increasing the pH under the mat until permissive conditions for C. *botulinum* growth are achieved. Growth of C. *botulinum* has been reported in acid media in which precipitated protein or meat particles are present, likely forming a permissive microenvironment "pocket" for C. *botulinum* proliferation (53). Upper limits for growth of C. *botulinum* are in the range of pH 8 to 9, but growth inhibition by alkalinization is not practical in foods.

Control of water activity, particularly by addition of NaCl, is of considerable utility for C. *botulinum* control. Food preservation by brining is largely due to the reduction of water activity, whereby sufficient free water is not available to the bacterium. Water activity (a_w) is defined as the vapor pressure of a food divided by the vapor pressure of pure water, or the equivalent of relative humidity/100. Brining is one of the most common practice for reducing water activity in food preservation. The percent brine in a food is defined as % NaCl × 100/% (H_2O + % NaCl). The brine values for inhibiting growth of group I and group II C. *botulinum* are 10 and 5%, respectively (53). Most group I and II strains are inhibited at slightly higher a_w. High concentrations of sucrose can also inhibit C. *botulinum* growth by reducing the a_w, but relatively high levels of 30 and 15% sucrose are required for inhibition of group I and group II strains, respectively. Similarly, glycerol, organic polymers, ions such as potassium, and other food components can bind free water and reduce the water activity, but the efficacy of inhibition of C. *botulinum* is relatively poor on a weight/percent basis compared to NaCl. Therefore, when these humectants are used as preservatives, it is essential to measure the water activity of a food during its formulation and to

Table 18.2 Primary physical treatments and antimicrobials used in formulation of botulism-safe foods[a]

Thermal treatment
pH and acidity
a_w and solute composition
Presence of antimicrobials and competitive microflora
    Organic acids
    Nitrites, sulfites, phenolic compounds
    Polyphosphates
    Fatty acids and esters
    Gas composition
    Naturally occurring antimicrobials
    Competitive microflora
    Indirect antimicrobials

[a] Modified from reference 44.

conduct challenge studies of the food (30, 44) during development and prior to commercialization.

Temperature is also commonly used to control *C. botulinum* in foods. The minimum temperature for growth of group I strains is 10°C (53, 58, 63), whereas group II strains can grow at temperatures as low as 3.3°C (44, 53, 104). In practice, these minimum temperatures do not apply to many *C. botulinum* strains. Growth at lower temperatures can require weeks to months due to slowing of metabolism and growth rate. The upper temperature limits for growth for group I and group II strains are 45 to 50°C and 40 to 45°C, respectively. Refrigeration alone has been considered for inhibition of *C. botulinum* under otherwise permissive conditions in foods, as many consumers have expressed preferences for "healthy" foods that have minimal processing and contain low levels of salt or other preservatives. Although certain minimally processed foods depend on refrigeration alone for botulism safety, this practice can result in *C. botulinum* growth and formation of botulinum toxin because of poor temperature control or temperature abuse in food stores, in food service establishments, during distribution, or by consumers. Temperature abuse is one of the most common mishandling practices that result in botulinum neurotoxin production and botulism outbreaks. Therefore, it is not recommended to rely solely on low temperature for botulism safety, and adequate processing and/or formulation by inclusion of secondary barriers such as antimicrobials is recommended (Table 18.2).

Since *C. botulinum* is a strict anaerobe, a permissive redox potential (E_h) is required for germination of spores and growth of vegetative cells. However, relatively high E_h levels of approximately +200 mV can allow germination and growth initiation of spores under certain conditions (53). Whiting and Naftulin (122) reported that the critical level of oxygen for germination and growth by *C. botulinum* is approximately 1 to 2%. The presence of carbon dioxide can enhance germination of botulinal spores (53). Although reduced oxygen and increased carbon dioxide levels may be permissive for *C. botulinum* growth and toxin production, increased safety risk depends not only on the gas environment but also on the product, storage temperature, packaging film used, and indigenous competitive flora.

Modified-atmosphere packaging and high-barrier films are used to effectively extend the shelf life of foods under refrigerated conditions. Reduced oxygen levels and increased concentrations of carbon dioxide in refrigerated modified-atmosphere packaged foods can decrease oxidative and chemical deterioration, as well as inhibit common aerobic spoilage microbes such as fungi and many bacteria (44, 92). Concerns have been raised

that these conditions may also select for anaerobic psychrotrophic bacteria such as group II *C. botulinum* and increase the botulism risk (53, 111, 112). The combination of high-barrier films and respiring foods such as vegetables may also lead to decreased oxygen levels for packaged foods in an ambient atmosphere. The reduced oxygen content, in turn, may prompt the growth of *C. botulinum*. In packaged mushrooms, a high-barrier plastic film reduced oxygen concentration through respiration of the mushrooms, resulting in botulinum toxin production (111, 112). Toxin was detected in prepackaged mushrooms after 3 days of storage, although mushrooms were still considered organoleptically acceptable (112). A simple solution to prevent botulinum toxin formation was to increase gas exchange by introducing two to four holes in the wrap covering the mushrooms. Similarly, packaging of garlic in oil and relying solely on refrigeration for botulism safety resulted in outbreaks and changes in regulatory formulation requirements (84). In summary, oxygen content alone cannot be relied on to inhibit botulinum growth and toxin production. The effect of food components, such as sulfhydryls or competitive microflora, may also reduce oxidative-reduction potential to a level at which *C. botulinum* can grow. Therefore, similar to temperature control, redox potential and enhanced oxygen transfer are not recommended as sole barriers to prevent growth of *C. botulinum*, and suitable processing conditions or inclusion of secondary barriers are recommended.

An important factor for prevention of *C. botulinum* growth and toxin formation in many foods is the presence of competitive microflora. In general, foods have low levels of *C. botulinum* spores, and the pathogen is often inhibited by competitive microflora such as lactic acid bacteria or yeasts that ferment sugars and other substrates in the food with production of inhibitory levels of organic acids, alcohols, and bacteriocins. Competitive microflora or antimicrobial fermentates derived from cultures have been included in food formulations to control growth of *C. botulinum* (44, 48, 68, 94).

For most foods, the use of a high level of a single inhibitor reduces the organoleptic acceptability of the food. Therefore, most foods are formulated to be organoleptically desirable by using combinations of inhibitory factors at sublethal levels. The individual actions often have cumulative or occasionally synergistic effects for inhibition of *C. botulinum*. An example of a low-acid food prepared to be botulism-safe by using a combination of treatments and antimicrobials is cured meats in which minimal heat processing is combined with the addition of sodium nitrite (53, 111). Another example is pasteurized process cheese, in which sublethal heat processing is combined with subinhibitory levels of moisture content and NaCl content, acid, and phosphate salts (44, 114). Process cheese and related products have an excellent safety record with regard to botulism (44). Extensive studies on process cheese at the Food Research Institute led to one of the first empirical models for botulism-safe product formulation (114). Strategies for controlling *C. botulinum* by using combinations of physical treatments and antimicrobials in various classes of foods have been described (10, 44, 53, 54, 79, 103).

The nutrient composition of a food is also important in the control of *C. botulinum*. The minimal nutrient requirements have been determined for growth of *C. botulinum* in chemically defined media (63, 96, 123). Certain nutrients are required in high quantities for growth of *C. botulinum*, such as arginine for group I *C. botulinum* and tryptophan for type E (62, 63, 96, 123). Increasing the arginine content also promoted growth in otherwise inhibitory foods, probably through active metabolism with release of ammonia and increase in pH (63). The presence of glutamic acid in medium increases salt tolerance in group I *C. botulinum* possibly by serving as a compatible solute or as a precursor to a protective osmolyte (63). Substrate utilization as well as protein and lipid degradation varies among the various botulinogenic clostridial groups (Table 18.3) (14, 42, 49, 63, 66, 123). All of the groups except *C. argentinense* (group IV) are able to utilize glucose as a primary carbon source. Depending on the group, certain strains ferment fructose, maltose, sucrose, and sorbitol to varying degrees. While *C. baratii* and *C. butyricum* utilize lactose, insignificant acid and growth from lactose are observed with *C. botulinum* groups I to IV. Similarly, the ability of the botulinogenic clostridia to digest proteins differs significantly among the groups (14, 42, 49, 63, 66). Strong proteolysis is observed by digestion of meat

particles or milk (casein as substrates), whereas gelatin is much more easily digested. Botulinogenic clostridia within groups I, III, and IV are proteolytic, as they digest milk, meat particles, and gelatin. Group II *C. botulinum* is nonproteolytic and does not digest casein or meat, but it does degrade gelatin. *C. butyricum* and *C. baratii* do not digest the complex proteins, nor do they utilize gelatin. Lipase and lecithinase reactions on plates are commonly used for characterization of clostridia (14, 49, 50, 66). Lipase hydrolyzes the breakdown of triglycerides into glycerol and fatty acids and is observed as a pearly lustrous film surrounding the colonies. Lecithinase mediates the breakdown of lecithin to diglyceride and phosphorylcholine and is seen as an opaque whitish halo of precipitation around the colony. Groups I to III of *C. botulinum* have a positive lipase reaction, whereas *C. argentinense*, *C. butyricum* type E, and *C. baratii* are negative in this activity. A summary of substrate characteristics is presented in Table 18.3.

USE OF PREDICTIVE MODELING AND CHALLENGE STUDIES FOR EVALUATION OF *C. BOTULINUM* NEUROTOXIN FORMATION

Predictive modeling can be useful in assessing the processing and formulation parameters for a botulism-safe food. Among the earliest applications of predictive modeling in food microbiology was the development of processing models to determine commercial sterility in canned foods (35; also see chapter 3). During the past 2 decades, advances in computing, software development, and the widespread availability of the Internet have led to significant advances in the field of predictive microbiology. These advances have led to the development and availability of predictive modeling tools (10, 35, 44, 53,

Table 18.3 Nutritional substrates metabolized by botulinogenic clostridia[a]

Organism	Glucose	Fructose	Sucrose	Maltose	Lactose	Gelatin	Milk	Meat	Lipase	Lecithinase
C. botulinum										
Types A, B, F (group I)	+	+/−	+/−	+/−	−	+	+	+	+	−
Types B, E, F (group II)	+	+	+/−	+/−	−	+	−	−	+	−
Types C, D (group III)	+	+/−	+/−	+/−	−	+	+/−	−	+	−
C. argentinense (group IV)	−	−	−	−	−	+	+	+/−	−	−
C. baratii toxigenic (group V)	+	+	+	+/−	+/−	−	−	−	−	+
C. butyricum toxigenic (group VI)	+	+	+	+	+	−	−	−	−	−

[a] Symbols: +, positive; −, negative; +/−, weak or variable.

59, 63, 79, 93, 103). These modeling approaches have provided valuable insights into the combined effects of biological and environmental factors on the growth, survival, and death of C. botulinum.

Predictive models provide valuable information for the development of botulism-safe foods. However, they should be validated for specific foods by challenge studies in a qualified laboratory (30, 44). The National Advisory Committee on the Microbiological Criteria for Foods developed guidelines for botulinum challenge studies on minimally processed refrigerated foods with extended shelf life (30). Similar approaches can be applied to shelf-stable low-acid foods with the additional recommendation that botulinum neurotoxin production should be determined for twice the expected shelf life rather than 1.5 times, as recommended for refrigerated foods. The challenge study should mimic as closely as possible the commercial process for production of the food and should utilize suitable strains of C. botulinum rather than related clostridia that have been proposed as surrogates. It is anticipated that as modeling procedures continue to advance, they will provide more in-depth information and guidelines to assist the food industry in the formulation, processing, production, and marketing of safe foods.

Laboratory Procedures

Since botulism is quite rare, it is recommended that toxin and organism diagnostic tests be performed in a suitable reference laboratory that has the necessary experience and reference cultures, toxins, and antitoxins for the procedures (16, 49, 77). A limited number of industry, government (local, state, and national), and academic laboratories have these capabilities. The primary laboratory in the United States is the National Botulism Surveillance and Reference Laboratory, Centers for Disease Control and Prevention, Atlanta, Ga. Information regarding botulism can be found on the websites http://www.cdc.gov and http://www.bt.gov/agent/botulism/index.asp, and the emergency telephone number to report botulism cases and request antitoxin is 770-488-7100. Emergency 24-h telephone numbers or Internet connections of qualified local and state public health laboratories (http://www.cdc.gov/other.htm#states or http://www.astho.org/state.html) can also be contacted (6, 16).

Detection of botulinum neurotoxin from representative specimens is a cornerstone of botulism diagnosis (16–18, 49, 96, 110, 111). The accepted procedure for botulinum neurotoxin detection is the mouse bioassay consisting of two essential steps: (i) determining whether a food substrate or an extract in gel-phosphate (0.1 M sodium phosphate −0.2% gelatin, pH 6.2) is lethal on intraperitoneal or intravenous injection into mice, and (ii) confirming the lethal agent as botulinum neurotoxin by neutralization with specific botulinum antitoxin (16, 49, 110, 111). Detailed procedures, required controls, and difficulties and pitfalls of the mouse bioassay have been described (16, 49, 109, 111). When botulinum neurotoxin of a nonproteolytic C. botulinum strain is tested, it is necessary to activate the toxin to the light and heavy chains by limited proteolysis using trypsin (16, 49, 110, 111). One problem often encountered when testing clinical and food samples is that these specimens may contain nonbotulinal substances that are lethal to mice and cause nonspecific deaths. These can be detected by evaluation of characteristic botulism symptoms, by onset time of symptoms in mice (generally >3 to 4 h for mice injected intraperitoneally), by "diluting out" these substances which generally have a lower toxicity than botulinum neurotoxin, and most definitely by use of serotype-specific antitoxins in a subset of the samples used for injection. Another difficulty encountered is that samples may contain more than one serotype of toxin due to contamination by C. botulinum producing different serotypes of toxins or by single strains that produce more than one serotype of toxin. These difficulties in toxin analysis have been described (16, 43, 49, 109). When delays in specimen collection and analysis occur or when mouse-lethal substances other than botulinum neurotoxin are present in samples, laboratory detection of botulinum neurotoxin may not be conclusive. In clinical cases of botulism, a positive test was reported to occur in only about 30% of samples obtained more than 2 days after human neurotoxin exposure (17). Therefore, it is important to process samples rapidly and to culture neurotoxigenic clostridia from the foods as described below. C. botulinum may produce lower quantities of toxin after repeated laboratory culturing, and clinical or wild isolates should be preserved in liquid nitrogen or at −80°C in oxygen-impermeable containers (63, 96). Toxin titers produced by strains and the identity of botulinal neurotoxin should be periodically confirmed using pure stock isolates and by mouse bioassay and neutralization by specific antitoxins (63, 96).

For many decades and to the present, the mouse bioassay has been and remains the most important and accepted laboratory method for detection and identification of botulinum neurotoxin (16, 49, 96, 111). However, due to certain drawbacks in the mouse assay as well as increased regulatory and ethical concerns in using animals for toxin determinations, there is considerable interest in assays not employing animals. The primary alternative assays employed have been based on immunological detection, particularly enzyme-linked immunosorbent assays and related immunological methods (16, 37, 51, 109). Since botulinum neurotoxins are zinc metallopeptidases with high specificity for their neuronal substrates, methods based on catalytic activity combined with sensitive detection methods, including high-throughput fluorogenic reporters and fluorescence resonance energy transfer sensing systems, have been developed (3, 47, 101). A promising methodology for detection of botulinum neurotoxins and their protein complexes is mass spectroscopy including high-resolution platforms to characterize the toxins and to detect reaction products from proteolytic cleavage of the neuronal substrates or detection of signature amino acid sequences of the neurotoxins (11, 56). It was reported that mass spectrometry could detect toxin equivalents of as little as 0.01 MLD_{50} and concentrations as low as 0.62 MLD_{50} per ml. Biosensor devices including microfluidic and nanofluidic platforms, synaptic chips, and other platforms are also in development (36, 83, 101). Currently most of these in vitro assays have the drawbacks that they are less sensitive than the mouse bioassay, have not been adequately validated for botulinum neurotoxin in foods and clinical samples, and do not measure each of the molecular steps required for intoxication and botulism (70, 82, 96, 101). Modifications of, and alternatives to, animal assays have also been considered for detection of antibodies to botulinum neurotoxin in the sera of patients who have been treated medicinally with the toxin (29, 98, 99). A nonanimal assay that depends on all the steps in the intoxication mechanism, including receptor binding, internalization, and catalytic cleavage of neuronal substrates such as cell-based assays (27, 28, 29, 46), would be ideal for determination of toxin titers and antibodies in patients' sera. Of particular promise is the use of neuronal cell cultures, since toxicity of these cells depends on all domains of botulinum neurotoxin and steps needed for intoxication (27–29, 46). However, most neuronal cell lines are fairly insensitive to botulinum neurotoxins. An approach to increase the sensitivity will be to introduce receptor proteins into the cell membrane, and this appears promising because the protein receptors and required ganglioside components are currently being elucidated (27, 28, 82).

Culturing Botulinogenic Clostridia

A food or clinical sample can be definitively shown to be the cause of botulism only if botulinal neurotoxin activity is demonstrated. However, an important aspect of investigations of botulism outbreaks is the culturing of the responsible botulinogenic bacteria (16, 41, 49, 51, 57, 65, 66, 109). The isolation, identification, and maintenance of pure cultures of botulinogenic bacteria present certain practical difficulties. Clostridia tend to grow as consortia, and pure cultures are often difficult to achieve and maintain (41, 63, 65, 66). The purity of botulinogenic clostridia must be ascertained by microscopy and by plating on nonselective media. Cultures should be routinely tested for botulinum neurotoxin formation by using the mouse bioassay and for spore formation. Since botulinogenic clostridia have complex nutrient requirements, rich media are commonly used for cultivation (16, 41, 49, 51, 65, 66, 109). Enrichment of neurotoxic clostridia is often carried out by heating the samples (e.g., 60 to 80°C for 10 min) or by treating the samples with ethanol to kill vegetative microbes and enrich for sporeformers (16, 49, 51, 65, 66, 109, 111). Non-heat-treated samples should also be cultured since spores may not be present in the samples. From a practical perspective, usually a small quantity of food and clinical specimens, ca. 1-g samples, are inoculated to 10 ml of cooked meat medium containing 0.5% glucose, or Trypticase-peptone-glucose-yeast extract broth (16, 49, 111). For isolation of *C. botulinum* Type E, the addition of trypsin to the medium may enhance recovery by the inactivation of bacteriocins that may be present (49, 111). When pure cultures are obtained, they are characterized by various tests, particularly by demonstration of botulinum neurotoxin by the mouse bioassay and specific neutralization by antibodies. Further characterization often includes lipase reaction, microscopic observation and Gram stain, proteolysis, and substrate utilization patterns (49, 57, 65, 66). Formerly, the volatile fatty acids produced were routinely determined by gas-liquid chromatography, but this is becoming less common, whereas molecular techniques such as the determination of the nucleotide sequence of genes encoding rRNA or interspacial regions and PCR analyses for toxin genes have become integral for identification (31, 41, 42, 65, 66). Reference works describe preparations of media, methods for anaerobic culture, and phenotypic and metabolic tests for identification.

Safety Precautions in Working with *C. botulinum* and Botulinum Neurotoxins

Botulinum neurotoxins are extremely toxic molecules and are considered the most potent poisons known (6, 63, 96). They have an estimated lethal human intravenous dose of 0.1 to 1 ng per kg of body weight and an oral lethal dose of 0.1 to 1 µg per kg (6, 95, 96). Because the consequences of an accidental intoxication are so severe, safety must be a primary concern of scientists studying these toxins (16, 49, 65, 77, 96). The CDC recommends biosafety level 3 primary containment and personnel precautions for facilities making large quantities of the botulinal neurotoxins and in working with high-producing strains (16, 65, 77). All personnel who work in the laboratory should be thoroughly educated on the hazards of working with *C. botulinum* and its toxins. They must have knowledge of spill control and toxin inactivation (0.1 to 0.5% hypochlorite or 0.1 M NaOH). Operations should be performed to prevent the formation of aerosols, e.g., use of closed containers during centrifugation, avoiding pressurized containers containing active toxin, and application of absorbent materials on spills prior to decontamination. Proper personal care protection must be used including eye protection and use of gloves and lab coats. Personnel who work with high levels of toxin (≥0.5 mg and/or high-producing cultures) should be immunized with pentavalent (A through E) toxoid, available from the CDC. Alternative toxoids are under preparation. A biosafety manual should be posted in the laboratory and should contain the proper emergency phone numbers and procedures for emergency response, spill control, and decontamination. When possible, culture and toxin handling and manipulation should be performed in a class II or III biological safety cabinet with appropriate respiratory protection. The use of needles and syringes for bioassays requires extreme caution. *C. botulinum* cultures and toxins are included in a group of select agents whose transfer is controlled by the CDC; restricted and secure working areas are mandatory (16, 78). To transfer select agents, both the laboratory and personnel sending and receiving cultures and toxin must be registered with the USDA and/or CDC and exchange the appropriate approval forms through their Biosafety Office. The laboratory and Principal Investigator should maintain frequent communication with the Biosafety Office and Responsible Official within the organization.

GENOMICS OF *C. BOTULINUM*

The genomic sequences of several *Clostridium* species such as *C. tetani*, *C. perfringens*, and *C. acetobutylicum* have recently been determined (31, 60) and others are being completed, including those of *C. difficile* and *C. botulinum*. The genome sequence of a single strain of each species has been determined, and it is not clear if the sequences will be representative of most strains, particularly in *C. botulinum* and *C. perfringens*, in which several distinct groups of organisms occur within the species. Additional genome sequencing is being performed with additional strains and types. The analyses of the genomes of the clostridia to date have revealed interesting features regarding the pathogenicity, spore formation, and metabolism of clostridia (31, 60, 88, 125). The availability of the genome sequence of *C. botulinum* type A (ATCC 3502) and eventually other serotypes and physiological groups will enable rational

approaches for identification of genes and proteins for development of inhibitors and novel antimicrobials that target the pathogen for enhancement of food safety.

The complete genome sequence of a type A strain of *C. botulinum* (strain Hall, ATCC 3502) is being determined by the Sanger Centre in England and its fully annotated sequence is anticipated to be published soon. The genome size is 3,886,916 bp and the G+C content is 28.2%. The strain contains a 16.34-kb plasmid pBOT3502, has a G+C content of 26.8%, and possesses a gene for a bacteriocin analogous to a boticin gene. The genome contains ~3,620 open reading frames, ~10 rRNA operons, and ~35 putative two-component regulators. The genes *botulinal neurotoxin/A* and *botr/A* are located on the chromosome. *C. botulinum* ATCC 3502 harbors a single toxin gene cluster, coding for two polycistronic mRNAs. The toxin gene cluster is flanked by defective insertion elements and transposases that are apparently nonfunctional because they contain several mutations. Genes for two putative hemolysins were also present, and further analysis of the genome will undoubtedly reveal additional genes for putative virulence factors. Currently, the genome information for *C. botulinum* Hall A (ATCC 3502) is available on the Sanger Centre website (http://www.sanger.ac.uk/projects/C_botulinum/) and it is expected to be published with a detailed analysis of genomic features in 2006.

The genome appears to be similar to that of *C. tetani* in that there is no obvious evidence of foreign DNA acquisition (31). Interestingly, analyses of the genomes for clostridia showed that genes involved in sporulation in *Bacillus* sp., including *spo0B*, *spo0F*, *spoVf*, *spoVM*, *kinA*, *kinE*, and additional genes of the phosphorelay system, and *spoI ger* were lacking or nonfunctional in *C. perfringens*, *C. acetobutylicum*, *C. tetani*, and probably in *C. botulinum* (11, 31, 57, 65, 88, 125). These results suggest that the mechanism of sporulation and spore germination is significantly different from that of *Bacillus*. Since *C. botulinum* is a very diverse species, it should be valuable to obtain the genome sequences of other strains and types and also to elucidate the proteome or the total proteins expressed by the bacterium. The availability of the genomic sequences of *C. botulinum* will also enable the development of microarrays for hybridization analysis of genes involved in pathogenesis as well as genes in important processes including sporulation, stress responses, and genes essential for growth in the intestinal tract, in wounds, and in foods.

Botulinum Neurotoxin and Food Bioterrorism

Botulinum neurotoxin is considered a potential biological warfare agent that could be administered in aerosols, foods, or water (6), and some history supports this consideration (6, 15). Botulinum neurotoxin is absorbed through mucous membranes, and three cases of botulism were documented in laboratory workers who apparently inhaled the toxin (16, 120). Botulinum neurotoxin is labile to many environmental conditions and chemicals, and the preparation of an aerosol weapon would be difficult. Immunization is not feasible for protection of human populations from botulism owing to the rarity of the disease, but pentavalent (A through E) toxoid is used for immunization of researchers and for military personnel who may be exposed to botulinum neurotoxin in warfare (6). Since immunization provides only partial protection against intoxication with large quantities of toxin, immunized researchers must still follow scrupulous laboratory practices in working with botulinum neurotoxin, including avoidance of aerosols, handling toxin in the biological safety cabinets, and use of closed containers during centrifugation and other procedures (16, 60, 77). Considerable efforts in the United States and some other countries are being devoted to the development of heptavalent vaccines, small molecule inhibitors, and other countermeasures.

CONCLUSIONS

Remarkable advances have been achieved during the past decade in elucidating the biochemistry, structure, and pharmacological mechanisms of botulinum neurotoxins. Structural and biochemical studies of these potent neurotoxins have provided much insight into the mechanisms of substrate catalysis, neurospecific binding, and trafficking of botulinum neurotoxin to their neuronal targets. These advances have certainly contributed to the remarkable success of botulinum neurotoxin as a pharmacological agent for the treatment of various neuronal diseases and may lead to improved vaccines and countermeasures. The availability of genomic sequences and comparative genomic analyses, together with the development of genetic tools such as gene replacement and vectors for controlled gene expression, will be invaluable in elucidating pathogenic mechanisms of botulinogenic clostridia.

Botulism is a rare disease, but its occurrence from consumption of foods can have great economic impact on the food industry, as well as tremendous negative exposure by the media. Consequently, inactivation of spores in foods or the prevention of growth by preservation methods and food formulation is an important goal of the food industry. Considerable information is available regarding the microbiological features of botulinogenic clostridia and preservation and formulation strategies for their control in foods. Nonetheless, botulism continues to occur through consumption of foods, and new technologies and research are needed to enhance control. Newer processing procedures such as pulsed electric fields, ohmic heating, high pressure, and light and sound require careful evaluation before they are widely applied in the food industry. Research on control of botulinogenic clostridia in food systems has contributed to fundamental and applied knowledge in the food industry, and it is anticipated that more discoveries will arise from the study of botulinogenic clostridia and their neurotoxins.

Research in my laboratory has been supported by the NIH, FDA, USDA, University of Wisconsin, and industry sponsors of the Food Research Institute, University of Wisconsin-Madison. I am grateful to my laboratory personnel over the years, and to collaborators and mentors on various projects involving neurotoxigenic clostridia.

References

1. **Alberts, B.** 2005. Modeling attacks on the food supply. *Proc. Natl. Acad. Sci. USA* **102:**9737–9738.

2. **Angulo, F. J., J. Getz, J. P. Taylor, K. A. Hendricks, C. L. Hatheway, S. S. Barth, H. M. Solomon, A. E. Larson, E. A. Johnson, L. N. Nickey, and A. A. Ries.** 1998. A large outbreak of botulism: the hazardous baked potato. *J. Infect. Dis.* **178:**172–177.

3. **Anne, C., F. Cornille, and C. Lenoir.** 2001. High-throughput fluorogenic assay for determination of botulinum type B protease activity. *Anal. Biochem.* **291:**253–261.

4. **Arnon, S. S.** 2004. Infant botulism, p. 1758–1766. *In* R. D. Feigen and J. D. Cherry (ed.), *Textbook of Pediatric Infectious Diseases*, 5th ed. W. B. Saunders, Philadelphia, Pa.

5. **Arnon, S. S., K. Damus, and J. Chin.** 1981. Infant botulism: epidemiology and relation to sudden infant death syndrome. *Epidemiol. Rev.* **3:**45–66.

6. **Arnon, S. S., R. Schechter, T. V. Inglesby, D. A. Henderson, J. G. Bartlett, M. Ascher, E. Eitzen, A. D. Fine, J. Hauer, M. Layton, S. Lillibridge, M. T. Osterholm, E. O'Toole, G. Parker, T. M. Perl, P. K. Russell, D. L. Swerdlow, K. Tonat, and the Working Group on Civilian Biodefense.** 2001. Botulinum toxin as a biological weapon. Medical and public health management. *JAMA* **285:**1059–1070.

7. **Arnon, S. S., R. Schechter, S. E. Maslanka, N. P. Jewell, and C. L. Hatheway.** 2006. Human botulism immune globulin for the treatment of infant botulism. *N. Engl. J. Med.* **354:**462–471.

8. **Aureli, P., M. Di Cunto, A. Maffei, G. De Chiara, G. Franciosa, L. Accorinti, A. M. Gambardella, and D. Greco.** 2000. An outbreak in Italy of botulism associated with a dessert made with mascarpone cream cheese. *Eur. J. Epidemiol.* **16:**913–918.

9. **Baird-Parker, A. C., and B. Freame.** 1967. Combined effect of water activity, spores, and temperature on the growth of *Clostridium botulinum* from spores and vegetative cell inocula. *J. Appl. Bacteriol.* **30:**420–429.

10. **Baranyi, J., and T. A. Roberts.** 2000. Principles and applications of predictive modeling of the effects of preservative factors on microorganisms, p. 342–358. *In* B. M. Lund, T. C. Baird-Parker, and G. W. Gould (ed.), *The Microbiological*

Safety and Quality of Foods, vol. 1. Aspen Publishers, Gaithersburg, Md.

11. **Barr, J., H. Moura, A. E. Boyer, A. R. Wollfitt, S. R. Kalb, A. Pavlopoulos, L. G. McWilliams, J. G. Schmidt, R. A. Martinez, and D. L. Ashley.** 2005. Botulinum neurotoxin detection and differentiation by mass spectroscopy. *Emerg. Infect. Dis.* **10:**1578–1583.

12. **Bell, C., and A. Kyriades.** 2000. Clostridium botulinum. *A Practical Approach to Its Control in Foods.* Blackwell Science, Oxford, United Kingdom.

13. **Bradshaw, M., S. S. Dineen, N. D. Maks, and E. A. Johnson.** 2004. Regulation of neurotoxin complex expression in *Clostridium botulinum* strains 62A, Hall A-*hyper*, and NCTC 2916. *Anaerobe* **10:**321–333.

14. **Cato, E. P., W. L. George, and S. M. Finegold.** 1986. Genus *Clostridium*, p. 1141–1200. *In* P. H. A. Sneath, N. S. Mair, M. E. Sharpe, and J. G. Holt (ed.), *Bergey's Manual of Systematic Bacteriology*, vol. 2. The Williams & Wilkins Co., Baltimore, Md.

15. **Caya, J. G.** 2001. *Clostridium botulinum* and the ophthalmologist: a review of botulism, including biological warfare ramifications of botulinum toxin. *Surv. Ophthamol.* **56:**25–34.

16. **Centers for Disease Control and Prevention.** 1998. Botulism in the United States, 1899–1996. *In Handbook for Epidemiologists, Clinicians, and Laboratory Workers.* Centers for Disease Control and Prevention, Atlanta, Ga.

17. **Cherington, M.** 1998. Clinical spectrum of botulism. *Muscle Nerve* **21:**701–710.

18. **Cherington, M.** 2004. Botulism: update and review. *Semin. Neurol.* **24:**155–163.

19. **DasGupta, B. R.** 1989. Structure of botulinum neurotoxin, p. 53–67. *In* L. L. Simpson (ed.), *Botulinum Neurotoxin and Tetanus Toxin.* Academic Press, San Diego, Calif.

20. **Dickson, E. C.** 1918. Botulism. A clinical and experimental study. *Rockefeller Inst. Med. Res. Monog.* No. 8:1–117.

21. **Dineen, S. S., M. Bradshaw, and E. A. Johnson.** 2003. Neurotoxin gene clusters in *Clostridium botulinum* type A strains: sequence comparison and evolutionary implications. *Curr. Microbiol.* **46:**345–352.

22. **Dineen, S. S., M. Bradshaw, C. Karasek, and E. A. Johnson.** 2004. Nucleotide sequence and transcriptional analysis of the type A2 neurotoxin gene cluster in *Clostridium botulinum. FEMS Microbiol. Lett.* **235:**9–16.

23. **Dixit, A., R. K. Dhaked, S. I. Alam, and L. Singh.** 2005. Characterization of *Clostridium* sp. RKD producing botulinum-like neurotoxin. *Syst. Appl. Microbiol.* **28:**405–414.

24. **Dodds, K. L.** 1993. *Clostridium botulinum* in the environment, p. 21–51. *In* A. H. W. Hauschild and K. L. Dodds (ed.), Clostridium botulinum: *Ecology and Control in Foods.* Marcel Dekker, New York, N.Y.

25. **Dodds, K. L.** 1993. *Clostridium botulinum* in foods, p. 53–68. *In* A. H. W. Hauschild and K. L. Dodds (ed.), Clostridium botulinum: *Ecology and Control in Foods.* Marcel Dekker, New York, N.Y.

26. **Dolman, E. C.** 1964. Botulism as a world health problem. *In* K. H. Lewis and K. Cassel (ed.), *Botulism.* U.S. Public Health Service, Washington, D.C.

27. **Dong, M., D. A. Richards, M. C. Goodnough, W. H. Tepp, E. A. Johnson, and E. R. Chapman.** 2003. Synaptotagmins I and II mediate entry of botulinum neurotoxin B into cells. *J. Cell Biol.* **162:**1293–1303.

28. **Dong, M., F. Yeh, W. H. Tepp, C. Dean, E. A. Johnson, R. Janz, and E. R. Chapman.** 2006. Receptor for botulinum neurotoxin type A. *Science* **312:**592–596.

29. **Dong, M., W. H. Tepp, E. A. Johnson, and E. R. Chapman.** 2004. Using fluorescent sensors to detect botulinum neurotoxin activity *in vitro* and in living cells. *Proc. Natl. Acad. Sci. USA* **101:**14701–14706.

30. **Doyle, M. P.** 1991. Evaluating the potential risk from extended-shelf-life refrigerated foods by *Clostridium botulinum* inoculation studies. *Food Technol.* **45**(4):154–156.

31. **Dürre, P. (ed.).** 2005. *Handbook of Clostridia.* CRC Press, Boca Raton, Fla.

32. **Eklund, M. W., D. I. Wieler, and F. T. Poysky.** 1967. Outgrowth and toxin production of non-proteolytic type B *Clostridium botulinum* at 3.3 to 5.6°C. *J. Bacteriol.* **93:**1461–1462.

33. **Eklund, M. W., and V. R. Dowell. Jr. (ed.).** 1987. *Avian Botulism: An International Perspective.* Charles C Thomas, Springfield, Ill.

34. **Eleopra, R., V. Tugnoli, O. Rossetto, D. De Grandis, and C. Monteucco.** 1998. Different time courses in recovery after poisoning neurotoxin serotypes A and E in humans. *Neurosci. Lett.* **256:**135–138.

35. **Esty, J. R., and K. F. Meyer.** 1922. The heat resistance of the spores of *B. botulinus* and allied anaerobes. XI. *J. Infect. Dis.* **31:**650–653.

36. **Ferracci, G., R. Miquelis, S. Kozaki, M. Seagar, and C. Leveque.** 2005. Synaptic vesicle chips to assay botulinum neurotoxin. *Biochem. J.* **391:**659–666.

37. **Ferreira, J. L., S. Maslanka, E. Johnson, and M. Goodnough.** 2003. Detection of botulinal neurotoxins A, B, E, and F by amplified enzyme-linked immunosorbent assay: collaborative assay. *J. Assoc. Off. Agric. Chem. Int.* **86:**314–331.

38. **Foran, P. G., N. Mohammed, G. O. Lisk, S. Nagwaney, G. W. Lawrence, E. Johnson, L. Smith, K. R. Aoki, and J. O. Dolly.** 2003. Evaluation of the therapeutic usefulness of botulinum neurotoxin B, C1, E, and F compared with the long lasting type A-Basis for distinct durations of inhibition of exocytosis in central neurons. *J. Biol. Chem.* **278:**1363–1371.

39. **Foster, E. M.** 1997. Historical overview of key issues in food safety. *Emerg. Infect. Dis.* **3:**481–482.

40. **Fox, C. K., C. A. Keet, and J. B. Strober.** 2004. Recent advances in infant botulism. *Pediatr. Neurol.* **32:**149–154.

41. **Franciosa, G., J. L. Ferreira, and C. L. Hatheway.** 1994. Detection of type-A, type-B, and type-E botulism neurotoxin genes in *Clostridium botulinum* and other *Clostridium* species by PCR-evidence of unexpressed type-B toxin genes in type-A toxigenic organisms. *J. Clin. Microbiol.* **32:**1911–1917.

42. **Franciosa, G., P. Aureli, and R. Schechter.** 2003. *Clostridium botulinum*, p. 61–89. *In* M. D. Miliotis and J. W. Bier (ed.), *International Handbook of Foodborne Pathogens.* Marcel Dekker, New York, N.Y.

43. Giménez, D. F., and J. A. Giménez. 1993. Serological subtypes of botulinal neurotoxins, p. 421–431. *In* B. R. Dasgupta (ed.), *Botulism and Tetanus Neurotoxins: Neurotransmission and Biomedical Aspects*. Plenum Press, New York, N.Y.

44. Glass, K. A., and E. A. Johnson. 2002. Formulating low-acid foods for botulinal safety, p. 323–350. *In* V. K. Juneja and J. N. Sofos (ed.), *Control of Foodborne Microorganisms*. Marcel Dekker, New York, N.Y.

45. Gupta, A., C. J. Sumner, M. Castor, S. Maslanka, and J. Sobel. 2005. Adult botulism type F in the United States, 1981–2002. *Neurology* **65:**1694–1700.

46. Hall, Y. H. J., J. A. Chaddock, H. J. Moulsdale, E. R. Kirby, F. C. G. Alexander, J. D. Marks, and K. A. Foster. 2004. Novel application of an in vitro technique to the detection and quantification of botulinum neurotoxin antibodies. *J. Immunol. Methods* **288:**55–60.

47. Hallis, B., B. A. F. James, and C. C. Shone. 1996. Development of novel assays for botulinum type A and B neurotoxins based on their endopeptidase activity. *J. Clin. Microbiol.* **34:**1934–1938.

48. Hammes, W. P., and P. S. Tichaczek. 1994. The potential of lactic acid bacteria for the production of safe and wholesome food. *Z. Lebensm. Unters. Forsch.* **198:**193–201.

49. Hatheway, C. L. 1988. Botulism, p. 111–133. *In* A. Balows, et al. (ed.), *Laboratory Diagnosis of Infectious Diseases: Principles and Practice*. Springer-Verlag, New York, N.Y.

50. Hatheway, C. L. 1993. *Clostridium botulinum* and other organisms that produce botulinum neurotoxin, p. 3–20. *In* A. H. W. Hauschild and K. L. Dodds (ed.), Clostridium botulinum: *Ecology and Control in Foods*. Marcel Dekker, New York, N.Y.

51. Hatheway, C. L., and E. A. Johnson. 1998. *Clostridium*: the spore-bearing anaerobes, p. 731–782. *In* L. Collier, A. Balows, and M. Sussman (ed.), *Topley & Wilson's Microbiology and Microbial Infections*, 9th ed., vol. 2. *Systematic Bacteriology*. Arnold, London, United Kingdom.

52. Hauschild, A. H. W. 1993. Epidemiology of foodborne botulism, p. 69–104. *In* A. H. W. Hauschild and K. L. Dodds (ed.), Clostridium botulinum: *Ecology and Control in Foods*. Marcel Dekker, New York, N.Y.

53. Hauschild, A. H. W. 1989. *Clostridium botulinum*, p. 111–189. *In* M. P. Doyle (ed.), *Foodborne Bacterial Pathogens*. Marcel Dekker, New York, N.Y.

54. Hauschild, A. H. W., and K. L. Dodds (ed.). 1993. Clostridium botulinum: *Ecology and Control in Foods*. Marcel Dekker, New York, N.Y.

55. Hauschild, A. H. W., R. Hilsheimer, K. F. Weiss, and R. B. Burke. 1988. *Clostridium botulinum* in honey, syrups, and dry infant cereals. *J. Food Prot.* **51:**892–894.

56. Hines, H. B., F. Lebeda, M. Hale, and E. E. Brueggemann. 2005. Characterization of botulinum progenitor toxins by mass spectrometry. *Appl. Environ. Microbiol.* **71:**4478–4486.

57. Holdeman, L. V., E. P. Cato, and W. E. C. Moore. 1979. *Anaerobe Laboratory Manual*, 4th ed. Virginia Polytechnic Institute and State University, Blacksburg.

58. International Commission on Microbiological Specifications for Foods. 1996. *Microorganisms in Foods. 5. Characteristics of Microbial Pathogens*. Blackie Academic & Professional, London, United Kingdom.

59. Ito, K. A., D. J. Seslar, W. A. Mercern, and K. F. Meyer. 1967. The thermal and chlorine resistance of *Cl. botulinum* types A, B, and E. spores, p. 108–122. *In* M. Ingram and T. A. Roberts (ed.), *Botulism 1966*. Chapman and Hall, London, United Kingdom.

60. Johnson, E. A. 2005. Clostridial neurotoxins. *In* P. Dürre (ed.), *Handbook of Clostridia*. CRC Press, Inc., Boca Raton, Fla.

61. Johnson, E. A. 2005. *Clostridium botulinum* and *Clostridium tetani*, p. 1035–1088. *In* P. Murray (ed.), *Topley and Wilson's Microbiology and Microbial Infections*, 8th ed. Hodder Arnold, London, United Kingdom.

62. Johnson, E. A., and M. Bradshaw. 2001. *Clostridium botulinum*: a metabolic and cellular perspective. *Toxicon* **39:**1703–1722.

63. Johnson, E. A., and M. C. Goodnough. 1998. Botulism, p. 723–741. *In* L. Collier, A. Balows, and M. Sussman (ed.), *Topley & Wilson's Microbiology and Microbial Infections*, 9th ed., vol. 2. Systematic Bacteriology. Arnold, London, United Kingdom.

64. Johnson, E. A., W. H. Tepp, M. Bradshaw, R. J. Gilbert, P. E. Cooke, and E. D. G. McIntosh. 2005. Characterization of *Clostridium botulinum* strains associated with an infant botulism case in the United Kingdom. *J. Clin. Microbiol.* **43:**2602–2607.

65. Johnson, E. A., P. San, and S. M. Finegold. 2007. *Clostridium*, p. 889–910. *In* P. R. Murray, E. J. Baron, J. H. Jorgensen, M. L. Landry, and M. A. Pfaller (ed.), *Manual of Clinical Microbiology*, 9th ed. ASM Press, Washington, D.C.

66. Jousimies-Somer, H. R., P. Summanen, D. M. Citron, E. J. Baron, H. M. Wexler, and S. M. Finegold. 2002. *Anaerobic Laboratory Manual*, 6th ed. Star Publishing Company, Belmont, Calif.

67. Kalluri, P., C. Crowe, M. Reller, L. Gaul, J. Hayslett, S. Barth, S. Eliasvberg, J. Ferreira, K. Holt, S. Bengston, K. Hendricks, and J. Sobel. 2003. An outbreak of foodborne botulism associated with food sold at a salvage store in Texas. *Clin. Infect. Dis.* **37:**1490–1495.

68. Kautter, D. A., S. M. Harmon, R. Y. Lynt, and T. Lilly. 1966. Antagonistic effect on *Clostridium botulinum* by organisms resembling it. *Appl. Microbiol.* **14:**616–622.

69. Kempner, W. 1897. Further contributions to the knowledge of meat poisoning. The antitoxin to botulism. *Z. Hyg. Infektionskrankh.* **26:**481–500. (In German.)

70. Lacy, D. B., W. Tepp, A. C. Cohen, B. R. DasGupta, and R. C. Stevens. 1998. Crystal structure of botulinum neurotoxin type A and implications for toxicity. *Nat. Struct. Biol.* **5:**898–902.

71. Lalli, G., S. Bohnert, K. Deinhardt, C. Verastegui, and G. Schiavo. The journey of tetanus and botulinum neurotoxins in neurons. *Trends Microbiol.* **11:**431–437.

72. Larson, A. E., and E. A. Johnson. 1999. Evaluation of botulinal toxin production in packaged fresh-cut cantaloupe and honeydew melons. *J. Food Prot.* **62:**948–952.

73. Lindler, L. E., F. J. Lebeda, and G. W. Korch (ed.). 2005. *Biological Weapons Defense. Infectious Diseases and Counterbioterrorism*. Humana Press, Totowa, N.J.

74. Lindstrom, M., M. Nevas, K. Kurki, R. Sauna-Aho, A. Latvala-Kiesila, I. Polonen, and H. Korkeala. 2004. Type C botulism due to toxic feed affecting 52,000 farmed foxes and minks in Finland. *J. Clin. Microbiol.* **42**:4718–4725.

75. Lynt, R. K., D. A. Kautter, and R. B. Read, Jr. 1975. Botulism in commercially canned foods. *J. Milk Food Technol.* **38**:546–550.

76. MacDonald, K. L., M. L. Cohen, and P. A. Blake. 1986. The changing epidemiology of adult botulism in the United States. *Am. J. Epidemiol.* **124**:794–799.

77. Malizio, C. J., M. C. Goodnough, and E. A. Johnson. 2000. Purification of botulinum type A neurotoxin. *Methods Mol. Biol.* **145**:27–39.

78. Margosch, D., M. A. Ehrmann, M. G. Gänzle, and R. F. Vogel. 2004. Comparison of pressure and heat resistance of *Clostridium botulinum* and other endospores in mashed carrots. *J. Food Prot.* **67**:2530–2537.

79. McKellar, R. C., and X. Lu (ed.). 2004. *Modeling Microbial Responses in Foods.* CRC Press, Boca Raton, Fla.

80. Meyer, K. F. 1956. The status of botulism as a world health problem. *Bull. W. H. O.* **15**:281–298.

81. Meyer, K. F., and B. Eddie. 1950. *Fifty Years of Botulism in the United States and Canada.* George Williams Hooper Foundation, San Francisco, Calif.

82. Montecucco, C., and G. Schiavo. 1995. Structure and function of tetanus and botulinum neurotoxins. *Q. Rev. Biophys.* **28**:423–472.

83. Moorthy, J., G. A. Mensing, D. Kim, S. Mohanty, D. T. Eddington, W. H. Tepp, E. A. Johnson, and D. J. Beebe. 2004. Microfluidic tectonics platform: a colorimetric, disposable botulinum toxin enzyme-linked immunosorbent assay system. *Electrophoresis* **25**:1705–1713.

84. Morse, D. L., L. K. Pickard, J. J. Buzewich, B. D. Devine, and M. Sharyegani. 1990. Garlic-in-oil associated botulism: episode leads to product modification. *Am. J. Public Health* **80**:1372–1373.

85. Nevas, M., M. Lindström, A. Virtanen, S. Hielm, M. Kuusi, S. S. Arnon, E. K. Vuori, and H. Korkeala. 2005. Infant botulism acquired from household dust presenting as sudden infant death syndrome. *J. Clin. Microbiol.* **43**:511–513.

86. NFPA/CMI Container Integrity Task Force, Microbiological Assessment Group Report. 1984. Botulism risk from post-processing contamination of commercially canned foods in metal containers. *J. Food Prot.* **47**:801–816.

87. Odlaug, T. E., and I. J. Pflug. 1978. *Clostridium botulinum* and acid foods. *J. Food Prot.* **41**:566–573.

88. Paredes, C. J., K. V. Alsaker, and E. T. Papoutsakis. 2005. A comparative genomic view of clostridial sporulation and physiology. *Nat. Rev. Microbiol.* **3**:969–978.

89. Pitt, M. L. M., and R. D. LeClaire. 2005. Pathogenesis by aerosol, p. 65–78. *In* L. E. Lindler, F. J. Lebeda, and G. W. Korch (ed.), *Biological Weapons Defense. Infectious Diseases and Counterbioterrorism.* Humana Press, Totowa, N.J.

90. Reddy, N. R., H. M. Solomon, G. A. Fingerhut, and E. J. Rhodehamel. 1999. Inactivation of *Clostridium botulinum* spores by high pressure processing. *J. Food Safety* **19**:277–288.

91. Reddy, N. R., H. M. Solomon, R. C. Tetzloff, and E. J. Rhodehamel. 2003. Inactivation of *Clostridium botulinum* type A spores by high-pressure processing at elevated temperatures. *J. Food Prot.* **66**:1402–1407.

92. Rhodehamel, E. J., N. R. Reddy, and M. D. Pierson. 1992. Botulism: the causative agent and its control in foods. *Food Control* **3**:125–143.

93. Roberts, T. A. 1997. Maximizing the usefulness of food microbiology research. *Emerg. Infect. Dis.* **3**:523–528.

94. Rodgers, S. 2004. Novel approaches in controlling safety of cook-chill meals. *Trends Food Sci. Technol.* **15**:366–372.

95. Sakaguchi, G. 1983. *Clostridium botulinum* toxins. *Pharmacol. Ther.* **19**:165–194.

96. Schantz, E. J., and E. A. Johnson. 1992. Properties and use of botulinum toxin and other microbial neurotoxins in medicine. *Microbiol. Rev.* **56**:80–99.

97. Schmidt, R. H., and G. E. Rodrick (ed.). 2003. *Food Safety Handbook.* Wiley-Interscience, Hoboken, N.J.

98. Sesardic, D. 2002. Alternatives in testing of bacterial toxins and antitoxins. *Dev. Biol.* (Basel) **111**:101–108.

99. Sesardic, D., R. G. Jones, T. Leung, T. Alsop, and R. Tierney. 2004. Detection of antibodies against botulinum toxins. *Mov. Disord.* **19**(Suppl. 8):S85–S91.

100. Shapiro, R. L., C. L. Hatheway, and D. L. Swerdlow. 1998. Botulism in the United States: a clinical and epidemiologic review. *Ann. Intern. Med.* **129**:221–228.

101. Sharma, S. K., and R. C. Whiting. 2005. Methods for detection of *Clostridium botulinum* toxin in foods. *J. Food Prot.* **68**:1256–1263.

102. Siegel, L. S. 1992. Destruction of botulinum toxins in food and water, p. 323–342. *In* A. H. W. Hauschild and K. L. Dodds (ed.), Clostridium botulinum. *Ecology and Control in Foods.* Marcel Dekker, Inc., New York, N.Y.

103. Smeldt, J. P. P. M., P. C. Hellemons, P. C. Wouters, and S. J. C. van Gerwen. 2002. Physiological and mathematical aspects in setting criteria for decontamination of foods by physical means. *Int. J. Food Microbiol.* **78**:57–77.

104. Smith, L. D. S. 1977. *Botulism. The Organism, Its Toxins, the Disease.* Charles C Thomas, Springfield, Ill.

105. Smith, L. D. S., and H. Sugiyama. 1988. *Botulism. The Organism, Its Toxins, the Disease,* 2nd ed. Charles C. Thomas, Springfield, Ill.

106. Smith, T. J., J. Lou, I. N. Geren, C. M. Forsyth, R. Tsai, S. L. LaPorte, W. H. Tepp, M. Bradshaw, E. A. Johnson, L. A. Smith, and J. D. Marks. 2005. Sequence variation with botulinum neurotoxin serotypes impacts antibody binding and neutralization. *Infect. Immun.* **73**:5450–5457.

107. Sobel, J., N. Tucker, A. Sulka, J. McLaughlin, and S. Maslanka. 2004. Foodborne botulism in the United States, 1990–2004. *Emerg. Infect. Dis.* **10**:1606–1611.

108. Solomon, H. A., and D. A. Kautter. 1988. Outgrowth and toxin production by *Clostridium botulinum* in bottled, chopped garlic. *J. Food Prot.* **51**:862–865.

109. Solomon, H. M., E. A. Johnson, D. T. Bernard, S. S. Arnon, and J. L. Ferreira. 2001. *Clostridium botulinum* and its toxins, p. 317–324. *In* F. P. Downes and K. Ito (ed.), *Compendium for the Microbiological Examination of Foods,* 4th ed. American Public Health Association, Washington, D.C.

110. Sugiyama, H. 1980. *Clostridium botulinum* neurotoxin. *Microbiol. Rev.* **44**:419–448.

111. Sugiyama, H. 1990. Botulism, p. 107–125. *In* D. O. Cliver (ed.), *Foodborne Diseases.* Academic Press, San Diego, Calif.

112. **Sugiyama, H., and K. S. Rutledge.** 1978. Failure of *Clostridium botulinum* to grow in fresh mushrooms packaged in plastic film overwraps with holes. *J. Food Prot.* 41:348–350.

113. **Swaminathan, S., S. Eswaramoorthy, and D. Kumaran.** 2004. Structure and activity of botulinum neurotoxins. *Mov. Disord.* 19(Suppl. 8):S17–S22.

114. **Tanaka, N., E. Traisman, P. Plantinga, L. Finn, W. Flom, L. Meske, and J. Guggisberg.** 1986. Evaluation of factors involved in antibotulinal properties of pasteurized process cheese spreads. *J. Food Prot.* 49:526–531.

115. **Townes, J. M., P. R. Cieslak, C. L. Hatheway, H. M. Solomon, J. T. Holloway, M. P. Baker, C. F. Keller, L. M. McCroskey, and P. M. Griffin.** 1996. An outbreak of type A botulism associated with a commercial cheese sauce. *Ann. Intern. Med.* 125:558–563.

116. **Tsukamoto, K., M. Mikamoto, T. Kohda, H. Ihara, S. Wang, T. Maegawa, S. Nakamura, T. Karasawa, and S. Kozaki.** 2002. Characterization of *Clostridium butyricum* neurotoxin associated with food-borne botulism. *Microb. Pathog.* 33:177–184.

117. **U.S. Food and Drug Adminstration.** 2001. *Bacteriological Analytical Manual.* [Online.] http://www.cfsan.fda.gov/~ebam/bam-toc.html. Accessed 10 February 2006.

118. **van Ermengem, E.** 1979. Classics in infectious disease. A new anaerobic bacillus and its relation to botulism. *Rev. Infect. Dis.* 1:701–719. (Originally published in 1897 as: Ueber einen neuen anaeroben Bacillus und seine Beziehungen zum Botulismus. *Z. Hyg. Infektionskr.* 26:1–56.)

119. **Villar, R. G., R. L. Shapiro, S. Busto, C. Riva-Posse, G. Verdejo, M. I. Farace, R. Rossetti, J. A. San Juan, C. M. Julia, J. Becher, S. E. Maslanka, and D. L. Swerdlow.** 1999. Outbreak of type A botulism and development of a botulism surveillance and antitoxin release system in Argentina. *JAMA* 281:1334–1338.

120. **von Holzer, E.** 1962. Botulismus durch Inhalation. *Med. Klin.* (München) 57:1735–1738.

121. **Wein, L. M., and Y. Liu.** 2005. Analyzing a bioterror attack on the food supply: the case of botulinum toxin in milk. *Proc. Natl. Acad. Sci. USA* 102:9984–9989.

122. **Whiting, R. C., and K. A. Naftulin.** 1992. Effect of headspace oxygen concentration on growth and toxin production by proteolytic strains of *Clostridium botulinum.* *J. Food Prot.* 55:23–27.

123. **Whitmer, M. E., and E. A. Johnson.** 1988. Development of improved defined media for *Clostridium botulinum* serotypes A, B, and E. *Appl. Environ. Microbiol.* 54:753–759.

124. **Woodruff, B. A., P. M. Griffin, L. M. McCroskey, J. F. Smart, R. B. Wainright, R. G. Bryant, L. C. Hutwagner, and C. L. Hatheway.** 1992. Clinical and laboratory features of botulism from toxin types A, B, and E in the United States, 1975–1988. *J. Infect. Dis.* 166:1281–1286.

125. **Wörner, K., H. Szurmant, C. Chiang, and J. A. Hoch.** 2005. Phosphorylation and functional analysis of the sporulation initiation factor Spo0A from *Clostridium botulinum.* *Mol. Microbiol.* 59:1000–1012.

Food Microbiology: Fundamentals and Frontiers, 3rd Ed.
Edited by M. P. Doyle and L. R. Beuchat
© 2007 ASM Press, Washington, D.C.

Bruce A. McClane

Clostridium perfringens

19

Clostridium perfringens was initially recognized as an important cause of foodborne disease in the 1940s and 50s (34). It later became apparent that *C. perfringens* causes two quite different human foodborne diseases, i.e., *C. perfringens* type A food poisoning and necrotic enteritis (also known as Darmbrand or Pig-Bel). Since foodborne necrotic enteritis is rare in industrialized societies, this chapter focuses mainly on *C. perfringens* type A food poisoning; details regarding necrotic enteritis are available in a review article (36).

CHARACTERISTICS OF THE BACTERIUM

C. perfringens is a gram-positive, rod-shaped, encapsulated, nonmotile bacterium that causes a spectrum of human and veterinary diseases (34, 41). The virulence of *C. perfringens* largely results from its prolific toxin-producing ability, including several toxins (e.g., *C. perfringens* enterotoxin [CPE] and β-toxin) with activity on the human gastrointestinal (GI) tract.

In addition to producing GI-active toxins, *C. perfringens* possesses several other characteristics favoring its ability to cause foodborne disease. First, its short doubling time (<10 min for some isolates [34]) allows *C. perfringens*

to rapidly multiply in food. Second, *C. perfringens* forms spores tolerant to environmental stresses such as radiation, desiccation, and heat (34) (Fig. 19.1), which allows this bacterium to survive in incompletely cooked or inadequately warmed foods (see "Virulence Factors Contributing to *C. perfringens* Type A Food Poisoning" below).

C. perfringens is considered to be an anaerobe because it does not produce colonies on agar plates continuously exposed to air (34). However, this bacterium tolerates moderate exposure to oxygen. Compared to most other anaerobes, *C. perfringens* requires only relatively modest reductions in oxidation-reduction potential (E_h) for growth (34).

Classification: Toxin Typing of *C. perfringens*

At least 14 different *C. perfringens* toxins have been identified (41). However, individual *C. perfringens* cells carry only defined subsets of this toxin gene repertoire, providing the basis for a toxin typing classification system for *C. perfringens* isolates (41). This scheme assigns isolates to one of five types (A through E), depending on their ability to express the four (alpha, beta, epsilon, and iota) "typing" toxins (Table 19.1). Toxin typing of

Bruce A. McClane, Department of Molecular Genetics and Biochemistry, University of Pittsburgh School of Medicine, E1240 Biomedical Science Tower, Pittsburgh, PA 15261–2072.

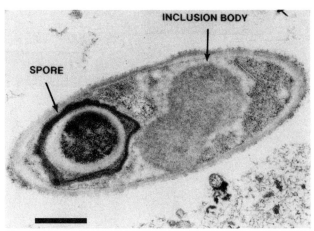

Figure 19.1 Electron micrograph of thin sections of a sporulating cell of *C. perfringens*. Magnification, ×40,000. Bar, 0.5 μm. Arrows indicate the endospore and a CPE-containing inclusion body in the cytoplasm of the mother cell. Reproduced with permission from (19).

C. perfringens traditionally involved laborious toxin antiserum neutralization tests in mice, but multiplex PCR-based toxin genotyping of *C. perfringens* isolates has now greatly simplified this process.

A distinct toxin type is associated with each of the two foodborne diseases caused by *C. perfringens*. Necrotic enteritis, a life-threatening illness, is usually caused by type C isolates, with β-toxin being considered the primary virulence factor responsible for this illness (36). *C. perfringens* type A food poisoning is generally self-limiting and (as implied by its name) is almost always associated with type A isolates producing CPE. CPE, the toxin responsible for the characteristic symptoms of *C. perfringens* type A food poisoning (see "Virulence Factors" below), is sometimes also produced by isolates belonging to other *C. perfringens* types. However, most CPE-producing *C. perfringens* isolates are of type A (15), which helps explain why those isolates are so strongly associated with *C. perfringens* type A food poisoning.

Table 19.1 Typing of *C. perfringens* based on toxins produced

C. perfringens type	Toxins produced			
	Alpha	Beta	Epsilon	Iota
A	+	−	−	−
B	+	+	+	−
C	+	+	−	−
D	+	−	+	−
E	+	−	−	+

Influence of Intrinsic and Extrinsic Factors of Foods on *C. perfringens*

C. perfringens growth in food is affected by a variety of environmental factors, including temperature, E_h, pH, and water activity (a_w), which are addressed below.

Temperature

Spore heat resistance in part contributes to the ability of *C. perfringens* to cause food poisoning by enabling this bacterium to survive in undercooked foods. The heat resistance of *C. perfringens* spores is influenced by both environmental and genetic factors. An example of an environmental influence is the medium in which *C. perfringens* spores are heated, as spores of some *C. perfringens* isolates can survive boiling for 1 h or longer in a protective medium such as cooked meat medium (34). The variation in heat resistance of spores of different *C. perfringens* isolates (see reference 34 for decimal reduction values for spores made by representative *C. perfringens* strains) represents genetic variation that influences the bacterium's thermal resistance. Of particular note, spores of food poisoning isolates are typically much more heat resistant than spores of *C. perfringens* isolates obtained from other sources (34, 54); the extreme heat resistance of food poisoning isolate spores may contribute to foodborne virulence (see "Virulence Factors" below). It is also noteworthy that incomplete cooking of foods may not only fail to kill *C. perfringens* spores in foods but also actually favors development of *C. perfringens* type A food poisoning by inducing spore germination.

Vegetative cells of *C. perfringens* have an optimal growth temperature of 43 to 45°C and often grow at temperatures of up to 50°C (34). Vegetative cells of food poisoning isolates also are typically more heat resistant than *C. perfringens* vegetative cells not associated with foodborne illness (54).

However, vegetative *C. perfringens* cells are not notably tolerant of cold temperatures, including either refrigeration or freezing conditions (34). Growth rates of *C. perfringens* rapidly decrease at temperatures below ~15°C, with no growth occurring at 6°C (34). In contrast, *C. perfringens* spores are cold resistant (34). Food poisoning can then occur if viable spores present in refrigerated or frozen foods germinate when contaminated food is warmed for serving.

Other Environmental Factors

C. perfringens growth in food is also affected by a_w, E_h, pH, and (likely) the presence of curing agents (34). The lowest a_w supporting growth of *C. perfringens* is 0.93 when other growth conditions are near optimal (34).

For an anaerobe, *C. perfringens* does not require an extremely reduced environment to grow. Provided the environmental E_h is suitably low for initiating growth (the exact E_h value needed to initiate *C. perfringens* growth depends on environmental factors such as pH), *C. perfringens* can produce reducing molecules such as ferredoxin to modify the E_h of its environment and create favorable growth conditions (34). As a practical guide for food microbiologists, the E_h of many common foods such as raw meats and gravies is often sufficient to support the growth of *C. perfringens* (34). Growth of *C. perfringens* is also pH sensitive, with optimal growth occurring at pH 6 to 7. *C. perfringens* grows poorly, if at all, at pH values of ≤5 and ≥8.3 (34).

The effectiveness of curing agents at commercially applicable concentrations in limiting *C. perfringens* growth in foods is an unsettled subject (34). Early studies revealed that curing-salt concentrations necessary for significantly affecting *C. perfringens* growth exceed commercially relevant levels, e.g., growth inhibition of *C. perfringens* reportedly requires at least 6 to 8% NaCl and 10,000-ppm $NaNO_3$ or 400-ppm $NaNO_2$. In contrast, later studies determined that curing salts can be at least partially effective at preventing *C. perfringens* growth in food, even when used at commercially acceptable levels. For example, (i) coapplication with other preservation factors such as heating and acidic pH can increase *C. perfringens* sensitivity to curing salts, (ii) simultaneous use of curing agents with other antimicrobials can exert a synergistic effect to inhibit *C. perfringens* growth, and (iii) foods often contain lower initial populations of *C. perfringens* cells and spores than those used in laboratory studies evaluating the effectiveness of curing agents for inhibiting *C. perfringens* growth. One practical argument supporting the ability of curing agents to influence *C. perfringens* growth in foods is the relatively rare association of commercially cured meat products with *C. perfringens* type A food poisoning outbreaks (34).

Preservation factors such as pH, a_w, and likely curing agents also control *C. perfringens* populations in foods by inhibiting the outgrowth of germinating *C. perfringens* spores (34). However, ungerminated spores can remain viable in foods containing preservation factors that prevent cell growth; those spores may later undergo germination and outgrowth if the growth-inhibiting factor(s) is removed during food preparation.

RESERVOIRS OF *C. PERFRINGENS* TYPE A FOOD POISONING STRAINS

C. perfringens is widely distributed throughout the natural environment (34), including soil (at levels of 10^3 to 10^4 CFU/g), foods (e.g., approximately 50% of raw or

frozen meat contain *C. perfringens*), dust, and the intestinal tract of humans and domestic animals (e.g., human feces usually contain 10^4 to 10^6 *C. perfringens* organisms/g). The widespread natural occurrence of *C. perfringens* is an important factor contributing to the frequent occurrence of *C. perfringens* type A food poisoning. However, it is now understood that <5% of global *C. perfringens* isolates carry the enterotoxin gene (*cpe*) necessary for causing *C. perfringens* type A food poisoning. Therefore, simply determining where any *C. perfringens* isolates reside in nature clearly has limited significance for understanding the specific reservoirs of *C. perfringens* type A food poisoning isolates.

Even determining where *cpe*-positive type A isolates exist in the environment is insufficient for identifying *C. perfringens* type A food poisoning reservoirs. While the *cpe* gene can be located on either the chromosome or a plasmid (11, 12, 49), nearly all human food poisoning isolates carry their *cpe* gene on the chromosome (Fig. 19.2). Therefore, identifying the specific reservoirs

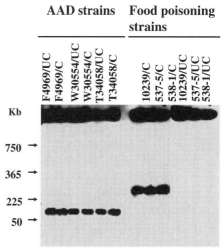

Figure 19.2 Pulsed-field gel electrophoresis Southern blots distinguish *C. perfringens* type A isolates carrying a chromosomal *cpe* gene from those with a plasmid *cpe* gene. Because of its large size, *cpe*-containing DNA is unable to enter pulsed-field gels when present on the chromosome unless that DNA is digested with a restriction enzyme. However, due to the smaller size of plasmid DNA, some *cpe*-containing DNA does enter pulsed-field gels when an isolate carries its *cpe* gene on a plasmid. Furthermore, the restriction enzyme I-CeuI cuts chromosomal DNA but not plasmid DNA. Therefore, a change in migration of *cpe*-containing DNA after I-CeuI treatment is indicative of a chromosomal *cpe* gene. Shown are pulsed-field gel Southern blot results, using an internal *cpe* probe, for three food poisoning isolates carrying a chromosomal *cpe* gene (NCTC10239, 537-5, and 538-1) and three nonfoodborne human GI disease isolates carrying a plasmid *cpe* gene (F4969, W30554, and T34058). C or UC indicates the sample was or was not, respectively, digested with I-CeuI prior to electrophoresis and Southern blotting. Reproduced with permission from reference 60.

for type A food poisoning isolates requires an understanding of the ecology of type A isolates carrying a chromosomal *cpe* gene. To that end, one study recently identified the presence of type A chromosomal *cpe* isolates (but no type A plasmid *cpe* isolates) in some raw meats sold at retail stores in the United States (66). Determining that the *cpe* gene is chromosomal in most, if not all, of the *cpe*-positive isolates present in common food poisoning vehicles provides a second reason, in addition to heat resistance, for why chromosomal *cpe* isolates are so closely associated with food poisoning. Since these chromosomal *cpe* food isolates were also capable of producing CPE, these findings also indicate that *C. perfringens* isolates with full food poisoning potential can be present in U.S. retail meats at the time of purchase. Additional studies are under way to address critical unanswered questions regarding the ecology of, and reservoirs for, *C. perfringens* food poisoning isolates, such as the following. Are chromosomal *cpe* isolates ever present in healthy human carriers, i.e., is there a human reservoir for these isolates? Are isolates with full food poisoning potential present in food animals? Do these isolates contaminate foods during processing? Do these isolates contaminate foods primarily during final handling, cooking, or holding, i.e., through food handler or environmental contamination?

Current knowledge of the principal reservoir(s) of *C. perfringens* type A food poisoning isolates is deficient. This is unfortunate relative to controlling or reducing *C. perfringens* type A food poisoning. By understanding how and when food becomes contaminated with *C. perfringens* food poisoning isolates, it may become possible to develop a focused strategy to intervene at specific locations in the food-processing chain where contamination with type A chromosomal *cpe* isolates occurs.

C. PERFRINGENS TYPE A FOOD POISONING OUTBREAKS

Incidence
The most recent statistics from the Centers for Disease Control and Prevention (CDC) (50) rank *C. perfringens* type A food poisoning as the third most commonly reported bacterial cause of foodborne disease outbreaks in the United States. From 1993 to 1997, 40 *C. perfringens* type A food poisoning outbreaks (representing 4.1% of total bacterial foodborne disease outbreaks) occurred in the United States, involving 2,772 cases (6.3% of total cases of bacterial foodborne diseases). Since most *C. perfringens* type A food poisoning cases are not identified, the official CDC statistics understate the true incidence and

impact of this foodborne disease. Conservative estimates suggest that 250,000 cases of *C. perfringens* type A foodborne illness occur annually in the United States, causing an average of 7 deaths per year (43). The economic costs associated with *C. perfringens* type A food poisoning exceed $120 million/year in 1989 dollars (63).

Identified *C. perfringens* type A food poisoning outbreaks are usually large (the average outbreak size is ~50 to 100 cases) and often occur in institutionalized settings (50). The large size of most recognized *C. perfringens* type A foodborne disease outbreaks is attributable to two factors: (i) foods at large institutions are often prepared in advance and then held for later serving, thereby allowing growth of *C. perfringens* if those foods are temperature abused; and (ii) the relatively mild and nondistinguishing symptoms of most *C. perfringens* type A food poisoning cases cause public health officials to become involved in investigating and reporting this foodborne illness only when large numbers of people become ill. *C. perfringens* type A food poisoning occurs throughout the year but is slightly more common during the summer (50), perhaps because higher ambient temperatures facilitate temperature abuse of foods during cooling and holding.

Food Vehicles for *C. perfringens* Type A Food Poisoning
The most recent CDC statistics indicate that the leading food vehicles for *C. perfringens* type A poisoning in the United States are meats, notably beef and poultry (50). Meat-containing products, e.g., gravies and stews, and Mexican foods also represent important vehicles for *C. perfringens* type A foodborne illness.

Contributing Factors Leading to *C. perfringens* Type A Food Poisoning
C. perfringens type A food poisoning usually results from temperature abuse during the cooking, cooling, or holding of foods. The CDC reports (50) that improper storage or holding temperatures contributed to 100% of recent *C. perfringens* type A food poisoning outbreaks for which contributing factors were identified. Improper cooking was identified as contributing to ~30% of these outbreaks, whereas use of contaminated equipment was a factor in ~15% of outbreaks.

The importance of temperature abuse in *C. perfringens* type A food poisoning is not surprising considering the relative heat tolerance of vegetative cells from type A chromosomal *cpe* isolates and, likely more importantly, the considerable heat resistance of spores produced by those isolates (see "Virulence Factors" below). Undercooking increases the likelihood of illness by inducing germination of, but not killing, *C. perfringens* spores present in foods (34). Vegetative cells resulting from these germinated spores can rapidly multiply to the populations necessary to cause foodborne illness if the foods are improperly cooled or stored (see "Infectious Dose and Susceptible Populations for *C. perfringens* Type A Foodborne Illness" below).

Prevention and Control of *C. perfringens* Type A Food Poisoning

Thorough cooking of food is important to prevent or control *C. perfringens* type A food poisoning. This is particularly true for large roasts and turkeys, for which, due to their size, it is difficult to obtain the high internal temperatures needed to kill *C. perfringens* spores. The difficulty of cooking large roasts and whole poultry carcasses to such high internal temperatures explains, in part, why those foods are such common vehicles for *C. perfringens* type A food poisoning. A second, perhaps even more important, step in preventing *C. perfringens* type A foodborne illness is to rapidly cool cooked foods and then store or serve them at nonpermissive conditions for *C. perfringens* growth (e.g., either at refrigeration temperature or at temperatures >70°C).

Example of a Recent *C. perfringens* Type A Food Poisoning Outbreak

In late November 2001, a *C. perfringens* type A food poisoning outbreak occurred in an Oklahoma residential care facility for the mentally challenged (4). This outbreak sickened seven residents, each of whom had consumed a Thanksgiving meal at the facility. In this outbreak, symptoms (with their frequency shown in parentheses) appeared after a median incubation period of 18 h and included diarrhea (100%), abdominal cramps (100%), vomiting (71%), and fever (14%). Three of the seven victims of this outbreak developed necrotizing enteritis, resulting in two fatalities. *C. perfringens* was cultured from the stool of each patient; multiplex PCR assays identified these fecal isolates as type A isolates carrying a chromosomal *cpe* gene. Western blotting confirmed that all these isolates could express CPE. Immunohistochemistry demonstrated the presence of CPE bound to the colonic epithelium in one fatality. The colonic epithelium of the other fatality was completely destroyed.

This outbreak illustrates several typical features of a *C. perfringens* type A food poisoning outbreak. Although no food remained for conclusive testing, turkey was implicated as the food vehicle responsible for this outbreak by case-control studies and investigators uncovered that improper food handling practices, i.e.,

improper storage and preparation, were used for the large number of turkeys served by the facility for the Thanksgiving meal. Difficulties in maintaining proper temperatures for thawing, cooking, and holding make turkey a common food poisoning vehicle for *C. perfringens* type A food poisoning outbreaks, which typically involve temperature-abused food. The involvement of an institutional setting for this Oklahoma outbreak is also typical for *C. perfringens* type A food poisoning outbreaks and may have been a contributing factor to this particular outbreak, i.e., the need to prepare and serve a large number of turkeys for a holiday dinner may have contributed to improper food handling practices. Finally, the initial symptomology and incubation period associated with this recent outbreak are characteristic of *C. perfringens* type A food poisoning outbreaks, as is the involvement of type A isolates carrying the chromosomal *cpe* gene (discussed below).

However, this Oklahoma outbreak was unusual for a *C. perfringens* type A food poisoning outbreak with respect to its nearly 30% fatality rate. This high mortality rate was not due to advanced patient age (the median patient age was only 50), immunocompromised health status, or physical debility (patients were generally healthy except for being mentally challenged). Instead, the unusual severity of this outbreak is attributable to the development of necrotizing enteritis in three patients (including the two who later died). While *C. perfringens* type C isolates are a well-known cause of foodborne necrotizing enteritis, the involvement of *cpe*-positive type A isolates in foodborne necrotizing enteritis had not been previously reported. It is likely that some Oklahoma patients developed necrotizing enteritis because of their psychiatric drug therapy, which can produce constipation and fecal impaction as side effects. Fecal impaction prevents the expulsion of CPE through the usual protective effects of diarrhea, allowing CPE more time to damage the GI tract of these patients. Consistent with this explanation, severe colonic necrosis was observed at the autopsy of the deceased patients.

Identification of *C. perfringens* Type A Food Poisoning Outbreaks

Public health agencies use descriptive criteria, such as incubation time, symptom and food vehicle history (e.g., is temperature-abused meat or poultry involved?), and food consumption based on case-control comparisons, to help identify *C. perfringens* type A food poisoning outbreaks. However, clinical features alone are not sufficient for identifying these outbreaks given the similarities between the onset times and symptomology of *C. perfringens* type A food poisoning and certain other foodborne illnesses (particularly the diarrheal form of *Bacillus cereus* food poisoning). In response, public health agencies usually include laboratory analyses for more reliable identification of *C. perfringens* type A food poisoning outbreaks. Bacteriologic criteria used by CDC to identify these outbreaks include demonstrating the presence of either (i) 10^5 *C. perfringens* organisms/g of stool from two or more ill persons, or (ii) 10^5 *C. perfringens* organisms/g of epidemiologically implicated food (34).

Simply determining the presence of *C. perfringens* in suspect food or feces is not sufficient to unequivocally identify an outbreak of *C. perfringens* type A food poisoning. CPE-negative *C. perfringens* isolates are widely distributed in the environment, including a presence, sometimes at high levels, in foods or feces from healthy people. Hence, CDC and the U.S. Food and Drug Administration (FDA) also use fecal CPE detection as a diagnostic criterion for identifying *C. perfringens* type A food poisoning outbreaks. Several commercially available serologic assays are available for fecal CPE detection, including a reverse-passive latex agglutination assay (Oxoid) and a rapid enzyme-linked immunosorbent assay (Tech Lab). CPE is also present in the feces of some people suffering from nonfoodborne GI diseases (such as antibiotic-associated diarrhea), so detecting CPE in the feces from a single ill individual is insufficient to establish a *C. perfringens* type A food poisoning outbreak. However, demonstrating the presence of CPE in feces from several epidemiologically associated ill individuals provides strong evidence for a *C. perfringens* type A food poisoning outbreak, particularly when those individuals consumed a common food and developed illness within the normal incubation period (see the next section) for this food poisoning. A limitation to the use of fecal CPE detection approaches for identifying *C. perfringens* food poisoning outbreaks is the need for rapid collection of fecal samples after the onset of food poisoning symptoms, as CPE can be labile in feces (1).

In theory, demonstrating the presence of CPE-positive *C. perfringens* food poisoning isolates in foods or feces represents an alternative or supplemental approach for diagnosing *C. perfringens* type A food poisoning outbreaks. However, several factors complicate the real-world usefulness of this approach. First, while CPE serologic detection assays can be used to evaluate the enterotoxigenic potential of a food or fecal isolate in vitro, this represents a challenge because an isolate must sporulate in laboratory medium to demonstrate CPE expression (CPE expression is sporulation associated) and in vitro *C. perfringens* sporulation is often difficult to achieve (32). Second, even if in vitro sporulation is

achieved, demonstrating CPE expression does not establish whether the isolate carries the chromosomal *cpe* gene which is strongly associated with food poisoning (discussed below). Internal *cpe* PCR assays also fail to identify whether an isolate carries the chromosomal *cpe* gene associated with food poisoning. In response, differences in the chromosomal *cpe* locus versus plasmid *cpe* loci (see "Genetics of CPE" below) have been exploited to develop multiplex PCR assays that can specifically identify chromosomal *cpe* isolates with the greatest food poisoning potential (Fig. 19.3).

However, the limitations discussed above can be overcome by coupling several microbiologic approaches. For example, investigators of the recent Oklahoma outbreak (discussed above) conclusively identified that outbreak as *C. perfringens* type A food poisoning by (i) using a multiplex toxin genotyping PCR to demonstrate that fecal isolates obtained from the outbreak victims are *cpe* positive and belong to genotype A; (ii) using a *cpe* genotyping PCR assay, similar to the one shown in Fig. 19.3, to establish that those type A disease isolates carry the chromosomal *cpe* gene typical of food poisoning isolates; and (iii) using Western blots to confirm that those disease isolates express CPE when induced to sporulate.

In summary, the laboratory plays an increasingly important role in identifying *C. perfringens* type A food poisoning outbreaks. Within the proper epidemiologic and clinical context, demonstrating the presence of CPE

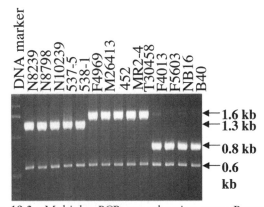

Figure 19.3 Multiplex PCR *cpe* subtyping assay. Representative results obtained with this assay are shown for culture lysates from type A isolates known to carry a chromosomal *cpe* gene (lanes 2 to 6 from the left), a plasmid *cpe* gene with an associated IS*1470*-like sequence (lanes 7 to 11 from the left), or a plasmid *cpe* gene with an associated IS*1151* sequence (lanes 12 to 15 from the left). The migration positions of molecular size markers are shown on the left. The sizes of expected PCR products are shown on the right. Used with permission from reference 49.

in fecal specimens obtained from several ill individuals provides compelling evidence for a *C. perfringens* type A food poisoning outbreak. Newer genetic approaches are also becoming increasingly useful epidemiologic tools for identifying these outbreaks.

CHARACTERISTICS OF *C. PERFRINGENS* TYPE A FOOD POISONING

Symptoms of *C. perfringens* type A food poisoning generally develop about 8 to 18 h after ingestion of contaminated food and then resolve spontaneously within the next 12 to 24 h (34, 42). As illustrated by the recent Oklahoma outbreak described above, victims of *C. perfringens* type A food poisoning almost always suffer diarrhea and abdominal cramps, with vomiting and fever being more variable symptoms. While death rates from *C. perfringens* type A food poisoning are usually

low, fatalities can occur in people who are debilitated, elderly, or medicated.

The typical pathogenesis of *C. perfringens* type A food poisoning is illustrated in Fig. 19.4. Initially, a food becomes contaminated with vegetative cells of *C. perfringens* type A isolates that carry a chromosomal *cpe* gene. If the food is temperature abused, those bacteria rapidly multiply until they are consumed with the contaminated food. Many of the ingested *C. perfringens* vegetative cells die when exposed to stomach acidity, but if the food vehicle was contaminated with $>10^6$ *C. perfringens* organisms/g, some ingested bacteria survive passage through the stomach and remain viable when entering the small intestine, where they multiply and sporulate. CPE is expressed by these sporulating *C. perfringens* cells and is eventually released into the intestinal lumen, where the sporulating cells lyse to release their endospores. Once released, CPE quickly binds to intestinal

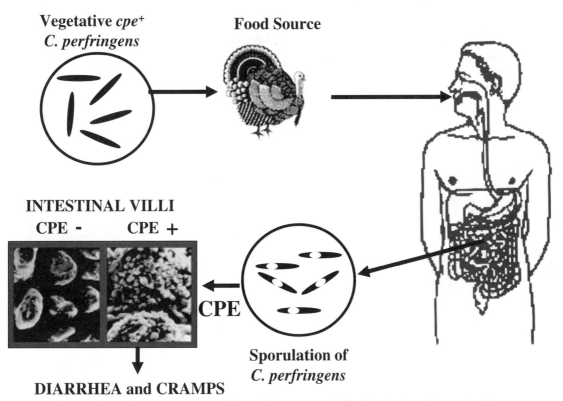

Figure 19.4 The pathogenesis of *C. perfringens* type A food poisoning. Vegetative *C. perfringens* cells multiply rapidly in contaminated food (usually a meat or poultry product) and, after ingestion, sporulate in the small intestine. Sporulating *C. perfringens* cells produce an enterotoxin (CPE) that causes morphologic damage to the small intestine, resulting in diarrhea and abdominal cramps. Reproduced with permission from reference 38.

epithelial cells and exerts its action, which induces intestinal tissue damage (Fig. 19.5). This CPE-induced intestinal tissue damage initiates intestinal fluid loss, which clinically manifests as diarrhea.

Two factors probably help to explain why most cases of *C. perfringens* type A food poisoning are relatively mild and self-limited: (i) CPE preferentially affects villus tip cells (55), which are the oldest intestinal cells and can be rapidly replaced in young, healthy individuals by normal turnover of intestinal cells, and (ii) the diarrhea associated with *C. perfringens* type A food poisoning probably helps mitigate the severity of illness by flushing unbound CPE and many sporulating *C. perfringens* cells containing intracellular CPE from the small intestine.

In rare situations where food poisoning victims do not develop diarrhea, e.g., due to severe constipation from drug therapy, the intestines endure prolonged exposure to CPE and possibly other toxins, which can result in necrotizing enteritis.

INFECTIOUS DOSE AND SUSCEPTIBLE POPULATIONS FOR *C. PERFRINGENS* TYPE A FOODBORNE ILLNESS

As mentioned earlier, *C. perfringens* cells are susceptible to killing by stomach acidity; hence, cases of *C. perfringens* type A food poisoning usually develop only after consumption of a heavily contaminated food, i.e., $>10^6$

SM101, FTG

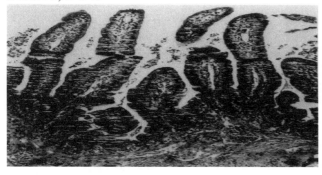

SM101, DS

MRS101, DS

MRS101(pJRC200), DS

Figure 19.5 Fulfilling molecular Koch's postulates demonstrates that CPE is important for the GI virulence of *C. perfringens* type A food poisoning isolates. Tissue specimens shown were collected from rabbit ileal loops treated with either concentrated vegetative (FTG) or sporulating (DS) culture lysates of *C. perfringens* strain SM101, an electroporatable derivative of food poisoning strain NCTC 8798; MRS101, which is a *cpe* knockout mutant of SM101; or MRS101(pJRC 200), which is the MRS101 mutant complemented with a shuttle plasmid carrying the cloned, wild-type *cpe* gene. Note that (i) tissue specimens treated with concentrated FTG lysates of SM101 (or its derivatives), which do not contain CPE, were indistinguishable from control ileal loop specimens and (ii) fluid accumulation was observed only in loops treated with MDS culture lysates of SM101 or MRS(pJRC 200), both of which were shown to contain CPE. Reprinted with permission from reference 53.

to 10^7 C. perfringens vegetative cells/g of food (34). CPE, the toxin responsible for disease symptoms of C. perfringens type A food poisoning, is usually produced in vivo when C. perfringens isolates sporulate in the intestines; therefore, this illness is typically an infection rather than an intoxication. Rare reports describe the onset of early symptoms that are consistent with preformed CPE in foods occasionally contributing to food poisoning. Nevertheless, the typically long incubation period of this food poisoning, despite the quick action of CPE, indicates that the involvement of preformed CPE in C. perfringens type A food poisoning symptoms must be uncommon (34).

Everyone is susceptible to C. perfringens type A food poisoning; however, this illness tends to be more serious in elderly, debilitated, or medicated individuals. Many individuals develop a serum antibody response to CPE following illness (3), but there is no evidence to indicate that previous exposure to this type of food poisoning provides significant future protection.

VIRULENCE FACTORS CONTRIBUTING TO C. PERFRINGENS TYPE A FOOD POISONING

Heat Resistance of Spores and Vegetative Cells of Food Poisoning Isolates

Most food poisoning cases are caused by C. perfringens type A isolates carrying a chromosomal cpe gene, while type A isolates carrying a plasmid-borne cpe gene cause human nonfoodborne GI diseases. The specific association between type A chromosomal cpe isolates and food poisoning was initially puzzling considering that C. perfringens isolates carrying chromosomal or plasmid cpe genes produce similar amounts of an identical CPE protein (10). A recent study (66) suggests that one reason for the strong association of chromosomal cpe isolates with food poisoning is that their spores and vegetative cells are especially heat resistant. Vegetative cells of chromosomal cpe isolates are, at 55°C, approximately twofold more heat resistant than vegetative cells of plasmid cpe isolates. Spore heat resistance differences are even more dramatic, with spores of chromosomal cpe isolates being ~60-fold more heat resistant than spores of plasmid cpe isolates.

Since cooked meat products are the most common food vehicle for C. perfringens type A food poisoning outbreaks (50), the greater heat resistance properties of chromosomal cpe isolates should favor their survival in foods that are incompletely cooked or inadequately held, which are the two major factors contributing to C. perfringens type A food poisoning.

CPE

Evidence that CPE Is Involved in C. perfringens Type A Food Poisoning

Epidemiologic studies provided strong initial evidence that CPE plays a pivotal role in C. perfringens type A foodborne illness. Those studies revealed the following:

1. A strong positive correlation exists between illness and the fecal presence of CPE; depending on the sensitivity of the assay used and how quickly a fecal sample was collected after the onset of symptoms, 80 to 100% of feces from individuals ill with C. perfringens type A food poisoning tested CPE positive, whereas virtually no feces from well individuals tested CPE positive (1, 3).
2. CPE is often present in the feces of food poisoning victims at levels known to cause significant GI effects in experimental animals (55).
3. Human volunteers fed highly purified CPE developed the characteristic GI symptoms of C. perfringens type A food poisoning (34, 61).
4. CPE-positive C. perfringens food poisoning isolates are considerably more effective than CPE-negative C. perfringens isolates at inducing fluid accumulation in rabbit ileal loops or diarrhea in human volunteers (61).
5. Rabbit ileal loop effects induced by CPE-positive isolates can be neutralized with CPE-specific antisera (25).

More recently, the importance of CPE for the GI pathogenesis of C. perfringens food poisoning isolates received compelling support from experiments (53) fulfilling molecular Koch's postulates. Those experiments (Fig. 19.5) revealed that sporulating (but not vegetative) culture lysates of SM101, a transformable derivative of a wild-type CPE-positive food-poisoning isolate, can induce both fluid accumulation and histopathologic damage in rabbit ileal loops, which is consistent with CPE (whose expression is sporulation associated) being necessary for SM101's GI activity. However, neither vegetative nor sporulating culture lysates of an isogenic SM101 cpe knockout mutant induced intestinal fluid accumulation or histopathologic damage. The cpe knockout mutant's loss of GI virulence could be specifically attributed to inactivation of its cpe gene because full GI virulence was restored by complementing the mutant with a shuttle plasmid carrying the wild-type cpe gene.

C. perfringens type A food poisoning is not the only GI disease involving CPE. CPE-positive C. perfringens type A isolates also cause several nonfoodborne human GI illnesses, including antibiotic-associated diarrhea and sporadic diarrhea, as well as some veterinary diarrheas (40). However, cpe-positive C. perfringens type A isolates causing nonfoodborne GI diseases are genotypically distinct from those causing food poisoning, i.e., cpe is typically located on the chromosome of food poisoning isolates but is present on a plasmid in nonfoodborne disease isolates (11, 12, 60). Molecular Koch's postulate studies using a cpe knockout mutant of F4969, a nonfoodborne human GI disease isolate, have also confirmed that CPE plays an important role in the pathogenesis of nonfoodborne human GI disease caused by type A isolates carrying a plasmid-borne cpe gene (53). However, it is likely that the pathogenesis of nonfoodborne GI diseases often involves additional toxins besides CPE based on recent studies (17) showing that F4969 is a relatively atypical type A nonfoodborne human GI isolate that lacks the cpb2 gene encoding beta2 toxin. In contrast, most type A nonfoodborne (17) GI disease isolates carry both a plasmid-borne cpe gene and a cpb2 gene. Since beta2 toxin can damage CaCo-2 cells, a human colon carcinoma cell line, it is possible that beta2 toxin also plays a role in CPE-associated nonfoodborne human GI diseases (17).

Genetics of CPE

Cloning and sequencing of the cpe gene (14) provided tools, e.g., cpe-specific primers, that substantially increased knowledge of cpe genetics. For example, it was learned that the cpe gene is present in only ~1 to 5% of all C. perfringens isolates, most of which classify as type A (15, 32). Discovering that only a small percentage of the overall C. perfringens population is cpe positive suggested that the cpe gene might be associated with mobile genetic elements. Early evidence supporting that association was provided by Southern blot analyses of pulsed-field gel electrophoresis gels, which demonstrated that the cpe gene in nonfoodborne human GI disease isolates and animal isolates is plasmid borne (11, 12, 60). Subsequent studies (6) demonstrated that the cpe plasmid of isolate F4969 (and possibly other nonfoodborne human GI disease isolates) can transfer by conjugation to cpe-negative C. perfringens isolates. Recent studies (17) have revealed that two distinct cpe-carrying plasmids exist in type A nonfoodborne human GI disease isolates, with one of those cpe plasmids also encoding beta2 toxin.

The chromosomal cpe gene found in nearly all type A food poisoning isolates also appears to be associated with a mobile genetic element. Specifically, the discovery of IS1470 sequences upstream and downstream of the chromosomal cpe open reading frame (ORF) in NCTC8239 and other food poisoning isolates (Fig. 19.6)

led to the proposal (7) that the chromosomal cpe gene is present on a 6.3-kb transposon (named Tn5565) with terminal IS1470 elements. Some evidence (5) suggests that this putative cpe-carrying transposon has a circular intermediate form, but actual excision and reintegration of Tn5565 have not yet been experimentally demonstrated.

Interestingly, no upstream insertion element and two different downstream insertion elements, i.e., IS1151 or a defective IS1470-like element, are associated with the plasmid cpe gene of type A isolates (Fig. 19.6). These cpe locus variations permitted development of a multiplex PCR assay capable of discriminating between type A isolates carrying a plasmid-borne versus chromosomal cpe gene (Fig. 19.3). These PCR cpe subtyping assays are now proving to be valuable laboratory tools for epidemiologic investigations.

Expression and Release of CPE

The expression and release of CPE from C. perfringens isolates have at least three interesting features: (i) CPE production is tightly regulated, i.e., this toxin is expressed during sporulation but not during vegetative growth; (ii) during sporulation, many CPE-positive isolates produce extremely large amounts of this toxin; and (iii) CPE is not actually secreted by sporulating C. perfringens cells but is instead released into the intestines when the mother cell lyses upon the completion of sporulation.

Regulation of CPE Synthesis

Duncan's classic studies first established a relationship between CPE expression and sporulation, e.g., it was shown that C. perfringens mutants blocked at stage 0 of sporulation completely lose their ability to produce CPE (44). Later, Western blot studies (14) confirmed a relationship between sporulation and CPE expression by revealing 1,500-fold-greater CPE expression by sporulating cells than vegetative cells of cpe-positive type A food poisoning isolate NCTC8239.

Resembling natural cpe-positive type A isolates, naturally cpe-negative C. perfringens types A, B, and C isolates also exhibit sporulation-associated CPE expression if transformed with a cpe-containing shuttle plasmid (13). This observation suggests that most C. perfringens isolates produce the regulatory factor(s) involved in modulating sporulation-associated CPE expression. The apparent widespread distribution of this regulator among C. perfringens isolates suggests that it may also control other C. perfringens genes during sporulation. Consistent with that suggestion, recent knockout studies have implicated two global regulatory proteins, i.e., Spo0A and CcpA, in the regulation of CPE production (26, 64).

I. Chromosomal *cpe* Locus in Type A Strain NCTC8239

II. Plasmid *cpe* Locus in Type A Isolate F4969

III. Plasmid *cpe* Locus in Type A Isolate F5603

IV. Plasmid Locus with Silent *cpe* Gene in Type E Strain NCIB10748

Figure 19.6 Comparison of *cpe* locus arrangement in various *C. perfringens* isolates. (I) Arrangement of the chromosomal *cpe* locus in food poisoning isolate NCTC8239. The chromosomal *cpe* locus appears to be similarly arranged in most other food poisoning isolates. (II and III) Arrangement of the plasmid *cpe* locus in type A human nonfoodborne GI disease isolates F4969 and F5603, respectively. Note that the *cpe* plasmid of F5603, like other type A isolates with a similar *cpe* locus, also encodes beta2 toxin. (IV) Arrangement of the silent *cpe* locus in type E isolate NCIB10748. Many other type E isolates appear to carry a similar silent *cpe* locus; note the presence of function iota toxin genes (*ibp* and *iap*) immediately upstream of the silent *cpe* locus in type E isolates. Compiled from references 2, 7, 17, 48, and 49.

Synthesis of CPE

CPE synthesis begins shortly after sporulation is induced and then progressively increases for the next 6 to 8 h (44). Late in sporulation, CPE can represent up to 15% of the total cell protein present inside a *C. perfringens* sporulating cell (14). In general, the better a *C. perfringens* isolate sporulates, the more CPE it produces; however, this correlation is not absolute (10).

Why do some CPE-positive *C. perfringens* type A strains produce so much enterotoxin during sporulation? This high-level CPE expression is not related to *cpe* gene location, as type A isolates carrying a chromosomal versus plasmid *cpe* gene produce similar levels of CPE (10–12). Nor is this strong enterotoxin expression due to a gene dosage effect, since most, if not all, *cpe*-positive isolates carry a single copy of the *cpe* gene.

RNA slot blot and Northern blot analyses (13, 45, 69) have demonstrated regulation of CPE production at the transcriptional level, with significant amounts of *cpe* mRNA expressed during sporulation, but little or no *cpe* mRNA produced during vegetative growth of *C. perfringens*. Northern blot studies (13) revealed that *cpe* mRNA is transcribed as a monocistronic message of ~1.2 kb, which is consistent with initial primer extension analysis studies (45) indicating that *cpe* mRNA transcription starts ~200 bp upstream of the *cpe* ORF translation start site. Subsequent primer extension, RNase T2 protection, and deletion mutagenesis studies (69) identified at least three start sites (i.e., P1, P2, and P3) for the initiation of *cpe* mRNA transcription. Therefore, it is possible that the presence of multiple promoters is a major contributor to strong CPE expression. Interestingly, P1 shares some homology with SigK-dependent promoters, whereas P2 and P3 share some homology with SigE-dependent promoters. Those homologies are interesting because SigE and SigK are sporulation-associated sigma factors that are active in mother cells during sporulation of *B. subtilis*, i.e., alternate sigma factors probably play a role in regulating CPE expression.

Posttranscriptional effects may also help regulate CPE expression levels. For example, the functional half-life of *cpe* mRNA in sporulating *C. perfringens* cells is reportedly 58 min (35), which is unusually long for a bacterial message. Such exceptional message stability could contribute to the abundant CPE expression noted for sporulating cells of some *C. perfringens* strains. Given that stem-loop structures can contribute to message stability, the putative stability of *cpe* mRNA could result from a stem-loop structure present 36 bp downstream of the 3′ end of the *cpe* ORF (14). This stem-loop structure is followed by an oligo(dT) tract, suggesting that it also functions as a rho-independent transcriptional terminator (14), which would be consistent with the previously described transcriptional start sites identified 200 bp upstream of the *cpe* initiation codon and the 1.2-kb size of *cpe* mRNA observed in Northern blot studies (13).

Release of CPE from *C. perfringens*

Unlike most *C. perfringens* toxins, CPE is not secreted via a classic bacterial transport system (14). Consistent with this, *cpe* does not encode the 5′ signal peptide associated with many secreted exotoxins (14). Instead, newly synthesized CPE accumulates in the cytoplasm of the mother cell, sometimes reaching sufficiently high concentrations to induce formation of cytoplasmic CPE-containing paracrystalline inclusion bodies (Fig. 19.1). This intracellular CPE is eventually released into the intestines upon the completion of sporulation, i.e., when the mother cell lyses to free its mature spore. This dependency upon mother cell lysis for CPE release helps explain why, despite CPE's quick intestinal action, *C. perfringens* type A food poisoning symptoms develop 8 to 24 h after ingestion of contaminated foods, i.e., sporulating *C. perfringens* cells need at least 8 to 12 h to complete sporulation and then release CPE into the intestine (34).

CPE Biochemistry

Studies conducted in the early 1970s (42) characterized CPE as a single polypeptide of ~35,000 Da, with a pI of 4.3. Later *cpe* ORF sequencing studies (14) revealed that the CPE protein is comprised of 319 amino acids, with a precise M_r of 35,317. Additional *cpe* ORF sequencing studies (10) determined that the CPE amino acid sequence is highly conserved among most, if not all, CPE-positive *C. perfringens* type A isolates. The consensus CPE sequence in type A isolates lacks homology with other proteins, except for some limited similarity, of unknown significance, with a nonneurotoxic protein produced by *Clostridium botulinum* (44).

CPE is a heat-labile protein; its biologic activity can be inactivated by heating for 5 min at 60°C (42). The enterotoxin is also quite sensitive to pH extremes (pH of <6 or >8) but is relatively resistant to some proteolytic treatments (42). In fact, limited trypsin or chymotrypsin treatment actually increases CPE activity about two- to threefold (41), suggesting that intestinal proteases could activate CPE during food poisoning. However, direct in vivo evidence supporting this hypothesis has not yet been presented.

CPE Action

Studies of the in vivo and in vitro effects of CPE have indicated that CPE has the novel mechanism of action described below.

CPE Effects on the GI Tract

CPE is classified as an enterotoxin because it induces fluid and electrolyte losses from the GI tract of many mammalian species (42). In rabbits, the principal target organ for CPE is the small intestine, with the ileum being particularly sensitive (42). Interestingly, the rabbit colon is relatively insensitive to CPE, despite the strong binding of this enterotoxin to rabbit colonic cells (42). Recent ex vivo studies (16) confirmed that the human ileum is also very sensitive to CPE. However, human colonic tissue had only a mild histopathologic response to CPE treatment, at least under the conditions used in ex vivo studies.

The biologic activity of CPE is readily distinguishable from such classical enterotoxins as cholera and *Escherichia coli* heat-labile enterotoxins (42): (i) CPE does not increase intestinal cAMP levels; (ii) CPE inhibits glucose absorption; and (iii) CPE quickly induces histopathologic damage, including epithelial desquamation and severe villus shortening, in the small intestine. While some other bacterial enterotoxins, e.g., Shiga toxin and *Clostridium difficile* toxins, also cause intestinal tissue damage, CPE is unique with respect to its ability to induce intestinal damage within as little as 15 to 30 min (55).

Two observations strongly suggest that CPE-induced tissue damage (Fig. 19.4) plays a major role in initiating CPE-induced fluid or electrolyte intestinal transport alterations. First, fluid transport alterations develop concurrently with tissue damage in CPE-treated rabbit ileum (55). Second, only those CPE doses capable of inducing tissue damage are able to induce intestinal fluid and electrolyte transport alterations in the rabbit ileum (42). Based on these two observations, CPE apparently initiates its intestinal effects during *C. perfringens* type A food poisoning by causing tissue damage that disrupts villus integrity and induces a breakdown of the normal intestinal secretion-absorption equilibrium.

CPE may also affect the paracellular permeability properties of the intestinal epithelium (59); those paracellular permeability alterations could also contribute to intestinal fluid and electrolyte secretion in CPE-treated intestinal tissue (discussed below). Additionally, the ability of CPE to induce significant release of some proinflammatory cytokines (33, 65) and to kill cells by the proinflammatory process of oncosis (discussed below) raises the possibility that inflammation could be another contributor to CPE-induced intestinal effects, particularly late in illness.

Overview of the Cellular Action of CPE
The cytotoxic action of CPE is responsible for the tissue damage that initiates CPE-induced intestinal fluid and electrolyte alterations. A model describing the cytotoxic

action of CPE is presented in Fig. 19.7. This model indicates that CPE acts on the intestine via a multistep cytotoxic process involving several early events. CPE first binds to a protein receptor(s). This binding localizes CPE in a 90-kDa small complex, which entraps CPE on the membrane surface. Third, a CPE-containing small complex interacts with other proteins to form several large complexes in mammalian plasma membranes. Finally, the formation of an ~155-kDa CPE-containing large complex alters normal plasma membrane permeability characteristics, possibly because that complex has pore-like properties. These plasma membrane permeability alterations lead to calcium influx into the CPE-treated cell, causing cell death. Meanwhile, the development of cellular morphologic damage due to calcium influx promotes formation of an ~200-kDa large complex, a slowly developing effect that could

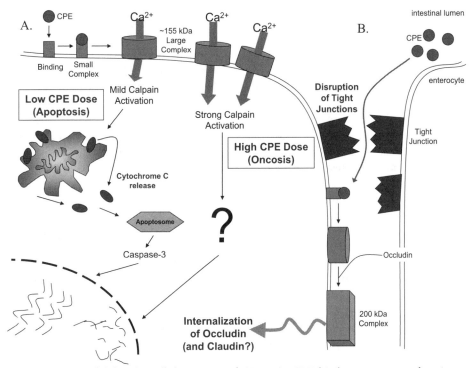

Figure 19.7 Model for the cellular action of CPE. (A) CPE binds to receptors, forming a small complex. At 37°C, the small complex interacts with other proteins to form an ~155-kDa large complex. The ~155-kDa complex is a pore (or portion of a pore) that triggers membrane permeability alterations, including Ca^{2+} influx. With high CPE doses, a massive Ca^{2+} influx occurs, which triggers oncosis; with low CPE doses, there is more moderate Ca^{2+} influx, which triggers apoptosis. Activation of either cell death pathway causes morphologic damage that allows CPE access to receptors on the basolateral surface of the intoxicated cell and adjacent cells. This results in the additional formation of the ~155-kDa large complex and promotes the formation of an ~200-kDa complex containing occludin. Formation of those two large CPE complexes triggers internalization of TJ proteins, which causes damage to the TJ and leads to paracellular permeability alterations that contribute to CPE-induced diarrhea. Reproduced with permission from reference 39.

induce tight junction (TJ) rearrangements that could produce paracellular permeability changes in the CPE-treated intestinal epithelium.

This model emphasizes the uniqueness of CPE cellular and molecular activity by predicting that CPE closely interacts with eucaryotic proteins at every step in its action. No other known membrane-active toxin interacts so closely with eucaryotic proteins throughout its action.

Cellular Action of CPE: Early Events

Step no. 1: receptor binding. A recent study (24) revealed that CPE can induce cation channel formation in protein-free artificial membranes, which indicates that a protein receptor is not absolutely required for CPE to cause biologic effects. However, this channel formation in artificial membranes required very high CPE concentrations and, even then, channels developed more slowly than occurs in CPE-treated mammalian cells containing receptors for this enterotoxin.

Ample evidence indicates that, at pathophysiologic CPE concentrations, the cytotoxic effects of this enterotoxin are mediated through receptor binding (37). For example, CPE interactions with sensitive mammalian cells exhibit the specificity and saturability expected of a receptor-mediated process. CPE-specific binding is also rapid and temperature sensitive, with less CPE binding occurring at 4°C than at 37°C. The small intestines of all tested mammalian species can specifically bind CPE, helping to explain why CPE induces intestinal fluid and electrolyte losses in many mammalian species (42).

There is even conclusive evidence that, at pathophysiologic concentrations, the cytotoxic action of CPE requires a receptor. For example, mouse fibroblasts naturally show no response to challenge with pathophysiologic CPE concentrations (27, 28, 57). In the late 1990s, Katahira et al. exploited this observation by conducting expression-cloning studies in mouse fibroblast transfectants that identified certain members of the claudin family of TJ proteins as functional CPE receptors (27, 28). Claudins, which are ~22-kDa proteins, play a critical role in maintaining normal TJ structure and function. More than 20 different claudins, which vary in their C-terminal cytoplasmic tail sequences, have been identified. Katahira's studies revealed that fibroblast transfectants expressing either claudin-3 or claudin-4 gained the ability to bind and respond to pathophysiologic CPE concentrations. In follow-up studies (18), those investigators determined that claudin-3, -4, -6, -7, -8, and -14, but not claudin-5 and -10, can bind and respond to CPE. CPE binding to claudin-3 apparently involves the second extracellular loop of this TJ protein (18).

Identification of certain claudins as functional CPE receptors by Katahira et al. was consistent with early studies indicating that the CPE receptor(s) is proteinaceous (37, 42). Those early studies demonstrated that protease pretreatment of sensitive mammalian cells or isolated intestinal brush border membranes destroys their CPE binding ability. Interestingly, biochemical approaches (68) had implicated an ~45- to 50-kDa mammalian membrane protein as the CPE receptor(s) in brush border membranes and Vero cells (a CPE-sensitive cell line). For example, immunoprecipitation analysis (68) of two CPE-sensitive cell lines revealed that CPE physically associates with an ~45- to 50-kDa eucaryotic membrane protein upon, or soon after, binding.

What might explain the apparent conflict between biochemical results and expression cloning data regarding the identity of CPE receptors? One possibility could be the existence of coreceptors for CPE binding, i.e., perhaps CPE binds simultaneously both to a claudin and to an ~45- to 50-kDa protein? This coreceptor hypothesis is supported by studies revealing that antibodies directed against a FLAG epitope present on recombinant claudin-3 can immunoprecipitate an ~45- to 50-kDa protein from lysates of mouse L cells expressing the tagged claudin-3; this observation is consistent with the ~45- to 50-kDa protein interacting with certain claudins prior to CPE treatment (27). Alternatively, the ~45- to 50-kDa protein identified by Wieckowski et al. (68) either could be a claudin aggregate or might simply associate with claudin receptors but play no direct role in CPE binding.

Step no. 2: small complex formation. Upon binding to sensitive mammalian cells, CPE is localized in an ~90-kDa, CPE-containing small complex (68). Formation of this small complex may simply reflect CPE binding to a claudin, which itself interacts with other proteins, e.g., other claudins or an ~45- to 50-kDa protein. Alternatively, CPE could first bind to a claudin and then subsequently interact with other proteins, such as the ~45- to 50-kDa protein described in the preceding section. Formation of this small CPE complex appears important for CPE action, as it forms in all CPE-sensitive cells examined to date (68).

When sequestered in the small complex, CPE largely remains exposed on the membrane surface. For example, it has been shown that CPE localized in a small complex is fully accessible to externally applied antibodies and proteases (31, 67). Despite its surface exposure, CPE localized in a small complex exhibits only limited dissociation from membranes or cells. In addition, kinetic studies (37) determined that CPE association with membranes at 4°C (a temperature at which all bound CPE is localized in a small complex) is a two-step process, i.e., CPE binding is rapidly followed by a second event. Collectively, these observations suggest that small complex formation entraps CPE on the membrane surface.

Step no. 3: formation of large CPE complexes. Under physiologic conditions, i.e., at 37°C, the ~90-kDa small complex rapidly transitions into a larger (~155-kDa) complex (Fig. 19.8). The small CPE complex appears to be a direct precursor for this ~155-kDa CPE-containing large complex. For example, if Vero cells treated with CPE at 4°C, a temperature at which all bound toxin is present in the small complex, are shifted to 37°C, formation of the ~155-kDa complex occurs almost immediately (68). At 37°C, the ~155-kDa complex is the predominant CPE-containing species present when Vero cells or CaCo-2 (CPE-sensitive, polarized human colon carcinoma cells) cells are briefly (e.g., for 20 min) treated with CPE (57). However, if CaCo-2 or Vero cell monolayers are CPE treated at 37°C for longer periods (e.g., for ~2 h), a second CPE-containing large complex of ~200 kDa (sometimes along with a third

CPE large complex of ~135 kDa) becomes discernible (Fig. 19.8). The small ~90-kDa CPE complex is easily distinguished from the large CPE complexes on the basis of its size, sensitivity to sodium dodecyl sulfate (SDS) (all large complexes are stable in SDS, whereas the small complex dissociates in the presence of SDS), and temperature sensitivity (formation of large CPE complexes is blocked at 4°C, whereas the small complex forms at low temperatures).

The non-CPE proteins present in the large CPE complexes are not clearly understood, with a single exception. Western blotting, coimmunoprecipitation, and electroelution studies (57) have revealed that the ~65-kDa TJ protein, occludin, is present in the ~200-kDa CPE complex. Those studies also established that occludin is not present in either the small CPE complex or the ~155-kDa large CPE complex. Less definitively, studies by Katahira et al. (27) have suggested that claudins could also be present in some large CPE complexes. Additionally, studies with affinity chromatography suggested that an ~45- to 50-kDa membrane protein, presumably the same protein(s) associated with the small CPE complex, may be present in some large CPE complexes (37).

Considerable evidence now implicates the ~155-kDa large complex in CPE-induced cytotoxicity. For example, there exists a close kinetic correlation (56) between the ~155-kDa complex formation and the onset of the membrane permeability alterations responsible for the death of CPE-treated CaCo-2 cell monolayers. Studies (56) have also shown that those two effects preceded the ~200-kDa CPE complex formation in CPE-treated CaCo-2 cell monolayers. Furthermore, CPE point and deletion mutants unable to form the ~155-kDa complex are also completely nontoxic (29, 58), whereas CPE deletion fragments with enhanced cytotoxicity, relative to native CPE, are more toxic than the native enterotoxin at ~155-kDa CPE complex formation (30).

Time of CPE Treatment (min)

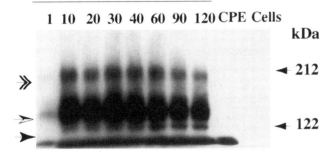

CPE Antibody

Figure 19.8 Kinetics of formation of CPE-containing large complexes in CaCo-2 cells. CPE was added to a suspension of CaCo-2 cells for the indicated times at 37°C. After removal of unbound enterotoxin, the cells were lysed with SDS, and cell lysates were analyzed by SDS-polyacrylamide gel electrophoresis (no sample boiling) using 4% acrylamide gels, followed by Western blotting with either CPE antibodies or occludin antibodies, as indicated. The time (in minutes) of CPE treatment is shown above the gel, whereas the migrations of myosin (212 kDa) and β-galactosidase (122 kDa) markers are indicated in the center space between these two blots. The double, open, and closed arrows indicate the location of the ~200-kDa large complex, the ~155-kDa large complex, and an ~135-kDa intermediate complex, respectively. Note that the formation of the ~200 kDa complex develops much more slowly than shown in this figure if CaCo-2 cell monolayers are CPE treated. Reproduced with permission from reference 57.

Step no. 4: plasma membrane permeability alterations. CPE rapidly affects the plasma membrane permeability properties of Vero cells and CaCo-2 cells (37). Initially, these CPE-induced plasma membrane permeability alterations are restricted to small molecules (<200 Da), suggesting that the initial CPE-induced membrane "lesion" is ~0.5 to 1 nm². This CPE-induced membrane lesion is neither directional nor very discriminating, i.e., CPE induces rapid increases in both influx and efflux of many small molecules, including ions (preferably cations) and amino acids. These small-molecule permeability alterations severely perturb the levels of ions and other small molecules present in a CPE-treated cell, leading to cell death. The kinetic development of these membrane permeability alterations is CPE dose dependent, but initial effects can develop within 5 min.

The molecular mechanism by which CPE changes small-molecule membrane permeability properties is becoming more evident. The ~155-kDa CPE-containing large complex forms concurrently with the onset of CPE-induced small molecule permeability alterations (56), consistent with this CPE complex causing CPE effects on membrane permeability properties. Involvement of the ~155-kDa large complex in CPE-induced membrane permeability changes receives further support from studies (29, 58) indicating that CPE mutants unable to form this large complex are also deficient in causing membrane permeability alterations.

How might the ~155-kDa large complex formation induce plasma membrane permeability changes? Electrophysiology studies (23) detected "pore" formation in apical membranes of CaCo-2 cells treated with CPE under physiologic conditions allowing ~155-kDa large complex formation. Other observations also support the ~155-kDa CPE complex representing part or all of a membrane pore. For example, protease challenge studies (67) revealed that, in contrast to CPE in small complex, CPE becomes very closely associated with membranes when present in the ~155-kDa complex; this observation is consistent with CPE present in the ~155-kDa complex being inserted into membranes to form a pore or channel. However, further studies are needed to formally prove that the ~155-kDa CPE complex corresponds to a complete or partial pore.

Cellular Action of CPE: Consequences of CPE-Induced Membrane Permeability Alterations

Cell death. Recently, considerable progress has been achieved in understanding how CPE kills sensitive mammalian cells (9). CaCo-2 cells treated with low CPE doses (1 μg/ml) exhibited morphologic alterations, DNA cleavage into a distinct fragment ladder, mitochondrial membrane depolarization, cytochrome *c* release from mitochondria, and caspase 3/7 activation. Very late events, e.g., DNA cleavage and morphologic damage, resulting from this low-CPE-dose treatment could be blocked by a caspase-3 inhibitor, but not by a caspase-1 inhibitor or by glycine, an oncosis inhibitor. In contrast, higher CPE doses (10 μg/ml) were shown to induce morphological damage and random DNA shearing without mitochondrial membrane depolarization, cytochrome *c* release, or caspase-3/7 activation. The morphological damage and random DNA shearing induced by this higher CPE dose could be blocked by glycine, but not by inhibitors of

caspase-1 or caspase-3/7. Collectively, these studies indicate that lower CPE doses kill mammalian cells via a classical caspase 3/7 apoptosis, whereas mammalian cells treated with higher CPE doses die from oncosis.

This identification of CPE dose-dependent variations in cell death pathway activation may have direct in vivo relevance. Evidence of inflammation has been observed in some, but not all, CPE-treated rabbit ileal loops (8, 9). Generally, inflammatory cell infiltration is more prominent in CPE-treated ileal loops treated with higher CPE doses. The recent CaCo-2 studies described above now suggest that this putative association between high CPE doses and inflammation may reflect the ability of high CPE doses to induce oncosis, which, unlike classical apoptosis, is proinflammatory. Since (i) inflammation contributes to the pathogenesis of several other foodborne GI illnesses and (ii) high CPE doses similar to those causing oncosis in vitro have been detected in the feces of some *C. perfringens* type A food poisoning victims (1), the proinflammatory effects of oncosis could contribute to intestinal pathology in some cases of CPE-induced GI disease. When lower CPE concentrations are present in the intestinal lumen, human intestinal epithelial cells probably die from apoptosis and inflammation plays a lesser (or no) role in pathology.

How do membrane permeability alterations trigger death of CPE-treated cells and why is cell death pathway activation CPE dose dependent? Recent studies (8) established that CPE-induced membrane permeability alterations result in Ca^{2+} influx (probably through a CPE-containing pore), which then serves as the direct trigger for cell death. Specifically, high CPE doses were determined to cause a rapid and massive change in cytoplasmic Ca^{2+} levels that result in oncosis, whereas lower CPE doses caused a slower developing and more modest elevation in cytoplasmic Ca^{2+} levels that produce a classical caspase 3/7-dependent apoptosis. Interestingly, both CPE-induced apoptosis and oncosis involve the cytoplasmic proteins calmodulin and calpain. However, high CPE doses induce a more rapid and stronger activation of calpain proteolytic activity than do low CPE doses. Further studies of CPE-induced cell death pathways are under way.

CPE-induced changes in TJs. A C-terminal CPE fragment, which is noncytotoxic and does not induce formation of any large CPE complex, can still bind to certain claudins and cause TJ rearrangements in MDCK cells (59). Those TJ rearrangements apparently lead to paracellular permeability changes in CPE-treated MDCK cells (59). Conceivably, native CPE could produce similar TJ rearrangements and paracellular permeability alterations in the intestinal epithelium. If so, those paracellular permeability effects might contribute to CPE-induced intestinal secretion, i.e., diarrhea, during *C. perfringens* type A food poisoning.

Native CPE is likely much better than C-terminal CPE fragments at inducing intestinal TJ rearrangements. Supporting this view, brief (1-h) treatment with native CPE (but not a C-terminal CPE fragment) is sufficient to cause TJ rearrangements in rat liver (52). Since (i) formation of the ~200-kDa complex coincides with internalization of the TJ protein occludin (57) and (ii) N-terminal CPE sequences are important for forming the large CPE complexes (58), the ability of native CPE (versus C-terminal CPE fragments) to more rapidly induce TJ structural rearrangements could implicate the large CPE complexes, particularly the ~200-kDa complex(?), in inducing or contributing to CPE-induced TJ changes and, by extension, intestinal paracellular permeability changes.

Although CPE-TJ interactions may contribute to CPE's intestinal activity, available data overwhelmingly indicate that the primary pathologic effect of CPE involves, via its cytotoxic action, the induction of histopathologic damage in the ileum. In support of this view, the onsets of histopathologic damage fluid and electrolyte losses closely correlate in CPE-treated rabbit ileal loops (55). Furthermore, CPE-induced TJ rearrangements in polarized monolayers of MDCK cells or rat liver are observed only when enterotoxin is applied to the basal side of those monolayers (51, 59). Since CPE is present in the intestinal lumen, it must initially interact with the apical side of epithelial cells. Therefore, the observation that CPE causes TJ rearrangements only when applied to the basolateral surface of polarized monolayers suggests that TJ rearrangements develop in the intestinal epithelium after CPE's cytotoxic effects produce sufficient histopathologic damage to provide the enterotoxin with access to the basolateral surface of epithelial cells. In this regard, it is important that studies (56) have revealed that the apical surface of polarized CaCo-2 cells is responsive to CPE-induced cytotoxicity,

although those cells are even more sensitive when CPE is applied to their basal surface. This difference in apical versus basal surface sensitivity is attributable to the presence of greater numbers of CPE receptors on the basolateral than apical surface of CaCo-2 cells.

CPE Structure-Function Relationships

Structure-function studies (29, 30, 41, 58) have provided considerable insights into how the CPE protein mediates the pathophysiologic effects described above. In particular, studies using CPE fragments and point mutants have localized several important functional regions on the CPE protein. An overview of CPE functional regions, as currently understood, is depicted in Fig. 19.9.

These mapping studies localized receptor-binding activity to the extreme C terminus of the CPE protein. For example, deleting only the five C-terminal amino acids from native CPE completely abrogated the enterotoxin's ability to bind to mammalian cells or isolated plasma membranes (30). Furthermore, either C-terminal CPE fragments or a synthetic peptide corresponding to amino acids 290 to 319 of native CPE, i.e., the 30 C-terminal CPE amino acids, are as equally efficient as native CPE at competing against ^{125}I-CPE binding to cells or isolated membranes (41).

Despite their binding ability, C-terminal CPE fragments are not themselves cytotoxic, i.e., they do not induce small-molecule permeability alterations in mammalian cells (41). This finding indicates that mere occupancy of the CPE receptor is insufficient to trigger CPE-induced cytotoxicity. It also supports the importance of postbinding steps for CPE activity and implicates sequences in the N-terminal half of the CPE molecule as being necessary for cytotoxicity, i.e., like most bacterial toxins, CPE segregates its receptor binding and activity regions.

However, the entire N-terminal half of CPE is not needed for cytotoxicity. For example, deletion mutagenesis studies (30) revealed that removing the first 36 or 44 N-terminal amino acids of CPE increases cytotoxic

Figure 19.9 Linear map of CPE functional regions. CPE regions involved in large complex formation and cytotoxicity are shown in black, with residues D48 and I51 (asterisks) being required for both those events to occur. The C-terminal region of CPE, which also reacts with MAb 3C9, is depicted in dark gray. The extreme N-terminal sequences of CPE (light gray) are unnecessary for cytotoxicity, and some of these sequences may be removed during disease by intestinal proteases (see the text). Compiled from references 29, 30, 37, and 58.

activity two- to threefold. This effect may have some in vivo relevance. Intestinal proteases such as trypsin and chymotrypsin can remove limited N-terminal amino acids from native CPE to produce an activated enterotoxin (20–22). This observation suggests that CPE may be proteolytically activated in the intestines during *C. perfringens* type A foodborne illness.

Why does limited removal of N-terminal sequences activate CPE's cytotoxicity? Deletion mutagenesis studies (30) have revealed that activated CPE fragments, i.e., CPE fragments lacking the extreme N-terminal sequences of native CPE, bind to receptors and form a small complex in a fashion similar to that of native enterotoxin. However, relative to native enterotoxin, these activated CPE fragments exhibit an enhanced ability to form the ~155-kDa CPE complex. Besides explaining why activated CPE fragments are more cytotoxic, this observation provides additional support for the important role of the ~155-kDa complex in CPE-induced cytotoxicity.

It has also been demonstrated that removing eight additional N-terminal amino acids from a CPE_{45-319} fragment eliminates all cytotoxic activity (30). This elimination of cytotoxic activity is specifically due to the CPE_{53-319} fragment being blocked for ~155-kDa large CPE complex formation (30). Collectively, CPE deletion fragment studies (30) map a region responsible for ~155-kDa complex formation to amino acids 45 to 53 of the native enterotoxin. This conclusion was supported by later studies (29) using random CPE point mutants, e.g., a CPE Gly49Asp random mutant, which binds and forms a small complex, was determined to be defective for forming the ~155-kDa CPE-containing complex and to be noncytotoxic. Recent alanine scanning mutagenesis studies (58) identified the core CPE sequence responsible for ~155-kDa complex formation as extending from residues Gly47 to Ile51, with residues 48 and 51 playing particularly important roles in CPE-induced cytotoxicity and ~155-kDa complex formation. Follow-up saturation mutagenesis studies (58) then revealed that the charge and size of the CPE Asp48 residue and the length of the aliphatic side chain on the CPE Ile51 residue are critical for ~155-kDa large complex formation and cytotoxicity.

Which CPE region is involved in inducing TJ rearrangements and paracellular permeability changes? The noncytotoxic C-terminal half of CPE slowly induces TJ rearrangements (59), likely because C-terminal CPE fragments can interact with claudin receptors, which are also major TJ structural proteins. TJ rearrangements are induced more rapidly by native CPE than by C-terminal CPE fragments, suggesting that N-terminal CPE sequences are important for normal CPE-induced TJ rearrangements and, possibly, for causing paracellular permeability changes. The ability of native CPE to more rapidly induce TJ rearrangements than C-terminal CPE fragments may involve the same N-terminal amino acid 45 to 53 region of native CPE that mediates cytotoxicity. Consistent with this hypothesis, CPE mutants containing alanine substitutions at either residue 48 or 51 are deficient for formation of the ~200-kDa complex which contains occludin, as well as for ~155-kDa complex formation (59). The ability of CPE, but not C-terminal CPE fragments that are fully capable of binding claudins, to associate with occludin indicates that CPE plays some role in localizing occludin in the ~200-kDa large complex, i.e., formation of the ~200-kDa complex does not simply result from interactions between claudin and occludin. Collectively, these findings indicate that native CPE disrupts TJs by simultaneously interacting with both claudins via its C-terminal region and occludin either directly or indirectly via its N-terminal amino acid 45 to 53 region.

Ongoing studies continue to more precisely map each CPE functional region and are also attempting to elucidate CPE's three-dimensional structure. Once this information becomes available, it will be possible to determine precisely how the enterotoxin exerts its cytotoxic and TJ-disrupting effects.

CPE Epitopes: Is a CPE Vaccine Possible?

CPE fragments prepared during structure-function mapping studies (described above) were reacted (22) with a series of CPE-specific monoclonal antibodies (MAbs). These epitope mapping studies identified at least four or five disparate regions scattered throughout the enterotoxin primary sequence that are involved in the presentation of CPE epitopes.

Of greatest significance, the CPE epitope recognized by MAb 3C9 mapped to the extreme C terminus of the enterotoxin protein. Since MAb 3C9 is a neutralizing monoclonal antibody that blocks CPE binding to cells (22), the presence of the MAb 3C9 epitope in the C terminus of CPE provides additional evidence for this CPE region having receptor binding activity. Moreover, identifying the presence of a neutralizing linear epitope in C-terminal CPE fragments, which are not themselves cytotoxic, suggests that those fragments could be potential CPE vaccine candidates. The possible use of C-terminal CPE fragments for developing a CPE vaccine was explored by preparing a 30-mer synthetic peptide corresponding to the extreme C-terminal CPE sequence and then chemically conjugating that peptide to a thyroglobulin carrier (47). When the resultant conjugate was administered intravenously to mice, the conjugate-immunized mice developed very high titers of serum antibodies capable of neutralizing the cytotoxicity of native CPE (47).

Effective immunity against *C. perfringens* type A food poisoning may require a secretory immunoglobulin A response in the intestinal lumen. Therefore, if a CPE vaccine becomes desirable, it may be necessary to pursue approaches that can specifically stimulate the development of intestinal immunoglobulin A immunity against CPE. Furthermore, since noncytotoxic C-terminal CPE fragments can sometimes induce, although slowly, TJ rearrangements (59), use of C-terminal CPE sequences for vaccine purposes may require fine-mapping the MAb 3C9 epitope to identify peptide sequences containing the MAb 3C9 neutralizing epitope, but lacking any claudin-binding activity.

Possible Medical Applications of CPE

Pancreatic, breast, and prostate cancer cells often exhibit upregulated expression of claudins that are capable of serving as CPE receptors (62). This observation opened the intriguing possibility of using CPE as an anticancer agent, and initial results to this end appear promising. For example, when CPE was injected into a Panc-1 (human pancreatic cancer cell) tumor xenograft growing on the back of a mouse, tumor necrosis and shrinkage occurred without harming the mouse (46). However, many questions must be addressed before CPE can be employed therapeutically in humans. For example, do cancer cells readily develop resistance to CPE? Will toxicity limit use of CPE to treating solid tumors by direct injection? Will immune responses against CPE limit its effectiveness? Nevertheless, the possibility of harnessing this toxic protein for medical applications is exciting.

CONCLUDING REMARKS

Since the publication of the first edition of this book, considerable progress has been made towards understanding the unique pathogenesis of *C. perfringens* type A food poisoning. Despite those advances, many challenges remain for CPE researchers, e.g., identifying the reservoirs of the chromosomal *cpe C. perfringens* isolates responsible for foodborne illness; determining how *C. perfringens* regulates CPE production during sporulation; defining the composition and stoichiometry of the small and large CPE complexes and elucidating precisely how each of those complexes contribute to CPE's intestinal activity; dissecting the importance of paracellular permeability effects versus cytotoxic effects of CPE-induced intestinal fluid and electrolyte loss; and unraveling the three-dimensional structure of CPE and mapping all important functional regions within that structure.

Answering these and other questions through basic research should lead to practical and effective approaches for controlling CPE-associated diseases, including *C. perfringens* type A food poisoning. For example, identifying the specific reservoirs of *C. perfringens* foodborne illness isolates would enable public health agencies to design specific hygienic measures to reduce food contamination with these bacteria. Additional studies may also lead to the development of agents capable of blocking CPE expression or activity. Finally, continued research on the mechanism of CPE activity may lead to the development of potent new anticancer agents.

Preparation of this chapter was supported by Public Health Service grant R37 AI19877-23 from the National Institute of Allergy and Infectious Disease and by National Research Initiative Competitive Grant 2005 35201 15387 from the USDA Cooperative State Research, Education, and Extension Service. I thank Sameera Sayeed and James G. Smedley III for their help in preparing figures for this chapter.

ADDENDUM IN PROOF

During the preparation of this chapter, our laboratory has reported that, in addition to their heat resistance properties, the spores and vegetative cells of chromosomal *cpe* isolates are also unusually tolerant of both freezing and refrigeration (J. Li and B. A. McClane, *Appl. Environ. Microbiol.* 72:4561–4568, 2006). Additionally, we have recently reported that chromosomal *cpe* isolates are more resistant than plasmid *cpe* isolates to osmotic stress and nitrites (J. Li and B. A. McClane, *Appl. Environ. Microbiol.*, 2006 Oct. 13, epub ahead of print). Collectively, these recent findings indicate that chromosomal *cpe* isolates are exceptionally resistant to a number of food preservation approaches, helping to explain their common association with food poisoning.

References

1. **Bartholomew, B. A., M. F. Stringer, G. N. Watson, and R. J. Gilbert.** 1985. Development and application of an enzyme-linked immunosorbent assay for *Clostridium perfringens* type A enterotoxin. *J. Clin. Pathol.* 38:222–228.

2. **Billington, S. J., E. U. Wieckowski, M. R. Sarker, D. Bueschel, J. G. Songer, and B. A. McClane.** 1998. *Clostridium perfringens* type E animal enteritis isolates with highly conserved, silent enterotoxin sequences. *Infect. Immun.* 66:4531–4536.

3. **Birkhead, G., R. L. Vogt, E. M. Heun, J. T. Snyder, and B. A. McClane.** 1988. Characterization of an outbreak of *Clostridium perfringens* food poisoning by quantitative fecal culture and fecal enterotoxin measurement. *J. Clin. Microbiol.* 26:471–474.

4. **Bos, J., L. Smithee, B. A. McClane, R. F. Distefano, F. Uzal, J. G. Songer, S. Mallonee, and J. M. Crutcher.** 2005. Fatal necrotizing enteritis following a foodborne outbreak of enterotoxigenic *Clostridium perfringens* type A infection. *Clin. Infect. Dis.* 15:e78–e83.

5. Brynestad, S., and P. E. Granum. 1999. Evidence that Tn*5565*, which includes the enterotoxin gene in *Clostridium perfringens*, can have a circular form which may be a transposition intermediate. *FEMS Microbiol. Lett.* **170:**281–286.

6. Brynestad, S., M. R. Sarker, B. A. McClane, P. E. Granum, and J. I. Rood. 2001. The enterotoxin (CPE) plasmid from *Clostridium perfringens* is conjugative. *Infect. Immun.* **69:**3483–3487.

7. Brynestad, S., B. Synstad, and P. E. Granum. 1997. The *Clostridium perfringens* enterotoxin gene is on a transposable element in type A human food poisoning strains. *Microbiology* **143:**2109–2115.

8. Chakrabarti, G., and B. A. McClane. 2005. The importance of calcium influx, calpain, and calmodulin for the activation of CaCo-2 cell death pathways by *Clostridium perfringens* enterotoxin. *Cell. Microbiol.* **7:**129–146.

9. Chakrabartys activated in CaCo-2 cells by *Clostridium perfringens* enterotoxin. *Infect. Immun.* **71:**4260–4270.

10. Collie, R. E., J. F. Kokai-Ki, G., X. Zhou, and B. A. McClane. 2003. Death pathwaun, and B. A. McClane. 1998. Phenotypic characterization of enterotoxigenic *Clostridium perfringens* isolates from non-foodborne human gastrointestinal diseases. *Anaerobe* **4:**69–79.

11. Collie, R. E., and B. A. McClane. 1998. Evidence that the enterotoxin gene can be episomal in *Clostridium perfringens* isolates associated with nonfoodborne human gastrointestinal diseases. *J. Clin. Microbiol.* **36:**30–36.

12. Cornillot, E., B. Saint-Joanis, G. Daube, S. Katayama, P. E. Granum, B. Carnard, and S. T. Cole. 1995. The enterotoxin gene (*cpe*) of *Clostridium perfringens* can be chromosomal or plasmid-borne. *Mol. Microbiol.* **15:** 639–647.

13. Czeczulin, J. R., R. E. Collie, and B. A. McClane. 1996. Regulated expression of *Clostridium perfringens* enterotoxin in naturally *cpe*-negative type A, B, and C isolates of *C. perfringens*. *Infect. Immun.* **64:**3301–3309.

14. Czeczulin, J. R., P. C. Hanna, and B. A. McClane. 1993. Cloning, nucleotide sequencing, and expression of the *Clostridium perfringens* enterotoxin gene in *Escherichia coli*. *Infect. Immun.* **61:**3429–3439.

15. Daube, G., P. Simon, B. Limbourg, C. Manteca, J. Mainil, and A. Kaeckenbeeck. 1996. Hybridization of 2,659 *Clostridium perfringens* isolates with gene probes for seven toxins (α, β, ε, ι, θ, μ and enterotoxin) and for sialidase. *Am. J. Vet. Res.* **57:**496–501.

16. Fernandez-Miyakawa, M. E., V. Pistone-Creydt, F. Uzal, B. A. McClane, and C. Ibarra. 2005. *Clostridium perfringens* enterotoxin damages the human intestine in vitro. *Infect. Immun.* **73:**8407–8410.

17. Fisher, D. J., K. Miyamoto, B. Harrision, S. Akimoto, M. R. Sarker, and B. A. McClane. 2005. Association of beta2 toxin production with *Clostridium perfringens* type A human gastrointestinal disease isolates carrying a plasmid enterotoxin gene. *Mol. Microbiol.* **56:**747–762.

18. Fujita, K., J. Katahira, Y. Horiguchi, N. Sonoda, M. Furuse, and S. Tsukita. 2000. *Clostridium perfringens* enterotoxin binds to the second extracellular loop of claudin-3, a tight junction membrane protein. *FEBS Lett.* **476:**258–261.

19. Garcia-Alvarado, J. S., R. G. Labbe, and M. A. Rodriguez. 1992. Sporulation and enterotoxin production by *Clostridium perfringens* type A at 37 and 43 degrees C. *Appl. Environ. Microbiol.* **58:**1411–1414.

20. Granum, P. E., and M. Richardson. 1991. Chymotrypsin treatment increases the activity of *Clostridium perfringens* enterotoxin. *Toxicon* **29:**445–453.

21. Granum, P. E., J. R. Whitaker, and R. Skjelkvale. 1981. Trypsin activation of enterotoxin from *Clostridium perfringens* type A. *Biochim. Biophys. Acta* **668:**325–332.

22. Hanna, P. C., E. U. Wieckowski, T. A. Mietzner, and B. A. McClane. 1992. Mapping functional regions of *Clostridium perfringens* type A enterotoxin. *Infect. Immun.* **60:**2110–2114.

23. Hardy, S. P., M. Denmead, N. Parekh, and P. E. Granum. 1999. Cationic currents induced by *Clostridium perfringens* type A enterotoxin in human intestinal CaCo-2 cells. *J. Med. Microbiol.* **48:**235–243.

24. Hardy, S. P., C. Ritchie, M. C. Allen, R. H. Ashley, and P. E. Granum. 2001. *Clostridium perfringens* type A enterotoxin forms mepacrine-sensitive pores in pure phospholipid bilayers in the absence of putative receptor proteins. *Biochim. Biophys. Acta* **1515:**38–43.

25. Hauschild, A. H., L. Niilo, and W. J. Dorward. 1971. The role of enterotoxin in *Clostridium perfringens* type A enteritis. *Can. J. Microbiol.* **17:**987–991.

26. Huang, I. H., M. Waters, R. R. Grau, and M. R. Sarker. 2004. Disruption of the gene (*spo0A*) encoding sporulation transcription factor blocks endospore formation and enterotoxin production in enterotoxigenic *Clostridium perfringens* type A. *FEMS Microbiol. Lett.* **233:**233–240.

27. Katahira, J., N. Inoue, Y. Horiguchi, M. Matsuda, and N. Sugimoto. 1997. Molecular cloning and functional characterization of the receptor for *Clostridium perfringens* enterotoxin. *J. Cell Biol.* **136:**1239–1247.

28. Katahira, J., H. Sugiyama, N. Inoue, Y. Horiguchi, M. Matsuda, and N. Sugimoto. 1997. *Clostridium perfringens* enterotoxin utilizes two structurally related membrane proteins as functional receptors *in vivo*. *J. Biol. Chem.* **272:**26652–26658.

29. Kokai-Kun, J. F., K. Benton, E. U. Wieckowski, and B. A. McClane. 1999. Identification of a *Clostridium perfringens* enterotoxin region required for large complex formation and cytotoxicity by random mutagenesis. *Infect. Immun.* **67:**6534–6541.

30. Kokai-Kun, J. F., and B. A. McClane. 1997. Deletion analysis of the *Clostridium perfringens* enterotoxin. *Infect. Immun.* **65:**1014–1022.

31. Kokai-Kun, J. F., and B. A. McClane. 1996. Evidence that region(s) of the *Clostridium perfringens* enterotoxin molecule remain exposed on the external surface of the mammalian plasma membrane when the toxin is sequestered in small or large complex. *Infect. Immun.* **64:**1020–1025.

32. Kokai-Kun, J. F., J. G. Songer, J. R. Czeczulin, F. Chen, and B. A. McClane. 1994. Comparison of Western immunoblots and gene detection assays for identification of potentially enterotoxigenic isolates of *Clostridium perfringens*. *J. Clin. Microbiol.* **32:**2533–2539.

33. **Krakauer, T., B. Fleischer, D. L. Stevens, B. A. McClane, and B. G. Stiles.** 1997. *Clostridium perfringens* enterotoxin lacks superantigenic activity but induces an interleukin-6 response from human peripheral blood mononuclear cells. *Infect. Immun.* **65:**3485–3488.

34. **Labbe, R. G.** 1989. *Clostridium perfringens,* p. 192–234. *In* M. P. Doyle (ed.), *Foodborne Bacterial Pathogens.* Marcel Decker, New York, N.Y.

35. **Labbe, R. G., and C. L. Duncan.** 1977. Evidence for stable messenger ribonucleic acid during sporulation and enterotoxin synthesis by *Clostridium perfringens* type A. *J. Bacteriol.* **129:**843–849.

36. **Lawrence, G. W.** 1997. The pathogenesis of enteritis necroticans, p. 198–207. *In* J. I. Rood, B. A. McClane, J. G. Songer, and R. W. Titball (ed.), *The Clostridia: Molecular Genetics and Pathogenesis.* Academic Press, London, United Kingdom.

37. **McClane, B. A.** 1994. *Clostridium perfringens* enterotoxin acts by producing small molecule permeability alterations in plasma membranes. *Toxicology* **87:**43–67.

38. **McClane, B. A.** 1992. *Clostridium perfringens* enterotoxin: structure, action and detection. *J. Food Safety* **12:**237–252.

39. **McClane, B. A.** 1984. Osmotic stabilizers differentially inhibit permeability alterations induced in Vero cells by *Clostridium perfringens* enterotoxin. *Biochim. Biophys. Acta* **777:**99–106.

40. **McClane, B. A., D. M. Lyerly, J. S. Moncrief, and T. D. Wilkins.** 2000. Enterotoxic clostridia: *Clostridium perfringens* type A and *Clostridium difficile,* p. 551–562. *In* V. A. Fischetti, R. P. Novick, J. J. Ferretti, D. A. Portnoy, and J. Rood (ed.), *Gram-Positive Pathogens.* ASM Press, Washington, D.C.

41. **McClane, B. A., and J. I. Rood.** 2001. Clostridial toxins involved in human enteric and histotoxic infections, p. 169–209. *In* H. Bahl and P. Duerre (ed.), *Clostridia: Biotechnology and Medical Applications.* Wiley-VCH, Weinheim, Germany.

42. **McDonel, J. L.** 1986. Toxins of *Clostridium perfringens* types A, B, C, D, and E, p. 477–517. *In* F. Dorner and H. Drews (ed.), *Pharmacology of Bacterial Toxins.* Pergamon Press, Oxford, United Kingdom.

43. **Mead, P. S., L. Slutsker, V. Dietz, L. F. McCaig, J. S. Bresee, C. Shapiro, P. M. Griffen, and R. V. Tauxe.** 1999. Food-related illness and death in the United States. *Emerg. Infect. Dis.* **5:**607–625.

44. **Melville, S. B., R. E. Collie, and B. A. McClane.** 1997. Regulation of enterotoxin production in *Clostridium perfringens,* p. 471–487. *In* J. I. Rood, B. A. McClane, J. G. Songer, and R. Titball (ed.), *The Clostridia: Molecular Genetics and Pathogenesis.* Academic Press, London, United Kingdom.

45. **Melville, S. B., R. Labbe, and A. L. Sonenshein.** 1994. Expression from the *Clostridium perfringens cpe* promoter in *C. perfringens* and *Bacillus subtilus. Infect. Immun.* **62:**5550–5558.

46. **Michl, P., M. Buchholz, M. Rolke, S. Kunsch, M. Lohr, B. McClane, S. Tsukita, G. Leder, G. Adler, and T. M. Gress.** 2001. Claudin-4: a new target for pancreatic cancer treatment using *Clostridium perfringens* enterotoxin. *Gastroenterology* **121:**678–684.

47. **Mietzner, T. A., J. F. Kokai-Kun, P. C. Hanna, and B. A. McClane.** 1992. A conjugated synthetic peptide corresponding to the C-terminal region of *Clostridium perfringens* type A enterotoxin elicits an enterotoxin-neutralizing antibody response in mice. *Infect. Immun.* **60:**3947–3951.

48. **Miyamoto, K., G. Chakrabarti, Y. Morino, and B. A. McClane.** 2002. Organization of the plasmid *cpe* locus of *Clostridium perfringens* type A isolates. *Infect. Immun.* **70:**4261–4272.

49. **Miyamoto, K., Q. Wen, and B. A. McClane.** 2004. Multiplex PCR genotyping assay that distinguishes between isolates of *Clostridium perfringens* type A carrying a chromosomal enterotoxin gene (*cpe*) locus, a plasmid *cpe* locus with an IS*1470*-like sequence, or a plasmid *cpe* locus with an IS*1151* sequence. *J. Clin. Microbiol.* **42:**1552–1558.

50. **Olsen, S. J., L. C. MacKinon, J. S. Goulding, N. H. Bean, and L. Slutsker.** 2000. Surveillance for foodborne-disease outbreaks—United States, 1993–97. *Morb. Mortal. Wkly. Rep.* **49:**1–51.

51. **Rahner, C., L. Mitic, and J. Anderson.** 2001. Heterogeneity in expression and subcellular localization of claudins 2, 3, 4, and 5 in the rat liver, pancreas, and gut. *Gastroenterology* **120:**411–422.

52. **Rahner, C., L. L. Mitic, B. A. McClane, and J. M. Anderson.** 1999. *Clostridium perfringens* enterotoxin impairs bile flow in the isolated perfused rat liver and induces fragmentation of tight junction fibrils. *Hepatology* **30:**326A.

53. **Sarker, M. R., R. J. Carman, and B. A. McClane.** 1999. Inactivation of the gene (*cpe*) encoding *Clostridium perfringens* enterotoxin eliminates the ability of two *cpe*-positive *C. perfringens* type A human gastrointestinal disease isolates to affect rabbit ileal loops. *Mol. Microbiol.* **33:**946–958.

54. **Sarker, M. R., R. P. Shivers, S. G. Sparks, V. K. Juneja, and B. A. McClane.** 2000. Comparative experiments to examine the effects of heating on vegetative cells and spores of *Clostridium perfringens* isolates carrying plasmid versus chromosomal enterotoxin genes. *Appl. Environ. Microbiol.* **66:**3234–3240.

55. **Sherman, S., E. Klein, and B. A. McClane.** 1994. *Clostridium perfringens* type A enterotoxin induces concurrent development of tissue damage and fluid accumulation in the rabbit ileum. *J. Diarrhoeal Dis. Res.* **12:**200–207.

56. **Singh, U., L. L. Mitic, E. Wieckowski, J. M. Anderson, and B. A. McClane.** 2001. Comparative biochemical and immunochemical studies reveal differences in the effects of *Clostridium perfringens* enterotoxin on polarized CaCo-2 cells versus Vero cells. *J. Biol. Chem.* **276:**33402–33412.

57. **Singh, U., C. M. Van Itallie, L. L. Mitic, J. M. Anderson, and B. A. McClane.** 2000. CaCo-2 cells treated with *Clostridium perfringens* enterotoxin form multiple large complex species, one of which contains the tight junction protein occludin. *J. Biol. Chem.* **275:**18407–18417.

58. **Smedley, J. G., III, and B. A. McClane.** 2004. Fine-mapping of the N-terminal cytotoxicity region of *Clostridium perfringens* enterotoxin by site-directed mutagenesis. *Infect. Immun.* **72:**6914–6923.

59. **Sonoda, N., M. Furuse, H. Sasaki, S. Yonemura, J. Katahira, Y. Horiguchi, and S. Tsukita.** 1999. *Clostridium perfringens* enterotoxin fragments remove specific claudins from tight junction strands: evidence for direct

involvement of claudins in tight junction barrier. *J. Cell Biol.* **147**:195–204.

60. **Sparks, S. G., R. J. Carman, M. R. Sarker, and B. A. McClane.** 2001. Genotyping of enterotoxigenic *Clostridium perfringens* isolates associated with gastrointestinal disease in North America. *J. Clin. Microbiol.* **39**:883–888.

61. **Strong, D. H., C. L. Duncan, and G. Perna.** 1971. *Clostridium perfringens* type A food poisoning. II. Response of the rabbit ileum as an indication of enteropathogenicity of strains of *Clostridium perfringens*. *Infect. Immun.* **3**:171–178.

62. **Swisshelm, K., R. Macek, and M. Kubbies.** 2005. Role of claudins in tumorigenesis. *Adv. Drug Deliv. Rev.* **57**:919–928.

63. **Todd, E. C. D.** 1989. Preliminary estimates of costs of foodborne disease in the United States. *J. Food Prot.* **52**:595–601.

64. **Varga, J., V. L. Stirewalt, and S. B. Melville.** 2004. The CcpA protein is necessary for efficient sporulation and enterotoxin gene (*cpe*) regulation in *Clostridium perfringens*. *J. Bacteriol.* **186**:5221–5229.

65. **Wallace, F. M., A. S. Mach, A. M. Keller, and J. A. Lindsay.** 1999. Evidence for *Clostridium perfringens* enterotoxin inducing a mitogenic and cytokine response *in vitro* and a cytokine response *in vivo*. *Curr. Microbiol.* **38**:96–100.

66. **Wen, Q., and B. A. McClane.** 2004. Detection of enterotoxigenic *Clostridium perfringens* type A isolates in American retail foods. *Appl. Environ. Microbiol.* **70**:2685–2691.

67. **Wieckowski, E., J. F. Kokai-Kun, and B. A. McClane.** 1998. Characterization of membrane-associated *Clostridium perfringens* enterotoxin following pronase treatment. *Infect. Immun.* **66**:5897–5905.

68. **Wieckowski, E. U., A. P. Wnek, and B. A. McClane.** 1994. Evidence that an ~50kDa mammalian plasma membrane protein with receptor-like properties mediates the amphiphilicity of specifically-bound *Clostridium perfringens* enterotoxin. *J. Biol. Chem.* **269**:10838–10848.

69. **Zhao, Y., and S. B. Melville.** 1998. Identification and characterization of sporulation-dependent promoters upstream of the enterotoxin gene (*cpe*) of *Clostridium perfringens*. *J. Bacteriol.* **180**:136–142.

Food Microbiology: Fundamentals and Frontiers, 3rd Ed.
Edited by M. P. Doyle and L. R. Beuchat
© 2007 ASM Press, Washington, D.C.

Per Einar Granum

Bacillus cereus

20

The *Bacillus cereus* group consists of six different species: *Bacillus anthracis, B. cereus, B. mycoides, B. pseudomycoides, B. thuringiensis,* and *B. weihenstephanensis.* These species are so closely related they could be within one species, with the exception of *B. anthracis* (which is not addressed here), which possesses large virulence plasmids. *B. cereus* can cause two different types of foodborne illness: (i) the diarrheal type, first recognized after a hospital outbreak associated with vanilla sauce in Oslo, Norway, in 1948 (35), and (ii) the emetic type, described about 20 years later after several outbreaks associated with fried rice in London (50). The diarrheal type is caused by enterotoxin(s) produced during vegetative growth of *B. cereus* in the small intestine (26), whereas the emetic toxin is produced (preformed) by growing cells in the food (42). For both types of foodborne illness the food involved has usually been heat treated, and surviving spores are the source of the food poisoning. *B. cereus* is not a competitive microorganism but grows well after cooking and cooling (<48°C). The heat treatment causes spore germination, and in the absence of competing flora, *B. cereus* grows well, with a generation time as low as 12 min under optimal conditions (13). Members of the *B.*

cereus group are common soil saprophytes and are easily spread to many types of foods, especially of plant origin (rice and pasta), but are also frequently isolated from meat, eggs, and dairy products (42). Increased numbers of psychrotolerant strains (mainly *B. weihenstephanensis*), specifically in the dairy industry, have led to increased surveillance of *B. cereus* in recent years (16, 29, 32, 43, 63).

B. cereus foodborne illness is likely highly underreported, as both types of illness are relatively mild and usually last for less than 24 h (42). However, occasional reports of more severe forms of the diarrheal type of *B. cereus* foodborne illness have been described (26), including three deaths caused by a newly discovered necrotic enterotoxin (46). Deaths due to intake of large amounts of the emetic toxin have also been reported (19, 49).

CHARACTERISTICS OF THE ORGANISM

The aerobic endosporeforming bacteria have traditionally all been placed in the genus *Bacillus.* Over the past 3 decades, this genus has been expanded to accommodate more than 100 species (6), and the family *Bacillaceae* has been divided into nine different genera (see *Bergey's*

Per Einar Granum, Dept. of Food Safety and Infection Biology, Norwegian School of Veterinary Science, Oslo, Norway.

Manual 2001, "Taxonomic Outline" [http://www.cme.msu.edu/Bergeys/]). The first genus is *Bacillus*, which includes members of the *B. cereus* group: *B. anthracis*, *B. cereus*, *B. mycoides*, *B. thuringiensis*, and, more recently, *B. pseudomycoides* (51) and *B. weihenstephanensis* (43). These bacteria have highly similar 16S and 23S rRNA sequences, indicating that they have diverged from a common evolutionary line relatively recently (6, 7). Although strains of *B. anthracis* have been related to the other species within the *B. cereus* group based on rRNA sequences, it is the most distinctive member of this group, both taxonomically and in its highly virulent pathogenicity. But even strains of *B. cereus* have caused anthrax-like symptoms (37), making it even more difficult to identify definitive criteria for differentiating the two species. In addition, extensive genomic studies of DNA (including full-genome sequencing) from strains of *B. cereus* and *B. thuringiensis* have revealed that there is no taxonomic basis for separate species status (15). Nevertheless, the name *B. thuringiensis* is retained for those strains that synthesize a crystalline inclusion (Cry protein) or δ-endotoxin that is highly toxic to specific insects. The *cry* genes are usually located on plasmids, and loss of the relevant plasmid(s) makes the bacterium indistinguishable from *B. cereus*.

Cells of the six species in the *B. cereus* group are large (cell width, >0.9 μm) and produce central to terminal ellipsoid or cylindrical spores that do not distend the sporangia (17, 43, 51). Bacteria of these *Bacillus* species sporulate on most media easily after 2 to 3 days. Both *B. cereus* and *B. thuringiensis* lose their motility during the early stages of sporulation. Criteria for differentiating the six species are shown in Table 20.1.

RESERVOIRS

Bacillus cereus is widespread in nature and frequently isolated from soil and growing plants (42). From this natural environment it is easily spread to foods, especially those of plant origin. It is frequently present in raw materials and ingredients used in the food industry such as vegetables, starch, and spices (30% of samples with 10^2 to 10^5 CFU/g). Through cross-contamination it may then be spread to other foods, such as meat products (42). High numbers of *B. cereus* have also been detected in cattle feces, through which these bacteria can be spread directly to meat. Milk and milk products are often contaminated by *B. cereus* by contact of soil and grass with udders of cows and subsequently into raw milk. *B. cereus* spores survive milk pasteurization and, after germination, the cells are free from competition from most other vegetative cells (5). According to the classic literature (17), *B. cereus* is unable to grow below 10°C and cannot grow in milk and milk products stored at temperatures between 4 and 8°C. However, the psychrotolerant strains (mainly *B. weihenstephanensis*) can grow at temperatures as low as 4 to 6°C (16, 29, 63). In addition to rice, pasta, and spices, dairy products are among the most common food vehicles for *B. cereus*. Most strains of *B. weihenstephanensis* are unable to cause food poisoning and can often outcompete other microbes in foods stored at temperatures below 8°C.

The closely related *B. thuringiensis* can produce enterotoxins (18, 23a, 52) and caused foodborne illness when administered to human volunteers (52). Spraying of the organism onto crops to protect against insect infestations has become a common practice in several countries and may become a public health issue. A recent study revealed that 31 of 40 randomly selected *B. cereus*-like strains isolated from foods were identified as *B. thuringiensis* (54). *B. thuringiensis* has reportedly caused an outbreak of foodborne illness (41). However, because the procedures normally used for identification of *B. cereus* do not differentiate these two species (Table 20.1), previous outbreaks caused by *B. thuringiensis* may have been unrecognized. Hence, the use of *B. thuringiensis* as a herbicide should be restricted to strains that do not produce enterotoxins.

Table 20.1 Criteria to differentiate members of the *B. cereus* group[a]

Species	Colony morphology	Hemolysis	Motility	Susceptible to penicillin	Parasporal crystal inclusion
B. cereus	White	+	+	−	−
B. anthracis	White	−	−	+	−
B. thuringiensis	White/gray	+	+	−	+
B. mycoides	Rhizoid	(+)	−	−	−

[a] *B. weihenstephanensis* can be differentiated from *B. cereus* based on growth at <7°C and not at 43°C and can be identified rapidly using rRNA or *cspA* (cold shock protein A)-targeted PCR (43). *B. pseudomycoides* is not distinguishable from *B. mycoides* by physiological and morphological characteristics but is clearly differentiated based on fatty acid composition and 16S RNA sequences (51).

FOODBORNE OUTBREAKS

The number of outbreaks of *B. cereus* foodborne illness is grossly underreported largely because the relatively short duration of both types of illness (usually <24 h) and the occurrence of rapid and complete recovery do not trigger epidemiologic investigations in many countries. The dominant type of *B. cereus* disease differs among countries. In Japan, the emetic type is reported about 10 times more frequently than the diarrheal type (59), whereas in Europe and North America the diarrheal type is most frequently reported (42, 56). This is likely associated with eating habits, although contaminated milk was reported to cause at least one large outbreak of the emetic type in Japan (58). Some patients experience both types of *B. cereus* foodborne illness concurrently (42). In addition, about 5% of *B. cereus* strains can produce both types of toxins (42).

Surveillance of foodborne illnesses differs greatly among countries. Hence, it is not possible to directly compare the incidence of outbreaks reported by different countries. The percentage of outbreaks and cases attributed to *B. cereus* in Japan, North America, and Europe varies from approximately 1 to 47% for outbreaks and from approximately 0.5 to 33% for cases (based on reports from different periods between 1960 and 2005) (42, 56). The greatest numbers of reported *B. cereus* outbreaks and cases are from Iceland, The Netherlands, and Norway. There are relatively few outbreaks of salmonellosis and campylobacter enteritis in Norway and Iceland, which are the two most frequently reported causes of foodborne illness in most of Europe and the United States. *B. cereus* in The Netherlands, between 1993 and 1998, was responsible for 12% of outbreaks for which the causative agent was identified. However, the actual incidence of *B. cereus* was only 2% of the total number of cases because most cases of foodborne illness were of unknown etiology (56). Examples of foods involved in foodborne outbreaks are shown in Table 20.2.

CHARACTERISTICS OF DISEASE

There are two types of *B. cereus* foodborne illness. The first type, which is caused by an emetic toxin, results in vomiting, whereas the second type, which is caused by enterotoxin(s), results in diarrhea (42). In a small number of cases both types of symptoms occur (42), likely due to the production of both types of toxin. There has been some debate about whether enterotoxin(s) can be preformed in foods. In reviewing the literature, it appears

Table 20.2 Examples of types of foods involved in *B. cereus* foodborne illness and outbreaks

Type of food	Country	No. of people involved	Type of syndrome[a]
Barbecued chicken	Many countries	—[b]	E, D
Cooked noodles	Spain	13	D
Cream cake	Norway	5	D
Fish soup	Norway	20	D
Hibachi steak	United States	11	E, D
Lobster pâté	United Kingdom	—	D
Meat loaf	United States	—	D
Meat with rice	Denmark	>200	D
Milk	Many countries	—	E, D
Milkshake	United States	36	?
Pea soup	The Netherlands	—	D
Sausages	Ireland, China	—	D
School lunch	Japan	1,877	E
Scrambled egg	Norway	12	D
Several rice dishes	Many countries	—	E, D
Stew	Norway	152	D
Turkey	United Kingdom, United States	—	D
Vanilla sauce	Norway (many countries)	>200	D
Vegetable sprouts	United States	3	E, D
Wheat flour dessert	Bulgaria	—	D

[a] E, emetic syndrome; D, diarrheal syndrome.
[b] —, not available.

that the incubation time (>6 h; average, 12 h) is too long for the diarrheal illness to be caused by preformed enterotoxin (42), and model experiments revealed that enterotoxin(s) is degraded as it proceeds to the ileum (26). However, there is no doubt that enterotoxins can be preformed in food, but the number of *B. cereus* cells in such food is also at least 2 orders of magnitude higher than the numbers necessary for causing food poisoning (16, 26). Usually foods with such large populations of *B. cereus* are no longer organoleptically acceptable to the consumer, although food containing >10^7 *B. cereus* organisms/ml may not always appear spoiled. The characteristics of the two types of *B. cereus* foodborne illness are described in Table 20.3.

Some recent reports have raised concern for *B. cereus* foodborne illness because of severe consequences, including fatal outcomes. An outbreak in Norway associated with eating stew containing approximately 10^4 to 10^5 *B. cereus* organisms per serving affected 17 people, of whom three were hospitalized, including one for 3 weeks. In another case of *B. cereus* foodborne illness, the emetic toxin, cereulide, was responsible for the death of a 17-year-old Swiss boy due to fulminant liver failure (49). A large amount of cereulide was detected in residue in the pan used to reheat the implicated food (pasta) and in the boy's liver and bile. Similarly, a 7-year-old girl died in Belgium only 13 h after ingesting a pasta salad that had been stored for several days in a refrigerator at 14°C. Her 2-year-old brother, who had only a small amount of the salad, also became ill and was hospitalized for 8 days (19). The newly discovered cytotoxin K (CytK) is similar to the β-toxin of *Clostridium perfringens* (and other related toxins) and was the cause of a severe outbreak of *B. cereus* foodborne illness in France in 1998

(46). Several people in this outbreak developed bloody diarrhea, and three died. This was the first recorded outbreak of *B. cereus* necrotic enteritis, although it is not nearly as severe as *C. perfringens* type C foodborne illness (14).

INFECTIVE DOSE AND SUSCEPTIBLE POPULATIONS

After the first recognized diarrheal outbreak of *B. cereus* foodborne illness in Oslo (due to a vanilla sauce), S. Hauge isolated the causative agent, grew it to 4×10^6 organisms/ml, and drank 200 ml of the culture (35). Approximately 13 h later he developed abdominal pain and watery diarrhea that lasted for approximately 8 h. The cell population of *B. cereus* ingested was approximately 8×10^8. Counts of *B. cereus* ranging from 200 to 10^9/g (or ml) (26, 28, 35, 42) have been reported in foods incriminated in outbreaks, indicating that the total dose ranged from approximately 5×10^4 to 10^{11}. The total number of *B. cereus* organisms required to be ingested to produce illness is likely in the range of 10^5 to 10^8 viable cells or spores. For the lower dose it is likely that only spores, which all survive the stomach acid barrier, can cause disease. Still, the wide range of infective dose is also in part due to differences in the amount of enterotoxin produced by different strains (26). Hence, food containing more than 10^3 *B. cereus* organisms/g (spores) cannot be considered completely safe for consumption. Little is known about susceptible populations, but the more severe types of the illness have occasionally involved young athletes (<19 years) or the elderly (>60 years) (27, 28, 46). More studies are needed to address this important question.

Table 20.3   Characteristics of the two types of foodborne illness caused by *B. cereus*[a]

Characteristics	Diarrheal syndrome	Emetics syndrome
Dose causing illness	10^5–10^7 (total)	10^5–10^8 (cells per g)
Toxin produced	In the small intestine of the host	Preformed in foods
Type of toxin	Protein; enterotoxin(s)	Cyclic peptide; emetic toxin
Incubation period	8–16 h (occasionally >24 h)	0.5–5 h
Duration of illness	12–24 h (occasionally several days)	6–24 h
Symptoms	Abdominal pain, watery diarrhea, and occasionally nausea	Nausea, vomiting, and malaise (sometimes followed by diarrhea, due to production of enterotoxin)
Foods most frequently implicated	Meat products, soups, vegetables, puddings, sauces, and milk or milk products	Fried and cooked rice, pasta, pastry, and noodles

[a] Based on references 26, 42, and 58.

VIRULENCE FACTORS AND MECHANISMS OF PATHOGENICITY

The two types of *B. cereus* foodborne illness are caused by very different types of toxins. The emetic toxin, causing vomiting, had been isolated and characterized (2), whereas the diarrheal disease is caused by one or more enterotoxins (9, 12, 26, 27, 42, 46).

The Emetic Toxin

The emetic toxin (Table 20.4) causes emesis (vomiting) only, and its structure was for many years a mystery because the only detection system involved living primates (39, 42). However, the discovery that the toxin could be detected by HEp-2 cells (vacuolation activity) (39) has led to its isolation and structure determination (2, 3). Although there has been some doubt as to whether the emetic toxin and the vacuolating factor are the same constituent (58, 60), there is no longer any doubt that they are identical (2, 3, 59). The emetic toxin has been named cereulide, and consists of a ring structure of three repeats of four amino acids and/or oxyacids: (D-O-Leu-D-Ala-L-O-Val-L-Val)$_3$. This ring structure (dodecadepsipeptide) has a molecular mass of 1.2 kDa and acts as a K$^+$ ionophore, like valinomycin (2). It causes inhibition of mitochondrial activity (inhibition of fatty acid oxidation). Emetic toxin stimulates the vagus afferent through binding to the 5-HT$_3$ receptor. A nonribosomal peptide synthetase gene is responsible for cereulide production in emetic strains (22), and this gene is located on a large plasmid (38). Neither of the characterized emetic strains grows below 10°C, nor does either one produce Hbl enterotoxin (21). The greatest amounts of emetic toxin are produced at 20 to 25°C.

For many years it was speculated that the emetic toxin was of a lipid structure (26, 42). Probably because of the hydrophobic nature of cereulide, it is associated with lipid environments and, hence, the reason for this belief. The emetic toxin is resistant to heat, pH, and proteolysis and is not antigenic (42) (Table 20.4). The recent death of a 17-year-old boy from liver failure caused by the emetic toxin has stimulated research on other aspects of cereulide besides food poisoning. Histopathological studies of mice injected intraperitoneally with large doses of synthetic cereulide revealed massive degeneration of hepatocytes (64). General recovery from pathological changes and regeneration of hepatocytes were observed after 4 weeks.

Enterotoxins

It is now apparent through cloning and sequencing studies that *B. cereus* produces at least three different proteins (or protein complexes), referred to as enterotoxins (30, 36, 46), that may cause foodborne illness (Table 20.5). Two of the enterotoxins are multicomponent and structurally related, whereas the third (cytK) is a single protein of 34 kDa. The three-component hemolysin (Hbl; consisting of three proteins: B, L$_1$, and L$_2$) with enterotoxin activity was the first to be fully characterized (11, 12). This toxin also has dermonecrotic and vascular permeability activities and causes fluid accumulation in ligated rabbit ileal loops. Hbl has been suggested to be a primary virulence factor in diarrhea caused by *B. cereus* (9), but several strains without the genes for this toxin have caused food poisoning. Evidence has shown that all three components of Hbl are necessary for maximal enterotoxin activity (9). It was first suggested that the B protein is the component that binds Hbl to the target cells and that L$_1$ and L$_2$ have lytic functions (8). However, more recently another model for the action of Hbl has been proposed, suggesting that the components of Hbl bind to target cells independently and then constitute a membrane-attacking complex resulting in a colloid osmotic lysis mechanism (10). More research is needed to elucidate the mode of action of Hbl. A 1:1:1 ratio of the three components appears to provide the greatest biological activity (9). There is substantial heterogeneity among the components of Hbl, and individual strains produce different combinations of single or

Table 20.4 Properties of the emetic toxin cereulide[a]

Trait	Property or activity
Molecular mass	1.2 kDa
Structure	Ring-shaped peptide
Isoelectric point	Uncharged
Antigenic	No
Biological activity in living primates	Vomiting
Receptor	5-HT$_3$ (stimulation of the vagus afferent)
Ileal loop tests (rabbit, mouse)	None
Cytotoxic	No
HEp-2 cells	Vacuolation activity
Stability to heat	90 min at 121°C
Stability to pH	Stable at pH 2–11
Effect of proteolysis (trypsin, pepsin)	None
Conditions under which toxin is produced	In food: rice and milk at 16–32°C
Mechanisms of production	Produced by a nonribosomal peptide synthetase

[a] Based on references 2, 3, 22, 42, 58, and 59.

Table 20.5 Toxins produced by *B. cereus*

Toxin	Type and size	Food poisoning	Reference(s)
Hemolysin BL (Hbl)	Protein, 3 components	Probably	9, 10, 36, 55
Nonhemolytic enterotoxin (Nhe)	Protein, 3 components	Yes	31, 45, 47, 48
Cytotoxin K (CytK)	Protein, 1 component, 34 kDa	Yes, 3 deaths	46
Emetic toxin (cereulide)	Cyclic peptide, 1.2 kDa	Yes, 2 deaths	2, 3, 19, 49

multiple bands of each component (57). This is likely due to sequence variation within multiple genes of *hbl*, but this must be validated by genetic studies. About 50% of *B. cereus* strains harbor *hbl* genes.

A nonhemolytic three-component enterotoxin (Nhe) was characterized by Lund and Granum (47, 48). The three components of this toxin are different from the components of Hbl, although there are similarities. More than 99% of *B. cereus* strains are capable of producing this enterotoxin complex, although not all can do it at 32 to 37°C. The enterotoxin is most active when the ratio of NheA, NheB, and NheC is 10:10:1. NheB binds to the receptors (unknown), and NheC appears to catalyze binding between NheA and NheB (45). However, NheC is apparently transcribed in even lower ratios in vivo (20) than the optimal ratio. This may be the reason why NheC is not detected on two-dimensional sodium dodecyl sulfate gels (24). If higher ratios than the one that is optimal for NheC are used in experiments, the biological activity is rapidly reduced (45), and at a 1:1:1 ratio very little activity remains.

The third characterized *B. cereus* enterotoxin involved in food poisoning is CytK. This toxin belongs to a family of β-barrel toxins and is similar to the β-toxin of *C. perfringens* and was the cause of severe consequences in an outbreak of *B. cereus* foodborne illness in which several people developed bloody diarrhea and three died (46). This was an outbreak of *B. cereus* necrotic enteritis, although it was not nearly as severe as the necrotic enteritis caused by *C. perfringens* type C (14). This strain is one of a few that do not produce Nhe or Hbl (although it harbors an Nhe operon that is quite different from the original sequence). This may explain why this strain was so aggressive, since the normal rice water diarrhea was not observed (possibly flushing out CytK). Another explanation is that this particular strain (together with only one other strain to date) produces a more virulent type of CytK than most other strains (*cytK* is present in about 35 to 40% of strains). This type of CytK (CytK-1) is 89% identical in amino acid sequence to the more commonly

present gene transcribing CytK-2 (23). CytK-1 is about five times more toxic on epithelial cells than CytK-2, but their activity on erythrocytes is very similar (23). As with other members of the β-barrel toxin family, CytK is inserted into membranes as a heptamer and produces a pore of 7 Å in diameter (34).

There is substantial sequence identity between the three proteins of Nhe and between the Nhe and Hbl proteins. Identity is greatest in the N-terminal third of the proteins. The most pronounced similarities occur between *nheA* and *hblC*, *nheB* and *hblD*, and *nheC* and *hblA* as determined not only by direct comparison of the sequences but also by predicted transmembrane helices of the six proteins. NheA and HblC have no predicted transmembrane helices, whereas NheB and HblD have two each and NheC and HblA each have one in the same position of the two proteins (31).

Although recent studies have revealed that the total toxicity of *B. cereus* supernatants, for most strains, is largely due to Nhe (49a), we cannot conclude that the two other toxins do not contribute to diarrhea. Furthermore, CytK-1 can be fatal when produced alone.

Gene Organization of the Enterotoxins

All three proteins of the Hbl are transcribed from one operon (*hbl*) (56), and Northern blot analysis has revealed an RNA transcript of 5.5 kb (Fig. 20.1). *hblC* (transcribing L_2) and *hblD* (transcribing L_1) are separated only by 37 bp and encode proteins of 447 amino acids (aa) and 384 aa. L_2 has a signal peptide of 32 aa, and L_1 has a signal peptide of 30 aa. The B-protein, transcribed from *hblA*, consists of 375 aa, with a signal peptide of 31 aa (36). The spacing between *hblD* and *hblA* is 36 bp, and that between *hblA* and *hblB* is 381 bp. The size of the putative HblB is 466 aa. The function of this putative protein is not yet known. The *hbl* operon is mapped to the unstable part of the *B. cereus* chromosome.

The *nhe* operon contains three open reading frames (ORFs) (Fig. 20.1): *nheA*, *nheB*, and *nheC* (31). The first two gene products have been addressed previously

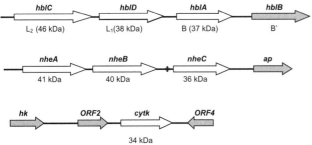

Figure 20.1 Genetic organization of the three enterotoxins *hbl*, *nhe*, and *cytK* (31, 36, 46, 56). The enterotoxins are all regulated by PlcR-PapR, which binds upstream of *hblC*, *nheA*, and *cytK*. There is an inverted repeat of 13 bp between *nheB* and *nheC*. An aminopeptidase gene (*ap*) is found about 600 bp downstream of *nheC*. *cytK* has an ORF (*ORF2*) upstream, encoding a protein of unknown function, and a histidine kinase (*hk*) further upstream. *ORF4* encodes a long-chain fatty acid coenzyme A ligase (orientated in the direction opposite to that of *cytK*). The arrowheads indicate the orientation of the genes. The gaps between the different genes are indicated by a solid line.

as the 45- and 39-kDa proteins, respectively. The last transcript from the *nhe* operon (*nheC*) is transcribed in small amounts and has been purified only after cloning into *Escherichia coli*, although transcription in *B. cereus* has been shown using antibodies made from a synthetic peptide based on the C-terminal part of NheC (20). The three proteins transcribed from the *nhe* operon have properties described in Table 20.6. The correct sizes of the 45-kDa protein and 39-kDa proteins are 41 and 39.8 kDa, respectively. The size of NheC is 36.5 kDa. There is a gap of 40 bp between *nheA* and *nheB*, and a gap of 109 bp between *nheB* and *nheC*. In the 109 bp between *nheB* and *nheC*, there is an inverted repeat of 13 bp (31). This structure may result in little production of NheC compared to that of NheA and NheB. This may be the reason for difficulties encountered in isolating the NheC protein.

The *B. cereus* strain producing CytK-1 did not transcribe other *B. cereus* enterotoxins. CytK is a protein of 34 kDa and is also hemolytic. *cytK-1* is genetically organized as shown in Fig. 20.1. The full genome sequence of

Table 20.6 Properties of the Nhe proteins[a]

Protein	Signal peptide (aa)	Active protein (aa)	Mol wt (active protein)	PI
NheA	26	360	41.019	5.13
NheB	30	372	39.820	5.61
NheC	30	329	36.481	5.28

[a] Based on reference 31.

this strain (NVH 0391-88) will soon be available (INRA, Paris, France).

Gene Regulation of Enterotoxin Production

Hbl, Nhe, and CytK are positively regulated by the pleiotropic regulator PlcR (1). A number of excreted proteins encoded by chromosomal genes in bacteria in the *B. cereus* group are under control of PlcR, and *plcR* has so far been found only in this group, although it is inactive in *B. anthracis*. The transcriptional activator PlcR (34 kDa) was first identified as a required activator of the *plcA* gene at the onset of stationary phase in *B. thuringiensis*. PlcR was shown to positively regulate its own transcription (1), by binding to a palindromic sequence with the consensus sequence TATGNAN₄TNCATA.

Further knowledge of the function and control of PlcR has been gained during the last few years. The transition state regulator in *B. thuringiensis*, Spo0A, controls PlcR expression at the onset of stationary phase in a growth medium-dependent manner. In the transition from exponential growth to stationary phase, triggered by depletion of nutrients, phosphorylation of the Spo0A protein is the key step, activating transcription of sporulation-specific sigma factors. The phosphorylated protein acts as a repressor of other genes. The concentration of Spo0A~P likely determines whether the bacterium sporulates or not in this phase.

In search of positive regulators of *plcR* transcription, oligopeptide permease (Opp), which is required for transport of small peptides into the cell, was determined to be necessary for hemolytic activity and insect virulence in *B. thuringiensis*, and therefore, *plcR* expression appears to be activated by uptake of peptide in a quorum-sensing system (25). The requirement for Opp was, however, independent of Spo0A.

The presence of an ORF, *orf2*, encoding a putative 48-aa peptide, 70 bp downstream from *plcR* was identified in 1996 (44), but only later was the exact nature of its indispensable role in the autoregulation of PlcR elucidated. It was demonstrated that the PlcR regulon was not expressed and that virulence was decreased when *orf2* was disrupted in *B. thuringiensis*. Furthermore, synthetic peptides corresponding to the C terminus of the *orf2*-encoded polypeptide activated expression of the PlcR regulon. The name PapR (mnemonic for peptide activating PlcR) was suggested for the gene product of *orf2*. PapR is secreted via the SecA pathway, processed, and then reimported via Opp. The processed form (presumably a pentapeptide) of PapR facilitates the binding of PlcR to the palindromic PlcR box in the promoter (61).

It has been suggested that the active peptide increases PlcR's affinity for the DNA target by inducing a conformational change. Furthermore, the PapR-PlcR activation complex was determined to be strain specific. The amino acid sequence of PapR was predicted after sequencing of *papR* from strains of *B. thuringiensis*, *B. cereus* (ATCC 14579), *B. anthracis*, and *B. mycoides*. Gene sequences in the C terminus of the peptide, which is the active component, differ between bacterial strains. The strain specificity of the system, however, lies not solely in the PapR C-terminal sequence. An analysis of secondary structure in PlcR predicted a helix-turn-helix DNA-binding domain in residues 18 to 55 in *B. cereus* strain 569, and on alignment of 18 PlcR sequences, the DNA-binding domain, located to the N-terminal part of the protein, was the most conserved region. The C-terminal region was most variable, and the 18 strains could be divided into two groups based on their C-terminal sequences. For each group, there was a consensus matching set of PapR C-terminal pentapeptide sequences. Additional studies revealed four classes of PlcR-PapR pairs, found by in vivo complementation assays and sequence comparisons, to further elucidate the basis of strain specificity of the system (62).

It is interesting that most *B. weihenstephanensis* strains also produce enterotoxins, but usually at temperatures below 30°C (L. P. S. Arnesen and P. E. Granum, unpublished results). This regulation is not only under the control of the above-described (PlcR-PapR) system; however, the actual mechanism has not been determined.

The Spore

The spore of *B. cereus* is an important factor in foodborne illness. First, the *B. cereus* spore is more hydrophobic than any other *Bacillus* spp. spores, which enables it to adhere to several types of surfaces (40). Hence, it is difficult to remove during cleaning and is a difficult target for disinfectants. *B. cereus* spores also contain appendages and/or pili (4, 40) that are, at least in part, involved in adhesion (40). Not only can these properties of the *B. cereus* spore enable them to withstand sanitation, and hence remain present on surfaces for subsequent contamination of foods, but also they aid in adherence to epithelial cells. Studies have revealed that the spores of many strains associated with foodborne outbreaks (4) can adhere to Caco-2 cells in culture and that these properties are associated with hydrophobicity and possibly to the spore's appendages (5).

Commercial Methods for Detection of *B. cereus* Toxins

No commercial kit for detection of the emetic toxin (cereulide) is yet available; however, with the structure of cereulide now known (2), it is likely that a kit will be available in the near future. Since a large portion of the genes transcribing the nonribosomal peptide synthetase responsible for cereulide production is known (22), PCR can be used for detection of such genes. Screening for cereulide production by *B. cereus* can also easily be done by using a sperm motility assay (33).

Neither of the two commercial immunoassays available for enterotoxin detection can quantify the toxicity of *B. cereus* enterotoxins. The Oxoid assay measures the presence of the HblC (L_2) component, whereas the Tecra assay detects the NheA (45-kDa) component (29, 30, 48). However, if one or both of the commercial kits react positively with proteins from *B. cereus* supernatant fluids, it is likely that the strain is enterotoxin positive. If culture supernatant fluids are also cytotoxic, the strains can be regarded as enterotoxin positive. At present there is no commercial method available for detecting CytK.

CONCLUDING REMARKS

B. cereus is a normal inhabitant of soil and is frequently isolated from a variety of foods, including vegetables, dairy products, and meat. It causes an emetic or a diarrheal type of food-associated illness that is becoming increasingly important in developed countries. The diarrheal type of illness is most prevalent in the Western Hemisphere, whereas the emetic type is most prevalent in Japan. Desserts, meat dishes, and dairy products are most frequently associated with diarrheal illness, whereas rice and pasta are the most common vehicles of emetic illness.

A *B. cereus* emetic toxin has been isolated and characterized. Three types of *B. cereus* enterotoxins involved in outbreaks of foodborne illness have been identified. Two of these enterotoxins possess three components and are related, whereas the third is a one-component protein (CytK). Deaths have been caused by the emetic toxin and by a strain producing only CytK.

Some strains of the *B. cereus* group are able to grow at refrigeration temperature. These variants raise concern regarding the safety of cooked, refrigerated foods with extended shelf lives. *B. cereus* spores are adhesive to many surfaces and survive normal cleaning and disinfection (except for hypochlorite and UVC) procedures. *B. cereus* foodborne illness is likely highly underreported and may not receive adequate attention from epidemiologists because of its relatively mild symptoms, which are of short duration. However, consumer interest in precooked, chilled food products with long shelf lives may lead to the development of food products well suited for *B. cereus* survival and growth. Such foods could increase the prominence of *B. cereus* as a foodborne pathogen.

References

1. **Agaisse, H., M. Gominet, O. A. Okstad, A. B. Kolsto, and D. Lereclus.** 1999. PlcR is a pleiotropic regulator of extracellular virulence factor gene expression in *Bacillus thuringiensis*. *Mol. Microbiol.* **32:**1043–1053.

2. **Agata, N., M. Mori, M. Ohta, S. Suwan, I. Ohtani, and M. Isobe.** 1994. A novel dodecadepsipeptide, cereulide, isolated from *Bacillus cereus* causes vacuole formation in HEp-2 cells. *FEMS Microbiol. Lett.* **121:**31–34.

3. **Agata, N., M. Ohta, M. Mori, and M. Isobe.** 1995. A novel dodecadepsipeptide, cereulide, is an emetic toxin of *Bacillus cereus*. *FEMS Microbiol. Lett.* **129:**17–20.

4. **Andersson, A., P. E. Granum, and U. Rönner.** 1998. The adhesion of *Bacillus cereus* spores to epithelial cells might be an additional virulence mechanism. *Int. J. Food Microbiol.* **39:**93–99.

5. **Andersson, A., U. Rönner, and P. E. Granum.** 1995. What problems does the food industry have with the sporeforming pathogens *Bacillus cereus* and *Clostridium perfringens*? *Int. J. Food Microbiol.* **28:**145–156.

6. **Ash, C., and M. D. Collins.** 1992. Comparative analysis of 23S ribosomal RNA gene sequence of *Bacillus anthracis* and emetic *Bacillus cereus* determined by PCR-direct sequencing. *FEMS Microbiol. Lett.* **73:**75–80.

7. **Ash, C., J. A. Farrow, M. Dorsch, E. Steckebrandt, and M. D. Collins.** 1991. Comparative analysis of *Bacillus anthracis*, *Bacillus cereus*, and related species on the basis of reverse transcriptase sequencing of 16S rRNA. *Int. J. Syst. Bacteriol.* **41:**343–346.

8. **Beecher, D. J., and J. D. Macmillan.** 1991. Characterization of the components of hemolysin BL from *Bacillus cereus*. *Infect. Immun.* **59:**1778–1784.

9. **Beecher, D. J., J. L. Schoeni, and A. C. L. Wong.** 1995. Enterotoxin activity of hemolysin BL from *Bacillus cereus*. *Infect. Immun.* **63:**4423–4428.

10. **Beecher, D. J., and A. C. L Wong.** 1997. Tripartite hemolysin BL from *Bacillus cereus*. Hemolytic analysis of component interaction and model for its characteristic paradoxical zone phenomenon. *J. Biol. Chem.* **272:**233–239.

11. **Beecher, D. J., and A. C. L. Wong.** 1994. Identification and analysis of the antigens detected by two commercial *Bacillus cereus* diarrheal enterotoxin immunoassay kits. *Appl. Environ. Microbiol.* **60:**4614–4616.

12. **Beecher, D. J., and A. C. L. Wong.** 1994. Improved purification and characterization of hemolysin BL, a hemolytic dermonecrotic vascular permeability factor from *Bacillus cereus*. *Infect. Immun.* **62:**980–986.

13. **Borge, G. A., M. Skeie, T. Langsrud, and P. E. Granum.** 2001. Growth and toxin profiles of *Bacillus cereus* isolated from different food sources. *Int. J. Food Microbiol.* **69:**237–246.

14. **Brynestad, S., and P. E. Granum.** 2002. *Clostridium perfringens* and foodborne infections. *Int. J. Food Microbiol.* **74:**195–202.

15. **Carlson, C. R., D. A. Caugant, and A.-B. Kolstø.** 1994. Genotypic diversity among *Bacillus cereus* and *Bacillus thuringiensis* strains. *Appl. Environ. Microbiol.* **60:**1719–1725.

16. **Christiansson, A., A. S. Naidu, I. Nilsson, T. Wadström, and H.-E. Pettersson.** 1989. Toxin production by *Bacillus cereus* dairy isolates in milk at low temperatures. *Appl. Environ. Microbiol.* **55:**2595–2600.

17. **Claus, D., and R. C. W. Berkeley.** 1986. Genus *Bacillus*, p. 1105–1139. *In* P. H. A. Seneath (ed.), *Bergeys Manual of Systematic Bacteriology*, vol. 2. The Williams and Wilkins Co., Baltimore, Md.

18. **Damgaaerd, P. H., H. D. Larsen, B. M. Hansen, J. Bresciani, and K. Jørgensen.** 1996. Enterotoxin-producing strains of *Bacillus thuringiensis* isolated from food. *Lett. Appl. Microbiol.* **23:**146–150.

19. **Dierick, K., E. Van Coillie, I. Swiecicka, G. Meyfroidt, H. Devlieger, A. Meulemans, G. Hoedemaekers, L. Fourie, M. Heyndrickx, and J. Mahillon.** 2005. Fatal family outbreak of *Bacillus cereus*-associated food poisoning. *J. Clin. Microbiol.* **43:**4277–4279.

20. **Dietrich, R., M. Moravek, C. Bürk, P. E. Granum, and E. Märtlbauer.** 2005. Production and characterization of antibodies against each of the three subunits of the *Bacillus cereus* nonhemolytic enterotoxin complex. *Appl. Environ. Microbiol.* **71:**8214–8220.

21. **Ehling-Schulz, M., B. Svensson, M.-H. Guinebritiere, T. Linback, M. Andersson, A. Schulz, A. Christiansson, P. E. Granum, E. Märtelbauer, C. Nguyen-The, M. Salkinoja-Salonen, and S. Scherer.** 2005. Emetic toxin formation of *Bacillus cereus* is restricted to a single evolutionary lineage of closely related strains. *Microbiology* **151:**183–197.

22. **Ehling-Schulz, M., N. Vukov, A. Schulz, R. Shaheen, M. Andersson, E. Martlbauer, and S. Scherer.** 2005. Identification and partial characterization of the nonribosomal peptide synthetase gene responsible for cereulide production in emetic *Bacillus cereus*. *Appl. Environ. Microbiol.* **71:**105–113.

23. **Fagerlund, A., O. Ween, T. Lund, S. P. Hardy, and P. E. Granum.** 2004. Genetic and functional analysis of the cytK family of genes in *Bacillus cereus*. *Microbiology* **150:**2689–2697.

23a. **Rivera, A. M. Gaviria, P. E. Granum, and F. G. Priest.** 2000. Common occurrence of enterotoxin genes and enterotoxicity in *Bacillus thuringiensis*. *FEMS Microbiol. Lett.* **190:**151–155.

24. **Gohar, M., O. A. Okstad, N. Gilois, V. Sanchis, A. B. Kolsto, and D. Lereclus.** 2002. Two-dimensional electrophoresis analysis of the extracellular proteome of *Bacillus cereus* reveals the importance of the PlcR regulon. *Proteomics* **2:**784–791.

25. **Gominet, M., L. Slamti, N. Gilois, M. Rose, and D. Lereclus.** 2001. Oligopeptide permease is required for expression of the *Bacillus thuringiensis* plcR regulon and for virulence. *Mol. Microbiol.* **40:**963–975.

26. **Granum, P. E.** 1994. *Bacillus cereus* and its toxins. *J. Appl. Bacteriol. Symp. Suppl.* **76:**61S–66S.

27. **Granum, P. E., A. Andersson, C. Gayther, M. C. te Giffel, H. D. Larsen, T. Lund, and K. O'Sullivan.** 1996. Evidence for a further enterotoxin complex produced by *Bacillus cereus*. *FEMS Microbiol. Lett.* **141:**145–149.

28. **Granum, P. E., and T. C. Baird-Parker.** 2000. *Bacillus* spp., p. 1029–1039. *In* B. Lund, T. Baird-Parker, and G. Gould (ed.), *The Microbiological Safety and Quality of Food.* Aspen Publishers, Gaithersburg, Md.

29. **Granum, P. E., S. Brynestad, and J. M. Kramer.** 1993. Analysis of enterotoxin production by *Bacillus cereus* from dairy

products, food poisoning incidents and non-gastrointestinal infections. *Int. J. Food Microbiol.* **17:**269–279.

30. **Granum, P. E., and T. Lund.** 1997. *Bacillus cereus* enterotoxins. *FEMS Microbiol. Lett.* **157:**223–228.

31. **Granum, P. E., K. O'Sullivan, and T. Lund.** 1999. The sequence of the non-haemolytic enterotoxin operon from *Bacillus cereus. FEMS Microbiol. Lett.* **177:**225–229.

32. **Griffiths, M. W.** 1990. Toxin production by psychrotrophic *Bacillus* spp. present in milk. *J. Food Prot.* **53:**790–792.

33. **Haggblom, M. M., C. Apetroaie, M. A. Andersson, and M. S. Salkinoja-Salonen.** 2002. Quantitative analysis of cereulide, the emetic toxin of *Bacillus cereus*, produced under various conditions. *Appl. Environ. Microbiol.* **68:**2479–2483.

34. **Hardy, S. P., T. Lund, and P. E. Granum.** 2001. CytK toxin of *Bacillus cereus* forms pores in planar lipid bilayers and is cytotoxic to intestinal epithelia. *FEMS Microbiol. Lett.* **197:**47–51.

35. **Hauge, S.** 1955. Food poisoning caused by aerobic spore forming bacilli. *J. Appl. Bacteriol.* **18:**591–595.

36. **Heinrichs, J. H., D. J. Beecher, J. M. MacMillan, and B. A. Zilinskas.** 1993. Molecular cloning and characterization of the *hblA* gene encoding the B component of hemolysin BL from *Bacillus cereus. J. Bacteriol.* **175:**6760–6766.

37. **Hoffmaster, A. R., J. Ravel, D. A. Rasko, G. D. Chapman, M. D. Chute, C. K. Marston, B. K. De, C. T. Sacchi, C. Fitzgerald, L. W. Mayer, M. C. Maiden, F. G. Priest, M. Barker, L. Jiang, R. Z. Cer, J. Rilstone, S. N. Peterson, R. S. Weyant, D. R. Galloway, T. D. Read, T. Popovic, and C. M. Fraser.** 2004. Identification of anthrax toxin genes in a *Bacillus cereus* associated with an illness resembling inhalation anthrax. *Proc. Natl. Acad. Sci. USA* **101:**8449–8454.

38. **Hoton, F. M., L. Andrup, I. Swiecicka, and J. Mahillon.** 2005. The cereulide genetic determinants of emetic *Bacillus cereus* are plasmid-borne. *Microbiology* **151:**2121–2124.

39. **Hughes, S., B. Bartholomew, J. C. Hardy, and J. M. Kramer.** 1988. Potential application of a HEp-2 cell assay in the investigation of *Bacillus cereus* emetic-syndrome food poisoning. *FEMS Microbiol. Lett.* **52:**7–12.

40. **Husmark, U.** 1993. Adhesion mechanisms of bacterial spores to solid surfaces. Ph.D. thesis, Department of Food Science, Chalmers University of Technology and SIK, The Swedish Institute for Food Research, Göteborg, Sweden.

41. **Jackson, S. G., R. B. Goodbrand, R. Ahmed, and S. Kasatiya.** 1995. *Bacillus cereus* and *Bacillus thuringiensis* isolated in a gastroenteritis outbreak investigation. *Lett. Appl. Microbiol.* **21:**103–105.

42. **Kramer, J. M., and R. J. Gilbert.** 1989. *Bacillus cereus* and other *Bacillus* species, p. 21–70. *In* M. P. Doyle (ed.), *Foodborne Bacterial Pathogens.* Marcel Dekker, New York, N.Y.

43. **Lechner, S., R. Mayr, K. P. Francic, B. M. Prub, T. Kaplan, E. Wieber-Gunkel, G. A. S. B. Stewart, and S. Scherer.** 1998. *Bacillus weihenstephanensis* sp. nov. is a new psychrotolerant species of the *Bacillus cereus* group. *Int. J. Syst. Bacteriol.* **48:**1373–1382.

44. **Lereclus, D., H. Agaisse, M. Gominet, S. Salamitou, and V. Sanchis.** 1996. Identification of a *Bacillus thuringiensis*

gene that positively regulates transcription of the phosphatidylinositol-specific phospholipase C gene at the onset of the stationary phase. *J. Bacteriol.* **178:**2749–2756.

45. **Lindbäck, T., A. Fagerlund, M. S. Rødland, and P. E. Granum.** 2004. Characterization of the *Bacillus cereus* Nhe enterotoxin. *Microbiology* **150:**3959–3967.

46. **Lund, T., M. L. De Buyser, and P. E. Granum.** 2000. A new cytotoxin from *Bacillus cereus* that may cause necrotic enteritis. *Mol. Microbiol.* **38:**254–261.

47. **Lund, T., and P. E. Granum.** 1996. Characterisation of a non-haemolytic enterotoxin complex from *Bacillus cereus* isolated after a foodborne outbreak. *FEMS Microbiol. Lett.* **141:**151–156.

48. **Lund, T., and P. E. Granum.** 1997. Comparison of biological effect of the two different enterotoxin complexes isolated from three different strains of *Bacillus cereus. Microbiology* **143:**3329–3336.

49. **Mahler, H., A. Pasi, J. M. Kramer, P. Schulte, A. C. Scoging, W. Bar, and S. Krahenbuhl.** 1997. Fulminant liver failure in association with the emetic toxin of *Bacillus cereus. N. Engl. J. Med.* **336:**1142–1148.

49a. **Moravek, M., R. Dietrich, C. Buerk, V. Broussole, M. H. Guinebretiere, P. E. Granum, C. Nguyen-the, and E. Maertlbauer.** 2006. Determination of the toxic potential of *Bacillus cereus* isolates by quantitative enterotoxin analyses. *FEMS Microbiol. Lett.* **257:**293–298.

50. **Mortimer, P. R., and G. McCann.** 1974. Food poisoning episodes associated with *Bacillus cereus* in fried rice. *Lancet* **i:**1043–1045.

51. **Nakamura, L. K.** 1998. *Bacillus pseudomycoides* sp. nov. *Int. J. Syst. Bacteriol.* **48:**1031–1035.

52. **Ray, D. E.** 1991. Pesticides derived from plants and other organisms, p. 585–636. *In* W. J. Hayes and E. R. Laws, Jr. (ed.), *Handbook of Pesticide Toxology.* Academic Press, Inc., New York, N.Y.

53. (Reference deleted.)

54. **Rosenquist, H., L. Smidt, S. R. Andersen, G. B. Jensen, and A. Wilcks.** 2005. Occurrence and significance of *Bacillus cereus* and *Bacillus thuringiensis* in ready-to-eat food. *FEMS Microbiol. Lett.* **250:**129–136.

55. **Ryan, P. A., J. M. Macmillan, and B. A. Zilinskas.** 1997. Molecular cloning and characterization of the genes encoding the L_1 and L_2 components of hemolysin BL from *Bacillus cereus. J. Bacteriol.* **179:**2551–2556.

56. **Schmidt, K. (ed.).** 2001. *WHO Surveillance Programme for Control of Foodborne Infections and Intoxications in Europe. Seventh Report.* FAO/WHO Collaborating Centre for Research and Training in Food Hygiene and Zoonoses, Berlin, Germany.

57. **Schoeni, J. L., and A. C. L. Wong.** 1999. Heterogeneity observed in the components of hemolysin BL, an enterotoxin produced by *Bacillus cereus. Int. J. Food Microbiol.* **53:**159–167.

58. **Shinagawa, K.** 1993. Serology and characterization of *Bacillus cereus* in relation to toxin production. *Bull. Int. Dairy Fed.* **287:**42–49.

59. **Shinagawa, K., H. Konuma, H. Sekita, and S. Sugii.** 1995. Emesis of rhesus monkeys induced by intragastric

administration with the HEp-2 vacuolation factor (cereulide) produced by *Bacillus cereus*. *FEMS Microbiol. Lett.* **130:**87–90.

60. **Shinagawa, K., S. Otake, N. Matsusaka, and S. Sugii.** 1992. Production of the vacuolation factor of *Bacillus cereus* isolated from vomiting-type food poisoning. *J. Vet. Med. Sci.* **54:**443–446.

61. **Slamti, L., and D. Lereclus.** 2002. A cell-cell signaling peptide activates the PlcR virulence regulon in bacteria of the *Bacillus cereus* group. *EMBO J.* **21:**4550–4559.

62. **Slamti, L., and D. Lereclus.** 2005. Specificity and polymorphism of the PlcR-PapR quorum-sensing system in the *Bacillus cereus* group. *J. Bacteriol.* **187:**1182–1187.

63. **van Netten, P., A. van de Moosdijk, P. van Hoensel, D. A. A. Mossel, and I. Perales.** 1990. Psychrotrophic strains of *Bacillus cereus* producing enterotoxin. *J. Appl. Bacteriol.* **69:**73–79.

64. **Yokoyama, K., M. Ito, N. Agata, M. Isobe, K. Shibayama, T. Horii, and M. Ohta.** 1999. Pathological effect of synthetic cereulide, an emetic toxin of *Bacillus cereus*, is reversible in mice. *FEMS Immunol. Med. Microbiol.* **24:**115–120.

Food Microbiology: Fundamentals and Frontiers, 3rd Ed.
Edited by M. P. Doyle and L. R. Beuchat
© 2007 ASM Press, Washington, D.C.

Bala Swaminathan
Didier Cabanes
Wei Zhang
Pascale Cossart

Listeria monocytogenes

21

Listeriosis has emerged as a major foodborne disease during the past 25 years, after a 1981 outbreak of listeriosis in Nova Scotia, Canada, was traced to contaminated coleslaw (254). However, the causative agent of listeriosis, *Listeria monocytogenes*, was discovered more than 70 years ago by E. G. D. Murray and James Pirie, who were working independently of each other and provided exquisitely detailed descriptions of listeriosis in small animals (241). A fascinating account of the discovery of *L. monocytogenes* has been compiled by Jim McLauchlin (201). The first documented human listeriosis case was a soldier who suffered from meningitis at the end of World War I. However, there is a suggestion in the literature that listeriosis may have been the cause of Queen Anne's 17 unsuccessful pregnancies (252).

Between 1930 and 1950, a few human listeriosis cases were reported. However, there are now hundreds of cases reported every year (242). The emergence of listeriosis is the result of complex interactions between various factors reflecting changes in social patterns. These factors include

- improvements during the past 50 years in medicine, public health, sanitation, and nutrition that have resulted in increased life expectancy, particularly in developed countries
- the ongoing epidemic of AIDS, as well as widespread use of immunosuppressive medications for the treatment of malignancies and management of organ transplantations, that have greatly expanded the immunocompromised population at increased risk for listeriosis
- changes in food production practices, particularly the high degree of centralization and consolidation of food production and processing, the ever-expanding national and international distribution of foods, and increased use of refrigeration as a primary means of preservation of foods
- changes in food habits (increased consumer demand for convenience food that has a fresh-cooked taste, that can be purchased ready-to-eat, refrigerated, or frozen and be prepared rapidly, and that requires essentially little cooking before consumption) and changes in handling and preparation practices

Bala Swaminathan, Centers for Disease Control and Prevention, 1600 Clifton Rd., MS-C03, Atlanta, GA 30333. **Didier Cabanes,** Group of Molecular Microbiology, Institute for Molecular and Cell Biology, Rua do Campo Alegre 823, 4150-180 Porto, Portugal. **Wei Zhang,** National Center for Food Safety and Technology, Illinois Institute of Technology, 6502 South Archer Road, Summit, IL 60501.
Pascale Cossart, Institut Pasteur, Unité des Interactions Bactéries-Cellules, Paris, F-75015 France; Inserm, U604, Paris, F-75015 France; INRA, USC2020, Paris, F-75015 France.

• improved diagnostic methods and enhanced public health surveillance.

Listeriosis is an atypical foodborne illness of major public health concern because of the severity of the disease (meningitis, septicemia, and abortion), a high case-fatality rate (approximately 20 to 30% of cases), a long incubation time, and a predilection for individuals who have an underlying condition that leads to impairment of T-cell-mediated immunity. *L. monocytogenes* differs in many respects from most other foodborne pathogens in that it is widely distributed, resistant to diverse environmental conditions including low pH and high NaCl concentrations, and microaerobic and psychrotrophic. The various ways the bacterium can enter into food-processing plants, its ability to survive for long periods of time in the environment (soil, plants, and water), on foods, and in food-processing plants, and its ability to grow at very low temperature (2 to 4°C) and to survive in or on food for prolonged periods under adverse conditions have made this bacterium a major concern for the agrifood industry during the last decade. The significance of *L. monocytogenes* as a foodborne pathogen is complex. The severity and case-fatality rate of the disease require appropriate preventive measures, but the characteristics of the microorganism are such that it is unrealistic to expect all food to be *Listeria*-free. This dilemma has generated considerable debate about strategies for prevention of listeriosis and the regulation of *L. monocytogenes* in foods. Further, it has prompted research in various areas, such as how the bacterium behaves in food, how it gets into food-processing plants, the reasons for its persistence in food-processing plants and equipment, and how to eliminate *Listeria* from food-processing establishments and prevent contamination of ready-to-eat (RTE) foods. Epidemiologic investigations of outbreaks have helped identify the vehicles of transmission. Basic research on the genetics and molecular biology of *L. monocytogenes* has provided detailed insights into the virulence characteristics of this fascinating pathogen.

CHARACTERISTICS OF THE ORGANISM

Classification—The Genus *Listeria*
Listeria belongs to the *Clostridium* subbranch together with *Staphylococcus*, *Streptococcus*, *Lactobacillus*, and *Brochothrix*. This phylogenetic position of *Listeria* is consistent with its low G+C% DNA content (36 to 42%) (243).

L. monocytogenes is one of six species in the genus *Listeria*. The other species are *L. ivanovii*, *L. innocua*, *L. seeligeri*, *L. welshimeri*, and *L. grayi* (241). Two subspecies of *L. ivanovii* have been described: *L. ivanovii* subsp.

ivanovii and *L. ivanovii* subsp. *londoniensis* (25). *L. murrayi*, which was a separate species in the genus *Listeria*, has been reclassified as a subspecies of *L. grayi*. Based on results of DNA-DNA hybridization, multilocus enzyme analysis, and 16S rRNA sequencing, the six species in the genus *Listeria* are divided on two lines of descent: (i) *L. monocytogenes* and its closely related species, namely, *L. innocua*, *L. ivanovii* (subspecies *ivanovii* and subspecies *londoniensis*), *L. welshimeri*, and *L. seeligeri* and (ii) *L. grayi*. Within the genus *Listeria*, only *L. monocytogenes* and *L. ivanovii* are considered to be pathogenic, as evidenced by their 50% lethal dose in mice and their ability to grow in mouse spleen and liver. *L. monocytogenes* is a human pathogen of high public health concern; *L. ivanovii* primarily is an animal pathogen.

Methodology for the isolation of *L. monocytogenes* from foods has been recently reviewed (103), and methods for its isolation from clinical specimens have been described (22). The identification of *Listeria* species is based on a limited number of biochemical markers, among which hemolysis is used to differentiate between *L. monocytogenes* and the most frequently encountered nonpathogenic *Listeria* species, *L. innocua* (265). The biochemical tests useful for discriminating between the species are acid production from D-xylose, L-rhamnose, alpha-methyl-D-mannoside, and D-mannitol (Fig. 21.1).

Further Characterization and Subtyping of *L. monocytogenes*
L. monocytogenes isolates are often characterized below the species level for the purposes of public health surveillance and to assist in outbreak investigations. Serotyping has proven its value over many years. There are 13 serotypes of *L. monocytogenes* which can cause disease, but more than 90% of human isolates belong to three serotypes: 1/2a, 1/2b, and 4b (249, 266). Because of the low discriminatory power of this method, phage-typing systems were developed and were the only means to distinguish between strains of the same serotype before the introduction of molecular typing methods. Since 1989, various molecular typing methods have been applied to *L. monocytogenes*, including multilocus enzyme electrophoresis, ribotyping, DNA microrestriction (DNA fragments generated with high-frequency cutting enzymes and separated by conventional agarose gel electrophoresis) and macrorestriction (large DNA fragments generated with infrequently cutting enzymes and separated by pulsed-field gel electrophoresis [PFGE]), and random amplification of polymorphic DNA (128). Due to their ability to type all strains and the high discriminatory power of some of them, these methods have become invaluable tools for epidemiologic investigations. In addition, unlike

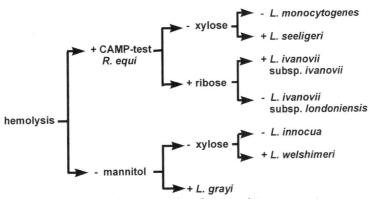

Figure 21.1 Phenotypic identification of *Listeria* species.

serotyping and phage typing that require specialized reagents (typing sera and bacteriophages) and thus are available only in a few reference laboratories, these methods can be performed in any reasonably equipped laboratory. In the United States, the Centers for Disease Control and Prevention has established a network (PulseNet) of public health and food regulatory laboratories that are routinely subtyping foodborne pathogenic bacteria to rapidly detect foodborne disease clusters that may have a common source. PulseNet laboratories use highly standardized protocols for subtyping of bacteria by PFGE and are able to quickly compare PFGE patterns of foodborne pathogens from different locations within the country via the Internet. *L. monocytogenes* was added to PulseNet in 1999, and a standardized PFGE protocol is available (127). Serotyping is still useful for first-level discrimination between isolates before more sensitive subtyping methods are applied, but serotyping reagents are generally available only in national reference laboratories. Recently, a PCR-based method has been developed and validated for grouping *L. monocytogenes* isolates on the basis of their serotype (74).

During the past 5 years, a plethora of new molecular methods have been proposed for subtyping *L. monocytogenes*. These include multilocus sequence typing (207), multi-virulence-locus sequence typing (51, 307), DNA sequencing-based subtyping (42), and macroarray analyses (27, 28, 73). Whether any of these approaches offer significant advantages over current molecular subtyping methods and whether they can be used routinely in public health laboratories for subtyping clinical isolates of *L. monocytogenes* remain to be determined.

L. monocytogenes serotype 4b strains are responsible for 33 to 50% of sporadic human cases worldwide and caused all major foodborne outbreaks in Europe and North America in the 1980s. In contrast, isolates recovered from foods in many countries belong mostly to the serotypes 1/2a and 1/2c. Although the reasons for this have not been fully elucidated, some recent observations provide intriguing clues. DNA from several epidemic-associated strains of *L. monocytogenes* 4b were resistant to restriction by Sau3A and other restriction enzymes known to be sensitive to cytosine methylation at 5′ GATC 3′ sites. This modification of Sau3A restriction appears to be host mediated (308). A putative restriction-modification system (85M, 85R, and 85S) has been identified in these strains (305). Also, they contain certain unique genetically unlinked DNA sequences. Epidemic-associated strains exhibiting these unique characteristics have been designated epidemic clone I (ECI). Also, a novel gene (*gtcA*) involved in the incorporation of galactose and glucose to the cell wall teichoic acid appears to be present in only serotype 4b and other serotype 4 isolates (234).

Evans et al. (84) examined several outbreak strains of *L. monocytogenes* 4b and classified them as ECI and II. The ECI isolates were involved in outbreaks prior to 1998, whereas ECII isolates were involved in multistate outbreaks in the United States in 1998 and 2002. ECII isolates have diverged in the serotype-specific region of the genome compared with other serotype 4b strains.

Wiedmann et al. (302) classified *L. monocytogenes* isolates into three distinct lineages based on the combination of their ribotype patterns and allelic analysis of two virulence genes (*hlyA* and *actA*). Lineage I was primarily composed of human clinical isolates associated with sporadic and outbreak-associated disease; lineage 2 contained human and animal isolates but none from epidemic outbreaks; and lineage III contained only animal isolates. Lineage I isolates were of serotypes 1/2b and 4b, whereas lineage II isolates were serotype 1/2a and 1/2c. The predominance of human clinical isolates in lineage I was confirmed in subsequent studies from the Wiedmann laboratory (99, 129, 150, 222). Zhang et al. (306) used DNA microarray analysis to identify lineage-specific and serotype-specific differences in genome content of lineage I and II strains and reported that 47 genes in 16 different contiguous segments of the genome that were present in serotype 1/2a strains of lineage II were absent in lineage I strains and nine additional genes were altered exclusively in 4b strains. Ward et al. (299) targeted the *prfA* virulence gene cluster to probe the intraspecific phylogeny of *L. monocytogenes* and subsequently classified outbreak-associated 4b isolates in their lineages I and III. They suggested that the low frequency of association between lineage III strains and human disease was likely to be due to rarity of exposure and not reduced virulence of lineage III strains. Liu et al. (184) observed that the serotype 4b strains in lineage III appeared to lack some virulence determinants found in serotype 4b strains clustered in lineage I.

Susceptibility to Physical and Chemical Agents

L. monocytogenes is able to initiate growth in the temperature range of 0 to 45°C, with growth occurring more slowly at lower temperatures. The average generation times for 39 *L. monocytogenes* strains were 43, 6.6, and 1.1 h at 4, 10, and 37°C, respectively, and the respective lag times were 151, 48, and 7.3 h (14). Temperatures below 0°C preserve or moderately inactivate the bacterium. Survival and injury during frozen storage depend on the substrate and the rate of freezing. *L. monocytogenes* is inactivated by exposure to temperatures above 50°C.

Zheng and Kathariou (308) cloned three genes (*ltrA*, *ltrB*, and *ltrC*) of *L. monocytogenes* that are essential for low-temperature growth. When a 1.2-kb internal fragment of *ltrB* was used as a probe in Southern hybridizations of HindIII-digested *L. monocytogenes* DNA, a 9.5-kb DNA fragment that hybridized with the probe was found to be unique to outbreak-associated serotype 4b strains of *L. monocytogenes*.

The pH range for the growth of *L. monocytogenes* was thought to be 5.6 to 9.6, although the bacterium can initiate growth in laboratory media at pH values as low as 4.4. Growth at low pH values is influenced by the incubation temperature and the type of acid. At pH values below 4.3, listeriae may survive but do not multiply. Experimentally, the presence of up to 0.1% acetic, citric, and lactic acids in tryptose broth inhibits the growth of *L. monocytogenes*, with inhibition increasing as the incubation temperature decreases. The antilisterial activity of these acids is related to their degree of dissociation, with citric and lactic acids being less detrimental for the pathogen at an equivalent pH than acetic acid.

L. monocytogenes grows optimally at water activity (a_w) values of >0.97. For most strains, the minimum a_w for growth is 0.93, but some strains may grow at a_w values as low as 0.90. Further, the bacterium may survive for long periods at a_w values as low as 0.83 (267). Also, an inverse relationship exists between the thermal resistance of *L. monocytogenes* and the a_w of the medium in which it is suspended (285), which must be addressed by food manufacturers who rely on low water activity and thermal treatment for preservation of their food products.

L. monocytogenes is able to grow in the presence of 10 to 12% sodium chloride and can grow to high populations in moderate salt concentrations (6.5%). The bacterium survives for long periods in high salt concentrations; the survival in high-salt environments is significantly increased by lowering the temperature.

The inoculum level of *L. monocytogenes* affects the ability of listeriae to grow in adverse environmental conditions (temperature, pH, and a_w). For example, at 25°C and a_w of 0.997, the minimum pH at which growth can be initiated is 4.45 at a cell concentration of 7.3 CFU/ml, whereas growth can be observed at a pH as low as 3.9 when the cell concentration is more than 6×10^6 CFU/ml (165).

LISTERIOSIS AND RTE FOODS

Certain RTE processed foods are high-risk vehicles for transmitting listeriosis for susceptible populations as determined by active surveillance for sporadic listeriosis and epidemiologic investigation of listeriosis outbreaks. These foods are usually preserved by refrigeration and offer an appropriate environment for the multiplication of *L. monocytogenes* during manufacture, aging, transportation, and storage. The foods in this category include unpasteurized milk and products prepared from unpasteurized milk, soft unfermented cheeses, unreheated frankfurters, certain delicatessen meats and poultry products, and some seafoods. In Canada, the regulatory policy directs inspection and compliance action on RTE foods that can support the growth of *L. monocytogenes*. The highest priority is given to those foods that have caused listeriosis and those that have greater than 10 days of shelf life (86). Several risk assessments have been conducted for *L. monocytogenes* in RTE foods (4, 6, 7, 49, 204). In 2003, the U.S. Food and Drug Administration, in collaboration with the U.S. Department of Agriculture Food Safety and Inspection Service and the Centers for Disease Control and Prevention, released the results of a risk assessment to predict the potential relative risk of listeriosis from eating certain RTE foods among three age-based groups of people: perinatal (16 weeks after fertilization to 30 days after birth), elderly (60 years of age and older), and intermediate age (general population, less than 60 years of age). This assessment evaluated foods within 23 categories considered to be principal potential sources of *Listeria* (Table 21.1). Deli meats were categorized in the "very high risk" category.

From the exposure models and "what-if scenarios" used in the risk assessment, it was determined that the following five factors affected consumer exposure to *L. monocytogenes* at the time of food consumption: (i) amount and frequency of consumption of a food, (ii) frequency and levels of *L. monocytogenes* in RTE food, (iii) the likelihood of the growth of *L. monocytogenes* in a food during refrigerated storage, (iv) refrigerated storage temperature, and (v) duration of refrigerated storage of a food before consumption. The risk assessment model was used to estimate the likely impact of control strategies by changing one or two input parameters and measuring the change in the model outputs. For example, one "what-if" scenario determined that the predicted number of listeriosis cases would be reduced by 69% if all home refrigerators were consistently operating at or below 7.2°C. Another scenario determined that reducing the maximum storage time of deli meats from 28 to 14 days would reduce the median number of cases in the elderly population by 13.6%.

Fluid Milk Products

Raw milk is a well-documented source of *L. monocytogenes*. In the United States, a Mexican-style cheese-associated outbreak in California in 1985 and a 2000

Table 21.1 Relative risk ranking and predicted median cases of listeriosis for the total U.S. population on a per serving and per annum basis[a]

Relative risk ranking	Predicted median cases of listeriosis for 23 food categories					
	Per serving basis[b]			Per annum basis[c]		
	Risk level	Food	No. of cases	Risk level	Food	No. of cases
1	High	Deli meats	7.7×10^{-8}	Very high	Deli meats	1,598.7
2		Frankfurters, not reheated	6.5×10^{-8}	High	Pasteurized fluid milk	90.8
3		Pâté and meat spreads	3.2×10^{-8}		High-fat and other dairy products	56.4
4		Unpasteurized fluid milk	7.1×10^{-9}		Frankfurters, not reheated	30.5
5		Smoked seafood	6.2×10^{-9}	Moderate	Soft unripened cheese	7.7
6		Cooked ready-to-eat crustaceans	5.1×10^{-9}		Pâté and meat spreads	3.8
7	Moderate	High-fat and other dairy products	2.7×10^{-9}		Unpasteurized fluid milk	3.1
8		Soft unripened cheese	1.8×10^{-9}		Cooked ready-to-eat crustaceans	2.8
9		Pasteurized fluid milk	1.0×10^{-9}		Smoked seafood	1.3
10	Low	Fresh soft cheese	1.7×10^{-10}	Low	Fruits	0.9
11		Frankfurters, reheated	6.3×10^{-11}		Frankfurters, reheated	0.4
12		Preserved fish	2.3×10^{-11}		Vegetables	0.2
13		Raw seafood	2.0×10^{-11}		Dry/semidry fermented sausages	<0.1
14		Fruits	1.9×10^{-11}		Fresh soft cheese	<0.1
15		Dry/semidry fermented sausages	1.7×10^{-11}		Semisoft cheese	<0.1
16		Semisoft cheese	6.5×10^{-12}		Soft ripened cheese	<0.1
17		Soft ripened cheese	5.1×10^{-12}		Deli-type salads	<0.1
18		Vegetables	2.8×10^{-12}		Raw seafood	<0.1
19		Deli-type salads	5.6×10^{-13}		Preserved fish	<0.1
20		Ice cream and other frozen dairy products	4.9×10^{-14}		Ice cream and other frozen dairy products	<0.1
21		Processed cheese	4.2×10^{-14}		Processed cheese	<0.1
22		Cultured milk products	3.2×10^{-14}		Cultured milk products	<0.1
23		Hard cheese	4.5×10^{-15}		Hard cheese	<0.1

[a] Table adapted from reference 6.
[b] Food categories were classified as high risk (>5 cases per billion servings), moderate risk (≤5 but ≥1 case per billion servings), and low risk (<1 case per billion servings).
[c] Food categories were classified as very high risk (>100 cases per annum), high risk (>10 to 100 cases per annum), moderate risk (≥1 to 10 cases per annum), and low risk (<1 case per annum).

outbreak in North Carolina were likely caused by the use of unpasteurized milk for cheesemaking (183, 192). The prevalence of *L. monocytogenes* in bulk tank raw milk varies from 1 to 13%, while its prevalence in milk-processing plants ranges from 7 to 28% (225). Therefore, raw milk and food products made from raw milk could be potential sources of *L. monocytogenes*. Recent surveys of raw milk for the presence of *L. monocytogenes* in several different countries showed that overall <5% of raw milk samples were positive for *L. monocytogenes* (64, 247). However, the ability of *L. monocytogenes* to survive and proliferate in raw dairy products stored at refrigeration temperatures makes this bacterium a particular concern for the dairy industry. Low initial contamination levels (e.g., <1 CFU/25 g) may increase to high numbers that could pose a human health hazard if dairy products are subject to extended refrigeration storage (247).

Most thermal-inactivation studies of *L. monocytogenes* in milk have shown that cells of *L. monocytogenes* suspended in milk were effectively inactivated under high temperature-short time pasteurization (HTST) conditions (71°C for 15 s or equivalent). Recent multistate surveys showed very low frequency (1 of 5,519 samples) of isolation of *L. monocytogenes* in commercial pasteurized fluid milk products sold in the United States (96). When thermal inactivation of freely suspended *L. monocytogenes* cells was compared with that of cells that were internalized within phagocytic leukocytes, differences were observed. The physiological state of the bacterial cells (actively growing cells versus cells in stationary phase) and growth of bacterial cells at elevated temperatures (infected cows that may have developed fever) before exposure to pasteurization conditions are some of the potential factors that may have caused the cells to become more heat resistant. Lou and Yousef (186) state that the following factors should be considered in evaluating the effect of pasteurization on *L. monocytogenes*: (i) the safety margin of pasteurization for inactivation of the bacterium may be lower than previously thought; (ii) the level of contamination in pooled milk is lower than that used in most inoculation studies; (iii) homogenization of milk destroys the integrity of phagocytic cells in milk, thus removing any protection offered to bacterial cells; and (iv) thermoduric spoilage microorganisms surviving HTST are likely to outcompete *L. monocytogenes*. Hence, there appears to be general agreement with the observations of the World Health Organization (WHO) informal study group, which concluded that "pasteurization is a safe process which reduces the number of *L. monocytogenes* in raw milk to levels that do not pose an appreciable risk to human health" (5). Recognizing the

small margin of safety offered by the HTST process, most raw milk processors have adopted processes that employ temperatures well above the minimum legal requirements for pasteurized milk. *L. monocytogenes* grows in pasteurized milk, with the numbers increasing 10-fold in 7 days at 4°C; also, listeriae grow more rapidly in pasteurized milk than in raw milk when incubated at 7°C. Therefore, fluid milk that is contaminated after pasteurization and stored under refrigeration may attain very high populations of *L. monocytogenes* after 1 week; temperature abuse may further enhance the multiplication of bacterial cells as was evidenced in a chocolate milk-associated outbreak in Illinois (60). Additional details may be obtained from a review on the prevalence, growth, and survival of *L. monocytogenes* in fluid milk and unfermented dairy products (96, 247).

Cheeses

L. monocytogenes can survive the cheese manufacturing and ripening process because of its relative hardiness to temperature fluctuations, ability to multiply at refrigeration temperature, and salt tolerance. Its growth in cheese milk is retarded but not completely inhibited by lactic starter cultures. During the manufacturing process, *L. monocytogenes* is primarily concentrated in the cheese curd, with only a very small proportion of cells appearing in whey. The behavior of listeriae in the curd is influenced by the type of cheese, ranging from growth in feta cheese to significant inactivation during cottage cheese manufacture. During ripening of the cheese, the numbers of *L. monocytogenes* cells may increase (Camembert), decrease gradually (Cheddar or Colby), or decrease rapidly during early ripening and then stabilize (blue cheese). Consumption of soft cheeses by susceptible persons is a risk factor for sporadic and epidemic listeriosis in North America and Europe. A decade ago, 49% of sporadic cases of listeriosis in France were attributed to the consumption of soft cheeses, but prevention efforts have reduced the incidence of sporadic cases of listeriosis there by 68% from 1987 to 1997 (122). The first documented foodborne listeriosis outbreak in Japan resulted in 86 cases of infection (mainly gastroenteritis) and was traced to *L. monocytogenes* serotype 1/2a contamination in washed-type cheese (193). In the United States, the Centers for Disease Control and Prevention recommends that persons at high risk for listeriosis should avoid the consumption of cheeses made from unpasteurized milk (47).

Meat and Poultry Products

The multiplication potential for *L. monocytogenes* in meat and poultry products depends on the type of meat, pH, and the type and cell populations of competitive flora. A recent survey in the United States revealed that 3.5% of commercial ground beef samples were positive for *L. monocytogenes* (250). Poultry supports the growth of *L. monocytogenes* better than other meats, whereas properly fermented sausage does not support *Listeria* growth following fermentation. Contamination of animal muscle tissue may occur either from symptomatic or asymptomatic carriage of *L. monocytogenes* by the food animal before slaughter or contamination of the carcass after slaughter. Because *L. monocytogenes* tends to concentrate and multiply in the kidney, mesenteric and mammary lymph nodes, and liver and spleen of infected animals, eating organ meat may be more hazardous than eating muscle tissue (89). Regardless of the route of contamination of meat, *L. monocytogenes* attaches strongly to the surface of raw meats and is difficult to remove or inactivate. *L. monocytogenes* multiplies readily in meat products, including vacuum-packaged beef, at pH values near 6.0, whereas there is very little or no multiplication at a pH of approximately 5.0 (89, 112). RTE meat products that have received a heat treatment followed by cooling in brine before packaging may provide a particularly conducive environment for multiplication of *L. monocytogenes* because of the reduction in competitive flora and the high salt tolerance of the bacterium.

A monitoring program for *L. monocytogenes* in cooked RTE meat products instituted by the U.S. Department of Agriculture revealed the following prevalence rates for 1993–96: beef jerky, 0 to 2.2%; cooked sausages, 1.0 to 5.3%; salads and spreads, 2.2 to 4.7%; and sliced ham and sliced luncheon meats, 5.1 to 81%. In Canada, the prevalence of *L. monocytogenes* in domestic RTE, cooked meat products was 24% in 1989–90 but declined to 3% or less during 1991–92. However, specimens obtained from the environments of establishments producing *Listeria*-positive foods remained constant at 12% between 1989 and 1992 (89). Lunden et al. (190) surveyed three meat-processing plants and one poultry-processing plant in Finland and found persistent and transient *L. monocytogenes* strains in all of them. Processing equipment and heat-treated products were frequently contaminated with persistent strains. These observations suggest that once *L. monocytogenes* is established in a food-processing plant environment, it may persist in that location for several years even if the products produced at that location are *Listeria* free for many months.

Cooked, RTE meat and poultry products have been implicated as the source of sporadic and outbreak-associated listeriosis on several occasions in North America and Europe. Consumption of unreheated frankfurters and undercooked chicken was identified as a risk factor for sporadic cases of listeriosis in a study conducted in the United States (264). A contaminated turkey frankfurter product was identified as the source of sporadic *L. monocytogenes* infection in a cancer patient in 1989 (45). Multistate outbreaks of listeriosis in the United States in 1998, 2000, and 2002 were linked to contaminated frankfurters and turkey deli meat (118, 206, 226).

Seafoods

The role of seafoods in human listeriosis was reviewed in 2000 (244). Shrimp, smoked mussels, and imitation crabmeat have been implicated as sources of human listeriosis. Gravad rainbow trout and cold-smoked rainbow trout were implicated as the sources of infection in two outbreaks in Sweden and Finland, respectively (83, 213). In the United States, *L. monocytogenes* has been isolated from both domestic and imported fresh, frozen, and processed seafood products, including crustaceans, molluscan shellfish, and finfish (153). A U.S. Food and Drug Administration survey of domestic and imported refrigerated or frozen cooked crab meat in 1987–88 determined contamination levels of 4.1% for domestic products and 8.3% for imported products (82). Crab and smoked fish samples analyzed by the U.S. Food and Drug Administration between 1991 and 1996 revealed that 7.5 and 13.6% of the samples, respectively, were contaminated with *L. monocytogenes*.

Huss et al. (143) have classified the following seafoods as potential high-risk foods for listeriosis: (i) molluscs, including fresh and frozen mussels, clams, oysters in shell or shucked; (ii) raw fish; (iii) lightly preserved fish products including salted, marinated, fermented, cold-smoked, and gravad fish; and (iv) mildly heat-processed fish products and crustaceans. Hoffman et al. (142) compared ribotypes of *L. monocytogenes* strains isolated from smoked-fish-processing environments and raw fish and identified unique ribotypes on raw fish samples which were not found in the plant environments, indicating that raw fish were possibly contaminated from a source different from that of environments within the plant. In Finland, the prevalence of *Listeria* spp. and *L. monocytogenes* in pooled unprocessed fresh rainbow trout was on average 35 and 14.6%, respectively. Most (95.6%) of *L. monocytogenes* organisms isolated were from fish gills, whereas only 4.4% were obtained from skin or viscera (212).

Seafood consumption in the United States is still much less than that of meats and cheeses. Also, seafood products are produced on a much smaller scale than meat and cheese products. This may help explain why large outbreaks of listeriosis have not been attributed to seafood consumption and case-control studies have not identified the consumption of seafoods as a contributing factor for listeriosis (244).

Effects of Newer Methods of Food Preservation

New trends in food preservation have recently been developed, including the use of biopreservatives, vacuum-packaging, and modified atmospheres. The antagonistic effect of nisin on *L. monocytogenes* has been demonstrated, and its activity in foods is strongly dependent on the chemical composition of the food to which it is added. Pediocins (from *Pediococcus pentosaceus* and *Pediococcus acidilactici*) inhibit *L. monocytogenes* growth. Bacteriocin from *Lactobacillus bavaricus* transiently affects *L. monocytogenes* in various beef systems, especially at low temperature. Meatborne lactic acid bacteria can effectively inhibit the growth of *L. monocytogenes* by rapid production of lactic acid and bacteriocin and can be used as potential biopreservatives in cooked meat products (295). A similar effect was observed with a bacteriocin from *Carnobacterium piscicola* in broth and skimmed milk (200). However, a drawback to the use of bacteriocins is the emergence of resistant mutants, as has been observed with bavaricin A and nisin (63, 168). Modi et al. (216) reported that a nisin-resistant strain of *L. monocytogenes* grew in the presence of nisin; it was more sensitive to heat than the wild-type strain. In vitro studies have also demonstrated antibacterial activity of essential oils against *L. monocytogenes* by partitioning the lipids of the cell membrane and mitochondria, rendering membranes more permeable and leading to leakage of cell contents. Examples of essential oils include carvacrol, thymol, eugenol, perillaldehyde, cinnamaldehyde, and cinnamic acid, which have MICs of 0.05 to 5 µl/ml in vitro (38).

A survey of vacuum-packaged processed meat in retail stores in the early 1990s revealed that 53% of the samples tested were contaminated with *L. monocytogenes* and that 4% contained more than 1,000 CFU/g (124). This observation corroborates experimental evidence that the growth of *L. monocytogenes* (a facultative anaerobe) is not significantly affected by vacuum packaging. There has been considerable interest in modified-atmosphere packaging of meat products (low oxygen and high carbon dioxide concentrations) over recent years because of the increasing demand for refrigerated convenience foods with extended shelf lives. Studies with meat juice, raw chicken, and precooked chicken nuggets revealed that such atmospheres do not significantly affect the growth of *L. monocytogenes*.

L. monocytogenes is sensitive to low doses of irradiation. In addition, heating *L. monocytogenes* to 5°C or above, freezing, or various other treatments can injure a substantial portion of surviving cells. Heat-stressed *L. monocytogenes* may be considerably less pathogenic than nonstressed cells.

On nonselective agar, injured cells can repair the damage induced by stress and grow, but they are subject to additional stress in selective agar and variable recovery is observed. Hence, the presence of sublethally damaged cells in food samples may lead to differences in cell recovery rates.

RESERVOIRS

The role of silage in the transmission of animal disease was bacteriologically documented in 1960, and *L. monocytogenes* strains were isolated from natural decaying vegetation in 1968 and thereafter (144, 300). The bacterium can survive and grow in soil and water. *Listeria* species have been detected in various aqueous environments, including surface water of canals and lakes, ditches of polders in the Netherlands, freshwater tributaries draining into a California bay, and sewage (52, 69, 133). Alfalfa plants and other crops grown on soil treated with sewage sludge are contaminated with *Listeria* (2). One-half of the radish samples grown in soil inoculated with *L. monocytogenes* were confirmed positive 3 months later (292). Similarly, the presence of *L. monocytogenes* in pasture grasses and grass silages has often been documented (144). The widespread presence of *L. monocytogenes* in soil is likely due to contamination by decaying plant and fecal material, with the soil providing a cool, moist environment and the decaying material providing the nutrients (90).

L. monocytogenes has been isolated from the feces of many healthy animals and birds; listeriosis in many animal species has been recorded. Humans exhibiting symptoms of listeriosis and asymptomatic carriers shed the organism in their feces. Figure 21.2 illustrates the many ways in which *L. monocytogenes* can be spread from the environment to animals and humans and back to the environment.

Food-Processing Plants

The entry of *L. monocytogenes* into food-processing plants occurs through soil on workers' shoes and clothing and on transport equipment, animals which excrete the bacterium or have contaminated hides or surfaces, raw plant tissue, raw food of animal origin, and possibly healthy human carriers. Growth of listeriae is favored by high humidity and the presence of nutrients. *L. monocytogenes* is most often detected in moist areas such as floor drains, condensed and stagnant water, floors, residues, and processing equipment (57). *L. monocytogenes* can attach to various kinds of surfaces (including stainless steel, glass, and rubber), and biofilms have been described in meat- and dairy-processing environments (152). *Listeria* spp. can survive on fingers after hand washing and

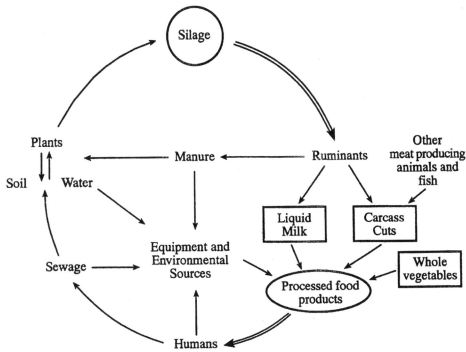

Figure 21.2 Potential routes of transmission of *L. monocytogenes*. Adapted from reference 90. Circles or ovals indicate areas of greatest risk of *L. monocytogenes* multiplication. Boxes indicate where direct consumption of minimally processed products (e.g., whole fresh vegetables, cooked carcass cuts of meat and fish, and effectively pasteurized milk) presents a low risk. Double arrows indicate consumer at risk.

in aerosols. The presence of *L. monocytogenes* in the food-processing chain is evidenced by the widespread distribution of the listeriae in processed products. Contaminated effluents from food-processing plants increase the spread of *L. monocytogenes* in the environment. Sources of *L. monocytogenes* in dairy-processing plants include the environment (floors and floor drains, especially in areas in and around coolers or places subject to outside contamination) and raw milk. Efforts to ensure that milk is safe from *L. monocytogenes* contamination should focus on promoting appropriate methods of pasteurization and on identifying and eliminating sources of postpasteurization contamination (298). The Pasteurized Milk Ordinance has been designed and recommended for dairy processors to prevent *Listeria* contamination in the processing environment and ensure the safety and quality of dairy products at risk.

The presence of *L. monocytogenes* on carcasses is usually attributed to contamination by fecal matter during slaughter. A high percentage (11 to 52%) of animals are healthy fecal carriers. Up to 45% of pigs harbor *L. monocytogenes* in tonsils and 24% of cattle have contaminated internal retropharyngeal nodes (37, 274).

L. monocytogenes has been recovered from both unclean and clean zones (especially on workers' hands) in slaughterhouses, with the most heavily contaminated working areas being cow dehiding and pig stunning and hoisting. Studies in turkey and poultry slaughterhouses failed to detect *L. monocytogenes* in feather samples, scalding tank water overflow, neck skin, livers, hearts, ceca, or large intestines. In contrast, *L. monocytogenes* was recovered from feather plucker drip water, chill water overflow, recycling water for cleaning gutters, and in mechanically deboned meat. These findings demonstrate the importance of the defeathering machine, chillers, and recycled water in product cross-contamination (109).

Postprocessing contamination is the most likely route of contamination of processed foods by *L. monocytogenes*. To date, there is no evidence to indicate that *L. monocytogenes* can survive validated heat-processing protocols used to render foods safe. *L. monocytogenes* is a particularly difficult bacterium to eliminate from the food-processing plant because it has the propensity to adhere to food contact surfaces and form biofilms, which makes it difficult to implement effective sanitation procedures (125). For example, *L. monocytogenes* can persistently survive on food contact surfaces such as stainless steel surfaces of dicing machines and repeatedly contaminate RTE meats (182, 189). The refrigerated, moist environments in food-processing plants provide a good growth environment for *L. monocytogenes*. Because *L. monocytogenes* is a frequent contaminant of raw materials used in food-processing plants, there are ample opportunities for reintroduction of listeriae into food-processing facilities (75). Extensive information on the problem of *L. monocytogenes* in various food-processing environments and approaches to control has been provided by Gravani (125). Tompkin (289) outlined a six-step *Listeria* control program for food-processing environments. These are (i) prevention of the establishment and growth of *Listeria* species in niches or other sites that can lead to contamination of RTE foods; (ii) implementation of a sampling program to assess how well the control program is working; (iii) rapid and effective response when the sampling program yields positive results for *Listeria* species; (iv) verification by follow-up sampling to ensure that the source of contamination has been identified and corrected; (v) short-term assessment of the last four to eight samplings to facilitate early detection of problems and trends; and (vi) long-term assessment at appropriate intervals (quarterly, annually, etc.) to identify widely scattered contamination events and to measure overall progress towards continuous improvement. Longitudinal studies based on various molecular subtyping analyses assist in identifying contamination patterns of *L. monocytogenes* in food-processing plants and critical validation of intervention strategies (167, 190). The design of the environmental sampling program and the response to positive findings by the sampling program determine the overall effectiveness of the *Listeria* control program in food-processing environments (289).

Surveys of Foods for Prevalence of *L. monocytogenes* and the Regulatory Status in Different Countries

Several surveys confirm that *L. monocytogenes* contamination of food is widespread in many parts of the world. A comprehensive qualitative and quantitative survey of retail RTE foods for *L. monocytogenes* was conducted by investigators at the National Food Processors Association in the United States (116). Product categories examined included luncheon meats, deli salads, Mexican-style cheeses, packaged salads, blue-veined and soft-ripened cheeses, smoked seafood, and seafood salads. Of 31,705 samples examined, 1.82% (the range by sample category varied from 0.17 to 4.7%) were positive for *L. monocytogenes*. Cell numbers of the pathogen in the positive foods varied from <0.3 CFU/g to 1.5×10^5 CFU/g. A

dose-response model was developed by combining the food survey data with concurrent data on illness in the population that consumed the foods surveyed. Results from this model revealed that a control strategy focused on foods that yielded higher cell numbers of *L. monocytogenes* is likely to have greater public health impact than the current zero-tolerance policy (50).

In a survey conducted in Denmark in 1994–1995, *L. monocytogenes* was isolated from 14.2 and 30.9% of raw fish and raw meats, respectively. Preserved, but not heat-treated, fish and meat products were more frequently contaminated (10.8 and 23.5%, respectively) than heat-treated meat products (5%). *L. monocytogenes* was present at levels exceeding 100 CFU/g in 1.3% of preserved (not heat-treated and heat-treated) fish and meat products packed under vacuum or modified atmospheres for extended shelf life. In contrast, a survey of the same type of products that were not packed under vacuum or modified atmospheres revealed significantly lower contamination levels (0.3 to 0.6%) (221). In a survey of retail foods in Japan, *L. monocytogenes* was isolated from 12, 20, 37, and 25% of minced beef, minced pork, minced chicken, and minced pork-beef mixture, respectively, and only five chicken samples had cell numbers greater than 100 CFU/g. *L. monocytogenes* was isolated from 5.4% of smoked salmon samples and 3.3% of RTE uncooked seafoods (145).

A survey conducted in Barcelona, Spain, revealed *L. monocytogenes* contamination in 9.3% of RTE foods and 2.9% of foods intended to be cooked before consumption (66). In another investigation, *L. monocytogenes* was isolated from 22% of retail smoked fish samples (72). *L. monocytogenes* has been isolated from many vegetables, including bean sprouts, cabbage, cucumbers, leafy vegetables, potatoes, prepackaged salads, radishes, salad vegetables, and tomatoes in North America, Europe, and Asia (9).

A recent survey in New York State revealed that many cases of human listeriosis were in fact associated with persistent *L. monocytogenes* strains in retail environments (249).

Because of the frequent occurrence of *L. monocytogenes* in foods and in the food-processing environment, food regulatory agencies in many countries have accepted the argument that it is impossible to produce *L. monocytogenes*-free foods and have established tolerance levels for *L. monocytogenes*. For example, the Canadian government policy directs inspection and compliance action to RTE foods that are capable of supporting the growth of *L. monocytogenes*. Foods are divided into three risk categories, with products causally linked to human listeriosis being placed in Category 1 and regulated more stringently than foods in Categories 2 and 3 (88). The French position directs that foods (25-g samples) should be *L. monocytogenes*-free, but when this is not possible, one must try to obtain the lowest level possible. In France, foods are divided into three groups as follows: (i) food for populations "at risk," (ii) foods heated in their wrapping or aseptically conditioned after treatment, and (iii) raw foods or foods susceptible to recontamination after treatment. In contrast, the United Kingdom and the United States, although they acknowledge the widespread distribution of *L. monocytogenes* in the food supply and the difficulties in producing *L. monocytogenes*-free RTE foods, have decided not to adopt tolerance levels for *L. monocytogenes* in these foods. Both countries argue that any "acceptable" levels for *L. monocytogenes* would require knowledge of the number of listeriae unlikely to cause human infection. Because there are no scientific data to establish such tolerance levels, these two countries continue a "zero-tolerance" policy for *L. monocytogenes* in RTE foods (268).

Human Carriage

Asymptomatic fecal carriage of *L. monocytogenes* has been studied in a variety of human populations, including healthy people, pregnant women, outpatients with gastroenteritis, slaughterhouse workers, laboratory workers handling *Listeria*, food handlers, and patients undergoing renal transplantation or hemodialysis (277). *L. monocytogenes* was isolated from 2 to 6% of fecal samples from healthy people. Patients with listeriosis often excrete high cell numbers of *L. monocytogenes*, e.g., specimens from 21% of patients had $\geq 10^4$ *L. monocytogenes* isolates/g of feces, and 18% of household contacts of patients with listeriosis fecally shed the same serotype and multilocus enzyme type of *L. monocytogenes* as the corresponding index case (151, 259). Among household contacts of 18 pregnant women, 8.3% asymptomatically shed *L. monocytogenes*, whereas no listeriae were isolated from 30 household contacts of age-, sex-, and hospital-matched controls (199). Results of an investigation of an outbreak in California in 1985 revealed that community-acquired outbreaks might be amplified through secondary transmission by fecal carriers. The very low prevalence of *L. monocytogenes* in human stools and short duration of fecal shedding argue against the appropriateness of routine stool screening of persons with a likelihood of work-related exposure to the pathogen (e.g., dairy workers) as a tool for prevention of listeriosis (131, 251). *L. monocytogenes* isolated from asymptomatic carriers may carry the full complement of virulence genes or may be attenuated by truncated virulence-associated genes such as *inlA* and *actA* (149, 223, 224).

L. monocytogenes has not been isolated from oropharyngeal samples of healthy people, and the presence of listeriae in cervicovaginal specimens may be associated with pregnancy-related listeriosis. The role of healthy carriers in the epidemiology of listeriosis is unclear and warrants further study.

FOODBORNE OUTBREAKS

Foodborne transmission of listeriosis was suggested early in the medical literature but was first definitively documented in 1981 during the investigation of an outbreak in Canada with the simultaneous use of a case-control study and strain typing (254). Since 1981, epidemiologic investigations have repeatedly revealed that the consumption of contaminated food is a primary mode of transmission of listeriosis. Food has been identified as the vehicle of several major (>30 cases) outbreaks of listeriosis investigated since 1981.

An outbreak of listeriosis in Massachusetts in 1979 may have been caused by raw produce, but the food source was not positively identified. Twenty patients with *L. monocytogenes* serotype 4b infection were hospitalized during a 2-month period; only nine cases had been detected in the previous 26 months. Ten of the patients were immunosuppressed adults and five died. Fifteen patients are thought to have acquired the infection in the hospital. Consumption of tuna fish, chicken salad, or cheese was associated with illness, but no specific brand was implicated. It was postulated that the raw celery and lettuce, served as a garnish with the three foods, may have been contaminated with *L. monocytogenes* (140). Although the source of infection was not definitively identified in this outbreak, consumption of cimetidine or antacids was implicated as a risk factor for listeriosis. Decreased gastric acidity might have increased the survival of *L. monocytogenes* cells as they passed through the stomach.

The first confirmed foodborne outbreak of listeriosis occurred in 1981 in Nova Scotia, Canada. Thirty-four pregnancy-associated cases and seven cases in nonpregnant adults occurred during a 6-month period. A case-control study implicated a locally prepared coleslaw as the vehicle, and the epidemic strain was subsequently isolated from an unopened package of this product. Cabbage fertilized with manure from sheep suspected to have had *Listeria* meningitis was the probable source. Harvested cabbage was stored over the winter and spring in an unheated shed, providing a definite growth advantage for the psychrotrophic *L. monocytogenes* (254). Pasteurized milk was identified as the most likely source of infection in another large outbreak of listeriosis in Boston, Mass., in 1983. Forty-nine cases occurred during a

2-month period, 42 in immunosuppressed adults and 7 in pregnant women; the overall case-fatality rate was 29% (92); a case-control study implicated pasteurized milk with 2% fat as the vehicle. Multiple serotypes of *L. monocytogenes* were isolated from raw milk at the implicated dairy, but none was the epidemic strain; no deviations from approved pasteurization process were noted at the dairy, suggesting that the contamination occurred after pasteurization of milk.

In 1985 in California, an outbreak of listeriosis with 142 cases during an 8-month period occurred in Los Angeles County, Calif., and was traced to contaminated Mexican-style cheese (183). Pregnant women accounted for 93 cases, and the remaining 49 were nonpregnant adults; 48 of 49 nonpregnant adults had a predisposing condition for listeriosis. Among pregnancy-associated cases, 87% occurred in Hispanic women. The case-fatality rates were 32% for perinatal cases and 32% for nonpregnant adults. Inadequate pasteurization of the milk that was used to prepare the cheese and mixing of raw milk with pasteurized milk likely resulted in the contaminated cheese.

L. monocytogenes-contaminated soft cheese was responsible for a 4-year (1983–1987) outbreak of 122 cases in Switzerland (21), and a contaminated paté was the vehicle of a 300-case outbreak in the United Kingdom in 1989–1990 (202). Contaminated pork tongue in aspic was the principal vehicle of 279 cases of listeriosis in 10 months in France in 1992 (147), potted pork ("rillettes") was associated with 39 cases in 1993, and soft cheese was the vehicle of 33 cases in 1995 (123). Recalling the implicated food, advising the general population through the mass media to avoid eating contaminated product, and taking appropriate action to prevent *L. monocytogenes* contamination at product-processing and -handling facilities terminated these large outbreaks.

A large multistate outbreak of listeriosis occurred in the United States between August 1998 and March 1999. A total of 101 outbreak-associated cases (15 were perinatal cases) were identified in 22 states. Fifteen adult deaths and six miscarriages or stillbirths were associated with this outbreak. A case-control study implicated turkey franks manufactured in one facility of a large food manufacturing company. *L. monocytogenes* serotype 4b of the epidemic PFGE subtype was isolated from open and unopened packages of frankfurters from the implicated factory (126, 206).

Between December 1998 and February 1999, an increase in cases of listeriosis due to *L. monocytogenes* serotype 3a was recognized in Finland. A total of 25 cases, most of whom were hematological or organ transplant patients, were identified as part of the outbreak; six

patients died from the *Listeria* infection. Butter served at the tertiary care hospital was implicated as the source of infection. The epidemic strain was isolated from all of 13 butter samples obtained from the hospital kitchen and from several lots from the dairy and wholesale store. One sample contained 11,000 *L. monocytogenes* isolates/g, but the other samples had lower counts (5 to 60 CFU/g) (191).

Between May and November 2000, a listeriosis outbreak was identified in the United States in 11 states. When subtyped, the *L. monocytogenes* isolates from these cases were all serotype 1/2a and were indistinguishable from each other by PFGE. Eight perinatal and 21 nonperinatal cases were reported. Among the 21 nonperinatal case patients, the median age was 65 years (range, 29 to 92 years); 13 (62%) were female. This outbreak resulted in four deaths and three miscarriages and stillbirths. A case-control study implicated sliced processed turkey meat from a delicatessen. A traceback investigation identified a single food-processing plant as the source of this outbreak and led to the recall of 16 million pounds of processed turkey meat. The same plant had been identified in a *Listeria* contamination event that had occurred 1 decade previously (226).

An outbreak of listeriosis, identified in North Carolina in 2000, affected 13 people, of whom 12 were young (18- to 34-year-old) Hispanic females. This outbreak resulted in five stillbirths and was caused by Mexican-style cheese made from contaminated raw milk at a local dairy (192). Another outbreak of listeriosis in Texas was also associated with legally imported but illegally distributed cheese prepared in Mexico. Five of six case patients reported eating the implicated cheese. These outbreaks underscore the need for educating Hispanic women about food safety considerations during pregnancy and enforcing existing laws that regulate the sale of raw milk and dairy products made by unlicensed manufacturers.

In July-October 2002, another multistate outbreak of listeriosis was recognized in Pennsylvania and eight other states in the United States (117). The outbreak affected 54 persons. There were eight deaths and three fetal deaths associated with this outbreak. Case-control studies implicated delicatessen turkey meats as the source of the outbreak. In traceback studies, the outbreak strain, which was related to the 1998–99 outbreak strain by PFGE, was found in the environment of one turkey-processing plant and in turkey-products produced by another plant. Together, the two plants recalled more than 30 million pounds of turkey meat products (Table 21.2).

The strains responsible for the major outbreaks between 1981 and 1992 (Canada, 1981; California, 1985; Switzerland, 1983–87; and France, 1992) were all of serotype 4b

Table 21.2 Invasive listeriosis outbreaks, 1990–2003

Year	Geographic location	No. of persons affected	No. of deaths/no. of fetal deaths	Vehicle	Serotype	Reference
1992	France	279	85	Pork tongue in jelly	4b	147
1994	Illinois	3	0	Chocolate milk	1/2b	233
1996	Ontario, Canada	2	0	Imitation crab (?)	1/2b	87
1998–99	U.S., multiple states	108	14/4	Frankfurters	4b	206
1998–99	Finland	11	4	Butter	3a	191
1999–2000	France	26	7	Pork tongue in jelly	4b	68
2000	U.S., multiple states	30	4/3	Turkey deli meat	1/2a	226
2000	North Carolina	13	0/5	Homemade Mexican-style cheese	4b	192
2002	U.S., multiple states	54	8/3	Delicatessen turkey meat	4b	118
2002	Quebec, Canada	17	0	Cheese made from raw milk		104
2003	Texas	12	2	Mexican-style cheese	4b	CDC, unpublished data

and belonged to a small number of closely related clones, as evidenced by ribotyping, multilocus enzyme electrophoresis, and DNA macrorestriction pattern analysis (35). This group of closely related, epidemic-associated strains was designated ECI. The 1998–99 and 2002 multistate outbreaks in the United States were caused by strains of a novel epidemic-associated lineage, ECII (84).

Also, in the past decade, several outbreaks of febrile gastroenteritis caused by *L. monocytogenes* have been documented (Table 21.3). These gastroenteritis outbreaks differ from the invasive outbreaks described above in several respects. They affect persons with no known predisposing risk factors for listeriosis. The infectious dose appears to be much higher (1.9×10^5 to 1.6×10^9 CFU/g or ml) than that for typical invasive listeriosis in the susceptible population. Finally, the symptoms appear within a day or less (typically within 18 to 27 h) of exposure (similar to other bacterial enteric infections) in gastrointestinal listeriosis in contrast to the several weeks of incubation observed for invasive listeriosis.

CHARACTERISTICS OF DISEASE

Human disease caused by *L. monocytogenes* occurs predominantly in certain well-defined high-risk groups that include pregnant women, neonates, immunocompromised adults, and the elderly but also occurs in persons with no predisposing underlying conditions. In nonpregnant adults, *L. monocytogenes* primarily causes septicemia, meningitis, and meningoencephalitis, with a mortality rate of 20 to 30%. Other infrequent manifestations of listeriosis in this population include endocarditis in persons with underlying cardiac lesions (including prosthetic or porcine valves) and various types of focal infections, including endoophthalmitis, septic arthritis, osteomyelitis, pleural infection, and peritonitis (277). Clinical conditions known to predispose persons to the serious manifestations of listeriosis include malignancy, organ transplants, immunosuppressive therapy, infection with the human immunodeficiency virus, and advanced age. Although pregnant women, particularly in the third trimester of pregnancy, may experience only mild flu-like symptoms (fever and/or myalgias with or without diarrhea) as a result of *L. monocytogenes* infection, the infection has serious consequences for the fetus, leading to stillbirth or abortion. In neonates who are less than 7 days old, sepsis and pneumonia are predominant syndromes, whereas in neonates older than 7 days, the infections manifest as meningitis and sepsis.

Several recent investigations of listeriosis outbreaks show that *L. monocytogenes* also causes febrile gastroenteritis in normal hosts (Table 21.2). Interestingly, Murray and Pirie had clearly stated in their original descriptions of cases from the 1920s that diarrhea was a common feature of listeriosis in small animals (201). The most compelling evidence for the gastrointestinal manifestation of *L. monocytogenes* infections comes from outbreak investigations by Dalton et al. (60) and Aureli et al. (11). Fever and diarrhea are the most consistent symptoms in gastrointestinal listeriosis. Also, the incubation time for the enteric form is rather short, usually in the range of 18 to 27 h. However, the incubation time varied between 24 and 144 h in the 2001 Japan outbreak, possibly because the level of contamination of the implicated cheese spanned a large range (Table 21.2). In the five investigations in which attempts were made to quantify

Table 21.3 Outbreaks of gastrointestinal manifestation of listeriosis in humans with no known predisposing condition

Year	Geographic location	Incubation time (h)	No. of persons affected	Symptoms	Vehicle	Serotype	Contamination level (CFU/g or ml)	Reference
1986–87	Pennsylvania	Unknown	Unknown	Fever, vomiting, and diarrhea in the week before positive culture	Unknown	Multiple serotypes	NA[a]	239
1994	Italy	18	18	Fever, diarrhea	Foods served at dinner	1/2b	NA	248
1996	Illinois	20	80	Fever, diarrhea	Chocolate milk, temperature-abused	1/2b	10^9	60
1998?	Finland	<27	5	Nausea, abdominal cramps, diarrhea, fever	Cold-smoked rainbow trout	1/2a	1.9×10^5	213
1997	Northern Italy	24	1,566	Headache, abdominal pain, diarrhea, fever	Cold salad of corn and tuna	4b	10^6	11
2000	Winnipeg, Canada	NA	>20	Gastrointestinal illness	Defective cans of aerosol-dispensed whipping cream	NA	NA	
2001	Los Angeles, Calif.	NA	16	Body aches, fever, headache, diarrhea, vomiting	Precooked, sliced turkey	1/2a	1.6×10^9	97
2001	Sweden	31	120	Diarrhea, fever, stomach cramps, vomiting	Fresh, raw milk cheese	1	6.3×10^7	62
2001	Japan	24–144	38	Gastroenteritis or flu-like symptoms	Locally manufactured cheese	1/2b	$<0.3 \times 10^7$ to 4.6×10^7	193

[a] NA, not available.

the numbers of *L. monocytogenes* in the implicated foods, the contamination level was very high (range, 1.9×10^5 to 1.6×10^9 CFU). From the data available thus far for gastrointestinal listeriosis, it appears that infection requires a high dose of *L. monocytogenes*. It is not known whether the strains involved in the gastrointestinal forms of listeriosis possess additional virulence factors similar to those of common enteric pathogens.

While listeriosis outbreaks attract the most attention, most cases of human listeriosis occur sporadically. However, some of these sporadic cases may be unrecognized common-source clusters. The source and route of infection of most of these cases remain unknown, although foodborne transmission is demonstrated in some cases. Many cases have not been associated with food because of the difficulties of prospectively investigating sporadic cases of the disease. Long incubation times (up to 5 weeks) make accurate food histories difficult to obtain and examination of incriminated foodstuffs difficult to accomplish because they are usually consumed or discarded. Understanding the epidemiology of sporadic cases is critical to the development of effective control strategies. An active surveillance for listeriosis conducted in the United States in a population base of 19 million people between 1989 and 1991 yielded an estimated incidence of 7.9 cases per million. A case-control study of dietary risk factors conducted during the same time revealed that foodborne transmission could be responsible for about one-third of cases and that patients with listeriosis were more likely than controls to have eaten soft cheeses or food purchased from store delicatessen counters (230, 260). By 1993, the incidence of sporadic cases of listeriosis in the active surveillance sites had declined to 4.4 cases per million (286), and the decrease was attributed to enhanced efforts by the food industry and the food regulatory agencies to prevent contamination of processed foods by *L. monocytogenes*. The most recent data from CDC's foodborne diseases active surveillance program (FoodNet) indicate that the incidence of listeriosis in the United States had declined to 3 cases per million population (46).

The reported incidence of listeriosis in 16 countries in Europe varied between 0.3 and 7.5 cases per million per year with a mean of 3.4 per million persons (67). These figures probably reflect differences in the sensitivity of the surveillance systems in addition to true differences in the incidence of disease among the countries. The mean rate of listeriosis in Australia was 3.0 cases per million between 1998 and 2000 (61). In Japan, the incidence of listeriosis is estimated to be 0.65 case per million people (145). Listeriosis is reported mainly from industrialized countries, and the prevalence in other areas is unknown

or low. Whether this reflects different consumption patterns and dietary habits, different host susceptibility, different food processing and storage technologies, or lack of awareness or laboratory facilities is not known.

Although exposure to *L. monocytogenes* is common, invasive listeriosis is rare. It is unclear whether this is due to early acquired protection, to intrinsic resistance of the average healthy host, or to most strains being only weakly virulent. The pathogenesis of human listeriosis is poorly understood. Many (2 to 6%) healthy individuals are asymptomatic fecal carriers of *L. monocytogenes*. The risk of clinical disease in those intestinal carriers of *L. monocytogenes* is unknown but must be very low given the rarity of the diagnosis of the disease. Endogenous infection by *L. monocytogenes* in the gut is plausible, especially in patients receiving immunosuppressive therapy, which not only impairs resistance to infection but also can result in alterations of intestinal defense mechanisms favoring listerial invasion. Nevertheless, asymptomatic fecal carriage has been observed in pregnant women who proceed to normal birth at term and women who have given birth to infected infants do not necessarily suffer the same problem in later pregnancies. Similarly, recent transplant recipients may harbor *L. monocytogenes* in the gut without developing the disease.

Epidemiologic studies since 1981 have focused on the role of contaminated food in transmission of listeriosis. However, two unusual transmission routes have been described. Hospital-acquired listeriosis is sporadically described, mainly in nursery mates, with equipment serving as the vehicle. Amniotic fluid during intrauterine infections could contain as many as 10^8 *L. monocytogenes* organisms/ml (203). Mineral oil has been implicated in an outbreak of neonatal listeriosis (261). Primary cutaneous infections without systemic involvement have been observed as an occupational disease in veterinarians and farmers, and most cases are caused by manipulation of presumably infected bovine fetuses or cows.

INFECTIOUS DOSE AND SUSCEPTIBLE POPULATIONS

Infectious Dose

The infectious dose of *L. monocytogenes* depends on many factors, including the immunological status of the host. In addition to host factors and exposure to particular foods, it is likely that microbial characteristics are important risk factors for disease. The occurrence and the course of infection may depend on virulence factors and infective dose. The severity of the disease is such that tests with human volunteers are impossible. Studies with monkeys and with mice suggest that reducing levels of exposure reduces clinical disease (85). However, these experiments do not help to determine the minimal infective dose for humans. When pregnant rhesus monkeys were exposed to *L. monocytogenes* at the beginning of the third trimester, they showed an increased risk of delivering a stillborn infant with pathology similar to that of humans, including acute inflammation, placentitis, fetal liver necrosis, and isolation of *Listeria* from the placental and fetal tissues (280).

Published data indicate that the populations of *L. monocytogenes* in contaminated food responsible for epidemic and sporadic foodborne cases were usually more than 100 CFU/g. However, the frankfurters implicated in the 1998 listeriosis outbreak in the United States contained less than 0.3 CFU/g (206). However, because enumeration procedures are not fully reliable and the time between consumption and analysis of the contaminated food can enable growth or death of *L. monocytogenes*, results may not always be indicative of the numbers consumed. More epidemiologic information is needed for an accurate assessment of the infectious dose.

Susceptible Populations

Most human cases of listeriosis occur in individuals who have a predisposing disease that leads to impairment of their T-cell-mediated immunity. The percentage of patients suffering from a known underlying condition varies greatly between studies, accounting for 70 to 85% of the cases in some surveys and for nearly all cases in others (36, 273). The most commonly affected populations include those of extreme ages (neonates and the elderly), pregnant women, and those persons who are immunosuppressed by medication (corticosteroids or cytotoxic drugs) especially after organ transplantation or illness (hematologic malignancies, such as leukemia, lymphoma, and myeloma, as well as solid malignancies). Listeriosis is 300 times more frequent in people with AIDS than in the general population (262). In addition to T-cell immunity impairment, a small percentage of listeriosis patients suffer from chronic diseases not usually associated with immunosuppression, such as congestive heart failure, diabetes, cirrhosis, alcoholism, and systemic lupus erythematosus, alone or in association with known predisposing diseases. Mortality in listeriosis is almost exclusively associated with predisposing diseases and conditions (110).

A concurrent infection could influence susceptibility to listeriosis. This was exemplified by a cluster of cases in 1987 in Philadelphia characterized by a number of different strains isolated from patients. A single food vehicle could not be identified because of the diversity of strains. Clinical and epidemiologic investigations

suggested that individuals who were previously asymptomatic for listerial infection but whose gastrointestinal tract harbored *L. monocytogenes* became symptomatic, possibly because of a coinfecting agent (262).

VIRULENCE FACTORS AND MECHANISMS OF PATHOGENICITY

In the early 1980s, *L. monocytogenes* was appearing to be an attractive model system to study intracellular parasitism. This was due to five main reasons: (i) *L. monocytogenes* can grow well in culture media, (ii) it can be genetically manipulated, (iii) it belongs to a genus that contains pathogenic and nonpathogenic species, (iv) it can infect mammalian cells in tissue culture, and (v) several laboratory animals are susceptible to *Listeria*. The *L. monocytogenes* genome sequence was determined in 2001, paving the way to the discovery of a series of new virulence factors (81, 160, 294). *Listeria* now appears to be one of the most documented intracellular pathogens (55).

Pathogenicity of *L. monocytogenes*

Many tests for addressing *L. monocytogenes* pathogenicity have been developed, including tissue culture assays and tests using laboratory animals, particularly mice (173, 231). Animal models are routinely infected intravenously (i.v.) or intragastrically, and virulence is evaluated either by comparing the 50% lethal dose (LD_{50}) or by enumerating bacteria in infected target organs, in particular the spleen and liver. Heterogeneity in the virulence of *L. monocytogenes* has been observed in several in vitro and in vivo studies (LD_{50} values of *L. monocytogenes* strains range from 10^3 to 10^7 CFU) (33, 291), but in most cases, no clear correlation between the level of virulence and the origin (human, animal, category of food, or environment) or the strain characteristics (serotype, phage type, multilocus enzyme type, or DNA micro- or macrorestriction patterns) could be established. However, only 3 of the 13 known serovars of *L. monocytogenes*, 1/2a, 1/2b, and 4b, account for more than 90% of human and animal cases of listeriosis (187, 262). Among the listeriosis-associated serotypes, 4b strains cause a large proportion of listeriosis cases worldwide, but serotypes 1/2a, 1/2b, and 1/2c predominate in food isolates (23, 24, 243). A number of observations suggest that there may be differences in tropism between *L. monocytogenes* strains. For example, a possible tropism of serovar 4b strains for humans was reported (302). This tropism for humans may explain why serovar 4b was most commonly isolated from human cases. Although rare nonpathogenic or weakly pathogenic *L. monocytogenes* isolates have been reported, all strains of *L. monocytogenes* are considered to be potentially capable of causing human disease.

Experimental Infection and Cell Biology of the Infectious Process

Experimental infection of rodents with *L. monocytogenes* has been and still is widely used for the study of cell-mediated immunity. When mice are exposed to a sublethal inoculum of *L. monocytogenes*, the ensuing infection follows a well-defined course, lasting for approximately 1 week. Routinely, mice are injected i.v. and bacterial growth kinetics are monitored in the spleen and liver. Within 10 min after i.v. injection, 90% of the inoculum is taken up by the liver and 5 to 10% is taken up by the spleen. During the first 6 h, the number of viable listeriae in the liver decreases 10-fold, indicating rapid destruction of most of the bacteria. Surviving listeriae then multiply within susceptible macrophages and grow exponentially in the spleen and liver for the next 48 h, peaking at day 2 or 3 postinfection (9). Rapid inactivation ensues during the next 3 to 4 days, indicating recovery of the host. Convalescent mice are resistant to challenge and have a delayed-type hypersensitivity characterized by swelling of the footpads injected with crude cell preparations of *L. monocytogenes*.

However, the natural route of infection with *L. monocytogenes* in humans is via the gastrointestinal tract. *L. monocytogenes* infects intestinal epithelial cells in a process that requires the interaction of internalin A, expressed at the bacterial surface, with epithelial cadherin (E-cadherin), expressed at the surface of epithelial cells (211) (see below). Although *L. monocytogenes* has a relatively broad host range, the efficiency with which intestinal epithelial cells from different mammalian species are infected is variable. For example, mice are relatively resistant to intestinal infection with *L. monocytogenes* because of a single amino acid difference between human and mouse E-cadherin (176). To circumvent this problem, a transgenic mouse expressing human E-cadherin in enterocytes was generated, allowing the study of orally acquired listeriosis in mice (177). Recent studies using different *L. monocytogenes* strains report that, in contrast to strains from other serovars, serotype 4b epidemic strains appear to be able to cause systemic infection in mice infected orally, suggesting that there might be serovar-specific virulence factors playing a role in mouse susceptibility to orally acquired listeriosis (58). Alternatively, as originally shown by Racz et al. (236) the guinea pig is also a good model to monitor the *L. monocytogenes* infectious process after intragastric inoculation because the guinea pig, like humans, has the appropriate E-cadherin receptor for internalin A.

From the intestinal lumen, bacteria traverse the epithelial-cell layer and disseminate via the bloodstream to other organs, such as the spleen and liver, where they are internalized by splenic and hepatic macrophages, in which they can survive and replicate (Fig. 21.3). They are subsequently transported via the blood to regional

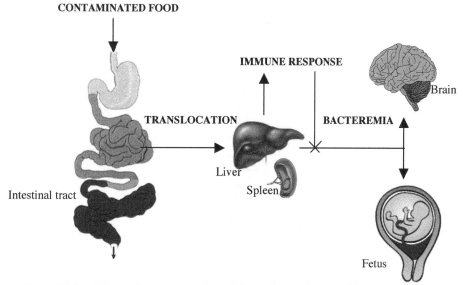

Figure 21.3   Schematic representation of the pathophysiology of *Listeria* infection.

lymph nodes. When they reach the liver and the spleen, most *Listeria* organisms are rapidly killed. In the initial phase of infection, infected hepatocytes are the target for neutrophils and later for mononuclear phagocytes that are responsible for control and resolution of infection. Depending on the level of T-cell response induced in the first days following initial infection, further dissemination via the blood to the brain or in the pregnant animal, the placenta, may subsequently occur.

L. monocytogenes possesses the ability to cross the maternofetal barrier and leads to placental abscesses, chorioamnionitis, and finally infection of the fetus. As the highest concentrations of *L. monocytogenes* are encountered in the gut and in the lung, it is thought that infection might be amplified through ingestion of contaminated amniotic fluid rather than solely as a consequence of the hematogenous transplacental route (185, 255). Several animal models have been developed for the study of pregnancy-associated (1, 13, 175, 217) and pulmonary *L. monocytogenes* infections (217).

L. monocytogenes has, in addition to its ability to cross initially the intestinal barrier and the maternofetal barrier, the capacity to cross the blood-brain barrier and reach the central nervous system (CNS) and cause meningitis, encephalitis, and brain abscesses. Invasion

of brain cell endothelial cells seems a prerequisite for meningeal pathogens that penetrate the CNS. It has been proposed that *L. monocytogenes* utilizes penetration of human brain microvascular endothelial cells (HBMEC) as a means of crossing the blood-brain barrier. *L. monocytogenes* adheres to HBMEC through the microvilli, and then infects them by an InlB-dependent mechanism (130).

Infection is not localized at the portal of entry into the host but involves dissemination and multiplication in a wide variety of cell types and tissues. In most infected tissues, *Listeria* has an intracellular location due to its capacity to induce its own phagocytosis into cells that are normally nonphagocytic. In vitro, *L. monocytogenes* invades many cell lines of different types (macrophages, fibroblasts, hepatocytes, and epithelial cells) (55). It has evolved specific strategies allowing entry into different cell types, but retaining some tropism for particular organs. Detailed analysis of infected tissue-cultured cells reveals a complex series of interactions between the bacteria and the cell (Fig. 21.4). Host cell infection begins with the adhesion and internalization of the bacteria either by phagocytosis in the case of macrophages or by induced phagocytosis (invasion) in the case of normally nonphagocytic cells. Bacterial invasion starts

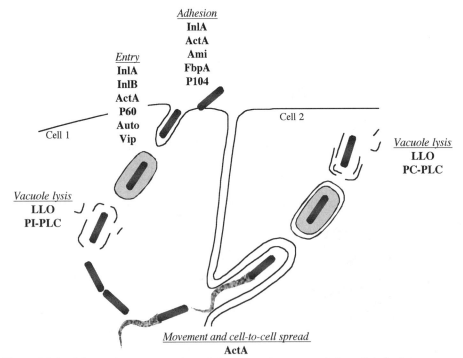

Figure 21.4   Schematic representation of the successive steps of the cell infectious process. Factors implicated in the different steps are indicated.

by a close contact with the plasma membrane that progressively enwraps the bacterium. This process is usually referred to as the "zipper" mechanism in contrast to the "trigger" mechanism used by *Salmonella* or *Shigella* (55). Following internalization, bacteria reside within membrane-bound vacuoles for ~30 min before lysing the membrane. When free in the cytosol, they replicate and concomitantly become covered with actin filaments. These filaments rearrange within 2 h into long comet tails (up to 40 μm in length) left behind in the cytosol while the bacteria move ahead at a speed of ~0.3 μm/s (59, 287, 288). When moving bacteria contact the plasma membrane, they induce the formation of bacterium-containing protrusions. Contact between these protrusions and neighboring cells results in the internalization of the protrusion. In the newly infected cell, the bacterium is surrounded by two plasma membranes that must be lysed to initiate a new cycle of multiplication and movement. The entire cycle is completed in about 5 h.

Once *Listeria* has entered the cytoplasm, it can disseminate directly from cell to cell, circumventing host defenses such as circulating antibodies and complement. This ability to disseminate in tissues by cell-to-cell spreading provides an explanation for the early observation that antibodies are not protective and that immunity to *Listeria* is T cell mediated (172).

Genetic Approaches to the Study of Virulence Factors

Genetic manipulation of *L. monocytogenes* in vitro is now possible. Conjugative transposons from the Tn*1545*-Tn*916* family or Tn*917* transposons were initially exploited to generate mutant libraries allowing the identification of several *L. monocytogenes* virulence factors (56, 100, 159, 232). More recently, Tn*1545*- or Tn*917*-derived tagged transposons were also used for signature-tagged mutagenesis in *L. monocytogenes* (12, 77, 194). Plasmid vectors originating from *Bacillus subtilis* or *Escherichia coli* are used for genetic studies in *L. monocytogenes*, including allelic exchange of chromosomal DNA, cloning, gene expression, or reporter gene fusion (94). For allelic exchange, in-frame deletion, or site-specific mutagenesis, thermosensitive vectors such as pKSV7 (279) or pAUL-A (253) are used. A novel thermosensitive plasmid, pMAD, harboring the β-galactosidase gene, facilitates the screening and the generation of allelic exchanges (8). For complementation and gene expression, shuttle plasmids replicating in *E. coli* and *L. monocytogenes*, pAT18, pMK4, and pAM401 are mostly used (284, 290, 303). Complementation of deletion mutants can be now achieved using a new integrative phage-derived plasmid, pPL2, that allows a single copy integration at a specific location in the *L. monocytogenes* chromosome (171). Listeriae are not naturally competent; however, transformation with plasmid DNA can be obtained on protoplasts or by electroporation (188, 296). Until recently, transduction was not possible in *L. monocytogenes*. It has now been successfully performed using bacteriophages that grow at 30°C, opening the way to new genetic approaches to modify the *Listeria* genome (141).

Adhesion to Mammalian Cells

Invasion of host cells and internalization of *L. monocytogenes* into mammalian cells have received considerable attention. Adhesion of *L. monocytogenes* to mammalian cells by itself may be an important and critical aspect of pathogen-host interaction. Adhesion may promote colonization of the gastrointestinal tract, direct the bacteria to appropriate target cells or tissues such as the CNS or the placenta, and even facilitate cell invasion by activating cell signal transduction pathways or triggering the synthesis of a target cell receptor required for invasion. Adherence involves a number of surface proteins, including InlA, ActA, Ami, p104, and FbpA. InlA is the first member of the internalin multigene family, a family characterized by the presence of a leucine-rich repeats (LRR) region (99). InlA promotes adherence and entry into cells expressing its receptor, the adhesion protein E-cadherin (211) (see below). ActA, the surface protein required for actin-based motility (see below), may also promote attachment via host cell proteoglycans (3, 282). An ActA-deficient mutant is significantly impaired in attachment and entry into IC-21 murine macrophages and Chinese hamster ovary epithelial cells. Ami is a bacterial surface amidase that presents an N-terminal domain similar to the amidase domain of the Atl autolysin of *Staphylococcus aureus* and a C-terminal, cell wall-anchoring domain made up of eight GW modules (30). Besides its lytic activity on *Listeria* cell walls (205), Ami is implicated in bacteria adhesion to target cells (215). Ami mutants are attenuated, indicating that Ami plays a role in the virulence of *L. monocytogenes*. The involvement of a cell surface protein of 104 kDa (p104) in the adhesion of *L. monocytogenes* to the human intestinal cell line Caco-2 has also been proposed (227). Finally, FbpA is a fibronectin binding protein shown to be required for intestinal and liver colonization after oral infection of transgenic mice expressing human E-cadherin. FbpA binds to immobilized human fibronectin and increases adherence of wild-type *L. monocytogenes* to HEp-2 cells. Despite the lack of conventional secretion/anchoring signals, FbpA is detected on the bacterial surface and was shown to be a substrate of the SecA2 pathway (77). In addition, FbpA behaves as a chaperone for two important virulence factors, LLO and InlB, probably preventing their degradation.

Entry into Mammalian Cells

InlA and InlB were the first listerial factors identified as mediating bacterial invasion into different target cell types. Their cellular receptors have also been identified, and the molecular signaling cascades triggered during *Listeria* entry into host cells are being characterized in detail. Recent findings suggest that other molecules are also necessary for internalization (P60, ActA, Auto, and Vip), revealing a complex dialogue between *Listeria* and eukaryotic cells during the early phases of the infection cycle.

Internalins

InlA belongs to a group of proteins, the internalin family, characterized by the presence of LRRs (39). InlA (800 amino acids) displays a classical signal sequence followed by an LRR region comprising 15 repeats of 22 amino acids. An interrepeat region (IR) separates the LRR region from a second repeat region (called the B repeat region). The carboxy terminus displays a cell wall-sorting motif, LPTTG, which is the substrate for the enzyme sortase A and allows covalent linkage of the protein to the peptidoglycan. The LRR and IR regions are necessary and sufficient to promote *Listeria* entry into human epithelial cells (176). The InlA receptor is E-cadherin (211). E-cadherin belongs to the cadherin superfamily of cell adhesion molecules that are transmembrane glycoproteins located at adherens junctions and allow cell-cell adhesion. *Listeria* seems to exploit the whole molecular junctional complex involved in adherens junctions for inducing its entry into target cells (281) (see above). The interaction between InlA and E-cadherin is species specific and involves recognition of a proline at position 16 (Pro16) of the E-cadherin molecule (174). InlA is truncated in some naturally occurring strains, and a recent epidemiologic survey indicates that clinical strains express a full-length InlA far more frequently than strains recovered from food products (148, 156), revealing that InlA is critical for the pathogenesis of human listeriosis.

InlB (630 amino acids) is another member of the internalin family involved in entry of *Listeria* into a broad range of cell lines including hepatocytes and nonepithelial cells (76). It contains a signal peptide, seven LRR repeats, one IR region, and one B repeat. The C-terminal domain presents three tandem repeats (GW modules) that mediate attachment of InlB to the bacterial cell wall through noncovalent interaction with lipoteichoic acids (155). The LRR region of InlB is sufficient to confer invasiveness to noninvasive *L. innocua* or to latex

beads (32). The crystal structures of the LRR domain (196), of the LRR/IR domains (258), and of the full-length InlB (195) indicate that the protein can accommodate several ligands. Indeed, InlB interacts with several cellular receptors, which likely cooperate to promote bacterial uptake. gC1qR was the first InlB receptor identified (31). gC1qR was initially identified as the receptor for C1q, the first component of the complement cascade (228). InlB-dependent entry is blocked by anti-gC1qR antibodies and by C1q, and transfection with human gC1qR enhances cell invasion (31). However, gC1qR has no transmembrane domain or GPI (glycosylphosphatidylinositol)-anchored domain, suggesting that it likely behaves as a coreceptor for a signaling protein. The signaling receptor was later identified as Met, the receptor for the hepatocyte growth factor. Met is a transmembrane protein that mediates several signaling pathways triggered by InlB. The interaction with the receptor Met occurs through the LRR domain of InlB (270). In addition, InlB directly binds to cellular glycosaminoglycans (GAGs) through its GW modules. InlB-dependent entry into epithelial cells is strongly affected by depletion of the cellular plasma membrane GAGs. It was proposed that interaction of InlB with GAGs through its GW modules leads to its detachment from the bacterial surface, allowing its clustering at the cellular surface through binding to Met by its LRR domain, and favoring the local activation of the signaling pathway downstream of Met (157).

P60

For a long time, Δ*iap* (invasion-associated protein) mutants could not been obtained, suggesting that the protein was essential for bacterial viability. Therefore, the role of P60 (encoded by *iap*) was first evaluated in spontaneous rough mutants expressing lower levels of P60 and forming long filamentous structures composed of bacterial chains (166). The rough mutants are less virulent and enter less efficiently in certain eukaryotic cells, suggesting a role for P60 in bacterial invasion (134, 139, 166). A viable Δ*iap* mutant was recently obtained, allowing more precise studies of the role of P60. As for rough mutants, it also had a defect in septum formation and in virulence after i.v. infection of mice. In addition, the Δ*iap* mutant is impaired in bacterial movement and spreading from cell to cell due to an improper localization of the ActA at the surface of *L. monocytogenes* (180, 229). However, in contrast to what is seen in studies using rough mutants, the invasiveness of this Δ*iap* mutant in mouse fibroblasts and human epithelial cells is only slightly diminished compared with that of wild-type bacteria (180, 229). P60 is an autolysin both secreted and associated with the bacterial cell wall (166, 246, 304). Secretion is mediated as for FbpA by the recently identified auxiliary secretion system SecA2, which mediates the secretion of at least 17 secreted and surface proteins of *L. monocytogenes* (180, 229). In addition, P60 plays an important role in the immune response against *L. monocytogenes* infection. Antibodies specific for P60 can act as opsonins and may play a role in preventing systemic infections in immunocompetent individuals (164). Moreover, P60 is a major protective antigen that induces both T-CD8 and Th1 protective immune responses (107, 108, 137).

ActA

ActA is a surface protein implicated in the attachment to cells (see above) and responsible for the actin-based motility of *Listeria* (see below). Moreover, expression of ActA in *L. innocua* is sufficient to promote bacterial entry in some epithelial cell lines (282).

Auto

A novel autolysin-encoding gene, *aut*, was identified recently (40). It is the only autolysin gene that is absent from the nonpathogenic species *L. innocua*. The *aut* gene encodes a surface protein, Auto, with an autolytic activity, as expected from the presence of a domain harboring homologies with autolysin-encoding genes, especially *N*-acetylglucosaminidases. The protein Auto possesses a C-terminal cell wall-anchoring domain made up of four GW modules, similar to those observed in the other autolysin, Ami, and in InlB (39). The morphology of an *aut* deletion mutant is similar to that of the wild type, with no defect in septation and cell division, suggesting no role for Auto in these functions (40). Auto is required for entry of *L. monocytogenes* into different nonphagocytic eukaryotic cell lines. An *aut* deletion mutant has a reduced virulence following i.v. inoculation of mice and oral infection of guinea pigs, which correlates with its low invasiveness. However, the autolytic activity of Auto by itself, rather than an invasive ability, might be critical for virulence. Indeed, Auto may control the general surface architecture exposed to the host by *L. monocytogenes* and/or the composition of the surface products released by the bacteria, hence affecting the host response to infection (40).

Vip

The gene *vip* is a new *L. monocytogenes* virulence gene that encodes an LPXTG surface protein absent from nonpathogenic *Listeria* species (41). *vip* is positively regulated by PrfA, the transcriptional activator of the major *Listeria* virulence factors. Vip is anchored to the *Listeria* cell wall by sortase A and is required for entry into some mammalian cells. Using a ligand overlay approach, a Vip cellular receptor was identified. It is the endoplasmic reticulum resident protein Gp96 reported to also interact with toll-like receptors (181). The Vip-Gp96 interaction is critical for bacterial entry into some cells. Comparative infection studies using oral and i.v. inoculation of nontransgenic and transgenic mice expressing human E-cadherin, demonstrated a role for Vip in *Listeria* virulence, not only at the intestine level but also in late stages of the infectious process. Vip thus appears to be a new virulence factor exploiting Gp96 as a receptor for cell invasion, and/or signaling events that may interfere with the host immune response in the course of the infection (41).

Sortases

The genome of *L. monocytogenes* (strain EGD-e) contains 41 genes that encode LPXTG proteins including InlA and Vip (39). LPXTG proteins are anchored to the peptidoglycan by transpeptidases named sortases (256). *srtA* encodes a sortase responsible for the anchoring of InlA, Vip, and at least 13 other proteins to the peptidoglycan (20, 41, 235). *L. monocytogenes* mutants lacking *srtA* are defective in internalizing into human enterocyticlike Caco-2 cells and hepatocytic HepG2 cells, as well as in colonizing the liver and spleen of mice infected orally. The *Listeria* genome encodes another sortase, SrtB, which has a limited number of substrates containing a C-terminal NXXTN sorting motif. SrtB is not required for the infectious process (19).

Escape from the Phagosome and Intracellular Growth

After internalization, *Listeria* resides in a vacuole that is lysed by listeriolysin O (LLO) in concert with two phospholipases C (PLCs). Once in the target cell cytoplasm, *Listeria* then multiplies, acquiring nutrients from the cytosol.

LLO

Nonhemolytic *L. monocytogenes* strains can be found in the environment but are avirulent in experimentally infected mice. This observation stimulated the study of the role of hemolysin in virulence. By use of conjugative transposons, nonhemolytic mutants that were avirulent in mice were obtained (100, 159, 232). Both wild-type and nonhemolytic strains were able to enter mammalian cells; however, nonhemolytic mutants are impaired in their ability to grow intracellularly. Electron microscopy revealed that nonhemolytic mutants, although invasive, were unable to escape from the phagocytic vacuole (100). The locus containing the transposon was cloned and named *hly* (209). It encodes LLO. Once internalized in a vacuole,

L. monocytogenes rapidly escapes into the cytosol, where it multiplies. LLO is a pore-forming toxin that enables the bacteria to escape into the cytosol, where they replicate. LLO accomplishes this by binding to cholesterol and then oligomerizing into large complex pores (146). LLO is a 60-kDa secreted protein involved in *Listeria* escape from primary and also secondary vacuoles (54, 78, 159, 232). LLO is a member of the pore-forming, cholesterol-dependent cytolysin family. These thiol-activated toxins are produced by several gram-positive bacteria, including *Streptococcus*, *Clostridium*, and *Bacillus* species. They are active only on cholesterol-containing membranes, with cholesterol likely acting as the receptor in the membrane. After an initial step of binding, LLO monomers diffuse laterally to form ring-shaped, ion-permeable pores of 30 nm diameter, without disrupting the plasma membrane (238). Unlike all other thiol-activated toxins, LLO has a maximal activity at pH 5 and is inactive at pH 7, thus impairing its deleterious effect on cellular membranes when the bacterium is free in the cytosol (115). In addition, optimal pore formation by LLO occurs between pH 5.5 and 6.0 (the pH of the early phagosome) (15). A PEST-like sequence (P, Pro; E, Glu; S, Ser; T, Thr) has been detected in the LLO molecule (65). This motif was first thought to target LLO for degradation when present in the cytosol, restricting its function to the vacuolar environment and inhibiting the lysis of host cells. Recent findings indicate that LLO is denaturized in the cytosol (263) and that the PEST sequence does not mediate proteasomal degradation by the host (257).

PLCs

Two phospholipases are produced and secreted by *L. monocytogenes*, a phosphoinositide-phospholipase (PI-PLC) and a broad-range phospholipase (PC-PLC) (114). The two phospholipases have a membrane-damaging activity and are involved in bacterial escape from primary and/or secondary phagosomes (44, 117, 132, 198, 293). Each of the two *L. monocytogenes* phospholipases is important for virulence since mutants deficient in either PI-PLC or PC-PLC are attenuated (43, 278). More importantly, double mutants deficient in both phospholipases are 500 times less virulent than single mutants, highlighting the importance and the complementarity of these factors in listeriosis (278). The PC-PLC is synthesized as a proenzyme matured by the metalloprotease encoded by *mpl* (70, 208). It hydrolyzes a wide variety of phospholipids including sphingomyelin. Both phospholipases act synergistically with LLO in lysing primary and secondary vacuoles (44, 105). In the absence of LLO, the PC-PLC can also promote lysis of primary vacuoles in human epithelial cell lines (132, 198). In synergy with

LLO, the PI-PLC induces hydrolysis of PI and production of diacylglycerol in macrophages (115, 271), leading to mobilization of protein kinase Cδ and subsequent elevation of intracellular calcium levels (297).

Once free in the cytosol, *L. monocytogenes* starts multiplying, with an approximate doubling time of 1 h (288). Several genes encoding virulence or metabolic determinants are induced during *L. monocytogenes* intracellular life, including those involved in phagosomal lysis, actin-based motility, and cell-to-cell spreading (34, 95, 161). In contrast to most bacteria, *L. monocytogenes* replicates into the cytosol when it is directly microinjected into cells (113). The cytosol permissiveness for *L. monocytogenes* growth is likely due to the bacterium's ability to use a variety of cytosolic nutrients, as suggested by the fact that intracellular multiplication of several auxotrophic mutants is not affected (197). Little is known about the precise nutritional requirements or the mechanisms used by *Listeria* to obtain nutrients from the host cell cytosol. Nevertheless, intracytosolic growth of *L. monocytogenes* is highly dependent on the *hpt* gene, which encodes a sugar uptake system (113), and is tightly regulated by PrfA (214).

Intracytoplasmic Movement and Cell-to-Cell Spreading

Induced phagocytosis, escape from the phagosome, and intracellular multiplication are essential steps for infection of individual cells but are not sufficient to achieve infection in mice. *L. monocytogenes* must also efficiently infect tissues by direct cell-to-cell spreading. Direct cell-to-cell spreading is the result of intracellular movement, protrusion formation, phagocytosis of the protrusion by a neighboring cell, formation of a double-membrane-bound vacuole, and lysis of the two-membrane vacuole. Intracytoplasmic movement is strictly coupled to continuous actin assembly that provides the force for bacterial propulsion (Color Plate 1). Actin polymerization requires expression of the *actA* gene. *actA* mutants are invasive, escape from the phagosome, replicate, and form intracellular microcolonies, but are not covered with actin, do not move intracellularly, and do not spread from cell to cell. ActA is a 639-amino-acid protein that presents a signal sequence at its aminoterminal domain and a transmembrane motif at its carboxyl-terminal domain that anchors the molecule to the bacterial surface (71, 163). A central repeat domain containing four proline-rich repeats stimulates the *Listeria* actin-based motility (169). This domain binds members of the enabled/vasodilator-stimulated phosphoprotein (Ena/VASP) family of proteins (48, 220), which modulate bacterial speed and directionality (10, 106). Immunolocalization of the ActA protein in infected cells has revealed that ActA is asymmetrically distributed on the bacterial surface, with an increasing concentration towards the opposite pole of the bacterial cell, which is the site of comet tail formation. The amino-terminal region of ActA alone can induce bacterial movement (170). It binds and activates Arp2/3, a seven-protein host complex that induces actin polymerization and the generation of a dendritic array of actin filaments (53). By doing so, ActA mimics WASP family proteins (29, 272). Actin filaments in *Listeria* tails, as in *Shigella* tails, are branched, in contrast to those in *Rickettsia conorii* tails, which display longer and unbranched actin filaments (119–121).

When the actin comet tail targets the bacterium to the host's plasma membrane, the membrane extends, forming long protrusions into the extracellular space. When these bacterium-containing membrane protrusions from one host cell enter a neighboring host cell, intercellular spread occurs. It has been suggested that bacteria could take advantage of normal paracytophagic behavior to favor the intercellular passage of bacteria (240). At this stage, the bacterial cell is enclosed in a double-membrane secondary vacuole, one membrane coming from the donor cell and the other from the recipient cell. Lysis of the two-membrane vacuoles requires at least four virulence factors, including LLO, a metalloprotease encoded by *mpl*, PI-PLC, and PC-PLC (see above).

Coregulation of Virulence Factors

Infection at the cell level involves the following steps: (i) adhesion to host cells, (ii) bacterially induced uptake, (iii) lysis of the phagosome, (iv) replication in the host cytoplasm, (v) actin-based intracellular movement and propulsion, and (vi) intercellular spread. Most *Listeria* virulence genes involved in these infectious steps (i.e., *prfA*, *plcA*, *hly*, *mpl*, *actA*, *plcB*, *inlA*, *inlB*, *inlC*, *hpt*, *bsh*, and *vip*) are regulated by the transcriptional activator PrfA (Color Plate 2) (41, 79, 179, 210, 214). PrfA is a protein comprising 233 amino acids that belongs to the Crp/Fnr family. Genes regulated by PrfA contain a binding site in the –41 region, the PrfA box, consisting of a 14-bp palindromic sequence to which the putative helix-turn-helix motif of PrfA binds in vitro (269). Expression of *prfA* is a complex phenomenon. *prfA* is the second gene of the *plcA-prfA* operon whose primary promoter is regulated by *prfA*. Hence, *prfA* autoregulates (activates) its own synthesis. *prfA* can also be transcribed from its own promoter region. It has been determined in a *prfA* mutant that transcription at the *prfA* specific promoter is upregulated (93). PrfA-regulated virulence factors such as LLO, PlcA, PlcB, and Vip are produced at higher levels as temperature increases to 37°C, the temperature of the host (41, 178). At a temperature lower than 30°C, the low

levels of expression are correlated to undetectable levels of PrfA, although its transcript is synthesized (237). At low temperatures, the untranslated leader region of the *prfA* mRNA forms a stable secondary structure masking the Shine-Dalgarno sequence (154). Consequently, PrfA is not translated. At high temperatures, this structure, named "thermosensor," melts. Mutations destabilizing the thermosensor at 30°C mimic the effect of an increase in temperature, i.e., they unmask the ribosome-binding site and lead to virulence gene expression. Knowledge of the *L. monocytogenes* genome sequence (111) allowed analysis of the complete PrfA regulon. A whole-genome array based on the sequence of the *L. monocytogenes* EGDe strain was constructed to study the PrfA regulon (214). This transcriptomic analysis revealed that PrfA could act both as an activator and as a repressor. It led to the identification of many new PrfA-regulated genes in addition to genes previously known to be part of the PrfA regulon. However, in many cases, the effect of PrfA is probably indirect.

Stress Proteins of *L. monocytogenes*

L. monocytogenes has evolved a series of adaptive responses in order to cope with a large variety of stresses to survive and/or multiply under harsh environmental conditions, outside as well as inside the host. The LisRK two-component signal transduction system is implicated in virulence, acid and ethanol tolerance, and oxidative stress (158). *virR*, a gene encoding the response regulator of another two-component system, was shown to regulate a key regulon in *L. monocytogenes*, controlling virulence by a global regulation of surface component modifications (194). The DnaK heat shock chaperone protein is required for survival of *L. monocytogenes* under high temperatures and acidic conditions as well as for efficient phagocytosis with macrophages (135, 136). The major heat shock chaperones, GroES and GroEL, are induced at high temperature, at low pH, and during cell infection (98, 136). The ClpC stress protein is required for survival under iron deprivation, high temperature, or osmolarity, in bone marrow macrophages, and in organs of mice (245). Two other proteases have been reported to be involved in stress response and virulence, namely, ClpE and ClpP (101, 219). The *fri* gene, which encodes ferritin, is necessary for optimal growth in minimal medium in both the presence and absence of iron, as well as after cold and heat shock. Ferritin also provides protection against reactive oxygen species and is essential for full virulence of *L. monocytogenes*. A comparative proteomic analysis revealed an effect of the *fri* deletion on the levels of LLO and several stress proteins (80, 138). Adaptation to osmotic stress depends on the intracellular accumulation of osmolytes and thus on uptake systems that include BetL, GbuABC, OpuC and OpuB, and OppA (26, 162, 275, 276, 301). Bile tolerance of *L. monocytogenes* involves the bile salt hydrolase, Bsh, an enzyme that deconjugates bile salts and that is required for both intestinal and hepatic phases of listeriosis (79). The transporter BtlA (17) and other systems such as the putative transporter of the glutamate decarboxylase, GadE, and the zinc uptake regulator, ZurR, contribute to tolerance to bile or various other stresses, e.g., low pH, salt, ethanol, detergent, and antibiotic (16). The alternative sigma factor B contributes to the ability of *L. monocytogenes* to survive and/or multiply under stressful conditions outside the host (acid, osmotic or oxidative stresses, low temperature, or carbon starvation) (91) but also to persist within the host during the infectious process. It has been demonstrated that sigma B contributes to survival following exposure to bile salts (18), but also to transcription of the virulence gene activator PrfA (218). Characterization of the sigma B-dependent general stress regulon confirmed the broad role of this sigma factor (283).

CONCLUDING REMARKS

Public health surveillance, outbreak investigations, and applied and basic research conducted during the past 25 years have helped to characterize the disease listeriosis, define the magnitude of its public health problem and its impact on the food industry, identify the risk factors associated with disease, and develop appropriate and targeted control strategies. The food-processing industry has made progress in reducing the prevalence of *L. monocytogenes* in processing plant environments and in high-risk foods, and preventive measures have been developed and implemented for persons at increased risk of infection. In the United States, the incidence of sporadic listeriosis continues to decline; in 2005, listeriosis incidence was 0.3 case per 100,000 population, which is very close to the national 2010 target of 0.25 case per 100,000 persons. Major factors that have contributed to this decline include enhanced surveillance and outbreak investigation efforts, enhanced regulatory control, and implementation of preventive strategies by the food industry.

Nevertheless, there are still many unanswered questions. Are all *L. monocytogenes* strains equally pathogenic for humans, and are all of them capable of causing outbreaks? What is the infectious dose of *L. monocytogenes* for nonpregnant adults, persons with immune system deficiencies, and pregnant women? Do the infectious doses vary significantly between strains? Are only a few strains capable of causing febrile gastroenteritis in humans, or are all strains equally capable of causing

this gastrointestinal syndrome, if they are present in high enough numbers? Recently, Garner et al. (102) reported that the general stress-responsive alternative sigma factor σ^B plays a critical role in the regulation of stress response and virulence genes that are important in the gastrointestinal stage of listeriosis.

Because *L. monocytogenes* is primarily a soil microorganism and is widely distributed in the environment, is it practical to expect all RTE foods to be *Listeria*-free? Is a "zero-tolerance" policy essential for maximum public health protection from listeriosis? How do we ensure that RTE prepackaged salads are not contaminated with *L. monocytogenes*? How do we prevent *Listeria* contamination of other RTE foods such as frankfurters, soft cheeses, and smoked fish? Are some strains of *L. monocytogenes* better able to form biofilms, and if yes, are these strains more likely to persist in food-processing environments for a long time? Finally, does the recurrence of Mexican-style cheese-associated listeriosis cases in the U.S. Hispanic population mirror a larger problem in the developing world, where milk is not routinely pasteurized?

A multidisciplinary approach has enabled the identification and characterization of several virulence factors. Research during the past 25 years has led to the (i) identification of the internalin receptor of mammalian cells; (ii) elucidation of the role of internalin multigene functions; (iii) understanding of the function of *actA*; (iv) understanding of *prfA* and the global regulation of virulence; and (v) the modulation of host cell signaling by the pathogen. *L. monocytogenes* has become one of the best understood intracellular pathogens at the cellular and molecular levels. The immune response to this pathogen continues to be a topic of intensive investigations.

Use of trade names is for identification only and does not imply endorsement by the Centers for Disease Control and Prevention or by the U.S. Department of Health and Human Services.

References

1. **Abram, M., D. Schluter, D. Vuckovic, B. Wraber, M. Doric, and M. Deckert.** 2003. Murine model of pregnancy-associated *Listeria monocytogenes* infection. *FEMS Immunol. Med. Microbiol.* 35:177–182.

2. **Al-Ghazali, M., and S. Al-Azawi.** 1990. *Listeria monocytogenes* contamination of crops grown on soil treated with sewage sludge cake. *J. Appl. Bacteriol.* 69:642–674.

3. **Alvarez-Dominguez, C., J. A. Vazquez-Boland, E. Carrasco-Marin, P. Lopez-Mato, and F. Leyva-Cobian.** 1997. Host cell heparan sulfate proteoglycans mediate attachment and entry of *Listeria monocytogenes*, and the listerial surface protein ActA is involved in heparan sulfate receptor recognition. *Infect. Immun.* 65:78–88.

4. **Anonymous.** 2002. Final Assessment Report: Proposal 239—Listeria *Risk Assessment & Risk Management*

Strategy. Food Standards Australia New Zealand. [Online.] www.foodstandards.gov.au/_srefiles/P239_DAR.021002. pdf. Accessed 24 May 2006.

5. **Anonymous.** 1988. *Food Listeriosis—Report of the WHO Informal Working Group.* WHO/EHE/FOS/88.5. World Health Organization, Geneva, Switzerland.

6. **Anonymous.** 2003. *Interpretive Summary: Quantitative Assessment of the Relative Risk to Public Health from Foodborne* Listeria monocytogenes *among Selected Categories of Ready-To-Eat Meats.* Center for Food Safety and Applied Nutrition, Food and Drug Administration, U.S. Department of Health and Human Services; Food Safety and Inspection Service, U.S. Department of Agriculture, Washington, D.C. [Online.] www.foodsafety.gov/~dms/lmr2toc.html. Accessed 24 May 2006.

7. **Anonymous.** 2004. *Microbiological Risk Assessment Series.* World Health Organization, Geneva, Switzerland.

8. **Arnaud, M., A. Chastanet, and M. Debarbouille.** 2004. New vector for efficient allelic replacement in naturally nontransformable, low-GC-content, gram-positive bacteria. *Appl. Environ. Microbiol.* **70:**6887–6891.

9. **Audurier, A., P. Pardon, J. Marly, and F. Lantier.** 1980. Experimental infection of mice with *Listeria monocytogenes* and *L. innocua. Ann. Microbiol.* **131B:**47–57.

10. **Auerbuch, V., J. J. Loureiro, F. B. Gertler, J. A. Theriot, and D. A. Portnoy.** 2003. Ena/VASP proteins contribute to *Listeria monocytogenes* pathogenesis by controlling temporal and spatial persistence of bacterial actin-based motility. *Mol. Microbiol.* **49:**1361–1375.

11. **Aureli, P., G. C. Fiorucci, D. Caroli, G. Marchiaro, O. Novara, L. Leone, and S. Salmaso.** 2000. An outbreak of febrile gastroenteritis associated with corn contaminated by *Listeria monocytogenes. N. Engl. J. Med.* **342:**1236–1241.

12. **Autret, N., I. Dubail, P. Trieu-Cuot, P. Berche, and A. Charbit.** 2001. Identification of new genes involved in the virulence of *Listeria monocytogenes* by signature-tagged transposon mutagenesis. *Infect. Immun.* **69:**2054–2065.

13. **Bakardjiev, A. I., B. A. Stacy, S. J. Fisher, and D. A. Portnoy.** 2004. Listeriosis in the pregnant guinea pig: a model of vertical transmission. *Infect. Immun.* **72:**489–497.

14. **Barbosa, W. B., L. Cabedo, H. J. Wederquist, J. N. Sofos, and G. R. Schmidt.** 1994. Growth variation among species and strains of *Listeria monocytogenes. J. Food Prot.* **57:**765–769.

15. **Beauregard, K. E., K. D. Lee, R. J. Collier, and J. A. Swanson.** 1997. pH-dependent perforation of macrophage phagosomes by listeriolysin O from *Listeria monocytogenes. J. Exp. Med.* **186:**1159–1163.

16. **Begley, M., C. G. Gahan, and C. Hill.** 2002. Bile stress response in *Listeria monocytogenes* LO28: adaptation, cross-protection, and identification of genetic loci involved in bile resistance. *Appl. Environ. Microbiol.* **68:**6005–6012.

17. **Begley, M., C. Hill, and C. G. Gahan.** 2003. Identification and disruption of btlA, a locus involved in bile tolerance and general stress resistance in *Listeria monocytogenes. FEMS Microbiol. Lett.* **218:**31–38.

18. **Begley, M., R. D. Sleator, C. G. Gahan, and C. Hill.** 2005. Contribution of three bile-associated loci, *bsh, pva,* and *btlB,* to gastrointestinal persistence and bile tolerance of *Listeria monocytogenes. Infect. Immun.* **73:**894–904.

19. **Bierne, H., C. Garandeau, M. G. Pucciarelli, C. Sabet, S. Newton, F. Garcia-del Portillo, P. Cossart, and A. Charbit.** 2004. Sortase B, a new class of sortase in *Listeria monocytogenes. J. Bacteriol.* **186:**1972–1982.

20. **Bierne, H., S. K. Mazmanian, M. Trost, M. G. Pucciarelli, G. Liu, P. Dehoux, L. Jansch, F. Garcia-del Portillo, O. Schneewind, P. Cossart, and The European Listeria Genome Consortium.** 2002. Inactivation of the *srtA* gene in *Listeria monocytogenes* inhibits anchoring of surface proteins and affects virulence. *Mol. Microbiol.* **43:**869–881.

21. **Bille, J.** 1989. Presented at the Foodborne Listeriosis Symposium, Wiesbaden, Germany, 7 September 1988.

22. **Bille, J., J. Rocourt, and B. Swaminathan.** 2003. *Listeria* and *Erysipelothrix,* p. 461–471. *In* P. R. Murray, E. J. Baron, J. H. Jorgensen, M. A. Pfaller, and R. H. Yolken (ed.), *Manual of Clinical Microbiology,* 8th ed, vol. 1. ASM Press, Washington, D.C.

23. **Boerlin, P., F. Boerlin-Petzold, E. Bannerman, J. Bille, and T. Jemmi.** 1997. Typing *Listeria monocytogenes* isolates from fish products and human listeriosis cases. *Appl. Environ. Microbiol.* **63:**1338–1343.

24. **Boerlin, P., and J. Piffaretti.** 1991. Typing of human, animal, food, and environmental isolates of *Listeria monocytogenes* by multilocus enzyme electrophoresis. *Appl. Environ. Microbiol.* **57:**1624–1629.

25. **Boerlin, P., J. Rocourt, F. Grimont, P. A. D. Grimont, C. Jacquet, and J. C. Piffaretti.** 1992. *Listeria ivanovii* subspecies *londoniensis. Int. J. Syst. Bacteriol.* **15:**42–46.

26. **Borezee, E., E. Pellegrini, and P. Berche.** 2000. OppA of *Listeria monocytogenes,* an oligopeptide-binding protein required for bacterial growth at low temperature and involved in intracellular survival. *Infect. Immun.* **68:**7069–7077.

27. **Borucki, M. K., S. H. Kim, D. R. Call, S. C. Smole, and F. Pagotto.** 2004. Selective discrimination of *Listeria monocytogenes* epidemic strains by a mixed-genome DNA microarray compared to discrimination by pulsed-field gel electrophoresis, ribotyping, and multilocus sequence typing. *J. Clin. Microbiol.* **42:**5270–5276.

28. **Borucki, M. K., J. Reynolds, D. R. Call, T. J. Ward, B. Page, and J. Kadushin.** 2005. Suspension microarray with dendrimer signal amplification allows direct and high-throughput subtyping of *Listeria monocytogenes* from genomic DNA. *J. Clin. Microbiol.* **43:**3255–3259.

29. **Boujemaa-Paterski, R., E. Gouin, G. Hansen, S. Samarin, C. Le Clainche, D. Didry, P. Dehoux, P. Cossart, C. Kocks, M. F. Carlier, and D. Pantaloni.** 2001. Listeria protein ActA mimics WASp family proteins: it activates filament barbed end branching by Arp2/3 complex. *Biochemistry* **40:**11390–11404.

30. **Braun, L., S. Dramsi, P. Dehoux, H. Bierne, G. Lindahl, and P. Cossart.** 1997. InlB: an invasion protein of *Listeria monocytogenes* with a novel type of surface association. *Mol. Microbiol.* **25:**285–294.

31. **Braun, L., B. Ghebrehiwet, and P. Cossart.** 2000. gC1q-R/p32, a C1q-binding protein, is a receptor for the InlB invasion protein of *Listeria monocytogenes. EMBO J.* **19:**1458–1466.

32. **Braun, L., F. Nato, B. Payrastre, J. C. Mazie, and P. Cossart.** 1999. The 213-amino-acid leucine-rich repeat region of

the *Listeria monocytogenes* InlB protein is sufficient for entry into mammalian cells, stimulation of PI 3-kinase and membrane ruffling. *Mol. Microbiol.* **34:**10–23.

33. Brosch, R., B. Catimel, G. Milon, C. Burchrieser, E. Vindel, and J. Rocourt. 1993. Virulence heterogeneity of *Listeria monocytogenes* strains from various sources (food, human, animal) in immunocompetent mice and its association with typing characteristics. *J. Food Prot.* **56:**296–301.

34. Bubert, A., Z. Sokolovic, S. K. Chun, L. Papatheodorou, A. Simm, and W. Goebel. 1999. Differential expression of *Listeria monocytogenes* virulence genes in mammalian host cells. *Mol. Gen. Genet.* **261:**323–336.

35. Buchrieser, C., R. Brosch, B. Catimel, and J. Rocourt. 1993. Pulsed-field electrophoresis applied for comparing *Listeria monocytogenes* strains involved in outbreaks. *Can. J. Microbiol.* **39:**395–401.

36. Bula, C., J. Bille, and M. Glauser. 1995. An epidemic of food-borne listeriosis in Western Switzerland: description of 57 cases involving adults. *Clin. Infect. Dis.* **20:**66–72.

37. Buncie, S. 1991. The incidence of *Listeria monocytogenes* in slaughtered animals, in meat, and in meat products in Yugoslavia. *Int. J. Food Microbiol.* **12:**173–180.

38. Burt, S. 2004. Essential oils: their antibacterial properties and potential applications in foods—a review. *Int. J. Food Microbiol.* **94:**223–253.

39. Cabanes, D., P. Dehoux, O. Dussurget, L. Frangeul, and P. Cossart. 2002. Surface proteins and the pathogenic potential of *Listeria monocytogenes*. *Trends Microbiol.* **10:**238–245.

40. Cabanes, D., O. Dussurget, P. Dehoux, and P. Cossart. 2004. Auto, a surface associated autolysin of *Listeria monocytogenes* required for entry into eukaryotic cells and virulence. *Mol. Microbiol.* **51:**1601–1614.

41. Cabanes, D., S. Sousa, A. Cebria, M. Lecuit, F. Garcia-del Portillo, and P. Cossart. 2005. Gp96 is a cellular receptor for a novel *Listeria monocytogenes* virulence factor, Vip, a surface protein. *EMBO J.* **24:**2827–2838.

42. Cai, S., D. Y. Kabuki, A. Y. Kuaye, T. G. Cargioli, M. S. Chung, R. Nielsen, and M. Wiedmann. 2002. Rational design of DNA sequence-based strategies for subtyping *Listeria monocytogenes*. *J. Clin. Microbiol.* **40:**3319–3325.

43. Camilli, A., H. Goldfine, and D. A. Portnoy. 1991. *Listeria monocytogenes* mutants lacking phosphatidylinositol-specific phospholipase C are avirulent. *J. Exp. Med.* **173:**751–754.

44. Camilli, A., L. G. Tilney, and D. A. Portnoy. 1993. Dual roles of PlcA in *Listeria monocytogenes* pathogenesis. *Mol. Microbiol.* **8:**143–157.

45. Centers for Disease Control and Prevention. 1989. Listeriosis associated with consumption of turkey franks. *Morb. Mortal. Wkly. Rep.* **38:**267–268.

46. Centers for Disease Control and Prevention. 2006. Preliminary FoodNet data on the incidence of infection with pathogens transmitted commonly through food—10 States, United States, 2005. *Morb. Mortal. Wkly. Rep.* **55:**392–395.

47. Centers for Disease Control and Prevention. 1996. *Preventing Foodborne Illness: Listeriosis*. U.S. Department of Health and Human Services, Public Health Service, Centers for Disease Control and Prevention, National Center for Infectious Diseases, Atlanta, Ga.

48. Chakraborty, T., F. Ebel, E. Domann, K. Niebuhr, B. Gerstel, S. Pistor, C. J. Temm-Grove, B. M. Jockusch, M. Reinhard, and U. Walter. 1995. A focal adhesion factor directly linking intracellularly motile *Listeria monocytogenes* and *Listeria ivanovii* to the actin-based cytoskeleton of mammalian cells. *EMBO J.* **14:**1314–1321.

49. Chen, Y., W. H. Ross, M. J. Gray, M. Wiedmann, R. C. Whiting, and V. N. Scott. 2006. Attributing risk to *Listeria monocytogenes* subgroups: dose response in relation to genetic lineages. *J. Food Prot.* **69:**335–344.

50. Chen, Y., W. H. Ross, V. N. Scott, and D. E. Gombas. 2003. *Listeria monocytogenes*: low levels equal low risk. *J. Food Prot.* **66:**570–577.

51. Chen, Y., W. Zhang, and S. J. Knabel. 2005. Multivirulence-locus sequence typing clarifies epidemiology of recent listeriosis outbreaks in the United States. *J. Clin. Microbiol.* **43:**5291–5294.

52. Colburn, K., C. Kaysner, C. Abeyta, Jr., and M. Wekell. 1990. *Listeria* species in a California estuarine environment. *Appl. Environ. Microbiol.* **56:**2007–2011.

53. Cossart, P. 2000. Actin-based motility of pathogens: the Arp2/3 complex is a central player. *Cell. Microbiol.* **2:**195–205.

54. Cossart, P., and J. Mengaud. 1989. *Listeria monocytogenes*: a model system for the molecular study of intracellular parasitism. *Mol. Biol. Med.* **6:**463–474.

55. Cossart, P., and P. J. Sansonetti. 2004. Bacterial invasion: the paradigms of enteroinvasive pathogens. *Science* **304:**242–248.

56. Cossart, P., M. F. Vicente, J. Mengaud, F. Baquero, J. C. Perez-Diaz, and P. Berche. 1989. Listeriolysin O is essential for virulence of *Listeria monocytogenes*: direct evidence obtained by gene complementation. *Infect. Immun.* **57:**3629–3636.

57. Cox, L., T. Kleiss, J. Cordier, C. Cordellana, P. Konkel, C. Pedrazzini, R. Beumer, and A. Siebenga. 1989. *Listeria* spp. in food processing, non-food and domestic environments. *Food Microbiol.* **6:**49–61.

58. Czuprynski, C. J., N. G. Faith, and H. Steinberg. 2003. A/J mice are susceptible and C57BL/6 mice are resistant to *Listeria monocytogenes* infection by intragastric inoculation. *Infect. Immun.* **71:**682–689.

59. Dabiri, G. A., J. M. Sanger, D. A. Portnoy, and F. S. Southwick. 1990. *Listeria monocytogenes* moves rapidly through the host-cell cytoplasm by inducing directional actin assembly. *Proc. Natl. Acad. Sci. USA* **87:**6068–6072.

60. Dalton, C. B., C. C. Austin, J. Sobel, P. S. Hayes, W. F. Bibb, L. M. Graves, B. Swaminathan, M. E. Proctor, and P. M. Griffin. 1997. An outbreak of gastroenteritis and fever due to *Listeria monocytogenes* in milk. *N. Engl. J. Med.* **336:**100–105.

61. Dalton, C. B., J. Gregory, M. D. Kirk, R. J. Stafford, R. Givney, E. Kraa, and D. Gould. 2004. Foodborne disease outbreaks in Australia, 1995 to 2000. *Commun. Dis. Intell.* **28:**211–224.

62. Danielsson-Tham, M. L., E. Eriksson, S. Helmersson, M. Leffler, L. Ludtke, M. Steen, S. Sorgjerd, and W. Tham. 2004. Causes behind a human cheese-borne outbreak of gastrointestinal listeriosis. *Foodborne Pathog. Dis.* **1:**153–159.

63. Davies, E., and M. Adams. 1994. Resistance of *Listeria monocytogenes* to the bacteriocin nisin. *Int. J. Food Microbiol.* **21:**341–347.

64. De Buyser, M. L., B. Dufour, M. Maire, and V. Lafarge. 2001. Implication of milk and milk products in food-borne diseases in France and in different industrialised countries. *Int. J. Food Microbiol.* **67:**1–17.

65. Decatur, A. L., and D. A. Portnoy. 2000. A PEST-like sequence in listeriolysin O essential for *Listeria monocytogenes* pathogenicity. *Science* **290:**992–995.

66. de Simon, M., and M. D. Ferrer. 1998. Initial numbers, serovars and phagevars of *Listeria monocytogenes* isolated in prepared foods in the city of Barcelona (Spain). *Int. J. Food Microbiol.* **44:**141–144.

67. de Valk, H., C. Jacquet, V. Goulet, V. Vaillant, A. Perra, F. Simon, J. C. Desenclos, P. Martin, and Listeria Surveillance Feasibility Study Participants. 2005. Surveillance of *Listeria* infections in Europe. *Euro Surveill.* **10:**251–255.

68. de Valk, H., V. Vaillant, C. Jacquet, J. Rocourt, F. Le Querrec, F. Stainer, N. Quelquejeu, O. Pierre, V. Pierre, J. C. Desenclos, and V. Goulet. 2001. Two consecutive nationwide outbreaks of listeriosis in France, October 1999-February 2000. *Am. J. Epidemiol.* **154:**944–950.

69. Dijkstra, R. 1982. The occurrence of *Listeria monocytogenes* in surface water of canals and lakes, in ditches of one big polder and in the effluents and canals of a sewage treatment plant. *Zentbl. Bakteriol. Hyg. Abt. 1 Orig. Reihe B* **176:**202–205.

70. Domann, E., M. Leimeister-Wachter, W. Goebel, and T. Chakraborty. 1991. Molecular cloning, sequencing, and identification of a metalloprotease gene from *Listeria monocytogenes* that is species specific and physically linked to the listeriolysin gene. *Infect. Immun.* **59:**65–72.

71. Domann, E., J. Wehland, M. Rohde, S. Pistor, M. Hartl, W. Goebel, M. Leimeister-Wachter, M. Wuenscher, and T. Chakraborty. 1992. A novel bacterial virulence gene in *Listeria monocytogenes* required for host cell microfilament interaction with homology to the proline-rich region of vinculin. *EMBO J.* **11:**1981–1990.

72. Dominguez, C., I. Gomez, and J. Zumalacarregui. 2001. Prevalence and contamination levels of *Listeria monocytogenes* in smoked fish and pate sold in Spain. *J. Food Prot.* **64:**2075–2077.

73. Doumith, M., C. Cazalet, N. Simoes, L. Frangeul, C. Jacquet, F. Kunst, P. Martin, P. Cossart, P. Glaser, and C. Buchrieser. 2004. New aspects regarding evolution and virulence of *Listeria monocytogenes* revealed by comparative genomics and DNA arrays. *Infect. Immun.* **72:**1072–1083.

74. Doumith, M., C. Jacquet, P. Gerner-Smidt, L. M. Graves, S. Loncarevic, T. Mathisen, A. Morvan, C. Salcedo, M. Torpdahl, J. A. Vazquez, and P. Martin. 2005. Multicenter validation of a multiplex PCR assay for differentiating the major *Listeria monocytogenes* serovars 1/2a, 1/2b, 1/2c, and 4b: toward an international standard. *J. Food Prot.* **68:**2648–2650.

75. Doyle, M. 1988. Effect of environmental and processing conditions on *Listeria monocytogenes*. *Food Technol.* **42:**169–171.

76. Dramsi, S., I. Biswas, L. Braun, E. Maguin, P. Mastroeni, and P. Cossart. 1995. Entry into hepatocytes requires expression of the *inlB* gene product. *Mol. Microbiol.* **16:**251–261.

77. Dramsi, S., F. Bourdichon, D. Cabanes, M. Lecuit, H. Fsihi, and P. Cossart. 2004. FbpA, a novel multifunctional *Listeria monocytogenes* virulence factor. *Mol. Microbiol.* **53:**639–649.

78. Dramsi, S., and P. Cossart. 2002. Listeriolysin O: a genuine cytolysin optimized for an intracellular parasite. *J. Cell Biol.* **156:**943–946.

79. Dussurget, O., D. Cabanes, P. Dehoux, M. Lecuit, C. Buchrieser, P. Glaser, and P. Cossart. 2002. *Listeria monocytogenes* bile salt hydrolase is a PrfA-regulated virulence factor involved in the intestinal and hepatic phases of listeriosis. *Mol. Microbiol.* **45:**1095–1106.

80. Dussurget, O., E. Dumas, C. Archambaud, I. Chafsey, C. Chambon, M. Hebraud, and P. Cossart. 2005. *Listeria monocytogenes* ferritin protects against multiple stresses and is required for virulence. *FEMS Microbiol. Lett.* **250:**253–261.

81. Dussurget, O., J. Pizarro-Cerda, and P. Cossart. 2004. Molecular determinants of *Listeria monocytogenes* virulence. *Annu. Rev. Microbiol.* **58:**587–610.

82. Elliott, E. L., and J. E. Kvenberg. 2000. Risk assessment used to evaluate the U.S. position on *Listeria monocytogenes* in seafood. *Int. J. Food Microbiol.* **62:**253–260.

83. Ericsson, H., A. Eklow, M. L. Danielsson-Tham, S. Loncarevic, L. O. Mentzing, I. Persson, H. Unnerstad, and W. Tham. 1997. An outbreak of listeriosis suspected to have been caused by rainbow trout. *J. Clin. Microbiol.* **35:**2904–2907.

84. Evans, M. R., B. Swaminathan, L. M. Graves, E. Altermann, T. R. Klaenhammer, R. C. Fink, S. Kernodle, and S. Kathariou. 2004. Genetic markers unique to *Listeria monocytogenes* serotype 4b differentiate epidemic clone II (hot dog outbreak strains) from other lineages. *Appl. Environ. Microbiol.* **70:**2383–2390.

85. Farber, J., E. Coates, N. Beausoleil, and J. Fournier. 1991. Feeding trials of *Listeria monocytogenes* with a nonhuman primate model. *J. Clin. Microbiol.* **29:**2606–2608.

86. Farber, J. M. 2000. Present situation in Canada regarding *Listeria monocytogenes* and ready-to-eat seafood products. *Int. J. Food Microbiol.* **62:**247–251.

87. Farber, J. M., E. M. Daley, M. T. MacKie, and B. Limerick. 2000. A small outbreak of listeriosis potentially linked to the consumption of imitation crab meat. *Lett. Appl. Microbiol.* **31:**100–104.

88. Farber, J. M., and J. Harwig. 1996. The Canadian position on *Listeria monocytogenes* in ready-to-eat foods. *Food Control* **7:**253–258.

89. Farber, J. M., and P. I. Peterkin. 1999. Incidence and behavior of *Listeria monocytogenes* in meat products, p. 505–564. *In* E. T. Ryser and E. H. Marth (ed.), *Listeria, Listeriosis and Food Safety*. Marcel Dekker, New York, N.Y.

90. Fenlon, D. R. 1999. *Listeria monocytogenes* in the natural environment, p. 21–37. *In* E. T. Ryser and E. H. Marth (ed.), *Listeria, Listeriosis and Food Safety*. Marcel Dekker, New York, N.Y.

91. Ferreira, A., D. Sue, C. P. O'Byrne, and K. J. Boor. 2003. Role of *Listeria monocytogenes* sigma(B) in survival of

lethal acidic conditions and in the acquired acid tolerance response. *Appl. Environ. Microbiol.* **69**:2692–2698.

92. **Fleming, D., S. Cochi, K. MacDonald, J. Brondum, P. Hayes, B. Plikaytis, M. Holmes, A. Audurier, C. Broome, and A. Reingold.** 1985. Pasteurized milk as a vehicle of infection in an outbreak of listeriosis. *N. Engl. J. Med.* **312**:404–407.

93. **Freitag, N., L. Rong, and D. Portnoy.** 1993. Regulation of the *prfA* transcriptional activator of *Listeria monocytogenes*: multiple promoter elements contribute to intracellular growth and cell-to-cell spread. *Infect. Immun.* **61**:2537–2544.

94. **Freitag, N. E.** 2000. Genetic tools for use with *Listeria monocytogenes*, p. 488–498. *In* V. A. Fischetti, R. P. Novick, J. J. Ferretti, D. A. Portnoy, and J. I. Rood (ed.), *Gram-Positive Pathogens.* ASM Press, Washington, D.C.

95. **Freitag, N. E., and K. E. Jacobs.** 1999. Examination of *Listeria monocytogenes* intracellular gene expression by using the green fluorescent protein of *Aequorea victoria*. *Infect. Immun.* **67**:1844–1852.

96. **Frye, C., and C. W. Donnelly.** 2005. Comprehensive survey of pasteurized fluid milk produced in the United States reveals a low prevalence of *Listeria monocytogenes*. *J. Food Prot.* **68**:973–979.

97. **Frye, D. M., R. Zweig, J. Sturgeon, M. Tormey, M. LeCavalier, I. Lee, L. Lawani, and L. Mascola.** 2002. An outbreak of febrile gastroenteritis associated with delicatessen meat contaminated with *Listeria monocytogenes*. *Clin. Infect. Dis.* **35**:943–949.

98. **Gahan, C. G., J. O'Mahony, and C. Hill.** 2001. Characterization of the *groESL* operon in *Listeria monocytogenes*: utilization of two reporter systems (*gfp* and *hly*) for evaluating in vivo expression. *Infect. Immun.* **69**:3924–3932.

99. **Gaillard, J., P. Berche, C. Frehel, E. Gouin, and P. Cossart.** 1991. Entry of *L. monocytogenes* into cells is mediated by internalin, a repeat protein reminiscent of surface antigens from Gram-positive cocci. *Cell* **65**:1127–1141.

100. **Gaillard, J. L., P. Berche, and P. Sansonetti.** 1986. Transposon mutagenesis as a tool to study the role of hemolysin in the virulence of *Listeria monocytogenes*. *Infect. Immun.* **52**:50–55.

101. **Gaillot, O., E. Pellegrini, S. Bregenholt, S. Nair, and P. Berche.** 2000. The ClpP serine protease is essential for the intracellular parasitism and virulence of *Listeria monocytogenes*. *Mol. Microbiol.* **35**:1286–1294.

102. **Garner, M. R., B. L. Njaa, M. Wiedmann, and K. J. Boor.** 2006. Sigma B contributes to *Listeria monocytogenes* gastrointestinal infection but not to systemic spread in the guinea pig infection model. *Infect. Immun.* **74**:876–886.

103. **Gasanov, U., D. Hughes, and P. M. Hansbro.** 2005. Methods for the isolation and identification of *Listeria* spp. and *Listeria monocytogenes*: a review. *FEMS Microbiol. Rev.* **29**:851–875.

104. **Gaulin, C., D. Ramsay, L. Ringuette, and J. Ismail.** 2003. First documented outbreak of *Listeria monocytogenes* in Quebec, 2002. *Can. Commun. Dis. Rep.* **29**:181–186.

105. **Gedde, M., D. Higgins, L. Tilney, and D. Portnoy.** 2000. Role of listeriolysin O in cell-to-cell spread of *Listeria monocytogenes*. *Infect. Immun.* **68**:999–1003.

106. **Geese, M., J. J. Loureiro, J. E. Bear, J. Wehland, F. B. Gertler, and A. S. Sechi.** 2002. Contribution of Ena/VASP proteins to intracellular motility of *Listeria* requires phosphorylation and proline-rich core but not F-actin binding or multimerization. *Mol. Biol. Cell* **13**:2383–2396.

107. **Geginat, G., M. Lalic, M. Kretschmar, W. Goebel, H. Hof, D. Palm, and A. Bubert.** 1998. Th1 cells specific for a secreted protein of *Listeria monocytogenes* are protective in vivo. *J. Immun.* **160**:6046–6055.

108. **Geginat, G., T. Nichterlein, M. Kretschmar, S. Schenk, H. Hof, M. Lalic-Multhaler, W. Goebel, and A. Bubert.** 1999. Enhancement of the *Listeria monocytogenes* p60–specific CD4 and CD8 T cell memory by nonpathogenic *Listeria innocua*. *J. Immun.* **162**:4781–4789.

109. **Genigeorgis, C., D. Dutulescu, and J. Fernandez Garayzabal.** 1989. Prevalence of *Listeria* spp. in poultry meat at the supermarket and slaughterhouse level. *J. Food Prot.* **52**:618–624.

110. **Gerner-Smidt, P., S. Ethelberg, P. Schiellerup, J. J. Christensen, J. Engberg, V. Fussing, A. Jensen, C. Jensen, A. M. Petersen, and B. G. Bruun.** 2005. Invasive listeriosis in Denmark 1994–2003: a review of 299 cases with special emphasis on risk factors for mortality. *Clin. Microbiol. Infect.* **11**:618–624.

111. **Glaser, P., L. Frangeul, C. Buchrieser, C. Rusniok, A. Amend, F. Baquero, P. Berche, H. Bloecker, P. Brandt, T. Chakraborty, A. Charbit, F. Chetouani, E. Couve, A. de Daruvar, P. Dehoux, E. Domann, G. Dominguez-Bernal, E. Duchaud, L. Durant, O. Dussurget, K. D. Entian, H. Fsihi, F. G. Portillo, P. Garrido, L. Gautier, W. Goebel, N. Gomez-Lopez, T. Hain, J. Hauf, D. Jackson, L. M. Jones, U. Kaerst, J. Kreft, M. Kuhn, F. Kunst, G. Kurapkat, E. Madueno, A. Maitournam, J. M. Vicente, E. Ng, H. Nedjari, G. Nordsiek, S. Novella, B. de Pablos, J. C. Perez-Diaz, R. Purcell, B. Remmel, M. Rose, T. Schlueter, N. Simoes, A. Tierrez, J. A. Vazquez-Boland, H. Voss, J. Wehland, and P. Cossart.** 2001. Comparative genomics of *Listeria* species. *Science* **294**:849–852.

112. **Glass, K. A., and M. P. Doyle.** 1989. Fate of *Listeria monocytogenes* in processed meat products during refrigerated storage. *Appl. Environ. Microbiol.* **55**:1565–1569.

113. **Goetz, M., A. Bubert, G. Wang, I. Chico-Calero, J. A. Vazquez-Boland, M. Beck, J. Slaghuis, A. A. Szalay, and W. Goebel.** 2001. Microinjection and growth of bacteria in the cytosol of mammalian host cells. *Proc. Natl. Acad. Sci. USA* **98**:12221–12226.

114. **Goldfine, H., and S. J. Wadsworth.** 2002. Macrophage intracellular signaling induced by *Listeria monocytogenes*. *Microbes Infect.* **4**:1335–1343.

115. **Goldfine, H., S. J. Wadsworth, and N. C. Johnston.** 2000. Activation of host phospholipases C and D in macrophages after infection with *Listeria monocytogenes*. *Infect. Immun.* **68**:5735–5741.

116. **Gombas, D. E., Y. Chen, R. S. Clavero, and V. N. Scott.** 2003. Survey of *Listeria monocytogenes* in ready-to-eat foods. *J. Food Prot.* **66**:559–569.

117. **Gonzalez-Zorn, B., G. Dominguez-Bernal, M. Suarez, M. T. Ripio, Y. Vega, S. Novella, and J. A. Vazquez-Boland.** 1999. The smcL gene of *Listeria ivanovii* encodes a sphingomyelinase C that mediates bacterial escape from the phagocytic vacuole. *Mol. Microbiol.* **33**:510–523.

118. Gottlieb, S. L., E. C. Newbern, P. M. Griffin, L. M. Graves, R. M. Hoekstra, N. L. Baker, S. B. Hunter, K. G. Holt, F. Ramsey, M. Head, P. Levine, G. Johnson, D. Schoonmaker-Bopp, V. Reddy, L. Kornstein, M. Gerwel, J. Nsubuga, L. Edwards, S. Stonecipher, S. Hurd, D. Austin, M. A. Jefferson, S. D. Young, K. Hise, E. D. Chernak, and J. Sobel. 2006. Multistate outbreak of listeriosis linked to turkey deli meat and subsequent changes in US regulatory policy. *Clin. Infect. Dis.* **42:**29–36.

119. Gouin, E., C. Egile, P. Dehoux, V. Villiers, J. Adams, F. Gertler, R. Li, and P. Cossart. 2004. The RickA protein of *Rickettsia conorii* activates the Arp2/3 complex. *Nature* **427:**457–461.

120. Gouin, E., H. Gantelet, C. Egile, I. Lasa, H. Ohayon, V. Villiers, P. Gounon, P. J. Sansonetti, and P. Cossart. 1999. A comparative study of the actin-based motility of the pathogenic bacteria *Listeria monocytogenes*, *Shigella flexneri* and *Rickettsia conorii*. *J. Cell Sci.* **112:**1697–1708.

121. Gouin, E., M. D. Welch, and P. Cossart. 2005. Actin-based motility of intracellular pathogens. *Curr. Opin. Microbiol.* **8:**35–45.

122. Goulet, V., H. de Valk, O. Pierre, F. Stainer, J. Rocourt, V. Vaillant, C. Jacquet, and J. C. Desenclos. 2001. Effect of prevention measures on incidence of human listeriosis, France, 1987–1997. *Emerg. Infect. Dis.* **7:**983–989.

123. Goulet, V., C. Jacquet, V. Vaillant, I. Rebiere, E. Mouret, E. Lorente, F. Steiner, and J. Rocourt. 1995. Listeriosis from consumption of raw milk cheese. *Lancet* **345:**1581–1582.

124. Grau, F., and P. Vanderlinde. 1992. Occurrence, numbers and growth of *Listeria monocytogenes* on some vacuum-packaged processed meats. *J. Food Prot.* **55:**4–7.

125. Gravani, R. 1999. *Listeria* in food processing facilities, p. 657–709. *In* E. T. Ryser and E. H. Marth (ed.), *Listeria, Listeriosis and Food Safety.* Marcel Dekker, New York, N.Y.

126. Graves, L. M., S. B. Hunter, A. R. Ong, D. Schoonmaker-Bopp, K. Hise, L. Kornstein, W. E. DeWitt, P. S. Hayes, E. Dunne, P. Mead, and B. Swaminathan. 2005. Microbiological aspects of the investigation that traced the 1998 outbreak. *J. Clin. Microbiol.* **43:**2350–2355.

127. Graves, L. M., and B. Swaminathan. 2005. PulseNet's step-by-step laboratory protocol for molecular subtyping of *Listeria monocytogenes* by macrorestriction and pulsed-field gel electrophoresis, p. 57–72. *In* C. Adley (ed.), *Food-Borne Pathogens,* vol. 21. Humana Press, Totowa, N.J.

128. Graves, L. M., B. Swaminathan, and S. Hunter. 1999. Subtyping *Listeria monocytogenes,* p. 279–297. *In* E. T. Ryser and E. H. Marth (ed.), *Listeria, Listeriosis, and Food Safety,* 2nd ed. Marcel Dekker, New York, N.Y.

129. Gray, M. J., R. N. Zadoks, E. D. Fortes, B. Dogan, S. Cai, Y. Chen, V. N. Scott, D. E. Gombas, K. J. Boor, and M. Wiedmann. 2004. *Listeria monocytogenes* isolates from foods and humans form distinct but overlapping populations. *Appl. Environ. Microbiol.* **70:**5833–5841.

130. Greiffenberg, L., W. Goebel, K. S. Kim, J. Daniels, and M. Kuhn. 2000. Interaction of *Listeria monocytogenes* with human brain microvascular endothelial cells: an electron microscopic study. *Infect. Immun.* **68:**3275–3279.

131. Grif, K., G. Patscheider, M. P. Dierich, and F. Allerberger. 2003. Incidence of fecal carriage of *Listeria monocytogenes* in three healthy volunteers: a one-year prospective stool survey. *Eur. J. Clin. Microbiol. Infect. Dis.* **22:**16–20.

132. Grundling, A., M. D. Gonzalez, and D. E. Higgins. 2003. Requirement of the *Listeria monocytogenes* broad-range phospholipase PC-PLC during infection of human epithelial cells. *J. Bacteriol.* **185:**6295–6307.

133. Guenich, H., H. Muller, A. Schrettenbrunner, and H. Seeliger. 1985. The occurrence of different *Listeria* species in municipal waste water. *Zentbl. Bakteriol. Hyg. Abt. 1 Orig. Reihe B* **181:**563–565.

134. Gutekunst, K. A., L. Pine, E. White, S. Kathariou, and G. M. Carlone. 1992. A filamentous-like mutant of *Listeria monocytogenes* with reduced expression of a 60-kilodalton extracellular protein invades and grows in 3T6 and Caco-2 cells. *Can. J. Microbiol.* **38:**843–851.

135. Hanawa, T., M. Fukuda, H. Kawakami, H. Hirano, S. Kamiya, and T. Yamamoto. 1999. The *Listeria monocytogenes* DnaK chaperone is required for stress tolerance and efficient phagocytosis with macrophages. *Cell Stress Chaperones* **4:**118–128.

136. Hanawa, T., T. Yamamoto, and S. Kamiya. 1995. *Listeria monocytogenes* can grow in macrophages without the aid of proteins induced by environmental stresses. *Infect. Immun.* **63:**4595–4599.

137. Harty, J. T., and E. G. Pamer. 1995. CD8 T lymphocytes specific for the secreted p60 antigen protect against *Listeria monocytogenes* infection. *J. Immunol.* **154:**4642–4650.

138. Hebraud, M., and J. Guzzo. 2000. The main cold shock protein of *Listeria monocytogenes* belongs to the family of ferritin-like proteins. *FEMS Microbiol. Lett.* **190:**29–34.

139. Hess, J., I. Gentschev, G. Szalay, C. Ladel, A. Bubert, W. Goebel, and S. H. Kaufmann. 1995. *Listeria monocytogenes* p60 supports host cell invasion by and in vivo survival of attenuated *Salmonella typhimurium*. *Infect. Immun.* **63:**2047–2053.

140. Ho, J. L., K. N. Shands, G. Friedland, P. Eckind, and D. W. Fraser. 1986. An outbreak of type 4b *Listeria monocytogenes* infection involving patients from eight Boston hospitals. *Arch. Intern. Med.* **146:**520–524.

141. Hodgson, D. A. 2000. Generalized transduction of serotype 1/2 and serotype 4b strains of *Listeria monocytogenes*. *Mol. Microbiol.* **35:**312–323.

142. Hoffman, A. D., K. L. Gall, D. M. Norton, and M. Wiedmann. 2003. *Listeria monocytogenes* contamination patterns for the smoked fish processing environment and for raw fish. *J. Food Prot.* **66:**52–60.

143. Huss, H. H., A. Reilly, and P. K. Ben Embarek. 2000. Prevention and control of safety hazards in cold smoked salmon production. *Food Control* **11:**149–156.

144. Husu, J., S. Sivela, and A. Rauramaa. 1990. Prevalence of *Listeria* species as related to chemical quality of farm-ensiled grass. *Grass Forage Sci.* **45:**309–314.

145. Inoue, S., A. Nakama, Y. Arai, Y. Kokubo, T. Maruyama, A. Saito, T. Yoshida, M. Terao, S. Yamamoto, and S. Kumagai. 2000. Prevalence and contamination levels of *Listeria monocytogenes* in retail foods in Japan. *Int. J. Food Microbiol.* **59:**73–77.

146. Jacobs, T., A. Darji, N. Frahm, M. Rohde, J. Wehland, T. Chakraborty, and S. Weiss. 1998. Listeriolysin O: cholesterol inhibits cytolysis but not binding to cellular membranes. *Mol. Microbiol.* **28**:1081–1089.

147. Jacquet, C., B. Catimel, R. Brosch, C. Buchrieser, P. Dehaumont, V. Goulet, V. Lepoutre, P. Veit, and J. Rocourt. 1995. Investigations related to the epidemic strain involved in the French listeriosis outbreak in 1992. *Appl. Environ. Microbiol.* **61**:2242–2246.

148. Jacquet, C., M. Doumith, J. I. Gordon, P. M. Martin, P. Cossart, and M. Lecuit. 2004. A molecular marker for evaluating the pathogenic potential of foodborne *Listeria monocytogenes*. *J. Infect. Dis.* **189**:2094–2100.

149. Jacquet, C., E. Gouin, D. Jeannel, P. Cossart, and J. Rocourt. 2002. Expression of ActA, Ami, InlB, and listeriolysin O in *Listeria monocytogenes* of human and food origin. *Appl. Environ. Microbiol.* **68**:616–622.

150. Jeffers, G. T., J. L. Bruce, P. L. McDonough, J. Scarlett, K. J. Boor, and M. Wiedmann. 2001. Comparative genetic characterization of *Listeria monocytogenes* isolates from human and animal listeriosis cases. *Microbiology* **147**:1095–1104.

151. Jensen, A. 1993. Excretion of *Listeria monocytogenes* in faeces after listeriosis: rate, quantity and duration. *Med. Microbiol. Lett.* **2**:176–182.

152. Jeong, D., and J. Frank. 1994. Growth of *Listeria monocytogenes* at 10°C in biofilms with microorganisms isolated from meat and dairy processing environments. *J. Food Prot.* **57**:576–586.

153. Jinneman, K. C., M. M. Wekell, and M. W. Eklund. 1999. Incidence and behavior of *Listeria monocytogenes* in fish and seafood, p. 601–630. *In* E. T. Ryser and E. H. Marth (ed.), *Listeria, Listeriosis and Food Safety.* Marcel Dekker, New York, N.Y.

154. Johansson, J., P. Mandin, A. Renzoni, C. Chiaruttini, M. Springer, and P. Cossart. 2002. An RNA thermosensor controls expression of virulence genes in *Listeria monocytogenes*. *Cell* **110**:551–561.

155. Jonquieres, R., H. Bierne, F. Fiedler, P. Gounon, and P. Cossart. 1999. Interaction between the protein InlB of *Listeria monocytogenes* and lipoteichoic acid: a novel mechanism of protein association at the surface of Grampositive bacteria. *Mol. Microbiol.* **34**:902–914.

156. Jonquieres, R., H. Bierne, J. Mengaud, and P. Cossart. 1998. The *inlA* gene of *Listeria monocytogenes* LO28 harbors a nonsense mutation resulting in release of internalin. *Infect. Immun.* **66**:3420–3422.

157. Jonquieres, R., J. Pizarro-Cerda, and P. Cossart. 2001. Synergy between the N- and C-terminal domains of InlB for efficient invasion of non-phagocytic cells by *Listeria monocytogenes*. *Mol. Microbiol.* **42**:955–965.

158. Kallipolitis, B. H., and H. Ingmer. 2001. *Listeria monocytogenes* response regulators important for stress tolerance and pathogenesis. *FEMS Microbiol. Lett.* **204**:111–115.

159. Kathariou, S., P. Metz, H. Hof, and W. Goebel. 1987. Tn916-induced mutations in the hemolysin determinant affecting virulence of *Listeria monocytogenes*. *J. Bacteriol.* **169**:1291–1297.

160. Khelef, N., M. Lecuit, C. Buchrieser, D. Cabanes, O. Dussurget, and P. Cossart. 2004. *Listeria monocytogenes* and the Genus *Listeria*. *In* M. Dworkin, S. Falkow, E. Rosenberg, K.-H. Schleifer, and E. Stackebrandt (ed.), *The Prokaryotes: An Evolving Electronic Resource for the Microbiological Community*, 3rd ed. Springer, New York, N.Y. [Online.] http://141.150.157.117:8080/prokPUB/index.htm. Accessed 31 December 2005.

161. Klarsfeld, A., P. Goossens, and P. Cossart. 1994. Five *Listeria monocytogenes* preferentially expressed in mammalian cells. *Mol. Microbiol.* **13**:585–597.

162. Ko, R., and L. T. Smith. 1999. Identification of an ATP-driven, osmoregulated glycine betaine transport system in *Listeria monocytogenes*. *Appl. Environ. Microbiol.* **65**:4040–4048.

163. Kocks, C., E. Gouin, M. Tabouret, P. Berche, H. Ohayon, and P. Cossart. 1992. *Listeria monocytogenes* induced actin assembly requires the ActA gene product, a surface protein. *Cell* **68**:521–531.

164. Kolb-Maurer, A., S. Pilgrim, E. Kampgen, A. D. McLellan, E. B. Brocker, W. Goebel, and I. Gentschev. 2001. Antibodies against listerial protein 60 act as an opsonin for phagocytosis of *Listeria monocytogenes* by human dendritic cells. *Infect. Immun.* **69**:3100–3109.

165. Koutsoumanis, K. P., and J. N. Sofos. 2005. Effect of inoculum size on the combined temperature, pH and aw limits for growth of *Listeria monocytogenes*. *Int. J. Food Microbiol.* **104**:83–91.

166. Kuhn, M., and W. Goebel. 1989. Identification of an extracellular protein of *Listeria monocytogenes* possibly involved in intracellular uptake by mammalian cells. *Infect. Immun.* **57**:55–61.

167. Lappi, V. R., J. Thimothe, K. K. Nightingale, K. Gall, V. N. Scott, and M. Wiedmann. 2004. Longitudinal studies on *Listeria* in smoked fish plants: impact of intervention strategies on contamination patterns. *J. Food Prot.* **67**:2500–2514.

168. Larsen, A., and B. Norrung. 1993. Inhibition of *Listeria monocytogenes* by bavaricin A, a bacteriocin produced by *Lactobacillus bavaricus* Ml401. *Lett. Appl. Microbiol.* **17**:132–134.

169. Lasa, I., V. David, E. Gouin, J. B. Marchand, and P. Cossart. 1995. The amino-terminal part of ActA is critical for the actin-based motility of *Listeria monocytogenes*; the central proline-rich region acts as a stimulator. *Mol. Microbiol.* **18**:425–436.

170. Lasa, I., E. Gouin, M. Goethals, K. Vancompernolle, V. David, J. Vandekerckhove, and P. Cossart. 1997. Identification of two regions in the N-terminal domain of ActA involved in the actin comet tail formation by *Listeria monocytogenes*. *EMBO J.* **16**:1531–1540.

171. Lauer, P., M. Y. Chow, M. J. Loessner, D. A. Portnoy, and R. Calendar. 2002. Construction, characterization, and use of two *Listeria monocytogenes* site-specific phage integration vectors. *J. Bacteriol.* **184**:4177–4186.

172. Lecuit, M., and C. Cossart. 2001. *Listeria monocytogenes*, p. 1437–1462. *In* S. Sussman (ed.), *Molecular Medical Microbiology*, vol. 2. Academic Press, London, United Kingdom.

173. Lecuit, M., and P. Cossart. 2002. Genetically-modified-animal models for human infections: the *Listeria* paradigm. *Trends Mol. Med.* **8:**537–542.

174. Lecuit, M., S. Dramsi, C. Gottardi, M. Fedor-Chaiken, B. Gumbiner, and P. Cossart. 1999. A single amino acid in E-cadherin responsible for host specificity towards the human pathogen *Listeria monocytogenes*. *EMBO J.* **18:**3956–3963.

175. Lecuit, M., D. M. Nelson, S. D. Smith, H. Khun, M. Huerre, M. C. Vacher-Lavenu, J. I. Gordon, and P. Cossart. 2004. Targeting and crossing of the human maternofetal barrier by *Listeria monocytogenes*: role of internalin interaction with trophoblast E-cadherin. *Proc. Natl. Acad. Sci. USA* **101:**6152–6157.

176. Lecuit, M., H. Ohayon, L. Braun, J. Mengaud, and P. Cossart. 1997. Internalin of *Listeria monocytogenes* with an intact leucine-rich repeat region is sufficient to promote internalization. *Infect. Immun.* **65:**5309–5319.

177. Lecuit, M., S. Vandormael-Pournin, J. Lefort, M. Huerre, P. Gounon, C. Dupuy, C. Babinet, and P. Cossart. 2001. A transgenic model for listeriosis: role of internalin in crossing the intestinal barrier. *Science* **292:**1722–1725.

178. Leimeister-Wachter, M., E. Domann, and T. Chakraborty. 1992. The expression of virulence genes in *Listeria monocytogenes* is thermoregulated. *J. Bacteriol.* **174:**947–952.

179. Leimeister-Wachter, M., C. Haffner, E. Domann, W. Goebel, and T. Chakraborty. 1990. Identification of a gene that positively regulates expression of listeriolysin, the major virulence factor of *Listeria monocytogenes*. *Proc. Natl. Acad. Sci. USA* **87:**8336–8340.

180. Lenz, L. L., S. Mohammadi, A. Geissler, and D. A. Portnoy. 2003. SecA2-dependent secretion of autolytic enzymes promotes *Listeria monocytogenes* pathogenesis. *Proc. Natl. Acad. Sci. USA* **100:**12432–12437.

181. Li, Z., J. Dai, H. Zheng, B. Liu, and M. Caudill. 2002. An integrated view of the roles and mechanisms of heat shock protein gp96-peptide complex in eliciting immune response. *Front. Biosci.* **7:**731–751.

182. Lin, C. M., K. Takeuchi, L. Zhang, C. B. Dohm, J. D. Meyer, P. A. Hall, and M. P. Doyle. 2006. Cross-contamination between processing equipment and deli meats by *Listeria monocytogenes*. *J. Food Prot.* **69:**71–79.

183. Linnan, M., L. Mascola, X. Lou, V. Goulet, S. May, C. Salminen, D. Hird, M. Yonekura, P. Hayes, R. Weaver, A. Audurier, B. Plikaytis, S. Fannin, A. Kleks, and C. Broome. 1988. Epidemic listeriosis associated with Mexican-style cheese. *N. Engl. J. Med.* **319:**823–828.

184. Liu, D., M. L. Lawrence, L. Gorski, R. E. Mandrell, A. J. Ainsworth, and F. W. Austin. 2006. *Listeria monocytogenes* serotype 4b strains belonging to lineages I and III possess distinct molecular features. *J. Clin. Microbiol.* **44:**214–217.

185. Lorber, B. 1997. Listeriosis. *Clin. Infect. Dis.* **24:**1–9.

186. Lou, Y., and A. E. Yousef. 1999. Characteristics of *Listeria monocytogenes* important to food processors, p. 131–224. *In* E. T. Ryser and E. H. Marth (ed.), *Listeria, Listeriosis, and Food Safety*, 2nd ed. Marcel Dekker, New York, N.Y.

187. Low, J. C., F. Wright, J. McLauchlin, and W. Donachie. 1993. Serotyping and distribution of Listeria isolates from cases of ovine listeriosis. *Vet. Rec.* **133:**165–166.

188. Luchansky, J. B., P. M. Muriana, and T. R. Klaenhammer. 1988. Application of electroporation for transfer of plasmid DNA to *Lactobacillus, Lactococcus, Leuconostoc, Listeria, Pediococcus, Bacillus, Staphylococcus, Enterococcus* and *Propionibacterium*. *Mol. Microbiol.* **2:**637–646.

189. Lunden, J. M., T. J. Autio, and H. J. Korkeala. 2002. Transfer of persistent *Listeria monocytogenes* contamination between food-processing plants associated with a dicing machine. *J. Food Prot.* **65:**1129–1133.

190. Lunden, J. M., T. J. Autio, A. M. Sjoberg, and H. J. Korkeala. 2003. Persistent and nonpersistent *Listeria monocytogenes* contamination in meat and poultry processing plants. *J. Food Prot.* **66:**2062–2069.

191. Lyytikainen, O., T. Autio, R. Maijala, P. Ruutu, T. Honkanen-Buzalski, M. Miettinen, M. Hatakka, J. Mikkola, V. J. Anttila, T. Johansson, L. Rantala, T. Aalto, H. Korkeala, and A. Siitonen. 2000. An outbreak of *Listeria monocytogenes* serotype 3a infections from butter in Finland. *J. Infect. Dis.* **181:**1838–1841.

192. MacDonald, P. D., R. E. Whitwam, J. D. Boggs, J. N. MacCormack, K. L. Anderson, J. W. Reardon, J. R. Saah, L. M. Graves, S. B. Hunter, and J. Sobel. 2005. Outbreak of listeriosis among Mexican immigrants as a result of consumption of illicitly produced Mexican-style cheese. *Clin. Infect. Dis.* **40:**677–682.

193. Makino, S. I., K. Kawamoto, K. Takeshi, Y. Okada, M. Yamasaki, S. Yamamoto, and S. Igimi. 2005. An outbreak of food-borne listeriosis due to cheese in Japan, during 2001. *Int. J. Food Microbiol.* **104:**189–196.

194. Mandin, P., H. Fsihi, O. Dussurget, M. Vergassola, E. Milohanic, A. Toledo-Arana, I. Lasa, J. Johansson, and P. Cossart. 2005. VirR, a response regulator critical for *Listeria monocytogenes* virulence. *Mol. Microbiol.* **57:**1367–1380.

195. Marino, M., M. Banerjee, R. Jonquieres, P. Cossart, and P. Ghosh. 2002. GW domains of the *Listeria monocytogenes* invasion protein InlB are SH3-like and mediate binding to host ligands. *EMBO J.* **21:**5623–5634.

196. Marino, M., L. Braun, P. Cossart, and P. Ghosh. 1999. Structure of the InlB leucine-rich repeats, a domain that triggers host cell invasion by the bacterial pathogen *L. monocytogenes*. *Mol. Cell* **4:**1063–1072.

197. Marquis, H., H. Bouwer, D. Hinrichs, and D. Portnoy. 1993. Intracytoplasmic growth and virulence of *Listeria monocytogenes* auxotrophic mutants. *Infect. Immun.* **61:**3756–3760.

198. Marquis, H., V. Doshi, and D. A. Portnoy. 1995. The broad-range phospholipase C and a metalloprotease mediate listeriolysin O-independent escape of *Listeria monocytogenes* from a primary vacuole in human epithelial cells. *Infect. Immun.* **63:**4531–4534.

199. Mascola, L., F. Sorvillo, V. Goulet, B. Hall, R. Weaver, and M. Linnan. 1992. Fecal carriage of *Listeria monocytogenes*—observations during a community wide, common-source outbreak. *Clin. Infect. Dis.* **15:**557–558.

200. Matthieu, F., M. Michel, A. Lebrihi, and G. Lefebvre. 1994. Effect of the bacteriocin carnocin CP5 and of the producing strain *Carnobacterium piscicola* CP5 on the viability of *Listeria monocytogenes* ATCC 15313 in salt solution, broth and skimmed milk, at various incubation temperatures. *Int. J. Food Microbiol.* **22:**155–172.

201. McLauchlin, J. 1997. The discovery of *Listeria. PHLS Microbiol. Dig.* **14:**76–78.

202. McLauchlin, J., S. Hall, S. Velani, and R. Gilbert. 1991. Human listeriosis and pate—a possible association. *Br. Med. J.* **303:**773–775.

203. McLauchlin, J., and P. Hoffman. 1989. Neonatal cross-infection from *Listeria monocytogenes. Commun. Dis. Rep.* **6:**3–4.

204. McLauchlin, J., R. T. Mitchell, W. J. Smerdon, and K. Jewell. 2004. *Listeria monocytogenes* and listeriosis: a review of hazard characterisation for use in microbiological risk assessment of foods. *Int. J. Food Microbiol.* **92:**15–33.

205. McLaughlan, A. M., and S. J. Foster. 1998. Molecular characterization of an autolytic amidase of *Listeria monocytogenes* EGD. *Microbiology* **144:**1359–1367.

206. Mead, P. S., E. F. Dunne, L. Graves, M. Wiedmann, M. Patrick, S. Hunter, E. Salehi, F. Mostashari, A. Craig, P. Mshar, T. Bannerman, B. D. Sauders, P. Hayes, W. Dewitt, P. Sparling, P. Griffin, D. Morse, L. Slutsker, and B. Swaminathan. 2005. Nationwide outbreak of listeriosis due to contaminated meat. *Epidemiol. Infect.* **134:**744–751.

207. Meinersmann, R. J., R. W. Phillips, M. Wiedmann, and M. E. Berrang. 2004. Multilocus sequence typing of *Listeria monocytogenes* by use of hypervariable genes reveals clonal and recombination histories of three lineages. *Appl. Environ. Microbiol.* **70:**2193–2203.

208. Mengaud, J., C. Braun-Breton, and P. Cossart. 1991. Identification of phosphatidylinositol-specific phospholipase C activity in *Listeria monocytogenes*: a novel type of virulence factor? *Mol. Microbiol.* **5:**367–372.

209. Mengaud, J., J. Chenevert, C. Geoffroy, J. L. Gaillard, and P. Cossart. 1987. Identification of the structural gene encoding the SH-activated hemolysin of *Listeria monocytogenes*: listeriolysin O is homologous to streptolysin O and pneumolysin. *Infect. Immun.* **55:**3225–3227.

210. Mengaud, J., S. Dramsi, E. Gouin, J. Vasquez-Boland, G. Milon, and P. Cossart. 1991. Pleiotropic control of *Listeria monocytogenes* virulence factors by a gene which is autoregulated. *Mol. Microbiol.* **5:**2273–2283.

211. Mengaud, J., H. Ohayon, P. Gounon, R. M. Mege, and P. Cossart. 1996. E-cadherin is the receptor for internalin, a surface protein required for entry of *L. monocytogenes* into epithelial cells. *Cell* **84:**923–932.

212. Miettinen, H., and G. Wirtanen. 2005. Prevalence and location of *Listeria monocytogenes* in farmed rainbow trout. *Int. J. Food Microbiol.* **104:**135–143.

213. Miettinen, M. K., A. Siitonen, P. Heiskanen, H. Haajanen, K. J. Bjorkroth, and H. J. Korkeala. 1999. Molecular epidemiology of an outbreak of febrile gastroenteritis caused by *Listeria monocytogenes* in cold-smoked rainbow trout. *J. Clin. Microbiol.* **37:**2358–2360.

214. Milohanic, E., P. Glaser, J. Y. Coppee, L. Frangeul, Y. Vega, J. A. Vazquez-Boland, F. Kunst, P. Cossart, and C. Buchrieser. 2003. Transcriptome analysis of *Listeria monocytogenes* identifies three groups of genes differently regulated by PrfA. *Mol. Microbiol.* **47:**1613–1625.

215. Milohanic, E., R. Jonquieres, P. Cossart, P. Berche, and J. L. Gaillard. 2001. The autolysin Ami contributes to the adhesion of *Listeria monocytogenes* to eukaryotic cells via its cell wall anchor. *Mol. Microbiol.* **39:**1212–1224.

216. Modi, K., M. Chikindas, and T. Montville. 2000. Sensitivity of nisin-resistant *Listeria monocytogenes* to heat and the synergistic action of heat and nisin. *Lett. Appl. Microbiol.* **30:**249–253.

217. Munder, A., A. Zelmer, A. Schmiedl, K. E. Dittmar, M. Rohde, M. Dorsch, K. Otto, H. J. Hedrich, B. Tummler, S. Weiss, and T. Tschernig. 2005. Murine pulmonary infection with *Listeria monocytogenes*: differential susceptibility of BALB/c, C57BL/6 and DBA/2 mice. *Microbes Infect.* **7:**600–611.

218. Nadon, C. A., B. M. Bowen, M. Wiedmann, and K. J. Boor. 2002. Sigma B contributes to PrfA-mediated virulence in *Listeria monocytogenes. Infect. Immun.* **70:**3948–3952.

219. Nair, S., C. Frehel, L. Nguyen, V. Escuyer, and P. Berche. 1999. ClpE, a novel member of the HSP100 family, is involved in cell division and virulence of *Listeria monocytogenes. Mol. Microbiol.* **31:**185–196.

220. Niebuhr, K., F. Ebel, R. Frank, M. Reinhard, E. Domann, U. D. Carl, U. Walter, F. B. Gertler, J. Wehland, and T. Chakraborty. 1997. A novel proline-rich motif present in ActA of *Listeria monocytogenes* and cytoskeletal proteins is the ligand for the EVH1 domain, a protein module present in the Ena/VASP family. *EMBO J.* **16:**5433–5444.

221. Norrung, B., J. K. Andersen, and J. Schlundt. 1999. Incidence and control of *Listeria monocytogenes* in foods in Denmark. *Int. J. Food Microbiol.* **53:**195–203.

222. Norton, D. M., J. M. Scarlett, K. Horton, D. Sue, J. Thimothe, K. J. Boor, and M. Wiedmann. 2001. Characterization and pathogenic potential of *Listeria monocytogenes* isolates from the smoked fish industry. *Appl. Environ. Microbiol.* **67:**646–653.

223. Olier, M., D. Garmyn, S. Rousseaux, J. P. Lemaitre, P. Piveteau, and J. Guzzo. 2005. Truncated internalin A and asymptomatic *Listeria monocytogenes* carriage: in vivo investigation by allelic exchange. *Infect. Immun.* **73:**644–648.

224. Olier, M., F. Pierre, S. Rousseaux, J. P. Lemaitre, A. Rousset, P. Piveteau, and J. Guzzo. 2003. Expression of truncated Internalin A is involved in impaired internalization of some *Listeria monocytogenes* isolates carried asymptomatically by humans. *Infect. Immun.* **71:**1217–1224.

225. Oliver, S. P., B. M. Jayarao, and R. A. Almeida. 2005. Foodborne pathogens in milk and the dairy farm environment: food safety and public health implications. *Foodborne Pathog. Dis.* **2:**115–129.

226. Olsen, S. J., M. Patrick, S. B. Hunter, V. Reddy, L. Kornstein, W. R. MacKenzie, K. Lane, S. Bidol, G. A. Stoltman, D. M. Frye, I. Lee, S. Hurd, T. F. Jones, T. N. LaPorte, W. Dewitt, L. Graves, M. Wiedmann, D. J. Schoonmaker-Bopp, A. J. Huang, C. Vincent, A. Bugenhagen, J. Corby,

E. R. Carloni, M. E. Holcomb, R. F. Woron, S. M. Zansky, G. Dowdle, F. Smith, S. Ahrabi-Fard, A. R. Ong, N. Tucker, N. A. Hynes, and P. Mead. 2005. Multistate outbreak of *Listeria monocytogenes* infection linked to delicatessen turkey meat. *Clin. Infect. Dis.* **40:**962–967.

227. Pandiripally, V. K., D. G. Westbrook, G. R. Sunki, and A. K. Bhunia. 1999. Surface protein p104 is involved in adhesion of *Listeria monocytogenes* to human intestinal cell line, Caco-2. *J. Med. Microbiol.* **48:**117–124.

228. Peerschke, E. I., K. B. Reid, and B. Ghebrehiwet. 1994. Identification of a novel 33-kDa C1q-binding site on human blood platelets. *J. Immunol.* **152:**5896–5901.

229. Pilgrim, S., A. Kolb-Maurer, I. Gentschev, W. Goebel, and M. Kuhn. 2003. Deletion of the gene encoding p60 in *Listeria monocytogenes* leads to abnormal cell division and loss of actin-based motility. *Infect. Immun.* **71:**3473–3484.

230. Pinner, R., A. Schuchat, B. Swaminathan, P. Hayes, K. Deaver, R. Weaver, B. Plikaytis, M. Reeves, C. Broome, and J. Wenger. 1992. Role of foods in sporadic listeriosis. 2. Microbiologic and epidemiologic investigation. *JAMA* **267:**2046–2050.

231. Pizarro-Cerda, J., M. Lecuit, and P. Cossart. 2002. Measuring and analysing invasion of mammalian cells by bacterial pathogens: the *Listeria monocytogenes* system. *Methods Microbiol.* **31:**161–177.

232. Portnoy, D. A., P. S. Jacks, and D. J. Hinrichs. 1988. Role of hemolysin for the intracellular growth of *Listeria monocytogenes*. *J. Exp. Med.* **167:**1459–1471.

233. Proctor, M. E., R. Brosch, J. W. Mellen, L. A. Garrett, C. W. Kaspar, and J. B. Luchansky. 1995. Use of pulsed-field gel electrophoresis to link sporadic cases of invasive listeriosis with recalled chocolate milk. *Appl. Environ. Microbiol.* **61:**3177–3179.

234. Promadej, N., F. Fiedler, P. Cossart, S. Dramsi, and S. Kathariou. 1999. Cell wall teichoic acid glycosylation in *Listeria monocytogenes* serotype 4b requires *gtcA*, a novel, serogroup-specific gene. *J. Bacteriol.* **181:**418–425.

235. Pucciarelli, M. G., E. Calvo, C. Sabet, H. Bierne, P. Cossart, and F. Garcia-Del Portillo. 2005. Identification of substrates of the *Listeria monocytogenes* sortases A and B by a non-gel proteomic analysis. *Proteomics* **5:**4808–4817.

236. Racz, P., K. Tenner, and E. Mero. 1972. Experimental Listeria enteritis. I. An electron microscopic study of the epithelial phase in experimental listeria infection. *Lab. Investig.* **26:**694–700.

237. Renzoni, A., A. Klarsfeld, S. Dramsi, and P. Cossart. 1997. Evidence that PrfA, the pleiotropic activator of virulence genes in *Listeria monocytogenes*, can be present but inactive. *Infect. Immun.* **65:**1515–1518.

238. Repp, H., Z. Pamukci, A. Koschinski, E. Domann, A. Darji, J. Birringer, D. Brockmeier, T. Chakraborty, and F. Dreyer. 2002. Listeriolysin of *Listeria monocytogenes* forms Ca2+-permeable pores leading to intracellular Ca2+ oscillations. *Cell. Microbiol.* **4:**483–491.

239. Riedo, F., R. Pinner, M. Tosca, M. Carter, L. Graves, M. Reeves, R. Weaver, B. Plikaytis, and C. Broome. 1994. A point-source foodborne listeriosis outbreak: documented incubation period and possible mild illness. *J. Infect. Dis.* **170:**693–696.

240. Robbins, J., A. Barth, H. Marquis, E. de Hostos, W. Nelson, and J. Theriot. 1999. *Listeria monocytogenes* exploits normal host cell processes to spread from cell to cell. *J. Cell Biol.* **146:**1333–1349.

241. Rocourt, J. 1999. The genus *Listeria* and *Listeria monocytogenes*: phylogenetic position, taxonomy, and identification, p. 1–20. *In* E. T. Ryser and E. H. Marth (ed.), *Listeria, Listeriosis, and Food Safety*, 2nd ed. Marcel Dekker, New York, N.Y.

242. Rocourt, J., and R. Brosch. 1992. *Human Listeriosis—1990.* WHO/HPP/FOS/92.3. World Health Organization, Geneva, Switzerland.

243. Rocourt, J., and C. Jacquet. 1994. Epidémiologie des infections humaines à *Listeria monocytogenes* en 1994: certitudes et interrogations. *Ann. Inst. Pasteur* **5:**168–174.

244. Rocourt, J., C. Jacquet, and A. Reilly. 2000. Epidemiology of human listeriosis and seafoods. *Int. J. Food Microbiol.* **62:**197–209.

245. Rouquette, C., C. de Chastellier, S. Nair, and P. Berche. 1998. The ClpC ATPase of *Listeria monocytogenes* is a general stress protein required for virulence and promoting early bacterial escape from the phagosome of macrophages. *Mol. Microbiol.* **27:**1235–1245.

246. Ruhland, G. J., M. Hellwig, G. Wanner, and F. Fiedler. 1993. Cell-surface location of *Listeria*-specific protein p60—detection of Listeria cells by indirect immunofluorescence. *J. Gen. Microbiol.* **139:**609–616.

247. Ryser, E. T. 1999. Incidence and behavior of *Listeria monocytogenes* in unfermented dairy products, p. 359–409. *In* E. T. Ryser and E. H. Marth (ed.), *Listeria, Listeriosis and Food Safety.* Marcel Dekker, New York, N.Y.

248. Salamina, G., E. Dalle Donne, A. Niccolini, G. Poda, D. Cesaroni, M. Bucci, R. Fini, M. Maldini, A. Schuchat, B. Swaminathan, W. Bibb, J. Rocourt, N. Binkin, and S. Salmaso. 1996. A foodborne outbreak of gastroenteritis involving *Listeria monocytogenes*. *Epidemiol. Infect.* **117:**429–436.

249. Salova, N., N. N. Filatov, E. V. Sizykh, A. N. Gerasimov, and L. A. Riapis. 2005. [Pulse-electrotypes of *Listeria monocytogenes* strains, isolated in Moscow]. *Zh. Mikrobiol. Epidemiol. Immunobiol.* **2005:**19–22. (In Russian.)

250. Samadpour, M., M. W. Barbour, T. Nguyen, T. M. Cao, F. Buck, G. A. Depavia, E. Mazengia, P. Yang, D. Alfi, M. Lopes, and J. D. Stopforth. 2006. Incidence of enterohemorrhagic *Escherichia coli*, *Escherichia coli* O157, *Salmonella*, and *Listeria monocytogenes* in retail fresh ground beef, sprouts, and mushrooms. *J. Food Prot.* **69:**441–443.

251. Sauders, B. D., D. Pettit, B. Currie, P. Suits, A. Evans, K. Stellrecht, D. M. Dryja, D. Slate, and M. Wiedmann. 2005. Low prevalence of *Listeria monocytogenes* in human stool. *J. Food Prot.* **68:**178–181.

252. Saxbe, W. B., Jr. 1972. *Listeria monocytogenes* and Queen Anne. *Pediatrics* **49:**97–101.

253. Schaferkordt, S., and T. Chakraborty. 1995. Vector plasmid for insertional mutagenesis and directional cloning in Listeria spp. *BioTechniques* **19:**720–722.

254. Schlech, W. F., III, P. Lavigne, R. Bortolussi, A. Allen, E. Haldane, A. Wort, A. Hightower, S. Johnson, S. King,

E. Nicholls, and C. V. Broome. 1983. Epidemic listeriosis—evidence for transmission by food. *N. Engl. J. Med.* 308:203–206.

255. Schlech, W. F., III. 2000. Foodborne listeriosis. *Clin. Infect. Dis.* 31:770–775.

256. Schneewind, O., P. Model, and V. A. Fischetti. 1992. Sorting of protein A to the staphylococcal cell wall. *Cell* 70:267–281.

257. Schnupf, P., D. A. Portnoy, and A. L. Decatur. 2006. Phosphorylation, ubiquitination and degradation of listeriolysin O in mammalian cells: role of the PEST-like sequence. *Cell. Microbiol.* 8:353–364.

258. Schubert, W. D., G. Gobel, M. Diepholz, A. Darji, D. Kloer, T. Hain, T. Chakraborty, J. Wehland, E. Domann, and D. W. Heinz. 2001. Internalins from the human pathogen *Listeria monocytogenes* combine three distinct folds into a contiguous internalin domain. *J. Mol. Biol.* 312:783–794.

259. Schuchat, A., K. Deaver, P. Hayes, L. Graves, L. Mascola, and J. Wenger. 1993. Gastrointestinal carriage of *Listeria monocytogenes* in household contacts of patients with listeriosis. *J. Infect. Dis.* 167:1261–1262.

260. Schuchat, A., K. Deaver, J. Wenger, B. Plikaytis, L. Mascola, R. Pinner, A. Reingold, and C. Broome. 1992. Role of foods in sporadic listeriosis. 1. Case-control study of dietary risk factors. *JAMA* 267:2041–2045.

261. Schuchat, A., C. Lizano, C. Broome, B. Swaminathan, C. Kim, and K. Win. 1991. Outbreak of neonatal listeriosis associated with mineral oil. *Pediatr. Infect. Dis. J.* 10:183–189.

262. Schuchat, A., B. Swaminathan, and C. V. Broome. 1991. Epidemiology of human listeriosis. *Clin. Microbiol. Rev.* 4:169–183.

263. Schuerch, D. W., E. M. Wilson-Kubalek, and R. K. Tweten. 2005. Molecular basis of listeriolysin O pH dependence. *Proc. Natl. Acad. Sci. USA* 102:12537–12542.

264. Schwartz, B., D. Hexter, C. Broome, A. Hightower, R. Hischorn, J. Porter, P. Hayes, W. Bibb, B. Lorber, and D. Faris. 1989. Investigation of an outbreak of listeriosis: new hypotheses for the etiology of epidemic *Listeria monocytogenes* infections. *J. Infect. Dis.* 159:680–685.

265. Seeliger, H., and D. Jones. 1986. *Listeria*, p. 1235–1245. *In* P. H. A. Sneath, N. S. Mair, M. E. Sharpe, and J. G. Holt (ed.), *Bergey's Manual of Systematic Bacteriology*, vol. 2. Williams & Wilkins, Baltimore, Md.

266. Seeliger, H. P. R., and K. Hohne. 1979. Serotyping of *Listeria monocytogenes* and related species. *Methods Microbiol.* 13:31–49.

267. Shahamat, M., A. Seaman, and M. Woodbine. 1980. Survival of *Listeria monocytogenes* in high salt concentrations. *Zentbl. Bakteriol. Hyg. Abt. I. Orig. A* 246:506–511.

268. Shank, F., E. L. Elliott, I. K. Wachsmuth, and M. E. Losikoff. 1996. U.S. position on *Listeria monocytogenes* in foods. *Food Control* 7:229–234.

269. Sheehan, B., A. Klarsfeld, R. Ebright, and P. Cossart. 1996. A single substitution in the putative helix-turn-helix motif of the pleiotropic activator PrfA attenuates *Listeria monocytogenes* virulence. *Mol. Microbiol.* 20:785–797.

270. Shen, Y., M. Naujokas, M. Park, and K. Ireton. 2000. InIB-dependent internalization of *Listeria* is mediated by the Met receptor tyrosine kinase. *Cell* 103:501–510.

271. Sibelius, U., T. Chakraborty, B. Krogel, J. Wolf, F. Rose, R. Schmidt, J. Wehland, W. Seeger, and F. Grimminger. 1996. The listerial exotoxins listeriolysin and phosphatidylinositol-specific phospholipase C synergize to elicit endothelial cell phosphoinositide metabolism. *J. Immunol.* 157:4055–4060.

272. Skoble, J., D. A. Portnoy, and M. D. Welch. 2000. Three regions within ActA promote Arp2/3 complex-mediated actin nucleation and *Listeria monocytogenes* motility. *J. Cell Biol.* 150:527–538.

273. Skogberg, K., J. Syrjanen, M. Jahkola, O. Renkonen, J. Paavonen, J. Ahonen, S. Kontiainen, P. Ruutu, and V. Valtonen. 1992. Clinical presentation and outcome of listeriosis in patients with and without immunosuppressive therapy. *Clin. Infect. Dis.* 14:815–821.

274. Skovgaard, N., and B. Norrung. 1989. The incidence of *Listeria* spp. in faeces of Danish pigs and in minced pork meat. *Int. J. Food Microbiol.* 8:59–63.

275. Sleator, R. D., C. G. Gahan, T. Abee, and C. Hill. 1999. Identification and disruption of BetL, a secondary glycine betaine transport system linked to the salt tolerance of *Listeria monocytogenes* LO28. *Appl. Environ. Microbiol.* 65:2078–2083.

276. Sleator, R. D., C. G. Gahan, and C. Hill. 2003. A postgenomic appraisal of osmotolerance in *Listeria monocytogenes*. *Appl. Environ. Microbiol.* 69:1–9.

277. Slutsker, L., and A. Schuchat. 1999. Listeriosis in humans, p. 75–95. *In* E. T. Ryser and E. H. Marth (ed.), *Listeria, Listeriosis and Food Safety*, 2nd ed. Marcel Dekker, New York, N.Y.

278. Smith, G. A., H. Marquis, S. Jones, N. C. Johnston, D. A. Portnoy, and H. Goldfine. 1995. The two distinct phospholipases C of *Listeria monocytogenes* have overlapping roles in escape from a vacuole and cell-to-cell spread. *Infect. Immun.* 63:4231–4237.

279. Smith, K., and P. Youngman. 1992. Use of a new integrational vector to investigate compartment-specific expression of the *Bacillus subtilis* spoIIM gene. *Biochimie* 74:705–711.

280. Smith, M. A., K. Takeuchi, R. E. Brackett, H. M. McClure, R. B. Raybourne, K. M. Williams, U. S. Babu, G. O. Ware, J. R. Broderson, and M. P. Doyle. 2003. Nonhuman primate model for *Listeria monocytogenes*-induced stillbirths. *Infect. Immun.* 71:1574–1579.

281. Sousa, S., D. Cabanes, C. Archambaud, F. Colland, E. Lemichez, M. Popoff, S. Boisson-Dupuis, E. Gouin, M. Lecuit, P. Legrain, and P. Cossart. 2005. ARHGAP10 is necessary for alpha-catenin recruitment at adherens junctions and for *Listeria* invasion. *Nat. Cell Biol.* 7:954–960.

282. Suarez, M., B. Gonzalez-Zorn, Y. Vega, I. Chico-Calero, and J. A. Vazquez-Boland. 2001. A role for ActA in epithelial cell invasion by *Listeria monocytogenes*. *Cell. Microbiol.* 3:853–864.

283. Sue, D., D. Fink, M. Wiedmann, and K. J. Boor. 2004. SigmaB-dependent gene induction and expression in *Listeria monocytogenes* during osmotic and acid stress

conditions simulating the intestinal environment. *Microbiology* **150:**3843–3855.

284. **Sullivan, M. A., R. E. Yasbin, and F. E. Young.** 1984. New shuttle vectors for *Bacillus subtilis* and *Escherichia coli* which allow rapid detection of inserted fragments. *Gene* **29:**21–26.

285. **Sumner, S. S., T. M. Sandros, M. Harmon, V. N. Scott, and D. T. Bernard.** 1991. Heat resistance of *Salmonella typhimurium* and *Listeria monocytogenes* in sucrose solutions of various water activities. *J. Food Sci.* **56:**1741–1743.

286. **Tappero, J., A. Schuchat, K. Deaver, L. Mascola, and J. Wenger.** 1995. Reduction in the incidence of human listeriosis in the United States. Effectiveness of prevention efforts. *JAMA* **273:**1118–1122.

287. **Theriot, J. A., T. J. Mitchison, L. G. Tilney, and D. A. Portnoy.** 1992. The rate of actin-based motility of intracellular *Listeria monocytogenes* equals the rate of actin polymerization. *Nature* **357:**257–260.

288. **Tilney, L. G., and D. A. Portnoy.** 1989. Actin filaments and the growth, movement, and spread of the intracellular bacterial parasite, *Listeria monocytogenes*. *J. Cell Biol.* **109:**1597–1608.

289. **Tompkin, R. B.** 2002. Control of *Listeria monocytogenes* in the food-processing environment. *J. Food Prot.* **65:**709–725.

290. **Trieu-Cuot, P., C. Carlier, C. Poyart-Salmeron, and P. Courvalin.** 1991. Shuttle vectors containing a multiple cloning site and a lacZ alpha gene for conjugal transfer of DNA from *Escherichia coli* to gram-positive bacteria. *Gene* **102:**99–104.

291. **Van Langendonck, N., E. Bottreau, S. Bailly, M. Tabouret, J. Marly, P. Pardon, and P. Velge.** 1998. Tissue culture assays using Caco-2 cell line differentiate virulent from non-virulent *Listeria monocytogenes* strains. *J. Appl. Microbiol.* **85:**337–346.

292. **Van Renterghem, B., F. Huysman, R. Rygole, and W. Verstraete.** 1991. Detection and prevalence of *Listeria monocytogenes* in the agricultural ecosystem. *J. Appl. Bacteriol.* **71:**211–217.

293. **Vazquez-Boland, J. A., C. Kocks, S. Dramsi, H. Ohayon, C. Geoffroy, J. Mengaud, and P. Cossart.** 1992. Nucleotide sequence of the lecithinase operon of *Listeria monocytogenes* and possible role of lecithinase in cell-to-cell spread. *Infect. Immun.* **60:**219–230.

294. **Vazquez-Boland, J. A., M. Kuhn, P. Berche, T. Chakraborty, G. Dominguez-Bernal, W. Goebel, B. Gonzalez-Zorn, J. Wehland, and J. Kreft.** 2001. *Listeria* pathogenesis and molecular virulence determinants. *Clin. Microbiol. Rev.* **14:**584–640.

295. **Vermeiren, L., F. Devlieghere, and J. Debevere.** 2006. Co-culture experiments demonstrate the usefulness of *Lactobacillus sakei* 10A to prolong the shelf-life of a model cooked ham. *Int. J. Food Microbiol.* **108:**68–77.

296. **Vicente, M. F., F. Baquero, and J. C. Perez-Diaz.** 1987. A protoplast transformation system for *Listeria* sp. *Plasmid* **18:**89–92.

297. **Wadsworth, S. J., and H. Goldfine.** 2002. Mobilization of protein kinase C in macrophages induced by *Listeria monocytogenes* affects its internalization and escape from the phagosome. *Infect. Immun.* **70:**4650–4660.

298. **Walker, R., L. Jensen, H. Kinde, A. Alexander, and L. Owen.** 1991. Environment survey for *Listeria* species in frozen milk plants in California. *J. Food Prot.* **54:**178–182.

299. **Ward, T. J., L. Gorski, M. K. Borucki, R. E. Mandrell, J. Hutchins, and K. Pupedis.** 2004. Intraspecific phylogeny and lineage group identification based on the *prfA* virulence gene cluster of *Listeria monocytogenes*. *J. Bacteriol.* **186:**4994–5002.

300. **Weis, J., and H. Seeliger.** 1975. Incidence of *Listeria monocytogenes* in nature. *Appl. Microbiol.* **30:**29–32.

301. **Wemekamp-Kamphuis, H. H., J. A. Wouters, R. D. Sleator, C. G. Gahan, C. Hill, and T. Abee.** 2002. Multiple deletions of the osmolyte transporters BetL, Gbu, and OpuC of *Listeria monocytogenes* affect virulence and growth at high osmolarity. *Appl. Environ. Microbiol.* **68:**4710–4716.

302. **Wiedmann, M., J. L. Bruce, C. Keating, A. E. Johnson, P. L. McDonough, and C. A. Batt.** 1997. Ribotypes and virulence gene polymorphisms suggest three distinct *Listeria monocytogenes* lineages with differences in pathogenic potential. *Infect. Immun.* **65:**2707–2716.

303. **Wirth, R., F. Y. An, and D. B. Clewell.** 1986. Highly efficient protoplast transformation system for *Streptococcus faecalis* and a new *Escherichia coli-S. faecalis* shuttle vector. *J. Bacteriol.* **165:**831–836.

304. **Wuenscher, M., S. Kohler, A. Bubert, U. Gerike, and W. Goebel.** 1993. The *iap* gene of *Listeria monocytogenes* is essential for cell viability, and its gene product, p60, has bacteriolytic activity. *J. Bacteriol.* **175:**3491–3501.

305. **Yildirim, S., W. Lin, A. D. Hitchins, L. A. Jaykus, E. Altermann, T. R. Klaenhammer, and S. Kathariou.** 2004. Epidemic clone I-specific genetic markers in strains of *Listeria monocytogenes* serotype 4b from foods. *Appl. Environ. Microbiol.* **70:**4158–4164.

306. **Zhang, C., M. Zhang, J. Ju, J. Nietfeldt, J. Wise, P. M. Terry, M. Olson, S. D. Kachman, M. Wiedmann, M. Samadpour, and A. K. Benson.** 2003. Genome diversification in phylogenetic lineages I and II of *Listeria monocytogenes*: identification of segments unique to lineage II populations. *J. Bacteriol.* **185:**5573–5584.

307. **Zhang, W., B. M. Jayarao, and S. J. Knabel.** 2004. Multi-virulence-locus sequence typing of *Listeria monocytogenes*. *Appl. Environ. Microbiol.* **70:**913–920.

308. **Zheng, W., and S. Kathariou.** 1997. Host-mediated modification of *Sau*3AI restriction in *Listeria monocytogenes*: prevalence in epidemic-associated strains. *Appl. Environ. Microbiol.* **63:**3085–3089.

Food Microbiology: Fundamentals and Frontiers, 3rd Ed.
Edited by M. P. Doyle and L. R. Beuchat
© 2007 ASM Press, Washington, D.C.

Keun Seok Seo
Gregory A. Bohach

Staphylococcus aureus

22

Staphylococcal food poisoning (SFP) is among the most prevalent causes of gastroenteritis worldwide. It results from ingestion of one or more preformed staphylococcal enterotoxins (SEs) in staphylococcus-contaminated food. The etiological agents of SFP are members of the genus *Staphylococcus*, predominantly *Staphylococcus aureus*. This form of food poisoning is considered an intoxication and does not involve infection by, and growth of, the bacteria in the host.

The association of staphylococci with foodborne illness was made more than a century ago. Barber, in 1914, was the first to implicate a toxin in SFP (8). He reported that repeated ingestion of contaminated milk produced symptoms typical of the illness. Barber cultured the milk, demonstrated the presence of a putative causative staphylococcal agent, and provided the first evidence that a soluble toxin was responsible for the disease. The next major advance in understanding SFP etiology was reported in 1930 by Dack et al. (33), who voluntarily consumed supernatants from cultures of "a yellow hemolytic staphylococcus" grown from contaminated sponge cake. Upon ingestion of the filtrates, Dack became ill with vomiting, abdominal cramps, and diarrhea. At that time, the only other foodborne toxin that had been recognized was botulinum toxin.

However, staphylococcal toxin, which exerts an effect on the gastrointestinal tract, was the first true enterotoxin to be described. It was particularly unique in comparison to botulinum toxin because its activity was "not entirely destroyed by heating even for 30 minutes at 100°C."

S. aureus has been extensively characterized. This bacterium produces a variety of extracellular products. Many of these, including the SEs, are virulence factors which have been implicated in diseases of humans and animals. As a group, the SEs elaborate a set of biological properties that enable staphylococci to cause at least two common human diseases, toxic shock syndrome (TSS) and SFP. This chapter will address primarily SFP; however, in regard to the SEs, there is significant overlap in the natural histories of both diseases. Hence, TSS also will be discussed in some sections in which this overlap is most relevant.

CHARACTERISTICS OF THE ORGANISM

Nomenclature, Characteristics, and Distribution of SE-Producing Staphylococci

The term staphylococci describes a group of small, spherical, gram-positive bacteria. Depending on the species and culture conditions, their cells have diameters ranging from

Keun Seok Seo and Gregory A. Bohach, Department of Microbiology, Molecular Biology, and Biochemistry, University of Idaho, Moscow, ID 83844.

approximately 0.5 to 1.5 µm. They are catalase-positive chemoorganotrophs with a DNA composition of 30 to 40 mol% guanine-plus-cytosine content. Staphylococci have a typical gram-positive cell wall containing peptidoglycan and teichoic acids. Except for clinical isolates such as some community-acquired methicillin-resistant *S. aureus* strains (129) and strains exposed to antimicrobial therapy, most staphylococci are sensitive to β-lactams, tetracyclines, macrolides, lincosamides, novobiocin, and chloramphenicol but are resistant to polymyxin and polyene. Some differential characteristics of *S. aureus* and several other selected species of staphylococci are summarized in Table 22.1.

There have been many useful schemes for classification of the staphylococci. According to *Bergey's Manual of Systematic Bacteriology* (121), staphylococci are classified in the family *Micrococcaceae*. This family includes the genera *Micrococcus*, *Staphylococcus*, *Stomatococcus*, and *Planococcus*. The genus *Staphylococcus* is further subdivided into 32 species and subspecies. Many of these are present in food as a result of human, animal, or environmental contamination. Several species of *Staphylococcus*, including both coagulase-negative and coagulase-positive isolates, can produce SEs. Although several species, including some coagulase-negative staphylococci, have the potential to cause gastroenteritis (16), nearly all cases of SFP are attributed to *S. aureus*. This is a reflection of the relatively high incidence of SE production by *S. aureus* in comparison to that by other staphylococcal species. Although the reason for this distinction is unknown, the SEs are superantigens (SAgs) and therefore are potential immunomodulating agents (see below). Hence, SE production is proposed to provide a selective advantage to *S. aureus*, a species that is common to both humans and animals, the two most common sources of food contamination.

Enterotoxigenic strains of staphylococci have been well characterized on the basis of a number of genotypic and phenotypic characteristics. An extensive phage-typing system is available for *S. aureus*. Most SE-producing isolates belong to phage groups I or III or are nontypeable. Although SE production by other phage groups is less common, it has been documented. Hajek and Marsalek (49) developed a classification scheme based largely on the animal host of origin. They were able to differentiate *S. aureus* into at least six biotypes. By far, SE production was most prevalent among human isolates within biotype A. SE production by other biotypes is rare except for that by biotype C bovine and ovine mastitis isolates. SEs may also be produced by *S. intermedius* and *S. hyicus* (formerly *S. aureus* biotypes E and F, respectively) and coagulase-negative staphylococci, albeit considerably less frequently than by *S. aureus*.

Currently known toxins in the SE family, with their physical and functional properties, are summarized in Table 22.2. Structures, functions, and mechanisms of pathogenicity of the SEs are discussed later in this chapter. This section will introduce the nomenclature and evolution of the SE family of toxins.

Classification Scheme Based on Antigenicity

Major advances in the characterization of SEs were made approximately two decades after Dack and colleagues (33) associated SFP with an exotoxin. Bergdoll and coworkers were the first investigators to produce purified SE preparations and develop specific antisera. They and others, using purified or partially purified toxins, induced protective antibodies in several species of animals. Immunity was strain specific and did not provide protection against strains other than the one that was used to induce the initial immune response (12). It soon became apparent that *S. aureus* could produce multiple toxins with similar molecular weights, as well as similar biological and physicochemical properties.

Initially, differentiation between antigenic forms of SE was based on the observation that many food isolates

Table 22.1 General characteristics of selected *Staphylococcus* species[a]

Characteristic	*S. aureus*	*S. chromogenes*	*S. hyicus*	*S. intermedius*	*S. epidermidis*	*S. saprophyticus*
Coagulase	+	−	+	+	−	−
Thermostable nuclease	+	−	+	+	+/−	−
Clumping factor	+	−	−	+	−	−
Yellow pigment	+	+	−	−	−	+/−
Hemolytic activity	+	−	−	+	+/−	−
Phosphatase	+	+	+	+	+/−	−
Lysostapin	Sensitive	Sensitive	Sensitive	Sensitive	Slightly sensitive	ND
Hyaluronidase	+	−	+	−	+/−	ND
Mannitol fermentation	+	+/−	−	+/−	−	+/−
Novobiocin resistance	−	−	−	−	−	+

[a] ND, not determined; +/−, variable or weak reaction.

Table 22.2 Biochemical and functional properties of SEs

SE or SEl[a]	Mass (kDa)	Zinc site(s)	Binding to MHC α/β-chain	Human TCR interactions[c]	Reference(s)
SEA	27.1	Cleft, domain 2	α/β	1.1, 5.3, 6.3, 6.4, 6.9, 7.3, 7.4, 9, 16, 18, 21.3, 22.1, 23	91
SED	26.9	Cleft, domain 2	α/β	1, 5.3, 6.9, 7.4, 8, 12	5, 70
SEE	26.8	Domain 2	α/β	5.1, 6.3, 6.4, 8.1, 8.2, 13.1, 18, 21.6	91
SElJ	31.2	ND[b]	?/β	ND	152
SElP	27.1	ND	ND	ND	77
SElN	26.1	ND	ND	9	64
SElO	26.8	ND	ND	5, 7, 22	64
SElH	25.1	Domain 2	ND/β	Vα10	106
SEB	28.4		α/–	1, 3.2, 6.4, 12, 14, 15, 17, 20	34
SEC1	27.5	Cleft	α/–	3.2, 5, 6.4, 6.9, 12, 13.2, 15, 17, 20	34
SEC2	26.6	Cleft	α/–	5, 12, 13.1, 13.2, 14, 15, 17, 20	34
SEC3	26.6	Cleft	α/–	2, 3, 5, 12, 13.1, 13.2, 14, 17, 20	34
SElU	27.2	ND	ND	ND	79
SEG	27.0		α/–	3, 12, 13.6, 14	64
SElR	27.0	ND	ND	3, 11, 12, 13.2, 14	100
SEI	24.9	ND	ND/β	1, 5, 5.3, 6, 23	64
SElK	30.0	ND	ND/β	5.1, 5.2, 6.7	102
SElL	26.8	ND	ND/β	5, 7, 22	64
SElM	24.8	ND	ND/β	6, 8, 9, 18, 21.3	64
SElQ	25.0	ND	ND	2, 5.1, 21.3	64
TSST-1	21.9		α/–	2, 8	30

[a] Nomenclature used in this table follows the new nomenclature rule of the INCSS (81).
[b] ND, not determined.
[c] Numbers indicate Vβ elements unless otherwise indicated for SEl.

produce one common antigenic type of toxin, tentatively designated the F type toxin. Most other enterotoxigenic strains, such as those from enteritis patients, also produce a second antigenic form that was classified as E toxin. The discovery of additional isolates that did not conform to this pattern prompted adoption of an improved nomenclature system. A committee assembled in 1962 established an alphabetical nomenclature (25) which, with some modification (81), is still currently recommended. Accordingly, SEs are sequentially assigned a letter of the alphabet in the order of their discovery. The F and E type toxins were designated SEA and SEB, respectively. Between 1962 and 1972, three additional classical SE serotypes (SEC, SED, and SEE) were reported (11). Initially, the designation SEF was used in reference to an exotoxin commonly produced by isolates of *S. aureus* associated with TSS. This designation was dropped when the toxin was confirmed to lack emetic activity. The term SEF was retired from use in the SE nomenclature system, and the toxin is now referred to as TSS toxin 1 (TSST-1) (15).

Compared to the rate of identification of the classical SEs (SEA through SEE), modern genomic analysis has greatly increased the rate at which previously unrecognized SEs and related proteins are discovered. As a result, classification based solely on antigenic properties is not practical. Instead, nomenclature is now based predominantly on the molecular relatedness of the primary sequences. Furthermore, it is not common to confirm the enterotoxigenic activities of putative SEs prior to publication. As a result, some toxins originally reported to be SEs have later been found to lack emetic activity. To ensure an orderly assignment for newly identified confirmed or putative toxins, investigators should contact the International Nomenclature Committee for Staphylococcal SAgs (INCSS) for guidance prior to publication (81). The INCSS also recommends that proteins related to the SEs but not confirmed to exhibit emetic activity in the monkey feeding assay (see below) be designated SE-like (SEl) until their enterotoxic activity can be confirmed. The primary sequences and other properties of currently reported SEs and SEls are summarized in Fig. 22.1 and Table 22.2 (5, 13, 32, 57, 69, 84, 116, 123, 131).

The incidence of SE involvement in SFP appears to change with time. Prior to 1971, SEA was the predominant toxin identified in cases of SFP, followed in frequency by SED and SEC. SEB was associated only rarely with SFP. In some cases, SEA was also identified as the causative agent in combination with SEC or SED (88). In a study examining more recent outbreaks of SFP from 1977 through 1981, SEA remained the most common toxin implicated (54). However, in contrast to results of previous surveys, SEB was the only other

SE identified. Holmberg and Blake (54) suggested that the observed decrease in cases attributed to other SEs was due to improved conditions for the processing and storage of milk, which had been commonly contaminated by SEC- and SED-producing strains of *S. aureus* in the past. In a more recent report, SEE was the classical SE least commonly associated with SFP (65). Of the newly described staphylococcal toxins, SEG, SEI, and other toxins corresponding to the *egc* operon (described below) (64) appear to be very common and widely distributed among staphylococcal human and animal isolates (127).

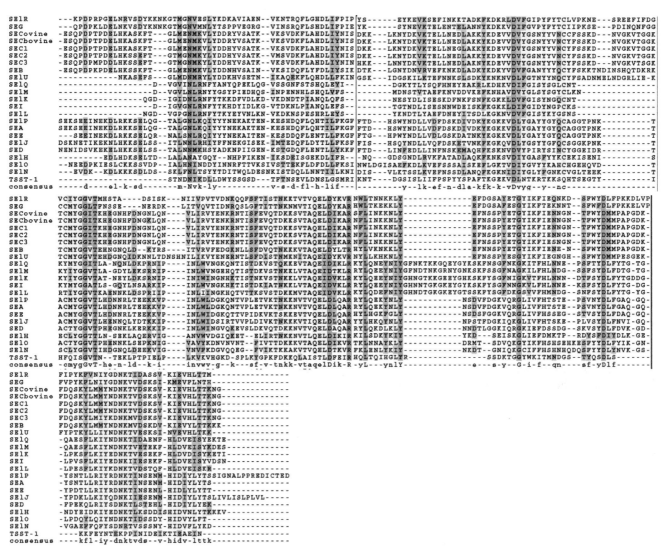

Figure 22.1 Alignment of primary sequences of mature SEs and SEIs according to the current literature. Also shown are the consensus sequences (at the bottom) and dashes to indicate gaps in the sequences made by alignment. Sequence alignment and output were conducted by using the CLUSTAL W program (139).

SE Antigenic Subtypes and Molecular Variants

Designation of SEs based on serological typing has been useful. However, sequence analysis and detailed immunological studies have produced some examples in which the antigenic characteristics of the proteins do not reflect their molecular or biological uniqueness. The best documented examples are with SEC. It had been noted for some time that the SEC serological variant can be further divided into at least three subtypes (SEC1, SEC2, and SEC3) based on minor differences in immunological reactivities. However, within each subtype, significant sequence variability may occur. For example, the SECs produced by strains FRI-909 and FRI-913 were both designated SEC3 according to their immunological reactivities. However, it was later shown that the sequences of the toxins differ by nine residues (84). The SEC variants produced by bovine and ovine isolates of *S. aureus* have very similar sequences and are apparently indistinguishable from SEC1 in immunological assays. In contrast, they behave differently from SEC1 in biological assays. For example, although SECbovine differs from SEC1 by only three residues, the potencies of the two toxins differ by several orders of magnitude in lymphocyte proliferation assays (84). Examples of heterogeneity among both the classical SEs and newly described toxins are becoming more common with the rapid rate of sequence determinations.

Staphylococcal Genetics and Evolutionary Aspects of SE Production

SEs Are SAgs and Belong to a Large PT Family

In discussing the genetics and evolution of the SEs, one must also consider other staphylococcal toxins, plus some toxins produced by other bacteria, especially group A streptococci. SEs are part of a large family of related toxins produced by *S. aureus* and *Streptococcus pyogenes* (17). This family of toxins has been termed the pyrogenic toxin (PT) family. Members of this family are grouped together based on shared biological and biochemical properties. The one feature in common to all PTs, including SEs, is their unique ability to act as SAgs (85).

SAgs are molecules that have the ability to stimulate an exceptionally high percentage of T cells. The mechanism by which this occurs distinguishes them from mitogens and conventional antigens (Ags). In regard to T-cell stimulation, SAgs are bifunctional molecules that interact with major histocompatibility complex class II (MHCII) molecules on Ag-presenting cells (APCs). Unlike interactions involving conventional Ags, this interaction does not require

processing and occurs outside of the MHCII peptide-binding groove (Fig. 22.2). The MHCII molecule-SAg complex interacts with the T-cell receptor (TCR). The interaction with the TCR is also nonconventional and relatively nonspecific; for most SAgs, binding occurs at a variable (V) location on the TCR β-chain (the Vβ region). Since SAgs bind outside the area on the TCR used for Ag recognition, they activate a much higher percentage of T cells than can be activated by conventional Ags. However, in contrast to mitogens which stimulate T cells in an indiscriminate manner, there is some degree of specificity in SAg action because only certain Vβ (or rarely Vα) sequences are recognized. Hence, not all T cells are stimulated.

The staphylococcal and streptococcal PTs are prototype microbial SAgs that exert a variety of immunomodulatory effects leading to shock, immunosuppression, and other systemic abnormalities associated with TSS. While SEs are included with the PTs, they have the unique distinction of possessing an additional ability to induce an emetic response upon oral ingestion and are thus solely responsible for SFP. It is generally agreed that many of the toxins in this family, including members of the SE family, arose from a common ancestral gene which crossed the genus barrier and became stably introduced in both *Staphylococcus* and *Streptococcus* genera. Evidence for

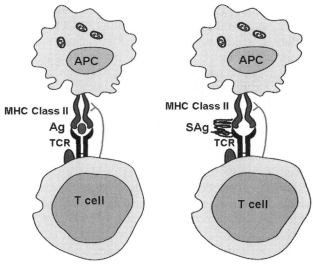

Figure 22.2 Interactions between APCs and T cells facilitated by conventional Ags and SAgs. Following processing by the APC, conventional Ags are presented to highly specific TCRs in association with the Ag-binding groove of the MHCII molecule. SAgs interact with MHCII molecules (without processing) outside the Ag-binding groove. The SAg-MHCII bimolecular complex binds to the TCR through specificity determined only by the variable region of the receptor α- or β-chain.

this idea is strongly supported by the observation that the structural genes for some SEs and related streptococcal PTs are carried on discrete genetic elements (see below).

Role of Genetic Elements in Generation and Dissemination of the SEs and SEls

Some staphylococcal and streptococcal PTs are encoded by structural genes located on bacteriophage genomes. Although the streptococcal toxins have been best studied, the genetic mobility of SEA and SEE appears to be similar to that of toxins of group A streptococci. Betley and Mekalanos confirmed that the SEA structural gene (*sea*) is carried by a lysogenic phage. *sea* was cloned directly from the induced bacteriophage genome (14). Both the *sea* and *see* genes map near the *att* site on their respective phage genomes. In a high percentage of cases, the toxin gene-carrying phages cannot be induced into a lytic cycle and appear to be defective. The most likely explanation for these observations is that the toxin genes were located originally on the bacterial genome but were subsequently obtained by the phage upon abnormal excision from the chromosome. This phenomenon is documented for some bacterial toxin genes that are transferred by lysogenic conversion into other genera of bacteria such as *Streptococcus* and *Corynebacterium* (68).

The role of plasmids has received considerable attention in relation to the transmission of PT genes. Of the PT genes, the *sed* gene was the first to be demonstrated to be plasmid carried. Bayles and Iandolo reported that in more than 20 characterized *sed*⁺ isolates, the *sed* structural gene is localized to a stable 27.6-kb plasmid (pIB485), which also carries genes for penicillin and cadmium resistance (5). Subsequent studies demonstrated that pIB485-related plasmids also harbor the genes for SElJ and SElR (99, 152).

The literature also includes reports associating SEC and SEB genes with transmissible penicillin resistance plasmids (96). However, it is now generally agreed that both SEC and SEB are chromosomally encoded and that the genes are harbored on pathogenicity islands in most strains. The best characterized staphylococcal pathogenicity island (SaPI) is SaPI1, which harbors the genes for SElK and SElL in addition to that for TSST-1 (96). This SaPI and another TSST-1 gene-carrying island, SaPI2, can be mobilized to different extents by superinfection. Early reports suggested that *seb* is located on a DNA element (67) that has been recently sequenced and designated SaPI3. The *sec* gene has been identified on SaPI4, which is highly related to SaPI1 (but lacks *tst*) and SaPIbov, an island common to bovine isolates (expresses SEC, TSST-1, and SElK) (39). In another molecular variation, the genes for SEG and SEI are clustered on an operon within

the staphylococcal chromosome which also contains genes for SElM, SElN, and SElO and several pseudogenes (39, 64). This enterotoxin gene cluster (designated *egc*) has had several variations identified and is proposed to be a nursery of new toxin genes.

Some staphylococcal toxin genes are insertion sites for genetic elements carrying other virulence determinants (13). The expression of several SEs is affected by this feature. For example, SEB synthesis and TSST-1 synthesis are mutually exclusive in *S. aureus*. At least in some isolates, this is due to the fact that SaPI3 and SaPI1 share a common insertion site (96). Likewise, the phage which harbors *sea* utilizes the beta-toxin locus, *hlb*, as its insertion site so that *sea*⁺ isolates do not produce β-toxin.

Mechanisms and Rationale for Generation of SE Diversity

Based on amino acid sequences, currently known members of the SE family are divided into three groups (Fig. 22.3). Group 1 contains SEB, the SEC subtypes and molecular variants, SEG, and SElR. Toxins in this group are highly related to one another and to several streptococcal PTs. Group 2 contains the highly related SEA, SElP, SEE, SElJ,

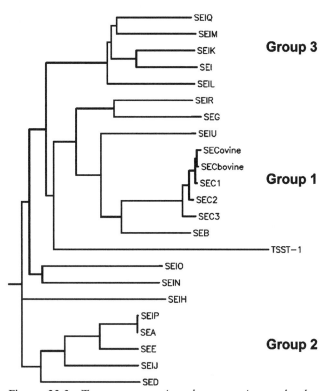

Figure 22.3 Tree representation demonstrating molecular relatedness of the currently known members of the SE family and TSST-1. This tree was created with the clustering feature of the PHYLIP program (37).

and SED, which is more distantly related. Group 3 contains SEI, SElL, SElK, SElM, and SElQ.

Most staphylococcal and streptococcal PTs, even those with no significant overall homology, contain four highly conserved stretches of primary sequence (53). This finding suggests that there is a selective advantage in host-parasite interactions for the bacteria to maintain certain toxin characteristics. At the same time, modification of selected regions of the proteins may allow the bacteria to broaden their host range. This molecular diversity may explain how a group of toxins with the same function but different host specificities may have arisen for the purpose of exploiting a broader repertoire of receptors.

Sequence comparisons of members within the PT family have provided several examples where diversity among the toxins appears to have arisen through gene duplication and/or homologous recombination. For example, SEC1 is most similar to SEC2 and SEC3. However, residues 14 through 26 of SEC1 (Fig. 22.1) are identical to those in the analogous region of SEB but significantly different from those of the other SEC subtypes (13). Genetic recombination between *seb* and *sec* in a strain producing both SEB and SEC2 (or SEC3) may explain the generation of SEC1.

Additional minor variabilities have resulted through point mutations; even closely related SEs display some sequence differences. This may reflect fine-tuning of the toxin sequences for interacting with cells from a variety of hosts, as demonstrated with the SEC subtype variants. The sequences of SEC toxins produced by strains of *S. aureus* isolated from humans differ slightly (>95% identity) from sequences of SEC variants produced by bovine and ovine isolates (84).

Staphylococcal Regulation of SE Expression

General Considerations

S. aureus is a widely distributed bacterium. It can exist in harsh environments as well as in various animals and humans. This versatile survival ability relies on a complex network of its virulence factors and adaptability. More than 50 genes involved in pathogenesis produce proteins that are either cell free or expressed on the bacterial surface. These enable the bacterium to evade host defenses, adhere to cells and the tissue matrix, propagate within the host, and destroy cells and tissues. Expression of virulence factors is temporarily controlled in response to cell density, nutrient availability, and environmental signals so that these factors are produced when required. During growth, expression of surface proteins is up-regulated early whereas that of cell-free secreted proteins is up-regulated postexponentially. These changes in expression

are regulated by a complex network of regulatory genes described below. In addition, several environmental signals also affect the production of extracellular proteins, such as high salt conditions, pH, and subinhibitory concentrations of antibiotics or glycerol monolaurate (52).

SEs are produced in extremely low quantities throughout most of the exponential growth phase (9, 94). There is generally a large increase in expression during the late exponential or early stationary phase of growth, with SEA and SED accumulating somewhat earlier than other SEs (13). Production of SE is dependent on de novo synthesis within the cell. The quantity of toxin produced is strain dependent, with SEB and SEC generally being produced in the highest quantities, up to 350 µg/ml. Even though there is a lower level of production of SEA, SED, and SEE, these toxins are generally detectable by gel diffusion assays which detect as little as 100 ng of SE per ml of culture. Some strains produce very low levels of toxins that require more sensitive analytical methods for detection. Some studies have revealed that SED and SEIJ are the toxins most likely to be undetected in *S. aureus* cultures, followed by SEC, SEA, and SEB in decreasing order.

Molecular and Environmental Regulation of SE Production

Several loci in *S. aureus* are involved in regulation of virulence factor expression. *S. aureus* genomic analyses reveal at least 16 two-component regulatory systems, of which *agrCA* (accessory gene regulator), *saeRS* (*S. aureus* exoprotein expression), *lytRS*, *arlRS* (autolysis-related locus), *srrAB* (staphylococcal respiratory response), and *yccFG* have been characterized (21, 40, 43, 109). In addition to SarA, a family of proteins homologous to SarA including products of *rot* and *xpr* (exoprotein regulator) has been characterized (28, 86, 125).

The ability of *S. aureus* to respond to environmental changes involves the density-sensing *agr* system (95), with some properties typical of bacterial two-component sensors and regulators (130). *agr* expression coincides temporally with expression of most SEs during the bacterial growth cycle. All are maximally expressed during late exponential and postexponential growth. SEA, which is not regulated by *agr*, is produced earlier (13). The *agr* locus maps at approximately 4 o'clock on the staphylococcal genome, near *purA*, *bla*, and *sea* on the standard map of *S. aureus*. It contains two divergent operons separated by approximately 120 bp (Fig. 22.4).

Transcription can be initiated from three promoters (P1, P2, and P3). P1 is weakly constitutive and transcribes *agrA*. P2 and P3 are induced strongly during late exponential and early stationary phases but are only weakly

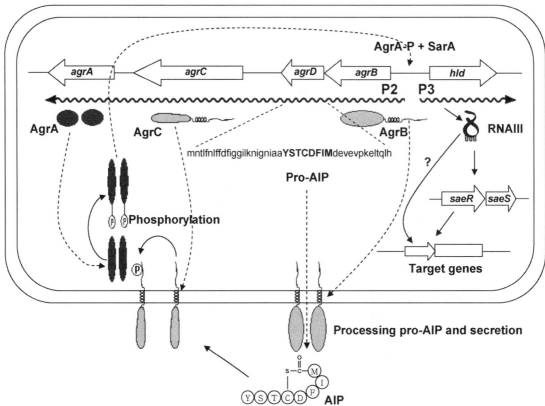

Figure 22.4 General characteristics of the *agr* locus in *S. aureus*. This physical map shows the relative locations of genes within the locus and other interacting regulatory genes and gene products (not drawn to scale).

expressed earlier. The P2 transcript, RNAII, encodes four proteins designated AgrA, AgrB, AgrC, and AgrD.

A model for signal transduction through the *agr* system is shown in Fig. 22.4. Together the four proteins form a quorum-sensing system enabling the bacterium to respond to its environment. A unique cyclic thiolactone pheromone peptide (autoinducing peptide [AIP]) derived from AgrD residues 46 to 53 is the activating molecule for the *agr* system. AIP is processed from AgrD by AgrB (151). Increased levels of AIP are recognized by the membrane-bound receptor AgrC which in turn initiates a signal transduction pathway (151). Its homology to conserved domains of histidine protein kinases, in particular a conserved histidine autophosphorylation site (130), further suggests that the 423-residue *agrC* gene product is likely to be the sensor. Autophosphorylation of the *agrC* product at the end of the exponential phase is thought to allow phosphorylation of AgrA, the response regulator, by phosphorylation of an aspartic acid residue, thereby activating expression of both RNAII and RNAIII (97). AgrA has homology to activator proteins

in other systems such as OmpR. AgrA and other typical response regulators exhibit a highly conserved N terminus of approximately 120 residues in length. Three amino acids, two aspartic acids, and a lysine are conserved among all response regulators.

Mutations in the *agr* locus result in decreased expression of several SEs and other exoproteins. Regulation of gene expression by *agr* can be transcriptional or translational. *agr* regulates alpha-hemolysin at both the transcriptional and translational levels, and SEB and SEC are regulated at the transcriptional level. Not all SEs are regulated by *agr*. SEA expression is not affected by *agr* mutations (143). At least 24 genes are under *agr* control; hence, it is a global regulator (95).

Strains with mutations in AgrA, AgrB, AgrC, or AgrD have an Agr mutant phenotype and do not initiate transcription from either P2 or P3. The 514-nucleotide transcript from P3, designated RNAIII, encodes the 26-residue staphylococcal delta-hemolysin and contains a significant amount of untranslated sequence. Interestingly, Agr mutant phenotypes can be complemented with plasmids carrying RNAIII under the control of an inducible promoter, even when the delta-hemolysin gene, *hld*, has been inactivated (75). Therefore, it appears that RNAIII is a diffusible element that plays a key role in the *agr* regulation of exoprotein structural genes, including those for the SEs (62, 75, 95). This RNA molecule can replace the regulatory function of the entire *agr* locus and acts reciprocally, up-regulating transcription of most of the extracellular protein genes and down-regulating many surface protein genes at the transcriptional level (95). The exact mechanism by which transcriptional regulation occurs has not been determined.

The *sae* locus is a positive effector of cell-free beta- and alpha-hemolysins, coagulase, nuclease, and protein A but apparently does not affect SEA, protease, lipase, staphylokinase, or cell-bound protein A. Mutations in *sae* had no effect on the production of RNAIII (95). These results suggest that the *sae* locus is distinct from other previously identified regulatory loci and may not act completely at the level of transcription.

The SarA protein family is a family of proteins homologous to SarA. Included in this family are SarA homologs (SarA, R, S, T, U, and V), MgrA (a global regulator of autolysis and virulence), and Rot (the repressor of toxin) (28, 61, 117). Sequence alignment and crystal structure studies demonstrated that SarA, R, and S are typical "winged-helix DNA binding proteins" with helix-turn-helix motifs (80, 82, 83). SarA binds to several target promoters, including *agr*, *hla*, *spa*, and *fnbA*. SarA binds to an intergenic region between *agr* promoters P2 and P3 which up-regulates expression of RNAIII, hence influencing expression of *agr*-regulated genes (29). Transcriptional gene fusion analysis revealed that *sarA* up-regulates both *tst* and *seb* (26). Mutation in the *sar* locus causes increased expression of alpha-toxin and decreased expression of delta- and beta-toxins (27).

Rot is a member of the SarA family of transcriptional factors. Rot acts as both a negative and a positive regulator of gene expression (117). Inactivation of *rot* was able to partially restore alpha-toxin and protease-positive phenotypes in an *agr*-null mutant (86). Rot also regulates expression of SEB (144). Deletion of the upstream promoter element to position −59 resulted in a dramatic reduction in the *seb* mRNA level. The sequence between −93 and −53 contains a Rot-binding site which negatively regulates transcription of *seb*. Even though the transcription of *rot* was not affected by *agr*, the activity of Rot is negatively regulated by *agr*. With the induction of RNAIII during the post-exponential growth phase, the activity of Rot as a repressor of exotoxin expression is inactivated (86). However, the exact mechanism of Rot regulation by the Agr system remains to be elucidated.

σ factors may also affect temporal expression of SEs. RNA polymerase purified from exponential-phase cultures of *S. aureus* contains a σ-70-related factor similar to that of *Escherichia coli* (112). The holoenzyme containing this σ factor transcribed *sea* more efficiently than *sec* or *agr* P2, suggesting additional complexity in the differential expression of certain SEs. While *agr* and most exoprotein genes, including *sec*, are expressed as the cells enter the stationary phase, *sea* is expressed earlier, during the exponential phase, simultaneously with the σ-70-like factor. Hence, differential SE expression at specific points of the bacterial growth phase may coincide with the availability of compatible σ factors.

The alternative σ factor (σ^B) is the major response to environmental stimuli. The *sigB* operon is constitutively transcribed under the control of σ^A and carries the SigB (σ^B) gene and the anti-*sigB* factor gene *rsbW*, the anti-repressor gene *rsbV*, and *rsbU*. RsbW usually binds to σ^B, which phosphorylates RsbV (76). Under environmental stresses such as high temperature, alkaline pH, high levels of NaCl, and the presence of catabolites (glucose, galactose, sucrose, glycerol, and maltose), phosphorylated RsbV is dephosphorylated by RsbU or RsbP and then binds RsbW, subsequently releasing and activating σ^B (13, 90). σ^B recognizes a unique promoter [GTTT(N_{14-17}) GGGTAT], which has been identified for 23 different *S. aureus* genes including the *sar* locus (44). This finding suggests that environmental signals activate σ^B which, in turn, activates global virulence regulation. SEB expression is negatively regulated by σ^B. When σ^B is activated by stress-induced conditions, expression of SEB is reduced (124). In contrast, expression of SEB is increased in a *sigB* mutant; however, the promoter of *seb* differs from the unique promoter for σ^B. This would suggest that expression of SEB is not directly regulated by σ^B.

SEC expression is affected by glucose through at least two different mechanisms. First, the metabolism of glucose indirectly influences SEC production through *agr* by reducing pH (115). Since *agr* is maximally expressed at neutral pH, growth in a nonbuffered environment containing glucose lowers pH levels, which directly reduces *agr* expression. Consequently, expression of *sec* and other *agr* target genes is affected correspondingly (115). Glucose also reduces *sec* expression in *agr* mutant strains. This observation suggests the existence of a second glucose-dependent mechanism for reduction of SE expression, independent of *agr* and apparently not involving pH (see below).

Other Relevant Molecular Aspects of SE Expression

Although many chemical and physical factors selectively inhibit SE expression (42, 59, 71), their effects on regulation and signal transduction are only beginning to be elucidated. *agr* is not the only signal transduction mechanism for *S. aureus*, as revealed by investigations into the inhibitory effect of glucose. The negative regulatory effect of glucose described above cannot be attributed entirely to lower pH levels in cultures grown on glucose because cultures containing the carbohydrate produce less SE, even when the pH is stably maintained. Although this effect has some attributes of catabolite repression, there are major differences between the glucose inhibitory effect in *S. aureus* and catabolite repression in *E. coli*. For example, inhibition by glucose cannot be reversed by adding cyclic AMP to staphylococcal cultures. The significance of these differences is still unclear. Even the catabolite-repressible staphylococcal *lac* operon has features that are significantly different from those of the analogous operon in *E. coli* and is apparently unresponsive to cyclic AMP.

S. aureus is an osmotolerant bacterium (see below). Although it is able to survive and grow in environments of low water activity, production of some SEs, especially that of SEB and SEC, is reduced when the bacterium is grown under osmotic stress. In experiments performed with SEC-producing strains, levels of *sec* mRNA and SEC protein are both reduced in response to high NaCl concentrations. However, addition of osmoprotectants reverses the effect. This reduced expression is also seen in *agr* mutant strains, indicating that the signal transduction pathway used in this mechanism occurs through an alternative pathway.

Low concentrations of the commonly used emulsifier glycerol monolaurate (GML) inhibit transcription of many exoprotein genes, including *sea*, without inhibiting *S. aureus* growth (110). Inhibition of SE production is not associated with a simultaneous effect on *agr* transcription and occurs in *agr* mutants as well as wild-type strains. These results, plus the finding that constitutive expression of some genes is not affected, suggest that GML interferes with non-*agr*-mediated signal transduction. It has been proposed that GML and a variety of related food additives exert this effect by insertion into the staphylococcal membrane, altering the membrane protein conformation and thereby interfering with signal transduction.

RESERVOIRS

Sources of Staphylococcal Food Contamination

Humans are the main reservoir for staphylococci involved in human disease, including *S. aureus*. Although most species are considered to be normal inhabitants of the external regions of the body, *S. aureus* is also a leading human pathogen. Colonized individuals may be carriers and are an important source for dissemination of staphylococci to others and to food. In humans, the anterior nares are the predominant site of colonization, although *S. aureus* can be present on other sites such as the skin or perineum. Dissemination of *S. aureus* among humans and from humans to food can occur through direct contact, indirectly through skin fragments, or through respiratory tract droplet nuclei.

Today, most sources of SFP are traced to humans who contaminate food during preparation. In addition to contamination by food preparers who are carriers, *S. aureus* may also be introduced into food by contaminated equipment used in food processing such as meat grinders, knives, storage utensils, cutting blocks, and saw blades. A survey of over 700 foodborne disease outbreaks revealed the following conditions most often associated with food poisoning: (i) inadequate refrigeration; (ii) preparation of foods far in advance; (iii) poor personal hygiene, e.g., not washing either hands or instruments properly; (iv) inadequate cooking or heating of food; and (v) prolonged use of warming plates when serving foods, a practice that promotes staphylococcal growth and SE production (22).

Animals, also an important source of *S. aureus*, are often heavily colonized with staphylococci. Predisposing factors which facilitate the survival of the bacterium are major concerns in the maintenance and processing of domestic animals and their products. For example, one very serious problem for the dairy industry is mastitis, an infectious disease of mammary tissue often caused by *S. aureus*. The combined losses and expenses associated with bovine mastitis make it the single most costly disease of animal agriculture in the United States. Colonization of animals by *S. aureus* is also a public health concern because it may result in contamination of milk and dairy products with *S. aureus* prior to or during processing.

It is not always possible to trace the source of staphylococcal food contamination to human or animal origin. Regardless of its source, many studies have demonstrated the common presence of *S. aureus* in many types of food products (Table 22.3). The numbers of cells of staphylococci are often low initially. However, the widespread presence of these cells provides a potential source of bacteria capable of causing SFP if conditions appropriate for SE expression are provided.

Resistance to Adverse Environmental Conditions

Some unique resistance properties of *S. aureus* facilitate its contamination of and growth in food. Outside the body, *S. aureus* is one of the most resistant non-spore-forming

Table 22.3 Prevalence of *S. aureus* in several foods

Product	No. of samples tested	% Positive for *S. aureus*	No. of *S. aureus* CFU/g[a]	Reference
Ground beef	74	57	≥100	135
	1,830	8	≥1,000	24
	1,090	9	≥100	107
Big-game meat	112	46	≥10	126
Pork sausage	67	25	100	135
Ground turkey	50	6	≥10	47
	75	80	>3.4	48
Salmon steaks	86	2	>3.6	36
Oysters	59	10	>3.6	36
Blue crabmeat	896	52	≥3	148
Peeled shrimp	1,468	27	≥3	137
Lobster tail	1,315	24	≥3	137
Assorted cream pies	465	1	≥25	141
Tuna pot pie	1,290	2	≥10	149
Delicatessen salads	517	12	≥3	104

[a] Determined by either direct plate count or most-probable-number techniques.

human pathogens and can survive for extended periods in a dry state. Its survival is facilitated by organic material, which is likely to be associated with the staphylococci from an inflammatory lesion. Isolation of staphylococci from air, dust, sewage, and water is relatively easy, and environmental sources of food contamination have been documented in several outbreaks of SFP.

S. aureus is known for acquiring genetic resistance to heavy metals and antimicrobial agents used in clinical medicine. However, generally the resistance of this bacterium to common food preservative methods is unremarkable. One noteworthy exception is its osmotolerance, which permits growth in media containing the equivalent of 3.5 M NaCl and survival at water activities of less than 0.86. This is especially problematic because other bacteria with which *S. aureus* does not compete efficiently are likely to be inhibited under these conditions.

The molecular basis for staphylococcal osmotolerance has been investigated in recent years, although systems for responding to osmotic stress have been more intensively studied with less tolerant bacteria. Considering the unique resistance of staphylococci, it would not be surprising to find that they have developed a highly efficient osmoprotectant system. As in other bacteria, several compounds accumulate in the cell or enhance staphylococcal growth under osmotic stress. Glycine betaine appears to be the most important osmoprotectant for *S. aureus*. To various degrees, other compounds, including L-proline, proline betaine, choline, and taurine, can

also act as compatible solutes for this bacterium. In *S. aureus*, intracellular levels of proline and glycine betaine accumulate to very high levels in response to increased concentrations of NaCl in the environment. Although the signal transduction pathway is not known for staphylococci, in other bacteria it involves a loss in turgor pressure in the cell and activation of required transport systems. High-affinity and low-affinity transport systems operate in *S. aureus* for both proline and glycine betaine (142, 147). The low-affinity systems are primarily stimulated by osmotic stress, have broad substrate specificity, and may be the same transporter shared by both osmoprotectants. By itself, the demonstration of a stress response system in *S. aureus* does not explain the unusual staphylococcal osmotolerance. Other, less tolerant microorganisms possess mechanisms for counteracting osmotic stress. The efficiency of the staphylococcal system may reflect an unusually high endogenous level of intracellular K^+ and a lack of need for de novo transporter synthesis. For example, in other well-studied systems such as that of *E. coli*, changes in osmotic stress activate K^+ transport systems. Elevated intracellular K^+ levels that result are required for induction of *proU* and eventually lead to the synthesis of transporters for glycine betaine and other osmoprotectants. In *S. aureus*, K^+ levels are high in unstressed cells. Therefore, the transport system is constitutively present in this bacterium and is preformed when high salt conditions are encountered. The net result is a very rapid and efficient response. *S. aureus* cells accumulate a 21-fold increase in proline after less than 3 min of exposure to high salt concentrations (142).

FOODBORNE OUTBREAKS

Incidence of SFP

SFP occurs as either isolated cases or outbreaks affecting a large number of people. Since SFP is usually self-limiting, the incentive to report cases has not been as great as that for other foodborne illnesses. Although there is a national surveillance system for SFP, it is not an officially reportable disease. It has been estimated that only 1 to 5% of all SFP cases are reported in the United States, usually at the state health department level. Most of these occur within highly publicized outbreaks. Isolated cases occurring in the home are not usually reported. Staphylococci account for an estimated 14% of the total of foodborne disease outbreaks within the United States (54). A study conducted by the USDA Economic Research Service found that more than an estimated 1.5 million cases of SFP occurred in a single year (1993), resulting in 1,210 deaths and a cost of $1.2 billion (23).

There has been an average of approximately 25 major outbreaks of SFP reported annually within the United States. The occurrence of SFP is cyclical, with the highest incidence typically occurring in the late summer when temperatures are warm and food is more likely to be stored improperly (54). A second peak occurs in November and December. Approximately one-third of these cases are associated with leftover holiday food.

SFP may be a leading cause of foodborne illness worldwide, although reporting in other countries is even less complete than that in the United States. In one study, 40 percent of outbreaks of foodborne gastroenteritis in Hungary were found to be due to SFP (8). The percentage is slightly lower in Japan (approximately 20 to 25%), where contamination of rice balls during preparation is a potential problem (8). Outbreaks due to improper manufacturing of canned corned beef have been reported in England, Brazil, Argentina, Malta, northern Europe, and Australia. Cases in Great Britain have been attributed to contaminated milk and cheese produced from milk from mastitic sheep (65). In some countries, ice cream has been a major vehicle of SFP. Despite increased knowledge of the epidemiology of SFP and attempts to educate the public, large outbreaks continue to represent a significant health hazard. For example, in 1998, approximately 4,000 individuals acquired SFP at a church event in Brazil (35). More recently, a massive outbreak in Osaka, Japan, in 2000 resulted in more than 10,000 cases from SEA and SEH in reconstituted milk (60).

Characteristics of a Recent Large Typical SFP Outbreak

The following is a summary of an outbreak of SFP reported by the U.S. Food and Drug Administration (4). Many of the aspects of this outbreak such as type of food involved, mechanism of contamination, inadequate safety measures in food handling, and clinical manifestations of illness are typical. This particular outbreak originated from one meal that was fed to 5,824 elementary school children at 16 sites in Texas. Of all the children exposed, a total of 1,364 developed clinical signs of SFP. Investigation into the source of the illness revealed that 95% of the children who became ill had eaten chicken salad from which large numbers of S. aureus cells were cultured.

The series of events leading up to the outbreak were as follows. Preparation of the meals was performed in a centralized kitchen facility and began on the preceding day. Frozen chickens used for the salad were boiled for 3 h. After cooking, the chickens were deboned, cooled to room temperature with a fan, ground into small pieces, placed into 30.5-cm-deep aluminum pans, and stored overnight in a walk-in refrigerator at 5.6 to 7.2°C. On the following morning, the remaining ingredients of the salad were added and the mixture was blended with an electric mixer. The food was placed in thermal containers and transported by truck to the various schools between 9:30 and 10:30 a.m. It was held at room temperature until being served between 11:30 a.m. and noon. It is believed that chicken became contaminated after cooking when it was deboned. Most likely, the storage of the warm chicken in the deep aluminum pans did not permit rapid cooling and provided an environment favorable for staphylococcal growth and SE production. Further growth of the bacteria probably occurred during the period when the food was kept in the warm classrooms. Prevention of the incident would have entailed more rapid cooling of the chicken and refrigeration of the salad after preparation.

CHARACTERISTICS OF DISEASE

Symptoms of SFP in Humans

SFP is usually a self-limiting illness presenting with emesis following a short incubation period. However, vomiting is not the only symptom that is commonly observed. Likewise, a significant number of patients with SFP do not vomit. Other common symptoms include nausea, abdominal cramps, diarrhea, headaches, muscular cramping, and/or prostration. In a summary of clinical symptoms involving 2,992 patients diagnosed with SFP, 82% complained of vomiting, 74% felt nauseated, 68% had diarrhea, and 64% exhibited abdominal pain (54).

In all cases of diarrhea, vomiting was always present. Diarrhea is usually watery but may contain blood as well. The absence of high fever is consistent with a lack of infection or significant toxemia in this type of food poisoning, although some patients present with low-grade fever. Other symptoms can include general weakness, dizziness, chills, and perspiration.

Symptoms typically develop within 6 h after ingestion of contaminated food. In one report, 75% of the exposed individuals exhibited symptoms of SFP within 6 to 10 h postingestion (54). The mean incubation rate is 4.4 h although incubation periods as short as 1 h have been reported. In another outbreak, symptoms lasted for 1 to 88 h with a mean of 26.3 h. Death, usually from severe dehydration or electrolyte imbalance, occurs in a small percentage of patients, with fatality rates ranging from 0.03% for the general public to 4.4% for more susceptible populations such as children and the elderly (54). Approximately 10% patients with confirmed SFP seek medical attention. Treatment is minimal in most cases, although administration of fluids is indicated when diarrhea and vomiting are severe.

TOXIC DOSE AND SUSCEPTIBLE POPULATIONS

Numbers of Staphylococci Required

Since many variables affect the amount of SE produced, one cannot predict with certainty the number of *S. aureus* cells in food required to cause SFP. Factors contributing to toxin concentrations have been extensively studied and include environmental conditions such as food composition, temperature, other physical and chemical parameters, and the presence of antimicrobial growth inhibitors. Also, bacterial factors to be considered include potential differences in the types, amounts, and numbers of different SEs the strain involved has the physiological ability to produce. It is likely that these combined conditions are unique for each isolated case or outbreak of SFP. Despite this variability, there are several general guidelines that are useful for assessing general risk. According to the U.S. Food and Drug Administration, doses of SE causing illness result when populations of greater than 10^5 *S. aureus* CFU per g of contaminated food are present (4). In other studies, 10^5 to 10^8 *S. aureus* CFU were observed to represent the typical range, despite the fact that lower levels were sometimes implicated (54).

Toxin Dose Required

Many investigations have been conducted to assess SE potency and the amount of toxin in food required to initiate SFP symptoms. Some of the most useful information

in this regard has come from the analysis of food recovered in outbreaks of the illness. Although SEs are quite potent, the amount of SE required to induce symptoms is relatively large compared to those of many other exotoxins that are acquired through contaminated food. A basal level of approximately 1 ng of SE per g of contaminated food is sufficient to cause symptoms associated with SFP. Although 1 to 5 µg of ingested toxin is typically associated with many outbreaks, the actual levels of detectable SE were considerably lower (<0.01 µg) in 16 SFP outbreaks (45). One of the most useful studies for predicting the minimal oral dose of SE required to induce SFP in humans was a well-documented investigation of an outbreak associated with ingestion of contaminated chocolate milk (36). In that study, the minimum dose of SEA required to cause SFP in schoolchildren was 144 ± 50 ng.

Many factors contribute to the likelihood of developing symptoms and their severity. The most important include susceptibility of the individual to the toxin, the total amount of food ingested, and the overall health of the affected person. The toxin type may also influence the likelihood and severity of disease. Though SFP outbreaks attributed to ingestion of SEA are much more common than those associated with other SEs, individuals exposed to SEB exhibit more severe symptoms. Forty-six percent of 2,291 individuals exposed to SEB exhibited SFP symptoms severe enough to be admitted to the hospital (54). Only 5% of 1,813 individuals exposed to SEA required such treatment. It is possible that these observations reflect differences in levels of toxin present because SEB is generally produced at higher concentrations than SEA.

Human volunteers and several species of macaque monkeys have been used to determine the minimal amount of purified SEs required to induce emesis when administered orally. Generally, monkeys are less susceptible to SE-induced enterotoxicity than humans. The study of purified SEs has provided useful comparative information and has important research applications, but its direct relevance to SFP is uncertain because potential stabilization of unpurified SEs by food is an important consideration. Based on a study in which human volunteers ingested partially purified toxin, Raj and Bergdoll estimated that 20 to 25 µg of SEB (0.4 µg/kg of body weight) is sufficient to cause vomiting in humans (111). In the rhesus monkey model, the 50% emetic dose is between 5 and 20 µg per animal (or approximately 1 µg/kg) when administered intragastrically. In our investigations, the minimal emetic dose of SEC1 for pigtail monkeys (*Macaca nemestrina*) is consistently between 0.1 and 1.0 µg/kg.

VIRULENCE FACTORS AND MECHANISMS OF PATHOGENICITY

SE Structure-Function Associations

Basic Structural and Biophysical Features

SEs and SEls are globular proteins of approximately 25 kDa. SE sequence analysis, relatedness, and diversity have been discussed above (Fig. 22.1 and 22.3 and Table 22.2). Although detailed studies have not been performed with every toxin, as a group they are stable molecules in many respects. Their recognition as heat-stable toxins arose from early studies in which enterotoxicity and antigenicity were not completely destroyed upon boiling of crude preparations. Furthermore, less extreme temperature treatments such as those used for pasteurization of milk have little or no effect on SE toxicity. The heat resistance of the SEs has been extensively studied. The general conclusion from this combined work is that SEs are difficult to inactivate by heating and have increased stability when present in high concentrations or in crude states such as in the environment of food. Since temperatures required to inactivate SEs are much higher than those needed to kill *S. aureus* under the same environmental conditions, toxic food involved in many cases of SFP is devoid of viable staphylococci at the time of serving. One additional SE property that has potential significance toward development of SFP is the resistance of SEs to inactivation by proteases present in the gastrointestinal tract. Resistance to pepsin, especially in a relatively low pH environment, is a key requirement for SE activity in vivo. Each of the SEs tested has some resistance to pepsin, a property not shared by at least one nonemetic staphylococcal PT, TSST-1. SEB is susceptible to degradation by pepsin at very low pHs, but partial neutralization of gastric acidity by ingested food is presumed to temporarily provide a protective environment for the toxin (8).

SEs may be cleaved by other common proteases; however, unless the fragments generated are separated in the presence of denaturing agents, proteolysis alone may not be sufficient to cause a loss of biological activity. This is apparently representative of inherent SE molecular stability which can be demonstrated by renaturing studies. For example, denaturation occurs only under strong denaturing conditions with high concentrations of urea or guanidine hydrochloride. If the denaturing conditions are removed, SEs may spontaneously renature and regain biological activity (128). Differences in stability exist among the toxins. For example, SEB and SEC1 are approximately 50-fold more stable under denaturing conditions than SEA (145). It has been suggested that the SE disulfide bond is responsible for the

inherent molecular stability. Several investigators have determined that the closed disulfide bond does contribute at least some degree of conformational stabilization but its disruption has only minimal effects on the overall stability and activity of the molecule (55) (see below).

Three-Dimensional Structure

Despite the sequence diversity of SEs, with only partial sequence conservation, structural studies revealed that the three-dimensional topologies of SEs are remarkably similar (136). The molecules have an overall ellipsoidal shape and are folded into two domains, the amino- and carboxy-terminal domains, which are connected by α-helices in the center (Fig. 22.5). The amino-terminal domain, domain 1, contains residues near the N terminus but not the N-terminal residues themselves. The folding conformation of this domain may have potential significance for the function of the toxin. Its topology, in which a β-barrel structure is capped at one end by an α-helix, is known as the oligonucleotide-oligosaccharide-binding (OB) fold. The internal portion of its β-barrel is rich in hydrophobic residues, and the potential oligomer-binding surface is covered with mainly hydrophilic residues. This OB folding pattern is found in staphylococcal nuclease and the B subunit of AB_5 heat-labile enterotoxins such as cholera toxin, pertussis toxin, and Shiga toxin (87), which share the feature of exerting their activity of interacting with either oligosaccharides or oligonucleotides. The other prominent feature of domain 1 is that in most toxins of this family it contains two cysteine residues responsible for forming the disulfide linkage characteristic of most SEs. This bond and the cysteine loop are located at the end of the domain opposite its α-helix cap. Crystallographic data for SEA, SEB, and SEC3 indicate that the loop regions of all three toxins are quite flexible (20, 119, 136).

The larger, carboxy-terminal domain (domain 2) can be described as a five-strand antiparallel β-sheet wall overlaid with a group of α-helices forming a β-grasp motif. It has structural similarity to the immunoglobulin-binding motifs of streptococcal proteins G and L, ubiquitin, and 2F2-2S ferredoxin and the Ras-binding domains of the Ser/Thr-specific protein kinase Raf-1 (46, 150). The structural similarities among SEs and other bacterial proteins suggest that SEs evolved through the recombination of these proteins (89).

Binding of Zinc by SEs

Biochemical and structural studies on SEs have revealed that some SEs are dependent on zinc ions to be functional and to be able to properly bind MHCII molecules. Fraser et al. (41) determined that SEA and SEE bind

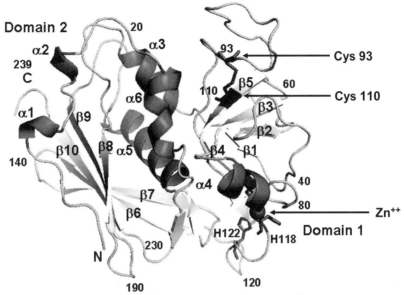

Figure 22.5 Schematic diagrams of the SEC3 crystal structure illustrating major structural features. Numerical designations defining the locations of select residues and each α-helix and β-strand are shown within the two major domains. Also indicated are the N and C termini. The intramolecular disulfide linkage between Cys residues 93 and 110 (arrows) connects the disulfide loop to the β5-strand containing the conserved residues (see Fig. 22.7) potentially important for emesis. The zinc atom bound by SEC3 faces toward the back of the SEC3 molecule between domains 1 and 2 and is coordinated by D83, H118, and H122. In contrast, the high-affinity zinc-binding site in SEA is positioned on the opposite edge of domain 2. The conformational topology of domain 1 is the same as those of the OB domains of several other proteins described in the text.

zinc via a single site with a dissociation constant of 1 to 2 µM. The binding site was subsequently predicted by mutagenesis of SEA to comprise a nonlinear stretch of residues (H187, H225, and D227) in the concave β-sheet within the C-terminal domain and N-terminal serine in the metal coordination (Fig. 22.6) (119). Crystallographic analysis of SEA crystals soaked in 10 mM $ZnCl_2$ revealed a second zinc-binding site, similar to that of the SEC2 site (see below). It is coordinated by D86 and H114 (corresponding to D83 and H118 in SEC2), a water molecule replaces the second histidine residue in SEC2 (H122), and E39 is analogous to D9 in SEC2 (Fig. 22.5) (118).

SEC2 and SEC3 bind zinc through a low-affinity mechanism (19, 20). The zinc ion is tetrahedrally coordinated by D83, H118, and H122 from one molecule and D9 from a neighboring molecule. The zinc-binding site is located in a classical motif (H-E-X-X-H) at the cleft in the α5-groove of SEC3 between the two domains. This zinc-binding motif is typically found in the catalytic site of metalloenzymes such as thermolysin (51) and in certain other bacterial protease-dependent toxins such

as botulinum and tetanus neurotoxins (73). However, none of the SEs is known to possess protease activity. Although the function of the zinc atom in SEC is not certain, it has been suggested that it plays a minor role in stabilizing the toxin structure.

Crystallographic studies revealed that SED binds two zinc ions and forms a homodimer in a zinc-dependent manner (134). One zinc ion is tetrahedrally coordinated by D182, H220, and D222 from one molecule and H218 from the other molecule which allows it to bind both MHCII α- and β-chains (see below). The second zinc-binding site is similar to that of SEC2.

Binding to MHCII Molecules
Despite the structural similarities of SEs, the mechanisms by which various SEs bind MHCII molecules are diverse. SEs interact with MHCII molecules in three ways: (i) binding a single α-chain of MHCII, (ii) binding a single β-chain of MHCII, or (iii) cross-linking two MHCII molecules.

The first characterized type of SE binding is to the α-chain of MHCII. SEB, SEC1, SEC2, SEC3, and TSST-1

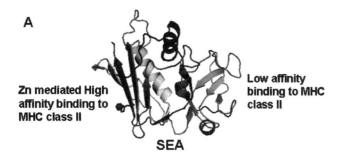

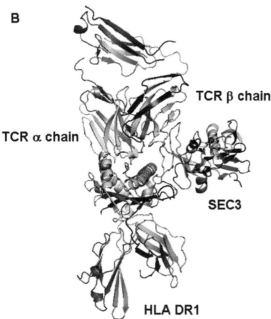

Figure 22.6 (A) Schematic diagram of the SEA crystal structure. SEA has two MHCII-binding sites. Relatively low affinity MHCII binding occurs at a generic binding site which is conserved in most SEs. A high-affinity MHCII-binding site is located on the external surface of domain 2. This includes the high-affinity zinc-binding site, formed by His187, His225, and Asp227. The zinc ion mediates cross-linking of SEA with MHCII molecules and is crucial for maximal B- and T-cell activation. (B) Hypothetical model of MHCII-SEC-TCR complex based on the modeling predicted from the crystal structures of the SEC3-HLA-DR1 (low-affinity binding site) and SEB-Vβ complexes.

bind in this manner and share a common MHCII-binding site at domain 1 (referred to as the generic site) with a comparatively low affinity (0.4 to 0.7 μM). The crystal structure of the SEB-HLA-DR1 complex showed that one major interaction is mediated by a salt bridge provided by Glu67 of SEB and Lys39 on HLA-DR1 (63). Although SEC2 and SEC3 are able to bind zinc, crystal structure data and site-directed mutagenesis studies

revealed that the zinc ion is not involved in the MHCII interaction (133).

The second type of interaction is with the β-chain of MHCII. A crystal structure of SElH complexed with HLA-DR1 revealed that toxin interacts with the β-chain of the receptor and binding requires its bound zinc ion (105). The zinc ion cross-linking plays an important role in interaction of SElH and HLA-DR1 by providing cross-linking through His81 on HLA-DR1 and His206 and Asp208 on SElH.

The third type of interaction involves SE cross-linking of two MHCII molecules. SEA has a low-affinity MHCII α-chain-binding site which overlaps that of SEB and TSST-1 plus a high-affinity site on the outside of domain 2 near the N terminus (58). The SEA high-affinity site involves coordination of zinc through three toxin residues (His187, His225, and Asp227) and His81 of the MHCII β-chain. Two MHCII-binding sites per toxin allow two MHCII molecules to be cross-linked (140) (Fig. 22.6). Similar modes of action, including formation of homodimers by some toxins, have been proposed for other SEs with high-affinity zinc-binding sites on the outer face of domain 2 (SED and SEE) (3, 134).

Binding to the TCR

SEs and other SAgs interact with a characteristic repertoire of TCR sequences (Table 22.1). The TCR specificity of each SE is determined by toxin residues in the shallow cavity at the top of the molecule (34). Crystallographic analysis of a molecular complex between SEC3 and a portion of the murine TCR β-chain (38) indicated that the toxin binds to the CDR1, CDR2, and HV4 loops of Vβ. Modeling permitted the deduction of a trimolecular complex model containing the SAg bridging the T cell and APC. Using SEC3 and SEB, this model shows that these toxins form a "wedge" between the TCR and the MHCII, orienting the peptide-binding cleft away from the TCR (Fig. 22.6B). Although different SAgs are predicted to bind with variations of this theme, in general, SAg binding results in receptor interactions significantly different from that in typical Ag presentation. The overall affinity of the entire complex determines the effectiveness of the stimulation (78), and binding by toxins with low affinity for the TCR can be compensated for by stronger binding to MHCII and vice versa. SElH appears to have the most divergent mechanism of binding and T-cell stimulation. This toxin binds to MHCII in a mechanism that does not allow efficient interactions between SElH and the TCR Vβ or between MHCII and Vβ. In one study (106), SElH stimulated T cells by interacting with the TCR Vα (Vα10) and Vβ-specific expansion could not be demonstrated.

Molecular Regions of SEs Responsible for Enterotoxicity

The structural aspects of SEs that enable them to survive degradation by pepsin and other enzymes in the gastrointestinal tract are required for the toxins to induce SFP. However, stability alone is not sufficient. SEs must also be able to interact with the appropriate target, leading to emesis, diarrhea, and other gastrointestinal tract symptoms. Initial attempts to define molecular regions responsible for enterotoxicity involved testing biological activities of protease-generated fragments derived from SEA, SEB, or SEC1 (128). Three main conclusions were drawn from this work. First, only large toxin fragments containing central and C-terminal portions of the SEs retain enough of the native structure to cause emesis. Second, N-terminal residues of the SEs are not required for emesis. SEC1 modified by removal of the 59 N-terminal residues retains the ability to cause emesis in monkeys. Smaller toxin fragments from SEC1 and other SEs are inactive. The third conclusion was that emesis seems to require preservation of structure in the area of the conserved SE disulfide loop.

The disulfide bond is a structural feature that is characteristic of most, but not all, SEs. The disulfide bond has long been associated with emesis and is not found uniformly in nonemetic staphylococcal and streptococcal exotoxins. With a few known exceptions, SEs contain exactly two cysteine residues which may potentially form an intramolecular linkage and a spacer disulfide loop. SEG and the SEC molecular variant produced by isolates from sheep (SECovine) deviate from this pattern and possess an additional third cysteine (Fig. 22.7). There is speculation regarding the importance in emesis of the disulfide bond, located near the center of every classical and most other SEs. Based on data acquired from analysis of recently identified SEs and SEls in the monkey feeding assay, evidence suggests that a stable disulfide bond facilitates an emetic response in vivo. Specifically, toxins lacking one or both cysteine residues are generally nonemetic or weakly emetic (7, 93, 101–103, 138). Potential structural contributions from the Cys-Cys bond that may contribute to the SE conformation necessary for emetic activity may include one or more of the following features: (i) proper positioning of cysteine residues upon formation of the disulfide bond, (ii) exposure and/orientation of crucial residues in the loop formed between the two linked cysteines, (iii) exposure and/or orientation of residues immediately adjacent to the disulfide linkage but not contained within the cysteine loop, and (iv) contributions to the overall SE conformation by the linkage of the two cysteine residues. Each of these possibilities has been considered. The cysteine residues and the loop probably do not play a

	Cysteine loop or analogous region	Conserved downstream sequences
Confirmed emetic		
SEA	CAGGTPNK----------TAC	MYGGVTLH
SEE	CAGGTPNK---------TAC	MYGGVTLH
SED	CYGGEIDR---------TAC	TYGGVTPH
SEI	CKS---------------KY	MYGGATLS
SEC1	CYFSSKD---NVGKVTGGKTC	MYGGITKH
SEB	CYFSKKTNDINSHQTDKRKTC	MYGGVTEH
SEG	CIIPKSE---PDINQNFGGCC	MYGGLTFN
Non-emetic		
SElQ	CNT--------------KY	MYGGITLA
SElL	SNS--------------RT	IYGGVTKA
Not reported		
SECovine	CCFSSKD---NVGKVTGGKTC	MYGGITKH
SElJ	CSGGKPNK---------TAC	MYGGVTLH
SElH	CEKISENI---------SEC	LYGGTTLN
SElO	CHGEHQVD---------TAC	TYGGVTPH
SElN	CVGLTEEK---------TSC	LYGGVTIH
SElM	CLKN------------KY	IYGGVTLA
SElK	CNT--------------KY	IYGGVTAT
SElR	CLVPKNE---SREEFIFDGVC	IYGGVTMH

Figure 22.7 Comparison of cysteine loop and adjacent sequences for SEs and the analogous regions of the SEls. Evidence suggests that proper positioning of the critical downstream residues by a stable disulfide bond is required for emesis. Toxins designated as emetic are those reported as inducing emesis in the monkey feeding assay (7, 93, 101–103, 138). SEI was reported to be weakly emetic (93).

direct role in the emetic response. It has been possible to replace the cysteine residues in several SEs by site-directed mutagenesis and show that neither of these two residues is absolutely critical. Although most of the SEs have a cysteine loop or analogous feature, the lengths of and compositions of residues within the loops of different SE types vary greatly. This lack of consistency among SE loop properties suggests that the loops are unlikely to have a shared enterotoxigenic function (Fig. 22.7). Furthermore, proteolytic nicking of toxins in their loops has no effect on their ability to cause emesis (128). In regard to overall protein conformation, Warren et al. (146) determined that the disulfide bond contributes only minimally. Presumably, then, if the disulfide linkage is important in emesis, the effect is likely to provide a particular orientation of residues near the disulfide linkage but not in the loop. The most convincing evidence in support of this possibility has been provided by mutagenesis of SEC1 in which its cysteine residues were replaced with either serine or alanine (55). It was found that mutants with serine substitutions were emetic, whereas the analogous mutants with alanine substitutions were nonemetic. Although serine and alanine are both considered to be conservative substitutions for cysteine, one difference between these two amino acids is that serine has the ability to hydrogen bond. Thus, hydrogen bonding by serine may be able to replace the disulfide linkage stabilization of local structure.

Which critical local residues require proper orientation by the disulfide bond (or hydrogen bonding at the same positions) in order for the SEs to induce emesis is unknown. Possible candidates are those within a highly conserved stretch of residues directly adjacent to, and downstream from, the disulfide loop (Fig. 22.7). In SEC3, these residues are located on the β5-strand. An attractive hypothesis is that in addition to stability in the gut, two other structural requirements need to be met for enterotoxicity. First, the appropriate conserved residues must be present in the toxin. Many PTs, emetic and nonemetic, have a similar set of highly conserved residues in a location analogous to the β5-strand of SEC3. Second, they must be positioned properly for interacting with their target in the gut. The unique SE disulfide bond may serve this function. Of the entire PT family, two non-SE toxins produced by *Streptococcus pyogenes* (streptococcal pyrogenic exotoxin and streptococcal SAg) may potentially form a disulfide linkage (114). However, the presence of more than two cysteines in these toxins suggests that the structure in this area and the degree of local stabilization by their putative disulfide linkage may not be identical to those of the SEs.

Additional studies on SEA revealed that single-site substitutions of residues close to the N terminus also influence emesis (50). This was especially the case for mutants constructed by replacement with glycine. For example, mutagenesis of residues 25, 47, and 48 causes significant reduction in the emetic potency of SEA. Although these residues are far from the disulfide bond in the SEA primary sequence, they are located near or within domain 1 of SEA and may potentially influence the area near the disulfide bond.

SE Antigenic Epitopes

The need for reagents that can detect SEs in food and clinical samples, plus a desire to differentiate antigenically different toxins in the toxin family, has been the impetus for considerable effort directed toward epitope characterization and mapping. Although the increased number of identified SEs and putative SEs is beginning to make immunological detection of SEs obsolete, several commercial reagents relying on this technique are still widely used. However, detection based on antigenicity is gradually being replaced by molecular techniques, especially multiplex PCR (92, 98).

Individually, each classical SE type and subtype has a sufficient degree of antigenic distinctness to allow its differentiation from other PTs by using highly specific polyclonal antisera and monoclonal antibodies. However, some degree of cross-reactivity among several SEs can often be demonstrated. The level of cross-reactivity generally correlates with shared primary sequences. The type C SE subtypes and their molecular variants exhibit a substantial amount of cross-reactivity, as do SEA and SEE, the two major serological types with the greatest sequence relatedness. For these toxins, cross-reactivity can be demonstrated even in relatively insensitive assays such as immunodiffusion assays in which these two toxins produce lines of partial identity to those of the heterologous antiserum. SEB and the SEC subtypes and molecular variants (see above) are also highly related at the amino acid level. Cross-reactivity between SEB and SEC may be demonstrated occasionally by immunodiffusion but more consistently by the use of sensitive methods such as radioimmunoassays and immunoblotting. Generally, it has not been possible to produce useful antibodies that cross-react among less-related SEs. Although one investigator has produced a monoclonal antibody that cross-reacts with all five major SE antigenic types A through E, this antibody has low affinity and cross-reacts with other staphylococcal proteins (10). The two most distantly related SEs determined to be recognized by a common epitope are SEA and SED (11).

The mapping of conserved and specific antigenic epitopes on SEs and their differentiation from potentially toxic regions have potential applications toward rational development of nontoxic vaccines. Considering the array of SE antigenic types, the most efficient toxoid would presumably contain one or more epitopes that are shared by multiple toxins. Several approaches have been used to partially localize antigenic epitopes on the SEs and differentiate them from toxic regions. One of the earlier methods used for this purpose was to identify protease-generated toxin fragments from several SEs that bind to cross-reactive antibodies (18). These studies revealed that both N- and C-terminal toxin fragments contain cross-reactive epitopes, but they were unable to define shorter stretches of residues.

There is some evidence that immunization with short, highly conserved peptides may have merit. For example, the use of synthetic peptides from highly conserved stretches of primary sequence has resulted in the production of neutralizing antibody for several of the SEs. Immunization with synthetic peptides corresponding to residues 130 to 160 of SEB or the same region of SEC1 (residues 148 to 162) induces antibodies that neutralize both native toxins (53, 66). The highly conserved SE sequence K-K-X-V-T-X-Q-E-L-D (Fig. 22.1), encompassed by both peptides, may represent part of an epitope that could be useful for protective immunity. It is yet to be determined if major epitopes identified on other toxins such as SEA also show promise (108).

SE Mode of Action in Induction of Emesis and Other Symptoms Related to SFP

SE-Induced Emesis Requires Nerve Stimulation

Except for rare SFP cases, in which massive doses of SEs are consumed, systemic dissemination of the toxins does not contribute significantly to illness. When fed to rodents, SEs do enter the circulation but are rapidly removed by the kidneys (6). Most studies using the simian model indicate that the SE site of action following ingestion is the abdominal viscera. Early studies into the mechanism of action of SEs tested the emetic responsiveness of animals to the toxins after disruption of well-defined neural systems or after visceral deafferation. The characteristic emetic response was determined to result from stimulation of local neural receptors in the abdomen (132) which transmit impulses through the vagus and sympathetic nerves, ultimately stimulating the medullary emetic center.

Cellular Histopathology in the Gastrointestinal Tract

Information on the histological effects of oral doses of SEs in humans is extremely limited. Most of what is known has been derived from information obtained from experiments with rhesus monkeys. Upon ingestion of the toxin, pathological changes compatible with a definition of gastroenteritis are observed in several parts of the gastrointestinal tract (72).

The primate stomach becomes hyperemic and is marked by lesions which begin with the influx of neutrophils into the lamina propria and epithelium. A mucopurulent exudate in the gastric lumen is also typically observed. Also characteristic are mucus-filled surface cells which eventually release their contents. Later in the illness, neutrophils become replaced by macrophages. Eventually, upon resolution of the symptoms, the cellular infiltrate clears.

A similar cellular infiltrate and lumen exudate occur in the small intestine, although they decrease in severity in sections taken from lower, compared to those from upper, portions of the intestine. Clearly evident in the jejunum are extension of crypts, disruption or loss of the brush border, and an extensive infiltrate of neutrophils and macrophages into the lamina propria. Changes in the colon are minimal in the monkey model. Only a mild cellular exudate and mucus depletion are evident. The only other significant effect in monkeys is acute lymphadenitis in the mesenteric lymph nodes.

Search for the Gastrointestinal Tract Target

The specific cells that SEs interact with in the abdomen have not been clearly identified, nor has their receptor. Evidence suggests that interaction of SEs with their target directly or indirectly causes production of inflammatory mediators that induce SFP symptoms. Jett et al. (66) determined that oral administration of SEB produces elevated levels of arachidonic acid cascade products. Specifically, they observed significant increases in prostaglandin E_2, leukotriene B_4, and 5-hydroxyeicosatetraenoic acid. These three compounds are potent vasoactive inflammatory mediators that can also act as chemoattractants for neutrophils. Both of these activities are consistent with the histopathology described above for the SFP monkey model.

Scheuber et al. (120) could not demonstrate an effect of prostanoid inhibitors on SEB-induced emesis but were able to correlate gastrointestinal symptoms with cysteinyl leukotriene generation. Intoxication of animals with SEB resulted in a 10-fold increase in levels of leukotriene E_4 in bile, plus those of an unidentified leukotriene in the urine. This group of investigators suggested a role for mast cells in the pathogenesis of SFP. Although induction of histamine production by SEB was responsible for some secondary nonenteric immediate hypersensitivity skin reactions, it did not correlate with emesis.

Evidence suggests that ingestion of SEs causes a stimulation of mast cells and possibly other inflammatory cells in the abdomen. Thus far, the abdominal receptor has not been identified. Experiments using anti-idiotype antibodies in binding assays and protection assays have provided circumstantial evidence for its existence on monkey mast cells (113). Komisar et al. showed that SEB stimulates rodent peritoneal mast cells as well and provided evidence for a protein receptor (74). It is possible that SEs act directly on their receptor and circumvent the typical two-stage mast cell immunoglobulin E antibody-Ag interaction (120).

Despite these observations, Alber et al. (2) were unable to directly stimulate monkey or human skin and intestinal mast cells with SEB to release inflammatory mediators. It was suggested that stimulation of mast cells in vivo occurs through a nonimmunological mechanism requiring the generation of neuropeptides released from peripheral terminals of primary sensory nerves. At least one putative mast cell-stimulating peptide, substance P, was implicated in SEB-induced toxicity by use of antibodies and a variety of inhibitors. The attractiveness of this explanation is that it is consistent with earlier predictions of neural involvement in the pathogenesis of SFP.

Is There a Relationship between Superantigenicity, TSS, and SFP?

The discovery of the mechanism of SAg action and the unique properties of this class of proteins provided an explanation for the multiple systemic effects seen in TSS patients. The massive cellular stimulation induced by superantigenic PTs explained the long-recognized fact that TSS patients have elevated serum cytokine and other mediator levels which mediate many symptoms of the disease. The realization that at least some of the pathogenesis of TSS could be attributed to immune cell stimulation led to the prediction by some investigators that SFP may also be a reflection of SAg function. Consistent with this prediction was the fact that TSS patients often have a gastrointestinal tract component characterized by vomiting and diarrhea. Also, patients with endotoxin shock have elevated cytokine levels and similarly display vomiting and diarrhea. If superantigenicity is responsible for SFP, the SEs presumably act directly on T cells and APCs in the gut. Although some SEs enter the circulation, they appear to be rapidly cleared by the kidneys so that significant systemic concentrations are unlikely to be achieved (6). This, plus the fact that TSS-associated symptoms (e.g., shock and fever) are not observed in SFP patients, suggests that SEs do not mediate SFP through systemic cytokines.

Despite their similarities and the evidence cited above, several lines of evidence suggest that the partial overlap between SFP and TSS symptoms is probably coincidental and that superantigenicity is not directly responsible for SFP. First, as discussed above, nonimmunological mast cell stimulation has been linked to the release of inflammatory mediator-affecting nerve interactions. The second line of evidence has come through mutagenesis of several SEs. Studies have revealed that T-cell stimulation and induction of emesis are separable functions and are determined by distinct portions of the SE molecules (1, 50, 55). It has been possible to construct SEA, SEB, and SEC1 mutants that are deficient in T-cell-stimulatory activity but that retain the ability to induce emesis and vice versa. Finally, although all PTs have been reported to have SAg function, only the SEs are emetic when ingested. The lack of emesis-inducing ability of some nonenterotoxic PTs has been attributed to instability in the gastrointestinal tract. However, at least one nonemetic PT, streptococcal pyrogenic exotoxin A, is very stable in gastric fluid (122).

If superantigenicity does not explain SFP, how do PTs cause vomiting and diarrhea in TSS? Several possibilities may explain how SEs and other PTs act on the gastrointestinal tract if they are not consumed through the oral route. First, the toxins' ability to induce TSS symptoms may be limited to the systemic circulation. If so, cytokines would need to enter the abdomen or gain access to the central nervous system from the circulation and mediate gastroenteritis pathogenesis. Alternatively, the PTs could enter the gut from the circulation and act directly at the local level as SAgs or through other mechanisms as proposed for SFP. The latter possibility is less likely since even nonenterotoxic PTs, including those that are susceptible to degradation in the gut, are known to cause gastrointestinal symptoms when they are associated with TSS.

These complex and unresolved issues are relevant to interpreting models for studying the enterotoxic activity of SEs. Oral administration of SEs, culture filtrates, or suspected food to monkeys is the preferred method for detecting enterotoxic properties of the SEs. However, intravenous administration of SEs to primates and other animals such as cats has been used as an alternative to feeding (31). Considering that systemic exposure to SEs mimics the situation leading to TSS symptoms, intravenous administration may not accurately reflect the pathogenesis and toxin properties required to induce emesis in SFP. Recently, other models that demonstrate emesis following administration of toxins to shrews and piglets have been proposed as alternatives to the "gold standard," the monkey feeding assay (56, 138). Additional studies will be required to determine whether these alternatives can replace the primate feeding assay.

CONCLUDING REMARKS

Progress has been made toward understanding the molecular aspects relevant to SFP. *S. aureus* has some unique properties that promote its ability to produce foodborne illness. However, further understanding of the molecular aspects of these unique properties should facilitate the implementation of more efficient ways to selectively inhibit staphylococcal survival, growth, and SE production in food. For example, exploitation of properties of the global regulation of virulence factors could be used to block exotoxin production and could have potential applications in the food industry.

One issue that continues to puzzle the scientific community is the questionable rationale for staphylococci to produce an emetic toxin. Unlike that of enteric pathogens that inhabit the gut, the ability to induce emesis and diarrhea as a mechanism to promote exit from the host and dissemination does not appear to be important for staphylococci. One may ask a similar question regarding the ability of SEs to induce lethal shock in TSS, since it is generally agreed that lethality is not advantageous to microorganisms. Instead, long-term host-pathogen coexistence usually relies upon adaptations that allow the microorganism to survive for extended periods without harming the host or being cleared by the immune response.

Based on what is now known regarding the superantigenic properties and proposed host-specific molecular adaptation of the SEs, one may propose that SEs are produced for the purpose of immunomodulating the host. Exposure of animals and peripheral blood mononuclear cell cultures to SEs and other SAgs has repeatedly been shown to induce at least a transient immunosuppression. Similarly, TSS patients often fail to produce a significant immune response to causative superantigenic PTs and remain susceptible to subsequent toxigenic illnesses. Hence, one may speculate that the harmful effects on the host (in SFP and TSS) are merely secondary effects of the staphylococci's attempting to affect immune cell function to allow the bacteria to survive and persistently colonize their many animal hosts.

Our efforts in preparation of the manuscript were supported by grants from the PHS (AI28401, U54AI57141, RR15587, and RR00166), the U.S. Department of Agriculture (99-35201), and the Idaho Agricultural Experiment Station. We also thank Yong Ho Park, Patrick Schlievert, Richard Novick, Amy Wong, Cynthia Stauffacher, Sibyl Munson, Brian Wilkinson, Lynn Jablonski, and Scott Minnich for comments, suggestions, and critical opinions in certain topics covered in this publication.

References

1. Alber, G., D. K. Hammer, and B. Fleischer. 1990. Relationship between enterotoxic- and T lymphocyte-stimulating activity of staphylococcal enterotoxin B. *J. Immunol.* **144**:4501–4506.

2. Alber, G., P. H. Scheuber, B. Reck, B. Sailer-Kramer, A. Hartmann, and D. K. Hammer. 1989. Role of substance P in immediate-type skin reactions induced by staphylococcal enterotoxin B in unsensitized monkeys. *J. Allergy Clin. Immunol.* **84**:880–885.

3. Al-Daccak, R., K. Mehindate, F. Damdoumi, P. Etongue-Mayer, H. Nilsson, P. Antonsson, M. Sundstrom, M. Dohlsten, R. P. Sekaly, and W. Mourad. 1998. Staphylococcal enterotoxin D is a promiscuous superantigen offering multiple modes of interactions with the MHC class II receptors. *J. Immunol.* **160**:225–232.

4. Anonymous. 1992. *Foodborne Pathogenic Microorganisms and Natural Toxins.* Center for Food Safety and Applied Nurition, U.S. Food and Drug Administration, Rockville, Md.

5. Bayles, K. W., and J. J. Iandolo. 1989. Genetic and molecular analyses of the gene encoding staphylococcal enterotoxin D. *J. Bacteriol.* **171**:4799–4806.

6. Beery, J. T., S. L. Taylor, L. R. Schlunz, R. C. Freed, and M. S. Bergdoll. 1984. Effects of staphylococcal enterotoxin A on the rat gastrointestinal tract. *Infect. Immun.* **44**:234–240.

7. Bergdoll, M. S. 1988. Monkey feeding test for staphylococcal enterotoxin. *Methods Enzymol.* **165**:324–333.

8. Bergdoll, M. S. 1985. The staphylococcal enterotoxin—an update, p. 247–254. *In* J. Jeljaszewicz (ed.), *The Staphylococci.* Gustav Fischer Verlag, Stuttgart, Germany.

9. Bergdoll, M. S. 1979. Staphylococcal intoxications, p. 443–494. *In* H. Riemann and F. L. Bryan (ed.), *Foodborne Infections and Intoxications.* Academic Press, Inc., New York, N.Y.

10. Bergdoll, M. S. 1989. *Staphylococcus aureus*, p. 463–523. *In* M. P. Doyle (ed.), *Foodborne Bacterial Pathogens.* Marcel Dekker, Inc., New York, N.Y.

11. Bergdoll, M. S., C. R. Borja, R. N. Robbins, and K. F. Weiss. 1971. Identification of enterotoxin E. *Infect. Immun.* **4**:593–595.

12. Bergdoll, M. S., M. J. Surgalla, and G. M. Dack. 1959. Staphylococcal enterotoxin. Identification of a specific precipitating antibody with enterotoxin-neutralizing property. *J. Immunol.* **83**:334–338.

13. Betley, M. J., D. W. Borst, and L. B. Regassa. 1992. Staphylococcal enterotoxins, toxic shock syndrome toxin and streptococcal pyrogenic exotoxins: a comparative study of their molecular biology. *Chem. Immunol.* **55**:1–35.

14. Betley, M. J., and J. J. Mekalanos. 1985. Staphylococcal enterotoxin A is encoded by phage. *Science* **229**:185–187.

15. Betley, M. J., P. M. Schlievert, M. S. Bergdoll, G. A. Bohach, J. J. Iandolo, S. A. Khan, P. A. Pattee, and R. R. Reiser. 1990. Staphylococcal gene nomenclature. *ASM News* **56**:182.

16. Blaiotta, G., D. Ercolini, C. Pennacchia, V. Fusco, A. Casaburi, O. Pepe, and F. Villani. 2004. PCR detection of staphylococcal enterotoxin genes in *Staphylococcus* spp. strains isolated from meat and dairy products. Evidence for new variants of SEG and SEI in *S. aureus* AB-8802. *J. Appl. Microbiol.* **97**:719–730.

17. Bohach, G. A., D. J. Fast, R. D. Nelson, and P. M. Schlievert. 1990. Staphylococcal and streptococcal pyrogenic toxins involved in toxic shock syndrome and related illnesses. *Crit. Rev. Microbiol.* **17**:251–272.

18. **Bohach, G. A., C. J. Hovde, J. P. Handley, and P. M. Schlievert.** 1988. Cross-neutralization of staphylococcal and streptococcal pyrogenic toxins by monoclonal and polyclonal antibodies. *Infect. Immun.* 56:400–404.

19. **Bohach, G. A., L. M. Jablonski, C. F. Deobald, Y. I. Chi, and C. V. Stauffacher.** 1995. Functional domains of staphylococcal enterotoxin, p. 339–356. *In* M. Ecklund, J. L. Richard, and M. Mise (ed.), *Molecular Approaches to Food Safety; Issues Involving Toxic Microorganisms.* Alaken, Fort Collins, Co.

20. **Bohach, G. A., C. V. Stauffacher, D. H. Ohlendorf, Y. I. Chi, G. M. Vath, and P. M. Schlievert.** 1996. The staphylococcal and streptococcal pyrogenic toxin family. *In* B. R. Singh and A. T. Tu (ed.), *Natural Toxins II.* Plenum, New York, N.Y.

21. **Brunskill, E. W., and K. W. Bayles.** 1996. Identification of LytSR-regulated genes from *Staphylococcus aureus. J. Bacteriol.* 178:5810–5812.

22. **Bryan, F. L.** 1976. *Staphylococcus aureus*, p. 12–128. *In* M. P. deFigueiredo and D. F. Splittstoesser (ed.), *Food Microbiology: Public Health and Spoilage Aspects.* AVI, Westport, Conn.

23. **Buzby, J. C., T. Roberts, C. T. Lin, and J. M. McDonald.** 1996. Bacterial foodborne disease medical costs and productively losses. *Agric. Econ. Rep.* **147.**

24. **Carl, K. E.** 1975. Oregon's experience with microbiological standards for meat. *J. Milk Food Technol.* 38:483–486.

25. **Casman, E. P., M. S. Bergdoll, and J. Robinson.** 1963. Designation of staphylococcal enterotoxins. *J. Bacteriol.* 85:715–716.

26. **Chan, P. F., and S. J. Foster.** 1998. Role of SarA in virulence determinant production and environmental signal transduction in *Staphylococcus aureus. J. Bacteriol.* 180:6232–6241.

27. **Cheung, A. L., Y. T. Chien, and A. S. Bayer.** 1999. Hyperproduction of alpha-hemolysin in a *sigB* mutant is associated with elevated SarA expression in *Staphylococcus aureus. Infect. Immun.* 67:1331–1337.

28. **Cheung, A. L., and G. Zhang.** 2002. Global regulation of virulence determinants in *Staphylococcus aureus* by the SarA protein family. *Front. Biosci.* 7:d1825–d1842.

29. **Chien, Y., and A. L. Cheung.** 1998. Molecular interactions between two global regulators, sar and agr, in *Staphylococcus aureus. J. Biol. Chem.* 273:2645–2652.

30. **Choi, Y., J. A. Lafferty, J. R. Clements, J. K. Todd, E. W. Gelfand, J. Kappler, P. Marrack, and B. L. Kotzin.** 1990. Selective expansion of T cells expressing V beta 2 in toxic shock syndrome. *J. Exp. Med.* 172:981–984.

31. **Clark, W. G., and J. S. Page.** 1968. Pyrogenic responses to staphylococcal enterotoxins A and B in cats. *J. Bacteriol.* 96:1940–1946.

32. **Couch, J. L., M. T. Soltis, and M. J. Betley.** 1988. Cloning and nucleotide sequence of the type E staphylococcal enterotoxin gene. *J. Bacteriol.* 170:2954–2960.

33. **Dack, G. M., W. E. Cary, O. Woolper, and H. Wiggers.** 1930. An outbreak of food poisoning proved to be due to a yellow hemolytic *Staphylococcus. Can. J. Microbiol.* 4:167–175.

34. **Deringer, J. R., R. J. Ely, S. R. Monday, C. V. Stauffacher, and G. A. Bohach.** 1997. V beta-dependent stimulation of bovine and human T cells by host-specific staphylococcal enterotoxins. *Infect. Immun.* 65:4048–4054.

35. **Do Carmo, L. S., C. Cummings, V. R. Linardi, R. S. Dias, J. M. De Souza, M. J. De Sena, D. A. Dos Santos, J. W. Shupp, R. K. Pereira, and M. Jett.** 2004. A case study of a massive staphylococcal food poisoning incident. *Foodborne Pathog. Dis.* 1:241–246.

36. **Everson, M. L., M. W. Hinds, R. S. Bernstein, and M. S. Bergdoll.** 1988. Estimation of human dose of staphylococcal enterotoxin A from a large outbreak of staphylocccal food poisoning involving chocolate milk. *Int. J. Food Microbiol.* 7:311–316.

37. **Felsenstein, J.** 1989. PHYLIP-Phylogeny Inference Package (Version 3.2). *Cladistics* 5:164–166.

38. **Fields, B. A., E. L. Malchiodi, H. Li, X. Ysern, C. V. Stauffacher, P. M. Schlievert, K. Karjalainen, and R. A. Mariuzza.** 1996. Crystal structure of a T-cell receptor beta-chain complexed with a superantigen. *Nature* 384:188–192.

39. **Fitzgerald, J. R., S. R. Monday, T. J. Foster, G. A. Bohach, P. J. Hartigan, W. J. Meaney, and C. J. Smyth.** 2001. Characterization of a putative pathogenicity island from bovine *Staphylococcus aureus* encoding multiple superantigens. *J. Bacteriol.* 183:63–70.

40. **Fournier, B., and A. Klier.** 2004. Protein A gene expression is regulated by DNA supercoiling which is modified by the ArlS-ArlR two-component system of *Staphylococcus aureus. Microbiology* 150:3807–3819.

41. **Fraser, J. D., S. Lowe, M. J. Irwin, N. R. Gascoigne, and K. R. Hudson.** 1993. Structural model of staphylococcal enterotoxin A interaction with MHC class II antigens, p. 7–30. *In* B. T. Huber and E. Palmer (ed.), *Superantigens: a Pathogen's View of the Immune System.* Cold Spring Harbor Laboratory Press, Plainview, N.Y.

42. **Friedman, M. E.** 1966. Inhibition of staphylococcal enterotoxin B formation in broth cultures. *J. Bacteriol.* 92:277–278.

43. **Fukuchi, K., Y. Kasahara, K. Asai, K. Kobayashi, S. Moriya, and N. Ogasawara.** 2000. The essential two-component regulatory system encoded by *yycF* and *yycG* modulates expression of the *ftsAZ* operon in *Bacillus subtilis. Microbiology* 146(Pt. 7):1573–1583.

44. **Gertz, S., S. Engelmann, R. Schmid, A. K. Ziebandt, K. Tischer, C. Scharf, J. Hacker, and M. Hecker.** 2000. Characterization of the sigma(B) regulon in *Staphylococcus aureus. J. Bacteriol.* 182:6983–6991.

45. **Gilbert, R. J., and A. A. Wieneke.** 1973. Staphylococcal food poisoning with special reference to the detection of enterotoxin in food, p. 273–285. *In* B. C. Hobbs and J. H. Christian (ed.), *The Microbiological Safety of Food.* Academic Press, New York, N.Y.

46. **Gronenborn, A. M., D. R. Filpula, N. Z. Essig, A. Achari, M. Whitlow, P. T. Wingfield, and G. M. Clore.** 1991. A novel, highly stable fold of the immunoglobulin binding domain of streptococcal protein G. *Science* 253:657–661.

47. **Guthertz, L. S., J. T. Fruin, R. L. Okoluk, and J. L. Fowler.** 1977. Microbial quality of frozen comminuted turkey meat. *J. Food Sci.* 42:1344–1477.

48. **Guthertz, L. S., J. T. Fruin, D. Spicer, and J. L. Fowler.** 1976. Microbial quality of fresh comminuted turkey meat. *J. Milk Food Technol.* 39:823–829.

49. Hajek, V., and E. Marsalek. 1973. The occurrence of entero-toxigenic *Staphylococcus aureus* strains in hosts of different animal species. *Zentbl. Bakteriol. Orig. A* **223:**63–68.

50. Harris, T. O., and M. J. Betley. 1995. Biological activities of staphylococcal enterotoxin type A mutants with N-terminal substitutions. *Infect. Immun.* **63:**2133–2140.

51. Hase, C. C., and R. A. Finkelstein. 1993. Bacterial extracellular zinc-containing metalloproteases. *Microbiol. Rev.* **57:**823–837.

52. Herbert, S., P. Barry, and R. P. Novick. 2001. Subinhibitory clindamycin differentially inhibits transcription of exoprotein genes in *Staphylococcus aureus*. *Infect. Immun.* **69:**2996–3003.

53. Hoffmann, M. L., L. M. Jablonski, K. K. Crum, S. P. Hackett, Y. I. Chi, C. V. Stauffacher, D. L. Stevens, and G. A. Bohach. 1994. Predictions of T-cell receptor- and major histocompatibility complex-binding sites on staphylococcal enterotoxin C1. *Infect. Immun.* **62:**3396–3407.

54. Holmberg, S. D., and P. A. Blake. 1984. Staphylococcal food poisoning in the United States. New facts and old misconceptions. *JAMA* **251:**487–489.

55. Hovde, C. J., J. C. Marr, M. L. Hoffmann, S. P. Hackett, Y. I. Chi, K. K. Crum, D. L. Stevens, C. V. Stauffacher, and G. A. Bohach. 1994. Investigation of the role of the disulphide bond in the activity and structure of staphylococcal enterotoxin C1. *Mol. Microbiol.* **13:**897–909.

56. Hu, D. L., K. Omoe, Y. Shimoda, A. Nakane, and K. Shinagawa. 2003. Induction of emetic response to staphylococcal enterotoxins in the house musk shrew (*Suncus murinus*). *Infect. Immun.* **71:**567–570.

57. Huang, I. Y., and M. S. Bergdoll. 1970. The primary structure of staphylococcal enterotoxin B. 3. The cyanogen bromide peptides of reduced and aminoethylated enterotoxin B, and the complete amino acid sequence. *J. Biol. Chem.* **245:**3518–3525.

58. Hudson, K. R., R. E. Tiedemann, R. G. Urban, S. C. Lowe, J. L. Strominger, and J. D. Fraser. 1995. Staphylococcal enterotoxin A has two cooperative binding sites on major histocompatibility complex class II. *J. Exp. Med.* **182:**711–720.

59. Iandolo, J. J., and W. M. Shafer. 1977. Regulation of staphylococcal enterotoxin B. *Infect. Immun.* **16:**610–616.

60. Ikeda, T., N. Tamate, K. Yamaguchi, and S. Makino. 2005. Mass outbreak of food poisoning disease caused by small amounts of staphylococcal enterotoxins A and H. *Appl. Environ. Microbiol.* **71:**2793–2795.

61. Ingavale, S., W. van Wamel, T. T. Luong, C. Y. Lee, and A. L. Cheung. 2005. Rat/MgrA, a regulator of autolysis, is a regulator of virulence genes in *Staphylococcus aureus*. *Infect. Immun.* **73:**1423–1431.

62. Janzon, L., and S. Arvidson. 1990. The role of the delta-lysin gene (*hld*) in the regulation of virulence genes by the accessory gene regulator (*agr*) in *Staphylococcus aureus*. *EMBO J.* **9:**1391–1399.

63. Jardetzky, T. S., J. H. Brown, J. C. Gorga, L. J. Stern, R. G. Urban, Y. I. Chi, C. Stauffacher, J. L. Strominger, and D. C. Wiley. 1994. Three-dimensional structure of a human class II histocompatibility molecule complexed with superantigen. *Nature* **368:**711–718.

64. Jarraud, S., M. A. Peyrat, A. Lim, A. Tristan, M. Bes, C. Mougel, J. Etienne, F. Vandenesch, M. Bonneville, and G. Lina. 2001. egc, a highly prevalent operon of enterotoxin gene, forms a putative nursery of superantigens in *Staphylococcus aureus*. *J. Immunol.* **166:**669–677.

65. Jay, J. M. 2000. *Modern Food Microbiology*, 6th ed., p. 441–455. Aspen Publishers, Gaithersburg, Md.

66. Jett, M., R. Neill, C. Welch, T. Boyle, E. Bernton, D. Hoover, G. Lowell, R. E. Hunt, S. Chatterjee, and P. Gemski. 1994. Identification of staphylococcal enterotoxin B sequences important for induction of lymphocyte proliferation by using synthetic peptide fragments of the toxin. *Infect. Immun.* **62:**3408–3415.

67. Johns, M. B., Jr., and S. A. Khan. 1988. Staphylococcal enterotoxin B gene is associated with a discrete genetic element. *J. Bacteriol.* **170:**4033–4039.

68. Johnson, L. P., and P. M. Schlievert. 1983. A physical map of the group A streptococcal pyrogenic exotoxin bacteriophage T12 genome. *Mol. Gen. Genet.* **189:**251–255.

69. Jones, C. L., and S. A. Khan. 1986. Nucleotide sequence of the enterotoxin B gene from *Staphylococcus aureus*. *J. Bacteriol.* **166:**29–33.

70. Kappler, J., B. Kotzin, L. Herron, E. W. Gelfand, R. D. Bigler, A. Boylston, S. Carrel, D. N. Posnett, Y. Choi, and P. Marrack. 1989. V beta-specific stimulation of human T cells by staphylococcal toxins. *Science* **244:**811–813.

71. Katsuno, S., and M. Kondo. 1973. Regulation of staphylococcal enterotoxin B synthesis and its relation to other extracellular proteins. *Jpn. J. Med. Sci. Biol.* **26:**26–29.

72. Kent, T. H. 1966. Staphylococcal enterotoxin gastroenteritis in rhesus monkeys. *Am. J. Pathol.* **48:**387–407.

73. Klimpel, K. R., N. Arora, and S. H. Leppla. 1994. Anthrax toxin lethal factor contains a zinc metalloprotease consensus sequence which is required for lethal toxin activity. *Mol. Microbiol.* **13:**1093–1100.

74. Komisar, J., J. Rivera, A. Vega, and J. Tseng. 1992. Effects of staphylococcal enterotoxin B on rodent mast cells. *Infect. Immun.* **60:**2969–2975.

75. Kornblum, J., B. Kreiswirth, S. J. Projan, H. Ross, and R. P. Novick. 1990. agr: a polycistronic locus regulating exoprotein synthesis *Staphylococcus aureus*, p. 373–402. *In* R. P. Novick and R. Skurray (ed.), *Molecular Biology of the Staphylococci*. VCH Publishers, New York, N.Y.

76. Kullik, I. I., and P. Giachino. 1997. The alternative sigma factor sigmaB in *Staphylococcus aureus*: regulation of the *sigB* operon in response to growth phase and heat shock. *Arch. Microbiol.* **167:**151–159.

77. Kuroda, M., T. Ohta, I. Uchiyama, T. Baba, H. Yuzawa, I. Kobayashi, L. Cui, A. Oguchi, K. Aoki, Y. Nagai, J. Lian, T. Ito, M. Kanamori, H. Matsumaru, A. Maruyama, H. Murakami, A. Hosoyama, Y. Mizutani-Ui, N. K. Takahashi, T. Sawano, R. Inoue, C. Kaito, K. Sekimizu, H. Hirakawa, S. Kuhara, S. Goto, J. Yabuzaki, M. Kanehisa, A. Yamashita, K. Oshima, K. Furuya, C. Yoshino, T. Shiba, M. Hattori, N. Ogasawara, H. Hayashi, and K. Hiramatsu. 2001. Whole genome sequencing of methicillin-resistant *Staphylococcus aureus*. *Lancet* **357:**1225–1240.

78. Leder, L., A. Llera, P. M. Lavoie, M. I. Lebedeva, H. Li, R. P. Sekaly, G. A. Bohach, P. J. Gahr, P. M. Schlievert,

K. Karjalainen, and R. A. Mariuzza. 1998. A mutational analysis of the binding of staphylococcal enterotoxins B and C3 to the T cell receptor beta chain and major histocompatibility complex class II. *J. Exp. Med.* **187:**823–833.

79. Letertre, C., S. Perelle, F. Dilasser, and P. Fach. 2003. Identification of a new putative enterotoxin SEU encoded by the *egc* cluster of *Staphylococcus aureus*. *J. Appl. Microbiol.* **95:**38–43.

80. Li, R., A. C. Manna, S. Dai, A. L. Cheung, and G. Zhang. 2003. Crystal structure of the SarS protein from *Staphylococcus aureus*. *J. Bacteriol.* **185:**4219–4225.

81. Lina, G., G. A. Bohach, S. P. Nair, K. Hiramatsu, E. Jouvin-Marche, and R. Mariuzza. 2004. Standard nomenclature for the superantigens expressed by *Staphylococcus*. *J. Infect. Dis.* **189:**2334–2336.

82. Liu, Y., A. Manna, R. Li, W. E. Martin, R. C. Murphy, A. L. Cheung, and G. Zhang. 2001. Crystal structure of the SarR protein from *Staphylococcus aureus*. *Proc. Natl. Acad. Sci. USA* **98:**6877–6882.

83. Liu, Y., A. C. Manna, C. H. Pan, I. A. Kriksunov, D. J. Thiel, A. L. Cheung, and G. Zhang. 2006. Structural and function analyses of the global regulatory protein SarA from *Staphylococcus aureus*. *Proc. Natl. Acad. Sci. USA* **103:**2392–2397.

84. Marr, J. C., J. D. Lyon, J. R. Roberson, M. Lupher, W. C. Davis, and G. A. Bohach. 1993. Characterization of novel type C staphylococcal enterotoxins: biological and evolutionary implications. *Infect. Immun.* **61:**4254–4262.

85. Marrack, P., and J. Kappler. 1990. The staphylococcal enterotoxins and their relatives. *Science* **248:**1066.

86. McNamara, P. J., K. C. Milligan-Monroe, S. Khalili, and R. A. Proctor. 2000. Identification, cloning, and initial characterization of *rot*, a locus encoding a regulator of virulence factor expression in *Staphylococcus aureus*. *J. Bacteriol.* **182:**3197–3203.

87. Merritt, E. A., and W. G. Hol. 1995. AB5 toxins. *Curr. Opin. Struct. Biol.* **5:**165–171.

88. Merson, M. H. 1973. The epidemiology of staphylococcal foodborne disease, p. 20–37. *In* Proceedings Staphylococci in Foods. Pennsylvania State University Press, University Park, Pa.

89. Mitchell, D. T., D. G. Levitt, P. M. Schlievert, and D. H. Ohlendorf. 2000. Structural evidence for the evolution of pyrogenic toxin superantigens. *J. Mol. Evol.* **51:**520–531.

90. Miyazaki, E., J. M. Chen, C. Ko, and W. R. Bishai. 1999. The *Staphylococcus aureus rsbW (orf159)* gene encodes an anti-sigma factor of SigB. *J. Bacteriol.* **181:**2846–2851.

91. Monday, S. R., and G. A. Bohach. 2001. Genes encoding staphylococcal enterotoxins G and I are linked and separated by DNA related to other staphylococcal enterotoxins. *J. Nat. Toxins* **10:**1–8.

92. Monday, S. R., and G. A. Bohach. 1999. Use of multiplex PCR to detect classical and newly described pyrogenic toxin genes in staphylococcal isolates. *J. Clin. Microbiol.* **37:**3411–3414.

93. Munson, S. H., M. T. Tremaine, M. J. Betley, and R. A. Welch. 1998. Identification and characterization of staphylococcal enterotoxin types G and I from *Staphylococcus aureus*. *Infect. Immun.* **66:**3337–3348.

94. Noleto, A. L., L. M. Malburg, Jr., and M. S. Bergdoll. 1987. Production of staphylococcal enterotoxin in mixed cultures. *Appl. Environ. Microbiol.* **53:**2271–2274.

95. Novick, R. P. 2003. Autoinduction and signal transduction in the regulation of staphylococcal virulence. *Mol. Microbiol.* **48:**1429–1449.

96. Novick, R. P. 2003. Mobile genetic elements and bacterial toxinoses: the superantigen-encoding pathogenicity islands of *Staphylococcus aureus*. *Plasmid* **49:**93–105.

97. Novick, R. P., S. J. Projan, J. Kornblum, H. F. Ross, G. Ji, B. Kreiswirth, F. Vandenesch, and S. Moghazeh. 1995. The *agr* P2 operon: an autocatalytic sensory transduction system in *Staphylococcus aureus*. *Mol. Gen. Genet.* **248:**446–458.

98. Omoe, K., D. L. Hu, H. Takahashi-Omoe, A. Nakane, and K. Shinagawa. 2005. Comprehensive analysis of classical and newly described staphylococcal superantigenic toxin genes in *Staphylococcus aureus* isolates. *FEMS Microbiol. Lett.* **246:**191–198.

99. Omoe, K., D. L. Hu, H. Takahashi-Omoe, A. Nakane, and K. Shinagawa. 2003. Identification and characterization of a new staphylococcal enterotoxin-related putative toxin encoded by two kinds of plasmids. *Infect. Immun.* **71:**6088–6094.

100. Omoe, K., K. Imanishi, D. L. Hu, H. Kato, H. Takahashi-Omoe, A. Nakane, T. Uchiyama, and K. Shinagawa. 2004. Biological properties of staphylococcal enterotoxin-like toxin type R. *Infect. Immun.* **72:**3664–3667.

101. Orwin, P. M., J. R. Fitzgerald, D. Y. Leung, J. A. Gutierrez, G. A. Bohach, and P. M. Schlievert. 2003. Characterization of *Staphylococcus aureus* enterotoxin L. *Infect. Immun.* **71:**2916–2919.

102. Orwin, P. M., D. Y. Leung, H. L. Donahue, R. P. Novick, and P. M. Schlievert. 2001. Biochemical and biological properties of staphylococcal enterotoxin K. *Infect. Immun.* **69:**360–366.

103. Orwin, P. M., D. Y. Leung, T. J. Tripp, G. A. Bohach, C. A. Earhart, D. H. Ohlendorf, and P. M. Schlievert. 2002. Characterization of a novel staphylococcal enterotoxin-like superantigen, a member of the group V subfamily of pyrogenic toxins. *Biochemistry* **41:**14033–14040.

104. Pace, P. J. 1975. Bacteriological quality of delicatessen foods. *J. Milk Food Technol.* **38:**347–353.

105. Petersson, K., M. Hakansson, H. Nilsson, G. Forsberg, L. A. Svensson, A. Liljas, and B. Walse. 2001. Crystal structure of a superantigen bound to MHC class II displays zinc and peptide dependence. *EMBO J.* **20:**3306–3312.

106. Petersson, K., H. Pettersson, N. J. Skartved, B. Walse, and G. Forsberg. 2003. Staphylococcal enterotoxin H induces V alpha-specific expansion of T cells. *J. Immunol.* **170:**4148–4154.

107. Pivnick, H., I. E. Erdman, D. Collins-Thompson, G. Roberts, M. A. Johnston, D. R. Conley, G. Lachapelle, U. T. Purvis, R. Foster, and M. Milling. 1976. Proposed microbiological standards for ground beef based on a Canadian survey. *J. Milk Food Technol.* **39:**408–412.

108. Pontzer, C. H., J. K. Russell, and H. M. Johnson. 1989. Localization of an immune functional site on staphylococcal enterotoxin A using the synthetic peptide approach. *J. Immunol.* **143:**280–284.

109. Pragman, A. A., J. M. Yarwood, T. J. Tripp, and P. M. Schlievert. 2004. Characterization of virulence factor regulation by SrrAB, a two-component system in *Staphylococcus aureus*. *J. Bacteriol.* **186:**2430–2438.

110. Projan, S. J., S. Brown-Skrobot, P. M. Schlievert, F. Vandenesch, and R. P. Novick. 1994. Glycerol monolaurate inhibits the production of beta-lactamase, toxic shock toxin-1, and other staphylococcal exoproteins by interfering with signal transduction. *J. Bacteriol.* **176:**4204–4209.

111. Raj, H. D., and M. S. Bergdoll. 1969. Effect of enterotoxin B on human volunteers. *J. Bacteriol.* **98:**833–834.

112. Rao, L., R. K. Karls, and M. J. Betley. 1995. In vitro transcription of pathogenesis-related genes by purified RNA polymerase from *Staphylococcus aureus*. *J. Bacteriol.* **177:**2609–2614.

113. Reck, B., P. H. Scheuber, W. Londong, B. Sailer-Kramer, K. Bartsch, and D. K. Hammer. 1988. Protection against the staphylococcal enterotoxin-induced intestinal disorder in the monkey by anti-idiotypic antibodies. *Proc. Natl. Acad. Sci. USA* **85:**3170–3174.

114. Reda, K. B., V. Kapur, J. A. Mollick, J. G. Lamphear, J. M. Musser, and R. R. Rich. 1994. Molecular characterization and phylogenetic distribution of the streptococcal superantigen gene (*ssa*) from *Streptococcus pyogenes*. *Infect. Immun.* **62:**1867–1874.

115. Regassa, L. B., J. L. Couch, and M. J. Betley. 1991. Steady-state staphylococcal enterotoxin type C mRNA is affected by a product of the accessory gene regulator (*agr*) and by glucose. *Infect. Immun.* **59:**955–962.

116. Ren, K., J. D. Bannan, V. Pancholi, A. L. Cheung, J. C. Robbins, V. A. Fischetti, and J. B. Zabriskie. 1994. Characterization and biological properties of a new staphylococcal exotoxin. *J. Exp. Med.* **180:**1675–1683.

117. Said-Salim, B., P. M. Dunman, F. M. McAleese, D. Macapagal, E. Murphy, P. J. McNamara, S. Arvidson, T. J. Foster, S. J. Projan, and B. N. Kreiswirth. 2003. Global regulation of *Staphylococcus aureus* genes by Rot. *J. Bacteriol.* **185:**610–619.

118. Schad, E. M., A. C. Papageorgiou, L. A. Svensson, and K. R. Acharya. 1997. A structural and functional comparison of staphylococcal enterotoxins A and C2 reveals remarkable similarity and dissimilarity. *J. Mol. Biol.* **269:**270–280.

119. Schad, E. M., I. Zaitseva, V. N. Zaitsev, M. Dohlsten, T. Kalland, P. M. Schlievert, D. H. Ohlendorf, and L. A. Svensson. 1995. Crystal structure of the superantigen staphylococcal enterotoxin type A. *EMBO J.* **14:**3292–3301.

120. Scheuber, P. H., C. Denzlinger, D. Wilker, G. Beck, D. Keppler, and D. K. Hammer. 1987. Staphylococcal enterotoxin B as a nonimmunological mast cell stimulus in primates: the role of endogenous cysteinyl leukotrienes. *Int. Arch. Allergy Appl. Immunol.* **82:**289–291.

121. Schleifer, K. H. 1986. Gram positive cocci, p. 999–1100. *In* P. A. Sneath (ed.), *Bergey's Manual of Systematic Bacteriology*, 1st ed., vol. 2. Williams & Wilkins Co., Baltimore, Md.

122. Schlievert, P. M., L. M. Jablonski, M. Roggiani, I. Sadler, S. Callantine, D. T. Mitchell, D. H. Ohlendorf, and G. A. Bohach. 2000. Pyrogenic toxin superantigen site specificity in toxic shock syndrome and food poisoning in animals. *Infect. Immun.* **68:**3630–3634.

123. Schmidt, J. J., and L. Spero. 1983. The complete amino acid sequence of staphylococcal enterotoxin C1. *J. Biol. Chem.* **258:**6300–6306.

124. Schmidt, K. A., N. P. Donegan, W. A. Kwan, Jr., and A. Cheung. 2004. Influences of sigmaB and *agr* on expression of staphylococcal enterotoxin B (*seb*) in *Staphylococcus aureus*. *Can. J. Microbiol.* **50:**351–360.

125. Smeltzer, M. S., M. E. Hart, and J. J. Iandolo. 1993. Phenotypic characterization of *xpr*, a global regulator of extracellular virulence factors in *Staphylococcus aureus*. *Infect. Immun.* **61:**919–925.

126. Smith, F. C., R. A. Field, and J. C. Adams. 1974. Microbiology of Wyoming big game meat. *J. Milk Food Technol.* **37:**129–131.

127. Smyth, D. S., P. J. Hartigan, W. J. Meaney, J. R. Fitzgerald, C. F. Deobald, G. A. Bohach, and C. J. Smyth. 2005. Superantigen genes encoded by the *egc* cluster and SaPIbov are predominant among *Staphylococcus aureus* isolates from cows, goats, sheep, rabbits and poultry. *J. Med. Microbiol.* **54:**401–411.

128. Spero, L., B. Y. Griffin, J. L. Middlebrook, and J. F. Metzger. 1976. Effect of single and double peptide bond scission by trypsin on the structure and activity of staphylococcal enterotoxin C. *J. Biol. Chem.* **251:**5580–5588.

129. Stefani, S., and P. E. Varaldo. 2003. Epidemiology of methicillin-resistant staphylococci in Europe. *Clin. Microbiol. Infect.* **9:**1179–1186.

130. Stock, J. B., A. J. Ninfa, and A. M. Stock. 1989. Protein phosphorylation and regulation of adaptive responses in bacteria. *Microbiol. Rev.* **53:**450–490.

131. Su, Y. C., and A. C. Wong. 1995. Identification and purification of a new staphylococcal enterotoxin, H. *Appl. Environ. Microbiol.* **61:**1438–1443.

132. Sugiyama, H., and T. Hayama. 1965. Abdominal viscera as site of emetic action for staphylococcal enterotoxin in the monkey. *J. Infect. Dis.* **115:**330–336.

133. Sundberg, E. J., M. W. Sawicki, S. Southwood, P. S. Andersen, A. Sette, and R. A. Mariuzza. 2002. Minor structural changes in a mutated human melanoma antigen correspond to dramatically enhanced stimulation of a CD4+ tumor-infiltrating lymphocyte line. *J. Mol. Biol.* **319:**449–461.

134. Sundstrom, M., L. Abrahmsen, P. Antonsson, K. Mehindate, W. Mourad, and M. Dohlsten. 1996. The crystal structure of staphylococcal enterotoxin type D reveals Zn2+-mediated homodimerization. *EMBO J.* **15:**6832–6840.

135. Surkiewicz, B. F., M. E. Harris, and R. W. Johnston. 1973. Bacteriological survey of frozen meat and gravy produced at establishments under federal inspection. *Appl. Microbiol.* **26:**574–576.

136. Swaminathan, S., W. Furey, J. Pletcher, and M. Sax. 1992. Crystal structure of staphylococcal enterotoxin B, a superantigen. *Nature* **359:**801–806.

137. Swartzentruber, A., A. H. Schwab, A. P. Duran, B. A. Wentz, and R. B. Read, Jr. 1980. Microbiological quality of frozen shrimp and lobster tail in the retail market. *Appl. Environ. Microbiol.* **40:**765–769.

138. **Taylor, S. L., L. R. Schlunz, J. T. Beery, D. O. Cliver, and M. S. Bergdoll.** 1982. Emetic action of staphylococcal enterotoxin A on weanling pigs. *Infect. Immun.* **36:**1263–1266.

139. **Thompson, J. D., D. G. Higgins, and T. J. Gibson.** 1994. CLUSTAL W: improving the sensitivity of progressive multiple sequence alignment through sequence weighting, position-specific gap penalties and weight matrix choice. *Nucleic Acids Res.* **22:**4673–4680.

140. **Tiedemann, R. E., and J. D. Fraser.** 1996. Cross-linking of MHC class II molecules by staphylococcal enterotoxin A is essential for antigen-presenting cell and T cell activation. *J. Immunol.* **157:**3958–3966.

141. **Todd, E. C., G. A. Jarvis, K. F. Weiss, G. W. Riedell, and S. Charbonneau.** 1983. Microbiological quality of frozen cream-type pies sold in Canada. *J. Food Prot.* **46:**34–40.

142. **Townsend, D. E., and B. J. Wilkinson.** 1992. Proline transport in *Staphylococcus aureus*: a high-affinity system and a low-affinity system involved in osmoregulation. *J. Bacteriol.* **174:**2702–2710.

143. **Tremaine, M. T., D. K. Brockman, and M. J. Betley.** 1993. Staphylococcal enterotoxin A gene (*sea*) expression is not affected by the accessory gene regulator (*agr*). *Infect. Immun.* **61:**356–359.

144. **Tseng, C. W., and G. C. Stewart.** 2005. Rot repression of enterotoxin B expression in *Staphylococcus aureus. J. Bacteriol.* **187:**5301–5309.

145. **Warren, J. R.** 1977. Comparative kinetic stabilities of staphylococcal enterotoxin types A, B, and C1. *J. Biol. Chem.* **252:**6831–6834.

146. **Warren, J. R., L. Spero, and J. F. Metzger.** 1974. Stabilization of native structure by the closed disulfide loop of staphylococcal enterotoxin B. *Biochim. Biophys. Acta* **359:**351–363.

147. **Wengender, P. A., and K. J. Miller.** 1995. Identification of a PutP proline permease gene homolog from *Staphylococcus aureus* by expression cloning of the high-affinity proline transport system in *Escherichia coli. Appl. Environ. Microbiol.* **61:**252–259.

148. **Wentz, B. A., A. P. Druan, A. Swartzentruber, A. H. Schwab, and R. B. Read, Jr.** 1983. Microbiological quality of fresh blue crabmeat, clams and oysters. *J. Food Prot.* **46:**978–981.

149. **Wentz, B. A., A. P. Duran, A. Swartzentruber, A. H. Schwab, and R. B. Read, Jr.** 1984. Microbiological quality of frozen onion rings and tuna pies. *J. Food Prot.* **47:**58–60.

150. **Wikstrom, M., T. Drakenberg, S. Forsen, U. Sjobring, and L. Bjorck.** 1994. Three-dimensional solution structure of an immunoglobulin light chain-binding domain of protein L. Comparison with the IgG-binding domains of protein G. *Biochemistry* **33:**14011–14017.

151. **Zhang, L., L. Gray, R. P. Novick, and G. Ji.** 2002. Transmembrane topology of AgrB, the protein involved in the post-translational modification of AgrD in *Staphylococcus aureus. J. Biol. Chem.* **277:**34736–34742.

152. **Zhang, S., J. J. Iandolo, and G. C. Stewart.** 1998. The enterotoxin D plasmid of *Staphylococcus aureus* encodes a second enterotoxin determinant (*sej*). *FEMS Microbiol. Lett.* **168:**227–233.

Food Microbiology: Fundamentals and Frontiers, 3rd Ed.
Edited by M. P. Doyle and L. R. Beuchat
© 2007 ASM Press, Washington, D.C.

Craig W. Hedberg

Epidemiology of Foodborne Diseases

23

The epidemiology of foodborne diseases is constantly changing (24). As new foods and new sources of foods become available, new opportunities for foodborne disease transmission often follow. For example, the development of export production of raspberries in Guatemala was closely followed by widespread outbreaks of diarrheal illnesses caused by the coccidian parasite *Cyclospora cayetanensis* in the United States and Canada (30). Diets change in response to tastes and concerns about health. Americans are being challenged to eat five servings of fresh fruits and vegetables every day to help prevent cancer, heart disease, obesity, and diabetes (54, 67). As a nation, we have increased consumption of fruits and vegetables but continue to fall short of this goal (40, 64). However, the number and variety of outbreaks of foodborne illness associated with fresh fruits and vegetables have increased (58) (Fig. 23.1). Moreover, the U.S. population itself is changing. As the population ages, the number and proportion of people at increased risk for serious complications of foodborne illnesses are increasing. Thus, even if the incidence of foodborne illnesses remains constant, the public health burden of these illnesses will increase.

Not all change is bad, however. During the early 1990s, eating hamburgers at fast food restaurants was identified as an important risk factor for *Escherichia coli* O157:H7 infections (59). Following a large foodborne outbreak of *E. coli* O157:H7 infections in the Pacific Northwest in 1993, public health recommendations and industry actions addressed this problem (2). Subsequently, a Foodborne Diseases Active Surveillance Network (FoodNet) case control study of sporadic *E. coli* O157:H7 infections, conducted during 1996 and 1997, revealed that eating hamburgers at fast food restaurants was not associated with illness (37). Further regulatory and industry activities have led to a reduction in the incidence of *E. coli* O157:H7 infections to levels meeting the National Health Objectives for 2010 (18) (Fig. 23.2).

Does the constantly changing epidemiology of foodborne diseases mean food is getting safer or less safe? When we get beyond looking at specific pathogens or foods, it is difficult to say. Food safety is the product of complex interactions between environmental, cultural, and socioeconomic factors (52). Before FoodNet was established as part of the Centers for Disease Control and Prevention's Emerging Infection Program, there was no systematic basis for even estimating the magnitude of foodborne illnesses in the United States (46). Although the estimate by

Craig W. Hedberg, Division of Environmental and Occupational Health, School of Public Health, University of Minnesota, Minneapolis, MN 55455.

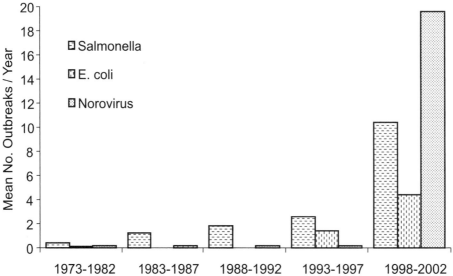

Figure 23.1 Confirmed foodborne outbreaks associated with fresh fruits and vegetables, United States, 1973–2002 (12b, 51).

Mead and colleagues (46) that 76 million foodborne illnesses occur each year in the United States was larger than previously reported estimates, it has become an internationally recognized standard and serves as the baseline against which future trends will be measured.

This chapter will provide an introduction to epidemiology and epidemiologic methods as they are applied to problems of foodborne diseases. An understanding of

epidemiology is important, because despite all of our best efforts to prevent foodborne diseases, humans remain the ultimate bioassay for low-level or sporadic contamination of our food supply (24). Epidemiologic methods of foodborne disease surveillance are needed to detect outbreaks, identify their causes, and assess the effectiveness of control measures. Epidemiologic data are also important in establishing food safety priorities, allocating food

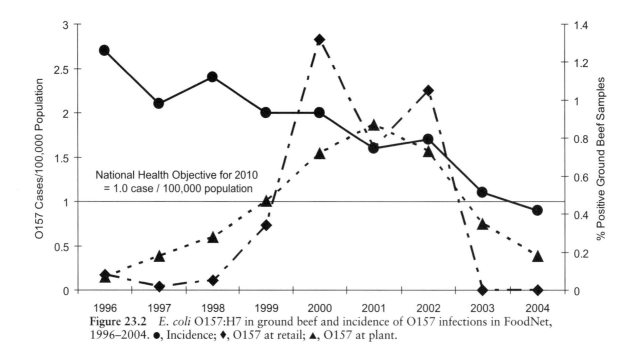

Figure 23.2 *E. coli* O157:H7 in ground beef and incidence of O157 infections in FoodNet, 1996–2004. ●, Incidence; ◆, O157 at retail; ▲, O157 at plant.

safety resources, stimulating public interest in food safety issues, establishing risk reduction strategies and public education campaigns, and evaluating the effectiveness of food safety programs (65). The examples provided in this chapter were drawn from the relevant literature and my own experiences as a foodborne disease epidemiologist. They were chosen to be illustrative, not necessarily representative. Epidemiologic data relevant to individual foodborne diseases are presented in their respective chapters.

EPIDEMIOLOGY

Epidemiology is the study of events in populations. Events are usually thought of in terms of diseases. For example, FoodNet was established, in part, to determine actively how many laboratory-confirmed infections caused by *Campylobacter, E. coli* O157:H7, and *Salmonella* occur in the populations under surveillance. A second major FoodNet activity has been a survey to determine how many diarrheal illnesses occur in the same populations that are under active surveillance for specific foodborne pathogens (7). By comparing the incidence and proportion of specific diseases attributable to foodborne transmission with the incidence of occurrence and medical evaluation for clinical syndromes, such as diarrhea, it becomes possible to determine what proportion of the diarrheal illnesses may be foodborne. This was the basis of the approach taken by Mead and colleagues in determining that 76 million foodborne illnesses occur each year in the United States (46).

Epidemiology also includes the study of factors associated with the occurrence of illnesses. The same population survey that FoodNet conducts to assess the frequency of diarrheal illness has also surveyed the population to find out how often people have potential exposures, such as eating undercooked hamburger, eating runny eggs, failing to wash their hands after handling raw chicken, or making a salad on the same cutting board they used to cut up raw chicken. Many of these same factors are likely to be assessed in the context of outbreak investigations to enable identification of the factors actually contributing to the occurrence of an outbreak.

The careful description of events in populations is generally called descriptive epidemiology. This process of determining characteristics of person, place, and time associated with illness occurrence forms the basis of all surveillance systems. The defining example of this process was the work of John Snow during the London cholera epidemic of 1860 (53). He located cases on a map of London, saw a cluster around the Broad Street pump, and pulled the pump handle. The contaminated water source was shut down, and the outbreak ended. Such dramatic examples of consequential epidemiology are not common. However, surveillance of foodborne disease, based on the principles of descriptive epidemiology, remains a cornerstone on which other food safety activities are built.

Surveillance is one function of public health agencies. In mid-September 1994, a microbiologist at the Minnesota Department of Health's Public Health Laboratory noted an increase in the number of *Salmonella enterica* serovar Enteritidis isolates being identified. Minnesota Department of Health epidemiologists began tracking the age, gender, and county of residence of the affected persons. After new cases were tracked for a couple of weeks, it was apparent that cases were clustered in southeastern Minnesota. The conclusion was made that a regional outbreak was occurring (28).

In contrast to the mere description of events, analytical epidemiology is the comparison of different rates at which events occur among groups. This involves determining a measure of association and a measure of the variability, or uncertainty, in the measurement of the point estimate of that association. For etiologic studies to determine the cause of an outbreak, the critical measure is the difference in risks of illness among those exposed to the contaminated food and those who were not exposed. This information is customarily presented as an odds ratio or risk ratio, depending on the form of the epidemiologic study design.

The classic tool of analytical epidemiology for foodborne diseases is the use of a case control study to identify the source of an outbreak. Case control studies are frequently used in foodborne outbreak investigations, because in most investigations it is easier to identify ill persons (cases) than it is to identify the entire group at risk. Outbreak investigations initiated from pathogen-specific surveillance will typically include case control studies. After concluding that a regional outbreak was occurring in southeastern Minnesota, epidemiologists constructed a questionnaire to ascertain what individuals had eaten in the 5 days before they became ill. The same questionnaire was used to interview healthy individuals of about the same ages from the same communities. By comparing the responses of the ill persons to the responses of the healthy individuals, who served as controls, it was determined that the outbreak was due to consumption of commercially manufactured ice cream (28). Ten (67%) of 15 infected individuals, but only 2 (13%) of 15 controls, had eaten brand A ice cream (matched odds ratio, 10.0; 95% confidence interval, 1.4–434). The magnitude of the odds ratio is a measure of the strength of the association between illness and exposure. The strength of this association left little doubt that the ice cream was the source of the outbreak. A public health intervention was initiated days before *Salmonella* was actually isolated from the suspected product. In addition, results of the case control study suggested that the problem likely involved multiple ice cream flavors and days of production. In response, a nationwide product recall was initiated. Further investigation led to the determination that the ice cream premix had been contaminated by raw egg residues in tanker trailers used to haul the premix from two other suppliers. Ultimately, what initially appeared to be a regional outbreak in southeastern Minnesota involved an estimated 224,000 illnesses, with cases reported from 41 states (28).

Recent developments in molecular subtyping (discussed below) make it possible to compare cases caused by an outbreak-associated strain with cases caused by unrelated strains (45). Because the occurrence of an outbreak implies a common source, the cases caused by unrelated strains with many different likely sources of exposure make a very efficient control group. This type of case-case comparison study has been critical for investigating *Listeria* infection outbreaks, where the risk for illness is dependent on underlying health conditions (21). However, the same methods could also be applied to outbreaks of *E. coli* O157:H7 or *Salmonella* infections, in which molecular subtyping can discriminate outbreak-associated from unrelated cases and public health resources may not be adequate to recruit and interview healthy control subjects.

Another closely related tool of analytical epidemiology is the cohort study. In outbreak investigations, retrospective cohort studies are conducted when an entire group (or population) with a common source of exposure can be identified. Individuals can be interviewed without prior knowledge of whether they were ill or not. Identifying groups based on their exposure status rather than their illness status allows for directly calculating a risk ratio for specific exposures. The risk ratio is the percentage of exposed persons who were ill divided by the percentage of unexposed persons who were ill. For example, in an outbreak of *Campylobacter* infections that was identified in a school in Madrid, Spain, 81 of 253 people interviewed developed a diarrheal illness consistent with campylobacteriosis, for an overall attack rate of 32% (35). The epidemic curve suggested a common source of exposure on a particular day. Information on foods eaten at the school on that day was available for 199 persons. Of these, 171 people ate a custard dessert and 77 became ill, for an attack rate of 45%. However, only 4 of 28 people who did not eat the custard

became ill, for an attack rate of 14% (35). Thus, the risk ratio for eating custard was 3.2, meaning that exposed individuals (i.e., persons who ate the custard) were three times more likely to become ill than were individuals who were not exposed.

Typical settings for cohort studies involve banquets, wedding receptions, and events where a list of all persons attending can be obtained. For specific exposures, such as those from a particular food item, the percentage of persons who became ill after eating the food can be determined. Thus, in addition to implicating the food item, it is possible to determine the attack rate among persons who ate it. Information on attack rates may also help in evaluating the factors contributing to the occurrence of the outbreak. For example, for many bacterial foodborne agents the attack rate may be related to the exposure dose. A high attack rate may imply high levels of contamination, such as may occur after prolonged temperature abuse.

In both case control and cohort studies, much attention is paid to the precision of the calculated estimate, as measured by the P value or 95% confidence interval. Most conventions treat a P value of <0.05 as significant. Similarly, 95% confidence intervals are expected to exclude 1.0 when a significant exposure occurs. Although these criteria have served as useful guides to data analysis over the years, it must always be noted that the main contributor to the P value and confidence interval is the size of the sample. The larger the sample size, the more likely that a statistically significant result may be obtained. Investigators should not discard potentially important associations just because they fail to achieve a P value of <0.05. Conversely, "highly significant" results may be totally spurious. A greater concern is that bias in sampling may have introduced reasons other than contamination of food for the observed difference in risks of illness. When potential sources of bias are properly controlled and accounted for, epidemiologic methods have demonstrated an impressive ability to accurately identify the sources and contributing causes of foodborne outbreaks.

As powerful as epidemiologic methods are, they do have limitations. In an outbreak setting where everyone ate a contaminated food item, there is no opportunity to demonstrate a difference in risks of illness between those who were exposed and those who were not. Alternative approaches, such as considering the amount eaten or when it was eaten, may allow some epidemiologic discrimination, but results are not likely to be "significant." For example, in the outbreaks of shigellosis associated with parsley, parsley was associated with illness at one restaurant in Minnesota but not at a second restaurant.

Chopped parsley was used on so many dishes at the second restaurant that most of the cases and the controls ate it (47).

In epidemiologic studies, the absence of an association does not necessarily mean that the food item or exposure in question is safe. For example, in the FoodNet case control study of sporadic *E. coli* O157:H7 infections, there was no association between sporadic infections and consumption of unpasteurized apple cider or attendance at day care (37). Both have been associated with outbreaks of disease and remain important potential sources of infection. They were not identified as the primary sources of sporadic *E. coli* O157:H7 infections, but it would be misleading to interpret the results to imply that there is no risk. Similarly, in a FoodNet case control study of *S. enterica* serovar Enteritidis infections, eating chicken prepared outside the home was the only significant risk factor associated with illness in a multivariate analysis (38). However, a U.S. Department of Agriculture risk assessment estimated that 182,000 egg-associated *S. enterica* serovar Enteritidis infections occurred in the United States during 2000 (57).

The concepts and methods of epidemiology can be used to examine the relationships between disease and all levels of food safety, from production and distribution to preparation and consumption.

AGENTS AND TRANSMISSION

Information about specific agents is contained in their respective chapters. However, in discussing the epidemiology of foodborne diseases it is important to keep in mind the chain of infection. This includes the agent, the reservoir that contains the agent, a means of escape from the reservoir, a mode of transmission to a susceptible host, and a means of entry into the host. These elements define the agent-host environment that results in the occurrence of illness.

Specific agents vary with respect to the types of illness they cause, from self-limited gastroenteritis associated with bacterial enterotoxins to invasive and life-threatening diseases associated with *Listeria monocytogenes*. Some agents, such as norovirus, are rapidly cleared from the gastrointestinal tract, and others, such as *Giardia*, may cause prolonged carriage. Infectious doses may range from a few *E. coli* O157:H7 CFU to hundreds of thousands of *Clostridium perfringens* CFU. Incubation periods range from a few hours for staphylococcal intoxication to weeks for viral hepatitis type A. Some agents such as *Shigella* infect only humans, whereas *Salmonella*, *Campylobacter*, and *E. coli* O157:H7 primarily originate from animal sources. Many agents are associated with typical food vehicles associated with their animal reservoirs, such as *E. coli* O157:H7 with ground beef and *S. enterica* serovar Enteritidis with eggs.

The characteristics of specific agents affect the potential sources of contamination during food production and preparation and form the basis for hazard analysis for specific foods. They also help shape the conduct of foodborne disease investigations and allow the investigation of source and transmission to be developed around the particular features of the disease.

FOODBORNE DISEASE SURVEILLANCE

Surveillance involves the systematic collection of data with analysis and dissemination of results. Surveillance systems may be passive or active, national or regional in scope, or based on a sentinel system of individual sites designed to provide estimates that can be generalized. Types of systems and their uses and limitations are described below.

Traditionally, there have been four purposes for conducting foodborne disease surveillance.

1. Identify, control, and prevent outbreaks of foodborne illness.
2. Determine the causes of foodborne illness.
3. Monitor trends in the occurrence of foodborne illness.
4. Quantify the magnitude of foodborne illness.

"Pulling the pump handle" is the goal of all public health epidemiologists conducting foodborne disease surveillance. Thus, it is necessary to identify, control, and prevent outbreaks of foodborne illness. For products with a long shelf life, such as frankfurters, ice cream, and toasted oats cereal, even a relatively long investigation can identify a contaminated product quickly enough to remove it from the marketplace in time to prevent many illnesses (9, 21, 28). For highly perishable products, such as fresh fruits and vegetables, often the outbreak will have run its course before the food item can be implicated (26, 47, 58).

Even if it is impossible to directly intervene to control the outbreak, there is still value in determining the causes of foodborne illness. Hazard analysis and critical control point (HACCP) systems for protecting the safety of foods rely on accurate knowledge of potential hazards (48). Many of these hazards, either in terms of specific agents, specific food ingredients, or various agent-food interactions, were originally determined as a result of foodborne illness surveillance. Because food sources and foodborne disease agents are constantly changing, hazard analysis is an ongoing process that requires continuous support from public health surveillance of foodborne diseases. Additionally, Mead and colleagues determined that the cause of most foodborne illnesses is unknown and much work remains to be done in this area (46). As the experience with *Cyclospora* outbreaks from Guatemalan raspberries reveals, foodborne disease surveillance may be the only way to identify both new causes of foodborne disease and potential foodborne hazards.

A third traditional purpose for conducting foodborne disease surveillance is to monitor trends in the occurrence of foodborne illness. This is important to identify priorities for new food safety activities and to monitor the effectiveness of existing programs. For example, during the 1980s, the increased occurrence of sporadic *S. enterica* serovar Enteritidis infections and outbreaks of infections in New England led to the identification of a new problem with *S. enterica* serovar Enteritidis contamination of grade A shell eggs (61). In the United States, the U.S. Department of Agriculture and the Food and Drug Administration have worked with the egg industry to develop and implement a number of control strategies (31). The incidence of *S. enterica* serovar Enteritidis infections recorded at FoodNet sites declined 46% from 1996 to 1999, suggesting that these control strategies have had a positive impact (44). Importantly, the incidence decreased by 71% in Connecticut, the FoodNet state with the highest incidence of *S. enterica* serovar Enteritidis infections (44).

As part of a national food safety initiative, FoodNet was established, in part, to quantify the magnitude of foodborne illnesses in the United States. Surveys of the population, physicians, and clinical laboratories linked to active surveillance of laboratory-confirmed infections provide the most comprehensive data available from which to estimate the magnitude of foodborne diseases. For the first time, there is a generally recognized estimate to be used as a basis for setting public health priorities and for evaluating the effectiveness of prevention measures. However, significant gaps remain in our knowledge base. Eighty-two percent of estimated foodborne illnesses cannot be attributed to known agents (46). And national surveillance for outbreaks of infections with norovirus, which is now recognized as the leading cause of outbreaks in which agents can be identified, has only recently been established (69) (Table 23.1).

Methods of Foodborne Disease Surveillance

Foodborne disease surveillance requires close collaboration between acute-disease epidemiologists, public health laboratories, and environmental health specialists to determine the likely sources of exposure and routes of food contamination. There are four primary components of foodborne disease surveillance (Table 23.2).

The first is identification and reporting of outbreaks associated with events and establishments, usually by ill persons who were part of the outbreak. There is a bias towards detecting outbreaks at events, such as weddings, where a large group gathers with a common experience and has some reason to discuss it later. People eating at a restaurant may never know their illness was also experienced by other restaurant patrons, who are unknown to them.

Outbreaks associated with events and establishments require prompt and thorough investigation to identify both the agent and the source. These outbreaks are usually recognized because of the occurrence of common symptoms, such as diarrhea and vomiting, and are

Table 23.1 Leading known causes of foodborne disease in the United States and ranking based on available surveillance systems[a]

| Agent | Estimated no. of cases | Rank by surveillance system | | |
		Active, 1999/2003[b]	Passive, 1999/2003[c]	Outbreak, 1999/2003[d]
Norovirus	9,200,000	NA[e]/NA	NA/NA	NA/1
Campylobacter spp.	1,963,141	1/2	2/NA	7/3
Salmonella, nontyphoidal	1,341,873	2/1	1/1	1/2
Clostridium perfringens	248,520	NA/NA	NA/NA	3/5
Giardia lamblia	200,000	3/NA	7/3	NA/NA

[a] Adapted from reference 43.
[b] Information from reference 12.
[c] Information from reference 12a. Does not distinguish route of transmission.
[d] Information from reference 12b.
[e] NA, not available.

Table 23.2 Primary components of foodborne disease surveillance

1. Investigation of outbreaks associated with events and establishments.
2. Pathogen-specific surveillance to identify clusters of cases caused by the same microorganism.
3. Determination of risk factors for sporadic cases of infection with common foodborne pathogens.
4. Population surveillance to determine the frequency of gastrointestinal illnesses, health care-seeking behavior, food consumption, and personal prevention measures.

reported to public health officials before a specific diagnosis has been established. Unfortunately, in many outbreak investigations, laboratory testing is not sought, is not available, or is not adequate to identify the causative agent. Of 2,751 outbreaks of foodborne disease reported to the Centers for Disease Control and Prevention from 1993 to 1997, 1,873 (68%) were classified as having unknown etiologies (51). However, many of these likely were outbreaks of norovirus gastroenteritis, which is not detected by clinical laboratory testing. With the development and use of norovirus testing by many state public health laboratories, the proportion of confirmed foodborne outbreaks in the United States increased from 1% in 1991 to 12% in 2000. However, 76% of these outbreaks were reported from only 11 states (69). In Minnesota, outbreaks whose etiologies could not be confirmed by laboratory testing have been characterized by the clinical and epidemiologic features of the reported illnesses. From 1981 to 1998, 120 (41%) of 295 confirmed foodborne outbreaks had clinical and epidemiologic features consistent with noroviruses and distinct from those of outbreaks caused by known bacterial infections or toxins (14). These include a median incubation period of between 24 and 48 h, 12- to 60-h duration of symptoms, and a relatively high proportion of ill persons experiencing vomiting (36).

Such epidemiologic profiling of outbreaks is useful to guide laboratory testing as well. Clinical laboratories, and many public health laboratories, do not test for the presence of enterotoxigenic *E. coli*. However, if a high proportion of individuals became ill several days after a common exposure and experienced diarrhea and cramps with little vomiting or fever, the investigator should seek specialized laboratory testing for enterotoxigenic *E. coli* (13). Just such a clinical and epidemiologic profile led to the identification of an atypical enteropathogenic strain of *E. coli* serogroup O39 as the cause of a foodborne outbreak associated with a restaurant (25). Similarly,

investigation of a series of event-associated outbreaks with unusually long median incubation periods of a week led to the recognition of *Cyclospora* as a foodborne pathogen (30).

The investigation of outbreaks associated with events is the oldest form of foodborne disease surveillance and is frequently overlooked in discussions of how to improve foodborne disease surveillance. However, it remains the primary way to identify "new" foodborne pathogens, such as *Cyclospora* or diarrheogenic *E. coli*, which are not routinely identified by clinical laboratories.

A major focus of public health surveillance for foodborne illnesses has been to improve pathogen-specific surveillance (63). Serotype-specific surveillance of *Salmonella* has identified many large, multistate outbreaks of salmonellosis; many were associated with previously unrecognized vehicles, most notably fresh fruits and vegetables (26, 57). Outbreaks of salmonellosis caused by cantaloupes, tomatoes, and alfalfa sprouts were all recognized because of unusual temporal clusters of cases involving an uncommon serotype. Epidemiologic investigations of these cases identified the sources.

Cases of *Salmonella* infections are electronically reported to the Centers for Disease Control and Prevention from state public health laboratories through the Public Health Laboratory Information System. An automated *Salmonella* outbreak detection algorithm (SODA) has been developed to look for unusual case clusters based on the 5-year mean number of cases from the same geographic area and week of the year (33). Similar schemes have been developed in Europe and Australia (60). Detection of clusters by SODA helped confirm the existence of multistate outbreaks of *S. enterica* serovar Stanley infections associated with alfalfa sprouts in 1995 and *S. enterica* serovar Agona infections associated with toasted oats cereal in 1998 (9, 42). However, both outbreaks were initially recognized by individual state health departments. Because it compares numbers of current cases to 5-year means, SODA appears to be most effective at detecting case clusters involving uncommon serotypes. As with other applications of serotype-specific surveillance, SODA is not likely to be especially sensitive to detect outbreaks caused by common serotypes, such as *S. enterica* serovar Typhimurium or *S. enterica* serovar Enteritidis. For these serovars, phage typing has been used to increase strain discrimination and facilitate outbreak detection. For example, outbreaks of *S. enterica* serovar Enteritidis infections associated with raw almonds in Canada and the United States and raw-milk cheese in France were detected because they were caused by uncommon phage types (23, 34).

Molecular subtyping schemes, such as pulsed-field gel electrophoresis (PFGE), can improve the investigation of outbreaks by distinguishing unrelated sporadic cases from those involving the main outbreak-associated strain. This distinction improves the specificity of the case definition and makes it more likely that the source of the outbreak can be identified. PFGE has been used in outbreaks of *E. coli* O157:H7, *Salmonella*, and *Shigella* infections (8, 9, 47). Molecular subtyping can also facilitate outbreak detection by changing serotype-specific surveillance into molecular subtype-specific surveillance. The public health utility of incorporating molecular subtyping into routine surveillance of *E. coli* O157:H7 infections at the state level has been demonstrated (3). However, many subtype cluster investigations initiated on a national basis fail to identify a common source of exposure (20). This failure may reflect the difficulties in conducting timely multistate outbreak investigations or uncertainties in the distributions of individual strains of these foodborne pathogens.

The ability to distinguish specific subtypes among relatively common pathogens, such as *E. coli* O157: H7 and *S. enterica* serovar Typhimurium, is the basis of the National Molecular Subtyping Network (PulseNet). PulseNet takes advantage of the combined revolutions in molecular biology and information technology. Highly reproducible PFGE patterns are generated under standardized conditions, and the PFGE patterns can be transmitted electronically between participating laboratories. For example, an outbreak of *E. coli* O157:H7 infections in Colorado was associated with consumption of a nationally distributed ground beef product. Within days, it was possible to compare the PFGE patterns of the outbreak strain to PFGE patterns of *E. coli* O157: H7 isolates throughout the United States (8). This ability gives PulseNet the potential to be the backbone of a public health surveillance system that can provide truly national surveillance for a variety of foodborne pathogens in a manner timely enough to provide an early warning system for outbreaks of foodborne illness.

Molecular subtyping also can be used to enhance the investigation of outbreaks associated with events or establishments, such as when multiple, apparently independent outbreaks of *Shigella sonnei* infections across the United States and Canada were shown to have been caused by a common strain (47). Although the independent investigations did not identify the cause, the linkage of the outbreaks allowed investigators to identify parsley that had been imported from Mexico as the common source.

All outbreak investigations require close collaboration with environmental health specialists and field investigators to identify and trace the sources of food items that may have been a vehicle for the causative agent. Although product tracebacks are frequently started only after a food item has been implicated, detailed product information is actually critical to epidemiologic analysis. In the outbreak of *S. enterica* serovar Enteritidis infections associated with brand A ice cream, for example, if ill persons and controls had initially been asked only if they had eaten ice cream, investigators would never have identified the source of the outbreak (28). Thirteen affected individuals (87%) and 10 controls (67%) ate some type of ice cream (odds ratio, 2.5; 95% confidence interval, 0.4–26.3). These data would not have raised suspicions about ice cream's being the source of the outbreak. Thus, collection of product source information should begin as early in the investigation as possible.

Determining risk factors for sporadic cases of infection with common foodborne pathogens is a third major component of foodborne disease surveillance. This can help identify targets for intervention and provide a basis to evaluate the intervention's effectiveness. Case control studies of sporadic *S. enterica* serovar Enteritidis infections in the United States and Europe helped establish the role of grade A shell eggs in the epidemiology of these infections. Case control studies of *Campylobacter* infections helped establish chicken as a primary source of *Campylobacter* (62). Similarly, case control studies of sporadic *E. coli* O157:H7 infections confirmed the importance of ground beef, particularly undercooked ground beef, as a source of the pathogen (39, 58).

FoodNet has conducted population-based case control studies for *E. coli* O157:H7, *Salmonella* serogroup B and D, *Campylobacter*, *Cryptosporidium*, and *Listeria monocytogenes* infections. These studies have provided updated and original estimates of the proportion of these infections that are attributable to specific food items. Results of the *E. coli* O157:H7 case control study demonstrated that eating hamburgers at a fast food restaurant was not associated with illness, as it had been in previous case control studies (37). This finding appears to be due to the fact that fast food restaurants are thoroughly cooking the hamburgers they serve. At home or at table service restaurants, where consumers have a choice to order an undercooked hamburger, eating undercooked hamburgers was associated with illness (37). Case control studies of *Campylobacter*, *S. enterica* serovar Enteritidis, and *S. enterica* serovar Heidelberg infections also demonstrated associations with eating out (29, 38, 55). Other interesting findings included associations between *S. enterica* serovar Enteritidis infection and chicken consumption, *S. enterica* serovar Heidelberg infection and egg consumption, and both *Campylobacter* and *Cryptosporidium* infections and foreign travel (55).

The final important component of foodborne disease surveillance is population surveillance to determine the frequency of gastrointestinal illnesses, health care-seeking behavior, food consumption, and personal prevention measures. This type of syndrome-specific surveillance is not usually conducted by state or local health departments. However, these measures formed the basis of the Centers for Disease Control and Prevention's recent estimates that 76 million foodborne diseases occur each year in the United States. These same data also are useful for understanding the results of active surveillance. By describing when and why people seek medical care when they are ill, these data can help to estimate how many illnesses occur for each one that gets a specific diagnosis. Finally, the population-based food consumption data are being used to estimate the proportion of specific diseases attributable to specific exposures in the FoodNet case control studies.

Surveillance for Food Hazards

Risk assessments for specific pathogens such as *E. coli* O157:H7 in ground beef and *S. enterica* serovar Enteritidis in shell eggs require making measurements of or assumptions about many parameters. At every step from farm to table, potential sources of contamination can be identified, measured, and modeled to determine the relative contribution to the overall risk of foodborne illness.

For many foodborne pathogens of public health importance, contamination of a finished product is a sporadic and rare occurrence. Thus, end-product testing is difficult to justify as a preventative measure. Given the scale of modern agriculture and food production systems, contamination levels below the statistical and microbiological threshold can still be large public health problems. For example, internal contamination of eggs with *S. enterica* serovar Enteritidis is rare; an estimated 1 in 20,000 eggs is contaminated on a national basis in the United States, with a range from 2.5 to 62.5 per 10,000 eggs from environmentally positive flocks (32). Despite this sporadic occurrence, consumption of contaminated shell eggs is estimated to cause more than 660,000 illnesses per year (32). For this reason, surrogate measures of contamination, such as identifying infected hens or environmental contamination in egg-laying houses, are more efficient and effective than testing of individual eggs.

Alternatively, measuring a relatively common organism such as commensal strains of *E. coli* may provide surveillance for control of operations that can be independently associated with a degree of risk for disease transmission. For example, in a series of baseline surveys of beef slaughter plants and ground beef conducted by

the U.S. Department of Agriculture in preparation for the introduction of HACCP regulations into the meat industry, 8.2% of steer and heifer carcasses and 15.8% of cow and bull carcasses were found to be contaminated by commensal *E. coli* strains (15, 16). However, only 4 (0.2%) of 2,081 steer and heifer carcasses, and none of 2,112 cow and bull carcasses, were contaminated with *E. coli* O157:H7. Of 563 finished ground beef samples, 78.6% were contaminated by commensal *E. coli* strains, but none were contaminated by *E. coli* O157:H7 (17). Thus, microbiological testing for commensal *E. coli* strains can provide useful information for monitoring of HACCPs, even though preventing contamination by *E. coli* O157:H7 is the desired outcome.

In addition to microbiological surveys of food products conducted by the U.S. Department of Agriculture and the Food and Drug Administration, the U.S. Department of Agriculture implemented the National Animal Health Monitoring System in 1983 to conduct national studies and compile data from industry sources to address emerging issues such as the association between calf management practices and the presence of *E. coli* O157:H7 and *Salmonella* in cattle herds (19, 41). The veterinary Diagnostic Laboratory Reporting System compiles and analyzes reports from state veterinary diagnostic laboratories to assess trends in infectious diseases among food animals (56).

Behavioral risk factors for foodborne disease range from choosing to eat alfalfa sprouts or undercooked ground beef to failing to wash hands between using a toilet and making a salad. These factors are evaluated and can be identified during the course of outbreak investigations. The frequency of these behaviors in the general population can be assessed through surveys, such as that being conducted as part of FoodNet. For example, 7.7% of respondents to the FoodNet population survey conducted from 1998 to 1999 reported eating alfalfa sprouts during the 7 days before the interview (10). However, 10.5% of California and Oregon residents reported eating sprouts. Thus, it is not entirely surprising that sprout-associated outbreaks of salmonellosis have been more common on the West Coast than in the rest of the United States (66). Among other potentially important food exposures, 25% of persons who ate eggs consumed runny eggs, 11% of persons who ate hamburgers ate hamburgers that were pink, 4.4% of respondents drank unpasteurized apple juice or apple cider, 3.4% drank unpasteurized milk, and 2.5% ate fresh oysters (10). Among persons who handled raw chicken, 88.6% reported washing their hands with soap and water after touching the raw chicken but only 48.3% of respondents had a meat thermometer in the home (10). Information of this type can provide important perspectives for understanding the occurrence of foodborne disease outbreaks and sporadic infections. It also serves as the basis for developing and modifying public education campaigns to improve food safety, such as the Partnership for Food Safety Education's national public education campaign, Fight Bac! (5). For example, the 2002–2003 FoodNet population survey revealed a decline in alfalfa sprout consumption of 51% overall and 39% in California and Oregon (11).

Lists of factors contributing to the occurrence of foodborne disease outbreaks are compiled by the Centers for Disease Control and Prevention from national surveillance data and by individual states. National data are typically published in summaries covering 5-year periods. These include detailed tables, by year, but very little in the way of synthesis or even trend analysis. For example, in outbreaks for which contributing factors were reported between 1988 and 1997, the two most commonly reported contributing factors were improper holding temperatures (60%) and poor personal hygiene (34%) (1, 51). During the first 5 years of this period, the incidence of poor personal hygiene declined as a contributing factor, associated with 44% of outbreaks in 1988 and only 29% of outbreaks in 1992. During the second 5-year period, the percentage of outbreaks involving poor personal hygiene increased from 27% of outbreaks in 1993 to 38% of outbreaks in 1997. It would be useful to know whether these changes represent actual trends in the occurrence of foodborne disease outbreaks or trends in the way outbreaks were investigated and reported or if they are mere variations in passively collected data that do not correlate with the actual occurrence of foodborne diseases. Unfortunately, contributing factors were reported for only 58% of all outbreaks reported to the Centers for Disease Control and Prevention, and not all outbreaks were reported (1, 51). For example, there was a 24% decline in the number of outbreaks reported between 1995 and 1996. There were 20 outbreaks reported from Minnesota in 1995 and one in 1996. Similarly, New York reported 89 outbreaks in 1995 and 9 in 1996. These apparent declines represent a lack of reporting to the passive outbreak surveillance system rather than an actual change in the occurrence of foodborne disease outbreaks in these states. Bias in what is reported is the largest systematic problem for interpreting national foodborne disease outbreak surveillance data.

The New York State Health Department has been a leader in attempting to quantify environmental and behavioral factors contributing to foodborne disease outbreaks. In contrast to the five categories of contributing factors historically tracked by the Centers for Disease Control and Prevention, the New York State

Department of Health's Bureau of Community Sanitation and Food Protection developed an expanded list of 19 contributing factors and began to distinguish foods based separately on methods of preparation and significant ingredients (68). The list of contributing factors has subsequently been further expanded and divided into 14 factors that may contribute to contamination of foods, 5 factors associated with pathogen survival or lack of inactivation, and 12 factors associated with pathogen proliferation and amplification (4). The Centers for Disease Control and Prevention has now incorporated this scheme into the national foodborne illness outbreak reporting system; however, much information will be submitted on an optional basis.

These refined summaries should provide better information on conditions found to contribute to the occurrence of individual outbreaks. However, they will not provide an estimate of the proportion of outbreaks attributable to the contributing factor identified. For example, the Environmental Health Specialists Network evaluated outbreaks associated with restaurants occurring within selected FoodNet sites (27). Norovirus accounted for more than 40% of outbreaks, with handling of food by an infected food worker and bare-hand contact with food identified as contributing factors in 65 and 35% of outbreaks, respectively (27). However, to put this in perspective it would be useful to know how often food workers in restaurants, generally, work while ill or have bare-hand contact with ready-to-eat foods. For comparison, 5% of food workers responding to a FoodNet population survey reported that they had worked while ill with vomiting or diarrhea (22). In a 2003 survey on the occurrence of foodborne illness risk factors in various food service settings, the Food and Drug Administration reported that 57% of observations on the prevention of hand contact with ready-to-eat foods in full service restaurants were out of compliance (6). Thus, the proportion of norovirus outbreaks attributable to these contributing factors is not simply the sum of all outbreaks in which these contributing factors were identified. These indirect comparisons suggest that the presence of an infected food worker is by far the most important risk factor for outbreaks of norovirus infections associated with restaurants.

This example highlights a critical need for addressing likely food hazards, first to determine the proportion of illnesses and outbreaks attributable to the hazard and then to estimate the predictive value of the hazard for causing disease. This is part of the process of risk assessment and is an important application of foodborne disease epidemiology (49). In a restaurant setting, we may know how many times bare-hand contact with foods or failure to wash hands is identified as a contributing factor in an outbreak of foodborne illness, but what is the frequency of this failure in restaurants? As noted above, the Food and Drug Administration began to address these data needs in 1997 with the creation of the FDA Retail Food Program Database of Foodborne Illness Risk Factors. This was followed by a second survey in 2003, with a third planned in 2008 to assess trends (6). The Environmental Health Specialists Network is also working to evaluate restaurants that have not experienced an outbreak by using tools similar to those for the outbreak evaluations (27). These efforts will help provide data to address important environmental questions using epidemiologic methods applied to surveillance for the hazards.

EVALUATING FOOD SAFETY SYSTEMS

Public health surveillance of foodborne disease is critical to the performance of food safety systems that are based on HACCP plans. Surveillance is required to identify new hazards. It also provides the ultimate feedback on the efficacy of HACCP plans. In the outbreak of salmonellosis due to commercial ice cream, cross contamination during transportation was identified as a hazard that had not been addressed in the manufacturer's HACCP plan. Epidemiologic data can also measure the success of food safety interventions. After implementation of the nationwide recall of brand A ice cream, a survey of brand A customers in Georgia revealed that 50% had not heard about the recall until 5 days after it was issued and 9% did not hear about it at all (43). Furthermore, 26% of persons who heard about the recall did not initially believe that the products were unsafe and many subsequently ate the ice cream. Of an estimated 11,000 cases of illness in Georgia associated with this outbreak, 16% were due to exposure after the recall (43).

As noted earlier, case control studies of sporadic *E. coli* O157:H7 infections conducted before and after the Jack-in-Box outbreak of 1993 demonstrated a change in the risk of illness associated with eating hamburgers at fast food restaurants. Recommendations that followed the outbreak indicated a reduction in the risk of eating hamburgers in fast food restaurants. This reduction appears to be due to thorough cooking of hamburger in these restaurants, since overall incidence rates for *E. coli* O157:H7 infections did not decline during this time period and the improved sensitivity of testing for *E. coli* O157:H7 provided a more complete picture of its abundance (Fig. 23.2). *E. coli* O157:H7 was isolated from fewer than 1 in 1,000 samples of ground beef sampled at the retail level by the U.S. Department of Agriculture from 1995 to 1998 (18). Following the introduction of

more sensitive culture methods late in 1999, the level of *E. coli* O157:H7-positive samples increased to 3.2 per 1,000. In 2000, more than 1% of samples tested positive. This increase was likely due primarily to the use of more sensitive testing methods rather than more frequent contamination of the finished product (50). However, following the 2002 USDA Food Safety Inspection Service notice to manufacturers of raw ground beef products that they must consider *E. coli* O157:H7 a potential hazard in their HACCP plans, the percentage of positive samples at the retail level decreased markedly and the incidence of *E. coli* O157:H7 infections decreased as well (18, 50) (Fig. 23.2). However, available microbiological and epidemiologic data continue to support the need for terminal control treatments, such as cooking or irradiation, to further reduce the risks of *E. coli* O157:H7 infections.

The FoodNet surveillance network was established, in part, to provide a tool to evaluate the public health impact of the U.S. Department of Agriculture's implementation of meat-processing plant HACCP. Surveillance data from the five original FoodNet sites provide consistent, population-based measures of the incidence of several important foodborne pathogens. Analysis of trends reveals declines in the occurrence of *Campylobacter*, *E. coli* O157:H7, *L. monocytogenes*, *Salmonella*, and *Yersinia* infections, with no change in the occurrence of *Shigella* infections and increases in the occurrence of *Vibrio* infections. It has been noted that these declining incidence rates occurred at the same time as the implementation of changes in meat- and poultry-processing plants, revisions to the Food and Drug Administration Food Code for restaurants, and increased attention to "Good Agricultural Practices" for fresh fruits and vegetables and eggs on farms (12). These are the types of effects FoodNet was designed to measure. However, determining cause and effect is not as simple as measuring the trends.

To begin with, there is considerable regional variability in the incidence of specific infections among the FoodNet sites. For example, in 1996, the reported incidence of *Campylobacter* infections ranged from 58 per 100,000 in California to 14 per 100,000 in Georgia (7). Similarly, the incidence of *E. coli* O157:H7 infections ranged from 5 per 100,000 in Minnesota to 0.6 per 100,000 in Georgia (7). Although the rates of *Salmonella* infections did not vary much between sites, the distribution of serotypes did; *S. enterica* serovar Enteritidis was the most common serotype in Connecticut but was rare in Georgia.

Overall, the incidence of *Campylobacter* infections reported in FoodNet declined from 25 per 100,000 in 1996 to 12.9 per 100,000 in 2004 (Fig. 23.3). However,

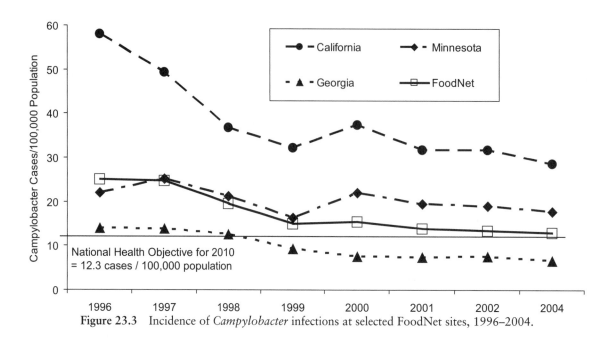

Figure 23.3 Incidence of *Campylobacter* infections at selected FoodNet sites, 1996–2004.

a large portion of that overall trend was due to marked declines in the incidence of *Campylobacter* infections in California, from 58 per 100,000 in 1996 to 28.6 per 100,000 in 2004 (12). Both the incidence and trend data continue to indicate considerable regional variability that cannot be readily explained. Thus, it is important to remember that the underlying factors that contribute to the regional variability of these infections may also lead to regional differences in how they respond to prevention and control measures.

SUMMARY

Humans remain the ultimate bioassay for low-level or sporadic contamination of our food supply (24). Epidemiologic methods of foodborne disease surveillance are powerful tools because they take advantage of events that are occurring throughout the population. This population-based lens, focused by advances in molecular subtyping and information technology available to public health laboratories, is particularly well suited to dealing with foodborne diseases associated with mass-produced and widely distributed food products.

Epidemiologic methods of foodborne disease surveillance are needed to detect outbreaks. In outbreak settings, rapid ascertainment of detailed information on food source and production is needed to confirm the cause of the outbreak and identify contributing factors. Epidemiologic data are also very important to establish food safety priorities, allocate food safety resources, stimulate

public interest in food safety issues, establish risk reduction strategies and public education campaigns, and evaluate the effectiveness of food safety programs (61). The principles and methods of epidemiologic surveillance can be applied to surveillance for food hazards as well as to surveillance for foodborne diseases.

Although this chapter focuses on experiences in the United States, the same methods of observation and analysis should form the basis of foodborne disease surveillance in developed and developing countries throughout the world. Since national food supplies are rapidly becoming global in origin, the need exists for an international system for foodborne disease surveillance as well. Models such as PulseNet provide opportunities to conduct multinational surveillance for at least the major bacterial foodborne disease agents. Because foodborne disease problems imported into countries such as the United States may represent endemic disease problems in the producing country, growing awareness of these problems may stimulate investment in interventions like safe water distribution and sanitary waste disposal systems that would benefit both the local population and consumers thousands of miles away.

References

1. **Bean, N. H., J. S. Goulding, C. Lao, and F. J. Angulo.** 1996. Surveillance for foodborne-disease outbreaks— United States, 1988–1992. *Morb. Mortal. Wkly. Rep.* **45** (SS-5):1–55.

2. Bell, B. P., M. Goldoft, P. M. Griffin, M. A. Davis, D. C. Gordon, P. I. Tarr, C. A. Bertelson, J. H. Lewis, T. J. Barrett, J. G. Wells, R. Baron, and J. Kobayashi. 1994. A multistate outbreak of *Escherichia coli* O157:H7-associated bloody diarrhea and hemolytic uremic syndrome from hamburgers: the Washington experience. *JAMA* 272:1349–1353.

3. Bender, J. B., C. W. Hedberg, J. M. Besser, D. J. Boxrud, K. L. MacDonald, and M. T. Osterholm. 1997. Surveillance for *Escherichia coli* O157:H7 infections in Minnesota by molecular subtyping. *N. Engl. J. Med.* 337:388–394.

4. Bryan, F. L., J. J. Guzewich, and E. C. D. Todd. 1997. Surveillance of foodborne disease. III. Summary and presentation of data on vehicles and contributory factors; their value and limitations. *J. Food Prot.* 60:701–714.

5. Carr, C. J., and F. C. Lu. 1998. Partnership for food safety education—"Fight BAC!" *Regul. Toxicol. Pharmacol.* 27:281–282.

6. Center for Food Safety and Applied Nutrition, Office of Compliance, Food and Drug Administration. 14 September 2004, posting date. *FDA Report on the Occurrence of Foodborne Illness Risk Factors in Selected Institutional Foodservice, Restaurant, and Retail Food Store Facility Types.* [Online.] Food and Drug Adminstration, Washington, D.C. http://www.cfsan.fda.gov/~dms/retrsk2.html.

7. Centers for Disease Control and Prevention. 1997. Foodborne diseases active surveillance network, 1996. *Morb. Mortal. Wkly. Rep.* 46:258–261.

8. Centers for Disease Control and Prevention. 1997. *Escherichia coli* O157:H7 infections associated with eating a nationally distributed commercial brand of frozen ground beef patties and burgers—Colorado, 1997. *Morb. Mortal. Wkly. Rep.* 46:777–778.

9. Centers for Disease Control and Prevention. 1998. Multistate outbreak of *Salmonella* serotype Agona infections linked to toasted oats cereal—United States, April–May, 1998. *Morb. Mortal. Wkly. Rep.* 47:462–464.

10. Centers for Disease Control and Prevention. 1999. *Foodborne Diseases Active Surveillance Network (FoodNet): Population Survey Atlas of Exposures, 1998–1999.* Centers for Disease Control and Prevention, Atlanta, Ga.

11. Centers for Disease Control and Prevention. October 25, 2006, posting date. *Foodborne Diseases Active Surveillance Network (FoodNet): Population Survey Atlas of Exposures, 2002.* [Online.] Centers for Disease Control and Prevention, Atlanta, Ga. http://www.cdc.gov/foodnet/surveys/pop/2002/2002Atlas.pdf.

12. Centers for Disease Control and Prevention. 2005. Preliminary FoodNet data on the incidence of infection with pathogens transmitted commonly through food—10 sites, United States, 2004. *Morb. Mortal. Wkly. Rep.* 54:352–356.

12a.Centers for Disease Control and Prevention. 2005. Summary of notifiable diseases—United States, 2003. *Morb. Mortal. Wkly. Rep.* 52(54):1–85.

12b.Centers for Disease Control and Prevention. 2006. Outbreak surveillance data: Annual Listing of Foodborne Disease Outbreaks, United States, 1990–2004. [Online.] Centers for Disease Control and Prevention, Atlanta, Ga. http://www.cdc.gov/foodborneoutbreaks/outbreak_data.htm. Accessed November 21, 2006.

13. Dalton, C. B., E. D. Mintz, J. G. Wells, C. A. Bopp, and R. V. Tauxe. 1999. Outbreaks of enterotoxigenic *Escherichia coli* infection in American adults: a clinical and epidemiologic profile. *Epidemiol. Infect.* 123:9–16.

14. Deneen, V. C., J.M Hunt, C. R. Paule, R. I. James, R. G. Johnson, M. J. Raymond, and C. W. Hedberg. 2000. The impact of foodborne calicivirus disease: the Minnesota experience. *J. Infect. Dis.* 181(S2):S281–S283.

15. Food Safety and Inspection Service, U. S. Department of Agriculture. 1994. *Nationwide Beef Microbiological Data Collection Program: Steers and Heifers.* U.S. Department of Agriculture, Washington, D.C.

16. Food Safety and Inspection Service, U. S. Department of Agriculture. 1996. *Nationwide Beef Microbiological Data Collection Program: Cows and Bulls.* U.S. Department of Agriculture, Washington, D.C.

17. Food Safety and Inspection Service, U. S. Department of Agriculture. 1996. *Nationwide Federal Plant Ground Beef Microbiological Survey.* U.S. Department of Agriculture, Washington, D.C.

18. Food Safety and Inspection Service, U. S. Department of Agriculture. April 12, 2006, posting date. Microbiological results of raw ground beef products analyzed for *Escherichia coli* O157:H7, summarized by calendar year. [Online.] U.S. Department of Agriculture, Washington, D.C. http://www.fsis.usda.gov/Science/Ecoli_O157_Summary_Tables/index.asp. Accessed November 21, 2006.

19. Garber, L. P., S. J. Wells, D. D. Hancock, M. P. Doyle, J. Tuttle, J. A. Shere, and T. Zhao. 1995. Risk factors for fecal shedding of *Escherichia coli* O157:H7 in dairy calves. *J. Am. Vet. Med. Assoc.* 207:46–49.

20. Gerner–Smidt, P., J. Kincaid, K. Kubota, K. Hise, S. B. Hunter, M. A. Fair, D. Norton, A. Woo–Ming, T. Kurzynski, M. J. Sotir, M. Head, K. Holt, and B. Swaminathan. 2005. Molecular surveillance of Shigatoxigenic *Escherichia coli* O157 by PulseNet USA. *J. Food Prot.* 68:1926–1931.

21. Graves, L. M., S. B. Hunter, A. R. Ong, D. Schoonmaker-Bopp, K. Hise, L. Kornstein, W. E. DeWitt, P. S. Hayes, E. Dunne, P. Mead, and B. Swaminathan. 2005. Microbiological aspects of the investigation that traced the 1998 outbreak of listeriosis in the United States to contaminated hot dogs and establishment of molecular subtyping-based surveillance for *Listeria monocytogenes* in the PulseNet network. *J. Clin. Microbiol.* 43:2350–2355.

22. Green, L., C. Selman, A. Banerjee, R. Marcus, C. Medus, F. J. Angulo, V. Radke, S. Buchanan, and the EHS-Net Working Group. 2005. Food service workers' self-reported food preparation practices: an EHS-Net study. *Int. J. Hyg. Environ. Health* 208:27–35.

23. Haeghebaert, S., P. Sulem, L. Deroudille, E. Vannerov-Adenot, O. Bagnis, P. Bouvet, F. Grimont, A. Brisabois, F. Le Querrec, C. Hervy, E. Espie, H. de Valk, and V. Vaillant. 2003. Two outbreaks of *Salmonella enteritidis* phage type 8 linked to the consumption of Cantal cheese made with raw milk, France, 2001. *Euro Surveill.* 8:151–156.

24. Hedberg, C. W., K. L. MacDonald, and M. T. Osterholm. 1994. Changing epidemiology of foodborne disease: a Minnesota perspective. *Clin. Infect. Dis.* 18:671–682.

25. Hedberg, C. W., S. J. Savarino, J. M. Besser, C. J. Paulus, V. M. Thelen, L. J. Meyers, D. N. Cameron, T. J. Barrett,

J. B. Kaper, M. T. Osterholm, and the Investigation Team. 1997. An outbreak of foodborne illness caused by *Escherichia coli* O39:NM, an agent not fitting into the existing scheme for classifying diarrheogenic *E. coli*. *J. Infect. Dis.* **176:**1625–1628.

26. Hedberg, C. W., F. J. Angulo, K. E. White, C. W. Langkop, W. L. Schell, M. G. Stobierski, A. Schuchat, J. M. Besser, S. Dietrich, L. Helsel, P. M. Griffin, J. W. McFarland, M. T. Osterholm, and the Investigation Team. 1999. Outbreaks of salmonellosis associated with eating uncooked tomatoes: implications for public health. *Epidemiol. Infect.* **122:**385–393.

27. Hedberg, C. W., S. J. Smith, E. Kirkland, V. Radke, T. F. Jones, C. A. Selman, and the EHS-Net Working Group. 2006. Systematic environmental evaluations to identify differences between outbreak and nonoutbreak restaurants. *J. Food Prot.* **69:**2697–2702.

28. Hennessy, T. W., C. W. Hedberg, L. Slutsker, K. E. White, J. M. Besser-Wiek, M. E. Moen, J. Feldman, W. W. Coleman, L. M. Edmonson, K. L. MacDonald, M. T. Osterholm, and the Investigation Team. 1996. A national outbreak of *Salmonella enteritidis* infections from ice cream. *N. Engl. J. Med.* **334:**1281–1286.

29. Hennessy, T. W., L. R. Cheng, H. Kassenborg, S. D. Ahuja, J. Mohle-Boetani, R. Marcus, B. Shiferaw, F. Angulo, and the FoodNet Working Group. 2004. Egg consumption is the principal risk factor for sporadic *Salmonella* serotype Heidelberg infections: a case-control study in FoodNet sites. *Clin. Infect. Dis.* **38:**S237–S243.

30. Herwaldt, B. L., M. L. Ackers, and the *Cyclospora* Working Group. 1999. An outbreak in 1996 of cyclosporiasis associated with imported raspberries. *N. Engl. J. Med.* **336:**1548–1556.

31. Hogue, A., P. White, J. Guard-Petter, W. Schlosser, R. Gast, E. Ebel, J. Farrar, T. Gomez, J. Madden, M. Madison, A. M. McNamara, R. Morales, D. Parham, P. Sparling, W. Sutherlin, and D. Swerdlow. 1997. Epidemiology and control of egg-associated *Salmonella enteritidis* in the United States of America. *Rev. Sci. Tech.* **16:**542–553.

32. Hope, B. K., R. Baker, E. D. Edel, A. T. Hogue, W. D. Schlosser, R. Whiting, R. M. McDowell, and R. A. Morales. 2002. An overview of the *Salmonella enteritidis* risk assessment for shell eggs and egg products. *Risk Anal.* **22:**203–218.

33. Hutwagner, L. C., E. K. Maloney, N. H. Bean, L. Slutsker, and S. M. Martin. 1997. Using laboratory-based surveillance data for prevention: an algorithm for detecting *Salmonella* outbreaks. *Emerg. Infect. Dis.* **3:**395–400.

34. Isaacs, S., J. Aramini, B. Ciebin, J. A. Farrar, R. Ahmed, D. Middleton, A. U. Chandran, L. J. Harris, M. Howes, E. Chan, A. S. Pichette, K. Campbell, A. Gupta, L. Y. Lior, M. Pearce, C. Clark, F. Rogers, F. Jamieson, I. Brophy, A. Ellis, and the Salmonella Enteritidis PT30 Outbreak Investigation Working Group. 2005. An international outbreak of salmonellosis associated with raw almonds contaminated with a rare phage type of *Salmonella enteritidis*. *J. Food Prot.* **68:**191–198.

35. Jimenez, M., P. Soler, J. D. Venanzi, C. Varela, and F. Martinez-Navarro. 2005. An outbreak of *Campylobacter jejuni* enteritis in a school of Madrid, Spain. *Euro Surveill.* **10:**118–121.

36. Kaplan, J. E., G. W. Gary, R. C. Baron, N. Singh, L. B. Schonberger, R. Feldman, and H. B. Greenberg. 1982. The epidemiology of Norwalk gastroenteritis and the role of Norwalk virus in outbreaks of acute nonbacterial gastroenteritis. *Ann. Intern. Med.* **96:**756–761.

37. Kassenborg, H. D., C. W. Hedberg, M. Hoekstra, M. C. Evans, A. E. Chin, R. Marcus, D. J. Vugia, K. Smith, S. D. Ahuja, L. Slutsker, and P. M. Griffin. 2004. Farm visits and undercooked hamburgers as major risk factors for sporadic *Escherichia coli* O157:H7 infection: data from a case-control study in 5 FoodNet sites. *Clin. Infect. Dis.* **38(S3):**S271–S278.

38. Kimura, A. C., V. Reddy, R. Marcus, P. R. Cieslak, J. C. Mohle-Boetani, H. D. Kassenborg, S. D. Segler, F. P. Hardnett, T. Barrett, D. L. Swerdlow, and the Emerging Infections Program FoodNet Working Group. 2004. Chicken consumption is a newly identified risk factor for sporadic *Salmonella* enterica serotype Enteritidis infections in the United States: a case-control study in FoodNet sites. *Clin. Infect. Dis.* **38(S3):**S285–S296.

39. Le Saux, N., J. S. Spika, B. Friesen, I. Johnson, D. Mlnychuck, C. Anderson, R. Dion, M. Rahman, and W. Tostowarky. 1993. Ground beef conumption in non-commercial settings is a risk factor for sporadic *Escherichia coli* O157:H7 infection in Canada. *J. Infect. Dis.* **167:**500–502.

40. Li, R., M. Serdula, S. Bland, A. Mokdad, B. Bowman, and D. Nelson. 2000. Trends in fruit and vegetable consumption among adults in 16 U.S. states: behavioral risk factor surveillance system. *Am. J. Public Health* **90:**777–781.

41. Losinger, W. C., S. J. Wells, L. P. Garber, H. S. Hurd, and L. A. Thomas. 1995. Management factors related to *Salmonella* shedding by dairy heifers. *J. Dairy Sci.* **78:**2464–2472.

42. Mahon, B. E., A. Ponka, W. N. Hall, K. Komatsu, S. E. Dietrich, A. Siitonen, G. Cage, P. S. Hayes, M. A. Lambert-Fair, N. H. Bean, P. M. Griffin, and L. Slutsker. 1997. An international outbreak of *Salmonella* infections caused by alfalfa sprouts grown from contaminated seeds. *J. Infect. Dis.* **175:**876–882.

43. Mahon, B. E., L. Slutsker, L. Hutwagner, C. Drenzek, K. Maloney, K. Toomey, and P. M. Griffin. 1999. Consequences in Georgia of a nationwide outbreak of *Salmonella* infections: what you don't know might hurt you. *Am. J. Public Health* **89:**31–35.

44. Marcus R., T. Rabatsky-Ehr , J. C. Mohle-Boetani, M. Farley, C. Medus, B. Shiferaw, M. Carter, S. Zansky, M. Kennedy, T. J. Van Gilder, and J. L. Hadler. 2004. Dramatic decrease in the incidence of Salmonella serotype Enteritidis infections in 5 FoodNet sites: 1996–1999. *Clin. Infect. Dis.* **38(S3):**S135–S141.

45. McCarthy, N., and J. Giesecke. 1999. Case-case comparisons to study causation of common infectious diseases. *Int. J. Epidemiol.* **28:**764–768.

46. Mead, P. S., L. Slutsker, V. Dietz, L. F. McCaig, J. S. Bresee, C. Shapiro, P. M. Griffin, and R. V. Tauxe. 1999. Food-related illness and death in the United States. *Emerg. Infect. Dis.* **5:**607–625.

47. Naimi, T. S., J. H. Wicklund, S. J. Olsen, G. Krause, J. G. Wells, J. M. Bartkus, D. J. Boxrud, M. Sullivan, H. Kassenborg, J. M. Besser, E. D. Mintz, M. T. Osterholm, and C. W. Hedberg. 2003. Concurrent outbreaks of *Shigella sonnei* and enterotoxigenic *Escherichia coli* infections associated with parsley: implications for surveillance and control of foodborne illness. *J. Food Prot.* **66:**535–541.

48. National Advisory Committee on Microbiological Criteria for Foods. 1998. Hazard analysis and critical control point principles and application guidelines. *J. Food Prot.* **61:**762–765.

49. National Advisory Committee on Microbiological Criteria for Foods. 1998. Principles of risk assessment for illness caused by foodborne biological agents. *J. Food Prot.* **61:**1071–1074.

50. Naugle, A. L., K. G. Holt, P. Levine, and R. Eckel. 2005. Food Safety and Inspection Service regulatory testing program for *Escherichia coli* O157:H7 in raw ground beef. *J. Food Prot.* **68:**462–468.

51. Olsen, S. J., L. C. MacKinon, J. S. Goulding, N. H. Bean, and L. Slutsker. 2000. Surveillance for foodborne-disease outbreaks—United States, 1993–1997. *Morb. Mortal. Wkly. Rep.* **49**(SS-1):1–51.

52. Potter, M. E., S. Gonzalez Ayala, and N. Silarug. 1997. Epidemiology of foodborne diseases p. 376–390. *In* M. P. Doyle, L. R. Beuchat, and T. J. Montville (ed.), *Food Microbiology: Fundamentals and Frontiers.* ASM Press, Washington, D.C.

53. Richardson, B. W., and W. H. Frost. 1936. *Snow on Cholera.* The Commonwealth Fund, New York, N.Y.

54. Rolls, B. J., and E. A. Bell. 2000. Dietary approaches to the treatment of obesity. *Med. Clin. N. Am.* **84:**401–418.

55. Roy, S. L., S. M. DeLong, S. A. Stenzel, B. Shiferaw, J. M. Roberts, A. Khalakdina, R. Marcus, S. D. Segler, D. D. Shaw, S. Thomas, D. J. Vugia, S. M. Zansky, V. Dietz, M. J. Beach, and the Emerging Infections Program Food-Net Working Group. 2004. Risk factors for sporadic cryptosporidiosis among immunocompetent persons in the United States from 1999 to 2001. *J. Clin. Microbiol.* **42:**2944–2955.

56. Salman, M. D., G. R. Frank, D. W. MacVean, J. S. Reif, J. K. Collins, and R. Jones. 1988. Validation of disease diagnoses reported to the National Animal Health Monitoring System from a large Colorado beef feedlot. *J. Am. Vet. Med. Assoc.* **192:**1069–1073.

57. Schroeder, C. M., A. L. Naugle, W. D. Schlosser, A. T. Hogue, F. J. Angulo, J. S. Rose, E. D. Ebel, W. T. Disney, K. G. Holt, and G. P. Goldman. 2005. Estimate of illnesses from *Salmonella enteritidis* in eggs, United States, 2000. *Emerg. Infect. Dis.* **11:**113–115.

58. Sivapalasingam, S., C. R. Friedman, L. Cohen, and R. V. Tauxe. 2004. Fresh produce: a growing cause of outbreaks of foodborne illness in the United States, 1973 through 1997. *J. Food Prot.* **67:**2342–2353.

59. Slutsker, L., A. A. Ries, K. Maloney, J. G. Wells, K. D. Greene, and P. M. Griffin. 1998. A nationwide case-control study of *Escherichia coli* O157:H7 infection in the United States. *J. Infect. Dis.* **177:**962–966.

60. Stern, L., and D. Lightfoot. 1999. Automated outbreak detection: a quantitative retrospective analysis. *Epidemiol. Infect.* **122:**103–110.

61. St. Louis, M. E., D. L. Morse, M. E. Potter, T. M. Demelfig, J. J. Guzewich, R. V. Tauxe, P. A. Blake, and the *Salmonella enteritidis* Working Group. 1988. The emergence of Grade A eggs as a major source of *Salmonella enteritidis* infections: new implications for the control of salmonellosis. *JAMA* **259:**2103–2107.

62. Tauxe, R. V. 1992. Epidemiology of *Campylobacter jejuni* infections in the United States and other industrialized nations, p. 9–19. *In* I. Nachamkin, M. J. Blaser, and L. S. Tompkins (ed.), Campylobacter jejuni: *Current Status and Future Trends.* ASM Press, Washington, D.C.

63. Tauxe, R. V. 1998. New approaches to surveillance and control of emerging foodborne diseases. *Emerg. Infect. Dis.* **4:**455–456.

64. Thompson, B., W. Demark-Wahnefried, G. Taylor, J. W. McClelland, G. Stables, S. Havas, Z. Feng, M. Topor, J. Heimendinger, K. D. Reynolds, and N. Cohen. 1999. Baseline fruit and vegetable intake among adults in seven 5 a day study centers located in diverse geographic areas. *J. Am. Diet. Assoc.* **99:**1241–1248.

65. Todd, E. C. D. 1996. Worldwide surveillance of foodborne disease: the need to improve. *J. Food Prot.* **59:**82–92.

66. Van Beneden, C. A., W. E. Keene, R. A. Strang, D. H. Werker, A. S. King, B. Mahon, K. Hedberg, A. Bell, M. T. Kelly, V. K. Balan, W. R. MacKenzie, and D. Fleming. 1999. Multinational outbreak of *Salmonella enterica* Newport infections due to contaminated alfalfa sprouts. *JAMA* **281:**158–162.

67. van't Veer, P., M. C. Jansen, M. Klerk, and F. J. Kok. 2000. Fruits and vegetables in the prevention of cancer and cardiovascular disease. *Public Health Nutr.* **3:**103–107.

68. Weingold, S. E., J. J. Guzewich, and J. K. Fudala. 1994. Use of foodborne disease data for HACCP risk assessment. *J. Food. Prot.* **57:**820–830.

69. Widdowson, M. A., A. Sulka, S. N. Bulens, R. S. Beard, S. S. Chaves, R. Hammond, E. D. Salehi, E. Swanson, J. Totaro, R. Woron, P. S. Mead, J. S. Bresee, S. S. Monroe, and R. I. Glass. 2005. Norovirus and foodborne disease, United States, 1991–2000. *Emerg. Infect. Dis.* **11:**95–102.

Mycotoxigenic Molds

Food Microbiology: Fundamentals and Frontiers, 3rd Ed.
Edited by M. P. Doyle and L. R. Beuchat
© 2007 ASM Press, Washington, D.C.

Ailsa D. Hocking

Toxigenic *Aspergillus* Species

24

Aspergillus was first described almost 300 years ago and is an important fungal genus associated with foods. Although a few species are used in production of food (e.g., *Aspergillus oryzae* in soy sauce manufacture), most *Aspergillus* species occur in foods as spoilage or biodeterioration fungi. They are extremely common in stored commodities such as grains, nuts, and spices and occur more frequently in tropical and subtropical than in temperate climates (93).

Mycotoxins were brought to the attention of scientists in the Western world in the early 1960s with the outbreak of turkey "X" disease in England, which killed about 100,000 turkeys and other farm animals. The cause of this disease was traced to peanut meal in the feed, which was heavily contaminated with *A. flavus*. Analysis revealed that a group of fluorescent compounds, later named aflatoxins, were responsible for this outbreak, the deaths of large numbers of ducklings in Kenya, and widespread hepatoma in hatchery-reared trout in California which occurred more or less simultaneously (3, 6, 108).

The genus *Aspergillus* contains a number of species capable of producing mycotoxins other than aflatoxins, and so correct identification of isolates from foods and knowledge of the ecology of these molds are of paramount importance.

TAXONOMY

Aspergillus contains more than 100 recognized species, most of which grow well in laboratory culture. There are a number of teleomorphic (ascosporic) genera which have *Aspergillus* conidial states (anamorphs), but the only two of real importance in foods are the xerophilic genus *Eurotium* (previously known as the *A. glaucus* group) and *Neosartorya* species which produce heat-resistant ascospores and cause spoilage in heat-processed foods, mainly fruit products (93).

The most widely used taxonomy for *Aspergillus* is that of Raper and Fennell (102), although their taxonomy has been updated (105) and many new species have since been described. A nomenclaturally correct classification for species within the genus *Aspergillus* was proposed by Gams et al. (45), grouping species into six subgenera which are subdivided into sections. A recent taxonomy for the most common *Aspergillus* species, including those important in foods, is provided by Klich (59).

Ailsa D. Hocking, Food Science Australia, Riverside Life Sciences Centre, North Ryde, New South Wales 2113, Australia.

In addition to morphological taxonomic techniques, secondary metabolites (39) and molecular techniques (e.g., see references 80, 82, 109, 119, and 128) have been used to clarify relationships within the genus *Aspergillus*.

ISOLATION, ENUMERATION, AND IDENTIFICATION

Techniques for the isolation and enumeration of *Aspergillus* species in foods have previously been described in detail (93, 107). Antibacterial media containing compounds to inhibit or reduce mold colony spreading, such as dichloran rose bengal chloramphenicol (DRBC) agar and dichloran–18% glycerol (DG18) agar, are recommended for enumerating fungi in foods (56, 93, 106).

The most complete taxonomy is that of Raper and Fennell (102), but key identification and descriptions of the most common foodborne *Aspergillus* species can be found elsewhere (59, 93, 107). Identification of *Aspergillus* species requires growth on media developed for this purpose, including Czapek agar, a derivative such as Czapek yeast extract agar, and malt extract agar. Growth on Czapek yeast extract–20% sucrose agar can be a useful aid in identifying species of *Aspergillus* (93).

Aspergillus species are conveniently color coded, and the color of the conidia can be a very useful starting point in identification, at least to the series level. As well as conidial color, microscopic morphology is important in identification. Phialides (cells producing conidia) may be produced directly from the swollen apex (vesicle) of long stalks (stipes), or there may be an intermediate row of supporting cells (metulae). Correct identification of *Aspergillus* species is an essential prerequisite for assessing the potential for mycotoxin contamination in a commodity, food, or feedstuff.

SIGNIFICANT *ASPERGILLUS* MYCOTOXINS

The *Aspergillus* mycotoxins of greatest significance in foods and feeds are aflatoxins (produced in foods by *A. flavus* and *A. parasiticus*); ochratoxin A from *A. ochraceus* and related species and from *A. carbonarius* and occasionally *A. niger*; sterigmatocystin, produced primarily by *A. versicolor* but also by *Emericella* species; and cyclopiazonic acid (*A. flavus* is the primary source, but this toxin is also reported to be produced by *A. tamarii*) (41). Citrinin, patulin, and penicillic acid may also be produced by certain *Aspergillus* species, and tremorgenic toxins are produced by *A. terreus* (territrems), *A. fumigatus* (fumitremorgens), and *A. clavatus* (tryptoquivaline) (37, 63, 110).

Aspergillus species, like *Penicillium* species, produce toxins that exhibit a wide range of toxicities, with the most significant effects being long term. Aflatoxin B_1 is perhaps the most potent liver carcinogen known for a wide range of animal species, including humans. Ochratoxin A and citrinin both affect kidney function. Cyclopiazonic acid has a wide range of effects, and tremorgenic toxins such as territrems affect the central nervous system. Table 24.1 lists the most significant toxins produced by *Aspergillus* species and their toxic effects.

A. FLAVUS AND A. PARASITICUS

The most important toxigenic aspergilli in foods are the aflatoxigenic molds, *A. flavus* and *A. parasiticus*, along with a recently described but much less common species, *A. nomius*, all of which are classified in *Aspergillus* section *Flavi* (45). Although these three species are closely related and show many similarities, a number of characteristics may be used in their differentiation (Table 24.2) and the suite of toxins produced is species specific (60, 64). *A. flavus* can produce aflatoxins B_1 and B_2 and cyclopiazonic acid, but only a proportion of isolates are toxigenic. *A. parasiticus* produces aflatoxins B_1, B_2, G_1, and G_2 but not cyclopiazonic acid, and almost all isolates are toxigenic. *A. nomius* is morphologically similar to *A. flavus*, but like *A. parasiticus*, it produces B and G aflatoxins without cyclopiazonic acid, and because this species appears to be uncommon, it is probably of lesser significance in the food supply. *A. flavus* and *A. parasiticus* are closely related to *A. oryzae* and *A. sojae*, species that are used in the manufacture of fermented foods and do not produce toxins (30). A number of other, less common species of *Aspergillus* and *Emericella* have also been reported to produce aflatoxins but are not found in foods (41).

Detection and Identification

The most effective medium for rapid detection of aflatoxigenic molds is *Aspergillus flavus* and *parasiticus* agar (97), a medium formulated specifically for this purpose. Under the incubation conditions specified (30°C for 48 to 72 h), *A. flavus*, *A. parasiticus*, and *A. nomius* produce a bright orange-yellow colony reverse which is readily recognized and diagnostic for the species. *A. flavus* and related species also grow well on general purpose yeast and mold enumeration media such as DRBC agar and DG18 agar and, with practice, are relatively easily recognized.

A. flavus and *A. parasiticus* (Fig. 24.1) are easily distinguished from other *Aspergillus* species by using appropriate media and methods (93, 107). However, differentiating between these two species and *A. nomius* is more difficult but is important because of the different mycotoxin profiles and toxin-producing potentials of these species. The texture of the conidial walls is a reliable

Table 24.1 Significant mycotoxins produced by *Aspergillus* species and their toxic effects

Mycotoxin(s)	Toxicity	Species
Aflatoxin B_1 and B_2	Acute liver damage; cirrhosis; carcinogenic (liver), teratogenic, and immunosuppressive	*A. flavus, A. parasiticus, A. nomius*
Aflatoxin G_1 and G_2	Effects similar to those of B aflatoxins; G_1 toxicity less than that of B_1 but greater than that of B_2	*A. parasiticus, A. nomius*
Cyclopiazonic acid	Degeneration and necrosis of various organs; tremorgenic; low oral toxicity	*A. flavus, A. tamarii*
Ochratoxin A	Kidney necrosis (especially in pigs); teratogenic, immunosuppressive, and probably carcinogenic	*A. ochraceus* and related species, *A. carbonarius, A. niger* (occasional)
Sterigmatocystin	Acute liver and kidney damage; carcinogenic (liver)	*A. versicolor, Emericella* spp.
Fumitremorgens	Tremorgenic (rats and mice)	*A. fumigatus*
Territrems	Tremorgenic (rats and mice)	*A. terreus*
Tryptoquivalines	Tremorgenic	*A. clavatus*
Cytochalasins	Cytotoxic	*A. clavatus*
Echinulins	Feed refusal (pigs)	*E. chevalieri, E. amstelodami*

differentiating feature: conidia of *A. flavus* (Fig. 24.1A and B) are usually smooth to finely roughened, and those of *A. parasiticus* (Fig. 24.1C and D) are clearly rough when observed under an oil immersion lens. Screening of isolates for aflatoxin production can also be used to differentiate the species. Cultures can be grown on coconut-cream agar and observed under UV light (29), or a simple agar plug technique coupled with thin-layer chromatography (36) can be used to screen cultures for aflatoxin production as an aid to identification. The combination of characteristics most useful in differentiation among the three aflatoxigenic species is summarized in Table 24.2.

Aflatoxins

Aflatoxins are difuranocoumarin derivatives. Aflatoxins B_1, B_2, G_1, and G_2 are produced in nature by the molds discussed above. The letters B and G refer to the fluorescent colors (blue and green, respectively) observed under long-wave UV light, and the subscripts 1 and 2 refer to the toxins' separation patterns on thin-layer-chromatography plates. Aflatoxins M_1 and M_2 are produced from the respective B aflatoxins by hydroxylation in lactating animals and are excreted in milk at a rate of approximately 1.5% of ingested B aflatoxins (43).

Aflatoxins are synthesized through the polyketide pathway through a series of steps which are now well understood (100). G-group aflatoxins are formed from the same substrate as B-group aflatoxins, namely *O*-methylsterigmatocystin, but by an independent pathway (135). Most of the genes involved in the biosynthesis of aflatoxins are contained within a single gene cluster in the genomes of *A. flavus* and *A. parasiticus*, and the regulation and expression of these genes are now relatively well understood (74, 89, 100).

Table 24.2 Distinguishing features of *A. flavus*, *A. parasiticus*, and *A. nomius*

Species	Conidia	Sclerotia	Toxins
A. flavus	Smooth to moderately roughened; variable in size	Large, globose	Aflatoxins B_1 and B_2, cyclopiazonic acid
A. parasiticus	Conspicuously roughened; little variation in size	Large, globose	Aflatoxins B and G
A. nomius	Similar to those of *A. flavus*	Small, elongated (bullet shaped)	Aflatoxins B and G

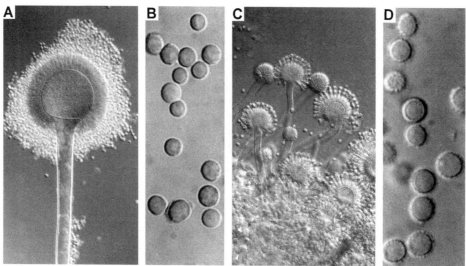

Figure 24.1 Aflatoxigenic fungi. (A) *A. flavus* head (magnification, ×215); (B) *A. flavus* conidia (magnification, ×1,350); (C) young *A. parasiticus* heads (magnification, ×215); (D) *A. parasiticus* conidia (magnification, ×1,350).

Toxicity

Aflatoxins are both acutely and chronically toxic in animals and humans, producing acute liver damage, liver cirrhosis, tumor induction, and teratogenesis (134). Perhaps of greater significance to human health are the immunosuppressive effects of aflatoxins, either alone or in combination with other mycotoxins (134). Immunosuppression can increase susceptibility to infectious diseases, particularly in populations where aflatoxin ingestion is chronic (23); can interfere with production of antibodies in response to immunization; and may accelerate the progress of human immunodeficiency virus/AIDS (92, 134).

Acute aflatoxicosis in humans is infrequent (134); however, several outbreaks have been reported. In 1967, 26 people in two farming communities in Taiwan became ill with apparent food poisoning. Nineteen were children, three of whom died. Rice from affected households contained about 200 µg of aflatoxin B$_1$/kg, which was probably responsible for the outbreak. An outbreak of hepatitis in India in 1974 which affected 400 people, 100 of whom died, almost certainly was caused by aflatoxins (65). The outbreak was traced to corn heavily contaminated with *A. flavus* and containing up to 15 mg of aflatoxins/kg. It was calculated that affected adults may have consumed 2 to 6 mg on a single day, implying that the acute lethal dose for adult humans is of the order of 10 mg. More recently, deaths of 13 Chinese children in the northwestern Malaysian state of Perak were reported (78), apparently due to ingestion of contaminated noodles. The presence of aflatoxins in postmortem tissue samples was confirmed. A case of systemic aspergillosis caused by an aflatoxin-producing strain of *A. flavus* has been reported in which aflatoxins B$_1$, B$_2$, and M$_1$ were detected in lung lesions and were considered to have played a role in damaging the immune system of the patient (81).

Undoubtedly, the greatest direct impact of aflatoxins on human health is their potential to induce liver cancer (77, 134). Human liver cancer has a high incidence in central Africa and parts of Southeast Asia, and studies in several African countries and Thailand have shown a correlation between aflatoxin intake and the occurrence of primary liver cancer (127). No such correlation could be demonstrated for populations in rural areas of the United States, despite the occurrence of considerable amounts of aflatoxins in corn (117). This apparent anomaly may be explained by the relationship between the roles of hepatitis B virus and aflatoxins in induction of human liver cancer. Hepatitis B virus interferes with the body's ability to detoxify aflatoxins (134), greatly increasing the probability of human liver cancer. However, evidence supports the hypothesis that high aflatoxin intakes are causally related to high incidences of cancer, even in the absence of hepatitis B virus (52, 90, 134).

In animals, aflatoxins have been shown to cause various syndromes, including immunosuppression and cancer of the liver, colon, and kidneys (31, 134). Regular low-level intake of aflatoxins reduces growth rates and other measures of productivity in animals (134).

Mechanism of Action

Aflatoxin B$_1$ is metabolized by cytochrome P450 in the liver, leading to the formation of highly reactive intermediates, one of which is 2,3-epoxy-aflatoxin B₁ (112). Binding of these reactive intermediates to DNA results in disruption of transcription (G-to-T transversion) and abnormal cell proliferation, leading to mutagenesis or carcinogenesis (112). Aflatoxins also inhibit oxygen uptake in the tissues by acting on the electron transport chain and inhibiting various enzymes, resulting in decreased production of ATP (31).

Toxin Detection

Chemical detection of aflatoxins has become fairly standardized (5). Samples are extracted with organic solvents such as chloroform and methanol in combination with small amounts of water. The presence of fats, lipids, or pigments in extracts reduces the efficiency of the separation, and solvents such as hexane may be used to partition these components from the extract (31). Extracts are further cleaned up, concentrated, and then separated by thin-layer chromatography or high-performance liquid chromatography. If thin-layer chromatography is used, aflatoxins are visualized under UV light and quantified by visual comparison with known concentrations of standards or by fluorimetry. High-performance liquid chromatography with detection by UV light or absorption spectrometry provides more readily quantifiable (although not necessarily more accurate or sensitive) results.

Immunoassay techniques including enzyme-linked immunosorbent assays (ELISA) and dipstick tests for aflatoxin detection have been developed (91, 120), and a number of kits are now commercially available. Immunoaffinity and fluorimetric detection have been combined in developing biosensors able to detect levels of aflatoxin down to 50 µg/kg in as little as 2 min (16, 17). Aflatoxigenic fungi can be detected in foods by using PCR targeting aflatoxin biosynthetic genes (46, 113, 137).

Occurrence of Aflatoxigenic Molds and Aflatoxins

A. flavus is widely distributed in nature, but *A. parasiticus* is probably less widespread, determination of the actual extent of its occurrence being complicated by the tendency for both species to be reported indiscriminately as *A. flavus*. In a wide-ranging survey of the mycoflora of commodities in Thailand (98, 99), *A. flavus* was one of the most commonly occurring molds in nuts and oilseeds, soybeans, mung beans, sorghum, and other commodities, but *A. parasiticus* was rarely encountered. *A. nomius* was reported from peanuts and corn, the first published report of the occurrence of this species in food (98).

A. flavus and *A. parasiticus* have a strong affinity with nuts and oilseeds. Corn, peanuts, and cottonseed are the most important crops invaded by these molds, and in many instances, invasion takes place before harvest, not during storage as was once believed. Peanuts are invaded while still in the ground if the crop suffers drought stress or related factors (21, 96). In corn, insect damage to developing kernels allows entry of aflatoxigenic molds, but invasion can also occur through the silks of developing ears (71). Cottonseeds are invaded through the nectaries (61).

Cereals and spices are common substrates for *A. flavus* (93), but aflatoxin production in these commodities is almost always a result of poor drying, handling, or storage, and aflatoxin levels are rarely significant. Significant amounts of aflatoxins can occur in peanuts, corn, and other nuts and oilseeds, particularly in some tropical countries where crops may be grown under marginal conditions and where drying and storage facilities are limited (4, 24, 77).

Factors Affecting Growth and Toxin Production

A. flavus and *A. parasiticus* have similar growth patterns. Both grow at temperatures ranging from 10 to 43°C, with an optimum from 30 to 33°C (7, 51), and aflatoxins are produced at temperatures from 12 to 40°C (25, 62, 86). The optimum water activity (a_w) for growth is near 0.996 (51), with minima reported as a_ws of 0.80 (8) to 0.82 to 0.83 (86, 95). Aflatoxins are generally produced in greater quantity at higher a_w (0.98 to 0.99), with toxin production apparently ceasing at or near an a_w of 0.85 (25, 62, 86). Although growth can take place over the pH range from just above 2.0 up to 10.5 (*A. parasiticus*) or 11.2 (*A. flavus*) (132), aflatoxin production has been reported for *A. parasiticus* only between pH 3.0 and 8.0, with an optimum near pH 6.0 (15). Reduction of available oxygen by modified-atmosphere packaging of foods in barrier film or with oxygen scavengers can inhibit aflatoxin formation by *A. flavus* and *A. parasiticus* (32).

Control and Inactivation

Control of aflatoxins in commodities generally relies on screening techniques which separate affected nuts, grains, or seeds. For corn, cottonseed, and figs, screening for aflatoxins can be done by examination under UV light: those particles which fluoresce may be contaminated. All peanuts fluoresce when exposed to UV light, so peanuts containing aflatoxins are segregated by electronic color-sorting machines which detect discolored kernels.

Aflatoxins can be partially destroyed by various chemical treatments. Oxidizing agents such as ozone and hydrogen peroxide have been demonstrated to remove aflatoxins from contaminated peanut meals (131). Although ozone was effective in the removal of aflatoxins B₁ and G₁ under the conditions applied

(100°C for 2 h), there was no effect on aflatoxin B_2, and the treatment decreased the lysine content of the meal (26). Treatment with hydrogen peroxide resulted in 97% destruction of aflatoxin in defatted peanut meal (115). The most practical chemical method of aflatoxin destruction appears to be the use of anhydrous ammonia gas at elevated temperatures and pressures, with a 95 to 98% reduction in total aflatoxin in peanut meal reported (131). This technique is used commercially for detoxification of animal feeds in Senegal, France, and the United States (131).

The ultimate method of control of aflatoxins in commodities, particularly peanuts, is to prevent the plants from becoming infected with aflatoxigenic strains of molds. A biological control strategy, biocompetitive inhibition, promotes early infection of plants with nontoxigenic strains of *A. flavus* to prevent the subsequent entry of toxigenic strains (18, 22, 27, 94). afla-guard, developed by the USDA Peanut Research Laboratory, has recently been registered in the United States for this purpose (123).

Aflatoxins are one of the few mycotoxins covered by legislation. Statutory limits are imposed by many countries on the amount of aflatoxin that can be present in particular foods. The limit imposed by most Western countries is 5 to 20 μg of aflatoxin B_1/kg in several foods, including peanuts and peanut products. The amount allowed in animal feeds varies, but up to 300 μg/kg is allowed in feedstuffs for beef cattle and sheep in the United States (13). Aflatoxins are now also regulated in food and animal feeds in many countries throughout Africa, Asia, and Oceania (38, 126).

Cyclopiazonic Acid

Cyclopiazonic acid is produced by *A. flavus* and has also been reported to be produced by *A. tamarii* (41, 49). Maximum production of cyclopiazonic acid by *A. flavus* occurs at 0.98 a_w and 20°C, with minimum production reported at 0.90 a_w and 30°C (50). The toxin is an indole, tetramic acid, which can occur in naturally contaminated agricultural commodities and compounded animal feeds. It is acutely toxic to rats and other test animals, causing severe gastrointestinal and neurological disorders (84). Degenerative changes and necrosis may occur in the digestive tract, liver, kidney, and heart (85). Cyclopiazonic acid may have been responsible for many of the symptoms observed in turkey "X" disease in the 1960s, originally attributed to aflatoxins (14, 19). A wide range of techniques are available for detection of cyclopiazonic acid, including ELISA, high-performance liquid chromatography, and liquid chromatography-mass spectrometry (LC-MS) (54, 76, 111, 139).

A. OCHRACEUS

A. ochraceus (Fig. 24.2A) is a widely distributed mold, particularly common on dried foods (93). It is the most commonly occurring species in what was known as the *A. ochraceus* group by Raper and Fennell (102), now correctly known as *Aspergillus* section *Circumdati* (45). Ochratoxin A was first isolated from *A. ochraceus* in 1965 (124). *A. ochraceus* also produces penicillic acid, a mycotoxin of lower toxicity and of uncertain importance in human health. Other reported toxic metabolites are xanthomegnin and viomellein (41).

Recent molecular studies (40) indicate that a new species, *A. westerdijkiae*, may be a more important ochratoxin A producer than *A. ochraceus*. Other closely related *Aspergillus* species can also produce ochratoxin A: *A. sclerotiorum*, *A. alliaceus*, *A. melleus*, and *A. sulphureus* have all been reported to produce ochratoxins (41), but these are of lesser significance in foods.

Isolation and Identification

A. ochraceus and other species in *Aspergillus* section *Circumdati* are xerophilic and are best isolated on reduced-a_w media such as DG18 agar (93, 106), although DRBC agar should also give satisfactory results. Colonies of *A. ochraceus* and related species are relatively deep ochre-brown to yellow-brown, with long stipes bearing radiate *Aspergillus* heads. The vesicles are spherical, bearing densely packed metulae and phialides with small, smooth, pale brown conidia.

Growth and Toxin Production

A. ochraceus is widely distributed in dried foods such as nuts, beans, dried fruit, dried meat, and dried fish (93). It is a xerophile, capable of growth at an a_w as low as 0.79, with an optimum a_w near 0.99 (93). *A. ochraceus* grows at temperatures from 8 to 37°C (87, 88) and within a wide pH range (2.2 to 10.3) (132). Ochratoxin A is produced optimally at 0.98 to 0.96 a_w (101) and at quite high temperatures (between 25 and 30°C) compared with the optimum temperatures for penicillic acid production, which are 10 to 20°C (9, 87, 88, 101).

Ochratoxin A contamination of foods is of great concern in Scandinavia and possibly in the Baltic states, where the source is *Penicillium verrucosum*. Ochratoxin A has also been found in figs (28). The toxic effects of ochratoxin A are discussed in chapter 25 of this volume.

Ochratoxins can be assayed by using routine chromatographic methods (thin-layer chromatography, high-performance liquid chromatography, and liquid chromatography) with detection of green fluorescence under UV light at 333 nm (121) as well as more sophisticated techniques such as LC-MS (8). Immunological assay

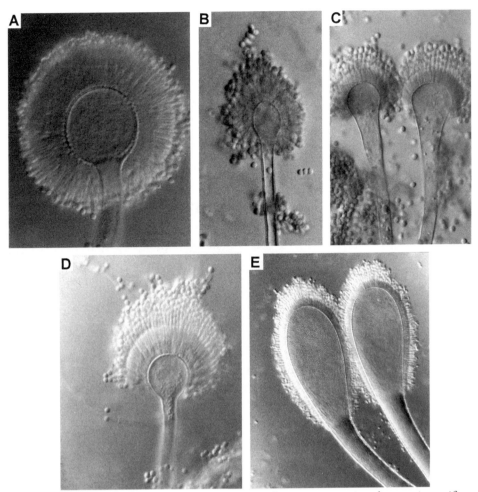

Figure 24.2 Some common mycotoxigenic *Aspergillus* species. (A) *A. ochraceus* (magnification, ×540); (B) *A. versicolor* (magnification, ×540); (C) *A. fumigatus* (magnification, ×540); (D) *A. terreus* (magnification, ×540); (E) *A. clavatus* (magnification, ×215).

techniques have been developed for detection of ochratoxin, including the use of immunoaffinity columns (129) and monoclonal antibodies (53) and ELISA-based methods (140).

A. CARBONARIUS AND *A. NIGER*

Aspergillus species in section *Nigri* have been recognized as a source of ochratoxin A only since the mid-1990s. Production of ochratoxin A by *A. niger* var. *niger* was reported in 1994 (1). However, *A. carbonarius* is the main producer of ochratoxin A in section *Nigri* (55, 58, 70, 133).

Isolation and Identification

A. niger and *A. carbonarius* can be isolated from foods by using the media recommended for *A. ochraceus*. Species in section *Nigri* are easily recognized in culture by the production of rapidly growing dark brown to black

colonies (93). *A. carbonarius* is distinguished from *A. niger* by having larger (7- to 10-μm-diameter) conidia which are generally much blacker and have heavy, tuberculate walls (59).

Growth and Toxin Production

A. niger grows optimally near 33 to 35°C with a minimum a_w for growth near 0.80 (93). *A. carbonarius* grows between 10 and 42°C, with an optimum near 30 to 35°C (12, 69). The optimum a_w range for growth is 0.93 to 0.98, depending on the strain (12). However, the optimum temperature for ochratoxin A production is lower than that for growth, around 15 to 20°C (33, 68).

A. carbonarius and *A. niger* also share habitats, and because of their high resistance to sunlight both have been isolated from sun-dried fruits such as sultanas, raisins, and figs (28, 69). Ochratoxin A has been reported to be present in these commodities, as well as in coffee and wine (11, 28, 57, 79, 104, 125, 141).

A. VERSICOLOR

A. versicolor (Fig. 24.2B) is the most important food spoilage and toxigenic species in Aspergillus section Versicolores (45), previously known as the A. versicolor group (102). A. versicolor is the major producer of sterigmatocystin, a carcinogenic dihyrofuranoxanthone which is a precursor of the aflatoxins (20), but aflatoxins are not produced by this species. Sterigmatocystin is also produced by members of the Aspergillus section Nidulantes, including A. nidulans and a number of Emericella species (42).

A. versicolor is a xerophile, with a minimum a_w for growth of 0.74 to 0.78 (93). It is very widely distributed in foods, particularly stored cereals, cereal products, nuts, spices, and dried meat products (93). DG18 agar is recommended for isolation. Small, gray-green colonies showing pinkish or reddish colors in the mycelium and/ or reverse and exhibiting mop-like heads are indicative of A. versicolor. The reported minimum temperature for growth is 9°C at 0.97 a_w, and the maximum temperature is 39°C at 0.87 a_w, with optimum growth at 27°C at 0.97 a_w (114). Little is known about factors affecting production of sterigmatocystin.

Occurrence, Toxicity, and Detection of Sterigmatocystin

The natural occurrence of sterigmatocystin in rice in Japan, wheat and barley in Canada, cereal-based products in the United Kingdom (138), and Ras cheese (2) has been reported. Sterigmatocystin has also been found in mold-affected buildings and building materials (83, 122). Although the toxin has not been detected in significant quantities, its acute and chronic toxicity are such that it should be regarded as a major potential hazard in foods and the environment (118). Sterigmatocystin has low acute oral toxicity because it is relatively insoluble in water and gastric juices. However, even low doses can cause tumors in mice (44) and pathological changes in rat livers (118). As a liver carcinogen, sterigmatocystin appears to be only about 1/150 as potent as aflatoxin B_1 but is still much more potent than most other known liver carcinogens. Doses as low as 15 µg/day fed continuously or a single dose of 10 mg caused liver cancer in ≥30% of male Wistar rats (118).

Sterigmatocystin can be detected by thin-layer chromatography, high-performance liquid chromatography, LC-MS, gas chromatography-mass spectrometry, and ELISA systems. Detection limits with chromatographic systems are generally of the order of 20 µg/kg (110).

A. FUMIGATUS

A. fumigatus, section Fumigati (Fig. 24.2C), is best recognized as a human pathogen, causing aspergillosis of the lung (63). It is thermophilic, with a temperature range for growth of 10 to 55°C and an optimum between 40 and 42°C. It is one of the least xerophilic of the common aspergilli, with a minimum a_w for growth of 0.85 (66). Its prime habitat is decaying vegetation, in which it causes spontaneous heating. A. fumigatus is isolated frequently from foods, particularly stored commodities, but is not regarded as a serious spoilage mold (93).

A. fumigatus is capable of producing several toxins which affect the central nervous system, causing tremors. Fumitremorgens A, B, and C are toxic cyclic dipeptides which are produced by A. fumigatus and A. caespitosus (41). When fed to laboratory rats and mice, fumitremorgens caused tremors and death in 70% of animals tested (48). Verruculogen, also produced by these species (20, 67), has a structure similar to that of the fumitremorgens. Verruculogen is tremorgenic when fed to mice and day-old chicks (48) and appears to cause inhibition of the alpha motor cells of the anterior horn (20). A. fumigatus also produces gliotoxin, a toxin with immunosuppressive activity which causes DNA damage and may play a role in the pathogenesis associated with this species (10, 47, 103, 116).

A. TERREUS

A. terreus, section Terrei (Fig. 24.2D), produces rapidly growing pale brown colonies, with Aspergillus heads bearing densely packed metulae and phialides with small conidia borne in long columns. A. terreus occurs commonly in soil and in foods, particularly stored commodities, but is not regarded as an important spoilage mold (93). It is probably thermophilic, growing much more strongly at 37°C than at 25°C (93), but little information has been published on its physiology. The reported minimum a_w for growth is 0.78 at 37°C (6).

A. terreus can produce a group of tremorgenic toxins known as territrems (73), distinctive compounds which do not contain nitrogen. Territrems are produced optimally at 28°C (34). Territrems are acutely toxic, with an intraperitoneal dose of 1 mg of territrem B injected into Swiss mice (body weight, 20 g) causing whole-body tremors within 5 min and other neurological symptoms within 20 to 30 min, all of which subside within 1 h (72). A 2-mg dose can cause death after convulsions and apnea. Territrem B appears to act by blocking acetylcholinesterase activity (72).

Territrems can be detected in chloroform extracts by thin-layer chromatography and by high-pressure liquid chromatography, exhibiting blue fluorescence under UV light (35). Natural contamination of food or feeds with territrems has not been reported (72).

A. CLAVATUS

A. clavatus (Fig. 24.2E) is found in soil and decomposing plant materials. It is the most common member of the section *Clavati*, subgenus *Clavati*, and is easily recognizable by its large, blue-green clavate (club-shaped) heads (59, 93, 102). Although *A. clavatus* has been reported to be present in various stored grains (93), it is especially common in malting barley, an environment particularly suited to its growth and sporulation (37).

A. clavatus produces patulin, cytochalasins, and the tremorgenic mycotoxins tryptoquivaline, tryptoquivalone, and related compounds. Patulin and cytochalasins may be formed in barley during the malting process (75), and outbreaks of *A. clavatus*-associated mycotoxicoses have been reported for stock fed on culms from distillery maltings in Europe, the United Kingdom, and South Africa (37).

A. clavatus is a recognized health hazard to workers in the malting industry. Inhalation of large numbers of highly allergenic spores can cause respiratory diseases such as bronchitis, emphysema, and malt worker's lung, a serious occupational extrinsic allergic alveolitis (37).

EUROTIUM SPECIES

The genus *Eurotium* is an ascomycete genus characterized by the formation of bright yellow cleistothecia, often enmeshed in yellow, orange, or red hyphae, overlayed by the gray-green (glaucous) *Aspergillus* heads of the anamorphic state. Members of this genus were (and often still are) referred to as the *A. glaucus* group (102). All *Eurotium* species are xerophilic; thus, DG18 agar is recommended for their isolation from and enumeration in foods (93). They are important spoilage molds in all types of stored commodities, often being the primary invaders of stored grains, spices, nuts, and animal feeds. The four most common species are *Eurotium chevalieri*, *E. repens*, *E. rubrum*, and *E. amstelodami*. Identification to the species level is based on colony color and ascospore morphology as observed on Czapek yeast extract–20% sucrose agar (59, 93).

E. chevalieri and *E. amstelodami* have been reported to produce toxic alkaloid metabolites called echinulin and neoechinulins (20). Echinulin was identified as being responsible for refusal by swine of moldy feed containing high populations of *E. chevalieri* and *E. amstelodami* (130). It has also been reported to be toxic to rabbits

(110). Neoechinulin from *E. repens* has been reported to have strong antioxidant properties (136).

CONCLUSION

Aspergillus is one of the most important genera in the spoilage of foods and animal feeds, particularly in warm-temperate climates and the tropics. The genus contains a number of highly mycotoxigenic molds, the aflatoxigenic and ochratoxigenic species clearly being the most important from the point of view of human health, with the list of foods and beverages contaminated by these toxins expanding. There is still much to learn about the interactions of toxigenic *Aspergillus* species with the plants they infect preharvest. The real significance of many of the other mycotoxins produced by *Aspergillus*, particularly their roles in cancer induction and immunosuppression, is still being elucidated. The interactive effects of naturally occurring mixtures of mycotoxins, e.g., aflatoxins and *Fusarium* toxins (particularly fumonisins) in corn, are poorly understood and may be of much greater significance to human and animal health than we yet realize.

References

1. Abarca, M. L., M. R. Bragulat, G. Castellá, and F. J. Cabañes. 1994. Ochratoxin production by strains of *Aspergillus niger* var. *niger*. *Appl. Environ. Microbiol.* 60:2650–2652.

2. AbdaAlla, E. A. M., M. M. Metwally, A. M. Mehriz, and Y. H. AbuSree. 1996. Sterigmatocystin: incidence, fate and production by *Aspergillus versicolor* in Ras cheese. *Nahrung* 40:310–313.

3. Allcroft, R., and R. B. A. Carnaghan. 1963. Toxic products in groundnuts—biological effects. *Chem. Ind.* 1963:50–53.

4. Arim, R. H. 1995. Present status of the aflatoxin situation in the Philippines. *Food Addit. Contam.* 12:291–296.

5. Association of Official Analytical Chemists International. 2000. *Official Methods of Analysis of the AOAC International*, 17th ed., vol. II, p. 1–64. Association of Official Analytical Chemists International, Washington, D.C.

6. Austwick, P. K. C., and G. Ayerst. 1963. Groundnut mycoflora and toxicity. *Chem. Ind.* 1963:55–61.

7. Ayerst, G. 1969. The effects of moisture and temperature on growth and spore germination in some fungi. *J. Stored Prod. Res.* 5:127–141.

8. Bacaloni, A., C. Cavaliere, A. Faberi, E. Pastorini, R. Samperi, and A. Lagana. 2005. Automated on-line solid-phase extraction-liquid chromatography-electrospray tandem mass spectrometry method for the determination of ochratoxin A in wine and beer. *J. Agric. Food Chem.* 53:5518–5525.

9. Bacon, C. W., J. G. Sweeney, J. D. Robbins, and D. Burdick. 1973. Production of penicillic acid and ochratoxin A on poultry feed by *Aspergillus ochraceus*: temperature and moisture requirements. *J. Food Prot.* 44:450–454.

10. Belkacemi, L., R. C. Barton, V. Hopwood, and E. G. V. Evans. 1999. Determination of optimum growth conditions for

gliotoxin production by *Aspergillus fumigatus* and development of a novel method for gliotoxin detection. *Med. Mycol.* 37:227–233.

11. Bellí, N., S. Marín, A. Duaigues, A. J. Ramos, and V. Sanchis. 2004. Ochratoxin A in wines, musts and grape juices from Spain. *J. Sci. Food Agric.* 84:591–594.

12. Bellí, N., A. J. Ramos, I. Coronas, V. Sanchis, and S. Marín. 2005. *Aspergillus carbonarius* growth and ochratoxin A production on a synthetic grape medium in relation to environmental factors. *J. Appl. Microbiol.* 98:839–844.

13. Bhatnagar, D., G. A. Payne, T. E. Cleveland, and J. F. Robens. 2004. Mycotoxins: current issues in USA, p. 17–47. *In* D. Barug, H. van Egmond, R. López-García, T. van Osenbruggen, and A. Visconti (ed.), *Meeting the Mycotoxin Menace*. Wageningen Academic Publishers, Wageningen, The Netherlands.

14. Bradburn, N., R. D. Coker, and G. Blunden. 1994. The aetiology of turkey "X" disease. *Phytochemistry* 35:817.

15. Buchanan, R. L., and J. C. Ayres. 1976. Effect of sodium acetate on growth and aflatoxin production in *Aspergillus parasiticus* NRRL 2999. *J. Food Sci.* 41:128–132.

16. Carlson, M. A., C. B. Bargeron, R. C. Benson, A. B. Fraser, T. E. Phillips, J. T. Velky, J. D. Groopman, P. T. Strickland, and H. W. Ko. 2000. An automated handheld biosensor for aflatoxin. *Biosens. Bioelectron.* 14:841–848.

17. Carman, A. S., S. S. Kuan, G. M. Ware, P. P. Umrigar, K. V. Miller, and H. G. Guerrero. 1996. Robotic automated analysis of foods for aflatoxin. *J. AOAC Int.* 79:456–464.

18. Chourasia, H. K., and R. K. Sinha. 1994. Potential of the biological control of aflatoxin contamination in developing peanut (*Arachis hypogaea* L.) by atoxigenic strains of *Aspergillus flavus*. *J. Food Sci. Technol. Mysore* 31:362–366.

19. Cole, R. J. 1986. Etiology of turkey "X" disease in retrospect: a case for the involvement of cyclopiazonic acid. *Mycotoxin Res.* 2:3–7.

20. Cole, R. J., and R. H. Cox. 1981. *Handbook of Toxic Fungal Metabolites*. Academic Press, New York, N.Y.

21. Cole, R. J., R. A. Hill, P. D. Blankenship, T. H. Sanders, and H. Garren. 1982. Influence of irrigation and drought on invasion of *Aspergillus flavus* in corn kernels and peanut pods. *Dev. Ind. Microbiol.* 23:299–326.

22. Cotty, P. J. 1994. Influence of field application of an atoxigenic strain of *Aspergillus flavus* on the populations of *A. flavus* infecting cotton bolls and on aflatoxin content of cottonseed. *Phytopathology* 84:1270–1277.

23. Denning, D. W., S. C. Quiepo, D. G. Altman, K. Makarananda, G. E. Neal, E. L. Camellere, M. R. A. Morgan, and T. E. Tupasi. 1995. Aflatoxin and outcome from acute lower respiratory infection in children in the Philippines. *Ann. Trop. Paediatr.* 15:209–216.

24. Dhavan, A. S., and M. R. Choudary. 1995. Incidence of aflatoxins in animal feedstuff: a decade's scenario in India. *J. AOAC Int.* 78:693–698.

25. Diener, U. L., and N. D. Davis. 1967. Limiting temperature and relative humidity for growth and production of aflatoxins and free fatty acids by *Aspergillus flavus* in sterile peanuts. *J. Am. Oil Chem. Soc.* 44:259–263.

26. Dollear, F. G., G. E. Mann, L. P. Codifer, H. K. Gardner, S. P. Koltun, and H. L. E. Vix. 1986. Elimination of aflatoxin from peanut meal. *J. Am. Oil Chem. Soc.* 45:862–865.

27. Dorner, J. W., R. J. Cole, and D. W. Wicklow. 1999. Aflatoxin reduction in corn through field application of competitive fungi. *J. Food Prot.* 62:650–656.

28. Doster, M. A., T. J. Michailides, and D. P. Morgan. 1996. *Aspergillus* species and mycotoxins in figs from California. *Plant Dis.* 80:484–489.

29. Dyer, S. K., and S. McCammon. 1994. Detection of toxigenic isolates of *Aspergillus flavus* and related species on coconut cream agar. *J. Appl. Bacteriol.* 76:75–78.

30. Egel, D. S., P. J. Cotty, and K. S. Elias. 1994. Relationships among isolates of *Aspergillus* sect. *Flavi* that vary in aflatoxin production. *Phytopathology* 84:906–912.

31. Ellis, W. O., J. P. Smith, B. K. Simpson, and J. H. Oldham. 1991. Aflatoxins in food: occurrence, biosynthesis, effects on organisms, detection, and methods of control. *Crit. Rev. Food Sci. Nutr.* 30:403–439.

32. Ellis, W. O., J. P. Smith, B. K. Simpson, H. Ramaswamy, and G. Doyon. 1994. Novel techniques for controlling growth of and aflatoxin production by *Aspergillus parasiticus* in packaged peanuts. *Food Microbiol.* 11:357–368.

33. Esteban, A., M. L. Abarca, M. R. Bragalat, and F. J. Cabañes. 2004. Effects of temperature and incubation time on production of ochratoxin A by black aspergilli. *Res. Microbiol.* 155:861–866.

34. Fang, S.-Y., and F.-C. Peng. 2003. Factors affecting territrem B production by *Aspergillus terreus* CCRC 32111 in potato-dextrose medium. *J. Food Drug Anal.* 11:27–31.

35. Fang, S.-Y., Q.-K. Wei, and F.-C. Peng. 2001. Determination of territrem B in rice media with different strains of *Aspergillus terreus* by chromatography. *J. Food Drug Anal.* 10:159–163.

36. Filtenborg, O., J. C. Frisvad, and J. A. Svendsen. 1983. Simple screening method for molds producing intracellular mycotoxins in pure culture. *Appl. Environ. Microbiol.* 45:581–585.

37. Flannigan, B., and A. R. Pearce. 1994. *Aspergillus* spoilage: spoilage of cereals and cereal products by the hazardous species *A. clavatus*, p. 115–127. *In* K. A. Powell, A. Renwick, and J. F. Peberdy (ed.), *The Genus* Aspergillus. *From Taxonomy and Genetics to Industrial Application*. Plenum Press, New York, N.Y.

38. Food and Agriculture Organization. 2004. Worldwide regulation for mycotoxins in food and feed in 2003. FAO Food and Nutrition papers, 81. Food and Agriculture Organization, Rome, Italy. [Online.] ftp://ftp.fao.org/docrep/fao/007/y5499e/y5499e00.pdf.

39. Frisvad, J. C. 1989. The connection between the penicillia and aspergilli and mycotoxins with species emphasis on misidentified isolates. *Arch. Environ. Contam. Toxicol.* 18:452–467.

40. Frisvad, J. C., J. M. Frank, J. A. M. P. Houbraken, A. F. A. Kuijpers, and R. A. Samson. 2004. New ochratoxin A producing species of *Aspergillus* section *Circumdati*. *Stud. Mycol.* (Utrecht) 50:23–43.

41. Frisvad, J. C., U. Thrane, R. A. Samson, and J. I. Pitt. 2006. Important mycotoxins and the fungi which produce them,

p. 3–31. *In* A. D. Hocking, J. I. Pitt, R. A. Samson, and U. Thrane (ed.), *Advances in Food Mycology.* Springer, New York, N.Y.

42. Frisvad, J. C., K. F. Nielsen, and R. A. Samson. 2006. Recommendations concerning the chronic problem of misidentification of mycotoxigenic fungi associated with foods and feeds, p. 33–46. *In* A. D. Hocking, J. I. Pitt, R. A. Samson, and U. Thrane (ed.), *Advances in Food Mycology.* Springer, New York, N.Y.

43. Frobish, R. A., B. D. Bradley, D. D. Wagner, P. E. Long-Bradley, and H. Hairston. 1986. Aflatoxin residues in milk of dairy cows after ingestion of naturally contaminated grain. *J. Food Prot.* **49:**781–785.

44. Fujii, K., H. Kurata, S. Odashirna, and Y. Hatsuda. 1976. Tumour induction by a single subcutaneous injection of sterigmatocystin in newborn mice. *Cancer Res.* **36:**1615–1618.

45. Gams, W., M. Christensen, A. H. S. Onions, J. I. Pitt, and R. A. Samson. 1985. Intrageneric taxa of *Aspergillus*, p. 55–62. *In* R. A. Samson and J. I. Pitt (ed.), *Advances in Penicillium and Aspergillus Systematics.* Plenum Press, New York, N.Y.

46. Geisen, R. 1996. Multiplex polymerase chain reaction for the detection of potential aflatoxin and sterigmatocystin producing fungi. *Syst. Appl. Microbiol.* **19:**388–392.

47. Golden, M. C., S. J. Hahm, R. E. Elessar, S. Saksonov, and J. J. Steinberg. 1998. DNA damage by gliotoxin from *Aspergillus fumigatus.* An occupational and environmental propagule: adduct detection as measured by ^{32}P DNA radiolabelling and two-dimensional thin-layer chromatography. *Mycoses* **41:**97–104.

48. Golinski, P. 1991. Secondary metabolites (mycotoxins) produced by fungi colonizing cereal grain in store—structure and properties, p. 355–403. *In* J. Chelkowski (ed.), *Cereal Grain. Mycotoxins, Fungi and Quality in Drying and Storage.* Elsevier, Amsterdam, The Netherlands.

49. Goto, T., D. T. Wicklow, and Y. Ito. 1996. Aflatoxin and cyclopiazonic acid production by a sclerotium-producing *Aspergillus tamarii* strain. *Appl. Environ. Microbiol.* **62:**4036–4038.

50. Gqaleni, N., J. E. Smith, J. Lacey, and G. Gettinby. 1996. The production of cyclopiazonic acid by *Penicillium commune* and cyclopiazonic acid and aflatoxins by *Aspergillus flavus* as affected by water activity and temperature on maize grains. *Mycopathologia* **136:**103–108.

51. Gqaleni, N., J. E. Smith, J. Lacey, and G. Gettinby. 1997. Effects of temperature, water activity, and incubation time on production of aflatoxins and cyclopiazonic acid by an isolate of *Aspergillus flavus* in surface agar culture. *Appl. Environ. Microbiol.* **63:**1048–1053.

52. Groopman, J. D., L. G. Cain, and T. W. Kensler. 1988. Aflatoxin exposure in human populations: measurements and relation to cancer. *Crit. Rev. Toxicol.* **19:**113–145.

53. Gyongyosihorvath, A., I. Barnavetro, and L. Solti. 1996. A new monoclonal antibody detecting ochratoxin A at the picogram level. *Lett. Appl. Microbiol.* **22:**103–105.

54. Hayashi, Y., and T. Yoshizawa. 2005. Analysis of cyclopiazonic acid in corn and rice by a newly developed method. *Food Chem.* **93:**215–221.

55. Heenan, C. N., K. J. Shaw, and J. I. Pitt. 1998. Ochratoxin A production by *Aspergillus carbonarius* and *A. niger* and detection using coconut cream agar. *J. Food Mycol.* **1:**67–72.

56. Hocking, A. D., J. I. Pitt, R. A. Samson, and U. Thrane (ed.). 2006. *Advances in Food Mycology.* Springer, New York, N.Y.

57. Hocking, A. D., P. Varelis, J. I. Pitt, S. Cameron, and S. L. Leong. 2003. Occurrence of ochratoxin A in Australian wine. *Aust. J. Grape Wine Res.* **9:**72–78.

58. Horie, Y. 1995. Productivity of ochratoxin A of *Aspergillus carbonarius* in *Aspergillus* section *Nigri. Nippon Kingakukai Kaiho* **36:**73–76.

59. Klich, M. A. 2002. *Identification of Common Aspergillus Species.* Centraalbureau voor Schimmelcultures, Utrecht, The Netherlands.

60. Klich, M. A., and J. I. Pitt. 1988. Differentiation of *Aspergillus flavus* from *A. parasiticus* and closely related species. *Trans. Br. Mycol. Soc.* **91:**99–108.

61. Klich, M. A., S. H. Thomas, and J. E. Mellon. 1984. Field studies on the mode of entry of *Aspergillus flavus* into cotton seeds. *Mycologia* **76:**665–669.

62. Koehler, P. E., L. R. Beuchat, and M. S. Chinnan. 1985. Influence of temperature and water activity on aflatoxin production by *Aspergillus flavus* in cowpea (*Vigna unguiculata*) seeds and meal. *J. Food Prot.* **48:**1040–1043.

63. Kozakiewicz, Z. 1994. *Aspergillus*, p. 575–616. *In* Y. H. Hui, J. R. Gorham, K. D. Murrell, and D. O. Cliver (ed.), *Foodborne Disease Handbook*, vol. 2. Marcel Dekker, New York, N.Y.

64. Kozakiewicz, Z. 1994. *Aspergillus* toxins and taxonomy, p. 303–311. *In* K. A. Powell, A. Renwick, and J. F. Peberdy (ed.), *The Genus Aspergillus. From Genetics to Industrial Application.* Plenum Press, New York, N.Y.

65. Krishnamachari, K. A.V., R. V. Bhat, V. Nagarajan, and T. B. G. Tilak. 1975. Investigations into an outbreak of hepatitis in parts of Western India. *Indian J. Med. Res.* **63:**1036–1048.

66. Lacey, J. 1994. Aspergilli in feeds and seeds, p. 73–92. *In* K. A. Powell, A. Renwick, and J. F. Peberdy (ed.), *The Genus Aspergillus. From Taxonomy and Genetics to Industrial Application.* Plenum Press, New York, N.Y.

67. Land, C. J., H. Lundstrom, and S. Werner. 1993. Production of tremorgenic mycotoxins by isolates of *Aspergillus fumigatus* from sawmills in Sweden. *Mycopathologia* **124:**87–93.

68. Leong, S. L. 2005. Black *Aspergillus* species: implications for ochratoxin A in Australian grapes and wine. Ph.D. thesis. University of Adelaide, Adelaide, South Australia.

69. Leong, S. L., A. D. Hocking, and J. I. Pitt. 2004. Occurrence of fruit rot fungi (*Aspergillus* section *Nigri*) on some drying varieties of irrigated grapes. *Aust. J. Grape Wine Res.* **10:**83–88.

70. Leong, S. L., A. D. Hocking, J. I. Pitt, B. A. Kazi, R. W. Emmett, and E. S. Scott. 2006. Australian research on ochratoxigenic fungi and ochratoxin A. *Int. J. Food Microbiol.* **111S1:**S10–S17.

71. Lillehoj, E. B., W. F. Kwolek, E. S. Horner, N. W. Widstrom, L. M. Josephson, A. O. Franz, and E. A. Catalano. 1980. Aflatoxin contamination of preharvest corn: role of *Aspergillus flavus* inoculum and insect damage. *Cereal Chem.* **57:**255–257.

72. Ling, K. H. 1994. Territrems, tremorgenic mycotoxin isolated from *Aspergillus terreus*. *J. Toxicol. Toxin Rev.* **13**:243–252.

73. Ling, K. H., C. K. Yang, and F. T. Peng. 1979. Territrems, tremorgenic mycotoxins of *Aspergillus terreus*. *Appl. Environ. Microbiol.* **37**:355–357.

74. Liu, B. H., and F. S. Chu. 1998. Regulation of *aflR* and its product, AflR, associated with aflatoxin biosynthesis. *Appl. Environ. Microbiol.* **64**:3718–3723.

75. Lopez-Diaz, T. M., and B. Flannigan. 1997. Production of patulin and cytochalasin E by *Aspergillus clavatus* during malting of barley and wheat. *Int. J. Food Microbiol.* **35**:129–136.

76. Losito, I., L. Monaci, A. Aresta, and C. G. Zambonin. 2002. LC-ion trap electrospray MS-MS for the determination of cyclopiazonic acid in milk samples. *Analyst* **127**:499–502.

77. Lubulwa, A. S. G., and J. S. Davis. 1994. Estimating the social costs of the impacts of fungi and aflatoxins, p. 1017–1042. *In* E. Highley, E. J. Wright, H. J. Banks, and B. R. Champ (ed.), *Stored Product Protection. Proceedings of the 6th International Working Conference on Stored-Product Protection.* CAB International, Wallingford, Oxford, United Kingdom.

78. Lye, M. S., A. A. Ghazali, J. Mohan, N. Alwin, and R. C. Nair. 1995. An outbreak of acute hepatic encephalopathy due to severe aflatoxicosis in Malaysia. *Am. J. Trop. Med. Hyg.* **53**:68–72.

79. Ministry of Agriculture, Fisheries and Food. 1999. *1998 Survey of Retail Products for Ochratoxin A.* Ministry of Agriculture, Fisheries, and Food, food surveillance information sheet 185:1–36. HMSO Publications Centre, London, United Kingdom.

80. Moody, S. F., and B. M. Tyler. 1990. Use of DNA restriction length polymorphisms to analyse the diversity of the *Aspergillus flavus* group, *Aspergillus flavus, Aspergillus parasiticus,* and *Aspergillus nomius. Appl. Environ. Microbiol.* **56**:2453–2461.

81. Mori, T., M. Matsumura, K. Yamada, S. Irie, K. Oshimi, K. Suda, T. Oguri, and M. Ichinoe. 1998. Systemic aspergillosis caused by an aflatoxin-producing strain of *Aspergillus flavus. Med. Mycol.* **36**:107–112.

82. Mullaney, E. J., and M. A. Klich. 1990. A review of molecular biological techniques for systematic studies of *Aspergillus* and *Penicillium*, p. 301–307. *In* R. A. Samson and J. I. Pitt (ed.), *Modern Concepts in Penicillium and* Aspergillus *Classification.* Plenum Press, New York, N.Y.

83. Nielsen, K. F., S. Gravesen, P. A. Nielsen, B. Andersen, U. Thrane, and J. C. Frisvad. 1999. Production of mycotoxins on artificially infested building materials. *Mycopathologia* **145**:43–56.

84. Nishie, K., R. J. Cole, and F. W. Dorner. 1985. Toxicity and neuropharmacology of cyclopiazonic acid. *Food Chem. Toxicol.* **23**:831–839.

85. Norred, W. P., R. E. Morrisey, R. T. Riely, R. J. Cole, and L. W. Dorner. 1985. Distribution, excretion and skeletal muscle effects of the mycotoxin (^{14}C) cyclopiazonic acid in rats. *Food Chem. Toxicol.* **23**:1069–1076.

86. Northolt, M. D., H. P. van Egmond, and W. E. Paulsch. 1977. Differences in *Aspergillus flavus* strains in growth and aflatoxin B$_1$ production in relation to water activity and temperature. *J. Food Prot.* **40**:778–781.

87. Northolt, M. D., H. P. van Egmond, and W. E. Paulsch. 1979. Ochratoxin A production by some fungal species in relation to water activity and temperature. *J. Food Prot.* **42**:485–490.

88. Northolt, M. D., H. P. van Egmond, and W. E. Paulsch. 1979. Penicillic acid production by some fungal species in relation to water activity and temperature. *J. Food Prot.* **42**:476–484.

89. Payne, G. A., and M. P. Brown. 1998. Genetics and physiology of aflatoxin biosynthesis. *Annu. Rev. Phytopathol.* **36**:329–362.

90. Peers, F., X. Bosch, J. Kaldor, A. Linsell, and M. Pluumen. 1987. Aflatoxin exposure, hepatitis B virus infection and liver cancer in Swaziland. *Int. J. Cancer* **39**:545–553.

91. Pestka, J. J., M. N. Abouzied, and Sutikno. 1995. Immunological assays for mycotoxin detection. *Food Technol.* **49**:120–128.

92. Pier, A. C. 1991. The influence of mycotoxins on the immune system, p. 489–497. *In* J. E. Smith and R. S. Henderson (ed.), *Mycotoxins and Animal Foods.* CRC Press, Boca Raton, Fla.

93. Pitt, J. I., and A. D. Hocking. 1997. *Fungi and Food Spoilage,* 2nd ed. Blackie Academic and Professional, London, United Kingdom.

94. Pitt, J. I., and A. D. Hocking. 2006. Mycotoxins in Australia: biocontrol of aflatoxin in peanuts. *Mycopathologia* **162**:233–243.

95. Pitt, J. I., and B. F. Miscamble. 1995. Water relations of *Aspergillus flavus* and closely related species. *J. Food Prot.* **58**:86–90.

96. Pitt, J. I., S. K. Dyer, and S. McCammon. 1991. Systemic invasion of developing peanut plants by *Aspergillus flavus. Lett. Appl. Microbiol.* **13**:16–20.

97. Pitt, J. I., A. D. Hocking, and D. R. Glenn. 1983. An improved medium for the detection of *Aspergillus flavus* and *A. parasiticus. J. Appl. Bacteriol.* **54**:109–114.

98. Pitt, J. I., A. D. Hocking, K. Bhudhasamai, B. F. Miscamble, K. A. Wheeler, and P. Tanboon-Ek. 1993. The normal mycoflora of commodities from Thailand. 1. Nuts and oilseeds. *Int. J. Food Microbiol.* **20**:211–226.

99. Pitt, J. I., A. D. Hocking, K. Bhudhasamai, B. F. Miscamble, K. A. Wheeler, and P. Tanboon-Ek. 1994. The normal mycoflora of commodities from Thailand. 2. Beans, rice, small grains and other commodities. *Int. J. Food Microbiol.* **23**:35–53.

100. Price, M. S., and G. A. Payne. 2005. Genomics and gene regulation of the aflatoxin biosynthetic pathway, p. 387–404. *In* Z. An (ed.), *Handbook of Industrial Mycology.* Marcel Dekker, New York, N.Y.

101. Ramos, A. J., N. la Bernia, S. Marin, V. Sanchis, and N. Magan. 1998. Effect of water activity and temperature on growth and ochratoxin production by three strains of *Aspergillus ochraceus* on a barley extract medium and on barley grains. *Int. J. Food Microbiol.* **44**:133–140.

102. Raper, K. B., and D. I. Fennell. 1965. *The Genus* Aspergillus. The Williams and Wilkins Co., Baltimore, Md.

103. Richards, J. L. 1997. Gliotoxin, a mycotoxin associated with cases of avian aspergillosis. *J. Nat. Toxins* **6**:11–18.

104. Rosa, C. A. R., C. E. Magnoli, M. E. Fraga, A. M. Dalcero, and D. M. N. Santana. 2004. Occurrence of ochratoxin A in wine and grape juice marketed in Rio de Janeiro, Brazil. *Food Addit. Contam.* **21:**358–364.

105. Samson, R. A., and J. I. Pitt. 1990. *Modern Concepts in Penicillium and Aspergillus Classification.* Plenum Press, New York, N.Y.

106. Samson, R. A., A. D. Hocking, J. I. Pitt, and A. D. King (ed.). 1992. *Modern Methods in Food Mycology.* Elsevier, Amsterdam, The Netherlands.

107. Samson, R. A., E. S. Hoekstra, and J. C. Frisvad. 2004. *Introduction to Food- and Airborne Fungi,* 7th ed. Centraalbureau voor Schimmelcultures, Utrecht, The Netherlands.

108. Sargeant, K., R. B. A. Carnaghan, and R. Allcroft. 1963. Toxic products in goundnuts—chemistry and origin. *Chemy Ind.* **1963:**53–55.

109. Schmidt, H., M. Ehrmann, R. E. Vogel, M. H. Taniwaki, and L. Niessen. 2003. Molecular typing of *Aspergillus ochraceus* and construction of species specific SCAR-primers based on AFLP. *Syst. Appl. Microbiol.* **26:**138–146.

110. Scott, P. M. 2004. Other mycotoxins, p. 406–440. *In* N. Magan and M. Olsen (ed.), *Mycotoxins in Food: Detection and Control.* Woodhead Publishing Limited, Cambridge, United Kingdom.

111. Scudamore, K. A. 2005. Principles and applications of mycotoxin analysis, p. 157–185. *In* D. E. Diaz (ed.), *The Mycotoxin Blue Book.* Nottingham University Press, Nottingham, United Kingdom.

112. Seo, J.-A., and J.-H. Yu. 2004. Toxigenic fungi and mycotoxins, p. 689–721. *In* Z. An (ed.), *Handbook of Industrial Mycology.* Marcel Dekker, New York, N.Y.

113. Shapira, R., N. Paster, O. Eyal, M. Menasherov, A. Mett, and R. Salomon. 1996. Detection of aflatoxigenic molds in grains by PCR. *Appl. Environ. Microbiol.* **62:**3270–3273.

114. Smith, S. L., and S. T. Hill. 1982. Influence of temperature and water activity on germination and growth of *Aspergillus restrictus* and *A. versicolor. Trans. Br. Mycol. Soc.* **79:**558–460.

115. Sreenivasamurthy, V., H. A. B. Parpia, S. Srikanta, and A. Shankarmurti. 1967. Detoxification of aflatoxin in peanut meal by hydrogen peroxide. *J. Assoc. Off. Anal. Chem.* **50:**350–354.

116. Stanzani, M., E. Orciuolo, R. Lewis, D. P. Kontoyiannis, S. L. R. Martins, L. S. St. John, and K. V. Komanduri. 2005. *Aspergillus fumigatus* suppresses the human cellular immune response via gliotoxin-mediated apoptosis of monocytes. *Blood* **105:**2258–2265.

117. Stoloff, L. 1983. Aflatoxin as a cause of primary liver-cell cancer in the United States: a probability study. *Nutr. Cancer* **5:**165–168.

118. Terao, K. 1983. Sterigmatocystin—a masked potent carcinogenic mycotoxin. *J. Toxicol. Toxin Rev.* **2:**77–100.

119. Tran-Dinh, N., J. I. Pitt, and D. A. Carter. 1999. Molecular genotype analysis of natural toxigenic and nontoxigenic isolates of *Aspergillus flavus* and *A. parasiticus. Mycol. Res.* **103:**1485–1490.

120. Trucksess, M. W., and M. E. Stack. 1994. Enzyme-linked immunosorbent assay of total aflatoxins B$_1$, B$_2$ and G$_1$

in corn: follow-up collaborative study. *J. AOAC Int.* **77:**655–658.

121. Trucksess, M. W., J. Giler, K. Young, K. D. White, and S. W. Page. 1999. Determination and survey of ochratoxin A in wheat, barley and coffee. *J. AOAC Int.* **82:**85–98.

122. Tuomi, T., K. Reijula, T. Johnsson, K. Hemminki, E. L. Hintikka, O. Lindroos, S. Kalso, P. Koukila-Kahkola, H. Mussalo-Rauhamaa, and T. Haahtela. 2000. Mycotoxins in crude building materials from water-damaged buildings. *Appl. Environ. Microbiol.* **66:**1899–1904.

123. USDA Agricultural Research Service. 6 August 2004, posting date. *afla-guard Receives EPA Registration.* [Online.] USDA Agricultural Research Service, Washington, D.C. http://www.ars.usda.gov/News/news.htm?modecode=66-04-00-00&newsid=2692.

124. van der Merwe, K. J., P. S. Steyn, L. Fourie, D. B. Scott, and J. J. Theron. 1965. Ochratoxin A, a toxic metabolite produced by *Aspergillus ochraceus* Wilh. *Nature* **205:**1112–1113.

125. van der Stegen, G., U. J. Jörissen, A. Pittet, M. Saccon, W. Steiner, M. Vincenzi, M. Winkler, J. Zapp, and C. Schlatter. 1997. Screening of European coffee final products for occurrence of ochratoxin A. *Food Addit. Contam.* **14:**211–216.

126. Van Egmond, H. P., and M. A. Jonker. 2004. Current regulations governing mycotoxin limits in food, p. 49–68. *In* N. Magan and M. Olsen (ed.), *Mycotoxins in Food: Detection and Control.* Woodhead Publishing Limited, Cambridge, United Kingdom.

127. van Rensburg, S. J. 1977. Role of epidemiology in the elucidation of mycotoxin health risks, p. 699–711. *In* J. V. Rodricks, C. W. Hesseltine, and M. A. Mehlman (ed.), *Mycotoxins in Human and Animal Health.* Pathatox Publishers, Park Forest South, Ill.

128. Varga, J., A. Juhasz, F. Kevei, and Z. Kozakiewicz. 2004. Molecular diversity of agriculturally important *Aspergillus* species. *Eur. J. Plant Pathol.* **110:**627–640.

129. Vargas, E. A., E. A. dos Santos, and A. Pittet. 2005. Determination of ochratoxin A in green coffee by immunoaffinity column cleanup and liquid chromatography: collaborative study. *J. AOAC Int.* **88:**773–779.

130. Vesonder, R. F., R. Lambert, D. T. Wicklow, and M. L. Biehl. 1988. *Eurotium* spp. and echinulin in feed refusal by swine. *Appl. Environ. Microbiol.* **54:**830–831.

131. Weng, C. Y., A. J. Martinez, and D. L. Park. 1994. Efficacy and permanency of ammonia treatment in reducing aflatoxin levels in corn. *Food Addit. Contam.* **11:**649–658.

132. Wheeler, K. A., B. F. Hurdman, and J. I. Pitt. 1991. Influence of pH on the growth of some toxigenic species of *Aspergillus, Penicillium* and *Fusarium. Int. J. Food Microbiol.* **12:**141–150.

133. Wicklow, D. T., P. F. Dowd, A. A. Alfatafta, and J. B. Gloer. 1996. Ochratoxin A: an antiinsectan metabolite from the sclerotia of *Aspergillus carbonarius* NRRL 369. *Can. J. Microbiol.* **42:**1100–1103.

134. Williams, J. H., T. D. Phillips, P. E. Jolly, J. K. Stiles, C. M. Jolly, and D. Aggarwal. 2004. Human aflatoxicosis in developing countries: a review of toxicology, exposure, potential health consequences, and interventions. *Am. J. Clin. Nutr.* **80:**1106–1122.

135. Yabe, K., M. Nakamura, and T. Hamasaki. 1999. Enzymatic formation of G-group aflatoxins and biosynthetic relationship between G- and B-group aflatoxins. *Appl. Environ. Microbiol.* **65:**3867–3872.

136. Yagi, R., and M. Doi. 1999. Isolation of an antioxidative substance produced by *Aspergillus repens. Biosci. Biotechnol. Biochem.* **63:**932–933.

137. Yang, Z.-Y., W.-B. Shim, J.-H. Kim, S.-J. Park, S.-J. Kang, B.-S. Nam, and D.-H. Chung. 2004. Detection of aflatoxin-producing molds in Korean fermented foods and grains by multiplex PCR. *J. Food Prot.* **67:**2622–2626.

138. Yoshizawa, T. 1991. Natural occurrence of mycotoxins in small grain cereals (wheat, barley, rye, oats, sorghum, millet, rice), p. 301–324. *In* J. E. Smith and R. S. Henderson (ed.), *Mycotoxins and Animal Foods.* CRC Press, Boca Raton, Fla.

139. Yu, W., and F. S. Chu. 1998. Improved direct competitive enzyme-linked immunosorbent assay for cyclopiazonic acid in corn, peanuts and mixed feed. *J. Agric. Food Chem.* **46:**1012–1017.

140. Zheng, Z. M., J. Hanneken, D. Houchins, R. S. King, P. Lee, and J. L. Richard. 2005. Validation of an ELISA test kit for the detection of ochratoxin A in several food commodities by comparison with HPLC. *Mycopathologia* **159:**265–272.

141. Zimmerli, B., and R. Dick. 1996. Ochratoxin A in table wine and grape-juice: occurrence and risk assessment. *Food Addit. Contam.* **13:**655–668.

Food Microbiology: Fundamentals and Frontiers, 3rd Ed.
Edited by M. P. Doyle and L. R. Beuchat
© 2007 ASM Press, Washington, D.C.

John I. Pitt

Toxigenic *Penicillium* Species

25

Experimental evidence that common microfungi can produce toxins is popularly believed to date only from about 1960. However, the study of mycotoxicology began 100 years ago. In 1891, in Japan, Sakaki demonstrated that an ethanol extract from moldy, unpolished "yellow" rice was fatal to dogs, rabbits, and guinea pigs, with symptoms indicating paralysis of the central nervous system (71). In consequence, the sale of yellow rice was banned in Japan in 1910. In 1913, an extract from a *Penicillium puberulum* culture isolated from moldy corn in Nebraska was found to be toxic to animals when injected at 200 to 300 mg/kg of body weight (1). This was the first reliable account of toxin production by a mold in pure culture. This careful study, aimed to resolve the question of whether common molds or mold products could have an injurious effect on animals, produced positive results and was far ahead of its time. Such direct evidence that common molds could be toxic was largely ignored.

The discovery of penicillin in 1929 gave impetus to a search for other *Penicillium* metabolites with antibiotic properties and, ultimately, to the recognition of citrinin, patulin, and griseofulvin as "toxic antibiotics" or, later, mycotoxins.

The literature on toxigenic penicillia is now quite vast. In a comprehensive review of the literature on fungal metabolites, about 120 common mold species were found to be demonstrably toxic to higher animals (14). Forty-two were reported to be produced by one or more *Penicillium* species. No fewer than 85 *Penicillium* species were listed as toxigenic (14). The literature on this subject has accumulated in a rather random fashion, with emphasis in the majority of papers on chemistry or toxicology rather than mycology. The impression gained from the literature is that toxin production by penicillia lacks species specificity, i.e., most toxins are produced by a variety of species. This viewpoint is commonly accepted. For example, citrinin has been reported to be produced by at least 22 species (14, 61), but perhaps as few as 3 species are actual producers. Explanations for this discrepancy include the fact that the taxonomy of toxigenic *Penicillium* species has been uncertain and that many reports on the toxigenicity of particular species have been based on misidentifications.

Recent developments in *Penicillium* taxonomy have profoundly changed the accuracy of identification. Improvements in morphological classification (47), the development of simple screening methods for secondary metabolites and the recognition that such compounds are of taxonomic value (23, 34), an emphasis on accurate identification and mycotoxin production (51), and

John I. Pitt, Food Science Australia, P. O. Box 52, North Ryde, New South Wales 1670, Australia.

the use of electrophoretic fingerprinting of certain iso-enzymes (3, 15) have all assisted in developing a more definitive taxonomy (57). Molecular studies to date have mainly confirmed the more traditional findings (63) and permitted refinement of phylogeny (37, 45). A classification system for the most common species has been published (49), and an accurate picture of the important species-mycotoxin relationships has been developed (25, 51). A complete list of accepted names and important synonyms for *Penicillium* and related genera was recently published (55).

TAXONOMY

Penicillium is a large genus. One hundred fifty species were recognized in the most recent complete taxonomic study (47), but subsequent studies indicated that this number is conservative. At least 50 species are of common occurrence (49). All common species grow and sporulate well on synthetic or semisynthetic media and are usually readily recognizable to the genus level.

Classification within *Penicillium* is based primarily on microscopic morphology (Fig. 25.1). The genus is divided into subgenera based on the number and arrangement of phialides (elements producing conidia) and of metulae and rami (elements supporting phialides) on the main stalk cells (stipes). The classification of Pitt (47) includes the following four subgenera: *Aspergilloides*, where phialides are borne directly on the stipes without intervening supporting elements; *Furcatum* and *Biverticillium*, where phialides are supported by metulae; and *Penicillium*, where both metulae and rami are usually present (Fig. 25.1). The majority of important toxigenic and food spoilage species are found in the subgenus *Penicillium*.

ENUMERATION

General enumeration procedures suitable for foodborne molds are effective for enumerating all common *Penicillium* species. Many antibacterial enumeration media can be expected to give satisfactory results. However, some *Penicillium* species grow rather weakly on very dilute media, such as potato dextrose agar. Dichloran–rose Bengal–chloramphenicol (DRBC) agar and dichloran–18% glycerol (DG18) agar (50) are recommended (28, 58).

IDENTIFICATION

For a comprehensive taxonomy of *Penicillium*, see the work of Pitt (47); for keys and descriptions to common species, see the work of Pitt (49); for foodborne species,

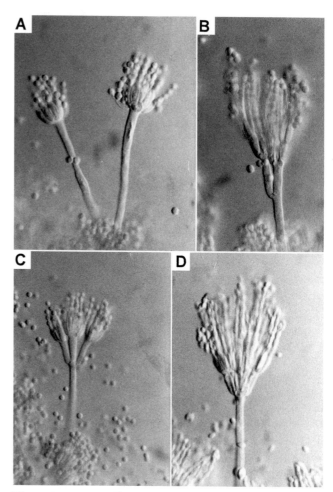

Figure 25.1 Penicilli representative of the four *Penicillium* subgenera. (A) *Penicillium* subgenus *Aspergilloides* (*P. glabrum*); (B) *Penicillium* subgenus *Penicillium* (*P. expansum*); (C) *Penicillium* subgenus *Furcatum* (*P. citrinum*); (D) *Penicillium* subgenus *Biverticillium* (*P. variabile*). Magnification, ×550.

see the work of Pitt and Hocking (50) and Samson et al. (58); and for all species in *Penicillium* subgenus *Penicillium*, see the work of Samson and Frisvad (56).

Identification of *Penicillium* isolates to the species level is not easy and is preferably carried out under carefully standardized conditions of media, incubation time, and temperature. In addition to microscopic morphology, gross physiological features, including colony diameters and colors of conidia and colony pigments, are used to distinguish species (46, 49, 56).

SIGNIFICANT *PENICILLIUM* MYCOTOXINS

As already noted, more than 80 *Penicillium* species have been reported to be toxin producers. In assessing the relevance of these reports, the following points need to be kept in mind: (i) many of the identifications reported in the literature are incorrect, and this has resulted in confusion; (ii) a number of genuinely toxic compounds are produced by species not usually occurring in commodities, foods, or feeds—such toxins are of little practical importance and are not discussed here; (iii) some compounds often cited as mycotoxins in the literature are of low toxicity and, again, are of little practical importance; and (iv) the growth of mold does not always indicate the production of toxin. The conditions under which toxins are produced are often narrower than the conditions for growth. Appropriate information is provided where this is known. Taking these factors into account, 27 mycotoxins were judged to have demonstrated toxicity to humans or domestic animals (51). Of these, a recent review described 17 as important in foods and as having the potential to affect human health (25).

The range of mycotoxin classes produced by *Penicillium* is broader than that for any other fungal genus. Moreover, molecular compositions are diverse in the extreme. Patulin is an unsaturated lactone with a molecular weight of 150, while penitrem A is a polycyclic ether with nine contiguous rings and a molecular weight of >650 (14, 16).

Toxicities due to *Penicillium* species are also very diverse. However, most toxins can be placed into two broad groups, viz, those that affect liver and kidney function and those that are neurotoxins. In general terms, the *Penicillium* toxins which affect liver or kidney function are asymptomatic or cause generalized debility in man or animals. In contrast, the toxicity of the neurotoxins in animals is often characterized by sustained trembling.

Table 25.1 lists nine mycotoxins, produced by 17 *Penicillium* species, which I consider the most significant for human health. Relative toxicities of the various compounds are shown for comparative purposes. For this reason, toxicities in a standard test animal (mouse) and by a standard route of administration (oral) have been provided where possible. The list of species shown as producing each toxin results from my own experience and is conservative. Some general notes follow on these toxins and the species which produce them.

Ochratoxin A

Ochratoxin A is undoubtedly the most important toxin produced by a *Penicillium* species. The International Agency for Research on Cancer has classified this compound as a possible human carcinogen (group 2B), based on sufficient evidence of carcinogenicity in experimental animal studies and inadequate evidence in humans (30). The target organ of toxicity in all mammalian species tested is the kidney, in which lesions can be produced by both acute and chronic exposure.

The source of ochratoxin A in nature is relatively complex. Ochratoxin A was originally described as a metabolite of *Aspergillus ochraceus*; recently, it was shown to be produced commonly by *Aspergillus carbonarius* and rarely by the related species *Aspergillus niger* (70). However, ochratoxin A is also produced by *Penicillium verrucosum* and the closely related *species Penicillium nordicum* (25). A majority of *P. verrucosum* isolates produce ochratoxin A, and sometimes citrinin as well (25).

P. verrucosum grows in barley and wheat crops in cold climates, especially Scandinavia, central Europe, and western Canada. Ochratoxin A plays a major role in the etiology of nephritis (kidney disease) in pigs in Scandinavia (33), and indeed in much of northern Europe. In an extensive survey of 70 Danish barley samples from farms where pigs were suffering from nephritis, 67 contained large populations of *P. verrucosum* and 66 contained ochratoxin A (23). It seems certain that *P. verrucosum* is the major source of ochratoxin A in Scandinavia and other cool temperate-zone areas.

The major source of ochratoxin A in foods is bread made from barley or wheat in which *P. verrucosum* has grown. Because ochratoxin A is fat soluble and not readily excreted, it also accumulates in the depot fat of pigs eating feeds containing ochratoxin A, and from there is ingested by humans eating pork. In consequence, ochratoxin A has been found to be widespread in the blood (5) and breast milk (6, 42) of Europeans. Although clear evidence of human disease is still elusive, such levels indicate the widespread occurrence of ochratoxin A in Europe.

It has been suggested that ochratoxin A is a causal agent of Balkan endemic nephropathy, a kidney disease with a high mortality rate in certain areas of Bulgaria, Yugoslavia, and Romania. The symptoms in people are sufficiently similar to those of ochratoxin A toxicity in animals that a role for ochratoxin A in this disease has been postulated on a number of occasions (44, 46). However, no conclusive epidemiological evidence has emerged to support this hypothesis, and its relevance to studies on ochratoxin A toxicity remains doubtful (40). Other toxins or contributing factors are probably involved (69).

Taxonomy

Described originally in 1901, *P. verrucosum* was ignored until revived by Samson et al. in 1976 (59). Their concept was very broad, encompassing several species previously considered distinct. *P. verrucosum*

Table 25.1 Significant mycotoxins produced by *Penicillium* species

Mycotoxin	Toxicity (animal, LD$_{50}$, route of administration)[a]	Species producing toxin[b]
Citreoviridin	Mice, 7.5 mg/kg, i.p.	*P. citreonigrum* Dierckx
	Mice, 20 mg/kg, oral	*E. ochrosalmoneum* Scott and Stolk
Citrinin	Mice, 35 mg/kg, i.p.	*P. citrinum* Thom
	Mice, 110 mg/kg, oral	*P. expansum* Link
		P. verrucosum Dierckx
Cyclopiazonic acid	Rats, 2.3 mg/kg, i.p.	*P. camemberti* Thom
	Male rats, 36 mg/kg, oral	*P. commune* Thom
	Female rats, 63 mg/kg, oral	*P. chrysogenum* Thom
		P. griseofulvum Dierckx
		P. hirsutum Dierckx
		P. viridicatum Westling
Ochratoxin A	Young rats, 22 mg/kg, oral	*P. verrucosum*
Patulin	Mice, 5 mg/kg, i.p.	*P. expansum*
	Mice, 35 mg/kg, oral	*P. vulpinum* (Cooke and Massee) Seifert and Samson
		P. griseofulvum
		P. roqueforti Thom
Penitrem A	Mice, 1 mg/kg, i.p.	*P. crustosum* Thom
		P. glandicola (Oudem.) Seifert and Samson
PR toxin	Mice, 6 mg/kg, i.p.	*P. roqueforti*
	Rats, 115 mg/kg, oral	
Roquefortine C	Mice, 340 mg/kg, i.p.	*P. roqueforti*
		P. chrysogenum
		P. crustosum
Secalonic acid D	Mice, 42 mg/kg, i.p.	*P. oxalicum* Currie and Thom

[a] References 14 and 51. LD$_{50}$, 50% lethal dose; i.p., intraperitoneal injection.
[b] Reference 25 and personal observations.

was restricted to a much narrower concept by Pitt (47). With the recognition that this concept is linked to ochratoxin production, it has now been widely accepted. *P. verrucosum* is classified in subgenus *Penicillium* section *Penicillium*, which includes many mycotoxigenic species of common occurrence in foods. *P. verrucosum* (Fig. 25.2A) is distinguished by slow growth on Czapek yeast extract agar (CYA) and malt extract agar (MEA) at 25°C (17 to 24 mm and 10 to 20 mm, respectively, after 7 days), bright green conidia, a clear to pale yellow exudate, and rough stipes (49). It is similar in general appearance to *Penicillium viridicatum*, differing most obviously by slower growth, and to *Penicillium solitum*, from which it differs by having green rather than blue conidia (49).

Enumeration and Identification
The media specified above for general enumeration of *Penicillium* species are effective for *P. verrucosum*. On dichloran–rose bengal–yeast extract–sucrose agar, a selective medium for the enumeration of *P. verrucosum*

and *P. viridicatum*, *P. verrucosum* produces a violet-brown reverse coloration (22). Isolation and identification of *P. verrucosum* in pure culture are essential for confirmation.

Ecology
P. verrucosum has been reported almost exclusively for grain from temperate zones. It is associated with barley and wheat from Scandinavia and Canada. It has also been isolated quite frequently from meat products in Germany and other European countries. It is uncommon elsewhere (50).

P. verrucosum grows most strongly at relatively low temperatures, down to 0°C, with a maximum near 31°C. It is xerophilic, growing down to a water activity (a$_w$) of 0.80. Maximum ochratoxin A production occurs at about 20°C and is possible down to an a$_w$ of about 0.85 (31). Cereals in storage only marginally above levels considered safe from mold growth appear especially vulnerable.

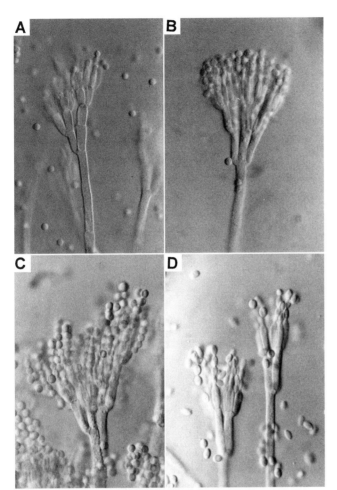

Figure 25.2 Some toxigenic *Penicillium* species. (A) *P. verrucosum*; (B) *P. crustosum*; (C) *P. roqueforti*; (D) *P. oxalicum*. Magnification, ×550.

Citrinin

Citrinin is a significant renal toxin affecting monogastric domestic animals such as pigs and dogs (21). It is also an important toxin in domestic birds, where it produces watery diarrhea, increased water consumption, and reduced weight gain due to kidney degeneration (65). The importance of citrinin in human health is currently difficult to assess. When ingested in the absence of other toxins, significant effects appear unlikely. However, citrinin is produced in nature along with ochratoxin A by some isolates of *P. verrucosum* (50) and with patulin by some isolates of *Penicillium expansum* (51). The possibility of synergy between these toxins should not be discounted.

Primarily recognized as a metabolite of *Penicillium citrinum*, citrinin has been reported from more than 20 other fungal species (14, 22, 61). However, apart from *P. citrinum*, only *P. expansum* and *P. verrucosum* are certain producers (51). These three species are among the

most commonly occurring penicillia, so citrinin is probably the most widely occurring *penicillium* toxin.

Taxonomy
Classified in subgenus *Furcatum* section *Furcatum* (47), *P. citrinum* is a very well-circumscribed species, accepted without controversy for many years. The most distinctive feature of *P. citrinum* is its penicillus (Fig. 25.1C), which consists of a cluster of three to five divergent metulae, which are usually apically swollen. Under a stereomicroscope, the phialides from each metula bear conidia as long columns, producing a distinctive pattern which can be of diagnostic value. Colonies of this species on CYA and MEA are of moderate size (25 to 30 mm and 14 to 18 mm, respectively), with the smaller size on MEA also being a distinctive feature. Growth normally occurs at 37°C, but colonies seldom exceed 10 mm after 7 days (49).

Enumeration and Identification
P. citrinum can be enumerated effectively on any of the enumeration media mentioned above. In all cases, confirmation requires isolation and identification by standard methods. No selective media have been developed for this species.

Ecology
P. citrinum is a ubiquitous mold and has been isolated from nearly every kind of food surveyed for fungi. The most common sources are cereals, especially rice, wheat, and corn, milled grains, and flour (50). Toxin production is also likely a common occurrence.

P. citrinum grows at temperatures from 5 to 7°C to 40°C, with an optimum near 30°C. It is xerophilic, with a minimum a_w for growth near 0.82 (50). Citrinin is produced over most of the temperature growth range, but the effect of a_w is unknown (31).

Patulin

Patulin is a lactone and produces teratogenic effects in rodents, as well as neurological and gastrointestinal effects (65). Early reports of carcinogenicity have not been substantiated, but mutagenicity in a cell line was recently reported (2). Like most other mycotoxins, patulin has no obvious immunological effects (36).

The most important *Penicillium* species producing patulin is *P. expansum*, which is best known as a fruit pathogen but also has widespread occurrence in other fresh and processed foods (50, 66). *P. expansum* produces patulin as it rots apples and pears. The use of rotting fruit in juice or cider manufacture can result in high concentrations of patulin, with a concentration of 350 μg/liter (4) or even 630 μg/liter (75) having been found in the resultant juice.

Levels in commercial practice are usually much lower than this, and given that patulin appears to lack chronic toxic effects in humans, low levels in juices are perhaps of little concern (43). However, because apple juice is widely consumed by children as well as adults, some countries have set an upper limit of 50 µg/liter for patulin in apple juice and other apple products (65, 73). It is also important as an indicator of the use of poor-quality raw materials in juice manufacture.

Patulin is produced by several other *Penicillium* species besides *P. expansum* (Table 25.1), but the potential for the production of unacceptable levels in foods appears to be much lower.

Taxonomy
One of the oldest described *Penicillium* species, *P. expansum* has been established as the principal cause of spoilage of pome fruits throughout this century. It belongs to subgenus *Penicillium* section *Penicillium* (47). *P. expansum* colonies usually grow quite rapidly on CYA and MEA (30 to 40 mm in diameter in 7 days) and show an orange-brown or cinnamon color in exudate, soluble pigment, and reverse (49). Penicilli are terverticillate, with smooth stipes (Fig. 25.1B). Destructive rots of pomaceous fruits are almost always due to this species.

Enumeration and Identification
This species can be enumerated effectively on DRBC agar or DG18 agar. No selective media have been developed; however, isolation from decaying apples and pears provides presumptive identification.

Ecology
The major habitat for *P. expansum* is as a destructive pathogen of pome fruits, but in reality it is a broad-spectrum pathogen of fruits, including avocados, grapes, tomatoes, and mangoes (66). Growth occurs from ca. −3 to 35°C, with an optimum near 25°C, and down to an a_w of 0.82. Patulin can be produced over the range of 0 to 25°C, but not at 31°C, with the optimum temperature being 25°C. The minimum a_w for patulin production by *P. expansum* is 0.95 at 25°C (31). Patulin is quite stable in apple juice during storage; pasteurization at 90°C for 10 s caused <20% reduction (50).

Cyclopiazonic Acid
Cyclopiazonic acid is a highly toxic compound that causes fatty degeneration and hepatic cell necrosis in the livers and kidneys of domestic animals. Chickens are particularly susceptible (18). When the toxin is administered by injection, central nervous system dysfunction occurs, and high doses may result in death of experimental animals (65). Mammals may be less affected (74).

Cyclopiazonic acid is produced by *Aspergillus flavus* and at least six *Penicillium* species (Table 25.1). Along with *A. flavus*, *Penicillium commune*, a common cause of cheese spoilage (50), is probably the most common source of cyclopiazonic acid in foods. It has been shown that *P. commune* is the wild ancestor of *Penicillium camemberti*, used in the production of Camembert-type cheeses. Nearly all isolates of *P. camemberti* produce cyclopiazonic acid, but not, according to some reports, under commercial cheese manufacturing conditions (60). Since others do not agree (62), the search for nontoxigenic *P. camemberti* strains suitable for use as starter cultures continues (34).

Taxonomy
The *Penicillium* species producing cyclopiazonic acid are all classified in *Penicillium* subgenus *Penicillium* (Table 25.1).

Enumeration and Identification
All of the species producing cyclopiazonic acid can be enumerated effectively on the media discussed in the introduction. No selective media are available for any of these species. Identification of any of them requires specialized methods and information (49, 58).

Ecology
The principal *Penicillium* species producing cyclopiazonic acid, *P. commune*, is a common cause of cheese spoilage around the world (27, 39). Other species producing this toxin have a wide range of habitats.

All of the *Penicillium* species producing cyclopiazonic acid have similar physiology, including the ability to grow at or near 0°C, with optima around 25°C and maxima at or below 37°C. Most are able to grow at or below an a_w of 0.85 (50).

Citreoviridin
The role of citreoviridin in the human disease acute cardiac beriberi has been well documented (72). Acute cardiac beriberi was a common disease in Japan in the second half of the 19th century. Symptoms were heart distress, labored breathing, nausea, and vomiting, followed by anguish, pain, restlessness, and sometimes maniacal behavior. In extreme cases, progressive paralysis leading to respiratory failure occurred. It is notable that victims of this disease were often young, healthy adults. In 1910, the incidence of acute cardiac beriberi suddenly decreased. This coincided with the implementation of a government inspection scheme which dramatically reduced the sale of moldy rice in Japan (71, 72).

The major source of citreoviridin is *Penicillium citreonigrum* (synonym *P. citreoviride*), a species which usually occurs in rice and occurs less commonly in other cereals and rarely in other foods (47, 50). Recent studies (53, 54) failed to find more than occasional infections of *P. citreonigrum* in Southeast Asian rice. The threat of citreoviridin toxicosis, at least in advanced Southeast Asian countries, now appears to be very low. However, the possibility cannot be discounted that *P. citreonigrum* still occurs in less developed African or Asian countries where adequate rice drying systems and controls are not yet in place.

Citreoviridin is also produced by *Eupenicillium ochrosalmoneum* (anamorph *Penicillium ochrosalmoneum*), which is a relatively uncommon, though widespread, species usually associated with cereals (47). The reported formation of citreoviridin by this species in corn in the United States is potentially of great concern (77, 78), but it has not been reported from Southeast Asian corn (52, 54). The ecology of this mold-commodity-mycotoxin relationship remains to be fully elucidated.

Taxonomy

When grown on standard identification media, *P. citreonigrum* is a distinctive species. Colonies grow quite slowly (after 7 days at 25°C, 20 to 28 mm in diameter on CYA and 22 to 26 mm in diameter on MEA; 0 to 10 mm at 37°C), produce small numbers of pale gray-green conidia, and exhibit yellow mycelial, soluble pigment and reverse colors. Penicilli consist of small clusters of phialides only; stipes are slender and not apically enlarged, and conidia are spherical, smooth walled, and tiny (47, 49).

E. ochrosalmoneum forms slowly growing colonies (usually of <30 mm in diameter) which are bright yellow on both CYA and MEA at 25°C. Growth on CYA at 37°C is similar to that at 25°C. Penicilli are biverticillate, with few, divergent metulae (49).

Enumeration and Identification

Little specific information exists on techniques for enumerating *P. citreonigrum* and *E. ochrosalmoneum*. Both would be expected to grow satisfactorily on media such as DRBC or DG18 agar. No selective media have been developed, however.

Ecology

The physiology of *P. citreonigrum* and of citreoviridin production has been little studied. Cardinal temperatures are below 5°C, 20 to 24°C, and 37 to 38°C (31). The minimum a_w for growth is not known, but this species is undoubtedly xerophilic (50). Citreoviridin is produced at temperatures from 10 to 37°C, with a maximum near 20°C (31).

Penitrem A

Chemicals capable of inducing a tremorgenic (trembling) response in vertebrate animals are regarded as rare, except for mold metabolites, as at least 20 such compounds have been reported (13). Tremorgens are neurotoxins; in low doses, they appear to cause no adverse effects on animals, which are able to feed and function more or less normally while sustained trembling occurs, even over periods as long as 18 days (32). However, relatively small increases in dosage (5- to 20-fold) can be rapidly lethal (29). The presence of tremorgens is exceptionally difficult to diagnose postmortem because visible pathological effects are not produced.

Several tremorgenic mycotoxins are produced by *Penicillium* species, with the most important being the highly toxic penitrem A. Verruculogen, which is equally toxic, is not produced by species of common occurrence in foods. Less toxic compounds include fumitremorgen B, paxilline, verrucosidin, and janthitrems (51). At doses around 1 mg/kg, penitrem A causes brain disturbance, brain damage, and death in rats, but the mechanism remains unclear (7, 10). The response in humans is also unclear: the evidence suggests that penitrem A produces an emetic effect, rendering serious toxic effects less likely (35).

The classification of *Penicillium* species producing penitrem A was particularly confusing, with penitrem production by several species described in early literature. It was later concluded that this confusion was due to misidentification and that the only common foodborne species producing penitrem A is *Penicillium crustosum* (48). Nearly all isolates of *P. crustosum* produce penitrem A at high levels, so the presence of this species in food or feed is a warning signal (20, 50).

Taxonomy

The validity of *P. crustosum* as a species was in doubt until 1980 (46). However, it has since been accepted without any confusion. A member of *Penicillium* subgenus *Penicillium*, *P. crustosum* produces large penicilli, with rami, metulae, and phialides characteristic of this subgenus (Fig. 25.2B). It is one of the faster growing species in section *Penicillium*, producing dull green colonies with a granular texture on both CYA and MEA. Microscopically, *P. crustosum* is characterized by large, rough-walled stipes and smooth-walled, usually spherical, conidia. However, the most distinctive feature of typical isolates is the production of enormous numbers of conidia on MEA, which become detached from the colony when the petri dish is jarred (46, 49).

Enumeration and Identification

P. crustosum grows relatively rapidly, so enumeration is best carried out using a modern medium such as DRBC agar, which inhibits spreading of colonies (76). Confirmation of this species requires isolation and growth on standard identification media. No selective media have been developed for this species.

Ecology

P. crustosum is a ubiquitous spoilage mold, occurring almost universally in cereal and animal feed samples (50). *P. crustosum* causes spoilage of corn, processed meats, nuts, cheese, and fruit juices and is a weak pathogen on pomaceous fruits and cucurbits (50, 66). The occurrence of penitrem A in animal feeds is well documented (29, 51). Its occurrence in human foods appears equally certain.

Little information about growth and toxin production by *P. crustosum* has been published. Like nearly all species in *Penicillium* subgenus *Penicillium*, *P. crustosum* does not grow at 37°C. Penitrem A appears to be produced only at high a_w (31).

PR Toxin and Roquefortine

Treated together here are toxins produced by *Penicillium roqueforti*, a species used in cheese manufacture which can also be a spoilage mold. Roquefortine has a relatively high 50% lethal dose (Table 25.1), but it has been reported to be the cause of the deaths of dogs in Canada, with symptoms similar to strychnine poisoning (38). PR toxin is apparently much more toxic than roquefortine (Table 25.1), but it has not been implicated in animal or human disease.

Taxonomy

P. roqueforti is classified in subgenus *Penicillium* section *Penicillium* (Fig. 25.2C). This species has been shown to comprise two distinct varieties: *P. roqueforti* var. *roqueforti* is widely used as a starter culture for mold-ripened cheeses, while *P. roqueforti* var. *carneum* occurs in meats and silage (24). *P. roqueforti* var. *roqueforti* produces PR toxin, *P. roqueforti* var. *carneum* produces patulin, and both varieties produce roquefortine (24).

P. roqueforti grows very rapidly on CYA and MEA and produces green reverse colors on one or both media. Stipes are often very rough; conidia are large, spherical, and smooth walled (49).

Enumeration and Identification

Like other similar species, *P. roqueforti* can be enumerated on DRBC or DG18 agar. A valuable property in some circumstances is the fact that *P. roqueforti* is more tolerant of acetic acid than are other *Penicillium* species. In consequence, *P. roqueforti* can be selectively enumerated on a medium such as malt acetic agar (MEA plus 0.5% glacial acetic acid) (50).

Ecology

P. roqueforti and the other two *Penicillium* species producing roquefortine are common in foods. As well as growing rapidly at refrigeration temperatures, *P. roqueforti* is able to grow in oxygen concentrations below 0.5%, even in the presence of up to 20% carbon dioxide (68). Hence, it is a common cause of spoilage in chilled meats, cheese, and other products (9, 50, 68). The presence of roquefortine in cheese and other foods is to be expected. However, human health effects from PR toxin or roquefortine in association with cheese have not been reported (50).

Because of their potential occurrence in staple foods, PR toxin and roquefortine are of considerable significance from the public health viewpoint. PR toxin is unstable in cheese in storage, but roquefortine has been isolated from finished products (61). Extensive searches for nontoxic strains for use as cheese starter cultures have so far been largely unsuccessful (34, 41).

Like other species in subgenus *Penicillium*, *P. roqueforti* is psychrotrophic, growing vigorously at temperatures as low as 2°C. It is notably tolerant of weak acid preservatives, such as sorbic acid, and is able to grow in 0.5% acetic acid or more. This species also has the lowest oxygen requirement for growth of any *Penicillium* species, being capable of growth in >0.5% oxygen in the presence of 20% carbon dioxide (68).

Secalonic Acid D

Secalonic acids are dimeric xanthones produced by a range of taxonomically distant molds (14). Secalonic acid D, the only secalonic acid produced by *Penicillium* species, has significant animal toxicity (11, 67). Secalonic acid D is produced as a major metabolite of *Penicillium oxalicum*. It has been found in grain dusts at levels of up to 4.5 mg/kg (19). The possibility that such concentrations can be toxic to grain handlers by inhalation should not be ignored. However, the role of secalonic acid D in human or animal disease remains unclear.

Taxonomy

P. oxalicum is a distinctive species in subgenus *Furcatum* section *Furcatum* (Fig. 25.2D). It grows rapidly on CYA at 25 and 37°C, forming a continuous layer of conidia which, under a low-power microscope, can be seen to lie in closely packed, readily fractured sheets with a uniquely shiny appearance. Penicilli are terminal and biverticillate, with long phialides producing large ellipsoidal conidia (49).

Enumeration and Identification

This species can be enumerated without difficulty on standard media such as DRBC and DG18 agars. Selective media have not been developed, however.

Ecology

A major habitat for *P. oxalicum* is corn at harvest, where it is the most common *Penicillium* species isolated (50). The occurrence of secalonic acid D in corn, and hence in grain dusts, is a potential hazard (19). *P. oxalicum* grows at temperatures from about 8°C to about 40°C. The minimum a_w for conidial germination is 0.86 (50).

QUANTIFYING MYCOTOXINS

Quantifying most *Penicillium* mycotoxins is relatively straightforward, but procedures vary with commodity and toxin. Since mycotoxins are usually present in minute amounts, best practice is essential to obtain good results. Scrupulous attention to safety is necessary, in analyst training, in laboratory equipment and procedures, and in toxin containment and disposal.

Basic procedures for mycotoxin assays are sampling and subsampling, extraction and cleanup, and detection, quantification, and confirmation. Each of these areas is described in general terms below.

Sampling and Subsampling

The incidence of mycotoxins in a food commodity is usually heterogeneous, so sampling plays an important part in assay accuracy. Adequate (representative) sample sizes depend on particle size. While a 500-g sample is adequate for oils or milk powder, 3 kg of flour or peanut butter is necessary. To be representative, corn samples should be 10 kg, and peanuts require 20 kg per batch (65).

Samples of raw materials should be taken with online samplers during processing or be composites from several sites in raw material lots. Much attention has been paid to sample plans for aflatoxins in peanuts (8, 12, 76), corn (17), and figs (64). Such detailed studies have not been carried out for sampling for other mycotoxins, but the aflatoxin systems should have broad applicability.

To obtain a representative subsample for analysis, samples of particulate foods should be ground or blended. After thorough mixing, subsamples (20 to 50 g) are taken. To check on the adequacy of sampling procedures, assays are often performed in duplicate.

Extraction and Cleanup

To release toxins from the ground food material, extraction in a solvent system is necessary. This is usually carried out by blending the subsample with a suitable solvent in an explosion-proof blender for 1 to 3 min or shaking the subsample on a wrist-action shaker (30 min). Various solvent systems have been used, but the preferred system today is methanol and acidified water (26).

The extract usually requires some form of cleanup procedure, in particular to remove lipids and pigments which would interfere with analysis. Dialysis, precipitation, chromatographic columns, solvent partitioning, and specific antibody layers are all in use (65). Concentration of the purified extract is then carried out using a steam bath or vacuum rotary evaporator.

Detection and Quantification

One of the earliest and still most common techniques for mycotoxin assays is thin-layer chromatography (TLC). Concentrated extracts are spotted (1 to 10 μl) on glass or aluminum plates coated with a thin layer of activated silica gel and then developed in tanks containing suitable solvents designed to separate toxins from interfering chemicals of all types. For some substrates, two-dimensional TLC (developing the plate twice at right angles) is of value. Appropriate standards are also run at the same time.

After development, toxins are visualized by comparison of spots on the chromatogram with standard spots. Many mycotoxins, including aflatoxins, fluoresce under UV light and are readily visualized. Others require reaction with spray reagents, such as sulfuric acid or aluminum chloride. Comparison of the R_f values and colors of unknown spots with those of standards enables separation of a wide variety of toxins. The intensities of spots compared with those of suitable standards provide quantification. Confirmation is important and usually involves use of a spray reagent to produce a different color.

In recent years, high-performance liquid chromatography (HPLC) has increasingly become the analysis method of choice. Passage of extracts through a long column packed with an inert layer and a suitable adsorbent causes separation of molecules in a manner similar to that for TLC. Visualization is done by spectrophotometric detectors. Sometimes, compounds must be derivatized before analysis to improve the test sensitivity. HPLC is more sensitive than TLC and is more suited to automation.

Rapid Methods of Analysis

Assay techniques such as TLC and HPLC are effective but slow. The search for more rapid methods continues. Observation of cracked corn kernels under UV light has long been used as a screening technique for aflatoxins. This method is also of value for aflatoxins in figs but is ineffective for peanuts, which autofluoresce under UV. Peanuts are screened for aflatoxins by using minicolumns, in which aflatoxin is bound to an absorbent material in a small tube and then assayed by UV light.

The newest approach to rapid assays involves the use of antibodies developed to specific toxins. Immunoassays using either spot or minicolumn tests have been developed for several major toxins (65).

Specific assay methods exist for most *Penicillium* toxins, based on the general principles outlined above (26, 65). However, the literature for the less common and less well-known toxins is widely scattered.

CONCLUSIONS

Penicillium species produce a very wide range of toxic compounds. The role of one toxin, citreoviridin, in a historical human health problem is well established. Of the other toxins discussed here, only ochratoxin is currently regarded as a serious threat to human health. However, *Penicillium* species are so widespread and abundant in foods and feeds that they must be considered a potential hazard to both human and animal health. Many species make several compounds known to be toxic, and the possibility of synergy also cannot be ignored. Much more research is needed to improve detection methods, to understand the ecology of toxigenic *Penicillium* species, and to evaluate the significance of *Penicillium* toxins in human health.

References

1. **Alsberg, C. L., and O. F. Black.** 1913. Contributions to the study of maize deterioration: biochemical and toxicological investigations of *Penicillium puberulum* and *Penicillium stoloniferum*. *Bull. Bur. Anim. Ind. U. S. Dept. Agric.* 270:1–47.

2. **Alves, I., N. G. Oliviera, A. Laires, A. S. Rodrigues, and J. Rueff.** 2000. Induction of micronuclei and chromosomal aberrations by the mycotoxin patulin in mammalian cells: role of ascorbic acid as a modulator of patulin clastogenicity. *Mutagenesis* 15:229–234.

3. **Banke, S., J. C. Frisvad, and S. Rosendahl.** 1997. Taxonomy of *Penicillium chrysogenum* and related xerophilic species, based on isozyme analysis. *Mycol. Res.* 101:617–624.

4. **Brackett, R. E., and E. H. Marth.** 1979. Patulin in apple juice from roadside stands in Wisconsin. *J. Food Prot.* 42:862–863.

5. **Breitholtz, A., M. Olsen, A. Dahlbäck, and K. Hult.** 1991. Plasma ochratoxin A levels in three Swedish populations surveyed using an ion-pair HPLC technique. *Food Addit. Contam.* 8:183–192.

6. **Breitholtz-Emanuelsson, A., M. Olsen, A. Oskarsson, I. Palminger, and K. Hult.** 1993. Ochratoxin A in cow's milk and human milk with corresponding human blood samples. *J. AOAC Int.* 76:842–846.

7. **Breton, P., J. C. Bizot, J. Buee, and I. Delamanche.** 1998. Brain neurotoxicity of penitrem A: electrophysiological, behavioral and histopathological changes. *Toxicon* 36:645–655.

8. **Brown, G. H.** 1982. Sampling for "needles in haystacks." *Food Technol. Aust.* 34:224–227.

9. **Bullerman, L. B.** 1981. Public health significance of molds and mycotoxins in fermented dairy products. *J. Dairy Sci.* 64:2439–2452.

10. **Cavanagh, J. B., J. L. Holton, C. C. Nolan, D. E. Ray, J. T. Naik, and P. G. Mantle.** 1998. The effects of the tremorgenic mycotoxin penitrem A on the rat cerebellum. *Vet. Pathol.* 35:53–63.

11. **Ciegler, A., A. W. Hayes, and R. F. Vesonder.** 1980. Production and biological activity of secalonic acid D. *Appl. Environ. Microbiol.* 39:285–287.

12. **Coker, R. D.** 1989. Control of aflatoxin in groundnut products with emphasis on sampling, analysis, and detoxification, p. 123–132. *In Aflatoxin Contamination of Groundnut: Proceedings of the International Workshop, 6–9 October, 1987, ICRISAT Centre, India. ICRISAT, Patancheru, India.*

13. **Cole, R. J.** 1981. Tremorgenic mycotoxins: an update, p. 17–33. *In* R. L. Ory (ed.), *Antinutrients and Natural Toxicants in Foods.* Food and Nutrition Press, Westport, Conn.

14. **Cole, R. J., and R. H. Cox.** 1981. *Handbook of Toxic Fungal Metabolites.* Academic Press, New York, N.Y.

15. **Cruickshank, R. H., and J. I. Pitt.** 1987. Identification of species in *Penicillium* subgenus *Penicillium* by enzyme electrophoresis. *Mycologia* 79:614–620.

16. **De Jesus, A. E., P. S. Steyn, F. R. van Heerden, R. Vleggaar, and P. L. Wessels.** 1981. Structure and biosynthesis of the penitrems A-F, six novel tremorgenic mycotoxins from *Penicillium crustosum*. *J. Chem. Soc. Chem. Commun.* 1981:289–291.

17. **Dickens, J. W., and T. B. Whitaker.** 1983. Sampling, BGYF, and aflatoxin analysis in corn, p. 35–37. *In* U. L. Diener, R. L. Asquith, and J. W. Dickens (ed.), *Aflatoxin and Aspergillus flavus in Corn.* Alabama Agricultural Experiment Station, Auburn University, Auburn, Ala.

18. **Dorner, J. W., R. J. Cole, L. G. Lomax, H. S. Gosser, and U. L. Diener.** 1983. Cyclopiazonic acid production by *Aspergillus flavus* and its effect on broiler chickens. *Appl. Environ. Microbiol.* 46:698–703.

19. **Ehrlich, K. C., L. S. Lee, A. Ciegler, and M. S. Palmgren.** 1982. Secalonic acid D: natural contaminant of corn dust. *Appl. Environ. Microbiol.* 44:1007–1008.

20. **El-Banna, A. A., and L. Leistner.** 1988. Production of penitrem A by *Penicillium crustosum* from foodstuffs. *Int. J. Food Microbiol.* 7:9–17.

21. **Friis, P., E. Hasselager, and P. Krogh.** 1969. Isolation of citrinin and oxalic acid from *Penicillium viridicatum* Westling and their nephrotoxicity in rats and pigs. *Acta Pathol. Microbiol. Scand.* 77:559–560.

22. Frisvad, J. C. 1983. A selective and indicative medium for groups of *Penicillium viridicatum* producing different mycotoxins in cereals. *J. Appl. Bacteriol.* **54**:409–416.

23. Frisvad, J. C. 1986. Taxonomic approaches to mycotoxin identification, p. 415–457. *In* R. J. Cole (ed.), *Modern Methods in the Analysis and Structural Elucidation of Mycotoxins.* Academic Press, Orlando, Fla.

24. Frisvad, J. C., and O. Filtenborg. 1989. Terverticillate penicillia: chemotaxonomy and mycotoxin production. *Mycologia* **81**:837–861.

25. Frisvad, J. C., U. Thrane, R. A. Samson, and J. I. Pitt. 2006. Important mycotoxins and the fungi which produce them, p. 3–31. *In* A. D. Hocking, J. I. Pitt, R. A. Samson, and U. Thrane (ed.), *Advances in Food Mycology.* Springer, New York, N.Y.

26. Helrich, K. (ed). 1990. *Official Methods of Analysis of the Association of Official Analytical Chemists,* 15th ed. Association of Official Analytical Chemists, Arlington, Va.

27. Hocking, A. D., and M. Faedo. 1992. Fungi causing thread mould spoilage of vacuum packaged Cheddar cheese during maturation. *Int. J. Food Microbiol.* **16**:123–130.

28. Hocking, A. D., J. I. Pitt, R. A. Samson, and U. Thrane. 2006. Recommended methods for food mycology, p. 343–348. *In* A. D. Hocking, J. I. Pitt, R. A. Samson, and U. Thrane (ed.), *Advances in Food Mycology.* Springer, New York, N.Y.

29. Hou, C. T., A. Ciegler, and C. W. Hesseltine. 1971. Tremorgenic toxins from penicillia. II. A new tremorgenic toxin, tremortin B, from *Penicillium palitans. Can. J. Microbiol.* **17**:599–603.

30. International Agency for Research on Cancer. 1994. Ochratoxin A, p. 489–521. *In IARC Monographs on the Evaluation of Carcinogenic Risks to Humans,* vol. 56. *Some Naturally Occurring Substances: Food Items and Constituents, Heterocyclic Aromatic Amines and Mycotoxins.* International Agency for Research on Cancer, Lyon, France.

31. International Commission on Microbiological Specifications for Foods. 1996. Toxigenic fungi: *Penicillium,* p. 397–413. *In Microorganisms in Foods. 5. Characteristics of Food Pathogens.* Blackie Academic and Professional, London, United Kingdom.

32. Jortner, B. S., M. Ehrich, A. E. Katherman, W. R. Huckle, and M. E. Carter. 1986. Effects of prolonged tremor due to penitrem A in mice. *Drug Chem. Toxicol.* **9**:101–116.

33. Krogh, P., B. Hald, P. Englund, L. Rutqvist, and O. Swahn. 1974. Contamination of Swedish cereals with ochratoxin A. *Acta Pathol. Microbiol. Scand. Sect. B* **82**:301–302.

34. Leistner, L., R. Geisen, and J. Fink-Gremmels. 1989. Mould-fermented foods of Europe: hazards and developments, p. 145–154. *In* S. Natori, K. Hashimoto, and Y. Ueno (ed.), *Mycotoxins and Phytotoxins '88.* Elsevier Science, Amsterdam, The Netherlands.

35. Lewis, P. R., M. B. Donohue, A. D. Hocking, L. Cook, and L. V. Granger. 2005. Tremor syndrome associated with a fungal toxin: sequelae of food contamination. *Med. J. Aust.* **182**:582–584.

36. Llewellyn, G. C., J. A. McCay, R. D. Brown, D. L. Musgrove, L. F. Butterworth, A. E. Munson, and K. L. White. 1998. Immunological evaluation of the mycotoxin patulin in female B6C3F(1) mice. *Food Cosmet. Toxicol.* **36**:1107–1115.

37. LoBuglio, K. F., J. I. Pitt, and J. W. Taylor. 1994. Independent origin of the synnematous *Penicillium* species, *P. duclauxiii, P. clavigerum* and *P. vulpinum,* as assessed by two ribosomal DNA regions. *Mycol. Res.* **98**:250–256.

38. Lowes, N. R., R. A. Smith, and B. E. Beck. 1992. Roquefortine in the stomach contents of dogs suspected of strychnine poisoning. *Can. Vet. J.* **33**:535–538.

39. Lund, F., O. Filtenborg, and J. C. Frisvad. 1995. Associated mycoflora of cheese. *Food Microbiol.* **12**:173–180.

40. Mantle, P. G., and K. M. Macgeorge. 1991. Nephrotoxic fungi in a Yugoslav community in which Balkan nephropathy is hyperendemic. *Mycol. Res.* **95**:660–664.

41. Medina, M., P. Gaya, and M. Nunez. 1985. Production of PR toxin and roquefortine by *Penicillium roqueforti* isolates from Cabrales blue cheese. *J. Food Prot.* **48**:118–121.

42. Miraglia, M., C. Brera, and M. Colatosti. 1996. Application of biomarkers to assessment of risk to human health from exposure to mycotoxins. *Microchem. J.* **54**:472–477.

43. Mortimer, D. N., I. Parker, M. J. Shepherd, and J. Gilbert. 1985. A limited survey of retail apple and grape juices for the mycotoxin patulin. *Food Addit. Contam.* **2**:165–170.

44. Pavlovic, M., R. Plestina, and P. Krogh. 1979. Ochratoxin A contamination of foodstuffs in an area with Balkan (endemic) nephropathy. *Acta Pathol. Microbiol. Scand. Sect. B* **87**:243–246.

45. Peterson, S. W. 2000. Phylogenetic analysis of *Penicillium* species based on ITS and LSU-rDNA nucleotide sequences, p. 163–188. *In* R. A. Samson and J. I. Pitt (ed.), *Integration of Modern Taxonomic Methods for Penicillium and Aspergillus Classification.* Harwood Academic Publishers, Amsterdam, The Netherlands.

46. Pfohl-Leszkowicz, A., T. Petkova-Bocharova, I. N. Chernozemsky, and M. Castegnaro. 2002. Balkan endemic nephropathy and associated urinary tract tumours: a review on aetiological causes and the potential role of mycotoxins. *Food Addit. Contam.* **19**:282–302.

47. Pitt, J. I. 1979. *The Genus Penicillium and Its Teleomorphic States Eupenicillium and Talaromyces.* Academic Press, London, United Kingdom.

48. Pitt, J. I. 1979. *Penicillium crustosum* and *P. simplicissimum,* the correct names for two common species producing tremorgenic mycotoxins. *Mycologia* **71**:1166–1177.

49. Pitt, J. I. 2000. *A Laboratory Guide to Common Penicillium Species,* 3rd ed. CSIRO Food Science Australia, North Ryde, NSW, Australia.

50. Pitt, J. I., and A. D. Hocking. 1997. *Fungi and Food Spoilage,* 2nd ed. Aspen Publishers, Gaithersburg, Md.

51. Pitt, J. I., and L. Leistner. 1991. Toxigenic *Penicillium* species, p. 91–99. *In* J. E. Smith and R. S. Henderson (ed.), *Mycotoxins and Animal Foods.* CRC Press, Boca Raton, Fla.

52. Pitt, J. I., A. D. Hocking, K. Bhudhasamai, B. F. Miscamble, K. A. Wheeler, and P. Tanboon-Ek. 1993. The normal mycoflora of commodities from Thailand. 1. Nuts and oilseeds. *Int. J. Food Microbiol.* **20**:211–226.

53. Pitt, J. I., A. D. Hocking, K. Bhudhasamai, B. F. Miscamble, K. A. Wheeler, and P. Tanboon-Ek. 1994. The normal

mycoflora of commodities from Thailand. 2. Beans, rice and other commodities. *Int. J. Food Microbiol.* **23:**35–53.

54. Pitt, J. I., A. D. Hocking, B. F. Miscamble, O. S. Dharmaputra, K. R. Kuswanto, E. S. Rahayu, and Sardjono. 1998. The mycoflora of food commodities from Indonesia. *J. Food Mycol.* **1:**41–60.

55. Pitt, J. I., R. A. Samson, and J. C. Frisvad. 2000. List of accepted names and their synonyms in the family *Trichocomaceae*, p. 9–49. *In* R. A. Samson and J. I. Pitt (ed.), *Integration of Taxonomic Methods for* Penicillium *and* Aspergillus *Classification.* Harwood Academic Publishers, Amsterdam, The Netherlands.

56. Samson, R. A., and J. C. Frisvad. 2004. *Penicillium* subgenus *Penicillium*: new taxonomic schemes and mycotoxins and other extrolites. *Stud. Mycol.* **49:**1–260.

57. Samson, R. A., and J. I. Pitt (ed.). 1990. *Modern Concepts in* Penicillium *and* Aspergillus *Classification.* Plenum Press, New York, N.Y.

58. Samson, R. A., E. S. Hoekstra, J. C. Frisvad, and O. Filtenborg. 1995. *Introduction to Food-Borne Fungi*, 4th ed. Centraalbureau voor Schimmelcultures, Baarn, The Netherlands.

59. Samson, R. A., A. C. Stolk, and R. Hadlok. 1976. Revision of the subsection Fasciculata of *Penicillium* and some allied species. *Stud. Mycol.* (Baarn) **11:**1–47.

60. Schoch, U., J. Lüthy, and C. Schlatter. 1984. Mutagenitätsprüfung industriell verwendeter *Penicillium camemberti*- und *P. roqueforti*-Stämme. *Z. Lebensm.-Unters. Forsch.* **178:**351–355.

61. Scott, P. M. 1977. *Penicillium* mycotoxins, p. 283–356. *In* T. D. Wyllie and L. G. Morehouse (ed.), *Mycotoxic Fungi, Mycotoxins, Mycotoxicoses, an Encyclopedic Handbook*, vol. 1. *Mycotoxigenic Fungi.* Marcel Dekker, New York, N.Y.

62. Scott, P. M. 1981. Toxins of *Penicillium* species used in cheese manufacture. *J. Food Prot.* **44:**702–710.

63. Scouboe, P., J. C. Frisvad, J. W. Taylor, D. Lauritsen, M. Boysen, and L. Rossen. 1999. Phylogenetic analysis of nucleotide sequences from the ITS region of terverticillate *Penicillium* species. *Mycol. Res.* **103:**873–881.

64. Sharman, M., S. Macdonald, A. J. Sharkey, and J. Gilbert. 1994. Sampling bulk consignments of dried figs for aflatoxin analysis. *Food Addit. Contam.* **11:**17–23.

65. Smith, J. E., C. W. Lewis, J. G. Anderson, and G. L. Solomons. 1994. *Mycotoxins in Human Nutrition and Health. Report EUR 16048 EN.* European Commission Directorate-General XII, Brussels, Belgium.

66. Snowdon, A. L. 1990. *A Colour Atlas of Post-Harvest Diseases and Disorders of Fruits and Vegetables.* 1. *General Introduction and Fruits.* Wolfe Scientific, London, United Kingdom.

67. Sorenson, W. G., F. H. Y. Green, V. Vallyathan, and A. Ciegler. 1982. Secalonic acid D toxicity in rat lung. *J. Toxicol. Environ. Health* **9:**515–525.

68. Taniwaki, M., A. D. Hocking, J. I. Pitt, and G. H. Fleet. 2001. Growth of fungi and mycotoxin production on cheese under modified atmospheres. *Int. J. Food Microbiol.* **68:**125–133.

69. Tatu, C. A., W. H. Orem, R. B. Finkelman, and G. L. Feder. 1998. The etiology of Baltic endemic nephropathy: still more questions than answers. *Environ. Health Perspect.* **106:**689–700.

70. Téren, J., J. Varga, Z. Hamari, E. Rinyu, and É. Kevei. 1996. Immunochemical detection of ochratoxin A in black *Aspergillus* strains. *Mycopathologia* **134:**171–176.

71. Ueno, Y., and I. Ueno. 1972. Isolation and acute toxicity of citreoviridin, a neurotoxic mycotoxin of *Penicillium citreo-viride* Biourge. *Jpn. J. Exp. Med.* **42:**91–105.

72. Uraguchi, K. 1969. Mycotoxic origin of cardiac beriberi. *J. Stored Prod. Res.* **5:**227–236.

73. Van Egmond, H. P. 1989. Current situation on regulations for mycotoxins. Overview of tolerances and status of standard methods of sampling and analysis. *Food Addit. Contam.* **6:**139–188.

74. Van Rensburg, S. J. 1984. Subacute toxicity of the mycotoxin cyclopiazonic acid. *Food Chem. Toxicol.* **22:**993–998.

75. Watkins, K. L., G. Fazekas, and M. V. Palmer. 1990. Patulin in Australian apple juice. *Food Aust.* **42:**438–439.

76. Whitaker, T. B., F. E. Dowell, W. M. Hagler, F. G. Giesbrecht, and J. Wu. 1994. Variability associated with sampling, sample preparation, and chemical testing for aflatoxin in farmers' stock peanuts. *J. Assoc. Off. Anal. Chem. Int.* **77:**107–116.

77. Wicklow, D. T., and R. J. Cole. 1984. Citreoviridin in standing corn infested by *Eupenicillium ochrosalmoneum. Mycologia* **76:**959–961.

78. Wicklow, D. T., R. D. Stubblefield, B. W. Horn, and O. L. Shotwell. 1988. Citreoviridin levels in *Eupenicillium ochrosalmoneum*-infested maize kernels at harvest. *Appl. Environ. Microbiol.* **54:**1096–1098.

Food Microbiology: Fundamentals and Frontiers, 3rd Ed.
Edited by M. P. Doyle and L. R. Beuchat
© 2007 ASM Press, Washington, D.C.

Lloyd B. Bullerman

26

Fusaria and Toxigenic Molds Other than Aspergilli and Penicillia

Toxigenic molds in genera other than *Aspergillus* and *Penicillium* are most often found as contaminants of plant-derived foods, especially cereal grains. As such, these molds and their toxic metabolites (mycotoxins) find their way into animal feeds and human foods. Animals, both food-producing animals and pets, are more often affected by the toxins of these molds than are humans. Nevertheless, there is evidence that grains and processed human foods are occasionally contaminated with mycotoxins produced by molds other than aspergilli and penicillia (33). Also, exposure of food-producing animals to mycotoxigenic molds may have an impact on the human food supply by causing deaths of animals, reducing their rate of growth, or depositing toxins in meats, milk, and eggs. Diseases in animals caused by mycotoxins may also suggest that similar conditions can occur in humans.

The most important group of mycotoxigenic molds other than *Aspergillus* and *Penicillium* species are species of the genus *Fusarium* (47). Many *Fusarium* species are plant pathogens, while others are saprophytic; most can be found in the soil (16, 17). In terms of human foods, *Fusarium* species are most often encountered as contaminants of cereal grains, oil seeds, and beans. Corn (maize), wheat, barley, and products made from these grains are most commonly contaminated. However, rye, triticale, millet, and oats can also be contaminated.

Another mold genus other than *Aspergillus* and *Penicillium* that can produce mycotoxins is *Alternaria* (86). *Alternaria* species are widely distributed in the environment and can be found in soil, decaying plant materials, and dust. Both plant pathogenic and saprophytic species occur. Toxigenic molds also belong to several other genera. These include species of *Acremonium, Chaetomium, Claviceps, Diplodia maydis, Myrothecium, Phoma herbarum, Phomopsis, Pithomyces chartarum, Rhizoctonia, Rhizopus, Stachybotrys,* and *Trichothecium roseum* (39).

DISEASES OF HUMANS ASSOCIATED WITH *FUSARIUM*

Alimentary Toxic Aleukia

During World War II, a very severe human disease occurred in the former Soviet Union, particularly in the Orenburg area of Russia. The disease, known as alimentary toxic aleukia, is believed to be caused by T-2 and

Lloyd B. Bullerman, Department of Food Science and Technology, University of Nebraska, 322 Food Industry Complex, East Campus, Lincoln, NE 68583-0919.

HT-2 toxins produced by *Fusarium sporotrichioides* and *Fusarium poae* of the *Sporotrichiella* section of *Fusarium* (35, 43). Because of the war, there was such a shortage of farm workers that much grain was not harvested in the fall and was overwintered in the field. By spring, near-famine conditions existed and people were forced to eat overwintered cereal grains that were milled into flour and made into bread. This resulted in a severe toxicosis manifested as alimentary toxic aleukia. The disease developed over several weeks, with increasingly severe symptoms with continued consumption of the toxic grain.

There are actually four stages to the disease (34, 35). The first stage is characterized by symptoms that appear a short time after ingestion of the toxic grain. These symptoms include a burning sensation in the mouth, tongue, esophagus, and stomach (Table 26.1). The stomach and intestinal mucosa become inflamed, resulting in vomiting, diarrhea, and abdominal pain. The first stage may appear and disappear rather quickly, within 3 to 9 days. During the second stage, the individual experiences no outward signs of the disease and feels well. However, during this stage the hematopoietic system is being damaged or destroyed by progressive leukopenia, granulopenia, and lymphocytosis. The blood-making capacity of the bone marrow is being destroyed, the platelet count decreases, and anemia develops. The leukocyte count decreases, and secondary bacterial infections occur. There are also disturbances in the central and autonomic nervous systems. The third stage develops suddenly and is marked by petechial hemorrhage on the skin and mucous membranes. The hemorrhaging becomes more severe, with bleeding from the nose and gums and hemorrhaging in the stomach and intestines. Necrotic lesions also develop in the mouth, gums, mucosa, larynx, and vocal cords. At this stage, the disease is highly fatal, with up to 60% mortality. If death does not occur, the fourth stage of the disease is recovery or convalescence. It takes 3 to 4 weeks of treatment for the necrotic lesions, hemorrhaging, and bacterial infections to clear up and 2 months or more for the blood-making capacity of the bone marrow to return to normal. Alimentary toxic aleukia has been reproduced in cats and monkeys administered T-2 toxin isolated from the strain of *F. sporotrichioides* involved in the fatal human outbreaks of the disease in Russia. The related compound HT-2 toxin may also contribute to the disease.

Urov or Kashin-Beck Disease

Urov or Kashin-Beck disease has been observed among the Cossack people of eastern Russia for well over 100 years (32). The disease has been endemic in areas along the Urov river in Siberia and was studied extensively by two Russian scientists, Kashin and Beck, thus the name Urov or Kashin-Beck disease (35). The disease is a deforming bone-joint osteoarthrosis that is manifested by a shortening of long bones and a thickening deformation of joints, plus muscular weakness and atrophy (Table 26.1). The disease most commonly occurs in preschool and school-age children, and the disease has been reproduced in puppies and rats by feeding them *F. poae* isolates obtained only from regions where the

Table 26.1 Human diseases which have been associated with *Fusarium* species and toxins

Disease	Food	Mold	Toxin	Symptoms or effects
Alimentary toxic aleukia	Cereal grains, wheat, rye, bread	*F. sporotrichioides*, *F. poae*	Possibly T-2 toxin	Burning sensation in mouth and throat, vomiting, diarrhea, abdominal pain, bone marrow destruction, hemorrhaging, death
Urov/Kashin-Beck disease	Cereal grains	*F. poae*	Unknown	Osteoarthritis, shortened long bones, deformed joints, muscular weakness
Drunken bread	Cereal grains, wheat, bread	*F. graminearum*	Unknown	Headache, dizziness, tinnitus, trembling, unsteady gait, abdominal pain, nausea, diarrhea
Akakabi-byo (scabby grain intoxication)	Cereal grains, wheat, barley, noodles	*F. graminearum*	Unknown, possibly deoxynivalenol	Anorexia, nausea, vomiting, headache, abdominal pain, diarrhea, chills, giddiness, convulsions
Foodborne illness outbreaks	Cereal grains, wheat, barley, corn, bread	*F. graminearum*	Deoxynivalenol, acetyldeoxynivalenol, nivalenol, T-2 toxin	Irritation of throat, nausea, headaches, vomiting, abdominal pain, diarrhea
Esophageal cancer	Corn	*F. verticillioides*	Unknown, possibly fumonisins and other toxins	Precancerous and cancerous lesions in the esophagus

disease is endemic (42). The same disease has also been reported in North Korea and northern China.

Drunken Bread

Another human mycotoxicosis reported in the former Soviet Union is known as "drunken bread" (32). This syndrome is apparently caused by the consumption of bread made from rye grain infected with *Fusarium graminearum*. The illness is milder than alimentary toxic aleukia and is a nonfatal, self-limiting disorder. Symptoms associated with this disease are headache, dizziness, tinnitus, trembling, and shaking of the extremities, with an unsteady or stumbling gait, hence the name (Table 26.1). There are also flushing of the face and gastrointestinal symptoms, including abdominal pain, nausea, and diarrhea. Victims may appear to be euphoric and confused. The duration of the illness is 1 to 2 days after consumption of the toxic food has ceased. The infected grain may seem normal or appear shriveled and light in weight with a white to pinkish coloration suggestive of fusarium head blight, or scab.

Akakabi-Byo

Akakabi-byo is also called scabby grain intoxication or red mold disease. It has been observed in Russia, Japan, and China (42) and is associated with eating bread made from scab-infested wheat, barley, or other grains infected with *F. graminearum*. Symptoms of this illness are anorexia, nausea, vomiting, headache, abdominal pain, diarrhea, chills, giddiness, and convulsions (Table 26.1). Clinical signs are similar to those of the so-called drunken bread syndrome discussed above. No specific mycotoxins have been shown to cause this illness, but deoxynivalenol and nivalenol occur naturally in scabby grain from the regions where the illness is endemic.

Foodborne Illness Outbreaks

Outbreaks of foodborne illness associated with *Fusarium* species have involved foods made from wheat or barley infected mainly with *F. graminearum*. These outbreaks resemble scabby grain intoxication or red mold disease and may be essentially the same thing. Foodborne illnesses have been reported in Japan, Korea, and China that have involved foods, particularly noodles, made from scabby wheat (83). The onset of illness is usually rapid, occurring in 5 to 30 min, suggesting the presence of a preformed toxin, most likely deoxynivalenol. The most common symptoms included nausea, vomiting, abdominal pain, and diarrhea, with headache, fever, chills, and throat irritation in some victims (Table 26.1). An outbreak occurred in India which was characterized by an onset time of 15 to 60 min, abdominal pain, irritation

of the throat, vomiting, and diarrhea, with blood in the stools in some cases (7). In addition, some victims had a facial rash, nausea, and flatulence. The food involved was bread made from molded wheat. Apparently, flour millers had mixed infected wheat with sound wheat and milled it into flour. *Fusarium* species were isolated from the wheat and flour, and trichothecene mycotoxins, including deoxynivalenol, acetyldeoxynivalenol, nivalenol, and T-2 toxin, were detected in many samples.

Outbreaks of foodborne illness that have occurred in China were reviewed by Kuiper-Goodman (37). These have involved corn and wheat contaminated with *Fusarium* species, deoxynivalenol, and zearalenone. An outbreak of precocious pubertal changes in thousands of young children in Puerto Rico in which zearalenone or its derivatives or other exogenous estrogenic substances were the suspected cause has also been reported (63, 64). Affected persons experienced premature pubarche, prepubertal gynecomastia, and precocious pseudopuberty. Zearalenone or a derivative was found in the blood of some of the patients, and food was believed to be the source of the estrogenic substances.

Esophageal Cancer

Fusarium verticillioides (*moniliforme*) has been associated with high rates of esophageal cancer in certain parts of the world, particularly the Transkei region of South Africa, northeastern Italy, and northern China (27, 37, 51, 57). In these regions, corn is a dietary staple and is the main or only food consumed. Corn and corn-based foods from these regions may contain significant amounts of fumonisins and possibly other metabolites of this fungus.

Immunotoxic Effects of *Fusarium* Toxins

Fusarium species produce a diverse array of biologically active compounds, and it is believed that many of these compounds, especially the trichothecenes, can affect the immune system by suppressing immune functions. T-2 toxin, for example, is known to be highly immunosuppressive (28). There are reports of so-called sick houses, i.e., houses where individuals have contracted diseases such as leukemia, where *Fusarium* species have been detected in dust (43). In one report, a husband and wife both contracted leukemia while living in a house that contained dust-associated spores of *Fusarium equiseti* (91). In another report, *Fusarium* spores were detected in a house in which four cases of leukemia occurred (90). *F. equiseti* is a known producer of diacetoxyscirpenol, a trichothecene that also has known immunosuppressive properties (43). Other *Fusarium* isolates from the house were toxic to ducklings, hamsters, and mice.

The conclusion was that mycotoxigenic *Fusarium* species, including *F. equiseti*, were possibly involved in the development of leukemia by their immunosuppressive effects (91).

Immunotoxicity can cause two general types of adverse effects on the immune system (54). In the first type, a toxin or chemical suppresses one or more functions of the immune system. This can result in increased susceptibility to infection or neoplastic disease. In the second type of immunotoxicity, the toxin or chemical may stimulate an immune function, resulting in autoimmune types of disorders. Trichothecenes are known to inhibit protein and DNA synthesis and to interact with cell membranes, causing weakening and damage. Exposure to trichothecenes can cause damage to the bone marrow, spleen, thymus, lymph nodes, and intestinal mucosa (54, 71). T-2 toxin and deoxynivalenol have been shown to affect B-cell and T-cell mitogen responses in lymphocytes. Dietary exposures to deoxynivalenol at concentrations as low as 2 μg/g of body weight for 5 weeks or 5 μg/g for 1 week cause decreased mitogen responses (54). T-2 toxin and diacetoxyscirpenol cause increased susceptibility to *Candida* infections, as well as to *Listeria*, *Salmonella*, *Mycobacterium*, and *Cryptococcus* infections, in experimental animals (54).

Dietary deoxynivalenol has been shown to stimulate immunoglobin production, causing elevated immunoglobin A levels in mice. Among the harmful effects of this stimulation is kidney damage very similar to a common human kidney condition known as glomerulonephritis or immunoglobulin A nephropathy (54). While the cause of this condition is unknown, there is an association with grain-based diets. Other *Fusarium* mycotoxins, e.g., zearalenone and fumonisins, may also have immunotoxic effects, but less information is available about these toxins. Of all the harmful effects of mycotoxins,

immunotoxicity or immunomodulation may have the most significant impact on human health. It appears that relatively low levels of the toxins can cause these responses.

TOXIGENIC *FUSARIUM* SPECIES AND THEIR TOXINS

According to Nelson et al. (53), there are 12 sections, or groupings, within the genus *Fusarium*. Only four sections, containing the most common toxic species, are discussed here, including *Sporotrichiella* (*F. sporotrichioides* and *F. poae*), *Gibbosum* (*F. equiseti*), *Discolor* (*F. graminearum* and *Fusarium culmorum*), and *Liseola* [*F. verticillioides* (*moniliforme*), *Fusarium proliferatum*, and *Fusarium subglutinans*]. The major mycotoxins produced by these species are summarized in Table 26.2. The work of Nelson et al. (53) should be consulted for identifying *Fusarium* species, although changes in *Fusarium* taxonomy are occurring. The key for the work of Nelson et al. (53) is based on the morphological species concept. Newer methods based on biological species concepts that employ mating types and on phylogenetic species concepts using DNA sequences are receiving intensive study (70, 77). The book by Samson et al. (65) can be used to identify most isolates encountered in food, and the book by Pitt and Hocking (56) is also a good general reference.

The genus *Fusarium* is characterized by the production of septate hyphae that generally range in color from white to pink, red, purple, or brown due to pigment production. The most common characteristic of the genus is the production of large septate, crescent-shaped, fusiform, or sickle-shaped spores known as macroconidia. The macroconidia exhibit a foot-shaped basal cell and a beak-shaped or snout-like apical cell (Fig. 26.1A). The macroconidia are produced from phialides in a stroma

Table 26.2 Major mycotoxins that may be produced by *Fusarium* species of importance in cereal grains and grain-based foods

Section	Species	Potential mycotoxin(s)
Sporotrichiella	*F. poae*	Type A trichothecenes, T-2 toxin, diacetoxyscirpenol
	F. sporotrichioides	T-2 toxin
Gibbosum	*F. equiseti*	Unknown
Discolor	*F. graminearum*	Deoxynivalenol (vomitoxin), 3-acetyldeoxynivalenol, 15-acetyldeoxynivalenol, zearalenone, possibly others
	F. culmorum	Nivalenol, zearalenone
Liseola	*F. verticillioides*	Fumonisins and others
	F. proliferatum	Fumonisins, moniliformin, and others
	F. subglutinans	Moniliformin and others

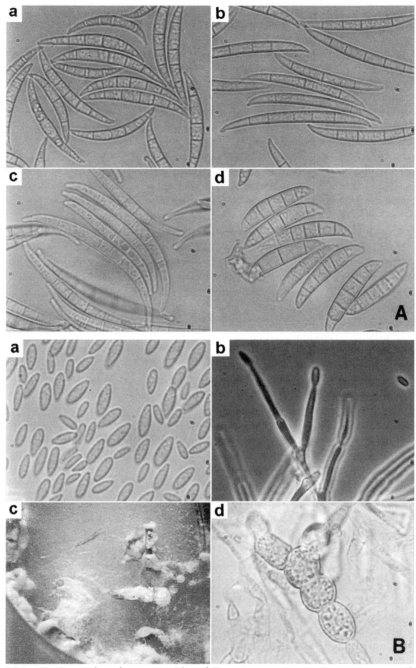

Figure 26.1 (A) Examples of macroconidia of *Fusarium* species. (a) *F. graminearum*; (b) *F. verticillioides*; (c) *F. equiseti*; (d) *F. culmorum*. Magnification, ×1,000. (B) Examples of micro- and macroscopic structures of *Fusarium*. (a) Microconidia; (b) monophialides; (c) sporodochia; (d) chlamydospores. Magnification, ×1,000 (a, b, and d) and ×10 (c).

known as a sporodochium or in mucoid or slimy masses known as pionnotes. Macroconidia can also be produced in the hyphae, but these are less typical and more variable. Some species also produce smaller one- or two-celled conidia known as microconidia (Fig. 26.1B). Some species also produce swollen, thick-walled chlamydospores in the hyphae or in the macroconidia (Fig. 26.1B). *Fusarium* species are highly variable because of their genetic makeup and can undergo mutations and morphological changes in culture after isolation.

Section *Sporotrichiella*

F. poae

F. poae is widespread in soils of temperate climate regions and is found on grains such as wheat, corn, and barley (42, 56). It exists primarily as a saprophyte but may be weakly parasitic, and it is most commonly found in temperate regions of Russia, Europe, Canada, and the northern United States; *F. poae* has also been found in warmer regions, such as Australia, India, Iraq, and South Africa. Its optimum growth temperature is 22 to 27°C, but it can grow at temperatures as low as 2 to 3°C. *F. poae* can be isolated from overwintered grain. Diseases that have been associated with *F. poae* include alimentary toxic aleukia, a hemorrhagic syndrome, and Urov or Kashin-Beck disease. *F. poae* produces rapid, profuse, white-to-pink mycelial growth on potato dextrose agar (PDA), with a red to very deep carmine red reverse coloration (53). The most distinctive feature of *F. poae* is its production of abundant globose to oval, almost pyriform (pear-shaped) microconidia, with few macroconidia. Conidia are produced on branched or unbranched monophialides. Chlamydospores are produced infrequently (53). The teleomorphic state has not been observed in *F. poae*. *F. poae* primarily produces type A trichothecene mycotoxins such as T-2 toxin and diacetoxyscirpenol.

F. sporotrichioides

F. sporotrichioides is found in soil and a wide variety of plant materials. This mold is found in the temperate to colder regions of the world, including Russia, northern Europe, Canada, northern United States, and Japan (42, 43). It can grow at low to very low temperatures, e.g., at −2°C on grain overwintering in the field. Its optimum growth temperature is 22 to 27°C. Diseases that have been associated with *F. sporotrichioides* include alimentary toxic aleukia, a hemorrhagic syndrome, and Akakabi-byo. *F. sporotrichioides* produces T-2 toxin, diacetoxyscirpenol, zearalenone, and fusarin C.

F. sporotrichiodes produces dense white to pink or brown mycelia on PDA, with a deep red reverse coloration. The most distinctive feature of *F. sporotrichioides* is the production of branched and unbranched polyphialides that produce two kinds of microconidia, i.e., oval to pear shaped and multiseptate spindle shaped, resembling macroconidia. Macroconidia are also produced, but on monophialides (Fig. 26.2A). *F. sporotrichioides* produces abundant chlamydospores singly and in chains and bunches (53, 56). The teleomorphic state has not been observed in *F. sporotrichioides*. *Fusarium tricinctum* and *Fusarium chlamydosporum* are also members of the *Sporotrichiella* section, and both have been reported to be toxigenic, producing moniliformin (53).

Section *Gibbosum*

F. equiseti

F. equiseti is a very cosmopolitan mold found in the soil. It is particularly common in tropical and subtropical areas but also occurs in temperate regions (42, 56). For the most part, *F. equiseti* is saprophytic, but it may be pathogenic to plants such as bananas, avocados, and curcubits. *F. equiseti* has been found in soils from Alaska to tropical regions and has been isolated from cereal grains and overwintered cereals in Europe, Russia, and North America. It has been suggested that *F. equiseti* may contribute to leukemia in humans by affecting the immune system (42, 90, 91).

Growth of *F. equiseti* on PDA is rapid, resulting in dense, cottony aerial mycelia that fill the petri dish. The color of the colony is white, with a pale salmon to almost brown reverse (53, 56). As cultures age, orange sporodochia may be produced. The most distinctive characteristic of *F. equiseti* is the shape of the macroconidia, which are long, slender, and curved, with five to seven septa. The apical cell is elongated, and the basal cell has a distinctive foot shape. Microconidia and chlamydospores are also produced (53, 56). The teleomorph of *F. equiseti* is *Gibberella intricans*, but its occurrence in nature is rare (53, 56). Other species in section *Gibbosum* include *Fusarium scirpi*, *Fusarium acuminatum*, and *Fusarium longipes*, with *F. acuminatum* also being reported to be toxigenic (53).

Section *Discolor*

F. graminearum

F. graminearum is a plant pathogen found worldwide in the soil and is the most widely distributed toxigenic *Fusarium* species (33). It causes various diseases of cereal grains, including gibberella ear rot in corn and fusarium head blight or scab in wheat and other small grains (42, 47). These two diseases are important to food microbiology and food safety because the mold and two of its main toxic metabolites, deoxynivalenol and zearalenone, may contaminate grain and subsequent food products made from the grain. Some confusion may surround the taxonomy of *F. graminearum*. In some of the older literature, the name *Fusarium roseum* was used for *F. graminearum* and other species, e.g., *F. roseum* Graminearum and *F. roseum* var. *graminearum*. These names are no longer valid. The major mycotoxins produced by *F. graminearum* are deoxynivalenol and zearalenone. In addition, 3-acetyldeoxynivalenol, 15-acetyldeoxynivalenol, diacetyldeoxynivalenol, butenolide, diacetoxyscirpenol, fusarenon-X (4-acetylnivalenol), monoacetoxyscirpenol, neosolaniol, nivalenol, or T-2 toxin may be produced by some strains (42).

Growth of *F. graminearum* on PDA is rapid, with dense aerial mycelia formed that fill the petri dish (53, 56). The mycelia are grayish with yellow and brown tinges and white margins. The reverse is usually a deep carmine red. The formation of sporodochia on PDA is sparse and may take 30 days to form. Sporodochia appear red-brown to orange. The macroconidia of *F. graminearum* are distinctive of the species, in that they are almost cylindrical, with the central dorsal and ventral surfaces parallel (Fig. 26.2B). They most often have five septa. The ends of the macroconidia are slightly and unequally curved, and the apical cell is cone shaped and slightly bent. The basal cell has a distinct foot shape. Microconidia are not formed, and chlamydospores, when formed, most often occur in the macroconidia, with some in the mycelia (53, 56). The teleomorphic state of *F. graminearum* is *Gibberella zeae*.

F. culmorum

F. culmorum is also widely distributed in the soil and causes diseases of cereal grains, of which ear rot in corn is important in food microbiology and food safety, since the mold and its toxins may contaminate corn-based foods. *F. culmorum* is similar in many ways to *F. graminearum*, and in the past, it has been lumped with *F. graminearum* in the nonexistent species *F. roseum*, e.g., *F. roseum* Culmorum and *F. roseum* var. *culmorum* (53, 56). Again, these names are not valid. Mycotoxins reported to be produced by *F. culmorum* include deoxynivalenol, zearalenone, and acetyldeoxynivalenol (42, 47).

Growth of *F. culmorum* on PDA is rapid, with the formation of dense aerial mycelia that are white with some tinges of yellow and brown. The reverse of the colony is a deep carmine red (53). Sporodochia that are orange to red-brown may be produced in older cultures. While *F. culmorum* may resemble *F. graminearum* in macroscopic appearance when growing on PDA, the macroconidia are quite different. Macroconidia of *F. culmorum* are very short and stout compared to those of other members of the section, and they have three to five septa (53, 56). The macroconida have curved dorsal and ventral surfaces, and the basal cell varies from being foot shaped to having a notched appearance. Chlamydospores are formed readily in the macroconidia and mycelia. Microconidia are not formed. No teleomorphic state of *F. culmorum* has been observed. Other species in section *Discolor* include *Fusarium heterosporum*, *Fusarium reticulatum*, *Fusarium sambucinum*, and *Fusarium crookswellense*. Of these, *F. heterosporum*, *F. crookswellense*, and *F. sambucinum* have been reported to form toxins (53).

Section *Liseola*

F. verticillioides (moniliforme)

The mold that has been called *F. moniliforme* in much of the scientific literature is now divided into two species, *F. verticillioides* and *Fusarium thapsinum* (44, 77). Within section *Liseola*, there exist six distinct mating types (A through F), of which mating types A and F were within the old *F. moniliforme* anamorph. These mating types were very distinct and differed in the ability to produce the mycotoxin fumonisin (38, 51). Mating type A is found in corn and is capable of producing high levels of fumonisins. The anamorph of this mating type is now designated *F. verticillioides* and is synonymous with *F. moniliforme* from corn. Mating type F is found in sorghum and produces little or no fumonisin. The anamorph of this mating type is now designated *F. thapsinum* and is synonymous with *F. moniliforme* from sorghum. The mating populations of organisms in section *Liseola*, with their anamorphs and teleomorphs, are summarized in Table 26.3. *Fusarium fujikuroi* is also an old synonym for *F. verticillioides* (*moniliforme*).

F. verticillioides is a soilborne plant pathogen that is found in corn growing in all regions of the world. It is the most prevalent mold associated with corn. It often produces symptomless infections of corn plants but may infect the grain as well, and it has been found worldwide on food- and feed-grade corn. The presence of *F. verticillioides* in corn grain is often not discernible, e.g., the grain does not appear infected, yet it is not uncommon to find complete lots of shelled corn with 100% internal kernel infection (42). The presence of *F. verticillioides* in corn is a major concern in food microbiology and food safety because of the possible widespread contamination of corn and corn-based foods with its toxic metabolites, especially the fumonisins.

F. verticillioides has long been suspected of being involved in animal and human diseases. In the 1880s, a mold found on corn in Italy and called *Oospora verticillioides* was associated with pellegra (43). In the United States, *F. moniliforme* (*verticillioides*) growing on corn was linked to diseases of farm animals in Nebraska and other parts of the Midwest (55, 72). In more recent years, animal diseases associated with *F. verticillioides* have included equine leukoencephalomalacia, a liquefactive necrosis of the brains of horses and other equines (36), pulmonary edema and hydrothorax in swine (31), and experimental liver cancer in rats (26). In addition, *F. verticillioides* has been associated with abnormal bone development in chicks and pigs, manifested as leg deformities and rickets-like diseases (42). Experimental toxicity has been induced in animals by feeding them culture

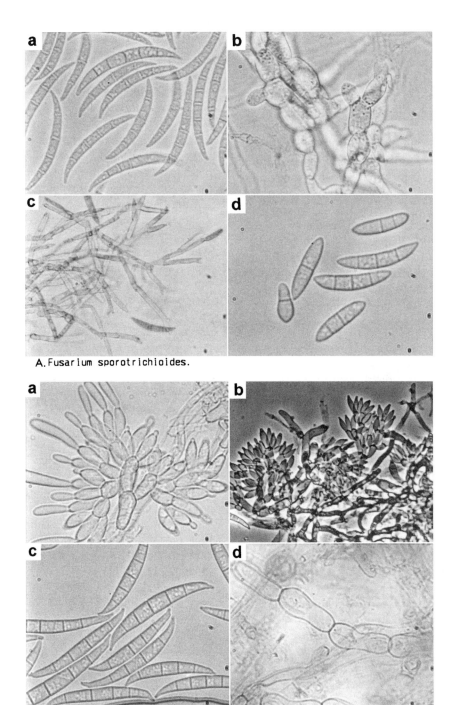

A. Fusarium sporotrichioides.

B. Fusarium graminearum.

Figure 26.2 Microscopic structures. (A) *F. sporotrichioides*. (a) Macroconidia; (b) chlamydospores; (c) phialides; (d) microconidia. Magnification, ×1,000 (a, b, and d) and ×550 (c). (B) *F. graminearum*. (a) Conidiophores (monophialides); (b) monophialides in sporodochia; (c) macroconidia; (d) chlamydospores. Magnification, ×1,000 (a, c, and d) and ×550 (b). (C) *F. verticillioides*. (a) Microconidia in chains; (b) microconidia; (c) monophialides producing microconidia; (d) macroconidia. Magnification, ×165 (a), ×550 (b), and ×1,000 (c and d). (D) *F. proliferatum*. (a) Microconidia in chains; (b) microconidia; (c) polyphialides; (d) macroconidia. Magnification, ×165 (a), and ×1,000 (b, c, and d).

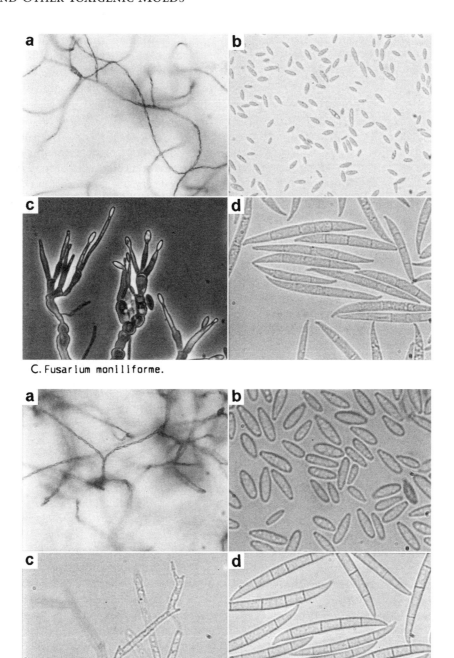

C. Fusarium moniliforme.

D. Fusarium proliferatum.

Figure 26.2 *(Continued)*

material of *F. verticillioides*. These include baboons, in which acute congestive heart failure and cirrhosis of the liver were observed (42); monkeys, in which atherogenic and hypercholesterolemic responses occurred (23); and chickens, donkeys, ducklings, geese, horses, mice, pigeons, pigs, rabbits, rats, and sheep (42).

The main human disease associated with *F. verticillioides* is esophageal cancer. Several studies have linked the presence of *F. verticillioides* and fumonisins in corn to high incidences of esophageal cancer in humans in certain regions of the world, including the Transkei region of South Africa, northeastern Italy, northern China,

Table 26.3 Mating populations of section *Liseola* of the genus *Fusarium*, with corresponding anamorphic and teleomorphic names[a]

Mating population	Anamorph (*Fusarium*)	Teleomorph (*Gibberella*)
A	*F. verticillioides*	*G. moniliformis*
B	*F. sacchari*	*Gibberella* sp.
C	*F. fujikuroi*	*G. fujikuroi*
D	*F. proliferatum*	*G. intermedia*
E	*F. subglutinans*	*G. subglutinans*
F	*F. thapsinum*	*G. thapsina*

[a] Data from reference 44. The teleomorph of mating population B (anamorph *F. sacchari*) is an unnamed species of *Gibberella*. The teleomorphs of *F. globosum* and *F. anthophilum* have not been reported and are unknown.

and an area around Charleston, South Carolina, in the United States (25, 27, 41, 45, 57). Mycotoxins that have been associated with *F. verticillioides* include fumonisins, fusaric acid, fusarins, and fusariocins.

F. verticillioides grows rapidly and produces dense white mycelia that might be tinged with purple on PDA (53). The reverse of the colony can range from colorless to purple. Macroconidia are long, slender, and almost straight to slightly curved, especially near the ends (Fig. 26.2C). Macroconidia have three to five septa, a snout-shaped apical cell, and a foot-shaped basal cell. The most distinctive microscopic feature of *F. verticillioides* is the formation of long chains of oval, single-celled microconidia on monophialides (Fig. 26.2C). Microconidia can also be formed in false heads. Chlamydospores are not formed. *F. verticillioides* can grow over a wide temperature range, from 2.5°C to 37°C, with an optimum of 22 to 27°C, and at water activities above 0.87 (56). Besides corn, *F. moniliforme* (*verticillioides* and/or *thapsinum*) has also been isolated from rice, sorghum, yams, hazelnuts, pecans, and cheeses (56, 77).

F. proliferatum

F. proliferatum is closely related to *F. verticillioides*, yet less is known about this species, possibly because of its frequent misidentification as *F. moniliforme* (42). It is also frequently isolated from corn, where it probably occurs in much the same way as *F. verticillioides*. *F. proliferatum* is capable of producing fumonisins but, as yet, has not been associated with animal or human diseases (42). In addition to producing fumonisins, *F. proliferatum* has been reported to produce moniliformin, fusaric acid, fusarin C, beauvericin, and fusaproliferin (47).

On PDA, *F. proliferatum* grows rapidly to produce heavy white aerial mycelia that may become tinged with purple. Macroconidia are produced abundantly and are

long, thin, septate (three to five septa), and only slightly curved to almost straight (Fig. 26.2D). The basal cell is foot shaped. Microconidia are single celled with a flattened base and are produced in short to varying-length chains or in false heads from polyphialides (more than one opening), which distinguishes *F. proliferatum* from *F. verticillioides* (53). Chlamydospores are not produced. *F. proliferatum* is widely distributed in the soil and may contaminate several types of food grains. *F. proliferatum* equates with mating population D, and the teleomorphic state is *Gibberella intermedia*. *F. proliferatum* appears to be a significant producer of fumonisins (38, 51).

F. subglutinans

F. subglutinans is very similar to *F. verticillioides* as well as *F. proliferatum*. It is widely distributed on corn and other grains. Little information is available about this mold, again probably because of its misidentification as *F. moniliforme* (49, 56). *F. subglutinans* has not been associated specifically with any reported animal or human diseases but has been found in corn from regions with high incidences of human esophageal cancer. Cultures of *F. subglutinans* have been shown to be acutely toxic to ducklings and rats and dermatoxic to rabbit skin (42). However, *F. subglutinans* does not produce fumonisins (52, 79). In previous studies, isolates of *F. subglutinans* from the United States, Mexico, Nigeria, and South Africa were examined. Thus, toxicity attributed to *F. subglutinans* must be due to other toxic metabolites, such as moniliformin, beauvericin, and fusaproliferin (44).

F. subglutinans grows rapidly on PDA, forming white aerial mycelia that may be tinged with purple. The reverse of the colony may range from colorless to dark purple (53, 56). Macroconidia are long, slender, and almost straight to slightly curved, with a foot-shaped basal cell and three to five septa. Microconidia are oval and usually single celled (but may be septate [one to three septa]) and are produced only in false heads. The teleomorphic state is *Gibberella subglutinans*, with the older synonym *Gibberella fujikuroi* var. *subglutinans*; synonyms of the anamorphic state are *F. moniliforme* var. *subglutinans*, *F. moniliforme* Subglutinans, and *Fusarium sacchari* (38, 53). *F. sacchari* is now the anamorph of mating population B, and *F. subglutinans* is the anamorph of mating population E (44). Another species in the *Liseola* section is *Fusarium anthophilum*.

DETECTION, ISOLATION, AND IDENTIFICATION OF FUSARIUM SPECIES

Fusarium species are most often associated with cereal grains, seeds, milled cereal products such as flour and corn meal, barley malt, animal feeds, and necrotic plant tissue. These substrates may also contain or be colonized by many other microorganisms, and *Fusarium* species may be present in small numbers. To isolate *Fusarium* species from these products, it is necessary to use selective media. The basic techniques for detection and isolation of *Fusarium* employ plating techniques, either plate counts of serial dilutions of products or the placement of seeds or kernels of grain directly on the surfaces of agar media in petri dishes, i.e., direct plating (12).

Several culture media have been used to detect and isolate *Fusarium* species. These include Nash-Snyder medium (50), modified Czapek Dox (MCZ) agar (53), Czapek iprodione-dichloran (CZID) agar (2), potato dextrose-iprodione-dichloran (PDID) agar (80), and dichloran-chloramphenicol-peptone agar (DCPA) (3, 56). Nash-Snyder medium and modified Czapek Dox agar contain pentachloronitrobenzene, a known carcinogen, and are not favored for routine use in food microbiology laboratories. However, these media can be useful for evaluating samples that are heavily contaminated with bacteria and other fungi. CZID agar is becoming a regularly used medium for isolating *Fusarium* from foods, but rapid identification of *Fusarium* isolates to the species level is difficult, if not impossible, on this medium. Isolates must be subcultured on other media, such as carnation leaf agar (CLA), for identification (53). However, CZID agar is a good selective medium for *Fusarium*. While some other molds may not be inhibited completely on CZID agar, most are, and *Fusarium* species can be distinguished readily. Thrane et al. (80) reported that PDID agar is as selective as CZID agar for *Fusarium* species, with the advantage that it supports *Fusarium* growth with morphological and cultural characteristics that are the same as those on PDA, which facilitates more rapid identification, since various monographs and manuals for *Fusarium* identification describe characteristics of colonies grown on PDA. Thrane et al. (80) compared several media for their suitability for supporting colony development by *Fusarium* and found that PDID and CZID agars were better than DCPA. Growth rates were much higher on DCPA, making colony counts more difficult. Conner (19), however, modified DCPA by adding 0.5 μg of crystal violet/ml and reported increased selectivity by *Aspergillus* and *Penicillium* species but not by *Fusarium* species.

Identification of *Fusarium* species is based largely on the production and morphology of macroconidia and microconidia. The identification keys described by Nelson et al. (53) rely heavily on the morphology of conidia and conidiophores, the structures on which conidia are produced. *Fusarium* species do not readily form conidia on all culture media, and conidia formed on high-carbohydrate

media, such as PDA, are often more variable and less typical. A medium that supports abundant and consistent spore production is CLA. Carnation leaves from actively growing, disbudded young carnation plants free of pesticide residues are cut into small pieces (5 mm^2), dried in an oven at 45 to 55°C for 2 h, and sterilized by irradiation (53). CLA is prepared by placing a few pieces of carnation leaf on the surface of 2.0% water agar (24, 53). *Fusarium* isolates are then inoculated onto the agar-leaf interface, where they form abundant and typical conidia and conidiophores in sporodochia rather than mycelia. CLA is low in carbohydrates and rich in other complex naturally occurring substances that apparently stimulate spore production.

Since many *Fusarium* species are plant pathogens and all are found in fields where crops are grown, these molds respond to light. Growth, pigmentation, and spore production are most typical when cultures are grown with alternating light and dark cycles of 12 h each. Fluorescent light or diffuse sunlight from a north-facing window is best. Fluctuating temperatures, such as 25°C during the day and 20°C at night, also enhance growth and sporulation. For identification keys, refer to the work of Nelson et al. (53), Samson et al. (65), Summerell et al. (77), Marasas et al. (43), and Marasas (40).

DETECTION AND QUANTITATION OF *FUSARIUM* TOXINS

Fusarium species produce several toxic or biologically active metabolites. The trichothecenes are a group of closely related compounds that are esters of sesquiterpene alcohols that possess a basic trichothecene skeleton and an epoxide group (76). The trichothecenes are divided into the following three groups: the type A trichothecenes, which include diacetoxyscirpenol, T-2 toxin, HT-2 toxin, and neosolaniol; the type B trichothecenes, which include deoxynivalenol, 3-acetydeoxynivalenol, 15-acetyldeoxynivalenol, nivalenol, and fusarenon-X; and the type C or so-called macrocyclic trichothecenes known as satratoxins. Of these, the toxin most commonly found in cereal grains or most often associated with human illness is deoxynivalenol (47). Other *Fusarium* toxins associated with diseases are zearalenone and the fumonisins. T-2 toxin occurs rarely in grain in the United States but was associated with alimentary toxic aleukia in Russia in the 1940s and earlier. Moniliformin, fusarin C, and fusaric acid are also of interest and concern but have not been shown to commonly occur or be specifically associated with diseases.

If present, *Fusarium* toxins are usually found at low levels in cereal grains and processed grain-based foods.

Their concentrations may range from less than nanogram to microgram quantities per gram of grain (parts per billion to parts per million, respectively). *Fusarium* toxins vary in their chemical structures and properties, making it difficult to develop a single method for quantitating all toxins. The basic steps involved in detection of *Fusarium* mycotoxins are similar to those for other mycotoxins. These include sampling, size reduction and mixing, subsampling, extraction, filtration, cleanup, concentration, separation of components, detection, quantification, and confirmation (9, 76).

The first problem encountered in the analysis of grains for *Fusarium* toxins is the same as that for other mycotoxins, i.e., sampling. Obtaining a representative sample from a large lot of cereal grain can be very difficult if the toxin is present in a relatively small percentage of the kernels, which may be the case with toxins such as deoxynivalenol and zearalenone. On the other hand, fumonisins appear to be more evenly distributed in grains such as corn. Processed grain-based foods may contain a more even distribution of toxins due to grinding and mixing. Samples are usually ground and mixed further, and a subsample of 50 to 100 g is taken for extraction. *Fusarium* toxins, like all mycotoxins, must be extracted from the matrix in which they are found.

Most mycotoxins are more soluble in slightly polar organic solvents than in water. The most commonly used extraction solvents consist of combinations of water and organics, such as methanol, acetone, and acetonitrile. Following extraction, the extract is filtered to remove solids and subjected to a cleanup step to remove interfering substances. Cleanup can be done in several ways, but the most common method used for *Fusarium* toxins is to pass the extract through a column packed with sorbent packing materials. In recent years, the use of small prepacked commercially available disposable columns or cartridges, such as Sep-Pak, Bond Elut, and MycoSep, has become common. After the extract has been cleaned, the sample may need to be concentrated before analysis in order to detect the toxin. This may be accomplished by mild heating, such as in a water bath, heating block, or rotary evaporator under reduced pressure or a stream of nitrogen. Detection and quantification of the toxins are done after they are separated from other components by chromatographic means. The most common chromatographic separation techniques used are thin-layer chromatography (TLC) and high-performance liquid chromatography (HPLC). Gas chromatography (GC) also has some applications, particularly when coupled with mass spectrometry.

A commonly used method for quantitating deoxynivalenol is TLC (22, 58, 82). GC is more sensitive than TLC but is also more laborious. While HPLC methods employing UV absorbance at 219 nm for detection are fairly sensitive, they require purification of deoxynivalenol, using high-capacity activated-charcoal columns (14). The method of choice for quantitation of zearalenone is HPLC with fluorescence detection (6, 87). A TLC method for zearalenone has been tested collaboratively and is useful as a screening method (74). The methods most commonly used for T-2 toxin are GC methods (15, 18, 59). Because type A trichothecenes lack a UV chromophore and are not fluorescent, TLC and HPLC methods are unsuitable, resulting in reliance on GC methods. The most widely used analytical methods for fumonisins are HPLC methods involving the formation of fluorescent derivatives (58). Methods have been developed using derivatizing agents, such as o-phthaldehyde (60, 61, 73), fluorescamine (88), and naphthalene dicarboxaldehyde (5). A TLC method for fumonisins has also been developed but is used mainly for screening (58, 62). For discussions of the methods most commonly used for analyses of *Fusarium* and other mycotoxins, see the work of Richard et al. (58) and Steyn et al. (76).

Immunoassays have been developed for *Fusarium* toxins. Enzyme-linked immunosorbent assay kits for *Fusarium* toxins are commercially available (Neogen Corporation, Lansing, Mich.). Qualitative kits for screening as well as kits for quantitative analyses are available for deoxynivalenol, zearalenone, T-2 toxin, and fumonisins. A rapid screening TLC kit for deoxynivalenol is available from Romer Labs (Union, Mo.). This method uses a special cleanup column that requires only 10 s per sample. An antibody-based affinity column for fumonisins is also available (Vicam, Watertown, Mass.) for quantitation as well as for use as a cleanup tool.

OCCURRENCE OF *FUSARIUM* TOXINS IN FOODS

Fusarium toxins, particularly deoxynivalenol and fumonisins, have been found in finished human food products. Various *Fusarium* toxins, including zearalenone, have been found naturally occurring in numerous cereal grains, but most of these grains have been destined for animal feed (30). Deoxynivalenol is the most common trichothecene found in commodity grains; therefore, the greatest potential exists for it to occur in finished foods (48). Food ingredients and products such as wheat flour bread, pasta, corn, cornmeal, breakfast cereals, corn chips, snack foods, popcorn, and beer have been found to be contaminated with at least trace amounts of deoxynivalenol, at up to 19 µg/g (1, 8, 67, 69, 81). Thus, there is evidence that deoxynivalenol is a contaminant of processed human food products and that levels sometimes exceed the U.S. government guideline of 1.0 µg/g for finished food products. Deoxynivalenol is quite heat stable and probably tolerates most thermal processes to some degree (67, 68).

Fumonisins have also been found in processed or finished food products. Food products that have been examined include cornmeal, corn grits, corn-based breakfast cereals, tortillas, tortilla chips, corn chips, popcorn, and hominy corn (10, 11, 29, 75, 78, 79). The most consistently contaminated products with the greatest amounts of fumonisins are those foods which receive only physical processing, such as milled products, e.g., cornmeal and corn muffin mixes. More highly processed corn products, such as corn flakes and corn pop cereals, corn chips, and corn tortilla chips, have been found to be negative for fumonisins, and only very low levels have been found in tortillas, popcorn, and hominy. Fumonisins have been detected in processed corn products in Germany, Italy, Japan, Spain, and Switzerland (21, 66, 84, 85, 92). Fumonisins and moniliformin have been found to co-occur in food-grade corn and corn-based foods in the United States (29).

OTHER TOXIC MOLDS

Other potentially toxic molds, aside from *Aspergillus*, *Penicillium*, and *Fusarium* species, that may contaminate foods include species of the genera *Acremonium*, *Alternaria*, *Byssochlamys*, *Chaetomium*, *Cladosporium*, *Claviceps*, *Myrothecium*, *Neosartorya*, *Phomopis*, *Rhizoctonia*, and *Rhizopus*. Molds such as *D. maydis*, *Phoma herbarum*, *Pithomyces chartarum*, *Stachybotrys chartarum*, and *T. roseum* are also potentially toxic (39). However, most of these molds are more likely to be present in animal feeds, and their significance to food safety may be minimal. *Stachybotrys* species seem to be of increasing importance as airborne contaminants of buildings and houses (65). Some have been shown to produce toxic secondary metabolites in vitro which have yet to be found to occur naturally.

The ergot mold, *Claviceps purpurea*, is the cause of the earliest recognized human mycotoxicosis, ergotism (4). Ergotism has been reported in sporadic outbreaks in Europe since 857, with near-epidemic outbreaks occurring in the Middle Ages, when it was known as St. Anthony's fire. Ergot is a disease of rye in which the rye grains are replaced by ergot sclerotia that contain toxic alkaloids. If sclerotia are not removed when the rye is milled into flour, the flour, and subsequent bread made from the flour, becomes contaminated. Ergotism can be manifested as a convulsive condition or a necrotic

gangrenous condition of the extremities. During the Middle Ages, necrotic gangrenous ergotism was characterized by swollen limbs and alternating cold and burning sensations in fingers, hands, and feet, hence the term "fire" in St. Anthony's fire. The main ergot alkaloid, ergotamine, has vasoconstrictive properties that cause these symptoms. In severe cases, the extremities, such as feet in humans and hooves in animals, were sloughed off. Convulsive ergotism may have contributed to the Salem witchcraft trials of 1692 in Salem, Massachusetts (13, 46). In more recent times, outbreaks of ergotism have occurred in Russia in 1926, Ireland in 1929, France in 1953, India in 1958, and Ethiopia in 1973 (4).

Alternaria species infect plants in the field, such as wheat, sorghum, and barley (20), and various fruits and vegetables, including apples, pears, citrus fruits, peppers, tomatoes, and potatoes. *Alternaria* species can also cause spoilage of these foods in refrigerated storage. Several *Alternaria* toxins have been described, including alternariol, alternariol monomethyl ether, altenuene, tenuazonic acid, and the altertoxins. Relatively little is known about the toxicity of these toxins; however, cultures of *Alternaria* that have been grown on corn or rice and fed to rats, chicks, turkey poults, and ducklings have been shown to be quite toxic. *Alternaria* was also implicated in the alimentary toxic aleukia toxicoses in Russia in the 1940s (34). The toxicity of various *Alternaria* toxins has not been studied extensively, but there is evidence that the toxins may have synergistic activity, e.g., mixtures of *Alternaria* toxins or culture extracts are more toxic than the individual toxins. Species of *Alternaria* that produce large amounts of tenuazonic acid and one or more of the other toxins include *A. tenuis*, *A. alternata*, *A. citri*, and *A. solani*. *A. alternata* f. sp. *lycopersici*, which is a pathogen of tomatoes, produces a host-specific phytotoxin known as AAL toxin which is nearly identical in structure to fumonisins, and it has been shown that fumonisins can cause lesions in tomatoes identical to those caused by AAL toxin and that AAL toxin is toxic to animal cells in tissue culture (89).

References

1. **Abbouzied, M. M., J. I. Azcona, W. E. Braselton, and J. J. Pestka.** 1991. Immunochemical assessment of mycotoxins in 1989 grain foods: evidence for deoxynivalenol (vomitoxin) contamination. *Appl. Environ. Microbiol.* **57**:672–677.

2. **Abildgren, M. P., F. Lund, U. Thrane, and S. Elmholt.** 1987. Czapek-Dox agar containing iprodione and dichloran as a selective medium for the isolation of *Fusarium* species. *Lett. Appl. Microbiol.* **5**:83–86.

3. **Andrews, S., and J. I. Pitt.** 1986. Selective medium for isolation of *Fusarium* species and dematiaceous hyphomycetes from cereals. *Appl. Environ. Microbiol.* **51**:1235–1238.

4. Beardall, J. M., and J. D. Miller. 1994. Diseases in humans with mycotoxins as possible causes, p. 487–539. *In* J. D. Miller and H. L. Trenholm (ed.), *Mycotoxins in Grain. Compounds Other than Aflatoxin.* Eagan Press, St. Paul, Minn.

5. Bennett, G. A., and J. L. Richard. 1992. High performance liquid chromatographic method for naphthalene dicarbox-aldehyde derivative of fumonisins, p. 143. *In Proceedings of the AOAC International Meeting, Cincinnati, Ohio.* AOAC International, Gaithersburg, Md.

6. Bennett, G. A., O. L. Shotwell, and W. F. Kwolek. 1985. Liquid chromatographic determination of a zearalenol and zearalenone in corn: collaborative study. *J. Assoc. Off. Anal. Chem.* 68:958–962.

7. Bhat, R. V., S. R. Beedu, Y. Ramakrisna, and K. L. Munshi. 1989. Outbreak of trichothecene mycotoxicosis associated with consumption of mould-damaged wheat products in Kashmir Valley, India. *Lancet* i:35–37.

8. Brumley, W. C., M. W. Trucksess, S. H. Adler, C. K. Cohen, K. D. White, and J. A. Sphon. 1985. Negative ion chemical ionization mass spectrometry of deoxynivalenol (DON): application to identification of DON in grains and snack foods after quantitation/isolation by thin-layer chromatography. *J. Agric. Food Chem.* 33:326–330.

9. Bullerman, L. B. 1987. Methods for detecting mycotoxins in foods and beverages, p. 571–598. *In* L. R. Beuchat (ed.), *Food and Beverage Mycology,* 2nd ed. Van Nostrand Reinhold Company Inc., New York, N.Y.

10. Bullerman, L. B. 1996. Occurrence of *Fusarium* and fumonisins on food grains and in foods, p. 27–38. *In* L. Jackson, J. DeVries, and L. Bullerman (ed.), *Fumonisins in Foods.* Plenum Publishing Corp., New York, N.Y.

11. Bullerman, L. B., and W. Y. J. Tsai. 1994. Incidence and levels of *Fusarium moniliforme, Fusarium proliferatum* and fumonisins in corn and corn-based foods and feeds. *J. Food Prot.* 57:541–546.

12. Burgess, L. W., B. A. Summerell, S. Bullock, K. P. Gott, and D. Backhouse. 1994. *Laboratory Manual for Fusarium Research,* 3rd ed. University of Sydney, Sydney, Australia.

13. Caporeal, L. R. 1976. Ergotism: the Satan loosed in Salem? *Science* 192:21–26.

14. Chang, H. L., J. W. Devries, P. A. Larson, and H. H. Patel. 1984. Rapid determination of deoxynivalenol (vomitoxin) by liquid chromatography using modified Romer column clean-ups. *J. Assoc. Off. Anal. Chem.* 67:52–54.

15. Chaytor, J. P., and M. J. Saxby. 1982. Development of a method for the analysis of T-2 toxin in maize by gas-chromatography-mass spectrometry. *J. Chromatogr.* 237:107–111.

16. Chelkowski, J. 1989. Mycotoxins associated with corn cob fusariosis, p. 53–62. *In* J. Chelkowski (ed.), *Fusarium. Mycotoxins, Taxonomy and Pathogenicity.* Elsevier Science Publishing Company, New York, N.Y.

17. Chelkowski, J. 1989. Formation of mycotoxins produced by fusaria in heads of wheat triticale and rye, p. 63–84. *In* J. Chelkowski (ed.), *Fusarium. Mycotoxins, Taxonomy and Pathogenicity.* Elsevier Science Publishing Company, New York, N.Y.

18. Cohen, H., and M. Lapointe. 1984. Capillary gas chromatographic determination of T-2 toxin, HT-2 toxin and diacetoxyscirpenol in cereal grains. *J. Assoc. Off. Anal. Chem.* 67:1105–1109.

19. Conner, D. E. 1992. Evaluation of methods for selective enumeration of *Fusarium* species in feedstuffs, p. 299–302. *In* R. A. Samson, A. D. Hocking, J. I. Pitt, and A. D. King (ed.), *Modern Methods in Food Mycology.* Elsevier Scientific Publishers, Amsterdam, The Netherlands.

20. Coulombe, R. A. 1991. *Alternaria* toxins, p. 425–433. *In* R. P. Sharma and D. K. Salunke (ed.), *Mycotoxins and Phytoalexins.* CRC Press, Inc., Boca Raton, Fla.

21. Doko, M. B., and A. Visconti. 1994. Occurrence of fumonisins B_1 and B_2 in corn and corn-based human foodstuffs in Italy. *Food Addit. Contam.* 11:433–439.

22. Eppley, R. M., M. W. Trucksess, S. Nesheim, C. W. Thorpe, and A. E. Pohland. 1986. Thin layer chromatographic method for detection of deoxynivalenol in wheat: collaborative study. *J. Assoc. Off. Anal. Chem.* 69:37–40.

23. Fincham, J. E., W. F. O. Marasas, J. J. F. Taljaard, N. P. J. Kriek, C. J. Badenhorst, W. C. A. Gelderblom, J. V. Seier, C. M. Smuts, M. Faber, M. J. Weight, W. Slazus, C. W. Woodroof, M. J. van Wyk, M. Kruger, and P. G. Thiel. 1992. Atherogenic effects in a non-human primate of *Fusarium moniliforme* cultures added to a carbohydrate diet. *Atherosclerosis* 94:13–25.

24. Fisher, N. L., L. W. Burgess, T. A. Toussoun, and P. E. Nelson. 1982. Carnation leaves as a substrate and for preserving cultures of *Fusarium* species. *Phytopathology* 72:151–153.

25. Franceschi, S., E. Bidoli, A. E. Baron, and C. LaVecchia. 1990. Maize and the risk of cancers of the oral cavity, pharynx and esophagus in northeastern Italy. *J. Natl. Cancer Inst.* 82:1407–1411.

26. Gelderblom, W. C. A., N. P. J. Kriek, W. F. O. Marasas, and P. G. Thiel. 1991. Toxicity and carcinogenicity of the *Fusarium moniliforme* metabolite fumonisin B_1 in rats. *Carcinogenesis* 12:1247–1251.

27. Gelderblom, W. C. A., W. F. O. Marasas, R. Vleggaar, P. G. Thiel, and M. E. Cawood. Fumonisins: isolation, chemical characterization and biological effects. *Mycopathologia* 117:11–16.

28. Graveson, S., J. C. Frisvad, and R. A. Samson. 1994. *Microfungi.* Munks-Gaard International Publishers Ltd., Copenhagen, Denmark.

29. Gutema, T., C. Munimbazi, and L. B. Bullerman. 2000. Occurrence of fumonisins and moniliformin in corn and corn-based food products of U.S. origin. *J. Food Prot.* 63:1732–1737.

30. Hagler, W. M., Jr., N. R. Towers, C. J. Mirocha, R. M. Eppley, and W. L. Bryden. 2001. Zearalenone: mycotoxin or mycoestrogen?, p. 321–331. *In* B. A. Summerell, J. F. Leslie, D. Backhouse, W. L. Bryden, and L. W. Burgess (ed.), *Fusarium. Paul E. Nelson Memorial Symposium.* APS Press, St. Paul, Minn.

31. Harrison, L. R., B. M. Colvin, J. T. Greene, L. E. Newman, and J. R. Cole. 1990. Pulmonary edema and hydrothorax in swine produced by fumonisin B1, a toxic metabolite of *Fusarium moniliforme. J. Vet. Diagn. Investig.* 2:217–221.

32. Hayes, A. W. 1981. Involvement of mycotoxins in animal and human health, p. 11–40. *In Mycotoxin Teratogenicity and Mutagenicity.* CRC Press, Inc., Boca Raton, Fla.

33. **International Agency for Research on Cancer.** 1993. *Some Naturally Occurring Substances: Food Items and Constituents, Heterocyclic Aromatic Amines and Mycotoxins. Monograph 56.* International Agency for Research on Cancer, Lyon, France.

34. **Joffee, A. Z.** 1960. The mycoflora of overwintered cereals and its toxicity. *Bull. Res. Counc. Isr.* **90:**101–126.

35. **Joffee, A. Z.** 1986. Effects of fusariotoxins in humans, p. 225–298. *In* A. Z. Joffee (ed.), Fusarium *Species: Their Biology and Toxicology.* John Wiley & Sons, New York, N.Y.

36. **Kellerman, T. S., W. F. O. Marasas, P. G. Thiel, W. C. A. Gelderblom, M. Cawood, and J. A. W. Coetzer.** 1990. Leukoencephalomalacia in two horses induced by oral dosing of fumonisin B$_1$. *Onderstepoort J. Vet. Res.* **57:**269–275.

37. **Kuiper-Goodman, T.** 1994. Prevention of human mycotoxicoses through risk assessment and risk management, p. 439–469. *In* J. D. Miller and H. L. Trenholm (ed.), *Mycotoxins in Grain. Compounds Other than Aflatoxin.* Eagan Press, St. Paul, Minn.

38. **Leslie, J. F., R. D. Plattner, A. E. Desjardins, and C. J. R. Klittich.** 1992. Fumonisin B$_1$ production by strains from different mating populations of *Gibberella fujikuroi* (*Fusarium* section Liseola). *Phytopathology* **82:**341–345.

39. **Mantle, P. G.** 1991. Miscellaneous toxigenic fungi, p. 141–152. *In* J. E. Smith and R. S. Henderson (ed.), *Mycotoxins and Animal Foods.* CRC Press, Inc., Boca Raton, Fla.

40. **Marasas, W. F. O.** 1991. Toxigenic fusaria, p. 119–139. *In* J. E. Smith and R. S. Henderson (ed.), *Mycotoxins and Animal Foods.* CRC Press, Inc., Boca Raton, Fla.

41. **Marasas, W. F. O., J. D. Miller, R. T. Riley, and A. Visconti.** 2001. Fumonisins—occurrence, toxicology, metabolism and risk assessment, p. 332–359. *In* B. A. Summerell, J. F. Leslie, D. Backhouse, W. L. Bryden, and L. W. Burges (ed.), *Fusarium. Paul E. Nelson Memorial Symposium.* APS Press, St. Paul, Minn.

42. **Marasas, W. F. O., P. E. Nelson, and T. A. Tousson.** 1984. *Toxigenic* Fusarium *Species: Identity and Mycotoxicology.* The Pennsylvania State University Press, University Park, Pa.

43. **Marasas, W. F. O., P. E. Nelson, and T. A. Tousson.** 1985. Taxonomy of toxigenic fusaria, p. 3–14. *In* J. Lacey (ed.), *Trichothecenes and Other Mycotoxins. Proceedings of the International Mycotoxin Symposium,* Sydney, Australia, 1984. John Wiley and Sons, New York, N.Y.

44. **Marasas, W. F. O., and J. P. Reeder.** 2000. Sections *Liseola, Sporotrichiella* and *Arthrosporiella. Fusarium* Laboratory Workshop, June 11–16. Kansas State University, Manhattan, Kans.

45. **Marasas, W. F. O., F. C. Wehner, S. J. van Rensberg, and D. J. van Schalkwyk.** 1981. Mycoflora of corn produced in human esophageal cancer areas in Transkei, Southern Africa. *Phytopathology* **71:**792–796.

46. **Matossian, M. K.** 1982. Ergot and the Salem witchcraft affair. *Am. Sci.* **70:**355–357.

47. **Miller, J. D.** 1995. Fungi and mycotoxins in grain: implications for stored product research. *J. Stored Prod. Res.* **31:**1–16.

48. **Miller, J. D., J. W. ApSimon, B. A. Blackwell, R. Greenhalgh, and A. Taylor.** 2001. Deoxynivalenol: a 25 year perspective on a trichothecene of agricultural importance, p. 310–320. *In* B. A. Summerell, J. F. Leslie, D. Backhouse, W. L. Bryden, and L. W. Burgess (ed.), *Fusarium. Paul E. Nelson Memorial Symposium.* APS Press, St. Paul, Minn.

49. **Mills, J. T.** 1989. Ecology of mycotoxigenic *Fusarium* species on cereal seeds. *J. Food Prot.* **52:**737–742.

50. **Nash, S. M., and W. C. Snyder.** 1962. Quantitative estimations by plate counts of propagules of the bean root rot *Fusarium* in field soils. *Phytopathology* **52:**567–572.

51. **Nelson, P. E., A. E. Desjardins, and R. D. Plattner.** 1993. Fumonisins, mycotoxins produced by *Fusarium* species: biology, chemistry and significance. *Annu. Rev. Phytopathol.* **31:**233–252.

52. **Nelson, P. E., R. D. Plattner, D. D. Shackelford, and A. E. Desjardins.** 1992. Fumonisin B$_1$ production by *Fusarium* species other than *F. moniliforme* in section Liseola and by some related species. *Appl. Environ. Microbiol.* **58:**984–989.

53. **Nelson, P. E., T. A. Tousoun, and W. F. O. Marasas.** 1983. Fusarium *Species. An Illustrated Manual for Identification.* The Pennsylvania State University Press, University Park, Pa.

54. **Pestka, J. J., and G. S. Bondy.** 1994. Immunotoxic effects of mycotoxins, p. 339–359. *In* J. D. Miller and H. L. Trenholm (ed.), *Mycotoxins in Grain. Compounds Other than Aflatoxin.* Eagan Press, St. Paul, Minn.

55. **Peters, A. T.** 1904. A fungus disease in corn. *Agric. Exp. Stn. Nebraska Annu. Rep.* **17:**13–22.

56. **Pitt, J. I., and A. D. Hocking.** 1997. *Fungi and Food Spoilage,* 2nd ed. Aspen Publishers, Gaithersburg, Md.

57. **Rheeder, J. P., W. F. O. Marasas, P. G. Thiel, E. W. Sydenham, G. S. Shepard, and D. J. van Schalkwyk.** 1992. *Fusarium moniliforme* and fumonisins in corn in relation to human esophageal cancer in Transkei. *Phytopathology* **82:**353–357.

58. **Richard, J. L., G. A. Bennett, P. F. Ross, and P. E. Nelson.** 1993. Analysis of naturally occurring mycotoxins in feedstuffs and foods. *J. Anim. Sci.* **71:**2563–2574.

59. **Romer, T. R., T. M. Boling, and J. L. McDonald.** 1978. Gas-liquid chromatographic determination of T-2 toxin and diacetoxyscirpenol in corn and mixed feeds. *J. Assoc. Off. Anal. Chem.* **61:**801–805.

60. **Ross, P. F., P. E. Nelson, J. L. Richard, G. D. Osweiler, L. G. Rice, R. D. Plattner, and T. M. Wilson.** 1990. Production of fumonisins by *Fusarium moniliforme* and *Fusarium proliferatum* isolates associated with equine leukoencephalomalacia and a pulmonary edema syndrome in swine. *Appl. Environ. Microbiol.* **56:**3225–3226.

61. **Ross, P. F., L. G. Rice, R. D. Plattner, G. O. Osweiler, T. M. Wilson, D. L. Owens, P. A. Nelson, and J. L. Richard.** 1991. Concentrations of fumonisin B$_1$ in feeds associated with animal health problems. *Mycopathologia* **114:**129–135.

62. **Rottinghaus, G. E., C. F. Coatney, and H. C. Minoir.** 1992. A rapid, sensitive thin layer chromatography procedure for the detection of fumonisin B$_1$ and B$_2$. *J. Vet. Diagn. Investig.* **4:**326–329.

63. **Saenz de Rodriguez, C. A.** 1984. Environmental hormone contamination in Puerto Rico. *N. Engl. J. Med.* **310:**1741–1742.

64. **Saenz de Rodriguez, C. A., A. M. Bongiovanni, and L. Conde de Borrego.** 1985. An epidemic of precocious development in Puerto Rican children. *J. Pediatr.* **107:**393–396.

65. Samson, R. A., E. S. Hoekstra, and J. C. Frisvad (ed.). 2004. *Introduction to Food- and Airborne Fungi*, 7th ed. Centraalbureau voor Schimmelcultures, Baarn, The Netherlands.

66. Sanchis, V., M. Abadias, L. Oncins, N. Sala, I. Vinas, and R. Canela. 1994. Occurrence of fumonisins B$_1$ and B$_2$ in corn-based products from the Spanish market. *Appl. Environ. Microbiol.* **60:**2147–2148.

67. Scott, P. M. 1984. Effects of food processing on mycotoxins. *J. Food Prot.* **47:**489–499.

68. Scott, P. M., S. R. Kanhere, P.-Y. Lau, J. E. Dexter, and R. Greenhalgh. 1983. Effects of experimental flour milling and bread baking on retention of deoxynivalenol (vomitoxin) in hard red spring wheat. *Cereal Chem.* **60:**421–424.

69. Scott, P. M., S. R. Kanhere, and D. Weber. 1993. Analysis of Canadian and imported beers for *Fusarium* mycotoxins by gas chromatography-mass spectrometry. *Food Addit. Contam.* **10:**381–389.

70. Seifert, K. A. 2001. *Fusarium* and anamorphic generic concepts, p. 15–28. *In* B. A. Summerell, J. F. Leslie, D. Backhouse, W. L. Bryden, and L. W. Burgess (ed.), *Fusarium. Paul E. Nelson Memorial Symposium.* APS Press, St. Paul, Minn.

71. Sharma, R. P., and Y. W. Kim. 1991. Trichothecenes, p. 339–359. *In* R. P. Sharma and D. K. Salunkhe (ed.), *Mycotoxins and Phytoalexins.* CRC Press, Inc., Boca Raton, Fla.

72. Sheldon, J. L. 1904. A corn mold (*Fusarium moniliforme* n. sp.). *Agric. Exp. Stn. Nebraska Annu. Rep.* **17:**23–43.

73. Shepard, G. W., E. W. Sydenham, P. G. Thiel, and W. C. A. Gelderblom. 1990. Quantitative determination of fumonisins B$_1$ and B$_2$ by high performance liquid chromatography with fluorescence detection. *J. Liquid Chromatogr.* **13:**2077–2087.

74. Shotwell, O. L., M. L. Goulden, and G. A. Bennett. 1976. Determination of zearalenone in corn: collaborative study. *J. Assoc. Off. Anal. Chem.* **59:**666–669.

75. Stack, M. E., and R. M. Eppley. 1992. Liquid chromatographic determination of fumonisins B$_1$ and B$_2$ in corn and corn products. *J. AOAC Int.* **75:**834–837.

76. Steyn, P. S., P. G. Thiel, and D. W. Trinder. 1991. Detection and quantification of mycotoxins by chemical analysis, p. 165–221. *In* J. E. Smith and R. S. Henderson (ed.), *Mycotoxins and Animal Foods.* CRC Press, Inc., Boca Raton, Fla.

77. Summerell, B. A., B. Salleh, and J. F. Leslie. 2003. A utilitarian approach to *Fusarium* identification. *Plant Dis.* **87:**117–128.

78. Sydenham, E. W., G. S. Shephard, P. G. Thiel, W. F. O. Marasas, and S. Stockenstrom. 1991. Fumonisin contamination of commercial corn-based human foodstuffs. *J. Agric. Food Chem.* **25:**767–771.

79. Thiel, P. G., W. F. O. Marasas, E. W. Sydenham, G. S. Shepard, W. C. A. Gelderblom, and J. J. Nieuwenhuis. 1991. Survey of fumonisin production by *Fusarium* species. *Appl. Environ. Microbiol.* **57:**1089–1093.

80. Thrane, U., O. Filtenborg, F. C. Frisvad, and F. Lund. 1992. Improved methods for the detection and identification of toxigenic *Fusarium* species, p. 285–291. *In* R. A. Samson, A. D. Hocking, J. I. Pitt, and A. D. King (ed.), *Modern Methods in Food Mycology.* Elsevier Science Publishers, New York, N.Y.

81. Trucksess, M. W., M. T. Flood, and S. W. Page. 1986. Thin layer-chromatography determination of deoxynivalenol in processed grain products. *J. Assoc. Off. Anal. Chem.* **69:**35–36.

82. Trucksess, M. W., S. Nesheim, and R. M. Eppley. 1984. Thin layer chromatographic determination of deoxynivalenol in wheat and corn. *J. Assoc. Off. Anal. Chem.* **67:**40–44.

83. Ueno, Y. 1983. Toxicoses, natural occurrence and control, p. 195–307. *In Trichothecenes. Chemical, Biological and Toxicological Aspects. Developments in Food Science.* Elsevier Science Publishing Company, Inc., New York, N.Y.

84. Ueno, Y., S. Aoyama, Y. Sugiura, D. S. Wang, U. S. Lee, E. Y. Hirooka, S. Hara, T. Karki, G. Chen, and S. Z. Yu. 1993. A limited survey of fumonisins in corn and corn-based products in Asian countries. *Mycotoxin Res.* **9:**27–34.

85. Usleber, E., M. Straka, and G. Terplan. 1994. Enzyme immunoassay for fumonisin B$_1$ applied to corn-based food. *J. Agric. Food Chem.* **42:**1392–1396.

86. Visconti, A., and A. Sibilia. 1994. *Alternaria* toxins, p. 315–336. *In* J. D. Miller and H. L. Trenholm (ed.), *Mycotoxins in Grain. Compounds Other than Aflatoxin.* Eagan Press, St. Paul, Minn.

87. Ware, G. M., and C. W. Thorp. 1978. Determination of zearalenone in corn by high-pressure liquid chromatography and fluorescence detection. *J. Assoc. Off. Anal. Chem.* **61:**1058–1061.

88. Wilson, T. M., P. F. Ross, L. G. Rice, G. D. Osweiler, H. A. Nelson, D. L. Owens, R. D. Plattner, C. Reggiardo, T. H. Noon, and J. W. Pickrell. 1990. Fumonisin B$_1$ levels associated with an epizootic of equine leukoencephalomalacia. *J. Vet. Diagn. Investig.* **2:**213–216.

89. Winter, C. K., D. G. Gilchrist, M. B. Dickman, and C. J. Jones. 1996. Chemistry and biological activity of AAL toxins, p. 307–316. *In* L. Jackson, J. DeVries, and L. Bullerman (ed.), *Fumonisins in Foods.* Plenum Publishing Corporation, New York, N.Y.

90. Wray, B. B., and K. G. O'Steen. 1975. Mycotoxin-producing fungi from a house associated with leukemia. *Arch. Environ. Health* **30:**571–573.

91. Wray, B. B., E. J. Rushings, R. C. Boyd, and A. M. Schindel. 1979. Suppression of phytohemagglutinin response by fungi from a "leukemia" house. *Arch. Environ. Health* **34:**350–353.

92. Zoller, O., F. Sager, and B. Zimmerli. 1994. Occurrence of fumonisins in foods. *Mitt. Geb. Lebensm. Hyg.* **85:**81–99.

Viruses

Food Microbiology: Fundamentals and Frontiers, 3rd Ed.
Edited by M. P. Doyle and L. R. Beuchat
© 2007 ASM Press, Washington, D.C.

Doris H. D'Souza
Christine L. Moe
Lee-Ann Jaykus

Foodborne Viral Pathogens

27

It is now recognized that human enteric viruses, the most significant of which include hepatitis A virus (HAV) and the noroviruses (NoVs), are responsible for a large proportion of the overall foodborne disease burden worldwide. For example, Mead et al. (152) in 1999 reported that an estimated 30.9 million (80%) of the estimated annual 38.6 million foodborne illnesses in the United States may be caused by viruses, a number much higher than previously believed. Guzewich and Ross (88), in their review of the published literature on foodborne disease outbreaks occurring between 1975 and 1998, determined that HAV and the NoVs accounted for 60% of all outbreaks reported. Widdowson et al. (224) estimated that 50% of all foodborne disease outbreaks would be found to be associated with NoVs if all specimens implicated in disease were screened for viruses. Transmitted primarily by the fecal-oral route, but also by vomitus, these viruses reach consumers through person-to-person spread, through the consumption of contaminated food or water, or through contact with virus-contaminated inanimate objects (fomites). However, the relative importance of the foodborne transmission route cannot be underestimated. For example, Fankhauser et al. (67), who reviewed 233 outbreaks of nonbacterial gastroenteritis reported to the U.S. Centers for Disease Control and Prevention from 1997 to 2000, noted that in 57% of these outbreaks, the mode of transmission was foodborne; for an additional 24%, the route of transmission was undetermined. Similar data have been reported from Europe (140).

Human enteric viruses have properties that are unique from those of bacterial foodborne pathogens. Viruses in general are species specific and tissue tropic, meaning that the human enteric viruses are believed to infect only humans, although there is some recent evidence that this may not always be the case. Nonetheless, they are usually not transmitted by animal sources. Since these viruses must resist the enzymatic conditions and extremes of pH encountered in the gastrointestinal tract, they tend to be quite resistant to a wide range of commonly used food-processing, preservation, and storage treatments. They are also notably persistent in foods and the environment, frequently surviving for days to weeks without substantial loss in infectivity. Although frequently present in low numbers in contaminated foods, their infectious doses are also low, meaning that any level of contamination may pose a human health risk.

Doris H. D'Souza, Dept. of Food Science and Technology, Institute of Agriculture, University of Tennessee, Knoxville, TN 37996-4591. **Lee-Ann Jaykus,** Dept. of Food Science, College of Agriculture and Life Sciences, North Carolina State University, Raleigh, NC 27695-7624. **Christine L. Moe,** Hubert Department of Global Health, Rollins School of Public Health, Emory University, Atlanta, GA 30322.

HUMAN ENTERIC VIRUSES OF EPIDEMIOLOGIC SIGNIFICANCE AND THEIR DISEASES

There is a wide variety of enteric viruses that can cause disease in humans, including the NoVs, HAV, human rotaviruses, human enteroviruses, astroviruses, hepatitis E virus, parvoviruses, and other relatively uncharacterized small, round viruses. From an epidemiologic perspective, the NoVs and HAV are the two most important enteric virus groups transmitted by foodborne routes.

The Norwalk virus was first reported in 1972 by Kapikian and colleagues who identified by electron microscopy a small, round-structured virus (SRSV) of 27 nm in diameter when viewing fecal material obtained in a 1968 outbreak of gastroenteritis (84, 114). Since that time, many similar so-called SRSVs have been reported. Virtually all of these viruses are now included in the *Caliciviridae* (which means "cup-like" shape) virus family. There have been significant changes during the past decade in *Caliciviridae* taxonomy, and current classifications include four genera: (i) *Vesivirus*, represented by vesicular exanthema of swine virus and feline calicivirus; (ii) *Lagovirus*, represented by rabbit hemorrhagic disease virus and European brown hare syndrome virus; (iii) *Norovirus* (members of which

were previously referred to as Norwalk-like viruses and which has the Norwalk virus as the prototype strain); and (iv) *Sapovirus* (members of which were previously referred to as the Sapporo-like viruses and which is represented by the Sapporo virus) (217). In general, the vesiviruses and lagoviruses are considered to be animal caliciviruses, posing no known human disease risk, whereas the sapoviruses and particularly the NoVs are the genera most often responsible for epidemic gastroenteritis (159, 179).

The genomes of the viruses within the *Caliciviridae* family comprise a single strand of positive-sense RNA ranging in size from 7.4 to 8.3 kb. The Norwalk virus genome sequence is 7,642 nucleotides (nt) in length, excluding the 3′ poly(A) tail, and has a base composition of 48% G+C. The genome includes three open reading frames: ORF1 (nt 146 to 5359) is the largest (corresponding to ~1,700 amino acids), encodes a nonstructural polyprotein, and contains the genes for nucleoside triphosphatase, a genome-linked virion protein, protease, and RNA polymerase; ORF2 (nt 5346 to 6935) encodes the viral capsid protein (550 amino acids; molecular weight, ~56,600); and ORF3 (nt 6938 to 7573) encodes a small basic structural protein of unknown function (17, 40) (Fig. 27.1). A large amount of sequence data has been

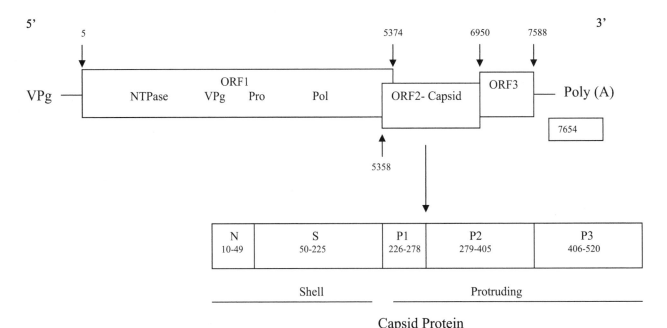

Figure 27.1 Norwalk virus genome organization (adapted from Green et al. [84] with permission). VPg, genome-linked virion protein. ORF1 encodes polyprotein posttranslationally cleaved into nonstructural proteins helicase, with a predicted nucleoside triphosphate-binding domain (NTPase); proteinase (Pro); and RNA-dependent RNA polymerase (Pol). ORF2 encodes capsid protein, which is translated into major and minor structural proteins which consist of shell and protruding domains. N, NH$_2$-terminal arm; S, shell. ORF3 encodes a basic protein of unknown function.

collected for the NoVs over the last decade. Currently, this genus comprises five genogroups: genogroup I (GI; prototype, Norwalk virus), GII (prototype, Snow Mountain virus), GIII (prototype, bovine enteric calicivirus), GIV (prototypes, Alphatron and Ft. Lauderdale viruses), and GV (prototype, murine NoV) (111). These five genogroups can be further subdivided into about 20 different genetic groups or genotypes that are defined as having greater than 80% amino acid similarity in the complete sequence corresponding to ORF2. The emergence of predominant NoV strains with additional RGD motifs or recombinant strains appears to be associated with advantages that either make the viruses more virulent or allow them to better evade the host immune response (7, 139). The sapoviruses are genetically related to the NoVs and occasionally cause viral gastroenteritis in humans, although from an epidemiologic perspective the NoVs are more important.

First discovered in 1972, HAV is classified as a member of the genus *Hepatovirus* within the family *Picornaviridae* (78, 97, 180). The virus is a naked, round particle with a diameter of 27 to 32 nm. The virion genome is a linear, single-stranded, 7.5-kb positive-sense RNA molecule that is enclosed in an icosahedral capsid that consists of three major proteins designated VP1, VP2, and VP3. The genome can be divided into three regions: a long 5′ terminal untranslated region of about 735 nt (nt 1 to 735), a large open reading frame encoding a polyprotein of 2,227 amino acids (nt 741 to 7415), and a short 3′ untranslated region (nt 7416 to 7478) with a poly(A) tail (Fig. 27.2). The single open reading frame is divided into three regions designated P1, P2, and P3. P1 codes for the capsid proteins, and its product is co- and posttranslationally cleaved into four smaller structural proteins (VP1, VP2, VP3, and VP4) and nonstructural proteins (2A, 2B, 2C, 3A, 3B, 3C, and 3D) by a virus-carried proteinase (78, 97, 180). The approximate genome locations corresponding to these various proteins have been defined (58, 78, 97, 180).

Human isolates of HAV comprise a single serotype, and monoclonal antibodies raised to different isolates fail to distinguish the isolates from one another. Nonetheless, there is substantial sequence heterogeneity within select genome regions that encode the putative VP1-2A junction of HAV. By using this region as the basis for typing, HAV isolates can be differentiated into seven unique genotypes such that a genotype is defined as a group of viruses that differ from one another in sequence homology by no more than 15% (15, 186). Of these seven genotypes (designated I to VII), only genotypes I, II, III, and VII are associated with human disease, whereas genotypes IV, V, and VI have been isolated from simians (15, 186). Genotypes I and III are the predominant human

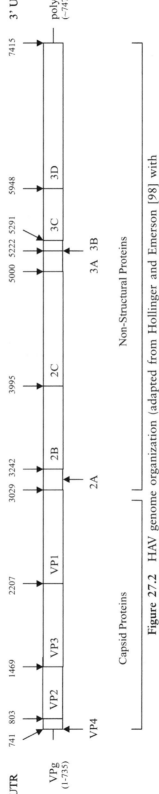

Figure 27.2 HAV genome organization (adapted from Hollinger and Emerson [98] with permission). 5′UTR, 5′ untranslated region (624 to 1,199 nt long); VPg, genome-linked virion protein, 22 to 24 amino acids. The single open reading frame encodes polyprotein posttranslationally cleaved into VP1, VP2, and VP3, which form the capsid; VP4 is the inner surface capsid protein. 2A, unknown function; 2B, RNA synthesis and cell membrane permeability; 2C, RNA replication; 3A and 3B, RNA replication proteins, cofactor for 3D; 3C, viral proteinases; 3D, RNA-dependent RNA polymerase. Poly (A) tail, about 35 to 100 nt long. 3′UTR, 3′ untranslated region (47 to 125 nt long).

genotypes and are further divided into subtypes A and B, whereas genotypes II and VII have only one human isolate each (15).

There are other human enteric viruses that can be transmitted via contaminated foods, although their epidemiologic significance is less well characterized. The human enteroviruses, for example, are smooth, round, nonenveloped particles of 27 nm in diameter with single-stranded, positive-sense RNA. These viruses are members of the *Picornaviridae* family (poliovirus is the prototype) and historically have been transmitted through the consumption of contaminated water and unpasteurized milk (210). Although poliomyelitis can occasionally be problematic in developing countries, outbreaks of foodborne poliomyelitis do not occur in developed countries due to extensive and effective vaccination. Other human enteroviruses such as coxsackieviruses and echoviruses have also been associated occasionally with foodborne disease (42, 43, 164). The symptoms of enteroviral infection are diverse and virus specific but may include gastrointestinal, respiratory, neurological, and skin manifestations.

The rotaviruses usually are transmitted by either waterborne or person-to-person routes, but they can be food borne (29). This virus group is the leading cause of infantile diarrhea worldwide (140) and is responsible for up to 130 million illnesses and 600,000 to 870,000 deaths per year, with the vast majority of rotavirus-associated deaths occurring in developing countries (152). The rotaviruses are 70 to 75 nm in diameter and appear roughly spherical in electron micrographs (42). They consist of 11 segmented, double-stranded RNA molecules encased in a double-layered protein coat. The 5' and 3' ends of the double-stranded RNA are highly conserved, with molecule sizes ranging from 667 bp (segment 11) to 3,302 bp (segment 1) and totaling 6,120 kDa, or 18,555 bp (56). All rotavirus genes are monocistronic except for segments 9 and 11. There are six structural viral proteins (termed VP1, VP2, VP3, VP4, VP6, and VP7) and five nonstructural proteins (termed NSP5 through NSP5). There are at least seven different rotavirus groups (A through G) according to the viruses' VP6 reactivity with monoclonal antibodies

Other significant human enteric viruses include the parvoviruses, astroviruses, and hepatitis E virus. The parvoviruses are perhaps the smallest of the enteric viruses, with diameters of 20 to 26 nm (13, 42). They are single-stranded-DNA viruses with a smooth protein coat. The astroviruses are SRSVs that are star shaped and have single-stranded, positive-sense RNA that is 6.8 to 7.2 kb in length. These viruses cause a diarrheal disease with an incubation period of 1 to 3 days and symptoms

lasting for 1 to 4 days. Hepatitis E virus is a small, 30-nm-diameter virus with single-stranded, positive-sense RNA with three open reading frames. Recently classified in the genus *Hepevirus*, hepatitis E virus appears to have a single serotype and at least four major genotypes (112). Hepatitis E virus causes a disease similar to that caused by HAV but with severe manifestations in pregnant women. It is much more prevalent in developing countries than in developed ones and is transmitted predominantly through sewage-contaminated water and person-to-person contact (42, 49).

FOODBORNE TRANSMISSION OF HUMAN ENTERIC VIRUSES

Enteric viruses can be transmitted directly via person-to-person contact or indirectly via consumption of contaminated food or water or contact with fomites (Fig. 27.3). The usual source of enteric virus contamination is human fecal matter, and up to 10^8 virions/g (up to 10^{11} virions/g for rotavirus) can be shed in the feces of infected individuals (123). However, the role of vomitus cannot be overlooked, particularly for the NoVs, as more than 30 million virus particles can be liberated in one vomiting episode (32). Aerosolization of vomitus can result in infection of exposed subjects who inhale and subsequently swallow the aerosolized virus (146); it can also provide a source of virus to contaminate nearby surfaces. From a foodborne disease standpoint, three types of commodities are commonly associated with viral disease outbreaks, namely, (i) molluscan shellfish contaminated by feces-impacted growing waters; (ii) fresh produce items contaminated by human feces during production or packing, usually through workers' hands or contact with contaminated water; and (iii) ready-to-eat (RTE) and prepared foods contaminated by infected food handlers as a result of poor personal hygiene (reviewed in reference 189; 65, 123, 168). Food items specifically associated with recent outbreaks of viral disease include raw molluscan shellfish, ice, fresh produce (green onions and raspberries), frozen produce (strawberries), and various RTE products, including salads, sandwiches, deli ham, and bakery items.

Contamination of marine waters with human sewage is the critical factor associated with virus contamination of molluscan shellfish (e.g., mussels, clams, cockles, and oysters). Sources of contamination in shellfish harvest waters include the illegal dumping of human waste, failing septic systems along shorelines, sewage treatment plants overloaded with storm water, and discharges of treated and untreated municipal wastewater and sludge (105). Molluscan shellfish are of particular concern

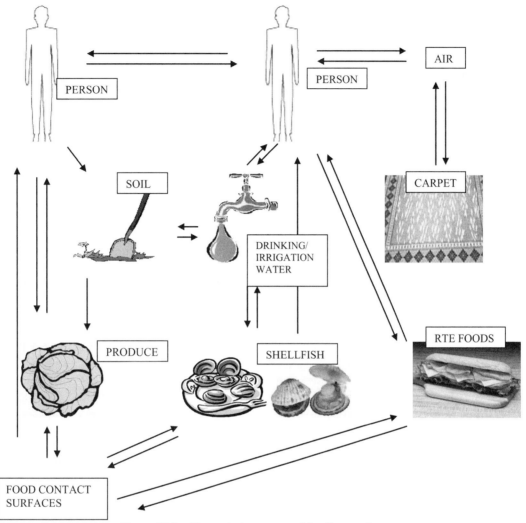

Figure 27.3 Transmission routes of foodborne viruses.

because they are filter feeders, meaning that they filter and even concentrate microorganisms in their gut during the process of feeding. This situation is further complicated by the facts that viruses tend to be environmentally persistent and that the animals are usually consumed either raw or only lightly cooked. Furthermore, there is no significant relationship between the levels of coliform and fecal coliform indicator bacteria, which are used in the United States to classify molluscan shellfish harvesting waters, and the presence of enteric viruses, so there is no reliable means of screening to prevent virally contaminated shellfish from reaching the marketplace (105).

The contamination of produce items usually occurs before the product reaches food service establishments (123). Produce may become contaminated with viruses when grown in fields irrigated with wastewater or fertilized and conditioned with improperly decontaminated sewage effluent. In fact, common sewage sludge treatments such as drying, pasteurization, anaerobic digestion, and composting reduce but do not necessarily eliminate enteric viruses (154). Viruses can survive in contaminated soil for long periods of time depending upon factors such as growing season, soil composition, temperature, sunlight, moisture, rainfall, resident microflora, and virus type (197, 228). Use of wastewater for spray irrigation is particularly risky as this may facilitate virus attachment to produce surfaces (184). Since produce items are usually harvested and processed several days before consumption, it is often difficult to identify the point at which contamination occurred (100); however, the role of infected food handlers who harvest, wash, and pack produce cannot be overlooked. Although washing produce may reduce microbial contamination, it will not eliminate viral or bacterial contaminants (47, 123, 145).

Poor personal hygienic practices of infected food handlers are the most important contributor to the spread of viral foodborne disease. The Centers for Disease Control and Prevention estimates that 50 to 95% of viral foodborne disease outbreaks from 1988 to 1992 were attributable to poor personal hygiene of infected food handlers (21), and similar data are cited for Europe (140). There is strong evidence suggesting that contaminated hands frequently play a role in the transmission of enteric viruses, serving as either virus donors or recipients. Nonenveloped viruses, such as the NoVs and HAV, survive better on skin than do enveloped ones, such as herpes and influenza viruses (206). Hands usually become contaminated by direct contact with any virus-containing fluid from oneself or others; they may also become contaminated by contact with virus-contaminated surfaces or objects. The extent of such contamination will vary depending on a variety of factors, including the virus load, the degree of discharge from the host, the hand-washing habits of the infected person, and the efficiency with which the virus is transferred and persists. Food handlers may transmit viruses to foods from contaminated hands, a contaminated surface, or other food items. Technically, any RTE food handled by a virus carrier, symptomatic or asymptomatic, can become contaminated.

The period during which an infected food handler sheds viruses in the feces is an important consideration. Hepatitis A, with an incubation period of 15 to 50 days, appears to be more readily transmitted during the latter half of the incubation period, meaning that food workers in retail settings with acute HAV infection can readily contaminate RTE products if they do not practice adequate personal hygiene (43). Although historically adults are considered infectious only in the first few days of a NoV infection, it has been determined that they can shed viruses in their feces for up to 3 weeks after the onset of disease (169, 187, 223, 229). Further complicating the issue is evidence of presymptomatic viral excretion from food handlers incubating NoV infection and viral excretion from food handlers with asymptomatic NoV infection (136, 199, 229).

EPIDEMIOLOGY OF FOODBORNE VIRAL DISEASE

Although many different types of enteric viruses can contaminate bivalve molluscs, only a few (HAV, NoVs, and astroviruses) have been epidemiologically linked to shellfish-associated viral disease (85). The most commonly implicated bivalves are oysters, followed by clams (177), and it is now estimated that human enteric viruses are the most common disease agents transmitted by molluscan shellfish (71, 125). The NoVs are most frequently identified as the causative agents in viral gastroenteritis outbreaks associated with bivalve-mollusc consumption (125). The first documented epidemiologic linkage between the NoVs and shellfish-associated gastroenteritis was in the mid-1970s, and many more outbreaks have occurred since then. For example, three large NoV infection outbreaks associated with improper discharge of untreated human waste material occurred in Louisiana in the 1990s alone (23, 121). Similarly, the first documented outbreaks of shellfish-associated hepatitis in the United States occurred in the early 1960s, and cases continue to occur today. Interestingly, a major outbreak of hepatitis A in Shanghai, China, in 1988, linked to the consumption of shellfish harvested from a site contaminated by human sewage pollution, resulted in an estimated 300,000 cases and currently ranks as the largest foodborne disease outbreak ever reported (91).

In recent years, many high-profile hepatitis A outbreaks have occurred in association with the consumption of contaminated produce items. For example, in the early months of 1997, a total of 258 cases of hepatitis A were reported in Michigan and Maine, of which >90% occurred among school employees and students (100). The epidemic curve suggested a common source, and subsequent case control studies in both states identified food items containing frozen strawberries as the likely vehicles of infection. Sporadic cases in other states were later identified, and nucleic acid sequencing of amplified regions of the viral genome confirmed that these additional sporadic cases were related to the outbreak cases and suggested a common source of exposure. The strawberries implicated in this outbreak were grown in Mexico, processed and frozen at a California facility, and distributed commercially through U.S. Department of Agriculture school lunch programs. Regulatory agency inspection of both processing and school facilities did not reveal a likely source of contamination, but inspection of three Mexican growing sites identified inadequate hygiene and toilet facilities (100).

In 2003, there were several large hepatitis A outbreaks associated with the consumption of green onions. As described in one report, an outbreak which involved 601 illnesses, with 3 deaths and at least 124 hospitalizations, occurred among patrons of a single Pennsylvania restaurant. The amplified sequences of HAV from the 170 patients tested were identical, indicating a common source of exposure. The case control study implicated foods containing green onions, particularly mild salsa, as the most likely vehicle of infection. All of the restaurant workers were screened for HAV infection, but as none were identified as the source, the conclusion was that the green onions, which originated from Mexico, were likely contaminated before arrival at the restaurant (222). A series of smaller outbreaks occurred among restaurant patrons in three other states during the months preceding this outbreak (8). Separate case control studies also implicated green onions grown in northern Mexico in two of these outbreaks, with post-harvest transmission by an infected food handler unlikely in both cases. Using molecular surveillance methods, the investigators compared gene sequences of HAV strains recovered in these outbreaks and the large Pennsylvania outbreak and found substantial sequence similarity between the epidemic U.S. HAV strains and the endemic Mexican strains. These series of hepatitis A outbreaks illustrate the importance of fresh produce as a vehicle for foodborne viral disease outbreaks.

Contamination of RTE and prepared foods most frequently results from poor hand-washing practices of infected food handlers after toilet use, as fecal material can be left on hands or even under nails which then can come in contact with food products. Indeed, handling cooked products with bare hands can serve as a major factor for pathogen transfer to RTE foods. Many hepatitis A outbreaks occur as the result of a single infected food handler at a single food establishment, and many of these outbreaks may be unrecognized and unreported. Nonetheless, hepatitis A outbreaks linked to poor personal hygiene of infected food handlers have been associated with lettuce, salads, sandwiches, hamburgers, spaghetti, and bakery products (41, 68, 188).

NoVs have been associated with many outbreaks caused by poor hygienic practices of food handlers, including outbreaks involving the consumption of contaminated salads and sandwiches (22, 136, 169) and bakery products (9, 73, 172). In recent outbreak investigations, scientists have combined molecular and immunological methods to support epidemiologic evidence indicating a common food source for virus transmission (9, 73, 169, 171) (Table 27.1).

The importance of person-to-person spread and environmental contamination in the propagation of viral gastroenteritis outbreaks cannot be overemphasized. The most recent examples include cruise ship outbreaks which have sporadically occurred during the past two decades. Isakbaeva et al. (102) investigated one such NoV gastroenteritis outbreak that affected six consecutive cruises despite extensive sanitation measures. Using genetic sequence analysis, the investigators documented strain persistence with likely foodborne, environmental, and person-to-person transmission (33). In a particularly interesting case, a GI NoV which initially infected athletes on a single college football team who consumed a common food item was later passed on to the rival team after contact during a football game in which the first team's members were clearly ill (22). Patterson et al. (172) reported on the role of a kitchen assistant who vomited into a sink used to prepare a potato salad that was subsequently identified as the vehicle of a NoV

Table 27.1 Representative viral foodborne disease outbreaks

Viral agent	Commodity	Sample type	Method(s)[a] of detection	Contamination source(s) or comment	Reference(s)
HAV	Green onions	Clinical	Serological testing	Distribution system or preharvest or postharvest processes	34
	Blueberries	Clinical and food	RT-PCR; sequencing	Food handlers or polluted groundwater	30
NoV	Deli meats	Clinical and food	RT-PCR (nested); sequencing	Food and clinical samples had identical amplicon sequences	51, 196
	Boxed lunch	Clinical	RT-PCR; EM	Person-to-person transmission during a football game	22
	Oysters	Clinical and food	IgG antibody detection; EM; RT-PCR; sequencing	Overboard disposal of sewage from handlers	23
	Raspberries	Clinical and food	RT-PCR; sequencing	Contaminated water	176
	Deli sandwiches	Clinical	RT-PCR (single and nested); sequencing	Asymptomatic food handler	169
	Potato salad	Clinical	RT-PCR; EM	Airborne transmission; kitchen assistant vomited in sink	172

[a] EM, electron microscopy.

infection outbreak. Marks et al. (146) described an outbreak of NoV gastroenteritis following a meal in a large hotel during which one of the diners vomited. In this case, attack rates were inversely proportional to the distances between persons with secondary cases and the individual with the primary case who vomited, consistent with airborne spread of the virus and infection by inhalation and subsequent ingestion of virus particles.

DISEASE, PATHOGENESIS, AND TREATMENT

NoVs

The disease caused by the NoVs is self-limiting and characterized by nausea, vomiting, diarrhea, and abdominal pain with occasional headache and low-grade fever (42, 43, 84, 114). Data from outbreak investigations and human challenge studies indicate that the incubation period ranges from 12 to 51 h, with a mean of 24 h, and illness lasts no more than 48 to 72 h (84, 114). In a community-based cohort study of NoV infection in the Netherlands, Rockx et al. (187) reported a median duration of symptoms of 5 days and found that diarrhea was more prevalent in NoV-infected children of <1 year of age whereas infected children of ≥5 years of age were more likely to vomit. Some outbreak investigations have noted less frequent vomiting in older patients (31). Severe illness or hospitalization is uncommon except in children, the elderly, and the immunocompromised, for whom rehydration therapy may be necessary. Sometimes, hospitalization for severe dehydration is necessary, though this situation is not frequent, and some deaths in elderly patients occur (84, 114). Outbreaks associated with NoV are characterized by a short incubation period, vomiting in >50% of the cases, average duration of symptoms of 12 to 72 h, a high attack rate, and stool specimens that are negative for bacterial pathogens. Average attack rates are high, typically at 30 to 45% but in some outbreaks up to 80% (31, 123, 213). Asymptomatic NoV infections have been documented in outbreaks (74) and human challenge studies (80, 134). Viruses can be shed in stools both before symptoms occur and for at least 3 weeks after symptoms have abated (75, 169, 187). Viruses can also be shed in high numbers in vomitus (84, 114).

Host susceptibility to infection appears to be due both to genetic determinants and to acquired immunity. Human challenge studies with Norwalk virus demonstrated that some volunteers remained uninfected even after exposure to high doses of the virus (110, 134). Further investigation revealed that these subjects lacked the H type 1 antigen, a histo-blood group antigen present in saliva and also on the surfaces of epithelial cells that

likely serves as the receptor for Norwalk virus binding. In vitro assays provide additional evidence that recombinant Norwalk virus capsid protein binds to H type 1 antigen (93). Synthesis of the H type 1 antigen depends on the alpha-1,2-fucosyltransferase enzyme that is encoded by the *FUT2* gene and determines the "secretor" status of an individual. Individuals who are homozygous recessive for the *FUT2* gene are secretor negative and appear to have innate genetic resistance to Norwalk virus infection. About 20% of Europeans are secretor negative, and this proportion is higher among Asian and African ethnic groups. In addition to secretor status, blood type also plays a role in susceptibility to Norwalk virus infection, as group O individuals in Norwalk virus human challenge studies appear to be more susceptible to Norwalk virus infection than group A or B individuals (101, 134). In contrast to those with Norwalk virus, a human challenge study with Snow Mountain virus, a GII.2 NoV, found no relationship between Snow Mountain virus infection and blood group secretor status (133). NoVs in GII.4 appear to bind to all histo-blood group antigens (not just type 1) (99) and thus are likely to correspond to a greater pool of susceptible individuals in the population. This characteristic may explain the greater frequency of GII.4 infection outbreaks than of outbreaks involving other NoV strains.

There is also evidence of acquired immunity to NoV infection, although the mechanisms are not well understood. Some human challenge studies have revealed that volunteers who became ill after a primary NoV challenge do not become ill when rechallenged with the same virus 6 to 14 weeks after the primary challenge, but do become ill when rechallenged with the same virus 24 to 42 months after the primary challenge (171, 227). This finding suggests the development of short-term protective immunity but perhaps not long-term immunity. Sera collected in human challenge studies and NoV outbreak investigations indicate that NoV-specific humoral antibody titers (immunoglobulin G [IgG], IgM, and IgA) rise in response to NoV infection. Serum IgG produced in response to infection with one NoV strain often cross-reacts with antigens from other NoV strains, and common epitopes within and between genogroups have been identified (170). IgA and IgM responses appear to be more type specific than that of IgG; however, it is not clear that serum antibodies provide any protection against subsequent NoV infection, and these antibodies may be only markers of susceptibility to NoV infection. Results of in vitro experiments suggest that serum IgG made in response to Norwalk virus infection is capable of blocking Norwalk virus attachment to synthetic H type 1 antigen (93). However, volunteers with anti-Norwalk virus IgG in their prechallenge sera were more likely to become infected after Norwalk virus challenge than those who did not have anti-Norwalk virus IgG (134). The high attack rates observed in NoV infection outbreaks also suggest that most of the population does not have protective immunity against NoV infection, either because the protective immunity is so short-lived or because there are many strains of NoV and protective immunity is not cross-reactive.

One recent Norwalk virus human challenge study revealed that 62% of secretor-positive volunteers became infected, suggesting that some mechanism exists to protect genetically susceptible individuals from Norwalk virus infection. Examination of Norwalk virus-specific salivary IgA levels indicated that the secretor-positive volunteers who did not become infected mounted an early (before 5 days postchallenge) Norwalk virus-specific salivary IgA response to the virus compared to infected secretor-positive volunteers, who did not mount a Norwalk virus-specific salivary IgA response until after 5 days postchallenge. The role of the mucosal response in protective immunity to Norwalk virus infection is not understood. It is possible that this early, specific salivary IgA response may represent a memory response that allows rapid production of neutralizing antibodies that may block virus binding to cell receptors. Why some secretor-positive volunteers demonstrate this salivary IgA response and whether it provides long-term or short-term immunity are not known. However, this observation of a protective response does suggest that induction of a mucosal immune response should be an important part of a vaccine strategy for NoV infection. Vaccine development efforts are under way and have been reviewed by Harrington et al. (92). Researchers are exploring various vaccination strategies, including the use of recombinant NoV capsid protein to stimulate immunity in oral vaccination strategies and Venezuelan equine encephalitis virus replicon-based vaccines (92). Because NoV infection outbreaks continue to occur in countries with high standards of sanitation and hygiene, vaccination of high-risk subgroups (such as food handlers) may be the only effective way to control epidemic NoV infection.

Patients who suffer from NoV illness have broadened and blunted villi of the proximal small intestine with infiltration of mononuclear cells and cytoplasmic vacuolization (84, 114). Biopsy specimens collected during human challenge studies indicate that histopathological changes appear within 24 h of virus challenge and usually persist for up to 2 weeks (5). Histopathological testing revealed that the intestinal mucosa appear intact and that viruses are not detected in the epithelial cells of the mucosa by electron microscopy. Also, transient maladsorption of fat,

D-xylose, and lactose is observed along with decreased levels of small intestine brush border enzymes (trehalase and alkaline phosphatase). However, adenylate cyclase activities of the jejunum are not elevated. Abnormal gastric motor function is thought to be responsible for the nausea and vomiting caused by NoVs.

NoV infection is self-limiting, and if necessary, oral fluid and electrolyte replacement therapy is usually adequate for replenishing fluid loss. Symptoms of the disease can also be reduced by the oral administration of bismuth subsalicylate (84, 114). In cases of severe vomiting and diarrhea, parenteral administration of fluids may be necessary. Infrequently, hospitalization for severe dehydration is necessary; likewise, death in elderly patients has been documented (97, 98). Repeated diarrheal episodes may promote intestinal mucosal damage, resulting in eventual malnutrition (84, 114). The development of antiviral drugs that block NoV binding to host cells may be valuable as a prophylactic approach in settings where NoV infection outbreaks are common.

HAV

The incubation period for HAV infection is 2 to 7 weeks, with an average of 28 days. The classic presentation of infectious hepatitis consists of four phases. The first phase is characterized by viral replication but a lack of symptoms. In the second, so-called preicteric, or prodromal, phase, which usually lasts 5 to 7 days, patients may experience anorexia, nausea, vomiting, alterations in taste, arthralgias, malaise, fatigue, urticaria, and pruritus. When seen by a health care provider during this phase, patients are often diagnosed as having gastroenteritis or a generalized viral syndrome. The third or icteric phase presents as darkening of the urine, pale-colored stools, jaundice, and right-upper-quadrant pain with hepatomegaly (97, 98, 180). Fecal shedding and viremia are maximal at the onset of the icteric phase and subside during its 7- to 28-day duration (207). During the last, or convalescent, phase, symptoms resolve and liver enzymes return to normal. Although icteric disease occurs in fewer than 10% of children younger than 6 years of age, it occurs in 40 to 50% of older children and in 70 to 80% of adults. In most cases, infection is mild and self-limiting; however, in older patients, more severe disease is possible. Complications such as acute liver failure, cholestatic hepatitis, and relapsing hepatitis occur but are rare. A variety of extrahepatic manifestations occasionally occur in patients with acute hepatitis A, including hemolysis, acalculous cholecystitis, acute renal failure, pleural or pericardial effusion, acute reactive arthritis, and pancreatitis; on occasion, neurological syndromes have also been reported (50). The overall mortality rate for HAV infection is approximately 0.01%. Immunity after infection is complete and considered to be lifelong in duration (207).

Hepatitis A infection begins with ingestion of the virus, which reaches the gastrointestinal tract in the bile and replicates in the hepatocyte (97, 98). Approximately 10^4 virions/ml of blood can be detected in the viremic phase of the disease (97, 98). Studies are under way to elucidate the nature of the host cell receptor that determines the tissue tropism of HAV to the hepatocyte. The virus may also circulate in the blood enclosed in lipid-associated membrane fragments, enabling it to be protected from neutralizing antibodies. Possible "enterohepatic" cycling of HAV may occur via ingestion of infected material, absorption from the stomach or small intestine, replication in the liver, secretion into bile, and excretion in stool or reabsorption (50). Enlargement of Kupffer cells is the first evidence of infection with HAV. For example, in experimental studies with owl monkeys, viral antigens were first detected in Kupffer cells of the liver at 14 days and in hepatocytes at 21 days after oral administration of the HM-175 strain of HAV (18).

Cardinal pathological features of acute hepatitis include the presence of hepatocellular degeneration, characterized as either ballooning or acidophilic (apoptotic) change, together with various degrees of portal and lobular inflammation and hepatocyte regeneration (50). Interestingly, although HAV targets the liver cells, it usually does not kill those cells (49). Rather, immune-mediated lysis, particularly involving natural killer and cytotoxic T cells, is recognized as the most probable cause of hepatic inflammation (97, 98). In addition, human gamma interferon produced by HAV-specific T cells may participate in pathogenesis and probably promotes clearance of HAV-infected hepatocytes, as increased levels of interferon have been detected in the sera of infected patients (143, 216). Biochemical changes associated with active HAV infection include bilirubin excretion in the urine and elevated alkaline phosphatase, alanine aminotransferase, and aspartate aminotransferase or serum glutamic-oxaloacetic transaminase levels, all of which persist until recovery from illness (97, 98).

The immune response against HAV infection occurs on two levels, humoral and cellular. In the humoral response, IgM, IgG, and IgA antibodies are directed against conformational epitopes, mostly the surface proteins VP1 and VP3 and precursor protein VP0 of the HAV particle. The IgA and IgM responses precede the IgG response, but the latter is long lived and provides immunity (50). Because immune-mediated injury appears to be the cause of hepatic inflammation, cytotoxic T cells also play a significant role in the immune response. Since detection of

virus particles, viral antigen, or RNA is complicated, the method of choice for diagnosis of infection is immunoassay, which detects the presence of IgM anti-HAV antibody in the blood (207).

Historically, Igs at doses as low as 0.01 to 0.04 ml/kg of body weight have been used for HAV postexposure prophylaxis. These can be effective in controlling both the incidence and the severity of disease as long as they are administered within 2 weeks of exposure (97, 98, 173). The prophylactic administration of Ig is recommended for patrons and food handlers provided that all of the following criteria exist: (i) the infected worker was responsible for handling RTE foods and was not wearing gloves; (ii) the infected worker demonstrated poor hygiene practices or already had diarrheal symptoms; and (iii) the patrons can be identified and treated within 2 weeks of exposure (44, 97, 98). As Ig provides only short-term protection against HAV infection, its administration is not recommended for preexposure prophylaxis (69). For that sort of protection, two Food and Drug Administration (FDA)-licensed formalin-killed whole-virus vaccines are available, generally administered as a single primary immunization followed by a booster dose after 6 to 12 months (131). The efficacy of the vaccine is 94 to 100%, and protection should last for about 20 years (69). Routine vaccination of all food handlers and restaurant employees is usually cost prohibitive, even during a hepatitis A epidemic (153). However, HAV vaccination can and perhaps should be offered to high-risk populations such as health care workers in infectious diseases and pediatrics, medical staff in laboratories handling stool samples, and staff in sewage treatment plants (96). Recently, HAV vaccination has been recommended for children up to 2 years of age (35).

ENVIRONMENTAL PERSISTENCE AND TRANSFER OF ENTERIC VIRUSES

Epidemiologic evidence from numerous, prolonged NoV infection outbreaks on cruise ships and in hotels and hospitals indicates that NoVs are very stable in the environment (33, 83, 102, 136), and this stability undoubtedly contributes to their success as agents of foodborne disease. For example, environmental sampling during an investigation of a prolonged NoV infection outbreak in a hotel in England revealed the detection of NoV RNA on the carpet where guests had vomited, even after the carpet had been cleaned with detergent and vacuumed (38). In a NoV infection outbreak at an elder care residential hostel in Australia, it was hypothesized that virus deposited on furniture and carpets remained infectious even after these items were professionally cleaned (135).

Likewise, surfaces such as carpets, toilet seats, lockers, curtains, and commodes have been implicated in virus transmission in hospital-acquired outbreaks (83).

A few laboratory-based studies have addressed the survival, persistence, removal, and transfer of enteric viruses on hands, in foods, and in food-processing and preparation environments. Virtually all of these studies have been done using enteric viruses for which mammalian cell culture hosts are available. For example, the vaccine strain of poliovirus is readily cultivatable and cell culture-adapted strains exist for HAV and the rotaviruses. However, the human NoV strains cannot be cultured in vitro nor does there exist an animal model for their propagation. Because of this limitation, studies focused on NoV behavior in the environment have used culturable members of the *Caliciviridae* family as surrogates, most notably the feline calicivirus, or have made inferences about NoV behavior based on RNA detection.

Many factors influence the stability of enteric viruses in the environment, the most important of which are relative humidity, temperature, degree of inoculum drying, type of suspending medium (fecal or otherwise), virus type, and the type of surface contaminated (1, 115, 148, 221). For instance, Mbithi et al. (148) reported that the half-life of HAV ranges from more than 7 days at relatively low humidity and 5°C to about 2 h at high humidity and 35°C. Using the feline calicivirus model, Doultree et al. (63) reported virus persistence on glass surfaces for more than 60 days at 4°C, with complete inactivation after 14 to 28 days at 25°C and after 1 to 10 days at 37°C. With respect to the stability of viruses on hands, Mbithi et al. (149) reported a biphasic pattern of HAV decay on finger pads, with 16 to 30% of the initially recoverable virus remaining detectable after 4 h. Similar results have been observed for rotavirus (11). To date, we do not have a clear understanding of the stability of the NoVs on human skin.

Enteric viruses are also readily transferred from one surface to another. Mbithi et al. (149) were the first to evaluate the transfer of HAV between hands and inanimate surfaces, finding that, on average, 25% of the initial virus inoculum could be transferred by touch. Dried virus suspensions were much less effective for tactile transmission, and there were significant differences in virus transfer levels related to pressure and friction. More recently, Bidawid et al. (27), using a pressure of 0.3 kg/cm² (2,700 g/9 cm²), determined that on average 9% of dried HAV inoculum was transferred from artificially contaminated finger pads to lettuce. These investigators systematically looked at cross contamination during food handling by using the feline calicivirus model (28), reporting that 13 to 46% of the virus inoculum was transferred from

finger pads to surfaces (food or environmental) and that transfer was less efficient (6 to 14%) when occurring in the opposite direction (i.e., from surfaces to hands). Both of these studies (27, 28) revealed that virus transfer is effectively interrupted if contaminated hands are rinsed liberally in water or decontaminated with ethanol (60 to 75%) and subsequently rinsed prior to touching the recipient surface.

It is generally recognized that enteric viruses are quite stable in foods as well. With respect to temperature, investigators found no significant reduction in virus titers in oysters artificially inoculated with model human enteroviruses after storage at refrigeration temperatures for more than 1 month (212). Kingsley and Richards (117) determined that, for oysters allowed to accumulate HAV by exposure to contaminated water, infectious virus could be detected for 3 weeks whereas HAV RNA could be detected by reverse transcription-PCR (RT-PCR) for up to 6 weeks after contamination. When three model enteroviruses were added to uncooked vegetables and stored under household conditions, infectivity was retained for 1 week and 1 month at room temperature and 10°C, respectively (175). Similar results were obtained for refrigerated soft fruit and salad vegetables on which poliovirus titers remained relatively stable, with D values of 14.2 days and frequently insignificant inactivation (124). Croci et al. (47) determined product-specific survival rates for HAV inoculated onto lettuce, fennel, and carrots followed by product refrigeration. In general, enteric viruses are stable for months to years when held at freezing temperatures. For example, HAV survived freezing for up to 2 years in frozen strawberries associated with a multistate outbreak (165), and in laboratory-based studies, there was only a 1-$\log_{10}$ reduction in poliovirus titer within the first 9 days of freezing (124).

Enteric viruses are also resistant to extremes of pH, which makes sense since they must survive gastric acidity in order to reach target sites in the small intestine. It has in fact been demonstrated that cell culture-adapted HAV can remain infectious after being held at pH 1 for 5 h and that human enteroviruses retain stability at pH 3 (166). Hewitt and Greening (95) determined that HAV remained infectious in acidic marinade (pH ~3.75) when held at 4°C for 4 weeks. Epidemiologic evidence indicates that the NoVs are similarly resistant to low pH. For example, Norwalk virus was reported to cause infection in human volunteers after incubation for 3 h at pH 2.7 (59). Another example of the resistance of human NoVs to prolonged acidic pH was seen in a large gastroenteritis outbreak in which orange juice (pH ~3.5) was implicated as the vehicle of infection (70). By contrast, the commonly used NoV surrogate, feline calicivirus, appears to be quite unstable at lower pH values (66, 95), calling into question its universal suitability as a NoV surrogate.

CONTROL OF VIRUSES IN FOODS AND THE FOOD PRODUCTION, PROCESSING, AND PREPARATION ENVIRONMENT

Molluscan Shellfish

Since the role of molluscan shellfish in the transmission of human enteric viruses has been recognized for decades, significant efforts have gone into the design of pre- and postharvest controls for this food commodity. Preharvest controls include prevention of human sewage pollution of shellfish harvesting waters, controlled purification, and the establishment of alternative microbiological indicators. Prevention of sewage pollution can be complicated because of the diversity of sources and unpredictable nature of a contamination event. Oyster beds close to sewage plants may need special consideration due to concerns about the potential for sewage overflow, particularly during times of excessive rainfall (37). Prevention of illegal dumping of human waste in marine water is critical (23).

Preventing human sewage contamination of molluscan shellfish is complicated by the lack of a reliable microbiological indicator whose presence and/or levels are correlated with the presence of enteric virus contamination. Consequently, extensive investigation has gone into the identification of alternative microbiological indicators. Bacteriophages, especially F-specific coliphages, somatic coliphages, and phages of *Bacteroides fragilis*, have long been proposed as possible indicators of viral contamination of the environment. Of these, the distribution and survival of F-specific coliphages (also known as FRNA phage, or male-specific coliphage) are considered to be more similar to those of human enteric viruses in the environment (62, 202). For instance, in a recent study examining the comparative survival of feline calicivirus, *Escherichia coli*, and FRNA phage in dechlorinated water at 4, 25, and 37°C, a correlation was found between the survival of the phage and that of feline calicivirus (6). Formiga-Cruz et al. (72) studied FRNA phages, somatic coliphages, and bacteriophages of *B. fragilis* as indicators of viral contamination in shellfish, demonstrating that FRNA phages are better predictors of NoV contamination than the other indicators evaluated. Others have sought to use FRNA phages as a surrogate for evaluating virus elimination during controlled purification, as these are more persistent than bacterial models (60, 61, 162). However, although one study examining the

distribution of FRNA phages in shellfish-harvesting areas suggested that their numbers in shellfish are consistently higher than those of *E. coli* bacteria (62), other studies show that shellfish may contain pathogenic viruses even though they lack FRNA phages (163).

Somatic coliphages have also been suggested as reliable indicators of the efficiency of shellfish depuration (161, 162, 178); however, others have debated these findings (39, 128). Evidence that phages infecting *B. fragilis* RYC2056 may be a suitable group of indicators for viral pollution in shellfish has recently been provided, but further research is needed to develop the appropriate methodology (161, 162). Likewise, the feasibility of using human adenoviruses as indicators of human enteric viruses in environmental and shellfish samples has been investigated (71, 174), although questions remain regarding the correlation between the presence of adenoviruses and that of human enteroviruses and NoVs in estuarine waters (25). In conclusion, a reliable alternative indicator for enteric virus contamination of shellfish and their harvesting waters has yet to be validated.

The two methods of controlled purification are depuration and relaying, and these are based on the principle that shellfish can purge at least some of their contaminants by extended feeding in clean water, provided that the feeding conditions (primarily temperature, salinity, and dissolved oxygen level) are favorable (184). In depuration, freshly harvested shellfish are placed for several days in a controlled environment, usually in tanks provided with a supply of clean, disinfected seawater under specific operating conditions (204). The water used in shellfish depuration can be disinfected by using UV light, chlorine, or ozone (52). In relaying, a method which relies on natural biological cleansing, shellfish are transferred from contaminated growing areas to approved, or naturally unpolluted (pristine), areas (204). Relaying usually takes 10 days to 2 weeks or more, and it is subject to more environmental fluctuation than is depuration (184).

Relaying and depuration are both influenced by many factors, including the type of shellfish, individual specimen variation in feeding rates, initial level of virus contamination, temperature, turbidity, availability of particulate matter, salinity, pH, and oxygen availability (45, 204). Although many depuration and relaying studies have been completed, the scientific consensus is that enteric viruses may not be eliminated during controlled purification. In fact, HAV appears to be more resistant to depuration than the human enteroviruses, which explains the fact that hepatitis A outbreaks associated with the consumption of depurated shellfish have occurred (182). A recent study revealed that in areas where current European microbiological standards characterizing waters as suitable for harvesting shellfish are in place, the infectivity of HAV in mussels after depuration is reduced by only 98.7% (2). In one of the few studies investigating the elimination of the NoVs by depuration, Schwab et al. (194) found only a 7% reduction of NoVs after 48 h of depuration treatment.

Postharvest controls for viral contamination of shellfish include heating and several alternative food-processing technologies. Although the human enteric viruses are not exceptionally resistant to heat, comparatively speaking, HAV is more thermally resistant than the NoVs. However, thermal inactivation of viruses in molluscan shellfish presents a worst-case scenario, since the viruses are sequestered in the gastrointestinal tracts of the animals and since heat penetration is slowed by dense shellfish tissues. Indeed, concern about the survival of human enteric viruses after heating has persisted after documented outbreaks of HAV and viral gastroenteritis linked to the consumption of cooked shellfish (14, 151, 160, 205). Early thermal inactivation studies with steamed clams revealed that it takes 4 to 6 min for complete inactivation of HAV at 100°C (120). When artificially contaminated cockles were immersed for 1 min in water at 85, 90, or 95°C or were steamed for 1 min, only partial reduction in HAV titers was achieved, but when the internal temperature of the shellfish reached 85 to 90°C for 1 min (155), HAV could be inactivated. Croci et al. (48) determined that heat treatments at 60°C for 30 min, 80°C for 10 min, and 100°C for 1 min were insufficient to completely inactivate HAV in contaminated mussels; however, heat treatments at 100°C for 2 min completely inactivated the virus. Unfortunately, elevated temperatures such as these tend to produce unacceptable sensory changes in the product. The European heat processing recommendations (internal temperature of 90°C for 1.5 min) for HAV elimination in cockles can also successfully eliminate feline calicivirus in shellfish (63, 203).

Alternative processing methods such as ionizing radiation and high pressure have also been suggested for inactivation of viruses in shellfish. In general, enteric viruses are extremely resistant to irradiation. Mallet et al. (144) demonstrated that treatment of artificially contaminated shellfish (oysters and clams) with a 3-kGy radiation dose could reduce HAV and rotavirus SA-11 populations by less than 2 $\log_{10}$ and also resulted in significant adverse organoleptic changes in the product. In model studies with high hydrostatic pressure, it was determined that HAV and feline calicivirus suspended in tissue culture medium were inactivated after exposure to 450 MPa for 5 min and 275 MPa for 5 min, respectively (118). However, pressures as high as 600 MPa for 15 min failed to inactivate poliovirus (226). Even though the process produces a "raw-like" product and the initial data on high-pressure processing are promising, more research is needed to definitively establish the efficacy of high pressure in inactivating human enteric viruses in raw molluscan shellfish.

Produce

Recent foodborne viral disease outbreaks associated with the consumption of fresh and frozen produce indicate the importance of this at-risk food group. Viral contamination of produce can occur during planting, irrigation, harvesting, processing, or shipping either through an infected worker or through contaminated water. Many produce items are particularly prone to virus contamination because they require extensive human handling, especially during harvest (55). Preharvest measures directed at reduction of viral contamination of produce are considered in the Good Agricultural Practices described by the FDA (214, 215). One important consideration in preventing viral contamination is the quality of irrigation water, manure, fertilizer, and compost. There are, however, little conclusive data regarding the efficacy of sewage treatment in virus inactivation or the degree of virus persistence in treated sewage or sludge. The virucidal efficacy of sewage disinfection can often be limited due to virus aggregation and association with particulate matter. Estimates of the efficacy of secondary sewage treatment and disinfection for enteric viruses are variable and depend upon many factors (193). Sewage spills, storm-related contamination of surface waters, illicit discharge of waste, and residential septic system failures are widely recognized as the leading sources of surface water and groundwater contamination, which may impact fruit and vegetable production (209). Since many produce items are subject to extensive human handling during harvesting, preharvest food safety strategies should also focus on food handlers. These strategies should address issues such as ready access to clean, on-site toilet and hand-washing facilities; restriction of the presence children and ill workers in production fields; and adequate employee education (69, 123, 214, 215).

Most decontamination efforts applied to fresh produce have focused on washing. The water used for washing produce should be free of fecal contamination. Ice used for the transport and storage of produce should be produced under hygienic conditions. For produce washing, waters are frequently treated with sanitizers, the most common of which is chlorine, although others such as chlorine dioxide, organic acids, and ozone have been used (197). Washing fresh produce can reduce but not eliminate foodborne pathogens. In some instances, simple home use remedies can be somewhat effective. For example, Lukasik et al. (141) found that water immersion and hand rubbing of virus-contaminated berries at 22 or 43°C removed or inactivated 41 to 79 and 60 to 90% of the input virus (bacteriophages and poliovirus type 1), respectively. Hand rubbing of produce in water held at a higher temperature (43°C) facilitated virus removal. The same investigators reported that automatic dishwashing detergent (0.05%) and 10% vinegar are more effective than 0.05% liquid dishwashing detergent, 2% NaCl, or commercially available produce washes (for domestic use) in removing viruses from strawberries immersed in lukewarm water. In this case, supplementing wash water with vinegar and automatic dishwashing detergent reduced virus contamination by 95 to >99% (141).

Several recent studies have compared chlorine to alternative surface disinfectants that might be more relevant for commercial use. Gulati et al. (87) compared the antiviral activities of commonly used antimicrobials, including 5.25% sodium hypochlorite, quaternary ammonium compounds, and 15% peroxyacetic acid–11% hydrogen peroxide, for rinsing produce. Only peroxyacetic acid–hydrogen peroxide formulations were proven to effectively reduce feline calicivirus titers by 3 $\log_{10}$, although this reduction occurred only at concentrations four times higher than those permitted by the FDA. Lukasik et al. (141) compared trisodium phosphate (1%), cetylpyridinium chloride, 0.5% hydrogen peroxide, and chlorine (0.3 to 300 ppm) as disinfectants in rinse water for washing of strawberries artificially contaminated with representative bacteriophages and poliovirus type 1. Hydrogen peroxide was somewhat effective but caused bleaching of the product. Trisodium phosphate, a common household cleaner, had inactivation rates ranging from 97 to >99% (141). Cetylpyridinium chloride was less effective (85 to 97% virus inactivation), whereas chlorine reduced virus levels by 70.4 to 99.5%, with free chlorine at 200 ppm considered optimal (141). HAV was inactivated by less than 70% when experimentally contaminated strawberries were washed with a 2 ppm chlorine dioxide (ClO_2) solution; similar HAV inactivation was achieved when artificially contaminated strawberry wash water was treated with a similar level of chlorine dioxide for a 30-s contact time (145).

UV radiation has been suggested as an alternative to chlorine for water disinfection. Feline calicivirus and poliovirus type 1 are highly susceptible to UV radiation, with 3-$\log_{10}$ reductions achieved with treatment of contaminated water with doses of 23 and 40 mJ/cm^2, respectively (77, 211). Bench-scale treatment of water with 0.37 mg of ozone/liter at pH 7 and 5°C for 10 s reduced NoV RNA and poliovirus type 1 by at least 3 $\log_{10}$ (201).

These promising technologies may someday prove to be viable alternatives to chlorine for the disinfection of produce wash water. As with shellfish, ionizing radiation is not a promising technology for inactivation of viruses in produce, as high doses are relatively ineffective and likely to negatively impact product quality.

Currently available treatments for chemical disinfection of food surfaces are not very promising. Not only is efficacy relatively poor, but some of the agents (such as hydrogen peroxide) proposed for decontamination cause unacceptable organoleptic changes to the product. Product-specific surface morphology and physiological characteristics make disinfection complicated; leafy vegetables can be more difficult to decontaminate because of their rough or wrinkled surfaces, and small fruits like raspberries and blackberries have many porous and irregular surfaces that can entrap virus particles (184). Furthermore, disinfectants incorporated in wash water may not be effective in removing or inactivating viruses that have penetrated through the skin of produce or those that might have entered tissues through cuts and abrasions.

RTE Foods

Adherence to strict hygienic practices when handling and preparing foods is critical to control viral contamination of RTE food products. The first line of defense is employee education and regular supervision. The importance of controlling viral contamination is further illustrated by several modifications to the most recent (2005) revised *Food Code*, including (i) exclusion and restriction of ill employees, (ii) emphasis on hand-washing procedures, and (iii) prohibition of bare-hand contact with RTE foods (214, 215). Restricting food service personnel from working directly with food or food equipment and providing a sick leave policy that allows workers to stay home while ill are to be encouraged. Virus shedding by individuals with asymptomatic infections, however, is still a concern. Glove use (and changing of gloves if damaged or soiled), frequent hand washing, and prevention of cross contamination during handling and preparation of foods are essential. One of the most important considerations in hand hygiene is contamination of fingernails, as virus-containing fecal material can be harbored here. Epidemiologic evidence indicates that long, artificial fingernails are a risk factor for NoV infection outbreaks (158). Lin et al. (132) determined that viruses are most effectively removed from artificially inoculated fingernails (natural and artificial) by scrubbing the nails with soap and a nailbrush. Frequently handled objects such as taps and doorknobs should be decontaminated, and bleach-sensitive items should be cleaned with detergent and hot water (36).

Hand disinfection remains an important issue. Reliable hand decontamination relies on three important factors, i.e., (i) an effective disinfecting agent, (ii) adequate use instructions, and (iii) regular compliance (191). Compliance is a consistent concern for the industry. Although hand washing helps prevent and control the transfer of viruses, hand-washing agents differ in their abilities to inactivate viruses (as reviewed in reference 168). Because hands are believed to play an important role in virus transmission, the efficiency of several hand-washing agents has been investigated. Mbithi et al. (150) found a medicated liquid soap containing 0.3% triclosan to be the most effective hand-washing agent to inactivate HAV and poliovirus on finger pads, reducing populations by 92 and 98%, respectively. Although alcohols are not as effective as hand-washing agents, these same investigators did find products containing high levels of alcohol to be effective in preventing the transfer of HAV and poliovirus from finger pads to stainless steel surfaces. Recent studies revealed that contact of 1-propanol or ethanol solutions with hands for 30 s reduces feline calicivirus titers by as much as 4 $\log_{10}$ (76). The same investigators found that an increase in alcohol concentration does not correlate with an increase in disinfection effectiveness. Bidawid et al. (28) reported that ethanol-based hand rubs contribute to the reduction of feline calicivirus spread. However, when considered collectively, the data for inactivation of viruses by alcohol are quite variable, and presently no hand disinfectant appears to be ideal and many are not even as effective as water and soap alone (27, 28). Some investigators suggest that alcohol-based hand washes are more useful for decontaminating hands between hand-washing events (28). Residual moisture on hands after hand washing can promote the transfer of viruses (206), suggesting that the means by which hands are dried can be a critical consideration for virus removal, especially if the hand-washing agents are not very effective (12, 41). Many believe that there is a need to establish new standards for the selection of effective formulations for hand-washing agents with particular reference to antiviral activities, particularly with respect to labeling (150, 192).

For control of cross contamination involving virus-contaminated surfaces, retailers rely on surface disinfection. There are data suggesting that most chemical disinfectants used in both institutional and domestic environments do not effectively inactivate HAV or the NoVs (reviewed in reference 168) (Table 27.2). Studies on several commercial disinfectants used in the food industry revealed that only products containing glutaraldehyde and sodium hypochlorite are effective in HAV and rotavirus inactivation and that their efficacy improves when the compounds are used at high concentrations for a relatively long time and at approximately room temperature (3). Of the 20 formulations tested by Mbithi et al. (147), only 2% gluteraldehyde, a quaternary ammonium compound containing 23% HCl, and sodium hypochlorite with free chlorine in excess of 5,000 ppm had demonstrable virucidal efficacy against HAV. Likewise, among a variety of disinfectants used at manufacturer-recommended concentrations, only sodium hypochlorite with available chlorine at 5,000 ppm was effective in reducing the feline calicivirus titer by more than 3 $\log_{10}$ (87). In other feline calicivirus studies, domestic bleach (5%

Table 27.2 Representative studies on enteric viral disinfection

Contact surface(s)	Viral agent(s)[a]	Chemical(s) studied	Contact time (min)	Log reduction	Reference
Stainless steel	HAV	2% Glutaraldehyde, 0.4% quaternary ammonium compound (with 23% HCl), 6% sodium hypochlorite (with chlorine at >5,000 ppm)	1	>4	147
Polystyrene	HAV, rotavirus	Sodium chlorite (effective at concn recommended by manufacturer)	1 at 28°C	~3	3
Stainless steel	FCV	5.25% Sodium hypochlorite (effective only with chlorine at 5,000 ppm), 4.75% o-benzyl p-chlorophenol and 4.75% o-phenylphenol or 5% o-benzyl p-chlorophenol and 10.5% o-phenylphenol (effective only at concn 4 times that recommended by manufacturer)	10 at 22°C	>3 >5	87
Aluminum, stainless steel, copper, polyvinyl chloride, high-density polyethylene	HAV	Sodium hypochlorite at 3,000 ppm	5 at 22°C	>5	109

[a] FCV, feline calicivirus.

sodium hypochlorite or available chlorine at 1,000 ppm) was the most effective surface disinfectant and also had the shortest required contact time (1 min) of the disinfectants tested (63, 190). However, the specific $\log_{10}$ reduction achieved by using sodium hypochlorite varies with virus type, surface type, and study design (63, 87, 147, 190). In addition, the presence of fecal material increases virus survival by protecting viruses from the action of disinfectants (94, 113). As with hand disinfection, there is no agent currently available that can make specific antiviral claims with respect to inactivation of the nonenveloped enteric viruses.

DETECTION OF VIRAL CONTAMINATION

A potential control strategy, albeit in developmental stages, would be the detection of viral contamination in foods. Unfortunately, the development of methods to detect enteric viruses in foods has lagged behind that of methods for clinical detection. Initial work directed at the diagnosis of the epidemiologically relevant foodborne virus infections from clinical specimens (e.g., stool or acute or convalescent sera) used techniques of electron microscopy, immune electron microscopy, or enzyme immunoassay. These were successful, largely because individuals in the acute phases of disease shed large amounts of virus. In the 1990s, the first molecular amplification methods were reported, and these quickly became the methods of choice for detection of the nonculturable human NoVs. For hepatitis A, most of the clinical diagnostic methods involve detection of IgM anti-HAV antibodies in sera (210), although some molecular amplification methods are available.

Unfortunately, the detection of viral contamination in foods remains a challenge. Unlike those in clinical samples, viruses in contaminated foods are typically present at low levels. Early virus detection approaches relied on concentrating and purifying the viruses from the food commodity, followed by detection based on infectivity in susceptible, live laboratory hosts. Most mammalian cell culture work was done to detect human enteroviruses in contaminated shellfish, as these viruses are easy to propagate on secondary cell lines of primate origin. However, both the NoVs and wild-type HAV cannot be cultivated in vitro, lacking both animal and mammalian cell culture hosts. As a result, in vitro virus amplification, much like traditional cultural enrichment for the detection of foodborne bacterial pathogens, cannot be used. The general approach applied to the detection of viruses in foods begins with separating and concentrating viruses from the food matrix. This step is then followed by molecular amplification, in which case the viral nucleic acid (RNA for HAV and the NoVs) is exponentially amplified in a sensitive and specific reaction. After amplification, amplicons are detected and their identity is confirmed. The major steps for the detection of viruses in foods can be described as follows: (i) virus concentration and purification, (ii) nucleic acid extraction, (iii) detection of amplicons, and (iv) confirmation of amplicon identity.

Virus Concentration and Purification

Specific challenges to virus concentration and purification include high sample volumes in relation to small amplification volumes, low levels of contamination, and the presence of residual food components that can later compromise detection (reviewed in references 64, 65, and 104). Consequently, when designing virus extraction methods, the analyst attempts to decrease sample volume and eliminate matrix-associated interfering substances or toxic compounds while simultaneously recovering a large proportion of the viruses present in the food sample. In order to achieve these goals, sample manipulations are undertaken that capitalize on the behavior of enteric viruses to act as proteins in solutions, to cosediment by simple centrifugation when adsorbed to larger particles, and to remain infectious at extremes of pH or in the presence of organic solvents. Early virus concentration and purification methods were applied to shellfish, and recent endeavors have included a broader range of at-risk food items.

Virus extraction methods are frequently preceded by crude filtration, such as that through cheesecloth, the purpose of which is to remove large matrix-associated particulates. Almost all virus concentration and purification methods incorporate the principles of adsorption and elution. By manipulating pH and ionic conditions, one can favor virus adsorption to, or elution from, the food matrix. This step is then followed by simple centrifugation ($<5,000 \times g$), after which the virus-containing phase (either precipitate or eluate) is recovered for further purification.

The suspension resulting from filtration, adsorption, and/or elution is usually of reduced volume and contains less of the initial food matrix. Further virus concentration can be achieved by precipitation steps using a number of alternative approaches. For instance, lowering the pH (so-called acid precipitation) to a value approximating the virus's isoelectric point will result in precipitation of some of the residual matrix, along with the virus. Precipitation can also be achieved by the addition of protein-precipitating agents such as polyethylene glycol. Organic flocculation, a method that results in the formation of a gelatinous "floc" to which the viruses adsorb, can also be used to promote virus precipitation. In all cases, the virus-containing solid materials can be readily separated from other matrix components by sedimenting with centrifugation at relatively low centrifugal forces ($<10,000 \times g$).

Another method to remove portions of the food matrix, particularly lipid materials, is organic solvent extraction. In this case, the virus-containing suspension is mixed with an equal volume of organic solvents such as chloroform, trichlorotrifluoroethane (Freon), or more environmentally friendly solvents such as Vertrel (Dupont, New Britain, Conn.); mixed; and centrifuged, with recovery of the aqueous phase. Alternative commercial virus purification agents (106, 127) as well as the cationic detergent cetyltrimethylammonium bromide have been used to remove matrix-associated polysaccharides (20, 106), and Sephadex, cellulose, and Chelex can remove salts and proteins (53, 225). Ultrafiltration has been used by some as a final concentration step, but the filters are prone to clogging. Some investigators have incorporated an immunocapture step to concentrate viruses from foods (26, 54, 119, 138). Figure 27.4 is an illustration of a prototype virus concentration method as applied to foods; there are many others. It should be noted that an effective method usually combines multiple steps and the choice and sequence of virus extraction steps depend on the food product being processed.

After virus extraction, sample volume reductions from 10- to 1,000-fold can be obtained and, theoretically, a 25-g food sample can be reduced to 25 µl to 2.5 ml in volume. The recovery of infectious viruses ranges from as low as 1 to 2% to as high as 90% (reviewed by Jaykus [104]). However, recovery efficiency is both virus and matrix specific. For example, HAV recovery tends to be low in comparison to relatively high recovery of human enteroviruses (e.g., poliovirus); simple sample matrices such as lettuce tend to be easier to work with than more complex matrices such as sandwiches and RTE salads. All methods have their own advantages and disadvantages. For example, immunocapture methods tend to be simple, requiring fewer sample manipulations with lower chances of virus loss and the recovery of infectious virus; however, they may be hindered by high specificity since broadly reactive antisera are not available for the NoVs. Presently, there is no universal extraction method that can be applied to all foods, especially when virus contamination levels are low.

Nucleic Acid Extraction

An efficient nucleic acid extraction step is critical because amplification efficiency is dependent on both the purity of the target template and the quantity of target molecules

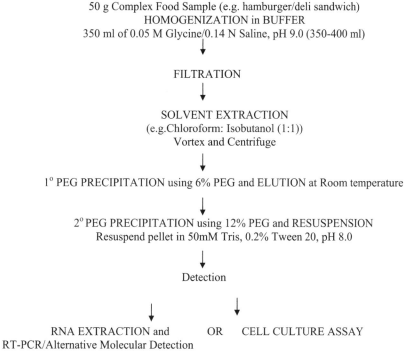

50 g Complex Food Sample (e.g. hamburger/deli sandwich)
HOMOGENIZATION in BUFFER
350 ml of 0.05 M Glycine/0.14 N Saline, pH 9.0 (350-400 ml)

↓

FILTRATION

↓

SOLVENT EXTRACTION
(e.g.Chloroform: Isobutanol (1:1))
Vortex and Centrifuge

↓

1° PEG PRECIPITATION using 6% PEG and ELUTION at Room temperature

↓

2° PEG PRECIPITATION using 12% PEG and RESUSPENSION
Resuspend pellet in 50mM Tris, 0.2% Tween 20, pH 8.0

↓

Detection

↓   ↓

RNA EXTRACTION and   OR   CELL CULTURE ASSAY
RT-PCR/Alternative Molecular Detection

Figure 27.4 Representative processing scheme for the concentration of viruses from complex food matrices (modification of method reported by Leggitt and Jaykus [127]). PEG, polyethylene glycol.

obtained from the sample. Nucleic acid extraction also provides additional sample concentration, usually by 10- to 100-fold. Over the last decade, guanidinium thiocyanate has emerged as the best agent for RNA extraction because it is effective at deproteinizing and purifying nucleic acids and also protects RNA against destruction by native RNases. Unfortunately, even after RNA extraction, matrix-associated inhibitors frequently remain, in which case the RNA can be further purified by binding to silica particles, glass powder, or magnetic poly(dT) beads, with subsequent RNA elution (57, 78, 86, 116, 118, 126). Others have used the metal-chelating agent Chelex-100 or Sephadex column chromatography for further RNA purification (210). Studies have reported various degrees of performance for different RNA extraction methods (16, 90, 181, 189). A few investigators have attempted to bypass upstream sample preparation by applying direct nucleic acid extraction to contaminated foods without prior virus concentration and purification steps (196). Although this method is easy and well suited to compositionally simple food matrices, residual amplification inhibitors frequently remain.

Detection of Amplicons

Because most of the enteric viruses have an RNA genome, the most commonly used detection method is RT-PCR. More recently, nucleic acid sequence-based amplification, a novel RNA amplification method, also has been applied to the detection of viruses in foods (86, 107, 108). The sensitivity and the specificity of molecular amplification are dependent on the efficiency of the upstream processing methods, the purity and yield of RNA, and the choice of primers. This is particularly important because the levels of viral contamination in foods are much lower than those in clinical matrices. Furthermore, even the best virus extraction and concentration methods frequently leave behind residual matrix-associated material. The result is a high degree of nonspecific amplification, which decreases both the sensitivity and the specificity of the molecular amplification assay. To alleviate this problem, the use of primers with low degeneracy and high melting temperatures is recommended. Indeed, a single primer pair that will detect all the epidemiologically relevant viruses within a single genus is ideal. This has been achieved for HAV detection, for which most primers are broadly reactive, targeting the viral VP1-2A junction (reviewed by D'Souza et al. [64, 65]).

However, the genetic diversity of the NoVs makes universal primer design quite challenging. Early NoV primers were designed to amplify conserved regions in the RNA-dependent RNA polymerase gene of the virus and tended to be strain specific (53, 156, 157). With the increased availability of sequence data from a variety of NoV strains, more broadly reactive primers have been designed. Some of these target the RNA polymerase genes (10, 81, 129), whereas others target the viral capsid region (84, 122, 218). Many, if not most, of these are degenerate. The interested reader is urged to consult D'Souza et al. (65) for a full description of candidate NoV primers and probes. Few studies have been done to compare primer performance with clinical specimens (89) and even fewer to compare performance in food concentrates (189). For epidemiologic investigations, primer sets targeting both RNA polymerase and capsid genes, namely, SR33/46 and Mon381/383 (51, 198, 199, 200), respectively, are frequently applied. Nested amplifications are almost always required to achieve detection limits necessary for naturally contaminated products (82, 89).

Confirmation of Amplicon Identity

In most cases, the products of molecular amplification are visualized by agarose gel electrophoresis. However, since nonspecific products of amplification are a constant concern, confirmation of amplicon sequence is highly recommended. The most widely applied confirmatory tool is Southern hybridization using specific oligoprobes internal to the amplicon. These probes can be enzyme labeled for colorimetric, luminescent, or fluorescent detection. Descriptions of various modifications of these protocols have been published, including those of DNA enzyme immunoassay (107, 195) and liquid electrochemiluminescence hybridization (108). For the genetically diverse NoVs, an oligonucleotide array dot blot format has been reported which facilitates amplicon confirmation and strain genotyping in a single assay (219, 220). Other amplicon confirmation methods include "nested" reactions (79, 89, 208) and restriction endonuclease digestion of RT-PCR products (79, 89). Definitive epidemiologic association now relies on direct sequencing of amplicons obtained from clinical and environmental or food samples (reviewed in references 65 and 189).

Within the last several years, real-time molecular amplification methods have come to the forefront (142). The term "real time" refers to the simultaneous detection and confirmation of amplicon identity as the amplification progresses. Real-time RT-PCR can be designed for the detection of double-stranded DNA (SYBR green 1-based methods) or for the detection of specific sequences within the amplicon (TaqMan or molecular beacons). In both cases, fluorescent dyes are used (142). Prototype real-time PCR and RT-PCR amplification technologies have been developed for the detection of HAV (46) and the NoVs (111, 137, 156, 167, 185), although only a few of these have been applied to the detection of viruses in food or environmental samples (4, 111, 137, 163) or to the simultaneous detection of enteric viruses by using multiplex real-time RT-PCR (24, 167).

Although significant strides have been made in the development of methods to detect viruses in foods, these methods are still not ideal (reviewed by Richards [183]). Indeed, these protocols are applied infrequently and usually only in response to known or suspected foodborne disease outbreaks. The most important reasons for their limited use include (i) the inability of molecular amplification methods to discriminate between infectious and inactivated virus; (ii) the lack of widely accepted, collaboratively tested methods, as available methods have been used only to a limited extent (19, 130, 219); (iii) the requirement that most methods be product specific, meaning that universal approaches do not exist; and (iv) the cost of and need for highly trained personnel. Although virus concentration from foods is likely to remain product dependent, research is still needed to develop and refine the prototype methods into collaboratively tested protocols and to do the necessary technology transfer to provide more universal access to the methods.

CONCLUSIONS

Epidemiologic evidence indicates that foods play a critical role in the transmission of human enteric viruses. This is in large part due to improvements in the ability to detect viral pathogens in clinical specimens, mostly related to recent advances in molecular techniques. Nonetheless, because both HAV and the NoVs can be transmitted by a variety of nonfoodborne routes, the relative importance of foods is poorly understood. Unfortunately, widespread epidemiologic surveillance for viral gastroenteritis is complicated by the fact that the disease, while miserable, is usually short-lived and recovery is complete, so many cases are not detected by health care professionals. In addition, tests for the detection of NoVs in stool specimens are not routinely applied. At the current time, sporadic cases of viral gastroenteritis usually go unrecognized and only relatively widespread outbreaks are investigated. When outbreaks are investigated, rarely is the implicated food tested for viral contamination, and when it is, detection limits are less than optimal. Improved and more widespread reporting and investigation of foodborne viral disease outbreaks, and targeted epidemiologic studies to identify the risk factors for viral gastroenteritis, would improve our understanding of attribution.

Even if better detection of viral disease were implemented, the availability of effective controls remains a stumbling block. For molluscan shellfish, the absence of a reliable indicator system for the classification of harvesting waters, along with the relative failure of conventional control strategies (depuration, relaying, and mild heat), suggests that occasional viral contamination of this commodity will continue to occur. High-pressure processing, a most promising control, should be further investigated. Fresh produce can become contaminated with viruses through a variety of routes, but water and workers' hands are the most important concerns. Water contamination and subsequent produce contamination are likely to occur sporadically, and implementing effective measures to improve personal hygiene at the farm level is complicated. More accessible, well-maintained hand-washing stations and mandatory hand-washing breaks at regular intervals may help reduce hand contamination of agricultural workers. This is also true for RTE foods, as a poorly paid, unskilled labor force results in high turnover and has little incentive for change. The lack of truly effective antiviral disinfectant agents for both hands and surfaces is a further complication. In point of fact, multiple control strategies will be needed, all of which will require significant resources. Perhaps the best place to start is with education. As more people understand how and why enteric viruses spread, then day-to-day changes such as routine hand washing and hand and surface sanitation will become a way of life and lead to better control of this substantial food safety and public health problem.

References

1. Abad, F. X., R. M. Pinto, and A. Bosch. 1994. Survival of enteric viruses on environmental fomites. *Appl. Environ. Microbiol.* **60:**3704–3710.

2. Abad, F. X., R. M. Pinto, R. Gajardo, and A. Bosch. 1997. Viruses in mussels: public health implications and depuration. *J. Food Prot.* **60:**677–681.

3. Abad, F. X., R. M. Pinto, and A. Bosch. 1997. Disinfection of human enteric viruses on fomites. *FEMS Microbiol. Lett.* **156:**107–111.

4. Abd El Galil, K. H., M. A. El Sokkary, S. M. Kheira, A. M. Salazar, M. V. Yates, W. Chen, and A. Mulchandani. 2004. Combined immunomagnetic separation-molecular beacon-reverse-transcription-PCR assay for detection of hepatitis A virus from environmental samples. *Appl. Environ. Microbiol.* **70:**4371–4374.

5. Agus, S. G., R. Dolin, R. G. Wyatt, A. J. Tousimis, and R. S. Northup. 1973. Acute infectious nonbacterial gastroenteritis: intestinal histopathology. Histologic and enzymatic alternations during illness produced by the Norwalk agent in man. *Ann. Intern. Med.* **79:**18–25.

6. Allwood, P. B., Y. S. Malik, C. W. Hedberg, and S. M. Goyal. 2003. Survival of F-specific RNA coliphage, feline calicivirus, and *Escherichia coli* in water: a comparative study. *Appl. Environ. Microbiol.* **69:**5707–5710.

7. **Ambert-Balay, K. F. Bon, F. Le Guyader, P. Pothier, and E. Kohli.** 2005. Characterization of new recombinant noroviruses. *J. Clin. Microbiol.* **43:**5179–5186

8. **Amon, J. J., R. Devasia, G. Xia, O. V. Nainan, S. Hall, B. Lawson, J. S. Wolthuis, P. D. M. MacDonald, C. W. Shepard, I. T. Williams, G. L. Armstrong, J. A. Gabel, P. Erwin, L. Sheeler, W. Kuhnert, P. Patel, G. Vaughan, A. Weltman, A. S. Craig, B. P. Bell, and A. Fiore.** 2005. Molecular epidemiology of foodborne hepatitis A outbreaks in the United States, 2003. *J. Infect. Dis.* **192:**1323–1330.

9. **Anderson, A. D., V. D. Garrett, J. Sobel, S. S. Monroe, R. L. Fankhauser, K. J. Schwab, J. S. Bresee, P. S. Mead, C. Higgins, J. Campana, and R. Glass.** 2001. Multistate outbreak of Norwalk-like virus gastroenteritis associated with a common caterer. *Am. J. Epidemiol.* **154:**1013–1019.

10. **Ando, T., S. S. Monroe, J. R. Gentsch, Q. Jin, D. C. Lewis, and R. I. Glass.** 1995. Detection and differentiation of antigenically distinct small round-structured viruses (Norwalk-like viruses) by reverse transcription-PCR and Southern hybridization. *J. Clin. Microbiol.* **33:**64–71.

11. **Ansari, S. A., S. A. Sattar, V. S. Springthorpe, G. A. Wells, and W. Tostowaryk.** 1988. Rotavirus survival on human hands and transfer of infectious virus to animate and nonporous inanimate surfaces. *J. Clin. Microbiol.* **26:**1513–1518.

12. **Ansari, S. A., S. A. Sattar, V. S. Springthorpe, G. A. Wells, and W. Tostowaryk.** 1989. In vivo protocol for testing efficacy of hand-washing agents against viruses and bacteria: experiments with rotavirus and *Escherichia coli*. *Appl. Environ. Microbiol.* **55:**3113–3118.

13. **Appleton, H.** 1994. Norwalk virus and the small round viruses causing foodborne gastroenteritis, p. 57–79. *In* Y. H. Hui, J. R. Gorham, K. D. Murrell, and D. O. Cliver (ed.), *Foodborne Disease Handbook,* vol. 2. *Diseases Caused by Viruses, Parasites, and Fungi.* Marcel Dekker, New York, N.Y.

14. **Appleton, H., and M. S. Pereira.** 1977. A possible virus aetiology in outbreaks of food poisoning from cockles. *Lancet* i(8015):780–781.

15. **Arauz-Ruiz, P., L. Sundqvist, Z. Garcia, L. Taylor, K. Visona, H. Norder, and L. O. Magnius.** 2001. Presumed common source outbreaks of hepatitis A in an endemic area confirmed by limited sequencing within the VP1 region. *J. Med. Virol.* **65:**449–456.

16. **Arnal, C., V. Ferre-Aubineau, B. Besse, B. Mignotte, L. Schwartzbrod, and S. Billaudel.** 1999. Comparison of 7 RNA extraction methods on stool and shellfish samples prior to hepatitis A virus amplification. *J. Virol. Methods* **77:**17–26.

17. **Asanaka, M., R. L. Atmar, V. Ruvolo, S. E. Crawford, F. H. Neill, and M. K. Estes.** 2005. Replication and packaging of Norwalk virus RNA in cultured mammalian cells. *Proc. Natl. Acad. Sci. USA* **102:**10327–10332.

18. **Asher, L. V., L. N. Binn, T. L. Mensing, R. H. Marchwicki, R. A. Vassell, and G. D. Young.** 1995. Pathogenesis of hepatitis A in orally inoculated owl monkeys (Aotus trivirgatus). *J. Med. Virol.* **47:**260–268.

19. **Atmar, R. L., F. H. Neill, C. M. Woodley, R. Manger, G. S. Fout, W. Burkhardt, L. Leja, E. R. McGovern, F. LeGuyader, T. G. Metcalf, and M. K. Estes.** 1996. Collaborative evaluation of a method for the detection of Norwalk virus in shellfish tissues by PCR. *Appl. Environ. Microbiol.* **62:**254–258.

20. **Atmar, R. L., F. H. Neill, J. L. Romalde, F. Le Guyader, C. M. Woodley, T. G. Metcalf, and M. K. Estes.** 1995. Detection of Norwalk virus and hepatitis A virus in shellfish tissues with the PCR. *Appl. Environ. Microbiol.* **61:**3014–3018.

21. **Bean, N. H., J. S. Goulding, M. T. Daniels, and F. J. Angulo.** 1997. Surveillance for foodborne disease outbreaks—United States, 1988–1992. *J. Food Prot.* **60:**1265–1286.

22. **Becker, K. M., C. L. Moe, K. L. Southwick, and J. N. MacCormack.** 2000. Transmission of Norwalk virus during a football game. *N. Engl. J. Med.* **343:**1223–1227.

23. **Berg, D. E., M. A. Kohn, T. A. Farley, and L. M. McFarland.** 2000. Multistate outbreaks of acute gastroenteritis traced to fecal-contaminated oysters harvested in Louisiana. *J. Infect. Dis.* **181**(Suppl. 2):S381–S386.

24. **Beuret, C.** 2004. Simultaneous detection of enteric viruses by multiplex real-time RT-PCR. *J. Virol. Methods* **115:**1–8.

25. **Beuret, C., A. Baumgartner, and J. Schluep.** 2003. Virus-contaminated oysters: a three-month monitoring of oysters imported to Switzerland. *Appl. Environ. Microbiol.* **69:**2292–2297.

26. **Bidawid, S., J. M. Farber, and S. A. Sattar.** 2000. Rapid concentration of detection of hepatitis A virus from lettuce and strawberries. *J. Virol. Methods* **88:**175–185.

27. **Bidawid, S., J. M. Farber, and S. A. Sattar.** 2000. Contamination of foods by food handlers: experiments on hepatitis A virus transfer to food and its interruption. *Appl. Environ. Microbiol.* **66:**2759–2763.

28. **Bidawid, S., N. Malik, O. Adegbunrin, S. A. Sattar, and J. M. Farber.** 2004. Norovirus cross-contamination during food handling and interruption of virus transfer by hand antisepsis: experiments with feline calicivirus as a surrogate. *J. Food Prot.* **67:**103–109.

29. **Bresee, J. S., M. A. Widdowson, S. S. Monroe, and R. I. Glass.** 2002. Foodborne viral gastroenteritis: challenges and opportunities. *Clin. Infect. Dis.* **35:**748–753.

30. **Calder, L., G. Simmons, C. Thornley, P. Taylor, K. Pritchard, G. Greening, and J. Bishop.** 2003. An outbreak of hepatitis A associated with consumption of raw blueberries. *Epidemiol. Infect.* **131:**745–751.

31. **Carrique-Mas, J., Y. Andersson, B. Petersen, K.-O. Hedlund, N. Sjogren, and J. Giesecke.** 2003. A Norwalk-like virus waterborne community outbreak in a Swedish village during peak holiday season. *Epidemiol. Infect.* **131:**737–744.

32. **Caul, E. O.** 1994. Small round structured viruses: airborne transmission and hospital control. *Lancet* **343:**1240–1241.

33. **Centers for Disease Control and Prevention.** 2003. Outbreaks of gastroenteritis associated with noroviruses on cruise ships—United States, 2002. *JAMA* **289:**167–169.

34. **Centers for Disease Control and Prevention.** 2003. Hepatitis A outbreak associated with green onions at a restaurant—Monaca, Pennsylvania. *Morb. Mortal. Wkly. Rep.* **52:**1–5.

35. **Centers for Disease Control and Prevention.** 2005. Hepatitis A vaccination coverage among children aged 24–35 months—United States, 2003. *Morb. Mortal. Wkly. Rep.* **54:**141–144.

36. Chadwick, P. R., G. Beards, D. Brown, E. O. Caul, J. Cheesbrough, I. Clarke, A. Curry, S. O'Brien, K. Quigley, J. Sellwood, and D. Westmoreland. 2000. Management of hospital outbreaks of gastroenteritis due to small round structured viruses. *J. Hosp. Infect.* **45:** 1–10.

37. Chalmers, J. W. T., and J. H. McMillan. 1995. An outbreak of viral gastroenteritis associated with adequately prepared oysters. *Epidemiol. Infect.* **115:**163–167.

38. Cheesbrough, J. S., J. Green, C. I. Gallimore, P. A. Wright, and D. W. G. Brown. 2000. Widespread environmental contamination with Norwalk-like viruses (NLV) detected in a prolonged hotel outbreak of gastroenteritis. *Epidemiol. Infect.* **125:**93–98.

39. Chung, H., L. Jaykus, G. Loveleace, and M. D. Sobsey. 1998. Bacteriophages and bacteria as indicators of enteric viruses in oysters and their harvest waters. *Water Sci. Technol.* **38:**37–44.

40. Clarke, I. N., and P. R. Lambden. 2000. Organization and expression of calicivirus genes. *J. Infect. Dis.* **181**(Suppl. 2): S309–S316.

41. Cliver, D. O. 1997. Virus transmission via food. *Food Technol.* **51**(4):71–78.

42. Cliver, D. O. 2001. Foodborne viruses, p. 501–513. *In* M. P. Doyle, L. R. Beuchat, and T. J. Montville (ed.), *Food Microbiology: Fundamentals and Frontiers*, 2nd ed. ASM Press, Washington, D.C.

43. Cliver, D. O., and S. M. Matsui. 2002. Viruses, p. 161–175. *In* D. O. Cliver and H. P. Reimann (ed.), *Foodborne Diseases*, 2nd ed. Academic Press, Amsterdam, The Netherlands.

44. Committee on Infectious Diseases. 1991. Hepatitis A, p. 234–237. *In* G. Peter (ed.), *Report of the Committee on Infectious Diseases*. American Academy of Pediatrics, Elk Grove Village, Ill.

45. Cook, D. W., and R. D. Ellender. 1986. Relaying to decrease the concentration of oyster-associated pathogens. *J. Food Prot.* **49:** 196–202.

46. Costa-Mattioli, M., S. Monpoeho, E. Nicand, M.-H. Aleman, S. Billaudel, and V. Ferré. 2002. Quantification and duration of viraemia during hepatitis A infection as determined by real-time RT-PCR. *J. Viral Hepat.* **9:** 101–106.

47. Croci, L., D. De Medici, G. Morace, A. Fiore, C. Scalfaro, F. Beneduce, and L. Toti. 2002. The survival of hepatitis A virus in fresh produce. *Int. J. Food Microbiol.* **73:**29–34.

48. Croci, L., M. Ciccozzi, D. De Medici, S. Di Pasquale, A. Fiore, A. Mele, and L. Toti. 1999. Inactivation of hepatitis A virus in heat-treated mussels. *J. Appl. Microbiol.* **87:**884–888.

49. Cromeans, T., O. V. Nainan, H. A. Fields, M. O. Favorov, and H. S. Margolis. 1994. Hepatitis A and E viruses, p. 1–56. *In* Y. H. Hui, J. R. Gorham, K. D. Murrell, and D. O. Cliver (ed.), *Foodborne Disease Handbook*, vol. 2. *Diseases Caused by Viruses, Parasites, and Fungi*. Marcel Dekker, New York, N.Y.

50. Cuthbert, J. A. 2002. Hepatitis A: old and new. *Clin. Microbiol. Rev.* **14:**38–58.

51. Daniels, N. A., D. A. Bergmire-Sweat, K. J. Schwab, K. A. Hendricks, S. Reddy, S. M. Rowe, R. L. Fankhauser, S. S. Monroe, R. L. Atmar, R. I. Glass, and P. Mead. 2000. A foodborne outbreak of gastroenteritis associated with Norwalk-like viruses: first molecular traceback to deli sandwiches contaminated during preparation. *J. Infect. Dis.* **181:**1467–1470.

52. De Leon, R., and C. P. Gerba. 1990. Viral disease transmission by seafood, p. 639–662. *In* J. O. Nriagu and M. S. Simmons (ed.), *Food Contamination from Environmental Sources*. Wiley and Sons, New York, N.Y.

53. De Leon, R., S. M. Matsui, R. S. Baric, J. E. Herrmann, N. R. Blacklow, H. B. Greenberg, and M. D. Sobsey. 1992. Detection of Norwalk virus in stool specimens by reverse transcriptase-polymerase chain reaction and nonradioactive oligoprobes. *J. Clin. Microbiol.* **30:**3151–3157.

54. Deng, M. Y., S. P. Day, and D. O. Cliver. 1994. Detection of hepatitis A virus in environmental samples by antigen-capture PCR. *Appl. Environ. Microbiol.* **60:**1927–1933.

55. Dentinger, C. M., W. A. Bower, O. V. Nainan, S. M. Cotter, G. Myers, L. M. Dubusky, S. Fowler, E. D. Salehi, and B. P. Bell. 2001. An outbreak of hepatitis A associated with green onions. *J. Infect. Dis.* **183:**1273–1276.

56. Desselberger, U. 2000. Rotaviruses: methods and protocols, p. 1–8. *In* J. Gray and U. Desselberger (ed.), *Methods in Molecular Medicine*, vol. 34. Humana Press, Totowa, N.J.

57. DiPinto, A., V. T. Forte, G. M. Tantillo, V. Terio, and C. Buonavoglia. 2003. Detection of hepatitis A virus in shellfish (*Mytilus galloprovincialis*) with RT-PCR. *J. Food Prot.* **66:**1681–1685.

58. Divizia, M., R. Gabrielli, A. Macaluso, B. Bagnato, L. Palombi, E. Buonomo, F. Cenko, L. Leno, S. Bino, A. Basha, and A. Pana. 2005. Nucleotide correlation between HAV isolates from human patients and environmental samples. *J. Med. Virol.* **75:**8–12.

59. Dolin, R., N. R. Blacklow, H. DuPont, R. F. Buscho, R. G. Wyatt, J. A. Kasel, R. Hornick, and R. M. Chanock. 1972. Biological properties of Norwalk agent of acute infectious nonbacterial gastroenteritis. *Proc. Soc. Exp. Biol. Med.* **140:**578–583.

60. Dore, W. J., and D. N. Lees. 1995. Behavior of *Escherichia coli* and male-specific bacteriophage in environmentally contaminated bivalve mollusks before and after depuration. *Appl. Environ. Microbiol.* **61:**2830–2834.

61. Dore, W. J., K. Henshilwood, and D. N. Lees. 2000. Evaluation of F-specific RNA bacteriophage as a candidate human enteric virus indicator for bivalve molluscan shellfish. *Appl. Environ. Microbiol.* **66:**1280–1285.

62. Dore, W. J., M. Mackie, and D. N. Lees. 2003. Levels of male-specific RNA bacteriophage and *Escherichia coli* in molluscan bivalve shellfish from commercial harvesting areas. *Lett. Appl. Microbiol.* **36:**92–96.

63. Doultree, J. C., J. D. Druce, C. J. Birch, D. S. Bowden, and J. A. Marshall. 1999. Inactivation of feline calicivirus, a Norwalk virus surrogate. *J. Hosp. Infect.* **41:**51–57.

64. D'Souza, D. H., and L. Jaykus. 2006. Molecular approaches for the detection of foodborne viral pathogens, p. 91–117. *In* J. Maurer (ed.), *PCR Methods for Food Microbiology*. Springer Science, New York, N.Y.

65. D'Souza, D. H., J. Jean, and L. Jaykus. 2006. Methods for detection of viral and parasitic pathogens in food, p. 188-1–188-23. *In* Y. H. Hui (ed.), *Handbook of Food*

Science, Technology and Engineering, vol. 4. CRC Press, Boca Raton, Fla.

66. Duizer, E., P. Bijkerk, B. Rockx, A. de Groot, F. Twisk, and M. Koopmans. 2004. Inactivation of caliciviruses. *Appl. Environ. Microbiol.* **70:**4538–4543.

67. Fankhauser, R. L., S. S. Monroe, J. S. Noel, C. D. Humphrey, J. S. Bresee, U. D. Parashar, T. Ando, and R. I. Glass. 2002. Epidemiologic and molecular trends of the "Norwalk-like viruses" associated with outbreaks of gastroenteritis in the United States. *J. Infect. Dis.* **186:**1–7.

68. Feinstone, S. M. 1996. Hepatitis A: epidemiology and prevention. *Eur. J. Gastroenterol. Hepatol.* **8:**300–305.

69. Fiore, A. E. 2004. Hepatitis A transmitted by food. *Clin. Infect. Dis.* **38:**705–715.

70. Fleet, G. H., P. Heiskanen, I. Reid, and K. A. Buckle. 2000. Foodborne viral illness—status in Australia. *Int. J. Food Microbiol.* **59:**127–136.

71. Formiga-Cruz, M., G. Tofino-Quesada, S. Bofill-Mas, D. N. Lees, K. Henshilwood, A. K. Allard, A.-C. Conden-Hansson, B. E. Hernroth, A. Vantarakis, A. Tsibouxi, M. Papapetropoulou, M. D. Furones, and R. Girones. 2002. Distribution of human virus contamination in shellfish from different growing areas in Greece, Spain, Sweden, and the United Kingdom. *Appl. Environ. Microbiol.* **68:**5990–5998.

72. Formiga-Cruz, M., A. K. Allard, A. C. Conden-Hansson, K. Henshilwood, B. E. Hernroth, J. Jofre, D. N. Lees, F. Lucena, M. Papapetropoulou, R. E. Rangdale, A. Tsibouxi, A. Vantarakis, and R. Girones. 2003. Evaluation of potential indicators of viral contamination in shellfish and their applicability to diverse geographical areas. *Appl. Environ. Microbiol.* **69:**1556–1563.

73. Friedman, D. S., D. Heisey-Grove, F. Argyros, E. Berl, J. Nsubuga, T. Stiles, J. Fontana, R. S. Beard, A. Moneow, M. E. McGrath, H. Sutherby, R. C. Dicker, A. DeMaria, and B. T. Matyas. 2005. An outbreak of norovirus gastroenteritis associated with wedding cakes. *Epidemiol. Infect.* **133:**1057–1063.

74. Gallimore, C. I., D. Cubitt, N. du Plessis, and J. J. Gray. 2004. Asymptomatic and symptomatic excretion of noroviruses during a hospital outbreak of gastroenteritis. *J. Clin. Microbiol.* **42:**2271–2274.

75. Gaulin, C., M. Frigon, D. Poirier, and C. Fournier. 1999. Transmission of calicivirus by a foodhandler in the presymptomatic phase of illness. *Epidemiol. Infect.* **123:**475–478.

76. Gehrke, C., J. Steinmann, and P. Goroncy-Bermes. 2004. Inactivation of feline calicivirus, a surrogate of norovirus (formerly Norwalk-like viruses), by different types of alcohol in vitro and in vivo. *J. Hosp. Infect.* **56:**49–55.

77. Gerba, C. P., D. M. Gramos, and N. Nwachuku. 2002. Comparative inactivation of enteroviruses and adenovirus 2 by UV light. *Appl. Environ. Microbiol.* **68:**5167–5169.

78. Goswami, B. B., M. Kulka, D. Ngo, P. Istafanos, and T. A. Cebula. 2002. A polymerase chain reaction-based method for the detection of hepatitis A virus in produce and shellfish. *J. Food Prot.* **65:**393–402.

79. Gouvea, V., N. Santos, M. Carmo-Timenetsky, and M. K. Estes. 1994. Identification of Norwalk virus in artificially seeded shellfish and selected foods. *J. Virol. Methods* **48:**177–187.

80. Graham, D. Y., X. Jiang, T. Tanaka, A. R. Opekun, H. P. Madore, and M. K. Estes. 1994. Norwalk virus infection of volunteers: new insights based on improved assays. *J. Infect. Dis.* **170:**34–43.

81. Green, J., C. I. Gallimore, J. P. Norcott, D. Lewis, and D. W. G. Brown. 1995. Broadly reactive reverse transcriptase polymerase chain reaction (RT-PCR) for the diagnosis of SRSV-associated gastroenteritis. *J. Med. Virol.* **47:**392–398.

82. Green, J., K. Henshilwood, C. I. Gallimore, D. W. G. Brown, and D. N. Lees. 1998. A nested reverse transcriptase PCR assay for detection of small round-structured viruses in environmentally contaminated molluscan shellfish. *Appl. Environ. Microbiol.* **64:**858–863.

83. Green, J., P. A. Wright, C. I. Gallimore, O. Mitchell, P. Morgan-Capner, and D. W. G. Brown. 1998. The role of environmental contamination with small round structured viruses in a hospital outbreak investigated by reverse-transcriptase polymerase chain reaction assay. *J. Hosp. Infect.* **39:**39–45.

84. Green, K. Y., R. M. Chanock, and A. Z. Kapikian. 2001. Human caliciviruses, p. 841–874. *In* D. M. Knipe, P. M. Howley, D. E. Griffin, R. A. Lamb, M. A. Martin, B. Roizman, and S. E. Straus (ed.), *Fields Virology.* Lippincott, Williams, and Wilkins, Philadelphia, Pa.

85. Green, S. M., P. R. Lambden, O. Caul, and I. N. Clarke. 1997. Capsid sequence diversity in small structured viruses from recent UK outbreaks of gastroenteritis. *J. Med. Virol.* **52:**14–19.

86. Greene, S. R., C. L. Moe, L.-A. Jaykus, M. Cronin, L. Grosso, and P. van Aarle. 2003. Evaluation of the NucliSens Basic Kit assay for the detection of Norwalk virus RNA in stool specimens. *J. Virol. Methods* **108:**123–131.

87. Gulati, B. R., P. B. Allwood, C. W. Hedberg, and S. M. Goyal. 2001. Efficacy of commonly used disinfectants for the inactivation of calicivirus on strawberry, lettuce, and a food-contact surface. *J. Food Prot.* **64:**1430–1434.

88. Guzewich, J., and M. P. Ross. 6 Oct. 1999, revision date. *Evaluation of Risks Related to Microbiological Contamination of Ready-to-Eat Foods by Food Preparation Workers and the Effectiveness of Interventions to Minimize Those Risks.* [Online.] Food and Drug Administration, Washington, D.C. http://vm.cfsan.fda.gov/~ear/rterisk.html.

89. Hafliger, D., M. Gilgen, J. Luthy, and P. Hubner. 1997. Seminested RT-PCR systems for small round-structured viruses in faecal specimens. *J. Virol. Methods* **57:**195–201.

90. Hale, A. D., J. Green, and D. W. G. Brown. 1996. Comparison of 4 RNA extraction methods for the detection of small round structured viruses in faecal specimens. *J. Virol. Methods* **57:**195–201.

91. Halliday, M. L., L. Y. Kang, T. K. Zhou, M. D. Hu, Q. C. Pan, T. Y. Fu, Y. S. Huang, and S. L. Hu. 1991. An epidemic of hepatitis A attributable to the ingestion of raw clams in Shanghai, China. *J. Infect. Dis.* **164:**852–859.

92. Harrington, P. R., L. Lindesmith, B. Yount, C. L. Moe, J. LePendu, and R. S. Baric. 2003. Norovirus attachment, susceptibility and vaccine design. *Recent Res. Dev. Virol.* **5:**19–44.

93. **Harrington, P. R., L. Lindesmith, B. Yount, C. L. Moe, and R. S. Baric.** 2002. Binding of Norwalk virus-like particles to ABH histo-blood group antigens is blocked by antisera from infected human volunteers or experimentally vaccinated mice. *J. Virol.* **76:**12335–12343.

94. **Hejkal, T. W., F. M. Wellings, P. A. Larock, and A. L. Lewis.** 1979. Survival of poliovirus within organic solids during chlorination. *Appl. Environ. Microbiol.* **38:**114–118.

95. **Hewitt, J., and G. E. Greening.** 2004. Survival and persistence of norovirus, hepatitis A virus, and feline calicivirus in marinated mussels. *J. Food Prot.* **67:**1743–1750.

96. **Hofmann, F., G. Wehrle, H. Berthold, and D. Koster.** 1992. Hepatitis A as an occupational hazard. *Vaccine* **10:**82–84.

97. **Hollinger, F. B., and J. R. Ticehurst.** 1996. Hepatitis A virus, p. 735–774. *In* B. N. Fields, D. M. Knipe, P. M. Howley, M. Chanock, J. L. Melnick, and T. P. Monath (ed.), *Fields Virology*, 3rd ed. Lippincott-Raven Publishers, Philadelphia, Pa.

98. **Hollinger, F. B., and S. U. Emerson.** 2001. Hepatitis A virus, p. 799–840. *In* D. M. Knipe, P. M. Howley, D. E. Griffin, R. A. Lamb, M. A. Martin, B. Roizman, and S. E. Straus (ed.), *Fields Virology*. Lippincott, Williams, and Wilkins, Philadelphia, Pa.

99. **Huang, P., T. Farkas, S. Marionneau, W. Zhong, N. Ruvoen-Clouet, A. L. Morrow, M. Altaye, L. K. Pickering, D. S. Newburg, J. LePendu, and X. Jiang.** 2003. Noroviruses bind to human ABO, Lewis and secretor histo-blood group antigens: identification of 4 distinct strain-specific patterns. *J. Infect. Dis.* **188:**19–31.

100. **Hutin, Y. J., V. Pool, E. H. Cramer, O. V. Nainan, J. Weth, I. T. Williams, S. T. Goldstein, K. F. Gensheimer, B. P. Bell, C. N. Shapiro, M. J. Alter, and H. S. Margolis.** 1999. A multistate foodborne outbreak of hepatitis A. *N. Engl. J. Med.* **340:**595–601.

101. **Hutson, A. M., R. L. Atmar, D. Y. Graham, and M. K. Estes.** 2002. Norwalk virus infection and disease is associated with ABO histo-blood group type. *J. Infect. Dis.* **185:**1335–1337.

102. **Isakbaeva, E. T., M.-A. Widdowson, R. S. Beard, S. N. Bulens, J. Mullins, S. S. Monroe, J. Bresee, P. Sassano, E. H. Cramer, and R. I. Glass.** 2005. Norovirus transmission on cruise ship. *Emerg. Infect. Dis.* **11:**154–158.

103. **Jansen, R. W., G. Siegl, and S. M. Lemon.** 1990. Molecular epidemiology of human hepatitis A virus defined by antigen-capture polymerase chain reaction. *Proc. Natl. Acad. Sci. USA* **87:**2867–2871.

104. **Jaykus, L.** 2000. Detection of human enteric viruses in foods, p. 137–163. *In* S. Sattar (ed.), *Foodborne Diseases Handbook*, 2nd ed., vol. 2. *Viruses, Parasites, Pathogens and HACCP.* Marcel Dekker, New York, N.Y.

105. **Jaykus, L. A., M. T. Hemard, and M. D. Sobsey.** 1994. Human enteric pathogenic viruses, p. 92–153. *In* C. R. Hackney and M. D. Pierson (ed.). *Environmental Indicators and Shellfish Safety.* Chapman and Hall, New York, N.Y.

106. **Jaykus, L. A., R. De Leon, and M. D. Sobsey.** 1996. A virion concentration method for detection of human enteric viruses in oysters by PCR and oligoprobe hybridization. *Appl. Environ. Microbiol.* **62:**2074–2080.

107. **Jean, J., B. Blais, A. Darveau, and I. Fliss.** 2002. Simultaneous detection and identification of hepatitis A virus and rotavirus by multiplex Nucleic Acid Sequence-Based Amplification (NASBA) and microtiter plate hybridization system. *J. Virol. Methods* **105:**123–132.

108. **Jean, J., D. H. D'Souza, and L. Jaykus.** 2004. Multiplex nucleic acid sequence-based amplification (NASBA) for the simultaneous detection of enteric viruses in ready-to-eat food. *Appl. Environ. Microbiol.* **70:**6603–6610.

109. **Jean, J., J.-F. Vachon, O. Moroni, A. Darveau, I. Kukavica-Ibrulj, and I. Fliss.** 2003. Effectiveness of commercial disinfectants for inactivating hepatitis A virus on agri-food surfaces. *J. Food Prot.* **66:**115–119.

110. **Johnson, P. C., J. J. Mathewson, H. L. DuPont, and H. B. Greenberg.** 1990. Multiple-challenge study of host susceptibility to Norwalk gastroenteritis in US adults. *J. Infect. Dis.* **161:**18–21.

111. **Jothikumar, N., T. L. Cromeans, B. H. Robertso, X. J. Meng, and V. R. Hill.** 2006. A broadly reactive one-step real-time RT-PCR assay for rapid and sensitive detection of hepatitis E virus. *J. Virol. Methods* **131:**65–71.

112. **Jothikumar, N., J. A. Lowther, K. Henshilwood, D. N. Lees, V. R. Hill, and J. Vinje.** 2005. Rapid and sensitive detection of noroviruses by using TaqMan-based one-step reverse transcription-PCR assays and application to naturally contaminated shellfish samples. *Appl. Environ. Microbiol.* **71:**1870–1875.

113. **Kampf, G., D. Grotheer, and J. Steinmann.** 2005. Efficacy of three ethanol-based hand rubs against feline calicivirus, a surrogate virus for norovirus. *J. Hosp. Infect.* **60:**140–149.

114. **Kapikian, A. Z., M. K. Estes, and R. M. Chanock.** 1996. Norwalk group of viruses, p. 783–810. *In* B. N. Fields, D. M. Knipe, P. M. Howley, M. Chanock, J. L. Melnick, and T. P. Monath (ed.), *Fields Virology*, 3rd ed. Lippincott-Raven Publishers, Philadelphia, Pa.

115. **Keswick, B., L. K. Pickering, H. L. DuPont, and W. E. Woodward.** 1983. Survival and detection of rotaviruses on environmental surfaces in day care centers. *Appl. Environ. Microbiol.* **46:**813–816.

116. **Kingsley, D. H., and G. P. Richards.** 2001. Rapid and efficient extraction method for reverse transcription-PCR detection of hepatitis A and Norwalk-like viruses in shellfish. *Appl. Environ. Microbiol.* **67:**4152–4157.

117. **Kingsley, D. H., and G. P. Richards.** 2003. Persistence of hepatitis A virus in oysters. *J. Food Prot.* **66:**331–334.

118. **Kingsley, D. H., D. G. Hoover, E. Papafragkou, and G. P. Richards.** 2002. Inactivation of hepatitis A virus and a calicivirus by high hydrostatic pressure. *J. Food Prot.* **65:**1605–1609.

119. **Kobayashi, S., K. Natori, N. Takeda, and K. Sakae.** 2004. Immunomagnetic capture RT-PCR for detection of norovirus from foods implicated in a foodborne outbreak. *Microbiol. Immunol.* **48:**201–204.

120. **Koff, R. S., and H. S. Sear.** 1967. Internal temperature of steamed clams. *N. Engl. J. Med.* **276:**737–739.

121. **Kohn, M. A., T. A. Farley, T. Ando, M. Curtis, S. A. Wilson, Q. Jin, S. S. Monroe, R. C. Baron, L. M. McFarland, and R. I. Glass.** 1995. An outbreak of Norwalk virus

gastroenteritis associated with eating raw oysters. *JAMA* **273**:466–471.

122. Kojima, S., T. Kageyama, S. Fukushi, F. B. Hoshino, M. Shinohara, K. Uchida, K. Natori, N. Takeda, and K. Katayama. 2002. Genogroup-specific PCR primers for detection of Norwalk-like viruses. *J. Virol. Methods* **100**:107–114.

123. Koopmans, M., and E. Duizer. 2004. Foodborne viruses: an emerging problem. *Int. J. Food Microbiol.* **90**:23–41.

124. Kurdziel, A. S., N. Wilkinson, S. Langton, and N. Cook. 2001. Survival of poliovirus on soft fruit and salad vegetables. *J. Food Prot.* **64**:706–709.

125. Lees, D. 2000. Viruses and bivalve shellfish. *Int. J. Food Microbiol.* **59**:81–116.

126. Lees, D. N., K. Henshilwood, J. Green, C. I. Gallimore, and D. W. Brown. 1995. Detection of small round-structured viruses in shellfish by reverse transcription-PCR. *Appl. Environ. Microbiol.* **61**:4418–4424.

127. Leggitt, P. R., and L.-A. Jaykus. 2000. Detection methods for human enteric viruses in representative foods. *J. Food Prot.* **63**:1738–1744.

128. Legnani, P., E. Leoni, D. Lev, R. Rossi, G. C. Villa, and P. Bisbini. 1998. Distribution of indicator bacteria and bacteriophages in shellfish and shellfish-growing waters. *J. Appl. Microbiol.* **85**:790–798.

129. LeGuyader, F., F. H. Neill, M. K. Estes, S. S. Monroe, T. Ando, and R. L. Atmar. 1996. Detection and analysis of a small round-structured virus strain in oysters implicated in an outbreak of acute gastroenteritis. *Appl. Environ. Microbiol.* **62**:4268–4272.

130. LeGuyader, F., L. Haugarreau, L. Miossec, E. Dubois, and M. Pommepuy. 2000. Three-year study to assess human enteric viruses in shellfish. *Appl. Environ. Microbiol.* **66**:3241–3248.

131. Lemon, S. M. 1997. Type A hepatitis: epidemiology, diagnosis, and prevention. *Clin. Chem.* **43**:1494–1499.

132. Lin, C. M., F. M. Wu, H. K. Kim, M. P. Doyle, B. S. Michaels, and L. K. Williams. 2003. A comparison of hand washing techniques to remove *Escherichia coli* and caliciviruses under natural and artificial fingernails. *J. Food Prot.* **66**:2296–2301.

133. Lindesmith, L., C. Moe, J. LePendu, J. A. Frelinger, J. Treanor, and R. Baric. 2005. Cellular and humoral immunity following Snow Mountain Virus challenge. *J. Virol.* **79**:2900–2909.

134. Lindesmith, L., C. Moe, S. Marionneau, N. Ruvoen, X. Jiang, J. Lindblad, P. Stewart, J. LePendu, and R. Baric. 2003. Human susceptibility and resistance to Norwalk virus infection. *Nat. Med.* **9**:548–553.

135. Liu, B., P. Maywood, L. Gupta, and B. Campbell. 2003. An outbreak of Norwalk-like virus gastroenteritis in an aged-care residential hostel. *N. S. W. Public Health Bull.* **14**:105–109.

136. Lo, S. V., A. M. Connolly, S. R. Palmer, D. Wright, P. D. Thomas, and D. Joynson. 1994. The role of the pre-symptomatic food handler in a common source outbreak of food-borne SRSV gastroenteritis in a group of hospitals. *Epidemiol. Infect.* **113**:513–521.

137. Loisy, F., R. L. Atmar, P. Guillon, P. Le Cann, M. Pommepuy, and F. Le Guyader. 2005. Real-time RT-PCR for norovirus screening in shellfish. *J. Virol. Methods* **123**:1–7.

138. Lopez-Sabater, E. I., M. Y. Deng, and D. O. Cliver. 1997. Magnetic immunoseparation PCR assay (MIPA) for detection of hepatitis A virus (HAV) in American oyster (*Crassostrea virginica*). *Lett. Appl. Microbiol.* **24**:101–104.

139. Lopman, B., H. Vennema, E. Kohli, P. Pothier, A. Sanchez, A. Negredo, J. Buesa, E. Schreier, M. Reacher, D. Brown, J. Gray, M. Iturriza, C. Gallimore, B. Bottiger, K. O. Hedlund, M. Torven, C. H. von Bonsdorff, L. Maunula, M. Poljsak-Prijatelj, J. Zimsek, G. Reuter, G. Szucs, B. Melegh, L. Svennson, Y. van Duijnhoven, and M. Koopmans. 2004. Increase in viral gastroenteritis outbreaks in Europe and epidemic spread of new norovirus variant. *Lancet* **363**:682–688.

140. Lopman, B. A., M. H. Reacher, Y. van Duijnhoven, F. X. Hanon, D. Brown, and M. Koopmans. 2003. Viral gastroenteritis outbreaks in Europe, 1995–2000. *Emerg. Infect. Dis.* **9**:90–96.

141. Lukasik, J., M. L. Bradley, T. M. Scott, M. Dea, A. Koo, W.-Y. Hsu, J. A. Bartz, and S. R. Farrah. 2003. Reduction of poliovirus 1, bacteriophages, *Salmonella montevideo*, and *Escherichia coli* O157:H7 on strawberries by physical and disinfectant washes. *J. Food Prot.* **66**:188–193.

142. Mackay, I. M., K. E. Arden, and A. Nitsche. 2002. Real-time PCR in virology. *Nucleic Acids Res.* **30**:1292–1305.

143. Maier, K., P. Gabriel, E. Koscielniak, Y.-D. Stierhof, K. H. Wiedmann, K. B. Flehmig, and A. Vallbracht. 1998. Human gamma interferon production by cytotoxic T lymphocytes sensitized during hepatitis A virus infection. *J. Virol.* **62**:3756–3763.

144. Mallet, J. C., L. E. Beghian, T. G. Metcalf, and J. D. Kaylor. 1991. Potential of irradiation technology for improving shellfish sanitation. *J. Food Safety* **11**:231–245.

145. Mariam, T. W., and D. O. Cliver. 2000. Hepatitis A virus control in strawberry products. *Dairy Food Environ. Sanit.* **20**:612–616.

146. Marks, P. J., I. B. Vipond, D. Carlisle, D. Deakin, R. E. Fey, and E. O. Caul. 2000. Evidence for airborne transmission of Norwalk-like virus (NLV) in a hotel restaurant. *Epidemiol. Infect.* **124**:481–487.

147. Mbithi, J. N., V. S. Springthorpe, and S. A. Sattar. 1990. Chemical disinfection of hepatitis A virus on environmental surfaces. *Appl. Environ. Microbiol.* **56**:3601–3604.

148. Mbithi, J. N., V. S. Springthorpe, and S. A. Sattar. 1991. Effect of relative humidity and air temperature on survival of hepatitis A virus on environmental surfaces. *Appl. Environ. Microbiol.* **57**:1394–1399.

149. Mbithi, J. N., V. S. Springthorpe, J. R. Boulet, and S. A. Sattar. 1992. Survival of hepatitis A virus on human hands and its transfer on contact with animate and inanimate surfaces. *J. Clin. Microbiol.* **30**:757–763.

150. Mbithi, J. N., V. S. Springthorpe, and S. A. Sattar. 1993. Comparative in vivo efficiencies of hand-washing agents

against hepatitis A virus (HM-175) and poliovirus type 1 (Sabin). *Appl. Environ. Microbiol.* **59:**3463–3469.

151. **McDonnell, S., K. B. Kirkland, W. G. Hlady, C. Aristeguieta, R. S. Hopcins, S. S. Monroe, and R. I. Glass.** 1997. Failure of cooking to prevent shellfish-associated viral gastroenteritis. *Arch. Intern. Med.* **157:**111–116.

152. **Mead, P. S., L. Slutsker, V. Dietz, L. F. McCaig, J. S. Bresee, C. Shapiro, P. M. Griffin, and R. V. Tauxe.** 1999. Food-related illness and death in the United States. *Emerg. Infect. Dis.* **5:**607–625.

153. **Meltzer, M. I., C. N. Shapiro, E. E. Mast, and C. Arcari.** 2001. The economics of vaccinating restaurant workers against hepatitis A. *Vaccine* **19:**2138–2145.

154. **Metcalf, T. G., J. L. Melnick, and M. K. Estes.** 1995. Environmental virology: from detection of virus in sewage and water by isolation to identification by molecular biology—a trip of over 50 years. *Annu. Rev. Microbiol.* **49:**461–487.

155. **Millard, J., H. Appleton, and J. V. Parry.** 1987. Studies on heat inactivation of hepatitis A virus with special reference to shellfish. *Epidemiol. Infect.* **98:**397–414.

156. **Miller, I., R. Gunson, and W. F. Carman.** 2002. Norwalk like virus by light cycler PCR. *J. Clin. Virol.* **25:**231–232.

157. **Moe, C., J. Gentsch, T. Ando, G. Grohmann, S. S. Monroe, X. Jiang, J. Wang, M. K. Estes, Y. Seto, C. Humphrey, S. Stine, and R. I. Glass.** 1994. Application of PCR to detect Norwalk virus in fecal specimens from outbreaks of gastroenteritis. *J. Clin. Microbiol.* **31:**2866–2872.

158. **Moe, C. L.** 2003. Next time you have vomiting patients in your office—consider Noroviruses. *Atlanta Med. J. Med. Assoc. Atlanta* **77:**13–19.

159. **Mohamed, N., S. Belak, K.-O. Hedlund, and J. Blomberg.** 2006. Experience from the development of a diagnostic single tube real-time PCR for human caliciviruses, Norovirus genogroups I and II. *J. Virol. Methods* **132:**69–76. (First published 9 Nov. 2005; doi:10.1016/j.jviromet.2005.09.006.)

160. **Morse, D. L., J. J. Guzewich, H. P. Hanrahan, R. Stricof, M. Shayegani, R. Deibel, J. Grabau, N. A. Nowak, J. E. Herrman, G. Cukor, and N. R. Blacklow.** 1986. Widespread outbreaks of clam and oyster-associated gastroenteritis-role of Norwalk virus. *N. Engl. J. Med.* **314:**678–681.

161. **Muniain-Mujika, I., M. Calvo, F. Lucena, and R. Girones.** 2003. Comparative analysis of viral pathogens and potential indicators in shellfish. *Int. J. Food Microbiol.* **83:**75–85.

162. **Muniain-Mujika, I., R. Girones, G. Tofino-Quesada, M. Calvo, and F. Lucena.** 2002. Depuration dynamics of viruses in shellfish. *Int. J. Food Microbiol.* **77:**125–133.

163. **Myrmel, M. E., M. M. Berg, E. Rimstad, and B. Grinde.** 2004. Detection of enteric viruses in shellfish from the Norwegian coast. *Appl. Environ. Microbiol.* **70:**2678–2684.

164. **New York Department of Health.** 1989. *A Review of Foodborne Disease Outbreaks in New York State, 1988.* Bureau of Community Sanitation and Food Protection, Albany, N.Y.

165. **Niu, M .T., L. B. Polish, B. H. Robertson, B. K. Khanna, B. A. Woodruff, C. N. Shapiro, M. A. Miller, J. D. Smith, J. K. Gedrose, M. J. Alterand, and H. S. Margolis.** 1992. Multistate outbreak of hepatitis A associated with frozen strawberries. *J. Infect. Dis.* **166:**518–524.

166. **Pallansch, M. A., and R. P. Roos.** 2001. Enteroviruses: polioviruses, coxsackieviruses, echoviruses, and newer enteroviruses, p. 723–775. *In* D. M. Knipe and P. M. Howley (ed.), *Fields Virology.* Lippincott Williams and Wilkins, Philadelphia, Pa.

167. **Pang, X. L., J. K. Preiksaitis, and B. Lee.** 2005. Multiplex real time RT-PCR for the detection and quantitation of norovirus genogroups I and II in patients with acute gastroenteritis. *J. Clin. Virol.* **33:**168–171.

168. **Papafragkou, E., D. H. D'Souza, and L. Jaykus.** 2006. Foodborne viruses: prevention and control, p. 289–330. *In* S. M. Goyal (ed.), *Food Virology.* Kluwer Academic/Plenum Publishers, New York, N.Y.

169. **Parashar, U. D., L. Dow, R. L. Fankhauser, C. D. Humphrey, I. J. Miller, T. Ando, K. S. Williams, C. R. Eddy, J. S. Noel, T. Ingram, J. S. Bresee, S. S. Monroe, and R. I. Glass.** 1998. An outbreak of viral gastroenteritis associated with consumption of sandwiches: implications for the control of transmission by food handlers. *Epidemiol. Infect.* **121:**615–621.

170. **Parker, T. D., N. Kitamoto, T. Tanaka, A. M. Hutson, and M. K. Estes.** 2005. Identification of Genogroup I and Genogroup II broadly reactive epitopes on the norovirus capsid. *J. Virol.* **79:**7402–7409.

171. **Parrino, T. A., D. S. Schreiber, J. S. Trier, A. Z. Kapikian, and N. R. Blacklow.** 1977. Clinical immunity in acute gastroenteritis caused by Norwalk agent. *N. Engl. J. Med.* **297:**86–89.

172. **Patterson, W., P. Haswell, P. T. Fryers, and J. Green.** 1997. Outbreak of small round structured virus gastroenteritis arose after kitchen assistant vomited. *Comm. Dis. Rep.* **7:**R101–R103.

173. **Pavia, A. T., L. Nielson, L. Armington, D. J. Thurman, E. Tierney, and C. R. Nichols.** 1990. A community-wide outbreak of hepatitis A in a religious community: impact of mass adminstration of immune globulin. *Am. J. Epidemiol.* **131:**1085–1093.

174. **Pina, S., M. Puig, F. Lucena, J. Jofre, and R. Girones.** 1998. Viral pollution in the environment and in shellfish: human adenovirus detection by PCR as an index of human viruses. *Appl. Environ. Microbiol.* **64:**3376–3382.

175. **Pirtle, E. C., and G. W. Beran.** 1991. Virus survival in the environment. *Rev. Sci. Tech.* **10:**733–748.

176. **Ponka, A., L. Maunula, C. H. Von Bonsdorff, and O. Lyytikainen.** 1999. Outbreak of calicivirus gastroenteritis associated with eating frozen raspberries. *Euro Surveill.* **4:**66–69.

177. **Potasman, I., A. Paz, and M. Odeh.** 2002. Infectious outbreaks associated with bivalve shellfish consumption: a worldwide perspective. *Clin. Infect. Dis.* **35:**921–928.

178. **Power, U. F., and J. K. Collins.** 1989. Differential depuration of poliovirus, *Escherichia coli*, and a coliphage by the common mussel, *Mytilus edulis. Appl. Environ. Microbiol.* **55:**1386–1390.

179. **Pringle, C. R.** 1999. Virus taxonomy—1999. The universal system of virus taxonomy, updated to include the new

proposals ratified by the International Committee on Taxonomy of Viruses during 1998. *Arch. Virol.* **144:** 421–429.

180. **Racaniello, V. R.** 2001. Picornaviridae: the viruses and their replication, p. 529–566. *In* D. M. Knipe and P. M. Howley (ed.), *Fundamental Virology.* Lippincott, Williams and Wilkins, Phildadephia, Pa.

181. **Ribao, C., I. Torrado, M. L. Vilarino, and J. L. Romalde.** 2004. Assessment of different commercial RNA-extraction and RT-PCR kits for detection of hepatitis A virus in mussel tissues. *J. Virol. Methods* **115:**177–182.

182. **Richards, G. P.** 1985. Outbreaks of shellfish-associated enteric virus illness in the United States: requisite for development of viral guidelines. *J. Food Prot.* **48:**815–823.

183. **Richards, G. P.** 1999. Limitations of molecular biological techniques for assessing the virological safety of foods. *J. Food Prot.* **62:**691–697.

184. **Richards, G. P.** 2001. Food-borne pathogens: enteric virus contamination of foods through industrial practices: a primer on intervention strategies. *J. Ind. Microbiol. Biotechnol.* **27:**117–125.

185. **Richards, G. P., M. A. Watson, R. L. Fankhauser, and S. S. Monroe.** 2004. Genogroup I and II noroviruses detected in stool samples by real-time reverse transcription-PCR using highly degenerate universal primers. *Appl. Environ. Microbiol.* **70:**7179–7184.

186. **Robertson, B. H., R. W. Jansen, B. Khanna, A. Totsuka, O. V. Nainan, G. Siegel, A. Widell, H. S. Margolis, S. Isomura, K. Ito, T. Ishizu, Y. Moritsugu, and S. M. Lemon.** 1992. Genetic relatedness of hepatitis A virus strains recovered from different geographical regions. *J. Gen. Virol.* **73:**1365–1377.

187. **Rockx, B., M. deWit, H. Vennema, J. Vinje, E. de Bruin, Y. van Duynhoven, and M. Koopmans.** 2002. Natural history of human calicivirus infection: a prospective cohort study. *Clin. Infect. Dis.* **35:**246–253.

188. **Rosenblum, L. S., I. R. Mirkin, D. T. Allen, S. Safford, and S. C. Hadler.** 1990. A multifocal outbreak of hepatitis A traced to commercially distributed lettuce. *Am. J. Public Health* **80:**1075–1079.

189. **Sair, A. I., D. H. D'Souza, and L. Jaykus.** 2002. Human enteric viruses as causes of foodborne disease. *Compr. Rev. Food Sci. Food Safety* **1:**73–89.

190. **Sattar, S. A.** 2004. Microbicides and the environmental control of nosocomial viral infections. *J. Hosp. Infect.* **56:**564–569.

191. **Sattar, S. A., V. S. Springthorpe, J. Tetro, R. Vashon, and B. Keswick.** 2002. Hygienic hand antiseptics: should they not have activity and label claims against viruses? *Am. J. Infect. Control* **30:**355–372.

192. **Sattar, S. A., and S. A. Ansari.** 2002. The fingerpad protocol to assess hygienic hand antiseptics against viruses. *J. Virol. Methods* **103:**171–181.

193. **Schaub, S. A., and R. K. Oshiro.** 2000. Public health concerns about caliciviruses as waterborne contaminants. *J. Infect. Dis.* **181**(Suppl. 2):374–380.

194. **Schwab, K. J., F. H. Neill, M. K. Estes, T. G. Metcalf, and R. L. Atmar.** 1998. Distribution of Norwalk virus within shellfish following bioaccumulation and subsequent depuration by detection using RT-PCR. *J. Food Prot.* **61:**1674–1680.

195. **Schwab, K. J., F. H. Neill, F. Le Guyader, M. K. Estes, and R. L. Atmar.** 2001. Development of a reverse transcription-PCR-DNA enzyme immunoassay for detection of "Norwalk-like" viruses and hepatitis A virus in stool and shellfish. *Appl. Environ. Microbiol.* **67:**742–749.

196. **Schwab, K. J., F. H. Neill, R. L. Fankhauser, N. A. Daniels, S. S. Monroe, D. A. Bergmire-Sweat, M. K. Estes, and R. L. Atmar.** 2000. Development of methods to detect "Norwalk-like viruses" (NLVs) and hepatitis A virus in delicatessen foods: application to a food-borne NLV outbreak. *Appl. Environ. Microbiol.* **66:**213–218.

197. **Seymour, I. J., and H. Appleton.** 2001. Foodborne viruses and fresh produce. *J. Appl. Microbiol.* **91:**759–773.

198. **Shieh, Y. C., K. R. Calci, and R. S. Baric.** 1999. A method to detect low levels of enteric viruses in contaminated oysters. *Appl. Environ. Microbiol.* **65:**4709–4714.

199. **Shieh, Y. C., S. S. Monroe, R. L. Frankhauser, G. W. Langlois, W. Burkhardt III, and R. S. Baric.** 2000. Detection of Norwalk-like virus in shellfish implicated in illness. *J. Infect. Dis.* **181**(Suppl. 2):S360–S366.

200. **Shieh, Y. C., R. S. Baric, J. W. Woods, and K. R. Calci.** 2003. Molecular surveillance of enterovirus and Norwalk-like virus in oysters relocated to a municipal-sewage-impacted gulf estuary. *Appl. Environ. Microbiol.* **69:**7130–7136.

201. **Shin, G.-A., and M. D. Sobsey.** 2003. Reduction of Norwalk virus, poliovirus 1, and bacteriophage MS2 by ozone disinfection of water. *Appl. Environ. Microbiol.* **69:**3975–3978.

202. **Sinton, L. W., C. H. Hall, P. A. Lynch, and R. J. Davies-Colley.** 2002. Sunlight inactivation of fecal indicator bacteria and bacteriophages from waste stabilization pond effluent in fresh and saline waters. *Appl. Environ. Microbiol.* **68:**1122–1131.

203. **Slomka, M. J., and H. Appleton.** 1998. Feline calicivirus as a model system for heat inactivation studies of small round structured viruses in shellfish. *Epidemiol. Infect.* **121:**401–407.

204. **Sobsey, M. D., and L. A. Jaykus.** 1991. Human enteric viruses and depuration of bivalve molluscs, p. 71–114. *In* W. S. Otwell, G. E. Rodrick, and R. E. Martin (ed.), *Molluscan Shellfish Depuration.* CRC Press, Boca Raton, Fla.

205. **Sockett, P. N., P. A. West, and M. Jacob.** 1985. Shellfish and public health. *PHLS Microbiol. Dig.* **2:**29–35.

206. **Springthorpe, S., and S. Sattar.** 1998. Handwashing: what can we learn from recent research? *Infect. Control Today* **2:**20–28.

207. **Stapleton, J. T.** 1995. Host immune response to hepatitis A virus. *J. Infect. Dis.* **171**(Suppl. 1):S9–S14.

208. **Sugieda, M., K. Nakajima, and S. Nakajima.** 1996. Outbreaks of Norwalk-like virus-associated gastroenteritis traced to shellfish: coexistence of two genotypes in one specimen. *Epidemiol. Infect.* **116:**339–346.

209. **Suslow, T. V., M. P. Oria, L. R. Beuchat, E. H. Garrett, M. E. Parish, L. J. Harris, J. N. Farber, and F. F. Busta.** 2003. Production practices as risk factors in microbial food safety of fresh and fresh-cut produce. *Compr. Rev. Food Sci. Food Safety* **2**(Suppl.):1–40.

210. Svensson, L. 2000. Diagnosis of foodborne viral infections in patients. *Int. J. Food Microbiol.* **59:**117–126.

211. Thurston-Enriquez, J. A., C. N. Haas, J. Jacangelo, K. Riley, and C. P. Gerba. 2003. Inactivation of feline calicivirus and adenovirus type 40 by UV radiation. *Appl. Environ. Microbiol.* **69:**577–582.

212. Tierney, J. T., R. Sullivan, J. T. Peeler, and E. P. Larkin. 1982. Persistence of polioviruses in shellstock and shucked oysters stored at refrigeration temperature. *J. Food Prot.* **45:**1135–1137.

213. Tseng, F., J. S. Leon, J. N. MacCormack, J. M.Maillard, and C. L. Moe. Molecular epidemiology of gastroenteritis outbreaks in North Carolina, United States, 1995–2000. *J. Med. Virol.,* in press.

214. U.S. Food and Drug Administration. 21 Sept. 2005, revision date. *Guidance for Industry: Guide To Minimize Microbial Food Safety Hazards for Fresh Fruits and Vegetables.* [Online.] U.S. Food and Drug Administration, Washington, D.C. http://www.foodsafety.gov/~dms/prodguid.html.

215. U.S. Food and Drug Administration, Center for Food Safety and Applied Nutrition. 10 Oct. 2005, revision date. *Food Code.* [Online.] U.S. Food and Drug Administration, Washington, D.C. http://www.cfsan.fda.gov/~dms/fc05-toc.html.

216. Vallbracht, A., B. Fleischer, and F. W. Busch. 1993. Hepatitis A: hepatotropism and influence on myelopoiesis. *Intervirology* **35:**33–139.

217. van Regenmortel, M. H.V., C. M. Fauquet, and D. H. L. Bishop. 2000. *Virus Taxonomy: Classification and Nomenclature of Viruses,* p. 725–739. Academic Press, San Diego, Calif.

218. Vinje, J., R. A. Hamidjaja, and M. D. Sobsey. 2004. Development and application of a capsid VP1 (region D) based reverse transcription PCR assay for genotyping of genogroup I and II noroviruses. *J. Virol. Methods* **116:**109–117.

219. Vinjé, J., H. Vennema, L. Maunula, C.-H. van Bonsdorff, M. Hoehne, E. Schreier, A. Richards, J. Green, D. Brown, S. S. Beard, S. S. Monroe, E. de Bruin, L. Svensson, and M. P. G. Koopmans. 2003. International collaborative study to compare reverse transcriptase PCR assays for detection and genotyping of noroviruses. *J. Clin. Microbiol.* **41:**1423–1433.

220. Vinje, J., and M. P. Koopmans. 2000. Simultaneous detection and genotyping of "Norwalk-like viruses" by oligonucleotide array in a reverse line blot hybridization format. *J. Clin. Microbiol.* **38:**2595–2601.

221. Ward, R. L., D. I. Bernstein, D. R. Knowlton, J. R. Sherwood, E. C. Young, T. M. Cusack, J. R. Rubino, and G. M. Schiff. 1991. Prevention of surface-to-human transmission of rotaviruses by treatment with disinfectant spray. *J. Clin. Microbiol.* **29:**1991–1996.

222. Wheeler, C., T. M. Vogt, G. L. Armstrong, G. Vaughan, A. Weltman, O. V. Nainan, V. Dato, G. Xia, K. Waller, J. Amon, T. M. Lee, A. Highbaugh-Battle, C. Hembree, S. Evenson, M. A. Ruta, I. T. Williams, A. E. Fiore, and B. P. Bell. 2005. An outbreak of hepatitis A associated with green onions. *N. Engl. J. Med.* **353:**890–897.

223. White, K. E., M. T. Osterholm, J. A. Mariotti, J. A. Korlath, D. H. Lawrence, T. L. Ristinen, and H. B. Greenberg. 1986. A foodborne outbreak of Norwalk virus gastroenteritis: evidence for post recovery transmission. *Am. J. Epidemiol.* **124:**120–126.

224. Widdowson, M.-A., A. Sulka, S. N. Bulens, R. S. Beard, S. S. Chaves, R. Hammond, E. D. P. Salehi, E. Swanson, J. Totaro, R. Woron, P. S. Mead, J. S. Bresee, S. S. Monroe, and R. I. Glass. 2005. Norovirus and foodborne disease, United States, 1991–2000. *Emerg. Infect. Dis.* **11:**95–102. [Online.]

225. Wilde, J., J. Eiden, and R. Yolken.1990. Removal of inhibitory substances from human fecal specimens for detection of group A rotaviruses by reverse transcriptase and polymerase chain reactions. *J. Clin. Microbiol.* **28:**1300–1307.

226. Wilkinson, N., A. S. Kurdziel, S. Langton, E. Needs, and N. Cook. 2001. Resistance of poliovirus to inactivation by high hydrostatic pressures. *Innov. Food Sci. Emerg. Technol.* **2:**95–98.

227. Wyatt, R. G., R. Dolin, N. R. Blacklow, H. L. DuPont, R. F. Buscho, T. S. Thornhill, A. Z. Kapikian, and R. M. Chanock. 1974. Comparison of three agents of acute infectious nonbacterial gastroenteritis by cross-challenge in volunteers. *J. Infect. Dis.* **129:**709–714.

228. Yates, M. V., C. P. Gerba, and L. M. Kelley. 1985. Virus persistence in groundwater. *Appl. Environ. Microbiol.* **49:**778–781.

229. Yotsuyanagi, H., K. Koike, K. Yasuada, K. Moriya, Y. Shintani, H. Fujie, K. Kurokawa, and S. Iino. 1996. Prolonged fecal excretion of hepatitis A virus in adult patients with hepatitis A as determined by polymerase chain reaction. *Hepatology* **24:**10–13.

Prions

Food Microbiology: Fundamentals and Frontiers, 3rd Ed.
Edited by M. P. Doyle and L. R. Beuchat
© 2007 ASM Press, Washington, D.C.

Paul Brown
Linda A. Detwiler

28

Bovine Spongiform Encephalopathy: Consequences for Human Health

Bovine spongiform encephalopathy (BSE), widely known as "mad cow disease," is a chronic, degenerative disease affecting the central nervous systems (CNS) of cattle. Worldwide, there have been more than 190,000 cases in cattle since the disease was first recognized in 1986, of which more than 186,000 have been in the United Kingdom. In addition to the officially reported and confirmed cases, it is estimated that as many as 3 to 5 million animals were infected and may have entered the food and feed chains without being detected (24). BSE has also been reported in most countries of Europe, as well as in Canada, Israel, Japan, and the United States, and it should be noted that the absence of reported cases in a country may not indicate so much the absence of disease as a lack of effective surveillance.

BSE is a member of the family of diseases known as the transmissible spongiform encephalopathies (TSEs), which affect both animals and humans. The animal diseases also include scrapie of sheep and goats, chronic wasting disease (CWD) of cervids, feline spongiform encephalopathy, and transmissible mink encephalopathy. Human TSEs include the prototypic disease kuru, which is limited to Papua New Guinea and is now virtually extinct; Creutzfeldt-Jakob disease (CJD) together with its variant form (vCJD) due to BSE infection; fatal familial insomnia; and the Gerstmann-Sträussler-Scheinker syndrome (Fig. 28.1). All TSEs share many common characteristics, including the following:

- incubation periods ranging from years to decades
- illnesses of weeks to months, with invariable progression to death
- accumulation in the brain (and other tissues) of fibrillar amyloid protein aggregates
- pathological changes confined to the CNS
- the absence of any detectable agent-specific immune response
- transmissibility by either natural or experimental means

Although these features define the fundamental biologic identity of TSEs, some important differences occur in pathogenesis, routes of natural transmission, and distributions of infectivity in tissues, and these differences must be taken into account for diagnosis, prevention, and control.

Paul Brown, National Institutes of Health, Bethesda, MD 20814. **Linda A. Detwiler,** Center for Public and Corporate Veterinary Medicine, Virginia-Maryland Regional College of Veterinary Medicine, 8075 Greenmead Dr., College Park, MD 20742.

ANIMALS

HUMANS

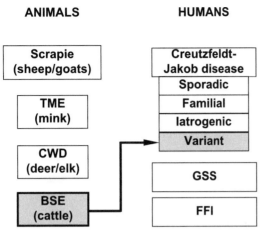

Figure 28.1 The family of transmissible spongiform encephalopathies. TME, transmissible mink encephalopathy; GSS, Gerstmann-Sträussler-Scheinker disease; FFI, fatal familial insomnia.

CHARACTERISTICS OF THE BSE AGENT

The unusual features of TSE have been recognized ever since it became apparent in the 1930s that the incubation period for scrapie in sheep exceeds 2 years and that no pathogenic agent can be seen by light microscopy (16), leading to the description of the agent as a "slow" or "unconventional" virus. The mystery deepened in the 1960s when the results of radiation experiments suggested that infectivity does not depend on nucleic acids (1). Building on this heretical observation, further research in the 1980s led to the discovery that a host-carried (rather than foreign) glycoprotein consisting of 253 amino acids and having a molecular mass of 35 kDa is closely associated with infectivity (4, 33). A quarter century of investigation following this discovery has focused on defining a mechanism by which a protein could "replicate" and so mimic the behavior of a nucleic acid-containing pathogen, in particular, how it could be compatible with the existence of different pathogenic agent strains.

The explanation is still imperfect, but a wealth of accumulating evidence has led to the conclusions that (i) a misfolded form of the protein (PrP^TSE), known as a prion, acts as a template to induce normal protein molecules to cascade into the same misfolded configuration; (ii) maximal infectivity is associated with an aggregate (or polymer) of 14 to 28 misfolded protein molecules; (iii) an as yet unidentified host molecule (chaperone) is probably necessary as a cofactor in replication; (iv) the degree of similarity in the primary structures of the protein in different species influences the ease with which the protein can induce intraspecies disease; and (v) the entire

process appears to occur spontaneously in the sporadic form of the disease but can be initiated (i.e., transmitted) by the introduction of tissue from a diseased host into a healthy host, as is likely to have happened when humans consumed BSE-contaminated meat products.

The agent of BSE has some further interesting features, first among which is the fact that almost all cases of the disease appear to be caused by a single strain that, even when experimentally transmitted to nonbovine species such as sheep and laboratory rodents, "breeds true" and is recognizable by a unique biological fingerprint. When tissue from an infected host is subjected to electrophoresis and Western blotting (using a labeled antibody to visualize the protein), the resulting three-band pattern reveals a combination of characteristics not seen in any other form of TSE: the molecular mass is slightly lower and the topmost (diglycosylated) protein band is more intense than those in sporadic and other forms of disease (Fig. 28.2) (27). The BSE agent can also be recognized by a distinctive combination of the incubation period length and the brain lesion profile when the agent is inoculated into a panel of several different mouse strains (12).

STABILITY OF THE INFECTIOUS AGENT

The agent of BSE shares with other TSE agents the property of unusual resistance to destruction (Table 28.1). None of the standard disinfection methods is effective, including irradiation and exposure to various chemical disinfectants. Even harsher conditions that are capable of inactivating all other known pathogens (including bacterial spores), such as heating under pressure at

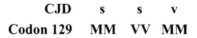

Figure 28.2 Western blots of misfolded prion protein (PrP^TSE) extracted from the brains in two cases of sporadic CJD(s) (types 1 and 2A) and a case of vCJD (v) (type 2B). The combination of the lower molecular mass and the comparatively higher density of the topmost protein band (asterisk) distinguishes vCJD from all other TSEs. M, methionine; V, valine. Courtesy of Mark Head, CJD Surveillance Unit, Western General Hospital, Edinburgh, Scotland.

Table 28.1 Categories of efficacy of various chemical and physical disinfection methods

Method type	Ineffective methods	Partially effective methods	Effective methods
Chemical	Alcohol	Chlorine dioxide	Hypochlorite (5%)
	Ammonia	Gluteraldehyde	NaOH (1–2 N)
	β-Propiolactone	Hypochlorite (0.5%)	Formic acid (100%)
	Detergents	Iodophores	
	Ethylene oxide	Guanidine thiocyanate (4 M)	
	Formaldehyde	Sodium dichloroisocyanurate	
	Hydrochloric acid	Sodium metaperiodate	
	Hydrogen peroxide	Urea (6–8 M)	
	Peracetic acid		
	Permangenate		
	Phenolics		
Physical methods	Boiling (100°C)	Steam heat (121°C)	Steam heat (134°C)
	Microwave radiation	Dry heat (300°C)	Dry heat (>600°C)
	UV radiation	Ionizing radiation (≥50 kG)	

121°C, exposure to dry heat at 600°C, and immersion in 0.1 N NaOH or 0.5% bleach, cannot ensure complete inactivation (7–9, 39, 42). The only procedures known to ensure sterility are exposure to dry heat at 1,000°C (7), immersion in either 1 N NaOH or fresh undiluted bleach (8), and steam heating under pressure at 132°C (9). The preferred method is sequential exposure to both physical and chemical inactivation treatments (38, 40, 60).

It is evident that foodstuffs cannot be subjected to these conditions and remain organoleptically acceptable; hence, the problem of processing food, including beef that may contain the BSE agent, would appear to be insoluble. However, exposure of beef or beef products to high temperature (132°C) under ultrahigh pressure (≥690 MPa = 1,000 atm) retains beef quality and reduces the very low level of infectivity that could realistically be present in commercial products to undetectable or non-transmissible levels (Tables 28.2 and 28.3) (11, 14).

Table 28.2 Proteinase-resistant protein (PrP^{TSE}) and infectivity reductions under various pressure, temperature, and exposure time conditions[a]

Pressure (MPa)[b]	Temperature (°C)	Cooking time (min)	Log$_{10}$ PrP^{TSE} reduction	Log LD$_{50}$ reduction
690	125	3	1.5	2.8
690	120	10	1.5	3.0
1,000	135	3	≥3.0	3.8
1,000	135	10	≥3.0	5.7
1,200	135	3	≥3.5	5.8
1,200	135	10	≥3.5	5.6

[a] A scrapie agent (263K)-infected hamster brain was homogenized in hot dog. The pretreatment (input) infectivity was 8 log LD$_{50}$/g.
[b] 690 MPa = 100,000 lb/in².

BODILY DISTRIBUTION OF INFECTIVITY

To appreciate the various levels of risk inherent in human exposure to BSE, it is necessary to understand the distribution of PrP^{TSE} and levels of infectivity in bovine tissues (Table 28.4). The CNS and its associated tissues (e.g., retinal and ganglionic tissues) invariably have high levels of infectivity, usually in the range of 10^4 to 10^6 mean lethal doses (LD$_{50}$)/g of tissue. The distribution of infectivity in the lymphoreticular system (including tonsils and lymphoid tissue of the third eyelid and distal ileum) is limited and is present in much lower levels, in the range of 10^1 to 10^2 LD$_{50}$/g (20, 50 52; G. A. H. Wells, personal communication). Until recently, no other tissue has been shown to have detectable infectivity (50), which is unlike the situation with sheep scrapie, CWD in deer and elk, and CJD in humans, in which a variety of peripheral nonlymphoid tissues have transmitted disease to experimental animals (6, 26).

However, a new study has revealed the presence of infectivity in an ever-expanding variety of tissues, including peripheral nerves and possibly muscle. It is important that these observations have been made in animals during the clinical stage of disease (13) and, moreover, that the level of infectivity is much lower than in the brain. To evaluate the risk from oral exposure to potentially infectious tissue in the most rigorous manner, it is necessary to perform an infectivity bioassay with healthy, normal animals infected with realistically possible doses given by the oral route. The fact that pigs can be infected by intracerebral inoculation but not by even large oral doses illustrates this point (51). The issue is especially relevant to recent studies in which PrP^{TSE} has been detected in muscle tissue of various species infected either naturally or experimentally with various strains of TSE (2, 3, 45, 47). Only a

Table 28.3 Misfolded-protein (PrP^{TSE}) reductions in various meat products spiked with 263K scrapie agent-infected hamster brain and then cooked for 5 min at 130°C under 690 MPa of pressure

Food product	Time relative to treatment	Weight equivalent of brain (µg) in loaded sample[a]							
		330	100	33	10	3.3	1	0.3	0.1
Hot dog	Before					+	+	±	−
	After	−	−	−	−	−			
Hamburger	Before					+	+	+	−
	After	−	−	−	−	−			
Corned beef	Before				+	+	+	+	−
	After	−	−	−	−	−			
Beef paté	Before				+	+	±	−	
	After	−	−	−	−	−			
Baby food	Before				+	+	±	−	
	After	−	−	−	−	−			
Cat food	Before				+	+	+	±	−
	After	−	−	−	−	−			

[a] The weight of the loaded sample is inversely related to the concentration of PrP^{TSE}. No PrP^{TSE} was detected in any of the treated samples, and thus all samples showed at least a 2.5- to 3-log reduction of PrP^{TSE} after cooking under high pressure.

few such studies have included parallel bioassays, such as intracerebral inoculation of supersensitive transgenic mice that overexpress the protein.

Bovine tissues can enter the human food chain in many ways, both obvious and subtle (Table 28.5). Examples of obvious food tissues are two staples of the human diet, muscle (meat) and milk. Although PrP^{TSE} and/or infectivity has been detected in muscle tissue of humans with vCJD (47) and that of animals with either natural or experimental infections with scrapie (2), CWD of deer (3), or even hamster-adapted BSE (45), muscle from naturally or experimentally infected cattle with BSE has not been shown to harbor the abnormal form of the protein or infectivity, with one exception (13, 20, 52). As methods of detection become more sensitive, additional tissues may be found to have PrP^{TSE} or infectivity. In the event that infectivity is definitively found to be present in muscle from cattle infected with BSE, available scientific and epidemiological evidence to date suggests that it does not appear to exist in amounts sufficient to increase the risk to human health in countries with adequate prevention and control measures.

Evidence that infectivity is not associated with milk includes temporospatial epidemiologic observations failing to detect evidence of significant maternal transmission, clinical observations of calves nursed by infected cows wherein calves have not developed BSE (23, 54), and experimental observations that milk from infected cows does not transmit disease when administered intracerebrally or orally to mice (32, 41).

An example of a not-so-obvious bovine tissue consumed by humans is tallow (and tallow derivatives) that is used in many different food and drug products such as emulsifiers and binders. Like muscle and milk, tallow does not appear to pose a risk to human health. Studies of tallow have shown it not to be infectious even though meat and bone meal (MBM) processed from the same animals was infectious (43, 44). Moreover, should infectivity ever be detected in tallow, its processing into tallow derivatives includes exposure to steam heat and NaOH under conditions that inactivate the BSE agent.

FEED-BORNE AND FOODBORNE OUTBREAKS

Cattle

BSE probably arose sometime in the 1970s or early 1980s, and although the species of origin of BSE is unknown, there is ample epidemiologic evidence to indicate that once cattle were infected, the epidemic was perpetuated by the feeding of TSE agent-contaminated MBM to cattle (56, 57). The disease was then spread to other countries by the exportation of contaminated feed and/or cattle incubating BSE.

It was initially thought that BSE was the result of an increase in the amount of scrapie infectivity entering the animal feed system coincident with a change in the United Kingdom rendering process. This hypothesis has the merit of explaining both the timing and geographic location of the epidemic. However, with time, information

Table 28.4 Distribution of tissue infectivity and PrP[TSE] in naturally occurring human and animal TSE diseases[a]

Tissue(s)	Human TSEs				BSE (cattle)		Scrapie sheep and goats	
	vCJD		Other TSEs					
	Infectivity	PrP[TSE]	Infectivity	PrP[TSE]	Infectivity	PrP[TSE]	Infectivity	PrP[TSE]
CNS tissues that attain a high titer of infectivity in the later stages of TSE and certain tissues anatomically associated with the CNS								
Brain	+	+	+	+	+	+	+	+
Spinal cord	+	+	+	+	+	+	+	+
Retina	NT	+	+	+	+	NT	NT	+
Optic nerve	NT	+	NT	+	+	NT	NT	+
Spinal ganglia	+	+	NT	+	+	NT	NT	+
Trigeminal ganglia	+	+	NT	+	+	NT	NT	+
Pituitary gland	NT	+	+	+	−	NT	+	NT
Dura mater	NT	−	+	−	NT	NT	NT	NT
Peripheral tissues that have tested positive for infectivity and/or PrP[TSE] in at least one form of TSE								
Peripheral nervous system								
Peripheral nerves	+	+	(−)	+	+	+	+	+
Enteric plexuses	NT	+	NT	(−)	NT	+	NT	+
Lymphoreticular tissues								
Spleen	+	+	+	+	−	−	+	+
Lymph nodes	+	+	+	−	−	−	+	+
Tonsil	+	+	NT	−	+	−	+	+
Thymus	NT	+	NT	−	−	NT	+	NT
Alimentary tract								
Esophagus	NT	−	NT	−	−	NT	NT	+
Stomach	NT	−	NT	NT	−	NT	NT	+
Duodenum	NT	−	NT	NT	−	NT	NT	+
Jejunum	NT	+	NT	−	−	NT	NT	+
Ileum	NT	+	NT	−	+	+	+	+
Appendix	−	+	NT	−	NA	NA	NA	NA
Large intestine	+	+	NT	−	−	NT	+	+
Other tissues								
Lung	NT	−	+	−	−	NT	−	−
Liver	NT	−	+	−	−	NT	+	NT
Kidney	NT	−	+	−	−	−	−	−
Adrenal tissue	NT	+	−	−	NT	NT	+	NT
Pancreas	NT	−	NT	−	−	NT	+	NT
Bone marrow	−	−	(−)	−	(+)	NT	+	NT
Skeletal muscle	NT	+	(−)	+	(+)	NT	−	+
Tongue	NT	−	NT	−	−	NT	NT	+
Blood vessels	NT	+	NT	+	−	NT	NT	+
Body fluids								
Cerebrospinal fluid	−	−	+	−	−	NT	+	NT
Blood	+	?	−	?	−	?	+	?
Tissues with no detected infectivity								
Musculoskeletal tissues								
Bone	NT	NT	NT	NT	−	NT	NT	NT
Heart/pericardium	NT	−	−	−	−	NT	−	NT
Tendon	NT	NT	NT	NT	−	NT	NT	NT
Other tissues								
Skin	NT	−	NT	−	−	NT	−	NT
Adipose tissue	NT	−	(−)	−	−	NT	NT	NT
Body fluids, secretions, and excretions								
Milk	NT	NT	(−)	NT	−	−	−	NT
Colostrum	NT	NT	(−)	NT	(−)	−	−	NT
Cord blood	NT	NT	(−)	NT	−	NT	NT	NT
Saliva	NT	−	−	NT	NT	NT	−	NT
Sweat	NT	NT	−	NT	NT	NT	NT	NT
Tears	NT	NT	−	NT	NT	NT	NT	NT
Nasal mucus	NT	−	−	NT	NT	NT	NT	NT
Bile	NT	NT	NT	NT	NT	NT	NT	NT
Urine	NT	NT	−	−	−	NT	NT	NT
Feces	NT	NT	−	NT	−	NT	−	NT

[a] The table is based on data from a WHO Consultation on Tissue Infectivity Distribution in Transmissible Spongiform Encephalopathies held in September 2005 (61). NT, not tested; NA, not applicable; ?, controversial results; ( ), limited data.

Table 28.5 Consumable bovine materials used by humans

Used directly (or after minimal processing)

Meat on the bone (T-bone, ox tail)

Deboned meat

Offals[a] (e.g., liver, lung, heart, kidney, thymus, and brain)

Fat (suet)

Bone (soup and broth)

Brain and endocrine powders in some unregulated "health food" supplements

Used after processing

Milk and milk products (e.g., butter, lactose, and casein)

Rennet (chymosin and pepsin derived from abomasum) used in the production of cheese, whey, and whey products such as lactose

Bone and skin to make gelatin, gelatin derivatives, and collagen

Meat, including tongue, meat extracts used as flavoring, and mechanically recovered meat[b]

Tripe (forestomachs)

Tallow[c] used for frying, used in food as shortening, and used to make tallow derivatives

Fat (suet), beef drippings

Blood and blood products (e.g., hemoglobin to clarify wine)

Intestines (duodenum to rectum for natural sausage casings)

[a] SRMs including offals as well as skull, ganglia, eyes, tonsils, and spinal cord in most countries are compulsorily removed for destruction.

[b] Production of mechanically recovered meat from ruminant animals is banned in many countries. Where used, its inclusion in meat products usually does not exceed 5 to 10% by weight.

[c] Tallow (bovine rendered fat) is one of the two major end products from rendering bovine carcass waste. Tallow derivatives include glycerol, fatty acids and their esters, stearates, polysorbates, and sorbitan esters. They are produced from tallow by hydrolysis at temperatures and pressures that have been shown to sterilize TSE infectivity.

has emerged that has challenged this assumption; specifically, the BSE strain is not identical to any known scrapie strain (12) and at least thus far the disease produced in cattle that are experimentally infected with scrapie does not resemble naturally occurring BSE (17, 18). Hence, although the origin of BSE may never be identified with certainty, current evidence is consistent with the following hypothesized origins:

- sheep or goats with an uncharacterized scrapie strain or a strain that was modified in the course of its adaptation to cattle (30)
- cattle that developed TSE as a consequence of a gene mutation (30, 35)
- other species such as wild ungulates infected with TSE that entered into the feed system (15, 30)

As a consequence of recycled ruminant tissue's having caused the BSE epidemic, changes in feeding practices have led to a precipitous decline in BSE cases in cattle worldwide. For example, there were approximately

1,000 new BSE cases in cattle reported each week in the United Kingdom during 1992 to 1993, whereas there were fewer than 250 confirmed cases during the entire year in 2005. Cases of BSE that continued to emerge after the 1988 feed ban in the United Kingdom (and after the initial feed bans in the European Union) likely were due to cross feeding and cross contamination on farms (22, 28, 29, 37).

Rendering may reduce but does not eliminate infectivity (36, 43, 44). Given that BSE can be orally transmitted to cattle with just 1 mg of infected tissue (21), a very low degree of contamination would be sufficient to recycle the disease. This was (and is) especially true in countries which do not have dedicated lines and equipment to manufacture and process feed for ruminants and nonruminants. Cross feeding is the practice of feeding meal for poultry, pigs, or pet food (which had legally contained ruminant MBM) to cattle on the same farm, usually due to simple human error or negligence. The continuance of BSE cases due to cross contamination and cross feeding of ruminant MBM has resulted in extended feed bans which have prohibited feeding of all mammalian or animal MBM to any animals used for human food.

All evidence indicates that BSE does not spread horizontally among cattle. The question of maternal transmission between an infected cow and her calf remains unanswered. Although it appears that this risk is small to nonexistent, the possibility has not been entirely eliminated (29, 55).

Humans

Human infection with BSE results in a variant form of CJD (vCJD) that was first recognized in the United Kingdom in 1996, approximately 10 years after the BSE epidemic began (58, 59) (Fig. 28.3). As of 1 December 2006, the number of cases had increased to 164, with 21 additional cases in France, 2 cases in Ireland, and single cases in several other European countries and Japan (all of which had also experienced BSE outbreaks). Humans most likely became infected with the agent that causes BSE through the consumption of beef products contaminated by CNS tissue, such as mechanically recovered meat that was pressure extracted from carcasses and often contained spinal cord tissue and paraspinal ganglia in addition to residual muscle shards (10, 48).

The period of maximum exposure to BSE in the United Kingdom occurred in the mid to late 1980s before the ban on high-risk tissues was instituted, and the delay in the peak occurrence of vCJD clearly delineates an average incubation period of about 12 to 15 years, although it is likely that additional cases with longer incubation periods will continue to occur in the future. The delay in the appearance of cases outside the United Kingdom reflects a delay in exposure to BSE in other countries as

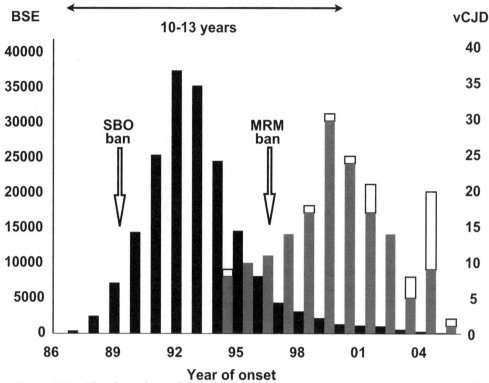

Figure 28.3 The chronology of BSE (black) in the United Kingdom compared to those of vCJD in the United Kingdom (gray) and non-United Kingdom countries (white). The BSE scale is at the left, and the vCJD scale is at the right. SBO, specified bovine offals; MRM, mechanically recovered meat.

a result of importation from the United Kingdom of contaminated beef products and cattle. In particular, France may experience more cases in the future because of the magnitude of its United Kingdom imports. Also, because all of the apparent foodborne cases have so far shown homozygosity, at codon 129 of the prion gene, there is a possibility that heterozygous individuals may have a significantly longer incubation period than the already affected homozygotes and may account for a second wave of cases that has not yet begun.

In three instances, patients with orally acquired disease have transmitted infection via blood transfusions, causing concern in the blood donor and recipient communities (31, 34, 46). Most countries have programs of blood donor deferrals of individuals whose histories include specified lengths of residence in the United Kingdom (and in some cases other European countries) or who have received blood or blood products from individuals in the United Kingdom. Plasma products are considered to pose a negligible risk because experimental evidence indicates large reductions of infectivity by one or more of the processing steps used in their manufacture (25).

CHARACTERISTICS OF DISEASE

Cattle

Affected animals develop a progressive degeneration of the nervous system. They may display changes in temperament, abnormalities of posture and movement, and changes in sensation, including signs of apprehension; nervousness or aggression; incoordination, especially hind-limb ataxia; tremor and difficulty in rising; and hyperesthesia to sound and touch. In addition, many animals have decreased milk production and loss of body condition despite continued appetite. Only a small proportion of affected cattle exhibit what would be considered typical "mad cow" signs. Many suspects show several, but not all, of the signs described above if they are closely observed.

BSE can be mistaken for other conditions or go unnoticed due to the subtlety of the signs. Neurologic, metabolic, or other kinds of disease that affect coordination and gait often predispose an animal to injuries such as broken limbs or soft-tissue damage. If the animal then becomes recumbent, because of a broken leg or torn ligament, the injury may be the prominent or sole presenting sign, and without a complete diagnostic workup and history of disease progression, the true underlying BSE cause of the nonambulatory condition may be overlooked. Even more troubling is the occurrence of a significant number of cases in older, recumbent cows in the absence of any preceding abnormalities.

It is thought that an animal usually becomes infected within the first year of life. The average incubation period of natural BSE is estimated to range from 2 to 8 years. Following the onset of clinical signs, the animal's condition gradually deteriorates until the animal becomes recumbent, dies, or is destroyed. The clinical progression of BSE may last from 2 weeks to 6 months. Most cases in the United Kingdom have occurred in dairy cows between 3 and 6 years of age (53), with the youngest confirmed victim being 20 months of age and the oldest over 22 years of age (19).

Humans

BSE infection in humans (vCJD) has certain clinical and neuropathological features that set it apart from other forms of CJD (10). The most distinctive feature is the young age at the onset of illness (Fig. 28.4), with many adolescents afflicted but only occasional cases in adults older than 50 years of age. This pattern contrasts with the peak occurrence of sporadic CJD in patients between 50 and 70 years of age. The clinical presentation is usually characterized by some form of psychiatric disturbance and complaints of sensory symptoms, particularly limb pain. As the illness progresses, however, the clinical distinction between the sporadic and variant forms of illness becomes progressively blurred. The combination of psychiatric and sensory symptoms in an adolescent or young adult is sufficient to raise a suspicion of vCJD in patients who reside or have resided in countries in which BSE has occurred. The average duration of illness is 14 months, or about twice as long as that of sporadic CJD (Fig. 28.5).

One puzzling feature of vCJD is the predisposition of individuals with a homozygous methionine-coding genotype at polymorphic codon 129 of the prion protein gene. However, alternative genotypes may conceivably be associated with longer incubation periods and only begin to appear in future cases, as suggested by the asymptomatic (pre- or subclinical) infection in a heterozygous recipient of blood from a patient with vCJD. Neuropathological examination, without which a definitive diagnosis cannot be made, reveals diffuse spongiform changes that are especially severe in the basal ganglia, posterior thalamus, and cerebellum, together with a feature unique to vCJD: myriad amyloid plaques surrounded by halos of vacuolation, the so-called "florid" or "daisy" plaques (Fig. 28.6).

GOVERNMENTAL REGULATORY MEASURES

Bovine products and by-products are widely used for both food and pharmaceuticals and hence require the highest level of safety. Because of the hardy nature of the BSE agent and its high potential for cross contamination, the most effective approach to protect bovine products

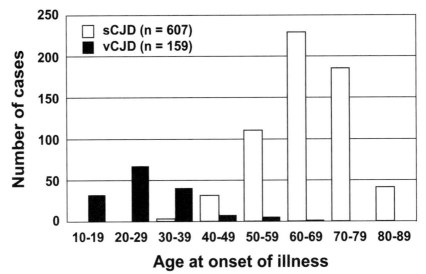

Figure 28.4 Age at onset in United Kingdom cases of vCJD and sporadic CJD, 1994 to 2005. Courtesy of Robert Will, CJD Surveillance Unit, Western General Hospital, Edinburgh, Scotland.

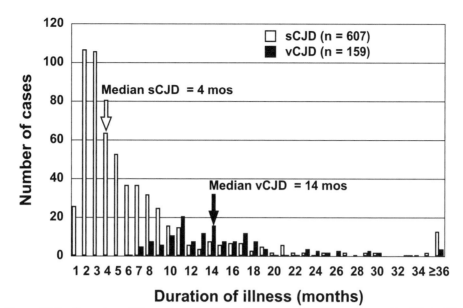

Figure 28.5 Duration of illness in United Kingdom cases of vCJD and sporadic CJD (sCJD), 1994 to 2005. Courtesy of Robert Will, CJD Surveillance Unit, Western General Hospital, Edinburgh, Scotland.

and bovine-derived materials from contamination by BSE is to ensure that infected animals or carcasses never enter processing plants. Because there are presently no diagnostic tools sensitive enough for detection of the disease during its long preclinical incubation, governments must rely primarily on measures to prevent exposure through feed (Table 28.6). These measures may be applied to the national herd in a BSE-free country or to the next generation of cattle in a country that has already identified the disease.

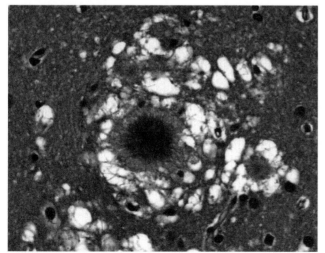

Figure 28.6 The pathognomonic vCJD daisy plaque consisting of a core of amyloid protein surrounded by vacuolar petals. Courtesy of James Ironside, CJD Surveillance Unit, Edinburgh, Scotland.

In a country in which BSE has been identified in cattle, or in which there has been substantial exposure, specific measures must also be taken to protect human health.

Protection of Animal Health

Countries should conduct a risk assessment to evaluate possible exposures from both external and internal sources of BSE and potential recycling of the agent within the cattle production system.

A country with no known exposure to BSE can best protect its national herd by prohibiting the importation of ruminant MBM, including MBM from other species if there is any possibility of cross contamination from BSE, and by prohibiting the importation of cattle from countries with BSE or with high risk factors for BSE. Imported live cattle pose a risk if they are eventually slaughtered, rendered, and incorporated into MBM.

Given that the primary, if not sole, route of BSE transmission is through the feeding of contaminated MBM to cattle, countries with any risk factors need to implement feed controls. The level of restriction is usually dependent upon the amount of contamination thought to be in the system. There are three main factors that can increase the stability of a national feed production system:

1. Feed bans. These regulations can range from the basic prohibition of feeding of ruminant MBM back to ruminants to prohibiting most animal proteins from being fed to all animals used for food production (including fish).

Table 28.6 Principal governmental measures taken to protect human and animal health

Precaution(s)[a]	Date(s) of implementation in:		
	Great Britain[b]	European Union[b]	United States
Ban on ruminant protein in ruminant feed	**July 1988**		
Ban on export of United Kingdom cattle born before July 1988 feed ban		**July 1989**	
Ban on export of United Kingdom cattle of >6 mos of age		Mar. 1990	
BSE active surveillance initiated			May 1990
Ban on SBO[c] in animal and human nutrition	**Sept. 1989/90**		
Ban on export of SBO and feed containing SBO to European Union countries	**Sept. 1990**		
High-risk waste to be rendered at 133°C/3 bar/20 min		Nov. 1990	
Ban on export from United Kingdom of SBO and feed containing SBO to non-European Union countries	July 1991		
Food and Drug Administration initiates series of guidelines to ensure freedom from BSE contamination of source materials for numerous products used by humans, including gelatin, cosmetics, vaccines, and blood and tissue donations			Dec. 1993 through Nov. 1999
Ban on mammalian MBM[d] in ruminant feed		**June 1994**	
Ban on mammalian protein in ruminant feed[e]	**Nov. 1994**		
Rendering methods must sterilize BSE		**Jan. 1995**	
SBO ban broadened to include the bony skull	Aug. 1995		
MRM[f] from bovine vertebral column banned and export prohibited	**Dec. 1995**		
Removal of lymph nodes and visible nervous tissue from United Kingdom bovine meat of >30 mos of age exported to the European Union	Jan. 1996		
Ban on export of all United Kingdom cattle and cattle products except milk		**Mar. 1996**	
Slaughtered and dead cattle >30 mos of age (except certain beef cattle >42 mos of age) ruled unfit for any use (hides for leather excluded from ban)	**Mar. 1996**	Mar. 1996	
Mammalian MBM prohibited from all animal feed and fertilizer	**Mar.–Apr. 1996**		
Mammalian MBM and MBM-containing feed recalled	**June 1996**		
Mammalian waste (except fat) to be rendered at 133°C/3 bars/20 min		July 1996	
BSE cohort cattle in United Kingdom ordered slaughtered and destroyed	Jan. 1997		
Ban on feeding of most mammalian protein to ruminants			**Aug. 1997**
Ban on import of live ruminants and most ruminant products from Europe			Dec. 1997
Replace human plasma and plasma products for use in the United Kingdom with imported sources	Aug. 1998		
Slaughter and destruction of offspring born to BSE-affected United Kingdom cattle after July 1996	Jan. 1999		
Leukodepletion of whole-blood donations from United Kingdom residents	July/Nov. 1999		
Ban on cattle and sheep SRM[g] throughout the European Union		**July 2000**	
Ban on MRM production from any part of cattle, sheep, and goats	**Jan. 2001**		
Ban on mammalian protein in all livestock feed		**Jan. 2001**	
Ban on slaughter techniques that could contaminate cattle carcasses with brain emboli (e.g., pithing and using pneumatic stun guns)		Jan. 2001	Jan. 2004
Immunologic brain examination on all slaughtered cattle of >30 mos of age	**Jan. 2001**		
All downers and SRMs banned from human food			**Jan. 2004**

[a] The most important measures are shown in boldface type.

[b] In Northern Ireland and Scotland, dates of implementation were sometimes different from those shown for England and Wales; also, individual European Community countries often adopted different measures on different dates.

[c] SBO, specified bovine offals (brain, spinal cord, thymus, tonsil, spleen, and intestines from cattle of >6 mos of age).

[d] MBM, high-protein residue produced by rendering.

[e] Some exemptions were made, e.g., milk, blood, and gelatin.

[f] MRM, mechanically recovered meat (residual meat derived from bones, including vertebral column with dorsal root ganglia and possibly spinal cord in situ).

[g] SRM, all tissues shown to be infectious in cattle, sheep, or goats; where infectivity is limited to animals over a certain age, the ban applies to animals over that age. The definition of SRM changes as new information is acquired.

2. Specified risk materials (SRMs) ban. This ban requires that high-infectivity-level tissues such as bovine brain and spinal cord be removed from both the food and feed chains and be destroyed. The intent of this control is to remove the primary source of infectivity from the entire system to prevent the possibility of cross contamination.

3. Regulation of rendering. Although no rendering process can completely remove all detectable infectivity, some are more effective than others. The best procedure studied to date requires a 20-min cooking exposure at 133°C under 3 bars of pressure (36).

Experience in countries that have spent considerable effort to eliminate BSE has underlined the need for an extremely high level of compliance with feed controls in order to remove the agent from the system and prevent new infections in cattle. There can be no complacency.

Protection of Human Health

Standard cooking temperatures do not inactivate the BSE agent, and there is no screening test to guarantee that an infected animal does not enter a processing plant. Therefore, the primary public health protection measure is to remove SRMs from the food supply and to mandate procedures to prevent the possibility of cross contamination between SRMs and edible tissue.

The basic list of SRMs includes brain, spinal cord, trigeminal ganglia, dorsal root ganglia, eye, skull, vertebral column, small intestine, and tonsil. Some countries have also included additional tissues that are known to be infectious in scrapie, even though they have not been shown to be infectious in BSE-affected cattle.

Processing can increase the BSE risk in edible products via cross contamination, especially considering that TSE agents tend to adhere to surfaces and are unusually durable. Standard measures that are used in slaughter plants to reduce the level of microbial contamination, e.g., dipping in 82.2°C water, do not inactivate the BSE agent. The time, temperature, and chemical treatments that would reduce levels of the TSE agent are extremely caustic and could corrode food-processing equipment and adversely affect food quality. However, certain, less onerous practices can at least reduce the risk of cross contamination, including the following:

- The use of dedicated equipment (e.g., knives, hooks, steels, etc.) for SRM removal. Such equipment should never to be used for edible tissues. Color coding of such equipment increases awareness and reduces the opportunity for misuse (Fig. 28.7).
- Removal of SRMs from the slaughter floor. This prevents contamination of the area where fabrication occurs and the edible product that could be contaminated and also provides an opportunity for quality and safety checks

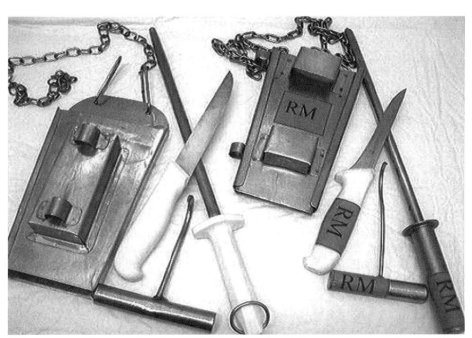

Figure 28.7 Red "RM" identification labels are used to differentiate between slaughterhouse tools used to handle specified risk materials (SRMs) and those for edible product. Courtesy of Ana Carolina Alonso Simplicio de Oliveira, Frigoalta, Brazil.

to ensure that the SRMs have been removed prior to further processing.

- Employee training on proper removal of SRMs and the significance of the procedure

Although testing will not guarantee identification of all infected animals, many countries have used testing at slaughter to reduce the level of infectivity in the system by eliminating the carcasses of animals that test positive. In addition, carcasses that may have been contaminated by close association with the positive carcass can also be eliminated.

In 1996, the sale of beef for human consumption from most cattle over the age of 30 months was prohibited in the United Kingdom. This was done because studies revealed that the levels of infectivity in SRM tissue were very high after this age. The decline of BSE cases in the United Kingdom epidemic in combination with greater practicality of BSE testing has led to a recent regulation change to allow the use of such animals if they test negative for BSE.

The tenacious nature of the agent makes the complete elimination of risk in countries with BSE extremely difficult. Hence, it is imperative for the government and the industry to reduce the avenues for contamination of food or pharmaceutical products by having multiple effectively implemented safeguards in place.

METHODS OF DETECTION

Postmortem

Historically, the diagnosis of BSE relied on the occurrence of clinical signs of the disease confirmed by postmortem histopathological examination of brain tissue (49). A diagnosis could also be made by electron microscopy detection of fibrils in denatured brain extracts, called scrapie-associated fibrils, or SAF, because they were first observed in the brains of scrapie-infected sheep.

In the 1990s, the development of tests to detect the pathognomonic PrP^{TSE} greatly enhanced the diagnostic capabilities for BSE and other TSEs, both because of the tests' improved sensitivity and the fact that they could be used on frozen or partially autolyzed tissue. Two types of tests for the detection of PrP^{TSE} have been internationally approved for the confirmatory diagnosis of BSE: immuno-histochemistry and Western blotting. In addition, a number of rapid immunoassays have been developed and approved by governments for use as screening tests, with positive results subjected to confirmatory Western blotting.

Premortem

There is at present no test that has been rigorously and reproducibly proven capable of detecting BSE (or any other TSE) during the preclinical phase of disease.

However, in the past 2 years a widespread effort to develop such a test has led to a number of different strategies with surprisingly good preliminary results. It is likely that a reliable test or tests will be available in the near future (5). Methods under investigation employ one of three approaches: (i) the use of antibodies that are specific for the misfolded form of the protein and thus eliminate background noise from the presence of normal protein, (ii) the use of polypeptide or synthetic ligands that take advantage of exposed epitopes unique to the misfolded form of the protein, and (iii) in vitro amplification of the protein to vastly increase test sensitivity. A point has been reached where the sensitivity of detection of the pathogenic protein may have surpassed that of infectivity bioassays, the historic "gold standard." Several laboratories have reported very promising results in a comparatively limited number of tests to detect scrapie in sheep, BSE in cattle, and CJD in humans, and data are accumulating for the detection of protein in the blood of presymptomatic animals, i.e., a truly practical screening test for infection.

BSE PREVALENCE IN THE UNITED STATES

Three cows with BSE have been identified in the United States through a PrP^{TSE} testing program initiated in 1990 that targeted cattle that had neurologic symptoms, were recumbent ("downers"), or died for unexplained reasons. One was detected in December 2003 that had been imported several years before from Canada, another was detected in November 2004, and the third was detected in March 2006, the latter two having been born and raised in the United States, and thus indigenously infected. A question thus arises as to the actual prevalence of BSE in the United States, which is made exceedingly difficult to answer by virtue of a widely varying rate of testing during the life of the program.

By using the number of tested animals as the statistical population, analyses were performed for two time periods: (i) from the inception of the program in 1990 through March 2006 and (ii) from January 2002 through March 2006, a period in which the test rate was much greater than that during the 1990s and that still included all of the positive tests. Both analyses yielded very similar results: a prevalence of approximately 2.7 per million (95% confidence interval of 1 per 100,000 to 3 per million) if only the two indigenously infected positives were counted and a prevalence of 4 per million (95% confidence interval of 1 per 100,000 to 0.8 per million) if the imported cow was included (which from the standpoint of both public health and U.S. trading partners would be the more important analysis).

EVALUATION OF POSSIBLE SPORADIC BSE

If in addition to orally acquired disease, BSE were to occur in a sporadic form (like CJD in humans), it would be necessary to continue indefinitely the prohibition of recycling contaminated tissues via supplemental feeding practices. Testing for sporadic disease would need to be done in countries with large national herds that are still free from orally acquired BSE, such as Argentina and Australia.

The criteria for selection of cattle to answer the question about spontaneous BSE are different from those for orally acquired BSE. Most importantly, we do not know at what age spontaneous cases of BSE may occur. It is unlikely to be the 3- to 5-year-old age group in which orally acquired BSE is most prevalent, and if the age distribution for spontaneous disease in cattle were to mimic that for sporadic (spontaneous) CJD in humans, the incidence of which peaks in the 50- to 75-year-old age group, the incidence of disease would not peak until 14 to 20 years of age (the last third of the 20-odd-year natural life span of a cow). Significant numbers of such older cattle do not exist, and thus it may never be possible to state with assurance that spontaneous BSE does not occur.

However, approximately 10% of sporadic CJD cases occur in patients 25 to 50 years of age, in whom the calculated incidence of CJD is 1 case per 10 million per year. This age bracket approximates the middle third of a cow's normal life span, or 7 to 13 years of age, and if the age distribution of sporadic BSE followed the sporadic CJD pattern, negative tests in a total of approximately 3 million animals would allow us to be 95% confident that sporadic BSE is not present at a prevalence higher than 1 per million and 30 million negative tests would lower the maximum prevalence to 1 per 10 million cattle.

CONCLUSIONS

Modifications of feed production and practices have resulted in a steady decline in numbers of cases of BSE around the world (62). Likewise, precautionary measures to exclude high-risk cattle and SRMs from all cattle from entering the human food chain have reduced the number of new cases of vCJD. Although this is an extremely positive trend, governments and industry cannot become complacent about measures to minimize animal and human risk from this family of diseases. There are several important issues about BSE, vCJD, and other TSEs that need to be monitored during the next several years:

- Secondary cases of human-to-human transmission of vCJD through blood transfusion. The costs associated with a loss of blood donation sources, leukodepletion

of the blood supply, and the extreme restrictions on the use of surgical equipment are significant. The extent to which latent BSE infection may be present in the population of the United Kingdom and other countries with vCJD is unknown (especially in view of recent evidence indicating that susceptibility is not limited to individuals who are homozygous for methionine at codon 129 of the prion protein gene) (3a, 30a), and there is at present no validated screening test to detect inapparent infection. Given the broader public health implications from human-to-human transmissions, it is imperative for countries to prevent primary foodborne transmission.

- Countries where BSE may be present but not detected. Many countries have imported vast amounts of MBM from countries with BSE-infected cattle, some of which do not have adequate surveillance programs and have not implemented policies to prevent contamination of animal feed and human food chains. These countries may still serve as a source of the disease.

- A wider distribution of infectivity in bovine tissues from BSE-infected cattle may have occurred than has been realized. Results of collaborative studies by the United Kingdom and Japan support the findings by Buschmann and Groschup (13) that BSE has a wider peripheral tissue distribution than was previously realized. As testing methodologies become more sensitive, the definition of risk material and requirements for feed controls may change.

- The emergence of new strains or species adaptation of existing strains. The origin of BSE has not been identified with certainty, and its emergence should warn us that TSEs may occur in other species and have unpredictable characteristics that will provide new challenges.

Finally, it is important that regulatory policies be modified in accord with advances in experimental and epidemiologic knowledge to minimize adverse consequences to both public and animal health. In particular, the development of preclinical diagnostic tests may vastly improve the precision of proactive measures to minimize risks to human and animal health.

We are grateful to Lisa McShane, Biometrics Research Branch, National Cancer Institute, National Institutes of Health, for statistical analyses relating to BSE.

References

1. **Alper, T., W. A. Cramp, D. A. Haig, and M. C. Clarke.** 1967. Does the agent of scrapie replicate without nucleic acid? *Nature* **214:**764–766.

2. **Andreoletti, O., S. Simon, C. Lacroux, N. Morel, G. Tabouret, A. Chabert, S. Lugan, F. Corbiere, P. Ferre, G. Foucras, H. Laude, F. Eychenne, J. Grassi, and F. Schelcher.** 2004. PrP^{Sc} accumulation in myocytes from sheep incubating natural scrapie. *Nat. Med.* **10:**591–593.

3. Angers, R. C., S. R. Browning, T. S. Seward, C. J. Sigurdson, M. W. Miller, E. A. Hoover, and G. C. Telling. 2006. Prions in skeletal muscles of deer with chronic wasting disease. *Science* **311**:1117.

3a. Bishop, M. T., P. Hart, L. Aitchison, H. N. Baybutt, C. Plinston, V. Thomson, N. L Tuzi, M. W. Head, J. W. Ironside, R. G. Will, and J. C. Manson. 2006. Predicting susceptibility and incubation time of human-to-human transmission of vCJD. *Lancet Neurol.* **5**:393–398.

4. Bolton, D. C., M. P. McKinley, and S. P. Prusiner. 1982. Identification of a protein that purifies with the scrapie prion. *Science* **218**:1309–1311.

5. Brown, P. 2005. Blood infectivity, processing and screening tests in transmissible spongiform encephalopathies. *Vox Sanguinis* **89**:63–70.

6. Brown, P., C. J. Gibbs Jr., P. Rodgers-Johnson, D. M. Asher, M. P. Sulima, A. Bacote, L. G. Goldfarb, and D. C. Gajdusek. 1994. Human spongiform encephalopathy: the National Institutes of Health series of 300 cases of experimentally transmitted disease. *Ann. Neurol.* **35:** 513–529.

7. Brown, P., E. H. Rau, P. Lemieux, B. K. Johnson, A. Bacote, and D. C. Gajdusek. 2004. Infectivity studies of both ash and air emissions from simulated incineration of scrapie-contaminated tissues. *Environ. Sci. Technol.* **38:** 6155–6160.

8. Brown, P., R. G. Rohwer, and D. C. Gajdusek. 1986. Newer data on the inactivation of scrapie virus or Creutzfeldt-Jakob disease virus in brain tissue. *J. Infect. Dis.* **153**:1145–1148.

9. Brown, P., R. G. Rohwer, E. M. Green, and D. C. Gajdusek. 1982. Effect of chemicals, heat, and histopathologic processing on high infectivity hamster-adapted scrapie virus. *J. Infect. Dis.* **145**:683–687.

10. Brown, P., R. G. Will, R. Bradley, D. M. Asher, and L. Detwiler. 2001. Bovine spongiform encephalopathy and variant Creutzfeldt-Jakob disease: background, evolution, and current concerns. *Emerg. Infect. Dis.* **7**:6–16.

11. Brown, P., R. Meyer, F. Cardone, and M. Pocchiari. 2003. Ultra-high-pressure inactivation of prion infectivity in processed meat: a practical method to prevent human infection. *Proc. Natl. Acad. Sci. USA* **100**:6093–6097.

12. Bruce, M. E., R. G. Will, J. W. Ironside, I. McConnell, D. Dummond, A. Suttie, L. McCardle, A. Chree, J. Hope, C. Birkett, S. Cousens, H. Fraser, and C. J. Bostock. 1997. Transmissions to mice indicate that "new variant" CJD is caused by the BSE agent. *Nature* **389**:498–501.

13. Buschmann, A., and M. Groschup. 2005. Highly bovine spongiform encephalopathy-sensitive transgenic mice confirm the essential restriction of infectivity to the nervous system in clinically diseased cattle. *J. Infect. Dis.* **192:** 934–942.

14. Cardone, F., P. Brown, R. Meyer, and M. Pocchiari. 2006. Inactivation of transmissible spongiform encephalopathy agents by ultra high pressure-temperature treatment. *Biochim. Biophys. Acta* **1764:** 558–562.

15. Colchester, A. C. F., and N. T. H. Colchester. 2005. The origin of bovine spongiform encephalopathy: the human prion disease hypothesis. *Lancet* **366:** 856–861.

16. Cuillé, J., and P. I. Chelle. 1936. La maladie dite "tremblante" de mouton; est-elle inoculable? *Compte Rendu Acad. Sci.* **203**:1552.

17. Cutlip, R. C., J. M. Miller, R. E. Race, A. L. Jenny, J. B. Katz, H. D. Lehmkuhl, B. M. DeBey, and M. M. Robinson. 1994. Intracerebral transmission of scrapie to cattle. *J. Infect. Dis.* **169**:814–820.

18. Cutlip, R. C., J. M. Miller, A. N. Hamir, J. Peters, M. M. Robinson, A. L. Jenny, H. D. Lehmkuhl, W. D. Taylor, and F. D. Bisplinghoff. 2001. Resistance of cattle to scrapie by the oral route. *Can. J. Vet. Res.* **65**:131–132.

19. Department for Environment, Food and Rural Affairs. 4 May 2006, revision date. BSE: statistics—youngest and oldest cases by year of onset (for passive surveillance cases) and on year of slaughter (for active surveillance cases). [Online.] Department for Environment, Food and Rural Affairs, London, United Kingdom. http://www.defra.gov.uk/animalh/bse/statistics/bse/yng-old.html.

20. Department for Environment, Food and Rural Affairs. 25 Oct. 2005, rivision date. *Distribution of Infectivity.* [Online.] Department for Environment, Food and Rural Affairs, London, United Kingdom. http://www.defra.gov.uk/animalh/bse/science-research/pathog.html#distribute.

21. Department for Environment, Food and Rural Affairs. 25 Oct. 2005, revision date. *Infectious Dose Experiment.* [Online.] Department for Environment, Food and Rural Affairs, London, United Kingdom. http://www.defra.gov.uk/animalh/bse/science-research/pathog.html#dose.

22. Doherr, M. G., A. R. Hett, J. Rufenacht, A. Zurbriggen, and D. Heim. 2002. Geographical clustering of cases of bovine spongiform encephalopathy (BSE) born in Switzerland after the feed ban. *Vet. Rec.* **151**:467–472.

23. Donnelly, C. A. 1998. Maternal transmission of BSE: interpretation of the data on the offspring of BSE-affected pedigree suckler cows. *Vet. Rec.* **142**:579–580.

24. Donnelly, C. A., N. M. Ferguson, A. C. Ghani, and R. M. Anderson. 2002. Implications of bovine spongiform encephalopathy (BSE) screening data for the scale of the British BSE epidemic and current European infection levels. *Proc. R. Soc. Lond. Ser. B* **269**:2179–2190.

25. Flan, B., and J.-T. Aubin. 2005. Evaluation de l'efficacité des procédés de purification des protéines plasmatiques à éliminer les agents transmissibles non conventionnels. *Virologie* **9**:S45–S56.

26. Hadlow, W. J., R. C. Kennedy, and R. E. Race. 1982. Natural infection of Suffolk sheep with scrapie virus. *J. Infect. Dis.* **146**:657–664.

27. Hill, A. F., M. Desbruslais, S. Joiner, K. C. L. Sidle, I. Gowland, J. Collinge, L. J. Doey, and P. Lantos. 1997. The same prion strain causes vCJD and BSE. *Nature* **389:** 448–450.

28. Hoinville, L. J. 1994. Decline in the incidence of BSE in cattle born after the introduction of the "feed ban." *Vet. Rec.* **134**:274–275.

29. Hoinville, L. J., J. W. Wilesmith, and M. S. Richards. 1995. An investigation of risk factors for cases of bovine spongiform encephalopathy born after the introduction of the "feed ban." *Vet. Rec.* **136**:312–318.

30. Horn, G., M. Bobrow, M. E. Bruce, M. Goedert, A. McLean, and J. Webster. 2001. *Review of the Origin of BSE.* Stationery Office, London, United Kingdom.

30a. Ironside, J. W., M. T. Bishop, K. Connelly, D. Hegazy, S. Lowrie, M. LeGrice, D. L. Ritchie, L. M. McCardle, and D. A. Hilton. 2006. Variant Creutzfeldt-Jakob disease: prion protein genotype analysis of positive appendix tissue samples from a retrospective prevalence study. *BMJ* **332**: 1186–1188.

31. Llewelyn, C. A., P. E. Hewitt, R. S. G. Knight, K. Amar, S. Cousens, J. Mackenzie, and R. G. Will. 2004. Possible transmission of variant Creutzfeldt-Jakob disease by blood transfusion. *Lancet* **363**:417–421.

32. Middleton, D. J., and R. M. Barlow. 1993. Failure to transmit bovine spongiform encephalopathy to mice by feeding them with extraneural tissues of affected cattle. *Vet. Rec.* **132**:545–547.

33. Oesch, B., D. Westaway, M. Walchli, M. P. McKinley, S. B. H. Kent, R. Aebersold, R. A. Barry, P. Tempst, D. B. Teplow, L. E. Hood, S. B. Prusiner, and C. Weismann. 1985. A cellular gene encodes scrapie PrP 27–30 protein. *Cell* **40**:735–746.

34. Peden, A., M. W. Head, D. L. Ritchie, J. E. Bell, and J. W. Ironside. 2004. Preclinical vCJD after blood transfusion in a *PRNP* codon 129 heterozygous patient. *Lancet* **364**:527–529.

35. Philips, Lord of Worth Matravers, J. Bridgeman, and M. Ferguson-Smith. 2000. *The BSE Inquiry.* [Online.] Stationery Office, London, United Kingdom. http://www.bseinquiry.gov.uk/pdf/index.htm.

36. Schreuder, B. E. C., R. E. Geertsma, L. J. M. van Keulen, J. A. A. M. van Asten, P. Enthoven, R. C. Oberthür, A. A. de Koeijer, and A. D. M. E. Osterhaus. 1998. Studies on the efficacy of hyperbaric rendering procedures in inactivating bovine spongiform encephalopathy (BSE) and scrapie agents. *Vet. Rec.* **142**:474–480.

37. Stevenson, M. A., J. W. Wilesmith, J. B. M. Ryan, R. S. Morris, A. B. Lawson, D. U. Pfeiffer, and D. Lin. 2000. Descriptive spatial analysis of the epidemic of bovine spongiform encephalopathy in Great Britain to June 1997. *Vet. Rec.* **147**:379–384.

38. Taguchi, F., Y. Tamai, K. Uchida, R. Kitajima, H. Kojima, T. Kawaguchi, Y. Ohtani, and S. Miura. 1991. Proposal for a procedure for complete inactivation of the Creutzfeldt-Jakob disease agent. *Arch. Virol.* **119**:297–301.

39. Taylor, D. M. 2000. Inactivation of transmissible spongiform encephalopathy agents: a review. *Vet. J.* **159**:10–17.

40. Taylor, D. M. 2004. Resistance of transmissible spongiform encephalopathy agents to decontamination. *Contrib. Microbiol.* **11**:136–145.

41. Taylor, D. M., C. E. Ferguson, C. J. Bostock, and M. Dawson. 1995. Absence of disease in mice receiving milk from cows with bovine spongiform encephalopathy. *Vet. Rec.* **136**:592.

42. Taylor, D. M., H. Fraser, I. McConnell, D. A. Brown, K. L. Brown, K. A. Lamza, and G. R. A. Smith. 1999. Decontamination studies with the agents of bovine spongiform encephalopathy and scrapie. *Arch. Virol.* **139**: 313–326.

43. Taylor, D. M., S. L. Woodgate, and M. J. Atkinson. 1995. Inactivation of the bovine spongiform encephalopathy agent by rendering procedures. *Vet. Rec.* **137**:605–610.

44. Taylor, D. M., S. L. Woodgate, A. J. Fleetwood, and R. J. G. Cawthorne. 1997. Effect of rendering procedures on the scrapie agent. *Vet. Rec.* **141**:643–649.

45. Thomzig, A., F. Cardone, D. Krugr, M. Pocchiari, P. Brown, and M. Beekes. 2006. Pathological prion protein in muscles of hamsters and mice infected with rodent-adapted BSE or vCJD. *J. Gen. Virol.* **87**:251–254.

46. United Kingdom Health Protection Agency. 9 Feb. 2006, posting date. New Case of Variant CJD Associated with Blood Transfusion. [Online.] United Kingdom Health Protection Agency, London, United Kingdom. http://www.hpa.org.uk/hpa/news/articles/press_releases/2006/060209_cjd.htm.

47. Wadsworth, J. D. F., S. Joiner, A. F. Hill, T. A. Campbell, M. Desbruslais, P. J. Luthert, and J. Collinge. 2001. Tissue distribution of protease resistant prion protein in variant Creutzfeldt-Jakob disease using a highly sensitive immuno-blotting assay. *Lancet* **358**:171–180.

48. Ward, H. J. T., D. Everington, S. Cousens, B. Smith-Bathgate, M. Leitch, S. Cooper, C. Heath, R. G. Knight, P. Smith, and R. G. Will. 2006. Risk factors for variant Creutzfeldt-Jakob disease: a case control study. *Ann. Neurol.* **59**: 111–120.

49. Wells, G. A. H., A. C. Scott, C. T. Johnson, R. F. Gunning, R. D. Hancock, M. Jeffrey, M. Dawson, and R. Bradley. 1987. A novel progressive spongiform encephalopathy in cattle. *Vet. Rec.* **121**:419–420.

50. Wells, G. A. H., J. Spiropoulos, S. A. C. Hawkins, and S. J. Ryder. 2005. Pathogenesis of experimental bovine spongiform encephalopathy: preclinical infectivity in tonsil and observations on the distribution of lingual tonsil in slaughtered cattle. *Vet. Rec.* **156**:401–407.

51. Wells, G. A. H., S. A. C. Hawkins, A. R. Austin, S. J. Ryder, S. H. Done, R. B. Green, I. Dexter, M. Dawson, and R. H. Kimberlin. Studies of the transmissibility of the agent of bovine spongiform encephalopathy to pigs. *J. Gen. Virol.* **84**:1021–1031.

52. Wells, G. A. H., S. A. Hawkins, R. B. Green, A. R. Austin, I. Dexter, Y. I. Spencer, M. J., Chaplin, M. J. Stack, and M. Dawson. 1998 Preliminary observations on the pathogenesis of experimental bovine spongiform encephalopathy (BSE): an update. *Vet. Rec.* **142**:103–106.

53. Wilesmith, J. W., and J. B. Ryan. 1992. Bovine spongiform encephalopathy: recent observations on the age-specific incidences. *Vet. Rec.* **130**:491–492.

54. Wilesmith, J. W., and J. B. M. Ryan. 1997. Absence of BSE in the offspring of pedigree suckler cows affected by BSE in Great Britain. *Vet. Rec.* **141**:250–251.

55. Wilesmith, J. W., G. A. H. Wells, J. B. M. Ryan, D. Gavier-Widen, and M. M. Simmons. 1997. A cohort study to examine maternally associated risk factors for bovine spongiform encephalopathy. *Vet. Rec.* **141**:239–243.

56. Wilesmith J. W., G. A. H. Wells, M. P. Cranwell, and J. B. M. Ryan. 1988. Bovine spongiform encephalopathy: epidemiological studies. *Vet. Rec.* **123**:638–644.

57. Wilesmith, J. W., J. B. Ryan, and W. D. Hueston. 1992. Bovine spongiform encephalopathy: case-control studies of calf feeding practices and meat and bonemeal inclusion in proprietary concentrates. *Res. Vet. Sci.* **52**:325–331.

58. **Will, R. G., J. W. Ironside, M. Zeidler, S. N. Cousens, K. Estibeiro, A. Alperovitch, S. Poser, M. Pocchiari, A. Hofman, and P. G. Smith.** 1996. A new variant of Creutzfeldt-Jakob disease in the UK. *Lancet* 347:921–925.

59. **Will, R. G., M. Zeidler, G. E. Stewart, J. A. Macleod, J. W. Ironside, S. N. Cousens, J. Mackenzie, K. Estibeiro, A. J. E. Green, and R. S. G. Knight.** 2000. Diagnosis of new variant Creutzfeldt-Jakob disease. *Ann. Neurol.* 47:575–582.

60. **World Health Organization.** 1999. *WHO Infection Control Guidelines for Transmissible Spongiform Encephalopathies. Report of a WHO Consultation.* WHO/CDS/ CSR/APH/2000.3 World Health Organization, Geneva, Switzerland.

61. **World Health Organization.** 2003. *WHO Guidelines on Transmissible Spongiform Encephalopathies in Relation to Biological and Pharmaceutical Products.* World Health Organization, Geneva, Switzerland.

62. **World Organization for Animal Health.** 22 May 2006, revision date. Bovine spongiform encephalopathy (BSE): geographical distribution of countries that reported BSE confirmed cases since 1989. [Online.]. OIE, Paris, France. http://www.oie.int/eng/info/en_esb.htm.

Foodborne and Waterborne Parasites

Food Microbiology: Fundamentals and Frontiers, 3rd Ed.
Edited by M. P. Doyle and L. R. Beuchat
© 2007 ASM Press, Washington, D.C.

H. Ray Gamble
Dante S. Zarlenga
Charles W. Kim

Helminths in Meat

29

Foodborne parasites have undoubtedly had an impact on human health throughout history. They are important from the standpoint of their direct effect on the well-being of humans, who almost universally consume animal meat as a source of protein and other nutrients. They also serve as a trade obstacle for countries where a high prevalence of zoonotic parasites in livestock may prevent trade with countries where these parasites are rare. There are four meat-borne helminths of medical significance: *Trichinella* spp., *Taenia solium*, and *Taenia asiatica*, which occur primarily in pork, and *Taenia saginata*, which is found in beef.

Despite the availability of sensitive, specific diagnostic tests, veterinary public health programs (meat inspection), and effective chemotherapeutic agents for human tapeworm carriers, these parasites continue to be a threat to public health throughout the world. There are a variety of reasons for this, including animal management systems which perpetuate infection, inadequate or poorly enforced inspection requirements for slaughtered animals, new sources of infection, and demographic changes in human populations that introduce new culinary practices of preparing meats. Thus, current control

and preventive procedures are often inadequate, and more effective control measures are needed to ensure safe meat for human consumption.

TRICHINELLOSIS

The history of trichinellosis is fascinating, going back to a period when infection with *Trichinella* species was presumed more than proven. Whether the commandment in the Bible (Leviticus 11 and Deuteronomy 14) not to eat the flesh of cloven-footed animals (swine) was due in part to the potential danger of contracting trichinellosis is only speculative. In the seventh century A.D., Mohammed prohibited the eating of pork. Hence, to this day, trichinellosis is rare among Jews and Muslims. The earliest known case of trichinellosis may be evidenced by the mummy of a person who probably lived near the Nile River ca. 1200 B.C., an observation made only in 1980 when *Trichinella* larvae were presumptively identified in the mummy's intercostal muscle (86).

Trichinella spiralis was first observed by Paget in 1835 in the muscles of a man during postmortem dissection (16). A landmark in the history of clinical trichinellosis

H. Ray Gamble, National Research Council, 500 Fifth St., NW, Washington, DC 20001. **Dante S. Zarlenga,** USDA Agricultural Research Service, 10300 Baltimore Ave., Beltsville, MD 20705. **Charles W. Kim,** Center for Infectious Diseases, Health Sciences Center, State University of New York at Stony Brook, Stony Brook, NY 11794.

was Zenker's demonstration in 1860 that encapsulated larvae in the arm muscle caused the illness and death of a young woman. Even after 1860, many cases of trichinellosis were undoubtedly not diagnosed because of the difficulty in recognizing the infection clinically.

Species of *Trichinella*

Trichinellosis is worldwide in its distribution because the etiologic agents, nematodes of the genus *Trichinella*, are ubiquitous in animals, both domestic and wild. Historically, it was assumed that there was only one species of *Trichinella*, *T. spiralis*. However, our understanding of the composition of this genus has changed remarkably in recent years. A large volume of information is now available on species within the genus and the molecular and biological characteristics that distinguish them (62, 95, 108, 109).

Initial knowledge of variation within the genus *Trichinella* was based primarily on observed biological differences. In 1972, new species were proposed for an isolate of *Trichinella* from arctic carnivores, designated *T. nativa* (11); an isolate first obtained from a hyena in Kenya, designated *T. nelsoni* (11); and a nonencapsulating isolate from a raccoon dog in Russia, designated *T. pseudotropicalis* (48). The initial distinguishing characteristic of *T. nativa* was its resistance to freezing; further, in contrast to *T. spiralis*, *T. nativa* has low infectivity for pigs (62). *T. nelsoni* occurs in equatorial Africa and is found in a wide range of wild animals, the primary hosts being the bush pig and the warthog (96). Like *T. nativa*, *T. nelsoni* has low infectivity for the domestic pig, and it develops a capsule slowly compared with other encapsulating *Trichinella* species (62). *T. pseudospiralis*, believed to be primarily a parasite of carnivorous birds, is smaller than the other species and, more importantly, does not form a capsule in host musculature (48). Since its description, this species has been found in various birds and mammals throughout the world (62, 109) and it has been implicated in human disease (1, 61).

The systematics of *Trichinella* have evolved rapidly within the past 30 years, with the use of biochemical, and then molecular, techniques to further delineate the phylogenetic relationships of an increasing number of isolates from various animals, including humans. Molecular differentiation has supported observed biological (62) and biochemical (73) differences, and as a result, new species and types of *Trichinella* have emerged. Presently, the genus *Trichinella* includes eight named species and three related genotypes, broadly grouped into those species that form capsules in the host and those that do not form a capsule (109). Those species forming a capsule include *T. spiralis*, *T. nativa*, and the related

genotype T-6, *T. britovi* and the related genotypes T-8 and T-9, *T. murrelli*, and *T. nelsoni*. The nonencapsulating species include *T. pseudospiralis*, *T. papuae*, and *T. zimbabwensis*.

Encapsulated species of *Trichinella* have been found only in mammalian hosts. Of these, *T. spiralis* is distributed in temperate regions worldwide and remains the species most closely associated with infections in domestic swine and synanthropic rodents and human infections resulting from the ingestion of pork. The freeze-resistant species *T. nativa* occurs in the Holarctic region worldwide and has a broad host distribution (108). Due to the ability of *T. nativa* to persist in frozen muscle tissue, this species poses a public health threat to hunters and consumers of game meats in regions where trichinellosis is endemic. Human trichinellosis resulting from walrus meat infected with *T. nativa* has been well documented (42, 74), leading public health officials in Canada to develop an inspection program for walruses killed for human consumption (117). The other freeze-resistant genotype of *Trichinella*, designated T-6, is found in North America. It has low infectivity for pigs, is found in a variety of wild mammals, and has been implicated in human disease (28).

T. britovi and related genotypes occur in temperate regions of Europe, Asia (T-9 from Japan), and Africa (T-8 from South Africa and Namibia). The distribution of *T. britovi* has recently been extended to West Africa (114). *T. britovi* has some intermediate characteristics of other species, including some tolerance to low temperatures, moderate infectivity for swine, and slow capsule formation (larvae have been confused for nonencapsulating species in some cases) (108). *T. britovi* has been the most common species implicated in human infections resulting from the ingestion of infected horsemeat in Europe. *T. murrelli* (110) is a North American species found in wildlife and occasionally horses and humans. It has low infectivity for domestic pigs but poses a risk to humans who eat game meats. *T. nelsoni* appears to have a distribution limited to the eastern part of the Afrotropical region from Kenya to South Africa (114).

The three nonencapsulating species of *Trichinella* are found in a variety of hosts, including mammals, birds, and reptiles (115). *T. pseudospiralis* has a cosmopolitan distribution and has been isolated from carnivorous and carrion-eating birds and a variety of mammals, including humans. *T. papuae* (113) has been detected in several regions of Papua New Guinea, and *T. zimbabwensis* has been reported to occur in Zimbabwe and Mozambique (112). Recent studies have highlighted the diverse host ranges of the nonencapsulating species. Both *T. papuae* and *T. zimbabwensis* have been shown to complete their life cycles in several species of reptiles (112). It is noteworthy that these two *Trichinella* species develop at host body temperatures of 26 to 32°C; encapsulating species inoculated into the same hosts were unable to develop at these temperatures. In contrast to *T. papuae* and *T. zimbabwensis*, *T. pseudospiralis* does not develop in reptiles but does develop in chickens.

As molecular techniques are refined and comparative studies are performed on *Trichinella* isolates, further taxonomic resolution should be applied to this genus. Several excellent reviews of the current status of *Trichinella* species are available (62, 95, 108, 115).

Life Cycles of *Trichinella* Species

All species of *Trichinella* complete their life cycle within one host; no intermediate hosts or extrinsic development is required. When larvae encysted in raw or inadequately cooked meat are ingested, the muscle fibers and capsules that enclose the parasites are digested in the stomach. In the intestine, the liberated larvae burrow into the lamina propria of the villi in the jejunum and ileum. Four molts occur within 48 h, and by the third day, the worms are sexually mature. The small tapered head of the adult worm has a round, unarmed mouth that opens into a tubular esophagus. Approximately one-third of the anterior portion of the body is composed of the stichosome, consisting of stacks of discoid stichocyte cells. The secreted contents of these cells are important for serological detection of infection and also contain antigens which confer protective resistance to the host. Immediately below the esophagus and the stichosome lies a thin-walled intestine, the hind portion of which terminates in the rectum, a muscular tube lined with chitin. The female worm measures about 3.5 mm in length and possesses a vulval opening about one-fifth the body length from the anterior end (Fig. 29.1). The male measures 1.3 to 1.6 mm in length and possesses a single testis that originates in the posterior portion of the body and extends anteriorly to near the posterior end of the esophagus, where it turns posteriorly to form the vas deferens, which becomes the enlarged vesicula seminalis. The vesicula seminalis becomes the ejaculatory duct to join the copulatory tube in the cloaca. The copulatory tube forms the copulatory bell that is extruded during copulation (Fig. 29.2). There are two ventrally located copulatory appendages on each side of the cloacal opening which possibly serve to clasp the female in copula. Between these appendages lie four tubercles, or papillae. Variations in size and differences in the cuticles have been noted for different species and genotypes. For example, *T. pseudospiralis* is as much as one-fourth to one-third smaller than other isolates.

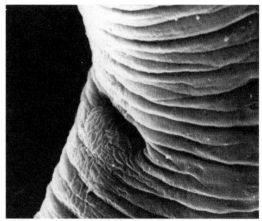

Figure 29.1 Scanning electron micrograph of female adult worm of *T. spiralis* with its prominent vulval opening (×2,450).

In the host, sexually mature adult worms reenter the lumen of the small intestine, where copulation takes place. The adult males die shortly after copulation. The female worms reburrow into the mucosa and begin to larviposit, usually into the central lacteals of the villi, about 7 days after infection and may continue to do so for a period of up to a few weeks. Each female worm bears approximately 1,500 newborn larvae, but this number varies based on *Trichinella* species and host.

The tiny newborn larvae (100 by 6 μm) are carried from the intestinal lymphatic vessels to the regional lymph nodes and into the thoracic duct and the venous blood, passing through the right side of the heart, through the pulmonary capillaries back to the left side of the heart, and into peripheral circulation. During migration, the larvae are known to enter many tissues, including those of the myocardium, brain, and other sites, but here they

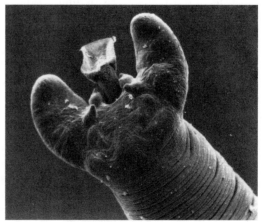

Figure 29.2 Scanning electron micrograph of male adult of *T. spiralis* with its copulatory bell (×1,400).

either are destroyed or reenter the bloodstream. Generally, only larvae that reach striated muscles are able to continue development. They penetrate the sarcolemma of the fibers, where they mature, reaching approximately 1,250 pm in length. They become coiled within the fibers and, in the case of most species, are encapsulated as a result of the host's cellular response. This host-parasite complex, called the nurse cell, is capable of supporting the infective larvae for months or even years. An increased vascular supply to the nurse cell provides nutrients and oxygen vital to the parasite's survival. The encapsulated cyst eventually becomes calcified, and as a result the larva dies.

Infection in humans typically represents a dead end in the parasite's life cycle. However, in animal hosts, the carcass serves as a source of infection. Typically, infected animals are refractory to subsequent infection. However, molecular evidence has documented dual infection in animals (108), and it is possible that these dual infections result from multiple exposures.

Epidemiology

Trichinellosis, human disease resulting from infection with species of the genus *Trichinella*, is considered a zoonosis because infection occurs as a result of ingestion of raw or poorly cooked meat from infected animals. Sources of human infection include domestic livestock, primarily pigs and horses, and wild mammals. The transmission patterns of *Trichinella* spp. can be divided, based on hosts, into syanthropic (domestic-animal) and sylvatic (wildlife) cycles (108). Humans can become involved in both of these cycles.

The domestic cycle is defined by habitat and, by definition, must involve domesticated (farm) animals. There are various patterns of transmission within the domestic cycle. Pigs may become infected through the deliberate feeding of uncooked or undercooked animal flesh containing infective larvae. Infected pigs may then serve as a source of infection to other pigs through cannibalism. Within the domestic cycle, rats may serve as a reservoir host, acquiring infection from wildlife, carrion, or pigs. Rats may transmit infection directly to pigs that ingest them. Pigs may acquire infection directly from wildlife when allowed to wander under unrestricted conditions.

The species of *Trichinella* most frequently associated with the domestic cycle in pigs is *T. spiralis*. However, there is an increasing number of reports of *T. britovi* as a cause of human infection caused by meat from domestic pigs (111). Lower infectivity of other *Trichinella* species diminishes their importance in the domestic cycle. Because cooking, freezing, and other processing methods kill trichinae in meat, most human infections have resulted from situations where meat preparation was not adequate. Pork products such as fresh sausage, summer sausage, and dried or smoked sausage have been implicated as vehicles of human trichinellosis in the United States. In recent years, only 10 to 20 cases of human trichinellosis have been reported annually in the United States (87, 121). Of these, fewer than one-half have been attributed to pork or pork products. However, pork and pork products continue to serve as a major source of infection in many parts of the world (25, 100, 130).

Another domesticated animal which has emerged as a major source of human trichinellosis is the horse. Outbreaks in 1975 and 1985 in the southern suburbs of Paris were attributed to the consumption of raw horsemeat in the form of steak tartare (9). It was initially difficult to believe that human infections resulted from ingestion of meat of an herbivorous animal, but a series of large outbreaks over the past three decades have documented the role of horsemeat as a cause of human trichinellosis. To date, more than 3,200 cases of human trichinellosis have been reported in France and Italy as a result of ingesting raw or undercooked horsemeat. Other countries where horsemeat is consumed have not had human cases as meat is cooked thoroughly before consumption. We know that three species have contributed to human trichinellosis from horsemeat, *T. spiralis*, *T. britovi*, and *T. murrelli* (6). Speculation following the first outbreaks was that the horsemeat was contaminated with meat from other species (4). However, the recovery of larvae from suspected meat (53) and identification of naturally infected horses (3) have solidified this source of human exposure. The route by which horses become infected has been the subject of considerable speculation. The majority of horses linked to human outbreaks in Italy and France originated from countries or regions where a high incidence of *Trichinella* infection is known to occur in pigs, rats, and other animal species. A recent epidemiological study reported that the intentional feeding of animal products and kitchen waste is a common occurrence among horse owners in parts of Eastern Europe and that horses are more willing to consume meat than previously realized (93).

Other domestic sources of trichinellosis have been reported sporadically. An outbreak in China was reported to be due to the consumption of mutton (18). An unusual outbreak in northern Germany was attributed to the consumption of air-dried meat known as pastyrma that had been purchased as camel meat in Egypt (8). It was not certain whether this delicacy was indeed camel meat, since pastyrma is usually made from beef.

If beef was the source, there may have been adulteration with pork. Dog meat has been implicated as a source of human trichinellosis in Thailand and China (19, 21, 58). Dog meat is prepared as a special dish and is favored by the Vietnamese, Chinese, and hill tribe Thais. Most recently, farmed crocodile flesh was reported to harbor *Trichinella* larvae (90), but no human cases of trichinellosis from this source have been reported.

The sylvatic cycle of *Trichinella* involves more than 100 species of mammals, and many of these species pose a risk to humans who eat raw or undercooked game meats. In the sylvatic cycle, wild carnivorous or omnivorous animals scavenge the carrion of dead animals or eat meat from prey animals. Higher infection rates typically occur in animals near the top of the food chain. For example, in the arctic environment, *Trichinella* is quite prevalent in polar bears, grizzly bears, foxes, and wolves. Transmission in the arctic environment is further facilitated by the survival of larvae in frozen meat. Trichinous polar bear meat may conceivably have been responsible for the deaths of three Swedish explorers on an expedition to the North Pole in 1897. More than 50 years later, laboratory examination verified the presence of larvae in minute particles of bear meat still remaining on the explorers' equipment (120). In Alaska, all human cases of trichinellosis have been traced to the consumption of bear or walrus meat and the etiological agent of these infections is *T. nativa* (42). Polar bear meat is considered a delicacy among the Eskimos. Among the Inuit Eskimos in northeastern Canada, there have been outbreaks resulting from the consumption of undercooked walrus meat (132) and probably other mammals (81). The repeated outbreaks of trichinellosis caused by the consumption of infected walrus meat have led to a program of inspection of harvested walruses prior to the distribution of meat to Inuit communities (117). In the former Soviet Union, up to 96% of human trichinellosis cases have been traced to the consumption of wild animals, particularly bears (5). Human infection resulting from badger meat has been reported in Korea (128) as well as Russia (101) and other countries of the former Soviet Union. In the United States, bear meat is the most commonly incriminated game meat, and wild pig is second (121).

Prevalence of Human Disease

Human trichinellosis is essentially nonexistent in some countries of the world but remains a significant problem in many others. While many thousands of clinical cases are reported each year throughout the world, it is impossible to measure true prevalence in humans.

In the United States, a National Institutes of Health report published in 1943 found 16.2% of the population to be infected with *Trichinella* (142). This type of information led to considerable publicity on the dangers of eating pork and was responsible for strict federal control of methods used to prepare ready-to-eat pork products. Numbers of cases of clinical trichinellosis reported to the Centers for Disease Control and Prevention declined from about 500/year in the 1940s to fewer than 50/year in the 1980s and fewer than 10/year from 1996 through 2000. Further, many of these cases resulted from nonpork sources such as bear and other game meats. Although there has been a general decline in incidence in the United States, trends in incidence and transmission patterns have changed over the years, especially with a large influx of people from Southeast Asia (65). Historically, trichinellosis was more common in individuals of German, Italian, and Polish descent because of culinary preferences for raw or undercooked pork. Between 1975 and 1984, trichinellosis among Southeast Asians, who prepare some dishes containing essentially raw pork, was 25 times more frequent than that among the general population (129). Of the 1,260 cases of trichinellosis reported to the Centers for Disease Control and Prevention during this period, 60 (4.8%) were among refugees from Southeast Asia. One of the largest outbreaks ever reported in the United States occurred in 1990 when Southeast Asian refugees from six states and Canada developed trichinellosis after eating pork sausage at a wedding held in Iowa (84).

In countries of the European Union, where slaughter inspection of pigs for *Trichinella* infection has been required for more than 100 years, the parasite has been eradicated from domestic swine, and in some of these countries human trichinellosis has not been reported for several decades. Other Western European countries report sporadic outbreaks often due to game meats or imported meats. In contrast, trichinellosis remains a serious public health problem in many parts of the world and may be considered an emerging or reemerging disease in some countries. Regions where trichinellosis continues to be a problem include parts of Eastern Europe, Asia, and some countries of Central and South America. Wang and Cui (136, 137) reported 23,004 cases of trichinellosis in China between 1964 and 1999. Almost all cases (95.8%) were linked to pork. The same authors found up to 5.6% of pork samples from local markets to be infected. In addition to outbreaks associated with pork, nine outbreaks occurring between 1974 and 1998 were associated with the ingestion of dog meat (19). In Argentina, 5,217 cases of human trichinellosis were reported to occur between 1990 and 1999, mostly in metropolitan areas near Buenos Aires (7). Similar situations with respect to the emergence or reemergence of trichinellosis are reported from Serbia, Romania, and other Eastern European countries (25). The latter regions, where *Trichinella* infection in pigs is highly endemic, are also the source of many of the horses implicated in outbreaks in human trichinellosis in France and Italy (6).

In most cases, poor hygienic conditions for raising pigs and inadequate veterinary control at slaughter are the reasons for high rates of human infection. Pigs raised outdoors, fed raw garbage, or raised in contact with large rodent populations are at risk of acquiring infection. Inadequate veterinary inspection may result from insensitive or improperly performed testing methods or from pigs that are butchered outside of normal inspection channels (backyard pigs). All of these factors are involved in the resurgence of swine and human infections as discussed above (25).

In addition to gaps in inspection programs, culinary habits (improper cooking) impact the risk of human exposure, and this impact is exemplified in some Asian countries where epidemics have been reported, including Thailand, Laos, Japan, China, and Hong Kong. A dish known as nahm is a common source among the inhabitants of northern Thailand (63). There are also other Thai dishes, such as lu (lahb [raw spiced meat] mixed with fresh blood) and satay (small pieces of spiced pork grilled rare to medium on a bamboo skewer), that are sources of infection. Sometimes satay is eaten raw. In both dishes, the larvae may remain alive. Outbreaks of trichinellosis in Laos have been due to the consumption of pork dishes known as som-mou, lap mou, and lap leuat. Laotian immigrants in the United States still prepare these traditional dishes, resulting in outbreaks (17).

Pathogenesis and Pathology

Damage caused by *Trichinella* infection varies with the intensity of the infection and the tissues invaded. The in-and-out movements of the adult worms, especially the females in intestinal tissue, cause an acute inflammatory response and petechial hemorrhages. The cellular response consists primarily of neutrophils with eosinophils. This response is followed by an infiltration of lymphocytes, plasma cells, and macrophages that peaks at about 12 days after infection, gradually declining thereafter. Thus, lesions in the intestine are due to the host's response to the adult worms or their protein products.

The newborn larvae cause an acute inflammatory response as they pass through or become lodged in various tissues and organs. The infiltration consists of lymphocytes, neutrophils, and especially eosinophils. Although there is myocarditis, viable larvae are more numerous in the pericardial fluid than in the myocardium. Pulmonary

hemorrhage and bronchopneumonia may be observed during this stage of larval migration through the capillaries. Rarely, encephalitis may result if the larvae migrate through cerebral capillaries.

When larvae encyst in striated muscle fibers, there is an immediate tissue response consisting of inflammation of the sarcolemma of the involved muscle fibers. The disturbance of ultrastructure and metabolic processes in muscle fibers results in basophilic transformation. This is followed by destruction of the muscle fibers and the eventual formation of a capsule. The larvae gradually die, provoking an intense granulomatous reaction or foreign body cellular response that culminates in calcification. The most heavily parasitized striated muscles include the diaphragm, intercostals, and ocular and masseter muscles. Those larvae that enter tissues other than striated muscles disintegrate and are eventually absorbed.

Clinical Manifestations

The vast majority of individuals infected with *Trichinella* spp. are asymptomatic, probably because low numbers of larvae are ingested. Classical trichinellosis is usually described as a febrile disease with gastrointestinal symptoms, periorbital edema, myalgia, petechial hemorrhage, and eosinophilia (68). During the intestinal stage of infection, gastrointestinal symptoms such as nausea, vomiting, and "toxic" diarrhea or dysentery and fever (over 38°C) and sweating may be observed. Malaise can be severe and lasts longer than fever. The onset of intestinal symptoms occurs usually within 72 h after infection, and symptoms may last for 2 weeks or longer.

Generally, from the second week, when the newborn larvae are migrating and can reach almost any tissue within the body, a characteristic edema is noted around the eyes and in some cases around the sides of the nose, at the temples, and even on the hands. Periorbital edema is present in about 85% of patients and is often complicated by subconjunctival and subungual splinter hemorrhages which disappear within 2 weeks. Eosinophilia greater than 50% is the most consistent manifestation. In severe cases, however, the number of eosinophils may be low. As the larvae enter striated muscle fibers, a striking myositis and muscular pain are evident. In addition to inflammation and pain, eosinophilia is observed during this phase. Serum creatine phosphokinase and transaminase levels can be elevated as a result of leakage of these enzymes from muscle fibers into the serum. Higher levels of enzymes in serum may not correlate well with the severity of the clinical condition. Hypoalbuminemia, which is believed to be due to a demand for protein by the parasite, is usually accompanied by hypopotassemia. Muscle atrophy and contractures may also occur.

Respiratory symptoms, including dyspnea, cough, and hoarseness, may either be due to myositis of respiratory muscles or be secondary to pulmonary congestion. Neurologic manifestations are rare but may result from invasion of the brain by migrating larvae. In severe cases, death may result following cardiac decompensation or respiratory failure, cyanosis, and coma.

Intimate contact between the parasite and host tissues stimulates the production of antibodies that can be demonstrated in the serum. The exact role of antibodies in acquired immunity is not entirely clear, but antibodies appear to be involved in destruction of newborn larvae. On the basis of experimental animal data, it would be safe to assume that in most cases in humans, the severity of infection is reduced considerably by the development of immunity from previous subclinical infections. Local gut immunity in mice has been shown to be T-cell dependent (52).

Recently it has been reported that different species of *Trichinella* present somewhat different clinical pictures in humans. For example, Serhir et al. (126) described a diarrheic syndrome in outbreaks caused by *T. nativa* in Arctic regions of North America. This syndrome, characterized by persistent diarrhea and little edema or fatigue, differs from the traditional myopathic syndrome characterized by edema, fatigue, fever, and rash. These distinct syndromes occur concurrently in outbreaks; following the accumulation of serological and epidemiological evidence, it was concluded that the diarrheic syndrome is seen in patients who have preexisting immunity to *T. nativa* from prior exposure (126). Similarly, infection with *T. murrelli* results in a greater number of patients' developing a rash (44%) and fewer patients' developing facial edema (58%) (27). Only a few outbreaks of human trichinellosis have been attributed to nonencapsulating species of *Trichinella* (12, 118). The only observed clinical difference in these cases was a prolonged period of symptoms (fever and myalgias), which lasted twice as long as those reported for encapsulating species.

Diagnosis

It is difficult to clinically diagnose trichinellosis because it mimics so many other infections as a result of its dissemination throughout the body. Thus, individual or small groups of cases may frequently go undiagnosed, whereas common-source outbreaks make diagnosis easier (12).

Diagnosis is based on a history of eating infected meat, symptoms, laboratory findings (including serology), and recovery of larvae from muscles (44). Classical symptoms of trichinellosis include fever, myalgia, and periorbital edema, although, as discussed previously, these symptoms may vary depending on the species of *Trichinella*.

The most characteristic feature of laboratory test results is marked eosinophilia. Although eosinophilia is not restricted to trichinellosis, its presence in 30 to 85% of infected individuals is a constant and important diagnostic aid. Eosinophilia is initiated in the second week of infection, reaching its peak by about day 20. It may be absent not only in patients with very severe and fatal cases but also in individuals with secondary infection. Muscle enzymes, including creatine phosphokinase and lactate dehydrogenase, are frequently elevated in trichinellosis patients, as are overall leukocyte counts (12, 26).

The most definitive diagnostic method is muscle biopsy to detect encapsulated or nonencapsulated larvae. The diagnostic success of biopsy depends on the chance distribution of larvae in the particular striated muscle that is sampled. Gastrocnemius, pectoralis major, deltoid, or bicep muscles are commonly used because of easy accessibility. The muscle strip is compressed tightly between two microscope slides and examined for the presence of larvae (Fig. 29.3). Part of the biopsy sample can be digested or fixed and then sectioned, stained, and examined. The presence of active larvae following digestion with artificial gastric juice indicates a recent infection.

There are numerous immunological tests available for the diagnosis of trichinellosis (64, 66, 92). Recommendations for the use of these tests have been summarized by the International Commission on Trichinellosis (46). The enzyme-linked immunosorbent assay (ELISA), using an excretory-secretory (ES) antigen collected from muscle larvae or a synthetic tyvelose antigen containing a dominant epitope, is the test of choice for human trichinellosis based on a high degree of sensitivity and specificity. Both ES and tyvelose antigens are preferable to crude extracts of *T. spiralis* muscle larvae due to a risk of cross-reactions with other helminth infections. Sensitivity of nearly 100% has been obtained for humans infected with *T. spiralis*

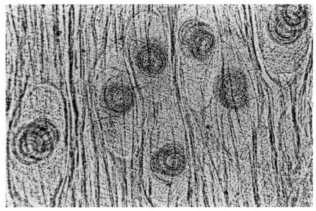

Figure 29.3 Pressed muscle containing *T. spiralis* larvae.

when immunoglobulin G (IgG) is measured by ELISA (13). Serological tests measuring other classes of antibodies result in lower sensitivity (92); however, detection of specific IgM or IgA is indicative of a recent infection. The use of antigens from other stages (adults, newborn larvae) has not resulted in comparable serodiagnosis. Western blotting to determine the presence of antibodies to ES antigens in the range of 40 to 70 kDa has been used as a confirmatory test. The sandwich ELISA has also been used to detect circulating antigens in sera (59).

Trichinellosis patients typically become seropositive between the second and fifth week of infection, although this time will vary with the species of *Trichinella*. The time required for seroconversion is inversely correlated with the infective dose. Thus, it is advisable to take multiple serum samples at intervals of several weeks in order to demonstrate seroconversion in patients whose sera were initially negative or to detect rising titers. Once patients become seropositive, antibody levels do not correlate with the severity of the disease (92). Patients typically remain seropositive for several years, but seropositivity has been reported to last for up to 35 years following infection (27, 69).

Treatment

The efficacy of treatment of trichinellosis depends on the intensity of infection, the species of *Trichinella*, the stage of infection, and the character and intensity of the host response. The purpose of treatment during the intestinal phase is to destroy adult worms and to interfere with the production of newborn larvae. The drug of choice is mebendazole (Vermox) at dosages of 200 to 400 mg three times a day for 3 days and then 400 to 500 mg three times a day for 10 days (44, 68). Mebendazole is also believed to be active against developing and encysted larvae but at dosages higher than those used against adult worms. Mebendazole is administered in the first week of infection before newborn larvae migrate, in moderate or severe infections in combination with corticosteroids, and in infections with *T. nativa*, which responds poorly to treatment with nonbenzimidazole compounds. Mebendazole is not recommended for women in the first trimester of pregnancy. Thiabendazole (Mintezol) is no longer used because of adverse reactions. Albendazole (Valbazen and Zentel) is better absorbed and is probably as active as or even more active than mebendazole (68). Pyrantel (Antiminth) at 10 mg/kg of body weight per day for 4 days and levamisole at 2.5 mg/kg/day (maximum of 1 to 50 mg) are active only against adult worms in the intestine.

To minimize hypersensitivity, it is recommended that corticosteroids be given in combination with anthelmintic drugs. Corticosteriods are recommended for acute severe trichinellosis not only for antiallergic action but also for anti-inflammatory and antishock actions. Supportive therapy, such as bed rest, is very important.

Prevention and Control

Prevention of human trichinellosis resulting from pork and pork products is accomplished by one or more methods including preventing infection on the farm, detecting infected carcasses at slaughter, processing meat to inactivate worms, and proper cooking. The recommended application of these methods for the control of trichinellosis in pork has been summarized by the International Commission on Trichinellosis (45).

There are several measures that can be directed against the principal sources of human trichinellosis (pigs) before they are slaughtered for the market. In the United States, Public Law 96-468 of 1980 requires treatment of meat-containing waste if it is to be fed to pigs (76). The Swine Health Protection Act of 1983 stipulates that all such garbage must be boiled for 30 min in a licensed facility before being fed to pigs (91). The other, more desirable alternative is to strictly prohibit feeding of garbage to pigs, a measure adopted by many states in the United States. Unfortunately, these preventive measures do not apply to the feeding of raw scraps of wild animal meat or scraps of raw household garbage to pigs. Some countries, including the United States, have established Good Production Practices for pigs which preclude exposure to *Trichinella* from all identified sources (45, 85). These Good Production Practices include elements of biosecurity (i.e., eliminating exposure to wildlife and rats) and general good hygiene.

In the United States, as in many countries, animals slaughtered for human consumption are examined by a federal or state inspector to determine if the animal is healthy and suitable for use as human food. Notably, all member countries of the European Union test each pig carcass for *Trichinella* infection by using one of several approved direct methods of inspection. These methods have historically included trichinoscope examination, in which a piece of muscle is compressed between two glass slides and scanned under a microscope, along with several artificial digestion methods, where a muscle sample is digested in acidified pepsin and the product is examined for larvae. Because the trichinoscope method does not adequately detect infection with nonencapsulating species, this method has been abandoned by the European Union. By applying rigorous slaughter testing using artificial digestion methods, *Trichinella* infection has been virtually eliminated from domestic pigs in some countries. However, this effort is not without cost, with an estimated $570 million spent annually by the European Union for *Trichinella* inspection (107). In the United States, pigs are not inspected individually for larvae because of the large numbers of animals that are produced (94). In some countries, testing programs are in place, but not all pigs are processed through official slaughter facilities. These gaps in veterinary control are one reason that trichinellosis remains a significant public health problem.

In the United States, pork products that are considered ready to eat, such as cold-smoked sausage, must be processed to kill all larvae by heating pork muscle to at least 58°C, freezing to −35°C, or freeze-drying (76, 91). Curing procedures are specified in the U.S. Department of Agriculture (USDA) Code of Federal Regulations (9CFR, §318.10) for most classes of pork products (76). The effectiveness of curing depends on a combination of salt concentration, temperature, and time (70).

The most effective measure to prevent trichinellosis caused by fresh pork from areas where the disease is endemic is education of the public. The responsibility lies with consumers to ensure that larvae are killed before the pork is eaten. The USDA recommends that fresh pork be cooked to an internal temperature of 160°F (71°C) before consumption in the home (92). Rapid cooking methods, such as the use of microwave ovens, may not heat the pork uniformly or for sufficient time to destroy the larvae (71). Commercial cooking of pork products requires lower temperatures to ensure inactivation of *Trichinella*, presuming that temperature is more closely controlled by commercial meat processors. The USDA Code of Federal Regulations (9CFR, §318.10) requires that pork be frozen for 20 days at 5°F (−15°C), 10 days at −10°F (−23°C), or 6 days at −20°F (−29°C) to kill the larvae, provided that the meat is less than about 6 in. (15 cm) thick (92). These freezing temperature and time requirements are not effective for killing the arctic isolate *T. nativa* or other freeze-resistant types of *Trichinella*.

TAENIASIS

Until the mid 1980s, it was believed that taeniasis was the result of ingesting tissue cysts of either *T. saginata* from cows or *T. solium* from pigs; however, shortly thereafter, Fan (30) identified and characterized a third human taeniid with adult morphological characteristics similar to those of *T. saginata* and larval characteristics more in line with those of *T. solium*. Pigs were putatively identified as the intermediate host for this unique taeniid that had a predilection for liver tissue in a multitude of experimental hosts. Since that time, controversy has arisen over the classification and therefore the naming of this third human taeniid. For this reason, both *Taenia*

asiatica (29) and *Taenia saginata asiatica* (37) have appeared and continue to appear in the literature as the scientific names identifying this organism. However, sufficient morphological, biological, and genetic differences exist to warrant use of the name *T. asiatica*, which better defines this organism as a unique lineage and eliminates confusion in comparisons of the disparate epidemiological information and clinical manifestations associated with this organism and *T. saginata*. Recent sequencing of the entire mitochondrial genome of *T. asiatica* (60) offers further genetic support for its unique classification. For these reasons, the term *T. asiatica* will be used to define this species here (57).

T. saginata Taeniasis

The history of taeniids has been described in the ancient literature dating back to 1500 B.C., with intriguing theories as to the parasites' origin and nature. However, it was not until 1782 that Goeze (49) identified a difference between the bovine and swine parasites, which was later confirmed in 1857 by Küchenmeister based on the morphologies of the scolices (72). The relationship between the adult worm in humans and the larval bladder worm in cattle was first established in the mid 1800s.

Life Cycle of *T. saginata*

The life cycle of *T. saginata* begins with cattle ingesting *T. saginata* eggs during grazing. The eggs hatch in the duodenum, liberating a six-hooked embryo that penetrates mesenteric venules or the lymphatics and reaches skeletal muscles or the heart, where it develops into the cysticercus larva. The cysticercus is essentially a miniature scolex and neck invaginated into a fluid-filled bladder that measures about 10 by 6 mm. The bladder larva becomes infective within 8 to 10 weeks following ingestion and can remain infective for more than a year.

A person eating raw or poorly prepared beef harboring viable larval cysts (referred to as *Cysticercus bovis*) is subject to infection whereby the larvae are released from their surrounding muscle tissues by digestion in the small intestine. The scolex of the cysticercus evaginates from the vesicle or bladder and attaches to the mucosa of the jejunum, where the larva develops into a mature adult worm in 8 to 10 weeks. A mature adult worm is characterized by a scolex with four hemispherical suckers of 0.7 to 0.8 mm in diameter situated at the four angles of the head. The entire strobila can grow to 17 m in length and possess 1,000 to 2,000 proglottids or segments immediately following the neck in a symmetric series of immature, mature, and gravid proglottids as they proceed posteriorly. The mature and gravid proglottids contain both male and female reproductive organs resulting in self-fertilization when distal proglottids come together by folding of the worm's body. The presence of only a single worm per definitive host suggests that the mechanism of genetic variation results from other than crossbreeding among adult worms.

The scolex attaches to the mucosal surface of the upper jejunum by means of the four suckers, but because of its length, the entire worm might extend down to the terminal ileum. The developing proglottids extend down the small intestine where the most distal gravid proglottids (20 mm long and 5 to 7 mm wide) detach singly from the rest of the strobila and independently migrate through the rectum to the outside. Each gravid proglottid contains about 80,000 eggs, which are expressed from the proglottids and deposited on the perianal skin. When gravid proglottids come to rest on the ground, eggs are extruded. The eggs may also be present as a result of indiscriminate, promiscuous defecation. Cattle ingest *T. saginata* eggs during grazing to complete the cycle.

Epidemiology

Taeniasis results when raw or inadequately cooked beef containing tissue cysts is consumed. Raw or rare beef is popular in many countries, particularly in the form of beef tartare (basterma) in the Middle East, the equivalent of steak tartare (kitfo) in parts of Africa, especially Ethiopia, shish kebab in India, lahb in Thailand, and yuk hoe in Korea. Although the meat of reindeer in northern Siberia (67), as well as meat of other herbivorous animals, has been reported to contain cysticerci of *T. saginata* and may have been responsible for taeniasis in humans, the only evidence of wild intermediate hosts in the Americas has been a single report at the turn of the century involving llama and pronghorn antelope (104). Thus, beef is the primary source of *T. saginata* taeniasis in humans.

With respect to the prevalence of taeniasis in humans, geographic areas can be classified into three groups: countries or regions where infection is highly endemic, with a prevalence in the human population exceeding 10%; countries with moderate infection rates; and countries with a very low prevalence, below 0.1% or even no infections caused by the parasite. The areas of high endemicity are Central and East African countries, such as Ethiopia, Kenya, and Zaire; Caucasian South Central Asian republics of the former Soviet Union; Middle Eastern countries, such as Syria and Lebanon; and parts of the former Yugoslavia. A high incidence was recently identified in Bali, Indonesia (135), where prevalence among three disparate villages was highly variable (1.1 to 27.5%). In older studies within the Asian republics of the former Soviet Union, the prevalence rate was shown to reach 45% (106). The incidence is low in the United States, where only 443 patients were reportedly treated for taeniasis due to *T. saginata* in 1981 (94). The predominant cestode found in Europe is *T. saginata*, especially in areas such as Poland (133, 134). *Taenia solium* is much less represented in this region of the world (15).

Chronic persistence of and periodic increases in human *T. saginata* infections correlate with factors that govern infections in cattle. Increased risk of infection to cattle is directly associated with environmental factors and access to pastures contaminated with human feces or sewage effluent (15), as well as feed, irrigation ditches, or cattle pens contaminated by farm workers. A recent outbreak on a feedlot in Alberta, Canada, resulted in the killing of nearly 3,000 animals, of which 67 were infected with *T. saginata* (75). In this instance, the water supply was implicated as the likely source of infection (124). Eggs of *T. saginata* are capable of surviving long periods in the environment and are resistant to moderate desiccation, disinfectants, and low temperatures (4 to 5°C). A longevity of 71 days in liquid manure, 33 days in river water, and 154 days on pasture has been reported (127). The eggs can also survive many sewage treatment processes and can remain viable in fluid effluent or dried sludge. Reviews (15, 24) summarize both biological and nonbiological treatments of sludge to reduce egg viability; however, a recent study in France (89) demonstrated that cows raised on pastures fertilized with liquid sludge did not acquire live cysticerci provided a 6-week delay occurred between application and grazing.

Despite an apparently low rate of human taeniasis in many developed countries, the incidence of bovine cysticercosis seems not to have decreased in recent years. Among affluent Americans, infection has been attributed to greater freedom of international travel in addition to a general preference for exotic diets. The penchant for raw beef and steak tartare is not limited to affluent Americans and Europeans but is also found, for example, among Ethiopians who can afford beef. A combination of poor sanitation and culinary preference for raw and rare beef in many parts of the world continues to contribute to a high risk of infection.

Pathogenesis and Pathology

The scolex of the adult worm generally is lodged in the upper part of the jejunum. Usually, only a single worm is present, although multiple infections have been reported. Recent data suggest that the multiple infections reported for *T. saginata* may have been due to the incorrect identification of *T. asiatica*, which is well documented as having the ability to multiply infect human hosts (80). The adult tapeworm of *T. saginata* does not cause pathologic changes; however, mucosal biopsies have shown minimal inflammatory reactions suggesting that the worm can have irritative action that results in bowel distension or spasm. Migration to unusual sites is rare, but complications such as appendicitis, invasion of pancreatic or bile ducts, intestinal obstruction, and perforation and vomiting of proglottids with aspiration have been reported.

Clinical Manifestations

Although most cases of taeniasis are asymptomatic, up to one-third of patients complain of nausea or abdominal "hunger" pain that is often relieved by eating. Epigastric pain may be accompanied by weakness, weight loss, increased appetite, headache, constipation, dizziness, and diarrhea. The patient usually becomes aware of the infection when a proglottid is passed in the stool or is found on the perianal area or even on underclothing.

The adult worm is weakly immunogenic, as manifested by moderate eosinophilia and increased levels of IgE in serum. Allergic reactions, such as urticaria and pruritus, may be due to the worm and its metabolites. The adult worms induce the production of antibodies. The persistence for years of a large, actively growing worm reflects lack of protective immunity to resident worms but is consistent with concomitant immunity observed in other helminth infections.

Diagnosis

Definitive diagnosis is based on the proglottid, since the eggs of *T. saginata* cannot be distinguished from those of other species of *Taenia* or those of *Multiceps* and *Echinococcus* species. The gravid proglottid of *T. saginata* has 15 to 20 lateral branches of the uterus on each side of the main uterine stem, a characteristic feature (Fig. 29.4); however, a similar range of uterine branches has been identified in *T. asiatica* as well, suggesting that the geographical source of the infection may play a role in diagnosis if uterine branches are used in the analysis. If the gravid proglottid is treated with 10% formaldehyde and injected with India ink, the uterine branches are very prominent. Uterine branches also can be seen by gently pressing the proglottid between two microscope slides and holding them in front of a bright light. If the scolex is present, the four characteristic hookless suckers can be used as a distinguishing feature for identification.

The egg is nearly spherical in shape, measuring 30 to 40 μm in diameter, has characteristic radial striations on its thick shell, and contains a hexacanth embryo with delicate lancet-shaped hooklets (Fig. 29.5). Rather than looking for the eggs in the stool, it is better to use the commercial Scotch tape method to obtain the eggs or proglottids from the perianal region. Since detection of eggs is nonspecific in terms of species, DNA probes that

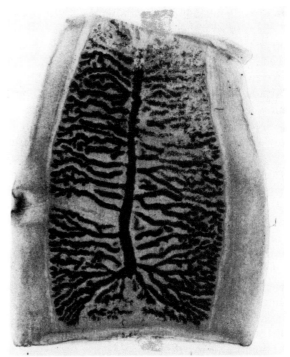

Figure 29.4   Gravid proglottid of *T. saginata*. Note at least 16 lateral uterine branches.

allow differentiation of tapeworm species by hybridization techniques have been developed (41, 54). PCR has been used to discriminate among various *Taenia* species. The amplification of a 0.55-kb DNA fragment provides a unique *T. saginata* genomic DNA template (51). *T. saginata* can also be easily delineated from *T. asiatica* by differential amplification of external transcribed rRNA gene

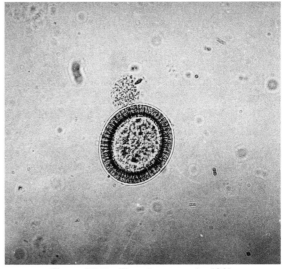

Figure 29.5   *T. saginata* egg (×590).

repeat fragments (144) and multiplex PCR (50). Current technologies will also allow the amplification of DNA from eggs or proglottids followed by DNA sequencing and comparison to sequences from other common taeniids for confirmatory diagnosis.

Although an antibody response is induced by the adult worm, specific serological tests such as those used for the diagnosis of trichinellosis have not been routinely used for diagnosis of taeniasis. However, preliminary data have shown that a <12-kDa antigen prepared from the cyst fluid of *Taenia hydatigena* may be capable of differentiating between the types of human taeniasis (55).

Treatment

The drug of choice in treating taeniasis is niclosamide (Niclocide and Yomesan), a taeniacide that is effective in damaging the worm to such an extent that a purge following therapy will often produce the scolex. Cure rates are very high. A single oral dose of 10 mg of praziquantel (Biltricide) per kg of body weight also has been reported to be highly effective (123).

Prevention and Control

In addition to effective drug therapy against the adult *Taenia* worms, there are measures that can be taken to prevent cysticercosis in cattle. The most important of these measures is adequate sewage disposal so that tapeworm eggs are unavailable to cattle. Since humans are the only definitive hosts of *T. saginata* and thus the only disseminators of eggs, which can be shed at a rate of 480,000 to 720,000 daily, the success of this measure depends on educating livestock producers and their employees about modes of transmission, examining workers' stools for *Taenia* species, and providing adequate toilet facilities in cattle feeding with establishments. Where toilet facilities are available, overloading of systems should be avoided. In addition to education and monitoring, prevention includes sanitary protection of cattle feed as well. Bovine cysticercosis resulting in extensive carcass condemnation has resulted from feeding with cottonseed harvested from an area where an overtaxed municipal water treatment facility and hard rains combined to contaminate the feed with *T. saginata* eggs.

Currently, meat inspection is used to detect cysticercosis in slaughtered cattle and thereby reduce taeniasis. Although practiced extensively, this methodology has limitations. Even routine examination of the heart and masseter muscles can miss a significant percentage of infected cattle. It is estimated that as few as 25% of infected cattle are detected by methods currently employed for meat inspection (119). This is due to assumptions regarding predilection sites in host tissues, the ability to use the heart as a representative site for whole-body infections, and the generally low number of cysts that occur in any animal subject to low-level, environmentally derived infections. In the United States, the percentage of cattle inspected by state and municipal authorities varies. Cattle from small farms where sanitary conditions can be poor are frequently slaughtered in local abattoirs without inspection requirements. Clearly, all infected carcasses should be condemned; however, approximately 70% of infected carcasses are trimmed of visible cysts and then passed unrestricted for consumption (125). Currently, the U.S. Code of Federal Regulations (9CFR, §318.10) requires infected carcasses to be frozen for a minimum of 10 days at 15°F (−9.4°C) or 5 days at 0°F (−18°C) to kill the organism (105). Nonetheless, the combination of the practice of trimming the cysts with the less than 100% animal inspection rate and the poor rate of identification at the abattoir is consistent with a low level of positive identification.

An alternative approach for assessing cysticercosis in cattle may be to use a sensitive serological method (122). ELISA techniques have been evaluated in recent years (22, 23), but the efficacy of these method has not been proven to be superior to that of physical inspection. The association between direct and serological methods is poor primarily because of the disparate immune responses among the hosts (138) and the lack of correlation between the level of parasitism and the level of the immune response. Serological testing might be useful for epidemiological purposes, especially if cysticercosis is suspected in a cattle herd. However, no large-scale study has been performed to validate the diagnostic efficacy of native or recombinant antigens with respect to sensitivity and specificity for detection of infection.

The development of an effective vaccine for use in areas of high endemicity would not only prevent initial infection but also protect cattle against challenge infections. Recombinant vaccines for bovine cysticercosis and their orthologous antigens for swine cysticercosis have been tested with remarkable success (78, 79); however, economic and logistical issues have thus far deterred application of vaccines as a prophylactic measure for bovine cysticercosis.

On the part of the consumer, public health education concerning the risks of eating raw or inadequately cooked beef is important. The cysticerci in meat are inactivated by freezing meat at −10°C for 10 days or −18°C for 5 days, heating to an internal temperature of 56°C, or salt curing under appropriate conditions. Major obstacles to this approach have been a reluctance by consumers to modify preferred culinary habits and to break with long-established cultural traditions. Thus, attempts to eradicate *T. saginata* taeniasis have been unsuccessful.

T. solium Taeniasis

Taeniasis caused by *T. solium* was known at the time of Hippocrates. The Greeks described the larval stage in the tongue of swine as resembling a hailstone. In contrast to *T. saginata, T. solium*, the pork tapeworm possesses an armed rostellum, has a smaller number of lateral uterine branches, and lacks a vaginal sphincter (116). Its phylogenetic classification as a unique species in 1758 is attributed to Linnaeus.

Life Cycle of *T. solium*

When pork containing a viable cyst of the larval stage of *T. solium* (often referred to as *Cysticercus cellulosae*) is ingested, the head of the larva evaginates from the fluid-filled milky white bladder. The scolex bears four suckers and an apical crown of hooklets. The cysticerci are referred to as pork measles and are larger (5 to 20 mm in diameter) than *Cysticercus bovis*. They attach to the wall of the small intestine and mature into adult worms in 5 to 12 weeks. The adult worm measures up to 8 m (usually 1.5 to 5 m) and has a scolex that is roughly quadrate, possessing a conspicuous rounded rostellum armed with a double row of large and small hooklets, numbering 22 to 32. A short cervical region is anterior to a series of proglottids or segments. Immature proglottids are broader than they are long, mature ones being nearly square, while gravid ones are longer than they are broad. The total number of distinct proglottids is less than 1,000. The terminal proglottids become separated from the rest of the strobila and migrate out of the anus or are passively expelled in the stool. A single gravid proglottid contains fewer than 50,000 eggs. Upon escape from the ruptured uterus of the gravid proglottid and after deposition on the soil, the eggs may remain viable for many weeks. The eggs of *T. solium* are more apt than those of *T. saginata* to appear in the stool.

In the normal cycle, the eggs are ingested by pigs, the usual intermediate host. The hexacanth embryo hatches in the duodenum, migrates through the intestinal wall to reach the blood and lymphatic channels, and is carried to the skeletal muscles and the myocardium. The embryo develops in 8 to 10 weeks into a cysticercus or *Cysticercus cellulosae*. In the abnormal cycle, humans can serve as intermediate, but terminal, hosts and harbor cysticerci acquired by accidental ingestion of eggs or by the autochronus cycle (an infection which results from the movement of eggs or gravid proglottids from the intestine back into the stomach by reverse peristalsis). The cysticercus develops most commonly in striated muscles and subcutaneous tissues but also in the brain, eye, heart, lung, and peritoneum.

Epidemiology

Humans are the only known natural definitive host for *T. solium* taeniasis, which has a worldwide distribution and is important in all countries where pork is consumed, especially those with suboptimal practices of sanitation and pig husbandry. It is very difficult to evaluate the prevalence of human infection because coproscopical methods cannot differentiate between infections caused by *T. saginata* and those caused by *T. solium*. Serological methods have been developed for this purpose but have not yet been tested in large field studies.

T. solium taeniasis has much more serious consequences than *T. saginata* taeniasis because the larval or cysticercus stage of *T. solium*, which infects pigs, can also infect humans. Cysticercosis in humans is an important public health problem in developing countries where hygiene and sanitary conditions are deficient or nonexistent and where swine-rearing practices are primitive (102). The relatively high prevalence of *T. solium* taeniasis in the Bantu population of South Africa has been attributed to a widespread practice of using tapeworm proglottids as a component of muti, a medication used by native herbalists for the treatment of intestinal worms (56). Infection of humans with larvae results from inadvertent ingestion of eggs in contaminated food, e.g., vegetables fertilized with night soil (human excrement) or contaminated water, from indirect contact with hands of individuals already infected with the adult worm (external autoinfection), or from direct contact with a tapeworm carrier. Another important mode of transmission involves patients infected with adult tapeworms, in which the eggs are carried by reversed peristalsis back to the duodenum or stomach, where they are stimulated to hatch and subsequently invade extraintestinal tissues to become cysticerci (internal autoinfection).

The prevalence of human neurocysticercosis is increasing in some parts of the world, and a considerable body of literature is devoted to the diagnosis, treatment, and prevention of this disease (47, 139). Human infection is practically nonexistent in Muslim countries (106) but is quite common in Mexico, Central and South America, Central and South Africa, Asia, non-Islamic Southeast Asia, Slavic countries, and Southern and Eastern Europe. Prevalence rates of cysticercosis have been reported to be 6.1% in Pondicherry, South India (103), 7 to 21% among different provinces of Madagascar (2), 6 to 9% based upon viable cysts or 4 to 36% based upon serology in three rural Venezuelan communities (38), 0.2 to 7.2%

in central and northern provinces of Vietnam (141), 21% in rural Peruvian communities (88), and 0.1 to 8% in Western and Central Africa (145). It should be noted that most of these studies, as well as other similar studies, have used indirect serology-based assays which are prone to false-positive and false-negative results. In general, specific habits of eating raw or undercooked pork are linked to foci of taeniasis; however, in some regions there is no link between porcine and human cysticercosis where non-pork eaters have as great a chance of contracting the disease as pork eaters (82). An interesting development is that reports worldwide are showing an association between the presence of neurocysticercosis and the incidence of epilepsy, suggesting that many infected patients are being misdiagnosed as epileptics (14, 97, 98).

Pathogenesis and Pathology

As in *T. saginata* infection, there is usually only a single adult worm, and pathological changes are similar. There is mild local inflammation of the intestinal mucosa as a result of attachment by the suckers and especially the hooklets. However, because of its smaller size, *T. solium* is less likely to cause intestinal obstruction. Rare instances of intestinal perforation with secondary peritonitis and gallbladder infection have been reported.

Pathological changes due to cysticerci can be serious, depending on the tissue invaded and the number of cysticerci that become established. Damage results from pressure caused by encapsulated larvae on surrounding tissue, since the cysticerci produce occupying lesions. There may be no prodromal symptoms or only slight muscular pain and a mild fever during invasion of the muscles and subcutaneous tissues. In ocular cysticercosis, which accounts for about 20% of neurocysticercosis cases, there may be loss of vision. Invasion of the meninges, cortex, cerebral substance, and ventricles evokes tissue reactions leading to focal epileptic attacks or other motor or sensory involvement. The reasons for the predilection of cysticerci for the central nervous system are still obscure.

Clinical Manifestations

Since only a single adult worm is usually present in *T. solium* taeniasis, there are no symptoms of epigastric fullness. Patients are asymptomatic and become aware of the infection only when they find proglottids in their stools or on perianal skin. However, there may be vague abdominal discomfort, hunger pains, anorexia, and nervous disorders. As noted above, rare instances of intestinal perforation with secondary peritonitis and gallbladder infection have been reported. Eosinophilia, as high as 28%, and leukopenia can occur. The persistence for years of large,

actively growing tapeworms does not appear to be consistent with the development of protective immunity. As mentioned earlier, seizures which often accompany forms of neurocysticercosis are to a high degree misdiagnosed as indicative of epilepsy primarily because of the inability to adequately diagnose the infection.

Diagnosis

Diagnosis is based on examination of stool specimens and perianal scrapings. The sensitivity of both methods is increased by three examinations daily. Since *T. solium* eggs cannot be distinguished from those of *T. saginata*, specific diagnosis is based on the identification of the gravid proglottid, which has fewer lateral branches (7 to 12) on each side of the main uterine stem (Fig. 29.6) than that of *T. saginata*. If the scolex is obtained, it possesses hooklets in addition to four suckers. The development of DNA probes (54), and more recently the use of PCR (83, 99) and multiplex PCR (143), have made it possible to distinguish *T. solium* from *T. saginata*.

The sensitivity of serological tests for taeniasis and cysticercosis varies, depending on the particular method and the clinical form of infection. Efforts have been made to improve the sensitivity of serology-based modern methods (39, 40, 131, 140), although emphasis on the use of serology has shifted primarily to the diagnosis of cysticercosis rather than taeniasis. A highly specific immunoassay for diagnosis of cysticercosis, using the <12-kDa *T. hydatigena* antigen, was reported (55) to be so specific as to distinguish between human clinical cases of cysticercosis and taeniasis.

Treatment

Niclosamide is the drug of choice because of its effectiveness against the scolex and the proliferative zone of the strobila. Single doses of praziquantel and mebendazole

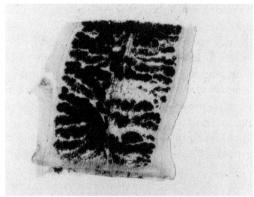

Figure 29.6 Gravid proglottid of *T. solium*. Note fewer lateral uterine branches than in *T. saginata* in Fig. 29.4.

have also been reported to be effective. It is imperative to treat patients harboring the adult worm, since cysticercosis can occur from internal autoinfection. The major concerns in treating patients with the adult worm are to prevent vomiting and to ensure rapid expulsion of disintegrated proglottids from the intestine.

Prevention and Control

Prevention and eradication of *T. solium* adults in the intestine and of cysticerci in various tissues in humans and in pigs is a concern in areas of endemicity where economic, social, and sanitary conditions are substandard. Needed changes are monumental in scope. Without fundamental changes, the most scrupulous personal hygiene and eating habits will not prevent or eradicate infection in underdeveloped or developing countries. In developed countries, infection can be avoided by adherence to modern animal husbandry practices. The best preventive measure in interrupting transmission from humans to animals is to introduce and maintain proper sanitary facilities to dispose of contaminated feces. Even with proper toilet facilities, measures must be taken to make sure that sewage treatment is adequate to kill the eggs.

Since one infected individual can infect literally thousands of pigs via contaminated feedlots, management personnel and employees must be educated regarding the parasite and means of avoiding transmission. Prospective employees who will come in contact with animals should be examined for tapeworm infection before employment and semiannually during employment, and chemical toilets should be installed and properly maintained at convenient locations on slaughtering premises. The most important practice is to keep pigs in enclosures or indoors to prevent access to human fecal matter.

Meat inspection programs are only partially effective for identifying and condemning infected carcasses. Meat inspection for measly pork is performed in the United States for animals intended for interstate commerce, but inspection can miss lightly infected carcasses. Thus, the consumer should make sure that meat is cooked to an internal temperature of 56°C or higher or is frozen at −10°C for at least 14 days to kill cysticerci (94). Other means of inactivating cysticerci in pork, such as irradiation, have not been commercially applied. The best preventive measure is to avoid eating raw or uninspected pork.

With the idea that eliminating swine cysticercosis will greatly assist in eradicating human taeniasis and neurocysticercosis, efforts to immunize pigs with recombinant antigens have met with remarkable success both in laboratory studies and field trials (77, 79). Unfortunately, vaccination as a means of prevention and control is subject to problems other than a lack of efficacy. Economic and regulatory issues as well as dissemination and stability of the vaccine also come into play. As a result, a swine vaccine has yet to be made commercially available.

Asian Taeniasis

Asian taeniasis was first described in the aboriginal population in the mountainous regions of Taiwan (30). Initially, the etiologic agent was considered to be *T. saginata* because of morphologic similarities among the adult worms, although notable differences, including shorter length, fewer number of proglottids, wider diameter of the scolex, and fewer number of testes in the mature proglottid of the agent, were found (30). Cloned rRNA fragments and sequence amplification by PCR showed that *T. asiatica* is similar but genetically distinct from *T. saginata* (144). Also, sequence variation in the 28S rRNA and mitochondrial cytochrome *c* oxidase I genes and restriction fragment length polymorphism pattern variation in the cytochrome c oxidase I and rRNA internal transcribed spacer I have been used to identify *T. asiatica* as genetically distinct from other taeniid cestodes (10). Most recently, PCR and multiplex PCR have been used to unequivocally differentiate *T. saginata* from *T. asiatica* (50). Although further studies are needed, the fact that *T. asiatica* is closely related to *T. saginata* suggests that it is unlikely to be an important cause of human cysticercosis (43).

Epidemiology

T. asiatica infection has been found in several Asian countries, notably the mountainous regions of Taiwan, Cheju Island of Korea, and Samosir Island of Indonesia. It is estimated that public health costs in these areas alone exceed $35 million annually (33). In Taiwan, of 1,661 aboriginal cases of Asian taeniasis, the overall clinical infection rate was 76% among nine aboriginal tribes in 10 counties in mountainous areas (35). Multiple studies since that time would suggest that the infection rate is quite variable among the aboriginal populations of Taiwan but nonetheless remains comparatively high. Pigs, cattle, goats, wild boars, and monkeys can serve as intermediate hosts, and this idea has been validated with experimentally infected animals (31). Of these, the wild boar appears to be the probable natural intermediate host in Taiwan. This may be true also in Indonesia, where people have become infected, presumably from consuming pork (20).

The cysticercus of *T. asiatica* is armed with tiny rostellar hooklets like that of *T. solium* and develops in a period shorter than that required for either *T. saginata* or *T. solium*. Interestingly, it is found mainly in the parenchyma of the liver (30), whereas cysticerci of *T. saginata* and *T. solium* are found primarily in the muscles of cattle and pigs, respectively. Thus, the custom of eating the viscera, especially the liver, of fresh-killed animals appears to be a major factor contributing to transmission.

Clinical Manifestations

Infected individuals pass proglottids in their feces, even for 30 years or more, suggesting that the life span of this form of *Taenia* is very long (35). Common clinical manifestations include pruritis ani, nausea, abdominal pain, and dizziness. Abdominal pain is usually localized on the midline of the epigastrium or in the umbilical region and varies in intensity from a dull aching, gnawing, or burning to intense, colic-like, sharp pain. There may be either an increased appetite or a lack of appetite, headache, or diarrhea, as well as constipation. In contrast to human infection with *T. solium* and *T. saginata*, humans have been reported to be capable of harboring multitudes of adult worms of *T. asiatica* (80).

Treatment

Clinical trials performed some time ago in Taiwan have shown that a single dose (150 mg) of praziquantel was highly effective against *T. asiatica* infection (32, 34). This treatment regime remains current today.

Prevention and Control

The best preventive measure is to avoid eating raw, uninspected pork or other meat that contains the cysticerci. Since many cases of Asian taeniasis have been reported to result from the consumption of infected pork (36), the meat should be cooked thoroughly or should be frozen for at least 4 days at −10°C to kill the cysticerci, the same procedures used to prevent *T. solium* taeniasis.

References

1. **Andrews, J. R. H., R. Ainsworth, and D. Abernethy.** 1994. *Trichinella pseudospiralis* in humans: description of a case and its treatment. *Trans. R. Soc. Trop. Med. Hyg.* 88:200–203.

2. **Andriantsimahavandy, A., V. E. Ravaoalimalala, P. Rajaonarison, P. Ravoniarimbinina, M. Rakotondrazaka, N. Raharilaza, D. Rakotoarivelo, M. Ratsitorahina, L. P. Rabarijaona, C. E. Ramarokoto, P. Leutscher, and R. Migliani.** 2003. The current epidemiological situation of cysticercosis in Madagascar. *Arch. Inst. Pasteur Madagascar* 69:46–51.

3. **Arriaga, C., L. Yépez-Mulia, N. Viveros, L. A. Adame, D. S. Zarlenga, J. R. Lichtenfels, E. Benitez, and M. G. Ortega-Pierres.** 1995. Detection of *Trichinella spiralis* muscle larvae in naturally infected horses. *J. Parasitol.* 81:781–783.

4. **Bellani, L., A. Mantovani, S. Pampiglione, and I. Fillippini.** 1978. Observations on an outbreak of human trichinellosis in Northern Italy, p. 535–539. *In* C. W. Kim and Z. S. Pawlowski (ed.), *Trichinellosis.* University Press of New England, Hanover, N.H.

5. **Bessonov, A. S.** 1981. Changes in the epizootic and epidemic situation of trichinellosis in the USSR, p. 365–368. *In* C. W. Kim, E. J. Ruitenberg, and J. S. Tepperna (ed.), *Trichinellosis.* Reedbooks, Chertsey, England.

6. **Boireau, P., I. Vallee, T. Roman, C. Perret, L. Mingyuan, H. R. Gamble, and A. A. Gajadhar.** 2000. *Trichinella* in horses: a low frequency infection with high human risk. *Vet. Parasitol.* 93:309–320.

7. **Bolpi, J., and R. Boffi.** 2001. Human trichinellosis in Argentina. Review of the casuistry registered from 1990 to 1999. *Parasite* 8:78–80.

8. **Bommer, W., H. Kaiser, W. Mannweiler, H. Mergerian, and G. Pottkamper.** 1985. An outbreak of trichinellosis in northern Germany caused by imported air-dried meat from Egypt, p. 314–317. *In* C. W. Kim (ed.), *Trichinellosis.* State University of New York Press, Albany, N.Y.

9. **Bouree, P., J. L. Leymarie, and C. Aube.** 1989. Epidemiological study of two outbreaks of trichinosis in France, due to horse meat, p. 382–386. *In* C. E. Tanner, A. R. Martinez-Fernandez, and F. Bolas-Fernandez (ed.), *Trichinellosis.* CSIC Press, Madrid, Spain.

10. **Bowles, J., and D. P. McManus.** 1994. Genetic characterization of the Asian *Taenia*, a newly described taeniid cestode of humans. *Am. J. Trop. Med. Hyg.* 50:33–34.

11. **Britov, V. A., and S. N. Boev.** 1972. Taxonomic rank of various strains of *Trichinella* and their circulation in nature. *Vestn. Akad. Nauk. SSSR* 28:27–32.

12. **Bruschi, F., and K. D. Murrell.** 2002. New aspects of human trichinellosis: the impact of new *Trichinella* species. *Postgrad. Med. J.* 78:15–22.

13. **Bruschi, F., A. Moretti, D. Wassom, and D. Piergili-Fioretti.** 2001. The use of a synthetic antigen for the serological diagnosis of human trichinellosis. *Parasite* 8:141–143.

14. **Bucardo, F., A. Meza-Lucas, F. Espinoza, R. C. Garcia-Jeronimo, R. Garcia-Rodea, and D. Correa.** 2005. The seroprevalence of *Taenia solium* cysticercosis among epileptic patients in Leon, Nicaragua, as evaluated by ELISA and western blotting. *Ann. Trop. Med. Parasitol.* 99: 41–45.

15. **Cabaret, J., S. Geerts, M. Madeline, C. Ballandonne, and D. Barbier.** 2002. The use of urban sewage sludge on pastures: the cysticercosis threat. *Vet. Res.* 33:575–597.

16. **Campbell, W. C. (ed.).** 1983. *Trichinella and Trichinosis*, p. 1–30. Plenum, New York, N.Y.

17. **Centers for Disease Control.** 1982. Common-source outbreaks of trichinosis—New York City; Rhode Island. *Morb. Mortal. Wkly. Rep.* 31:161–164.

18. **Coordinating Group for Prevention and Treatment of Trichinosis, Harbin City.** 1981. *A Survey of Trichinosis Due to Eating Scalded Mutton.* WHO/HELM/82.5. World Health Organization, Geneva, Switzerland.

19. **Cui, J., and Z. Q. Wang.** 2001. Outbreaks of human trichinellosis caused by consumption of dog meat in China. *Parasite* 8:74–77.

20. **Depary, A. A., and M. L. Kosman.** 1990. Taeniasis in Indonesia with special reference to Samosir Island, North Sumatra. *Southeast Asian J. Trop. Med. Public Health* 22:239–241.

21. Dissamarn, R., and P. Indrakamhang. 1985. Trichinosis in Thailand during 1962–1983. *Int. J. Zoonoses* **12:**257–266.

22. Dorny, P., I. Phiri, S. Gabriel, N. Speybroeck, and J. Vercruysse. 2002. A sero-epidemiological study of bovine cysticercosis in Zambia. *Vet. Parasitol.* **104:**211–215.

23. Dorny, P., F. Vercammen, J. Brandt, W. Vansteenkiste, D. Berkvens, and S. Geerts. 2000. Sero-epidemiological study of *Taenia saginata* cysticercosis in Belgian cattle. *Vet. Parasitol.* **88:**43–49.

24. Dumontet, S., H. Dinel, and B. Baloda. 1997. Pathogen reduction in biosolids by composting and other biological treatments: a literature review, p. 251–295. *In Proceedings of the 10th International Congress on Environmental and Sanitary Problems in the Mediterranean Areas*, vol. 1. Maratea, Italy, 10–13 October 1997.

25. Dupouy-Camet, J. 2000. Trichinellosis: a worldwide zoonosis. *Vet. Parasitol.* **93:**191–200.

26. Dupouy-Camet, J., W. Kociecka, F. Bruschi, F. Bolas-Fernandez, and E. Pozio. 2002. Opinion on the diagnosis and treatment of human trichinellosis. *Expert Opin. Pharmacother.* **3:**1117–1129.

27. Dupouy-Camet, J., A. Paugam, G. De Pinieux, V. Lavarde, and A. Vieillefond. 2001. *Trichinella murrelli:* pathological features in human muscles at different delays after infection. *Parasite* **8:**176–179.

28. Dworkin, M. S., H. R. Gamble, D. S. Zarlenga, and P. O. Tennican. 1996. Outbreak of trichinellosis associated with eating cougar jerky. *J. Infect. Dis.* **174:**663–666.

29. Eom, K. S., and H. J. Rim. 1993. Morphologic descriptions of *Taenia asiatica* sp. *Korean J. Parasitol.* **31:**1–6.

30. Fan, P. C. 1988. Taiwan *Taenia* and taeniasis. *Parasitol. Today* **4:**86–88.

31. Fan, P. C. 1990. Asian *Taenia saginata*: species or strain? *Southeast Asian J. Trop. Med. Public Health* **22:**245–250.

32. Fan, P. C. 1995. Review of taeniasis in Asia. *Chung Hua Min Kuo Wei Sheng Wu Chi Mien I Hsueh Tsa Chih* **28:**79–94.

33. Fan, P. C., and W. C. Chung. 1997. Sociocultural factors and local customs related to taeniasis in East Asia. *Kao Hsiung I Hsueh Ko Hsueh Tsa Chih* **13:**647–652.

34. Fan, P. C., W. C. Chung, C. H. Chan, Y. A. Chen, F. Y. Cheng, and M. C. Hsu. 1986. Studies on taeniasis in Taiwan. V. Field trial on evaluation of therapeutic efficacy of mebendazole and praziquantel against taeniasis. *Southeast Asian J. Trop. Med. Public Health* **17:**82–90.

35. Fan, P. C., W. C. Chung, C. Y. Lin, and C. H. Chan. 1992. Clinical manifestations of taeniasis in Taiwan aborigines. *J. Helminthol.* **66:**118–123.

36. Fan, P. C., W. C. Chung, C. T. Soh, and M. L. Kosman. 1992. Eating habits of east Asian people and transmission of taeniasis. *Acta Trop.* **50:**305–315.

37. Fan, P. C., C. Y. Lin, C. C. Chen, and W. C. Chung. 1995. Morphological description of *Taenia saginata asiatica* (Cyclophyllidea: Taeniidae) from man in Asia. *J. Helminthol.* **69:**299–303.

38. Ferrer, E., Z. Cabrera, G. Rojas, M. Lares, A. Vera, B. A. de Noya, I. Fernandez, H. U. Romero, L. J. Harrison, R. M. Parkhouse, and M. M. Cortez. 2003. Evidence for high seroprevalence of *Taenia solium* cysticercosis in individuals from three rural communities in Venezuela. *Trans. R. Soc. Trop. Med. Hyg.* **97:**522–526.

39. Ferrer, E., M. M. Cortez, Z. Cabrera, G. Rojas, I. Davila, B. Alarcon de Noya, H. A. Perez, I. Fernandez, H. U. Romero, L. J. Harrison, R. M. Parkhouse, and T. Garate. 2005. Oncospheral peptide-based ELISAs as potential seroepidemiological tools for *Taenia solium* cysticercosis/ neurocysticercosis in Venezuela. *Trans. R. Soc. Trop. Med. Hyg.* **99:**568–576.

40. Fleury, A., C. Beltran, E. Ferrer, T. Garate, L. J. Harrison, R. M. Parkhouse, E. Garcia, G. Fragoso, J. Costa-Cruz, G. Biondi, S. Agapejev, and E. Sciutto. 2003. Application of synthetic peptides to the diagnosis of neurocysticercosis. *Trop. Med. Int. Health* **12:**1124–1130.

41. Flisser, A., A. Reid, E. Garcia-Zepeda, and D. P. McManus. 1988. Specific detection of *Taenia saginata* eggs by DNA hybridisation. *Lancet* **ii:**1429–1430.

42. Forbes, L. B. 2000. The occurrence and ecology of *Trichinella* in marine mammals. *Vet. Parasitol.* **93:**321–334.

43. Galan-Puchades, M. T., and M. V. Fuentes. 2000. The Asian *Taenia* and the possibility of cysticercosis. *Korean J. Parasitol.* **38:**1–7.

44. Gamble, H. R., and K. D. Murrell. 1988. Trichinellosis, p. 1018–1024. *In* W. Balows (ed.), *Laboratory Diagnosis of Infectious Disease: Principles and Practice.* Springer-Verlag, New York, N.Y.

45. Gamble, H. R., A. S. Bessonov, K. Cuperlovic, A. A. Gajadhar, F. van Knapen, K. Noeckler, H. Schenone, and X. Zhu. 2000. International Commission on Trichinellosis: recommendations on methods for the control of *Trichinella* in domestic and wild animals intended for human consumption. *Vet. Parasitol.* **93:**393–408.

46. Gamble, H. R., E. Pozio, F. Bruschi, K. Nöckler, C. M. O. Kapel, and A. A. Gajadhar. 2004. International Commission on Trichinellosis: recommendations on the use of serological tests for the detection of *Trichinella* infection in animals and man. *Parasite* **11:**3–13.

47. Garcia, H. H., and O. H. Del Brutt. 2000. *Taenia solium* cysticercosis. *Infect. Dis. Clin. N. Am.* **14:**97–119.

48. Garkavi, B. L. 1972. Species of *Trichinella* from wild carnivores. *Veterinariya* (Moscow) **49:**90–101.

49. Goeze, J. A. E. 1782. *Versuch einer Naturgeschichte der Eingeweidewürmer thierischer Körper.* P. Pape, Blankenberg, Germany.

50. Gonzalez, L. M., E. Montero, N. Morakote, S. Puente, J. L. Diaz De Tuesta, T. Serra, R. Lopez-Velez, D. P. McManus, L. J. Harrison, R. M. Parkhouse, and T. Garate. 2004. Differential diagnosis of *Taenia saginata* and *Taenia saginata asiatica* taeniasis through PCR. *Diagn. Microbiol. Infect. Dis.* **49:**183–188.

51. Gottstein, B., P. Deplazes, I. Tanner, and J. S. Skaggs. 1991. Diagnostic identification of *Taenia saginata* with the polymerase chain reaction. *Trans. R. Soc. Trop. Med. Hyg.* **85:**248–249.

52. Grencis, R. K., and D. Wakelin. 1985. Analysis of lymphocyte subsets involved in mediation of intestinal immunity to *Trichinella spiralis* in the mouse, p. 26–30. *In* C. W. Kim (ed.), *Trichinellosis.* State University of New York Press, Albany, N.Y.

53. Haeghebaert, S., and E. Maillot. 1999. *Community-Wide Outbreak of Trichinellosis: Tarn et Garonne, Haute Garonne, Tarn Districts, January–March 1998*. Réseau National de Santé Publique, Rennes, France.

54. Harrison, L. J. S., J. Delgado, and R. M. E. Parkhouse. 1990. Differential diagnosis of *Taenia saginata* and *Taenia solium* with DNA probes. *Parasitology* 100:459–461.

55. Hayunga, E. G., M. P. Sumner, M. L. Rhoads, K. D. Murrell, and R. S. Isenstein. 1991. Development of a serologic assay for cysticercosis, using an antigen isolated from *Taenia* spp. cyst fluid. *Am. J. Vet. Res.* 52:462–470.

56. Heinz, H., and G. Macnab. 1965. Cysticercosis in the Bantu of Southern Africa. *S. Afr. J. Med. Sci.* 30:19–31.

57. Hoberg, E. P. 2006. Phylogeny of *Taenia*: species definitions and origins of human parasites. *Parasitol. Int.* 55:23–30.

58. Hou, H. W., et al. 1983. *A Survey of an Outbreak of Trichinosis Caused by Eating Roasted Dog Meat*. WHO/HELM/84.15. World Health Organization, Geneva, Switzerland.

59. Ivanoska, D., K. Cuperlovic, H. R. Gamble, and K. D. Murrell. 1989. Comparative efficacy of antigen and antibody detection tests for human trichinellosis. *J. Parasitol.* 75:38–41.

60. Jeon, H. K., K. H. Lee, K. H. Kim, U. W. Hwang, and K. S. Eom. 2005. Complete sequence and structure of the mitochondrial genome of the human tapeworm, *Taenia asiatica* (Platyhelminthes; Cestoda). *Parasitology* 130:717–726.

61. Jongwutiwes, S., N. Chantachum, P. Kraivichian, P. Siriyasatien, C. Putaporntip, A. Tamburrini, G. La Rosa, C. Sreesunpasirikul, P. Yingyourd, and E. Pozio. 1998. First outbreak of human trichinellosis caused by *Trichinella pseudospiralis*. *Clin. Infect. Dis.* 26:111–115.

62. Kapel, C. M. O. 2000. Host diversity and biological characteristics of the *Trichinella* genotypes and their effect on their transmission. *Vet. Parasitol.* 93:263–278.

63. Khamboonruang, C. 1990. The present status of trichinellosis in Thailand. *Southeast Asian J. Trop. Med. Public Health* 22:312–315.

64. Kim, C. W. 1975. The diagnosis of parasitic diseases. *Prog. Clin. Pathol.* 6:267–288.

65. Kim, C. W. 1991. The significance of changing trends in trichinellosis. *Southeast Asian J. Trop. Med. Public Health* 22:316–320.

66. Kim, C. W. 1994. A decade of progress in trichinellosis, p. 35–47. *In* W. C. Campbell (ed.), *Trichinellosis*. Istituto Superiore di Sanita Press, Rome, Italy.

67. Kirichek, V. S., M. N. Belousov, and A. S. Nikitin. 1984. New data on the epidemiology of taeniasis in the extreme north of the USSR (data of investigations made in the Yamalo-Nenets autonomous region). *Med. Parazitol.* (Mosk) 1984(6):27–33. (In Russian.)

68. Kocieka, W. 2000. Human disease, pathology, diagnosis and treatment. *Vet. Parasitol.* 93:365–383.

69. Kociecka, W., F. Bruschi, C. Marini, B. Mrozewicz, and L. Pielok. 2001. Clinical appraisal of patients and detection of serum antibodies by ELISA and CIA tests in late periods of Trichinella sp. invasion. *Parasite* 8:147–151.

70. Kotula, A. W. 1983. Postslaughter control of *Trichinella spiralis*. *Food Technol* 37:91–94.

71. Kotula, A. W., K. D. Murrell, L. Acosta-Stein, L. Lamb, and L. Douglass. 1983. Destruction of *Trichinella spiralis* during cooking. *J. Food Sci.* 48:765–768.

72. Küchenmeister, F. 1857. *Animal and Vegetable Parasites*. The Sydenham Society, London, United Kingdom.

73. La Rosa, G., G. Marucci, and E. Pozio. 2003. Biochemical analysis of encapsulated and non-encapsulated species of *Trichinella* (Nematoda, Trichinellidae) from cold- and warm-blooded animals reveals a high genetic divergence in the genus. *Parasitol. Res.* 91:462–466.

74. Leclair, D., L. B. Forbes, S. Suppa, J. F. Proulx, and A. A. Gajadhar. 2004 A preliminary investigation on the infectivity of *Trichinella* larvae in traditional preparations of walrus meat. *Parasitol. Res.* 93:507–509.

75. Lees, W., J. Nightingale, D. Brown, B. Scandrett, and A. Gajadhar. 2002. Outbreak of Cysticercus bovis (*Taenia saginata*) in feedlot cattle in Alberta. *Can. Vet. J.* 43:227–228.

76. Leighty, J. C. 1983. Regulatory action to control *Trichinella spiralis*. *Food Technol.* 37:95–97.

77. Lightowlers, M. W. 2003. Vaccines for prevention of cysticercosis. *Acta Trop.* 87:129–135.

78. Lightowlers, M. W., R. Rolfe, and C. G. Gauci. 1996. *Taenia saginata*: vaccination against cysticercosis in cattle with recombinant oncosphere antigens. *Exp. Parasitol.* 84:330–338.

79. Lightowlers, M. W., and C. G. Gauci. 2001. Vaccines against cysticercosis and hydatidosis. *Vet. Parasitol.* 101:337–352.

80. Liu, H.-Y., D. Chao, and P.-C. Fan. 1981. Prevalence and chemotherapy of taeniasis among the aborigines in Nanao district, I-lan County, Northeastern Taiwan. *Proc. Natl. Sci. Counc. A.* 5:188–195.

81. MacLean, J. D., J. Viallet, C. Law, and M. Staudt. 1989. Trichinosis in the Canadian arctic: report of five outbreaks and a new clinical syndrome. *J. Infect. Dis.* 160:513–520.

82. Mafojane, N. A., C. C. Appleton, R. C. Krecek, L. M. Michael, and A. L. Willingham III. 2003. The current status of neurocysticercosis in Eastern and Southern Africa. *Acta Trop.* 87:25–33.

83. Mayta, H., A. Talley, R. H. Gilman, J. Jimenez, M. Verastegui, M. Ruiz, H. H. Garcia, and A. E. Gonzalez. 2000. Differentiating *Taenia solium* and *Taenia saginata* infections by simple hematoxylin-eosin staining and PCR-restriction enzyme analysis. *J. Clin. Microbiol.* 38:133–137.

84. McAuley, J. B., M. K. Michelson, A. W. Hightower, S. Engeran, L. A. Wintermeyer, and P. M. Schantz. 1992. A trichinosis outbreak among Southeast Asian refugees. *Am. J. Epidemiol.* 135:1404–1410.

85. Miller, L. E., H. R. Gamble, and B. Lautner. 1997. Use of risk factor evaluation and ELISA testing to certify herds for trichinae status in the United States. p. 687–690. *In Proceedings of the 9th International Conference on Trichinellosis* (M. Ortega-Pierres, H. R. Gamble, D. Wakelin, and F. van Knapen (ed.), Centro de Investigacion y Estudios Avanzados del Instituto Politecnico National Mexico, Mexico City, Mexico.

86. Millet, N. B., G. D. Hart, T. A. Reyman, M. R. Zimmerman, and P. K. Lewin. 1980. ROM I: mummification for the common people, p. 71–84. *In* A. Cockburn and E. Cockburn (ed.), *Mummies, Disease and Ancient Cultures*. Cambridge University Press, Cambridge, United Kingdom.

87. Moorhead, A., P. E. Grunenwald, V. J. Diet, and P. M. Schantz. 1999. Trichinellosis in the United States, 1991–1996: declining but not gone. *Am. J. Trop. Med. Hyg.* 60:66–69.

88. Moro, P. L., L. Lopera, N. Bonifacio, R. H. Gilman, B. Silva, M. Verastegui, A. Gonzales, H. H. Garcia, L. Cabrera, et al. 2003. *Taenia solium* infection in a rural community in the Peruvian Andes. *Ann. Trop. Med. Parasitol.* 97:373–379.

89. Moussavou-Boussougou, M. N., S. Geerts, M. Madeline, C. Ballandonne, D. Barbier, and J. Cabaret. 2005. Sewage sludge or cattle slurry as pasture fertilizers: comparative cysticercosis and trichostrongylosis risk for grazing cattle. *Parasitol. Res.* 97:27–32.

90. Mukaratirwa, S., and C. M. Foggin. 1999. Infectivity of *Trichinella* sp. isolated from *Crocodylus niloticus* to the indigenous Zimbabwean pig. *Int. J. Parasitol.* 29:1129–1131.

91. Murrell, K. D. 1985. Strategies for the control of trichinosis transmitted by pork. *Food Technol.* 39:65–68, 110–111.

92. Murrell, K. D., and F. Bruschi. 1994. Clinical trichinellosis. *Prog. Clin. Parasitol.* 4:117–150.

93. Murrell, K. D., M. Djordjevic, K. Cuperlovic, L. Sofronic, M. Savic, M. Djordjevic, and S. Damjanovic. 2004. Epidemiology of *Trichinella* infection in the horse: the risk from animal product feeding practices. *Vet. Parasitol.* 123:223–233.

94. Murrell, K. D., R. Fayer, and J. P. Dubey. 1986. Parasitic organisms. *Adv. Meat Res.* 2:311–377.

95. Murrell, K. D., J. R. Lichtenfels, D. S. Zarlenga, and E. Pozio. 2000. The systematics of the genus *Trichinella* with a key to species. *Vet. Parasitol.* 93:293–307.

96. Nelson, G. S., and J. Mikundi. 1963. A strain of *Trichinella spiralis* from Kenya of low infectivity to rats and domestic pigs. *J. Helminthol.* 37:329–338.

97. Nicoletti, A., A. Bartoloni, V. Sofia, F. Bartalesi, J. R. Chavez, R. Osinaga, F. Paradisi, J. L. Dumas, V. C. Tsang, A. Reggio, and A. J. Hall. 2005. Epilepsy and neurocysticercosis in rural Bolivia: a population-based survey. *Epilepsia* 46:1127–1132.

98. Nsengiyumva, G., M. Druet-Cabanac, B. Ramanankandrasana, B. Bouteille, L. Nsizabira, and P. M. Preux. 2003. Cysticercosis as a major risk factor for epilepsy in Burundi, east Africa. *Epilepsia* 44:950–955.

99. Nunes, C. M., A. K. Dias, F. E. Dias, S. M. Aoki, H. B. de Paula, L. G. Lima, and J. F. Garcia. 2005. *Taenia saginata*: differential diagnosis of human taeniasis by polymerase chain reaction-restriction fragment length polymorphism assay. *Exp. Parasitol.* 110:412–415.

100. Ortega Pierres, M. G., C. Arriaga, and L. Yepez-Mulia. 2000. Epidemiology of trichinellosis in Central and South America. *Vet. Parasitol.* 93:201–225.

101. Ozeretskovskaya, N. N., L. G. Mikhailova, T. P. Sabgaida, and A. S. Dovgalev. 2005. New trends and clinical patterns of human trichinellosis in Russia at the beginning of the XXI century. *Vet. Parasitol.* 132:167–171.

102. Pal, D. K., A. Carpio, and J. W. Sander. 2000. Neurocysticercosis and epilepsy in developing countries. *J. Neurol. Neurosurg. Psychiatry* 68:137–143.

103. Parija, S. C., and P. S. Sahu. 2003. A serological study of human cysticercosis in Pondicherry, South India. *J. Commun. Dis.* 35:283–289.

104. Pawlowski, Z., and M. G. Schultz. 1972. Taeniasis and cysticercosis (*Taenia saginata*). *Adv. Parasitol.* 10:269–343.

105. Pawlowski, Z. S. 1982. Taeniasis and cysticercosis, p. 313–348. *In* J. H. Steele (ed.), *Parasitic Zoonosis*, vol. 1. CRC I Handbook Series in Zoonoses. CRC Press, Boca Raton, Fla.

106. Pawlowski, Z. S. 1990. Cestodiasis, p. 490–504. *In* K. S. Warren and A. A. F. Mahmoud (ed.), *Tropical and Geographic Medicine*, 2nd ed. McGraw-Hill, New York, N.Y.

107. Pozio, E. 1998. Trichinellosis in the European Union: epidemiology, ecology and economic impact. *Parasitol. Today* 14:35–38.

108. Pozio, E. 2000. Factors affecting the flow among domestic, synanthropic and sylvatic cycles of *Trichinella*. *Vet. Parasitol.* 93:241–262.

109. Pozio, E. 2005. The broad spectrum of *Trichinella* hosts: from cold- to warm-blooded animals. *Vet. Parasitol.* 132:3–11.

110. Pozio, E., and G. La Rosa. 2000. *Trichinella murrelli* n. sp: etiological agent of sylvatic trichinellosis in temperate areas of North America. *J. Parasitol.* 86:134–139.

111. Pozio, E., G. La Rosa, and M. A. Gomez Morales. 2001. Epidemiology of human and animal trichinellosis in Italy since its discovery in 1887. *Parasite* 8:106–108.

112. Pozio, E., G. Marucci, A. Casulli, L. Sacchi, S. Mukaratirwa, C. M. Foggin, and G. La Rosa. 2004. *Trichinella papuae* and *Trichinella zimbabwensis* induce infection in experimentally infected varans, caimans, pythons and turtles. *Parasitology* 128:333–342.

113. Pozio, E., I. L. Owen, G. La Rosa, L. Sacchi, P. Rossi, and S. Corona. 1999. *Trichinella papuae* n. sp. (Nematoda), a new non-encapsulated species from domestic and sylvatic swine of Papua New Guinea. *Int. J. Parasitol.* 29:1825–1839.

114. Pozio, E., P. Pagani, G. Marucci, D. S. Zarlenga, E. P. Hoberg, D. DeMeneghi, G. La Rosa, and L. Rossi. 2005. *Trichinella britovi* etiological agent of sylvatic trichinellosis in the Republic of Guinea (West Africa) and a re-evaluation of geographical distribution for encapsulated species in Africa. *Int. J. Parasitol.* 35:955–960.

115. Pozio, E., and D. S. Zarlenga. 2005. Recent advances on the taxonomy, systematics and epidemiology of *Trichinella*. *Int. J. Parasitol.* 35:1191–1204.

116. Proctor, E. M. 1972. Identification of tapeworms. *S. Afr. Med. J.* 46:234–238.

117. Proulx, J. F., J. D. MacLean, T. W. Gyorkos, D. Leclair, A. K. Richter, B. Serhir, L. Forbes, and A. A. Gajadhar. 2002. Novel prevention program for trichinellosis in inuit communities. *Clin. Infect. Dis.* 34:1508–1514.

118. Ranque, S., B. Faugere, E. Pozio, G. La Rosa, A. Tamburrini, J.-F. Pellissier, and P. Brouqui. 2000. *Trichinella pseudospiralis* outbreak in France. *Emerg. Infect. Dis.* 6:543–547.

119. Rhoads, M. L., K. D. Murrell, G. W. Dilling, M. M. Wong, and N. F. Baker. 1985. A potential diagnostic reagent for bovine cysticercosis. *J. Parasitol.* 71:779–787.

120. Roberts, D. 1986. The last trace. *Am. Photogr.* **16**:64–68.

121. Roy, S. L., A. S. Lopez, and P. M. Schantz. 2003. Trichinellosis surveillance—United States, 1997–2001. *MMWR Surveill. Summ.* **52**:1–8.

122. Ruitenberg, E. J., F. Van Knapen, and J. W. Weiss. 1979. Foodborne parasitic infections—a review. *Vet. Parasitol.* **5**:1–10.

123. Ruiz-Perez, A., M. Santana-Ane, B. Villaverde-Ane, F. Bandera-Tirado, and N. Santana-Santos. 1995. The minimum dosage of praziquantel in the treatment of *Taenia saginata*, 1986–1993. *Rev. Cubana Med. Trop.* **47**:219–220.

124. Scandrett, W. B., and A. A. Gajadhar. 2004. Recovery of putative taeniid eggs from silt in water associated with an outbreak of bovine cysticercosis. *Can. Vet. J.* **45**:758–760.

125. Schultz, M. G., J. A. Hermos, and J. H. Steele. 1970. Epidemiology of beef tapeworm infection in the United States. *Public Health Rep.* **85**:169–176.

126. Serhir, B., J. D. MacLean, S. Healey, B. Segal, and L. Forbes. 2001. Outbreak of trichinellosis associated with arctic walruses in northern Canada, 1999. *Can. Commun. Dis. Rep.* **27**:31–36

127. Snyder, G. R., and K. D. Murrell. 1983. Bovine cysticercosis, p. 161–170. *In* G. Woods (ed.), *Practices in Veterinary Health and Preventive Medicine*. Iowa State University Press, Ames, Iowa.

128. Sohn, W.-M., H.-M. Kim, D.-I. Chung, and S. T. Yee. 2000. The first human case of *Trichinella spiralis* infection in Korea. *Korean J. Parasitol.* **38**:111–115.

129. Stehr-Green, J. K., and P. M. Schantz. 1986. Trichinosis in Southeast Asian refugees in the United States. *Am. J. Public Health* **76**:1238–1239.

130. Takahashi, Y., L. Mingyuan, and J. Waikagul. 2000. Epidemiology of trichinellosis in Asia and the Pacific rim. *Vet. Parasitol.* **93**:227–239.

131. Verastegui, M., R. H. Gilman, H. H. Garcia, A. E. Gonzalez, Y. Arana, C. Jeri, I. Tuero, C. M. Gavidia, M. Levine, V. C. Tsang, et al. 2003. Prevalence of antibodies to unique *Taenia solium* oncosphere antigens in taeniasis and human and porcine cysticercosis. *Am. J. Trop. Med. Hyg.* **69**:438–444.

132. Viens, P., and P. Auger. 1981. Clinical and epidemiological aspects of trichinosis in Montreal, p. 275–277. *In* C. W. Kim, E. J. Ruitenberg, and J. S. Teppema (ed.), *Trichinellosis*. Redbooks, Chertsey, England.

133. Waloch, M. 2004. Cestode infections in Poland in 2002. *Przegl. Epidemiol.* **58**:165–169. (In Polish.)

134. Waloch, M. 2005. Cestode infections in Poland in 2003. *Przegl. Epidemiol.* **59**:331–335. (In Polish.)

135. Wandra, T., P. Sutisna, N. S. Dharmawan, S. S. Margono, R. Sudewi, T. Suroso, P. S. Craig, and A. Ito. 2006. High prevalence of *Taenia saginata* taeniasis and status of *Taenia solium* cysticercosis in Bali, Indonesia, 2002–2004. *Trans. R. Soc. Trop. Med. Hyg.* **100**:346–353.

136. Wang, Z. Q., and J. Cui. 2001. Epidemiology of swine trichinellosis in China. *Parasite* **8**:67–70.

137. Wang, Z. Q., and J. Cui. 2001. The epidemiology of human trichinellosis in China during 1964–1999. *Parasite* **8**:63–66.

138. Wanzala, W., J. A. Onyango-Abuje, E. K. Kang'ethe, H. Ochanda, and L. J. Harrison. 2002. Serodiagnosis of bovine cysticercosis by detecting live *Taenia saginata* cysts using a monoclonal antibody-based antigen-ELISA. *J. S. Afr. Vet. Assoc.* **73**:201–206.

139. White, A. C., Jr. 2000. Neurocysticercosis: updates on epidemiology, pathogenesis, diagnosis, and management. *Ann. Rev. Med.* **51**:187–206.

140. Wilkins, P. P., J. C. Allan, M. Verastegui, M. Acosta, A. G. Eason, H. H. Garcia, A. E. Gonzalez, R. H. Gilman, and V. C. Tsang. 1999. Development of a serologic assay to detect *Taenia solium* taeniasis. *Am. J. Trop. Med. Hyg.* **60**:199–204.

141. Willingham, A. L., III, N. V. De, N. Q. Doanh, D. Congle, T. V. Dung, P. Dorny, P. D. Cam, and A. Dalsgaard. 2003. Current status of cysticercosis in Vietnam. *Southeast Asian J. Trop. Med. Public Health* **34**:35–50.

142. Wright, W. H., K. B. Kerr, and L. Jacobs. 1943. Studies on trichinosis. XV. Summary of the findings of *Trichinella spiralis* in a random sampling and other samplings of the population of the United States. *Public Health Rep.* **58**:1293–1313.

143. Yamasaki, H., J. C. Allan, M. O. Sato, M. Nakao, Y. Sako, K. Nakaya, D. Qiu, W. Mamuti, P. S. Craig, and A. Ito. 2004. DNA differential diagnosis of taeniasis and cysticercosis by multiplex PCR. *J. Clin. Microbiol.* **42**:548–553.

144. Zarlenga, D. S., D. P. McManus, P. C. Fan, and J. H. Cross. 1991. Characterization and detection of a newly described Asian Taeniid using cloned ribosomal DNA fragments and sequence amplification by the polymerase chain reaction. *Exp. Parasitol.* **72**:174–183.

145. Zoli, A., O. Shey-Njila, E. Assana, J. P. Nguekam, P. Dorny, J. Brandt, and S. Geerts. 2003. Regional status, epidemiology and impact of *Taenia solium* cysticercosis in Western and Central Africa. *Acta Trop.* **87**:35–42.

Food Microbiology: Fundamentals and Frontiers, 3rd Ed.
Edited by M. P. Doyle and L. R. Beuchat
© 2007 ASM Press, Washington, D.C.

Eugene G. Hayunga

30

Helminths Acquired from Finfish, Shellfish, and Other Food Sources

A variety of human helminthic infections can be acquired through the consumption of food products from infected animals and plants, through the accidental ingestion of infected invertebrates in foodstuffs or drinking water, or through inadvertent fecal contamination by humans or animals (Table 30.1). Effective prevention involves exploiting parasite vulnerabilities (36) and requires a sound understanding of parasite life cycles and modes of transmission. Unlike bacteria, the infective stages of helminths generally do not propagate. As a result, the critical control point for foodborne helminthiases is initial food preparation, not subsequent storage, reheating, or processing. Although these helminthic infections can readily be prevented, the reality is that safe water supplies, adequate sanitation, and reliable food handling simply do not exist for much of the world's population, a fact generally not appreciated by tourists "who explore tropical countries with a zeal undamped by any knowledge of preventive medicine" (47).

Foodborne helminths, although taxonomically diverse, share the common characteristic of requiring more than one host to complete their life cycles. Termed biohelminths by Kisielewska (44), their transmission requires close behavioral contact between hosts. Typically, the definitive

host of a biohelminth occupies the highest trophic level of the food chain. Prevention of biohelminth infections can be accomplished by avoiding the intermediate hosts or by adequately cooking foods. In contrast, helminths with eggs or free-living stages that can survive a certain length of time in the external environment, termed geohelminths (44), are typically transmitted via contaminated water or foods and are best controlled by improved sanitation. In addition, any parasite capable of penetrating the skin can also penetrate the buccal epithelium.

HELMINTHS ACQUIRED FROM FINFISH AND SHELLFISH

Anisakis and Related Roundworms

Several related nematodes of the genera *Anisakis*, *Pseudoterranova*, and *Contracaecum* may be acquired by eating raw fish or squid in seafood dishes such as sushi, sashimi, seviche, and lomi-lomi (61). The noninvasive form of anisakiasis is generally asymptomatic, resulting in "tingling throat syndrome" when worms are released from seafood following digestion and migrate up the esophagus into the pharynx, where they subsequently may be expectorated (62). In the invasive form, worms

Eugene G. Hayunga, Office of Extramural Policy, National Institute of Child Health and Human Development, National Institutes of Health, Bethesda, MD 20892-7510.

649

Table 30.1 Sources of infection with some foodborne helminths of humans

Helminth	Beef or pork[a]	Finfish	Shellfish	Other invertebrates	Fruits or vegetables	Other food sources	Fecal contamination
Alaria americana						X[b]	
Angiostrongylus sp.			X	X	X		
Anisakis sp.		X	X[c]				
Ascaris lumbricoides							X
Baylisascaris procyonis							X
Capillaria philippinensis		X					
Clonorchis sinensis		X	X				
Dicrocoelium dendriticum				X	X[d]		
Diphyllobothrium sp.		X					
Dipylidium caninum				X			
Dracunculus medinensis				X			
Echinococcus granulosus							X
Echinococcus multiocularis							X
Echinostomum sp.		X		X		X[e]	
Eustrongylides sp.		X					
Fasciola hepatica					X	X[f]	
Fasciolopsis buski					X		
Gnathostoma spinigerum	X	X		X		X[g]	
Heterophyes heterophyes		X					
Hymenolepis diminuta				X			
Hymenolepis nana							X
Ligula intestinalis		X					
Macracanthorhynchus hirudinaceus				X			
Metagonimus yokogawai		X					
Moniliformis moniliformis				X			
Multiceps multiceps							X
Nanophyetus salmincola		X					
Nybelinia surmenicola			X[c]				
Opisthorchis sp.		X	X				
Paragonimus westermani			X			X[h]	
Phaeneropsolos bonnei				X			
Philometra sp.		X					
Prosthodendrium molenkampi				X			
Spirometra sp.	X	X		X		X[i]	
Strongyloides sp.						X[j]	
Taenia saginata	X						
Taenia solium	X						X
Toxocara canis							X
Trichinella spiralis	X					X[k]	
Trichostrongylus sp.							X

[a] Described in chapter 29.
[b] Frog, raccoon, and opossum.
[c] Squid.
[d] Acquired by ingestion of ants on unwashed herbs and vegetables.
[e] Frog and tadpole.
[f] The condition halzoun occurs when adult worms in uncooked sheep liver attach to the pharynx.
[g] Pork, chicken, duck, frog, eel, snake, and rat.
[h] Wild boar.
[i] Acquired by ingestion of procercoids in copepods and plerocercoids in frogs, tadpoles, lizards, snakes, birds, and mammals; infection has also been reported to occur from eating undercooked pork.
[j] Transmammary infection has been reported.
[k] Reported to be acquired from a variety of animals including bear, walrus, and horse.

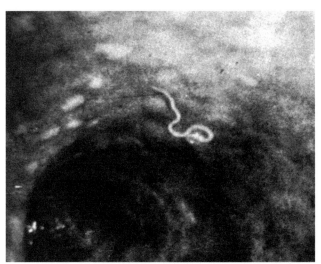

Figure 30.1 *Anisakis* embedded in the human gastric mucosa as visualized by gastroscopy. (Photograph contributed by Tomoo Oshima; illustration courtesy of the Armed Forces Institute of Pathology, Washington, D.C., AFIP 76-2118.)

typically penetrate the mucosa or submucosa of the stomach or small intestine (Fig. 30.1), resulting in epigastric pain, nausea, vomiting, and diarrhea, usually 12 h after consumption of the infected seafood. Chronic anisakiasis may mimic peptic ulcers, appendicitis, enteritis, Crohn's disease, or gastric carcinomas. The only effective treatment is surgical removal of the worms.

The life cycles of these helminths are not completely known. The adult worms are intestinal parasites of dolphins, whales, seals, and sea lions. Eggs passed in the feces of marine mammals embryonate in seawater and develop into larvae that are eaten by krill. The infected crustaceans are next eaten by fish or squid, and the larvae develop further. The life cycle is completed in marine mammals, but when fish or squid are eaten by the unsuitable human host the parasites do not develop further or reproduce. Anisakidae larvae have been found in rockfish, herring, cod, halibut, mackerel, and salmon.

The prominent reddish brown larvae of *Pseudoterranova* are readily visible in contrast to the whitish fish tissue, but the smaller, lighter-colored *Anisakis* larvae are more difficult to detect (61). Fish raised in commercial pens are less likely to be infected than wild-caught fish (15). Comparison of salmon steaks with salmon fillets indicates predilection sites for the larvae and suggests that certain cuts may pose a lower risk for infection (19). In some fishes, most of the juvenile larvae are found in the viscera (16), suggesting that immediate evisceration after catching would prevent postmortem migration of

larvae into the musculature. However, only thorough cooking or prolonged freezing will kill the parasites and completely eliminate the risk of infection. The U.S. Food and Drug Administration recommends that all finfish and shellfish intended to be eaten raw, partly cooked, or marinated be blast frozen to $-35°C$ or below for 15 h or regularly frozen to $-20°C$ or below for 7 days (74). Cold smoking and most methods of brining of fish are not reliable preventive measures (66).

Most human infections with the so-called sushi parasite have been reported from Japan and The Netherlands. In Japan, hypochlorhydria or achlorhydria has been found in more than half of anisakiasis patients and may predispose them to infection (23). In the United States, fewer than 10 cases of anisakiasis are diagnosed annually (74). Considering the increasing popularity of raw seafood dishes and the proliferation of sushi restaurants, the low rate of reported infection is remarkable. The majority of human infections acquired in the United States have been associated with dishes prepared at home. The source of commercial fish harvesting with regard to the distribution of reservoir hosts may also contribute to the epidemiology of this disease, as geographic variation has been reported for rates of infection of fish from different localities (17, 21).

Diphyllobothrium latum

The broad tapeworm *Diphyllobothrium latum* occurs in northern temperate regions of the world where raw or undercooked freshwater fishes are eaten; prevalent since neolithic times (45), infection is closely related to dietary and cultural practices in food preparation (22). Infective larvae may be found in whitefish, trout, pike, and salmon. Cases of infection have been reported throughout Europe, particularly in the Baltic countries; the Great Lakes region of the United States; Canada; Japan; South America; and Australia. *Diphyllobothrium latum* was introduced into the United States by European immigrants in the middle of the 19th century (17). Although it is the largest of the human tapeworms, measuring up to 9 m in length, infections may be mild or asymptomatic. Nausea, abdominal pain, diarrhea, and weakness are common manifestations. *Diphyllobothrium latum* may also cause pernicious anemia and vitamin B_{12} deficiency because the worm is highly efficient in competing with its host for available vitamins.

Diphyllobothrium latum eggs shed in feces require approximately 12 days to embryonate, at which time the ciliated coracidium exits the egg through the operculum. When eaten by a *Cyclops* or *Diaptomus* copepod, the coracidium penetrates the body cavity of the freshwater crustacean and develops into the procercoid stage of the parasite. When small fish such as minnows eat infected crustaceans, the procercoid penetrates into the viscera or muscles of the fish and develops into another unsegmented stage called the plerocercoid. The carnivorous fish that eat these fish are termed paratenic hosts because, although the plerocercoid penetrates into their viscera or muscles, it does not develop beyond this stage. Further development from the plerocercoid into the adult worm requires ingestion by the human definitive host.

Diagnosis of infection is based on the presence of eggs in the feces. The distinctive *Diphyllobothrium latum* egg is ovoid, approximately 45 by 70 µm, with a prominent operculum at one end and a characteristic knob at the abopercular end; proglottids may occasionally be found in the feces. Infection is best prevented by adequate cooking or freezing of fish before consumption. Improved sanitation measures can also help reduce prevalence by interrupting the parasite's life cycle. Praziquantel is the recommended anthelminthic (1). *Diphyllobothrium dendriticum* and *Ligula intestinalis*, tapeworms of piscivorous birds, and *Diphyllobothrium pacificum*, a tapeworm of seals, have also been found in humans.

Clonorchis sinensis

Although the correct scientific name for the Chinese liver fluke is *Opisthorchis sinensis*, its original name, *Clonorchis sinensis*, remains more commonly accepted. Widespread throughout Asia, the parasite is typically acquired by eating infected freshwater fishes; the infective stage of *Clonorchis sinensis* has also been found in crayfish (24). Reservoir hosts include dogs, cats, foxes, pigs, rats, mink, badgers, and tigers.

Adult worms, measuring approximately 1.2 to 2.4 cm in length and 0.3 to 0.5 cm in width, reside in the bile duct. When eggs passed in the feces are eaten by *Parafossarulus manchouricus* or other hydrobiid snails such as *Bulimus*, *Semisulcospira*, *Alocinma*, or *Melanoides* species, the miracidium is released into the digestive tract and penetrates into the hemocoel, where it develops into first the sporocyst and then the redia stage. The redia gives rise to free-swimming cercariae that leave the snail and penetrate the second intermediate host, a cyprinid fish, where they encyst as metacercariae in the gills, fins, or muscles or under the skin. When the definitive host eats an infected fish, metacercariae excyst in the duodenum, migrate into the bile duct, and develop into adult worms.

Clonorchis sinensis may live in the human host for as long as 25 to 30 years, and massive infections with as many as 500 to 1,000 parasites have been reported. The severity of symptoms is related to the intensity and duration of infection. Diarrhea, epigastric pain, and anorexia are typical manifestations of acute clonorchiasis. The adult worm produces localized tissue damage that may result in hyperplasia or metaplasia of the bile duct epithelium, duct thickening, fibrosis, bilary stasis, and secondary bacterial infection. Pancreatitis may occur when worms enter the pancreatic duct. An association between cholangiocarcinoma and *Clonorchis sinensis* infection has also been reported (39). Diagnosis is made by identifying eggs in the feces. The operculate *Clonorchis sinensis* eggs measure approximately 27 by 16 µm and are characterized by distinctive opercular shoulders and a small spinelike process at the abopercular end. Infection is best prevented by avoiding raw, undercooked, or pickled finfish and shellfish. Snail control and improved sanitation can also help reduce transmission. Praziquantel and albendazole are effective anthelminthics.

Two closely related bile duct flukes of dogs and cats may also be acquired by eating uncooked cyprinid fishes. *Opisthorchis felineus* occurs throughout Eastern Europe and portions of the former USSR, and *Opisthorchis viverrini* has been reported to occur in Southeast Asia. The clinical picture for human infection with these species is very similar to that for clonorchiasis.

The Human Lung Fluke *Paragonimus westermani*

Paragonimus westermani is the best known and most widely distributed lung fluke, although other species of this genus may parasitize humans (78). *Paragonimus westermani* infection is common throughout the Far East and occurs to a lesser extent in parts of Africa and the Indian subcontinent. *Paragonimus westermani* is a common parasite of mink in Canada and the eastern United States. Other reservoir hosts include a variety of animals that may eat crustaceans. Infection in dogs, cats, tigers, and cattle has been reported.

The reddish brown, thick-bodied flatworms are found encapsulated in cystic structures adjacent to the bronchi or bronchioles. They measure approximately 0.8 to 1.6 cm in length by 0.4 to 0.8 cm in width and are 3.0 to 0.5 cm thick. Eggs are released by the parasite into the bronchioles, where they may be expectorated or swallowed and passed in the feces. Eggs require several weeks in an aqueous environment for development. Upon hatching, the free-swimming miracidium penetrates a snail of the genus *Brotia*, *Semisulcospira*, *Tarebia*, or *Thiara*, where it undergoes further development into a sporocyst, and then two generations of rediae. Approximately 11 weeks after the snail is infected, cercariae are shed and then penetrate any of a variety of freshwater crabs and crayfish, where they become encysted in the muscles, gills, and other organs as metacercariae. Important shellfish hosts include the freshwater and brackish water crabs *Eriocheir*, *Potamon*, and *Sundathelphusa* and the crayfish *Procambarus*.

Humans become infected by eating raw or improperly cooked freshwater crabs or crayfish. Dishes such as raw crayfish salad, jumping salad (live shrimp), drunken crab (live crabs in wine), and crayfish curd are popular throughout the Orient. Pickling does not kill the parasite. The infective metacercariae can be killed by boiling the crabs for several minutes until the meat has congealed and turned opaque (42). Infection may also be acquired from shellfish juices used in food dishes or folk remedies, from food prepared by using contaminated utensils or chopping blocks, or from drinking water contaminated with metacercariae released from dead or injured crustaceans.

Migration of parasites through host tissues produces localized hemorrhage and infiltration of lymphocytes. Pulmonary symptoms include dyspnea, chronic cough, chest pain, night sweats, hemoptysis, and persistent rales. The severity of symptoms appears to be related to the number of parasites present. Pleural effusion and fibrosis may occur in long-standing infections, although there is also evidence that lesions may resolve without treatment. Neurological complications result from migration of the parasite into the spinal cord or brain. *Paragonimus westermani* has also been found in the intestinal wall, peritoneum, pleural cavity, and testes. *Paragonimus szechuanensis* and *Paragonimus hueitungensis* cause cutaneous larva migrans characterized by eosinophilia, anemia, and low-grade fever. Diagnosis is based on the presence of eggs in expectorant or feces. The characteristic golden-brown eggs are 80 to 120 µm long by 48 to 60 µm wide. Chest X rays may show nodular shadows or calcified spots. Praziquantel and bithionol are effective anthelminthics.

Capillaria philippinensis

Capillaria philippinensis, a nematode found in freshwater fishes, causes a severe and potentially fatal infection in humans (9). In 1967 and 1968, the disease reached epidemic proportions in The Philippines with over 1,000 confirmed cases and more than 100 deaths (11). The infection appears to be acquired as a result of eating raw fish that contain infective larvae. The freshwater fishes *Ambassis miops*, *Eleotris melanosoma*, and *Hypseleotris bipartita* have been implicated in experimental infections, but the life cycle of *Capillaria philippinensis* has not been fully elucidated nor have natural reservoir hosts been identified.

Both larvae and adult worms may be found embedded in the intestine (Fig. 30.2). Diagnosis of infection is made

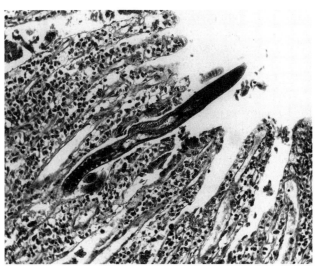

Figure 30.2 Transverse section of larval *Capillaria philippinensis* embedded in human intestinal glands. Magnification, ×95. (Illustration courtesy of the Armed Forces Institute of Pathology, AFIP 69-1066.)

by identifying eggs in the feces. Symptoms of capillariasis include borborygmus, abdominal pain, nausea, vomiting, diarrhea, and anorexia during the acute phases. If the infection is untreated, intestinal malabsorption and intractable diarrhea lead to cachexia and possibly death. Treatment consists of a regimen of mebendazole with electrolyte and protein supplementation.

Gnathostoma Roundworms

Roundworms of the genus *Gnathostoma* reside in the stomach walls of a variety of carnivorous mammals. The life cycle of this parasite typically involves two intermediate hosts, a copepod and a freshwater fish; however, a variety of animals may serve as paratenic or transport hosts. Unable to mature in human hosts, the parasite wanders aimlessly through the tissues, causing a severe larva migrans that may persist for years. Gnathostomiasis has been reported to occur throughout the world, particularly in Thailand, and is considered to be an emerging infection in several parts of Latin America.

Gnathostoma spinigerum has been found in tigers, leopards, lions, domestic cats, minks, and dogs. The stout, reddish, female worm ranges in length from 25 to 54 mm and is characterized by spines covering the anterior half of the body and by a prominent cephalic bulb covered by rows of sharp hooklets. Male worms are about half as long as females. Human infection typically results from eating raw, marinated, or poorly cooked freshwater fish (66). Recent cases in Mexico have been attributed to marinated freshwater tilapia (60). In addition, infection can be acquired by eating pork, chicken, duck, frog, eel, snake, or rat or by accidentally ingesting infected copepods in drinking water. It is also possible that larvae may penetrate the skin during food handling (13). Larvae can be killed by cooking or by immersion in strong vinegar for 5 h or longer; immersion in lime juice and chilling at 4°C for 1 month are not effective (4). Raw foods, particularly fish and chicken, should be avoided in areas where infection is endemic, and drinking water should be filtered before consumption.

Nausea, abdominal pain, and vomiting usually develop between 24 and 48 h following the ingestion of infected meat or fish. Symptoms of larva migrans or creeping eruption include pruritus, urticaria, tenderness, and painful subcutaneous swelling. Invasion of the central nervous system may result in meningitis and neuropathy. Diagnosis is difficult, and chemotherapy is of questionable value. Albendazole and ivermectin have been reported as alternatives to surgical removal of the larvae (1).

Other Helminths Associated with Seafood

Heterophyes heterophyes, an intestinal fluke acquired from the mullet, has been found in Egypt, Israel, and throughout Asia. The parasite may cause nausea, diarrhea, and abdominal pain, but light infections are often asymptomatic. There have been reports of fatal myocarditis and neurological complications when helminth eggs penetrate the intestine and enter the circulatory system. The closely related trematode *Metagonimus yokagawai*, acquired from salmonid fishes, has been found in Asia, the Balkans, Israel, Spain, and portions of the former USSR. Fish-eating mammals and birds, such as dogs, cats, and pelicans, serve as reservoir hosts. Infection with species of another trematode, *Echinostomum*, has been reported in Japan and has been attributed to eating raw freshwater fish, in particular sashimi (34). Diagnosis of infection is made by identifying eggs passed in the feces. Intestinal symptoms are mild and depend upon the number of worms present. Praziquantel is an effective anthelminthic.

Salmon poisoning, a severe and frequently fatal disease of dogs in the Pacific Northwest, is associated with the intestinal trematode *Nanophyetus* (*Troglotrema*) *salmincola*. Disease in dogs occurs because *Nanophyetus salmincola* serves as a vector for the rickettsial pathogen *Neorickettsia helmintheca*. Human *Nanophyetus salmincola* infection, characterized by nausea, diarrhea, and intestinal discomfort, may be acquired by eating uncooked salmonid fishes and has also been attributed to handling freshly killed coho salmon (35).

The nematode *Eustrongylides* is a common parasite of piscivorous birds. In addition to fish, reptiles and amphibians may also serve as intermediate hosts and thus contain infective larvae. An infection in New York City was attributed to eating raw fish prepared at home (77). Cases in Maryland and New Jersey were reported to occur in fishermen who had eaten live bait (8, 27). As in *Anisakis* infections, surgical removal of worms is the only effective treatment.

Philometra, a nematode closely related to *Dracunculus*, was acquired by a fisherman in Hawaii as a result of filleting a carangid fish (18). Infection with the tapeworm *Nybelinia surmenicola* has been attributed to eating raw squid (43). Sparganosis may be transmitted by a variety of animals, including fish. Although typically associated with snails, *Angiostrongylus* may also be acquired from freshwater prawns and land crabs.

HELMINTHS ACQUIRED FROM VEGETATION

The Sheep Liver Fluke *Fasciola hepatica*

Sheep liver rot, caused by the digenetic trematode *Fasciola hepatica*, was recognized as early as the 14th century (20). *Fasciola hepatica* was the first trematode for which a complete life cycle was elucidated (46). Sheep, cattle, and other herbivores acquire the infection by eating metacercariae encysted on aquatic plants. *Fasciola hepatica* has been found in goats, horses, deer, rabbits, camels, vicuña, swine, dogs, and squirrels. Human infection is prevalent in sheep-raising areas throughout the world. In the United States, human fascioliasis has been reported in California (53) and Puerto Rico (37).

Fasciola hepatica is a large fluke, measuring approximately 3 cm in length by 1.5 cm in width, and is readily identified by its characteristic "cephalic zone," a distinct conical projection at the anterior end. Adult worms reside in the biliary passages and gall bladder. Eggs passed with feces into the water require 9 to 15 days to mature but may remain viable for several months in soil if they remain moist. Upon hatching, the free-swimming miracidium penetrates a lymnaeid snail of the genus *Lymnaea*, *Succinea*, *Fossaria*, or *Practicolella*, where it develops further. Free-swimming cercariae are shed from the snail and attach to aquatic vegetation, where they encyst as metacercariae. The metacercariae are susceptible to drying but can survive over winter (55).

Humans become infected by ingesting infested freshwater plants or free metacercariae in drinking water. Human cases are frequently traced to watercress, *Nasturtium officinalis* (41). Transmission can be controlled by using molluscicides, by draining ponds, and by protecting crops and water supplies from contact with livestock.

Some inflammation is associated with migration of the parasite, but mild infections are often asymptomatic. Tissue destruction occurs when worms penetrate the liver. In sheep, liver rot causes massive damage. There may be mechanical obstruction of bile ducts, hyperplasia of the biliary epithelium, and proliferation of the connective tissue. Worms may erode the walls of the bile ducts and invade the liver parenchyma. Secondary bacterial infection and portal cirrhosis have been reported, but liver calcification appears to be rare. A pharyngeal form of disease, called halzoun, may occur following the ingestion of raw liver from infected animals when worms present in the liver attach to the pharynx. Pain, bleeding, and edema of the face and neck are associated with halzoun.

Diagnosis is based on the presence of eggs in the feces. The relatively large, operculate *Fasciola hepatica* eggs are approximately 130 to 150 µm long by 63 to 90 µm wide. Ingestion of liver from infected sheep or cattle may result in spurious infection when eggs present in the food are passed in feces. Such false-positive findings can be ruled out by subsequent stool examinations. A lengthy regimen of bithionol has been the treatment of choice, although triclabendazole offers the advantage of a single dosage.

The Giant Intestinal Fluke *Fasciolopsis buski*

Fasciolopsis buski, the largest trematode parasite of humans, is endemic throughout Asia. The pig is an important reservoir host, but infection in dogs and rabbits has also been reported. Although only 0.8 to 3.0 mm thick, *Fasciolopsis buski* can grow to 7.5 cm in length by 2 cm in width. Adult worms attach to the bowel walls, primarily along the duodenum and jejunum. Eggs passed in the feces are unembryonated and require 3 to 7 weeks in freshwater to develop. Upon hatching, the free-swimming miracidium penetrates a snail of the genus *Segmentina* or *Hippeutis*, where it develops further. Approximately 4 to 7 weeks after the snail is infected, cercariae are shed into the water, and then attach to vegetation, where they encyst as metacercariae.

Humans become infected by eating infested water chestnuts, bamboo, caltrop, or lotus. Individuals may also acquire the parasite by peeling the hulls of plants with their teeth. The metacercariae excyst in the small intestine, attach to the mucosa, and develop into adult worms in about 3 months. Drying or cooking the plants before eating kills the metacercariae (5). Immersing of vegetables in boiling water for a few seconds, or even peeling and washing them in clear water, is sufficient to preclude infection (4). Poor sanitation and the use of human and swine feces as fertilizer are major factors in disease transmission (59).

Adult worms feed not only on intestinal contents but also on the intestinal epithelium, leading to local ulceration and hemorrhage. Nausea, abdominal pain, diarrhea, and hunger pangs are common. In heavy infections, stools are profuse and light yellow in color, suggestive of malabsorption; intestinal obstruction and ascites have been reported. Diagnosis is based on the presence of eggs in the feces. The yellow-brown eggs have a small operculum and are approximately 130 to 140 µm long by 80 to 85 µm wide. Adult worms are seen in feces only following chemotherapy or purgation. Praziquantel is the anthelminthic of choice.

Other Helminths Associated with Vegetation

Fresh vegetables grown in areas where night soil (human waste) is used as fertilizer are frequently contaminated and, thus, may facilitate transmission of any of a number of geohelminths. Human infection with *Dicrocoelium* is explained by the accidental ingestion of ants on vegetation, and *Angiostrongylus costaricensis* is thought to be acquired by eating raw fruits and vegetables on which snails have left larvae in mucus deposits or by accidentally ingesting infected snails on unwashed vegetation. *Trichostrongylus* and *Echinococcus granulosus* infections have also been attributed to contaminated vegetation (4).

HELMINTHS ACQUIRED FROM INVERTEBRATES IN DRINKING WATER

The Guinea Worm *Dracunculus medinensis*

The long, threadlike roundworm *Dracunculus* has plagued humans since antiquity. It is believed to be the "fiery serpent" described in the Bible (Numbers 21: 4–9) and is symbolically depicted in the caduceus, the insignia of the medical profession. As a result of an aggressive and highly effective eradication program, the worldwide annual incidence of dracunculiasis has been reduced by 99% from that estimated in 1986 and the disease has been completely eliminated from many countries; it now occurs primarily in the Sudan and several other African countries (38, 64).

Although infection is rarely fatal, this parasite causes considerable discomfort and disability. The female nematode is almost a meter in length but less than 2 mm thick; males are inconspicuous and only 2 cm long. Worms develop to maturity in the body cavity or deep connective tissues. Females then migrate to the subcutaneous tissues, become gravid, and stimulate the formation of a blister that eventually ruptures to expose part of the worm. Upon contact with freshwater, the worm bursts to release larvae, which are then ingested by copepods of the genus *Cyclops*, *Mesocyclops*, or *Thermocyclops*.

Humans become infected by swallowing infected copepods present in drinking water (Fig. 30.3). The larvae penetrate the digestive tract and take about a year to reach maturity. Transmission is clearly related to poverty, the quality of drinking water, and water contact by infected individuals. A measure as simple as sieving drinking water through a piece of cloth will remove copepods and prevent infection. The provision of safe drinking-water supplies in a village is usually followed by disappearance of the disease (51, 57). *Dracunculus* infections in dogs in China and the former USSR have been reported, but it is not clear whether canines play any significant role as reservoir hosts.

The traditional treatment of winding the worm around a small stick and slowly extruding it can still be effective provided asepsis is maintained. Surgical removal of *Dracunculus* has been practiced in India and Pakistan.

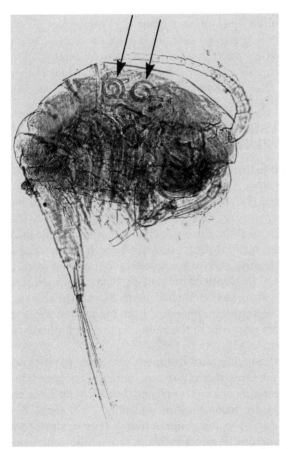

Figure 30.3 Arrows show *Dracunculus* larvae within the body cavity of the intermediate host, *Cyclops*. Magnification, ×60. (Specimen contributed by E. L. Schiller, Johns Hopkins University School of Public Health, Baltimore, Md.; illustration courtesy of the Armed Forces Institute of Pathology, AFIP 68-4629.)

Although not curative, metronidazole decreases inflammation and facilitates removal of the worm, but this compound may be mutagenic at the dosage recommended; mebendazole has been reported to kill the parasite directly, but the treatment of choice remains slow extraction of the worm combined with wound care (1).

Other Helminths Acquired from Copepods
There is evidence that the tissue-dwelling nematode *Gnathostoma spinigerum* may be acquired directly from copepods (12). Sparganosis, although typically acquired from a variety of vertebrate hosts, may also result from ingesting infected copepods.

HELMINTHS ACQUIRED FROM OTHER INVERTEBRATES

Snails
The rodent lungworm *Angiostrongylus cantonensis* is a slender roundworm about 25 mm in length that typically resides in the pulmonary arteries of rodents. The adult nematodes lay eggs that hatch in the lungs. First-stage larvae migrate to the trachea, are swallowed, and pass in the feces, where they are ingested by any of a variety of slugs, land snails, or planarians. When rodents eat infected molluscs, the larvae migrate to the brain, developing into adults in about 4 weeks, and then to the pulmonary arteries where they begin to lay eggs after an additional 2 weeks. In humans, migration of the parasite through the brain typically presents as eosinophilic meningitis but may also result in ocular involvement and even fatal consequences (58, 71). Cases of infection have been reported from Taiwan, Thailand, India, The Philippines, Japan, Hawaii, several Pacific islands, Cuba, and the Ivory Coast.

Foodborne infections have been attributed to the consumption of raw or undercooked molluscs such as the freshwater snail *Pila* or the giant African land snail *Achatina fulica*. The parasite has been found in other potential food sources such as the land crab *Cardisoma hirtipes* and the coconut crab *Birgus latro* (2). Freshwater prawns, which are frequently eaten raw in Thailand, Vietnam, and Tahiti, have been implicated in human infection (58), although shrimp, crabs, and frogs serve as paratenic hosts. In addition, the parasite can be acquired by the accidental ingestion of slugs on lettuce, helminth larvae left by snails in mucus deposits on fruits and vegetables, and contaminated drinking water. *Angiostrongylus cantonensis* infection has also been reported in the United States in a child who had deliberately consumed a raw snail as a result of peer pressure (52). Transmission can be prevented by thorough cooking and appropriate attention to food-handling practices, particularly the careful washing of fruits and vegetables and washing of hands after exposure to molluscs while gardening.

Angiostrongylus costaricensis is a closely related roundworm found in the mesenteric arteries of the cotton rat *Sigmodon hispidus*. Human abdominal angiostrongyliasis in Costa Rica, Honduras, Panama, Mexico, Brazil, and Venezuela has been reported. The veronicellid slug *Vaginulus plebius* has been implicated in the transmission of *Angiostrongylus costaricensis*. Human infection is acquired by accidental ingestion of slugs or by contact with contaminated fruits, vegetables, or grass.

Several species of the intestinal trematode *Echinostomum* may be acquired by eating metacercariae encysted in snails. Human echinostomiasis occurs primarily throughout Asia (34), although there is archeological evidence of human infection in the New World (67). Infections are generally mild and often asymptomatic. Adult worms are found in a variety of domestic animals and birds, and the Norway rat is believed to be an important reservoir host. Diagnosis of infection is based on identification of unembryonated, operculate eggs measuring approximately 83 to 116 µm in length by 58 to 69 µm in width in stool specimens. Praziquantel appears to be an effective anthelminthic.

Ants

Dicrocoelium dendriticum is a trematode found in the biliary passages of sheep, deer, and other herbivores. The intermediate hosts are land snails and ants. Parasites are shed from snails in slime balls that are deposited on the grass and eaten by ants; the mammalian host becomes infected by ingestion of infected ants. Reports of human infection are frequently spurious, as ingestion of liver from infected sheep can result in a false-positive diagnosis when eggs present in the food are passed in feces. However, genuine human cases have been reported in Europe, Asia, and Africa. Transmission of *Dicrocoelium* can best be prevented by careful washing of herbs and vegetables to remove ants.

Fleas

Dipylidium caninum is a common tapeworm of dogs and cats throughout the world. Tapeworm eggs are ingested by the flea *Ctenocephalides*, where they develop into the cysticercoid stage. Human infection appears to be limited to young children and results from accidental ingestion of infected fleas. Infection can best be prevented by control of fleas on pets and by periodic worming of animals when necessary.

Beetles, Cockroaches, and Other Insects

Hymenolepis diminuta is a tapeworm of rats, mice, and other rodents. Eggs passed in rodent feces are ingested by flour beetles (*Tribolium*, *Tenebrio*), cockroaches (*Blattella*, *Periplaneta*), or fleas, where they develop into the cysticercoid stage. Human infection, acquired by the accidental ingestion of infected insects, is typically asymptomatic, although nausea, abdominal pain, and diarrhea may occur. Preventive measures include protection of grains and foodstuffs from insects and rodent control.

The acanthocephalan *Moniliformis moniliformis* is an intestinal parasite of rats that also utilizes cockroaches and beetles as intermediate hosts. As is the case with *Hymenolepis diminuta*, human infection occurs usually in young children. The giant leechlike acanthocephalan *Macracanthorhynchus hirudenaceus* is a cosmopolitan parasite of swine. The spiny proboscis embeds itself in the intestinal mucosa; female worms measure up to 65 cm in length. Human infections are rare but have been directly attributed to eating raw beetles. The spirurid nematode *Gongylonema* is typically acquired by ingestion of cockroaches; an infection in an adult resident of New York City was reported (26). The trematodes *Prosthodendrium molenkampi* and *Phaeneropsolos bonnei* may be acquired from ingesting dragonfly and damselfly aquatic larvae.

HELMINTHS ACQUIRED FROM OTHER FOOD SOURCES

Tapeworms Causing Sparganosis

Sparganosis is infection with the plerocercoid stage of the diphyllobothrid tapeworm *Sparganum* or *Spirometra*. Human infection may be acquired by the accidental ingestion of copepods infected with the procercoid stage of the tapeworm. Alternatively, humans may serve as paratenic hosts by eating any of a variety of animals infected with the plerocercoid stage, such as frogs, tadpoles, lizards, snakes, birds, and mammals. Foodborne transmission has been attributed to eating raw pork (10). Infection may also result from the folk medicine practice of applying poultices of frog or snake to the eyes, skin, or vagina, as spargana are capable of migrating out of the infected animal flesh and penetrating the lesion.

The sparganum is a wrinkled, ivory-white, ribbonlike flatworm that can grow up to 30 cm in length but is only about 3 mm wide (Fig. 30.4). Histological sections reveal parenchymal tissue typical of undifferentiated cestode plerocercoids. Although usually found in subcutaneous tissue, spargana are highly motile and may migrate into muscle, internal organs, brain, and eye tissue. *Sparganum proliferum* is an especially hyperplastic species,

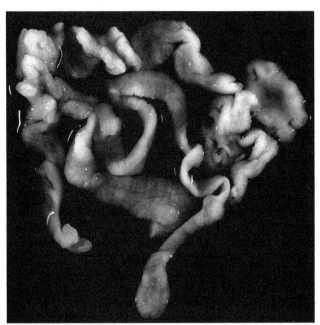

Figure 30.4 A sparganum removed from a subcutaneous nodule in the inguinal region. Magnification, ×1.7. (Illustration courtesy of the Armed Forces Institute of Pathology, AFIP 70-7392.)

exhibiting extensive branching and the capability of asexual reproduction by budding into separate organisms.

Migration of spargana is characterized by a painful, localized inflammatory reaction. Excessive lacrimation, periorbital edema, and swelling of the eyelids are associated with ocular sparganosis. Subcutaneous lesions may develop into abscesses; if the lesion ulcerates, the sparganum may be mistaken for a guinea worm. Diagnosis of infection is presumptive and usually not confirmed until the worm is removed and identified through histological sectioning. Surgical removal of spargana is the treatment of choice. Praziquantel has been proven to be an effective anthelminthic for laboratory animals.

Other Biohelminths

Alaria americana is a strigeid trematode typically found in the intestines of dogs and foxes. The normal life cycle involves sporocysts in snails and mesocercariae (a nonencysted stage) in tadpoles and frogs. Human infection has been attributed to eating inadequately cooked frog legs, raccoon, and opossum. Parasites have been recovered from the eye and from intradermal lesions, and at least one massive, fatal infection has been reported with mesocercariae disseminated into the lungs and other internal organs (30). Human infection with the trematode *Echinostomum* has also been attributed to eating crustaceans, tadpoles, and frogs (34).

Pentasomids, or "tongue worms," are larval arthropods that have been recovered from liver, spleen, lungs, and eye. Human infection has been attributed to eating raw snake, lizard, goat, and sheep. Ocular involvement probably results from direct contact with pentastomid eggs in water. Most infections are asymptomatic, although respiratory discomfort and intestinal obstruction have been reported (48). Hoarseness and coughing due to young worms attached to the pharynx, a condition known as halzoun in the Near East, have been attributed to pentastomid infection (5). Halzoun may also be caused by *Fasciola hepatica* in sheep liver. Ingestion of liver from sheep infected with adult *Fasciola hepatica* or *Dicrocoelium dentriticum* worms may result in false-positive diagnoses of infections when parasite eggs are released into the alimentary canal. Although typically associated with fish and copepods, *Gnathostoma* may also be acquired from pork, chicken, duck, frog, eel, snake, or rat.

Trichinella spiralis, described in chapter 29 of this volume, is typically associated with ingestion of raw or undercooked pork. However, human infection has also been acquired from sources as diverse as bear and walrus. An outbreak of trichinellosis in France was traced to horsemeat imported from the United States (3). Wild boar, a paratenic host for *Paragonimus westermani*, has been implicated as another potential source of human infection with that species (50). Prenatal and transmammary transmission of infective larval stages of helminths is more common than generally realized and in some instances may be the major route of infection. It has been observed for cestodes, trematodes, and most frequently, nematodes (49). Larvae of *Strongyloides fuelleborni* have been recovered from human milk (6), and *Strongyloides stercoralis* may be similarly transmitted (4). Strongyloidiasis is a common cause of infant death in Papua New Guinea (76).

HELMINTHS ACQUIRED FROM FECAL CONTAMINATION

Inadequate washing of produce or poor hygiene among food handlers can result in a variety of helminthic infections (63). Personal hygiene is critical because helminth eggs are often adherent and contamination may be found not only on hands but also under fingernails, on clothing, and in wash water (54). Parasites can be acquired from both animal and human waste, and the use of night soil to fertilize crops may be a major source of infection. In addition to the direct transmission of zoonoses from animal to man, there is evidence that birds and dogs may play an important ancillary role in disseminating helminth eggs found in untreated human waste (68, 72).

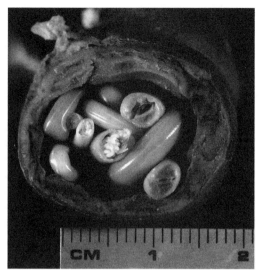

Figure 30.5 Numerous adult *Ascaris lumbricoides* worms obstructing the jejunum of a 13-year-old Zairian. Magnification, ×2.5. (Illustration courtesy of the Armed Forces Institute of Pathology, AFIP 72-13204.)

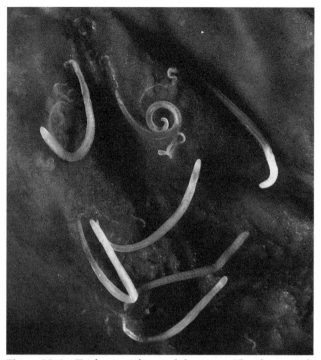

Figure 30.6 *Trichuris trichiura* adult worms. The photograph shows the slender anterior ends threaded beneath the colonic epithelium. Magnification, ×3.7. (Illustration courtesy of the Armed Forces Institute of Pathology, AFIP 69-3583.)

A variety of geohelminth species utilize the fecal-oral route for person-to-person and animal-to-person transmission. *Ascaris lumbricoides*, the largest intestinal roundworm of humans (Fig. 30.5), has been known since antiquity. *Ascaris* eggs have been found in mummified remains in Egypt and in archeological artifacts from the sites of Roman legion encampments. In parts of Central and South America, infection rates have been reported to approach 45% (4). It has been estimated that, at one time, in China alone 18,000 tons of *Ascaris* eggs were produced each year (70). Unembryonated eggs shed in the feces are resistant to desiccation, moderate freezing temperatures, and chemical treatment of sewage and may remain dormant in the soil for years (7). The swine ascarid, *Ascaris suum*, can also infect humans but does not develop into maturity (14).

The whipworm *Trichuris trichiura* (Fig. 30.6) occurs throughout the world but is most common in the tropics and in regions where sanitation is poor. In the United States, infection now occurs primarily in the southeast, but archeological evidence reveals that distribution was once far more widespread (28). *Trichuris* eggs can survive in the soil for several years. There is no reservoir host for *Trichuris trichiura*, but *Trichuris suis* and *Trichuris vulpis*, parasites of swine and dogs, respectively, have been found in humans.

Trichostrongylus is a common intestinal nematode of herbivores that resembles the hookworm. Human trichostrongyliasis, caused by the accidental ingestion of larvae on contaminated vegetation or in drinking water,

occurs primarily in Asia but has been reported throughout the world.

Oesophagostomum is a common intestinal nematode of primates, swine, and domestic animals throughout Asia, Africa, and South America. In humans, accidentally ingested larvae produce nodular lesions and abscesses, approximately 1 to 2 cm in diameter, in the intestinal wall and sometimes other organs (33). As in anisakiasis, there is no effective anthelminthic, and surgical removal of the parasite is the only effective treatment. Definitive diagnosis is based upon identifying worms in surgical or biopsy specimens. In livestock, severe diarrhea may kill the animal and parasitic nodules render the intestine unsuitable for making sausage casings, thus representing a serious economic loss for farmers (25).

The dwarf tapeworm *Hymenolepis nana* is the only human tapeworm that does not require an intermediate host; its life cycle is maintained by person-to-person contact and by autoinfection, although rodents may serve as reservoir hosts. Distributed throughout the world, *Hymenolepis nana* is common in southern Europe, the former USSR, India, and Latin America. Young children are most frequently infected. Eggs shed in feces are fully embryonated and immediately infective but have poor resistance outside the host. Internal autoinfection may occur when eggs hatch in the small intestine and penetrate the villi to repeat the developmental cycle without leaving the host. Eggs ingested by fleas and flour beetles develop into cysticercoids, but insects do not appear to play a major role in transmission. Mild infections are typically asymptomatic, and even large numbers of *Hymenolepis nana* worms are well tolerated.

Toxocara canis, an intestinal roundworm of dogs and foxes, appears to be the principal cause of visceral larva migrans in the United States (56). *Toxocara cati* is a related species in domestic felines. Following the accidental ingestion of soil contaminated with embryonated *Toxocara* eggs, the larvae migrate throughout the body and may cause serious complications when they invade the central nervous system or the eye. Eggs may survive for years in the environment, and the typically high rates of contamination in municipal parks, playgrounds, and schoolyards pose a serious potential risk for young children (73). Appropriate preventive measures include animal control, particularly leash laws and waste disposal regulations, the periodic worming of pets, and scrupulous encouragement of hand washing before meals. Ocular involvement is a serious complication, and tragically, there are reports that eyes with benign *Toxocara* inflammatory lesions have been unnecessarily enucleated because of suspected retinoblastoma (31). Ocular lesions may be more likely to occur in mild infections (32).

Baylisascaris procyonis, an ascarid of raccoons, can cause severe neurological complications and even death when this aggressive species migrates through the tissues of an unsuitable host. Fatal human infections have been reported in children believed to have been exposed to *Baylisascaris* eggs present on pieces of firewood or in raccoon feces deposited by animals nesting in unused hearths and chimneys (29, 40); fatal infections have also been reported for animals (65).

Echinococcus granulosus is a tapeworm of dogs and other canids. Human infection following the accidental ingestion of eggs in canine feces is common in sheep-raising areas throughout the world and among indigenous populations such as Eskimos, where there is a close ecological relationship between humans and dogs. *Echinococcus multilocularis* is a closely related species producing an alveolar cyst that grossly resembles an invading neoplasm. The course of the disease resembles that of a growing carcinoma, and the disease is among the most lethal of all helminthic infections (4). Risk factors for *Echinococcus* infection include exposure to infected dogs; consumption of contaminated water, ice, or snow; contact with infected foxes or their skins; and ingestion or handling of contaminated strawberries, huckleberries,

cranberries, dandelions, or other vegetation (69, 75). Increases in grassland rodent and red fox populations in some parts of Europe, attributed to changes in land use as a result of agricultural policy, may further contribute to the risk of human infection (75).

Multiceps multiceps develops into a similar proliferative stage called a coenurus. In sheep, coenuriasis of the brain or spinal cord is relatively common and is known as gid or staggers. Human infections may occur in any organ but frequently involve subcutaneous tissue, brain, and eye. Cysticercosis, described in chapter 29, results from the accidental ingestion of *Taenia solium* eggs. Racemose cysticercosis has been described as infection with an aberrant cysticercus form of *Taenia solium* but may represent coenurus infection.

I thank Judy A. Sakanari, University of California, San Francisco, for sharing research findings about Anisakis *and other fishborne parasites and John H. Cross, Uniformed Services University of the Health Sciences, Bethesda, Md., for providing background information about capillariasis and other helminthic diseases in the Far East. I am especially indebted to John S. Mackiewicz, State University of New York at Albany, for encouraging attention to the importance of ecological factors when considering parasitic diseases. Illustrations for this chapter were obtained from the extensive slide collection of the Armed Forces Institute of Pathology, Washington, D.C.*

References

1. **Abramowicz, M. (ed.).** 2004. Drugs for parasitic infections. *Med. Lett. Drugs Ther.* **1189:**1–12.

2. **Alicata, J. E.** 1965. Notes and observations on murine angiostrongylosis and eosinophilic meningoencephalitis in Micronesia. *Can. J. Zool.* **43:**667–672.

3. **Ancelle, T., J. Dupouy-Camet, M. E. Bougnoux, V. Fourestie, H. Petit, G. Mougeot, J. P. Nozais, and J. LaPierre.** 1988. Two outbreaks of trichinosis caused by horsemeat in France in 1985. *Am. J. Epidemiol.* **127:**1302–1311.

4. **Beaver, P. C., R. C. Jung, and E. W. Cupp.** 1984. *Clinical Parasitology*, 9th ed. Lea & Febiger, Philadelphia, Pa.

5. **Bogitsh, B. J., and T. C. Cheng.** 1990. *Human Parasitology.* Saunders College Publications, Holt, Rinehart & Winston, New York, N.Y.

6. **Brown, R. C., and M. H. F. Girardeau.** 1977. Transmammary passage of *Strongyloides* sp. larvae in the human host. *Am. J. Trop. Med. Hyg.* **26:**215–219.

7. **Bryan, F. D.** 1977. Diseases transmitted by foods contaminated by waste water. *J. Food Prot.* **40:**45–56.

8. **Centers for Disease Control and Prevention.** 1982. Intestinal perforation caused by larval *Eustrongylides*—Maryland. *Morb. Mortal. Wkly. Rep.* **31:**383–389.

9. **Chitwood, M. B., C. Valesquez, and N. G. Salazar.** 1968. *Capillaria philippinensis* sp. n. (Nematoda:Trichinellida) from the intestine of man in the Philippines. *J. Parasitol.* **54:**368–371.

10. **Corkum, K. C.** 1966. Sparganosis in some vertebrates of Louisiana and observations on a human infection. *J. Parasitol.* **52:**444–448.

11. Cross, J. H. 1992. Intestinal capillariasis. *Clin. Microbiol. Rev.* **5:**120–129.

12. Daengsvang, S. 1971. Infectivity of *Gnathostoma spinigerum* larvae in primates. *J. Parasitol.* **57:**476–578.

13. Daengsvang, S., B. Sermswatsri, P. Youngyi, and D. Guname. 1970. Development of adult *Gnathostoma spinigerum* in the definitive host (cat and dog) by skin penetration of the advanced third-stage larvae. *Southeast Asia J. Trop. Med. Public Health* **1:**187–192.

14. Davies, N. J., and J. M. Goldsmid. 1978. Intestinal obstruction due to *Ascaris suum* infection. *Trans. R. Soc. Trop. Med. Hyg.* **72:**107.

15. Deardorff, T. L., and M. L. Kent. 1989. Prevalence of larval Anisakis simplex in pen-reared and wild-caught salmon (Salmonidae) from Puget Sound, Washington. *J. Wildl. Dis.* **25:**416–419.

16. Deardorff, T. L., and R. M. Overstreet. 1981. Larval *Hysterothylacium* (=*Thynnascaris*) (Nematoda: Anisakidae) from fishes and invertebrates in the Gulf of Mexico. *Proc. Helminthol. Soc. Wash.* **48:**113–126.

17. Deardorff, T. L., and R. M. Overstreet. 1990. Seafood-transmitted zoonoses in the United States: the fishes, the dishes, and the worms, p. 211–265. *In* D. R. Ward and D. R. Hackney (ed.), *Microbiology of Marine Food Products*. Van Nostrand Reinhold, New York, N.Y.

18. Deardorff, T. L., R. M. Overstreet, M. Okihiro, and R. Tam. 1986. Piscine adult nematode invading an open lesion in a human hand. *Am. J. Trop. Med. Hyg.* **35:**827–830.

19. Deardorff, T. L., and R. Throm. 1988. Commercial blast freezing of third-stage Anisakis simplex larvae encapsulated in salmon and rockfish. *J. Parasitol.* **74:**600–603.

20. de Brie, J. 1897. Le Bon Berger ou le Vray Reegime et Gouvenement de Bergers et Bergeres: Compose par le Rustique Jehan de Brie le Bon Berger. Isidor Liseux, Paris, France.

21. Delaware Sea Grant. 2000, posting date. Eating raw finfish: what are the risks, the benefits? [Online.] University of Delaware, Newark, Del. http://www.ocean.udel.edu/mas/seafood/raw.html.

22. Desowitz, R. S. 1981. *New Guinea Tapeworms and Jewish Grandmothers*. Norton and Company, New York, N.Y.

23. Dooley, J. R., and R. C. Neafie. 1976. Anisakiasis, p. 475–481. *In* C. H. Binford and D. H. Connor (ed.), *Pathology of Tropical and Extraordinary Diseases*, vol. 2. Armed Forces Institute of Pathology, Washington, D.C.

24. Dooley, J. R., and R. C. Neafie. 1976. Clonorchiasis and opisthorchiasis, p. 509–516. *In* C. H. Binford and D. H. Connor (ed.), *Pathology of Tropical and Extraordinary Diseases*, vol. 2. Armed Forces Institute of Pathology, Washington, D.C.

25. Dooley, J. R., and R. C. Neafie. 1976. Oesophagostomiasis, p. 440–445. *In* C. H. Binford and D. H. Connor (ed.), *Pathology of Tropical and Extraordinary Diseases*, vol. 2. Armed Forces Institute of Pathology, Washington, D.C.

26. Eberhard, M. L., and C. Busillo. 1999. Human *Gongylonema* infection in a resident of New York City. *Am. J. Trop. Med. Hyg.* **61:**51–52.

27. Eberhard, M. L., H. Hurwitz, A. M. Sun, and D. Coletta. 1989. Intestinal perforation caused by larval *Eustrongylides* (Nematode:Dioctophymatoidae) in New Jersey. *Am. J. Trop. Med. Hyg.* **40:**648–650.

28. Faulkner, C. T., S. E. Cowie, P. E. Martin. S. R. Martin, C. S. Mayes, and S. Patton. 2000. Archeological evidence of parasitic infection from the 19th century company town of Fayette, Michigan. *J. Parasitol.* **86:**846–849.

29. Fox, A. S., K. R. Kazacos, N. S. Gould, P. T. Heydemann, C. Thomas, and K. M. Boyer. 1985. Fatal eosinophilic meningoencephalitis and visceral larva migrans caused by the raccoon ascarid *Baylisascaris procyonis*. *N. Engl. J. Med.* **312:**1619–1623.

30. Freeman, R. S., P. F. Stuart, J. B. Cullen, A. C. Ritchie, A. Mildon, B. J. Fernandes, and R. Bonin. 1976. Fatal human infection with mesocercariae of the trematode *Alaria americana*. *Am. J. Trop. Med. Hyg.* **25:**803–807.

31. Glickman, L. T. 1984. Toxocariasis. *In* K. S. Warren and A. A. F. Mahmoud (ed.), *Tropical and Geographical Medicine*. McGraw-Hill, New York, N.Y.

32. Glickman, L. T., P. M. Schantz, and R. H. Cypress. 1979. Canine and human toxocariasis: review of transmission, pathogenesis and clinical disease. *J. Am. Vet. Med. Assoc.* **175:**1265–1269.

33. Gordon, J. A., C. M. D. Ross, and H. Affleck. 1969. Abdominal emergency due to an oesophagostome. *Ann. Trop. Med. Parasitol.* **63:**161–164.

34. Graczyk, T. K., and B. Fried. 1998. Echinostomiasis: a common but forgotten food-borne disease. *Am. J. Trop. Med. Hyg.* **58:**501–504.

35. Harrell, L. W., and T. L. Deardorff. 1990. Human nanophyetiasis: transmission by handling naturally infected coho salmon (*Oncorhynchus kisutch*). *J. Infect. Dis.* **161:**146–148.

36. Hayunga, E. G. 1989. Parasites and immunity: tactical considerations in the war against disease—or, how did the worms learn about Clausewitz? *Perspect. Biol. Med.* **32:**349–370.

37. Hillyer, G. V. 1981. Fascioliasis in Puerto Rico: a review. *Bol. Assoc. Med. P. R.* **73:**94–101.

38. Hopkins, D. R., E. Ruiz-Tiben, N. Diallo, P. C. Withers, Jr., and J. H. Maguire. 2002. Dracunculiasis eradication: and now, Sudan. *Am. J. Trop. Med. Hyg.* **57:**415–422.

39. Hou, P. C. 1965. Hepatic clonorchiasis and carcinoma of the bile duct in a dog. *J. Pathol. Bacteriol.* **89:**365.

40. Huff, D. S., R. C. Neafie, M. J. Binder, G. A. DeLeon, L. W. Brown, and K. R. Kazacos. 1984. Case 4, the first fatal *Baylisascaris* infection in humans: an infant with eosinophilic meningoencephalitis. *Pediatr. Pathol.* **2:**345–352.

41. Jones, E. A., J. M. Key, H. P. Milligan, and D. Owens. 1977. Massive infection with *Fasciola hepatica* in man. *JAMA* **63:**836–842.

42. Katz, M., D. D. Despommier, and R. W. Gwadz. 1982. *Parasitic Diseases*. Springer-Verlag, New York, N.Y.

43. Kikuchi, Y., T. Takenouchi, M. Kamiya, and H. Ozake. 1981. Trypanorhynchiid cestode larva found on the human palatine tonsil. *Jpn. J. Parasitol.* **30:**3497–3499.

44. Kisielewska, K. 1970. Ecological organization of intestinal helminth groupings in *Clethrionomys glareolus* (Schreb.) (Rodentia). I. Structure and seasonal dyamics of helminth groupings in a host population in the Bialowieza National Park. *Acta Parasitol. Polonica* **18:**121–147.

45. Le Bailly, M., U. Leuzinger, H. Schlichtherie, and F. Bouchet. 2005. Diphyllobothrium: Neolithic parasite? *J. Parasitol.* **91:**957–959.

46. Leuckart, K. G. 1882. Zur Entwickelungsgeschichte des Leberegels. Zweite Mittheilung. *Zool. Anz.* **5:**524–528.

47. MacArthur, W. P. 1933. Cysticercosis as seen in the British Army, with special reference to the production of epilepsy. *Trans. R. Soc. Trop. Med. Hyg.* **27:**343–363.

48. Meyers, W. M., R. C. Neafie, and D. H. Connor. 1976. Pentastomiasis, p. 546–550. *In* C. H. Binford and D. H. Connor (ed.), *Pathology of Tropical and Extraordinary Diseases*, vol. 2. Armed Forces Institute of Pathology, Washington, D.C.

49. Miller, G. C. 1981. Helminths and the transmammary route of infection. Trends and perspectives in parasitology. *Parasitology* **82:**335–342.

50. Miyazaki, I., and S. Habe. 1976. A newly recognized mode of human infection with the lung fluke, *Paragonimus westermani* (Kerbert 1978). *J. Parasitol.* **62:**646–648.

51. Muller, R. 1971. *Dracunculus* and dracunculiasis. *Adv. Parasitol.* **9:**73–151.

52. New, D., M. D. Little, and J. Cross. 1995. *Angiostrongylus cantonensis* infection from eating raw snails. *N. Engl. J. Med.* **332:**1105–1106.

53. Norton, R. A., and L. Monroe. 1961. Infection by *Fasciola hepatica* acquired in California. *Gastroenterology* **41:**46–48.

54. Ockert, G., and J. Obst. 1973. Ausstreuung umhüllter Onkosphären durch Bandwurmträger. *Monatsh. Veterinarmed.* **28:**97–98.

55. Ollerenshaw, C. B. 1980. Forecasting liver fluke disease, p. 33–52. *In* A. E. R. Taylor and R. Muller (ed.), *12th Annual Symposium of the British Society for Parasitology.* Blackwell, London, United Kingdom.

56. Paul, A. J., K. S. Todd, Jr., and J. A. Dipietro. 1988. Environmental contamination by eggs of *Toxocara* species. *Vet. Parasitol.* **26:**339–342.

57. Rao, C. K., R. C. Paul, and M. I. D. Sharma. 1981. Guinea worm disease in India—control status and strategy of its eradication. *J. Commun. Dis.* **13:**1–7.

58. Rosen, L., G. Loison, J. Laigret, and G. D. Wallace. 1967. Studies on eosinophilic meningitis. 3. Epidemiologic and clinical observations on Pacific islands and the possible etiologic role of *Angiostrongylus cantonensis*. *Am. J. Epidemiol.* **85:**17–44.

59. Sadun, E. H., and C. Maiphoom. 1953. Studies in the epidemiology of the human intestinal fluke, *Fasciolopsis buski* (Lankester) in Central Thailand. *Am. J. Trop. Med. Hyg.* **2:**1070–1084.

60. Sakanari, J. A., M. Moser, and T. L. Deadorff. 1995. *Fish Parasites and Human Health: Epidemiology of Human Helminthic Infections*. Report no. T-CSGCP-034. California Sea Grant College, University of California, La Jolla, Calif.

61. Sakanari, J. A., H. M. Loinaz, T. L. Deardorff, R. B. Raybourne, H. H. McKerrow, and J. G. Frierson. 1988. Intestinal anisakiasis: a case diagnosed by morphologic and immunologic methods. *Am. J. Clin. Pathol.* **90:**107–113.

62. Sakanari, J. A., and J. H. McKerrow. 1989. Anisakiasis. *Clin. Microbiol. Rev.* **2:**278–284.

63. Sanchez, J. L., I. Hernandez-Fragoso, C. Rios, and C. K. Ho. 1990. Parasitological evaluation of a foodhandler population cohort in Panama: risk factors for intestinal parasitism. *Mil. Med.* **156:**250–255.

64. Santibanez, T., L. Barker, J. Santoli, C. Bridges, G. Euler, and M. McCauley. 2004. Progress toward global eradication of dracunculiasis, 2002–2003. *Morb. Mortal. Wkly. Rep.* **53:**871–872.

65. Sato, H., Y. Une, S. Kawakami, E. Saito, H. Kamiya, N. Akao, and H. Furuoka. 2005. Fatal *Baylisascaris* larva migrans in a colony of Japanese macaques kept by a safari-style zoo in Japan. *J. Parasitol.* **91:**716–719.

66. Schantz, P. M. 1989. The dangers of eating raw fish. *N. Engl. J. Med.* **320:**1143–1145.

67. Sianto, L., K. J. Reinhard, M. Chame, S. Mendonca, M. L. C. Concalves, A. Fernandes, L. F. Ferreira, and A. Araujo. 2005. The finding of *Echinostoma* (Trematoda: Digenea) and hookworm eggs in coprolites collected from a Brazilian mummified body dated 600–1,200 years before present. *J. Parasitol.* **91:**972–975.

68. Silverman, P. H., and R. B. Griffiths. 1955. A review of methods of sewage disposal in Great Britain with special reference to the epizootiology of *Cysticercus bovis*. *Ann. Trop. Med. Parasitol.* **49:**436–450.

69. Stehr-Green, J. K., P. A. Stehr-Green, P. M. Schantz, J. F. Wilson, and A. Lanier. 1988. Risk factors for infection with *Echinococcus multilocularis* in Alaska. *Am. J. Trop. Med. Hyg.* **38:**380–385.

70. Stoll, N. R. 1947. This wormy world. *J. Parasitol.* **33:**1–18.

71. Toma, H., S. Matsumura, C. Oshiro, and Y. Sato. 2002. Ocular angiostrongyliasis without meningitis symptoms in Okinawa, Japan. *J. Parasitol.* **88:**211–213.

72. Traub, R. J., I. D. Robertson, P. Irwin, N. Mencke, and R. C. A. Thompson. 2002. The role of dogs in transmission of gastrointestinal parasites in a remote tea-growing community in northeastern India. *Am. J. Trop. Med. Hyg.* **67:**539–545.

73. Uga, S., and N. Kataoka. 1995. Measures to control *Toxocara* egg contamination in sandpits of public parks. *Am. J. Trop. Med. Hyg.* **52:**21–24.

74. U.S. Food and Drug Administration, Center for Food Safety and Applied Nutrition. 2003. *Foodborne Pathogenic Microorganisms and Natural Toxins Handbook (Bad Bug Book)*.[Online] U.S. Food and Drug Administration, Washington, D.C. http://vm.cfsan.fda.gov/~mow/chap25.html.

75. Viel, J. F., P. Giraudoux, V. Abrial, and S. Bresson-Hadni. 1999. Water vole (*Arvicola terrestris scherman*) density as risk factor for human alveolar echinococcosis. *Am. J. Trop. Med. Hyg.* **61:**559–565.

76. Vince, J. D., R. W. Ashford, M. J. Gratten, and J. Bana-Koiri. 1979. *Strongyloides* species infestation in young infants of Papua New Guinea: association with generalized oedema. *Papua New Guinea Med. J.* **22:**120–127.

77. Wittner, M., J. W. Turner, G. Jacquotte, L. R. Ash, M. P. Salgo, and H. B. Tanowitz. 1989. Eustrongylidiasis—a parasitic infection acquired by eating sushi. *N. Engl. J. Med.* **320:**1124–1126.

78. Yokogawa, M. 1969. *Paragonimus* and paragonimiasis. *Adv. Parasitol.* **7:**375–387.

Food Microbiology: Fundamentals and Frontiers, 3rd Ed.
Edited by M. P. Doyle and L. R. Beuchat
© 2007 ASM Press, Washington, D.C.

Ynes R. Ortega

Protozoan Parasites

31

Protozoan parasites have long been associated with food-borne and waterborne outbreaks of disease in humans. Difficulties arise with the inactivation of these organisms because of their resistance to environmental stresses. Groups of protozoan parasites transmitted via foods include Apicomplexa, microsporidia, flagellates, ciliates, and amoebae (Table 31.1).

A major characteristic of apicomplexan parasites is that a vertebrate host is required to complete the complex life cycle and produce infectious cysts. Of this group, *Cryptosporidium* species, *Cyclospora cayetanensis*, and *Isospora belli* inhabit the intestinal mucosa and produce diarrheal illnesses in humans. These coccidian parasites affect immunocompetent as well as immunocompromised individuals, causing more severe and prolonged symptoms in the latter. *Cryptosporidium* can infect animals, and some genotypes have been described as having host preferences or specificities. *C. cayetanensis* has been isolated exclusively from humans. Another apicomplexan, *Toxoplasma gondii*, infects human tissues other than the intestinal mucosa and can cause birth defects, blindness, and chorioretinitis. The life cycle stages of apicomplexan parasites are produced intracellularly in the host. For *Cyclospora*, *Toxoplasma*, and *Isospora*,

sporogony occurs typically outside the host, requiring the passage of time before oocysts are infective to a new host, but *Cryptosporidium* oocysts are excreted already sporulated and are infectious when shed.

Microsporidia, once considered to be protozoan parasites, have been reclassified as fungi based on recent phylogenetic analyses. Microsporidia have been implicated in human foodborne and waterborne diseases and are able to propagate and complete their life cycle intracellularly. Of the microsporidia, members from five genera have been implicated in human diseases: *Encephalitozoon*, *Enterocytozoon*, *Septata*, *Pleistophora*, and *Vittaforma*. None of these species are host, tissue, or organ specific, with the exception of *Enterocytozoon bienusii*, which appears to infect only the human intestinal tract.

Flagellates, ciliates, and amoebae can propagate in more than one host. All three types of parasites can cause disease in humans and animals. The infectious stage of this group is the cyst. Once ingested by the susceptible host, the trophozoites or motile forms will be released from the cyst and proceed to colonize the host's intestinal cells. In contrast with Apicomplexa and microsporidia, flagellates, ciliates, and amoebae do not present

Ynes R. Ortega, Center for Food Safety, University of Georgia, 1109 Experiment St., Griffin, GA 30223–1797.

Table 31.1 Protozoa of medical importance acquired from food and water

Phylum	Protozoa(n)	Infective stage
Apicomplexa	*Cryptosporidium parvum*	Oocyst
	Cyclospora cayetanensis	Oocyst
	Isospora belli	Oocyst
	Toxoplasma gondii	Oocyst or tissue cyst
	Sarcocystis hominis	Oocyst
	Sarcocystis suihominis	Oocyst
Ciliphora	*Balantidium coli*	Cyst
Microspora	*Enterocytozoon bieneusi*	Spore
	Septata intestinalis	Spore
Sarcomastigophora	*Acanthamoeba* spp.[a]	Trophozoite
	Dientamoeba fragilis	Trophozoite
	Entamoeba dispar[b]	Cyst
	Entamoeba histolytica	Cyst
	Giardia intestinalis	Cyst
	Naegleria fowleri[a]	Trophozoite

[a] Species of normally free-living amoebae that invade through the mucosa and can in some instances infect the central nervous system of humans. Amoebic meningitis is usually fatal.

[b] Commensal, morphologically similar to *E. histolytica*, and not normally capable of inducing disease.

intracellular stages but instead attach with specialized structures to the luminal surfaces of the intestinal cells and feed off the nutrients and cellular debris of the intestinal epithelium. *Giardia lamblia*, a flagellate originally considered to be a commensal organism, is well recognized as an etiological agent of acute or chronic diarrhea in humans. *Balantidium coli*, a ciliate, also causes diarrhea in humans, and if the infection is not treated, it can cause ulcerative colitis. Various species of amoebae can infect humans. Most of them are commensal, not producing disease; however, others can cause diarrhea, dysentery, or amebomas if the infection is not treated.

CRYPTOSPORIDIUM

Cryptosporidium was first isolated in 1910 from the intestines of mice. Various species of *Cryptosporidium* capable of infecting animals have since been described. It was not until 1976 that *Cryptosporidium parvum* infection in humans was first described. Since 1982, *C. parvum* has been one of the most frequently identified opportunistic pathogens associated with diarrhea and wasting syndrome in patients with AIDS. Cryptosporidiosis in immunocompromised patients produces a life-threatening, prolonged, cholera-like illness (156). In immunocompetent patients, *C. parvum* causes self-resolving, acute

diarrheal disease with variability in the severity of symptoms. Isolate variations and associations with the severity of disease have been subjects of intensive investigation for some time. Recently, *C. parvum* isolates have been separated into two genotypes: type I and type II. Genotype I has been isolated exclusively from humans. Genotype II has been isolated from cattle and humans that have been exposed to infected cattle and is readily infective to laboratory animals. Genotype differentiation is based on the sequencing of the thrombospondin-related adhesive protein (TRAP-C2) (132). Genotype I has now been renamed *C. hominis* and seems to be anthroponotic (118). Other *Cryptosporidium* species (*C. felis, C. canis, C. meleagridis, C. baileyi, C. muris,* and *C. parvum*) have also been identified as infecting humans, occurring most frequently in human immunodeficiency virus (HIV)-positive individuals (33, 54, 64, 75, 115, 131, 135).

Cryptosporidiosis is acquired after ingesting food or water contaminated with infective *Cryptosporidium* oocysts. Once ingested, the oocysts excyst (Fig. 31.1). The released sporozoites proceed to invade the enterocytes. A parasitophorous vacuole is formed as the parasite enters the cell. This structure contains the intracellular parasites and communicates with the host cell via a feeder organelle. The parasite goes through two asexual multiplication stages called merogony: type I meronts contain eight merozoites which are released and proceed to invade other enterocytes to form type II meronts, each containing four merozoites. Once merozoites from type II meronts are released, they infect enterocytes but differentiate into sexual stages identified as macrogametocytes (female) and microgametocytes (male). The union of a microgametocyte and a macrogametocyte produces a zygote (immature oocyst), which matures within its host into a fully sporulated oocyst (Fig. 31.2). Two types of oocysts are produced and excreted. Thin-walled oocysts excyst endogenously, resulting in autoinfection, and thick-walled oocysts, which are environmentally resistant, are shed in the feces and are immediately infectious to other hosts (46, 167). An additional life stage which may represent an extracellular trophozoite or gamont stage has been described for *C. andersoni*, a parasite that infects the gut of cattle. This stage has also been reported for *C. parvum*. This finding, along with results of phylogenetic analysis, indicates that *Cryptosporidium* has a closer phylogenetic affinity with the gregarines than with the coccidia (36, 90, 144).

To date, 15 species of *Cryptosporidium* have been identified in various animals, of which *C. canis* (dogs), *C. felis* (cats), *C. muris* (rodents), *C. parvum* (cattle, sheep, and goats), *C. meleagridis* (turkeys), *C. suis* (pigs), and *C. hominis* (humans) are known to infect humans.

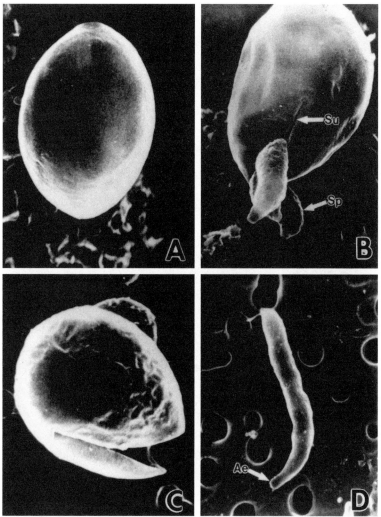

Figure 31.1 Scanning electron micrographs of oocysts and excysting sporozoites of *C. parvum*. (A) Intact oocyst prior to excystation (×11,200). (B) Three sporozoites (Sp) excysting from oocyst simultaneously via the cleaved suture (Su) (×11,200). (C) Empty oocyst (×11,200). (D) Excysted sporozoite; Ae, apical end (×9,800). From Fayer et al. (61a).

Generally, *C. parvum* infects the brush border of the intestinal epithelium and causes villous atrophy, though the exact mechanism for this pathological alteration is not yet fully understood. Immunocompetent patients develop a profuse watery diarrhea accompanied by epigastric cramping, nausea, and anorexia, which is usually self-limiting and lasts for about 15 days.

Extraintestinal dissemination has been observed in immunocompromised patients. *Cryptosporidium* may be found in epithelial cells from other organs such as those of the respiratory tract and the biliary tree (108). Immunocompromised patients (e.g., AIDS patients or those receiving immunosuppressive drugs) develop severe diarrhea (3 to 6 liters/day), which persists for several weeks to months or even years. These cases are the most severe and life threatening, with continuous shedding of oocysts. In patients with HIV infections, the CD4+-cell count is the best marker for the ability of the immune system to self-resolve the infection. In the developed world, patients with CD4-cell counts of 180 cells/mm^3 or more usually develop self-limiting cryptosporidiosis, while those with counts below 180 cells/mm^3 usually develop chronic and profuse diarrhea, which is exacerbated by the lack of effective therapy (69). The pathogenesis of the diarrhea is not clear, and the presence of a toxin has been suggested but not yet demonstrated.

The mean prevalence of *C. parvum* infection in Europe and the United States is 1 to 3%, and the prevalence is

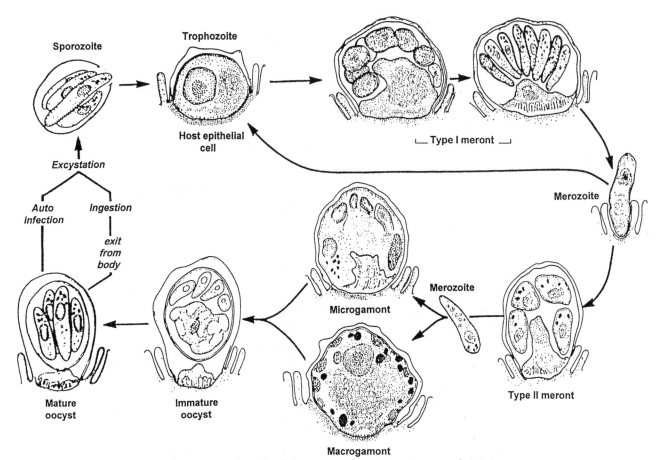

Figure 31.2 Life cycle of *Cryptosporidium*. From Fayer et al. (61a).

considerably higher in developing countries. Outbreaks of disease associated with *Cryptosporidium* have been reported in the United States as well as in other countries (167). *Cryptosporidium* infection is highly associated with travel abroad, exposure to farm animals, and person-to-person transmission in settings such as day care centers and medical institutions. A large number of waterborne outbreaks of cryptosporidiosis have been documented, and some of the most notable were those in Milwaukee, Wis. (111), and in Georgia (86), with a total of more than 400,000 and 13,000 cases reported, respectively. In Minnesota in 1995, *Cryptosporidium* was associated with cases of acute gastroenteritis experienced by attendees of a social event. This outbreak was epidemiologically associated with the consumption of contaminated chicken salad (9). In 1993 in central Maine and in 1996 in New York, apple cider was associated with outbreaks of cryptosporidiosis (12, 116). In 1998 in Spokane, Wash., another foodborne outbreak affected about 50 people, but this outbreak could not be traced to a specific type of food.

Cryptosporidium has been identified in produce intended for human consumption. In Norway, 26% of lettuce samples and 74% of mung bean sprout samples examined were *Cryptosporidium* positive. *Giardia* was identified in dill (20%), lettuce (20%), mung bean sprout (30%), radish (10%), and strawberry (20%) samples (142). The risk of acquiring cryptosporidiosis and giardiasis based on the frequency and level of mung bean sprout consumption was assessed to be 20 cases per 100,000 (142, 143). Analysis of produce obtained from a regional market in Peru revealed that 14.5% contained *Cryptosporidium* oocysts (126). The immunofluorescence assay specific to *Cryptosporidium* has been used as a gold standard assay. An alternative method for identifying *Cryptosporidium* from fresh produce has been developed by using polyclonal sera raised to a recombinant viral capsid protein of an RNA virus that is a symbiont of *Cryptosporidium* (98).

Although there is no effective and definite therapeutic agent for the treatment of cryptosporidiosis, spiramycin has been reported to decrease diarrhea in the early stages of infections (119, 134, 146); however, it is not effective in advanced infections. Azithromycin, nitazoxanide, and paromomycin have been reported to be effective in AIDS patients with cryptosporidiosis (28, 48, 71, 76). Alternative experimental therapies that have been evaluated for humans include the use of passive immunotherapies such as the administration of hyperimmune bovine colostrum. Monoclonal antibodies and polyclonal hyperimmune hen yolk against *Cryptosporidium* have also been tested in animal models with some success (35, 62, 133).

Cryptosporidium oocysts are 4 to 6 μm in diameter. Because of their size, they can be overlooked in examinations of feces or confused with yeast cells. *Cryptosporidium* can be identified by light microscopy by using acid-fast staining. Immunoassays (fluorescent-antibody immunoassay or enzyme immunoassay) are available and are very sensitive and specific diagnostic procedures (15, 16, 37, 159, 160). *Cryptosporidium* can be mistaken for *Cyclospora*, which is also positive in acid-fast staining but larger (Color Plate 3). It is important that laboratories carefully measure the diameters of cells, particularly if the cells appear to be larger than those of *Cryptosporidium*. New molecular diagnostic tests such as PCR are being developed and may improve the sensitivity in detecting *Cryptosporidium* oocysts in produce (20, 45, 51, 78, 97, 98, 101, 141). It has been demonstrated that shellfish can concentrate large volumes of *Cryptosporidium* cells along with other particulate matter (63, 77, 80, 83, 84, 112). *Cryptosporidium* has also been identified in shellfish organs by using conventional diagnostic assays, and molecular tools (multiplex nested PCR) have been examined to optimize parasite identification (77–79, 81). The use of natural grassland buffers may contribute to the control of the parasite load that may potentially be deposited in water banks and oceans as a consequence of rangeland runoff, thus contaminating shellfish (18). To date shellfish have not been associated with outbreaks of *Cryptosporidium* infections.

Although *Cryptosporidium* oocysts are highly resistant to chlorine and chlorine dioxide, treatment with 1,000 μg/ml for 1 min resulted in 0.5- and 2-log reductions, respectively, according to the viability assay (in vitro excystation or tissue culture infectivity) (38). More than 90% inactivation was achieved when *Cryptosporidium* oocysts were treated with ozone at 1 μg/ml for 5 min, chlorine dioxide at 1.3 μg/ml for 1 h, and chlorine and monochloramine for 80 μg/ml for 90 min (100).

CYCLOSPORA

Cyclospora was probably first reported to occur in humans in 1979 by Ashford, who described it as an *Isospora*-like coccidian affecting humans in Papua New Guinea (17).

Thereafter, other investigators found similar structures in fecal samples of patients with diarrhea, but because of the morphologies of the unsporulated oocysts and their autofluorescence, they were considered to be CLBs (coccidian-like bodies or cyanobacterium-like bodies) (103, 104, 151). In 1993, conclusive identification was made and the CLBs were fully characterized as coccidian parasites and placed into the genus *Cyclospora* with the proposed species name *cayetanensis* (124, 127).

Cyclospora belongs to the family Eimeriidae, subphylum Apicomplexa. *Cyclospora* species infect moles, rodents, insectivores, snakes, and humans. In 1998, a *Cyclospora* species which morphologically resembles *Cyclospora* species isolated from humans was isolated from nonhuman primates (58, 107, 155). However, phylogenetic analysis has demonstrated that the isolate is a different species (55, 107).

Even before their true identity was established, epidemiological information on the intriguing CLBs was being collected worldwide. In Nepal, the prevalence of *Cyclospora* infection is highest in adult expatriates (152), whereas in areas of Peru, where infection is endemic, children under 10 years of age are the most susceptible to infection, though most are asymptomatic (113). Adults from areas of endemicity do not develop symptoms of the infection, but adults of medium to high socioeconomic status living outside of the areas of endemicity, as well as foreign travelers, present with clinical disease. These observations suggest that prior exposure to the parasite may result in protective immunity.

Newly shed oocysts (Fig. 31.3A) of *C. cayetanensis* require 2 weeks to sporulate and become infectious (Fig. 31.3B) under optimal laboratory conditions (127). The requirement of this period for the oocyst to become infectious suggests that contamination of produce occurs with oocysts that are fully sporulated or almost fully sporulated. Otherwise, unsporulated oocysts would require optimal time and environmental conditions to induce oocyst sporulation, while produce would still remain edible.

In studies performed in areas of endemicity, *Cyclospora* oocysts have been isolated from produce by washing thoroughly with distilled water. The washes are concentrated by centrifugation, and pellets are fixed and preserved in 10% formalin. Samples are examined directly by using epifluorescence microscopy and phase-contrast microscopy (127). Nested PCR targeted to amplify the 18S rRNA gene is also used with these preparations; however, it should be noted that further testing using restriction fragment length polymorphism is required, since the described PCR cross-reacts with *Eimeria* species. *Eimeria* parasites are infectious to animals but not to humans and can be readily found in the environment (95, 150).

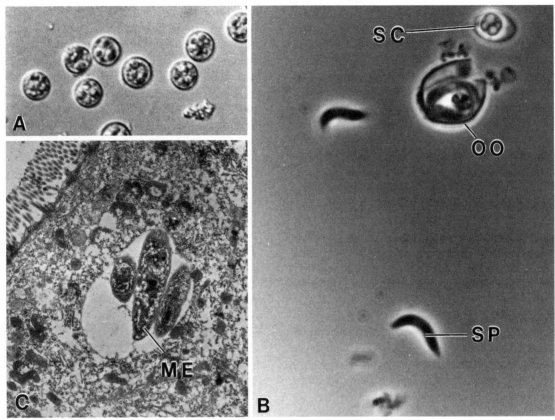

Figure 31.3 *C. cayetanensis* oocysts. (A) Phase-contrast microscopy of unsporulated oocysts. (B) Oocysts (OO) in the process of excystation. Note the two sporozoites (SP) free of the sporocyst (SC). (C) Transmission electron microscopy of human small intestine showing *Cyclospora* intracellular stages. ME, merozoite.

Cyclosporiasis is characterized by mild to severe nausea, anorexia, abdominal cramping, mild fever, and watery diarrhea. Diarrhea alternating with constipation has been commonly reported. Some patients present with flatulent dyspepsia and, less frequently, joint pain and night sweats. The onset of illness is usually sudden, and symptoms persist an average of 7 weeks (42, 125).

C. cayetanensis infects epithelial cells of the duodenum and jejunum of humans. Merogony and gametogony occur intracytoplasmically within the parasitophorous vacuoles in intestinal cells (Fig. 31.3C) (125, 158). Duodenal and jejunal biopsy specimens from patients with cyclosporiasis show various degrees of jejunal villous blunting, atrophy, and crypt hyperplasia (42, 125). Extensive lymphocytic infiltration into the surface epithelium is present, especially at the tips of the shortened villi. The reactive inflammatory response of the host does not correlate with the number of intracellular parasites present in the tissues (125).

Routes of transmission for *Cyclospora* are undocumented, although the fecal-oral route, either directly or via

water and food, is probably the major one. In the United States, epidemiological evidence suggests that water has been responsible for sporadic cases of cyclosporiasis. In Utah, a man became infected after cleaning his basement that was flooded with runoff from a nearby farm following heavy rains. In 1990, an outbreak involving residents in a physicians' dormitory in a Chicago hospital was epidemiologically associated with tap water from unprotected reservoir tanks that served the building and which had a broken water pump (13, 93). In Pokhara, Nepal, British Gurkha soldiers were confirmed to have cyclosporiasis; oocysts were isolated from drinking water. The water, consisting of a mixture of river and municipal water, was routinely chlorinated and served the houses in the camp where the soldiers were stationed (136).

A few reports have described the isolation of *C. cayetanensis* oocysts from animals (chickens, ducks, and dogs) (39, 74, 164, 165). To date, after several experimental studies attempting to infect these and additional animal species, it seems that *Cyclospora* species are host specific (59).

Since the early 1990s, *Cyclospora* has been linked to waterborne and foodborne disease outbreaks. Foodborne outbreaks have been epidemiologically associated with the consumption of contaminated fresh produce such as raspberries, lettuce, basil, and snow peas (10, 11, 14, 26, 87, 88, 92). In 1996, a large foodborne outbreak occurred following a wedding reception at a restaurant in Boston, resulting in 57 guests having cyclosporiasis. Berries were implicated as the vehicle of transmission for *Cyclospora* oocysts (68). In 1999, another foodborne outbreak among attendees of two events was reported. Sixty-two cases of infection were documented, and the illness was associated with the consumption of chicken pasta salad and tomato basil salad. The most likely vehicle of illness was fresh basil, grown either in Mexico or in the United States, which was included in both salads (105). However, *Cyclospora* oocysts were not recovered from or detected in the produce associated with these outbreaks. In the United States, most cases of cyclosporiasis have been reported in the months of April to August, suggesting some possible seasonality.

In the United States in 1996 and 1997, *Cyclospora* infections were associated with imported raspberries (87–89). It was speculated that contamination of produce could have occurred when raspberries were sprayed with insecticide possibly diluted with contaminated surface water. Analysis of irrigation water demonstrated the presence of *Cyclospora* oocysts (26). In Peru, *C. cayetanensis* oocysts have been isolated from vegetables obtained from markets in areas of endemicity. In studies where vegetables were experimentally inoculated with *C. cayetanensis* oocysts, it was demonstrated that washing with water does not remove all the oocysts (126). *Cyclospora* has also been isolated from produce from Nepal (149). Neither the minimum infectious dose of oocysts nor the sporulation and survival rates under different environmental conditions are known, although human volunteer studies have been attempted without success (61). Because of the potential low number of oocysts present in foods, molecular assays and methodologies to improve parasite recovery have been studied previously (95, 96, 123).

Analysis of the intervening transcribed spacer-1 (ITS-1) region sequences of *Cyclospora* isolates from the 1996 outbreak demonstrated that all are identical, suggesting that a single source of contamination caused the infection (2). Another study examined *C. cayetanensis* and *C. papionis* isolates. Although high sequence variability is present, conserved species-specific ITS-1 sequences were identified. This consistent and remarkable diversity among *Cyclospora* species ITS-1 sequences argues for polyparasitism and simultaneous transmission of multiple strains (121).

There is strong evidence to suggest seasonality of *Cyclospora* infections. In Peru, more than 6 years worth of prospective epidemiological studies investigating endemic *Cyclospora* infections found that nearly all infections occurred between December and July (113). Rarely were infections documented at any other time of the year. In the United States, major cases have occurred during the period from May to July. In Nepal, infection and illness occur most frequently from May to August. The specific reasons for this marked seasonality have not been defined.

The only successful antimicrobial treatment for *Cyclospora* infection is trimethoprim-sulfamethoxazole (TMP-SMX) (113, 128, 162). AIDS patients appear to have higher parasite infestation levels than immunocompetent individuals infected with *Cyclospora*. However, the prevalence of *Cyclospora* among HIV patients is not higher than that among immunocompetent populations. This is probably due to the frequent use of TMP-SMX for *Pneumocystis carinii* prophylaxis among HIV patients (128). Ciprofloxacin has also been examined as an alternative treatment for sulfa-sensitive patients (162).

ISOSPORA

Isospora is a coccidian parasite which infects humans. It is frequently identified in AIDS patients (21, 40). *Isospora* can be acquired by ingestion of contaminated food or water. Oocysts excyst in the intestine, and sporocysts are released. The sporocysts in turn release sporozoites which infect intestinal epithelial cells. Asexual and sexual life cycle stages occur in the cytoplasm of enterocytes. Unsporulated *Isospora* oocysts require 12 to 48 h to mature and become infectious outside the host.

Isospora can be detected in stools from infected patients by observing unsporulated oocysts shed in feces. *I. belli* oocysts (10 to 19 by 20 to 30 μm) can be detected in direct wet mount preparations in heavy infections during ovum and parasite examination (Fig. 31.4). However, most infections are not heavy, and rates of shedding of oocysts may be variable; therefore, it is necessary to examine a series of samples and perform modified acid-fast staining (Color Plate 3C). *I. belli* infects the entire intestine and produces severe intestinal disease (41). Deaths from overwhelming infections have been reported, especially in immunocompromised patients. Symptoms include diarrhea, nausea, steatorrhea, headache, and weight loss. The disease may persist for months and even years (50).

Isospora infection is rare in immunocompetent people but occurs in 0.2 to 0.3% of immunocompromised AIDS

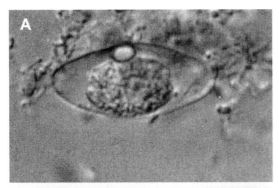

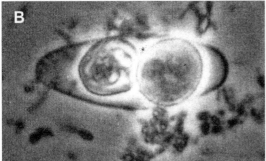

Figure 31.4 Bright-field photomicrograph of *I. belli* oocysts unsporulated (A) and sporulated (B).

patients in the United States and 8 to 20% of AIDS patients in Africa and Haiti. *Isospora* infection is endemic in many parts of Africa, Asia, and South America. The treatment of choice is TMP-SMX. In HIV patients, recurrence is common after discontinuation of therapy (129).

TOXOPLASMA

T. gondii is a coccidian parasite that infects a variety of warm-blooded hosts. Cats are the definitive hosts, and other warm-blooded animals can serve as intermediate hosts. Cats excrete oocysts in their feces. Oocysts are environmentally resistant and can survive several years in moist, shaded conditions. Infections are acquired principally by ingestion of food and water containing oocysts, ingestion of animal tissues containing cystic forms (bradyzoites), or transplacental transmission (55).

The unsporulated oocysts require 24 h outside the host to differentiate and become infectious. When oocysts are ingested by the intermediate host, the oocyst walls are ruptured and the sporozoites are released (Fig. 31.5). They invade epithelial cells and rapidly multiply asexually, producing tachyzoites. Tachyzoites multiply by endodyogeny, a process in which the mother tachyzoite is consumed by the formation of two daughter zoites (Fig. 31.6). Eventually bradyzoites, which are slowly multiplying forms, develop, forming tissue cysts. Cysts persist for the duration of the life of the host. By encysting, the

parasite evades the host's immune response and ensures its viability (52, 56).

When animal tissues containing cysts are ingested by cats, proteolytic enzymes digest the cyst walls and the bradyzoites are released. They infect and multiply in the intestinal epithelial cells, transforming into tachyzoites. These in turn disperse via blood and lymph nodes. When tissues of other animal species are ingested by felines, tachyzoites or bradyzoites begin the enteroepithelial cycle and sexual multiplication also occurs (Fig. 31.5). Macro- and microgametocytes are produced. After fertilization, the zygote differentiates into oocysts which are passed in the feces (56).

Although most *Toxoplasma* infections occur by ingestion of contaminated meat, other foods, and water, infection can also be acquired by organ transplantation or by blood transfusion (22, 32, 43, 138). Disseminated toxoplasmosis may occur in patients who have received organ transplants and are receiving immunosuppressive therapy. Chorioretinitis is frequently observed in adults who acquire the infection. In immunosuppressed patients, toxoplasmosis can reactivate from latent infections (53, 139).

Toxoplasmosis can also be acquired vertically by transplacental transmission when a pregnant woman becomes infected. After multiplying in the placenta, tachyzoites spread into the fetal tissues. Infection can occur at any stage of the pregnancy, but the fetus is affected the most when infection occurs during the first months of pregnancy. Most infected children do not show any signs of the disease until later in life, when they may present with chorioretinitis and mental retardation (3, 6, 31, 49, 94).

The overall prevalence in humans and animals varies according to eating habits and lifestyle. The prevalence of *T. gondii* is highest in swine with outdoor access. Confined housing reduces the exposure to cat feces and infected rodents (8, 24).

Toxoplasmosis can be acquired by ingestion of lamb, poultry, horse, and wild game animals. Cooking, freezing, or gamma irradiation will kill the *Toxoplasma* cysts and oocysts. Temperatures of 61°C or higher for 3.6 min will inactivate the parasites, and freezing at −13°C will result in death of cysts (57, 110). Pyrimethamine in combination with folinic acid or trisulfapyrimidine is the treatment of choice for acute infections. TMP-SMX is effective and frequently used to prevent recurrence of acute infections in AIDS patients (70).

MICROSPORIDIA

Most microsporidium infections have been reported to occur in AIDS patients (39). Five genera of microsporidia have been associated with human infections:

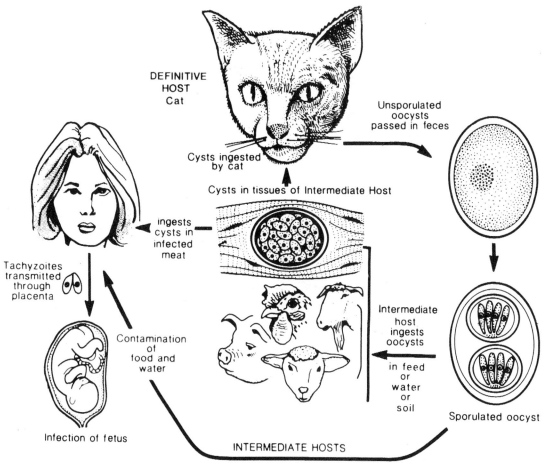

Figure 31.5 Life cycle of *T. gondii*. From Dubey and Beattie (55).

Enterocytozoon, Septata, Pleistophora, Encephalitozoon, and *Vittaforma*. *E. bienusii*, the only species which seems to be tissue specific, is associated only with human enteric infections. Microsporidium spores, which are highly resistant environmental forms, vary in size among the different species. They are ovoid or piriform and 1 to 2 μm in diameter. *Encephalitozoon intestinalis* spores are bigger than those of *Enterocytozoon bienusii*. These spores contain a polar filament which is used to eject the sporoplasm and penetrate the host cell cytoplasm. After entering the cell, the parasite multiplies asexually and eventually forms spores, lysing the host cell and invading neighboring cells. Although the mechanisms of transmission are not clear, it is believed that the infection can be acquired by ingesting spores in contaminated water and produce (114).

Infection with microsporidia is characterized by watery and large-volume stools. Microsporidia are occasionally associated with biliary tract disease and may potentially cause cholangiopathy, as observed in AIDS patients (23).

Microsporidium spores can be histologically identified in tissue, but identification in fecal samples requires special staining processes. Microsporidia can be identified in tissues by using various conventional stains such as hematoxylin-eosin, and the Gram and Giemsa stains. Detection in fecal samples can be achieved by using nonspecific staining techniques such as the use of calcofluor white and a modified trichrome with Chromotrope 2R; however, small structures and yeast spores are also stained by these procedures (25, 109, 147). Identification of microsporidia in the environment presents an additional challenge. Most species of the animal kingdom can be parasitized by species of microsporidia, and noninfectious species can be confused with those infectious to humans. Definitive identification is limited to transmission electron microscopy (19, 29, 30, 34, 47). New tests using specific monoclonal antibodies and molecular tools such as PCR are being developed to aid in the detection of microsporidia in tissues and in fecal and environmental samples (65, 97, 120, 154).

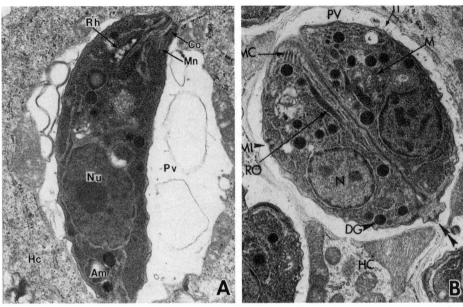

Figure 31.6 Transmission electron micrographs of *T. gondii*. (A) Sporozoite in parasitophorous vacuole (Pv) of host cell (Hc) at 24 h after inoculation; Am, amylopectin granule; Co, conoid; Mn, microneme; Nu, nucleus of sporozoite; Rh, rhoptry. (B) Final stage of endodyogeny to form two daughter tachyzoites that are still attached to the posterior ends (arrowheads); DG, dense granules; HC, host cell; IT, intravacular tubules; M, mitochondrion; MC, microneme; MI, micropore; N, nucleus; RO, rhoptry. From Dubey and Beattie (55).

Infections caused by *Septata* can be treated with albendazole or with metronidazole and atovaquone. To date, there is no effective treatment for infections caused by *E. bienusii* (44). However, nitazoxanide was reported to resolve the infection in an AIDS patient (27).

GIARDIA

Giardia is a protozoan flagellate that belongs to the phylum Sarcomastigophora. Initially thought to be a commensal organism in humans, *Giardia* is now clearly recognized as a common cause of diarrhea and malabsorption. *Giardia* infects millions of people throughout the world in both epidemic and sporadic forms. Most human infections result from ingestion of contaminated water or food or direct fecal-oral transmission such as would occur in person-to-person contact in child care centers and in male homosexual activity (1).

Three species of *Giardia* have been described based on differences discernible in cysts and trophozoites by light microscopy: *G. agilis* from amphibians; *G. muris* from rodents, birds, and reptiles; and *G. lamblia* (also called *G. intestinalis* or *G. duodenalis*) from various mammals, including humans. Two additional species which are indistinguishable from *G. lamblia* by light

microscopy, *G. ardeae* from herons and *G. psittaci* from psittacine birds, have been identified based on ultrastructural morphologic differences (60).

G. lamblia does not appear to be host restricted, and wild animals such as beavers and muskrats have been implicated in waterborne outbreaks of giardiasis in humans. More recently, molecular classification using small-subunit RNA has placed *Giardia* as one of the most primitive eukaryotic organisms (161). Two major molecular groups or assemblages (A and B) of *G. lamblia* have been recognized as infecting humans. Other assemblages infect animals exclusively: dogs (C and D), cats (F), rats (G), and cattle, goats, sheep, and pigs (E) (117).

Giardia can be observed in two forms: the trophozoite and the cyst. The cyst is the infectious form and is relatively inert and environmentally resistant. After cysts are ingested, excystation occurs in the duodenum after exposure to the acidic gastric pH and the pancreatic enzymes chemotrypsin and trypsin. Each cyst releases two vegetative trophozoites. The trophozoites replicate in the crypts of the duodenum and upper jejunum and reproduce asexually by binary fission. Some of the trophozoites then encyst in the ileum, possibly as a result of exposure to bile salts or cholesterol starvation (1). The trophozoites and cysts are excreted in the feces.

Cysts are round or oval. Each cyst measures 4 to 11 by 7 to 10 μm, has four nuclei, and contains axonemes and median bodies. Trophozoites have the shape of a teardrop (viewed dorsally or ventrally) and measure 10 to 20 μm in length by 5 to 15 μm in width. The trophozoite has a concave sucking disk with four pairs of flagella, two axonemes, two median bodies, and two nuclei. The ventral disks act as suction cups, allowing mechanical attachment to the surface of the intestine (Fig. 31.7) (102). Infections may result from the ingestion of 10 or fewer *Giardia* cysts (145). Boiling is very effective to inactivate *Giardia* cysts, but cysts can survive after freezing for a few days.

Ozone is an excellent disinfectant and can even be used to inactivate microorganisms such as protozoa, which are very resistant to conventional disinfectants. The apparent activation energy for the inactivation of protozoa is 80 kJ/mol. By-products of ozonation include bromate, iodate, and chlorate, which may be of concern depending on the chemical composition of the water to be disinfected. A 2-log reduction in active *G. lamblia* and *G. muris* cysts with ozone required 2.4 times longer than the contact time recommended by the Surface Water Treatment Rule.

A surrogate organism used as an indicator for protozoan parasite inactivation is *Bacillus subtilis* because it has reported values similar to those for *Cryptosporidium*. Chlorine dioxide at the dose of 2.5 μg/ml reduces the number of active spores by ca. 2.0 and 0.5 logs at water temperatures of 23.2 and 5.2°C, respectively. This chlorine concentration is below the maximum treatment concentration set by the U.S. Environmental Protection Agency (66, 137, 140, 163).

G. lamblia is prevalent worldwide and is especially common in areas where poor sanitary conditions and insufficient water treatment facilities prevail. Seasonality of giardiasis has been reported during late summer in the United Kingdom, the United States, and Mexico. The majority of *Giardia* infections are asymptomatic, but they can present as chronic diarrhea. Travelers to areas of endemicity are at high risk for developing symptomatic giardiasis. A study in Leningrad (now St. Petersburg), Russia, reported that 95% of travelers developed symptomatic giardiasis. Hikers and campers are also at increased risk, since *Giardia* cysts, often of animal origin, can be found in freshwater lakes and streams. The prevalence of *Giardia* infection can be as high as 35% in children attending child care centers. Although these children are frequently asymptomatic, they may infect other family members who may develop symptomatic giardiasis (7, 67, 106).

In waterborne outbreaks of diarrhea in which the etiological agent is identified, *Giardia* has been the most common agent. Waterborne transmission is commonly a result of inadequate water treatment or sewage contamination of drinking, well, or surface water. Giardiasis has also been associated with exposure to contaminated recreational water such as that in swimming pools. *Giardia* cysts are susceptible to inactivation by ozone and halogens; however, the concentration of chlorine used for drinking water may not cause inactivation. Inactivation by chlorine requires prolonged contact time, and filtration is the recommended means for purifying water (4, 99, 130).

Symptomatic patients present with loose, foul-smelling stools and increased levels of fat and mucus in fecal samples. Flatulence, abdominal cramps, bloating, and nausea are common, as are anorexia, malaise, and weight loss. Blood is not present in stools. Fever is occasionally present at the beginning of the infection. In contrast with most other forms of acute infectious diarrhea, *G. lamblia* infection results in prolonged symptoms. Although giardiasis may resolve spontaneously, the illness frequently lasts for several weeks and sometimes for months if left untreated. Those with chronic giardiasis have profound malaise and diffuse epigastric and abdominal discomfort. Although diarrhea may persist, it may be replaced by constipation or even by normal bowel habits (1).

Malabsorption associated with giardiasis may be responsible for substantial weight loss. Even in asymptomatic infections, malabsorption of fats, carbohydrates, and vitamins may occur. Reduced intestinal disaccharidase activity may persist even after *Giardia* is eradicated.

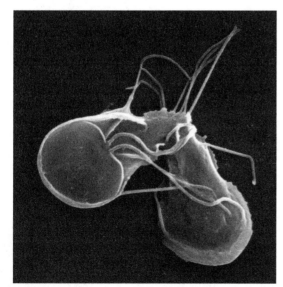

Figure 31.7 Scanning electron micrograph of *G. lamblia* trophozoites. One trophozoite shows the dorsal surface and the other shows the ventral surface with sucking disk and flagella.

Lactase deficiency is the most common residual deficiency and occurs in 20 to 40% of cases (153).

Villous blunting, lymphocytic infiltration, and malabsorption are observed in biopsy samples in symptomatic cases. No tissue invasion is observed, and high numbers of trophozoites are sometimes present in the crypts without obvious pathology. To date, the presence of a toxin has not been demonstrated and no other potential mechanisms by which *Giardia* causes diarrhea have been identified.

Giardia infection can be diagnosed by finding cysts or, less commonly, trophozoites in fecal specimens (Fig. 31.8A). *Giardia* in feces can be detected by enzyme immunoassays, indirect and direct immunofluorescent assays using monoclonal antibodies, or PCR. All of these procedures are highly sensitive and specific for environmental and stool samples. In some patients with chronic diarrhea and malabsorption, results of stool examinations are repeatedly negative despite ongoing suspicion of giardiasis (5, 85, 166). Giardia cysts were found in 41.5% of nondepurated mussels from the Galician Coast that were destined for human consumption (82).

Effective treatment for patients with symptomatic giardiasis is mainly a single treatment course with metronidazole. In refractory cases, multiple or combination courses have occasionally been required. Tinidazole is widely used throughout the world, and a single dose is effective for the treatment of giardiasis (91, 157).

BALANTIDIUM

Balantidium coli is a ciliate parasite that, although found worldwide, is not highly prevalent. It is a commensal parasite of pigs. The trophozoites reside in the large intestine and multiply by binary fission. In humans, *B. coli* can cause ulcerative colitis and diarrhea. The ulcers differ from those caused by *Entamoeba histolytica* in that the epithelial surface is damaged but with lesions more superficial than those caused by amoebae. The parasite encysts and is excreted in the feces. The cyst, which is the environmentally resistant form, is large and oblong, 45 to 65 μm in diameter (Fig. 31.8B). Both cysts and trophozoites contain two nuclei. Trophozoites move via their cilia and rotate on their longitudinal axis (72). Treatment of infection is preferentially done with tetracycline and, alternatively, with iodoquinol and metronidazole.

AMOEBIASIS

Various amoebae can infect humans as commensals; however, *E. histolytica* is pathogenic to humans. *Entamoeba dispar*, which is morphologically similar to *E. histolytica*, is not pathogenic. Cysts and trophozoites of *E. histolytica* are excreted in the feces of infected individuals (Fig. 31.8C), with cysts being environmentally resistant. Once cysts are ingested, the trophozoite excysts, colonizes the large intestine, and multiplies by binary fission followed by encystation.

Most patients are asymptomatic even when shedding cysts in their feces. In other instances, parasites can invade the mucosa and cause an ulceration that goes from the luminal surface of the intestine through the lamina propria and to the muscularis mucosa. The parasite then spreads laterally, forming a flask-shaped ulcer. The trophozoites feed on cell debris and red blood cells (73). Infection can progress, producing ulcerative amoebic colitis and causing perforation of the intestinal wall. Patients complain of diarrhea with stools containing blood and mucus, back pain, tenesmus, dehydration, and abdominal tenderness. Fulminant colitis is characterized by severe bloody diarrhea, fever, and abdominal tenderness due to transmural necrosis of the bowel. Another presentation of amoebiasis is ameboma, which resembles carcinoma and is not necessarily associated with pain. Amoebae can also disseminate to the liver. Complications with amoebiasis are observed when the liver parenchyma is gradually replaced with necrotic debris, inflammatory cells, and trophozoites. Patients present with hepatomegaly, weight loss, and anemia (73, 148).

Asymptomatic amoebiasis can be treated with iodoquinol, paromomycin, or diloxanide. If mild to severe infection and hepatic abcesses are present, amoebiasis can be treated with metronidazole or tinidazole followed with iodoquinol to treat asymptomatic amoebiasis.

In summary, foodborne disease outbreaks caused by parasites are being documented with increased frequency.

SUMMARY

The consumption of fresh fruits and vegetables in developed countries has increased, and local production cannot meet this demand. Thus, importation of produce and other foods from countries where sanitary standards are minimal or lacking makes it even more difficult to control foodborne diseases (122). In addition, inactivation of these parasites has been a challenging task.

We face new challenges to ensure the safety of produce and other foods. Enrichment media are not available for cultivating parasites; therefore, isolation and identification methodologies need to be extremely sensitive and specific to detect small numbers of various forms of parasites that may be present in foods. Molecular tools such as PCR, restriction fragment length polymorphism, and variations of these techniques are being developed

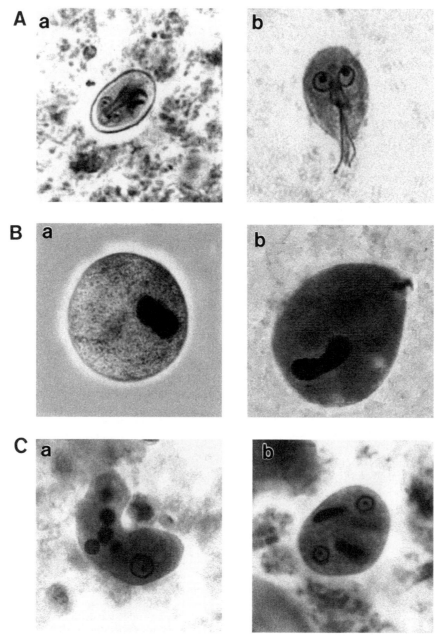

Figure 31.8 Hematoxylin and eosin staining. (A) *G. lamblia* cyst (a) with two nuclei and trophozoite (b) showing two nuclei and median body visible at one pole. (B) *B. coli* cyst (a) showing large macronucleus and cilia beneath cyst wall and trophozoite (b) showing oval macronucleus. (C) *E. histolytica* trophozoite (a) showing a nucleus and few red blood cells in the cytoplasm and cyst (b) showing two of four nuclei and rod-shaped inclusion bodies with rounded ends. (Courtesy of Lynne S. Garcia.)

to improve the sensitivity and specificity of detection and identification processes.

Treatment of drinking water by chlorination at permissible concentrations is relatively ineffective at inactivating spores, cysts, and oocysts of parasites and cannot be recommended as a sole method for water treatment. Boiling can inactivate cysts, oocysts, and spores. Irradiation of produce to inactivate pathogens that may be present has also been examined. This methodology effectively inactivates *Toxoplasma* cyst forms and oocysts. There is an increase in consumers' acceptance of purchasing irradiated produce (57), thereby providing additional incentive to use irradiation as a treatment to reduce or eliminate at least some genera of parasites in food and water.

References

1. Adam, R. D. 2001. Biology of *Giardia lamblia*. *Clin. Microbiol. Rev.* **14:**447–475.

2. Adam, R. D., Y. R. Ortega, R. H. Gilman, and C. R. Sterling. 2000. Intervening transcribed spacer region 1 variability in *Cyclospora cayetanensis*. *J. Clin. Microbiol.* **38:**2339–2343.

3. Ades, A. E. 1991. Evaluating the sensitivity and predictive value of tests of recent infection: toxoplasmosis in pregnancy. *Epidemiol. Infect.* **107:**527–535.

4. Adin, A. 2004. Particle count and size alteration for membrane fouling reduction in non-conventional water filtration. *Water Sci. Technol.* **50:**273–278.

5. Aldeen, W. E., D. Hale, A. J. Robison, and K. Carroll. 1995. Evaluation of a commercially available ELISA assay for detection of *Giardia lamblia* in fecal specimens. *Diagn. Microbiol. Infect. Dis.* **21:**77–79.

6. Amos, C. S. 1977. The ocular manifestations of congenital infections produced by *Toxoplasma* and cytomegalovirus. *J. Am. Optom. Assoc.* **48:**532–538.

7. Ang, L. H. 2000. Outbreak of giardiasis in a daycare nursery. *Commun. Dis. Public Health* **3:**325–327.

8. Anonymous. 1970. Transmission of toxoplasmosis. *Br. Med. J.* **1:**126–127.

9. Anonymous. 1996. Foodborne outbreak of diarrheal illness associated with *Cryptosporidium parvum*—Minnesota, 1995. *Morb. Mortal. Wkly. Rep.* **45:**783–784.

10. Anonymous. 1997. Outbreak of cyclosporiasis—northern Virginia-Washington, D.C.-Baltimore, Maryland, Metropolitan area, 1997. *Can. Commun. Dis. Rep.* **23:**166–168.

11. Anonymous. 1997. Outbreaks of cyclosporiasis—United States, 1997. *Morb. Mortal. Wkly. Rep.* **46:**451–452.

12. Anonymous. 1997. Outbreaks of *Escherichia coli* O157:H7 infection and cryptosporidiosis associated with drinking unpasteurized apple cider—Connecticut and New York, October 1996. *Morb. Mortal. Wkly. Rep.* **46:**4–8.

13. Anonymous. 2000. Outbreaks of diarrhea associated with Cyanobacteria (blue-green-algae)-like-bodies—Chicago and Nepal, 1989 and 1990. *Morb. Mortal. Wkly. Rep.* **40:**325–327.

14. Anonymous. 2004. Outbreak of cyclosporiasis associated with snow peas—Pennsylvania, 2004. *Morb. Mortal. Wkly. Rep.* **53:**876–878.

15. Anusz, K. Z., P. H. Mason, M. W. Riggs, and L. E. Perryman. 1990. Detection of *Cryptosporidium parvum* oocysts in bovine feces by monoclonal antibody capture enzyme-linked immunosorbent assay. *J. Clin. Microbiol.* **28:**2770–2774.

16. Arrowood, M. J., and C. R. Sterling. 1989. Comparison of conventional staining methods and monoclonal antibody-based methods for *Cryptosporidium* oocyst detection. *J. Clin. Microbiol.* **27:**1490–1495.

17. Ashford, R. W. 1979. Occurrence of an undescribed coccidian in man in Papua New Guinea. *Ann. Trop. Med. Parasitol.* **73:**497–500.

18. Atwill, E. R., K. W. Tate, M. D. Pereira, J. Bartolome, and G. Nader. 2006. Efficacy of natural grassland buffers for removal of *Cryptosporidium parvum* in rangeland runoff. *J. Food Prot.* **69:**177–184.

19. Awadalla, H. N., I. F. el Naga, M. M. el-Temsahi, and A. Y. Negm. 1998. Detection of Microsporidia by different staining techniques. *J. Egypt. Soc. Parasitol.* **28:**729–738.

20. Balatbat, A. B., G. W. Jordan, Y. J. Tang, and J. Silva, Jr. 1996. Detection of *Cryptosporidium parvum* DNA in human feces by nested PCR. *J. Clin. Microbiol.* **34:**1769–1772.

21. Ballal, M., T. Prabhu, A. Chandran, and P. G. Shivananda. 1999. *Cryptosporidium* and *Isospora belli* diarrhoea in immunocompromised hosts. *Indian J. Cancer* **36:**38–42.

22. Barcan, L. A., M. L. Dallurzo, L. O. Clara, A. Valledor, S. Macias, E. Zorkin, S. Gerona, and B. Livellara. 2002. *Toxoplasma gondii* pneumonia in liver transplantation: survival after a severe case of reactivation. *Transplant. Infect. Dis.* **4:**93–96.

23. Beaugerie, L., M. F. Teilhac, A. M. Deluol, J. Fritsch, P. M. Girard, W. Rozenbaum, Q. Y. Le, and F. P. Chatelet. 1992. Cholangiopathy associated with Microsporidia infection of the common bile duct mucosa in a patient with HIV infection. *Ann. Intern. Med.* **117:**401–402.

24. Beazley, D. M., and R. S. Egerman. 1998. Toxoplasmosis. *Semin. Perinatol.* **22:**332–338.

25. Berlin, O. G. W., L. R. Ash, C. N. Conteas, and J. B. Peter. 1999. Rapid, hot chromotrope stain for detecting microsporidia. *Clin. Infect. Dis.* **29:**209.

26. Bern, C., B. Hernandez, M. B. Lopez, M. J. Arrowood, M. A. de Mejia, A. M. De Merida, A. W. Hightower, L. Venczel, B. L. Herwaldt, and R. E. Klein. 1999. Epidemiologic studies of *Cyclospora cayetanensis* in Guatemala. *Emerg. Infect. Dis.* **5:**766–774.

27. Bicart-See, A., P. Massip, M. D. Linas, and A. Datry. 2000. Successful treatment with nitazoxanide of *Enterocytozoon bieneusi* microsporidiosis in a patient with AIDS. *Antimicrob. Agents Chemother.* **44:**167–168.

28. Bissuel, F., L. Cotte, M. Rabodonirina, P. Rougier, M. A. Piens, and C. Trepo. 1994. Paromomycin: an effective treatment for cryptosporidial diarrhea in patients with AIDS. *Clin. Infect. Dis.* **18:**447–449.

29. Blanshard, C., W. S. Hollister, C. S. Peacock, D. G. Tovey, D. S. Ellis, E. U. Canning, and B. G. Gazzard. 1992. Simultaneous infection with two types of intestinal microsporidia in a patient with AIDS. *Gut* **33:**418–420.

30. Boldorini, R., A. Tosoni, G. Mazzucco, M. Cernuschi, P. Caramello, E. Maran, G. Costanzi, and G. Monga. 1996. Intracellular protozoan infection in small intestinal biopsies of patients with AIDS. Light and electron microscopic evaluation. *Pathol. Res. Pract.* **192:**249–259.

31. Brady-McCreery, K. M., M. A. Hussein, and E. A. Paysse. 2003. Congenital toxoplasmosis with unusual retinal findings. *Arch. Ophthalmol.* **121:**1200–1201.

32. Brinkman, K., S. Debast, R. Sauerwein, F. Ooyman, J. Hiel, and J. Raemaekers. 1998. *Toxoplasma* retinitis/encephalitis 9 months after allogeneic bone marrow transplantation. *Bone Marrow Transplant.* **21:**635–636.

33. Caccio, S., E. Pinter, R. Fantini, I. Mezzaroma, and E. Pozio. 2002. Human infection with *Cryptosporidium*

felis: case report and literature review. *Emerg. Infect. Dis.* 8:85–86.

34. Cali, A., L. Weiss, P. Takvorian, H. Tanowitz, and M. Wittner. 1994. Ultrastructural identification of AIDS associated microsporidiosis. *J. Eukaryot. Microbiol.* 41:24S.

35. Cama, V. A., and C. R. Sterling. 1991. Hyperimmune hens as a novel source of anti-Cryptosporidium antibodies suitable for passive immune transfer. *J. Protozool.* 38: 42S–43S.

36. Carreno, R. A., D. S. Martin, and J. R. Barta. 1999. *Cryptosporidium* is more closely related to the gregarines than to coccidia as shown by phylogenetic analysis of apicomplexan parasites inferred using small-subunit ribosomal RNA gene sequences. *Parasitol. Res.* 85:899–904.

37. Casemore, D. P., M. Armstrong, and R. L. Sands. 1985. Laboratory diagnosis of cryptosporidiosis. *J. Clin. Pathol.* 38:1337–1341.

38. Chauret, C. P., C. Z. Radziminski, M. Lepuil, R. Creason, and R. C. Andrews. 2001. Chlorine dioxide inactivation of *Cryptosporidium parvum* oocysts and bacterial spore indicators. *Appl. Environ. Microbiol.* 67:2993–3001.

39. Chu, D. M., J. B. Sherchand, J. H. Cross, and P. A. Orlandi. 2004. Detection of *Cyclospora cayetanensis* in animal fecal isolates from Nepal using an FTA filter-base polymerase chain reaction method. *Am. J. Trop. Med. Hyg.* 71: 373–379.

40. Chui, D. W., and R. L. Owen. 1994. AIDS and the gut. *J. Gastroenterol. Hepatol.* 9:291–303.

41. Comin, C. E., and M. Santucci. 1994. Submicroscopic profile of *Isospora belli* enteritis in a patient with acquired immune deficiency syndrome. *Ultrastruct. Pathol.* 18: 473–482.

42. Connor, B. A., J. Reidy, and R. Soave. 1999. Cyclosporiasis: clinical and histopathologic correlates. *Clin. Infect. Dis.* 28:1216–1222.

43. Conrath, J., A. Mouly-Bandini, F. Collart, and B. Ridings. 2003. *Toxoplasma gondii* retinochoroiditis after cardiac transplantation. *Graefes Arch. Clin. Exp. Ophthalmol.* 241:334–338.

44. Costa, S. F., and L. M. Weiss. 2000. Drug treatment of microsporidiosis. *Drug Resist. Updat.* 3:384–399.

45. Coupe, S., C. Sarfati, S. Hamane, and F. Derouin. 2005. Detection of *Cryptosporidium* and identification to the species level by nested PCR and restriction fragment length polymorphism. *J. Clin. Microbiol.* 43:1017–1023.

46. Current, W. L., and P. H. Bick. 1989. Immunobiology of *Cryptosporidium* spp. *Pathol. Immunopathol. Res.* 8: 141–160.

47. Curry, A. 2003. Electron microscopy and the investigation of new infectious diseases. *Int. J. Infect. Dis.* 7:251–257.

48. Danziger, L. H., T. P. Kanyok, and R. M. Novak. 1993. Treatment of cryptosporidial diarrhea in an AIDS patient with paromomycin. *Ann. Pharmacother.* 27:1460–1462.

49. De, M. R., R. Ceccarelli, R. Frulio, C. Palmero, and P. Vittone. 2003. Retinochoroiditis associated with congenital toxoplasmosis in children: IgG antibody profiles demonstrating the synthesis of local antibodies. *Eur. J. Ophthalmol.* 13:74–79.

50. DeHovitz, J. A., J. W. Pape, M. Boncy, and W. D. Johnson, Jr. 1986. Clinical manifestations and therapy of *Isospora belli* infection in patients with the acquired immunodeficiency syndrome. *N. Engl. J. Med.* 315:87–90.

51. Deng, M. Q., and D. O. Cliver. 2000. Comparative detection of *Cryptosporidium parvum* oocysts from apple juice. *Int. J. Food Microbiol.* 54:155–162.

52. Derouin, F. 1992. Pathogeny and immunological control of toxoplasmosis. *Braz. J. Med. Biol. Res.* 25:1163–1169.

53. Derouin, F., D. Vittecoq, B. Beauvais, and A. Bussel. 1987. *Toxoplasma parasitaemia* associated with serological reactivation of chronic toxoplasmosis in a patient with the acquired immunodeficiency syndrome. *J. Infect.* 14:189–190.

54. Ditrich, O., L. Palkovic, J. Sterba, J. Prokopic, J. Loudova, and M. Giboda. 1991. The first finding of *Cryptosporidium baileyi* in man. *Parasitol. Res.* 77:44–47.

55. Dubey, J. P., and C. P. Beattie. 1988. *Toxoplasmosis of Animals and Man.* CRC Press, Inc., Boca Raton, Fla.

56. Dubey, J. P., N. L. Miller, and J. K. Frenkel. 1970. *Toxoplasma gondii* life cycle in cats. *J. Am. Vet. Med. Assoc.* 157:1767–1770.

57. Dubey, J. P., D. W. Thayer, C. A. Speer, and S. K. Shen. 1998. Effect of gamma irradiation on unsporulated and sporulated *Toxoplasma gondii* oocysts. *Int. J. Parasitol.* 28:369–375.

58. Eberhard, M. L., A. J. da Silva, B. G. Lilley, and N. J. Pieniazek. 1999. Morphologic and molecular characterization of new *Cyclospora* species from Ethiopian monkeys: *C. cercopitheci* sp. n., *C. colobi* sp. n., and *C. papionis* sp. n. *Emerg. Infect. Dis.* 5:651–658.

59. Eberhard, M. L., Y. R. Ortega, D. E. Hanes, E. K. Nace, R. Q. Do, M. G. Robl, K. Y. Won, C. Gavidia, N. L. Sass, K. Mansfield, A. Gozalo, J. Griffiths, R. Gilman, C. R. Sterling, and M. J. Arrowood. 2000. Attempts to establish experimental *Cyclospora cayetanensis* infection in laboratory animals. *J. Parasitol.* 86:577–582.

60. Erlandsen, S. L., and W. J. Bemrick. 1987. SEM evidence for a new species, *Giardia psittaci. J. Parasitol.* 73: 623–629.

61. Fano-Sobsey, E. M., M. L. Eberhard, J. R. Seed, D. J. Weber, K. Y. Won, E. K. Nace, and C. L. Moe. 2004. Human challenge pilot study with *Cyclospora cayetanensis. Emerg. Infect. Dis.* 10:726–728.

61a. Fayer, R., C. A. Spear, and J. P. Dubey. 1990. The general biology of *Crytosporidium*, p. 1–41. *In* R. Fayer (ed.), *Cryptosporidium and Cryptosporidiosis*, CRC Press, Boca Raton, Fla.

62. Fayer, R., M. Tilley, S. J. Upton, A. J. Guidry, D. W. Thayer, M. Hildreth, and J. Thomson. 1991. Production and preparation of hyperimmune bovine colostrum for passive immunotherapy of cryptosporidiosis. *J. Protozool.* 38:38S–39S.

63. Fayer, R., J. M. Trout, E. J. Lewis, M. Santin, L. Zhou, A. A. Lal, and L. Xiao. 2003. Contamination of Atlantic coast commercial shellfish with *Cryptosporidium. Parasitol. Res.* 89:141–145.

64. Fayer, R., J. M. Trout, L. Xiao, U. M. Morgan, A. A. Lai, and J. P. Dubey. 2001. *Cryptosporidium canis* n. sp. from domestic dogs. *J. Parasitol.* 87:1415–1422.

65. Fedorko, D. P., N. A. Nelson, and C. P. Cartwright. 1995. Identification of microsporidia in stool specimens by using PCR and restriction endonucleases. *J. Clin. Microbiol.* **33**:1739–1741.

66. Finch, G. R., E. K. Black, C. W. Labatiuk, L. Gyurek, and M. Belosevic. 1993. Comparison of *Giardia lamblia* and *Giardia muris* cyst inactivation by ozone. *Appl. Environ. Microbiol.* **59**:3674–3680.

67. Flanagan, P. A. 1992. *Giardia* —diagnosis, clinical course and epidemiology. A review. *Epidemiol. Infect.* **109**:1–22.

68. Fleming, C. A., D. Caron, J. E. Gunn, and M. A. Barry. 1998. A foodborne outbreak of *Cyclospora cayetanensis* at a wedding: clinical features and risk factors for illness. *Arch. Intern. Med.* **158**:1121–1125.

69. Furio, M. M., and C. J. Wordell. 1985. Treatment of infectious complications of acquired immunodeficiency syndrome. *Clin. Pharm.* **4**:539–554.

70. Gallant, J. E., R. D. Moore, and R. E. Chaisson. 1994. Prophylaxis for opportunistic infections in patients with HIV infection. *Ann. Intern. Med.* **120**:932–944.

71. Galvagno, G., G. Cattaneo, and E. Reverso-Giovantin. 1993. Chronic diarrhea due to *Cryptosporidium*: the efficacy of spiramycin treatment. *Pediatr. Med. Chir.* **15**:297–298. (In Italian.)

72. Garcia, L. S. 1993. Intestinal protozoa: flagellates and cilliates, p. 31–48. *In* L. S. Garcia and D. A. Bruckner (ed.), *Diagnostic Medical Parasitology*, 2nd ed. American Society for Microbiology, Washington, D.C.

73. Garcia, L. S. 2006. Intestinal protozoa: amoeba, p. 6–30. *In* L. S. Garcia and D. A. Bruckner (ed.), *Diagnostic Medical Parasitology*, 2nd ed. American Society for Microbiology, Washington, D.C.

74. Garcia-Lopez, H. L., L. E. Rodriguez-Tovar, and C. E. Medina-De la Garza. 1996. Identification of *Cyclospora* in poultry. *Emerg. Infect. Dis.* **2**:356–357.

75. Gatei, W., R. W. Ashford, N. J. Beeching, S. K. Kamwati, J. Greensill, and C. A. Hart. 2002. *Cryptosporidium muris* infection in an HIV-infected adult, Kenya. *Emerg. Infect. Dis.* **8**:204–206.

76. Giacometti, A., F. Burzacchini, O. Cirioni, F. Barchiesi, M. Dini, and G. Scalise. 1999. Efficacy of treatment with paromomycin, azithromycin, and nitazoxanide in a patient with disseminated cryptosporidiosis. *Eur. J. Clin. Microbiol. Infect. Dis.* **18**:885–889.

77. Gomez-Bautista, M., L. M. Ortega-Mora, E. Tabares, V. Lopez-Rodas, and E. Costas. 2000. Detection of infectious *Cryptosporidium parvum* oocysts in mussels (*Mytilus galloprovincialis*) and cockles (*Cerastoderma edule*). *Appl. Environ. Microbiol.* **66**:1866–1870.

78. Gomez-Couso, H., F. Freire-Santos, C. F. Amar, K. A. Grant, K. Williamson, M. E. res-Mazas, and J. McLauchlin. 2004. Detection of *Cryptosporidium* and *Giardia* in molluscan shellfish by multiplexed nested-PCR. *Int. J. Food Microbiol.* **91**:279–288.

79. Gomez-Couso, H., F. Freire-Santos, G. A. Hernandez-Cordova, and M. E. res-Mazas. 2005. A histological study of the transit of *Cryptosporidium parvum* oocysts through clams (*Tapes decussatus*). *Int. J. Food Microbiol.* **102**:57–62.

80. Gomez-Couso, H., F. Freire-Santos, J. Martinez-Urtaza, O. Garcia-Martin, and M. E. res-Mazas. 2003. Contamination of bivalve molluscs by *Cryptosporidium* oocysts: the need for new quality control standards. *Int. J. Food Microbiol.* **87**:97–105.

81. Gomez-Couso, H., F. Freire-Santos, M. R. Ortega-Inarrea, J. A. Castro-Hermida, and M. E. res-Mazas. 2003. Environmental dispersal of *Cryptosporidium parvum* oocysts and cross transmission in cultured bivalve molluscs. *Parasitol. Res.* **90**:140–142.

82. Gomez-Couso, H., F. Mendez-Hermida, J. A. Castro-Hermida, and E. res-Mazas. 2005. *Giardia* in shellfish-farming areas: detection in mussels, river water and waste waters. *Vet. Parasitol.* **133**:13–18.

83. Gomez-Couso, H., F. Mendez-Hermida, J. A. Castro-Hermida, and E. res-Mazas. 2006. *Cryptosporidium* contamination in harvesting areas of bivalve molluscs. *J. Food Prot.* **69**:185–190.

84. Graczyk, T. K., and K. J. Schwab. 2000. Foodborne infections vectored by molluscan shellfish. *Curr. Gastroenterol. Rep.* **2**:305–309.

85. Guerreiro, N. M., P. M. Herrera, L. de Escalona, V. G. de Yanes, O. de Febres, O. Naveda, and M. de Naveda. 1991. *Giardia lamblia*: comparison of two diagnostic methods and evaluation of response to treatment with metronidazole. *G. E. N.* **45**:105–110. (In Spanish.)

86. Hayes, E. B., T. D. Matte, T. R. O'Brien, T. W. McKinley, G. S. Logsdon, J. B. Rose, B. L. Ungar, D. M. Word, P. F. Pinsky, and M. L. Cummings. 1989. Large community outbreak of cryptosporidiosis due to contamination of a filtered public water supply. *N. Engl. J. Med.* **320**:1372–1376.

87. Herwaldt, B. L. 2000. *Cyclospora cayetanensis*: a review, focusing on the outbreaks of cyclosporiasis in the 1990s. *Clin. Infect. Dis.* **31**:1040–1057.

88. Herwaldt, B. L., M. L. Ackers, and the *Cyclospora* Working Group. 1997. An outbreak in 1996 of cyclosporiasis associated with imported raspberries. *N. Engl. J. Med.* **336**:1548–1556.

89. Herwaldt, B. L., M. J. Beach, et al. 1999. The return of *Cyclospora* in 1997: another outbreak of cyclosporiasis in North America associated with imported raspberries. *Ann. Intern. Med.* **130**:210–220.

90. Hijjawi, N. S., B. P. Meloni, U. M. Ryan, M. E. Olson, and R. C. Thompson. 2002. Successful in vitro cultivation of *Cryptosporidium andersoni*: evidence for the existence of novel extracellular stages in the life cycle and implications for the classification of Cryptosporidium. *Int. J. Parasitol.* **32**:1719–1726.

91. Hill, D. R. 1993. Giardiasis. Issues in diagnosis and management. *Infect. Dis. Clin. N. Am.* **7**:503–525.

92. Hoang, L. M., M. Fyfe, C. Ong, J. Harb, S. Champagne, B. Dixon, and J. Isaac-Renton. 2005. Outbreak of cyclosporiasis in British Columbia associated with imported Thai basil. *Epidemiol. Infect.* **133**:23–27.

93. Huang, P., J. T. Weber, D. M. Sosin, P. M. Griffin, E. G. Long, J. J. Murphy, F. Kocka, C. Peters, and C. Kallick. 1995. The first reported outbreak of diarrheal illness associated with *Cyclospora* in the United States. *Ann. Intern. Med.* **123**:409–414.

94. Hume, O. S. 1972. Toxoplasmosis and pregnancy. *Am. J. Obstet. Gynecol.* **114**:703–715.

95. Jinneman, K. C., J. H. Wetherington, W. E. Hill, A. M. Adams, J. M. Johnson, B. J. Tenge, N. L. Dang, R. L. Manger, and M. M. Wekell. 1998. Template preparation for PCR and RFLP of amplification products for the detection and identification of *Cyclospora* sp. and *Eimeria* spp. oocysts directly from raspberries. *J. Food Prot.* **61:**1497–1503.

96. Jinneman, K. C., J. H. Wetherington, W. E. Hill, C. J. Omiescinski, A. M. Adams, J. M. Johnson, B. J. Tenge, N. L. Dang, and M. M. Wekell. 1999. An oligonucleotide-ligation assay for the differentiation between *Cyclospora* and *Eimeria* spp. polymerase chain reaction amplification products. *J. Food Prot.* **62:**682–685.

97. Karasudani, T., S. Aoki, J. Takeuchi, M. Okuyama, M. Oseto, S. Matsuura, T. Asai, and H. Inouye. 2001. Sensitive detection of *Cryptosporidium* oocysts in environmental water samples by reverse transcription-PCR. *Jpn. J. Infect. Dis.* **54:**122–124.

98. Kniel, K. E., and M. C. Jenkins. 2005. Detection of *Cryptosporidium parvum* oocysts on fresh vegetables and 1 herbs using antibodies specific for a *Cryptosporidium parvum* viral antigen. *J. Food Prot.* **68:**1093–1096.

99. Kong, L. I., L. J. Swango, B. L. Blagburn, C. M. Hendrix, D. E. Williams, and S. D. Worley. 1988. Inactivation of *Giardia lamblia* and *Giardia canis* cysts by combined and free chlorine. *Appl. Environ. Microbiol.* **54:**2580–2582.

100. Korich, D. G., J. R. Mead, M. S. Madore, N. A. Sinclair, and C. R. Sterling. 1990. Effects of ozone, chlorine dioxide, chlorine, and monochloramine on *Cryptosporidium parvum* oocyst viability. *Appl. Environ. Microbiol.* **56:**1423–1428.

101. Laberge, I., M. W. Griffiths, and M. W. Griffiths. 1996. Prevalence, detection and control of *Cryptosporidium parvum* in food. *Int. J. Food Microbiol.* **32:**1–26.

102. Lanfredi-Rangel, A., J. A. Diniz, Jr., and W. de Souza. 1999. Presence of a protrusion on the ventral disk of adhered trophozoites of *Giardia lamblia*. *Parasitol. Res.* **85:**951–955.

103. Long, E. G., A. Ebrahimzadeh, E. H. White, B. Swisher, and C. S. Callaway. 1990. Alga associated with diarrhea in patients with acquired immunodeficiency syndrome and in travelers. *J. Clin. Microbiol.* **28:**1101–1104.

104. Long, E. G., E. H. White, W. W. Carmichael, P. M. Quinlisk, R. Raja, B. L. Swisher, H. Daugharty, and M. T. Cohen. 1991. Morphologic and staining characteristics of a cyanobacterium-like organism associated with diarrhea. *J. Infect. Dis.* **164:**199–202.

105. Lopez, A. S., D. R. Dodson, M. J. Arrowood, P. A. Orlandi, Jr., A. J. da Silva, J. W. Bier, S. D. Hanauer, R. L. Kuster, S. Oltman, M. S. Baldwin, K. Y. Won, E. M. Nace, M. L. Eberhard, and B. L. Herwaldt. 2001. Outbreak of cyclosporiasis associated with basil in Missouri in 1999. *Clin. Infect. Dis.* **32:**1010–1017.

106. Lopez, C. E., A. C. Dykes, D. D. Juranek, S. P. Sinclair, J. M. Conn, R. W. Christie, E. C. Lippy, M. G. Schultz, and M. H. Mires. 1980. Waterborne giardiasis: a communitywide outbreak of disease and a high rate of asymptomatic infection. *Am. J. Epidemiol.* **112:**495–507.

107. Lopez, F. A., J. Manglicmot, T. M. Schmidt, C. Yeh, H. V. Smith, and D. A. Relman. 1999. Molecular characterization of *Cyclospora*-like organisms from baboons. *J. Infect. Dis.* **179:**670–676.

108. Lopez-Velez, R., R. Tarazona, C. A. Garcia, E. Gomez-Mampaso, A. Guerrero, V. Moreira, and R. Villanueva. 1995. Intestinal and extraintestinal cryptosporidiosis in AIDS patients. *Eur. J. Clin. Microbiol. Infect. Dis.* **14:**677–681.

109. Luna, V. A., B. K. Stewart, D. L. Bergeron, C. R. Clausen, J. J. Plorde, and T. R. Fritsche. 1995. Use of the fluorochrome calcofluor white in the screening of stool specimens for spores of microsporidia. *Am. J. Clin. Pathol.* **103:**656–659.

110. Lunden, A., and A. Uggla. 1992. Infectivity of *Toxoplasma gondii* in mutton following curing, smoking, freezing or microwave cooking. *Int. J. Food Microbiol.* **15:**357–363.

111. MacKenzie, W. R., N. J. Hoxie, M. E. Proctor, M. S. Gradus, K. A. Blair, D. E. Peterson, J. J. Kazmierczak, D. G. Addiss, K. R. Fox, and J. B. Rose. 1994. A massive outbreak in Milwaukee of *Cryptosporidium* infection transmitted through the public water supply. *N. Engl. J. Med.* **331:**161–167.

112. MacRae, M., C. Hamilton, N. J. Strachan, S. Wright, and I. D. Ogden. 2005. The detection of *Cryptosporidium parvum* and *Escherichia coli* O157 in UK bivalve shellfish. *J. Microbiol. Methods* **60:**395–401.

113. Madico, G., J. McDonald, R. H. Gilman, L. Cabrera, and C. R. Sterling. 1997. Epidemiology and treatment of *Cyclospora cayetanensis* infection in Peruvian children. *Clin. Infect. Dis.* **24:**977–981.

114. Marshall, M. M., D. Naumovitz, Y. Ortega, and C. R. Sterling. 1997. Waterborne protozoan pathogens. *Clin. Microbiol. Rev.* **10:**67–85.

115. Matos, O., M. Alves, L. Xiao, V. Cama, and F. Antunes. 2004. *Cryptosporidium felis* and *C. meleagridis* in persons with HIV, Portugal. *Emerg. Infect. Dis.* **10:**2256–2257.

116. Millard, P. S., K. F. Gensheimer, D. G. Addiss, D. M. Sosin, G. A. Beckett, A. Houck-Jankoski, and A. Hudson. 1994. An outbreak of cryptosporidiosis from fresh-pressed apple cider. *JAMA* **272:**1592–1596.

117. Monis, P. T., R. H. Andrews, G. Mayrhofer, and P. L. Ey. 2003. Genetic diversity within the morphological species *Giardia intestinalis* and its relationship to host origin. *Infect. Genet. Evol.* **3:**29–38.

118. Morgan-Ryan, U. M., A. Fall, L. A. Ward, N. Hijjawi, I. Sulaiman, R. Fayer, R. C. Thompson, M. Olson, A. Lal, and L. Xiao. 2002. *Cryptosporidium hominis* n. sp. (Apicomplexa: Cryptosporidiidae) from Homo sapiens. *J. Eukaryot. Microbiol.* **49:**433–440.

119. Moskovitz, B. L., T. L. Stanton, and J. J. Kusmierek. 1988. Spiramycin therapy for cryptosporidial diarrhoea in immunocompromised patients. *J. Antimicrob. Chemother.* **22**(Suppl. B):189–191.

120. Muller, A., K. Stellermann, P. Hartmann, M. Schrappe, G. Fatkenheuer, B. Salzberger, V. Diehl, and C. Franzen. 1999. A powerful DNA extraction method and PCR for detection of microsporidia in clinical stool specimens. *Clin. Diagn. Lab. Immunol.* **6:**243–246.

121. Olivier, C., S. van de Pas, P. W. Lepp, K. Yoder, and D. A. Relman. 2001. Sequence variability in the first internal

transcribed spacer region within and among *Cyclospora* species is consistent with polyparasitism. *Int. J. Parasitol.* 31:1475–1487.

122. Orlandi, P. A., D. M. Chu, J. W. Bier, and G. J. Jackson. 2002. Parasites in the food supply. *Food Technol.* 56:72–81.

123. Orlandi, P. A., and K. A. Lampel. 2000. Extraction-free, filter-based template preparation for rapid and sensitive PCR detection of pathogenic parasitic protozoa. *J. Clin. Microbiol.* 38:2271–2277.

124. Ortega, Y. R., R. H. Gilman, and C. R. Sterling. 1994. A new coccidian parasite (Apicomplexa: Eimeriidae) from humans. *J. Parasitol.* 80:625–629.

125. Ortega, Y. R., R. Nagle, R. H. Gilman, J. Watanabe, J. Miyagui, H. Quispe, P. Kanagusuku, C. Roxas, and C. R. Sterling. 1997. Pathologic and clinical findings in patients with cyclosporiasis and a description of intracellular parasite life-cycle stages. *J. Infect. Dis.* 176:1584–1589.

126. Ortega, Y. R., C. R. Roxas, R. H. Gilman, N. J. Miller, L. Cabrera, C. Taquiri, and C. R. Sterling. 1997. Isolation of *Cryptosporidium parvum* and *Cyclospora cayetanensis* from vegetables collected in markets of an endemic region in Peru. *Am. J. Trop. Med. Hyg.* 57:683–686.

127. Ortega, Y. R., C. R. Sterling, R. H. Gilman, V. A. Cama, and F. Diaz. 1993. *Cyclospora* species—a new protozoan pathogen of humans. *N. Engl. J. Med.* 328:1308–1312.

128. Pape, J. W., R. I. Verdier, M. Boncy, J. Boncy, and W. D. Johnson, Jr. 1994. *Cyclospora* infection in adults infected with HIV. Clinical manifestations, treatment, and prophylaxis. *Ann. Intern. Med.* 121:654–657.

129. Pape, J. W., R. I. Verdier, and W. D. Johnson, Jr. 1989. Treatment and prophylaxis of *Isospora belli* infection in patients with the acquired immunodeficiency syndrome. *N. Engl. J. Med.* 320:1044–1047.

130. Payment, P. 1999. Poor efficacy of residual chlorine disinfectant in drinking water to inactivate waterborne pathogens in distribution systems. *Can. J. Microbiol.* 45:709–715.

131. Pedraza-Diaz, S., C. F. Amar, J. McLauchlin, G. L. Nichols, K. M. Cotton, P. Godwin, A. M. Iversen, L. Milne, J. R. Mulla, K. Nye, H. Panigrahl, S. R. Venn, R. Wiggins, M. Williams, and E. R. Youngs. 2001. *Cryptosporidium meleagridis* from humans: molecular analysis and description of affected patients. *J. Infect.* 42:243–250.

132. Peng, M. M., L. Xiao, A. R. Freeman, M. J. Arrowood, A. A. Escalante, A. C. Weltman, C. S. Ong, W. R. MacKenzie, A. A. Lal, and C. B. Beard. 1997. Genetic polymorphism among *Cryptosporidium parvum* isolates: evidence of two distinct human transmission cycles. *Emerg. Infect. Dis.* 3:567–573.

133. Perryman, L. E., and J. M. Bjorneby. 1991. Immunotherapy of cryptosporidiosis in immunodeficient animal models. *J. Protozool.* 38:98S–100S.

134. Portnoy, D., M. E. Whiteside, E. Buckley III, and C. L. MacLeod. 1984. Treatment of intestinal cryptosporidiosis with spiramycin. *Ann. Intern. Med.* 101:202–204.

135. Posada, G., D. Pizarro, and E. Mohs. 1987. Oral rehydration in children with *Cryptosporidium muris* diarrhea. *Bol. Med. Hosp. Infant. Mex.* 44:740–744. (In Spanish.)

136. Rabold, J. G., C. W. Hoge, D. R. Shlim, C. Kefford, R. Rajah, and P. Echeverria. 1994. *Cyclospora* outbreak associated with chlorinated drinking water. *Lancet* 344:1360–1361.

137. Radziminski, C., L. Ballantyne, J. Hodson, R. Creason, R. C. Andrews, and C. Chauret. 2002. Disinfection of *Bacillus subtilis* spores with chlorine dioxide: a bench-scale and pilot-scale study. *Water Res.* 36:1629–1639.

138. Raisanen, S. 1978. Toxoplasmosis transmitted by blood transfusions. *Transfusion* 18:329–332.

139. Reiter-Owona, I., H. Seitz, U. Gross, M. Sahm, J. K. Rockstroh, and H. M. Seitz. 2000. Is stage conversion the initiating event for reactivation of *Toxoplasma gondii* in brain tissue of AIDS patients? *J. Parasitol.* 86:531–536.

140. Rice, E. W., and J. C. Hoff. 1981. Inactivation of *Giardia lamblia* cysts by ultraviolet irradiation. *Appl. Environ. Microbiol.* 42:546–547.

141. Ripabelli, G., A. Leone, M. L. Sammarco, I. Fanelli, G. M. Grasso, and J. McLauchlin. 2004. Detection of *Cryptosporidium parvum* oocysts in experimentally contaminated lettuce using filtration, immunomagnetic separation, light microscopy, and PCR. *Foodborne Pathol. Dis.* 1:216–222.

142. Robertson, L. J., and B. Gjerde. 2001. Occurrence of parasites on fruits and vegetables in Norway. *J. Food Prot.* 64:1793–1798.

143. Robertson, L. J., J. D. Greig, B. Gjerde, and A. Fazil. 2005. The potential for acquiring cryptosporidiosis or giardiosis from consumption of mung bean sprouts in Norway: a preliminary step-wise risk assessment. *Int. J. Food Microbiol.* 98:291–300.

144. Rosales, M. J., G. P. Cordon, M. S. Moreno, and C. M. Sanchez. 2005. Extracellular like-gregarine stages of *Cryptosporidium parvum*. *Acta Trop.* 95:74–78.

145. Rose, J. B., C. N. Haas, and S. Regli. 1991. Risk assessment and control of waterborne giardiasis. *Am. J. Public Health* 81:709–713.

146. Saez-Llorens, X., C. M. Odio, M. A. Umana, and M. V. Morales. 1989. Spiramycin vs. placebo for treatment of acute diarrhea caused by *Cryptosporidium*. *Pediatr. Infect. Dis. J.* 8:136–140.

147. Schottelius, J., E. M. Kuhn, and R. Enriquez. 2000. Microsporidia and *Candida* spores: their discrimination by Calcofluor, trichrome-blue and methylene-blue combination staining. *Trop. Med. Int. Health* 5:453–458.

148. Sepulveda, B. 1982. Amebiasis: host-pathogen biology. *Rev. Infect. Dis.* 4:1247–1253.

149. Sherchand, J. B., J. H. Cross, M. Jimba, S. Sherchand, and M. P. Shrestha. 1999. Study of *Cyclospora cayetanensis* in health care facilities, sewage water and green leafy vegetables in Nepal. *Southeast Asian J. Trop. Med. Public Health* 30:58–63.

150. Shields, J. M., and B. H. Olson. 2003. PCR-restriction fragment length polymorphism method for detection of *Cyclospora cayetanensis* in environmental waters without microscopic confirmation. *Appl. Environ. Microbiol.* 69:4662–4669.

151. Shlim, D. R., M. T. Cohen, M. Eaton, R. Rajah, E. G. Long, and B. L. Ungar. 1991. An alga-like organism associated

with an outbreak of prolonged diarrhea among foreigners in Nepal. *Am. J. Trop. Med. Hyg.* **45**:383–389.

152. **Shlim, D. R., C. W. Hoge, R. Rajah, R. M. Scott, P. Pandy, and P. Echeverria.** 1999. Persistent high risk of diarrhea among foreigners in Nepal during the first 2 years of residence. *Clin. Infect. Dis.* **29**:613–616.

153. **Singh, K. D., D. K. Bhasin, S. V. Rana, K. Vaiphei, R. Katyal, V. K. Vinayak, and K. Singh.** 2000. Effect of *Giardia lamblia* on duodenal disaccharidase levels in humans. *Trop. Gastroenterol.* **21**:174–176.

154. **Sironi, M., C. Bandi, S. Novati, and M. Scaglia.** 1997. A PCR-RFLP method for the detection and species identification of human microsporidia. *Parasitologia* **39**: 437–439.

155. **Smith, H. V., C. A. Paton, R. W. Girdwood, and M. M. Mtambo.** 1996. *Cyclospora* in non-human primates in Gombe, Tanzania. *Vet. Rec.* **138**:528.

156. **Soave, R.** 1988. Cryptosporidiosis and isosporiasis in patients with AIDS. *Infect. Dis. Clin. N. Am.* **2**:485–493.

157. **Speelman, P.** 1985. Single-dose tinidazole for the treatment of giardiasis. *Antimicrob. Agents Chemother.* **27**:227–229.

158. **Sun, T., C. F. Ilardi, D. Asnis, A. R. Bresciani, S. Goldenberg, B. Roberts, and S. Teichberg.** 1996. Light and electron microscopic identification of *Cyclospora* species in the small intestine. Evidence of the presence of asexual life cycle in human host. *Am. J. Clin. Pathol.* **105**:216–220.

159. **Tee, G. H., A. H. Moody, A. H. Cooke, and P. L. Chiodini.** 1993. Comparison of techniques for detecting antigens of *Giardia lamblia* and *Cryptosporidium parvum* in faeces. *J. Clin. Pathol.* **46**:555–558.

160. **Ungar, B. L.** 1990. Enzyme-linked immunoassay for detection of *Cryptosporidium* antigens in fecal specimens. *J. Clin. Microbiol.* **28**:2491–2495.

161. **van Keulen, H., S. R. Campbell, S. L. Erlandsen, and E. L. Jarroll.** 1991. Cloning and restriction enzyme mapping of ribosomal DNA of *Giardia duodenalis*, *Giardia ardeae* and *Giardia muris*. *Mol. Biochem. Parasitol.* **46**:275–284.

162. **Verdier, R. I., D. W. Fitzgerald, W. D. Johnson, Jr., and J. W. Pape.** 2000. Trimethoprim-sulfamethoxazole compared with ciprofloxacin for treatment and prophylaxis of *Isospora belli* and *Cyclospora cayetanensis* infection in HIV-infected patients. A randomized, controlled trial. *Ann. Intern. Med.* **132**:885–888.

163. **Von, G. U.** 2003. Ozonation of drinking water: part II. Disinfection and by-product formation in presence of bromide, iodide or chlorine. *Water Res.* **37**:1469–1487.

164. **Yai, L. E., A. R. Bauab, M. P. Hirschfeld, M. L. de Oliveira, and J. T. Damaceno.** 1997. The first two cases of Cyclospora in dogs, Sao Paulo, Brazil. *Rev. Inst. Med. Trop. Sao Paulo* **39**:177–179.

165. **Zerpa, R., N. Uchima, and L. Huicho.** 1995. *Cyclospora cayetanensis* associated with watery diarrhoea in Peruvian patients. *J. Trop. Med. Hyg.* **98**:325–329.

166. **Zimmerman, S. K., and C. A. Needham.** 1995. Comparison of conventional stool concentration and preserved-smear methods with Merifluor *Cryptosporidium/Giardia* Direct Immunofluorescence Assay and ProSpecT *Giardia* EZ Microplate Assay for detection of *Giardia lamblia*. *J. Clin. Microbiol.* **33**:1942–1943.

167. **Zu, S. X., and R. L. Guerrant.** 1993. Cryptosporidiosis. *J. Trop. Pediatr.* **39**:132–136.

Preservatives and Preservation Methods

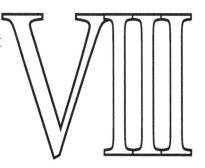

VIII

Food Microbiology: Fundamentals and Frontiers, 3rd Ed.
Edited by M. P. Doyle and L. R. Beuchat
© 2007 ASM Press, Washington, D.C.

József Farkas

Physical Methods of Food Preservation

32

When the values of environmental stress factors discussed in chapter 1 of this volume are outside the vital range for growth and survival of a specific microorganism, cellular damage occurs. Depending on the severity of this effect, growth may be inhibited (microbistatic effect, or reversible inhibition) or cells may be killed (microbicidal effect) (172). Food spoilage can be prevented by applying one or more of the following strategies: inhibition of microbial growth, destruction (irreversible inactivation) of microbial cells, and mechanical removal of micro organisms from the food.

Physical methods of food preservation are those that utilize physical treatments to inhibit, destroy, or remove undesirable microorganisms without involving antimicrobial additives or products of microbial metabolism as preservative factors (73). Microbial growth can be inhibited by physical dehydration processes (drying, freeze-drying, and freeze concentration), cool storage, or frozen storage. Microorganisms can be destroyed by established physical microbicide treatments such as heating (including microwave heat treatment), UV or ionizing radiation, and emerging methods of new nonthermal treatments, such as the use of high hydrostatic pressure, pulsed electric fields (PEFs), oscillating magnetic fields, photodynamic effects, and a combination of physical processes such as heat irradiation, dehydro-irradiation, and manothermosonication. Mechanical removal of microorganisms from food may be accomplished by membrane filtration of food liquids. This chapter discusses the microbiological fundamentals of the physical preservation methods outlined above, with the exception of mechanical removal. Wherever possible, the mechanisms and underlying principles are explained.

PHYSICAL DEHYDRATION PROCESSES

One of the oldest methods for preserving food is dehydration, and water is one of the most important factors controlling the rate of deterioration of food, by either microbial or nonmicrobial effects. Knowledge of the moisture content alone is not sufficient to predict the stability of foods, because it is not the total moisture content but rather the availability of water that determines the shelf life of a food. A proportion of the total water in food is strongly bound to specific sites.

József Farkas, Department of Refrigeration and Livestock Products' Technology, Faculty of Food Science, Corvinus University of Budapest, Ménesi út 45, H-1118 Budapest, Hungary.

The availability of water is measured by the water activity (a_w) of a food and is defined as the ratio of the vapor pressure of water in a food, P, to the vapor pressure of pure water, P_0, at the same temperature:

$$a_w = \frac{P}{P_0}$$

The relation of a_w to the solute concentration is expressed by Raoult's law for "ideal" solutions,

$$a_w = \frac{P}{P_0} = \frac{n_2}{n_1 + n_2}$$

where n_1 and n_2 are the numbers of moles of solute and solvent, respectively. An ideal solution may be defined as one in which the molecules of a solute are affected by their environment in exactly the same manner as they are in a pure state at the same temperature and pressure as the solution. In reality, there are no such solutions, because interactions between molecules may reduce the effective number of particles in a solution. On the other hand, dissociation may greatly increase the number. Departures from the ideal are large for nonelectrolytes at concentrations above 1 molal (molality is the number of gram moles per kilogram of water) and for electrolytes at all concentrations.

The movement of water vapor from a food to the surrounding air depends on the moisture content, the composition of the food, temperature, and the humidity of the air. At a constant temperature, the moisture content of food changes until it comes into equilibrium with water vapor in the surrounding air. This moisture level is called the equilibrium moisture content of the food. At the equilibrium moisture content, the food neither gains nor loses weight during storage under those conditions. The relative humidity of the surrounding air, expressed as a percentage, is then the equilibrium relative humidity (ERH). When different values of relative humidity versus equilibrium moisture contents are plotted, a curve known as a water sorption isotherm is obtained. ERH is related to a_w by the following expression: ERH (%) = $a_w \times 100$.

Fresh, raw, high-moisture foods such as fruits, vegetables, meats, and fish have a_w levels of 0.98 or higher. Food preservation by dehydration is based on the principle that microbial growth is inhibited if the available water required for the growth is removed, i.e., if the a_w is reduced (119). Technologies for drying solid foods remove water by using hot air; freeze-drying does so by sublimation after freezing. Liquid foods can be dried by freeze concentration, whereby freezing is followed by mechanical removal of frozen water.

Table 32.1 a_ws of various foods[a]

Foods	a_w
Fresh, raw fruits, vegetables, meat, and fish	>0.98
Cooked meat, bread	0.98–0.95
Cured meat products, cheeses	0.95–0.91
Sausages, syrups	0.91–0.87
Flours, rice, beans, peas	0.87–0.80
Jams, marmalades	0.80–0.75
Candies	0.75–0.65
Dried fruits	0.65–0.60
Dehydrated vermicelli, spices, milk powder	0.60–0.20

[a] Adapted from references 35 and 211.

These processes reduce the a_ws of both the food and the microbial cells to levels insufficient for microbial growth. a_ws of various raw materials and food products are listed in Table 32.1. The microbiological stability of many intermediate-moisture foods with a_w levels of 0.70 to 0.85 (jams, dry sausages, etc.) depends also on other preservative factors, e.g., reduced pH, chemical preservatives, and pasteurization; however, the reduced a_w in these foods is of major importance. Because of the chemical composition and water binding capacity of components, various foods may have very different moisture contents at the same a_ws, as is illustrated in Table 32.2.

The a_w requirements of various microorganisms vary significantly (98). In the vital range of growth, decreasing the a_w increases the lag phase of growth and decreases the growth rate. Foodborne microorganisms are grouped according to their minimal a_w requirements in Table 32.3.

In general, among bacteria, gram-negative species have the highest a_w requirements for growth. Important gram-negative spoilage bacteria such as *Pseudomonas* spp. and most members of the family *Enterobacteriaceae* can usually grow only at a_ws above 0.96 and 0.93,

Table 32.2 Moisture contents of various dry or dehydrated food products when their a_w is 0.70 at 20°C[a]

Food(s)	Moisture content (%)
Various seed grains	4–9
Milk powder	7–10
Cocoa powder	7–10
Whole egg powder	10–11
Skim milk powder	10–15
Dried, fat-free meat	10–15
Rice and legume seeds	12–15
Dehydrated vegetables	12–22
Dried soups	13–21
Dried fruits	18–25

[a] Adapted from references 31, 35, 146, 147, and 211.

Table 32.3 Minimal a_w levels required for growth of foodborne microorganisms at 25°C[a]

Group of microorganisms	Minimal a_w required
Most bacteria	0.91–0.88
Most yeasts	0.88
"Regular" molds	0.80
Halophilic bacteria	0.75
Xerotolerant molds	0.71
Xerophilic molds and osmophilic yeasts	0.62–0.60

[a] Adapted from references 35, 99, 146, and 211.

respectively. Less sensitive to reduced a_w are gram-positive, non-spore-forming bacteria. Although many members of the family *Lactobacillaceae* have minimum a_w requirements near 0.94, some representatives of the family *Micrococcaceae* are capable of growth at a_ws below 0.90. Staphylococci are unique among nonhalophilic bacteria in being capable of growing in substrates containing high levels of NaCl. The generally recognized minimum a_w for the growth of *Staphylococcus aureus* is about 0.86 (98); however, under otherwise ideal conditions, growth has been demonstrated at an a_w as low as 0.83. Production of staphylococcal enterotoxin in food slurries has not been observed at an a_w below 0.93 (98). Most spore-forming bacteria do not grow at an a_w below 0.93. Germination and outgrowth of spores of food-poisoning strains of *Bacillus cereus* are prevented at a_ws of 0.93 to 0.97, depending on the nature of the bulk solute (101). Various serological types of *Clostridium botulinum* differ in their a_w requirements. The lower a_w limits in salt-adjusted media for growth from spore inocula are 0.95 for type A, 0.94 for type B, and 0.97 for type E (159). Toxin production occurs at an a_w closely approaching these growth minima. The minimum a_w for the growth and spore germination of *Clostridium perfringens* is between 0.97 and 0.95 in complex media when sucrose

or NaCl is used to adjust a_w (108). For both *C. botulinum* and *C. perfringens*, growth proceeds at lower a_ws when the a_w is controlled by glycerol instead of by NaCl (98).

Several species of yeasts have a_w requirements much lower than those of bacteria. Salt-tolerant species such as *Debaryomyces hansenii*, *Hansenula anomala*, and *Candida pseudotropicalis* may grow well on cured meats and pickles at NaCl concentrations of up to 11% (a_w, 0.93). Some "osmophilic" species, e.g., *Zygosaccharomyces rouxii*, *Z. bailii*, and *Z. bisporus*, are able to grow in food with high sugar contents (jams, honey, syrups). Terms such as "xerotolerant" and "saccharotolerant" would be more appropriate for these osmophilic yeasts.

In general, molds have a_w requirements much lower than those of other groups of foodborne microorganisms (24, 174). The most common xerotolerant molds belong to the genus *Eurotium* (their asexual forms are members of the *Aspergillus glaucus* group). The minimal a_w for the growth of *Eurotium* species is in the range of 0.71 to 0.77, while the optimal a_w is 0.96. True xerophilic molds (173) such as *Monascus* (*Xeromyces*) *bisporus* do not grow at a_ws greater than 0.97 to 0.99. The relationship of a_w to mold growth and toxin formation is complex (190). It is of great public health importance that the minimal a_w requirement for mycotoxin production is generally higher than that for the growth of toxigenic molds (Table 32.4). That is, under reduced-a_w conditions in which, some toxigenic molds can grow, toxins are not produced.

Differences in a_w limits for growth reflect differences in osmoregulatory capacities (70). Mechanisms of tolerance to low a_w are different in bacteria and fungi. The key point, however, is that the cell osmoregulation mechanism operates to maintain homeostasis with respect to water content (24). In general, the strategy employed by microorganisms as protection against osmotic stress is the intracellular accumulation of compatible solutes, i.e.,

Table 32.4 Minimal a_w requirements for growth of and mycotoxin production by some toxigenic molds[a]

Mycotoxin(s)	Mold	Minimal a_w requirement for:	
		Growth	Toxin production
Aflatoxins	*Aspergillus flavus*	0.82	0.83–0.87
	Aspergillus parasiticus	0.82	0.87
Ochratoxin	*Aspergillus ochraceus*	0.77	0.85
	Penicillium cyclopium	0.82–0.85	0.87–0.90
Patulin	*Penicillium expansum*	0.81	0.99
	Pencillium patulum	0.81	0.95
Stachybotryn	*Stachybotrys atra*	0.94	0.94

[a] Adapted from references 31, 35, and 175.

such compounds that have the general property of binding water while interfering minimally with the metabolic activities of the cell. While some bacteria accumulate K^+ ions and amino acids such as proline as a response to low a_w (15), halotolerant and xerotolerant fungi concentrate polyols such as glycerol, erythritol, and arabitol (44). These polyols not only act as osmoregulators but also prevent inhibition or inactivation of enzymes and probably serve as nutrient reserves (16). In the case of halophilic bacteria, KCl is a requirement, and osmophilic yeasts have a high tolerance for high sugar concentrations (14).

Drying

Drying is a complicated process, simultaneously involving heat and mass transfer phenomena (151). During hot-air drying of vegetables, vermicelli, and similar solid foods, the air stream has two functions; it transfers heat to evaporate the water from the raw materials, and it carries away the vapor produced. Vacuum microwave drying of food has been reported as an alternative drying technique to improve the quality and retain sensitive compounds typically lost through thermal oxidative degradation (150, 179).

In the course of drying, microorganisms on the raw materials are affected by both the drying temperature and changes of a_w. Microbiological consequences of drying technology are influenced by a number of other factors, e.g., sizes and compositions of food pieces and time-temperature combinations. During the warm-up period of drying, the temperature is still low and the a_w is high. The length of this phase depends mainly on the size of food particles. If this phase is long, microorganisms may even grow during the slow increase of temperature in the range 20 to 40°C under the existing high a_w. During later phases of drying, the temperature exceeds 50 to 70°C. Although there is no opportunity for growth, heat destruction is not significant either, because the higher temperature develops parallel to decreased moisture content. Wet heat is more lethal than dry heat. With certain drying technologies, the surface temperature may reach even 100°C while the internal temperature remains lower. When microorganisms are mainly on the product surface, during the drying process their inactivation is enhanced. On the other hand, the moisture content of the food surface decreases rapidly with heating, thereby decreasing the heat sensitivity of microbial cells. Therefore, the lethal combination of high temperature and high humidity is rare in drying technologies. In fact, the major microbicidal effect is due to the high-temperature and low-humidity conditions which last for a long period of time. Loss

of viability of microorganisms continues during storage because reversibly damaged cells, unable to resuscitate at a low a_w, gradually die. Further viability losses may occur during rehydration of dried products, especially when the rehydration is rapid, causing large differences in osmotic pressure.

The a_w of dried foods is usually considerably lower than the a_w critical for microbial growth. Therefore, most dried products are stable microbiologically, provided that the relative humidity of the storage atmosphere is less than 70%. If the ERH increases or the temperature changes, conditions on the product surface may change and permit the growth of xerotolerant molds (211).

Freeze-Drying

Freeze-drying (lyophilization) is a combined method of food preservation. Its principle is dehydration of the food in the frozen state through vacuum sublimation of its ice content (66). This method of water removal is very gentle. In other methods for dehydrating solid foods, the distribution of solutes in the food changes due to a continuous transport of solutions toward the surface and increased solute loss when they reach the surface. During freeze-drying, the sublimation front moves toward the core and the ice sublimates at the location where it is formed. Thus, solutes remain inside the food at their original location, their loss is insignificant, and the food retains its original form and structure very well. As a result, rehydration of freeze-dried foods is very fast and almost complete, 90 to 95% of the original moisture contents being regained. Reduced amounts of oxygen in the vacuum chamber may also result in reduced oxidation of the product during freeze-drying (179). On the other hand, the very large surface areas of freeze-dried products make them vulnerable to other damage such as lipid oxidation. To prevent spontaneous rehydration, freeze-dried foods should be kept in vapor-impermeable packages. They should be packaged in an inert atmosphere to delay oxidation.

The effect of freeze-drying of foods on their microbial contaminants results from a combination of a decrease in a_w by freezing and further reduction of a_w by sublimation of the ice. The extents of cell damage that occur during these phases may be different, depending on the conditions (temperatures and rates) of freezing and sublimation. Microbial survival also depends on the composition of freeze-dried food. Carbohydrates, proteins, and collodial substances, in general, have a protective effect on microorganisms. During storage of products freeze-dried to 2 to 5% residual moisture content, a gradual, slow decrease in cell viability occurs, similar to that for conventionally dried foods, especially in the presence of atmospheric oxygen. Again, there may be additional microbial destruction during rehydration of freeze-dried products because large differences in solution concentrations develop suddenly.

Because of the mild manner of moisture removal, lyophilization conditions can be intentionally optimized for maximal survival of microbial cells, e.g., preservation of stock cultures. Low dehydration temperature, protective additives such as glycerol in the cell suspension, and storage of lyophilized cultures in a nitrogen atmosphere or in vacuo increase culture viability. Gram-positive bacteria survive freeze-drying better than gram-negative bacteria.

COOL STORAGE

Descriptions of the state of the art of food chilling and reviews on a number of key aspects of chilled foods have been published previously (67). Cool (or chill) storage generally refers to storage at temperatures above freezing, from about 16 down to −2°C. While pure water freezes at 0°C, most foods remain unfrozen until a temperature of about −2°C or lower is reached. One should bear in mind, however, that many fresh fruits and vegetables may suffer from a chilling injury when they are kept below critical temperatures of 4 to 12°C (223). Depending on the inherent storability of raw foods, the duration of cool storage may vary from a few days to several weeks (Table 32.5).

The preservation effect of cool storage (refrigeration) is based on the fact that reducing temperature reduces the rates of chemical reactions and the growth of microorganisms. Refrigeration temperatures allow the growth of psychrophilic and psychrotrophic microorganisms (189). If the initial population of such microorganisms is high, refrigerated foods can spoil in a short time. Since the temperature requirements for various microorganisms differ, refrigeration may considerably change the qualitative composition of the microbiota.

Because growth rates of microorganisms increase significantly with an increase in temperature of only a few degrees, variation of storage temperatures should be avoided. Psychrotrophic pathogens such as *Yersinia enterocolitica*, *Vibrio parahaemolyticus*, *Listeria monocytogenes*, and *Aeromonas hydrophila* are important, especially in "minimally processed, extended shelf-life" products (134).

There is some evidence that the minimum growth temperature for microorganisms is dictated by the inhibition of solute transport. In this regard, it has been proposed that the growth temperature range for a microorganism depends on how well the organism can regulate cellular lipid fluidity within a given range (85). Psychrotrophs contain increased amounts of unsaturated fatty acid residues in their lipids when grown at low temperatures. This increase in the degree of unsaturation of fatty acids leads to a decrease in the lipid melting point, suggesting that increased synthesis of unsaturated fatty acids at low temperatures acts to maintain the lipid in a liquid and mobile state, thereby allowing membrane proteins to continue to function (101). In addition, the transport permeases of psychrotrophs are apparently more active at low temperatures than those of mesophiles, and a cold-resistant transport system is characteristic of psychrotrophic bacteria.

Controlled-Atmosphere Storage

The minimal temperature for microbial growth is influenced by many factors and is lowest when other growth conditions are optimal. When other environmental conditions are unfavorable, the minimal temperature needed for microbial growth increases considerably. Very important environmental factors include the pH, a_w, and oxygen concentration. Therefore, when suboptimal values for these factors are combined with low temperature, the refrigeration stability of many commodities increases. Controlled-atmosphere storage of certain fruits and vegetables is widely used (2). In this operation, a reduced (2 to 5%) oxygen content and generally an increased (8 to 10%) carbon dioxide content are established and maintained in airtight, chilled storage rooms. This modified storage atmosphere is advantageous because it depresses the respiration and adverse changes in sensorial and textural qualities of stored fruits and vegetables while inhibiting the growth of certain spoilage microorganisms.

Growth inhibition is evident both in the extension of the lag phase of microorganisms and in the reduction in microbial populations. The inhibitory effect of car-

Table 32.5 Shelf-life extension for raw foods by cool storage[a]

Food	Avg useful storage life (days) at:	
	0°C (32°F)	22°C (72°F)
Meat	6–10	1
Fish	2–7	1
Poultry	5–18	1
Fruits	2–180	1–20
Leafy vegetables	3–20	1–7
Root crops	90–300	7–50

[a] Adapted from reference 182.

bon dioxide is due to several factors. It reduces after-ripening, thereby maintaining greater phytoimmunity of plant tissues. When dissolved in aqueous phase, carbon dioxide reduces the pH. But the primary mechanism of preservation is direct inhibition of microbial respiration. The simultaneous reduction of oxygen concentration contributes significantly to the inhibitory effect of carbon dioxide. Psychrotrophic spoilage bacteria such as those of the aerobic genera *Pseudomonas* and *Acinetobacter* are particularly sensitive to carbon dioxide, but lactic acid bacteria and yeasts are not (33).

Modified-Atmosphere Packaging

Modified-atmosphere packaging (MAP) of food has gained substantial success in the past decades, facilitated by developments in films and equipment in the packaging industry and the increasing unease among consumers toward chemical preservatives (25). MAP operates by a principle similar to that of controlled-atmosphere storage (139). Unpreserved bread and bakery products which are prone to mold spoilage are obvious candidates for MAP with elevated carbon dioxide and very low oxygen conentration in the package atmosphere (81). In the case of vacuum-packaged (VP) meat products, it is not the composition but the pressure of air that is changed. The residual air pressure is only 0.3×10^5 to 0.4×10^5 Pa instead of the 10^5 Pa for atmospheric air, thereby reducing the amount of available oxygen. This modified atmosphere is inhibitory to certain microorganisms and greatly increases the keeping quality of foods. However, MAP and VP are not without safety concerns (96). With regard to VP of chilled foods, one must remember that *C. botulinum*, as well as some other pathogens, grows well in the absence of oxygen. Thus, a maximum temperature of 3.3°C is advisable for some VP foods, especially pasteurized foods.

Almost any combination of carbon dioxide, nitrogen, and oxygen may be used in MAP to sustain the visual appearance and/or to extend shelf lives of meat and meat products (65, 156). In meats, aerobic spoilage bacteria such as those of the *Pseudomonas-Acinetobacter-Moraxella* group are inhibited (46). The lactic acid bacteria then become the dominant components of the microbiota, but they grow more slowly than the former group and produce less offensive sensory changes (65).

MAP for respiring ("live") products, e.g., fresh fruit and vegetables, is rather complicated because proper gas permeability of the packaging film is required to achieve an equilibrium atmosphere in the package while, at the same time, both the respiration rate and the gas permeability change with the temperature (111). Despite an increasing interest in the use of MAP to extend the shelf lives of many perishable products, the concern about the potential for growth of pathogenic bacteria at refrigeration temperatures remains the factor limiting the further expansion of the method. Therefore, several studies have examined the effects of MAP and VP on the growth and survival of psychrotrophic pathogens and the effects of physicochemical changes occurring in these foods (155).

The effects of epiphytic flora on the growth of *L. monocytogenes* were assessed with different types of ready-to-eat leafy vegetables under a modified atmosphere at 10°C (68). The highest growth rate was obtained for butterhead lettuce, and no growth occurred on lamb lettuce. The pathogen grew faster on young, yellow leaves than on older, green leaves of broad leaf chicory.

The survival and growth of *Bacillus cereus* and *L. monocytogenes* on mung bean sprouts and chicory endive have been studied previously (68). *B. cereus* proliferated during storage at 10°C under atmospheric pressure conditions, but the population diminished in a moderate vacuum system (reduced oxygen tension, 4×10^4 Pa of pressure). In the latter system, *L. monocytogenes* expired on mung bean sprouts but grew to significant populations on chicory endive.

Fish and fish products are the greatest concern for *C. botulinum* growth in MAP and VP products (75). The U.S. National Academy of Sciences recommended that fish not be packed under modified atmospheres until the safety of the system has been established. In European Community projects, the microbiological safety of MAP fishes (cod and rainbow trout) stored at 0, 5, and up to 12°C was studied. In no instance was the growth or survival of *Aeromonas* spp., *Y. enterocolitica*, and *Salmonella enterica* serovar Typhimurium greater under MAP than in the aerobically stored control, and frequently the growth was reduced under MAP (25). However, these investigations did not include studies of *C. botulinum*. Recent studies have shown that *C. botulinum* can grow and produce toxin in products such as MAP pizza and English-style crumpets while the products remain organoleptically acceptable (32).

FREEZING AND FROZEN STORAGE

In principle, of all the preservation processes, freezing provides the longest high-quality shelf life for high-moisture foods because crystallization of the freezable water together with the subzero temperature creates an environment unfavorable to microbial growth and most deteriorative physicochemical processes. Freezing normally lowers the temperature of a food to −18°C, and the food is stored at that temperature or below. Many convenience food products are made possible by this technology, which preserves foods without causing major changes in sensory quality. Freezing and frozen transportation and storage are, however, highly energy intensive. There are three basic freezing methods in commercial use: placing the product in cold air, placing the food or the food package in contact with a surface that is cooled by a refrigerant (indirect contact freezing), and submerging the food or spraying cold refrigerant liquid onto the food or package surface (182). Commercial freezing of cellular systems almost always causes some damage because of intercellular and intracellular ice crystal growth, and the damage is not reversed upon thawing. The freezing rate is recognized as a critical parameter in tissue damage. Slow freezing results in the formation of large ice crystals, whereas rapid cooling results in numerous smaller ice crystals and leads to better product quality.

The freezing of foods does not occur at one defined temperature but rather over a broad temperature range. Depending on the composition of the food, water starts freezing at -1 to $-3°C$. Freezing increases the solute concentration in the water not yet frozen, which decreases further the freezing point of the solution. At the so-called eutectic temperature, the concentration of solutes reaches the solubility limit (the point at which they precipitate) and the residual water freezes. A totally frozen state, which involves a complex system of ice crystals and crystallized soluble substances, results at -15 to $-20°C$ for fruits and vegetables and at less than $-40°C$ for meats.

As an effect of freezing, both the temperature and the a_w decrease. Thus, in frozen foods, only those microorganisms which are cold tolerant and xerotolerant can grow. Although the spoilage of refrigerated nonfrozen meat is due mainly to bacteria, on frozen meats molds that are able to grow at lower a_ws may be problematic.

During freezing, microorganisms suffer multiple types of damage which may cause inactivation immediately or at a later time (125). This damage is the outcome of several effects, the magnitude of which is influenced by, for example, the rate and temperature of freezing and the composition of the food. The rapid cooling of mesophilic bacteria from a normal growth temperature brings about immediate death to a proportion of the culture. Freezing also involves a temperature shock and may cause metabolic injury, presumably by virtue of damage to the plasma membrane. Gram-negative bacteria, particularly mesophiles, appear to be more susceptible to this cold shock than gram-positive bacteria (99). Rapid chilling results in membrane phase transitions from liquid crystal to gel status without allowing lateral-phase separation of phospholipid and protein domains. These rapid transitions in turn result in a loss of permeability control of cytoplasmic and outer membranes (130).

Cold shock sensitizes cells to various forms of oxidative stress (63, 126).

During the freezing of water, osmotic shock occurs, and intercellular ice crystal formation causes mechanical injury. The concentration of cellular liquids changes the pH and ionic strength, thereby inactivating enzymes, denaturing proteins, and hampering the functions of DNA, RNA, and cellular organelles. The main targets of injury are cell membranes. Not only do injured cells die gradually during frozen storage, but surviving cells are reexposed to detrimental osmotic effects during thawing. The injury of microbial cells may be reversible or irreversible. In the former case, cells are able to repair if nutrients, energy sources, and specific ions (mainly Mg^{2+} and phosphate) are available and metabolism can commence. The extent of injury and repair, as well as death rates, varies according to the conditions of freezing, frozen storage, and thawing.

Freezing rate is an important determinant of microbial viability. During slow freezing, crystallization occurs extracellularly as the cell cytoplasmic fluid becomes supercooled at -5 to $-10°C$. Because their vapor pressure is greater than that of ice crystals, cells lose water, which freezes extracellularly. As a result of crystallization, the concentration of the external solution increases, which also removes water from the cells. On the whole, microorganisms are exposed to osmotic effects for a relatively long time, which causes increased injury. With increased freezing rates, the duration of osmotic effects decreases, thereby increasing survival to a maximum. When freezing rates are too high, crystal formation occurs also intracellularly, injuring the cells drastically and decreasing survival rate.

The most important factor influencing the effect of freezing on microbial cells is the composition of the suspending medium. Certain compounds enhance, while others diminish, the lethal effects of freezing. Sodium chloride is very important in this regard because it reduces the freezing points of solutions, thereby extending the time during which cells are exposed to high solute concentrations before freezing occurs. Other compounds, e.g., glycerol, sucrose, gelatin, and proteins, generally have a cryoprotective effect.

At temperatures below ca. $-8°C$, there is, in theory, no microbial growth (97). The fates of microorganisms surviving freezing may vary during frozen storage. Often, the death rate for survivors is high initially and then lowers gradually, and finally the survival level stabilizes. During frozen storage, the death rate of microbial populations is usually lower than it is during the freezing process. Death during frozen storage is probably due to the unfrozen, very concentrated residual solution formed by freezing. The concentration and composition of this residual solution may change during the course of storage, and the size of the ice crystals may increase, especially at fluctuating storage temperatures. These processes also influence the survival of microorganisms. In general, there is less loss of viability during frozen storage when the temperature is stable rather than fluctuating. The lower the temperature of frozen storage, the lower is the death rate of survivors (97). Gram-positive microorganisms survive frozen storage better than gram-negative bacteria.

Although freezing and frozen storage reduce considerably the viable cell counts in food, freezing preservation cannot be considered a sterilization process; the extent of microbial destruction may be significantly reduced by the components of the food. Under some conditions, survivors can grow during the thawing of frozen food. Their populations may then equal or exceed the level prior to destruction during freezing and frozen storage. The cells and the tissue structures of foods are also damaged by freezing, frozen storage, and thawing. During thawing, a solution rich in nutrients is released from the food cells, the microorganisms can penetrate the damaged tissue more easily, and a liquid film forms on the surface of the product. These conditions favor microbial growth. Therefore, thawed products may be vulnerable to rapid microbial spoilage. Refreezing of thawed products may be a dangerous practice and should be avoided.

PRESERVATION BY HEAT TREATMENT

The most effective and most widely used method for destroying microorganisms and inactivating enzymes is heat treatment (170). Heat resistance is, therefore, one of the most important characteristics of microorganisms from an industrial point of view.

The heat processing method called pasteurization (named after Louis Pasteur) is a relatively mild heat treatment aimed at inactivating enzymes and destroying a large population (99 to 99.9%) of vegetative bacterial cells. The main objective of this treatment is to eliminate non-spore-forming pathogenic bacteria. Because a significant proportion of the spoilage microorganisms are also destroyed by pasteurization, pasteurized products have extended shelf lives. For safety and keeping quality, pasteurization must be complemented with packaging which prevents recontamination. Pasteurized food must be stored at low temperatures to prevent the growth of sporeformers whose spores survive the heat treatment. The shelf lives of pasteurized products depend on the type of food and the conditions of pasteurization and storage.

Sterilization refers to the elimination of viable microorganisms. Industrially, the emphasis is not on absolute sterility but rather on freedom from pathogenic microorganisms and shelf stability of hermetically packaged products. Shelf-stable, microbiologically safe products may contain low numbers of viable, but dormant, bacterial spores. These products are called "commercially sterile."

Technological Fundamentals

Heat preservation technologies can be divided into two broad categories: those that heat foods in their final containers and those that heat foods prior to packaging. The latter technology requires separate sterilization of the food and the packaging material before the food is placed aseptically (in a sterile environment) into the package (aseptic processing) (191). Commercial sterilization of food in hermetically closed containers (cans or bottles) is called canning, or appertization, named after Nicholas Appert, who invented the basic process in the early 1800s.

A major technological development in heat processing has been the introduction of ultrahigh-temperature (UHT), short-time treatments of uniform fluid products (rapid heating to temperatures of about 140°C, holding for several seconds in heat exchangers, and then rapid cooling). High-temperature, short-time processes provide possibilities for increasing the flexibility of packaging and product types as well as improving product quality and processing efficiency (120). The use of higher temperatures for shorter times results in less damage to important nutrients and functional ingredients. The thermobacteriology of UHT-processed foods has been reviewed previously (17). Aseptic processing is widely used for fruit juices, dairy products, creams, and sauces. Continuous heat treatment with heat exchangers enables an efficient process with improved product quality by comparison with canned foods. Significant progress has also been made in the adaptation of this technology to soups including particles of meat, fish, or vegetables (191).

There has been a recent expansion in the use of the combination of mild heating of packaged foods and well-controlled chill storage. Products undergoing this treatment are also known as REPFEDs (refrigerated processed foods of extended durability) (154), or cook-chill products, and include "sous vide" meals (foods mildly heated within vacuum packs) (148). For these chilled, pasteurized foods, specific attention is given to the control of nonproteolytic, psychrotrophic C. botulinum because of its ability to grow at temperatures as low as 3°C (106). To ensure safety from microbes, some form of demonstrable intrinsic preservation of products, i.e., an additional hurdle, is generally regarded as necessary (133).

Fundamentals of Thermobacteriology

Heat in the presence of water (so-called wet-heat treatment) kills microorganisms by the denaturation of nucleic acids, structural proteins, and enzymes. Identification of primary sites for lethal heat-induced damage is difficult because of many changes which heat brings about in cells. Nevertheless, there is substantial evidence that DNA damage, directly or indirectly, may be the key lethal event in heated cells (71). However, heat also damages many other, less critical sites. In general, the thermal stability of ribosomes corresponds to the maximal growth temperature of microorganisms (102), and cytoplasmic membranes seem to be major sites of injury caused by mild heating (58), with a consequent increase of sensitivity to environmental stress factors. However, degradation of rRNA in membrane-damaged, heated cells often precedes loss of viability. Therefore, it is not thought to be a prime cause of cell death (94). In bacterial spores, some part of the germination system (e.g., specific enzyme[s]) may be the least heat-resistant component and thus a key site of heat inactivation (74). Dry heat is less lethal and kills microorganisms by dehydration and oxidation. Dry heat necessitates higher temperatures and longer heating times to generate the same lethality as wet heat.

Traditionally, calculating the heat treatment required to kill a given number of bacteria is based on the assumption that the destruction of a pure bacterial culture by wet heat at a constant temperature (T) exhibits negative exponential kinetics of a first-order reaction:

$$N = N_0 \cdot e^{-k \cdot t}$$

where N is the actual number of surviving cells at time t, N_0 is the initial number of viable cells, k is the rate constant (time^{-1}) of heat destruction (the death rate), and t is the heat processing time. That is, equal periods at lethal temperatures result in equal percentages of destruction, regardless of the microbial population size. A theoretical treatment of this relationship is given in chapter 1.

The destruction rate is inversely related to the decimal reduction time (D value), $D_T = \log 10 / k_T$. The D value is defined as the time in minutes required to destroy 90% of the population, as shown in Fig. 32.1 (survival curve). Survival curves provide data on the rate of destruction of a specific microorganism in a specific medium or food at a specific temperature. D values for microorganisms are greatly affected by environmental factors such as pH and composition of the substrate.

Under given conditions, the rate of death is constant at any given temperature and independent of the initial viable cell number. Plotting the logarithms of D values

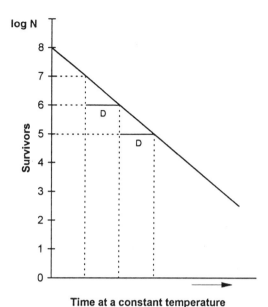

Figure 32.1 Bacterial survival curve showing logarithmic order of death and the concept of the D value.

against exposure temperatures gives the thermal death time curve (Fig. 32.2). The thermal death time (F) is the time necessary to kill a given number of microorganisms at a specified temperature.

While the D value represents a measure of heat resistance of the given microorganism at a given temperature,

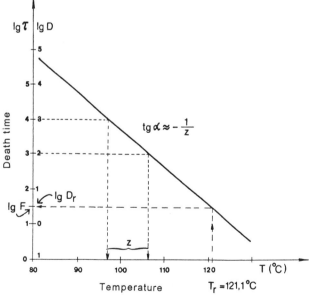

Figure 32.2 Thermal death time curve showing definition of z value and F value. T_r, reference temperature; τ, death time; lg, log; D_r, D value at the reference temperature; tgα, slope of the thermal death time curve.

the thermal death time curve indicates the relative resistance levels at different temperatures. This enables one to calculate the lethal effects of different temperatures. The slope of the thermal death time curve is expressed as the z value, which is the temperature increase required to reduce the thermal death time by a factor of 10.

Heat resistance levels (D values) of various microorganisms and the effect of temperature on the death rate are different and are influenced by many factors (41, 97). There are inherent differences among species, among strains within the same species, and between spores and vegetative cells. Bacterial spores are more heat resistant than vegetative cells of the same strain, some being capable of surviving for minutes at 120°C or for several hours at 100°C. Less heat resistant spores may have $D_{100°C}$ values of less than 1 min (143). Vegetative cells of bacteria, yeasts, and molds are killed after a few minutes at 70 to 80°C.

The ages of cells and their stage of growth, the temperature at which microorganisms are grown, and the composition of the growth medium also affect heat resistance. Physicochemical factors such as pH, a_w, salt content, and organic composition of the medium in which microorganisms are heated can also profoundly influence heat resistance (30, 143, 201).

Tables 32.6 and 32.7 illustrate selected heat resistance data for vegetative bacterial cells and endospores, respectively. Vegetative cells of spore-forming bacteria are as heat sensitive as other vegetative cells. The high heat resistance of spores results from their specific structure and composition and is basically the consequence of the dehydrated state of the spore core (71). The molecular basis of heat resistance in spores is covered more fully in chapter 3 of this volume.

Bacterial spores suspended in oils are more heat resistant than those suspended in aqueous systems. This increased heat resistance is ascribed to the low a_w of oils, and the difference in z values suggests that the mechanisms of inactivation differ for spores suspended in lipids and those suspended in aqueous systems (1).

Vegetative forms of yeasts can usually be killed by heating for 10 to 20 min at 55 to 60°C (183). Given the same period of heat treatment, some yeast ascospores require a temperature only 5 to 10°C higher than that required by vegetative cells to cause similar lethality (103). Vegetative propagules and conidiospores of most

Table 32.6 Relative levels of heat resistance by some vegetative bacteria[a]

Organism	Heating medium	D value (min) at: 70°C	65°C	60°C	55°C	z value (°C)
Escherichia coli	Ringer solution, pH 7.0				4	
Lactobacillus planatarum	TSB[b] plus sucrose (a_w = 0.95)		4.7–8.1			
	Tomato juice, pH 4.5	11.0				12.5
Pseudomonas aeruginosa	Nutrient agar				1.9	
Pseudomonas fluorescens	Nutrient agar				1–2	
Salmonella enterica serovar Senftenberg (775 W)	Skim milk			10.8		6.0
	Phosphate buffer (0.1 M), pH 6.5		0.29			
	Phosphate buffer plus sucrose (30:70, wt/vol)		1.4:4.3			
	Phosphate buffer plus sucrose (30:70, wt/vol)		2.0:17.0			
	Milk chocolate	440				18.0
Salmonella enterica serovar Senftenberg	Heart infusion broth (a_w = 0.99), pH 7.4			6.1		6.8
	Heart infusion broth plus broth plus NaCl (a_w = 0.90), pH 7.4			2.7		13
	Heart infusion broth plus sucrose (a_w = 0.90), pH 7.4			75.2		8.9
Salmonella enterica serovar Typhimurium	Phosphate buffer (0.1 M), pH 6.5		0.056			
	Milk chocolate	816				19.0
Staphylococcus aureus	Custard or pea soup			7.8		4.5

[a] Adapted from reference 210.
[b] TSB, tryptic soy broth.

molds can be inactivated by a 5- to 10-min wet-heat treatment at 60°C (40). Several mold species form sclerotia or ascospores that are much more heat resistant (42, 202). Spores of some molds can survive dry-heat treatment for 30 min at 120°C.

Calculating Heat Processes for Foods

Food in containers does not heat instantaneously, and since all temperatures above a minimum value contribute to microbial destruction, a method to determine the relative effect of changing temperature while a food is being heated and cooled during thermal processing is necessary.

For the determination of heat processes for canned foods, the F value was introduced. The F value is defined as the number of minutes at a specific temperature required to destroy a specific number of viable cells having a specific z value (Fig. 32.2). Since F values represent the number of minutes to diminish a homogeneous population with a specific z value at a specific temperature, and because z values as well as temperatures vary, it is necessary to designate a reference F value. This reference

is the F_0 value, which is the number of minutes at 121°C (250°F) required to destroy a specific number of cells whose z value is 10°C (18°F). The F_0 value of a heat treatment is called its sterilization value, and is a measure of the lethality of a given heat treatment. It is related to τ, the time required for identical lethality at temperature T, shown by the equation

$$\log F - \log \tau = \frac{T - 121.1}{z}$$

F_0 requirements of various foods differ. Since heat sensitivities of microorganisms and the characteristics of thermal death curves are affected by many factors, thermal death curves should be established for the specific food for which a heat process is designed. Because the spores of *C. botulinum* types A and B are the most heat-resistant spores produced by foodborne pathogens (Table 32.7), commercial sterilization of low-acid (pH $\geq$ 4.6) foods must be sufficient to destroy these spores. The targets for the heat processing of acid foods (pH < 4.6) are heat-resistant molds.

Table 32.7 Approximate heat resistance levels (D values) for some bacterial spores

Type of food and typical organism	D value (min) at:		z value (°C)
	121°C	100°C	
Low-acid foods (pH > 4.6)			
Thermophilic aerobe			
Bacillus stearothermophilus	4.0–4.5	3,000	7
Thermophilic anarobes			
Clostridium thermosaccharolyticum	3.0–4.0		12–18
Desulfotomaculum nigrificans	2.0–3.0		
Mesophilic anerobes			
Clostridium sporogenes	0.1–1.5		9–13
Clostridium botulinum types A and B	0.1–0.2	50	10
Clostridium perfringens		0.3–20	10–30
Mesophilic aerobes			
Bacillus licheniformis		13	6
Bacillus subtilis		11	7
Bacillus cereus		5	10
Bacillus megaterium		1	9
Acid foods (pH ≤ 4.6)			
Thermotolerant aerobe			
Bacillus coagulans	0.01–0.1		
Mesophilic aerobes			
Bacillus polymyxa		0.1–0.5	
Bacillus macerans		0.1–0.5	
Clostridium butyricum (or *Clostridium pasteurianum*)		0.1–0.5	

[a]Adapted from references 98 and 206.

A "botulinum cook" is a heat process that reduces the population of the most resistant *C. botulinum* spores by an arbitrarily established factor of 12 decimal values: the 12 D concept. This provides a substantial margin of safety in low-acid canned foods. In practice, 2.45 min is the 12 D value of botulinal spores in phosphate buffer (97). *Clostridium sporogenes* (PA 3679) has been used as a surrogate microorganism for *C. botulinum* in developing processes for sterilized food because of its nontoxic property and similarity in biochemical requirements (157). The botulinum cook will produce a product that is safe but not necessarily sterile. Other heat-resistant spores, which may cause spoilage but are not pathogenic, may remain and be viable (Table 32.7). Thus, their higher heat resistance often is the determining factor in establishing commercial heat processes. On the other hand, *C. botulinum* can be controlled by pasteurization in some types of cured meat products because of the products' reduced a_w and pH and the addition of nitrite. This multiple-hurdle preservation system facilitates stability.

Detailed guidance for calculating heat processes on the basis of semi-log-linear kinetics using the model of Bigelow (12) is provided in several monographs (e.g., reference 205). Kinetic models taking into account the concurrent activation of dormant spores and distinct inactivation of dormant and activated spores during UHT treatments have been developed for advanced process design (120, 196). In the case of vegetative cells, the death rate coefficient depends significantly on environmental factors. Reichart and Mohácsi-Farkas (187) developed several kinetic models for describing the combined effect of a_w, pH, and redox potential on the thermal destruction rate of seven foodborne spoilage microorganisms (*Lactobacillus plantarum*, *Lactobacillus brevis*, *Saccharomyces cerevisiae*, *Zygosaccharomyces bailii*, *Yarrowia lipolytica*, *Paecilomyces varioti*, and *Neosartoria fischeri*).

Assessment of the Thermal Process by Using Enzymatic Time-Temperature Integrators

The assessment of the thermal process with enzymatic time-temperature integrators is used to evaluate the process values $^{10°C}F_{121.1°C}$ inside particles of a solid-plus-liquid food on the basis of determining decreases in the enzyme activity (80).

Recent Developments in Thermobacteriology of Food

Previous concepts about the heat preservation of foods assumed implicitly that heat resistance values for microorganisms measured under isothermal conditions are constant and unaffected by the heating rate in the sublethal temperature range. It has been demonstrated, however, that the rate of temperature elevation affects the subsequent isothermal death of vegetative cells of some bacteria (127, 128). Exposure of cells to sublethal heat shock temperatures above their growth maxima induces resistance to heating at higher temperatures. These effects were observed with nutritionally rich media. The induction or enhanced synthesis of a family of proteins (heat shock proteins [hsps] in response to heat shock and also to other environmental stresses and chemical or mechanical treatments has been reported as a phenomenon that occurs in most, if not all, microorganisms (3, 4, 13). Many of the hsp inducers have in common the capacity to produce protein damage (87). It has been proposed that the signal for hsp induction is protein denaturation and an increase in the amount of unfolded protein in the cell (164). However, the mechanism responsible for the development of thermotolerance has not been completely elucidated, and the relationship between hsp induction and the acquisition of thermotolerance is unclear (4, 13). The lesions produced by various chemicals triggering thermotolerance are different from those produced by heat (13). Some investigators propose two states of tolerance to heat, one not requiring hsps and another induced by more severe damage and requiring hsps (116). These observations lead to the conclusion that environmental stresses imposed by industrial processes can induce protective responses in microorganisms (220). Thus, the possibility that heat resistance of vegetative cells may increase during heating must be taken into account when one is establishing safe heat treatments for foods that may be heated relatively slowly (43, 44, 107).

The traditional model of thermal inactivation kinetics, which assumes that the reduction in log numbers of survivors decreases in a linear manner with time, has served the food industry and regulatory agencies well for many years. Indeed, in situations where bacterial spores are the critical targets or death is rapid, a good fit to this hypothesis is generally observed. However, as a result of the introduction of milder thermal processes that use low processing temperatures under a wide range of environmental conditions, the food industry is increasingly requiring accurate predictions for death kinetics for vegetative bacteria. Under these conditions, significant deviations from log-linear kinetics are encountered, "shoulder" and/or "tailing" of survival curves occurs frequently, and methods incorporating D and z values cannot be relied upon. The theory underlying log-linear death kinetics assumes that inactivation results from random single hits of key targets within microbial cells. Explanations offered as to why variations in the straight-line survivor curve occur include (i) experimental errors and artifacts such as a mixed population of test microorganisms, flocculation and deflocculation of cells during the heating period, the requirement for heat activation for spore germination and aggregation and absorption of cells on the walls of the vessel, and the presence of cells in aerosol droplets above the surface of the suspension (82, 169); (ii) multiple-hit destruction mechanisms (71); and (iii) variability of heat sensitivities (resistance distribution) in the heated population (82, 135, 214) and heat adaptation of cells during treatment (82).

Mathematical models were developed to describe observed inactivation kinetics (29, 117) which allow for variability of heat sensitivity throughout a population of cells, i.e., normal distribution of heat sensitivity. A further, improved model describes the heat sensitivity of cells as being distributed by a logistic function, which allows for accurate predictions of thermal inactivation to be made at a relatively low heating rate (204) and even when parameters other than temperature, e.g., pH and NaCl concentration, are varied (29).

Microwave and Radio Frequency Heat Treatment

Electromagnetic heating of food using radio frequency (1 to 500 MHz) or microwave (500 MHz to 10 GHz) energy has the advantage of generating internal heat. Water molecules, because of their negatively charged oxygen atoms and positively charged hydrogen atoms, are electric dipoles. When a rapidly oscillating radio frequency or microwave field is applied to a biological tissue or food rich in water, the water molecules reorient with each change in the field direction. This reorientation creates intermolecular friction, which produces heat (149). Such electroheat technologies for food processing are reviewed, e.g., by Tempest (208).

Microwave heating modes may reduce process times and energy and water usage in some food-processing areas. Microwave energy is suited for heat processing of foods and for pest control in the food industry (36, 192). However, the geometry of the food product; its thermal, physical, and dielectric properties; its mass; the power input; and radiation frequency all influence the interaction of electromagnetic waves within a food (178). These factors can cause differential heating of different food constituents or different parts of a multicomponent meal. Oliveira and Franca (160) developed a model to simulate the temperature distribution in microwave-heated foods by using Maxwell's equation and Lambert's law. Thermal nonuniformity has limited commercial application of microwave heating for the purpose of microbial inactivation, even though microwave heating has gained widespread application in home food preparation. The production of heterogeneous temperatures within a food might create problems if it results in insufficient heating compared with conventional cooking and thereby allows survival of potentially pathogenic microorganisms (90). Therefore, metal shielding and other packaging designs which increase the heating uniformity of food are desirable (37, 88).

Radio frequency energy can penetrate dielectric materials more deeply than microwave energy, due to its having lower frequency and longer wavelengths than microwave energy (216). Recently, a novel thermal process was explored to produce shelf-stable egg products in polymeric trays with quality parameters better than those of retort-treated samples (124).

Regarding specific electromagnetic effects, electromagnetic fields cause ion shifts on cellular membranes, leading to changes in permeability, functional disturbances, and cell rupture (178). Theoretically, microbial cells in food may be differentially heated compared to the food matrix, which could result in specific inactivations. However, these theoretical considerations need to be validated with food. So far, experimental data on the specific effect of radio frequency or microwave frequency electromagnetic energies are contradictory, but in the majority of studies, no nonthermal effects have been detected (152).

A more fundamental approach in food-related research concerning microwave effects is needed, and more research should be conducted to determine these effects and utilize the advantages of various frequencies. The choice of the most widely used frequency, 2.45 GHz, was a compromise of various factors, and this frequency is not necessarily the most suitable frequency for a certain application (178). Sophisticated temperature registration methods and dielectric measurements can be important tools for better understanding interactions in studies of specific effects of electromagnetic fields.

Ohmic Heating

The use of direct electrical heating, known as ohmic (electrical resistance) heating, is another heating technology that can be used for pumpable food products containing particulate matter (197). An electric current is passed through the material as it is pumped through a tube. The heating rate is a function of the electrical conductance levels of the solid and liquid phases. If conductance levels of the two phases are about the same, both phases will heat at the same rate. The electrical current heats up the product quickly so that the final temperatures required to kill microorganisms are achieved with far less total cooking time, resulting in improved product quality as compared to retorted foods. In the limited thermobacteriological studies conducted, microbial death appeared to be caused solely by thermal effects (163). The role of electric resistance heating in aseptic processing was discussed by Fryer (60). As electric conductivities of all the constituents of the food product determine the rate of heating of the different constituents, ohmic heating often requires well-controlled pretreatments to eliminate air in food and to control salt levels (224). Advantages and limitations for electromagnetic and electric resistance heating methods have been summarized by Ohlsson (158).

ULTRASOUND

High-power ultrasound can kill microorganisms (176), but its use as a single variable to achieve inactivation of microbes is not practical for a number of reasons. However, sonic energy at lower power values can sensitize microorganisms to heat (20, 193) and when combined with heat may be an attractive approach for minimizing heat damage in some foods (19, 47). Although the heat-ultrasound synergism disappears at the boiling point of the menstruum (61), this disappearance can be prevented by overpressure, commonly no more than 1 MPa (mano-thermosonication), thereby offering the possibility of a new sterilization process for liquid foods (185).

PRESERVATION BY IRRADIATION

UV Radiation

The UV radiation wavelength range most destructive to microorganisms is between 240 and 280 nm. The great microbicidal efficiency of the so-called germicidal radiation is based on the fact that this radiation can most effectively damage nucleic acids. Inactivation is caused by the cross-linking of thymine dimers of the DNA, which prevents repair and reproduction.

Gram-negative bacteria are most easily killed by UV radiation, and bacterial endospores and molds are much more resistant. Viruses are more UV radiation resistant than bacterial cells (86, 207). UV radiation resistance of various microorganisms also depends on the presence of cellular pigments. Cocci that form colored colonies are less susceptible to UV radiation than those with colorless colonies. Dark-colored conidia of certain molds are highly UV radiation resistant (144).

The very low penetration of UV radiation and the difficulty in attaining an even exposure level over all points of the food surface are major technological problems which limit the feasibility of UV radiation in food preservation. Therefore, UV sources may be used mainly for disinfection of air, e.g., in aseptic filling of liquid foods such as milk, beer, and fruit juices (64), in packaging of sliced bread, and in the ripening rooms for cheese and dry sausages, or for disinfection of water when the use of chlorine is undesirable (21). One type of treatment chamber for pasteurization of juices utilizes turbulent flow to form a "continuously renewed surface" (64, 200). Possible uses of UV irradiation of packaging materials and preformed containers as a single treatment and in combination with hydrogen peroxide for aseptic packaging have been described by Narasimhan et al. (153) and Stannard et al. (203), respectively. The U.S. Food and Drug Administration has amended the food additive regulation to provide for the safe use of UV irradiation to inactivate human pathogens and other microorganisms in juice products (213).

High-Intensity Pulsed Light

During the past decade, high-intensity pulsed light has been developed into a commercially available process for decontamination of (food) surfaces with high-intensity xenon arc lamps (79). The killing effect is derived mainly from the UV component of the absorbed light or heat. On nonporous surfaces, the level of destruction of microorganisms is very high. When the surfaces are more complex, such as in the case of food materials, the microbial reduction is only 2 to 3 log cycles.

Ionizing Radiation

The potential for application of ionizing radiation to food products is based mainly on the fact that ionizing radiation very effectively damages DNA so that cell division is impaired. Given the right doses, this impairment is obtained without serious effects on the food itself. Therefore, microorganisms, insect gametes, and plant meristems can be prevented from reproducing or growing, thus rendering the food stable (Table 32.8). Irradiation offers a broad scope also as a quarantine treatment for fresh produce. There are many general reviews on various aspects, prospects, and problems of food irradiation (e.g., references 39, 50, 53, 105, 122, 140, 158, 161, 217, and 218).

In those parts of the world where the transport of food is difficult and where refrigerated storage of food is scarce and extremely expensive, the use of ionizing radiation may facilitate wider distribution of some food than would otherwise be feasible. In this way, a more varied and possibly nutritionally superior diet may become available to the residents.

Application of ionizing radiation to foods is limited to the use of high-energy electromagnetic radiation (gamma rays of ^{60}Co or X rays) with energies up to 5 MeV or electrons from electron accelerators with energies up to

Table 32.8 Preservative effects of ionizing radiation

Effect(s)	Result(s)
Inhibition of sprouting	Increased shelf lives of root and bulb crops; reduction of malting losses
Decrease of after-ripening and delay of senescence of some fruits and vegetables	Increased shelf lives of fruits and vegetables
Killing and sterilization of insects	Insect disinfestation of food
Prevention of growth and reproduction of parasites transmitted by food	Prevention of parasitic diseases
Reduction of microbial populations	Decreased contamination of food; increased shelf life of food; prevention of food poisoning

Table 32.9 Approximate doses of radiation needed to kill various organisms[a]

Organisms	Dose (kGy)
Higher animals	0.005–0.1
Insects	0.01–1
Non-spore-forming bacteria	0.5–10
Bacterial spores	10–50
Viruses	10–200

[a] Data from reference 212.

10 MeV (28). These types of radiation are chosen because (i) they produce the desired effects in foods, (ii) they do not induce radioactivity in foods or packaging materials, and (iii) they are available in quantities and at costs that allow commercial use of the process. Other kinds of ionizing radiation, in some respects, do not suit the needs of food irradiation.

The penetration depths of electron beams and X or gamma rays into matter are different. The practical usable depth limit for 10-MeV electrons in water-equivalent material is 3.9 cm, and the so-called half-thickness value for 5-MeV X rays is 23.0 cm. Except for different depths of penetration, electromagnetic radiation and electrons are equivalent in food irradiation and can be used interchangeably.

Microbiological Fundamentals

One of the main reasons for using ionizing radiation is to kill microorganisms that cause spoilage or are a health hazard to consumers. The DNA in the chromosomes is the most critical target of ionizing radiation. Effects on the cytoplasmic membrane appear to play an additional important role in radiation-induced damage of cells (78). Although changes caused by radiation are at the cellular level, the consequences of these changes vary with the microorganism. Radiation sensitivity is roughly inversely proportional to the size and complexity of an organism (Table 32.9). This correlation is related to genome size: the DNA in the nuclei of insect cells is a target much larger than the genomes of bacteria, and the cells of mammalian organisms containing DNA molecules that

must provide much more genetic information than those of insects are correspondingly larger and even more sensitive to radiation (39). Differences in radiation sensitivities within groups of similar organisms are related to differences in their chemical and physical structures and the ability to recover from radiation injury. The amount of radiation energy required to control microorganisms in foods, therefore, varies according to the resistance of the particular species and the number of viable cells present.

The relative radiation resistance of various foodborne microorganisms can be found in several reviews (e.g., references 39, 52, 54, and 142). Summarizing data from a large number of references, Table 32.10 presents ranges of radiation resistance levels of some foodborne microorganisms in fresh and frozen foods of animal origin. In general, gram-negative bacteria, including the common spoilage microorganisms in many foods (e.g., pseudomonads) and particularly enteric pathogens (salmonellae, shigellae, etc.), are more sensitive than vegetative gram-positive bacteria. *Bacillus* and *Clostridium* spores are more resistant. Rarely, even more highly resistant vegetative forms may be encountered, e.g., *Deinococcus* (formerly *Micrococcus*) *radiodurans* and the gram-negative, rod-shaped *Deinobacter* sp. (76). The radiation sensitivity of many molds is of the same order as that of vegetative bacteria. However, molds with melanized hyphae have a radiation resistance comparable to that of bacterial spores (194). Yeasts are as resistant as the more resistant vegetative bacterial cells. Viruses are highly radiation resistant (11, 219).

Ionizing radiation can affect DNA directly by the ionizing ray or indirectly by radiolytic products of water, particularly free radicals ·H, ·OH, and e_{aq}^-. The most important of these three radicals in causing DNA damage is the ·OH radical. The ·OH radicals formed in the hydration layer around the DNA molecule are responsible for 90% of DNA damage. Thus, in living cells, indirect radiation damage is predominant. The principal lesions

Table 32.10 Typical radiation resistance levels of some foodborne microorganisms in fresh and frozen foods of animal origin

Microorganism(s)	D_{10} (kGy) for:	
	Fresh food	Frozen food
Vibrio spp.	0.03–0.12	0.11–0.75
Yersinia enterocolitica	0.04–0.21	0.4
Pseudomonas putida	0.06–0.11	
Campylobacter jejuni	0.08–0.20	0.21–0.32
Shigella spp.		0.2–0.4
Aeromonas hydrophila	0.14–0.19	
Bacillus cereus (vegetative cells)	0.17	
Proteus vulgaris	0.2	
Escherichia coli	0.23–0.35	0.3–0.6
Escherichia coli O157:H7	0.24–0.27	0.31–0.44
Staphylococcus aureus	0.26–0.6	0.3–0.45
Brucella abortus	0.34	
Salmonella spp.	0.3–0.8	0.4–1.3
Listeria monocytogenes	0.27–1.0	0.52–1.3
Lactobacillus spp.	0.3–0.9	
Streptococcus faecalis	0.65–1.0	
Clostridium perfringens (vegetative cells)	0.59–0.83	
Moraxella phenylpyruvica	0.63–0.88	
Clostridium botulinum type E (spores)	1.25–1.40	
Bacillus cereus (spores)	1.6	
Clostridium sporogenes (spores)	1.5–2.2	
Clostridium botulinum types A and B (spores)	1.0–3.6	
Deinococcus radiodurans	5–3.1	
Deinobacter sp.	5.05	
Coxsackievirus		6.8–8.1

induced by ionizing radiation in intracellular DNA are chemical damage to the purine and pyrimidine bases and to deoxyribose sugar and physicochemcial damage resulting in a break in the phosphodiester backbone in one strand of the molecule (single-strand break) or in both strands at the same place (double-strand break) (145). Double-strand breaks are produced by ionizing radiation at about 5 to 10% of the rate for single-strand breaks. Most microorganisms can repair single-strand breaks. The more sensitive microorganisms, such as *Escherichia coli*, cannot repair double-strand breaks, but highly resistant species (e.g., *Deinococcus* sp.) are able to repair both single- and double-strand breaks (145).

The low water content of the spore protoplast is a major factor affecting the radiation resistance of bacterial spores. During germination, the water content of the spore protoplast increases and therefore the radiation resistance markedly decreases.

Radiation survival is conveniently represented by the logarithm of the number of surviving microorganisms plotted against the radiation dose. Typical radiation survival curves are shown in Fig. 32.3. The type 1 curve is an exponential survival curve, quite common for the more radiation sensitive microorganisms. The type 2 curve is characterized by an initial shoulder, indicating that equal increments of radiation are more effective at high doses than at low doses. The shoulder in the curve may be explained by a certain amount of repair of cells treated with low doses. The type 3 curve is characterized as concave with a resistant tail.

Similar to heat resistance, the radiation response in microbial populations can be expressed by the decimal reduction dose (D_{10} value). When the dose-survival curve is a straight line, the D_{10} is the reciprocal of its slope,

$$D_{10} = \frac{\text{radiation dose}}{\log N_0 - \log N}$$

where N_0 is the initial number of microorganisms and N is the number of microorganisms surviving the radiation dose. Survival curves of type 2 and type 3 are often best represented by quoting an inactivation dose, for example, the dosage to inactivate 90 or 99% of the initial viable cell count.

Several environmental factors influence the radiation resistance of microorganisms (34).

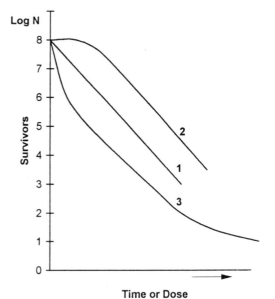

Figure 32.3 Typical radiation survival curves. N, survivors of radiation dose; N_0, original number of viable cells; 1, exponential survival curve; 2, curve characterized by an initial shoulder; 3, concave curve with a resistant tail.

(i) The composition of the medium surrounding the microorganisms plays an important role in the dose requirement necessary to achieve a given effect. Cells irradiated in phosphate buffer are more sensitive than those irradiated in foodstuffs. In general, the more complex the medium, the greater the competition of the medium components for the free radicals formed from water and the activated molecules produced by the radiation, thus protecting the microorganisms. Buchanan et al. (18) reported that prior growth at acidic conditions increased the resistance of *E. coli* O157:H7 to ionizing radiation at acidic pH. It is impossible to predict the foods in which particular bacterial cells will be more radiation sensitive or radiation resistant.

(ii) As in heat treatment, a reduction in the moisture content (reduction of a_w) of the food protects microorganisms against the lethal effect of ionizing radiation. Under drier conditions, the yield of free radicals produced from water by radiation is lower, and thus the indirect DNA damage that the free radicals may cause is decreased.

(iii) The temperature at which a product is treated may influence the radiation resistance of microorganisms. For vegetative cells, elevated-temperature treatments, generally those with temperatures in the sublethal range above 45°C, synergistically enhance the lethal effect of radiation. This enhancement is thought to occur because the repair systems, which normally operate at ambient and slightly above ambient temperatures, are damaged at higher temperatures.

(iv) Freezing causes a striking increase in the radiation resistance of vegetative cells. Microbial radiation resistance in frozen foods is about two- or threefold higher than that at ambient temperature. This increase is due to the immobilization of the free radicals and prevention of their diffusion when the medium is frozen. Thus, in the frozen state, the indirect DNA damage by ·OH radicals is nearly prevented. The change in resistance with temperature demonstrates the importance of the indirect action in high-moisture foods.

(v) The lethal effect of ionizing radiation on microbial cells increases in the presence of oxygen (209). In the total absence of oxygen, and in wet conditions, radiation resistance usually increases by a factor of 2 to 4. In dry conditions, the radiation resistance might increase by a factor of 8 to 17. Data reported for cell suspensions irradiated in sealed tubes can frequently be plotted as type 3 survival curves (Fig. 32.3). These tailing effects probably represent a shift to anaerobic conditions, i.e., irradiation of aerated water will consume all the oxygen after a dose of 500 Gy.

The foregoing discussion emphasizes the necessity for specifying the environmental conditions when measuring and citing the radiation sensitivity of a microorganism.

Technological Fundamentals

Generally, there is a minimum dose requirement to inactivate microorganisms. Whether every mass element of a food requires irradiation will depend on the purpose of the treatment. In some cases, irradiation of the surface will suffice. In others, the entire food must receive the minimum dose. Guidelines for dose requirements are given in Table 32.11. Detailed requirements must again be considered as specific for a given food.

Radurization is the term describing substantial reduction in the number of spoilage microorganisms in food, which thereby extends the refrigeration shelf life of the food. With relatively high doses, those approaching 5 kGy, the normal spoilage microorganisms are often eliminated and the more resistant and metabolically less active species, e.g., resistant *Moraxella* spp., lactic acid bacteria, and yeasts, remain. During chilled storage, this surviving biota grows less rapidly than the normal spoilage biota, and the storage life is correspondingly further extended to three or four times normal. As those of the biota in unirradiated foods, the nature and behavior of the spoilage biota are strongly influenced by the nature of the packaging. In anaerobic packages, lactic acid bacteria and yeasts predominate.

Radicidation is the term describing reduction in the number of specific viable non-spore-forming pathogenic microorganisms (other than viruses) and parasites, which thereby improves the hygienic quality of foods and reduces the microbiological safety risks in foods.

Table 32.11 Dose requirements for various applications of food irradiation

Application	Dose requirement (kGy)
Inhibition of sprouting of potatoes and onions	0.03–0.12
Insect disinfestation of seed products, flours, fresh and dried fruits, etc.	0.2–0.8
Parasite disinfestation of meat and other foods	0.1–3.0
Radurization of perishable food items (fruits, vegetables, meat, poultry, fish)	0.5–10
Radicidation of frozen meat, poultry, eggs, and other foods and feeds	3.0–10
Reduction or elimination of microbial populations in dry food ingredients (spices, starch, enzyme preparates, etc.)	3.0–10
Radappertization of meat, poultry, and fishery products	25–60

Radappertization is the term used to describe the application of ionizing radiation to prepackaged, enzyme-inactivated foods in doses sufficient to reduce the number and/or activity of viable microorganisms to such an extent that, in the absence of postprocessing contamination, no microbial spoilage or toxicity should occur, no matter how long or under what conditions the food is stored. The dose requirement for radappertization is determined by the most radiation resistant microorganism associated with the enzyme-inactivated food. For non-acid, low-salt foods, this is the *C. botulinum* type A spore. By analogy with thermal processing, the safety of the radappertization process is based on a 12 *D* reduction in the viability of botulinal spores. For non-acid, low-salt foods, the required dose is about 45 to 50 kGy (39, 219).

The usefulness of ionizing radiation in overcoming microbiological problems has been clearly demonstrated with a variety of foodstuffs and bacteria of public health significance, especially pathogenic bacteria such as salmonellae, *L. monocytogenes*, campylobacters, *Y. enterocolitica*, and *E. coli* O157:H7 (8, 26, 27, 52, 56, 59, 112). Radiation doses of 3 to 10 kGy are sufficient to reduce viable cell counts in spices, herbs, and enzyme preparations to satisfactory levels (49, 55).

Radiation treatment causes practically no temperature rise in the product. The largest dose likely to be employed is about 50 kGy. This amount of energy is equivalent to about 12 calories (50 J). Hence, if all the ionizing radiation degrades to heat, the temperature of a food will rise about 12°C. For this reason, irradiation has been termed a cold process.

Ionizing irradiation has the advantage of being able to be applied through any type of packaging material, including those materials which cannot withstand heat. Radiation can be applied after packaging, thus avoiding recontamination and reinfestation. Most standard packaging materials are satisfactory for use with irradiated foods (141).

One should keep in mind that the quality of the irradiated foods, as that of any other preserved food, is a function of the quality of the original material and that to obtain good results, good manufacturing practice is needed. The longest shelf life may be obtained if the quality of the raw material is good and proper hygienic conditions are maintained. In certain cases, the use of more thorough inspection and sorting to remove substandard-quality raw materials before treatment (or treatments such as washing, blanching, or chemical rinse) can be used to reduce the radiation dose required to produce a high-quality product. The benefits of irradiation should never be considered as a substitute for product quality or as compensation for poor handling and storage conditions.

The basic irradiation process applies a prescribed amount of ionizing radiation of either electrons or electromagnetic rays and uses certain other procedures which may be needed. These procedures will vary according to the food being treated and the purpose of the treatment. For radappertized foods intended to have shelf stability without refrigeration, packaging suitable to protect the food from microbial contamination after irradiation is essential, and inactivation of the radiation-resistant food enzymes by mild heat processing may be necessary. For radurized foods, enzyme inactivation is not necessary and protective packaging may or may not be needed. Protection to prevent reinfestation may be needed for products irradiated to control insects. Specific storage conditions may be required, depending on the nature of products treated.

Some products may require irradiation under special conditions, such as at low temperature or in an oxygen-free atmosphere, or combination treatments such as heat and irradiation (51). Sensitization of microorganisms with the aim of lowering the dose can be obtained by heating (77) and chemical means. The microbiota surviving irradiation is more sensitive to subsequent food-processing treatments than the microbiota of unirradiated products.

The actual dose employed is a balance between what is needed and what can be tolerated by the product without unwanted changes. High doses cause sensorial changes (off flavors and/or texture changes) in high-moisture food. These undesirable effects can be limiting factors. In the case of protein foods of animal origin, the maximum dose permitted usually is set by the level at which off flavor develops.

Radiation-induced changes in the texture of some foods, such as fresh fruits and vegetables, can be a limiting factor. Polysaccharides such as pectin can be degraded by irradiation. This degradation is often accompanied by a release of calcium, resulting in softening of the product. These observations, and also the observed increase in permeability, can be explained by the fact that the natural resistance of the irradiated plant tissues is often reduced, resulting in an accelerated spoilage when microorganisms are given a chance to grow after irradiation. However, this radiation-induced initial softening in several horticultural commodities is often compensated for by the fact that during extended postirradiation storage, this softening (after-ripening) takes place considerably more slowly than in the untreated plant tissues.

In some cases, the chemical and physical effects of irradiation can be regarded as improvements. For example, the radiation-induced softening and increase in permeability are advantageous when shortening of the cooking time (e.g., for dried vegetables) is required, and increased yield of juice can be obtained from irradiated fruits.

NEW PHYSICAL PRESERVATION TREATMENTS

During recent years, several other physical treatments such as the use of high hydrostatic pressure and electric and magnetic fields have attracted much interest as promising tools for food processing and preservation and for the creation of new types of food products.

High Hydrostatic Pressure for Food Preservation

One advantage of the application of high or ultrahigh hydrostatic pressure (300 to 1,000 MPa) is that the pressure is transferred throughout food instantly and uniformly (Pascal principle) and thus the treatment is independent of sample size and geometry. It can be applied at ambient or moderate temperatures. Because only noncovalent bonds are affected by the treatment, there is good retention of color, nutrients, and flavor (23). High-pressure-treated fruit juices and other fruit products have been marketed in Japan since 1990, and some commercial products have been available in the United States and Europe since 1996. Recent reviews (48, 92, 100, 115, 195) summarize the effects of high-pressure treatment on microorganisms, foods, and food ingredients and developments in high-pressure food-processing technologies. In addition to several batch systems, semicontinuous systems associated with aseptic packaging have been developed (200).

The chemical principle behind the effect of ultrahigh pressure is that noncovalent bonds, such as hydrogen bonds, and ionic and hydrophobic bonds in proteins and carbohydrates undergo changes which alter their molecular structures. Microbial inactivation is probably due to a number of factors including protein conformation changes and membrane perturbation leading to cell leakage (188). In general, gram-positive microorganisms are more resistant to pressure than gram-negative microorganisms (123). Microbial resistance to heat and resistance to pressure are often correlated. For example, more heat resistant strains of *L. monocytogenes* are more resistant to pressure, and pressure-treated cells have marked differences in ultrastructure (129). However, a heat-resistant strain of *S. enterica* serovar Senftenberg has been shown to be less resistant to pressure than other strains (136). Studies on the injury and survival of *A. hydrophila* and *Y. enterocolitica* caused by high hydrostatic pressure suggest that posttreatment growth occurs only under exaggerated

temperature abuse conditions (45). High-hydrostatic-pressure treatment of orange juice sensitizes cells of *E. coli* O157:H7 that survive treatment to the acidic conditions of orange juice, resulting in significantly lower survival rates during refrigerated storage (121). Pressure cycling, i.e., the application of consecutive, short pressure treatments interrupted by brief decompressions, strongly enhances the pressure sensitivity of *E. coli* in milk, particularly in combination with natural antimicrobial peptides, and minimizes problems associated with pressure-resistant mutants (62). Vegetative cells are relatively pressure sensitive, being readily killed by pressure in the range of 300 to 700 MPa, and bacterial spores are resistant to pressures at ambient temperatures (69, 186). However, the combination of elevated temperature and pressure of, e.g., 400 MPa is effective against bacterial spores (198). Spores can be inactivated by alternating pressure cycles which initiate spore germination under moderate pressure (e.g., 30 min at 80 MPa) with those that kill the germinated spores under high pressure (e.g., 30 min at 350 MPa) (69, 162). Pressure-induced germination of spores is affected by pH and ionic environment (69). More recent data on pressure inactivation of bacterial spores show several log cycles of inactivation of highly heat resistant spores of *Bacillus subtilis* and *Bacillus stearothermophilus* at 150 to 450 MPa and 70 to 95°C. This finding suggests that combining moderate pressures with pasteurization temperatures can lead to sterilization (83). The heat resistance of spores of various bacteria does not correlate with their pressure resistance (131). Another preservation approach for enhanced inhibition and inactivation of bacterial spores is to combine high hydrostatic pressure, mild heat, and chemical preservatives (199).

Ascospores of some yeasts are more pressure resistant than vegetative cells. Nevertheless, ascospores of *Saccharomyces cerevisiae* in fruit juices can be inactivated by high pressure, having *D* values between 8 sec and 10.8 min at 500 and 300 MPa, respectively (225).

Inactivation of foodborne viruses by high hydrostatic pressure has recently been investigated, with promising results (113). Pressure sensitivities of viruses vary substantially, but pressure inactivation of hepatitis A virus and a calicivirus seems to be feasible, and parasites are generally quite sensitive (79).

Application of high pressure to foods and food components can cause protein denaturation, altered gelation properties of biopolymers, increased susceptibility to enzymatic degradation, and increased drying rates of plant tissues as a result of permeabilization of cell walls (22). Activation and inactivation of enzymes can be controlled by using selected pressure and temperature conditions (184). However, various enzymes have considerable differences in barosensitivities (84) and further work is required to study the possible regeneration of enzyme activities in foods during subsequent storage (114).

Recently, the influence of combined pressure and temperature treatments on enterotoxins of various pathogenic bacteria has been studied (132). Individual toxins showed different responses to high pressure with or without heat treatment, but overall the results indicate that pressure application may increase inactivation by heat treatment. No epitope masking of bacterial toxins was observed after pressurization.

The efficiency of high-pressure preservation, just as that of other treatments, depends on the food composition and the pH. According to studies using short high-pressure pulses, achieving an end temperature above 105°C by adiabatic heating produces sterility even in low-acid foods (38, 137). Due to pressure, the product sterilization temperature is reduced by 5 to 15°C, and heat degradation is minimized since the time-temperature integral is very small compared to that of the conventional retort process.

There is much interest also in the application of high hydrostatic pressure to foods at low or subzero temperters (23). Combinations of high pressure and subzero temperature offer various processing opportunities such as thawing, freezing, cold storage without freezing, and pressurization in the frozen state. The findings of a study by Picart et al. (171) indicate that all of these processes may have a beneficial effect on the safety of foods from microbes.

Pulsed Electric Fields

A comprehensive analysis of the use of PEFs to inactivate microorganisms has been published by Barbosa-Cánovas et al. (7). High-voltage discharges can kill microorganisms (104). An external PEF charges cells in such a way that they behave as small dipoles and an electric potential develops over the cell membranes. When the induced electric potential exceeds some critical value, a reversible increase in membrane permeability results. For vegetative bacteria, this critical electric field strength is about 15 kV/cm. When the critical electric field strength is greatly exceeded, the ion channels and pores of membranes are irreversibly extended to such a degree that cell contents leak out and cells die (dielectric rupture theory). Much greater field strength is required for inactivation of ascospores and endospores (114).

The critical electric field is affected by pulse duration and resistivity of the suspension material (221). Because electric resistivity decreases as ionic strength increases, the microbicidal effect of the PEF is generally inversely proportional to the ionic strength of the suspension material, whereas reduced pH increases the inactivation ratio by affecting the cytoplasm when poration is completed

(181, 215). Microbial inactivation is proportional to the number and duration of pulses. Gram-positive bacteria are less sensitive to electric pulse treatment than gram-negative bacteria, and yeast cells, being larger than bacterial cells, are more sensitive to PEF processing (93). The PEF acts synergistically with nisin in reducing the viable count of vegetative cells of *B. cereus* (177, 178). PEF inactivation is also a function of the treatment temperature and the microbial growth stage. Cells in the logarithmic growth phase are more sensitive than cells in stationary and lag phases. Increasing the medium temperature decreases the transmembrane potential of the cell membrane in addition to causing thermal injury effects, thus resulting in a higher inactivation ratio (181).

Over the last decade, high-voltage PEF treatment applied across electrodes between which liquid foods can be pumped has been investigated as a potential nonthermal preservation treatment (6, 9, 104). Test results show that bipolar square-wave pulses are the most efficient in terms of microbial inactivation for commercial PEF pasteurization of fluid foods. The application of many rapid high-power pulses can significantly increase the temperature of the product unless this generated heat is removed.

Fruit juices can be pasteurized by PEF treatment without significant loss of flavor, color, and vitamin C. Both electrical stimulation and pulsed electricity significantly reduce the viability of microorganisms on beef surfaces (10). PEF treatment has a less pronounced effect on dry solid powders than on liquids (109, 110). Peleg (166) proposed kinetic models for PEF inactivation of microorganisms, calculating the relationship between survival rate, field strength, and the number of pulses. A continuous PEF processing system suitable for industrial application has yet to be developed.

Oscillating Magnetic Fields

Magnetic fields cause a change in the orientations of biomolecules and biomembranes to a direction parallel or perpendicular to the applied magnetic field and induce a change in the ionic drift across the plasma membrane (89, 180). Thus, magnetic fields may affect growth and reproduction of microorganisms. As in high-voltage pulse treatment, the total exposure time is very short. Oscillating magnetic fields with an intensity (flux density) of 5 to 50 T with short pulses of 25 μs to 10 ms and frequencies of 5 to 500 kHz can inactivate vegetative microorganisms and pasteurize food material of high electrical resistivity, packaged in plastic packaging, without significant temperature increase (180). Microwave cooking also involves application of magnetic fields but differs in that it produces a thermal effect.

The explanation of the microbicidal effect of oscillating magnetic field pulses suggests that the oscillating magnetic field channels enough energy into the magnetoactive parts (magnetic dipoles) of large biological molecules to break bonds in DNA or proteins critical for survival and reproduction of microorganisms. It has been suggested that magnetic energy may change membrane fluidity and alter ion fluxes across cell membranes (180). Results of inactivation, however, are inconsistent (95). Further research is required to overcome technical difficulties, understand the mechanisms of microbial inactivation, and establish the commercial feasibility of this treatment.

Recent Developments in Calculating Kinetics of Microbial Inactivation by Nonthermal Food-Processing Technologies

In most inactivation studies, D and z values are still used, based on the assumptions that microbial destruction is exponential and that the logarithmic relationship between microbial inactivation and processing time is linear. However, recent reviews of the nonlinearity of semilogarithmic survivor curves indicate that first-order kinetics, which dominate current microbial kinetic modeling, may not be appropriate for many nonthermal processes.

Peleg and Cole (168) proposed that the survival curve is a cumulative form of the resistance distribution of the exposed microbial population. Thus, the semilogarithmic survival curve can be of any shape, and the linear curve is a special case of the Weibull distribution of resistances. For a microbial resistance curve that has a sigmoidal shape, Fermi's equation, which is used to describe mechanical changes of biomaterials at and around their glass transition temperature, may be a better model to quantify the relationship (166). This model has been successfully applied to microorganisms exposed to PEFs. For mathematical models, the Bigelow model, the Arrhenius equation, Fermi's equation, and Weibull frequency distribution have been compared (91). Weibull distributions predict high-pressure inactivation with more accuracy and the Fermi model provides a better description of gamma irradiation inactivation kinetics than the traditional Bigelow model. A statistical approach to the interpretation of various types of microbial survival curves and the exploration of the possibility of expressing their characteristics, not in the form of reaction kinetics, but as cumulative forms of distributions of lethal events was attempted also by Peleg (167).

FUTURE WORK NEEDED

Like the established physical preservation methods discussed earlier in this chapter, the nonthermal physical treatments discussed above induce sublethal injury to microbial cells surviving the treatments. Injury makes these cells sensitive to physical and chemical environments to which normal cells are resistant. Thus, both conventional and newly developed physical treatments can be used in combination with other preservative factors to increase antimicrobial efficiency and enhance safety and shelf lives of foods (5, 57, 118, 138, 165, 222). Various newer nonthermal physical treatments appear to be promising tools for food preservation. However, conflicting observations about their effects on bacterial spores and enzymes, as well as conflicting kinetic data, need to be reconciled and optimal processing parameters should be established (73). A mechanistic understanding of the effects of these processes on foods, food components, and microbial cells, as well as on storage-dependent changes after nonthermal treatments, is limited. It is essential to determine the pathogens of greatest public health concern for each alternative processing technology and establish target or surrogate microorganisms (72, 74, 131). Commercial application of new technologies and combinations of preservation methods will also raise theoretical questions, e.g., what constitutes an equivalent treatment in processes when microbial mortality patterns are different. Further research should concentrate on answering these questions.

References

1. **Ababouch, L., and F. F. Busta.** 1987. Effect of thermal treatments in oils on bacterial spore survival. *J. Appl. Bacteriol.* 62:491–502.

2. **Agar, I. T., J. Streif, and F. Bangerth.** 1997. Effect of high CO_2 and controlled atmosphere (CA) on the ascorbic and dehydroascorbic acid content of some berry fruits. *Postharvest Biol. Technol.* 11:47–55.

3. **Arsene, F., T. Tomoyasu, and B. Bukan.** 2000. The heat shock response of *Escherichia coli*. *Int. J. Food Microbiol.* 55:3–9.

4. **Auffray, Y., E. Lecesne, A. Hartky, and P. Boulibounes.** 1995. Basic features of the *Streptococcus thermophilus* heat shock response. *Curr. Microbiol.* 30:87–91.

5. **Aymerich, T. A., A. Jofré, M. Garriga, and M. Hugas.** 2005. Inhibition of *Listeria monocytogenes* and *Salmonella* by natural antimicrobials and high hydrostatic pressure in sliced cooked ham. *J. Food Prot.* 68:173–177.

6. **Barbosa-Cánovas, G. V., M. D. Pierson, Q. H. Zhang, and D. W. Schaffner.** 2000. Pulsed electric fields. *J. Food Sci.* 65(Suppl.):65–79.

7. **Barbosa-Cánovas, G. V., M. S. Tapia, M. P. Cano, O. Martin-Beloso, and A. Martinez.** 2005. *Novel Food Processing Technologies*. CRC Press, Boca Raton, Fla.

8. **Bari, M. L., M. Nakauma, S. Todoriki, V. K. Juneja, K. Isshiki, and S. Kawamoto.** 2005. Effectiveness of irradiation treatments in inactivating *Listeria monocytogenes* on fresh vegetables at refrigeration temperature. *J. Food Prot.* 68:318–323.

9. Barsotti, I., P. Merle, and K. Cheftel. 1999. Food processing by pulsed electric fields. (Part I and II). *Food Rev. Int.* 15(2):163–213.

10. Bawcom, D. W., L. D. Thompson, M. F. Miller, and C. B. Ramsey. 1995. Reduction of microorganisms on beef surfaces utilizing electricity. *J. Food Prot.* 58:35–38.

11. Bidawid, S., J. M. Farber, and S. A. Sattar. 2000. Inactivation of hepatitis A virus (HAV) in fruits and vegetables by gamma irradiation. *Int. J. Food Microbiol.* 57:91–97.

12. Bigelow, W. D. 1921. The logarithmic nature of thermal death time curves. *J. Infect. Dis.* 29:528–536.

13. Boutibonnes, P., V. Bisson, B. Thammavongs, A. Hartke, J. M. Panoff, A. Benackour, and Y. Auffray. 1995. Induction of thermotolerance by chemical agents in *Lactococcus lactis* subsp. *lactis* IL 1403. *Int. J. Food Microbiol.* 25:83–94.

14. Brown, A. D. 1974. Microbial water relations: features of the intracellular composition of sugar-tolerant yeasts. *J. Bacteriol.* 118:769–777.

15. Brown, A. D. 1976. Microbial water stress. *Bacteriol. Rev.* 40:803–846.

16. Brown, A. D., and J. R. Simpson. 1972. Water relations of sugar-tolerant yeasts: the role of intracellular polyols. *J. Gen. Microbiol.* 72:589–591.

17. Brown, K. L., and C. A. Ayres. 1982. Thermobacteriology of UHT processed foods. *Dev. Food Microbiol.* 1:119–152.

18. Buchanan, R. L., S. G. Edelson, and G. Boyd. 1999. Effects of pH and acid resistance on the radiation resistance of enterohemorrhagic *Escherichia coli*. *J. Food Prot.* 62:219–228.

19. Cabeza, M. C., M. L. Garcia, L. de la Hoz, I. Cambero, and J. A. Ordonez. 2005. Thermo-ultrasonication eliminates salmonellae from intact eggshells without changing the functional properties. *J. Food Sci.* 70:M292–M295.

20. Cabeza, M. C., J. A. Ordonez, I. Cambero, L. de la Hoz, and M. L. Garcia. 2004. Effect of thermoultrasonication on *Salmonella enterica* serovar Enteritidis in distilled water and in intact shell eggs. *J. Food Prot.* 67:1886–1891.

21. Chang, J. C. H., S. F. Ossoff, D. C. Lobe, M. H. Dorfman, C. M. Dumais, R. G. Qualls, and J. D. Johnson. 1985. UV inactivation of pathogenic and indicator microorganisms. *Appl. Environ. Microbiol.* 49:1361–1365.

22. Cheftel, J. C. 1992. Effects of high hydrostatic pressure on food constituents: an overview, p. 195–209. *In* C. Balny, R. Hayashi, K. Heremans, and P. Masson (ed.), *High Pressure and Biotechnology*. John Libbey and Co. Ltd., London, United Kingdom.

23. Cheftel, J. C., M. Thiebaud, and E. Dumay. 2003. High pressure-low temperature processing of foods: a review, p. 327–340. *In* R. Winter (ed.), *Advances in High Pressure Bioscience and Biotechnology*. Springer, Berlin, Germany.

24. Christian, J. H. B. 2000. Drying and reduction of water activity, p. 146–174. *In* B. M. Lund, T. C. Baird-Parker, and G. W. Gould (ed.), *The Microbiological Safety and Quality of Food*, vol. I. Aspen Publishers, Inc., Gaithersburg, Md.

25. Church, N. 1994. Developments in modified-atmosphere packaging and related technologies. *Trends Food Sci. Technol.* 5:345–352.

26. Clardy, S., D. M. Foley, F. Caporaso, M. L. Calicchia, and A. Prakash. 2002. Effect of gamma irradiation on *Listeria monocytogenes* in frozen artificially contaminated sandwiches. *J. Food Prot.* 65:1740–1744.

27. Clavero, M. R. S., J. D. Monk, L. R. Beuchat, M. P. Doyle, and R. E. Brackett. 1994. Inactivation of *Escherichia coli* O157:H7, salmonellae and *Campylobacter jejuni* in raw ground beef by gamma irradiation. *Appl. Environ. Microbiol.* 60:2069–2075.

28. Codex Alimentarius Commission. 2003. *Codex General Standard for Irradiated Foods*. CODEX STAN 106–1983, Rev. 1–2003. Codex Alimentarius Commission, Rome, Italy.

29. Cole, M. B., K. W. Davies, G. Munro, C. D. Holyoak, and D. C. Kilsby. 1993. A vitalistic model to describe the thermal inactivation of *Listeria monocytogenes*. *J. Ind. Microbiol.* 12:232–239.

30. Condon, S., and F. J. Sala. 1992. Heat resistance of *Bacillus subtilis* in buffer and foods of different pH. *J. Food Prot.* 55:605–608.

31. Corry, J. E. L. 1987. Relationship of water activity to fungal growth, p. 51–99. *In* L. R. Beuchat (ed.), *Food and Beverage Mycology*. AVI Van Nostrand Reinhold Co., New York, N.Y.

32. Daifas, D. P., J. P. Smith, B. Blanchfield, and J. W. Austin. 1999. Growth and toxin production by *Clostridium botulinum* in English style crumpets packaged under modified atmosphere. *J. Food Prot.* 62:349–355.

33. Davies, A. D. 1995. Advances in modified atmosphere packaging, p. 301–320. *In* G. W. Gould (ed.), *New Methods of Food Preservation*. Blackie Academic and Professional, Glasgow, United Kingdom.

34. Davies, R. 1976. The inactivation of vegetative bacterial cells by ionizing radiation, p. 239–255. *In* F. A. Skinner and W. B. Hugo (ed.), *Inhibition and Inactivation of Vegetative Microbes*. Academic Press, London, United Kingdom.

35. Deák, T. 1991. Preservation by dehydration. p. 70–82. *In* E. Szenes and M. Oláh (ed.), *Handbook of Canning Industry*. (In Hungarian.) Integra Projekt, Budapest, Hungary.

36. Decereau, R. V. 1985. *Microwaves in the Food Processing Industry*. Academic Press, Orlando, Fla.

37. Decereau, R. V. 1992. *Microwave Foods: New Product Development*. Food & Nutrition Press, Trumbull, Conn.

38. De Heij, W., R. Van der Berg, L. Van Schepdael, H. Hoogland, and H. Bijmolt. 2005. Sterilisation—only better. *New Food* 2005(2):56, 58–60, 61.

39. Diehl, J. F. 1995. *Safety of Irradiated Foods*, 2nd ed., revised and expanded. Marcel Dekker, Inc., New York, N.Y.

40. Doyle, M. P., and E. H. Marth. 1975. Thermal inactivation of conidia from *Aspergillus flavus* and *Aspergillus parasiticus*. *J. Milk Food Technol.* 38:678.

41. Doyle, M. P., A. S. Mazzotta, T. Wang, D. W. Wieseman, and V. N. Scott. 2001. Heat resistance of *Listeria monocytogenes*. *J. Food Prot.* 64:410–429.

42. Eckardt, C., and E. Ahrens. 1977. Untersuchungen über *Byssochlamys fulva* Olliver & Smith als potentiellen Verderbniserreger in Erdbeerkonserven. II. Hitzeresistenz der Ascosporen von *Byssochlamys fulva*. *Chem. Mikrobiol. Technol. Lebensm.* 5:76–80.

43. Edelson-Mammel, S. G., R. C. Whiting, S. G. Joseph, and R. L. Buchanan. 2005. Effect of prior growth conditions on the thermal inactivation of 13 strains of *Listeria monocytogenes* in two heating menstrua. *J. Food Prot.* **68:**168–172.

44. Edgey, M., and A. D. Brown. 1978. Response of xerotolerant and nontolerant yeasts to water stress. *J. Gen. Microbiol.* **104:**343–345.

45. Ellenberg, L., and D. G. Hoover. 1999. Injury and survival of *Aeromonas hydrophila* 7965 and *Yersinia enterocolitica* 9610 from high hydrostatic pressure. *J. Food Safety* **19:**263–276.

46. Erichsen, I., and G. Molin. 1981. Microbial flora of normal and high pH beef stored at 4°C in different gas environments. *J. Food Prot.* **44:**866.

47. Ernshaw, R. G., J. Appleyard, and R. M. Hurst. 1995. Understanding physical inactivation processes: combined preservation opportunities using heat, ultrasound and pressure. *Int. J. Food Microbiol.* **28:**197–219.

48. Farkas, D. F., and D. G. Hoover. 2000. High pressure processing. *J. Food Sci.* **65**(Suppl.):47S–64S.

49. Farkas, J. 1988. *Irradiation of Dry Food Ingredients.* CRC Press, Inc., Boca Raton, Fla.

50. Farkas, J. 1989. Microbiological safety of irradiated foods: a review. *Int. J. Food Microbiol.* **9:**1–15.

51. Farkas, J. 1990. Combination of irradiation with mild heat treatment. *Food Control* **1:**223–229.

52. Farkas, J. 1998. Irradiation as a method for decontaminating food. A review. *Int. J. Food Microbiol.* **44:**189–204.

53. Farkas, J. 2000. Food irradiation, p. 785–812. *In* A. Mozumder and Y. Hatano (ed.), *Charged Particle and Photon Interactions with Matter. Chemical, Physicochemical, and Biological Consequences with Applications.* Marcel Dekker, Inc., New York, N.Y.

54. Farkas, J. 2000. Elimination of foodborne pathogens by ionising radiation, p. 157–176. *In* F. J. H. Smulders and J. D. Collins (ed.), *Food Safety Assurance and Veterinary Public Health,* vol. 2. *Safety Assurance during Food Processing.* Wageningen Academic Publishers, Wageningen, The Netherlands.

55. Farkas, J. 2001. Radiation decontamination of spices, herbs, condiments, and other dried food ingredients, p. 291–312. *In* R. Molins (ed.), *Food Irradiation: Principles and Applications.* Wiley-Interscience, New York, N.Y.

56. Farkas, J., T. Sáray, C. Mohácsi-Farkas, K. Horti, and E. Andrássy. 1997. Effect of low dose gamma radiation on shelf life and microbiological safety of pre-cut/prepared vegetables. *Adv. Food Sci.* **19:**111–119.

57. Fenice, M., R. Di Giambattista, J.-L. Leuba, and F. Federici. 1999. Inactivation of *Mucor plumbeus* by combined actions of chitinase and high hydrostatic pressure. *Int. J. Food Microbiol.* **52:**109–113.

58. Flowers, R. S., and S. E. Martin. 1980. Ribosome assembly during recovery of heat-injured *Staphylococcus aureus.* *J. Bacteriol.* **141:**645–651.

59. Foong, S. C., G. L. Gonzalez, and J. S. Dickson. 2004. Reduction and survival of *Listeria monocytogenes* in ready-to-eat meats after irradiation. *J. Food Prot.* **67:**77–82.

60. Fryer, P. 1995. Electrical resistance heating of foods, p. 205–235. *In* G. W. Gould (ed.), *New Methods of Food Preservation.* Blackie Academic and Professional, Glasgow, United Kingdom.

61. Garcia, M. L., J. Burgos, B. Sanz, and J. A. Ordonez. 1989. Effect of heat and ultrasonic waves on the survival of two strains of *Bacillus subtilis.* *J. Appl. Bacteriol.* **67:**619–628.

62. Garcia-Graells, C., B. Masschalek, and C. W. Michiels. 1999. Inactivation of *Escherichia coli* in milk by high-hydrostatic-pressure treatment in combination with antimicrobial peptides. *J. Food Prot.* **62:**1248–1254.

63. George, A. M., W. A. Cramp, and M. B. Yatvin. 1980. The influence of membrane fluidity on radiation induced changes in the DNA of *E. coli* K1060. *Int. J. Radiat. Biol. Relat. Stud. Phys. Chem. Med.* **38:**427–438.

64. Geveke, D. J. 2005. UV inactivation of bacteria in apple cider. *J. Food Prot.* **68:**1739–1742.

65. Gill, C. O., and G. Molin. 1991. Modified atmospheres and vacuum packaging, p. 172–199. *In* N. J. Russell and G. W. Gould (ed.), *Food Preservatives.* Blackie, Glasgow, United Kingdom.

66. Goldblith, S. A., L. Rey, and W. W. Rothmayr. 1975. *Freeze Drying and Advanced Food Technology.* Academic Press, London, United Kingdom.

67. Gormley, T. R. (ed.). 1990. *Chilled Foods. The State of the Art.* Elsevier Applied Science, London, United Kingdom.

68. Gorris, L. G. M. 1994. Improvement of the safety and quality of refrigerated ready-to-eat foods using novel mild preservation techniques, p. 57–72. *In* R. P. Singh and F. A. R. Oliveira (ed.), *Minimal Processing of Foods and Process Optimization. An Interface.* CRC Press, Boca Raton, Fla.

69. Gould, G. W. 1973. Inactivation of spores in food by combined heat and hydrostatic pressure. *Acta Aliment.* **2:**377–383.

70. Gould, G. W. 1989. Drying, raised osmotic pressure and low water activity, p. 97–117. *In* G. W. Gould (ed.), *Mechanisms of Action of Food Preservation Procedures.* Elsevier Applied Science, London, United Kingdom.

71. Gould, G. W. 1989. Heat induced injury and inactivation, p. 11–42. *In* G. W. Gould (ed.), *Mechanisms of Action of Food Preservation Procedures.* Elsevier Applied Science, London, United Kingdom.

72. Gould, G. W. 2000. Strategies for food preservation, p. 19–35. *In* B. M. Lund, T. C. Baird-Parker, and G. W. Gould (ed.), *The Microbiological Safety and Quality of Food,* vol. 1. Aspen Publishers, Inc., Gaithersburg, Md.

73. Gould, G. W. 2000. New and emerging physical methods of preservation, p. 277–293. *In* B. M. Lund, T. C. Baird-Parker, and G. W. Gould (ed.), *The Microbiological Safety and Quality of Food,* vol. 1. Aspen Publishers, Inc., Gaithersburg, Md.

74. Gould, G. W. 2000. Emerging technologies in food preservation and processing in the last 40 years, p. 1–11. *In* G. V. Barbosa-Cánovas and G. W. Gould (ed.), *Innovations in Food Processing.* Technomic Publishing Co., Inc., Lancaster, Pa.

75. Graham, A. F., D. R. Mason, F. J. Maxwell, and M. W. Peck. 1997. Effect of pH and NaCl on growth from spores of non-proteolytic *Clostridium botulinum* at chill temperatures. *Lett. Appl. Microbiol.* **24:**95–100.

76. Grant, I. R., and M. F. Patterson. 1989. A novel radiation-resistant *Deinobacter* sp. isolated from irradiated pork. *Lett. Appl. Microbiol.* **8:**21–24.

77. **Grant, I. R., and M. F. Patterson.** 1995. Combined effect of gamma radiation and heating on the destruction of *Listeria monocytogenes* and *Salmonella typhimurium* in cook-chill roast beef and gravy. *Int. J. Food Microbiol.* **27:**117–128.

78. **Grecz, N., D. B. Rowley, and A. Matsuyama.** 1983. The action of radiation on bacteria and viruses. *In* E. S. Josephson and M. S. Peterson (ed.), *Preservation of Food by Ionizing Radiation*, vol. 2. CRC Press, Boca Raton, Fla.

79. **Guan, D., and D. G. Hoover.** 2005. Emerging decontamination techniques for meat, p. 388–417. *In* J. N. Sofos (ed.), *Improving the Safety of Fresh Meat*. CRC Press, Boca Raton, Fla.

80. **Guiavarc'h, Y., F. Zuber, A. Van Loey, and M. Hendrickx.** 2005. Combined use of two single-component enzymatic time-temperature integrators: application to industrial continuous rotary processing of canned ravioli. *J. Food Prot.* **68:**375–383.

81. **Guynot, M. E., S. Marin, V. Sanchis, and A. J. Ramos.** 2003. Modified atmosphere packaging for prevention of mold spoilage of bakery products with different pH and water activity levels. *J. Food Prot.* **66:**1864–1872.

82. **Han, Y. W.** 1975. Death rates of bacterial spores: nonlinear survivor curves. *Can. J. Microbiol.* **21:**1464–1467.

83. **Heinz, V., and O. Knorr.** 1998. High pressure germination and inactivation kinetics of bacterial spores. *In* N. S. Isaacs (ed.), *High Pressure Food Science, Bioscience and Chemistry*. Royal Society of Chemistry, Cambridge, United Kingdom.

84. **Hendrickx, M., L. Ludikhuyze, I. Van den Broeck, and C. Weermaes.** 1998. Effects of high pressure on enzymes related to food quality. *Trends Food Sci. Technol.* **9:**197–203.

85. **Herbert, R. A., and J. P. Sutherland.** 2000. Chill storage, p. 101–121. *In* B. M. Lund, T. C. Baird-Parker, and G. W. Gould (ed.), *The Microbiological Safety and Quality of Food*, vol. 1. Aspen Publishers, Inc., Gaithersburg, Md.

86. **Hiatt, C. W.** 1964. Kinetics of the inactivation of viruses. *Bacteriol. Rev.* **28:**150–163.

87. **Hightower, L. E.** 1991. Heat shock, stress proteins, chaperones, and proteotoxicity. *Cell* **66:**191–197.

88. **Ho, Y. C., and K. L. Yam.** 1993. Effect of metal shielding on microwave heating uniformity of a cylindrical food model. *J. Food Process. Preserv.* **16:**337–359.

89. **Hofmann, G. A.** 1985. Inactivation of microorganisms by an oscillating magnetic field. U.S. patent 4,524,079.

90. **Hollywood, N. W., Y. Varabioff, and G. E. Mitchell.** 1991. The effect of microwave and conventional cooking on the temperature profiles and microbial flora of minced beef. *Int. J. Food Microbiol.* **14:**67–76.

91. **Hu, X., P. Maltikarjunan, J. Koo, L. S. Andrews, and M. L. Jahncke.** 2005. Comparison of kinetic models to describe high pressure and gamma irradiation used to inactivate *Vibrio vulnificus* and *Vibrio parahaemolyticus* prepared in buffer solution and whole oysters. *J. Food Prot.* **68:**292–295.

92. **Hugas, M., M. Garriga, and J. M. Monfort.** 2002. New mild technologies in meat processing: high pressure as a model. *Meat Sci.* **3:**359–371.

93. **Hülsheger, H., J. Potel, and E. G. Neumann.** 1983. Electric field effects on bacteria and yeast cells. *Radiat. Environ. Biophys.* **22:**149–162.

94. **Hurst, A.** 1984. Reversible heat damage, p. 303–318. *In* A. Hurst and A. Nasim (ed.), *Repairable Lesions in Microorganisms*. Academic Press, London, United Kingdom.

95. **Institute of Food Technologists.** 2000. Kinetics of microbial inactivation for alternative food processing technologies. A report of the Institute of Food Technologists for the Food and Drug Administration. *J. Food Sci.* **65:**1S–108S.

96. **Institute of Food Technologists.** 20 Feb. 2002, posting data. *IFT Expert Report on Emerging Microbiological Food Safety Issues: Implications for Control in the 21st Century.* [Online.] Institute of Food Technologists, Chicago, Ill. http://members.ift.org/IFT/Research/IFTExpertReports/microsfs_report.htm.

97. **International Commission on Microbiological Specifications for Foods.** 1980. Temperature, p. 1–37. *In* J. H. Silliker et al. (ed.), *Microbial Ecology of Foods*, vol. 1. *Factors Affecting Life and Death of Microorganisms*. Academic Press, New York, N.Y.

98. **International Commission on Microbiological Specifications for Foods.** 1996. *Microorganisms in Foods. 5. Microbiological Specifications of Food Pathogens.* Blackie Academic and Professional, London, United Kingdom.

99. **Ingram, M., and B. M. Mackey.** 1976. Inactivation by cold, p. 111–151. *In* F. A. Skinner and W. B. Hugo (ed.), *Society of Applied Bacteriology Symposia Series*, no. 5. *Inhibition and Inactivation of Vegetative Microbes*. Academic Press, New York, N.Y.

100. **Isaacs, N. S.** 1998. *High Pressure Food Science, Bioscience and Chemistry.* Royal Society of Chemistry, Cambridge, United Kingdom.

101. **Jakobsen, M., O. Filtenborg, and F. Bramsnaes.** 1972. Germination and outgrowth of the bacterial spore in the presence of different solutes. *Lebensm. Wiss. Technol.* **5:**159–162.

102. **Jay, J. M.** 1992. *Modern Food Microbiology*, 4th ed. Chapman & Hall, New York, N.Y.

103. **Jermini, M. F. G., and W. Schmidt-Lorenz.** 1987. Heat resistance of vegetative cells and asci of two *Zygosaccharomyces* yeasts in broth at different water activity values. *J. Food Prot.* **50:**835–841.

104. **Jeyamkondan, S., D. S. Jayas, and R. A. Holley.** 1999. Pulsed electric field processing of foods: a review. *J. Food Prot.* **62:**1088–1096.

105. **Josephson, E. S., and M. S. Peterson (ed.).** 1983. *Preservation of Foods by Ionizing Radiation*, vol. 1–3. CRC Press, Boca Raton, Fla.

106. **Juneja, V. K.** 2003. *Sous-vide* processed foods: safety hazards and control of microbial risk, p. 97–124. *In* J. S. Novak and V. K. Juneja (ed.), *Microbial Safety of Minimally Processed Foods*. CRC Press, Boca Raton, Fla.

107. **Juneja, V. K., P. G. Klein, and B. S. Marmer.** 1998. Heat shock and thermotolerance of *Escherichia coli* O157:H7 in a model beef gravy system and ground beef. *J. Appl. Microbiol.* **84:**677–684.

108. **Kang, C. K., M. Woodburn, A. Pagenkopf, and R. Cheney.** 1969. Growth, sporulation and germination of *Clostridium perfringens* in media of controlled water activity. *Appl. Microbiol.* **18:**798–805.

109. Keith, W. D., L. J. Harris, and M. W. Griffiths. 1997. Reduction of bacterial levels in flour by pulsed electric fields. *J. Food Process Eng.* **21**:263–269.

110. Keith, W. D., L. J. Harris, L. Hudson, and M. W. Griffiths. 1997. Pulsed electric fields as a processing alternative for microbial reduction in spice. *Food Res. Int.* **30**:185–191.

111. Keteleer, A., and P. P. Tobback. 1994. Modified atmosphere storage of respiring product, p. 59–64. *In* L. Leistner and L. G. M. Gorris (ed.), *Food Preservation by Combined Processes*. Final Report FLAIR Concerted Action No. 7, Subgroup B. EUR 15776 EN. European Commission, The Netherlands.

112. Khattak, M. K., N. Bibi, A. B. Khattak, and M. A. Chaudry. 2005. Effect of irradiation on microbial safety and nutritional quality of minimally processed bitter gourd (*Momordica charanta*). *J. Food Sci.* **70**:M255–M259.

113. Kingsley, D. H., D. Guan, and D. G. Hoover. 2005. Pressure inactivation of hepatitis A virus in strawberry puree and sliced green onions. *J. Food Prot.* **68**:1748–1751.

114. Knorr, D. 1994. Non-thermal processes for food preservation, p. 3–15. *In* R. P. Singh and F. A. R. Oliveira (ed.), *Minimal Processing of Foods and Process Optimization. An Interface.* CRC Press, Boca Raton, Fla.

115. Knorr, D. 2000. Process aspects of high-pressure treatment of food systems, p. 13–30. *In* G. V. Barbosa-Cánovas and G. W. Gould (ed.), *Innovations in Food Processing*. Technomic Publishing Co., Inc., Lancaster, Pa.

116. Laszlo, A. 1988. Evidence for two states of thermotolerance in mammalian cells. *Int. J. Hyperthermia* **4**:513–526.

117. Le Jean, G., G. Abraham, E. Debray, Y. Candau, and G. Piar. 1994. Kinetics of thermal destruction of *Bacillus stearothermophilus* spores using a two reaction model. *Food Microbiol.* **11**:229–241.

118. Leistner, L. 1995. Principles and application of hurdle technology, p. 1–21. *In* G. W. Gould (ed.), *New Methods of Food Preservation*. Blackie Academic and Professional, Glasgow, United Kingdom.

119. Leistner, L., and N. J. Russell. 1991. Solutes and low water activity, p. 111–134. *In* N. J. Russell and G. W. Gould (ed.), *Food Preservatives*. Blackie Academic and Professional, Glasgow, United Kingdom.

120. Lewis, M. 1993. UHT processing: safety and quality aspects. *Food Technol. Int. Eur.* **1993**:47–51.

121. Linton, M., J. M. J. McClements, and M. F. Patterson. 1999. Survival of food *Escherichia coli* O157:H7 during storage in pressure-treated orange juice. *J. Food Prot.* **62**:1038–1040.

122. Loaharanu, P., and P. Thomas (ed.). 2001. *Irradiation for Food Safety and Quality*. Technomic Publishing Co., Inc., Lancaster, Pa.

123. López-Caballero, M. E., J. Carballo, and F. Jiménez-Colmenero. 1999. Microbiological changes in pressurized, prepackaged sliced cooked ham. *J. Food Prot.* **62**:1411–1415.

124. Luechapattanaporn, K., Y. Wang, J. Wang, J. Tang, L. M. Hallberg, and C. P. Dunne. 2005. Sterilization of scrambled eggs in military polymeric trays by radiofrequency energy. *J. Food Sci.* **70**:288–294.

125. Lund, B. M. 2000. Freezing, p. 122–145. *In* B. M. Lund, T. C. Baird-Parker, and G. W. Gould (ed.), *The Microbiological Safety and Quality of Food*, vol. 1. Aspen Publishers, Inc., Gaithersburg, Md.

126. Mackey, B. M., and C. M. Derrick. 1986. Peroxide sensitivity of cold-shocked *Salmonella typhimurium* and *Escherichia coli* and its relationship to minimal medium recovery. *J. Appl. Bacteriol.* **60**:501–511.

127. Mackey, B. M., and C. M. Derrick. 1986. Elevation of heat resistance of *Salmonella typhimurium* by sublethal heat shock. *J. Appl. Bacteriol.* **61**:389–394.

128. Mackey, B. M., and C. M. Derrick. 1987. Changes in the heat resistance of *Salmonella typhimurium* during heating at rising temperatures. *Lett. Appl. Microbiol.* **4**:3–16.

129. Mackey, B. M., K. Forestiere, N. S. Isaacs, R. Stenning, and B. Brooker. 1994. The effect of high hydrostatic pressure on *Salmonella thompson* and *Listeria monocytogenes* examined by electron microscopy. *Lett. Appl. Microbiol.* **19**:429–432.

130. Macleod, R. A., and P. H. Calcott. 1976. Cold shock and freezing damage to microbes, p. 81–109. *In* T. R. G. Gray and J. R. Postgate (ed.), *The Survival of Vegetative Microbes. Twenty-Sixth Symposium of the Society for General Microbiology.* Cambridge University Press, Cambridge, United Kingdom.

131. Margosch, D., M. A. Ehrmann, M. G. Gaenzle, and R. V. Vogel. 2005. Comparison of pressure and heat resistance of *Clostridium botulinum* and other endospores in mashed carrots. *J. Food Prot.* **67**:2530–2537.

132. Margosch, D., M. Moravek, M. G. Gaenzle, E. Maertlbauer, R. F. Vogel, and A. Ehrmann. 2005. Effect of high pressure and heat on bacterial toxins. *Food Technol. Biotechnol.* **43**:211–217.

133. Martens, T. (ed.). 1999. *Harmonization of Safety Criteria for Minimally Processed Foods. Rational and Harmonization Report.* FAIR Concerted Action FAIR CT96-1020. Alma University Restaurants, Leuven, Belgium.

134. Marth, E. H. 1998. Extended shelf life refrigerated foods: microbiological quality and safety. *Food Technol.* **52**(12):57–62.

135. McKellar, R. C., and X. Lu. 2004. Primary models, p. 21–62. *In* R. C. McKellar and X. Lu (ed.), *Modeling Microbial Responses in Food*. CRC Press, Boca Raton, Fla.

136. Metrick, C., D. Hoover, and D. Farkas. 1989. Effects of high hydrostatic pressure on heat-resistant and heat-sensitive strains of *Salmonella*. *J. Food Sci.* **54**:1547–1549.

137. Meyer, R. S., K. L. Cooper, D. Knorr, and H. L. Lelifeld. 2000. High-pressure sterilization of foods. *Food Technol.* **54**(11):67–72.

138. Mohácsi-Farkas, C., G. Kiskó, J. Farkas, L. Mészáros, and T. Sáray. 1999. Assessment of antibacterial effects of essential oils by automated impedimetry and preliminary studies on their utility as biopreservatives, p. 279–281. *In* A. C. J. Tuijtelaars, R. A. Samson, F. M. Rombouts, and S. Notermans (ed.), *Food Microbiology and Food Safety into the Next Millennium*. Foundation Food Micro '99, Zeist, The Netherlands.

139. Molin, G. 2000. Modified atmosphere, p. 214–234. *In* B. M. Lund, T. C. Baird-Parker, and G. W. Gould (ed.), *The Microbiological Safety and Quality of Food*, vol. 1. Aspen Publishers, Inc., Gaithersburg, Md.

140. Molins, R. (ed.). 2001. *Food Irradiation: Principles and Applications*. Wiley-Interscience, New York, N.Y.

141. Molins, R. 2001. Global status of food irradiation in 2000, p. 443–455. *In* R. Molins (ed.), *Food Irradiation: Principles and Applications*. Wiley-Interscience, New York, N.Y.

142. Monk, D. J., L. R. Beuchat, and M. P. Doyle. 1995. Irradiation inactivation of food-borne microorganisms. *J. Food Prot.* 58:197–208.

143. Montville, T. J., and G. M. Sapers. 1981. Thermal resistance of spores from pH elevating strains of *Bacillus licheniformis*. *J. Food Sci.* 46:1710–1712.

144. Moreno, M. A., M. del Carmen Ramos, A. Gonzalez, and G. Suarez. 1987. Effect of ultraviolet light irradiation on viability and aflatoxin production by *Aspergillus parasiticus*. *Can. J. Microbiol.* 33:927–929.

145. Moseley, B. E. B. 1989. Ionizing radiation: action and repair, p. 43–70. *In* G. W. Gould (ed.), *Mechanisms of Action of Food Preservation Procedures*. Elsevier Applied Science, London, United Kingdom.

146. Mossel, D. A. A. 1975. Occurrence, prevention and monitoring of microbial quality loss of foods and dairy products. *Crit. Rev. Environ. Control* 5:1–139.

147. Mossel, D. A. A., and M. Ingram. 1955. The physiology of the microbial spoilage of foods. *J. Appl. Bacteriol.* 18:232–268.

148. Mossel, D. A. A., and C. B. Struijk. 1991. Public health implications of refrigerated pasteurized ("sous vide") foods. *Int. J. Food Microbiol.* 13:187–206.

149. Mudgett, R. E. 1985. Dielectrical properties of foods, p. 15–37. *In* R. V. Decareau (ed.), *Microwaves in the Food Processing Industry*. Academic Press, Orlando, Fla.

150. Mui, W. W. Y., T. D. Durance, and C. H. Seaman. 2002. Flavor and texture of banana chips dried by combination of hot air, vacuum, and microwave processing. *J. Agric. Food Chem.* 50:1883–1889.

151. Mulet, A., M. Blasco, J. Garcia-Reverter, and J. V. Garcia-Perez. 2005. Drying kinetics of *Curcuma longa* rhizomes. *J. Food Sci.* 70:318–323.

152. Mullin, J. 1995. Microwave processing, p. 112–134. *In* G. W. Gould (ed.), *New Methods of Food Preservation*. Blackie Academic and Professional, Glasgow, United Kingdom.

153. Narasimhan, R., M. M. Habibullah-Khan, J. Ernest, and K. Thangavel. 1989. Effect of ultraviolet radiation on the bacterial flora of the packaging materials of milk and milk products. *Cherion* 18(2):89–92.

154. Notermans, S., J. Dufrenne, and B. M. Lund. 1990. Botulism risk of refrigerated processed foods of extended durability. *J. Food Prot.* 53:1020–1024.

155. Nychas, G. J. E. 1994. Modified atmosphere packaging of meats, p. 417–436. *In* R. P. Singh and F. A. R. Oliveira (ed.), *Minimal Processing of Foods and Process Optimization. An Interface*. CRC Press, Boca Raton, Fla.

156. Nychas, G. J. E., and J. S. Arkoudelos. 1990. Microbiological and physico-chemical changes in minced meat under carbon dioxide, nitrogen or air at 3°C. *Int. J. Food Sci. Technol.* 25:389.

157. Ocio, M. J., T. Sanchez, P. S. Fernandez, M. Rodrigo, and A. Martinez. 1994. Thermal resistance characteristics of PA3679 in the temperature range of 110–121°C as affected by pH, type of acidulant and substrate. *Int. J. Food Microbiol.* 22:239–247.

158. Ohlsson, T. 2000. Minimal processing of foods with thermal methods, p. 141–148. *In* G. V. Barbosa-Cánovas and G. W. Gould (ed.), *Innovations in Food Processing*. Technomic Publishing Co., Inc., Lancaster, Pa.

159. Ohye, D. F., and J. H. B. Christian. 1967. Combined effects of temperature, pH and water activity on growth and toxin production by *C. botulinum* types A, B and E, p. 217–223. *In* M. Ingram and T. A. Roberts (ed.), *Botulism, 1966*. Chapman & Hall, London, United Kingdom.

160. Oliveira, M. E. C., and A. S. Franca. 2002. Microwave heating of foodstuffs. *J. Food Eng.* 53:347–349.

161. Olson, D. G. 1998. Irradiation of food. *Food Technol.* 52(1):56–62.

162. Paidhungat, M., B. Setlow, W. B. Daniels, D. Hoover, E. Papfragha, and P. Setlow. 2002. Mechanism of induction of germination of *Bacillus subtilis* spores by high pressure. *Appl. Environ. Microbiol.* 68:3172–3175.

163. Palaniappan, S., S. K. Sastry, and E. R. Richter. 1992. Effects of electroconductive heat treatment and electrical pretreatment on thermal death kinetics of selected microorganisms. *Biotechnol. Bioeng.* 39:225–232.

164. Parsell, D. A., and R. T. Sauer. 1989. Induction of heat shock-like response by unfolded protein in *Escherichia coli*: dependence on protein level not protein degradation. *Genes Dev.* 3:1226–1232.

165. Patterson, M. 2001. Combination treatments involving food irradiation, p. 313–327. *In* R. Molins (ed.), *Food Irradition: Principles and Applications*. Wiley-Interscience, New York, N.Y.

166. Peleg, M. 1995. A model of microbial survival after exposure to pulsed electric fields. *J. Sci. Food Agric.* 67:93–99.

167. Peleg, M. 2000. Modeling and simulating microbial survival in foods subjected to a combination of preservation methods, p. 163–181. *In* G. V. Barbosa-Cánovas and G. W. Gould (ed.), *Innovations in Food Processing*. Technomic Publishing Co., Inc., Lancaster, Pa.

168. Peleg, M., and M. B. Cole. 1998. Reinterpretation of microbial survival curves. *Crit. Rev. Food Sci. Nutr.* 38:353–380.

169. Perkin, A. G., F. L. Davies, P. Neaves, B. Jarvis, C. A. Ayres, K. L. Brown, W. C. Falloon, H. Dallyn, and P. G. Bean. 1980. Determination of bacterial spore inactivation at high temperatures, p. 173–188. *In* G. W. Gould and J. E. L. Corry (ed.), *Microbial Growth and Survival in Extreme Environments*. Academic Press, London, United Kingdom.

170. Pflug, I., E. Dumay, J.-P. Guiraud, and J. C. Cheftel. 2005. Heat treatment, p. 36–64. *In* B. M. Lund, T. C. Baird-Parker, and G. W. Gould (ed.), *The Microbiological Safety and Quality of Food*, vol. 1. Aspen Publishers, Inc., Gaithersburg, Md.

171. Picart, L., E. Dumay, J.-P. Guiraud, and J. C. Cheftel. 2005. Combined pressure-sub-zero temperature processing of smoked salmon mince: phase transition phenomena and inactivation of *Listeria innocua*. *J. Food Eng.* **68**:43–56.

172. Pirchardt, K. 1993. *Lebensmittelmikrobiologie*, 3rd ed. Springer-Verlag, Berlin, Germany.

173. Pitt, J. I. 1975. Xerophilic fungi and the spoilage of foods of plant origin, p. 273–307. *In* R. B. Duckworth (ed.), *Water Relations of Foods*. Academic Press, London, United Kingdom.

174. Pitt, J. I., and A. D. Hocking. 1997. *Fungi and Food Spoilage*. Blackie Academic and Professional, London, United Kingdom.

175. Pitt, J. I., and B. F. Miscamble. 1995. Water relations of *Aspergillus flavus* and closely related species. *J. Food Prot.* **58**:86–90.

176. Piyasena, P. E., E. Mohareb, and R. C. McKellar. 2003. Inactivation of microbes using ultrasound: a review. *Int. J. Food Microbiol.* **87**:207–216.

177. Pol, I. E., H. C. Mastwijk, P. V. Bartels, and E. J. Smid. 1999. Inactivation of *Bacillus cereus* by nisin combined with pulsed electric field, p. 287–290. *In* A. C. J. Tuijtelaars, R. A. Samson, F. M. Rombouts, and S. Notermans (ed.), *Food Microbiology and Food Safety into the Next Millennium*. Foundation Food Micro '99, Zeist, The Netherlands.

178. Ponne, C. T., and P. V. Bartels. 1995. Interaction of electromagnetic energy with biological material—relation to food processing. *Radiat. Phys. Chem.* **45**:591–607.

179. Popovich, D. G., C. Hu, T. D. Durance, and D. D. Kitts. 2005. Retention of ginsenoides in dried ginseng root: comparison of drying methods. *J. Food Sci.* **70**:355–358.

180. Pothakamury, U. R., U. R. Monsalve-Gonzalea, G. V. Barbosa-Canovas, and B. G. Swanson. 1993. Magnetic field inactivation of microorganisms and generation of biological changes. *Food Technol.* **47**(12):85–93.

181. Pothakmury, U. R., H. Vega, Q. Zhang, G. V. Barbosa-Canovas, and B Swanson. 1996. Effect of growth stage and processing temperature on the inactivation of *E. coli* by pulsed electric fields. *J. Food Prot.* **59**:1167–1191.

182. Potter, N. N. 1986. *Food Science*, 4th ed. AVI-Van Nostrand Reinhold, New York, N.Y.

183. Put, H. M. C., J. de Jong, F. E. M. Sand, and A. M. van Grinsven. 1976. Heat resistance studies on yeast spp. causing spoilage in soft drinks. *J. Appl. Bacteriol.* **40**:135–152.

184. Rapenau, G., A. Vanloey, C. Smout, and M. Hendrickx. 2005. Effect of pH on thermal and/or pressure inactivation of Victoria grape (*Vitis vinifera sativa*) polyphenol oxidase: a kinetic study. *J. Food Sci.* **70**:301–307.

185. Raso, J., S. Condon, and F. J. Sala Trepat. 1994. Mano-Thermo-Sonication: a new method of food preservation, p. 37–41. *In* L. Leistner and L. G. M. Gorris (ed.), *Food Preservation by Combined Process*. Final Report FLAIR Concerted Action No. 7, Subgroup B. EUR 15776 EN. European Commission, The Netherlands.

186. Reddy, N. R., H. M. Solomon, G. A. Fingerhut, E. J. Rhodehamel, V. M. Balasubramaniam, and P. Palaniappan. 1999. Inactivation of *Clostridium*

187. *botulinum* type E spores by high pressure processing. *J. Food Safety* **19**:277–288.

188. Reichart, O., and C. Mohácsi-Farkas. 1994. Mathematical modeling of the combined effect of water activity, pH and redox potential on the heat destruction. *Int. J. Food Microbiol.* **24**:103–112.

188. Ritz, M., J. L. Tolozan, M. Federighi, and M. F. Pilet. 2002. Physiological damages of *Listeria monocytogenes* treated by high hydrostatic pressure. *Int. J. Food Microbiol.* **79**:47–53.

189. Roberts, T. A., G. Hobbs, J. H. B. Christian, and N. Skovgaard. 1981. *Psychrotrophic Microorganisms in Spoilage and Pathogenicity*. Academic Press, London, United Kingdom.

190. Rockland, L. B., and L. R. Beuchat (ed.). 1987. *Water Activity: Theory and Applications to Food*. Marcel Dekker Inc., New York, N.Y.

191. Rose, D. 1995. Advances and potential for aseptic processing, p. 283–303. *In* G. W. Gould (ed.), *New Methods of Food Preservation*. Blackie Academic and Professional, Glasgow, United Kingdom.

192. Rosenberg, U., and W. Bogl. 1987. Microwave pasteurization, sterilization, blanching and pest control in the food industry. *Food Technol.* **6**:92–99.

193. Sala, F. J., J. Burgos, S. Condon, P. Lopez, and J. Raso. 1995. Effect of heat and ultrasound on microorganisms and enzymes, p. 176–204. *In* G. W. Gould (ed.), *New Methods of Food Preservation*. Blackie Academic and Professional, Glasgow, United Kingdom.

194. Saleh, Y. G., M. S. Mayo, and D. G. Ahearn. 1988. Resistance of some common fungi to gamma irradiation. *Appl. Environ. Microbiol.* **54**:2134–2135.

195. San Martin, M. F., G. V. Barbosa-Cánovas, and B. G. Swanson. 2002. Food processing by high hydrostatic pressure. *Crit. Rev. Food Sci. Nutr.* **42**:627–645.

196. Sapru, V., A. A. Teixeira, G. H. Smerage, and J. A. Lindsay. 1992. Predicting thermophilic spore population dynamics for U. H. T. sterilization processes. *J. Food Sci.* **57**:1248–1257.

197. Sastry, S. K., A. Yousef, H. Y. Cho, S. R. Unal, S. Salenghe, W. C. Wang, M. Lima, S. Kulshrstha, P. Wongsa-Ngasri, and I. Senoy. 2002. Ohmic heating and moderate electric field (MEF) processing, p. 785–793. *In* J. Welti-Chanes, G. V. Barbosa-Cánovas, and J. M. Aguilera (ed.), *Engineering and Food for the 21st Century*. CRC Press, Boca Raton, Fla.

198. Seyderhelm, I., and D. Knorr. 1992. Reduction of *Bacillus stearothermophilus* spores by combined high pressure and temperature treatments. *ZFL International Zeitschrift für Lebensmittel-Technik, Marketing, Verpackung und Analytik, EFS* **43**(4):17–20.

199. Shearer, A. E. H., C. P. Dunne, A. Sikes, and D. G. Hoover. 2000. Bacterial spore inhibition and inactivation in foods by pressure, chemical preservatives and mild heat. *J. Food Prot.* **63**:1503–1510.

200. Sizer, C. E., and V. M. Balsubramaniam. 1999. New intervention processes for minimally processed juices. *Food Technol.* **53**(10):64–67.

201. Splittstoesser, D. F., S. B. Leasor, and K. M. J. Swanson. 1986. Effect of food composition on the heat

resistance of yeast ascospores. *J. Food Sci.* **51**:1265–1267.

202. **Splittstoesser, D. F., and C. M. Splittstoesser.** 1977. Ascospores of *Byssochlamys fulva* compared with those of a heat resistant *Aspergillus. J. Food Sci.* **42**:685–688.

203. **Stannard, C. J., J. S. Abbiss, and J. M. Wood.** 1983. Combined treatment with hydrogen peroxide and ultra-violet irradiation to reduce microbial contamination levels in preformed packaging cartons. *J. Food Prot.* **46**:1060–1064.

204. **Stephens, P. J., M. B. Cole, and M. V. Jones.** 1994. Effect of heating rate on the thermal inactivation of *Listeria monocytogenes. J. Appl. Bacteriol.* **77**:702–708.

205. **Stoforos, N. G., J. Noronha, M. Hendrickx, and P. Tobback.** 1997. A critical analysis of mathematical procedures for the evaluation and design of in-container thermal processes for foods. *Crit. Rev. Food Sci. Nutr.* **37**:411–441.

206. **Stumbo, C. R., K. S. Purohit, T. V. Ramakrishnan, D. A. Evans, and F. J. Francis.** 1983. *Handbook of Lethality Guides for Low-Acid Canned Foods*, vol. 1 and 2. CRC Press, Boca Raton, Fla.

207. **Tartera, C., A. Bosch, and J. Jofre.** 1988. The inactivation of bacteriophages infecting *Bacteroides fragilis* by chlorine treatment and UV-irradiation. *FEMS Microbiol. Lett.* **56**:313–316.

208. **Tempest, P.** 1996. *Electroheat Technologies for Food Processing. Bulletin of APV Processed Food Sector.* Crawley, West Sussex, United Kingdom.

209. **Thayer, D. W., and G. Boyd.** 1999. Irradiation and modified atmosphere packaging for the control of *Listeria monocytogenes* on turkey meat. *J. Food Prot.* **62**:1136–1142.

210. **Tomlins, R. I., and Z. J. Ordal.** 1976. Thermal injury and inactivation in vegetative bacteria, p. 153–190. *In* F. A. Skinner and W. B. Hugo (ed.), Society of Applied Bacteriology Symposium Series, no. 5. *Inhibition and Inactivation of Vegetative Microbes.* Academic Press, New York, N.Y.

211. **Troller, J. A.** 1983. Effect of low moisture environments on the microbial stability of foods, p. 173–198. *In* A. H. Rose (ed.), *Economic Microbiology*, vol. 8. Academic Press, London, United Kingdom.

212. **Urbain, W. M.** 1986. *Food Irradiation.* Academic Press, Orlando, Fla.

213. **U.S. Food and Drug Administration.** 2000. Irradiation in the production, processing and handling of food. *Fed. Regist.* **65**:71056–71058.

214. **Vas, K., and G. Proszt.** 1957. Observations on the heat destruction of spores of *Bacillus cereus. J. Appl Bacteriol.* **21**:431–441.

215. **Vega-Mercado, H., U. P. Pothakamury, F. J. Chang, G. V. Barbosa-Canovas, and G. G. Swanson.** 1996. Inactivation of *Escherichia coli* by combining pH, ionic strength, and pulsed electric fields hurdles. *Food Res. Int.* **29**: 119–199.

216. **Wang, Y., T. D. Wing, J. Tang, and I. M Hallberg.** 2003. Sterilization of foodstuffs using radiofrequency heating. *J. Food Sci.* **68**:S39–S44.

217. **Wilkinson, V. M., and G. W. Gould.** 1996. *Food Irradiation. A Reference Guide.* Butterworth-Heinemann, Oxford, United Kingdom.

218. **World Health Organization.** 1994. *Safety and Nutritional Adequacy of Irradiated Food.* World Health Organization, Geneva, Switzerland.

219. **World Health Organization.** 1999. *High-Dose Irradiation: Wholesomeness of Food Irradiated with Doses above 10 kGy.* World Health Organization, Geneva, Switzerland.

220. **Yousef, A. E., and V. K. Juneja.** 2003. *Microbial Stress Adaptation and Food Safety.* CRC Press, Boca Raton, Fla.

221. **Zhang, Q., B.-L. Quin, G. V. Barbosa-Cánovas, and B. G. Swanson.** 1994. Growth stage and temperature affect the inactivation of *E. coli* by pulsed electric fields, p. 104. *In Proceedings of the Canadian Institute of Food Science Technology 37th Annual Conference.* Vancouver, British Columbia, Canada.

222. **Zhang, Y., and G. S. Mittal.** 2005. Inactivation of spoilage microorganisms in mango juice using low energy pulsed electric field in combination with antimicrobials. *Ital. J. Food Sci.* **17**:167–176.

223. **Zhuang, H., M. M. Barth, and T. R. Hankinson.** 2003. Microbial safety, quality, and sensory aspects of fresh-cut fruits and vegetables, p. 255–278. *In* J. S. Novak, G. M. Sapers, and V. K. Juneja (ed.), *Microbial Safety of Minimally Processed Foods.* CRC Press, Boca Raton, Fla.

224. **Zoltai, P., and P. Swearingen.** 1996. Product development consideration for ohmic heating. *Food Technol.* **50**(5):263.

225. **Zook, C. D., M. E. Parish, R. J. Braddock, and M. O. Balaban.** 1999. High pressure inactivation kinetics of *Saccharomyces cerevisiae* ascospores in orange and apple juices. *J. Food Sci.* **64**:533–553.

Food Microbiology: Fundamentals and Frontiers, 3rd Ed.
Edited by M. P. Doyle and L. R. Beuchat
© 2007 ASM Press, Washington, D.C.

P. Michael Davidson
T. Matthew Taylor

33

Chemical Preservatives and Natural Antimicrobial Compounds

The overall quality of food products diminishes from the time of harvest or slaughter until they are consumed. Quality loss may be due to microbiological, enzymatic, chemical, or physical changes. The consequences of microbiological changes include hazards to consumers because of the presence of microbial toxins or pathogenic microorganisms or economic losses through growth of spoilage microorganisms and resultant off-odors, off-flavors, texture problems, discoloration, slime, or haze. Food preservation technologies, some in use since ancient times, protect foods from the effects of microorganisms and inherent deterioration. Microorganisms may be inhibited or inactivated by heat, chilling, freezing, water activity reduction, nutrient restriction, acidification, modification of packaging atmosphere, fermentation, or nonthermal treatments (e.g., high pressure) or through addition of antimicrobial compounds.

Food antimicrobials are chemical compounds added to or present in foods that retard growth of or kill microorganisms. Most food antimicrobials are bacteriostatic or fungistatic at use concentrations and not bactericidal or fungicidal. Therefore, food antimicrobials do not preserve a food indefinitely. Food antimicrobials are often used in combination with other food preservation procedures. Targets for antimicrobials are pathogenic and spoilage microorganisms. Food antimicrobials are sometimes referred to as "preservatives." However, the term "preservatives" often includes food additives that function as antibrowning agents (e.g., citric acid) and antioxidants (e.g., butylated hydroxyanisole).

In this chapter, antimicrobial compounds are divided into two classes: traditional and naturally occurring. Antimicrobials are classified as traditional when they (i) have been used for many years, (ii) are approved by many countries for inclusion as antimicrobials in foods (e.g., lysozyme and lactoferrin, which are naturally occurring but regulatory-agency approved), or (iii) are produced by synthetic means (as opposed to natural extracts). Ironically, many synthetic traditional antimicrobials are found in nature. These include acetic acid from vinegar, benzoic acid from cranberries, and sorbic acid from mountain ash tree berries (rowanberries).

FACTORS AFFECTING ACTIVITY

The effectiveness of food antimicrobials depends on many factors associated with the food product, its storage environment, its handling, and the target microorganisms

P. **Michael Davidson** and T. **Matthew Taylor,** Department of Food Science and Technology, 2605 River Rd., University of Tennessee, Knoxville, TN 37996-4591.

themselves. Food preservation is best achieved when the microorganisms to be inhibited, antimicrobial type and concentration, storage time and temperature, food pH and buffering capacity, and the presence of other agents which may affect shelf life are known and taken into account. Gould (128) classified factors that affect the activity of antimicrobials into microbial, intrinsic, extrinsic, and process.

Microbial factors that affect antimicrobial activity include inherent resistance (vegetative cells versus spores; strain differences), initial number and growth rate, interaction with other microorganisms (e.g., antagonism), cellular composition (gram reaction), cellular status (injury), and ability to form biofilms. Intrinsic factors affecting activity are those associated with a food product and include nutrients, pH, buffering capacity, oxidation reduction potential, and water activity. Extrinsic factors affecting antimicrobial activity may include temperature of storage, atmosphere, and relative humidity. Time of storage is also an important factor. Processing factors include changes in food composition, shifts in microflora, changes in microbial populations, and changes in food microstructure. Most of these factors influence antimicrobial activity in an interactive manner.

pH is one of the most important factors influencing the effectiveness of many food antimicrobials. Because many regulatory-agency-approved food antimicrobials are weak acids, they are most effective in their undissociated form. This is because weak acids are able to penetrate the cytoplasmic membrane of a microorganism more effectively in the protonated form. Another major factor affecting activity is polarity. This relates both to the ionization of the molecule and to the contribution of any alkyl side groups or hydrophobic parent molecules. Antimicrobials must be at least partially hydrophobic to attach and pass through the cell membrane but also be at least partially soluble in the aqueous phase in which the microorganisms exist.

TRADITIONAL ANTIMICROBIALS

Organic Acids and Esters

Many organic acids are used as food additives, but not all have antimicrobial activity. Research suggests that the most active are acetic, lactic, propionic, sorbic, and benzoic acids. Others including citric, caprylic, malic, and fumaric acids have variable but usually limited activity. Esters of fatty acids are discussed here because they are derivatives of organic acids and likely have related mechanisms.

The antimicrobial effectiveness of organic acids is related to pH, and the undissociated form of the acid is primarily responsible for inhibition of microorganisms. Therefore, in selecting an organic acid for use as an antimicrobial food additive, both the product pH and the pK_a of the acid must be taken into account. The use of organic acids is generally limited to foods with pH <5.5, since most organic acids have pK_as of 3.0 to 5.0 (86).

The mechanisms of action of organic acids and their esters have some common elements. There is little evidence that the organic acids and related esters influence cell wall synthesis in prokaryotes or that they significantly interfere with protein synthesis or genetic mechanisms. As stated previously, in the undissociated form, organic acids can penetrate the cell membrane lipid bilayer more easily. Once inside the cell, the acid dissociates because the cell interior has a higher pH than the exterior (Fig. 33.1). Bacteria generally maintain internal pH near neutrality to prevent conformational changes to the cell structural proteins, enzymes, nucleic acids, and phospholipids. Protons generated from intracellular dissociation of the organic acid acidify the cytoplasm and must be extruded to the exterior. According to the chemiosmotic theory (213), the cytoplasmic membrane is impermeable to protons and they must be transported

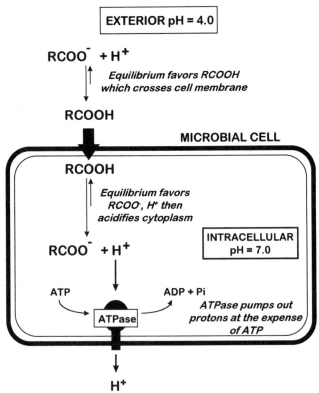

Figure 33.1 Fate of an organic acid (RCOOH) in a low pH environment in the presence of a microbial cell.

to the exterior. This proton extrusion creates an electrochemical potential across the membrane called the proton motive force (PMF). Lambert and Stratford (174) demonstrated that yeast cells pump excess protons out of the cell, utilizing energy in the form of ATP. However, the intracellular pH is eventually raised to a point that the cell may resume growth. The time it takes to accomplish this increase in intracellular pH is dependent on the extracellular pH and inhibitor concentration and is termed "lag time." Some researchers have suggested that only the undissociated form of weak organic acids has activity. Eklund (90), however, demonstrated that while the undissociated species has significantly greater activity, the anion does contribute slightly to antimicrobial activity. Paul and Hirshfield (248) suggested that perturbation of membrane function by organic acids, leading to interference with membrane-bound protein function, is also a growth-inhibiting mechanism. Another possible mechanism, proposed by Paul and Hirshfield (248), involves accumulation of inhibitory concentrations of anion in the cytoplasm of the cell. High concentrations of anions could lead to an increased osmolarity and to interference with metabolic processes.

Many environmental stress factors (e.g., heat, cold, starvation, low pH, and mild organic acids) can trigger resistance responses of microorganisms to subsequent stressors (32). Developed resistance may be termed tolerance, adaptation, or habituation depending on how the microorganism is exposed to the stress and the physiological conditions that lead to enhanced survival (38, 108). For example, *Escherichia coli* O157:H7, *Salmonella enterica* serovar Typhimurium, and *Listeria monocytogenes* can become more acid resistant and possibly resistant to other stresses (e.g., heat or osmotic stress) if subjected to relatively mild acidity before exposure to more acidic conditions (37, 38, 108, 113, 184, 185, 230, 248). In foods, adaptive alteration of cells might be of concern if the resistance is sustained by the adapted cells. However, the greater the degree of severity of the antimicrobial challenge, the less likely the microorganism is to survive for extended periods, even if it is adapted (72).

Acetic Acid and Acetates

Acetic acid (pK_a, 4.75; molecular mass, 60.05 Da) (Fig. 33.2), the primary component of vinegar, and its sodium, potassium, and calcium salts, sodium and calcium diacetate, and dehydroacetic acid (methylacetopyranone) are some of the oldest food antimicrobials. Acetic acid is more effective against yeasts and bacteria than against molds. Only *Acetobacter* species (microorganisms involved in vinegar production), lactic acid

CH₃COOH
Acetic Acid

COOH

Benzoic Acid

CH₃CHOHCOOH
Lactic Acid

CH₃CH₂COOH
Propionic Acid

CH₃CH=CH-CH=CHCOOH
Sorbic Acid

Figure 33.2 Organic acids used as antimicrobial food preservatives.

bacteria, and butyric acid bacteria are tolerant to acetic acid (86). Bacteria inhibited by acetic acid include *Bacillus* spp., *Campylobacter jejuni*, *Clostridium* spp., *E. coli*, *L. monocytogenes*, *Pseudomonas* spp., *Salmonella*, and *Staphylococcus aureus* (86, 94, 178, 212, 362). Molds and yeasts are more resistant to acetic acid than are bacteria. Yeasts and molds sensitive to acetic acid include *Aspergillus*, *Penicillium*, *Rhizopus*, and some strains of *Saccharomyces* (86, 169). The general mechanism by which acetic acid inhibits microorganisms is related to other organic acids as discussed previously.

Acetic acid and its salts have shown variable success as antimicrobials in food applications. Acetic acid can increase poultry shelf life when added to cut-up chicken parts in cold water at pH 2.5 (217). Addition of acetic acid at 0.1% to scald tank water used in poultry processing decreases the heat resistance of *Salmonella enterica* serovars Newport and Typhimurium and *C. jejuni* (235). In contrast, Lillard et al. (187) found that 0.5% acetic acid in the scald water has no significant effect on *Salmonella*, total aerobic bacteria, or *Enterobacteriaceae* on unpicked poultry carcasses. Acetic acid has shown variable effectiveness as an antimicrobial for use as a spray sanitizer on meat carcasses. Use of 2% acetic acid resulted in reductions in viable *E. coli* O157:H7 on beef after 7 days at 5°C (305). Acetic acid was the most effective antimicrobial in ground roasted beef slurries against *E. coli* O157:H7 growth in comparison with citric or lactic acid (2). Acetic acid added at 0.1% to bread dough inhibited growth of 6 log CFU of rope-forming *Bacillus subtilis* per g in wheat bread (pH 5.14) stored at 30°C for >6 days (283). In brain heart infusion broth, 0.2% acetic acid at pH 5.1 or 0.1% at pH 4.8 inhibited rope-forming strains of *B. subtilis* and *Bacillus licheniformis* for >6 days at 30°C (283).

Sodium acetate at 1.0% increases the shelf life of catfish fillets by 6 days stored at 4°C (167). Al-Dagal and Bazaraa (3) found that whole or peeled shrimp dipped in a 10% sodium acetate (wt/wt) solution for 2 min had extended microbiological and sensory shelf life compared to controls. Dehydroacetic acid has a high pK_a of 5.27 and is therefore active at higher pH values. It is inhibitory to bacteria at 0.1 to 0.4% and to fungi at 0.005 to 0.1% (86).

Sodium diacetate ($pK_a = 4.75$) is approved for use in processed meat and poultry products by the USDA Food Safety and Inspection Service (9 CFR 424.21) not to exceed 0.25% of the product formulation. It is most often used in combination with sodium or potassium lactate (see below) to inhibit *L. monocytogenes* in these products. Sodium diacetate is effective at 0.1 to 2.0% in inhibiting mold growth in cheese spread (86). At 32 mM (0.45%) in brain heart infusion broth (pH 5.4), sodium diacetate is inhibitory to *L. monocytogenes*, *E. coli*, *Pseudomonas fluorescens*, *Salmonella enterica* serovar Enteritidis and *Shewanella putrefaciens* but not *S. aureus*, *Yersinia enterocolitica*, *Pseudomonas fragi*, *Enterococcus faecalis*, or *Lactobacillus fermentis* after 48 h at 35°C (295). In addition, 21 to 28 mM sodium diacetate suppressed growth by the natural microflora of ground beef after storage at 5°C for up to 8 days (295). Degnan et al. (75) showed a 2.6 log decrease in viable *L. monocytogenes* in blue crab meat washed with 2M sodium diacetate after 6 days at 4°C. Sodium diacetate is useful in the baking industry because it has little effect on the yeast used in baking.

Acetic acid is used commercially in baked goods, cheeses, condiments and relishes, dairy product analogues, fats and oils, gravies and sauces, and meats. The sodium and calcium salts are used in breakfast cereals, cheeses, fats and oils, gelatin, hard candy, jams and jellies, meats, soft candy, snack foods, soup mixes, and sweet sauces. Sodium diacetate is used in baked goods, candy, cheese spreads, gravies, meats, sauces, and soup mixes.

Benzoic Acid and Benzoates

Benzoic acid (molecular mass, 122.12 Da) (Fig. 33.2) and sodium benzoate were the first antimicrobial compounds permitted in foods by the U.S. Food and Drug Administration (59). Benzoic acid occurs naturally in cranberries, plums, prunes, cinnamon, cloves, and most berries. Sodium benzoate is highly soluble in water (66.0 g/100 ml at 20°C), while benzoic acid is much less so (0.27% at 18°C). The undissociated form of benzoic acid ($pK_a = 4.19$) is the most effective antimicrobial agent. Therefore, the normal-use pH range is 2.5 to 4.5. Eklund (93) demonstrated that both the dissociated and undissociated forms of benzoic acid inhibit various bacteria and yeast but that the MICs for the undissociated acid are 15 to 290 times lower.

Benzoic acid and sodium benzoate are used primarily as antifungal agents. The inhibitory concentration of benzoic acid at pH <5.0 against most yeasts ranges from 20 to 700 μg/ml, while for molds it is 20 to 2,000 μg/ml (59). Some fungi, including *Byssochlamys nivea*, *Talaromyces flavus*, *Pichia membranaefaciens*, and *Zygosaccharomyces bailii*, are resistant to benzoic acid (313). *L. monocytogenes* is inhibited by 1,000 to 3,000 μg of benzoic acid/ml at pH 5.6, depending on incubation temperature (101, 367). Benzoic acid at 0.1% is effective in reducing viable *E. coli* O157:H7 in apple cider (pH 3.6 to 4.0) by 3 to 5 log CFU/ml in 7 days at 8°C (107). In combination with heat, 0.1% sodium benzoate reduced the time for a 3-log reduction of the ascospores of the heat-resistant mold, *Neosartorya fischeri*, from 205 to 210 min at 85°C to 85 to 123 min in mango and grape juice at pH 3.5 (262). Against *Acinetobacter* spp. isolated from various meat products, sodium benzoate was inhibitory to all species of the spoilage microbe at 0.5% at room temperature and 4°C (290). A gelatin coating containing benzoic acid extended the shelf life of tilapia fillets up to 7 days as measured by microbial counts, volatile basic nitrogen, and sensory properties (240).

Sodium benzoate (up to 0.1%) is used as an antimicrobial in carbonated and still beverages, syrups, cider, margarine, olives, pickles, relishes, soy sauce, jams, jellies, preserves, pie and pastry fillings, fruit salads, salad dressings, and in the storage of vegetables (59).

Lactic Acid and Lactates

Lactic acid ($pK_a = 3.79$; molecular mass, 90.08 Da) (Fig. 33.2) is produced naturally during fermentation of foods by lactic acid bacteria. Lactic acid and lactate salts act as antimicrobials, pH control agents, and flavorings in food products. Much research has been done investigating the efficacy of lactic acid as a sanitizer on meat and poultry carcasses to reduce or eliminate pathogens, such as *E. coli* O157:H7. In most cases, lactic acid sprays or dips at 0.2 to 2.5% are effective in reducing contamination on beef, veal, pork, poultry, and fish and, in some cases, improving shelf life (17, 22, 82, 153, 311, 312).

At pH 3.8, L-lactate is significantly more inhibitory than D-lactate to *E. coli* O157:H7 (180).

Sodium lactate (2.5 to 5.0%) inhibits *Clostridium botulinum*, *Clostridium sporogenes*, *L. monocytogenes*, *Salmonella*, *S. aureus*, *Y. enterocolitica*, and spoilage bacteria in various meat products (57, 146, 197, 208, 252, 298, 344, 352). Mixtures of sodium or calcium lactate (1.25 to 6%) and sodium diacetate (0.25 to 5%) have been demonstrated to be effective in inhibiting growth and/or causing slight reductions in numbers of *L. monocytogenes* in meat and seafood products, including bratwurst, comminuted beef, frankfurters, pork bologna, turkey ham, and smoked salmon, at 4 to 10°C (19, 23, 124, 205, 291, 366). Similarly, a mixture of sodium lactate (2.5%) and 0.25% sodium acetate inhibited growth of *L. monocytogenes* in sliced cooked ham and a sausage product stored for 5 weeks at 4°C (30). Uhart et al. (337) tested sodium lactate (6%) with sodium diacetate (3%) and pediocin (6,000 arbitrary units [AU]) against *L. monocytogenes* in beef wieners stored at 4°C. All three compounds in combination were able to reduce the target microorganism by 1.5 to 2.5 logs after 2 to 3 weeks of storage. Dipping wieners and bratwurst in up to 6% sodium lactate and 3% sodium diacetate was not effective against *L. monocytogenes* growth, whereas incorporation of the compounds at 1.32 to 3.4% and 0.1 to 0.25%, respectively, into the meat formulations prevented growth for 8 to 12 weeks depending on product and temperature (124). Sodium or calcium lactate (1 to 4.8%) has been shown to be effective alone and in combination with sodium acetate, sodium diacetate, sodium citrate or buffered sodium citrate in inhibiting germination and outgrowth of *Clostridium perfringens* during cooling or refrigeration of vacuum-packaged heated roast beef, turkey breast, and beef goulash (8, 159, 160, 287). Houtsma et al. (145) reported that toxin production by proteolytic *C. botulinum* was delayed at 15 and 20°C by sodium lactate concentrations of 2 and 2.5%, respectively, and complete inhibition of toxin production at 15, 20, and 30°C occurred at 3, 4, and >4%, respectively. Huang and Juneja (147) reported that sodium lactate at up to 4.5% had no effect on the heat resistance of *E. coli* O157:H7 in ground beef. Similarly, Murphy et al. (220) found that 4.8% sodium lactate in ground chicken meat had no effect on the *D* value of *Salmonella* and increased the *D* value of *L. monocytogenes* at 55 to 70°C. In contrast, Aymerich et al. (11) reported a synergistic effect of sodium lactate with high pressure in inactivating *L. monocytogenes* and *Salmonella*. At pH 7.0 with lower concentrations (0.18 to 0.45%) of lactate, growth of *Campylobacter* sp. in microbiological media is stimulated compared to growth in media not containing the organic acid salt (141).

Alakomi et al. (2a) reported that lactic acid caused permeabilization of the outer membrane of the gram-negative bacteria *E. coli* O157:H7, *Pseudomonas aeruginosa*, and *S. enterica* serovar Typhimurium. In addition, they suggested that lactic acid may act to potentiate other antimicrobial substances, such as lysozyme. Since lactate salts at ≥2.5% have minimal effect on product pH and most of the lactate remains in the less effective anionic form, there is some question about their mechanism of inhibition. Some have suggested that high concentrations of the salts reduce water activity (a_w) sufficiently to inhibit microorganisms (74). However, Chen and Shelef (57) and Weaver and Shelef (352) measured the a_w of cooked meat model systems and liver sausage, respectively, containing lactate salts up to 4% and concluded that a_w reduction is not sufficient to inhibit *L. monocytogenes*. It is most likely that at the high concentrations of lactate used, sufficient undissociated lactic acid is present, possibly in combination with a slightly reduced pH and a_w, to inhibit some microorganisms.

Propionic Acid and Propionates

Up to 1% propionic acid ($pK_a = 4.87$) (Fig. 33.2) is produced naturally in Swiss cheese by *Propionibacterium freudenreichii* subsp. *shermanii*. The activity of propionates depends on the pH of the substance to be preserved, with the undissociated acid being the most active form. Eklund (93) demonstrated that undissociated propionic acid is 11 to 45 times more effective than the dissociated form.

Propionic acid and sodium, potassium, and calcium propionates are used primarily against molds; however, some yeasts and bacteria are also inhibited. The microorganism in bread dough responsible for rope formation, *B. subtilis*, is inhibited by propionic acid at pH 5.6 to 6.0 (86). Propionates (0.1 to 5.0%) retard the growth of bacteria including *E. coli*, *S. aureus*, *Sarcina lutea*, *Salmonella*, *Proteus vulgaris*, *Lactobacillus plantarum*, and *L. monocytogenes* and of the yeasts *Candida* and *Saccharomyces cerevisiae* (86).

Propionic acid and propionates are used as antimicrobials in baked goods and cheeses. Propionates may be added directly to bread dough because they have no effect on the activity of baker's yeast. Propionic acid added at 0.1% to bread dough inhibited growth of 6 log CFU of rope-forming *B. subtilis* per g in wheat bread (pH 5.30) stored at 30°C for >6 days (283). There is no limit to the concentration of propionates allowed in foods, but amounts used are generally less than 0.4%.

Sorbic Acid and Sorbates

Sorbic acid (Fig. 33.2) was first identified in 1859 by A. W. van Hoffman, a German chemist, from the berries of the mountain ash tree (rowanberry) (318). Sorbic acid is a *trans-trans*, unsaturated monocarboxylic fatty acid which is slightly soluble in water (0.16 g/100 ml) at 20°C. The potassium salt of sorbic acid is readily soluble in water (58.2 g/100 ml at 20°C). As with other organic acids, the antimicrobial activity of sorbic acid is greatest when the compound is in the undissociated state. With a pK_a of 4.75, activity is greatest at pH less than 6.0 to 6.5. The undissociated acid is 10 to 600 times more effective than the dissociated form (90).

Sorbates are the best characterized of all food antimicrobials as to their spectrum of action. They inhibit fungi and certain bacteria. Food-related yeasts inhibited by sorbates include species of *Brettanomyces, Byssochlamys, Candida, Cryptococcus, Debaryomyces, Hansenula, Pichia, Rhodotorula, Saccharomyces, Sporobolomyces, Torulaspora,* and *Zygosaccharomyces* (112, 318). Food-related mold species inhibited by sorbates include species in the genera *Alternaria, Aspergillus, Botrytis, Cephalosporium, Fusarium, Geotrichum, Helminthosporium, Mucor, Penicillium, Pullularia (Aureobasidium), Sporotrichum,* and *Trichoderma* (318). Sorbates inhibit the growth of yeasts and molds in microbiological media, cheeses, fruits, vegetables and vegetable fermentations, sauces, and meats. Sorbates inhibit growth and mycotoxin production by the mycotoxigenic molds *Aspergillus flavus, Aspergillus parasiticus, B. nivea, Penicillium expansum,* and *Penicillium patulum* (39, 40, 181, 279, 286, 336). High initial mold populations can degrade sorbic acid in cheese. Some species of *Aspergillus, Fusarium, Geotrichum, Mucor, Penicillium, Saccharomyces,* and *Zygosaccharomyces* can grow in the presence of and degrade potassium sorbate (112, 318). Sorbates may be degraded through a decarboxylation reaction resulting in the formation of 1,3-pentadiene, a compound having a kerosene-like or hydrocarbon-like odor (186).

Bacteria inhibited by sorbate include *Acinetobacter, Aeromonas, Bacillus, Campylobacter, Clostridium, E. coli* O157:H7, *Lactobacillus, L. monocytogenes, Pseudomonas, Salmonella, Staphylococcus, Vibrio,* and *Y. enterocolitica* (150, 318, 336). Sorbic acid inhibits primarily catalase-producing bacteria, whereas, with some exceptions, catalase-negative lactic acid bacteria are generally resistant to sorbates. This allows use of sorbates in products fermented by lactic acid bacteria.

Sorbate inhibits the growth of many pathogenic bacteria in or on foods including *Aeromonas* spp. on sundried fish with 1.5% salt; *Salmonella* and *S. aureus* in sausage; *S. aureus* in bacon; *Vibrio parahaemolyticus* in seafood; *Salmonella, S. aureus,* and *E. coli* in poultry; *Y. enterocolitica* in pork; *S. enterica* serovar Typhimurium in milk and cheese; and *E. coli* O157:H7 in queso fresco

cheese (131, 164, 255, 318). Uljas and Ingham (340) utilized 0.1% sorbic acid in combination with freeze-thawing and storage at 25°C for 12 or 4 h at 35°C and storage without freeze-thawing at 35°C for 6 h to achieve a 5-log reduction of *E. coli* O157:H7 in apple cider at pH 4.1. In addition, the compound inhibits growth of the spoilage bacteria *Pseudomonas putrefaciens* and *P. fluorescens,* histamine production by *Proteus morganii* and *Klebsiella pneumoniae,* and listeriolysin O production by *L. monocytogenes* (206, 318).

Sorbates are effective anticlostridial agents in cured meats and other meat and seafood products (314, 318). The compound prevents spores of *C. botulinum* from germinating and forming toxin in beef, pork, poultry, soy protein frankfurters and emulsions, and bacon (318). Lopez et al. (194) showed that 0.1% potassium sorbate inhibited unheated strains of *Bacillus stearothermophilus* spores by ca. 50% and prevented growth of spores heated at 121°C for 1 min. Against other strains, 0.1% potassium sorbate had little effect on heated spores. Potassium sorbate is a strong inhibitor of both *Bacillus* and *Clostridium* spore germination at pH 5.7 but much less so at pH 6.7 (310).

One of the primary targets of sorbic acid in vegetative cells appears to be the cytoplasmic membrane. Sorbic acid inhibits amino acid uptake, which in turn was theorized to be responsible for eliminating the membrane proton motive force (PMF) through nutrient depletion (110, 299–301). Ronning and Frank (281) also showed that sorbic acid reduces the cytoplasmic membrane electrochemical gradient and consequently the PMF. They concluded that the sorbic-acid-induced loss of PMF inhibits amino acid transport, which could eventually result in the inhibition of many cellular enzyme systems. In contrast, Eklund (90) showed that while low concentrations of sorbic acid reduce the ΔpH of the PMF of *E. coli* vesicles, concentrations much greater than those required for inhibition reduce, but do not eliminate, the $\Delta\Psi$ component. Since the $\Delta\Psi$ component alone could energize active uptake of amino acids, the amino acid uptake inhibition theory does not entirely explain the mechanism of inhibition by sorbic acid (91). Stratford and Anslow (319) suggest that sorbic acid does not act as a classic weak acid inhibitor, i.e., by passing through the cytoplasmic membrane as an undissociated acid and inhibiting via cytoplasmic acidification. They point out that sorbic acid is inhibitory at a relatively high pH and does not release as many protons at the MIC. They theorize that sorbic acid acts primarily through membrane disruption. In a related study, Bracey et al. (33) found little relationship between inhibition and reduction in intracellular pH of *S. cerevisiae* by sorbic acid. Since sorbic acid caused an increase in ATP consumption, which was partially attributed to increased proton pumping by membrane H⁺-ATPase, they theorized that inhibition was due to induction of an energy-utilizing mechanism to compensate for a reduced intracellular pH. The mechanism by which sorbic acid inhibits microbial growth may also be partially due to its effect on enzymes. The mechanism of sorbate action against bacterial spores has been studied extensively. In some of the few reports involving effects on the cell wall, Gould (127) and Seward et al. (294) demonstrated that sorbic acid inhibits cell division of germinated spores of *Bacillus* and *C. botulinum* Type E. Sorbic acid at pH 5.7 competitively inhibited L-alanine and L-α-NH₂-*n*-butyric acid-induced germination of *Bacillus cereus* T spores and L-alanine- and L-cysteine-induced germination of *C. botulinum* 62A (310).

Sorbate is applied to foods by direct addition, dipping, spraying, dusting, or incorporation into packaging. Baked goods can be protected from yeast and molds through the use of 0.05 to 0.10% potassium sorbate applied either as a spray after baking or by direct addition. Food products in which sorbates are used and their typical use concentrations (as a percentage) include beverage syrups, 0.1; cakes and icings, 0.05 to 0.1; cheese and cheese products, 0.2 to 0.3; cider, 0.05 to 0.1; dried fruits, 0.02 to 0.05; fruit drinks, 0.025 to 0.075; margarine, 0.1; pie fillings, 0.05 to 0.1; salad dressings, 0.05 to 0.1; semimoist pet food, 0.1 to 0.3; and wine, 0.02 to 0.04 (7). In combination with sodium lactate (2%), sorbate washes applied to cantaloupe rinds have been shown to be effective at reducing *Salmonella* populations by up to 3 logs when used at 0.02% sorbate (338).

Miscellaneous Organic Acids

Many organic acids and their esters have been examined as potential antimicrobials, but most have little or no activity. They are used in foods as acidulants or flavoring agents rather than as preservatives. Fumaric acid is used to prevent the malolactic fermentation in wines (241) and as an antimicrobial agent in wines (257). A combination of 0.15% fumaric acid and 0.05% sodium benzoate caused a 5-log CFU/ml reduction of *E. coli* O157:H7 in three commercial cider products after products were held at refrigeration for 24 h (64). In microbiological media, sodium citrate was bactericidal only to *S. aureus* at 0.625 to 2.5%, while for *B. subtilis*, *E. faecalis*, *E. coli*, *P. aeruginosa*, and *Y. enterocolitica*, ≥25% was required for inhibition (178, 179). Citric acid is inhibitory to *Salmonella* on poultry carcasses and in mayonnaise, to *C. botulinum* growth and toxin production in shrimp and tomato products, and to *S. aureus* in microbiological medium (212, 259, 326, 363). Buffered sodium citrate

(≥1%) reduced the growth of *C. perfringens* on beef rounds and pork loins during prolonged chilling from 54.4 to 7.2°C compared to control products (324). Citric acid retards growth and toxin production by *A. parasiticus* and *Aspergillus versicolor*, but not *P. expansum* (269). Citric acid (0.156%, wt/vol) in mango juice (pH 3.5) reduced the heat resistance of *N. fischeri* spores 2.3 times at 85°C (262). Nakai and Siebert (227) reported MIC values for caproic and caprylic acid against *Listeria innocua* and *Listeria ivanovii* of 2.83 g/liter at pH 5.25; against *P. aeruginosa* the MICs of caproic acid and formic acid were 2.83 and 1.91 g/liter, respectively. Thus, there is potential use for other organic acids depending on the target microorganism and food product application.

Fatty Acid Esters

Certain fatty acid esters have antimicrobial activity in foods. One of the most effective is glyceryl monolaurate (monolaurin) (162). Monolaurin is active against gram-positive bacteria including *Bacillus*, *Lactococcus*, *L. monocytogenes*, *Micrococcus*, and *S. aureus* at concentrations of ≤100 µg/ml but is much less effective against gram-negative bacteria (231, 233, 264, 350). In combination with sodium benzoate (0.5 mg/ml), monolaurin has been reported to be inhibitory to *L. monocytogenes* at 5 mg/ml in broth at pH 7.0 (97). Monolaurin was shown to inhibit spores and vegetative cells of *Bacillus cereus* T, *C. botulinum* 62A, and *C. sporogenes* PA3679 at 0.075 to 0.18 mM (53). Spores were more sensitive than vegetative cells with the exception of *C. botulinum* 62A, for which the sensitivity of each was equivalent (53). The presence of EDTA expands the activity spectrum of monolaurin to include gram-negative bacteria such as *E. coli* O157:H7, *Pseudomonas*, *Salmonella*, and *Vibrio* and decreases the MICs against gram-positive strains (162). Razavi-Rohani and Griffiths (264) showed that EDTA could not be replaced by other chelators including sodium citrate. Monolaurin is effective against *L. monocytogenes* in Camembert cheese, cottage cheese, crawfish tail homogenate, some meat products, and yogurt (231, 232, 350). Monolaurin is inhibitory to some molds and yeasts including *Aspergillus*, *Alternaria*, *Candida*, *Cladosporium*, *Penicillium*, and *Saccharomyces* (162).

DMDC

Dimethyl dicarbonate (DMDC) (Fig. 33.3) is a colorless liquid which is slightly soluble in water (3.6%). The compound is very reactive with many substances including water, ethanol, alkyl and aromatic amines, and sulfhydryl groups (125). The primary target microorganisms for DMDC are yeasts including *Saccharomyces*, *Zygosaccharomyces*, *Rhodotorula*, *Candida*, *Pichia*, *Torulopsis*,

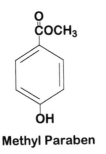

Methyl Paraben

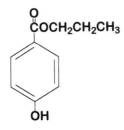

Propyl Paraben

Figure 33.3 Alkyl esters of *p*-hydroxybenzoic acid (parabens).

Torula, *Endomyces*, *Kloeckera*, and *Hansenula*. The compound is also bactericidal at 30 to 400 μg/ml to a number of species including *Acetobacter pasteurianus*, *E. coli*, *P. aeruginosa*, *S. aureus*, several *Lactobacillus* species, and *Pediococcus cerevisiae* (125). The compound was shown to be bactericidal at 0.025% and more effective than either sodium bisulfite or sodium benzoate against *E. coli* O157:H7 in apple cider at 4°C (107). Molds are generally more resistant to DMDC than yeasts or bacteria. The mechanism by which DMDC acts is most likely related to inactivation of enzymes. A related compound, diethyl dicarbonate, reacts with imidazole groups, amines, or thiols of proteins (89). In addition, diethyl dicarbonate readily reacts with histidyl groups of proteins. This can cause inactivation of the enzymes lactate dehydrogenase or alcohol dehydrogenase by reacting with the histidine in the active site (125).

Lactoferrin

In milk and colostrum, the primary iron-binding protein is lactoferrin. This protein is also found in other physiological fluids and polymorphonuclear leukocytes. Lactoferrin (molecular mass, 76.5 kDa) is a glycoprotein that exists in milk primarily as a tetramer with Ca^{2+} and has two iron binding sites per molecule. For each Fe^{3+} bound by lactoferrin, one bicarbonate (HCO_3^-) is required. Lactoferrin must be low in iron saturation and bicarbonate must be present for the compound to be an effective antimicrobial.

Lactoferrin is active in milk possibly because of the low iron concentration and the presence of bicarbonate (357). The exact biological role of lactoferrin is unknown; however, it may act as a barrier to infection of the nonlactating mammary gland and may help protect the gastrointestinal tract of the newborn against infection (323).

Lactoferrin is inhibitory to a number of microorganisms including *B. subtilis*, *B. stearothermophilus*, *L. monocytogenes*, *Micrococcus* species, *E. coli*, and *Klebsiella* species (171, 198, 239, 249, 270). Payne et al. (249) found that bovine lactoferrin had to be reduced to 18% iron saturation using dialysis (apo-lactoferrin) to have bacteriostatic activity against four strains of *L. monocytogenes* and an *E. coli* strain at concentrations of 15 to 30 mg/ml in ultrahigh-temperature-treated milk. At 2.5 mg/ml, the compound has no activity against *Salmonella* serovar Typhimurium or *P. fluorescens* and little activity against *E. coli* O157:H7 or *L. monocytogenes* VPHI (250). Branen and Davidson (35) studied the interactive antimicrobial effects of combinations of lactoferrin with monolaurin, nisin, or lysozyme against *L. monocytogenes*, *E. coli*, *Salmonella* serovar Enteritidis, and *P. fluorescens* in microbiological media and milk. In microbiological media, lactoferrin in combination with nisin inhibited *L. monocytogenes* and in combination with monolaurin inhibited *E. coli* O157:H7. None of the antimicrobial combinations with lactoferrin was effective in ultrahigh-temperature-treated milk held at 25°C, which was attributed to factors such as food component interactions and storage temperature.

Production of an iron-deficient environment by lactoferrin may be part of its mechanism of inhibition. Iron stimulates the growth of bacteria in many genera including *Clostridium*, *Escherichia*, *Listeria*, *Pseudomonas*, *Salmonella*, *Staphylococcus*, *Vibrio*, and *Yersinia*. Some gram-negative bacteria may be resistant because they adapt to low-iron environments by producing siderophores such as phenolates and hydroxamates (96). Microorganisms with a low-iron requirement, such as lactic acid bacteria, would not be inhibited by lactoferrin. There is also evidence that lactoferrin may have effects on gram-negative bacteria in addition to iron deprivation. Ellison et al. (100) found that both lactoferrin and EDTA, another chelator, cause release of anionic lipopolysaccharides from the outer membrane of *E. coli*. Lactoferrin causes this loss by chelation of cations, including magnesium, calcium, and iron, that stabilize lipopolysaccharides in the membrane. Since it is cationic, lactoferrin also may increase the outer membrane permeability to hydrophobic compounds, including other antimicrobials. According to Naidu and Bidlack (226), lactoferrin blocks adhesion of microorganisms to mucosal surfaces, inhibits expression of fimbria and other colonizing factors of enteric pathogens, such as *E. coli*, and inactivates lipopolysaccharides of gram-negative bacteria.

Lactoferricin B or hydrolyzed lactoferrin (HLF) is a small peptide produced by acid-pepsin hydrolysis of bovine lactoferrin (24). It contains 25 amino acid residues and has a molecular mass of ca. 3 kDa. Jones et al. (157) determined that the compound is inhibitory and cidal to *Shigella*, *Salmonella*, *Y. enterocolitica*, *E. coli* O157:H7, *S. aureus*, *L. monocytogenes*, and *Candida* at concentrations ranging from 1.9 to 125 μg/ml. *Proteus*, *Serratia*, and *Pseudomonas cepacia* are resistant to 500 μg of lactoferricin B per ml. In contrast, while HLF was effective against *L. monocytogenes*, enterohemorrhagic *E. coli*, and *S. enterica* serovar Enteritidis in peptone yeast extract glucose broth, it was not active in a more complex medium, Trypticase soy broth (34). The addition of EDTA enhanced the activity of HLF in Trypticase soy broth, indicating that the decreased activity of HLF may have been due, in part, to excess cations in the medium. Venkitanarayanan et al. (346) found that while 50 or 100 μg of lactoferricin B per ml reduced the number of viable *E. coli* O157:H7 organisms in 1% peptone, it was much less effective as an antimicrobial in ground beef. Min and Krochta (211) evaluated lactoferrin and hydrolyzed lactoferrin against the mold *Penicillium commune*. Both were effective at inhibiting the mold in 1% peptone at ≥10 mg/ml but demonstrated no inhibition in the more complex potato dextrose agar. In contrast to other studies, Murdock and Matthews (219) reported that pepsin-hydrolyzed lactoferrin at 8 mg/ml reduced populations of *E. coli* O157:H7 by ca. 1 to 2 logs CFU/ml of milk adjusted to pH 4.0 or 7.0 and stored at 4°C. *L. monocytogenes* was only reduced ca. 0.5 to 1.0 log under the same conditions but was reduced by around 2 logs at 35°C in milk adjusted to pH 4.0. In comparison, lactoferrin had no effect on either microorganism in milk at 4 or 35°C at pH 4.0 or 7.0 (219).

Naidu (225) developed an antimicrobial system in which lactoferrin is immobilized to food-grade polysaccharides and is dissolved in a citrate-bicarbonate buffer with NaCl. The resulting product ("activated lactoferrin") was reported to be effective at inhibiting attachment of *E. coli* O157:H7 on meats. The lactoferrin product is approved by the USDA Food Safety Inspection Service to reduce bacterial contamination by the target microorganism on beef carcasses (224).

Lysozyme

Lysozyme (peptidoglycan *N*-acetylmuramoyl hydrolase; EC 3.2.1.17) is a 14.6-kDa enzyme present in avian eggs, mammalian milk, tears and other secretions, insects, and fish. While tears contain the greatest concentration of lysozyme, dried egg white (3.5%) is the commercial source for foods (156). Lysozyme is the primary antimicrobial compound in egg albumen, but its activity is enhanced by ovotransferrin, ovomucoid, and alkaline pH. Lysozyme is relatively stable to heat (80°C, 30 min; 100°C, 1 to 2 min) at lower pHs but is inactivated at lower temperatures when the pH is increased. The optimum temperature for activity of the enzyme is 55 to 60°C, but it retains 50% activity at 10 to 25°C (151). The enzyme catalyzes hydrolysis of the α-1,4 glycosidic bonds between *N*-acetylmuramic acid and *N*-acetylglucosamine of the peptidoglycan of bacterial cell walls. This causes cell wall degradation and lysis in hypotonic solutions.

Lysozyme is most active against gram-positive bacteria, most likely because the peptidoglycan of the cell wall is more exposed. The enzyme shows greatest activity (lysis or inhibition) against *Bacillus coagulans*, *B. stearothermophilus*, *Clostridium thermosaccharolyticum*, *Clostridium tyrobutyricum*, and *Micrococcus* (156). Lysozyme has moderate antimicrobial activity or variable activity, depending on the strain, against *B. cereus*, *C. jejuni*, *C. botulinum*, *L. monocytogenes*, *Lactobacillus* spp., *P. aeruginosa*, and *Y. enterocolitica* (156). Those microorganisms that are normally not lysed or inhibited by lysozyme include *Aeromonas hydrophila*, *Brochothrix thermosphacta*, *C. perfringens*, *E. coli* O157:H7, *S. enterica* serovar Typhimurium, *Shigella* spp., *S. aureus*, and *Vibrio cholerae* (156). The MIC of lysozyme against fungi, including *Candida*, *Sporothrix*, *Penicillium*, *Paecilomyces*, and *Aspergillus*, was >9,530 μg/ml in potato dextrose agar at pH 5.6 (263). Only *Fusarium graminearum* (1,600 μg/ml) and *Aspergillus ochraceus* (3,260 μg/ml) were inhibited by less than the maximum concentration evaluated. However, when combined with an equivalent concentration of EDTA, lysozyme was inhibitory to most species of fungi tested at ≤500 μg/ml.

Variation in susceptibility of gram-positive bacteria is likely due to the presence of teichoic acids and other materials that bind the enzyme and the fact that certain species have greater proportions of 1,6 or 1,3 glycosidic linkages in the peptidoglycan, which are more resistant than the 1,4 linkage (331). Some strains of *L. monocytogenes* are not inhibited by lysozyme alone but are inhibited when EDTA is added (35, 148, 250). Hughey and Johnson (148) hypothesized that the peptidoglycan of the microorganism may be partially masked by other cell wall components and that EDTA enhances penetration of the lysozyme to the peptidoglycan. Growth conditions may also affect lysozyme activity. Johansen et al. (155) suggested that low pH (5.5) causes increased inhibition of *L. monocytogenes* by lysozyme because the organism has a slower growth rate, which allows enzymatic hydrolysis of the cell wall to exceed the cell proliferation rate. Incorporation of lysozyme into liposomes was also reported to enhance its activity against *L. monocytogenes* (355).

Lysozyme is less effective against gram-negative bacteria due to their reduced peptidoglycan content (5 to 10%) and presence of the outer membrane of lipopolysaccharide and lipoprotein (357). Gram-negative cell susceptibility can be increased by pretreatment with chelators (e.g., EDTA or phosphates) (31, 35). The chelators reportedly bind Ca^{2+} or Mg^{2+}, which are essential for maintaining integrity of the lipopolysaccharide layer (331). In addition, gram-negative cells may be sensitized to lysozyme if the cells are subjected to pH shock, heat shock, osmotic shock, drying, and freeze-thaw cycling (46, 331).

Lysozyme activity is influenced by food type as well as processing and storage conditions. The enzyme has shown potential for use as an antimicrobial with EDTA to control the growth of *L. monocytogenes* in vegetables but is less effective in refrigerated meat and soft cheese products (149). Surface application to ham and bologna of lysozyme and nisin (25.5 g/liter, 1:3) and EDTA (25.5 g/liter) in a gelatin coating had a bactericidal effect on *B. thermosphacta*, *Leuconostoc mesenteroides*, *L. monocytogenes*, and *Lactobacillus sakei* but was less effective against serovar Typhimurium and *E. coli* O157: H7 (123). Incorporation into ham or bologna sausages (500 mg/kg, 1:3; and 500 mg/kg, respectively) initially reduced *B. thermosphacta* and *L. mesenteroides* populations but had variable effects on the remaining microorganisms (120). Nattress and Baker (228) showed that a combination of lysozyme and nisin (ca. 260 µg/cm²) inhibited lactic acid bacteria and *B. thermosphacta* on vacuum-packaged pork loins stored at 2°C for 6 weeks. The efficacy of the treatment declined following preparation of the loins for retail display, and the populations of *Enterobacteriaceae* were higher on treated samples. The presence of higher levels of *Enterobacteriaceae* was attributed to reduced growth of the lactic acid bacteria (228). Samuelson et al. (292) found that EDTA plus lysozyme inhibited serovar Typhimurium on poultry. In contrast, no inhibition was demonstrated with up to 2.5 µg of EDTA/ml and 200 µg of lysozyme/ml in milk against either serovar Typhimurium or *P. fluorescens* in a study by Payne et al. (250). Lysozyme has been evaluated in several studies for its use as an antimicrobial packaging component. Min et al. (210) incorporated lysozyme into whey protein films and demonstrated inhibition of *L. monocytogenes* in microbiological media and on

smoked salmon at 4 and 10°C. Park et al. (246) showed that lysozyme added to chitosan films enhanced inhibition of *Streptococcus faecalis* and *E. coli*. There is a significant influence of the food product on activity of lysozyme and EDTA.

Lysozyme is approved for use to prevent gas formation ("blowing") in cheeses such as Edam and Gouda by *C. tyrobutyricum*. In the United States, egg white lysozyme was approved as a "direct food substance affirmed as generally recognized as safe" in 1998 (6). Lysozyme is also approved for use in casings for frankfurters and in cooked meat and poultry products. In Japan, lysozyme is used to a great extent to preserve seafood, vegetables, pasta, and salads.

Nitrites

Sodium nitrite ($NaNO_2$) and potassium nitrite (KNO_2) have a specialized use in cured meat products. Meat curing utilizes salt, sugar, spices, and ascorbate or erythorbate in addition to nitrite. Nitrites have many functions in cured meats as well as serving as antimicrobials. As nitric oxide, nitrite reacts with the meat pigment myoglobin to form nitrosomyoglobin and the characteristic cured meat color. It also contributes to the flavor and texture of cured meats and serves as an antioxidant. Meat curing is often combined with drying, heating, smoking, or fermentation as preservation adjuncts. At one time, sodium nitrate ($NaNO_3$) and potassium nitrate (KNO_3) were used extensively in cured meat production. Their use diminished when it was discovered that nitrate is converted to nitrite and that nitrite is the effective antimicrobial agent. The specific contribution of nitrite to the antimicrobial effects of curing salt was not recognized until the late 1920s, and evidence that nitrite is an effective antimicrobial agent came in the 1950s (328).

The primary use for sodium nitrite as an antimicrobial is to inhibit *C. botulinum* growth and toxin production in cured meats. In association with other components in the curing mix, such as salt, and reduced pH, nitrite exerts a concentration-dependent antimicrobial effect on the outgrowth of spores from *C. botulinum* and other clostridia. Nitrite inhibits bacterial sporeformers by inhibiting outgrowth of the germinated spore (87). Only very high nitrite concentrations significantly inhibit spore germination.

It was first suggested in the 1920s that nitrites were more effective at a lower pH (360), and the interaction of nitrite and reduced pH against bacteria is well established. Nitrite is more inhibitory under anaerobic conditions. Ascorbate and isoascorbate enhance the antibotulinal action of nitrite, probably by acting as reducing agents (278). Storage and processing temperatures, salt concentration, and initial inoculum size also significantly influence the antimicrobial effectiveness of nitrite. Roberts and Ingram (277) and Duncan and Foster (87) demonstrated that nitrite addition prior to heating does not increase inactivation of spores but inhibits outgrowth following heating.

Nitrite has variable effects on microorganisms other than *C. botulinum*. At 200 μg/ml and pH 5.0, nitrite completely inhibited growth of *E. coli* O157:H7 at 37°C (336). However, at pH 6.8 in broth, ~5,700 μg of nitrite/ml was required to reach the MIC of three *E. coli* O157:H7 strains (216). *C. perfringens* growth at 20°C is inhibited by 200 μg of nitrite/ml and 3% salt or 50 μg of nitrite/ml and 4% salt at pH 6.2 in a laboratory medium (117). *L. monocytogenes* growth is inhibited for 40 days at 5°C by 200 μg of sodium nitrite/ml with 5% NaCl in vacuum-packaged and film-wrapped smoked salmon (252). Gibson and Roberts (117, 118) found limited inhibition by nitrite (400 μg/ml) and salt (up to 6%) against fecal streptococci, *Salmonella*, and enteropathogenic *E. coli*. Certain strains of *Salmonella*, *Lactobacillus*, *C. perfringens*, and *Bacillus* are resistant to nitrite (48, 254, 274). In growth media, Gill and Holley (121) reported significant improvements in the inhibition of gram-negative pathogens such as *E. coli*, *Salmonella*, and *Shigella* when sodium nitrite (180 μg/ml) and EDTA (900 μg/ml) were combined or when nitrite (180 μg/ml) and NaCl (270 μg/ml) were combined.

The mechanism of nitrite inhibition has been studied for over 60 years; however, the likely targets of clostridial inhibition by nitrite have been elucidated only in the past 20 years. Woods et al. (360) showed that nitrite caused a reduction in intracellular ATP and excretion of pyruvate in cells of *C. sporogenes*. Since these cells oxidize pyruvate to acetate to produce ATP using the phosphoroclastic system, it was theorized that this enzyme system was being inhibited by nitrite. Two enzymes in the system, pyruvate-ferredoxin oxidoreductase (PFR) and ferridoxin, were suspected to be susceptible to nitrite. Inhibition was due to reaction of nitrite in the form of nitric oxide with the nonheme iron of the proteins. PFR was most susceptible. Later Woods and Wood (359) showed that the phosphoroclastic system of *C. botulinum* is also inhibited by nitrite. Carpenter et al. (47) confirmed that nitrite inhibits the phosphoroclastic system of *C. botulinum* and *C. pasteurianum* but that the iron-sulfur enzyme, ferredoxin, is more susceptible than PFR. McMindes and Siedler (207) reported that nitric oxide is the active antimicrobial principle of nitrite and that pyruvate decarboxylase may be an additional target for growth inhibition by nitrite. Roberts et al. (278) also confirmed inhibition of the phosphoroclastic system and found that ascorbate enhanced inhibition. In addition, they showed that other iron-containing enzymes of *C. botulinum*, including other oxidoreductases and the iron-sulfur protein, hydrogenase, are inhibited. It has been suggested that inhibition of clostridial ferridoxin and/or pyruvate-ferredoxin oxidoreductase is the ultimate mechanism of growth inhibition for clostridia (47, 328). These observations are supported by the fact that addition of iron to meats containing nitrite reduces the inhibitory effect of the compound (329).

The mechanism of inhibition against nonsporeforming bacteria may be different from that against sporeformers. Nitrite inhibits active transport, oxygen uptake, and oxidative phosphorylation of *P. aeruginosa* by oxidizing ferrous iron of an electron carrier, such as cytochrome oxidase, to the ferric form (285). Muhoberac and Wharton (218) and Yang (364) also found inhibition of cytochrome oxidase of *P. aeruginosa*. *E. faecalis* and *Lactococcus lactis*, which do not depend on active transport or cytochromes, are not inhibited by nitrite (285). Woods et al. (361) theorized that nitrites inhibit aerobic bacteria by binding the heme iron of cytochrome oxidase.

Meat products that may contain nitrites include bacon, bologna, corned beef, frankfurters, luncheon meats, ham, fermented sausages, shelf-stable canned cured meats, and perishable canned cured meat (e.g., ham). Nitrite is also used in a variety of fish and poultry products. The concentration used in these products is specified by governmental regulations but is generally limited to 156 μg/g for most products and 100 to 120 μg/ml in bacon. Sodium erythorbate or isoascorbate is required in products containing nitrites as a cure accelerator and as an inhibitor to the formation of nitrosamines, carcinogenic compounds formed by reactions of nitrite with secondary or tertiary amines. Sodium nitrate is used in certain European cheeses to prevent spoilage by *C. tyrobutyricum* or *C. butyricum* (328).

para-Hydroxybenzoic Acid Esters

Alkyl esters of *p*-hydroxybenzoic acid (parabens) (Fig. 33.4) were first reported to possess antimicrobial activity in the 1920s. Esterification of the carboxyl group of benzoic acid allows the molecule to remain undissociated up to pH 8.5, giving the parabens an effective range of pH 3.0 to 8.0 (1). In most countries, the methyl, propyl, and heptyl parabens are allowed for direct addition to foods as antimicrobials, while the ethyl and butyl esters are approved in some countries.

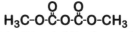

Figure 33.4 Dimethyldicarbonate (DMDC).

The antimicrobial activity of *p*-hydroxybenzoic acid esters is, in general, directly proportional to the chain length of the alkyl component. As the alkyl chain length of the parabens increases, inhibitory activity generally increases. Increasing activity with decreasing polarity is more evident against gram-positive than against gram-negative bacteria. Parabens are generally more active against molds and yeast than against bacteria (Table 33.1). Against bacteria, the parabens are more active against gram-positive genera (Table 33.2). Little research on the activity of the *n*-heptyl ester in foods seems to be available. Chan et al. (55) did show that this compound was very effective in inhibiting bacteria involved in the malolactic fermentation of wines.

The primary mechanism of the parabens likely involves interference with the function of the cytoplasmic membrane (71). Bredin et al. (36) demonstrated that propyl paraben induced potassium efflux that was related to the porin expression in the bacterial outer membrane of *E. coli* and concluded that this permeabilizing effect was related to paraben antibacterial properties. Eklund (92) postulated that since the parabens caused leakage of cellular components, they are capable of neutralizing chemical and electrical forces which establish a normal membrane gradient. He found that parabens eliminated the ΔpH of the cytoplasmic membrane. In contrast, the compounds did not significantly affect the $\Delta\Psi$ component of the PMF (91). Eklund concluded that neutralization of the PMF and subsequent transport inhibition could not be the only mechanism of inhibition for the parabens. Gram-negative bacteria are probably resistant to the parabens because of a screening effect by the cell wall lipopolysaccharide layer (92, 110). Bargiota et al. (18) examined the relationship between lipid composition of *S. aureus* and resistance to parabens. Paraben-resistant strains had a greater percentage of total lipid, higher relative percentage of phosphatidylglycerol and decreased cyclopropane fatty acids than sensitive strains. These changes could influence membrane fluidity. Juneja and Davidson (158) altered the lipid composition of *L. monocytogenes* by growth in the presence of added fatty acids. In the presence of $C_{14:0}$ or $C_{18:0}$ fatty acids the microorganism showed increased resistance to parabens, whereas growth in the presence of $C_{18:1}$ increased sensitivity to parabens. Results indicated that for *L. monocytogenes*, a correlation existed between lipid composition of the cell membrane and susceptibility to parabens.

Table 33.1 Concentration ranges of esters of *p*-hydroxybenzoic acid necessary for total inhibition of various fungi[a]

Fungus	Concn (μg/ml)				
	Methyl	Ethyl	Propyl	Butyl	Heptyl
Alternaria sp.			100		50–100
Aspergillus flavus	>608	330–500	180–360	388	
Aspergillus niger	1,000	400–500	200–250	125–200	
Aspergillus parasiticus	530–>608	415–500	270–360	388	
Byssochlamys fulva			200		
Candida albicans	1,000	500–1,000	125–250	125	
Debaryomyces hansenii		400			
Fusarium graminearum	530	330	270	194	
Fusarium moniliforme	>608	330	180	290	
Fusarium oxysporum	608	330	180	290	
Penicillium citrinum	608	250	180	195	
Penicillium digitatum	500	250	63	<32	
Penicillium chrysogenum	500–>608	250–330	125–270	63–388	
Rhizopus nigricans	500	250	125	63	
Saccharomyces bayanus	930		220		
Saccharomyces cerevisiae	1,000	500	125–200	32–200	25–100
Torula utilis			200		25
Zygosaccharomyces bailii		900			
Zygosaccharomyces bisporus		400			
Zygosaccharomyces rouxii		700			

[a] pH values and incubation temperatures and times vary. Data from references 1, 154, 161, 165, 204, and 325.

Table 33.2 Concentration ranges of esters of *p*-hydroxybenzoic acid necessary for total inhibition of growth of various bacteria[a]

Bacterium	Concn (μg/ml)				
	Methyl	Ethyl	Propyl	Butyl	Heptyl
Gram-positive organisms					
Bacillus cereus	1,000–2,000	830–1,000	125–400	63–400	12
Bacillus megaterium	1,000		320	100	
Bacillus subtilis	1,980–2,130	1,000–1,330	250–450	63–115	
Clostridium botulinum	1,000–1,200	800–1,000	200–400	200	
Lactococcus lactis			400		12
Listeria monocytogenes	1,430–1,600		512		
Micrococcus sp.		60–110	10–100		
Sarcina lutea	4,000	1,000	400–500	125	12
Staphylococcus aureus	1,670–4,000	1,000–2,500	350–540	120–200	12
Streptococcus faecalis		130	40		
Gram-negative organisms					
Aeromonas hydrophila	550		100		
Enterobacter aerogenes	2,000	1,000	1,000	4,000	
Escherichia coli	1,200–2,000	1,000–2,000	400–1,000	1,000	
Klebsiella pneumoniae	1,000	500	250	125	
Pseudomonas aeruginosa	4,000	4,000	8,000	8,000	
Pseudomonas fluorescens	1,310		670		
Pseudomonas fragi			4,000		
Pseudomonas putida	450				
Salmonella serovar Typhimurium			180–>300		
Vibrio parahaemolyticus			50–100		
Yersinia enterocolitica	350				

[a] pH values and incubation temperatures and time vary. Data from references 1, 18, 86, 88, 93, 95, 158, 161, 165, 177, 183, 214, 251, 265, and 266.

To take advantage of their respective solubility and increased activity, methyl and propyl parabens are normally used in a combination of 2 to 3:1 (methyl:propyl). The compounds may be incorporated into foods by dissolving in water, ethanol, propylene glycol, or the food product itself. The *n*-heptyl ester is used in fermented malt beverages (beers) and noncarbonated soft drinks and fruit-based beverages. Parabens are used in a variety of foods including baked goods, beverages, fruit products, jams and jellies, fermented foods, syrups, salad dressings, wine, and fillings.

Sulfites

Sulfur dioxide (SO_2) and its salts have been used as disinfectants since the time of the ancient Greeks and Romans (129, 242). Salts of sulfur dioxide include potassium sulfite (K_2SO_3), sodium sulfite (Na_2SO_3), potassium bisulfite ($KHSO_3$), sodium bisulfite ($NaHSO_3$), potassium metabisulfite ($K_2S_2O_5$), and sodium metabisulfite (NaS_2O_5). As antimicrobials, sulfites are used primarily in fruit and vegetable products to control three groups of microorganisms: spoilage and fermentative yeasts and molds on fruits and fruit products (e.g., wine), acetic acid bacteria, and malolactic bacteria (242). In addition to use as antimicrobials, sulfites act as antioxidants and inhibit enzymatic and nonenzymatic browning in a variety of foods.

The most important factor impacting the antimicrobial activity of sulfites is pH. Sulfur dioxide and its salts exist as a pH-dependent equilibrium mixture when dissolved in water: $SO_2 \cdot H_2O \leftrightarrows HSO_3^- + H^+ \leftrightarrows SO_3^{2-} + H^+$. Aqueous solutions of sulfur dioxide theoretically yield sulfurous acid (H_2SO_3); however, evidence indicates that the actual form is more likely $SO_2 \cdot H_2O$ (130). As the pH decreases, the proportion of $SO_2 \cdot H_2O$ increases and the bisulfite (HSO_3^-) ion concentration decreases. The pK_a values for sulfur dioxide, depending on temperature, are 1.76 to 1.90 and 7.18 to 7.20 (130, 242, 282). The inhibitory effect of sulfites is most pronounced when the acid or $SO_2 \cdot H_2O$ is in the undissociated form. Therefore, their most effective pH range is less than 4.0. $SO_2 \cdot H_2O$ is 1,000, 500, and 100 times more active than HSO_3^- or SO_3^{2-} against *E. coli*, yeasts, and *Aspergillus niger*, respectively (267). Increased effectiveness at low pH is likely due to the ability of the un-ionized sulfur dioxide to pass across the cell membrane (261, 282).

Sulfur dioxide is fungicidal even in low concentrations. The inhibitory concentration range of sulfur dioxide is 0.1 to 20.2 µg/ml for *Saccharomyces, Zygosaccharomyces, Pichia, Hansenula,* and *Candida* species (267). Roland and Beuchat (279) found that sulfur dioxide inhibits *B. nivea* growth and patulin production in apple juice.

Sulfites, especially as the bisulfite ion, are very reactive. These reactions not only determine the mechanism of action of the compounds but also influence antimicrobial activity. For example, sulfites form addition compounds (α-hydroxysulfonates) with aldehydes and ketones, which reduces or eliminates their antimicrobial activity compared to the free forms. The most likely targets for inhibition by sulfites include disruption of the cytoplasmic membrane, inactivation of DNA replication, protein synthesis, inactivation of membrane-bound or cytoplasmic enzymes, or reaction with individual components in metabolic pathways. With *S. cerevisiae, Saccharomycodes ludwigii,* or *Z. bailii,* sulfites gain access to the cell by free diffusion (256, 320, 321). Other fungi have active transport systems for the compound. Sulfites dissipate the PMF and inhibit solute active transport (282). Hinze and Holzer (142) did experiments to demonstrate that the most sensitive enzyme in the Embden-Meyerhoff-Parnas pathway of *S. cerevisiae* was glyceraldehyde-3-phosphate dehydrogenase. Inactivation of this enzyme led to a decrease in cellular ATP. It is still uncertain as to whether the enzyme is the target, its cofactor NAD, or the substrate glyceraldehyde-3-phosphate.

Sulfites may be used to inhibit acetic acid-producing bacteria, lactic acid bacteria, and spoilage bacteria in meat products (129). At 100 to 200 µg/ml, sulfites inhibit *Acetobacter* sp. that cause wine spoilage (242). The concentration of sulfur dioxide required to inhibit lactic acid bacteria varies significantly depending on conditions but can be 1 to 10 µg/ml in fruit products at pH 3.5 or less (356). Sulfur dioxide is more inhibitory to gram-negative rods than to gram-positive rods. Banks and Board (14) tested several genera of *Enterobacteriaceae* isolated from sausage for their metabisulfite sensitivity. The microorganisms tested and the concentrations of free sulfite (in micrograms per milliliter) necessary to inhibit their growth at pH 7.0 were *Salmonella,* 15 to 109; *E. coli,* 50 to 195; *Citrobacter freundii,* 65 to 136; *Y. enterocolitica,* 67 to 98; *Enterobacter agglomerans,* 83 to 142; *Serratia marcescens,* 190 to 241; and *Hafnia alvei,* 200 to 241.

Sulfur dioxide and sulfites are used to control the growth of undesirable microorganisms in fruits, fruit juices, wines, sausages, fresh shrimp, and acid pickles, and during extraction of starches. They are added at 50 to 100 µg/ml to expressed grape juices used for making wines to inhibit molds, bacteria, and undesirable yeasts (5). At appropriate concentrations, sulfur dioxide does not interfere with wine yeasts or with the flavor of wine. During fermentation, sulfur dioxide also serves as an antioxidant, clarifier, and dissolving agent. The optimum level of sulfur dioxide (50 to 75 µg/ml) is maintained to prevent postfermentation changes by microorganisms. Researchers have also demonstrated a 4.3 log CFU/g reduction of *Salmonella* on dried apple slices that were exposed to sodium metabisulfite (4.18%) for 10 min prior to dehydration and storage (84). Similarly, metabisulfite has been shown to be effective in the control of *L. monocytogenes* on fruit. A 10-min immersion in a 4.18% metabisulfite solution of peach slices, followed by dehydration for 6 h, reduced *L. monocytogenes* by 4.3 log CFU/g (85). In some countries, sulfites may be used to inhibit the growth of microorganisms on fresh meat and meat products. Sulfite or metabisulfite added in sausages is effective in delaying the growth of molds, yeasts, and salmonellae during storage at refrigerated or room temperature (14). Sulfur dioxide restores a bright color but may give a false impression of freshness.

NATURALLY OCCURRING COMPOUNDS AND SYSTEMS

Many food products contain compounds which have antimicrobial activity. In the natural state, these compounds may play a role in extending the shelf life of a food product. In addition, many of these naturally occurring compounds have been studied for their potential as direct food antimicrobials. There are many problems associated with the use of natural compounds as direct food additives. An ideal naturally occurring antimicrobial would be effective enough to be added as a whole food or as an edible component (e.g., an herb or spice). Few, if any, antimicrobials are present in foods at concentrations great enough to be antimicrobials without some purification. Often, even if purification of an antimicrobial is possible, adding it to another food may lead to undesirable sensory changes. The ultimate challenge is to find a naturally occurring antimicrobial which can be added to a "microbiologically sensitive" food product in a nonpurified form from another nonsensitive food. The nonpurified food would have to contain an antimicrobial which is completely nontoxic and is highly effective in controlling the growth of microorganisms. This may well be impossible. According to Beuchat and Golden (27), the challenge is to isolate, purify, stabilize, and incorporate natural antimicrobials into foods without adversely affecting sensory, nutritional, or safety characteristics. This must be done without significant increases in costs for formulation, processing, or marketing.

Animal Sources

LPS

Lactoperoxidase is a glycoprotein enzyme that occurs in raw milk, colostrum, saliva, and other biological secretions. It has one heme group that contains Fe^{3+}. It has 610 amino acids and a molecular mass of 78 kDa (96). Bovine milk naturally contains 10 to 60 mg of lactoperoxidase per liter (96, 357). Lactoperoxidase reacts with thiocyanate (SCN^-) in the presence of hydrogen peroxide and forms antimicrobial compound(s); this is termed the lactoperoxidase system (LPS). Thiocyanate is found in many biological secretions, including milk. It is formed by the detoxification of thiosulfates, by metabolism of sulfur-containing amino acids, and in the diet through metabolism of various glucosides (357). Fresh milk contains 1 to 10 µg of thiocyanate per ml, which is not always sufficient to activate the LPS. Wilkins and Board (357) recommended addition of 10 to 12 µg of thiocyanate/ml. Hydrogen peroxide, the third component of the LPS, is not present in fresh milk due to the action of natural catalase, peroxidase, or superoxide dismutase. Approximately 8 to 10 µg of hydrogen peroxide per ml of milk is required for LPS. This can be added directly, through the action of lactic acid bacteria, or through the enzymatic action of xanthine oxidase, glucose oxidase, or sulfhydryl oxidase. The amount of hydrogen peroxide to activate the LPS is much lower than that required to pasteurize raw milk (ca. 300 to 800 µg/ml). In the LPS reaction, thiocyanate is oxidized to the antimicrobial hypothiocyanate ($OSCN^-$) which also exists in equilibrium with hypothiocyanous acid ($pK_a = 5.3$) (115, 271).

The LPS is more effective against gram-negative bacteria, including pseudomonads, than gram-positive bacteria (29). However, it does inhibit both gram-positive and gram-negative foodborne pathogens including *Salmonella*, *S. aureus*, *L. monocytogenes*, and *C. jejuni* (28, 114, 163, 306). There is variable activity against catalase-negative microorganisms, including the lactic acid bacteria. In resistant strains, the enzyme NADH: $OSCN^-$ oxidoreductase oxidizes NADH with reduction of $OSCN^-$ (45). The LPS system can increase the shelf life of raw milk in countries that have a poorly developed refrigerated storage system (96). Barrett et al. (21) found that the keeping quality of milk pasteurized at 72°C for 15 s was better than that of milk heated at 80°C for 15 s. Lactoperoxidase was completely inactivated at 80°C whereas residual LPS activity was detected following treatment at 72°C. Since higher levels of hypothiocyanite were detected in milk treated at the lower temperature, it was theorized that lactoperoxidase may have a role in the keeping quality of pasteurized milk (21). LPS has been shown to reduce growth of *S. aureus*, *L. monocytogenes*, *E. coli* O157:H7, *Salmonella*, *Y. enterocolitica*, and *P. aeruginosa* in ground beef, beef cubes, tomato juice, carrot juice, milk, and liquid whole egg compared to controls (99, 166, 330).

There are several potential mechanisms for inhibition by the LPS. The thiocyanate ion can oxidize sulfhydryl groups to disulfides, sulfenylthiocyanates (-S-SCN), or sulfenic acid (-S-OH) (96). Therefore, the compound may react with cysteine side chains on proteins and inactivate enzymes with functionally important sulfhydryl groups. Enzymes that are inhibited in vitro by thiocyanate include aldolase, glyceraldehyde phosphate dehydrogenase, hexokinase, lactate dehydrogenase, and 6-phosphogluconate dehydrogenase (96). Thiocyanate may also oxidize NADH or NADPH to the corresponding NAD or NADP. Through reaction with NADH or interference with membrane proteins, thiocyanate may cause leakage of the membrane or loss of electrochemical potential. This eventually leads to inhibition of transport of amino acids and sugars.

Chitosan

Chitosan, (1-4)-2-amino-2-deoxy-beta-D-glucan, is a natural constituent of fungal cell walls and is produced commercially from chitin, a by-product of shellfish processing. Chitosan is actually a series of polymers with different ratios of glucosamine and N-acetylglucosamine. Chitosan inhibits growth of foodborne molds, yeasts, and bacteria including *A. flavus*, *Byssochlamys*, *Botrytis cinerea*, *Mucor racemosus*, *Rhizopus stolonifer*, *S. cerevisiae*, and *Z. bailii* and *E. coli*, *Lactobacillus fructivorans*, *L. monocytogenes*, *Salmonella*, *S. aureus*, and *Y. enterocolitica* (245, 280, 322, 349). Reported MICs for bacteria and yeasts vary widely depending on the molecular weight of the polymer, degree of acetylation, pH, temperature, and the presence of interfering substances such as proteins and fats (56, 273, 280, 322, 334, 335). In combination with other antimicrobials like benzoates, however, the MIC of chitosan can be reduced. Sagoo et al. (289) reported that chitosan produced reductions of ~2 to 3 logs of *Saccharomyces* and *Saccharomycodes* spp. over 2 h at 0.05% and pH 6.2. Upon mixing 0.005% chitosan with 0.025% sodium benzoate at pH 4 or 5, yeast populations were reduced by 2 to 4 logs in only 60 min (289). Zivanovic et al. (368) evaluated the antimicrobial efficiency of chitosan polysaccharide or oligosaccharide in oil-in-water emulsions against *L. monocytogenes* or *Salmonella* serovar Typhimurium. Chitosan polysaccharide in acetic acid was bactericidal to both *L. monocytogenes* and serovar Typhimurium at 25°C, but the reduction rate depended on the strain. Chitosan oligosaccharide was less effective against all tested strains and showed a concentration-dependent effect. Chitosan was added to mayonnaise inoculated with 5 to 6 log CFU of two spoilage microorganisms per g in dressings, *Z. bailii* or *Lactobacillus fructivorans*. Storage at 5 or 25°C resulted in total inactivation of *L. fructivorans* and an initial 1- to 2-log decrease in *Z. bailii* at 25°C (234, 280). One recent use of chitosan has been to generate films of the polymer with antimicrobials attached to provide intimate contact with the surface of the food product, effectively making the packaging a functional component of the whole food product. Chitosan films generated with garlic oil, sorbic acid, and nisin were shown to be highly inhibitory towards *E. coli*, *Salmonella*, *Staphylococcus*, *L. monocytogenes*, and *Bacillus* spp. without loss of acceptability with respect to mechanical and physical properties of the film (260). Park et al. (247) found that chitosan-based coatings applied on strawberries inhibited growth of inoculated *Cladosporium* and *Rhizopus* as well as the total aerobic count and coliforms and reduced weight loss of strawberries during storage. A coating of chitosan (9 mg/g) on shrimp in a salad reduced spoilage microflora by around 4 log CFU/g for 4 weeks at 5°C (280). Chitosan may directly affect the microbial cell by interaction with the anionic cell wall polysaccharides or components of the cytoplasmic membrane, resulting in altered permeability or prevention of transport (104, 334). Sagoo et al. (289) speculated that these interactions may allow for enhanced inhibition of yeasts by benzoic acid. It has also been theorized that chitosan may interfere with protein synthesis and/or may inhibit cells via chelation of metals, spore elements, or nutrients (81).

Avidin

Avidin is a glycoprotein present in egg albumen. The concentration varies with the hen's age, but the mean is 0.05% of the total egg albumen protein (331). The protein has a molecular mass of 66 to 69 kDa and four identical subunits of 128 amino acids each. It is stable to heat and a wide pH range. Avidin strongly binds the cofactor biotin at a ratio of four molecules of biotin per molecule of avidin. Biotin is a cofactor for enzymes in the tricarboxylic acid cycle and fatty acid biosynthesis. While the biological role of the protein is unknown, it may be secreted by macrophages and play a role in the immune system (331). Avidin inhibits the growth of some bacteria and yeasts that have a requirement for biotin, with the primary mechanism being

nutrient deprivation. However, Korpela (172) investigated another potential mechanism and found that avidin can bind porin proteins in the outer membrane of *E. coli*. This finding suggested that avidin may inhibit microorganisms in vitro by interfering with transport through porins.

Ovotransferrin

Another iron-binding molecule, ovotransferrin or conalbumin, occurs in egg albumen. This 77- to 80-kDa glycoprotein makes up 10 to 13% of the total egg white protein (357). Ovotransferrin has 49% sequence homology with lactoferrin (331). Each ovotransferrin molecule has two iron binding sites and, like lactoferrin, it binds an anion such as bicarbonate or carbonate with each ferric iron bound. The compound is 80% denatured by heating at 60°C for 5 min. To be inhibitory, ovotransferrin must be in stoichiometric excess of iron and the pH must be in the alkaline range (above 7.5) (332, 357). Ovotransferrin is inhibitory to both gram-positive and gram-negative bacteria, but the former are generally more sensitive. *Bacillus* and *Micrococcus* species are particularly sensitive (331). Some yeasts are also sensitive (345). As with lactoferrin, one of the primary mechanisms suggested for ovotransferrin is deprivation of iron from microorganisms that require the element. However, Tranter and Board (333) suggested that inhibition may be partially or completely related to the cationic nature of the protein. Certain gram-positive bacteria and yeasts are inhibited in egg albumen with or without iron (333).

Plant Sources

Spices and Their Essential Oils

Spices are roots, bark, seeds, buds, leaves, or fruit of aromatic plants added to foods as flavoring agents. However, it has been known since ancient times that spices and their essential oils have varying degrees of antimicrobial activity. The earliest report on use of spices as preservatives was around 1550 B.C., when the ancient Egyptians used spices for food preservation and embalming the dead. Cloves, cinnamon, oregano, and thyme and to a lesser extent sage and rosemary have demonstrated the greatest antimicrobial activity in most studies.

The major antimicrobial components of clove (*Syzygium aromaticum*) and cinnamon (*Cinnamomum zeylanicum*) are eugenol [2-methoxy-4-(2-propenyl)-phenol] (Fig. 33.5) and cinnamic aldehyde (3-phenyl-2-propenal) (Fig. 33.5), respectively. Cinnamon contains 0.5 to 10% volatile oil, of which 75% is cinnamic aldehyde and 8% is eugenol, while cloves contain 14 to 21% volatile oil, 95% of which is eugenol. Cinnamon and cinnamic aldehyde have demonstrated antimicrobial activity against

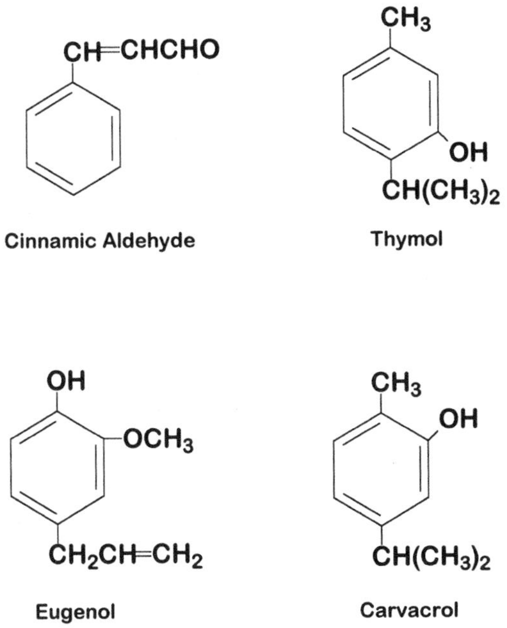

Figure 33.5 Examples of antimicrobial compounds in spice essential oils.

the bacteria *A. hydrophila*, *Bacillus* spp., *C. jejuni*, *E. coli* O157:H7 and other verotoxin-producing *E. coli* organisms, *Lactobacillus*, *L. monocytogenes*, salmonellae, *Shigella* spp., *S. aureus*, and *Streptococcus* and the fungi *Aspergillus*, *Candida*, *Penicillium*, and *Saccharomyces* (10, 52, 66, 70, 102, 106, 111, 122, 139, 229, 307, 308). Clove and eugenol are inhibitory to *A. hydrophila*, *Bacillus* spp., *C. jejuni*, *E. coli* O157:H7, *Lactobacillus*, *L. monocytogenes*, salmonellae, *Shigella* spp., *S. aureus*, and *Streptococcus* and the fungi *Aspergillus*, *Candida*, *Penicillium*, and *Saccharomyces* (10, 12, 25, 102, 111, 116, 122, 182, 192, 193, 215, 229, 307, 308). Mytle et al. (222) studied the inhibitory effect of clove oil on growth of *L. monocytogenes* on hot dogs. At 1 and 2% clove oil, *L. monocytogenes* growth was inhibited at 5 or 15°C. In contrast, Singh et al. (304) found little inhibition of *L. monocytogenes* inoculated on the surface of hot dogs of various fat percentages that were subsequently dipped in aqueous suspensions of 1 or 10 ml of clove oil per liter. Hao et al. (136, 137) applied eugenol to beef slices or cooked chicken, and the products were inoculated with *A. hydrophila* or *L. monocytogenes* and stored at 5 or 15°C. Eugenol inhibited growth of both microorganisms; however, *L. monocytogenes* was less sensitive. Stecchini et al. (317) showed that essential oil of clove at 500 μg/ml inhibits *A. hydrophila* growth in microbiological media and in vacuum-packaged and air-packed cooked pork. Ceylan et al. (52) studied the

effect of cinnamon against *E. coli* O157:H7 in apple juice at 8 and 25°C. With 0.3% cinnamon, *E. coli* O157:H7 was reduced by 1.6 and 2.0 log CFU/ml at 8 and 25°C, respectively. Azzouz and Bullerman (11a) evaluated 16 ground herbs and spices at 2% (wt/vol) against nine mycotoxin-producing *Aspergillus* and *Penicillium* species. The most effective antimicrobial spice evaluated was clove, which inhibited growth initiation at 25°C by all species for over 21 days. Cinnamon was the next most effective spice, inhibiting three *Penicillium* species for over 21 days. Bullerman (41) determined that 1.0% cinnamon in raisin bread inhibits growth and aflatoxin production by *A. parasiticus*.

The antimicrobial activities of oregano (*Origanum vulgare*) and thyme (*Thymus vulgaris*) have been attributed to their essential oils, which contain carvacrol (2-methyl-5-[1-methylethyl] phenol) (Fig. 33.5) and thymol (5-methyl-2-[1-methylethyl] phenol) (Fig. 33.5), respectively. The essential oils and isolated compounds have inhibitory activity against a number of bacterial species, molds, and yeasts including *Aeromonas* spp., *B. cereus*, *B. thermosphacta*, *C. jejuni*, *E. coli*, *E. faecalis*, *L. plantarum*, *L. monocytogenes*, *Pediococcus cerevisiae*, *Pseudomonas* spp., *Proteus* spp., salmonellae, *Shigella* spp., *S. aureus*, *V. parahaemolyticus*, *Y. enterocolitica*, and *Aspergillus*, *Candida*, *Geotrichum*, *Penicillium*, *Pichia* spp., *Rhodotorula*, *S. cerevisiae*, and *Schizosaccharomyces pombe* (4, 10, 12, 13, 25, 42–44, 54, 70, 73, 83, 98, 103, 111, 135, 139, 168, 170, 175, 192, 193, 201, 236, 237, 243, 253, 272, 284, 288, 293, 304, 341, 342).

Sage (*Salvia officinalis*) and rosemary (*Rosmarinus officinalis*) also have antimicrobial activity (42). Rosemary contains primarily borneol (endo-1,7,7-trimethylbicyclo[2.2.1] heptan-2-ol) along with pinene, camphene, and camphor, while sage contains thujone (4-methyl-1-[1-methylethyl]bicyclo[3.1.0]-hexan-3-one). At 2% in growth medium, sage and rosemary are more active against gram-positive than gram-negative bacterial strains (297). The inhibitory effect of these two spices at 0.3% was bacteriostatic, while at 0.5% they were bactericidal to gram-positive strains. Of 18 spices tested against *L. monocytogenes* in culture medium, Pandit and Shelef (244) reported the most effective compound to be rosemary. The most inhibitory fraction of rosemary was α-pinene. Smith-Palmer et al. (307) demonstrated that rosemary (0.02 to 0.05%) and sage (0.02 to 0.075%) were inhibitory to gram-positive *L. monocytogenes* and *S. aureus* but not to gram-negative bacteria. In refrigerated fresh pork sausage, *L. monocytogenes* Scott A growth was delayed by 0.5% ground rosemary or 1% rosemary essential oil (244). In contrast, others (98, 200, 288) found that essential oil of rosemary had little antimicrobial activity against bacteria and yeasts in agar diffusion assays. The lack of activity may be due to the assays utilized. Sensitivity of *B. cereus*, *S. aureus*, and *Pseudomonas* to sage is greatest in microbiological medium and significantly reduced in rice and chicken and noodles (296). It was theorized that loss of activity was due to solubilization of the antimicrobial fraction in the lipid of the foods. Hefnawy et al. (138) evaluated 10 herbs and spices against two strains of *L. monocytogenes* in tryptose broth. The most effective spice was sage, which at 1% decreased viable *L. monocytogenes* by 5 to 7 logs after 1 day at 4°C. In contrast, Araujo et al. (9) and Hammer et al. (135) reported little antibacterial or antifungal activity with sage extracts.

Sweet basil (*Ocimum basilicum*) essential oil has demonstrated antimicrobial activity, with linalool and methyl chavicol being the primary antimicrobial agents. Basil essential oil extract was active against certain fungi, including *Mucor* and *Penicillium* species, but had little activity against bacteria (173). Wan et al. (348) screened the essential oil components of sweet basil, linalool and methyl chavicol, against 35 strains of bacteria, yeasts, and molds. Again, the compounds demonstrated limited activity against most microorganisms except *Mucor* and *Penicillium*. In contrast, methyl chavicol (0.1%) in filter-sterilized fresh lettuce supernatant reduced viable *A. hydrophila* by 5 log CFU/g and, as a wash for lettuce leaves, the compound was as effective as 125-μg/ml chlorine (348). Smith-Palmer et al. (307) reported MICs for basil essential oil of 0.25, 0.25, 0.1, 0.05, and 0.1% for *C. jejuni*, *E. coli*, *Salmonella enterica* serovar Enteritidis, *L. monocytogenes*, and *S. aureus*, respectively. Yin and Cheng (365) detected no antifungal activity against *A. flavus* or *A. niger* with water extracts of basil or ginger in a paper disk assay.

Vanillin (4-hydroxy-3-methoxybenzaldehyde) is a major constituent of vanilla beans, the fruit of an orchid (*Vanilla planifola*, *Vanilla pompona*, or *Vanilla tahitensis*). Vanillin is most active against molds and nonlactic gram-positive bacteria. López-Malo et al. (190) prepared fruit-based agars containing mango, papaya, pineapple, apple, and banana with up to 2,000 μg of vanillin per ml and inoculated each with *A. flavus*, *A. niger*, *A. ochraceus*, or *A. parasiticus*. Vanillin at 1,500 μg/ml significantly inhibited all strains of *Aspergillus* in all media. Cerrutti and Alzamora (50) demonstrated complete inhibition of growth for 40 days at 27°C of *Debaryomyces hansenii*, *S. cerevisiae*, *Z. bailii*, and *Zygosaccharomyces rouxii* in laboratory media and apple puree at a_w of 0.99 and 0.95 by 2,000-μg/ml vanillin. In contrast, 2,000-μg/ml vanillin was not effective against the yeasts in banana puree. These researchers attributed this loss of activity to binding

of the phenolic vanillin by protein or lipid in the banana and mango, a phenomenon demonstrated for other antimicrobial phenolic compounds (276). López-Malo et al. (189, 191) studied the effect of vanillin concentration, pH, and incubation temperature on *A. flavus*, *A. niger*, *A. ochraceus*, and *A. parasiticus* growth on potato dextrose agar. Depending on the species, vanillin in combination with reduced pH had an additive or synergistic effect on growth of the molds. The most sensitive was *A. ochraceus* (MIC = 500 μg/ml at pH 3.0 and ≤25°C or pH 4.0 and ≤15°C), and the most resistant was *A. niger* (MIC = 1,000 μg/ml, pH 3.0, 15°C). Cerrutti et al. (51) utilized vanillin to produce a shelf-stable strawberry puree. A combination of 3,000-μg/ml vanillin, 1,000-μg/ml calcium lactate, and 500-μg/ml ascorbic acid preserved strawberry puree acidified with citric acid to pH 3.0 and adjusted to an a_w of 0.95 with sucrose for over 60 days against growth of natural microflora and inoculated yeasts. Delaquis et al. (76) studied the antimicrobial efficacy of vanillin and vanillic acid on *L. monocytogenes*, *L. innocua*, *L. grayi*, and *L. seeligeri* at various pHs. Vanillin was inhibitory at pH 5 to 8, but vanillic acid was effective only at pH 5.

Several other spice essential oils shown to have potential for antibacterial and antifungal activity include cilantro (coriander; *Coriandrum sativum*) (79, 119), fingerroot (*Boesenbergia pandurata*) (327), lemongrass (*Cymbopogon citratus*) (111), savory (*Satureja* sp.) (61, 67, 134, 284), and tea tree oil (*Melaleuca alternifolia*) (68). Limited or no activity has been demonstrated with anise, bay (laurel), black pepper, cardamom, cayenne (red pepper), celery seed, chili powder, cumin, curry powder, dill, fenugreek, ginger, juniper oil, mace, marjoram, nutmeg, orris root, paprika, sesame, spearmint, tarragon, turmeric, and white pepper (42, 73, 203). Because of their hydrophobicity, there is a general need to increase the concentration of essential oils when utilized in foods versus microbiological media (in vitro). This is because the essential oils interact with hydrophobic food components (lipids and hydrophobic areas on proteins) resulting in less available active components to act against microorganisms. For example, it has been reported that the high protein content in milk reduces the inhibitory activity of carvacrol and clove oil against *B. cereus* and *Salmonella* serovar Enteritidis, respectively (258, 309). Further, the physical structure of foods may potentiate or inhibit the action of essential oils; diffusion through food gels and emulsion characteristics have been demonstrated to retard inhibition of essential oils against pathogens such as *L. monocytogenes* and *Y. enterocolitica* (42). In contrast, Gaysinsky et al. (116) demonstrated that eugenol encapsulated in surfactant micelles was as effective as free eugenol but was soluble in the aqueous phase.

Most studies on the mechanism of action of spice essential oil components have focused on the effect of the compounds on the cytoplasmic membrane of the target microorganisms. For example, Ultee et al. (342, 343) determined that carvacrol depleted intracellular ATP, reduced membrane potential, and increased permeability of the cytoplasmic membrane of *B. cereus*, leading to inhibition and eventually cell death. Similarly, studies have shown tea tree oil and β-pinene to affect potassium and proton leakage in *E. coli* and in yeasts (42, 132). Other reported effects on the cytoplasmic membrane by spice essential oil components are dissipation of PMF components, interference with ATP generation, enzyme inhibition resulting in prevention of substrate utilization, and prevention of oxidative respiration (144). Gill and Holley (122) reported that eugenol inhibited glucose uptake or utilization by *L. monocytogenes* and *Lactobacillus sakei* possibly. They reported that eugenol could function by inhibition of membrane-bound ATPase activity, inhibition of cellular metabolic enzymes, or disruption of membranes. Further, they reported that cinnamaldehyde causes disruption of the cytoplasmic membrane so as to disperse the PMF.

Onions and Garlic

Probably the most well characterized antimicrobial system in plants is that found in the juice and vapors of onions (*Allium cepa*) and garlic (*Allium sativum*). Growth and toxin production of many microorganisms have been shown to be inhibited by onion and garlic including the bacteria *B. cereus*, *C. botulinum* type A, *E. coli*, *L. plantarum*, *Salmonella*, *Shigella*, and *S. aureus* and the fungi *A. flavus*, *A. parasiticus*, *Candida albicans*, *Cryptococcus*, *Rhodotorula*, *Saccharomyces*, *Torulopsis*, and *Trichosporon* (26, 65, 126).

Cavallito and Bailey (49) identified the major antimicrobial component of garlic as allicin (diallyl thiosulfinate; thio-2-propene-1-sulfinic acid-5-allyl ester) (Fig. 33.6). Allicin is formed by the action of the enzyme allinase on the substrate alliin (S-[2-propenyl]L-cysteine sulfoxide). The reaction occurs only when cells of the garlic are disrupted, releasing the enzyme to act on the substrate. A similar reaction occurs in onion, except that the substrate is (S-[1-propenyl]L-cysteine sulfoxide) and one of the major products is thiopropanal-S-oxide. The products apparently responsible for antimicrobial

$$CH_2{=}CH{-}CH_2{-}\underset{\underset{O}{\|}}{S}{-}S{-}CH_2{-}CH{=}CH_2$$

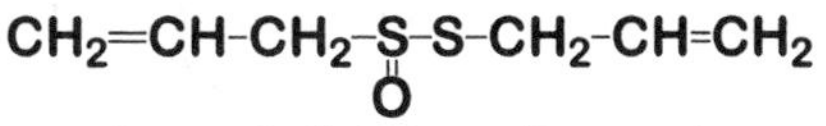

Figure 33.6 Allicin (diallyl thiosulfinate; thio-2-propene-1-sulfinic acid-5-allyl ester).

activity are also responsible for the flavor of onions and garlic. In addition to antimicrobial sulfur compounds, onions contain the phenolic compounds protocatechuic acid and catechol, which could contribute to their antimicrobial activity (347).

The mechanism of action of allicin is most likely inhibition of sulfhydryl-containing enzymes (26). Wills (358) found that 0.5 mM allicin inhibited many sulfhydryl (alcohol dehydrogenase, choline esterase, choline oxidase, glyoxylase, hexokinase, papain, succinic dehydrogenase, urease, and xanthine oxidase) and some nonsulfhydryl enzymes (lactic dehydrogenase, tyrosinase, and alkaline phosphatase). Barone and Tansey (20) theorized that allicin inactivates proteins by oxidizing thiols to disulfides and inhibiting the intracellular reducing activity of glutathione and cysteine.

Hops

Resin from the flowers of the hop vine (*Humulus lupulus* L.) is used in the brewing industry for imparting a desirable bitter flavor to beer. About 3 to 12% of the resin is composed of α-bitter acids including humulone (humulon), cohumulone, and adhumulone and β-bitter acids including lupulone (lupulon), colupulone, xanthohumol, and adlupulone (238). Both types of bitter acids possess principal antimicrobial activity against gram-positive bacteria and fungi at low water activity (26). Lactic acid bacteria that spoil beer, including species of *Lactobacillus* and *Pediococcus*, were resistant to the antimicrobial effect of humulone, colupulone, and *trans*-isohumulone, while the same species that did not spoil beer were sensitive (105, 303). *E. coli* and *B. subtilis* are naturally resistant to both the iso-α acid and β-acids, while strains of *Bacillus megaterium*, *Enterococcus salivarius*, *L. monocytogenes*, and *S. aureus* are susceptible (133, 176, 209). There is little definitive research on the antimicrobial mechanism of hop extract bitter acids. However, there is evidence that the mechanism may involve the cytoplasmic membrane since the activity is selectively increased against gram-positive bacteria and fungi and is reduced by lipid (176). In addition, hop bitter acids inhibit growth of beer spoilage bacteria (gram positives) by dissipating transmembrane pH gradient (302). However, Fernandez and Simpson (105) found that resistance to *trans*-isohumulone does not confer resistance to sorbic acid or benzoic acid, indicating a potential difference in targets for the mechanism of antimicrobial action.

Isothiocyanates

Isothiocyanates (R—N=C=S) are derivatives from glucosinolates in cells of plants of the Cruciferae or mustard family (cabbage, kohlrabi, Brussels sprouts, cauliflower, broccoli, kale, horseradish, mustard, turnips, and ruta-baga). These compounds are formed from the action of the enzyme myrosinase (thioglucoside glucohydrolase EC 3.2.3.1) on the glucosinolates when the plant tissue is injured or mechanically disrupted. In addition to the allyl side group, other isothiocyanate side groups include ethyl, methyl, benzyl, and phenyl. These compounds have been reported to be potent antimicrobial agents (77, 78).

Isothiocyanates are inhibitory to fungi, yeasts, and bacteria in the range of 16 to 110 ng/ml in the vapor phase (152) and 10 to 600 μg/ml in liquid media (202). Inhibition against bacteria varies, but generally gram-positive bacteria are less sensitive to allyl isothiocyanate (AIT) than gram-negative bacteria. Delaquis and Mazza (77) found a 1- to 5-log decrease in viable cells of *E. coli*, *L. monocytogenes*, and *Salmonella* serovar Typhimurium in the presence of 2,000 μg of AIT per ml of air. Delaquis and Sholberg (78) examined this effect further and showed that 1,000 μg of AIT per ml of air apparently decreased viable *E. coli* O157:H7, serovar Typhimurium, and *L. monocytogenes* by up to 6 logs. However, cells recovered to a large extent if they were exposed to air. *E. coli* O157:H7 was the most resistant. Growth of *P. expansum*, *A. flavus*, and *Botrytis cinerea* was inhibited at 100 μg of AIT per ml of air (78). Ward et al. (351) prepared horseradish essential oil distillate (~90% AIT) and applied it to the headspace of cooked roast beef or agar inoculated with *E. coli* O157:H7, *L. monocytogenes*, serovar Typhimurium, *S. aureus*, *Serratia grimesii*, and *Lactobacillus sake*. AIT at 2 μl/liter inhibited all the microorganisms on the agar, but 20 μl/liter was required to inhibit the pathogens and spoilage microorganisms on the beef. On beef, *L. monocytogenes*, *S. grimesii*, and *L. sake* were the most resistant (351). Delaquis et al. (80) added 20 μl of horseradish essential oil per liter with pre-cooked roast beef slices. The beef was stored for 28 days at 4°C, and inoculated spoilage bacteria were monitored. *Pseudomonas* and *Enterobacteriaceae* were inhibited to the greatest extent, while lactic acid bacteria were more resistant. The development of off-odors and flavors was delayed, and cooked meat color was preserved in the treated roasts. Nadarajah et al. (223) added AIT to fresh ground beef inoculated with *E. coli* O157:H7 and stored it under nitrogen at 4, 10, or −18°C. AIT reduced the pathogen to undetectable levels at 4 and −18°C from an initial 3.0 log CFU/g and from 1 (10 and −18°C) to >3 (4°C) logs from an initial 6 log CFU/g.

The mechanism by which isothiocyanates inhibit cells may be due to inhibition of enzymes by direct reaction with disulfide bonds or through thiocyanate (SCN⁻) anion reaction to inactivate sulfhydryl enzymes (77). Lin et al. (188) suggested that AIT inhibited *Salmonella enterica* serovar Montevideo and *E. coli* O157:H7 through disruption of the cell membrane and leakage of intracellular material. While these compounds have very low sensory thresholds, it has been suggested that they may be useful as food antimicrobials due to their low inhibitory concentrations.

Phenolic Compounds

Simple phenolic compounds include monophenols (e.g., *p*-cresol), diphenols (e.g., hydroquinone), and triphenols (e.g., gallic acid). Gallic acid occurs in plants as quinic acid esters or hydrolyzable tannins (tannic acid) (143). The use of simple phenols for preservation is found in the application of wood smoke. Smoking of foods such as meats, cheeses, fish, and poultry not only imparts a desirable flavor but also has a preservative effect through drying and the deposition of chemicals. While many chemicals are deposited on smoked foods, the major contributors of flavor and antioxidant and antimicrobial effects are phenol and cresol. Liquid smoke on the surface of Cheddar cheese inhibits growth of *Aspergillus oryzae*, *Penicillium camemberti*, and *Penicillium roqueforti* (354). Of the eight major phenolic compounds in liquid smoke, isoeugenol is the most effective antifungal compound followed by *m*-cresol and *p*-cresol.

The phenolic acids, including derivatives of *p*-hydroxybenzoic acid (protocatechuic, vanillic, gallic, syringic, and ellagic) and *o*-hydroxybenzoic acid (salicylic), may be found in plants and plant foods. Tannic acid is inhibitory to *A. hydrophila*, *E. coli*, *L. monocytogenes*, *Salmonella* serovar Enteritidis, *S. aureus*, and *E. faecalis* (63, 251). The intestinal microorganisms *Bacteroides fragilis*, *C. perfringens*, *E. coli*, and serovar Typhimurium strains were inhibited by both tannic acid and propyl gallate, while neither *Bifidobacterium infantis* nor *Lactobacillus acidophilus* was affected (62). Chung et al. (62) theorized that tannic acid inhibited by binding iron but that propyl gallate had a different inhibitory mechanism. Stead (315) demonstrated stimulation of growth of the lactic acid spoilage bacteria *Lactobacillus collinoides* and *Lactobacillus brevis* by gallic, quinic, and chlorogenic acids. He suggested that the compounds were being metabolized by the microorganisms and concluded that the presence of these compounds in fruit-based beverages may result in antagonistic reactions with other food grade antimicrobials.

Hydroxycinnamic acids include caffeic, *p*-coumaric, ferulic, and sinapic acids. They frequently occur as esters and less often as glucosides (143). Herald and Davidson (140) demonstrated that ferulic acid at 1,000 μg/ml and *p*-coumaric acid at 500 and 1,000 μg/ml inhibit the growth of *B. cereus* and *S. aureus*. The compounds were much less effective against *P. fluorescens* and *E. coli*. In contrast, alkyl esters of hydroxycinnamic acids including methyl caffeoate, ethyl caffeoate, propyl caffeoate, methyl *p*-coumarate, and methyl cinnamate were effective inhibitors of the growth of *P. fluorescens* (16). Lyon and McGill (195) tested the antimicrobial activity of caffeic, cinnamic, ferulic, salicylic, sinapic, and vanillic acids on *Erwinia carotovora* and demonstrated total inhibition at pH 6.0 in nutrient broth with 0.5 µg of cinnamic, ferulic, salicylic, or vanillic acid per ml. While effective as growth inhibitors of *E. carotovora*, none of these phenolic compounds inhibit the pectolytic enzymes, polygalacturonic acid lyase and polygalacturonase, produced by the organism except caffeic acid, which inhibits only the latter (196). Wen et al. (353) tested chlorogenic, caffeic, *p*-coumaric, and ferulic acids against *L. monocytogenes*. MICs were 0.20, 0.22, 0.27, and 0.29% (all wt/vol) for the cinnamic, *p*-coumaric, ferulic, and caffeic acids, respectively. Chlorogenic acid was ineffective at 1.0% (wt/vol). Mixtures of the acids demonstrated additive interactive inhibition. Caffeic acid in apple juice at concentrations of 0.2 to 1.0 g/liter reduced survival of stationary- or log-phase *E. coli* O157:H7 cells (268). Stead (316) determined the effects of caffeic, coumaric, and ferulic acids against the wine spoilage lactic acid bacteria *L. collinoides* and *Lactobacillus brevis*. At pH 4.8 in the presence of 5% ethanol, *p*-coumaric and ferulic acids were the most inhibitory compounds at 500 or 1,000 µg/ml. At 100 µg/ml, all three hydroxycinnamic acids stimulate growth of the microorganisms, suggesting that these compounds may play a role in initiating the malolactic fermentation of wines.

Many of the studies with hydroxycinnamic acids have involved their antifungal properties. Chipley and Uraih (60) showed that ferulic acid inhibited aflatoxin B_1 and G_1 production by *A. flavus* and *A. parasiticus* by up to 75%. Salicylic and *trans*-cinnamic acids totally inhibited aflatoxin production at the same level. Baranowski et al. (15) studied the effects of caffeic, chlorogenic, *p*-coumaric, and ferulic acids at pH 3.5 on the growth of *S. cerevisiae*. Caffeic and chlorogenic acids had little effect on the organism at 1,000 µg/ml. In the presence of *p*-coumaric acid, however, the organism was completely inhibited by 1,000 µg/ml. Ferulic acid was the most effective growth inhibitor tested. At 50 µg/ml, this compound extended the lag phase of *S. cerevisiae*, and at 250 µg/ml, growth of the organism was completely inhibited. The degree of inhibition is inversely related to the polarity of the compounds.

Furocoumarins are related to the hydroxycinnamates. These compounds, including psoralen and its derivatives, are phytoalexins in citrus fruits, parsley, carrots, celery, and parsnips. Purified psoralen and natural sources of the compound (e.g., cold-pressed lime oil and lime peel extract) have demonstrated antimicrobial activity

against *E. coli* O157:H7, *Erwinia carotovora*, *L. monocytogenes*, and *Micrococcus luteus* following irradiation with long-wave (365-nm) UV light (199, 339). These compounds inhibit growth by interfering with DNA replication. This occurs when the furocoumarin and DNA are exposed to UV light at 365 nm, which causes monoadducts of DNA and cross-linking of the furocoumarin and DNA (221).

The flavonoids include catechins, flavons, flavonols, and their glycosides (143). Proanthocyanidins or condensed tannins are polymers of favan-3-ol and are found in apples, grapes, strawberries, plums, sorghum, and barley (143). Cushnie and Lamb (69) reported that certain of the flavonoids had antifungal, antiviral, and antibacterial activity. Benzoic acid, proanthocyanidins, and flavonols account for 66% of cranberry microbial inhibition against *Saccharomyces bayanus*, with the last two being the most important (204).

Much of the research on the mechanism of phenolic compounds has focused on their effect on the cytoplasmic membrane. Simple phenols disrupt the cytoplasmic membrane and cause leakage of cells (71). Phenolics may also inhibit cellular proteins directly. Rico-Muñoz et al. (275) investigated the effect of the naturally occurring propyl gallate and *p*-coumaric, ferulic, and caffeic acids on the activity of membrane-bound ATPase of *S. aureus*. All of the compounds inhibited ATPase activity. It was suggested that compounds that inhibited ATPase interact directly with the enzyme, possibly through binding. The authors concluded that all phenolic compounds probably do not have the same mechanism of action and there may be several targets which lead to inhibition of microorganisms by these compounds. Chipley (58) found that 0.1 mM 2,4-dinitrophenol noncompetitively inhibits enzymes associated with the cell envelopes of both *E. coli* and *Salmonella* serovar Enteritidis. In contrast, with the same enzymes in a purified isolated state, no inhibition was demonstrated. These results suggested that a conformational change in the membrane of the vesicle causes inhibition of the enzymes and that there is no direct effect of dinitrophenol.

CONCLUSIONS

Traditional food antimicrobials are an important tool in preserving foods from the hazardous and detrimental effects of microorganisms. In the future, they will continue to play an important role in food preservation. However, it is doubtful that any newly discovered synthetic antimicrobials will be approved for use in foods by worldwide regulatory agencies. Therefore, the future of traditional antimicrobials lies in novel uses (e.g., new foods or in combination with other preservation procedures), in combinations designed on the basis of their mechanisms so as to obtain a wide spectrum of activity against microorganisms, and on systems for better and more efficient delivery of antimicrobials to their targets.

While extensive research on sources and activities of natural antimicrobials has taken place, more information is needed on activity spectra, mechanisms, targets, and interactions with environmental conditions and food components. Prior to their regulatory approval as food antimicrobials, they will have to be proven toxicologically safe. The latter point, along with potential negative sensory effects and cost, could be the greatest hurdle to their future use.

References

1. Aalto, T. R., M. C. Firman, and N. E. Rigler. 1953. p-Hydroxybenzoic acid esters as preservatives. I. Uses, antibacterial and antifungal studies, properties and determination. *J. Am. Pharm. Assoc.* **42**:449–458.

2. Abdul-Raouf, U. M., L. R. Beuchat, and M. S. Ammar. 1993. Survival and growth of *Escherichia coli* O157:H7 in ground, roasted beef as affected by pH, acidulants, and temperature. *Appl. Environ. Microbiol.* **59**:2364–2368.

2a. Alakomi, H.-L., E. Skyttä, M. Saarela, T. Mattila-Sandholm, K. Latva-Kala, and I. M. Helander. 1999. Lactic acid permeabilizes gram-negative bacteria by disrupting the outer membrane. *Appl. Environ. Microbiol.* **66**:2001–2005.

3. Al-Dagal, M. M., and W. A. Bazaraa. 1999. Extension of shelf life of whole and peeled shrimp with organic acid salts and bifidobacteria. *J. Food Prot.* **62**:51–56.

4. Aligiannis, N., E. Kalpoutzakis, S. Mitaku, and I. B. Chinou. 2001. Composition and antimicrobial activity of the essential oils of two *Origanum* species. *J. Agric. Food Chem.* **49**:4168–4170.

5. Amerine, M. A., and M. A. Joslyn. 1970. *Table Wines: the Technology of Their Production*, 2nd ed. Univ. Calif. Press, Berkeley.

6. Anonymous. 1998. Direct food substances affirmed as generally recognized as safe; egg white lysozyme. *Fed. Regist.* **63**:12421–12426.

7. Anonymous. 1999. Sorbic acid and potassium sorbate for preserving freshness. Public. ZS-1D. Eastman Chemical Co., Kingsport, Tenn.

8. Aran, N. 2001. The effect of calcium and sodium lactates on growth from spores of *Bacillus cereus* and *Clostridium perfringens* in a "sous-vide" beef goulash under temperature abuse. *Int. J. Food Microbiol.* **63**:117–123.

9. Araujo, C., M. J. Sousa, M. F. Ferreira, and C. Leao. 2003. Activity of essential oils from Mediterranean Lamiaceae species against food spoilage yeasts. *J. Food Prot.* **66**:625–632.

10. Aureli, P., A. Costantini, and S. Zolea. 1992. Antimicrobial activity of some plant essential oils against *Listeria monocytogenes*. *J. Food Prot.* **55**:344–348.

11. Aymerich, T., A. Jofre, M. Garriga, and M. Hugas. 2005. Inhibition of *Listeria monocytogenes* and *Salmonella* by natural antimicrobials and high hydrostatic pressure in sliced cooked ham. *J. Food Prot.* **68**:173–177.

11a. Azzouz, M. A., and L. B. Bullerman. 1982. Comparative antimycotic effects of selected herbs, spices, plant components and commercial fungal agents. *J. Food Prot.* **45**:1298–1301.

12. Bagamboula, C. F., M. Uyttendaele, and J. Debevere. 2003. Antimicrobial effect of spices and herbs on *Shigella sonnei* and *Shigella flexneri*. *J. Food Prot.* **66**:668–673.

13. Bagamboula, C. F., M. Uyttendaele, and J. Debevere. 2004. Inhibitory effect of thyme and basil essential oils, carvacrol, thymol, estragol, linalool and p-cymene towards *Shigella sonnei* and *S. flexneri*. *Food Microbiol.* **21**:33–42.

14. Banks, J. G., and R. G. Board. 1982. Sulfite-inhibition of *Enterobacteriaceae* including *Salmonella* in British fresh sausage and in culture systems. *J. Food Prot.* **45**:1292–1297.

15. Baranowski, J. D., P. M. Davidson, C. W. Nagel, and A. L. Branen. 1980. Inhibition of *Saccharomyces cerevisiae* by naturally-occurring hydroxycinnamates. *J. Food Sci.* **45**: 592–594.

16. Baranowski, J. D., and C. W. Nagel. 1983. Properties of alkyl hydroxycinnamates and effects on *Pseudomonas fluorescens*. *Appl. Environ. Microbiol.* **45**:218–222.

17. Barboza de Martinez, Y., K. Ferrer, and E. Marquez Salas. 2002. Combined effects of lactic acid and nisin solution in reducing levels of microbiological contamination in red meat carcasses. *J. Food Prot.* **65**:1780–1783.

18. Bargiota, E. E., E. Rico-Muñoz, and P. M. Davidson. 1987. Lethal effect of methyl and propyl parabens as related to *Staphylococcus aureus* lipid composition. *Int. J. Food Microbiol.* **4**:257–266.

19. Barmpalia, I. M., I. Geornaras, K. E. Belk, J. A. Scanga, P. A. Kendall, G. C. Smith, and J. N. Sofos. 2004. Control of *Listeria monocytogenes* on frankfurters with antimicrobials in the formulation and by dipping in organic acid solutions. *J. Food Prot.* **67**:2456–2464.

20. Barone, F. E., and M. R. Tansey. 1977. Isolation, purification, identification, synthesis and kinetics of the activity of the anticandidal component of *Allium sativum*, and a hypothesis for its mode of action. *Mycologia* **69**:793–825.

21. Barrett, N. E., A. S. Grandison, and M. J. Lewis. 1999. Contribution of the lactoperoxidase system to the keeping quality of pasteurized milk. *J. Dairy Res.* **66**:73–80.

22. Bautista, D. A., N. Sylvester, S. Barbut, and M. W. Griffiths. 1997. The determination of efficacy of antimicrobial rinses on turkey carcasses using response surface designs. *Int. J. Food Microbiol.* **34**:279–292.

23. Bedie, G. K., J. Samelis, J. N. Sofos, K. E. Belk, J. A. Scanga, and G. C. Smith. 2001. Antimicrobials in the formulation to control *Listeria monocytogenes* postprocessing contamination on frankfurters stored at 4 degrees C in vacuum packages. *J. Food Prot.* **64**:1949–1955.

24. Bellamy, W., M. Takase, H. Wakabayashi, K. Kawase, and M. Tomita. 1992. Antibacterial spectrum of lactoferrin B, a potent bactericidal peptide derived from the N-terminal region of bovine lactoferrin. *J. Appl. Bacteriol.* **73**:472–479.

25. Bennis, S., F. Chami, N. Chami, T. Bouchikhi, and A. Remmal. 2004. Surface alteration of *Saccharomyces cerevisiae* induced by thymol and eugenol. *Lett. Appl. Microbiol.* **38**:454–458.

26. Beuchat, L. R. 1994. Antimicrobial properties of spices and their essential oils, p. 167–179. *In* V. M. Dillon and R. G. Board (ed.), *Natural Antimicrobial Systems and Food Preservation.* CAB Intl., Wallingford, England.

27. Beuchat, L. R., and D. A. Golden. 1989. Antimicrobials occurring naturally in foods. *Food Technol.* **43:**134–142.

28. Beumer, R. R., A. Noomen, J. A. Marijs, and E. H. Kampelmacher. 1985. Antibacterial action of the lactoperoxidase system on *Campylobacter jejuni* in cow's milk. *Neth. Milk Dairy J.* **39:**107–114.

29. Björck, L. 1978. Antibacterial effect of the lactoperoxidase system on psychrotrophic bacteria in milk. *J. Dairy Res.* **45:**109–118.

30. Blom, H., E. Nerbink, R. Dainty, T. Hagtvedt, E. Borch, H. Nissen, and T. Nesbakken. 1997. Addition of 2.5% lactate and 0.25% acetate controls growth of *Listeria monocytogenes* in vacuum-packed, sensory-acceptable servelat sausage and cooked ham stored at 4°C. *Int. J. Food Microbiol.* **38:**71–76.

31. Boland, J. S., P. M. Davidson, and J. Weiss. 2003. Enhanced inhibition of *Escherichia coli* O157:H7 by lysozyme and chelators. *J. Food Prot.* **66:**1783–1789.

32. Bower, C. K., and M. A. Daeschel. 1999. Resistance responses of microorganisms in food environments. *Int. J. Food Microbiol.* **50:**33–44.

33. Bracey, D., C. D. Holyoak, and P. J. Coote. 1998. Comparison of the inhibitory effect of sorbic acid and amphotericin B on *Saccharomyces cerevisiae*: is growth inhibition dependent on reduced intracellular pH? *J. Appl. Microbiol.* **85:**1056–1066.

34. Branen, J., and P. M. Davidson. 2000. Activity of hydrolysed lactoferrin against foodborne pathogenic bacteria in growth media: the effect of EDTA. *Lett. Appl. Microbiol.* **30:**233–237.

35. Branen, J. K., and P. M. Davidson. 2004. Enhancement of nisin, lysozyme, and monolaurin antimicrobial activities by ethylenediaminetetraacetic acid and lactoferrin. *Int. J. Food Microbiol.* **90:**63–74.

36. Bredin, J., A. Davin-Regli, and J. M. Pages. 2005. Propyl paraben induces potassium efflux in *Escherichia coli*. *J. Antimicrob. Chemother.* **55:**1013–1015.

37. Brudzinski, L., and M. A. Harrison. 1998. Influence of incubation conditions on survival and acid tolerance response of *Escherichia coli* O157:H7 and non-O157:H7 isolates exposed to acetic acid. *J. Food Prot.* **61:**524–546.

38. Buchanan, R. L., and S. G. Edelson. 1999. pH-dependent stationary-phase acid resistance response of enterohemorrhagic *Escherichia coli* in the presence of various acidulants. *J. Food Prot.* **62:**211–218.

39. Bullerman, L. B. 1983. Effects of potassium sorbate on growth and aflatoxin production by *Aspergillus parasiticus* and *Aspergillus flavus*. *J. Food Prot.* **46:**940–942.

40. Bullerman, L. B. 1984. Effects of potassium sorbate on growth and patulin production by *Penicillium patulum* and *Penicillium roqueforti*. *J. Food Prot.* **47:**312–315.

41. Bullerman, L. B. 1974. Inhibition of aflatoxin production by cinnamon. *J. Food Sci.* **39:**1163–1165.

42. Burt, S. 2004. Essential oils: their antibacterial properties and potential applications in foods—a review. *Int. J. Food Microbiol.* **94:**223–253.

43. Burt, S. A., and R. D. Reinders. 2003. Antibacterial activity of selected plant essential oils against *Escherichia coli* O157:H7. *Lett. Appl. Microbiol.* **36:**162–167.

44. Burt, S. A., R. Vlielander, H. P. Haagsman, and E. J. A. Veldhuizen. 2005. Increase in activity of essential oil components carvacrol and thymol against *Escherichia coli* O157:H7 by addition of food stabilizers. *J. Food Prot.* **68:**919–926.

45. Carlsson, J., Y. Iwami, and T. Yamada. 1983. Hydrogen peroxide excretion by oral streptococci and effect of lactoperoxidase-thiocyanate-hydrogen peroxide. *Infect. Immun.* **40:**70–80.

46. Carneiro de Melo, A. M. S., C. A. Cassar, and R. J. Miles. 1998. Trisodium phosphate increases sensitivity of gram-negative bacteria to lysozyme and nisin. *J. Food Prot.* **61:**839–844.

47. Carpenter, C. E., D. S. A. Reddy, and D. P. Cornforth. 1987. Inactivation of clostridial ferredoxin and pyruvate-ferredoxin oxidoreductase by sodium nitrite. *Appl. Environ. Microbiol.* **53:**549–552.

48. Castellani, A. G., and C. F. Niven. 1955. Factors affecting the bacteriostatic action of sodium nitrite. *Appl. Microbiol.* **3:**154–159.

49. Cavallito, C. J., and J. H. Bailey. 1944. Allicin, the antibacterial principle of *Allium sativum*. I. Isolation, physical properties, and antibacterial action. *J. Am. Chem. Soc.* **16:**1950–1951.

50. Cerrutti, P., and S. M. Alzamora. 1996. Inhibitory effects of vanillin on some food spoilage yeasts in laboratory media and fruit purées. *Int. J. Food Microbiol.* **29:**379–386.

51. Cerrutti, P., S. M. Alzamora, and S. L. Vidales. 1997. Vanillin as an antimicrobial for producing shelf-stable strawberry purées. *J. Food Sci.* **62:**608–610.

52. Ceylan, E., D. Y. C. Fung, and J. R. Sabah. 2004. Antimicrobial activity and synergistic effect of cinnamon with sodium benzoate or potassium sorbate in controlling *Escherichia coli* O157:H7 in apple juice. *J. Food Sci.* **69:**M102-M106.

53. Chaibi, A., L. H. Ababouch, and F. F. Busta. 1996. Inhibition of bacterial spores and vegetative cells by glycerides. *J. Food Prot.* **59:**716–722.

54. Chami, N., S. Bennis, F. Chami, A. Aboussekhra, and A. Remmal. 2005. Study of anticandidal activity of carvacrol and eugenol in vitro and in vivo. *Oral Microbiol. Immunol.* **20:**106–111.

55. Chan, L., R. Weaver, and C. S. Ough. 1975. Microbial inhibition caused by *p*-hydroxybenzoate esters in wines. *Am. J. Enol. Viticult.* **26:**201–207.

56. Chen, C. S., W. Y. Liau, and G. J. Tsai. 1998. Antibacterial effects of N-sulfonated and N-sulfobenzoyl chitosan and application to oyster preservation. *J. Food Prot.* **61:**1124–1128.

57. Chen, N., and L. A. Shelef. 1992. Relationship between water activity, salts of lactic acid, and growth of *Listeria monocytogenes* in a meat model system. *J. Food Prot.* **55:**574–578.

58. Chipley, J. R. 1974. Effects of 2,4-dinitrophenol and N,N′-cyclohexylcarbodiimide on cell-envelope associated enzymes of *Escherichia coli* and *Salmonella enteriditis*. *Microbios* **10:**115.

59. **Chipley, J. R.** 2005. Sodium benzoate and benzoic acid, p. 11–49. *In* P. M. Davidson, J. N. Sofos, and A. L. Branen (ed.), *Antimicrobials in Foods*, 3rd ed. CRC Press, Boca Raton, Fla.

60. **Chipley, J. R., and N. Uraih.** 1980. Inhibition of *Aspergillus* growth and aflatoxin release by derivatives of benzoic acid. *Appl. Environ. Microbiol.* **40:**352.

61. **Chorianopoulos, N., E. Kalpoutzakis, N. Aligiannis, S. Mitaku, G. J. Nychas, and S. A. Haroutounian.** 2004. Essential oils of *Satureja*, *Origanum*, and *Thymus* species: chemical composition and antibacterial activities against foodborne pathogens. *J. Agric. Food Chem.* **52:**8261–8267.

62. **Chung, K. T., Z. Lu, and M. W. Chou.** 1998. Mechanism of inhibition of tannic acid and related compounds on the growth of intestinal bacteria. *Food Chem. Toxicol.* **36:**1053–1060.

63. **Chung, K. T., and C. A. Murdock.** 1991. Natural systems for preventing contamination and growth of microorganisms in foods. *Food Microstruct.* **10:**361–374.

64. **Comes, J. E., and R. B. Beelman.** 2002. Addition of fumaric acid and sodium benzoate as an alternative method to achieve a 5-log reduction of *Escherichia coli* O157:H7 populations in apple cider. *J. Food Prot.* **65:**476–483.

65. **Conner, D. E., L. R. Beuchat, R. E. Worthington, and H. L. Hitchcock.** 1984. Effects of essential oils and oleoresins of plants on ethanol production, respiration and sporulation of yeasts. *Int. J. Food Microbiol.* **1:**63–74.

66. **Corona, A., and P. Quaglio.** 2001. Sensitivity of *Escherichia coli* VTEC to herbs and spices. *Industrie Alimentari* **40:**738–740.

67. **Cox, S. D., J. E. Gustafson, C. M. Mann, J. L. Markham, Y. C. Liew, R. P. Hartland, H. C. Bell, J. R. Warmington, and S. G. Wyllie.** 1998. Tea tree oil causes K+ leakage and inhibits respiration in *Escherichia coli*. *Lett. Appl. Microbiol.* **26:**355–358.

68. **Cox, S. D., C. M. Mann, and J. L. Markham.** 2001. Interactions between components of the essential oil of *Melaleuca alternifolia*. *J. Appl. Microbiol.* **91:**492–497.

69. **Cushnie, T. P. T., and A. J. Lamb.** 2005. Antimicrobial activity of flavonoids. *Int. J. Antimicrob. Agents* **26:**343.

70. **Daferera, D. J., B. N. Ziogas, and M. G. Polissiou.** 2000. GC-MS analysis of essential oils from some Greek aromatic plants and their fungitoxicity on *Penicillium digitatum*. *J. Agric. Food Chem.* **48:**2576–2581.

71. **Davidson, P. M.** 2005. Parabens, p. 291–304. *In* P. M. Davidson, J. N. Sofos, and A. L. Branen (ed.), *Antimicrobials in Foods*, 3rd ed. CRC Press, Boca Raton, Fla.

72. **Davidson, P. M., and M. A. Harrison.** 2002. Resistance and adaptation to food antimicrobials, sanitizers, and other process controls. *Food Technol.* **56:**69–78.

73. **Davidson, P. M., and A. S. Naidu.** 2000. Phyto-phenols, p. 266–294. *In* A. S. Naidu (ed.), *Natural Food Antimicrobial Systems*. CRC Press, Boca Raton, Fla.

74. **Debevere, J. M.** 1989. The effect of sodium lactate on the shelf life of vacuum-packed coarse liver pate. *Fleischwirtschaft* **69:**223–224.

75. **Degnan, A. J., C. W. Kaspar, W. S. Otwell, M. L. Tamplin, and J. B. Luchansky.** 1994. Evaluation of lactic acid bacterium fermentation products and food-grade chemicals to control *Listeria monocytogenes* in blue crab (*Callinectes sapidus*) meat. *Appl. Environ. Microbiol.* **60:**3198–3203.

76. **Delaquis, P., K. Stanich, and P. Toivonen.** 2005. Effect of pH on the inhibition of *Listeria* spp. by vanillin and vanillic acid. *J. Food Prot.* **68:**1472–1476.

77. **Delaquis, P. J., and G. Mazza.** 1995. Antimicrobial properties of isothiocyanates in food preservation. *Food Technol.* **49:**73–84.

78. **Delaquis, P. J., and P. L. Sholberg.** 1997. Antimicrobial activity of gaseous allyl isothiocyanate. *J. Food Prot.* **60:**943–947.

79. **Delaquis, P. J., K. Stanich, B. Girard, and G. Mazza.** 2002. Antimicrobial activity of individual and mixed fractions of dill, cilantro, coriander and eucalyptus essential oils. *Int. J. Food Microbiol.* **74:**101–109.

80. **Delaquis, P. J., S. M. Ward, R. A. Holley, M. C. Cliff, and G. Mazza.** 1999. Microbiological, chemical and sensory properties of pre-cooked roast beef preserved with horseradish essential oil. *J. Food Sci.* **64:**519–524.

81. **Devlieghere, F., A. Vermeulen, and J. M. Debevere.** 2004. Chitosan: antimicrobial activity, interactions with food components and applicability as a coating on fruit and vegetables. *Food Microbiol.* **21:**703–714.

82. **Dickson, J. S., and M. E. Anderson.** 1992. Microbiological decontamination of animal carcasses by washing and sanitizing systems: a review. *J. Food Prot.* **55:**13–140.

83. **Di Pasqua, R., V. De Feo, F. Villani, and G. Mauriello.** 2005. In vitro antimicrobial activity of essential oils from Mediterranean Apiaceae, Verbenaceae and Lamiaceae against foodborne pathogens and spoilage bacteria. *Ann. Microbiol.* **55:**139–143.

84. **Di Persio, P. A., P. A. Kendall, M. Calicioglu, and J. N. Sofos.** 2003. Inactivation of *Salmonella* during drying and storage of apple slices treated with acidic or sodium metabisulfite solutions. *J. Food Prot.* **66:**2245–2251.

85. **Di Persio, P. A., P. A. Kendall, and J. N. Sofos.** 2004. Inactivation of *Listeria monocytogenes* during drying and storage of peach slices treated with acidic or sodium metabisulfite solutions. *Food Microbiol.* **21:**641–648.

86. **Doores, S.** 2005. Organic acids, p. 95–136. *In* P. M. Davidson, J. N. Sofos, and A. L. Branen (ed.), *Antimicrobials in Foods*, 3rd ed. Marcel Dekker, Inc., New York, N.Y.

87. **Duncan, C. L., and E. M. Foster.** 1968. Effect of sodium nitrite, sodium chloride, and sodium nitrate on germination and outgrowth of anaerobic spores. *Appl. Microbiol.* **16:**406–411.

88. **Dymicky, M., and C. N. Huhtanen.** 1979. Inhibition of *Clostridium botulinum* by *p*-hydroxybenzoic acid n-alkyl esters. *Antimicrob. Agents Chemother.* **15:**798–801.

89. **Ehrenberg, L., I. Fedorscsak, and F. Solymosy.** 1976. Diethyl pyrocarbonate in nucleic acid research. *Prog. Nucleic Acid Mol. Biol.* **16:**189.

90. **Eklund, T.** 1983. The antimicrobial effect of dissociated and undissociated sorbic acid at different pH levels. *J. Appl. Bacteriol.* **54:**383–389.

91. **Eklund, T.** 1985. The effect of sorbic acid and esters of *p*-hydroxybenzoic acid on the proton motive force in *Escherichia coli* membrane vesicles. *J. Gen. Microbiol.* **131:**73–76.

92. **Eklund, T.** 1980. Inhibition of growth and uptake processes in bacteria by some chemical food preservatives. *J. Appl. Bacteriol.* **48:**423–432.

93. **Eklund, T.** 1985. Inhibition of microbial growth at different pH levels by benzoic and propionic acids and esters

of p-hydroxybenzoic acid. *Int. J. Food Microbiol.* **2:** 159–167.

94. **Eklund, T.** 1989. Organic acids and esters, p. 161–200. *In* G. W. Gould (ed.), *Mechanisms of Action of Food Preservation Procedures.* Elsevier Applied Sci, London, England.

95. **Eklund, T., I. F. Nes, and Skjelkvåle.** 1981. Control of Salmonella at different temperatures by propyl paraben and butylated hydroxyanisole, p. 377. *In* T. A. Roberts, G. Hobbs, J. H. B. Christian, and N. Skovgaard (ed.), *Psychrotrophic Microorganisms in Spoilage and Pathogenicity.* Academic Press, London, England.

96. **Ekstrand, B.** 1994. Lactoperoxidase and lactoferrin, p. 15–63. *In* V. M. Dillon and R. G. Board (ed.), *Natural Antimicrobial Systems and Food Preservation.* CAB Intl., Wallingford, England.

97. **Elci, S., and N. O. Akpolat.** 2003. Effect of glycerol monolaurate and sodium benzoate at different concentrations, temperatures and pH on the growth of *Listeria monocytogenes. Biotechnol. Biotechnol. Equip.* **17:**123–127.

98. **Elgayyar, M., F. A. Draughon, D. A. Golden, and J. R. Mount.** 2001. Antimicrobial activity of essential oils from plants against selected pathogenic and saprophytic microorganisms. *J. Food Prot.* **64:**1019–1024.

99. **Elliot, R. M., J. C. McLay, M. J. Kennedy, and R. S. Simmonds.** 2004. Inhibition of foodborne bacteria by the lactoperoxidase system in a beef cube system. *Int. J. Food Microbiol.* **91:**73–81.

100. **Ellison, R. T., T. G. Giehl, and F. M. LaForce.** 1988. Damage of the outer membrane of enteric gram-negative bacteria by lactoferrin and transferrin. *Infect. Immun.* **56:**2774–2781.

101. **El-Shenawy, M. A., and E. H. Marth.** 1988. Sodium benzoate inhibits growth of or inactivates *Listeria monocytogenes. J. Food Prot.* **51:**525–530.

102. **Fabio, A., A. Corona, E. Forte, and P. Quaglio.** 2003. Inhibitory activity of spices and essential oils on psychrotrophic bacteria. *Microbiologica* **26:**115–120.

103. **Falcone, P., B. Speranza, M. A. Del Nobile, M. R. Corbo, and M. Sinigaglia.** 2005. Study on the antimicrobial activity of thymol intended as a natural preservative. *J. Food Prot.* **68:**1664–1670.

104. **Fang, S. W., C. J. Li, and D. Y. C. Shin.** 1994. Antifungal activity of chitosan and its preservative effect on low-sugar candied kumquat. *J. Food Prot.* **56:**136–140.

105. **Fernandez, J. L., and W. J. Simpson.** 1993. Aspects of the resistance of lactic acid bacteria to hop bitter acids. *J. Appl. Bacteriol.* **75:**315–319.

106. **Filoche, S. K., K. Soma, and C. H. Sissons.** 2005. Antimicrobial effects of essential oils in combination with chlorhexidine digluconate. *Oral Microbiol. Immunol.* **20:**221–225.

107. **Fisher, T. L., and D. A. Golden.** 1998. Survival of *Escherichia coli* O157:H7 in apple cider as affected by dimethyl dicarbonate, sodium bisulfite, and sodium benzoate. *J. Food Sci.* **63:**904–906.

108. **Foster, J. W.** 1995. Low pH adaptation and the acid tolerance response of *Salmonella* Typhimurium. *Crit. Rev. Microbiol.* **21:**215–237.

109. **Francois, J., E. Vam Scjaftingen, and H. G. Hers.** 1986. Effect of benzoate on the metabolism of fructose 2,6-biphosphate in yeast. *Eur. J. Biochem.* **154:**141–145.

110. **Freese, E., C. W. Sheu, and E. Galliers.** 1973. Function of lipophilic acids as antimicrobial food additives. *Nature* **241:**321–327.

111. **Friedman, M., P. R. Henika, and R. E. Mandrell.** 2002. Bactericidal activities of plant essential oils and some of their isolated constituents against *Campylobacter jejuni, Escherichia coli, Listeria monocytogenes,* and *Salmonella enterica. J. Food Prot.* **65:**1545–1560.

112. **Fujita, K.-I., and I. Kubo.** 2005. Naturally occurring antifungal agents against *Zygosaccharomyces bailii* and their synergism. *J. Agric. Food Chem.* **53:**5187–5191.

113. **Garren, D. M., M. A. Harrison, and S. M. Russell.** 1998. Acid tolerance and acid shock response of *Escherichia coli* O157:H7 and non-O157:H7 isolates provide cross protection to sodium lactate and sodium chloride. *J. Food Prot.* **61:**158–161.

114. **Gay, M., and A. Amgar.** 2005. Factors moderating *Listeria monocytogenes* growth in raw milk and in soft cheese made from raw milk. *Lait* **85:**153–170.

115. **Gaya, P., M. Medina, and M. Nunez.** 1991. Effect of the lactoperoxidase system on *Listeria monocytogenes* behavior in raw milk at refrigeration temperatures. *Appl. Environ. Microbiol.* **57:**3355–3360.

116. **Gaysinsky, S., P. M. Davidson, B. D. Bruce, and J. Weiss.** 2005. Stability and antimicrobial efficiency of eugenol encapsulated in surfactant micelles as affected by temperature and pH. *J. Food Prot.* **68:**1359–1366.

117. **Gibson, A. M., and T. A. Roberts.** 1986. The effect of pH, sodium chloride, sodium nitrite and storage temperature on the growth of *Clostridium perfringens* and faecal streptococci in laboratory medium. *Int. J. Food Microbiol.* **3:**195–210.

118. **Gibson, A. M., and T. A. Roberts.** 1986. The effect of pH, water activity, sodium nitrite and storage temperature on the growth of enteropathogenic *Escherichia coli* and salmonellae in laboratory medium. *Int. J. Food Microbiol.* **3:**183–194.

119. **Gill, A. O., P. Delaquis, P. Russo, and R. A. Holley.** 2002. Evaluation of antilisterial action of cilantro oil on vacuum packed ham. *Int. J. Food Microbiol.* **73:**83–92.

120. **Gill, A. O., and R. A. Holley.** 2000. Inhibition of bacterial growth on ham and bologna by lysozyme, nisin and EDTA. *Food Res. Int.* **33:**83–90.

121. **Gill, A. O., and R. A. Holley.** 2003. Interactive inhibition of meat spoilage and pathogenic bacteria by lysozyme, nisin and EDTA in the presence of nitrite and sodium chloride at 24 degrees C. *Int. J. Food Microbiol.* **80:**251–259.

122. **Gill, A. O., and R. A. Holley.** 2004. Mechanisms of bactericidal action of cinnamaldehyde against *Listeria monocytogenes* and of eugenol against *L. monocytogenes* and *Lactobacillus sakei. Appl. Environ. Microbiol.* **70:**5750–5755.

123. **Gill, A. O., and R. A. Holley.** 2000. Surface application of lysozyme, nisin, and EDTA to inhibit spoilage and pathogenic bacteria on ham and bologna. *J. Food Prot.* **63:**1338–1346.

124. Glass, K. A., D. A. Granberg, A. L. Smith, A. M. McNamara, M. Hardin, J. Mattias, K. Ladwig, and E. A. Johnson. 2002. Inhibition of *Listeria monocytogenes* by sodium diacetate and sodium lactate on wieners and cooked bratwurst. *J. Food Prot.* **65:**116–123.

125. Golden, D. A., R. W. Worobo, and C. S. Ough. 2005. Dimethyl dicarbonate and diethyl dicarbonate, p. 305–326. *In* P. M. Davidson, J. N. Sofos, and A. L. Branen (ed.), *Antimicrobials in Foods*, 3rd ed. CRC Press, Boca Raton, Fla.

126. González-Fandos, E., M. L. García-López, M. L. Sierra, and A. Otero. 1994. Staphylococcal growth and enterotoxins (A-D) and thermonuclease synthesis in the presence of dehydrated garlic. *J. Appl. Bacteriol.* **77:**549–552.

127. Gould, G. W. 1964. Effect of food preservatives on the growth of bacteria from spores, p. 17. *In* N. Molin (ed.), *Microbial Inhibitors in Food*. Almqvist and Miksell, Stockholm, Sweden.

128. Gould, G. W. (ed.). 1989. *Mechanisms of Action of Food Preservation Procedures*. Elsevier Applied Sci., London, England.

129. Gould, G. W. 2000. The use of other chemical preservatives: sulfite and nitrite, p. 200–213. *In* B. M. Lund, T. C. Baird-Parker, and G. W. Gould (ed.), *The Microbiological Safety and Quality of Food*. Aspen Publishers, Gaithersburg, Md.

130. Gould, G. W., and N. J. Russell. 1991. Sulphite, p. 72–88. *In* N. J. Russell and G. W. Gould (ed.), *Food Preservatives*. Blackie and Son Ltd., Glasgow, Scotland.

131. Gram, L. 1991. Inhibition of mesophilic spoilage *Aeromonas* spp. on fish by salt, potassium sorbate, liquid smoke, and chilling. *J. Food Prot.* **54:**436–442.

132. Griffin, S. G., S. G. Wyllie, J. L. Markham, and D. N. Leach. 1999. The role of structure and molecular properties of terpenoids in determining their antimicrobial activity. *Flavour Fragrance J.* **14:**322–332.

133. Haas, G. J., and R. Barsoumian. 1994. Antimicrobial activity of hop resins. *J. Food Prot.* **57:**59–61.

134. Hammer, K. A., C. F. Carson, and T. V. Riley. 2003. Antifungal activity of the components of *Melaleuca alternifolia* (tea tree) oil. *J. Appl. Microbiol.* **95:**853–860.

135. Hammer, K. A., C. F. Carson, and T. V. Riley. 1999. Antimicrobial activity of essential oils and other plant extracts. *J. Appl. Microbiol.* **86:**985–990.

136. Hao, Y. Y., R. E. Brackett, and M. P. Doyle. 1998. Efficacy of plant extracts in inhibiting *Aeromonas hydrophila* and *Listeria monocytogenes* in refrigerated, cooked poultry. *Food Microbiol.* **15:**367–378.

137. Hao, Y. Y., R. E. Brackett, and M. P. Doyle. 1998. Inhibition of *Listeria monocytogenes* and *Aeromonas hydrophila* by plant extracts in refrigerated cooked beef. *J. Food Prot.* **61:**307–312.

138. Hefnawy, Y. A., S. I. Moustafa, and E. H. Marth. 1993. Sensitivity of *Listeria monocytogenes* to selected spices. *J. Food Prot.* **56:**876–878.

139. Helander, I. M., H. L. Alakomi, K. Latva-Kala, T. Mattila-Sandholm, I. Pol, E. J. Smid, L. G. M. Gorris, and A. von Wright. 1998. Characterization of the action of selected essential oil components on gram-negative bacteria. *J. Agric. Food Chem.* **46:**3590–3595.

140. Herald, P. J., and P. M. Davidson. 1983. The antibacterial activity of selected hydroxycinnamic acids. *J. Food Sci.* **48:**1378–1379.

141. Hinton, A., Jr. 2006. Growth of *Campylobacter* in media supplemented with organic acids. *J. Food Prot.* **69:**34–38.

142. Hinze, H., and H. Holzer. 1985. Effect of sulfite or nitrite on the ATP content and the carbohydrate metabolism in yeast. *Z. Lebensm. Unters. Forsch.* **181:**87–91.

143. Ho, C. T. 1992. Phenolic compounds in food. An overview. *In* C. T. Ho, C. Y. Lee, and M. T. Huang (ed.), *Phenolic Compounds in Food and Their Effects on Health. Analysis, Occurrence, and Chemistry*, vol. Symp. Ser. 506. American Chemical Society, Washington, D.C.

144. Holley, R. A., and D. Patel. 2005. Improvement in shelf-life and safety of perishable foods by plant essential oils and smoke antimicrobials. *Food Microbiol.* **22:**273–292.

145. Houtsma, P. C., A. Heuvelink, J. Dufrenne, and S. Notermans. 1994. Effect of sodium lactate on toxin production, spore germination and heat resistance of proteolytic *C. botulinum* strains. *J. Food Prot.* **57:**327–330.

146. Houtsma, P. C., J. C. Wit, and F. M. Rombouts. 1996. Minimum inhibitory concentration (MIC) of sodium lactate and sodium chloride for spoilage organisms and pathogens at different pH values and temperatures. *J. Food Prot.* **59:**1300–1304.

147. Huang, L. H., and V. K. Juneja. 2003. Thermal inactivation of *Escherichia coli* O157:H7 in ground beef supplemented with sodium lactate. *J. Food Prot.* **66:**664–667.

148. Hughey, V. L., and E. A. Johnson. 1987. Antimicrobial activity of lysozyme against bacteria involved in food spoilage and food-borne disease. *Appl. Environ. Microbiol.* **53:**2165–2170.

149. Hughey, V. L., R. A. Wilger, and E. A. Johnson. 1989. Antibacterial activity of hen egg white lysozyme against *Listeria monocytogenes* Scott A in foods. *Appl. Environ. Microbiol.* **55:**631–638.

150. ICMSF. 1996. *Micoorganisms in Foods 5. Microbiological Specifications of Food Pathogens*. Blackie Academic and Professional, London, England.

151. Inovatech. 2000. Inovapure Product Description. Canadian Inovatech, Inc., Abbotsford, B.C., Canada.

152. Isshiki, K., K. Tokuora, R. Mori, and S. Chiba. 1992. Preliminary examination of allyl isothiocyanate vapor for food preservation. *Biosci. Biotechnol. Biochem.* **56:**1476–1477.

153. Jangho, K., and D. L. Marshall. 2001. Effect of lactic acid on *Listeria monocytogenes* and *Edwardsiella tarda* attached to catfish skin. *Food Microbiol.* **18:**589–596.

154. Jermini, M. F. G., and W. Schmidt-Lorenz. 1987. Activity of Na-benzoate and ethyl-paraben against osmotolerant yeasts at different water activity values. *J. Food Prot.* **50:**920–927.

155. Johansen, C., L. Gram, and A. S. Meyer. 1994. The combined inhibitory effect of lysozyme and low pH on growth of *Listeria monocytogenes*. *J. Food Prot.* **57:**561–566.

156. Johnson, E. A., and A. E. Larson. 2005. Lysozyme, p. 361–387. *In* P. M. Davidson, A. L. Branen, and J. N.

Sofos (ed.), *Antimicrobials in Food*, 3rd ed. CRC Press, Boca Raton, Fla.

157. **Jones, E. M., A. Smart, G. Bloomberg, L. Burgess, and M. R. Millar.** 1994. Lactoferrin, a new antimicrobial peptide. *J. Appl. Bacteriol.* **77:**208–214.

158. **Juneja, V. K., and P. M. Davidson.** 1992. Influence of altered fatty acid composition on resistance of *Listeria monocytogenes* to antimicrobials. *J. Food Prot.* **56:**302–305.

159. **Juneja, V. K., and H. Thippareddi.** 2004. Control of *Clostridium perfringens* in a model roast beef by salts of organic acids during chilling. *J. Food Safety* **24:**95–108.

160. **Juneja, V. K., and H. Thippareddi.** 2004. Inhibitory effects of organic acid salts on growth of *Clostridium perfringens* from spore inocula during chilling of marinated ground turkey breast. *Int. J. Food Microbiol.* **93:**155–163.

161. **Jurd, L., A. D. King, K. Mihara, and W. L. Stanley.** 1971. Antimicrobial properties of natural phenols and related compounds. I. Obtusastyrene. *Appl. Microbiol.* **21:**507–510.

162. **Kabara, J. J., and D. L. Marshall.** 2005. Medium-chain fatty acids and esters, p. 327–360. *In* P. M. Davidson, J. N. Sofos, and A. L. Branen (ed.), *Antimicrobials in Foods*, 3rd ed. CRC Press, Boca Raton, Fla.

163. **Kamau, D. N., S. Doores, and K. M. Pruitt.** 1990. Antibacterial activity of the lactoperoxidase system against *Listeria monocytogenes* and *Staphylococcus aureus* in milk. *J. Food Prot.* **53:**1010–1014.

164. **Kasrazadeh, M., and C. Genigeorgis.** 1995. Potential growth and control of *Escherichia coli* O157:H7 in soft hispanic type cheese. *Int. J. Food Microbiol.* **25:**289–300.

165. **Kato, A., and I. Shibasaki.** 1975. Combined effect of different drugs on the antibacterial activity of fatty acids and their esters. *J. Antibacterial Antifungal Agents* (Japan) **8:**355–361.

166. **Kennedy, M., A. L. O'Rourke, J. McLay, and R. Simmonds.** 2000. Use of a ground beef model to assess the effect of the lactoperoxidase system on the growth of *Escherichia coli* O157:H7, *Listeria monocytogenes* and *Staphylococcus aureus* in red meat. *Int. J. Food Microbiol.* **57:**147–158.

167. **Kim, J. M., M. R. Marshall, J. A. Cornell, J. F. Preston, and C. I. Wei.** 1995. Antibacterial activity of carvacrol, citral, and geraniol against *Salmonella typhimurium* in culture medium and on fish cubes. *J. Food Sci.* **60:**1364–1368.

168. **Kim, J. M., M. R. Marshall, and C. Wei.** 1995. Antibacterial activity of some essential oil components against five foodborne pathogens. *J. Agric. Food Chem.* **43:**2839–2845.

169. **Kirby, G. W., L. Atkin, and C. N. Frey.** 1973. Further studies on the growth of bread molds as influenced by acidity. *Cereal Chem.* **14:**865.

170. **Kisko, G., and S. Roller.** 2005. Carvacrol and *p*-cymene inactivate *Escherichia coli* O157: H7 in apple juice. *BMC Microbiol.* **5:**36.

171. **Korhonen, H.** 1978. Effect of lactoferrin and lysozyme in milk on the growth inhibition of *Bacillus stearothermophilus* in the Thermocult method. *Suomen Eläinlääkärilehti* **84:**255–267.

172. **Korpela, J.** 1984. Avidin, a high affinity biotin-binding protein, as a tool and subject of biological research. *Med. Biol.* **65:**5–26.

173. **Lachowicz, K. J., G. P. Jones, D. R. Briggs, F. E. Bienvenu, J. Wan, A. Wilcock, and M. J. Coventry.** 1998. The synergistic preservative effects of the essential oils of sweet basil (*Ocimum basilicum* L.) against acid-tolerant food microflora. *Lett. Appl. Microbiol.* **26:**209–214.

174. **Lambert, R. J., and M. Stratford.** 1999. Weak-acid preservatives: modeling microbial inhibition and response. *J. Appl. Microbiol.* **86:**157–164.

175. **Lambert, R. J. W., P. N. Skandamis, P. J. Coote, and G. J. E. Nychas.** 2001. A study of the minimum inhibitory concentration and mode of action of oregano essential oil, thymol and carvacrol. *J. Appl. Microbiol.* **91:**453–462.

176. **Larson, A. E., R. R. Yu, O. A. Lee, S. Price, G. J. Haas, and E. A. Johnson.** 1996. Antimicrobial activity of hop extracts against *Listeria monocytogenes* in media and in food. *Int. J. Food Microbiol.* **33:**195–207.

177. **Lee, J. S.** 1973. What seafood processors should know about *Vibrio parahaemolyticus*. *J. Milk Food Technol.* **36:**405–408.

178. **Lee, Y. L., T. Cesario, J. Owens, E. Shanbrom, and L. D. Thrupp.** 2002. Antibacterial activity of citrate and acetate. *Nutrition* **18:**665–666.

179. **Lee, Y. L., L. Thrupp, J. Owens, T. Cesario, and E. Shanbrom.** 2001. Bactericidal activity of citrate against Gram-positive cocci. *Lett. Appl. Microbiol.* **33:**349–351.

180. **Leitch, E. C. M., and C. S. Stewart.** 2002. *Escherichia coli* O157 and non-O157 isolates are more susceptible to L-lactate than to D-lactate. *Appl. Environ. Microbiol.* **68:**4676–4678.

181. **Lennox, J. E., and L. J. McElroy.** 1984. Inhibition of growth and patulin synthesis in *Penicillium expansum* by potassium sorbate and sodium propionate in culture. *Appl. Environ. Microbiol.* **48:**1031–1033.

182. **Leuschner, R. G. K., and J. Zamparini.** 2002. Effects of spices on growth and survival of *Escherichia coli* O157 and *Salmonella enterica* serovar Enteritidis in broth model systems and mayonnaise. *Food Control* **13:**399–404.

183. **Lewis, J. C., and L. Jurd.** 1972. Sporostatic action of cinnamylphenols and related compounds on *Bacillus megaterium*, p. 384–389. *In* H. O. Halvorson, R. Hewson, and L. L. Campbell (ed.), *Spores*, 5th ed. American Society for Microbiology, Washington, D.C.

184. **Leyer, G. J., and E. A. Johnson.** 1993. Acid adaptation induces cross-protection against environmental stresses in *Salmonella* Typhimurium. *Appl. Environ. Microbiol.* **59:**1842–1847.

185. **Leyer, G. J., L.-L. Wang, and E. A. Johnson.** 1995. Acid adaptation of *Escherichia coli* O157:H7 increases survival in acidic foods. *Appl. Environ. Microbiol.* **61:**3752–3755.

186. **Liewen, M. B., and E. H. Marth.** 1985. Growth of sorbate-resistant and -sensitive strains of *Penicillium roqueforti* in the presence of sorbate. *J. Food Prot.* **48:**525–529.

187. **Lillard, H. S., L. C. Blankenship, J. A. Dickens, S. E. Craven, and A. D. Shackelford.** 1987. Effect of acetic acid on the microbiological quality of scalded picked and unpicked broiler carcasses. *J. Food Prot.* **50:**112–114.

188. Lin, C. M., J. F. Preston, and C. I. Wei. 2000. Antibacterial mechanism of allyl isothiocyanate. *J. Food Prot.* **63:** 727–734.

189. López-Malo, A., S. M. Alzamora, and A. Argaiz. 1998. Vanillin and pH synergistic effects on mold growth. *J. Food Sci.* **63:**143–146.

190. López-Malo, A., S. M. Alzamora, and A. Argaiz. 1995. Effect of natural vanillin on germination time and radial growth of moulds in fruit-based agar systems. *Food Microbiol.* **12:**213–219.

191. López-Malo, A., S. M. Alzamora, and A. Argaiz. 1997. Effect of vanillin concentration, pH and incubation temperature on *Aspergillus flavus*, *Aspergillus niger*, *Aspergillus ochraceus*, and *Aspergillus parasiticus*. *Food Microbiol.* **14:**117–124.

192. López-Malo, A., S. M. Alzamora, and E. Palou. 2002. Aspergillus flavus dose-response curves to selected natural and synthetic antimicrobials. *Int. J. Food Microbiol.* **73:**213–218.

193. López-Malo, A., S. M. Alzamora, and E. Palou. 2005. *Aspergillus flavus* growth in the presence of chemical preservatives and naturally occurring antimicrobial compounds. *Int. J. Food Microbiol.* **99:**119–128.

194. Lopez, M., S. Martinez, J. Gonzalez, R. Martin, and A. Bernardo. 1998. Sensitization of thermally injured spores of *Bacillus stearothermophilus* to sodium benzoate and potassium sorbate. *Lett. Appl. Microbiol.* **27:**331–335.

195. Lyon, G. D., and F. M. McGill. 1988. Inhibition of *Erwinia carotovora in vitro* by phenolics. *Potato Res.* **31:**461–467.

196. Lyon, G. D., and F. M. McGill. 1989. Inhibition of polygalacturonase and polygalacturonic acid lyase from *Erwinia carotovora* subsp. *carotovora* by phenolics *in vitro*. *Potato Res.* **32:**267–274.

197. Maas, M. R., K. A. Glass, and M. P. Doyle. 1989. Sodium lactate delays toxin production by *Clostridium botulinum* in cook-in-bag turkey products. *Appl. Environ. Microbiol.* **55:**2226–2229.

198. Mandel, I. D., and S. A. Ellison. 1985. The biological significance of the non-immunoglobulin defense factors, p. 1–14. *In* K. M. Pruitt and J. O. Tenovuo (ed.), *The Lactoperoxidase System: Its Chemistry and Biological Significance.* Marcel Dekker, Inc., New York, N.Y.

199. Manderfeld, M. M., H. W. Schafer, P. M. Davidson, and E. A. Zottola. 1997. Isolation and identification of antimicrobial furocoumarins from parsley. *J. Food Prot.* **60:**72–77.

200. Mangena, T., and N. Y. O. Muyima. 1999. Comparative evaluation of the antimicrobial activities of essential oils of *Artemisia afra*, *Pteronia incana*, and *Rosmarinus officinalis* on selected bacteria and yeast strains. *Lett. Appl. Microbiol.* **28:**291–296.

201. Manohar, V., C. Ingram, J. Gray, N. A. Talpur, B. W. Echard, D. Bagchi, and H. G. Preuss. 2001. Antifungal activities of origanum oil against *Candida albicans*. *Mol. Cell. Biochem.* **228:**111–117.

202. Mari, M., R. Iori, O. Leoni, and A. Marchi. 1993. In vitro activity of glucosinolate derived isothiocyanates against postharvest fruit pathogens. *Ann. Appl. Biol.* **123:**155–164.

203. Marth, E. H. 1966. Antibiotics in foods—naturally occurring, developed, and added. *Residue Rev.* **12:**65–161.

204. Marwan, A. G., and C. W. Nagel. 1986. Microbial inhibitors of cranberries. *J. Food Sci.* **51:**1009–1013.

205. Mbandi, E., and L. A. Shelef. 2002. Enhanced antimicrobial effects of combination of lactate and diacetate on *Listeria monocytogenes* and *Salmonella* spp. in beef bologna. *Int. J. Food Microbiol.* **76:**191–198.

206. McKellar, R. C. 1993. Effect of preservatives and growth factors on secretion of listeriolysin O by *Listeria monocytogenes*. *J. Food Prot.* **56:**380–384.

207. McMindes, M. K., and A. J. Siedler. 1988. Nitrite mode of action: inhibition of yeast pyruvate decarboxylase (E.C. 4.1.1.1) and clostridial pyruvate:oxidoreductase (E.C. 1.2.7.1) by nitric oxide. *J. Food Sci.* **53:** 917–919.

208. Meng, J. H., and C. Genigeorgis. 1993. Modeling the lag phase of non-proteolytic *Clostridium botulinum* toxigenesis in cooked turkey and chicken breast as affected by temperature, sodium lactate, sodium chloride and spore inoculum. *Int. J. Food Microbiol.* **19:**109–122.

209. Millis, R. J., and M. J. Schendel. February 1994. Inhibition of food pathogens by hop acids. U.S. patent 5,286,506.

210. Min, S., L. J. Harris, J. H. Han, and J. M. Krochta. 2005. *Listeria monocytogenes* inhibition by whey protein films and coatings incorporating lysozyme. *J. Food Prot.* **68:**2317–2325.

211. Min, S., and J. M. Krochta. 2005. Inhibition of *Penicillium commune* by edible whey protein films incorporating lactoferrin, lactoferrin hydrolysate, and lactoperoxidase systems. *J. Food Sci.* **70:**87–94.

212. Minor, T. E., and E. H. Marth. 1970. Growth of *Staphylococcus aureus* in acidified pasteurized milk. *J. Milk Food. Technol.* **33:**516–520.

213. Mitchell, P., and J. Moyle. 1969. Estimation of membrane potential and pH difference across the cristae membrane of rat liver mitochondria. *Eur. J. Biochem.* **7:**471–484.

214. Moir, C. J., and M. J. Eyles. 1992. Inhibition, injury and inactivation of four psychrotrophic foodborne bacteria by the preservatives methyl p-hydroxybenzoate and potassium sorbate. *J. Food Prot.* **55:**360–366.

215. Moreira, M. R., A. G. Ponce, C. E. del Valle, and S. I. Roura. 2005. Inhibitory parameters of essential oils to reduce a foodborne pathogen. *Lebensm. Wiss. Technol.* **38:**565–570.

216. Morita, H., H. Yoshikawa, T. Suzuki, S. Hisamatsu, Y. Kato, R. Sakata, Y. Nagata, and T. Yoshimura. 2004. Anti-microbial action against verotoxigenic *Escherichia coli* O157:H7 of nitric oxide derived from sodium nitrite. *Biosci. Biotechnol. Biochem.* **68:**1027–1034.

217. Mountney, G. J., and J. O'Malley. 1985. Acids as poultry meat preservatives. *Poultry Sci.* **44:**582.

218. Muhoberac, B. B., and D. C. Wharton. 1980. EPR study of hemo-NO complexes of ascorbic acid-reduced *Pseudomonas* cytochrome oxidase and corresponding model complexes. *J. Biol. Chem.* **255:**8437–8442.

219. Murdock, C. A., and K. R. Matthews. 2002. Antibacterial activity of pepsin-digested lactoferrin on foodborne pathogens in buffered broth systems and ultra-high temperature milk with EDTA. *J. Appl. Microbiol.* **93:**850–856.

220. Murphy, R. Y., T. Osaili, L. K. Duncan, and J. A. Marcy. 2004. Effect of sodium lactate on thermal inactivation

of *Listeria monocytogenes* and *Salmonella* in ground chicken thigh and leg meat. *J. Food Prot.* **67:**1403–1407.

221. **Musajo, L., F. Bordin, and R. Bevilacqua.** 1967. Photoreactions at 3655A linking the 3–4 double bond of furocoumarins with pyrimidine bases. *Photochem. Photobiol.* **6:**927–931.

222. **Mytle, N., G. L. Anderson, M. P. Doyle, and M. A. Smith.** 2006. Antimicrobial activity of clove (*Syzgium aromaticum*) oil in inhibiting *Listeria monocytogenes* on chicken frankfurters. *Food Control* **17:**102.

223. **Nadarajah, D., J. H. Han, and R. A. Holley.** 2005. Use of mustard flour to inactivate *Escherichia coli* O157:H7 in ground beef under nitrogen flushed packaging. *Int. J. Food Microbiol.* **99:**257–267.

224. **Naidu, A. S.** 2002. Activated lactoferrin—a new approach to meat safety. *Food Technol.* **56:**40–45.

225. **Naidu, A. S.** January 2001. Immobilized lactoferrin antimicrobial agents and use. U.S. patent 6,172,040 B1.

226. **Naidu, A. S., and W. R. Bidlack.** 1998. Milk lactoferrin-natural microbial blocking agent (MBA) for food safety. *Environ. Nutr. Interact.* **2:**35–50.

227. **Nakai, S. A., and K. J. Siebert.** 2004. Organic acid inhibition models for *Listeria innocua*, *Listeria ivanovii*, *Pseudomonas aeruginosa*, and *Oenococcus oeni*. *Food Microbiol.* **21:**67–72.

228. **Nattress, F. M., and L. P. Baker.** 2003. Effects of treatment with lysozyme and nisin on the microflora and sensory properties of commercial pork. *Int. J. Food Microbiol.* **85:**259–267.

229. **Nielsen, P. V., and R. Rios.** 2000. Inhibition of fungal growth on bread by volatile components from spices and herbs, and the possible application in active packaging, with special emphasis on mustard essential oil. *Int. J. Food Microbiol.* **60:**219–229.

230. **O'Driscoll, B., C. G. M. Gahan, and C. Hill.** 1996. Adaptive acid tolerance response in *Listeria monocytogenes*: isolation of an acid-tolerant mutant which demonstrates increased virulence. *Appl. Environ. Microbiol.* **62:**1693–1698.

231. **Oh, D. H., and D. L. Marshall.** 1994. Enhanced inhibition of *Listeria monocytogenes* by glycerol monolaurate with organic acids. *J. Food Sci.* **59:**1258–1261.

232. **Oh, D. H., and D. L. Marshall.** 1995. Influence of packaging method, lactic acid and monolaurin on *Listeria monocytogenes* in crawfish tail meat homogenate. *Food Microbiol.* **12:**159–163.

233. **Oh, D. H., and D. L. Marshall.** 1993. Influence of temperature, pH, and glycerol monolaurate on growth and survival of *Listeria monocytogenes*. *J. Food Prot.* **56:** 744–749.

234. **Oh, H. I., Y. J. Kim, E. J. Chang, and J. Y. Kim.** 2001. Antimicrobial characteristics of chitosans against food spoilage microorganisms in liquid media and mayonnaise. *Biosci. Biotechnol. Biochem.* **65:**2378–2383.

235. **Okrend, A. J., R. W. Johnston, and A. B. Moran.** 1986. Effect of acetic acid on the death rates at 52°C of *Salmonella newport*, *Salmonella typhimurium*, and *Campylobacter jejuni* in poultry scald water. *J. Food Prot.* **49:**500–503.

236. **Olasupo, N. A., D. J. Fitzgerald, M. J. Gasson, and A. Narbad.** 2003. Activity of natural antimicrobial compounds against *Escherichia coli* and *Salmonella enterica* serovar Typhimurium. *Lett. Appl. Microbiol.* **37:**448–451.

237. **Olasupo, N. A., D. J. Fitzgerald, A. Narbad, and M. J. Gasson.** 2004. Inhibition of *Bacillus subtilis* and *Listeria innocua* by nisin in combination with some naturally occurring organic compounds. *J. Food Prot.* **67:**596–600.

238. **Omar, M. M.** 1992. Phenolic compounds in botanical extracts used in foods, flavors, cosmetics, and pharmaceuticals, p. 154–168. *In* C. T. Ho, C. Y. Lee, and M. T. Huang (ed.), *Phenolic Compounds in Food and Their Effects on Health. I. Analysis, Occurrence, and Chemistry*, vol. 506. American Chemical Society, Washington, D.C.

239. **Oram, J. D., and B. Reiter.** 1968. Inhibition of bacteria by lactoferrin and other iron chelating agents. *Biochim. Biophys. Acta* **170:**351–365.

240. **Ou, C.-Y., S.-F. Tsay, C.-H. Lai, and Y.-M. Weng.** 2002. Using gelatin-based antimicrobial edible coating to prolong shelf-life of tilapia fillets. *J. Food Qual.* **25:**213–222.

241. **Ough, C. S., and R. E. Kunkee.** 1974. The effect of fumaric acid on malolactic fermentation in wines from warm areas. *Am. J. Enol. Viticult.* **25:**188–190.

242. **Ough, C. S., and L. Were.** 2005. Sulfur dioxide and sulfites, p. 143–168. *In* P. M. Davidson, J. N. Sofos, and A. L. Branen (ed.), *Antimicrobials in Foods*, 3rd ed. CRC Press, Boca Raton, Fla.

243. **Ozkan, G., O. Sagdic, and M. Ozcan.** 2003. Inhibition of pathogenic bacteria by essential oils at different concentrations. *Food Sci. Technol. Int.* **9:**85–88.

244. **Pandit, V. A., and L. A. Shelef.** 1994. Sensitivity of *Listeria monocytogenes* to rosemary (*Rosmarinus officianalis* L.). *Food Microbiol.* **11:**57–63.

245. **Papineau, A. M., D. G. Hoover, D. Knorr, and D. F. Farkas.** 1991. Antimicrobial effect of water-soluble chitosans with high hydrostatic pressure. *Food Biotechnol.* **5:**45–57.

246. **Park, S. I., M. A. Daeschel, and Y. Zhao.** 2004. Functional properties of antimicrobial lysozyme-chitosan composite films. *J. Food Sci.* **69:**M215-M221.

247. **Park, S. I., S. D. Stan, M. A. Daeschel, and Y. Y. Zhao.** 2005. Antifungal coatings on fresh strawberries (Fragaria x ananassa) to control mold growth during cold storage. *J. Food Sci.* **70:**M202-M207.

248. **Paul, B., and I. Hirshfield.** 2003. The effect of acid treatment on survival and protein expression of a laboratory K-12 strain *Escherichia coli*. *Res. Microbiol.* **154:**115–121.

249. **Payne, K. D., P. M. Davidson, S. P. Oliver, and G. L. Christen.** 1990. Influence of bovine lactoferrin on the growth of *Listeria monocytogenes*. *J. Food Prot.* **53:**468–472.

250. **Payne, K. D., S. P. Oliver, and P. M. Davidson.** 1994. Comparison of EDTA and apo-lactoferrin with lysozyme on the growth of foodborne pathogenic and spoilage bacteria. *J. Food Prot.* **57:**62–65.

251. **Payne, K. D., E. Rico-Muñoz, and P. M. Davidson.** 1989. The antimicrobial activity of phenolic compounds against *Listeria monocytogenes* and their effectiveness in a model milk system. *J. Food Prot.* **52:**151–153.

252. **Pelroy, G. A., M. E. Peterson, R. Paranjpye, J. Almond, and M. W. Eklund.** 1994. Inhibition of *Listeria monocytogenes* in cold-process (smoked) salmon by sodium nitrite and packaging method. *J. Food Prot.* **57:**114–119.

253. Penalver, P., B. Huerta, C. Borge, R. Astorga, R. Romero, and A. Perea. 2005. Antimicrobial activity of five essential oils against origin strains of the Enterobacteriaceae family. *APMIS* 113:1–6.

254. Perigo, J. A., and T. A. Roberts. 1968. Inhibition of clostridia by nitrite. *J. Food Technol.* 3:91–94.

255. Pierson, M. D., L. A. Smoot, and N. J. Stern. 1979. Effect of potassium sorbate on growth of *Staphylococcus aureus* in bacon. *J. Food Prot.* 42:302–304.

256. Pilkington, B. J., and A. H. Rose. 1988. Reactions of *Saccharomyces cerevisiae* and *Zygosaccharomyces bailii* to sulphite. *J. Gen. Microbiol.* 134:2823–2830.

257. Pilone, G. J. 1975. Control of malo-lactic fermentation in table wines by addition of fumaric acid, p. 121–138. *In* J. G. Carr, C. V. Cutting, and G. C. Whiting (ed.), *Lactic Acid Bacteria in Beverages and Foods.* Academic Press, London, England.

258. Pol, I. E., W. G. C. van Arendonk, H. C. Mastwijk, J. Krommer, E. J. Smid, and R. Moezelaar. 2001. Sensitivities of germinating spores and carvacrol-adapted vegetative cells and spores of *Bacillus cereus* to nisin and pulsed-electric-field treatment. *Appl. Environ. Microbiol.* 67:1693–1699.

259. Post, F. J., T. L. Amoroso, and M. Solberg. 1985. Inhibition of *Clostridium botulinum* type E in model acidified food systems. *J. Food Sci.* 50:966–968.

260. Pranato, Y., S. K. Rakshit, and V. M. Salokhe. 2005. Enhancing antimicrobial activity of chitosan films by incorporating garlic oil, potassium sorbate and nisin. *Food Sci. Technol.* 38:859–865.

261. Rahn, O., and J. E. Conn. 1944. Effect of increase in acidity on antiseptic efficiency. *Ind. Eng. Chem.* 36:185.

262. Rajashekhara, E., E. R. Suresh, and S. Ethiraj. 1998. Thermal death of ascospores of *Neosartorya fischeri* ATCC 200957 in the presence of organic acids and preservatives in fruit juices. *J. Food Prot.* 61:1358–1362.

263. Razavi-Rohani, S. M., and M. W. Griffiths. 1999. The antifungal activity of butylated hydroxyanisole and lysozyme. *J. Food Safety* 19:97–108.

264. Razavi-Rohani, S. M., and M. W. Griffiths. 1994. The effect of mono and polyglycerol laurate on spoilage and pathogenic bacteria associated with foods. *J. Food Safety* 14:131–151.

265. Reddy, N. R., and M. D. Pierson. 1982. Influence of pH and phosphate buffer on inhibition of *Clostridium botulinum* by antioxidants and related phenolic compounds. *J. Food Prot.* 45:925–927.

266. Reddy, N. R., M. D. Pierson, and R. V. Lechowich. 1982. Inhibition of *Clostridium botulinum* by antioxidants, phenols and related compounds. *Appl. Environ. Microbiol.* 43:835–839.

267. Rehm, H. J., and H. Wittman. 1962. Beitrag Zur Kenntnis der antimikrobiellen wirkung der Schwefligen Saure I. Ubersicht uber einflussnehmende Factoren auf die antimikrobielle wirking der Schwefligen Saure. *Z. Lebensm. Unters. Forsch.* 118:413–429.

268. Reinders, R. D., S. Biesterveld, and P. G. H. Bijker. 2001. Survival of *Escherichia coli* O157: H7 ATCC 43895 in a model apple juice medium with different concentrations of proline and caffeic acid. *Appl. Environ. Microbiol.* 67:2863–2866.

269. Reiss, J. 1976. Prevention of the formation of mycotoxins in whole wheat bread by citric acid and lactic acid. *Experientia* 32:168.

270. Reiter, B. 1978. Review of the progress of dairy science: antimicrobial systems in milk. *J. Dairy Res.* 45:131–147.

271. Reiter, B., and B. G. Harnulv. 1984. Lactoperoxidase antibacterial system: natural occurrence, biological functions and practical applications. *J. Food Prot.* 47:724–732.

272. Rhayour, K., T. Bouchikhi, A. Tantaoui-Elaraki, K. Sendide, and A. Remmal. 2003. The mechanism of bactericidal action of oregano and clove essential oils and of their phenolic major components on *Escherichia coli* and *Bacillus subtilis*. *J. Essential Oil Res.* 15:356–362.

273. Rhoades, J., and S. Roller. 2000. Antimicrobial actions of degradated and native chitosan against spoilage organisms in laboratory media and foods. *Appl. Environ. Microbiol.* 66:80–86.

274. Rice, K. M., and M. D. Pierson. 1982. Inhibition of *Salmonella* by sodium nitrite and potassium sorbate in frankfurters. *J. Food Sci.* 47:1615–1617.

275. Rico-Muñoz, E., E. E. Bargiota, and P. M. Davidson. 1987. Effect of selected phenolic compounds on the membrane-bound adenosine triphosphatase of *Staphylococcus aureus*. *Food Microbiol.* 4:239–249.

276. Rico-Muñoz, E., and P. M. Davidson. 1983. The effect of corn oil and casein on the antimicrobial activity of phenolic antioxidants. *J. Food Sci.* 48:1284–1288.

277. Roberts, T. A., and M. Ingram. 1966. The effect of sodium chloride, potassium nitrate and sodium nitrite on the recovery of heated bacterial spores. *J. Food Technol.* 1:147–163.

278. Roberts, T. A., L. F. J. Woods, M. J. Payne, and R. Cammack. 1991. Nitrite, p. 89–111. *In* N. J. Russell and G. W. Gould (ed.), *Food Preservatives.* Blackie and Son Ltd., Glasgow, Scotland.

279. Roland, J. O., and L. R. Beuchat. 1984. Biomass and patulin production by *Byssochlamys nivea* in apple juice as affected by sorbate, benzoate, SO₂ and temperature. *J. Food Sci.* 49:402–406.

280. Roller, S., and N. Covill. 2000. The antimicrobial properties of chitosan in mayonnaise and mayonnaise-based shrimp salads. *J. Food Prot.* 63:202–209.

281. Ronning, I. E., and H. A. Frank. 1987. Growth inhibition of putrefactive anaerobe 3679 caused by stringent-type response induced by protonophoric activity of sorbic acid. *Appl. Environ. Microbiol.* 53:1020–1027.

282. Rose, A. H., and B. J. Pilkington. 1989. Sulphite, p. 201–224. *In* G. W. Gould (ed.), *Mechanisms of Action of Food Preservation Procedures.* Elsevier Applied Sci., London, England.

283. Rosenquist, H., and Å. Hansen. 1998. The antimicrobial effect of organic acids, sour dough and nisin against *Bacillus subtilis* and *B. lichenformis* isolated from wheat bread. *J. Appl. Microbiol.* 85:621–631.

284. Rota, C., J. J. Carraminana, J. Burillo, and A. Herrera. 2004. In vitro antimicrobial activity of essential oils from aromatic plants against selected foodborne pathogens. *J. Food Prot.* 67:1252–1256.

285. Rowe, J. J., J. M. Yabrough, J. B. Rake, and R. G. Eagon. 1979. Nitrite inhibition of aerobic bacteria. *Curr. Microbiol.* 2:51.

286. Rusul, G., and E. H. Marth. 1987. Growth and aflatoxin production by *Aspergillus parasiticus* NRRL 2999 in the presence of potassium benzoate or potassium sorbate at different initial pH values. *J. Food Prot.* 50:820–825.

287. Sabah, J. R., H. Thippareddi, J. L. Marsden, and D. Y. C. Fung. 2003. Use of organic acids for the control of *Clostridium perfringens* in cooked vacuum-packaged restructured roast beef during an alternative cooling procedure. *J. Food Prot.* 66:1408–1412.

288. Sacchetti, G., S. Maietti, M. Muzzoli, M. Scaglianti, S. Manfredini, M. Radice, and R. Bruni. 2005. Comparative evaluation of 11 essential oils of different origin as functional antioxidants, antiradicals and antimicrobials in foods. *Food Chem.* 91:621–632.

289. Sagoo, S. K., R. G. Board, and S. Roller. 2002. Chitosan potentiates the antimicrobial action of sodium benzoate on spoilage yeasts. *Lett. Appl. Microbiol.* 34:168–172.

290. Saha, S. C., and B. A. Chopade. 2002. Effect of food preservatives on *Acinetobacter* genospecies isolated from meat. *J. Food Sci. Technol.* 39:26–32.

291. Samelis, J., J. N. Sofos, M. L. Kain, J. A. Scanga, K. E. Belk, and G. C. Smith. 2001. Organic acids and their salts as dipping solutions to control *Listeria monocytogenes* inoculated following processing of sliced pork bologna stored at 4 degrees C in vacuum packages. *J. Food Prot.* 64:1722–1729.

292. Samuelson, K. J., J. H. Rupnow, and G. W. Froning. 1985. The effect of lysozyme and ethylenediaminetetraacetic acid on *Salmonella* on broiler parts. *Poultry Sci.* 64:1488–1490.

293. Seaberg, A. C., R. G. Labbe, and K. Shetty. 2003. Inhibition of *Listeria monocytogenes* by elite clonal extracts of oregano (*Origanum vulgare*). *Food Biotechnol.* 17:129–149.

294. Seward, R. A., R. H. Dielbel, and R. C. Lindsay. 1982. Effects of potassium sorbate and other antibotulinal agents on germination and outgrowth of *Clostridium botulinum* Type E spores in microculture. *Appl. Environ. Microbiol.* 44:1212–1221.

295. Shelef, L. A., and L. Addala. 1994. Inhibition of *Listeria monocytogenes* and other bacteria by sodium diacetate. *J. Food Safety* 14:103–115.

296. Shelef, L. A., E. K. Jyothi, and M. A. Bulgarelli. 1984. Growth of enteropathogenic and spoilage bacteria in sage-containing broth foods. *J. Food Sci.* 49:737–740.

297. Shelef, L. A., O. A. Naglik, and D. W. Bogen. 1980. Sensitivity of some common foodborne bacteria to the spices sage, rosemary, and allspice. *J. Food Sci.* 45:1042–1044.

298. Shelef, L. A., and Q. Yang. 1991. Growth suppression of *Listeria monocytogenes* by lactates in broth, chicken, and beef. *J. Food Prot.* 54:283–287.

299. Sheu, C. W., and E. Freese. 1972. Effects of fatty acids on growth and envelope proteins of *Bacillus subtilis*. *J. Bacteriol.* 111:516–524.

300. Sheu, C. W., W. N. Konings, and E. Freese. 1972. Effects of acetate and other short-chain fatty acids on sugars and amino acid uptake of *Bacillus subtilis*. *J. Bacteriol.* 111:525–530.

301. Sheu, C. W., D. Salomon, J. L. Simmons, T. Sreevalsan, and E. Freese. 1975. Inhibitory effects of lipophilic acids and related compounds on bacteria and mammalian cells. *Antimicrob. Agents Chemother.* 7:349–363.

302. Simpson, W. J. 1993. Studies on the sensitivity of lactic acid bacteria to hop bitter acids. *J. Inst. Brew.* 99:405–411.

303. Simpson, W. J., and A. R. W. Smith. 1992. Factors affecting antibacterial activity of hop compounds and their derivatives. *J. Appl. Bacteriol.* 72:327–334.

304. Singh, A., R. K. Singh, A. K. Bhunia, and N. Singh. 2003. Efficacy of plant essential oils as antimicrobial agents against *Listeria monocytogenes* in hotdogs. *Lebensm. Wiss. Technol.* 36:787–794.

305. Siragusa, G. R., and J. S. Dickson. 1993. Inhibition of *Listeria monocytogenes*, *Salmonella typhimurium* and *Escherichia coli* O157:H7 on beef muscle tissue by lactic or acetic acid contained in calcium alginate gels. *J. Food Safety* 13:147–158.

306. Siragusa, G. R., and M. G. Johnson. 1989. Inhibition of *Listeria monocytogenes* growth by the lactoperoxidase-thiocyanate-H_2O_2 antimicrobial system. *Appl. Environ. Microbiol.* 55:2802–2805.

307. Smith-Palmer, A., J. Stewart, and L. Fyfe. 1998. Antimicrobial properties of plant essential oils and essences against five important food-borne pathogens. *Lett. Appl. Microbiol.* 26:118–122.

308. Smith-Palmer, A., J. Stewart, and L. Fyfe. 2004. Influence of subinhibitory concentrations of plant essential oils on the production of enterotoxins A and B and alpha-toxin by *Staphylococcus aureus*. *J. Med. Microbiol.* 53:1023–1027.

309. Smith-Palmer, A., J. Stewart, and L. Fyfe. 2001. The potential application of plant essential oils as natural food preservatives in soft cheese. *Food Microbiol.* 18:463–470.

310. Smoot, L. A., and M. D. Pierson. 1981. Mechanisms of sorbate inhibition of *Bacillus cereus* T and *Clostridium botulinum* 62A spore germination. *Appl. Environ. Microbiol.* 42:477–483.

311. Smulders, F. J. M., P. Barendsen, J. G. van Longtestjin, D. A. A. Mossel, and G. M. Van der Marel. 1986. Lactic acid: considerations in favour of its acceptance as a meat decontaminant. *J. Food Technol.* 21:419–436.

312. Snijders, J. M. A., J. G. van Longtestjin, D. A. A. Mossel, and F. J. M. Smulders. 1985. Lactic acid as a decontaminant in slaughter and processing procedures. *Vet. Q.* 7:277–282.

313. Sofos, J. N., L. R. Beuchat, P. M. Davidson, and E. A. Johnson. 1998. Naturally occurring antimicrobials in food. *Regul. Toxicol. Pharm.* 28:71–72.

314. Sofos, J. N., F. F. Busta, and C. E. Allen. 1979. Botulism control by nitrite and sorbate in cured meats: a review. *J. Food Prot.* 42:739–770.

315. Stead, D. 1994. The effect of chlorogenic, gallic and quinic acids on the growth of spoilage strains of *Lactobacillus collinoides* and *Lactobacillus brevis*. *Lett. Appl. Microbiol.* 18:112–114.

316. Stead, D. 1993. The effect of hydroxycinnamic acids on the growth of wine-spoilage lactic acid bacteria. *J. Appl. Bacteriol.* 75:135–141.

317. Stecchini, M. L., I. Sarais, and P. Giavedoni. 1993. Effect of essential oils on *Aeromonas hydrophila* in a culture medium and in cooked pork. *J. Food Prot.* **56**:406–409.

318. Stopforth, J. D., J. N. Sofos, and F. F. Busta. 2005. Sorbic acid and sorbates, p. 49–90. *In* P. M. Davidson, J. N. Sofos, and A. L. Branen (ed.), *Antimicrobials in Foods*, 3rd ed. CRC Press, Boca Raton, Fla.

319. Stratford, M., and P. A. Anslow. 1998. Evidence that sorbic acid does not inhibit yeast as a classic "weak acid preservative". *Lett. Appl. Microbiol.* **27**:203–206.

320. Stratford, M., P. Morgan, and A. H. Rose. 1987. Sulphur dioxide resistance in *Saccharomyces cerevisiae* and *Saccharomyces ludwigii*. *J. Gen. Microbiol.* **133**:2173–2179.

321. Stratford, M., and A. H. Rose. 1986. Transport of sulphur dioxide by *Saccharomyces cerevisiae*. *J. Gen. Microbiol.* **132**:1–6.

322. Sudarshan, N. R., D. G. Hoover, and D. Knorr. 1992. Antibacterial action of chitosan. *Food Biotechnol.* **6**:257–272.

323. Teraguchi, S., K. Shin, T. Ogata, M. Kingaku, A. Kaino, H. Miyauchi, Y. Fukuwatari, and S. Shimamura. 1995. Orally administered bovine lactoferrin inhibits bacterial translocation in mice fed bovine milk. *Appl. Environ. Microbiol.* **61**:4131–4134.

324. Thippareddi, H., V. K. Juneja, R. K. Phebus, J. L. Marsden, and C. L. Kastner. 2003. Control of *Clostridium perfringens* germination and outgrowth by buffered sodium citrate during chilling of roast beef and injected pork. *J. Food Prot.* **66**:376–381.

325. Thompson, D. P. 1994. Minimum inhibitory concentrations of esters of *p*-hydroxybenzoic acid (paraben) combinations against toxigenic fungi. *J. Food Prot.* **57**:133–135.

326. Thomson, J. E., G. J. Banwart, D. H. Sanders, and A. J. Mercuri. 1967. Effect of chlorine, antibiotics, β-propiolactone, acids and washing on *Salmonella typhimurium* on eviscerated fryer chickens. *Poultry Sci.* **46**:146.

327. Thongson, C., P. M. Davidson, W. Mahakarnchanakul, and P. Vibulsresth. 2005. Antimicrobial effect of Thai spices against *Listeria monocytogenes* and *Salmonella typhimurium* DT104. *J. Food Prot.* **68**:2054–2058.

328. Tompkin, R. B. 2005. Nitrite, p. 169–236. *In* P. M. Davidson, J. N. Sofos, and A. L. Branen (ed.), *Antimicrobials in Foods*, 3rd ed. CRC Press, Boca Raton, Fla.

329. Tompkin, R. B., L. N. Christiansen, and A. B. Shaparis. 1978. The effect of iron on botulinal inhibition in perishable canned cured meat. *J. Food Technol.* **13**:521–527.

330. Touch, V., S. Hayakawa, S. Yamada, and S. Kaneko. 2004. Effects of a lactoperoxidase-thiocyanate-hydrogen peroxide system on *Salmonella enteritidis* in animal or vegetable foods. *Int. J. Food Microbiol.* **93**:175–183.

331. Tranter, H. S. 1994. Lysozyme, ovotransferrin and avidin, p. 65–97. *In* V. M. Dillon and R. G. Board (ed.), *Natural Antimicrobial Systems and Food Preservation*. CAB Intl., Wallingford, England.

332. Tranter, H. S., and R. G. Board. 1984. Influence of incubation temperature and pH on the antimicrobial properties of hen egg albumen. *J. Appl. Bacteriol.* **56**:53–61.

333. Tranter, H. S., and R. G. Board. 1982. The antimicrobial defense of avian eggs: biological perspective and chemical basis. *J. Appl. Biochem.* **4**:295–338.

334. Tsai, G. J., and W. H. Su. 1999. Antibacterial activity of shrimp chitosan against *Escherichia coli*. *J. Food Prot.* **62**: 239–243.

335. Tsai, G. J., Z. Y. Wu, and W. H. Su. 2000. Antibacterial activity of chitooligosaccharide mixture prepared by cellulase digestion of shrimp chitosan and its application to milk preservation. *J. Food Prot.* **63**:747–752.

336. Tsai, S., and C. Chou. 1996. Injury, inhibition and inactivation of *Escherichia coli* O157:H7 by potassium sorbate and sodium nitrite as affected by pH and temperature. *J. Sci. Food Agric.* **71**:10–12.

337. Uhart, M., S. Ravishankar, and N. D. Maks. 2004. Control of *Listeria monocytogenes* with combined antimicrobials on beef franks stored at 4 degrees C. *J. Food Prot.* **67**:2296–2301.

338. Ukuku, D. O., and W. F. Fett. 2004. Effect of nisin in combination with EDTA, sodium lactate, and potassium sorbate for reducing *Salmonella* on whole and fresh-cut cantaloupe. *J. Food Prot.* **67**:2143–2150.

339. UlateRodriguez, J., H. W. Schafer, E. A. Zottola, and P. M. Davidson. 1997. Inhibition of *Listeria monocytogenes*, *Escherichia coli* O157:H7, and *Micrococcus luteus* by linear furanocoumarins in a model food system. *J. Food Prot.* **60**:1050–1054.

340. Uljas, H. E., and S. C. Ingham. 1999. Combinations of intervention treatments resulting in a 5-$\log_{10}$-unit reductions in numbers of *Escherichia coli* O157:H7 and *Salmonella typhimurium* DT104 organisms in apple cider. *Appl. Environ. Microbiol.* **65**:1924–1929.

341. Ultee, A., L. G. M. Gorris, and E. J. Smid. 1998. Bactericidal activity of carvacrol towards the food-borne pathogen *Bacillus cereus*. *J. Appl. Microbiol.* **85**:211–218.

342. Ultee, A., and E. J. Smid. 2001. Influence of carvacrol on growth and toxin production by *Bacillus cereus*. *Int. J. Food Microbiol.* **64**:373–378.

343. Ultee, A. E., P. W. Kets, and E. J. Smid. 1999. Mechanisms of action of carvacrol on the food-borne pathogen *Bacillus cereus*. *Appl. Environ. Microbiol.* **65**:4606–4610.

344. Unda, J. R., R. A. Mollins, and H. W. Walker. 1991. *Clostridium sporogenes* and *Listeria monocytogenes*: survival and inhibition in microwave-ready beef roasts containing selected antimicrobials. *J. Food Sci.* **56**:198–205.

345. Valenti, P., P. Visca, G. Antonini, and N. Orsi. 1985. Antifungal activity of ovotransferrin towards genus *Candida*. *Mycopathology* **89**:165–175.

346. Venkitanarayanan, K. S., T. Zhao, and M. P. Doyle. 1999. Antibacterial effect of lactoferrin B on *Escherichia coli* O157:H7 in ground beef. *J. Food Prot.* **62**:747–750.

347. Walker, H. W., and M. A. Stahmann. 1955. Chemical nature of disease resistance in plants. *Ann. Rev. Plant Physiol.* **6**:351–366.

348. Wan, J., A. Wilcock, and M. J. Coventry. 1998. The effect of essential oils of basil on the growth of *Aeromonas hydrophila* and *Pseudomonas fluorescens*. *J. Appl. Microbiol.* **84**:152–158.

349. Wang, G. H. 1992. Inhibition and inactivation of five species of foodborne pathogens by chitosan. *J. Food Prot.* **55**:916–919.

350. **Wang, L.-L., and E. A. Johnson.** 1997. Control of *Listeria monocytogenes* by monoglycerides in foods. *J. Food Prot.* **60:**131–138.

351. **Ward, S. M., P. J. Delaquis, R. A. Holley, and G. Mazza.** 1998. Inhibition of spoilage and pathogenic bacteria on agar and pre-cooked roast beef by volatile horseradish distillates. *Food Res. Int.* **31:**19–26.

352. **Weaver, R., and L. A. Shelef.** 1993. Antilisterial activity of sodium, potassium or calcium lactate in pork liver sausage. *J. Food Safety* **13:**133–146.

353. **Wen, A. M., P. Delaquis, K. Stanich, and P. Toivonen.** 2003. Antilisterial activity of selected phenolic acids. *Food Microbiol.* **20:**305–311.

354. **Wendorff, W. L., W. E. Riha, and E. Muehlenkamp.** 1993. Growth of molds on cheese treated with heat or liquid smoke. *J. Food Prot.* **56:**963–966.

355. **Were, L. M., B. Bruce, P. M. Davidson, and J. Weiss.** 2004. Encapsulation of nisin and lysozyme in liposomes enhances efficacy against *Listeria monocytogenes*. *J. Food Prot.* **67:**922–927.

356. **Wibowo, D., R. Eschenbruch, C. R. Davis, G. H. Fleet, and T. H. Lee.** 1985. Occurrence and growth of lactic acid bacteria in wine. *Am. J. Enol. Viticult.* **36:** 302–313.

357. **Wilkins, K. M., and R. G. Board.** 1989. Natural antimicrobial systems, p. 285–362. *In* G. W. Gould (ed.), *Mechanisms of Action of Food Preservation Procedures*. Elsevier Applied Sci., London, England.

358. **Wills, E. D.** 1956. Enzyme inhibition by allicin, the active principal of garlic. *Biochem. J.* **63:**514–520.

359. **Woods, L. F. J., and J. M. Wood.** 1982. The effect of nitrite inhibition on the metabolism of *Clostridium botulinum*. *J. Appl. Bacteriol.* **52:**109–110.

360. **Woods, L. F. J., J. M. Wood, and P. A. Gibbs.** 1981. The involvement of nitric oxide in the inhibition of the phosphoroclastic system in *Clostridium sporogenes* by sodium nitrite. *J. Gen. Microbiol.* **125:**399–406.

361. **Woods, L. F. J., J. M. Wood, and P. A. Gibbs.** 1989. Nitrite, p. 225–246. *In* G. W. Gould (ed.), *Mechanisms of Action of Food Preservation Procedures*. Elsevier Applied Sci., London, England.

362. **Woolford, M. K.** 1975. Microbiological screening of the straight chain fatty acids (C_1-C_{12}) as potential silage additives. *J. Sci. Food Agric.* **26:**219–228.

363. **Xiong, R., G. Xie, and A. S. Edmondson.** 1999. The fate of *Salmonella enteritidis* PT4 in home-made mayonnaise prepared with citric acid. *Lett. Appl. Microbiol.* **28:**36–40.

364. **Yang, T.** 1985. Mechanism of nitrite inhibition of cellular respiration in *Pseudomonas aeruginosa*. *Curr. Microbiol.* **12:**35–40.

365. **Yin, M.-C., and W.-S. Cheng.** 1998. Inhibition of *Aspergillus niger* and *Aspergillus flavus* by some herbs and spices. *J. Food Prot.* **61:**123–125.

366. **Yoon, K. S., C. S. Burnette, K. A. Abou-Keid, and R. C. Whiting.** 2004. Control of growth and survival of *Listeria monocytogenes* on smoked salmon by combined potassium lactate and sodium diacetate and freezing stress during refrigeration and frozen storage. *J. Food Prot.* **67:**2465–2471.

367. **Yousef, A. E., M. A. El-Shenawy, and E. H. Marth.** 1989. Inactivation and injury of *Listeria monocytogenes* in a minimal medium as affected by benzoic acid and incubation temperature. *J. Food Sci.* **54:**650–652.

368. **Zivanovic, S., C. C. Basurto, S. Chi, P. M. Davidson, and J. Weiss.** 2004. Molecular weight of chitosan influences antimicrobial activity in oil-in-water emulsions. *J. Food Prot.* **67:**952–959.

Food Microbiology: Fundamentals and Frontiers, 3rd Ed.
Edited by M. P. Doyle and L. R. Beuchat
© 2007 ASM Press, Washington, D.C.

Thomas J. Montville
Michael L. Chikindas

Biopreservation of Foods

34

This chapter provides an overview of the biologically based preservation technologies termed "biopreservation." Biopreservation is defined as the use of microorganisms (including bacteriophages), their metabolic products, or both to preserve foods that are not generally considered fermented. The preservative, nutritional, and functional properties of fermented foods are covered in chapters 35 to 42.

Acid production by lactic acid bacteria (LAB) in temperature-abused foods (controlled acidification) is covered in the first part of the chapter. Some LAB produce antimicrobial proteins, called bacteriocins, which inhibit spoilage and pathogenic bacteria without changing (e.g., through acidification, protein denaturation, and other processes) the physicochemical nature of the food. The largest section of this chapter deals with bacteriocins. The chapter closes by examining the use of bacteriophages as biocontrol agents.

BIOPRESERVATION BY CONTROLLED ACIDIFICATION

Organic acids inhibit microbial growth, as discussed more fully in chapter 33. While organic acids are usually added to foods, LAB can produce lactic acid in situ.

The controlled production of acid in situ is an important form of biopreservation. Many factors determine the effectiveness of in situ acidification. These include the product's initial pH, its buffering capacity, target microorganism characteristics, the type and concentration of fermentable carbohydrate, ingredients that influence the viability and growth rate of LAB, and the growth rates of LAB and target pathogens at refrigerated and abuse temperatures (66). Clearly, such applications require customization. The production of bacteriocins, diacetyl, and hydrogen peroxide may also contribute to inhibition caused by culture fermentates. For example, Microgard is a cultured milk product added to much of the cottage cheese produced in the United States as a Generally Recognized as Safe (GRAS) food preservative. It is made by fermenting milk using *Propionibacterium shermanii* to produce acetate, propionic acid, and low-molecular-weight proteins. Although Microgard contains a bacteriocin, propionic acid plays a major role in its activity (4, 82).

The idea of using LAB to prevent botulinal toxigenesis through in situ acid production dates back to the 1950s. This technology relies on the inability of *Clostridium*

Thomas J. Montville and Michael L. Chikindas, Department of Food Science, Rutgers—The State University of New Jersey, New Brunswick, NJ 08901-8520.

botulinum to grow at pH < 4.6 as a defense against temperature abuse. LAB and a fermentable carbohydrate are added to the food. The LAB grow and produce acid only under conditions of temperature abuse. Under proper refrigeration, LAB cannot grow and no acid is formed. Saleh and Ordal (120) used a "normal cheese culture," *Lactobacillus bulgaricus*, or *Lactococcus lactis* in experiments designed to prevent toxin production in chicken à la king containing a fermentable carbohydrate inoculated with spores of *C. botulinum*. When incubated at 30°C in the absence of LAB, 16 of 16 samples rapidly became toxic. In the presence of the normal cheese culture, fewer than three samples became toxic after 5 days at 30°C. In all samples to which the LAB had been added, including those positive for toxin, the pH was reduced to <4.5. There were no significant differences between cultures, and the inhibition was attributed to acid production.

The discovery that carcinogenic nitrosamines are formed from nitrites in cured meats initiated a search for nitrite substitutes. Tanaka et al. (131) used controlled acidification to reduce nitrite concentrations in bacon. When bacon was inoculated with 3 log botulinal spores/g and incubated at 28°C, toxin was produced in 58% of the conventional bacon samples prepared with 120-μg/g nitrite but no sucrose or starter culture. When bacon was prepared with 80- or 40-μg/g nitrite, 0.7% sucrose, and starter cultures, ≤2% became toxic. The U.S. Department of Agriculture (USDA) approved the "Wisconsin Process" for bacon manufacture in 1986.

BACTERIOCINS

Bacteriocins are ribosomally synthesized antimicrobial peptides that are not lethal to the host. Interest in bacteriocins produced by LAB has grown dramatically and is documented by many major reviews and books (for examples, see references 36, 55, 61, 88, 99, 112, and 130). Many bacteriocins inhibit bacteria such as *Listeria monocytogenes*, which is recalcitrant to traditional preservation methods. In addition, bacteriocinogenic LAB are associated with, and are used as, starter cultures. The use of bacteriocins, of the bacteria which produce them, or of both is attractive to the food industry due to increasing consumer demand for natural products and increasing concern about foodborne diseases. However, this interest is tempered by regulatory uncertainty and concerns that the development of bacteriocin-resistant pathogens might render the technology ineffective.

General Characteristics

Bacteriocins produced by LAB are a heterogeneous group of ribosomally synthesized small proteins. They normally act against closely related bacteria, but not against the producer. "Closely related" can cover a wide range of gram-positive bacteria. Chelating agents, hydrostatic pressure, or injury can render gram-negative bacteria bacteriocin sensitive (72, 73, 127) or act synergistically to enhance bacteriocin activity against gram-positive bacteria. Degradative enzymes such as lysozyme are not considered bacteriocins. However, in addition to acting on the membrane of the target cell, some bacteriocins, e.g., colicins produced by *Escherichia coli*, inhibit protein synthesis, degrade RNA, or have other biological functions (77).

The seven characteristics of colicins cited in the Tagg et al. (130) review were initially applied to bacteriocins. Few LAB bacteriocins meet all of these criteria, which would restrict the designation "bacteriocins" to only those plasmid-mediated proteins which are bactericidal to a narrow range of closely related bacteria having specific binding sites for that bacteriocin and whose biosynthesis ultimately kills the producing cell.

Bacteriocins differ in their spectra of activity, biochemical characteristics, and genetic determinants (74, 99). Most bacteriocins are small (3 to 10 kDa), have a high isoelectric point, and contain both hydrophobic and hydrophilic domains. Klaenhammer (74) further classified bacteriocins into four major groups.

Group I bacteriocins contain the unusual amino acids dehydroalanine, dehydrobutyrine, lanthionine, and β-methyllanthione (68). These amino acids are produced by posttranslational modification of serine and threonine to their dehydro forms. The dehydro amino acids react with cysteine to form thioether (single-sulfhydryl) lanthionine rings. Bacteriocins containing these lanthionine rings are commonly referred to as lantibiotics. There are many structurally similar lantibiotics. Nisin, the best characterized LAB bacteriocin (133), is produced in at least two related forms. Nisin A contains histidine at position 27, whereas nisin Z has an asparagine (97). Subtilin, produced by *Bacillus subtilis*, also contains five lanthionine rings and a conformation similar to that of nisin but has other amino acid substitutions including a carboxy terminus two amino acids shorter than nisin. The 12 amino acids at the amino terminus of nisin and epidermin are similar, but epidermin lacks the central lanthionine ring common to nisin and subtilin and has a cyclized carboxy terminus. Lacticin 481 (108), lactocin S (96), and carnocin (129) are other lantibiotics produced by LAB.

Group II bacteriocins are small heat-stable proteins with a consensus leader sequence containing a Gly-Gly^{-1}-Xaa^{+1} cleavage site important for processing the prebacteriocin during export. Group II is subdivided into three subgroups. Bacteriocins active against *L. monocytogenes* and having a -Tyr-Gly-Asn-Gly-Val-Xaa-Cys amino-terminal consensus sequence are classified in the subgroup IIa. Pediocin PA-1/AcH, sakacins A and P, leucocin A, bavaricin MN, and curvacin A are members of this subgroup (36). Subgroup IIb (98) contains bacteriocins such as lactococcin G, lactococcin M, lactacin F, and plantaricins EF and JK, which require two different peptides for activity (3, 92). Bacteriocins in subgroup IIc, such as lactacin B, require reduced cysteine for activity.

Bacteriocins in Groups III and IV differ markedly from other bacteriocins (74). The larger (>30-kDa) heat-labile antimicrobial proteins such as helveticins J and V and lactacins A and B are classified as Group III bacteriocins. Group IV bacteriocins such as leuconocin S, lactocin 27, and pediocin SJ-1 have lipid or carbohydrate moieties. The composition and function of the nonprotein portions are largely unknown.

Methodological Considerations

Bacteriocin-producing bacteria are easy to isolate from foods. The methods used for their initial isolation and characterization are relatively simple (86). The most common method for demonstrating bacteriocin production (the Bac⁺ phenotype) is to overlay a colony of the putative bacteriocin producer with an agar medium containing the bacterium being tested for sensitivity. An inhibition zone in the confluent growth of the target bacterium is presumptive evidence for bacteriocin production. Such zones can also be produced by acid, bacteriophage, hydrogen peroxide, or other nonspecific inhibitors. Negative-control experiments are required to confirm the production of bacteriocin. A positive control, confirming the protease sensitivity of the inhibitor, is equally important and can often be done on the same petri dish.

Research beyond the initial isolation of the Bac⁺ bacteria and characterization of the inhibitor as a bacteriocin is more difficult. Some isolates produce large inhibition zones on agar media but have no detectable activity in broth. High (10,000 arbitrary units [AU]/ml to >50,000 AU/ml) bacteriocin activities (see below for definitions) in the culture supernatants facilitate bacteriocin purification. The purification usually involves salting out the protein followed by some combination of gel filtration, ion-exchange, and affinity and hydrophobic-interaction chromatography. Amino acid sequences are then determined from the electrophoretically pure protein. As more bacteriocins are being purified to the sequence level, it is becoming common to discover that independently purified bacteriocins are identical.

The lack of recognized standards for bacteriocin activity is a major impediment to progress in this field. Only nisin has an international unit of activity. One gram of a commercially available nisin preparation (Nisaplin) contains 25 mg of pure nisin and is defined as having 1,000,000 International Units (earlier known as Reading Units) of activity. Nisin activity is measured by the well diffusion assay of Tramer and Fowler (134) using *Micrococcus luteus* ATCC 10420 as the sensitive organism. *M. luteus* is also used to measure the activity of other peptide antimicrobials such as scorpion defensins (21). Investigators frequently use other indicator strains in assays for other bacteriocins if these strains are more sensitive (generate larger zone sizes) than *M. luteus*.

Bacteriocin activity is estimated from the size of the inhibition zones produced by the diffusion of the bacteriocin in confluent lawns of bacteriocin-sensitive bacteria. The zone sizes obtained in diffusion assays are proportional to the log of the bacteriocin activity. Results are easily misinterpreted when zone sizes are reported without considering the log-linear nature of the assay.

Arbitrary Units (AU) are more useful than zone sizes as quantitative measurements of bacteriocin activity and are usually determined by assaying twofold serial dilutions of the sample. The reciprocal of the highest dilution producing inhibition becomes the number of AU. This can be divided by the sample volume to yield the number of arbitrary units per milliliter. The arbitrary nature of this system for measuring activity cannot be overemphasized. By virtue of the twofold dilutions, the assay is insensitive to activity differences that are less than twofold. The assumption of linearity with volume used to calculate the number of arbitrary units per milliliter is rarely verified. The choice of indicator bacterium, assay technique, length of incubation time, assay medium, and other factors are so idiosyncratic as to make it virtually impossible to compare the AU derived from different procedures. There is no easy solution to this problem. However, by assaying a known concentration of nisin under the same experimental conditions, an approximation of the relationship between AU and equivalent nisin International Units can be generated.

Some investigators use conceptually different assays to generate other "Arbitrary Units" unrelated to those described above. For example, the ability to reduce growth rate, optical density, or viability by 50% can be a measure of bacteriocin activity (59). These assays do not require the 12- to 24-h incubation period of the diffusion assays.

Methods of bacteriocin quantification that do no rely on biological activity are more quantitative since they do not depend on the sensitivity of an arbitrarily chosen indicator bacterium. Enzyme-linked immunosorbent assays work well if antibodies to the bacteriocin are available (for examples, see references 8 and 83). A recombinant construction consisting of the *gusA* gene under the control of the nisin-regulated *nisA* promoter (for examples, see references 107 and 29a) can also determine nisin concentration. Nisin induces activity of the *nisA* promoter and, subsequently, production of β-glucuronidase, whose activity can be measured. A bioluminescence assay for nisin quantification (103) is based on a recombinant construct consisting of the nisin-regulated promoter and bioluminescence genes *luxAB* from *Xenorhabdus luminescens*. *Lc. lactis* MG1614 cells transformed with such a recombinant DNA can detect as little as 0.0125 ng of nisin per ml (145).

Bacteriocin Applications in Foods

There are many different ways to use bacteriocins in foods. The first is to add bacteriocins directly to the food for the purpose of inhibiting spoilage or pathogenic bacteria. Only nisin is commercially available for this purpose. It is sold as a partially purified product of dairy or nondairy fermentation (146) and marketed by companies under the names of Nisaplin and Novasin (25a), among others. The efficacy of pediocin addition has also been demonstrated. The second way to use bacteriocins is to add bacteriocinogenic cultures to the food (17, 35) or use them as starter cultures that produce the bacteriocin in situ. A third way to use bacteriocins is to facilitate the use of defined starter cultures in fermented foods. The use of defined starter cultures rather than indigenous bacteria improves product quality and consistency. However, unless the indigenous bacteria can be inactivated, they usually predominate over the inocula. Because of this, the benefits of defined starter cultures are most pronounced in the dairy industry, where indigenous bacteria are inactivated during pasteurization. The use of bacteriocin-producing starter cultures in fermented meats and vegetables might inactivate the indigenous bacteria and facilitate the use of defined starter cultures.

Addition of Nisin

Nisin is added to milk, cheese, and dairy products, a variety of canned foods, mayonnaise, and baby foods throughout the world (133). It is GRAS (38) in a variety of applications and is used commercially as an antimastitis teat dip (123). Nisin has potential as a treatment for ulcers, in personal hygiene, and as a general sanitizing agent. When nisin is absorbed onto surfaces, it inhibits listerial growth (24) and prevents biofilm formation (9). Nisin also sensitizes spores to heat, so that the thermal treatment can be reduced. This application is not approved in the United States.

Many nisin applications target the inhibition of *C. botulinum*. Spores of *C. botulinum* are much less nisin sensitive than are vegetative cells of *L. monocytogenes*. While nisin at concentrations as low as 200 IU/ml can reduce *L. monocytogenes* viability by 6 logs, concentrations of up to 10,000 IU/ml are required to obtain similar results with botulinal spores (94).

Temperature is the major determinant of the inhibitory action of nisin. Nisin is much less effective at elevated temperatures than at refrigeration temperatures against *C. botulinum* 56A spores in a model food system (117). Eventually, growth in this system occurs when nisin activity falls below some threshold level, which is lower at decreasing temperatures. For example, botulinal growth occurred when residual nisin concentrations fell below 154 IU/ml at 35°C, but not until the nisin concentration was <12 IU/ml at 15°C.

Many other factors influence the sporostatic efficacy of nisin (122, 133). In studies used to support the GRAS affirmation of nisin in pasteurized process cheese, nisin concentrations between 500 and 2,000 IU/ml inhibited botulinal spore outgrowth by 50% in broth, but concentrations of up to 10,000 IU/ml were ineffective in cooked meat medium (121). Nisin at 100 to 250 μg/g allows some salt reduction or increased moisture content in pasteurized process cheese spreads without elevating the risk of botulism (126). Specific phospholipids also decrease the antibotulinal efficacy of nisin (117), and fats in general are considered antagonistic due to hydrophobicity (133).

In most applications, nisin serves as one part of a multiple-barrier inhibitory system. Nisin may be a useful adjunct to modified-atmosphere storage. Nisin increases shelf life and delays toxin production by type E strains of *C. botulinum* in fresh fish packaged in a carbon dioxide atmosphere. However, toxin can sometimes be detected before samples are obviously spoiled (132). The combination of nisin and modified atmosphere to prevent *L. monocytogenes* growth in pork is more effective than either used alone (37). Nisin (5 μg/ml) added to liquid whole egg prior to pasteurization extends refrigerated shelf life from 6 to 11 days to 17 to 20 days (29).

Most research has focused on defining conditions under which botulinal spores are inhibited by nisin. Inhibition of spores from other bacterial species also depends on many factors. The ability of nisin to inhibit thermally stressed *Bacillus* spores is influenced by the time-temperature combination used to affect the thermal stress, subsequent incubation temperature, pH, and even the type of acidulant used (106). In general, nisin is more effective at lower temperatures, against lower spore populations, and under acidic conditions.

Addition of Pediocin

While inactive against bacterial spores, pediocins also inhibit vegetative cells of *L. monocytogenes*. European patents cover the use of pediocin PA-1 as a dried powder or culture liquid to extend the shelf life of salads and salad dressing (49) and as an antilisterial agent in foods such as cream, cottage cheese, meats, and salads (133, 140).

Pediocins are more effective than nisin in inhibiting bacteria in meat and are even more effective in dairy products. Dipping meat in 5,000 AU of crude pediocin PA-1 per ml decreases the viability of attached *L. monocytogenes* 100- to 1,000-fold. Pretreating meat with pediocin reduces subsequent *L. monocytogenes* attachment (100). Pediocin AcH at 1,350 AU/ml reduces *Listeria* populations in ground beef, sausage, and other products by 1 to 7 logs. Pediocin AcH is more effective at 4°C than at 25°C against *L. monocytogenes* in wiener (frankfurter or hot dog) exudate. Emulsifiers such as Tween 80 or the entrapment of the pediocin in multilamellar vesicles increases pediocin effectiveness in fatty foods (26–28). In most cases, the bacteriocin rapidly reduces the viability of *Listeria* and delays growth of the survivors.

Addition of Bacteriocin-Producing Bacteria to Nonfermented Foods

Many antilisterial applications involve the addition of bacteriocinogenic culture rather than the bacteriocin. In wieners held at 4°C, *L. monocytogenes* grows after a 20- to 30-day lag period and increases from 4 to 6 log CFU/g by the end of 60 days (6). At 7 CFU/g, Bac$^+$ and Bac$^-$ derivative strains of *Pediococcus acidilactici* inhibit *L. monocytogenes* growth in wieners for 60 days. Inhibition by the Bac$^-$ strain was not due to acidification or hydrogen peroxide. However, at low inoculum populations, the Bac$^+$ strain, but not its Bac$^-$ derivative, extends the lag period of *Listeria*. The degree of inhibition increases with decreasing temperatures and, in this case, is greater under anaerobic conditions than aerobic conditions.

Lactobacillus bavaricus MN (which produces bavaricin MN) inhibits listerial growth in a model gravy system at 4°C, even in the absence of a fermentable carbohydrate (148, 149). The addition of a fermentable carbohydrate, reduction of incubation temperatures, and increased *Lactobacillus*:*Listeria* inoculation ratios all increase the inhibition. These variables also influence the success of this technology in *sous vide* beef cubes (148, 149). *Lb. bavaricus* causes *L. monocytogenes* viability to decline 10-fold over 6 weeks at 4°C under the most favorable conditions (4°C, high inoculation ratio, and beef packed in gravy which contained a fermentable carbohydrate).

Under the least favorable conditions (10°C, low inoculation ratio, beef cubes without gravy), listerial growth is inhibited by the bavaricin for 1 week but increases 100-fold in beef without the *Lb. bavaricus*.

Carnobacterium piscicola LK5 is more effective against *L. monocytogenes* at 5°C than at 19°C in ultra-high-temperature-treated milk, dog food made from beef, pasteurized crab meat, creamed corn, and wieners (15). *C. piscicola* LK5 inhibits *L. monocytogenes* Scott A, even when present at populations 100-fold higher than *C. piscicola*.

Use of Bacteriocin-Producing Starter Cultures To Improve Safety of Fermented Foods

If a food is to be fermented by LAB, the use of bacteriocin-producing starter cultures can add value to the product. For example, the presence of a nisin producer among the strains used to make Cheddar cheese increases the shelf life of pasteurized processed cheese in which it is used by 14 to 87 days at 22°C (115).

Bacteriocinogenic pediococci appear to be especially effective in fermented meats. In situ pediocin production by *P. acidilactici* PAC 1.0 during the manufacture of fermented dry sausage reduces *L. monocytogenes* viability >10-fold relative to the acid-induced decrease caused by a nonbacteriocinogenic control (44). When wild-type Bac$^+$ *P. acidilactici* H is used to ferment summer sausage, concentrations of pediocin up to 5,000 AU/g are produced, reducing *L. monocytogenes* viability by 3.4 log CFU/g (68).

Use of Bacteriocinogenic Starter Cultures To Direct the Fermentation of Fermented Foods

The use of undefined indigenous bacteria in some types of fermented foods compromises quality, makes process control difficult, and introduces an uncontrolled manufacturing variable. These problems can be overcome by the use of bacteriocinogenic starter cultures. Their ability to outgrow the indigenous microbiota (58, 109, 110, 144), thereby facilitating the use of defined starter cultures in unpasteurized foods, is a promising application.

There are several novel applications of nisin or nisin-producing strains in fermented foods. The use of a nisin-producing *Lc. lactis* paired with nisin-resistant *Leuconostoc mesenteroides* in sauerkraut fermentations retards *Lactobacillus plantarum* growth (58). This allows *Leuconostoc mesenteroides* to establish itself and results in a higher-quality product. The LAB found in wine are very sensitive to nisin and can be inhibited without inhibiting yeasts or influencing the taste (109, 110). At 100 IU/ml, nisin inhibits the bacteria which, by their malolactic fermentation, spoil wine. If the malolactic fermentation is desired, nisin-resistant *Leuconostoc oenos* can be added with nisin. This promotes malolactic fermentation and suppresses the indigenous LAB (23).

Genetics of LAB Bacteriocins

Location of Bacteriocin Genes

The genetic information encoding bacteriocin production and immunity can be located on plasmids, on the chromosome, or both. For instance, the gene coding for enterocin AS-48RJ, a bacteriocin very similar to the cyclic peptide enterocin AS-48, is chromosomal, whereas genes coding for enterocin AS-4 and related bacteriocins are plasmid mediated (2). The nisin structural gene is located on the chromosome of *Lc. lactis* (16, 32, 111). The nisin gene resides within a 70-kb conjugative transposon and is genetically linked to genes coding for sucrose fermentation. Integration of this transposable element into the chromosome of a Nis⁻ strain was observed after conjugation by probing the digested total DNA of transconjugants for different regions of the nisin-coding transposon (62). Other examples of chromosomally located LAB bacteriocins are helveticin J and lactacin B (74). At the same time, many of the class IIa bacteriocins (including the best-characterized pediocin PA-1/AcH) have their genes located on plasmids (45, 89, 142).

Organization of Bacteriocin Operons: a Generic Operon

The DNA sequence and amino acid composition of many bacteriocins are known (for a review, see reference 119). Therefore, a general picture of the genetic organization of bacteriocin genes is emerging. The structural gene encoding bacteriocin production appears to be located in an operon-like structure (74). The organization of a generic operon is shown in Fig. 34.1. It carries the genes coding for the bacteriocin prepeptide (structural gene), immunity, maturation, processing, and export of the bacteriocin molecules as well as genes encoding products involved in the regulation of bacteriocin biosynthesis. The "real" operons are different from the "generic" model. Not all specific bacteriocin operons contain all of the genes incorporated into the generic bacteriocin gene cluster. The structural gene usually encodes a prepeptide which comprises the precursor of the mature bacteriocin preceded by an N-terminal extension or leader sequence. The secondary structure of this N-terminal extension is predicted to be an α-helix that is cleaved during the maturation or export process. Class II bacteriocins encode a prepeptide containing a consensus sequence (Gly^{-2}-Gly^{-1}) at the cleavage site and present strong homology in their hydrophobicity profile. These leader peptides play a significant role in the recognition of the precursor by the ABC transporters which affect the proteolytic

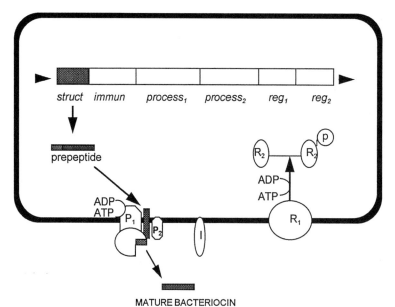

Figure 34.1 A generic bacteriocin operon. The structural gene (*struct*) codes for a prepropeptide which is modified and excreted by the processing gene products (P1 and P2) and may be regulated by a signal transduction pathway coded for by *reg₁* and *reg₂*. For additional abbreviations and explanations, see the text.

cleavage of the leader peptides and the export of the "maturated" (processed) bacteriocins across the membrane (138). Most bacteriocin leader sequences, except for a few such as divergicin A (150), lactococcin 972 (84), propionicin T1 (40), and some enterocins (20), do not exhibit characteristics of *sec*-dependent export proteins. Rather, they may be exported by a *sec*-independent pathway involving a transport protein encoded within the operon. In addition, the N-terminal extension may help neutralize bacteriocin activity within the producer strain, since a nonprocessed bacteriocin is much less antimicrobial than a mature peptide (113).

The structural genes of many bacteriocins produced by LAB have been described and reviewed (29a, 55, 74). These include the structural genes encoding for nisin (*nisA*), pediocin PA-1 (*pedA*), lactococcins A and B (*lcnA* and *lcnB*), sakacin A (*sakA*), leucocin A (*leuA*), lactacin F (*lafA* and *lafX*), and lactococcin M (*lcnM* and *lcnN*).

The immunity gene makes Bac$^+$ cells immune to their own bacteriocin. Immunity is specific and usually coordinated with bacteriocin production. Several immunity genes have been sequenced, and their ability to confer immunity to the producer strains has been proven (1, 60, 78, 136, 141). In the case of nisin, the *nisI* gene product is predicted to be an extracellular lipoprotein which, by anchoring to the membrane through its lipid moiety, confers immunity to the producer cells. Additional genes (*nisE*, *F*, and *G*) thought to be involved in immunity are in the nisin gene cluster (125). Class IIa operons for bacteriocins code for 11-kDa immunity proteins, which protect the producer cells. These immunity proteins are highly specific four-helix bundle cytosolic molecules which establish immunity only to their own bacteriocins or to very closely related ones. The domain that recognizes the C-terminal membrane-penetrating region of class IIa bacteriocins is located in the C-terminal part of immunity proteins (for a review, see reference 43).

Processing and export genes code for at least two proteins that ensure that the mature bacteriocin is formed and exported from the cytoplasm. One of these proteins is an ATP-binding cassette translocator of the hemolysin B (HlyB) subfamily. These translocators (also known as ABC transporters) utilize ATP hydrolysis as an energy source for protein translocation and are characterized by the presence of two domains: a membrane spanning domain and a cytoplasmic ATP-binding domain (39). Examples of genes coding for ATP-binding transporter proteins in LAB bacteriocins are *nisT* (nisin), *lcnC* (lactococcin A), and *pedD* (pediocin PA-1) (34, 85, 128). The second protein of the secretion apparatus is a structural homologue of the hemolysin D (HlyD), an accessory

protein which may facilitate transport. The HlyD homologous protein does not seem to be present in the nisin gene cluster.

Some of the ABC transporters found in bacteria that produce Class II bacteriocins have similar N-terminal amino acid sequences. This domain may contain the proteolytic cleavage site for the N-terminal extension of these bacteriocins. In contrast, the nisin gene cluster contains the *nisP* gene. Its gene product, a serine protease, is thought to be involved in cleavage of the N-terminal extension of nisin. In addition, the nisin operon contains the genes *nisB* and *nisC*, which encode two membrane-associated proteins involved in catalyzing posttranslational modifications of precursor molecules (34, 141).

Quorum Sensing in Regulation of Bacteriocin Production

Some microbial genes are regulated to be expressed when the cell density has reached a certain population size, or quorum. This process is called quorum sensing or cell-to-cell signaling and involves synthesis and detection of small diffusible pheromones (5). The observation that some bacteria produce bacteriocins in a cell-density-dependent manner led to a discovery of quorum-sensing involvement in the synthesis of these peptides. Originally it was noticed that when diluted in fresh media, some strains would stop producing bacteriocins. However, the synthesis would reoccur if filter-sterilized medium from the same strain cells of high density was added (33).

Regulation of Lantibiotics

Class I bacteriocins undergo extensive posttranslational modification prior to their secretion from the cell. Biosynthesis of at least some lantibiotics is regulated by quorum sensing. All genes necessary for nisin A production are arranged in one gene cluster. Expression of *nisA-BTCIP* is regulated by the P$_{nisA}$ promoter. This operon includes the structural gene for nisin precursor peptide (*nisA*) and genes necessary for maturation and export of, and immunity to, nisin (*nisBT*, *nisCP*, and *nisI*, respectively). Regulatory genes (*nisRK*) and the rest of the immunity genes (*nisFEG*) are under control of promoters P$_{nisR}$ and P$_{nisF}$, respectively. NisK and NisR are a two-component signal transduction system. When a mature nisin A molecule binds to NisK, the signal is transduced to a response regulator NisR, which subsequently is able to bind to P$_{nisA}$ and P$_{nisF}$ (75). Subtilin has an almost identical regulatory system, with minor differences.

Regulation of Pediocin-Like Bacteriocins

The synthesis of many class IIa bacteriocins such as plantaricin from *Lb. plantarum* is regulated by peptide pheromones (30). The signal transduction is conveyed via a standard two-component system. This system is sometimes called a three-component system when the structural gene for pheromone itself is included (33).

The quorum-sensing regulation of sakacin K has been studied in detail (13). The pheromone binds to a sensor-kinase protein which activates the response regulator. The response regulator interacts with promoters upstream of the regulatory and transport operons as well as with the promoter of the structural gene itself. The structural gene for bacteriocin is under the stringent control of this system, while regulatory and transport genes appear to be less responsive to the regulation because the pheromone is exported by the very same ABC transporter (33).

Mechanism of Action against Vegetative Cells

Bacteriocins produced by LAB disrupt the integrity of the cytoplasmic membrane, increasing its permeability to low-molecular-weight compounds. The addition of bacteriocins to vegetative cells results in a rapid and nonspecific efflux of preaccumulated ions, amino acids, and in some cases (nisin, las 5) but not others (pediocin PA-1), ATP molecules (18, 19, 137, 143, 147). This increased flux of compounds rapidly dissipates chemical and electrical gradients across the membrane. The proton motive force, which serves as the major energy source for many vital processes, is dissipated within minutes after contact with bacteriocin (11, 12). Bacteriocin-treated cells have decreased levels of intracellular ATP. The loss of ATP can be caused by ATP efflux or hydrolysis. While efflux may be a direct result of membrane disruption, in other cases, such as with pediocin PA-1, hydrolysis can result from a shift in the ATP equilibrium due to P_i efflux or as a futile attempt of the cell to regenerate the proton motive force (1, 147). Ultimately, these changes in permeability render the cell unable to protect its cytoplasm from the environment. This leads to cell inhibition and often death.

Given the diversity in the biochemical attributes of bacteriocin molecules and genetic determinants, it is surprising that they all act by the common mechanism of membrane permeabilization. A structural motif similar to other antimicrobial peptides occurring in nature (see below) may explain their interaction with membranes (105). In this regard, most bacteriocins with known sequences are amphiphilic cationic peptides showing α-helix, β-sheet, or in the case of lantibiotics, screw-like, secondary structures. Based on the amphiphilic characteristics of bacteriocins, there are at least two different mechanisms which may explain their membrane-permeabilization action (Fig. 34.2). Bacteriocins may

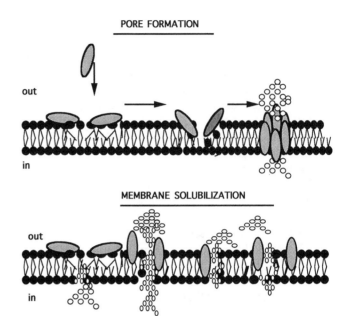

Figure 34.2 Models for pore formation and detergent-like mechanisms of bacteriocin action.

act by a multiple-step poration complex in which bacteriocin monomers bind, insert, and oligomerize in the cytoplasmic membrane to form a pore with the hydrophilic residues facing inward and the hydrophobic ones facing outward. The small size of bacteriocins makes it is unlikely that one molecule could form a pore. Alternatively, bacteriocins may disrupt the membrane integrity by a detergent-like membrane solubilization action. The action of nisin, pediocin PA-1, lactococcins A and B, and lactacin F in vivo is concentration and time dependent, supporting a poration complex mechanism (1, 19, 136, 143, 147). A generalized solubilization mechanism would result in an all-or-none lysis of the cells, with a sudden collapse of the bioenergetic parameters and no saturation kinetics. The addition of bacteriocins to membrane vesicles or liposomes loaded with different-size probes does not always result in leakage of the probe. Rather, leakage depends on both the size of the probe and the amount of bacteriocin added (19, 47).

Bacteriocins act similarly in that they permeabilize the cytoplasmic membrane. However, they may require different conditions in the target membrane to establish a successful interaction. In vivo, the lantibiotic bacteriocins act on energized membranes (46). The N-terminal part of the nisin molecule is involved in its insertion into the lipid bilayer, while the C-terminal moiety is involved in the initial interaction with the membrane surface (80). Prior to forming pores, nisin has to bind to the peptidoglycan precursor lipid II, which serves as a nisin-docking molecule in the membrane structure (10). Alternatively, most nonlantibiotic bacteriocins act on nonenergized membranes but seem to require a membrane receptor protein (19, 136, 143). Mutation in the *rpoN* gene, which codes for σ⁵⁴ in *Enterococcus faecalis* JH2-2, causes resistance to group IIa bacteriocins (25). This suggests a possible role for the *rpoN* gene product in the docking and/or receptor recognition of the target cell by a type IIa bacteriocin. Nissen-Meyer et al. (102) examined the sequences of four pediocin-like bacteriocins with different inhibitory spectra to identify regions that determine inhibitory specificity. The specificity of inhibitory activity correlates with the C-terminal sequences (30), suggesting that it determines specificity (reviewed in reference 31). Later, the same group (42) showed that a 15-residue pediocin PA-1 C-terminal derivative inhibits activity of the intact molecule. This suggests that the truncated peptide contains a receptor-binding motif that competes with the native pediocin PA-1 for the same surface receptor on the sensitive cells.

The activity of some bacteriocins, including lactacin F, lactococcins M and G, and plantaricins EF and JK, occurs through the action of two peptides. These two-component bacteriocins are also postulated to form pores in the cytoplasmic membrane (74, 76, 77).

Mechanism of Action Against Spores

Most information about the mechanisms of bacteriocin action against spores pertains to nisin. Spore germination and outgrowth is a multistep process detailed in chapter 3. Nisin allows spores to germinate and may act as a progerminant (88) but inhibits outgrowth of the preemergent spore. Heat resistance and refractility under phase microscopy are lost at this point. The growth of vegetative botulinal cells is inhibited at much lower nisin concentrations than those required to inhibit outgrowth of spores. *Bacillus* species spore coats opened by mechanical pressure are much more sensitive to nisin than are spores of species with coats that are opened by lysis (133). At a molecular level, nisin modifies the sulfhydryl groups in the envelopes of germinated spores, presumably because the dehydro residues of nisin act as electron acceptors (53, 54).

Similarity to Other Antimicrobial Proteins

The antimicrobial proteins of LAB are not unique in structure or function. Many higher organisms produce similar cytolytic pore-forming proteins (105) including the cecropins, mellittins, magainins, and defensins. In many cases, these proteins have been better studied than the bacteriocins produced by LAB.

A brief review of these antimicrobial proteins provides an opportunity to compare and contrast them with the LAB bacteriocins (93). Colicins produced by *E. coli* have molecular weights of 35 to 70 kDa and form ion-permeable channels in the cytoplasmic membrane of sensitive bacteria. Cecropins are small (ca. 25-amino-acid) peptides of the moth *Hyalophora cecropia* which are active against gram-negative and gram-positive bacteria. Cecropins contain amphiphilic helices consistent with the structural requirements for membrane insertion. Mellittin is the major toxic component of bee venom and causes membrane lysis, permeabilization, and inhibition of membrane-bound enzyme systems. It contains 26 amino acids, with the N-terminal region rich in hydrophobic residues and the C-terminal end predominantly hydrophilic. Magainins are antimicrobial peptides of the frog *Xenopus laevis* and contain approximately 23 amino acids which exert microbicidal action by permeabilizing sensitive membranes. Defensins are small antimicrobial peptides found in human, rabbit, guinea pig, and rat phagocytes. These small (29- to 34-amino-acid) peptides also exert their antimicrobial activity by permeabilization or disruption of cell membranes. Clearly, antimicrobial proteins are common in the biosphere and are used by many organisms to protect themselves. While these molecules are the prehistoric immune system for bacteria, they remain the first line of defense for eucaryotes.

Resistance to Bacteriocins

While nisin-resistant starter cultures can be advantageous, the appearance of nisin-resistant pathogens can undermine the use of nisin as an antimicrobial. Genetically stable nisin-resistant *L. monocytogenes* can be isolated at a frequency of 10⁻⁶ (57, 90). Nisin-resistant isolates are generated from vegetative cells of *Staphylococcus aureus*, *Bacillus licheniformis*, *B. subtilis*, *Bacillus cereus* (90), and *C. botulinum* (87) at similar frequencies. The use of multiple bacteriocins to overcome this problem has been suggested (38, 56) but is effective only if resistance to each bacteriocin is conferred by different mechanisms. *L. monocytogenes* resistant to mesentericin 52, curvaticin 13, or plantaricin C19 can be isolated at frequencies of 10⁻³ to 10⁻⁸. Strains resistant to any of these bacteriocins are cross-resistant to the other two. When all three bacteriocins are used together, resistant strains are isolated at frequencies similar to those obtained when the bacteriocins are used alone (114). Isolates resistant to all three of these bacteriocins, however, are not resistant to nisin.

Some mechanisms of bacteriocin resistance parallel mechanisms of antibiotic resistance (95) (Table 34.1). This comparison is made to provide a conceptual context for bacteriocin resistance and does not suggest that bacteriocin-resistant pathogens are cross-resistant to anti-

Table 34.1 Parallel mechanisms of antibiotic and bacteriocin resistance[a]

Mechanism of resistance	Example		
	Specificity	Antibiotic	Bacteriocins
Destruction	Specific or general	β-Lactamases	Protease, specific bacteriocinase
Modification	Specific	Methylation of aminoglycosides	Dehydroreductase (70)
Altered receptors	Specific	Penicillin-binding proteins	Probable, but not reported to date
Membrane composition	General	Altered membranes in resistant *E. coli* and bacilli	Demonstrated for nisin resistance (88, 90)

[a] Adapted from reference 95.

biotics. Mechanisms characterized as "specific" have a biochemical specificity to a particular antimicrobial and cannot generate cross-resistance. General mechanisms of resistance, such as changes in membrane permeability, can be viewed as affecting the intrinsic resistance of the bacteria and might cause cross-resistance to other food preservatives. Membranes in nisin-resistant *L. monocytogenes* have more straight-chain fatty acids, resulting in a higher phase transition temperature (Tc) than that of the wild-type strain (90). Presumably, this lack of fluidity hinders nisin insertion into the membrane. Membrane fluidity plays an important role in resistance of *Listeria* to other antimicrobials (71). *L. monocytogenes* grown in the presence of C14:0 or C18:0 fatty acids has a higher Tc and increased resistance to four common antimicrobials compared to cells grown in the presence of C18:1, which have a lower Tc and are more sensitive to other antimicrobials. A different nisin-resistant mutant (51) showed increased expression of a protein with strong homology to the glycosyltransferase domain of penicillin-binding proteins, a histidine protein kinase, a protein of unknown function, and ClpB (putative functions from homology). Increased expression of the putative penicillin-binding protein may affect the cell wall composition and thereby alter the sensitivity to the cell wall targeted by bacteriocin. This mutant was also characterized by increased sensitivity to β-lactam antibiotics and by a slight decrease in sensitivity to another lantibiotic, mersacidin.

For type IIa bacteriocins, decreased mannose-specific phosphoenolpyruvate-dependent phosphotransferase system gene expression, a more positive cell surface, increased lysinylation of membrane phospholipids, and increased membrane fluidity due to unsaturated phosphatidylglycerol contribute to bacteriocin resistance (135).

The issue of potential cross-resistance between bacteriocins and antibiotics has been considered. Bacteriocins

are different from antibiotics. They have different mechanisms of action. Therefore, cells respond to different stresses in different ways. This statement is confirmed by the sensitivity of multidrug-resistant bacteria to nisin (124). In order to survive, microorganisms can develop resistance to any antimicrobial substance used to preserve food. However, the use of intelligently designed multiple-hurdle technology not only allows inhibition of bacterial growth but also makes bacteriocin-resistant cells sensitive again (91). In addition, bacteriocin-resistant bacteria may be even more sensitive to other hurdles used in food preservation (101).

Our understanding of bacteriocin resistance is still incomplete. Under actual application conditions, the frequency of bacteriocin resistance may be much lower than that obtained in optimal laboratory media (57, 87). Nonetheless, it is prudent to conduct additional research on bacteriocin resistance so that it might be advantageously manipulated. Resistance management is already a component in other antimicrobial applications ranging from the use of antibiotics in hospitals to the use of *Bacillus thuringiensis* toxin as an agricultural insecticide.

Regulatory Status
LAB are GRAS for the production of fermented foods. GRAS status, which is conferred by the U.S. Food and Drug Administration (FDA), is especially desirable because it allows a compound to be used in a specific application without additional regulatory approval. The linkage of GRAS status to a specific application is often overlooked. Thus, the GRAS status of LAB for the production of fermented foods does not automatically make LAB, or their metabolic products, GRAS for uses such as the preservation of foods which are not fermented. Nisin is the only bacteriocin that has GRAS affirmation. The 1988 GRAS affirmation for the use of nisin in pasteurized processed cheese (41) was supported by

toxicological data. This affirmation is the foundation for additional GRAS affirmations.

Bacteriocins produced by GRAS organisms are not automatically GRAS themselves. Bacteriocins that are not GRAS are regulated as food additives (133) and require premarket approval by the FDA. A food fermented by bacteriocin-producing starters can be used as an ingredient in a second food product. Its use as an ingredient might coincidentally extend the shelf life of the product without necessitating preservative declarations. However, if the ingredient is added for the purpose of extending shelf life, the FDA would undoubtedly consider it an additive and require both premarket clearance and label declaration. Purified bacteriocins used as preservatives definitely require premarket approval by the FDA.

GRAS status for bacteriocins can be based on documented use prior to 1958, a consensus of scientific opinion, or a formal GRAS affirmation from the FDA. The presence of bacteriocins in foods prior to 1958 might be inferred from the ease with which bacteriocinogenic bacteria are isolated from a variety of foods. This suggests that they are long-standing members of the natural microbiota of foods. Furthermore, strains that produce nisin (69, 81,116) and strains that produce pediocin PA-1/AcH (7, 22, 48, 58) have been independently isolated from foods in different parts of the world. This demonstrates widespread occurrence of bacteriocinogenic bacteria in nature and suggests that bacteriocins have been consumed for decades. While these arguments are reasonable, the FDA might nonetheless require both isolation of the bacteriocinogenic organism and its bacteriocin from food produced prior to 1958 before accepting them as GRAS under the prior-use clause.

The international regulation of bacteriocins is complex and beyond the scope of this chapter. Nisin is the only bacteriocin approved internationally for use in foods. The Joint Food and Agriculture/World Health Organization accepted nisin as a food additive in 1969 and set the maximum intake level as 33,000 IU per kg body weight. Based on this, many countries allow nisin in a variety of products, sometimes with no restrictions as to maximum level. In addition to milk, cheese, and other dairy products, these uses include canned tomatoes, canned soups, other canned vegetables, mayonnaise, and baby food (139).

USE OF BACTERIOPHAGES FOR BIOPRESERVATION

General Characteristics

American pharmaceutical companies marketed bacteriophage therapies in the 1930s. The Germans and Soviets used phage therapy to cure dysentery during World War II (14). With the advent of antibiotics in the 1940s, interest in bacteriophage therapies waned. The specter of widespread antibiotic resistance has now revived medical interest, and food scientists are also revisiting bacteriophage biopreservation. Indeed, the Institute of Food Technologists' white paper, *Analysis and Evaluation of Preventive Control Measures for the Control and Reduction/Elimination of Microbial Hazards on Fresh and Fresh-Cut Produce* (67), contains a section on bacteriophage biocontrol. Reviews by Greer (52) and Hudson et al. (65) present different perspectives on the topic in greater detail than can be covered here.

Bacteriophages are viruses whose only hosts are bacteria. Lytic bacteriophages have a cycle of cell infection and death that starts when they attach to sensitive bacterial cells and release endolysins. These endolysins degrade the cell wall from the outside to allow the phage DNA (or RNA) to enter the bacterium. If the phage DNA escapes the restriction/modification (R/M) system of the bacterium, it seizes the growing cell's metabolic machinery to produce phage components. The phage components self-assemble and are released during a lytic burst which kills the bacterial host. The released bacteriophages adsorb to additional bacteria to continue the lytic cycle. If high concentrations of bacteriophages attach to the cells, they may release enough endolysin to lyse the cell independent of the lytic cycle. This is called "lysis from without."

Bacteriophages are natural components of food microbiota and are routinely consumed as part of our diet (14, 52, 66). They are stable over a wide pH range but inactivated at 60 to 75°C (66). Their pre- and postharvest uses are being championed by companies to reduce animal carriage of *Salmonella* and *E. coli* O157:H7 and as a food additive for control of *L. monocytogenes* (52).

Scientific Issues To Be Considered for Bacteriophage Biopreservation of Food

Using bacteriophages to control foodborne bacteria is fundamentally different from their use to control human infections. Human infections are characterized by a large population of a single bacterial species actively growing in a well-defined ecosystem. Treatment is temporally limited and does not create selective pressure in the pathogen reservoir. In foods, there can be low populations of several nongrowing pathogens amidst large populations of other bacteria. This is problematic since bacteriophage therapy in humans requires a threshold population (10^5 to 10^6 CFU) of actively growing bacteria (66). Continued bacteriophage use in foodborne pathogen environments can create selective pressure and generate bacteriophage resistance.

Bacteriophage characteristics affect the efficacy of their application. These include multiplicity of infection (MOI), ability to self-replicate, requirement that the host be growing, a limited host range, and hosts becoming resistant. The MOI is the ratio of bacteriophage particles to pathogen cells. Higher MOIs enhance efficacy. Bacteriophages are self-replicating and can increase the MOI during the lytic cycle. When the MOI is very high (10^6 to 10^8 phage/bacterium), the extracellular action of endolysins (lysis from without) can obviate the need for host growth. The mechanism is then strictly enzymatic rather than the result of the lytic cycle, and the MOI does not increase.

The requirement for growth of the target pathogen makes bacteriophages unlikely to kill low populations of nongrowing pathogens. A more promising application might be controlling growth of spoilage bacteria when a product is temperature abused, in a fashion similar to that of controlled acidification.

Lytic bacteriophages have a limited host range. Phage sensitivity is determined by the ability of the bacteriophage to attach to cell receptors and insensitivity to the host R/M system. A restricted host range can be advantageous because pathogens could be killed without destroying desirable fermentative or probiotic bacteria, normal spoilage microbiota, or resident human microbiota. However, multiple strains of bacteriophage would be required to kill multiple pathogens in the same food, or even multiple strains on the same pathogen. For example, 30 different O serotypes would have to be recognized to lyse the majority of enteropathogenic and enterotoxigenic E. coli (14). Changes in surface receptors or R/M systems can make sensitive cells bacteriophage resistant, undermining the efficacy of bacteriophage biocontrol. Resistance might be controlled through the use of multiple-strain cocktails, rotation of the bacteriophage in the cocktails, or through use of multiple hurdles to decrease the generation of phage-resistant pathogens.

Issues as Addressed in Research Literature

An evaluation of bacteriophage control of E. coli O157: H7 (104) illustrates many of the issues raised above. Single- and three-strain phage cocktails reduce viable counts up to 5 log in a broth system at 30°C. At 12°C, there is no effect because the E. coli growth required for phage replication does not occur. At 37°C, phage biocontrol is initially effective but then fails as phage-resistant E. coli organisms begin to dominate the population. These bacteriophage-insensitive mutants occur at a frequency of 10^{-6}, even in multiphage cocktails, but revert to sensitivity after 50 generations. Thus, if occasional

bacteriophage-insensitive mutants of E. coli are generated where they are not recycled into the pathogen reservoir (e.g., at the slaughterhouse rather than the farm), resistance should not reduce the efficacy of the control step. When meat inoculated with E. coli O157:H7 (100 CFU/g) was spotted with a phage cocktail (10^8 PFU), seven of nine samples were negative for E. coli O157:H7 after enrichment. At this high MOI (2×10^6), the lethality may be caused by lysis from without.

The MOI was also a key factor in bacteriophage biocontrol of Salmonella and Campylobacter on chicken skin (50). When applied at an MOI of 1, bacteriophage caused less than a 1-$\log_{10}$ reduction in the pathogen population relative to untreated chicken skin. At an MOI of 100 to 1,000, the reduction is up to 100-fold. At an MOI of 10^5, no pathogens were recovered. Since MOIs of 10^7 eliminate even Salmonella strains that can adsorb bacteriophages, but are "phage resistant" due to their R/M systems, the mechanism is probably lysis from without.

A study on bacteriophage biopreservation of produce (79) illustrates the variability caused by the food system and potential use of bacteriophages in multiple-hurdle systems. Intralytix bacteriophage mixtures applied to honeydew melon cause a 2- to 4-log reduction of L. monocytogenes but are ineffective against the pathogen on apple slices. The combined use of bacteriophages and nisin at 16,000 IU/ml has no effect in apple slices but increases the reduction of L. monocytogenes to up to 5.8 log in honeydew. Nisin alone gives only a 3.2-log reduction. The authors attributed the synergy to pore-forming action of nisin, but other factors might be important. Since nisin preparation contains mostly milk solids and salts, but only 2.5% pure nisin, the synergy could be due to the other material in the preparation. For example, the salts might increase bacteriophage efficiency by increasing attachment.

Applications in Foods

"Undue skepticism and unfounded optimism are both misplaced as regards the discovery of phage therapy" (14) and their application to food. While there is a clear potential for this technology, a number of fundamental scientific issues need to be addressed before bacteriophages can be used as components of food preservation. Like bacteriocins, bacteriophages are not "silver bullets" but need to be used from a perspective that considers the microbial ecology of the food. The unique interplay of the specific bacteriophage with a specific pathogen in a specific food caused Hudson et al. (65) to conclude that "each combination of food matrix and host will require specific optimization of biocontrol parameters."

Table 34.2 Analogies between the use of insecticides in production agriculture and the use of antimicrobials for food safety

Control of insects in crops	Control of bacteria in foods
Chemical pesticides	Chemical preservatives
Integrated pest management.	Microbial competition
Bioinsecticides	Bacteriocins
Insecticidal plants	Antimicrobial foods

CONCLUSIONS AND OUTLOOK FOR THE FUTURE

The biological methods of food preservation covered here mark only the beginning of the biopreservation era in the food industry. Controlled acidification is conceptually straightforward, but its successful application depends on a variety of product-specific factors. This has limited both its commercial use and academic interest in controlled acidification. The use of antimicrobial proteins, in one form or another, is sure to increase. In production agriculture, *B. thuringiensis* insecticidal proteins have been applied to plants for the last 20 years. The genes for this protein are now being cloned into the plant itself. By analogy (Table 34.2), the day may come when we can genetically engineer pathogen resistance into microbially sensitive foods.

Research in our laboratory and preparation of the manuscript were supported by state appropriations, U.S. Hatch Act Funds, the U.S. Israel Bilateral Research and Development Fund, and grants from the U.S Department of Agriculture CSREES NRI Food Safety Program. We acknowledge the contribution of Karen Winkowski as coauthor of earlier versions of this chapter, which appeared in the first and second editions of this book.

References

1. Abee, T., F. M. Rombouts, J. Hugenholtz, G. Guihard, and L. Letellier. 1994. Mode of action of nisin Z against *Listeria monocytogenes* Scott A grown at high and low temperatures. *Appl. Environ. Microbiol.* **60:**1962–1968.

2. Abriouel, H., R. Lucas, N. Ben Omar, E. Valdivia, M. Maqueda, M. Martinez-Canamero, and A. Galvez. 2005. Enterocin AS-48RJ: a variant of enterocin AS-48 chromosomally encoded by *Enterococcus faecium* RJ16 isolated from food. *Syst. Appl. Microbiol.* **28:**383–397.

3. Allison, G., C. Fremaux, C. Ahn, and T. R. Klaenhammer. 1994. Expansion of bacteriocin activity and host range upon complementation of two peptides encoded within the lactacin F operon. *J. Bacteriol.* **176:**2235–2241.

4. Al-Zoreky, N., J. C. Ayres, and W. Sandine. 1991. Antimicrobial activity of Microgard® against food spoilage and pathogenic microorganisms. *J. Dairy Sci.* **74:**758–763.

5. Bassler, B. L. 2002. Small talk. Cell-to-cell communication in bacteria. *Cell* **109:**421–424.

6. Berry, E. D., R. W. Hutkins, and R. Mandigo. 1991. The use of bacteriocin producing *Pediococcus acidilactici* to control post processing *Listeria monocytogenes* contamination of frankfurters. *J. Food Prot.* **54:**681–686.

7. Bhunia, A. K., and M. C. Johnson. 1992. Monoclonal antibody-colony immunoblot method specific for isolation of *Pediococcus acidilactici* from foods and correlation with pediocin (bacteriocin) production. *Appl. Environ. Microbiol.* **58:**2315–2320.

8. Bouksaim, M., C. Lacroix, R. Bazin, and R. E. Simard. 1999. Production and utilization of polyclonal antibodies against nisin in an ELISA and for immuno-location of nisin in producing and sensitive bacterial strains. *J. Appl. Microbiol.* **87:**500–510.

9. Bower, C. K., J. McGuire, and M. A. Daeschel. 1995. Suppression of *Listeria monocytogenes* colonization following adsorption of nisin onto silica surfaces. *Appl. Environ. Microbiol.* **61:**992–997.

10. Breukink, E., I. Wiedemann, C. van Kraaij, O. P. Kuipers, H. Sahl, and B. de Kruijff. 1999. Use of the cell wall precursor lipid II by a pore-forming peptide antibiotic. *Science* **286:**2361–2364.

11. Bruno, M. E. C., A. Kaiser, and T. J. Montville. 1992. Depletion of proton motive force by nisin in *Listeria monocytogenes* cells. *Appl. Environ. Microbiol.* **58:**2255–2259.

12. Bruno, M. E. C., and T. J. Montville. 1993. Common mechanistic action of bacteriocins from lactic acid bacteria. *Appl. Environ. Microbiol.* **59:**3003–3010.

13. Brurberg, M. B., I. F. Nes, and V. G. Eijsink. 1997. Pheromone-induced production of antimicrobial peptides in *Lactobacillus*. *Mol. Microbiol.* **26:**347–360.

14. Brüsson, H. 2005. Phage therapy: the *Escherichia coli* experience. *Microbiology* **151:**2133–2140.

15. Buchanan, R. L., and L. A. Klawitter. 1992. Effectiveness of *Carnobacterium piscicola* LK5 for controlling the growth of *Listeria monocytogenes* Scott A in refrigerated foods. *J. Food Safety* **12:**217–224.

16. Buchman, G. W., S. Banergee, and J. N. Hansen. 1988. Structure, expression and evolution of a gene encoding the precursor of nisin, a small protein antibiotic. *J. Biol. Chem.* **263:**16260–16266.

17. Buyong, N., J. Kok, and J. B. Luchansky. 1998. Use of a genetically enhanced, pediocin-producing starter culture, *Lactococcus lactis* subsp. *lactis* MM217, to control *Listeria monocytogenes* in cheddar cheese. *Appl. Environ. Microbiol.* **64:**4842–4845.

18. Chen, Y., and T. J. Montville. 1995. Efflux of ions and ATP depletion induced by pediocin PA-1 are concomitant with cell death in *Listeria monocytogenes* Scott A. *J. Appl. Bacteriol.* **79:**684–690.

19. Chikindas, M. L., M. J. Garcia-Garcera, A. J. M. Driessen, A. M. Ledeboer, J. Nissen-Meyer, I. F. Nes, T. Abee, W. N. Konings, and G. Venema. 1993. Pediocin PA-1, a bacteriocin from *Pediococcus acidilactici* PAC1.0, forms hydrophilic pores in the cytoplasmic membrane of target cells. *Appl. Environ. Microbiol.* **59:**3577–3584.

20. Cintas, L. M., P. Casaus, C. Herranz, L. S. Havarstein, H. Holo, P. E. Hernandez, and I. F. Nes. 2000. Biochemical and genetic evidence that *Enterococcus faecium* L50

produces enterocins L50A and L50B, the *sec*-dependent enterocin P, and a novel bacteriocin secreted without an N-terminal extension termed enterocin Q. *J. Bacteriol.* **182:**6806–6814.

21. **Cociancich, S., M. Goyffon, F. Bontems, P. Bulet, F. Bouet, A. Menez, and J. Hoffman.** 1993. Purification and characterization of a scorpion defensin, a 4kDa antibacterial peptide presenting structural similarity with insect defensins and scorpion toxins. *Biochem. Biophys. Res. Commun.* **194:**17–22.

22. **Daba, H., C. Lacroix, J. Huang, R. E. Simard, and L. Lemieux.** 1994. Simple method of purification and sequencing of a bacteriocin produced by *Pediococcus acidilactici* UL5. *J. Appl. Bacteriol.* **77:**682–698.

23. **Daeschel, M. A.** 1990. Controlling wine malolactic fermentation with nisin and nisin-resistant strains of *Leuconostoc oenos*. *Appl. Environ. Microbiol.* **51:**601–603.

24. **Daeschel, M. A., J. McGuire, and H. Al-Makhlafi.** 1992. Antimicrobial activity of nisin adsorbed to hydrophilic and hydrophobic silicon surfaces. *J. Food Prot.* **55:**731–735.

25. **Dalet, K., C. Briand, Y. Cenatiempo, and Y. Hechard.** 2000. The *rpoN* gene of *Enterococcus faecalis* directs sensitivity to subclass IIa bacteriocins. *Curr. Microbiol.* **41:**441–443.

25a. **Danisco.** 2005. Products and services. Nisaplin and Novasin antimicrobials. [Online.] http://www.danisco.com/cms/connect/corporate/products%20and%20services/product%20range/antimicrobials/antimicrobial%20ingredients/nisaplin%20and%20novasin%20antimicrobials/nisaplin_and_nocasin_en.htm. Accessed 23 January 2006.

26. **Degnan, A. J., N. Buyong, and J. B. Luchansky.** 1993. Antilisterial activity of pediocin AcH in model food systems in the presence of an emulsifier or encapsulated within liposomes. *Int. J. Food Microbiol.* **18:**127–138.

27. **Degnan, A. J., and J. B. Luchansky.** 1992. Influence of beef tallow and muscle on the antilisterial activity of pediocin AcH and liposome-encapsulated pediocin AcH. *J. Food Prot.* **55:**552–554.

28. **Degnan, A. J., A. E. Yousef, and J. B. Luchansky.** 1992. Use of *Pediococcus acidilactici* to control *Listeria monocytogenes* in temperature-abused vacuum-packaged wieners. *J. Food Prot.* **55:**98–103.

29. **Delves-Broughton, J., G. C. Williams, and S. Williamson.** 1992. The use of the bacteriocin, nisin, as a preservative in pasteurized white egg. *Lett. Appl. Microbiol.* **15:**133–136.

29a. **de Ruyter, P. G., O. P. Kuipers, and W. M. de Vos.** 1996. Controlled gene expression systems for *Lactococcus lactis* with the food-grade inducer nisin. *Appl. Environ. Microbiol.* **62:**3662–3667.

30. **Diep, D. B., L. S. Havarstein, J. Nissen-Meyer, and I. F. Nes.** 1994. The gene encoding plantaricin A, a bacteriocin from *Lactobacillus plantarum* C11, is located on the same transcription unit as an *agr*-like regulatory system. *Appl. Environ. Microbiol.* **60:**160–166.

31. **Diep, D. B., and I. F. Nes.** 2002. Ribosomally synthesized antibacterial peptides in Gram-positive bacteria. *Curr. Drug Targets* **3:**107–122.

32. **Dodd, H. M., N. Horn, and M. J. Gasson.** 1990. Analysis of the genetic determinant for the production of the peptide antibiotic nisin. *J. Gen. Microbiol.* **136:**555–556.

33. **Eijsink, V. G., L. Axelsson, D. B. Diep, L. S. Havarstein, H. Holo, and I. F. Nes.** 2002. Production of class II bacteriocins by lactic acid bacteria; an example of biological warfare and communication. *Antonie Leeuwenhoek* **81:**639–654.

34. **Engelke, G., Z. Gutowski-Eckel, M. Hammelmann, and K. D. Entian.** 1992. Biosynthesis of the lantibiotic nisin: genomic organization and membrane localization of the NisB protein. *Appl. Environ. Microbiol.* **58:**3730–3743.

35. **Ennahar, S., O. Assobhel, and C. Hasselmann.** 1998. Inhibition of *Listeria monocytogenes* in a smear-surface soft cheese by *Lactobacillus plantarum* WHE 92, a pediocin AcH producer. *J. Food Prot.* **61:**186–191.

36. **Ennahar, S., T. Sashihara, K. Sonomoto, and A. Ishizaki.** 2000. Class IIa bacteriocins: biosynthesis, structure and activity. *FEMS Microbiol. Rev.* **24:**85–106.

37. **Fang, T. J., and L. W. Lin.** 1994. Inactivation of *Listeria monocytogenes* on raw pork treated with modified atmosphere packaging and nisin. *J. Food Drug Anal.* **2:** 189–200.

38. **Farber, J. M.** 1993. Current research on *Listeria monocytogenes* in foods: an overview. *J. Food Prot.* **56:**640–643.

39. **Fath, M. J., and R. Kolter.** 1993. ABC transporters: bacterial exporters. *Microbiol. Rev.* **57:**995–1017.

40. **Faye, T., T. Langsrud, I. F. Nes, and H. Holo.** 2000. Biochemical and genetic characterization of propionicin T1, a new bacteriocin from *Propionibacterium thoenii*. *Appl. Environ. Microbiol.* **66:**4230–4236.

41. **Federal Register.** 1988. Nisin preparation: affirmation of GRAS status as a direct human food ingredient. 21CFR Part 184. **53:**11247–11251.

42. **Fimland, G., O. R. Blingsmo, K. Sletten, G. Jung, I. F. Nes, and J. Nissen-Meyer.** 1996. New biologically active hybrid bacteriocins constructed by combining regions from various pediocin-like bacteriocins: the C-terminal region is important for determining specificity. *Appl. Environ. Microbiol.* **62:**3313–3318.

43. **Fimland G., L. Johnsen, B. Dalhus, and J. Nissen-Meyer.** 2005. Pediocin-like antimicrobial peptides (class IIa bacteriocins) and their immunity proteins: biosynthesis, structure, and mode of action. *J. Pept. Sci.* **11:**688–696.

44. **Foegeding, P. M., A. B. Thomas, D. H. Pinkerton, and T. R. Klaenhammer.** 1992. Enhanced control of *Listeria monocytogenes* by in situ-produced pediocin during dry fermented sausage production. *Appl. Environ. Microbiol.* **58:**884–890.

45. **Fremaux, C., Y. Hechard, and Y. Cenatiempo.** 1995. Mesentericin Y105 gene clusters in *Leuconostoc mesenteroides* Y105. *Microbiology* **141:**1637–1645.

46. **Gao, F. H., T. Abee, and W. N. Konings.** 1991. The mechanism of action of the peptide antibiotic nisin in liposomes and cytochrome C oxidase proteoliposomes. *Appl. Environ. Microbiol.* **57:**2164–2170.

47. **Garcia-Garcera, M. J. G., G. L. Elferink, J. M. Driessen, and W. N. Konings.** 1993. In vitro pore-forming activity of the lantibiotic nisin. Role of PMF force and lipid composition. *Eur. J. Biochem.* **212:**417–422.

48. **Garver, K. I., and P. M. Muriana.** 1993. Detection, identification and characterization of bacteriocin-producing lactic

acid bacteria from retail food products. *Int. J. Food Microbiol.* 19:241–258.

49. **Gonzalez, C. F.** 1988. Method for inhibiting bacterial spoilage and composition for this purpose. European Patent Application 88101624.

50. **Goode, D., V. M. Allen, and P. A. Barrow.** 2003. Reduction of experimental *Salmonella* and *Campylobacter* contamination of chicken skin by application of lytic bacteriophage. *Appl. Environ. Microbiol.* 69:5032–5036.

51. **Gravesen, A., K. Sorensen, F. M. Aarestrup, and S. Knochel.** 2001. Spontaneous nisin-resistant *Listeria monocytogenes* mutants with increased expression of a putative penicillin-binding protein and their sensitivity to various antibiotics. *Microb. Drug Resist.* 7:127–135.

52. **Greer, G. G.** 2005. Bacteriophage control of foodborne bacteria. *J. Food Prot.* 68:1102–1111.

53. **Gross, E., and J. L. Morell.** 1967. The presence a of dehydroalanine in the antibiotic nisin and its relationship to activity. *J. Am. Chem. Soc.* 89:2791–2792.

54. **Gross, E., and J. L. Morell.** 1971. The structure of nisin. *J. Am. Chem. Soc.* 93:4634–4635.

55. **Guder, A., I. Wiedemann, and H. G. Sahl.** 2000. Posttranslationally modified bacteriocins—the lantibiotics. *Biopolymers* 2000 55:62–73.

56. **Hanlin, M. B., N. Kalchayan, P. Ray, and B. Ray.** 1993. Bacteriocins of lactic acid bacteria in combination have greater antibacterial activity. *J. Food Prot.* 56:252–255.

57. **Harris, L. J., H. P. Fleming, and T. R. Klaenhammer.** 1991. Sensitivity and resistance of *Listeria monocytogenes* ATCC 19115 Scott A and VAL 500 to nisin. *J. Food Prot.* 54:836–840.

58. **Harris, L. J., H. P. Fleming, and T. R. Klaenhammer.** 1992. Novel paired starter culture system for sauerkraut, consisting of a nisin-resistant *Leuconostoc mesenteroides* strain and a nisin-producing *Lactococcus lactis* strain. *Appl. Environ. Microbiol.* 58:1484–1489.

59. **Herranz, C., Y. Chen, H. J. Chung, L. M. Cintas, P. E. Hernandez, T. J. Montville, and M. L Chikindas.** 2001. Enterocin P selectively dissipates the membrane potential of *Enterococcus faecium* T136. 2001. *Appl. Environ. Microbiol.* 67:1689–1692.

60. **Holo, H., O. Nissen, and I. F. Nes.** 1991. Lactococcin A, a new bacteriocin from *Lactococcus lactis* subsp. *cremoris*: isolation and characterization of the protein and its gene. *J. Bacteriol.* 173:3879–3887.

61. **Hoover, G., and L. R. Steenson.** 1993. *Bacteriocins of Lactic Acid Bacteria.* Academic Press, New York, N.Y.

62. **Horn, N., S. Swindell, H. Dodd, and M. Gasson.** 1991. Nisin biosynthesis genes are encoded by a novel conjugative transposon. *Mol. Gen. Genet.* 228:129–135.

63. Reference deleted.

64. Reference deleted.

65. **Hudson, J. A. C. Billington, G. Carey-Smith, and G. Greening.** 2005. Bacteriophages as biocontrol agents in food. *J. Food Prot.* 68:426–437.

66. **Hutton, M. T., P. A. Chehak, and J. H. Hanlin.** 1991. Inhibition of botulism toxin production by *Pediococcus acidilactici* in temperature abused refrigerated foods. *J. Food Safety* 11:255–267.

67. **Institute of Food Technologists.** 2001. Analysis and evaluation of preventive control measures for the control and reduction/elimination of microbial hazards on fresh and fresh-cut produce. *Compr. Rev. Food Sci. Food Safety* 2(Suppl. 1):13–37.

68. **Jack, R. W., and G. Jung.** 2000. Lantibiotics and microcins: polypeptides with unusual chemical diversity. *Curr. Opin. Chem. Biol.* 4:310–317.

69. **Jager, K., and S. Harlander.** 1992. Characterization of a bacteriocin from *Pediococcus acidilactici* PC and comparison of bacteriocin-producing strains using molecular typing procedures. *Appl. Microbiol. Biotechnol.* 37:631–637.

70. **Jarvis, B., and J. Farr.** 1971. Partial purification, specificity and mechanism of the nisin-inactivating enzyme from *Bacillus cereus. Biochim. Biophys. Acta* 227:232–240.

71. **Juneja, V. K., and P. M. Davidson.** 1993. Influence of altered fatty acid composition on resistance of *Listeria monocytogenes* to antimicrobials. *J. Food Prot.* 56:302–305.

72. **Kalchayanand, N., M. B. Hanlin, and B. Ray.** 1992. Sublethal injury makes Gram-negative and resistant Gram-positive bacteria sensitive to the bacteriocins, pediocin AcH and nisin. *Lett. Appl. Microbiol.* 15:239–243.

73. **Kalchayanand, N., T. Sikes, C. P. Dunne, and B. Ray.** 1994. Hydrostatic pressure and electroporation have increased bactericidal efficiency in combination with bacteriocins. *Appl. Environ. Microbiol.* 60:4174–4177.

74. **Klaenhammer, T. R.** 1993. Genetics of bacteriocins produced by lactic acid bacteria. *FEMS Microbiol. Rev.* 12:39–86.

75. **Kleerebezem, M.** 2004. Quorum sensing control of lantibiotic production; nisin and subtilin autoregulate their own biosynthesis. *Peptides* 25:1405–1414.

76. **Kok, J., H. Holo, M. J. van Belkum, A. J. Haandrikman, and I. F. Nes.** 1993. Nonnisin bacteriocins in lactococci: biochemistry, genetics and mode of action, p. 121–150. *In* D. G. Hoover and L. R. Steenson (ed.), *Bacteriocins of Lactic Acid Bacteria.* Academic Press, New York, N.Y.

77. **Konisky, J.** 1982. Colicins and other bacteriocins with established modes of action. *Annu. Rev. Microbiol.* 36:125–144.

78. **Kuipers, O. P., M. M. Beerthuyzen, R. J. Siezen, and W. M. de Vos.** 1993. Characterization of the nisin gene cluster *nisABTCIPR* of *Lactococcus lactis*: requirement of expression of the *nisA* and *nisI* gene for producer immunity. *Eur. J. Biochem.* 216:281–292.

79. **Leverentz, B., W. S. Conway, M. J. Camp, W. J. Janisiewcz, T. Abuladze, M. Yang, R. Saftner, and A. Sulakvelidze.** 2003. Control of *Listeria monocytogenes* on fresh-cut produce by treatment with lytic bacteriophages and a bacteriocin. *Appl. Environ. Microbiol.* 69:4519–4526.

80. **Lins, L., P. Ducarme, E. Breukink, and R. Brasseur.** 1999. Computational study of nisin interaction with model membrane. *Biochim. Biophys. Acta* 1420:111–120.

81. **Luchansky, J. B., K. A. Glass, K. D. Harrsono, A. J. Degnan, N. G. Faith, B. Cauvin, G. Bascus-Taylor, K. Arihara, B. Bater, A. J. Maurer, and R. G. Cassers.** 1992. Genomic analysis of *Pediococcus* starter cultures used to control *Listeria monocytogenes* in turkey summer sausage. *Appl. Environ. Microbiol.* 58:3053–3059.

82. Lyon, W. J., J. E. Sethi, and B. A. Glatz. 1993. Inhibition of psychrotrophic organisms by propionicin PLG-1, a bacteriocin produced by *Propionibacterium thoenii*. *J. Dairy Sci.* **76:**1506–1513.

83. Martinez, J. M., M. I. Martinez, C. Herranz, A. Suarez, M. F. Fernandez, L. M. Cintas, J. M. Rodriguez, and P. E. Hernandez. 1999. Antibodies to a synthetic 1-9-N-terminal amino acid fragment of mature pediocin PA-1: sensitivity and specificity for pediocin PA-1 and cross-reactivity against Class IIa bacteriocins. *Microbiology* **145:**2777–2787.

84. Martinez, B., M. Fernandez, J. E. Suarez, and A. Rodriguez. 1999. Synthesis of lactococcin 972, a bacteriocin produced by *Lactococcus lactis* IPLA 972, depends on the expression of a plasmid-encoded bicistronic operon. *Microbiology* **145:**3155–3161.

85. Marugg, J. D., C. F. Gonzalez, B. S. Kunka, A. M. Ledeboer, M. J. Pucci, M. Y. Toonen, S. A. Walker, L. C. M. Zoetmulder, and P. A. Vandenbergh. 1992. Cloning, expression, and nucleotide sequence of genes involved in production of pediocin PA-1, a bacteriocin from *Pediococcus acidilactici* PAC 1.0. *Appl. Environ. Microbiol.* **58:**2360–2367.

86. Mayr-Harting, A., A. J. Hedges, and R. C. W. Beerkley. 1972. Methods for studying bacteriocins. *Methods Microbiol.* **7A:**313–342.

87. Mazzotta, A. S., A. D. Crandall, and T. J. Montville. 1997. Nisin resistance in *Clostridium botulinum* spores and vegetative cells. *Appl. Environ. Microbiol.* **63:**2654–2659.

88. Mazzotta, A. S., and T. J. Montville. 1999. Characterization of fatty acid composition, spore germination, and thermal resistance in a nisin-resistant mutant of *Clostridium botulinum* 169B and in the wild-type strain. *Appl. Environ. Microbiol.* **65:**659–664.

89. Miller, K. W., P. Ray, T. Steinmetz, T. Hanekamp, and B. Ray. 2005. Gene organization and sequences of pediocin AcH/PA-1 production operons in *Pediococcus* and *Lactobacillus* plasmids. *Lett. Appl. Microbiol.* **40:**56–62.

90. Ming, X., and M. A. Daeschel. 1993. Nisin resistance of foodborne bacteria and the specific resistance responses of *Listeria monocytogenes* Scott A. *J. Food Prot.* **11:**944–948.

91. Modi, K. D., M. L. Chikindas, and T. J. Montville. 2000. Sensitivity of nisin-resistant *Listeria monocytogenes* to heat and the synergistic action of heat and nisin. *Lett. Appl. Microbiol.* **30:**249–253.

92. Moll, G. N., E. van den Akker, H. H. Hauge, J. Nissen-Meyer, I. F. Nes, W. N. Konings, and A. J. Driessen. 1999. Complementary and overlapping selectivity of the two-peptide bacteriocins plantaricin EF and JK. *J. Bacteriol.* **181:**4848–4852.

93. Montville, T. J., and M. E. C. Bruno. 1994. Evidence that dissipation of proton motive force is a common mechanism of action for bacteriocins and other antimicrobial proteins. *Int. J. Food Microbiol.* **24:**53–74.

94. Montville, T. J., A. M. Rogers, and A. Okereke. 1992. Differential sensitivity of *Clostridium botulinum* strains to nisin. *J. Food Prot.* **56:**444–448.

95. Montville, T. J., K. Winkowski, and R. D. Ludescher. 1995. Models and mechanisms for bacteriocin action and application. *Int. Dairy J.* **5:**797–815.

96. Mortvedt, C. I., J. Nissen-Meyer, K. Sletten, and I. F. Nes. 1991. Purification and amino acid sequence of lactocin S, a bacteriocin produced by *Lactobacillus sake* L45. *Appl. Environ. Microbiol.* **57:**1829–1834.

97. Mulders, J. W., I. J. Boerrigter, H. S. Rollema, R. J. Siezen, and W. M. de Vos. 1991. Identification and characterization of the lantibiotic nisin Z, a natural nisin variant. *Eur. J. Biochem.* **201:**581–584.

98. Nes, I. F., and H. Holo. 2000. Class II antimicrobial peptides from lactic acid bacteria. *Biopolymers* **55:**50–61.

99. Nettles, C. G., and S. F. Barefoot. 1993. Biochemical and genetic characteristics of bacteriocins of food-associated lactic acid bacteria. *J. Food Prot.* **56:**338–356.

100. Nielsen, J. W., J. S. Dickson, and J. D. Crouse. 1990. Use of a bacteriocin produced by *Pediococcus acidilactici* to inhibit *Listeria monocytogenes* associated with fresh meat. *Appl. Environ. Microbiol.* **56:**2142–2145.

101. Nilsson, L., Y. Chen, M. L. Chikindas, H. H. Huss, L. Gram, and T. J. Montville. 2000. Carbon dioxide and nisin act synergistically on *Listeria monocytogenes*. *Appl. Environ. Microbiol.* **66:**769–774.

102. Nissen-Meyer, J., H. Holo, L. S. Havarstein, K. Sletten, and I. F. Nes. 1992. A novel lactococcal bacteriocin whose activity depends on the complementary action of two peptides. *J. Bacteriol.* **174:**5686–5692.

103. Nussbaum, A., and A. Cohen. 1988. Use of a bioluminescence gene reporter for the investigation of red-dependent and gam-dependent plasmid recombination in *E. coli* K12. *J. Mol. Biol.* **203:**391–402.

104. O'Flynn, G., R. P. Ross, G. F. Fitzgerald, and A. Coffey. 2004. Evaluation of a cocktail of three bacteriophages for control of *Escherichia coli* O157:H7. *Appl. Environ. Microbiol.* **70:**3417–3424.

105. Ojcius, D. M., and J. D. E. Young. 1991. Cytolytic pore-forming proteins and peptides: is there a common structural motif? *Trends Biochem. Sci.* **16:**225–229.

106. Oscroft, C. A., J. G. Banks, and S. McPhee. 1990. Inhibition of thermally-stressed *Bacillus* spores by combinations of nisin, pH and organic acids. *Lebensm. Wiss. Technol.* **23:**538–544.

107. Pavan, S., P. Hols, J. Delcour, M. C. Geoffroy, C. Grangette, M. Kleerebezem, and A. Mercenier. 2000. Adaptation of the nisin-controlled expression system in *Lactobacillus plantarum*: a tool to study in vivo biological effects. *Appl. Environ. Microbiol.* **66:**4427–4432.

108. Piard, J. C., O. P. Kuipers, H. S. Rollema, M. J. Desmazeaud, and W. M. de Vos. 993. Structure, organization, and expression of the *lct* gene for lacticin 481, a novel lantibiotic produced by *Lactococcus lactis*. *J. Biol. Chem.* **268:**16361–16368.

109. Radler, F. 1990. Possible use of nisin in winemaking. I. Action of nisin against lactic acid bacteria and wine yeasts in solid and liquid media. *Am. J. Enol. Vitic.* **41:**1–6.

110. Radler, F. 1990. Possible use of nisin in winemaking. II. Experiments to control lactic acid bacteria in the production of wine. *Am. J. Enol. Vitic.* **41:**7–11.

111. Rauch, P. J. G., and W. M. deVos. 1992. Characterization of the novel nisin-sucrose conjugative transposon

TN5276 and its insertion in *Lactococcus lactis*. *J. Bacteriol.* **174:**1280–1287.

112. **Ray, B., and M. A. Daeschel.** 1992. *Food Biopreservation of Microbial Origin.* CRC Press, Boca Raton, Fla.

113. **Ray, B., R. Schamber, and K. W. Miller.** 1999. The pediocin AcH precursor is biologically active. *Appl. Environ. Microbiol.* **65:**2281–2286.

114. **Rekhif, N., A. Atrih, and G. Lefebvre.** 1995. Selection and properties of spontaneous mutants of *Listeria monocytogenes* ATCC 15313 resistant to different bacteriocins produced by lactic acid bacteria strains. *Curr. Microbiol.* **230:**827–853.

115. **Roberts, R. E., and E. A. Zottola.** 1993. Shelf-life of pasteurized process cheese spreads made from cheddar cheese manufactured with a nisin producing starter culture. *J. Dairy Sci.* **76:**1830–1836.

116. **Rodriguez, J. M., L. M. Cintas, P. Casaus, N. Horn, H. M. Dodd, P. E. Hernandez, and M. J. Gasson.** 1995. Isolation of nisin-producing *Lactococcus lactis* strains from dry fermented sausages. *J. Appl. Bacteriol.* **78:**109–115.

117. **Rogers, A. M., and T. J. Montville.** 1994. Quantification of factors influencing nisin's inhibition of *Clostridium botulinum* 56A in a model food system. *J. Food Sci.* **59:**663–668, 686.

119. **Sablon, E., B. Contreras, and E. Vandamme.** 2000. Antimicrobial peptides of lactic acid bacteria: mode of action, genetics and biosynthesis. *Adv. Biochem. Eng. Biotechnol.* **68:**21–60.

120. **Saleh, M. A., and Z. J. Ordal.** 1955. Studies on growth and toxin production of *Clostridium botulinum* in precooked frozen food. II. Inhibition by lactic acid bacteria. *Food Res.* **20:**340–346.

121. **Scott, V. N., and S. L. Taylor.** 1981. Effect of nisin on outgrowth of *Clostridium botulinum* spores. *J. Food Sci.* **46:**117–120.

122. **Scott, V. N., and S. L. Taylor.** 1981. Temperature, pH, and spore load on the ability of nisin to prevent the outgrowth of *Clostridium botulinum* spores. *J. Food Sci.* **46:**121–126.

123. **Sears, P. M., B. S. Smith, W. K. Stewart, R. Gonzalez, S. O. Rubino, S. A. Gusik, E. S. Kulisek, S. J. Projan, and P. Blackburn.** 1992. Evaluation of a nisin-based germicidal formulation on teat skin of live cows. *J. Dairy Sci.* **75:**3185–3190.

124. **Severina, E., A. Severin, and A. Tomasz.** 1998. Antibacterial efficacy of nisin against multidrug-resistant Gram-positive pathogens. *J. Antimicrob. Chemother.* **41:**341–347.

125. **Siegers, K., and K. D. Entian.** 1995. Genes involved in immunity to the lantibiotic nisin produced by *Lactococcus lactis* 6F3. *Appl. Environ. Microbiol.* **61:**1082–1089.

126. **Somers, E. B., and S. L. Taylor.** 1987. Antibotulinal effectiveness of nisin in pasteurized process cheese spreads. *J. Food Prot.* **50:**842–848.

127. **Stevens, K. A., B. W. Sheldon, N. A. Klapes, and T. R. Klaenhammer.** 1992. Effect of treatment conditions on nisin inactivation of Gram-negative bacteria. *J. Food Prot.* **55:**763–767.

128. **Stoddard, G. W., J. P. Petzel, M. J. van Belkum, J. Kok, and L. L. McKay.** 1992. Molecular analyses of the lactococcin A gene cluster from *Lactococcus lactis* subsp. *lactis* biovar *diacetylactis* WM4. *Appl. Environ. Microbiol.* **58:**1952–1961.

129. **Stoffels, G., J. Nissen-Meyer, A. Gudmundsdottir, K. Sletten, H. Holo, and I. F. Nes.** 1992. Purification and characterization of a new bacteriocin isolated from a *Carnobacterium* sp. *Appl. Environ. Microbiol.* **58:**1417–1422.

130. **Tagg, J. R., A. S. Dajani, and L. W. Wannamaker.** 1976. Bacteriocins of Gram-positive bacteria. *Bacteriol. Rev.* **40:**722–756.

131. **Tanaka, N. E., E. Traisman, M. H. Lee, and R. Casses.** 1980. Inhibition of botulism toxin formation in bacon by acid development. *J. Food Prot.* **43:**450–452.

132. **Taylor, L. Y., O. O. Cann, and B. J. Welch.** 1990. Antibotulinal properties of nisin in fresh fish packaged in an atmosphere of carbon dioxide. *J. Food Prot.* **53:**953–957.

133. **Thompson, L. V., and J. Delves-Broughton.** 2005. Nisin, p. 239–275. *In* P. M. Davidson, J. N. Sofos, and A. L. Branen (ed.), *Antimicrobials in Foods*, 3rd ed. CRC Press, Boca Raton, Fla.

134. **Tramer, J., and G. G. Fowler.** 1964. Estimation of nisin in foods. *J. Sci. Food Agric.* **15:**522–528.

135. **Vadyvaloo, V., S. Arous, A. Gravesen, Y. Héchard, R. Chauhan-Haubrock, J. W. Hasting, and M. Rautenbach.** 2004. Cell-surface alterations in class IIa bacteriocin-resistant *Listeria monocytogenes* strains. *Microbiology* **150:**3025–3033.

136. **van Belkum, M. J., B. J. Hayema, R. E. Jeeninga, J. Kok, and G. Venema.** 1991. Organization and nucleotide sequence of two lactococcal bacteriocin operons. Cloning of two bacteriocin genes from a lactococcal bacteriocin plasmid. *Appl. Environ. Microbiol.* **57:**492–498.

137. **van Belkum, M. J., J. Kok, G. Venema, H. Holo, I. F. Nes, W. N. Konings, and T. Abee.** 1991. The bacteriocin lactococcin A specifically increases the permeability of lactococcal cytoplasmic membranes in a voltage-independent, protein-mediated manner. *J. Bacteriol.* **173:**7934–7941.

138. **van Belkum, M. J., R. W. Worobo, and M. E. Stiles.** 1997. Double-glycine-type leader peptides direct secretion of bacteriocins by ABC transporters: colicin V secretion in *Lactococcus lactis*. *Mol. Microbiol.* **23:**1293–1301.

139. **Vandenbergh, P. A.** 1993. Lactic acid bacteria, their metabolic products and interference with microbial growth. *FEMS Microbiol. Rev.* **12:**221–238.

140. **Vandenbergh, P. A., M. J. Pucci, B. S. Kunka, and E. B. Vedamuthu.** 1989. Method for inhibiting *Listeria monocytogenes* using a bacteriocin. European Patent Application 89101126.6.

141. **van der Meer, J. R., J. Polman, M. M. Beerthuyzen, R. J. Siezen, O. P. Kuipers, and W. M. de Vos.** 1993. Characterization of the *Lactococcus lactis* nisin A operon genes *nisP*, encoding a subtilisin-like serine protease involved in precursor processing, and *nisR*, encoding a regulatory protein involved in nisin biosynthesis. *J. Bacteriol.* **175:**2578–2588.

142. **Van Reenen, C. A., M. L. Chikindas, W. H. Van Zyl, and L. M. Dicks.** 2003. Characterization and heterologous expression of a class IIa bacteriocin, plantaricin 423 from *Lactobacillus plantarum* 423, in *Saccharomyces cerevisiae*. *Int. J. Food Microbiol.* **81:**29–40.

143. **Venema, K., T. Abee, A. J. Haandrikman, K. J. Leenhouts, J. Kok, W. N. Konings, and G. Venema.** 1993. Mode of action of lactococcin B, a thiol-activated bacteriocin from *Lactococcus lactis. Appl. Environ. Microbiol.* **59:**1041–1048.

144. **Vogel, R. F., B. S. Pohle, P. S. Tichaczek, and W. Hammes.** 1993. The competitive advantage of *Lactobacillus curvatus* LTH 1174 in sausage fermentations is caused by formation of curvacin. *Syst. Appl. Microbiol.* **16:**457–462.

145. **Wahlstrom, G., and P. E. Saris.** 1999. A nisin bioassay based on bioluminescence. *Appl. Environ. Microbiol.* **65:**3742–3745.

146. **Whole Foods Market.** 2000–2006. Health Info. Reference library: ingredients: nisin. [Online.] http://www.wholefoods.com/healthinfo/nisin.html. Accessed 23 January 2006.

147. **Winkowski, K., M. E. C. Bruno, and T. J. Montville.** 1994. Correlation of bioenergetic parameters with cell death in *Listeria monocytogenes* cells exposed to nisin. *Appl. Environ. Microbiol.* **60:**4186–4187.

148. **Winkowski, K., A. D. Crandall, and T. J. Montville.** 1993. Inhibition of *Listeria monocytogenes* by *Lactobacillus bavaricus* MN in meat systems at refrigeration temperatures. *Appl. Environ. Microbiol.* **59:**2552–2557.

149. **Winkowski, K., and T. J. Montville.** 1992. Use of a meat isolate, *Lactobacillus bavaricus* MN, to inhibit *Listeria monocytogenes* growth in a model meat gravy system. *J. Food Safety* **13:**19–31.

150. **Worobo, R. W., M. J. van Belkum, H. Sailer, K. L. Roy, J. C. Vederas, and M. E. Stiles.** 1995. A signal peptide secretion-dependent bacteriocin from *Carnobacterium divergens. J. Bacteriol.* **177:**3143–3149.

Fermentations and Beneficial Microorganisms

Food Microbiology: Fundamentals and Frontiers, 3rd Ed.
Edited by M. P. Doyle and L. R. Beuchat
© 2007 ASM Press, Washington, D.C.

Mark E. Johnson
James L. Steele

Fermented Dairy Products

35

The primary microflora used in the production of fermented milk products are the homofermentative lactic acid bacteria (LAB). Additionally, yeasts, molds, and several other species of bacteria, including heterofermentative LAB, may be added to specific products; however, their purpose is not for acid development but for the production of flavor components or carbon dioxide (Table 35.1). It is also common today to use traditional starter bacteria such as *Lactobacillus helveticus* strains, not as acid producers as they are in Swiss, Parmesan, and mozzarella cheeses but as flavor enhancers (adjuncts) in cheeses such as Cheddar cheese and lower-fat versions of several varieties of cheese.

While their main function is to ferment lactose to lactic acid, it would be wrong to assume that LAB are added only to produce acid. They are also involved in the development of flavor of fermented dairy products, but their importance varies with the individual product. Starter bacteria include both mesophilic (optimal growth at 25 to 30°C) and thermophilic (optimal growth at 37 to 42°C) species. Mesophilic LAB include *Lactococcus lactis* subsp. *lactis* and *Lactococcus lactis* subsp. *cremoris*. Thermophilic LAB include *Streptococcus thermophilus*, *Lactobacillus delbrueckii* subsp.

bulgaricus, and *Lb. helveticus*. A protocooperative relationship exists between *S. thermophilus* and starter lactobacilli which results in increased rate and extent of acid production. While this may be desirable for manufacturers of mozzarella cheese, it may be detrimental in the manufacture of Swiss cheese, for which a slow rate of acid development is desired to prevent the formation of slits. *Lactobacillus helveticus* strains are being replaced with less-stimulatory *Lactobacillus lactis* or *Lactobacillus casei* strains. Traditionally the thermophiles were paired (cocci and rod) to get maximum rates of acid development. Today, single strains of *S. thermophilus* are used alone in the manufacture of mozzarella cheese. *S. thermophilus* is also being paired with mesophiles as starters in Cheddar cheese to increase the rate of acid development during the warm (38°C) cooking procedure. Mesophiles are also being paired with thermophiles or are used in cheeses in which thermophiles were the traditional starters. In addition to acid development, starters or adjuncts play a role in the breakdown of casein (peptides), which can stimulate the growth of other bacteria. Lactobacilli used in Swiss cheese enhance growth and gas development of propionibacteria (71), and they also can increase

Mark E. Johnson, Center for Dairy Research, Department of Food Science, University of Wisconsin–Madison, Madison, WI 53706–1565. **James L. Steele,** Department of Food Science, University of Wisconsin–Madison, Madison, WI 53706–1565.

Table 35.1   Microorganisms involved in the manufacture of cheeses and fermented milks

Product	Principal acid producer	Intentionally introduced secondary microflora
Cheeses		
Colby, Cheddar, cottage, cream	*Lc. lactis* subsp. *cremoris/lactis*	*Lactobacillus* sp. as adjuncts
Gouda, Edam, Havarti	*Lc. lactis* subsp. *cremoris/lactis*	*Leuconostoc* sp., Cit⁺ *Lc. lactis* subsp. *lactis*
Brick, Limburger	*Lc. lactis* subsp. *cremoris/lactis*	*Geotrichum candidum*, *B. linens*, *Micrococcus* sp., *Arthrobacter*
Camembert	*Lc. lactis* subsp. *cremoris/lactis*	*P. camemberti*, sometimes *B. linens*, *Arthrobacter*
Blue	*Lc. lactis* subsp. *cremoris/lactis*	Cit⁺ *Lc. lactis* subsp. *lactis*, *G. candidum*, *P. roqueforti*
Mozzarella, Provolone, Romano, Parmesan	*S. thermophilus*, *Lb. delbrueckii* subsp. *bulgaricus*, *Lb. helveticus*	*Lactobacillus* sp. as adjuncts; animal lipases added to Romano and Provolone for picante or rancid flavor
Swiss	*S. thermophilus*, *Lb. helveticus*, *Lb. delbrueckii* subsp. *bulgaricus* and *lactis*	*P. freudenreichii* subsp. *shermanii*
Fermented milks		
Yogurt	*S. thermophilus*, *Lb. delbrueckii* subsp. *bulgaricus*	None
Buttermilk	*Lc. lactis* subsp. *cremoris/lactis*	*Leuconostoc* sp., Cit⁺ *Lc. lactis* subsp. *lactis*
Sour cream	*Lc. lactis* subsp. *cremoris/lactis*	None

the level of succinic acid and flavor development in Swiss and other cheese varieties. The contribution of succinic acid to cheese flavor is, however, not well characterized.

In some fermented dairy products, additional bacteria, often referred to as secondary microflora (but essential to flavor development), are added to influence flavor and alter texture of the final product. Two LAB, *Leuconostoc* species and strains of *Lc. lactis* subsp. *lactis* capable of metabolizing citric acid (Cit⁺), are added to produce aroma compounds and carbon dioxide in cultured buttermilk and certain cheeses (Gouda, Edam, blue, and Havarti). When used together, these bacteria make up 10 to 20% of the total starter culture, with *Leuconostoc* species being present at about three times the population of Cit⁺ *Lc. lactis* subsp. *lactis*. Heterofermentative lactobacilli (*Lactobacillus brevis*, *Lactobacillus fermentum*, and *Lactobacillus kefir*) are part of the varied microflora (including several yeast species) found in the more exotic cultured milks such as kefir and koumiss, in which they produce ethanol, carbon dioxide, and lactic acid. They are not used in other fermented dairy products because of the copious quantities of carbon dioxide produced. *Propionibacterium freudenreichii* subsp. *shermanii* is added to Swiss-type cheeses, in which it metabolizes L-lactic acid to propionic acid, acetic acid, and carbon dioxide. While these acids are

not responsible for the distinctive flavor of Swiss cheese, other metabolic activities of the propionibacteria most certainly contribute to the development of the desired flavor components. The carbon dioxide forms the "eyes" in Swiss-type cheeses. Propionibacteria also ferment citric acid to glutamic acid (19).

Other types of secondary microflora include undefined mixtures of yeasts, molds, and bacteria. These microorganisms can be added directly to the milk or are smeared, sprayed, or rubbed onto the cheese surface. This group of microorganisms has extremely varied and complex metabolic activities, their main function being to produce unique flavors. The use of these secondary microflora is usually limited to surface-ripened and mold-ripened cheeses. Yeasts (*Debaryomyces*, *Candida*, *Yersinia*, and *Geotrichum* species) and bacteria (*Brevibacterium linens*, *Arthrobacter* species, and *Micrococcus* species) are employed in the aging of surface-ripened cheeses such as Limburger, Danbo, and Grùyere. However, they are also being used in the production of specialty cheeses (including Cheddar), especially by farmstead cheese makers. Molds (*Penicillium camemberti* and *Penicillium roqueforti*) are used in Camembert and blue-veined cheeses, respectively.

Fermented dairy products are not commercially produced in an environment free of contaminating microorganisms; however, cheese makers, in a sense, are sometimes very glad for contamination from the environment, and at other times it can be extremely detrimental for the quality of the cheese. Indeed, metabolism of contaminants, e.g., lactobacilli, is thought to be essential for the development of characteristic flavor in aged cheeses. Rapid acid development by starter within 4 to 8 h lowers the pH to less than 5.3 in cheese and to less than 4.6 in fermented milk products. After fermentation is complete, only acid-tolerant bacteria can grow. However, if acid development is slow or if the pH does not decrease sufficiently, contaminants which otherwise would have been inhibited may be able to grow. In some cheeses the pH can increase during ripening and permit the growth of previously inhibited bacteria. Dairy products may contain yeasts, molds, and many different genera of bacteria whose metabolic activities destroy quality. Spoilage of dairy products is described in chapter 7.

IMPORTANCE OF THE STARTER CULTURE

The key to commercial development of fermented milk products is the consistent and predictable rate of acid development by LAB. The rate and extent of pH decrease during manufacture and in the finished product are critical. pH has profound effects on moisture control during cheese manufacture, retention of coagulants, loss of minerals, hydration of proteins, and electrochemical interactions between protein molecules. These, in turn, have consequences for the development of flavor and physical properties (body and texture) of cheese and fermented milks. The reader is referred to other texts (26, 37) for discussions of these complex and interrelated phenomena.

In the past, antibiotic residues and over mature starters have been causes of inconsistent acid production. However, these problems have been overcome through monitoring of the milk supply and the use of improved starter media and preparation (78). Today bacteriophage infection is the most common cause of inconsistent acid development, causing significant loss of revenue to the cheese industry.

In cultured milks, desired flavors are derived directly from the metabolism of starter cultures and deliberately added aroma-producing secondary microflora. Thus, the desired flavor of the product dictates the choice of microorganisms. With cultured milks and some cheeses such as mozzarella, cream cheese, and cottage cheese, the short time between processing and consumption (1 day to 4 weeks), coupled with refrigerated storage, precludes the development of flavor other than that produced by starter and secondary added microflora. In other cheeses, the choice of microorganism(s) depends primarily on the

manufacturing protocol such as the temperature to which the product will be subjected, desired rate and extent of acid development, and desired physical properties of the finished product.

There is considerable debate as to the exact contribution of the starter culture to flavor development in cheese, especially in cheeses to which no secondary microflora is added. In these cheeses, nonstarter LAB, particularly lactobacilli, are the dominant adventitious microflora during ripening (60), and it is generally believed that they play the most significant role in the ripening of these cheeses. The use of nonstarter LAB, especially lactobacilli, as adjunct flavor cultures is a burgeoning research area and is practiced commercially. Unfortunately, the selection of appropriate cultures is a trial and error process since agreement on specific compounds that contribute to desired cheese flavor is often lacking. Differences in descriptions of desired flavor arise from the sheer complexity of cheese flavor as well as individual flavor perceptions and taste preferences. It should be understood that flavor development in cheese is a dynamic process and occurs in an environment that is constantly changing. It is not just that cheese develops stronger flavor with age, but that the very nature of the flavor is evolving. The development of flavor in different cheese varieties has been described (27).

LACTOSE METABOLISM

Energy transduction and fermentation pathways for carbohydrate metabolism in LAB have been described in detail (28, 61). Lactose, a disaccharide composed of glucose and galactose, is the only free-form sugar present in milk (45 to 50 g/liter). Lactococci translocate lactose into the cell by a phosphoenol-pyruvate phosphotransferase system. The lactose is phosphorylated during translocation and then cleaved by phospho-β-galactosidase into glucose and galactose-6-phosphate (Fig. 35.1). The glucose moiety enters the glycolytic pathway, and galactose-6-P is converted into tagatose-6-phosphate via the tagatose pathway. Both sugars are cleaved by specific aldolases into triose phosphates which are converted to pyruvic acid at the expense of NAD$^+$. For continued energy production, NAD$^+$ must be regenerated. This is usually accomplished by reducing pyruvic acid to lactic acid.

S. thermophilus and some thermophilic lactobacilli transport lactose via a lactose-galactose antiport system driven by an electrochemical proton gradient (61). Lactose is not phosphorylated but is cleaved by β-galactosidase to yield glucose and galactose. The glucose moiety enters the glycolytic pathway, but galactose is excreted

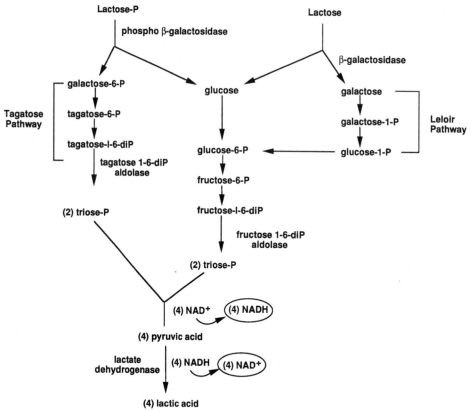

Figure 35.1 Lactose metabolism in homofermentative LAB.

from the cells and accumulates in milk or cheese. Thermophilic lactobacilli that do not excrete galactose and *Lb. helveticus* strains able to transport excreted galactose utilize the Leloir pathway to metabolize galactose. *Lb. delbrueckii* subsp. *bulgaricus* and most strains of *S. thermophilus* cannot metabolize galactose. This presents a problem in cheese manufacture since residual sugar can be metabolized heterofermentatively by other bacteria. Rapid production of carbon dioxide by heterofermentative bacteria causes cheese to crack and packages to swell. Residual sugar can also react with amino groups and form pink or brown pigments, i.e., Maillard-browning reaction products. Thermophiles may not metabolize the residual sugar during cold storage (4 to 7°C) of the cheese. Therefore, lactococci are sometimes included in the starter to ensure that all residual sugar is fermented. In pasta filata cheese manufacture, the curd is heated (52 to 66°C) and molded. The heat treatment may inactivate starter bacteria and prevent further sugar metabolism.

It is not known how lactose is transported in cells by *Leuconostoc* species or heterofermentative lactobacilli;

however, lactose is known to be hydrolyzed by β-galactosidase (33). The galactose moiety is transformed into glucose-6-phosphate (Leloir pathway) and, together with glucose, is metabolized through the phosphoketolase pathway (Fig. 35.2). Heterofermentative LAB lack aldolase, but through a dehydrogenation-decarboxylation system, a pentose sugar (xylulose-5-phosphate) and carbon dioxide are formed. Xylulose-5-phosphate is then cleaved by phosphoketolase to yield glyceraldehyde and acetylphosphate. Lactic acid and ethanol, respectively, are formed from these intermediates, facilitating the regeneration of NAD$^+$. However, during cometabolism of lactose and citric acid, *Leuconostoc* species convert acetylphosphate into acetic acid and generate ATP.

Two enzymes, L-lactic acid dehydrogenase (L-LDH) and D-lactic acid dehydrogenase (D-LDH), are responsible for the conversion of pyruvate to L-lactic acid and D-lactic acid, respectively. Lactococci produce only L-lactic acid, while *Lb. delbrueckii* subsp. *bulgaricus* and *Leuconostoc* species form only D-lactic acid. Other lactobacilli possess both enzymes and produce both D- and L-lactic acid.

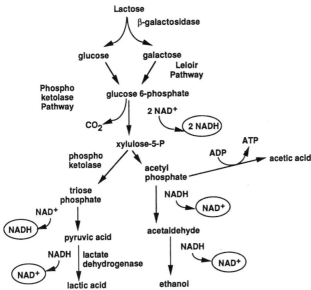

Figure 35.2 Lactose metabolism in heterofermentative LAB.

The key to end product formation in lactose metabolism lies in the regeneration or reducing equivalents. Lactococci have the enzymatic potential to produce compounds other than lactic acid to regenerate NAD$^+$, but these activities are not usually expressed under aerobic conditions (25). Oxygen in milk is used as an electron acceptor by LAB through the activity of oxidases and peroxidases (16, 67). As a consequence, hydrogen from NADH is transferred to oxygen to produce hydrogen peroxide, and NAD$^+$ is regenerated. However, under anaerobic conditions and low sugar levels, lactococci (7, 25, 72) produce formic acid and ethanol (Fig. 35.3) to regenerate NAD$^+$. However, Jensen et al. (38) demonstrated that even slight aeration of an otherwise anaerobic fermentation decreases synthesis of pyruvate formate lyase. This results in a shift away from production of ethanol and formic acid to acetic acid and carbon dioxide. Heterofermentative LAB convert acetylphosphate into acetic acid rather than ethanol under aerobic conditions and regenerate NAD(P)$^+$ through NAD(P)H oxidases (52, 76).

Ethanol has a very high flavor threshold value and would not be expected to contribute directly to flavor in the amounts produced by homofermentative LAB. However, subsequent esterification of ethanol with short-chain fatty acids yields esters with very low flavor thresholds. These compounds are responsible for the fruity flavor defects of Cheddar cheese (8). Short-chain fatty acids are probably generated by nonstarter LAB or exogenous sources, since starter bacteria have limited lipase activity (40).

The potential for production of diacetyl and carbon dioxide from lactose metabolism in lactococci with reduced LDH activity has been described (54). Formation of ethanol, acetic acid, and formic acid would regenerate NAD$^+$ (Fig. 35.3).

As a result of lactose metabolism (and oxidase activity), the environment becomes anaerobic and the oxidation-reduction potential is reduced. In cheese, further metabolism by nonstarter LAB may be needed to maintain this low potential. It has been postulated that a low oxidation-reduction potential is necessary for the production and stability of reduced sulfur-containing compounds thought to be vital for the development of certain cheese flavors (31, 53). Sugar and citric acid metabolism may result in the formation of α-dicarbonyls such as glyoxal, methylglyoxal, and diacetyl. These compounds readily react with amino acids and produce a myriad of compounds that contribute to cheese flavor (32).

Although lactic acid is commonly thought of as the end of fermentation, this is not always the case. Co-metabolism of citric acid and lactic acid by facultative lactobacilli produces carbon dioxide and causes blowing of packaged cheese (29). Propionibacteria metabolize lactic acid to acetic and propionic acids. Clostridia also metabolize lactic acid to acetic acid and carbon dioxide.

In some cases lactose is not fermented but utilized in the production of exopolysaccharides. Under stress, e.g., low pH or low water activity, some lactococci strains do not ferment the galactose moiety of lactose but form exopolysaccharides containing methyl pentoses and galactose (51). *S. thermophilus* also converts lactose to an exopolysaccharide during stationary phase of growth or low temperatures (30).

PRODUCTION OF AROMA COMPOUNDS

Although lactic acid is the main metabolic end product of lactose metabolism in cultured dairy products and is responsible for the acid taste, it is nonvolatile and odorless and does not contribute to aroma. The main volatile flavor components of fermented milks are acetic acid, acetaldehyde, and diacetyl. In yogurt, these volatile compounds are formed by the lactic acid starter bacteria, *S. thermophilus*, and *Lb. delbrueckii* subsp. *bulgaricus*. *Leuconostoc* species and Cit$^+$ *Lc. lactis* subsp. *lactis* are added to produce aroma compounds in buttermilk and some cheese varieties. Literature describing the production of aroma compounds in fermented dairy products has been compiled by Imhof and Bosset (36).

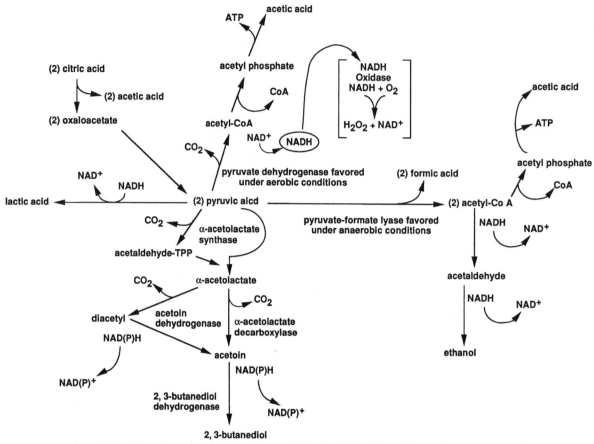

Figure 35.3   Pyruvic acid and citric acid metabolism in LAB. Abbreviations: CoA, coenzyme A; Tpp, thiamine pyrophosphate.

Diacetyl Production

The production of diacetyl, acetic acid, and carbon dioxide from citric acid by LAB has been reviewed by Hugenholtz (34) and is shown in Fig. 35.3. The carbon dioxide produced is responsible for the holes (eyes) in Gouda and Edam cheeses and the effervescent quality of buttermilk. Milk contains 0.15 to 0.2% citric acid, but not all LAB can metabolize it. However, *Leuconostoc* species, Cit⁺ *Lc. lactis* subsp. *lactis*, and facultative heterofermentative lactobacilli (49, 59) metabolize citric acid. *Leuconostoc* species and Cit⁺ *Lc. lactis* subsp. *lactis* strains utilize citric acid and lactose simultaneously and under certain conditions can derive energy via metabolism of citric acid. Citric acid is transported into the cell by a citric acid permease, which is plasmid encoded in lactococci and *Leuconostoc* (42, 74), and metabolized to pyruvic acid without generation of NADH. The result is an excess of pyruvic acid which does not have to be reduced to lactic acid to regenerate NAD⁺; therefore, it is

available for other reactions. Citrate metabolism in *Leuconostoc* sp. and *Lc. lactis* generates an electrochemical proton motive force-generating process (4). Ramos et al. (62) appear to have resolved conflicting reports on the pathway leading to the formation of diacetyl. Using ¹³C nuclear magnetic resonance, they verified that diacetyl formation involves nonenzymatic decarboxylative oxidation of α-acetolactate (an unstable intermediate derived from two molecules of pyruvic acid) and that the alternative suggested pathway, via a diacetyl synthase, is highly unlikely. However, it has been suggested that not all the diacetyl produced by Cit⁺ *Lc. lactis* subsp. *lactis* strains can be explained by spontaneous decarboxylation of the α-acetolactate (3). α-Acetolactate can be produced only when pyruvic acid accumulates within the cell. *Leuconostoc* species metabolize citric acid during growth but do not form diacetyl until the pH is below 5.4. α-Acetolactate synthase is inhibited at pH 5.4 or higher by many intermediates of lactose metabolism; however, the inhibition is relieved at lower pH values (15, 63). When diacetyl is not formed, *Leuconostoc* species form lactic acid from pyruvic acid derived from citric acid and regenerate NAD$^+$. Since NAD$^+$ is regenerated, there is less demand to form ethanol from the acetylphosphate that is generated via the phosphoketolase pathway. Consequently, acetic acid is formed with the generation of ATP (Fig. 35.3) and growth is enhanced (14, 65).

Diacetyl can be reduced by 2,3-butanediol dehydrogenases (17) to acetoin and 2,3-butanediol, both flavorless compounds. The presence of citric acid inhibits these reactions, but reduction begins when citric acid is exhausted. To ensure residual levels of citric acid in cultured milks, the ratio of starter culture to Cit$^+$ bacteria must be controlled (22). The cultured milk is stirred when the pH is reduced to 4.5 or below. This introduces oxygen and helps to increase and maintain the desired diacetyl content. The introduction of oxygen is required for nonenzymatic oxidative decarboxylation of α-acetolactate to diacetyl. In addition, LAB produce NADH oxidase, which transfers hydrogen to oxygen and regenerates NAD$^+$. The NADH oxidase activity replaces the role of the 2,3-butanediol dehydrogenases in regenerating NAD$^+$ and allows the accumulation of diacetyl, rather than its reduction to acetoin and 2,3-butanediol (5). NADH oxidase is more active at lower temperatures, whereas dehydrogenases are less active (6). For rapid acidification, non-citrate-metabolizing lactococci must be the dominant acid producers. If not, cometabolism of citric acid and lactose by Cit$^+$ *Lc. lactis* subsp. *lactis* would quickly consume the citric acid and result in the reduction of diacetyl to acetoin and 2,3-butanediol before pH 4.6 is reached. Hugenholtz (34) described the use of genetic engineering to construct strains of lactococci with elevated levels of diacetyl.

Acetaldehyde Production

There are several metabolic pathways in LAB that can lead to the formation of acetaldehyde (46, 47). This has resulted in some controversy over the primary pathway utilized by LAB. Cleavage of threonine by threonine aldolase to glycine and acetaldehyde has been suggested to be the most important mechanism for acetaldehyde production in yogurt and buttermilk (47, 81). However, using radiolabeled threonine, Wilkins et al. (79) demonstrated that only 2% of the acetaldehyde produced by mixed cultures of *Lb. delbrueckii* subsp. *bulgaricus* and *S. thermophilus* originated from threonine, even though both bacteria possess threonine aldolase. Bongers et al. (9) developed a genetically modified *Lc. lactis* strain (added pyruvate decarboxylase activity and over-expressed NADH oxidase activity) that produced a high

level of acetaldehyde. Chaves et al. (12) demonstrated that the main pathway for acetaldehyde formation in *S. thermophilus* is catalyzed by serine hydroxymethyltransferase displaying threonine aldolase activity. Acetaldehyde is also formed by Cit$^+$ *Lc. lactis* subsp. *lactis* strains (41). When the ratio of diacetyl to acetaldehyde in fermented milks is lower than 3:1, a yogurt or green apple flavor defect is observed (50). The defect is due to excess metabolic activity by Cit$^+$ *Lc. lactis* subsp. *lactis*. Excessive acetaldehyde in yogurt is the result of over-ripening and is always associated with high acid content. Prevention of an excessive amount of acetaldehyde may be accomplished by the use of *Leuconostoc*, which metabolizes acetaldehyde to ethanol. Obviously, to prevent overripening the product must be cooled rapidly and stored at lower temperatures. The trend for faster acid development and larger fermentation vessels may limit the ability of the manufacturer to cool the product fast enough to prevent overripening.

PROTEOLYTIC SYSTEMS IN LAB

Proteolytic systems in LAB contribute to their ability to grow in milk and are necessary for the development of flavor in ripened cheeses. LAB are amino acid auxotrophs typically requiring several amino acids for growth. Well-characterized examples include strains of lactococci and lactobacilli which require 6 to 15 amino acids, respectively. The quantities of free amino acids present in milk are not sufficient to support the growth of these bacteria to high cell density; therefore, they require a proteolytic system capable of utilizing the peptides present in milk and hydrolyzing milk proteins (α_{S1}-, α_{S2}-, β-, and γ-caseins) to obtain essential amino acids. Peptides and amino acids formed by proteolysis may impart flavor directly or serve as flavor precursors in fermented dairy products. Additionally, the resulting flavors may have either positive or negative impacts. The production of high-quality fermented dairy products is dependent on the proteolytic systems of LAB.

Proteolytic Systems and Their Physiological Role

Since the mid 1980s, significant progress has been made in defining the biochemistry and genetics of proteolytic systems from LAB. This is especially true for the lactococcal system, in which many of the enzymes have been purified and characterized, and numerous genes encoding components of proteolytic systems have been sequenced. For extensive reviews of the proteolytic enzyme systems of LAB, see Kunji et al. (45) and Christensen et al. (13). A model of the lactococcal proteolytic enzyme system is

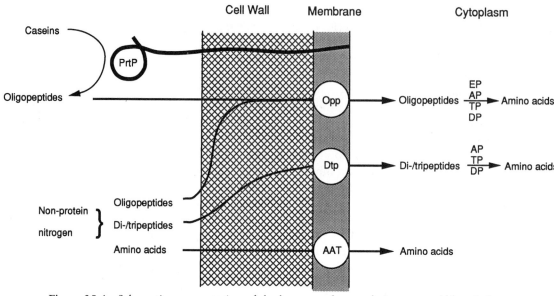

Figure 35.4 Schematic representation of the lactococcal proteolytic system. Abbreviations: PrtP, cell-envelope-associated proteinase; Opp, oligopeptide transport system; Dtp, di/tripeptide transport systems; AAT, amino acid transport systems; EP, endopeptidases; AP, aminopeptidases; TP, tripeptidases; DP, dipeptidases.

presented in Fig. 35.4. While growing in milk, LAB obtain essential amino acids in a variety of ways. They first utilize nonprotein nitrogen sources such as free amino acids and small peptides. Caseins, which compose 80% of all proteins present in milk, become the primary nitrogen source after nonprotein nitrogen is depleted. Proteolytic systems in LAB can be divided into three components, viz, enzymes outside the cytoplasmic membrane, transport systems, and intracellular enzymes.

Extensive investigations have revealed that a cell envelope-associated proteinase, designated PrtP, is the only extracellular proteolytic enzyme present in lactococci. The enzyme is a serine-protease which is expressed as a preproproteinase. A signal peptidase removes the signal peptide upon transport across the cytoplasmic membrane. Subsequently, a lipoprotein maturase (PrtM) is thought to cause a conformational change in the proproteinase, resulting in release of the proregion via autoproteolysis, and an active PrtP. The activated enzyme remains associated with the cell due to the presence of a C-terminal membrane anchor sequence. Genes encoding lactococcal proteinases with different substrate specificities have been sequenced, and the amino acid residues involved in substrate binding and catalysis have been determined. A critical feature of the enzyme is its broad cleavage specificity, which results in the release of more than 100 oligopeptides from soluble β-casein, 20% of

which are small enough to be transported by the oligopeptide transport system (39). Loss of PrtP, which is typically plasmid encoded in lactococcal strains, results in derivatives capable of reaching only about 10% of the final cell density of the parental strain, indicating that PrtP is essential for growth of lactococci to high cell density in milk.

Transport of nitrogenous compounds across the lactococcal cytoplasmic membrane takes place via group-specific amino acid transport systems, di- and tripeptide transport systems, and an oligopeptide transport system (Opp). Of these systems, Opp is of greatest importance during growth of lactococci in milk. Growth studies with Opp derivatives have shown that Opp is essential for the uptake of PrtP-generated peptides from β-casein and oligopeptides present in the nonprotein nitrogen component of milk (39). This system, which is organized in an operon, is composed of two ATP-binding proteins, two integral membrane proteins, and a substrate-binding protein. Peptides from 4 to 18 residues can be transported with little specificity for particular side chains (21).

Once inside the cell, peptides are hydrolyzed by peptidases. Peptidase classes which have been identified in lactococci include exopeptidases and endopeptidases. The greatest variety of enzymes are from the exopeptidase class, which includes aminopeptidases, tripeptidases, and

Table 35.2 Peptidases purified and characterized from lactococci

Peptidase	Abbreviation	Specificity[a]
X-prolyl dipeptidyl aminopeptidase	PepX	X-Pro$\updownarrow$Y-…
Aminopeptidase N	PepN	X$\updownarrow$Y-Z…
Aminopeptidase C	PepC	X$\updownarrow$Y-Z…
Aminopeptidase A	PepA	Asp(Glu)$\updownarrow$Y-Z…
Pyrrolidone carboxylyl peptidase	PCP	pGlu$\updownarrow$Y-Z…
Prolyl iminopeptidase	PepI	Pro$\updownarrow$Y-Z…
Dipeptidase	PepV	X$\updownarrow$Y
Prolidase	PepQ	X$\updownarrow$YPro
Tripeptidase	PepT	X$\updownarrow$Y-Z
Endopeptidases	PepO, PepF	…W-X$\updownarrow$Y-Z…

[a] ($\updownarrow$) The vertical double-ended arrow indicates which peptide bond is hydrolyzed.

dipeptidases. No carboxypeptidases have been detected in lactococci. Peptidases identified in lactococci and their cleavage specificity are summarized in Table 35.2. This combination of endopeptidases, aminopeptidases, tripeptidases, and dipeptidases converts the transported peptides into free amino acids, which lactococci require for growth. Examination of the growth of strains lacking a single peptidase in milk has revealed that only the PepN single mutant grows significantly slower. Characterization of lactococcal derivatives lacking more than one peptidase has indicated that, in general, the more peptidases that are inactivated the slower the strain grows in milk. The only possible exception to this general rule is PepX, which has not been observed to have a significant effect on growth rate in milk (55). The most direct interpretation of these results is that the multiple peptidase mutants have a reduced ability to obtain essential amino acids from casein-derived oligopeptides, thereby limiting the availability of amino acids for new protein biosynthesis. However, it is also possible that the reduced growth rate in milk observed with multiple peptidase mutants is related to altered regulation of the proteolytic system or a reduced ability to turn over cellular proteins (13).

Proteolytic systems of other LAB have not been as extensively studied. However, numerous components of the proteolytic enzyme systems of *Lb. helveticus* and *Lb. delbrueckii* have been characterized. In general, these components have homologs in the lactococcal system (13).

Proteolysis and Cheese Flavor Development

While flavor development in various types of cheese remains a poorly defined process, it is generally agreed that proteolysis is essential for flavor development in bacterially ripened cheeses (27, 75). Proteolytic enzymes present in this group of products include chymosin, plasmin, and proteolytic enzymes from starter cultures, adjunct cultures, and nonstarter LAB. The specificities and relative activities of the proteolytic enzymes present in the cheese matrix determine which peptides and amino acids accumulate and, hence, how flavor develops.

Free amino acids and peptides in the cheese matrix can contribute to cheese flavor either directly or indirectly and with positive or negative effects. Cheese flavor development has been the subject of comprehensive reviews (27, 68, 73). A major negative effect of proteolytic products is bitterness, which is believed to be caused by hydrophobic peptides ranging in length from 3 to 27 amino residues (48). These peptides are believed to be generated from casein principally by the joint action of chymosin and the LAB proteinases (11) and can be hydrolyzed to nonbitter peptides and amino acids by LAB peptidases. Therefore, the accumulation of bitter peptides is dependent on the relative rates of their formation and hydrolysis. A variety of volatile compounds can be derived from catabolism of amino acids (2, 13, 69, 77, 80). Numerous sulfur-containing compounds, particularly methanethiol, are thought to be important in cheese flavor. Methionine is believed to be the precursor to methanethiol; a number of enzymes and/or pathways have been identified in LAB capable of converting methionine to methanethiol (77). Alternatively, amino acid catabolism can give rise to compounds which have a negative impact on cheese flavor. For example, catabolism of aromatic amino acids can give rise to compounds such as indole and skatole, which contribute to "unclean" flavors in cheese (13). Overall, proteolysis is believed to be essential for the development of characteristic flavor compounds in bacterially ripened cheeses; however, the mechanism(s) by which products of proteolysis give rise to beneficial flavor compounds remains unknown. Additionally, other than bitterness, the mechanisms by which proteolysis impacts on the development of undesirable flavor compounds are unknown.

BACTERIOPHAGES AND BACTERIOPHAGE RESISTANCE

Bacteriophage infection may lead to a decrease or complete inhibition of lactic acid production by the starter culture. This has a major impact on the manufacture of fermented dairy products, as lactic acid synthesis is required to produce these products. Additionally, slow acid production disrupts manufacturing schedules and typically results in products which are downgraded to lower economic value. The severe consequences of bacteriophage infection have led to extensive investigations into bacteriophages and the mechanisms by which LAB resist infection. Bacteriophage infection of LAB was first described in lactococcal starter cultures in the 1930s. Since then, the dairy industry has employed improved sanitation regimes, utilized sophisticated starter culture propagation vessels, developed starter culture systems to minimize the impact of phage infection, and isolated and constructed starter strains with enhanced bacteriophage resistance. However, bacteriophage infection of the starter culture has remained a significant problem in the dairy industry. Klaenhammer and Fitzgerald (44) listed four reasons why dairy product fermentations are particularly susceptible to bacteriophage infection:

1. Phage contamination can occur when fermentations are not protected from environmental contaminants and in a nonsterile fluid medium, pasteurized milk.
2. Processing efficiency is easily disrupted because batch culture fermentations occur under increasingly stringent manufacturing schedules.
3. Increasing reliance on specialized strains limits the number and diversity of available dairy starter cultures.
4. Continuous use of defined cultures provides an ever-present host for bacteriophage attack.

Historically, most bacteriophage-related problems have occurred with lactococcal starter cultures; however, problems with starter systems which employ *S. thermophilus* and *Lb. delbrueckii* subsp. *bulgaricus* have been increasing. It is principally bacteriophage infection of *S. thermophilus* strains that has been the problem (56). For more extensive reviews of bacteriophages and bacteriophage resistance in LAB, readers are referred to other reports (1, 20, 24, 56, 70).

Bacteriophages of LAB

Significant progress in the characterization of bacteriophages from LAB has been made. All of the bacteriophages examined contain double-stranded linear DNA genomes with either cohesive or circularly permuted terminally redundant ends. Both lytic and temperate bacteriophages have been characterized. A major outcome of this characterization has been a clear classification, based on DNA homology and morphological studies, of LAB bacteriophages. Complete nucleotide sequence information is now available for representative isolates of most industrially important LAB bacteriophage species (20). This information has enhanced our understanding of how bacteriophage and host interactions evolve over

time and how to construct novel bacteriophage defense mechanisms.

Lactococcal Bacteriophage Resistance Mechanisms

Selective environmental pressure placed on lactococci by bacteriophages over thousands of years has resulted in strains which contain numerous bacteriophage defense mechanisms. The best-characterized bacteriophage-resistant strain is *Lc. lactis* ME2. This strain has been shown to contain at least five distinct phage defense loci, including one which interferes with bacteriophage adsorption, two restriction-modification systems, and two abortive infection mechanisms (44). These defense loci are encoded by plasmids capable of conjugal transfer, which suggests that genetic exchange between starter cultures has had an important role in the development of bacteriophage-resistant starter cultures. Recombinant DNA techniques have also been employed to construct lactococcal strains with enhanced bacteriophage resistance. These include the use of antisense RNA derived from conserved bacteriophage genes and the cloning of bacteriophage origins of replication on multicopy plasmids. The latter approach is thought to titrate bacteriophage replication factors and result in a bacteriophage defense mechanism similar to abortive bacteriophage infection. The remainder of this section covers the previously mentioned three naturally occurring bacteriophage defense mechanisms.

Interference with Bacteriophage Adsorption

The adsorption of a bacteriophage to a host cell is determined by bacteriophage specificity, physicochemical properties of the cell envelope, accessibility and density of bacteriophage receptor material, and electrical potential across the cytoplasmic membrane (66). The complexity of this interaction has facilitated the isolation of bacteriophage-resistant starter cultures by exposing them to bacteriophages and isolating resistant variants. Mutants isolated in such a fashion typically have a reduced capacity to adsorb the bacteriophage(s) used in the challenge and are referred to as bacteriophage-insensitive mutants. This is a common practice for deriving starter cultures with reduced bacteriophage sensitivity.

Researchers have begun to characterize host components required for bacteriophage adsorption. It is now thought that bacteriophages initially interact reversibly with cell envelope-associated polysaccharide and then interact irreversibly with cell membrane protein(s) (57). The best-characterized example of a lactococcal mechanism which interferes with adsorption of bacteriophages has been described in *Lc. lactis* subsp. *cremoris* SK110.

Reduction in the ability of bacteriophage to adsorb to cells of this strain is due to masking of the phage receptors by a galactose-containing lipoteichoic acid, rather than the absence of receptors.

Inhibition of adsorption of lactococcal bacteriophages due to the lack of a membrane-bound protein has also been reported. Characterization of phage-resistant mutants of *Lc. lactis* subsp. *lactis* C2 has revealed that these mutants bind normally with bacteriophages but no plaques are formed. Subsequently, it was demonstrated that a 32-kDa membrane protein essential for bacteriophage infection was lacking in these mutants. Similarly, it has been suggested that a membrane protein is involved in bacteriophage adsorption to *Lc. lactis* subsp. *lactis* ML3. It remains to be determined if this protein also has a role in the injection of phage DNA into the cytoplasm (24).

R/M Systems

Restriction and modification (R/M) systems are widely distributed in lactococci and are often plasmid encoded. In fact, it is not uncommon for strains to contain two or more R/M systems, and at least 23 plasmids which encode these systems have been identified. The two components of an R/M system are a site-specific modifying enzyme and a corresponding site-specific restriction endonuclease. This system enables the cell to differentiate bacteriophage DNA from its own DNA and inactivate foreign DNA by hydrolysis. The typical end result of bacteriophage infection of a culture containing an R/M system is a reduction in the number of progeny bacteriophage produced. The extent of reduction is dependent on both the activity of the R/M system and the number of unmodified restriction endonuclease sites on the bacteriophage genome. It is important to note that bacteriophages which escape restriction give rise to modified progeny phage that are immune to the corresponding restriction endonuclease. Therefore, while this is an important and widely distributed mechanism of bacteriophage defense, it is also very fragile.

Two mechanisms by which bacteriophages have evolved resistance to R/M systems have been identified. Characterization of lactococcal bacteriophages has revealed that they contain far fewer restriction endonuclease sites than expected for genomes of their size, suggesting that evolutionary pressure has selected for bacteriophages with few restriction endonuclease sites. This view is supported by the observation that lactococcal bacteriophages which have recently emerged in the dairy industry are more sensitive to R/M systems. The characterization of bacteriophages which have evolved to overcome a specific R/M system has revealed that they have acquired a functional copy of the modification enzyme from that system. These examples illustrate that bacteriophage-host interactions are continually changing, with bacterial strains acquiring new defense mechanisms and bacteriophages evolving mechanisms to overcome them.

Abortive Bacteriophage Infection

Like R/M systems, abortive bacteriophage infection (Abi) systems are widely distributed in lactococci and are frequently plasmid encoded, although some are also encoded by episomes. By definition, these mechanisms inhibit bacteriophage infection following adsorption, DNA penetration, and the early stages of the bacteriophage lytic cycle. Few infections successfully release viable progeny, and those which are successful result in fewer progeny bacteriophages being released. The end result for the host, even one that does not release viable bacteriophages, is death.

All of the characterized Abi systems have an unusually low G+C content, suggesting that they have recently been acquired by horizontal gene transfer. Frequently, Abi genes are associated with systems R/M. These systems work well together; the R/M system reduces the number of viable bacteriophage genomes, and the Abi system reduces the number of progeny bacteriophages released from infected cells which evade the R/M system.

GENETICS OF LAB

Research on the genetics of LAB began in the early 1970s. Initially, research focused on plasmids and natural gene transfer systems in lactococci. Interest in the genetics of LAB has expanded rapidly, and there are now hundreds of researchers active in this area. This interest has resulted in the development of a relatively detailed understanding of the basic genetics of these bacteria, natural gene transfer systems, and the development of tools required for application of recombinant DNA techniques. For a more complete description of the genetics of LAB, the reader is referred to other publications (10, 43, 58) and chapters 42 and 46.

General Genetics of LAB

Genetic elements of LAB which have been characterized include chromosomes, introns, transposable elements, and plasmids. Chromosomes of LAB are relatively small compared to those of other eubacteria, ranging from 1.8 to 3.4 Mbp (18). As of the time of writing this chapter, there were 25 completed or ongoing genome sequencing projects (43). The value of genomic sequences of LAB to basic and applied research on these organisms cannot be overstated. For example, this information should allow for the development of a comprehensive view of the metabolic potential of these organisms. Group I and group II introns have been identified in LAB. These elements are ribozymes that catalyze a self-splicing reaction from mRNAs which contain the intron. To date, group I introns in LAB have been identified only in bacteriophage genomes. The only group II intron identified in LAB is associated with the plasmid-encoded gene required for conjugal transfer (23). Transposable elements, genetic elements capable of moving as discrete units from one site to another in the genome, have been described in LAB (10, 58). Insertion sequences, the simplest of transposable elements, are widely distributed in bacteria and have also been identified in numerous LAB. Their ability to mediate molecular rearrangements and affect gene regulation has had both positive and negative implications for dairy product fermentations. In one case, the incorporation of an insertion sequence into a prophage of *Lb. casei* resulted in the conversion of a temperate bacteriophage into a virulent bacteriophage. Alternatively, insertion sequence-mediated cointegration has played a pivotal role in the dissemination of many beneficial characteristics via conjugation. More complex transposable elements, such as self-transmissible conjugal transposons which code for the production of the bacteriocin nisin, the ability to metabolize sucrose, and a bacteriophage defense mechanism, have also been described in detail (10, 58). Plasmids, i.e., autonomous replicating extrachromosomal circular DNA molecules, have been identified in several LAB. These are of particular importance in lactococci, where they encode numerous characteristics essential for dairy product fermentations, including lactose metabolism, proteinase activity, oligopeptide transport, bacteriophage resistance mechanisms, bacteriocin production and immunity, bacteriocin resistance, exopolysaccharide production, and citric acid utilization (58).

Genetic Modification of LAB by Using Recombinant DNA Techniques

Experiments employing recombinant DNA techniques have led to most of the recent advances in the understanding of physiology and genetics of LAB. The power of recombinant DNA approaches is that strains can be constructed which differ in a single defined genetic alternation, e.g., inactivation of a specific gene. By comparing the wild-type culture to its isogenic derivative, the role of that gene in the phenotype being examined can be unequivocally determined. This general approach

has resulted in a detailed understanding of how these bacteria utilize lactose, obtain essential amino acids, produce diacetyl, and resist bacteriophage infection. Additionally, recombinant DNA approaches have been used to construct novel bacteriophage resistance mechanisms and to overproduce enzymes of interest in dairy fermentations. In the future, it is likely that numerous commercial strains will be constructed utilizing recombinant DNA techniques. Readers interested in more comprehensive reviews on the physiology and genetics of LAB are referred to other publications (10, 35, 64).

References

1. **Allison, G. E., and T. R. Klaenhammer.** 1998. Phage defense mechanisms in lactic acid bacteria. *Int. Dairy J.* 8:207–226.

2. **Ardö, Y.** 2006. Flavour formation by amino acid catabolism. *Biotechnol. Adv.* 24:238–242.

3. **Aymes, F., C. Monnet, and G. Corrieu.** 1999. Effect of alpha-acetolactate decarboxylase inactivation on alpha-acetolactate and diacetyl production by *Lactococcus lactis* subsp. *lactis* biovar *diacetylactis*. *J. Biosci. Bioeng.* 87:87–92.

4. **Bandel, M., M. E. Lhotte, C. Marty-Teysset, A. Veyrat, H. Prevost, V. Dartois, C. Divies, W. N. Konings, and J. S. Lolkema.** 1998. Mechanism of the citrate transporters in carbohydrate and citrate cometabolism in *Lactococcus* and *Leuconostoc* species. *Appl. Environ. Microbiol.* 64:1594–1600.

5. **Bassit, N., C. Y. Boquien, D. Picque, and G. Corrieu.** 1993. Effect of initial oxygen concentration on diacetyl and acetoin production by *Lactococcus lactis* subsp. *lactis* biovar diacetylactis. *Appl. Environ. Microbiol.* 59:1893–1897.

6. **Bassit, N., C. Y. Boquien, D. Picque, and G. Corrieu.** 1995. Effect of temperature on diacetyl and acetoin production by *Lactococcus lactis* subsp. *lactis* biovar *diacetylactis* CNRZ 483. *J. Dairy Res.* 62:123–129.

7. **Bills, D. D., and E. A. Day.** 1966. Dehydrogenase activity of lactic streptococci. *J. Dairy Sci.* 49:1473–1477.

8. **Bills, D. D., M. E. Morgan, L. M. Libby, and E. A. Day.** 1965. Identification of compounds responsible for fruity flavor defect of experimental cheeses. *J. Dairy Sci.* 48:1168–1173.

9. **Bongers, R. S., M. H. N. Hoefnagel, and M. Kleerebezem.** 2005. High-level acetaldehyde production in *Lactococcus lactis* by metabolic engineering. *Appl. Environ. Microbiol.* 71:1109–1113.

10. **Broadbent, J. R.** 2001. Genetics of lactic acid bacteria, p. 243–299. *In* E. H. Marth and J. L. Steele (ed.), *Applied Dairy Microbiology*, 2nd ed. Marcel Dekker, Inc., New York, N.Y.

11. **Broadbent, J. R., M. Strickland, B. C. Weimer, M. E. Johnson, and J. L. Steele.** 1998. Small peptide accumulation and bitterness in Cheddar cheese made from single strain *Lactococcus lactis* starters with distinct proteinase specificities. *J. Dairy Sci.* 81:327–337.

12. Chaves, A. C. S. D., M. Ferenandez, A. L. S. Lerayer, I. Mierau, M. Kleerebezem, and J. Hugenholtz. 2002. Metabolic engineering of acetaldehyde production by *Streptococcus thermophilus*. *Appl. Environ. Microbiol.* 68:5656–5662.

13. Christensen, J. E., E. G. Dudley, J. R. Pederson, and J. L. Steele. 1999. Peptidases and amino acid catabolism in lactic acid bacteria. *Antonie Leeuwenhoek* 76:217–246.

14. Cogan, T. M. 1987. Co-metabolism of citrate and glucose by *Leuconostoc* spp.: effects on growth, substrates and products. *J. Appl. Bacteriol.* 63:551–558.

15. Cogan, T. M., R. J. Fitzgerald, and S. Doonan. 1984. Acetolactate synthase of *Leuconostoc lactis* and its regulation of acetoin production. *J. Dairy Res.* 51:597–604.

16. Condon, S. 1987. Responses of lactic acid bacteria to oxygen. *FEMS Microbiol. Rev.* 46:269–280.

17. Crow, V. L. 1990. Properties of 2,3-butanediol dehydrogenases from *Lactococcus lactis* subsp. *lactis* in relation to citrate fermentation. *Appl. Environ. Microbiol.* 56:1656–1665.

18. Davidson, B. E., N. Kordias, M. Dobos, and A. J. Hillier. 1996. Genomic organization of lactic acid bacteria. *Antonie Leeuwenhoek* 70:161–183.

19. Deborde, C., D. B. Rolin, A. Bondon, J. D. De Certaines, and P. Boyaval. 1998. In vivo nuclear magnetic resonance study of citrate metabolism in *Propionibacterium freudenreichii* subsp. *shermanii*. *J. Dairy Res.* 65:503–514.

20. Desiere, F., S. Lucchini, C. Canchaya, M. Ventura, and H. Brussow. 2002. Comparative genomics of phages and prophages in lactic acid bacteria. *Antonie Leeuwenhoek* 82:73–91.

21. Detmers, F. J. M., E. R.S. Kunji, F. C. Lanfermeijer, B. Poolman, and W. N. Konings. 1998. Kinetics and specificity of peptide uptake by the oligopeptide transport system of *Lactococcus lactis*. *Biochemistry* 37:16671–16679.

22. Driesson, F. M., and Z. Puhan. 1988. Technology of mesophilic fermented milks. *Int. Dairy Fed. Bull.* 227:75–81.

23. Dunny, G. M., and L. L. McKay. 1999. Group II introns and expression of conjugative transfer functions in lactic acid bacteria. *Antonie Leeuwenhoek* 76:77–88.

24. Forde, A., and G. F. Fitzgerald. 1999. Bacteriophage defense systems in lactic acid bacteria. *Antonie Leeuwenhoek* 76:89–113.

25. Fordyce, A. M., V. L. Crow, and T. D. Thomas. 1984. Regulation of product formation during glucose or lactose limitation in nongrowing cells of *Streptococcus lactis*. *Appl. Environ. Microbiol.* 48:332–337.

26. Fox, P. F. (ed.). 1993. *Cheese: Chemistry, Physics and Microbiology*, vol. 1 and 2. Chapman and Hall, Ltd., London, United Kingdom.

27. Fox, P. F., J. Law, P. L. H. McSweeney, and J. Wallace. 1993. Biochemistry of cheese ripening, p. 389–438. *In* P. F. Fox (ed.), *Cheese: Chemistry, Physics and Microbiology*, vol. 1. Chapman and Hall, Ltd., London, United Kingdom.

28. Fox, P. F., J. A. Lucey, and T. M. Cogan. 1990. Glycolysis and related reactions during cheese manufacture and ripening. *Food Sci. Nutr.* 29:237–253.

29. Fryer, T. F., M. E. Sharpe, and B. Reiter. 1970. Utilization of milk citrate by lactic acid bacteria and "blowing" of film wrapped cheese. *J. Dairy Sci.* 37:17–28.

30. Gancel, F., and G. Novel. 1994. Exopolysaccharide production by *Streptococcus salivarius* ssp. *thermophilus* cultures. 1. Conditions of production. *J. Dairy Sci.* 77:685–688.

31. Green, M. L., and D. J. Manning. 1982. Development of texture and flavor in cheese and other fermented products. *J. Dairy Res.* 49:737–748.

32. Griffith, R., and E. G. Hammond. 1989. Generation of Swiss cheese flavor components by the reaction of amino acids with carbonyl compounds. *J. Dairy Sci.* 72:604–613.

33. Huang, D. Q., H. Prévost, and C. Diviès. 1995. Principal characteristics of β-galactosidase from *Leuconostoc* spp. *Int. Dairy J.* 5:29–43.

34. Hugenholtz, J. 1993. Citrate metabolism in lactic acid bacteria. *FEMS Microbiol. Rev.* 12:165–178.

35. Hutkins, R. W. 2001. Metabolism of starter cultures, p. 207–241. *In* E. H. Marth and J. L. Steele (ed.), *Applied Dairy Microbiology*, 2nd ed. Marcel Dekker, Inc., New York, N.Y.

36. Imhof, R., and J. O. Bosset. 1994. Review: relationships between micro-organisms and formation of aroma compounds in fermented dairy products. *Lebensm. Unters Forsch.* 198:267–276.

37. International Dairy Federation. 1988. *Fermented Milks: Science and Technology*. International Dairy Federation, Brussels, Belgium.

38. Jensen, N. B. S., C. R. Melchiorsen, K. V. Jokumsen, and J. Villadsen. 2001. Metabolic behavior of *Lactococcus lactis* MG1363 in microaerobic continuous cultivation at a low dilution rate. *Appl. Environ. Microbiol.* 67:2677–2682.

39. Juillard, V., D. Le Bars, E. R. S. Kunji, W. N. Konings, J-.C. Gripon, and J. Richard. 1995. Oligopeptides are the main source of nitrogen for *Lactococcus lactis* during growth in milk. *Appl. Environ. Microbiol.* 61:3024–3030.

40. Kamaly, M. K., and E. H. Marth. 1989. Enzyme activities of lactic streptococci and their role in maturation of cheese: a review. *J. Dairy Sci.* 72:1945–1966.

41. Keenan, T. W., R. C. Lindsay, M. E. Morgan, and E. A. Day. 1966. Acetaldehyde production by single strain lactic streptococci. *J. Dairy Sci.* 49:10–14.

42. Kempler, G. M., and L. L. McKay. 1981. Biochemistry and genetics of citrate utilization in *Streptococcus lactis* spp. *diacetylactis*. *J. Dairy Sci.* 64:1527–1539.

43. Klaenhammer, T., E. Altermann, F. Arigoni, A. Bolotin, F. Breidt, J. Broadbent, R. Cano, S. Chaillou, J. Reutscher, M. Gasson, M. van de Guchte, J. Guzzo, T. Hawkins, P. Hols, R. Hutkins, M. Kleerebezem, J. Kok, O. Kuipers, M. Lubbers, E. Maguin, L. McKay, D. Mills, A. Nauta, R. Overbeek, H. Pel, D. Pridmore, M. Saier, D. van Sinderen, A. Sorokin, J. Steele, D. O'Sullivan, W. de Vos, B. Weimer, M. Zagorec, and R. Siezen. 2002. Discovering lactic acid bacteria by genomics. *Antonie Leeuwenhoek* 82:29–58.

44. Klaenhammer, T. R., and G. F. Fitzgerald. 1994. Bacterio-phages and bacteriophage resistance. *In* M. J. Gasson and W. M. de Vos (ed.), *Genetics and Biotechnology of Lactic Acid Bacteria.* Chapman and Hall, Ltd., London, United Kingdom.

45. Kunji, E. R. S., I. Mierau, A. Hagting, B. Poolman, and W. N. Konings. 1996. The proteolytic systems of lactic acid bacteria. *Antonie Leeuwenhoek* 70:187–221.

46. Lees, G. J., and G. R. Jago. 1978. Role of acetaldehyde in metabolism: a review. 1. Enzymes catalyzing reactions involving acetaldehyde. *J. Dairy Sci.* 61:1205–1215.

47. Lees, G. J., and G. R. Jago. 1978. Role of acetaldehyde in metabolism: a review. 2. The metabolism of acetaldehyde in cultured dairy products. *J. Dairy Sci.* 61:1216–1224.

48. Lemieux, L., and R. E. Simard. 1992. Bitter flavour in dairy products. II. A review of bitter peptides from caseins: their formation, isolation and identification, structure masking and inhibition. *Lait* 72:335–382.

49. Lindgren, S. E., and L. T. Axelsson. 1990. Anaerobic L-lactate degradation by *Lactobacillus plantarum. FEMS Microbiol. Lett.* 66:209–214.

50. Lindsay, R. C., E. A. Day, and W. E. Sandine. 1965. Green fla-vor defect in lactic starter cultures. *J. Dairy Sci.* 48:863–869.

51. Liu, S. Q., R. V. Asmundson, P. K. Gopal, R. Holland, and V. L. Crow. 1998. Influence of reduced water activ-ity on lactose metabolism by *Lactococcus lactis* subsp. *cremoris* at different pH values. *Appl. Environ. Microbiol.* 64:2111–2116.

52. Lucey, C. A., and S. Condon. 1986. Active role of oxygen and NADH oxidase in growth and energy metabolism of *Leuconostoc. J. Gen. Microbiol.* 132:1789–1796.

53. Manning, D. J. 1979. Chemical production of essential Cheddar flavor compounds. *J. Dairy Res.* 46:531–537.

54. McKay, L. L., and K. A. Baldwin. 1974. Altered metabo-lism in a *Streptococcus lactis* C2 mutant deficient in lactate dehydrogenase. *J. Dairy Sci.* 57:181–186.

55. Mierau, I., E. R. S. Kunji, K. J. Leenhouts, M. A. Hellen-doorn, A. J. Haandrikman, B. Poolman, W. N. Konings, G. Venema, and J. Kok. 1996. Multiple-peptidase mutants of *Lactococcus lactis* are severely impaired in their ability to grow in milk. *J. Bacteriol.* 178:2794–2803.

56. Moineau, S. 1999. Applications of phage resistance in lac-tic acid bacteria. *Antonie Leeuwenhoek* 76:377–382.

57. Montville, M. R., B. Ardestani, and B. L. Geller. 1994. Lac-tococcal bacteriophage require a host cell wall carbohydrate and a plasma membrane protein for adsorption and ejec-tion of DNA. *Appl. Environ. Microbiol.* 60:3204–3211.

58. Morelli, L., F. K. Vogensen, and A. von Wright. 2004. Genetics of lactic acid bacteria, p. 249–293. *In* S. Salminen, A. von Wright, and A. Ouwehand (ed.), *Lactic Acid Bacte-ria: Microbiology and Functional Aspects,* 3rd ed. Marcel Dekker, Inc., New York, N.Y.

59. Palles, T., T. Beresford, S. Condon, and T. M. Cogan. 1998. Citrate metabolism in *Lactobacillus casei* and *Lac-tobacillus plantarum. J. Appl. Microbiol.* 85:147–154.

60. Peterson, S. D., and R. T. Marshall. 1990. Nonstarter lactobacilli in Cheddar cheese: a review. *J. Dairy Sci.* 73:1395–1410.

61. Poolman, B. 1993. Energy transduction in lactic acid bac-teria. *FEMS Microbiol. Rev.* 12:125–148.

62. Ramos, A., K. N. Jordan, T. M. Cogan, and H. Santos. 1994. ^{13}C Nuclear magnetic resonance studies of citrate and glucose cometabolism by *Lactococcus lactis. Appl. Environ. Microbiol.* 60:1739–1748.

63. Ramos, A., J. S. Lolkema, W. N. Konings, and H. Santos. 1995. Enzyme basis for pH regulation of citrate and pyru-vate metabolism by *Leuconostoc oenos. Appl. Environ. Microbiol.* 61:1303–1310.

64. Salminen, S., A. von Wright, and A. Ouwehand (ed.). 2004. *Lactic Acid Bacteria: Microbiology and Functional Aspects,* 3rd ed. Marcel Dekker, Inc., New York, N.Y.

65. Schmitt, P., and C. Divies. 1991. Co-metabolism of citrate and lactose by *Leuconostoc mesenteroides* subsp. *cremo-ris. J. Ferment. Bioeng.* 71:72–74.

66. Sijtsma, L., N. Jansen, W. C. Hazeleger, J. T. M. Wouters, and K. J. Hellingwerf. 1990. Cell surface characteristics of bacteriophage-resistant *Lactococcus lactis* subsp. *cremo-ris* SK110 and its bacteriophage-sensitive variant SK112. *Appl. Environ. Microbiol.* 56:3230–3233.

67. Smart, J. B., and T. D. Thomas. 1987. Effect of oxygen on lactose metabolism in lactic streptococci. *Appl. Environ. Microbiol.* 53:533–541.

68. Smit, G., B. A. Smit, and W. J. M. Engels. 2005. Flavour formation by lactic acid bacteria and biochemical fla-vor profiling of cheese products. *FEMS Microbiol. Rev.* 29:591–610.

69. Smit, G., A. Verheul, R. van Kranenburg, E. Ayad, R. Siezen, and W. J. M. Engels. 2000. Cheese flavour devel-opment by enzymatic conversions of peptides and amino acids. *Food Res. Int.* 33:153–160.

70. Sturino, J. M., and T. R. Klaenhammer. 2004. Bacterio-phage defense systems and strategies for lactic acid bacte-ria. *Adv. Appl. Microbiol.* 56:331–378.

71. Thierry, A., D. Salvat-Brunaud, and J.-L. Maubois. 1999. Influence of thermophilic lactic acid bacteria strains on propionibacteria growth and lactate consumption in an Emmental juice-like medium. *J. Dairy Res.* 66:105–113.

72. Thomas, T. D., D. C. Ellwoos, and V. M. C. Longyear. 1979. Change from homo- to heterolactic fermentation by *Streptococcus lactis* resulting from glucose limitation in anaerobic chemostat cultures. *J. Bacteriol.* 138:109–117.

73. Urbach, G. 1995. Contribution of lactic acid bacteria to flavor compound formation in dairy products. *Int. Dairy J.* 5:877–903.

74. Vaughan, E. E., S. David, A. Harrington, C. Daly, G. F. Fitzgerald, and W. M. De Vos. 1995. Characterization of plasmid-encoded citrate permease (*citP*) genes from *Leu-conostoc* species reveals high sequence conservation with the *Lactococcus lactis citP* gene. *Appl. Environ. Microbiol.* 61:3172–3176.

75. Visser, S. 1993. Proteolytic enzymes and their relation to cheese ripening and flavor: an overview. *J. Dairy Sci.* 76:329–350.

76. Warriner, K. S. R., and J. G. Morris. 1995. The effects of aeration on the bioreductive abilities of some hetero-fermentative lactic acid bacteria. *Lett. Appl. Microbiol.* 20:322–327.

77. Weimer, B., K. Seefeldt, and B. Dias. 1999. Sulfur metabolism in bacteria associated with cheese. *Antonie Leeuwenhoek* 76:247–261.

78. Whitehead, W. E., J. W. Ayres, and W. E. Sandine. 1993. A review of starter media for cheese making. *J. Dairy Sci.* 76:2344–2353.

79. Wilkins, D. W., R. H. Schmidt, R. B. Shireman, K. L. Smith, and J. J. Jezeski. 1986. Evaluating acetaldehyde synthesis from L-[^{14}C(U)] threonine by *Streptococcus thermophilus* and *Lactobacillus bulgaricus*. *J. Dairy Sci.* 69:1219–1224.

80. Yvon, M., and L. Rijnen. 2001. Cheese flavor formation by amino acid catabolism. *Int. Dairy J.* 11:185–201.

81. Zourari, A., J. P. Accolas, and M. J. Desmazeaud. 1992. Metabolism and biochemical characteristics of yogurt bacteria. A review. *Lait* 72:1–34.

Food Microbiology: Fundamentals and Frontiers, 3rd Ed.
Edited by M. P. Doyle and L. R. Beuchat
© 2007 ASM Press, Washington, D.C.

Frederick Breidt, Jr.
Roger F. McFeeters
Ilenys Díaz-Muñiz

Fermented Vegetables

36

Vegetable fermentation by lactic acid bacteria (LAB) in a salt brine began as a way to preserve foods for out-of-season use and for long journeys, especially by sea. It is believed that cucumbers were first pickled 4,500 years ago in Mesopotamia. Although originally used to preserve foods, pickling is frequently done because people enjoy the resulting flavor. The current market for pickled vegetables in the United States is $2 billion or more and is increasing in size. The two primary retail fermented vegetable products produced in the United States are cucumber pickles and sauerkraut. Currently, pickled cucumbers are the most popular pickled foods. The word "pickle" by itself usually refers to a pickled cucumber. The retail market for cucumber pickles in the United States is dominated by acidified, pasteurized, and refrigerated products, not fermented pickles (40). Commercial sale of hamburger dill pickle slices to the food service industry makes up the majority of the market for fermented cucumber pickles. Fermented cabbage, i.e., sauerkraut, was introduced in the United States by immigrants from Germany and other European countries. Although the popularity of sauerkraut in Europe and the United States continues today, consumption has declined in

the United States. Fermented olives are also produced in the United States, mostly in California, but the majority of olives sold in the United States are produced in southern Spain. There are niche markets for other pickled vegetables, including deli-style, half-sour cucumber pickles, which are fermented only slightly prior to consumption (30), and Korean kimchi (40). Other acidified vegetable products familiar to consumers include peppers, okra, cauliflower, and green tomatoes.

Fermented pickle products in the United States typically have a pH value of about 3.7, with acetic acid added in addition to the lactic acid which is produced during fermentation. Fermented foods are considered acid foods and are defined in the U.S. Code of Federal Regulations (21 CFR part 114) (101) as foods that naturally have a pH below 4.6. Acidified vegetables are defined in 21 CFR part 114 as low-acid (pH above 4.6) foods to which acid or acid food ingredients have been added to attain a pH value of 4.6 or below. Most acidified pickles have a pH of about 3.7 and an acetic acid concentration of 1 to 2% and contain 2 to 4% NaCl. Acidified pickles, which are not fermented, are typically heated (pasteurized) to prevent spoilage by LAB. Some pickled pepper products

---

Frederick Breidt, Jr., Roger F. McFeeters, and Ilenys Díaz-Muñiz, Dept. of Food Science, Agricultural Research Service, United States Department of Agriculture, 322 Schaub Hall, Box 7624, North Carolina State University, Raleigh, NC 27695-7624.

are acidified with 2 to 3% acetic acid to reach a pH of 3.3 or below. These products may not be heat processed because the concentration of acetic acid is high enough to prevent growth of bacteria and there is better retention of product texture. The high acid/low pH is sufficient to ensure at least a 5-log reduction of acid-tolerant pathogenic bacteria.

Sweet cucumber pickles are often produced from small (<20 mm in diameter) cucumbers that have been fermented. In these products, the lactic acid and salt from the fermentation are removed by washing the fermented cucumbers with water. The cover liquor of sweet pickles is similar to that of sour pickles, but contains 25 to 30% sugar (from corn syrup). Traditional genuine dill pickles, containing lactic acid (up to 2%) from fermentation and salt (4% or more), are a minor market in the United States. Sauerkraut and green table olives, however, are often packed with the naturally present lactic acid and brine from the fermentation. Typically, sauerkraut sold in the United States contains both lactic acid and acetic acid, which are produced during fermentation (up to 2% acid), and about 2% NaCl, although acid and salt levels may be adjusted prior to packaging.

HISTORY OF VEGETABLE FERMENTATION RESEARCH IN THE UNITED STATES

Prior to the 1920s, research in the United States on pickled vegetables was primarily focused on product surveys (73, 74, 90) and descriptions of brining methods (1, 20, 49). Reports on the microbiology and biochemistry of vegetable fermentations appeared in the literature between 1918 and 1920 (21, 51, 59, 77). Initial reports of the "lactic bacilli" and organic acids in cucumber pickles, sauerkraut, and olives were further advanced by a series of studies done by E. B. Fred at the University of Wisconsin. During the decade of the 1920s, his group studied the sauerkraut fermentation, including the growth of yeasts (46, 85), microbial ecology (92, 93), and chemistry of the fermentation (47, 86).

In the early 1930s, Fabian began studies on fermented vegetables at Michigan State University. An initial report described the study of softening and spoilage of fermented cucumbers (34). Fabian et al. (35) and Costilow and Fabian (15–17) published extensively in the field of pickled vegetables. Pederson, at Cornell University, studied sauerkraut fermentation from the 1930s to the early 1970s. His group reported on various aspects of the microbiology (78–81) and nutritional and biochemical properties (83, 84, 103) of the sauerkraut fermentation, which culminated in a comprehensive review (82). Stamer and other researchers subsequently published a series of articles on the microflora of sauerkraut (95, 96), including the development of red or pink discoloration in sauerkraut (95), a continuation of the work done by Fred and Peterson in the early 1920s (46).

Etchells and Veldhuis were the founding scientists of a U.S. Department of Agriculture (USDA) laboratory in Raleigh, N.C., in the mid 1930s, where research on pickled vegetables from 1938 (27, 102) to the present has been ongoing. Included in these studies were the development of pasteurization methods (32, 33, 54), investigations of the yeasts that are responsible for spoilage of cucumber pickle products (28, 31), and a preservation-prediction chart to describe the storage stability of sweet pickles based on salt and sugar concentrations (3). Bell determined the role of pectin-degrading enzymes in cucumber softening in the 1970s (2, 4, 5). Fleming and coworkers, also at the Raleigh USDA laboratory, later developed purging technology for commercial cucumber fermentations (29, 37, 44). Costilow and others have done further research using this technology (14, 18, 19). Other developments include an investigation of the antimicrobial properties of oleuropein in olives and its degradation products (45), the role of malolactic enzymes in the production of carbon dioxide by LAB during cucumber fermentation (72), a *Lactobacillus plantarum* strain that lacks the ability to carry out the malolactic reaction (22), and the use of calcium to improve the texture of pickled vegetables (70, 71).

As we advance into the second century of research on acid and acidified foods, researchers are building on the solid foundation laid by the researchers noted above as well as others. Current research on pickled vegetables includes the genomics of LAB (8), mathematical modeling of bacterial growth and competition (25), the molecular ecology of vegetable fermentations (60, 88), closed-tank fermentation technology to reduce salt waste (38), the use of clays to filter brines for recycling (12), sensory perception of pickled vegetable products (53), and the safety of acidified foods (6, 9).

COMMERCIAL FERMENTATIONS

In the United States, commercial cucumber (*Cucumis sativus*) fermentations are commonly done in 8,000- to 10,000-gallon, open-top, plastic or fiberglass tanks that are located out-of-doors so the brine surface is exposed to sunlight. The UV radiation in sunlight is relied upon to kill aerobic surface yeasts which can metabolize lactic acid produced by the fermentation. Cucumbers are covered with a salt brine and held below the brine surface with wooden headboards. Fermentations are typically carried out in brine equilibrated at about 6% NaCl. Calcium chloride (0.1 to 0.4%, equilibrated) is added to the cover brine to maintain the firm, crisp texture of the fermented cucumbers during storage (41). Cucumber fermentations typically undergo a homolactic acid fermentation, which does not result in production of carbon dioxide from sugars (glucose and fructose, about 1% each). However, carbon dioxide may be generated from respiration of cucumbers when they are submerged in brine and by the decarboxylation of malate during the fermentation. Some LAB have an inducible malolactic enzyme which converts malate to lactate and carbon dioxide. The malolactic enzyme reaction occurs intracellularly and results in the uptake of a proton, raising internal cell pH. Cucumber fermentations are purged with air during the active fermentation period to remove carbon dioxide from the tank (14). Purging prevents bloating damage in the cucumbers. Potassium sorbate (350 μg/ml) or 0.16% acetic acid can be used to limit the growth of aerobic microorganisms, particularly molds and yeasts, which can grow as a consequence of dissolved oxygen in the fermentation brine resulting from air purging (40, 89). Excessive growth of aerobic microorganisms which can cause spoilage problems (48) is also controlled by stopping air purging for several hours each day. Fermentation is carried out primarily by *Lb. plantarum*, a homofermentative, acid-tolerant LAB. After fermentation, cucumbers may be stored in the fermentation tanks for 1 year or more. In the northern states, the concentration of NaCl is often increased during storage to as high as 10 or 15% to minimize freezing damage and maintain the desirable texture of fermented cucumbers.

Currently, starter cultures of *Lb. plantarum* are rarely used by the industry. Most commercial cucumber fermentations rely upon growth of the epiphytic LAB which occur naturally on the surface of cucumbers. These bacteria effectively control the microbial ecology of the fermentation by consuming the glucose and fructose present, producing lactic acid, and lowering the brine pH. The initial pH of brined cucumbers is about 6.5. In practice, commercial fermentations may use recycled brine or acetic acid may be added to brine solutions, so the initial pH of the fermentation can vary significantly. In addition to lactic acid, LAB produce a variety of metabolites, e.g., bacteriocins, peroxides, and peptides, that can be inhibitory to other bacteria (24). At the end of the fermentation, there may be up to 2% lactic acid, a pH of 3.1 to 3.5, and little or no residual sugar. In this anaerobic, acidic, high-salt environment lacking sugar, very few microorganisms are capable of growing or surviving, effectively preserving the cucumbers.

Commercial production of sauerkraut (41) from cabbage (*Brassica oleracea*) in the United States is also done in bulk fermentation tanks that may contain 100 tons or more of shredded or chopped cabbage. The cabbage for these fermentations consists of large heads, typically 8 to 10 lb (3.6 to 4.5 kg). The outer leaves and woody core of the cabbage are removed prior to shredding. The shredded cabbage is dry salted on belts as it is conveyed to indoor cement fermentation tanks. This process typically results in an equilibrated NaCl concentration of about 2%. The cabbage is manually distributed in the tanks to create a concave surface and covered with a plastic liner. Water is added on top of the plastic to hold down the liner on the cabbage and allow anaerobic conditions to develop under the plastic. After the first 24 to 48 h, carbon dioxide gas may accumulate under the liner. To prevent heaving of the cabbage, some of the brine formed from the salted cabbage may be drained from the bottom of the tanks during the first week of fermentation. A salt concentration of 2% and temperature of 18°C help ensure a normal fermentation by the naturally present LAB (82). Cabbage typically contains 4 to 5% sugar, consisting of about 2.5% glucose and 2% fructose with smaller amounts of sucrose and other sugars (42). The primary fermentation is initiated by naturally occurring, heterofermentative LAB, primarily *Leuconostoc mesenteroides*, although *Weissella* species are also present and sometimes dominate the initial stage of fermentation (82, 87). The initial heterolactic stage of the fermentation results in production of both lactic and acetic acids. The volatile acetic acid makes an important contribution to the flavor and aroma of the final product. Heterofermentative microorganisms also use fructose as an electron acceptor, converting it to mannitol (68). After about 1 week of fermentation, the heterofermentative LAB, which may reach 9 log CFU/ml or greater, die off. They are replaced by the more acid-tolerant homofermentative LAB. This stage of the fermentation process is usually dominated by *Lb. plantarum*, presumably because it is the most acid-resistant microorganism present (64). The fermentation end products resulting from both stages of the fermentation can include mannitol and acetic acid (about 1% each) and lactic acid, which may exceed 2%, depending on how long the homolactic fermentation is allowed to continue. For most manufacturers in the United States, this may be up to 1 year because sauerkraut is stored in the fermentation tank until it is processed for food service or retail sale. Some European manufacturers package sauerkraut at the end of the heterolactic fermentation stage (about 1 week after the start of fermentation) to produce a product with mild acid flavor (40). Spices, wines, and other ingredients may be added to the sauerkraut to augment flavor.

Since cucumbers are fermented in brine containing salt at a concentration too high to be used in products for consumption, salt is reduced to about 2% prior to packaging for distribution and sale. This results in a waste stream with high concentrations of salt plus a high biological oxygen demand from the lactic acid and other organic components that diffuse out of the cucumbers during the desalting process. The brine from the fermentation tanks is usually recycled and may be used in subsequent fermentations (69). Prior to recycling, fermentation brines may be processed to remove "softening enzymes," primarily polygalacturonases, which can degrade pectic substances in the cucumber cell wall and soften the fruits (12). Waste brine is also generated from sauerkraut manufacture when draining excess brine after the initiation of fermentation.

Other vegetable fermentation processes, including olive fermentation, represent smaller markets in the United States, with Spain being the largest producer and exporter of black olives and green table olives. Commercial production of kimchi, a Korean fermented cabbage product, is now widespread in South Korea. Like cucumber and cabbage fermentation in the United States, both olive and kimchi fermentation practices are based on traditional methods. There are several methods used for processing olives (36). Green table olives are treated with lye (NaOH) and then washed prior to being brined and fermented. Olives may also be aerated during lye treatment to allow blackening (oxidation) of olives for the manufacture of canned black olives (36). Both of these processes can generate NaOH waste, in addition to the salt waste generated from fermentation brines. For commercial production of kimchi, Chinese cabbage (*Brassica campestris* subsp. *pekinesis*) is used. The cabbage is dry salted or brined, and then washed to remove excess salt prior to fermentation and packing with spices (41). Although recycling is used extensively in the vegetable fermentation industry, particularly for cucumber fermentation brine, the need to dispose of NaCl waste is a continuing problem.

BIOCHEMISTRY OF VEGETABLE FERMENTATIONS

There is continuing research interest in fermentation and storage of vegetables, particularly cucumbers, with reduced salt. Chloride waste from vegetable fermentations could be greatly reduced if the salt required for fermentation and storage could be reduced sufficiently to eliminate the need for a desalting step prior to conversion into final products. Lu et al. (61) investigated the effect of replacement of NaCl with different anions and cations on the sugar fermentation in cucumber juice. Interestingly, it

was clear from these experiments that fructose was the preferred sugar for *Lb. plantarum*, since more fructose was fermented in almost every experiment. Sugar utilization decreased as cation or anion concentrations increased with the addition of different salts. Divalent cations (Ca^{2+} and Mg^{2+}) reduced the extent of fermentation at lower concentrations than monovalent cations (Na^+ and K^+).

A number of the volatile components in cucumbers fermented with *Lb. plantarum* in 2% NaCl were identified by Zhou and McFeeters (104). Thirty-seven volatile compounds were identified, though most showed little change as a result of fermentation. The most notable effect of fermentation on cucumber volatiles was the inhibition of production of (E, Z)-2,6-nonadienal and 2-nonenal, the two most important odor impact compounds in fresh cucumbers. Marsili and Miller (62) identified *trans*- and *cis*-4-hexenoic acid as the most potent odorants that define the characteristic brine aroma of cucumbers fermented commercially in about 6% NaCl. Zhou et al. (105) exposed fermented cucumber slurries with 2% NaCl to oxygen and observed nonenzymatic formation of hexanal plus a series of *trans* unsaturated aldehydes with 5 to 8 carbon atoms that correlated with the development of oxidized odor intensity of the fermented cucumber tissue. Calcium disodium EDTA at a concentration of at least 100 μg/ml protects nonfermented pickles against lipid oxidation and bleaching of pigments in the presence of light (11). However, there was some reduction in firmness retention in pickles when this compound was used.

Retention of firmness is a key quality issue in the fermentation and storage of cucumbers and peppers. It has not been possible to ensure the retention, in cucumbers fermented in reduced salt, of firmness equivalent to that which can be achieved by fermenting in 6% NaCl and storage in 6% or greater NaCl concentrations (39). However, in the past several years there has been some increased understanding of the cucumber tissue softening. Fleming et al. (43) showed that calcium is beneficial in maintaining the firmness of fermented cucumbers. Nonenzymatic softening of blanched, acidified cucumber tissue was found to follow first-order kinetics (70). This kinetic behavior made it possible to determine the entropy and enthalpy of activation for nonenzymatic softening of cucumbers, even though the chemical reactions responsible for softening were not known. Both the enthalpy and entropy of activation were high at pH 3.0 in the presence of 1.5 M NaCl. Calcium inhibited cucumber softening because it reduced the entropy of activation so much that the overall free energy of activation was reduced (71). This thermodynamic behavior is more like that which occurs when polymers change conformation, such as occurs in protein denaturation. It is very different from the characteristics observed for acid hydrolysis of pectin (58). Krall and McFeeters (58) found that the rate of acid hydrolysis of pectin was too slow to be the cause of nonenzymatic cucumber tissue softening. McFeeters et al. (66) determined the combined effects of temperature, salt, and calcium concentration on the rate of softening of fermented cucumber tissue. The kinetics of softening for fermented cucumbers did not follow a simple first-order reaction.

As with many other plant tissues, cucumbers contain enzymes that can degrade components of the plant cell walls, which may result in changes in texture. Pectinesterase (5), exopolygalacturonase (91), and endopolygalacturonase (67) activities have been found in cucumbers. Pectinesterase removes methyl groups from pectin when cucumbers are fermented or acidified (52, 65, 98). However, it has not been determined if enzymatic hydrolysis of pectin by cucumber polygalacturonases is a significant factor in softening of fermented cucumbers. Commercially important enzymatic softening of fermented cucumbers has been associated with the introduction of fungal polygalacturonases into fermentation tanks, particularly on flowers attached to small cucumbers (4). Buescher and Burgin (10) developed a sensitive diffusion plate assay to measure polygalacturonase activity in fermentation brines and determined that an aluminosilicate clay can adsorb and remove polygalacturonase activity from fermentation brines that are recycled.

In addition to enzymes that degrade pectin, enzymes that may degrade other cell wall polysaccharides in cucumbers have been investigated to a very limited extent. Meurer and Gierschner (75) reported endo-β-1,4-gluconase activity in the cucumber that is inactivated below pH 4.8 and an endoglucomannan-splitting enzyme that retains activity down to pH 4.0 but is inactivated during fermentation. They detected six enzymes in fresh cucumbers, which hydrolyze p-nitrophenylglycosides of α-D-galactose, α-D-galactose, α-D-glucose, β-D-xylose, α-D-mannose, and α-L-arabinose, which were inactivated during fermentation. Enzymes capable of hydrolyzing these synthetic substrates are common in plants, e.g., most of the same enzymatic activities have been found in pears (26), olives (50), and Semillon grapes (97). Maruvada (63) observed the same p-nitrophenylglycosidases observed by Meurer and Gierschner (75) in cucumbers. She found that the concentration of all of these enzymes declined to nondetectable levels during the first week of fermentation in 2% NaCl brines. Fleming et al. (39) combined blanching of fresh cucumbers to partially inactivate enzymes, calcium addition, and rapid fermentation with a malolactic-negative $Lb.$ plantarum strain in order to ferment cucumbers and

maintain desirable texture with the NaCl concentration reduced to 4%.

Cabbage contains a group of glucosinolates which have received considerable attention in recent years due to potential health benefits of some of the degradation products formed during the processing of cabbage. A recent report indicates that a high intake of sauerkraut correlates with a reduced incidence of breast cancer in women (94). Daxenbichler et al. (23), who were concerned about potentially toxic compounds derived from glucosinolates, did an excellent study of glucosinolates and their degradation products in three cabbage cultivars that were fermented and stored for up to 28 weeks. They reported the presence of 11 glucosinolates in fresh cabbage and found that all glucosinolates disappeared within 2 weeks after the start of fermentation. Tolonen et al. (99) found isothiocyanates and allyl cyanide to be the predominant degradation products of glucosinolates in sauerkraut fermented with and without a starter culture. Only minor amounts of goitrin, a toxic compound, and the beneficial phytochemical sulforaphane nitrile were found in sauerkraut. Tolonen et al. (100) found greater amounts of glucosinolate degradation products in sauerkraut fermented with Lactobacillus sakei than in sauerkraut made with starter cultures consisting of other LAB. Ciska and Pathak (13) reported that ascorbigen, a compound formed from the reaction of a degradation product of indole glucosinolate (glucobrassicin) and ascorbic acid, is the dominant glucosinolate degradation product in sauerkraut. A large fraction of glucoraphinin present in fresh cabbage was converted to sulforphorane during fermentation, though sulforphorane was only a minor glucosinolate degradation product in fermented cabbage.

There is some concern about the formation of biogenic amines in sauerkraut. Kalač et al. (55) reported that tyramine was the major biogenic amine formed in sauerkraut stored for up to 12 months. Only trace levels of histamine, tryptamine, and spermine were detected. These results were confirmed in a survey of vegetables (76) showing that the content of tyramine was 4.9 mg/100 g in canned sauerkraut, virtually the same concentration reported by Kalač et al. (55). These biogenic amine levels would not represent a health risk, with the possible exception of individuals taking medications containing monamine oxidase inhibitors.

GENOMICS RESEARCH ON LAB IN VEGETABLE FERMENTATIONS

Bolotin et al. (7) reported the first genome sequence for a lactic acid bacterium, Lactococcus lactis IL1403. They observed that the metabolic potential of this microorganism is more extensive than previously considered. The

Lc. lactis genome sequence contains putative genes for unexpected functionalities, including aerobic respiration, and biosynthesis pathways for all amino acids. Release of the *Lc. lactis* genome sequence was followed by sequencing of several other LAB genomes (55). Among these genome sequences are those belonging to the predominant bacteria present in fermented vegetables, *Lb. plantarum*, *L. mesenteroides*, *Pediococcus pentosaceus*, and *Lactobacillus brevis*. The *Lb. plantarum* WCFS1 genome sequence was published in 2002 by Kleerebezem et al. (57). Genome sequencing efforts by the Joint Genome Institute (Walnut Creek, Calif.) and the Lactic Acid Bacteria Genome Consortium (LABGC, USA) have contributed draft sequences for 10 genomes. These genomes include those for *L. mesenteroides* ATCC 8293, *P. pentosaceus* ATCC 25745, and *Lb. brevis* ATCC 367.

The genome sequences for the *L. mesenteroides*, *Lb. brevis*, and *P. pentosaceus* consist of approximately 2,000,000 bp. The *Lb. plantarum* genome consists of approximately 3,000,000 bp. The 2,000,000- or 3,000,000-bp genomes have approximately 2,000 or 3,000 open reading frames (ORFs), respectively. About 75% of the ORFs in the *L. mesenteroides*, *Lb. plantarum*, *Lb. brevis*, and *P. pentosaceus* sequences have been assigned functions in metabolic pathways based on BLAST (Basic Local Alignment Search Tool) predicted protein sequence similarities. These ORFs have been classified according to clusters of orthologous groups (COG) categories. Each COG represents proteins or sets of paralogs from a representative number of lineages or an ancient conserved domain. Table 36.1 shows a comparison of COG categories among the four LAB genome sequences noted above. About 50% of the ORFs have been assigned functions relating to the metabolism or transport of amino acids, carbohydrates, and inorganic ion, the proteins involved in transcription and translation, including ribosomal structures and biogenesis, replication, recombination, and repair, and cell wall and membrane biogenesis. Only about 7% of these ORFs are dedicated to energy production by fermentation of sugars.

Table 36.1 Distribution of ORFs among COG functional categories[a]

COG categories	*Lb. brevis* ATCC 367	*L. mesenteroides* ATCC 8293	*P. pentosaceus* ATCC 25745	*Lb. plantarum* WCFS1
Energy production and conversion	6.8	6.0	6.7	8.8
Cell cycle control, cell division, chromosome partitioning	2.4	2.2	1.8	2.2
Amino acid transport and metabolism	13.6	17.3	10.4	16.1
Nucleotide transport and metabolism	7.2	8.3	8.8	7.5
Carbohydrate transport and metabolism	14.4	12.8	17.0	21.9
Coenzyme transport and metabolism	5.5	7.1	4.9	7.8
Lipid transport and metabolism	3.8	3.9	3.9	4.3
Translation, ribosomal structure, and biogenesis	17.4	15.9	17.5	15.4
Transcription	15.7	11.3	14.0	20.2
Replication, recombination, and repair	16.2	11.7	13.2	12.8
Cell wall/membrane/envelope biogenesis	10.8	9.4	11.8	13.2
Cell motility	0.2	0.1	0.3	0.3
Posttranslational modification, protein turnover, chaperones	5.4	5.5	5.2	4.9
Inorganic ion transport and metabolism	10.6	9.0	8.5	11.2
Secondary metabolite biosynthesis, transport, and catabolism	3.1	2.3	1.4	2.6
General function prediction only	32.2	26.9	28.4	38.6
Function unknown	20.5	16.4	18.1	20.0
Signal transduction mechanisms	7.8	6.2	6.5	8.4
Intracellular trafficking, secretion, and vesicular transport	2.4	2.2	2.9	2.1
Defense mechanisms	4.1	3.0	4.1	5.4
Extracellular structures	0.0	0.1	0.1	0.3
Other ORF without a category	0.0	0.0	0.0	0.0

[a] Values are in percentages.

An average of 16% of the ORFs identified in the four genome sequences have putative functions related to carbohydrate transport and metabolism. The *Lb. plantarum* sequence contains the largest number of putative phosphotransferase systems (PTS) genes, which may encode 25 complete complexes and several incomplete complexes (57). *P. pentosaceus* has a slightly reduced number of putative PTS genes compared to *Lb. plantarum*. *Leuconostoc mesenteroides* contains at least five putative phosphotransferase transport systems and several incomplete systems. *Lb. brevis* seems to be at a disadvantage regarding PTSs. There are only six putative PTS-related genes in the *Lb. brevis* genome, which may potentially encode two complete PTSs at the most. Although PTS genes are evidently present in the *L. mesenteroides*, *Lb. plantarum*, *Lb. brevis*, and *P. pentosaceus* genomes, genes for phospho-β-galactosidase have not been found; nonetheless, β-galactosidase genes are apparently present in all four genomes.

Intracellular glucose may be converted to pyruvic acid via glycolysis or to pentoses via the pentose phosphate or phosphogluconate pathway. Putative genes coding for phosphofructokinase are present in *Lb. plantarum* and *P. pentosaceus*. As expected, phosphofructokinase genes have not been found in the sequences for the heterofermentative bacteria *Lb. brevis* and *L. mesenteroides*. All four bacteria, *L. mesenteroides*, *Lb. plantarum*, *Lb. brevis*, and *P. pentosaceus*, have predicted genes for enzymes involved in catabolic functions, including glucose-6-phosphate dehydrogenase, lactonase, 6-phosphogluconate dehydrogenase, ribose-5-phosphate isomerase, ribulose-phosphate 3-epimerase, and phosphopentose

isomerase. Putative genes for phosphoketolases, which catalyze the initial step in the conversion of D-xylulose-5-phosphate to ethanol and lactic acid, have been found in *Lb. plantarum*, *L. mesenteroides*, and *P. pentosaceus*. These genes have assigned functions based on sequence similarity; therefore, further research will be needed to determine which genes are expressed.

Predicted gene products related to pyruvic acid catabolism in *L. mesenteroides*, *Lb. plantarum*, *Lb. brevis*, and *P. pentosaceus* are noted in Table 36.2. These data suggest that *Lb. plantarum* and *L. mesenteroides* contain more pyruvate-dissipating enzymes than *Lb. brevis* and *P. pentosaceus*. The absence of an oxaloacetate decarboxylase in these four LAB was unexpected. Putative genes for lactate dehydrogenases are present in all four bacteria.

As expected, none of the predominant LAB in fermented vegetables have predicted genes for a complete citric acid cycle. However, putative genes coding for enzymes involved in the citric acid cycle reductive route are present in most of the available genomes. Additionally, all four microorganisms have putative genes for the malolactic enzyme.

The genome sequences of *L. mesenteroides*, *Lb. plantarum*, *Lb. brevis*, and *P. pentosaceus* contain multiple copies of the rRNA clusters, which display minimal or no intracellular polymorphism. The 16S, 5S, and 23S rRNA sequence copies in each bacterium have 99 or 100% identity over their entire length. The *L. mesenteroides* genome sequence contains four putative rRNA clusters, which are located close to the origin of replication. The *Lb. plantarum*, *Lb. brevis*, and *P. pentosaceus* genome

Table 36.2 Pyruvic-acid-dissipating enzymes present in the predominant LAB in fermented vegetables[a]

Pyruvate catabolism-related enzymes	*Lb. brevis* ATCC 367	*L. mesenteroides* ATCC 8293	*P. pentosaceus* ATCC 25745	*Lb. plantarum* WCFS1
Pyruvate oxidase	X	X	X	X
Pyruvate dehydrogenase	X	X	X	X
Pyruvate formate lyase		X	X	X
Acetolactate synthase	X	X		X
Lactate dehydrogenases	X	X	X	X
Hydroxyisocaproate dehydrogenase		X		X
Malic enzyme (E.C. 1.1.1.38)				X
Malate dehydrogenase (E.C. 1.1.1.40)	X	X	X	
Pyruvate kinase	X	X	X	X
Oxaloacetate decarboxylate				
Pyruvate carboxylase		X	X	X

[a] X, enzyme is present.

sequences contain five putative rRNA clusters distributed around the genome. The *Lb. brevis*, *Lb. plantarum*, and *P. pentosaceus* 16S, 5S, and 23S rRNA sequences are at least 95% identical to each other.

With the exception of the cysteine tRNA, most tRNAs are present in the four bacterial genomes in multiple copies. Ten amino acids (asparagine, aspartate, cysteine, histidine, isoleucine, methionine, phenylalanine, trypsine, tyrosine, and valine) seem to be uniquely encoded by a single codon in these bacteria. The strains analyzed contained several plasmids. *Lb. plantarum* WCFS1 contains two small plasmids of approximately 2,000 bp and a larger plasmid of about 36,000 bp (57). Functions assigned to the ORFs in these plasmids include conjugal plasmid transfer. *L. mesenteroides* ATCC 8293 contains a plasmid of approximately 37,000 bp, which codes for several mobile genetic elements. Although plasmids encoding bacteriocin production, lactose utilization, and citric acid utilization related genes have been isolated from several *Leuconostoc* species, none of these functions appear to be present in the sequenced plasmid. Only the *P. pentosaceus* genome sequence data do not reveal the presence of plasmids. The overall G+C content of these plasmids tends to be lower than the 40% chromosomal G+C content. Most of the ORFs identified in these plasmids have no assigned functions.

The recent advances in genomics, molecular microbial ecology, analytical biochemistry, plant breeding, and fermentation technology suggest a bright future for vegetable fermentation science and applications, enhancements of existing products, and new processing techniques. Principal areas for the application of this technology may include the development of commercial low-salt fermentations to reduce salt wastes. Current industrial fermentation practices are in large part based on traditional practices, which have been adapted to a larger scale. The development of low-salt fermentations and storage of fermented vegetables for commercial use present significant technological hurdles, including the potential need for starter cultures (and the impact of bacteriophage on starter cultures) and for new product handling equipment. Future products may have nutritional properties superior to those of current products and incorporate nontraditional vegetables. The reasons for developing these products will be the same as they have been for centuries. Fermented vegetable products are microbiologically safe and nutritious. They have appealing sensory characteristics and can be conveniently stored for extended periods without refrigeration.

Paper no. FSR06-07 of the Journal Series of the Department of Food Science, North Carolina State University, Raleigh, N.C. 27695-7624. Mention of a trademark or proprietary product does not constitute a guarantee or warranty of the product by the U.S. Department of Agriculture or North Carolina Agricultural Research Service, nor does it imply approval to the exclusion of other products that may be suitable.

References

1. **Baumann, A.** 12 September 1911. Pickling green cucumbers. U.S. patent. 1,003,320.

2. **Bell, T. A.** 1951. Pectolytic enzyme activity in various parts of the cucumber plant and fruit. *Bot. Gaz.* 113(2):216–221.

3. **Bell, T. A., and J. L. Etchells.** 1952. Sugar and acid tolerance of spoilage yeasts from sweet-cucumber pickles. *Food Technol.* 6:468–472.

4. **Bell, T. A., J. L. Etchells, and I. D. Jones.** 1950. Softening of commercial cucumber salt-stock in relation to polygalacturonase activity. *Food Technol.* 4:157–163.

5. **Bell, T. A., J. L. Etchells, and I. D. Jones.** 1951. Pectinesterase in the cucumber. *Arch. Biochem. Biophys.* 31:431–441.

6. **Bjornsdottir, K., F. Breidt, Jr., and R. F. McFeeters.** 2006. Protective effects of organic acids on survival of *Escherichia coli* O157:H7. *Appl. Environ. Microbiol.* 72:660–664.

7. **Bolotin, A., P. Wincker, S. Mauger, O. Jaillon, K. Malarme, J. Weissenbach, S. D. Ehrlich, and A. Sorokin.** 2001. The complete genome sequence of the lactic acid bacterium *Lactococcus lactis* ssp. *lactis* IL1403. *Gen. Res.* 11:731–753.

8. **Breidt, F., Jr.** 2004. A genomic study of *Leuconostoc mesenteroides* and the molecular ecology of sauerkraut fermentations. *J. Food Sci.* 69:30–32.

9. **Breidt, F., J. S. Hayes, and R. F. McFeeters.** 2004. The independent effects of acetic acid and pH on the survival of *Escherichia coli* O157:H7 in simulated acidified pickle products. *J. Food Prot.* 67:12–18.

10. **Buescher, R. W., and C. Burgin.** 1992. Diffusion plate assay for measurement of polygalacturonase activities in pickle brines. *J. Food Biochem.* 16:59–68.

11. **Buescher, R., and C. Hamilton.** 2000. Protection of cucumber pickle quality by CaNa2EDTA. *J. Food Qual.* 23:429–441.

12. **Buescher, R., and C. Hamilton.** 2002. Adsorption of polygalacturonase from recycled cucumber pickle brines by Pure-Flo B80 clay. *J. Food Biochem.* 26:153–156.

13. **Ciska, E., and D. R. Pathak.** 2004. Glucosinolate derivatives in stored fermented cabbage. *J. Agric. Food Chem.* 52:7938–7943.

14. **Costilow, R. N., C. L. Bedford, D. Mingus, and D. Black.** 1977. Purging of natural salt-stock pickle fermentations to reduce bloater damage. *J. Food Sci.* 42:234–240.

15. **Costilow, R. N., and F. W. Fabian.** 1953. Availability of essential vitamins and amino acids for *Lactobacillus plantarum* in cucumber fermentations. *Appl. Microbiol.* 1:320–326.

16. **Costilow, R. N., and F. W. Fabian.** 1953. Effect of various microorganisms on the vitamin and amino acid content of cucumber brines. *Appl. Microbiol.* 1:327–329.

17. **Costilow, R. N., and F. W. Fabian.** 1953. Microbiological studies of cucumber fermentations. *Appl. Microbiol.* 1:314–319.

18. Costilow, R. N., K. Gates, and C. L. Bedford. 1981. Air purging of commercial salt-stock pickle fermentations. *J. Food Sci.* **46:**278–282.

19. Costilow, R. N., and M. Uebersax. 1982. Effects of various treatments on the quality of salt-stock pickles from commercial fermentations purged with air. *J. Food Sci.* **47:**1866–1868.

20. Cruess, W. V. 26 February 1918. Pickling olives. U.S. patent 1,257,584.

21. Cruess, W. V., and J. R. Zion. 1919. Olive investigations. *USDA Exp. Sta. Rec.* **42:**805–819.

22. Daeschel, M. A., R. F. McFeeters, H. P. Fleming, T. R. Klaenhammer, and R. B. Sanozky. 1984. Mutation and selection of *Lactobacillus plantarum* strains that do not produce carbon dioxide from malate. *Appl. Environ. Microbiol.* **47:**419–420.

23. Daxenbichler, M. E., C. H. VanEtten, and P. H. Williams. 1980. Glucosinolate products in commercial sauerkraut. *J. Agric. Food Chem.* **28:**809–811

24. DeVuyst, L., and E. J. Vandamme. 1994. Antimicrobial potential of lactic acid bacteria, p. 91–142. *In* L. DeVuyst and E. J. Vandamme (ed.), *Bacteriocins of Lactic Acid Bacteria.* Blackie Academic and Professional, London, United Kingdom.

25. Dougherty, D. P., F. Breidt, Jr., R. F. McFeeters, and S. R. Lubkin. 2002. Energy-based dynamic model for variable temperature batch fermentation by *Lactococcus lactis. Appl. Environ. Microbiol.* **68:**2468–2478.

26. El-Rayah-Ahmed, A., and J. M. Labavitch. 1980. Cell wall metabolism in ripening fruit. II. Changes in carbohydrate-degrading enzymes in ripening 'Bartlett' pears. *Plant Physiol.* **65:**1014–1016.

27. Etchells, J. L. 1938. Rate of heat penetration during the pasteurization of cucumber pickle. *Fruit Prod. J.* **18:**68–70.

28. Etchells, J. L. 1950. Salt-tolerant yeasts from commercial cucumber brines. *Texas Rep. Biol. Med.* **8:**103–104.

29. Etchells, J. L., T. A. Bell, H. P. Fleming, R. E. Kelling, and R. L. Thompson. 1973. Suggested procedure for the controlled fermentation of commercially brined pickling cucumbers—the use of starter cultures and reduction of carbon dioxide accumulation. *Pickle Pak Sci.* **3:**4–14.

30. Etchells, J. L., T. A. Bell, and W. R. Moore, Jr. 1976. Refrigerated dill pickles, questions and answers. *Pickle Pak Sci.* **5:**1–20.

31. Etchells, J. L., A. F. Borg, and T. A. Bell. 1961. Influence of sorbic acid on populations and species of yeasts occurring in cucumber fermentations. *Appl. Microbiol.* **9:**139–144.

32. Etchells, J. L., and I. D. Jones. 1942. Pasteurization of pickle products. *Fruit Prod.* **21:**330–332.

33. Etchells, J. L., and I. D. Jones. 1943. Bacteriological changes in cucumber fermentation. *Food Ind.* **15:**54–56.

34. Fabian, F. W. 1930. Pickle manufacture. *Fruit Prod. J. Am. Vinegar Ind.* **9:**219–220, 231.

35. Fabian, F. W., C. S. Bryan, and J. L. Etchells. 1932. *Experimental Work on Cucumber Fermentation.* Technical Bulletin no. 126. Michigan Agricultural Experiment Station, East Lansing, Mich.

36. Fernández, A. G., P. G. Garcia, and M. B. Balbuena. 1995. Olive fermentations, p. 593–630. *In* H.-J. Rehm and G. Reed (ed.), *Biotechnology.* VCH Publishers, New York, N.Y.

37. Fleming, H. P., J. L. Etchells, R. L. Thompson, and T. A. Bell. 1975. Purging of CO_2 from cucumber brines to reduce bloater damage. *J. Food Sci.* **40:**1304–1310.

38. Fleming, H. P., E. G. Humphries, O. O. Fasina, R. F. McFeeters, R. L. Thompson, and F. Breidt, Jr. 2002. Bag-in-box technology: pilot system for process-ready, fermented cucumbers. *Pickle Pak Sci.* **8:**1–8.

39. Fleming, H. P., E. G. Humphries, R. L. Thompson, and R. F. McFeeters. 2002. Bag-in-box technology: storage stability of process-ready, fermented cucumbers. *Pickle Pak Sci.* **8:**14–18.

40. Fleming, H. P., K.-H. Kyung, and F. Breidt, Jr. 1995. Vegetable fermentations, p. 629–661. *In* H.-J. Rehm and G. Reed (ed.), *Biotechnology.* VCH Publishers, New York, N.Y.

41. Fleming, H. P., L. C. McDonald, R. F. McFeeters, R. L. Thompson, and E. G. Humphries. 1995. Fermentation of cucumbers without sodium chloride. *J. Food Sci.* **60:**312–315, 319.

42. Fleming, H. P., R. F. McFeeters, and E. G. Humphries. 1988. A fermentor for study of sauerkraut fermentation. *Biotechnol. Bioeng.* **31:**189–197.

43. Fleming, H. P., R. L. Thompson, T. A. Bell, and L. H. Hontz. 1978. Controlled fermentation of sliced cucumbers. *J. Food Sci.* **43:**888–891.

44. Fleming, H. P., R. L. Thompson, J. L. Etchells, R. E. Kelling, and T. A. Bell. 1973. Bloater formation in brined cucumbers fermented by *Lactobacillus plantarum. J. Food Sci.* **38:**499–503.

45. Fleming, H. P., W. M. Walter, and J. L. Etchells. 1973. Antimicrobial properties of oleuropein and products of its hydrolysis from green olives. *Appl. Microbiol.* **26:**777–782.

46. Fred, E. B., and W. H. Peterson. 1922. The production of pink sauerkraut by yeasts. *J. Bacteriol.* **7:**257–269.

47. Fred, E. B., and W. H. Peterson. 1924. Factors determining quality in kraut. *Can. Age* **5:**161–165.

48. Gates, K., and R. N. Costilow. 1981. Factors influencing softening of salt-stock pickles in air-purged fermentations. *J. Food Sci.* **46:**274–277, 282.

49. Hasbrouck, F. F. 1910. Salting and curing cucumber pickles. *Pure Prod.* **6:**509–514.

50. Heredia, A., R. Guillén, A. Jiménez, and J. Fernández-Bolanos. 1993. Activity of glycosidases during development and ripening of olive fruit. *Z. Lebensm. Unters. Forsch.* **196:**147–151.

51. Hilts, R. W., and R. S. Hollingshead. 1920. A Chemical Study of the Ripening and Pickling of California Olives. *USDA Bulletin* no. 803. USDA, Washington, D.C.

52. Hudson, J. M., and R. W. Buescher. 1986. Relationship between degree of pectin methylation and tissue firmness of cucumber pickles. *J. Food Sci.* **51:**138–140, 149.

53. Johanningsmeier, S. D., H. P. Fleming, R. L. Thompson, and R. F. McFeeters. 2005. Chemical and sensory properties of sauerkraut produced with *Leuconostoc mesenteroides* starter cultures of differing malolactic phenotypes. *J. Food Sci.* **70:**S343–S349.

54. Jones, I. D., J. L. Etchells, M. K. Veldhuis, and O. Veerhoff. 1941. Pasteurization of genuine dill pickles. *Fruit Prod.* **20:**304–305, 316, 325.

55. Kalač, P., J. Spicka, M. Krizek, and T. Pelikanova. 2000. Changes in biogenic amine concentrations during sauerkraut storage. *Food Chem.* **69:**309–314.

56. Klaenhammer, T., E. Altermann, F. Arigoni, A. Bolotin, F. Breidt, J. Broadbent, R. Cano, S. Chaillou, J. Deutscher, M. Gasson, M. van de Guchte, J. Guzzo, A. Hartke, T. Hawkins, P. Hols, R. Hutkins, M. Kleerebezem, J. Kok, O. Kuipers, M. Lubbers, E. Maguin, L. McKay, D. Mills, A. Nauta, R. Overbeek, H. Pel, D. Pridmore, M. Saier, D. van Sinderen, A. Sorokin, J. Steele, D. O'Sullivan, W. de Vos, B. Weimer, M. Zagorec, and R. Siezen. 2002. Discovering lactic acid bacteria by genomics. *Antonie Leeuwenhoek* **82**:29–58.

57. Kleerebezem, M., J. Boekhorst, R. van Kranenburg, D. Molenaar, O. P. Kuipers, R. Leer, R. Tarchini, S. A. Peters, H. M. Sandbring, M. W. E. J. Fiers, W. Stickema, R. M. K. Lankhorst, P. A. Bron, S. M. Hoffer, M. N. M. Groot, R. Kerkhoven, M. de Vries, B. Ursing, W. M. de Vos, and R. J. Siezen. 2003. Complete genome sequence of *Lactobacillus plantarum* WCFS1. *Proc. Natl. Acad. Sci. USA* **100**:1990–1995.

58. Krall, S. M., and R. F. McFeeters. 1998. Pectin hydrolysis: effect of temperature, degree of methylation, pH, and calcium on hydrolysis rates. *J. Agric. Food Chem.* **46**:1311–1315.

59. LeFevre, E., and L. A. Round. 1919. Preliminary report on some haliphylic bacteria. *J. Bacteriol.* **4**:177–182.

60. Lu, Z., F. Breidt, Jr., V. Plengvidhya, and H. P. Fleming. 2003. Bacteriophage ecology in commercial sauerkraut fermentations. *Appl. Environ. Microbiol.* **69**:3192–3202.

61. Lu, Z., H. P. Fleming, and R. F. McFeeters. 2002. Effects of fruit size on fresh cucumber composition and the chemical and physical consequences of fermentation. *J. Food Sci.* **67**:2934–2939.

62. Marsili, R. T., and N. Miller. 2000. Determination of major aroma impact compounds in fermented cucumbers by solid-phase microextraction-gas chromatography-mass spectrometry-olfactometry detection. *J. Chromatogr. Sci.* **38**:307–314.

63. Maruvada, R. 2005. Evaluation of the importance of enzymatic and non-enzymatic softening in low salt cucumber fermentations. M.S. Thesis. North Carolina State University, Raleigh.

64. McDonald, L. C., H. P. Fleming, and H. M. Hassan. 1990. Acid tolerance of *Leuconostoc mesenteroides* and *Lactobacillus plantarum*. *Appl. Environ. Microbiol.* **56**: 2120–2124.

65. McFeeters, R. F., and S. A. Armstrong. 1984. Measurement of pectin methylation in plant cell walls. *Anal. Biochem.* **139**:212–217.

66. McFeeters, R. F., M. Balbuena, M. Brenes, and H. P. Fleming. 1995. Softening rates of fermented cucumber tissue: effects of pH, calcium, and temperature. *J. Food Sci.* **60**:786–788.

67. McFeeters, R. F., T. A. Bell, and H. P. Fleming. 1980. An endo-polygalacturonase in cucumber fruit. *J. Food Biochem.* **4**:1–16.

68. McFeeters, R. F., and K.-H. Chen. 1986. Utilization of electron acceptors for anaerobic mannitol metabolism by *Lactobacillus plantarum*. Compounds which serve as electron acceptors. *Food Microbiol.* **3**:73–81.

69. McFeeters, R. F., W. Coon, M. P. Palnitkar, M. Velting, and N. Fehringer. 1978. *Reuse of Fermentation Brines in the Cucumber Pickling Industry*, p. 1–115. EPA-600/2-78-207. U.S. Environmental Protection Agency, Washington, D.C.

70. McFeeters, R. F., and H. P. Fleming. 1989. Inhibition of cucumber tissue softening in acid brines by multivalent cations: inadequacy of the pectin "egg box " model to explain textural effects. *J. Agric. Food Chem.* **37**:1053–1059.

71. McFeeters, R. F., and H. P. Fleming. 1990. Effect of calcium ions on the thermodynamics of cucumber tissue softening. *J. Food Sci.* **55**:446–449.

72. McFeeters, R. F., H. P. Fleming, and R. L. Thompson. 1982. Malic acid as a source of carbon dioxide in cucumber fermentations *J. Food Sci.* **47**:1862–1865.

73. McGill, A. 1909. *Pickles*. Bulletin no. 163(9). Lab. Inland Rev. Dept., Ottawa, Canada.

74. McGill, A. 1913. *Bottled Pickles*. Bulletin no. 249(11). Inland Rev. Dept., Ottawa, Canada.

75. Meurer, P., and K. Gierschner. 1992. Occurrence and effect of indigenous and eventual microbial enzymes in lactic acid fermented vegetables. *Acta Aliment.* **21**:171–188.

76. Moret, S., D. Smela, T. Populin, and L. S. Conte. 2005. A survey on free biogenic amine content of fresh and preserved vegetables. *Food Chem.* **89**:355–361.

77. Nelson, V. E., and A. J. Beck. 1918. By-product of the fermentation of cabbage. *J. Am. Chem. Soc.* **40**:1001–1005.

78. Pederson, C. S. 1930. *Effect of Pure Culture Inoculation on the Quality and Chemical Composition of Sauerkraut*. Technical Bulletin no. 169. N.Y. State Agricultural Experiment Station, Geneva, N.Y.

79. Pederson, C. S. 1932. Floral changes in the fermentation of sauerkraut. *Zentbl. Bakteriol. Parasitenkd. Infektion. Hyg. Naturwiss. Allgemeine Landwirtschaftliche Technische Mikrobiol.* **85**(Abt. II):215–223.

80. Pederson, C. S. 1935. The effect of inoculation on the quality, chemical composition and bacterial flora of sauerkraut. *Zentbl. Bakteriol. Parasitenkd.* **92**(Abt. II):342–348.

81. Pederson, C. S., and M. N. Albury. 1954. The influence of salt and temperature on the microflora of sauerkraut fermentation. *Food Technol.* **8**:1–5.

82. Pederson, C. S., and M. N. Albury. 1969. *The Sauerkraut Fermentation*. Technical Bulletin no. 824. N.Y. State Agricultural Experiment Station, Geneva, N.Y.

83. Pederson, C. S., G. L. Mack, and W. L. Athawes. 1939. Vitamin C content of sauerkraut. *Food Res.* **4**:31–45.

84. Pederson, C. S., J. Whitcombe, and W. B. Robinson. 1956. The ascorbic acid content of sauerkraut. *Food Technol.* **10**:365–367.

85. Peterson, W. H., and E. B. Fred. 1923. An abnormal fermentation of sauerkraut. *Centr. Bakt. Parasit.* **58**(Abt. II): 199–204.

86. Peterson, W. H., E. B. Fred, and J. A. Viljoen. 1926. Variations in the chemical composition of cabbage and sauerkraut. *USDA Expt. Sta. Rec.* **54**:803.

87. Plengvidhya, V. 2003. Microbial ecology of sauerkraut fermentation and genome analysis of lactic acid bacterium *Leuconostoc mesenteroides* ATCC 8293. Ph.D. Thesis. North Carolina State University, Raleigh.

88. Plengvidhya, V., F. Breidt, Jr., and H. P. Fleming. 2004. Use of RADP-PCR as a method to follow the progress of

starter cultures in sauerkraut fermentations. *Int. J. Food Microbiol.* **93**:287–296.

89. **Potts, E. A., and H. P. Fleming.** 1979. Changes in dissolved oxygen and microflora during fermentation of aerated, brined cucumbers. *J. Food Sci.* **44**:429–434.

90. **Prescott, A. B.** 1880. Contributions from the chemical laboratory of the University of Michigan. *J. Amer. Chem. Soc.* **2**:333–340.

91. **Pressey, R., and J. K. Avants.** 1975. Cucumber polygalacturonase. *J. Food Sci.* **40**:937–939.

92. **Preuss, L. M., W. H. Peterson, and E. B. Fred.** 1928. Gas production in the making of sauerkraut. *J. Ind. Eng. Chem.* **20**:1187–1190.

93. **Priem, L. A., W. H. Peterson, and E. B. Fred.** 1927. Studies of commercial sauerkraut with special reference to changes in the bacterial flora during fermentation at low temperatures. *J. Agric. Res.* **34**:79–95.

94. **Rybaczyk-Pathak, D.** 2005. Joint association of high cabbage/sauerkraut intake at 12–13 years of age and adulthood with reduced breast cancer risk in Polish migrant women: results from the U.S. Component of the Polish Women's Health Study (PWHS), abstr. 3697. *Abstr. Am. Assoc. Cancer Res. 4th Annu. Frontiers Cancer Prevention Res.*, Baltimore, Md., November 2, 2005.

95. **Stamer, J. R., G. Hrazdina, and B. O. Stoyla.** 1973. Induction of red color formation in cabbage juice by *Lactobacillus brevis* and its relationship to pink sauerkraut. *Appl. Microbiol.* **26**:161–166.

96. **Stamer, J. R., and B. O. Stoyla.** 1978. Stability of sauerkraut packaged in plastic bags. *J. Food Prot.* **41**:525–529.

97. **Takayanagi, T., T. Okuda, and K. Yokotsuka.** 1997. Changes in glycosidase activity in grapes during development. *J. Inst. Enol. Vitic. Yamanashi Univ.* **32**:1–4.

98. **Tang, H. C. L., and R. F. McFeeters.** 1983. Relationships among cell wall constituents, calcium and texture during cucumber fermentation and storage. *J. Food Sci.* **48**: 66–70.

99. **Tolonen, M., T., T. Marianne, V. Britta, P. Juha-Matti, K. Hannu, and R. Eeva-Liisa.** 2002. Plant-derived biomolecules in fermented cabbage. *J. Agric. Food Chem.* **50**:6798–6803.

100. **Tolonen, M., S. Rajaniemi, J.-M. Pihlava, T. Johansson, P. E. J. Saris, and E.-L. Ryhanen.** 2004. Formation of nisin, plant-derived biomolecules and antimicrobial activity in starter culture fermentations of sauerkraut. *Food Microbiol.* **21**:167–179.

101. **United States Code of Federal Regulations.** 2006. *Acidified Foods.* 21 CFR Part 114. U.S. Department of Health and Human Services, Washington, D.C.

102. **Veldhuis, M. K.** 1938. The preservation of brine samples for chemical analysis. *Fruit Prod. J.* **18**:6–7.

103. **Vorbeck, M. L., F. A. Lee, L. R. Mattick, and C. S. Pederson.** 1961. Volatile flavor of sauerkraut—gas chromatographic identification of a volatile acidic off-odor. *J. Food Sci.* **26**:569–572.

104. **Zhou, A., and R. F. McFeeters.** 1998. Volatile compounds in cucumbers fermented in low-salt conditions. *J. Agric. Food Chem.* **46**:2117–2122.

105. **Zhou, A., R. F. McFeeters, and H. P. Fleming.** 2000. Development of oxidized odor and volatile aldehydes in fermented cucumber tissue exposed to oxygen. *J. Agric. Food Chem.* **48**:193–197.

Food Microbiology: Fundamentals and Frontiers, 3rd Ed.
Edited by M. P. Doyle and L. R. Beuchat
© 2007 ASM Press, Washington, D.C.

Steven C. Ricke
Irene Zabala Diaz
Jimmy T. Keeton

Fermented Meat, Poultry, and Fish Products

37

Fermented meat products are defined as meats that are deliberately inoculated during processing to ensure sufficient control of microbial activity to alter the product characteristics (8). If fresh meat is not preserved or cured in some manner, it spoils rapidly owing to the growth of indigenous gram-negative bacteria and subsequent putrefaction resulting from their metabolic activities (31). Although some manufacturers still depend upon naturally occurring microflora to ferment meat, most use starter cultures consisting of a single species or multiple-species combinations of lactic acid bacteria (LAB) and/or staphylococci that have been selected for metabolic activities especially suited for fermentation in meat ecosystems. Understanding the technological, microbiological, and biochemical processes that occur during meat, poultry, and fish fermentation is essential for ensuring safe, palatable products.

MANUFACTURE OF FERMENTED MEAT AND POULTRY PRODUCTS

Sausage Categories and Meat Fermentation

Dry and semidry sausages represent the largest category of fermented meat products, with many present-day processing practices having their origin in the Mediterranean region. Traditionally, dry sausages acquired their particular sensory characteristics from exposure to salt, indigenous gram-positive microorganisms such as LAB, staphylococci, micrococci (to be consistent with earlier literature, this group is referred to collectively as micrococci in this chapter, although phylogenic and chemotaxonomic analyses have indicated sufficient heterogeneity among individual members to warrant reclassification of *Micrococcaceae* into several distinct genera [97]), and yeasts residing on the meat, and the rapid drying conditions existing in the warm, dry Mediterranean climate. These products were heavily seasoned and stuffed into sausage casings that excluded air, which favored the growth of gram-positive bacteria. Typically, they were not smoked and preserved after fermentation and the accumulation of lactic acid and, to a lesser degree, other organic acids, carbon dioxide, and alcohols. They derived their name, sausage, from the Latin term "salsus," meaning salted. Sausage processing practices later spread to northern Europe, and by the Middle Ages, hundreds of varieties of dry and semidry sausages were manufactured across the continent. In contrast to

Steven C. Ricke, Department of Food Science, University of Arkansas, 2650 North Young Avenue, Fayetteville, AR 72704-5690. **Irene Zabala Diaz**, Departmento Experimental de Biologia, F.E.C.-L.U.Z., Av. Goajira, Bloque A-1, Maracaibo, Edo. Zulia, 4001, Venezuela. **Jimmy T. Keeton**, Department of Animal Science, Room 338, Kleberg Animal and Food Science Center, Texas A&M University, College Station, TX 77843-2471.

Table 37.1 Categories and origins of selected dry and semidry sausages[a]

Category	Origin	Description and unique characteristics
Dry sausages		
Salamis (Genoa, Milano, Siciliano)	Italy	Lean pork (coarse), some with wine or cured beef (fine), garlic, large-diameter casing (beef or hog bungs) with flax twine, some with coating of white mold
Lombardia salami	Italy	Coarse cut, high fat content, brandy added, twine wrap over casing
Cappicola	Italy	Boneless pork shoulder butt combined with red hot or sweet peppers, mildly cured
Mortadella	Italy	Finely chopped, cured beef and pork with added cubes of backfat, mildly spiced, smoked, encased in beef bladders
Pepperoni	Italy	Cured pork, some beef, cubed fat, red peppers, small-diameter casing
Chorizos	Spain or Portugal	Pork (coarse), highly spiced, hot, small-diameter casing
D'Arles	France	Similar to Italian salamis, coarse, large-diameter casings (hog bungs)
Lyons	France	All pork (fine), diced fat (fine), spices and garlic, cured, large-diameter casing
Alesandri and Alpino	United States	Similar to Italian salami
Katenrauchwurst, Dauerwurst, Plockwurst, Zerevelat	Germany	Dry sausages, smoked, air dried
Semidry sausages		
Summer sausages (cervelat, farmer cervelat)	Generic	Mildly seasoned soft cervelat, beef and pork (coarse), no garlic, cured, small-diameter casing
Holsteiner cervelat	Germany	Similar to farmer cervelat, packed in ring-shaped casing
Thuringer cervelat	Germany	Medium dry to soft, tangy flavor, mildly spiced, smoked, some beef and veal added
Gothaer cervelat	Germany	Very lean pork, finely chopped, cured, soft texture, mild seasoning
Goteborg cervelat, medwurst	Sweden	Coarse, salty, soaked in brine before being heavily smoked
Landjaeger cervelat	Switzerland	Small diameter (frankfurter size), pressed flat, smoked, flavored with garlic and caraway seeds
Teewurst, frische Mettwurst	Germany	Undried, spreadable, smoked
Lebanon bologna	United States	Smoked, coarse, large diameter, very acidic

[a] From references 5, 8, 76, 83, and 85.

the Mediterranean variety, northern European sausages were prepared during the cold winter months and stored until summer and thus were called summer sausages. These sausages contained more water than their Mediterranean counterparts and were lightly spiced, heavily smoked at cool temperatures, and less susceptible to spoilage owing to lower ambient temperatures. Summer sausages are similar to present-day semidry sausages. Examples of the most common dry and semidry sausage varieties are listed in Table 37.1 and are categorized as salamis or cervelats.

Product Categories and Compositional End-point Characteristics

Compositional characteristics of some dry and semidry sausage categories produced in the United States, as well as some processing criteria (moisture-to-protein ratios, pH limits, and ingredients), are defined by the

U.S. Department of Agriculture Food Safety Inspection Service in the *Food Standards and Labeling Policy Book* (106). For example, shelf-stable dry sausage must have a moisture-to-protein ratio (MPR) of 1.9:1 or less, unless specified otherwise. Nonrefrigerated, semidry, shelf-stable sausage must have an MPR of 3.1:1 or less and a pH of 5.0 or less, unless it is commercially sterilized or unless specified otherwise. Alternatively, nonrefrigerated, semidry, shelf-stable sausages are those that (i) are fermented to a pH of 4.5 or lower (or the pH may be as high as 4.6 if combined with a water activity [a_w] of no higher than 0.91); (ii) are in an intact form or, if sliced, are vacuum packed; (iii) have an internal brine concentration of no less than 5%; (iv) are cured with nitrite or nitrate; and (v) are smoked with wood. Examples of the compositional characteristics of dry and semidry fermented meat products are given in Table 37.2.

Table 37.2 Compositions of two types of fermented sausages

Parameter	Value (% [wt/wt])[a]	
	Summer sausage/ Thuringer cervelat	Pepperoni/hard salami
Moisture	50	30
Fat	24	39
Protein	21	21
Salt	3.4	4.2
PH	4.9	4.7
Total acidity	1.0	1.3
Yield	90	64

[a] Except pH. Adapted from reference 101.

Manufacturing Procedures and Processing Conditions

Dry and semidry sausage manufacture using starter cultures involves the following basic steps: (i) reducing the particle size of high-quality raw meat trimmings; (ii) incorporation of salt, nitrate (Europe, mostly) or nitrite (United States and Europe), glucose, spices, seasonings, and a specific inoculum selected on the basis of the incubation temperature optimum and level of lactic acid desired (various strains of lactobacilli, leuconostocs, pediococci, and streptococci [9]); (iii) uniformly blending all ingredients and further reducing the particle size; (iv) vacuum stuffing the meat into a semipermeable casing to minimize the presence of oxygen; (v) incubation (ripening) at or near the temperature optimum of the starter culture until a specific pH end point is achieved or until carbohydrate utilization is complete; (vi) heating (usually, but product dependent) of the product to inactivate the inoculum and ensure pathogen destruction; and (vii) drying (aging) the product to the required MPR end point. A generic scheme for manufacturing dry and semidry sausages using starter cultures is shown in Table 37.3.

Factors Affecting Color, Texture, Flavor, and Appearance of Fermented Meats

Raw Meat Tissues

Bacterial contamination of carcasses occurs during the slaughter process, with most microorganisms originating from the hide, skin, and/or intestinal contents. These include predominantly gram-negative bacteria, such as enteric Enterobacteriaceae (including Escherichia coli and Salmonella), Pseudomonas, gram-positive LAB, and staphylococci associated with humans, animals, and the environment (9). Fresh or frozen raw meats to be used

for fermented sausages should be chilled to <4.5°C, have small microbial populations, be free of physical and chemical defects, and meet the compositional specifications for the product being manufactured. The predominant bacteria that develop in meats not held under vacuum are typically gram-negative, oxidase-positive, aerophilic rods composed of psychrotrophic pseudomonads along with psychrotrophic Enterobacteriaceae (61). Spoilage is usually associated with gram-negative, proteolytic bacteria that produce putrid, "rotten egg-like" odors and flavors (9). Small numbers of salt-tolerant LAB and other gram-positive bacteria are initially present in meat and become the dominant microflora if oxygen is excluded, as in the case of vacuum-packaged or vacuum-encased meat products. Initial populations of LAB may range from 3 to 4 log CFU/g, but they increase during fermentation to 7 to 8 log CFU/g. Micrococcaceae and Staphylococcaceae (containing Micrococcus and Staphylococcus spp., respectively) also occur in fermented meats and tend to decline as the pH decreases. Some strains are lipolytic and proteolytic and contribute flavor as a consequence of breakdown products. Micrococci and staphylococci reduce nitrate to nitrite to generate nitric oxide, which reacts with myoglobin to produce the characteristic cured color of fermented meats (9). Some LAB cause greening (Lactobacillus viridescens) due to hydrogen peroxide, while others cause gas production (Lactobacillus brevis and Leuconostoc mesenteroides) and souring (Brochrothrix), with consequent off flavors and odors.

Pathogenic bacteria pose a serious health threat if not controlled on raw materials and finished fermented products. Organisms that are of greatest concern include Staphylococcus aureus, E. coli O157:H7, Salmonella, Listeria monocytogenes, Campylobacter, and the nematode Trichinella spiralis (9). S. aureus produces a heat-stable enterotoxin when cell numbers reach 6 log CFU/g. Thus, its growth must be controlled early during processing, using proper sanitation and starter culture dominance to reduce the pH to <5.3 as rapidly as possible. E. coli O157:H7 is somewhat acid tolerant but is inactivated by reducing the pH quickly, followed by heat treatment. Nonheated products rely on reductions in a_w, combined with salt and temperature. Salmonellae are acid sensitive, heat sensitive, and competitively inhibited by starter cultures; however, they become more heat resistant with drying. Moist heat early during processing, combined with good sanitation, is effective for controlling Salmonella. L. monocytogenes can grow at refrigeration temperatures, but it is not tolerant of acidic conditions. Processing procedures that control other pathogens can also control L. monocytogenes.

Table 37.3 Generic manufacturing scheme for dry and semidry fermented sausages with starter cultures[a]

Processing parameter	Procedure and conditions		
	Semidry sausage	Dry sausage (North America)	Dry sausage (Europe)
Meat tissue selection	Fresh/frozen meats with low microbial populations, no discoloration, no off odors, limited age, no DFD tissue, trimmed free of blood clots, glands, sinews, gristle, and bruises, refrigerated to <4.5°C (<40°F)	Same as semidry	Same as semidry
Comminution, grinding, blending ingredients; inoculum level at ca. 10^7 CFU/g	Fresh/tempered −3°C (26°F) meats, coarse grind/chop/mince lean (1/4 to 1/2 in., 6.35 to 12.7 mm) and fat (1/2 to 1 in., 12.7 to 25.4 mm) meats separately, combine to a specified fat end point; blend with seasonings and cure ingredients to uniformly distribute ingredients; avoid overmixing and excessive protein extraction or fat smearing	Same as semidry	Same as semidry
Ingredients used Salt (2.5 to 3%) Glucose (0.4 to 0.8%) Nitrite (<150 mg/kg) Sodium erythorbate (550 mg/kg) Antioxidants (natural or synthetic) Spices (sterilized)	Rehydrate frozen or lyophilized culture with nonchlorinated, distilled water at ambient temperature <1 h before use; tap water is usually acceptable; if antimicrobial compounds are present (i.e., chlorine) and are in excess, distilled water should be used; add inoculum (high-temperature optimum for smokehouse incubation) to meat batch; blend, but avoid excessive mixing (causes fat smearing and coating of lean particles); fine grind/chop (1/8 to 3/16 in., 3.2 to 4.8 mm) to specified particle size	Same as semidry, with the following exceptions: use high-temperature inoculum (>32.5°C [90°F]) for smokehouse incubation and low-temperature inoculum (21.3°C [70°F]) for "green" or "ripening" room incubation	Same as semidry, with the following exceptions: use nitrate (200 to 600 µg/g) alone or in combination with nitrite (nitrate used mostly in Europe; used only for Lebanon bologna and country-style hams in the United States; low-temperature fermentation (21.3°C [70°F]) is used most often
Starter culture (frozen, dehydrated)	*Lb. plantarum* (21 to 38°C [70 to 100°F]), *P. acidilactici* (32 to 46°C [90 to 115°F]), *P. pentosaceus* (21 to 38°C [0 to 100°F]), *S. carnosus* (21 to 38°C [0 to 100°F])		*Lb. sake, Lb. curvatus, S. carnosus, S. xylosus, Lb. plantarum, P. pentosaceus, P. acidilactici, Penicillium* spp., *Debaryomyces* spp. (20 to 24°C [68 to 75°F])
Vacuum encasing (stuffing)	Keep at 2°C (ca. 34°F); subject to vacuum to remove oxygen and encase in fibrous or natural casing; oxygen exclusion accelerates anaerobic fermentation and favors LAB growth and color and flavor development	Same as semidry	Same as semidry
Incubation and fermentation (ripening)	High-temperature smokehouse incubation at 32.5 to 38.1°C (90 to 100°F) and 90% relative humidity for 18+ h (dependent upon sausage diameter) to an end-point pH of <4.7; air movement, >1 m/s; smoke at the end of fermentation	Low-temperature smokehouse incubation or "ripening" room at 15 to 26°C (60 to 78°F) and 90% relative humidity for ca. 72 h to an end-point pH of <4.7; air movement, >1 m/s; smoke application at the end of fermentation if desired	Low-temperature "ripening" at 26°C (78°F) and 88% relative humidity for 3 days to an end-point pH of 4.7 to 4.8; chamber relative humidity held at 5 to 10% lower than relative humidity within the sausage, or use the following schedule: a. 22.2 to 23.9°C (72 to 75°F), 94 to 95% relative humidity for 24 h b. 20 to 22.2°C (68 to 72°F), 90 to 92% relative humidity for 24 h c. 18.3 to 20°C (65 to 68°F), 85 to 88% relative humidity for 24 h

(Continued)

Table 37.3 *(Continued)*

Processing parameter	Procedure and conditions		
	Semidry sausage	Dry sausage (North America)	Dry sausage (Europe)
Drying (aging)	Drying chamber at 12.9 to 15.7°C (55 to 60°F) and 65 to 70% relative humidity for 12+ days (dependent upon sausage diameter) to specified moisture-to-protein ratio	Drying chamber at 10 to 11.2°C (50 to 52°F) and 68 to 72% relative humidity for 21+ days (dependent upon sausage diameter) to specified moisture-to-protein ratio	Remains in the ripening room or moved to drying chamber at 20°C (68°F) and 88% relative humidity for 10 days and then 15.7°C (60°F) and 82% relative humidity for 14 days to a specified end point, or hold at 11.7 to 15.0°C (53 to 59°F) at 75 to 80% relative humidity to specified end point

[a] Adapted from references 8 and 61.

Following good manufacturing practices (3) and using established fermentation temperature/time guidelines ensure the inhibition of pathogen growth.

Postrigor pH and the residual glycogen concentration in muscle tissues influence the quality of fermented sausages. ATP levels in muscle tissues average 1 μm/g 24 h after slaughter, while pH values range from 5.5 to 5.7 for beef, 5.7 to 5.9 for pork, and 5.8 to 6.0 for poultry. Pork meat sometimes exhibits pale, soft, and exudative characteristics and has tissue pH values of 5.3 to 5.5, a wet or watery meat surface, and a pale pink color. However, these tissues may be incorporated into dry sausages at concentrations of up to 50% of the meat block without impairing sensory qualities (102). Use of 100% pale, soft, and exudative meats in sausages, however, will likely result in products with a pale yellowish cured color, poor water-holding capacity, low a_w, increased susceptibility to oxidative rancidity, and poor (soft, grainy, and noncohesive) textural characteristics. Meats exhibiting the dark, firm, and dry (DFD) condition are characterized by a dry surface, dark red color, and high pH (>6.0 to 6.2). This condition occurs more frequently in beef than in pork. DFD trimmings are not suitable for dry sausage because of their excessive water-binding capacity and potential for accelerated microbial spoilage. Dark red meat from more mature animals, however, may be desirable because of its contribution to product appearance.

The use of lamb and mutton meats in fermented sausages is limited (2), but studies indicate that acceptable sausage products can be produced when the meats are combined with appropriate seasonings and limited amounts of mutton fat (13, 113). Fat tissues from beef, lamb, and pork have high proportions of saturated fatty acids and yield products that are firmer and have a more desirable texture than products containing poultry and turkey fats, which have a predominance of polyunsaturated fatty acids. Polyunsaturated fatty acids are more susceptible to auto-oxidation and rancidity, which can lead to the development of off flavors. Thus, poultry meats may be a less desirable source of fat for fermented sausage formulations because they contain higher levels of polyunsaturated fatty acids.

Lean poultry meat, if used, is often supplemented with pork or beef meat to avoid sausages that appear too light in color and to ensure appropriate textural attributes. Poultry sausages are usually lower in fat (15%) than red meat sausages (23 to 45%), initially have a higher pH, and contain more moisture that can affect product uniformity (diameter) during drying. Incorporation of turkey thigh meat into sausages can result in darker red products owing to a higher myoglobin content in the lean tissue, but turkey and chicken have higher proportions of polyunsaturated fatty acids and moisture than do red meats. Slightly larger amounts of fermentable carbohydrate should be used in formulas containing higher-pH meats, such as poultry. Heating to >68.9°C (155°F), thermal processing as outlined by the U.S. Department of Agriculture (103, 105), or heating for a specified time-temperature combination (104) may be required to ensure the destruction of *Salmonella* in poultry tissues. Mechanically deboned poultry meat is an acceptable meat source for fermented dry sausages when limited to 10% of the meat block (40), as sausages tend to become soft when larger amounts are used.

Ingredients

Incorporation of sodium chloride, sodium or potassium nitrite and/or nitrate, glucose, and homofermentative lactic acid starter cultures (Table 37.3) in sausage formulas dramatically alters the ecology of the culture environment and the chemical characteristics of finished products (Table 37.4). Incubation of sausages at the optimum temperature for growth of LAB in a reduced-oxygen environment causes a reduction in pH by the rapid conversion of glucose to lactic acid as a consequence of the exponential growth and subsequent suppression of indigenous microflora, such as psychrotrophic pseudomonads, *Enterobacteriaceae*, and most pathogens. *Pseudomonas* species are sensitive to salt, nitrite, elevated incubation temperatures (>37°C), and reduced oxygen tension, while the competitiveness of the *Enterobacteriaceae* is restricted at reduced oxygen levels, low pH, and the presence of salt.

The glucose content of postrigor meats (4.5 to 7 μmol/g) is not sufficient to significantly reduce the pH; therefore, 0.4 to 0.8% fermentable carbohydrate in the form of glucose, corn syrup, sucrose, lactose, maltodextrins, or starch is added to sausage formulas to enable reduction of the pH to 4.6 to 5.0. About 1 oz of glucose per 100 lb of meat (0.62 g of glucose/kg of meat) is required to reduce the pH 0.1 unit. Carbohydrates such as lactose, raffinose, trehalose, dextrins, and maltodextrins used in place of glucose yield less lactic acid owing to incomplete utilization of the substrate (52). In the United States, fermented sausages having final pH values of 4.8 to 5.0 contain approximately 25 g of lactic acid per kg (dry weight), while some Italian and Hungarian salamis that are fermented with limited amounts of carbohydrate may undergo a decrease in pH of only 0.5 unit.

Nitrates and nitrites (40 to 50 μg/g [minimum]) in fermented sausages react with the heme moiety of myoglobin to facilitate the development of cured color, retard lipid oxidation, inhibit the growth of *Clostridium botulinum* (through a synergistic relationship with salt), and enhance cured flavor. In the United States, sodium chloride (2.5 to 3.0%) or a combination of sodium and potassium chloride is used with sodium or potassium nitrite (156 μg/g [maximum] ingoing into the product,

Table 37.4 Chemical characteristics of selected fermented sausage products

Category	Final pH	Lactic acid (%)	Moisture/ protein ratio	Moisture loss (%)	Moisture[a] (%)	Comments
Dry sausages[b]	5.0–5.3	0.5–1.0	<2.3:1	25–50 (30)[c]	<35	Heat processed (optional[d]); dried or aged after fermentation for moisture loss; may be smoked
Cervelat			1.9:1		32–38	
Cappicola	4.7–4.8		1.3:1		23–29	Shelf stable
German "Dauerwurst"	4.7–4.8		1.1:1		25–27	
German salami	4.5–4.8		1.6:1		34–35	
Pepperoni		0.8–1.2	1.6:1	35	25–32	
Italian salami, hard or dry	4.9		1.9:1	30	32–38	
Genoa salami	4.9	0.79	2.3:1	28	33–39	
Thuringer, dry		1.0	2.3:1	28	46–50	
Semidry sausages[b]	4.7–5.1	0.5–1.3	>2.3:1–3.7:1	8–15 (15)	45–50	Heat processed[d]; typically smoked; packaged after processing and chilling
Lebanon bologna	4.7	1.0–1.3	2.5:1	10–15	56–62	Keep refrigerated
Cervelat, soft			2.6:1	10–15		
Salami, soft			2.3:1–3.7:1	10–15	41–51	
Summer sausage	<5.0	1.0	3.1:1	10–15	41–52	
Thuringer, soft					46–50	
Other (for comparison)						
Dried beef			2.04:1	29		
Beef jerky			0.75:1	>50	28–30	
Air-dried sausage			2.1:1			

[a] Water activity ranges for dry and semidry sausages are <0.85 to 0.91 and 0.90 to 0.94, respectively. The European Economic Directive 77/99 requests a_w of <0.91 or pH of <4.5 for dry sausages to be shelf stable or a combined a_w and pH of <0.95 and <5.2, respectively.

[b] Data from references 1, 8, 54, 74, 82, 85, 99, and 107.

[c] Values in parentheses are averages.

[d] USDA-FSIS Title 9 CFR may be amended to require specified time-temperature heating combinations after fermentation or verification that processing conditions destroy all pathogenic microorganisms.

not to exceed 200 μg/g in the final product). Nitrite serves as the primary curing agent in fermented sausages, but sodium or potassium nitrate may be added legally to dry sausages at a maximum level of 1,718 μg/g, calculated as sodium nitrate. In Europe, sodium or potassium nitrate (200 to 600 μg/g), in combination with nitrate-reducing bacteria such as *Micrococcus varians* and *Staphylococcus carnosus*, is utilized during low-temperature fermentation (ripening) to enable metabolism by these acid-sensitive, nitrate-reducing bacteria. Sodium ascorbate is often used in combination with nitrates and nitrites in dry sausages, at concentrations of up to 550 and 600 μg/g in the United States and Europe, respectively. Ascorbates, isoascorbates, and erythorbates accelerate the reduction of nitrous acid to nitric oxide, thus enhancing color development, reducing residual nitrite, and retarding the formation of *N*-nitrosamines, a class of potent carcinogens.

Ground pepper, paprika, garlic, mace, pimento, cardamon, red pepper, and mustard are commonly used in fermented sausages, but as with all spices, they should be sterilized to avoid wild fermentations. Red pepper and mustard are known to stimulate lactic acid formation, possibly because of the available manganese in these spices, which enhances the glycolytic enzyme fructose-1,6-diphosphate aldolase (61). Garlic, rosemary, and sage contain antioxidant and antimicrobial compounds and may assist in preserving the flavor, color, and microbial shelf life of fermented sausages. Glucono-delta-lactone (GDL), a chemical acidulant, is hydrolyzed to gluconic acid, which is then converted to lactic acid and acetic acid by indigenous lactobacilli. The acidulant is sometimes used at concentrations ranging from 0.25 to 0.5%, although up to 1% is allowed in the United States. At a concentration of 0.25% GDL (40), dry sausage color and consistency are important in conjunction with a rapid decline in the pH caused by natural starter culture fermentation. At higher concentrations, GDL can inhibit the growth of lactobacilli, produce undesirable aromas, and impart a sweet flavor from the unfermented sugar.

Smoke contains phenols, carbonyls, and organic acids, which may act to preserve sausage products. Phenols are effective antioxidants and microbial inhibitors, while carbonyls contribute a desirable amber color by combining with free amino groups to form brown furfural compounds. Organic acids, such as formic, acetic, propionic, butyric, and isobutyric acids, in smoke inhibit the growth of microorganisms on the surfaces of sausages and promote coagulation of surface proteins.

Fermentations fail for various reasons, especially if a product does not reach the appropriate pH within a specified time period. Some of the most common causes include the following: starter culture not added or mishandled, nonfermentable or insufficient amount of sugar, antimicrobial agents incorporated into the formulation, antibiotic residues in meat trimmings, or processing temperature/humidity fluctuations.

MANUFACTURE OF FERMENTED FISH PRODUCTS

Fermented fish products include a variety of fish sauces, fish pastes, and fish-vegetable blends that have been salted, packed whole in layers, or ground into small particles and then fermented in their own "pickle." These products are eaten as a proteinaceous staple or condiment in Southeast Asia but are consumed as a condiment in northern Europe (14). Fish fermentation involves minimal bacterial conversion of carbohydrates to lactic acid but entails extensive tissue degradation by proteolytic and lipolytic enzymes derived from viscera and muscle tissues. Low-molecular-weight compounds from fish tissue degradation are the primary contributors to aroma and flavor characteristics of sauces. Indigenous microorganisms do contribute to aroma and flavor but are limited to species tolerant of high salt concentrations (10 to 20%) in the curing brine. Partial tissue hydrolysis is responsible for the unique textural attributes of pastes and fish-vegetable blends.

Fish Sauces

Fish sauces, such as nuoc-mam (Vietnam), patis (Philippines), nam-pla (Thailand), budu (Malaysia), nuoc-mam-nuoc (Thailand), and shottsuru (Japan), are liquids consumed as a condiment with rice and vary in color from clear brown to yellow-brown. Sauces have a predominantly salty taste and are derived from decanting or pressing fermented fish or shrimp after a 9-month to 1-year fermentation (69). Products fermented over a 1- to 2-year period have a distinctive sharp, meaty aroma and may range in protein content from 9.6 to 15.2%. Commercial fish sauce production begins with layering seine-netted fish, shrimp, or shellfish with salt in concrete vats at an approximate ratio of 3:1 (fish to salt), sealing the vats, allowing supernatant liquor to develop, and carefully decanting this liquid. Enzymatic digestion or fermentation may range from 6 months for small fish to 18 months for larger species. The first liquid removed from the fermenting fish contains an abundance of peptides, amino acids, ammonia, and volatile fatty acids and is considered the highest-quality product. Extracts may be supplemented with caramel, caramelized sugar, molasses, roasted corn, or roasted barley to enhance the color and keeping qualities. For some products, the sauce is ripened in the sun for 1 to 3 months and blended with bacterial by-products from the manufacture of monosodium glutamate.

The nitrogen content of the supernatant liquor increases as a result of proteolysis during fermentation. Initially, salt penetrates the tissues by osmosis (0 to 25 days), a protein-rich liquid develops through autolysis (80 to 120 days), and ultimately, the fish tissue is transformed into a nitrogen-containing liquid (140 to 200 days). Proteases, such as cathepsin B and trypsin-like enzymes, have been shown to increase the soluble protein content of the liquor during the first 2 months, but their activity gradually decreases through an inhibition feedback mechanism with the buildup of amino acids and polypeptides. New polypeptide formation occurs during the last fermentation stage as the level of free amino acids increases. Aseptically produced fish sauces do not have a typical aroma (15), suggesting that some microbial involvement is required for aroma development. Bacterial populations in raw fish, predominantly facultative anaerobes, have been shown to be high (2.7×10^4 CFU/g) initially but decline (2×10^3 CFU/g) over a 6- to 9-month period. *Bacillus*, *Lactococcus*, *Micrococcus*, and *Staphylococcus* have been confirmed as indigenous microflora, but as fermentation progresses, populations and the number of species decline with the changing brine environment. Crisan and Sands (23) identified *Bacillus cereus* and *Bacillus licheniformis* as the dominant bacteria in nam-pla, but after fermentation for 7 months, *B. licheniformis*, *Bacillus megaterium*, and *Bacillus subtilis* were the dominant species. *Micrococcus copoyenes*, *M. varians*, *Bacillus pumilus*, and *Candida clausenii* are other halotolerant microorganisms, i.e., those capable of growth in 10% salt brine but not in 20% brine, that have been isolated from sauces.

The characteristic aroma and flavor of fish sauce are complex and cannot be attributed to specific volatile fatty acids, peptides, or amino acids derived from bacterial fermentation alone. One theory suggests a complex interaction of enzymatic activity and oxidation during fermentation; however, some evidence exists for bacterial production of volatile fatty acids in fresh fish before salting and in the brief period following salting (16). Dougan and Howard (30) characterized nam-pla aroma as being ammoniacal-trimethylaminic, cheesy-ethanoic, and *n*-butanoic and having other aromas attributable to low-molecular-weight volatile fatty acids, meaty ketones, keto acids, and amino acids, especially glutamic acid. The flavor of fish sauce has a strong salt component, with contributions from a combination of mono-amino acids and possibly aspartic and glutamic acids. Other factors,

such as the pH of the brine, fermentation temperature, and salt concentration, also affect flavor.

Fish Paste

Fish pastes, which are more widely produced than sauces, are consumed raw or cooked as a condiment with rice and vegetables. In some countries, fish pastes may be the primary source of dietary protein for low-income families. A wide variety of fermented fish pastes are produced, and they require shorter processing periods than sauces. These include bagoong, tinabal, balbakwa (Philippines), pra-hoc (Cambodia), padec (Laos), blachan or bleachon, trassiudang (Malaysia), sidal (Pakistan), and shiokara (Japan) (14). Paste production consists of mixing cleaned, eviscerated whole or ground fish, shrimp, plankton, or squid with salt at a ratio of 3:1 (fish to salt) and then placing them in vats to ferment. Proteolytic enzymes from visceral tissue and, to some extent, bacteria break down the tissue until it attains a pasty consistency. Aging in hermetically sealed containers may follow, or the paste may be hand kneaded and then aged. Pickle (liquid exudates), which forms as a result of the osmotic differential of the brine solution and the fish tissue, is decanted, aged, and consumed as fish sauce. When pickle no longer forms, the fish paste is ready for use or aged further. Typical paste has a salty, cheese-like aroma, but other characteristics vary depending upon the method and region of production.

Bacteria do not appear to play a major role in the proteolysis of fish paste but may contribute to aroma and flavor or to spoilage. Populations of 6.5×10^3 CFU/g, representing 40 bacterial species, have been reported to occur in bagoong (20% salt), among which *Bacillus*, *Micrococcus*, and *Moraxella* species were dominant (34). Eighteen strains of bacteria, some of which were halophilic, were found in shiokara (15, 114), and *Micrococcus* species appeared to be responsible for ripening. Halophilic *Vibrio* and *Achromobacter* species were also isolated and were likely responsible for spoilage.

Proteolytic enzymes from fish viscera, stomach, pancreas, and intestine are more active than muscle enzymes and are most often responsible for the release of free amino acids and polypeptides, which are believed to undergo further microbial synthesis to yield specific flavor compounds. Some of the compounds derived from proteolytic degradation include volatile fatty acids (formic, ethanoic, propanoic, isobutanoic, and *n*-pentanoic acids), ammonia, trimethylamine, and mono- and dimethylamines. Aminobutane and 2-methylpropylamine are thought to result from microbial action. If carbohydrates are added to the raw materials and fermented to alcohol, this may suppress proteolytic and microbial activity. Bogoong is the residue of partially hydrolyzed fish or shrimp (69) having a pH of 5.2 and containing 65 to 68% moisture, 13 to 15% protein, 2% fat, 20 to 30% salt, 1.45% nitrogen, 0.18% volatile nitrogen, 0.015% trimethylamine, 0.011% hydrogen sulfide, and a 35% solids base. Pra-hoc, used in dishes such as soups, may contain as much as 24% protein and 17% salt. Shiokara, a Japanese fish or squid product with or without malted rice, is texturally between a sauce and paste and is characterized as a dark brown liquid containing lumps of solid tissue. In its final form, the squid product contains 74.2% water, 7.8% salt, 11.6% protein, and 8.7% ash.

Fermented Rice and Shrimp or Rice and Fish

Balao balao, a traditional food of the Philippines, is a cooked rice and shrimp product which is fermented at room temperature for 7 to 10 days in a 20% salt brine (69). A similar Filipino fermented product in which fish is substituted for shrimp is known as burong isda, while the Japanese version is called naresushi or funasushi. Bacterial isolates involved in the fermentation of burong have been demonstrated to be capable of starch hydrolysis and are identified (69) as having characteristics similar to those of *Lactobacillus plantarum* and *Lactobacillus coryneformis* subsp. *coryneformis*, but with the capacity to convert oligosaccharides and reducing sugars to lactic acid. An enzyme with a pH optimum of 4.0 has been isolated from these bacteria and found to hydrolyze amylose to oligosaccharides. However, during the initial stages of fermentation, other bacteria may hydrolyze starch for use by LAB.

STARTER CULTURES IN MEATS

The use of starter cultures in fermented meat products is a relatively recent practice compared with their use in fermented dairy foods and alcoholic beverages (57). The rationale for the use of a starter culture in meat fermentation is similar in concept to that for the use of starter cultures in dairy products. In the United States, lactobacilli, pediococci, and staphylococci are the predominant culture microorganisms, while in Europe these genera are most often used in combination with staphylococci and *Micrococcus* spp. (9, 10, 57). Proper inoculation and incubation procedures are among the most critical steps for the production of safe, flavorful, uniform, and wholesome fermented sausages. Inoculation is accomplished by one of the following three methods: natural fermentation, which relies on indigenous microflora in the meat to serve as the inoculum; back inoculation, which involves transfer of a portion of raw meat from

a previous batch of sausage to the present batch, i.e., transfer from a "mother batch" of raw sausage; and the use of starter cultures, i.e., inoculation of unfermented meat with a pure strain or strains of LAB. Cultures come in three basic forms, namely, frozen liquid, frozen pellets, or freeze-dried form. Frozen liquid has been the most popular form in the United States, but the use of frozen pellets is increasing. Frozen pellets and dry cultures are used more outside the United States. The primary reasons for using starter cultures are to ensure an immediate dominance of lactic acid microflora to inhibit other microorganisms, especially pathogens, and to provide metabolic efficiency for the production of lactic acid or nitrate reduction (9). Use of commercial starter cultures is the predominant method of inoculation in the United States.

Development of Commercial Starter Cultures

In fresh meat, LAB are a minor component of the microflora, but when meat is packaged and stored under vacuum, the resulting microenvironment facilitates their growth (51). Commercial application of starter cultures in the cheese industry during the 1930s led investigators to identify potential starter cultures for fermented meats in the 1940s. Use of pure cultures in sausages began after several studies in the 1950s demonstrated that LAB are responsible for lactic acid production (25, 71–73). An essential requirement for starter cultures is that they can be produced and preserved in a viable and metabolically active form suitable for commercial distribution. Additional requirements of homofermentative, catalase-positive starter cultures include the production of adequate quantities of lactic acid, the ability to grow at salt concentrations of up to 6%, and the capacity to enhance the flavor of finished sausages without production of biogenic amines and slimes. In the United States, *Lactobacillus* species were first used as starter cultures for meat fermentations in the temperature range of 20 to 25°C. Original starter cultures were logically derived from the predominant microflora of fermented meat products, but when attempts were made to preserve these cultures via lyophilization, as was routinely done with dairy starter cultures, these strains invariably died. Deibel et al. (26) were able to surmount this problem by using *Pediococcus cerevisiae* (now called *Pediococcus acidilactici*), the first commercially available meat starter culture. Although *P. cerevisiae* was not a predominant bacterium in naturally fermented meats, it survived lyophilization, possessed characteristic lactic acid fermentation properties, had a higher optimal growth temperature, and was tolerant of salt concentrations of up to at least 6.5% (8,

10, 26). When Deibel et al. (26) developed a lyophilization procedure for maintaining starter cultures consisting of pediococci, it was considered the best method for distributing a viable culture in a reliable and economical manner (12). Eventually, lyophilization proved commercially problematic because rehydration procedures were time-consuming and introduced unacceptable variation (8). Also, lyophilized cultures exhibited inordinately long lag phases, thus lengthening fermentation times. Implementation of frozen culture technology in the late 1960s, along with improvements in conditions for storing, handling, and shipping, led to commercial acceptance of frozen culture concentrates (12, 32). This approach eliminated the need for a rehydration step and provided larger numbers of viable cells than did lyophilization (80). Widespread use of pure starter culture strains in the United States occurred during the late 1970s and was a consequence of the development of efficient, high-volume dry sausage production technologies that required short ripening times and produced consistently uniform products with low defect levels.

Foodborne illness outbreaks associated with coagulase-positive staphylococci also focused attention on the need to control the fermentation process and ensure the production of safe products. In addition to solving these problems, the introduction of frozen culture concentrates also renewed interest in lactobacilli as starter cultures (8, 11). In 1974, *Lb. plantarum* was patented as a starter culture to be used alone or in combination with *P. acidilactici* for dry and semidry sausages (8, 33).

Currently, the predominant genera of bacteria used either singly or as mixed starter cultures in the United States and Europe are *Pediococcus*, *Lactobacillus*, *Micrococcus*, and *Staphylococcus* (8, 41). These cultures are available either as frozen liquid, frozen pellets, or a freeze-dried form or in a low-temperature stabilized liquid form (syrup). The frozen liquid and pellet forms are used most often in the United States, while pellets and dry cultures are used more worldwide (9). *P. acidilactici*, *Pediococcus pentosaceus*, *Lb. plantarum*, and *S. carnosus* (formerly known as *M. varians* or *Micrococcus halobius*) (9, 24) are favored in the United States for their rapid and nearly complete (>90%) conversion of glucose to lactic acid (pH 4.6 to 5.1) at high temperatures (ca. 32°C). Culture type selection in the United States is based on fermentation temperature and the final pH of the product. The final desired pH of fermented meat products depends on the "activity" of the culture, which generally refers to the ability of a culture to reduce the pH of the product under a set of defined conditions, especially the desired temperature of the fermentation. The rate of pH decline increases with increasing culture

activity. Highly active meat starter cultures are able to decrease the pH in a designed meat system from 5.6 to 4.8 within 8 h of fermentation. *P. acidilactici* is used in high-temperature fermentation (35 to 46.1°C) and enables rapid pH decline. For low-temperature fermentation (21 to 35°C) and rapid pH decline, *P. pentosaceus* is preferred, while for low temperatures (21 to 35°C) and slow pH decline (slow fermentation), *Lb. plantarum* strains are preferred. In addition, selected *Staphylococcus* spp. are added for color and flavor development. To improve the textural properties of fermented sausages, control pathogens, and expand the temperature range for fermentation, mixtures of meat starter cultures can be used.

In Europe, less fermentative (pH 5.2 to 5.6) bacteria, e.g., *Staphylococcus xylosus*, *S. carnosus*, *Staphylococcus simulans*, *Staphylococcus saprophyticus*, and to a lesser extent, *Micrococcus* (64), which have a lower growth temperature optimum (<24°C), are preferred for flavor development and red color enhancement. Micrococci, which are commonly employed as starter cultures in the manufacture of European dry sausage, reduce nitrate and nitrite via reductase enzymes. Chemical reduction of nitrate/nitrite to nitric oxide via nitrate and nitrite reductase, followed by reaction with the singular heme moiety of myoglobin, forms dinitrosylhemochrome, which develops into a characteristic pink cure color when heated (75). Some lactobacilli (*Lactobacillus delbrueckii* subsp. *lactis*, *Lactobacillus sake*, *Lactobacillus farciminis*, *Lb. brevis*, *Lactobacillus buchneri*, and *Lactobacillus suebicus*) reduce nitrite to nitric oxide in vitro (112). Red color development in meat can also be attributed to bacterial consumption of oxygen. Increased oxygen consumption by rapidly growing facultative anaerobes such as lactobacilli could reduce oxygen tension on meat surfaces. *Lactobacillus fermentum* forms a physical barrier that would limit access of oxygen to the meat surface underneath. In theory, bacteria initially consume oxygen and reduce the oxygen pressure to levels that allow for metmyoglobin formation (4). However, further consumption of oxygen establishes a low-oxygen environment and allows for metmyoglobin reduction to bright red myoglobin derivatives. Bacterial metabolites or intracellular components from bacterial cells are presumed responsible for metmyoglobin conversion, but the exact mechanism of oxymyoglobin oxidation is not fully understood. *P. pentosaceus* and *P. acidilactici* have temperature optima of 32 and 35°C, respectively, while lactobacilli have lower optima, in the range of 21 to 24°C. Rapidly fermented products, such as beef sticks and pepperoni, utilize *P. acidilactici* and *P. pentosaceus* alone or in combination with *Lactobacillus* species or *Micrococcus* species, which permits the use of elevated fermentation temperatures ranging from 32 to 46°C. *Lactobacillus curvatus*, *Lb. sake*, *Lb. plantarum*, *P. pentosaceus*, and *P. acidilactici* produce bacteriocins, which may find broader application in starter cultures for meat fermentations in the future.

Characteristics of Commercial Meat Starter Cultures

Bacterial starter cultures for meat and poultry products in the United States have been selected on the basis of being homofermentative, capable of rapidly converting glucose or sucrose to DL-lactic acid anaerobically, with sustained growth to pH 4.5, tolerant of salt brines (NaCl and KCl) up to 6%, and capable of growth in the presence of sodium or potassium nitrate or nitrite in the range of 600 µg/g (nitrate) to 150 µg/g (nitrite). These cultures are aciduric, capable of growth in the range of 21 to 43°C, nonproteolytic, nonlipolytic, inactivated at 60.5 to 63.2°C, resistant to phage infection and mutation, and able to outgrow and/or suppress pathogens. Specific advantages favoring the use of commercial starter cultures over natural fermentations, i.e., back inoculation, are consistency of the inoculum, reduced risk of bacterial cross-contamination, uniformity of lactic acid development, production of desirable flavor components, predictability of the pH end point (as regulated by the carbohydrate level and incubation temperature), and reduced risk of proteolytic and pathogenic bacterial outgrowth. Other benefits include the acceleration of fermentation to increase commercial plant throughput and the reduction of product defects, such as off flavors, lack of tangy flavor, excessive softness, crumbly texture, gas pockets, and pinholes, all of which can be attributed to heterofermentative bacteria.

An inoculum population of 7 log CFU/g of raw product is sufficient for rapid lactic acid development within 6 to 18 h under controlled temperature, air flow, and relative humidity conditions (9), but the incubation time is also dependent on the carbohydrate type and concentration, spice composition, exclusion of oxygen, and product diameter or thickness. In the United States, commercial processors inoculate meats with homofermentative, gram-positive bacteria that have been isolated from and adapted to specific meat products. These inocula (starter cultures) may consist of a single species of *Lb. plantarum*, *Lactobacillus pentosus*, *Lb. sake*, *Lb. curvatus*, *M. varians*, *P. acidilactici*, or *P. pentosaceus* or, more commonly, a combination of these bacteria. Use of starter cultures ensures the dominance of desirable bacteria, production of acids consisting of >90% lactic acid, and modulation of fermentation based on combinations of inoculum level, glucose concentration, and

incubation time/temperature intervals (8, 9). Thus, the combined effects of low pH, increased acidity, concomitant loss of moisture during drying, reduction of a_w, concentration of curing salts, such as sodium chloride and sodium nitrite, bacterial inhibition of spoilage or pathogenic microorganisms, and heat processing (if applied) preserve fermented meat and poultry products against spoilage by inactivating indigenous tissue and bacterial enzymes.

Mold and yeast starter cultures are not widely used in the United States, with the exception of dry sausages produced in the San Francisco area. White or gray molds are typical on casing surfaces of sausages produced in Hungary, Italy, Spain, Greece, Yugoslavia, Romania, Slovakia, and the Czech Republic. Their effect is primarily cosmetic, but it has been reported that the mycelial coat can reduce moisture loss and facilitate uniform drying (8). Catalase produced by molds may serve as an antioxidant by reacting with surface oxygen to prevent it from entering the product, while nitrate reductase promotes the development of a red surface (64). Green mold on the surface of sausage is typically the result of excessive humidity and sporulation, but *Penicillium chrysogenum* on French sausage is noted for its ability to produce a preferred ebony color. Molds are capable of decomposing lactic acid, which increases the pH and results in a milder flavor (38). *Penicillium nalgiovense* is most commonly used for this purpose, but application of *P. chrysogenum* and *Penicillium camemberti* may also be used. In Germany, only *Penicillium candidum*, *P. nalgiovense*, and *Penicillium roqueforti* are approved for application to sausages. The use of nontoxigenic molds may reduce the risk of mycotoxin production by other molds (38, 56).

Yeasts such as *Candida famata* and *Debaryomyces hansenii* are used alone or in combination with bacterial cultures to produce a powdery surface or a fruity/alcoholic aroma which may be construed as spoiled if uncontrolled (8). These cultures are added as a population of 6 log CFU/g and are characterized by a high salt tolerance and growth at low a_w, e.g., as low as 0.87 (42). *Streptomyces griseus* subsp. *butter* produces a cellar-ripened sausage aroma and enhances the color because of its nitrate reductase and catalase activities (64).

Classification of Bacterial Starter Cultures for Meat

The genera most commonly used for meat starter cultures are *Lactobacillus*, *Pediococcus*, *Micrococcus*, and *Staphylococcus* (8, 41, 99); *Pediococcus*, *Lactobacillus*, and other LAB are preferred when acid production is of primary importance. Specific strains of *Micrococcus* and coagulase-negative staphylococci are used in meat curing when lower incubation temperatures and less acid production are required for flavor development, as found in many European sausages (8, 99). Eight genera (*Lactobacillus, Leuconostoc, Pediococcus, Streptococcus, Carnobacterium* [formerly called *Lactobacillus*], *Enterococcus, Lactococcus,* and *Vagococcus* [the last three genera were formerly called *Streptococcus*]) are most commonly used as starter cultures (46). *Lactobacillus hordniae* and *Lactobacillus xylosus* are now in the genus *Lactobacillus*, while *Streptococcus diacetilactis* has been classified as a citrate-utilizing strain of *Lactococcus lactis* subsp. *lactis*.

Lactobacilli and pediococci are gram-positive, nonsporulating rods and cocci, respectively, which produce lactic acid as a major end product during the fermentation of carbohydrates (7, 98). Originally classified on the basis of morphology, glucose fermentation pathway, optimal growth temperature, and stereoisomer of lactic acid produced, this phenotypic grouping has remained largely intact but may change considerably as molecular characterization is completed. Ongoing research on oligonucleotide cataloging and 16S and 23S rRNA sequencing indicate that all gram-positive bacteria cluster in 1 of the 11 major eubacterial phyla and can be further divided into two main groups or clusters (7, 99, 111). One cluster, designated the *Actinomycetes* subdivision, is comprised of bacteria possessing a mol% G+C of DNA above 55% and would include meat fermentation genera such as *Micrococcus*, while the low-mol% G+C cluster, designated the *Clostridium* subdivision, would include meat fermentation genera such as the staphylococci (7, 98, 99, 111). LAB are thought to form a large related cluster which lies phylogenetically between strictly anaerobic species, such as the clostridia, and facultative or strictly aerobic staphylococci and bacilli (7, 48, 49, 99).

At the species level, lactobacilli appear to fall into three clusters which do not appear to correlate with current classification schemes (7, 99). Most of the species involved or associated with meat fermentations fall within the *Lactobacillus casei/Pediococcus* subgroup, comprised of obligately homofermentative, some heterofermentative, and all facultative heterofermentative lactobacilli (7). The pediococci are gram-positive cocci and are the only LAB capable of dividing in two planes. Consequently, they can appear as pairs, tetrads, or other formations (35, 80, 99). The latest taxonomic classification places pediococci as members of the gram-positive, facultatively anaerobic phylogenetic cluster, which contains 15 genera, including the staphylococci, streptococci, micrococci, and leuconostocs (89). This designation will probably change because comparison studies using 16S rRNA and nucleotide sequencing have indicated that pediococci are more closely aligned with the phylogenetic cluster that includes the lactobacilli and leuconostocs (22, 96, 109). Most of the strains designated *P. cerevisiae* that were used as meat starters have been reclassified as *P. acidilactici*, and now *P. acidilactici, Pediococcus damnosus, Pediococcus parvulus,* and *P. pentosaceus* form an evolutionary group with the *Lb. pentosus, Lb. brevis,* and *Lb. buchneri* complex (99).

The lactobacilli associated with meat fermentations are members of the genus *Lactobacillus*. They are characterized as gram-positive, non-spore-forming rods which are catalase negative on media not containing blood, are usually nonmotile, usually reduce nitrate, and ferment glucose (43, 50). The genus *Lactobacillus* currently consists of more than 50 species (99). DNA hybridization and sequencing comparisons indicate that despite the numerous species, as a whole the lactobacilli comprise a well-defined group of bacteria (43). Part of the reason for the large number of species is that the genus *Lactobacillus* has been studied exhaustively, not only from the perspective of taxonomy and identification but also from the standpoint of nutritional requirements for application in studies of biochemistry and metabolism (90). Most research on meat fermentations has emphasized *Lb. plantarum* as the predominant species (8, 99). However, attempts to identify lactobacilli isolated from meat are usually less than successful because most of the documented descriptions and schemes of identification are based on isolates from other food sources (51, 82, 87, 91). A series of investigations on atypical lactic acid streptococci isolated from fermented meats resulted in their identification as *Lb. sake* and *Lb. curvatus* (45 [as cited in reference 41]). These species outnumbered the typical lactobacilli, identified as *Lb. plantarum, Lb. brevis, Lactobacillus alimentarius, Lb. casei, Lb. farciminis, Weissella virdescens,* unspecified leuconostocs, and pediococci, 1,000-fold and were typically psychrotrophic and less acid tolerant (41, 81, 99). Efforts to accurately classify and identify these strains are becoming more important as various isolates of LAB become more commonly used as starter cultures.

Starter Culture Metabolism

The term "fermentation" was first defined by Pasteur as life in environments devoid of oxygen. However, meat fermentations are more accurately defined as bacterial or microbial metabolic processes in which carbohydrates and related compounds are oxidized, with the release of energy, in the absence of oxygen as a final electron acceptor. Thus, by definition, fermentative microorganisms cannot use oxygen as a terminal electron acceptor to generate ATP and must either conduct anaerobic respiration, using more electronegative electron acceptors (carbon dioxide, sulfate, nitrate, or fumarate), or ferment intermediate metabolites formed from the substrate as electron acceptors (37). The bacteria responsible for fermentation are either facultative or obligate anaerobes.

As a collective group, the gram-positive acidogenic bacteria (predominantly LAB) include the genera *Lactobacillus*, *Streptococcus*, *Pediococcus*, *Leuconostoc*, *Lactococcus*, and *Enterococcus*, which can metabolize a large number of mono- and oligosaccharides, polyalcohols, aliphatic compounds, mono-, di-, and tricarboxylic acids, and some amino acids, although individual species characteristically have a limited range of carbon and energy sources (59). During fermentation of sausages, members of this group of bacteria are responsible for two basic microbiological processes that occur simultaneously and are interdependent, viz, the production of nitric oxide by nitrate- and nitrite-reducing bacteria and the decrease in pH as a result of anaerobic glycolysis. These two activities are synergistic due to the pH dependency of nitrite and nitrate reduction. The following discussion details reactions that are important in meat, poultry, and fish fermentations and is based on several excellent reviews (7, 48, 57).

Carbohydrate Fermentation Pathways

Acidogenic bacteria ferment indigenous and added carbohydrates to form primarily lactic acid. The formation of lactic acid, which is an anaerobic process, helps to create a reduced pH environment in the meat, and this in turn contributes to the development of cured meat color. The formation of lactic acid also causes coagulation of meat proteins and, in combination with drying, gives sausages their characteristic firm texture. There are three potential pathways that LAB may use to form lactic acid from carbohydrates. The homofermentative pathway yields two moles of lactic acid per mole of glucose; the heterofermentative pathway yields one mole of lactic acid, one mole of ethanol, one mole of acetic acid, and one mole of carbon dioxide per mole of glucose; and the bifidum pathway (which is not discussed at length here, since it is generally not found in the bacteria important for meat starter cultures) yields three moles of acetic acid and two moles of lactate per two moles of glucose (37). The homofermentative pathway, or glycolysis, is the primary means of generating lactic acid in meat fermentations. Both *P. acidilactici* and *P. pentosaceus* are microaerophilic and, under anaerobic growth conditions, homofermentative, yielding DL-lactic acid (35, 80). Glucose and most other monosaccharides are fermented, and unlike other pediococci, both species can ferment pentoses (35). It is probable that *P. acidilactici* and *P. pentosaceus* transport

glucose via the phosphoenolpyruvate phosphotransferase system and derive ATP using the Embden-Meyerhof pathway (glycolysis). The Embden-Meyerhof pathway features the formation of fructose-1,6-diphosphate (FDP), which is cleaved by FDP aldolase to form dihydroxyacetonephosphate and glyceraldehyde-3-phosphate. These three-carbon intermediates are converted to pyruvate, which also energetically favors ATP formation via substrate-level phosphorylation at two separate metabolic transformation steps (two ATP per glucose). Because the resulting NADH formed during glycolysis requires oxidation to regenerate NAD^+ and maintain the oxidation-reduction balance, pyruvate is reduced to lactate by an NAD^+-dependent lactate dehydrogenase. Since lactic acid is virtually the only end product, the fermentation is referred to as a homolactic fermentation (7).

Heterofermentation is also characterized by dehydrogenation steps, yielding 6-phosphogluconate, followed by decarboxylation and formation of a pentose-5-phosphate. The pentose is cleaved by phosphoketolase to form glyceraldehyde-3-phosphate, which eventually becomes lactic acid through glycolysis, and acetyl phosphate (which is required to maintain electron balance) is reduced to ethanol (7, 37). In theory, homofermentative and heterofermentative bacteria can be distinguished by the pattern of fermentation end products and the presence or absence of FDP aldolase and phosphoketolase (48). Thus, obligately homofermentative species possess a constitutive FDP aldolase and lack phosphoketolase, while the reverse is true of obligately heterofermentative species (7, 48, 50). Species of LAB used to ferment meat can be regarded essentially as facultative heterofermenters because they not only possess a constitutive FDP aldolase and, consequently, use glycolysis for hexose fermentation but also possess a pentose-inducible phosphoketolase. The type of fermentation is dependent on growth conditions. In an ecosystem such as meat, with a wide variety of complex substrates serving as sources of pentoses, organic acids, and other fermentable compounds in addition to hexoses, it is possible for LAB to be homofermentative when using hexoses but heterofermentative when using pentoses and other substrates. Lactic acid is the primary metabolite derived from both homo- and heterofermentation, but during heterofermentation the production of mixtures of lactic acid, acetic acid, ethanol, and carbon dioxide also occurs.

ATP Generation, Pyruvate Metabolism, and Energy Recycling

The most important energy requirements of the bacterial cell are macromolecular synthesis and the transport of essential solutes against a concentration gradient (7).

Fermentation that is characteristic of LAB is essentially the oxidation of a substrate to generate energy-rich intermediates that are used in turn for ATP production by substrate-level phosphorylation. Although LAB can tolerate more minimal and broad ranges of internal pHs, there still must be a substantial amount of ATP generated to keep the cytoplasmic pH above a threshold level to offset the massive external acid production (7).

Bacteria have considerable flexibility in the pathways they can use to generate ATP and regulate its distribution. In aerobic bacteria, which rely on an electron transport chain, large quantities of ATP can be generated via a proton motive force across the electron transport chain coupled to a H^+-translocating ATP synthase (7). Even though LAB do not possess ATP-generating capabilities via electron transport, lactobacilli do have some metabolic flexibility in altering pyruvate pathways and conserving energy by end-product efflux. Some LAB have alternative pathways of pyruvate metabolism other than direct reduction to lactic acid. Oxidation involves the formation of NADH from NAD^+, which requires regeneration back to NADH, and pyruvate serves as the key intermediate for this by acting as the electron acceptor (7). Pyruvate-formate lyase generates formate and acetyl Coenzyme A (CoA) from pyruvate and CoA, and the acetyl CoA can be used either as a precursor for substrate-level phosphorylation, for direct reduction to ethanol, or both. This alteration in pyruvate metabolism has been demonstrated in studies where lactobacilli have been shown to shift fermentation patterns and, concomitantly, the amount of potential ATP formed as the growth rate changes (29). In these studies, mixed acid-forming strains of lactobacilli produced less lactic acid and more acetic acid with decreasing growth rates, and for every mole of acetic acid, an extra mole of ATP (two versus one mole of ATP produced per mole of lactic acid) was potentially formed (29). During fermentation, LAB form sufficient end products in the cytoplasm to result in a very high internal-to-external cell gradient (77). It has been suggested that when a fermentation product, such as lactic acid, exceeds the electrochemical proton gradient, additional metabolic energy is gained by product efflux through carrier-mediated transport in symport with protons (63, 77, 100). Metabolic energy is conserved because a product gradient is essentially a proton gradient which can be used to directly produce ATP by proton-driven ATPase, and it helps to maintain a proton motive force without consumption of ATP (7, 77).

Nitrate/Nitrite Reduction

Cured meat derives its characteristic pink color from the reaction of myoglobin (Fe^{2+}) or metmyoglobin (Fe^{3+}) with nitric oxide to ultimately form nitric oxide myoglobin, an unstable, bright pink compound (75). Nitric oxide myoglobin is formed directly from the reaction of purple-red myoglobin with nitric oxide. Another proposed route of formation is through the oxygenation of myoglobin to bright red oxymyoglobin (Fe^{2+}) and subsequent oxidation to brown metmyoglobin (Fe^{3+}). Nitric oxide reduces metmyoglobin to gray-brown nitric oxide metmyoglobin (Fe^{2+}), which transmutates to nitric oxide myoglobin with the loss of oxygen. Nitric oxide myoglobin, when heated, forms nitrosohemochrome (Fe^{2+}), the stable pink pigment present in commercially processed cured meats. Nitrate/nitrite-reducing bacteria transform added nitrates into nitrites and eventually into more reduced forms for reaction with the dominant myoglobin or metmyoglobin pigment forms. *Micrococcaceae* possess nitrate reductase to convert nitrate to nitrite, which in turn is converted to nitric oxide via nitrous acid under acidic conditions. Nitric oxide is a strong electron donor and rapidly reacts with the heme moiety of myoglobin to form the characteristic pink meat pigment, dinitrosylhemochrome. During fermentations, some strains of LAB may produce metabolites, such as carbon dioxide, acetate, formate, succinate, and acetoin, that cause off flavors; however, nitrate and nitrite prevent the formation of certain off flavors from compounds such as formate by their inhibition of pyruvate-formate lyase activity in lactobacilli (42). The formation of nitric oxide from the reduction of nitrates and nitrites is crucial because it reduces nitrosometmyoglobin to nitrosomyoglobin and induces red color formation (57). The advantage of microbial activity is the reduction of nitrates, thus removing excess nitrate/nitrite from the meat.

Metabolic Activities Important in Commercial Starter Cultures

Strains of *P. acidilactici* were initially selected for use as starter cultures because they ferment carbohydrates rapidly to lactic acid at high temperatures and effectively lower the pH of the meat, from a range of 5.6 to 6.2 down to a range of 4.7 to 5.2 (12). Traditionally, semidry sausages (summer sausage, thuringer, and beef sticks) and some types of pepperoni are allowed to undergo greening, i.e., fermentation at 16 to 38°C, which favors the development of indigenous microflora and extends fermentation times (12). Although indigenous microflora have the advantage of imparting a unique flavor(s) and other desirable sensory properties to sausages, there is also sufficient variation in lactic acid production to be a considerable disadvantage for large-scale production of sausage. The use of starter cultures consisting of pediococci allows for the inoculation of large numbers of one microorganism into the raw meat, followed by a uniform fermentation. More importantly, meat inoculated with *P. acidilactici* can be incubated at higher temperatures (43 to 50°C), which precludes the growth of most indigenous microorganisms so that flavor and pH characteristics are predictable (12). Consequently, commercial production is enhanced since flavor development can be controlled and the fermentation process hastened.

Strains of *P. acidilactici* that have been selected and developed for use as commercial starter cultures are well suited for the production of semidry sausages which are fermented and/or smoked at higher temperatures (26 to 50°C). However, when used for production of dry sausages, which require lower fermentation temperatures (15 to 27°C), these strains generally produce lactic acid at a much lower rate (12). Since *P. pentosaceus* has a lower optimum temperature than that of *P. acidilactici* for growth (28 to 32°C versus 40°C) (35), it has been promoted as an effective meat starter culture. *P. pentosaceus* is also more attractive because it has a 25% lower Arrhenius energy of activation for fermentation than that of *P. acidilactici* (79). Thus, *P. pentosaceus* has a lower optimum growth temperature and a higher capacity for rapid production of lactic acid than does *P. acidilactici*.

Metabolic Contributions of Starter Cultures to Sausage Sensory Qualities

Fermentation of sausages most often involves inoculation of ground meats with a starter culture, followed by an incubation period to allow enzymatic conversion of available carbohydrate to approximately equimolar concentrations of DL-lactic acid. Fermentable carbohydrates in the form of residual muscle glucose (0.1 mg/g) and added glucose or sucrose are the primary sources of energy available for metabolism. The decline in pH to 4.5 to 4.7 with the buildup of lactic acid results in a dramatic loss of water-holding capacity as the pH nears the isoelectric point (pI, ~5.1) of myofibrillar proteins. Product texture becomes firmer, density increases with dehydration during aging, and partial denaturation of the myofibrillar proteins occurs due to the presence of lactic acid. Thus, the tangy flavor and chewy texture of fermented sausages are consequences of the dominant fermentation metabolite, lactic acid, and dehydration.

Lactic acid is the dominant flavor component in fermented sausage, but spices, salt, sugar, sodium nitrite reaction products, smoke, and meat components, such as fatty acids, amino acids, and peptides, are also contributors to the flavor profile (76). Secondary flavor contributions from metabolites of lactobacilli and pediococci are minor in fermented sausages produced in the United States; however, natural heterofermentative lactobacilli (*Lb. brevis* and *Lb. buchneri*) from "back inoculations," combined with staphylococci and micrococci, give many European sausages their characteristic flavor as a result of volatile acids, alcohols, and carbon dioxide. *Micrococcaceae*, for example, possess lipolytic and proteolytic enzymes which generate aldehydes, ketones, and short-chain fatty acids that give characteristic aromas and flavors to sausages (61). They also produce catalase, which decomposes hydrogen peroxide. *Leuconostoc mesenteroides* and *Lb. brevis* produce ethanol, acetic acid, lactic acid, pyruvic acid, acetoin (which imparts a nutty flavor and aroma), and carbon dioxide to give an effervescent sensation, but fermentation conditions must be controlled to avoid excessive pinholes, gas pockets, and off flavors.

In natural fermentations as well as sausages inoculated with lactobacilli, *Lb. sake* and *Lb. curvatus* have been observed to be the dominant bacteria at temperatures of 25°C and below, while *Lb. plantarum* tends to dominate at higher fermentation temperatures (47). *Lb. sake*, *Lb. curvatus*, and to a lesser extent, *Lb. plantarum*, form DL-lactic acid and have flavin-dependent oxidases capable of forming hydrogen peroxide. Tolerance to peroxide may be one of the factors contributing to their dominance during fermentation. Hydrogen peroxide, if not decomposed, can induce oxidation of unsaturated fatty acids, leading to rancid flavor development, and/or oxidize the heme component of myoglobin, leading to fading or the formation of green, yellow, gray, or other off-color pigments (64). *Lb. curvatus* does not possess pseudocatalase or true catalase and may permit the accumulation of hydrogen peroxide. *Lb. sake*, *Lb. plantarum*, and pediococci, on the other hand, exhibit catalase activity and can therefore decompose hydrogen peroxide formed during fermentation. Sensory analysis of unspiced fermented dry sausages produced using various starter culture combinations (17) has revealed that sausage flavor is influenced by culture composition. Berdague et al. (17) reported that the butter odor of dry sausages was largely dependent upon the degradation of sugars by way of pyruvic acid and that curing and rancid odors were correlated with compounds resulting from lipid oxidation. Their work indicated that *S. saprophyticus* and *Staphylococcus warneri* were most associated with butter odor, which in turn was correlated with the presence of acetoin, diacetyl, 1,3-butanediol, and 2,3-butanediol. Distinctive curing odors were associated with combinations of *S. carnosus* and *P. acidilactici*, *S. carnosus* and *Lb. sake*, and *S. carnosus* and *P. pentosaceus* and were correlated with 2-pentanone, 2-hexanone, 2-heptanone, and an unknown compound.

A combination of *S. saprophyticus* and *Lb. sake*, which produced the most acetic acid, was associated with fruity odors derived from esters and with a less intense rancid odor. Meat proteins are hydrolyzed by endogenous and microbial proteases to yield peptides and amino acids, which in turn are degraded to ammonia and amines, causing a slight rise in the pH. Demeyer and Samejima (27), however, suggested that the major protease activity in fermented meat is derived from the meat enzymes. Amino acid degradation products and nucleotide inosine monophosphate intensify meat flavors and contribute to the overall sausage flavor. Thus, it appears that compounds produced via endogenous proteolysis in combination with starter culture fermentation play a significant role in the nonacidic flavor and aroma of fermented sausages.

Genetics and Biotechnology of Meat Starter Cultures

To utilize biotechnological approaches with meat starter cultures, genetic transfer systems and mobile genetic elements for carrying a potentially important gene(s) must be identified. The standard approach to gene cloning in lactic acid bacteria is to develop plasmid vectors based on either indigenous cryptic plasmids or heterologous plasmids resistant to a broad range of antibiotics (36). Since Chassy et al. (21) first demonstrated the presence of plasmids in lactobacilli, significant progress has been made, particularly in elucidating genetic systems in dairy lactobacilli and *Lc. lactis* (36, 43).

Plasmids have been detected in many strains of *P. pentosaceus*, *P. cerevisiae*, and *P. acidilactici*, and some may be transferable between strains. Some properties that are characteristic of a particular species of *Pediococcus* have been shown to be plasmid linked. The production of bacteriocin and bacteriocin-like substances and the ability to ferment some sugars have been shown to be linked to or encoded by plasmids in strains of *P. cerevisiae*, *P. pentosaceus*, and *P. acidilactici*. Numerous plasmids have been isolated from meat lactobacilli, and some have been characterized functionally. Nes (70) observed that 8 of 10 strains of *Lb. plantarum* examined for the presence of extrachromosomal DNA contained one to six plasmids. Six strains containing plasmids were commonly used as starter cultures in dry sausage. Metabolic functions in *Lb. plantarum* and four atypical *Lactobacillus* species isolated from fresh meat (58) that have been shown to be plasmid linked include maltose utilization and cysteine metabolism (92). When Schillinger and Lücke (88) screened 221 strains of lactobacilli from meat and meat products for antibacterial compounds, they linked

immunity and production of a bacteriocin in a strain of *Lb. sake* with an 18-MDa plasmid by showing that cured variants no longer possessed either trait. *Lb. sake* also produces a bacteriocin, lactocin S, which inhibits species of *Lactobacillus*, *Pediococcus*, and *Leuconostoc*. Both bacteriocin production by *Lb. sake* and its immunity factor have been shown to be associated with an unstable 50-kb plasmid (66–68).

Genetic transfer has been demonstrated for several species of lactobacilli and lactococci, but limited studies have been performed with species associated directly with meat fermentations. In vivo gene transfer systems have involved either conjugation, which requires cell-to-cell contact (bacterial mating) for transfer of genetic material, or transduction (transfection or phage-mediated genetic exchange), while in vitro physiological transformation requires uptake of naked DNA by the recipient cell (44, 108). Intergenic conjugation of an antibiotic-resistant streptococcal plasmid was demonstrated in *Lb. plantarum* (110), and successful inter- and intragenic conjugation of the same plasmid has also been reported for *Lb. plantarum* (93) as well as a strain of *Lactobacillus* isolated from fermented sausage (84). Conjugative transfer of an *Escherichia-Streptococcus* nonconjugative shuttle plasmid has also been achieved in *Lb. plantarum* by cointegration (94). Further genetic analysis and manipulation of meat lactobacilli have been handicapped by a lack of natural competence and transformation procedures (43, 65). This will change with the emergence of electroporation methods, which have been used successfully to transform plasmids in *Lb. plantarum* (6, 19, 60, 78).

Although shuttle vectors have been used to construct and introduce heterologous genes into lactobacilli, they generally contain antibiotic resistance markers, which greatly facilitate the process but may not receive regulatory approval for use in foods (43). Furthermore, considerable instability and loss of function occur during replication because these plasmids, for the most part, belong to a group which replicates via single-stranded DNA intermediates or rolling circle replication (36, 39). This form of replication is a potential source of plasmid segregational instability. For meat lactobacilli, vectors containing replicons from *Lactobacillus*, *Staphylococcus*, or *E. coli* sequences have all been shown to contribute to this instability (55, 78). Therefore, plasmids containing a *Lactobacillus* replicon can be stabilized under selective conditions and used to express genes from multiple copies of the plasmid, but under nonselective conditions they are frequently segregationally unstable and are lost (55, 78). Consequently, efforts have focused on gene cloning strategies that enable heterologous genes to integrate in a stable fashion into the bacterial chromosome (36, 45). Recombination in chromosomes can occur as general recombination, when chromosomal DNA is transferred from one bacterium to the next via conjugation, or transduction, which necessitates extensive sequence homology between externally introduced DNA and chromosomal DNA as well as specific bacterial recombination factors (108). Recombination can also be mediated by transposable genetic elements which involve short target sequences of 10 nucleotide base pairs or less and occur independent of bacterial host function (108). Transposons have been introduced into *Lb. curvatus* (53) and *Lb. plantarum* (6), and integration of plasmids by single-crossover events within regions of homology can be accomplished by constructing plasmid suicide vectors which lack the ability to replicate in lactic acid bacteria (36). Scheirlinck et al. (86) used a suicide plasmid to integrate *Bacillus stearothermophilus* α-amylase genes and *Clostridium thermocellum* endoglucanase genes into *Lb. plantarum*. This homologous insertion into the chromosome was made possible by generating plasmid

constructs containing an unknown portion of a *Lactobacillus* sequence as part of the suicide vector to facilitate site-specific integration. Complete replacement of a gene(s) can be achieved with homologous double-crossover recombination events. An alternative to using homologous sequences is to locate insertion sequences already present on the chromosome that can transpose chromosomal DNA to plasmids and to use these in the construction of integrative vectors (44). Only a few such elements have been identified in meat lactobacilli, and only one, IS*1163* from *Lb. sake*, has been isolated, sequenced, and described in any detail (95).

Development of genetically engineered strains of lactobacilli that have large-scale use in fermented meats or other foods has remained elusive. Part of the problem is that transfer of structural genes to a new host does not guarantee that the genes will be expressed (44). Optimal expression of cloned genes requires not just cloning of the structural gene but also efficient promoters, ribosome-binding sites, and termination sites, all of which must be identified, isolated, cloned, and sequenced (44). Integration systems that have been used do not allow for high enough levels of heterologous gene expression for

Table 37.5 Research needs for meat starter cultures[a]

Improving current starter culture technology
Substrate specificity of starter culture microorganisms
Immobilized cells for meat fermentation
Increased rate of lactic acid production (reduction of lag phase)
Control of pathogens without heating
Use of starter cultures in combination with acidulants, e.g., GDL
Identification and testing of nitrite substitutes (development of cured color with *C. botulinum* outgrowth protection)
New cryoprotectant (antifreeze) solutions for frozen starter cultures
Development of new pediocins that are more soluble and are not bound by fat, subject to proteolysis, or inactivated by other meat components
Production of biogenic amine-free (negative) meat starter culture
Development of meat cultures to be grown at higher salt concentrations
Production of starter cultures for pathogen and yeast control
Rapid detection and identification using molecular detection techniques
Improvement of starter cultures for higher level of nitrite reduction and better red color development
Probiotic LAB (*Lactobacillus acidophilus*) applied to meat fermentation
Use of cultures to enhance meat product nutrition and quality
Enhancement of nutritional quality by in situ production of critical dietary nutrients for selected populations
Greater utilization of nontraditional meat sources via fermentation
Accelerated curing of traditional meat products (nonnitrite hams, bacon, comminuted meats, meat snacks, natural acidification against *C. botulinum*)
Generation of natural antioxidants in situ (α-tocopherol production by starter cultures)
Production of antimycotic agents for mold inhibition
Development of starter cultures unique to fish pastes and sauces

[a] Based on comments from reference 24.

practical application to fermented meats, and they give unsatisfactory stabilization of the foreign gene (45). One approach to solving this problem essentially involves using the structural gene (*amyL*) that encodes the production and secretion of α-amylase from *B. licheniformis* as a reporter gene for the cloning of expression-secretion sequences from *Lb. plantarum* (45). The key requirement for this approach is that the *amyL* gene does not possess a specific and/or regulated promoter that is operational in any gram-positive bacterium outside its genus of origin, *Bacillus*. Thus, the high expression and secretion of over 90% of the *Bacillus* extracellular amylase in *Lb. plantarum* serve as an expression reporter to indicate that replacement of the *Bacillus* promoter by an *Lb. plantarum* promoter has taken place. The plasmid containing the silent *amyL* coding frame could subsequently be used as a probe to locate expression (transcription and translation) or expression-secretion regions on the chromosome of *Lb. plantarum* strains expressing and secreting α-amylase.

Newly developed techniques for genetic manipulations in LAB have been introduced. Inducible systems for gene expression that utilize inducible promoters linked to promoterless reporter genes have been described for *Lactococcus*. The applicability of such expression systems in other LAB has been demonstrated (28). The recent publication of the entire genome sequence of *Lc. lactis* will provide novel genetic tools for the genetic manipulation of LAB, including new insertion sequence elements, prophages, and competence genes that have been found after analysis of the entire genome sequence (18). These findings should lead to the development of procedures that will enhance the understanding of gene regulation and genetic engineering of LAB. Numerous opportunities are available for advances in meat starter culture research. Potential research areas are summarized in Table 37.5. The number of groups conducting research on genetic systems, particularly in lactobacilli, has multiplied severalfold in recent years. Earlier predictions (20, 62) for biotechnological approaches to optimize fermentations are nearing reality.

References

1. Acton, J. C. 1977. The chemistry of dry sausages. *Proc. Recip. Meat Conf.* 30:49–62.
2. Al-Sheddy, I. A., D. Y. C. Fung, and C. L. Kastner. 1995. Microbiology of fresh and restructured lamb meat: a review. *Crit. Rev. Microbiol.* 21:31–52.
3. American Meat Institute Foundation. 1997. *Good Manufacturing Practices for Fermented Dry and Semi-Dry Sausage Products*. American Meat Institute Foundation, Washington, D.C.
4. Arihara, K., H. Kushida, Y. Kondo, M. Itoh, J. B. Luchansky, and R. G. Cassens. 1993. Conversion of metmyoglobin to bright red myoglobin derivatives by *Chromobacterium violaceum*, *Kurthia* sp., and *Lactobacillus fermentum* JCM1173. *J. Food Sci.* 58:38–42.
5. Association of Food and Drug Officials. 1999. *Safe Practices for Sausage Production—Distance Learning Course Manual*, version 1. United States Department of Agriculture, Food Safety and Inspection Service, Washington, D.C. Accessed 12 October 2005. [Online.] www.aamp.com/links/documents/Sausage.pdf.
6. Aukrist, T., and I. F. Nes. 1988. Transformation of *Lactobacillus plantarum* with the plasmid pTV1 by electroporation. *FEMS Microbiol. Lett.* 52:127–132.
7. Axelsson, L. T. 1993. Lactic acid bacteria: classification and physiology, p. 127–159. *In* S. Salminen and A. von Wright (ed.), *Lactic Acid Bacteria*. Marcel Dekker, Inc., New York, N.Y.
8. Bacus, J. N. 1986. Fermented meat and poultry products. *Adv. Meat Res.* 2:123–164.
9. Bacus, J. N. 11 May 2005, posting date. *Microbiology—Shelf-Stable Dried Meats*. Accessed 12 October 2005. [Online.] www.fsis.usda.gov/PDF/FSRE_SS_5Microbiology Dried.pdf.
10. Bacus, J. N., and W. L. Brown. 1981. Use of microbial cultures: meat products. *Food Technol.* 35:74–78, 83.
11. Bacus, J. N., and W. L. Brown. 1985. The lactobacilli: meat products, p. 58–71. *In* S. E. Gilliland (ed.), *Bacterial Starter Cultures for Foods*. CRC Press, Inc., Boca Raton, Fla.
12. Bacus, J. N., and W. L. Brown. 1985. The pediococci: meat products, p. 86–95. *In* S. E. Gilliland (ed.), *Bacterial Starter Cultures for Foods*. CRC Press, Inc., Boca Raton, Fla.
13. Bartholomew, D. R., and C. I. Osuala. 1986. Acceptability of flavor, texture, and appearance of mutton processed meat products made by smoking, curing, spicing, adding starter cultures and modifying fat source. *J. Food Sci.* 51:1560–1562.
14. Beddows, C. G. 1985. Fermented fish and fish products, p. 1–39. *In* B. J. B. Wood (ed.), *Microbiology of Fermented Foods*, vol. II. Elsevier Applied Science Publishers, London, United Kingdom.
15. Beddows, C. G., A. G. Ardeshir, and W. Johari bin Daud. 1979. Biochemical changes occurring during the manufacture of budu. *J. Sci. Food Agric.* 30:1097–1103.
16. Beddows, C. G., A. G. Ardeshir, and W. Johari bin Daud. 1980. Development and origin of the volatile fatty acids in budu. *J. Sci. Food Agric.* 31:86–92.
17. Berdague, J. L., P. Monteil, M. C. Montel, and R. Talon. 1993. Effects of starter cultures on the formation of flavor compounds in dry sausage. *Meat Sci.* 35:275–287.
18. Bolotin, A., S. Mauger, K. Malarme, S. D. Ehrlich, and A. Sorokin. 1999. Low-redundancy sequencing of the entire *Lactococcus lactis* IL 1403 genome. *Antonie Leeuwenhoek* 76:27–76.
19. Bringle, F., and J.-C. Hubert. 1990. Optimized transformation by electroporation of *Lactobacillus plantarum* strains by plasmid vectors. *Appl. Microbiol. Biotechnol.* 33:664–670.
20. Chassy, B. M. 1987. Prospect for the genetic manipulation of lactobacilli. *FEMS Microbiol. Lett.* 46:297–312.

21. Chassy, B. M., E. Gibson, and A. Giuffrida. 1976. Evidence for extrachromosomal elements in *Lactobacillus*. *J. Bacteriol.* **127:**1576–1578.

22. Collins, M. D., A. M. Williams, and S. Wallbanks. 1990. The phylogeny of *Aerococcus* and *Pediococcus* as determined by 16 rRNA sequence analysis: description of *Tetragenococcus* gen. nov. *FEMS Microbiol. Lett.* **70:**255–262.

23. Crisan, E. V., and A. Sands. 1975. The microbiology of four fermented fish sauces. *Appl. Microbiol.* **29:**106.

24. Curtis, S. I. 1995. Personal communication.

25. Deibel, R. H., and C. F. Niven, Jr. 1957. *Pediococcus cerevisiae*: a starter culture for summer sausage. *Bacteriol. Proc.* **1957:**14–15.

26. Deibel, R. H., G. D. Wilson, and C. F. Niven, Jr. 1961. Microbiology of meat curing. IV. A lyophilized *Pediococcus cerevisiae* starter culture for fermented sausages. *Appl. Microbiol.* **9:**239–243.

27. Demeyer, D., and K. Samejima. 1991. Animal biotechnology and meat processing, p. 127–143. *In* L. O. Fiems, B. G. Cottyn, and D. I. Demeyer (ed.), *Animal Biotechnology and the Quality of Meat Production*. Elsevier, Amsterdam, The Netherlands.

28. de Vos, W. M. 1999. Gene expression systems for lactic acid bacteria. *Curr. Opin. Microbiol.* **2:**289–295.

29. de Vries, W., W. M. C. Kapteijn, E. G. van der Beek, and A. H. Stouthamer. 1970. Molar growth yields and fermentation balance of *Lactobacillus casei* L3 in batch cultures and continuous cultures. *J. Gen. Microbiol.* **63:**333–345.

30. Dougan, J., and G. E. Howard. 1975. Some flavouring constituents of fermented fish sauces. *J. Sci. Food Agric.* **26:**887–894.

31. Egan, A. F. 1983. Lactic acid bacteria of meat and meat products. *Antonie Leeuwenhoek* **49:**327–336.

32. Everson, C. W., W. E. Danner, and P. A. Hammes. 1970. Bacterial starter cultures in sausage products. *J. Agric. Food Chem.* **18:**570–571.

33. Everson, C. W., W. E. Danner, and P. A. Hammes. June 1974. Process for curing dry and semidry sausages. U.S. patent 3,814,817.

34. Fujii, T., S. D. Basuki, and H. Tozawa. 1980. Microbiological studies on the ageing of fish sauce; chemical composition and microflora of fish sauce produced in the Philippines. *Nippon Suissan Gakkaishi* **46:**1235–1240.

35. Garvie, E. I. 1986. Genus *Pediococcus*, p. 1075–1079. *In* P. H. A. Sneath, N. S. Mair, M. E. Sharpe, and J. G. Holt (ed.), *Bergey's Manual of Systematic Bacteriology*, vol. 2. The Williams and Wilkins Co., Baltimore, Md.

36. Gasson, M. J. 1993. Progress and potential in the biotechnology of the lactic acid bacteria. *FEMS Microbiol. Rev.* **12:**3–20.

37. Gottschalk, G. 1986. *Bacterial Metabolism*, 2nd ed. Springer-Verlag, New York, N.Y.

38. Grazia, L., P. Romano, A. Bagni, D. Roggiani, and G. Guglielmi. 1986. The role of moulds in the ripening process of salami. *Food Microbiol.* **3:**19–25.

39. Gruss, A., and S. D. Ehrlich. 1989. The family of highly interrelated single-stranded deoxyribonucleic acid plasmids. *Microbiol. Rev.* **53:**231–241.

40. Hammer, G. F. 1987. Meat processing: ripened products. *Fleischwirtschaft* **67:**71–74.

41. Hammes, W. P., A. Bantleon, and S. Min. 1990. Lactic acid bacteria in meat fermentation. *FEMS Microbiol. Lett.* **87:**165–174.

42. Hammes, W. P., and H. J. Knauf. 1994. Starters in the processing of meat products. *Meat Sci.* **36:**155–168.

43. Hammes, W. P., N. Weiss, and W. Holzapfel. 1992. The genera *Lactobacillus* and *Carnobacterium*, p. 1535–1594. *In* A. Balows, H. G. Trüper, M. Dworkin, W. Harder, and K.-H. Schleifer (ed.), *The Prokaryotes—a Handbook on the Biology of Bacteria: Ecophysiology, Isolation, Identification, Applications*, 2nd ed. Springer-Verlag, New York, N.Y.

44. Harlander, S. K. 1992. Genetic improvement of microbial starter cultures, p. 20–26. *In Applications of Biotechnology to Traditional Fermented Foods*. National Academy Press, Washington, D.C.

45. Hols, P., T. Ferain, D. Garmyn, N. Bernard, and J. Delcour. 1994. Use of homologous expression-secretion signals and vector-free stable chromosomal integration in engineering of *Lactobacillus plantarum* for α-amylase and levanase expression. *Appl. Environ. Microbiol.* **60:**1401–1413.

46. Jay, J. M. 1992. Fermented foods and related products, p. 371–409. *In Modern Food Microbiology*, 4th ed. Chapman and Hall, New York, N.Y.

47. Kagermeier, A. 1981. *Taxonomie und Vorkommen von Milchsaurebakterien in Fleischprodukten*. Dissertation. Fakultat fur Biologie, Ludwig-Maximilian-Universität München, Munich, Germany.

48. Kandler, O. 1983. Carbohydrate metabolism in lactic acid bacteria. *Antonie Leeuwenhoek* **49:**209–224.

49. Kandler, O. 1984. Current taxonomy of lactobacilli. *Dev. Ind. Microbiol.* **25:**109–123.

50. Kandler, O., and N. Weiss. 1986. Regular, non-sporing gram-positive rods, p. 1208–1234. *In* P. H.A. Sneath, N. S. Mair, M. E. Sharpe, and J. G. Holt (ed.), *Bergey's Manual of Systematic Bacteriology*, vol. 2. The Williams and Wilkins Co., Baltimore, Md.

51. Kitchell, A. G., and B. G. Shaw. 1975. Lactic acid bacteria in fresh and cured meat, p. 209–220. *In* J. G. Carr, C. V. Cutting, and G. C. Whiting (ed.), *Lactic Acid Bacteria in Beverages and Food*. Academic Press, Inc., New York, N.Y.

52. Klettner, P.-G., and D. List. 1980. Beitrag zum Einfluss der Kohlenhydratart auf den Verlauf der Rohwurstreifung. *Fleischwirtschaft* **60:**1589–1593.

53. Knauf, H. J., R. F. Vogel, and W. P. Hammes. 1989. Introduction of the transposon Tn919 into *Lactobacillus curvatus*. *FEMS Microbiol. Lett.* **65:**101–104.

54. Languer, H. J. 1972. Aromastoffe in der Rohwurst. *Fleischwirtschaft* **52:**1299–1306.

55. Leer, R. J., N. van Luijk, M. Posno, and P. H. Pouwels. 1992. Structural and functional analysis of two cryptic plasmids from *Lactobacillus pentosus* MD353 and *Lactobacillus plantarum* ATCC 8014. *Mol. Gen. Genet.* **234:**265–274.

56. Leistner, L. 1986. Mould-ripened foods. *Fleischwirtschaft* **66:**1385–1388.

57. Liepe, H. U. 1983. Starter cultures in meat production, p. 400–424. *In* H.-J. Rehm and G. Reed (ed.), *Biotechnology, Food and Feed Production with Microorganisms*, vol. 5. Verlag Chemie, Weinheim, Germany.

58. Liu, M.-L., J. K. Kondo, M. B. Barnes, and D. T. Bartholomew. 1988. Plasmid-linked maltose utilization in *Lactobacillus* spp. *Biochimie* **70**:351–355.

59. London, J. 1990. Uncommon pathways of metabolism among lactic acid bacteria. *FEMS Microbiol. Lett.* **87**:103–112.

60. Luchansky, J. B., P. M. Muriana, and T. R. Klaenhammer. 1988. Application of electroporation for transfer of plasmid DNA to *Lactobacillus, Lactococcus, Leuconostoc, Listeria, Pediococcus, Bacillus, Staphylococcus, Enterococcus* and *Propionibacterium*. *Mol. Microbiol.* **2**:637–646.

61. Lücke, F.-K. 1985. Fermented sausages, p. 41–83. *In* B. J. B. Wood (ed.), *Microbiology of Fermented Foods*, vol. 2. Elsevier Applied Science Publishing Co., Inc., London, United Kingdom.

62. McKay, L. L., and K. A. Baldwin. 1990. Applications for biotechnology: present and future improvements in lactic acid bacteria. *FEMS Microbiol. Lett.* **87**:3–14.

63. Michels, P. A. M., J. P. J. Michels, J. Boonstra, and W. N. Konings. 1979. Generation of electrochemical proton gradient in bacteria by the extrusion of metabolic end products. *FEMS Microbiol. Lett.* **5**:357–364.

64. Mogensen, G. 1993. Starter cultures, p. 1–22. *In* J. Smith (ed.), *Technology of Reduced-Additive Foods*. Blackie Academic and Professional, Chapman and Hall, New York, N.Y.

65. Morelli, L., P. S. Cocconcelli, V. Bottazzi, G. Damiani, L. Ferretti, and V. Sgaramella. 1987. *Lactobacillus* protoplast transformation. *Plasmid* **17**:73–75.

66. Mortvedt, C. I., and I. F. Nes. 1989. Bacteriocin production by a *Lactobacillus* strain isolated from fermented meat. *Eur. Food Chem. Proc.* **1**:336–341.

67. Mortvedt, C. I., and I. F. Nes. 1990. Plasmid-associated bacteriocin production by a *Lactobacillus sake* strain. *J. Gen. Microbiol.* **136**:1601–1607.

68. Mortvedt, C. I., J. Nissen-Meyer, K. Sletten, and I. F. Nes. 1991. Purification and amino acid sequence of lactocin S, a bacteriocin produced by *Lactobacillus sake* L45. *Appl. Environ. Microbiol.* **57**:1829–1834.

69. National Research Council. 1992. *Applications of Biotechnology to Traditional Fermented Foods*, p. 121–149. National Academy Press, Washington, D.C.

70. Nes, I. F. 1984. Plasmid profiles of ten strains of *Lactobacillus plantarum*. *FEMS Microbiol. Lett.* **21**:359–361.

71. Niinivaara, F. P. 1955. The influence of pure cultures of bacteria on the maturing and reddening of raw sausage. *Acta Agric. Fenn.* **85**:95–101.

72. Niven, C. F., Jr. 1951. Sausage discolorations of bacterial origin. *American Meat Institute Foundation Bulletin*, no. 13. American Meat Institute Foundation, Chicago, Ill.

73. Niven, C. F., Jr., R. H. Deibel, and G. D. Wilson. 1958. *The AMIF Sausage Starter Culture*. Circular no. 41. American Meat Institute Foundation, Chicago, Ill.

74. Palumbo, S. A., and J. L. Smith. 1977. Lebanon bologna processing. *Proc. Recip. Meat Conf.* **30**:63–68.

75. Pearson, A. M., and W. F. Tauber. 1996. *Processed Meats*, 3rd ed. Chapman and Hall, New York, N.Y.

76. Pederson, C. S. 1979. Fermented sausage, p. 210–234. *In Microbiology of Food Fermentations*. AVI Publishing Co., Inc., Westport, Conn.

77. Poolman, B. 1993. Energy transduction in lactic acid bacteria. *FEMS Microbiol.* **12**:125–148.

78. Posno, M., R. J. Leer, N. Van Luijk, M. J. F. van Giezen, P. T. H. M. Heuvelmans, B. C. Lokman, and P. H. Pouwels. 1991. Incompatibility of *Lactobacillus* vectors with replicons derived from small cryptic *Lactobacillus* plasmids and segregational instability of the introduced vectors. *Appl. Environ. Microbiol.* **57**:1822–1828.

79. Raccach, M. 1984. Method for selection of lactic acid bacteria and determination of minimum temperature for meat fermentations. *J. Food Prot.* **47**:670–671.

80. Raccach, M. 1987. Pediococci and biotechnology. *Crit. Rev. Microbiol.* **14**:291–309.

81. Reuter, G. 1975. Classification problems, ecology and some biochemical activities of lactobacilli in meat products, p. 221–229. *In* J. G. Carr, C. V. Cutting, and G. C. Whiting (ed.), *Lactic Acid Bacteria in Beverages and Food*. Academic Press, Inc., New York, N.Y.

82. Rogosa, M., and M. E. Sharpe. 1959. An approach to the classification of the lactobacilli. *J. Appl. Bacteriol.* **22**:329–340.

83. Romans, J. R., W. J. Costello, C. W. Carlson, M. L. Greaser, and K. W. Jones. 1994. Sausages, p. 773–886. *In The Meat We Eat*. Interstate Publishers, Inc., Danville, Ill.

84. Romero, D. A., and L. L. McKay. 1986. Isolation and plasmid characterization of a *Lactobacillus* species involved in the manufacture of fermented sausage. *J. Food Prot.* **48**:1028–1035.

85. Rust, R. E. 1976. *Sausage and Processed Meats Manufacturing*. American Meat Institute, Washington, D.C.

86. Scheirlinck, T., J. Mahillon, H. Joos, P. Dhaese, and F. Michiels. 1989. Integration and expression of α-amylase and endoglucanase genes in the *Lactobacillus plantarum* chromosome. *Appl. Environ. Microbiol.* **55**:2130–2137.

87. Schillinger, U., and F.-K. Lücke. 1987. Identification of lactobacilli from meat and meat products. *Food Microbiol.* **4**:199–208.

88. Schillinger, U., and F.-K. Lücke. 1989. Antibacterial activity of *Lactobacillus sake* isolated from meat. *Appl. Environ. Microbiol.* **55**:1901–1906.

89. Schleifer, K. H. 1986. Gram-positive cocci, p. 999–1002. *In* P. H. A. Sneath, N. S. Mair, M. E. Sharpe, and J. G. Holt (ed.), *Bergey's Manual of Systematic Bacteriology*, vol. 2. The Williams and Wilkins Co., Baltimore, Md.

90. Sharpe, M. E. 1981. The genus *Lactobacillus*, p. 1653–1679. *In* M. P. Starr, H. G. Trüper, A. Balows, and H. G. Schlegel (ed.), *The Prokaryotes—a Handbook on Habitats, Isolation, and Identification of Bacteria*, vol. II (1st ed.). Springer-Verlag, New York, N.Y.

91. Sharpe, M. E., T. F. Fryer, and D. G. Smith. 1966. Identification of the lactic acid bacteria. *In* B. M. Gibbs and F. A. Skinner (ed.), *Identification Methods for Microbiologists*, part A. Academic Press, Ltd., London, United Kingdom.

92. Shay, B. J., A. F. Egan, M. Wright, and P. J. Rogers. 1988. Cysteine metabolism in an isolate of *Lactobacillus sake*:

plasmid composition and cysteine transport. *FEMS Microbiol. Lett.* **56:**183–188.

93. **Shrago, A. W., B. M. Chassy, and W. J. Dobrogosz.** 1986. Conjugal plasmid transfer (pAMb1) in *Lactobacillus plantarum. Appl. Environ. Microbiol.* **52:**574–576.

94. **Shrago, A. W., and W. J. Dobrogosz.** 1988. Conjugal transfer of group B streptococcal plasmids and comobilization of *Escherichia coli-Streptococcus* shuttle plasmids to *Lactobacillus plantarum. Appl. Environ. Microbiol.* **54:**824–826.

95. **Skaugen, M., and I. F. Nes.** 1994. Transposition in *Lactobacillus sake* and its abolition of lactocin S production by insertion of IS*1163*, a new member of the IS*3* family. *Appl. Environ. Microbiol.* **60:**2818–2825.

96. **Stackebrandt, E., V. J. Fowler, and C. R. Woese.** 1983. A phylogenetic analysis of lactobacilli, *Pediococcus pentosaceus* and *Leuconostoc mesenteroides. Syst. Appl. Microbiol.* **4:**326–337.

97. **Stackebrandt, E., C. Koch, O. Gvozdiak, and P. Schumann.** 1995. Taxonomic dissection of the genus *Micrococcus: Kocuria* gen. nov., *Nesterenkonia* gen. nov., *Kytococcus* gen. nov., *Dermacoccus* gen. nov., and *Micrococcus* Cohn 1872 gen. emend. *Int. J. Syst. Bacteriol.* **45:**682–692.

98. **Stackebrandt, E., and M. Teuber.** 1988. Molecular taxonomy and phylogenetic position of lactic acid bacteria. *Biochimie* **70:**317–324.

99. **Stiles, M. E., and W. H. Holzapfel.** 1997. Lactic acid bacteria of foods and their current taxonomy. *Int. J. Food Microbiol.* **36:**1–29.

100. **ten Brink, R. Otto, U. P. Hansen, and W. N. Konings.** 1985. Energy recycling by lactate efflux in growing and nongrowing cells of *Streptococcus cremoris. J. Bacteriol.* **162:**383–390.

101. **Terrell, R. N., G. C. Smith, and Z. L. Carpenter.** 1977. Practical manufacturing technology for dry and semi-dry sausage. *Proc. Recip. Meat Conf.* **30:**39–44.

102. **Townsend, W. E., C. E. Davis, and C. E. Lyon.** 1978. Some properties of fermented dry sausage prepared from PSE and normal pork. *In Kongressdokumentation.* 24th Europäischer Fleischforscher-Kongress, Kulmbach, Germany.

103. **U.S. Department of Agriculture, Food Safety Inspection Service.** 1995. Prescribed treatment for pork and products containing pork to destroy trichinae, part 318.10. *Code of Federal Regulations, Title 9.* Office of the Federal Register, Washington, D.C.

104. **U.S. Department of Agriculture, Food Safety Inspection Service.** 1995. Requirements for the production of cooked beef, roast beef, and cooked corn beef, part 318.17. *Code of Federal Regulations, Title 9.* Office of the Federal Register, Washington, D.C.

105. **U.S. Department of Agriculture, Food Safety and Inspection Service.** June 1999, revision date. Appendix A. Compliance guidelines for meeting lethality performance standards for certain meat and poultry products. [Online.] www.fsis.usda.gov/OPPDE/rdad/FRPubs/95-033F/95-033F_Appendix%20A.htm.

106. **U.S. Department of Agriculture, Food Safety and Inspection Service.** May 2003, revision date. *The Food Standards and Labeling Policy Book.* Office of Policy Program Development, Washington, D.C. [Online.] www.fsis.usda.gov/oppde/larc/policies/policybook.pdf.

107. **Vandekerckhove, P., and D. Demeyer.** 1975. Die Zusammernstzung belgischer Rohwurst (Salami). *Fleischwirtschaft* **55:**680–682.

108. **von Wright, A., and M. Sibakov.** 1993. Genetic modification of lactic acid bacteria, p. 161–198. *In* S. Salminen and A. von Wright (ed.), *Lactic Acid Bacteria.* Marcel Dekker, Inc., New York, N.Y.

109. **Weiss, N.** 1992. The genera *Pediococcus* and *Aerococcus,* p. 1502–1507. *In* A. Balows, H. G. Trüper, M. Dworkin, W. Harder, and K.-H. Schleifer (ed.), *The Prokaryotes—A Handbook on the Biology of Bacteria: Ecophysiogy, Isolation, Identification, Applications,* 2nd ed. Springer-Verlag, New York, N.Y.

110. **West, C. A., and P. J. Warner.** 1985. Plasmid profiles and transfer of plasmid-encoded antibiotic resistance in *Lactobacillus plantarum. Appl. Environ. Microbiol.* **50:**1319–1321.

111. **Woese, C. R.** 1987. Bacterial evolution. *Microbiol. Rev.* **51:**221–271.

112. **Wolf, G., E. K. Aarendt, U. Pfahler, and W. P. Hammes.** 1990. Heme-dependent and heme-independent nitrite reduction by lactic acid bacteria results in different N-containing products. *Int. J. Food Microbiol.* **10:**323–330.

113. **Wu, W. J., D. C. Rule, J. R. Busboom, R. A. Field, and B. Ray.** 1991. Starter culture and time/temperature of storage influences on quality of fermented mutton sausage. *J. Food Sci.* **56:**919–925.

114. **Zenitani, B.** 1955. Studies on fermented fish products. I. On the aerobic bacteria in "Shiokara." *Bull. Jpn. Soc. Sci. Fish.* **21:**280–283.

Food Microbiology: Fundamentals and Frontiers, 3rd Ed.
Edited by M. P. Doyle and L. R. Beuchat
© 2007 ASM Press, Washington, D.C.

M. J. Robert Nout
Prabir K. Sarkar
Larry R. Beuchat

Indigenous Fermented Foods

38

A food is considered fermented when one or more of its constituents have been acted upon by selected microorganisms or their enzymes to produce a significantly altered final product desirable for human consumption. Most fermentations are caused by molds, yeasts, or bacteria, either singularly or in combination. Indigenous or traditional fermented foods have been prepared and consumed for hundreds of years and are strongly linked to cultures and traditions of millions of people around the world, especially in rural communities. The origins of most fermentation technologies have been lost in the mists of history. Some products and practices no doubt fell by the wayside; those that remain today have survived the test of time. Fermented food products are important components of the diet as staples, adjuncts to staples, condiments, and beverages.

This oldest form of food biotechnology originated as a necessity for enhancing the keeping quality of diverse plant and animal food materials through organic acid, alcoholic, and alkaline fermentations. Fermentation, a relatively efficient low-energy preservation process, also improves digestibility, flavor, appearance, nutrient contents, and other quality attributes and reduces antinutritional components of the substrates and cooking time. Many fermented foods are now receiving global attention for their health-promoting or disease-preventing or -curing effects. Whereas a considerable number of food fermentation processes have been scaled up for commercial purposes, most types of fermented foods are still produced on a home scale. Such products often contain mixed genera and populations of microorganisms because of the lack of controlled processing facilities. In view of larger-scale industrialized food fermentation, microbial ecology and metabolic activities of functional microorganisms must be investigated. Present-day developments in molecular methods enable accurate characterization of strains and development of tailor-made fermented food products.

A variety of fermented foods can be found widespread over the world. Following the sequence in Table 38.1, some of them will be described in this chapter, mainly to illustrate the complexity of biochemical, nutritional, and sensorial changes that result from an array of microbial activities in a range of raw materials.

M. J. Robert Nout, Dept. of Agrotechnology and Food Sciences, Wageningen University, Bomenweg 2, 6703HD Wageningen, The Netherlands. **Prabir K. Sarkar**, Dept. of Botany, University of North Bengal, Siliguri 734013, India. **Larry R. Beuchat**, Center for Food Safety, University of Georgia, 1109 Experiment St., Griffin, GA 30223-1797.

Table 38.1 Some important indigenous fermented foods

Product(s)	Country(ies) and/or area(s)	Substrate(s)	Functional microflora[a]	Type(s) of fermentation[b]	Description and usage
Cereal and starch crop products					
Ang-kak	China, Southeast Asia	Rice	Molds	SSF, TS	Dry purple-red powder; colorant
Banku	Egypt	Maize or cassava	Yeasts, LAB	SSF, N	Dumpling; staple
Ben-saalga	Burkina Faso	Pearl millet	Bacteria, LAB	SmF, N	Sour, thin gruel; breakfast staple and infant food
Bhatura	India, Pakistan	Wheat	LAB, yeasts	SSF, TS	Deep-fried, flat, leavened bread; snack
Bouza	Egypt	Wheat	Yeasts, LAB	SmF, TS	Pale yellow, thick, sour alcoholic drink
Breads (leavened yeast and sour-dough breads)	Worldwide	Wheat and/or rye	Bakers' yeast or yeast-LAB mixed cultures	SSF	Baked leavened dough; staple
Busaa	Kenya, Uganda	Maize, finger millet	Yeasts, LAB	SmF, N	Sour alcoholic drink
Chicha	South America	Maize	Molds, yeasts, LAB, AAB	SmF, N	Clear, yellowish, effervescent, sour alcoholic drink
Deguè	Burkina Faso	Pearl millet	Bacteria, molds	SSF, N	Balls diluted with milk or water to make porridge
Gari	Nigeria, West Africa	Cassava	LAB, yeasts	SSF, N	Granulate; precooked instant porridge; breakfast staple
Injera	Ethiopia, Sudan	Teff (or corn or sorghum)	LAB, yeasts	SSF, N	Sour, soft, steam-baked, flat pancake; staple
Jalebi	India	Wheat	LAB, yeasts	SSF, N	Pretzel-like syrupy confection
Jnard	India, Nepal, Bhutan	Finger millet	Molds, yeasts, LAB	SSF, TS	Sweet-sour alcoholic drink
Kenkey	Ghana	Maize	LAB, yeasts	SSF, N	Sour dumpling; cooked; staple
Lafun	Nigeria, West Africa	Cassava	LAB	SmF, N	White flour made into a stiff porridge; staple
Lao-chao	China, Indonesia	Rice	Molds, yeasts	SmF, TS	Sweet-sour, juicy, alcoholic snack
Mahewu	South Africa	Maize	LAB	SmF, N	Sour, nonalcoholic drink
Mawè	Benin, Togo	Maize	LAB, yeasts	SSF, N	Sourdough made into porridge or gruel; staple
Merissa	Sudan	Sorghum	LAB, yeasts	SSF-SmF, N	Thick, sour alcoholic drink
Minchin	China, Thailand	Wheat (or rice)	Molds	SSF, N	Thin strips or noodles; staple
Munkoyo	Zambia, Zaire	Maize	LAB, yeasts	SmF, N	Sweet-sour alcoholic drink
Naan	Afghanistan, Iran, Pakistan, India	Wheat	Yeasts, LAB	SSF, N	Flat, baked bread; staple
Ogi	Nigeria, West Africa	Maize or sorghum or millet	Molds, yeasts, AAB, LAB	SmF, N	Sour gruel; staple
Pito	Nigeria, Ghana	Sorghum or maize	Molds, LAB, yeasts	SmF, TS	Sweet-sour alcoholic drink
Poi	Hawaii	Taro corm	LAB, yeasts	SSF, N	Sour porridge; staple or condiment
Poto-poto	Congo	Maize	LAB, yeasts	SSF, N	Sourdough balls made into porridge or gruel; staple
Pozol	Mexico	Nixtamal[c]	Molds, yeasts, LAB, other bacteria	SSF, N	Balls diluted with water to make sour, nonalcoholic porridge
Puto	Philippines	Rice	LAB, yeasts	SSF, N	Spongy, steamed cake; snack
Ruou nep than	Vietnam	Rice	Molds, yeasts	SSF-SmF, TS	Alcoholic drink

(Continued)

Table 38.1 (*Continued*)

Product(s)	Country(ies) and/or area(s)	Substrate(s)	Functional microflora[a]	Type(s) of fermentation[b]	Description and usage
Saké	Japan	Rice	Molds, yeasts, bacteria, LAB	SmF, TS	Alcoholic drink
Tapé	Indonesia	Cassava or rice	Molds, yeasts	SSF, TS	Sweet-sour alcoholic snack
Tapuy	Philippines	Rice	Yeasts, LAB	SSF, TS	Sour-sweet alcoholic drink
Yakju and Takju	Korea	Rice	Molds, yeasts, LAB, other bacteria	SSF-SmF, TS	Alcoholic drink
Legume products					
Daddawa	West Africa, Nigeria	African locust bean	Bacteria	SSF, N	Flavoring agent; soup and stew ingredient
Inyu	Taiwan, China, Hong Kong	Black soybeans	Molds, LAB, yeasts	SmF, N	Syrup; flavor enhancer
Kecap asin	Indonesia	Soybeans	Molds, LAB, yeasts	SSF-SmF, TS	Thin, transparent, light brown salty liquid; condiment
Kecap manis	Indonesia	Soybeans, palm sugar, herbs	Molds, LAB, yeasts	SSF-SmF, TS	Thick, dark brown syrup; sweet condiment
Kinema	India, Nepal	Soybeans	Bacteria	SSF, N	Paste made into thick curry; side dish
Meitauza	China, Taiwan	Soybean press cake	Molds	SSF, N	Cake, fried or cooked; side dish
Meju	Korea	Black soybeans	Molds, LAB, yeasts	SSF-SmF, N	Syrup; seasoning agent
Natto	Japan	Soybeans	Bacteria	SSF, PS	Mucilaginous snack
Oncom	Indonesia	Peanut press cake	Molds	SSF, TS	Cake, deep fried or roasted; side dish or soup ingredient
Papad	India, Pakistan, Bangladesh	Black gram	Yeasts	SSF, N	Deep-fried or roasted snack or condiment
Sufu	China, Taiwan	Soybean curd	Molds	SSF, N	Paste; condiment
Tempeh	Indonesia	Soybeans	Molds, bacteria	SSF, TS	White, mold-penetrated and covered cake, stewed or deep fried; side dish, snack, or soup ingredient
Wadi	India, Pakistan, Bangladesh	Black gram	LAB, yeasts	SSF, N	Balls or cones; condiment
Cereal-legume mixture products					
Dhokla	India	Rice, Bengal gram	LAB, yeasts	SSF, N	Steamed, soft cake; snack
Idli	India, Sri Lanka	Rice, black gram	LAB, yeasts	SSF, N	Steamed, spongy cake; snack
Miso	Japan	Soybeans, rice	Molds, yeasts, LAB	SSF, TS	Paste; soup base or seasoning agent
Soy sauces	East and Southeast Asia	Soybeans, wheat	Molds, LAB, yeasts	SSF-SmF, TS	Brown, salty liquid; seasoning agent
Taoco	Indonesia	Soybeans, cereals	Molds, LAB, yeasts	SSF-SmF, TS	Yellow paste; seasoning agent
Vegetable products					
Gundruk	India, Nepal	Mustard leaves	LAB	SSF, N	Shreds; soup ingredient or pickle
Soibum, Mesu, Naw-Mai-Dong	India, Nepal, Thailand	Young bamboo shoot	LAB, yeasts	SSF, N	Cubes; consumed as a pickle or made into curry

(*Continued*)

Table 38.1 Some important indigenous fermented foods *(Continued)*

Product(s)	Country(ies) and/or area(s)	Substrate(s)	Functional microflora[a]	Type(s) of fermentation[b]	Description and usage
Kanji	India, Pakistan, Israel	Carrot or beet	Yeasts	SmF, N	Deep purple, sour, alcoholic drink
Kimchi	Korea	Cabbage (or radish taproot), garlic	LAB	SSF, N or TS	Sour, carbonated vegetable; staple
Sauerkraut	Europe, Russia, United States	White cabbage	LAB	SSF-SmF, N or PS	Sour shreds; consumed raw or cooked with meat or sausages
Sinki	India, Nepal	Radish taproot	LAB	SSF, N	Sour shreds; pickle or soup ingredient
Dairy products					
Gorgonzola, Blue Stilton, Roquefort cheese	Italy, United Kingdom, France	Cow's or sheep's milk	LAB, molds	SSF, TS or PS	Blue-veined cheese; strong-flavored side dish or cooking ingredient
Camembert cheese	France	Cow's milk	LAB, molds	SSF, TS or PS	Mold surface-ripened cheese; soft-texture side dish with gradually developing strong flavor
Dahi	India, Pakistan, Bangladesh, Sri Lanka	Cow's or buffalo's milk	LAB	SmF, N or PS	Thick gel; dessert
Gouda cheese	The Netherlands	Cow's milk	LAB	SSF, TS or PS	Small-eyed or blind cheese; multipurpose protein food
Kefir	Scandinavia, Russia	Goat's, sheep's, or cow's milk	Yeasts, LAB	SmF, TS	Effervescent, sour, mild alcoholic drink
Koumiss	Russia	Mare's milk	LAB, yeasts	SmF, N	Effervescent, cloudy, sour, alcoholic drink
Lassi	India	Cow's or buffalo's milk	LAB	SmF, N	Sour drink
Yogurt	Europe, worldwide	Cow's milk	LAB	SmF, TS or PS	Viscous or thick gel; dessert or side dish
Fish products					
Bagoong	Philippines	Fish or shrimp or oyster	Bacteria	SSF, N	Brown paste; condiment
Izushi	Japan	Fish, rice, vegetable	LAB	SSF, N	Pickle
Katsuobushi	Japan	Bonito or skipjack tuna	Molds	SSF, N	Strips, dried; seasoning agent
Som-fak	Thailand	Fish fillet, rice, garlic	LAB	SSF, N	Served raw or cooked; main course or snack
Meat products					
Country-cured ham	Europe, United States	Pork	Bacteria, LAB, molds	SSF, N	Cured meat; ham slices consumed raw or cooked
Nem	Vietnam	Pork, garlic	Bacteria, LAB	SSF, N or TS	Meat cubes, fried; side dish
Nham	Thailand	Pork, cooked rice	LAB	SSF, N	Sour slices, deep fried, crispy; snack
Salami	Europe, United States	Pork and beef	LAB	SSF, N or PS	Sausage

(Continued)

Table 38.1 *(Continued)*

Product(s)	Country(ies) and/or area(s)	Substrate(s)	Functional microflora[a]	Type(s) of fermentation[b]	Description and usage
Miscellaneous products					
Balao balao	Philippines	Rice, shrimp	LAB	SSF, N	Main dish or sauce
Basi	Philippines	Sugarcane juice	Yeasts, LAB	SmF, TS	Sweet-sour, effervescent, cloudy, alcoholic drink
Bongkrek	Indonesia	Coconut press cake	Molds	SSF, TS	Bars, roasted or fried; snack or soup ingredient
Kishk	Egypt, Syria, Lebanon, Jordan, Iraq, North Africa	Wheat, milk	LAB, yeasts, other bacteria	SSF, N	Brownish, sour, dried balls; snack or soup ingredient
Kombucha	Japan, Indonesia, China, Russia	Tea liquor, sugar	AAB, yeasts	SmF, TS	Sour, mildly alcoholic drink
Miang, or Leppet-So	Myanmar, Thailand	Tea leaves	LAB	SSF, N	Sour-bitter tasting soft snack
Tarhana, or Trahana	Turkey, Greece	Tomatoes, wheat flour	Yeasts	SSF, N or PS	Tomato dough, dehydrated granulate; soup ingredient
Palm wines	All tropical palm-growing countries	Sap of coconut, date, palmyra, oil, nipa, raphia, or kithul palm	Yeasts, LAB, bacteria, AAB	SmF, N	Sweet-sour alcoholic drink
Ugba, or Ogiri	Nigeria, West and Central Africa	African oil bean or castor oil beans or melon or sesame seeds	Bacteria	SSF, N	Dark brown balls; salad ingredient or flavoring agent in soups, stews, and sauces

[a] AAB, acetic acid bacteria; LAB, lactic acid bacteria.

[b] SSF, solid-state fermentation; SmF, submerged fermentation; SSF-SmF, solid-state fermentation followed by submerged fermentation; N, natural and/or backslopped fermentation; TS, traditional undefined starter; PS, pure culture starter.

[c] Corn grains cooked in alkaline water.

CEREAL AND STARCH CROP PRODUCTS

Bakery Products

Bread, in various forms, has been a staple in the diets of many population groups for many centuries. The history of bread traces back to about 3,000 B.C. The development of cereal foods has proceeded through several stages, from roasted grain to gruels to flat breads and finally to leavened bread loaves. Early Egyptians developed the use of fermentation for breads and constructed baking ovens. We will focus here on leavened breads owing their sensorial properties, at least in part, to fermentative activities of microorganisms.

In principle, bread is made from dough that is fermented and baked. The essential ingredients are wheat or rye flour, salt, water, and a leavening agent (26). Usually some fat, sugar, milk solids, and bread-improving emulsifiers are added, but these are not essential. The function of water (50 to 60% of flour weight) is to hydrate the starch and gluten (extensible and elastic proteins in wheat), enabling the mixing and kneading of a viscoelastic dough that retains the carbon dioxide gas formed during fermentation. The most commonly used leavening agent is bakers' yeast, *Saccharomyces cerevisiae* (27), which is commercially available as dehydrated granules (instant dry yeast), fresh yeast cake, or yeast cream (a suspension). Dry yeast must first be reactivated in a flour-water suspension for about 20 min. Yeast cream and cake have the advantage that no activation is required, but they are prone to spoilage by lactic acid bacteria (LAB) and thus have limited shelf lives. Based on flour weight, about 1 to 6% yeast dry matter is required. The function of salt (1 to 2%) is to moderate the fermentation rate in order to obtain a steady production of gas that can be adequately retained in the dough. After fermentation for several hours at 25 to 30°C, doughs are remixed to obtain a homogenous distribution of gas cells. The dough is portioned to the required weight or size and is molded and put into baking pans. After another period of fermentation, the dough has at least doubled its volume and is baked in a hot-air or steam oven for 20 to 40 min at temperatures ranging from 180 to 230°C.

Sourdough bread is slightly acidic because of the leavening agent, sourdough. In contrast to the pure-culture bakers' yeast, sourdough comprises a stable, mixed microflora containing 10^7 to 10^9 LAB CFU/g, predominantly *Lactobacillus sanfransiscensis* and occasionally *Lactobacillus pontis*, *Lactobacillus panis*, *Lactobacillus frumenti*, or *Lactobacillus reuteri*, and 10^5 to 10^7 yeast CFU (predominantly *Candida milleri*)/g, obtained by repeated propagation of sourdough fermentations by reinoculation. Long-term propagation of a sourdough during the last seven decades has been documented (16), and anecdotal reports exist of sourdoughs maintained over several centuries. Commercial sourdough starters have been developed and are available as dehydrated granules and (semi)dried preferments. Sourdoughs are required for rye breads to achieve bakeability, and they are widely used in rye and wheat breads because of the high sensory quality they impart to these breads. Sourdough contributes to the characteristic flavor (61), improves texture, and delays staling and microbial spoilage of bread (60).

Naan (nan) is made by mixing white wheat flour with sugar, salt, backslop (called khamira), and water. The hand-kneaded dough is left in an earthen jar to ferment for 12 to 24 h. After maturity, the leavened dough is made into balls, placed on a smooth surface sprinkled with flour, and flattened by a wooden rolling pin. Smoothly flattened round dough is transferred onto a circular pad of cotton cloth and is slapped onto the inner wall of the clay-clad brick oven, called the tandoor, where it sticks for baking at 120 to 150°C until the dough is puffed off and light brown. The bread is speared with a skewer and removed from the oven wall to be served hot, usually along with meat or chicken preparations. From a new dough (pH 5.9) for making naan, 10^5 CFU of yeasts/g and 10^2 CFU of LAB/g can be obtained compared with respective counts of 10^8 and 10^9 CFU/g from ripe, fermented dough (pH 4.8) (5). *S. cerevisiae* is the predominant yeast. Presently, bakers' yeast and dahi are added to shorten the fermentation period.

Kenkey

Kenkey is a dense, sour-tasting, cooked mass, served as thick slices at breakfast combined with tea, sardines, or other foods. Cleaned whole corn (maize, *Zea mays*) kernels are soaked in water for 2 days; during this period, the kernels soften (43), which is essential during the next operation, i.e., coarse wet milling. The resulting wet grits are kneaded into a stiff dough, which is covered and left to ferment at ambient temperature (25 to 30°C) for 2 to 4 days. Dominant microorganisms are obligate heterofermentative lactobacilli, e.g., *Lactobacillus fermentum*, and yeasts, mainly *Candida krusei* and *S. cerevisiae*. When fermented according to local preference for odor and acidity (28), the dough is divided into two equal portions, and one portion is cooked, with the addition of some water, while being kneaded continuously with a cooking stick into an elastic gelatinized paste called aflata. The aflata is then mixed through the remaining uncooked dough. The resulting mass is molded by hand into units of 200 to 400 g and wrapped in banana leaves (Fanti kenkey) or corn sheaths (Ga kenkey). The packages are cooked by immersion in boiling water for a few hours. The function of the aflata is twofold: it acts as an adhesive, keeping the mixture in shape, and it carries the water needed for the swelling (gelatinization) of the uncooked, gritty dough. During the fermentation, the level of available lysine (and thus protein quality) and nutrient bioavailability increase, and flavor compounds (2,3-butanediol, butanoic acid, lactic acid, 3-methylbutanoic acid, octanoic acid, 2-phenylethanol, and propanoic acid) are formed (28).

Mawè

Mawè is an intermediate product used for the preparation of, e.g., ablo, a steam-cooked corn bread, and porridge, e.g., aklui (23). To prepare mawè, cleaned, dry, whole corn kernels are milled into grits, partly reground to obtain a fine grind, mixed with water, kneaded into a dough, covered, and allowed to ferment naturally during 2 to 4 days at ambient temperature (30°C). The pH decreases to 3.7 to 3.8, and an attractive freshly sour flavor is formed due mainly to heterofermentative lactobacilli (*Lb. fermentum* and *Lactobacillus cellobiosus*) and *C. krusei*.

Ogi

To prepare ogi, kernels of corn are soaked in warm water for 1 to 3 days, after which they are wet milled and sieved with water through a screen to remove fiber, hulls, and much of the germ. The filtrate is fermented to yield a sour, white, starchy sediment. Fermentation is by lactobacilli (*Lb. fermentum*, *Lb. cellobiosus*, *Lactobacillus brevis*, and *Lactobacillus plantarum*) originating from the environment, although other bacteria (*Enterobacter sakazakii* and *Corynebacterium* spp.) and yeasts (*C. krusei*, *Candida kefyr*, and *Rhodotorula* spp.) are also involved (42). Ogi may be diluted in water to 8 to 10% solids and boiled into a pap or cooked and turned into a stiff gel (eko) before eating.

Ogi is a major breakfast cereal for adults and a traditional food for weaning babies. As a result of the preparation method, significant (40%) losses of protein occur but the digestibility of the remaining protein is improved by 20% (42). In Nigeria, industrialization of ogi manufacture has taken place (54), enabling better control of quality and hygiene. Based on upgraded village technology, the final product is packaged and distributed as a long-shelf-life dehydrated powder, obtained by rotary drying or spray drying of the fermented wet cake. The nutritional value of ogi can be improved by enrichment with soybeans to obtain a 15% protein content (54).

Pozol

Pozol, which dates back to the Aztec period, is made from nixtamal, which consists of corn kernels that have been boiled in lime water containing about 10% calcium hydroxide. This treatment, which probably evolved from the use of naturally occurring alkaline water of volcanic origin, facilitates the swelling of the corn and removal of pericarps (decortication). The resulting cooked endosperms are washed, drained, and milled to obtain masa, a coarse paste which is molded into balls (51), wrapped in banana leaves, and left to ferment for 1 to 2 weeks at ambient temperature (22 to 27°C). During the process, the pH increases to about 7.5 after nixtamilization and then gradually decreases to 3.8 to 4.0 after 1 week due to fermentative acidification, dominated by LAB. *Streptococcus* spp. account for 25 to 50% of the microflora, and *Lb. plantarum* and *Lb. fermentum*, together with *Leuconostoc* and *Weissella* species, are the other dominant microorganisms (4). Yeasts, including *Candida* spp. and *Trichosporon cutaneum*, are encountered in combination with the LAB (51). When left to ferment for longer periods, yeasts and molds (*Geotrichum candidum* and *Rhizopus* spp.) develop on the surface (51), imparting a musty flavor. In addition to the development of desired flavor, fermentation also contributes to the digestibility and increased riboflavin, niacin, and tryptophan contents of the product.

Ang-Kak

Red kojic rice (ankak, or anka) is made by solid-state fermentation of cooked rice with the ascomycetous molds *Monascus* spp. Rice is washed, steamed for about 1 h, cooled to 36°C, inoculated with starter, heaped to ferment until the temperature rises to 42°C, and then spread and shelved. It is used in the fermentation industry for coloring red rice wine and foods such as sufu, fish sauce, and red soybean curd. The azaphilone pigments produced by *Monascus ruber*, *Monascus pilosus*, and *Monascus purpureus* include the orange rubropunctatin and monascorubrin, purple rubropunctamin and monascorubramin,

and yellow ankaflavin and monascin (59), which are heat stable over a wide range of pHs and thus of interest as biocolorants in foods. Several other secondary metabolites are produced, including xanthomonasins, monascumic acid, monascusones, monacolins, and γ-aminobutyric acid. The pleasant flavor of ang-kak is derived from alcohols, aldehydes, ketones, esters, and terpenoid compounds.

Of recent interest are the potential health-promoting effects of ang-kak, such as cholesterol-lowering ability due to mevonolin (monacolin K), hypotensive effects due to γ-aminobutyric acid, and anti-inflammatory effects (3). The optimum cultural conditions for the production of pigments by a *Monascus* sp. isolated from the solid koji of Kaoliang liquor are reported to be pH 6.0 for a 3-day incubation at 32°C. Among the carbon sources tested, starch, maltose, and galactose are suitable for pigment production; a starch content of 3.5% (5% rice powder) and a sodium or potassium nitrate content of 0.5% gave maximum yield of pigment in laboratory media. Zinc may act as a growth inhibitor of *Monascus purpureus* and concomitantly as a stimulant for glucose uptake and the synthesis of secondary metabolites such as pigments.

Ragi

The Indonesian word ragi refers to a starter or inoculum, and the name following ragi indicates the intended use of the starter, e.g., ragi-tempe, ragi-tapé, ragi-peuyeum, and ragi-tapai. Similar starters are Indian bakhar, Nepalese murcha, Thai loog-pang, Vietnamese men, Philippine bubod, Malaysian jui-paing, Chinese chu, Japanese tane koji, and Korean nuruk. To prepare ragi, rice flour is mixed with a variety of herbs, spices, and water to make dough which is inoculated by dusting with powdered ragi from a previous batch, flattened into cakes (about 3 cm in diameter and 1 cm thick), placed on a bamboo tray, covered with leaves or a cloth, incubated for 2 to 5 days at ambient temperature (20 to 30°C), air or sun dried, and preserved until needed.

A widely used type of ragi and ragi-like starters combines three groups of microorganisms, mucoraceous fungi, yeasts, and LAB (45). Ragi contains molds, namely *Amylomyces rouxii* and *Aspergillus*, *Mucor*, and *Rhizopus* spp. *Amylomyces* reproduces through thick-walled chlamydospores which ensure survival when the starter cakes are dried and stored prior to being used. Among the diverse yeast species in ragi, *Saccharomycopsis fibuligera* and *Pichia anomala* are the principal amylolytic and ethanol-producing yeasts, respectively. *Pediococcus pentosaceus*, *Weissella* spp., *Lactobacillus curvatus*, and *Enterococcus faecium* form the LAB component of ragi microflora (70). The microflora of ragi varies with the location and additives used. The molds and several yeasts convert starchy materials into fermentable sugars, which are subsequently converted into ethanol by the yeasts and organic acids by the LAB and molds.

Puto

To make puto, rice grains are soaked, ground to a semi-paste consistency (called galapong), mixed with starter (called lebadura), and fermented at two different stages, during which time the volume and lactic acid content increase 3- and 20-fold, respectively. The fermented batter is poured into molds and steam cooked for 15 to 30 min to make puto. The predominant microorganism in the fermenting batter is always *Leuconostoc mesenteroides* (30), followed by *Enterococcus faecalis* and then *S. cerevisiae* and *Pediococcus dextrinicus*. The yeasts produce low levels of ethanol and, along with *L. mesenteroides*, leaven the batter, rendering a spongy texture to the product.

Rice Beers

Although the term "rice wine" is also in use for a rice-fermented alcoholic drink, the term "rice beer" is technically correct because, like beer, rice beer is produced from grain rather than fruit and it undergoes a two-stage fermentation process wherein starches in the rice are broken down into sugars which are then converted into alcohol. While most of these processes still follow indigenous technology, significant development in the manufacturing process has been made in Japan, China, and Korea; modern Japanese saké manufacture is highly

sophisticated. The manufacture of rice beers can be characterized as a biotechnological process which includes steaming, inoculation with starter, mashing, and fermentation. The microorganisms involved in the fermentations of some rice beers are listed in Table 38.2. Depending on the fermentation performance, the ethanol content varies and can reach up to 15% (vol/vol) (14, 15).

Injera

To prepare injera, teff flour and water are combined with irsho, a fermented yellow fluid saved from a previous batch. The resultant thin, watery paste is generally incubated for 1 to 3 days. A portion of the fermented paste is then mixed with 3 parts water and boiled to give a product called absit, which is in turn mixed with a portion of the original fermented flour to yield a thin injera. Thick injera (aflegna) is a teff paste that has undergone only minimal fermentation (12 to 24 h) and is characterized by a sweet flavor and a reddish color. A third type of injera (komtata) is made from overfermented paste and, consequently, has a sour taste, probably due to extensive growth of LAB (17). Although the microflora compositions responsible for fermentation of the sweeter types of injera have not been fully determined, *Candida guilliermondii* is apparently a primary yeast in this process. The carbon sources for fermentation originate from the grain. Initially, a rise in free sugars, mainly sucrose, takes place, followed by a decline due to microbial assimilation (72). Regardless of the method used to prepare injera,

Table 38.2 Microorganisms involved in the production of Asian rice beers

Beer(s)	Starter	Functional microorganisms
Brem	Ragi-tapé	*Amylomyces* spp., *Mucor* spp., *Rhizopus* spp., *Saccharomyces cerevisiae*, *Candida glabrata*, *Pichia anomala*, *Issatchenkia orientalis*
Ruou nep than	Men	*Amylomyces rouxii*, *Rhizopus* spp., *Saccharomyces cerevisiae*, *Pichia anomala*
Saké, Mirin	Tane-koji	*Aspergillus oryzae*, *Saccharomyces sake*, *Pichia anomala*, *Lactobacillus sakei*
Sato, Ou, Nam-Khao	Loog-pang	*Mucor* spp., *Rhizopus* spp., *Candida* spp., *Saccharomyces* spp.
Shaoshing	Chu	*Aspergillus oryzae*, *Rhizopus* spp., *Saccharomyces cerevisiae*
Takju, Yakju	Nuruk	*Aspergillus* spp., *Mucor* spp., *Rhizopus* spp., *Saccharomyces cerevisiae*, *Pichia anomala*, *Hansenula subpelliculosa*, *Candida* spp., *Debaryomyces polymorphus*, *Lactobacillus plantarum*, *Leuconostoc mesenteroides*
Tapai	Jui-piang	*Amylomyces rouxii*, *Rhizopus oryzae*, *Mucor* spp., *Saccharomycopsis fibuligera*, *Pichia anomala*
Tapuy	Bubod	*Rhizopus oryzae*, *Amylomyces rouxii*, *Saccharomycopsis fibuligera*, *Rhodotorula glutinis*, *Debaryomyces hansenii*, *Candida* spp., *Lactobacillus plantarum*, *Leuconostoc* spp.

the fermented dough with batter consistency is baked on a hot, oiled clay griddle for a few minutes, resulting in a large, pancake-like bread injera.

Gari

Fermented root of the cassava plant (*Manihot esculenta*) is known as gari in the rain forest belt of West Africa. To prepare gari, the corky outer peel and the thick cortex are removed and the inner portion of the root is grated. The pulp is then packed into jute bags, and weights are applied to express some of the juice. After 3 to 4 days of fermentation, cassava is sieved and heated while constantly turning over a hot steel pan. This process is known as garification. The final product contains 10 to 15% moisture, 80 to 85% starch, 0.1% fat, 1 to 1.5% crude protein, and 1.5 to 2.5% crude fiber. Palm oil may be added as a colorant just before or after drying. For the production of 1 ton of gari, 4 tons of cassava roots are required. Cassava is a highly perishable crop once harvested; garification is a clever approach to achieve a safe, shelf-stable product. In Nigeria, gari production has been industrialized (55).

Fresh cassava roots of bitter varieties contain cyanogenic glycosides, viz., linamarin and lotaustralin, that decompose during the fermentation of gari with the liberation of gaseous hydrocyanic acid. The hydrolysis of cyanogenic glycosides is due mainly to endogenous linamarinase, reducing cyanide levels from 300 mg (initial) to 10 to 20 mg of HCN/kg of product (55). *Lb. plantarum* and other LAB contribute significantly to decreasing the pH (74). The acid condition favors fungal growth, mainly that of *Galactomyces candidum*, which contributes to the characteristic aroma and flavor by its aldehydes and esters.

Tapé

Tapé ketan (rather similar to lao-chao) and tapé ketella are prepared by fermenting rice and cassava, respectively. Glutinous rice or peeled and chopped cassava root is soaked, steam cooked until soft, spread in thin layers onto bamboo trays, inoculated with powdered ragi-tapé (starter), and left to ferment under cover for 1 to 3 days at 27 to 30°C to produce a soft, white mass (65). The essential biochemical changes, caused by *Amylomyces rouxii* and *Pichia burtonii*, are the hydrolysis of starch into maltose and glucose and the conversion of a part of the sugars into alcohol and organic acids, thereby imparting a sweet-sour taste to the product.

Poi

To prepare poi, the corm of the taro plant (*Colocasia esculenta*) is cooked for several hours by baking or steaming, peeled, ground, combined with water to make a smooth, sticky paste, and stored airtight for 2 to 3 days at ambient temperature (8). *Lactobacillus delbrueckii* and *Lactococcus lactis* predominate in the early stages of fermentation, lowering the pH to 3.8 to 4.0. These bacteria, along with *Saccharomyces kefyr*, produce a large amount of lactic acid and moderate amounts of acetic, propionic, succinic, and formic acids. *Candida vini* and *Geotrichum candidum*, which are prevalent in the later stages of fermentation, impart a pleasant fruity aroma to mature poi.

LEGUME PRODUCTS

Daddawa

Daddawa (dawadawa) preparation is still largely a family art practiced at home. Seeds of the African locust bean tree (*Parkia biglobosa*) are washed to remove yellow powdery pulp, leaving black beans which are then boiled in water in a covered container for 18 to 24 h, with occasional replenishing of water to swell the seeds and soften the very tough seed coats, which are then removed by pounding and rinsing. The cotyledons are reboiled for 30 min to 2 h when a native softening agent called kuru (containing mostly potash) is added. The cotyledons are drained, heaped (10 to 15 cm deep) in calabash trays or dumped in a hole in the ground, and covered with locally available leaves and sackcloth. Beans are left to ferment for 2 to 4 days at 25 to 35°C, during which time they become dark brown and covered in a sticky mucilaginous layer and develop a pungent odor. The bean mass is air dried in the sun or hot shade, where the beans darken further, and the beans are then used loose or shaped into balls or pyramids (9, 63).

Although daddawa is dominated and produced mainly by *Bacillus subtilis*, it contains several other species of *Bacillus* and *Leuconostoc* (52, 63). During fermentation, the temperature of 25°C and the pH of 7.0 of the beans increase to 45°C and pH 8.1 at 36 h. The content of free amino acids increases fivefold due to extensive proteolysis. Certain antinutritional factors such as oligosaccharides, phytic acid, and oxalate decrease during fermentation (52).

Kinema

Kinema resembles natto except that, although in natto intact whole soybeans are used, in kinema the beans are crushed to form grits about half the size of cotyledons. Kinema is a naturally fermented product, containing *Bacillus* spp., enterococci, and yeasts. However, *B. subtilis* is the principal bacterium in the microflora and is largely responsible for the production of kinema. Spores of *B. subtilis*, which are normally present on soybeans, survive the cooking treatment to initiate and carry out the fermentation. Strong proteolytic activity causes an increase in pH from an initial pH of 6.9 to a pH of 8.6 at the end of fermentation (62, 63).

Meitauza

Meitauza is prepared from okara (insoluble carbohydrate residue that is left over after the production of soy milk or tofu) which is ground, steeped, strained, and formed into round cakes 10 to 14 cm in diameter and 2 to 3 cm thick at the middle and 1 to 1.3 cm thick at the edges. The cakes are placed in a vessel and left to ferment with moderate aeration until, after 10 to 15 days, they are covered with white mycelium of *Actinomucor elegans*. The molded cakes are then sun dried. Meitauza is served either fried in vegetable oil or cooked with vegetables as a flavoring agent (56).

Natto

Natto, a popular breakfast and dinner item in Japan, is usually eaten with rice along with soy sauce and spicy mustard. It is the only food in the category of alkaline fermentations that has been industrialized (62). With the use of whole soybeans, three types of natto are prepared. Yukiwari-natto and hama-natto are koji (*Aspergillus oryzae*)-based products, while the more common itohiki-natto is a *Bacillus*-fermented product. Itohiki-natto, generally referred to as natto, is popular in the eastern Kanto region (Tokyo). Natto soybeans are small (up to 5.5 mm in diameter) with a clear hilum, thin seed coat, and high carbohydrate content. Smaller beans are preferred, as the fermentation process reaches the center of the beans easier. To prepare natto traditionally, soybeans are washed, soaked overnight, boiled until tender (approximately 15 min), drained, partially air dried for 20 min over bamboo trays, and put into shallow paper containers covered with wax paper. The containers are stacked in large wooden boxes, covered with straw mats, and left near an oven at approximately 36°C to ferment for 1 day. Intentional inoculation is not necessary because straw contains the fermenting microorganism, *B. subtilis*. However, not all strains of *B. subtilis* are suitable for making good natto (53).

Yukiwari-natto is made by mixing itohiki-natto with rice koji and salt and then aging at 25 to 30°C for about 2 weeks. To prepare hama-natto, washed soybeans are soaked in water for 4 h and steamed for 1 h. After cooling, the beans are inoculated with koji, fermented for about 20 h, dried to a moisture content of 12%, submerged in brine, and aged for 6 to 12 months.

Natto is prized for its high nutritional value and improved digestibility, both resulting from fermentation. The nature of the free amino acid profile of natto is similar to that of kinema (44). Natto has a characteristic pungent but pleasant aroma. Sulfur-containing compounds deriving from the cooked soybeans and pyrazines formed during fermentation are the main contributors to the characteristic natto odor. The sulfur compounds include 4-ethyl-2-methylthiazole, 3,5-dimethyl-1,2,4-trithiolane, and thialdine. The pyrazines present at the highest concentrations include tetramethyl, trimethyl, and 2,5-dimethyl derivatives (57). Natto is also characterized by the presence of a sticky paste on its surface. When stirred, the paste increases in volume and becomes stickier and is held together like a spider web by gossamer-like threads. Natto mucin contains 22% fructan and 78% poly-DL-glutamic acid with a γ-peptide linkage (γ-PGA) which has a high viscoelasticity and is spinnable due to the formation of network structures of randomly coiled γ-PGA through intermolecular H bondings in the presence of fructan (25). Production of γ-PGA in the natto strains of *B. subtilis* is regulated by the *comQXPA* quorum-sensing system and is genetically unstable because of the translocation of IS4*Bsu1* into the *comP* gene at a high frequency (41). The IS4*Bsu1* is widely distributed among *B. subtilis* strains in other similar soybean-fermented foods, such as kinema, Thai thua-nao, Chinese douchi, Korean chungkuk-jang, and Burmese chine pepoke (24). Natto mucin can absorb 5,000 times its weight in water, and this remarkable property has been put to use in cosmetics and wrappings of food products.

Oncom

Oncom is a by-product of peanut oil pressing, produced by soaking peanut (*Arachis hypogaea*) press cake for 1 day, mixing with starchy ingredients such as cassava residues, steaming for about 1 h, cooling, and inoculating with pregrown fungal mycelium, usually *Neurospora sitophila* or *Neurospora intermedia*. The inoculated dough is molded to form brick-shaped pieces that are incubated for a few days in banana leaves at ambient temperature (25 to 30°C) (31). Oncom hitam (black oncom) and oncom merah (yellow-red oncom) contain different mycofloras. The merah type contains mainly *Neurospora*, whereas hitam is dominated by *Rhizopus* spp. *Rhizopus* sporangiospores are black due to melanoids, and carotenoids form the basis of the orange-yellow color of *Neurospora*.

Oncom flavor has been described as fruity and somewhat alcoholic; after frying, mincemeat or almond flavors can be detected. The enzymatic activities (lipases, proteases) provoke an increase in free fatty acids and degradation of proteins, resulting in improved protein digestibility which is relevant for consumers with digestive disorders. Recently, experiments with oncom-miso made from soybeans and oncom demonstrated increased anti-oxidative and antimutagenic activity associated with the enzymatic release of isoflavone-aglycones (37). Instead of dry, spore-based starters, starters used for oncom are propagated and maintained by mycelial growth in a kind of fed-batch solid-state fermentation kept active by the processors. Although very little controlled experimentation has been done on this fermentation, it is presumed that a method of vegetative propagation is needed because the *Neurospora* spores have limited viability when stored dry and have poor germination ability.

Sufu

Actinomucor elegans (7, 21) and *Actinomucor taiwanensis* (11) are used as pure-culture starters in the manufacture of Chinese fu-ru, or sufu. The process of preparing sufu starts with the production of soy milk by soaking dehulled soybeans, grinding, sieving, and cooking the watery extract, the latter to inactivate trypsin inhibitors and reduce some of the undesirable beany flavor. Next, a coagulation step is carried out, by adding salts (calcium or magnesium sulfate) or acid, in order to obtain a precipitate of mainly soy protein and entrapped lipids. This precipitate is collected and pressed to obtain sheets of tofu (soybean curd) of the required moisture content and firmness. After cutting of the tofu into cubes (dices), the tofu is inoculated with a suspension of mold spores. Incubation for a few days usually results in luxuriant mycelial development, giving the dices a fluffy appearance. These are now called pehtze, containing about 74% water, 12% protein, and 4.3% lipid. After flattening of the mycelium to form a protective skin on the cubes, they are submerged in a maturation mix and stored for several months to develop into a flavorsome, soft, cheese-like product. The main functions of the maturation mix are preservation, flavoring, and coloring. Preservation of sufu is achieved by a combination of salt and alcohol (rice beer may be used), whereas ang-kak and other ingredients impart specific flavor and color to the product (21). Depending upon the desired flavor and color, pehtzes may be submerged in salted, fermented rice or soybean mash, fermented soybean paste, or a solution containing 5 to 12% sodium chloride, red rice, and 10% ethanol. Red rice and soybean mash impart a red color to sufu. Use of brine containing high levels of ethanol results in sufu with a marked alcoholic bouquet. The major functions of the molds in this process are the formation of a protective layer of mycelial biomass surrounding the pehtze cubes and, most importantly, the release of several enzymes that are responsible for the partial degradation of the protein, fiber, and lipid fractions in pehtze during the maturation. This degradation results in softening of the texture, solubilization of constituents, and accumulation of flavor-enhancing compounds, such as glycine and glutamic acid (20, 36). In view of the optimization of industrial sufu-making processes, the response of *Actinomucor elegans* to temperature, salt, and alcohol has been studied. The higher the salt and alcohol levels during the maturation, the slower the enzymatic reactions take place, thus requiring longer maturation times. With the objective of accelerating the maturation, the salt and alcohol levels could be lowered. This is feasible to a level of about 10% alcohol in combination with 6% salt; at lower levels the product is susceptible to spoilage by LAB, however, as well as survival of pathogens and enterotoxin formation by *Staphylococcus aureus*.

Tempeh

Tempeh (the Indonesian spelling is "tempe") is made from cooked seeds (those of soybeans, cereals, or others) or seed-processing by-products by solid-state fungal fermentation (47, 48). Tempeh is an attractive nonmeat protein food that can be used as an ingredient in a large variety of traditonal Indonesian dishes as well as in Western-style spreads, snacks, and burgers.

Soybeans are soaked in water at ambient temperature overnight or until hulls (testae) can be easily removed by hand. LAB and yeasts predominant in water in which soybeans have been soaked are *Lactobacillus casei*, *Lactococcus* spp., *Pichia burtonii*, *Candida diddensiae*, and *Rhodotorula mucilaginosa* (40). Fermentative acidification during the soaking stage has been shown to suppress the growth of spoilage and pathogenic bacteria (48).

After removal of the hulls from the soaked soybeans, cotyledons are cooked for 30 to 60 min, drained, and cooled. In the traditional tempeh process, simple methods are employed for the inoculation of the cooked beans. In principle, it is possible to use some previously made tempeh as inoculum (32); however, as tempeh contains a considerable population of bacteria other than those desired for fermentation, the reuse of tempeh as an inoculum incurs the risk of fermentation failure due to bacterial overgrowth. Therefore, professional tempeh manufacturers use traditional mold spore concentrates. These are, e.g., harvested from cooked rice on which selected strains of *Rhizopus oligosporus* have been cultured or cooked soybeans that have been held between leaves of *Hibiscus tiliaceus* (the waru tree). The latter type of widely used starter, made by specialized households, is available in the public markets in Indonesia.

The inoculated beans are then spread onto bamboo frames, wrapped in a punctured plastic sheet or between banana leaves, and allowed to ferment at ambient temperature (25 to 30°C) for 1 to 2 days. At this point, the soybeans are covered with white *Rhizopus oligosporus* mycelium and bound together as a cake (32). Aerobic mold growth requires oxygen and produces heat and carbon dioxide. Care should be taken that the beans do not dehydrate and that no overheating (>40°C) occurs. This is achieved by allowing only restricted access of air to the beans and by limiting the thickness of the bean layers or packages (47).

Several factors may limit the acceptability and shelf life. These include the production of black sporangia and spores, indicating inadequate fermentation conditions, which results in an undesirable gray color, and enzymatic browning, comparable to the browning of cut apples, which is initiated by prolonged storage or mechanical abuse. Whereas freshly fermented tempeh has an attractive mushroom-like flavor, prolonged storage may lead to yeasty off-odors or ammoniacal odors resulting from excessive protein and amino acid degradation.

During fermentation, carbohydrases, lipases, proteases, phytases, and other enzymes degrade macromolecular substrates, resulting in very significant increases in water-soluble nutrients for enhanced digestion, biosynthesis of B vitamins such as folate (18), and transformation of soy isoflavones into health-promoting antioxidant compounds (47).

Wadi and Papad

Wadi (wari) is prepared by soaking dal (dehusked split beans), generally of black gram (*Phaseolus mungo*), in water for 6 to 12 h, draining, grinding into a smooth soft dough, and fermenting for 1 to 3 days at 20 to 27°C, with or without spices but with salt and backslop added. In an alternative method, the dough is combined with shredded waxgourd (*Benincasa hispida*) and whisked vigorously until it becomes light and fluffy due to the incorporation of air. The fermented or whisked dough is hand-molded into cones or balls (3 to 8 cm in diameter), deposited onto bamboo or palm mats smeared with oil, and sun dried for 4 to 8 days (6). The surface of the cones or balls becomes covered with a mucilaginous coating which retains the gas formed during fermentation within them. The wadis look hollow with many air pockets and yeast spherules in the interior and have a characteristic surface crust.

Initially the microflora includes LAB, *Bacillus* spp., flavobacteria, and yeasts. Gradually, a domination by gas-producing *L. mesenteroides*, *Lb. fermentum*, *S. cerevisiae*, and *Trichosporon cutaneum* is achieved. *Candida vartiovaarae* and *Kluyveromyces marxianus* are also often found. Summer is more favorable for the prevalence of bacteria, and winter is more favorable for the

yeasts (67). The production of acid and gas results in a decrease of pH from 5.6 to 3.2, an increase in total acid (as lactic acid) from 0.5 to 1.5%, and a twofold increase in the volume of the dough. The LAB are mainly responsible for acidification of dough, a condition which favors the growth of yeasts and leavening. Fermentation brings about a significant increase in soluble solids, nonprotein nitrogen, soluble nitrogen, free amino acids, and B vitamins. Most of these changes cause improvement in digestibility and nutritional value. Increase in total acidity during fermentation helps to enhance the shelf life of the product (66).

Papad (papadam, or appalam) is a thin, usually circular, wafer-like product used to prepare curry or eaten as a crackly snack or appetizer with meals after roasting or deep frying. In the indigenous method of preparation, black gram flour alone or blended with Bengal gram (*Cicer arietinum*), lentil (*Lens culinaris*), red gram, or green gram (*Phaseolus aureus*) flour is hand-kneaded with a small quantity of peanut oil, common salt (about 8%), papad khar (a natural additive), and water and then beaten or pounded into a stiff paste. The paste may be seasoned with spices. The dough (sometimes with backslop added) is left to ferment for 1 to 6 h and then formed into long cylinders and cut and shaped into small balls which are rolled into thin, circular flat sheets (10 to 24 cm in diameter, 0.2 to 1.2 mm thick) by using a wooden rolling pin and generally dried under shade to 12 to 17% moisture content (6, 64). *C. krusei* and *S. cerevisiae* are involved in fermenting the dough, presumably resulting in modest leavening.

CEREAL-LEGUME MIXTURE PRODUCTS

Idli

Idli is a classical example of cereal-legume mixture food that provides an improved balance of carbohydrates and proteins. Because of its appealing sour flavor, spongy texture, nutritional quality, and easy digestibility, idli is also fed to infants as a complementary food and is used as a main dish in diets provided to patients in hospitals (49).

The substrates used in preparing idli are white polished rice and black gram dal (1:4 to 4:1), which are washed and soaked in water separately at ambient temperature for 5 to 10 h. While rice is coarsely ground, the dal is ground into a smooth, mucilaginous paste. The two slurries are combined generally in the ratio of 2:1 and stirred well with added salt (0.8%) to form a thick batter which is put in a closed container and left in a warm place (25 to 35°C) to ferment overnight or longer (14 to 24 h). The fermentation period must allow a definite leavening (two- to threefold increase in volume) of the batter and development of a pleasant acid flavor. The fermented batter is poured into small cups (8 to 10 cm in diameter) and steamed in a covered pan to yield a soft, spongy product (39). The open texture is attributed to the protein (globulin) and polysaccharide (arabinogalactan) in black gram (71).

Although bacteria and yeasts that participate in the fermentation are generally introduced by the substrates, it is often the practice to add backslop to the newly ground substrates. With the progress of fermentation, both bacterial and yeast cell numbers increase significantly, with a concomitant decrease in pH and an increase in the volume of the batter and its amylase and protease activities. *L. mesenteroides* is the most commonly encountered bacterium, followed by *Lb. fermentum*, *Enterococcus faecalis*, and *Pediococcus dextrinicus* (39). During fermentation, along with *L. mesenteroides*, yeasts such as *S. cerevisiae*, *Debaryomyces hansenii*, *Pichia anomala*, and *Trichosporon pullulans* are predominant, and *Trichosporon cutaneum* develops subsequently. *S. cerevisiae* is the only yeast that eventually persists (68).

The major functions of the fermentation of idli include the leavening of batter and improvement of flavor and nutritional value. The role of LAB is to reduce the pH of the batter from an initial 6.0 to an optimum level (4.1 to 4.5) for yeast activity. The LAB may also play a role in the breakdown of phytate present in black gram. *L. mesenteroides* isolated from soy idli secretes β-N-acetylglucosaminidase and α-D-mannosidase, which are involved in the hydrolysis of hemagglutinin. Yeasts help in the degradation of starch, a process that cannot be carried out by *L. mesenteroides*, into maltose and glucose by producing extracellular amylolytic enzymes. Yeasts also produce carbon dioxide and play a significant role in leavening. Fermentation of batter by inoculating the ingredients with yeasts individually and in combination with *L. mesenteroides* has revealed that yeasts contribute not only to gas production, resulting in good texture, but also to sensory qualities. The higher activity of amylases and levels of B vitamins and free amino acids attained in yeast-enriched fermentations represent positive contributions of yeasts (73).

Miso

Fermented soybean pastes are known as miso in Japan, chiang in China, jang or doenjang in Korea, taoco in Indonesia, and tao chieo in Thailand. In addition to soybeans and salt, most of these products contain cereals such as rice or barley (38). Miso is fermented using *Aspergillus oryzae* and a yeast, *Zygosaccharomyces rouxii*. Sometimes, *Tetragenococcus halophila* and *Enterococcus faecalis* are also involved in the fermentation. Heat-treated rice and/

or soybeans are used to prepare shinshu or rice-soybean miso. After the initial solid-state fermentation dominated by *Aspergillus oryzae*, salt (38% of the original weight of dry soybeans) is added to the koji and mixed thoroughly. The mixture is inoculated with *Z. rouxii*; traditionally, sound miso from a previous batch is used to inoculate the koji-soybean-salt mixture prior to fermentation. Although halophilic yeasts such as *Torulopsis versatilis* may be present, only *Z. rouxii* produces the desired metabolites for an acceptable product. The mixture, known as green miso, is packed into vats or tanks to undergo anaerobic fermentation and aging at 25 to 30°C. White miso takes about 1 week, salty miso takes 1 to 3 months, and soybean miso takes over 1 year. White miso contains 4 to 8% salt, which permits rapid fermentation, and yellow or brown misos contain 11 to 13% salt. Moisture content ranges from 44 to 52%, protein content ranges from 8 to 19%, carbohydrate content ranges from 6 to 30%, and fat content ranges from 2 to 10%, depending on the ratio of soybeans, rice, and barley used as ingredients.

During fermentation and aging, soybean protein is digested by proteases produced by *Aspergillus oryzae* in the koji. Amino acids and their salts, particularly sodium glutamate, contribute to flavor. The addition of commercial enzyme preparations to enhance fermentation has met with some success. The relative amount of carbohydrates in miso is a reflection of the amount of rice in the product. Starch is extensively saccharified by koji amylases to yield glucose and maltose, some of which is utilized as a source of energy by the microorganisms responsible for fermentation. Miso contains 0.6 to 1.5% acids, mainly lactic, succinic, and acetic. Esters of ethyl and higher alcohols with fatty acids in soybean lipid are important in giving miso its characteristic aroma. Up to 35% of the initial lipid content is degraded into fatty acids; the extent of maturation can be conveniently monitored by the levels of fatty acid ethyl esters (75). Furanones HEMF [4-hydroxy-2(or 5)-ethyl-5(or 2)-methyl-3(2H)-furanone] and HDMF [4-hydroxy-2, 5-dimethyl-3(2H)-furanone] produced by *Z. rouxii* have been identified as important flavor components in miso. Miso also contains B vitamins (riboflavin and cyanocobalamin) as a result of yeast fermentation. Miso is considered to be a health-promoting, functional food, offering protection against gastrointestinal disorders; cancers of the breast, stomach, and colon; and cholesterol-associated and degenerative diseases (38).

Soy Sauces

Soy sauces are light to dark brown liquids with meat-like salty flavor used in cooking and as a table condiment. Traditionally made in the Orient, they are now also produced in Europe and the Americas. In Japan, there are several distinguished types, including Koikuchi, Usukuchi, Tamari, Saishikomi, and Shiro soy sauces, all having characteristic colors and flavors. All contain relatively high levels (17 to 19%) of salt and are used to enhance the flavors of meats, seafoods, and vegetables. Typical ranges in other characteristics are as follows: pH 4.6 to 4.8 and 0.5 to 2.5 g of total nitrogen, 0.2 to 1.1 g of formol nitrogen, 3.8 to 2.0 g of reducing sugar, and traces to 2.2 ml of ethanol (per 100 ml).

There are two specific fermentation stages involved in soy sauce production, the first being an aerobic koji fermentation. Seed (tane) koji is produced by culturing single or mixed strains of *Aspergillus oryzae* or *Aspergillus sojae* on either steamed, polished rice or a mixture of wheat bran and soybean flour. Seed koji is added to a soybean-wheat mixture at a concentration of 0.1 to 0.2% and fermented into what is then simply called koji. The second stage is an anaerobic moromi or salt mash which undergoes LAB and yeast (*Z. rouxii*) fermentations.

The two main groups of enzymes produced by *Aspergillus oryzae* during koji fermentation are carbohydrases (α-amylases, amyloglucosidase, maltase, sucrase, pectinase, β-galactosidase, cellulase, hemicellulase, and pentosan-degrading enzymes) and proteinases, although lipase activity has also been reported. These enzymes hydrolyze carbohydrates to sugars and proteins to produce amino acids and low-molecular-weight peptides. These soluble products are essential for yeast and bacterial activities during the moromi fermentation (2, 12). In the moromi

fermentation, *Tetragenococcus halophila* initially proliferates and produces lactic acid, which lowers the pH to 5.5 or less. This is followed by the growth of acid-tolerant yeasts, notably *Z. rouxii*, which produce about 3% alcohol and several compounds that contribute to the characteristic aroma of soy sauce.

Although *Z. rouxii* is the dominant moromi yeast, other yeasts such as *Candida versatilis* and *Candida etchellsii* also produce phenolic compounds which contribute to soy sauce aroma. Nearly 300 flavor compounds have been identified in Japanese soy sauce (50); major categories are summarized in Table 38.3. *Z. rouxii* produces flavor compounds, including alcohols, glycerol, esters, and furanones. Of the latter, HEMF produced by *Z. rouxii* and *Candida* spp. gives Japanese-type soy sauce its characteristic flavor (22). This compound is also reported to have antitumor and antioxidative properties (33). Notwithstanding their high salt contents, soy sauces require pasteurization and adequate bottling for preservation.

VEGETABLE PRODUCTS

Kimchi

Kimchi is a generic term used to denote a group of fermented vegetable foods in Korea. More specific names are used for these products, depending on the substrate, processing method, season, and locality. Each family has its own recipe handed down from generation to generation. T'ong baegu'u-kimchi, the most common kimchi, is

---

Table 38.3 Major flavor components in soy sauce

Type	Component(s)
Produced by *Zygosaccharomyces rouxii*	
Alcohols	Benzyl alcohol, 1-butanol, ethanol, glycerol
Esters	Bornyl acetate, ethyl acetate, ethyl lactate, ethyl myristate
Furanones	4-Hydroxy-5-methyl-3(3*H*)-furanone (HMMF), 4-hydroxy-2(or 5)-ethyl-5(or 2)-methyl-3(2*H*)-furanone (HEMF)
Produced by *Candida versatilis* and *Candida etchellsii*	
Phenolic compounds	2-Methoxy-4-ethylphenol, 2-methoxyphenol
Higher alcohols	Isobutyl alcohol, isoamylalcohol, 2-phenylethanol
Miscellaneous	
Aldehydes	Acetaldehyde, benzaldehyde, 2-methylpropanal
Volatile carboxylic acids	Acetic acid, butanoic acid, 3-methylbutanoic acid
Ketones	2-Hexanone, 3-hydroxyl-2-butanone
Furans	2-Acetyl furan
Pyrroles	2-Acetylpyrrole
Pyrazines	2,3-Dimethylpyrazine, 3-ethyl-2,5-dimethylpyrazine, 2-methylpyrazine

made by cutting Chinese-type cabbage into two or four parts and adding salt (approximately 3%), followed by rinsing; draining; packing in layers with premixed garlic, onion, ginger, hot pepper powder, and salt; and fermenting for 2 to 3 days at around 20°C. The flavor of kimchi is dependent on the ingredients, fermentation conditions, and LAB involved in the fermentation (34). There is a succession of microfloras during kimchi production. *L. mesenteroides* is a major bacterium found in the initial to the middle stage of fermentation. *Weissella confusa* and *Leuconostoc citreum* present in the raw materials also initiate and remain present throughout the fermentation period, indicating their importance in kimchi production. *Lb. brevis* and *Lb. plantarum*, which appear at the middle stage of fermentation, are overtaken by *Lactobacillus sakei* and *Lb. curvatus*, which dominate the late stage of fermentation when the pH decreases to 3.9 (35). They produce various constituents such as lactic acid, acetic acid, ethanol, carbon dioxide, mannitol, and dextran that impart a good flavor to kimchi. Kimchi closely resembles sauerkraut. The optimal range in salt concentration in sauerkraut is 0.7 to 3.0%, and that in kimchi is 3.0 to 5.0%, and secondly, kimchi is less sour than sauerkraut but carbonated.

FISH PRODUCTS

Basically salt-fermented fish products are popular in the Orient (34). Where large quantities (>20%) of salt are added, fermentation by microorganisms is largely suppressed and enzymes from fish flesh and gut cause proteolysis of the fish. To allow fermentation to gain more importance, a source of carbohydrate is added so that LAB prevail and ferment the carbohydrate into organic acids, which reduces the high buffering capacity of the fish to result in a rapid decrease in pH, aiding preservation and production of a tangy odor (58). To prepare low-salt fermented products like Korean sik-hae (side dish), whole fish is degutted; stripped; mixed with salt (6 to 7%); cured overnight; drained; mixed with cooked millet (7 to 8%), minced garlic (3 to 4%), and red pepper powder (8 to 9%); and packed in an earthen jar for fermentation at 20°C for 1 week. The jar is then kept in a cool place for 3 to 6 weeks to develop optimum flavor. The pH decreases rapidly during the first 3 to 5 days from 6.5 to <5.0, and softening begins after 3 days of fermentation. The amino-N concentration increases steadily up to 14 days, and this increase coincides with the attainment of optimum flavor. The important fermenting organisms are *L. mesenteroides* and *Lb. plantarum* (34). Garlic affects the microflora by inhibiting the growth of gram-negative bacteria, in particular, due to its allicin content and by stimulating the growth of LAB due to the supply of fermentable fructose from its inulin reserve.

MISCELLANEOUS PRODUCTS

Kombucha

For kombucha, black tea leaves are steeped in boiling water, combined with sugar (5 to 15%), cooled, inoculated, and left for 7 to 14 days, during which a film grows on the surface. When sufficiently thick (ca. 2.5 cm), the film falls to the bottom and further film develops on the surface. This biofilm is used as a starter for new batches; the liquid (pH of ca. 2.5) is claimed to benefit health (19). The cellulosic pellicle is a result of symbiotic associations of acetic acid bacteria, chiefly *Acetobacter aceti* subsp. *xylinus*, and various yeasts, e.g., *Brettanomyces*, *Zygosaccharomyces*, and *Saccharomyces* spp. Among them, *Zygosaccharomyces kombuchaensis* is the dominant species. Ethanol-producing yeasts are succeeded by acid-producing bacteria.

Palm Wines

The sap obtained from decapitated inflorescence of palm is fermented to produce wine. There is an art in binding the flower spathes, pounding to cause the sap to flow properly by cutting the spathe tip, and collecting the sap. The fermentation starts as soon as the sap flows into a container, and within a few hours it becomes reasonably high in alcohol content (up to 4%). The palm wine fermentation is always an alcoholic-lactic-acetic one, involving yeasts, LAB, and acetic acid bacteria. The microorganisms responsible are mainly *S. cerevisiae*, *Schizosaccharomyces pombe*, *Lb. plantarum*, and *L. mesenteroides*. The LAB are responsible for the consistency and soluble white coloration of palm wines through the production of glucans (29).

NUTRITION AND HEALTH-PROMOTING ASPECTS

Many traditional fermented foods are staples in the diets of vast populations of people who would otherwise have less than minimum intakes of protein and/or calories. Although the quality or quantity of proteins in vegetable-based fermented foods generally is not dramatically increased over that in raw substrates, digestibility may be improved through fermentation. Antinutritional factors in plant materials, e.g., protease inhibitors and lectins in leguminous seeds and phytic acid in cereals and seeds, may actually be reduced by fermentation. Factors that inhibit digestive activity, or that form indigestible complexes with proteins and minerals, are degraded during fermentation processes by endogenous (plant) and microbial enzymes. Naturally occurring toxic components, e.g., cyanogenic glycosides and glucosinolates, can be degraded substantially in fermented foods, rendering inedible materials into wholesome foods. Food components such as soybean isoflavones may be transformed into potent antioxidant aglycones, which have potential for reducing hypertension, cardiovascular ailments, and cancer, as a result of food fermentation. A significant synthesis of vitamins, especially B vitamins such as thiamine, niacin, folate, and riboflavin, contributes to the bioenrichment of foods (69).

The presence of large numbers of microorganisms (e.g., LAB) in fermented foods such as yogurt may have probiotic effects, i.e., supporting a healthy balance of the gut microbiota and offering enhanced colonization resistance and immunoresistance to the host (46). The positive contribution of traditional fermented foods to the nutritional well-being and health of those who consume them on a regular basis is widely recognized.

FOOD SAFETY ASPECTS

Of interest to food microbiologists and sanitarians is the possibility of microorganisms' producing toxic substances or of pathogenic microorganisms' surviving during fermentation or storage of indigenous fermented foods. An investigation of the aflatoxin-forming ability of aspergilli used in the Japanese food industry revealed that although fluorescent compounds were formed, none produced aflatoxin. Recently it was found that *Aspergillus oryzae* and *Aspergillus sojae*, typical "domesticated" industrial molds that have been used for centuries in the production of koji for soy sauce and miso, are incapable of forming aflatoxins. The *aflR* gene (aflatoxin pathway-specific regulatory gene) is impaired in its ability to activate transcription of aflatoxin biosynthetic genes and is unable to interact with *aflJ* (coactivator gene) (10). The inability of potentially toxigenic molds to form mycotoxins in food fermentations may be caused by altered gene expression, unfavorable food environment, or competition with other microorganisms.

A variety of pathogenic bacteria, viruses, and parasites may occur in raw ingredients or in fermented foods due to postprocessing contamination. Whereas many gram-negative bacteria cannot survive when exposed to acidic and/or elevated salt, alcohol, or sugar conditions, many other bacteria survive such stress conditions. Viruses and parasitic cysts can be eradicated only by adequate heat treatments. Safety of fermented foods must be evaluated and controlled on the basis of safety management

approaches, i.e., the hazard analysis and critical control point system (1) (see chapter 46 in this volume). Although traditional fermented foods have a history of safe use, adequate measures for hygienic processing and protection against contamination must be implemented.

MODERN TOOLS IN RESEARCH AND DEVELOPMENT OF FERMENTED FOODS

The food industry is in continuous transition, adopting an innovative and market-oriented position (13). Within this trend, there are ample opportunities for genomics-based approaches to be applied in the fermented food industry. These include transcriptomics, proteomics, and other postgenomics approaches and are important new tools that may be applied to optimize processes and to gain a better insight into mechanisms. In view of the protection of origin (a.o.c., or certified origin of production), an unequivocal characterization of traditional fermented foods and their microflora will be required; this could be based on combinations of food compositional analysis and metabolomics (46).

In addition, these so-called "-omics" approaches offer opportunities to develop novel products and diversify processes. Innovative processes using nontraditional fermentation conditions, e.g., immobilized cells or agitated solid-state fermentors, or using pure-culture inoculation instead of multistrain natural fermentations, may invoke changes in secondary metabolite production. In view of maintaining the character of the food, as well as safeguarding safety for the consumer, the impact of novel processing should be investigated, understood, and possibly controlled. -Omics approaches will also offer opportunities to advance indigenous fermented food quality assurance and allow running cost to be reduced to more affordable levels. Finally, in the field of probiotics, -omics offer analytical approaches to substantiate health claims.

References

1. **Adams, M. R., and M. J. R. Nout (ed.).** 2001. *Fermentation and Food Safety.* Aspen Publishers, Gaithersburg, Md.
2. **Aidoo, K. E., J. E. Smith, and B. J. B. Wood.** 1994. Industrial aspects of soy sauce fermentations using *Aspergillus*, p. 155–169. *In* K. A. Powell, A. Renwick, and J. F. Peberdy (ed.), *The Genus* Aspergillus: *from Taxonomy and Genetics to Industrial Application.* Plenum Press, New York, N.Y.
3. **Akihisa, T., H. Tokuda, K. Yasukawa, M. Ukiya, A. Kiyota, N. Sakamoto, T. Suzuki, N. Tanabe, and H. Nishino.** 2005. Azaphilones, furanoisophthalides, and amino acids from the extracts of *Monascus pilosus*-fermented rice (red-mold rice) and their chemopreventive effects. *J. Agric. Food Chem.* 53:562–565.

4. **Ampe, F., N. Ben-Omar, C. Moizan, C. Wacher, and J. P. Guyot.** 1999. Polyphasic study of the spatial distribution of microorganisms in Mexican pozol, a fermented maize dough, demonstrates the need for cultivation-independent methods to investigate traditional fermentations. *Appl. Environ. Microbiol.* **65:**5464–5473.

5. **Batra, L. R.** 1981. Fermented cereals and grain legumes of India and vicinity, p. 547–553. *In* M. Moo-Young and C. W. Robinson (ed.), *Advances in Biotechnology*, vol. 2. *Fuels, Chemicals, Foods and Waste Treatment.* Pergamon Press, New York, N.Y.

6. **Batra, L. R., and P. D. Millner.** 1974. Some Asian fermented foods and beverages and associated fungi. *Mycologia* **66:**942–950.

7. **Benjamin, C. R., and C. W. Hesseltine.** 1957. The genus *Actinomucor. Mycologia* **49:**240–249.

8. **Brown, A. C., and A. Valiere.** 2004. The medicinal uses of poi. *Nutr. Clin. Care* **7:**69–74.

9. **Campbell-Platt, G.** 1980. African locust bean (*Parkia* species) and its West African fermented food product, dawadawa. *Ecol. Food Nutr.* **9:**123–132.

10. **Chang, P. K.** 2004. Lack of interaction between AFLR and AFLJ contributes to nonaflatoxigenicity of *Aspergillus sojae. J. Biotechnol.* **107:**245–253.

11. **Chou, C. C., F. M. Ho, and C. S. Tsai.** 1988. Effects of temperature and relative humidity on the growth of and enzyme production by *Actinomucor taiwanensis* during sufu pehtze preparation. *Appl. Environ. Microbiol.* **54:**688–692.

12. **Chou, C.-C., and J.-H. Rwan.** 1995. Mycelial propagation and enzyme production in koji prepared with *Aspergillus oryzae* on various rice extrudates and steamed rice. *J. Ferment. Bioeng.* **79:**509–512.

13. **De Vos, W. M.** 2005. Frontiers in food biotechnology—fermentations and functionality. *Curr. Opin. Biotechnol.* **16:**187–189.

14. **Dung, N. T. P.** 2004. Defined fungal starter granules for purple glutinous rice wine. Ph.D. thesis. Wageningen University, Wageningen, The Netherlands.

15. **Fleet, G. H.** 1998. The microbiology of alcoholic beverages, p. 217–262. *In* B. J. B. Wood (ed.), *Microbiology of Fermented Foods*, 2nd ed., vol. 1. Blackie Academic & Professional, London, United Kingdom.

16. **Gänzle, M. G., and R. F. Vogel.** 2002. Contribution of reutericyclin production to the stable persistence of *Lactobacillus reuteri* in an industrial sourdough fermentation. *Int. J. Food Microbiol.* **80:**31–45.

17. **Gashe, B. A.** 1985. Involvement of lactic acid bacteria in the fermentation of tef (*Eragrostis tef*), an Ethiopian fermented food. *J. Food Sci.* **50:**800–801.

18. **Ginting, E., and J. Arcot.** 2004. High-performance liquid chromatographic determination of naturally occurring folates during tempe preparation. *J. Agric. Food Chem.* **52:**7752–7758.

19. **Greenwalt, C. J., K. H. Steinkraus, and R. A. Ledford.** 2000. Kombucha, the fermented tea: microbiology, composition, and claimed health effects. *J. Food Prot.* **63:**976–981.

20. **Han, B. Z., F. M. Rombouts, and M. J. R. Nout.** 2004. Amino acid profiles of sufu, a Chinese fermented soybean food. *J. Food Comp. Anal.* **17:**689–698.

21. **Han, B. Z., F. M. Rombouts, and M. J. R. Nout.** 2001. A Chinese fermented soybean food. *Int. J. Food Microbiol.* **65:**1–10.

22. **Hanya, Y., and T. Nakadai.** 2003. Yeasts and soy products, p. 413–428. *In* T. Boekhout and V. Robert (ed.), *Yeasts in Food: Beneficial and Detrimental Aspects.* B. Behr's Verlag GmbH & Co. KG, Hamburg, Germany.

23. **Hounhouigan, D. J., M. J. R. Nout, C. M. Nago, J. H. Houben, and F. M. Rombouts.** 1993. Changes in the physico-chemical properties of maize during natural fermentation of mawe. *J. Cereal Sci.* **17:**291–300.

24. **Inatsu, Y., K. Kimura, and Y. Itoh.** 2002. Characterization of *Bacillus subtilis* strains isolated from fermented soybean foods in Southeast Asia: comparison with *B. subtilis* (natto) starter strains. *Jpn. Agric. Res. Q.* **36:**169–175.

25. **Ishikawa, H., K. Okubo, and T. Oki.** 1972. Characteristic spinnability of a natto mucin solution. *Nippon Kagaku Kaishi* **11:**2171–2177.

26. **Jakobsen, M., M. D. Cantor, and L. Jespersen.** 2001. Production of bread, cheese and meat, p. 1–22. *In* H. D. Osiewacz (ed.), *The Mycota, Industrial Applications*, vol. 10. Springer-Verlag, New York, N.Y.

27. **Jenson, I.** 1998. Bread and bakers' yeast, p. 172–198. *In* B. J. B. Wood (ed.), *Microbiology of Fermented Foods*, 2nd ed., vol. 1. Blackie Academic & Professional, London, United Kingdom.

28. **Jespersen, L., M. Halm, K. Kpodo, and M. Jakobsen.** 1994. Significance of yeasts and moulds occurring in maize dough fermentation for "kenkey" production. *Int. J. Food Microbiol.* **24:**239–248.

29. **Joshi, V. K., D. K. Sandhu, and N. S. Thakur.** 1999. Fruit based alcoholic beverages, p. 647–744. *In* V. K. Joshi and A. Pandey (ed.), *Biotechnology: Food Fermentation*, vol. 2. Educational Publishers, Ernakulam, India.

30. **Kelly, W. J., R. V. Asmundson, G. L. Harrison, and C. M. Huang.** 1995. Differentiation of dextran-producing *Leuconostoc* strains from fermented rice cake (puto) using pulsed-field gel electrophoresis. *Int. J. Food Microbiol.* **26:**345–352.

31. **Ko, S. D.** 1986. Indonesian fermented foods not based on soybeans, p. 67–84. *In* C. W. Hesseltine and H. L. Wang (ed.), *Mycologia Memoir*, no. 11. *Indigenous Fermented Food of Non-Western Origin.* J. Cramer, Berlin, Germany.

32. **Ko, S. D., and C. W. Hesseltine.** 1979. Tempe and related foods, p. 115–140. *In* A. H. Rose (ed.), *Microbial Biomass*, vol. 4. Academic Press, London, United Kingdom.

33. **Koga, T., K. Moro, and T. Matsudo.** 1998. Antioxidative behaviors of 4-hydroxy-2,5-dimethyl-3(2H)-furanone and 4-hydroxy-2(or 5)-ethyl-5(or 2)-methyl-3(2H)-furanone against lipid peroxidation. *J. Agric. Food Chem.* **46:**946–951.

34. **Lee, C.-H.** 1996. Lactic fermented foods and their benefits in Asia. *Food Biotechnol.* **5:**187–197.

35. **Lee, J.-S., G.-Y. Heo, J. W. Lee, Y.-J. Oh, J. A. Park, Y.-H. Park, Y.-R. Pyun, and J. S. Ahn.** 2005. Analysis of kimchi microflora using denaturing gradient gel electrophoresis. *Int. J. Food Microbiol.* **102:**143–150.

36. **Liu, Y.-H., and C. C. Chou.** 1994. Contents of various types of proteins and water soluble peptides in sufu during ageing and the amino acid composition of taste oligopeptides. *J. Chin. Agric. Chem. Soc.* **32:**276–283.

37. **Matsuo, M.** 2004. Low-salt O-miso produced from koji fermentation of oncom improves redox state and cholesterolemia in rats more than low-salt soybean-miso. *J. Nutr. Sci. Vitaminol.* (Tokyo) **50:**362–366.

38. **Minamiyama, Y., and S. Okada.** 2003. Miso: production, properties, and benefits to health, p. 277–286. *In* E. R. Farnworth (ed.), *Handbook of Fermented Functional Foods.* CRC Press, Boca Raton, Fla.

39. **Mukherjee, S. K., M. N. Albury, C. S. Pederson, A. G. Van Veen, and K. H. Steinkraus.** 1965. Role of *Leuconostoc mesenteroides* in leavening the batter of idli, a fermented food of India. *Appl. Microbiol.* **13:**227–231.

40. **Mulyowidarso, R. K., G. H. Fleet, and K. A. Buckle.** 1989. The microbial ecology of soybean soaking for tempe production. *Int. J. Food Microbiol.* **8:**35–46.

41. **Nagai, T., L.-S. Phan Tran, Y. Inatsu, and Y. Itoh.** 2000. A new IS*4* family insertion sequence, IS*4Bsu*1, responsible for genetic instability of poly-gamma-glutamic acid production in *Bacillus subtilis. J. Bacteriol.* **182:**2387–2392.

42. **Nago, M. C., J. D. Hounhouigan, N. Akissoe, E. Zanou, and C. Mestres.** 1998. Characterization of the Beninese traditional ogi, a fermented maize slurry: physicochemical and microbiological aspects. *Int. J. Food Sci. Technol.* **33:**307–315.

43. **Nche, P. F., G. T. Odamtten, M. J. R. Nout, and F. M. Rombouts.** 1996. Soaking determines the quality of aflata for kenkey production. *J. Cereal Sci.* **24:**291–297.

44. **Nikkuni, S., T. B. Karki, K. S. Vilkhu, T. Suzuki, K. Shindoh, C. Suzki, and N. Okada.** 1995. Mineral and amino acid contents of kinema, a fermented soybean food prepared in Nepal. *Food Sci. Technol. Int.* **1:**107–111.

45. **Nout, M. J. R., and K. E. Aidoo.** 2002. Asian fungal fermented food, p. 23–47. *In* H. D. Osiewacz (ed.), *The Mycota: Industrial Applications,* vol. 10. Springer-Verlag, New York, N.Y.

46. **Nout, M. J. R., W. M. De Vos, and M. H. Zwietering (ed.).** 2005. *Food Fermentation.* Wageningen Academic Publishers, Wageningen, The Netherlands.

47. **Nout, M. J. R., and J. L. Kiers.** 2005. Tempe fermentation, innovation and functionality: up-date into the 3rd millennium. *J. Appl. Microbiol.* **98:**789–805.

48. **Nout, M. J. R., and F. M. Rombouts.** 1990. Recent developments in tempe research. *J. Appl. Bacteriol.* **69:**609–633.

49. **Nout, M. J. R., and P. K. Sarkar.** 1999. Lactic acid food fermentation in tropical climates. *Antonie Leeuwenhoek* **76:**395–401.

50. **Nunomura, N., and M. Sasaki.** 1992. Japanese soy sauce flavour with emphasis on off-flavours, p. 287–312. *In* G. Charalambous (ed.), *Off-flavours in Foods and Beverages.* Elsevier, Amsterdam, The Netherlands.

51. **Nuraida, L., M. C. Wacher, and J. D. Owens.** 1995. Microbiology of pozol, a Mexican fermented maize dough. *World J. Microbiol. Biotechnol.* **11:**567–571.

52. **Odunfa, S. A.** 1988. Review: African fermented foods, from art to science. *MIRCEN J.* **4:**259–273.

53. **Ohta, T.** 1986. Natto, p. 85–93. *In* N. R. Reddy, M. D. Pierson, and D. K. Salunkhe (ed.), *Legume-based Fermented Foods.* CRC Press, Boca Raton, Fla.

54. **Onyekwere, O. O., I. A. Akinrele, and O. A. Koleoso.** 1989. Industrialization of ogi fermentation, p. 329–362. *In* K. H. Steinkraus (ed.), *Industrialization of Indigenous Fermented Foods.* Marcel Dekker, New York, N.Y.

55. **Onyekwere, O. O., I. A. Akinrele, O. A. Koleoso, and G. Heys.** 1989. Industrialization of gari fermentation, p. 363–410. *In* K. H. Steinkraus (ed.), *Industrialization of Indigenous Fermented Foods.* Marcel Dekker, New York, N.Y.

56. **O'Toole, D.** 1999. Characteristics and use of okara, the soybean residue from soy milk production—a review. *J. Agric. Food Chem.* **47:**363–371.

57. **Owens, J. D., N. Allagheny, G. Kipping, and J. M. Ames.** 1997. Formation of volatile compounds during *Bacillus subtilis* fermentation of soya beans. *J. Sci. Food Agric.* **74:**132–140.

58. **Owens, J. D., and L. S. Mendoza.** 1985. Enzymatically hydrolysed and bacterially fermented fishery products. *J. Food Technol.* **20:**237–293.

59. **Pastrana, L., P. J. Blanc, A. L. Santerre, M. O. Loret, and G. Goma.** 1995. Production of red pigments by *Monascus ruber* in synthetic media with a strictly controlled nitrogen source. *Process Biochem.* **30:**333–341.

60. **Pepe, O., G. Blaiotta, G. Moschetti, T. Greco, and F. Villani.** 2003. Rope-producing strains of *Bacillus* spp. from wheat bread and strategy for their control by lactic acid bacteria. *Appl. Environ. Microbiol.* **69:**2321–2329.

61. **Ruiz, J. A., J. Quilez, M. Mestres, and J. Guasch.** 2003. Solid-phase microextraction method for headspace analysis of volatile compounds in bread crumb. *Cereal Chem.* **80:**255–259.

62. **Sarkar, P. K.** 2005. Microbiology of traditional alkaline fermented foods, p. 899–915. *In* T. Satyanarayana and B. N. Johri (ed.), *Microbial Diversity: Current Perspectives and Potential Applications.* I. K. International, New Delhi, India.

63. **Sarkar, P. K., B. Hasenack, and M. J. R. Nout.** 2002. Diversity and functionality of *Bacillus* and related genera isolated from spontaneously fermented soybeans (Indian kinema) and locust beans (African soumbala). *Int. J. Food Microbiol.* **77:**175–186.

64. **Shurpalekar, S. R.** 1986. Papads, p. 191–217. *In* N. R. Reddy, M. D. Pierson, and D. K. Salunkhe (ed.), *Legume-Based Fermented Foods.* CRC Press, Boca Raton, Fla.

65. **Siebenhandl, S., L. N. Lestario, D. Trimmel, and E. Berghofer.** 2001. Studies on tape ketan—an Indonesian fermented rice food. *Int. J. Food Sci. Nutr.* **52:**347–357.

66. **Soni, S. K., and D. K. Sandhu.** 1990. Biochemical and nutritional changes associated with Indian Punjabi warri fermentation. *J. Food Sci. Technol.* **27:**82–85.

67. **Soni, S. K., and D. K. Sandhu.** 1999. Fermented cereal products, p. 895–949. *In* V. K. Joshi and A. Pandey (ed.), *Biotechnology: Food Fermentation,* vol. 2. Educational Publishers, Ernakulam, India.

68. **Soni, S. K., and D. K. Sandhu.** 1991. Role of yeast domination in Indian idli batter fermentation. *World J. Microbiol. Biotechnol.* **7:**505–507.

69. **Steinkraus, K. H.** 1994. Nutritional significance of fermented foods. *Food Res. Int.* **27:**259–267.

70. **Sujaya, I. N., S. Amachi, K. Saito, A. Yokota, K. Asano, and F. Tomita.** 2002. Specific enumeration of lactic acid bacteria in ragi tape by colony hybridization with specific oligonucleotide probes. *World J. Microbiol. Biotechnol.* **18:**263–270.

71. **Susheelamma, N. S., and M. V. L. Rao.** 1979. Functional role of the arabinogalactan of black gram (*Phaseolus mungo*) in the texture of leavened foods (steamed puddings). *J. Food Sci.* **44:**1309–1312, 1316.

72. **Umeta, M., and R. M. Faulks.** 1988. The effect of fermentation on the carbohydrates in tef (*Eragrostis tef*). *Food Chem.* **27:**181–189.

73. **Venkatasubbaiah, P., C. T. Dwarakanath, and V. S. Murthy.** 1985. Involvement of yeast flora in idli batter fermentation. *J. Food Sci. Technol.* **22:**88–90.

74. **Westby, A., and D. R. Twiddy.** 1992. Characterization of gari and fu-fu preparation procedures in Nigeria. *World J. Microbiol. Biotechnol.* **8:**175–182.

75. **Yamabe, S., K. Kaneko, H. Inoue, and T. Takita.** 2004. Maturation of fermented rice-koji miso can be monitored by an increase in fatty acid ethyl ester. *Biosci. Biotechnol. Biochem.* **68:**250–252.

Food Microbiology: Fundamentals and Frontiers, 3rd Ed.
Edited by M. P. Doyle and L. R. Beuchat
© 2007 ASM Press, Washington, D.C.

Sterling S. Thompson
Kenneth B. Miller
Alex S. Lopez

Cocoa and Coffee

39

Cocoa and coffee are two of the many foods that rely on a microbial curing process or fermentation for flavor development. The popularity and worldwide appeal of these products are due primarily to their unique flavors and aromas. Although a primary curing process is conducted in the preparation of each product before marketing, fermentation of cocoa is absolutely essential for flavor development whereas with coffee the curing process is less crucial to flavor and more important for the removal of pulp. Consequently, this chapter will focus mainly on the more comprehensive role of fermentation in cocoa curing and to a lesser extent on the role of fermentation in the production of coffee.

COCOA PROCESSING

Commercial cocoa is derived from the seeds (beans) of the ripe fruit (pods) of the plant *Theobroma cacao*, which is native to the Amazon region of South America. It has been used by the Amerindians to produce a beverage since time immemorial and was introduced to Europe in the 15th century by Cortez during the period of discovery and colonization of the Americas. Its popularity and demand led to the establishment and spread of rootstock

to virtually all of the European colonies located between 15° north and 15° south of the equator with climates that could support cocoa production.

Of the *Theobroma* species, only *T. cacao* produces beans suitable for chocolate manufacturing. Immediately following harvesting of the ripe fruit, or following a brief storage period, the seeds are removed and subjected sequentially to fermentation and drying processes often referred to as "curing" which are carried out on farms, estates, or cooperatives in the producing countries. The origin of these processes has been lost in antiquity, but it was believed at one time that fermentation was conducted simply to aid in removing the mucilaginous pulp surrounding the seed so as to facilitate drying and storage, as in the case of coffee. This in fact is one purpose of fermentation, but the main reason for fermentation of cocoa is to induce biochemical transformations within the beans that lead to the formation of color, aroma, and flavor precursors of chocolate. Without this step, cocoa beans are excessively bitter and astringent and when processed do not develop the flavor that is characteristic of chocolate. The characteristics and strength of chocolate flavor are governed primarily by the genetics of the cocoa variety, and the fermentation process releases and

Sterling S. Thompson and Kenneth B. Miller, Microbiology Research Technical Center, Hershey Foods Corporation, 1025 Reese Ave., Hershey, PA 17033-0805. **Alex S. Lopez,** 6621 Creeping Thyme St., Las Vegas, NV 89148.

develops this flavor potential (36). The inherited characteristics of the bean therefore set a limit to what can be achieved by fermentation. It is impossible to improve genetically inferior material by superior processing techniques; on the other hand, it is quite easy to ruin good-quality cocoa by inadequate curing.

The cocoa fruit varies among varieties in size, shape, external color, and appearance. These characteristics have often been used in classifying cocoa, but as far as the flavor quality is concerned, the only really important morphological differences are those that distinguish between the white-seeded Criollo variety of South and Central America and the purple-seeded Forastero variety of the Amazon. The former type is the source of the original "fine" cocoa and has almost disappeared from the market because of its susceptibility to disease, its lower productivity, and its replacement by the hardier, more prolific Forastero variety and varietal crosses which now account for over 95% of the world production. Hence, the following discussion refers primarily to the processing of Forastero cocoa.

Flowers are produced seasonally from cushions that emerge on the bark of the trunk and stems. Fertilized flowers bear fruit 170 days after pollination, a period during which the fruit grows to maturity and changes color from green or dark red-purple to yellow, orange, or red, depending on the variety. The mature fruits are thick walled and contain 30 to 40 beans, each enveloped in a sweet, white, mucilaginous pulp and loosely attached to an axial placenta. Only the beans are used in chocolate manufacturing. For the purpose of describing the curing process, the bean may be envisaged as comprising two main parts, namely, the testa (seed coat), together with the attached sugary, mucilaginous pulp that surrounds it, and the embryo, or the cotyledons, contained within. The mucilage, containing sugars and citric acid (64), serves as a substrate for the microorganisms that are involved in the natural fermentation process; the cotyledons, referred to as the nib in the cured bean, are used in chocolate manufacturing.

Processing begins with the harvesting of healthy ripe fruits, an operation carried out over a period of 3 to 4 days at a frequency which varies according to the size of the farm and yield. Fermentation is a batch type process, and harvesting is conducted to allow for the accumulation of sufficient material for each batch while taking precautions that, in the process, pods do not overripen and the seeds within do not germinate. The pods are usually collected in piles in the field and broken open on site or at the processing plant (fermentary) at the end of the harvesting operation. The beans are removed manually or mechanically on some large estates in West African countries, Mexico,

and Brazil where pod breaking and bean extraction are mechanized (41). Once removed from the pods, the seeds are aggregated in heaps or in receptacles of one kind or another and left to ferment for a period of 2 to 8 days. During this interval, microorganisms which are transferred to the seeds from laborers' hands, fruit surfaces, and containers used in transporting and fermentation degrade the cells of the mucilage that surround the bean (45). The collective microbial activity resulting from the accidental inoculation by a multitude of microorganisms is referred to as fermentation or sweating. This process results in the liberation of pulp juices from which alcohols and acids are produced with the evolution of heat. During fermentation, concentrations of ethanol and lactic and acetic acids sequentially increase and decrease. This can result in an excess of acids that remains in the bean at the end of the bean fermentation (65). Together, these factors provoke changes and affect the curing of the bean. The two principal objectives of fermentation are to remove mucilage, thus provoking aeration during fermentation of the beans and facilitating drying later on, and to provide heat and acetic acid necessary for inhibiting germination, which ensures proper curing of the beans (37).

Methods of Fermentation

The manner in which cacao is fermented varies considerably from country to country, and in many instances even adjacent farms may adopt different curing methods. A substantial amount of research has been devoted to fermentation practices. Today, many of the primitive methods such as fermentation in banana leaf-lined holes in the ground, in derelict canoes, and in makeshift banana and bamboo frames are the exception rather than the rule. In general, large farms with adequate production of cacao will opt for permanent facilities specifically constructed for this purpose. In such instances, fermentation is carried out in batteries of wooden or fiberglass boxes. However, most of the world's cocoa is produced on small farm holdings under very rural conditions, and the relatively small volumes produced do not always merit permanent processing facilities. In this case, cocoa beans are fermented in any convenient receptacle such as fruit boxes, baskets, plastic buckets, and fertilizer bags, or when these are not readily available, the beans are simply piled on a sheet and covered with any handy material. On the whole, however, the majority of the world's cocoa is fermented on drying platforms, in heaps covered with banana leaves, in baskets, or in an assortment of wooden boxes (17, 31, 56). Approximately one-half of the world crop is fermented in some type of box, and the remaining half is fermented by using heaps or other primitive methods.

Fermentation on Drying Platforms

Fermentation on drying platforms is practiced in parts of Central America where Criollo cocoa was once grown. Wet cocoa beans are spread directly onto drying platforms where they ferment and dry during the day and are heaped into piles each night to conserve heat and retard the growth of surface molds. Criollo cocoa requires only a short fermentation for flavor development, about 2 to 3 days. Forastero varieties require a fermentation time of 5 to 8 days for the development of flavor (18). Although Criollo cocoa has been largely replaced by Forastero hybrids in these countries, in many instances the old method of fermentation still persists (52). This practice preserves the fine flavor characteristics of Criollo beans; however, it is inappropriate for Forastero varieties which require longer fermentation times for optimal flavor development. Fermenting cocoa on the drying floor is convenient, but unless properly managed, the process tends to produce underfermented cocoa with the added danger of undesirable mold growth, which will cause the development of off-flavor.

Fermentation in Heaps

Fermentation in heaps is a popular method among small holding farmers in Ghana and many other African cocoa-producing countries. It is also observed sporadically in the Amazon region of Brazil. This method does not require a permanent structure and is well suited to family holdings with a small production. Judging from Ghanaian cocoa, fermentation in heaps can produce good-quality products. Varying quantities of cocoa beans from 25 to 1,000 kg are heaped in the field on banana or plantain leaves and covered with the same material. The beans are mixed (turned) periodically to ensure even fermentation and to decrease the potential for mold growth. This is often done daily or every other day by forming another heap. Mixing is laborious, and small heaps may not be turned at all. The duration of fermentation can be from 4 to 8 days.

Fermentation in Baskets

Fermentation in baskets is practiced principally by small-scale producers in Nigeria, the Amazon region, the Philippines, and some parts of Ghana. Small lots of cocoa are placed into woven baskets lined with plantain leaves. The surface is covered with plantain leaves, and the leaves are weighted down to hold them in place. The turning procedure and the fermentation are similar to those used for small heaps. Basket fermentation is often used when fermentation in heaps leaves beans liable to predial larceny.

Fermentation in Boxes

Fermentation in boxes is considered to be an improvement over other methods. This batch process requires a fixed volume of beans and is the method of choice on large estates. The sizes of the containers vary from region to region, but the design and function are standard. The container, or "sweat box," may be a single unit or one of a number of compartments within a large box created by subdividing the space into units measuring approximately 1 by 1 by 1 m with either fixed or movable internal partitions. These boxes hold between 600 and 700 kg of freshly harvested (wet) cocoa beans. The box is always raised above ground level and placed over a drain which carries away the pulp juices (sweatings) liberated by the degradation of the mucilage during fermentation. The wooden floor of the box generally has holes or spaces between the boards or slats to facilitate drainage and aeration. Sweat boxes vary considerably in size from that of a small fruit box (0.4 by 0.4 by 0.5 m) to boxes measuring 7 by 5 by 1 m used on some Malaysian estates (28). Large estates and cooperatives often have batteries of 20 to 30 sweat boxes arranged in tiers in three to seven rows, one below the other to facilitate mixing or turning. Mixing is achieved by simply removing a dividing wall and shoveling the beans into the next box or, in the case of the tier design, into the box below. On some Malaysian estates, boxes are built on pallets and a forklift is used to transfer the contents into an empty box. Variations occur not only in the size of the sweat box but also in the type of wood used in construction, in methods of drainage and aeration, and in duration of fermentation. The recommendation is to ferment a 1-m^3 volume for 6 to 7 days with two to three mixings during this period. In the majority of cases, boxes are filled to within 10 cm of the top and the surface is covered with a padding of banana leaves or jute sacking to help retain the heat and prevent the surface beans from drying.

In some countries, fermentation norms have been modified in an attempt to overcome problems such as acidity by varying the prefermentation treatment and the depth of beans in the sweat boxes (36). In Malaysia, for instance, harvested cocoa pods are stored up to 15 days before breaking and removal of the beans or the beans are pressed or predried to reduce the pulp volume before fermentation (6, 13).

The progress of the fermentation is assessed by the odor and the external and internal color changes in the beans. When the process is judged to be complete, the beans are dried in the sun or in mechanical dryers.

Microbiology of Cocoa Fermentation

Fermentation begins immediately after beans are removed from the pods, as they become inoculated with a variety of microorganisms from the pod surface, knives, laborers' hands, containers used to transport the beans to the fermentary, dried mucilage on surfaces of the fermentation box (tray, platform, or basket) from the previous fermentation, insects, and banana or plantain leaves (22, 23, 31, 42, 45, 54, 61). It is the pulp surrounding the bean, not the cocoa bean, that undergoes microbial fermentation. Chemical changes take place within the bean as a result of the fermentation of the pulp. The testa of the bean acts as a natural barrier between microbial fermentation activities outside the bean and chemical reactions within the bean. However, there is a migration of ethanol, acetic acid, lactic acid, and water of microbial origin from the outside to the inside of the bean. After the bean dies, soluble bean components leach through the skin and are lost in the drainings. The pulp consists of about 85% water, 2.7% pentosans, 0.7% sucrose, 10% glucose and fructose, 0.6% protein, 0.7% acids, and 0.8% inorganic salts (25), making it a rich substrate for microbial growth. The concentrations of sucrose, glucose, and fructose in the bean are influenced by the age of the pod (62).

The initial microbial population is variable in number and type; however, the key groups active during fermentation are yeasts, lactic acid bacteria, and acetic acid bacteria (34, 45, 54, 68). Climatic conditions may influence the sequence of microorganisms involved in the fermentation (68). It is theorized that *Bacillus* species play an important role during the latter stages of the fermentation in Brazil and become the dominant group during drying (68). More research is needed to confirm this theory. More than 100 aerobic spore-forming bacteria were isolated from cocoa bean fermentations in Brazil. Bacteria were identified as *Bacillus subtilis*, *B. licheniformis*, *B. firmus*, *B. coagulans*, *B. pumilus*, *B. marcerans*, *B. polymyxa*, *B. laterosporus*, *B. stearothermophilus*, *B. circulans*, *B. pasteurii*, *B. megaterium*, *B. brevis*, and *B. cereus*. *B. subtilis*, *B. circulans*, and *B. licheniformis* were encountered more frequently than the other *Bacillus* species during fermentation (68).

During fermentation, yeasts, lactic acid bacteria, and acetic acid bacteria develop in succession. Species of microorganisms that have been detected in cocoa during fermentation in Ghana, Malaysia, and Belize are listed in Table 39.1. At the onset of fermentation, a pH of 3.4 to 4.0, a sugar content of 10 to 12%, and a low oxygen tension favor the growth of yeasts (62, 67). Yeasts utilize the carbohydrates in the pulp under aerobic and anaerobic conditions and may account for 40 to 65% of the microflora when the fermentation begins (12, 46). The yeast phase lasts 24 to 48 h, during which populations may increase to 90% of the total microflora. Yeast populations have been determined in several investigations (34, 42, 45, 62).

Some yeasts produce various pectinolytic enzymes that degrade the cocoa pulp, thereby aiding in the drainage of juices (20, 54). In addition to metabolizing sugar to produce ethanol, yeasts utilize citric acid, causing the pH to increase (54). All yeast species that contribute to fermentation are not present simultaneously, but the species follow a succession which is influenced by the turning step (aeration) and the fact that fermenting bean masses are not homogeneous (34). Several genera of yeasts are involved in fermentation (Table 39.1). In one study (12), of the 142 yeast genera detected in fermenting bean masses, 105 were asporogenous and 37 were ascosporogenous. Between 48 and 72 h of fermentation, the yeast population begins to decrease so that, by the third day, it is reduced to 10% of the total microbial population (6). Three factors are responsible for the rapid decline in the dominance of yeasts. First, yeasts rapidly metabolize sucrose, glucose, and fructose in the pulp to form carbon dioxide and ethanol, causing a reduction in energy source. Second, the production of ethanol produces a toxic environment that suppresses yeast growth. For example, Schwan et al. (68) reported that a decline in the population of *Kloeckera apiculata* was associated with an ethanol concentration of greater than 4%. A small amount of heat is developed simultaneously with ethanol production (4). Third, acetic acid, which is produced from ethanol by the acetic acid bacteria, is also toxic to yeasts. The acetic acid concentration may reach 1 to 2% (8).

The anaerobic conditions created by yeasts make the environment suitable for lactic acid bacteria (46). Lactic acid bacteria prefer a low oxygen concentration or, if oxygen is present, a high concentration of carbon dioxide (34). Such an environment develops as the pulp collapses and the yeast population decreases. The population of lactic acid bacteria increases rapidly, but large numbers may be present for only a brief period (36, 54, 56). Studies have demonstrated that the population of lactic acid bacteria may reach 10^6 to 10^7 CFU/g in a typical fermentation in Belize and a high of 20% of the total microflora after 1.5 days in fermentations in Trinidad (61). In Brazil, the population of lactic acid bacteria is about 65% of the total microflora after 14 h of fermentation. The population of lactic acid bacteria remains high for up to 3 days, at which time it decreases to less than 10% of the total microflora (46).

Both homofermentative and heterofermentative lactic acid bacteria occur in cocoa fermentations; however, the majority are homofermentative (68). Lactic acid bacteria detected in traditional box fermentations in Brazil were isolated and characterized as homofermentative and heterofermentative. The homofermentative species included

Table 39.1 Microorganisms isolated from fermenting cocoa beans

Type of microorganism	Microorganisms isolated in:		
	Ghana[a]	Malaysia[a]	Belize[b]
Lactic acid bacteria	*Lactobacillus plantarum*	*Lactobacillus plantarum*	*Lactobacillus plantarum*
	Lactobacillus mali	*Lactobacillus collinoides*	*Lactobacillus fermentum*
	Lactobacillus collinoides		*Lactobacillus brevis*
	Lactobacillus fermentum		*Lactobacillus buchneri*
			Lactobacillus cellobiosus
			Lactobacillus casei pseudoplantarum
			Lactobacillus delbrueckii
			Lactobacillus fructivorans
			Lactobacillus kandleri
			Lactobacillus gasseri
			Leuconostoc mesenteroides
			Leuconostoc paramesenteroides
			Leuconostoc oenos
Acetic acid bacteria	*Acetobacter rancens*	*Acetobacter rancens*	*Acetobacter* spp.
	Acetobacter ascendens	*Acetobacter lovaniensis*	*Gluconobacter oxydans*
	Acetobacter xylinum	*Acetobacter xylinum*	
	Gluconobacter oxydans	*Gluconobacter oxydans*	
Yeasts	*Candida* spp.	*Candida* spp.	*Brettanomyces claussenii*
	Hansenula spp.	*Debaryomyces* spp.	*Candida* spp.
	Kloeckera spp.	*Hanseniaspora* spp.	*Candida boidinii*
	Pichia spp.	*Hansenula* spp.	*Candida cocoai*
	Saccharomyces spp.	*Kloeckera* spp.	*Candida intermedia*
	Saccharomycopsis spp.	*Rhodotorula* spp.	*Candida guilliermondii*
	Schizosaccharomyces spp.	*Saccharomyces* spp.	*Candida krusei*
	Torulopsis spp.	*Torulopsis* spp.	*Candida reukaufii*
			Kloeckera apis
			Kloeckera javanica
			Pichia membranaefaciens
			Saccharomyces cerevisiae
			Saccharomyces chevalieri
			Schizosaccharomyces spp.
			Schizosaccharomyces malidevorans

[a] Data from Carr et al. (8).
[b] Unpublished data.

Lactobacillus plantarum, L. casei, L. delbrueckii, L. acidophilus, Pediococcus cerevisiae, P. acidilactici, and *Streptococcus (Lactococcus) lactis.* The heterofermentative species included *Leuconostoc mesenteroides* and *Leuconostoc brevis* (47). Citric acid is metabolized either into acetic acid, carbon dioxide, and lactic acid by heterofermentative species or into acetylmethylcarbinol and carbon dioxide by homofermentative lactics. Lactic acid bacteria may be more important in cocoa fermentation in Brazil where, during the first 48 h of the fermentation, their population was consistently larger than the yeast population (46). This population dynamic differs from those of most other fermentations, where yeasts are the dominant microorganism during the first 48 h. If

lactic acid bacteria remain a high percentage of the total microbial population during the fermentation, high concentrations of lactic acid will be produced. Since lactic acid is not volatile, it will remain in the chocolate after manufacturing, producing an undesirable chocolate (58, 71, 80).

As the beans are turned to aerate the mass, more of the pulp is metabolized and conditions become aerobic. The population of lactic acid bacteria decreases and the population of acetic acid bacteria increases with increased aeration. The population of acetic acid bacteria generally reaches 10^5 to 10^6 CFU/g in a typical fermentation in Belize. Members of two genera of acetic acid bacteria, *Acetobacter* and *Gluconobacter*, have been isolated from fermenting cocoa beans. *Acetobacter* species occur more frequently than *Gluconobacter* species (8, 9, 45). The population of acetic acid bacteria has been observed to make up 80 to 90% of the total microflora after 2 days in fermentations in Trinidad (61). Acetic acid bacteria oxidize ethanol into acetic acid exothermally (4), causing the temperature of the bean mass to rise to 45 to 50°C. Turning the beans periodically facilitates oxidation of the ethanol into acetic acid and conserves the high temperature of the bean mass. When all of the ethanol is oxidized into acetic acid and then carbon dioxide and water, the fermentation subsides and the temperature of the bean mass decreases quickly.

During the later stages of fermentation and drying, aerobic spore-forming *Bacillus* species develop and may become dominant (8, 45). *Bacillus* species are present during the first 72 h of the fermentation, but during this early stage, their population remains constant. They become dominant later in the fermentation, making up over 80% of the microbial population (67, 69). Development of *Bacillus* species in the bean mass is favored by increased aeration, an increased pH (3.5 to 5.0) of the pulp, and an increase in temperature to 45 to 50°C (8, 45, 69).

Bacillus species can produce several compounds that may contribute to the acidity and off-flavors of fermented cocoa. The C_3, C_4, and C_5 free fatty acids that are present in the bean mass during the aerobic phase of fermentation may contribute to the development of some of the off-flavors of chocolate (39, 50). The importance of *Bacillus* species in cocoa bean fermentation is not well established, but they are reputed to produce acetic and lactic acids, 2,3-butanediol, and tetramethylpyrazine which can affect the flavor of chocolate (39, 69, 81).

A key factor that must be considered when deciding on a fermentation scheme is when to remove the beans from their fermentation environment and begin drying. Extending the fermentation can result in undesirable microbial activity, leading to putrefaction and the production of compounds such as butyric and valeric acids that contribute to off-flavors (40). Forsyth and Quesnel (17) suggested that the following factors may collectively indicate when fermentation is optimum: (i) external color of the beans, (ii) time schedule, (iii) decrease in temperature, (iv) bean cut test using the internal color as a criterion, (v) aroma of the fermenting mass, and (vi) plumping or swelling of the beans.

A more desirable measurement of optimum fermentation would be a chemical method that is relatively rapid, inexpensive, and easy to perform and interpret. We have observed that the end point of fermentation can be determined by the pH of the beans, provided that a normal temperature curve is established. The minimum pH that gives acceptable cocoa liquor is 5.2; however, the actual fermentation pH may be slightly lower. Other methods to measure the qualitative and quantitative changes that occur during cocoa fermentation have been reviewed by Shamsuddin and Dimick (70).

Fermentation Using Pure-Culture Seeding

Fermentation of cocoa beans continues to be a natural process that is required for the development of cocoa aroma and chocolate flavor. However, the fermentation step is unpredictable because cocoa beans will vary in their degrees of fermentation. Cocoa beans can be underfermented, overfermented, or optimally fermented with the desired sensory attributes. Consequently, the natural process yields beans of various sensory qualities. If the fermentation process is well understood, it may be possible to manipulate the fermentation by using a defined starter culture, as is used in the dairy industry, to produce the desired fermentation. Over 40 species of microorganisms are known to be involved in the natural process. However, most of these microorganisms are probably not necessary to achieve the natural process. Studies have been conducted to determine the potential application of a defined microbial inoculum to improve the natural fermentation (14, 62, 63, 65). The isolates used in these studies were obtained and identified from previous fermentations. A seeding study was conducted using *Saccharomyces chevalieri*, *Candida zeylaniodes*, and *Kluyveromyces fragilis* isolates. Of the three yeasts used, *S. chevalieri* produced the best quality cocoa (63). *Torulopsis candida* (ATCC 20031), *C. norvegensis* (ATCC 22971), *K. fragilis* (ATCC 8601), and *S. chevalieri* were evaluated for their abilities to increase the yield of the cocoa sweatings for use in the development of products, such as soft drinks, jams, and marmalades. The defined yeast fermentations did not negatively affect the physicochemical properties of the cocoa sweatings or the quality of the chocolate (14). A defined inoculum cocktail consisting of *S. chevalieri*, *Lactobacillus lactis*, *L. plantarum*, *Acetobacter aceti*, and *Gluconobacter oxydans* subsp. *suboxydans* was successfully used to produce an acceptable-quality chocolate (65). Based on the fermentation process and the quality of the chocolate made by using a defined cocktail inoculum, specific characteristics are required of the microorganisms used. The yeast species should be highly fermentative, e.g., *S. chevalieri*, and produce pectinolytic enzymes. The lactic acid bacteria should be members of the homolactic group, e.g., *L. lactis* and *L. plantarum*. The acetic acid bacteria must be able to grow in an environment of up to 6% ethanol and tolerate 45°C and pH 3.5, e.g., *A. aceti* and *G. oxydans* subsp. *suboxydans* (65). Several benchtop seeding studies were conducted using an inoculum mixture isolated in natural fermentation studies in Belize. The mixture contained 11 microorganisms consisting of yeasts and lactic and acetic acid bacteria. The final chocolate was judged to be acceptable by a trained sensory panel. If it could be demonstrated that initiating the fermentation with the desired starter culture would consistently reduce the fermentation time and yield beans with desirable sensory attributes, this would represent a scientific step forward. Use of a defined inoculum, if appropriate steps are taken, would eliminate natural contamination by spoilage microorganisms, thereby yielding a more desirable fermented, dried bean. Several issues and questions must be resolved to determine whether a starter culture approach can supplement or replace the traditional fermentation. To scale up the seeding of cocoa with a defined inoculum to the farm level, whether applied to small farm holdings or large estates, would be a significant challenge, especially as it relates to cost, culture maintenance, and seeding method.

Biochemistry of Cocoa Fermentation

The actual production of chocolate flavor precursors occurs within the cocoa bean and is primarily the result of biochemical changes that take place during fermentation and drying. The mode of fermentation and the microbial environment during these stages of cocoa production provide the necessary conditions for complex biochemical reactions to occur. Although flavor compounds such as lactic and acetic acids are produced outside the bean by microbial activity, chocolate flavor development is largely dependent on the enzymatic formation of flavor precursors within the cotyledon that are unique to cocoa. Such classes of compounds include free amino acids, peptides, reducing sugars, and polyphenols. When fermented, dried cocoa beans containing these flavor precursors are subjected to roasting during chocolate manufacture, a necessary step in flavor development, a series of complex, nonenzymatic browning reactions occurs to produce flavor and color compounds characteristic of chocolate (27, 35, 57, 59, 60). However, if unfermented cocoa beans lacking these precursor compounds are roasted, very little chocolate flavor is produced. It is, therefore, important that these flavor precursor compounds are formed inside the cocoa bean during fermentation.

The initiation of fermentation also corresponds to an incipient germination phase which is necessary for mobilization of the enzymes and hydration of bean components in preparation for growth. However, the germination phase is undesirable in cocoa beans used to make chocolate.

Bean Death

Bean death is a critical event during cocoa fermentation which allows the biochemical reactions responsible for flavor development to occur within the cocoa bean. Although rising temperatures and increasing acetic acid concentrations during fermentation have been implicated in causing seed death (48), more recent data (33) indicate that the production of ethanol during the anaerobic yeast growth phase correlates very closely with the death of the seed. Total inability of the seed to germinate occurs about 24 h after maximum concentrations of ethanol are attained within the cotyledon (Fig. 39.1). As a result, events associated with germination and certain quality defects, e.g., the utilization of valuable seed components such as cocoa butter and the opening of the testa by hypocotyl extension, will not occur. This produces a more stable, desirable end product.

From a flavor perspective, events associated with the death of the seed also cause cellular membranes to leak and permit enzymes and substrates to react to form flavor precursor compounds important to chocolate flavor development. As shown in Fig. 39.1, the activity of these enzymes in the cotyledon results in significant increases in free amino acid and reducing-sugar (glucose and fructose) contents. While the temperature of beans increases, the concentration of organic acids increases, causing a decrease in pH. All of these factors influence the biochemistry within the bean and have an impact on cocoa flavor and quality.

Environmental Factors

There are several environmental factors, viz., pH, temperature, and moisture, in the fermenting mass that influence cocoa bean enzyme reactions. Each enzyme has an optimum pH at which it is most active, and within a defined range, an enzyme reaction accelerates as temperature increases. In addition, a certain amount of moisture is necessary to allow enzymes and their substrates to react to form products. Significant changes in pH, temperature, and moisture occur during cocoa fermentation and drying processes (Fig. 39.1) that influence the type and quantity of flavor precursor compounds produced by enzymatic action.

Moisture content within the cotyledon during fermentation is usually more than 35% and will permit adequate migration of enzymes and substrates for enzymatic activity. However, once the drying process begins, moisture content gradually decreases, making it increasingly difficult for enzymes and substrates to react. When a moisture content of 6 to 8% is achieved, virtually all enzyme activity ceases.

The pH of the unfermented cotyledon is about 6.5 and may decrease to as low as 4.5 by the end of the

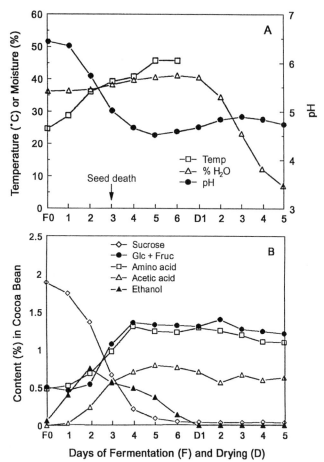

Figure 39.1 Physical (A) and chemical (B) changes in cocoa beans during fermentation and drying in Belize. Fermentation was conducted with 2,000 lb of wet cocoa beans from ripe pods in wooden boxes that were turned daily. Drying was conducted in flat-bed dryers indirectly heated with hot air. Data represent results from an average of 11 fermentation trials using composite samples collected daily. (A) Temperature was measured in the whole bean mass. Moisture (%) and pH analyses are based on shell-free cotyledons. (B) Sucrose, glucose (Glc), fructose (Fruc), total amino acid, acetic acid, and ethanol contents (%) were determined by analysis of water extracts from shell-free cotyledon samples. Data are taken from Lehrian (33).

fermentation. This lowering of pH occurs after seed death and is due primarily to diffusion into the bean of organic acids produced by lactic acid and acetic acid bacteria. It is the growth of these and other microorganisms that also contributes to the increasing temperature of the mass of fermenting beans. Typically, temperature will rise from 25 to about 50°C, followed by a slight decrease as bacterial growth subsides. An increase in temperature of more than 20°C during fermentation can have a profound impact on enzyme activity. If very little change in temperature occurs, enzyme activity is reduced, resulting in fewer

flavor precursors and poor chocolate flavor. Likewise, if appropriate amounts of organic acids are not produced during fermentation, the pH of the cotyledon will not be suitable for optimal enzyme activity and the flavor profile of the resulting cocoa will be affected. However, too much acid will produce excessive sourness that can mask the chocolate flavor.

Consequently, there is a delicate balance among the length of fermentation, environmental factors, and microbial activity that influences enzyme activity within the cotyledon. Hydrolytic and oxidative enzymes play a major role in reactions that produce flavor precursors. A summary of cocoa bean enzymes and their substrates and optimum pHs is given in Table 39.2 (36). More recently, Hansen et al. have studied cocoa bean enzymes and their activities during the fermentation and drying process in an effort to understand their impact on cocoa flavor and quality (24).

Hydrolytic Enzyme Reactions

Hydrolytic enzymes such as invertase, glycosidases, and proteases have highest activity during the anaerobic phase of cocoa fermentation. The products of these enzyme activities during cocoa fermentation fall into three basic categories: sugars, amino acids and peptides, and cyanidins. Sugars and amino acids and peptides participate in nonenzymatic browning reactions during roasting to form important chocolate flavor precursors, whereas the cyanidins have more of an impact on color development and some minor flavor components.

Sucrose is the major sugar in unfermented cocoa beans. It is not a reducing sugar and therefore does not participate in nonenzymatic browning reactions that occur during roasting to contribute to chocolate flavor. However, sucrose is converted to glucose and fructose by invertase during the fermentation process. These reducing sugars account for more than 95% of the total reducing monosaccharides in cocoa beans, and their concentrations increase almost threefold during fermentation while sucrose is depleted (Fig. 39.1).

Another class of hydrolytic enzymes within the cocoa bean that contributes to both flavor and color during fermentation is the glycosidases. The substrates for these enzymes are the purple anthocyanins located in specialized vacuoles within the cotyledon that are responsible for the characteristic deep purple color of the unfermented bean. The actions of specific glycosidase enzymes begin at seed death and are responsible for cleaving the sugar moieties, galactose, and arabinose attached to the anthocyanins. This results in a bleaching of the purple color of the beans as well as the release of reducing sugars that can participate in flavor precursor reactions

Table 39.2 Characteristics of the principal enzymes active during the curing of the cocoa bean[a]

Enzyme(s)	Location	Substrate(s)	Products	pH	Temp(s) (°C)	Reference(s)
Invertase	Testa	Sucrose	Glucose and fructose	4.0	52	38
				5.25	37	
Glycosidases (β-galactosidase)	Bean	Glycosides (3-β-D-galactosidyl cyanidin and 3-α-L-arabinosidyl cyanidin)	Cyanidin and sugars	3.8–4.5	45	16
Proteases	Bean	Proteins	Peptides and amino acids	4.7	55	7, 49
Polyphenol oxidases	Bean	Polyphenols (epicatechin)	σ-Quinones and σ-diquinones	6.0	31.5, 34.5	51

[a] Information taken from Lopez (36).

during roasting (16). Pigments themselves do not carry any flavor potential (16, 39, 55). Cocoa beans that still contain significant purple color are considered to have been poorly fermented and are less desirable.

Although there are small amounts of free amino acids present in unfermented cocoa beans, the total free amino acid pool increases significantly during fermentation due to the actions of both endo- and exoproteases on cocoa bean proteins. After seed death occurs, these proteolytic enzymes are free to act on protein substrates within the bean and their activity becomes dependent on pH and temperature. A vicilin-like globular storage protein within the cotyledon is the primary target of these proteolytic enzymes, and ratios of free amino acids and peptides that are unique to cocoa are produced (78). These flavor precursor compounds contribute to the development of cocoa flavor when roasted in the presence of reducing sugars.

Oxidative Enzyme Reactions

Significant oxidative enzyme activity also occurs, being most prevalent late in the aerobic phase of fermentation but continuing well into the drying of cocoa. Polyphenol oxidase is the major oxidase in cocoa and is responsible for much of the brown color that occurs during fermentation as well as some flavor modifications. This enzyme becomes active during the aerobic phase of the fermentation as a result of oxygen's permeating the cotyledon. Events that contribute to activity include seed death, the subsequent breakdown of cellular membranes, reduction in the amount of seed pulp, and aeration of the bean mass by agitation. Oxygen continues to penetrate the beans during the drying process, enabling polyphenol oxidase activity to continue until rising temperatures and insufficient moisture become inhibiting factors.

Catachins and leucocyanidins are the major classes of polyphenols that are subject to oxidation in cocoa beans. Epicatechin makes up more than 90% of the total catechin fraction and is the major substrate of polyphenol

oxidase (21). Oxidation of epicatechin during the aerobic phase of fermentation and drying is largely responsible for the characteristic brown color of fermented cocoa beans. Polyphenols in the dihydroxy configuration are oxidized to form quinones which in turn can polymerize with other polyphenols or form complexes with amino acids and proteins to yield characteristic colored compounds and high-molecular-weight insoluble material. This formation of complexes also has an impact on flavor. The formation of these less soluble polyphenolic complexes reduces astringency and bitterness associated with native polyphenols present in unfermented cocoa (18, 53). In addition, the ability of polyphenols to form complexes with proteins results in the reduction of off-flavors associated with the roasting of peptide and protein material (26, 79).

Despite the beneficial effects of fermentation on the color and flavor of cocoa, the biochemical changes during fermentation have been shown to decrease the native polyphenol and antioxidant contents (76). In recent years, much attention has been devoted to the potential health benefits of polyphenols and their antioxidant activity in cocoa and chocolate products (32). The nonfat portion of the cocoa bean is a rich source of certain classes of polyphenols, and it has been shown that the cocoa solid contents of chocolate products are highly correlated with the polyphenol and antioxidant contents of these products (29). Therefore, the flavor-enhancing properties of cocoa fermentation must be balanced against the polyphenol- and antioxidant-lowering effects of fermentation. The availability of cocoa beans with various degrees of fermentation provides the chocolate manufacturer with the opportunity to produce chocolate products that meet the diverse needs of the modern consumer.

Flavor and Quality Implications

The ultimate goal of biochemical changes during fermentation is to produce cocoa beans with desirable flavor and color characteristics. Good chocolate flavor potential is achieved by the production of specific amino acids, peptides, and sugars through the action of proteases and invertases on cocoa bean substrates. Proper control of fermentation conditions and microflora ensures that concentrations of organic acids are maintained at reasonable levels to minimize sour and putrid off-flavors while still developing the pH and temperature environment for enzyme-substrate reactions that produce chocolate flavor precursors. Enzyme-mediated conversion of polyphenol materials during fermentation and drying processes reduces astringency and bitterness and produces the desirable brown color typical of properly fermented cocoa beans. The drying process will then preserve the flavor and color characteristics of the beans until they are made into chocolate.

Although cocoa has been successfully produced for centuries, flavor characteristics of specific varieties of cocoa are becoming diluted due to the prevalence of genetic hybrids. While these hybrids are being selected for high crop yield and disease resistance, certain flavor attributes are being lost. The actual identities of the flavor precursors responsible for chocolate flavor have not yet been confirmed. However, recent work has focused on characterizing cocoa bean enzymes and proteins that yield breakdown products unique to cocoa during the fermentation process (77). This work needs to continue in order to understand flavor development and maintain the high quality of chocolate flavor that the consumer expects.

Drying

After the beans are fermented, they have a moisture content of about 40 to 50% which must be reduced to 6 to 8% for safe storage. A higher final moisture content will result in mold growth during storage. The drying process relies on air movement to remove water. This environment favors aerobic microorganisms which proliferate at rates that decrease with moisture loss. Sun drying is the preferred method, but in regions where harvesting coincides with frequent rainfall, some form of artificial drying is necessary and desirable. In general, sun drying is employed on small farms, whereas large estates may resort to both natural and artificial drying. Sun drying allows a slow migration of moisture throughout the bean, which transports flavor precursors that were formed during fermentation.

During sun drying, beans are placed on wooden platforms, mats, polypropylene sheets, or concrete floors in layers ranging from 5 to 7 cm thick. The beans are constantly mixed to promote uniform drying, to break agglomerates that may form, and to discourage mold growth. Under sunny conditions, the beans dry in about a week, but under cloudy or rainy conditions, drying times may be prolonged

to 3 or 4 weeks, increasing the risk of mold development and spoilage.

Various types of artificial dryers employed to overcome the dependence on weather conditions have been described by McDonald et al. (43). Hot air dryers of one form or another, fueled by wood or oil as a source of cheap, readily available energy, are generally employed. The beans may be heated by direct contact with the flue gases; however, the preferred method relies on indirect heating via heat exchangers. Improperly used or poorly maintained heating systems present the danger of contamination with smoke which results in smoky or hammy off-flavors characteristic of beans from some countries. Platforms, trays, and rotary dryers of various designs, coupled to furnaces, are used, but in every instance, the initial drying must be slow and frequent mixing must be done to obtain uniform removal of water. This method results in volatilization of acids and sufficient time for oxidative, biochemical reactions to occur. For this reason, temperatures should not exceed 60°C and drying should take at least 48 h. Elevated temperatures also tend to produce cocoa with brittle shells and cotyledons which crumble during handling. In short, the drying rate should be controlled so as to remove moisture at a rate that will avoid case hardening (rapid drying on the bean surface with moisture retention inside the bean) or excessive mold growth while still allowing sufficient time for biochemical oxidative reactions and loss of acid to occur.

Storage

Due to marketing practices and manufacturing procedures, fermented, cured, dried beans may be stored for periods of 3 to 12 months, in warehouses on farms, on wharfs in exporting and receiving countries, and at factories before being processed into chocolate. The efficiency of the drying process will determine the shelf life of the product. Uniformly dried beans with a moisture content of 7 to 8%, which are stored at a relative humidity of 65 to 70%, will generally maintain that moisture, resist mold growth and insect infestation, and not require repeated fumigation. The cocoa quality can change during storage, depending on temperature, relative humidity, and ventilation conditions. Slow oxidation and acid loss continue to enhance product quality somewhat, but prolonged storage results in a noticeable staling (37).

COFFEE PROCESSING

Coffee of various tonnages is produced in Brazil, Central America, Colombia, Ethiopia, Guatemala, Hawaii, India, Indonesia, Mexico, Vietnam, and other, smaller countries. Coffee beans also undergo a fermentation step to prepare the fruit for commercial use (5, 10, 72). The fermentation of coffee causes the breakdown of the pulp layer surrounding the bean in order to aid in the processing of the fruit into a desirable finished product. Natural fermentation, which involves several different microorganisms, is the primary method used to remove the mucilage. Natural fermentation involves molds, yeasts, several species of lactic acid bacteria, coliforms, and other gram-negative bacteria. These microorganisms originate from the surface of the fruit and the soil (1, 19, 74). Three nonmicrobial methods, addition of enzymes, chemical treatment (sodium hydroxide), and mechanical force, may also be used to remove the mucilage (30). The coffee fruit is a fleshy berry approximately the size of a small cherry (10). The coffee bean is composed of six parts: skin, pulp, mucilage, parchment (outer layer around the bean), sliver skin (inner layer), and bean. The pulp layer is approximately 29% of the dry weight of the fruit. It consists of 76% water, 10% protein, 2% fiber, 8% mineral salts, and 4% soluble and insoluble components, such as tannins, pectin, reducing and nonreducing sugars, caffeine, chlorogenic and caffeic acids, cellulose, hemicellulose, lignin, and amino acids (15). The mucilage is 5% of the dry weight of the bean. It contains water; pectin; pectic acid; small quantities of arabinose, galactose, xylose, and rhamrose (2); organic acids (15); and hydrolytic and oxidative enzymes (2, 3). Coffee beans are produced by the genus *Coffea*. Over 40 species are known, but only a few are used to produce coffee. *Coffea arabica*, *C. canephora*, *C. robusta*, *C. liberica*, and *C. excelsa* are the important species. *C. arabica* and *C. canephora* are the two predominant species on the world market.

It takes approximately 1 year from the flowering stage for the fruit to reach maturity. As the fruit ripens, it changes color from green to cherry red. The ripe, red coffee cherry consists of two green beans surrounded by the pulp layer that is enclosed in parchment. It is necessary to first remove the parchment and pulp layer to obtain the green coffee beans which are then dried, roasted, milled, and used for making coffee beverages. Removal of the pulp can be accomplished by either a natural drying step that yields "natural coffee" or a wet fermentation step that yields "washed coffee." The percent yield of dried coffee obtained from ripe cherries varies among species; *C. arabica* produces 12 to 18%, *C. robusta* produces 17 to 22%, and *C. liberica* produces 10%.

The role of fermentation in coffee production is less critical in the development of flavor than that in cocoa production, although improperly fermented fruit will result in off-flavors. Since pectic substances are a large portion of the mucilage (30), the microorganisms responsible for colonization and utilization of this material must be capable of producing pectinases. Coffee fermentation studies have demonstrated that the highest microbial activity occurs during the first 12 to 24 h after the beans are harvested (75). Most of the microorganisms detected during this early stage belong to the genera *Aerobacter* (*Enterobacter*) and *Escherichia*. Populations of these bacteria increase from 10^2 to 10^9 CFU/g during the first 24 h. Pectinolytic species of *Bacillus*, *Fusarium*, *Penicillium*, and *Aspergillus* have also been detected. Only a few yeast species have been identified. Fermentation studies conducted with Kona coffee beans demonstrated that *Erwinia dissolvens* is the main cause of mucilage decomposition (19).

The pH of unfermented coffee beans can range from 5.4 to 6.4. During fermentation, the pH may decrease to 3.7 (1, 44). Low levels of ethanol have been detected under anaerobic fermentation conditions (30). The duration of a natural fermentation will depend on climate conditions, regional factors, coffee variety, the degree of anaerobiosis of the ferment, and the microorganisms present (30). An optimal fermentation time is considered to be 16 to 24 h (73). Extending the fermentation time will result in the development of flavor defects and bean discoloration (44, 52). Once the fermentation is complete, any remaining mucilage is removed by washing.

Dry Processing

The natural drying process relies on partial dehydration of the pulp layer while the coffee fruit is ripening on the tree. This method is more likely to be used in Brazil and Ethiopia and for robusta coffee (66). The ripe fruit is harvested (hand or machine stripped) and subjected to additional drying. The mature berries may be spread onto platforms, soil, concrete, or tarmac in layers of up to 10 cm thick. Berries are heaped up at night and respread each morning. It can take from 10 to 25 days of sun drying. During this time, a natural microbial fermentation occurs; enzymes are secreted that break down the pulp and mucilage (30). Bacteria, yeasts, and pectinolytic molds are present during the fermentation. However, bacteria are more prevalent than yeasts or molds. Unlike that in cocoa fermentation, there is no defined microbial succession that occurs with coffee maturation and fermentation. Fermentation of the sugars produces alcohol and acetic, lactic, butyric, and other carboxylic acids. Some mucilage will penetrate the coffee bean during drying, and for this reason the natural process produces a light brown bean instead of the blue-green color of wet processed coffee. The resulting dry berry is free of pulp and mucilage but is still surrounded by dry skins which are mechanically removed. Molds may grow during slow drying, producing off-flavors typical of a butyric fermentation. Coffee should have a final moisture content of 11 to 12%.

Wet Processing

Colombia, Central America, and Hawaii are more likely to use the wet process on arabica coffees (66). Two factors are the key to obtaining a good-quality final product, viz., using a uniform population of mature berries and rapid processing (2, 30). After harvesting, the mature beans are mechanically depulped. This is followed by a fermentation step which is used to convert the remaining mucilage into water-soluble products that are removed by washing prior to final drying. The microorganisms involved in the fermentation secrete pectinase enzymes which break down the pectinacious mucilage. Although a large number of microorganisms have been isolated from fermentation processes, the specific importance of yeasts or bacteria is not well understood. Generally, the berries are fermented in wooden or concrete bins. The white, sticky, partially depulped beans are held under water for 12 to 60 h, depending on environmental conditions, the ripeness of the berries, and the variety being processed. Microbial by-products are periodically washed away during the fermentation step to avoid the development of excessive off-flavors. Once the sticky pulp layer is converted into water-soluble products, the beans are washed with water and dried to a moisture content of 11 to 12%. Overfermentation of the berries will cause spoilage and the development of taints (11, 30).

CONCLUSION

Although serving somewhat different purposes, microbial fermentation plays a critical role in the production of both cocoa and coffee. A common goal of fermentation of both of these products is the breakdown and removal of fruit pulp. This process aids in the proper drying of coffee and, in the case of cocoa, provides a suitable environment for flavor and color development. As in most natural curing processes, there is a delicate balance between environmental factors and conditions enabling microbial activities, all of which influence the biochemical changes that take place in processed foods. Coffee and cocoa are no exceptions, and it is the proper control of the fermentation process that largely determines the color and flavor qualities of the final products. Consequently, understanding the microbiology and biochemistry of cocoa and coffee production, as well as the factors that influence them, is critical to quality control.

References

1. **Agate, A. D., and J. V. Bhat.** 1966. Role of pectinolytic yeasts in the degradation of the mucilage layer of *Coffee* and *Robusta* cherries. *Appl. Microbiol.* **14:**256–260.

2. **Amorium, H. V., and V. L. Amorium.** 1977. Coffee enzymes and coffee quality, p. 27–55. *In* R. L. Ory and A. J. St. Angelo (ed.), *ACS Symposium Series*, vol. 47. *Enzymes in Food and Beverage Processing.* American Chemical Society, Washington, D.C.

3. **Amorium, H. V., and M. Mello.** 1991. Significance of enzymes in non-alcoholic coffee beverage, p. 189–209. *In* P. F. Fox (ed.), *Food Enzymology*, vol. 2. Elsevier, Amsterdam, The Netherlands.

4. **Anonymous.** 1979. *The Fermentation of Cocoa.* National Union of Cocoa Producers, Villahemosa, Mexico.

5. **Arunga, R.** 1982. Coffee, p. 259–274. *In* A. H. Rose (ed.), *Economic Microbiology*, vol. 7. *Fermented Foods.* Academic Press, New York, N.Y.

6. **Biehl, B., B. Meyer, G. Crone, L. Pollman, and M. B. Said.** 1984. Chemical and physical changes in the pulp during ripening and post-harvest storage of cocoa pods. *J. Sci. Food Agric.* **48:**189–208.

7. **Biehl, B., U. Passern, and D. Passern.** 1977. Subcellular structures in fermenting cocoa beans—effect of aeration and temperature during seed and fragment incubation. *J. Sci. Food Agric.* **28:**41–52.

8. **Carr, J. G., P. A. Davies, and J. Dougan.** 1979. *Cocoa Fermentation in Ghana and Malaysia.* University of Bristol Research Station, Long Ashton, Bristol and Tropical Products Institute, London, United Kingdom.

9. **Carr, J. G., P. A. Davies, and J. Dougan.** 1980. *Cocoa Fermentation in Ghana and Malaysia*, vol. II. University of Bristol Research Station, Long Ashton, Bristol and Tropical Products Institute, London, United Kingdom.

10. **Castelein, J., and H. Verachtert.** 1983. Coffee fermentation, p. 587–615. *In* H. J. Rehm and G. Reed (ed.), *Biotechnology*, vol. 5. Chemie Verlag, Weinheim, Germany.

11. **Clarke, R. J., and R. Macrae.** 1987. *Coffee Technology*, vol. 2. Elsevier Applied Science, New York, N.Y.

12. **de Camargo, R. J., J. Leme, and A. M. Filho.** 1963. General observations on the microflora of fermenting cocoa beans (*Theobroma cocoa*) in Bahia (Brazil). *Food Technol.* **17:**116–118.

13. **Duncan, R. J. E., G. Godfrey, T. N. Yap, G. L. Pettipher, and T. Tharumarajah.** 1989. Improvement of Malaysian cocoa bean flavor by modification of harvesting, fermentation and drying methods—the Sime-Cadbury process. *Planter* **65:**157–173.

14. **Dzogbefia, V. P., R. Buamah, and J. H. Oldham.** 1999. The controlled fermentation of cocoa (*Theobroma cacao* L) using yeasts: enzymatic process and associated physicochemical changes in cocoa sweatings. *Food Biotechnol.* **13:**1–12.

15. **Elias, L. G.** 1978. Composición guimica de la pulpa de café y otros subproductos, p. 19–29. *In* J. E. Braham and R. Bressan (ed.), *Pulpa de Café: Composicion, Tecnologia y Utilicatión.* INCAP, Panamiá, Panama.

16. **Forsyth, W. G. C., and V. C. Quesnel.** 1957. Cocoa glycosidase and colour changes during fermentation. *J. Sci. Food Agric.* **8:**505–509.

17. **Forsyth, W. G. C., and V. C. Quesnel.** 1957. Variations in cocoa preparation, p. 157–168. *In Sixth Meeting of the InterAmerican Cocoa Committee.* InterAmerican Cocoa Conference, Bahia, Brazil.

18. **Forsyth, W. G. C., and V. C. Quesnel.** 1963. Mechanisms of cocoa curing. *Adv. Enzymol.* **25:**457–492.

19. **Frank, H. A., N. A. Jum, and A. S. Dela Cruz.** 1965. Bacteria responsible for mucilage-layer deposition in Kona coffee cherries. *Appl. Microbiol.* **13:**201–207.

20. **Gauthier, B., J. Guiraud, J. C. Vincent, J. P. Porvais, and P. Galzy.** 1977. Components on yeast flora from the traditional fermentation of cocoa in the Ivory Coast. *Rev. Ferment. Ind. Aliment.* **32:**160–163.

21. **Griffiths, L. A.** 1957. Detection of the substrate of enzymatic browning in cocoa by a post-chromatographic enzymatic technique. *Nature* **180:**1373–1374.

22. **Grimaldi, J.** 1954. *The Cocoa Fermentation Process in the Cameroon,* vol. V. 5RTIC/DOC 24. Bioq. 12. Reunion del Comite Técnico Interamericano del Cocoa.

23. **Grimaldi, J.** 1978. The possibilities of improving techniques of pod breaking and fermentation in the traditional process of the preparation of cocoa. *Café Cacao* **22:**303–316.

24. **Hansen, C. E., M. del Olmo, and C. Burri.** 1998. Enzyme activities in cocoa beans during fermentation. *J. Sci. Food Agric.* **77:**273–281.

25. **Hardy, F.** 1960. *Cocoa Manual,* p. 350. Inter American Institute of Agricultural Science, Turrialba, Costa Rica.

26. **Hardy, F., and C. Rodrigues.** 1953. *A Report on Cocoa Research, 1945–1951,* p. 89–91. The Imperial College of Tropical Agriculture, St. Augustine, Trinidad.

27. **Hodge, J.** 1953. Chemistry of browning reactions in model systems. *J. Agric. Food Chem.* **1:**928–943.

28. **Hoi, O. K.** 1977. Cocoa bean processing—a review. *Planter* (Kuala Lumpur) **53:**507–530.

29. **Hurst, W. J., K. B. Miller, J. Apgar, D. A. Stuart, C. Y. Lee, N. McHale, and B. Ou.** 2005. The determination of polyphenols and the correlation to cocoa content in selected U.S. confectionery products. Poster at Cornell Institute of Food Science Symposium on Functional Foods, Bioactive Compounds, and Human Health, 21 May 2005, Ithaca, N.Y.

30. **Jones, K. L., and S. E. Jones.** 1984. Fermentations involved in the production of cocoa, coffee and tea, p. 433–446. *In* M. E. Bushell (ed.), *Modern Applications of Traditional Biotechnologies.* Elsevier, Amsterdam, The Netherlands.

31. **Knapp, A. W.** 1937. *Cocoa Fermentation. A Critical Survey of Its Scientific Aspects.* John Bale Sons and Curnow, London, United Kingdom.

32. **Kris-Etherton, P. M., and C. L. Keen.** 2002. Evidence that the antioxidant flavonoids in tea and cocoa are beneficial for cardiovascular health. *Curr. Opin. Lipidol.* **13:**41–49.

33. **Lehrian, D. W.** 1989. Recent developments in the chemistry and technology of cocoa processing, p. 22–33. *In* Y. B. Che Man, M. N. B. Abdul Karim, and B. A. Amabi (ed.), *Food Processing Issues and Prospects.* Faculty of Food Science and Biotechnology, University Pertanian Malaysia, Serdang, Selangor, Malaysia.

34. **Lehrian, D. W., and G. R. Patterson.** 1983. Cocoa fermentation, p. 529–575. *In* H. J. Rehm and G. Reed (ed.), *Biotechnology,* vol. 5. Chemie Verlag, Weinheim, Germany.

35. **Lopez, A. S.** 1972. The development of chocolate aroma from non-volatile precursors, p. 640–646. *In Proceedings of Fourth International Conference on Cocoa Research (Trinidad and Tobago).*

36. **Lopez, A. S.** 1986. Chemical changes occurring during the processing of cacao, p. 19–53. *In* P. S. Dimick (ed.), *Proceedings of the Cocoa Biotechnology Symposium.* Department of Food Science, The Pennsylvania State University, University Park, Pa.

37. **Lopez, A. S., and P. S. Dimick.** 1995. Cocoa fermentation, p. 562–577. *In* H. J. Rehm and G. Reed (ed.), *Biotechnology,* vol. 9. Chemie Verlag, Weinheim, Germany.

38. **Lopez, A. S., D. W. Lehrian, and L. V. Lehrian.** 1978. Optimum temperature and pH of invertase of the seeds of *Theobroma cocoa* L. *Rev. Theobromo* (Brazil) **8:**105–112.

39. **Lopez, A. S., and V. C. Quesnel.** 1971. An assessment of some claims relating to the production and composition of chocolate aroma. *Int. Chocolate Rev.* **26:**19–24.

40. **Lopez, A. S., and V. C. Quesnel.** 1973. Volatile fatty acid production in cocoa fermentation and the effect on chocolate flavor. *J. Sci. Food Agric.* **24:**319–326.

41. **Lozano, A. R.** 1958. Mechanical pod breakers. *Cocoa* (Turrialba) **3(14):**35.

42. **Martelli, H. L., and H. F. K. Dittmar.** 1961. Cocoa fermentation. V. Yeasts isolated from cocoa beans during the curing process. *Appl. Microbiol.* **9:**370–371.

43. **McDonald, C. R., R. A. Lass, and A. S. Lopez.** 1981. Cocoa drying—a review. *Cocoa Growers Bull.* **31:**5–39.

44. **Menchu, J. F., and C. Rolz.** 1973. Coffee fermentation technology. *Café Cacao* **17:**53–61.

45. **Ostovar, K., and P. G. Keeney.** 1973. Isolation and characterization of microorganisms involved in the fermentation of Trinidad's cocoa beans. *J. Food Sci.* **38:**611–617.

46. **Passos, F. M. L., A. S. Lopez, and D. O. Silva.** 1984. Aeration and its influence on the microbial sequence in cocoa fermentations in Bahia with emphasis on lactic acid bacteria. *J. Food Sci.* **49:**1470–1474.

47. **Passos, F. M. L., D. O. Silva, A. S. Lopez, C. L. L. F. Ferreira, and W. V. Guimaraes.** 1984. Characterization and distribution of lactic acid bacteria from traditional cocoa bean fermentations in Bahia. *J. Food Sci.* **49:**205–208.

48. **Quesnel, V. C.** 1965. Agents inducing the death of the cocoa seeds during fermentation. *J. Sci. Food Agric.* **16:**441–447.

49. **Quesnel, V. C.** 1972. *Annual Report of Cocoa Research,* p. 48. University of the West Indies, St. Augustine, Trinidad.

50. **Quesnel, V. C.** 1972. Cacao curing in retrospect and prospect, p. 602–606. *In Proceedings of Fourth International Conference on Cocoa Research (Trinidad and Tobago).*

51. **Quesnel, V. C., and K. Jugmohunsingh.** 1970. Browning reaction in drying cocoa. *J. Sci. Food Agric.* **21:**537.

52. **Rodriquez, D. B., and H. A. Frank.** 1969. Acetaldehyde as a possible indicator of spoilage in coffee Kona (Hawaiian coffee). *J. Sci. Food Agric.* **20:**15–19.

53. **Roelofsen, P. A.** 1953. Polygalacturonase activity of yeast, *Neurospora* and tomato extract. *Biochim. Biophys. Acta* **10:**410–413.

54. **Roelofsen, P. A.** 1958. Fermentation, drying, and storage of cocoa beans. *Adv. Food Res.* **8:**225–296.

55. **Rohan, T. A.** 1957. Cocoa preparation and quality. *West Afr. Cocoa Res. Inst.* **1956–1957:**76.

56. **Rohan, T. A.** 1963. *FAO Agriculture Studies,* no. 60. *Processing of Raw Cocoa from the Market.* Food and Agriculture Organization, Rome, Italy.

57. **Rohan, T. A.** 1969. The flavor of chocolate: its precursors and a study of their reactions. *Gordian* **69**:443–447, 500–501, 542–544, 587–589.

58. **Rohan, T. A., and T. Stewart.** 1965. The precursors of chocolate aroma: the distribution of free amino acids in different commercial varieties of cocoa beans. *J. Food Sci.* **30**:416–419.

59. **Rohan, T. A., and T. S. Stewart.** 1967. The precursors of chocolate aroma: production of free amino acids during fermentation of cocoa beans. *J. Food Sci.* **32**:396–398.

60. **Rohan, T. A., and T. S. Stewart.** 1967. The precursors of chocolate aroma: production of reducing sugars during fermentation of cocoa beans. *J. Food Sci.* **32**:399–402.

61. **Rombouts, J. E.** 1952. Observations on the microflora of fermenting cocoa beans in Trinidad. *Trinidad Proc. Soc. Appl. Bacteriol.* **15**:103–111.

62. **Rombouts, J. E.** 1953. Critical review of the yeast species previously described from cocoa. *Trop. Agric.* (Trinidad) **30**:34–41.

63. **Sanchez, J., G. Daguenet, J. P. Guiraud, J. C. Vincent, and P. Galzy.** 1985. A study of the yeast flora and the effect of pure culture seeding during the fermentation process of cocoa beans. *Lebensm. Wiss. Technol.* **18**:69–76.

64. **Saposhnikova, K.** 1952. Changes in the acidity and carbohydrates during ripening of the cocoa fruit: variations of acidity and weight of seeds during fermentation of cocoa in Venezuela. *Agron. Trop.* (Maracay) **2**:185–195.

65. **Schwan, R. F.** 1998. Cocoa fermentations conducted with a defined microbial cocktail inoculum. *Appl. Environ. Microbiol.* **64**:1477–1483.

66. **Schwan, R. F., and A. E. Wheals.** 2003. Mixed microbial fermentations of chocolate and coffee, p. 429–449. *In* T. Boekhout and V. Robert (ed.), *Yeasts in Food: Beneficial and Detrimental Aspects.* Woodhead Publishing Ltd., Cambridge, England.

67. **Schwan, R. F., A. S. Lopez, D. O. Silva, and M. C. D. Vanetti.** 1990. Influência da frequência e intervalos de revolvimentos sobre a fermentacao de cocoa e qualidade do chocolate. *Rev. Agratrop.* **2**:22–31.

68. **Schwan, R. F., A. H. Rose, and R. G. Board.** 1995. Microbial fermentation of cocoa beans, with emphasis on enzymatic degradation of the pulp. *J. Appl. Bacteriol.* **79**:96–107.

69. **Schwan, R. F., M. C. D. Vanetti, D. O. Silva, A. S. Lopez, and C. A. deMoraes.** 1986. Characterization and distribution of aerobic spore-forming bacteria from cocoa fermentations in Bahia. *J. Food Sci.* **51**:1583–1584.

70. **Shamsuddin, S. B., and P. S. Dimick.** 1986. Qualitative and quantitative measurements of cocoa bean fermentation, p. 55–78. *In* P. S. Dimick (ed.), *Proceedings of the Cacao Biotechnology Symposium.* Department of Food Science, The Pennsylvania State University, University Park, Pa.

71. **Sieki, K.** 1973. Chemical changes during cocoa fermentation using the tray method in Nigeria. *Int. Chocolate Rev.* **28**:38–42.

72. **Sivetz, M., and N. Desrosier.** 1979. *Coffee Technology,* p. 74–116. AVI Publishing Company, Inc., Westport, Conn.

73. **Suryakantha Raju, K., S. Vishveshwara, and C. S. Srinivasan.** 1978. Association of some characters with cup quality in *Coffea canephora* × *Coffea arabica* hybrids. *Indian Coffee* **1978**:195–197.

74. **Van Pee, W., and J. M. Castelein.** 1972. Study of the pectinolytic microflora, particularly the Enterobacteriaceae, from fermenting coffee in the Congo. *J. Food Sci.* **37**:171–174.

75. **Vaughn, R. H., R. DeCamargo, H. Falanghe, G. Mello-Ayres, and A. Serzedello.** 1958. Observations on the microbiology of the coffee fermentation in Brazil. *Food Technol.* **12**:57.

76. **Villeneuve, F., E. Cros, J. C. Vincent, and J. J. Macheix.** 1989. Search for a cocoa fermentation index III—changes in the flavan-3-ols in the bean. *Café Cacao* **33**:165–170.

77. **Voigt, J., and B. Beihl.** 1995. Precursors of the cocoa specific aroma components are derived from the vicilin class (7S) globulin of the cocoa seeds by proteolytic processing. *Bot. Acta* **108**:283–289.

78. **Voigt, J., D. Wrann, H. Heinrichs, and B. Biehl.** 1994. The proteolytic formation of essential cocoa-specific aroma precursors depends on particular chemical structures of the vicilin-class globulin of the cocoa seeds lacking in the globular storage proteins of coconuts, hazelnuts and sunflower seeds. *Food Chem.* **51**:197–205.

79. **Wadsworth, R. V.** 1955. The preparation of cocoa, p. 131–142. *In Cocoa Conference.* Report of the 1955 London Cocoa Conference, Grosvenor House, 13–15 September 1955. London, United Kingdom.

80. **Weissberger, W., T. E. Kavanagh, and P. G. Keeney.** 1971. Identification and quantification of several non-volatile organic acids of cocoa beans. *J. Food Sci.* **36**:877–879.

81. **Zak, D. K., K. Ostovar, and P. G. Keeney.** 1972. Implication of *Bacillus subtilis* in the synthesis of tetramethylpyrazine during fermentation of cocoa beans. *J. Food Sci.* **37**:967–968.

Food Microbiology: Fundamentals and Frontiers, 3rd Ed.
Edited by M. P. Doyle and L. R. Beuchat
© 2007 ASM Press, Washington, D.C.

Iain Campbell

Beer

40

This chapter is a general overview of the scientific principles of the brewing industry. More detailed information on the science and technology of beer production is available in textbooks (3, 9, 10, 12, 13). Detailed information on the fermentation aspects of brewing can be found in recent texts (1, 18, 22).

The legal definition of beer varies among countries. The strictest definition, as in Germany, limits the ingredients to hops, yeast, water, and malt, not necessarily of barley, although that cereal is understood if no other is specified. In many other countries, sugars or unmalted cereals are permitted as adjuncts. Normally, the upper limit is about 30% of the grist, but in some tropical countries where barley and its malt are unavailable, acceptable beers can be brewed with 100% unmalted cereal plus enzymes. Figure 40.1 shows a general outline of the traditional brewing process.

Archeological evidence has shown that beer has been produced for at least 5,000 years, perhaps as a fortuitous discovery from the baking of bread. Brewing yeast is unable to ferment the starch of cereal grains, so a preliminary germination is required in which the starch and protein are hydrolyzed enzymatically into simple sugars and amino acids which provide the main nutrients for the yeasts to carry out the fermentation. This may have happened accidentally with moist grain, and naturally occurring fermentative yeasts would have produced a primitive beer. In subsequent developments over the millennia, barley became the principal cereal for beer production because of its husk, which provides an excellent natural filter for clarification of the extracted wort.

MALTING

The first stage of the brewing process is the production of malt, now almost always by specialist maltsters. It is a common mistake to think that during malting the grains convert starch into simple sugars which the yeast can metabolize. In fact, that happens during mashing, and the changes during malting are limited to the breakdown of cell walls and the protein matrix in which the starch granules are embedded, but such modification of the grain is necessary for hydrolysis of the starch during mashing.

Not all barley is suitable for malting. One important property is the nitrogen content. Too low a level of nitrogen will restrict yeast growth in the subsequent fermentation, but normally the problem is too high a level of

Iain Campbell, International Centre for Brewing and Distilling, Heriot-Watt University, Edinburgh EH14 4AS, United Kingdom.

Malting

    Conversion of barley starch to fermentable sugars

    (glucose, maltose, maltotriose) and protein to free amino nitrogen

Milling, Mashing

    Extraction of sugars, amino acids, and other yeast nutrients and

    enzymes with hot water to yield sweet wort

Wort Boiling

    Boiling with hops to extract aroma and bittering compounds,

    then sterilization to yield hopped wort

Fermentation

    Conversion by *Saccharomyces cerevisiae* of fermentable sugars

    to ethanol and carbon dioxide

Post-Fermentation Treatments

    Maturation (improvement of flavor), clarification, packaging,

    pasteurization

Figure 40.1 The brewing process.

nitrogen in the barley, more than is necessary for yeast growth, with the resulting excess nitrogen encouraging microbial, particularly bacterial, spoilage of the final beer. Also, grain intended for malting must be carefully dried and stored to avoid the risk of either dormancy (delayed germination) or death of the barley (3, 7, 14).

The malting process occurs in three stages: steeping, germination, and kilning. Steeping in water over 24 to 48 h,

usually in two stages with an "air rest" between, stimulates the growth of the embryo plant, as would occur if it had been planted in soil. Germination is controlled by temperature, aeration, and humidity to the point where the stem is just about to emerge from the grain, by which time rootlets have already formed. By that stage, the zone of modification of the starch has extended almost to the end of the grain opposite from the embryo. In the traditional malting process, grain transferred from the steeping vessel was spread onto the floor as a layer about 0.8 m deep and manually turned daily to prevent the developing rootlets from binding together over the 2 weeks of germination. Since late in the 19th century, numerous types of mechanical malting devices have been developed to speed up the process and reduce the labor requirements. Figure 40.2 shows one of the first, invented by Saladin, which still is operated in essentially the same design. The rotating helical turners carried on a gantry run along the full length of the germination vessel to turn the germinating grain, which is aerated by moist air passing through the slotted floor.

At the end of germination the grain has a moisture content of about 50%, so drying to <6% moisture is required for storage. However, kilning is more complex than simply drying the grain. Malt is not only a source of fermentable sugars and other yeast nutrients, it is also a source of amylolytic and proteolytic enzymes for hydrolysis of any additional unmalted cereal in the recipe. Although these enzymes are moderately heat resistant, they are increasingly inactivated by drying temperatures above 70°C. Also, higher temperatures give darker-colored malt because of browning reactions, first explained by Maillard, between the sugars and nitrogen components of the grain. For some beers, darker malts are desirable, but with the penalty of reduced enzymatic activity for hydrolysis of cereal adjuncts.

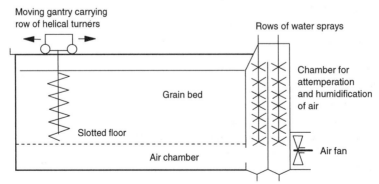

Figure 40.2 Saladin malting system. The row of helical turners moves from one end of the grain bed to the other at intervals during germination to turn the malt and prevent matting of the rootlets.

The production of malt is more the concern of the botanist and biochemist, but there are important microbiological aspects. First, malt or malt extracts are food products, and high standards of hygiene are required. Although some bacterial contamination is inevitable, potentially the most troublesome contaminants are fungi. Certain fungi have specific spoilage or toxigenic effects: the production of mycotoxins or, in the case of some species of *Fusarium*, of a polypeptide "gushing factor" which, by creating nuclei for development of carbon dioxide bubbles, causes violent frothing of the beer when the bottle or can is opened. However, since the growth of any mold on barley gives an obvious weathered appearance and a moldy aroma which persists to the final beer, simple sensory assessment is sufficient for the maltster to decide on acceptance or rejection of a barley sample. In fact, the maltster normally has no alternative to such an immediate assessment when deciding to accept a load of grain from a waiting truck, since conventional microbiological analysis would require at least 1 week.

MASHING AND WORT PRODUCTION

Wort is the sugary solution prepared by steeping malt in warm water, either alone or with sugar (e.g., glucose, sucrose, or maltose crystals or syrups) or unmalted cereal adjunct if appropriate, and the resulting grist is extracted with more warm water. Details of the process vary among breweries, and it is impracticable to deal with all of the possibilities here.

In traditional milling, although the contents of the grains must be ground finely to maximize the yield of fermentable extract, it is important that the outermost layer, the husk, is only cracked rather than ground to powder. In a later stage of the process, the husk functions as a natural filter. Barley has become the dominant cereal of the brewing industry largely because its husk structure is ideal for this purpose.

The traditional decoction mashing process for Bavarian and Czech beers has origins predating the invention of the thermometer. Consistent brewing requires reproducible conditions, and consistent temperatures were achieved by mixing the grist with a measured volume of well water, heating a measured proportion to boiling, and returning the boiled slurry to the mash tun. Well water is of constant temperature throughout the year, and water always boils at the same temperature at a given altitude, so with the same volumes each time, mashing temperatures would always be the same. Experience showed that several steps of increasing temperature were required, which we now know are the optima of α- and β-amylases and protease of the malt. This stepwise increase was achieved by a succession of accurately measured decoctions. While double- or triple-decoction mashing undoubtedly produces a good-quality wort, even from poorly modified malts, the process is expensive to operate and now is used only for prestigious products for which the traditional process has sales appeal. Infusion mashing is also a traditional process, associated particularly with British ales and similar beers of Belgium and northern Germany. At its simplest, the mash tun contains a floating bed of grist, suspended on entrapped air, with wort drawn off at the base continuously replaced by a top spray of hot sparge water until further extraction of sugar is impracticable. Originally, infusion mashing operated at a constant temperature of about 70°C (traditionally, the temperature at which the brewer's reflection was just obscured by the steam beginning to rise from the brewing liquor), and well-modified malt was essential (3).

The majority of breweries now use a mashing vessel in which the grist is mixed with warm water, heated in steps by using a steam jacket rather than decoction through the optimum temperatures of the various mashing enzymes, and finally filtered in the lauter tun (German: lauter, clear) with a horizontal slotted metal base on which the bed of husk material functions as a filter, since all cereal solids must be removed before the next stage of the process. When unmalted cereal forms part of the recipe, the ground cereal is heated in a cereal cooker at least to the gelatinization temperature of the starch of the cereal (70 to 75°C in the case of maize, a common ingredient), but often to boiling, and then transferred to the mashing vessel to allow hydrolysis of its starch and protein by the malt enzymes.

In recent years, many breweries have replaced the lauter tun with a mash filter, in reality a type of filter press (13). As in the lauter tun, the cereal solids constitute the filter bed, but on a support of vertical polypropylene sheets with numerous fine perforations, much smaller than the slots in the bottom plates of the lauter tun. Therefore, the malt can be ground more finely than for a lauter tun, giving better extraction of nutrients, but the main advantage of the mash filter is that in each of the filter chambers the mash can be squeezed to expel the wort. The process is faster than filtration in a lauter tun, and a much smaller amount of sparge liquor is required. Therefore, stronger worts can be produced and more quickly.

HOPS AND HOP BOILING

Over brewing history, many different herbs have been used as flavoring for beer, but for the past five centuries hops have been preferred, almost certainly for microbiological reasons. It is now recognized that hops have an important antimicrobial, particularly antibacterial, effect, and it is presumed that the medieval brewers realized that hopped beers maintained their quality for longer periods of time than did beers with other flavorings.

Hops require a temperate climate to grow, but in practice their cultivation is restricted to certain areas, e.g., the county of Kent in Britain and Washington state in the United States. The different hop varieties in common use vary in their content of the bitter acids, resins, and oils which contribute to beer flavor and aroma and in whether they contain seeds. Many brewers believe that hop seeds impart a harsh flavor to beer. The selection of a particular variety, or blend of varieties, is an important part of a beer recipe. Flavor is extracted by boiling, with the incidental advantages of sterilizing and concentrating the wort, isomerizing hop α- and β-acids (see below), and purging the wort of harsh, grainy flavors.

In appearance, hops resemble small pine cones but have a softer texture. Cone hops are still used by some traditional breweries, but the majority of breweries now use processed hops, either pellets prepared from ground cone material or hop extract. Modern hop extraction technology uses liquid carbon dioxide to avoid potential problems of residual organic solvent. Also, various preisomerized hop products are available that can be added during the last few minutes of boiling to reduce losses. Thus, it is interesting that the boiling stage, originally intended for extraction of flavors, is now in many breweries simply a process of biochemical and microbiological stabilization of the wort.

To the brewer, the most important hop components are the resins, tannins, and essential oils (21). Typically, these constitute approximately 14, 4, and 0.5%, respectively, of the weight of the mature hop cones. Bitterness of beer is due to the α-acid (humulone) and β-acid (lupulone) components of the resins, but not directly: these complex acids are isomerized during hop boiling into the bitter iso-α-acid and iso-β-acid isohumulone and hulupone, respectively (10, 12, 21). Within these general groups of iso-acids are numerous analogs according to the acyl side chains on the resin molecule. Essential oils contribute aroma to the beer, but most of the oil is lost during boiling. Therefore, aroma hops, as distinct from high-α-acid bitterness hops, are added late in the boil to ensure maximum aroma effect. Also, if sugar crystals or syrup are included in the recipe, this is the appropriate stage for addition. Hop tannins have little direct influence on flavor but react with malt protein to form "hot break," a precipitate of protein and tannin complexes and insoluble calcium salts and phosphates. The hot break must be removed before fermentation, since the particulate material would adversely affect fermentation and flavor development by the yeast.

After traditional hop boiling with cone hops, the hot break is removed from the wort by filtration through the settled bed of spent hops. Pellets or extract does not provide a suitable filter medium, so wort prepared with these products must be clarified by centrifugation. In modern plants, the centrifugal effect is provided by a whirlpool hop separator: the hopped wort is run tangentially into a circular tank, often with a dished or slightly coned base. Hop debris and hot break collect at the center of the resulting vortex while clear wort is drawn from the side of the vessel.

Obviously the wort, still at approximately 100°C, has to be cooled to 20°C or lower before "pitching" with the yeast inoculum. Less obvious, but nevertheless essential, is the requirement, explained below, to aerate the wort to contain 6 to 8 μg of dissolved oxygen/ml. After cooling and aeration, the wort is ready for the fermentation stage of the process.

FERMENTATION

Brewing Yeast (*Saccharomyces cerevisiae*)

Formerly, the actively fermenting yeasts of the fermentation industries, both culture yeasts and common contaminant "wild yeasts," were classified as different species of *Saccharomyces*. *Saccharomyces cerevisiae* of the traditional northern European ales and similar beers was collected (skimmed) as "yeast head" or "top yeast" from the surface of an active fermentation, and *S. carlsbergensis*, "bottom yeast" which did not form such a head, had to be collected from the bottom of the vessel at the end of fermentation (4). Originally, bottom yeast was used only for Czech, Bavarian, and Danish lager beers, but now non-head-forming yeasts are more widely used to maximize the useful capacity of enclosed fermentation vessels. *S. bayanus* and *S. uvarum*, which are important culture yeasts in the wine industry, are also possible brewery contaminants, as is *S. diastaticus*, an amylolytic yeast. Most of these species are now classified officially by yeast taxonomists as a single species, *S. cerevisiae* (4), but still it is convenient in the brewing industry to distinguish the different types by their former specific names.

It is important to realize that *S. cerevisiae* is not a facultative anaerobe like *Escherichia coli* (18, 22). Although brewing yeast changes between oxidative and fermentative metabolism according to environmental aerobic or anaerobic conditions, it cannot grow indefinitely under anaerobic conditions. As with all eukaryotic cells, yeast cell membranes contain unsaturated fatty acids and sterols, which can be synthesized only in the presence of atmospheric oxygen. The amounts of unsaturated fatty acids in malt and sterols naturally present in wort are too small to support yeast growth, hence the requirement for initial aeration of the wort to allow the yeast to synthesize these compounds.

Pitching yeast in satisfactory condition must be added in the correct amount to inoculate the fermentation vessel: usually to 1×10^7 to 2×10^7 cells/ml, although measurement of yeast cake or slurry by weight is more convenient on a production scale. In the course of a typical fermentation, the yeast population increases by a factor of about eight, i.e., only three successive cell divisions. Subsequent multiplication is inhibited by the lack of oxygen, unsaturated fatty acids, and sterols, and since aeration late in the fermentation would cause flavor problems, no further cell growth is possible.

A brewery fermentation is essentially the lag, log, and early stationary phases of the yeast growth curve (Fig. 40.3). Brewing is unusual in modern biotechnology in reusing the culture from the previous fermentation, although a new culture is prepared at regular intervals, e.g., after 15 successive fermentations, although the number varies among breweries. In the lag phase of 6 to 12 h, the yeast utilizes the dissolved oxygen in the wort to restore its unsaturated fatty acid and sterol supply and to adjust from the anaerobic, acidic, alcoholic conditions at the end of the previous fermentation to the very different environment of fresh wort.

Yeast nutrients, including sugars, are taken up from the wort to generate new yeast cells; fermentable sugars

also provide the energy for this process. A traditional-strength wort of 10° Plato, i.e., 10% solute, has a specific gravity of 1.040, which is largely due to the sugars in solution. Amino acids, vitamins, and inorganic salts account for less than 1% of the dissolved solids. Typically, the approximate sugar composition of such a wort would be as follows: maltose, 4%; glucose and maltotriose, each 2%; and higher dextrins, which are not utilized by brewing yeast, about 2%. However, the various fermentable sugars are not utilized simultaneously: this is related to the transport of nutrients into the cells and varies with different strains. Normally, brewing yeasts transport and metabolize the simplest sugar first, i.e., glucose, and maltose transport begins only after much of the glucose has been metabolized. For a similar reason, utilization of maltotriose begins late in the fermentation, when most of the maltose has been consumed. Note also in Fig. 40.3 that fermentation of sugars continues after yeast growth ceases, due to the continuing action of the relevant enzymes, but more slowly than before. However, without yeast growth and protein synthesis, there is no further requirement for amino nitrogen; in fact, there may be some release of amino nitrogen from the small percentage of dead cells.

Ethanol and carbon dioxide are the main products of yeast metabolism, but small amounts of other compounds make an important contribution to the flavor of the beer (1, 18). By-products of the Embden-Meyerhof pathway, the major metabolic route in the anaerobic conditions of fermentation, contribute to flavor, but a major effect is the nitrogen metabolism of yeast. *S. cerevisiae* has a limited range of amino acid permeases, and most amino acids for protein synthesis have to be synthesized by the yeast cell. Keto acids, which are intermediates in the biosynthetic pathway, may be converted to higher alcohols (Fig. 40.4), which make a significant contribution to the flavor of beer and indeed all fermented beverages. Also related to transport is a decrease in pH during the fermentation, a result of excretion of H^+ during uptake of nutrients. The pH of wort is normally 5.1 to 5.2 and decreases during fermentation to 3.8 to 4.0, mainly during the period of yeast growth.

Progress of Fermentation

Traditional fermentation vessels are open rectangular tanks of 2 to 3 m in depth (10). Such vessels are particularly useful for top yeasts, brewing strains which rise to the surface of the fermenting beer, from where they are skimmed off as the inoculum for the next fermentation. Since about 1970 the cylindroconical fermentation vessel (CCFV) (Fig. 40.5) has become the preferred type (12, 13). These vessels are enclosed to reduce the risk of

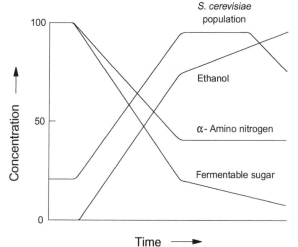

Figure 40.3 Theoretical progress of a typical brewery fermentation, showing changes in the population of *S. cerevisiae* and concentrations of fermentable sugar, amino nitrogen, and ethanol. The graph shows yeasts in suspension and the start of settling of cells from the beer at the end of fermentation. The time axis is not calibrated, since fermentation rates differ widely among breweries. Other variables are shown as percentages of the initial or final value, expressed as 100%.

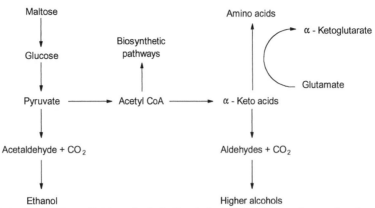

Figure 40.4 Formation of higher alcohols (fusel alcohols) as by-products of amino acid biosynthesis. Note the similarity of the reactions pyruvate → acetaldehyde → ethanol and α-keto acid → aldehyde → higher alcohols; both are important for redox balance under anaerobic conditions.

contamination and facilitate recovery of carbon dioxide and have a conical lower section to facilitate recovery of bottom yeast, now almost universally used, which settles out late in the fermentation. The shape of the vessel also encourages a vigorous mixing of yeast in the wort,

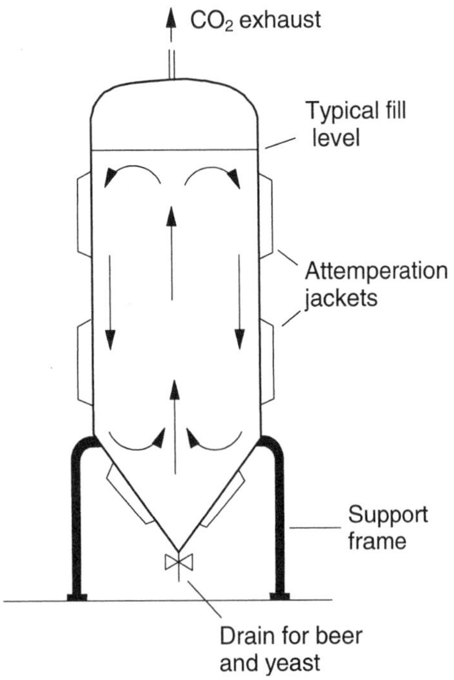

CO₂ exhaust

Typical fill level

Attemperation jackets

Support frame

Drain for beer and yeast

Figure 40.5 CCFV. Vigorous circulation of fermenting wort is created by a central upward flow of CO_2 bubbles and a downward flow in contact with the cooling jackets. Cooling of the cone section is optional, depending on whether the brewery stores settled yeast before the next fermentation.

giving a fast fermentation. Central upward movement with the bubbles of carbon dioxide and peripheral downward movement by the cooling panels on the walls occur in the fermenting wort.

In recent years some of the larger brewery companies have installed very large CCFVs, of capacities of up to 6,000 hl, for large-scale production of their principal brands. Although all CCFVs operate in a similar way, an important difference is that the very large vessels may be four to five times the size of the brewhouse and must be filled in installments. The first batch of wort must be inoculated immediately, as it is dangerously bad microbiological practice to store uninoculated wort. The wort is aerated and the yeast begins to grow, and the second, third, fourth, and possibly even fifth batches of wort are added as available. Unfortunately, there is a biochemical problem. The wort must be aerated since yeast needs oxygen to grow efficiently. An intermediate state of oxygen deficiency causes excessive production of esters, especially ethyl acetate. However, oxygen added late in the fermentation encourages the production of unwanted diacetyl. Therefore, the solution is to add with the first batch of wort most of the yeast and dissolved oxygen required, perhaps three times as much as that volume of wort would normally require. The rest of the pitching yeast and dissolved oxygen are added with the second batch of wort, and the later wort additions are fermented by the yeast already present.

An important property of brewing yeasts is the ability to flocculate, i.e., spontaneously aggregate into clumps, late in the fermentation. Too early flocculation stops the fermentation as the yeasts settle out of partly fermented beer, but nonflocculent yeasts are also troublesome;

they have to be removed by filtration or centrifugation. Currently, the preferred explanation of flocculation is a lectin-like activity of the cell wall which develops during fermentation, but other factors are also involved, e.g., divalent ions, particularly Ca^{2+}, decreasing amounts of flocculation-inhibiting fermentable sugar, and increasing amounts of stimulatory ethanol (19).

POSTFERMENTATION TREATMENTS

Conditioning

"Green beer" from the fermentation vessel contains acetaldehyde, diacetyl, and other unwanted by-products of yeast metabolism, which must be removed (10, 18). Although present in only small amounts, these by-products have a marked effect on the flavor or, more correctly, the aroma of the beer. In traditional cask conditioning, the beer, still with about 1% fermentable sugar, undergoes a secondary fermentation in casks, absorbing these undesirable flavor compounds and generating sufficient carbon dioxide for dispense. Similar changes are associated with the traditional low-temperature (0 to 2°C) secondary fermentation over several months for pilsner and other lager beers (German: lager, store). It is now known that such storage at low temperature is not essential; the beneficial flavor changes can be achieved more rapidly by a few days' storage of the beer in contact with 10^5 to 10^6 yeast cells/ml at 15°C, followed by cooling to 0°C only long enough to precipitate "chill haze" material and yeast sediment. It is important to avoid accidental access of air at this stage, since diacetyl is formed from an acetolactate precursor by spontaneous chemical reactions in the presence of oxygen and yeast is no longer present to remove the buttery off-flavor of diacetyl.

Filtration

With the exception of cask-conditioned beers, which are clarified by the addition of fining agents, it is common practice now to filter beer to complete clarity. Various designs of filters are in common use, most using either cellulose fibers or diatomite or pumice powder as the filter medium (11). Such materials, which adsorb microbial and inert haze-forming material in the depths of the filter, do not sterilize the beer since sufficiently fine filters would cause unacceptably low flow rates. Typically, the yeast count is reduced to <100 cells/liter, whereas at least 10^8 yeast cells/liter are required to form a visible haze.

However, membrane filtration, which does sterilize beer, is becoming increasingly popular. Pasteurization is unnecessary for sterile-filtered beer, so membrane filtration avoids the substantial energy cost and the potential flavor defects caused by heating the beer. With two filters used in series—a rough prefilter to remove as much particulate material as possible and a second filter, the membrane filter, to sterilize—an acceptable flow rate and long filter life can be achieved.

Pasteurization and Packaging

Draft beer is pasteurized by passage through a heat exchanger at about 70°C before filling into already cleaned and sterilized kegs. Beer for sale in bottles or cans is pasteurized after filling. In both systems, the heat treatment is equal in terms of pasteurization units (PU; 1 PU = 60°C for 1 min), but the "tunnel" pasteurizer for bottles and cans uses a lower temperature (typically 62°C) for a longer time. Individual breweries have their own preferred pasteurization treatment. In theory, 5 PU is sufficient to kill the small numbers of brewing yeasts likely to pass through a cellulose filter, but to eliminate the slightly more heat-resistant bacterial or wild-yeast contaminants which may occasionally be present, up to 30 PU may be applied. Since increasing heat treatment may adversely affect flavor, the choice of PU value is a compromise between potential risks of oxidized flavors and microbial spoilage (10, 13).

HIGH-GRAVITY BREWING

The basic principle of high-gravity brewing is that it is theoretically possible to double the production of a brewery without the expense of additional brewhouse or fermentation capacity by fermenting double-strength wort. The resulting double-strength beer would be diluted to normal strength immediately before packaging. The economic advantages are obvious. At first sight, the addition of water to beer might seem rather questionable, but that water would have been used anyway in the first stage of the standard process. Unfortunately, the idea is very difficult in practice. In order to match the flavor of high-gravity beer with that of the normal product, the concentrations of everything must be doubled: malt, hops, yeast, and even the dissolved oxygen in the wort.

Dealing with the nonmicrobiological aspects first: with standard mash and lauter tuns, doubling the weight of malt does not yield double the extract. Extraction becomes less efficient at higher concentrations, and efficient sparging of the grain bed in the lauter tun dilutes the wort excessively. Therefore, until the development of a mash filter which can produce high-gravity wort from an all-malt grist, it was necessary to use sugar syrup adjuncts to increase the gravity, illegal in some countries. There is a similar problem with hops, since extraction becomes progressively less efficient from larger quantities of hops or hop pellets. That particular problem was solved by the use of hop extracts, especially preisomerized extracts which may be added either late in hop boiling or at the stage of dilution of high-gravity beer to normal strength.

Although the brewer can easily add twice the previous amount of yeast to the double-strength wort, it is difficult to achieve twice the dissolved oxygen concentration. A maximum concentration of 6 to 8 μg of dissolved oxygen/ml is routinely achieved by injecting air into wort between the cooler (after hop boiling) and the fermentation vessel. In the high-gravity process, twice the amount of yeast would require 12 to 16 μg of dissolved oxygen/ml, achieved only by injecting pure oxygen rather than air. As noted earlier, reduced oxygen levels cause excess ester production.

A standard wort of specific gravity 1.040 is termed 10° Plato, i.e., it contains 10% sugar. Double-strength wort of 20% sugar is likely to retard yeast growth by osmotic stress. A more serious situation occurs at the end of fermentation, when the yeast is recovered from beer of 8% instead of the previous 4% alcohol by volume and its viability may be too low for successful repitching in the next fermentation. A brewery which already has a yeast culture plant, previously used to replace the yeast culture occasionally, could easily prepare a new yeast culture for each fermentation. Since the plant is already available, the additional expense is small in comparison with the savings from high-gravity operation. Finally, the dilution water must be deoxygenated to a level of <0.1 μg of dissolved oxygen/ml to prevent any flavor problems from staling reactions or the development of diacetyl. The essential deaeration plant is the only equipment which is required specifically for high-gravity operations.

The example described above of brewing double-strength wort has only recently been successfully accomplished in practice. Previously, 1.5 times normal strength was the practical limit, because of flavor problems. Without sufficiently efficient oxygenation of the wort in the first stages of the fermentation, the yeast is unable to grow normally and produces excessive amounts of esters, as indicated in the section above on very large cylindroconical vessels. With recent developments in oxygenation, however, twice the normal strength is now practicable.

MICROBIAL CONTAMINANTS OF THE BREWING PROCESS

In comparison with other foods and drinks, beer is a relatively stable product, protected by its alcohol and carbon dioxide contents, anaerobiosis, low pH, and the antibacterial properties of hops, in addition to the heat treatments of wort boiling and beer pasteurization. Even

so, beer production carries a risk of microbial contamination at all stages of the process. On barley growing in the field, and on germinating barley during malting, the most important contaminants are molds. The field fungi which develop during cultivation are seldom a problem, and then only after a particularly wet growing season which allows *Fusarium* to develop sufficiently to cause a problem of gushing (6). However, storage fungi, principally *Aspergillus* and *Penicillium* spp. which develop on improperly dried barley after harvest, can have serious effects, as mentioned earlier, in forming mycotoxins.

In the brewery itself, yeasts and bacteria are potentially the troublesome contaminants. In beer, its acidic (usually about pH 3.9), alcoholic, anaerobic characteristics, with the additional antibacterial effects of carbon dioxide and hop acids and oils, restrict the range of potential spoilage microorganisms. Yeasts on grain are mainly aerobes which are unable to grow in fermenting wort or beer, but some species of the genera *Brettanomyces, Candida, Debaryomyces, Pichia, Saccharomyces, Torulaspora,* and *Zygosaccharomyces* are potential spoilage yeasts (4). Presumably their natural habitat and original source are plants, but once established in a brewery they are difficult to eradicate and are often transferred from one fermentation to the next in the pitching yeast, in increasing numbers each time. Members of the fermentative genera *Saccharomyces, Torulaspora,* and *Zygosaccharomyces,* and equivalent non-spore-forming *Candida* spp. may contaminate the fermentation or persist through filtration and pasteurization to produce turbidity and off-flavors in the beer. Most of these contaminants have diameters smaller than the 5- to 7-μm diameters of culture yeasts, so filters designed to retain brewing yeasts are less efficient with contaminants. *Dekkera* and its non-spore-forming anamorph *Brettanomyces* cause acetic acid spoilage and turbidity. *Debaryomyces, Pichia,* and equivalent aerobic *Candida* spp. are oxidative yeasts and so are limited to the early stages of fermentation or to beer to which air has gained access postfermentation. These yeasts cause turbidity and yeasty or estery off-flavors and in bottled or canned beer often form a surface film fragmenting into flaky particles or a deposit.

Only a few species of the lactic bacteria *Lactobacillus* and *Pediococcus* are sufficiently resistant to the antibacterial properties of beer to grow at any stage of the process. Hop components are particularly effective antibacterial agents. If able to grow, these bacteria cause turbidity and off-flavor, often caused by diacetyl (15). Diacetyl is a strongly flavored minor by-product of their metabolism which is highly valued in the dairy industry as the flavor of butter but is generally considered to be a spoilage fault of beer. Superattenuation by starch-fermenting lactobacilli and slime formation by pediococci are other possible faults. *Zymomonas*, like the lactic acid bacteria, is capable of growth under anaerobic conditions and grows at all stages of the process, but its most usual effect is in packaged beer, causing further fermentation with turbidity, off-flavors, and often "rotten apple" or dimethyl sulfide aromas.

Other possible spoilage bacteria are limited by their properties to specific stages of the process (20). *Aceto-bacter* and *Gluconobacter* are strict aerobes and grow, with characteristic acetic acid production, only in beer accidentally exposed to air. Enterobacteria (including *Obesumbacterium*, the most important of that group in the brewery environment) cause turbidity and off-flavor and often produce indole, phenols, diacetyl, hydrogen sulfide, and dimethyl sulfide, but grow well in the early stages of fermentation until inhibited by the falling pH and increasing ethanol content. Even so, they survive to be transferred with the yeast recovered for the next fermentation. Finally, *Megasphaera* (cocci) and *Pectinatus* (rods) species are recently discovered strictly anaerobic gram-negative bacteria which form acetic, butyric, and propionic acids, hydrogen sulfide, dimethyl sulfide, and turbidity and have become troublesome only because of modern advances in maintaining very low dissolved oxygen levels in beer.

Other bacteria are occasionally isolated, e.g., *Bacillus* and *Micrococcus* spp., but are unable to grow in wort or beer. Their main nuisance is the further laboratory testing required to confirm that they are not the more troublesome *Lactobacillus* or *Pediococcus*, which are of similar appearance on a Gram-stained film (15). A simple distinguishing test is that the lactic acid bacteria do not form the enzyme catalase (Fig. 40.6).

Isolation of Microbial Contaminants

Culture media for microbiological quality control can be either nonselective or selective. All brewing yeasts and bacteria should grow on nonselective media, usually either malt extract agar or Wallerstein Laboratories nutrient agar, a semisynthetic medium of similar nutrient value but more consistent composition. These media are useful for examination, by plating or membrane filtration, of samples in which no microorganisms should be found or in which they should be present only in low numbers, e.g., pasteurized beer or swabs or rinsings from recently cleaned and sterilized plants. Any microorganism, whether bacterium, culture yeast, or wild yeast, can be regarded as a contaminant in these situations (5, 15, 20).

Selective media are intended to suppress the growth of culture yeasts but allow the growth of any contaminants

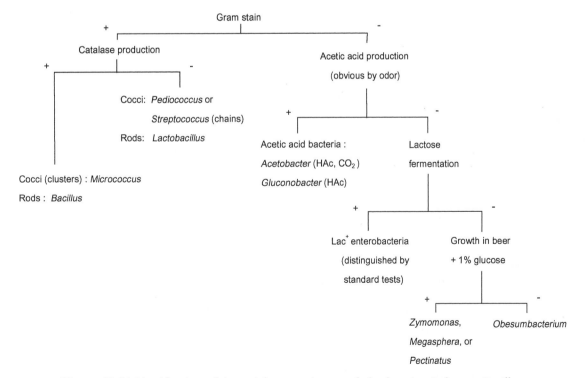

Figure 40.6 Identification of bacterial contaminants of the brewing industry. *Bacillus*, *Micrococcus*, and *Streptococcus* may be present but are unlikely to grow in beer, and *Megasphera* and *Pectinatus* grow only under strictly anaerobic conditions. HAc, acetic acid.

which may be present. Such media should normally be used to examine samples of pitching yeast or samples taken during or immediately after fermentation in which culture yeast is present. A commonly used selective medium is lysine agar, a synthetic medium of glucose, inorganic salts, vitamins, and L-lysine as a sole nitrogen source. *Saccharomyces* spp. are unable to grow, but yeasts of other genera do grow by using the NH$_2$ groups of lysine for nitrogen requirements. Unfortunately, wild strains of *S. cerevisiae* and other *Saccharomyces* spp. are particularly likely contaminants but cannot be detected on lysine agar. However, diastatic yeasts, including the former *S. diastaticus*, can be isolated on a medium containing starch, salts, vitamins, and ammonium sulfate, which all yeasts can utilize as a nitrogen source.

The most useful selective agent for detection or counting of bacterial contaminants is the antifungal antibiotic cycloheximide (actidione), which can be added to any suitable culture medium but is usually added to Wallerstein Laboratories nutrient agar as "actidione agar" (5). Some wild yeasts are sufficiently resistant to grow, but in general, a more effective way to recover wild *Saccharomyces* contaminants relies on their sporulation. Sporulation, which is stimulated by starvation conditions, is presumably advantageous in nature, but most brewery culture yeasts, after innumerable generations in rich media, have lost the ability to form viable spores. Yeast spores, unlike bacterial spores, are only marginally more heat resistant than the vegetative cells, but the difference is sufficient to recover contaminants on nonselective media after heat treatment. Practical details of microbiological analyses are fully explained in analytical manuals (5).

In general, complete identification of recovered contaminants is unnecessary. Bacteria and non-*Saccharomyces* wild yeasts need to be identified to the genus level only, as indicated in Fig. 40.6 and 40.7. Only fermentative yeasts need further identification, to determine whether or not they are *Saccharomyces* and, if so, *S. cerevisiae*. Even identification to the species level in this case is insufficient. It is important to determine whether it is the culture yeast itself, possibly persisting on improperly cleaned fermentor or other equipment surfaces or surviving through filtration and pasteurization. Remedial action in that situation would obviously be different from that required by the discovery of a wild *S. cerevisiae* strain as a contaminant. A quick and simple test to distinguish industrial yeast strains is their sensitivity to a range of antifungal compounds. In theory, a small-scale fermentation can distinguish the culture yeast from others, but the delay in obtaining useful results is unacceptably long. With recent advances in genetic methods,

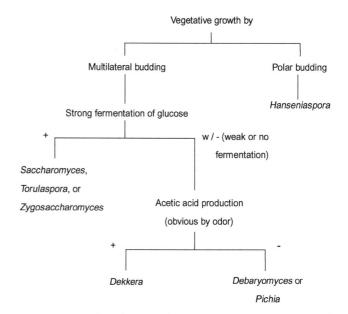

Figure 40.7 Identification of common yeast contaminants of the brewing industry. All genera listed are teleomorphic, i.e., they form spores. Anamorphic (non-spore-forming) forms of *Dekkera* and *Hanseniaspora* are *Brettanomyces* and *Kloeckera*, respectively. The anamorph of all other genera listed is *Candida*. *S. cerevisiae* ferments glucose, sucrose, maltose, and raffinose but not lactose.

DNA fingerprinting is a potentially useful rapid method to distinguish different strains, but at present it is likely to be used only in the laboratories of the largest brewing companies (8, 15, 16).

Sterilization

In most industrial fermentation processes, steam is the usual method of sterilization of the plant. In the brewery, however, chemical sterilants are preferred, mainly because they can be used at ambient temperature and avoid attemperation problems with adjacent working fermentors (17). Modern breweries use automatic in-place cleaning and sterilization equipment to remove deposits (of yeast and organic soil derived from beer foam) by a powerful jet of water containing detergent, followed by a spray of chemical sanitizer, and finally by a spray of sterile rinse water. Until recently, caustic sterilants containing 2% NaOH and additives were widely used, but acid sterilants based on phosphoric or peracetic (CH$_3$COOOH) acids have become more popular, not least because they are unaffected by carbon dioxide and can be used in closed vessels without the long delay and expense of draining off that valuable gas. Chlorine-based sanitizers are also used, but not widely because of the possibility of flavor problems: residual chlorine reacts with unknown compounds in beer to produce a strong medicinal flavor of organohalogen compounds. Quaternary ammonium compounds and biguanides are also regarded with suspicion. These compounds are widely used in other food industries because of their protective persistence on treated surfaces, even after rinsing, but this is unacceptable in the brewing industry because of the possible adverse effects of residual sanitizer on beer foam.

MODERN DEVELOPMENTS

For most of its history, brewing was a small-scale operation, either a domestic enterprise or one associated with monasteries. Developments over the past 300 years have coincided with increasingly large-scale commercial production and with general improvements in scientific knowledge and technology. Research in the brewing industry has improved our understanding of the processes involved, and better-quality malts and beers can be produced more rapidly and efficiently. For example, in 1900 it took at least 14 days to produce a batch of malt, mashing required up to 8 h, ale and lager fermentations could last 7 and 14 days, respectively, and subsequent lager maturation could take many weeks longer. Now these times have been halved, with improvement in consistency and quality, although some traditionalist consumers may disagree.

New products offer possibilities for increasing market share. One recent example is diet beer or light beer of lower carbohydrate content and, by implication, lower calorie content. Up to 20% of the carbohydrates of wort can be dextrins, which standard brewing yeast does not utilize and which thus remain in the beer. Even in countries where the addition of hydrolytic enzymes is not permitted, dextrin-fermenting yeasts can legally be used to ferment that carbohydrate. One possible source of such strains is hybridization between *S. diastaticus* and brewing yeasts, and successful results have indeed been achieved in that way. Recent genetic engineering research has also produced potentially useful diastatic yeasts, but at present commercial brewers, in Europe at any rate, are reluctant to risk the public disapproval of their use. Another obvious use of diastatic yeasts is for beers of higher ethanol content by fermentation of the dextrins (8). In the currently popular "ice beer," the ethanol content is increased by selectively freezing out part of the water content of the beer, with the incidental benefit of improved flavor and stability due to the chilling process.

Another recent development is the sale of certain brands of beer in clear bottles rather than the brown or green glass bottles which were traditionally used as protection against photochemical reactions which create a "light-struck" stale flavor, even over relatively short periods of storage. Although there are other possible staling reactions, e.g., those associated with excessive pasteurization, light-struck hop degradation compounds are particularly important because of their low flavor threshold. New types of hop iso-acids, often tetrahydro isomers, are more stable to light and allow brewers to satisfy the current popular demand for clear glass bottles (2, 21). As with other isomerized hop products, these can be added at any stage between the last few minutes of hop boiling and final packaging of the beer.

Continuous fermentation is not actually a modern development, but recently there has been renewed interest in continuous fermentation of beer. Between about 1955 and 1970 there was considerable research on continuous operation of industrial microbiological processes in general, and those in the brewing industry were no exception. At that time some breweries built continuous fermentation systems based on a succession of two or three stirred vessels, in each of which the homogeneous contents replicated the successive stages of batch fermentation: aerobic growth of the yeast, anaerobic fermentation, and in some plants, a third anaerobic vessel for flavor development and yeast separation. Other breweries preferred a single-vessel system, an unstirred tower in which the different levels represented the stages of batch fermentation, and fermented beer overflowed from a yeast separator at the top. Although undoubtedly faster and more economical than previous batch operations, a serious disadvantage of both types of continuous systems was the high cost of construction and operation in comparison with that of the cylindroconical vessels which were introduced in their modern form during the 1960s. Another problem was the difficulty of replicating in a different type of fermentation the flavors of traditional batch beers.

Recently, however, there has been renewed interest in continuous fermentation, now using immobilized-cell technology. Although very successful on a laboratory scale, full-size fermentors suffer from damage to the immobilized-cell support by the expansion of bubbles of carbon dioxide as they rise up the necessarily tall column. New types of support currently under development avoid this difficulty, but the greatest success of immobilized-cell systems has been in postfermentation conditioning or the production of low-alcohol beer, processes with less vigorous evolution of gas than active fermentation.

There is no doubt that yet more novel fermentation methods and products will be developed in the brewing industry in the future. We can also be sure that there will be a continuing demand for traditional beers, and the small breweries specializing in such products are certainly enjoying much success at the present time.

References

1. **Boulton, C. A., and D. E. Quain.** 2001. *Brewing Yeast and Fermentation.* Blackwell, London, United Kingdom.

2. **Bradley, L. L.** 1997. Uses of iso-alpha acids and chemically modified products. *Ferment* **10:**48–50.

3. **Briggs, D. E., J. S. Hough, R. Stevens, and T. W. Young.** 1981. *Malting and Brewing Science,* 2nd ed., vol. I. *Malt and Sweet Wort.* Chapman and Hall, London, United Kingdom.

4. **Campbell, I.** 2003. Wild yeasts in brewing and distilling, p. 247–266. *In* F. G. Priest and I. Campbell (ed.), *Brewing Microbiology,* 3rd ed. Kluwer, New York, N.Y.

5. **Campbell, I.** 2003. Microbiological methods in brewing analysis, p. 367–392. *In* F. G. Priest and I. Campbell (ed.), *Brewing Microbiology,* 3rd ed. Kluwer, New York, N.Y.

6. **Flannigan, B.** 2003. The microbiota of barley and malt, p. 113–180. *In* F. G. Priest and I. Campbell (ed.), *Brewing Microbiology,* 3rd ed. Kluwer, New York, N.Y.

7. **Gibson, G.** 1989. Malting plant technology, p. 279–325. *In* G. H. Palmer (ed.), *Cereal Science and Technology.* Aberdeen University Press, Aberdeen, United Kingdom.

8. **Hammond, J. R. M.** 2003. Yeast genetics, p. 67–112. *In* F. G. Priest and I. Campbell (ed.), *Brewing Microbiology,* 3rd ed. Kluwer, New York, N.Y.

9. **Hardwick, W. A. (ed.).** 1995. *Handbook of Brewing.* Dekker, New York, N.Y.

10. **Hough, J. S., D. E. Briggs, R. Stevens, and T. W. Young.** 1982. *Malting and Brewing Science,* 2nd ed., vol. II. *Hopped Wort and Beer.* Chapman and Hall, London, United Kingdom.

11. **Leeder, G.** 1998. Design of a state-of-the-art filter cellar. *Ferment* **11:**108–121.

12. **Lewis, M. J., and T. W. Young.** 1995. *Brewing.* Chapman and Hall, London, United Kingdom.

13. **Moll, M.** 1994. *Beers and Coolers.* Intercept, Andover, United Kingdom.

14. **Palmer, G. H.** 1989. Cereals in malting and brewing, p. 61–242. *In* G. H. Palmer (ed.), *Cereal Science and Technology.* Aberdeen University Press, Aberdeen, United Kingdom.

15. **Priest, F. G.** 2003. Gram-positive brewery bacteria, p. 181–217. *In* F. G. Priest and I. Campbell (ed.), *Brewing Microbiology,* 3rd ed. Kluwer, New York, N.Y.

16. **Schofield, M. A., S. M. Rowe, J. R. M. Hammond, S. W. Molzahn, and D. E. Quain.** 1995. Differentiation of yeast strains by DNA fingerprinting. *J. Inst. Brewing* **101:**75–78.

17. **Singh, M., and J. Fisher.** 2003. Cleaning and disinfection in the brewing industry, p. 337–366. *In* F. G. Priest and I. Campbell (ed.), *Brewing Microbiology,* 3rd ed. Kluwer, New York, N.Y.

18. **Slaughter, J. C.** 2003. Biochemistry and physiology of yeast growth, p. 19–66. *In* F. G. Priest and I. Campbell (ed.), *Brewing Microbiology,* 3rd ed. Kluwer, New York, N.Y.

19. **Stratford, M.** 1992. Yeast flocculation: a new perspective. *Adv. Microb. Physiol.* **33:**1–71.

20. **Van Vuuren, H. J. J., and F. G. Priest.** 2003. Gram-negative brewery bacteria, p. 219–245. *In* F. G. Priest and I. Campbell (ed.), *Brewing Microbiology,* 3rd ed. Kluwer, New York, N.Y.

21. **Verzele, M., and D. C. de Keukeleire.** 1991. *Developments in Food Science,* vol. 27. *Chemistry and Analyses of Hop and Beer Bitter Acids.* Elsevier, Amsterdam, The Netherlands.

22. **Walker, G. M.** 1998. *Yeast Physiology and Biotechnology.* Wiley, Chichester, United Kingdom.

Food Microbiology: Fundamentals and Frontiers, 3rd Ed.
Edited by M. P. Doyle and L. R. Beuchat
© 2007 ASM Press, Washington, D.C.

Graham H. Fleet

Wine

41

Winemaking is a bioprocess that has its origins in antiquity. Scientific understanding of the process commenced with the studies of Louis Pasteur, who demonstrated that wines were the product of alcoholic fermentation of grape juice by yeasts (3). Since then, winemaking has developed into a modern, multinational industry with a strong research and development base in the disciplines of viticulture and enology. Viticulture concerns the study of grapes and grape cultivation, and enology covers postharvest processing of the grapes, from crushing through fermentation to packaging and retailing of the wine. The sensory and health sciences are also relevant to the modern wine industry. The appeal of wine is intimately linked to human perceptions of aroma, flavor, and color and increasing evidence that its consumption, in moderation, is beneficial to health (22, 167).

Microorganisms are fundamental to the winemaking process. To understand their contribution, it is necessary to know (i) the taxonomic identities of the species and strains associated with the process, (ii) the kinetics of their growth and survival throughout the entire production chain, (iii) the biochemical, physiological, and genomic responses of these species and their effects on the physical and chemical properties of the wine, (iv) the

influence of winemaking practices upon the microbial response, and (v) the linkage between microbial action, sensory quality, and consumer acceptability of the wine (83). This chapter will focus on the occurrence, growth, and significance of microorganisms in winemaking. It covers wines produced only from grapes and includes table wines, sparkling wines, and fortified wines.

THE PROCESS OF WINEMAKING

Details of the process of winemaking are beyond the scope of this chapter and are described elsewhere (25, 175, 181). Figure 41.1 outlines the main steps in the production of white and red table wines.

Grapes

Numerous varieties of the grape, *Vitis vinifera*, are used in winemaking, with the particular variety determining the fruity or floral characteristics of the final product. Some main varieties used in white-wine making are Riesling, Traminer, Muller-Thurgau, Chardonnay, Semillon, and Sauvignon Blanc, and those used in red-wine making include Cabernet Sauvignon, Merlot, Cabernet Franc, Pinot Noir, Shiraz, Gamay, Grenache, and Barbera.

Graham H. Fleet, School of Chemical Sciences and Engineering, The University of New South Wales, Sydney, New South Wales 2052, Australia.

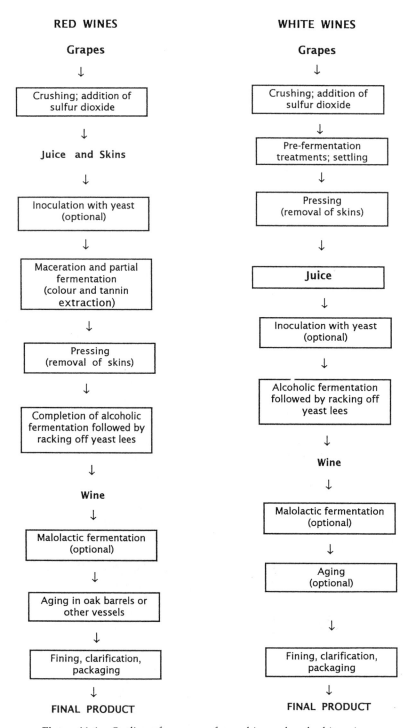

Figure 41.1 Outline of processes for making red and white wines.

The grapes are harvested at an appropriate stage of maturity which determines the chemical composition of the juice extracted from them. Particularly important are the concentrations of sugars and acids which are the major constituents of the juice (Table 41.1) and impact its fermentation properties. Other preharvest conditions that affect the chemical composition of the grape and its juice include climate, sunlight exposure, soil, use of fertilizers, availability of water, vine age, and use of fungicides and insecticides. Traditionally, grapes were

Table 41.1 Main components of grape juice[a]

Substance(s)	Concn
Glucose	75–150 mg/ml
Fructose	75–150 mg/ml
Pentose sugars	0.8–2 mg/ml
Pectin	0.1–1 mg/ml
Tartaric acid	2–10 mg/ml
Malic acid	1–8 mg/ml
Citric acid	0.1–0.5 mg/ml
Ammonia	5–150 μg/ml
Amino acids (total)	150–2,500 μg/ml
Protein	10–100 μg/ml
Vitamins	Varies
Anthocyanins	0.5 mg/ml
Flavonoids and nonflavonoids	0.1–1.0 mg/ml

[a] Data obtained from references 25, 175, and 181.

harvested manually, but now there is increasing use of mechanical harvesters which are often operated at night to minimize the temperature of the berries at the time of crushing (45).

Crushing and Prefermentation Treatments

For white wines, the grapes are mechanically destemmed and crushed and the juice is drained away from the skins. If required, clarification of the juice is done by cold settling, filtration, centrifugation, or combinations of these methods. Cold settling is generally done at 5 to 10°C for 24 to 48 h with the addition of pectolytic enzymes to help break down grape material. The juice is then transferred into the fermentation tank, where the fermentation may commence naturally or may be initiated by inoculation with selected yeasts (75).

Red grapes are mechanically destemmed and crushed, and the juice plus skins (must) are directly transferred into the fermentation tank. Fermentation begins either naturally or after inoculation, and during the first few days, the skins rise to the top of the juice to form a cap. Throughout this early stage, often described as maceration, juice is regularly pumped over the cap. This step extracts purple and red anthocyanin pigments, as well as other phenolic substances, from the grape skins to give color and tannic and astringent character to the wine. The extraction process is assisted by the production of ethanol during this preliminary fermentation. When sufficient extraction has been achieved, the partially fermented wine is drained and pressed from the skins into another tank for completion of the fermentation. Some variations of this process include thermovinification and carbonic maceration. In thermovinification, the juice plus skins are heated to 45 to 55°C with pumping over to accelerate color and

tannin extraction, whereafter the juice is separated from the skins and transferred into the fermentation tank. In carbonic maceration, uncrushed grapes are placed in a tank which is gassed with carbon dioxide. The temperature is maintained at 25 to 35°C for several days, during which the grapes undergo endogenous metabolism that extracts color and phenolics from the skin. After 8 to 10 days, the grapes are pressed to yield a partially fermented juice (1 to 1.5% ethanol) that is transferred into a tank for subsequent fermentation (24, 25).

Other pretreatments of the juice or must include the adjustment of pH and sugar concentration (where permitted), addition of diammonium phosphate or other nutrients to assist yeast growth, addition of sulfur dioxide (50 to 75 μg/ml) as an antioxidant and antimicrobial, and addition of ascorbic acid or erythorbic acid as an antioxidant.

Alcoholic Fermentation

Traditionally, fermentation of the juice was conducted in large wooden barrels or concrete tanks, but most wineries now use stainless steel tanks with facilities for temperature control (principally cooling) and cleaning in place and other modern features for process management (25, 61). However, some premium-quality white wines (e.g., Chardonnay) may be fermented in wooden (oak) barrels. White wines are generally fermented at 10 to 18°C for 7 to 14 days or more, where the lower temperature and lower fermentation rate favor the retention of desirable, volatile flavor compounds. Red wines are fermented for about 7 days at 20 to 30°C, where the higher temperature is necessary to extract color from the grape skins.

The alcoholic fermentation can be conducted either as an indigenous, or wild, fermentation or as an induced, or seeded, fermentation. With indigenous fermentation, yeasts resident in the grape juice initiate and complete the fermentation. With seeded fermentation, selected stains of yeasts, generally those of *Saccharomyces cerevisiae* or *Saccharomyces bayanus*, are inoculated into the juice at initial populations of 10^6 to 10^7 cells/ml. Such yeasts are commercially available as active dry preparations and are used extensively throughout the world (52). The advantages and disadvantages of indigenous and seeded fermentations have been well discussed previously (3, 60, 86, 111, 165). Essentially, seeded fermentations are more rapid and predictable, and indigenous fermentations have more varied outcomes, with the potential of failures but with the prospect of wines with more interesting characters due to contributions from a greater diversity of yeast species and strains.

Alcoholic fermentation is considered complete when the fermentable sugars, glucose and fructose, of the juice are completely utilized. The wine is then drained or pumped (racked) from the sediment of yeast and grape material (lees) and transferred into stainless steel tanks or wooden barrels for malolactic fermentation, if desired, and ageing. Clarification by filtration or centrifugation may be done at this stage. Leaving the wine in contact with the lees for long periods is not encouraged because the yeast cells autolyze, with the potential of adversely affecting wine flavor and providing nutrients for the subsequent growth of spoilage bacteria (84).

Malolactic Fermentation

It has been known since the early 1900s that wines frequently undergo another biochemical transformation which has been termed the malolactic fermentation (25, 48, 114, 130, 223). The malolactic fermentation commences naturally about 2 to 3 weeks after completion of the alcoholic fermentation and lasts 2 to 4 weeks. Lactic acid bacteria resident in the wine are responsible for the malolactic fermentation, but many winemakers now encourage this reaction by inoculation with commercial cultures of *Oenococcus oeni*, formerly known as *Leuconostoc oenos* (58). The main reaction is decarboxylation of L-malic acid into L-lactic acid, giving a decrease in acidity of the wine and an increase in its pH by about 0.3 to 0.5 unit. Wines produced from grapes cultivated in cool climates tend to have higher concentrations of malic acid (e.g., >5 g/liter, pH 3.0 to 3.5), which can mask their varietal character. A decrease in acidity by malolactic fermentation gives a wine with a softer, mellower taste. Also, growth of malolactic bacteria in wine contributes additional metabolites that may confer complex and interesting flavor characteristics. Apart from flavor considerations, there are practical reasons for having wines complete malolactic fermentation. Wines that have not undergone malolactic fermentation before bottling risk the natural onset of this reaction at some later stage in the bottle. If this happens, the wine becomes gassy and cloudy and is spoiled. There is also a view that wines with completed malolactic fermentation have greater microbiological stability and are less prone to spoilage by other species of lactic acid bacteria. Fewer nutrients are available for microbial growth, and bacteriocin production by malolactic bacteria may be a further inhibitory factor.

Malolactic fermentation is not necessarily beneficial to all wines. Wines produced from grapes grown in warmer climates have less malic acid (e.g., <2 g/liter, pH >3.5). Further reduction in acidity by malolactic fermentation is deleterious to overall sensory balance, and also, it increases the pHs of the wines to levels at which spoilage bacteria are more likely to grow. However,

preventing the natural occurrence of malolactic fermentation in these wines (as might occur after bottling) is an extra technical burden. Consequently, many winemakers prefer to encourage the malolactic fermentation and later adjust wine acidity, if necessary. Nevertheless, there are winemakers who prefer not to have the malolactic fermentation occur in their wines.

Postfermentation Processes

Most white wines are not stored for lengthy periods after fermentation. If storage is necessary, it is generally done in stainless steel tanks. Some white wines (e.g., Chardonnay) may be aged in wooden barrels. Most red wines are aged for periods of 1 to 2 years by storage in wooden (generally oak) barrels. During this time, chemical reactions that contribute to flavor development occur between wine constituents and components extracted from the wood of the barrels (24, 25, 175, 181). Critical points for control during storage and ageing are the exclusion of oxygen and the addition of sulfur dioxide to free levels of 20 to 25 μg/ml. These controls are necessary to prevent the growth of spoilage bacteria and yeasts and to prevent unwanted oxidation reactions.

Just before bottling, the wines may be cold stored at 5 to 10°C to precipitate excess tartarate and then clarified by application of one or more processes which include the addition of fining agents (bentonite, albumen, isinglass, gelatin), centrifugation, pad filtration, and membrane filtration. For some white wines with residual sugar, potassium sorbate at up to 100 to 200 μg/ml may be added to control yeast growth (75, 175).

Wine Flavor

The distinctive flavors of wine originate from grape constituents and the processing operations which include alcoholic fermentation, malolactic fermentation, and ageing (41, 181, 208). The grapes contribute many volatile components (e.g., terpenes) that give wines their distinctive varietal, fruity characters. In addition, they contribute nonvolatile acids (tartaric and malic) that impact flavor and tannins (flavanoid phenols) that give bitterness and astringency. The fermentation steps, especially alcoholic fermentation, increase the chemical and flavor complexity by assisting extraction of compounds from the grapes, modifying some grape-derived substances, and producing a vast array of volatile and nonvolatile metabolic end products. Further chemical alterations occur during ageing, and enzymes derived from the grapes and excreted by yeasts and malolactic bacteria, as well as those added at prefermentation, might be expected to participate in chemical-flavor transformations. Thus, the final flavor represents contributions from many compounds and cannot be attributed to any one "impact substance" (89).

YEASTS

Yeasts are significant in winemaking because they carry out the alcoholic fermentation and they can cause spoilage of the wine. Their autolytic products may affect sensory quality and influence the growth of malolactic and spoilage bacteria (84). By interacting with grape anthocyanins, they can modify wine color (143). Some yeast species associated with grapes (e.g., *Metschnokowia*, *Candida*, and *Cryptococcus* species) have antifungal activity and could naturally control the development of spoilage and mycotoxigenic molds (84, 85). About 25 yeast species are commonly isolated from grapes and wines, with *Saccharomyces* and *Hanseniaspora* species being the most prevalent. Cultural methods for their isolation, enumeration, and identification are well described elsewhere (82, 90, 132). However, there is increasing evidence that, at certain stages during the production chain, some yeasts may enter a viable but nonculturable state and require culture-independent molecular methods, such as PCR-denaturing gradient gel electrophoresis, for their detection (62, 146, 147, 162). Nucleic acid-based, molecular methods are widely used to characterize and differentiate wine yeasts (14, 62, 132, 190). Based on new DNA sequence-phylogenesis criteria, yeast taxonomy is in a phase of transition and revision, and this process includes classification of many wine yeasts (126).

Origin

Wine yeasts originate from any of three sources, namely, the surfaces of grapes, the surfaces of winery equipment (crushers, presses, fermenters, tanks, pipes, pump barrels, filtration units), and inoculum cultures (85, 88).

Grapes are a primary source of yeasts that enter the winery environment. These yeasts probably come from wind-blown soil and vegetation, as well as attacks from insects and birds, but the processes by which grape berries become contaminated and colonized by yeasts throughout their cultivation in the vineyard are not well understood (88). Very few yeasts (10 to 10^3 CFU/g) are found on immature grape berries, but the numbers increase to 10^4 to 10^6 CFU/g as the berries mature to harvest. Unripe grapes harbor a predominance of *Rhodotorula*, *Sporobolomyces*, *Cryptococcus*, and *Candida* species, along with the yeast-like fungus *Aureobasidium pullulans*. Most of these species also occur on mature grapes, but at the mature stage, species of *Hanseniaspora* (*Kloeckera*) and *Metschnikowia* often predominate (43, 88,

123, 162). Damaged grapes with increased availability of fermentable substrates have increased populations of *Hanseniaspora* and *Metschnikowia* species as well as other yeasts, including *Saccharomyces* and *Zygosaccharomyces*. *Saccharomyces* yeasts are rarely isolated from healthy, mature grapes by plate culture methods, raising questions about their origin in natural, unseeded wine fermentations (88, 140). Nevertheless, they must be present on grapes at low populations because they are readily isolated by enrichment culture. Moreover, aseptically harvested and crushed grapes naturally ferment and give a dominance of *Saccharomyces*. Using such methods, followed by molecular characterization of the isolates, many studies have demonstrated significant diversity in the strains of *S. cerevisiae* obtained from various vineyards and even the same vineyard over several vintages. Some studies suggest that certain strains are unique to particular geographical regions and may contribute to the regional characters, or "terroirs," of wines, but data supporting this view are not consistent (30, 54, 168, 178, 215, 218). Explanations for these variations probably reside in the facts that the grape surface is a unique phyllospheric habitat and numerous factors are likely to affect its yeast ecology. Such factors include the surface chemistry of the grape and its ability to support microbial attachment and growth; microbial exposure to the natural stresses of temperature, sunlight, irradiation, rainfall, and desiccation; tolerance to chemical inhibitors from the grape itself and from the application of insecticides and fungicides; and interactions with other species (other yeasts, bacteria, and filamentous fungi) (85, 88, 215).

The surfaces of winery equipment that come into contact with grape juice and wine are locations for development of the so-called residential or winery yeast flora. The extent of this development depends upon the nature of the surface and the effectiveness of cleaning and sanitizing operations. These surfaces are considered to be the main source of *Saccharomyces* in wine fermentations. Such sources harbor multiple strains of *S. cerevisiae* that have accumulated from the grapes and starter cultures used in previous years and carry over into subsequent vintages (39, 44, 54, 82, 108, 189).

Starter cultures, if used to inoculate the juice, will be a principal source of yeasts. Presently, various strains of *S. cerevisiae* and *Saccharomyces bayanus* are used, but the future may see the development and use of other species such as those in the genera *Hanseniaspora*, *Candida*, and *Pichia* (29, 36, 37, 111, 202). Commercial yeast preparations used for inoculation are not necessarily pure and may contain a proportion of species other than *S. cerevisiae* (52).

Growth during Fermentation

Many studies have described the yeast populations that grow during alcoholic fermentation. Early research gave qualitative descriptions of the main species isolated from different stages (early, middle, and final) of fermentation. Subsequent studies monitored the growth of individual species throughout the entire course of fermentation. Recently, molecular techniques have made it possible to follow the development of particular strains throughout fermentation. Virtually all ecological studies show that *S. cerevisiae* and, to a lesser extent, *Saccharomyces bayanus* are the principal wine yeasts and predominate during the middle to final stages of fermentation. Nevertheless, there is an important biodiversity of other yeasts that contribute to most fermentations and, indeed, can predominate under some conditions (60, 86, 125).

Figure 41.2 gives a general representation of the growth of yeasts during the fermentation of grape juice, whether it be conducted through indigenous or inoculated processes (see "Alcoholic Fermentation" above). Freshly extracted grape juice harbors a yeast population of 10^3 to 10^5 CFU/ml, comprising mostly *Hanseniaspora* (*Kloeckera*) species, but species of *Candida*, *Metschnikowia*, *Pichia*, *Hansenula*, *Kluyveromyces*, and *Rhodotorula* also occur. These species are often referred to as the non-*Saccharomyces* yeasts, or wild flora. The juice will also contain low populations of indigenous *Saccharomyces* species, depending on the extent of their occurrence on grapes and contamination from equipment used to process the juice. Fermentation is initiated by the growth of various species of the non-*Saccharomyces* yeasts (e.g.,

Hanseniaspora uvarum, *Kloeckera apiculata*, *Candida stellata*, *Candida colliculosa*, and *Metschnikowia pulcherrima*) as well as *Saccharomyces* species. The growth of the non-*Saccharomyces* species is generally limited to the first 2 to 4 days of fermentation, after which they die off. Nevertheless, they achieve maximum populations of 10^6 to 10^7 CFU/ml before death, and such growth is metabolically significant in terms of substrates utilized (hence not available to *S. cerevisiae*) and end products released into the wine (184). Also, the dead cells of these yeasts become part of the total yeast pool for subsequent autolysis. Their death is attributed to an inability to tolerate the increasing concentrations of ethanol, which is produced largely by *S. cerevisiae*, but other factors may be involved. After 4 days or so, the fermentation is continued and completed by *S. cerevisiae*, which reaches final populations of about 10^8 CFU/ml. In some cases, *Saccharomyces bayanus*, *Saccharomyces paradoxus*, and *Saccharomyces uvarum*, and not *S. cerevisiae*, may be the dominant species, especially for fermentations at lower temperatures (54, 178, 199).

This basic ecological profile has been found in wineries throughout the world by using both culture (42, 44, 87, 92, 101, 160, 163, 191) and, now, culture-independent (38, 147) analytical methods and is readily demonstrated in experimental fermentations (110, 161). It is determined by the relative abilities of the different species to survive, grow, and interact under the stresses initially imposed by the chemical and physical properties of the juice (e.g., low pH and high sugar concentration) and the stresses of the changing environment as fermentation progresses (e.g., increasing ethanol level and anaerobiosis) (11, 33, 110). Cell-cell interactions and quorum-sensing molecules may also be involved (153, 201). In addition to the successional growth of different species throughout fermentation, further underlying ecological complexity is demonstrated by the successional development of strains within a species. This revelation became evident with the use of molecular techniques that have enabled strain differentiation and recognition. As many as 100 genetically distinct strains of *S. cerevisiae* have been found in some fermentations (39, 92, 163, 188, 189, 200). Moreover, strain diversity throughout fermentation has been reported for non-*Saccharomyces* species (160, 163, 191).

Factors Affecting Yeast Growth during Fermentation

Many intrinsic and extrinsic variables determine the rate and extent of the growth of yeasts during fermentation (Table 41.2) (28, 56, 60, 86, 125). Yeast growth is best measured by plate counts, but carbon dioxide

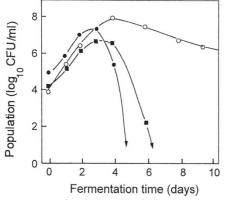

Figure 41.2 Generalized growth of yeast species during alcoholic fermentation of wine. ○, *S. cerevisiae*; ●, *Kloeckera* and *Hanseniaspora* species; ■, *Candida* species. Variations will occur in the initial and maximum populations for each species; for fermentations inoculated with *S. cerevisiae*, the initial population is approximately 10^6 CFU/ml (86).

Table 41.2 Factors affecting the growth of yeasts during alcoholic fermentation

Grape juice characteristics
    Sugar concentration
    Level of assimilable nitrogen
    pH
    Amt of fungicide and pesticide residues
    Content of dissolved oxygen
    Accumulation of toxic metabolites

Processing factors
    Addition of sulfur dioxide
    Extent of settling and clarification of juice
    Addition of yeast nutrients
    Inoculation with selected yeasts
    Temperature control
    Pumping over

Biological factors
    Influences of grape fungi and bacteria
    Population and composition of indigenous yeasts
    Presence of killer yeasts

production, as measured by the loss of culture weight, and the utilization of juice sugars are also used to monitor fermentation kinetics (81, 82, 90, 161).

Juice Composition

In most circumstances, grape juices provide all the nutrients and conditions necessary for vigorous and complete fermentation. However, chemical and physical properties of the juice vary according to grape variety, climatic influences, viticulture practices, and maturity at harvest. Relevant properties include the sugar concentration, the amount of nitrogenous substances, concentrations of vitamins, the dissolved oxygen content, the amount of soluble solids, the presence of fungicide and pesticide residues, pH, and the presence of any yeast-inhibitory or -stimulatory substances produced by the growth of molds and bacteria on the grapes (56, 86).

The concentrations of fermentable hexoses in grape juice vary between 150 and 300 mg/ml (Table 41.1) and may be as high as 400 mg/ml in juice prepared from grapes infected with *Botrytis cinerea* ("pourriture noble") (63). The initial sugar concentration will affect growth rates of the different species and strains of yeasts and the extent to which they contribute to the overall fermentation. The growth of *Candida stellata* and *Torulaspora delbrueckii* may be favored in juices with higher sugar concentrations (33).

Free amino acids and ammonium ions (Table 41.1) are the principal nitrogen sources used by yeasts during fermentation. Most juices contain sufficient nitrogen substrates (>150 mg of assimilable nitrogen/liter) to allow rapid and complete fermentation, but heavily processed or clarified juices may be nitrogen deficient (16, 115). Nitrogen availability in vineyard soils and the use of nitrogen fertilizers can affect the concentration of assimilable nitrogen in the juice and subsequent yeast growth (16). The nitrogen demand by yeasts increases significantly with increasing sugar concentration in the juice and varies with the strain of *S. cerevisiae*. Consequently, supplementation of juices with various yeast foods or diammonium phosphate is a common practice to ensure that nitrogen availability is not a factor which limits yeast growth. Most studies on the nitrogen requirements of wine yeasts have been conducted with *S. cerevisiae*, and little is known about the nitrogen demands of non-*Saccharomyces* species or their ability to remove specific nitrogen substrates from the juice before the growth of *S. cerevisiae* (16, 102). The ability of wine yeasts to utilize grape juice proteins as a source of nitrogen requires further consideration. Strains of *S. cerevisiae* generally do not produce extracellular proteolytic enzymes (81), but some non-*Saccharomyces* wine yeasts are proteolytic (e.g., *Kloeckera apiculata* and *Metschnikowia pulcherrima*) (29, 34, 81, 92, 206).

Grape juices generally contain enough vitamins (inositol, thiamine, biotin, pantothenic acid, and nicotinic acid) to permit maximum growth of *S. cerevisiae*. Vitamin losses may occur in heavily processed juices, where supplementation can improve yeast growth. Species of non-*Saccharomyces* are more demanding of vitamins than *S. cerevisiae*, and vitamin availability may be a factor which limits their contribution to fermentation.

The pH of grape juice varies between 2.8 and 4.0, depending on the concentrations of tartaric and malic acids. Although growth and fermentation rates for *S. cerevisiae* are decreased as the pH is decreased from 3.5 to 3.0 (33, 113), it is not fully understood how juice pH affects the relative growth rates of the non-*Saccharomyces* yeasts and their potential to influence alcoholic fermentation.

Treatment of grapes with fungicides and pesticides before harvest can give juices that contain residues of these substances. Depending on their concentration and chemical nature, these residues may decrease yeast growth and even change the ecology, thereby leading to slow or incomplete fermentations (21, 26, 86).

Conditions which stimulate yeast growth and fermentation include aeration of the juice before or during the early stages of fermentation and the presence of grape solids and particulate materials (21, 28, 56). Different yeast species may selectively adsorb to such particles to form a biofilm of immobilized biomass.

Clarification of Grape Juice

The procedures used to clarify juices, especially for white wine fermentations, will influence the populations of indigenous yeasts in the juice and their potential contribution to the fermentation. Centrifugation and filtration remove yeast cells, thereby decreasing or eliminating the contribution of indigenous species to the fermentation. In contrast, clarification by cold settling presents opportunities for the growth of indigenous yeasts, especially those species or strains that grow well at low temperatures (e.g., *Kloeckera apiculata*) (25, 33, 150, 180).

Sulfur Dioxide

The addition of sulfur dioxide to grapes or juice for controlling oxidation reactions and restricting the growth of indigenous microflora is a well-established practice (185). Nevertheless, strong growth of non-*Saccharomyces* species occurs during the early stages of most commercial fermentations in which the usual amounts of sulfur dioxide (50 to 100 μg/ml) have been added (86). These findings question the efficacy of sulfur dioxide in controlling indigenous yeasts and challenge one of the reasons for using sulfur dioxide in winemaking. Good comparative data on the responses of wine yeasts to this agent under the conditions of winemaking are lacking (70).

Temperature of Fermentation

Temperature control is an important practice in modern winemaking. It affects the growth rates and metabolic activities of yeasts (33), their ability to tolerate ethanol (93), and their contribution to the fermentation. Fastest yeast growth and fermentation occur at 25 to 30°C, and the ecology of the fermentation follows the pattern outlined in Fig. 41.2. However, when the temperature is decreased below 20°C, there is an increased contribution of the non-*Saccharomyces* species to the fermentation. Species such as *Kloeckera apiculata* and *Candida stellata* exhibit increased tolerance to ethanol and do not die off as shown in Fig. 41.2. They can produce maximum populations of 10^7 to 10^8 CFU/ml which remain viable until the end of fermentation (72, 113). Moreover, they may have higher growth rates than *S. cerevisiae* at low temperatures (33). The impact of such ecological shifts on the chemical compositions and sensory qualities of wines has yet to be determined. *Saccharomyces bayanus* is more cryophilic than *S. cerevisiae* and is more likely to be found in fermentations conducted at lower temperatures (199).

Inoculation of the Juice

Perhaps the most significant technological innovation in winemaking during the last 50 years has been the seeding (inoculation) of the juice with selected strains of *S. cerevisiae* or *Saccharomyces bayanus*. These strains have been selected according to criteria that enhance the efficiency of the process and product quality (Table 41.3) (52, 111, 116). Seeding of the fermentation is undertaken with the assumption that the inoculated strain will outcompete and dominate over indigenous strains of *S. cerevisiae* and non-*Saccharomyces* yeasts. Although much evidence shows that inoculated strains dominate at the end of fermentation, the view that early growth of the indigenous species is suppressed or insignificant cannot be supported. Growth of the indigenous non-*Saccharomyces* species, according to Fig. 41.2, still occurs (112), and moreover, indigenous strains of *S. cerevisiae* may grow despite massive competition from the seeded strain. Indeed, if conditions in the juice do not favor the growth of the seeded strain, indigenous *S. cerevisiae* may dominate the fermentation, and this domination can be verified by molecular techniques that allow the differentiation of yeast strains. Although there is high probability that inoculated *S. cerevisiae* will dominate the fermentation, seeding does not guarantee the dominance

Table 41.3 Some properties used to select yeasts for application in wine fermentations[a]

Desirable properties
  Complete and rapid fermentation of sugars
  High tolerance of alcohol
  Resistance to sulfur dioxide
  Fermentation at low temperatures
  Production of good flavor and aroma profiles
  Production of glycerol
  Production of β-glycosidases
  Malic acid degradation
  Killer phenomenon
  Good sedimentation properties
  Tolerance of pesticides and fungicides
  Fermentation under pressure
  Suitability for mass culture, freeze-drying distribution, and
    rehydration

Undesirable properties
  Production of sulfur dioxide
  Production of hydrogen sulfide
  Production of volatile acidity
  High level of formation of acetaldehyde, pyruvate, and esters
  Foaming properties
  Formation of ethyl carbamate precursors
  Production of polyphenol oxidase (affects wine color)
  Inhibition of malolactic fermentation
  Production of mousy and other taints

[a] From references 52 and 116.

of any particular strain or its exclusive contribution to the fermentation (44, 70, 92, 108, 171, 189, 191). Significant factors that affect this outcome are the population of indigenous yeasts already in the juice and the extent to which they have adapted to grow in that juice (161).

Interactions with Other Microorganisms

Various species of molds, acetic acid bacteria, and lactic acid bacteria naturally occur on grapes and on winery equipment. Conditions which allow their proliferation on the grape or in the juice have the potential to affect yeast growth during alcoholic fermentation and are discussed in later sections.

Killer yeasts are certain strains which produce extracellular proteins or glycoproteins, termed killer toxins (zymocins), that destroy other yeasts (196). Usually, strains of one species kill only strains within that species, but killer interactions between different species also occur. Killer toxin-producing strains of *S. cerevisiae* and killer-sensitive strains of *S. cerevisiae* occur as part of the natural flora of wine fermentations. In some wineries, killer strains of *S. cerevisiae* predominate at the end of fermentation, suggesting that they have asserted their killer property and taken over the fermentation. Killer strains of wine isolates of *Candida*, *Pichia*, *Hanseniaspora*, and *Hansenula* occur and can assert their killer action against wine strains of *S. cerevisiae* (107, 226, 228). Expression of the killer phenomenon during wine fermentations is affected by many factors which include the ethanol concentration, pH, temperature, amount of assimilable nitrogen, presence of fining agents, and relative populations of killer and killer-sensitive strains (86). There are several implications of killer yeasts in winemaking. First, inoculated strains of *S. cerevisiae* may be destroyed by indigenous killer strains of *S. cerevisiae* or non-*Saccharomyces* species, leading to sluggish or stuck fermentation, slower fermentation, or completion of the fermentation by a less desirable species. Second, there may be advantage in conducting the fermentation with selected or genetically engineered killer strains of *S. cerevisiae* for the purposes of controlling the growth of less desired indigenous species. Moreover, strains could be selected or constructed to produce stable, broad-spectrum killer toxins that would protect the wine from infection by spoilage yeasts. Finally, strains might be selected to have immunity against the killing action of indigenous yeasts, thereby giving them a greater chance of dominating the fermentation (196).

Stuck or Sluggish Fermentations

A sporadic but serious problem is the premature cessation of yeast growth and alcoholic fermentation, giving wine with residual, unfermented sugar (>2 to 4 g/liter) and a lower-than-expected concentration of ethanol. Such fermentations are referred to as being stuck or sluggish if they take longer than normal to give low residual sugar (21, 115). Factors considered to cause this problem include excessive clarification and processing of the juice; fermentation temperatures that are too high; juice deficiency in nutrients or growth factors; the presence of fungicide residues; influences from other microorganisms such as molds, acetic acid bacteria, and killer yeasts; ethanol toxicity; and accumulation of medium chain-length fatty acids such as octanoic and decanoic acids that can become toxic to yeast growth. Another consideration is failure in the transport of grape juice sugars into the yeast cell and the multitude of factors that affect expression of the genes responsible for this activity (20, 21). Initiatives to overcome stuck fermentations include the addition of nitrogen-containing yeast foods, controlled aeration of the juice or wine, and the addition of yeast cell wall hulls or other bioadsorbents to remove toxic substances (21).

Biochemistry, Physiology, and Genomics

Yeasts utilize grape juice constituents as substrates for their growth, thereby generating metabolic end products that are excreted into the wine (25, 60, 116). The main products are carbon dioxide and ethanol and, to a lesser extent, glycerol and succinic acid. In addition, many hundreds of volatile and nonvolatile secondary metabolites are produced in small amounts that, collectively, contribute to the sensory quality of the wine. These substances include a vast range of organic acids, higher alcohols, esters, aldehydes, ketones, sulfur compounds, and amines. The chemical identities of individual substances and their flavor or aroma sensations, sensory thresholds, and concentrations in wines are well documented (41, 89), and the biochemistry of their formation in *S. cerevisiae*, at least, is well known (17, 127, 208). However, further studies are needed to determine the metabolic characteristics of the non-*Saccharomyces* yeasts. The production of these metabolites varies considerably depending on the yeast strain, yeast species, and conditions of fermentation (111, 184). Table 41.4 shows some of the main metabolites produced by yeasts associated with wine fermentations.

During wine fermentation, *S. cerevisiae* responds to the changing environment by sequential expression and regulation of many genes associated with carbohydrate, nitrogen, and sulfur metabolism and genes required to tolerate the stresses of high sugar concentration, low pH, the presence of ethanol, and nutrient deficiency (5, 136, 187, 217, 231). A natural process of yeast strain adaptation and selection probably occurs in wine ecosystems (172).

Table 41.4 Some principal compounds produced by yeasts during alcoholic fermentation of wine[a]

Compound	Concn in wine (mg/liter)[b]	Sensory description
Ethanol	100–150 g/liter	Burning
Propanol	9–68	Pungent, harsh
Isobutanol	9–174	Solvent, bitter
Isoamyl alcohol	6–490	Malt, burnt, nail polish
2-Phenylethyl alcohol	4–197	Floral, rose
Ethyl acetate	23–64	Pineapple, varnish, nail polish
Isoamyl acetate	0.1–3.4	Banana, pear
2-Phenylethyl acetate	0–18.5	Rose, fruity, honey
Acetic acid	100–115	Vinegary, sour
Succinic acid	0.5–1.7	Salty, bitter
Acetaldehyde	10–75	Nutty, pungent, green apple
Diacetyl	0.1–5	Buttery
Glycerol	5–15 g/liter	Slightly sweet

[a] Adapted from references 89, 111, 127, and 208.
[b] Unless otherwise noted.

The first stage of the alcoholic fermentation involves gene expression under conditions of exponential growth in an environment of low pH and relatively high sugar concentration. The second stage involves significant metabolic activity of stationary-phase cells under the added stress of the presence of ethanol and possible nutrient limitation. Many of the secondary metabolites that impact wine flavor are produced during this second stage and reflect the yeast response to stress conditions (11, 116).

Carbohydrates

Glucose and fructose in juice are metabolized by the glycolytic pathway into pyruvate, which is decarboxylated by pyruvate decarboxylase into acetaldehyde. Acetaldehyde is reduced to ethanol by alcohol dehydrogenase. Although most of the pyruvate is converted to ethanol and carbon dioxide, small proportions are converted to secondary metabolites (11, 20, 116, 208). Glycerol, which imparts desirable smoothness and viscosity to the wine, is produced during glycolysis. Its production is increased by the presence of sulfur dioxide, higher incubation temperature, and increased sugar concentration, but there are significant strain and species influences. Transport of sugars into the cell involves several transporter genes, and factors which affect their activity are important rate-limiting steps in the fermentation process (127, 217). Wine yeasts vary in their ability to take up glucose and fructose, and this variation can impact the residual sweetness of the wine (18).

The potential for wine yeasts to degrade grape pectins and enhance juice extraction needs more consideration, given that some strains of *S. cerevisiae* and non-*Saccharomyces* may produce pectolytic enzymes (34, 81, 173, 206).

Nitrogen Compounds

Wine yeasts metabolize ammonium ions and amino acids in the juice as sources of nitrogen (16, 115). However, genes for the transport and assimilation of amino acids are not expressed until after the ammonium ions are utilized (136, 187). *S. cerevisiae* does not have extracellular proteolytic activity, but some wine species of *Hanseniaspora*, *Kloeckera*, *Candida*, and *Metschnikowia* produce extracellular proteases that may break down juice proteins (29, 34, 81, 206).

The metabolism of nitrogen substrates by yeasts has important implications in winemaking (16, 21, 115, 127, 136). Juices that are limiting in nitrogen content can give stuck or sluggish fermentations and wines with unacceptably high contents of hydrogen sulfide. Metabolism of arginine (the most predominant amino acid in grape juices) by *S. cerevisiae* can lead to the production of urea which is able to react with ethanol to form ethyl carbamate, a suspected carcinogen. The amount of urea produced depends on many factors, including the concentration of arginine relative to those of other nitrogen substances in the juice and the strain of *S. cerevisiae*. Metabolism of amino acids by decarboxylation, transamination, reduction, and deamination reactions produces higher alcohols, fatty acids, esters, and carbonyl compounds that impact wine flavor (16, 145, 208, 222).

Yeasts also release amino acids into the wine. This occurs during the final stages of alcoholic fermentation by mechanisms not fully understood and later when the cells have died and there is autolytic degradation of yeast proteins (35, 117). These amino acids can serve as nutrients for the growth of malolactic bacteria or spoilage bacteria (1, 84).

Sulfur Compounds

Yeasts produce a range of volatile sulfur compounds which, depending on threshold concentrations, have a positive or negative impact on wine flavor (127, 176, 208). The most predominant compounds are sulfur dioxide (sulfite), hydrogen sulfide, and dimethyl sulfide, with lesser amounts of other organic sulfites, mercaptans, and thioesters. The production of sulfite and hydrogen sulfide in *S. cerevisiae* is linked to the biosynthesis of cysteine and methionine by the sulfate reduction pathway (5, 136, 204). The formation of sulfite by *S. cerevisiae* depends on the strain. Most strains produce less than 10 µg/ml, but some give levels up to 100 µg/ml (185). High-sulfite-producing strains are avoided in winemaking because of the negative effect of sulfite on wine quality and the potential of sulfite to cause allergic reactions in some consumers and to inhibit malolactic bacteria. At concentrations exceeding 50 µg/liter, hydrogen sulfide gives an unpleasant "rotten egg" aroma to wine. Many chemical and biological factors affect the production of hydrogen sulfide during wine fermentation, the most significant of which is the strain of *S. cerevisiae*. Some strains produce hydrogen sulfide at concentrations exceeding 1 µg/ml. Hydrogen sulfide production is genetically based, but it is also influenced by factors such as the composition of the grape juice and fermentation conditions (204). Elemental sulfur used as a fungicide on grapes prior to harvest, metabisulfite added to the grapes and juice at crushing, and sulfate which occurs naturally in the juice are all significant precursors of hydrogen sulfide in the wine. Deficiencies in assimilable nitrogen and vitamins in the juice can cause hydrogen sulfide production by *S. cerevisiae* (16, 115, 222). Under these conditions, the intracellular pool of cysteine and methionine is low, allowing the sulfate reduction pathway to operate with consequent production of hydrogen sulfide. If the juice contains an adequate supply of assimilable nitrogen (e.g., ammonium ions, amino acids), cysteine and methionine are produced at concentrations which, through feedback inhibition, decrease the activity of the sulfate reduction pathway and the production of hydrogen sulfide. There is also a view that, under nitrogen-limiting conditions, *S. cerevisiae* degrades intracellular proteins to provide essential amino acids including cysteine and methionine from which hydrogen sulfide is formed (115).

The potential for the non-*Saccharomyces* yeasts to produce sulfur compounds during wine fermentations requires study; some species produce hydrogen sulfide at levels similar to those produced by *S. cerevisiae* (144).

The volatile thiols 4-mercapto-4 methylpentan-2-one, 3-mercaptohexan-1-ol, and 3-mercapto-hexyl acetate have recently been found in some wines. They have very low perception thresholds (3 to 50 ng/liter) and, at low levels, impart most desirable passion fruit, guava, grapefruit, or box tree aromas to some wines. These compounds occur in grape juice as nonvolatile conjugates to cysteine and are deconjugated into the volatile form by *S. cerevisiae* and *Saccharomyces bayanas* during alcoholic fermentation (121, 122, 211).

Organic Acids

Of the numerous organic acids produced in wine by yeasts, succinic and acetic acids are the most significant (174, 208). Succinic acid has a bitter, salty taste and is produced by *S. cerevisiae* at concentrations up to 2.0 mg/ml, depending on the strain. Lower concentrations are produced by non-*Saccharomyces* species. The production of succinic acid is not associated with any major defects in wine quality. In contrast, acetic acid becomes detrimental to wine flavor at concentrations exceeding 1.5 mg/ml and may lead to stuck fermentations (21). Most strains of *S. cerevisiae* produce only small amounts (<0.75 mg/ml) of this acid, but some can produce more than 1.0 mg/ml and are unsuitable for winemaking (116). Factors which limit yeast growth such as low temperature, high sugar concentrations, low pH, deficiency in available nitrogen, and excessive clarification cause increased acetic acid production by *S. cerevisiae* (116, 208). *Candida*, *Kloeckera*, and *Hanseniaspora* species may produce larger amounts of acetic acid than *S. cerevisiae*, but there is substantial strain variation in this property (184, 195).

The production of lactic acid by wine yeasts is considered insignificant (<0.1 mg/ml), but some species of *Saccharomyces*, *Kluyveromyces*, and *Candida* can produce this acid in amounts of 5 to 10 mg/ml (174, 208). Such strains could be used to increase the acidity of some wines. Although tartaric acid is prevalent in grape juice and wine, it is not metabolized by wine yeasts (174). However, malic acid is partially (5 to 50%) metabolized by *S. cerevisiae* and other wine yeasts. It is completely degraded by some species of *Schizosaccharomyces* and some strains of *Zygosaccharomyces bailii* (95, 174), by which it is oxidatively decarboxylated into pyruvate which is then converted to ethanol. The possibility of using species of *Schizosaccharomyces* to deacidify wines in place of the malolactic fermentation has attracted considerable interest, but such use must be carefully controlled as these yeasts can produce off-flavors (95, 197).

Yeasts produce small amounts (1 to 15 µg/ml) of free fatty acids in wines (174). Of special note are hexanoic, octanoic, and decanoic acids which, on accumulation, may become toxic to *S. cerevisiae* and *Oenococcus oeni* and contribute to stuck fermentations (1, 21, 84).

Monoterpenes

Monoterpenes are important aroma compounds that occur in grapes and confer a variety of desirable, fruity characters (e.g., grape, raspberry, and passion fruit). Most of these monoterpenes occur as nonvolatile precursors that are glycosidically conjugated to glucose or disaccharides. Their potent volatility and fragrance are released on hydrolysis of this linkage. Glycosidase activity produced by yeasts during alcoholic fermentation catalyzes this transformation and can significantly impact wine flavor. The extent of this activity depends on the yeast species and strain, with the non-*Saccharomyces* yeasts having stronger glycosidase activity than *S. cerevisiae* (29, 34, 78, 80, 183). Moreover, it is possible that wine yeasts may also synthesize monoterpenes de novo (32).

Autolysis

The autolytic degradation of yeast cells at the end of alcoholic fermentation and during cellar storage of the wines is often underestimated as a significant biochemical event. During autolysis, yeast proteins, nucleic acids, and lipids are extensively degraded, releasing peptides, amino acids, nucleotides, bases, and free fatty acids into the wine. These products impact wine flavor and serve as nutrients for the growth of bacteria (1, 84). In addition to *S. cerevisiae*, the non-*Saccharomyces* species would be involved in autolytic reactions (35, 117, 170, 230).

Yeasts and Flavor Diversity

Individual yeasts produce an array of several hundred flavor metabolites (organic acids, higher alcohols, esters, aldehydes, ketones, amines, sulfur compounds, etc.) which, collectively, contribute to the individuality of wine character. This array varies with the yeast species, strains within a species, and conditions of fermentation (73, 111, 127, 58, 184, 208). Moreover, the metabolite profile may vary, depending on whether the yeasts grow in single or mixed cultures (120). Wines produced by indigenous, nonseeded fermentations are often perceived as having more diverse and interesting sensory characters than those produced by inoculation with strains of *S. cerevisiae* (86, 165). These differences may be attributed to the greater ecological and metabolic spectrum of yeasts associated with nonseeded fermentations (127). Also, the regional characters of some wines have been ascribed, in part, to influences of the local yeast flora. However, such relationships are not always consistent or predictable and reflect the challenges that remain in linking yeast ecology and wine chemistry with sensory perception (41, 89).

Spoilage Yeasts

Yeasts can spoil wine at several stages during the process: during alcoholic fermentation and bulk storage in the winery and after packaging (25, 60, 67, 132, 205, 209). Growth of inappropriate species or strains of yeasts during alcoholic fermentation can give an inferior wine (e.g., high in content of esters, acetic acid, and hydrogen sulfide), and the product is spoiled. Wine that is exposed to air (e.g., as that in incompletely filled barrels or tanks) quickly develops a film or surface flora of weakly fermentative or oxidative yeasts of the genera

Candida, *Pichia*, and *Hansenula*. Particularly significant is *Pichia membranifaciens*. These species oxidize ethanol, glycerol, and acids, giving wines with unacceptably high levels of acetaldehyde, esters, and acetic acid. Fermentative species that grow in wines during cellar storage and in packaged wines include *Zygosaccharomyces bailii*, *Brettanomyces* (*Dekkera*) species, and *Saccharomycodes ludwigii*. In addition to causing excessive carbonation, sediments, and haze, these species produce acid and estery off-flavors. *Brettanomyces* species also cause unpleasant mousy or medicinal taints due to the formation of tetrahydropyridines and volatile phenolic substances such as 4-ethyl guaiacol and 4-ethyl phenol (67, 205). Packaged wines that contain residual sugars are prone to refermentation, especially by *S. cerevisiae*, causing swelling and explosion of the container (209). Good quality control and assurance practices, including end product specifications (e.g., <1 to 2 viable yeast cells in 200 ml), are essential to prevent spoilage problems (132, 209). Real-time PCR methods are available to quickly detect wine spoilage yeasts (40, 141, 155).

Genetic Improvement of Wine Yeasts

A vast biodiversity of wine yeasts naturally occurs in vineyards and wineries, and selection from these reservoirs has generally met the needs of winemaking. Nevertheless, the process of strain selection and development can be accelerated and more specifically targeted through the use of classical and modern genetic improvement technologies (116, 165, 167). The sequence of the *S. cerevisiae* genome has been determined, and the functions of many of its genes are known (166). These advances have made it possible to genetically engineer desirable characteristics into wine strains of *S. cerevisiae*, and many targets have been identified and investigated to improve wine-processing efficiency, wine quality, and wine appeal (Table 41.5). There is little doubt that successful, technologically advanced strains can be engineered, but their commercialization will depend on safety and environmental approval by government authorities and consumer acceptance of foods and beverages modified by modern biotechnologies.

LACTIC ACID BACTERIA

Lactic acid bacteria are significant in winemaking mainly because they cause spoilage and are responsible for the malolactic fermentation. They also release autolytic products, which may impact flavor, and they may contribute to the levels of biogenic amines and ethyl carbamate in wines. Wine lactic acid bacteria have the unique ability to tolerate the stresses of the wine environment, namely,

Table 41.5 Directions for the genetic improvement of wine yeasts

Characteristics for improved wine flavor and quality
  More balanced production of specific flavor volatiles (esters, higher alcohols)
  Increased glycerol production
  Increased release of grape terpenoids and volatile thiols
  Decreased production of acetic acid, hydrogen sulfide, and phenolic off-flavors
  Decreased production of biogenic amines and ethyl carbamate
  Decreased ethanol production for low-alcohol wines

Characteristics for improved process efficiency
  More efficient utilization of grape juice sugars and amino acids
  Increased ethanol tolerance
  Malolactic activity
  Antimicrobial action (killer toxins bacteriocins against spoilage yeasts and bacteria)
  Enzymes (protease, pectinase) to assist clarification
  Enhanced sedimentation properties

low pH, the presence of ethanol and sulfur dioxide, low temperature, anaerobiosis, and dilute concentrations of nutrients. Their occurrence, growth, and significance in wines have been reviewed previously (8, 12, 48, 114, 129, 130, 223). The relevant species are members of the genera *Oenococcus*, *Leuconostoc*, *Pediococcus*, and *Lactobacillus* and include *Oenococcus oeni*, *Leuconostoc mesenteroides*, *Pediococcus parvulus*, *Pediococcus pentosaceus*, *Pediococcus damnosus* (formerly *Pediococcus cerevisiae*), and various species of *Lactobacillus* (e.g., *Lactobacillus brevis*, *Lactobacillus plantarum*, *Lactobacillus fermentum*, *Lactobacillus buchneri*, *Lactobacillus hilgardii*, and *Lactobacillus trichodes*). *Oenococcus oeni* is found uniquely in wine ecosystems and is intimately associated with the malolactic fermentation. Consequently, the taxonomy, biochemistry, physiology, and genomics of this species have been extensively studied (8, 130, 148, 216). Wine lactic acid bacteria are isolated and enumerated by culture on plates of MRS agar supplemented with 10 to 15% (vol/vol) tomato juice or apple juice (82, 90). They are identified by an extensive array of biochemical and physiological tests (31, 58, 66). However, molecular methods based on rRNA gene sequencing, multilocus sequence typing, specific PCR primers, restriction fragment length polymorphism, random amplified polymorphic DNA, and nucleic acid probes are now routinely used for species identification and strain differentiation (23, 53, 104, 157, 229).

Ecology

The lactic acid bacteria of wines originate from the grapes and winery equipment, but inoculation of selected species to conduct the malolactic fermentation is widely practiced. Freshly extracted grape juice contains lactic acid bacteria at populations of 10^3 to 10^4 CFU/ml, but the bacteria undergo little or no growth during the alcoholic fermentation and tend to die off because of competition from yeasts (Fig. 41.3). Nevertheless, these bacteria are capable of abundant growth in the juice, and if yeast growth is delayed, they grow and spoil the juice or cause stuck alcoholic fermentation. About 1 to 3 weeks after completion of the alcoholic fermentation, the surviving lactic acid bacteria commence vigorous growth to conduct the malolactic fermentation. Final populations of 10^6 to 10^8 CFU/ml are produced. The onset, duration, and ecology of this growth are determined by many factors which include the properties of the wine, vinification variables, and influences of other microorganisms. Consequently, the natural occurrence of malolactic fermentation and its completion by the preferred species, *Oenococcus oeni*, can be unpredictable (114, 130). Moreover, multiple strains of *Oenococcus oeni* may be associated with the one fermentation (179). Strains of *Oenococcus oeni* obtained from different locations exhibit significant genetic diversity (53).

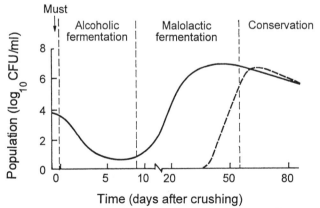

Figure 41.3 Growth of lactic acid bacteria during vinification of red wines, pH 3.0 to 3.5. The solid line shows the growth of *Oenococcus oeni*, often the only species present. Occasionally, species of *Lactobacillus* and *Pediococcus* develop towards the end of malolactic fermentation or at later stages during conservation (broken line). For wines of pH 3.5 to 4.0, a similar growth curve is obtained but there may be slight growth and death of lactic acid bacteria during the early stages of alcoholic fermentation. Also, there is a greater chance that species of *Lactobacillus* and *Pediococcus* will grow and conduct malolactic fermentation.

Of the wine properties, pH, the concentration of ethanol, and the concentration of sulfur dioxide have strong influences on the growth of lactic acid bacteria. Different species, and even strains within species, show different responses to these properties (50, 90, 223). Wines of low pH (e.g., pH 3.0), high ethanol content (>12%, vol/vol), and high total sulfur dioxide level (>50 µg/ml) are less likely to support the growth of lactic acid bacteria and may not undergo successful malolactic fermentation. Strains of *Oenococcus oeni* are more tolerant of low pH than those of *Leuconostoc*, *Pediococcus*, and *Lactobacillus* species and generally predominate in wines of pH 3.0 to 3.5. Wines with pHs exceeding 3.5 tend to have mixed microfloras consisting of *Oenococcus oeni* and various species of *Pediococcus* and *Lactobacillus*. Species of *Pediococcus* and *Lactobacillus* are more tolerant of higher concentrations of sulfur dioxide than is *Oenococcus oeni* and are more likely to occur in wines with larger amounts of this substance (50, 51). Thus, winemaker management of pH and sulfur dioxide content is important if it is desired to have malolactic fermentation conducted solely by *Oenococcus oeni*.

Other factors that affect the growth of *Oenococcus oeni* in wine and successful completion of malolactic fermentation include excessive growth of molds and acetic acid bacteria on grapes, yeast species and strains responsible for the alcoholic fermentation, and bacteriophages. Substances produced by the growth of molds or acetic acid bacteria on damaged grapes may either stimulate or inhibit malolactic fermentation (130, 223). During alcoholic fermentation and subsequent autolysis, yeasts release nutrients that encourage the growth of lactic acid bacteria (1, 84). However, some strains of *S. cerevisiae* produce high concentrations of sulfur dioxide, proteins, and fatty acids (hexanoic, octanoic, decanoic, and dodecanoic) that inhibit malolactic bacteria (1, 84, 130). It is not known how non-*Saccharomyces* species may impact the development of the malolactic fermentation. Bacteriophages active against *Oenococcus oeni* occur in wines and can interrupt and delay the malolactic fermentation (49, 130). Lysogeny of *Oenococcus oeni* is common, and bacteriophage-resistant strains have been described previously (50, 159).

The fate of lactic acid bacteria after malolactic fermentation depends on the wine and winemaking practices, but they may survive in wine for long periods. Because wine pH is increased by malolactic fermentation, the wine becomes a more favorable environment for bacterial growth, and spoilage microorganisms such as *Pediococcus* and *Lactobacillus* species may develop, especially in wines of pH 3.5 to 4.0. As noted already, wines after malolactic fermentation may have better microbiological

stability. Nevertheless, *Oenococcus oeni* and various species of *Lactobacillus* and *Pediococcus* can reestablish growth in such wines (224). The death of *Oenococcus oeni* in wines subsequent to the growth of *Pediococcus* and *Lactobacillus* species can occur (51). Strains of *Pediococcus*, *Lactobacillus*, and *Oenococcus oeni* produce bacteriocins that are active against one another, and this property probably contributes to the bacterial ecology at this stage (12, 13, 151). Strains of *Oenococcus oeni* that produce bacteriocins with broad-spectrum action against pediococci and lactobacilli would have obvious value in controlling spoilage by these bacteria.

Biochemistry, Physiology, and Genomics

Lactic acid bacteria change the chemical composition of wine by utilizing some constituents for growth and generating metabolic end products. Depending upon the species and strains that develop, such changes will have a positive or negative impact on wine quality. Although detailed studies which correlate the growth of lactic acid bacteria with changes in wine composition are few (114), significant experimental research has been done to understand the biochemistry, physiology, and genomics of this growth (8, 12, 114, 129, 130). The genome of *Oenococcus oeni* PSU-1 has been sequenced (148).

Carbohydrates

Wines contain residual amounts of glucose and fructose (0.5 to 1 mg/ml) and smaller amounts (<0.5 mg/ml) of other hexose and pentose sugars. Concentrations of these sugars decrease with the growth of lactic acid bacteria, but consistent trends have not emerged (51). This finding is consistent with the ability of these bacteria to ferment a wide range of hexose and pentose sugars and the substantial variation among species and strains in conducting these reactions (50, 129, 156, 216). The pathways utilized by lactic acid bacteria for sugar transport and metabolism are well described (129), but it is not known how wine conditions of low pH and high ethanol concentration impact these reactions. *Pediococcus* species as well as some species of *Lactobacillus* ferment hexoses by the Embden-Meyerhof-Parnas pathway (homofermenters), while *Oenococcus oeni* and some other species of *Lactobacillus* use the hexose monophosphate or phosphoketolase pathway (heterofermenters). The latter pathway is used by all species to metabolize pentose sugars. Fructose is also metabolized into mannitol by *Oenococcus oeni* (182).

It is not known if wine lactic acid bacteria produce enzymes that hydrolyze grape juice pectins or other polysaccharides, but such properties occur in other lactic acid bacteria (142). Some strains of *Oenococcus oeni* produce β-1,3-glucanase, capable of lysing the glucans in yeast cell walls (106).

Nitrogen Compounds

The metabolism of nitrogen compounds by lactic acid bacteria can impact wine flavor, protein stability, and public health safety.

The concentrations of some wine amino acids (e.g., arginine and histidine) decrease with the growth of lactic acid bacteria, suggesting their utilization as a nitrogen source (51, 114). Arginine is metabolized into ornithine, ammonia, and carbon dioxide by the arginine deiminase pathway (129, 148). Wine lactic acid bacteria, including strains of *Oenococcus oeni*, produce citrulline as an intermediate of this metabolism. Citrulline chemically reacts with ethanol to produce the carcinogen ethyl carbamate. Consequently, there is increasing concern about the levels of ethyl carbamate in wines and the role of malolactic bacteria in contributing to its production (4, 129, 214). The ability to decarboxylate amino acids into amines (e.g., histamine, putrescine, and tyramine) occurs widely throughout wine lactic acid bacteria, including *Oenococcus oeni*. The contribution of these bacteria to the content of biogenic amines in wines is an emerging public health issue. The histidine decarboxylating genes of some wine lactic acid bacteria have been sequenced, and DNA probes for the rapid detection of amine-producing strains have been developed (105, 131).

The production of extracellular proteases and peptidases by wine lactic acid bacteria requires more detailed study. Extracellular protease production by *Oenococcus oeni* has been described previously, and this property may be of nutritional advantage to the organism, as well as decreasing the levels of proteins that contribute to haze and sediments in wine (77).

Organic Acids

The decarboxylation of L-malic acid into L-lactic acid is one of the most significant metabolic reactions conducted by lactic acid bacteria in wines. These bacteria possess mechanisms for the transport of malic acid into the cell and the efflux of lactic acid and also possess the malolactic enzyme for decarboxylation. Purification and properties of the malolactic enzyme, as well as the bioenergetics and genomics of the reaction, have been described previously for several species (8, 114, 130, 148). However, factors which regulate the expression of malolactic activity are not completely understood. Glucose-induced inhibition has been reported previously for some strains of *Oenococcus oeni* (149).

The metabolism of tartaric acid is not usually encountered during or after malolactic fermentation, but when it occurs the wine is spoiled (67, 205). Citric acid, which occurs in wine at concentrations up to 0.7 g/liter, can be completely or partially metabolized during malolactic fermentation, depending on the wine pH and species of lactic acid bacteria. Its degradation is frequently correlated with small increases in the concentrations of acetic acid and diacetyl, and this observation is consistent with the action of citrate lyase, which is produced by some but not all species of wine lactic acid bacteria (9, 51, 130). The concentrations of fumaric, gluconic, and pyruvic acids can decrease during malolactic fermentation. Gluconic acid, which is significantly increased in wines made from grapes infected with *Botrytis cinerea*, is metabolized by most lactic acid bacteria except the pediococci (223). Sorbic acid, which may be added to wines in some countries to control yeast growth, can be metabolized by *Oenococcus oeni* to form 3,5-hexadien-2-ol and 2-ethoxyhexa-3-diene that cause geranium-like off-flavors (67, 90, 205).

Stress Response

Malolactic bacteria are required to grow and perform under stressful environmental conditions, as mentioned previously. Their ability to tolerate the combined effects of low pH (3.0 to 3.5) and high ethanol concentrations (up to 14 to 15%) is particularly noteworthy (91), and the molecular mechanisms underlying the cellular response to these conditions are attracting significant research (8, 109, 148). Explanations are far from complete but involve a multiple of adaptive reactions, including the generation of a proton motive force through the malolactic reaction, activation of proton-extruding ATPase, synthesis of various stress proteins, and modification of membrane structures. Progress in understanding these mechanisms should lead to strain development and management strategies that promote more effective and efficient malolactic fermentation.

Flavor Enhancement

Much has been written about the impact of malolactic fermentation on wine flavor (48, 114, 130, 208). Malolactic fermentation not only affects the taste of wine through deacidification but it also contributes other flavor characteristics (often described as buttery, nutty, fruity, or vegetative) that may enhance or detract from overall acceptability (57). Such changes are related to the wine constituents metabolized by the malolactic bacteria and the nature, concentrations, and flavor thresholds of the products generated. Autolysis of malolactic bacteria may also impact flavor. Connecting sensory impression to flavor substances produced by particular species or strains of malolactic bacteria has proved elusive because of confounding influences of grape variety, yeasts, and vinification variables.

Mechanisms advanced to explain the impact of malolactic fermentation on wine flavor include the production of diacetyl, esters, higher alcohols, and sulfur volatiles and the release of glycosydically linked terpenes and other compounds (208). Diacetyl production, which is linked to citrate and sugar metabolism, is well documented. Diacetyl contributes buttery or butterscotch aromas that have a positive impact at concentrations of less than 5 mg/liter, but higher levels are detrimental (9). The concentrations of various esters, organic acids, and higher alcohols change during malolactic fermentation, suggesting a role of bacterial esterases and lipases, but these observations and linkages are not definitive and require more detailed study (48, 142). Volatile thiols, responsible for vegetable and fruit aromas in wines, are associated with very low concentrations of substances such as methanethiol, demethysuflide, 3-(methysulfanyl) propan-1-ol, 3-(methylsulfanyl)-propionic acid, 4-mercapto-4-methylpentan-2-one, and 3-mercapthexanol. There is preliminary research suggesting that *Oenococcus oeni* and other lactic acid bacteria produce these substances through the metabolism of methionine and cysteine-conjugated precursors in grapes (169). In a similar context, there are increasing data demonstrating that *Oenococcus oeni* and other species possess various glycosidase activities that liberate desirable flavor volatiles from nonvolatile, glycosylated, terpenoid precursors in grapes (59, 103, 135, 213).

Spoilage Reactions

Uncontrolled growth of lactic acid bacteria during or after malolactic fermentation results in wine spoilage (25, 60, 67, 90, 205, 223). Wines containing high concentrations of residual glucose and fructose are more likely to support bacterial growth, with the production of unacceptable amounts of acetic acid, D-lactic acid, and carbon dioxide. Mannitol taint is caused by some strains of heterofermentative lactobacilli (e.g., *Lactobacillus brevis*) due to enzymatic reduction of fructose to mannitol. Excessive production of diacetyl (>5 μg/ml) gives overpowering buttery flavors (9). Certain strains of *Leuconostoc mesenteroides*, *Oenococcus oeni*, *Lactobacillus hilgardii*, *Lactobacillus brevis*, and *Lactobacillus cellobiosus* have been implicated in the formation of mousy taints due to the production of acetyl-tetrahydropyridines from lysine and ornithine metabolism (46). The degradation of glycerol, especially by species of *Pediococcus* and *Lactobacillus*, gives acrolien and associated bitterness (208). Spoilage arising from the degradation of tartaric and sorbic acids has been mentioned already.

However, the ability to metabolize glycerol and tartaric acid is not widespread among the lactic acid bacteria. The production of extracellular polysaccharides by some strains of *Oenococcus oeni* and *Pediococcus damnosus* gives unsightly ropiness and increased viscosity that retards processing by filtration (130). Glucan synthase genes responsible for this property have been identified previously (221).

Control of Malolactic Fermentation

Control of malolactic fermentation is one of the challenges of modern winemaking (48, 114). Because the natural occurrence of malolactic fermentation can be unpredictable, commercial cultures of malolactic bacteria have become available for inoculation into the wine to induce this reaction. Generally, these are strains of *Oenococcus oeni* that have been selected for a range of desirable criteria (Table 41.6) (52, 114). Many factors affect the successful induction of malolactic fermentation by inocula, the most important of which include selection of the appropriate strain, proper reactivation and preculture of the freeze-dried concentrate, the level of inoculum, and the timing of inoculation (152). Generally, wines are inoculated to give 10^6 to 10^7 CFU/ml of bacteria just after alcoholic fermentation is completed. Arguments have been advanced for inoculating malolactic bacteria into the juice either before or simultaneously with the yeast culture. Under these conditions, bacterial cells are not exposed to the stresses of high ethanol concentration, giving a higher probability of successful growth and malic acid degradation. Risks associated with early inoculation of malolactic bacteria include their metabolism of grape juice sugars to yield unacceptable concentrations of

Table 41.6 Desirable properties of bacteria for use in malolactic fermentation

Strong malolactic activity under wine conditions

Strong ability to grow in wines, including those of low pH (3.0–3.2) and high ethanol content (14%) and those containing sulfur dioxide (50 μg/ml total), and at low temperatures (15–20°C)

Resistance to destruction by bacteriophages; nonlysogenic

Resistance to fungicide and pesticide residues

Production of desirable flavors and no off-flavors

Release of conjugated terpenes and volatile thiols

Production of bacteriocins effective against spoilage bacteria

Nonproduction of biogenic amines and precursors of ethyl carbamate

Nonproduction of yeast inhibitory factors if used before alcoholic fermentation

Suitability for mass culture, freeze-drying, distribution, and rehydration acetic acid and interference with yeast growth and alcoholic fermentation (97, 179, 186, 192).

Even under optimum conditions, inoculation with malolactic bacteria does not ensure successful completion of the fermentation. In some instances, the particular properties of the wine may not be suitable for growth of the bacterial strain or the mixture of strains inoculated (97). Several biotechnological innovations have been considered to overcome this problem. Bioreactor systems consisting of high concentrations of malolactic bacteria immobilized in beads of alginate, carrageenan, or other support have been developed (48, 61, 124, 134). An alternative system uses high densities (10^9 to 10^{10} CFU/ml) of *Oenococcus oeni* retained within a cell recycle membrane bioreactor (94). Under these conditions, cells of malolactic bacteria act as biocatalysts and, in the absence of growth, rapidly convert malic acid into lactic acid in wine that is passed through the reactor on a continuous basis. Such technologies give rapid, continuous deacidification of wines but have not proved commercially successful, mainly because of instability of the malolactic activity of cells in the reactor.

The availability of strains of *S. cerevisiae* that could carry out malolactic fermentation simultaneously with alcoholic fermentation would be most attractive to winemakers. With the use of recombinant DNA technology, strains of *S. cerevisiae* expressing the malolactic gene of *Lactococcus lactis* and the malate transport gene of *Schizosaccharomyces pombé* have been constructed and reported to give good malolactic activity in wine (219, 220).

Complete prevention of malolactic fermentation is an option preferred by some winemakers. Wines of low pH (<3.2), high ethanol content (>14%), and high sulfur dioxide level (>50 mg/ml) are less prone to malolactic fermentation. Bacteriocin, nisin, and the antibacterial enzyme lysozyme are effective in preventing the growth of lactic acid bacteria in wine and could be added if permitted (47, 67, 76, 96).

ACETIC ACID BACTERIA

Acetic acid bacteria cause the vinegary spoilage of wines through the oxidation of ethanol into acetaldehyde and acetic acid. In addition, their growth on grapes before fermentation can produce substances that not only affect wine flavor but interfere with the growth of yeasts, leading to stuck fermentations (64, 67, 205). The taxonomy of acetic acid bacteria is complex and has undergone many revisions. Currently, these bacteria are classified into eight genera: *Acetobacter*, *Gluconobacter*, *Gluconoacetobacter*, *Acidomonas*, *Asaia*, *Kozakia*, *Swaminathania*, and *Saccharibacter* (203, 212). The main isolates from grapes

and wine have been *Gluconobacter oxydans*, *Acetobacter pasteurianus*, and *Acetobacter aceti*, but recently *Gluconoacetobacter hansenii* and *Gluconoacetobacter liquefaciens* have been found (68, 98, 99), and additional species are likely to occur. These bacteria are isolated by culturing on glucose-yeast extract, calcium carbonate agar, and WL-nutrient agar to which 1 to 2% (wt/vol) of ethanol has been added (10, 64). Some isolates give very weak growth and are difficult to maintain, and there is evidence that they enter a viable but nonculturable state in wine ecosystems (10, 69). The species can be differentiated by key phenotypic tests (64, 68, 69), but PCR-based methods are now used for species and strain differentiation (98, 99, 212).

Ecology

Sound, unspoiled grapes harbor low populations of acetic acid bacteria, generally less than 10^2 CFU/g, with *Gluconobacter oxydans* being the predominant species. Damaged, spoiled grapes and those infected with the mold *Botrytis cinerea* have higher populations (>10^6 CFU/g) that are characterized by a mixture of *Gluconobacter*, *Acetobacter*, and *Gluconoacetobacter* species (7, 68, 98, 99).

In the absence of yeasts, acetic acid bacteria quickly grow in grape juice, reaching populations of 10^6 to 10^8 CFU/ml. The extent of their growth during alcoholic fermentation depends on their initial populations relative to those of yeasts, juice pH, and the concentration of added sulfur dioxide. In juice prepared from sound grapes where the yeast population is 10 to 100 times greater than that of acetic acid bacteria, there appears to be little growth or influence of these bacteria on the alcoholic fermentation (65, 68, 98, 99). However, when initial populations of these bacteria exceed about 10^4 CFU/ml, they may grow in conjunction with yeasts during the early stages of fermentation. Populations as high as 10^8 CFU/ml can develop but die off due to the combined influences of ethanol and anaerobiosis caused by yeast growth. Nevertheless, acetic acid and other substances produced by acetic acid bacteria become inhibitory to the yeast, causing premature cessation of the alcoholic fermentation (64, 65).

There is a notable evolution of species, and even strains within species, throughout vinification, reflecting different organism responses to the combined stresses of ethanol, low pH, sulfur dioxide, and oxygen availability. *Gluconobacter oxydans* quickly dies off at the beginning of fermentation, giving way to a prevalence of *Acetobacter pasteurianus*, *Acetobacter aceti*, and *Gluconoacetobacter* species in the later stages (68, 98, 99). At the end of alcoholic fermentation, the population of acetic acid bacteria is generally less than 10^2 to 10^3 CFU/ml. However, subsequent transfer of the wine from fermentation tanks to other storage vessels may produce sufficient agitation and aeration to encourage the growth of the survivors to a level of 10^5 CFU/ml or higher. It is not uncommon to isolate *Acetobacter pasteurianus* or *Acetobacter aceti* from wines during bulk storage in barrels or tanks kept under anaerobic conditions (64, 82, 225). The mechanisms by which these aerobic bacteria survive for long periods under apparently anaerobic or semianaerobic conditions require explanation. Their growth is activated by exposure of the wine to air, and the wine is quickly spoiled (69). Even after bottling, ingress of oxygen through defective corks can initiate their growth and wine spoilage (10). Wine spoilage by acetic acid bacteria is best controlled by good hygienic practices and prevention of wine exposure to air.

Biochemistry

Oxidation of ethanol into acetic acid is a key reaction of acetic acid bacteria. Ethanol is first oxidized by ethanol dehydrogenase into acetaldehyde which is oxidized by acetaldehyde dehydrogenase into acetic acid. Species of *Acetobacter* further oxidize acetic acid into carbon dioxide and water by the tricarboxylic acid (TCA) cycle, but this reaction is inhibited in the presence of ethanol. Species of *Gluconobacter* do not have a fully functional TCA cycle and are unable to completely oxidize acetic acid (55). Sulfur dioxide, which is generally present in wine, can chemically trap acetaldehyde, causing an accumulation of this intermediate at the expense of its further oxidation into acetic acid. Ethanol concentrations above 10% become increasingly inhibitory to the growth of acetic acid bacteria and their ability to oxidize ethanol. Aldehyde dehydrogenase is less stable than ethanol dehydrogenase at high concentrations of ethanol, and such conditions give an increased accumulation of acetaldehyde. Lower oxygen concentrations also favor the accumulation of acetaldehyde. Ethyl acetate is another end product that is significant in the vinegary spoilage of wines (64, 65, 67, 90).

Acetobacter and *Gluconobacter* lack a functional Embeden-Meyerhoff-Parnas pathway and metabolize hexose and pentose sugars by the hexose monophosphate pathway into acetic and lactic acids. However, *Acetobacter* species give only weak metabolism of sugars due to their decreased ability to phosphorylate these substrates. At pHs below 3.5 or glucose concentrations above 5 to 15 mM, as occur in grapes or grape juice, metabolism of glucose by the hexose monophosphate pathway is inhibited and glucose is directly oxidized into gluconic and ketogluconic acids (7). Consequently, grapes heavily infected with *Gluconobacter oxydans* give juices with high concentrations (50 to 70 mg/ml) of gluconic acid. Strains of *Acetobacter aceti* and *Acetobacter pasteurianus* also produce gluconic acid in grape juice but to a lesser extent than *Gluconobacter oxydans* (65).

Strains of *Gluconobacter oxydans* and *Acetobacter aceti* oxidize glycerol into dihydroxyacetone (7). Glycerol is not normally present in grape juice, but its concentration may be significant (up to 20 mg/ml) in juices prepared from grapes infected with *Botrytis cinerea* or other molds. The metabolism of glycerol by acetic acid bacteria, either on grapes, during the early stages of alcoholic fermentation, or in wine, leads to significant production of dihydroxyacetone that may affect wine quality as well as bind added sulfur dioxide (7, 64, 205).

Species of *Acetobacter* use the TCA cycle for the metabolism of organic acids, causing decreases in concentrations of citric, succinic, malic, tartaric, and lactic acids in wines. Moreover, lactate may be oxidized into acetoin. The metabolism of amino acids and protein by wine acetic acid bacteria requires study. Some acetic acid bacteria produce extracellular cellulose and other polysaccharides that cause ropiness and filtration difficulties (64, 67).

OTHER BACTERIA

Most bacteria cannot survive the high ethanol contents and low pHs of wines. Nevertheless, there are occasional reports that *Bacillus* species and other bacterial species can survive and grow in wines and contribute to spoilage (25, 67, 82, 90). *Bacillus coagulans*, *Bacillus circulans*, and *Bacillus subtilis* were isolated from spoiled sweet wines, and *Bacillus megaterium* was isolated from spoiled brandy. *Bacillus coagulans* and *Bacillus badius* isolated from wine corks had limited ability to grow in wines (82, 128, 205). *Bacillus thuringiensis* is prevalent on wine grapes because of its use in vineyards as a bioinsecticide. It carries over into grape juice and wine, where it remains viable but is unable to grow. It has no impact on the ability of *S. cerevisiae* to conduct alcoholic fermentation but is inhibitory to *Oenococcus oeni* in agar culture but not liquid culture (6). Juices and wines of high pH (e.g., 4.0) have, on rare occasions, been spoiled by the growth of *Clostridium butyricum* and have elevated concentrations of butyric, isobutyric, propionic, and acetic acids (205). Species of *Actinomyces* and *Streptomyces* have been isolated from wines, corks, and wooden barrels and may contribute to the musty, earthy, or corky taints sometimes found in wines (2, 82, 128).

MOLDS

Molds (filamentous fungi) impact wine production at several stages during the chain of operations. Their main influence occurs during grape cultivation, when they cause grape spoilage and loss of yield, produce metabolites that adversely impact wine quality, and produce mycotoxins. They also contaminate wooden barrels and corks (82, 84). Although molds are usual contaminants of grape juice, the conditions of anaerobiosis, increasing ethanol concentration, and the presence of sulfur dioxide prevent their growth during fermentation and conservation and in the final product.

Grapes and vines harbor a range of mold species, the populations and diversity of which depend on grape variety, degree of berry maturity and physical damage, climatic conditions, and viticultural practices. Infections of the vine by species such as *Plasmopara viticola* (downy mildew), *Uncinula necator* (powdery mildew), *Phomopsis viticola* (cane and leaf spot), and *Eutypa lata* (dieback) have major effects on the quality and yield of grapes and, eventually, can kill the vine (71, 90). Rotting and spoilage of grape berries before harvest are caused by a variety of species, the principal one being *Botrytis cinerea*, which causes bunch rot (63). Other species include those of *Penicillium*, *Aspergillus*, *Rhizopus*, *Mucor*, *Alternaria*, and *Cladosporum*, whose presence varies with grape maturity and climatic conditions (133, 193, 194). Grapes that are heavily infected with molds have altered chemical compositions and mold enzymes that adversely affect wine flavor, color, and filterability and the growth of yeasts during alcoholic fermentation (27, 84, 207). Fungal contamination of vines and grapes is controlled by the application of fungicides. However, such use must be carefully managed to minimize residues in the juice and their potential inhibitory effects on the alcoholic and malolactic fermentations (26). Biocontrol of grape molds with appropriate, antagonistic yeasts is a novel direction for minimizing the use of chemical fungicides (85).

Ochratoxin A is a mycotoxin with carcinogenic and immunosuppressive properties. Its recent discovery in wines has focused attention on grape molds and their public health significance. Ochratoxin presence in wine is linked to the growth of *Aspergillus carbonarius* and *Aspergillus niger* on grapes (118, 133, 193). Other mycotoxins of potential concern are patulin and tricothecene (194). However, mycotoxin production on grapes does not necessarily mean that mycotoxins will occur in the wine, because they may be inactivated during the fermentation processes or adsorbed onto the cell surfaces of yeasts (15).

Although *Botrytis cinerea* is well known for causing bunch rot, its controlled development on grapes gives the distinctive and highly prized botrytized sweet wines, Sauternes, Trockenbeerenauslese, and Tokay (63, 181). Under certain climatic and viticultural conditions, this species parasitizes grape berries without significantly disrupting the general integrity of their skins, causing the so-called pourriture noble, or noble rot. Mold growth on and in the berry leads to its dehydration and the concentration of its constituents. This concentration effect, along with fungal metabolism of grape sugars and acids (especially tartaric), gives a juice that has increased sugar concentration (300 to 400 mg/ml); high concentrations of glycerol (3 mg/ml), other polyols, and gluconic acid; and less tartaric acid. Fermentations of such juices require particular attention as they are prone to become stuck, possibly due to nitrogen deficiency combined with the higher sugar content, and also there is evidence that the mold secretes antiyeast substances (84). *Botrytis cinerea* also produces various phenolic oxidases and glycosidases that can affect wine color and flavor, and it produces extracellular soluble glucans that block membranes during filtration processes (63).

Mold contamination of wine corks and wooden barrels can cause earthy, moldy, or corky taints in wines and rejection of the product (128, 198). Molds use the cork or wood as a growth substrate, generating potent aroma compounds such as trichloroanisoles, 1-octen-3-one, geosmin, and guaiacol that are subsequently leached into the wine to cause the taint. Molds isolated from these sources include species of *Penicillium*, *Aspergillus*, *Trichoderma*, *Cladosporum*, *Paecilomyces*, and *Monilia* (2, 198).

SPARKLING WINES OR CHAMPAGNE

Sparkling wines contain dissolved carbon dioxide which is produced by secondary fermentation of a base wine in a closed system, such as a bottle or a tank (19, 25, 119, 175). The microbial ecology and biochemistry of base wine production are essentially the same as those of table wine production. Three types of secondary fermentation processes are used. The traditional *méthode champenoise* requires fermentation in the bottle in which the wine is sold. A base wine (usually white wine) is selected to which sugar or sugar syrup is added to give a final concentration of about 2.4 mg/ml. Yeast nutrients (diammonium phosphate and vitamins) may also be added. A suitable strain of *S. cerevisiae* is inoculated into the wine, which is thoroughly mixed and transferred into the bottles (leverage, or tirage). The bottles are sealed with crown caps or corks and then stored at 12 to 15°C for fermentation. The bottles are specially made to withstand the high pressure (about 600 kPa) of carbon dioxide produced during fermentation. Fermentation is completed in 3 to 6 months, after which the wine is aged in the bottles in contact with the yeast lees for at least 6 to 12 months or longer for premium-quality wines. After storage, the bottles are restacked with their neck downwards and periodically twisted or shaken to facilitate settling of the yeast sediment into the neck. This process is called riddling, or remuage. Finally, the sedimented yeast plug is removed from the bottle neck by a process termed disgorgement. The yeast plug is frozen by dipping the neck of the bottle into calcium chloride solution at −24°C, and then the cork or cap is removed, allowing the internal pressure to force out the yeast plug with little loss of wine. The lost wine is replaced by the addition of base wine (dosage), which may contain a small amount of sugar to give a desired amount of sweetness, and sulfur dioxide. Remuage, disgorgement, and dosage are automated processes in the modern winery. Following these operations, the bottles are corked and an agraffe (wire clamp) is applied.

Variations in the *méthode champenoise* include the transfer method and bulk fermentation, or the Charmat process. In the transfer process, fermentation is conducted in the bottle as already described, followed by a short maturation on lees. The bottle contents are chilled and emptied into a pressurized tank. Dosage is added to the wine, which is filtered to remove the yeasts and bottled. In the Charmat process, the base wine, with added sugar and yeast, is fermented in a pressurized tank. After fermentation, the wine is chilled and clarified by centrifugation before transfer into a second (pressurized) tank containing the dosage. The wine is finally filtered and bottled (19, 119).

Special strains of *S. cerevisiae* are required for the secondary fermentation. Criteria for these strains include the following: give complete sugar fermentation under conditions of low temperature (10 to 15°C), relatively high ethanol concentration (8 to 12%), low pH (as low as 3.0), the presence of up to 20 μg of free sulfur dioxide/ml, low nutrient availability, and increasing pressure of carbon dioxide (up to 600 kPa); flocculate and sediment to facilitate the riddling process; give good flavor; and undergo autolysis during ageing on the lees (138, 139). Usually, the yeast is inoculated into the wine base at a population of 1×10^6 to 5×10^6 CFU/ml. The secondary fermentation is completed over the next 30 to 40 days, during which time the yeast grows to a maximum population of about 10^7 CFU/ml. Subsequently, yeast cells slowly die, and generally, viable cells cannot be detected after 100 days. Autolysis of the yeast then commences (35, 210).

Chemical changes occur in two phases, namely, during the secondary fermentation and during the period of ageing and yeast autolysis. During secondary fermentation, the added sugar is utilized with the production of carbon dioxide and about 1% ethanol. In the early stages, amino nitrogen is consumed, but after exhaustion of the sugar, yeast cells give a small efflux of amino nitrogen into the wine. Minor changes in the concentrations of glycerol, some organic acids, and some esters have been noted, but the data are not consistent. The yeast cells are operating under extreme environmental conditions, especially increasing concentrations of carbon dioxide and increasing physical pressure, and these circumstances are likely to impact their metabolic behavior.

Yeast autolysis is characterized by the degradation of cellular macromolecules and the release of their degradation products, as well as other cell constituents, into the external medium. This process correlates with gradual increases in concentrations of proteins, amino acids, nucleic acids, lipids, free fatty acids, and mannoproteins (originating from the yeast cell wall) in the wine, as well as changes in the concentrations of various esters, higher alcohols, carbonyl compounds, terpenes, and lactones. These changes influence sparkling wine aroma, flavor, and bubble properties (35, 79, 117, 138, 139, 164, 170, 230).

Innovations to increase the efficiency of the secondary fermentation and ageing processes include the use of immobilized yeast cells (227), killer yeasts (210), and autolytic mutant yeasts (138).

FORTIFIED WINES

Fortified wines such as sherry, port, and Madeira have ethanol concentrations of 15 to 22%. This higher ethanol concentration is achieved by the addition of ethanol (usually derived from the distillation of wine products) at certain stages during the process. Details of fortified wine production are given elsewhere (25, 100, 177).

With sherry-style wines, an essentially dry white wine base is produced from particular grape varieties. The yeast ecology and the biochemistry of the fermentation are similar to those described already. At the completion of alcoholic fermentation, the wine is fortified and transferred into oak casks for ageing and maturation. In the case of oloroso sherries, the wine is fortified to contain 18 to 19% ethanol, and this fortification stops any further microbiological processes. Consequently, subsequent maturation is a chemical process. With the finos sherries, the wine is fortified to contain 15 to 16% ethanol and aged in a series of oak casks by the solera system, in which portions of the wine in each cask are systematically removed and replaced with an amount of younger wine of the same style so that the wine is continuously blended and emerges with a consistent character. The frequency of transfers and amount of wine removed from each cask vary according to the producer. This process encourages the natural formation of a surface film or velum of yeast growth (flor) at the air-wine interface. Essentially, the velum is a wrinkled layer of yeast biomass about 5 to 10 mm thick. It represents a unique ecological niche where yeasts (and probably bacteria as well) have adapted to survive and metabolize in the presence of high concentrations of ethanol. The velum consists mainly of a mixture of distinct, hydrophobic strains of *S. cerevisiae*, but strains of *Torulaspora delbrueckii*, *Dekkera bruxellensis*, *Candida cantorelli*, and *Zygosaccharomyces* may also be present (74, 137). There appears to be a definitive evolution of species and strains during the process of velum development, and the ecologies may be different at the top (aerobic) and bottom (anaerobic) locations. The oxidative metabolism of the velum decreases the concentrations of wine acids, glycerol, and alcohol and substantially increases the concentration of acetaldehyde. Amino acids (including proline) are assimilated, with the consequent production of higher alcohols. These changes, as well as many other less quantitative reactions conducted by the flor yeasts, contribute to the unique flavor of finos sherries (137, 154).

Port-style wines are prepared from a red wine base, and the Madeira-style wines are produced from a blend of red and white wines. In these processes, fortification with ethanol is done at an appropriate stage during the alcoholic fermentation so that the fermentation is arrested to give a wine with residual unfermented sugar and a desired amount of sweetness. These wines are then subjected to ageing and maturation which do not involve any secondary fermentations because of the high concentration of ethanol (100, 177).

Despite their high ethanol concentrations, fortified wines may undergo spoilage from ethanol-tolerant strains of yeasts (*Zygosaccharomyces bisporus*, *S. cerevisiae*, and *Rhodotorula*, *Candida*, and *Brettanomyces* species) and lactic acid bacteria (*Lactobacillus hilgardii* and other species) (205, 209).

References

1. Alexandre, H., P. J. Costello, F. Remize, J. Guzzo, and M. Guilloux-Benatier. 2004. *Saccharomyces cerevisiae*-*Oenococcus oeni* interactions in wine: current knowledge and perspectives. *Int. J. Food Microbiol.* 93:141–154.
2. Alvarez-Rodriguez, M. L., C. Belloch, M. Villa, F. Uruburu, G. Larriba, and J.-J. R. Coque. 2003. Degradation of vanillic acid and production of guaiacol by microorganisms isolated from cork samples. *FEMS Microbiol. Lett.* 220:49–55.
3. Amerine, M. A. 1985. Winemaking, p. 67–81. *In* H. Koprowski and S. A. Plotin (ed.), *World's Debt to Pasteur*. Alan R. Liss Incorporated, New York, N.Y.
4. Arena, M. E., and M. C. Manca de Nadra. 2005. Influence of ethanol and low pH on arginine and citrulline metabolism in lactic acid bacteria from wine. *Res. Microbiol.* 156:858–864.
5. Backhus, L. E., J. de Risi, P. O. Brown, and L. F. Bisson. 2001. Functional geonomic analysis of a commercial wine strain of *Saccharomyces cerevisiae* under differing nitrogen conditions. *FEMS Yeast Res.* 1:111–125.
6. Bae, S., G. H. Fleet, and G. M. Heard. 2004. Occurrence and signficiance of *Bacillus thuringiensis* on wine grapes. *Int. J. Food Microbiol.* 94:301–312.
7. Barbe, J. C., G. de Revel, A. Joyeux, A. Bertrand, and A. Lonvaud-Funel. 2001. Role of botrytized grape microorganisms in SO$_2$-binding phenomena. *J. Appl. Microbiol.* 90:34–42.
8. Bartowsky, E. 2005. *Oenococcus oeni* and malolactic fermentation—moving into the molecular arena. *Aust. J. Grape Wine Res.* 11:174–187.
9. Bartowsky, E., and P. A. Henschke. 2004. The buttery attribute of wine—diacetyl—desirability, spoilage and beyond. *Int. J. Food Microbiol.* 96:235–252.
10. Bartowsky, E. J., D. Xia, R. L. Gibson, G. H. Fleet, and P. A. Henschke. 2003. Spoilage of bottled red wine by acetic acid bacteria. *Lett. Appl. Microbiol.* 36:307–314.
11. Bauer, F. F., and I. S. Pretorius. 2000. Yeast stress response and fermentation efficiency: how to survive the making of wine—a review. *S. Afr. J. Enol. Viticult.* 21:27–51.
12. Bauer, R., and L. M. T. Dicks. 2004. Control of malolactic fermentation in wine—a review. *S. Afr. J. Enol. Viticult.* 25:74–88.
13. Bauer, R., H. A. Nel, and L. M. T. Dicks. 2003. Pediocin PD-1 as a method to control growth of *Oenococcus oeni* in wine. *Am. J. Enol. Viticult.* 54:86–91.
14. Beh, A. L., G. H. Fleet, C. Prakitchaiwattana, and G. M. Heard. 2006. Evaluation of molecular methods for the analysis of yeasts in foods and beverages, p. 69–106. *In* A. Hocking (ed.), *Advances in Food Mycology*. Springer, Berlin, Germany.
15. Bejaoui, H., F. Mathieu, P. Taillandier, and A. Lebrihi. 2004. Ochratoxin A removal in synthetic and natural grape juices by selected enological *Saccharomyces* strains. *J. Appl. Microbiol.* 97:1038–1044.
16. Bell, S.-J., and P. A. Henschke. 2005. Implications of nitrogen nutrition for grapes, fermentation and wine. *Aust. J. Grape Wine Res.* 11:242–295.
17. Berry, D. R., and J. C. Slaughter. 2003. Alcoholic beverages fermentation, p. 25–29. *In* A. G. H. Lea and J. Piggott (ed.), *Fermented Beverage Production*, 2nd ed. Kluwer Academic, New York, N.Y.
18. Berthels, N. J., R. R. Cordero Oteso, F. F. Bauer, J. M. Thevelein, and I. S. Pretorius. 2004. Discrepancy in glucose and fructose utilization during fermentation by *Saccharomyces cerevisiae* wine yeast strains. *FEMS Yeast Res.* 4:683–689.
19. Bidan, P., M. Feuillat, and J. Moulin. 1986. Rapport de la France. Les vins Mousseux. *Bull. Off. Int. Vin* 59:563–626.
20. Bisson, L. F. 1993. Yeasts—metabolism of sugars, p. 55–75. *In* G. H. Fleet (ed.), *Wine Microbiology and Biotechnology*. Harwood Academic Publishers, Chur, Switzerland.
21. Bisson, L. F. 1999. Stuck and sluggish fermentations. *Am. J. Enol. Viticult.* 50:107–119.
22. Bisson, L. F., A. L. Waterhouse, S. E. Ebeler, M. Andrew-Walker, and J. T. Lapsley. 2002. The present and the future of the international wine industry. *Nature* 418:696–699.

23. Blasco, L., S. Ferrer, and I. Pardo. 2003. Development of specific fluorescent oligonucleotide probes for insitu identification of wine lactic acid bacteria. *FEMS Microbiol. Lett.* **225**:115–123.

24. Boulton, R. 2003. Red wines, p. 107–108. *In* A. G. H. Lea and J. R. Piggott (ed.), *Fermented Beverage Production*, 2nd ed. Kluwer Academic, New York, N.Y.

25. Boulton, R. B., V. L. Singleton, L. F. Bisson, and R. E. Kunkee. 1995. *Principles and Practices of Winemaking*. Chapman and Hall, New York, N.Y.

26. Cabras, R., and A. Angioni. 2000. Pesticide residues in grapes, wines and their processing products. *J. Agric. Food Chem.* **48**:967–973.

27. Calonnec, A., P. Carolaro, C. Poupot, D. Dubourdieu, and P. Dariet. 2004. Effects of *Uncinula necator* on the yield and quality of grapes (*Vitis vinifera*) and wine. *Plant Pathol.* **53**:434–445.

28. Cantarelli, C. 1989. Factors affecting the behaviour of yeast in wine fermentation, p. 127–151. *In* C. Cantarelli and G. Lanzarini (ed.), *Biotechnology Applications in Beverage Production*. Elsevier Applied Science, London, United Kingdom.

29. Capece, A., C. Fiore, A. Maraz, and P. Romano. 2005. Molecular and technological approaches to evaluate strain biodiversity in *Hanseniaspora uvarum* of wine origin. *J. Appl. Microbiol.* **98**:136–144.

30. Cappello, M. S., G. Bleve, F. Grieco, F. Dellaglio, and G. Zacheo. 2004. Characterization of *Saccharomyces cerevisiae* strains isolated from must of grape grown in experimental vineyard. *J. Appl. Microbiol.* **97**:1274–1280.

31. Carr, F. J., D. Chill, and N. Maida. 2002. The lactic acid bacteria: a literature survey. *Crit. Rev. Microbiol.* **28**: 281–370.

32. Carrau, F. M., K. Medina, E. Boido, L. Farina, C. Gaggero, E. Dellacassa, G. Versini, and P. A. Henschke. 2005. De novo synthesis of monoterpenes by *Saccharomyces cerevisiae* wine yeasts. *FEMS Microbiol. Lett.* **243**:107–115.

33. Charoenchai, C., G. H. Fleet, and P. Henschke. 1998. Effects of temperature, pH and sugar concentration on the growth rates and cell biomass of wine yeast. *Am. J. Enol. Viticult.* **49**:283–288.

34. Charoenchai, C., G. H. Fleet, P. Henschke, and B. E. N. Todd. 1997. Screening of non-*Saccharomyces* wine yeasts for the presence of extracellular hydrolytic enzymes. *Aust. J. Grape Wine Res.* **3**:2–8.

35. Charpentier, C., and M. Feuillat. 1993. Yeast autolysis, p. 225–242. *In* G. H. Fleet (ed.), *Wine Microbiology and Biotechnology*. Harwood Publishers, Chur, Switzerland.

36. Ciani, M., F. Fatichenti, and I. Mannazzu. 2002. Yeasts in winemaking biotechnology, p. 111–123. *In* M. Ciani (ed.), *Biodiversity and Biotechnology of Wine Yeasts*. Research Signpost, Kerala, India.

37. Clemente-Jimenez, J. M., L. Mingorance-Cazorla, S. Martinez-Rodriguez, F. J. L. Heras-Vazquez, and F. Rodiriguez-Vico. 2005. Influence of sequential yeast mixtures on wine fermentation. *Int. J. Food Microbiol.* **98**:301–308.

38. Cocolin, L., L. Bisson, and D. A. Mills. 2000. Direct profiling of the dynamics in wine fermentations. *FEMS Microbiol. Lett.* **189**:81–87.

39. Cocolin, L., V. Pepe, F. Comitini, G. Comi, and M. Ciani. 2004. Enological and genetic traits of *Saccharomyces cerevisiae* isolated from former and modern wineries. *FEMS Yeast Res.* **5**:237–246.

40. Cocolin, L., K. Rantsiou, L. Iacumin, R. Zirconi, and G. Comi. 2004. Molecular detection and identification of *Brettanomyces/Dekkera bruxellensis* and *Brettanomyces/Dekkera anomalus* in spoiled wine. *Appl. Environ. Microbiol.* **70**:1347–1355.

41. Cole, V. C., and A. C. Noble. 2003. Flavor chemistry, p. 393–412. *In* A. G. H. Lea and J. R. Piggott (ed.), *Fermented Beverage Production*, 2nd edition. Kluwer Academic, New York, N.Y.

42. Combina, M., A. Elia, L. Mercado, C. Catania, A. Ganga, and C. Martinez. 2003. Dynamics of yeast populations during spontaneous fermentations of wines from Mendonza, Argentina. *Int. J. Food Microbiol.* **99**:237–243.

43. Combina, M., L. Mercado, P. Borgo, A. Elia, V. Jofre, A. Ganga, and C. Martinez. 2005. Yeasts associated to Malbec grape berries from Mendoza, Argentina. *J. Appl. Microbiol.* **98**:1055–1061.

44. Constanti, M., M. Poblet, L. Arola, A. Mas, and J. M. Guillamón. 1997. Analysis of yeast populations during alcoholic fermentation in a newly established winery. *Am. J. Enol. Viticult.* **48**:339–344.

45. Coombe, B. G., and P. R. Dry. 2004. *Viticulture*, 2nd ed., vol. 1. *Resources*. Winetitles, Adelaide, Australia.

46. Costello, P. J., and P. A. Henschke. 2002. Mousy off-flavour of wine. Precursors and biosynthesis of the causative N-heterocycles 2-ethyl tetrahydropyridine, 2-acetyltetrahydropyridine and 2-acetyl-1-pyrroline by *Lactobacillus hilgardii* DSM 20176. *J. Agric. Food Chem.* **50**:7079–7087.

47. Daeschel, M. A., D. S. Jung, and B. T. Watson. 1991. Controlling malolactic fermentation with nisin and nisin-resistant strains of *Leuconostoc oenos*. *Appl. Environ. Microbiol.* **57**:601–603.

48. Davis, C., D. Wibowo, R. Eschenbruch, T. H. Lee, and G. H. Fleet. 1985. Practical implications of malolactic fermentation—a review. *Am. J. Enol. Viticult.* **36**:209–301.

49. Davis, C. R., N. F. A. Silveira, and G. H. Fleet. 1985. Occurrence and properties of bacteriophages of *Leuconstoc oenos* in Australian wines. *Appl. Environ. Microbiol.* **50**:872–876.

50. Davis, C. R., D. Wibowo, G. H. Fleet, and T. H. Lee. 1988. Properties of wine lactic acid bacteria: their potential enological significance. *Am. J. Enol. Viticult.* **39**:137–142.

51. Davis, C. R., D. Wibowo, T. H. Lee, and G. H. Fleet. 1986. Growth and metabolism of lactic acid bacteria during and after malolactic fermentation of wines at different pH. *Appl. Environ. Microbiol.* **51**:539–545.

52. Degré, R. 1993. Selection and cultivation of wine yeast and bacteria, p. 421–447. *In* G. H. Fleet (ed.), *Wine Microbiology and Biotechnology*. Harwood Academic Publishers, Chur, Switzerland.

53. de las Rivas, B., A. Marcobal, and R. Munoz. 2004. Allelic diversity and population structure in *Oenococcus oeni* as determined from sequence analysis of housekeeping genes. *Appl. Environ. Microbiol.* **70**:7210–7219.

54. Demuyter, C., M. Lollier, J.-L. Legras, and C. Le Jeune. 2004. Predominance of *Saccharomyces uvarum* during spontaneous alcoholic fermentation for three consecutive years in an Alsation winery. *J. Appl. Microbiol.* 97:1140–1148.

55. Deppenmeier, U., M. Hoffmeister, and C. Prost. 2002. Biochemistry and biotechnological applications of *Gluconobacter* strains. *Appl. Microbiol. Biotechnol.* 60:233–242.

56. Dequin, S., J. M. Salmon, H.-V. Nguyen, and B. Blondin. 2003. Wine yeasts, p. 389–411. *In* T. Boekhout and V. Robert (ed.), *Yeasts in Foods: Beneficial and Detrimental Aspects.* Behr's-Verlag, Hamburg, Germany.

57. de Revel, G., N. Martin, L. Pripis-Nicolau, A. Lonvaud-Funel, and A. Bertrand. 1999. Contribution to the knowledge of malolactic fermentation influence on wine aroma. *J. Agric. Food Chem.* 47:4003–4008.

58. Dicks, L. M. T., F. Dellaglio, and M. D. Collins. 1995. Proposal to reclassify *Leuconostoc oenos* as *Oenococcus oeni* (corrig.) gen. nov. *Int. J. Syst. Bacteriol.* 45:395–397.

59. D'Inecco, N., E. J. Bartwosky, S. Kassara, A. Lante, P. Spettoli, and P. A. Henschke. 2004. Release of glycosidically bound flavour compounds of Chardonnay by *Oenococcus oeni* during malolactic fermentation. *Food Microbiol.* 21:257–265.

60. Dittrich, H. H. 1995. Wine and brandy, p. 464–503. *In* G. Reed and T. W. Nagodawithana (ed.), *Biotechnology,* 2nd ed., vol. 9. *Enzymes, Biomass, Food and Feed.* VCH, Weinheim, Germany.

61. Divies, C. 1993. Bioreactor technology and wine fermentation, p. 449–475. *In* G. H. Fleet (ed.), *Wine Microbiology and Biotechnology.* Harwood Academic Publishers, Chur, Switzerland.

62. Divol, B., and A. Lonvaud-Funel. 2005. Evidence for viable but non-culturable yeasts in botrytis-affected wine. *J. Appl. Microbiol.* 99:85–93.

63. Doneche, B. 1993. Botrytized wines, p. 327–351. *In* G. H. Fleet (ed.), *Wine Microbiology and Biotechnology.* Harwood Academic Publishers, Chur, Switzerland.

64. Drysdale, G. S., and G. H. Fleet. 1988. Acetic acid bacteria in winemaking—a review. *Am. J. Enol. Viticult.* 39:143–154.

65. Drysdale, G. S., and G. H. Fleet. 1989. The effect of acetic acid bacteria upon the growth and metabolism of yeasts during the fermentation of grape juice. *J. Appl. Bacteriol.* 67:471–481.

66. du Plessis, H. W., L. M. T. Dicks, I. S. Pretorius, M. G. Lambrechts, and M. du Toit. 2004. Identification of lactic acid bacteria isolated from South African brandy base wines. *Int. J. Food Microbiol.* 91:19–29.

67. du Toit, M., and I. S. Pretorius. 2000. Microbial spoilage and preservation of wine: using weapons from nature's own arsenal—a review. *S. Afr. J. Enol. Viticult.* 21:74–96.

68. du Toit, W. J., and M. G. Lambrechts. 2002. The enumeration and identification of acetic acid bacteria from South African red wine fermentations. *Int. J. Food Microbiol.* 74:57–64.

69. du Toit, W. J., I. S. Pretorius, and A. Lonvaud-Funel. 2005. The effect of sulphur dioxide and oxygen on the viability and culturability of a strain of *Acetobacter pasteurianus* and a strain of *Brettanomyces bruxellensis* isolated from wine. *J. Appl. Microbiol.* 98:862–871.

70. Egli, C. M., W. D. Edinger, C. M. Mitrakul, and T. Henick-Kling. 1998. Dynamics of indigenous and inoculated yeast populations and their effect on the sensory character of Riesling and Chardonnay wines. *J. Appl. Microbiol.* 85:779–789.

71. Emmet, R. W., A. R. Harris, R. H. Taylor, and J. K. McGechan. 1988. Grape diseases and vineyard protection, p. 232–278. *In* B. G. Coombe and P. R. Dry (ed.), *Viticulture,* vol. 2. Winetitles, Adelaide, Australia.

72. Erten, H. 2002. Relations between elevated temperatures and fermentative behaviour of *Kloeckera apiculata* and *Saccharomyces cerevisiae* associated with winemaking in mixed cultures. *World J. Microbiol. Biotechnol.* 18:373–378.

73. Estevez, P., M. Luisa Gil, and E. Falque. 2004. Effects of seven yeast strains on the volatile composition of Palomino wines. *Int. J. Food Sci. Technol.* 39:61–69.

74. Esteve-Zarzoso, B., M. J. Peris-Toran, E. Garcia-Maiquez, F. Uruburu, and A. Querol. 2001. Yeast population dynamics during the fermentation and biological aging of sherry wines. *Appl. Environ. Microbiol.* 67:2056–2061.

75. Ewart, A. 2003. White wines, p. 89–106. *In* A. G. H. Lea and J. R. Piggott (ed.), *Fermented Beverage Production,* 2nd ed. Kluwer Academic, New York, N.Y.

76. Faia, A. M., and F. Radler. 1990. Investigation of the bactericidal effect of nisin on lactic acid bacteria of wine. *Vitis* 29:233–238.

77. Farias, M. E., and M. C. Manca de Nadra. 2000. Purification and partial characterization of *Oenococcus oeni* protease. *FEMS Microbiol. Lett.* 185:263–266.

78. Fernandez-Gonzalez, M., R. di Stefano, and A. Briones. 2003. Hydrolysis and transformation of terpene glycosides from Muscat must by different yeast species. *Food Microbiol.* 20:35–41.

79. Feuillat, M. 2003. Yeast macromolecules: origin, composition and enological interest. *Am. J. Enol. Viticult.* 54:211–213.

80. Fia, G., G. Giovani, and I. Rosi. 2005. Study of β-glucosidase production by wine-related yeasts during alcoholic fermentation. A new rapid fluorimetric method to determine enzymatic activity. *J. Appl. Microbiol.* 99:509–517.

81. Fleet, G. H. 1992. Spoilage yeasts. *Crit. Rev. Biotechnol.* 12:1–44.

82. Fleet, G. H. 1993. The microorganisms of winemaking—isolation, enumeration and identification, p. 1–26. *In* G. H. Fleet (ed.), *Wine Microbiology and Biotechnology.* Harwood Academic Publishers, Chur, Switzerland.

83. Fleet, G. H. 1999. Microorganisms in food ecosystems. *Int. J. Food Microbiol.* 50:101–117.

84. Fleet, G. H. 2003. Yeast interactions and wine flavour. *Int. J. Food Microbiol.* 86:11–22.

85. Fleet, G. H. 2003. Yeasts in fruit and fruit products, p. 267–288. *In* T. Boeckhout and V. Robert (ed.), *Yeasts in Food: Beneficial and Detrimental Aspects.* Behr's-Verlag, Hamburg, Germany.

86. Fleet, G. H., and G. M. Heard. 1993. Yeasts—growth during fermentation, p. 27–54. *In* G. H. Fleet (ed.), *Wine Microbiology and Biotechnology.* Harwood Academic Publishers, Chur, Switzerland.

87. Fleet, G. H., S. Lafon-Lafourcade, and P. Ribereau-Gayon. 1984. Evolution of yeasts and lactic acid bacteria during fermentation and storage of Bordeaux wines. *Appl. Environ. Microbiol.* **48:**1034–1038.

88. Fleet, G. H., C. Prakitchaiwattana, A. L. Beh, and G. Heard. 2002. The yeast ecology of wine grapes, p. 1–17. *In* M. Ciani (ed.), *Biodiversity and Biotechnology of Wine Yeasts.* Research Signpost, Kerala, India.

89. Francis, J. L., and J. L. Newton. 2005. Determining wine aroma from compositional data. *Aust. J. Grape Wine Res.* **11:**114–126.

90. Fugelsang, K. C. 1997. *Wine Microbiology.* Chapman & Hall, New York, N.Y.

91. G-Alegria, E., I. Lopez, J. I. Ruiz, J. Saenz, E. Fernandez, M. Zarazaga, M. Dizy, C. Torres, and F. Ruiz-Larrea. 2004. High tolerance of wild *Lactobacillus plantarum* and *Oenococcus oeni* strains to lyophilisation and stress environmental conditions of acid pH and ethanol. *FEMS Microbiol. Lett.* **230:**53–61.

92. Ganga, M. A., and C. Martinez. 2004. Effect of wine yeast monoculture practice in the biodiversity of non-*Saccharomyces* yeasts. *J. Appl. Microbiol.* **96:**76–83.

93. Gao, C., and G. H. Fleet. 1988. The effects of temperature and pH on the ethanol tolerance of the wine yeasts, *Saccharomyces cerevisiae, Candida stellata* and *Kloeckera apiculata. J. Appl. Bacteriol.* **65:**405–410.

94. Gao, C., and G. H. Fleet. 1995. Cell-recycle membrane bioreactor for conducting continuous malolactic fermentation. *Aust. J. Grape Wine Res.* **1:**32–38.

95. Gao, C., and G. H. Fleet. 1995. Degradation of malic and tartaric acids by high density cell suspensions of wine yeats. *Food Microbiol.* **12:**65–71.

96. Gao, Y. C., G. Zhang, S. Krentz, S. Darius, J. Power, and G. Lagarde. 2002. Inhibition of spoilage lactic acid bacteria by lysozyme during wine alcoholic fermentation. *Aust. J. Grape Wine Res.* **8:**76–83.

97. Gockowiak, H., and P. A. Henschke. 2003. Interaction of pH, ethanol concentration and wine matrix on induction of malolactic fermentation with commercial "direct inoculation" starter cultures. *Aust. J. Grape Wine Res.* **9:**200–209.

98. Gonzalez, A., N. Hierro, M. Poblet, A. Mas, and J. M. Guillamon. 2005. Application of molecular methods to demonstrate species and strain evolution of acetic acid bacteria population during wine production. *Int. J. Food Microbiol.* **102:**295–304.

99. Gonzalez, A., H. M. Poblet, N. Rozes, A. Mas, and J. M. Guillamon. 2004. Application of molecular methods for the differentiation of acetic acid bacteria in a red wine fermentation. *J. Appl. Microbiol.* **96:**853–860.

100. Goswell, R. W., and R. E. Kunkee. 1977. Fortified wines, p. 478–533. *In* A. H. Rose (ed.), *Economic Microbiology,* vol. 1. Academic Press, London, United Kingdom.

101. Granchi, L., M. Bosco, A. Messini, and M. Vincenzini. 1999. Rapid detection and quantification of yeast species during spontaneous wine fermentation by PCR–RFLP analysis of the rDNA ITS region. *J. Appl. Microbiol.* **87:**949–956.

102. Granchi, L., D. Gannucci, C. Viti, L. Giovannetti, and M. Vincenzini. 2003. *Saccharomyces cerevisiae* biodiversity in spontaneous commercial fermentations of grape musts with "adequate" and "inadequate" assimilable-nitrogen content. *Lett. Appl. Microbiol.* **36:**54–58.

103. Grimaldi, A., E. Bartowsky, and V. Jiranek. 2005. A survey of glycosidase activities of commercial wine strains of *Oenococcus oeni. Int. J. Food Microbiol.* **105:**233–244.

104. Guerrini, S., A. Bastianini, G. Blaiotta, L. Granchi, G. Moschetti, S. Coppola, P. Romano, and M. Vincenzini. 2003. Phenotypic and genotypic characterization of *Oenococcus oeni* strains isolated from Italian wines. *Int. J. Food Microbiol.* **83:**1–14.

105. Guerrini, S., S. Mangani, L. Granchi, and M. Vincenzini. 2002. Biogenic amine production by *Oenococcus oeni. Curr. Microbiol.* **44:**374–378.

106. Guilloux-Benatier, M., M. O. Pageault, A. Man, and M. Feuillat. 2000. Lysis of yeast cells by *Oenococcus oeni* enzymes. *J. Ind. Microbiol. Biotechnol.* **25:**193–197.

107. Gutiérrez, A. R., S. Epifanio, P. Garijo, P. Lopez, and P. Santamaria. 2001. Killer yeasts: incidence in the ecology of spontaneous fermentation. *Am. J. Enol. Viticult.* **52:**352–356.

108. Gutiérrez, A. R., P. Santamaria, S. Epifania, P. Garijo, and R. Lopez. 1999. Ecology of spontaneous fermentation in one winery during 5 consecutive years. *Lett. Appl. Microbiol.* **29:**411–415.

109. Guzzo, J., M. P. Jobin, F. Delmas, L. C. Fortier, D. Garmyn, R. Tourdot-Marechal, B. Lee, and C. Divies. 2000. Regulation of stress response in *Oenococcus oeni* as a function of environmental changes and growth phase. *Int. J. Food Microbiol.* **55:**27–31.

110. Hansen, E. H., P. Nissen, P. Sommer, J. C. Nielsen, and N. Arneborg. 2001. The effect of oxygen on the survival of non-*Saccharomyces* yeasts during mixed culture fermentations of grape juice with *Saccharomyces cerevisiae. J. Appl. Microbiol.* **91:**541–547.

111. Heard, G. M. 1999. Novel yeasts in winemaking—looking to the future. *Food Aust.* **51:**347–352.

112. Heard, G. M., and G. H. Fleet. 1985. Growth of natural yeast flora during the fermentation of inoculated wines. *Appl. Environ. Microbiol.* **50:**727–728.

113. Heard, G. M., and G. H. Fleet. 1988. The effects of temperature and pH on the growth of yeast species during the fermentation of grape juice. *J. Appl. Bacteriol.* **65:**23–28.

114. Henick-Kling, T. 1993. Malolactic fermentation, p. 289–326. *In* G. H. Fleet (ed.), *Wine Microbiology and Biotechnology.* Harwood Academic Publishers, Chur, Switzerland.

115. Henschke, P., and V. Jiranek. 1993. Yeasts—metabolism of nitrogen compounds, p. 77–164. *In* G. H. Fleet (ed.), *Wine Microbiology and Biotechnology.* Harwood Academic Publishers, Chur, Switzerland.

116. Henschke, P. A. 1997. Wine yeast, p. 527–560. *In* F. K. Zimmermann and K. D. Entian (ed.), *Yeast Sugar Metabolism.* Technomic Publishing Co., Lancaster, Pa.

117. Hernawan, T., and G. H. Fleet. 1995. Chemical and cytological changes during the autolysis of yeasts. *J. Ind. Microbiol.* **14:**440–450.

118. Hocking, A. D., P. Varelis, J. I. Pitt, S. F. Cameron, and S. L. Leong. 2003. Occurrence of ochratoxin A in Australian wine. *Aust. J. Grape Wine Res.* **9:**72–78.

119. Howe, P. 2003. Sparkling wines, p. 139–156. *In* A. G. H. Lea and J. R. Piggot (ed.), *Fermented Beverage Production*, 2nd ed. Kluwer Academic, New York, N.Y.

120. Howell, K. S., D. Cozzolino, E. J. Bartowsky, G. H. Fleet, and P. A. Henschke. 2006. Metabolic profiling as a tool for revealing *Saccharomyces* interactions during wine fermentation. *FEMS Yeast Res.* 6:91–101.

121. Howell, K. S., M. Klein, J. H. Swiegers, Y. Hayasaka, G. M. Elsey, G. H. Fleet, P. B. Hoj, I. S. Pretorius, and M. A. de Barros Lopes. 2005. Genetic determinants of volatile-thiol release by *Saccharomyces cerevisiae* during wine fermentation. *Appl. Environ. Micrbiol.* 71:5420–5426.

122. Howell, K. S., J. H. Swiegers, G. M. Elsey, T. E. Siebert, E. J. Bartowsky, G. H. Fleet, I. S. Pretorius, and M. A. de Barros Lopes. 2004. Variation in 4-mercapto-4-methyl-pentan-2-one release by *Saccharomyces cerevisiae* commercial wine strains. *FEMS Microbiol. Lett.* 240:125–129.

123. Jolly, N. P., O. P. H. Augustyn, and I. S. Pretorius. 2003. The occurrence of non-*Saccharomyces cerevisiae* yeast species over three vintages in four vineyards and grape must from four production regions of the Western Cape, South Africa. *S. Afr. J. Enol. Viticult.* 24:35–42.

124. Kourkoutas, Y., A. Bekatorou, I. A. Banat, R. Marchant, and A. A. Koutinas. 2004. Immobilization technologies and support materials suitable in alcohol beverages production—a review. *Food Microbiol.* 21:377–397.

125. Kunkee, R. E., and L. Bisson. 1993. Wine-making yeasts, p. 69–128. *In* A. H. Rose and J. S. Harrison (ed.), *The Yeasts*, 2nd ed., vol. 5. *Yeast Technology*. Academic Press, London, United Kingdom.

126. Kurtzman, C. P. 2003. Phylogenetic circumscription of *Saccharomyces*, *Kluyveromyces* and other members of *Saccharomycetaceae* and the proposal of the new genera *Lachancea*, *Nakaseomyces*, *Vanderwaltozyma* and *Zygotorulaspora*. *FEMS Yeast Res.* 4:232–245.

127. Lambrechts, M. G., and I. S. Pretorius. 2000. Yeast and its importance in wine aroma—a review. *S. Afr. J. Enol. Viticult.* 21:97–129.

128. Lee, T. H., and R. F. Simpson. 1993. Microbiology and chemistry of cork taints in wine, p. 353–372. *In* G. H. Fleet (ed.), *Wine Microbiology and Biotechnology*. Harwood Academic Publishers, Chur, Switzerland.

129. Liu, S. Q. 2002. Malolactic fermentation in wine—beyond deacidificiation. *J. Appl. Microbiol.* 92:589–601.

130. Lonvaud-Funel, A. 1999. Lactic acid bacteria in the quality improvement and depreciation of wine. *Antonie Leeuwenhoek* 76:317–331.

131. Lonvaud-Funel, A. 2001. Biogenic amines in wines: role of lactic acid bacteria. *FEMS Microbiol. Lett.* **199:**9–13.

132. Loureiro, V., and M. Malfeito-Ferreira. 2003. Spoilage yeasts in the wine industry. *Int. J. Food Microbiol.* 86:23–50.

133. Magnoli, C., M. Violante, M. Combina, G. Palacio, and A. Dalcero. 2003. Mycoflora and ochratoxin-producing strains of *Aspergillus* section Nigri in wine grapes in Argentina. *Lett. Appl. Microbiol.* 37:179–184.

134. Maicas, S. 2001. The use of alternative technologies to develop malolactic fermentation in wine. *Appl. Microbiol. Biotechnol.* 56:35–39.

135. Mansfield, A. K., B. Zoecklein, and R. S. Whiton. 2002. Quantification of glycosidase activity in selected strains of *Brettanomyces bruxellensis* and *Oenococcus oeni*. *Am. J. Enol. Viticult.* 53:303–307.

136. Marks, V. D., G. K. van der Merwe, and H. J. J. van Vuuren. 2003. Transcriptional profiling of wine yeast in fermenting grape juice: regulatory effect of diammonium phosphate. *FEMS Yeast Res.* 3:269–287.

137. Martinez, P., M. J. Valcarcel, L. Perez, and T. Benitez. 1995. Metabolism of *Saccharomyces cerevisiae* flor yeasts during fermentation and biological ageing of fino sherry: by-products and aroma compounds. *Am. J. Enol. Viticult.* 49:240–250.

138. Martinez-Rodriguez, A., A. V. Carrascosa, J. M. Barcenilla, M. A. Pozo-Bayon, and M. Carmen Polo. 2001. Autolytic capacity and foam analysis as additional criteria for the selection of yeast strains for sparkling wine production. *Food Microbiol.* 18:183–191.

139. Martinez-Rodriguez, A. J., M. C. Polo, and A. V. Carrascosa. 2001. Structural and ultrastructural changes in yeast cells during autolysis in a model wine system and in sparkling wines. *Int. J. Food Microbiol.* 71:45–51.

140. Martini, A., M. Ciani, and G. Scorzetti. 1996. Direct enumeration and isolation of wine yeasts from grape surfaces. *Am. J. Enol. Viticult.* 47:435–440.

141. Martorell, P., A. Querol, and M. T. Fernandez-Espinar. 2005. Rapid identification and enumeration of *Saccharomyces cerevisiae* cells in wine by real-time PCR. *Appl. Environ. Microbiol.* 71:6823–6830.

142. Matthews, A., A. Grimaldi, M. Walker, E. Bartowsky, P. Grbin, and V. Jiranek. 2004. Lactic acid bacteria as a potential source of enzymes for use in vinification. *Appl. Environ. Microbiol.* 70:5715–5731.

143. Medina, K., E. Boido, E. Dellacassa, and F. Carrau. 2005. Yeast interactions with anthocyanins during red wine fermentation. *Am. J. Enol. Viticult.* 56:104–109.

144. Mendes-Ferreira, A., A. Mendes-Faia, and C. Leao. 2002. Survey of hydrogen sulphide production by wine yeasts. *J. Food Prot.* 65:1033–1037.

145. Mendes-Ferrerira, A., A. Mendes-Faia, and C. Leao. 2004. Growth and fermentation patterns of *Saccharomyces cerevisiae* under different ammonium concentrations and its implications in winemaking industry. *J. Appl. Microbiol.* 97:540–545.

146. Millet, V., and A. Lonvaud-Funel. 2000. The viable but non-culturable state of wine microorganisms during storage. *Lett. Appl. Microbiol.* 30:136–141.

147. Mills, D., J. Eric, and L. Cocolin. 2002. Yeast diversity and persistence in *Botrytis*-affected wine fermentation. *Appl. Environ. Microbiol.* 68:4884–4893.

148. Mills, D. A., H. Rawsthorne, C. Parker, D. Tamir, and K. Makarova. 2005. Genomic analysis of *Oenococcus oeni* PSU-1 and its relevance to winemaking. *FEMS Microbiol. Rev.* 29:465–475.

149. Miranda, M., A. Ramos, M. Veiga-da-Cunha, M. C. Loureiro-Dias, and H. Santos. 1997. Biochemical basis for glucose-induced inhibition of malolactic fermentation in *Leuconostoc oenos*. *J. Bacteriol.* 179:5345–5354.

150. Mora, J., and A. Mulet. 1991. Effects of some treatments of grape juice on the population and growth of yeast species during fermentation. *Am. J. Enol. Viticult.* **42**:133–136.

151. Navarro, L., M. Zarazaga, J. Saenz, F. Ruiz-Larrea, and C. Torres. 2000. Bacteriocin production by lactic acid bacteria isolated from Rioja wines. *J. Appl. Microbiol.* **88**:44–51.

152. Nielsen, J. C., C. Prahl, and A. Lonvaud-Funel. 1996. Malolactic fermentation in wine by direct inoculation with freeze-dried *Leuconostoc oenos* cultures. *Am. J. Enol. Viticult.* **47**:42–48.

153. Nissen, P., D. Nielsen, and N. Arneborg. 2003. Viable *Saccharomyces cerevisiae* cells at high concentrations cause early growth arrest of non-*Saccharomyces* yeasts in mixed cultures by a cell-cell contact mediated mechanism. *Yeast* **20**:331–341.

154. Peinado, R. A., J. A. Mauricio, and J. Moreno. 2006. Aromatic series in sherry wines with gluconic acid subjected to different biological aging conditions by *Saccharomyces cerevisiae* var *capensis*. *Food Chem.* **94**:232–239.

155. Phister, T. G., and D. A. Mills. 2003. Real-time PCR assay for detection and enumeration of *Dekkera bruxellensis* in wine. *Appl. Environ. Microbiol.* **69**:7430–7434.

156. Pimentel, M. S., M. H. Silva, I. Cortes, and A. M. Faia. 1994. Growth and metabolism of sugar and acids of *Leuconostoc oenos* under different conditions of temperature and pH. *J. Appl. Bacteriol.* **76**:42–48.

157. Pinzani, P., L. Bonciani, M. Pazzagli, C. Orlando, S. Guerrini, and L. Granchi. 2004. Rapid detection of *Oenococcus oeni* in wine by real-time quantitative PCR. *Lett. Appl. Microbiol.* **38**:118–124.

158. Plata, C., C. Millan, J. C. Mauricio, and J. M. Ortega. 2003. Formation of ethyl acetate and isoamyl acetate by various species of wine yeasts. *Food Microbiol.* **20**:217–224.

159. Poblet-Icart, M., A. Bordons, and A. Lonvaud-Funel. 1998. Lysogeny of *Oenococcus oeni* (syn. *Leuconostoc oenos*) and study of their induced bacteriophages. *Curr. Microbiol.* **36**:365–369.

160. Pohve-Jemec, K., N. Cadez, T. Zagorc, V. Bubic, A. Zupec, and P. Raspor. 2001. Yeast population dynamics in five spontaneous fermentations of Malvasia must. *Food Microbiol.* **18**:247–259.

161. Pohve-Jemec, K., and P. Raspor. 2005. Initial *Saccharomyces cerevisiae* concentration in single or composite cultures dictates bioprocess kinetics. *Food Microbiol.* **22**:293–300.

162. Prakitchaiwattana, C. J., G. H. Fleet, and G. M. Heard. 2004. Application and evaluation of denaturing gradient gel electrophoresis to analyse the yeast ecology of wine grapes. *FEMS Yeast Res.* **4**:865–877.

163. Pramateftaki, P. V., P. Lanaridis, and M. A. Typas. 2000. Molecular identification of wine yeasts as species or strain level: a case study with strains from two vine-growing areas of Greece. *J. Appl. Microbiol.* **89**:236–248.

164. Presa-Owens, C. D. L., P. Schlich, H. D. Davies, and A. C. Noble. 1998. Effect of *Methode Champenoise* process on aroma of four *V. vinifera* varieties. *Am. J. Enol. Viticult.* **49**:289–294.

165. Pretorius, I. 2000. Tailoring wine yeast for the new millennium: novel approaches to the ancient art of winemaking. *Yeast* **16**:675–729.

166. Pretorius, I. S., and E. F. Bauer. 2002. Meeting the consumer challenge through genetically customized wine yeast strains. *Trends Biotechnol.* **20**:426–432.

167. Pretorius, I. S., and P. B. Hoj. 2005. Grape and wine biotechnology: challenges, opportunities and potential benefits. *Aust. J. Grape Wine Res.* **11**:83–108.

168. Pretorius, I. S., T. J. van der Westhuizen, and O. P. H. Augustyn. 1999. Yeast biodiversity in vineyards and wineries and its importance to the South African wine industry. A review. *S. Afr. J. Enol. Viticult.* **20**:61–76.

169. Pripis-Nicolau, L., G. de Revel, A. Bertrand, and A. Lonvaud-Funel. 2004. Methionine catabolism and production of volatile sulphur compounds by *Oenococcus oeni*. *J. Appl. Microbiol.* **96**:1176–1184.

170. Pueyo, E., A. Martinez-Rodriguez, M. Polo, G. Santa-Maria, and B. Bartolome. 2000. Release of lipids during yeast autolysis in a model wine system. *J. Agric. Food Chem.* **48**:116–122.

171. Querol, A., E. Barrio, T. Huerta, and D. Ramon. 1992. Molecular monitoring of wine fermentations conducted by active dry yeast strains. *Appl. Environ. Microbiol.* **58**:2948–2953.

172. Querol, A., M. T. Fernandez-Espinar, M. L. del Olmo, and E. Barrio. 2003. Adaptive evolution of wine yeast. *Int. J. Food Microbiol.* **86**:3–10.

173. Radio, F., M. Kishida, and H. Kawasaki. 2005. Endopolygalacturonase in *Saccharomyces* wine yeasts: effect of carbon source on enzyme production. *FEMS Yeast Res.* **5**:663–668.

174. Radler, F. 1993. Yeasts—metabolism of organic acids, p. 165–182. *In* G. H. Fleet (ed.), *Wine Microbiology and Biotechnology*. Harwood Academic Publishers, Chur, Switzerland.

175. Rankine, B. L. 1989. *Making Good Wine. A Manual of Winemaking Practices for Australia and New Zealand*. Sun Books, Melbourne, Australia.

176. Rauhut, D. 1993. Yeasts—production of sulfur compounds, p. 183–223. *In* G. H. Fleet (ed.), *Wine Microbiology and Biotechnology*. Harwood Academic Publishers, Chur, Switzerland.

177. Reader, H. P., and M. Dominguez. 2003. Fortified wines: sherry, port and madeira, p. 157–194. *In* A. G. H. Lea and J. R. Piggott (ed.), *Fermented Beverage Production*. Kluwer Academic, New York, N.Y.

178. Redzepovic, S., S. Orlic, S. Sikora, A. Majdak, and I. S. Pretorius. 2002. Identification and characterization of *Saccharomyces cerevisiae* and *Saccharomyces pardoxus* strains isolated from Croatian vineyards. *Lett. Appl. Microbiol.* **35**:305–310.

179. Reguant, C., R. Carrete, N. Ferrer, and A. Bordons. 2005. Molecular analysis of *Oenococcus oeni* population dynamics and the affect of aeration and temperature during alcoholic fermentation on malolactic fermentation. *Int. J. Food Sci. Technol.* **40**:451–459.

180. Regueiro, L. A., C. L. Costas, and J. E. L. Rubio. 1993. Influence of viticultural and enological practices on the

development of yeast populations during winemaking. *Am. J. Enol. Viticult.* **44**:405–408.

181. Ribéreau-Gayon, P., D. Dubourdieu, B. Donéche, and A. Lonvaud. 2000. *Handbook of Enology*, vol. 1. *The Microbiology of Wine and Vinifications.* John Wiley & Sons, Chichester, United Kingdom.

182. Richter, H., I. Hamann, and G. Unden. 2003. Use of the mannitol pathway in fructose fermentation of *Oenococcus oenii* due to limiting redox regeneration capacity of the ethanol pathway. *Arch. Microbiol.* **179**:227–233.

183. Rodriguez, M. E., C. A. Lopes, M. Broock, S. Valles, D. Ramon, and A. C. Caballero. 2004. Screening and typing of Patagonian wine yeasts for glycosidase activities. *J. Appl. Microbiol.* **96**:84–95.

184. Romano, P., C. Fiore, M. Paraggio, M. Caruso, and A. Capece. 2003. Function of yeast species and strains in wine flavour. *Int. J. Food Microbiol.* **86**:169–180.

185. Romano, P., and G. Suzzi. 1993. Sulfur dioxide and wine microorganisms, p. 373–394. *In* G. H. Fleet (ed.), *Wine Microbiology and Biotechnology.* Harwood Academic Publishers, Chur, Switzerland.

186. Rosi, I., G. Fia, and V. Canuti. 2003. Influence of pH values and inoculation time on the growth and malolactic activity of a strain of *Oenococci oeni. Aust. J. Grape Wine Res.* **9**:194–199.

187. Rossignol, T., L. Dulau, A. Julien, and B. Blondin. 2003. Genome-wide monitoring of wine yeast gene expression during alcoholic fermentation. *Yeast* **20**:1369–1385.

188. Sabate, J., J. Cano, A. Querol, and J. M. Guillamon. 1998. Diversity of *Saccharomyces* strains in wine fermentation: analysis for two consecutive years. *Lett. Appl. Microbiol.* **26**:452–455.

189. Santamaria, P., P. Garijo, R. Lopez, C. Tenorio, and A. R. Gutierrez. 2005. Analysis of yeast population during spontaneous alcoholic fermentation: effect of age of the cellar and practice of inoculation. *Int. J. Food Microbiol.* **103**:49–56.

190. Schuller, D., E. Valero, S. Dequin, and M. Casal. 2003. Survey of molecular methods for typing wine yeast strains. *FEMS Microbiol. Lett.* **231**:19–26.

191. Schutz, M., and J. Gafner. 1994. Dynamics of the yeast strain population during spontaneous alcoholic fermentation determined by CHEF gel electrophoresis. *Lett. Appl. Microbiol.* **19**:253–259.

192. Semon, M. J., C. G. Edwards, D. Forsyth, and C. O. Dinn. 2001. Inducing malolactic fermentation in Chardonnay musts and wines using different strains of *Oenococcus oeni. Aust. J. Grape Wine Res.* **7**:52–59.

193. Serra, R., L. Abrunhosa, Z. Kozakiewicz, and A. Venancio. 2003. Black *Aspergillus* species as ochratoxin A producers in Portuguese wine grapes. *Int. J. Food Microbiol.* **88**:63–68.

194. Serra, R., A. Braga, and A. Venancio. 2005. Mycotoxin-producing and other fungi isolated from grapes for wine production, with particular emphasis on ochratoxin A. *Res. Microbiol.* **156**:515–521.

195. Shimazu, Y., and M. Watanabe. 1981. Effects of yeast strains and environmental conditions on formation of organic acids in must during fermentation. *J. Ferment. Technol.* **59**:27–32.

196. Shimizu, K. 1993. Killer yeasts, p. 243–264. *In* G. H. Fleet (ed.), *Wine Microbiology and Biotechnology.* Harwood Academic Publishers, Chur, Switzerland.

197. Silva, S., F. Ramon Portugal, P. Andrade, M. de Fatima Texeira, and P. Strehaiano. 2003. Malic acid consumption by dry immobilized cells of *Schizosaccharomyces pombe. Am. J. Enol. Viticult.* **54**:50–55.

198. Silva-Pereira, C., J. J. Figueiredo Marques, and M. V. Ramao. 2000. Cork taint in wine: scientific knowledge and public perception—a critical review. *Crit. Rev. Microbiol.* **26**:147–162.

199. Sipiczki, M. 2002. Taxonomic and physiological diversity of *Saccharomyces bayanus*, p. 53–69. *In* M. Ciani (ed), *Biodiversity and Biotechnology of Wine Yeasts.* Research Signpost, Kerala, India.

200. Sipiczki, M., P. Romano, A. Capece, and M. Parragio. 2004. Genetic segregation of natural *Saccharomyces cerevisiae* strains derived from spontaneous fermentation of Aglianico wine. *J. Appl. Microbiol.* **96**:1169–1175.

201. Smith, J. L., P. M. Fratamico, and J. S. Novak. 2004. Quorum sensing: a primer for food microbiologists. *J. Food Prot.* **67**:1053–1070.

202. Soden, A., I. L. Francis, H. Oakey, and P. A. Henschke. 2000. Effects of co-fermentation with *Candida stellata* and *Saccharomyces cerevisiae* on the aroma and composition of Chardonnay wine. *Aust. J. Grape Wine Res.* **6**:21–30.

203. Sokollek, S. J., C. Hertel, and W. P. Hammes. 1998. Cultivation and preservation of vinegar bacteria. *J. Biotechnol.* **60**:195–206.

204. Spiropoulos, A., J. Tanaka, I. Flerianos, and L. F. Bisson. 2000. Characterisation of hydrogen sulfide fermentation in commercial and natural wine isolates of *Saccharomyces. Am. J. Enol. Viticult.* **51**:233–248.

205. Sponholz, W. R. 1993. Wine spoilage by microorganisms, p. 395–420. *In* G. H. Fleet (ed.), *Wine Microbiology and Biotechnology.* Harwood Academic Publishers, Chur, Switzerland.

206. Straus, M. L. A., N. P. Jolly, M. G. Lambrechts, and P. van Rensburg. 2001. Screening for the production of extracellular hydrolytic enzymes by non-*Saccharomyces* wine yeasts. *J. Appl. Microbiol.* **91**:182–190.

207. Stumner, B. E., I. L. Francis, T. Zanker, K. A. Lattey, and E. S. Scott. 2005. Effects of powdery mildew on the sensory properties and composition of Chardonnay juice and wine when grape sugar ripeness is standardized. *Aust. J. Grape Wine Res.* **11**:66–76.

208. Swiegers, J. H., E. J. Bartowsky, P. A. Henschke, and I. S. Pretorius. 2005. Yeast and bacterial modulation of wine aroma and flavour. *Aust. J. Grape Wine Res.* **11**:139–173.

209. Thomas, S. D. 1993. Yeasts as spoilage organisms in beverages, p. 517–562. *In* A. H. Rose and J. S. Harrison (ed.), *The Yeasts*, 2nd ed., vol. 5. *Yeast Technology.* Academic Press, London, United Kingdom.

210. Todd, B. E. N., G. H. Fleet, and P. A. Henschke. 2000. Promotion of autolysis through the interaction of killer and sensitive yeasts: potential application in sparkling wine production. *Am. J. Enol. Viticult.* **51**:65–72.

211. Tominaga, T., R. Baltenweck-Guyot, C. Peyrot des Gachon, and D. Dubourdieu. 2000. Contribution of volatile thiols to the aromas of white wines made from several *Vitis vinifera* grape varieties. *Am. J. Enol. Viticult.* **51**:178–181.

212. Trcek, J. 2005. Quick identification of acetic acid bacteria based on nucleotide sequences of the 16S-23S DNA internal transcribed spacer region and of the PQQ-dependent alcohol dehydrogenase gene. *Syst. Appl. Microbiol.* **28**:735–745.

213. Ugliano, M., A. Genovese, and L. Moio. 2003. Hydrolysis of wine aroma precursors during malolactic fermentation with four commercial starter cultures of *Oenococcus oeni*. *J. Agric. Food Chem.* **51**:5073–5078.

214. Uthurry, C. A., J. A. Suarez Lepe, J. Lombardero, and J. R. Garcia Del Hierro. 2006. Ethyl carbamate production by selected yeast and lactic acid bacteria in red wine. *Food Chem.* **94**:262–270.

215. Van der Westhuizen, T. J., O. P. H. Augustyn, W. Khan, and I. S. Pretorius. 2000. Seasonal variation of indigenous *Saccharomyces cerevisiae* strains isolated from vineyards of the Western Cape in South Africa. *S. Afr. J. Enol. Viticult.* **21**:10–16.

216. van Vuuren, H. J. J., and L. M. T. Dicks. 1993. *Leuconostoc oenos*—a review. *Am. J. Enol. Viticult.* **44**:99–112.

217. Varela, C., J. Cardenas, F. Melo, and E. Agosin. 2005. Quantitative analysis of wine yeast gene expression profiles under winemaking conditions. *Yeast* **22**:369–383.

218. Versavaud, A., P. Courcoux, C. Roulland, L. Dulau, and J. N. Hallet. 1995. Genetic diversity and geographical distribution of wild *Saccharomyces cerevisiae* strains from the wine-producing area of Charentes, France. *Appl. Environ. Microbiol.* **61**:3521–3529.

219. Volschenk, H., M. Viljoen, J. Grobler, F. Bauer, A. Lonvaud-Funel, M. Denayrolles, R. E. Subden, and H. J. J. van Vuuren. 1997. Malolactic fermentation in grape must by a genetically engineered strain of *Saccharomyces cerevisiae*. *Am. J. Enol. Viticult.* **48**:193–197.

220. Volschenk, H., M. Viljoen-Bloom, R. E. Subden, and H. J. J. van Vurren. 2001. Malo-ethanolic fermentation in grape must by recombinant strains of *Saccharomyces cerevisiae*. *Yeast* **18**:693–670.

221. Walling, E., E. Gindreau, and A. Lonvaud-Funel. 2004. A putative glucan synthase gene *dps* detected in exopolysaccharide producing *Pediococcus damnosus* and *Oenococcus oeni* strains isolated from wine and cider. *Int. J. Food Microbiol.* **98**:53–62.

222. Wang, X. D., J. C. Bohlscheid, and C. G. Edwards. 2003. Fermentative activity and production of volatile compounds by *Saccharomyces* grown in synthetic grape juice media deficient in assimilable nitrogen and/or pantothenic acid. *J. Appl. Microbiol.* **94**:349–359.

223. Wibowo, D., R. Eschenbruch, C. Davis, G. H. Fleet, and T. H. Lee. 1985. Occurrence and growth of lactic acid bacteria in wine—a review. *Am. J. Enol. Viticult.* **36**:302–313.

224. Wibowo, D., G. H. Fleet, T. H. Lee, and R. E. Eschenbruch. 1988. Factors affecting the induction of malolactic fermentation in red wines with *Leuconostoc oenos*. *J. Appl. Bacteriol.* **64**:421–428.

225. Wilkes, K. L., and M. R. Dharmadhikari. 1997. Treatment of barrel wood infected with acetic acid bacteria. *Am. J. Enol. Viticult.* **48**:516–520.

226. Yap, N. A., M. de Barros Lopes, P. Langridge, and P. A. Henschke. 2000. The incidence of killer activity of non-*Saccharomyces* yeasts towards indigenous yeast species of grape must: potential application in wine fermentation. *J. Appl. Microbiol.* **89**:381–389.

227. Yokotsuka, K., M. Yajima, and T. Matsudo. 1997. Production of bottle-fermented sparkling wine using yeast immobilised in double-layer gel beads or strands. *Am. J. Enol. Viticult.* **48**:471–481.

228. Zagorc, T., A. Maraz, N. Cadez, K. Povhe Jemec, G. Peter, M. Resnik, J. Nemanic, and P. Raspor. 2001. Indigenous wine killer yeasts and their application as a starter culture in wine fermentation. *Food Microbiol.* **18**:441–451.

229. Zavaleta, A. I., A. J. Martinez-Murcia, and F. Rodriguez-Valera. 1997. Intraspecific genetic diversity of *Oenococcus oeni* as derived from DNA fingerprinting and sequence analysis. *Appl. Environ. Microbiol.* **63**:1261–1267.

230. Zhao, J., and G. H. Fleet. 2003. Degradation of DNA during the autolysis of *Saccharomyces cerevisiae*. *J. Ind. Microbiol. Biotechnol.* **30**:175–182.

231. Zuzuarregui, A., P. Carrasco, A. Palacios, A. Julien, and M. del Olmo. 2005. Analysis of the expression of some stress induced genes in several commercial wine yeasts strains at the beginning of vinification. *J. Appl. Microbiol.* **98**:299–307.

Food Microbiology: Fundamentals and Frontiers, 3rd Ed.
Edited by M. P. Doyle and L. R. Beuchat
© 2007 ASM Press, Washington, D.C.

Todd R. Klaenhammer

Probiotics and Prebiotics

42

THE PROBIOTIC CONCEPT

Many beneficial bacteria within our food supply inhabit the gastrointestinal tract (GIT) of humans, animals, and birds. Probiotic (which means "for life") bacteria have long been considered to influence general health and well-being via their commensal association with the GIT and its normal microbiota. The concept of probiotics was first popularized at the turn of the century by the Russian Nobel laureate, Elie Metchnikoff (Fig. 42.1). He proposed that a normal, healthy, gastrointestinal microbiota in humans and animals provided resistance to "putrefactive" intestinal pathogens (4). His theorized that the intestinal flora influences the incidence and severity of enteric infections and either enhances or slows atrophy and aging processes. During that period, it was reported that infants subsisting on breast milk develop a characteristic intestinal flora of rod- and bifid-shaped bacteria, then called *Bacillus bifidus* (now known to represent *Lactobacillus* and, primarily, *Bifidobacterium* species). The fermentative bacteria constituting this flora produced lactic acid, were associated with fermented milk products, and did not produce putrefactive compounds or toxins. Metchnikoff proceeded to isolate a *Lactobacillus* culture from a fermented milk consumed by Bulgarian peasants who were renowned for living long and healthy lives. The culture produced large amounts of lactic acid and survived during intestinal implantation studies. From these and many observations made during his work on the intestinal microbiota, Metchnikoff suggested that by transforming the "wild population of the intestine into a cultured population . . . the pathological symptoms may be removed from old age, and . . . in all probability, the duration of life of man may be considerably increased" (as related by Bibel [4]). The value of consuming fermented dairy products (yogurt, kefir, sour cream) became popular in Europe, with the Pasteur Institute supporting the manufacture of a product called "Le Ferment," based on the Bulgarian *Lactobacillus*. Later, Minoru Shirota (ca. 1930 [58]) selected from human feces a *Lactobacillus* culture, *Lactobacillus casei* Shirota, that survived passage through the GIT. The fermented milk Yakult is manufactured using a pure culture of strain Shirota. The product remains a dietary staple in the Japanese and Korean societies and is now available in Europe.

THE NORMAL MICROBIOTA

Hundreds of microbial species live as commensals, largely on mucosal tissues of the nose, mouth, GIT, and vagina (Fig. 42.2). Bacterial populations found throughout the

Todd R. Klaenhammer, Dept. of Food Science, 339 Schaub Hall, Box 7624, North Carolina State University, Raleigh, NC 27695-7624.

Figure 42.1 Probiotic pioneers: Eli Mechnikoff (1845–1916), *Lactobacillus casei* strain Shirota, and *Bifidobacterium* species. From references 12 and 58, with permission.

human body are estimated at 10^{14} cells, 10-fold more cells than the 10^{13} comprising the human body itself. It is well established that the composition of the microbiota is complex, dynamic, and specific to each host and can change markedly with diet, age, and lifestyle (Fig. 42.3) (40, 64). Comparative studies with colonized and germfree gnotobiotic animals have established that the normal gastrointestinal microbiota is responsible for many important properties that affect nutrition (digestion and feed conversion), pathogen sensitivity, enterohepatic circulation (bile salt hydrolases and glucuronidases), and immunomodulation via the muscosal immunity system (39, 64). Microbial interference and colonization resistance are properties of the normal microbiota that reflect the ability of the existing flora to competitively occupy available sites and establish an immunity profile through which host and nonhost microorganisms are recognized. The collective microbiotas of humans, animals, and birds vary considerably with the GIT architecture (64). Groups of microorganisms are located at different locations throughout the GIT and include strains that are either harmful or beneficial, and those that may act positively or negatively depending on the circumstances and specific strains involved (Fig. 42.4) (17). For example, there is a clear distinction between *Escherichia coli* and the fecal coliforms comprising the normal flora and the varieties that are pathogenic and/or enterotoxigenic, such as *E. coli* O157:H7. General effects on the host depend on the many factors that direct the composition of the normal microbiota, including diet, age, exposure to different microorganisms, and the genetic or physiological condition of the host tissues themselves. The digestive tract contains four major microbial population categories (13, 65, 67) defined as:

- autochthonous microbiota: populations of microbes that are present at high levels and permanently colonize the host
- normal microbiota: microorganisms that are frequently present but can vary in number and be sporadically absent

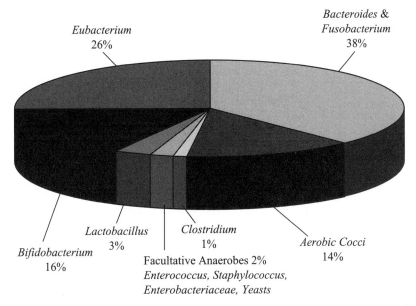

Figure 42.2 Composition of the culturable microflora in the human colon. Reprinted from reference 47 with permission.

• true pathogens: microorganisms that are periodically acquired but can persist, causing infection and disease
• allochthonous microbiota: microbes of another origin that are present temporarily (most probiotic cultures are allochthonous [67])

Molecular technologies (69) have revealed that individual humans carry their own unique collection of autochthonous strains. While the exact microorganisms vary considerably among individuals, it is now clear that each person has a persistent flora that can be recovered repeatedly over extended periods. Severe disturbances to the GIT, such as antibiotic therapy, enteric infections, or dietary stress, can temporarily disrupt the autochthonous and normal microbiota. However, these populations appear to quickly recover as the host returns to the normal state. Of the predominant colonic microorganisms, both *Lactobacillus* and *Bifidobacterium* species are considered

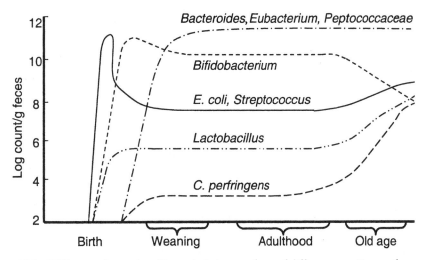

Figure 42.3 Differences in species of bacteria in human feces of different ages. From reference 40.

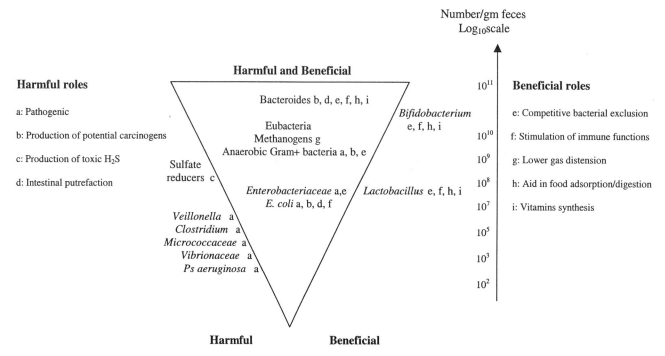

Figure 42.4 Predominant colonic microorganisms categorized into potentially harmful or beneficial groups. Adapted from reference 17.

exclusively beneficial and elicit no harmful or pathogenic responses within their normal ecological niches (Fig. 42.4). As a result, they are the two major species around which the probiotic and prebiotic concepts have evolved. From this platform, the question of probiotics and their ability to positively or negatively impact the autochthonous, normal, or pathogenic microbiota of the GIT can be discussed.

PROBIOTICS, ABIOTICS, AND PREBIOTICS

In contrast to "antibiotic," the term "probiotic" was first coined to describe a substance produced by one microorganism that stimulates the growth of another microorganism (35). Fuller (15) defined probiotics "as a live microbial feed supplement which beneficially affects the host animal by improving its intestinal balance." The definition has evolved in recent years to consider the many types of probiotic cultures (mono- and mixed-strain cultures, multiple probiotic species), applications (gastrointestinal versus topical), and mechanisms of probiotic activity (live cells, dead cells, and cellular components). Therefore, the following definitions are applied in this text:

Probiotics: live microbial cell preparations that, when applied or ingested in certain numbers, exert a

beneficial effect on health and well-being (adapted from definitions in references 14, 20, 38, and 50)

Abiotics: nonviable probiotic organisms or cellular components thereof that exert beneficial effects on health or well-being (adapted from the definition in reference 59)

In addition to the multiple impacts of probiotics and abiotics, benefits to intestinal health can be realized by feeding ingredients that selectively promote the growth of the existing beneficial microbiota. Growth factors in human breast milk, which stimulate the growth of *Bifidobacterium*, have long been recognized to change the composition of the intestinal microbiota. Bifidogenic factors are typically complex carbohydrates (e.g., galactosyllactose in breast milk) which are not metabolized by the host or microbiota residing in the upper GIT. As a consequence, these factors reach the colon, where they are preferentially metabolized by the residing bifidobacteria. Stimulation of the natural population levels of beneficial bacteria in this manner decreases colonic pH, stimulates mucosal immunity, and retards enteric infections (11). Expansion of this core concept toward the development of food ingredients that are metabolized selectively by one or more groups of beneficial intestinal bacteria has been championed by Gibson and Roberfroid (17) and has led to the following definitions:

Prebiotic: a nondigestible food ingredient that beneficially affects the host by selectively stimulating the growth and/or activity of one or a limited number of bacteria in the colon

Synbiotic: combination of a prebiotic ingredient with a probiotic culture

Prebiotics

Considering the specificity and uniqueness of the autochthonous and normal microbiota, the use of prebiotics to stimulate the growth and activity of beneficial bacteria in an individual's intestinal microbiota is a logical and effective approach to extend probiotic benefits. Food ingredients classified as prebiotics generally exhibit the following characteristics:

- limited hydrolysis and absorption in the upper GIT
- selective growth stimulation of beneficial bacteria in the colon
- potential to repress pathogens and limit virulence via a number of processes including attenuation of virulence, immunostimulation, and stimulation of a beneficial flora that promotes colonization resistance

The best-known prebiotics are fructo-oligosaccharides derived from food sources. The largest natural source is inulin recovered from the chicory root by water extraction. Inulin is composed of a glucose-$n$fructose polymer, with a degree of polymerization (DP, i.e., the number of fructose residues, n) that varies from 2 to 60 (average DP = 10). The bonds are β1-2 osidic linkages. Inulin can also be found in edible plants such as onions, asparagus, bananas, wheat, and Jerusalem artichokes. Oligofructoses that contain glucose-$n$fructose moieties, plus polymeric fructose chains with DP values from 2 to 10, can be synthesized or generated as hydrolysis products of inulin. Most prebiotic compounds are bifidogenic in nature (Table 42.1). However, there are some exceptions. Notable among these is the finding by Kaplin and Hutkins (25) that some lactobacilli residing in the small intestine also use fructo-oligosaccharides. Studies with a variety of bifidogenic factors and prebiotics are providing evidence of health-related effects occurring via the prebiotic and its stimulated microbiota in the areas of colonization resistance, reduction of colon cancer markers (enzymes and aberrant crypt foci) in animals, reduction of serum triglyceride levels, and enhanced adsorption of minerals (calcium, magnesium, iron, and zinc) (reviewed in reference 11).

The current potential for prebiotics rests largely with compounds that can be extracted from foods and investigation of their impact on the residing intestinal microbiota and health-based markers. One area of

Table 42.1 Prebiotic compounds influencing members of the intestinal microfora[a]

Prebiotic factor	Origin	Microbes stimulated	Effects
Oligosaccharides	Onion, garlic, chicory root, burdock, asparagus, Jerusalem artichoke, soybean, wheat bran	*Bifidobacterium* species	Increase in number of bifidobacteria, suppression of putrefactive bacteria, prevention of constipation and diarrhea
Fructo-oligosaccharides, inulin, oligofructose	See above	*Bifidobacterium* species, *Lactobacillus acidophilus*, *Lactobacillus casei*, *Lactobacillus plantarum*	Growth of bifidobacteria and acid promotion
Fructans	Ash-free white powder from tubers of Jerusalem artichoke	*Bifidobacterium* species	Growth promotion of *Bifidobacterium*
Human kappa casein and derived glycomacropeptide	Human milk: chymotrypsin and pepsin hydrolysate	*Bifidobacterium bifidum*	Growth promotion
Stachyose and raffinose	Soybean extract	*Bifidobacterium* species	Growth factor
Casein macropeptide	Bovine milk	*Bifidobacteritum* species	Growth promotion
Lactitol (4-O-β-D-glactopyranosyl)-D-glucitol	Synthetic sugar alcohol of lactose	*Bifidobacteritum* species	Growth promotion
Lactulose (4-O-β-D-glactopyranosyl)-D-fructose	Synthetic derivative of lactose	*Bifidobacteritum* species	Growth promotion

[a] Compiled from references 18, 51, and 63.

exciting research is the generation of designer prebiotics that may offer multiple activities in retarding undesirable microorganisms, better promoting the native desirable microbiota, or stimulating the growth or activity of synbiotic cultures (16). The following aspects are good targets for development of prebiotics:

- expand avenues for incorporation into appropriate food vehicles
- improved stimulation of bacterial floras
- antipathogenic and antiadhesive properties
- low-dosage forms
- synthesis from dietary polysaccharides
- noncariogenic properties
- good preservative and drying characteristics
- low caloric value
- controllable viscosity

Probiotic Cultures

Probiotic microorganisms designed for delivery in food or dairy products, via supplementation or fermentation, are usually members of the *Lactobacillus* or *Bifidobacterium* genus (Table 42.2). Nonpathogenic microorganisms that occupy important niches in the host gut or tissues, such as yeasts, enterococci, and members of the *Enterobacteriaceae*, are also used as human and animal probiotics.

Considerable effort has been directed to the development of probiotic cultures for animals. Antibiotics have been used extensively in animal feeds since the 1950s to improve growth and feed conversion. Concern over the development, transmission, and spread of antibiotic resistance determinants through the common practice of feeding antibiotics to animals has revived interest in developing probiotic cultures for animals and poultry (15). Over 30 years ago, Nurmi and Rantala (43) discovered that newly hatched chicks could be protected from *Salmonella* infection by exposure to a suspension of gut contents from the adult chicken. This competitive

exclusion concept has been revived by the development of a probiotic mixture composed of 29 species of nonpathogenic bacteria isolated from the chicken gut. The probiotic mixture is sprayed over the chicks so that as they preen, their intestinal tract is seeded with microbes which occupy this ecological niche to provide colonization resistance against *Salmonella*, *Campylobacter*, and *Listeria* (22, 42, 49). Ironically, this "seeding" occurs naturally in small flocks, where mothers are in close contact with hatchlings. Similarly, Zhao et al. (75) specifically targeted colonization of cattle by *E. coli* O157:H7 by creating a probiotic cocktail of 17 isolates of *E. coli* and 1 *Proteus mirabilis* strain. These were isolated from cattle and selected on the basis of their ability to inhibit *E. coli* O157:H7 in vitro. In challenge studies with *E. coli* O157:H7, the "probiotic mixture" reduced the level of pathogen carriage in most of the probiotic-treated animals. The primary health targets for animal probiotics are enhancement of animal growth, attainment of weight gains, and reduction in the number of enteric pathogens. This is of significant interest in food microbiology because control of enteric pathogens in foods, at the farm level, can greatly alter the risk of foodborne illness. Microorganisms used as probiotics for farm animals (Table 42.3) are pure cultures, such as *Lactobacillus reuteri* (7), or multiple-strain cultures which act more broadly in multiple hosts under varied conditions (15).

Bacterial probiotics are generally effective in chickens, pigs, and preruminant calves, whereas fungal probiotics have shown better results in adult ruminants. In the animal probiotic field, host specificity, age of the animal, and targeted benefit are critical in selecting cultures for specific probiotic applications.

Health Benefits of Probiotic Cultures

The important role of the intestinal microbiota in health and resistance to disease is well recognized. Clinical investigators using well-characterized cultures are slowly accumulating evidence which supports the

Table 42.2 Examples of human probiotic species and strains with research documentation[a]

Lactobacillus acidophilus (NCFM, SBT-2062, DDS1)	*Lactobacillus salivarius* (UCC118)
Lactobacillus casei (Shirota, CRL431, DN014 001, immunitas)	*Streptococcus thermophilus* (1131)
Lactobacillus johnsonii (La1, Lj1)	*Bifidobacterium lactis* (BB12)
Lactobacillus fermentum	*Bifidobacterium longum* (SBT-2928, BB536)
Saccharomyces boulardii	*Bifidobacterium breve* Yakult
Lactobacillus rhamnosus (GG, 271, GR1)	*Lactobacillus paracasei* (CLR431, F19)
Lactobacillus plantarum (299v)	*Lactobacillus delbrueckii* subsp. *bulgaricus* (2038)
Lactobacillus reuteri (SD2112)	

[a] Compiled from reference 54.

Table 42.3 Probiotic cultures used in farm animals[a]

Genus	Probiotic strains
Lactobacillus species	*L. acidophilus, L. delbrueckii* subsp. *bulgaricus, L. casei, L. rhamnosus, L. reuteri, L. plantarum, L. fermentum, L. brevis, L. helveticus*
Bifidobacterium species	*B. animalis, B. bifidum, B. pseudolongum*
Bacillus species	*B. brevis, B. thermophilus, B. subtilis, B. cereus, B. toyoi, B. natto, B. mesentericus, B. licheniformis*
Others	*Clostridium botulinum, Pediococcus pentosaceus, Lactococcus lactis, Saccharomyces cerevisiae, Aspergillus oryzae, Candida pintolopesii,* Preempt (29 species), cocktail of *E. coli* (17 strains) and *Proteus mirabilis* (1 strain)

[a] Adapted from reference 15.

primary claim that probiotics exert a beneficial influence on the intestinal ecosystem and may also offer some protection against gastrointestinal infections and inflammatory bowel disease. Proposed health outcomes and their suspected mechanisms are compiled in Table 42.4. For details of the research supporting each of the above clinical areas, the reader is referred to the Institute of Food Technologists scientific status summary (53) and Tannock's books on this topic (65, 68).

Some of these health claims remain controversial. However, well-designed studies using carefully selected and defined probiotic cultures are increasingly supporting these health outcomes (65). General examination of these clinical studies indicates three major avenues through which probiotic cultures appear to carry out beneficial activities in the GIT via probiotic and/or abiotic mechanisms (Fig. 42.5).

First, probiotic cultures appear to affect the microbial composition and the associated metabolic and enzymatic activities of resident harmful or developing pathogenic bacteria. Alterations of the microbiota are likely prerequisites to lowering the levels of procarcinogenic enzymes, limiting the production of carcinogens, and altering the concentrations of secondary bile acids.

Second, probiotic cultures, via their intimate association with the intestinal mucosa, their cellular components, and their effects on the associated microbiota,

appear to improve immunological function in the GIT and the integrity of the mucosal barrier. Modulation and enhancement of the immune system appear to be focal mechanisms that impact many of the effects shown in Table 42.2, such as inflammatory responses, anticarcinogenic activity, and resistance to colonization (23).

Third, abiotic components, which are dead cells, their cellular constituents (enzymes and cell wall components), and their growth by-products (bioactive peptides) can play important roles in eliciting probiotic-type benefits. Clear examples are the delivery of lactase by dead cells (60, 70), immune system modulation by cellular components (21, 37, 45), and antihypertensive effects by angiotensin-converting enzyme tripeptides (36).

It is difficult to draw overarching conclusions on the benefits of probiotics because the study results can vary with the strains used, the population level of probiotic cells delivered, the health marker targeted, and the number, age, and condition of the subjects evaluated. Also, little is known about the minimal dose of probiotic and the frequency of consumption required to elicit a physiological effect. Levels of cells delivered can range from 10^6 to 10^9 CFU/ml or CFU/g depending on the food carrier and method of elevating the probiotic bacterial concentration. For example, in the manufacture of Sweet Acidophilus milk, *L. acidophilus* NCFM cells are concentrated to 10^{10} CFU/ml in frozen culture concentrates and then added directly to 2% fluid milk to yield a target concentration approaching 10^7 CFU/ml. Alternatively, as is the case in some yogurt fermentations, probiotic cultures are added as inoculants and allowed to grow during the fermentation phase, reaching levels of 10^8 to 10^9 CFU/ml. The proposed target level for daily consumption of a probiotic culture is 10^8 to 10^9 CFU/day (up to 10^9 to 10^{10} CFU/day if losses are expected during stomach transit) (54). Achieving these levels of viable probiotic cultures in foods can be a challenging technological task. Substantial effort is required, on a strain-to-strain-dependent basis, to design the fermentation, concentration, and storage (dried/frozen) conditions that will enhance the viability and functional activity of probiotic cultures. Stress preconditioning can enhance the storage stability of probiotic cultures in harsh environments (21).

MOLECULAR TAXONOMY

There are a multitude of potential bacterial strains that can be used as probiotics, and not all strains elicit the same technological properties or probiotic characteristics. Proper identification of probiotic strains and cocktails of multiple probiotic cultures is the critical first step

Table 42.4 Proposed health benefits and mechanisms of probiotics[a]

Proposed health outcome	Suspected mechanisms
Promotion of lactose digestion in lactose-intolerant individuals	Lactase activity from probiotic cultures Lactase released from transient bile-sensitive lactic acid bacteria that lyse in the small intestine
Resistance to enteric pathogens	Colonization resistance Systemic immunity Shortened duration and enhanced recovery from diarrhea Adjuvant increasing secretory antibody production Alteration to unfavorable or antagonistic conditions for pathogens (lowering pH, production of bacteriocins and short-chain fatty acids)
Anticarcinogenic	Antimutagenic activity (binding of mutagens) Lowering procarcinogenic enzyme activities (nitroreductase, β-glucuronidase, azoreductase) of colonic bacteria in humans and animals Reduction in the number of aberrant crypt foci in colon of animals Stimulation of immune function Influence on secondary bile salt concentrations
Reduction of toxic impacts of small bowel bacterial overgrowth	Decrease in the production of toxic metabolites, e.g., dimethyamine, by colonic microbiota
Immune system modulation	Strengthening of nonspecific and antigen-specific defenses against infections and tumors Adjuvant effect on antigen-specific immune responses Lowering of inflammatory responses Amelioration of atopic eczema in infants Influence on cytokine production and developmental pathways for Th1/Th2 development
Treatment of blood lipids, heart disease	Alteration of bile salt hydrolase activity Reduction of cholesterol (mechanism unknown)
Anti-hypertensive effects	Bacterial peptidase action on milk proteins produces a tripeptide that acts as an angiotensin 1-converting enzyme inhibitor to lower blood pressure in hypertensive animals
Reduction of ulcerative *Helicobacter pylori* activity	Production of inhibitors by lactic acid bacteria, in fermented milks, against *H. pylori*
Treatment of hepatic encephalopathy	Competitive exclusion or inhibition of urease-producing gut flora
Reduction of urogenital infections	Adhesion to urinary and vaginal cells Competitive exclusion Production of inhibitors (biosurfactants, hydrogen peroxide)
Alteration of gut motility	Unknown

[a] Compiled from references 53 to 55 and 65.

to establishing the cause-and-effect relationships responsible for health benefits under investigation. This has been a historical problem for the probiotic field. Strains under investigation were often misidentified or incorrectly named, and health claims were inappropriately inferred on the basis of name recognition. Underlying these problems were two basic issues: (i) reliance on phenotypic criteria for taxonomic identification and (ii) the existence of multiple closely related species that could not be distinguished by phenotypic criteria or gross genetic analysis (e.g., % G+C content). Therefore, the genus and species names used in the probiotic literature, prior to 1995, should be viewed with caution.

Phylogenetic analysis is now widely recognized as the most powerful tool for taxonomic classification of bacterial cultures (76). The availability of molecular tools to

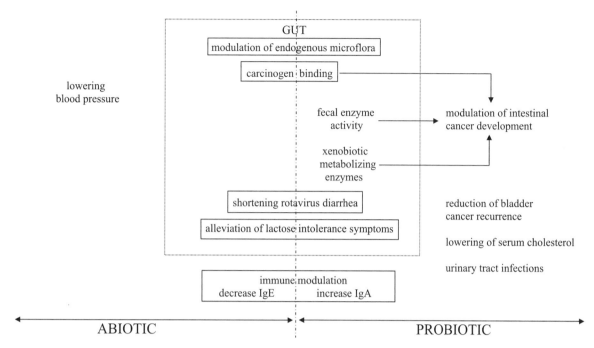

Figure 42.5 Heath benefits and suspected mechanisms of probiotics versus abiotics. IgE, IgA, immunoglobulins E and A.

properly identify probiotic species and individual strains over the last decade has helped dispel confusion over strain identity and ancestry. The accumulating sequence information on rRNAs provides a growing resource for comparative identification of probiotic cultures, both established candidates and potentially new candidates. At this juncture, phylogenetic analysis has recognized 54 species of *Lactobacillus*, 18 of which are considered to be of some interest in probiotics, and 31 species of *Bifidobacterium*, 11 of which have been detected in human feces (66). Phylogenetic analysis can be conducted to various degrees and combined with other characteristics (phenotypes) as needed to make definitive taxonomic classifications. A number of DNA sequence-based typing systems have been used to analyze conserved regions of the rRNA operon or other conserved genes in probiotic cultures (44):

- PCR amplification and sequencing of ~1,500 bp of the 16S rRNA gene
- PCR amplification and sequencing of ~450 bp of the internal transcribed spacer region between the 16S and 23S rRNA genes (Fig. 42.6)
- PCR amplification and sequencing of ~50 bp of the variable region of 16S rRNA to identify members of the *L. acidophilus* complex (33)
- PCR amplification and sequencing of alternative genes that are universally present and highly conserved, e.g., the *recA* gene of bifidobacteria (31)

The availability of efficient and cost-effective commercial sequencing facilities has made bacterial species identification by phylogenetic analysis commonplace. The ability to properly identify probiotic species provides the first giant step toward eliminating any confusion over strain identity and ancestry. The practice is expected to also uncover new probiotic species that have been hidden below the surface of traditional taxonomic descriptions. DNA sequence analysis of the rRNA operons of strains loosely classified as *Lactobacillus acidophilus* reveals six closely related species that compose the "Acidophilus" complex: *L. acidophilus*, *L. crispatus*, *L. amylovorus*, *L. gallinarum*, *L. gasseri*, and *L. johnsonii* (Fig. 42.6) (24, 29, 34). All of these are considered to have probiotic potential, but it remains to be determined what specific roles and benefits are exerted by each species exclusively or collectively by the group.

MOLECULAR APPROACHES TO THE INVESTIGATION OF PROBIOTIC BACTERIA

The primary scientific barrier to the acceptance of probiotics is the inability to determine how a probiotic influences an individual's complex GIT microbiota and then determine the impact of those specific changes on factors that could affect general health and well-being.

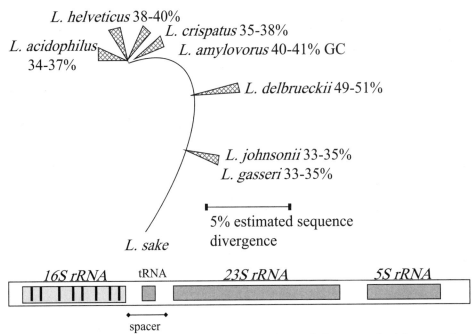

Figure 42.6 Phylogenic relationships among members of the *L. acidophilus* complex. Adapted from references 27 and 57.

This has led to scientific criticism of the probiotic field and a charge to demonstrate benefits with the rigor of pharmaceutical studies, applying new molecular technologies to investigate direct cause-and-effect relationships (65). The recent introduction of molecular technologies to microbial ecology provides a set of powerful tools for the analysis of the complex, variable, and dynamic communities of the GIT and the impact of probiotic cultures on those communities. These technologies do not rely on culture-based techniques and therefore provide a more accurate picture of the microbial composition, diversity, and fluctuation within a community (69). The power of this technology was illustrated by Suau et al. (62), who sequenced 284 PCR-generated 16S rRNA amplicons cloned from one human fecal sample and reported that only 24% of the sequences correlated with known (culturable) organisms. This means that 76% of the sequences represent previously unidentified microorganisms. Our understanding of the impact of probiotics on the intestinal microbiota will be enhanced considerably by the use of molecular methods that can more accurately reflect those changes occurring within the entire gastrointestinal microbiota.

Powerful technologies include genetics-based molecular techniques used to fingerprint specific DNA patterns that are characteristic for a single strain. Methods applied to probiotic cultures have been reviewed by O' Sullivan (44) and include:

- ribotyping
- analysis of restriction fragment length polymorphisms of genomic DNA using pulse-field gel electrophoresis (Fig. 42.7)
- randomly amplified polymorphic DNA
- 16S rDNA sequencing
- in situ PCR and fluorescent in situ hybridization allowing in situ visualization of a specific probiotic culture, or indicator flora, within mixed microbial populations

The molecular tools available for genomic fingerprinting can unequivocally link a fed probiotic culture with the strain recovered in the GIT or feces (2, 67, 71). These technologies have demonstrated that when feeding is stopped, the probiotic strain is no longer recovered after about 14 to 25 days. It is thought that probiotics (allochthonous flora) are not likely to permanently colonize the GIT. Instead, continuous delivery would be required to maintain their presence (65).

Analysis of complex microbial communities has been revolutionized by the analysis of rRNA genes and the use of in situ probes or denaturing-gradient gel electrophoresis (DGGE) to distinguish between different species that may be present. A sample from the GIT is used in a PCR amplification with universal primers designed to amplify

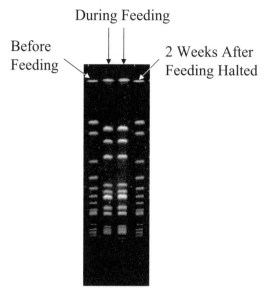

Figure 42.7 DNA fingerprint of the predominant *Lactobacillus* culture isolated from human feces before feeding with a probiotic, after feeding with *L. acidophilus* NCFM, and 2 weeks after feeding was halted. SmaI-digested DNA fragments prepared from individual *Lactobacillus* colonies were separated on a pulsed-field electrophoresis gel.

a variable region of the 16S rRNA that is flanked by sequences conserved among all bacteria (Fig. 42.6). The amplicons are identical in length but vary in their melting properties due to variations in the internal 16S rRNA sequence and % G+C content of the organism's DNA. Analysis under denaturing conditions using temperature or chemicals yields a specific banding profile, with each band representing a 16S rRNA amplicon from a different bacterium that was originally present in the GIT sample. These DNA amplicon bands can be extracted from the gel and sequenced to reveal the identity of the organisms present. The presence and intensity of the bands may also reflect the relative population levels of various species in the sample. Profiles of organisms from human fecal samples show some common bands among different individuals, reflecting universal and dominant species. There are also unique bands characteristic of an individual's microbiota at that moment in time (76). DGGE analysis of human fecal samples after feeding *L. salivarius* revealed a markedly stable flora that was not modified by probiotic treatment (69). Additional work of this type is needed and may require targeting of specific microbial populations in the GIT that are expected to be impacted by introduction of probiotic cultures. The documented ability of foods and food-bearing probiotics to modify the intestinal flora will be instrumental to proving or refuting probiotic claims on a mechanistic basis.

The developments in molecular techniques over the past decade have removed many of the key issues which previously hindered scientific progress in probiotics. Exacting methods for the identification, tracking, and analysis of probiotic cultures within complex microbial ecosystems now promise to revolutionize our understanding of the functional roles and in vivo effects associated with probiotic bacteria.

CRITERIA FOR SELECTION OF PROBIOTIC CULTURES

Different probiotic species and even strains within a species exhibit distinctive properties that can markedly affect their survival in foods, fermentation characteristics, and other probiotic properties. Strain selection becomes, therefore, a critical parameter to ensure the culture's fermentation or probiotic performance. Desirable traits for selection of functional probiotics are summarized under four major categories in Table 42.5.

These criteria are based on experience with microbial selection in vitro, propagation (viability and technological suitability), and safe use of lactic acid bacteria in foods (nonpathogenic, nontoxic, genetically stable, normal inhabitant of target species, and viability). Some criteria (i.e., survival during transit, retention in microbiota, anti-inflammatory, and immunostimulation) require in vivo analysis by human or animal studies. These studies can be lengthy and expensive, and they become more complicated as multiple probiotic strains, singly or in cocktails, are evaluated for functional properties. As a result, additional efforts are needed to develop model systems in animals, or additional in vitro systems, that better predict probiotic performance in vivo. Some recent examples include the development of a transit tolerance test (9) that evaluates probiotic survival in simulated gastric juice (pH 2) containing pepsin and sodium chloride. The majority (14 of 15 strains) of probiotic cultures evaluated lost 90% viability during the gastric portion of the test but survived well under conditions mimicking those of the small intestine. Gastric transit can be detrimental to delivery of viable probiotic cells to the small intestine. Efforts to protect cultures with dairy foods or components that act as buffering agents can improve survival (9). A more dynamic in vitro system for screening probiotic cultures has been developed by Marteau et al. (38), with a simulated GIT model composed of major components (stomach, small intestine, and colon), chemical constituents (pH, bile, pepsin, nutrients, and water), and peristaltic movement. In this model in vitro system, the survival of lactobacilli and bifidobacteria was similar to the results obtained in human in vivo survival studies.

Table 42.5 Desirable selection criteria for probiotic strains[a]

Appropriateness
  Taxonomic identification known by phylogenetic analysis and rRNA sequencing
  Origin: normal inhabitant of the species targeted and isolated from a healthy individual
  Safety: nontoxic, nonpathogenic, generally-recognized-as-safe status

Technological suitability
  Amenable to mass production and storage: adequate growth, recovery, concentration, freezing, dehydration, storage, and distribution
  Viability at high populations (preferred at 10^7–10^9 CFU/g)
  Stability of desired characteristics during culture preparation, storage, and delivery
  Desirable organoleptic qualities (or no undesirable qualities) when included in foods or fermentation processes
  Genetically stable, maintains phenotypic properties
  Genetically accessible for potential modification

Competitiveness
  Capable of survival, proliferation, and metabolic activity at the target site in vivo
  Resistant to bile
  Resistant to acid
  Able to compete with the normal microbiota, including the same or closely related species; potentially resistant to bacteriocins, acid, and other antimicrobials produced by resident microbiota
  Adherence, colonization, and retention evaluated

Performance and functionality
  Able to exert one or more clinically documented health benefits
  Antagonistic toward pathogenic or cariogenic bacteria
  Production of antimicrobial substances (bacteriocins, hydrogen peroxide, organic acids, or other inhibitory compounds)
  Immunostimulatory
  Anti-inflammatory
  Antimutagenic
  Anticarcinogenic
  Production of bioactive compounds (enzymes, vaccines, and peptides)

[a] Adapted from reference 27.

Selection criteria that address competitiveness (e.g., adherence, antimicrobials, and bacteriocins) and performance issues (e.g., immunostimulation and anticarcinogenic effects) remain more complicated and controversial because the underlying mechanisms by which probiotics exert functional effects in vivo remain to be elucidated. Therefore, it is impossible to precisely define the microbial feature or collection of characteristics that dictate some of the above selection criteria. Establishing the mechanisms that link in vitro phenotypic criteria to in vivo functionality remains one of the major scientific challenges for probiotics in the coming decade. In this regard, this field is perfectly poised to apply the recent progress in sequencing capacity and functional genomics to the investigation of these bacteria and their probiotic capabilities (26, 28). There remain a myriad of possible probiotic strains, representing a diverse set of phenotypes, which are being linked increasingly to a variety of benefits. Defining and screening genetic traits important for functional probiotic activities promises to identify superior strains and allow the construction of strain combinations that elicit unique or multiple effects.

GENOMICS

For probiotic lactic acid bacteria, the complete genomes have now been published for *B. longum* (56), *L. acidophilus* (3), *L. johnsonii* (46), *L. plantarum* (30), and *L. salivarius* (10). Partial genome information is available for *L. gasseri*, *L. casei*, and a second strain of *B. longum* from the Joint Genome Institute website (http://genome.jgi-psf.org/mic_home.html). The genomic analysis has revealed some fascinating differences and similarities within these probiotic bacteria (5, 28). First, these bacteria are rich in transporters for uptake of sugars and amino acids, emphasizing their dependence on importing, rather than synthesizing, nutrients from the GIT. In this regard, members of the *L. acidophilus* complex (*L. acidophilus*, *L. gasseri*, and *L. johnsonii*) are largely deficient in their ability to synthesize amino acids and other essential nutrients. This is consistent with the evolution of these bacteria within the small intestine, where nutrients are abundant and can be acquired via transporters. In contrast, *B. longum* encodes complete pathways for synthesis of all amino acids and nucleotides, reflecting the more nutrient-competitive environment of the colon where bifidobacteria naturally reside. In addition, *B. longum* harbors a surprisingly large number of genes predicted to encode proteins required for catabolism of complex carbohydrates and plant polymers (56). Such compounds, like fructo-oligosaccharides, are considered prebiotics which can promote the growth of bifidobacteria and some lactobacilli.

For the lactobacilli, evolution within the nutritionally rich environments of milk and the GIT has resulted in genome reduction and loss of some biosynthetic capabilities. A comparative genomic analysis highlighted the biosynthetic deficiencies of the members of the *L. acidophilus* complex compared with the expanded metabolic capacity of *L. plantarum* (5). The most metabolically diverse bacterium among the probiotic lactobacilli is *L. plantarum*. Its larger genome (3 Mbp) encodes a relatively large number of carbohydrate transport and catabolism pathways which were clustered in a 600-kb "lifestyle adaptation region." This metabolic capacity and flexibility explains why this bacterium is found naturally in the GIT and can also dominate microbial populations in fermenting plant materials.

Complete genome annotation and functional analysis support the identification of key genes that are expected to direct functions important in probiotic activity. For example, within the *L. acidophilus* genome (Color Plate 4), a series of surface-associated proteins (e.g., proteins associated with mucin binding and fibronectin binding) have been identified and correlated with the ability of the bacterium to adhere to intestinal epithelial cells in vitro (6). Across a number of *Lactobacillus* species, a number of gene regions have already been investigated and implicated in probiotic activity. Among them are genetic loci linked to bile salt hydrolase activity, acid tolerance, bacteriocin activity, oligosaccharide utilization, exopolysaccharide production, and communication with immunomodulatory cells of the host (19, 28).

Comparative genomic analysis is a powerful tool for recognizing the relationships between microbes and identifying key similarities and differences. Alignment of whole genomes has illustrated the overall synteny between the closely related members of the *L. acidophilus* complex and their substantial differences from *L. plantarum* (5, 29). Moreover, genomic content can be compared between strains of the same species, allowing for an inventory of genetic content or its deficiencies. Molenaar et al. (41) compared the genomic content of 20 strains using DNA microarrays based on the *L. plantarum* WCFS1 genome. Remarkably, considerable variation was observed between strains, notably in properties considered important for probiotic activities such as exopolysaccharide synthesis, bacteriocin production, and carbohydrate utilization.

As a result of genomics, correlations between gene, function, and mechanisms of probiotic activity are now rapidly emerging in this field. Both the potential functionality and safety of probiotic cultures can now be rationally considered based on genome content.

DELIVERY VEHICLES FOR BIOACTIVE COMPOUNDS

Lactic acid bacteria, such as lactobacilli and lactococci, have been consumed at concentrations of 10^8 to 10^9 for centuries in fermented foods, notably in dairy products.

Their generally-recognized-as-safe status, acid tolerance, and capacity for safe consumption at elevated levels provide a unique opportunity to exploit these beneficial bacteria as live vehicles for the delivery of biological molecules to the small intestine. Oral ingestion of proteins (vaccines or enzymes) results in denaturation, degradation, and loss of biological activity. In contrast, bioactive molecules can be protected by viable or nonviable cells during passage through the stomach and then released into the small intestine. Bioactive molecules targeted for delivery by probiotic lactic acid bacteria include vaccine antigens against viral and bacterial pathogens, digestive enzymes for humans, and growth-promoting enzymes in animals. Wells et al. (73) have pioneered the development and use of lactic acid bacteria as vaccine delivery vehicles. In 2000 it was demonstrated that *Lactococcus lactis* secreting the cytokine interleukin-10 could decrease the GIT inflammatory responses of mice with colitis (61). More recently, researchers have developed recombinant *Lactococcus* and *Lactobacillus* species to secrete prototype viricidal compounds and antibodies that interfere with the infectivity of human immunodeficiency virus type 1 in vitro (8, 48). There are many exciting opportunities to employ probiotic bacteria as vehicles to deliver bioactive molecules to targeted locations in the GIT. This is likely to be one of the most important areas for practical application of probiotic cultures that are derived through recombinant DNA technology (32).

SAFETY

Probiotic lactic acid bacteria have been consumed in various forms, at high concentrations, for centuries. As a consequence, they enjoy a generally-recognized-as-safe status in U.S. regulatory statutes. They are not considered "pathogenic" in any capacity, and recent reviews conclude that the infection potential of lactobacilli and bifidobacteria is low (1, 52). This conclusion is based on the widespread presence of these probiotic bacteria in foods, their commensal relationship with the host, and their extremely low correlation with infections (54). However, rare incidents of lactic acid bacteria being recovered from infected tissues are documented, but the bacteria occur only as opportunists (often in endocarditis) in immunocompromised and elderly individuals. Some intestinal isolates, primarily *L. rhamnosus*, have warranted increased surveillance due to a higher correlation rate with secondary infections. When considered in context, these incidents are extremely rare and speak more to the susceptibility of the individual rather than the infective potential of the probiotic. However, it remains an important and critical question whether any lactic acid bacterium isolated from the GIT or feces can be considered inherently safe. The history of safe use suggests that the answer is yes, but it is widely accepted that new isolates should be evaluated for pathogenicity and toxicity in animal studies before being included in any commercial probiotic product for humans. Ongoing genome-sequencing projects have not revealed any pathogenic potential for probiotic cultures. However, genome-sequencing projects have identified genes in probiotic species that annotate as antibiotic resistance genes or multidrug resistance transporters (30) (http://genome.jgi-psf.org/mic_home.html). This is not surprising, given the widespread use of antibiotics in agriculture and medicine and the apparent widespread distribution of antibiotic resistance genes among commensal and food bacteria (72). It will be important to determine whether any antibiotic resistance determinant identified in probiotic cultures is functional or clinically relevant and whether it has any reasonable probability for genetic transfer to other microbes.

CONCLUSIONS

The probiotic field has offered considerable promise since the early observations by Metchnikoff of the importance of the GIT microbiota in limiting the development of putrefactive organisms and the potential role of lactic acid bacteria in establishing a healthy microbiota. Today, microbiological and molecular methods are providing critical new insights into probiotic bacteria, the normal microbiota, and the interactions between the two that may be responsible for eliciting beneficial outcomes realized in health and well-being. Proving or refuting the probiotic concept will be a worthy challenge to the most talented of scientists, due to the complexity of the host, the dynamic interactions within microbial ecosystems, and the multifaceted impacts of our associated microorganisms on health and well-being. Some of the key issues and important challenges for food microbiologists working on probiotics in the years ahead will be:

- taxonomic identification of all probiotic cultures used in research, clinical trials, and commercial products
- definition of the active principles responsible for probiotic and abiotic activities
- correlation of genus, species, strain, phenotype, and genotype to specific probiotic functions
- determination of the impact of probiotics on the normal microbiota and associated host tissues
- stabilization of cultures and components for delivery in probiotic applications
- use of probiotic lactic acid bacteria for targeted delivery of novel bioactive compounds

- science-driven implementation of findings in commercial development
- assured safety of probiotic cultures, components, and food carriers
- identification of physiologically relevant biomarkers that can assess parameters of probiotic effectiveness (strain, dose, growth, and colonization potential)
- epidemiological and long-term studies in humans and animals to investigate the impact of probiotic cultures on diseases of longevity, e.g. colon cancer, heart disease, and inflammatory bowel diseases

This paper is from the Journal Series of the Department of Food Science, NCSU, Raleigh, NC 27695-7624. Research at NC State University on probiotic lactobacilli is supported by the N.C. Dairy Foundation, Danisco USA, Inc., and Dairy Management, Inc. Thanks are extended to Rodolphe Barrangou-Poueys for his creative input into Fig. 42.4 and 42.5 and to W. Michael Russell and Michael Callanan for their critical review of the text.

References

1. **Adams, M. R., and P. Marteau.** 1995. On the safety of lactic acid bacteria from food. *Int. J. Food Microbiol.* **27:**263–264.

2. **Alander, M., R. Satokari, R. Korpela, M. Saxelin, T. Vilpponen-Salmela, T. Mattila-Sandholm, and A. von Wright.** 1999. Persistence of colonization of human colonic mucosa by a probiotic strain, *Lactobacillus rhamnosus* GG, after oral consumption. *Appl. Environ. Microbiol.* **65:** 351–354.

3. **Altermann, E., W. M. Russell, M. A. Azcarate-Peril, R. Barrangou, B. L. Buck, O. McAuliffe, N. Souther, A. Dobsen, T. Doung, M. Callanan, S. Lick, A. Hamrick, R. Cano, and T. R. Klaenhammer.** 2005. Complete genome sequence of the probiotic lactic acid bacterium *Lactobacillus acidophilus* NCFM. *Proc. Natl. Acad. Sci. USA* **102:**3906–3912.

4. **Bibel, D. J.** 1988. Elie Metchnikoff's bacillus of long life. *ASM News* **54:**661–665.

5. **Boekhorst, J., R. J. Siezen, M. C. Zwahlen, D. Vilanova, R. D. Pridmore, A. Mercenier, M. Kleerebezem, W. M. de Vos, H. Brussow, and F. Desiere.** 2004. The complete genomes of *Lactobacillus plantarum* and *Lactobacillus johnsonii* reveal extensive differences in chromosome organization and gene content. *Microbiology* **150:**3601–3611.

6. **Buck, B. L., E. Altermann, T. Svingerud, and T. R. Klaenhammer.** 2005. Functional analysis of putative adhesion factors in *Lactobacillus acidophilus* NCFM. *Appl. Environ. Microbiol.* **71:**8344–8351.

7. **Cacas, I. A.** 1998. *Lactobacillus reuteri*: an effective probiotic for poultry and other animals. p. 475–516. *In* S. Salminen and A. von Wright (ed.), *Lactic Acid Bacteria: Microbiology and Functional Aspects*. Marcel Dekker, New York, N.Y.

8. **Chang, T. L., C. H. Chang, D. A. Simpson, Q. Xu, P. K. Martin, L. A. Lagenaur, G. K. Schoolnik, D. D. Ho, S. L. Hillier, M. Holodniy, J. A. Lewicki, and P. P. Lee.** 2003. Inhibition of HIV infectivity by a natural human isolate of *Lactobacillus jensenii* engineered to express functional

two-domain CD4. *Proc. Natl. Acad. Sci. USA* **100**:11672–11677.

9. **Charteris, W. P., P. M. Kelly, L. Morelli, and J. K. Collins.** 1998. Development and application of an in vitro methodology to determine the transit tolerance of potentially probiotic *Lactobacillus* and *Bifidobacterium* species in the upper human gastrointestinal tract. *J. Appl. Bacteriol.* **84**:759–768.

10. **Claesson, M. J., Y. Li, S. Leahy, C. Canchaya, J. P. van Pijkeren, A. M. Cerdeno-Tarraga, J. Parkhill, S. Flynn, G. C. O'Sullivan, J. K. Collins, D. Higgins, F. Shanahan, G. E. Fitzgerald, D. van Sinderen, and P. W. O'Toole.** 2006. Multireplicon genome architecture of *Lactobacillus salivarius. Proc. Natl. Acad. Sci. USA* **103**:6718–6723.

11. **Crittenden, R. G.** 1999. Prebiotics, p. 141–156. *In* G. W. Tannock (ed.), *Probiotics: a Critical Review.* Horizon Scientific Press, Norwich, United Kingdom.

12. **Danone.** 1997. Bifidobacteria. *Danone World Newsl.* no. 16. http://www.danonenewsletter.fr.

13. **Dubos, R., R. W. Schaedler, R. Costello, and P. Hoet.** 1965. Indigenous, normal and autochthonous flora of the gastrointestinal tract. *J. Exp. Med.* **122**:67–76.

14. **Fuller, R.** 1989. Probiotics in man and animals. *J. Appl. Bacteriol.* **66**:365–378.

15. **Fuller, R.** 1999. Probiotics for farm animals, p. 15–22. *In* G. W. Tannock (ed.), *Probiotics: a Critical Review.* Horizon Scientific Press, Norwich, United Kingdom.

16. **Gibson, G. R.** 1998. Dietary modulation of the human gut microflora using prebiotics. *Br. J. Nutr.* **80**(Suppl. 2): S209–S212.

17. **Gibson, G. R., and M. B. Roberfroid.** 1995. Dietary modulation of the human colonic microbiota: introducing the concept of prebiotics. *J. Nutr.* **125**:1401–1412.

18. **Gomes, A. M. P., and F. X. Malcata.** 1999. *Bifidobacterium* spp. and *Lactobacillus acidophilus*: biological, biochemical, technological and therapeutical perperties relevant for use as probiotics. *Trends Food Sci. Technol.* **10**:139–157.

19. **Grangette, G., S. Nutten, E. Palumbo, S. Morath, C. Hermann, J. Dewulf, J., B. Pot, T. Hartung, P. Hols, and A. Mercenier.** 2005. Enhanced anti-inflammatory capacity of a *Lactobacillus plantarum* mutant synthesizing modified teichoic acids. *Proc. Natl. Acad. Sci. USA* **102**:10321–10326.

20. **Guarner, F., and G. J. Schaafsma.** 1998. Probiotics. *Int. J. Food Microbiol.* **39**:237–238.

21. **Hosono, A., J. Lee, A. Ametani, M. Natsume, M. Hirayama, T. Adachi, and S. Kaminogawa.** 1997. Characterization of a water-soluble polysaccharide fraction with immunopotentiating activity from *Bifidobacterium adolescentis* M101-4. *Biosci. Biotechnol. Biochem.* **61**:312–316.

22. **Hume, M. E., D. E., Corrier, D. J. Nisbet, and J. R. DeLoach.** 1998. Early *Salmonella* challenge time and reduction in chick cecal colonization following treatment with a characterized competitive exclusion culture. *J. Food Prot.* **61**:673–676.

23. **Isolauri, E., E. Salminen, and S. Salminen.** 1998. Lactic acid bacteria and immune modulation, p. 255–268. *In* S. Salminen and A. von Wright (ed.), *Lactic Acid Bacteria: Microbiology and Functional Aspects.* Marcel Dekker, New York, N.Y.

24. **Johnson, J. L., C. F. Phelps, C. S. Cummins, J. London, and F. Gasser.** 1980. Taxonomy of the *Lactobacillus acidophilus* group. *Int. J. Syst. Bacteriol.* **30**:53–68.

25. **Kaplan, H., and R. W. Hutkins.** 2000. Fermentation of fructooligosaccharides by lactic acid bacteria and bifidobacteria. *Appl. Environ. Microbiol.* **66**:2682–2684.

26. **Klaenhammer, T. R.** 1998. Functional activities of *Lactobacillus* probiotics: genetic mandate. *Int. Dairy J.* **8**:497–506.

27. **Klaenhammer, T. R., and M. J. Kullen.** 1999. Selection and design of probiotics. *Int. J. Food Microbiol.* **50**:45–58.

28. **Klaenhammer, T. R., and W. M. Russell.** 2000. Species of the *Lactobacillus acidophilus* complex, p. 1151–1157. *In* R. K. Robinson, C. Batt, and P. D. Patel (ed.), *Encyclopedia of Food Microbiology*, vol. 2. Academic Press, Inc., San Diego, Calif.

29. **Klaenhammer, T. R., R. Barrangou, B. L. Buck, M. A. Azcarate-Peril, and E. Altermann.** 2005. Genomic features of lactic acid bacteria effecting bioprocessing and health. *FEMS Microbiol. Rev.* **29**:393–409.

30. **Kleerebezem, M., J. Boekhorst, R. van Kranenburg, D. Molenaar, O. P. Kuipers, R. Leer, R. Tarchini, S. A. Peters, H. M. Sandbrink, M. W. Fiers, W. Stiekema, R. M. Lankhorst, P. Bron, S. M. Hoffer, M. N. Groot, R. Kerkhoven, M. de Vries, B. Ursing, W. M. de Vos, and R. Siezen.** 2003. Complete genome sequence of *Lactobacillus plantarum* WCFS1 *Proc. Natl. Acad. Sci. USA* **100**:1990–1995.

31. **Kullen, M. J., L. J. Brady, and D. J. O'Sullivan.** 1997. Evaluation of using a short region of the *recA* gene for rapid and sensitive speciation of dominant bifidobacteria in the human large intestine. *FEMS Microbiol. Lett.* **154**:377–383.

32. **Kullen, M. J., and T. R. Klaenhammer.** 1999. Genetic modification of intestinal lactobacilli and bifidobacteria, p. 65–84. *In* G. W. Tannock (ed.), *Probiotics: a Critical Review.* Horizon Scientific Press, Norwich, United Kingdom.

33. **Kullen, M. J., R. B. Sanozky-Dawes, D. C. Crowell, and T. R. Klaenhammer.** 2000. Use of DNA sequence of variable regions of the 16S rRNA gene for rapid and accurate identification of bacteria in the *Lactobacillus acidophilus* complex. *J. Appl. Microbiol.* **89**:511–518.

34. **Lauer, E., C. Helming, and O. Kandler.** 1980. Heterogeneity of the species *Lactobacillus acidophilus* (Moro) Hansen and Moquot as revealed by biochemical characteristics and DNA-DNA hybridization. *Zentbl. Bakteriol. Mikrobiol. Hyg. 1 Abt. Orig.* **C1**:150–168.

35. **Lilly, D. M., and R. H. Stillwell.** 1965. Probiotics: growth promoting factors produced by microorganisms. *Science* **147**:747–748.

36. **Maeno, M., N. Tamamoto, and T. Takano.** 1996. Identification of antihypertensive peptides from casein hydrolysate produced by a proteinase from *Lactobacillus helveticus* CP790. *J. Dairy Sci.* **73**:1316–1321.

37. **Marin, M. L., J. H. Lee, J. Murtha, Z. Ustunol, and J. J. Pestka.** 1997. Differential cytokine production in clonal macrophage and T-cell lines cultured with bifidobacteria. *J. Dairy Sci.* **80**:2713–2720.

38. **Marteau, P., M. Minekus, R. Havenaar, and J. H. H. Huis in't Veld.** 1997. Survival of lactic acid bacteria in a dynamic model of the stomach and small intestine: validation and the effect of bile. *J. Dairy Sci.* **80**:1031–1037.

39. McCracken, V. J., and H. R. Gaskins. 1999. Probiotics and the immune system, p. 85–111. *In* G. W. Tannock (ed.), *Probiotics: a Critical Review.* Horizon Scientific Press, Norwich, United Kingdom.

40. Mitsuoka, T. 1992. The human gastrointestinal tract, p. 69–114. *In* B. J. B. Wood (ed.), *The Lactic Acid Bacteria*, vol. 1. *The Lactic Acid Bacteria in Health and Disease.* Elsevier Applied Science, London, United Kingdom.

41. Molenaar, D. F., F. H. Bringel, W. M. Schuren, W. M. de Vos, R. J. Siezen, and M. Kleerebezem. 2005. Exploring *Lactobacillus plantarum* genome diversity by using microarrays. *J. Bacteriol.* 187:6128–6136.

42. Nisbet, D. J., G. I. Tellez, V. K. Lowry, R. C. Anderson, G. Garcia, G. Nava, M. H. Kogut, D. E. Corrier, and L. H. Stanker. 1998. Effect of a commercial competitive exclusion culture (Preempt) on mortality and horizontal transmission of *Salmonella gallinarum* in broiler chickens. *Avian Dis.* 42:651–656.

43. Nurmi, E., and M. Rantala. 1973. New aspects of *Salmonella* infection in broiler production. *Nature* (London) 241:210–211.

44. O'Sullivan, D. J. 1999. Methods for the analysis of the intestinal microflora, p. 23–44. *In* G. W. Tannock (ed.), *Probiotics: a Critical Review.* Horizon Scientific Press, Norwich, United Kingdom.

45. Perdigon, G., M. E. Nader de Macias, S. Alvarez, G. Oliver, and A. A. Pesce de Ruiz Holgado. 1986. Effect of perorally administered lactobacilli on macrophage activation in mice. *Infect. Immun.* 53:404–410.

46. Pridmore, R. D., B. Berger, F. Desiere, D. Vilanova, C. Barretto, A. C. Pittet, M. C. Zwahlen, M. Rouvet, E. Altermann, R. Barrangou, B. Mollet, A. Mercenier, T. R. Klaenhammer, F. Arigoni, and M. A. Schell. 2004. The genome sequence of the probiotic intestinal bacterium *Lactobacillus johnsonii* NCC 533. *Proc. Natl. Acad. Sci. USA* 101:2512–2517.

47. Puhan, Z. 1999. Effect of probiotic fermented dairy products in human nutrition. *Ind. Latte* 35(3–4):3–11.

48. Pusch, O., D. Boden, S. Hannify, F. Lee, L. D. Tucker, M. R. Boyd, J. M. Wells, and B. Ramratnam. 2005. Bioengineering lactic acid bacteria to secrete the HIV-1 virucide cyanovirin. *J. Acquir. Immune Defic. Syndr.* 40:512–520.

49. Radloff, J. 1998. Spray guards chicks from infections. *Science News* 153:196.

50. Salminen, S., A. Ouwehand, Y. Benno, and Y. K. Lee. 1999. Probiotics: how should they be defined? *Trends Food Sci. Technol.* 10:107–110.

51. Salminen, S., M. Roberfroid, P. Ramos, and R. Fonden. 1998. Prebiotic substrates and lactic acid bacteria, p. 343–358. *In* S. Salminen and A. von Wright (ed.), *Lactic Acid Bacteria: Microbiology and Functional Aspects.* Marcel Dekker, New York, N.Y.

52. Salminen, S., A. Von Wright, L. Morelli, P. Marteau, D. Brassart, W. M. de Vos, R. Fonden, M. Saxelin, K. Collins, G. Mogensen, S.-E. Birkeland, and T. Mattila-Sandholm. 1998. Demonstration of safety of probiotics—a review. *Int. J. Food Microbiol.* 44:93–106.

53. Sanders, M. E. 1999. Probiotics—scientific status summary. *Food Technol.* 53:67–77.

54. Sanders, M. E., and J. Huis in't Veld. 1999. Bringing a probiotic-containing functional food to market: microbiological, product, regulatory, and labeling issues. *Antonie Leeuwenhoek* 76:293–315.

55. Sandholm, T. M., S. Blum, J. K. Collins, R. Crittenden, W. de Vos, C. Dunne, R. Fonden, G. Grenov, E. Isolauri, B. Kiely, P. Marteau, L. Morelli, A. Ouwehand, R. Reniero, M. Saarela, S. Salminen, M. Saxelin, E. Schiffrin, F. Shanahan, E. Vaughan, and A. von Wright. 1999. Probiotics: towards demonstrating efficacy. *Trends Food Sci. Technol.* 10:393–399.

56. Schell, M. A., M. Karmirantzou, B. Snel, D. Vilanova, B. Berger, G. Pessi, M. C. Zwahlen, F. Desiere, P. Bork, M. Delley, R. D. Pridmore, and F. Arigoni. 2002. The genome sequence of *Bifidobacterium longum* reflects its adaptation to the human gastrointestinal tract. *Proc. Natl. Acad. Sci. USA* 99:14422–14427.

57. Schleifer, K. H., and W. Ludwig. 1995. Phylogenetics for the genus *Lactobacillus* and related genera. *Syst. Appl. Microbiol.* 18:461–467.

58. Shortt, C. 1998. Living it up for dinner. *Chem. Ind.* 20:300–303.

59. Shortt, C. 1999. The probiotic century: historical and current perspectives. *Trends Food Sci. Technol.* 10:411–417.

60. Somkuti, G. A., M. E. Dominiecki, and D. H. Steinberg. 1998. Permeabilization of *Streptococcus thermophilus* and *Lactobacillus delbrueckii* subsp. *bulgaricus* with ethanol. *Curr. Microbiol.* 36:202–206.

61. Steidler, L., W. Hans, L. Schotte, S. Neirynck, F. Obermeier, W. Falk, W. Fiers, and E. Remaut. 2000. Treatment of murine colitis by *Lactococcus lactis* secreting IL-10. *Science* 289:1352–1355.

62. Suau, A., R. Bonnet, M. Sutren, J.-J. Godon, G. R. Gibson, M. D. Collins, and J. Dore. 1999. Direct analysis of genes encoding 16S rRNA from complex communities reveals many novel molecular species. *Appl. Environ. Microbiol.* 65:4799–4807.

63. Tamine, A. Y. 1997. Bifidobacteria—an overview of physiological, biochemical, and technological aspects, p. 9. *In* R. Harmink (ed.), *Non-digestible Oligosaccharides: Healthy Food for the Colon.* Krukkerij Modern, Bennekonm, The Netherlands.

64. Tannock, G. W. 1995. *The Normal Microflora.* Chapman & Hall, London, United Kingdom.

65. Tannock, G. W. 1999. *Probiotics: a Critical Review.* Horizon Scientific Press, Norwich, United Kingdom.

66. Tannock, G. W. 1999. A fresh look at the intestinal microflora, p. 5–14. *In* G. W. Tannock (ed.), *Probiotics: a Critical Review.* Horizon Scientific Press, Norwich, United Kingdom.

67. Tannock, G. W. 1999. Analysis of the intestinal microflora: a renaissance. *Antonie Leeuwenhoek* 76:265–278.

68. Tannock, G. W. 2005. *Probiotics & Prebiotics.* Caister Academic Press, Norwich, United Kingdom.

69. Vaughan, E. E., H. G. H. J. Heilig, E. G. Zoetendal, R. Satokari, K. Collins, A. D. L. Ackermans, and W. M. de Vos. 1999. Molecular approaches to study probiotic bacteria. *Trends Food Sci. Technol.* 10:400–404.

70. Vesa, T. H., P. Parteau, S. Zidi, F. Briet, P. Pochart, and J. C. Rambaud. 1996. Digestion and tolerance of lactose from yoghurt and different semi-solid fermented dairy products containing *Lactobacillus acidophilus* and bifidobacteria in lactose maldigesters—is bacterial lactase important? *Eur. J. Clin. Nutr.* **50:**730–733.

71. Walter, J., G. W. Tannock, A. Tilsala-Timisjarvi, S. Rodtong, D. M. Loach, K. Munro, and T. Alatossava. 2000. Detection and identification of gastrointestinal *Lactobacillus* species by using denaturing gradient gel electrophoresis and species-specific PCR primers. *Appl. Environ. Microbiol.* **66:**297–303.

72. Wang, H. H., M. Manuzon, M. Lehman, K. Wan, H. Luo, T. E. Wittum, A. Yousef, and L. O. Bakaletz. 2006. Food commensal microbes as a potentially important avenue in transmitting antibiotic resistance genes. *FEMS Microbiol. Lett.* **254:**226–231.

73. Wells, J. M., K. Robinson, L. M. Chamberlain, K. M. Schofield, and R. W. LePage. 1996. Lactic acid bacteria as vaccine delivery vehicles. *Antonie Leeuwenhoek* **70:**317–330.

74. Woese, C. R. 1987. Bacterial evolution. *Microbiol. Rev.* **51:**221–271.

75. Zhao, T., M. P. Doyle, B. G. Harmon, C. A. Brown, P. O. Mueller, and A. H. Parks. 1998. Reduction of carriage of enterohemorrhagic *Escherichia coli* O157:H7 in cattle by inoculation with probiotic bacteria. *J. Clin. Microbiol.* **36:**641–647.

76. Zoetendal, E. G., A. D. L. Akkermans, and W. M. de Vos. 1998. Temperature gradient gel electrophoresis analysis of 16S rRNA from human fecal samples reveals sand host-specific communities of active bacteria. *Appl. Environ. Microbiol.* **64:**3854–3859.

Advanced Techniques in Food Microbiology

Food Microbiology: Fundamentals and Frontiers, 3rd Ed.
Edited by M. P. Doyle and L. R. Beuchat
© 2007 ASM Press, Washington, D.C.

Peter Feng

43

Rapid Methods for the Detection of Foodborne Pathogens: Current and Next-Generation Technologies

In the second edition of *Food Microbiology: Fundamentals and Frontiers*, the chapter on rapid methods entitled "Development and Impact of Rapid Methods for the Detection of Food-Borne Pathogens" (19) gave an overview of rapid methods, including broad aspects about food testing such as historical perspectives, the traditional microbiological testing procedures, the various problems involved in food analysis, the approaches used to try to resolve these problems, and how this process was affected by the development of rapid methods. That chapter also covered the impact of biotechnology on the origin and development of rapid methods, the definitions of what is a "rapid method," the various rapid-method groups, the assay formats in each group, and a brief discussion of next-generation technologies. Also included in the previous chapter was a discussion of method validation, the impact of rapid methods on the food industry, and the advantages and limitations of rapid methods on the problems associated with food testing and how these affected rapid methods as well. Although some of these previously discussed issues continue to be of relevance and many of the same problems continue to plague food-testing methods, the information from the previous chapter is only summarized here and updated to include more recent information. Similarly, the groups of rapid methods previously classified based on technology still remain applicable and many of the various assay formats within each group continue to exist; hence, these formats are briefly presented here. However, this chapter focuses on the modifications that were made to these assay formats and emphasizes the technologies that have seen significant increases in assay development. Also, the tables that list the many assays based on technology have been updated and the tests that are no longer available commercially have been deleted from the tables. In addition, many bacterial-toxin assays have been introduced and so toxin detection methods are now listed in a separate table that includes all assay formats.

The history of biotechnology (Fig. 43.1) reveals that scientific discoveries lead to technological advances and these are usually followed by a phase of research and development to find applications for the technologies (29). In the 1990s, a phase of application development occurred where two technologies, antibody- and nucleic acid-based assays, which were not predicted by the Delphi Forecast of 1981 (27) to have an impact, ended up dominating food microbiology diagnostics and were pivotal in the emergence of "rapid methods." Future

Peter Feng, Division of Microbiological Studies, U.S. Food and Drug Administration, College Park, MD 20740-3835.

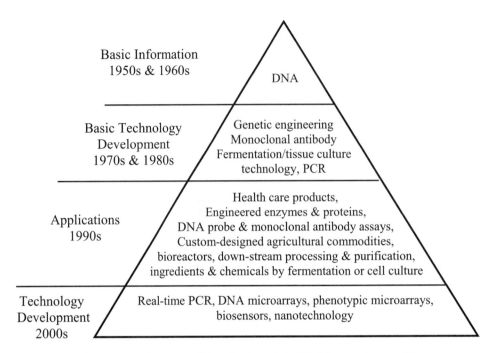

Figure 43.1 History of biotechnology. Modified from reference 29.

testing trends remain difficult to predict since technological advancements are continuing at an even faster pace and constantly altering the formats and platforms used in diagnostic assays. This is especially evident in nucleic acid-based assays, where advances in automated sequencing coupled with powerful bioinformatics computer software have enabled the sequencing and analysis of tens of thousands of nucleotide bases in a few weeks instead of the months or years that it used to take. As a result, there has been a tremendous increase in the number of commercially available DNA-based assays. Also, not only is the time from scientific discovery to development of applications getting shorter, but also these phases are becoming less well defined as applications often develop simultaneously with discovery. We are currently in another application phase, where new technologies such as real-time PCR (rtPCR), DNA microarrays, and biosensors are providing very sophisticated tools for use in diagnostics. This chapter therefore include discussions on these "next-generation technologies," which will have an impact on the way we test for pathogens and toxins in foods.

PROBLEMS IN THE MICROBIOLOGICAL TESTING OF FOODS

Microbiological analysis of foods for pathogens and their toxins continues to be a difficult and labor-intensive task, requiring multiple steps and incubations over several days

(Fig. 43.2). As discussed in the rapid-methods chapter of the second edition of this book (19), the complexity of food matrices and composition, as well as the heterogeneous distribution of low levels of pathogens that may be stress-injured by food processing, not only is problematic in testing but also causes major problems in sampling, especially in obtaining statistically reproducible samples. In addition, the presence of bacteria from the normal flora, especially at very high levels in raw foods, can interfere with assays; coupled with the required sensitivity to detect 1 cell/25 g of foods or to establish the absence of toxins in foods, this continues to present formidable challenges to all assays and technologies, including rapid methods. The most common strategy to overcome these problems has been extraction and concentration to test for toxins and enrichment to test for pathogens (Fig. 43.2). Although enrichment is effective, it is a time-consuming and labor-intensive procedure; however, virtually all rapid methods require some enrichment step(s) prior to testing (Fig. 43.2; see Tables 43.3, 43.4, and 43.6). As a result, the performance of the assay itself may be completed within minutes or hours, but analysis of a food sample requires days. The continued reliance on these time-consuming preparation steps is a limitation, but it is gradually being reduced by the development of faster and more sensitive assays. However, the complexity of food matrices remains the major obstacle to the development of effective sampling and rapid testing methods.

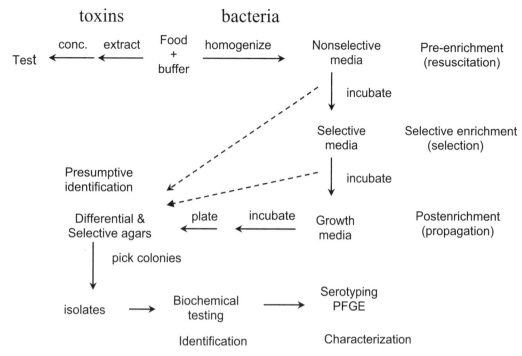

Figure 43.2 Conventional microbiological analysis of foods.

DEVELOPMENT OF RAPID METHODS

Scientists have always strived to develop faster and simpler assays, and so "rapid-methods" development has been an ongoing process. However, as illustrated by the pyramid of biotechnology (Fig. 43.1) (29), advances in basic research gave rise to biotechnology which led to the emergence of rapid diagnostic methods in the 1980s. Rapid-methods development has continued at a fast pace, as is evident when comparing the assay listings in the tables in this chapter with those in the equivalent chapter in the previous edition (19). Some of the notable changes include increases in the use of ATP-based assays, specialty media with chromogenic or fluorogenic substrates, toxin assays, DNA-based assays, and new detection targets. As expected, assay development trends have continued to parallel incidences of outbreaks of foodborne illness. Hence, the types of assays for *Salmonella* and *Listeria*, foodborne pathogens of constant concern, have become steadily more common. Similarly, the number of types of assays for *Campylobacter* has increased as this pathogen continues to be a major cause of human infections and is of great concern to the poultry industry. However, for other pathogens, like *Escherichia coli* O157:H7, which continues to be of concern and for which the number of types of assays has increased in the last few years, assay development may gradually decline as a result of recent Centers for Disease Control and Prevention (CDC) statistics (9, 10) which showed a decline in the incidence of *E. coli* O157:H7 infections compared to 5 or 6 years ago. Other changes in detection targets include the introduction of assays for emerging pathogens such as *Enterobacter sakazakii*, which has been implicated in infant meningitis associated with contaminated dried infant formulas (49). There have also been increases in the number of types of assays for *Salmonella enterica* serovar Enteritidis as well as for Shiga toxin-producing *E. coli* (STEC) other than O157:H7. The increase in the number of STEC assays may be attributed not only to increasing incidences of STEC infections worldwide but perhaps also to changes implemented by the CDC in 2001 to include all STEC strains in the list of notifiable diseases. Trends in future assay development, however, are as difficult to predict as incidences of foodborne illnesses. Furthermore, as we have transitioned into another phase of technological advancements, which are providing tools to develop more rapid and sensitive assays, this may cause a resurgence of assay development for all foodborne pathogens and toxins.

RAPID METHODS—EXISTING TECHNOLOGIES

The term "rapid" in "rapid methods" is subject to interpretation, so a nonstringent definition of "rapid methods" would encompass a broad group of assays and technologies. In the rapid-methods chapter of this book's second edition (19), rapid methods were loosely categorized into four groups or technologies: (i) miniaturized biochemical and other identification systems, (ii) modifications of assays and specialized media, (iii) antibody-based methods, and (iv) DNA-based assays. Although this grouping scheme remains applicable, some formats within the groups have changed or no longer exist. Also, the group delineations are becoming less well defined as assays are being developed using overlapping technologies. Some of these changes are addressed in the following sections.

The assay listings were updated on the basis of computer searches to ensure that the methods and their manufacturers still exist and/or PubMed searches for publications on the methods. If an assay was not found by these searches, it was not included in the listing. The AOAC test kit website (http://www.aoac.org/testkits/testedmethods.html) was also searched, and in some instances a method was listed on this website but no additional information could be obtained about the assay or the manufacturer. In these cases, those methods were included in the tables. Also, the official validation status of the methods was obtained from the AOAC website; however, the tables provide only "Official Method" designations of AOAC International but not of the Performance Tested validation status of the AOAC Research Institute. The distinction between these programs is discussed in the Validation section of this chapter. The rapid-methods market is large and competitive, so that assays and manufacturers are often sold or acquired; therefore, the tables provide only partial listings and the information may not be entirely current. Also, some assays are manufactured by one company but distributed by another, and so some listings may be duplicated.

Miniaturized Biochemical and Other Identification Assays

All identification assays require a pure culture of the unknown bacteria, which is then identified most often by its biochemical characteristics. It is a lengthy, labor-intensive, and media-consuming process. The introduction of miniaturized biochemical kits, which provide a quick biochemical profile of bacteria at a great savings in cost, labor, and time, has simplified the process of bacterial identification and continues to be important in regulatory testing of foods.

Most of the miniaturized kits (Table 43.1) are manual tests and can be labor-intensive when large numbers of isolates are to be tested. However, identification assays are amenable to automation, so many simple and convenient automated identification assays have been developed. Most of these systems continue to use biochemical

traits, but some also use fatty acid or carbon oxidation profiles. Recently, there has been an increase in the number of automated assays that identify bacteria on the basis of changes in conductance or impedance due to the metabolism of species-specific substrates; this has also been used in various plating media to tentatively identify bacteria (see below).

The databases for identification assays continue to expand and improve with use; as a result, some systems have undergone major modifications in response to the expanded databases. For instance, the Vitek assay has expanded from a 30-tests/card format to a 64-tests/card format in Vitek2 to improve the analytical capability of the system and to allow the testing of additional phenotypes. Other assays have also been improved or expanded to include additional identification targets (Table 43.1). It should be mentioned that although most of these tests work well in identifying commonly isolated bacterial species, their efficiency can vary among microorganisms (50), especially with the more difficult-to-identify bacteria such as the *Vibrio* species (53).

Modifications and Specialized Substrates and Media

The group of assays based on specialized substrates and media continues to serve as a catch-all for a wide variety of assays that provide simplicity and savings in labor, time, and materials. It includes special filtration assays such as hydrophobic grid membrane filters that provide expanded counting capacity; disposable hydratable media gel cards such as Petrifilm; and convenient, easy-to-use swab devices for surface or environmental testing. The greatest strides in assay development in this group have been in the use of chromogenic and fluorogenic substrates and the use of ATP in total bacterial analysis.

Previously, most specialty substrate media were developed for the detection of coliforms and *E. coli*. Commonly used chromogens like o-nitrophenyl-β-D-galactoside (ONPG) and 5-bromo-4-chloro-3-indolyl-β-D-galactopyranoside (X-Gal) have been used to measure β-galactosidase activity, which is indicative of coliforms. Similarly, the chromogen 5-bromo-4-chloro-3-indolyl-β-D-glucuronide (X-Gluc) and the fluorogen 4-methyl-umbelliferyl-β-D-glucuronide (MUG) have widely been used to detect the β-glucuronidase activity of *E. coli* (20). Many of these substrates or combinations thereof have also been used to develop special plating media for enterohemorrhagic *E. coli* including serotype O157:H7. More recently, chromogenic media have been introduced for the presumptive identification of other foodborne pathogens (Table 43.2). There are several media for *Salmonella* detection in which chromogens are used

Table 43.1 Partial list of miniaturized and automated identification kits for bacteria[a]

Assay name	Format	Manufacturer	Target bacteria
API[b]	Biochemical	bioMérieux	*Enterobacteriaceae, Listeria, Staphylococcus, Campylobacter,* nonfermenting bacteria, anaerobic bacteria
Rapid ID 32E	Biochemical	bioMérieux	*Enterobacteriaceae*
Micro-ID[c]	Biochemical	REMEL	*Enterobacteriaceae, Listeria*
RapID	Biochemical	REMEL	*Enterobacteriaceae*
Enterotube II	Biochemical	Becton Dickinson	*Enterobacteriaceae*
BBL Crystal	Biochemical	Becton Dickinson	*Enterobacteriaceae, Vibrionaceae,* nonfermenting bacteria, anaerobic bacteria
Microgen	Biochemical	Microgen	Gram-negative bacteria, *Listeria, Bacillus*
GNID30	Biochemical	TECRA	Gram-negative bacteria
LIDID10	Biochemical	TECRA	*Listeria*
Vitek[c]	Biochemical[b]	bioMérieux	*Enterobacteriaceae,* gram-negative bacteria
Vitek-2	Biochemical[b]	bioMérieux	Gram-positive bacteria, total bacteria
Microlog	Carbon oxidation[b]	Biolog	*Enterobacteriaceae,* gram-negative bacteria, gram-positive bacteria
MIS[c]	Fatty acid[b]	Microbial-ID	*Enterobacteriaceae, Listeria, Bacillus, Staphylococcus, Campylobacter*
Walk/Away	Biochemical[b]	Dade MicroScan	*Enterobacteriaceae, Listeria, Bacillus, Staphylococcus, Campylobacter*
Replianalyzer	Biochemical[b]	Oxoid	*Enterobacteriaceae, Listeria, Bacillus, Staphylococcus, Campylobacter*
Riboprinter	Nucleic acid[b]	Qualicon	*Salmonella, Staphylococcus, Listeria, Escherichia coli*
Cobas	Biochemical[b]	Becton Dickinson	*Enterobacteriaceae,* gram-negative bacteria
Cobas Micro-ID	Biochemical[b]	Becton Dickinson	Nonfermenting bacteria
Malthus[b]	Conductance[b]	Malthus	*Salmonella, Listeria, Campylobacter, E. coli, Pseudomonas,* coliforms
RABIT	Conductance[b]	Don Whitley	Bacteria, *Enterobacteriaceae, E. coli,* coliforms, *Salmonella, Staphylococcus aureus*
BactoScan	Conductance[b]	Foss Electric	Total bacteria
Bactometer	Impedance[b]	bioMérieux	*Salmonella,* bacteria, coliforms, *E. coli*
MicroFoss	Growth dye[b]	BioSysFoss	Bacteria, *Enterobacteriaceae,* coliforms, *E. coli, E. coli* O157
Bac Trac	Impedance[b]	Sy Lab	Bacteria, coliforms, *E. coli, Listeria, Enterobacteriaceae, Salmonella, Staphylococcus aureus*

[a] Updated from Table 38.1 in the second edition of this book (19).
[b] Automated systems.
[c] Selected systems have AOAC Official First or Final Action.

to detect esterase activity, α-galactosidase production, or combinations of X-Gal and glucuronic acid or X-Gal and propylene glycol fermentation (56). There are media for *Listeria* spp. in which the base Oxford formulation has been modified by replacing esculin with the chromogen 3,4-cyclohexenoesculetin-β-D-glucoside to detect

esculin hydrolysis (61). Chromogenic media have also been developed for differentiating phospholipase C-producing *L. monocytogenes* and *L. ivanovii* from other *Listeria* spp. (16); for differentiating *Staphylococcus aureus* from other staphylococci; and for differentiating *V. parahaemolyticus, V. vulnificus,* and *V. cholerae* from

Table 43.2 Partial list of commercially available assays and specialty substrate media for detection of foodborne bacteria[a]

Target bacteria	Assay name	Format[b]	Manufacturer
Bacteria	Redigel[c]	Media	3M
	ISOgrid[c]	HGMF	Neogen
	Enliten	ATP	Promega
	Profile-1	ATP	New Horizon
	Uni-Lite	ATP	Biotrace
	Lightning	ATP	BioControl
	Celsis Digital	ATP	Celsis
	MicroStar RMS	ATP	Millipore
	Swabcheck	ATP	S & S
	Microbial ATP	ATP	BioThema
	Ultrasnap	ATP	Hygiena
	Petrifilm[c]	Media-film	3M
	SimPlate	Media	BioControl
	BactiFlow	Cytometry	Chemunex
Campylobacter	SimPlate	Media	BioControl
Coliforms/*E. coli*	ISOgrid[c]	HGMF/MUG	Neogen
	Petrifilm[c]	Media-film	3M
	SimPlate[c]	Media	BioControl
	Redigel	Media	RCR Scientific
	ChromoCULT	Media	Merck
	ColiQuik[d]	MUG/ONPG	Hach
	ColiBlue[d]	Media	Hach
	Colilert[c,d]	MUG/ONPG	Idexx
	LST-MUG[c]	MPN media	Becton Dickinson
	ColiComplete[c]	MUG–X-Gal	BioControl
	Colitrak[c]	MPN-MUG	BioControl
	ColiGel & E*Colite[d]	MUG–X-Gal	Charm Sciences
	MicroFoss	Growth-dye	Foss
	Path-Chek	Swab/media	Microgen
	MUG disc	MUG	Remel
	CHROMagar	Media	Becton Dickinson
	Rapid'E.coli2	Media	Bio Rad
	EC X-gluc	Media	Biolife
	RIDA COUNT	Media-sheet	R-Biopharm
Enterobacter sakazakii	ESIA	Media	AES Chemunex
EHEC[e]/O157:H7	Rainbow agar	Media	Biolog
	BCMO157:H7	Media	Biosynth
	Fluorocult O157:H7	Media	Merck
	CHROMagar	Media	Becton Dickinson
	O157:H7 ID	Media	bioMérieux
	EC O157	Media	Biolife
	ISOgrid[c]	HGMF/SD39	Neogen
Listeria spp.	Swabcheck	ATP	S & S
	ALOA	Media	AES Chemunex
	CHROMagar	Media	Becton Dickinson
	PDX-LIB	Media	Paradigm Diagnostics

(Continued)

Table 43.2 *(Continued)*

Target bacteria	Assay name	Format[b]	Manufacturer
Listeria monocytogenes	BCM	Media	Biosynth
	Rapid′L.mono	Media	Bio-Rad
Salmonella	ISOgrid[c]	HGMF	Neogen
	OSRT	Medium/motility	Unipath (Oxoid)
	CHROMagar	Media	Becton Dickinson
	ABC	Media	Lab M
	Compass	Media	Biokar Diagnostics
	SM ID2	Media	bioMérieux
	Biosector	Media	Biolife
	ASAP	Media	AES Chemunex
	SMS	Media	AES Chemunex
	MUCAP	C$_8$esterase	Biolife
	XLT-4	Media	Becton Dickinson
	Path-Chek	Swab/media	Microgen
	RIDA COUNT	Media-sheet	R-Biopharm
Staphylococcus aureus	Petrifilm[c]	Media-film	3M
	CHROMagar	Media	Becton Dickinson
	RIDA COUNT	Media-sheet	R-Biopharm
Vibrio parahaemolyticus	Path-Chek	Swab/media	Microgen
	CHROMagar[f]	Media	Becton Dickinson

[a] Updated from Table 38.2 in the second edition of this book (19).
[b] Abbreviations: HGMF, hydrophobic grid membrane filter; MPN, most probable number.
[c] AOAC Official First or Final Action.
[d] Application for water analysis.
[e] EHEC, enterohemorrhagic *Escherichia coli*.
[f] Also detects *V. vulnificus* and *V. cholerae*.

other *Vibrio* spp. Many of the assays with these specialty media use proprietary technologies and substrates, and so the precise mechanisms are not well known.

Bioluminescence assays that use the enzyme luciferase to measure bacterial ATP levels are useful for a quick check of the sanitary quality of foods and food-processing surfaces. In the presence of ATP, the organic compound luciferin is oxidized to oxyluciferin; the photons released as a by-product can be quantified with a luminometer and are directly proportional to the bacterial load. ATP assays do not require culture enrichment and can be performed in minutes; most of them are packaged in self-contained, luminometer-ready disposable tubes, containing a swab and all the reagents including lysis buffer to release bacterial ATP. These assays are very easy to perform, but, because ATP is present in all living cells, assays used to sample organically complex environments such as animal carcasses require a pretreatment step to remove nonbacterial cells prior to testing. Furthermore, although ATP assays are useful for analysis of the total bacterial load, they cannot determine the presence of specific bacteria.

The AK-Phage test (Table 43.3) has made ATP assays more specific by using antibodies and immunomagnetic separation to selectively capture target microorganisms prior to testing. In addition, species-specific bacteriophages instead of chemicals are used to lyse cells to release ATP, thereby providing additional assay specificity.

Nucleic Acid-Based Assays

When the rapid-methods chapter of this book's second edition was published, approximately 12 nucleic acid-based assays were commercially available, and they were greatly outnumbered by antibody-based assays. However, advances in DNA technology have stimulated a major surge in the development of DNA-based assays, including assays for additional pathogens such as *Enterobacter sakazakii* (Table 43.3).

DNA-based assay formats consist of probes, bacteriophages, and PCR. Of these, the only commercially available phage assay, Bacterial Ice Nucleation Diagnostic, which used P22 phages cloned with the *ina* gene to detect *Salmonella*, is no longer available because the technology

Table 43.3 Partial list of commercially available nucleic acid-based assays for detection of foodborne bacterial pathogens[a]

Target bacteria	Assay name	Format[b]	Manufacturer
Bacillus cereus	B. cereus	PCR	Biotecon Diagnostics
Campylobacter	GENETRAK	Probe	Neogen
	AccuProbe	Probe	Gen-Probe
	Probelia	PCR	Sanofi Diagnostic Pasteur
	Campylobacter	PCR	Biotecon Diagnostics
Clostridium botulinum	Probelia	PCR	Sanofi Diagnostic Pasteur
Clostridium perfringens	C. perfringens	PCR	Biotecon Diagnostics
Enterobacter sakazakii	BAX	PCR	Qualicon
Escherichia coli	GENETRAK	Probe	Neogen
	Genevision	PCR	Warnex
Coliforms	coliforms	PCR	Biotecon Diagnostics
E. coli O157:H7	BAX	PCR	Qualicon
	Probelia	PCR	Probelia
	AssuranceGDS[c]	PCR	BioControl
	Genevision	PCR	Warnex
	TaqMan	PCR	Perkin Elmer
	foodProof	PCR	Roche
	GENETRAK	Probe	Neogen
	GeneQuence	Probe	Neogen
	AK-Phage	IMS/ATP	Alaska Food Diagnostics
Listeria spp.	GENETRAK[c]	Probe	Neogen
	GeneQuence	Probe	Neogen
	BAX	PCR	Qualicon
	OligoScan	Probe	Micro-Tech LLC
	Genevision	PCR	Warnex
	foodProof	PCR	Roche
	AK-Phage	IMS/ATP	Alaska Food Diagnostics
L. monocytogenes	Probelia	PCR	Sanofi Diagnostic Pasteur
	BAX	PCR	Qualicon
	AccuProbe	Probe	Gen-Probe
	Foodproof	PCR	Biotecon Diagnostics
	GENETRAK[c]	Probe	Neogen
	GeneQuence	Probe	Neogen
	Genevision	PCR	Warnex
	AK-Phage	IMS/ATP	Alaska Food Diagnostics
	LightCycler	PCR/probe	Roche
	iQ-Check	PCR	Bio-Rad
Salmonella	GENETRAK[c]	Probe	Neogen
	GeneQuence	Probe	Neogen
	BAX	PCR	Qualicon

(Continued)

Table 43.3 *(Continued)*

Target bacteria	Assay name	Format[b]	Manufacturer
	Probelia	PCR	Sanofi Diagnostic Pasteur
	Genevision	PCR	Warnex
	TaqMan	PCR	Perkin Elmer
	foodProof	PCR	Roche
	iQ-Check	PCR	Bio-Rad
	fastrAK	IMS/ATP	Alaska Food Diagnostics
Shigella	*E. coli*/Shigella	PCR	Biotecon Diagnostics
Staphylococcus aureus	GENETRAK	Probe	Neogen
	AccuProbe	Probe	Gen-Probe
	Genevision	PCR	Warnex
	S. aureus	PCR	Biotecon Diagnostics
Yersinia enterocolitica	GENETRAK	Probe	Neogen

[a] Updated from Table 38.3 in the second edition of this book (19).
[b] Probe – hybridization; PCR includes both conventional and rtPCR assays.
[c] AOAC Official First or Final Action.

was sold and eventually discontinued. However, the utility of cloned bacteriophages for use in the diagnostics of pathogens in foods continues to be explored. Bioluminescence assays using *lux* genes cloned into phages have been used to detect *Salmonella* in eggs (11). Phages have been used as specific lysis agents in ATP assays for *L. monocytogenes* and *Salmonella* (Table 43.3) and have also been coupled with immunomagnetic separation (IMS) to detect *S. enterica* serovar Enteritidis and *E. coli* O157:H7 in foods (18). In the latter assay, bacteria captured by IMS are used to propagate species-specific phage and the resulting phage titer is indicative of the presence of the specifically targeted bacteria.

DNA probe-based kits still exist, but GeneTrak assays, which pioneered the use of DNA probes for pathogen in foods testing, is now marketed by Neogen. In addition, the original manual tests have been coupled with instrumentation to produce automated versions of the same assays (GeneQuence). There are few other companies that use the probe hybridization format for diagnostics, but oligonucleotides are increasingly being used as primers or incorporated as detection probes in rtPCR or in DNA microarrays to provide an additional level of target specificity.

PCR is an extremely powerful tool that uses enzymes to exponentially amplify a single copy of DNA by 10^6-fold in a few hours. It has exquisite sensitivity in being able to differentiate single-nucleotide polymorphisms (SNP) and can be designed as multiplex tests to simultaneously detect several targets. For instance, multiplex PCR has been used to simultaneously detect 10 *Staphylococcus aureus* toxin genes (46) or 13 foodborne bacterial

pathogens (69). Multiplex assays are very useful for identification and characterization of bacterial isolates.

In the rapid-methods chapter published in the second edition of this book (19), PCR assays for pathogen testing were just becoming commercially available, so only a few kits existed. However, despite their advantages and the extensive research undertaken to develop such assays, PCR for food testing did not become as widely adopted as anticipated. This may be partly because, like other assays, PCR is adversely affected by the complex composition of foods, so that the extreme sensitivity achievable with pure cultures is greatly reduced when testing foods. One notable problem is the presence of PCR inhibitors in foods, which can affect amplification efficiency or primer binding, thereby causing false reactions (58, 68). This interference occurs more frequently in foods with high protein and fat contents but has also been observed in produce samples such as berries and bean sprouts. As a result, most PCR assays used in testing foods continue to require a brief culture enrichment step; hence, the potential of PCR to eliminate the need for culture enrichment has not been realized.

In addition to the need for culture enrichment, many PCR assays require DNA extraction procedures (41) to purify DNA templates in order to further reduce the effects of inhibitors and to improve amplification efficiency. Because of all the factors and variables that can potentially affect PCR results (39), it is critical that PCR assays used for testing foods be designed to include internal amplification controls (30), which consists of primers specific for a nontarget DNA sequence like 16S rDNA, so that it is coamplified with the target DNA. With the inclusion of such controls, a PCR result showing only the 16S rDNA amplicon would indicate that the extraction was effective in yielding amplifiable DNA and that amplification worked accordingly and no target DNA was present. Conversely, a PCR result with neither the 16S rDNA nor the target-specific amplicon could mean that the sample did not have amplifiable DNA, perhaps due to poor extraction, but could also be indicative of reaction failure, which may be caused by a number of factors (39) including the presence of inhibitors (58).

Aside from PCR assays, other amplification technologies such as nucleic acid-based sequence amplification (NASBA) are being used to develop assays for detecting bacterial pathogens (24) and enteric viruses in ready-to-eat foods (33). Unlike PCR, NASBA is an isothermal amplification system, and so it does not require temperature cycling. Also, it can be applied to RNA targets, therefore offering the advantage of detecting only viable cells.

One drawback of earlier commercial PCR assays was the use of gel-based detection of amplicons that approximated the expected size. Agarose gel electrophoresis is very useful in research but labor-intensive in diagnostics, especially when large numbers of samples are to be tested. This limitation, however, has been eliminated by the introduction of rtPCR, which provides "real-time" results during amplification. Many rtPCR assays, using various assay formats and technologies (see "Next-Generation Technologies" below), have been developed for pathogen testing (Table 43.3).

Antibody-Based Assays
The simplicity and specificity of antibody-antigen interaction have facilitated the design of many assays and formats, so that antibody assays continue to constitute the largest group of rapid methods used in food testing. The five basic antibody-based assay formats described in the previous edition of this book (19) continue to be used, but many have undergone changes. In addition, antibodies are increasingly being used as the specificity component of very sensitive and rapid biosensor assays.

Latex Agglutination Assays and Reverse Passive Latex Agglutination Assays
Latex agglutination (LA) assays use antibody-coated colored latex beads or colloidal gold particles that clump or agglutinate in the presence of a specific antigen. LA assays need pure cultures for testing and are not very sensitive (see Table 43.6), but they are fast and easy to perform and are very useful for quick identification of isolates based on serology. Many LA assays can type O157 antigen and a few can also type H7 antigen in the analysis for *E. coli* O157:

H7 in foods (Table 43.3). However, many O157 tests are not specific solely for the O157:H7 serotype and therefore can also detect O157 non-H7 strains that usually do not carry virulence factors and are nonpathogenic. Also, due to structural similarities in lipopolysaccharide, some anti-O157 antibodies can cross-react with other enteric bacteria. Since STEC serotypes other than O157:H7 are increasingly causing foodborne illness worldwide (5, 15), LA kits for typing other STEC serotypes are now available, as are kits for quick serological identification of *Salmonella*, *S. aureus*, *Campylobacter*, and other foodborne pathogens (Table 43.4).

Reverse passive (RPLA) is a variation of LA, the main difference being that in LA the antigens (bacterial cells) are insoluble so agglutinate (clump) with the antibody for a positive test whereas in RPLA the antigens (toxins) are soluble and so the formation of a diffuse lattice network or layer with the antibody is considered to indicate a positive test. RPLA is usually done on a microplate and used for testing bacterial toxins in food extracts or in pure culture suspensions. Many RPLA assays are available for testing a variety of bacterial toxins (Table 43.5).

Gel Immunodiffusion
Gel immunodiffusion is a simple antibody assay format that is used only in the *Salmonella* 1-2 test, one of the earliest rapid methods used in food testing. The L-shaped device in which the assay is performed consists of a semisolid agar-filled motility chamber that is inoculated with an antibody to *Salmonella* flagella and a horizontal chamber containing selective medium inoculated with an aliquot of sample pre-enrichment culture. *Salmonella* growth is amplified in the selective enrichment, and the bacteria migrate into the motility chamber, where they contact the diffusing antibody to form a visible line of precipitation. Atypical, nonmotile salmonellae are not detected in this assay. This assay format never became widely used, but the 1-2 test is an AOAC official method and continues to be marketed worldwide (Table 43.5).

Immunomagnetic Separation
Pathogen testing has consistently been hampered by the complex composition of foods (4, 62). The assay efficiency can be greatly improved if the target pathogen can be specifically selected for or concentrated before the sample is tested. For many years, selective enrichment culture has been used for that purpose, but it was time-consuming and not always effective, as the harsh reagents and antibiotics used were not entirely target specific and potentially caused stress-related injury of the target itself. Hence, IMS, which uses antibodies coupled to magnetic beads to specifically capture the target, became an attractive alternative. The use of antibodies not only provides specificity but also

Table 43.4 Partial list of commercially available antibody-based assays for the detection of foodborne pathogens[a]

Target bacteria	Assay name	Format[b]	Manufacturer
Campylobacter	Campyslide	LA	Becton Dickinson
	Meritec	LA	Meridian
	Microscreen	LA	Microgen
	Dryspot	LA	Oxoid
	Pathatrix	IMS	Matrix Microscience
	Assurance Gold	ELISA	BioControl
	Transia Plate	ELISA	Diffchamb AB
	Alert	ELISA	Enojen
	VIA	ELISA	TECRA
	Unique	Cap.EIA	TECRA
	CampyVIA	ELISA	Biotrace
	EIAFoss	ELISA	FOSS
	VIDAS	ELFA	bioMérieux
	Singlepath	Ab-ppt	Merck
	BV	ECL	BioVeris
	DIA/PRO	biosensor	UMEDIK
	DETEX	ElectroIA	Molecular Circuitry
Enterobacter sakazakii	Pathatrix	IMS	Matrix Microscience
Escherichia coli			
O103	DryspotSeroCheck	LA	Oxoid
	Dynabeads	IMS	Invitrogen
O111	O111-F	LA	Denka Seiken
	DryspotSeroCheck	LA	Oxoid
	E. coli O111	LA	Mast
	Dynabeads	IMS	Invitrogen
O128	DryspotSeroCheck	LA	Oxoid
O145	DryspotSeroCheck	LA	Oxoid
	Dynabeads	IMS	Invitrogen
O26	O26-F	LA	Denka Seiken
	DryspotSeroCheck	LA	Oxoid
	E. coli O26	LA	Mast
	Dynabeads	IMS	Invitrogen
O91	DryspotSeroCheck	LA	Oxoid
H7	RIM	LA	Remel
	Wellcolex	LA	Murex
O157	RIM	LA	Remel
	Dryspot	LA	Oxoid
	Prolex	LA	Pro-Lab
	Wellcolex	LA	Murex
	O157-AD	LA	Denka Seiken
	Captivate	LA	IDG/LabM Ltd
	Microscreen	LA	Microgen
	ANI *E. coli* O157	LA	ANI Biotech
	E. coli O157	LA	Mast
	Pathatrix	IMS	Matrix Microscience
	Dynabeads	IMS	Invitrogen
	VIP[c]	Ab-ppt	BioControl
	Reveal[c]	Ab-ppt	Neogen
	QUIX	Ab-ppt	Universal Health Watch

(Continued)

Table 43.4 Partial list of commercially available antibody-based assays for the detection of foodborne pathogens[a] *(Continued)*

Target bacteria	Assay name	Format[b]	Manufacturer
	ImmunoCardSTAT	Ab-ppt	Meridian Diagnostics
	PATH-STIK	Ab-ppt	Celsis
	TransiaCard	Ab-ppt	Diffchamb AB
	RapidChek	Ab-ppt	Strategic Diagnostics Int.
	Single Path	Ab-ppt	Merck
	O157 Antigen	Ab-ppt	Morningstar Diagnostics
	SMART-II	Ab-ppt	New Horizon
	O157 Coli-Strip	Ab-ppt	Coris BioConcept
	Envisio O157	IMS-Ab-ppt	Centrus International
	PetrifilmHEC	BlotEIA	3M
	EZcoli	TubeEIA	Becton Dickinson
	Assurance[c]	ELISA	BioControl
	HECO157	ELISA	3M Canada
	ECOVIA	ELISA	TECRA
	E.coli O157	ELISA	LMD
	PremierO157	ELISA	Meridian
	Transia Plate O157	ELISA	Diffchamb AB
	Ridascreen	ELISA	rBiopharma
	Colortrix	ELISA	Matrix Microscience
	EIAFoss	ELISA	FOSS
	VIDAS	ELFA	bioMérieux
	VIDAS ICE	ELFA	bioMerieux
	BV	ECL	BioVeris
	DETEX	ElectroIA	Molecular Circuitry
	DIA/PRO	Biosensor	UMEDIK
	RBD3000	Biosensor	AATI
	BioFlash	Biosensor	Innovative Biosensors
Listeria spp.	Microscreen	LA	Microgen
	ListerTest	IMS	VICAM
	Dynabeads	IMS	Invitrogen
	Pathatrix	IMS	Matrix Microscience
	Singlepath	Ab-ppt	Merck
	VIP[c]	Ab-ppt	BioControl
	Reveal	Ab-ppt	Neogen
	RapidTest/Clearview	Ab-ppt	Oxoid
	RapidChek	Ab-ppt	Strategic Diagnostics Int.
	VIA[c]	ELISA	TECRA
	Assurance[c]	ELISA	BioControl
	Transia Plate	ELISA	Diffchamb AB
	VIDAS LIS[c]	ELFA	bioMérieux
	EIAFoss	ELISA	FOSS
	LOCATE	ELISA	R-Biopharm
	Unique	Cap.EIA	TECRA
	BV	ECL	BioVeris
	RBD3000	Biosensor	AATI
	DIA/PRO	Biosensor	UMEDIK
	DETEX	ElectroIA	Molecular Circuitry
Listeria monocytogenes	VIDAS LMO	ELFA	bioMérieux
	TransiaPlate	ELISA	Diffchamb AB

(Continued)

Table 43.4 *(Continued)*

Target bacteria	Assay name	Format[b]	Manufacturer
Salmonella	Bactigen	LA	Wampole
	Spectate	LA	R-Biopharm
	Microscreen	LA	Microgen
	Wellcolex	LA	Remel
	Serobact	LA	REMEL
	RapidTest	LA	Unipath
	ANI Salmonella	LA	ANI Biotech
	Salmonella Verify	LA	VICAM
	Salmonella Seiken	LA	Denka Seiken
	Salmonella Latex	LA	Oxoid
	Slidex	LA	bioMérieux
	Pastorex	LA	Sanofi Diagnostics
	Dynabeads	IMS	Invitrogen
	Salmonella Screen	IMS	VICAM
	Pathatrix	IMS	Matrix Microscience
	VIA[c]	ELISA	TECRA
	Ultima[c]	ELISA	TECRA
	Assurance[c]	ELISA	BioControl
	Salmonella Screen	ELISA	GEM Biomedical
	LOCATE	ELISA	R-Biopharm
	Colortrix	ELISA	Matrix Microscience
	Salmonella	ELISA	Bioline/Mast Diagnostics
	Transia Plate Gold	ELISA	Diffchamb AB
	Salmotype	ELISA	Labor Diagnostics Leipzig
	EIAFoss	ELISA	FOSS
	Mastazyme	ELISA	Mast
	VIDAS SLM[c]	ELFA	bioMérieux
	VIDAS ICS[c]	ELFA	bioMerieux
	BV	ECL	BioVeris
	DIA/PRO	Biosensor	UMEDIK
	RBD3000	Biosensor	AATI
	BactiFlow	Biosensor	AES Chemunex
	Salmonella 1-2[c]	Ab-diff.	BioControl
	Unique[c]	Cap.EIA	TECRA
	PATH-STIK	Ab-ppt	Celsis
	Reveal	Ab-ppt	Neogen
	VIP[c]	Ab-ppt	BioControl
	Clearview	Ab-ppt	Unipath
	TransiaCard	Ab-ppt	Diffchamb AB
	Singlepath	Ab-ppt	Merck
	SMART-II	Ab-ppt	New Horizon
	RapidCheck	Ab-ppt	Strategic Diagnostics Int.
S. enterica serovar Enteritidis	S. enteritidis Verify	LA	VICAM
	CHEKIT	ELISA	Bommeli Diagnostics
	FlockChek	ELISA	IDEXX
	SE Screen	IMS	VICAM
Shigella	Wellcolex	LA	Remel
Staphylococcus aureus	Staphyloslide	LA	Becton Dickinson
	StaphyTect plus	LA	Oxoid

(Continued)

Table 43.4 Partial list of commercially available antibody-based assays for the detection of foodborne pathogens[a] *(Continued)*

Target bacteria	Assay name	Format[b]	Manufacturer
	Microscreen	LA	Microgen
	ANI S. aureus	LA	ANI Biotech
	Slidex	LA	bioMérieux
	Bactident	LA	Merck
	BACTi Staph	LA	Remel
	Staphaurex	LA	Remel
	STAVIA	ELISA	TECRA
Vibrio cholerae	V. cholerae 01-AD	LA	Denka Seiken
Yersinia enterocolitica	ANI Y.enterocolitica	LA	ANI Biotech
	Microscreen	LA	Microgen

[a] Updated from Table 38.4 in the second edition of this book (19).
[b] ELFA, enzyme-linked fluorescence assay; Ab-ppt, immunoprecipitation; ECL, electrochemiluminescence; ElectroIA, electroimmunoassay; blot EIA, blot enzyme immunoassay; Cap.EIA, capture EIA; Ab-diff., antibody diffusion.
[c] AOAC Official First or Final Action.

minimizes cell damage since no harsh reagents are used. IMS is also much faster than selective enrichment, since it can capture target cells in minutes instead of requiring overnight incubation. Also, IMS is very versatile and can be coupled to almost any type of assay, including biosensors and rtPCR. However, it is important to recognize that although IMS reduces the cell numbers of nontarget bacteria from enrichment cultures, it does not yield a pure culture of the target microbe. Hence, it is not a stand-alone test and needs to be coupled to other tests for more definitive detection and identification results.

In most instances, IMS improves selective isolation of target microbes from mixed microbial populations or improves the detection efficiency of assays. Hence, many IMS kits, including automated systems (63), have been developed for selective capture of pathogens in foods (Table 43.5). However, IMS is equally susceptible to inhibitors or interference by food components, and so its effectiveness can vary with the foods being tested (62). In one study, IMS bound 50% of a pure culture of *E. coli* O157:H7, but the binding efficiency was greatly reduced in ground-beef suspensions (21). Others have found IMS to be less efficient than selective enrichment in the recovery of *Salmonella* from broiler carcass rinses (59). There are other examples of limitations in the application of IMS binding in foods (4).

Enzyme-Linked Immunosorbent Assay

The enzyme-linked immunosorbent assay (ELISA) was used in one of the earliest commercially available rapid methods for testing of foods for pathogens and was quickly adopted by other manufacturers for their assays. Most ELISAs are "sandwich" assays, in which an antibody-coated solid matrix is used to capture the antigen (bacteria or toxin) from enrichment culture and a second antibody, conjugated with an enzyme (often alkaline phosphatase or horseradish peroxidase), is added to form an antibody-antigen-conjugate "sandwich." A substrate is then added, which is cleaved by the enzyme to produce a colored product that can be read visually or with a spectrophotometer and recorded. Most ELISAs originally used breakable 96-well microtiter plates, which allowed the flexibility of testing few to large numbers of samples but was labor-intensive and required several manipulations to add reagents and to thoroughly wash wells in order to avoid false reactions. To eliminate this inconvenience, many ELISAs have been coupled to autoanalyzers and involve disposable devices containing the wash buffers and reagents; they can be completed in less that 1 h postenrichment. There are many variations of the ELISA format, and some include signal-enhancing modifications such as the enzyme-linked coagulation assay and the enzyme-linked fluorescence assay, which uses a fluorogenic substrate to enhance sensitivity. Others couple ELISA with IMS and use quantum dots as fluorescence markers (64) or use chemiluminescence (CL) to enhance the signal detected. An enzyme-linked IMS-CL assay was able to detect 10^3 cells of *E. coli* O157:H7/ml of buffer in 75 min or 10 cells seeded in ground beef in 7 h, including a 5.5-h enrichment step (25). The principle of ELISA can be adapted to many solid matrices, including dipsticks, membranes, and pipette tips. The

Table 43.5 Partial listing of rapid methods for the detection of bacterial toxins[a]

Bacteria	Toxin	Assay	Format[b]	Company
Bacillus cereus	Diarrheal enterotoxin	BDEVIA	ELISA	TECRA
		BCET	RPLA	Denka Seiken
Clostridium botulinum	A, B, E, F	ELCA	ELISA	Elcatech
		Bot Toxin	ELISA	METAbiologics
		SMART-II	Ab-ppt	New Horizon
		BTA	Ab-ppt	Alexeter Tech
		BV	ECL	BioVeris
Clostridium perfringens	Enterotoxin	PET	RPLA	Denka Seiken
Escherichia coli	Shiga toxin	Verotest	ELISA	Antex Biologics
		PremierEHEC	ELISA	Meridian Diagnostics
		VTEC	RPLA	DenkaSeiken
		VTECScreen	LA	DenkaSeiken
		TaqMan	rtPCR	Perkin Elmer
		RidaScreen	ELISA	rBiopharma
		Duopath	Ab-ppt	Merck KgaA
		ProSpecT	ELISA	Remel
		Transia Plate	ELISA	Diffchamb AB
		AssuranceGDS[c]	PCR	BioControl
	Heat-labile toxin	VET	RPLA	DenkaSeiken
	Heat-stable toxin	COLIST	ELISA	DenkaSeiken
		E.coli ST	ELISA	Oxoid
Staphylococcus aureus	Enterotoxin	SET	RPLA	Denka Seiken
		VIA[c]	ELISA	TECRA
		VIAID	ELISA	TECRA
		Unique	Cap.EIA	TECRA
		TransiaTube	ELISA	Diffchamb AB
		TransiaPlate	ELISA	Diffchamb AB
		Ridascreen	ELISA	rBiopharma
		VIDAS SET	ELFA	bioMérieux
	SEA, SEB	BV	ECL	BioVeris
	SEB	SMART	Ab-ppt	New Horizon
		BTA	Ab-ppt	Alexeter Tech
Vibrio cholerae	CT	VET	RPLA	Denka Seiken
Vibrio parahaemolyticus	Hemolysin	KAP	RPLA	Denka Seiken

[a] Table modified from reference 19.
[b] Ab-ppt, immunoprecipitation; ELFA, enzyme-linked fluorescence assay; Cap.EIA, capture EIA.
[c] AOAC Official First or Final Action.

Unique test, which can be done manually or with automation (Unique Plus), uses an antibody-coated paddle to selectively capture the target bacteria; this is followed by ELISA detection.

The simplicity and convenience of automation has recently caused the demise of a few manual ELISA kits. *Salmonella*-Tek, which not only pioneered the use of rapid methods for pathogen testing in foods but, along with *Listeria*-Tek, both being Official AOAC methods, has been discontinued and is no longer available. Many ELISA methods available for testing for pathogens and toxins are listed in Tables 43.4 and 43.5.

Immunoprecipitation

Also known as immunochromatography or lateral flow, the immunoprecipitation assay has a format based on technology developed for use in home pregnancy tests. It is also an antibody "sandwich" test, but instead of an enzyme conjugate, the detection antibody is labeled with colored latex beads or colloidal gold. Conceptually, these disposable plastic devices or dipsticks consist of chambers above a series of absorbent pads that are coated in sequence with (i) latex-bound target-specific antibody (inoculation chamber), (ii) a band of target-specific capture antibody (test chamber), and (iii) another band of capture antibody that is specific not to the target but to the latex-bound antibody (control chamber). A 0.1-ml aliquot of enrichment broth is added to the inoculation chamber, and if the target pathogen is present, it binds the latex antibody and the complex is wicked across the device by the absorbent pads. At the test chamber, the antigen-latex antibody "complex" is captured by the band of target-specific antibody to form a visible band of precipitation. The excess "complex" is then wicked to the control chamber, where it is bound by the capture antibody specific to the latex-antibody to form another line of precipitation. The control chamber ensures that the sample has wicked across the entire device so that only the appearance of bands in both the test and control chamber constitutes a valid positive test. These assays are extremely simple, require no washing or manipulations, and can be done in 10 min after enrichment. This format is used in many antibody-based assays for detecting pathogens (Table 43.4) and has also been adapted for testing Shiga toxins in bacterial cultures (55) (Table 43.5). The sensitivities of these assays range from 10^7 to 10^8 cells/ml or 3 to 20 ng, depending on the toxin type (67) (Table 43.6). Another modification uses immunomagnetic beads instead of antibody-latex, and the specific bacteria or toxin captured in the antigen-antibody-magnetic bead complex is determined with a sensitive magnetic scanner rather than visually (Envisio test).

NEXT-GENERATION TECHNOLOGIES

Recently, experts picked seven technologies credited with having the greatest impact on life science research. These included (i) the automated DNA sequencer, (ii) the Basic Local Alignment Search Tool (BLAST) software, (iii) the DNA microarray, (iv) the yeast two-hybrid assay,

Table 43.6 Approximate detection sensitivity of various formats

Type of assay	No. of bacteria (CFU/g)	Amt of toxin (ng/ml)
ATP	10^4	NA[a]
LA	10^7	NA
RPLA	NA	0.5–4.0
ELISA	10^4–10^5	0.01–1
Immunomagnetic separation	$<10^3$	NA
Immunodiffusion	10^5–10^6	NA
Immunoprecipitation for Stx	10^4–10^5 10^7–10^8	NA 6–20
DNA probe	10^4–10^6	NA
PCR	10^1–10^2	NA
Biosensor	10^1–10^2	NA

[a] NA, information not available or applicable.

(v) matrix-assisted laser desorption ionization-time of flight (MALDI-TOF) mass spectrometry, (vi) microfluidics, and (vii) the optical trap or tweezers (23). Of these, the automated sequencer and the BLAST algorithm have probably had the greatest impact on nucleic acid-based assays, since they enabled the genomic sequencing of virtually all major foodborne pathogens, including virulence factors (1), as well as allowing fast and efficient analysis and comparison of the large amounts sequence data. It was estimated that at the end of 2004, the GenBank DNA database had over 40 million sequences, composed of 44.5 billion nucleotides, most of which can be searched within minutes by BLAST (28). These tools enabled the design of highly specific primers and probes for use in diagnostic assays such as rtPCR and, coupled with advances in microfluidic technology, were also keys to the development of DNA microarrays and biosensors.

rtPCR

The term "real-time" may be misinterpreted to infer the amplification process, but, in actuality, it is the results that are obtained in real time. There are many rtPCR platforms using various heating mechanisms, illumination sources, and detectors (40), and all provide very fast temperature ramping, so the amplification cycle times are much shorter than for conventional PCR. These assays also use a variety of detection formats that provide "real-time" results, thereby eliminating the need for gel-based detection. The PCR assays listed in Table 43.3 include many rtPCR assays.

The simplest and most economical rtPCR assays use SYBR Green I, which is a fluorescent dye that specifically binds double-stranded DNA. As the amount of double-stranded amplicon increases with increasing cycles, so does the intensity of the SYBR Green I fluorescence, which is monitored in real time during amplification. Since SYBR Green I does not exhibit sequence specificity, the assay specificity is dependent solely on primer sequences; however, the T_m, which is characteristic of the length, sequences, and G+C content of an amplicon, can be obtained by melting-curve analyses postamplification to give some assurance of assay specificity. One SYBR Green-based test was used to detect 17 species of food- and waterborne pathogens in stools, and the organisms were identified by T_m. This assay had a detection limit of 10^5 cells/g but needed overnight enrichment culture to detect lower levels (22). Another SYBR Green I assay used IMS to capture *Salmonella* in milk, ground beef, and alfalfa sprouts prior to PCR and detected as few as 1.5 cells/25 g in 13 h, including enrichment culture (45).

One limitation of the SYBR Green I assay is the lack of sequence specificity, which means that the identification of amplicons based solely on primers and T_m can be inconsistent due to mispriming and genetic variations. These problems can be overcome by the use of specific probes in conjunction with rtPCR.

The 5′ exonuclease or TaqMan format uses a sequence-specific probe that is colabeled with a fluorescent reporter and adjacently with a quencher dye. The probe binds to a specific sequence downstream from the amplification primer, but it is cleaved during amplification by the 5′ exonuclease activity of *Taq* polymerase, separating the reporter from quencher. Therefore, as the number of amplicons increases, so does the intensity of fluorescence, which is monitored in real time. This format has been used to detect staphylococcal enterotoxins A, B, C1, and D as well as a few antibiotic resistance genes, including that for methicillin (36). It was also used to detect stx_2 and to genetically serotype for the O157 antigen in O157:H7 cells seeded in feces, milk, apple juice, and ground beef, with detection limits of 10^4 to 10^9 cells/g (31). A TaqMan assay was also coupled with IMS to detect 10^2 cells of O157:H7/ml of culture or 10^4 cells/g of ground beef (21). Some TaqMan assays have also been modified with the use of minor-groove binder probes, which stabilize duplexes and improve the detection of SNP. One such assay (34) was developed to detect both stx_1 and stx_2 genes as well as the +93 *uid*A SNP that is highly conserved in *E. coli* O157:H7 (48).

Molecular beacon is another rtPCR assay that uses sequence-specific probes. The probes consist of a hairpin or stem-loop oligonucleotide, which is coupled with a fluorophore and a quencher at the 3′ and 5′ ends,

respectively. Hybridization of the probe to the specific target separates the reporter from the quencher, resulting in the emission of fluorescence. Like other rtPCR assays, the signal intensity, which is monitored in real time, increases with amplification (6). This format was used to develop an assay for *E. coli* O157:H7 in milk (44) and also is used in commercially available PCR assays for other pathogens.

Fluorescence resonance energy transfer (FRET) is another rtPCR probe assay. FRET uses two sequence-specific probes, in which one probe is 3′ labeled with a FRET donor dye that becomes excited by an energy source and, on deactivation, transfers its excited energy to a compatible acceptor which is 5′ labeled in the adjacent probe. The distance separating the donor and acceptor molecules is critical for efficient energy transfer and is in the range of 1 to 3 nm. Hence, the two specific probes must be designed to be about 1 to 5 bp apart (13). A FRET assay was able to detect *L. monocytogenes* seeded at 10^3 to 10^4 cells/25 ml of nonfat dry milk (37), and the format is also used in commercially available rtPCR assays for *Salmonella* and *Listeria*.

The rtPCR assays that use specific probes are extremely sensitive and well suited for the detection of SNP, but the probes can be costly because they must be custom synthesized with the specific reporter and/or quenchers. Melting-point analysis can also be performed with these assays as an added confirmation of the identity of the amplicons, except for the TaqMan assay, where the reporter dye has been dissociated from the probe by exonuclease activity. Many multiplex rtPCR assays are able to detect several gene targets; however, in designing multiplex tests, the fluorophores used must be carefully selected to ensure adequate separation in emission spectra. Unfortunately, there are substantial wavelength overlaps among the commonly used fluorophores, so that it is only practical to distinguish at most four targets in a single multiplex reaction (17).

Immunosensors or Biosensors

A biosensor consists of a biological component, such as an antibody, coupled to a physicochemical transducer, which can convert minute signal fluctuations resulting from biological interactions into measurable, digital electronic readings. There are many biosensor formats, and several have been explored for pathogen testing in foods.

Fiber-optic biosensors using an antibody sandwich format on fiber waveguides detect the capture of antigens via fluorescently tagged conjugates. One optic biosensor in 24 h, including enrichment, detected 10^3 cells of *L. monocytogenes*/ml of culture medium, 10^3 to

10^7 cells/ml in mixed cultures, and 10 to 10^3 cells/g of hot dogs or bologna (26). RAPTOR, a commercially available fiber-optic sensor, was used to develop an assay for *S. enterica* serovar Typhimurium and was able to detect 50 cells/ml of spent sprout irrigation water (38). Surface plasmon resonance (SPR) is another technology that is being explored for foodborne-pathogen testing. SPR assays use antibodies immobilized on a gold electrode surface and measure minuscule changes in resonance frequency caused by antigen-antibody binding. One piezoelectric SPR assay detected 10^3 to 10^8 cells of *E. coli* O157:H7/ml within 30 to 50 min (65), whereas another was able to detect 10^2 to 10^9 cells of *S. enterica* serovar Typhimurium/ml (51). BIAcore 3000, a commercially available SPR sensor, was used to develop an assay that detected 10^5 cells of *L. monocytogenes*/ml in <30 min (43). Multiplex SPR tests have also been developed to simultaneously detect *E. coli* O157:H7, *S. enterica* serovar Typhimurium, *Legionella pneumophila*, and *Yersinia enterocolitica* (52).

Many evanescent-wave biosensors are also sandwich antibody assays, but they use the surface-specific characteristics of evanescent waves to selectively excite fluorophore-bound antibody conjugates with a laser. Evanescent-wave sensors reportedly can detect bacterial targets in turbid and heterogenous samples with little interference from other components. One assay simultaneously detected staphylococcal enterotoxin B (SEB) and botulinum toxoid A (BotA) in canned tomatoes, sweet corn, green beans, mushrooms, and tuna, with a detection limit of 20 to 500 ng/ml for BotA and 0.1 to 0.5 ng/ml for SEB (60). Evanescent-wave sensors were used to detect ricin, SEB, and cholera toxin, and also used to detect *L. monocytogenes* and *S. enterica* serovar Typhimurium in chicken excreta, with detection limits of 10^3 to 10^4 cells/ml (66).

Other technologies used in biosensors include IMS-immunoliposomes in which sulfohodamine B-encapsulated liposomes coated with anti-O157 antibodies were able to detect 1 cell of O157:H7 in water, apple juice, and cider within 8 h (14). The RBD3000 biosensor (AATI) uses flow cytometry and an antibody-fluorescent tag to detect bacteria with a limit of 10 cells/ml within 20 min; it is capable of handing 42 samples. Another biosensor, known as CANARY (for "cellular analysis and notification of antigen risk and yields"), uses B lymphocytes engineered to express aquorin, a calcium-sensitive bioluminescent protein from jellyfish. By cross-linking antibodies to the cell membrane, the specific binding of pathogens causes elevation in calcium levels, which results in light emission by aquorin (57). Innovative Biosensors uses CANARY in its *E. coli* O157:H7

assay, which can detect 50 cells in pure culture in 3 min or 1,000 cells in ground beef.

The potential application of biosensor technology to pathogen testing in foods offers many attractive features, including portability, multi-analyte testing, speed, and sensitivity. However, like most assays, the exquisite sensitivity achievable with cultures does not translate to food testing due to interference (66). Also, many sensors use sophisticated microfuidics that are susceptible to clogging by food components; therefore, adequate sample preparation techniques are another important consideration in developing biosensor assays for foods. Presently, this technology is being explored commercially for use in food safety (2).

DNA Microarrays

Also known as gene array, DNA chip, biochip, gene chip, or genome chip technology, DNA microarray analysis is a very powerful technology that enables the simultaneous analysis of thousands of genes. Chips are fabricated using a variety of solid supports, but glass or nylon filters are most often used and some of these have been silanized or derivatized with other agents to enhance binding (12). Microarrays can be classified into amplicon and oligonucleotide arrays. Amplicon arrays uses cDNA fragments, typically between 100 and >1,000 bp, that are obtained by PCR and then covalently attached or adsorbed to a derivatized support. Due to the use of longer probes, amplicon arrays offer a broad range of hybridization conditions and are very useful for examining gene relatedness among large populations, where exact gene sequences are not known (42). However, they also tend to exhibit less specificity compared to oligonucleotide probes and so are not as effective in detecting SNPs. Oligonucleotide arrays are made by immobilizing presynthesized oligonucleotides (<100 bp) onto the support or, as done on gene chips, by direct synthesis of oligonucleotides onto the support via methods such as photolithography (7, 42). Oligonucleotide probes are synthesized; therefore, this type of array has the advantage of customizing probes for optimal performance and designing probes to detect very specific targets such as unique SNPs.

Once an array has been synthesized, it can be used for direct detection without amplification if the target is abundant, as in genome analysis. In such applications, the target DNA is extracted, labeled, and hybridized directly to the array. However, for low-copy targets, it must be amplified by PCR and labeled before it can be used for array analysis. Approaches used in PCR-microarray assays include amplification using arbitrary or random primers, primers to universal genes such as rRNA, and primers to specific genetic markers (42).

Many microarrays have been developed for the detection and identification of pathogens. An oligonucleotide array was used to specifically identify 18 pathogenic prokaryotes, eukaryotes, and viruses. This assay was able to detect as little as 10 fg of *Bacillus anthracis* DNA (70). Another microarray for *Campylobacter jejuni* and *C. coli* gave better detection of these pathogens in chicken feces than did other methods (35). An oligonucleotide microarray based on multiple trait and virulence markers from *V. cholerae*, *V. vulnificus*, and *V. parahaemolyticus* was used for the identification of pathogenic *Vibrio* species. It was determined to have a detection sensitivity of 10^2 to 10^3 cells/ml of pure culture or 1 cell/g of oyster tissue homogenates after 5 h of enrichment culture (54). Microarray assays are often coupled with IMS to selectively capture target microbes prior to analysis. A microarray-IMS assay was able to detect 55 cells of *E. coli* O157:H7/ml of chicken rinsate without the need for enrichment culture (8). Another microarray-automated IMS assay detected 10 *E. coli* O157:H7 cells in river water samples or 100 cells each of *E. coli* O157:H7, *Salmonella*, and *Shigella* as a multiplex assay (63). Microarrays are also well suited for characterization of pathogens, and one system was used to analyze for 25 virulence and 23 antibiotic resistance genes in *Salmonella* and pathogenic *E. coli* (72).

Despite the tremendous potential of microarray technology, few products have been commercialized for the detection of pathogens in foods. Some reports suggest that arrays are not as sensitive as PCR and do not have sufficient sensitivity to detect low levels of microbial contamination (42). Also, the difficulties inherent in food testing apply to microarrays as well, since the sensitivity of these assays decreases when testing foods. Hence, factors such as adequate sample preparation (71) and the need for confirmation of results (7) must be considered in designing microarrays for food testing.

ADVANTAGES AND LIMITATIONS

The number of rapid methods available for pathogen and toxin testing of foods is impressive. However, the complexities of food composition, such as inhibitors to pathogen determination, present a variety of challenges. There are many advantages and limitations to using rapid methods for food testing.

Need for Enrichment

The method is merely one component in the overall scheme of foodborne pathogen analysis; therefore, even though most rapid methods can be completed in a few minutes or hours, the total analysis time can require several days due to the need for culture enrichment. This continued reliance on enrichment diminishes the impact of rapid methods, but this limiting factor is gradually becoming less influential with the introduction of more rapid and sensitive methods. However, it is important to recognize that enrichment provides many essential benefits for pathogen detection, such as diluting the effects of inhibitors, providing target cell growth to detectable numbers, repairing stressed or injured cells, and, in the case of PCR, enabling the differentiation of viable from nonviable cells. Hence, it would be difficult to completely eliminate enrichment culture from the process of pathogen detection in foods.

Confirmation Requirement

Rapid methods for foodborne-pathogen and toxin testing are useful screening tools, well suited for use by the food industry or testing laboratories to enable a more rapid release of products into the market. As a screening tool, negative results by rapid methods are considered definitive but positive results are regarded only as presumptive; therefore, rapid-method kit inserts include instructions for confirmation, which is often done by standard culture-based microbiological methods. Recently some rapid methods have been introduced and validated that do not require confirmation. While some new assays are more sensitive and specific than traditional methods, it is important not to underestimate the complexities of food testing, the potential for antibody cross-reactivity or nonspecific reactions, and the possibility of inadvertent cross-contamination of sample with the positive control strain. For in-food testing, the definitive proof of contamination remains to be an "isolate in hand"; confirmation is therefore highly recommended and perhaps should be required. Although confirmation extends the analysis time by a few days, this may not be an imposing burden since negative results are most often encountered in food testing.

Interference and Food-Dependent Efficiency

Some rapid methods perform better in some foods than others, so that their efficiencies can be food-dependent. This can be attributed to the complex composition of foods but can also be due to the different technologies used in the methods. For example, a food component may interfere with antigen-antibody binding but may not affect DNA assays, while, conversely, another component can inhibit DNA hybridization or *Taq* polymerase but may not affect antibodies. In addition, the detection limits of methods vary widely (Table 43.6), so that some methods can detect lower levels of contaminants than others. Because of the many influences that can affect assay efficiency, comparative testing is recommended to ensure that a particular assay is effective for the intended application.

Lack of Gene Expression

Most DNA assays use short oligonucleotides to obtain exquisite specificity, capable of detecting single-nucleotide differences. However, a positive result by a DNA assay indicates only that the complementary sequence is present, and additional conclusions should not be drawn without further testing. For example, a positive result with a toxin gene-specific PCR assay indicates only that bacteria with those gene sequences are present and that they are potentially toxigenic. It should not be assumed that the gene is actually expressed or that the toxin, if produced, is functional. Bacteria are well known to carry genes that are not expressed due to physiological factors or the presence of genetic mutations (47, 48). This provides an additional reason to confirm results, as emphasized above. Another concern regarding DNA assays is the issue of cell viability, since assays such as PCR can theoretically amplify 1 copy of DNA regardless of its viability. This concern is often resolved by the inclusion of a brief enrichment culture step prior to PCR or by using RNA templates for amplification (63). As a final point, with foodborne intoxications such as those caused by *C. botulinum* or *S. aureus*, in which illness is caused by the ingestion of preformed toxins, DNA assays can detect the presence only of potentially toxigenic cells, not the presence of toxins, in foods.

Single-Target Design

Most rapid methods are designed to detect a single target, which makes them well suited for use in quality control programs to screen foods for the presence of specific pathogens. However, in outbreak investigations or in food surveillance, where multiple targets are tested, the use of single-target kits can be logistically complicated and costly. The trend of single-target assays is changing, and there are currently many technologies that enable multi-analyte testing. Many rtPCR methods are multiplex tests that detect several genes or species. In addition, biosensor assays are available that can detect several bacteria and toxins simultaneously (66), and DNA microarrays can perform thousands of tests on a single chip.

Need for Validation

It is evident from the list of assays in the tables in this chapter that there is an abundance of rapid methods available for testing foodborne pathogens and toxins. New rapid methods are frequently being introduced, and often there are no data published in peer-reviewed

journals describing their performance. Since the efficiencies of these methods can be food dependent, it is critical that new methods be comparatively evaluated against existing assays, preferably through validation studies to ensure their performance.

The microbiological safety of foods is of primary concern to public health, and so the food industry and regulatory agencies have implemented programs to monitor foods for the presence of pathogens and toxins. As a screening tool, negative results by rapid methods are accepted but positive results are regarded as only presumptive and must be confirmed. This creates a situation in which false-positive results are identified during confirmation but false-negative results are not unless other assays are used simultaneously during the evaluation. False-negative results are of food safety concern, with potentially serious ramifications such as consumption of contaminated foods that may cause foodborne illnesses.

The level of sensitivity of detection methods continues to improve. While this may be desirable, it can also present a challenging situation for the food industry and regulatory agencies. For instance, the current specification for ready-to-eat foods is "absence" of pathogens and toxins. But "absence" actually means "not detected" by the method used for food testing. The challenge with increased sensitivity of a method is that it may present situations in which pathogens were found to be "absent" from foods previously analyzed by an accepted method but it no longer meets specifications if more sensitive methods are used. Such situations have already occurred, in which low levels of bacterial toxins were not detected in contaminated foods by a standard method but were detected by more sensitive rapid methods. This situation raises several questions. Must manufacturers upgrade their processing or quality control programs to comply with greater test sensitivity? What is the health risk of the presence of lower than previously detectable levels of pathogens and toxins? What are the ultimate specification limits as regulations become more stringent when more sensitive methods are developed?

New assays should be validated before commrcialization by comparative testing with standard methods. In the United States, assays used in clinical testing must be reviewed and approved by the Food and Drug Administration (FDA) Center for Devices and Radiological Health, but no such requirements exist for pathogen and toxin testing methods used in foods. Hence, most validations for foodborne-pathogen assays are done by independent organizations such as the Association of Official Analytical Chemists International (AOAC). The agencies which have oversight of food safety, the FDA and U.S. Department of Agriculture (USDA), both have active roles and interests in validation of food-testing assays since such methods are fundamental tools in regulatory assessment of food safety. Since these federal agencies regulate different food commodities, methods and procedures used to test for the same target pathogens or toxin can be very different. As a result, each agency has internal evaluation procedures for the methods, and these procedures can vary depending on urgency or need. The procedures validate preferred methods which are subsequently published in the FDA *Bacteriological Analytical Manual* (BAM) or the USDA-Food Safety and Inspection Service *Microbiology Laboratory Guidebook* (MLG).

The BAM was first drafted in 1965 to provide a document of standardized methods for use within the FDA testing laboratories. The BAM contains many methods that are official AOAC methods, but it also contains methods that are developed and evaluated internally by FDA and have not been subjected to collaborative studies (32). The BAM is available online (http://www.cfsan. fda.gov/~ebam/bam-toc.html) for ease of accessibility.

Likewise, the MLG is a manual of internally validated methods for use in the microbiological testing of meat, poultry, eggs, and other commodities regulated by the USDA. The MLG is also available online (http://www. fsis.usda.gov/Science/Microbiological_Lab_Guidebook/ index.asp).

AOAC International has three validation programs, which are the most widely accepted method validation programs in the United States. The Collaborative Study program and the Peer-Verified programs are both administered by the AOAC, whereas the Test Kit Performance Tested program is administered by the AOAC Research Institute (RI), an affiliate of AOAC (3). These three programs were described in the rapid-methods chapter of this book's second edition (19) and in other reviews (3) and so are not presented in detail here. However, it is important to clarify the apparent confusion that exists between the AOAC Collaborative Study program and the Performance Tested program of AOAC RI.

The Collaborative Study program is the backbone of AOAC validation programs and has been in existence for many years. Relying strictly on volunteer scientists to manage, review, and perform the studies, a Collaborative Study validation is an extensive, multilayered study involving many laboratories, food samples, and bacterial inoculation levels to thoroughly compare a new test to a standard method. The end result of this rigorous process is that methods meeting requirements and validated by the Collaborative Study qualify as Official methods. These Official methods must be performed exactly as specified in the procedure without deviations or modifications.

The Test Kit Performance Tested methods program of AOAC RI was introduced in the early 1990s with the sole purpose of validating performance claims of commercially available rapid methods. It is a much simpler program to complete, in which the manufacturer submits to AOAC RI the assay claims and performance data, which are then evaluated by an independent laboratory and two technical reviewers. The independent laboratory tests the assay and compares its results with data submitted by the manufacturer. If the claims are substantiated by the results, it is designated Performance Tested, and this status can be maintained by payment of a yearly fee. It is important to recognize that Performance Tested methods are not Official methods but are often confused with them.

The AOAC, in part to alleviate the confusion, has incorporated the Performance Tested program into the Collaborative Study program, so that all assays with proprietary technologies must first be submitted to AOAC RI for performance testing. Assays can then remain at Performance Tested status, or the data generated during performance testing can be used to satisfy part of the requirements of a Collaborative Study and proceed with this type of study to obtain Official method status.

A thorough validation is a lengthy and laborious process. Recent concerns regarding bioterrorism and food security have prompted the development of an emergency situation whereby extensive validation of a method is not feasible and more abbreviated validation procedures are needed to quickly evaluate a method prior to use. To address this, the AOAC, with the support of USDA-FSIS, developed the e-CAM system, which consists of electronic compilations of analytical methods worldwide, based on the degree to which the method has been validated. The e-CAM system classifies methods into five categories, including Regulatory (REG), Harmonized Collaborative Validation (HCV), Multiple-Laboratory Validation (MLV), Single-Laboratory Validation (SLV), and Research Non-Validated (RNV). REG methods are those used or specified by national and international regulatory agencies for enforcement purposes. HCV methods are those that have been validated through a full collaborative study that meets the standards of AOAC or International Standards Organization (ISO) and must report valid data from at least 8 and 10 collaborating laboratories for quantitative and qualitative methods, respectively. MLV methods include those that have been validated by several laboratories according to AOAC and ISO standards but do not meet the requirement for the minimum number of laboratories required for HCV. MLV methods must report valid data from at least two laboratories, one of which may be the originating laboratory requesting the validation.

Hence, MLV resembles the Performance Tested program of AOAC RI. SLV methods include those that were comparatively evaluated by only one laboratory in accordance with AOAC and ISO standards; SLV is equivalent to an "in-house" validation study, which is one of the requirements in a Collaborative Study validation. Lastly, RNV methods are experimental or state-of the-art methods that appear to be useful or applicable in a testing situation; however, they have not been fully characterized, optimized, or validated. RNV methods are usually accompanied by a disclaimer stating their nonvalidated status. The AOAC e-CAM is an Internet-based system, and its intent is to harmonize testing methods worldwide and enable the international analytical community to electronically access all types of analytical methods that may be suitable for their particular testing situations.

Advancements in technology continue to have a great impact on the evolution of rapid methods. Since publication of the second edition of this book (19), several commercial rapid methods kits or systems have been discontinued, others have been modified, and many methods using new technologies have been introduced or are being developed for the detection of both existing and emerging foodborne pathogens. Many of the new methods use next-generation technologies such as rtPCR, biosensors, and DNA chips, which may be more rapid, more sensitive, and capable of multitarget testing and hence are well suited for use in screening large numbers of samples in compliance or food security surveillance programs. However, the complexity of foods continues to present challenges to testing methods, and all assays remain susceptible to interferences by food ingredients and inhibitors and the presence of the normal microflora. Hence, culture enrichment or adequate sample preparation procedures are required. Moreover, since rapid-method assays use different technologies which have different detection sensitivities that often provide presumptive results, additional testing is usually needed to confirm the results. Furthermore, comparative evaluation by with standard methods or validation of rapid methods is critical to document their efficacy in detecting foodborne pathogens and toxins.

References

1. **Abee, T., W. van Schaik, and R. J. Siezen.** 2004. Impact of genomics on microbial food safety. *Trends Biotechnol.* **22:**653–660.

2. **Alocilja, E. C., and S. M. Radke.** 2003. Market analysis of biosensors for food safety. *Biosens. Bioelectron.* **18:**841–846.

3. **Andrews, W. H.** 1996. AOAC INTERNATIONAL'S three validation programs for methods used in the microbiological analysis of foods. *Trends Food Sci. Technol.* **7:**147–151.

4. **Benoit, P. W., and D. W. Donahue.** 2003. Methods for rapid separation and concentration of bacteria in food that bypass time-consuming cultural enrichment. *J. Food Prot.* **66:**1935–1948.

5. **Brooks, J. T., E. G. Sowers, J. G. Wells, K. D. Greene, P. M. Griffin, R. M. Hoekstra, and N. A. Strockbine.** 2005. Non-O157 Shiga toxin-producing *Escherichia coli* infections in the United States, 1983–2002. *J. Infect. Dis.* **192:**1422–1429.

6. **Broude, E. N.** 2005. Molecular beacons and other hairpin probes, p. 846–850. *In* J. Fuchs and M. Podda (ed.), *Encyclopedia of Diagnostic Genomics and Proteomics.* Marcel Dekker, New York, N.Y.

7. **Call, D. R.** 2005. Challenges and opportunities for pathogen detection using DNA microarrays. *Crit. Rev. Microbiol.* **31:**91–99.

8. **Call, D. R., F. J. Brockman, and D. P. Chandler.** 2001. Detecting and genotyping *Escherichia coli* O157:H7 using multiplexed PCR and nucleic acid microarrays. *Int. J. Food Microbiol.* **67:**71–80.

9. **Centers for Disease Control and Prevention.** 2005. Summary of notifiable diseases—United States, 2003. *Morb. Mortal. Wkly. Rep.* **52:**1–85.

10. **Centers for Disease Control and Prevention.** 2004. Preliminary FoodNet data on the incidences of infection with pathogens transmitted commonly through foods-selected sites United States, 2003. *Morb. Mortal. Wkly. Rep.* **53:** 338–343.

11. **Chen, J., and M. W. Griffiths.** 1996. *Salmonella* detection in eggs using *lux*⁺ bacteriophages. *J. Food Prot.* **59:**908–914.

12. **Cunningham, M. J., and D. D. Dao.** 2005. Microarray-fabricating, p. 819–823. *In* J. Fuchs and M. Podda (ed.), *Encyclopedia of Diagnostic Genomics and Proteomics.* Marcel Dekker, New York, N.Y.

13. **De Angelis, D. A.** 2005. Fluorescence resonance energy transfer (FRET), p. 491–495. *In* J. Fuchs and M. Podda (ed.), *Encyclopedia of Diagnostic Genomics and Proteomics.* Marcel Dekker, New York, N.Y.

14. **DeCory, T. R., R. A. Durst, S. J. Zimmerman, L. A. Garringer, G. Paluca, H. H. DeCory, and R. A. Montagna.** 2005. Development of an immunomagnetic bead-immunoliposome fluorescence assay for rapid detection of *Escherichia coli* O157:H7 in aqueous samples and comparison of the assay with a standard microbiological method. *Appl. Environ. Microbiol.* **71:**1856–1864.

15. **Eklund, M., J. P. Nuorti, P. Ruutu, and A. Siitonen.** 2005. Shiga toxigenic *Escherichia coli* (STEC) infections in Finland during 1998–2002: a population-based surveillance study. *Epidemiol. Infect.* **133:**845–852.

16. **El Marrakchi, A., N. Boum'handi, and A. Hamama.** 2005. Performance of a new chromogenic plating medium for the isolation of *Listeria monocytogenes* from marine environments. *Lett. Appl. Microbiol.* **40:**87–91.

17. **Exner, M. M.** 2005. Multiplex real-time PCR, p. 855–859. *In* J. Fuchs and M. Podda (ed.), *Encyclopedia of Diagnostic Genomics and Proteomics.* Marcel Dekker, New York, N.Y.

18. **Favrin, S. J., A. J. Sabah, and M. W. Griffiths.** 2001. Development and optimization of a novel immunomagnetic

separation-bacteriophage assay for detection of *Salmonella enterica* serovar Enteritidis in broth. *Appl. Environ. Microbiol.* **67:**217–224.

19. Feng, P. 2001. Development and impact of rapid methods for detection of food-borne pathogens, p. 775–796. *In* M. P. Doyle, L. R. Beuchat, and T. J. Montville (ed.). *Food Microbiology: Fundamentals and Frontiers*, 2nd ed. ASM Press, Washington, D.C.

20. Feng, P., and P. A. Hartman. 1982. Fluorogenic assays for immediate confirmation of *Escherichia coli*. *Appl. Environ. Microbiol.* **43:**1320–1329.

21. Fu, Z., S. Rogelj, and T. L. Kieft. 2005. Rapid detection of *Escherichia coli* O157:H7 by immunomagnetic separation and real-time PCR. *Int. J. Food Microbiol.* **97:**47–57.

22. Fukushima, H., Y. Tsunomori, and R. Seki. 2003. Duplex real-time SYBR Green PCR assays for the detection of 17 species of food- or waterborne pathogens in stools. *J. Clin. Microbiol.* **41:**5134–5146.

23. Gallagher, R., and J. Perkel. 2005. Seven cheers for technology. *Scientist* **19:**6.

24. Gasanov, U., D. Hughes, and P. M. Hansbro. 2005. Methods for the isolation and identification of *Listeria* spp. and *Listeria monocytogenes*: a review. *FEMS Microbiol. Rev.* **29:**851–875.

25. Gehring, A. G., P. L. Irwin, S. A. Reed, S. I. Tu, P. E. Andreotti, H. Akhavan-Tafti, and R. S. Handley. 2004. Enzyme-linked immunomagnetic chemiluminescent detection of *Escherichia coli* O157:H7. *J. Immunol. Methods* **293:**97–106.

26. Geng, T., M. T. Morgan, and A. K. Bhunia. 2004. Detection of low levels of *Listeria monocytogenes* by using a fiber-optic immunosensor. *Appl. Environ. Microbiol.* **70:**6138–6146.

27. Gutteridge, C. S., and M. L. Arnott. 1989. Rapid methods: an over the horizon view, p. 297–319. *In* M. R. Adams and C. F. A. Hope (ed.), *Rapid Methods in Food Microbiology: Progress in Industrial Microbiology*. Elsevier, New York, N.Y.

28. Harding, A. 2005. BLAST. *Scientist* **19:**21–25.

29. Harlander, S. 1989. Food biotechnology: yesterday, today and tomorrow. *Food Technol.* **43:**196–206.

30. Hoorfar, J., N. Cook, B. Malorny, M. Wagner, D. DeMedici, A. Abdulmawjood, and P. Fach. 2003. Making internal amplification control mandatory for diagnostic PCR. *J. Clin. Microbiol.* **41:**5835.

31. Hsu, C. F., T. Y. Tsai, and T. M. Pan. 2005. Use of the duplex TaqMan PCR system for detection of Shiga-like toxin-producing *Escherichia coli* O157. *J. Clin. Microbiol.* **43:**2668–2673.

32. Jackson, G. J., and I. K. Wachsmuth. 1998. The US Food and Drug Administration's selection and validation of tests for foodborne microbes and microbial toxins. *Food Control* **7:**37–39.

33. Jean, J., D. H. D'Souza, and L. A. Jaykus. 2004. Multiplex nucleic acid sequence-based amplification for simultaneous detection of several enteric viruses in model ready-to-eat foods. *Appl. Environ. Microbiol.* **70:**6603–6610.

34. Jinneman, K. C., K. J. Yoshitomi, and S. D. Weagant. 2003. Multiplex real-time PCR method to identify Shiga toxin genes *stx*1 and *stx*2 and *Escherichia coli* O157:H7/H– serotype. *Appl. Environ. Microbiol.* **69:**6327–6333.

35. Keramas, G., D. D. Bang, M. Lund, M. Madsen, H. Bunkenborg, P. Telleman, and C. B. Christensen. 2004. Use of culture, PCR analysis and DNA microarray for detection of *Campylobacter jejuni* and *Campylobacter coli* from chicken feces. *J. Clin. Microbiol.* **42:**3985–3991.

36. Klotz, M., S. Opper, K. Heeg, and S. Zimmermann. 2003. Detection of *Staphylococcus aureus* enterotoxins A to D by real-time fluorescence PCR assay. *J. Clin. Microbiol.* **41:**4683–4687.

37. Koo, K., and L. A. Jaykus. 2003. Detection of *Listeria monocytogenes* from a model food by fluorescence resonance energy transfer-based PCR with an asymmetric fluorogenic probe set. *Appl. Environ. Microbiol.* **69:**1082–1088.

38. Kramer, M. F., and D. V. Lim. 2004. A rapid and automated fiber optic-based biosensor assay for the detection of *Salmonella* in spent irrigation water used in the sprouting of sprout seeds. *J. Food Prot.* **67:**46–52.

39. Krupp, G., and U. Spengler. 2005. Polymerase chain reaction, p. 1052–1059. *In* J. Fuchs and M. Podda (ed.), *Encyclopedia of Diagnostic Genomics and Proteomics*. Marcel Dekker, New York, N.Y.

40. Kubista, M., and N. Zoric. 2005. Real-time PCR platforms, p. 1126–1130. *In* J. Fuchs and M. Podda (ed.), *Encyclopedia of Diagnostic Genomics and Proteomics*. Marcel Dekker, New York, N.Y.

41. Lantz, P.-G., B. Hahn-Hagerdal, and P. Radstrom. 1994. Sample preparation methods in PCR-based detection of food pathogens. *Trends Food Sci. Technol.* **5:**384–389.

42. Lemarchand, K., L. Masson, and R. Brousseau. 2004. Molecular biology and DNA microarray technology for microbial quality monitoring of water. *Crit. Rev. Microbiol.* **30:**145–172.

43. Leonard, P., S. Hearty, J. Quinn, and R. O'Kennedy. 2004. A generic approach for the detection of whole *Listeria monocytogenes* cells in contaminated samples using surface plasmon resonance. *Biosens. Bioelectron.* **19:**1331–1335.

44. McKillip, J. L., and M. Drake. 2000. Molecular beacon polymerase chain reaction detection of *Escherichia coli* O157:H7 in milk. *J. Food Prot.* **63:**855–859.

45. Mercanoglu, B., and M. W. Griffiths. 2005. Combination of immunomagnetic separation with real-time PCR for rapid detection of *Salmonella* in milk, ground beef, and alfalfa sprouts. *J. Food Prot.* **68:**557–561.

46. Monday, S. R., and G. A. Bohach. 1999. Use of multiplex PCR to detect classical and newly described pyrogenic toxin genes in staphylococcal isolates. *J. Clin. Microbiol.* **37:**3411–3414.

47. Monday, S. R., S. A. Minnich, and P. C. H. Feng. 2004. A 12-base-pair deletion in the flagellar master control gene *flhC* causes non-motility of the pathogenic German sorbitol-fermenting *Escherichia coli* O157:H– strains. *J. Bacteriol.* **186:**2319–2327.

48. Monday, S. R., T. S. Whittam, and P. Feng. 2001. Genetic and evolutionary analysis of insertions in the *gusA* gene, which caused the absence of glucuronidase activity in *Escherichia coli* O157:H7. *J. Infect. Dis.* **184:**918–921.

49. Nazarowec-White, M., and J. M. Farber. 1997. *Enterobacter sakazakii*: a review. *Int. J. Food Microbiol.* **34**:103–113.

50. Odumeru, J. A., M. Stele, L. Fruhner, C. Larkin, J. Jiang, E. Mann, and W. B. McNab. 1999. Evaluation of accuracy and repeatability of identification of food-borne pathogens by automated bacterial identification systems. *J. Clin. Microbiol.* **37**:944–949.

51. Oh, B. K., Y. K. Kim, K. W. Park, W. H. Lee, and J. W. Choi. 2004. Surface plasmon resonance immunosensor for the detection of *Salmonella typhimurium*. *Biosens. Bioelectron.* **19**:1497–1504.

52. Oh, B. K., W. Lee, B. S. Chun, Y. M. Bae, W. H. Lee, and J. W. Choi. 2005. The fabrication of protein chip based on surface plasmon resonance for detection of pathogens. *Biosens. Bioelectron.* **20**:1847–1850.

53. O'Hara, C. M., E. G. Sowers, C. A. Bopp, S. B. Duda, and N. A. Strockbine. 2003. Accuracy of six commercially available systems for identification of members of the family *Vibrionaceae*. *J. Clin. Microbiol.* **41**:5654–5659.

54. Panicker, G., D. R. Call, M. J. Krug, and A. K. Bej. 2004. Detection of pathogenic *Vibrio* spp. in shellfish by using multiplex PCR and DNA microarrays. *Appl. Environ. Microbiol.* **70**:7436–7444.

55. Park, C. H., H. J. Kim, D. L. Hixon, and A. Bubert. 2003. Evaluation of the Duopath verotoxin test for detection of shiga toxins in cultures of human stools. *J. Clin. Microbiol.* **41**:2650–2653.

56. Perez, J. M., P. Cavalli, C. Roure, R. Renac, Y. Gille, and A. M. Freydiere. 2003. Comparison of four chromogenic media and Hektoen agar for detection and presumptive identification of *Salmonella* strains in human stools. *J. Clin. Microbiol.* **41**:1130–1134.

57. Rider, T. H., M. S. Petrovick, F. E. Nargi, J. D. Harper, E. D. Schwoebel, R. H. Mathews, D. J. Blanchard, L. T. Bortolin, A. M. Young, J. Chen, and M. A. Hollis. 2003. A B cell-based sensor for rapid identification of pathogens. *Science* **301**:213–215.

58. Rossen, L., P. Norskov, K. Holmstrom, and O. F. Rasmussen. 1992. Inhibition of PCR by components of food samples, and DNA-extraction solutions. *Int. J. Food Microbiol.* **17**:37–45.

59. Rybolt, M. L., R. W. Wills, J. A. Byrd, T. P. Doler, and R. H. Bailey. 2004. Comparison of four *Salmonella* isolation techniques in four different inoculated matrices. *Poult. Sci.* **83**:1112–1116.

60. Sapsford, K. E., C. R. Taitt, N. Loo, and F. S. Ligler. 2005. Biosensor detection of botulinum toxoid A and staphylococcal enterotoxin B in food. *Appl. Environ. Microbiol.* **71**:5590–5592.

61. Smith, P. A., D. Mellors, A. Holroyd, and C. Gray. 2001. A new chromogenic medium for the isolation of *Listeria* spp. *Lett. Appl. Microbiol.* **32**:78–82.

62. Stevens, K. A., and L. A. Jaykus. 2004. Bacterial separation and concentration from complex sample matrices: a review. *Crit. Rev. Microbiol.* **30**:7–24.

63. Straub, T. M., B. P. Dockendorff, M. D. Quinonez-Diaz, C. O. Valdez, J. L. Shutthanandan, B. J. Tarasevich, J. W. Grate, and C. J. Bruckner-Lea. 2005. Automated methods for multiplexed pathogen detection. *J. Microbiol. Methods* **62**:303–316.

64. Su, X. L., and Y. Li. 2004. Quantum dot biolabeling coupled with immunomagnetic separation for detection of *Escherichia coli* O157:H7. *Anal. Chem.* **76**:4806–4810.

65. Su, X. L., and Y. Li. 2004. A self-assembled monolayer-based piezoelectric immunosensor for rapid detection of *Escherichia coli* O157:H7. *Biosens. Bioelectron.* **19**:563–574.

66. Taitt, C. R., J. P. Golden, Y. S. Shubin, L. C. Shriver-Lake, K. E. Sapsford, A. Rasooly, and F. S. Ligler. 2004. A portable array biosensor for detecting multiple analytes in complex samples. *Microb. Ecol.* **47**:175–185.

67. Teel, L. D., B. R. Steinberg, N. E. Aronson, and A. D. O'Brien. 2003. Shiga toxin-producing *Escherichia coli*-associated kidney failure in a 40-year-old patient and late diagnosis by novel bacteriologic and toxin detection methods. *J. Clin. Microbiol.* **41**:3438–3440.

68. Vaneechoute, M., and J. Van Eldere. 1997. The possibilities and limitations of nucleic acid amplification technology in diagnostic microbiology. *J. Med. Microbiol.* **46**:188–194.

69. Wang, R. F., W. W. Cao, and C. E. Cerniglia. 1997. A universal protocol for PCR detection of 13 species of foodborne pathogens in foods. *J. Appl. Microbiol.* **83**:727–736.

70. Wilson, W. J., C. L. Strout, T. Z. DeSantis, J. L. Stilwell, A. V. Carrano, and G. L. Andersen. 2002. Sequence-specific identification of 18 pathogenic microorganisms using microarray technology. *Mol. Cell. Probes* **16**:119–127.

71. Yoo, S. Y., K. C. Keum, S. Y. Yoo, J. Y. Choi, K. H. Chang, N. C. Yoo, W. M. Yoo, J. M. Kim, D. Lee, and S. Y. Lee. 2004. Development of DNA microarray for pathogen detection. *Biotechnol. Bioprocess Eng.* **9**:93–99.

72. Zhao, C. S., P. F. McDermott, C. M. Schroeder, D. G. White, and J. H. Meng. 2005. A DNA microarray for identification of virulence and antimicrobial resistance genes in *Salmonella* and *Escherichia coli*. *Mol. Cell. Probes* **19**:195–201.

Food Microbiology: Fundamentals and Frontiers, 3rd Ed.
Edited by M. P. Doyle and L. R. Beuchat
© 2007 ASM Press, Washington, D.C.

Todd R. Klaenhammer
Erika Pfeiler
Tri Duong

44

Genomics and Proteomics of Foodborne Microorganisms

Foods are teeming with microorganisms that may be innocuous, pathogenic threats, spoilage agents, or beneficial microbes driving fermentations or acting as biocontrol agents. Historically, the overarching priorities in this field have been the destruction and control of undesirable organisms in foods and/or promotion of the growth and activity of desirable microbes. Recent recommendations from the 2005 American Academy of Microbiology report, *Research Opportunities in Food and Agricultural Microbiology,* have emphasized the importance of studying the distribution of microorganisms and microbial communities in foods; understanding the nature, specificity, and adaptation of microorganisms to food environments and their human, animal, and plant hosts; and investigating the impact of production and processing practices on the evolution, persistence, resistance, and flow of microbes in foods. More generally stated, the recommendation promotes an understanding of the interaction of microbes with our foods, hosts, and processing and storage environments created by mankind.

In response to this challenge, genomics and proteomics now underlie a renaissance in food microbiology that will impact every facet of how we understand and investigate microorganisms associated with our foods. Food microbiologists investigating pathogens, commensals, or starter cultures will rely extensively on DNA and protein sequence information to define genetic content, construct metabolic pathways, and predict microbial substrates and products. In addition, microarray and gene expression profiles can now be exploited to investigate the responses of microbes to food environments, processing conditions, host ecosystems, and the presence of other organisms.

This chapter outlines the basic concepts underlying genomics, proteomics, and microarray technologies. Furthermore, selected examples are illustrated where these approaches have already provided an in-depth understanding of important microbial properties, behavior, processes, or interactions within food environments.

DNA SEQUENCING

The heart of all genomics research lies in DNA sequencing. With few exceptions, all genome-scale sequencing has been done using the chain termination technique developed by Fredrick Sanger in 1974, which, along with his work in sequencing a phage genome, earned him the Nobel prize in chemistry in 1980. This method utilizes normal DNA replication with a template strand, a primer, DNA

Todd R. Klaenhammer, Erika Pfeiler, and Tri Duong, Genomic Sciences Program, Department of Food Science, North Carolina State University, Raleigh, NC 27695-7624.

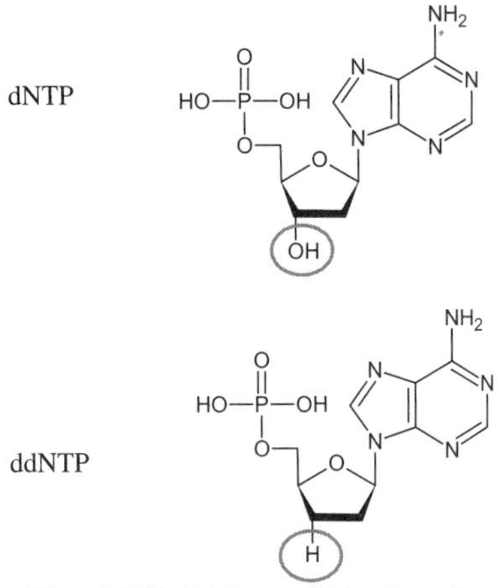

Figure 44.1 The dNTP (A) has a 3′ hydroxyl present on the deoxyribose that the ddNTP (B) does not. This stops DNA strand growth because DNA polymerase no longer has a way to connect the bases in the growing strand.

polymerase, and a mix of deoxynucleotide triphosphates (dNTPs). Dideoxynucleotide triphosphates (ddNTPs) are added to the reaction mixture because the absence of a 3′ hydroxyl in these nucleotides stops strand growth (Fig. 44.1). Therefore, four separate reactions are set using the same template DNA and the four ddNTPs, A, T, G, or C. When the products of these reactions are run on an acrylamide gel, they are separated by size, and the sequence can then be determined by reading the bands up the gel. While manual sequencing works for short pieces of DNA, sequencing an entire genome requires a more high-throughput approach, and in the 1990s, such a setup was made possible. First, the development of ddNTPs which each have a different fluorescent label made it possible to sequence from a single reaction instead of four. Second, capillary electrophoresis was introduced, whereby the reaction products are run through a capillary tube through which a scanner reads the fluorescence wavelength coming from each amplicon fragment that passes by. Software programs are then used to correlate the wavelength reading with the appropriate nucleotide and to assemble the approximately 600-bp lengths of sequence into an entire genome. This is accomplished by matching overlapping sequences in different contigs and assembling those to create longer and fewer contigs until the genome is closed. Genomes of important foodborne microorganisms that have been sequenced are listed in Table 44.1.

BIOINFORMATICS

Once the genome sequence is complete, the next task is to find genes in the sequence and annotate them. Elucidating the content of assembled DNA sequences belongs in part to the field of bioinformatics. While bioinformatics is defined as a "scientific discipline that encompasses all aspects of biological information acquisition, processing, storage, distribution, analysis and interpretation" (54), the focus here is on bioinformatic tools and the process of genome annotation. Genome annotation is the process of analyzing the raw DNA sequence and determining its biological significance. Genome annotation can be divided into three stages: nucleotide-level, protein-level, and process-level annotation (59).

Nucleotide annotation is the first step in the complete annotation of a genome sequence. Genomic features are located and identified; this includes the identification of genes, untranslated RNAs, and any markers from preexisting physical or genetic maps. The most visible part of this phase is gene calling. In prokaryotic genomes, gene calling is primarily the identification of long open reading frames (ORFs), or nucleic acid sequences uninterrupted by stop codons. Simplified, "ORF calling" is the process of performing a six-frame translation of the DNA sequence and identifying all ORFs longer than a chosen threshold. Several software algorithms are available for both eukaryotic and prokaryotic gene calling, including GLIMMER (22), Genezilla (44), NCBI ORF finder, GENEMARK.hmm (43), and Genie (40). In addition to gene calling, software packages exist to identify other genomic features such as Rho-independent terminators and bacterial ribosome-binding sites upstream of the initiation codon of ORFs (e.g., TransTerm and RBSfinder [26]).

After genes are found, protein-level annotation attempts to define the protein complement of an organism, name the proteins, and assign them putative functions. Most approaches to computational protein annotation rely on sequence similarity comparisons with sequences previously annotated in preexisting databases. A typical protein annotation pipeline would search for similarities by using BLASTP (2) against several databases available through the National Center for Biotechnology Information (http://www.ncbi.nlm.nih.gov). Another approach is to search against a database of functional domains. One commonly used database is PFAM (9), which contains alignments and Hidden Markov Model profiles for almost 8,000 protein families. Other databases available include those available through the Expert Protein Analysis System (ExPASy) Proteomics Server such as Prosite (33) and SWISS-PROT (8).

Finally, process-level annotation concerns itself with relating the genome to biological processes. Several online databases are available to aid researchers in placing genes and gene products into a cellular context. However, the wide variation in terminology can make

Table 44.1 Microorganisms of foodborne significance whose genomes have been sequenced

Organism	Importance	NCBI accession no.	Genome size	Plasmids
Bacillus cereus 10987	Diarrheal and emetic illness, both infective and intoxicant	NC_003909	5.2 Mb	pBc10987 (NC_005707)
Bacillus cereus ATCC 14579		NC_004722	5.4 Mb	pBClin15 (NC_004721)
Bacillus cereus Zk		NC_006274	5.3 Mb	pE33L9 NC_007107) pE33L8 (NC_007106) pE33L54 (NC_007105) pE33L5 (NC_007104) pE33L466 (NC_007103)
Bacillus licheniformis ATCC 14580	Spoilage	NC_006270	4.2 Mb	pBL63 (NC_006270)
Bacillus subtilis subsp. *subtilis* 168	Spoilage	NC_000964	4.2 Mb	
Bifidobacterium longum NCC2705	Probiotic	NC_004943	2.3 Mb	pBLO1 (NC_004307)
Campylobacter jejuni NCTC11168	Diarrheal illness, infective	NC_002163	1.6 Mb	pTet (NC_007141)
Campylobacter jejuni RM1221		NC_003912	1.7 Mb	
Clostridium perfringens 13	Diarrheal illness, intoxicant	NC_003366	3.0 Mb	PCp13 (NC_003042)
Clostridium perfringens ATCC 13124		NC_008261	3.3 Mb	
Clostridium perfringens SM101		NC_008262	2.9 Mb	NC_008263 NC_008264
Enterococcus faecalis V583	Indicator of water quality	NC_004668	3.2 Mb	pTEF1 (NC_004669) pTEF2 (NC_004671) pTEF3 (NC_004670)
Escherichia coli O157:H7 EDL933	Diarrheal illness, hemolytic uremic syndrome; infective	NC_002655	5.5 Mb	pO157 (NC_007414)
Escherichia coli O157:H7 VT2-Sakai		NC_002695	5.4 Mb	pO157 (NC_002128) pOSAK1 (NC_002127)
Helicobacter pylori 26695	Gastric ulcers, infective	NC_000915	1.6 Mb	
Helicobacter pylori J99		NC_000921	1.6 Mb	
Helicobacter pylori HPAG1		NC_008086	1.6 Mb	pHPAG1 (NC008087)
Lactobacillus acidophilus NCFM	Probiotic	NC_006814.1	1.9 Mb	
Lactobacillus brevis ATCC 367	Starter culture	NC_008497	2.3 Mb	NC_008498 NC_008499
Lactobacillus casei ATCC 334	Starter culture	NC_008526	2.9 Mb	NC_008502
Lactobacillus delbrueckii subsp. *bulgaricus* ATCC 11842	Starter culture	NC_008054	1.9 Mb	
Lactobacillus delbrueckii subsp. *bulgaricus* ATCC BAA-365	Starter culture	NC_008529	1.9 Mb	
Lactobacillus gasseri ATCC 33323	Probiotic	NC_008530	1.9 Mb	
Lactobacillus johnsonii NCC533	Probiotic	NC_005362	1.9 Mb	

(Continued)

Table 44.1 Microorganisms of foodborne significance whose genomes have been sequenced *(Continued)*

Organism	Importance	NCBI accession no.	Genome size	Plasmids
Lactobacillus plantarum WCFS1	Vegetable fermentation, probiotic	NC_004567	3.3 Mb	pWCFS101 (NC_006375) pWCFS102 (NC_006376) pWCFS103 (NC_006377)
Lactobacillus sakei subsp. *sakei* 23k	Starter culture	NC_007576	1.9 Mb	
Lactobacillus salivarius subsp. *salivarius* UCC118	Probiotic	NC_007929	1.8 Mb	pSF118-20 (NC_006529) pSF118-44 (NC_006530) pMP118 (NC_007930)
Lactococcus lactis subsp. *cremoris* SK11	Cheese production	NC_008527	2.4 Mb	NC_008503 NC_008504 NC_008505 NC_008506 NC_008507
Lactococcus lactis subsp. *lactis* IL1403	Milk fermentation	NC_002662	2.3 Mb	
Leuconostoc mesenteroides subsp. *mesenteroides* ATCC 8293	Starter culture	NC_008531	2.0 Mb	pLEUM1 (NC_008496)
Listeria monocytogenes 4b F2365	Septicemia, meningitis (or meningoencephalitis), encephalitis, and intrauterine or cervical infections in pregnant women, infective	NC_002973	2.9 Mb	
Listeria monocytogenes 4b H7858		NZ_AADR00000000	2.9 Mb	
Listeria monocytogenes EGD-e		NC_003210	2.9 Mb	
Pediococcus pentosaceus ATCC 25745	Starter culture	NC_008525	1.8 Mb	
Salmonella enterica subsp. *enterica* serovar Choleraesuis SC-B67	Diarrheal illness, infective	NC_006905	4.8 Mb	pSCV50 (NC_006855) pSC138 (NC_006856)
Salmonella enterica serovar Paratyphi ATCC 9150		NC_006511	4.5 Mb	
Salmonella enterica serovar Typhi CT18		NC_003198	4.8 Mb	pHCM1 (NC_003384) pHCM2 (NC_003385)
Salmonella enterica serovar Typhi Ty2		NC_004631	4.7 Mb	
Salmonella enterica serovar Typhimurium LT2 SGSC1412	Diarrheal illness, infective	NC_003197	4.8 Mb	PSLT (NC_003277)
Shigella boydii Sb227	Diarrheal and emetic illness, infective	NC_007613	4.5 Mb	pSB4-227 (NC_007608)
Shigella dysenteriae Sd197	Dysentery	NC_007606	4.4Mb	pSD1_197 (NC_007607)
Shigella flexneri 2a 2457T	Diarrheal and emetic illness, infective	NC_004741	4.5 Mb	
Shigella flexneri 2a 301		NC_004337	4.6 Mb	pCP301 (NC_004851)
Shigella flexneri 5 8401		NC_008258	4.6 Mb	
Shigella sonnei Ss046	Diarrheal and emetic illness, infective	NC_007384	4.8 Mb	pSS (NC_007385)

(Continued)

Table 44.1   *(Continued)*

Organism	Importance	NCBI accession no.	Genome size	Plasmids
Staphylococcus aureus MW2	Diarrheal and emetic illness, infective and intoxicant	NC_003923	2.8 Mb	
Staphylococcus aureus Mu50		NC_002758	2.8 Mb	
Staphylococcus aureus N315		NC_002745	2.8 Mb	
Staphylococcus aureus subsp. aureus COL		NC_002951	2.8 Mb	pT181 (NC_006629)
Staphylococcus aureus subsp. aureus MRSA252		NC_002952	2.9 Mb	
Staphylococcus aureus subsp. aureus MSSA476		NC_002953	2.7 Mb	pSAS (NC_005951)
Streptococcus thermophilus CNRZ1066	Starter culture	NC_006449	1.8 Mb	
Streptococcus thermophilus LMD-9		NC_008532	1.9 Mb	NC_008500 NC_008501
Streptococcus thermophilus LMG 18311		NC_006448	1.8 Mb	
Vibrio cholerae El Tor N16961 chromosome I	Chlolera, infective	NC_002505	2.9 Mb	
Vibrio cholerae El Tor N16961 chromosome II		NC_002506	1.1 Mb	
Vibrio vulnificus CMCP6 chromosome I	Gastroenteritis and septicemia, infective	NC_004459	3.2 Mb	
Vibrio vulnificus CMCP6 chromosome II		NC_004460	1.8 Mb	
Vibrio vulnificus YJ016 chromosome I		NC_005139	3.3 Mb	pYJ016 (NC_005128)
Vibrio vulnificus YJ016 chromosome II		NC_005140	3.4 Mb	
Xanthomonas campestris pv. campestris 8004	Xanthan gum production	NC_007086	5.2 Mb	
Xanthomonas campestris pv. campestris ATCC 33913		NC_003902	5.1 Mb	
Xanthomonas campestris pv. vesicatoria 85–10		NC_007508	5.2 Mb	pXCV2 (NC_007504) pXCV19 (NC_007505) pXCV38 (NC_007506) pXCV183 (NC_007507)
Saccharomyces cerevisiae chromosomes 1–16	Fermentations of alcohol and food	NC_001133 through NC_001148	12 Mb	
Hepatitis A virus	Fever, abdominal pain, and jaundice	NC_001489	7.4 kb	
Hepatitis E virus	Fever, abdominal pain, and jaundice	NC_001434	7.2 kb	
Norwalk virus	Acute gastroenteritis	NC_001959	7.6 kb	
Cryptosporidium parvum chromosomes 1–8	Diarrheal and emetic illness, infective	NC_006980 through NC_006987	9.1 Mb	

effective searching difficult, whether by computers or by people. The Gene Ontology (GO) project is a standardized vocabulary for describing the functions of genes and their products (7). The GO attempts to describe proteins on three levels: molecular function, biological process, and cellular component. Molecular function describes activities such as catalytic or binding activity. A biological process is a series of events carried out by one or more groups of molecular functions. Cellular components describe gene products in terms of the cellular structures that they comprise. The Kyoto Encyclopedia of Genes and Genomes (KEGG) (http://www.genome.ad.jp/kegg/) is an online resource that provides knowledge to help researchers tie together genomic information with pathways and chemical information (35). For example, Fig. 44.2 illustrates the known metabolic pathways for folate metabolism and identifies which genes (denoted by shaded boxes) are predicted within the genome of *Lactococcus lactis* (12). Using this type of analysis, this genome revealed components for both folate (vitamin B_{11}) and riboflavin (vitamin B_2) biosynthetic pathways. By knowing the steps of these pathways, Sybesma et al. (60) were able to apply genetic approaches for metabolic engineering of these pathways and simultaneously overproduce both vitamins.

Genomes can also be displayed graphically using tools that present either intrinsic DNA information, such as GC-Skew, or annotation data, such as ORF location and deduced function. Color Plate 5 shows the genome map of *Lactobacillus plantarum* WCFS1 with the predicted origin of replication at the top (38). The outer two circles show ORFs encoded on the positive strand in red and ORFs encoded on the negative strand in blue. Additionally, GC-Skew is shown in green and the G+C content is shown in black. Finally, the innermost rings show deduced gene function such as prophage-related functions in green, *IS*-like elements in purple, rDNA operons in black, and tRNA-encoding genes in red. Annotation of the features of a genome can also reveal many interesting structural properties that may be unique to a genome. For example, the *Lactobacillus acidophilus* genome contains a SPIDR (SPacers Interspersed Direct Repeats) locus (1). Within this 1.5-kb region were 32 nearly perfect repeats of 29 bp separated by unique 32-bp spacers. SPIDR loci constitute a novel family of repeat sequences that are present in bacteria and archaea. SPIDR loci have now been identified in more than 40 microorganisms, and they can provide a unique molecular signature that can be used in strain identification and tracking.

COMPARATIVE GENOMICS

With the number of complete bacterial genomes increasing rapidly, one powerful approach to defining unique or conserved gene content is comparative genomics, via an in silico analysis. This was beautifully illustrated when, after genome sequencing of *Listeria monocytogenes* and *Listeria innocua*, a comparative analysis of the two genomes revealed extensive colinearity (31). It was found that 270 genes (9.5%) were specific to the *L. monocytogenes* genome whereas only 149 (5%) were specific to *L. innocua*. Many of the genes specific to *L. monocytogenes* were key to survival in the mammalian gut, including those encoding bile salt hydrolases and acid tolerance via glutamic acid decarboxylases, as well as a virulence locus responsible for infection and motility during pathogenesis (Color Plate 6). Similar comparative genomic analysis has revealed a number of pathogenicity islands in other foodborne pathogens. The genomic analysis also revealed that pathogenic *L. monocytogenes* and nonpathogenic *L. innocua* probably were derived from a common ancestor, with the latter subsequently losing most of the genes directing pathogenesis. Therefore, in silico comparative analysis of complete genomes can reveal important clues to a microorganism's genetic content, capabilities, and relationship to other microorganisms. A number of software tools are available for these comparisons, including the Artemis Comparison Tool (ACT; Sanger Center) and MUMmer (41).

Genome sequence information has also provided key insights into the evolution and adaptation of microorganisms to specific foods (11). *Streptococcus thermophilus* is a lactic acid bacterium used worldwide for the manufacture of yogurt and some hard cheeses; its estimated global market value is around $40 billion. The natural habitat for this bacterium is raw milk and artisanal dairy starter cultures that have been propagated by cheesemakers for centuries. Sequencing of the genome of two yogurt strains of *S. thermophilus* and comparison to other streptococci revealed that the bacterium had undergone extensive rounds of genome decay, with over 10% of the genes characterized as nonfunctional "pseudogenes." Notably, with the exception of lactose metabolism, many of the other gene regions associated with carbohydrate transport and metabolism were also degraded. Genes commonly associated with pathogenic streptococci (e.g., *S. agalactiae* and *S. pneumoniae*) were either lost or inactivated due to mutations or transposon insertions. Bolotin et al. (11) concluded that this bacterium has evolved specifically to reside in milk and that in the adaptation process it eliminated or degraded much of its genomic complement that was not required for growth in this environment.

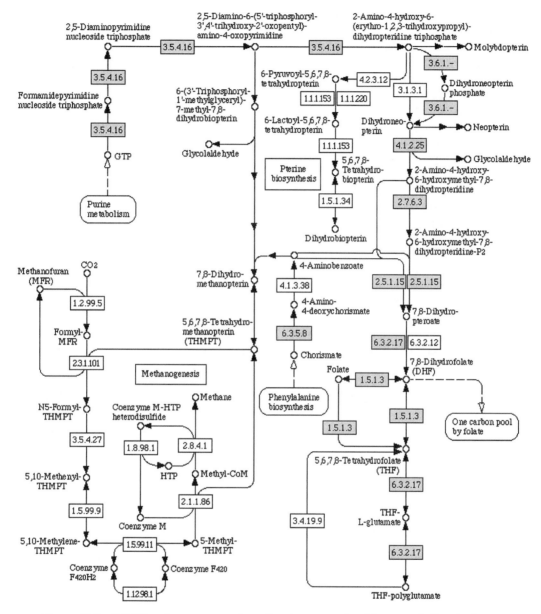

Figure 44.2 KEGG pathway map and predicted enzymes for folate metabolism in *Lactococcus lactis*. Pathway intermediates and reaction products are shown with EC numbers for enzymes which catalyze these reactions. Shaded boxes indicate catalytic activities encoded in the *L. lactis* genome.

FUNCTIONAL GENOMICS

Once a genome sequence is generated and annotated, an important step is to identify or confirm the function of the genes found therein. While annotating genes on the basis of homology is a good way to begin this task, a number of genes (30 to 40%) in each bacterial genome have no homology to any genes found in the databases. Also, it is possible that a gene with a known phenotype in one microorganism may play a different role in another organism. The discipline of functional genomics deals with defining the roles of genes in their appropriate micro-organisms. This task is usually carried out through two approaches: forward genetics and reverse genetics. For-ward genetics deals with identifying the genotype of a microbe with a known phenotype or feature. Routinely, attempts are made to create random mutations and screen for derivatives that exhibit a change in the phenotype under investigation. Mutant organisms can be created in different ways. Point mutations, or changes to a single nucleotide, can be created by using mutagens such as ethyl methanesulfonate. This approach is effective but often leads to secondary mutations that cause additional, unrelated changes in phenotype. Another method to cre-ate random mutations is via transposable elements, i.e., DNA sequences that have the ability to randomly insert into the genome. The advantage of using these elements for mutagenesis is that they can be subsequently located in the genome and the precise position of the insertion mutation can be mapped, facilitating the identification of the genetic loci encoding the phenotype.

In contrast, reverse genetics utilizes the known genome sequence of the microorganism, or at the very least the sequence of the gene of interest. Reverse genetics begins with a directed change to the genotype of an organism and determination of the resulting phenotype. Muta-tions are directed into the targeted DNA sequence by using gene inactivation or replacement strategies. Most are based on integration via homologous recombina-tion of a cloning vector that replicates conditionally (for example, a plasmid that replicates only at a certain tem-perature range) in the bacterial cell. The vector encodes a selective marker such as an antibiotic resistance gene and a region of homology that is unique to the targeted gene (Fig. 44.3). Once the vector is transformed into the bacterial cell, homologous recombination events in the population are expected to occur at frequencies of 10^{-6} to 10^{-8}. This event can be detected by selecting for antibiotic-resistant derivatives under conditions where plasmid replication cannot occur. Once a potential inte-grant is isolated, the position of the integration event is confirmed by detecting the junction fragments by PCR or DNA-DNA hybridization (Southern blotting). Once confirmed, integrants or deletion mutants can be charac-terized for any changes in phenotype that may elucidate the role of the gene or operon under study. Care must be taken to minimize the downstream or polar effects of integrants. Therefore, gene deletions or replacements are the preferred approach and require double crossover events (Fig. 44.4). The advantage of this method is that genes are deleted or inactivated and the resulting deriv-ative no longer contains any vector DNA or selection markers (e.g., antibiotic resistance genes). Using targeted

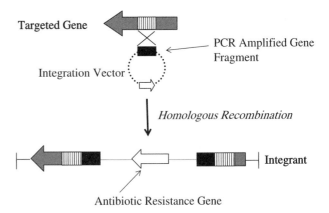

Figure 44.3 Plasmid insertion into a gene through homol-ogous recombination for inactivation of gene function. The phenotype of the mutant can then be analyzed to investigate the function of the gene.

knockout or deletion strategies to correlate phenotypes also demands one more step. Using genetic complementa-tion, the original gene is reintroduced into the mutant, in *trans*, and recovery of the original phenotype is expected (30, 50, 57).

With knowledge of the sequence of a gene of interest, a gene or operon of interest can also be cloned and over-expressed on high-copy-number plasmids with or without strong promoters. The use of gene expression or inacti-vation strategies to elucidate or confirm gene function is a very powerful approach to understanding microbial metabolism, physiology, and interactions. Genomics provides the information and tools to understand, con-trol, and, where appropriate, exploit the benefits of microorganisms in foods.

DNA MICROARRAYS

DNA microarray analysis is a relatively new technol-ogy that allows investigators to take a genome-wide approach to biological systems. Their power comes from the exploitation of sequence complementarity between two strands of duplex DNA, providing both sensitivity and specificity (24, 58). While many different microarray platforms are available, all operate under the same principle: sequence-specific nucleic acid probes are immobilized onto a solid substrate and are then used, via hybridization, to probe a pool of labeled target nucleic acid for the presence of these specific sequences. The abundance of the target sequence is then determined from the amount of label detected at the location of a specific probe.

The principles behind DNA microarray technol-ogy make it very applicable to many different uses that

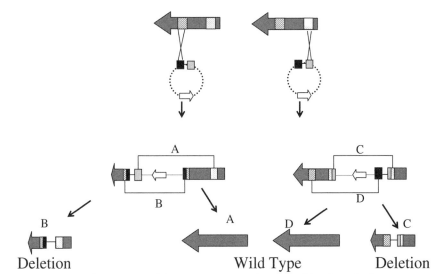

Figure 44.4 A replacement or deletion mutant can be created by first cloning two noncontiguous portions of a gene into an integration vector. The vector integrates into a targeted gene within one region of homology (black or light gray regions). Excision of the plasmid from the integrant structure can occur in a manner that either resolves the original gene (wild type) or leaves the deleted version. Points of resolution at steps A, B, C, or D result in various combinations, as illustrated.

include comparative genomics and global gene expression analysis. Microarrays can be used to determine the presence or absence of similar sequences between two different genomes. In this case, one genome serves as a reference and is spotted onto the array while a second test genome is then hybridized against the array. Currently, the main large-scale application for microarrays is in gene expression analysis (58), and different microarray solutions exist. In this overview, the "home-baked" cDNA microarray and Affymetrix GeneChip platforms are discussed.

The most flexible and accessible platform for academic users seems to be the home-baked cDNA microarray (Color Plate 7). The fabrication of these arrays begins with probe selection. In general, probes are PCR amplified from clones from an EST (for "expressed sequence tags") library or, in the case of microbes with complete DNA sequence information, using specific primers to create gene-specific amplicons. These probes are then purified and spotted individually onto glass slides. In the simplest form of a microarray experiment, the array is used to probe pools of fluorescently labeled cDNA reverse transcribed from total RNAs isolated from reference cells and test cells under some defined condition, e.g., cells grown in glucose (reference) versus cells grown in lactose (test). Once the targets have been hybridized to the array, it is scanned using a laser scanner, generating an image for each dye used. A common example is to label the reference sample with Cy3, which is represented

in a color image as red spots, and to label the test sample with Cy5, which is represented in a color image as green. When the two images are overlaid, genes which are expressed more in the reference sample show up as red spots whereas those that are expressed more in the test sample show up as green spots. A gene that is equally expressed in both samples will show up as a yellow spot. One can then determine the relative message abundance between the two samples based on the signal generated (17, 24).

The Affymetrix GeneChip platform is the most widely used commercial microarray solution. It is based on 20-base oligonucleotides synthesized in situ on a glass surface using photolithography and solid-phase DNA synthesis. An important feature of the GeneChip is the amount of redundancy on the array: for each gene there are 20 probes representing different portions of the gene. Additionally, for each probe there is a single-base mismatch probe which is used to subtract any cross hybridization effects (42). Three major limitations exist when this system is compared to home-baked microarrays:

1. The Affymetrix system is a single-dye system, which excludes the ability to perform competitive hybridization experiments using target cDNA from two conditions.
2. The system is proprietary. Every component required for use of the system, including the reagents, the fluidics station, and the scanner, must be purchased.

3. The system is limited to research involving the major model systems. Custom probe array production is available but is prohibitively expensive.

The major advantage of microarrays is the amount of throughput. Microarrays have given researchers the ability to assay for the expression of all the genes in a particular genome rather than using a gene-by-gene approach. Using traditional hybridization-based methods such as Northern blots, researchers were able to assay for expression of only one gene during a particular hybridization experiment. Slot/dot blots were able to increase the number of genes that could be assayed, but researchers were still severely limited in the number of probes that could be blotted on a membrane. With the advent of computer-controlled precision robotics, researchers were able to miniaturize the slot/dot blots while increasing the number of probes arrayed up to 1,000-fold (14).

Genome content between closely related strains or species can also be compared by hybridizations to whole-genome arrays representing the microorganism where the complete genome sequence is available. Doumith et al. (23) used this approach to create a DNA array of the "flexible" part of *Listeria* genomes that had been sequenced. Probing that array with the DNA from 113 *L. monocytogenes* strains, it was found that 93 of the strains harbored all the previously identified virulence factors. In addition, the method identified 30 genes that were *L. monocytogenes* specific, which may prove useful for tracking strains in listeriosis outbreaks.

PROTEOMICS

Microarray technologies provide vast quantities of global gene expression data at the level of mRNA expression. However, it has been shown that in many cases, the abundance of mRNA transcripts does not directly correlate to the final amount of protein expressed (3). Proteomic techniques are required to provide function and quantitative data to complement genomic data. Proteomics can be defined as "the use of quantitative protein-level measurements of gene expression to characterize biological processes and decipher the mechanisms of gene expression control" (4).

Proteomics has been traditionally associated with the display of a large number of proteins from a cell line or microorganism by using two-dimensional polyacrylamide gel electrophoresis (2D-PAGE). 2D-PAGE is a technique where proteins are separated according to isoelectric point by isoelectric focusing in the first dimension and according to molecular weight (Fig. 44.5) by sodium dodecyl sulfate-PAGE in the second dimension (51). 2D-PAGE

has the ability to resolve a large number of proteins and discern changes in charge and mass due to posttranslational modification. However, it was not until the advent of biological mass spectrometry (MS) that the identity of the resolved proteins could be easily elucidated. Following separation on 2D gels, proteins are stained to allow visualization, excised from the gel as gel fragments, and then digested into peptides with sequence-specific proteases such as trypsin. The resulting peptides can be extracted and sequenced using MS-based methods.

There are two main approaches to MS-based protein identification. In peptide mass mapping, the mass spectrum of the digested peptide mixture is acquired using matrix-assisted laser desorption ionization (MALDI). The resulting mass spectrum serves as a peptide mass fingerprint for the protein, which can then be identified by searching against protein databases (13). This approach can be readily scaled up and automated so that the mass spectra of hundreds of proteins can be acquired and searched against databases. The tandem mass spectrometric (MS/MS) method is based on the fragmentation and amino acid sequencing of individual peptides in the protein digest. Using this approach, the peptide mixture is ionized directly from the liquid phase by electrospray ionization. The ionized peptides are sprayed into a tandem mass spectrometer, which can then resolve individualized peptides in the mixture. Single peptides are isolated and further resolved into amino- or carboxy-terminal-containing fragments. While the MS/MS method is more complex and less scalable than mass peptide fingerprinting, the sequence information acquired using this technique is more specific for the identification of an individual protein rather than a listing of peptide masses. MS/MS data can also be searched against both protein and nucleic acid databases.

The primary use of these 2D-PAGE/MS methods is in protein expression profiling. Analogous to gene expression profiling using microarrays, proteins from an organism exposed to different conditions are resolved on separate 2D gels. The gels are subsequently analyzed using computer software to identify changes in protein expression by bacterial cells exposed to different conditions (19, 52).

Conventional 2D-PAGE technology relies on comparisons of protein spots from at least two different gels. Due to gel-to-gel variation, it is not possible to directly superimpose the images of two separate gels, thus adding further complexity to the analysis of these comparisons. The introduction of differential gel electrophoresis (DIGE) has addressed these pitfalls of 2D-PAGE. Analogous to microarrays, 2D-DIGE makes use of fluorescent dyes that covalently label all proteins in a protein sample. By labeling protein samples with different fluorescent

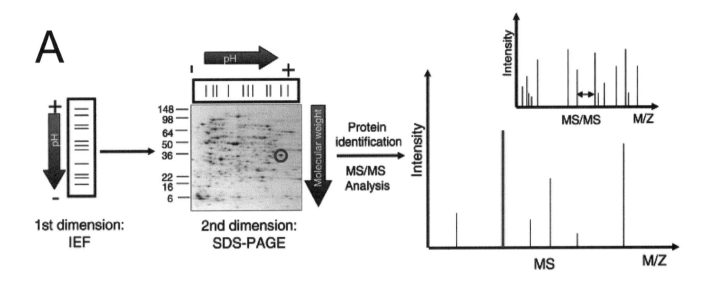

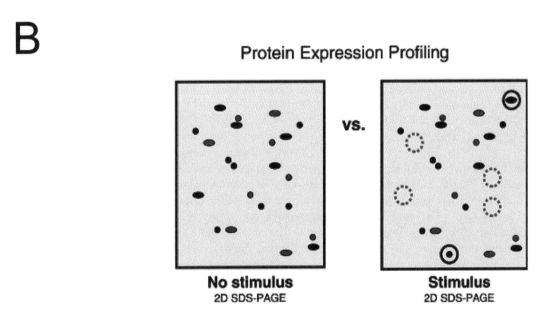

Figure 44.5 Proteomic methods. (A) 2D-PAGE/MS. In the first dimension, proteins are separated based on isolectric point (pI) by using isoelectric focusing (IEF). The proteins migrate along a pH gradient until they reach their pI, at which point they carry no net charge and stop migrating. In the second dimension, proteins are further separated according to molecular weight using sodium dodecyl sulfate-PAGE (SDS-PAGE). Gels are stained to identify protein bands. Individual spots (circles) are excised from the gel, trypsin digested, and sequenced using MS/MS. (B) Protein expression profiling. By overlaying images of 2D-PAGE gels, comparisons can be made between the proteomes of different organisms or differences in protein expression of a single organism in different conditions. Downregulated (dotted circles) and upregulated (solid circles) proteins can be visualized. Adapted from reference 53.

dyes that have nonoverlapping excitation and emission profiles, it is possible to compare the protein expression from several conditions on the same gel. This method solves many of the issues associated with run-to-run variation in conventional 2D-PAGE (61).

Alternative gel-free methods based on liquid chromatography (LC) have been developed, including direct LC separation of digested peptides coupled to tandem mass spectrometry (LC/MS/MS) and multidimensional LC coupled to mass spectrometry (multidimensional protein identification technology [MudPIT]) (21, 47, 65). These methods perform the protein separations using LC-based methods, dispensing with issues such as resolvability of proteins and sensitivity of detection in gel-based methods. Additionally, with the availability of robotics for LC methods, automation and increased throughput can be introduced.

GENOMIC AND PROTEOMIC ADVANCES IN FOOD MICROBIOLOGY

One use of genomic approaches to food microbiology has been the identification and characterization of genes that confer resistance to stressors such as acid or bile. These genes can be important in understanding how both foodborne pathogens such as *E. coli* O157:H7 and commensals such as some *Lactobacillus* spp. survive the rigors of the intestinal tract or how they survive in acidified foods. Bacteria are able to survive in these environments because they induce an acid tolerance response that allows them to adapt. While this physiological phenomenon was identified before the advent of bacterial genomics (29), the availability of sequenced bacterial genomes allows the use of transcriptional (microarrays) and translational (proteomic techniques) analyses to characterize microbial responses to acid and pH and elucidate which genes are expressed and how they are regulated.

Probably one of the best characterized acid resistance mechanisms is the Gad system of *E. coli*. This system is present in all strains of *E. coli* that are of public health significance, including O157:H7 (32). The acid resistance of this system is contributed by three proteins encoded by the *gadA*, *gadB*, and *gadC* genes. The *gadA* and *gadBC* genes exist in different parts of the genome (Fig. 44.6). The GadA and GadB proteins catalyze the exchange of the α-carboxyl of the amino acid glutamate for a proton in the environment. This reaction leads to the creation of a molecule of carbon dioxide and a molecule of γ-amino butyric acid (GABA). Consumption of an H⁺ ion from the cytoplasm contributes to an increase in the intracellular pH. GadC is a membrane transporter known as an antiporter that expels GABA from the cell and imports fresh glutamate (15, 56). These genes were first discovered before the sequence of *E. coli* was published in 1997, but an understanding of how they are regulated and which other microorganisms use them has come about using the tools of genomics.

While the molecular mechanisms of this defense system are simple, the regulatory system that controls them is quite complex and not entirely understood. The main player in the activation of the acid resistance system is

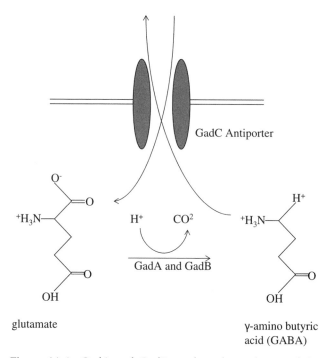

glutamate   γ-amino butyric
                                               acid (GABA)

Figure 44.6 GadA and GadB catalyze the exchange of the α-carboxyl of glutamate for a proton in the environment, leading to the creation of a molecule of carbon dioxide and one molecule of GABA. GadC is an antiporter that expels GABA from the cell and imports fresh glutamate.

the GadE protein, a transcriptional activator that binds upstream of *gadA* and *gadBC* and directs the binding of RNA polymerase to transcribe these genes (38). What controls *gadE*, however, is a question that many researchers are trying to answer. The first regulatory circuit discovered, the EvgAS circuit, was first characterized using microarrays. Many known acid resistance genes, including the Gad genes, were induced when the regulator EvgA was overexpressed in *E. coli* cells. Also, strains overexpressing the EvgA protein showed greater acid resistance than did wild-type cells during the exponential phase of growth. EvgA is part of a two-component regulatory system in which the bacterial cell senses specific environmental changes (such as a drop in pH) and responds to them by upregulating genes designed to deal with that particular type of change. A histidine protein kinase (EvgS) is the sensing part of this system, and once it receives a signal, it "turns on" the response regulator (EvgA), which transcriptionally activates specific genes (Fig. 44.7). Use of a microarray to compare EvgA overexpression strains to an *evgA* deletion strain identified several other genes upregulated by EvgA. These genes were *ydeO*, *gadA*, *gadB*, *gadC*, and *gadE*. Based on a highly conserved DNA sequence found upstream of the genes upregulated by EvgA, the DNA-binding site for

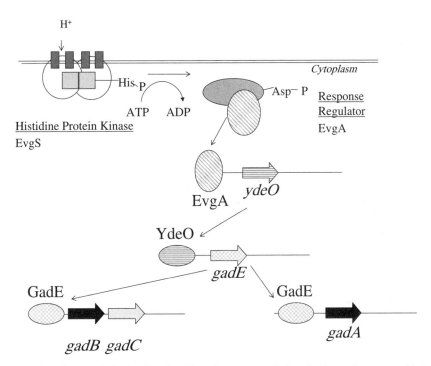

Figure 44.7 The EvgA/S circuit of acid resistance regulation is dependent on a histidine protein kinase (EvgS) which senses an environmental change, causing it to activate its corresponding response regulator (EvgA), which is then able to act as a transcriptional regulator. The regulation follows a pathway to produce GadE, which ultimately induces the transcription of GadA and GadBC (see Fig. 44.6).

this protein was determined and the rest of the *E. coli* genome was then examined for potential binding sites for EvgA. This binding site was present upstream of *ydeO*, indicating that EvgA acts as a transcriptional activator for this gene. Thus, the first regulatory circuit for the Gad system was discovered using genomic data, microarrays, and bioinformatics to identify key genes and their regulatory sequences (45, 46).

While the details of the Gad system were described primarily by experiments with *E. coli*, genomics and bioinformatics have enabled researchers to identify and study the effects of these genes in other microorganisms. In *L. monocytogenes*, for example, the *gadA* and *gadBC* genes were identified based on similarity to these genes in *E. coli*, *Lactococcus lactis*, and *Mycobacterium tuberculosis*. After creating deletion mutants of these genes and testing their survival in synthetic and porcine gastric fluids, it was determined that *L. monocytogenes* absolutely depends on the Gad system to maintain pH homeostasis (20).

In addition to the data generated by using HCl to lower the pH, the *gadA* and *gadBC* genes have been shown to be upregulated in *E. coli* in the presence of acetic acid, an organic acid typical of those encountered in acidified and fermented foods (6). *E. coli* O157:H7 has been known to

survive in low-acid food products, including unpasteurized apple cider, where the pH can be below 4.0 (34). Glutamate, the proton-accepting amino acid that is central to this system, is used extensively in food products as a flavor enhancer or as a way to adjust acidity. It is also present in protein-rich foods such as meat and milk. Many vegetables contain large amounts of bound glutamate, and their fermented counterparts show increased levels of free glutamate (20). In this regard, it is likely that foods and food components can precondition some foodborne pathogens for improved survival and pathogenicity, once ingested.

While the ability to survive the acidic conditions in the stomach or food products is important to food-associated microbes, infection in the intestinal tract requires that bacteria possess the ability to withstand additional types of stresses, such as those imposed by bile. Bile acts as a detergent that can dissolve the phospholipid cytoplasmic membranes of bacteria, leading to death of the organisms (18). *L. monocytogenes* is able to withstand the presence of high concentrations of bile. In fact, it can withstand levels so high that it can colonize the gallbladder, which is the human body's primary repository for bile (10). While the mechanisms through which bacteria tolerate bile are not entirely understood, some genes have been identified that play important roles. One of the most intensively studied groups are those encoding bile salt hydrolases (BSH). Bile salts exist primarily as steroid rings attached through an amide bond to an amino acid group, either glycine or taurine (Fig. 44.8). Bile salt hydrolases catalyze the hydrolysis of the conjugated salt to release the amino acid. Disruption of the *bsh* gene in *L. monocytogenes* leads to greater sensitivity to bile (10). Three mechanisms through which BSHs contribute to bile tolerance have been proposed: (i) conferring a nutritional advantage on the cell by liberating amino acids; (ii) facilitating incorporation of the cholesterol moiety into the cell membrane, strengthening it against the detergent action of bile; and (iii) acting as a detoxification mechanism against bile salts (10). While these hypotheses are still under investigation, a comparison between the genomes of *L. monocytogenes* and the nonpathonogenic *L. innocua* showed that only the pathogen carried the gene for BSH. Bioinformatic analysis of the *bsh* promoter region indicated that *bsh* is under the control of the transcriptional regulator PrfA, which controls expression of other known virulence genes in *L. monocytogenes* (25). A study comparing wild-type *L. monocytogenes* and a derivative without the *bsh* gene (Δ*bsh*) in mice revealed that the wild-type strain caused significantly more deaths than did the Δ*bsh* strain. These studies indicate that BSH contributes to the virulence of *L. monocytogenes*, but the mechanism remains undefined. The contribution of genome sequences and functional genomics has greatly facilitated our understanding of the pathogenicity of *L. monocytogenes*.

The transcriptional regulator PrfA is also regulated by another factor, sigma B, which is the analogue to RpoS found in *E. coli* and other gram-negative bacteria. Sigma B is the general stress response sigma factor in gram-positive bacteria and controls stress responses in *L. monocytogenes*, including salt tolerance and growth at low temperatures (27, 36, 37, 49). Tolerance to both of these stressors is mediated in much the same way via the uptake of small organic compounds known as compatible solutes that act to increase osmolarity within the cell without disrupting cellular functions. The osmolytes utilized by *Listeria* are glycine-betaine, carnitine, and proline, with proline being the only osmolyte that *Listeria* can synthesize on its own. The cell relies on osmolyte transporters to import these molecules: Gbu, a transporter for glycine-betaine; OpuC, the carnitine transporter; and BetL, another glycine-betaine uptake system. Studies of mutants with mutations in each of these genes have shown that the mutants have a decreased ability to take up osmolytes and to grow at low temperatures or tolerate salt (39, 48, 55, 63, 64). Of particular interest is OpuC, which, when mutated, leads to a decrease in virulence of *Listeria*. Microarray studies have shown that OpuC is also the only osmolyte transporter regulated by sigma B, again illustrating that sigma B-controlled genes such as *opuC* and *bsh* are associated with both stress tolerance and virulence (5, 16, 28, 55).

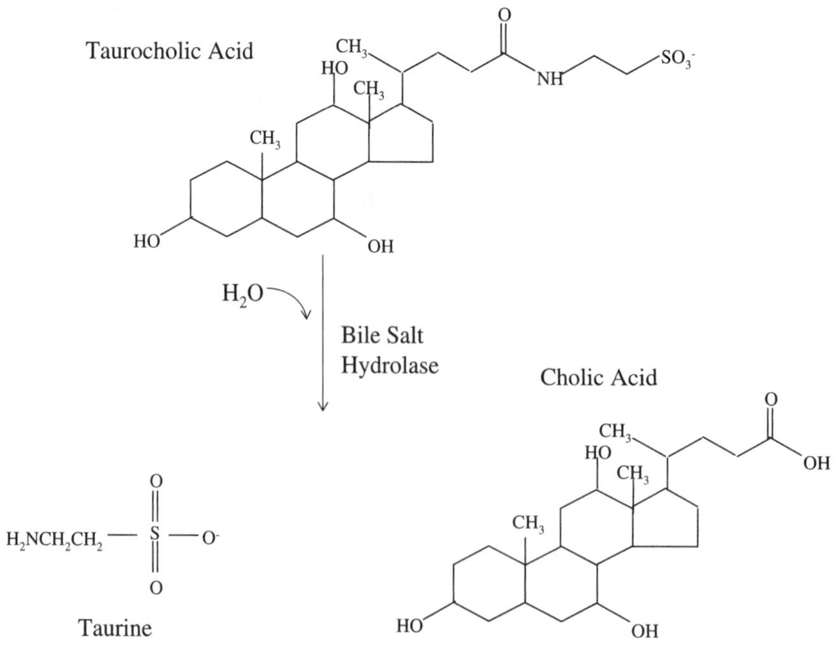

Figure 44.8 Bile salt hydrolases catalyze the hydrolysis of the peptide bond between the amino acid and the cholesterol-derived backbone of the bile salt.

Trost et al. (62) were also able to characterize the secretory proteome of *L. monocytogenes* by using the complementary techniques of 2D-PAGE/MS and LC/MS/MS (62). The researchers identified 105 proteins in the culture supernatant, including the 8 proteins known to be involved in virulence. Comparison of the secretory proteomes of *L. monocytogenes* and *L. innocua* identified a number of additional proteins unique to *L. monocytogenes* which may be involved in pathogenicity and/or adaptation to its lifestyle. Additionally, by combining in silico comparative genomic analysis with these proteomic techniques, the authors determined that the differences between the pathogenic and nonpathogenic *Listeria* species appear most strongly in the secretory proteome.

CONCLUSIONS

The rise of "omic" technologies has led to an explosion of DNA, protein, and metabolic information about microorganisms associated with our foods. With this information, it is now possible to understand the complete genetic complement of pathogenic, spoilage, and bioprocessing organisms and determine how environmental conditions affect the expression of key genetic traits. Pathogens can be better tracked in foods and outbreaks because of their signature sequences revealed in genomes. Pathogenicity islands have exposed the key gene combinations underlying the survival, infection, and disease properties of foodborne pathogens. Different bioprocessing and probiotic strains can be compared for their genetic content, and their beneficial capabilities can be reliably predicted. Foods and and their microenvironments could potentially be better designed and formulated to minimize the expression of undesirable pathogenic traits (e.g., acid tolerance, virulence, and toxin formation) or to optimize the expression of beneficial properties in desirable microorganisms (e.g., cryoprotection, acidification rates, and adherence to intestinal tissues). The nature of food microbiology has changed dramatically from its historical emphasis on microbial phenotypic properties and behavior to a new perspective dominated by genomic and comparative genomic information. The food microbiologists of the future will become increasingly reliant on genomics and the other omics technologies in their efforts to understand and control microorganisms associated with our foods.

We gratefully acknowledge support by the Genomic Science, Microbiology, and Food Science programs at North Carolina State University.

References

1. Altermann, E., W. M. Russell, M. A. Azcarate-Peril, R. Barrangou, B. L. Buck, O. McAuliffe, N. Souther, A. Dobson, T. Duong, M. Callanan, S. Lick, A. Hamrick, R. Cano, and T. R. Klaenhammer. 2005. Complete genome sequence of the probiotic lactic acid bacterium *Lactobacillus acidophilus* NCFM. *Proc. Natl. Acad. Sci. USA* **102**:3906–3912.

2. Altschul, S. F., W. Gish, W. Miller, E. W. Myers, and D. J. Lipman. 1990. Basic local alignment search tool. *J. Mol. Biol.* **215**:403–410.

3. Anderson, L., and J. Seilhamer. 1997. A comparison of selected mRNA and protein abundances in human liver. *Electrophoresis* **18**:533–537.

4. Anderson, N. L., and N. G. Anderson. 1998. Proteome and proteomics: new technologies, new concepts, and new words. *Electrophoresis* **19**:1853–1861.

5. Angelidis, A. S., L. T. Smith, L. M. Hoffman, and G. M. Smith. 2002. Identification of OpuC as a chill-activated and osmotically activated carnitine transporter in *Listeria monocytogenes*. *Appl. Environ. Microbiol.* **68**:2644–2650.

6. Arnold, C. N., J. McElhanon, A. Lee, R. Leonhart, and D. A. Siegele. 2001. Global analysis of *Escherichia coli* gene expression during the acetate-induced acid tolerance response. *J. Bacteriol.* **183**:2178–2186.

7. Ashburner, M., C. A. Ball, J. A. Blake, D. Botstein, H. Butler, J. M. Cherry, A. P. Davis, K. Dolinski, S. S. Dwight, J. T. Eppig, M. A. Harris, D. P. Hill, L. Issel-Tarver, A. Kasarskis, S. Lewis, J. C. Matese, J. E. Richardson, M. Ringwald, G. M. Rubin, G. Sherlock, and The Gene Ontology Consortium. 2000. Gene ontology: tool for the unification of biology. *Nat. Genet.* **25**:25–29.

8. Bairoch, A., B. Boeckmann, S. Ferro, and E. Gasteiger. 2004. Swiss-Prot: juggling between evolution and stability. *Brief Bioinform.* **5**:39–55.

9. Bateman, A., L. Coin, R. Durbin, R. D. Finn, V. Hollich, S. Griffiths-Jones, A. Khanna, M. Marshall, S. Moxon, E. L. Sonnhammer, D. J. Studholme, C. Yeats, and S. R. Eddy. 2004. The Pfam protein families database. *Nucleic Acids Res.* **32**:D138–D141.

10. Begley, M., C. G. M. Gahan, and C. Hill. 2005. The interaction between bacteria and bile. *FEMS Microbiol. Rev.* **29**:625–651.

11. Bolotin, A., B. Quinquis, P. Renault, A. Sorokin, S. D. Erlich, S. Kulakauskas, B. Purnelle, D. Prozzi, K. Ngui, D. Masuy, F. Hancy, S. Burteau, M. Boutry, J. Delcour, A. Goffeau, and P. Hols. 2004. Complete sequence and comparative analysis of the dairy bacterium *Streptococcus thermophilus*. *Nat. Biotechnol.* **22**:1554–1558.

12. Bolotin, A., P. Wincker, S. Mauger, O. Jaillon, K. Malarme, J. Weissenbach, S. D. Ehrlich, and A. Sorokin. 2001. The complete genome sequence of the lactic acid bacterium *Lactococcus lactis* ssp. lactis IL1403. *Genome Res.* **11**:731–753.

13. Bonk, T., and A. Humeny. 2001. MALDI-TOF-MS analysis of protein and DNA. *Neuroscientist* **7**:6–12.

14. Brown, P. O., and D. Botstein. 1999. Exploring the new world of the genome with DNA microarrays. *Nat. Genet.* **21**:33–37.

15. Castanie-Cornet, M.-P., T. A. Penfound, D. Smith, J. F. Elliott, and J. W. Foster. 1999. Control of acid resistance in *Escherichia coli*. *J. Bacteriol.* **181**:3525–3535.

16. Cetin, M. S., C. Zhang, R. W. Hutkins, and A. K. Benson. 2004. Regulation of transcription of compatible solute transporters by the general stress sigma factor, sigma B, in *Listeria monocytogenes*. *J. Bacteriol.* **186**:794–802.

17. Cheung, V. G., M. Morley, F. Aguilar, A. Massimi, R. Kucherlapati, and G. Childs. 1999. Making and reading microarrays. *Nat. Genet.* **21**:15–19.

18. Coleman, R., P. J. Lowe, and D. Billington. 1980. Membrane lipid composition and susceptibility to bile salt damage. *Biochim. Biophys. Acta* **559**:294–300.

19. Cordwell, S. J., A. S. Nouwens, and B. J. Walsh. 2001. Comparative proteomics of bacterial pathogens. *Proteomics* **1**:461–472.

20. Cotter, P. D., C. G. M. Gahan, and C. Hill. 2001. A glutamate decarboxylase system protects *Listeria monocytogenes* in gastric fluid. *Mol. Microbiol.* **40**:465–475.

21. Delahunty, C., and J. R. Yates III. 2005. Protein identification using 2D-LC-MS/MS. *Methods* **35**:248–255.

22. Delcher, A. L., D. Harmon, S. Kasif, O. White, and S. L. Salzberg. 1999. Improved microbial gene identification with GLIMMER. *Nucleic Acids Res.* **27**:4636–4641.

23. Doumith, M., C. Cazalet, N. Simoes, L. Frangeul, C. Jacquet, F. Kunst, P. Martin, P. Cossart, P. Glaser, and C. Buchrieser. 2004. New aspects regarding evolution and virulence of *Listeria monocytogenes* revealed by comparative genomics and DNA arrays. *Infect. Immun.* **72**:1072–1083.

24. Duggan, D. J., M. Bittner, Y. Chen, P. Meltzer, and J. M. Trent. 1999. Expression profiling using cDNA microarrays. *Nat. Genet.* **21**:10–14.

25. Dussurget, O., D. Cabanes, P. Dehoux, M. Lecuit, C. Buchrieser, P. Glaser, and P. Cossart. 2002. *Listeria monocytogenes* bile salt hydrolase is a PrfA-regulated virulence factor involved in the intestinal and hepatic phases of listeriosis. *Mol. Microbiol.* **45**:1095–1106.

26. Ermolaeva, M. D., H. G. Khalak, O. White, H. O. Smith, and S. L. Salzberg. 2000. Prediction of transcription terminators in bacterial genomes. *J. Mol. Biol.* **301**:27–33.

27. Ferreira, A., D. Sue, C. P. O'Byrne, and K. J. Boor. 2003. Role of *Listeria monocytogenes* sigma B in survival of lethal acidic conditions and in the acquired acid tolerance response. *Appl. Environ. Microbiol.* **69**:2692–2698.

28. Fraser, K. R., D. Sue, M. Wiedmann, K. Boor, and C. P. O'Byrne. 2003. Role of sigma B in regulating the compatible solute uptake systems of *Listeria monocytogenes*: osmotic induction of *opuC* is sigma B dependent. *Appl. Environ. Microbiol.* **69**:2015–2022.

29. Gale, E. 1946. The bacterial amino acid decarboxylases. *Adv. Enzymol.* **6**:1–32.

30. Gibson, G., and S. Muse. 2002. *A Primer of Genome Science*. Sinauer Associates, Sunderland, Mass.

31. Glaser, P., L. Frangeul, C. Buchrieser, C. Rusniok, A. Amend, F. Baquero, P. Berche, H. Bloecker, P. Brandt, T. Chakraborty, A. Charbit, F. Chetouani, E. Couve, A. de Daruvar, P. Dehoux, E. Domann, G. Dominguez-Bernal, E. Duchaud, L. Durant, O. Dussurget, K.-D. Entian, H. Fsihi, F. G.-D. Portillo, P. Garrido, L. Gautier, W. Goebel, N. Gomez-Lopez, T. Hain, J. Hauf, D. Jackson, L.-M. Jones, U. Kaerst, J. Kreft, M. Kuhn, F. Kunst, G. Kurapkat, E. Madueno, A. Maitournam, J. M. Vicente, E. Ng, H. Nedjari, G. Nordsiek, S. Novella, B. de Pablos, J.-C. Perez-Diaz, R. Purcell, B. Remmel, M. Rose, T. Schlueter,

N. Simoes, A. Tierrez, J.-A. Vazquez-Boland, H. Voss, J. Wehland, and P. Cossart. 2001. Comparative genomics of *Listeria* species. *Science* **294**:849–852.

32. Grant, M. A., S. D. Weagant, and P. Feng. 2001. Glutamate decarboxylase genes as a prescreening marker for detection of pathogenic *Escherichia coli* groups. *Appl. Environ. Microbiol.* **67**:3110–3114.

33. Hulo, N., C. J. Sigrist, V. Le Saux, P. S. Langendijk-Genevaux, L. Bordoli, A. Gattiker, E. De Castro, P. Bucher, and A. Bairoch. 2004. Recent improvements to the PROSITE database. *Nucleic Acids Res.* **32**:D134–D137.

34. Jay, J. M. 2000. *Modern Food Microbiology*, 6th ed. Aspen, Gaithersburg, Md.

35. Kanehisa, M., and S. Goto. 2000. KEGG: Kyoto encyclopedia of genes and genomes. *Nucleic Acids Res.* **28**:27–30.

36. Kazmierczak, M. J., S. C. Mithoe, K. J. Boor, and M. Wiedmann. 2003. *Listeria monocytogenes* sigma B regulates stress response and virulence functions. *J. Bacteriol.* **185**:5722–5734.

37. Kim, H., K. J. Boor, and H. Marquis. 2004. *Listeria monocytogenes* sigma B contributes to invasion of human intestinal epithelial cells. *Infect. Immun.* **72**:7374–7378.

38. Kleerebezem, M., J. Boekhorst, R. van Kranenburg, D. Molenaar, O. P. Kuipers, R. Leer, R. Tarchini, S. A. Peters, H. M. Sandbrink, M. W. Fiers, W. Stiekema, R. M. Lankhorst, P. A. Bron, S. M. Hoffer, M. N. Groot, R. Kerkhoven, M. de Vries, B. Ursing, W. M. de Vos, and R. J. Siezen. 2003. Complete genome sequence of *Lactobacillus plantarum* WCFS1. *Proc. Natl. Acad. Sci. USA* **100**:1990–1995.

39. Ko, R., L. Tombras Smith, and G. M. Smith. 1994. Glycine betaine confers enhanced osmotolerance and cryotolerance on *Listeria monocytogenes*. *J. Bacteriol.* **176**:426–431.

40. Kulp, D., D. Haussler, M. G. Reese, and F. H. Eeckman. 1996. A generalized hidden Markov model for the recognition of human genes in DNA. *Proc. Int. Conf. Intell. Syst. Mol. Biol.* **4**:134–142.

41. Kurtz, S., A. Phillippy, A. Delcher, M. Smoot, M. Shumway, C. Antonescu, and S. Salzberg. 2004. Versatile and open software for comparing large genomes. *Genome Biol.* **5**:R12.

42. Lipshutz, R. J., S. P. Fodor, T. R. Gingeras, and D. J. Lockhart. 1999. High density synthetic oligonucleotide arrays. *Nat. Genet.* **21**:20–24.

43. Lukashin, A. V., and M. Borodovsky. 1998. GeneMark.hmm: new solutions for gene finding. *Nucleic Acids Res.* **26**:1107–1115.

44. Majoros, W. H., M. Pertea, and S. L. Salzberg. 2004. TigrScan and GlimmerHMM: two open source ab initio eukaryotic gene-finders. *Bioinformatics* **20**:2878–2879.

45. Masuda, N., and G. M. Church. 2002. *Escherichia coli* gene expression responsive to levels of the response regulator EvgA. *J. Bacteriol.* **184**:6225–6234.

46. Masuda, N., and G. M. Church. 2003. Regulatory network of acid resistance genes in *Escherichia coli*. *Mol. Microbiol.* **48**:699–712.

47. McCormack, A. L., D. M. Schieltz, B. Goode, S. Yang, G. Barnes, D. Drubin, and J. R. Yates III. 1997. Direct analysis and identification of proteins in mixtures by LC/MS/MS and database searching at the low-femtomole level. *Anal. Chem.* **69**:767–776.

48. Mendum, M. L., and L. T. Smith. 2002. Characterization of glycine betaine porter I from *Listeria monocytogenes* and its roles in salt and chill tolerance. *Appl. Environ. Microbiol.* **68:**813–819.

49. Nadon, C. A., B. M. Bowen, M. Wiedmann, and K. J. Boor. 2002. Sigma B contributes to PrfA-mediated virulence in *Listeria monocytogenes*. *Infect. Immun.* **70:**3948–3952.

50. Nelson, D., and C. Michael. 2000. *Lehninger Principles of Biochemistry*, 3rd ed. Worth Publishers, New York, N.Y.

51. O'Farrell, P. H. 1975. High resolution two-dimensional electrophoresis of proteins. *J. Biol. Chem.* **250:**4007–4021.

52. Pandey, A., and M. Mann. 2000. Proteomics to study genes and genomes. *Nature* **405:**837–846.

53. Phillips, C. I., and M. Bogyo. 2005. Proteomics meets microbiology: technical advances in the global mapping of protein expression and function. *Cell. Microbiol.* **7:**1061–1076.

54. Rajpal, D. K. 2005. Understanding biology through bioinformatics. *Int. J. Toxicol.* **24:**147–152.

55. Sleator, R. D., C. G. M. Gahan, and C. Hill. 2001. Identification and disruption of the *proBA* locus in *Listeria monocytogenes*: role of proline biosynthesis in salt tolerance and murine infection. *Appl. Environ. Microbiol.* **67:**2571–2577.

56. Smith, D., T. Kassam, B. Singh, and J. F. Elliott. 1992. *Escherichia coli* has two homologous glutamate decarboxylase genes that map to distinct loci. *J. Bacteriol.* **174:**5820–5826.

57. Snyder, L., and W. Champness. 2003. *Molecular Genetics of Bacteria*, 2nd ed. ASM Press, Washington, D.C.

58. Southern, E., K. Mir, and M. Shchepinov. 1999. Molecular interactions on microarrays. *Nat. Genet.* **21:**5–9.

59. Stein, L. 2001. Genome annotation: from sequence to biology. *Nat. Rev. Genet.* **2:**493–503.

60. Sybesma, W., C. Burgess, M. Starrenburg, D. van Sinderen, and J. Hugenholtz. 2004. Multivitamin production in *Lactococcus lactis* using metabolic engineering. *Metab. Eng.* **6:**109–115.

61. Tonge, R., J. Shaw, B. Middleton, R. Rowlinson, S. Rayner, J. Young, F. Pognan, E. Hawkins, I. Currie, and M. Davison. 2001. Validation and development of fluorescence two-dimensional differential gel electrophoresis proteomics technology. *Proteomics* **1:**377–396.

62. Trost, M., D. Wehmhoner, U. Karst, G. Dieterich, J. Wehland, and L. Jansch. 2005. Comparative proteome analysis of secretory proteins from pathogenic and nonpathogenic *Listeria* species. *Proteomics* **5:**1544–1557.

63. Wemekamp-Kamphuis, H. H., R. D. Sleator, J. A. Wouters, C. Hill, and T. Abee. 2004. Molecular and physiological analysis of the role of osmolyte transporters BetL, Gbu, and OpuC in growth of *Listeria monocytogenes* at low temperatures. *Appl. Environ. Microbiol.* **70:**2912–2918.

64. Wemekamp-Kamphuis, H. H., J. A. Wouters, R. D. Sleator, C. G. M. Gahan, C. Hill, and T. Abee. 2002. Multiple deletions of the osmolyte transporters BetL, Gbu, and OpuC of *Listeria monocytogenes* affect virulence and growth at high osmolarity. *Appl. Environ. Microbiol.* **68:**4710–4716.

65. Wolters, D. A., M. P. Washburn, and J. R. Yates III. 2001. An automated multidimensional protein identification technology for shotgun proteomics. *Anal. Chem.* **73:**5683–5690.

Food Microbiology: Fundamentals and Frontiers, 3rd Ed.
Edited by M. P. Doyle and L. R. Beuchat
© 2007 ASM Press, Washington, D.C.

R. C. Whiting
R. L. Buchanan

Progress in Microbiological Modeling and Risk Assessment

45

Modeling techniques are commonly applied to the design and data analyses of microbiological experiments. The value of models to describe experimental results and make quantitative predictions, especially to interpolate results for values not explicitly included in the experimental design, is widely recognized. Progress in modeling was described in earlier editions of this text (123) and recently in extensive reviews (76, 79).

The U.S. Department of Agriculture (USDA) Pathogen Modeling Program (PMP; www.ars.usda.gov/main/main.htm) has been periodically updated and remains available for free downloading (109). Version 7.0 of this program (Jan. 2006) contains growth models for nine foodborne pathogens; most include both aerobic and anaerobic conditions. The program contains survival models for four pathogens, thermal-death models for three pathogens in meats, inactivation-by-irradiation models for four pathogens in meats, time-to-turbidity models for spores of proteolytic and nonproteolytic *Clostridium botulinum*, and models for the growth of *C. botulinum* and *Clostridium perfringens* during cooling. The growth and survival models are based on data from microorganisms in broth media. Experience has shown that although these models include the most important factors affecting growth rates (temperature, pH, and salt concentration or water activity), many foods have additional factors that slow microbial growth, thereby making the models predict more rapid growth than will be found in the foods. Lag phases, in particular, were observed to differ between broth models and foods. Reasons for this are discussed further in following sections. These models are valuable tools for microbiologists, but they must be validated for the food of interest and used with appropriate judgment.

More recently, microbiological growth data were collected in a searchable database called Combase (3). This database (www.combase.cc), a collaborative project between the USDA Agricultural Research Service and the United Kingdom Institute of Food Research, contains over 30,000 records of microbial responses to food environments. Microbiologists can specify the microorganism, temperature, matrix, and other factors that are applicable to a specific situation and rapidly retrieve the relevant literature. Users can readily obtain the relevant data and use them to create a model explicitly for their intended purpose and situation.

Use of risk assessments to analyze the safety of food processes and systems has likewise increased dramatically

R. C. Whiting and R. L. Buchanan, Center for Food Safety and Applied Nutrition, Food and Drug Administration, 5100 Paint Branch Pkwy., College Park, MD 20740.

since the previous edition of this book. The overall paradigm of risk analyses with risk assessment, risk management, and risk communication; the four components of risk assessment; and the relationship of risk analysis to food safety and the hazard analysis and critical control point (HACCP) system were described in documents from the Codex Alimentarius Committee on Food Hygiene (12–14), the Food and Agriculture Organization and the World Health Organization (34, 35, 37, 38), and the Institute of Medicine, National Research Council (57), and in various texts (15, 123). The extensive literature describing advances in both microbiological modeling and risk assessment has made it impossible to provide a comprehensive review in a single chapter. Therefore, a general overview and evaluation of both fields will be provided, concentrating on selected areas that we feel are of particular importance or have demonstrated significant advancement. Regretfully, many noteworthy papers could not be included in this review, but this chapter will be a starting point for searching the literature. It is also important to note the positive contribution that predictive microbiology has made to the acquisition and archiving of systematically collected data on the behavior of microorganisms of food safety significance. The abundance of data acquired has led to major advances in the ability to study food safety microbiology in a quantitative manner.

MODELING

Microbiological models are mathematical equations that describe a microorganism's response over time and the way that response changes with changes in the environment. The standard approach is to consider each processing step as a unit operation where microorganisms grow, survive, or are inactivated during the time they are in that step. Mixing, partitioning, removal, and cross-contamination are additional operations that can affect either the cell population or the probability that the food is contaminated (85). The primary-level model describes how numbers of the microorganism change during the time course of that step. Secondary-level models estimate how the parameters of the primary-level model change with different or changing environmental conditions (temperature, pH, water activity). At the end of the step, the new number of microorganisms is calculated to become the input for the next step. A series of process steps can conclude with an estimate of the number of pathogenic microorganisms consumed. Dose-response models estimate how likely it is that the consumption of a specified number of microorganisms by a specific population (children, the elderly, immunocompromised individuals) will cause illness. When estimates of microbial numbers

throughout the food-processing chain are linked to the likelihood of illness, a risk assessment has been created.

The most frequently used primary-level models for microbial growth are the Gompertz equation, the Baranyi model, and the two- or three-phase log-linear model. Microbial survival is calculated by either log-linear inactivation or logistic-based models. Microbial inactivation models are typically log-linear D and z value models; however, the Weibull equation is receiving interest for nonlinear inactivation curves which are frequently observed (see below). Secondary growth models usually employ the square root model when temperature, pH, and water activity are the environmental factors or polynomial regression equations, especially when additional environmental factors are included. Secondary-level models for calculating survival and inactivation are typically regression equations when more than a single environmental factor is considered. Some models such as the square root model are based on microbiological behavior even if they are not strictly mechanistic models. Polynomial regression models have no underlying biological basis.

Tertiary-level models apply user-friendly software that combines the primary and secondary models to calculate predicted microbial behavior. Confidence intervals and other metrics (e.g., time for 4-log increases or decreases) may also be calculated. The PMP is the most widely used tertiary program. The PMP models microorganisms in a fixed environment; changes in environments are represented by creating a sequence of small steps, each of which is based on static conditions. The PMP cooling models are examples of this approach. Dynamic modeling procedures are beginning to replace this simpler approach (see below).

Several authors have proposed changes in the mathematical basis used for modeling in certain situations, such as those in which data may not meet the assumption of being normally distributed (82). Residuals for growth data were observed to have a Gaussian distribution; hence, a robust nonlinear regression method was employed. An approach that combines microbiological knowledge with polynomial regression (including artificial neural networks) to create secondary models was proposed that avoids some of the problems with overparameterization and areas of poor fitting which were often found with polynomial regression models (45). The three steps were to (i) compile available knowledge about the effects of the factors on microbial behavior from the literature and specific experimental data, (ii) translate this behavior into the form of partial derivatives, and (iii) determine the parameter values via a flexible optimization technique such as polynomial regression. A fuzzy logic approach to secondary modeling of high-pressure inactivation was demonstrated (62). Fuzzy logic uses expert knowledge by setting up a series of classifications, i.e., pressure was described as "high" if it was between 300 and 600 MPa. Rules are established for the effects of the input parameters on the output parameters. Numerical values and weighting factors then lead to the predictions.

Structured Foods

The environmental factors (pH, water activity) are usually considered to be uniform within a food, and models are frequently created from observations with broth cultures. However, it is recognized that foods have microenvironments with characteristics that may critically differ from the bulk characteristics. Different structures which may be present within a food were evaluated, and an approach to their modeling was devised (128). Individual microorganisms may be planktonic in liquids, constrained to colonies in gels or highly viscous foods, or growing on a surface (solid-air boundary) or interface (oil-water emulsion). Compared to planktonic growth, the growth of cells in the other structures may be restricted because of limited availability or depletion of nutrients, accumulation of metabolic end products, and cell interactions (quorum sensing). Use of broth-based models for nonbroth foods is frequently considered "fail safe" because the broth-based models predict more rapid growth; however, the development of applicable models with better accuracy for nonbroth food matrices would provide a better situation.

As water activities in low-viscosity broths are decreased with increasing concentrations of glucose or sucrose, lag times increase and exponential growth rates for *Listeria monocytogenes* decrease (102). Increasing the viscosity of broth with polyvinylpyrrolidone decreases listerial growth rates. Studies on the growth of *L. monocytogenes* on surfaces of iceberg lettuce and agar plates revealed that growth is slower than that predicted by the PMP, which is based on growth in broths (63).

Dynamic Conditions

Initial predictive models were based on fixed or static conditions, both for environmental conditions (temperature, pH) and for space (agitated or homogeneous broths, entire solid food represented by the coldest point). However, if a primary-level model is differentiable, then the model can be integrated over the time course of the processing step with the time or spatial dependence of the environmental conditions taken into account. Many of the primary-level inactivation and survival models were evaluated for their ability to model dynamic conditions (44), and the concepts behind dynamic modeling for

growth and inactivation were explained (5). Dynamic situations of *C. perfringens* growth in cooked ground beef with different cooling patterns were successfully simulated (56). A dynamic version of the Baranyi and square root models quantified the influence of temperature on the growth of *Bacillus cereus* and *Enterobacter cloacae* in liquid whole-egg product (48).

Foods have thermal gradients when they are experiencing shifts in environmental temperatures. Finite element analyses can calculate the transient temperatures at different locations within the food. These gradients can then be coupled to dynamic microbial models to calculate microbial growth or inactivation at every point in the food, not just at the center or coldest point. Models for growth of *L. monocytogenes* in nonagitated milk under fluctuating storage temperatures (1), growth of *Salmonella enterica* serovar Enteritidis in pasteurized, intact eggs (30), and growth of *C. perfringens* during cooling of cooked ham (2) illustrate this type of modeling. Predictive modeling for microbial inactivation in a food undergoing hot-air heating was developed by using one-dimensional heat transfer via convection, radiation, and evaporation; nonlinear thermal inactivation data for *Escherichia coli* K-12; and microbial considerations including possible growth before reaching the inactivation temperatures (112). Spatial distributions may also apply to diffusion of nutrients and metabolic by-products within a food. A two-species competition model with limited diffusion across an agar medium leads to dynamic but spatially segregated growth (29). The model calculated local microbial growth as a function of the local environment and the transfer of biomass across the medium. Significant differences in the growth-no growth boundaries for *L. monocytogenes* in broth and on agar were modeled (65).

Strain Variation

Microbiologists have frequently treated differences among strains of the same species as a relatively minor factor, and much of the previous microbiological research used single strains or cocktails (which do not differentiate between the strains but instead measure the growth or survival kinetics of the most efficient or resistant strain) to determine growth or inactivation characteristics. Recent studies have revealed that the range in these characteristics within a species can be very significant and that judgments on the safety of food processes must consider these ranges (25). The potential impact on safety was shown by a comparison of growth rates of four clinical and four seafood isolates of *L. monocytogenes* during acid and osmotic stress (118). Clinical isolates had significantly shorter lag phases and were more adaptable to changing conditions, whereas the generation times were not different. Another study of the thermal inactivation, growth, and survival parameters of three genotypic lineages of *L. monocytogenes* revealed some differences among lineages but also extensive strain-to-strain differences within lineages (22). In the case of thermal resistance, the responses of *L. monocytogenes* strains were further confounded by differential responses to preadaptation (33). Strains of *L. monocytogenes* collected during the same time period and in the same locations from foods and clinical cases were genotyped (11). By comparing the frequencies and numbers consumed with the frequencies of illness associated with the different genotypes, the virulence of certain genotypes was found to be nearly 2 orders of magnitude greater than the average virulence.

Growth rates and heat resistance levels of *Salmonella* isolates were also found to have serotype-specific differences (61). Differences in growth rates were relatively small, but there were significant differences among serotypes; the latter coefficient of variation was approximately 20%. Seventeen strains of *E. coli* O157:H7 were compared for growth, survival, thermal inactivation, and toxin production (124). The ratio of the standard deviation to the mean was smallest for the exponential growth rate but was approximately 0.40 for lag phase durations, survival times, and thermal inactivation *D* values. A tissue culture verotoxin assay revealed more than a 2-log difference in toxin production levels between the lower and higher toxin-producing strains. Twelve strains of *Enterobacter sakazakii* were thermally inactivated at 58°C (32), and the *D* values ranged from 30.5 to 591.9 s, with the strains falling into two distinct heat resistance phenotypes that appear to be attributable to the presence or absence of a single protein (127).

Lag Times

Most studies addressing lag phase durations used inocula of stationary-phase cells that were grown in a high-nutrient, favorable-environment broth and considered the lag phase durations to be related to growth conditions. Growth models based on these inocula frequently were poor predictors of the lag phase in foods (77). The lag phase is a complex phenomenon that is highly dependent on the cells' cultural history and their biological characteristics at the time of environmental change. Attempts at modeling the lag phase have been reviewed previously (107). The previous and present environments in which the cells are exposed, the magnitude of the environmental shift, and the rate at which the cells can adapt are all factors in the length of the lag phase. Exponential growth and stationary phase are two important physiological states, but other factors include acid adaptation, heat

or cold shock, nutrient starvation, desiccation, freezing, and injury.

A few papers describe how initial cell divisions can be modeled in a manner closer to the cells' biological processes (27, 28). A cell cycle is described as having three phases, beginning with an increase in cell mass in which the cell increases its contents of RNA, enzymes, and proteins. The next phase is the replication of DNA, and the last period is cell division. Appropriate models have been created that fit data from cells grown during temperature transitions.

Realization that the lag time observed for a culture is not the same as those of individual cells has led to several different approaches to modeling the lag phase. In a culture, the relatively few cells with the shortest lag phases will become the parent cells of virtually all of the daughter cells in the culture. This has led to studies of the distribution of the lag phase duration among individual cells and the realization that the inoculum size can affect the lag phase. A stochastic lag-growth model (77) revealed that the observed culture growth curves had less variability and shorter times with a 500-cell inoculum than with a 50-cell inoculum. Cultures inoculated with only 5 cells had highly variable lag phases; some were as short as those of the 500-cell-inoculum cultures, but most were longer. The standard practice when conducting inoculated-pack studies and other growth experiments is to use inocula of 10^3 to 10^4 cells. When natural contamination levels are less than 100 cells per package, as frequently occurs with many foodborne pathogens, the lag times will frequently be longer than those determined in the studies. Although models based on higher inoculum levels are conservative, more accurate predictions of lag phase durations and their variations at different inoculum levels would be preferable.

A mathematical model with deterministic and stochastic components was created for growth from single cells (81). A flow chamber microscopic system was used to observe the time for daughter cells to be released by the parent cell. The time for the first division was longer and more varied than those for the second and third generations, indicating that the lag phase adjustments may not be completed in the first generation. Individual cell sizes were measured over time, and a stochastic model was proposed interrelating cell lengths, lag times, and cell division times for unstressed and heat-shocked cells (67).

Lag time distributions have been determined by several investigators. This determination was done by sequentially diluting cultures on microtiter plates until single cells per well were obtained (42). Growth curves were obtained by measuring the optical density with time, and the lag phase was back calculated by using the exponential growth rate. It was shown that individual lag phases of *L. monocytogenes* were affected by culture temperature and pH. In low-stress culture conditions, the range of the lag time distributions was compressed towards zero time, but under high-stress conditions, the distributions shifted towards longer times and greater variability. The preferred distribution for moderate-stress conditions was the gamma distribution, and for harsher conditions it was the Weibull distribution.

However, consensus on the "best distribution" to describe the lag period has not been reached. For example (75), the lag phases for individual cells of *E. coli* O157:H7 were best fitted with a log normal distribution or other relationships that provided for substantial right-hand skewing of the data. Times for germination and growth of spores were observed and modeled with a two-component lag period (4). The observed population lag time was not the simple average of the individual spore germination and outgrowth times and was dependent on the inoculum size, particularly when fewer than 10 spores were present.

Studies of the lag times of microbial populations have shown marked effects from changes in temperature, pH, and other factors. Sudden increases in temperature for exponentially growing *E. coli* K-12 cells were modeled, and the amplitude of the temperature rise played a major role in the length of the observed lag (108). The greater the magnitude of the temperature shifts in both directions, the greater the lag phase duration for *L. monocytogenes* (23, 122). Exponential-phase cells had the shortest lag times, followed by stationary-phase and starved cells, and then frozen and finally desiccated cells.

Growth-No Growth Boundary

Growth rates of microorganisms become more variable as the limits of growth are approached. However, knowing the growth limits for combinations of temperatures, pHs, and other factors is very important for assessing the likelihood of growth when formulating a food. This growth boundary is a mathematical expression of the hurdle concept as the model shows the areas of growth, no growth, or uncertainty in the multidimensional space of the environmental factors (71). The growth-no growth boundary model is a probability-type model and provides a clear demarcation in these situations where the kinetic model has severe limitations. The modeling method described in the early literature was the logit(p) approach, where logit(p) = ln($p/[1-p]$) and p is the probability of growth occurring. Independent factors affecting the probability were fitted by polynomial regression (80). The logit(p) approach was successfully adapted to fit by nonlinear logistic regression (96, 111).

A model for growth of *Staphylococcus aureus* in medium with different combinations of pHs and sorbate concentrations and glycerol to control the relative humidity has been based on the geometric mean of the time of the last no-turbidity measurement and polynomial regression (103). The growth boundary at different storage times was determined for *L. monocytogenes* in ready-to-eat cooked meat products with different amounts of salt, moisture, lactate, and diacetate (70). The time for growth of 1 log of cells was estimated from a growth curve, and the log of this time was modeled by polynomial regression. In this model, the boundary is not fixed; rather, it moves towards less favorable conditions as the storage time lengthens.

A probability model for the boundary between survival and death of *E. coli* O157:H7 in mayonnaise was developed with temperature and acetic acid and sucrose concentrations being the compositional factors (78). The various test media were observed visually for growth, and nongrowth samples were tested to determine whether any surviving cells were present. The logit(p) as a probability of survival was used by fitting it to a polynomial regression model. Undissociated acetic acid concentration was a factor in this model.

Competition between Microorganisms: the Spoilage Flora

Competition with the indigenous spoilage microorganisms has long been recognized as affecting the growth of pathogens in a food and is one reason that broth-based models of pure cultures often predict rates or levels of growth that are higher than those observed in foods, especially naturally contaminated foods. Therefore, some modelers have included different numbers of cells of spoilage microorganisms in the modeling. The Seafood Spoilage Predictor is an example of a tertiary software program to predict nonpathogen populations during storage and the remaining product shelf life (20). A Food and Agriculture Organization-World Health Organization risk assessment team modeled the growth of *L. monocytogenes* in smoked fish with several estimates of the suppression of the growth rate by lactic acid bacteria (38). Models predicting the interactive growth of *Lactobacillus curvatus* and *Enterobacter cloacae* have included temperature, pH, and glucose and lactate concentrations as environmental factors (74).

Identifying the appropriate factors that control the growth of different microorganisms is not always straightforward. Some microorganisms produce acids and decrease the pH; others can produce bacteriocins. A competitive model for *L. monocytogenes* and spoilage microorganisms in cold-smoked salmon was based on the fact that lactic acid bacteria, being the dominant flora, suppress growth of *L. monocytogenes* primarily by limiting its maximum population (46). This factor was added to an *L. monocytogenes* growth model by including $[1 - (LAB_t/LAB_{max})]$, where LAB_t is the number of lactic acid bacterial cells at time t and LAB_{max} is the maximum number of lactic acid bacterial cells. Without the addition of the spoilage microorganisms, *L. monocytogenes* grew to a population of 10^8 CFU/g, but in the presence of the spoilage flora it grew to a population of 10^2 to 10^4 CFU/g. This approach was used as a basis to expand models that considered temperature, water phase salt concentration, and phenolic content (i.e., degree of smoking) as additional factors (17).

A dynamic model was proposed that included the effects of nutrient exhaustion and/or metabolic waste products (116). This approach was extended to microbial interactions and also structured foods (93). The coculture microorganism, a lactic acid bacterium, was incorporated into the model based on pH reduction and production of undissociated lactic acid.

Cross Contamination

The importance of environmental contamination, recontamination of heat-processed foods, and the numerous potential pathways for cross contamination has long been recognized; however, few investigators have attempted to quantify or model cross contamination (91, 97). A general systematic approach for modeling contamination of foods from surfaces by air, hands, or liquids was developed whereby transfer coefficients from the literature were collected and generalized models were described (26). Rates of bacterial transfer between hands and other food preparation surfaces were measured, and a normal distribution was determined to describe the variability of the logarithm of the percent transfer rates (10). Representative means and standard deviations (log of percentage values) were 0.94 ± 0.68 for chicken to hand, 0.90 ± 0.59 for cutting board to lettuce, 0.36 ± 0.90 for faucet spigot to hand, and -0.80 ± 1.09 for hand to spigot.

A process risk assessment for cross contamination in a smoked-fish plant modeled *L. monocytogenes* contamination among food, food contact surfaces, gloves, and the environment (59). Some of the parameters modeled included the total number of gloves in the production stage, the probability of contact between a food and a glove, the duration that a pair of gloves was worn, the prevalence of contaminated raw fish, the relative reduction in the prevalence of contaminated fish due to smoking, and the prevalence of contaminated food contact surfaces at the end of the previous shift. The median estimate was that 10.7% of the product was contaminated.

Sensitivity analysis revealed that the frequency of contact of employees' gloves with food and food contact surfaces and the frequency of changing of gloves were the most significant factors influencing cross contamination. Cross contamination with *Salmonella* and *Campylobacter* on kitchen surfaces has been modeled, and a risk assessment was done to estimate the importance of pathogen prevalence on chicken carcasses and kitchen hygienic practices for the probabilities of illness (66).

Nonlinear Inactivation: Weibull Model

Heat inactivation of vegetative cells and spores as well as death from exposure to an unfavorable environment has traditionally been modeled by first-order kinetics (88, 89). This produces a linear plot on semilogarithm graphs with resulting D and z values. However, frequent departures from linear population decreases are observed with shoulders, upwards or downwards concavity, or tailing. These curves have been attributed to cell clusters, heterogeneity in thermal resistance levels among the cell population, the existence of a more resistant subpopulation, experimental artifacts, and other contributing factors. During thermal inactivation, nonlinearity is more frequently observed at lower inactivation temperatures. Nonthermal inactivation methods (e.g., acidification) and survival conditions are also prone to nonlinear decreases. Numerous nonlinear and multiple-phase models have been used to describe inactivation kinetics (52, 113), with the Weibull model being used by an increasing number of modelers.

The basic form of the Weibull model is as follows:

$$\log (N_t/N_0) = -bt^n$$

where N_t is the cell population at time t, N_0 is the initial population, b is the rate parameter, and n is the shape parameter. When n is 1, the model simplifies to the familiar log-linear thermal death model and $b = 1/D$. When n is <1, the curve is concave upwards, and when n is >1, the curve is concave downwards. As the b parameter increases, the overall rate of inactivation increases. The Weibull model is a primary model describing cell population changes with time. How the b and n parameters change with different environments can be described by secondary-level models. These primary and secondary models are derived under static conditions and can be solved numerically to estimate the inactivation curve during changing conditions (90).

A primary and secondary set of models based on the Weibull distribution was used to describe the interactions of temperature and pH in the thermal inactivation of *Bacillus cereus* spores in carrot juice (16). The investigators determined that temperature and pH do not have a significant effect on the shape parameter (n). The temperature was modeled for the b parameter by using the z concept, and the pH value was inversely related to the b parameter. Thermal inactivation studies of *Bacillus pumilus* spores revealed that different heating temperatures, heating-medium pHs, and recovery-medium pHs do not change the shape parameter (n) (19). However, the rate parameter (b) is affected by all three and can be modeled with the traditional z value approach.

Thermal inactivation of *L. monocytogenes* cells in a medium at different pHs follows log-linear decreases, whereas that of *Pseudomonas aeruginosa* cells is fitted to the Weibull model (50). For *P. aeruginosa*, the b parameter is dependent on both temperature and pH whereas the n parameter is dependent only on pH.

The Weibull function has also been used to model other lethal processes for vegetative cells. Pulsed-electric-field inactivation of *E. coli* in carrot juice (99) and high-hydrostatic-pressure inactivation of *Listeria innocua* in peptone broth (8) are recent examples for which several nonlinear models, including the Weibull, were fitted and compared. The pulsed-electric-field inactivation data were best fitted by the Weibull model, whereas high-hydrostatic-pressure data were better fitted by a log-logistic model than the Weibull model, but both models were superior to the log-linear model.

The highly flexible nature of the Weibull function has provided an important new tool for modeling microbiological inactivation and survival. However, it also increases the number of parameters that must be fitted when considering experimental data, which in turn may lead to wider confidence intervals when the data are applied to microbiological risk assessments. It would be beneficial for future comparisons of inactivation models to evaluate the gains in "accuracy" achieved with the Weibull function compared to simpler models versus the potential increased uncertainty obtained when the Weibull function is used. An accepted but critical assumption for the log-linear inactivation model is that D values obtained from tests with high initial numbers can be applied to the lower cell numbers more typically present in foods. The corresponding applicability of parameter values of nonlinear models has not been unequivocally demonstrated.

RISK ANALYSIS

Making decisions about the safety of a food always involves some form of risk assessment. Usually these assessments have a qualitative or semiquantitative design with both quantitative data, e.g., D values and growth rates, and subjective information, e.g., the maximum

number of contaminating microorganisms, the virulence of the microorganism, and consumer handling practices. Risk analysis improved this process by formalizing the structure and management of the analysis, quantifying all of the data steps and calculations, evaluating the ranges of possible values, considering the quality of the data, linking all steps of the food process chain to consumption and the likelihood of illness, and linking HACCP plans and critical control points to public health (number of illnesses). The conceptual framework for risk analysis has undergone significant advances, much by international bodies, and numerous risk assessments have been completed and used for regulatory decision making (see www.foodrisk.org for risk assessments conducted by the U.S. Food and Drug Administration, the USDA Food Safety and Inspection Service, and other organizations).

Risk assessments may be done to estimate the risk from consuming a serving of a food, to determine the total number of illnesses in a population, to evaluate the rates of illnesses in different human populations, to compare or rank different classes of foods, to determine which steps in a food process contribute the most to the final pathogen population, to identify where poor-quality or absent data make analysis of the process highly uncertain, to evaluate the effectiveness of mitigations at different points in the food-processing and distribution chain, and to compare the unavoidable trade-offs between different risks (43). A series of risk analysis principles were developed by the Codex Alimentarious Commission (120).

1. Microbial risk assessment must be soundly based upon science.
2. There should be a functional separation between risk assessment and risk management.
3. Microbiological risk assessment should be conducted according to a structured approach that includes hazard identification, hazard characterization, exposure assessment, and risk characterization.
4. A microbiological risk assessment should clearly state the purpose of the exercise, including the form of risk estimate that will be the output.
5. A microbiological risk assessment should be transparent. This requires full and systematic documentation, a statement of assumptions and value judgments and rationale, and a formal record.
6. Any constraints that impact the risk assessment, such as cost, resources, and time, should be identified, and their possible consequences should be described.

7. The risk estimate should contain a description of the uncertainty and where during the risk assessment process the uncertainty arose.

8. Data should be such that uncertainty in the risk estimate can be determined; data and data collection systems should, as far as possible, be of sufficient quality and precision so that uncertainty in the risk estimate is minimized.

9. A microbiological risk assessment should explicitly consider the dynamics of microbiological growth, survival, and death in foods and the complexity of the interaction (including sequelae) between humans and the agent following consumption as well as the potential for further spread.

10. Wherever possible, risk estimates should be reassessed over time by comparison with independent human illness data.

11. A microbiological risk assessment may need reevaluation when new relevant information becomes available.

Risk analysis is an integral part of U.S. and international decision making in food safety. The World Trade Organization via the Application of Sanitary and Phytosanitary Measures Agreement and the Agreement on Technical Barriers to Trade specifies methods of risk assessment, among other criteria, that a member nation must use to ensure that its measures achieve the appropriate level of protection and are not more trade restrictive than required (55). The executive branch of the U.S. government currently works under Executive Order 12866 (1993), which articulates the need for federal agencies to base new regulations on appropriate assessments of risk to support the impact and effectiveness of any proposed regulation.

Economic analyses have not traditionally been considered part of the risk assessment process. However, decisions that risk managers frequently need to make include an economic, cost-benefit, or most-cost-effective-mitigation component (58, 126). When an economic analysis follows a risk assessment, it is important that the risk assessment questions and outputs be appropriately specified in the beginning of the risk assessment process in order that the results will be suitable for economic analysis.

Managing the Risk Analysis Process

A risk analysis has three interactive components: risk assessment, risk management, and risk communication. Risk managers are the people who need the risk assessment to assist in making a decision about the safety of a food, the estimated impact of mitigation strategies, or other safety-related questions. Risk managers pose specific questions to the risk assessment team. The risk managers in most organizations also assemble the risk assessment team and provide staff, funds, and time for the risk assessors to conduct the risk assessment. Risk communication refers to the sharing of goals, information, and data among risk assessors and managers in addition to other stakeholders such as the public and industry. Transparency is an important feature of a risk assessment, meaning that the data, modeling, and interpretation of results are clearly understandable and communicated to all parties.

A risk assessment is a process of collecting and evaluating data, constructing models, calculating the requested values, and interpreting the results. Risk analysis is frequently a cyclic process whereby risk managers decide whether a process or product is unsafe and request a risk assessment to evaluate the process or product and potential corrections. Risk assessors conduct the assessment and present their analysis back to the risk managers. The risk managers then decide on a course of action (which may include making no changes), organize the response, and order a follow-up evaluation of the effectiveness of the action. If the process or product is still considered unsafe, the cycle can be repeated based on changes made in the process or product.

Experience has shown that regular communication between the risk managers and risk assessors is necessary to ensure that the risk assessment will have maximum value to the risk managers. Frequently, issues from the available data or questions that need refining arise that are within the realm of the risk managers. For example, in the Food and Drug Administration-Food Safety and Inspection Service *L. monocytogenes* risk assessment (41), different types of cheeses were among the food categories compared. Cheeses could be grouped by type (feta, cheddar, cottage cheese) or by their moisture levels. Risk managers determined that grouping by moisture level was most appropriate because the classification of cheeses for regulatory purposes is by moisture level. However, it is also important that risk managers respect the risk assessors' role in evaluating scientific data, creating distributions and models, and conducting the calculations and not impose any bias upon the risk assessment. The risk management process also includes managing communication and transparency by creating outside expert panels, conducting public meetings, requesting independent reviews of the finished risk assessment, and disseminating the results in appropriate forums. The latter may include preparing an interpretive summary to accompany the detailed risk assessment, and disseminating the risk assessment through publications and public presentations. Additional information on principles of risk analysis and managing risk assessments can be found elsewhere (6, 7, 34, 35, 94).

Microbial Risk Assessments

Numerous papers have described the basic approaches and methods for conducting microbial risk assessments (54, 120, 123). The initial step is to determine the purpose of the risk assessment, i.e., what is the question to be answered and the calculation to be made. A risk profile may be drafted to describe the food process, review the safety questions, and evaluate the available data. The risk profile may be sufficient by itself, or it may show that critical data gaps exist that must be filled before a risk assessment can be conducted that will be sufficiently precise to provide guidance for the risk managers.

A risk assessment has four components: hazard identification, exposure assessment, hazard characterization, and risk characterization. The hazard identification is the collection of information on the pathogen, food process, and illness that is relevant to the risk assessment. The exposure assessment is the determination of the probability of consuming the pathogen and the cell numbers consumed. For a food process, this includes the initial contamination frequency and pathogen cell numbers, the effect of any inactivation process, the amount of growth during transportation and storage, the frequency of recontamination, and handling practices of the food handler or preparer. For microbial hazards, modeling the changes in growth, survival, or inactivation of populations during each step from raw material to consumption typically is the largest and most complex part of the risk assessment.

The hazard characterization or dose-response relationship describes the health impact on individuals of consuming a specified number of pathogen cells. It includes consideration of the virulence of the pathogen strain and the susceptibility of humans to illness. Many foodborne pathogens are opportunistic and affect individuals with impaired immune systems or other vulnerabilities. Children, the elderly, pregnant women, individuals with immune system diseases, those undergoing medical treatments that suppress the immune system, and those with other underlying diseases are vulnerable to infectious microorganisms (e.g., *Salmonella* and *L. monocytogenes*) and infectious-toxigenic microorganisms (*E. coli* O157:H7). The food matrix may also affect the dose response by making a pathogen acid adapted or by neutralizing stomach acidity and thereby allowing more pathogens to enter the intestinal tract.

The dose-response relationship usually shows a sigmoid curve with a plot of the log dose versus the probability of illness (53, 83). This is commonly interpreted as being indicative of the existence of a threshold; however, this is not a valid assumption because threshold and nonthreshold models can produce a sigmoid curve when plotted on a log-dose-versus-probability-of-illness scale (37). Data for the dose response come from animal studies, human volunteer feeding studies, and epidemiological investigations (60). Correlating animal and healthy human responses to the responses of the susceptible human population has considerable uncertainties. Volunteer studies are not possible with the highly virulent pathogens. An epidemiological study of an outbreak of *L. monocytogenes* in a Finnish hospital illustrates the collection of data necessary to determine human susceptibilities (72). The pathogen populations in the food when consumed, the amount of food consumed, the health statuses of individuals who became ill, and the numbers of people who became ill and did not become ill were determined. The probabilities of developing illness in outbreaks typically range from 0.10 to 1.0 (10 to 100% of those exposed become ill) and provide points for plotting a dose-response curve. Three models have received the most frequent use for determining dose response, but none are preeminent for all pathogens (53). These models include the exponential, beta-Poisson, and Weibull-gamma models and are one-, two-, and three-parameter models, respectively. All three are sigmoid-shaped, nonthreshold models but have significant differences when extrapolated to low doses. Unfortunately, it is extremely difficult to collect data for low doses and low probabilities to precisely fit a model. As nonthreshold models, each assumes that a single cell of a pathogen has some probability of causing illness. When all of these models are plotted on a log dose-log probability of illness basis to show the probability ranges at which an acceptable level of protection (ALOP) and food safety objective (FSO) would be set, the models are linear. This means that in the linear portion of the model, a 10-fold increase in dose will have a 10-fold increase in the probability of causing illness. On a population basis, increasing the frequency of contamination 10-fold (all else equal) will increase the probability of illness 10-fold.

Despite the difficulties in determining dose-response relationships for different pathogens and populations, these relationships are beginning to be described (64). Dose-response models have been published for *L. monocytogenes* (38, 41), *E. coli* O157:H7 (49, 95, 104), *Salmonella* (36, 69, 86), *Vibrio parahemolyticus* (40), and *Cryptosporidium* and *Giardia* species (98, 110).

The final component of a risk assessment is risk characterization. It combines the exposure assessment for the foods of concern with the specific dose-response relationship of the pathogen and population to calculate the various outputs, sensitivity analyses, and scenarios desired by risk managers.

A risk assessment links each of the processing steps to the total number of people exposed at consumption and then to public health (probability of illness). Prior to risk assessments, each process step, e.g., pasteurization, was evaluated in isolation from the rest of the food chain. Hence, it was difficult to evaluate the relative impact of different intervention strategies taken at different steps in the food process, for example, determining whether reducing the initial contamination levels or lowering a storage temperature would be more effective in reducing the number of illnesses. The linkage to public health sets a safety objective for the food process; otherwise, there are no criteria to determine whether a control step is adequate. Decisions of whether a pasteurization step requires inactivation of 5 or 7 log CFU of a pathogen/g or whether the air temperature in a cooling step should be 4.4 or 7.2°C (40 or 45°F), as examples, cannot be made without a risk assessment. Therefore, a risk assessment determines requirements for the performance of each control step; the critical control points of the HACCP plan are set to achieve these requirements (101).

Variation, Uncertainty, and Sensitivity Analysis

An important aspect of the risk assessment process is that it is a stochastic rather than a deterministic approach. A stochastic approach uses the distribution of values in a data set when making calculations, in contrast to a deterministic approach which uses the averages or the median to represent the data set. A data set has a distribution in its values from variability and uncertainty (84). Variability is the real differences that exist between different samples, e.g., different lots of raw materials are not all contaminated at the same level. Additional data or better measurements will not minimize or eliminate this variability. Uncertainty, on the other hand, is from sparse data and the errors in sampling and testing. Additional replications or more accurate and precise tests would reduce the uncertainty. Many data sets have a combination of variability and uncertainty. Mathematical functions can be fitted to describe the distribution in a data set (18, 119); however, selecting a normal or log normal function, a function that has a logical reason to be used, or a function with prior demonstration of its applicability is preferable to fitting a number of functions and using the function that fits that specific data set the best.

Another type of uncertainty is in the design of the process model. All models must be a simplification of the multiple pathways that a food may undergo. It is unknown if all the important steps have been put into the model.

There may be a source of recontamination that is important but that is not included in the calculation, thereby making the model inaccurate by an unknown amount.

Many of the data parameters that are needed to model a food process have not been measured in a survey or scientific study. Examples may be the time a product is being transported, the temperature in a refrigerated processing facility, and many consumer food-handling practices. For these parameters, expert opinion will be needed. This involves soliciting input on parameters from authorities on such topics. Various protocols for asking questions and estimating a parameter's distribution can be used to obtain as unbiased and accurate an estimate as possible (114, 119). For example, Bayesian methods have been used to estimate the uncertainty and variability in the prevalence of *B. cereus* spores (73) and *L. monocytogenes* (24) in specific food-related environments or processing conditions.

A risk assessment uses distributions for the input parameters in its calculations by an iterative process termed a Monte Carlo analysis (68, 92, 119). A value is selected from each distribution, and the output of the risk assessment is calculated. Then another set of values is selected, and the output is calculated again. This iterative process is repeated until the output distribution becomes evident (often several thousand iterations). For analysis of the output, the fraction of iterations that exceed a certain value or the level at which the upper 10% of the iterations (servings) lies is more important information than the average value of the output distribution. The ability to analyze the distribution in the output is a major advantage for stochastic models compared to the single-value output from a deterministic model.

A two-dimensional risk assessment separates variation from uncertainty by assigning an uncertainty distribution to the parameters of the variability distributions. Therefore, a nested set of distributions describes each factor, i.e., the initial distribution is the variation and each parameter is characterized by an uncertainty distribution. To conduct and analyze a two-dimensional risk assessment, a value for each uncertainty distribution is selected according to its distribution (selected randomly or by Latin hypercube sampling). With this specific set of uncertainty values defining the variation distributions, the risk assessment is conducted and an output distribution is obtained from the iterations. Then another set of uncertainty values is selected to define the variation distributions, and the risk assessment is rerun to obtain another output distribution. Figure 45.1 illustrates seven outputs of a risk analysis with each sigmoid curve being the cumulative plot of the variation in the output using one set of uncertainty values. The placement of the curve indicates the general level of risk;

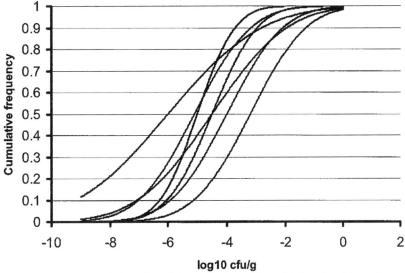

Figure 45.1 Cumulative plot of the variation and sensitivity of a risk analysis.

the 5, 50, and 95% levels can be readily observed. The median of the curves is at approximately −5 $\log_{10}$ CFU/g. The steepness of the curve reflects the variation; a relatively steep curve has low variation compared to a sloping curve. The variation in the risk assessment has values ranging from approximately −8 to −2 $\log_{10}$ CFU/g. The spread of the different curves is an indication of the impact of the uncertainties. If the uncertainty is relatively small, the curves are close together; if they are widely spread relative to the slopes of the individual curves, then the uncertainty is high compared to the variation. In fig. 45.1, the uncertainty is about 3 $\log_{10}$ CFU/g.

Various techniques can be used with the risk assessment to determine which of the many parameters make the greatest contribution to the average value, the variation, or the uncertainty of the output distributions. These techniques are termed sensitivity analyses and provide much insight for risk managers on why the output results occur and which of the many parameters have the greatest influence on the output (87, 129). Different scenarios can be analyzed by changing the input parameter values and observing changes in output distributions. These scenarios can be used to estimate the impact of regulatory or process changes or to improve the understanding of how various factors in a complex process interact. As examples, the impact of processing changes on public health has been estimated by scenario analysis for *Salmonella* in eggs (125) and *L. monocytogenes* in ready-to-eat foods (9).

To improve a food process so that all lots or servings meet a safety standard (see "FSO Paradigm" below), a

general order of evaluation would be to first reduce the uncertainty to determine whether the process is actually acceptable or not (R.C. Whiting, submitted for publication). When results from a risk assessment are analyzed, it may be apparent that the average lot or serving is acceptable but that a small percentage may exceed the standard, and the process would be judged unacceptable. Sensitivity analyses can identify the specific parameters whose uncertainty makes the greatest contribution to the overall uncertainty. Additional data can then be collected to refine the distributions, thereby reducing the uncertainty for those parameters. For example, the warehouse storage times for different lots of a refrigerated, ready-to-eat food may not be accurately known. With new information, the risk assessment can be recalculated and the process may be determined to be acceptable without changing the process. If a portion of the process lots are still unacceptable, then a sensitivity analysis for variation identifies the parameter(s) that significantly contributes to the output variation. When production schedules and warehouse inventory control are not optimized, for example, different lots may have different storage times. Better controls could reduce the distribution of storage times, particularly avoiding holding of lots for extensive periods of time. Reducing variation would likely improve the consistency of the sensory and nutritional qualities of the product as well as the microbial consistency. With variation reduced, the risk assessment is rerun. If the process still has an unacceptable portion of trials that exceed the standard, the process must be changed to lower the absolute value of the output distribution. In this example, production and warehousing schedules would need to be changed so that storage times are minimized for all lots. By this iterative process of risk assessments and sensitivity analyses, the design of a food process can be improved until it meets the desired safety standard.

Simplified Risk Assessments

The large risk assessments that regulatory agencies conduct frequently involve one or more years of work by a team of specialists and a major expense (38, 41). A risk assessment for a food process chain is not as complex, but it is still a significant project. A stepwise analysis from simple deterministic estimates to the complete stochastic risk assessment may find that insights from the initial, simpler analyses are adequate for the decisions that are needed (115).

Qualitative and semiquantitative risk assessments are one approach towards a faster analysis. A spreadsheet-based risk assessment with inputs expressed in qualitative terms has been developed and distributed (100). The probability of contamination of a raw product per serving, for example, is entered as "rare (1 in 1,000)," "infrequent (1%)," "sometimes (10%)," "common (50%)," or "all (100%)." The spreadsheet converts these qualitative values into numerical values to calculate risk estimates and a relative risk ranking score. This tool has been used to assess 10 seafood hazard combinations (106). Situations in which data may not be easily obtained might be treated with the combination of fuzzy math and risk assessment (21). A risk assessment related to the presence of *Enterobacter sakazakii* and *Salmonella* spp. in powdered infant formula was rapidly developed by a Food and Agriculture Organization-World Health Organization consultation by focusing on risk reductions that

could be achieved by different risk mitigations compared to an agreed-upon baseline scenario (39).

FSO Paradigm

To integrate HACCP and risk assessment concepts, a series of terms have been developed (13, 15, 35, 47, 51, 105, 117, 121). Figure 45.2 illustrates a simple food process in which an individual serving starts as raw ingredients with some frequency and level of contamination, is pasteurized to reduce pathogen numbers, is stored under conditions in which survivors may grow, is consumed, and may result in illness. The FSO terms and their points of application in the food process are indicated.

The paradigm begins with the ALOP. This is the public health goal for the food that the process is designed to meet. ALOPs can be expressed as the number of cases per year in a country or as the probability of illness per serving for a specified population. The ALOP is set by a societal process; it is a level that the public will tolerate in order to have the foods it desires. However, it is necessary to communicate to the public that risks cannot be zero and that ALOPs must be achievable by the food industry. The dose-response relationship for the selected susceptible human population will determine the number of cells of a pathogen consumed with a serving that has a risk equal to or below the ALOP. This maximum number of cells of a pathogen and/or the frequency of contamination is the FSO and becomes the microbial target that a food process should be designed and controlled to achieve. Factoring in the serving size will convert the FSO into a CFU-per-gram basis. Because it is unfeasible to determine or regulate microbial populations at the time of consumption, the maximum numbers of cells of the pathogen can be specified at earlier points in the processing chain. These are the performance objectives (POs). There can be several POs

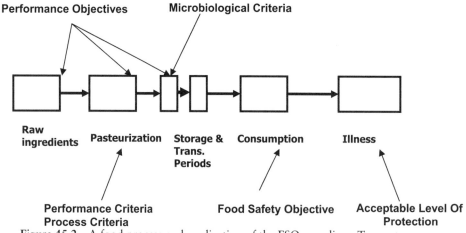

Figure 45.2 A food process and application of the FSO paradigm. Trans., transport.

in a food process, and the FSO can be viewed as the PO at consumption. For foods that support growth, allowances for growth between manufacture or retail and consumption are necessary. This requires designation of appropriate home and food service storage temperatures and times. These guidelines need to be based on reasonable levels of abuse as most foods that allow pathogen growth would not have counts below the FSO after extreme storage temperatures and/or times. The selection of an FSO (and corresponding ALOP) then allows calculation of microbial levels at various steps in the food process. A more stringent FSO will require comparable reduction in initial contamination or growth or an increase in an inactivation step. A case study from the Netherlands that explored the process of setting FSOs for microbiological hazards has been published (31). For an inactivation process step, the decrease that should be achieved is termed the performance criterion. Performance criteria could also be established for a storage period in which a 1-log increase, for example, would be allowed. Several inactivation processes could achieve a specified performance criterion. Thermal inactivation is the traditional inactivation process but high pressure, UV light, and pulsed electrical fields are examples of other processes for inactivation that could be employed with juices, for example. There are usually multiple combinations of parameters for each process that could achieve the performance criteria, e.g., different time-temperature combinations that achieve a 7-log reduction by thermal inactivation. The specific operating parameters chosen to achieve the performance criteria are the process criteria. These parameters then become the critical control points in the HACCP plan that controls the safety of the process. The application of risk assessment techniques to better link microbiological criteria to public health goals is an area that is actively being pursued by regulatory agencies, international intergovernmental organizations (e.g., Codex Alimentarius), and the food industry.

References

1. **Alavi, S. H., V. M. Puri, S. J. Knabel, R. H. Mohtar, and R. C. Whiting.** 1999. Development and validation of a dynamic growth model for *Listeria monocytogenes* in fluid whole milk. *J. Food Prot.* **62:**170–176.

2. **Amézquita, A., C. L. Weller, L. Wang, H. Thippareddi, and D. E. Burson.** 2005. Development of an integrated model for heat transfer and dynamic growth of *Clostridium perfringens* during the cooling of cooked boneless ham. *Int. J. Food Microbiol.* **101:**123–144.

3. **Baranyi, J., and M. L. Tamplin.** 2004. Combase: a common database on microbial response to food environments. *J. Food Prot.* **67:**1967–1971.

4. **Barker, G. C., P. K. Malakar, and M. W. Peck.** 2005. Germination and growth from spores: variability and uncertainty in the assessment of food borne hazards. *Int. J. Food Microbiol.* **100:**67–76.

5. **Bernaerts, K., E. Dens, K. Vereecken, A. H. Geeraerd, A. R. Standaert, F. Devlieghere, J. Debevere, and J. F. van Impe.** 2004. Concepts and tools for predictive modeling of microbial dynamics. *J. Food Prot.* **67:**2041–2052.

6. **Buchanan, R. L., and S. B. Dennis.** 2001. Microbial risk assessment—a tool for regulatory decision making. *J. Assoc. Food Drug Off.* **65:**36–46.

7. **Buchanan, R. L., S. B. Dennis, and M. Miliotis.** 2004. Initiating and managing risk assessments within a risk analysis framework: FDA/CFSAN's practical approach. *J. Food Prot.* **67:**2058–2062.

8. **Buzurl, S., and H. Alpas.** 2004. Modeling the synergistic effect of high pressure and heat on inactivation kinetics of *Listeria innocua*: a preliminary study. *FEMS Microbiol. Lett.* **238:**29–36.

9. **Carrington, C. D., S. B. Dennis, R. C. Whiting, and R. L. Buchanan.** 2004. Putting a risk assessment model to work: *Listeria monocytogenes* "what if" scenarios. *J. Assoc. Food Drug Off.* **68:**5–19.

10. **Chen, Y., K. M. Jackson, F. P. Chea, and D. W. Schaffner.** 2001. Quantification and variability analysis of bacterial cross-contamination rates in common food service tasks. *J. Food Prot.* **64:**72–80.

11. **Chen, Y., W. H. Ross, M. J. Gray, M. Wiedmann, R. C. Whiting, and V. N. Scott.** 2006. Attributing risk to *Listeria monocytogenes* subgroups: dose response in relation to genetic lineages. *J. Food Prot.* **69:**335–344.

12. **Codex Alimentarius Commission.** 2003. *Basic Texts on Food Hygiene*, 3rd ed., p. 45–52. Secretariat of the Joint Food and Agriculture Organization-World Health Organization Food Standards Programme, Food and Agriculture Organization, Rome, Italy.

13. **Codex Alimentarius Commission.** 2003. *Basic Texts on Food Hygiene*, 3rd ed., p. 53–62. Secretariat of the Joint Food and Agriculture Organization-World Health Organization Food Standards Programme, Food and Agriculture Organization, Rome, Italy.

14. **Codex Alimentarius Commission.** 2005. *Codex Alimentarius Procedural Manual*, 15th ed., p. 101–107. Secretariat of the Joint Food and Agriculture Organization-World Health Organization Food Standards Programme, Food and Agriculture Organization, Rome, Italy.

15. **Cole, M. B., and R. B. Tompkin.** 2005. Microbiological performance objectives and criteria, p. 673–695. *In* J. N. Sofos (ed.), *Improving the Safety of Fresh Meat*. Woodhead Publishing, Cambridge, United Kingdom.

16. **Collado, J., A. Fernández, L. M. Cunha, M. J. Ocio, and A. Martínez.** 2003. Improved model based on the Weibull distribution to describe the combined effect of pH and temperature on the heat resistance of *Bacillus cereus* in carrot juice. *J. Food Prot.* **66:**978–984.

17. **Cornu, M., A. Beaufort, S. Rudelle, L. Laloux, H. Bergis, N. Miconnet, T. Serot, and M. L. Delignette-Muller.** 2006. Effects of temperature, water-phase salt, and phenolic contents on *Listeria monocytogenes* growth rates on cold-smoked salmon and the evaluation of secondary models. *Int. J. Food Microbiol.* **106:**159–168.

18. Corradini, M. G., M. D. Normand, A. Nussinovitch, J. Horowitz, and M. Peleg. 2001. Estimating the frequency of high microbial counts in commercial food products using various distribution functions. *J. Food Prot.* **64:**674–681.

19. Couvert, O., S. Gaillard, N. Savy, P. Mafart, and I. Leguérinel. 2005. Survival curves of heated bacterial spores: effect of environmental factors on Weibull parameters. *Int. J. Food Microbiol.* **101:**73–81.

20. Dalgaard, P., P. Buch, and S. Silberg. 2002. Seafood spoilage predictor—development and distribution of a product specific application software. *Int. J. Food Microbiol.* **73:**343–349.

21. Davidson, V. J., and J. Ryks. 2003. Comparison of Monte Carlo and fuzzy math simulation methods for quantitative microbial risk assessment. *J. Food Prot.* **66:**1900–1910.

22. de Jesús, A. J., and R. C. Whiting. 2002. Thermal inactivation, growth, and survival studies of *Listeria monocytogenes* strains belonging to three distinct genotypic lineages. *J. Food Prot.* **66:**1611–1617.

23. Delignette-Muller, M. L., F. Baty, M. Cornu, and H. Bergis. 2005. Modelling the effect of a temperature shift on the lag phase duration of *Listeria monocytogenes. Int. J. Food Microbiol.* **100:**77–84.

24. Delignette-Muller, M. L., M. Cornu, R. Pouillot, and J.-B. Denis. 2006. Use of Bayesian modeling in risk assessment: application to growth of *Listeria monocytogenes* and food flora in cold-smoked salmon. *Int. J. Food Microbiol.* **106:**195–208.

25. Delignette-Muller, M. L., and L. Rosso. 2000. Biological variability and exposure assessment. *Int. J. Food Microbiol.* **58:**203–212.

26. den Aantrekker, E. D., R. M. Boom, M. H. Zwietering, and M. van Schothorst. 2003. Quantifying recontamination through factory environments—a review. *Int. J. Food Microbiol.* **80:**117–130.

27. Dens, E. J., K. Bernaerts, A. R. Standaert, J.-U. Kreft, and J. F. van Impe. 2005. Cell division theory and individual-based modeling of microbial lag, part II. Modeling lag phenomena induced by temperature shifts. *Int. J. Food Microbiol.* **101:**319–332.

28. Dens, E. J., K. Bernaerts, A. R. Standaert, and J. F. van Impe. 2005. Cell division theory and individual-based modeling of microbial lag, part I. The theory of cell division. *Int. J. Food Microbiol.* **101:**303–318.

29. Dens, E. J., and J. F. van Impe. 2001. On the need for another type of predictive model in structured foods. *Int. J. Food Microbiol.* **64:**247–260.

30. Denys, S., J. G. Pieters, and K. Dewettinck. 2005. Computational fluid dynamics analysis for process impact assessment during thermal pasteurization of intact eggs. *J. Food Prot.* **68:**366–374.

31. de Swarte, C., and R. A. Donker. 2005. Towards an FSO/ALOP based food safety policy. *Food Control* **16:**825–830.

32. Edelson-Mammel, S. G., and R. L. Buchanan. 2004. Thermal inactivation of *Enterobacter sakazakii* in rehydrated infant formula. *J. Food Prot.* **67:**60–63.

33. Edelson-Mammel, S. G., R. C. Whiting, S. W. Joseph, and R. L. Buchanan. 2005. Effect of prior growth conditions on the thermal inactivation of 13 strains of *Listeria monocytogenes* in two heating menstrua. *J. Food Prot.* **68:**168–172.

34. Food and Agriculture Organization-World Health Organization. 1997. *FAO Food and Nutrition Paper,* no. 65. *Risk Management and Food Safety: Report of a Joint FAO/WHO Consultation.* Food and Agriculture Organization, Rome, Italy.

35. Food and Agriculture Organization-World Health Organization. 2002. *Principles and Guidelines for Incorporating Microbiological Risk Assessment in the Development of Food Safety Standards: Guidelines and Related Texts. Report of a Joint FAO/WHO Consultation, Kiel, Germany.* ISBN 92 5 1048452. Food and Agriculture Organization, Rome, Italy.

36. Food and Agriculture Organization-World Health Organization. 2002. *Microbiological Risk Assessment Series,* no. 2. *Risk Assessment of Salmonella in Eggs and Broiler Chickens.* Food and Agriculture Organization, Rome, Italy.

37. Food and Agriculture Organization-World Health Organization. 2003. *Microbiological Risk Assessment Series,* no. 3. *Hazard Characterization for Pathogens in Food and Water.* Food and Agriculture Organization, Rome, Italy.

38. Food and Agriculture Organization-World Health Organization. 2004. *Microbiological Risk Assessment Series,* no. 5. Risk Assessment of Listeria monocytogenes *in Ready-To-Eat Foods.* ISBN 92 5 105127 5. Food and Agriculture Organization, Rome, Italy.

39. Food and Agriculture Organization-World Health Organization. 2004. *Microbiological Risk Assessment Series,* no. 6. Enterobacter sakazakii *and Other Microorganisms in Powdered Infant Formula.* Food and Agriculture Organization, Rome, Italy.

40. Food and Drug Administration. 19 July 2005, posting date. *Quantitative Risk Assessment on the Public Health Impact of Pathogenic Vibrio parahaemolyticus in Raw Oysters.* [Online.] Food and Drug Administration, Washington, D.C. www.cfsan.fda.gov/~dms/vpra-toc.html.

41. Food and Drug Administration-Food Safety and Inspection Service. October 2003, posting date. Updated 10 October 2005. *Quantitative Assessment of Relative Risk to Public Health from Foodborne Listeria monocytogenes among Selected Categories of Ready-To-Eat foods.* [Online.] Food and Drug Administration, Washington, D.C. www.foodsafety.gov/~dms/lmr2-toc.html.

42. Francois, K., F. Devlieghere, K. Smet, A. R. Standaert, A. H. Geeraerd, J. F. van Impe, and J. Debevere. 2005. Modelling the individual cell lag phase: effect of temperature and pH on the individual cell lag distribution of *Listeria monocytogenes. Int. J. Food Microbiol.* **100:**41–53.

43. Gaylor, D. W. 2005. Risk/benefit assessments of human diseases: optimum dose for intervention. *Risk Anal.* **25:**161–168.

44. Geeraerd, A. H., C. H. Herremans, and J. F. van Impe. 2000. Structural model requirements to describe microbial inactivation during a mild heat treatment. *Int. J. Food Microbiol.* **59:**185–209.

45. Geeraerd, A. H., V. P. Valdramidis, F. Devlieghere, H. Bernaert, J. Debevere, and J. F. van Impe. 2004. Development of a novel approach for secondary modeling in predictive microbiology: incorporation of microbiological

knowledge in black box polynomial modeling. *Int. J. Food Microbiol.* **91:**229–244.

46. **Giménez, B., and P. Dalgaard.** 2004. Modelling and predicting the simultaneous growth of *Listeria monocytogenes* and spoilage micro-organisms in cold-smoked salmon. *J. Appl. Microbiol.* **96:**96–109.

47. **Gorris, L. G. M.** 2005. Food safety objective: an integral part of food chain management. *Food Control* **16:**801–809.

48. **Grijspeerdt, K., and K. De Reu.** 2005. Practical application of dynamic temperature profiles to estimate the parameters of the square root model. *Int. J. Food Microbiol.* **101:** 83–92.

49. **Haas, C. N., A. Thayyar-Madabusi, J. B. Rose, and C. P. Gerba.** 2000. Development of a dose-response relationship for *Escherichia coli* O157:H7. *Int. J. Food Microbiol.* **56:**153–159.

50. **Hassani, M., I. Álvarez, J. Raso, S. Condón, and R. Pagán.** 2005. Comparing predicting models for heat inactivation of *Listeria monocytogenes* and *Pseudomonas aeruginosa* at different pH. *Int. J. Food Microbiol.* **100:**213–222.

51. **Havelaar, A. H., M. J. Nauta, and J. T. Jansen.** 2004. Fine-tuning food safety objectives and risk assessment. *Int. J. Food Microbiol.* **93:**11–29.

52. **Heldman, D. R., and R. L. Newsome.** 2003. Kinetics models for microbial survival during processing. *Food Technol.* **57(8):**40–46, 100.

53. **Holcomb, D. L., M. A. Smith, G. O. Ware, Y.-C. Hung, R. E. Brackett, and M. P. Doyle.** 1999. Comparison of six dose-response models for use with food-borne pathogens. *Risk Anal.* **19:**1091–1100.

54. **Hoornstra, E., and S. Notermans.** 2001. Quantitative microbiological risk assessment. *Int. J. Food Microbiol.* **66:**21–29.

55. **Horton, L. R.** 2001. Risk analysis and the law: international law, the World Trade Organization, Codex Alimentarius and national legislation. *Food Addit. Contam.* **18:**1057–1067.

56. **Huang, L.** 2003. Dynamic computer simulation of *Clostridium perfringens* growth in cooked ground beef. *Int. J. Food Microbiol.* **87:**217–227.

57. **Institute of Medicine, National Research Council.** 2003. *Scientific Criteria To Ensure Safe Food*, p. 401. ISBN 0-309-50920-3. The National Academies Press, Washington, D.C.

58. **Ivanek, R., Y. T. Gröhn, L. W. Tauer, and M. Wiedmann.** 2004. The cost and benefit of *Listeria monocytogenes* food safety measures. *Crit. Rev. Food Sci. Nutr.* **44:**513–523.

59. **Ivanek, R., Y. T. Gröhn, M. Wiedmann, and M. T. Wells.** 2004. Mathematical model of *Listeria monocytogenes* cross-contamination in a fish processing plant. *J. Food Prot.* **67:**2688–2697.

60. **Jones, R. C., S. I. Gerber, P. S. Diaz, L. L. Williams, S. B. Dennis, E. S. Parish, and W. S. Paul.** 2004. Intensive investigation of bacterial foodborne disease outbreaks: proposed guidelines and tools for the collection of dose-response data by local health departments. *J. Food Prot.* **67:**616–623.

61. **Juneja, V. K., H. M. Marks, and L. Huang.** 2003. Growth and heat resistance kinetic variation among various

isolates of *Salmonella* and its application to risk assessment. *Risk Anal.* **23:**199–213.

62. **Kilimann, K. V., C. Hartmann, A. Delagado, R. F. Vogel, and M. G. Gänzle.** 2005. A fuzzy logic-based model for the multistage high-pressure inactivation of *Lactococcus lactis* ssp. Cremoris MG 1363. *Int. J. Food Microbiol.* **98:**89–105.

63. **Koseki, S., and S. Isobe.** 2005. Growth of *Listeria monocytogenes* on iceberg lettuce and solid media. *Int. J. Food Microbiol.* **101:**217–225.

64. **Kothary, M. H., and U. S. Babu.** 2001. Infective dose of foodborne pathogens in volunteers: a review. *J. Food Safety* **21:**49–73.

65. **Koutsoumanis, K. P., P. A. Kendall, and J. N. Sofos.** 2004. A comparative study on growth limits of *Listeria monocytogenes* as affected by temperature, pH and a_w when grown in suspension or on a solid surface. *Food Microbiol.* **21:**415–422.

66. **Kusumaningrum, H. D., E. D. van Asselt, R. R. Beumer, and M. H. Zwietering.** 2004. A quantitative analysis of cross-contamination of *Salmonella* and *Campylobacter* spp. via domestic kitchen surfaces. *J. Food Prot.* **67:**1892–1903.

67. **Kutalik, Z., M. Razaz, A. Elfwing, A. Ballagi, and J. Baranyi.** 2005. Stochastic modeling of individual cell growth using flow chamber microscopy images. *Int. J. Food Microbiol.* **105:**177–190.

68. **Lammerding, A. M., and R. C. McKellar.** 2004. Predictive microbiology in quantitative risk assessment, p. 263–284. *In* R. C. McKellar and X. Lu (ed.), *Modeling Microbial Responses in Food.* CRC Press, Boca Raton, Fla.

69. **Latimer, H. K., L. Jaykus, R. A. Morales, P. Cowen, and D. Crawford-Brown.** 2001. A weighted composite dose-response model for human salmonellosis. *Risk Anal.* **21:**295–305.

70. **Legan, J. D., D. L. Seman, A. L. Milkowski, J. A. Hirschey, and M. H. Vandeven.** 2004. Modeling the growth boundary of *Listeria monocytogenes* in ready-to-eat cooked meat products as a function of the product salt, moisture, potassium lactate, and sodium diacetate concentrations. *J. Food Prot.* **67:**2195–2204.

71. **Le Marc, Y., C. Pin, and J. Baranyi.** 2005. Methods to determine the growth domain in a multidimensional environmental space. *Int. J. Food Microbiol.* **100:**3–12.

72. **Lyytikäinen, O., T. Autio, R. Maijala, P. Ruutu, T. Honkanen-Buzalski, M. Miettinen, M. Hatakka, J. Mikkola, V.-J. Anttila, T. Johansson, and L. Rantala.** 2000. An outbreak of *Listeria monocytogenes* serotype 3a infections from butter in Finland. *J. Infect. Dis.* **181:**1838–1841.

73. **Malakar, P. K., G. C. Barker, and M. W. Peck.** 2004. Modeling the prevalence of *Bacillus cereus* spores during the production of a cooked chilled vegetable product. *J. Food Prot.* **67:**939–946.

74. **Martens, D. E., C. Báal, P. Malakar, M. H. Zwietering, and K. van't Reit.** 1999. Modelling the interactions between *Lactobacillus curvatus* and *Enterobacter cloacae.* I. Individual growth kinetics. *Int. J. Food Microbiol.* **51:**53–65.

75. **McKellar, R. C., and A. Hawke.** 2006. Assessment of distributions for fitting lag times of individual cells in bacterial populations. *Int. J. Food Microbiol.* **106:**169–175.

76. **McKellar, R. C., and X. Lu.** 2004. *Modeling Microbial Responses in Food*, p. 343. CRC Press, Boca Raton, Fla.

77. McKellar, R. C., and X. Lu. 2005. Development of a global stochastic model relating the distribution of individual cell and population physiological states. *Int. J. Food Microbiol.* **100:**33–40.

78. McKellar, R. C., X. Lu, and P. J. Delaquis. 2002. A probability model describing the interface between survival and death of *Escherichia coli* O157:H7 in a mayonnaise model system. *Food Microbiol.* **19:**235–247.

79. McMeekin, T. A., J. Olley, D. A. Ratkowsky, and T. Ross. 2002. Predictive microbiology: towards the interface and beyond. *Int. J. Food Microbiol.* **73:**395–407.

80. McMeekin, T. A., K. Presser, D. Ratkowsky, T. Ross, M. Salter, and S. Tienungoon. 2000. Quantifying the hurdle concept by modeling the bacterial growth/no growth interface. *Int. J. Food Microbiol.* **55:**93–98.

81. Métris, A., Y. Le Marc, A. Elfwing, A. Ballagi, and J. Baranyi. 2005. Modelling the variability of lag times and the first generation times of single cells of *E. coli*. *Int. J. Food Microbiol.* **100:**13–19.

82. Miconnet, N., A. H. Geeraerd, J. F. van Impe, L. Rosso, and M. Cornu. 2005. Reflections of the use of robust and least-squares non-linear regression to model challenge tests conducted in/on food products. *Int. J. Food Microbiol.* **104:**161–177.

83. Moon, H., J. J. Chen, D. W. Gaylor, and R. L. Kodell. 2004. A comparison of microbial dose-response models fitted to human data. *Regul. Toxicol. Pharmacol.* **40:**177–184.

84. Nauta, M. J. 2000. Separation of uncertainty and variability in quantitative microbial risk assessment models. *Int. J. Food Microbiol.* **57:**9–18.

85. Nauta, M. J. 2002. Modelling bacterial growth in quantitative microbial risk assessment: is it possible? *Int. J. Food Microbiol.* **73:**297–304.

86. Oscar, T. 2004. Dose-response model for 13 strains of *Salmonella*. *Risk Anal.* **24:**41–49.

87. Patil, S. R., and H. C. Frey. 2004. Comparison of sensitivity analysis methods based on applications to a food safety risk assessment model. *Risk Anal.* **24:**573–585.

88. Peleg, M. 2000. Microbial survival curves—the reality of flat "shoulders" and absolute thermal death times. *Food Res. Int.* **33:**531–538.

89. Peleg, M., and J. B. Cole. 1998. Reinterpretation of microbial survival curves. *Crit. Rev. Food Sci.* **38:**353–380.

90. Peleg, M., and C. M. Penchina. 2000. Modeling microbial survival during exposure to a lethal agent with varying intensity. *Crit. Rev. Food Sci.* **40:**159–172.

91. Pin, C., J. P. Sutherland, and J. Baranyi. 1999. Validating predictive models of food spoilage organisms. *J. Appl. Microbiol.* **87:**491–499.

92. Poschet, F., A. H. Geeraerd, N. Scheerlinck, B. M. Nicolaï, and J. F. van Impe. 2003. Monte Carlo analysis as a tool to incorporate variation on experimental data in predictive microbiology. *Food Microbiol.* **20:**285–295.

93. Poschet, F., K. M. Vereecken, A. H. Geeraerd, B. M. Nicolaï, and J. F. van Impe. 2005. Analysis of a novel class of predictive microbial growth models and application to coculture growth. *Int. J. Food Microbiol.* **100:**107–124.

94. Powell, D. A. 2000. Food safety and the consumer—perils of poor risk communication. *Can. J. Anim. Sci.* **80:**393–404.

95. Powell, M. R., E. Ebel, W. Schlosser, M. Walderhaug, and J. Kause. 2000. Dose-response envelope for *Escherichia coli* O157:H7. *Quant. Microbiol.* **2:**141–163.

96. Ratkowsky, D. A. 2002. Some examples of, and some problems with, the use of nonlinear logistic regression in predictive food microbiology. *Int. J. Food Microbiol.* **73:**119–125.

97. Reij, M. W., E. D. den Aantrekker, and ILSI Europe Risk Analysis in Microbiology Task Force. 2004. Recontamination as a source of pathogens in processed foods. *Int. J. Food Microbiol.* **91:**1–11.

98. Robertson, L. J., J. D. Greig, B. Gjerde, and A. Fazil. 2005. The potential for acquiring cryptosporidiosis or giardiosis from consumption of mung bean sprouts in Norway: a preliminary step-wise risk assessment. *Int. J. Food Microbiol.* **98:**291–300.

99. Rodrigo, D., G. V. Barbosa-Cánovas, A. Martínez, and M. Rodrigo. 2003. Weibull distribution function based on an empirical mathematical model for inactivation of *Escherichia coli* by pulsed electric fields. *J. Food Prot.* **66:**1007–1012.

100. Ross, T., and J. Sumner. 2002. A simple, spreadsheet-based, food safety risk assessment tool. *Int. J. Food Microbiol.* **77:**39–53.

101. Serra, J. A., E. Domenech, I. Escriche, and S. Martorell. 1999. Risk assessment and critical control points from the production perspective. *Int. J. Food Microbiol.* **46:**9–26.

102. Stecchini, M. L., M. Del Torre, and E. Venir. 2004. Growth of *Listeria monocytogenes* as influenced by viscosity and water activity. *Int. J. Food Microbiol.* **96:**181–187.

103. Stewart, C. M., M. B. Cole, J. D. Legan, L. Slade, M. H. Vandeven, and D. W. Schaffner. 2001. Modeling the growth boundary of *Staphylococcus aureus* for risk assessment purposes. *J. Food Prot.* **64:**51–57.

104. Strachan, N. J. C., M. P. Doyle, F. Kasuga, O. Rotariu, and I. D. Ogden. 2005. Dose response modeling of *Escherichia coli* O157 incorporating data from foodborne and environmental outbreaks. *Int. J. Food Microbiol.* **103:**35–47.

105. Stringer, M. 2005. Food safety objectives—role in microbiological food safety management. *Food Control* **16:**775–794.

106. Sumner, J., and T. Ross. 2002. A semi-quantitative seafood safety risk assessment. *Int. J. Food Microbiol.* **77:**55–59.

107. Swinnen, I. A. M., K. Bernaerts, E. J. J. Dens, A. H. Geeraerd, and J. F. van Impe. 2004. Predictive modelling of the microbial lag phase: a review. *Int. J. Food Microbiol.* **94:**137–159.

108. Swinnen, I. A. M., K. Bernaerts, K. Gysemans, and J. F. van Impe. 2005. Quantifying microbial lag phenomena due to a sudden rise in temperature: asystematic macroscopic study. *Int. J. Food Microbiol.* **100:**85–96.

109. Tamplin, M. L. 2005. Modeling pathogen behavior in foods, p. 113–120. *In* P. M. Fratamico, A. K. Bhunia, and J. L. Smith (ed.), *Foodborne Pathogens: Microbiology and*

Molecular Biology. Academic Press, Wymondham, United Kingdom.

110. **Teunis, P. F. M., and A. H. Havelaar.** 2002. Risk assessment for protozoan parasites. *Int. Biodeterior. Biodegradation* **50:**185–193.

111. **Tienungoon, S., D. A. Ratkowsky, T. A. McMeekin, and T. Ross.** 2000. Growth limits of *Listeria monocytogenes* as a function of temperature, pH, NaCl, and lactic acid. *Appl. Environ. Microbiol.* **66:**4979–4987.

112. **Valdramidis, V. P., N. Belaubre, R. Zuniga, A. M. Foster, M. Havet, A. H. Geeraerd, M. J. Swain, K. Bernaerts, J. F. van Impe, and A. Kondjoyan.** 2005. Development of predictive modeling approaches for surface temperature and associated microbiological inactivation during hot dry air decontamination. *Int. J. Food Microbiol.* **100:**261–274.

113. **Valdramidis, V. P., A. H. Geeraerd, K. Bernaerts, F. Devlieghere, J. Debevere, and J. F. van Impe.** 2004. *Bulletin of the IDF,* no. 392. *Accurate Modeling of Non-Loglinear Survival Curves,* p. 97–110. International Dairy Federation, Brussels, Belgium.

114. **Van der Fels-Klerx, H. J., R. M. Cooke, M. N. Nauta, L. H. Goossens, and A. H. Havelaar.** 2005. A structured expert judgment study for a model of *Campylobacter* transmission during broiler-chicken processing. *Risk Anal.* **25:**109–124.

115. **van Gerwen, S. J. C., M. C. te Giffel, K. van't Riet, R. R. Beumer, and M. H. Zwietering.** 2000. Stepwise quantitative risk assessment as a tool for characterization of microbiological food safety. *J. Appl. Microbiol.* **88:**938–951.

116. **van Impe, J. F., F. Poschet, A. H. Geeraerd, and K. M. Vereecken.** 2005. Towards a novel class of predictive microbial growth models. *Int. J. Food Microbiol.* **100:**97–105.

117. **van Schothorst, M.** 2005. A proposed framework for the use of FSOs. *Food Control* **16:**811–816.

118. **Vialette, M., A. Pinon, E. Chasseignaux, and M. Lange.** 2003. Growth kinetics comparison of clinical and seafood *Listeria monocytogenes* isolates in acid and osmotic environment. *Int. J. Food Microbiol.* **82:**121–131.

119. **Vose, D.** 2000. *Risk Analysis,* 2nd ed. Wiley & Sons, New York, N.Y.

120. **Voysey, P. A., and M. Brown.** 2000. Microbiological risk assessment: a new approach to food safety control. *Int. J. Food Microbiol.* **58:**173–179.

121. **Walls, I., and R. L. Buchanan.** 2005. Use of food safety objectives as a tool for reducing foodborne listeriosis. *Food Control* **16:**795–799.

122. **Whiting, R. C., and L. K. Bagi.** 2002. Modeling the lag phase of *Listeria monocytogenes. Int. J. Food Microbiol.* **73:**291–295.

123. **Whiting, R. C., and R. L. Buchanan.** 2001. Predictive modeling and risk assessment, p. 813–831. *In* M. P. Doyle, L. R. Beuchat, and T. J. Montville (ed.), *Food Microbiology: Fundamentals and Frontiers,* 2nd ed. ASM Press, Washington, D.C.

124. **Whiting, R. C., and M. H. Golden.** 2002. Variation among *Escherichia coli* O157:H7 strains relative to their growth, survival, thermal inactivation, and toxin production in broth. *Int. J. Food Microbiol.* **75:**127–133.

125. **Whiting, R. C., A. Hogue, W. D. Schlosser, E. D. Ebel, R. Morales, A. Baker, and R. McDowell.** 2000. A risk assessment for *Salmonella* Enteritidis in shell eggs. *J. Food Sci.* **65:**864–869.

126. **Williams, R. A., and K. M. Thompson.** 2004. Integrated analysis: combining risk and economic assessments while preserving the separation of powers. *Risk Anal.* **24:**1613–1623.

127. **Williams, T. L., S. R. Monday, S. Edelson-Mammel, R. L. Buchanan, and S. M. Musser.** 2005. A top-down proteomics approach for differentiating thermal resistant strains of *Enterobacter sakazakii. Proteomics* **5:**4161–4169.

128. **Wilson, P. D. G., T. F. Brocklehurst, S. Arino, D. Thuault, M. Jakobsen, M. Lange, J. Farkas, J. W. T. Wimpenny, and J. F. van Impe.** 2002. Modelling microbial growth in structured foods: towards a unified approach. *Int. J. Food Microbiol.* **73:**275–289.

129. **Zwietering, M. H., and S. J. C. van Gerwen.** 2000. Sensitivity analysis in quantitative microbial risk assessment. *Int. J. Food Microbiol.* **58:**213–221.

Food Microbiology: Fundamentals and Frontiers, 3rd Ed.
Edited by M. P. Doyle and L. R. Beuchat
© 2007 ASM Press, Washington, D.C.

Dane Bernard
Virginia N. Scott

46

Hazard Analysis and Critical Control Point System: Use in Controlling Microbiological Hazards

Several chapters of this book have presented current information regarding consumer health problems associated with harmful microbes or their toxins in food or water. Many of these problems are being addressed by the food industry by application of a food safety management system, the Hazard Analysis and Critical Control Point (HACCP) system, that is specifically designed for minimizing food safety risks. At the same time, it has become accepted that the HACCP system works best in an environment where certain components of an overall system of food safety-related controls are supported as prerequisites to HACCP, including those functions normally classified as good manufacturing practices (GMPs) or good hygienic practices (4, 18).

It is well accepted that it is the responsibility of the food industry to produce, transport, process, and prepare foods that present a minimum level of risk from foodborne hazards. To accomplish this, the industry must identify and control the potential risks posed by pathogenic microorganisms, toxic chemicals, and hazardous physical agents. The controls applied to achieve this must be both practical and achievable within the limits of the available technology. To assist in identifying hazards and managing risks resulting from

exposure to the hazards, the HACCP system has been developed.

HACCP is regarded by many as the best system available for designing programs to assist food firms in producing and processing foods that are safe to consume. Recognizing this, the U.S. Food and Drug Administration (FDA) and the U.S. Department of Agriculture's Food Safety and Inspection Service (USDA FSIS) have published final rules (8, 9, 11) mandating the development and implementation of HACCP plans to cover a variety of domestic and imported seafoods, fruit and vegetable juices, and meat and poultry products. Regulations on HACCP for egg products are under development.

Paralleling these U.S. government initiatives is an international movement toward a food safety assurance and regulatory scheme based on the principles of HACCP. For example, in April 2004 the European Union adopted several new regulations on the hygiene of foods, including one (852/2004/EC) that mandates that all food business operators implement procedures based on HACCP principles. The requirement for HACCP also applies to imports. This directive, which became effective 1 Jan. 2006, replaces several product-specific ("vertical") directives and a "horizontal" food hygiene

Dane Bernard, Keystone Foods LLC, 5 Tower Bridge, 300 Barr Harbor Drive, Suite 600, W. Conshohocken, PA 19428. **Virginia N. Scott,** GMA/FPA, 1350 I St. N.W., Suite 300, Washington, DC 20005.

directive (93/43/EEC) that required consistency with HACCP principles for a wide range of food items. Other government authorities across the globe, including those in Canada, Australia, and Japan, have adopted or are adopting HACCP-based food safety control systems.

Agreements and treaties that support the World Trade Organization (WTO) illustrate the growing importance of HACCP as a tool to judge the acceptability of foods traded internationally. These agreements provide guidance to WTO member nations about how to establish sanitary and phytosanitary requirements without creating technical trade barriers (a prohibited practice that may result in sanctions against a country that is a member of the WTO). For the establishment of sanitary measures addressing food safety, these agreements reference the standards, guidelines, and recommendations developed and adopted by the Codex Alimentarius Commission as an acceptable scientific baseline for consumer protection (3).

In an effort to promote global food safety, the Codex Alimentarius Commission adopted a code of practice detailed in "Hazard Analysis and Critical Control Point (HACCP) System and Guidelines for Its Application." This document was most recently revised in 2003 to provide better guidance, especially to small, less developed businesses. The HACCP guidelines are appended to the code of hygienic practice (4), which serves to emphasize that, for all its merits, HACCP does not stand alone but depends upon a foundation of general good hygienic practices, a point that will be discussed later.

Although HACCP has now been in wide use for many years as a mandated program, issues have arisen that have yet to be fully resolved. Specifically, determining an appropriate separation between prerequisite programs and HACCP, as well as the treatment and enforcement of prerequisite programs and company-specific operating procedures, within a mandatory HACCP environment has become an item of some concern. Also, as HACCP is customized according to each individual facility, inspection by government authorities carries unique challenges. Much guidance has been developed, but the distinction between compliance requirements for HACCP and those for prerequisite programs and an appropriate government response to nonconformance for each category are as yet unclear. Resolution of these issues is of paramount importance.

ORIGIN OF HACCP

The acronym HACCP was first coined during collaborative work among the Pillsbury Company, the U.S. Army Natick Laboratories, and the National Aeronautics and Space Administration to describe the systematic

approach to food safety developed to ensure the safety of foods being produced for manned space flights. In order to approach the goal of 100% assurance that these foods would be safe to consume, a system had to be developed that went well beyond the limited effectiveness of sampling and analyzing finished goods for the presence of hazardous microorganisms, potentially injurious foreign materials, or toxic chemicals. Methods used by others, including "Haz-Ops" studies and "Modes of Failure Analysis," were considered in the search for new ways to ensure that foods were safe. Although the HACCP concept reflects much of this earlier thinking, these programs did not take into account the special needs of the food industry. From these approaches, the HACCP concept was distilled. The concept is essentially that if knowledge can be ascertained about how a product (in this case a food) may become unsafe if consumed, then control measures can be developed to prevent and/or detect such failures and keep food that presents an unacceptable risk from reaching consumers. This concept formed the foundation for development of the HACCP system of food safety control.

Howard Bauman presented the HACCP concept, along with the original three principles, to the public in 1971 at the first National Conference on Food Protection (19). Although the concept created some interest in the early 1970s, initial attempts at incorporating HACCP into food operations were not very successful. It is likely that a major contributor to these early unsuccessful attempts was a lack of description and understanding of HACCP concepts. Early HACCP plans typically had far more critical control points (CCPs) than were needed to ensure production of a safe food, were not "plant friendly," and were too cumbersome to be sustained. However, in 1985, a subcommittee of the Food Protection Committee of the National Academy of Sciences issued a report on microbiological criteria that included an endorsement of HACCP (17). Based on this recommendation, the National Advisory Committee on Microbiological Criteria for Foods (NACMCF) was formed in 1988 as an advisory group for the Secretaries of Agriculture, Commerce, Defense, and Health and Human Services. One of the activities of the NACMCF involved developing material to promote an understanding of HACCP and to encourage adoption of the HACCP approach to food safety. The NACMCF has been a leader in elaborating HACCP principles and in providing guidance for their application to food-processing operations and to the sponsoring agencies. In recognition that the ideas underlying the successful application of HACCP continue to evolve, the NACMCF has updated its 1989 and 1992 guidance documents. The committee's latest guidance document on HACCP, completed in 1997, was published in 1998 (18).

HACCP OVERVIEW

The systematic approach to food safety embodied by HACCP is based on seven principles. The HACCP principles describe a format for identifying and ensuring control of those factors (hazards) that are reasonably likely to cause a food to be unsafe for consumption. In the hazard analysis step (principle 1), potential hazards are identified and evaluated to determine which hazards are significant enough that they must be addressed in a HACCP plan. The remaining six HACCP principles describe the steps necessary to develop a management strategy to control the significant hazards. HACCP is based on a common-sense application of technical and scientific principles to the food production process from field to table. The process of developing a HACCP plan is analogous to conducting a qualitative assessment of risk posed by a particular hazard, coupled with selection and implementation of appropriate control measures.

The principles adopted by the NACMCF are used in this chapter. There are variations of the principles outlined by other groups and official bodies and even more variation in the interpretation of the principles and in the resulting guidance about how they are to be applied during development and implementation of HACCP plans. Nevertheless, there is surprising agreement in the basic concepts and in the overarching message that users of HACCP should not focus so tightly on the "doctrine" that the intended common-sense approach to food safety management is lost. In other words, one should use common sense in applying the seven principles.

The approach taken in this chapter is that HACCP plans should address only significant food safety hazards. It is these safety hazards that warrant the extensive monitoring, record-keeping, and verification activities required by the HACCP system of food safety management. The FDA, in its final rules on seafood and juice safety, and the USDA FSIS, in its regulation on HACCP and pathogen reduction for meat and poultry, also limit HACCP to issues of food safety. There is growing international acceptance of the safety-only philosophy of HACCP as well. This may be viewed as an unnecessarily narrow interpretation by some, as the concepts embodied in the seven principles for developing a HACCP plan can be applied to the control of quality factors and economic adulteration as well. However, extending the system to quality factors dilutes the focus on safety and results in too many CCPs that are not truly critical. The safety-only approach is increasingly viewed as a functional

necessity in order for HACCP to be globally successful. Production under a HACCP system is one factor used to judge the acceptability of products in domestic and international trade. Although quality is a negotiable item (especially when viewed on a global scale), safety should be nonnegotiable. That said, experience shows that even HACCP experts still have difficulty coming to agreement on the safety implications of certain hazards. This results in a situation in which it is sometimes difficult to obtain agreement on what is a good HACCP plan. This situation becomes even worse when quality factors are included in the mix. Thus, if the seven principles are applied to develop a control strategy for quality factors, this should not be labeled as a HACCP plan.

Most contemporary references on HACCP point to its applicability to all segments of the food industry, from basic agriculture through food preparation and handling, food processing, food packaging, food service, distribution systems, and consumer handling and use. Although in a broad sense this may be true, the ways in which the HACCP principles are interpreted and applied will differ depending on which segment in the food chain is being addressed. This subject is discussed further in a paper published by the Food and Agriculture Organization (7). However, for a sharper focus for this chapter, we have limited our examples of the application of HACCP to the food-manufacturing segment.

HACCP continues to evolve as a concept. As additional real-world experience is gained with the system, more techniques will be developed that will make it easier to understand and apply. The guidance documents about HACCP will also continue to be modified. Thus, each firm, besides following the legal requirements that pertain to its operation, should do what it feels is appropriate to develop a workable, plant-friendly, and effective HACCP plan. Remember that many of the substantial benefits of HACCP accrue because plans are operation and product (or product category) specific, therefore addressing the unique needs of each food facility. Each operation therefore should view the HACCP guidance available today as just that, guidance. Given this relatively traditional stand regarding the individuality of HACCP plans, the imposition of HACCP as a regulatory program brings with it unique challenges to this notion.

The basic premise underlying HACCP is to prevent hazards during production rather than relying solely on inspection or testing of finished goods to detect and correct problems. Although testing of finished products usually plays a role in a HACCP system (as a verification tool), the main focus of HACCP is on prevention through process control. As most food microbiologists will attest, results derived from sampling and testing support a particular claim about only the units tested. When test results are interpreted, the limitations of the testing methodology and the sampling plan must be considered, because statements about the nature of units not tested can be made only by inference within the context of these limitations. By contrast, if the hazards of concern are properly identified and effective control measures are available and followed, HACCP provides greater confidence in the safety of each unit produced than can be obtained through any practical sampling and testing program. The degree of control of a hazard should be key in determining the need for verification sampling and testing of finished goods. If the level of control over a hazard is high, as it may be with a fully cooked item, sampling and testing will have little benefit. For raw or raw, transformed foods, the lower level of control of hazards (in this case, enteric pathogens) means that the need for verification testing may be greater.

A CAUTIONARY NOTE

In the opinion of many HACCP authorities, the most important (and least understood) activity in the development of a HACCP plan is the hazard analysis step. Although this process will be discussed in more depth below, it must be emphasized that a HACCP plan will not be effective in minimizing risks associated with a food unless the hazard analysis is done correctly and effective control measures are adopted. A note of caution, however: HACCP is not a zero-defects program, even though this should always be the ultimate goal in food safety. When HACCP is applied to processes with a well-defined and controlled microbial kill step, such as canning, we closely approach the zero risk threshold. However, the application of HACCP to raw or raw, transformed products for which control steps are less well defined and less effective than those for cooked products is a risk reduction exercise that cannot offer the same level of protection provided for other processed foods. Unless and until more effective control measures are available and adopted in these parts of the industry, gains available through the adoption of HACCP alone will be marginal.

In addition, any hazard analysis will be limited by the information available to the experts conducting the analysis. Some risk factors, such as a pathogen that adapts to a new environment, cannot be anticipated and thus cannot be addressed in a HACCP plan. HACCP is not a magic bullet that will cure all food safety problems. However, the adoption of HACCP will put a system in place that has the potential to be more responsive to newly identified problems. Because the natural world is always presenting us with new hazards, it is essential to the success of a HACCP system that plans are reviewed in light of new scientific information. To be effective over time, the hazard analysis and resulting HACCP plans should be reviewed frequently.

PRELIMINARY STEPS

Certain preliminary steps must be undertaken before beginning the hazard analysis. The recommended steps that precede the hazard analysis include the following: (i) gain management support, (ii) assemble the HACCP team, (iii) describe the food and the method of its distribution, (iv) identify the intended use and consumers of the food, (v) develop a flow diagram, and (vi) verify the flow diagram.

Regarding point (i) from the above list, most who have attempted to develop HACCP programs agree that it is very difficult to put HACCP systems in place without firm support by company management. HACCP often requires basic changes in management styles and requires the commitment of resources to the process. Assembling a HACCP team [point (ii)] can be especially challenging for small firms that do not have the advantage of a large technical staff. Even though it may be a challenge for some firms, most will agree that individuals with the necessary expertise must be involved in the process. However, they need not necessarily be employed at the firm. All firms developing HACCP plans must find a way to have the requisite expertise reflected in their plans. The team should be multidisciplinary and include individuals from areas such as engineering, production, sanitation, quality assurance, and food microbiology. The team should also include local personnel who are involved in the operation, as they are more familiar with the variability and limitations of the operation. Points (iii) through (vi) from the list are necessary to allow for development of HACCP plans that are tailored to specific products and food operations. In order to address these preliminary steps, the food firm must have sufficient information about the ingredients, the processing methods, the distribution system, and the expected consumers so it can determine where and how significant food safety hazards may be introduced or enhanced.

ESTABLISHING PLAN-SPECIFIC OBJECTIVES AND PERFORMANCE CRITERIA

The topics of food safety objectives, performance objectives, and performance criteria have recently been introduced into discussions of the application of risk analysis in food safety systems (6, 15). Although these terms have been defined (5), how they are to be applied, especially with respect to the development and implementation of HACCP plans, remains to be more fully developed. Currently there are performance criteria (e.g., the FSIS requirement for a 6.5-log reduction of *Salmonella* in meat and a 7-log reduction in poultry [12]) that form the basis for process criteria (cooking times and temperatures) used in HACCP plans for meat and poultry products in the United States. In cases where these criteria apply, they are to be used as targets for the validation of the effectiveness of control measures. However, the relation of these criteria to specific food safety objectives (which apply at consumption) or performance objectives (which apply at other points in the food chain) has not been defined.

In the absence of defined food safety objectives, it may be useful for the HACCP team to identify plan-specific public health-related objectives as the plan is being developed. These may be in the form of a statement of the food safety-related results expected from implementation of the HACCP plan. These objectives will typically become clear as the team conducts its hazard analysis and determines the significant hazards that will be addressed in the plan. Where possible, performance criteria should be identified that lead to meeting the objective. An example of such an objective is the qualitative statement that canned foods should pose only a negligible risk from *Clostridium botulinum*. This can be accomplished by achieving or exceeding the performance criterion of a 12-log reduction (or equivalent) of the target organism. A similar qualitative objective would be to produce a pasteurized product that poses a negligible risk from vegetative pathogens such as *Salmonella*. In the production of refrigerated juice products, a 5-$\log_{10}$ reduction of the most resistant pathogen of public health significance likely to occur in the juice (e.g., *Salmonella*) is considered adequate to protect consumers from this hazard and has been set as a performance criterion by the FDA (9). Process steps designed to inactivate pathogenic microorganisms should be designed to meet criteria that have been developed to be protective of public health.

For raw products, defined and agreed-to objectives are more difficult to determine, and associated performance criteria are even more challenging. Currently, the objective of a HACCP plan may be to produce a product that presents a risk of illness from vegetative pathogens that is as low as reasonably achievable. This objective is usually accomplished by application of treatments designed to reduce numbers of microorganisms present and reduce the potential for cross-contamination and by control of conditions (e.g., preservatives) specifically designed to reduce or prevent the proliferation of pathogens. The

next step may well be to elaborate a target food safety objective based on a farm-to-table risk assessment and categorization of the risk relative to public health goals such as those outlined in *Healthy People 2010* (23). With this science-based food safety objective in hand, control measures may need to be developed or adapted to meet the objective. Appropriate performance objectives or criteria can then be established at appropriate points in the farm-to-table continuum to meet the food safety objective.

Similar public health-related objectives and criteria could be identified for chemical and physical hazards as well.

PRINCIPLE 1: CONDUCT A HAZARD ANALYSIS

After addressing the preliminary points discussed above, the HACCP team conducts a hazard analysis or ensures that an analysis is conducted. The purpose of the hazard analysis is to develop a list of hazards that are of such significance that they are reasonably likely to cause injury or illness if not effectively controlled. Hazards that are not reasonably likely to occur would not require further consideration within a HACCP plan. The NACMCF (18) defines a hazard (for HACCP) as "a biological, chemical, or physical agent that is reasonably likely to cause illness or injury in the absence of its control." The NACMCF (18) also provides a definition for a hazard analysis as "the process of collecting and evaluating information on hazards associated with the food under consideration to decide which are significant and must be addressed in the HACCP plan." Based on the results of the hazard analysis, the hazards that should be addressed in the HACCP plan, along with control measures for each hazard, are identified. For a hazard to be listed among the significant hazards that should be addressed in the HACCP plan, the hazard should be of such a nature that its prevention, elimination, or reduction to acceptable levels is essential to the production of a safe food.

On this last point, note that HACCP does not provide an excuse for food operators to ignore low-risk hazards or to ignore existing legal requirements, including those that define adulteration. Operations must be in compliance with all legal requirements. Low-risk hazards should not be dismissed as insignificant. It is expected that these will usually be addressed within other management systems or prerequisite programs, such as GMP programs, which are prerequisites for the HACCP system. It is often the proper functioning of this prerequisite program that ensures that a potential hazard is not reasonably likely to occur. Most authorities agree that if a food operation does not have sound prerequisite programs that comply with GMPs and other regulatory requirements, HACCP will have little chance for success. It is clear, however, that the NACMCF recognizes that not all biological, chemical, or physical properties that may cause a food to be unsafe for consumption are significant enough to warrant being addressed within a HACCP plan. Making the decision as to whether or not a hazard is reasonably likely to present a significant consumer safety problem is at the heart of the hazard assessment.

Conducting a Hazard Analysis

The NACMCF (18) advises that the process of conducting a hazard analysis involves two stages: hazard identification and hazard evaluation.

Hazard Identification

Hazard identification is sometimes described as a brainstorming session. During this stage, the HACCP team reviews the ingredients used in the product, the activities conducted at each step in the production process, and the equipment used to make the product. The team also considers the methods of storage and distribution and the intended use and consumers of the product.

Based on its review of the information just described, the team develops a list of potential biological, chemical, and physical hazards that may be introduced, increased (e.g., pathogen growth), or controlled at each step described on the flow diagram. During hazard identification, the HACCP team should focus on developing a list of potential hazards associated with each ingredient, each process step under direct control of the food operation, and the final product. Knowledge of any adverse health-related events historically associated with the product is important in this exercise. It is also important to be specific about the hazard; rather than listing the hazard as "pathogens," the team should list the specific pathogen (e.g., *Salmonella* or spores of *Clostridium perfringens*), especially when the controls may be different for different hazards, e.g., heat treatment versus refrigeration.

Potential Biological Hazards

A biological hazard is one that, if not properly controlled, is reasonably likely to result in foodborne illness. The primary microorganisms of concern are pathogenic bacteria, such as *C. botulinum*, *Listeria monocytogenes*, *Salmonella* species, and *Staphylococcus aureus*. It is also important to identify potential hazards from other biological agents, such as pathogenic viruses and parasites. The potential biological hazards associated with a product (and each ingredient) and process should be listed in preparation for evaluation.

Potential Chemical Hazards

Like biological hazards, the HACCP team must identify all potential chemical hazards associated with the production of the food commodity before evaluating the significance of each. Potential chemical hazards include toxic substances and any other compounds that may render a food unsafe for consumption, not only by the general public, but also by the small percentage of the population that may be particularly sensitive to a specific chemical. For example, sulfiting agents used to preserve fresh leafy vegetables, dried fruits, and wines have caused allergic-type reactions in sensitive individuals. Other chemical hazards to be considered include aflatoxin and other mycotoxins, fish and shellfish toxins, scombrotoxin (histamine) from the decomposition of certain types of fish, and allergenic ingredients such as tree nuts and shellfish.

Potential Physical Hazards

The HACCP team must identify any potential physical hazards associated with the finished product. Foreign objects that are capable of injuring the consumer represent potential physical hazards. Examples of such potential hazards include glass fragments, pieces of wood or metal, and plastic. During the hazard evaluation stage, differentiation may be made between foreign objects that are aesthetically unpleasant and those that are capable of causing injury. The HACCP team should define the performance criteria (size, shape, etc.) for physical hazards capable of causing injury. For example, a public health hazard analysis board at the FSIS concluded that bone particles less than 10 mm in size do not pose a significant hazard to consumers (2). In addition, the FDA's current *Compliance Policy Guide* states that products containing a hard or sharp foreign object that measures 7 to 25 mm in length should be considered adulterated (1). This guideline was based on a thorough review of 25 years of data by the FDA indicating that objects less than 7 mm in size pose little risk unless a food is intended for a special risk group such as infants or the elderly (20).

Hazard Evaluation

The second stage of the hazard analysis, the hazard evaluation, is conducted after the list of potential biological, chemical, or physical hazards is assembled. During the hazard evaluation, the HACCP team decides which of the potential hazards present a significant risk to consumers. Based on this evaluation, a determination is made as to which hazards must be addressed in the HACCP plan. During the evaluation of each potential hazard, the food; its methods of preparation, transportation, and storage; and the likely consumers of the product should be considered. The team must consider the influence of the likely food preparation practices and methods of storage and whether the intended consumers are particularly susceptible to a potential hazard.

According to the NACMCF (18), each potential hazard should be evaluated based on the severity of the potential hazard and its likely occurrence. Severity is the seriousness of the consequences (potential illness or injury) resulting from exposure to the hazard. Considerations of severity (e.g., the magnitude and duration of illness or injury and the potential for sequelae) can be helpful in understanding the public health impact of the hazard. Consideration of the likely occurrence of the hazard in the food as eaten is usually based upon a combination of experience, epidemiological data, and information in the technical literature. In conducting the hazard evaluation, it is helpful to consider the likelihood of occurrence and the severity of the potential consequences if the hazard is not properly controlled. Because the likelihood of occurrence and the severity of consequences of a hazard are the two main factors to be considered in a risk assessment, many have compared a hazard analysis to a qualitative assessment of risk to consumers as a result of a hazard in food. Although the process and output of a risk assessment are different from a hazard analysis, the identification of potential hazards and the hazard evaluation may be facilitated by using the same systematic approach.

One tool developed to help visualize the concept of evaluating hazards according to a qualitative estimate of risk is presented in Table 46.1. This table is an adaptation of a table developed by the NACMCF in 1997 (18) to help explain the two stages of conducting a hazard analysis. This table gives three examples using a logic sequence that parallels a qualitative risk assessment scheme for conducting a hazard analysis. The examples in this table relate to biological hazards, but chemical and physical hazards are equally important to consider.

Hazards identified as significant in one operation or facility may not be significant in another operation producing the same or a similar product. This may happen because of differences in sources of supply, product formulations, production methods, etc. For example, due to differences in equipment and/or an effective maintenance program, the probability of metal contamination may be high in one facility but remote in another. The key to this determination, however, is the proper conduct and documentation of a hazard analysis. Vast differences in HACCP plans should not be observed between two facilities that manufacture the same items in very similar manners.

When a team has determined that a potential hazard is not reasonably likely to occur, this decision should be made on the basis of all available evidence from the literature and, when available, from the plant's own historical records. Note that severe hazards may need to be considered in a HACCP plan even if they are considered not reasonably likely to occur, e.g., *C. botulinum* in a meat product. A summary of the HACCP team's deliberations and the rationale developed during the hazard analysis must be kept for future reference. This summary should include the conclusions developed during the hazard evaluation about the likelihood of occurrence and the severity of hazards. This information will be useful during validation, future reviews, and updates of the hazard analysis and the HACCP plan.

In our view, another factor may sometimes need to be considered in determining whether a hazard will be addressed within a HACCP construct. In the case where the only control measures are not particularly well defined or amenable to the establishment of CCPs with defined critical limits, hazards may be best addressed within a specific prerequisite program. This will be discussed in a later section on prerequisite programs as applied to the control of *L. monocytogenes* in certain ready-to-eat foods.

Identification of Control Measures

Upon completion of the hazard analysis, the hazards associated with each step in the production of the food should be listed along with any measure(s) that are used to control the hazard(s). The term control measure was substituted for preventive measure in the 1997 NACMCF HACCP document (18) because not all hazards can be prevented but control is usually possible. The NACMCF (18) defines a control measure as "any action or activity that can be used to prevent, eliminate or reduce a significant hazard." There may be several control measures available, but usually only one is selected as a CCP (e.g., a metal detector is often selected as a CCP, even though magnets and screens on the line may help prevent metal in the product). More than one control measure may be required to control a specific hazard; for example, control of pathogens in unpasteurized citrus juice may involve controls at the washing and brushing step and chemical or heat treatment of the unpeeled fruit. More than one hazard may be addressed by a specified control measure; a heat step controls pathogens that are present in raw materials and those that contaminate products during the processing steps.

As an example, consider the identification of an appropriate control strategy for enteric pathogens such as *Salmonella* in the production of frozen cooked beef patties, such as that described in Table 46.1. When making the selection of a control measure, the HACCP team considers the intrinsic properties of the food and its method of preparation compared with those physical parameters

Table 46.1 Examples of how the stages of hazard analysis are used to identify and evaluate hazards

Hazard analysis stage	Step	Evaluation of:		
		Frozen cooked beef patties produced in a manufacturing plant	Product containing eggs prepared for food service	Commercial frozen precooked, deboned chicken for further processing
Stage 1, hazard identification	Determine potential hazards associated with product	Enteric pathogens (e.g., *Escherichia coli* O157:H7 and *Salmonella*)	*Salmonella* in finished product	*Staphylococcus aureus* in finished product
Stage 2, hazard evaluation	Assess severity of health consequences if potential hazard is not properly controlled	Epidemiological evidence indicates that these pathogens cause severe health effects, including death, among children and the elderly. Undercooked beef patties have been linked to disease from these pathogens	*Salmonella* causes a foodborne infection that is a moderate to severe illness; salmonellosis can be caused by ingestion of only a few cells of *Salmonella*	Certain strains of *S. aureus* produce an enterotoxin, which can cause a moderate foodborne illness
	Determine likelihood of occurrence of potential hazard if not properly controlled	Likely occurrence of *E. coli* O157:H7 is low, whereas the likelihood of salmonellae is moderate in raw beef trimmings	Product is made with liquid eggs, which have been associated with past outbreaks of salmonellosis. Recent problems with *Salmonella* serotype Enteritidis in eggs cause increased concern. Probability of *Salmonella* in raw eggs cannot be ruled out. If hazard is not effectively controlled, some consumers are likely to be exposed to *Salmonella* from this food	Product may be contaminated with *S. aureus* due to human handling during deboning of cooked chicken. Enterotoxin capable of causing illness will occur only as *S. aureus* cells multiply to about 10^6/g. Operating procedures during deboning and subsequent freezing prevent growth of *S. aureus*; thus, the potential for enterotoxin formation is very low
	Using information described above, determine if this potential hazard is to be addressed in the HACCP plan	HACCP team decides that enteric pathogens are hazards for this product. Hazards must be addressed in the plan	HACCP team determines that if the potential hazard is not properly controlled, consumption of product is likely to result in an unacceptable health risk. Hazard must be addressed in the plan	HACCP team determines that the potential for enterotoxin formation is very low. However, it is still desirable to keep the initial number of *S. aureus* cells low. Employee practices that minimize contamination, rapid carbon dioxide freezing, and handling instructions have been adequate to control this potential hazard. Potential hazard does not need to be addressed in HACCP plan

that permit the growth and survival of enteric pathogens and with those conditions needed to inactivate them. Based on the HACCP team's review of control options for enteric pathogens, the team may determine that sufficient heat must be applied to kill *Salmonella*. If done correctly, the cooking step will reduce the likelihood of enteric pathogens in finished patties to an acceptable level. Hence, the control measure selected is cooking.

Influence of Prerequisite Programs on the Hazard Analysis

As noted earlier in this chapter, each operation must have a firm foundation in prerequisite programs before considering the implementation of a HACCP plan. One of the most difficult decisions that a HACCP team will face is determining whether a potential hazard can be managed under existing programs, such as GMP compliance or some other prerequisite program, or whether the hazard should be managed under the HACCP plan. Prerequisite programs typically include objectives other than food safety, and it may not be easy to associate performance of a prerequisite program element, e.g., pest control or chemical storage programs, with specific production lots or batches (22). Consequently, it is usually more effective to manage non-food safety issues and low-risk hazards within a quality system rather than include their performance and control as part of the HACCP plan. This is appropriate provided that uninterrupted adherence to the prerequisite program is not essential for food safety.

Nevertheless, prerequisite programs play an important role in controlling potential health hazards. Occasional deviation from a prerequisite program requirement would alone not be expected to create a food safety hazard or concern. For example, supplier control programs and chemical control programs can be used to minimize potential chronic health hazards such as those from mycotoxins or pesticides. Similarly, foreign material contamination in many food processes can be minimized by preventive maintenance programs and by upstream control devices such as sifters and magnets. Deviations from compliance in a prerequisite program usually do not result in action against the product (22). In contrast, potential acute health hazards such as the presence of *Salmonella* are usually controlled in a HACCP system where definitive CCPs are available to eliminate the hazard. Deviations from compliance in a HACCP system normally result in action against the product, such as evaluation of the product to determine appropriate disposition (rework, destroy, etc.). This is a key consideration that can aid in distinguishing between control points within prerequisite programs and CCPs that should be included in a HACCP plan.

The decision as to whether a hazard warrants control within a HACCP program will depend for the most part on the HACCP team's evaluation of the risk to consumers that may result from failure to control the hazard. Use of the logical evaluation of potential hazards as presented in Table 46.1, following the same systematic approach as a qualitative risk assessment, should help in this task.

Other considerations in deciding on prerequisite programs versus HACCP involve those hazards that are controlled, in part, by systemwide adherence to GMPs or sanitation standard operating procedures. Recontamination of cooked products by *L. monocytogenes* is an example of this type of situation. Control measures to prevent recontamination of cooked products by *L. monocytogenes* may be so diverse that management within a HACCP program will be extremely difficult. In addition, if the plant chooses to address the possibility of postprocessing contamination by *L. monocytogenes* within a HACCP plan, they may find it difficult to identify discrete CCPs and critical limits that are meaningful. (An exception would be the delivery of a lethal process such as postpackaging pasteurization with heat or high pressure to control *L. monocytogenes* recontamination.) More importantly, failure to conduct a single environmental control activity at the prescribed time is not likely to result in contamination by *L. monocytogenes*. This factor would call into question whether this is really a CCP. Additionally, many in the scientific community remain skeptical over the potential for small numbers of *L. monocytogenes* cells to present a health hazard even to immunocompromised individuals (16). In this example, as in all hazard analysis exercises, it is important that the HACCP team develop a scientific rationale for its decisions.

The use of prerequisite programs in addressing certain potential hazards has been recognized by regulatory agencies. Both the USDA FSIS and the FDA consider sanitation standard operating procedures to be prerequisite programs for products produced under the HACCP regulations. In 2003, the FSIS issued a directive on the presence of foreign material in meat or poultry products that notes that an establishment can determine that foreign material is a food safety hazard that is not reasonably likely to occur as the result of prerequisite programs (14). The FSIS also recognized that certain aspects of *L. monocytogenes* control may be addressed in a prerequisite program in its interim final rule on the control of *L. monocytogenes* in ready-to-eat meat and poultry products (13).

When a hazard is considered to be not reasonably likely to occur due to the operation of a prerequisite program, it is particularly important to document the HACCP team deliberations and the rationale developed during the hazard analysis. It is important to monitor the prerequisite program to ensure that it is carried out properly, as lack of compliance with the program can bring into question the validity of the hazard analysis.

PRINCIPLE 2: DETERMINE THE CRITICAL CONTROL POINTS

Once the significant hazards and control measures are identified, CCPs within the process scheme are identified. A control point is any point, step, or procedure at which biological, physical, or chemical factors can be controlled. A CCP is defined by the NACMCF (18) as "a step at which control can be applied and [where control] is essential to prevent or eliminate a food safety hazard or reduce it to an acceptable level." Each significant hazard identified during the hazard analysis should be addressed by at least one CCP. Examples of CCPs include cooking, chilling, product formulation control, application of a bactericidal rinse, a decontamination step, etc.

Although measures needed for adequate control of a hazard should be reflected in the HACCP plan, in order to keep HACCP programs plant friendly and sustainable, CCPs should not be redundant. Redundant CCPs typically add little to the margin of safety but will add to the record-keeping and administrative burden of a firm's management structure. They will also add significant cost without concomitant benefit. Experience has shown that HACCP plans that are unnecessarily cumbersome will not be supported over extended periods.

As an example of this concept, a firm may have multiple magnets and metal detectors in a line to protect production equipment from damage and consumers from harm. If the product is passed through a metal detection and reject device after it is in its final package, the last detector will typically be regarded as the CCP for this hazard, whereas upline metal detectors or magnets will be considered control points.

For the cooked beef patties referred to above, it was determined that cooking would be used to control the enteric pathogen hazard and that this would be the CCP, because it is the only step at which the hazard could be eliminated.

Considerations for Raw Products

The objective of a HACCP plan will be different for raw products than for products that are processed by cooking or applying some other bactericidal treatment. For products that receive a defined bactericidal treatment, the hazard from microbial pathogens can be reduced, often to negligible levels, through proper design and application of the treatment. For raw products, however, the risk associated with the product can, at best, only be minimized.

Decision Trees and Their Use

CCP decision trees have been developed to assist firms in identifying CCPs. Both the NACMCF and Codex documents include various versions of this tool. The HACCP team may utilize a decision tree to evaluate each of the points where food safety hazards can be prevented, eliminated, or reduced to acceptable levels. Each point should then be categorized as either a control point or a CCP. The results of this evaluation should be summarized and added to the material developed under principle 1.

The most common problem with using a decision tree is trying to apply the questions in a tree prior to the completion of the hazard analysis. By applying the questions to potential hazards that are not reasonably likely to cause illness or injury, the HACCP team will unintentionally identify CCPs that are not related to controlling product safety. Another common problem is that no decision tree will give correct answers for all applications. According to the NACMCF (18), "Although application of the CCP decision tree can be useful in determining if a particular step is a CCP for a previously identified hazard, it is merely a tool and not a mandatory element of HACCP. A CCP decision tree is not a substitute for expert knowledge." In our view, it is best to learn the logic flow exemplified in the decision tree(s) and apply this along with common sense rather than relying exclusively on a decision tree for guidance.

PRINCIPLE 3: ESTABLISH CRITICAL LIMITS

A critical limit is defined as "a maximum and/or minimum value to which a biological, chemical or physical parameter must be controlled at a CCP to prevent, eliminate or reduce to an acceptable level the occurrence of a significant food safety hazard" (18). Each CCP will have one or more control measures that must be properly applied to ensure prevention, elimination, or reduction of significant hazards to acceptable levels. Each control measure will have an associated critical limit(s) that serves as a safety boundary for the control measure(s) applied at each CCP and should be viewed as the dividing line between a product that is unacceptable (at least until evaluated further) and one that is acceptable. Critical limits may be set for such factors as temperature, time, physical dimension, humidity, moisture level, water activity, pH, titratable acidity, salt concentration, available chlorine, viscosity, and the presence or concentration of preservatives. As HACCP has evolved, it has been determined that attributes may also be established as critical limits, e.g., the critical limit for a metal detector may be specified as "on and functioning." The critical limit for delivery of the time component of a thermal process in a continuous flow system may be a specific flow rate of a liquid as defined by a pump setting and the correct hold tube length and diameter.

Setting Critical Limits

Each firm must ensure that the critical limits identified in its HACCP plan are adequate for the intended purpose. Hence, the firm or its process authority should base critical limits on a benchmark or performance standard which the treatment should achieve. There are many possible sources for these benchmarks. They may be derived from sources such as regulatory standards and guidelines (when they relate directly to a health hazard), literature surveys, experimental studies, and experts such as process authorities. As noted previously, for juice products U.S. regulatory authorities require a treatment targeted to achieve a 5-$\log_{10}$ reduction of the pathogen of concern. Hence, a heat treatment could be judged adequate if it achieves this performance standard. The firm should have adequate scientific studies to validate the critical limits it has selected and the proper delivery of the control measure to each unit of the product. The normal variation in process delivery should also be considered in establishing critical limits, and a firm should consider setting target or operational values such that critical limits are not routinely violated. Validation, which includes assurance that the critical limits are well founded as well as effectively and consistently delivered, will be discussed in more detail in "Principle 6: Establish Verification Procedures" below. For the beef patty example, the USDA FSIS is in the process of establishing a performance standard but currently specifies in regulations the time and temperature combinations that must be met to ensure a 5-$\log_{10}$ reduction of *Salmonella* (12). In this instance the firm would not need to establish the scientific basis for the critical limits but would select a process from the regulations that can be achieved with the equipment in the facility.

Microbiological Criteria as Critical Limits

Just as microbiological testing is not a good tool for monitoring, microbiological criteria are typically not useful as critical limits in a HACCP program. Because HACCP targets process control, factors that lend themselves to real-time monitoring and quick feedback should be identified as control measures. Thus, the critical limits used in a HACCP program do not typically relate to a criterion that is directly associated with microbiological testing of products or ingredients. However, there will be instances where the only option open to a processor is to hold an ingredient lot and perform microbiological testing before release. This situation may arise when a particular ingredient can be obtained only from sources where little control is exercised over factors that affect the contaminants or pathogens associated with the ingredient. In this case, a CCP may be located at receiving, where the incoming ingredient would be sampled for analysis and placed in controlled storage until the results of the analysis are available. In this instance, microbiological testing is the monitoring activity and a microbiological criterion would be the critical limit. When this approach is used, no products should be released until the results of all testing are finalized.

PRINCIPLE 4: ESTABLISH MONITORING PROCEDURES

Monitoring is a planned sequence of observations or measurements to assess whether a CCP is under control and to produce an accurate record for future use in verification. What is monitored will be a measurement or observation of the physical factor identified as the control (preventive) measure for a significant hazard. Examples of monitoring activities include the following:

- measuring temperature
- tracking elapsed time, e.g., time at a specific temperature
- sampling the product and determining pH
- determining moisture level or water activity

Monitoring serves three main purposes. First, monitoring tracks the system's operation in a manner essential to food safety management. If monitoring leads to the detection of a trend toward loss of control, that is, approaching a target level, then action can be taken to bring the process back into control before a deviation occurs. Second, monitoring is used to determine when a deviation occurs at a CCP, i.e., violation of the critical limit, which will initiate an appropriate corrective action. Third, it provides the basis for written documentation of process control for use in verifying that the HACCP plan has been followed.

When establishing monitoring activities, the firm should specify the following:

- what control measure(s) is being monitored
- how often the monitoring needs to be conducted
- what procedures will be followed to collect data and what methods and equipment will be used
- who will be responsible for performing each monitoring activity

The frequency at which activities are monitored must be consistent with the needs of the operation in relation to the variation inherent in the control step; it must also be adequate to ensure proper control. The frequency of monitoring should be integrated with a product-coding system which is designed to prevent excessive amounts of a product from being involved in a corrective action if problems arise and a critical limit is violated. For example, if a product must reach a certain temperature during a cooking step and that temperature is checked only once per hour, then if the temperature drops below the critical limit, all the product produced after the last check must be reworked or destroyed. Less of the product would be involved if checks were more frequent or if a method of continuous monitoring with automatic line shutdown or diversion was used.

In most instances, control measures that can be monitored on a continuous basis are preferred over those that do not provide this level of security. Those control measures that can be monitored continuously are typically amenable to continual recording and electronic monitoring with automated trend analysis. Such monitoring can provide advance warning to allow for correction of a problem before violation of a critical limit occurs. In the example above, constant monitoring of temperature with an alarm to warn of a trend toward violation of a critical limit may avoid a deviation or at least would minimize the amount of the product subject to corrective action.

Certain characteristics, however, cannot be measured on a continuous basis, or they may not need continuous monitoring. Products that are mixed in a batch, such as some acidified products, do not need continuous monitoring of pH. A control measure like maintaining a certain water activity in a food cannot easily be monitored continually. Such items will be monitored through a planned sequence of sampling and testing of appropriate frequency to document control.

One of the underlying concepts of HACCP is to promote awareness of food safety and individual responsibility to the line level in an operation. One way to promote this awareness is to involve line operators in monitoring activities. In many operations, monitoring activities are assigned to quality control personnel when they could be accomplished just as effectively by line operators, with the extra benefit of line worker involvement. The role of quality control personnel may be more appropriate in verification, or "checking the checker."

Those responsible for monitoring a control measure at a CCP must

- be trained in the appropriate monitoring techniques
- understand the importance of monitoring
- accurately report and record results of the monitoring activity
- immediately report deviations from critical limits so that corrective actions can be taken

Such personnel may often be given authority to shut down the line when a deviation occurs.

Returning again to our example of cooked beef patties, it was determined that the control measure for enteric pathogens such as *Salmonella* would be heat applied during cooking and that the CCP would be a cooking step. The parameters being monitored by the firm are temperature of the oven and the duration of the cooking cycle (as controlled by conveyor belt speed). Critical limits for these physical factors must be set such that a time and temperature combination specified in the regulations is met. If this firm chooses to use equipment that provides continuous monitoring (recording) of time and temperature, then the operator's duty is to ensure that the monitoring equipment is working and to obtain a manual reading periodically as a verification activity. If there is a deviation in either the time or the temperature component of the cooking step, the operator should initiate the proper corrective action.

PRINCIPLE 5: ESTABLISH CORRECTIVE ACTIONS

The HACCP system is designed to identify situations where health hazards are reasonably likely to occur and to establish strategies to prevent or minimize their occurrence. However, ideal circumstances do not always prevail. When there is a deviation from established critical limits, corrective actions must be taken. Corrective actions, in the HACCP context, are procedures to be followed when a deviation or failure to meet a critical limit occurs. Corrective actions should be developed and adopted which address how the firm will

- fix or correct the cause of noncompliance to ensure that the process is brought under control and the critical limits are being met
- hold the product produced when the deviation occurred and determine the disposition of noncompliant products, ensuring that no products that may be harmful are released
- take steps to prevent reoccurrence
- determine whether adjustments in the HACCP plan are needed

Because of the variations in CCPs for different foods and the diversity of possible deviations, corrective action plans must be developed that are specific to the particular operation. A corrective action plan should be developed for deviations that may occur at each CCP. Individual positions responsible for thoroughly understanding the process, product, and HACCP plan are to be assigned responsibility for ensuring that the appropriate corrective actions have been implemented. These corrective action plans are often general in nature but must take into account the points listed above. Records of the corrective actions that have been taken must be maintained. Long-term solutions should be sought when a particular critical limit appears to be violated routinely. Such repeated events should trigger a review of the plan, the CCP, the identified critical limit, and the method of process control to determine what improvements are needed to reduce the deviations.

Corrective actions for inadequately cooking the hamburger patties considered earlier may include correcting the equipment problem that caused the temperature or time error. In addition, any products involved in the deviation must not be released until they are evaluated to determine whether they are safe for consumption. In this case, microbiological testing is seldom appropriate; the number of samples that would need analysis to detect a low-level contaminant would be excessive. Instead, the preferred method would be to obtain an expert evaluation of the time and temperature actually observed during the deviation period. If these are determined to be adequate to inactivate the pathogens of concern, the product may be safely consumed. Alternatively, the product may be reprocessed, diverted for another use in which it will be rendered safe, or destroyed.

PRINCIPLE 6: ESTABLISH VERIFICATION PROCEDURES

There are several HACCP-related activities that will be conducted under the label of verification. Verification is the use of methods, procedures, or tests, in addition to those used in monitoring, to determine if the HACCP plan is being followed and whether the records of monitoring activities are accurate. Verification is a second level of review, beyond the primary review of line personnel who are actually conducting the monitoring. This secondary level of review is typically the responsibility of quality control personnel and will be aimed at verifying that records are being kept accurately and that monitoring activities are being properly conducted. This review may include observing monitoring activities or conducting an independent test. Verification sampling and testing are conducted where appropriate. In fact, most product testing for microorganisms is conducted as a verification activity. Verification often involves a third level of review where outside audit teams are used to review all aspects of the HACCP plan, monitoring procedures, record-keeping practices, etc.

In addition, the concept of validation has evolved as a separate and distinct activity to be conducted as a subset of verification. Validation is described as that element of verification focused on collecting and evaluating scientific and technical information to determine

if the HACCP plan, when properly implemented, will effectively control the significant hazards (18). Validation usually focuses on the process of ensuring that the plan reflects identification of all significant hazards, that the control measures identified are appropriate, that the designated critical limits are adequate to control the hazard, and that the rest of the features of the plan are satisfactory (21). Most now feel that validation should be viewed as a distinct function but one that is carried out in parallel with other activities. Plans need to be revalidated whenever significant changes occur in product formulations, equipment, or processing procedures. In addition, it is appropriate to reassess a HACCP plan periodically. U.S. regulations for seafood, juice, and meat and poultry require this reassessment to be done at least annually.

Validation of the control procedures for inactivation of a pathogen, e.g., *Salmonella*, in a product would include a determination of whether the parameters of the process are adequate to inactivate an appropriate number of *Salmonella* cells. In addition, validation information should include the rationale for the microbiological reduction that the cooking process is supposed to achieve. The determination of the appropriate time and temperature to achieve the appropriate level of kill would involve reviewing appropriate literature, consulting with experts in the field to determine the approximate thermal inactivation kinetics of the pathogen in the product, or conducting challenge studies (21). In many instances, appropriate times and temperatures will have been established and may be published in guidance documents that can serve as "safe harbors." For a juice product, validation of the control procedures for *Salmonella* could include a determination of whether the time and temperature selected for the pasteurization treatment would achieve the 5-$\log_{10}$ reduction of *Salmonella*; the rationale for the level of reduction is specified in the regulation. The FDA has published a safe harbor heat treatment—160°F for 3 s (6 s for apple juice)—in the juice hazards and controls guidance (10). Validation of the control procedures for *Salmonella* identified in the earlier example for cooked beef patties would require determining the heating rate of the product in the oven to be used for cooking so that the cooking time and temperature needed to achieve an adequate level of kill can be delivered. Uniform delivery of cooking in the oven must also be validated so that each unit of the product passing through the oven is assured of receiving at least the minimum treatment. Alternatively, cooking steps can be validated microbiologically by using "inoculated packs"—products inoculated with the pathogen of concern or a nonpathogenic simulator microorganism—if these are statistically designed to ensure that "worst-case" conditions are encountered and enough sample units are run.

Verification activities for thermal processes would include calibrations necessary to ensure that the time and temperature measurements and other data produced during monitoring are accurate. The firm should also review all HACCP records, especially the records for any lethal step, such as a heat treatment, before the final release of any lot of the product. Occasional microbiological testing of the finished product may be included as an additional verification step. Verification activities should take place on a regular and predetermined schedule. Some will be conducted daily; others will be weekly, monthly, or even less frequently. The need for accuracy and precision, as well as equipment variation, should be considered in establishing verification schedules.

PRINCIPLE 7: ESTABLISH RECORD-KEEPING AND DOCUMENTATION PROCEDURES

Record keeping is integral to maintaining a HACCP system. Without effective record keeping and review, the HACCP system will not be maintained as an ongoing practice. Records provide the basis for management assessment of the safety of products produced and for documentation of the safety of products for customers. Records provide a means to trace the production history of foods produced, to document that critical limits were met, and to prove that any needed corrective actions were carried out.

The HACCP plan and associated support documents should be on file at the food establishment. Generally, the records utilized in the HACCP system will include the following:

1. The HACCP plan

   - Listing of the HACCP team and assigned responsibilities
   - Description of the product and its intended use
   - Block flow diagram for the entire manufacturing process indicating CCPs
   - Hazards associated with each CCP and control measures
   - Critical limits for each CCP
   - Monitoring procedures (who, what, when, where, how)
   - Corrective action plans for deviations from critical limits
   - Record-keeping procedures
   - Procedures for verification of the HACCP system
   - Records of validation studies

2. Records obtained during the operation of the plan

   - Records of data collected by monitoring of control measures at CCPs

   - Corrective action records
   - Records of certain verification activities, e.g., observations of monitoring activities and calibration of equipment

Examples of operational records for a cooked product may include a computerized log of the operational details of the oven used for cooking. Such records would document the time the unit was turned on, when it reached the target temperature, when the product began to move through, the temperature and time during the production run, etc. A record of operator checks of a separate temperature-monitoring device, as well as confirmation of throughput or residence time in the oven, would also be expected. If any deviations occurred, a record of the time and extent of the deviation should be available, along with a description of actions needed to bring the cooker back into control and the disposition of any of the product produced while the deviation occurred. In this case, the processor may choose to automatically stop the line if a time or temperature parameter is violated. This eliminates the need to determine product disposition. Other operational records would include the calibration records for the temperature-monitoring equipment, time measurements, and documentation of any other practices needed to ensure the accuracy of the primary monitoring activities.

In addition to the above, there should be a HACCP master file which will contain a record of the deliberations of the HACCP team and supporting documentation for various aspects of the plan. This file should include justification for the critical limits identified, details on sampling plans and methods of analysis, standard operating procedures for monitoring activities and corrective actions, and other details that may need further review and modification to support the plan. The forms used, format, practices to be followed, and review procedures should be explained in the HACCP master file.

CONCLUSIONS

The HACCP concept provides a systematic, structured approach to ensuring the safety of food products. However, there is no blueprint or universal formula for putting together the specific details of a HACCP plan. The plan must be dynamic. It must allow for modifications to production, substitution of new materials and ingredients, and development of new products. In order for HACCP to work well, it should be a participatory endeavor at all levels of an organization, both in formulating and in managing the plan.

References

1. **Anonymous.** 23 March 1999, posting date. *FDA/ORA Compliance Policy Guides Manual 2006*, ch. 5, section 555.425. [Online.] Food and Drug Administration, Washington, D.C. http://www.fda.gov/ora/compliance_ref/cpg/cpgfod/cpg555-425.htm.

2. **Anonymous.** 1995. *Bone Particles and Foreign Material in Meat and Poultry Products: a Report to the Food Safety and Inspection Service.* Public Health Hazard Analysis Board, U.S. Department of Agriculture-Food Safety and Inspection Service, Washington, D.C.

3. **Anonymous.** *GATT 1994 World Trade Organization Agreement on Sanitary and Phytosanitary Measures.* [Online.] World Trade Organization, Geneva, Switzerland. http://www.wto.org/english/docs_e/legal_e/15sps_01_e.htm.

4. **Codex Alimentarius Commission.** 2003. *Recommended International Code of Practice: General Principles of Food Hygiene.* CAC/RCP 1-1969, revision 4 (2003). World Health Organization-Food and Agriculture Organization of the United Nations, Rome, Italy.

5. **Codex Alimentarius Commission.** 2004. *Procedural Manual*, 14th ed. World Health Organization-Food and Agriculture Organization of the United Nations, Rome, Italy.

6. **Codex Alimentarius Commission.** 2005. *Report of the Thirty-Seventh Session of the Codex Committee on Food Hygiene.* ALINORM 05/28/13 (appendix III). World Health Organization-Food and Agriculture Organization of the United Nations, Rome, Italy.

7. **Food and Agriculture Organization of the United Nations.** 1995. *Food and Nutrition Paper 58. The Use of Hazard Analysis and Critical Control Point (HACCP) Principles in Food Control: Report of an FAO Expert Technical Meeting, Vancouver, Canada.* Food and Agriculture Organization of the United Nations, Rome, Italy.

8. **Food and Drug Administration.** 1995. Procedures for the safe and sanitary processing and importing of fish and fishery products, final rule. *Fed. Regist.* **60:**65096–65202.

9. **Food and Drug Administration.** 2001. Hazard analysis and critical control point (HAACP) [*sic*]: procedures for the safe and sanitary processing and importing of juice, final rule. *Fed. Regist.* **66:**6138–6202.

10. **Food and Drug Administration.** 3 March 2004, posting date. *Juice HACCP Hazards and Controls Guidance*, 1st ed. [Online.] Food and Drug Administration, Washington, D.C. http://www.cfsan.fda.gov/~dms/juicgu10.html.

11. **Food Safety and Inspection Service, U.S. Department of Agriculture.** 1996. Pathogen reduction: hazard analysis and critical control point (HACCP) systems, final rule. *Fed. Regist.* **61:**38806–38989.

12. **Food Safety and Inspection Service, U.S. Department of Agriculture.** 1999. Performance standards for the production of certain meat and poultry products. *Fed. Regist.* **64:**732–749.

13. **Food Safety and Inspection Service, U.S. Department of Agriculture.** 2003. Control of *Listeria monocytogenes* in ready-to-eat meat and poultry products, final rule. *Fed. Regist.* **68:**34208–34254.

14. **Food Safety and Inspection Service, U.S. Department of Agriculture.** 30 May 2003, posting date. *FSIS Directive 7310.5. Presence of Foreign Material in Meat and Poultry Products.* [Online.] Food Safety and Inspection Service, U.S. Department of Agriculture, Washington, D.C. http://www.fsis.usda.gov/OPPDE/rdad/FSISDirectives/7310.5.pdf.

15. **International Commission on Microbiological Specifications for Foods.** 2002. *Microorganisms in Foods*, vol. 7. *Microbiological Testing in Food Safety Management.* Kluwer Academic/Plenum Publishers, New York, N.Y.

16. **International Life Sciences Institute Research Foundation-Risk Science Institute Expert Panel on *Listeria monocytogenes* in Foods.** 2005. Achieving continuous improvement in reductions in foodborne listeriosis—a risk-based approach. *J. Food Prot.* **68:**1932–1994.

17. **National Academy of Sciences-National Research Council.** 1985. *An Evaluation of the Role of Microbiological Criteria for Foods and Food Ingredients.* National Academy Press, Washington, D.C.

18. **National Advisory Committee on Microbiological Criteria for Foods.** 1998. Hazard analysis and critical control point principles and application guidelines. *J. Food Prot.* **61:**762–775.

19. **National Conference on Food Protection.** 1971. *Proceedings of the National Conference on Food Protection.* Department of Health Education and Welfare, Public Health Service, Washington, D.C.

20. **Olsen, Alan R.** 1998. Regulatory action criteria for filth and other extraneous materials. I. Review of hard or sharp foreign objects as physical hazards in food. *Regul. Toxicol. Pharmacol.* **28:**181–189.

21. **Scott, V. N.** 2005. How does industry validate elements of HACCP plans? *Food Control* **16:**497–503.

22. **Sperber, W. H., K. E. Stevenson, D. T. Bernard, K. E. Deibel, L. J. Moberg, L. R. Hontz, and V. N. Scott.** 1998. The role of prerequisite programs in managing a HACCP system. *Dairy Food Environ. Sanit.* **18:**418–423.

23. **U.S. Department of Health and Human Services.** 3 November 2000, posting date. *Healthy People 2010*, vol. 1, ch. 10. [Online.] U.S. Department of Health and Human Services, Washington, D.C. http://www.healthypeople.gov/Document/tableofcontents.htm#volume1.

Food Microbiology: Fundamentals and Frontiers, 3rd Ed.
Edited by M. P. Doyle and L. R. Beuchat
© 2007 ASM Press, Washington, D.C.

Timothy J. Barrett
Peter Gerner-Smidt

47

Molecular Source Tracking and Molecular Subtyping

Bacterial subtyping may be defined as the characterization of bacteria below the species (or subspecies) level. Such characterization will allow bacterial isolates to be placed into groups in which members are more or less similar to one another based on one or more (usually many) characteristics and may also suggest how such groups relate to one another. Subtyping can thus be used to study the population structure of a particular bacterial species, to determine the possible evolution of the subject organism, or to study the molecular epidemiology of an organism. The types of methods used for subtyping and the approaches to data analysis and interpretation may vary greatly with the reason for performing the subtyping. This chapter will focus almost entirely on subtyping for molecular epidemiology.

The term molecular epidemiology, in the context of foodborne bacteria, is usually applied to the subtyping of bacteria that cause foodborne disease and the ways in which such subtyping data contribute to understanding the transmission of those bacteria to humans. Molecular epidemiology can be applied to identifying the source of a particular outbreak or to a broader understanding of the role of certain foods or processes in outbreak-related or sporadic infections.

REASONS FOR PERFORMING MOLECULAR SUBTYPING

Perhaps the most easily appreciated reason for molecular subtyping is to facilitate the identification and investigation of foodborne disease outbreaks. By identifying isolates with the same molecular subtype, one can determine which isolates are most likely to have a common source. When isolates of foodborne pathogens with the same subtype are found to cluster temporally, especially if they also cluster geographically, it may indicate an outbreak of foodborne disease. Such cluster detection is the principal focus of the PulseNet system (104). PulseNet is described in more detail in a later section of this chapter. Once clusters are recognized, epidemiologic investigations can determine if a cluster actually represents an outbreak. Molecular subtyping can further facilitate epidemiologic investigations by determining which infections, of all those occurring at the same time as the outbreak, are most likely to be outbreak related (those involving isolates sharing the same or very similar subtypes). Subtyping contributes to tracking of the source of an outbreak in two ways: by facilitating the epidemiologic investigation and by matching the subtype of isolates from food products with that of isolates from patients.

Timothy J. Barrett and Peter Gerner-Smidt, Foodborne and Diarrheal Diseases Branch, Centers for Disease Control and Prevention, 1600 Clifton Rd., Atlanta, GA 30333.

Table 47.1 Properties of methods commonly used for molecular subtyping of foodborne pathogens

Method	Strain discrimination	Interlaboratory reproducibility	Supply costs	Specialized equipment	Automatable	Unambiguous output
Plasmid profile analysis	Moderate	Moderate	Low	No	No	No
Ribotyping	Moderate	Moderate	Low	No	No	No
Automated ribotyping (RiboPrinter)	Moderate	High	High	Yes	Yes	Yes
PFGE	High	Moderate	Moderate	Yes	No	No
RAPD analysis	High	Low	Low	No	No/Yes[a]	No/Yes[a]
REP or ERIC PCR	Moderate	Low	Low	No	No/Yes[a]	No/Yes[a]
PCR-RFLP analysis	Low-moderate	High	Low	No	No	No
AFLP analysis[b]	High	Moderate	High	Yes	Yes	Yes
MLST[b]	Moderate	High	High	Yes	Yes	Yes
MLVA[b]	High	High	High	Yes	Yes	Yes

[a] "No" refers to standard method; "yes" refers to Bacterial Barcodes system.
[b] Comments refer to method as performed using automated DNA sequencing equipment and software.

There are many reasons for performing molecular subtyping other than cluster detection and outbreak investigation. Public health laboratories use subtyping to identify new and emerging bacterial pathogens, such as *Salmonella enterica* serotype Typhimurium DT104 in the 1990s. Food-processing plants may use subtyping to track microorganisms through the plant to determine where they enter into or reside within the plant and to monitor efforts to eliminate them. The types of subtyping methods used depend on the microorganism under study and the specific application (Table 47.1). Subtyping methods will be described in detail in the next section, and specific applications will be described in following sections.

Regardless of the subtyping method being used, the importance of considering subtyping data in the context of other available data cannot be overemphasized. Factors such as the discriminative power of the subtyping method, the genetic diversity of the organism being analyzed, and the occurrence and distribution of common molecular subtypes must all be considered in the interpretation of subtyping results. Whenever possible, interpretation should be made in the context of epidemiologic and environmental investigations. Subtyping can only indicate which isolates are more likely or less likely to be related to an outbreak, food, or environmental source. Subtyping data alone cannot prove such a link.

SUBTYPING METHODS IN COMMON USE

Phenotypic Methods

Although the focus of this chapter is on molecular methods, it is important to consider them in the context of earlier phenotypic methods such as serotyping, phage

typing, biotyping, and antimicrobial susceptibility typing. Most of these phenotypic methods have long and successful histories of use in subtyping for the same purposes for which molecular methods are now used. In fact, most of them are still used prior to, or in conjunction with, molecular methods. Serotyping of *Salmonella* is an excellent example. Serotyping is so widely practiced that microbiologists tend to think of it more as identification than as subtyping, but serotyping is itself a powerful method of subtyping, with more than 2,500 named serotypes of *Salmonella*. With relatively rare serotypes, serotyping alone may provide a sufficient level of subtyping. For more common serotypes, phage-typing systems with good strain discrimination have been developed, but these systems are serotype specific and are available in few laboratories, so molecular methods are generally used to provide further strain discrimination.

Although molecular methods typically provide greater strain discrimination than phenotypic methods, this is not always the case, and it is only one reason why molecular methods are generally preferred. Perhaps the most important reason for using molecular methods is that they do not require specialized reagents or expertise and can thus be performed in almost any laboratory with basic molecular biology capabilities. This same ease of performance is also the most obvious drawback to molecular subtyping, as it allows every laboratory to perform and interpret subtyping according to their own criteria, making interlaboratory comparisons more difficult unless laboratories can agree to standardized protocols.

Plasmid Profile Analysis

One of the first molecular methods used for strain identification or source tracking is plasmid profile analysis. Plasmids are small pieces of extrachromosomal DNA that are typically circular and supercoiled. Plasmids range in size from a few hundred to several hundred thousand base pairs and are present in most bacterial species. Plasmid DNA is extracted by using a method that will separate plasmid DNA from chromosomal DNA, and the plasmids are then separated by agarose gel electrophoresis. How the plasmids travel in the gel depends on both size and conformation. Figure 47.1 shows a plasmid profile analysis of nine strains of ceftriaxone-resistant *Salmonella*. Sensitivity can be increased by restriction digestion of the plasmid DNA so that different plasmids of the same approximate sizes can be distinguished.

One of the first successful uses of plasmid profile analysis in epidemiologic investigations occurred in 1981, when exposure to marijuana was associated with a cluster of *Salmonella* serotype Muenchen infections (106). Early uses of plasmid profile analysis included identifying outbreak-associated strains of *Salmonella* serotype Typhimurium (40, 77) and linking patient and hamburger isolates of *Escherichia coli* O157 (88). Plasmid analysis continues to be useful today. During an outbreak of *Salmonella* serotype Typhimurium DT104 infections in England and Wales in 2000, all isolates analyzed by pulsed-field gel electrophoresis (PFGE) had the same profile but 67% of the isolates could be classified as outbreak associated based on plasmid analysis (42). One

major drawback of plasmid profiling is that bacteria may gain or lose plasmids during the course of an outbreak, resulting in different plasmid profiles and confounding interpretation.

RFLP Methods

DNA may be digested by restriction endonucleases, which are enzymes that cut the DNA at specific nucleotide sequences (the recognition sequence). The number and sizes of the resulting DNA fragments (restriction fragments) are unique to a particular strain. Other strains will yield different restriction fragment patterns. This is the principle behind all restriction fragment length polymorphism (RFLP)-based methods. In the initial methods, the DNA was digested with high-frequency-cutting enzymes like EcoRI and HinDIII and the resulting fragments were separated by standard agarose gel electrophoresis (100). The resulting restriction patterns were very complex due to the huge number of fragments obtained, resulting in poor resolution of individual fragments and difficulty in interpreting results. To overcome this problem, variations on the basic method were developed.

Ribotyping

Ribotyping is one of the first subtyping methods described that utilizes the transfer of electrophoretically separated genomic restriction fragments onto nitrocellulose or nylon membranes followed by hybridization with specific DNA or RNA probes. Genomic digests with frequently cutting enzymes yield extremely complex patterns that are difficult to interpret. The development of hybridization methods allowed the patterns to be made less difficult to interpret while maintaining sufficient complexity to provide useful discrimination between strains. The method was first described by Grimont and Grimont in 1986 (33) as a tool for taxonomic classification, but it was soon extended to molecular epidemiology of bacterial pathogens (101).

The original probe used for ribotyping was *E. coli* 16S and 23S rRNA sequences (33), but others have used rRNA itself (101) or cDNA obtained by reverse transcription of *E. coli* 16S and 23S rRNA (113). A variety of methods for labeling the probe have also been used. Regardless of the probe and label, all variations of ribotyping depend on the nucleotide sequence variability of the regions surrounding the highly conserved rRNA-encoding DNA sequences (and occasionally that within the intergenic spacer regions). The primary advantages to ribotyping include the facts that all eubacteria have rRNA-encoding DNA and the genes encoding rRNA (the *rrn* operon) are sufficiently conserved that one probe works for all bacteria. In addition, most bacteria have

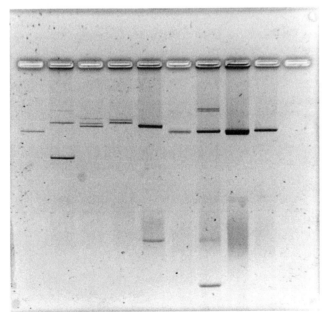

Figure 47.1 Example of plasmid profile analysis. Each lane contains plasmid DNA extracted from an isolate of ceftriaxone-resistant *Salmonella*. In this instance, plasmid DNA was separated by PFGE rather than standard gel electrophoresis for better separation of large plasmids.

several *rrn* operons, so reasonably complex patterns are obtained. However, different restriction enzymes will not necessarily yield comparable results, so optimization is still required for each specific application. Ribotyping has been successfully applied to the molecular subtyping of numerous foodborne bacteria including *Yersinia enterocolitica* (9), *Campylobacter jejuni* (19), *Salmonella* serotype Typhi (1), nontyphoidal *Salmonella* (39), and *Listeria monocytogenes* (20). The last reference is of particular note, because the authors made use of a commercially available automated method for performing and interpreting ribotyping.

The RiboPrinter (Qualicon, Wilmington, Del.) is an automated system that performs ribotyping and analysis with sample preparation and loading being the only steps that require operator involvement (12). DNA is extracted and restriction digested; then the fragments are separated by electrophoresis, transferred onto a membrane, and hybridized with a probe; and finally the resulting blot is analyzed automatically by the RiboPrinter software. Figure 47.2 shows an example of the RiboPrinter output, including an image of the actual hybridized membrane and the automatically generated identification and subtype output. This high level of automation allows technicians

with minimal training to perform ribotyping and removes many of the problems of intra- or interlaboratory variability. The biggest drawbacks to the RiboPrinter are the initial and maintenance costs, the cost of reagents, and the lack of substantial databases for enzymes other than EcoRI (80). The equipment and supply cost issues may be ameliorated by the reduced labor costs. Specific applications of ribotyping in general and automated ribotyping using the RiboPrinter are discussed below.

PFGE

The problem of excessively complex RFLP patterns was solved in a different way by the development of PFGE. When infrequently cutting enzymes are used, the actual number of fragments obtained can be reduced to a manageable number. Although this strategy is obvious, the sizes of the resulting fragments prevent them from being separated by standard gel electrophoresis. The development of PFGE in 1983 (97) eliminated this drawback. During the following few years, several different types of PFGE were devised, including transverse alternating field electrophoresis (31), field inversion gel electrophoresis (13), orthogonal field alternation gel electrophoresis (13), and contour-clamped homogeneous electric field

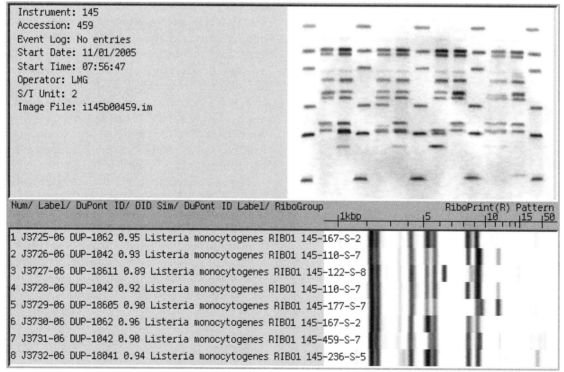

Figure 47.2 Automated ribotyping using DuPont RiboPrinter. Output shows appearance of blotted and hybridized gel (above) and software-normalized image (below). Identification and subtype (RiboGroup) are automatically generated.

electrophoresis (16). Although these methods differ considerably in the physical ways in which fragments are separated, they all depend on changing the direction of current flow (pulsing) over a gradient of time intervals between changes. The contour-clamped homogeneous electric field technology has become so widely used that many laboratories performing PFGE may not even be aware that other technologies have been described.

In recent years, PFGE has become the "gold standard" for molecular subtyping and source tracking for most foodborne bacteria, as well as many other bacterial agents. PFGE is so widely known and accepted that a lengthy list of applications would be superfluous. The application of PFGE to specific foodborne pathogens is described below. As with any subtyping method, PFGE can present problems in interlaboratory reproducibility and interpretation of results. Such problems can never be totally eliminated, but they can be minimized by using carefully standardized protocols. Potential problems and suggested approaches to solving them were recently described (5). Figure 47.3a provides an example of a PFGE gel, and Figure 47.3b provides a dendrogram of the same gel data created by using BioNumerics software.

(a)

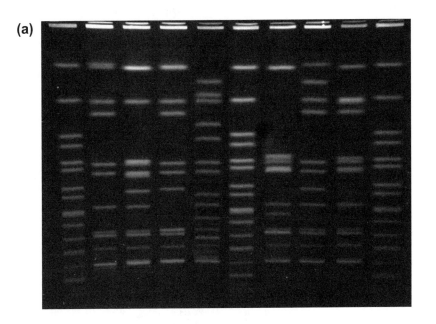

(b) Dice (Opt:1.50%) (Tol 1.5%-1.5%) (H>0.0% S>0.0%) [0.0%-100.0%]
PFGE-BlnI   **PFGE-BlnI**

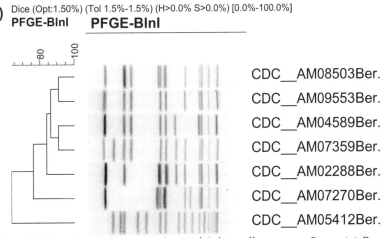

CDC__AM08503Ber.

CDC__AM09553Ber.

CDC__AM04589Ber.

CDC__AM07359Ber.

CDC__AM02288Ber.

CDC__AM07270Ber.

CDC__AM05412Ber.

Figure 47.3 PFGE analysis of seven isolates of *Salmonella* serotype Berta. (a) Raw data: lanes 2 through 4 and 6 through 9 contain BlnI digests of *Salmonella* serotype Berta genomic DNA; lanes 1, 5, and 10 contain XbaI digests of a molecular-weight-standard strain. (b) Analyzed data: dendrogram showing relatedness of *Salmonella* serotype Berta isolates. The dendrogram was produced by using BioNumerics software.

PCR-Based Methods

Although many different PCR-based subtyping methods have been reported, most methods fall into one of three categories. The first type of PCR-based subtyping method involves using primers homologous to known, conserved, multicopy sequences. In this case, differences in amplicon lengths are the basis of subtype comparisons. Amplification of the intergenic 16S-23S spacer region was reported to yield sufficient polymorphism to separate strains of *Pseudomonas cepacia* (51). Two more widely used methods, based on the presence of repeated elements, were reported by Versalovic et al. (110). These authors found that amplification of repetitive extragenic palindromic (REP) elements or enterobacterial repetitive intergenic consensus (ERIC) sequences could discriminate among laboratory strains of *E. coli*, suggesting that these methods could be used for subtyping of foodborne pathogens. Although REP and ERIC typing methods have not become as widely used as other typing methods such as PFGE, they do have the advantages of being relatively inexpensive and easy to perform. They have been used to track fecal pollution by subtyping of *E. coli* isolates from diverse settings such as dairy and swine production systems (63) and Apalachicola Bay (78). As other subtyping methods, successful source tracking can be improved with the use of appropriate statistical methods (37). The biggest drawback to REP and ERIC typing may be difficulties with interlaboratory reproducibility.

Reproducibility has clearly been the biggest drawback to the second type of PCR-based subtyping method, the arbitrary primer methods. Although the terms random amplified polymorphic DNA (RAPD) analysis and arbitrarily primed PCR (AP-PCR) are often used interchangeably, there are some differences between the methods. In both methods, primers are not chosen based on homology to known target sequences. Both methods depend on low stringency to allow imperfect matches. The RAPD method, as originally described by Williams et al. (116), used 10-mer primers and 45 PCR cycles at 36°C. AP-PCR, as originally described by Welsh and McClelland (115), used larger (20- and 34-mer) primers that were designed for other purposes and were arbitrarily chosen for subtyping by PCR. Two cycles at low stringency were followed by 40 cycles at high stringency. Both of these methods allow typing to be performed by creating a complex pattern of PCR products by using primers that are not specifically designed to match target sequences (knowledge of the target sequence is not necessary). The primary advantage to the RAPD and AP-PCR approaches is that they can be used with little knowledge of the target microorganism or even without first identifying the microbe. However, different primers work better with

different microorganisms, so choosing primers based on the target microorganism is preferable. The biggest drawback to these methods, and the primary reason that they have not been more widely accepted, is concerns about reproducibility, especially among laboratories.

There have been many reports of minor differences in PCR conditions resulting in different RAPD or AP-PCR profiles. Factors reported to affect profiles include primer concentration (65, 72), polymerase concentration and source (69, 95), $MgCl_2$ concentration (23), and even the model of thermal cycler used (65, 71). Although each of these factors can be carefully standardized within a laboratory, the tremendous sensitivity of these methods to minor variations in experimental conditions has thus far prevented them from being widely used for interlaboratory comparisons. A more thorough review of PCR-based methods, including applications and problems, was recently published by Rademaker and Savelkoul (82).

The third category of PCR-based methods combines PCR with restriction digestion: either PCR amplification followed by restriction digestion of the amplicon or restriction digestion of the genome followed by selective amplification of certain restriction fragments. In the first approach, the target locus is typically 1 to 2 kilobases and digestion results in several fragments that are then separated by gel electrophoresis, forming a characteristic RFLP profile. Southern blotting or probing is not necessary to obtain a PCR-RFLP profile. A specific example of the utility of PCR-RFLP in subtyping of *Campylobacter* is discussed in detail in "*Campylobacter*" below.

The second approach is exemplified by amplified fragment length polymorphism (AFLP) analysis, described by Vos et al. in 1995 (112). The most widely used version of AFLP analysis involves digestion of genomic DNA with two restriction enzymes, one of which cuts the DNA very frequently (MseI and TaqI are often used) and the other of which cuts with only average frequency (such as EcoRI). Specific double-stranded oligonucleotide adaptors allow ligation to create new fragments while preventing religation to recreate the original restriction sites. The resulting new fragments are then amplified by PCR with adaptor-specific primers. These primers contain extensions of 1 to 3 bp at the 3′ end. Each 1-nucleotide extension will be selective for approximately one-fourth of the newly ligated fragments. The number of fragments being generated in AFLP analysis can thus be adjusted by the number of nucleotides added to the adaptor-specific primers. A target range of 40 to 200 fragments is typically sought. Separation of the resulting fragments in agarose or acrylamide gels is possible, but the resulting patterns are complex and difficult to interpret. Most AFLP analysis performed for the purpose of bacterial subtyping is done using automated DNA sequencers. Applications of AFLP to specific foodborne bacteria are discussed below. Savelkoul et al. have written a general review of AFLP as applied to bacterial, plant, and animal genetics (94).

DNA Sequence-Based Methods

MLST

Multilocus sequence typing (MLST) was one of the first subtyping or strain-tracking methods to be based on DNA sequences rather than DNA fragment sizes. All of the RFLP-based methods described above suffer, to different degrees, from the same problems. In addition to the problems in interlaboratory variability in methods and interpretation, RFLP methods have additional shortcomings. Fragments from different test strains that appear to be the same size on a gel are not necessarily the same DNA fragments. Fragments that are very close in size may not be separated during electrophoresis and thus may not be recognized. The reasons for differences in patterns are not known, and strains with fewer differences are not necessarily more closely related phylogenetically than strains with more differences. DNA sequence-based subtyping yields unambiguous data that can be readily understood by any laboratory and which can also be used to quantify differences among test strains.

MLST was derived from the earlier technology of multilocus enzyme electrophoresis. Multilocus enzyme electrophoresis is a phenotypic method that detects changes in certain housekeeping loci by charting variation in the electrophoretic mobilities of their gene products (98). Only changes that result in a different electrophoretic mobility are detected. MLST is a more sensitive method that detects all nucleotide changes by direct sequencing. MLST is typically performed by sequencing 450- to 500-bp internal fragments of seven housekeeping genes. Since these housekeeping genes are under little selective pressure, the accumulation of changes occurring by chance is rather low, making MLST an ideal tool for understanding the evolution of microorganisms or strains. However, this same lack of rapid change in the housekeeping genes makes MLST a less ideal method for investigating outbreaks or conducting traceback studies. An MLST typing scheme has been described previously for *Campylobacter* (21), and MLST has been applied to other foodborne bacteria as described below.

VNTR-Based Typing (Including Multiple-Locus VNTR Analysis [MLVA])

The recent sequencing of several complete bacterial genomes has revealed that much of the bacterial genome consists of repeated short nucleotide sequences. The repeats may vary in size, location, and complexity. Repeats can be classified according to structure, such as inverted, dyad, and direct repeats. In bacteria, the unit of repeat is typically 1 to 10 nucleotides (although repeating units of up to 100 bp are recognized). Variability occurs both in the repeating unit and in the number of repeats. These repeating arrays are commonly called variable-number tandem repeats (VNTR). VNTR are often found in regions controlling gene expression but also within open reading frames (107, 108, 118).

The sequence of the repeat unit and the number of copies repeated are characteristic of each VNTR. For highly clonal microorganisms, there may be little variability for a given VNTR, and multiple loci must be examined. This approach has been termed MLVA. In some ways, MLVA may be analogous to MLST, where multiple loci must be sequenced to provide sufficient strain discrimination. As typically performed, MLVA determines the number of repeats at a VNTR locus by the size of a PCR amplicon rather than by the actual DNA sequence (56). In Fig. 47.4, the two isolates in panels a and b differ at three of the four sites shown, as determined by the fragment sizes shown in the chromatograms. The dendrogram in Fig. 47.4c shows the relationships among eight *E. coli* O157:H7 isolates based on alleles at seven sites.

One of the first, and perhaps one of the most important, applications of MLVA was the subtyping of the highly monomorphic potential agent of bioterrorism *Bacillus anthracis* (47). It has also proven useful for the subtyping of two other potential agents of bioterrorism for which other methods had not proven adequate, *Yersinia pestis* (54, 62) and *Francisella tularensis* (26, 46). Applications of MLVA to foodborne pathogens are discussed below.

APPLICATION OF MOLECULAR SUBTYPING METHODS TO SPECIFIC FOODBORNE BACTERIA

Salmonella

Bacteria of the genus *Salmonella* are so widespread in animals, humans, and the environment that virtually every commonly used subtyping method has been applied to the molecular epidemiology of these bacteria. Although it is not generally thought of as a subtyping method, serotyping of *Salmonella* is actually a very sensitive subtyping method, with more the 2,500 recognized serotypes. With rare serotypes, simply determining the serotype may be sufficient to at least suggest a source or epidemiologic connection. For example, an outbreak of *Salmonella*

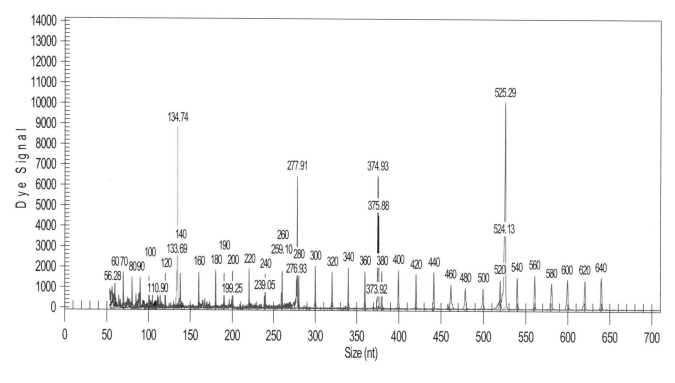

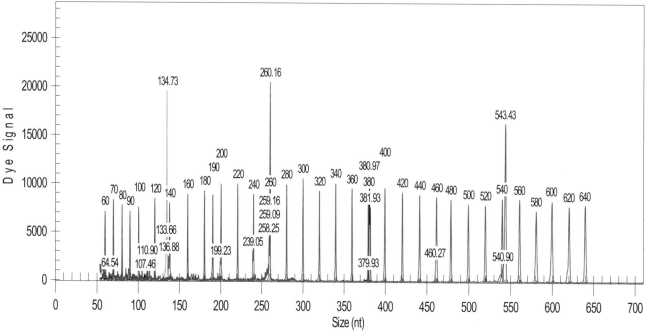

Figure 47.4 Example of MLVA of *E. coli* O157:H7 isolates. (Top and center) Chromatograms from automated DNA sequencer showing different fragment sizes at three of four sites. nt, nucleotides. (Bottom) Dendrogram showing relatedness of isolates based on MLVA data. vals, values.

VNTR_vals
MLVA_composit   **MLVA_composite**

11.0	8.0	16.0	15.0	6.0	13.0	2.0	7.0	7.0	K2127
11.0	8.0	16.0	15.0	6.0	13.0	2.0	7.0	7.0	K2128
11.0	8.0	16.0	15.0	6.0	13.0	2.0	7.0	7.0	K2130
11.0	8.0	16.0	15.0	6.0	13.0	2.0	7.0	7.0	K2132
11.0	8.0	16.0	15.0	6.0	13.0	2.0	7.0	7.0	K2164
11.0	8.0	16.0	15.0	6.0	13.0	2.0	7.0	7.0	K2366
11.0	8.0	16.0	12.0	6.0	13.0	2.0	7.0	7.0	K2129
18.0	9.0	16.0	21.0	4.0	5.0	4.0	7.0	7.0	K2131

Figure 47.4 *Continued.*

serotype Hartford infections occurred among visitors to a theme park in Orlando, Fla., in 1995 (18). Serotype Hartford is a sufficiently rare serotype that further subtyping was not needed to suggest an association with drinking orange juice at the park. The association was further strengthened by the isolation of another rare serotype, serotype Gaminara, from the implicated brand of orange juice as well as from a person with a serotype Hartford coinfection. Ironically, the PFGE subtypes of serotype Gaminara isolates from the patient and the orange juice were different, which could have led to erroneous conclusions if the epidemiologic evidence was not considered.

For the more common serotypes such as Enteritidis and Typhimurium, serotyping alone typically does not provide sufficient strain discrimination for source tracking. Some of the earliest descriptions of molecular strain typing involved the application of plasmid profile analysis to *Salmonella* serotypes Muenchen (106), Typhimurium (40), and Enteritidis (113). Though less widely used today, plasmid profile analysis of *Salmonella* isolates may still be useful (42).

The most widely used subtyping method for *Salmonella* today, and clearly the current gold standard, is PFGE. PFGE has been successfully used for identifying and investigating outbreaks of infection with serotype Typhimurium (7) and many other serotypes (Centers for Disease Control and Prevention, unpublished data). Ridley at al. found PFGE superior to plasmid analysis and ribotyping for characterization of serotype Enteritidis isolates (87). In addition to outbreak investigations, PFGE has been successfully applied to tracking of *Salmonella* in production environments. Liljebjelke et al. used PFGE to reveal that serotype Typhimurium and serotype Enteritidis isolates from processed poultry carcasses from one farm were indistinguishable from isolates obtained in the poultry house environment and at a company breeder farm, suggesting vertical as well as horizontal transmission (55).

Although PFGE has proven very useful in subtyping of *Salmonella*, other methods have been developed and evaluated. Kotetishvili et al. used MLST to subtype a collection of *Salmonella* isolates, finding a larger number of subtypes by MLST than by PFGE (53). This study appeared promising but was not designed to evaluate whether MLST could be used to distinguish between outbreak and nonoutbreak isolates of a specific serotype. Such a study was done by Fakhr et al., who found that MLST was not as discriminating as PFGE when applied only to serotype Typhimurium isolates (24). An interesting variant of MLST, based on sequences of stable temperate phages in serotype Typhimurium, was reported by Ross and Heuzenroeder (91). The initial report found this approach to provide greater strain discrimination than standard MLST. MLVA has been recently reported to be a promising method for subtyping of serotype Typhimurium (60, 83), but it has not yet been used in a sufficient number of different laboratories to determine whether it truly represents a viable alternative to PFGE.

E. coli O157:H7

One of the first applications of PFGE to foodborne disease investigations occurred during the large hamburger-associated outbreak of *E. coli* O157:H7 infections in the western United States in 1993 (4, 6). The demand for PFGE subtyping to support investigations of *E. coli* O157 infection outbreaks was the driving force behind the creation of PulseNet, the National Molecular Subtyping Network for Foodborne Disease Surveillance (104). PFGE continues to be the standard method for subtyping of *E. coli* O157 in PulseNet. The relative clonality of the bacterium limits the applicability of MLST (76), and methods such as AFLP analysis, though showing some potential utility, have received little attention (121). Foley at al. compared PFGE, MLST, and REP element PCR for typing of isolates from food, cattle, and humans and concluded that PFGE was the method of choice (29).

One new method that appears to have promise for subtyping of *E. coli* O157 is MLVA. Several authors have reported the identification of highly diverse loci (48) and the ability of MLVA to discriminate between apparently unrelated isolates with indistinguishable PFGE patterns (57, 61, 75). The future of MLVA in subtyping of *E. coli* O157 is discussed in "Future of Molecular Typing of Foodborne Pathogens" below.

L. monocytogenes

For the past several years, PFGE has also been the gold standard for molecular subtyping of *L. monocytogenes*. It is the standard method for PulseNet and has contributed to the identification of clusters of *L. monocytogenes* infections and tracking of the sources of outbreaks (32, 93). It has also been used to track *L. monocytogenes* strains in processing plants. Gudmundsdottir et al. used PFGE to identify potential sources of contamination of cold-smoked salmon in a processing plant in Iceland (34). Ribotyping, particularly automated ribotyping using the RiboPrinter, has been useful for detecting case clusters and tracking contamination, but it has not proven to be as sensitive as PFGE (30, 64, 93). At least one study found AFLP analysis to be similar to PFGE in discriminatory power (30). MLST was also reported to identify a larger number of subtypes of *L. monocytogenes* than PFGE, but this study did not determine the epidemiologic utility of the method (86).

Zhang et al. developed a novel approach to subtyping *L. monocytogenes* based on sequence differences in three virulence genes and three virulence-associated genes (119). They referred to the new method as multi-virulence-locus sequence typing (MVLST) because of its similarity to MLST. The MVLST method was able to differentiate strains that had indistinguishable PFGE patterns, but the epidemiologic significance was unclear. Zhang and Knabel later reported an abbreviated version of MVLST in which a multiplex PCR assay first identified *L. monocytogenes* serotypes 1/2a and 4b and then two PCR products were sequenced to provide interstrain discrimination (120). This method would presumably be less sensitive than the method based on sequencing of six loci, but it is faster and less expensive in both reagent cost and time. Although these and other methods may be promising, PFGE remains the method of choice.

C. jejuni

Unlike *Salmonella*, *E. coli* O157:H7, and *L. monocytogenes*, there is no consensus on the method of choice for subtyping of *C. jejuni*, and there is even some debate about whether subtyping is worth doing, at least for routine surveillance (38). PFGE was suggested as an epidemiologic tool for *C. jejuni* as long ago as 1991 (117). It was used to discriminate between outbreak and sporadic isolates of *C. jejuni* within a single serotype as early as 1994 (103) and continues to be used today (90). Despite these successes, several authors have expressed concern about the effects of genomic instability on PFGE typing. There is clear evidence that genomic rearrangement and genetic exchange between strains alter PFGE patterns in vivo in poultry (10, 114), although at least some strains appear to be genetically stable (66). This lack of stability makes it more difficult to use PFGE to trace the source of *C. jejuni* contamination or to establish links between animal sources and human infections. Unfortunately, PFGE is not the only method affected by genetic instability.

Another early molecular typing method for *C. jejuni* was based on amplification of the flagellin gene *flaA*, followed by restriction digestion of the amplicon and separation of the fragments by gel electrophoresis (73). The sensitivity and specificity of this method were improved by sequencing of the *flaA* gene, and the method was further simplified by focusing on the short variable region within the gene (67). Both of these approaches, the original method and the simplified method, are affected by recombination within the flagellin locus (36) and are generally not as sensitive as other available methods.

AFLP analysis has been a sensitive and useful method for epidemiologic investigations of *C. jejuni* and *Campylobacter coli* in several studies, typically being as sensitive as, or more sensitive than, PFGE (58, 59, 96, 99). It would be expected that AFLP analysis would also be affected by genomic instability, but to our knowledge, that has not yet been demonstrated. MLST has also been reported to be useful, especially when combined with *flaA* typing (17, 92). Other methods which may prove useful but have not been sufficiently evaluated include RFLP analysis of the *lps* gene cluster (49), sequencing of the *cmp* gene encoding the major outer membrane protein (44), and sequencing of *flaB* rather than *flaA* (68).

The plasticity of the *Campylobacter* genome and the relative lack of identifiable outbreaks have led some authors to question the utility of real-time molecular typing. Michaud et al. found that molecular typing was useful to confirm already suspected outbreaks, but no link could be found for most clusters of isolates with indistinguishable PFGE patterns (70). Despite such concerns, the continued use of PFGE and other methods shows a very clear need for molecular typing methods in understanding the epidemiology of *Campylobacter* infections and suggests that continued method development is warranted.

FUTURE OF MOLECULAR TYPING OF FOODBORNE PATHOGENS

With the increasing international trade of food and food animals, a foodborne infection in one country may have its origin in another country, even one on another continent. It is therefore crucial that molecular subtyping methods for foodborne pathogens be harmonized worldwide to facilitate the rapid comparison of strains isolated in different countries. This method harmonization and comparison are best done in the framework of surveillance networks. Such networking systems have been in place for the last decade in Europe in the Salm-Net and Enter-Net networks (27, 28) and are available in the United States in PulseNet (104) and most recently internationally in the extension of the PulseNet network (105). The World Health Organization Global Salm-Surv network also focuses on surveillance of foodborne infections internationally, building capacity through training workshops (3). The future success of these networks will depend in large measure on the development of new subtyping methods.

An ideal subtyping method should type all strains, discriminate between all isolates that are epidemiologically unrelated while identifying all related isolates, and be 100% reproducible, universally applicable, and inexpensive. The results generated should be definitive, fully portable, and easy to interpret. New methods should correlate with previously used methods so that data generated with the new methods can be related to historic data generated with the previous methods. It is evident that no currently used subtyping method fulfills all these requirements, and it is likely that none ever will. Nevertheless, new methods are constantly being developed and evaluated that bring the science closer to the ideal.

The methods used in surveillance epidemiology are often the same as those used in forensic medicine or by population geneticists and taxonomists, and most of the new methods currently under development and evaluation for surveillance purposes originated in these contexts.

In recent years, the main focus of method development has been on DNA sequencing-based methods. This section will describe some of the approaches that are currently being explored. Sequence-based methods have already been implemented in the surveillance of foodborne viral infections (2, 25, 50), but the development of sequence-based subtyping methods for foodborne bacterial pathogens has barely begun. Some methods involve determining one or more DNA sequences in the genome (such as MLST), others detect point mutations at specific sites (single-nucleotide polymorphisms [SNPs]) or variations in repeating regions on the genome (MLVA), others detect specific genes or open reading frames via genome microarrays, and still others combine the different approaches.

The prototype of a sequence-based method is MLST. In this method, the sequences of a number of housekeeping genes are determined and compared among different strains. Because housekeeping genes are typically rather conserved, this method seldom provides as much discrimination below the species level as may be achieved by other methods such as PFGE. For this reason, MLST is not especially useful for detection of clusters of infections with most bacterial foodborne pathogens (24, 52, 76, 102, 107) with the notable exception of thermotolerant *Campylobacter* spp. (92). However, MLST yields extremely useful information for phylogenetic analysis and may be the preferred method for addressing broader questions such as whether human and animal isolates of a given species represent the same populations.

Because of the often limited discrimination of traditional MLST, other similar systems have been developed targeting more rapidly evolving genes, such as virulence or virulence-associated genes, or genes introduced into the genome such as prophage genes. The results to date have been mixed. Foley et al. (29) compared PFGE using the enzyme XbaI, REP element PCR using a commercially available system (Bacterial Barcodes Inc., Athens, Ga.), and MLST targeting three housekeeping genes and four virulence-associated genes (*eaeA*, *hlyA*, *fliC*, and *uidA*) for the subtyping of Shiga toxin-producing *E. coli* O157 and found that PFGE by far was the most discriminatory. In contrast, a study of *E. coli* producing extended-spectrum beta-lactamases revealed MLST using eight housekeeping and two beta-lactamase genes to be slightly more discriminating than PFGE (74). With *L. monocytogenes*, an MLST-type approach targeting virulence or virulence-associated genes appears promising (86, 119).

Sequencing of nonhousekeeping genes also can provide useful information for subtyping of *Salmonella*. Ross and Heuzenroeder sequenced 25 loci in three prophages and compared the results with those of PFGE and MLST for subtyping of *Salmonella* serotype Typhimurium. The phage-sequencing method was superior to both the other methods (91). Even if the analysis was limited to determining the presence or absence of each of the 25 prophage loci by PCR, the method was still more discriminatory than PFGE. As one would expect, determination of presence or absence was not quite as discriminating as sequencing, because several of the loci were polymorphic. Although this prophage-based typing method appears very promising for serotype Typhimurium, it is not yet clear whether it would be broadly applicable to other *Salmonella* serotypes or other foodborne pathogens.

Such detection of the presence or absence of a sequence has also been used for subtyping of foodborne pathogens by using the array format. In DNA arrays, polymorphism in DNA segments or genes that are known or expected to be present in some isolates but absent from others (based on comparisons of published genome sequences or subtractive hybridization) is detected by hybridization of labeled genomic DNA to target probes on a slide or a nylon membrane, in microwells, or in another format. Borucki et al. (11) have described a 629-probe microarray from a shotgun library that was as discriminatory as PFGE for subtyping of *L. monocytogenes* in their initial testing. Another microarray derived from a comparative genomic study with 409 probes from the genomes of two *L. monocytogenes* and one *Listeria innocua* strain also shows promise (22). A similar approach has been used to study the phylogenomics of *C. jejuni* (14) and the diversity of *Salmonella* (81, 85). Although this *Campylobacter* array appeared to provide sufficient discrimination for its use in the surveillance of that pathogen, the *Salmonella* arrays did not. However, in another study where the probes were selected both from published polymorphic genome sequences and from sequences obtained from subtractive hybridization experiments, it was possible to create a microarray for subtyping of serotype Typhimurium with greater discriminatory power (79). Microarrays may also be constructed to target virulence and antimicrobial resistance genes, allowing for identification and subtyping of and virulence and resistance profile determinations for pathogens in a single step (15, 111).

A simplified approach to sequenced-based subtyping is the detection of SNPs. An SNP is a change in a single nucleotide in one sequence relative to another. SNPs are detected by PCR, hybridization, or sequencing of the short region containing the SNP. Several SNP systems have been proposed based on established MLST schemes (89). They offer the same phylogenetic information as the MLST schemes they are based on but with less discrimination. Such SNP schemes may be useful for distinguishing between outbreak and nonoutbreak strains but are generally not suitable for surveillance purposes (8, 35, 41). However, by selecting SNPs from more variable regions it has been possible to discriminate within highly clonal bacteria such as *B. anthracis* and *Mycobacterium tuberculosis* (35, 84). SNPs in intergenic regions that may be useful for subtyping purposes have been described previously for *Salmonella* serotype Typhimurium (43). With the increasing number of genomes that have been fully or partially sequenced, it is likely that SNP-based subtyping systems will be described for all foodborne pathogens in the near future.

The sequence-based method that is closest to implementation in the routine setting for foodborne pathogens is MLVA. As described earlier is this chapter, MLVA is based on the detection of differences in the sequences of repetitive DNA (109). Repeats in one locus may be organized in tandem, with the number of repeats varying among different strains. Such loci are designated VNTR loci. MLVA involves determination of the number of the repeat copy units in multiple loci by simple PCR amplification of the repeat regions and determination of their sizes. Thus, an MLVA type may be designated as an array of numbers, each representing the copy number at a VNTR locus. Three partially overlapping MLVA subtyping schemes, each with good discriminatory power, have been published for Shiga toxin-producing *E. coli* O157 (48, 57, 75). An adapted version of these methods has been validated internally at the Centers for Disease Control and Prevention and is currently under external validation within the PulseNet network system (45). The results to date are extremely encouraging. A promising MLVA subtyping scheme for *Salmonella* serotype Typhimurium has also been published previously (60). By using that protocol, it was possible to differentiate among strains of the highly clonal serotype Typhimurium DT104 complex that cannot readily be differentiated by PFGE (M. Torpdhal and E. M. Nielson, Multiple locus variable number of tandem repeats analysis [MLVA] used for surveillance of S. Typhimurium, *Abstr. 7th Int. Meet. Microb. Epidemiol. Markers*, abstr. 17A, 2005). Within PulseNet, MLVA subtyping protocols for *Salmonella* serotypes Enteritidis and Typhimurium and *L. monocytogenes* are under development.

Sequence-based approaches to subtyping of bacteria will likely be widely implemented in the surveillance of foodborne infections in the near future. They are attractive for several reasons. Sequence-based methods have the potential to generate not only information of utility to surveillance, but by targeting virulence or antimicrobial susceptibility genes they may be able to generate results that are also clinically useful.

Sequence-based methods may potentially be partially or fully automated, thus increasing the throughput, reducing the risk of human errors, and potentially lowering the cost of the surveillance. However, commercial availability of such methods is unlikely before the method formats have been agreed upon internationally, either through formal agreement or simply through widespread use. International subtyping networks such as PulseNet, Enter-Net, and Salm-Net will play central roles in establishing the preferred technology. As described above, work is currently under way within the framework of PulseNet International to harmonize

MLVA protocols for Shiga toxin-producing *E. coli* O157, *Salmonella*, and *L. monocytogenes*, and within PulseNet USA, a project is under way to identify SNPs through comparative sequencing of Shiga toxin-producing *E. coli* O157 with the aim of establishing an SNP-based subtyping system for this pathogen. Although most of the sequenced-based methods are currently more costly than the existing molecular subtyping methods and the equipment needed is present only in relatively few routine surveillance laboratories, the cost of sequencing is decreasing and it is anticipated that these methods may become affordable to a broader range of laboratories in the foreseeable future.

Although the move toward sequence-based subtyping is highly desirable, it must be remembered that even the newest methods are not without limitations. These methods generate definitive results in a portable format that is much less complex than most of the current molecular surveillance data, e.g., those from PFGE, but that does not necessarily mean that interpretation of the data becomes simpler. This issue is very important to consider when the targets for the new methods are chosen, and differentiation criteria will need to be defined during the process of method validation.

Although sequence-based data per se are definitive by nature, i.e., a sequence of characters, the reproducibility of

the sequencing is not 100%. Analysis should be attempted only with high-quality data, and even then the "correct" interpretation of raw data is not always obvious. Figure 47.5 (top) shows a short chromatogram from an automated DNA sequencer. The peaks between positions 33 and 40 (and thus the number of bases), for example, are far less clear than the peaks in other areas of the chromatogram. However, assuming correct interpretation of the raw data, sequencing does indeed provide unambiguous comparisons as shown in the sequence alignment in the bottom panel of Fig. 47.5. It must also be considered that stability of the subtypes is unlikely to be absolute. The loci or genes that are targeted by these methods need to be selected carefully, with consideration of not only their discriminatory power but also their stability. New methods must be validated with a panel of carefully selected historical strains with well-established epidemiologic relationships, and their value in real-time surveillance must be evaluated on a small scale. For example, in the PulseNet validation of MLVA for Shiga toxin-producing *E. coli* O157, it has become clear that not all the selected loci are fully stable and it is necessary to set differentiation criteria for each of the VNTR loci used in the system.

Finally, it is important to validate the new methods against those previously used so that information in existing databases, such as the PFGE data in the

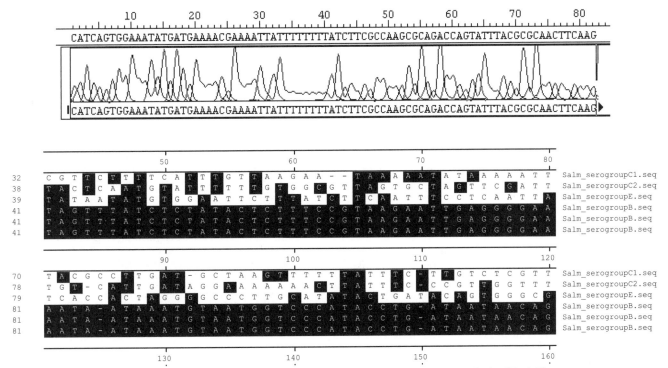

Figure 47.5 Example of output from automated DNA sequencing analysis. (Top) Chromatogram showing unanalyzed data (base calls). (Bottom) Analyzed data (aligned sequences).

PulseNet databases, is not lost. Validation of new methods is a time-consuming process, so the current standard methods will probably remain the standards for at least the next 5 years. The new methods will increasingly be introduced in smaller surveillance projects and gradually be implemented into the major surveillance networks, initially supplementing but possibly later replacing the older methods.

Predicting the future is always perilous, especially in a field where new technology can suddenly open unforeseen possibilities. Although it seems highly likely that some type of sequenced-based subtyping will ultimately replace all the current standards, it is difficult to predict exactly what the format of the new technology will be. Whatever advances new technology provides, it must be remembered that subtyping data will continue to be just one part of the picture and cannot alone replace epidemiologic and environmental investigations.

References

1. **Altwegg, M., F. W. Hickman-Brenner, and J. J. Farmer III.** 1989. Ribosomal RNA gene restriction patterns provide increased sensitivity for typing *Salmonella typhi* strains. *J. Infect. Dis.* **160:**145–149.

2. **Amon, J. J., R. Devasia, G. Xia, O. V. Nainan, S. Hall, B. Lawson, J. S. Wolthuis, P. D. Macdonald, C. W. Shepard, I. T. Williams, G. L. Armstrong, J. A. Gabel, P. Erwin, L. Sheeler, W. Kuhnert, P. Patel, G. Vaughan, A. Weltman, A. S. Craig, B. P. Bell, and A. Fiore.** 2005. Molecular epidemiology of foodborne hepatitis A outbreaks in the United States, 2003. *J. Infect. Dis.* **192:**1323–1330.

3. **Anonymous.** 2000. Global Salm-Surv on internet. *Wkly. Epidemiol. Rec.* **75:**236–237.

4. **Barrett, T. J., H. Lior, J. H. Green, R. Khakhria, J. G. Wells, B. P. Bell, K. D. Greene, J. Lewis, and P. M. Griffin.** 1994. Laboratory investigation of a multistate food-borne outbreak of *Escherichia coli* O157:H7 by using pulsed-field gel electrophoresis and phage typing. *J. Clin. Microbiol.* **32:**3013–3017.

5. **Barrett, T. J., E. Ribot, and B. Swaminathan.** 2004. Molecular subtyping for epidemiology: issues in comparability of patterns and interpretation of data, p. 259–266. *In* D. H. Persing, F. C. Tenover, J. Versalovic, Y.-W. Tang, E. R. Unger, D. A. Relman, and T. J. White (ed.), *Molecular Microbiology: Diagnostic Principles and Practices.* ASM Press, Washington, D.C.

6. **Bell, B. P., M. Goldoft, P. M. Griffin, M. A. Davis, D. C. Gordon, P. I. Tarr, C. A. Bartleson, J. H. Lewis, T. J. Barrett, J. G. Wells, et al.** 1994. A multistate outbreak of *Escherichia coli* O157:H7-associated bloody diarrhea and hemolytic uremic syndrome from hamburgers. The Washington experience. *JAMA* **272:**1349–1353.

7. **Bender, J. B., C. W. Hedberg, D. J. Boxrud, J. M. Besser, J. H. Wicklund, K. E. Smith, and M. T. Osterholm.** 2001. Use of molecular subtyping in surveillance for *Salmonella enterica* serotype Typhimurium. *N. Engl. J. Med.* **344:**189–195.

8. **Best, E. L., A. J. Fox, J. A. Frost, and F. J. Bolton.** 2005. Real-time single-nucleotide polymorphism profiling using Taqman technology for rapid recognition of *Campylobacter jejuni* clonal complexes. *J. Med. Microbiol.* **54:**919–925.

9. **Blumberg, H. M., J. A. Kiehlbauch, and I. K. Wachsmuth.** 1991. Molecular epidemiology of *Yersinia enterocolitica* O:3 infections: use of chromosomal DNA restriction fragment length polymorphisms of rRNA genes. *J. Clin. Microbiol.* **29:**2368–2374.

10. **Boer, P., J. A. Wagenaar, R. P. Achterberg, J. P. Putten, L. M. Schouls, and B. Duim.** 2002. Generation of *Campylobacter jejuni* genetic diversity in vivo. *Mol. Microbiol.* **44:**351–359.

11. **Borucki, M. K., S. H. Kim, D. R. Call, S. C. Smole, and F. Pagotto.** 2004. Selective discrimination of *Listeria monocytogenes* epidemic strains by a mixed-genome DNA microarray compared to discrimination by pulsed-field gel electrophoresis, ribotyping, and multilocus sequence typing. *J. Clin. Microbiol.* **42:**5270–5276.

12. **Bruce, J.** 1996. Automated system rapidly identifies and characterizes microorganisms in food. *Food Technol.* **50:**77–81.

13. **Carle, G. F., M. Frank, and M. V. Olson.** 1986. Electrophoretic separations of large DNA molecules by periodic inversion of the electric field. *Science* **232:**65–68.

14. **Champion, O. L., M. W. Gaunt, O. Gundogdu, A. Elmi, A. A. Witney, J. Hinds, N. Dorrell, and B. W. Wren.** 2005. Comparative phylogenomics of the food-borne pathogen *Campylobacter jejuni* reveals genetic markers predictive of infection source. *Proc. Natl. Acad. Sci. USA* **102:**16043–16048.

15. **Chen, S., S. Zhao, P. F. McDermott, C. M. Schroeder, D. G. White, and J. Meng.** 2005. A DNA microarray for identification of virulence and antimicrobial resistance genes in *Salmonella* serovars and *Escherichia coli.* *Mol. Cell. Probes* **19:**195–201.

16. **Chu, G., D. Vollrath, and R. W. Davis.** 1986. Separation of large DNA molecules by contour-clamped homogeneous electric fields. *Science* **234:**1582–1585.

17. **Clark, C. G., L. Bryden, W. R. Cuff, P. L. Johnson, F. Jamieson, B. Ciebin, and G. Wang.** 2005. Use of the oxford multilocus sequence typing protocol and sequencing of the flagellin short variable region to characterize isolates from a large outbreak of waterborne *Campylobacter* sp. strains in Walkerton, Ontario, Canada. *J. Clin. Microbiol.* **43:**2080–2091.

18. **Cook, K. A., T. E. Dobbs, W. G. Hlady, J. G. Wells, T. J. Barrett, N. D. Puhr, G. A. Lancette, D. W. Bodager, B. L. Toth, C. A. Genese, A. K. Highsmith, K. E. Pilot, L. Finelli, and D. L. Swerdlow.** 1998. Outbreak of *Salmonella* serotype Hartford infections associated with unpasteurized orange juice. *JAMA* **280:**1504–1509.

19. **de Boer, P., B. Duim, A. Rigter, J. van Der Plas, W. F. Jacobs-Reitsma, and J. A. Wagenaar.** 2000. Computer-assisted analysis and epidemiological value of genotyping methods for *Campylobacter jejuni* and *Campylobacter coli.* *J. Clin. Microbiol.* **38:**1940–1946.

20. **De Cesare, A., J. L. Bruce, T. R. Dambaugh, M. E. Guerzoni, and M. Wiedmann.** 2001. Automated ribotyping using different enzymes to improve discrimination of

Listeria monocytogenes isolates, with a particular focus on serotype 4b strains. *J. Clin. Microbiol.* 39:3002–3005.

21. Dingle, K. E., F. M. Colles, D. R. Wareing, A. Ure, A. J. Fox, F. E. Bolton, H. J. Bootsma, R. J. Willems, R. Urwin, and M. C. Maiden. 2001. Multilocus sequence typing system for *Campylobacter jejuni. J. Clin. Microbiol.* 39:14–23.

22. Doumith, M., C. Cazalet, N. Simoes, L. Frangeul, C. Jacquet, F. Kunst, P. Martin, P. Cossart, P. Glaser, and C. Buchrieser. 2004. New aspects regarding evolution and virulence of *Listeria monocytogenes* revealed by comparative genomics and DNA arrays. *Infect. Immun.* 72:1072–1083.

23. Ellsworth, D. L., K. D. Rittenhouse, and R. L. Honeycutt. 1993. Artifactual variation in randomly amplified polymorphic DNA banding patterns. *BioTechniques* 14:214–217.

24. Fakhr, M. K., L. K. Nolan, and C. M. Logue. 2005. Multilocus sequence typing lacks the discriminatory ability of pulsed-field gel electrophoresis for typing *Salmonella enterica* serovar Typhimurium. *J. Clin. Microbiol.* 43:2215–2219.

25. Fankhauser, R. L., S. S. Monroe, J. S. Noel, C. D. Humphrey, J. S. Bresee, U. D. Parashar, T. Ando, and R. I. Glass. 2002. Epidemiologic and molecular trends of "Norwalk-like viruses" associated with outbreaks of gastroenteritis in the United States. *J. Infect. Dis.* 186:1–7.

26. Farlow, J., K. L. Smith, J. Wong, M. Abrams, M. Lytle, and P. Keim. 2001. *Francisella tularensis* strain typing using multiple-locus, variable-number tandem repeat analysis. *J. Clin. Microbiol.* 39:3186–3192.

27. Fisher, I. S. 1999. The Enter-Net international surveillance network—how it works. *Euro Surveill.* 4:52–55.

28. Fisher, I. S. 1995. Salm-Net: a network for human *Salmonella* surveillance in Europe. *Euro Surveill.* 0:7–8.

29. Foley, S. L., S. Simjee, J. Meng, D. G. White, P. F. McDermott, and S. Zhao. 2004. Evaluation of molecular typing methods for *Escherichia coli* O157:H7 isolates from cattle, food, and humans. *J. Food Prot.* 67:651–657.

30. Fonnesbech Vogel, B., V. Fussing, B. Ojeniyi, L. Gram, and P. Ahrens. 2004. High-resolution genotyping of *Listeria monocytogenes* by fluorescent amplified fragment length polymorphism analysis compared to pulsed-field gel electrophoresis, random amplified polymorphic DNA analysis, ribotyping, and PCR-restriction fragment length polymorphism analysis. *J. Food Prot.* 67:1656–1665.

31. Gardiner, K., W. Laas, and D. Patterson. 1986. Fractionation of large mammalian DNA restriction fragments using vertical pulsed-field gradient gel electrophoresis. *Somat. Cell Mol. Genet.* 12:185–195.

32. Graves, L. M., S. B. Hunter, A. R. Ong, D. Schoonmaker-Bopp, K. Hise, L. Kornstein, W. E. DeWitt, P. S. Hayes, E. Dunne, P. Mead, and B. Swaminathan. 2005. Microbiological aspects of the investigation that traced the 1998 outbreak of listeriosis in the United States to contaminated hot dogs and establishment of molecular subtyping-based surveillance for *Listeria monocytogenes* in the PulseNet network. *J. Clin. Microbiol.* 43:2350–2355.

33. Grimont, F., and P. A. Grimont. 1986. Ribosomal ribonucleic acid gene restriction patterns as potential taxonomic tools. *Ann. Inst. Pasteur Microbiol.* 137B:165–175.

34. Gudmundsdottir, S., B. Gudbjornsdottir, H. L. Lauzon, H. Einarsson, K. G. Kristinsson, and M. Kristjansson. 2005. Tracing *Listeria monocytogenes* isolates from cold-smoked salmon and its processing environment in Iceland using pulsed-field gel electrophoresis. *Int. J. Food Microbiol.* 101:41–51.

35. Gutacker, M. M., J. C. Smoot, C. A. Migliaccio, S. M. Ricklefs, S. Hua, D. V. Cousins, E. A. Graviss, E. Shashkina, B. N. Kreiswirth, and J. M. Musser. 2002. Genome-wide analysis of synonymous single nucleotide polymorphisms in *Mycobacterium tuberculosis* complex organisms: resolution of genetic relationships among closely related microbial strains. *Genetics* 162:1533–1543.

36. Harrington, C. S., F. M. Thomson-Carter, and P. E. Carter. 1997. Evidence for recombination in the flagellin locus of *Campylobacter jejuni*: implications for the flagellin gene typing scheme. *J. Clin. Microbiol.* 35:2386–2392.

37. Hassan, W. M., S. Y. Wang, and R. D. Ellender. 2005. Methods to increase fidelity of repetitive extragenic palindromic PCR fingerprint-based bacterial source tracking efforts. *Appl. Environ. Microbiol.* 71:512–518.

38. Hedberg, C. W., K. E. Smith, J. M. Besser, D. J. Boxrud, T. W. Hennessy, J. B. Bender, F. A. Anderson, and M. T. Osterholm. 2001. Limitations of pulsed-field gel electrophoresis for the routine surveillance of *Campylobacter* infections. *J. Infect. Dis.* 184:242–244.

39. Hilton, A. C., and C. W. Penn. 1998. Comparison of ribotyping and arbitrarily primed PCR for molecular typing of *Salmonella enterica* and relationships between strains on the basis of these molecular markers. *J. Appl. Microbiol.* 85:933–940.

40. Holmberg, S. D., I. K. Wachsmuth, F. Hickman-Brenner, and M. L. Cohen. 1984. Comparison of plasmid profile analysis, phage typing and antimicrobial susceptibility testing in characterizing *Salmonella typhimurium* isolates from outbreaks. *J. Clin. Microbiol.* 19:100–104.

41. Hommais, F., S. Pereira, C. Acquaviva, P. Escobar-Paramo, and E. Denamur. 2005. Single-nucleotide polymorphism phylotyping of *Escherichia coli. Appl. Environ. Microbiol.* 71:4784–4792.

42. Horby, P. W., S. J. O'Brien, G. K. Adak, C. Graham, J. I. Hawker, P. Hunter, C. Lane, A. J. Lawson, R. T. Mitchell, M. H. Reacher, E. J. Threlfall, L. R. Ward, and P. O. I. Team. 2003. A national outbreak of multi-resistant *Salmonella enterica* serovar Typhimurium definitive phage type (DT) 104 associated with consumption of lettuce. *Epidemiol. Infect.* 130:169–178.

43. Hu, H., R. Lan, and P. R. Reeves. 2006. Adaptation of multilocus sequencing to study variation within a major clone: evolutionary relationships of *Salmonella enterica* serovar Typhimurium. *Genetics* 172:743–750.

44. Huang, S., T. Luangtongkum, T. Y. Morishita, and Q. Zhang. 2005. Molecular typing of *Campylobacter* strains using the *cmp* gene encoding the major outer membrane protein. *Foodborne Pathog. Dis.* 2:12–23.

45. Hyytia-Trees, E., S. C. Smole, P. I. Fields, B. Swaminathan, and E. Ribot. 2006. Second generation subtyping: a proposed PulseNet protocol for multiple-locus variable-number tandem repeat analysis of Shiga toxin-producing *Escherichia coli* O157 (STEC O157). *Foodborne Pathog. Dis.* 3:118–131.

46. Johansson, A., J. Farlow, P. Larsson, M. Dukerich, E. Chambers, M. Bystrom, J. Fox, M. Chu, M. Forsman, A. Sjostedt, and P. Keim. 2004. Worldwide genetic relationships among *Francisella tularensis* isolates determined by multiple-locus variable-number tandem repeat analysis. *J. Bacteriol.* **186:**5808–5818.

47. Keim, P., L. B. Price, A. M. Klevytska, K. L. Smith, J. M. Schupp, R. Okinaka, P. J. Jackson, and M. E. Hugh-Jones. 2000. Multiple-locus variable-number tandem repeat analysis reveals genetic relationships within *Bacillus anthracis*. *J. Bacteriol.* **182:**2928–2936.

48. Keys, C., S. Kemper, and P. Keim. 2005. Highly diverse variable number tandem repeat loci in the *E. coli* O157:H7 and O55:H7 genomes for high-resolution molecular typing. *J. Appl. Microbiol.* **98:**928–940.

49. Knudsen, K. N., D. D. Bang, E. M. Nielsen, and M. Madsen. 2005. Genotyping of *Campylobacter jejuni* strains from Danish broiler chickens by restriction fragment length polymorphism of the LPS gene cluster. *J. Appl. Microbiol.* **99:**392–399.

50. Koopmans, M., H. Vennema, H. Heersma, E. van Strien, Y. van Duynhoven, D. Brown, M. Reacher, and B. Lopman. 2003. Early identification of common-source foodborne virus outbreaks in Europe. *Emerg. Infect. Dis.* **9:** 1136–1142.

51. Kostman, J. R., T. D. Edlind, J. J. LiPuma, and T. L. Stull. 1992. Molecular epidemiology of *Pseudomonas cepacia* determined by polymerase chain reaction ribotyping. *J. Clin. Microbiol.* **30:**2084–2087.

52. Kotetishvili, M., A. Kreger, G. Wauters, J. G. Morris, Jr., A. Sulakvelidze, and O. C. Stine. 2005. Multilocus sequence typing for studying genetic relationships among *Yersinia* species. *J. Clin. Microbiol.* **43:**2674–2684.

53. Kotetishvili, M., O. C. Stine, A. Kreger, J. G. Morris, Jr., and A. Sulakvelidze. 2002. Multilocus sequence typing for characterization of clinical and environmental *Salmonella* strains. *J. Clin. Microbiol.* **40:**1626–1635.

54. Le Fleche, P., Y. Hauck, L. Onteniente, A. Prieur, F. Denoeud, V. Ramisse, P. Sylvestre, P. Benson, F. Ramisse, and G. Vergnaud. 2001. A tandem repeats database for bacterial genomes: application to the genotyping of *Yersinia pestis* and *Bacillus anthracis*. *BMC Microbiol.* **1:**2.

55. Liljebjelke, K. A., C. L. Hofacre, T. Liu, D. G. White, S. Ayers, S. Young, and J. J. Maurer. 2005. Vertical and horizontal transmission of *Salmonella* within integrated broiler production system. *Foodborne Pathog. Dis.* **2:**90–102.

56. Lindstedt, B. A. 2005. Multiple-locus variable tandem repeats analysis for genetic fingerprinting of pathogenic bacteria. *Electrophoresis* **26:**2567–2582.

57. Lindstedt, B. A., E. Heir, E. Gjernes, T. Vardund, and G. Kapperud. 2003. DNA fingerprinting of Shiga-toxin producing *Escherichia coli* O157 based on multiple-locus variable-number tandem-repeats analysis (MLVA). *Ann. Clin. Microbiol. Antimicrob.* **2:**12.

58. Lindstedt, B. A., E. Heir, T. Vardund, and G. Kapperud. 2000. Fluorescent amplified-fragment length polymorphism genotyping of *Salmonella enterica* subsp. *enterica* serovars and comparison with pulsed-field gel electrophoresis typing. *J. Clin. Microbiol.* **38:**1623–1627.

59. Lindstedt, B. A., E. Heir, T. Vardund, K. K. Melby, and G. Kapperud. 2000. Comparative fingerprinting analysis of *Campylobacter jejuni* subsp. *jejuni* strains by amplified-fragment length polymorphism genotyping. *J. Clin. Microbiol.* **38:**3379–3387.

60. Lindstedt, B. A., T. Vardund, L. Aas, and G. Kapperud. 2004. Multiple-locus variable-number tandem-repeats analysis of *Salmonella enterica* subsp. *enterica* serovar Typhimurium using PCR multiplexing and multicolor capillary electrophoresis. *J. Microbiol. Methods* **59:**163–172.

61. Lindstedt, B. A., T. Vardund, and G. Kapperud. 2004. Multiple-locus variable-number tandem-repeats analysis of *Escherichia coli* O157 using PCR multiplexing and multicolored capillary electrophoresis. *J. Microbiol. Methods* **58:**213–222.

62. Lowell, J. L., D. M. Wagner, B. Atshabar, M. F. Antolin, A. J. Vogler, P. Keim, M. C. Chu, and K. L. Gage. 2005. Identifying sources of human exposure to plague. *J. Clin. Microbiol.* **43:**650–656.

63. Lu, Z., D. Lapen, A. Scott, A. Dang, and E. Topp. 2005. Identifying host sources of fecal pollution: diversity of *Escherichia coli* in confined dairy and swine production systems. *Appl. Environ. Microbiol.* **71:**5992–5998.

64. Lukinmaa, S., K. Aarnisalo, M. L. Suihko, and A. Siitonen. 2004. Diversity of *Listeria monocytogenes* isolates of human and food origin studied by serotyping, automated ribotyping and pulsed-field gel electrophoresis. *Clin. Microbiol. Infect.* **10:**562–568.

65. MacPherson, J. M., P. E. Eckstein, G. J. Scoles, and A. A. Gajadhar. 1993. Variability of the random amplified polymorphic DNA assay among thermal cyclers, and effects of primer and DNA concentration. *Mol. Cell. Probes* **7:**293–299.

66. Manning, G., B. Duim, T. Wassenaar, J. A. Wagenaar, A. Ridley, and D. G. Newell. 2001. Evidence for a genetically stable strain of *Campylobacter jejuni*. *Appl. Environ. Microbiol.* **67:**1185–1189.

67. Meinersmann, R. J., L. O. Helsel, P. I. Fields, and K. L. Hiett. 1997. Discrimination of *Campylobacter jejuni* isolates by *fla* gene sequencing. *J. Clin. Microbiol.* **35:**2810–2814.

68. Mellmann, A., J. Mosters, E. Bartelt, P. Roggentin, A. Ammon, A. W. Friedrich, H. Karch, and D. Harmsen. 2004. Sequence-based typing of *flaB* is a more stable screening tool than typing of *flaA* for monitoring of *Campylobacter* populations. *J. Clin. Microbiol.* **42:**4840–4842.

69. Meunier, J. R., and P. A. D. Grimont. 1993. Factors affecting reproducibility of random amplified polymorphic DNA fingerprinting. *Res. Microbiol.* **144:**373–379.

70. Michaud, S., S. Menard, and R. D. Arbeit. 2005. Role of real-time molecular typing in the surveillance of *Campylobacter enteritis* and comparison of pulsed-field gel electrophoresis profiles from chicken and human isolates. *J. Clin. Microbiol.* **43:**1105–1111.

71. Micheli, M. R., R. Bova, E. Pascale, and E. D'Ambrosio. 1994. Reproducible DNA fingerprinting with the random amplified polymorphic DNA (RAPD) method. *Nucleic Acids Res.* **22:**1921–1922.

72. Muralidharan, K., and E. K. Wakeland. 1993. Concentration of primer and template qualitatively affects products

in random-amplified polymorphic DNA PCR. *BioTechniques* 14:362–364.

73. Nachamkin, I., K. Bohachick, and C. M. Patton. 1993. Flagellin gene typing of *Campylobacter jejuni* by restriction fragment length polymorphism analysis. *J. Clin. Microbiol.* 31:1531–1536.

74. Nemoy, L. L., M. Kotetishvili, J. Tigno, A. Keefer-Norris, A. D. Harris, E. N. Perencevich, J. A. Johnson, D. Torpey, A. Sulakvelidze, J. G. Morris, Jr., and O. C. Stine. 2005. Multilocus sequence typing versus pulsed-field gel electrophoresis for characterization of extended-spectrum beta-lactamase-producing *Escherichia coli* isolates. *J. Clin. Microbiol.* 43:1776–1781.

75. Noller, A. C., M. C. McEllistrem, A. G. Pacheco, D. J. Boxrud, and L. H. Harrison. 2003. Multilocus variable-number tandem repeat analysis distinguishes outbreak and sporadic *Escherichia coli* O157:H7 isolates. *J. Clin. Microbiol.* 41:5389–5397.

76. Noller, A. C., M. C. McEllistrem, O. C. Stine, J. G. Morris, Jr., D. J. Boxrud, B. Dixon, and L. H. Harrison. 2003. Multilocus sequence typing reveals a lack of diversity among *Escherichia coli* O157:H7 isolates that are distinct by pulsed-field gel electrophoresis. *J. Clin. Microbiol.* 41:675–679.

77. Olsvik, O., H. Sorum, K. Birkness, K. Wachsmuth, M. Fjolstad, J. Lassen, K. Fossum, and J. C. Feeley. 1985. Plasmid characterization of *Salmonella typhimurium* transmitted from animals to humans. *J. Clin. Microbiol.* 22:336–338.

78. Parveen, S., R. L. Murphree, L. Edmiston, C. W. Casper, K. M. Portier, and M. L. Tamplin. 1997. Association of multiple-antibiotic resistance profiles with point and non-point sources of *Escherichia coli* in Apalachicola Bay. *Appl. Environ. Microbiol.* 63:2607–2612.

79. Pelludat, C., R. Prager, H. Tschape, W. Rabsch, J. Schuchhardt, and W. D. Hardt. 2005. Pilot study to evaluate microarray hybridization as a tool for *Salmonella enterica* serovar Typhimurium strain differentiation. *J. Clin. Microbiol.* 43:4092–4106.

80. Pfaller, M. A., and R. J. Hollis. 2004. Automated ribotyping, p. 245–258. *In* D. H. Persing, F. C. Tenover, J. Versalovic, Y.-W. Tang, E. R. Unger, D. A. Relman, and T. J. White (ed.), *Molecular Microbiology: Diagnostic Principles and Practices.* ASM Press, Washington, D.C.

81. Porwollik, S., E. F. Boyd, C. Choy, P. Cheng, L. Florea, E. Proctor, and M. McClelland. 2004. Characterization of *Salmonella enterica* subspecies I genovars by use of microarrays. *J. Bacteriol.* 186:5883–5898.

82. Rademaker, J. L., and P. Savelkoul. 2004. PCR amplification-based microbial typing, p. 197–221. *In* D. H. Persing, F. C. Tenover, J. Versalovic, Y.-W. Tang, E. R. Unger, D. A. Relman, and T. J. White (ed.), *Molecular Microbiology: Diagnostic Principles and Practice.* ASM Press, Washington, D.C.

83. Ramisse, V., P. Houssu, E. Hernandez, F. Denoeud, V. Hilaire, O. Lisanti, F. Ramisse, J. D. Cavallo, and G. Vergnaud. 2004. Variable number of tandem repeats in *Salmonella enterica* subsp. *enterica* for typing purposes. *J. Clin. Microbiol.* 42:5722–5730.

84. Read, T. D., S. L. Salzberg, M. Pop, M. Shumway, L. Umayam, L. Jiang, E. Holtzapple, J. D. Busch, K. L. Smith, J. M. Schupp, D. Solomon, P. Keim, and C. M. Fraser. 2002. Comparative genome sequencing for discovery of novel polymorphisms in *Bacillus anthracis*. *Science* 296:2028–2033.

85. Reen, F. J., E. F. Boyd, S. Porwollik, B. P. Murphy, D. Gilroy, S. Fanning, and M. McClelland. 2005. Genomic comparisons of *Salmonella enterica* serovar Dublin, Agona, and Typhimurium strains recently isolated from milk filters and bovine samples from Ireland, using a *Salmonella* microarray. *Appl. Environ. Microbiol.* 71:1616–1625.

86. Revazishvili, T., M. Kotetishvili, O. C. Stine, A. S. Kreger, J. G. Morris, Jr., and A. Sulakvelidze. 2004. Comparative analysis of multilocus sequence typing and pulsed-field gel electrophoresis for characterizing *Listeria monocytogenes* strains isolated from environmental and clinical sources. *J. Clin. Microbiol.* 42:276–285.

87. Ridley, A. M., E. J. Threlfall, and B. Rowe. 1998. Genotypic characterization of *Salmonella enteritidis* phage types by plasmid analysis, ribotyping, and pulsed-field gel electrophoresis. *J. Clin. Microbiol.* 36:2314–2321.

88. Riley, L. W., R. S. Remis, S. D. Hlegerson, H. B. McGee, J. G. Wells, B. R. Davis, R. J. Hebert, E. S. Olcott, L. M. Johnson, N. T. Hargrett, P. A. Blake, and M. L. Cohen. 1983. Hemorrhagic colitis associated with a rare *Escherichia coli* serotype. *N. Engl. J. Med.* 308:681–685.

89. Robertson, G. A., V. Thiruvenkataswamy, H. Shilling, E. P. Price, F. Huygens, F. A. Henskens, and P. M. Giffard. 2004. Identification and interrogation of highly informative single nucleotide polymorphism sets defined by bacterial multilocus sequence typing databases. *J. Med. Microbiol.* 53:35–45.

90. Ronner, A. C., E. Borch, and B. Kaijser. 2005. Genetic profiling of *Campylobacter jejuni* strains from humans infected in Sweden or in Thailand, and from healthy Swedish chickens, studied by pulsed-field gel electrophoresis (PFGE). *Scand. J. Infect. Dis.* 37:579–584.

91. Ross, I. L., and M. W. Heuzenroeder. 2005. Discrimination within phenotypically closely related definitive types of *Salmonella enterica* serovar Typhimurium by the multiple amplification of phage locus typing technique. *J. Clin. Microbiol.* 43:1604–1611.

92. Sails, A. D., B. Swaminathan, and P. I. Fields. 2003. Utility of multilocus sequence typing as an epidemiological tool for investigation of outbreaks of gastroenteritis caused by *Campylobacter jejuni*. *J. Clin. Microbiol.* 41:4733–4739.

93. Sauders, B. D., E. D. Fortes, D. L. Morse, N. Dumas, J. A. Kiehlbauch, Y. Schukken, J. R. Hibbs, and M. Wiedmann. 2003. Molecular subtyping to detect human listeriosis clusters. *Emerg. Infect. Dis.* 9:672–680.

94. Savelkoul, P. H., H. J. Aarts, J. de Haas, L. Dijkshoorn, B. Duim, M. Otsen, J. L. Rademaker, L. Schouls, and J. A. Lenstra. 1999. Amplified-fragment length polymorphism analysis: the state of an art. *J. Clin. Microbiol.* 37:3083–3091.

95. Schierwater, B., and A. Ender. 1993. Different thermostable DNA polymerases may amplify different RAPD products. *Nucleic Acids Res.* 21:4647–4648.

96. Schouls, L. M., S. Reulen, B. Duim, J. A. Wagenaar, R. J. Willems, K. E. Dingle, F. M. Colles, and J. D. Van Embden. 2003. Comparative genotyping of *Campylobacter jejuni*

by amplified fragment length polymorphism, multilocus sequence typing, and short repeat sequencing: strain diversity, host range, and recombination. *J. Clin. Microbiol.* **41:**15–26.

97. **Schwartz, D. C., W. Saffran, J. Welsh, R. Haas, M. Goldenberg, and C. R. Cantor.** 1983. New techniques for purifying large DNAs and studying their properties and packaging. *Cold Spring Harbor Symp. Quant. Biol.* **47**(Pt. 1):189–195.

98. **Selander, R. K., D. A. Caugant, H. Ochman, J. M. Musser, M. N. Gilmour, and T. S. Whittam.** 1986. Methods of multilocus enzyme electrophoresis for bacterial population genetics and systematics. *Appl. Environ. Microbiol.* **51:**873–884.

99. **Siemer, B. L., E. M. Nielsen, and S. L. On.** 2005. Identification and molecular epidemiology of *Campylobacter coli* isolates from human gastroenteritis, food, and animal sources by amplified fragment length polymorphism analysis and Penner serotyping. *Appl. Environ. Microbiol.* **71:**1953–1958.

100. **Southern, E. M.** 1979. Analysis of restriction-fragment patterns from complex deoxyribonucleic acid species. *Biochem. Soc. Symp.* **44:**37–41.

101. **Stull, T. L., J. J. LiPuma, and T. D. Edlind.** 1988. A broadspectrum probe for molecular epidemiology of bacteria: ribosomal RNA. *J. Infect. Dis.* **157:**280–286.

102. **Sukhnanand, S., S. Alcaine, L. D. Warnick, W. L. Su, J. Hof, M. P. Craver, P. McDonough, K. J. Boor, and M. Wiedmann.** 2005. DNA sequence-based subtyping and evolutionary analysis of selected *Salmonella enterica* serotypes. *J. Clin. Microbiol.* **43:**3688–3698.

103. **Suzuki, Y., M. Ishihara, M. Saito, N. Ishikawa, and T. Yokochi.** 1994. Discrimination by means of pulsed-field gel electrophoresis between strains of *Campylobacter jejuni* Lior type 4 derived from sporadic cases and from outbreaks of infection. *J. Infect.* **29:**183–187.

104. **Swaminathan, B., T. J. Barrett, S. B. Hunter, and R. V. Tauxe.** 2001. PulseNet: the molecular subtyping network for foodborne bacterial disease surveillance, United States. *Emerg. Infect. Dis.* **7:**382–389.

105. **Swaminathan, B., P. Gerner-Smidt, L.-K. Ng, S. Likinmaa, K.-M. Kam, S. Rolando, E. P. Gutierrez, and N. Binsztein.** 2006. Building PulseNet International: an interconnected system of laboratory networks to facilitate timely public health recognition and response to foodborne disease outbreaks and emerging foodborne diseases. *Foodborne Pathog. Dis.* **3:**36–50.

106. **Taylor, D. N., I. K. Wachsmuth, Y. H. Shangkuan, E. V. Schmidt, T. J. Barrett, J. S. Schrader, C. S. Scherach, H. B. McGee, R. A. Feldman, and D. J. Brenner.** 1982. Salmonellosis associated with marijuana: a multistate outbreak traced by plasmid fingerprinting. *N. Engl. J. Med.* **306:**1249–1253.

107. **Torpdahl, M., M. N. Skov, D. Sandvang, and D. L. Baggesen.** 2005. Genotypic characterization of *Salmonella* by multilocus sequence typing, pulsed-field gel

electrophoresis and amplified fragment length polymorphism. *J. Microbiol. Methods* **63:**173–184.

108. **van Belkum, A.** 1999. Short sequence repeats in microbial pathogenesis and evolution. *Cell. Mol. Life Sci.* **56:**729–734.

109. **van Belkum, A., S. Scherer, L. van Alphen, and H. Verbrugh.** 1998. Short-sequence DNA repeats in prokaryotic genomes. *Microbiol. Mol. Biol. Rev.* **62:**275–293.

110. **Versalovic, J., T. Koeuth, and J. R. Lupski.** 1991. Distribution of repetitive DNA sequences in eubacteria and application to fingerprinting bacterial genomes. *Nucleic Acids Res.* **15:**7865–7876.

111. **Vora, G. J., C. E. Meador, M. M. Bird, C. A. Bopp, J. D. Andreadis, and D. A. Stenger.** 2005. Microarray-based detection of genetic heterogeneity, antimicrobial resistance, and the viable but nonculturable state in human pathogenic *Vibrio* spp. *Proc. Natl. Acad. Sci. USA* **102:**19109–19114.

112. **Vos, P., R. Hogers, M. Bleeker, M. Reijans, T. van de Lee, M. Hornes, A. Frijters, J. Pot, J. Peleman, M. Kuiper, et al.** 1995. AFLP: a new technique for DNA fingerprinting. *Nucleic Acids Res.* **23:**4407–4414.

113. **Wachsmuth, I. K., J. A. Kiehlbauch, C. A. Bopp, D. N. Cameron, N. A. Strockbine, J. G. Wells, and P. A. Blake.** 1991. The use of plasmid profiles and nucleic acid probes in epidemiologic investigations of foodborne, diarrheal diseases. *Int. J. Food Microbiol.* **12:**77–89.

114. **Wassenaar, T. M., B. Geilhausen, and D. G. Newell.** 1998. Evidence of genomic instability in *Campylobacter jejuni* isolated from poultry. *Appl. Environ. Microbiol.* **64:**1816–1821.

115. **Welsh, J., and M. McClelland.** 1990. Fingerprinting genomes using PCR with arbitrary primers. *Nucleic Acids Res.* **18:**7213–7218.

116. **Williams, J. G. K., A. R. Kubelik, K. J. Livak, J. A. Rafalski, and S. V. Tingey.** 1990. DNA polymorphisms amplified by arbitrary primers are useful as genetic markers. *Nucleic Acids Res.* **18:**6531–6535.

117. **Yan, W., N. Chang, and D. E. Taylor.** 1991. Pulsed-field gel electrophoresis of *Campylobacter jejuni* and *Campylobacter coli* genomic DNA and its epidemiologic application. *J. Infect. Dis.* **163:**1068–1072.

118. **Yeramian, E., and H. Buc.** 1999. Tandem repeats in complete bacterial genome sequences: sequence and structural analyses for comparative studies. *Res. Microbiol.* **150:**745–754.

119. **Zhang, W., B. M. Jayarao, and S. J. Knabel.** 2004. Multivirulence-locus sequence typing of *Listeria monocytogenes*. *Appl. Environ. Microbiol.* **70:**913–920.

120. **Zhang, W., and S. J. Knabel.** 2005. Multiplex PCR assay simplifies serotyping and sequence typing of *Listeria monocytogenes* associated with human outbreaks. *J. Food Prot.* **68:**1907–1910.

121. **Zhao, S., S. E. Mitchell, J. Meng, S. Kresovich, M. P. Doyle, R. E. Dean, A. M. Casa, and J. W. Weller.** 2000. Genomic typing of *Escherichia coli* O157:H7 by semi-automated fluorescent AFLP analysis. *Microbes Infect.* **2:**107–113.

Index